INTERNATIONAL ENERGY AGENCY
AGENCE INTERNATIONALE DE L'ENERGIE

W9-AGN-938

ENERGY STATISTICS OF NON-OECD COUNTRIES

1997 - 1998

STATISTIQUES DE L'ENERGIE DES PAYS NON-MEMBRES

 OECD

2000 Edition

INTERNATIONAL ENERGY AGENCY
9, RUE DE LA FÉDÉRATION, 75739 PARIS CEDEX 15, FRANCE

The International Energy Agency (IEA) is an autonomous body which was established in November 1974 within the framework of the Organisation for Economic Co-operation and Development (OECD) to implement an international energy programme.

It carries out a comprehensive programme of energy co-operation among twenty-four* of the OECD's twenty-nine Member countries. The basic aims of the IEA are:

- To maintain and improve systems for coping with oil supply disruptions;
- To promote rational energy policies in a global context through co-operative relations with non-member countries, industry and international organisations;
- To operate a permanent information system on the international oil market;
- To improve the world's energy supply and demand structure by developing alternative energy sources and increasing the efficiency of energy use;
- To assist in the integration of environmental and energy policies.

IEA Member countries: Australia, Austria, Belgium, Canada, Denmark, Finland, France, Germany, Greece, Hungary, Ireland, Italy, Japan, Luxembourg, the Netherlands, New Zealand, Norway, Portugal, Spain, Sweden, Switzerland, Turkey, the United Kingdom, the United States. The European Commission also takes part in the work of the IEA.

ORGANISATION FOR ECONOMIC CO-OPERATION AND DEVELOPMENT

Pursuant to Article 1 of the Convention signed in Paris on 14th December 1960, and which came into force on 30th September 1961, the Organisation for Economic Co-operation and Development (OECD) shall promote policies designed:

- To achieve the highest sustainable economic growth and employment and a rising standard of living in Member countries, while maintaining financial stability, and thus to contribute to the development of the world economy;
- To contribute to sound economic expansion in Member as well as non-member countries in the process of economic development; and
- To contribute to the expansion of world trade on a multilateral, non-discriminatory basis in accordance with international obligations.

The original Member countries of the OECD are Austria, Belgium, Canada, Denmark, France, Germany, Greece, Iceland, Ireland, Italy, Luxembourg, the Netherlands, Norway, Portugal, Spain, Sweden, Switzerland, Turkey, the United Kingdom and the United States. The following countries became Members subsequently through accession at the dates indicated hereafter: Japan (28th April 1964), Finland (28th January 1969), Australia (7th June 1971), New Zealand (29th May 1973), Mexico (18th May 1994), the Czech Republic (21st December 1995), Hungary (7th May 1996), Poland (22nd November 1996) and the Republic of Korea (12th December 1996). The Commission of the European Communities takes part in the work of the OECD (Article 13 of the OECD Convention).

AGENCE INTERNATIONALE DE L'ÉNERGIE
9, RUE DE LA FEDERATION, 75739 PARIS CEDEX 15, FRANCE

L'agence internationale de l'énergie (AIE) est un organe autonome institué en novembre 1974 dans le cadre de l'Organisation de coopération et de développement économiques (OCDE) afin de mettre en œuvre un programme international de l'énergie.

Elle applique un programme général de coopération dans le domaine de l'énergie entre vingt-quatre* des vingt-neuf pays Membres de l'OCDE. Les objectifs fondamentaux de l'AIE sont les suivants :

- tenir à jour et améliorer des systèmes permettant de faire face à des perturbations des approvisionnements pétroliers ;
- œuvrer en faveur de politiques énergétiques rationnelles dans un contexte mondial grâce à des relations de coopération avec les pays non membres, l'industrie et les organisations internationales ;
- gérer un système d'information continue sur le marché international du pétrole ;
- améliorer la structure de l'offre et de la demande mondiales d'énergie en favorisant la mise en valeur de sources d'énergie de substitution et une utilisation plus rationnelle de l'énergie ;
- contribuer à l'intégration des politiques d'énergie et d'environnement.

Pays Membres de l'AIE : Allemagne, Australie, Autriche, Belgique, Canada, Danemark, Espagne, États-Unis, Finlande, France, Grèce, Hongrie, Irlande, Italie, Japon, Luxembourg, Norvège, Nouvelle-Zélande, Pays-Bas, Portugal, Royaume-Uni, Suède, Suisse et Turquie. La Commission des Communautés européennes participe également aux travaux de l'AIE.

ORGANISATION DE COOPÉRATION
ET DE DÉVELOPPEMENT ÉCONOMIQUES

En vertu de l'article 1er de la Convention signée le 14 décembre 1960, à Paris, et entrée en vigueur le 30 septembre 1961, l'Organisation de Coopération et de Développement Économiques (OCDE) a pour objectif de promouvoir des politiques visant :

- à réaliser la plus forte expansion de l'économie et de l'emploi et une progression du niveau de vie dans les pays Membres, tout en maintenant la stabilité financière, et à contribuer ainsi au développement de l'économie mondiale ;
- à contribuer à une saine expansion économique dans les pays Membres, ainsi que les pays non membres, en voie de développement économique ;
- à contribuer à l'expansion du commerce mondial sur une base multilatérale et non discriminatoire conformément aux obligations internationales.

Les pays Membres originaires de l'OCDE sont : l'Allemagne, l'Autriche, la Belgique, le Canada, le Danemark, l'Espagne, les États-Unis, la France, la Grèce, l'Irlande, l'Islande, l'Italie, le Luxembourg, la Norvège, les Pays-Bas, le Portugal, le Royaume-Uni, la Suède, la Suisse et la Turquie. Les pays suivants sont ultérieurement devenus Membres par adhésion aux dates indiquées ci-après : le Japon (28 avril 1964), la Finlande (28 janvier 1969), l'Australie (7 juin 1971), la Nouvelle-Zélande (29 mai 1973), le Mexique (18 mai 1994), la République tchèque (21 décembre 1995), la Hongrie (7 mai 1996), la Pologne (22 novembre 1996) et la République de Corée (12 décembre 1996). La Commission des Communautés européennes participe aux travaux de l'OCDE (article 13 de la Convention de l'OCDE).

TABLE OF CONTENTS

PART I: METHODOLOGY

PART II: STATISTICAL DATA

ANNUAL TABLES 1997-1998

SUMMARY TABLES

OIL DEMAND BY PRODUCT

TABLE DES MATIERES

TABLEAUX RECAPITULATIFS

PRODUITS PETROLIERS

ABBREVIATIONS

Btu:	British thermal unit
GWh:	gigawatt hour
kcal:	kilocalorie
kg:	kilogramme
kJ:	kilojoule
Mt:	million tonnes
m^3:	cubic metre
t:	metric ton = tonne = 1000 kg
TJ:	terajoule
toe:	tonne of oil equivalent = 10^7 kcal
GDP	Gross Domestic Product
CHP:	combined heat and power
GCV:	gross calorific value
HHV:	higher heating value = GCV
LHV:	lower heating value = NCV
NCV:	net calorific value
PPP:	purchasing power parity
AfDB	African Development Bank
EU:	European Union
FAO	Food and Agriculture Organisation of the United Nations
IEA:	International Energy Agency
OECD:	Organisation for Economic Co-Operation and Development
OLADE:	Organización Latinoamericana de Energía
UN:	United Nations
IPCC	Intergovernmental Panel on Climate Change
ISIC	International Standard Industrial Classification
UNIPEDE	International Union of Producers and Distributors of Electrical Energy
-	not applicable, nil or not available

Note	**See multilingual pullout at the end of the publication.**
Attention	**Voir le dépliant en plusieurs langues à la fin du présent recueil.**
Achtung	**Aufklappbarer Text auf der letzten Umschlagseite.**
Attenzione	**Riferirsi al glossario poliglotta alla fine del libro.**
注意事項	**巻末の日本語の折り込みページを参照**
Nota	**Véase el glosario plurilingüe al final del libro.**
Примеч.	**Смотрите многоязычный словарь в конце книги.**

ABREVIATIONS

Btu:	British thermal unit
GWh:	gigawattheure
kcal:	kilocalorie
kg:	kilogramme
kJ:	kilojoule
Mt:	million de tonnes
m^3:	mètre cube
t:	tonne métrique = 1000 kg
tep:	tonne d'équivalent pétrole = 10^7 kcal
TJ:	térajoule
PCI:	pouvoir calorifique inférieur
PCS:	pouvoir calorifique supérieur
PPA:	parité de pouvoir d'achat
AIE:	Agence Internationale de l'Energie
BAfD:	Banque Africaine de développement
FAO:	Organisation des Nations Unies pour l'alimentation et l'agriculture
OCDE:	Organisation de Coopération et de Développement Economiques
OLADE:	Organización Latinoamericana de Energía
ONU:	Organisation des Nations Unies
UE:	Union Européenne
CITI	Classification Internationale Type par Industrie
GIEC	Groupe d'experts intergouvernemental sur l'évolution du climat
UNIPEDE	Union Internationale des Producteurs et Distributeurs d'Energie Electrique
-	sans objet, néant ou non-disponible

Attention	**Voir le dépliant en plusieurs langues à la fin du présent recueil.**
Note	**See multilingual pullout at the end of the publication.**
Achtung	**Aufklappbarer Text auf der letzten Umschlagseite.**
Attenzione	**Riferirsi al glossario poliglotta alla fine del libro.**
注意事項	巻末の日本語の折り込みページを参照
Nota	**Véase el glosario plurilingüe al final del libro.**
Примеч.	Смотрите многоязычный словарь в конце книги.

INTRODUCTION

This publication is intended for those involved in analytical and policy work related to international energy issues. It provides detailed statistics on production, trade and consumption for each source of energy in more than 100 non-OECD countries, and main regions, including developing countries, Central and Eastern European countries and the former USSR. Starting with this issue, the data for Togo and Eritrea will also be shown. The consistency and complementarity of OECD and non-OECD countries' statistics ensure an accurate picture of the global energy situation.

The data shown in this publication for the Member countries of the Economic Commission for Europe of the United Nations (UN-ECE) are based mostly on information provided in four annual questionnaires common to the OECD, the UN-ECE and the European Union: "Oil", "Natural Gas", "Solid Fuels, Wastes and Manufactured Gases", and "Electricity and Heat" completed by the national administrations. The commodity balances for the other countries are based on national energy data of heterogeneous nature which have been converted and adjusted to fit the IEA's format. This volume has been prepared in close collaboration with other international organisations including the Organizacíon Latino Americana De Energía (OLADE), the Asia Pacific Energy Research Centre, the Statistical Office of the United Nations and the Forestry Department of the Food and Agriculture Organisation of the United Nations. It draws upon and complements the extensive work of the United Nations in the field of world energy statistics.

While every effort is made to ensure the accuracy of the data, quality is not homogeneous throughout the publication. Special methodological issues arise in a number of countries. In some countries data are based on secondary sources, and where incomplete or unavailable, on estimates. In general, data are likely to be more accurate for production, trade and total consumption than for individual sectors in transformation or final consumption. Commodity balances are presented in two formats reflecting the degree of detail available. Moreover, the breakdown by fuel of electricity and heat production in the transformation sector provided in the *Energy Statistics of OECD Countries* is not shown in this publication. General issues of data quality, country notes and individual country data should be consulted when using regional aggregates.

A companion volume – *Energy Balances of Non-OECD Countries* – presents corresponding data in comprehensive balances expressed in a common unit, million tonnes of oil equivalent (Mtoe), with $1\ toe = 10^7\ kcal = 41.868$ gigajoules.

Energy data on OECD and non-OECD countries are collected by the team in the Energy Statistics Division (ESD) of the IEA Secretariat, headed by Mr. Jean-Yves Garnier. Non-OECD countries statistics are currently the responsibility of Mr. Yannis Yaxas, Ms. Yukimi Shimura, Ms. Stephane de la Rue du Can, and Ms. Rikke Jacobsen. Dr. Sohbet Karbuz has overall editorial responsibility. Secretarial support was supplied by Ms. Sharon Michel and Ms. Susan Stolarow.

We would like to thank our numerous contacts world-wide in national administrations, and in public and private companies for their helpful co-operation.

Complete supply and consumption data from 1971 to 1998 are available on diskettes suitable for use on IBM-compatible personal computers. An order form has been provided in the back of this publication.

In addition, a data service will be available on the internet some time in 2000. It will include unlimited access through an annual subscription as well as the possibility to obtain data or a pay-per-view basis. Details will be available at http://www.iea.org.

Enquiries about data, methodology, or comments and suggestions should be addressed to:

Dr. Sohbet Karbuz
Energy Statistics Division
International Energy Agency
9 rue de la Fédération, 75739 Paris Cedex 15, France

Telephone: (+33-1) 40-57-66-34
Fax: (+33-1) 40-57-66-49
E-mail: sohbet.karbuz@iea.org or wed@iea.org

INTRODUCTION

Ce recueil s'adresse aux lecteurs qui participent aux travaux d'analyse et d'étude des questions de fond concernant la situation énergétique internationale. Elle fournit, pour chaque source d'énergie, des statistiques détaillées sur la production, les échanges et la consommation dans plus de 100 pays ne faisant pas partie de l'OCDE ainsi que pour plusieurs régions qui comprennent des pays en développement, des pays d'Europe centrale et orientale et l'ex-URSS. Il convient de souligner que la présente édition a été enrichie des données relatives au Togo et à l'Erythrée. La compatibilité et la complémentarité des statistiques des pays membres de l'OCDE et des pays ne faisant pas partie de l'OCDE garantissent la qualité de l'image de la situation énergétique mondiale.

Les données publiées sur les pays membres de la Commission Economique pour l'Europe des Nations Unies (CEE-ONU) sont basées sur les informations recueillies dans quatre questionnaires annuels communs de l'OCDE, de la CEE-ONU et de l'Union Européenne : « Pétrole », « Gaz naturel », « Combustibles solides, déchets et gaz manufacturés » et « Electricité et chaleur », remplis par les administrations nationales. Les bilans par produit des autres pays sont basés sur des statistiques énergétiques nationales de nature hétérogène qui ont été converties et harmonisées afin d'être conformes au format de l'AIE. Ce document a été préparé en étroite collaboration avec d'autres organisations internationales dont l'Organización Latino Americana De Energía (OLADE), le Asia Pacific Energy Research Centre, le Bureau de Statistiques des Nations Unies, et le Département des forêts de l'Organisation des Nations Unies pour l'Alimentation et l'Agriculture. Il utilise et complète le travail considérable déjà accompli par les Nations Unies dans le domaine des statistiques énergétiques.

Bien que tout ait été mis en œuvre pour assurer l'exactitude de ces données, la qualité des chiffres de cette publication n'est pas toujours homogène. Des problèmes méthodologiques particuliers se posent dans un certain nombre de pays. Par ailleurs pour certains pays les données proviennent de sources secondaires et, lorsque les données manquent, sont basées sur des estimations. D'une façon générale, les chiffres sont sans doute plus exacts en ce qui concerne la production, les échanges et la consommation totale que pour les secteurs désagrégés de la transformation ou de la consommation finale. Les bilans par produit sont présentés en deux formats différents selon les détails disponibles. De plus, la ventilation par combustible de la production d'électricité et de chaleur du secteur transformation fournie dans les *Statistiques de l'énergie des pays de l'OCDE* n'est pas présentée dans la présente publication. Il convient, lors de l'utilisation des agrégats régionaux, de consulter également la note générale sur la qualité des données, les notes relatives aux différents pays et les données par pays.

Un recueil complémentaire - *Bilans énergétiques des pays non-membres 1997-1998* - contient des données équivalentes et exprimées sous la forme de bilans globaux dans une unité commune, à savoir en millions de tonnes d'équivalent pétrole (Mtep), sur la base de 1 tep = 10^7 kcal = 41,868 gigajoules.

Les données énergétiques sur les pays membres et non-membres de l'OCDE sont collectées par la Division des statistiques énergétiques (ESD) du Secrétariat de l'AIE, dirigée par M. Jean-Yves Garnier. M. Yannis Yaxas, Mlle Yukimi Shimura, Mlle Stéphane de la Rue du Can et Mlle Rikke Jacobsen sont actuellement responsables des statistiques des pays non-membres de l'OCDE.

Dr. Sohbet Karbuz est responsable de la publication. Mme Sharon Michel et Mme Susan Stolarow ont assuré le secrétariat d'édition et la saisie des textes.

Nous aimerions remercier les personnes des adminsitrations nationales avec qui nous sommes en contact pour leur aide précieuse.

Des données complètes sur l'offre et la demande sont disponibles, pour les années 1971 à 1998, sur disquettes exploitables par des ordinateurs personnels compatibles IBM. Un formulaire de commande est fourni à la fin de cet ouvrage.

En outre, un service de données sera disponible sur internet dans le courant de l'année 2000. Ce service comprendra une souscription annuelle pour un accès illimité ou bien la possibilité de payer uniquement pour des données sélectionnées. Pour plus de détails, veuillez consulter http://www.iea.org.

Les demandes de renseignements sur les données ou la méthodologie doivent être adressées au chef de la section des pays non-membres, division des statistiques de l'énergie :

Dr. Sohbet Karbuz
Division des statisitiques de l'énergie
Agence Internationale de l'Energie
9 rue de la Fédération, 75739 Paris Cedex 15, France

Téléphone : (+33-1) 40-57-66-34
Fax : (+33-1) 40-57-66-49
E-mail : sohbet.karbuz@iea.org ou wed@iea.org

PART I:
METHODOLOGY

PARTIE I :
METHODOLOGIE

1. ISSUES OF DATA QUALITY

A. Methodology

Considerable effort has been made to ensure that data presented in this publication adhere to the IEA definitions contained in the General Notes (Part I.2 and 3). These definitions are used by most of the international organisations that collect energy statistics. Nevertheless, the national energy statistics which are reported to international organisations are often collected using criteria and definitions which differ, sometimes considerably, from those employed by the international organisations. The extent to which the IEA Secretariat has identified these differences and, where possible, adjusted the data to meet international definitions, is outlined below. Recognized anomalies occurring in specific countries are presented in Part I.6, Country Notes and Sources. The Country Notes simply identify some of the more important and obvious deviations from IEA methodology in certain countries and are by no means a comprehensive list of anomalies by country.

B. Estimation

In addition to any adjustments undertaken to compensate for differences in definitions, estimations are sometimes required to complete major aggregates from which key statistics are missing. Examples may be found in the more detailed accounts given below. Except as concerns combustible renewables and waste (see E.5 below), it has been the Secretariat's aim to provide all of the elements of commodity balances down to the level of Final Consumption for all countries and years. This entails providing the elements of supply as well as inputs of primary fuels and outputs of secondary fuels to and from the main transformation activities such as oil refining and electricity generation. This has often required estimations prepared after consultation with national statistical offices, oil companies, electricity utilities and national energy experts.

C. Time Series and Political Changes

Commodity balances for the republics of the former USSR have been constructed since 1992. These balances have been constructed from official data and, where necessary, estimates have been made based on information obtained from industry sources and other international organisations. Summary tables and commodity balances for the country 'former USSR' are now the sum of individual former USSR republics. In previous editions of this publication, intra-USSR trade was excluded and average net calorific values were applied to the aggregate data instead of the country-specific factors. This is also the case for the country 'Former Yugoslavia'.

Commodity balances for the Slovak Republic have been constructed since 1971. For the years 1989 to 1998, the balances for the Slovak Republic have been constructed from official data. For the years 1971 to 1988, estimates have been made where necessary (see Country Notes and Sources).

Energy statistics for some countries undergo countinuous changes in their coverage or methodology so that, for example, it is possible to present more detailed energy accounts for China beginning in 1980. Consequently, "breaks in series" are considered to be unavoidable.

The IEA Secretariat reviews its databases each year. In the light of new assessments, important revisions are made to the time series of individual countries during the course of this review. Therefore, data in this publication have been substantially revised with respect to previous editions. Please see country notes on that.

D. Classification of Fuel Uses

National statistics sources often lack adequate information on the consumption of fuels in different categories of end use. Many countries do not conduct annual surveys of the fuel consumption in the main sectors of economic activity and consequently published data are based on out-of-date surveys. Sectoral disaggregation of consumption for individual countries should therefore be interpreted with caution.

Previous to the reforms undertaken in the 1990's the sectoral classification of fuel consumption in the transition economies (eastern Europe and the countries of the former USSR) and China differed greatly from that practiced in market economies. Sectoral consumption was defined according to the economic branch to which the user of the fuel belonged rather than according to the purpose or use of the fuel. Consumption of gasoline in the vehicle fleet of an enterprise attached to the economic branch 'iron and steel' was classified as industrial consumption of gasoline in the iron and steel industry. Where possible, the data have been adjusted to fit international classifications, for example, all gasoline is assumed to be consumed in the 'transport' sector and is reported in this publication as such. However, it has not been possible to reclassify products other than gasoline and jet fuel with the same facility, and few adjustments have been made to the remaining products.

E. Specific Issues by Fuel

1. Oil

The IEA Secretariat collects comprehensive statistics for oil supply and use, including refineries' own use of oil, oil delivered to ships' bunkers and oil for use as petrochemical feedstock. National statistics often do not report these amounts. Reported production of refined products frequently refers to net rather than gross refinery output and consumption of oil products may be limited to sales to domestic markets and may not include deliveries to international shipping or aircraft. Oil consumed as petrochemical feedstock in integrated refinery/petrochemical complexes is often not included in available official statistics.

Where possible, the Secretariat, in consultation with the oil industry, makes estimates of these unreported data. In the absence of any other indication of refinery fuel use the consumption has been estimated to be about 5 per cent of refinery throughput and has been divided equally between refinery gas and heavy fuel oil.

For a description of the nature of the adjustments made to the sectoral consumption of oil products, see above section 'Classification of Fuel Uses'.

2. Natural Gas

The IEA defines natural gas production as marketable production, i.e. that which is net of field losses, flaring, venting and reinjection. Furthermore, natural gas should be comprised mainly of methane and other gases such as ethane and higher hydrocarbons should be reported under the heading of 'oil'.

Readily available data for natural gas, however, often do not identify the separate elements of field losses, flaring, venting and reinjection. Moreover, reported data are frequently unaccompanied by adequate definitions so that it is difficult or impossible to identify gas at the different stages of its separation into dry gas (methane) and the heavier fractions in gas separation facilities.

Natural gas supply and demand statistics are normally reported in volumetric units and it is difficult to obtain accurate data on the calorific value of the gas. Where the heat value is unknown, a general gross calorific value of 38 TJ/million m^3 has been applied.

It should also be noted that reliable consumption data for natural gas at a disaggregated level are often difficult to locate. This is especially true of some of the larger natural gas consuming countries in the Middle East. Industrial use of natural gas for these countries is therefore frequently missing from data appearing in this publication.

3. Electricity

As defined in the General Notes (Part I.2), an autoproducer of electricity is an establishment which, in addition to its main activities, generates electricity wholly or partly for its own use. Data on the basis of this definition are frequently unknown in non-OECD countries. In such cases the fuels inputs for autoproduced electricity are reported in the appropriate end use sector.

When statistics of the production of electricity from inputs of Combustible Renewables and Waste are available, they are included in total electricity production. These data are not comprehensive; for example, much of the electricity generated from waste biomass in sugar refining remains unreported.

Inputs of fuels for electricity generation are estimated (when unreported) using information on electricity output, fuel efficiency and the type of generation capacity.

4. Heat

The transition economies (eastern Europe and countries of the former USSR) and China employed a methodology with respect to heat that set them apart from common practice in market economies. The approach taken was to allocate the transformation of primary fuels (coal, oil and gas) by industry into heat *for consumption on site* to the transformation activity *Heat Production*, not to industrial consumption as in the IEA methodology[1]. The transformation output of *Heat* was then allocated to the various end use sectors. The losses occurring in the transformation of fuels into heat in industry were not included in final consumption of industry. Although a number of countries have recently switched to the practice of the international organisations, this important distinction reduces the validity of cross-country comparisons of sectoral end use consumption between transition economies and market economies.

5. Combustible Renewables and Waste

The IEA publishes production and domestic supply of combustible renewables and waste for all non-OECD countries and all regions for the years 1974 to 1998 (1971 to 1998 on diskettes).

---

1 The international methodology restricts the inclusion of heat in transformation sector to that sold to third parties. See definition in Part I.2.

Data shown are often from secondary sources, inconsistent and may be of questionable quality, which makes comparisons between countries difficult. The historical data of many countries derive from surveys which were often irregular, irreconcilable and conducted at the local rather than national level making them incomparable between regions and time. Where historical series are incomplete or unavailable, they have been estimated using the methodology consistent with the projection framework of the IEA's 1998 edition of *World Energy Outlook* (September 1998). First, nationwide biomass and wastes domestic supply per capita was compiled or estimated for 1995. Secondly, per capita supply for the years 1971 to 1994 was estimated, using a log/log equation with either GDP per capita or the percentage of urban population as the exogenous variable, depending on the region. Thirdly, total biomass and waste supply after 1996 was estimated assuming a growth rate either constant, or equal to the population growth, or following the 1971-1994 estimation. However, these estimated time-series should be treated very cautiously.

The chart below attempts to provide a broad indication of the data quality and estimation methodology by region.

Region	Main source of data	Data Quality	Exogenous Variables
Africa	FAO database and AfDB	Low	% urban population
Latin America	National and OLADE	High	None
Asia	Surveys	High to Low	GDP per capita
Non-OECD Europe	Questionnaires and FAO	High to Medium	None
Former USSR	National and questionnaires and FAO	High to Medium	None
Middle East	FAO	Medium to Low	None

Although the methods used for estimations are consistent with those used for biomass in the *World Energy Outlook 1998*, minor numerical discrepancies occur for a number of reasons. The reasons include later data revisions and the level of country disaggregations employed.

For the years 1994 to 1998, balances down to final consumption by end-use for individual products or product categories have been compiled for all the countries. The data for the years 1997 and 1998 are shown in the Annual Tables. Charcoal production is shown in the Summary Tables. The figures confirm the importance of vegetal fuels in the energy sector of many developing countries.

The IEA hopes that the inclusion of these data will encourage national administrations and other agencies active in the field to enhance the level and quality of data collection and coverage. More details on the methodology used by country may be provided on request, and comments are welcome.

2. GENERAL NOTES

The tables include all "commercial" sources of energy, both primary (hard coal, brown coal/lignite, peat, natural gas, crude oil, NGL, hydro, geothermal/solar, wind, tide, etc. and nuclear power) and secondary (coal products, manufactured gases, petroleum products, electricity and heat). Data also include various sources of combustible renewables and waste, such as solid biomass and animal products, gas/liquids from biomass, municipal waste and industrial waste.

Each table is divided into three main parts: the first showing *supply* elements, the second showing the *transformation* and *energy* sectors, and the third showing *final consumption* broken down into the various end-use sectors.

The following description refers to the layout of the full commodity balance type presentation. In the case of the regional and abbreviated tables, the definition of the products is the same. However, sectoral breakdown has been restricted to main totals (Energy Sector, etc.) and to selected main categories (Iron and Steel, etc.).

A. Supply

The first part of the basic energy balance shows the following elements of supply:

   Production
+ *Inputs from other sources*
+ *Imports*
- *Exports*
- *International marine bunkers*
± *Stock changes*
= *Domestic supply*

1. Production

Production refers to the quantities of fuels extracted or produced, calculated after any operation for removal of inert matter or impurities (e.g. sulphur from natural gas).

2. Inputs from Other Sources

All inputs of origin other than primary energy sources explicitly recognised in the tables are listed under inputs *from other sources*, e.g. under crude oil: inputs of origin other than crude oil and NGL such as hydrogen, synthetic crude oil (including mineral oil extracted from bituminous minerals such as shales, bituminous sand, etc.); under additives: benzol, alcohol and methanol produced from natural gas; under refinery feedstocks: backflows from the petrochemical industry used as refinery feedstocks; under hard coal: recovered slurries, middlings, recuperated coal dust and other low-grade coal products that cannot be classified according to type of coal from which they are obtained; under gas works gas: natural gas, refinery gas, and LPG, that are treated or mixed in gas works (i.e. gas works gas produced from sources other than coal).

3. Imports and Exports

Imports and exports comprise amounts having crossed the national territorial boundaries of the country whether or not customs clearance has taken place.

a) Coal

Imports and exports comprise the amount of fuels obtained from or supplied to other countries, whether or not there is an economic or customs union between the relevant countries. Coal in transit should not be included.

b) Oil and Gas

Quantities of crude oil and oil products imported or exported under processing agreements (i.e. refining on account) are included. Quantities of oil in transit are excluded. Crude oil, NGL and natural gas are reported as coming from the country of origin; refinery feedstocks and oil products are reported as coming from the country of last consignment.

Re-exports of oil imported for processing within bonded areas are shown as an export of product from the processing country to the final destination.

c) Electricity

Amounts are considered as imported or exported when they have crossed the national territorial boundaries of the country. If electricity is "wheeled" or transited through a country, the amount is shown as both an import and an export

4. International Marine Bunkers

International marine bunkers cover those quantities delivered to sea-going ships of all flags, including warships. Consumption by ships engaged in transport in inland and coastal waters and by fishing vessels in all waters is not included. See definitions of transport (Section 2.C.2) and agriculture (Section 2.C.3).

5. Stock Changes

Stock changes reflect the difference between opening stock levels on the first day of the year and closing levels on the last day of the year of stocks on national territory held by producers, importers, energy transformation industries and large consumers. Oil and gas stock changes in pipelines are not taken into account. With the exception of large users mentioned above, changes in final users' stocks are not taken into account. A stock build is shown as a negative number, and a stock draw as a positive number.

6. Domestic Supply

Domestic supply is defined as production + inputs from other sources + imports - exports - international marine bunkers ± stock changes.

7. Transfers

Transfers comprise *interproduct transfers*, *products transferred* and *recycled products*.

Interproduct transfers result from reclassification of products either because their specification has changed or because they are blended into another product, e.g. kerosene may be reclassified as gasoil after blending with the latter in order to meet its winter diesel specification. The net balance of *interproduct transfers* is zero.

Products transferred is intended for petroleum products imported for further processing in refineries. For example, fuel oil imported for upgrading in a refinery is transferred to the feedstocks category.

Recycled products are finished products which pass a second time through the marketing network, **after** having been once delivered to final consumers (e.g. used lubricants which are reprocessed).

8. Statistical Differences

Statistical difference is defined as deliveries to final consumption + use for transformation and consumption within the energy sector + distribution losses – domestic supply – transfers. Statistical differences arise because the data for the individual components of supply are often derived from different data sources by the national administration. Furthermore, the inclusion of changes in some large consumers' stocks in the supply part of the balance introduces distortions which also contribute to the statistical differences.

B. Transformation and Energy Sectors

The *transformation sector* comprises the conversion of primary forms of energy to secondary and further transformation (e.g. coking coal to coke, crude oil to petroleum products, heavy fuel oil to electricity).

The *energy sector* comprises the amount of fuels used by the energy producing industries (e.g. for heating, lighting and operation of all equipment used in the extraction process, for traction and for distribution).

The following categories are distinguished in the *transformation* and *energy* sectors:

1. Transformation Sector

- *Electricity plants* (refers to plants which are designed to produce electricity only). If one or more units of the plant is a CHP unit (and the inputs and outputs can not be distinguished on a unit basis) then the whole plant is designated as a CHP plant. Both public[2] and autoproducer[3] plants are included here.

- *Combined heat and power plants* (refers to plants which are designed to produce both heat and electricity). UNIPEDE refers to these as co-generation power stations. If possible, fuel inputs and electricity/heat outputs are on a unit basis rather than on a plant basis. However, if data are not available on a unit basis, the convention for defining a CHP plant noted above should be adopted. Both public and autoproducer plants are included here. *Note that for autoproducer's CHP plants, all fuel inputs to electricity production are taken into account, while only the part of fuel inputs to heat **sold** is shown. Fuel inputs for the production of heat consumed within the autoproducer's establishment are **not** included here but are included with figures for the final consumption of fuels in the appropriate consuming sector.*

- *Heat plants* (refers to plants [including heat pumps and electric boilers] designed to produce heat only and who sell heat to a third party [e.g. residential, commercial or industrial consumers] under the provisions of a contract). Both public and autoproducer plants are included here.

- *Blast furnaces/Gas works* (covers the quantities of fuels used for the production of town gas, blast furnace gas and oxygen steel furnace gas). The production of pig-iron from iron ore in blast furnaces uses fuels for supporting the blast furnace charge and providing heat and carbon for the reduction of the iron ore. Accounting for the calorific content of the fuels entering the process is a complex matter as transformation (into blast furnace gas) and consumption (heat of combustion) occur simultaneously. Some carbon is also retained in the pig-iron; almost all of this reappears later in the oxygen steel furnace gas (or converter gas) when the pig-iron

is converted to steel. In principle, the quantities of all fuels (e.g. pulverised coal injection (PCI) coal, coke oven coke, natural gas and oil) entering blast furnaces and the quantity of blast furnace gas and oxygen steel furnace gas produced are collected. However, except for coke oven coke inputs, the data are often uncertain or incomplete. The Secretariat then needs to split these inputs into the transformation and consumption components. The transformation component is shown in the row *blast furnaces/gas works* in the column appropriate for the fuel, and the consumption component is shown in the row *iron and steel* in final consumption in the column appropriate for the fuel. Starting with 1999, the Secretariat decided to assume an 80/20 split of coke oven coke between the transformation and consumption components. This results in more of the energy inputs appearing in the *blast furnaces/gas works* row and less appearing in the *iron and steel* row than in the previous editions. This split has been estimated based on the results of the model developed for OECD countries.[4] It provides a more consistent balancing of the carbon input into and output from the blast furnaces when the IEA data are used to calculate CO_2 emissions from fuel combustion using the Intergovernmental Panel on Climate Change (IPCC) methodology as published in the *Revised 1996 IPCC Guidelines for National Greenhouse Gas Inventories.*[5]

- *Coke/Patent fuel/BKB plants* (covers the use of fuels for the manufacture of coke, coke oven gas, patent fuels and BKB).

- *Petroleum refineries* (covers the use of hydrocarbons for the manufacture of finished petroleum products).

- *Petrochemical industry* (covers backflows returned from the petrochemical sector). Note, backflows from oil products that are used for non-energy purposes (i.e. white spirit and lubricants) are not included here, but in non-energy use.

---

2   Public supply undertakings generate electricity and/or heat for sale to third parties, *as their primary activity*. They may be privately or publicly owned. Note that the sale need not take place through the public grid.

3   Autoproducer undertakings generate electricity and/or heat, wholly or partly for their own use as an activity which supports their primary activity. They may be privately or publicly owned.

---

4   Please refer to *Energy Statistics of OECD Countries* for more information on the blast furnace model used for the OECD countries.

5   The *Revised 1996 IPCC Guidelines for National Greenhouse Gas Inventories* are available from the IPCC National Greenhouse Gas Inventories Programme at http://www.ipcc-nggip.iges.or.jp.

- *Liquefaction* (includes diverse liquefaction processes, such as coal and natural gas liquefaction in South Africa).

- *Other transformation sector* (covers charcoal burners losses and other non-specified transformation).

2. Energy Sector

Energy producing industries' own use includes energy consumed by transformation industries for heating, pumping, traction and lighting purposes [ISIC[6] Divisions 10, 11, 12, 23 and 40]:

- *Coal mines* (hard coal and lignite);

- *Oil and gas extraction* (flared gas is not included);

- *Petroleum refineries*;

- *Electricity, CHP and heat plants*;

- *Pumped storage* (electricity consumed in hydro-electric plants);

- *Other energy sector* (including own consumption in patent fuel plants, coke ovens, gas works, BKB and lignite coke plants as well as the non-specified energy sector's use).

3. Distribution losses

Distribution losses include losses in gas distribution, electricity transmission, and coal transport. It may also include unaccounted for use of crude oil and petroleum products.

C. Final Consumption

The term *final consumption* (equal to the sum of end-use sectors' consumption) implies that energy used for transformation and for own use of the energy producing industries is excluded. Final consumption reflects for the most part deliveries to consumers (see note on stock changes in Section 2.A.5).

In final consumption, petrochemical feedstocks are covered under **industry** as an *of which* item under chemical industry for those oil products that are

principally used for energy purposes. Separated from these are the other oil products that are mainly used for non-energy purposes (see non-energy use in Section 2.C.4), which are shown in the rows for non-energy uses and included only in **total final consumption**. Backflows from the petrochemical industry are not included in final consumption (see Sections 2.A.2 and 2.B.1).

1. Industry Sector

Consumption of the *industry sector* is specified in the following sub-sectors (energy used for transport by industry is not included here but is reported under transport):

- *Iron and steel industry* [ISIC Group 271 and Class 2731];

- *Chemical industry* [ISIC Division 24];

- *of which: petrochemical feedstocks.* The petrochemical industry includes cracking and reforming processes for the purpose of producing ethylene, propylene, butylene, synthesis gas, aromatics, butadene and other hydrocarbon-based raw materials in processes such as steam cracking, aromatics plants and steam reforming. [Part of ISIC Group 241]; See feedstocks under Section 2.C.4 (Non-energy use).

- *Non-ferrous metals* basic industries [ISIC Group 272 and Class 2732];

- *Non-metallic mineral* products such as glass, ceramic, cement, etc. [ISIC Division 26];

- *Transport equipment* [ISIC Divisions 34 and 35];

- *Machinery.* Fabricated metal products, machinery and equipment other than transport equipment [ISIC Divisions 28, 29, 30, 31 and 32];

- *Mining (excluding fuels) and quarrying* [ISIC Divisions 13 and 14];

- *Food and tobacco* [ISIC Divisions 15 and 16];

- *Paper, pulp and print* [ISIC Divisions 21 and 22];

- *Wood and wood products* (other than pulp and paper) [ISIC Division 20];

- *Construction* [ISIC Division 45];

---

6 International Standard Industrial Classification of All Economic Activities, Series M, No. 4/Rev. 3, United Nations, New York, 1990.

- *Textile and leather* [ISIC Divisions 17, 18 and 19];

- *Non-specified* (any manufacturing industry not included above) [ISIC Divisions 25, 33, 36 and 37];

Note: Most countries have difficulties supplying an industrial breakdown for all fuels. In these cases, the *non-specified* industry row has been used. *Regional aggregates of industrial consumption should therefore be used with caution.*

2.   Transport Sector

Consumption in the *Transport sector* covers all transport activity in mobile engines regardless of the economic sector to which it is contributing [ISIC Divisions 60, 61 and 62], and is divided into the following sub-sectors:

- *Air*: Deliveries of aviation fuels to international civil aviation and to all domestic air transport, commercial, private, agricultural, military, etc. It also includes use for purposes other than flying, e.g. bench testing of engines, but not use of fuels for road transport by airline companies;

- *Road*: All fuels used in road vehicles (including military) as well as agricultural and industrial highway use. Excludes motor gasoline used in stationary engines, and diesel oil for use in tractors that are not for highway use;

- *Rail*: All quantities used in rail traffic, including industrial railways;

- *Pipeline transport*: Energy used for transport of materials by pipeline;

- *Internal navigation* (including small craft and coastal vessels not purchasing their bunker requirements under international marine bunker contracts). Fuel used for ocean, coastal and inland fishing should be included in agriculture;

- *Non-specified*.

- Note: Most countries have difficulties supplying an industrial breakdown for all fuels. In these cases, the *non-specified* industry row has been used. *Regional aggregates of industrial consumption should therefore be used with caution.* Please see Country Notes (Part I.5)

3.   Other Sectors

- *Agriculture*: Defined as all deliveries to users classified as agriculture, hunting and forestry by the ISIC, and therefore includes energy consumed by such users whether for traction (excluding agricultural highway use), power or heating (agricultural and domestic). Also includes fuels used for ocean, coastal and inland fishing. ISIC Divisions 01, 02 and 05;

- *Commercial and public services*: All activities coming into ISIC Divisions 41, 50, 51, 52, 55, 63, 64, 65, 66, 67, 70, 71, 72, 73, 74, 75, 80, 85, 90, 91, 92, 93 and 99;

- *Residential*: All consumption by households, excluding fuels used for transport. Includes households with employed persons (ISIC Division 95) which is a small part of total residential consumption;

- *Non-specified*: Includes all fuel use not elsewhere specified (e.g. military fuel consumption with the exception of transport fuels in international marine bunkers, the domestic air and road sectors and consumption in the above-designated categories for which separate figures have not been provided).

4.   Non-energy use

Non-energy use covers use of *other* petroleum products such as white spirit, paraffin waxes, lubricants, bitumen and other products (see Sections 3.C.10 and 3.C.11). It also includes the non-energy use of coal (excluding peat). They are shown separately in final consumption under the heading *non-energy use*. It is assumed that the use of these products is exclusively non-energy use. It should be noted that petroleum coke is shown as *non-energy use* only when there is evidence of such use, otherwise it is shown under energy use in industry or in other sectors.

Feedstocks for petrochemical industry are accounted for in industry under chemical industry and shown separately under: *of which: feedstocks*). This covers all oil and gas, including naphtha except the *other petroleum products* listed in Section 3.C.11.

3. NOTES ON ENERGY SOURCES

A. Coal

The heading *coal mines* refers only to coal which is used directly within the coal industry. It excludes coal burned in pithead power stations (included under *transformation* - electricity plants) and free allocations to miners and their families (considered as part of household consumption and therefore included under *other sectors* - residential).

1. Coking Coal

Coking coal refers to coal with a quality that allows the production of a coke suitable to support a blast furnace charge. Its gross calorific value is greater than 23 865 kJ/kg (5 700 kcal/kg) on an ash-free but moist basis.

2. Other Bituminous Coal and Anthracite

Other bituminous coal is used for steam raising and space heating purposes and includes all anthracite coals and bituminous coals not included under coking coal. Its gross calorific value is greater than 23 865 kJ/kg (5 700 kcal/kg), but usually lower than that of coking coal.

3. Sub-Bituminous Coal

Non-agglomerating coals with a gross calorific value between 17 435 kJ/kg (4 165 kcal/kg) and 23 865 kJ/kg (5 700 kcal/kg) containing more than 31 per cent volatile matter on a dry mineral matter free basis.

4. Lignite and Brown Coal

Lignite/brown coal is a non-agglomerating coal with a gross calorific value of less than 17 435 kJ/kg (4 165 kcal/kg), and greater than 31 per cent volatile matter on a dry mineral matter free basis.

Oil shale and tar sands produced and combusted directly are included in this category. Oil shale and tar sands used as inputs for other transformation processes are also included here. This includes the portion of the oil shale or tar sands consumed in the transformation process. Shale oil and other products derived from liquefaction are included in *from other sources* under crude oil (*other hydrocarbons*).

5. Peat

Combustible soft, porous or compressed, fossil sedimentary deposit of plant origin with high water content (up to 90 per cent in the raw state), easily cut, of light to dark brown colour. Peat used for non-energy purposes is not included.

6. Coke Oven Coke and Gas Coke

Coke oven coke is the solid product obtained from the carbonisation of coal, principally coking coal, at high temperature. It is low in moisture content and volatile matter. Also included are semi-coke, a solid product obtained from the carbonisation of coal at a low temperature, lignite coke and semi-coke made from lignite/brown coal, coke breeze and foundry coke. The heading *other energy sector* represents consumption at the coking plants themselves. Consumption in the iron and steel industry does not include coke converted into blast furnace gas. To obtain the total consumption of coke oven coke in the iron and steel industry, the quantities converted into blast furnace gas have to be added (these are shown under *blast furnaces/gas works*).

Gas coke is a by-product of hard coal used for the production of town gas in gas works. Gas coke is used for heating purposes. *Other energy sector* data represent consumption of gas coke at gas works.

7. Patent Fuel and Brown Coal / Peat Briquettes (BKB)

Patent fuel is a composition fuel manufactured from hard coal mines with the addition of a binding agent. The amount of patent fuel produced is, therefore, slightly higher than the actual amount of coal consumed in the transformation process. Consumption of patent fuels during the patent fuel manufacturing process is shown under *other energy sector*.

BKB are composition fuels manufactured from lignite/brown coal, produced by briquetting under high pressure. These figures include peat briquettes, dried lignite fines and dust. The heading *other energy sector* includes consumption by briquetting plants.

B. Crude Oil, NGL, Refinery Feedstocks

Petroleum refineries under *transformation* shows inputs of crude oil, NGL, refinery feedstocks, additives and other hydrocarbons into the refining process.

1. Crude Oil

Crude oil is a mineral oil consisting of a mixture of hydrocarbons of natural origin, being yellow to black in colour, of variable density and viscosity. It also includes lease condensate (separator liquids) which are recovered from gaseous hydrocarbons in lease separation facilities.

Other hydrocarbons, including synthetic crude oil, mineral oils extracted from bituminous minerals such as shales, bituminous sand, etc., and oils from coal liquefaction are included in the row *from other sources*. See Section 2.A.2.

Emulsified oils (e.g. orimulsion) are included here.

2. Natural Gas Liquids (NGL)

NGLs are the liquid or liquefied hydrocarbons produced in the manufacture, purification and stabilisation of natural gas. These are those portions of natural gas which are recovered as liquids in separators, field facilities, or gas processing plants. NGLs include but are not limited to ethane, propane, butane, pentane, natural gasoline and condensate. They may also include small quantities of non-hydrocarbons.

3. Refinery Feedstocks

A refinery feedstock is a product or a combination of products derived from crude oil and destined for further processing other than blending in the refining industry. It is transformed into one or more components and/or finished products. This definition covers those finished products imported for refinery intake and those returned from the petrochemical industry to the refining industry.

4. Additives

Additives are non-hydrocarbon substances added to or blended with a product to modify its properties, for example, to improve its combustion characteristics. Alcohols and ethers (MTBE, methyl tertiary-butyl ether) and chemical alloys such as tetraethyl lead are included here. Ethanol is not included here, but under *Gas/Liquids from Biomass*.

C. Petroleum Products

Petroleum products are any oil-based products which can be obtained by distillation and are normally used outside the refining industry. The exceptions to this are those finished products which are classified as refinery feedstocks above.

Production of petroleum products shows gross refinery output for each product.

Refinery fuel (row *petroleum refineries*, under *energy sector*) represents consumption of petroleum products, both intermediate and finished, within refineries, e.g. for heating, lighting, traction, etc.

1. Refinery Gas (not liquefied)

Refinery gas is defined as non-condensable gas obtained during distillation of crude oil or treatment of oil products (e.g. cracking) in refineries. It consists mainly of hydrogen, methane, ethane and olefins. It also includes gases which are returned from the petrochemical industry. Refinery gas production refers to gross production. Own consumption is shown separately under *petroleum refineries* in the *energy* sector.

2. Liquefied Petroleum Gases (LPG) and Ethane

These are the light hydrocarbons fraction of the paraffin series, derived from refinery processes, crude oil stabilisation plants and natural gas processing plants comprising propane (C_3H_8) and butane (C_4H_{10}) or a combination of the two. They are normally liquefied under pressure for transportation and storage.

Ethane is a naturally gaseous straight-chain hydrocarbon (C_2H_6). It is a colourless paraffinic gas which is extracted from natural gas and refinery gas streams.

3. Motor Gasoline

This is light hydrocarbon oil for use in internal combustion engines such as motor vehicles, excluding aircraft. Motor gasoline is distilled between 35°C and 215°C and is used as a fuel for land based spark ignition engines. Motor gasoline may include additives, oxygenates and octane enhancers, including lead compounds such as TEL (Tetraethyl lead) and TML (tetramethyl lead).

4. Aviation Gasoline

Aviation gasoline is motor spirit prepared especially for aviation piston engines, with an octane number suited to the engine, a freezing point of -60°C, and a distillation range usually within the limits of 30°C and 180°C.

5. Jet Fuel

This category comprises both gasoline and kerosene type jet fuels meeting specifications for use in aviation turbine power units.

a) Gasoline type jet fuel

This includes all light hydrocarbon oils for use in aviation turbine power units. They distil between 100°C and 250°C. It is obtained by blending kerosenes and gasoline or naphthas in such a way that the aromatic content does not exceed 25 per cent in volume, and the vapour pressure is between 13.7 kPa and 20.6 kPa. Additives can be included to improve fuel stability and combustibility.

b) Kerosene type jet fuel

This is medium distillate used for aviation turbine power units. It has the same distillation characteristics and flash point as kerosene (between 150°C and 300°C but not generally above 250°C). In addition, it has particular specifications (such as freezing point) which are established by the International Air Transport Association (IATA).

6. Kerosene

Kerosene comprises refined petroleum distillate intermediate in volatility between gasoline and gas/diesel oil. It is a medium oil distilling between 150°C and 300°C.

7. Gas/Diesel Oil (Distillate Fuel Oil)

Gas/diesel oil includes heavy gas oils. Gas oils are obtained from the lowest fraction from atmospheric distillation of crude oil, while heavy gas oils are obtained by vacuum redistillation of the residual from atmospheric distillation. Gas/diesel oil distils between 180°C and 380°C. Several grades are available depending on uses: diesel oil for diesel compression ignition (cars, trucks, marine, etc.), light heating oil for industrial and commercial uses, and other gas oil including heavy gas oils which distil between 380°C and 540°C and which are used as petrochemical feedstocks.

8. Heavy Fuel Oil (Residual)

This heading defines oils that make up the distillation residue. It comprises all residual fuel oils, including those obtained by blending. Its kinematic viscosity is above 10 cSt at 80°C. The flash point is always above 50°C and the density is always more than 0.90 kg/l.

9. Naphtha

Naphtha is a feedstock destined either for the petrochemical industry (e.g. ethylene manufacture or aromatics production) or for gasoline production by reforming or isomerisation within the refinery. Naphtha comprises material in the 30°C and 210°C distillation range or part of this range. Naphtha imported for blending is shown as an import of naphtha, then shown in the transfers row as a negative entry for naphtha and a positive entry for the corresponding finished product (e.g. gasoline).

10. Petroleum Coke

Petroleum coke is defined as a black solid residue, obtained mainly by cracking and carbonising of petroleum derived feedstocks, vacuum bottoms, tar and pitches in processes such as delayed coking or fluid coking. It consists mainly of carbon (90 to 95 per cent) and has a low ash content. It is used as a feedstock in coke ovens for the steel industry, for heating purposes, for electrode manufacture and for production of chemicals. The two most important qualities are "green coke" and "calcinated coke". This category also includes "catalyst coke" deposited on the catalyst during refining processes: this coke is not recoverable and is usually burned as refinery fuel.

11. Other Petroleum Products

The category *other petroleum products* groups together white spirit and SBP, lubricants, bitumen, paraffin waxes and others.

a) *White Spirit and SBP:*

White spirit and SBP are refined distillate intermediates with a distillation in the naphtha/ kerosene range.

They are sub-divided as:

i) Industrial Spirit (SBP): Light oils distilling between 30°C and 200°C, with a temperature difference between 5 per cent volume and 90 per cent volume distillation points, including losses, of not more than 60°C. In other words, SBP is a light oil of narrower cut than motor spirit. There are 7 or 8 grades of industrial spirit, depending on the position of the cut in the distillation range defined above.

ii) White Spirit: Industrial spirit with a flash point above 30°C. The distillation range of white spirit is 135°C to 200°C.

b) *Lubricants:*

Lubricants are hydrocarbons produced from distillate or residue; they are mainly used to reduce friction between bearing surfaces. This category includes all finished grades of lubricating oil, from spindle oil to cylinder oil, and those used in greases, including motor oils and all grades of lubricating oil base stocks.

c) *Bitumen:*

Solid, semi-solid or viscous hydrocarbon with a colloidal structure, being brown to black in colour, obtained as a residue in the distillation of crude oil, vacuum distillation of oil residues from atmospheric distillation. Bitumen is often referred to as asphalt

and is primarily used for surfacing of roads and for roofing material. This category includes fluidized and cut back bitumen.

d) *Paraffin Waxes:*

Saturated aliphatic hydrocarbons (with the general formula C_nH_{2n+2}). These waxes are residues extracted when dewaxing lubricant oils, and they have a crystalline structure with carbon number greater than 12. Their main characteristics are that they are colourless, odourless and translucent, with a melting point above 45°C.

e) *Others:*

Includes the petroleum products not classified above, for example: tar, sulphur, and grease. This category also includes aromatics (e.g. BTX or benzene, toluene and xylene) and olefins (e.g. propylene) produced within refineries.

D. Gases

The figures for these four categories of gas are all expressed in terajoules based on **gross calorific values**.

1. Natural Gas

Natural gas comprises gases, occurring in underground deposits, whether liquefied or gaseous, consisting mainly of methane. It includes both "non-associated" gas originating from fields producing only hydrocarbons in gaseous form, and "associated" gas produced in association with crude oil as well as methane recovered from coal mines (colliery gas).

Production is measured after purification and extraction of NGL and sulphur, and excludes re-injected gas, quantities vented or flared. It includes gas consumed by gas processing plants and gas transported by pipeline.

2. Gas Works Gas

Gas works gas covers all types of gas produced in public utility or private plants, whose main purpose is the manufacture, transport and distribution of gas. It includes gas produced by carbonisation (including gas produced by coke ovens and transferred to gas works), by total gasification (with or without enrichment with oil products), by cracking of natural gas, and by reforming and simple mixing of gases and/or air. This heading also includes substitute natural gas, which is a high calorific value gas manufactured by chemical conversion of a hydrocarbon fossil fuel.

3. Coke Oven Gas

Coke oven gas is obtained as a by-product of the manufacture of coke oven coke for the production of iron and steel.

4. Blast Furnace Gas

Blast furnace gas is produced during the combustion of coke in blast furnaces in the iron and steel industry. It is recovered and used as a fuel partly within the plant and partly in other steel industry processes or in power stations equipped to burn it. Also included here is oxygen steel furnace gas which is obtained as a by-product of the production of steel in an oxygen furnace and is recovered on leaving the furnace. The gas is also known as converter gas or LD gas or BOS gas.

E. Combustible Renewables and Waste

The figures for these four categories of fuels are all expressed in terajoules based on **net calorific values**.

1. Solid Biomass and Animal Products

Biomass is defined as any plant matter used directly as fuel or converted into other forms before combustion. Included are wood, vegetal waste (including wood waste and crops used for energy production), animal materials/wastes, sulphite lyes, also known as "black liquor" (an alkaline spent liquor from the digesters in the production of sulphate or soda pulp during the manufacture of paper where the energy content derives from the lignin removed from the wood pulp) and other solid biomass.

Charcoal produced from solid biomass is also included here. Since charcoal is a secondary product, its treatment is slightly different than that of the other primary biomass. Production of charcoal (an output in the transformation process) is offset by the inputs of primary biomass into the charcoal production process. The losses from this process are included in the row *other transformation sector.*

Other supply (e.g. trade and stock changes) as well as consumption are aggregated directly with the primary biomass. However, in some countries, only the primary biomass is reported. Reported production of charcoal is available separately in the Summary Tables at the end of the publication.

2. Gas/Liquids from Biomass

Biomass gases are derived principally from the anaerobic fermentation of biomass and solid wastes and combusted to produce heat and/or power. Included in this category are landfill gas and sludge gas (sewage gas and gas from animal slurries). Bio-additives such as ethanol are also included in this category.

3. Municipal Waste

Municipal waste consists of products that are combusted directly to produce heat and/or power and comprises wastes produced by the residential, commercial and public services sectors that are collected by local authorities for disposal in a central location. Hospital waste is included in this category.

4. Industrial Waste

Industrial waste consists of solid and liquid products (e.g. tyres) combusted directly, usually in specialised plants, to produce heat and/or power and that are not reported in the category *solid biomass and animal products*.

F. Electricity and Heat

1. Electricity

Gross electricity production is measured at the terminals of all alternator sets in a station; it therefore includes the energy taken by station auxiliaries and losses in transformers that are considered integral parts of the station.

The difference between gross and net production is generally calculated as 7 per cent for conventional thermal stations, 1 per cent for hydro stations, and 6 per cent for nuclear, geothermal and solar stations. Hydro stations' production includes production from pumped storage plants.

2. Heat

In recent years, the production of heat for sale has been increasing in importance. To reflect this, heat production represents all heat production from public CHP and heat plants as well as heat sold by autoproducer CHP and heat plants to third parties.

Corresponding fuels to produce quantities of heat for sale are being recorded in the transpormation sector under the rows *CHP plants* and *Heat plants*. The use of fuels for heat which is not sold is recorded under the sectors in which the fuel use occurs.

3. Hydro Power

Potential and kinetic energy of water converted into electricity in hydroelectric plants.

4. Geothermal Energy

Energy available as heat emitted from within the earth's crust, usually in the form of hot water or steam. It is exploited at suitable sites:

- for electricity generation using dry stream or high enthalpy brine after flashing

- directly as heat for district heating, agriculture, etc.

5. Solar Energy

Solar radiation exploited for hot water production and electricity generation, by:

- flat plate collectors, mainly of the thermosyphon type, for domestic hot water or for the seasonal heating of swimming pools

- photovoltaic cells

- solar thermal-electric plants

Passive solar energy for the direct heating, cooling and lighting of dwellings or other buildings is not included.

6. Tide/Wave/Ocean Energy

Mechanical energy derived from tidal movement or wave motion and exploited for electricity generation.

7. Wind Energy

Kinetic energy of wind exploited for electricity generation in wind turbines.

4. NOTES ON SUMMARY TABLES

Production of Charcoal

Charcoal is a secondary biomass product and refers to the solid residue, consisting mainly of carbon, derived from the distillation of wood or other biomass products in the absence of air. Please note that in the Annual Tables, charcoal is included under Solid biomass (Part I.3. Notes on Energy Sources).

Oil Demand by Product, in thousand tonnes and in million barrels per day

Oil demand is defined as the sum of domestic supply, transfers and statistical differences less inputs to refinery, petrochemical industry and liquefaction in the transformation sector, and refineries' own use in the energy sector. *Motor gasoline* includes additives and liquids from biomass.

Other covers crude oil, other hydrocarbons, refinery gas, petroleum coke, white spirit and SBP, lubricants, bitumen, paraffin waxes and others such as tar, sulphur, grease, as well as aromatics (e.g. BTX or benzene, toluene and xylene) and olefins (e.g. propylene) produced within refineries.

Refinery fuel shows petroleum refineries' own use of petroleum fuels for the operation of equipment, heating and lighting, which include mainly refinery gas, gas/diesel oil and heavy fuel oil.

Bunkers shows international marine bunkers consumption of liquid fuels, which include mainly gas/diesel oil and heavy fuel oil.

The following conversion factors are used, unless otherwise indicated, for all countries and all years.

Conversion Factors

	Barrels per tonne	
Refinery gas	8.00	
Ethane	16.85	
LPG	11.60	
Naphtha	8.50	(8.90 OECD Europe)
Aviation Gasoline	8.90	
Motor Gasoline	8.53	(8.45 OECD Europe)
Jet Gasoline	7.93	(8.25 OECD)
Jet Kerosene	7.93	(7.88 OECD Europe)
Other Kerosene	7.74	(7.88 OECD Europe)
Gas/Diesel Oil	7.46	
Heavy Fuel Oil	6.66	(6.45 OECD Europe)
White Spirit	7.00	(8.46 OECD)
Lubricants	7.09	
Bitumen	6.08	
Paraffin Waxes	7.00	(7.85 OECD)
Petroleum Coke	5.50	
Non Specified Products	7.00	(8.00 OECD)

Consumption of Electricity

Covers electricity use in transformation, energy sector and final consumption, but excludes distribution losses.

5. GEOGRAPHICAL COVERAGE

Africa includes Algeria, Angola, Benin, Cameroon, Congo, Democratic Republic of Congo, Côte d'Ivoire, Egypt, Eritrea, Ethiopia, Gabon, Ghana, Kenya, Libya, Morocco, Mozambique, Nigeria, Senegal, South Africa, Sudan, United Republic of Tanzania, Togo, Tunisia, Zambia, Zimbabwe and **Other Africa**.

Other Africa includes Botswana, Burkina Faso, Burundi, Cape Verde, Central African Republic, Chad, Djibouti, Equatorial Guinea, Gambia, Guinea, Guinea-Bissau, Lesotho, Liberia, Madagascar, Malawi, Mali, Mauritania, Mauritius, Niger, Rwanda, Sao Tome and Principe, Seychelles, Sierra Leone, Somalia, Swaziland and Uganda.

Middle East includes Bahrain, Islamic Republic of Iran, Iraq, Israel, Jordan, Kuwait, Lebanon, Oman, Qatar, Saudi Arabia, Syria, United Arab Emirates and Yemen.

Non-OECD Europe includes Albania, Bosnia and Herzegovina, Bulgaria, Croatia, Cyprus, Gibraltar, Former Yugoslav Republic of Macedonia (FYROM), Malta, Romania, Slovak Republic, Slovenia and Federal Republic of Yugoslavia.

Former USSR includes Armenia, Azerbaijan, Belarus, Estonia, Georgia, Kazakhstan, Kyrgyzstan, Latvia, Lithuania, Republic of Moldova, Russia, Tajikistan, Turkmenistan, Ukraine and Uzbekistan.

Latin America includes Argentina, Bolivia, Brazil, Chile, Colombia, Costa Rica, Cuba, Dominican Republic, Ecuador, El Salvador, Guatemala, Haiti, Honduras, Jamaica, Netherlands Antilles, Nicaragua, Panama, Paraguay, Peru, Trinidad and Tobago, Uruguay, Venezuela and **Other Latin America**.

Other Latin America includes Antigua and Barbuda, Bahamas, Barbados, Belize, Bermuda, Dominica, French Guiana, Grenada, Guadeloupe, Guyana, Martinique, St. Kitts and Nevis, Anguilla, Saint Lucia, St. Vincent and Grenadines and Suriname.

China includes the People's Republic of China and Hong Kong (China).

Asia includes Bangladesh, Brunei, Chinese Taipei, India, Indonesia, DPR of Korea, Malaysia, Myanmar, Nepal, Pakistan, Philippines, Singapore, Sri Lanka, Thailand, Vietnam and **Other Asia**.

Other Asia and Other Oceania includes Afghanistan, Bhutan, Fiji, French Polynesia, Kiribati, Maldives, New Caledonia, Papua New Guinea, Samoa, Solomon Islands and Vanuatu.

The **Organisation for Economic Co-Operation and Development (OECD)** includes Australia, Austria, Belgium, Canada, the Czech Republic, Denmark, Finland, France, Germany, Greece, Hungary, Iceland, Ireland, Italy, Japan, Korea, Luxembourg, Mexico, the Netherlands, New Zealand, Norway, Poland, Portugal, Spain, Sweden, Switzerland, Turkey, the United Kingdom and the United States.

Within OECD:

- **Australia** excludes the overseas territories;
- **Denmark** excludes Greenland and the Danish Faroes;
- **France** includes Monaco, and excludes overseas departments (French Polynesia, Guadeloupe, Guyane, Martinique, Nouvelle-Calédonie, La Réunion and St.-Pierre et Miquelon);
- **Germany** includes the new federal states of Germany;
- **Italy** includes San Marino and the Vatican;
- **Japan** includes Okinawa;

- **The Netherlands** excludes Suriname and the Netherlands Antilles;
- **Portugal** includes the Azores and Madeira;
- **Spain** includes the Canary Islands;
- **Switzerland** includes Liechtenstein;
- **United States** includes Puerto Rico, Guam, the Virgin Islands and the Hawaiian Free Trade Zone.

The Organisation of the Petroleum Exporting Countries (OPEC) includes Algeria, Indonesia, Iran, Iraq, Kuwait, Libya, Nigeria, Qatar, Saudi Arabia, the United Arab Emirates and Venezuela.

Please note that the following countries have not been considered due to lack of data:

- **Africa:** Comoros, Namibia, Saint Helena and Western Sahara;
- **America:** Aruba, British Virgin Islands, Cayman Islands, Falkland Islands, Montserrat, Saint Pierre-Miquelon and Turks and Caicos Islands;
- **Asia and Oceania:** American Samoa, Cambodia, Christmas Island, Cook Islands, Laos, Macau, Mongolia, Nauru, Niue, Palau, Tonga and Wake Island.

6. COUNTRY NOTES AND SOURCES

General References –

Annual Bulletin of Coal Statistics for Europe, Economic Commission for Europe (ECE), New York, 1994.

Annual Bulletin of Electric Energy Statistics for Europe, Economic Commission for Europe (ECE), New York, 1994.

Annual Bulletin of Gas Statistics for Europe, Economic Commission for Europe (ECE), New York, 1994.

Annual Bulletin of General Energy Statistics for Europe, Economic Commission for Europe (ECE), New York, 1994.

Annual Report July 1991- June 1992, South African Development Community (SADC), Gaborone, 1993.

Annual Statistical Bulletin 1998, Organization of Petroleum Exporting Countries (OPEC), Vienna, 1999.

APEC Energy Database, Tokyo,2000.

Arab Oil and Gas Directory 1999, Arab Petroleum Research Centre, Paris, 2000.

ASEAN Energy Review 1995 Edition, ASEAN-EC Energy Management Training and Research Centre (AEEMTRC), Jakarta, 1996.

Base CHELEM-PIB, Centre d'Etudes Prospectives et d'Informations Internationales (CEPII) Paris, 1999.

Eastern Bloc Energy, Tadcaster, various issues to May 1999.

Energy Indicators of Developing Member Countries, Asian Development Bank (ADB), Manila, 1994.

Energy Information Administration (EIA), Washington, D.C.,

Energy-Economic Information System (SIEE), Latin American Energy Organization (OLADE), Ecuador, 2000.

Energy Statistics Yearbook 1990, South African Development Community (SADC), Luanda, 1992.

Energy Statistics Yearbook 1996, United Nations, New York, 1998.

Food and Agriculture Organisation of the United Nations, 2000 Forestry Data, Rome, 2000.

Forests and Biomass Sub-sector in Africa, African Energy Programme of the African Development Bank, Abidjan, 1996.

International Coal Report, various issues to May 1999.

International Energy Annual 1990, 1991, 1992, 1993, Energy Information Administration (EIA), Washington, D.C., 1991-1994.

International Energy Data Report 1992, World Energy Council, London, 1993.

Les Centrales Nucléaires dans le Monde Commissariat à l'Énergie Atomique, Paris, 2000.

Middle East Economic Survey (MEES), Nicosia, various issues to June 1999.

Natural Gas in the World, 1998 Survey, Cedigaz, Paris, 2000.

Notes d'Information et Statistiques, Banque Centrale des Etats de l'Afrique de l'Ouest, Dakar, 1995.

Pétrole 1994, Comité Professionnel du Pétrole (CPDP), Paris, 1995.

PIW's Global Oil Stocks & Balances, New York, various issues to June 1995.

PlanEcon Energy Outlook for Eastern Europe and the Former Soviet Republics, Washington, October 1999, May 2000.

Prospects of Arab Petroleum Refining Industry, Organization of Arab Petroleum Exporting Countries (OAPEC), Kuwait, 1990.

Review of Wood Energy Data in RWEDP Member Countries, Regional Wood Energy Development Programme in Asia, Food and Agriculture Organisation of the United Nations, Bangkok, 1997.

Statistical Handbook 1993 - States of the Former USSR, The World Bank, Washington, 1993.

Statistical Yearbook of the Member States of the CMEA, Council of Mutual Economic Assistance (CMEA), Moscow, 1985 and 1990.

The United Nations Energy Statistics Database 1996, United Nations Statistical Office, New York, 1998.

World Development Indicators 2000 on CD-ROM, The World Bank, Washington, 2000.

Note:

- The OLADE database was used for some of the Latin American countries.
- For the period 1971 to 1997, the UN database was the only source of information for the eleven individual countries which are not listed below, and for the regions Other Africa, Other Latin America and Other Asia and Oceania. It was also used in a number of other countries as a complementary source.

Albania

Large quantities of oil widely reported to have moved through Albania into former Yugoslavia are not included in oil trade for 1993. Although estimated to represent up to 100 per cent of underlying domestic consumption, no reliable figures for this trade were available. Series have been revised since 1990.

Aide Memoire of World Bank Mission to Albania May/June 1991 and Secretariat estimates.

The UN Energy Statistics Database 1996, 1998

UN ECE Energy Questionnaires 1994 and 1995.

Sources for Combustible Renewables and Waste:

The UN Energy Statistics Database, 1997, UN ECE Energy Questionnaires and Secretariat estimates.

Algeria

Combustible Renewables and Waste historical series were entirely revised based on Ministry of Energy and Mines submission and Secretariat estimates.

Sources 1992-1998:

Direct communication to the Secretariat from the Ministry of Industry, and Energy, Information Systems Management Department, Algiers.

Sources up to 1991:

Bilan Energétique National, Gouvernement Algérien, Algiers, 1984.

Algérie Energie, No 6, Ministère de l'Energie et des Industries Chimiques et Pétrochimiques, Algiers, 1979 to 1983.

Annuaire Statistique de l'Algérie 1980-1984, Office National des Statistiques, Algiers, 1985.

Sources for Combustible Renewables and Waste:

The UN Energy Statistics Database 1996, 1998.

Angola

Historical series of crude oil and oil products have been revised from 1991 onwards, based on submissions from the oil refinery and national oil company.

Sources 1992-1998:

Direct communications to the Secretariat from oil industry sources.

Eskom Annual Statistical Yearbook 1993, 1994, 1995 (Johannesburg, 1994, 1995, 1996) citing Empresa Nacional de Electricidade, Luanda as source.

The UN Energy Statistics Database 1996, 1998.

Sources up to 1991:

Le Pétrole et l'Industrie Pétrolière en Angola en 1985, Poste d'Expansion Economique de Luanda, Luanda, 1985.

Sources for Combustible Renewables and Waste:

Secretariat estimates based on 1991 data from African Energy Programme of the African Development Bank, *Forests and Biomass Sub-sector in Africa,* Abidjan, 1996.

Argentina

Historical series have been considerably revised using *Balance Energético Nacional serie 1986-1998* and various editions of *Informe del Sector Eléctrico.*

Sources up to 1998:

Annuario Estadístico de La Republica Argentina, Instituto Nacional de Estadistica y Censos, Buenos Aires, September 1997.

Direct communication to the Secretariat from the Ministry of Economy and Public Services, Secretariat of Energy, Buenos Aires.

Other sources from the Secretaría de Energía, Buenos Aires.

Anuario de Combustibles, Ministerio de Obras y Servicios Públicos, Secretaria de Energía, Buenos Aires, 1980 to 1984, 1986, 1988 to 1992, 1995, 1997.

Combustibles Boletin Mensual, Ministerio de Obras y Servicios Públicos, Secretaria de Energía, Buenos Aires, various editions.

Natural Gas Projection up to 2000, Gas del Estado Argentina, Buenos Aires, 1970, 1984 to 1986.

Anuario Estadístico de la Republica Argentina 1970-1981, Instituto Nacional de Estadístico y Censos, Secretaria de Planificación, Buenos Aires, 1982.

Balance Energetico Nacional 1970-1985, Ministerio de Obras y Servicios Públicos, Secretaria de Energía, Buenos Aires, 1986.

Plan Energetico Nacional 1986-2000, Ministerio de Obras y Servicios Públicos, Secretaria de Energía, Subsecretaria de Planificación Energetica, Buenos Aires, 1985.

Anuario Estadístico, Yacimientos Petrolíferos Fiscales, Buenos Aires, 1984 to 1987.

Memoria Y Balance General, Yacimientos Petrolíferos Fiscales, Buenos Aires, 1984 to 1986.

Sources for Combustible Renewables and Waste:

SIEE, OLADE and *Balance Energético Nacional.*

Bahrain

Historical series were revised for natural gas and certain oil products from 1986 onwards.

Sources 1992-1998:

Statistical Abstract, 1994, 1998 Council of Ministers, Control Statistics Organisation, Bahrain, 1995, 1999.

Direct communication to the Secretariat from oil industry sources.

Sources up to 1991:

Statistical Abstract 1990, Council of Ministers, Central Statistics Organisation, Bahrain, 1991.

1986 Annual Report, Bahrain Monetary Agency, Bahrain, 1987.

B.S.C. Annual Report, Bahrain Petroleum Company, Bahrain, 1982, 1983 and 1984.

Foreign Trade Statistics, Council of Ministers, Central Statistics Organisation, Bahrain, 1985.

Bahrain in Figures, Council of Ministers, Central Statistics Organisation, Bahrain, 1983, 1984 and 1985.

Bangladesh

Energy statistics are reported for a fiscal year.

Combustible Renewables and Waste data series have been revised from 1971 onwards, following estimates provided by the Bangladesh Bureau of Statistics, Ministry of Planning.

Sources 1996-1998:

Statistical Yearbook of Bangladesh 1996, 1997, 1998 7th Edition, Ministry of Planning, Bangladesh Bureau of Statistics, Dhaka, 1997, 1998, 1999.

Direct communication to the Secretariat from oil and gas industry sources and electricity utility.

Sources 1992-1995:

Statistical Pocket Book of Bangladesh, Ministry of Planning, Bangladesh Bureau of Statistics, Dhaka, 1986 to 1996.

The UN Energy Statistics Database 1996, 1998.

Sources up to 1991:

Bangladesh Energy Balances 1976-1981, Government of Bangladesh, Dhaka, 1982.

Statistical Yearbook of Bangladesh 1991, Government of Bangladesh, Dhaka, 1976 to 1991.

Monthly Statistical Bulletin of Bangladesh, Ministry of Planning, Bangladesh Bureau of Statistics, Statistics Division, Dhaka, June 1986 and October 1989.

Benin

Sources up to 1998:

Direct communication to the Secretariat from the Direction de l'Energie, Cotonou, 1999, 2000.

Direct communication to the Secretariat from the Electricity utility 1998, 1999.

The UN Energy Statistics Database 1996, 1998.

Rapport sur l'Etat de l'Economie Nationale, Ministère de l' Economie, Cotonou, septembre 1993.

Sources for Combustible Renewables and Waste:

Secretariat estimates based on 1991 data from *Forests and Biomass Sub-sector in Africa,* African Energy Programme of the African Development Bank, Abidjan, 1996.

Bolivia

Sources 1992-1998:

Anuario Estadístico 1996-1997, 1998, Superintendencia de Electricidad, La Paz, 1998, 1999.

Direct communication to the Secretariat from the Superintendencia de Hidrocarburos, La Paz.

Informe Estadístico 1992, 1993, 1994, 1995, 1996 and 1997, Yacimientos Petrolíferos Fiscales Bolivianos, La Paz, 1993, 1994, 1995, 1996, 1997, 1998.

Memoria Anual 1992, Yacimientos Petrolíferos Fiscales Bolivianos, La Paz, 1993.

Sources up to 1991:

Boletin Estadístico 1973-1985, Banco Central de Bolivia, Division de Estudios Económicos, La Paz, 1986.

Diez Anos de Estadística Petrolera en Bolivia 1976-1986, Dirección de Planeamiento, Division de Estadística, La Paz, 1987.

Empresa Nacional de Electricidad S.A. 1986 Ende Memoria, Empresa Nacional de Electricidad, La Paz, 1987.

Sources for Combustible Renewables and Waste:

SIEE, OLADE.

Brazil

Historical series have been entirely revised. The amount for the humid gas has been removed from the natural gas this year.

Sources 1971-1998:

Direct communication to the Secretariat from Ministério de Minas e Energia, Brasilia.

Brunei

Sources 1992-1998:

Direct communication to the secretariat from the Office of the Prime Minister, Petroleum Unit 1999.

Direct communication to the secretariat from the Ministry of Development, Electrical Services Department 1999, 2000.

Brunei Statistical Yearbook, 1992 to 1994, Ministry of Finance, Statistics Section, Brunei, 1993, 1995.

Direct communication to the Secretariat from the UN Energy Statistics Unit.

Sources up to 1991:

Fifth National Development Plan 1986-1990, Ministry of Finance, Economic Planning Unit, Bandar Seri Bagawan, 1985.

Sources for Combustible Renewables and Waste:

The UN Energy Statistics Database 1996, 1998.

Bulgaria

See Part I.1, Issues of Data Quality above.

Sources 1992-1998:

UN ECE Energy Questionnaires.

Energy Balances, National Statistical Institute, Sofia, 1995.

Sources up to 1991:

Energy Development of Bulgaria, Government of Bulgaria, Sofia, 1980 and 1984.

Energy in Bulgaria, Government of Bulgaria, Sofia, 1980 to 1983.

General Statistics in the Republic of Bulgaria 1989/1990, Government of Bulgaria, Sofia, 1991.

Sources for Combustible Renewables and Waste:

The UN Energy Statistics Database 1996, 1997 and UN ECE Energy Questionnaires.

Cameroon

Crude oil and oil products data were revised from 1980 onwards following oil industry source submission.

Sources up to 1998:

Direct communication to the Secretariat from oil industry sources and the electricity utility.

The UN Energy Statistics Database 1996, 1998.

Sources for Combustible Renewables and Waste:

Secretariat estimates based on 1991 data from *Forests and Biomass Sub-sector in Africa,* African Energy Programme of the African Development Bank, Abidjan, 1996.

Chile

Historical series were revised for natural gas, LPG, NGL and electricity from 1971.

Sources 1992-1998:

Balance Nacional de Energía 1995, 1997, 1998 Comisión Nacional de Energía, Santiago, 1996, 1998, 1999.

Balance Nacional de Energía 1977-1996, Comisión Nacional de Energía, Santiago, September 1997.

Balance de Energía Preliminar 1993, Comisión Nacional de Energía, Santiago, 1994.

Balance de Energía 1973 - 1992, Comisión Nacional de Energía, Santiago, 1993, *1975-1994,* Santiago, 1995.

Sources up to 1991:

Compendio Estadístico Chile 1985, Ministerio de Economía, Fomento Y Reconstrucción, Instituto Nacional de Estadísticas, Santiago, 1986.

Sources for Combustible Renewables and Waste:

Comisión Nacional de Energía.

People's Republic of China

See Part I.1, Issues of Data Quality above.

Coal production statistics refer to unwashed and unscreened coal. IEA coal statistics normally refer to coal after washing and screening for the removal of inorganic matter.

Preliminary information seems to indicate that coal demand may have dramatically fallen in 1998-1999. However, due to the stocks available at many mines which shut down in 1998 and 1999, consumption may be much higher than prelimianry suggests.

It is known that much of the agricultural use of diesel in China is for transport purposes but the national data have not been adjusted by the IEA Secretariat to take account of this.

See Part I.1, Issues of Data Quality for further information.

Sources 1992-1998:

Energy Balances of China, provided to the Secretariat by the State Statistical Bureau for 1993, 1994, 1995, 1996, 1997 and 1998.

China Energy Databook, Lawrence Berkeley National Laboratory, Berkeley, 1996.

China Statistical Yearbook 1995, State Statistical Bureau of the People's Republic of China, Beijing, 1995.

1995 Energy Report of China, State Planning Commission of the People's Republic of China, Beijing, 1995.

China Petroleum Newsletter Monthly, China Petroleum Information Institute, Beijing, various issues to June 1995.

China OGP, China Oil Gas and Petrochemicals/Xinhua News Agency, Beijing, various issues to June 1995.

Petroleum Data Monthly, China Oil Gas and Petrochemicals/Xinhua News Agency, Beijing, various issues to June 1995.

Energy of China, China International Book Trading Co, Beijing, various issues to June 1995.

Energy Commodity Account of China, Asian Development Bank, Manila, 1994.

China's Customs Statistics, State Statistical Bureau, Economic Information & Agency, Beijing, various editions to 1995.

China's Customs Statistics, General Administration of Customs, PRC, Economic Information and Agency, Hong Kong, various editions from 1991 to 1995.

Statistical Yearbook of China, State Statistical Bureau of the People's Republic of China, Economic Information & Agency, Hong Kong, various editions from 1981 to 1994.

Statistical Communique, State Statistical Bureau of the People's Republic of China, Hong Kong, March 1995.

China's Downstream Oil Industry in Transition, Fesharaki Associates Consulting and Technical Services, Inc., Honolulu, 1993.

Sources up to 1991:

Electric Industry in China in 1987, Ministry of Water Resources and Electric Power, Department of Planning, Beijing, 1988.

Statistical Yearbook of China 1991, State Statistical Bureau of the People's Republic of China, Beijing, 1992.

Outline of Rational Utilization and Conservation of Energy in China, Bureau of Energy Conservation State Planning Commission, Beijing, June 1987.

China Coal Industry Yearbook, Ministry of Coal Industry, People's Republic of China, Beijing, 1983, 1984 and 1985.

Energy in China 1989, Ministry of Energy, People's Republic of China, Beijing, 1990.

China: A Statistics Survey 1975-1984, State Statistical Bureau, Beijing, 1985.

China Petro-Chemical Corporation (SINOPEC) Annual Report, SINOPEC, Beijing, 1987.

Almanac of China's Foreign Economic Relations and Trade, The Editorial Board of the Almanac, Beijing, 1986.

Sources for Combustible Renewables and Waste:

Secretariat estimates based on a per capita average consumption from various surveys and studies.

Chinese Taipei

Autoproducer electricity includes only the inputs and outputs of the iron and steel industry and waste disposal plants. Energy used by the other autoproducers (mostly industrial cogeneration) was counted as final consumption.

Sources 1992-1998:

Energy Balances in Taiwan, Ministry of Economic Affairs, Taipei, 1992 to 1998.

Yearbook of Energy Statistics, Ministry of Trade Industry and Energy, Taipei, 1996.

Direct communication to the Secretariat from the electricity utilities.

Sources up to 1991:

The Energy Situation in Taiwan, Ministry of Economic Affairs, Energy Committee, Taipei, 1986, 1987, 1988 and 1992.

Industry of Free China 1975-1985, Council for Economic Planning and Development, Taipei, 1986.

Taiwan Statistical Data Book 1954-1985, Council for Economic Planning and Development, Taipei, 1986.

Energy Policy for the Taiwan Area, Ministry of Economic Affairs, Energy Committee, Taipei, 1984.

Energy Balances in Taiwan, Ministry of Economic Affairs, Taipei, 1984 to 1991.

Energy Indicators Quarterly, Taiwan Area, Ministry of Economic Affairs, Taipei, 1986.

Taipower 1987 Annual Report, Taipower, Taipei, 1988.

Energy Data Report, 1986, WEC National Committee, Taipei, 1986.

Sources for Combustible Renewables and Waste:

The UN Energy Statistics Database 1996, 1998 and Secretariat estimates.

Colombia

Historical series have been revised based on national Energy Balance tables obtained from the Ministry of Energy and Mines web site for Coal products, Natural Gas, NGL, LPG, Fuel Oil, Electricity and Combustible Renewables and Waste.

Sources 1992-1998:

Direct communication to the Secretariat from the Ministry of Mines and Energy, Energy Information Department, Bogotá.

Sources up to 1991:

Boletin Minero-Energético, Ministerio de Minas y Energía, Bogota, December 1991.

Estadísticas Minero-Energéticas 1940-1990, Ministerio de Minas y Energía, Bogota, 1990.

Estadísticas Basicas del Sector Carbón, Carbocol, Oficina de Planeación, Bogotá, various editions from 1980 to 1988.

Colombia Estadística 1985, DANE, Bogotá, 1970 to 1983 and 1987.

Empresa Colombiana de Petróleos, Informe Anual, Empresa Colombiana de Petróleos, Bogotá, 1979, 1980, 1981 and 1985.

Estadísticas de la Industria Petrolera Colombiana Bogota 1979-1984, Empresa Colombiana de Petróleos, Bogotá, 1985.

Informe Estadístico Sector Eléctrico Colombiano, Government of Colombia, Bogotá, 1987 and 1988.

La Electrificacion en Colombia 1984-1985, Instituto Colombiano de Energía Electrica, Bogotá, 1986.

Balances Energéticos 1975-1986, Ministerio de Minas y Energía, Bogota, 1987.

Energía y Minas Para el Progreso Social 1982-1986, Ministerio de Minas y Energía, Bogota, 1987.

Sources for Combustible Renewables and Waste:

Ministry of Mines and Energy, Energy Information Department.

Congo

Sources up to 1998:

Annual Statistical Yearbook 1993, 1994, 1995, Eskom, Johannesburg, 1994, 1995, 1996, citing Empresa Nacional de Electricidade, Luanda as source.

L'Energie en Afrique, IEPE/ENDA, Paris, 1995, in turn sourced from the Direction des Etudes et de la Planification, Ministère des Mines et de l'Energie,

and the Société Congolaise de Raffinage, Brazzaville.

Direct communication to the Secretariat from the Ministère de l'Energie et de l'Hydraulique.

Direct communication to the Secretariat from the oil industry.

Sources for Combustible Renewables and Waste:

Secretariat estimates based on 1991 data from *Forests and Biomass Sub-sector in Africa,* African Energy Programme of the African Development Bank, Abidjan, 1996.

Democratic Republic of Congo

Sources up to 1998:

L'Energie en Afrique, IEPE/ENDA, Paris, 1995, in turn sourced from the *Annuaire Statistique Energétique 1990,* Communauté Economique des Pays des Grands Lacs, Bujumbura, 1990.

The UN Energy Statistics Database 1996, 1998.

Sources for Combustible Renewables and Waste:

Secretariat estimates based on 1991 data from *Forests and Biomass Sub-sector in Africa,* African Energy Programme of the African Development Bank, Abidjan, 1996.

Cote d'Ivoire

Historical series were revised from 1986 onwards based on submissions from the Ministry of Energy.

Sources 1996-1998:

Direct communication to the Secretariat from oil industry and the Ministry of Energy, Abidjan, July 1998, 2000.

La Côte d'Ivoire en chiffres, Ministère de l'Economie et des Finances, edition 1996-97.

Sources 1992-1995:

Direct communication to the Secretariat from the Bureau des Economies d'Energie, Abidjan.

L'Energie en Afrique, IEPE/ENDA, Paris, 1995, in turn sourced from the Ministère des Mines et de L'Energie, Abidjan.

The UN Energy Statistics Database 1996, 1998.

Sources up to 1991:

Etudes & Conjoncture 1982 - 1986, Ministère de l'Economie et des Finances, Direction de la Planification et de la Prévision, Abidjan, 1987.

Sources for Combustible Renewables and Waste:

Secretariat estimates based on 1991 data from *Forests and Biomass Sub-sector in Africa,* African Energy Programme of the African Development Bank, Abidjan, 1996.

Cuba

Historical series have been considerably revised using *Anuario Estadístico de Cuba 1998,* from Oficina Nacional de Estadísticas.

Sources up to 1998:

Annuario Estadistico de Cuba 1996, 1998, Oficina Nacional de Estadísticas, Edición 1998, 2000.

Compendio estadístico de energía de Cuba 1989, Comite Estatal de Estadísticas, Havana, 1989.

Anuario Estadístico de Cuba, Comite Estatal de Estadísticas, Havana, various editions from 1978 to 1987.

Sources for Combustible Renewables and Waste:

SIEE, OLADE.

Anuario Estadístico de Cuba 1996, Oficina Nacional de Estadísticas, Havana, 1998.

Cyprus

UN ECE Energy Questionnaires.

Electricity Authority of Cyprus Annual Report 1988, 1992, 1996, Electricity Authority of Cyprus, Nicosia, 1989, 1993, 1997.

Industrial Statistics 1988, Ministry of Finance, Department of Statistics, Nicosia, 1989.

Sources for Combustible Renewables and Waste:

UN ECE Energy Questionnaires and Secretariat estimates.

Ecuador

Sources 1996-1998:

Sector Energético Ecuatoriano, Edición No 7, Direccion de Planificacion, Ministerio de Energia y Minas, Quito, November 1997, and Secretariat estimates.

Sources 1992-1995:

Balance Energético Nacional 1995, Ministerio de Energia y Minas, Quito, December 1996.

Balances Energéticos 1988-1994, Ministerio de Energia y Minas, Quito, 1996.

Sources up to 1991:

Ministerio de Energia y Minas.

Cuentas Nacionales, Banco Central del Ecuador, Quito, various editions from 1982 to 1987.

Memoria 1980-1984, Banco Central del Ecuador, Quito, 1985.

Ecuadorian Energy Balances 1974-1986, Instituto Nacional de Energía, Quito, 1987.

Informacion Estadística Mensual, No. 1610, Instituto Nacional de Energía, Quito, 1988.

Plan Maestro de Electrificación de Ecuador, Ministerio de Energía y Minas, Quito, 1989.

Sources for Combustible Renewables and Waste:

Ministerio de Energia y Minas and *SIEE,* OLADE.

Egypt

Stock changes of crude oil may include some input to foreign refineries in Egypt. Final consumption of gas/diesel oil and heavy fuel oil may include international marine bunkers prior to 1985. **Historical series for electricity have been considerably revised**.

Sources 1992-1998:

Direct submission to the Secretariat from the Ministry of Petroleum.

Annual Report 1995, 1997, 1998, Ministry of Petroleum, Egyptian General Petroleum Corporation, Cairo, 1996, 1998, 1999.

Annual Report of Electricity Statistics 1996/1997, 1998/1999 Ministry of Electricity and Energy, Egyptian Electricity Authority, Cairo, 1998.

Arab Oil and Gas, The Arab Petroleum Research Center, Paris, October 1997.

Middle East Economic Survey, Middle East Petroleum and Economic Publications, Nicosia, February 1994, June 1996, March 1998.

A Survey of the Egyptian Oil Industry 1993, Embassy of the United States of America in Cairo, Cairo, 1994.

Sources up to 1991:

Annual Report of Electricity Statistics 1990/1991, Ministry of Electricity and Energy, Egyptian Electricity Authority, Cairo, 1992.

Statistical Yearbook of the Arab Republic of Egypt, Central Agency for Public Mobilisation and Statistics, Cairo, 1977 to 1986.

L'Electricité, l'Energie, et le Pétrole, République Arabe d'Egypte, Organisme Général de l'Information, Cairo, 1990.

Annual Report, The Egyptian General Petroleum Corporation, Cairo, 1985.

Sources for Combustible Renewables and Waste:

The UN Energy Statistics Database 1996, 1998, and Secretariat estimates.

El Salvador

Historical series for electricity have been considerably revised using *Boletin de Estadísticas Electricas No.1 1999,* from Superintendencia General de Electricidad y Telecomunicaciones.

Sources up to 1998:

Direct communication to the Secretariat from the Ministerio de Economia, Direccíon de Hidrocarburos y Minas, San Salvador, and OLADE.

Sources for Combustible Renewables and Waste:

SIEE, OLADE.

Eritrea

Data are available from 1992. Prior to that Eritrea data are included in Ethiopia.

Sources 1992-1998:

Direct Communication to the Secretariat from the Ministry of Energy and Mines, State of Eritrea.

Ethiopia

Ethiopia energy data includes Eritrea from 1971 to 1991. From 1992 the data are reported for the two countries separately, see country notes on

Eritrea. Geothermal production of electricity for Ethiopia has been discontinued from 1997 because of lack of reliable data.

Sources 1992-1998

Direct communication to the Secretariat from the Ministry of Economic Development and Co-Operation, Addis Ababa, 1998, 1999, 2000.

Direct communication to the Secretariat from the UN Energy Statistics Unit.

Sources up to 1991:

Ten Years of Petroleum Imports, Refinery Products, and Exports, Ministry of Mines & Energy, Addis Ababa, 1989.

Energy Balance for the Year 1984, Ministry of Mines & Energy, Addis Ababa, 1985.

1983 Annual Report, National Bank of Ethiopia, Addis Ababa, 1984.

Quarterly Bulletin, National Bank of Ethiopia, Addis Ababa, various editions from 1980 to 1985.

Sources for Combustible Renewables and Waste:

Secretariat estimates based on 1992 data from Eshetu, L. and Bogale, W., *Power Restructuring in Ethiopia,* AFREPREN, Nairobi, 1996.

Gabon

Sources 1992-1998:

Direct communication to the Secretariat from the Société Gabonaise de Raffinage, Port Gentil, 1997, 2000.

Tableau de Bord de l'Economie, Situation 1997, Perspectives 1998-1999, Direction Générale de l'Economie, Ministère des Finance, de l'Economie, du Budget et des participations, chargé de la privatisation, Mai 1998.

Rapport d'Activité, Banque Gabonaise de Développement, Libreville, 1985, 1990, 1992 and 1993.

The UN Energy Statistics Database 1996, 1998.

Sources up to 1991:

Tableau de Bord de l'Economie, Situation 1983 Perspective 1984-85, Ministère de l'Economie et des Finances, Direction Générale de l'Economie, Libreville, 1984.

Sources for Combustible Renewables and Waste:

Secretariat estimates based on 1991 data from *Forests and Biomass Sub-sector in Africa,* African Energy Programme of the African Development Bank, Abidjan, 1996.

Direct communication from the oil industry.

Ghana

Historical Series have extensively been revised following data submission from the Ministry of Energy and Mines.

Sources 1992-1998

National Energy Statistics, Ministry of Energy and Mines, Accra, 1997, 1999.

Quarterly Digest of Statistics, Government of Ghana, Statistical Services, Accra, March 1990, March 1991, March 1992, March 1995.

Sources up to 1991:

Energy Balances, Volta River Authority, Accra, various editions from 1970 to 1985.

Sources for Combustible Renewables and Waste:

Ministry of Mines and Energy, *the UN Energy Statistics Database 1996*, 1998 and Secretariat estimates.

Honduras

Direct communication to the Secretariat from the Empressa Nacional de Energia Eléctrica, Comayaguela.

Direct Communication to the Secretariat from Secretariat de Recursos Naturales y del Ambiente, Tegucigalpa.

Sources for Combustible Renewables and Waste:

SIEE, OLADE.

Hong Kong, China

Direct communication to the Secretariat from the electricity utilities.

Hong Kong Energy Statistics - Annual Report, Census and Statistics Department, Hong Kong, various editions to 1998.

Hong Kong Energy Statistics - Quarterly Supplement, Census and Statistics Department, Hong Kong, various editions to 1998.

Hong Kong Monthly Digest of Statistics, Census and Statistics Department, Hong Kong, various editions to 1994.

Sources for Combustible Renewables and Waste:

The UN Energy Statistics Database 1996, 1998 and Secretariat estimates.

India

Coal Products data were revised from 1978 based on The Annual Report published by the Ministry of Coal.

Natural gas supply and demand data are reported by fiscal year. Oil supply data are reported by calendar year, and the oil consumption data have been pro-rated using available fiscal year end use detail. Autoproducer's electricity has been assumed to be consumed in the industrial sector.

Sources 1992-1998:

Direct communication to the Secretariat from the Ministry of Petroleum and Natural Gas.

Annual Report 1994-1995, 1995-1996, 1998-1999, 1999-2000, Ministry of Coal, New Delhi, 1995, 1996, 1999, 2000.

Direct Communication to the Secretariat from the Ministry of Power.

Indian Oil and Gas, Ministry of Petroleum and Natural Gas, Economics & Statistics Division, New Delhi, January 1985 to March 1998.

Coal Directory of India, 1992-1993, 1993-1994, 1995-1996, 1996-1997, 1997-1998, Ministry of Coal, Coal Controller's Organization, Calcutta, 1994, 1995, 1996, 1997, 1998.

Annual Review of Coal Statistics, 1993-1994, 1995-1996, 1996-1997, 1997-1998, Ministry of Coal, Coal Controller's Organization, Calcutta, 1995, 1997, 1998, 1999.

Indian Petroleum and Natural Gas Statistics, Ministry of Petroleum and Natural Gas, Economics & Statistics Division, New Delhi, 1985 to 1997.

Energy Data Directory, Yearbook "TEDDY", and *Annual Report,* Tata Energy Research Institute "TERI", New Delhi, 1986, 1987, 1988, 1990, 1994-1995 and 1995-1996.

General Review, Public Electricity Supply, India Statistics, Central Electricity Authority, New Delhi, 1982 to 1985, 1995, 1996.

Monthly Abstract of Statistics, Ministry of Planning, Central Statistics Organisation, Department of Statistics, New Delhi, various editions from 1984 to March 1998.

Annual Report 1993-1994, 1998-1999 Ministry of Petroleum and Natural Gas, New Delhi, 1995, 1999.

India Monthly Coal Statistics, Ministry of Energy, Department of Coal, Coal Controllers Organisation, New Delhi, various editions to June 1996.

Annual Report 1994-1996, 1998-1999 Ministry of Energy, Department of Non-Conventional Energy, New Delhi, 1996, 1999.

India's Energy Sector, July 1995, Center for Monitoring Indian Economy PVT Ltd., Bombay, 1995.

Monthly Review of the Indian Economy, Center for Monitoring Indian Economy PVT Ltd., New Delhi, various issues from 1994 to June 1999.

Sources up to 1991:

Indian Oil Corporation Limited 1987-88 Annual Report, Indian Oil Corporation Limited, New Delhi, 1989-1992.

Report 1986-87, Ministry of Energy, Department of Coal, New Delhi, 1981 to 1987.

Annual Report 1986-1987, Ministry of Energy, Department of Non-Conventional Energy, New Delhi, 1987.

Economic Survey, Ministry of Finance, New Delhi, various editions from 1975 to 1986.

Statistical Outline of India, Ministry of Finance, New Delhi, 1983, 1984, 1986, and 1987.

Monthly Coal Bulletin, vol xxxvi no.2., Ministry of Labour, Directorate General of Mines Safety, New Delhi, February 1986.

Sources for Combustible Renewables and Waste:

Secretariat estimates based on a per capita average consumption from various surveys and studies.

Indonesia

Although it has been estimated that electricity output by autoproducers is approximately equal to that generated by the national electricity utility, since this amount is not known with any certainty, autoproducer output refers only to that amount sold to the public electricity grid.

Sources 1992-1998:

Direct communication to the Secretariat from the Ministry of Mines and Energy, Directorate General of Electricity and Energy Development.

Statistik Perminyakan Indonesia 1995 to 1998, Indonesia Oil and Gas Statistics, Directorate General of Oil and Gas.

Indonesian Coal Yearly Statistics 1999, Directorate of Coal, Directorate General of Mines Ministry of Energy and Mines.

The Petroleum Report Indonesia, U.S. Embassy in Jakarta, Jakarta, 1986 to 1996.

Oil Statistics of Indonesia, Direktorat Jenderal Minyak dan Gas Bumi, Jakarta, 1981 to December 1998.

Statistik Dan Informasi Ketenagalistrikan Dan Energi, Direktorat Jenderal Listrik Dan Pengembangan Energi, Jakarta, December 1998, December 1999.

Mining and Energy Yearbook, 1998, Ministry of Mines and Energy, Jakarta, 1998.

Sources up to 1991:

Indonesian Financial Statistics, Bank of Indonesia, Jakarta, 1982.

Indikator Ekonomi 1980-1985, Biro Pusat Statistik, Jakarta, 1986.

Statistical Yearbook of Indonesia, Biro Pusat Statistik, Jakarta, 1978 to 1984, 1992.

Statistik Pertambangan Umum, 1973 - 1985, Biro Pusat Statistik, Jakarta, 1986.

Energy Planning for Development in Indonesia, Directorate General for Power, Ministry of Mines and Energy, Jakarta, 1981.

Commercial Information, Electric Power Corporation, Perusahaan Umum Listrik Negara, Jakarta, 1984 and 1985.

Sources for Combustible Renewables and Waste:

The UN Energy Statistics Database 1996, 1998 and Secretariat estimates.

Islamic Republic of Iran

Energy statistics are reported on a fiscal year basis.

Sources 1992-1998:

Direct communication to the Secretariat from the Ministry of Energy, Office of Deputy Minister for Energy, Teheran 1998, 2000.

Direct communication to the Secretariat from the Ministry of Petroleum, 1999.

Electric Power in Iran, Ministry of Energy, Power Planning Bureau, Statistics Section, Teheran, 1992.

Sources up to 1991:

Electric Power in Iran, Ministry of Energy, Power Planning Bureau, Statistics Section, Teheran, 1967 to 1977, 1988, 1990, 1991.

Sources for Combustible Renewables and Waste:

The UN Energy Statistics Database 1996, 1998, *Forestry Statistics,* FAO, 2000, Rome, and Secretariat estimates.

Israel

For reasons of confidentiality the annual "*Energy in Israel*" only reports aggregated data for kerosene and jet kerosene.

Sources 1992-1998:

Direct communication to the Secretariat from the Ministry of Energy and Infrastructure, Jerusalem.

Energy in Israel, Ministry of Energy and Infrastructure, Central Bureau of Statistics, Jerusalem, 1992, 1993, 1994, 1995, 1996, 1997.

Statistical Report 1993, 1994, 1995, the Israel Electric Corporation, Haifa, April 1994, May 1995, April 1996.

Statistical Results 1992, the Israel Electric Corporation, Haifa, June 1993.

Central Bureau of Statistics (CBS), Israel 2000.

Sources up to 1991:

Energy in Israel, Ministry of Energy and Infrastructure, Central Bureau of Statistics, Jerusalem, 1975 to 1991.

Statistical Abstract of Israel, Ministry of Energy and Infrastructure, Central Bureau of Statistics, Jerusalem, 1985.

Supplement to Monthly Bulletin of Statistics, Ministry of Energy and Infrastructure, Central Bureau of Statistics, Jerusalem, various editions from 1984 to 1986.

Sources for Combustible Renewables and Waste:

Forestry Statistics, FAO, 2000, Rome, and Secretariat estimates.

Jamaica

Sources 1994-1998:

Direct communication to the Secretariat from the Statistical Institute of Jamaica, Kingston.

Sources 1992-1993:

Economic and Social Survey Jamaica 1992 to 1995, Planning Institute of Jamaica, Kingston, 1993, 1994, April 1995, April 1996.

Sources up to 1991:

National Energy Outlook 1985-1989, Petroleum Corporation of Jamaica, Economics and Planning Division, Kingston, 1985.

Energy and Economic Review, Petroleum Corporation of Jamaica, Energy Economics Department, Kingston, September 1986, December 1986 and March 1987.

Production Statistics 1988, Planning Institute of Jamaica, Kingston, 1989.

Statistical Digest, Research and Development Division, Bank of Jamaica, Kingston, 1984, 1985, 1986, 1989, 1990 and 1991.

Sources for Combustible Renewables and Waste:

SIEE, OLADE.

Jordan

Sources 1992-1998:

Direct communication to the Secretariat from the National Electric Power Company, Amman.

Annual Report 1995, 1996, 1998, National Electric Power Compan, Amman, 1996, 1997, 1999.

Annual Report 1992, 1993, Jordan Electricity Authority, Amman, 1993, 1994.

Energy and Electricity in Jordan 1992, 1993, 1994, 1995, Jordan Electricity Authority, Amman, 1993, 1994, 1995, 1996.

Statistical Yearbook, 1994, Department of Statistics, Amman, 1995.

Sources up to 1991:

Monthly Statistical Bulletin, Central Bank of Jordan, Department of Research Studies, Amman, various issues.

Statistical Yearbook, Department of Statistics, Amman, 1985, 1986 and 1988.

1986 Annual Report, Ministry of Energy and Mineral Resources, Amman, 1987.

1989 Annual Report, Ministry of Energy and Mineral Resources, Amman, 1990.

Sources for Combustible Renewables and Waste:

Forestry Statistics, FAO, Rome, 2000, and Secretariat estimates.

Kenya

Sources 1992-1998:

Economic Survey, 1995, 1996, 1997, 1998, 1999, Central Bureau of Statistics, Nairobi.

The UN Energy Statistics Database 1996, 1998.

Sources up to 1991:

Economic Survey, Government of Kenya, Nairobi, 1989.

Economic Survey 1991, Ministry of Planning and National Development, Central Bureau of Statistics, Nairobi, 1992.

Kenya Statistical Digest, Ministry of Planning and National Development, Central Bureau of Statistics, Nairobi, 1988.

Sources for Combustible Renewables and Waste:

Secretariat estimates based on 1991 data from *Forests and Biomass Sub-sector in Africa,* African Energy Programme of the African Development Bank, Abidjan, 1996.

Democratic People's Republic of Korea

Historical series have been entirely revised using *the UN Energy Statistics Database 1996*, 1998.

Sources for Combustible Renewables and Waste:

The UN Energy Statistics Database 1996, 1998 and Secretariat estimates.

Kuwait

Historical series have been entirely revised.

Data include 50 per cent of Neutral Zone output.

Sources 1992-1998:

Direct communication to the Secretariat from the Ministry of Oil, Safat, the Ministry of Planning and the Ministry of Electricity & Water, Kuwait City.

Monthly Digest of Statistics, Ministry of Planning, Central Statistical Office, Kuwait, 1998.

A Survey of the Kuwait Oil Industry, Embassy of the United States of America in Kuwait City, Kuwait, 1993.

Twelfth Annual Report 1991-1992, Kuwait Petroleum Corporation, Kuwait, 1993.

Sources up to 1991:

Quarterly Statistical Bulletin, Central Bank of Kuwait, Kuwait, various editions from 1986 and 1987.

The Kuwaiti Economy, Central Bank of Kuwait, Kuwait, various editions from 1980 to 1985.

Annual Statistical Abstract, Ministry of Planning, Central Statistical Office, Kuwait, 1986 and 1989.

Monthly Digest of Statistics, Ministry of Planning, Central Statistical Office, Kuwait, various editions from 1986 to 1990.

Economic and Financial Bulletin Monthly, Central Bank of Kuwait, Kuwait, various editions from 1983 to 1986.

Kuwait in Figures, The National Bank of Kuwait, Kuwait, 1986 and 1987.

Sources for Combustible Renewables and Waste:

Forestry Statistics, FAO, Rome, 2000, and Secretariat estimates.

Lebanon

There was no refinery production in 1993, 1994, 1995, 1996, 1997 and 1998.

Direct communication to the Secretariat from Association Libanaise pour la Maîtrise de l'Energie.

L'Energie au Liban, Les Bilans Energétiques en 1998, Association Libanaise pour la Maîtrise de l'Energie, Beirut, 1999.

L'Energie au Liban, le Défi, Association Libanaise pour la Maîtrise de l'Energie, Beirut, décembre 1996.

Les Bilans Energétiques au Liban, Association Libanaise pour la Maîtrise de l'Energie, Beirut, décembre 1994.

Sources for Combustible Renewables and Waste:

Forestry Statistics, FAO, Rome, 2000, and Secretariat estimates.

Libya

Historical series for natural gas have been considerably revised between 1980 and 1997. This may result in a break in series between 1979 and 1980.

Statistical Abstract of Libya, 19th vol, Government of Libya, Tripoli, 1983.

Sources for Combustible Renewables and Waste:

The UN Energy Statistics Database 1996, 1998 and Secretariat estimates.

Malaysia

Historical series have been entirely revised.

Sources up to 1998:

Direct communication to the Secretariat from the Ministry of Energy, Telecommunications and Posts, Kuala Lumpur.

Direct communication to the Secretariat from Petroliam Nacional Berhad, Kuala Lumpur, May 2000.

Malaysia Energy Centre, Kuala Lumpur, 2000.

Sources for Combustible Renewables and Waste:

The UN Energy Statistics Database, 1996, 1998 and *Forestry Statistics,* FAO, Rome, 2000.

Malta

UN ECE Questionnaire on Oil 1995 to 1998.

UN ECE Questionnaire on Coal 1994 and 1995.

UN ECE Questionnaire on Electricity and Heat, 1994 to 1998.

Morocco

Sources 1992-1998:

Direct communication to the Secretariat from Ministère de l'Energie et des Mines - Direction des Mines, Rabat.

Annuaire Statistique du Maroc, Ministère du Plan, Direction de la Statistique, Rabat, 1980, 1984, 1986 to 1999.

Electricity consumption by economic sector by direct communication from the Ministère du Plan, Rabat.

Sources up to 1991:

Rapport d'Activité 1992, Office National de l'Electricité, Casablanca, 1993.

Le Maroc en Chiffres 1986, Ministère du Plan, Direction de la Statistique, Rabat, 1987.

Rapport Annuel, Office National de Recherches et d'Exploitations Pétrolieres, Maroc, 1984.

Rapport d'Activité du Secteur Pétrolier 1983, Ministère de l'Energie et des Mines, Direction de l'Energie, Rabat, 1984.

Rapport sur les Données Energétiques Nationales 1979-1981, Ministère de l'Energie et des Mines, Rabat, 1982.

Sources for Combustible Renewables and Waste:

The UN Energy Statistics Database 1996, 1998 and Secretariat estimates.

Mozambique

Coal Historical series were revised based on Ministry of Energy and Mines submission and Secretariat estimates.

Sources 1992-1998:

Direct communication to the Secretariat from the Ministry of Energy and Mineral Resources, (Maputo June 1998, April 1999, November 1999) and the electricity utility.

Annual Statistical Yearbook 1993, 1994, 1995, Eskom, Johannesburg, 1994, 1995, 1996, citing Electricidade de Mozambique, Maputo, as source.

The UN Energy Statistics Database 1996, 1998.

Sources for Combustible Renewables and Waste:

Secretariat estimates based on 1991 data from *Forests and Biomass Sub-sector in Africa,* African Energy Programme of the African Development Bank, Abidjan, 1996.

Myanmar

Sources 1992-1998:

Direct communications to the Secretariat from the Ministry of Energy, Planning Department, Rangoon, 1996, 1997, 1998, 1999 and 2000.

Review of the Financial Economic and Social Conditions, Ministry of National Planning and Economic Development, Central Statistical Organization, Rangoon, 1995, 1996.

Statistical Yearbook, Ministry of National Planning and Economic Development, Central Statistical Organization, Rangoon, 1995, 1996.

The UN Energy Statistics Database 1996, 1998.

Sources up to 1991:

Sectoral Energy Demand in Myanmar, UNDP Economic and Social Commission for Asia and The Pacific, Bangkok, 1992.

Selected Monthly Economic Indicators, paper no. 3, Ministry of Planning and Finance, Central Statistical Organization, Rangoon, 1989.

Sources for Combustible Renewables and Waste:

Secretariat estimates based on 1990 data from *UNDP Sixth Country Programme Union of Myanmar,* World Bank, Programme Sectoral Review of Energy, by Sousing John, et. al., Washington, D.C., 1991.

Nepal

Historical series have been considerably revised based on direct communication from the Water and Energy Commission Secretariat.

Energy statistics are reported for a fiscal year.

Sources for Combustible Renewables and Waste:

Water and Energy Commission Secretariat (WECS), Ministry of Water Resources.

Netherlands Antilles

Between 1992 and 1996, data are based on the *UN Energy Statistics Database, 1*994, 1995, 1996, 1997, 1998. Secretariat partially estimates 1997 and 1998 data based on the direct communication with the Central Bureau of Statistics and the oil industry. This change in the sources may result in a break in series between 1996 and 1997.

Direct communication to the Secretariat from the Central Bureau of Statistics, Fort Amsterdam, Curaçao.

Direct communication to the Secretariat from the oil industry.

Nicaragua

Sources 1971-1998:

Direct communication to the Secretariat from the Instituto Nicaraguense de Energía, Dirección General de Hidrocarburos, Managua, 1999.

Direct communication to the Secretariat from the Electricity Utility.

Estatidisticas des Suministro de los Hidrocarburos, Instituto Nicaraguense de Energía, 1999.

Informe Annual 1996: Datos Estadisticos del Sector Electrico, INE, Managua, 1999.

Instituto Nicaraguense de Energía, Managua, 2000.

Sources for Combustible Renewables and Waste:

SIEE, OLADE.

Nigeria

Recent years have been revised based on Annual Statistical Bulletin 1998, OPEC.

Direct submission to the Secretariat from the Energy Commission of Nigeria.

Sources 1992-1998:

Annual Report and Statement of Accounts 1995, Central Bank of Nigeria, Lagos, 1996.

Direct communication from the oil industry. Statistical difference probably includes oil products smuggled into neighbouring countries for consumption.

Nigerian Petroleum News, Energy Publications, monthly reports, various issues up to May 1998.

Sources up to 1991:

Annual Report and Statement of Accounts, Central Bank of Nigeria, Lagos, various editions from 1981 to 1987.

Basic Energy Statistics for Nigeria, Nigerian National Petroleum Corporation, Lagos, 1984.

NNPC Annual Statistical Bulletin, Nigerian National Petroleum Corporation, Lagos, 1983 to 1987.

The Economic and Financial Review, Central Bank of Nigeria, Lagos, various editions.

Sources for Combustible Renewables and Waste:

Secretariat estimates based on 1991 data from *Forests and Biomass Sub-sector in Africa,* African Energy Programme of the African Development Bank, Abidjan, 1996.

Oman

Sources 1992-1998:

Direct communication to the Secretariat from the Ministry of Petroleum and Minerals, Muscat, 1997, 1998 and 1999.

Direct communication to the Secretariat from the Ministry of Electricity & Water, Office of the Under Secretary, Ruwi, 1998 and 1999.

Quarterly Bulletin December 1994, Central Bank of Oman, Muscat, 1995.

Annual Report 1992, Central Bank of Oman, Muscat, 1993.

Statistical Yearbook, 1994, 1995, 1996, 1997 Ministry of Development, Muscat, 1995, 1996, 1997 and 1998.

Sources up to 1991:

Quarterly Bulletin, Central Bank of Oman, Muscat, 1986, 1987, 1989 and 1995.

Annual Report to His Majesty the Sultan of Oman, Department of Information and Public Affairs, Petroleum Development, Muscat, 1981, 1982, and 1984.

Oman Facts and Figures 1986, Directorate General of National Statistics, Development Council, Technical Secretariat, Muscat, 1987.

Quarterly Bulletin on Main Economic Indicators, Directorate General of National Statistics, Muscat, 1989.

Statistical Yearbook, Directorate General of National Statistics, Development Council, Muscat, 1985, 1986, 1988 and 1992.

Pakistan

Energy statistics are reported on a fiscal year basis.

Sources 1992-1998:

Energy Year Book, Ministry of Petroleum and Natural Resources, Directorate General of New and Renewable Energy Resources, Islamabad, various editions from 1979 to 1999.

Pakistan Economic Survey 1994-1995, 1996, 1997, Government of Pakistan, Finance Division, Islamabad, 1995, 1997, 1998.

Statistical Supplement 1993/1994, Finance Division, Economic Adviser's Wing, Government of Pakistan, Islamabad, 1995.

Sources up to 1991:

As above.

Monthly Statistical Bulletin, no. 12, Federal Bureau of Statistics, Islamabad, December 1989.

1986 Bulletin, The State Bank of Pakistan, Islamabad, 1987.

Sources for Combustible Renewables and Waste:

Secretariat estimates based on 1991 data from "Household Energy Strategy Study (HESS)" of 1991.

Panama

Sources 1992-1994:

Direct communication to the Secretariat from the Electricity and Hydro Resources Institute, Panama.

Sources up to 1991:

Balance Energetico Nacional, Serie Historica 1970-80, CONADE Peica Programa Energético del Istmo Centro Americano, Comisión Nacional de Energía, Panama, 1981.

Sources for Combustible Renewables and Waste:

SIEE, OLADE and "Balance Energetico Nacional."

Paraguay

The Itaipu hydroelectric plant, operating since 1984 and located on the Paraná river (which forms the border of Brazil and Paraguay) was formed as a joint venture between Eletrobrás and the Paraguayan government. Production is shared equally between Brazil and Paraguay. Consumption in Paraguay accounts for less than 5 per cent of the total power produced, the remaining 45 per cent is exported to Brazil, and has been accounted as such.

Sources 1992-1998:

Direct communication to the Secretariat from the Non-conventional energy department of Ministerio de Obras Públicas y Comunicaciones.

Direct communication to the Secretariat from the Electricity utility.

Sources up to 1991:

Boletin Estadístico no. 316, Banco Central del Paraguay, Departamento de Estudios Económicos, Asuncion, 1984.

Importaciones Por Productos Principales 1964-1986, Banco Central del Paraguay, Asuncion, 1987.

Sources for Combustible Renewables and Waste:

SIEE, OLADE and Secretariat estimates.

Peru

Natural Gas data were revised from 1985 onwards based on submissions from the Ministry of Energy and Mines.

Sources 1992-1998:

Direct communication to the Secretariat from the Ministry of Energy and Mines, Lima.

Balance Nacional de Energía 1997, 1998 Ministerio de Energía y Minas, Oficina Técnica de Energía, República del Perú.

Sources up to 1991:

Resena Económica, Banco Central de Reserva del Peru, Lima, 1984 and 1985.

Balance Nacional de Energía, Ministerio de Energía y Minas, Oficina Sectorial de Planificación, Lima, various editions from 1978 to 1987.

Annual Report, Petróleos del Peru, Lima, 1983 and 1984.

Estadísticas de las Operaciones Exploración/ Producción 1985, Petróleos del Peru, Lima, 1986.

Sources for Combustible Renewables and Waste:

SIEE, OLADE.

Philippines

Sources 1996-1998:

Philippine Energy Bulletin 1997, 1998, Department of Energy, Metro Manila, 1998, 1999.

Sources 1992-1995:

Direct communication to the Secretariat from the Office of Energy Affairs, Metro Manila.

Philippine Statistical Yearbook 1977-1983, and *1993,* National Economic and Development Authority, Manila, 1978-1984, and 1994.

The APEC Energy Statistics 1994, Tokyo, October 1996.

The UN Energy Statistics Database, 1998.

Sources up to 1991:

1990 Power Developmemt Program (1990 - 2005), National Power Corporation, Manila, 1990.

Philippine Medium-term Energy Plan 1988 - 1992, Office of Energy Affairs, Manila, 1989.

1985 and *1989 Annual Report,* National Power Corporation, Manila, 1986, 1990.

Philippine Economic Indicators, National Economic and Development Authority, Manila, various editions of 1985.

Accomplishment Report: Energy Self-Reliance 1973-1983, Ministry of Energy, Manila, 1984.

Industrial Energy Profiles 1972-1979, vol. 1-4, Ministry of Energy, Manila, 1980.

National Energy Program, Ministry of Energy, Manila, 1982-1987 and 1986-1990.

Philippine Statistics 1974-1981, Ministry of Energy, Manila, 1982.

Energy Statistics, National Economic and Development Authority, Manila, 1983.

Quarterly Review, Office of Energy Affairs, Manila, various editions.

Sources for Combustible Renewables and Waste:

Department of Energy, National Economic and Development Authority and *the UN Energy Statistics Database 1996,* 1997.

Qatar

Historical series have been revised for LPG, following direct communications with the petrochemical industries.

Sources 1992-1998:

Direct communication to the Secretariat from the oil and petrochemical industries.

Annual Statistical Abstract, Presidency of the Council of Ministers, Central Statistical Office, Doha, 1994, 1995, 1996, 1997, 1998, 1999.

The UN Energy Statistics Database 1996, 1998.

Sources up to 1991:

Qatar General Petroleum Corporation 1981-1985, General Petroleum Corporation, Doha, 1986.

Economic Survey of Qatar 1990, Ministry of Economy and Commerce, Department of Economic Affairs, Doha, 1991.

Statistical Report 1987 Electricity & Water, Ministry of Electricity, Doha, 1988.

State of Qatar Seventh Annual Report 1983, Qatar Monetary Agency, Department of Research and Statistics, Doha, 1984.

Sources for Combustible Renewables and Waste:

Forestry Statistics, FAO, Rome, 2000, and Secretariat estimates.

Romania

See Part I.1, Issues of Data Quality, above.

Sources 1992-1998:

UN ECE Energy Questionnaires.

Buletin Statistic de Informare Publica, Comisia Nationala Pentru Statistica, Bucharest, various issues to June 1995.

Renel Information Bulletin, Romanian Electricity Authority, Bucharest, 1990, 1991, 1992, 1993, 1994.

Sources up to 1991:

Anuarul Statistic al Republicii Socialiste Romania, Comisia Nationala Pentru Statistica, Bucharest, 1984, 1985, 1986, 1990 and 1991.

Sources for Combustible Renewables and Waste:

UN ECE Energy Questionnaires and Secretariat estimates.

Saudi Arabia

The data include 50 per cent of Neutral Zone output.

Sources 1992-1998:

Direct submissions from oil industry sources.

A Survey of the Saudi Arabian Oil Industry 1993, Embassy of the United States of America in Riyadh, Riyadh, January 1994.

Electricity Growth and Development in the Kingdom of Saudi Arabia up to the year of 1416H. (1996G.), Ministry of Industry and Electricity, Riyadh, 1997.

Sources up to 1991:

Annual Reports, ARAMCO, various issues.

Petroleum Statistical Bulletin 1983, Ministry of Petroleum and Mineral Resources, Riyadh, 1984.

Achievement of the Development Plans 1970-1984, Ministry of Planning, Riyadh, 1985.

The 1st, 2nd, 3rd and 4th Development Plans, Ministry of Planning, Riyadh, 1970, 1975, 1980 and 1985.

Annual Report, Saudi Arabian Monetary Agency, Research and Statistics Department, Riyadh, 1984, 1985, 1986, 1988 and 1989.

Statistical Summary, Saudi Arabian Monetary Agency, Research and Statistics Department, Riyadh, 1986.

Sources for Combustible Renewables and Waste:

Forestry Statistics, FAO, Rome, 2000, and Secretariat estimates.

Senegal

Sources 1992-1998:

Direct communication to the Secretariat from the Ministère de l'Energie, des Mines et de l'Industrie, Direction de l'Energie, Dakar, 1998, 1998 and 1999.

Direct communication from oil industry sources.

Report of Senegal on the Inventory of Greenhouse Gases Sources, Ministère de l'Environnement et de la Protection de la Nature, Dakar, 1994.

Direct communication to the Secretariat from ENDA - Energy Program, Dakar, 1997.

The UN Energy Statistics Database 1996, 1998.

Sources up to 1991:

Situation Economique 1985, Ministère de l'Economie et des Finances, Direction de la Statistique, Senegal, 1986.

Sources for Combustible Renewables and Waste:

Secretariat estimates based on 1994 data from *Forests and Biomass Sub-sector in Africa,* African Energy Programme of the African Development Bank, Abidjan, 1996, and from direct communication with ENDA, Senegal.

Singapore

Official Singapore trade statistics do not show oil trade between Indonesia and Singapore. The quantity of this trade for crude oil has been estimated.

Sources 1992-1998:

Direct communication to the Secretariat from the Public Utilities Board and Secretariat estimates.

Direct submissions from oil industry sources.

Yearbook of Statistics Singapore 1993, 1995, 1996, Department of Statistics, Singapore, 1994, 1996, 1997.

The Strategist Oil Report, Singapore, various issues up to March 1999.

Singapore Trade Statistics, Department of Statistics, Singapore, various editions from 1985 to 1995.

Petroleum in Singapore 1993/1994, Petroleum Intelligence Weekly, Singapore, 1994.

AEEMTRC, 1996.

Sources up to 1991:

Monthly Digest of Statistics, Department of Statistics, Singapore, various editions from 1987 and 1989.

Yearbook of Statistics Singapore 1975/1985, Department of Statistics, Singapore, 1986.

Asean Oil Movements and Factors Affecting Intra-Asean Oil Trade, Institute of Southeast Asian Studies, Singapore, 1988.

The Changing Structure of the Oil Market and Its Implications for Singapore's Oil Industry, Institute of Southeast Asian Studies, Singapore, 1988.

Public Utilities Board Annual Report (1986 and 1989), Public Utilities Board, Singapore, 1987 and 1990.

Sources for Combustible Renewables and Waste:

The UN Energy Statistics Database 1996, 1998 and Secretariat estimates.

Slovak Republic

See Part I.1, Issues of Data Quality, above.

Direct submission from the Power Research Institute (EGU), Bratislava. Primary source: Statistical Office of the Slovak republic:

1989-1995: complete balances.

1980-1988: electricity and heat balances; primary balances for coal, oil, gas and electricity.

UN ECE Energy Questionnaires 1995 to 1998.

Statistical Yearbook of the Czechoslovak Socialist Republic, Federal Czech and Slovak Statistical Office, Prague, 1981 to 1991.

Sources for Combustible Renewables and Waste:

UN ECE Energy Questionnaires, *the UN Energy Statistics Database 1996,* 1997 and from the Slovak government as a result of IEA in-depth energy survey.

South Africa

In the second half of 1995, the IEA undertook an in-depth energy survey of South Africa and the results were published in *Energy Policies of South Africa,* IEA/OECD, Paris, 1996. As a result of the survey, much of the data have been revised and improved. Natural gas production began in 1993; all gas is liquefied and refined into petroleum products. Input of coal to coal liquefaction is shown in the coal column of the balance, and the crude produced appears as an output in the crude column; this crude is included in refinery input.

Sources 1992-1998:

Direct submissions to the Secretariat from the Department of Minerals and Energy and related institutions, June 2000.

Direct submission from the Institute for Energy Studies, Rand Afrikaans University, Pretoria, 1998, 1999.

Digest of South African Energy Statistics 1998, Department of Minerals and Energy, Pretoria, 1999.

Direct submissions from the Department of Mineral and Energy Affairs, Pretoria.

Direct submissions from the Energy Research Institute, University of Cape Town.

Eskom Annual Report, Electricity Supply Commission (ESKOM), South Africa, 1989 to 1994.

Statistical Yearbook, Electricity Supply Commission (ESKOM), South Africa, 1983 to 1994.

South Africa's Mineral Industry, Department of Mineral and Energy Affairs, Braamfontein, 1995.

South African Energy Statistics, 1950-1993, Department of Mineral and Energy Affairs, Pretoria, 1995.

Wholesale Trade Sales of Petroleum Products, Central Statistical Service, Pretoria, 1995.

South African Coal Statistics 1994, South African Coal Report, Randburg, 1995.

Energy Balances in South Africa 1970-1993, Energy Research Institute, Plumstead, 1995.

Sources up to 1991:

Statistical News Release 1981-1985, Central Statistical Service, South Africa, various editions from 1986 to 1989.

Annual Report Energy Affairs 1985, Department of Mineral and Energy Affairs, Pretoria, 1986.

Energy Projections for South Africa (1985 Balance), Institute for Energy Studies, Rand Afrikaans University, South Africa, 1986.

Sources for Combustible Renewables and Waste:

South African Energy Statistics 1950-1989, No. 1, National Energy Council, Pretoria, 1989 and Secretariat estimates.

Sri Lanka

Sources 1992-1998:

Sri Lanka Energy Balance 1996, 1997, 1998, Energy Conservation Fund, Colombo, 1997, 1999, 2000.

Annual Report 1993, Central Bank of Sri Lanka, Colombo, July 1994.

Direct communication to the Secretariat from the Ceylon Electricity Board, *Sri Lanka Energy Balances, 1994.*

Sources up to 1991:

Energy Balance Sheet 1991, 1992, Energy Unit, Ceylon Electricity Board, Colombo, 1992, 1993.

Bulletin 1989, Central Bank of Sri Lanka, Colombo, July 1989.

Bulletin (monthly), Central Bank of Sri Lanka, Colombo, May 1992.

Sectoral Energy Demand in Sri Lanka, UNDP Economic and Social Commission for Asia and The Pacific, Bangkok, 1992.

External Trade Statistics 1992, Government of Sri Lanka, Colombo, 1993.

Sources for Combustible Renewables and Waste:

Energy Conservation Fund and Ceylon Electricity Board.

Sudan

Historical Series have been revised for Combustible Renewables and Waste, based on direct communication and Secretariat estimates.

Sources 1992-1998:

Direct communication to the Secretariat from the Ministry of Energy and Mines, Khartoum, June 1998, April 1999, and March 2000.

Sources up to 1991:

Foreign Trade Statistical Digest 1990, Government of Sudan, Khartoum, 1991.

Sources for Combustible Renewables and Waste:

Secretariat estimates based on 1990 data from Bhagavan, M.R., Editor, *Energy Utilities and Institutions in Africa*, AFREPREN, Nairobi, 1996.

Syria

Sources 1992-1998:

Statistical Abstract 1992-1998, Office of the Prime Minister, Central Bureau of Statistics, Damascus, 1993, 1996, 1997, 1998, 1999.

The UN Energy Statistics Database 1996, 1998.

Sources up to 1991:

Quarterly Bulletin, Central Bank of Syria, Research Department, Damascus, 1984.

Sources for Combustible Renewables and Waste:

Forestry Statistics, FAO, Rome, 2000, and Secretariat estimates.

United Republic of Tanzania

Sources up to 1998:

Direct communication to the Secretariat from the electricity utility.

Tanzanian Economic Trends, Economic Research Bureau, University of Dar-es-Salaam, 1991.

Sources for Combustible Renewables and Waste:

Secretariat estimates based on 1990 data from *Energy Statistics Yearbook 1990,* SADC, Luanda, 1992.

Thailand

Electric Power in Thailand, Ministry of Science, Technology and Energy, National Energy Administration, Bangkok, 1985, 1986, 1988 to 1999.

Oil and Thailand, Ministry of Science, Technology and Energy, National Energy Administration, Bangkok, 1979 to 1999.

Thailand Energy Situation, Ministry of Science, Technology and Energy, National Energy Administration, Bangkok, 1978 to 1999.

Sources for Combustible Renewables and Waste:

Thailand Energy Situation, Ministry of Science, Technology and Energy, National Energy Administration.

Trinidad and Tobago

Natural Gas series have been revised from 1984 based on sources below:

Sources 1992-1998:

Direct communication to the Secretariat from the Ministry of Energy and Natural Resources, Port of Spain.

Annual Economic Survey 1994, 1995, Central Bank of Trinidad and Tobago, Port of Spain 1995, 1996, 1997, 1998.

Petroleum Industry Monthly Bulletin, Ministry of Energy and Natural Resources, Port of Spain, various issues to 1999.

Sources up to 1991:

Annual Statistical Digest, Central Statistical Office, Port of Spain, 1983 and 1984.

History And Forecast, Electricity Commission, Port of Spain, 1987.

Annual Report, Ministry of Energy and Natural Resources, Port of Spain, 1985 and 1986.

The National Energy Balances 1979-1983, Ministry of Energy and Natural Resources, Port of Spain, 1984.

Trinidad and Tobago Electricity Commission Annual Report, Trinidad and Tobago Electricity Commission, Port of Spain, 1984 and 1985.

Sources for Combustible Renewables and Waste:

SIEE, OLADE.

Tunisia

Sources 1992-1998:

Energy balance and electricity consumption and production data provided by direct submission from the Observatoire National de l'Energie, Agence pour la Maîtrise de l'Energie, Tunis.

Sources up to 1991:

Bilan Energétique de l'Année 1991, Banque Centrale de Tunisie, Tunis, September 1992.

Rapport d'Activité 1990, Observatoire National de l'Energie, Agence pour la Maîtrise de l'Energie, Tunis, 1991.

Rapport Annuel 1990, Banque Centrale de Tunisie, Tunis, 1991.

Activités du Secteur Pétrolier en Tunisie, Banque Centrale de Tunisie, Tunis, 1987.

Statistiques Financières, Banque Centrale de Tunisie, Tunis, 1986.

Entreprise Tunisienne d'Activités Pétrolières (ETAP), Tunis, 1987.

Annuaire Statistique de la Tunisie, Institut National de la Statistique, Ministère du Plan, Tunis, 1985 and 1986.

L'Economie de la Tunisie en Chiffres, Institut National de la Statistique, Tunis, 1984 and 1985.

Activités et Comptes de Gestion, Société Tunisienne de l'Electricité et du Gaz, Tunis, 1987.

Sources for Combustible Renewables and Waste:

Secretariat estimates based on 1991 data from *Analyse du Bilan de Bois d'Energie et Identification d'un Plan d'Action,* Ministry of Agriculture, Tunis, 1998.

United Arab Emirates

The series for natural gas, electricity and some petroleum products have been entirely revised.

Estimates of annual sales for marine bunkers in the facilities offshore of Fujairah in the UAE have been made in consultation with the oil industry.

Sources 1993-1998:

Statistical Yearbook 1995, 1996, 1998, Department of Planning, Abu Dhabi, November 1998, 1998, 2000.

Direct communications with the Ministry of Electricity and Water, Abu Dhabi, March 2000

Sources up to 1992:

Annual Report 1998, Ministry of Electricity & Water, Dubai.

Abu Dhabi National Oil Company, 1985 Annual Report, Abu Dhabi National Oil Company, Abu Dhabi, 1986.

United Arab Emirates Statistical Review 1981, Ministry of Petroleum and Mineral Resources, Abu Dhabi, 1982.

Annual Statistical Abstract, Ministry of Planning, Central Statistical Department, Abu Dhabi, various editions from 1980 to 1993.

Sources for Combustible Renewables and Waste:

Forestry Statistics, FAO, Rome, 2000, and Secretariat estimates.

Uruguay

The power produced from the Salto Grande hydroelectric plant, operating since 1980 and located on the Uruguay river (natural border of Argentina and Uruguay), is equally shared between the two countries. Electricity exports include power produced in Salto Grande and exported to Argentina.

Sources 1992-1998:

Balance Energetico Nacional, Ministerio de Industria, Energía y Mineria, Dirección Nacional de Energía, Montevideo, 1971 to 1998.

Sources up to 1991:

UTE Memoria Anual, Administración Nacional de Usinas y Trasmiones Eléctricas, Montevideo, 1978.

Boletín Estadístico 1975-1985, Banco Central del Uruguay, Departemento de Investigaciones Económicas, Montevideo, 1986.

Informaciones y Estadísticas Nacionales e Internationales no. 48, Centro de Estadísticas Nacionales y Comercio Internacional, Montevideo, 1989.

Boletín Mensual Energético, Ministerio de Industria y Energía, Dirección Nacional de Energía, Montevideo, various editions from 1986 to 1987.

Sources for Combustible Renewables and Waste:

Direccion Nacional de Energia and *SIEE,* OLADE.

Former USSR

Coal production statistics refer to unwashed and unscreened coal. IEA coal statistics normally refer to coal after washing and screening for the removal of inorganic matter. Also see notes under 'Classification of Fuel Uses' and 'Heat', above.

The energy balances presented for the former USSR include Secretariat estimates of fuel consumption in the main categories of transformation. These estimates are based on secondary sources and on isolated references in FSU literature.

Summary tables, commodity and energy balance tables for former USR are now the sum of individual former USSR republics after 1992. In previous editions of this publication, intra-USSR trade was excluded and average net calorific values were applied to the aggregate data instead of the country-specific factors. This is also the case for "Former Yugoslavia".

Sources 1991 to 1998 for NIS:

PlanEcon Energy Outlook for the Former Soviet Republics, Washington, June 1995 and 1996 and *PlanEcon Energy Outlook for Eastern Europe and the Former Soviet Republics,* Washington, September 1997, October 1998 and October 1999.

Foreign Scouting Service, Commonwealth of Independent States, IHS Energy Group – IEDS Petroconsultants, Geneva, April 1999.

Global E&P Service, Commonwealth of Independent States, IHS Energy Group - IEDS Petroconsultants, Geneva, April 2000.

Statistical Bulletin, various editions, The State Committee of Statistics of the CIS, Moscow, 1993 and 1994.

External Trade of the Independent Republics and the Baltic States in 1991, The State Committee of Statistics of the CIS, Moscow, 1992.

Statistical Yearbook, 1993, 1994 and 1995, The State Committee of Statistics of the CIS, Moscow, 1994, 1995 and 1996.

Sources up to 1990 for the Former USSR:

Statistical Yearbook, The State Committee for Statistics of the USSR, Moscow, various editions from 1980 to 1989.

External Trade of the Independent Republics and the Baltic States, 1990 and 1991, The State Committee of Statistics of the CIS, Moscow, 1992.

External Trade of the USSR, annual and quarterly, various editions, The State Committee of Statistics of the USSR, Moscow, 1986 to 1990.

CIR Staff Paper no. 14, 28, 29, 30, 32 and 36, Center for International Research, U.S. Bureau of the Census, Washington, 1986, 1987 and 1988.

Yearbook on Foreign Trade, The Ministry of Foreign Trade, Moscow, 1986.

Armenia

Direct communications to the Secretariat from the Ministry of Energy and Fuels, 1992 and 1996.

UN ECE Questionnaire on Electricity, 1992 to 1998.

UN ECE Questionnaire on Coal, 1995 and 1996.

UN ECE Questionnaire on Gas 1997 and 1998.

Sources for Combustible Renewables and Waste:

Forestry Statistics, FAO, Rome, 2000, and Secretariat estimates.

Azerbaijan

Direct communications to the Secretariat from the Ministry of Economics, June 1999.

Direct communications to the Secretariat from the State Committee of Statistics and Analysis, May 1993, June 1994, May 1996 and June 1998 .

UN ECE Energy Questionnaires, 1992 to 1998.

Sources for Combustible Renewables and Waste:

Forestry Statistics, FAO, Rome, 2000, and Secretariat estimates.

Belarus

Direct communication to the Secretariat from the Ministry of Statistics and Analysis, May 1996.

UN ECE Energy Questionnaires, 1990 to 1997.

Sources for Combustible Renewables and Waste:

UN ECE Energy Questionnaires and Secretariat estimates.

Estonia

Up to 1994, process heat from final consumption is included in the transformation sector.

UN ECE Energy Questionnaires, 1991 to 1998.

UN ECE Questionnaires (Summary), 1970 to 1990.

UN ECE Questionnaire on Coal, 1990.

Statistical Office of Estonia, Energy Balances, 1994.

Sources for Combustible Renewables and Waste:

UN ECE Energy Questionnaires and Secretariat estimates.

Georgia

Official Energy Balance of Georgia 1990-1998, Ministry of Economy and Ministry of Energy, Tbilissi.

Energy Outlook of Georgia 1998-2010, draft, Georgian Energy Research Institute, Tbilissi.

Energy outlook of Georgia 1993-2015, Georgian Public Institute Techinform, Tbilissi.

Direct communication to the Secretariat from the Socio-Economical Information Committee, 1992.

Sources for Combustible Renewables and Waste:

The UN Energy Statistics Database 1996, 1997.

Kazakhstan

Direct communication to the Secretariat from the Agency on Statistics of the Republic of Kazakhstan, April 2000.

UN ECE Energy Questionnaires, 1993, 1995, 1997 and 1998.

Direct communication to the Secretariat from the State Committee of Statistics and Analysis, August 1994.

Sources for Combustible Renewables and Waste:

Forestry Statistics, Rome, FAO, 2000.

Kyrgyzstan

UN ECE Energy Questionnaires, 1993 to 1998.

Direct communication to the Secretariat, 1992.

Sources for Combustible Renewables and Waste:

The UN Energy Statistics Database 1996, 1998.

Latvia

UN ECE Questionnaire on Natural Gas, 1992 to 1998.

UN ECE Questionnaire on Coal, 1993 to 1998.

UN ECE Questionnaire on Electricity and Heat, 1993 to 1998.

UN ECE Questionnaire on Oil 1995 to 1998.

Balance of Latvian Energy Sources 1991 and 1992, EC PHARE Project Implementation Unit, Riga, 1994.

Sources for Combustible Renewables and Waste:

UN ECE Energy Questionnaires and Secretariat estimates.

Lithuania

UN ECE Energy Questionnaires, 1992 to 1998.

Direct communications to the Secretariat, February 1994 and May 1996.

Balances of Electricity, Heat, Fuel and Energy in Lithuania 1991–1993, Lithuanian Ministry of Energy Agency, Vilnius.

Lithuania Power Demand and Supply Options, The World Bank, Washington, 1993.

Energy Conservation Potentials in Lithuania and Latvia, Riso National Laboratory, Roskilde, 1992.

Energy in Lithuania (Power, Heat and Fuel Balances 1980-1992), Lithuanian Energy Institute, Vilnius, 1993.

Sources for Combustible Renewables and Waste:

UN ECE Energy Questionnaires and Secretariat estimates.

Republic of Moldova

UN ECE Questionnaire on Electricity and Heat, 1991 to 1998.

UN ECE Questionnaire on Natural Gas, 1991 to 1998.

UN ECE Questionnaire on Coal, 1992 to 1998.

UN ECE Questionnaire on Oil, 1993 to 1998.

Direct communication to the Secretariat from the Ministry of Industry and Energy, July 1992.

Sources for Combustible Renewables and Waste:

The UN Energy Statistics Database 1996, 1997.

Russia

Process heat from final consumption is included in the transformation sector.

Direct communication to the Secretariat, June 1998, March 2000.

UN ECE Questionnaire on Coal, 1992 to 1998.

UN ECE Questionnaire on Natural Gas, 1991 to 1998.

UN ECE Questionnaire on Electricity and Heat, 1991 to 1998.

UN ECE Questionnaires on Oil, 1991 to 1998.

Oil and Natural Gas: Direct communication to the Secretariat from the State Committee of Statistics of Russia, May 1995.

Energy trade: Direct communication to the Secretariat from the State Committee of Statistics of Russia, July 1994.

Statistical Yearbook of Russia 1994. The State Committee of Statistics, Moscow, 1994.

The Russian Federation in 1992, Statistical Yearbook, The State Committee of Statistics of Russia, Moscow, 1993.

Russian Federation External Trade, annual and quarterly various editions, The State Committee of Statistics of Russia, Moscow.

Statistical Bulletin, various editions, The State Committee of Statistics of the CIS, Moscow, 1993 and 1994.

Statistical Bulletin n° 3, The State Committee of Statistics of Russia, Moscow, 1992.

Fuel and Energy Balance of Russia 1990, The State Committee of Statistics of Russia, Moscow, 1991.

Energetika, Energo-Atomisdat, Moscow, 1981 and 1987.

Sources for Combustible Renewables and Waste:

The State Committee of Statistics of Russian Federation and Secretariat estimates.

Tajikistan

UN ECE Questionnaire on Coal, 1994, 1995, 1997 and 1998.

UN ECE Questionnaire on Natural Gas, 1994, 1995, 1997 and 1998.

UN ECE Questionnaire on Electricity and Heat, 1994, 1995, 1997 and 1998.

UN ECE Questionnaire on Oil, 1994 to 1998.

Turkmenistan

Direct communication to the Secretariat from the National Institute on Statistics and Forecasting of Turkmenistan, November 1999.

Ukraine

UN ECE Questionnaire on Coal, 1991, 1992 and 1995 to 1998.

UN ECE Questionnaire on Oil, 1992 and 1995 to 1998.

UN ECE Questionnaire on Natural Gas, 1992 and 1995 to 1998.

UN ECE Questionnaire on Electricity and Heat, 1991, 1992, and 1994 to 1998.

Direct communication to the Secretariat from the Ministry of Statistics, the Coal Ministry, the National Dispatching Company, November 1995.

Coal: Direct communications to the Secretariat from the State Mining University of Ukraine, May 1995 and 1996.

Natural Gas: Direct communication to the Secretariat from Ukrgazprom, February 1995.

Direct communication to the Secretariat from the Ministry of Statistics of the Ukraine, July 1994.

Ukraine in 1992, Statistical Handbook, Ministry of Statistics of the Ukraine, Kiev, 1993.

Ukraine Power Demand and Supply Options, The World Bank, Washington, 1993.

Power Industry in Ukraine, Ministry of Power and Electrification, Kiev, 1994.

Energy Issues Paper, Ministry of Economy, March 1995.

Ukraine Energy Sector Statistical Review 1993, 1994, 1995, 1996 and 1997, World Bank Regional Office, Kiev, 1994, 1995, 1996, 1997 and 1998.

Global Energy Saving Strategy for Ukraine, Commission of the European Communities, TACIS, Madrid, July 1995.

Sources for Combustible Renewables and Waste:

Statistical Office in Kiev, World Bank and Secretariat estimates.

Uzbekistan

Up to 1994, crude oil includes NGL.

UN ECE Questionnaires 1995 to 1997.

Direct communications to the Secretariat from the Institute of Power Engineering and Automation, March 1994 and June 1996.

Venezuela

Historical series have entirely been revised for most products based on the National Energy Balance, provided by the Ministry of Energy and Mines.

Sources 1992-1998:

Direct communication to the Secretariat from the Ministry of Energy and Mines.

25 anos del Balance Energético, Ministerio de Energía y Minas, Abril de 1998.

Petróleo y Otros Datos Estadísticos, Dirección General Sectorial de Hidrocarburos, Dirección de Economía de Hidrocarburos, Caracas, 1993, 1994, 1995, 1997 and 1998.

Sources up to 1991:

Petróleo y Otros Datos Estadísticos, Dirección General Sectorial de Hidrocarburos, Dirección de Planificación y Economía de Hidrocarburos, Caracas, 1983 to 1991.

Balance Energetico Consolidado de Venezuela 1970-1984, Ministerio de Energía y Minas, Dirección General Sectorial de Energía, División de Programación Energetica, Caracas, 1986.

Compendio Estadístico del Sector Eléctrico, Ministerio de Energía y Minas, Dirección de Electricidad, Carbón y Otras Energías, Caracas, 1984, 1989, 1990 and 1991.

Memoria y Cuenta, Ministerio de Energía y Minas, Caracas, 1991.

Petróleos de Venezuela S.A. 1985 Annual Report, Petróleos de Venezuela, Caracas, 1991.

Sources for Combustible Renewables and Waste:

The UN Energy Statistics Database 1996, 1998.

Vietnam

Historical series between 1995 and 1998 have been considerably revised using the energy balance data provided by the Institute of Energy, Hanoi.

Sources 1971-1998:

Direct communications to the Secretariat from the Center for Energy-Environment Research and Development, Pathumthami, 1997, 1998 and 1999.

Data were supplied by RWEDP in Asia, Bangkok, Thailand.

Direct communication from the Oil Industry.

Sectoral Energy Demand in Vietnam, UNDP Economic and Social Commission for Asia and The Pacific, Bangkok, 1992.

Energy Commodity Account of Vietnam 1992, Asian Development Bank, Manila, 1994.

World Economic Problems No 2 (20), National Centre for Social Sciences of the S.R. Vietnam, Institute of World Economy, Hanoi, 1993.

Vietnam Energy Review, Institute of Energy, Hanoï, 1995, 1997, 1998.

Sources for Combustible Renewables and Waste:

Secretariat estimates based on 1992 data from *Vietnam Rural and Household Energy Issues and Options: Report No. 161/94*, World Bank, ESMAP, Washington, D.C., 1994.

Yemen

Sources 1992-1998:

Statistical Yearbook 1993 to 1998, Ministry of Planning and Development, Central Statistical Organization, Republic of Yemen, Yemen, 1994, 1995, 1996, 1997, 1998 and 1999.

Statistical Indicators in the Electricity Sector, Ministry of Planning and Development, Central Statistical Organization, Republic of Yemen, Yemen, 1993.

Sources up to 1991:

Statistical Yearbook, Government of Yemen Arab Republic, Yemen, 1988.

Sources for Combustible Renewables and Waste:

The UN Energy Statistics Database 1996, 1998.

Forestry Statistics, FAO, Rome, 2000, and Secretariat estimates.

Former Yugoslavia

From 1992 onwards, the energy balance for former Yugoslavia has been prepared by adding together the basic statistics of its component countries: Croatia, Slovenia, Former Yugoslav Republic of Macedonia, Bosnia-Herzegovina, and the Federal Republic of Yugoslavia. In the case of the latter two, the quality of data beyond basic production flows is questionable. Oil imports into Former Yugoslavia breaking the UN embargo of Serbia &

Montenegro are theoretically included, although the absolute levels of this trade are not known with any certainty. As such the balance for former Yugoslavia should be seen as an estimation that is necessary for completing regional aggregates.

Sources up to 1991:

Statisticki Godisnjak Yugoslavije, Socijalisticka Federativna Rebublika Jugoslavija, Savezni Zavod Za Statistiku, Beograd, 1985 to 1991.

Indeks, Socijalisticka Federativna Rebublika Jugoslavija, Beograd, 1990, 1991, 1992.

Bosnia and Herzegovina

UN ECE Energy Questionnaires 1993, 1994.

Sources for Combustible Renewables and Waste:

The UN Energy Statistics Database 1996, 1997.

Croatia

Direct communication to the Secretariat from the energy institute "Hrvoje Požar", May 2000.

UN ECE Energy Questionnaires 1993 to 1997.

Former Yugoslav Republic of Macedonia (FYROM)

UN ECE Energy Questionnaires, 1993 to 1995, 1998.

Sources for Combustible Renewables and Waste:

The UN Energy Statistics Database 1996, 1997 and *Forestry Statistics,* FAO, Rome, 2000.

Slovenia

UN ECE Energy Questionnaires 1993 to 1998.

Zambia

Sources 1992-1996:

Direction Communication to the Secretariat from oil sources.

Annual Statistical Yearbook 1993, 1994 and 1995 (*Consumption in Zambia 1978-1983),* Eskom, Lusaka, 1984.

Sources for Combustible Renewables and Waste:

Secretariat estimates based on 1991 data from *Forests and Biomass Sub-sector in Africa,* African Energy Programme of the African Development Bank, Abidjan, 1996.

Zimbabwe

Following submission from the Ministry of Environment and Tourism and the Secretariat estimates coal times series were revised from 1971.

Sources 1996-1998:

Direct communication to the Secretariat from the Ministry of Environment and Tourism, Harare, 1999, 2000.

Direct communication to the Secretariat from the electricity utility.

Electricity Statistics Information, Central Statistical Office, Causeway, February 1998.

Sources 1992-1995:

Eskom Annual Statistical Yearbook 1993, 1995 and 1995, Johannesburg, 1994, 1995, 1996, citing Zimbabwe Electricity Supply Authority, Harare as source.

The UN Energy Statistics Database 1996, 1998.

Sources up to 1991:

Zimbabwe Statistical Yearbook 1986, Central Statistical Office, Harare, 1990.

Quarterly Digest of Statistics, Central Statistical Office, Harare, 1990.

Zimbabwe Electricity Supply Authority Annual Report, Zimbabwe Electricity Supply Authority, Harare, 1986 to 1991.

Sources for Combustible Renewables and Waste:

Secretariat estimates based on 1991 data from *Forests and Biomass Sub-sector in Africa,* African Energy Programme of the African Development Bank, Abidjan, 1996.

Other Africa, Other Latin America, and Other Asia and Oceania

The series for these 'sum' countries are made up by simple addition of their component countries. Intra-component country trade is therefore included as part of total trade, although to truly represent trade to and from the grouping, intra-component country movements should be discounted. Trade is therefore likely to be overstated.

1. DE LA QUALITE DES DONNEES

A. Méthodologie

Des efforts considérables ont été déployés pour veiller à ce que les données présentées dans cette publication correspondent aux définitions de l'AIE figurant dans les Notes générales (parties I.2 et 3). Ces définitions sont utilisées par la plupart des organisations internationales qui recueillent des statistiques sur l'énergie. Toutefois, les statistiques nationales sur l'énergie communiquées à ces organisations sont souvent fondées sur des critères et définitions qui diffèrent, parfois sensiblement, de ceux qu'elles emploient. On trouvera ci-après des informations sur les différences repérées par le Secrétariat de l'AIE et sur les ajustements qu'il a pu, le cas échéant, opérer pour faire correspondre les données aux définitions internationales. Les différences notoires relevées dans certains pays sont signalées dans la partie I.6, Notes et sources par pays. Les notes par pays indiquent seulement quelques-unes des déviations les plus importantes et les plus évidentes constatées dans certains pays par rapport à la méthodologie de l'AIE, et ne constituent en aucun cas une liste exhaustive des différences et anomalies par pays.

B. Estimation

Outre les ajustements opérés pour compenser les différences de définition, des estimations se révèlent parfois nécessaires pour compléter certains agrégats importants pour lesquels il manque des statistiques essentielles. On en trouvera des exemples dans les notes plus détaillées proposées ci-dessous. Sauf en ce qui concerne les énergies combustibles renouvelables et les déchets (voir plus loin E.5), le Secrétariat s'est attaché à fournir tous les éléments des bilans par produit jusqu'à la Consommation finale pour tous les pays et toutes les années. Il a fallu pour cela indiquer les éléments de l'offre ainsi que les entrées de combustibles primaires et les sorties de combustibles secondaires correspondant aux principales activités de transformation telles que le raffinage du pétrole et la production d'électricité. Des estimations ont souvent dû être établies, après consultation d'offices statistiques nationaux, de compagnies pétrolières, de compagnies d'électricité et d'experts nationaux en matière d'énergie.

C. Séries chronologiques et changements politiques

Les bilans par produit des républiques de l'ex-URSS ont été établis à partir de 1992. Ces bilans ont été établis à l'aide de données officielles et, lorsque cela a été nécessaire, des estimations ont été effectuées à partir de données obtenues auprès de diverses industries et d'autres organisations internationales. Il est à noter que les tableaux récapitulatifs et les bilans par produit de la région "ex-URSS" sont à présent la somme des pays individuels de l'ex-URSS. Dans les éditions précédentes à cette publication, les échanges entre pays membres de l'ex-URSS n'étaient pas comptabilisés ; d'autre part, les valeurs calorifiques moyennes nettes étaient appliquées aux agrégats régionaux au lieu de facteurs spécifiques à chaque pays. Ceci est aussi le cas pour l'"ex-Yougoslavie".

Les bilans par produit de la République slovaque ont été établis à partir de 1971. Pour les années 1989 à 1998, les bilans de la République slovaque ont été établis à partir de données officielles. Pour les années 1971 à 1988, des estimations ont été faites lorsque c'était nécessaire (voir Notes et sources par pays).

Les statistiques énergétiques de certains pays ont subi diverses modifications soit au niveau des données prises en compte soit de la méthodologie utilisée, de sorte que des données énergétiques plus détaillées ont pu être présentées, par exemple, pour la Chine à partir de 1980. En conséquence, certaines "ruptures de séries" apparaissent inévitables.

Le Secrétariat de l'AIE réexamine ses bases de données chaque année. A la lumière des nouvelles évaluations, d'importantes révisions sont apportées aux séries chronologiques des pays au cours de cet examen. En conséquence, les données de la présente publication ont été sensiblement révisées par rapport aux précédentes éditions. Veuillez vous référer aux notes relatives aux différents pays.

D. Classification des utilisations des combustibles

Les sources statistiques nationales manquent souvent d'informations adéquates concernant la consommation de combustibles suivant les différentes catégories d'utilisations finales. Beaucoup de pays ne font pas d'enquêtes annuelles de la consommation de combustibles dans les principaux secteurs de l'économie, c'est pourquoi les données publiées sont fondées sur des enquêtes anciennes. La ventilation par secteurs de la consommation dans les différents pays doit donc être interprétée avec prudence.

Avant les réformes entreprises dans les années 1990, la classification sectorielle de la consommation de combustibles dans les économies en transition (d'Europe de l'Est et de l'ex-URSS) et en Chine était très différente de celle pratiquée dans les économies de marché. La consommation sectorielle était définie d'après la branche de l'économie à laquelle appartenait l'utilisateur du combustible et non d'après la vocation ou l'usage du combustible. Par exemple, la consommation d'essence des véhicules d'une entreprise appartenant au secteur de la sidérurgie faisait partie de la consommation industrielle d'essence du secteur de la sidérurgie. Les données ont été ajustées dans la mesure du possible pour correspondre aux classifications internationales: par exemple, la totalité de la consommation d'essence est supposée relever du secteur des transports et est reportée comme telle dans la présente publication. Cependant, il n'a pas été possible de reclasser tous les produits aussi facilement que l'essence et le carburéacteur, et peu d'ajustements ont été opérés sur les autres produits.

E. Problèmes particuliers par combustible

1. Pétrole

Le Secrétariat de l'AIE recueille des statistiques très complètes concernant l'offre et la consommation de pétrole, y compris l'autoconsommation des raffineries, les soutages maritimes et le pétrole utilisé en pétrochimie. Les statistiques nationales font rarement état de ces chiffres. La production indiquée de produits raffinés correspond souvent à la production nette, et non brute, des raffineries et la consommation de produits pétroliers se limite parfois aux ventes sur les marchés intérieurs et ne comprend pas les livraisons aux transports internationaux maritimes ou aériens. Le pétrole utilisé comme produit de base en pétrochimie dans des complexes intégrés de raffinage/pétrochimie n'est souvent pas comptabilisé dans les statistiques officielles disponibles.

Lorsque cela est possible, le Secrétariat procède, en consultation avec l'industrie pétrolière, à une estimation des données non communiquées. En l'absence de toute autre indication sur l'utilisation de combustibles par les raffineries, la consommation des raffineries a été estimée à environ 5 pour cent des entrées en raffinerie et divisée à parts égales entre le gaz de raffinerie et le fioul lourd.

Pour une description des ajustements apportés à la consommation sectorielle de produits pétroliers, se reporter au paragraphe ci-devant intitulé "Classification des utilisations des combustibles".

2. Gaz naturel

L'AIE définit la production de gaz naturel comme la production commercialisable, c'est-à-dire nette des pertes à l'extraction et des quantités brûlées à la torchère, rejetées et réinjectées. De plus, le gaz naturel doit être constitué essentiellement de méthane, et les autres gaz tels que l'éthane et les hydrocarbures à chaîne plus longue doivent être comptabilisés dans la rubrique "pétrole".

Toutefois, les données directement disponibles sur le gaz naturel n'indiquent souvent pas le détail des pertes à l'extraction ainsi que des quantités brûlées à la torchère, rejetées et réinjectées. Les données communiquées sont rarement accompagnées de définitions adéquates, c'est pourquoi il est difficile, voire impossible, d'identifier le gaz aux différents stades de sa séparation en gaz sec (méthane) et en fractions plus lourdes dans les installations de séparation.

Les statistiques concernant l'offre et la demande de gaz naturel sont généralement exprimées en unités de volume et il est difficile d'obtenir des données précises sur le pouvoir calorifique du gaz. Lorsque ce chiffre n'est pas connu, un pouvoir calorifique supérieur de 38 TJ/million de m^3 a été appliqué.

Signalons par ailleurs qu'il est souvent difficile d'obtenir des données fiables sur la ventilation de la consommation de gaz naturel. Cette constatation s'applique tout particulièrement à certains des plus grands pays consommateurs de gaz naturel du Moyen-Orient. Pour cette raison, la consommation industrielle de gaz naturel de ces pays n'est souvent pas indiquée dans les données figurant dans la présente publication.

3. Electricité

Comme l'indiquent les Notes générales (partie I.2), on appelle autoproducteur d'électricité une entreprise qui, en plus de ses activités principales, produit de l'énergie électrique destinée en totalité ou en partie à couvrir ses besoins propres. Les pays non membres de l'OCDE disposent rarement de données répondant à cette définition. Pour ces pays, les entrées de combustibles correspondant à l'autoproduction d'électricité sont indiquées dans la rubrique utilisation finale industrielle.

Lorsqu'il existe des statistiques concernant la production d'électricité à partir d'énergies Renouvelables Combustibles et de Déchets, elles sont inclues dans la production totale d'électricité. Ces données ne sont pas complètes : par exemple, l'électricité produite à partir des déchets végétaux des sucreries n'est généralement pas indiquée.

Les entrées de combustibles utilisés pour produire de l'électricité sont estimées (lorsqu'elles ne sont pas communiquées) à l'aide de données sur la production d'électricité, le rendement énergétique et le type d'installation de production.

4. Chaleur

Les économies en transition (d'Europe de l'Est et de l'ex-URSS) et la Chine utilisaient, s'agissant de la chaleur, une méthodologie qui les démarquait des pratiques communément adoptées dans les économies de marché. Cette méthodologie consistait à comptabiliser la transformation par l'industrie de combustibles primaires (charbon, pétrole et gaz) en chaleur destinée à la *consommation sur place* dans l'activité de transformation Production de chaleur et non dans la consommation industrielle, comme le prévoit la méthodologie de l'AIE[1] ; la production de Chaleur était ensuite répartie entre les différents secteurs d'utilisation finale. Les pertes survenant pendant la transformation des combustibles en chaleur par l'industrie n'étaient pas comptabilisées dans la consommation industrielle finale. Bien que plusieurs pays aient récemment adopté la pratique des organisations internationales, cette différence non négligeable limite la validité des comparaisons entre les pays en transition et les économies de marché pour ce qui est de la consommation finale par secteur.

---

1   La méthodologie internationale ne comptabilise la chaleur dans le secteur de transformation que dans la mesure où elle est vendue à des tiers. Voir la définition dans la partie I.2.

5. Energies renouvelables combustibles et déchets

L'AIE publie les données de production et d'approvisionnement intérieur en énergies renouvelables combustibles et en déchets de tous les pays non-membres de l'OCDE et toutes les régions pour les années 1974 à 1998 (1971 à 1998 sur disquettes).

Ces données proviennent souvent de sources secondaires, elles ne sont pas harmonisées et leur qualité est douteuse, ce qui rend difficile les comparaisons entre les pays. Les données historiques de nombreux pays proviennent d'enquêtes souvent irrégulières, non harmonisées et menées à un niveau local plutôt que national. Elles ne sont donc comparables ni entre régions ni dans le temps. Lorsque des séries chronologiques étaient incomplètes ou non disponibles, le Secrétariat a procédé à l'estimation des données selon une méthodologie compatible avec le cadre prévisionnel de l'édition 1998 de l'ouvrage de l'AIE intitulé *World Energy Outlook* (septembre 1998). Premièrement, l'approvisionnement intérieur par habitant de biomasse et de déchets a été calculé ou estimé pour 1995. Deuxièmement, l'approvisionnement par habitant a été estimé pour les années 1971 à 1994 en utilisant une équation log/log avec comme variable exogène soit le PIB par habitant, soit le taux d'urbanisation, selon la région. Troisièmement, l'approvisionnement total de biomasse et de déchets après 1996 a été estimé en supposant un taux de croissance soit constant, soit égal à la croissance de la population, soit conforme à l'estimation sur la période 1971-1994. Cependant, ces séries historiques estimées doivent être traitées avec prudence.

Le tableau ci-dessous fournit des indications sommaires sur la qualité des données et la méthode d'estimation par région.

Région	Principales sources de données	Qualité des données	Variable exogène
Afrique	base de données FAO et BAfD	Faible	taux d'urbanisation
Amérique latine	nationales et OLADE	Élevée	aucune
Asie	enquêtes	Élevée à faible	PIB per habitant
Europe non-OCDE	questionnaires et FAO	Élevée à moyenne	aucune
ex-USSR	nationales et questionnaires et FAO	Élevée à moyenne	aucune
Moyen-Orient	FAO	Moyenne à faible	aucune

Bien qu'un soin tout particulier ait été apporté à la compatibilité des estimations avec celles réalisées dans la partie biomasse du *World Energy Outlook 1998*, des différences mineures subsistent pour certaines raisons. Parmi ces raisons, il convient de souligner des révisions plus tardives des données utilisées ainsi que des niveaux différents de désagrégations par pays.

Pour les années 1994 à 1998, tous les éléments des bilans jusqu'à la consommation par usage ont été compilés par produit ou catégorie de produits pour tous les pays. Les données pour les années 1996 et 1998 sont fournies dans les Tableaux annuels. La production de charbon de bois est fournie dans les Tableaux récapitulatifs. Les valeurs confirment l'importance des énergies d'origine végétale dans le secteur énergétique de nombreux pays en voie de développement.

L'AIE espère que l'inclusion de ces données encouragera les administrations et les autres agences spécialisées à améliorer la qualité et étendre la couverture de la collecte des données. Plus de détails sur la méthode suivie par pays peuvent être obtenus sur demande, et les commentaires sont bienvenus.

2. NOTES GENERALES

Les tableaux couvrent l'ensemble des sources "commerciales" d'énergie, c'est-à-dire à la fois les sources primaires (houille, lignite, tourbe, gaz naturel, pétrole brut, liquides de gaz naturel (LGN) et énergies hydrauliques, géothermique/solaire, éolienne, marémotrice, etc. ainsi que l'énergie nucléaire) et les sources secondaires (dérivés du charbon, gaz manufacturés, produits pétroliers, électricité et chaleur). Des données sont également indiquées pour diverses sources d'énergies renouvelables combustibles et de déchets, telles que la biomasse solide et les produits d'origine animale, les gaz/liquides tirés de la biomasse, les déchets urbains et les déchets industriels.

Chaque tableau est divisé en trois parties principales: la première récapitulant les éléments de *l'approvisionnement*; la deuxième faisant état des secteurs *transformation* et *énergie*; et la troisième indiquant la *consommation finale* ventilée entre les divers secteurs d'utilisation finale.

La description ci-dessus concerne les bilans par produit complets. Dans le cas des bilans régionaux et des bilans agrégés les définitions des produits sont les mêmes que pour les bilans complets. Cependant, la ventilation par secteur est réduite aux principaux totaux (secteur énergie, etc.) et à des catégories de consommation choisies (sidérurgie, etc.).

A. Approvisionnement

La première partie du bilan énergétique de base fournit les éléments suivants de l'approvisionnement:

Production
+ *Apports d'autres sources*
+ *Importations*
- *Exportations*
- *Soutages maritimes internationaux*
± *Variations des stocks*
= *Approvisionnement intérieur*

1. Production

La *production* comprend les quantités de combustibles extraites ou produites, après extraction des matières inertes ou des impuretés (par exemple, après extraction du soufre contenu dans le gaz naturel).

2. Apports d'autres sources

La rubrique *apports d'autres sources* couvre tous les apports de produits dont l'origine ne correspond pas explicitement aux définitions des sources d'énergie primaire figurant dans les tableaux ; par exemple, pour la catégorie pétrole brut : les produits provenant d'autres sources que le pétrole brut ou les LGN, comme l'hydrogène, le pétrole brut de synthèse (y compris les huiles minérales extraites de minéraux bitumineux tels que les schistes, les sables asphaltiques, etc.) ; pour les additifs : le benzol, l'alcool et le méthanol produits à partir du gaz naturel ; pour les produits d'alimentation des raffineries : les quantités renvoyées par l'industrie pétrochimique et utilisées comme produits d'alimentation des raffineries ; pour la houille : les schlamms et les mixtes, la poussière récupérée et d'autres produits charbonniers de basse qualité, qui ne sont pas classables d'après le type de charbon d'origine ; pour le gaz d'usine à gaz : le gaz naturel, le gaz de raffinerie et le GPL, traités ou mélangés dans les usines à gaz (c'est-à-dire, gaz d'usine à gaz produit à partir d'autres sources que le charbon).

3. Importations et exportations

La rubrique *importations* et *exportations* désigne les quantités de produits ayant franchi les frontières du territoire national, que le dédouanement ait été effectué ou non.

a) Charbon

Les *importations* et *exportations* comprennent les quantités de combustibles obtenues d'autres pays ou fournies à d'autres pays, qu'il existe ou non une union économique ou douanière entre les pays en question. Le charbon en transit n'est pas pris en compte.

b) Pétrole et gaz

Cette rubrique comprend les quantités de pétrole brut et de produits pétroliers importées ou exportées au titre d'accords de traitement (à savoir, raffinage à façon). Les quantités de pétrole en transit ne sont pas prises en compte. Le pétrole brut, les LGN et le gaz naturel sont répertoriés en fonction de leur pays d'origine. Pour les produits d'alimentation des raffineries et les produits pétroliers, en revanche, c'est le dernier pays de provenance qui est pris en compte.

Les réexportations de pétrole importé pour raffinage en zone franche sont comptabilisées dans les exportations de produits pétroliers par le pays de raffinage vers le pays de destination finale.

c) Electricité

Les quantités sont considérées comme importées ou exportées lorsqu'elles ont franchi les limites territoriales du pays. Si l'électricité transite par un pays le montant correspondant est inclu à la fois dans les importations et les exportations.

4. Soutages maritimes internationaux

Les *soutages maritimes internationaux* correspondent aux quantités fournies aux navires de haute mer, y compris les navires de guerre, quel que soit leur pavillon. La consommation des navires assurant le transport par cabotage ou navigation intérieure et des navires de pêche n'est pas comprise. Voir ci-dessous les définitions des secteurs des transports (section 2.C.2) et de l'agriculture (section 2.C.3).

5. Variations des stocks

Les *variations des stocks* expriment la différence enregistrée entre le premier jour et le dernier jour de l'année dans le niveau des stocks détenus sur le territoire national par les producteurs, les importateurs, les entreprises de transformation de l'énergie et les gros consommateurs. Les variations des quantités de pétrole et de gaz stockées dans les oléoducs et les gazoducs ne sont pas prises en compte. Sauf chez les gros consommateurs susmentionnés, les variations des stocks des utilisateurs finals ne sont pas comptabilisées. Une augmentation des stocks est indiquée par un chiffre négatif, tandis qu'une diminution apparaît sous la forme d'un chiffre positif.

6. Approvisionnement intérieur

L'*approvisionnement intérieur* est ainsi défini : production + apports d'autres sources + importations - exportations - soutages maritimes internationaux ± variations des stocks.

7. Transferts

La rubrique *transferts* comprend les lignes *transferts entre produits*, *produits transférés* et *produits recyclés*.

Les *transferts entre produits* visent les produits dont le classement a changé soit parce que leurs spécifications ont été modifiées soit parce qu'ils ont été mélangés pour former un autre produit. Ainsi, le kérosène peut être reclassé comme gazole après mélange avec ce dernier produit pour obtenir un gazole conforme aux spécifications hivernales. Le solde net des *transferts entre produits* est nul.

Les *produits transférés* sont des produits pétroliers importés pour subir un traitement complémentaire dans des raffineries. Par exemple : le fioul importé pour conversion dans une raffinerie est transféré dans la catégorie des produits d'alimentation.

Les *produits recyclés* sont des produits finis qui sont remis dans le circuit commercial, **après** avoir été livrés une première fois au consommateur final (par exemple les lubrifiants usés qui sont régénérés).

8. Ecarts statistiques

L'écart statistique est défini comme les livraisons destinées à la consommation finale + les quantités utilisées pour la transformation et la consommation dans le secteur de l'énergie + les pertes de distribution – l'approvisionnement intérieur – les transferts. En effet, les écarts statistiques proviennent des données relatives aux différentes composantes de l'approvisionnement qui sont souvent tirées par l'administration nationale de sources différentes. En outre, la prise en compte des variations des stocks de certains gros consommateurs dans la partie approvisionnement du bilan crée des distorsions qui contribuent aux écarts statistiques.

B.   Secteurs transformation et énergie

Le *secteur transformation* englobe les activités de transformation des formes d'énergie primaire en énergie secondaire, et de transformation ultérieure (par exemple, celle du charbon à coke en coke, du pétrole brut en produits pétroliers, du fioul lourd en électricité).

Le *secteur énergie* englobe les quantités de combustibles utilisées par les industries productrices d'énergie (par exemple, pour le chauffage, l'éclairage et le fonctionnement de tous les équipements intervenant dans le processus d'extraction, ou encore pour la traction et la distribution).

Dans les secteurs *transformation* et *énergie* on distingue les catégories suivantes :

1.   Secteur Transformation

- *Centrales électriques* (désigne les centrales conçues pour produire uniquement de l'électricité). Si une unité ou plus de la centrale est une installation de cogénération (et que l'on ne peut pas comptabiliser séparément, sur une base unitaire, les combustibles utilisés et la production), elle est considérée comme une centrale de cogénération. Tant les centrales publiques[2] que les installations des autoproducteurs[3] entrent dans cette rubrique.

---

2   La production publique désigne les installations dont la *principale activité* est la production d'électricité et/ou de chaleur pour la vente à des tiers. Elles peuvent appartenir au secteur privé ou public. Il convient de noter que les ventes ne se font pas nécessairement par l'intermédiaire du réseau public.

3   L'autoproduction désigne les installations qui produisent de l'électricité et/ou de la chaleur, en totalité ou en partie pour leur

- *Centrales de cogénération chaleur/électricité* (désigne les centrales conçues pour produire de la chaleur et de l'électricité). L'UNIPEDE les appelle "installations de production combinée d'énergie électrique et de chaleur". Dans la mesure du possible, les consommations de combustibles et les productions de chaleur/électricité doivent être exprimées sur la base des unités plutôt que des centrales. Cependant, à défaut de données disponibles exprimées sur une base unitaire, il convient d'adopter la convention indiquée ci-dessus pour la définition d'une centrale de cogénération. *On notera que, dans le cas des installations de cogénération chaleur/électricité des autoproducteurs, sont comptabilisés tous les combustibles utilisés pour la production d'électricité, tandis que seule la partie des combustibles utilisés pour la production de chaleur **vendue** est indiquée. Les combustibles utilisés pour la production de la chaleur destinée à la consommation interne des autoproducteurs **ne sont pas** comptabilisés dans cette rubrique mais dans les données concernant la consommation finale de combustibles du secteur de consommation approprié.*

- *Centrales calogènes* (désigne les installations [pompes à chaleur et chaudières électriques comprises] conçues pour produire uniquement de la chaleur et qui en vendent à des tiers [par exemple, consommateurs des secteurs résidentiels, commerciaux ou industriels] selon les termes d'un contrat). Cette rubrique comprend aussi bien les centrales publiques que les installations des autoproducteurs.

- *Hauts-fourneaux et Usines à gaz* (couvrant les quantités de combustibles utilisées pour la production de gaz de ville, de gaz de haut fourneau et de gaz de convertisseur à oxygène). Dans la production de fonte brute à partir de minerai de fer dans les hauts-fourneaux, les combustibles sont utilisés pour la charge des hauts-fourneaux et pour l'apport de chaleur et de carbone nécessaire à la réduction du minerai. Comptabiliser le pouvoir calorifique des combustibles qui entrent dans ce procédé est une tâche complexe, car la transformation (en

---

consommation propre, en tant qu'activité qui contribue à leur activité principale. Elles peuvent appartenir au secteur privé ou public.

gaz de haut fourneau) et la consommation (de chaleur de combustion) interviennent simultanément. Il se produit aussi une rétention de carbone dans la fonte brute; la quasi-totalité de ce carbone réapparaît ultérieurement dans le gaz de convertisseur à oxygène (appelé aussi gaz de convertisseur ou gaz LD) lorsque la fonte brute est transformée en acier. En principe, les quantités de tous les combustibles (par exemple, charbon utilisé pour l'injection de charbon pulvérisé, coke de four à coke, gaz naturel et fioul) entrant dans les hauts-fourneaux et la quantité de gaz de haut fourneau et gaz de convertisseur à oxygène produite sont recueillies. Cependant, à l'exception des entrées de coke de four, les données sont souvent douteuses ou incomplètes. Le Secrétariat doit répartir ces apports entre les secteurs de transformation et de consommation. L'élément correspondant à la transformation est indiqué dans la ligne *hauts-fourneaux et usines à gaz* à la colonne appropriée pour le combustible dont il s'agit, et l'élément correspondant à la consommation est indiqué dans la ligne *sidérurgie,* à la consommation finale, à la colonne appropriée pour le combustible dont il s'agit. Avec la présente publication, le Secrétariat a décidé de supposer une ventilation des entrées de coke de four de 80% et 20% respectivement entre les secteurs de transformation et de consommation. Le résultat de ce changement, par rapport aux éditions précédentes, est une augmentation de la quantité d'énergie entrant à la ligne *hauts-fourneaux et usines à gaz* et une diminution à la ligne *sidérurgie.* Cette ventilation a été estimée à partir des résultats du modèle appliqué aux pays de l'OCDE.[4] Elle fournit un équilibrage plus cohérent entre les entrées de carbone dans les hauts fourneaux et les sorties correspondantes, lorsque les données de l'AIE sont utilisées pour calculer les émissions de CO_2 dues à la combustion, en utilisant la méthodologie du Groupe d'experts intergouvernemental sur l'évolution du climat (GIEC) telle qu'elle est publiée dans *Lignes directrices révisées en 1996 du GIEC pour les inventaires nationaux de gaz à effet de serre*[5].

- *Cokeries/Fabriques d'agglomérés/Fabriques de briquettes de lignite* (couvrant les combustibles utilisés pour la production de coke, de gaz de cokerie, d'agglomérés et de briquettes de lignite [BKB]).

- *Raffineries de pétrole* (couvrant les hydrocarbures utilisés pour la production de produits pétroliers finis).

- *Industrie pétrochimique* (couvrant les produits renvoyés aux raffineries par l'industrie pétrochimique). Il convient de noter que les retours en raffinerie des produits pétroliers utilisés à des fins non énergétiques (i.e. white spirit et lubrifiants) ne sont pas inclus sous cette rubrique, mais sous utilisations non énergétiques.

- La *liquéfaction* (comprend divers procédés de liquéfaction, notamment la liquéfaction de charbon et de gaz naturel en Afrique du Sud).

- *Secteur transformation - autres* (couvrant les pertes des charbonnières et les autres transformations non spécifiées).

2. Secteur énergie

La consommation propre du secteur de la production d'énergie comprend l'énergie consommée par les industries de transformation pour la chauffe, le pompage, la traction et l'éclairage [Divisions 10, 11, 12, 23 et 40 de la CITI[6]] :

- *mines de charbon* (houille et lignite) ;

- *extraction de pétrole et de gaz* (le gaz brûlé à la torche n'est pas compris) ;

- *raffineries de pétrole* ;

- *centrales électriques, centrales de cogénération et centrales calogènes* ;

- *énergie absorbée par le pompage* (électricité consommée dans les centrales hydrauliques) ;

- *secteur énergie-autres* (comprend la consommation propre des fabriques

---

4   Voir *Statistiques de l'énergie des pays de l'OCDE* pour plus d'informations sur le modèle d'estimation des hauts fourneaux appliqué aux pays de l'OCDE.

5   La publication *Lignes directrices révisées en 1996 du GIEC pour les inventaires nationaux de gaz à effet de serre* est

disponible auprès du Programme du GIEC sur les inventaires de gaz à effet de serre à http://www.ipcc-nggip.iges.or.jp.

6   Classification internationale type par industrie de toutes les branches d'activité économique, Série M, No. 4/Rév. 3, Nations Unies, New York, 1990.

d'agglomérés, des cokeries, des usines à gaz, des fabriques de briquettes et de coke de lignite, ainsi que les utilisations non spécifiées du secteur énergie).

3. Pertes de distribution

Les *pertes de distribution* incluent les pertes enregistrées lors de la distribution du gaz, du transport de l'électricité et du transport du charbon. Peut également inclure des usages non comptabilisés de pétrole brut et de produits pétroliers.

C. Consommation finale

Le terme *consommation finale* (qui correspond à la somme des consommations des secteurs d'utilisation finale) signifie que l'énergie utilisée pour la transformation et pour la consommation propre des industries productrices d'énergie est exclue. La consommation finale recouvre la majeure partie des livraisons aux consommateurs (voir la note sur les variations des stocks à la section 2.A.5).

Dans la consommation finale, les produits d'alimentation de l'industrie pétrochimique sont inclus dans le secteur **industrie** dans une sous-catégorie de la rubrique *industrie chimique* pour les produits pétroliers qui sont utilisés essentiellement à des fins énergétiques. En revanche, les produits pétroliers qui sont principalement utilisés à des fins non énergétiques (voir utilisations non énergétiques à la section 2.C.4), sont indiqués sous les rubriques *utilisations non énergétiques* et inclus uniquement dans la **consommation finale totale**. Les retours de produits de l'industrie pétrochimique ne sont pas pris en compte dans la consommation finale (voir les sections 2.A.2 et 2.B.1).

1. Secteur industrie

La consommation du *secteur industrie* est répartie entre les sous-secteurs suivants (l'énergie utilisée par l'industrie pour le transport n'est pas prise en compte ici mais figure dans la rubrique transports) :

- *Sidérurgie* [Groupe 271 et Classe 2731 de la CITI] ;

- *Industrie chimique* [Division 24 de la CITI] ; *dont* : produits d'alimentation de l'industrie pétrochimique. L'industrie pétrochimique

comprend les opérations de craquage et de reformage destinées à la production de l'éthylène, du propylène, du butylène, du gaz de synthèse, des aromatiques, du butadiène et d'autres matières premières à base d'hydrocarbures [partie du Groupe 241 de la CITI] ; voir produits d'alimentation, dans la section 2.C.4 (Utilisations non énergétiques) ;

- Industries de base des *métaux non ferreux* [Groupe 272 et Classe 2732 de la CITI] ;

- *Produits minéraux non métalliques* tels que verre, céramiques, ciment, etc. [Division 26 de la CITI] ;

- *Matériel de transport* [Divisions 34 et 35 de la CITI] ;

- *Construction mécanique*. Fabrication d'ouvrages en métaux, de machines et de matériels à l'exclusion du matériel de transport [Divisions 28, 29, 30, 31 et 32 de la CITI] ;

- *Industries extractives (à l'exception des combustibles)* [Divisions 13 et 14 de la CITI] ;

- *Industrie alimentaire et tabacs* [Divisions 15 et 16 de la CITI] ;

- *Papier, pâte à papier et imprimerie* [Divisions 21 et 22 de la CITI] ;

- *Bois et produits dérivés* (à exclusion de la pâte à papier et du papier) [Division 20 de la CITI] ;

- *Construction* [Division 45 de la CITI] ;

- *Textiles et cuir* [Divisions 17, 18 et 19 de la CITI] ;

- *Non spécifiés* (tout autre secteur industriel non spécifié précédemment) [Divisions 25, 33, 36 et 37 de la CITI].

Note : La plupart des pays éprouvent des difficultés pour fournir une ventilation par branche d'activité pour tous les combustibles. Dans ces cas, la rubrique *non spécifiés* a été utilisée. *Les agrégats régionaux de la consommation industrielle doivent donc être employés avec précaution.*

2. Secteur transports

La consommation dans le *secteur transports* couvre toutes les activités de transport (liées à des moteurs mobiles) quel que soit le secteur économique concerné [Divisions 60, 61 et 62 de la CITI], et elle est ventilée entre les différents sous-secteurs suivants :

- *Transport aérien* : Livraisons de carburants aviation à l'aviation civile internationale et pour toutes les activités de transport aérien intérieur, à savoir commerciales, privées, agricoles, militaires, etc. Comprend également les quantités utilisées à des fins autres que le vol proprement dit, par exemple, l'essai de moteurs au banc, mais non le carburant utilisé par les compagnies aériennes pour le transport routier ;

- *Transport routier* : La totalité des carburants utilisés dans les véhicules routiers (militaires compris) ainsi que le carburant consommé par les transports agricoles et industriels sur route. Ne tient pas compte de l'essence moteur employée dans les moteurs fixes, ni du gazole utilisé par les tracteurs ailleurs que sur route ;

- *Transport ferroviaire* : Toutes les quantités utilisées par le trafic ferroviaire, y compris par les chemins de fer industriels ;

- *Transport par conduites* : L'énergie utilisée pour le transport de substances par conduites ;

- *Navigation intérieure* (y compris la consommation des petites embarcations et des bateaux de cabotage n'achetant pas leur soutage aux termes de contrats de soutages maritimes internationaux). Le carburant utilisé pour la pêche en haute mer, le long du littoral et dans les eaux intérieures doit être comptabilisé dans le secteur agriculture ;

- *Non spécifiés.*

- Note : La plupart des pays éprouvent des difficultés à fournir une ventilation détaillée de la consommation industrielle pour tous les combustibles. Dans ces cas-là, la rubrique *non spécifiés* a été utilisée. *Les agrégats régionaux de la consommation industrielle doivent donc être employés avec précaution. Voir les notes relatives aux différents Pays (Section 6).*

3. Autres secteurs

- *Agriculture* : Cette rubrique couvre, par définition, toutes les livraisons aux usagers classés dans les rubriques agriculture, chasse et sylviculture de la CITI, et comprend donc les produits énergétiques consommés par ces usagers que ce soit pour la traction automobile (à l'exception des carburants utilisés par les engins agricoles sur route), pour la production d'énergie ou le chauffage (dans les secteurs agricoles ou résidentiels). Elle comprend aussi

les carburants utilisés pour la pêche en haute mer, le long du littoral et dans les eaux intérieures. Divisions 01, 02 et 05 de la CITI ;

- *Services marchands et publics* : Cette rubrique recouvre toutes les activités qui relèvent des Divisions 41, 50, 51, 52, 55, 63, 64, 65, 66, 67, 70, 71, 72, 73, 74, 75, 80, 85, 90, 91, 92, 93 et 99 ;

- *Résidentiel* : Cette rubrique couvre toutes les quantités consommées par les ménages, à l'exception des combustibles utilisés dans les transports. Elle comprend les ménages employant du personnel domestique (Division 95 de la CITI), ce qui représente une faible part de la consommation résidentielle totale ;

- *Non spécifiés* : Cette rubrique couvre toutes les quantités de combustibles consommées qui n'ont pas été précisées ailleurs (par exemple, la consommation de combustibles pour les activités militaires, à l'exclusion des carburants des soutages maritimes internationaux, du secteur du transport routier et du transport aérien intérieur, et la consommation dans les catégories précitées pour lesquelles des données ventilées n'ont pas été fournies).

4. Utilisations non énergétiques

Les utilisations non énergétiques comprennent la consommation des *autres* produits pétroliers, notamment white spirit, paraffines, lubrifiants, bitume et produits divers (voir les sections 3.C.10 et 3.C.11). Ils incluent également les utilisations non énergétiques du charbon (excepté pour la tourbe). Ces produits se trouvent ventilés à part, dans la consommation finale, sous la rubrique *utilisations non énergétiques*. Il est présumé que l'usage de ces produits est strictement non énergétique. Il convient de noter que le coke de pétrole ne figure sous la rubrique *utilisations non énergétiques* que si cette utilisation est prouvée ; dans le cas contraire ce produit est comptabilisé avec les utilisations énergétiques dans l'industrie ou dans les autres secteurs.

Les chiffres concernant les produits d'alimentation de l'industrie pétrochimique sont comptabilisés dans le secteur de l'industrie, au titre de l'industrie chimique et figurent séparément à la rubrique : *dont : produits d'alimentation.* Sont compris dans cette rubrique tous les produits pétroliers, y compris le naphta, à l'exception des *autres produits pétroliers* énumérés à la section 3.C.11.

3. NOTES CONCERNANT LES SOURCES D'ENERGIE

A. Charbon

La rubrique *mines de charbon* ne comprend que le charbon utilisé directement par les charbonnages. Elle exclut le charbon consommé par les centrales électriques minières (compris dans la rubrique *transformation* - centrales électriques) et les quantités de charbon allouées gratuitement aux mineurs et à leurs familles (considérées comme consommation des ménages et classées de ce fait dans la rubrique *autres secteurs* - résidentiel).

1. Charbon à coke

On appelle charbon à coke un charbon d'une qualité permettant la production d'un coke susceptible d'être utilisé dans les hauts-fourneaux. Son pouvoir calorifique supérieur dépasse 23 865 kJ/kg (5 700 kcal/kg), valeur mesurée pour un combustible exempt de cendres, mais humide.

2. Autres charbons bitumineux et anthracite

Les autres charbons bitumineux sont utilisés pour la production de vapeur et pour le chauffage des locaux. Cette catégorie comprend tous les charbons anthraciteux et bitumineux autres que les charbons à coke. Son pouvoir calorifique supérieur dépasse 23 865 kJ/kg (5 700 kcal/kg), mais est généralement inférieur à celui du charbon à coke.

3. Charbons sous-bitumineux

On appelle charbons sous-bitumineux les charbons non agglutinants d'un pouvoir calorifique supérieur compris entre 17 435 kJ/kg (4 165 kcal/kg) et 23 865 kJ/kg (5 700 kcal/kg), contenant plus de 31 pour cent de matières volatiles sur produit sec exempt de matières minérales.

4. Lignite et charbon sous-bitumineux

Le lignite/charbon sous-bitumineux est un charbon non agglutinant dont le pouvoir calorifique supérieur n'atteint pas 17 435 kJ/kg (4 165 kcal/kg), et qui contient plus de 31 pour cent de matières volatiles sur produit sec exempt de matières minérales. Les schistes bitumineux sont également inclus dans cette catégorie.

5. Tourbe

Sédiment fossile d'origine végétale poreux ou comprimé, combustible à haute teneur en eau (jusqu'à 90 pour cent sur brut), facilement rayé, de couleur brun clair à brun foncé. La tourbe utilisée à des fins non énergétiques n'est pas prise en compte.

6. Coke de four à coke (coke de cokerie) et coke d'usine à gaz

Le coke de cokerie est un produit solide obtenu par carbonisation à haute température du charbon, et surtout du charbon à coke ; la teneur en eau et en matières volatiles est faible. Le semi-coke, produit solide obtenu par carbonisation à basse température, le coke, le semi-coke de lignite/charbon sous-bitumineux, le poussier de coke et le coke de fonderie sont également inclus dans cette rubrique. La rubrique *secteur énergie-autres* représente la consommation interne des cokeries. La consommation de l'industrie sidérurgique ne comprend pas le coke transformé en gaz de haut fourneau. Pour obtenir la consommation totale de coke de cokerie de l'industrie sidérurgique, il faut ajouter les quantités de coke transformées en gaz de haut fourneau (elles apparaissent sous la rubrique *hauts-fourneaux et usines à gaz*).

Le coke d'usine à gaz est un sous-produit de la houille utilisée pour la production de gaz de ville dans les usines à gaz. Il est principalement utilisé pour le chauffage. La rubrique *secteur énergie-autres* couvre la consommation de coke de gaz dans les usines à gaz.

7. Agglomérés et briquettes de lignite (et de tourbe) (BKB)

Les agglomérés sont des combustibles composites fabriqués à partir de fines de charbon par moulage avec adjonction d'un liant. La quantité d'agglomérés produite est donc légèrement supérieure au tonnage de houille effectivement utilisé à cet effet. La consommation d'agglomérés durant le processus de fabrication des agglomérés apparaît sous la rubrique *secteur énergie-autres*.

Les briquettes de lignite (BKB) sont des combustibles composites fabriqués à partir du lignite/charbon sous-bitumineux, et agglomérés sous haute pression. Ces données couvrent les briquettes de tourbe, le lignite séché, la poussière de lignite et les fines de lignite. La consommation des usines de briquettes est comprise sous la rubrique *secteur énergie-autres*.

B. Pétrole brut, liquides de gaz naturel et produits d'alimentation des raffineries

Sous le titre *transformation*, la rubrique *raffineries de pétrole* indique les quantités de pétrole brut, de LGN, de produits d'alimentation des raffineries, d'additifs et d'autres hydrocarbures qui sont utilisées dans le processus de raffinage.

1. Pétrole brut

C'est une huile minérale, constituée d'un mélange d'hydrocarbures d'origine naturelle. Sa couleur va du jaune au noir, sa densité et sa viscosité sont variables. Cette catégorie comprend aussi les condensats (provenant des séparateurs) directement récupérés sur les périmètres d'exploitation des hydrocarbures gazeux dans les installations de séparation des phases liquide et gazeuse.

Les autres hydrocarbures, notamment le pétrole brut synthétique, les huiles minérales extraites des roches bitumineuses telles que schistes, sables asphaltiques, etc. ainsi que les huiles issues de la liquéfaction du charbon figurent à la ligne *autres sources*. Voir la section 2.A.2.

Les huiles émulsionnées (par exemple, l'orimulsion) sont prises en compte dans cette catégorie.

2. Liquides de gaz naturel (LGN)

Les LGN sont des hydrocarbures liquides ou liquéfiés obtenus pendant le traitement, la purification et la stabilisation du gaz naturel. Il s'agit des fractions de gaz naturel qui sont récupérées sous forme liquide dans les installations de séparation, dans les installations sur les gisements ou dans les usines de traitement du gaz. Les LGN comprennent l'éthane, le propane, le butane, le pentane, l'essence naturelle et les condensats, sans que la liste soit limitative. Ils peuvent aussi inclure certaines quantités de substances autres que des hydrocarbures.

3. Produits d'alimentation des raffineries

C'est un produit ou une combinaison de produits dérivés du pétrole brut et destinés à subir un traitement ultérieur autre qu'un mélange dans l'industrie du raffinage. Il est transformé en un ou plusieurs constituants et/ou produits finis. Cette définition recouvre les produits finis qui sont importés pour la consommation des raffineries et ceux qui sont renvoyés par l'industrie pétrochimique aux raffineries.

4. Additifs

Les additifs sont des substances autres que des hydrocarbures qui sont ajoutées ou mélangées à un produit afin de modifier ses propriétés, pour améliorer par exemple ses caractéristiques lors de la combustion. Les alcools et les éthers (MTBE ou méthyl tertio-butyl éther), ou des substances telles que le plomb tétraéthyle, sont comptabilisés dans cette rubrique. L'éthanol n'y figure pas, mais apparaît à la rubrique *gaz/liquides tirés de la biomasse*.

C. Produits pétroliers

Ce sont tous les produits dérivés du pétrole qui peuvent être obtenus par distillation et qui sont, en général, utilisés en dehors de l'industrie du raffinage. Les produits finis classés comme produits d'alimentation des raffineries (voir ci-dessus) n'entrent pas dans cette catégorie.

Les données sur la *production* de produits pétroliers font apparaître, pour chaque produit, la production brute des raffineries.

La consommation de combustibles des raffineries (figurant dans le *secteur énergie*, ligne *raffineries de pétrole*) représente leur consommation de produits pétroliers, qu'il s'agisse de produits intermédiaires ou de produits finis, utilisés par exemple pour le chauffage, l'éclairage, la traction, etc.

1. Gaz de raffinerie (non liquéfiés)

Cette catégorie couvre, par définition, les gaz non condensables obtenus dans les raffineries lors de la distillation du pétrole brut ou du traitement des produits pétroliers (par craquage, par exemple). Il s'agit principalement d'hydrogène, de méthane, d'éthane et d'oléfines. Sont compris également les gaz retournés aux raffineries par l'industrie pétrochimique. La production de gaz de raffinerie correspond à la production brute. La consommation propre des raffineries est comptabilisée séparément à la ligne *raffineries de pétrole* dans la rubrique *secteur énergie*.

2. Gaz de pétrole liquéfiés (GPL) et éthane

Il s'agit des fractions légères d'hydrocarbures paraffiniques qui s'obtiennent lors du raffinage ainsi que dans les installations de stabilisation du pétrole brut et de traitement du gaz naturel. Ce sont le propane (C_3H_8) et le butane (C_4H_{10}) ou un mélange de ces deux hydrocarbures. Ils sont généralement liquéfiés sous pression pour le transport et le stockage.

L'éthane (C_2H_6) est un hydrocarbure à chaîne droite, gazeux à l'état naturel. C'est un gaz paraffinique incolore que l'on extrait du gaz naturel et des gaz de raffinerie.

3. Essence moteur

C'est un hydrocarbure léger utilisé dans les moteurs à combustion interne, tels que ceux des véhicules à moteur, à l'exception des aéronefs. L'essence moteur est distillée entre 35°C et 215°C et utilisée comme carburant pour les moteurs terrestres à allumage commandé. L'essence moteur peut contenir des additifs, des composés oxygénés et des additifs améliorant l'indice d'octane, notamment des composés plombés comme le PTE (plomb tétraéthyle) et le PTM (plomb tétraméthyle).

4. Essence aviation

Il s'agit d'une essence spécialement préparée pour les moteurs à pistons des avions, avec un indice d'octane adapté au moteur, un point de congélation de -60°C et un intervalle de distillation habituellement compris entre 30°C et 180°C.

5. Carburéacteurs

Cette catégorie comprend les carburéacteurs type essence et les carburéacteurs type kérosène, qui répondent aux spécifications d'utilisation pour les turbomoteurs pour avion.

a) Carburéacteur type essence

Cette catégorie comprend tous les hydrocarbures légers utilisés dans les turbomoteurs pour avion. Ils distillent entre 100°C et 250°C. Ils sont obtenus par mélange de kérosène et d'essence ou de naphtas, de manière à ce que la teneur en aromatiques soit égale ou inférieure à 25 pour cent en volume, et que la pression de vapeur se situe entre 13,7 kPa et 20,6 kPa. Des additifs peuvent être ajoutés afin d'accroître la stabilité et la combustibilité du carburant.

b) Carburéacteur type kérosène

C'est un distillat moyen utilisé dans les turbomoteurs pour avion, qui répond aux même caractéristiques de distillation et présente le même point d'éclair que le kérosène (entre 150°C et 300°C, mais ne dépassant pas 250°C en général). De plus, il est conforme à des spécifications particulières (concernant notamment le point de congélation), définies par l'Association du transport aérien international (IATA).

6. Kérosène

Le kérosène comprend les distillats de pétrole raffiné dont la volatilité est comprise entre celle de l'essence et celle du gazole/carburant diesel. C'est une huile moyenne qui distille entre 150°C et 300°C.

7. Gazole/carburant diesel (Distillat de coupe intermédiaire)

Les gazoles/carburants diesel sont des huiles lourdes. Les gazoles sont extraits de la dernière fraction issue de la distillation atmosphérique du pétrole brut, tandis que les gazoles lourds sont obtenus par redistillation sous vide du résidu de la distillation atmosphérique. Le gazole/carburant diesel distille entre 180°C et 380°C. Plusieurs qualités sont disponibles, selon l'utilisation : gazole pour moteur diesel à allumage par compression (automobiles, poids lourds, bateaux, etc.), fioul léger pour le chauffage des locaux industriels et commerciaux, et autres gazoles, y compris les huiles lourdes distillant entre 380°C et 540°C utilisées comme produit d'alimentation dans l'industrie pétrochimique.

8. Fioul lourd (résiduel)

Ce sont les huiles lourdes constituant le résidu de distillation. La définition englobe tous les fiouls résiduels (y compris ceux obtenus par mélange). La viscosité cinétique est supérieure à 10 cSt à 80°C. Le point d'éclair est toujours supérieur à 50°C, et la densité est toujours supérieure à 0,9 kg/l.

9. Naphtas

Les naphtas sont un produit d'alimentation des raffineries destiné soit à l'industrie pétrochimique (par exemple, fabrication d'éthylène ou production de composés aromatiques) soit à la production d'essence par reformage ou isomérisation dans la raffinerie. Les naphtas correspondent aux fractions distillant entre 30°C et 210°C ou sur une partie de cette plage de température. Les naphtas importés pour mélange doivent être indiqués dans les importations, puis repris à la ligne *transferts*, affectés d'un signe négatif pour les naphtas, et d'un signe positif pour les produits finis correspondants (par exemple, essence).

10. Coke de pétrole

Le coke de pétrole est un résidu solide noir brillant, obtenu principalement par craquage et carbonisation de produits d'alimentation dérivés du pétrole, de produits de la distillation sous vide du pétrole brut, de goudrons et de poix, dans des procédés tels que la cokéfaction différée ou la cokéfaction fluide. Il se compose essentiellement de carbone (90 à 95 pour cent) et brûle en laissant peu de cendres. Il est employé comme produit d'alimentation dans les cokeries des usines sidérurgiques, pour la chauffe, pour la fabrication d'électrodes et pour la production de substances chimiques. Les deux qualités les plus importantes de coke sont le coke de pétrole et le coke de pétrole calciné. Cette catégorie comprend également le coke de catalyse, qui se dépose sur le catalyseur pendant les opérations de raffinage ; ce coke n'est pas récupérable, et il est en général brûlé comme combustible dans les raffineries.

11. Autres produits pétroliers

La catégorie *autres produits pétroliers* regroupe les white spirit et SBP, les lubrifiants, le bitume, les paraffines et d'autres produits.

a) White spirit et essences spéciales (SBP)

Ce sont des distillats intermédiaires raffinés, dont l'intervalle de distillation se situe entre celui des naphtas et celui du kérosène.

Ils se subdivisent en :

i) Essences spéciales (SBP) : Huiles légères distillant entre 30°C et 200°C et dont l'écart de température entre les points de distillation de 5 pour cent et 90 pour cent en volume, y compris les pertes, est inférieur ou égal à 60°C. En d'autres termes, il s'agit d'une huile légère, de coupe plus étroite que celle des essences moteur. On distingue 7 ou 8 qualités d'essences spéciales, selon la position de la coupe dans l'intervalle de distillation défini plus haut.

ii) White spirit : Essence industrielle dont le point d'éclair est supérieur à 30°C. L'intervalle de distillation du white spirit est compris entre 135°C et 200°C.

b) *Lubrifiants*

Les lubrifiants sont des hydrocarbures obtenus à partir de distillats ou de résidus ; ils sont principalement utilisés pour réduire les frottements entre surfaces d'appui. Cette catégorie comprend tous les grades d'huiles lubrifiantes, depuis les spindles jusqu'aux huiles à cylindres, et les huiles entrant dans les graisses, y compris les huiles moteur et tous les grades d'huiles de base pour lubrifiants.

c) *Bitume*

Hydrocarbure solide, semi-solide ou visqueux, à structure colloïdale, de couleur brune à noire ; c'est un résidu de la distillation du pétrole brut obtenu par distillation sous vide des huiles résiduelles de distillation atmosphérique. Le bitume est aussi souvent appelé asphalte, et il est principalement employé pour le revêtement des routes et pour les matériaux de toiture. Cette catégorie couvre le bitume fluidisé et le cutback.

d) *Paraffines*

Hydrocarbures aliphatiques saturés (dont la formule générale est C_nH_{2n+2}). Les paraffines sont des résidus du déparaffinage des huiles lubrifiantes ; elles présentent une structure cristalline avec $C > 12$ et, pour principales caractéristiques, d'être incolores, inodores et translucides ainsi que d'avoir un point de fusion supérieur à 45°C.

e) *Autres*

Tous les produits pétroliers qui ne sont pas classés ci-dessus, par exemple, le goudron, le soufre et la graisse. Cette catégorie comprend également les composés aromatiques (par exemple, BTX ou benzène, toluène et xylènes) et les oléfines (par exemple, propylène) produits dans les raffineries.

D. Gaz

Les valeurs relatives aux quatre catégories de gaz sont toutes exprimées en térajoules, sur la base du **pouvoir calorifique supérieur**.

1. Gaz naturel

Le gaz naturel est constitué de gaz, méthane essentiellement, sous forme liquide ou gazeuse, extraits de gisements naturels souterrains. Il peut s'agir aussi bien de gaz "non associé" provenant de gisements qui produisent uniquement des hydrocarbures sous forme gazeuse, que de gaz "associé" obtenu en même temps que le pétrole brut, ou de méthane récupéré dans les mines de charbon (grisou).

La production est mesurée après élimination des impuretés et extraction des LGN et du soufre. Elle exclut les quantités de gaz réinjectées et les quantités rejetées ou brûlées à la torchère. Elle comprend les quantités de gaz utilisées dans l'industrie gazière et le gaz transporté par gazoduc.

2. Gaz d'usine à gaz

Cette catégorie couvre tous les types de gaz produits dans les usines des entreprises publiques ou privées ayant pour principal objet la production, le transport et la distribution de gaz. Cette catégorie comprend le gaz produit par carbonisation (y compris le gaz produit dans les fours à coke et transféré aux usines à gaz), par gazéification totale avec ou sans enrichissement au moyen de produits pétroliers, par craquage du gaz naturel ou par reformage et simple mélange avec d'autres gaz et/ou de l'air. Cette rubrique recouvre également le gaz naturel de substitution dont le pouvoir calorifique est élevé, et qui est produit par conversion chimique d'hydrocarbures.

3. Gaz de cokerie

Le gaz de cokerie est un sous-produit de la fabrication du coke de cokerie utilisé en sidérurgie.

4. Gaz de haut fourneau

Le gaz de haut fourneau est obtenu lors de la combustion du coke dans les hauts-fourneaux de l'industrie sidérurgique. Il est récupéré et utilisé comme combustible, en partie dans l'usine même, et en partie pour d'autres procédés de l'industrie sidérurgique, ou encore dans des centrales électriques dotées d'équipements adaptés pour en brûler. Cette rubrique comprend également le gaz obtenu comme sous-produit lors de l'élaboration de l'acier dans les fours à oxygène ou convertisseurs basiques avec soufflage d'oxygène, qui est récupéré à la sortie du gueulard. Ce gaz est également appelé gaz de convertisseur ou gaz LD ou gaz BOS.

E. Energies renouvelables combustibles et déchets

Les données concernant ces quatre catégories de combustibles sont toutes exprimées en térajoules, sur la base du **pouvoir calorifique inférieur**.

1. Biomasse solide et produits d'origine animale

La biomasse est, par définition, toute matière végétale utilisée directement comme combustible ou transformée avant de la brûler sous une autre forme. Elle comprend le bois, les déchets végétaux (y compris les déchets de bois et les cultures destinées à la production d'énergie), les matières/déchets d'origine animale, les lessives sulfitiques (résidus de fabrication de la pâte à papier), également appelées "liqueur noire" (liqueur alcaline de rejet des digesteurs lors de la production de pâte au sulfate ou à la soude dans le procédé d'élaboration du papier, dont le contenu énergétique provient de la lignine extraite de la pâte chimique) et autre biomasse solide.

Le charbon de bois produit à partir de la biomasse solide est également inclus ici. Etant donné que le charbon de bois est un produit secondaire, son traitement est légèrement différent de celui de la biomasse primaire. La production de charbon de bois (un produit du processus de transformation) est compensée par les consommations correspondantes de biomasse primaire dans le processus de production. Les pertes lors de ce processus sont reportées dans la ligne *Secteur transformation – autres*. Les autres approvisionnements (par exemple le commerce et les variations de stocks) ainsi que la consommation de charbon de bois sont agrégés directement avec la biomasse primaire. Cependant, dans quelques pays, seule la biomasse primaire est reportée. La production de charbon de bois reportée est disponible séparement dans les Tableaux sommaires à la fin de la publication.

2. Gaz/liquides tirés de la biomasse

Ce sont les gaz qui sont produits principalement par fermentation anaérobie de biomasse et de déchets solides et brûlés pour produire de la chaleur et/ou de l'énergie électrique. Cette catégorie comprend les gaz de décharge et les gaz de digestion des boues (gaz issus des eaux usées et des lisiers). Les additifs tirés de la biomasse (ou bio-additifs) comme l'éthanol entrent également dans cette catégorie.

3. Déchets urbains et assimilés

Les déchets urbains correspondent aux produits brûlés directement pour produire de la chaleur et/ou de l'énergie électrique, dont notamment les déchets des secteurs résidentiel et commercial ainsi que du secteur des services publics, qui sont recueillis par les autorités municipales pour leur élimination dans des installations centralisées. Les déchets hospitaliers entrent dans cette catégorie.

4. Déchets industriels

Il s'agit de produits liquides et solides brûlés directement, généralement dans des installations spécialisées, pour produire de la chaleur et/ou de l'énergie électrique et qui ne sont pas notifiés dans la catégorie *biomasse solide et produits d'origine animale*.

F. Electricité et chaleur

1. Electricité

La production brute d'électricité est mesurée aux bornes de tous les groupes d'alternateurs d'une centrale. Elle comprend donc l'énergie absorbée par les équipements auxiliaires et les pertes dans les transformateurs qui sont considérés comme faisant partie intégrante de la centrale.

La différence entre production nette et brute est généralement évaluée à 7 pour cent dans les centrales thermiques classiques, à 1 pour cent dans les centrales hydroélectriques et à 6 pour cent dans les centrales nucléaires, géothermiques ou solaires. La production hydraulique comprend la production des centrales à accumulation par pompage (également appelées centrales de pompage).

2. Chaleur

La production de chaleur destinée à la vente acquiert une importance grandissante depuis quelques années. Pour tenir compte de cette évolution, la production de chaleur représente toute la production publique de chaleur provenant des centrales de cogénération chaleur/électricité et calogènes ainsi que la chaleur vendue à des tiers par les centrales de cogénération chaleur/électricité et calogènes des autoproducteurs. Les quantités correspondantes de combustibles utilisées pour produire la chaleur destinée à la vente sont indiquées, dans le secteur transformation, aux lignes *installations de cogénération* et *centrales calogènes*. Les quantités de combustibles utilisées pour produire la chaleur qui n'est pas vendue sont rapportées dans les secteurs où cette consommation a lieu.

3. Hydroélectricité

Énergie potentielle et cinétique des eaux transformée en électricité dans les centrales hydroélectriques.

4. Énergie géothermique

Énergie thermique provenant de l'intérieur de l'écorce terrestre, généralement sous forme d'eau chaude ou de vapeur. Elle est exploitée dans les sites qui s'y prêtent :

- pour la production d'électricité en mettant à profit la vapeur sèche ou la saumure naturelle de haute enthalpie après vaporisation instantanée

- directement sous forme de chaleur pour le chauffage urbain, l'agriculture, etc.

5. Énergie solaire

Rayonnement solaire exploité pour la production d'eau chaude et d'électricité, au moyen de :

- capteurs plans, qui fonctionnent essentiellement en thermosiphon, pour la production d'eau chaude sanitaire ou pour le chauffage saisonnier des piscines

- cellules photovoltaïques

- centrales thermo-hélio-électriques

L'énergie solaire passive pour le chauffage, la climatisation et l'éclairage directs des logements ou autres bâtiments n'est pas prise en compte.

6. Energies marémotrice/houlomotrice

Énergie mécanique résultant du mouvement des marées, de la houle ou des vagues exploitée pour la production d'électricité.

7. Énergie éolienne

Énergie cinétique du vent exploitée pour la production d'électricité au moyen d'aérogénérateurs.

4. NOTES CONCERNANT LES TABLEAUX RECAPITULATIFS

Production de charbon de bois

Le charbon de bois est un produit secondaire de la biomasse, qui se compose principalement de carbone, issu de la combustion lente et incomplète du bois ou d'autres produits végétaux en l'absence d'air. On notera que dans les tableaux annuels, le charbon de bois est inclus dans la biomasse solide (voir les Notes concernant les sources d'énergie, partie 3).

Demande de produits raffinés par groupes principaux, en milliers de tonnes et millions de barils-jour

La demande de produits raffinés par groupes principaux est définie comme la somme de l'approvisionnement intérieur, des transferts et des écarts statistiques moins les entrées dans les raffineries, les industries pétrochimique et de liquéfaction du secteur transformation, et la consommation propre du secteur énergie. *L'essence moteur* comprend les additifs et les liquides tirés de la biomasse. Les *autres* comprennent le pétrole brut, les autres hydrocarbures, le gaz de raffinerie, le coke de pétrole, le white spirit et les essences spéciales, les lubrifiants, le bitume, les paraffines et les autres produits tels que le goudron, le soufre, la graisse ainsi que les composés aromatiques (par exemple, BTX ou benzène, toluène et xylène) et les oléfines (par exemple, propylène) produits dans les raffineries.

Combustibles de raffinerie indique la consommation propre de combustibles pétroliers par les raffineries pour les opérations d'équipement, de chauffage et d'éclairage, qui comprennent principalement des gaz de raffinerie, du gazole-carburant diesel et du fioul lourd.

Les *soutages* maritimes internationaux de combustibles liquides comprennent principalement du gazole-carburant diesel et du fioul lourd.

Les coefficients de conversion suivants sont valables pour tous les pays et pour toutes les années.

Coefficients de Conversion

	Barils/tonne
Gaz de raffinerie	8.00
Ethane	16.85
GPL	11.60
Naphta	8.50 (8.90 OCDE Europe)
Essence aviation	8.90
Essence moteur	8.53 (8.45 OCDE Europe)
Carburéacteur type essence	7.93 (8.25 OCDE)
Carburéacteur type kérosène	7.93 (7.88 OCDE Europe)
Autre kérosène	7.74 (7.88 OCDE Europe)
Gazole/carburant diesel	7.46
Fioul lourd	6.66 (6.45 OCDE Europe)
White Spirit	7.00 (8.46 OCDE)
Lubrifiants	7.09
Bitume	6.08
Paraffines	7.00 (7.85 OCDE)
Coke de pétrole	5.50
Autres produits non spécifiés	7.00 (8.00 OCDE)

Consommation d'électricité

Comprend l'utilisation de l'électricité dans les secteurs de transformation, d'énergie et de consommation finale, mais ne comprend pas les pertes de distribution.

5. COUVERTURE GEOGRAPHIQUE

L'Afrique comprend l'Afrique du Sud, l'Algérie, l'Angola, le Bénin, le Cameroun, le Congo, la République démocratique du Congo, la Côte d'Ivoire, l'Egypte, l'Erythrée, l'Ethiopie, le Gabon, le Ghana, le Kenya, la Libye, le Maroc, le Mozambique, le Nigéria, le Sénégal, le Soudan, la République unie de Tanzanie, le Togo, la Tunisie, la Zambie, le Zimbabwe et les **autres pays d'Afrique.**

Les autres pays d'Afrique comprennent le Botswana, le Burkina Faso, le Burundi, le Cap-Vert, la République centrafricaine, Djibouti, la Gambie, la Guinée, la Guinée-Bissau, la Guinée équatoriale, le Lesotho, le Libéria, Madagascar, le Malawi, le Mali, la Mauritanie, Maurice, le Niger, l'Ouganda, le Rwanda, Sao Tomé et Principe, les Seychelles, la Sierra Leone, la Somalie, le Swaziland et le Tchad.

Le Moyen-Orient comprend l'Arabie saoudite, Bahreïn, les Emirats arabes unis, la République islamique d'Iran, l'Iraq, Israël, la Jordanie, le Koweït, le Liban, Oman, le Qatar, la Syrie et le Yémen.

La région Europe hors OCDE comprend l'Albanie, la Bosnie-Herzégovine, la Bulgarie, Chypre, la Croatie, Gibraltar, l'ex-République yougoslave de Macédoine (FYROM), Malte, la Roumanie, la Slovaquie, la Slovénie et la République fédérative de Yougoslavie.

L'ex-URSS comprend l'Arménie, l'Azerbaïdjan, la Biélorussie, l'Estonie, la Géorgie, le Kazakhstan, le Kirghizistan, la Lettonie, la Lituanie, la République de Moldovie, l'Ouzbékistan, la Fédération de Russie, le Tadjikistan, le Turkménistan et l'Ukraine.

L'Amérique latine comprend les Antilles néerlandaises, l'Argentine, la Bolivie, le Brésil, le Chili, la Colombie, le Costa Rica, Cuba, la République dominicaine, El Salvador, l'Equateur, le Guatemala, Haïti, le Honduras, la Jamaïque, le Nicaragua, Panama, le Paraguay, le Pérou, Trinité-et-Tobago, l'Uruguay, le Venezuela et les **autres pays d'Amérique latine.**

Les autres pays d'Amérique latine comprennent Antigua-et-Barbuda, les Bahamas, la Barbade, le Belize, les Bermudes, la Dominique, la Grenade, la Guadeloupe, le Guyana, la Guyane française, la Martinique, Saint-Kitts et Nevis, Anguilla, Sainte-Lucie, Saint-Vincent et les Grenadines et le Suriname.

La Chine comprend la République populaire de Chine et Hong Kong (Chine).

L'Asie comprend le Bangladesh, Brunei, la République populaire démocratique de Corée, l'Inde, l'Indonésie, la Malaisie, Myanmar, le Népal, le Pakistan, les Philippines, Singapour, Sri Lanka, le Taipei chinois, la Thaïlande, le Viet Nam et les **autres pays d'Asie**.

Les autres pays d'Asie comprennent l'Afghanistan, le Bhoutan, les Fidji, Kiribati, les Maldives, la Nouvelle-Calédonie, la Papouasie-Nouvelle-Guinée, la Polynésie française, le Samoa, les Iles Salomon et Vanuatu.

L'Organisation de coopération et de développement économiques (OCDE) comprend l'Allemagne, l'Australie, l'Autriche, la Belgique, le Canada, la Corée, le Danemark, l'Espagne, les Etats-Unis, la Finlande, la France, la Grèce, la Hongrie, l'Irlande, l'Islande, l'Italie, le Japon, le Luxembourg, le Mexique, la Norvège, la Nouvelle-Zélande, les Pays-Bas, la Pologne, le Portugal, la République tchèque, le Royaume-Uni, la Suède, la Suisse et la Turquie.

Dans la zone de l'OCDE :

• L'**Australie** ne comprend pas les territoires d'outre-mer.

- L'**Allemagne** tient compte des nouveaux Länder à partir de 1970.

- Le Groenland et les Iles Féroé danoises ne sont pas pris en compte dans les données relatives au **Danemark**.

- L'**Espagne** englobe les Iles Canaries.

- Les **Etats-Unis** englobent Porto-Rico, Guam et les Iles Vierges ainsi que la zone franche d'Hawaï.

- Dans les données relatives à la **France**, Monaco est pris en compte, mais non les départements d'outre-mer (à savoir la Guadeloupe, la Guyane, la Martinique, la Nouvelle-Calédonie, la Polynésie française, la Réunion et Saint Pierre et Miquelon).

- L'**Italie** englobe Saint-Marin et le Vatican.

- Le **Japon** englobe Okinawa.

- Ni le Suriname ni les Antilles néerlandaises ne sont pris en compte dans les données relatives aux **Pays-Bas**.

- Le **Portugal** englobe les Açores et l'Ile de Madère.

- La **Suisse** englobe le Liechtenstein.

L'Organisation des pays exportateurs de pétrole (OPEP) comprend l'Algérie, l'Indonésie, l'Iran, l'Irak, le Koweit, la Libye, le Nigéria, Qatar, l'Arabie saoudite, les Emirats arabes unis et le Venezuela.

On notera que les pays suivants n'ont pas été pris en compte par suite d'un manque de données :

- **Afrique** : Comores, Namibie, Sainte-Hélène et Sahara Occidental.

- **Amérique** : Aruba, Iles Vierges Britanniques, Iles Caïmanes, Iles Falkland, Montserrat, Saint-Pierre et Miquelon et les Iles Turks et Caïcos.

- **Asie et Océanie** : Samoa américaines, Cambodge, Ile Christmas, Iles Cook, Laos, Macao, Mongolie, Nauru, Nioué, Palaos, Tonga et Ile de Wake.

PART II:
STATISTICAL DATA

PARTIE II
DONNEES STATISTIQUES

INTERNATIONAL ENERGY AGENCY

ANNUAL TABLES

TABLEAUX ANNUELS

1997-1998

World / Monde : 1997

SUPPLY AND CONSUMPTION / APPROVISIONNEMENT ET DEMANDE	Coal / Charbon (1000 tonnes)							Oil / Pétrole (1000 tonnes)			
	Coking Coal / Charbon à coke	Other Bit. Coal / Autres charb. bit.	Sub-Bit. Coal / Charbon sous-bit.	Lignite / Lignite	Peat / Tourbe	Oven and Gas Coke / Coke de four/gaz	Pat. Fuel and BKB / Agg./briq. de lignite	Crude Oil / Pétrole brut	NGL / LGN	Feed-stocks / Produits d'aliment.	Additives / Additifs
Production	605421	2883090	463673	801861	22593	362544	26589	3254259	198384	-	10354
From Other Sources	150	3071	427	-	-	5	-	722	-	24547	1704
Imports	192724	331229	5328	7675	14	18973	1188	1846421	16747	49915	3105
Exports	-180797	-350257	-5833	-1290	-158	-24190	-1553	-1770474	-59067	-7425	-2003
Stock Changes	-2177	-7623	7334	2733	-573	2655	57	-28014	-21	-565	114
DOMESTIC SUPPLY	**615321**	**2859510**	**470929**	**810979**	**21876**	**359987**	**26281**	**3302914**	**156043**	**66472**	**13274**
Intl. Marine Bunkers	-	-	-	-	-	-	-	-	-	-	-
Transfers	-	-	-	-	-	-	-	-3369	-100837	44137	-698
Statistical Differences	-9222	-8111	7218	2345	-236	-16157	-1	9256	-1369	1680	333
TRANSFORMATION	**567267**	**1948854**	**438009**	**766435**	**17181**	**239631**	**1188**	**3287582**	**42106**	**112289**	**12902**
Electricity Plants	16479	1533603	396183	549905	5571	53	698	29605	280	-	-
CHP Plants	42	304806	39080	170896	6452	-	242	2609	-	-	-
Heat Plants	45	38528	792	16872	1630	193	110	5	-	-	-
Blast Furnaces/Gas Works	8951	25184	1214	7041	-	239189	138	-	219	-	-
Coke/Pat. Fuel/BKB Plants	541750	18967	740	21721	3528	196	-	-	-	-	-
Petroleum Refineries	-	-	-	-	-	-	-	3263755	41607	112289	12902
Petrochemical Industry	-	-	-	-	-	-	-	-	-	-	-
Liquefaction	-	27766	-	-	-	-	-	-7909	-	-	-
Other Transform. Sector	-	-	-	-	-	-	-	-483	-	-	-
ENERGY SECTOR	**213**	**76834**	**32**	**1289**	**154**	**1964**	**99**	**8483**	**1401**	**-**	**-**
Coal Mines	80	31460	31	370	96	526	49	13	-	-	-
Oil and Gas Extraction	-	3211	-	-	-	26	-	5901	1401	-	-
Petroleum Refineries	-	175	-	-	-	-	-	2568	-	-	-
Electr., CHP+Heat Plants	-	33730	1	47	-	75	-	1	-	-	-
Pumped Storage (Elec.)	-	-	-	-	-	-	-	-	-	-	-
Other Energy Sector	133	8258	-	872	58	1337	50	-	-	-	-
Distribution Losses	10	5572	33	4102	225	124	-	7305	689	-	-
FINAL CONSUMPTION	**38609**	**820139**	**40073**	**41498**	**4080**	**102111**	**24993**	**5431**	**9641**	**-**	**7**
INDUSTRY SECTOR	**38190**	**594728**	**35891**	**22216**	**1228**	**92392**	**3092**	**4748**	**9641**	**-**	**7**
Iron and Steel	-	-	-	-	-	-	-	-	-	-	-
Chemical and Petrochem.	-	-	-	-	-	-	-	-	-	-	-
of which: Feedstocks	-	-	-	-	-	-	-	-	-	-	-
Non-Ferrous Metals	-	-	-	-	-	-	-	-	-	-	-
Non-Metallic Minerals	-	-	-	-	-	-	-	-	-	-	-
Transport Equipment	-	-	-	-	-	-	-	-	-	-	-
Machinery	-	-	-	-	-	-	-	-	-	-	-
Mining and Quarrying	-	-	-	-	-	-	-	-	-	-	-
Food and Tobacco	-	-	-	-	-	-	-	-	-	-	-
Paper, Pulp and Print	-	-	-	-	-	-	-	-	-	-	-
Wood and Wood Products	-	-	-	-	-	-	-	-	-	-	-
Construction	-	-	-	-	-	-	-	-	-	-	-
Textile and Leather	-	-	-	-	-	-	-	-	-	-	-
Non-specified	-	-	-	-	-	-	-	-	-	-	-
TRANSPORT SECTOR	**8**	**14963**	**214**	**431**	**-**	**2**	**22**	**16**	**-**	**-**	**-**
Air	-	-	-	-	-	-	-	-	-	-	-
Road	-	-	-	-	-	-	-	-	-	-	-
Rail	-	-	-	-	-	-	-	-	-	-	-
Pipeline Transport	-	-	-	-	-	-	-	-	-	-	-
Internal Navigation	-	-	-	-	-	-	-	-	-	-	-
Non-specified	-	-	-	-	-	-	-	-	-	-	-
OTHER SECTORS	**-**	**-**	**-**	**-**	**-**	**-**	**-**	**-**	**-**	**-**	**-**
Agriculture	-	-	-	-	-	-	-	-	-	-	-
Comm. and Publ. Services	-	-	-	-	-	-	-	-	-	-	-
Residential	-	-	-	-	-	-	-	-	-	-	-
Non-specified	-	-	-	-	-	-	-	-	-	-	-
NON-ENERGY USE	**240**	**8354**	**-**	**428**	**49**	**1353**	**237**	**-**	**-**	**-**	**-**
in Industry/Trans./Energy	-	-	-	-	-	-	-	-	-	-	-
in Transport	-	-	-	-	-	-	-	-	-	-	-
in Other Sectors	-	-	-	-	-	-	-	-	-	-	-

World / Monde : 1997

SUPPLY AND CONSUMPTION / APPROVISIONNEMENT ET DEMANDE	Oil cont. / Pétrole cont. (1000 tonnes)										
	Refinery Gas / Gaz de raffinerie	LPG + Ethane / GPL + éthane	Motor Gasoline / Essence moteur	Aviation Gasoline / Essence aviation	Jet Fuel / Carbu-réacteurs	Kerosene / Kérosène	Gas/ Diesel / Gazole	Heavy Fuel Oil / Fioul lourd	Naphtha / Naphta	Petrol. Coke / Coke de pétrole	Other Prod. / Autres prod.
Production	101543	87325	824003	1619	204263	84557	967910	649536	161973	65658	210745
From Other Sources	-	3203	-	-	-	35	96	-	-	-	255
Imports	-	53429	87280	641	29970	16163	177042	154289	66291	18264	37579
Exports	-	-31007	-106320	-227	-38227	-16913	-207546	-212172	-57024	-23529	-36998
Stock Changes	5	-2625	-2026	76	-1638	-323	-5356	5535	-651	-809	-277
DOMESTIC SUPPLY	101548	110325	802937	2109	194368	83519	932146	597188	170589	59584	211304
Intl. Marine Bunkers	-	-	-	-	-14	-	-26840	-97529	-	-	-345
Transfers	100	85030	4534	196	-943	-2365	-674	-6434	-7962	-6	-6884
Statistical Differences	618	1614	-2106	5	-2141	-113	4106	1842	1125	208	3164
TRANSFORMATION	5311	4637	547	-	-	626	42418	250099	20872	8232	1711
Electricity Plants	1177	529	38	-	-	78	38853	190017	1136	6216	351
CHP Plants	3638	335	-	-	-	9	983	22440	31	248	542
Heat Plants	-	41	-	-	-	18	1225	33155	46	-	192
Blast Furnaces/Gas Works	262	2411	-	-	-	-	135	2772	1375	416	-
Coke/Pat. Fuel/BKB Plants	-	-	-	-	-	-	-	20	-	1287	-
Petroleum Refineries	-	-	-	-	-	-	-	-	-	-	-
Petrochemical Industry	234	1320	509	-	-	521	1222	1694	18284	65	626
Liquefaction	-	-	-	-	-	-	-	-	-	-	-
Other Transform. Sector	-	1	-	-	-	-	-	1	-	-	-
ENERGY SECTOR	92306	3148	1213	-	20	514	10501	53873	224	25635	5058
Coal Mines	-	2	-	-	-	63	1907	295	-	-	2
Oil and Gas Extraction	94	156	911	-	-	10	4593	2134	3	1599	25
Petroleum Refineries	90370	2822	279	-	20	382	2652	46647	217	24029	4575
Electr., CHP+Heat Plants	3	5	6	-	-	7	1071	1757	-	-	-
Pumped Storage (Elec.)	-	-	-	-	-	-	-	-	-	-	-
Other Energy Sector	1839	163	17	-	-	52	278	3040	4	7	456
Distribution Losses	65	353	213	-	17	7	268	173	278	-	169
FINAL CONSUMPTION	4584	188831	803392	2310	191233	79894	855551	190922	142378	25919	200301
INDUSTRY SECTOR	4582	92207	7475	2	77	9121	95964	146143	142378	15176	4193
Iron and Steel	-	-	-	-	-	-	-	-	-	-	-
Chemical and Petrochem.	-	-	-	-	-	-	-	-	-	-	-
of which: Feedstocks	-	-	-	-	-	-	-	-	-	-	-
Non-Ferrous Metals	-	-	-	-	-	-	-	-	-	-	-
Non-Metallic Minerals	-	-	-	-	-	-	-	-	-	-	-
Transport Equipment	-	-	-	-	-	-	-	-	-	-	-
Machinery	-	-	-	-	-	-	-	-	-	-	-
Mining and Quarrying	-	-	-	-	-	-	-	-	-	-	-
Food and Tobacco	-	-	-	-	-	-	-	-	-	-	-
Paper, Pulp and Print	-	-	-	-	-	-	-	-	-	-	-
Wood and Wood Products	-	-	-	-	-	-	-	-	-	-	-
Construction	-	-	-	-	-	-	-	-	-	-	-
Textile and Leather	-	-	-	-	-	-	-	-	-	-	-
Non-specified	-	-	-	-	-	-	-	-	-	-	-
TRANSPORT SECTOR	-	10128	786874	2247	190535	760	494235	17265	-	-	32
Air	-	-	27	2247	190535	642	-	-	-	-	-
Road	-	9966	782739	-	-	55	444859	382	-	-	32
Rail	-	-	-	-	-	-	-	-	-	-	-
Pipeline Transport	-	-	-	-	-	-	-	-	-	-	-
Internal Navigation	-	-	-	-	-	-	-	-	-	-	-
Non-specified	-	-	-	-	-	-	-	-	-	-	-
OTHER SECTORS	-	-	-	-	-	-	-	-	-	-	-
Agriculture	-	-	-	-	-	-	-	-	-	-	-
Comm. and Publ. Services	-	-	-	-	-	-	-	-	-	-	-
Residential	-	-	-	-	-	-	-	-	-	-	-
Non-specified	-	-	-	-	-	-	-	-	-	-	-
NON-ENERGY USE	-	-	-	-	-	-	-	-	-	10533	186407
in Industry/Trans./Energy	-	-	-	-	-	-	-	-	-	-	-
in Transport	-	-	-	-	-	-	-	-	-	-	-
in Other Sectors	-	-	-	-	-	-	-	-	-	-	-

World / Monde : 1997

SUPPLY AND CONSUMPTION / APPROVISIONNEMENT ET DEMANDE	Gas / Gaz (TJ)				Comb. Renew. & Waste / En. Re. Comb. & Déchets (TJ)				(GWh)	(TJ)
	Natural Gas / Gaz naturel	Gas Works / Usines à gaz	Coke Ovens / Cokeries	Blast Furnaces / Hauts fourneaux	Solid Biomass / Biomasse solide	Gas/Liquids from Biomass / Gaz/Liquides tirés de biomasse	Municipal Waste / Déchets urbains	Industrial Waste / Déchets industriels	Electricity / Electricité	Heat / Chaleur
Production	88916493	449038	1841709	2586937	42064256	576172	582949	684308	14048656	11549480
From Other Sources	-	22712	3842	-	2779	690	-	22009	-	-
Imports	20750657	-	-	-	20977	22662	-	-	418643	145
Exports	-20930138	-	-	-	-41845	-3756	-	-	-424194	-145
Stock Changes	177760	-	-	-	-2387	-40922	-	60	-	-
DOMESTIC SUPPLY	88914772	471750	1845551	2586937	42043779	554846	582949	706377	14043105	11549480
Intl. Marine Bunkers	-	-	-	-	-	-	-	-	-	-
Transfers	-	-	-	-	-	-6805	-	-	-	-
Statistical Differences	-47980	12376	-78695	-424	-7158	-15084	-16994	-21734	-4816	-1692
TRANSFORMATION	29835506	31272	359690	891477	3763326	117566	562155	324947	7471	-
Electricity Plants	14634471	4354	195935	651808	417167	59931	331514	61900	-	-
CHP Plants	11222748	13500	108278	227093	1467160	15340	158028	206792	-	-
Heat Plants	3012057	13418	35270	12429	120191	1820	72613	56240	-	-
Blast Furnaces/Gas Works	43724	-	20207	-	1884	-	-	-	-	-
Coke/Pat. Fuel/BKB Plants	3814	-	-	147	6279	-	-	15	-	-
Petroleum Refineries	315	-	-	-	105	40475	-	-	-	-
Petrochemical Industry	-	-	-	-	-	-	-	-	-	-
Liquefaction	446584	-	-	-	-	-	-	-	-	-
Other Transform. Sector	471793	-	-	-	1750540	-	-	-	7471	-
ENERGY SECTOR	8644577	41999	417771	171306	1620	1717	110	8372	1361174	933980
Coal Mines	9396	-	2590	-	9	-	-	-	104774	84340
Oil and Gas Extraction	6722292	-	-	-	-	-	-	-	134962	169813
Petroleum Refineries	1636373	13505	1862	-	-	-	-	163	163836	450904
Electr., CHP+Heat Plants	18860	11275	72	-	56	423	110	-	785704	104018
Pumped Storage (Elec.)	-	-	-	-	-	-	-	-	93436	-
Other Energy Sector	257656	17219	413247	171306	1555	1294	-	8209	78462	124905
Distribution Losses	870083	9246	4070	47051	73	-	60	20726	1155515	678830
FINAL CONSUMPTION	49516626	401609	985325	1476679	38269816	413674	3630	330598	11514129	9934978
INDUSTRY SECTOR	22385781	143925	907364	1476559	5367703	2229	420	308374	4934277	3999033
Iron and Steel	-	-	-	-	-	-	-	-	-	-
Chemical and Petrochem.	-	-	-	-	-	-	-	-	-	-
of which: Feedstocks	-	-	-	-	-	-	-	-	-	-
Non-Ferrous Metals	-	-	-	-	-	-	-	-	-	-
Non-Metallic Minerals	-	-	-	-	-	-	-	-	-	-
Transport Equipment	-	-	-	-	-	-	-	-	-	-
Machinery	-	-	-	-	-	-	-	-	-	-
Mining and Quarrying	-	-	-	-	-	-	-	-	-	-
Food and Tobacco	-	-	-	-	-	-	-	-	-	-
Paper, Pulp and Print	-	-	-	-	-	-	-	-	-	-
Wood and Wood Products	-	-	-	-	-	-	-	-	-	-
Construction	-	-	-	-	-	-	-	-	-	-
Textile and Leather	-	-	-	-	-	-	-	-	-	-
Non-specified	-	-	-	-	-	-	-	-	-	-
TRANSPORT SECTOR	1772010	17	-	-	3088	354253	-	595	225673	-
Air	-	-	-	-	-	-	-	-	-	-
Road	89423	-	-	-	-	354253	-	-	204	-
Rail	-	-	-	-	-	-	-	-	-	-
Pipeline Transport	-	-	-	-	-	-	-	-	-	-
Internal Navigation	-	-	-	-	-	-	-	-	-	-
Non-specified	-	-	-	-	-	-	-	-	-	-
OTHER SECTORS	-	-	-	-	-	-	-	-	-	-
Agriculture	-	-	-	-	-	-	-	-	-	-
Comm. and Publ. Services	-	-	-	-	-	-	-	-	-	-
Residential	-	-	-	-	-	-	-	-	-	-
Non-specified	-	-	-	-	-	-	-	-	-	-
NON-ENERGY USE	-	-	764	120	-	-	-	-	-	-
in Industry/Transf./Energy	-	-	-	-	-	-	-	-	-	-
in Transport	-	-	-	-	-	-	-	-	-	-
in Other Sectors	-	-	-	-	-	-	-	-	-	-

World / Monde : 1998

SUPPLY AND CONSUMPTION / APPROVISIONNEMENT ET DEMANDE	Coal / Charbon (1000 tonnes)							Oil / Pétrole (1000 tonnes)			
	Coking Coal / Charbon à coke	Other Bit. Coal / Autres charb. bit.	Sub-Bit. Coal / Charbon sous-bit.	Lignite / Lignite	Peat / Tourbe	Oven and Gas Coke / Coke de four/gaz	Pat. Fuel and BKB / Agg./briq. de lignite	Crude Oil / Pétrole brut	NGL / LGN	Feed-stocks / Produits d'aliment.	Additives / Additifs
Production	555488	2736304	491771	794158	12200	347097	23657	3317742	200740	-	10835
From Other Sources	-	2488	312	-	-	1	-	523	-	24993	2041
Imports	190477	342747	6507	5942	10	20486	953	1879531	17718	41186	3402
Exports	-174783	-370527	-6258	-548	-180	-25351	-1148	-1850797	-60431	-7649	-3421
Stock Changes	-83	-819	-3156	2385	7797	-997	158	-33132	-470	1624	-75
DOMESTIC SUPPLY	**571099**	**2710193**	**489176**	**801937**	**19827**	**341236**	**23620**	**3313867**	**157557**	**60154**	**12782**
Intl. Marine Bunkers	-	-	-	-	-	-	-	-	-	-	-
Transfers	-	-	-	-	-	-	-	-60	-101479	48675	-595
Statistical Differences	-18153	103326	-3332	1949	-125	-6846	-980	14649	-2205	727	1270
TRANSFORMATION	**515501**	**1991322**	**448066**	**760235**	**15382**	**233500**	**1078**	**3309925**	**42176**	**109556**	**13447**
Electricity Plants	15401	1579641	405350	559910	4288	99	705	26055	342	-	-
CHP Plants	38	309678	40076	162503	5930	-	155	3426	-	-	-
Heat Plants	35	31861	693	12890	1543	149	178	2	-	-	-
Blast Furnaces/Gas Works	11049	24330	1299	6870	-	232891	40	-	263	-	-
Coke/Pat. Fuel/BKB Plants	488978	18271	648	18062	3621	361	-	-	-	-	-
Petroleum Refineries	-	-	-	-	-	-	-	3288882	41571	109556	13447
Petrochemical Industry	-	-	-	-	-	-	-	-	-	-	-
Liquefaction	-	27541	-	-	-	-	-	-7953	-	-	-
Other Transform. Sector	-	-	-	-	-	-	-	-487	-	-	-
ENERGY SECTOR	**242**	**79561**	**30**	**1699**	**100**	**1717**	**49**	**9132**	**439**	**-**	**-**
Coal Mines	55	33650	30	377	37	451	11	13	-	-	-
Oil and Gas Extraction	-	1992	-	-	-	48	-	6199	439	-	-
Petroleum Refineries	-	141	-	-	-	-	-	2908	-	-	-
Electr., CHP+Heat Plants	-	33781	-	33	2	91	-	12	-	-	-
Pumped Storage (Elec.)	-	-	-	-	-	-	-	-	-	-	-
Other Energy Sector	187	9997	-	1289	61	1127	38	-	-	-	-
Distribution Losses	49	5268	40	3841	70	116	-	4499	693	-	-
FINAL CONSUMPTION	**37154**	**737368**	**37708**	**38111**	**4150**	**99057**	**21513**	**4900**	**10565**	**-**	**10**
INDUSTRY SECTOR	**36762**	**550941**	**34321**	**21707**	**1934**	**90389**	**3056**	**4299**	**10565**	**-**	**10**
Iron and Steel	-	-	-	-	-	-	-	-	-	-	-
Chemical and Petrochem.	-	-	-	-	-	-	-	-	-	-	-
of which: Feedstocks	-	-	-	-	-	-	-	-	-	-	-
Non-Ferrous Metals	-	-	-	-	-	-	-	-	-	-	-
Non-Metallic Minerals	-	-	-	-	-	-	-	-	-	-	-
Transport Equipment	-	-	-	-	-	-	-	-	-	-	-
Machinery	-	-	-	-	-	-	-	-	-	-	-
Mining and Quarrying	-	-	-	-	-	-	-	-	-	-	-
Food and Tobacco	-	-	-	-	-	-	-	-	-	-	-
Paper, Pulp and Print	-	-	-	-	-	-	-	-	-	-	-
Wood and Wood Products	-	-	-	-	-	-	-	-	-	-	-
Construction	-	-	-	-	-	-	-	-	-	-	-
Textile and Leather	-	-	-	-	-	-	-	-	-	-	-
Non-specified	-	-	-	-	-	-	-	-	-	-	-
TRANSPORT SECTOR	**7**	**14399**	**210**	**423**	**-**	**1**	**13**	**11**	**-**	**-**	**-**
Air	-	-	-	-	-	-	-	-	-	-	-
Road	-	-	-	-	-	-	-	-	-	-	-
Rail	-	-	-	-	-	-	-	-	-	-	-
Pipeline Transport	-	-	-	-	-	-	-	-	-	-	-
Internal Navigation	-	-	-	-	-	-	-	-	-	-	-
Non-specified	-	-	-	-	-	-	-	-	-	-	-
OTHER SECTORS	**-**	**-**	**-**	**-**	**-**	**-**	**-**	**-**	**-**	**-**	**-**
Agriculture	-	-	-	-	-	-	-	-	-	-	-
Comm. and Publ. Services	-	-	-	-	-	-	-	-	-	-	-
Residential	-	-	-	-	-	-	-	-	-	-	-
Non-specified	-	-	-	-	-	-	-	-	-	-	-
NON-ENERGY USE	**242**	**8058**	**-**	**421**	**7**	**1196**	**232**	**-**	**-**	**-**	**-**
in Industry/Trans./Energy	-	-	-	-	-	-	-	-	-	-	-
in Transport	-	-	-	-	-	-	-	-	-	-	-
in Other Sectors	-	-	-	-	-	-	-	-	-	-	-

World / Monde : 1998

SUPPLY AND CONSUMPTION / *APPROVISIONNEMENT ET DEMANDE*	Oil cont. / *Pétrole cont.* (1000 tonnes)										
	Refinery Gas / *Gaz de raffinerie*	LPG + Ethane / *GPL + éthane*	Motor Gasoline / *Essence moteur*	Aviation Gasoline / *Essence aviation*	Jet Fuel / *Carbu-réacteurs*	Kerosene / *Kérosène*	Gas/ Diesel / *Gazole*	Heavy Fuel Oil / *Fioul lourd*	Naphtha / *Naphta*	Petrol. Coke / *Coke de pétrole*	Other Prod. / *Autres prod.*
Production	103816	88252	838329	1637	205407	83504	967107	652293	165123	67893	216167
From Other Sources	-	3035	-	-	-	-	-	-	-	-	-
Imports	-	53279	90587	714	30387	14875	170109	163520	65530	16943	33770
Exports	-	-34349	-109564	-274	-39119	-15999	-200617	-208126	-55879	-20651	-35199
Stock Changes	4	-2235	-1056	-30	-291	510	-3289	1409	-930	-783	-111
DOMESTIC SUPPLY	**103820**	**107982**	**818296**	**2047**	**196384**	**82890**	**933310**	**609096**	**173844**	**63402**	**214627**
Intl. Marine Bunkers	-	-	-	-	-14	-	-28112	-99452	-	-	-353
Transfers	-31	86516	5276	97	-847	-2176	-2395	-7902	-5616	232	-14993
Statistical Differences	-22	2057	-3203	-13	-1287	243	4715	-3679	901	84	566
TRANSFORMATION	**5522**	**4955**	**556**	**-**	**-**	**651**	**39858**	**256024**	**20418**	**8698**	**1695**
Electricity Plants	1205	361	38	-	-	82	36596	195722	1138	6305	311
CHP Plants	3848	751	-	-	-	4	696	23215	47	231	507
Heat Plants	15	43	-	-	-	21	1155	32792	46	4	253
Blast Furnaces/Gas Works	258	2300	-	-	-	-	144	2287	1279	583	-
Coke/Pat. Fuel/BKB Plants	-	-	-	-	-	-	-	12	-	1575	-
Petroleum Refineries	-	-	-	-	-	-	-	-	-	-	-
Petrochemical Industry	196	1499	518	-	-	544	1267	1995	17908	-	624
Liquefaction	-	-	-	-	-	-	-	-	-	-	-
Other Transform. Sector	-	1	-	-	-	-	-	1	-	-	-
ENERGY SECTOR	**93833**	**2621**	**2018**	**-**	**38**	**495**	**9776**	**55049**	**196**	**26750**	**4732**
Coal Mines	-	1	3	-	-	60	2096	304	-	1	1
Oil and Gas Extraction	438	191	1070	-	-	22	3746	1758	2	1776	32
Petroleum Refineries	91420	2328	882	-	38	382	2649	48409	183	24946	4294
Electr., CHP+Heat Plants	-	-	7	-	-	4	973	1420	-	-	-
Pumped Storage (Elec.)	-	-	-	-	-	-	-	-	-	-	-
Other Energy Sector	1975	101	56	-	-	27	312	3158	11	27	405
Distribution Losses	71	178	126	-	16	1	272	170	335	-	1
FINAL CONSUMPTION	**4341**	**188801**	**817669**	**2131**	**194182**	**79810**	**857612**	**186820**	**148180**	**28270**	**193419**
INDUSTRY SECTOR	**4339**	**91408**	**6850**	**2**	**60**	**9141**	**98105**	**145356**	**148180**	**13730**	**4104**
Iron and Steel	-	-	-	-	-	-	-	-	-	-	-
Chemical and Petrochem.	-	-	-	-	-	-	-	-	-	-	-
of which: Feedstocks	-	-	-	-	-	-	-	-	-	-	-
Non-Ferrous Metals	-	-	-	-	-	-	-	-	-	-	-
Non-Metallic Minerals	-	-	-	-	-	-	-	-	-	-	-
Transport Equipment	-	-	-	-	-	-	-	-	-	-	-
Machinery	-	-	-	-	-	-	-	-	-	-	-
Mining and Quarrying	-	-	-	-	-	-	-	-	-	-	-
Food and Tobacco	-	-	-	-	-	-	-	-	-	-	-
Paper, Pulp and Print	-	-	-	-	-	-	-	-	-	-	-
Wood and Wood Products	-	-	-	-	-	-	-	-	-	-	-
Construction	-	-	-	-	-	-	-	-	-	-	-
Textile and Leather	-	-	-	-	-	-	-	-	-	-	-
Non-specified	-	-	-	-	-	-	-	-	-	-	-
TRANSPORT SECTOR	**-**	**10398**	**802555**	**2068**	**193464**	**809**	**502531**	**18005**	**-**	**-**	**30**
Air	-	-	24	2068	193464	695	-	-	-	-	-
Road	-	10232	798998	-	-	59	443267	140	-	-	30
Rail	-	-	-	-	-	-	-	-	-	-	-
Pipeline Transport	-	-	-	-	-	-	-	-	-	-	-
Internal Navigation	-	-	-	-	-	-	-	-	-	-	-
Non-specified	-	-	-	-	-	-	-	-	-	-	-
OTHER SECTORS	**-**	**-**	**-**	**-**	**-**	**-**	**-**	**-**	**-**	**-**	**-**
Agriculture	-	-	-	-	-	-	-	-	-	-	-
Comm. and Publ. Services	-	-	-	-	-	-	-	-	-	-	-
Residential	-	-	-	-	-	-	-	-	-	-	-
Non-specified	-	-	-	-	-	-	-	-	-	-	-
NON-ENERGY USE	**-**	**-**	**-**	**-**	**-**	**-**	**-**	**-**	**-**	**14307**	**182851**
in Industry/Trans./Energy	-	-	-	-	-	-	-	-	-	-	-
in Transport	-	-	-	-	-	-	-	-	-	-	-
in Other Sectors	-	-	-	-	-	-	-	-	-	-	-

World / Monde : 1998

SUPPLY AND CONSUMPTION *APPROVISIONNEMENT ET DEMANDE*	Gas / *Gaz* (TJ)				Comb. Renew. & Waste / *En. Re. Comb. & Déchets* (TJ)				(GWh) Electricity *Electricité*	(TJ) Heat *Chaleur*
	Natural Gas *Gaz naturel*	Gas Works *Usines à gaz*	Coke Ovens *Cokeries*	Blast Furnaces *Hauts fourneaux*	Solid Biomass *Biomasse solide*	Gas/Liquids from Biomass *Gaz/Liquides tirés de biomasse*	Municipal Waste *Déchets urbains*	Industrial Waste *Déchets industriels*		
Production	90716941	433338	1788929	2503986	42600190	562572	766119	651409	14403046	11405031
From Other Sources	-	25521	4152	-	1675	690	-	18377	-	-
Imports	21028560	746	-	-	22372	4154	-	-	416123	145
Exports	-21228761	-	-	-	-42498	-3009	-	-	-422032	-145
Stock Changes	-802500	-	-	-	-1607	3235	-	-210	-	-
DOMESTIC SUPPLY	**89714240**	**459605**	**1793081**	**2503986**	**42580132**	**567642**	**766119**	**669576**	**14397137**	**11405031**
Intl. Marine Bunkers	-	-	-	-	-	-	-	-	-	-
Transfers	-	-	-	-	-	-4273	-	-	-	-
Statistical Differences	-320174	32518	-71813	-20461	-5958	-18095	-21707	-21552	-17696	1
TRANSFORMATION	**30751689**	**38661**	**366671**	**887546**	**3708511**	**120647**	**736521**	**326957**	**6956**	**-**
Electricity Plants	15635173	9819	210598	677528	412263	61729	335655	70488	-	-
CHP Plants	11173598	14634	105085	197811	1413170	17316	332417	187452	-	-
Heat Plants	2952220	14208	32545	12111	114251	1426	68449	69009	-	-
Blast Furnaces/Gas Works	54890	-	18443	-	1675	-	-	-	-	-
Coke/Pat. Fuel/BKB Plants	3271	-	-	96	54523	-	-	8	-	-
Petroleum Refineries	278	-	-	-	116	40176	-	-	-	-
Petrochemical Industry	-	-	-	-	-	-	-	-	-	-
Liquefaction	382423	-	-	-	-	-	-	-	-	-
Other Transform. Sector	549836	-	-	-	1712513	-	-	-	6956	-
ENERGY SECTOR	**9018945**	**39300**	**382325**	**159726**	**1125**	**1434**	**110**	**16344**	**1355094**	**1016404**
Coal Mines	7734	-	2913	-	1	-	-	2	106085	83564
Oil and Gas Extraction	6937495	-	-	-	-	-	-	-	132221	185880
Petroleum Refineries	1757388	11614	2044	-	-	-	-	438	157484	476322
Electr., CHP+Heat Plants	23730	9502	20	-	59	404	110	-	790944	106664
Pumped Storage (Elec.)	-	-	-	-	-	-	-	-	94570	-
Other Energy Sector	292598	18184	377348	159726	1065	1030	-	15904	73790	163974
Distribution Losses	867686	5545	5691	47120	138	68	60	19489	1239137	723698
FINAL CONSUMPTION	**48755746**	**408617**	**966581**	**1389133**	**38842607**	**423125**	**7721**	**285234**	**11778254**	**9664930**
INDUSTRY SECTOR	**22208781**	**133350**	**888725**	**1388951**	**5523421**	**4614**	**420**	**263204**	**4969641**	**3790881**
Iron and Steel	-	-	-	-	-	-	-	-	-	-
Chemical and Petrochem.	-	-	-	-	-	-	-	-	-	-
of which: Feedstocks	-	-	-	-	-	-	-	-	-	-
Non-Ferrous Metals	-	-	-	-	-	-	-	-	-	-
Non-Metallic Minerals	-	-	-	-	-	-	-	-	-	-
Transport Equipment	-	-	-	-	-	-	-	-	-	-
Machinery	-	-	-	-	-	-	-	-	-	-
Mining and Quarrying	-	-	-	-	-	-	-	-	-	-
Food and Tobacco	-	-	-	-	-	-	-	-	-	-
Paper, Pulp and Print	-	-	-	-	-	-	-	-	-	-
Wood and Wood Products	-	-	-	-	-	-	-	-	-	-
Construction	-	-	-	-	-	-	-	-	-	-
Textile and Leather	-	-	-	-	-	-	-	-	-	-
Non-specified	-	-	-	-	-	-	-	-	-	-
TRANSPORT SECTOR	**1792554**	**17**	**-**	**-**	**1786**	**357699**	**-**	**3159**	**224456**	**-**
Air	-	-	-	-	-	-	-	-	-	-
Road	102856	-	-	-	-	357699	-	-	250	-
Rail	-	-	-	-	-	-	-	-	-	-
Pipeline Transport	-	-	-	-	-	-	-	-	-	-
Internal Navigation	-	-	-	-	-	-	-	-	-	-
Non-specified	-	-	-	-	-	-	-	-	-	-
OTHER SECTORS	**-**	**-**	**-**	**-**	**-**	**-**	**-**	**-**	**-**	**-**
Agriculture	-	-	-	-	-	-	-	-	-	-
Comm. and Publ. Services	-	-	-	-	-	-	-	-	-	-
Residential	-	-	-	-	-	-	-	-	-	-
Non-specified	-	-	-	-	-	-	-	-	-	-
NON-ENERGY USE	**-**	**-**	**668**	**182**	**-**	**-**	**-**	**-**	**-**	**-**
in Industry/Transf./Energy	-	-	-	-	-	-	-	-	-	-
in Transport	-	-	-	-	-	-	-	-	-	-
in Other Sectors	-	-	-	-	-	-	-	-	-	-

OECD Total / Total OCDE : 1997

SUPPLY AND CONSUMPTION / APPROVISIONNEMENT ET DEMANDE	Coal / Charbon (1000 tonnes)							Oil / Pétrole (1000 tonnes)			
	Coking Coal / Charbon à coke	Other Bit. Coal / Autres charb. bit.	Sub-Bit. Coal / Charbon sous-bit.	Lignite / Lignite	Peat / Tourbe	Oven and Gas Coke / Coke de four/gaz	Pat. Fuel and BKB / Agg./briq. de lignite	Crude Oil / Pétrole brut	NGL / LGN	Feed-stocks / Produits d'aliment.	Additives / Additifs
Production	257423	847511	437603	531322	14634	144303	10193	902462	105680	-	10171
From Other Sources	150	3071	427	-	-	1	-	318	-	24547	184
Imports	140622	226774	3671	3182	-	14675	963	1437443	16727	46758	2999
Exports	-169831	-132362	-5832	-129	-56	-11012	-1424	-386724	-21920	-7425	-457
Intl. Marine Bunkers	-	-	-	-	-	-	-	-	-	-	-
Stock Changes	-1324	6550	7268	3304	-763	2157	-12	-3557	-43	-487	80
DOMESTIC SUPPLY	**227040**	**951544**	**443137**	**537679**	**13815**	**150124**	**9720**	**1949942**	**100444**	**63393**	**12977**
Transfers	-	-	-	-	-	-	-	-	-65537	41387	-698
Statistical Differences	-4019	3796	7501	4603	-	-118	-2	-10116	-1122	1680	336
TRANSFORMATION	**217802**	**830435**	**434521**	**522505**	**11362**	**111916**	**883**	**1938878**	**31157**	**106460**	**12615**
Electricity Plants	16479	671054	394299	418220	5571	5	680	13501	41	-	-
CHP Plants	42	128293	37555	79916	4423	-	60	1465	-	-	-
Heat Plants	45	13060	713	213	745	187	5	-	-	-	-
Blast Furnaces/Gas Works	8844	12235	1214	6103	-	111529	138	-	-	-	-
Coke/Pat. Fuel/BKB Plants	192392	5793	740	18053	623	195	-	-	-	-	-
Petroleum Refineries	-	-	-	-	-	-	-	1924409	31116	106460	12615
Petrochemical Industry	-	-	-	-	-	-	-	-	-	-	-
Liquefaction	-	-	-	-	-	-	-	-14	-	-	-
Other Transform. Sector	-	-	-	-	-	-	-	-483	-	-	-
ENERGY SECTOR	**171**	**3108**	**7**	**909**	**-**	**338**	**51**	**922**	**359**	**-**	**-**
Coal Mines	67	2461	7	36	-	1	2	-	-	-	-
Oil and Gas Extraction	-	-	-	-	-	-	-	922	359	-	-
Petroleum Refineries	-	175	-	-	-	-	-	-	-	-	-
Electr., CHP+Heat Plants	-	302	-	1	-	2	-	-	-	-	-
Pumped Storage (Elec.)	-	-	-	-	-	-	-	-	-	-	-
Other Energy Sector	104	170	-	872	-	335	49	-	-	-	-
Distribution Losses	10	-	29	-	-	5	-	-	-	-	-
FINAL CONSUMPTION	**5038**	**121797**	**16081**	**18868**	**2453**	**37747**	**8784**	**26**	**2269**	**-**	**-**
INDUSTRY SECTOR	**4629**	**91357**	**12060**	**9649**	**1165**	**31920**	**2930**	**26**	**2269**	**-**	**-**
Iron and Steel	3446	9998	634	1	-	26105	12	-	-	-	-
Chemical and Petrochem.	28	12210	1323	1000	26	508	319	26	2269	-	-
of which: Feedstocks	-	-	-	-	-	-	-	-	*2269*	-	-
Non-Ferrous Metals	8	1683	2079	361	-	1320	50	-	-	-	-
Non-Metallic Minerals	18	41086	962	525	-	1102	1925	-	-	-	-
Transport Equipment	-	504	192	2	-	61	12	-	-	-	-
Machinery	7	1751	372	86	-	491	7	-	-	-	-
Mining and Quarrying	-	573	427	4	-	551	34	-	-	-	-
Food and Tobacco	183	7058	3288	1609	-	387	322	-	-	-	-
Paper, Pulp and Print	-	8448	939	199	-	36	162	-	-	-	-
Wood and Wood Products	-	470	107	49	-	3	9	-	-	-	-
Construction	425	1769	39	1366	-	72	-	-	-	-	-
Textile and Leather	1	1116	319	398	-	23	28	-	-	-	-
Non-specified	513	4691	1379	4049	1139	1261	50	-	-	-	-
TRANSPORT SECTOR	**7**	**17**	**211**	**1**	**-**	**1**	**-**	**-**	**-**	**-**	**-**
Air	-	-	-	-	-	-	-	-	-	-	-
Road	-	-	-	-	-	-	-	-	-	-	-
Rail	7	17	4	1	-	1	-	-	-	-	-
Pipeline Transport	-	-	-	-	-	-	-	-	-	-	-
Internal Navigation	-	-	207	-	-	-	-	-	-	-	-
Non-specified	-	-	-	-	-	-	-	-	-	-	-
OTHER SECTORS	**162**	**30082**	**3810**	**9065**	**1288**	**5216**	**5658**	**-**	**-**	**-**	**-**
Agriculture	12	2472	131	276	40	225	12	-	-	-	-
Comm. and Publ. Services	2	2812	1448	258	15	2549	953	-	-	-	-
Residential	148	24559	1445	8232	1233	2440	4682	-	-	-	-
Non-specified	-	239	786	299	-	2	11	-	-	-	-
NON-ENERGY USE	**240**	**341**	**-**	**153**	**-**	**610**	**196**	**-**	**-**	**-**	**-**
in Industry/Trans./Energy	3	341	-	153	-	610	186	-	-	-	-
in Transport	-	-	-	-	-	-	-	-	-	-	-
in Other Sectors	237	-	-	-	-	-	10	-	-	-	-

OECD Total / Total OCDE : 1997

SUPPLY AND CONSUMPTION / APPROVISIONNEMENT ET DEMANDE	Oil cont. / *Pétrole cont.* (1000 tonnes)										
	Refinery Gas / *Gaz de raffinerie*	LPG + Ethane / *GPL + éthane*	Motor Gasoline / *Essence moteur*	Aviation Gasoline / *Essence aviation*	Jet Fuel / *Carbu- réacteurs*	Kerosene / *Kérosène*	Gas/ Diesel / *Gazole*	Heavy Fuel Oil / *Fioul lourd*	Naphtha / *Naphta*	Petrol. Coke / *Coke de pétrole*	Other Prod. / *Autres prod.*
Production	73013	51907	603005	1160	139533	43917	576183	284484	92770	59312	134706
From Other Sources	-	-	-	-	-	-	-	-	-	-	-
Imports	-	36888	58023	137	17701	5534	83403	64667	58291	17313	27031
Exports	-	-7186	-61398	-180	-15308	-2536	-93056	-65180	-23111	-21942	-24588
Intl. Marine Bunkers	-	-	-	-	-14	-	-18934	-61443	-	-	-342
Stock Changes	2	-1697	-1730	88	-1606	-46	-4406	1924	-552	-749	-240
DOMESTIC SUPPLY	73015	79912	597900	1205	140306	46869	543190	224452	127398	53934	136567
Transfers	100	57493	-2605	196	-878	-2242	-689	-8981	-6537	-6	-8251
Statistical Differences	-51	1588	-2157	-1	-821	-832	3464	1293	1095	-97	1704
TRANSFORMATION	1848	4287	509	-	-	535	9292	119287	19781	7808	1119
Electricity Plants	713	522	-	-	-	4	7163	107147	1115	6191	-
CHP Plants	639	14	-	-	-	-	612	6810	31	248	493
Heat Plants	-	39	-	-	-	10	160	1467	-	-	-
Blast Furnaces/Gas Works	262	2392	-	-	-	-	135	2149	351	403	-
Coke/Pat. Fuel/BKB Plants	-	-	-	-	-	-	-	20	-	901	-
Petroleum Refineries	-	-	-	-	-	-	-	-	-	-	-
Petrochemical Industry	234	1320	509	-	-	521	1222	1694	18284	65	626
Liquefaction	-	-	-	-	-	-	-	-	-	-	-
Other Transform. Sector	-	-	-	-	-	-	-	-	-	-	-
ENERGY SECTOR	68980	1415	226	-	2	260	2347	22504	179	24665	4314
Coal Mines	-	2	-	-	-	-	564	24	-	-	2
Oil and Gas Extraction	-	37	-	-	-	3	194	231	3	1599	25
Petroleum Refineries	68980	1313	222	-	2	257	1397	21891	176	23059	4227
Electr., CHP+Heat Plants	-	-	3	-	-	-	53	292	-	-	-
Pumped Storage (Elec.)	-	-	-	-	-	-	-	-	-	-	-
Other Energy Sector	-	63	1	-	-	-	139	66	-	7	60
Distribution Losses	-	18	88	-	15	1	45	65	-	-	-
FINAL CONSUMPTION	2236	133273	592315	1400	138590	42999	534281	74908	101996	21358	124587
INDUSTRY SECTOR	2236	81812	3554	2	75	7548	56477	55417	101996	11854	1361
Iron and Steel	-	1021	-	-	-	266	1214	4447	-	259	-
Chemical and Petrochem.	2236	69688	1403	-	-	1343	8892	11672	101807	1070	1291
of which: Feedstocks	*996*	*69154*	*917*	-	-	*993*	*4003*	*1762*	*100438*	-	-
Non-Ferrous Metals	-	663	1	-	-	181	1960	2221	25	24	-
Non-Metallic Minerals	-	1225	4	-	-	96	4742	8113	3	7119	38
Transport Equipment	-	312	30	2	62	4	1194	735	-	-	-
Machinery	-	1239	40	-	13	382	3140	1589	-	214	27
Mining and Quarrying	-	417	1	-	-	90	3450	971	-	37	2
Food and Tobacco	-	761	9	-	-	10	5423	6510	-	-	-
Paper, Pulp and Print	-	485	3	-	-	117	2120	9253	-	131	-
Wood and Wood Products	-	242	1	-	-	2	2997	416	-	-	-
Construction	-	159	841	-	-	957	8402	547	-	-	2
Textile and Leather	-	364	1	-	-	24	1901	4128	-	29	-
Non-specified	-	5236	1220	-	-	4076	11042	4815	161	2971	1
TRANSPORT SECTOR	-	9486	584606	1398	137939	67	307108	8362	-	-	-
Air	-	-	27	1398	137939	-	-	-	-	-	-
Road	-	9425	581425	-	-	48	278940	84	-	-	-
Rail	-	-	-	-	-	12	16666	43	-	-	-
Pipeline Transport	-	-	-	-	-	-	-	11	-	-	-
Internal Navigation	-	1	3154	-	-	4	8973	7938	-	-	-
Non-specified	-	60	-	-	-	3	2518	297	-	-	-
OTHER SECTORS	-	41975	4155	-	576	35384	170696	11129	-	115	12
Agriculture	-	2569	2936	-	-	3464	44606	1802	-	10	-
Comm. and Publ. Services	-	8868	1082	-	-	13349	47295	6036	-	13	12
Residential	-	29714	110	-	-	18435	76709	3045	-	92	-
Non-specified	-	824	27	-	576	136	2086	246	-	-	-
NON-ENERGY USE	-	-	-	-	-	-	-	-	-	9389	123214
in Industry/Transf./Energy	-	-	-	-	-	-	-	-	-	9389	112174
in Transport	-	-	-	-	-	-	-	-	-	-	7795
in Other Sectors	-	-	-	-	-	-	-	-	-	-	3245

OECD Total / Total OCDE : 1997

SUPPLY AND CONSUMPTION / APPROVISIONNEMENT ET DEMANDE	Gas / Gaz (TJ)				Comb. Renew. & Waste / En. Re. Comb. & Déchets (TJ)				(GWh)	(TJ)
	Natural Gas / Gaz naturel	Gas Works / Usines à gaz	Coke Ovens / Cokeries	Blast Furnaces / Hauts fourneaux	Solid Biomass / Biomasse solide	Gas/Liquids from Biomass / Gaz/Liquides tirés de biomasse	Municipal Waste / Déchets urbains	Industrial Waste / Déchets industriels	Electricity / Electricité	Heat / Chaleur
Production	40511489	171057	1059629	1621612	5627725	208586	582949	318989	8905243	2236564
From Other Sources	-	14452	3842	-	-	690	-	22009	-	-
Imports	15576828	-	-	-	16572	3484	-	-	277053	145
Exports	-7246733	-	-	-	-119	-	-	-	-271421	-145
Intl. Marine Bunkers	-	-	-	-	-	-	-	-	-	-
Stock Changes	-259316	-	-	-	-87	-2014	-	57	-	-
DOMESTIC SUPPLY	48582268	185509	1063471	1621612	5644091	210746	582949	341055	8910875	2236564
Transfers	-	-	-	-	-	-6805	-	-	-	-
Statistical Differences	-210490	-1558	-17359	-30156	-2247	-15084	-16994	-3988	-	-
TRANSFORMATION	12501334	25291	269757	787831	1800240	117508	562155	256692	5454	-
Electricity Plants	7932765	-	161387	596969	326206	59931	331514	61900		-
CHP Plants	4285413	12045	91773	188980	1413755	15340	158028	139037		-
Heat Plants	233704	13246	2888	1735	51652	1762	72613	55740		-
Blast Furnaces/Gas Works	23994	-	13709	-	-	-	-	-		-
Coke/Pat. Fuel/BKB Plants	3814	-	-	147	6241	-	-	15		-
Petroleum Refineries	-	-	-	-	-	40475	-	-		-
Petrochemical Industry	-	-	-	-	-	-	-	-		-
Liquefaction	1406	-	-	-	-	-	-	-		-
Other Transform. Sector	20238	-	-	-	2386	-	-	-	5454	-
ENERGY SECTOR	3992084	7984	355224	151796	13	423	110	301	797368	143368
Coal Mines	9143	-	2338	-	-	-	-	-	40710	15930
Oil and Gas Extraction	2860546	-	-	-	-	-	-	-	55929	15
Petroleum Refineries	1036870	3282	1862	-	-	-	-	163	95887	55897
Electr., CHP+Heat Plants	1044	-	-	-	3	423	110	-	473012	24673
Pumped Storage (Elec.)	-	-	-	-	-	-	-	-	82714	-
Other Energy Sector	84481	4702	351024	151796	10	-	-	138	49116	46853
Distribution Losses	151809	7989	2228	21901	-	-	60	-	545910	175861
FINAL CONSUMPTION	31726551	142687	418903	629928	3841591	70926	3630	80074	7562143	1917335
INDUSTRY SECTOR	12945168	52038	417067	629808	1698573	2229	420	65423	3015987	584239
Iron and Steel	1333954	8299	386537	625453	61	-	-	3	343526	15684
Chemical and Petrochem.	4598640	11772	6542	2163	24412	28	-	20799	571959	193147
of which: Feedstocks	1006279	-	-	-	-	-	-	-	-	-
Non-Ferrous Metals	582751	2911	209	59	2420	156	-	2376	275179	4776
Non-Metallic Minerals	1230638	2838	12209	2133	21429	-	-	6889	158417	4687
Transport Equipment	293246	22	160	-	73	-	-	9	124491	20261
Machinery	850232	9223	6667	-	274	-	-	98	289457	15116
Mining and Quarrying	166982	1	2117	-	126	-	-	19	101494	4758
Food and Tobacco	1284064	6313	931	-	182865	903	-	299	206847	33815
Paper, Pulp and Print	1257787	3950	349	-	893520	157	-	2274	374885	54523
Wood and Wood Products	91491	-	-	-	488106	-	-	654	58664	6272
Construction	33157	-	414	-	651	1	-	47	13288	2038
Textile and Leather	350554	1199	140	-	4254	-	-	6	114510	18177
Non-specified	871672	5510	792	-	80382	984	420	31950	383270	210985
TRANSPORT SECTOR	1094919	-	-	-	-	62033	-	-	104985	-
Air	-	-	-	-	-	-	-	-	-	-
Road	23206	-	-	-	-	62033	-	-	-	-
Rail	-	-	-	-	-	-	-	-	89246	-
Pipeline Transport	1070672	-	-	-	-	-	-	-	4044	-
Internal Navigation	-	-	-	-	-	-	-	-	-	-
Non-specified	1041	-	-	-	-	-	-	-	11695	-
OTHER SECTORS	17686464	90649	1072	-	2143018	6664	3210	14651	4441171	1333096
Agriculture	239254	6	-	-	33117	62	15	-	76391	12500
Comm. and Publ. Services	5342545	31012	141	-	59236	3981	2453	770	2054653	297942
Residential	11560074	59579	931	-	1995328	10	742	-	2291754	904630
Non-specified	544591	52	-	-	55337	2611	-	13881	18373	118024
NON-ENERGY USE	-	-	764	120	-	-	-	-	-	-
in Industry/Transf./Energy	-	-	764	120	-	-	-	-	-	-
in Transport	-	-	-	-	-	-	-	-	-	-
in Other Sectors	-	-	-	-	-	-	-	-	-	-

OECD Total / Total OCDE : 1998

SUPPLY AND CONSUMPTION / APPROVISIONNEMENT ET DEMANDE	Coal / Charbon (1000 tonnes)							Oil / Pétrole (1000 tonnes)			
	Coking Coal / Charbon à coke	Other Bit. Coal / Autres charb. bit.	Sub-Bit. Coal / Charbon sous-bit.	Lignite / Lignite	Peat / Tourbe	Oven and Gas Coke / Coke de four/gaz	Pat. Fuel and BKB / Agg./briq. de lignite	Crude Oil / Pétrole brut	NGL / LGN	Feed-stocks / Produits d'aliment.	Additives / Additifs
Production	248030	816444	466948	537514	6459	139002	8182	895106	106126	-	10584
From Other Sources	-	2488	312	-	-	-	-	284	-	24572	159
Imports	142102	236200	4964	2647	-	16478	694	1468002	17524	39932	3277
Exports	-167400	-144242	-6257	-60	-66	-11267	-982	-393475	-22641	-7645	-1531
Intl. Marine Bunkers	-	-	-	-	-	-	-	-	-	-	-
Stock Changes	-551	-6613	-3104	1006	6741	-990	131	-4317	-427	1543	-102
DOMESTIC SUPPLY	**222181**	**904277**	**462863**	**541107**	**13134**	**143223**	**8025**	**1965600**	**100582**	**58402**	**12387**
Transfers	-	-	-	-	-	-	-	-	-64934	46503	-595
Statistical Differences	-4316	51268	-3297	4145	-	-543	-49	-2043	-2012	727	1286
TRANSFORMATION	**212648**	**843389**	**444524**	**527491**	**10041**	**107117**	**805**	**1962793**	**31215**	**105632**	**13078**
Electricity Plants	15401	687987	403292	428763	4288	5	705	9809	36	-	-
CHP Plants	38	126544	38657	78079	4365	-	60	1175	-	-	-
Heat Plants	34	10735	628	86	726	143	-	-	-	-	-
Blast Furnaces/Gas Works	8776	12324	1299	6267	-	106608	40	-	-	-	-
Coke/Pat. Fuel/BKB Plants	188399	5799	648	14296	662	361	-	-	-	-	-
Petroleum Refineries	-	-	-	-	-	-	-	1952296	31179	105632	13078
Petrochemical Industry	-	-	-	-	-	-	-	-	-	-	-
Liquefaction	-	-	-	-	-	-	-	-	-	-	-
Other Transform. Sector	-	-	-	-	-	-	-	-487	-	-	-
ENERGY SECTOR	**188**	**2946**	**5**	**1319**	**-**	**316**	**38**	**603**	**361**	**-**	**-**
Coal Mines	54	2269	5	29	-	1	1	-	-	-	-
Oil and Gas Extraction	-	-	-	-	-	-	-	603	361	-	-
Petroleum Refineries	-	141	-	-	-	-	-	-	-	-	-
Electr., CHP+Heat Plants	-	349	-	1	-	2	-	-	-	-	-
Pumped Storage (Elec.)	-	-	-	-	-	-	-	-	-	-	-
Other Energy Sector	134	187	-	1289	-	313	37	-	-	-	-
Distribution Losses	**10**	**-**	**27**	**-**	**-**	**5**	**-**	**-**	**-**	**-**	**-**
FINAL CONSUMPTION	**5019**	**109210**	**15010**	**16442**	**3093**	**35242**	**7133**	**161**	**2060**	**-**	**-**
INDUSTRY SECTOR	**4633**	**85347**	**11709**	**8834**	**1883**	**30568**	**2852**	**161**	**2060**	**-**	**-**
Iron and Steel	3771	10213	674	12	-	24374	12	-	-	-	-
Chemical and Petrochem.	19	10082	1273	953	22	522	328	161	2060	-	-
of which: Feedstocks	-	-	-	-	-	-	-	-	2060	-	-
Non-Ferrous Metals	8	1593	2094	355	-	1235	54	-	-	-	-
Non-Metallic Minerals	16	37530	944	310	-	1200	1815	-	-	-	-
Transport Equipment	-	586	207	-	-	38	3	-	-	-	-
Machinery	6	1227	317	73	-	533	4	-	-	-	-
Mining and Quarrying	-	693	353	-	-	483	47	-	-	-	-
Food and Tobacco	184	6432	3221	1461	-	596	315	-	-	-	-
Paper, Pulp and Print	-	8172	1017	191	1861	35	182	-	-	-	-
Wood and Wood Products	-	411	62	48	-	2	9	-	-	-	-
Construction	170	1246	21	1206	-	54	-	-	-	-	-
Textile and Leather	-	917	270	375	-	10	27	-	-	-	-
Non-specified	459	6245	1256	3850	-	1486	56	-	-	-	-
TRANSPORT SECTOR	**7**	**9**	**208**	**-**	**-**	**-**	**-**	**-**	**-**	**-**	**-**
Air	-	-	-	-	-	-	-	-	-	-	-
Road	-	-	-	-	-	-	-	-	-	-	-
Rail	7	9	4	-	-	-	-	-	-	-	-
Pipeline Transport	-	-	-	-	-	-	-	-	-	-	-
Internal Navigation	-	-	204	-	-	-	-	-	-	-	-
Non-specified	-	-	-	-	-	-	-	-	-	-	-
OTHER SECTORS	**137**	**23506**	**3093**	**7480**	**1210**	**4095**	**4090**	**-**	**-**	**-**	**-**
Agriculture	4	2097	45	199	43	240	14	-	-	-	-
Comm. and Publ. Services	5	2018	888	190	8	2050	623	-	-	-	-
Residential	128	19284	743	6888	1159	1803	3438	-	-	-	-
Non-specified	-	107	1417	203	-	2	15	-	-	-	-
NON-ENERGY USE	**242**	**348**	**-**	**128**	**-**	**579**	**191**	**-**	**-**	**-**	**-**
in Industry/Trans./Energy	3	348	-	128	-	579	182	-	-	-	-
in Transport	-	-	-	-	-	-	-	-	-	-	-
in Other Sectors	239	-	-	-	-	-	9	-	-	-	-

OECD Total / Total OCDE : 1998

| | \multicolumn{11}{c}{Oil cont. / *Pétrole cont.* (1000 tonnes)} | | | | | | | | | | |
| SUPPLY AND CONSUMPTION | Refinery Gas | LPG + Ethane | Motor Gasoline | Aviation Gasoline | Jet Fuel | Kerosene | Gas/ Diesel | Heavy Fuel Oil | Naphtha | Petrol. Coke | Other Prod. |
APPROVISIONNEMENT ET DEMANDE	*Gaz de raffinerie*	*GPL + éthane*	*Essence moteur*	*Essence aviation*	*Carbu- réacteurs*	*Kérosène*	*Gazole*	*Fioul lourd*	*Naphta*	*Coke de pétrole*	*Autres prod.*
Production	73212	51040	614081	1192	143057	42568	578801	289520	93375	61405	138905
From Other Sources	-	-	-	-	-	-	-	-	-	-	.
Imports	-	34915	61053	129	18895	3200	84447	67567	59073	16582	25440
Exports	-	-7320	-63821	-203	-17352	-2019	-94685	-67987	-23184	-19544	-22987
Intl. Marine Bunkers	-	-	-	-	-14	-	-19152	-63605	-	-	-350
Stock Changes	-	-1215	-1739	-29	-318	601	-5267	-583	-417	-714	-359
DOMESTIC SUPPLY	73212	77420	609574	1089	144268	44350	544144	224912	128847	57729	140649
Transfers	-31	56429	-1491	97	-828	-2065	-2427	-6900	-4986	232	-16104
Statistical Differences	15	1731	-3351	12	-925	-778	3977	1346	610	-279	-470
TRANSFORMATION	1766	4226	518	-	-	563	11332	122934	19354	7964	1055
Electricity Plants	712	331	-	-	-	7	9133	110305	1115	6272	-
CHP Plants	600	77	-	-	-	-	643	7282	47	231	431
Heat Plants	-	39	-	-	-	12	145	1413	-	4	-
Blast Furnaces/Gas Works	258	2280	-	-	-	-	144	1927	284	583	-
Coke/Pat. Fuel/BKB Plants	-	-	-	-	-	-	-	12	-	874	-
Petroleum Refineries	-	-	-	-	-	-	-	-	-	-	-
Petrochemical Industry	196	1499	518	-	-	544	1267	1995	17908	-	624
Liquefaction	-	-	-	-	-	-	-	-	-	-	-
Other Transform. Sector	-	-	-	-	-	-	-	-	-	-	-
ENERGY SECTOR	69196	1460	863	-	37	291	2295	23388	192	25530	3679
Coal Mines	-	1	-	-	-	1	578	10	-	-	1
Oil and Gas Extraction	-	42	-	-	-	16	184	218	2	1776	32
Petroleum Refineries	69196	1370	860	-	37	274	1360	22784	183	23727	3586
Electr., CHP+Heat Plants	-	-	3	-	-	-	51	312	-	-	-
Pumped Storage (Elec.)	-	-	-	-	-	-	-	-	-	-	-
Other Energy Sector	-	47	-	-	-	-	122	64	7	27	60
Distribution Losses	-	18	-	-	15	1	45	62	-	-	-
FINAL CONSUMPTION	2234	129876	603351	1198	142463	40652	532022	72974	104925	24188	119341
INDUSTRY SECTOR	2234	80069	3071	2	59	7425	55269	54038	104925	11014	1231
Iron and Steel	-	978	-	-	-	237	1127	5220	-	246	-
Chemical and Petrochem.	2234	67984	1020	-	-	1239	8939	10526	104721	879	1157
of which: Feedstocks	*779*	*67363*	*757*	-	-	*876*	*4232*	*1916*	*103298*	-	-
Non-Ferrous Metals	-	637	-	-	-	163	1981	2123	24	16	-
Non-Metallic Minerals	-	1176	2	-	-	102	4497	8208	-	7796	37
Transport Equipment	-	271	18	2	46	10	1164	1104	-	-	-
Machinery	-	1168	43	-	13	378	2936	1448	-	139	18
Mining and Quarrying	-	422	-	-	-	67	3154	1550	-	25	1
Food and Tobacco	-	770	7	-	-	23	5421	6249	-	-	15
Paper, Pulp and Print	-	461	2	-	-	114	2055	8180	-	106	-
Wood and Wood Products	-	211	2	-	-	1	3060	417	-	-	-
Construction	-	163	649	-	-	876	8177	500	-	-	1
Textile and Leather	-	368	1	-	-	47	1850	3986	-	29	-
Non-specified	-	5460	1327	-	-	4168	10908	4527	180	1778	2
TRANSPORT SECTOR	-	9789	596343	1196	141889	78	313076	9230	-	-	-
Air	-	-	24	1196	141889	-	-	-	-	-	-
Road	-	9726	593454	-	-	57	285970	66	-	-	-
Rail	-	-	-	-	-	12	15832	22	-	-	-
Pipeline Transport	-	-	-	-	-	-	20	-	-	-	-
Internal Navigation	-	3	2865	-	-	2	9048	8901	-	-	-
Non-specified	-	60	-	-	-	7	2206	241	-	-	-
OTHER SECTORS	-	40018	3937	-	515	33149	163677	9706	-	116	8
Agriculture	-	2499	2703	-	-	3345	43394	1765	-	46	-
Comm. and Publ. Services	-	8103	1097	-	-	11022	45652	5192	-	6	8
Residential	-	28878	110	-	-	18606	72819	2626	-	64	-
Non-specified	-	538	27	-	515	176	1812	123	-	-	-
NON-ENERGY USE	-	-	-	-	-	-	-	-	-	13058	118102
in Industry/Transf./Energy	-	-	-	-	-	-	-	-	-	13058	107283
in Transport	-	-	-	-	-	-	-	-	-	-	7972
in Other Sectors	-	-	-	-	-	-	-	-	-	-	2847

OECD Total / Total OCDE : 1998

	Gas / *Gaz* (TJ)				Comb. Renew. & Waste / *En. Re. Comb. & Déchets* (TJ)				(GWh)	(TJ)
SUPPLY AND CONSUMPTION *APPROVISIONNEMENT ET DEMANDE*	Natural Gas *Gaz naturel*	Gas Works *Usines à gaz*	Coke Ovens *Cokeries*	Blast Furnaces *Hauts fourneaux*	Solid Biomass *Biomasse solide*	Gas/Liquids from Biomass *Gaz/Liquides tirés de biomasse*	Municipal Waste *Déchets urbains*	Industrial Waste *Déchets industriels*	Electricity *Electricité*	Heat *Chaleur*
Production	40712269	159134	1030653	1569688	5737107	230248	766119	309580	9124708	2201337
From Other Sources	-	14499	4152	-	-	690	-	18377	-	-
Imports	16068707	-	-	-	18145	680	-	-	278927	145
Exports	-7420022	-	-	-	-198	-	-	-	-269480	-145
Intl. Marine Bunkers	-	-	-	-	-	-	-	-	-	-
Stock Changes	-608692	-	-	-	1559	-490	-	-206	-	-
DOMESTIC SUPPLY	**48752262**	**173633**	**1034805**	**1569688**	**5756613**	**231128**	**766119**	**327751**	**9134155**	**2201337**
Transfers	-	-	-	-	-	-4273	-	-	-	-
Statistical Differences	-375433	2402	-11166	-19484	-2542	-18095	-21707	-4820	-	-
TRANSFORMATION	**13256435**	**27519**	**276638**	**760475**	**1734064**	**120577**	**736521**	**265093**	**4921**	**-**
Electricity Plants	8618405	-	173269	613427	315833	61729	335655	70488	-	-
CHP Plants	4379945	13499	88670	145083	1358701	17316	332417	125620	-	-
Heat Plants	207743	14020	2845	1869	51873	1356	68449	68977	-	-
Blast Furnaces/Gas Works	26676	-	11854	-	-	-	-	-	-	-
Coke/Pat. Fuel/BKB Plants	3271	-	-	96	5396	-	-	8	-	-
Petroleum Refineries	-	-	-	-	-	40176	-	-	-	-
Petrochemical Industry	-	-	-	-	-	-	-	-	-	-
Liquefaction	-	-	-	-	-	-	-	-	-	-
Other Transform. Sector	20395	-	-	-	2261	-	-	-	4921	-
ENERGY SECTOR	**4147919**	**12707**	**332628**	**142190**	**12**	**404**	**110**	**589**	**810528**	**135997**
Coal Mines	7491	-	2212	-	-	-	-	-	39961	15693
Oil and Gas Extraction	2965083	-	-	-	-	-	-	-	55036	10
Petroleum Refineries	1104312	3979	2044	-	-	-	-	438	94947	52206
Electr., CHP+Heat Plants	1136	3	-	-	3	404	110	-	488588	22897
Pumped Storage (Elec.)	-	-	-	-	-	-	-	-	83615	-
Other Energy Sector	69897	8725	328372	142190	9	-	-	151	48381	45191
Distribution Losses	133057	4550	2682	21686	-	68	60	-	603702	176518
FINAL CONSUMPTION	**30839418**	**131259**	**411691**	**625853**	**4019995**	**87711**	**7721**	**57249**	**7715004**	**1888822**
INDUSTRY SECTOR	**12762923**	**47735**	**410011**	**625671**	**1791093**	**4614**	**420**	**44293**	**3041970**	**543428**
Iron and Steel	1309496	7200	379356	621371	181	-	-	1	341131	11663
Chemical and Petrochem.	4465482	11480	6335	2154	25233	21	-	18946	559670	187021
of which: Feedstocks	*988150*	-	-	-	-	-	-	-	-	-
Non-Ferrous Metals	579276	2527	227	49	2405	165	-	2184	285174	5009
Non-Metallic Minerals	1234051	2392	12516	2097	22955	-	-	7552	156885	4386
Transport Equipment	287800	15	164	-	-	-	-	9	124162	18506
Machinery	868257	8876	6963	-	307	1	-	81	293706	14311
Mining and Quarrying	157130	-	2116	-	4	-	-	8	103757	3038
Food and Tobacco	1270366	5939	1012	-	187529	3403	-	276	209454	33933
Paper, Pulp and Print	1227684	3548	379	-	911645	160	-	2281	376617	56766
Wood and Wood Products	91314	-	-	-	551888	-	-	464	57513	6310
Construction	33196	-	374	-	288	1	-	8	14374	1443
Textile and Leather	348614	979	152	-	4320	-	-	6	114198	18034
Non-specified	890257	4779	417	-	84338	863	420	12477	405329	183008
TRANSPORT SECTOR	**973572**	**-**	**-**	**-**	**-**	**73084**	**-**	**-**	**106207**	**-**
Air	-	-	-	-	-	-	-	-	-	-
Road	25023	-	-	-	-	73084	-	-	-	-
Rail	-	-	-	-	-	-	-	-	88795	-
Pipeline Transport	947463	-	-	-	-	-	-	-	5067	-
Internal Navigation	-	-	-	-	-	-	-	-	-	-
Non-specified	1086	-	-	-	-	-	-	-	12345	-
OTHER SECTORS	**17102923**	**83524**	**1012**	**-**	**2228902**	**10013**	**7301**	**12956**	**4566827**	**1345394**
Agriculture	240043	2	-	-	34972	67	15	-	77115	12325
Comm. and Publ. Services	5150656	29302	-	-	65015	7240	6380	683	2106673	303212
Residential	11152844	54196	1012	-	2071813	23	906	-	2362765	910907
Non-specified	559380	24	-	-	57102	2683	-	12273	20274	118950
NON-ENERGY USE	**-**	**-**	**668**	**182**	**-**	**-**	**-**	**-**	**-**	**-**
in Industry/Transf./Energy	-	-	668	182	-	-	-	-	-	-
in Transport	-	-	-	-	-	-	-	-	-	-
in Other Sectors	-	-	-	-	-	-	-	-	-	-

Non-OECD Total / Total non-OCDE : 1997

SUPPLY AND CONSUMPTION / APPROVISIONNEMENT ET DEMANDE	Coal / Charbon (1000 tonnes)							Oil / Pétrole (1000 tonnes)			
	Coking Coal / Charbon à coke	Other Bit. Coal / Autres charb. bit.	Sub-Bit. Coal / Charbon sous-bit.	Lignite / Lignite	Peat / Tourbe	Oven and Gas Coke / Coke de four/gaz	Pat. Fuel and BKB / Agg./briq. de lignite	Crude Oil / Pétrole brut	NGL / LGN	Feed-stocks / Produits d'aliment.	Additives / Additifs
Production	347998	2035579	26070	270539	7959	218241	16396	2351797	92704	-	183
From Other Sources	-	-	-	-	-	4	-	404	-	-	1520
Imports	52102	104455	1657	4493	14	4298	225	408978	20	3157	106
Exports	-10966	-217895	-1	-1161	-102	-13178	-129	-1383750	-37147	-	-1546
Intl. Marine Bunkers	-	-	-	-	-	-	-	-	-	-	-
Stock Changes	-853	-14173	66	-571	190	498	69	-24457	22	-78	34
DOMESTIC SUPPLY	**388281**	**1907966**	**27792**	**273300**	**8061**	**209863**	**16561**	**1352972**	**55599**	**3079**	**297**
Transfers	-	-	-	-	-	-	-	-3369	-35300	2750	-
Statistical Differences	-5203	-11907	-283	-2258	-236	-16039	1	19372	-247	-	-3
TRANSFORMATION	**349465**	**1118419**	**3488**	**243930**	**5819**	**127715**	**305**	**1348704**	**10949**	**5829**	**287**
Electricity Plants	-	862549	1884	131685	-	48	18	16104	239	-	-
CHP Plants	-	176513	1525	90980	2029	-	182	1144	-	-	-
Heat Plants	-	25468	79	16659	885	6	105	5	-	-	-
Blast Furnaces/Gas Works	107	12949	-	938	-	127660	-	-	219	-	-
Coke/Pat. Fuel/BKB Plants	349358	13174	-	3668	2905	1	-	-	-	-	-
Petroleum Refineries	-	-	-	-	-	-	-	1339346	10491	5829	287
Petrochemical Industry	-	-	-	-	-	-	-	-	-	-	-
Liquefaction	-	27766	-	-	-	-	-	-7895	-	-	-
Other Transform. Sector	-	-	-	-	-	-	-	-	-	-	-
ENERGY SECTOR	**42**	**73726**	**25**	**380**	**154**	**1626**	**48**	**7561**	**1042**	**-**	**-**
Coal Mines	13	28999	24	334	96	525	47	13	-	-	-
Oil and Gas Extraction	-	3211	-	-	-	26	-	4979	1042	-	-
Petroleum Refineries	-	-	-	-	-	-	-	2568	-	-	-
Electr., CHP+Heat Plants	-	33428	1	46	-	73	-	1	-	-	-
Pumped Storage (Elec.)	-	-	-	-	-	-	-	-	-	-	-
Other Energy Sector	29	8088	-	-	58	1002	1	-	-	-	-
Distribution Losses	-	5572	4	4102	225	119	-	7305	689	-	-
FINAL CONSUMPTION	**33571**	**698342**	**23992**	**22630**	**1627**	**64364**	**16209**	**5405**	**7372**	**-**	**7**
INDUSTRY SECTOR	**33561**	**503371**	**23831**	**12567**	**63**	**60472**	**162**	**4722**	**7372**	**-**	**7**
Iron and Steel	-	-	-	-	-	-	-	-	-	-	-
Chemical and Petrochem.	-	-	-	-	-	-	-	-	-	-	-
of which: Feedstocks	-	-	-	-	-	-	-	-	-	-	-
Non-Ferrous Metals	-	-	-	-	-	-	-	-	-	-	-
Non-Metallic Minerals	-	-	-	-	-	-	-	-	-	-	-
Transport Equipment	-	-	-	-	-	-	-	-	-	-	-
Machinery	-	-	-	-	-	-	-	-	-	-	-
Mining and Quarrying	-	-	-	-	-	-	-	-	-	-	-
Food and Tobacco	-	-	-	-	-	-	-	-	-	-	-
Paper, Pulp and Print	-	-	-	-	-	-	-	-	-	-	-
Wood and Wood Products	-	-	-	-	-	-	-	-	-	-	-
Construction	-	-	-	-	-	-	-	-	-	-	-
Textile and Leather	-	-	-	-	-	-	-	-	-	-	-
Non-specified	-	-	-	-	-	-	-	-	-	-	-
TRANSPORT SECTOR	**1**	**14946**	**3**	**430**	**-**	**1**	**22**	**16**	**-**	**-**	**-**
Air	-	-	-	-	-	-	-	-	-	-	-
Road	-	-	-	-	-	-	-	-	-	-	-
Rail	-	-	-	-	-	-	-	-	-	-	-
Pipeline Transport	-	-	-	-	-	-	-	-	-	-	-
Internal Navigation	-	-	-	-	-	-	-	-	-	-	-
Non-specified	-	-	-	-	-	-	-	-	-	-	-
OTHER SECTORS	**-**	**-**	**-**	**-**	**-**	**-**	**-**	**-**	**-**	**-**	**-**
Agriculture	-	-	-	-	-	-	-	-	-	-	-
Comm. and Publ. Services	-	-	-	-	-	-	-	-	-	-	-
Residential	-	-	-	-	-	-	-	-	-	-	-
Non-specified	-	-	-	-	-	-	-	-	-	-	-
NON-ENERGY USE	**-**	**8013**	**-**	**275**	**49**	**743**	**41**	**-**	**-**	**-**	**-**
in Industry/Trans./Energy	-	-	-	-	-	-	-	-	-	-	-
in Transport	-	-	-	-	-	-	-	-	-	-	-
in Other Sectors	-	-	-	-	-	-	-	-	-	-	-

Non-OECD Total / Total non-OCDE : 1997

SUPPLY AND CONSUMPTION / APPROVISIONNEMENT ET DEMANDE	Oil cont. / Pétrole cont. (1000 tonnes)										
	Refinery Gas / Gaz de raffinerie	LPG + Ethane / GPL + éthane	Motor Gasoline / Essence moteur	Aviation Gasoline / Essence aviation	Jet Fuel / Carbu-réacteurs	Kerosene / Kérosène	Gas/ Diesel / Gazole	Heavy Fuel Oil / Fioul lourd	Naphtha / Naphta	Petrol. Coke / Coke de pétrole	Other Prod. / Autres prod.
Production	28530	35418	220998	459	64730	40640	391727	365052	69203	6346	76039
From Other Sources	-	3203	-	-	-	35	96	-	-	-	255
Imports	-	16541	29257	504	12269	10629	93639	89622	8000	951	10548
Exports	-	-23821	-44922	-47	-22919	-14377	-114490	-146992	-33913	-1587	-12410
Intl. Marine Bunkers	-	-	-	-	-	-	-7906	-36086	-	-	-3
Stock Changes	3	-928	-296	-12	-32	-277	-950	3611	-99	-60	-37
DOMESTIC SUPPLY	**28533**	**30413**	**205037**	**904**	**54048**	**36650**	**362116**	**275207**	**43191**	**5650**	**74392**
Transfers	-	27537	7139	-	-65	-123	15	2547	-1425	-	1367
Statistical Differences	669	26	51	6	-1320	719	642	549	30	305	1460
TRANSFORMATION	**3463**	**350**	**38**	**-**	**-**	**91**	**33126**	**130812**	**1091**	**424**	**592**
Electricity Plants	464	7	38	-	-	74	31690	82870	21	25	351
CHP Plants	2999	321	-	-	-	9	371	15630	-	-	49
Heat Plants	-	2	-	-	-	8	1065	31688	46	-	192
Blast Furnaces/Gas Works	-	19	-	-	-	-	-	623	1024	13	-
Coke/Pat. Fuel/BKB Plants	-	-	-	-	-	-	-	-	-	386	-
Petroleum Refineries	-	-	-	-	-	-	-	-	-	-	-
Petrochemical Industry	-	-	-	-	-	-	-	-	-	-	-
Liquefaction	-	-	-	-	-	-	-	-	-	-	-
Other Transform. Sector	-	1	-	-	-	-	-	1	-	-	-
ENERGY SECTOR	**23326**	**1733**	**987**	**-**	**18**	**254**	**8154**	**31369**	**45**	**970**	**744**
Coal Mines	-	-	-	-	-	63	1343	271	-	-	-
Oil and Gas Extraction	94	119	911	-	-	7	4399	1903	-	-	-
Petroleum Refineries	21390	1509	57	-	18	125	1255	24756	41	970	348
Electr., CHP+Heat Plants	3	5	3	-	-	7	1018	1465	-	-	-
Pumped Storage (Elec.)	-	-	-	-	-	-	-	-	-	-	-
Other Energy Sector	1839	100	16	-	-	52	139	2974	4	-	396
Distribution Losses	65	335	125	-	2	6	223	108	278	-	169
FINAL CONSUMPTION	**2348**	**55558**	**211077**	**910**	**52643**	**36895**	**321270**	**116014**	**40382**	**4561**	**75714**
INDUSTRY SECTOR	**2346**	**10395**	**3921**	**-**	**2**	**1573**	**39487**	**90726**	**40382**	**3322**	**2832**
Iron and Steel	-	-	-	-	-	-	-	-	-	-	-
Chemical and Petrochem.	-	-	-	-	-	-	-	-	-	-	-
of which: Feedstocks	-	-	-	-	-	-	-	-	-	-	-
Non-Ferrous Metals	-	-	-	-	-	-	-	-	-	-	-
Non-Metallic Minerals	-	-	-	-	-	-	-	-	-	-	-
Transport Equipment	-	-	-	-	-	-	-	-	-	-	-
Machinery	-	-	-	-	-	-	-	-	-	-	-
Mining and Quarrying	-	-	-	-	-	-	-	-	-	-	-
Food and Tobacco	-	-	-	-	-	-	-	-	-	-	-
Paper, Pulp and Print	-	-	-	-	-	-	-	-	-	-	-
Wood and Wood Products	-	-	-	-	-	-	-	-	-	-	-
Construction	-	-	-	-	-	-	-	-	-	-	-
Textile and Leather	-	-	-	-	-	-	-	-	-	-	-
Non-specified	-	-	-	-	-	-	-	-	-	-	-
TRANSPORT SECTOR	**-**	**642**	**202268**	**849**	**52596**	**693**	**187127**	**8903**	**-**	**-**	**32**
Air	-	-	-	849	52596	642	-	-	-	-	-
Road	-	541	201314	-	-	7	165919	298	-	-	32
Rail	-	-	-	-	-	-	-	-	-	-	-
Pipeline Transport	-	-	-	-	-	-	-	-	-	-	-
Internal Navigation	-	-	-	-	-	-	-	-	-	-	-
Non-specified	-	-	-	-	-	-	-	-	-	-	-
OTHER SECTORS	**-**	**-**	**-**	**-**	**-**	**-**	**-**	**-**	**-**	**-**	**-**
Agriculture	-	-	-	-	-	-	-	-	-	-	-
Comm. and Publ. Services	-	-	-	-	-	-	-	-	-	-	-
Residential	-	-	-	-	-	-	-	-	-	-	-
Non-specified	-	-	-	-	-	-	-	-	-	-	-
NON-ENERGY USE	**-**	**-**	**-**	**-**	**-**	**-**	**-**	**-**	**-**	**1144**	**63193**
in Industry/Transf./Energy	-	-	-	-	-	-	-	-	-	-	-
in Transport	-	-	-	-	-	-	-	-	-	-	-
in Other Sectors	-	-	-	-	-	-	-	-	-	-	-

Non-OECD Total / Total non-OCDE : 1997

SUPPLY AND CONSUMPTION APPROVISIONNEMENT ET DEMANDE	Gas / Gaz (TJ)				Comb. Renew. & Waste / En. Re. Comb. & Déchets (TJ)				(GWh)	(TJ)
	Natural Gas Gaz naturel	Gas Works Usines à gaz	Coke Ovens Cokeries	Blast Furnaces Hauts fourneaux	Solid Biomass Biomasse solide	Gas/Liquids from Biomass Gaz/Liquides tirés de biomasse	Municipal Waste Déchets urbains	Industrial Waste Déchets industriels	Electricity Electricité	Heat Chaleur
Production	48405004	277981	782080	965325	36436531	367586	-	365319	5143413	9312916
From Other Sources	-	8260	-	-	2779	-	-	-	-	-
Imports	5173829	-	-	-	4405	19178	-	-	141590	-
Exports	-13683405	-	-	-	-41726	-3756	-	-	-152773	-
Intl. Marine Bunkers	-	-	-	-	-	-	-	-	-	-
Stock Changes	437076	-	-	-	-2300	-38908	-	3	-	-
DOMESTIC SUPPLY	40332504	286241	782080	965325	36399688	344100	-	365322	5132230	9312916
Transfers	-	-	-	-	-	-	-	-	-	-
Statistical Differences	162510	13934	-61336	29732	-4911	-	-	-17746	-4816	-1692
TRANSFORMATION	17334172	5981	89933	103646	1963086	58	-	68255	2017	-
Electricity Plants	6701706	4354	34548	54839	90961	-	-	-	-	-
CHP Plants	6937335	1455	16505	38113	53405	-	-	67755	-	-
Heat Plants	2778353	172	32382	10694	68539	58	-	500	-	-
Blast Furnaces/Gas Works	19730	-	6498	-	1884	-	-	-	-	-
Coke/Pat. Fuel/BKB Plants	-	-	-	-	38	-	-	-	-	-
Petroleum Refineries	315	-	-	-	105	-	-	-	-	-
Petrochemical Industry	-	-	-	-	-	-	-	-	-	-
Liquefaction	445178	-	-	-	-	-	-	-	-	-
Other Transform. Sector	451555	-	-	-	1748154	-	-	-	2017	-
ENERGY SECTOR	4652493	34015	62547	19510	1607	1294	-	8071	563806	790612
Coal Mines	253	-	252	-	9	-	-	-	64064	68410
Oil and Gas Extraction	3861746	-	-	-	-	-	-	-	79033	169798
Petroleum Refineries	599503	10223	-	-	-	-	-	-	67949	395007
Electr., CHP+Heat Plants	17816	11275	72	-	53	-	-	-	312692	79345
Pumped Storage (Elec.)	-	-	-	-	-	-	-	-	10722	-
Other Energy Sector	173175	12517	62223	19510	1545	1294	-	8071	29346	78052
Distribution Losses	718274	1257	1842	25150	73	-	-	20726	609605	502969
FINAL CONSUMPTION	17790075	258922	566422	846751	34428225	342748	-	250524	3951986	8017643
INDUSTRY SECTOR	9440613	91887	490297	846751	3669130	-	-	242951	1918290	3414794
Iron and Steel	-	-	-	-	-	-	-	-	-	-
Chemical and Petrochem.	-	-	-	-	-	-	-	-	-	-
of which: Feedstocks	-	-	-	-	-	-	-	-	-	-
Non-Ferrous Metals	-	-	-	-	-	-	-	-	-	-
Non-Metallic Minerals	-	-	-	-	-	-	-	-	-	-
Transport Equipment	-	-	-	-	-	-	-	-	-	-
Machinery	-	-	-	-	-	-	-	-	-	-
Mining and Quarrying	-	-	-	-	-	-	-	-	-	-
Food and Tobacco	-	-	-	-	-	-	-	-	-	-
Paper, Pulp and Print	-	-	-	-	-	-	-	-	-	-
Wood and Wood Products	-	-	-	-	-	-	-	-	-	-
Construction	-	-	-	-	-	-	-	-	-	-
Textile and Leather	-	-	-	-	-	-	-	-	-	-
Non-specified	-	-	-	-	-	-	-	-	-	-
TRANSPORT SECTOR	677091	17	-	-	3088	292220	-	595	120688	-
Air	-	-	-	-	-	-	-	-	-	-
Road	66217	-	-	-	-	292220	-	-	204	-
Rail	-	-	-	-	-	-	-	-	-	-
Pipeline Transport	-	-	-	-	-	-	-	-	-	-
Internal Navigation	-	-	-	-	-	-	-	-	-	-
Non-specified	-	-	-	-	-	-	-	-	-	-
OTHER SECTORS	-	-	-	-	-	-	-	-	-	-
Agriculture	-	-	-	-	-	-	-	-	-	-
Comm. and Publ. Services	-	-	-	-	-	-	-	-	-	-
Residential	-	-	-	-	-	-	-	-	-	-
Non-specified	-	-	-	-	-	-	-	-	-	-
NON-ENERGY USE	-	-	-	-	-	-	-	-	-	-
in Industry/Transf./Energy	-	-	-	-	-	-	-	-	-	-
in Transport	-	-	-	-	-	-	-	-	-	-
in Other Sectors	-	-	-	-	-	-	-	-	-	-

Non-OECD Total / Total non-OCDE : 1998

SUPPLY AND CONSUMPTION APPROVISIONNEMENT ET DEMANDE	Coal / *Charbon* (1000 tonnes)							Oil / *Pétrole* (1000 tonnes)			
	Coking Coal *Charbon à coke*	Other Bit. Coal *Autres charb. bit.*	Sub-Bit. Coal *Charbon sous-bit.*	Lignite *Lignite*	Peat *Tourbe*	Oven and Gas Coke *Coke de four/gaz*	Pat. Fuel and BKB *Agg./briq. de lignite*	Crude Oil *Pétrole brut*	NGL *LGN*	Feed-stocks *Produits d'aliment.*	Additives *Additifs*
Production	307458	1919860	24823	256644	5741	208095	15475	2422636	94614	-	251
From Other Sources	-	-	-	-	-	1	-	239	-	421	1882
Imports	48375	106547	1543	3295	10	4008	259	411529	194	1254	125
Exports	-7383	-226285	-1	-488	-114	-14084	-166	-1457322	-37790	-4	-1890
Intl. Marine Bunkers	-	-	-	-	-	-	-	-	-	-	-
Stock Changes	468	5794	-52	1379	1056	-7	27	-28815	-43	81	27
DOMESTIC SUPPLY	**348918**	**1805916**	**26313**	**260830**	**6693**	**198013**	**15595**	**1348267**	**56975**	**1752**	**395**
Transfers	-	-	-	-	-	-	-	-60	-36545	2172	-
Statistical Differences	-13837	52058	-35	-2196	-125	-6303	-931	16692	-193	-	-16
TRANSFORMATION	**302853**	**1147933**	**3542**	**232744**	**5341**	**126383**	**273**	**1347132**	**10961**	**3924**	**369**
Electricity Plants	-	891654	2058	131147	-	94	-	16246	306	-	-
CHP Plants	-	183134	1419	84424	1565	-	95	2251	-	-	-
Heat Plants	1	21126	65	12804	817	6	178	2	-	-	-
Blast Furnaces/Gas Works	2273	12006	-	603	-	126283	-	-	263	-	-
Coke/Pat. Fuel/BKB Plants	300579	12472	-	3766	2959	-	-	-	-	-	-
Petroleum Refineries	-	-	-	-	-	-	-	1336586	10392	3924	369
Petrochemical Industry	-	-	-	-	-	-	-	-	-	-	-
Liquefaction	-	27541	-	-	-	-	-	-7953	-	-	-
Other Transform. Sector	-	-	-	-	-	-	-	-	-	-	-
ENERGY SECTOR	**54**	**76615**	**25**	**380**	**100**	**1401**	**11**	**8529**	**78**	**-**	**-**
Coal Mines	1	31381	25	348	37	450	10	13	-	-	-
Oil and Gas Extraction	-	1992	-	-	-	48	-	5596	78	-	-
Petroleum Refineries	-	-	-	-	-	-	-	2908	-	-	-
Electr., CHP+Heat Plants	-	33432	-	32	2	89	-	12	-	-	-
Pumped Storage (Elec.)	-	-	-	-	-	-	-	-	-	-	-
Other Energy Sector	53	9810	-	-	61	814	1	-	-	-	-
Distribution Losses	39	5268	13	3841	70	111	-	4499	693	-	-
FINAL CONSUMPTION	**32135**	**628158**	**22698**	**21669**	**1057**	**63815**	**14380**	**4739**	**8505**	**-**	**10**
INDUSTRY SECTOR	**32129**	**465594**	**22612**	**12873**	**51**	**59821**	**204**	**4138**	**8505**	**-**	**10**
Iron and Steel	-	-	-	-	-	-	-	-	-	-	-
Chemical and Petrochem.	-	-	-	-	-	-	-	-	-	-	-
of which: Feedstocks	-	-	-	-	-	-	-	-	-	-	-
Non-Ferrous Metals	-	-	-	-	-	-	-	-	-	-	-
Non-Metallic Minerals	-	-	-	-	-	-	-	-	-	-	-
Transport Equipment	-	-	-	-	-	-	-	-	-	-	-
Machinery	-	-	-	-	-	-	-	-	-	-	-
Mining and Quarrying	-	-	-	-	-	-	-	-	-	-	-
Food and Tobacco	-	-	-	-	-	-	-	-	-	-	-
Paper, Pulp and Print	-	-	-	-	-	-	-	-	-	-	-
Wood and Wood Products	-	-	-	-	-	-	-	-	-	-	-
Construction	-	-	-	-	-	-	-	-	-	-	-
Textile and Leather	-	-	-	-	-	-	-	-	-	-	-
Non-specified	-	-	-	-	-	-	-	-	-	-	-
TRANSPORT SECTOR	**-**	**14390**	**2**	**423**	**-**	**1**	**13**	**11**	**-**	**-**	**-**
Air	-	-	-	-	-	-	-	-	-	-	-
Road	-	-	-	-	-	-	-	-	-	-	-
Rail	-	-	-	-	-	-	-	-	-	-	-
Pipeline Transport	-	-	-	-	-	-	-	-	-	-	-
Internal Navigation	-	-	-	-	-	-	-	-	-	-	-
Non-specified	-	-	-	-	-	-	-	-	-	-	-
OTHER SECTORS	**-**	**-**	**-**	**-**	**-**	**-**	**-**	**-**	**-**	**-**	**-**
Agriculture	-	-	-	-	-	-	-	-	-	-	-
Comm. and Publ. Services	-	-	-	-	-	-	-	-	-	-	-
Residential	-	-	-	-	-	-	-	-	-	-	-
Non-specified	-	-	-	-	-	-	-	-	-	-	-
NON-ENERGY USE	**-**	**7710**	**-**	**293**	**7**	**617**	**41**	**-**	**-**	**-**	**-**
in Industry/Trans./Energy	-	-	-	-	-	-	-	-	-	-	-
in Transport	-	-	-	-	-	-	-	-	-	-	-
in Other Sectors	-	-	-	-	-	-	-	-	-	-	-

Non-OECD Total / Total non-OCDE : 1998

	Oil cont. / *Pétrole cont.* (1000 tonnes)										
SUPPLY AND CONSUMPTION / *APPROVISIONNEMENT ET DEMANDE*	Refinery Gas / *Gaz de raffinerie*	LPG + Ethane / *GPL + éthane*	Motor Gasoline / *Essence moteur*	Aviation Gasoline / *Essence aviation*	Jet Fuel / *Carbu- réacteurs*	Kerosene / *Kérosène*	Gas/ Diesel / *Gazole*	Heavy Fuel Oil / *Fioul lourd*	Naphtha / *Naphta*	Petrol. Coke / *Coke de pétrole*	Other Prod. / *Autres prod.*
Production	30604	37212	224248	445	62350	40936	388306	362773	71748	6488	77262
From Other Sources	-	3035	-	-	-	-	-	-	-	-	-
Imports	-	18364	29534	585	11492	11675	85662	95953	6457	361	8330
Exports	-	-27029	-45743	-71	-21767	-13980	-105932	-140139	-32695	-1107	-12212
Intl. Marine Bunkers	-	-	-	-	-	-	-8960	-35847	-	-	-3
Stock Changes	4	-1020	683	-1	27	-91	1978	1992	-513	-69	248
DOMESTIC SUPPLY	30608	30562	208722	958	52102	38540	361054	284732	44997	5673	73625
Transfers	-	30087	6767	-	-19	-111	32	-1002	-630	-	1111
Statistical Differences	-37	326	148	-25	-362	1021	738	-5025	291	363	1036
TRANSFORMATION	3756	729	38	-	-	88	28526	133090	1064	734	640
Electricity Plants	493	30	38	-	-	75	27463	85417	23	33	311
CHP Plants	3248	674	-	-	-	4	53	15933	-	-	76
Heat Plants	15	4	-	-	-	9	1010	31379	46	-	253
Blast Furnaces/Gas Works	-	20	-	-	-	-	-	360	995	-	-
Coke/Pat. Fuel/BKB Plants	-	-	-	-	-	-	-	-	-	701	-
Petroleum Refineries	-	-	-	-	-	-	-	-	-	-	-
Petrochemical Industry	-	-	-	-	-	-	-	-	-	-	-
Liquefaction	-	-	-	-	-	-	-	-	-	-	-
Other Transform. Sector	-	1	-	-	-	-	-	1	-	-	-
ENERGY SECTOR	24637	1161	1155	-	1	204	7481	31661	4	1220	1053
Coal Mines	-	-	3	-	-	59	1518	294	-	1	-
Oil and Gas Extraction	438	149	1070	-	-	6	3562	1540	-	-	-
Petroleum Refineries	22224	958	22	-	1	108	1289	25625	-	1219	708
Electr., CHP+Heat Plants	-	-	4	-	-	4	922	1108	-	-	-
Pumped Storage (Elec.)	-	-	-	-	-	-	-	-	-	-	-
Other Energy Sector	1975	54	56	-	-	27	190	3094	4	-	345
Distribution Losses	71	160	126	-	1	-	227	108	335	-	1
FINAL CONSUMPTION	2107	58925	214318	933	51719	39158	325590	113846	43255	4082	74078
INDUSTRY SECTOR	2105	11339	3779	-	1	1716	42836	91318	43255	2716	2873
Iron and Steel	-	-	-	-	-	-	-	-	-	-	-
Chemical and Petrochem.	-	-	-	-	-	-	-	-	-	-	-
of which: Feedstocks	-	-	-	-	-	-	-	-	-	-	-
Non-Ferrous Metals	-	-	-	-	-	-	-	-	-	-	-
Non-Metallic Minerals	-	-	-	-	-	-	-	-	-	-	-
Transport Equipment	-	-	-	-	-	-	-	-	-	-	-
Machinery	-	-	-	-	-	-	-	-	-	-	-
Mining and Quarrying	-	-	-	-	-	-	-	-	-	-	-
Food and Tobacco	-	-	-	-	-	-	-	-	-	-	-
Paper, Pulp and Print	-	-	-	-	-	-	-	-	-	-	-
Wood and Wood Products	-	-	-	-	-	-	-	-	-	-	-
Construction	-	-	-	-	-	-	-	-	-	-	-
Textile and Leather	-	-	-	-	-	-	-	-	-	-	-
Non-specified	-	-	-	-	-	-	-	-	-	-	-
TRANSPORT SECTOR	-	609	206212	872	51575	731	189455	8775	-	-	30
Air	-	-	-	872	51575	695	-	-	-	-	-
Road	-	506	205544	-	-	2	157297	74	-	-	30
Rail	-	-	-	-	-	-	-	-	-	-	-
Pipeline Transport	-	-	-	-	-	-	-	-	-	-	-
Internal Navigation	-	-	-	-	-	-	-	-	-	-	-
Non-specified	-	-	-	-	-	-	-	-	-	-	-
OTHER SECTORS	-	-	-	-	-	-	-	-	-	-	-
Agriculture	-	-	-	-	-	-	-	-	-	-	-
Comm. and Publ. Services	-	-	-	-	-	-	-	-	-	-	-
Residential	-	-	-	-	-	-	-	-	-	-	-
Non-specified	-	-	-	-	-	-	-	-	-	-	-
NON-ENERGY USE	-	-	-	-	-	-	-	-	-	1249	64749
in Industry/Transf./Energy	-	-	-	-	-	-	-	-	-	-	-
in Transport	-	-	-	-	-	-	-	-	-	-	-
in Other Sectors	-	-	-	-	-	-	-	-	-	-	-

Non-OECD Total / Total non-OCDE : 1998

	Gas / *Gaz* (TJ)				Comb. Renew. & Waste / *En. Re. Comb. & Déchets* (TJ)				(GWh)	(TJ)
SUPPLY AND CONSUMPTION	Natural Gas	Gas Works	Coke Ovens	Blast Furnaces	Solid Biomass	Gas/Liquids from Biomass	Municipal Waste	Industrial Waste	Electricity	Heat
APPROVISIONNEMENT ET DEMANDE	*Gaz naturel*	*Usines à gaz*	*Cokeries*	*Hauts fourneaux*	*Biomasse solide*	*Gaz/Liquides tirés de biomasse*	*Déchets urbains*	*Déchets industriels*	*Electricité*	*Chaleur*
Production	50004672	274204	758276	934298	36863083	332324	-	341829	5278338	9203694
From Other Sources	-	11022	-	-	1675	-	-	-	-	-
Imports	4959853	746	-	-	4227	3474	-	-	137196	-
Exports	-13808739	-	-	-	-42300	-3009	-	-	-152552	-
Intl. Marine Bunkers	-	-	-	-	-	-	-	-	-	-
Stock Changes	-193808	-	-	-	-3166	3725	-	-4	-	-
DOMESTIC SUPPLY	40961978	285972	758276	934298	36823519	336514	-	341825	5262982	9203694
Transfers	-	-	-	-	-	-	-	-	-	-
Statistical Differences	55259	30116	-60647	-977	-3416	-	-	-16732	-17696	1
TRANSFORMATION	17495254	11142	90033	127071	1974447	70	-	61864	2035	-
Electricity Plants	7016768	9819	37329	64101	96430	-	-	-	-	-
CHP Plants	6793653	1135	16415	52728	54469	-	-	61832	-	-
Heat Plants	2744477	188	29700	10242	62378	70	-	32	-	-
Blast Furnaces/Gas Works	28214	-	6589	-	1675	-	-	-	-	-
Coke/Pat. Fuel/BKB Plants	-	-	-	-	49127	-	-	-	-	-
Petroleum Refineries	278	-	-	-	116	-	-	-	-	-
Petrochemical Industry	-	-	-	-	-	-	-	-	-	-
Liquefaction	382423	-	-	-	-	-	-	-	-	-
Other Transform. Sector	529441	-	-	-	1710252	-	-	-	2035	-
ENERGY SECTOR	4871026	26593	49697	17536	1113	1030	-	15755	544566	880407
Coal Mines	243	-	701	-	1	-	-	2	66124	67871
Oil and Gas Extraction	3972412	-	-	-	-	-	-	-	77185	185870
Petroleum Refineries	653076	7635	-	-	-	-	-	-	62537	424116
Electr., CHP+Heat Plants	22594	9499	20	-	56	-	-	-	302356	83767
Pumped Storage (Elec.)	-	-	-	-	-	-	-	-	10955	-
Other Energy Sector	222701	9459	48976	17536	1056	1030	-	15753	25409	118783
Distribution Losses	734629	995	3009	25434	138	-	-	19489	635435	547180
FINAL CONSUMPTION	17916328	277358	554890	763280	34822612	335414	-	227985	4063250	7776108
INDUSTRY SECTOR	9445858	85615	478714	763280	3732328	-	-	218911	1927671	3247453
Iron and Steel	-	-	-	-	-	-	-	-	-	-
Chemical and Petrochem.	-	-	-	-	-	-	-	-	-	-
of which: Feedstocks	-	-	-	-	-	-	-	-	-	-
Non-Ferrous Metals	-	-	-	-	-	-	-	-	-	-
Non-Metallic Minerals	-	-	-	-	-	-	-	-	-	-
Transport Equipment	-	-	-	-	-	-	-	-	-	-
Machinery	-	-	-	-	-	-	-	-	-	-
Mining and Quarrying	-	-	-	-	-	-	-	-	-	-
Food and Tobacco	-	-	-	-	-	-	-	-	-	-
Paper, Pulp and Print	-	-	-	-	-	-	-	-	-	-
Wood and Wood Products	-	-	-	-	-	-	-	-	-	-
Construction	-	-	-	-	-	-	-	-	-	-
Textile and Leather	-	-	-	-	-	-	-	-	-	-
Non-specified	-	-	-	-	-	-	-	-	-	-
TRANSPORT SECTOR	818982	17	-	-	1786	284615	-	3159	118249	-
Air	-	-	-	-	-	-	-	-	-	-
Road	77833	-	-	-	-	284615	-	-	250	-
Rail	-	-	-	-	-	-	-	-	-	-
Pipeline Transport	-	-	-	-	-	-	-	-	-	-
Internal Navigation	-	-	-	-	-	-	-	-	-	-
Non-specified	-	-	-	-	-	-	-	-	-	-
OTHER SECTORS	-	-	-	-	-	-	-	-	-	-
Agriculture	-	-	-	-	-	-	-	-	-	-
Comm. and Publ. Services	-	-	-	-	-	-	-	-	-	-
Residential	-	-	-	-	-	-	-	-	-	-
Non-specified	-	-	-	-	-	-	-	-	-	-
NON-ENERGY USE	-	-	-	-	-	-	-	-	-	-
in Industry/Transf./Energy	-	-	-	-	-	-	-	-	-	-
in Transport	-	-	-	-	-	-	-	-	-	-
in Other Sectors	-	-	-	-	-	-	-	-	-	-

Africa / Afrique

SUPPLY AND CONSUMPTION 1997	Coking Coal	Other Bit. Coal	Sub-Bit. Coal	Lignite	Peat	Oven and Gas Coke	Pat. Fuel and BKB	Crude Oil	NGL	Feed-stocks	Additives
	Coal (1000 tonnes)							Oil (1000 tonnes)			
Production	4243	221713	-	-	12	5607	-	341138	30315	-	-
Imports	2853	3144	-	-	-	287	85	29640	-	-	-
Exports	-	-64337	-	-	-	-424	-	-246742	-19649	-	-
Intl. Marine Bunkers	-	-	-	-	-	-	-	-	-	-	-
Stock Changes	-116	-2537	-	-	-	91	-	-7062	-	-	-
DOMESTIC SUPPLY	6980	157983	-	-	12	5561	85	116974	10666	-	-
Transfers and Stat. Diff.	351	9592	-	-	-	-	-	161	-8507	-	-
TRANSFORMATION	7331	134352	-	-	-	4311	-	115038	2036	-	-
Electricity and CHP Plants	-	101208	-	-	-	-	-	-	-	-	-
Petroleum Refineries	-	-	-	-	-	-	-	122289	2036	-	-
Other Transform. Sector	7331	33144	-	-	-	4311	-	-7251	-	-	-
ENERGY SECTOR	-	-	-	-	-	41	-	1884	-	-	-
DISTRIBUTION LOSSES	-	-	-	-	-	-	-	185	123	-	-
FINAL CONSUMPTION	-	33223	-	-	12	1209	85	28	-	-	-
INDUSTRY SECTOR	-	20863	-	-	-	1207	-	28	-	-	-
Iron and Steel	-	-	-	-	-	-	-	-	-	-	-
Chemical and Petrochem.	-	-	-	-	-	-	-	-	-	-	-
Non-Metallic Minerals	-	-	-	-	-	-	-	-	-	-	-
Non-specified	-	-	-	-	-	-	-	-	-	-	-
TRANSPORT SECTOR	-	33	-	-	-	1	-	-	-	-	-
Air	-	-	-	-	-	-	-	-	-	-	-
Road	-	-	-	-	-	-	-	-	-	-	-
Non-specified	-	-	-	-	-	-	-	-	-	-	-
OTHER SECTORS	-	-	-	-	-	-	-	-	-	-	-
Agriculture	-	-	-	-	-	-	-	-	-	-	-
Comm. and Publ. Services	-	-	-	-	-	-	-	-	-	-	-
Residential	-	-	-	-	-	-	-	-	-	-	-
Non-specified	-	-	-	-	-	-	-	-	-	-	-
NON-ENERGY USE	-	7617	-	-	-	-	-	-	-	-	-

APPROVISIONNEMENT ET DEMANDE 1998	Charbon à coke	Autres charb. bit.	Charbon sous-bit.	Lignite	Tourbe	Coke de four/gaz	Agg./briq. de lignite	Pétrole brut	LGN	Produits d'aliment.	Additifs
	Charbon (1000 tonnes)							Pétrole (1000 tonnes)			
Production	2454	228193	-	-	12	5031	-	332675	30719	-	-
Imports	3717	3486	-	-	-	259	80	33004	-	-	-
Exports	-	-66297	-	-	-	-424	-	-247853	-18882	-	-
Intl. Marine Bunkers	-	-	-	-	-	-	-	-	-	-	-
Stock Changes	14	-2343	-	-	-	105	-	-8735	-	-	-
DOMESTIC SUPPLY	6185	163039	-	-	12	4971	80	109091	11837	-	-
Transfers and Stat. Diff.	351	11166	-	-	-	1	-	2867	-9588	-	-
TRANSFORMATION	6536	138359	-	-	-	3836	-	109568	1969	-	-
Electricity and CHP Plants	-	106004	-	-	-	-	-	-	-	-	-
Petroleum Refineries	-	-	-	-	-	-	-	116761	1969	-	-
Other Transform. Sector	6536	32355	-	-	-	3836	-	-7193	-	-	-
ENERGY SECTOR	-	-	-	-	-	41	-	1941	-	-	-
DISTRIBUTION LOSSES	-	-	-	-	-	-	-	421	280	-	-
FINAL CONSUMPTION	-	35846	-	-	12	1095	80	28	-	-	-
INDUSTRY SECTOR	-	25316	-	-	-	1093	-	28	-	-	-
Iron and Steel	-	-	-	-	-	-	-	-	-	-	-
Chemical and Petrochem.	-	-	-	-	-	-	-	-	-	-	-
Non-Metallic Minerals	-	-	-	-	-	-	-	-	-	-	-
Non-specified	-	-	-	-	-	-	-	-	-	-	-
TRANSPORT SECTOR	-	48	-	-	-	1	-	-	-	-	-
Air	-	-	-	-	-	-	-	-	-	-	-
Road	-	-	-	-	-	-	-	-	-	-	-
Non-specified	-	-	-	-	-	-	-	-	-	-	-
OTHER SECTORS	-	-	-	-	-	-	-	-	-	-	-
Agriculture	-	-	-	-	-	-	-	-	-	-	-
Comm. and Publ. Services	-	-	-	-	-	-	-	-	-	-	-
Residential	-	-	-	-	-	-	-	-	-	-	-
Non-specified	-	-	-	-	-	-	-	-	-	-	-
NON-ENERGY USE	-	7555	-	-	-	-	-	-	-	-	-

Africa / Afrique

SUPPLY AND CONSUMPTION 1997	Oil cont. (1000 tonnes)										
	Refinery Gas	LPG + Ethane	Motor Gasoline	Aviation Gasoline	Jet Fuel	Kerosene	Gas/ Diesel	Heavy Fuel Oil	Naphtha	Petrol. Coke	Other Prod.
Production	1773	2940	20314	80	6592	5293	33142	35595	9043	122	3701
Imports	-	1738	3194	278	1529	1273	7885	2688	24	2	1097
Exports	-	-5558	-1440	-	-2243	-1106	-8128	-13293	-9636	-	-192
Intl. Marine Bunkers	-	-	-	-	-	-	-1224	-5846	-	-	-
Stock Changes	-	2	-54	-	-256	66	-10	500	-23	-	25
DOMESTIC SUPPLY	1773	-878	22014	358	5622	5526	31665	19644	-592	124	4631
Transfers and Stat. Diff.	-	6987	34	-3	54	56	59	-2382	1577	-	1195
TRANSFORMATION	-	-	-	-	-	69	2547	8970	-	-	-
Electricity and CHP Plants	-	-	-	-	-	69	2547	8970	-	-	-
Petroleum Refineries	-	-	-	-	-	-	-	-	-	-	-
Other Transform. Sector	-	-	-	-	-	-	-	-	-	-	-
ENERGY SECTOR	1709	-	-	-	-	-	223	1293	-	-	-
DISTRIBUTION LOSSES	-	106	-	-	-	-	-	-	-	-	-
FINAL CONSUMPTION	64	6003	22048	355	5676	5513	28954	6999	985	124	5826
INDUSTRY SECTOR	64	605	39	-	-	360	5389	5971	985	2	31
Iron and Steel	-	-	-	-	-	-	-	-	-	-	-
Chemical and Petrochem.	-	-	-	-	-	-	-	-	-	-	-
Non-Metallic Minerals	-	-	-	-	-	-	-	-	-	-	-
Non-specified	-	-	-	-	-	-	-	-	-	-	-
TRANSPORT SECTOR	-	112	21859	355	5676	11	15133	14	-	-	-
Air	-	-	-	355	5676	-	-	-	-	-	-
Road	-	112	21814	-	-	-	14542	-	-	-	-
Non-specified	-	-	-	-	-	-	-	-	-	-	-
OTHER SECTORS	-	-	-	-	-	-	-	-	-	-	-
Agriculture	-	-	-	-	-	-	-	-	-	-	-
Comm. and Publ. Services	-	-	-	-	-	-	-	-	-	-	-
Residential	-	-	-	-	-	-	-	-	-	-	-
Non-specified	-	-	-	-	-	-	-	-	-	-	-
NON-ENERGY USE	-	-	-	-	-	-	-	-	-	122	5728

APPROVISIONNEMENT ET DEMANDE 1998	Pétrole cont. (1000 tonnes)										
	Gaz de raffinerie	GPL + éthane	Essence moteur	Essence aviation	Carbu- réacteurs	Kérosène	Gazole	Fioul lourd	Naphta	Coke de pétrole	Autres prod.
Production	1701	2727	19335	22	6253	4866	31175	34309	8809	181	3408
Imports	-	1987	4360	384	1686	1638	8360	3285	24	2	1253
Exports	-	-6369	-1345	-57	-2287	-976	-6433	-10820	-9536	-	-145
Intl. Marine Bunkers	-	-	-	-	-	-	-1107	-5446	-	-	-
Stock Changes	-	1	-219	-	-228	-12	146	-328	-20	-	52
DOMESTIC SUPPLY	1701	-1654	22131	349	5424	5516	32141	21000	-723	183	4568
Transfers and Stat. Diff.	-	7977	143	-3	160	65	455	-1645	1708	-	1264
TRANSFORMATION	-	-	-	-	-	69	2867	9984	-	-	-
Electricity and CHP Plants	-	-	-	-	-	69	2867	9984	-	-	-
Petroleum Refineries	-	-	-	-	-	-	-	-	-	-	-
Other Transform. Sector	-	-	-	-	-	-	-	-	-	-	-
ENERGY SECTOR	1598	-	-	-	-	-	223	1247	-	-	-
DISTRIBUTION LOSSES	-	53	-	-	-	-	-	-	-	-	-
FINAL CONSUMPTION	103	6270	22274	346	5584	5512	29506	8124	985	183	5832
INDUSTRY SECTOR	103	583	36	-	-	433	5604	7101	985	2	31
Iron and Steel	-	-	-	-	-	-	-	-	-	-	-
Chemical and Petrochem.	-	-	-	-	-	-	-	-	-	-	-
Non-Metallic Minerals	-	-	-	-	-	-	-	-	-	-	-
Non-specified	-	-	-	-	-	-	-	-	-	-	-
TRANSPORT SECTOR	-	119	22079	346	5584	14	15493	12	-	-	-
Air	-	-	-	346	5584	-	-	-	-	-	-
Road	-	119	22028	-	-	-	14878	-	-	-	-
Non-specified	-	-	-	-	-	-	-	-	-	-	-
OTHER SECTORS	-	-	-	-	-	-	-	-	-	-	-
Agriculture	-	-	-	-	-	-	-	-	-	-	-
Comm. and Publ. Services	-	-	-	-	-	-	-	-	-	-	-
Residential	-	-	-	-	-	-	-	-	-	-	-
Non-specified	-	-	-	-	-	-	-	-	-	-	-
NON-ENERGY USE	-	-	-	-	-	-	-	-	-	181	5734

Africa / Afrique

SUPPLY AND CONSUMPTION 1997	Gas (TJ)				Comb. Renew. & Waste (TJ)				(GWh)	(TJ)
	Natural Gas	Gas Works	Coke Ovens	Blast Furnaces	Solid Biomass	Gas/Liquids from Biomass	Municipal Waste	Industrial Waste	Electricity	Heat
Production	4176004	30282	27152	43872	9803355	-	-	-	406395	-
Imports	29836	-	-	-	31	-	-	-	7892	-
Exports	-2203830	-	-	-	-10081	-	-	-	-10609	-
Intl. Marine Bunkers	-	-	-	-	-	-	-	-	-	-
Stock Changes	-	-	-	-	-	-	-	-	-	-
DOMESTIC SUPPLY	2002010	30282	27152	43872	9793304	-	-	-	403678	-
Transfers and Stat. Diff.	-10913	1	-5	-	-40	-	-	-	-262	-
TRANSFORMATION	901091	-	1126	-	1070374	-	-	-	-	-
Electricity and CHP Plants	779172	-	-	-	-	-	-	-	-	-
Petroleum Refineries	-	-	-	-	-	-	-	-	-	-
Other Transform. Sector	121919	-	1126	-	1070374	-	-	-	-	-
ENERGY SECTOR	501709	7	-	-	-	-	-	-	43158	-
DISTRIBUTION LOSSES	30588	-	668	2786	-	-	-	-	40663	-
FINAL CONSUMPTION	557709	30276	25353	41086	8722892	-	-	-	319595	-
INDUSTRY SECTOR	391539	28469	25353	41086	927538	-	-	-	156272	-
Iron and Steel	-	-	-	-	-	-	-	-	-	-
Chemical and Petrochem.	-	-	-	-	-	-	-	-	-	-
Non-Metallic Minerals	-	-	-	-	-	-	-	-	-	-
Non-specified	-	-	-	-	-	-	-	-	-	-
TRANSPORT SECTOR	31357	-	-	-	-	-	-	-	5187	-
Air	-	-	-	-	-	-	-	-	-	-
Road	-	-	-	-	-	-	-	-	8	-
Non-specified	-	-	-	-	-	-	-	-	-	-
OTHER SECTORS	-	-	-	-	-	-	-	-	-	-
Agriculture	-	-	-	-	-	-	-	-	-	-
Comm. and Publ. Services	-	-	-	-	-	-	-	-	-	-
Residential	-	-	-	-	-	-	-	-	-	-
Non-specified	-	-	-	-	-	-	-	-	-	-
NON-ENERGY USE	-	-	-	-	-	-	-	-	-	-

APPROVISIONNEMENT ET DEMANDE 1998	Gaz (TJ)				En. Re. Comb. & Déchets (TJ)				(GWh)	(TJ)
	Gaz naturel	Usines à gaz	Cokeries	Hauts fourneaux	Biomasse solide	Gaz/Liquides tirés de biomasse	Déchets urbains	Déchets industriels	Electricité	Chaleur
Production	4416420	31758	22972	39774	10055975	-	-	-	416267	-
Imports	30678	-	-	-	32	-	-	-	11656	-
Exports	-2377385	-	-	-	-10262	-	-	-	-13317	-
Intl. Marine Bunkers	-	-	-	-	-	-	-	-	-	-
Stock Changes	-	-	-	-	-	-	-	-	-	-
DOMESTIC SUPPLY	2069713	31758	22972	39774	10045745	-	-	-	414606	-
Transfers and Stat. Diff.	-18924	-	-	-	-1	-	-	-	-15046	-
TRANSFORMATION	931708	-	1323	-	1094542	-	-	-	-	-
Electricity and CHP Plants	828765	-	-	-	-	-	-	-	-	-
Petroleum Refineries	-	-	-	-	-	-	-	-	-	-
Other Transform. Sector	102943	-	1323	-	1094542	-	-	-	-	-
ENERGY SECTOR	511323	187	-	-	-	-	-	-	24987	-
DISTRIBUTION LOSSES	33544	-	1794	3752	-	-	-	-	43290	-
FINAL CONSUMPTION	574214	31571	19855	36022	8931204	-	-	-	331283	-
INDUSTRY SECTOR	415892	28711	19855	36022	950505	-	-	-	162708	-
Iron and Steel	-	-	-	-	-	-	-	-	-	-
Chemical and Petrochem.	-	-	-	-	-	-	-	-	-	-
Non-Metallic Minerals	-	-	-	-	-	-	-	-	-	-
Non-specified	-	-	-	-	-	-	-	-	-	-
TRANSPORT SECTOR	16356	-	-	-	-	-	-	-	5205	-
Air	-	-	-	-	-	-	-	-	-	-
Road	-	-	-	-	-	-	-	-	23	-
Non-specified	-	-	-	-	-	-	-	-	-	-
OTHER SECTORS	-	-	-	-	-	-	-	-	-	-
Agriculture	-	-	-	-	-	-	-	-	-	-
Comm. and Publ. Services	-	-	-	-	-	-	-	-	-	-
Residential	-	-	-	-	-	-	-	-	-	-
Non-specified	-	-	-	-	-	-	-	-	-	-
NON-ENERGY USE	-	-	-	-	-	-	-	-	-	-

Latin America / Amérique latine

SUPPLY AND CONSUMPTION 1997	Coal (1000 tonnes)							Oil (1000 tonnes)			
	Coking Coal	Other Bit. Coal	Sub-Bit. Coal	Lignite	Peat	Oven and Gas Coke	Pat. Fuel and BKB	Crude Oil	NGL	Feed-stocks	Additives
Production	1761	43084	-	-	-	10376	-	339020	9097	-	1520
Imports	14546	4434	-	-	-	1943	2	68547	-	664	-
Exports	-1101	-31584	-	-	-	-128	-	-183354	-123	-	-1546
Intl. Marine Bunkers	-	-	-	-	-	-	-	-	-	-	-
Stock Changes	-175	328	-	-	-	68	-	-20	3	-	36
DOMESTIC SUPPLY	15031	16262	-	-	-	12259	2	224193	8977	664	10
Transfers and Stat. Diff.	-173	161	-	-	-	-1	-	4965	-6471	368	-3
TRANSFORMATION	12568	10318	-	-	-	9476	-	222637	956	1032	-
Electricity and CHP Plants	-	10058	-	-	-	48	-	88	239	-	-
Petroleum Refineries	-	-	-	-	-	-	-	223193	717	1032	-
Other Transform. Sector	12568	260	-	-	-	9428	-	-644	-	-	-
ENERGY SECTOR	-	-	-	-	-	-	-	161	1004	-	-
DISTRIBUTION LOSSES	-	51	-	-	-	119	-	3576	-	-	-
FINAL CONSUMPTION	2290	6054	-	-	-	2663	2	2784	546	-	7
INDUSTRY SECTOR	2290	5733	-	-	-	2404	-	2678	546	-	7
Iron and Steel	-	-	-	-	-	-	-	-	-	-	-
Chemical and Petrochem.	-	-	-	-	-	-	-	-	-	-	-
Non-Metallic Minerals	-	-	-	-	-	-	-	-	-	-	-
Non-specified	-	-	-	-	-	-	-	-	-	-	-
TRANSPORT SECTOR	-	-	-	-	-	-	-	-	-	-	-
Air	-	-	-	-	-	-	-	-	-	-	-
Road	-	-	-	-	-	-	-	-	-	-	-
Non-specified	-	-	-	-	-	-	-	-	-	-	-
OTHER SECTORS	-	-	-	-	-	-	-	-	-	-	-
Agriculture	-	-	-	-	-	-	-	-	-	-	-
Comm. and Publ. Services	-	-	-	-	-	-	-	-	-	-	-
Residential	-	-	-	-	-	-	-	-	-	-	-
Non-specified	-	-	-	-	-	-	-	-	-	-	-
NON-ENERGY USE	-	-	-	-	-	259	-	-	-	-	-

APPROVISIONNEMENT ET DEMANDE 1998	Charbon (1000 tonnes)							Pétrole (1000 tonnes)			
	Charbon à coke	Autres charb. bit.	Charbon sous-bit.	Lignite	Tourbe	Coke de four/gaz	Agg./briq. de lignite	Pétrole brut	LGN	Produits d'aliment.	Additifs
Production	1478	46417	-	-	-	10333	-	346109	8016	-	1882
Imports	14731	4537	-	-	-	1834	2	70935	-	698	-
Exports	-660	-34649	-	-	-	-301	-	-182429	-113	-	-1890
Intl. Marine Bunkers	-	-	-	-	-	-	-	-	-	-	-
Stock Changes	-182	-1244	-	-	-	184	-	837	-2	-	32
DOMESTIC SUPPLY	15367	15061	-	-	-	12050	2	235452	7901	698	24
Transfers and Stat. Diff.	-249	1762	-	-	-	-18	-	-223	-6437	389	-14
TRANSFORMATION	12346	11032	-	-	-	9231	-	232132	1036	1087	-
Electricity and CHP Plants	-	10668	-	-	-	94	-	82	306	-	-
Petroleum Refineries	-	-	-	-	-	-	-	232810	730	1087	-
Other Transform. Sector	12346	364	-	-	-	9137	-	-760	-	-	-
ENERGY SECTOR	-	7	-	-	-	-	-	220	-	-	-
DISTRIBUTION LOSSES	39	74	-	-	-	111	-	94	-	-	-
FINAL CONSUMPTION	2733	5710	-	-	-	2690	2	2783	428	-	10
INDUSTRY SECTOR	2733	5532	-	-	-	2468	-	2677	428	-	10
Iron and Steel	-	-	-	-	-	-	-	-	-	-	-
Chemical and Petrochem.	-	-	-	-	-	-	-	-	-	-	-
Non-Metallic Minerals	-	-	-	-	-	-	-	-	-	-	-
Non-specified	-	-	-	-	-	-	-	-	-	-	-
TRANSPORT SECTOR	-	-	-	-	-	-	-	-	-	-	-
Air	-	-	-	-	-	-	-	-	-	-	-
Road	-	-	-	-	-	-	-	-	-	-	-
Non-specified	-	-	-	-	-	-	-	-	-	-	-
OTHER SECTORS	-	-	-	-	-	-	-	-	-	-	-
Agriculture	-	-	-	-	-	-	-	-	-	-	-
Comm. and Publ. Services	-	-	-	-	-	-	-	-	-	-	-
Residential	-	-	-	-	-	-	-	-	-	-	-
Non-specified	-	-	-	-	-	-	-	-	-	-	-
NON-ENERGY USE	-	-	-	-	-	222	-	-	-	-	-

Latin America / Amérique latine

SUPPLY AND CONSUMPTION 1997	Refinery Gas	LPG + Ethane	Motor Gasoline	Aviation Gasoline	Jet Fuel	Kerosene	Gas/ Diesel	Heavy Fuel Oil	Naphtha	Petrol. Coke	Other Prod.
					Oil cont. (1000 tonnes)						
Production	5896	6550	50567	143	11530	2335	66702	53207	6897	4459	12128
Imports	-	4817	5155	127	2197	446	15893	13040	3442	183	1172
Exports	-	-2912	-14072	-44	-5862	-96	-16480	-28253	-728	-1210	-3250
Intl. Marine Bunkers	-	-	-	-	-	-	-1542	-6079	-	-	-
Stock Changes	-	-121	-227	-6	-185	96	516	-208	-12	13	176
DOMESTIC SUPPLY	5896	8334	41423	220	7680	2781	65089	31707	9599	3445	10226
Transfers and Stat. Diff.	-6	6217	1053	9	-126	9	-125	2930	-1554	196	303
TRANSFORMATION	174	7	30	-	-	5	4768	12246	493	424	129
Electricity and CHP Plants	174	-	30	-	-	5	4768	12246	-	25	129
Petroleum Refineries	-	-	-	-	-	-	-	-	-	-	-
Other Transform. Sector	-	7	-	-	-	-	-	-	493	399	-
ENERGY SECTOR	5440	206	54	-	17	43	1222	3427	4	871	272
DISTRIBUTION LOSSES	63	2	2	-	-	-	139	65	35	-	29
FINAL CONSUMPTION	213	14336	42390	229	7537	2742	58835	18899	7513	2346	10099
INDUSTRY SECTOR	213	2854	710	-	-	484	6257	16084	7513	1775	1223
Iron and Steel	-	-	-	-	-	-	-	-	-	-	-
Chemical and Petrochem.	-	-	-	-	-	-	-	-	-	-	-
Non-Metallic Minerals	-	-	-	-	-	-	-	-	-	-	-
Non-specified	-	-	-	-	-	-	-	-	-	-	-
TRANSPORT SECTOR	-	110	40942	168	7497	-	41694	1228	-	-	-
Air	-	-	-	168	7497	-	-	-	-	-	-
Road	-	9	40857	-	-	-	39625	-	-	-	-
Non-specified	-	-	-	-	-	-	-	-	-	-	-
OTHER SECTORS	-	-	-	-	-	-	-	-	-	-	-
Agriculture	-	-	-	-	-	-	-	-	-	-	-
Comm. and Publ. Services	-	-	-	-	-	-	-	-	-	-	-
Residential	-	-	-	-	-	-	-	-	-	-	-
Non-specified	-	-	-	-	-	-	-	-	-	-	-
NON-ENERGY USE	-	-	-	-	-	-	-	-	-	476	8545

APPROVISIONNEMENT ET DEMANDE 1998	Gaz de raffinerie	GPL + éthane	Essence moteur	Essence aviation	Carbu- réacteurs	Kérosène	Gazole	Fioul lourd	Naphta	Coke de pétrole	Autres prod.
					Pétrole cont. (1000 tonnes)						
Production	7329	6900	53754	151	11914	2213	66060	56040	7178	4046	13040
Imports	-	5149	5311	104	2474	436	17117	12702	3520	56	893
Exports	-	-2986	-17948	-11	-5869	-164	-14141	-27543	-862	-859	-3465
Intl. Marine Bunkers	-	-	-	-	-	-	-1614	-5784	-	-	-
Stock Changes	-	-128	-161	-3	-256	128	-51	12	-193	21	311
DOMESTIC SUPPLY	7329	8935	40956	241	8263	2613	67371	35427	9643	3264	10779
Transfers and Stat. Diff.	1	6149	2304	-21	-271	23	-22	-939	-1523	405	628
TRANSFORMATION	187	8	30	-	-	6	5382	13512	443	734	94
Electricity and CHP Plants	187	-	30	-	-	6	5382	13512	-	33	94
Petroleum Refineries	-	-	-	-	-	-	-	-	-	-	-
Other Transform. Sector	-	8	-	-	-	-	-	-	443	701	-
ENERGY SECTOR	6860	104	22	-	-	26	1108	2575	4	1086	468
DISTRIBUTION LOSSES	69	33	3	-	-	-	146	64	84	-	1
FINAL CONSUMPTION	214	14939	43205	220	7992	2604	60713	18337	7589	1849	10844
INDUSTRY SECTOR	214	3113	367	-	-	432	6108	15858	7589	1478	1265
Iron and Steel	-	-	-	-	-	-	-	-	-	-	-
Chemical and Petrochem.	-	-	-	-	-	-	-	-	-	-	-
Non-Metallic Minerals	-	-	-	-	-	-	-	-	-	-	-
Non-specified	-	-	-	-	-	-	-	-	-	-	-
TRANSPORT SECTOR	-	112	42017	159	7952	-	44099	1246	-	-	-
Air	-	-	-	159	7952	-	-	-	-	-	-
Road	-	9	41936	-	-	-	41960	-	-	-	-
Non-specified	-	-	-	-	-	-	-	-	-	-	-
OTHER SECTORS	-	-	-	-	-	-	-	-	-	-	-
Agriculture	-	-	-	-	-	-	-	-	-	-	-
Comm. and Publ. Services	-	-	-	-	-	-	-	-	-	-	-
Residential	-	-	-	-	-	-	-	-	-	-	-
Non-specified	-	-	-	-	-	-	-	-	-	-	-
NON-ENERGY USE	-	-	-	-	-	-	-	-	-	371	9489

Latin America / Amérique latine

SUPPLY AND CONSUMPTION 1997	Gas (TJ) Natural Gas	Gas Works	Coke Ovens	Blast Furnaces	Comb. Renew. & Waste (TJ) Solid Biomass	Gas/Liquids from Biomass	Municipal Waste	Industrial Waste	(GWh) Electricity	(TJ) Heat
Production	3480467	29257	80145	107834	2949607	317000	-	-	691370	-
Imports	93276	-	-	-	404	19178	-	-	46894	-
Exports	-119636	-	-	-	-222	-3756	-	-	-46819	-
Intl. Marine Bunkers	-	-	-	-	-	-	-	-	-	-
Stock Changes	-	-	-	-	-6410	-38908	-	-	-	-
DOMESTIC SUPPLY	3454107	29257	80145	107834	2943379	293514	-	-	691445	-
Transfers and Stat. Diff.	-3923	-	-	-1	-1731	-	-	-	-663	-
TRANSFORMATION	969790	4354	9704	24520	362071	-	-	-	-	-
Electricity and CHP Plants	848058	4354	6144	24520	126158	-	-	-	-	-
Petroleum Refineries	-	-	-	-	105	-	-	-	-	-
Other Transform. Sector	121732	-	3560	-	235808	-	-	-	-	-
ENERGY SECTOR	869207	10690	17248	12272	1294	1294	-	-	19908	-
DISTRIBUTION LOSSES	70085	630	852	20354	-	-	-	-	112826	-
FINAL CONSUMPTION	1541102	13583	52341	50687	2576496	292220	-	-	558048	-
INDUSTRY SECTOR	1134072	683	52341	50687	1283266	-	-	-	258034	-
Iron and Steel	-	-	-	-	-	-	-	-	-	-
Chemical and Petrochem.	-	-	-	-	-	-	-	-	-	-
Non-Metallic Minerals	-	-	-	-	-	-	-	-	-	-
Non-specified	-	-	-	-	-	-	-	-	-	-
TRANSPORT SECTOR	57782	17	-	-	347	292220	-	-	2149	-
Air	-	-	-	-	-	-	-	-	-	-
Road	51017	-	-	-	-	292220	-	-	139	-
Non-specified	-	-	-	-	-	-	-	-	-	-
OTHER SECTORS	-	-	-	-	-	-	-	-	-	-
Agriculture	-	-	-	-	-	-	-	-	-	-
Comm. and Publ. Services	-	-	-	-	-	-	-	-	-	-
Residential	-	-	-	-	-	-	-	-	-	-
Non-specified	-	-	-	-	-	-	-	-	-	-
NON-ENERGY USE	-	-	-	-	-	-	-	-	-	-

APPROVISIONNEMENT ET DEMANDE 1998	Gaz (TJ) Gaz naturel	Usines à gaz	Cokeries	Hauts fourneaux	En. Re. Comb. & Déchets (TJ) Biomasse solide	Gaz/Liquides tirés de biomasse	Déchets urbains	Déchets industriels	(GWh) Electricité	(TJ) Chaleur
Production	3741891	32330	78354	104894	3001732	281455	-	-	717745	-
Imports	146182	-	-	-	543	3474	-	-	47862	-
Exports	-152608	-	-	-	-270	-3009	-	-	-47877	-
Intl. Marine Bunkers	-	-	-	-	-	-	-	-	-	-
Stock Changes	-	-	-	-	-6140	3725	-	-	-	-
DOMESTIC SUPPLY	3735465	32330	78354	104894	2995865	285645	-	-	717730	-
Transfers and Stat. Diff.	-35147	17	-57	1	-1458	-	-	-	748	-
TRANSFORMATION	1095213	9819	11874	23806	362741	-	-	-	-	-
Electricity and CHP Plants	931467	9819	8270	23806	134572	-	-	-	-	-
Petroleum Refineries	-	-	-	-	116	-	-	-	-	-
Other Transform. Sector	163746	-	3604	-	228053	-	-	-	-	-
ENERGY SECTOR	913175	8101	15918	11670	1030	1030	-	-	21473	-
DISTRIBUTION LOSSES	59964	699	925	20024	-	-	-	-	115299	-
FINAL CONSUMPTION	1631966	13728	49580	49395	2628841	284615	-	-	581706	-
INDUSTRY SECTOR	1206707	653	49580	49395	1342546	-	-	-	261750	-
Iron and Steel	-	-	-	-	-	-	-	-	-	-
Chemical and Petrochem.	-	-	-	-	-	-	-	-	-	-
Non-Metallic Minerals	-	-	-	-	-	-	-	-	-	-
Non-specified	-	-	-	-	-	-	-	-	-	-
TRANSPORT SECTOR	66864	17	-	-	302	284615	-	-	2240	-
Air	-	-	-	-	-	-	-	-	-	-
Road	57853	-	-	-	-	284615	-	-	169	-
Non-specified	-	-	-	-	-	-	-	-	-	-
OTHER SECTORS	-	-	-	-	-	-	-	-	-	-
Agriculture	-	-	-	-	-	-	-	-	-	-
Comm. and Publ. Services	-	-	-	-	-	-	-	-	-	-
Residential	-	-	-	-	-	-	-	-	-	-
Non-specified	-	-	-	-	-	-	-	-	-	-
NON-ENERGY USE	-	-	-	-	-	-	-	-	-	-

Asia (excluding China) / Asie (Chine non incluse)

SUPPLY AND CONSUMPTION 1997	Coking Coal	Other Bit. Coal	Sub-Bit. Coal	Lignite	Peat	Oven and Gas Coke	Pat. Fuel and BKB	Crude Oil	NGL	Feed-stocks	Additives
	Coal (1000 tonnes)							Oil (1000 tonnes)			
Production	53047	379185	23984	46916	-	21380	267	165837	10511	-	67
Imports	19910	46401	1041	1	-	534	132	193567	-	1715	96
Exports	-	-45616	-	-	-	-9	-	-77206	-1769	-	-
Intl. Marine Bunkers	-	-	-	-	-	-	-	-	-	-	-
Stock Changes	-333	-3275	-	-779	-	90	-	-3904	-	-	-1
DOMESTIC SUPPLY	72624	376695	25025	46138	-	21995	399	278294	8742	1715	162
Transfers and Stat. Diff.	1	77	-	-1	-	-	-	7995	-7321	2382	-
TRANSFORMATION	50896	243906	1259	40944	-	17595	13	285944	14	4097	162
Electricity and CHP Plants	-	243651	1259	40544	-	-	13	-	-	-	-
Petroleum Refineries	-	-	-	-	-	-	-	285944	14	4097	162
Other Transform. Sector	50896	255	-	400	-	17595	-	-	-	-	-
ENERGY SECTOR	-	3896	-	-	-	-	-	-	-	-	-
DISTRIBUTION LOSSES	-	-	-	-	-	-	-	67	552	-	-
FINAL CONSUMPTION	21729	128970	23766	5193	-	4400	386	278	855	-	-
INDUSTRY SECTOR	21729	126308	23766	5141	-	4400	124	278	855	-	-
Iron and Steel	-	-	-	-	-	-	-	-	-	-	-
Chemical and Petrochem.	-	-	-	-	-	-	-	-	-	-	-
Non-Metallic Minerals	-	-	-	-	-	-	-	-	-	-	-
Non-specified	-	-	-	-	-	-	-	-	-	-	-
TRANSPORT SECTOR	-	78	-	-	-	-	-	-	-	-	-
Air	-	-	-	-	-	-	-	-	-	-	-
Road	-	-	-	-	-	-	-	-	-	-	-
Non-specified	-	-	-	-	-	-	-	-	-	-	-
OTHER SECTORS	-	-	-	-	-	-	-	-	-	-	-
Agriculture	-	-	-	-	-	-	-	-	-	-	-
Comm. and Publ. Services	-	-	-	-	-	-	-	-	-	-	-
Residential	-	-	-	-	-	-	-	-	-	-	-
Non-specified	-	-	-	-	-	-	-	-	-	-	-
NON-ENERGY USE	-	-	-	2	-	-	-	-	-	-	-

APPROVISIONNEMENT ET DEMANDE 1998	Charbon à coke	Autres charb. bit.	Charbon sous-bit.	Lignite	Tourbe	Coke de four/gaz	Agg./briq. de lignite	Pétrole brut	LGN	Produits d'aliment.	Additifs
	Charbon (1000 tonnes)							Pétrole (1000 tonnes)			
Production	52683	383779	22893	43357	-	21065	266	165925	10638	-	65
Imports	18546	44443	1061	1	-	518	175	194405	-	170	100
Exports	-	-50227	-	-	-	-9	-	-77533	-2434	-	-
Intl. Marine Bunkers	-	-	-	-	-	-	-	-	-	-	-
Stock Changes	-266	-1303	-	417	-	-62	-	-4198	-	-	-7
DOMESTIC SUPPLY	70963	376692	23954	43775	-	21512	441	278599	8204	170	158
Transfers and Stat. Diff.	-	102	-	-1	-	-	-	6015	-7387	1568	-
TRANSFORMATION	48274	260643	1376	37894	-	17210	-	284441	11	1738	158
Electricity and CHP Plants	-	260113	1376	37494	-	-	-	-	-	-	-
Petroleum Refineries	-	-	-	-	-	-	-	284441	11	1738	158
Other Transform. Sector	48274	530	-	400	-	17210	-	-	-	-	-
ENERGY SECTOR	-	3711	-	-	-	-	-	-	-	-	-
DISTRIBUTION LOSSES	-	-	-	-	-	-	-	134	400	-	-
FINAL CONSUMPTION	22689	112440	22578	5880	-	4302	441	39	406	-	-
INDUSTRY SECTOR	22689	108799	22578	5829	-	4302	179	39	406	-	-
Iron and Steel	-	-	-	-	-	-	-	-	-	-	-
Chemical and Petrochem.	-	-	-	-	-	-	-	-	-	-	-
Non-Metallic Minerals	-	-	-	-	-	-	-	-	-	-	-
Non-specified	-	-	-	-	-	-	-	-	-	-	-
TRANSPORT SECTOR	-	47	-	-	-	-	-	-	-	-	-
Air	-	-	-	-	-	-	-	-	-	-	-
Road	-	-	-	-	-	-	-	-	-	-	-
Non-specified	-	-	-	-	-	-	-	-	-	-	-
OTHER SECTORS	-	-	-	-	-	-	-	-	-	-	-
Agriculture	-	-	-	-	-	-	-	-	-	-	-
Comm. and Publ. Services	-	-	-	-	-	-	-	-	-	-	-
Residential	-	-	-	-	-	-	-	-	-	-	-
Non-specified	-	-	-	-	-	-	-	-	-	-	-
NON-ENERGY USE	-	-	-	1	-	-	-	-	-	-	-

Asia (excluding China) / Asie (Chine non incluse)

SUPPLY AND CONSUMPTION 1997	Oil cont. (1000 tonnes)										
	Refinery Gas	LPG + Ethane	Motor Gasoline	Aviation Gasoline	Jet Fuel	Kerosene	Gas/ Diesel	Heavy Fuel Oil	Naphtha	Petrol. Coke	Other Prod.
Production	4650	6004	36907	8	19026	15655	87378	71697	15171	292	13232
Imports	-	3847	9571	53	4048	6765	41966	39889	3552	10	2635
Exports	-	-4071	-7800	-1	-8013	-760	-20197	-25243	-6704	-9	-5710
Intl. Marine Bunkers	-	-	-	-	-	-	-3832	-18034	-	-	-2
Stock Changes	-	-817	430	-2	70	-386	477	1024	-63	9	-187
DOMESTIC SUPPLY	4650	4963	39108	58	15131	21274	105792	69333	11956	302	9968
Transfers and Stat. Diff.	4	7005	110	-	-723	-253	607	-1045	-1286	7	186
TRANSFORMATION	9	-	-	-	-	-	8502	28986	-	-	-
Electricity and CHP Plants	9	-	-	-	-	-	8502	28979	-	-	-
Petroleum Refineries	-	-	-	-	-	-	-	-	-	-	-
Other Transform. Sector	-	-	-	-	-	-	-	7	-	-	-
ENERGY SECTOR	4642	46	2	-	1	29	1118	7468	41	15	12
DISTRIBUTION LOSSES	-	199	5	-	1	1	-	1	-	-	-
FINAL CONSUMPTION	3	11723	39211	58	14406	20991	96779	31833	10629	294	10142
INDUSTRY SECTOR	3	3349	43	-	-	254	13381	27705	10629	28	1061
Iron and Steel	-	-	-	-	-	-	-	-	-	-	-
Chemical and Petrochem.	-	-	-	-	-	-	-	-	-	-	-
Non-Metallic Minerals	-	-	-	-	-	-	-	-	-	-	-
Non-specified	-	-	-	-	-	-	-	-	-	-	-
TRANSPORT SECTOR	-	129	39006	58	14406	8	72810	1276	-	-	-
Air	-	-	-	58	14406	-	-	-	-	-	-
Road	-	129	39001	-	-	-	67880	84	-	-	-
Non-specified	-	-	-	-	-	-	-	-	-	-	-
OTHER SECTORS	-	-	-	-	-	-	-	-	-	-	-
Agriculture	-	-	-	-	-	-	-	-	-	-	-
Comm. and Publ. Services	-	-	-	-	-	-	-	-	-	-	-
Residential	-	-	-	-	-	-	-	-	-	-	-
Non-specified	-	-	-	-	-	-	-	-	-	-	-
NON-ENERGY USE	-	-	-	-	-	-	-	-	-	266	8986

APPROVISIONNEMENT ET DEMANDE 1998	Pétrole cont. (1000 tonnes)										
	Gaz de raffinerie	GPL + éthane	Essence moteur	Essence aviation	Carbu- réacteurs	Kérosène	Gazole	Fioul lourd	Naphta	Coke de pétrole	Autres prod.
Production	4714	6300	36591	4	16880	15938	88008	73958	15842	316	13236
Imports	-	3986	10101	50	2897	7626	36165	43575	1935	10	1973
Exports	-	-3741	-7471	-1	-6533	-755	-19403	-23832	-5218	-3	-4911
Intl. Marine Bunkers	-	-	-	-	-	-	-4190	-18643	-	-	-2
Stock Changes	-	-783	856	4	136	-272	-308	-604	-252	3	-87
DOMESTIC SUPPLY	4714	5762	40077	57	13380	22537	100272	74454	12307	326	10209
Transfers and Stat. Diff.	-	7224	145	-	-171	577	933	-4248	-434	5	-281
TRANSFORMATION	16	-	-	-	-	-	7579	29127	-	-	-
Electricity and CHP Plants	16	-	-	-	-	-	7579	29120	-	-	-
Petroleum Refineries	-	-	-	-	-	-	-	-	-	-	-
Other Transform. Sector	-	-	-	-	-	-	-	7	-	-	-
ENERGY SECTOR	4694	45	3	-	1	29	1140	7973	-	15	10
DISTRIBUTION LOSSES	-	53	2	-	1	-	-	1	-	-	-
FINAL CONSUMPTION	4	12888	40217	57	13207	23085	92486	33105	11873	316	9918
INDUSTRY SECTOR	4	3735	58	-	-	239	11760	28587	11873	29	1063
Iron and Steel	-	-	-	-	-	-	-	-	-	-	-
Chemical and Petrochem.	-	-	-	-	-	-	-	-	-	-	-
Non-Metallic Minerals	-	-	-	-	-	-	-	-	-	-	-
Non-specified	-	-	-	-	-	-	-	-	-	-	-
TRANSPORT SECTOR	-	109	39966	57	13207	8	69806	1445	-	-	-
Air	-	-	-	57	13207	-	-	-	-	-	-
Road	-	109	39964	-	-	-	64974	18	-	-	-
Non-specified	-	-	-	-	-	-	-	-	-	-	-
OTHER SECTORS	-	-	-	-	-	-	-	-	-	-	-
Agriculture	-	-	-	-	-	-	-	-	-	-	-
Comm. and Publ. Services	-	-	-	-	-	-	-	-	-	-	-
Residential	-	-	-	-	-	-	-	-	-	-	-
Non-specified	-	-	-	-	-	-	-	-	-	-	-
NON-ENERGY USE	-	-	-	-	-	-	-	-	-	287	8753

Asia (excluding China) / Asie (Chine non incluse)

SUPPLY AND CONSUMPTION 1997	Gas (TJ)				Comb. Renew. & Waste (TJ)				(GWh)	(TJ)
	Natural Gas	Gas Works	Coke Ovens	Blast Furnaces	Solid Biomass	Gas/Liquids from Biomass	Municipal Waste	Industrial Waste	Electricity	Heat
Production	7032881	26599	43386	196709	14413714	-	-	-	1054245	-
Imports	234808	-	-	-	600	-	-	-	2831	-
Exports	-2542654	-	-	-	-5367	-	-	-	-1923	-
Intl. Marine Bunkers	-	-	-	-	-	-	-	-	-	-
Stock Changes	-7035	-	-	-	-	-	-	-	-	-
DOMESTIC SUPPLY	4718000	26599	43386	196709	14408947	-	-	-	1055153	-
Transfers and Stat. Diff.	-4739	189	-3	1	1	-	-	-	-1356	-
TRANSFORMATION	1947774	-	6084	12367	443141	-	-	-	-	-
Electricity and CHP Plants	1872398	-	6084	12367	9083	-	-	-		
Petroleum Refineries	315	-	-	-	-	-	-	-		
Other Transform. Sector	75061	-	-	-	434058	-	-	-		
ENERGY SECTOR	1040002	-	-	-	-	-	-	-	60353	-
DISTRIBUTION LOSSES	92144	488	-	397	-	-	-	-	168905	-
FINAL CONSUMPTION	1633341	26300	37299	183946	13965807	-	-	-	824539	-
INDUSTRY SECTOR	1341655	2098	37299	183946	1417738	-	-	-	374136	-
Iron and Steel	-	-	-	-	-	-	-	-	-	-
Chemical and Petrochem.	-	-	-	-	-	-	-	-	-	-
Non-Metallic Minerals	-	-	-	-	-	-	-	-	-	-
Non-specified	-	-	-	-	-	-	-	-	-	-
TRANSPORT SECTOR	894	-	-	-	-	-	-	-	7850	-
Air	-	-	-	-	-	-	-	-	-	-
Road	236	-	-	-	-	-	-	-	-	-
Non-specified	-	-	-	-	-	-	-	-	-	-
OTHER SECTORS	-	-	-	-	-	-	-	-	-	-
Agriculture	-	-	-	-	-	-	-	-	-	-
Comm. and Publ. Services	-	-	-	-	-	-	-	-	-	-
Residential	-	-	-	-	-	-	-	-	-	-
Non-specified	-	-	-	-	-	-	-	-	-	-
NON-ENERGY USE	-	-	-	-	-	-	-	-	-	-

APPROVISIONNEMENT ET DEMANDE 1998	Gaz (TJ)				En. Re. Comb. & Déchets (TJ)				(GWh)	(TJ)
	Gaz naturel	Usines à gaz	Cokeries	Hauts fourneaux	Biomasse solide	Gaz/Liquides tirés de biomasse	Déchets urbains	Déchets industriels	Electricité	Chaleur
Production	7136348	26786	44252	190134	14529677	-	-	-	1107209	-
Imports	278753	-	-	-	392	-	-	-	3652	-
Exports	-2542459	-	-	-	-5411	-	-	-	-1898	-
Intl. Marine Bunkers	-	-	-	-	-	-	-	-	-	-
Stock Changes	1777	-	-	-	-	-	-	-	-	-
DOMESTIC SUPPLY	4874419	26786	44252	190134	14524658	-	-	-	1108963	-
Transfers and Stat. Diff.	-14380	-128	-120	-	34	-	-	-	2869	-
TRANSFORMATION	2016910	-	6727	12440	437774	-	-	-	-	-
Electricity and CHP Plants	2016632	-	6727	12440	8545	-	-	-		
Petroleum Refineries	278	-	-	-	-	-	-	-		
Other Transform. Sector	-	-	-	-	429229	-	-	-		
ENERGY SECTOR	1082073	-	-	-	-	-	-	-	66124	-
DISTRIBUTION LOSSES	82503	233	-	557	-	-	-	-	177970	-
FINAL CONSUMPTION	1678553	26425	37405	177137	14086918	-	-	-	867738	-
INDUSTRY SECTOR	1380168	2098	37405	177137	1400454	-	-	-	382349	-
Iron and Steel	-	-	-	-	-	-	-	-	-	-
Chemical and Petrochem.	-	-	-	-	-	-	-	-	-	-
Non-Metallic Minerals	-	-	-	-	-	-	-	-	-	-
Non-specified	-	-	-	-	-	-	-	-	-	-
TRANSPORT SECTOR	971	-	-	-	-	-	-	-	8509	-
Air	-	-	-	-	-	-	-	-	-	-
Road	233	-	-	-	-	-	-	-	-	-
Non-specified	-	-	-	-	-	-	-	-	-	-
OTHER SECTORS	-	-	-	-	-	-	-	-	-	-
Agriculture	-	-	-	-	-	-	-	-	-	-
Comm. and Publ. Services	-	-	-	-	-	-	-	-	-	-
Residential	-	-	-	-	-	-	-	-	-	-
Non-specified	-	-	-	-	-	-	-	-	-	-
NON-ENERGY USE	-	-	-	-	-	-	-	-	-	-

China / Chine : 1997

SUPPLY AND CONSUMPTION / APPROVISIONNEMENT ET DEMANDE	Coal / Charbon (1000 tonnes)							Oil / Pétrole (1000 tonnes)			
	Coking Coal / Charbon à coke	Other Bit. Coal / Autres charb. bit.	Sub-Bit. Coal / Charbon sous-bit.	Lignite / Lignite	Peat / Tourbe	Oven and Gas Coke / Coke de four/gaz	Pat. Fuel and BKB / Agg./briq. de lignite	Crude Oil / Pétrole brut	NGL / LGN	Feed-stocks / Produits d'aliment.	Additives / Additifs
Production	192969	1179851	-	-	-	136531	10017	160741	-	-	-
From Other Sources	-	-	-	-	-	-	-	-	-	-	-
Imports	-	7724	-	-	-	1	-	35470	-	-	-
Exports	-	-35331	-	-	-	-10581	-	-19829	-	-	-
Intl. Marine Bunkers	-	-	-	-	-	-	-	-	-	-	-
Stock Changes	-	-13405	-	-	-	361	54	-1386	-	-	-
DOMESTIC SUPPLY	192969	1138839	-	-	-	126312	10071	174996	-	-	-
Transfers	-	-	-	-	-	-	-	-	-	-	-
Statistical Differences	-	-13951	-	-	-	-16110	-3	-1152	-	-	-
TRANSFORMATION	192969	574535	-	-	-	63311	-	166865	-	-	-
Electricity Plants	-	495879	-	-	-	-	-	643	-	-	-
CHP Plants	-	62447	-	-	-	-	-	149	-	-	-
Heat Plants	-	-	-	-	-	-	-	-	-	-	-
Blast Furnaces/Gas Works	-	7326	-	-	-	63311	-	-	-	-	-
Coke/Pat. Fuel/BKB Plants	192969	8883	-	-	-	-	-	-	-	-	-
Petroleum Refineries	-	-	-	-	-	-	-	166073	-	-	-
Petrochemical Industry	-	-	-	-	-	-	-	-	-	-	-
Liquefaction	-	-	-	-	-	-	-	-	-	-	-
Other Transform. Sector	-	-	-	-	-	-	-	-	-	-	-
ENERGY SECTOR	-	69235	-	-	-	1549	-	3815	-	-	-
Coal Mines	-	24592	-	-	-	525	-	13	-	-	-
Oil and Gas Extraction	-	3211	-	-	-	26	-	3165	-	-	-
Petroleum Refineries	-	-	-	-	-	-	-	636	-	-	-
Electr., CHP+Heat Plants	-	33344	-	-	-	73	-	1	-	-	-
Pumped Storage (Elec.)	-	-	-	-	-	-	-	-	-	-	-
Other Energy Sector	-	8088	-	-	-	925	-	-	-	-	-
Distribution Losses	-	-	-	-	-	-	-	1884	-	-	-
FINAL CONSUMPTION	-	481118	-	-	-	45342	10068	1280	-	-	-
INDUSTRY SECTOR	-	328519	-	-	-	42304	-	926	-	-	-
Iron and Steel	-	7668	-	-	-	17071	-	24	-	-	-
Chemical and Petrochem.	-	84548	-	-	-	13557	-	602	-	-	-
of which: Feedstocks	-	12774	-	-	-	8834	-	482	-	-	-
Non-Ferrous Metals	-	7671	-	-	-	2126	-	11	-	-	-
Non-Metallic Minerals	-	120116	-	-	-	2690	-	105	-	-	-
Transport Equipment	-	6201	-	-	-	367	-	1	-	-	-
Machinery	-	19530	-	-	-	4258	-	8	-	-	-
Mining and Quarrying	-	4851	-	-	-	1281	-	124	-	-	-
Food and Tobacco	-	28589	-	-	-	306	-	14	-	-	-
Paper, Pulp and Print	-	15356	-	-	-	79	-	1	-	-	-
Wood and Wood Products	-	4694	-	-	-	39	-	1	-	-	-
Construction	-	3721	-	-	-	125	-	30	-	-	-
Textile and Leather	-	20195	-	-	-	115	-	5	-	-	-
Non-specified	-	5379	-	-	-	290	-	-	-	-	-
TRANSPORT SECTOR	-	14220	-	-	-	-	-	-	-	-	-
Air	-	-	-	-	-	-	-	-	-	-	-
Road	-	-	-	-	-	-	-	-	-	-	-
Rail	-	14220	-	-	-	-	-	-	-	-	-
Pipeline Transport	-	-	-	-	-	-	-	-	-	-	-
Internal Navigation	-	-	-	-	-	-	-	-	-	-	-
Non-specified	-	-	-	-	-	-	-	-	-	-	-
OTHER SECTORS	-	138379	-	-	-	3038	10027	354	-	-	-
Agriculture	-	18794	-	-	-	1247	-	-	-	-	-
Comm. and Publ. Services	-	14412	-	-	-	544	79	170	-	-	-
Residential	-	105173	-	-	-	1247	9924	-	-	-	-
Non-specified	-	-	-	-	-	-	24	184	-	-	-
NON-ENERGY USE	-	-	-	-	-	-	41	-	-	-	-
in Industry/Trans./Energy	-	-	-	-	-	-	41	-	-	-	-
in Transport	-	-	-	-	-	-	-	-	-	-	-
in Other Sectors	-	-	-	-	-	-	-	-	-	-	-

China / Chine : 1997

SUPPLY AND CONSUMPTION *APPROVISIONNEMENT ET DEMANDE*	Oil cont. / *Pétrole cont.* (1000 tonnes)										
	Refinery Gas *Gaz de raffinerie*	LPG + Ethane *GPL + éthane*	Motor Gasoline *Essence moteur*	Aviation Gasoline *Essence aviation*	Jet Fuel *Carbu- réacteurs*	Kerosene *Kérosène*	Gas/ Diesel *Gazole*	Heavy Fuel Oil *Fioul lourd*	Naphtha *Naphta*	Petrol. Coke *Coke de pétrole*	Other Prod. *Autres prod.*
Production	5341	6679	35102	77	4466	1304	49245	23112	17439	-	19488
From Other Sources	-	-	-	-	-	-	-	-	-	-	-
Imports	-	3748	554	-	3438	1450	17982	17785	531	-	4019
Exports	-	-410	-1905	-	-265	-740	-9663	-2838	-	-	-1680
Intl. Marine Bunkers	-	-	-	-	-	-	-852	-1975	-	-	-
Stock Changes	-	49	-568	-	28	-145	-1869	-85	-	-	-
DOMESTIC SUPPLY	5341	10066	33183	77	7667	1869	54843	35999	17970	-	21827
Transfers	-	-	-	-	-	-	-	-	-	-	-
Statistical Differences	728	161	280	-	1	802	2052	3943	-	-	-1
TRANSFORMATION	1039	12	8	-	-	-	6981	11896	531	-	-
Electricity Plants	280	7	8	-	-	-	6654	8777	-	-	-
CHP Plants	759	5	-	-	-	-	327	2502	-	-	-
Heat Plants	-	-	-	-	-	-	-	-	-	-	-
Blast Furnaces/Gas Works	-	-	-	-	-	-	-	617	531	-	-
Coke/Pat. Fuel/BKB Plants	-	-	-	-	-	-	-	-	-	-	-
Petroleum Refineries	-	-	-	-	-	-	-	-	-	-	-
Petrochemical Industry	-	-	-	-	-	-	-	-	-	-	-
Liquefaction	-	-	-	-	-	-	-	-	-	-	-
Other Transform. Sector	-	-	-	-	-	-	-	-	-	-	-
ENERGY SECTOR	3963	1395	-	-	-	155	3029	6790	-	-	-
Coal Mines	-	-	-	-	-	34	285	20	-	-	-
Oil and Gas Extraction	94	73	-	-	-	7	1758	1378	-	-	-
Petroleum Refineries	3866	1280	-	-	-	106	301	3838	-	-	-
Electr., CHP+Heat Plants	3	5	-	-	-	7	662	1419	-	-	-
Pumped Storage (Elec.)	-	-	-	-	-	-	-	-	-	-	-
Other Energy Sector	-	37	-	-	-	1	23	135	-	-	-
Distribution Losses	-	18	-	-	-	-	-	-	-	-	-
FINAL CONSUMPTION	1067	8802	33455	77	7668	2516	46885	21256	17439	-	21826
INDUSTRY SECTOR	1067	692	-	-	-	363	6622	13613	17439	-	-
Iron and Steel	1	7	-	-	-	8	399	3295	-	-	-
Chemical and Petrochem.	949	487	-	-	-	74	1205	5134	17439	-	-
of which: Feedstocks	*288*	*75*	-	-	-	*11*	*20*	*908*	*17439*	-	-
Non-Ferrous Metals	-	4	-	-	-	8	164	592	-	-	-
Non-Metallic Minerals	8	90	-	-	-	20	1072	2844	-	-	-
Transport Equipment	4	7	-	-	-	35	207	135	-	-	-
Machinery	6	49	-	-	-	70	702	598	-	-	-
Mining and Quarrying	-	-	-	-	-	6	276	25	-	-	-
Food and Tobacco	-	18	-	-	-	11	307	267	-	-	-
Paper, Pulp and Print	94	1	-	-	-	47	128	90	-	-	-
Wood and Wood Products	-	-	-	-	-	7	110	22	-	-	-
Construction	-	4	-	-	-	42	870	132	-	-	-
Textile and Leather	-	10	-	-	-	24	280	367	-	-	-
Non-specified	5	15	-	-	-	11	902	112	-	-	-
TRANSPORT SECTOR	-	16	33455	77	7668	-	20413	5752	-	-	-
Air	-	-	-	77	7668	-	-	-	-	-	-
Road	-	16	33455	-	-	-	12507	197	-	-	-
Rail	-	-	-	-	-	-	3776	26	-	-	-
Pipeline Transport	-	-	-	-	-	-	10	151	-	-	-
Internal Navigation	-	-	-	-	-	-	4067	5378	-	-	-
Non-specified	-	-	-	-	-	-	53	-	-	-	-
OTHER SECTORS	-	8094	-	-	-	2153	19850	1891	-	-	-
Agriculture	-	-	-	-	-	14	10757	29	-	-	-
Comm. and Publ. Services	-	604	-	-	-	1467	8815	1862	-	-	-
Residential	-	7490	-	-	-	672	278	-	-	-	-
Non-specified	-	-	-	-	-	-	-	-	-	-	-
NON-ENERGY USE	-	-	-	-	-	-	-	-	-	-	21826
in Industry/Transf./Energy	-	-	-	-	-	-	-	-	-	-	21826
in Transport	-	-	-	-	-	-	-	-	-	-	-
in Other Sectors	-	-	-	-	-	-	-	-	-	-	-

China / Chine : 1997

SUPPLY AND CONSUMPTION / APPROVISIONNEMENT ET DEMANDE	Gas / Gaz (TJ)				Comb. Renew. & Waste / En. Re. Comb. & Déchets (TJ)				(GWh)	(TJ)
	Natural Gas / Gaz naturel	Gas Works / Usines à gaz	Coke Ovens / Cokeries	Blast Furnaces / Hauts fourneaux	Solid Biomass / Biomasse solide	Gas/Liquids from Biomass / Gaz/Liquides tirés de biomasse	Municipal Waste / Déchets urbains	Industrial Waste / Déchets industriels	Electricity / Electricité	Heat / Chaleur
Production	983449	186112	424655	249475	8672689	50528	-	-	1163416	1175695
From Other Sources	-	-	-	-	-	-	-	-	-	-
Imports	107462	-	-	-	378	-	-	-	7965	-
Exports	-107462	-	-	-	-60	-	-	-	-7763	-
Intl. Marine Bunkers	-	-	-	-	-	-	-	-	-	-
Stock Changes	-	-	-	-	-	-	-	-	-	-
DOMESTIC SUPPLY	983449	186112	424655	249475	8673007	50528	-	-	1163618	1175695
Transfers	-	-	-	-	-	-	-	-	-	-
Statistical Differences	5938	13744	4172	29851	-	-	-	-	-4304	1
TRANSFORMATION	202372	-	34990	22910	-	-	-	-	-	-
Electricity Plants	199643	-	22320	16125	-	-	-	-	-	-
CHP Plants	2729	-	12670	6785	-	-	-	-	-	-
Heat Plants	-	-	-	-	-	-	-	-	-	-
Blast Furnaces/Gas Works	-	-	-	-	-	-	-	-	-	-
Coke/Pat. Fuel/BKB Plants	-	-	-	-	-	-	-	-	-	-
Petroleum Refineries	-	-	-	-	-	-	-	-	-	-
Petrochemical Industry	-	-	-	-	-	-	-	-	-	-
Liquefaction	-	-	-	-	-	-	-	-	-	-
Other Transform. Sector	-	-	-	-	-	-	-	-	-	-
ENERGY SECTOR	276860	23318	32277	-	-	-	-	-	185310	238198
Coal Mines	-	-	252	-	-	-	-	-	38097	1522
Oil and Gas Extraction	227955	-	-	-	-	-	-	-	31468	10498
Petroleum Refineries	42971	-	-	-	-	-	-	-	18241	178857
Electr., CHP+Heat Plants	5501	11275	72	-	-	-	-	-	96182	43032
Pumped Storage (Elec.)	-	-	-	-	-	-	-	-	-	-
Other Energy Sector	433	12043	31953	-	-	-	-	-	1322	4289
Distribution Losses	24041	-	-	-	-	-	-	-	83855	16647
FINAL CONSUMPTION	486114	176538	361560	256416	8673007	50528	-	-	890149	920851
INDUSTRY SECTOR	392937	56942	285435	256416	-	-	-	-	578360	711435
Iron and Steel	11869	-	262862	256416	-	-	-	-	93416	119049
Chemical and Petrochem.	340393	6488	15078	-	-	-	-	-	150683	305810
of which: Feedstocks	-	-	-	-	-	-	-	-	-	-
Non-Ferrous Metals	2642	11697	288	-	-	-	-	-	51352	41932
Non-Metallic Minerals	13255	11381	3846	-	-	-	-	-	60808	15248
Transport Equipment	1949	4185	162	-	-	-	-	-	20123	18985
Machinery	15205	20579	2803	-	-	-	-	-	52566	32996
Mining and Quarrying	780	-	-	-	-	-	-	-	19055	4712
Food and Tobacco	2556	120	126	-	-	-	-	-	29884	48197
Paper, Pulp and Print	390	-	18	-	-	-	-	-	19546	41442
Wood and Wood Products	-	-	-	-	-	-	-	-	7025	6480
Construction	43	-	-	-	-	-	-	-	11741	2071
Textile and Leather	3855	241	252	-	-	-	-	-	39807	62876
Non-specified	-	2251	-	-	-	-	-	-	22354	11637
TRANSPORT SECTOR	910	-	-	-	-	-	-	-	20200	-
Air	-	-	-	-	-	-	-	-	-	-
Road	910	-	-	-	-	-	-	-	-	-
Rail	-	-	-	-	-	-	-	-	-	-
Pipeline Transport	-	-	-	-	-	-	-	-	-	-
Internal Navigation	-	-	-	-	-	-	-	-	-	-
Non-specified	-	-	-	-	-	-	-	-	20200	-
OTHER SECTORS	92267	119596	76125	-	8673007	50528	-	-	291589	209416
Agriculture	-	-	-	-	-	-	-	-	63977	533
Comm. and Publ. Services	6844	17138	10118	-	-	-	-	-	94258	38403
Residential	85423	102458	66007	-	8672689	50528	-	-	133354	164322
Non-specified	-	-	-	-	318	-	-	-	-	6158
NON-ENERGY USE	-	-	-	-	-	-	-	-	-	-
in Industry/Transf./Energy	-	-	-	-	-	-	-	-	-	-
in Transport	-	-	-	-	-	-	-	-	-	-
in Other Sectors	-	-	-	-	-	-	-	-	-	-

China / Chine : 1998

SUPPLY AND CONSUMPTION *APPROVISIONNEMENT ET DEMANDE*	Coal / *Charbon* (1000 tonnes)							Oil / *Pétrole* (1000 tonnes)			
	Coking Coal *Charbon à coke*	Other Bit. Coal *Autres charb. bit.*	Sub-Bit. Coal *Charbon sous-bit.*	Lignite *Lignite*	Peat *Tourbe*	Oven and Gas Coke *Coke de four/gaz*	Pat. Fuel and BKB *Agg./briq. de lignite*	Crude Oil *Pétrole brut*	NGL *LGN*	Feed-stocks *Produits d'aliment.*	Additives *Additifs*
Production	156281	1057031	-	-	-	128991	9235	161000	-	-	-
From Other Sources	-	-	-	-	-	-	-	-	-	-	-
Imports	-	8688	-	-	-	-	-	27320	-	-	-
Exports	-	-32297	-	-	-	-11464	-	-15600	-	-	-
Intl. Marine Bunkers	-	-	-	-	-	-	-	-	-	-	-
Stock Changes	-	8813	-	-	-	-192	-3	449	-	-	-
DOMESTIC SUPPLY	156281	1042235	-	-	-	117335	9232	173169	-	-	-
Transfers	-	-	-	-	-	-	-	-	-	-	-
Statistical Differences	-	37333	-	-	-	-6226	-942	782	-	-	-
TRANSFORMATION	156281	580034	-	-	-	64801	-	166807	-	-	-
Electricity Plants	-	502116	-	-	-	-	-	744	-	-	-
CHP Plants	-	63199	-	-	-	-	-	245	-	-	-
Heat Plants	-	-	-	-	-	-	-	-	-	-	-
Blast Furnaces/Gas Works	-	6851	-	-	-	64801	-	-	-	-	-
Coke/Pat. Fuel/BKB Plants	156281	7868	-	-	-	-	-	-	-	-	-
Petroleum Refineries	-	-	-	-	-	-	-	165818	-	-	-
Petrochemical Industry	-	-	-	-	-	-	-	-	-	-	-
Liquefaction	-	-	-	-	-	-	-	-	-	-	-
Other Transform. Sector	-	-	-	-	-	-	-	-	-	-	-
ENERGY SECTOR	-	72017	-	-	-	1300	-	4280	-	-	-
Coal Mines	-	26928	-	-	-	450	-	13	-	-	-
Oil and Gas Extraction	-	1992	-	-	-	48	-	3400	-	-	-
Petroleum Refineries	-	-	-	-	-	-	-	855	-	-	-
Electr., CHP+Heat Plants	-	33287	-	-	-	89	-	12	-	-	-
Pumped Storage (Elec.)	-	-	-	-	-	-	-	-	-	-	-
Other Energy Sector	-	9810	-	-	-	713	-	-	-	-	-
Distribution Losses	-	-	-	-	-	-	-	1960	-	-	-
FINAL CONSUMPTION	-	427517	-	-	-	45008	8290	904	-	-	-
INDUSTRY SECTOR	-	305000	-	-	-	41692	-	694	-	-	-
Iron and Steel	-	7661	-	-	-	16806	-	68	-	-	-
Chemical and Petrochem.	-	78397	-	-	-	12537	-	430	-	-	-
of which: Feedstocks	-	*12485*	-	-	-	*6818*	-	*672*	-	-	-
Non-Ferrous Metals	-	8453	-	-	-	2084	-	9	-	-	-
Non-Metallic Minerals	-	109966	-	-	-	3023	-	119	-	-	-
Transport Equipment	-	5667	-	-	-	394	-	1	-	-	-
Machinery	-	16364	-	-	-	4573	-	15	-	-	-
Mining and Quarrying	-	4482	-	-	-	1273	-	-	-	-	-
Food and Tobacco	-	27481	-	-	-	319	-	9	-	-	-
Paper, Pulp and Print	-	14089	-	-	-	37	-	2	-	-	-
Wood and Wood Products	-	4687	-	-	-	31	-	-	-	-	-
Construction	-	6010	-	-	-	146	-	22	-	-	-
Textile and Leather	-	17815	-	-	-	182	-	19	-	-	-
Non-specified	-	3928	-	-	-	287	-	-	-	-	-
TRANSPORT SECTOR	-	13754	-	-	-	-	-	-	-	-	-
Air	-	-	-	-	-	-	-	-	-	-	-
Road	-	-	-	-	-	-	-	-	-	-	-
Rail	-	13754	-	-	-	-	-	-	-	-	-
Pipeline Transport	-	-	-	-	-	-	-	-	-	-	-
Internal Navigation	-	-	-	-	-	-	-	-	-	-	-
Non-specified	-	-	-	-	-	-	-	-	-	-	-
OTHER SECTORS	-	108763	-	-	-	3316	8249	210	-	-	-
Agriculture	-	18794	-	-	-	1400	-	-	-	-	-
Comm. and Publ. Services	-	15619	-	-	-	516	69	17	-	-	-
Residential	-	74350	-	-	-	1400	8156	-	-	-	-
Non-specified	-	-	-	-	-	-	24	193	-	-	-
NON-ENERGY USE	-	-	-	-	-	-	41	-	-	-	-
in Industry/Trans./Energy	-	-	-	-	-	-	41	-	-	-	-
in Transport	-	-	-	-	-	-	-	-	-	-	-
in Other Sectors	-	-	-	-	-	-	-	-	-	-	-

China / Chine : 1998

| | | | | | Oil cont. / Pétrole cont. (1000 tonnes) | | | | | | |
| SUPPLY AND CONSUMPTION | Refinery Gas | LPG + Ethane | Motor Gasoline | Aviation Gasoline | Jet Fuel | Kerosene | Gas/ Diesel | Heavy Fuel Oil | Naphtha | Petrol. Coke | Other Prod. |
APPROVISIONNEMENT ET DEMANDE	Gaz de raffinerie	GPL + éthane	Essence moteur	Essence aviation	Carbu- réacteurs	Kérosène	Gazole	Fioul lourd	Naphta	Coke de pétrole	Autres prod.
Production	5450	7474	34885	126	4264	1897	48977	21004	17412	-	21225
From Other Sources	-	-	-	-	-	-	-	-	-	-	-
Imports	-	4910	428	-	3238	1340	14060	20541	552	-	2378
Exports	-	-511	-1883	-	-255	-922	-6129	-3225	-	-	-2331
Intl. Marine Bunkers	-	-	-	-	-	-	-1484	-1716	-	-	-
Stock Changes	-	-76	1	-	27	231	1276	-149	-	-	-
DOMESTIC SUPPLY	**5450**	**11797**	**33431**	**126**	**7274**	**2546**	**56700**	**36455**	**17964**	**-**	**21272**
Transfers	-	-	-	-	-	-	-	-	-	-	-
Statistical Differences	-2	203	212	-	-	305	748	1888	-	-	-
TRANSFORMATION	**1170**	**30**	**8**	**-**	**-**	**-**	**2044**	**13787**	**552**	**-**	**-**
Electricity Plants	290	30	8	-	-	-	2044	10003	-	-	-
CHP Plants	880	-	-	-	-	-	-	3430	-	-	-
Heat Plants	-	-	-	-	-	-	-	-	-	-	-
Blast Furnaces/Gas Works	-	-	-	-	-	-	-	354	552	-	-
Coke/Pat. Fuel/BKB Plants	-	-	-	-	-	-	-	-	-	-	-
Petroleum Refineries	-	-	-	-	-	-	-	-	-	-	-
Petrochemical Industry	-	-	-	-	-	-	-	-	-	-	-
Liquefaction	-	-	-	-	-	-	-	-	-	-	-
Other Transform. Sector	-	-	-	-	-	-	-	-	-	-	-
ENERGY SECTOR	**3493**	**978**	**-**	**-**	**-**	**146**	**2565**	**6078**	**-**	**-**	**-**
Coal Mines	-	-	-	-	-	30	390	49	-	-	-
Oil and Gas Extraction	438	147	-	-	-	6	1171	1537	-	-	-
Petroleum Refineries	3055	792	-	-	-	106	352	3404	-	-	-
Electr., CHP+Heat Plants	-	-	-	-	-	4	573	1063	-	-	-
Pumped Storage (Elec.)	-	-	-	-	-	-	-	-	-	-	-
Other Energy Sector	-	39	-	-	-	-	79	25	-	-	-
Distribution Losses	-	13	-	-	-	-	-	-	-	-	-
FINAL CONSUMPTION	**785**	**10979**	**33635**	**126**	**7274**	**2705**	**52839**	**18478**	**17412**	**-**	**21272**
INDUSTRY SECTOR	**785**	**1254**	**-**	**-**	**-**	**517**	**11704**	**12531**	**17412**	**-**	**-**
Iron and Steel	-	4	-	-	-	33	590	3073	-	-	-
Chemical and Petrochem.	736	515	-	-	-	82	1664	3690	17412	-	-
of which: Feedstocks	343	78	-	-	-	1	70	914	17412	-	-
Non-Ferrous Metals	-	4	-	-	-	5	490	566	-	-	-
Non-Metallic Minerals	38	424	-	-	-	19	2278	3118	-	-	-
Transport Equipment	-	8	-	-	-	43	386	86	-	-	-
Machinery	7	125	-	-	-	79	1258	492	-	-	-
Mining and Quarrying	-	-	-	-	-	5	431	5	-	-	-
Food and Tobacco	-	26	-	-	-	4	571	255	-	-	-
Paper, Pulp and Print	3	5	-	-	-	68	255	142	-	-	-
Wood and Wood Products	-	2	-	-	-	6	170	27	-	-	-
Construction	-	57	-	-	-	35	1535	166	-	-	-
Textile and Leather	1	55	-	-	-	32	511	606	-	-	-
Non-specified	-	29	-	-	-	106	1565	305	-	-	-
TRANSPORT SECTOR	**-**	**5**	**33635**	**126**	**7274**	**-**	**22306**	**5656**	**-**	**-**	**-**
Air	-	-	-	126	7274	-	-	-	-	-	-
Road	-	5	33635	-	-	-	3287	-	-	-	-
Rail	-	-	-	-	-	-	-	-	-	-	-
Pipeline Transport	-	-	-	-	-	-	-	-	-	-	-
Internal Navigation	-	-	-	-	-	-	-	-	-	-	-
Non-specified	-	-	-	-	-	-	19019	5656	-	-	-
OTHER SECTORS	**-**	**9720**	**-**	**-**	**-**	**2188**	**18829**	**291**	**-**	**-**	**-**
Agriculture	-	-	-	-	-	16	11202	3	-	-	-
Comm. and Publ. Services	-	2028	-	-	-	1506	7188	288	-	-	-
Residential	-	7692	-	-	-	666	439	-	-	-	-
Non-specified	-	-	-	-	-	-	-	-	-	-	-
NON-ENERGY USE	**-**	**-**	**-**	**-**	**-**	**-**	**-**	**-**	**-**	**-**	**21272**
in Industry/Transf./Energy	-	-	-	-	-	-	-	-	-	-	21272
in Transport	-	-	-	-	-	-	-	-	-	-	-
in Other Sectors	-	-	-	-	-	-	-	-	-	-	-

China / Chine : 1998

SUPPLY AND CONSUMPTION / *APPROVISIONNEMENT ET DEMANDE*	Gas / *Gaz* (TJ)				Comb. Renew. & Waste / *En. Re. Comb. & Déchets* (TJ)				(GWh)	(TJ)
	Natural Gas / *Gaz naturel*	Gas Works / *Usines à gaz*	Coke Ovens / *Cokeries*	Blast Furnaces / *Hauts fourneaux*	Solid Biomass / *Biomasse solide*	Gas/Liquids from Biomass / *Gaz/Liquides tirés de biomasse*	Municipal Waste / *Déchets urbains*	Industrial Waste / *Déchets industriels*	Electricity / *Electricité*	Heat / *Chaleur*
Production	1008400	179496	424891	255344	8719168	50799	-	-	1197617	1270677
From Other Sources	-	-	-	-	-	-	-	-	-	-
Imports	100815	-	-	-	274	-	-	-	7777	-
Exports	-100815	-	-	-	-7	-	-	-	-7784	-
Intl. Marine Bunkers	-	-	-	-	-	-	-	-	-	-
Stock Changes	-	-	-	-	-	-	-	-	-	-
DOMESTIC SUPPLY	1008400	179496	424891	255344	8719435	50799	-	-	1197610	1270677
Transfers	-	-	-	-	-	-	-	-	-	-
Statistical Differences	9581	30227	-36	-1693	-	-	-	-	-8355	1
TRANSFORMATION	193255	-	35009	48575	-	-	-	-	-	-
Electricity Plants	190340	-	22332	26334	-	-	-	-	-	-
CHP Plants	2915	-	12677	22241	-	-	-	-	-	-
Heat Plants	-	-	-	-	-	-	-	-	-	-
Blast Furnaces/Gas Works	-	-	-	-	-	-	-	-	-	-
Coke/Pat. Fuel/BKB Plants	-	-	-	-	-	-	-	-	-	-
Petroleum Refineries	-	-	-	-	-	-	-	-	-	-
Petrochemical Industry	-	-	-	-	-	-	-	-	-	-
Liquefaction	-	-	-	-	-	-	-	-	-	-
Other Transform. Sector	-	-	-	-	-	-	-	-	-	-
ENERGY SECTOR	288795	18305	23555	-	-	-	-	-	179272	312073
Coal Mines	-	-	701	-	-	-	-	-	42358	9003
Oil and Gas Extraction	233738	-	-	-	-	-	-	-	29498	32870
Petroleum Refineries	49079	-	-	-	-	-	-	-	19232	210853
Electr., CHP+Heat Plants	3292	9499	18	-	-	-	-	-	85553	55755
Pumped Storage (Elec.)	-	-	-	-	-	-	-	-	-	-
Other Energy Sector	2686	8806	22836	-	-	-	-	-	2631	3592
Distribution Losses	23478	-	-	-	-	-	-	-	82811	10723
FINAL CONSUMPTION	512453	191418	366291	205076	8719435	50799	-	-	927172	947882
INDUSTRY SECTOR	410049	51228	290124	205076	-	-	-	-	593372	720904
Iron and Steel	13255	-	261875	205076	-	-	-	-	93663	117629
Chemical and Petrochem.	350529	5690	21361	-	-	-	-	-	145941	291231
of which: Feedstocks	-	-	-	-	-	-	-	-	-	-
Non-Ferrous Metals	3076	11065	54	-	-	-	-	-	53860	52709
Non-Metallic Minerals	15248	11004	3902	-	-	-	-	-	64910	8109
Transport Equipment	6151	3809	126	-	-	-	-	-	20155	16483
Machinery	14858	17613	2356	-	-	-	-	-	56056	31522
Mining and Quarrying	520	-	-	-	-	-	-	-	19041	7563
Food and Tobacco	2123	241	234	-	-	-	-	-	35766	60172
Paper, Pulp and Print	520	30	18	-	-	-	-	-	20804	46127
Wood and Wood Products	-	105	-	-	-	-	-	-	4480	3740
Construction	520	-	18	-	-	-	-	-	18884	1609
Textile and Leather	3249	181	180	-	-	-	-	-	39509	73895
Non-specified	-	1490	-	-	-	-	-	-	20303	10115
TRANSPORT SECTOR	1126	-	-	-	-	-	-	-	20180	-
Air	-	-	-	-	-	-	-	-	-	-
Road	1126	-	-	-	-	-	-	-	-	-
Rail	-	-	-	-	-	-	-	-	-	-
Pipeline Transport	-	-	-	-	-	-	-	-	-	-
Internal Navigation	-	-	-	-	-	-	-	-	-	-
Non-specified	-	-	-	-	-	-	-	-	20180	-
OTHER SECTORS	101278	140190	76167	-	8719435	50799	-	-	313620	226978
Agriculture	-	-	-	-	-	-	-	-	62349	533
Comm. and Publ. Services	13688	16151	10123	-	-	-	-	-	109711	33666
Residential	87590	124039	66044	-	8719168	50799	-	-	141560	187111
Non-specified	-	-	-	-	267	-	-	-	-	5668
NON-ENERGY USE	-	-	-	-	-	-	-	-	-	-
in Industry/Transf./Energy	-	-	-	-	-	-	-	-	-	-
in Transport	-	-	-	-	-	-	-	-	-	-
in Other Sectors	-	-	-	-	-	-	-	-	-	-

Non-OECD Europe / Europe non-OCDE

SUPPLY AND CONSUMPTION 1997	Coal (1000 tonnes)							Oil (1000 tonnes)			
	Coking Coal	Other Bit. Coal	Sub-Bit. Coal	Lignite	Peat	Oven and Gas Coke	Pat. Fuel and BKB	Crude Oil	NGL	Feed-stocks	Additives
Production	324	254	2066	118955	-	6211	1082	9482	488	-	104
Imports	9308	4982	616	2248	8	593	5	25384	-	77	1
Exports	-	-3	-1	-9	-	-172	-	-40	-	-	-
Intl. Marine Bunkers	-	-	-	-	-	-	-	-	-	-	-
Stock Changes	-387	42	66	-78	-	-16	5	119	-5	-4	-1
DOMESTIC SUPPLY	**9245**	**5275**	**2747**	**121116**	**8**	**6616**	**1092**	**34945**	**483**	**73**	**104**
Transfers and Stat. Diff.	-97	35	-283	265	-	72	4	-17	-318	-	-
TRANSFORMATION	**8760**	**3827**	**2229**	**113225**	**-**	**4997**	**135**	**34558**	**128**	**73**	**104**
Electricity and CHP Plants	-	3814	2150	110483	-	-	135	-	-	-	-
Petroleum Refineries	-	-	-	-	-	-	-	34557	128	73	104
Other Transform. Sector	8760	13	79	2742	-	4997	-	1	-	-	-
ENERGY SECTOR	**42**	**4**	**25**	**190**	**-**	**-**	**3**	**259**	**37**	**-**	**-**
DISTRIBUTION LOSSES	**-**	**1**	**4**	**300**	**-**	**-**	**-**	**109**	**-**	**-**	**-**
FINAL CONSUMPTION	**346**	**1478**	**206**	**7666**	**8**	**1691**	**958**	**2**	**-**	**-**	**-**
INDUSTRY SECTOR	**343**	**1193**	**65**	**4044**	**8**	**1591**	**7**	**2**	**-**	**-**	**-**
Iron and Steel	-	-	-	-	-	-	-	-	-	-	-
Chemical and Petrochem.	-	-	-	-	-	-	-	-	-	-	-
Non-Metallic Minerals	-	-	-	-	-	-	-	-	-	-	-
Non-specified	-	-	-	-	-	-	-	-	-	-	-
TRANSPORT SECTOR	**1**	**-**	**3**	**312**	**-**	**-**	**-**	**-**	**-**	**-**	**-**
Air	-	-	-	-	-	-	-	-	-	-	-
Road	-	-	-	-	-	-	-	-	-	-	-
Non-specified	-	-	-	-	-	-	-	-	-	-	-
OTHER SECTORS	**-**	**-**	**-**	**-**	**-**	**-**	**-**	**-**	**-**	**-**	**-**
Agriculture	-	-	-	-	-	-	-	-	-	-	-
Comm. and Publ. Services	-	-	-	-	-	-	-	-	-	-	-
Residential	-	-	-	-	-	-	-	-	-	-	-
Non-specified	-	-	-	-	-	-	-	-	-	-	-
NON-ENERGY USE	**-**	**14**	**-**	**-**	**-**	**-**	**-**	**-**	**-**	**-**	**-**

APPROVISIONNEMENT ET DEMANDE 1998	Charbon (1000 tonnes)							Pétrole (1000 tonnes)			
	Charbon à coke	Autres charb. bit.	Charbon sous-bit.	Lignite	Tourbe	Coke de four/gaz	Agg./briq. de lignite	Pétrole brut	LGN	Produits d'aliment.	Additifs
Production	192	250	1910	116975	4	5515	1152	9268	472	421	171
Imports	7290	5054	482	1796	10	558	2	25117	-	19	16
Exports	-	-	-1	-12	-	-183	-	-42	-	-	-
Intl. Marine Bunkers	-	-	-	-	-	-	-	-	-	-	-
Stock Changes	403	171	-52	615	-	-38	-	-16	-21	-14	2
DOMESTIC SUPPLY	**7885**	**5475**	**2339**	**119374**	**14**	**5852**	**1154**	**34327**	**451**	**426**	**189**
Transfers and Stat. Diff.	335	-25	-35	170	9	-60	11	428	-237	215	-2
TRANSFORMATION	**7740**	**4244**	**2166**	**112063**	**-**	**4404**	**78**	**34436**	**137**	**641**	**187**
Electricity and CHP Plants	-	4224	2101	108661	-	-	77	-	-	-	-
Petroleum Refineries	-	-	-	-	-	-	-	34435	137	641	187
Other Transform. Sector	7740	20	65	3402	-	4404	1	1	-	-	-
ENERGY SECTOR	**54**	**5**	**25**	**159**	**-**	**-**	**3**	**255**	**77**	**-**	**-**
DISTRIBUTION LOSSES	**-**	**1**	**13**	**249**	**-**	**-**	**-**	**64**	**-**	**-**	**-**
FINAL CONSUMPTION	**426**	**1200**	**100**	**7073**	**23**	**1388**	**1084**	**-**	**-**	**-**	**-**
INDUSTRY SECTOR	**426**	**700**	**34**	**4266**	**10**	**1333**	**8**	**-**	**-**	**-**	**-**
Iron and Steel	-	-	-	-	-	-	-	-	-	-	-
Chemical and Petrochem.	-	-	-	-	-	-	-	-	-	-	-
Non-Metallic Minerals	-	-	-	-	-	-	-	-	-	-	-
Non-specified	-	-	-	-	-	-	-	-	-	-	-
TRANSPORT SECTOR	**-**	**2**	**2**	**318**	**-**	**-**	**-**	**-**	**-**	**-**	**-**
Air	-	-	-	-	-	-	-	-	-	-	-
Road	-	-	-	-	-	-	-	-	-	-	-
Non-specified	-	-	-	-	-	-	-	-	-	-	-
OTHER SECTORS	**-**	**-**	**-**	**-**	**-**	**-**	**-**	**-**	**-**	**-**	**-**
Agriculture	-	-	-	-	-	-	-	-	-	-	-
Comm. and Publ. Services	-	-	-	-	-	-	-	-	-	-	-
Residential	-	-	-	-	-	-	-	-	-	-	-
Non-specified	-	-	-	-	-	-	-	-	-	-	-
NON-ENERGY USE	**-**	**-**	**-**	**-**	**-**	**-**	**-**	**-**	**-**	**-**	**-**

Non-OECD Europe / Europe non-OCDE

SUPPLY AND CONSUMPTION 1997	Oil cont. (1000 tonnes)										
	Refinery Gas	LPG + Ethane	Motor Gasoline	Aviation Gasoline	Jet Fuel	Kerosene	Gas/ Diesel	Heavy Fuel Oil	Naphtha	Petrol. Coke	Other Prod.
Production	1330	619	7167	3	497	228	10896	7952	2415	499	2537
Imports	-	146	1963	2	562	57	2516	5964	385	157	262
Exports	-	-180	-2885	-	-65	-23	-3587	-771	-245	-246	-252
Intl. Marine Bunkers	-	-	-	-	-	-	-167	-898	-	-	-1
Stock Changes	3	4	-341	-	54	-12	-257	-536	-8	-36	26
DOMESTIC SUPPLY	1333	589	5904	5	1048	250	9401	11711	2547	374	2572
Transfers and Stat. Diff.	-57	202	-1	-	-101	9	202	138	-329	102	804
TRANSFORMATION	65	14	-	-	-	-	66	7031	67	-	225
Electricity and CHP Plants	65	-	-	-	-	-	51	6318	21	-	41
Petroleum Refineries	-	-	-	-	-	-	-	-	-	-	-
Other Transform. Sector	-	14	-	-	-	-	15	713	46	-	184
ENERGY SECTOR	873	-	22	-	-	24	77	830	-	-	396
DISTRIBUTION LOSSES	-	2	96	-	-	4	69	39	243	-	140
FINAL CONSUMPTION	338	775	5785	5	947	231	9391	3949	1908	476	2615
INDUSTRY SECTOR	338	214	47	-	-	9	845	3318	1908	196	493
Iron and Steel	-	-	-	-	-	-	-	-	-	-	-
Chemical and Petrochem.	-	-	-	-	-	-	-	-	-	-	-
Non-Metallic Minerals	-	-	-	-	-	-	-	-	-	-	-
Non-specified	-	-	-	-	-	-	-	-	-	-	-
TRANSPORT SECTOR	-	78	5629	5	947	8	6177	250	-	-	11
Air	-	-	-	5	947	-	-	-	-	-	-
Road	-	78	5627	-	-	7	5607	-	-	-	11
Non-specified	-	-	-	-	-	-	-	-	-	-	-
OTHER SECTORS	-	-	-	-	-	-	-	-	-	-	-
Agriculture	-	-	-	-	-	-	-	-	-	-	-
Comm. and Publ. Services	-	-	-	-	-	-	-	-	-	-	-
Residential	-	-	-	-	-	-	-	-	-	-	-
Non-specified	-	-	-	-	-	-	-	-	-	-	-
NON-ENERGY USE	-	-	-	-	-	-	-	-	-	280	1920

APPROVISIONNEMENT ET DEMANDE 1998	Pétrole cont. (1000 tonnes)										
	Gaz de raffinerie	GPL + éthane	Essence moteur	Essence aviation	Carbu- réacteurs	Kérosène	Gazole	Fioul lourd	Naphta	Coke de pétrole	Autres prod.
Production	1370	690	7145	6	493	242	11171	7313	2569	771	2957
Imports	-	156	1822	5	596	63	2864	4919	366	176	537
Exports	-	-191	-2696	-2	-123	-22	-4147	-1088	-206	-215	-246
Intl. Marine Bunkers	-	-	-	-	-	-	-273	-929	-	-	-1
Stock Changes	4	-18	28	2	-8	-7	-104	268	7	-47	12
DOMESTIC SUPPLY	1374	637	6299	11	958	276	9511	10483	2736	685	3259
Transfers and Stat. Diff.	-36	143	-305	-1	-	-28	234	-716	-279	-47	83
TRANSFORMATION	36	17	-	-	-	-	78	5789	69	-	308
Electricity and CHP Plants	35	1	-	-	-	-	50	5135	23	-	63
Petroleum Refineries	-	-	-	-	-	-	-	-	-	-	-
Other Transform. Sector	1	16	-	-	-	-	28	654	46	-	245
ENERGY SECTOR	1009	8	60	-	-	2	55	758	-	18	507
DISTRIBUTION LOSSES	-	-	99	-	-	-	72	40	251	-	-
FINAL CONSUMPTION	293	755	5835	10	958	246	9540	3180	2137	620	2527
INDUSTRY SECTOR	293	186	70	-	-	6	886	2760	2137	210	461
Iron and Steel	-	-	-	-	-	-	-	-	-	-	-
Chemical and Petrochem.	-	-	-	-	-	-	-	-	-	-	-
Non-Metallic Minerals	-	-	-	-	-	-	-	-	-	-	-
Non-specified	-	-	-	-	-	-	-	-	-	-	-
TRANSPORT SECTOR	-	71	5680	10	958	2	6202	168	-	-	6
Air	-	-	-	10	958	-	-	-	-	-	-
Road	-	71	5679	-	-	2	5749	1	-	-	6
Non-specified	-	-	-	-	-	-	-	-	-	-	-
OTHER SECTORS	-	-	-	-	-	-	-	-	-	-	-
Agriculture	-	-	-	-	-	-	-	-	-	-	-
Comm. and Publ. Services	-	-	-	-	-	-	-	-	-	-	-
Residential	-	-	-	-	-	-	-	-	-	-	-
Non-specified	-	-	-	-	-	-	-	-	-	-	-
NON-ENERGY USE	-	-	-	-	-	-	-	-	-	410	1906

Non-OECD Europe / Europe non-OCDE

SUPPLY AND CONSUMPTION 1997	Gas (TJ)				Comb. Renew. & Waste (TJ)				(GWh)	(TJ)
	Natural Gas	Gas Works	Coke Ovens	Blast Furnaces	Solid Biomass	Gas/Liquids from Biomass	Municipal Waste	Industrial Waste	Electricity	Heat
Production	658039	418	41465	57383	204327	-	-	2050	206524	449662
Imports	770429	-	-	-	1505	-	-	-	14310	-
Exports	-	-	-	-	-777	-	-	-	-11556	-
Intl. Marine Bunkers	-	-	-	-	-	-	-	-	-	-
Stock Changes	288	-	-	-	-254	-	-	-	-	-
DOMESTIC SUPPLY	1428756	418	41465	57383	204801	-	-	2050	209278	449662
Transfers and Stat. Diff.	40937	-	603	-119	-1321	-	-	-	-285	-1693
TRANSFORMATION	454915	-	3835	8789	2694	-	-	1191	1803	-
Electricity and CHP Plants	328976	-	3835	8789	282	-	-	737	-	-
Petroleum Refineries	-	-	-	-	-	-	-	-	-	-
Other Transform. Sector	125939	-	-	-	2412	-	-	454	1803	-
ENERGY SECTOR	98688	-	11044	7238	27	-	-	-	26012	56485
DISTRIBUTION LOSSES	15557	3	322	1613	40	-	-	-	25349	49100
FINAL CONSUMPTION	900533	415	26867	39624	200719	-	-	859	155829	342384
INDUSTRY SECTOR	599907	145	26867	39624	16311	-	-	859	65070	65044
Iron and Steel	-	-	-	-	-	-	-	-	-	-
Chemical and Petrochem.	-	-	-	-	-	-	-	-	-	-
Non-Metallic Minerals	-	-	-	-	-	-	-	-	-	-
Non-specified	-	-	-	-	-	-	-	-	-	-
TRANSPORT SECTOR	628	-	-	-	412	-	-	-	4560	-
Air	-	-	-	-	-	-	-	-	-	-
Road	-	-	-	-	-	-	-	-	57	-
Non-specified	-	-	-	-	-	-	-	-	-	-
OTHER SECTORS	-	-	-	-	-	-	-	-	-	-
Agriculture	-	-	-	-	-	-	-	-	-	-
Comm. and Publ. Services	-	-	-	-	-	-	-	-	-	-
Residential	-	-	-	-	-	-	-	-	-	-
Non-specified	-	-	-	-	-	-	-	-	-	-
NON-ENERGY USE	-	-	-	-	-	-	-	-	-	-

APPROVISIONNEMENT ET DEMANDE 1998	Gaz (TJ)				En. Re. Comb. & Déchets (TJ)				(GWh)	(TJ)
	Gaz naturel	Usines à gaz	Cokeries	Hauts fourneaux	Biomasse solide	Gaz/Liquides tirés de biomasse	Déchets urbains	Déchets industriels	Electricité	Chaleur
Production	617323	418	36353	49841	194359	-	-	258	205387	432004
Imports	730862	-	-	-	1486	-	-	-	7746	-
Exports	-38	-	-	-	-396	-	-	-	-8732	-
Intl. Marine Bunkers	-	-	-	-	-	-	-	-	-	-
Stock Changes	3616	-	-	-	102	-	-	-4	-	-
DOMESTIC SUPPLY	1351763	418	36353	49841	195551	-	-	254	204401	432004
Transfers and Stat. Diff.	7987	-	140	715	-1539	-	-	-	-	-
TRANSFORMATION	471856	2	3744	8728	2291	-	-	119	1931	-
Electricity and CHP Plants	331774	-	3738	8711	149	-	-	119	-	-
Petroleum Refineries	-	-	-	-	-	-	-	-	-	-
Other Transform. Sector	140082	2	6	17	2142	-	-	-	1931	-
ENERGY SECTOR	83930	-	8248	5866	4	-	-	2	25112	42773
DISTRIBUTION LOSSES	24938	3	290	1101	114	-	-	-	25099	41591
FINAL CONSUMPTION	779026	413	24211	34861	191603	-	-	133	152259	347640
INDUSTRY SECTOR	461443	144	24202	34861	12946	-	-	133	61135	72618
Iron and Steel	-	-	-	-	-	-	-	-	-	-
Chemical and Petrochem.	-	-	-	-	-	-	-	-	-	-
Non-Metallic Minerals	-	-	-	-	-	-	-	-	-	-
Non-specified	-	-	-	-	-	-	-	-	-	-
TRANSPORT SECTOR	2037	-	-	-	227	-	-	-	4184	-
Air	-	-	-	-	-	-	-	-	-	-
Road	-	-	-	-	-	-	-	-	58	-
Non-specified	-	-	-	-	-	-	-	-	-	-
OTHER SECTORS	-	-	-	-	-	-	-	-	-	-
Agriculture	-	-	-	-	-	-	-	-	-	-
Comm. and Publ. Services	-	-	-	-	-	-	-	-	-	-
Residential	-	-	-	-	-	-	-	-	-	-
Non-specified	-	-	-	-	-	-	-	-	-	-
NON-ENERGY USE	-	-	-	-	-	-	-	-	-	-

Former USSR / Ex-URSS : 1997

SUPPLY AND CONSUMPTION *APPROVISIONNEMENT ET DEMANDE*	Coal / *Charbon* (1000 tonnes)							Oil / *Pétrole* (1000 tonnes)			
	Coking Coal *Charbon à coke*	Other Bit. Coal *Autres charb. bit.*	Sub-Bit. Coal *Charbon sous-bit.*	Lignite *Lignite*	Peat *Tourbe*	Oven and Gas Coke *Coke de four/gaz*	Pat. Fuel and BKB *Agg./briq. de lignite*	Crude Oil *Pétrole brut*	NGL *LGN*	Feed-stocks *Produits d'aliment.*	Additives *Additifs*
Production	94732	211327	20	104196	7947	36921	5030	353116	6027	-	12
From Other Sources	-	-	-	-	-	4	-	404	-	-	-
Imports	4887	28766	-	2244	6	938	1	31071	20	701	9
Exports	-9865	-41011	-	-1152	-102	-1864	-129	-144287	-589	-	-
Intl. Marine Bunkers	-	-	-	-	-	-	-	-	-	-	-
Stock Changes	158	4838	-	286	190	-96	10	-181	24	-74	-
DOMESTIC SUPPLY	**89912**	**203920**	**20**	**105574**	**8041**	**35903**	**4912**	**240123**	**5482**	**627**	**21**
Transfers	-	-	-	-	-	-	-	-49	-32	-	-
Statistical Differences	-5284	-7821	-	-2522	-236	-	-	-2454	-10	-	-
TRANSFORMATION	**75422**	**142842**	**-**	**89289**	**5819**	**27051**	**157**	**233661**	**5425**	**627**	**21**
Electricity Plants	-	281	-	11872	-	-	-	-	-	-	-
CHP Plants	-	113085	-	59294	2029	-	52	995	-	-	-
Heat Plants	-	25455	-	16366	885	-	105	4	-	-	-
Blast Furnaces/Gas Works	-	-	-	938	-	27050	-	-	219	-	-
Coke/Pat. Fuel/BKB Plants	75422	4021	-	819	2905	1	-	-	-	-	-
Petroleum Refineries	-	-	-	-	-	-	-	232662	5206	627	21
Petrochemical Industry	-	-	-	-	-	-	-	-	-	-	-
Liquefaction	-	-	-	-	-	-	-	-	-	-	-
Other Transform. Sector	-	-	-	-	-	-	-	-	-	-	-
ENERGY SECTOR	**-**	**591**	**-**	**190**	**154**	**36**	**45**	**1442**	**1**	**-**	**-**
Coal Mines	-	507	-	160	96	-	44	-	-	-	-
Oil and Gas Extraction	-	-	-	-	-	-	-	1442	1	-	-
Petroleum Refineries	-	-	-	-	-	-	-	-	-	-	-
Electr., CHP+Heat Plants	-	84	-	30	-	-	-	-	-	-	-
Pumped Storage (Elec.)	-	-	-	-	-	-	-	-	-	-	-
Other Energy Sector	-	-	-	-	58	36	1	-	-	-	-
Distribution Losses	-	5520	-	3802	225	-	-	1484	14	-	-
FINAL CONSUMPTION	**9206**	**47146**	**20**	**9771**	**1607**	**8816**	**4710**	**1033**	**-**	**-**	**-**
INDUSTRY SECTOR	**9199**	**20402**	**-**	**3382**	**55**	**8323**	**31**	**810**	**-**	**-**	**-**
Iron and Steel	-	-	-	-	-	-	-	-	-	-	-
Chemical and Petrochem.	-	-	-	-	-	-	-	-	-	-	-
of which: Feedstocks	-	-	-	-	-	-	-	-	-	-	-
Non-Ferrous Metals	-	-	-	-	-	-	-	-	-	-	-
Non-Metallic Minerals	-	-	-	-	-	-	-	-	-	-	-
Transport Equipment	-	-	-	-	-	-	-	-	-	-	-
Machinery	-	-	-	-	-	-	-	-	-	-	-
Mining and Quarrying	-	-	-	-	-	-	-	-	-	-	-
Food and Tobacco	-	-	-	-	-	-	-	-	-	-	-
Paper, Pulp and Print	-	-	-	-	-	-	-	-	-	-	-
Wood and Wood Products	-	-	-	-	-	-	-	-	-	-	-
Construction	-	-	-	-	-	-	-	-	-	-	-
Textile and Leather	-	-	-	-	-	-	-	-	-	-	-
Non-specified	-	-	-	-	-	-	-	-	-	-	-
TRANSPORT SECTOR	**-**	**615**	**-**	**118**	**-**	**-**	**22**	**16**	**-**	**-**	**-**
Air	-	-	-	-	-	-	-	-	-	-	-
Road	-	-	-	-	-	-	-	-	-	-	-
Rail	-	-	-	-	-	-	-	-	-	-	-
Pipeline Transport	-	-	-	-	-	-	-	-	-	-	-
Internal Navigation	-	-	-	-	-	-	-	-	-	-	-
Non-specified	-	-	-	-	-	-	-	-	-	-	-
OTHER SECTORS	**-**	**-**	**-**	**-**	**-**	**-**	**-**	**-**	**-**	**-**	**-**
Agriculture	-	-	-	-	-	-	-	-	-	-	-
Comm. and Publ. Services	-	-	-	-	-	-	-	-	-	-	-
Residential	-	-	-	-	-	-	-	-	-	-	-
Non-specified	-	-	-	-	-	-	-	-	-	-	-
NON-ENERGY USE	**-**	**382**	**-**	**273**	**49**	**484**	**-**	**-**	**-**	**-**	**-**
in Industry/Trans./Energy	-	-	-	-	-	-	-	-	-	-	-
in Transport	-	-	-	-	-	-	-	-	-	-	-
in Other Sectors	-	-	-	-	-	-	-	-	-	-	-

Former USSR / Ex-URSS : 1997

SUPPLY AND CONSUMPTION / APPROVISIONNEMENT ET DEMANDE	Oil cont. / *Pétrole cont.* (1000 tonnes)										
	Refinery Gas / *Gaz de raffinerie*	LPG + Ethane / *GPL + éthane*	Motor Gasoline / *Essence moteur*	Aviation Gasoline / *Essence aviation*	Jet Fuel / *Carbu- réacteurs*	Kerosene / *Kérosène*	Gas/ Diesel / *Gazole*	Heavy Fuel Oil / *Fioul lourd*	Naphtha / *Naphta*	Petrol. Coke / *Coke de pétrole*	Other Prod. / *Autres prod.*
Production	6202	5655	38307	43	10956	647	64053	84968	23	974	17225
From Other Sources	-	-	-	-	-	-	-	-	-	-	-
Imports	-	1873	5766	42	300	62	5068	5805	-	599	1003
Exports	-	-1056	-7804	-2	-230	-75	-29899	-31362	-23	-122	-828
Intl. Marine Bunkers	-	-	-	-	-	-	-38	-126	-	-	-
Stock Changes	-	-10	245	-	-15	7	152	-147	-	-46	11
DOMESTIC SUPPLY	6202	6462	36514	83	11011	641	39336	59138	-	1405	17411
Transfers	-	32	-	-	-	-	-	-	-	-	49
Statistical Differences	-	-	-	-	-	-	-	-	-	-	-
TRANSFORMATION	2176	317	-	-	-	17	1172	43017	-	-	238
Electricity Plants	-	-	-	-	-	-	80	2881	-	-	222
CHP Plants	2176	316	-	-	-	9	42	9161	-	-	8
Heat Plants	-	1	-	-	-	8	1050	30975	-	-	8
Blast Furnaces/Gas Works	-	-	-	-	-	-	-	-	-	-	-
Coke/Pat. Fuel/BKB Plants	-	-	-	-	-	-	-	-	-	-	-
Petroleum Refineries	-	-	-	-	-	-	-	-	-	-	-
Petrochemical Industry	-	-	-	-	-	-	-	-	-	-	-
Liquefaction	-	-	-	-	-	-	-	-	-	-	-
Other Transform. Sector	-	-	-	-	-	-	-	-	-	-	-
ENERGY SECTOR	3361	1	909	-	-	3	1695	3107	-	84	-
Coal Mines	-	-	-	-	-	-	6	2	-	-	-
Oil and Gas Extraction	-	-	909	-	-	-	1355	-	-	-	-
Petroleum Refineries	1522	-	-	-	-	-	-	609	-	84	-
Electr., CHP+Heat Plants	-	-	-	-	-	-	324	27	-	-	-
Pumped Storage (Elec.)	-	-	-	-	-	-	-	-	-	-	-
Other Energy Sector	1839	1	-	-	-	3	10	2469	-	-	-
Distribution Losses	2	8	22	-	1	1	15	3	-	-	-
FINAL CONSUMPTION	663	6168	35583	83	11010	620	36454	13011	-	1321	17222
INDUSTRY SECTOR	661	1414	3082	-	2	78	6033	8250	-	1321	24
Iron and Steel	-	-	-	-	-	-	-	-	-	-	-
Chemical and Petrochem.	-	-	-	-	-	-	-	-	-	-	-
of which: Feedstocks	-	-	-	-	-	-	-	-	-	-	-
Non-Ferrous Metals	-	-	-	-	-	-	-	-	-	-	-
Non-Metallic Minerals	-	-	-	-	-	-	-	-	-	-	-
Transport Equipment	-	-	-	-	-	-	-	-	-	-	-
Machinery	-	-	-	-	-	-	-	-	-	-	-
Mining and Quarrying	-	-	-	-	-	-	-	-	-	-	-
Food and Tobacco	-	-	-	-	-	-	-	-	-	-	-
Paper, Pulp and Print	-	-	-	-	-	-	-	-	-	-	-
Wood and Wood Products	-	-	-	-	-	-	-	-	-	-	-
Construction	-	-	-	-	-	-	-	-	-	-	-
Textile and Leather	-	-	-	-	-	-	-	-	-	-	-
Non-specified	-	-	-	-	-	-	-	-	-	-	-
TRANSPORT SECTOR	-	143	28772	83	11003	24	11527	378	-	-	21
Air	-	-	-	83	11003	-	-	-	-	-	-
Road	-	143	27955	-	-	-	6385	17	-	-	21
Rail	-	-	-	-	-	-	-	-	-	-	-
Pipeline Transport	-	-	-	-	-	-	-	-	-	-	-
Internal Navigation	-	-	-	-	-	-	-	-	-	-	-
Non-specified	-	-	-	-	-	-	-	-	-	-	-
OTHER SECTORS	-	-	-	-	-	-	-	-	-	-	-
Agriculture	-	-	-	-	-	-	-	-	-	-	-
Comm. and Publ. Services	-	-	-	-	-	-	-	-	-	-	-
Residential	-	-	-	-	-	-	-	-	-	-	-
Non-specified	-	-	-	-	-	-	-	-	-	-	-
NON-ENERGY USE	-	-	-	-	-	-	-	-	-	-	8204
in Industry/Transf./Energy	-	-	-	-	-	-	-	-	-	-	-
in Transport	-	-	-	-	-	-	-	-	-	-	-
in Other Sectors	-	-	-	-	-	-	-	-	-	-	-

Former USSR / Ex-URSS : 1997

SUPPLY AND CONSUMPTION APPROVISIONNEMENT ET DEMANDE	Gas / Gaz (TJ)				Comb. Renew. & Waste / En. Re. Comb. & Déchets (TJ)				(GWh)	(TJ)
	Natural Gas Gaz naturel	Gas Works Usines à gaz	Coke Ovens Cokeries	Blast Furnaces Hauts fourneaux	Solid Biomass Biomasse solide	Gas/Liquids from Biomass Gaz/Liquides tirés de biomasse	Municipal Waste Déchets urbains	Industrial Waste Déchets industriels	Electricity Electricité	Heat Chaleur
Production	25192190	5313	165277	310030	352769	58	-	363269	1236229	7687559
From Other Sources	-	8260			-	-	-	-		
Imports	3903842	-	-	-	122	-	-	-	61090	-
Exports	-8292146	-	-	-	-25217	-	-	-	-72930	-
Intl. Marine Bunkers	-	-	-	-		-	-	-		-
Stock Changes	443823	-	-	-	4364	-	-	3	-	-
DOMESTIC SUPPLY	21247709	13573	165277	310030	332038	58	-	363272	1224389	7687559
Transfers	-	-	-	-	-	-	-	-	-	-
Statistical Differences	113	-	-66103	-	-1820	-	-	-17746	-	-
TRANSFORMATION	10889916	1627	34194	35060	79024	58	-	67064	214	-
Electricity Plants	1136886		-	-	8356	-	-	-	-	
CHP Plants	6649061	1455	-	24366	487	-	-	67018		
Heat Plants	2652414	172	32382	10694	67873	58	-	46		
Blast Furnaces/Gas Works	-	-	1812	-	-	-	-	-		
Coke/Pat. Fuel/BKB Plants	-	-	-	-	-	-	-	-		
Petroleum Refineries	-	-	-	-	-	-	-	-		
Petrochemical Industry	-	-	-	-	-	-	-	-		
Liquefaction	-	-	-	-	-	-	-	-	-	
Other Transform. Sector	451555	-	-	-	2308	-	-	-	214	-
ENERGY SECTOR	577490	-	1978	-	286	-	-	8071	193056	495929
Coal Mines	-	-	-	-	9	-	-	-	19593	65400
Oil and Gas Extraction	513513	-	-	-	-	-	-	-	44173	159300
Petroleum Refineries	3000	-	-	-	-	-	-	-	13986	203477
Electr., CHP+Heat Plants	9454	-	-	-	26	-	-	-	91140	6197
Pumped Storage (Elec.)	-	-	-	-	-	-	-	-	663	-
Other Energy Sector	51523	-	1978	-	251	-	-	8071	23501	61555
Distribution Losses	392828	136	-	-	33	-	-	20726	146868	437222
FINAL CONSUMPTION	9387588	11810	63002	274970	250875	-	-	249665	884251	6754408
INDUSTRY SECTOR	3698557	3550	63002	274970	16488	-	-	242092	421874	2638315
Iron and Steel	-	-	-	-	-	-	-	-	-	-
Chemical and Petrochem.	-	-	-	-	-	-	-	-	-	-
of which: Feedstocks	-	-	-	-	-	-	-	-	-	-
Non-Ferrous Metals	-	-	-	-	-	-	-	-	-	-
Non-Metallic Minerals	-	-	-	-	-	-	-	-	-	-
Transport Equipment	-	-	-	-	-	-	-	-	-	-
Machinery	-	-	-	-	-	-	-	-	-	-
Mining and Quarrying	-	-	-	-	-	-	-	-	-	-
Food and Tobacco	-	-	-	-	-	-	-	-	-	-
Paper, Pulp and Print	-	-	-	-	-	-	-	-	-	-
Wood and Wood Products	-	-	-	-	-	-	-	-	-	-
Construction	-	-	-	-	-	-	-	-	-	-
Textile and Leather	-	-	-	-	-	-	-	-	-	-
Non-specified	-	-	-	-	-	-	-	-	-	-
TRANSPORT SECTOR	585520	-	-	-	2329	-	-	595	80742	-
Air	-	-	-	-	-	-	-	-	-	-
Road	14054	-	-	-	-	-	-	-	-	-
Rail	-	-	-	-	-	-	-	-	-	-
Pipeline Transport	-	-	-	-	-	-	-	-	-	-
Internal Navigation	-	-	-	-	-	-	-	-	-	-
Non-specified	-	-	-	-	-	-	-	-	-	-
OTHER SECTORS	-	-	-	-	-	-	-	-	-	-
Agriculture	-	-	-	-	-	-	-	-	-	-
Comm. and Publ. Services	-	-	-	-	-	-	-	-	-	-
Residential	-	-	-	-	-	-	-	-	-	-
Non-specified	-	-	-	-	-	-	-	-	-	-
NON-ENERGY USE	-	-	-	-	-	-	-	-	-	-
in Industry/Transf./Energy	-	-	-	-	-	-	-	-	-	-
in Transport	-	-	-	-	-	-	-	-	-	-
in Other Sectors	-	-	-	-	-	-	-	-	-	-

Former USSR / Ex-URSS : 1998

SUPPLY AND CONSUMPTION / APPROVISIONNEMENT ET DEMANDE	Coal / Charbon (1000 tonnes)							Oil / Pétrole (1000 tonnes)			
	Coking Coal / Charbon à coke	Other Bit. Coal / Autres charb. bit.	Sub-Bit. Coal / Charbon sous-bit.	Lignite / Lignite	Peat / Tourbe	Oven and Gas Coke / Coke de four/gaz	Pat. Fuel and BKB / Agg./briq. de lignite	Crude Oil / Pétrole brut	NGL / LGN	Feed-stocks / Produits d'aliment.	Additives / Additifs
Production	93370	204021	20	95868	5725	35797	4822	354756	5975	-	15
From Other Sources	-	-	-	-	-	1	-	239	-	-	-
Imports	3387	30503	-	1498	-	837	-	34936	194	367	9
Exports	-6723	-42804	-	-476	-114	-1703	-166	-160254	-775	-4	-
Intl. Marine Bunkers	-	-	-	-	-	-	-	-	-	-	-
Stock Changes	499	2052	-	347	1056	-4	30	-174	-20	95	-
DOMESTIC SUPPLY	90533	193772	20	97237	6667	34928	4686	229503	5374	458	24
Transfers	-	-	-	-	-	-	-	-60	-32	-	-
Statistical Differences	-14273	1720	-	-2365	-134	-	-	-2172	-10	-	-
TRANSFORMATION	69973	144338	-	82343	5341	25809	195	222627	5318	458	24
Electricity Plants	-	288	-	10573	-	-	-	-	-	-	-
CHP Plants	-	118893	-	58399	1565	-	18	2006	-	-	-
Heat Plants	-	21106	-	12330	817	-	177	1	-	-	-
Blast Furnaces/Gas Works	2166	-	-	603	-	25809	-	-	263	-	-
Coke/Pat. Fuel/BKB Plants	67807	4051	-	438	2959	-	-	-	-	-	-
Petroleum Refineries	-	-	-	-	-	-	-	220620	5055	458	24
Petrochemical Industry	-	-	-	-	-	-	-	-	-	-	-
Liquefaction	-	-	-	-	-	-	-	-	-	-	-
Other Transform. Sector	-	-	-	-	-	-	-	-	-	-	-
ENERGY SECTOR	-	875	-	221	100	60	8	1833	1	-	-
Coal Mines	-	730	-	193	37	-	7	-	-	-	-
Oil and Gas Extraction	-	-	-	-	-	-	-	1833	1	-	-
Petroleum Refineries	-	-	-	-	-	-	-	-	-	-	-
Electr., CHP+Heat Plants	-	145	-	28	2	-	-	-	-	-	-
Pumped Storage (Elec.)	-	-	-	-	-	-	-	-	-	-	-
Other Energy Sector	-	-	-	-	61	60	1	-	-	-	-
Distribution Losses	-	5193	-	3592	70	-	-	1826	13	-	-
FINAL CONSUMPTION	6287	45086	20	8716	1022	9059	4483	985	-	-	-
INDUSTRY SECTOR	6281	19888	-	2778	41	8660	17	700	-	-	-
Iron and Steel	-	-	-	-	-	-	-	-	-	-	-
Chemical and Petrochem.	-	-	-	-	-	-	-	-	-	-	-
of which: Feedstocks	-	-	-	-	-	-	-	-	-	-	-
Non-Ferrous Metals	-	-	-	-	-	-	-	-	-	-	-
Non-Metallic Minerals	-	-	-	-	-	-	-	-	-	-	-
Transport Equipment	-	-	-	-	-	-	-	-	-	-	-
Machinery	-	-	-	-	-	-	-	-	-	-	-
Mining and Quarrying	-	-	-	-	-	-	-	-	-	-	-
Food and Tobacco	-	-	-	-	-	-	-	-	-	-	-
Paper, Pulp and Print	-	-	-	-	-	-	-	-	-	-	-
Wood and Wood Products	-	-	-	-	-	-	-	-	-	-	-
Construction	-	-	-	-	-	-	-	-	-	-	-
Textile and Leather	-	-	-	-	-	-	-	-	-	-	-
Non-specified	-	-	-	-	-	-	-	-	-	-	-
TRANSPORT SECTOR	-	539	-	105	-	-	13	11	-	-	-
Air	-	-	-	-	-	-	-	-	-	-	-
Road	-	-	-	-	-	-	-	-	-	-	-
Rail	-	-	-	-	-	-	-	-	-	-	-
Pipeline Transport	-	-	-	-	-	-	-	-	-	-	-
Internal Navigation	-	-	-	-	-	-	-	-	-	-	-
Non-specified	-	-	-	-	-	-	-	-	-	-	-
OTHER SECTORS	-	-	-	-	-	-	-	-	-	-	-
Agriculture	-	-	-	-	-	-	-	-	-	-	-
Comm. and Publ. Services	-	-	-	-	-	-	-	-	-	-	-
Residential	-	-	-	-	-	-	-	-	-	-	-
Non-specified	-	-	-	-	-	-	-	-	-	-	-
NON-ENERGY USE	-	155	-	292	7	395	-	-	-	-	-
in Industry/Trans./Energy	-	-	-	-	-	-	-	-	-	-	-
in Transport	-	-	-	-	-	-	-	-	-	-	-
in Other Sectors	-	-	-	-	-	-	-	-	-	-	-

Former USSR / Ex-URSS : 1998

SUPPLY AND CONSUMPTION / APPROVISIONNEMENT ET DEMANDE	Oil cont. / Pétrole cont. (1000 tonnes)										
	Refinery Gas / Gaz de raffinerie	LPG + Ethane / GPL + éthane	Motor Gasoline / Essence moteur	Aviation Gasoline / Essence aviation	Jet Fuel / Carbu-réacteurs	Kerosene / Kérosène	Gas/ Diesel / Gazole	Heavy Fuel Oil / Fioul lourd	Naphtha / Naphta	Petrol. Coke / Coke de pétrole	Other Prod. / Autres prod.
Production	6649	5440	37393	34	11209	656	62694	78606	12	1174	15294
From Other Sources	-	-	-	-	-	-	-	-	-	-	-
Imports	-	1827	5018	40	390	186	4632	6334	-	117	912
Exports	-	-1775	-5813	-	-375	-84	-30309	-26632	-12	-30	-649
Intl. Marine Bunkers	-	-	-	-	-	-	-52	-107	-	-	-
Stock Changes	-	-9	-240	-	2	6	1085	469	-	-46	-68
DOMESTIC SUPPLY	6649	5483	36358	74	11226	764	38050	58670	-	1215	15489
Transfers	-	32	-	-	-	-	-	-	-	-	60
Statistical Differences	-	-	-	-	-	-	-	-	-	-	-
TRANSFORMATION	2347	674	-	-	-	13	1114	43420	-	-	238
Electricity Plants	-	-	-	-	-	-	80	2943	-	-	217
CHP Plants	2333	673	-	-	-	4	52	9752	-	-	13
Heat Plants	14	1	-	-	-	9	982	30725	-	-	8
Blast Furnaces/Gas Works	-	-	-	-	-	-	-	-	-	-	-
Coke/Pat. Fuel/BKB Plants	-	-	-	-	-	-	-	-	-	-	-
Petroleum Refineries	-	-	-	-	-	-	-	-	-	-	-
Petrochemical Industry	-	-	-	-	-	-	-	-	-	-	-
Liquefaction	-	-	-	-	-	-	-	-	-	-	-
Other Transform. Sector	-	-	-	-	-	-	-	-	-	-	-
ENERGY SECTOR	3592	-	1070	-	-	1	1593	3176	-	101	-
Coal Mines	-	-	-	-	-	-	9	3	-	-	-
Oil and Gas Extraction	-	-	1070	-	-	-	1272	-	-	-	-
Petroleum Refineries	1617	-	-	-	-	-	-	596	-	101	-
Electr., CHP+Heat Plants	-	-	-	-	-	-	307	24	-	-	-
Pumped Storage (Elec.)	-	-	-	-	-	-	-	-	-	-	-
Other Energy Sector	1975	-	-	-	-	1	5	2553	-	-	-
Distribution Losses	2	8	22	-	-	-	9	3	-	-	-
FINAL CONSUMPTION	708	4833	35266	74	11226	750	35334	12071	-	1114	15311
INDUSTRY SECTOR	706	1145	3248	-	1	73	5853	7780	-	997	53
Iron and Steel	-	-	-	-	-	-	-	-	-	-	-
Chemical and Petrochem.	-	-	-	-	-	-	-	-	-	-	-
of which: Feedstocks	-	-	-	-	-	-	-	-	-	-	-
Non-Ferrous Metals	-	-	-	-	-	-	-	-	-	-	-
Non-Metallic Minerals	-	-	-	-	-	-	-	-	-	-	-
Transport Equipment	-	-	-	-	-	-	-	-	-	-	-
Machinery	-	-	-	-	-	-	-	-	-	-	-
Mining and Quarrying	-	-	-	-	-	-	-	-	-	-	-
Food and Tobacco	-	-	-	-	-	-	-	-	-	-	-
Paper, Pulp and Print	-	-	-	-	-	-	-	-	-	-	-
Wood and Wood Products	-	-	-	-	-	-	-	-	-	-	-
Construction	-	-	-	-	-	-	-	-	-	-	-
Textile and Leather	-	-	-	-	-	-	-	-	-	-	-
Non-specified	-	-	-	-	-	-	-	-	-	-	-
TRANSPORT SECTOR	-	80	28949	74	11122	12	11525	243	-	-	24
Air	-	-	-	74	11122	-	-	-	-	-	-
Road	-	80	28416	-	-	-	6425	55	-	-	24
Rail	-	-	-	-	-	-	-	-	-	-	-
Pipeline Transport	-	-	-	-	-	-	-	-	-	-	-
Internal Navigation	-	-	-	-	-	-	-	-	-	-	-
Non-specified	-	-	-	-	-	-	-	-	-	-	-
OTHER SECTORS	-	-	-	-	-	-	-	-	-	-	-
Agriculture	-	-	-	-	-	-	-	-	-	-	-
Comm. and Publ. Services	-	-	-	-	-	-	-	-	-	-	-
Residential	-	-	-	-	-	-	-	-	-	-	-
Non-specified	-	-	-	-	-	-	-	-	-	-	-
NON-ENERGY USE	-	-	-	-	-	-	-	-	-	-	9221
in Industry/Transf./Energy	-	-	-	-	-	-	-	-	-	-	-
in Transport	-	-	-	-	-	-	-	-	-	-	-
in Other Sectors	-	-	-	-	-	-	-	-	-	-	-

Former USSR / Ex-URSS : 1998

SUPPLY AND CONSUMPTION / APPROVISIONNEMENT ET DEMANDE	Gas / Gaz (TJ)				Comb. Renew. & Waste / En. Re. Comb. & Déchets (TJ)				(GWh)	(TJ)
	Natural Gas / Gaz naturel	Gas Works / Usines à gaz	Coke Ovens / Cokeries	Blast Furnaces / Hauts fourneaux	Solid Biomass / Biomasse solide	Gas/Liquids from Biomass / Gaz/Liquides tirés de biomasse	Municipal Waste / Déchets urbains	Industrial Waste / Déchets industriels	Electricity / Electricité	Heat / Chaleur
Production	25798129	3416	151454	294289	320859	70	-	341571	1224475	7501013
From Other Sources	-	11022	-	-	-	-	-	-	-	-
Imports	3580613	746	-	-	129	-	-	-	57849	-
Exports	-8170728	-	-	-	-25952	-	-	-	-71885	-
Intl. Marine Bunkers	-	-	-	-	-	-	-	-	-	-
Stock Changes	-199201	-	-	-	2872	-	-	-	-	-
DOMESTIC SUPPLY	**21008813**	**15184**	**151454**	**294289**	**297908**	**70**	**-**	**341571**	**1210439**	**7501013**
Transfers	-	-	-	-	-	-	-	-	-	-
Statistical Differences	-	-	-60574	-	-452	-	-	-16732	-	-
TRANSFORMATION	**10593188**	**1321**	**31356**	**33522**	**71221**	**70**	**-**	**61745**	**104**	**-**
Electricity Plants	958621	-	-	-	7158	-	-	-	-	-
CHP Plants	6500731	1135	-	23297	475	-	-	61713	-	-
Heat Plants	2604947	186	29694	10225	61465	70	-	32	-	-
Blast Furnaces/Gas Works	-	-	1662	-	-	-	-	-	-	-
Coke/Pat. Fuel/BKB Plants	-	-	-	-	-	-	-	-	-	-
Petroleum Refineries	-	-	-	-	-	-	-	-	-	-
Petrochemical Industry	-	-	-	-	-	-	-	-	-	-
Liquefaction	-	-	-	-	-	-	-	-	-	-
Other Transform. Sector	528889	-	-	-	2123	-	-	-	104	-
ENERGY SECTOR	**656074**	**-**	**1976**	**-**	**79**	**-**	**-**	**15753**	**189752**	**525561**
Coal Mines	-	-	-	-	1	-	-	-	17634	58589
Oil and Gas Extraction	586121	-	-	-	-	-	-	-	44325	153000
Petroleum Refineries	3000	-	-	-	-	-	-	-	12942	205468
Electr., CHP+Heat Plants	9633	-	-	-	52	-	-	-	95696	6152
Pumped Storage (Elec.)	-	-	-	-	-	-	-	-	654	-
Other Energy Sector	57320	-	1976	-	26	-	-	15753	18501	102352
Distribution Losses	409947	60	-	-	24	-	-	19489	155421	494866
FINAL CONSUMPTION	**9349604**	**13803**	**57548**	**260767**	**226132**	**-**	**-**	**227852**	**865162**	**6480586**
INDUSTRY SECTOR	**3541185**	**2781**	**57548**	**260767**	**18088**	**-**	**-**	**218778**	**399074**	**2453931**
Iron and Steel	-	-	-	-	-	-	-	-	-	-
Chemical and Petrochem.	-	-	-	-	-	-	-	-	-	-
of which: Feedstocks	-	-	-	-	-	-	-	-	-	-
Non-Ferrous Metals	-	-	-	-	-	-	-	-	-	-
Non-Metallic Minerals	-	-	-	-	-	-	-	-	-	-
Transport Equipment	-	-	-	-	-	-	-	-	-	-
Machinery	-	-	-	-	-	-	-	-	-	-
Mining and Quarrying	-	-	-	-	-	-	-	-	-	-
Food and Tobacco	-	-	-	-	-	-	-	-	-	-
Paper, Pulp and Print	-	-	-	-	-	-	-	-	-	-
Wood and Wood Products	-	-	-	-	-	-	-	-	-	-
Construction	-	-	-	-	-	-	-	-	-	-
Textile and Leather	-	-	-	-	-	-	-	-	-	-
Non-specified	-	-	-	-	-	-	-	-	-	-
TRANSPORT SECTOR	**731628**	**-**	**-**	**-**	**1257**	**-**	**-**	**3159**	**77931**	**-**
Air	-	-	-	-	-	-	-	-	-	-
Road	18621	-	-	-	-	-	-	-	-	-
Rail	-	-	-	-	-	-	-	-	-	-
Pipeline Transport	-	-	-	-	-	-	-	-	-	-
Internal Navigation	-	-	-	-	-	-	-	-	-	-
Non-specified	-	-	-	-	-	-	-	-	-	-
OTHER SECTORS	**-**	**-**	**-**	**-**	**-**	**-**	**-**	**-**	**-**	**-**
Agriculture	-	-	-	-	-	-	-	-	-	-
Comm. and Publ. Services	-	-	-	-	-	-	-	-	-	-
Residential	-	-	-	-	-	-	-	-	-	-
Non-specified	-	-	-	-	-	-	-	-	-	-
NON-ENERGY USE	**-**	**-**	**-**	**-**	**-**	**-**	**-**	**-**	**-**	**-**
in Industry/Transf./Energy	-	-	-	-	-	-	-	-	-	-
in Transport	-	-	-	-	-	-	-	-	-	-
in Other Sectors	-	-	-	-	-	-	-	-	-	-

Middle East / Moyen-Orient

SUPPLY AND CONSUMPTION 1997	Coal (1000 tonnes)							Oil (1000 tonnes)			
	Coking Coal	Other Bit. Coal	Sub-Bit. Coal	Lignite	Peat	Oven and Gas Coke	Pat. Fuel and BKB	Crude Oil	NGL	Feed-stocks	Additives
Production	922	165	-	472	-	1215	-	982463	36266	-	-
Imports	598	9004	-	-	-	2	-	25299	-	-	-
Exports	-	-13	-	-	-	-	-	-712292	-15017	-	-
Intl. Marine Bunkers	-	-	-	-	-	-	-	-	-	-	-
Stock Changes	-	-164	-	-	-	-	-	-12023	-	-	-
DOMESTIC SUPPLY	1520	8992	-	472	-	1217	-	283447	21249	-	-
Transfers and Stat. Diff.	-1	-	-	-	-	-	-	6554	-12888	-	-
TRANSFORMATION	1519	8639	-	472	-	974	-	290001	2390	-	-
Electricity and CHP Plants	-	8639	-	472	-	-	-	15373	-	-	-
Petroleum Refineries	-	-	-	-	-	-	-	274628	2390	-	-
Other Transform. Sector	1519	-	-	-	-	974	-	-	-	-	-
ENERGY SECTOR	-	-	-	-	-	-	-	-	-	-	-
DISTRIBUTION LOSSES	-	-	-	-	-	-	-	-	-	-	-
FINAL CONSUMPTION	-	353	-	-	-	243	-	-	5971	-	-
INDUSTRY SECTOR	-	353	-	-	-	243	-	-	5971	-	-
Iron and Steel	-	-	-	-	-	-	-	-	-	-	-
Chemical and Petrochem.	-	-	-	-	-	-	-	-	-	-	-
Non-Metallic Minerals	-	-	-	-	-	-	-	-	-	-	-
Non-specified	-	-	-	-	-	-	-	-	-	-	-
TRANSPORT SECTOR	-	-	-	-	-	-	-	-	-	-	-
Air	-	-	-	-	-	-	-	-	-	-	-
Road	-	-	-	-	-	-	-	-	-	-	-
Non-specified	-	-	-	-	-	-	-	-	-	-	-
OTHER SECTORS	-	-	-	-	-	-	-	-	-	-	-
Agriculture	-	-	-	-	-	-	-	-	-	-	-
Comm. and Publ. Services	-	-	-	-	-	-	-	-	-	-	-
Residential	-	-	-	-	-	-	-	-	-	-	-
Non-specified	-	-	-	-	-	-	-	-	-	-	-
NON-ENERGY USE	-	-	-	-	-	-	-	-	-	-	-

APPROVISIONNEMENT ET DEMANDE 1998	Charbon (1000 tonnes)							Pétrole (1000 tonnes)			
	Charbon à coke	Autres charb. bit.	Charbon sous-bit.	Lignite	Tourbe	Coke de four/gaz	Agg./briq. de lignite	Pétrole brut	LGN	Produits d'aliment.	Additifs
Production	1000	169	-	444	-	1363	-	1052903	38794	-	-
Imports	704	9836	-	-	-	2	-	25812	-	-	-
Exports	-	-11	-	-	-	-	-	-773611	-15586	-	-
Intl. Marine Bunkers	-	-	-	-	-	-	-	-	-	-	-
Stock Changes	-	-352	-	-	-	-	-	-16978	-	-	-
DOMESTIC SUPPLY	1704	9642	-	444	-	1365	-	288126	23208	-	-
Transfers and Stat. Diff.	-1	-	-	-	-	-	-	8995	-13047	-	-
TRANSFORMATION	1703	9283	-	444	-	1092	-	297121	2490	-	-
Electricity and CHP Plants	-	9283	-	444	-	-	-	15420	-	-	-
Petroleum Refineries	-	-	-	-	-	-	-	281701	2490	-	-
Other Transform. Sector	1703	-	-	-	-	1092	-	-	-	-	-
ENERGY SECTOR	-	-	-	-	-	-	-	-	-	-	-
DISTRIBUTION LOSSES	-	-	-	-	-	-	-	-	-	-	-
FINAL CONSUMPTION	-	359	-	-	-	273	-	-	7671	-	-
INDUSTRY SECTOR	-	359	-	-	-	273	-	-	7671	-	-
Iron and Steel	-	-	-	-	-	-	-	-	-	-	-
Chemical and Petrochem.	-	-	-	-	-	-	-	-	-	-	-
Non-Metallic Minerals	-	-	-	-	-	-	-	-	-	-	-
Non-specified	-	-	-	-	-	-	-	-	-	-	-
TRANSPORT SECTOR	-	-	-	-	-	-	-	-	-	-	-
Air	-	-	-	-	-	-	-	-	-	-	-
Road	-	-	-	-	-	-	-	-	-	-	-
Non-specified	-	-	-	-	-	-	-	-	-	-	-
OTHER SECTORS	-	-	-	-	-	-	-	-	-	-	-
Agriculture	-	-	-	-	-	-	.	-	-	-	-
Comm. and Publ. Services	-	-	-	-	-	-	-	-	-	-	-
Residential	-	-	-	-	-	-	-	-	-	-	-
Non-specified	-	-	-	-	-	-	-	-	-	-	-
NON-ENERGY USE	-	-	-	-	-	-	-	-	-	-	-

Middle East / Moyen-Orient

SUPPLY AND CONSUMPTION 1997	Refinery Gas	LPG + Ethane	Motor Gasoline	Aviation Gasoline	Jet Fuel	Kerosene	Gas/ Diesel	Heavy Fuel Oil	Naphtha	Petrol. Coke	Other Prod.
Production	3338	10174	32634	105	11663	15213	80407	88521	18215	-	7983
Imports	-	372	3054	2	195	576	2329	4451	66	-	360
Exports	-	-9634	-9016	-	-6241	-11577	-26536	-45232	-16577	-	-498
Intl. Marine Bunkers	-	-	-	-	-	-	-251	-3128	-	-	-
Stock Changes	-	-35	219	-4	272	97	41	3063	7	-	-88
DOMESTIC SUPPLY	**3338**	**877**	**26891**	**103**	**5889**	**4309**	**55990**	**47675**	**1711**	**-**	**7757**
Transfers and Stat. Diff.	-	6959	5714	-	-490	-27	-2138	-488	197	-	291
TRANSFORMATION	**-**	**-**	**-**	**-**	**-**	**-**	**9090**	**18666**	**-**	**-**	**-**
Electricity and CHP Plants	-	-	-	-	-	-	9090	18666	-	-	-
Petroleum Refineries	-	-	-	-	-	-	-	-	-	-	-
Other Transform. Sector	-	-	-	-	-	-	-	-	-	-	-
ENERGY SECTOR	**3338**	**85**	**-**	**-**	**-**	**-**	**790**	**8454**	**-**	**-**	**64**
DISTRIBUTION LOSSES	**-**	**-**	**-**	**-**	**-**	**-**	**-**	**-**	**-**	**-**	**-**
FINAL CONSUMPTION	**-**	**7751**	**32605**	**103**	**5399**	**4282**	**43972**	**20067**	**1908**	**-**	**7984**
INDUSTRY SECTOR	**-**	**1267**	**-**	**-**	**-**	**25**	**960**	**15785**	**1908**	**-**	**-**
Iron and Steel	-	-	-	-	-	-	-	-	-	-	-
Chemical and Petrochem.	-	-	-	-	-	-	-	-	-	-	-
Non-Metallic Minerals	-	-	-	-	-	-	-	-	-	-	-
Non-specified	-	-	-	-	-	-	-	-	-	-	-
TRANSPORT SECTOR	**-**	**54**	**32605**	**103**	**5399**	**642**	**19373**	**5**	**-**	**-**	**-**
Air	-	-	-	103	5399	642	-	-	-	-	-
Road	-	54	32605	-	-	-	19373	-	-	-	-
Non-specified	-	-	-	-	-	-	-	-	-	-	-
OTHER SECTORS	**-**	**-**	**-**	**-**	**-**	**-**	**-**	**-**	**-**	**-**	**-**
Agriculture	-	-	-	-	-	-	-	-	-	-	-
Comm. and Publ. Services	-	-	-	-	-	-	-	-	-	-	-
Residential	-	-	-	-	-	-	-	-	-	-	-
Non-specified	-	-	-	-	-	-	-	-	-	-	-
NON-ENERGY USE	**-**	**-**	**-**	**-**	**-**	**-**	**-**	**-**	**-**	**-**	**7984**

APPROVISIONNEMENT ET DEMANDE 1998	Gaz de raffinerie	GPL + éthane	Essence moteur	Essence aviation	Carbu- réacteurs	Kérosène	Gazole	Fioul lourd	Naphta	Coke de pétrole	Autres prod.
Production	3391	10716	35145	102	11337	15124	80221	91543	19926	-	8102
Imports	-	349	2494	2	211	386	2464	4597	60	-	384
Exports	-	-11456	-8587	-	-6325	-11057	-25370	-46999	-16861	-	-465
Intl. Marine Bunkers	-	-	-	-	-	-	-240	-3222	-	-	-
Stock Changes	-	-7	418	-4	354	-165	-66	2324	-55	-	28
DOMESTIC SUPPLY	**3391**	**-398**	**29470**	**100**	**5577**	**4288**	**57009**	**48243**	**3070**	**-**	**8049**
Transfers and Stat. Diff.	-	8685	4416	-	-99	-32	-1578	-367	189	-	393
TRANSFORMATION	**-**	**-**	**-**	**-**	**-**	**-**	**9462**	**17471**	**-**	**-**	**-**
Electricity and CHP Plants	-	-	-	-	-	-	9462	17471	-	-	-
Petroleum Refineries	-	-	-	-	-	-	-	-	-	-	-
Other Transform. Sector	-	-	-	-	-	-	-	-	-	-	-
ENERGY SECTOR	**3391**	**26**	**-**	**-**	**-**	**-**	**797**	**9854**	**-**	**-**	**68**
DISTRIBUTION LOSSES	**-**	**-**	**-**	**-**	**-**	**-**	**-**	**-**	**-**	**-**	**-**
FINAL CONSUMPTION	**-**	**8261**	**33886**	**100**	**5478**	**4256**	**45172**	**20551**	**3259**	**-**	**8374**
INDUSTRY SECTOR	**-**	**1323**	**-**	**-**	**-**	**16**	**921**	**16701**	**3259**	**-**	**-**
Iron and Steel	-	-	-	-	-	-	-	-	-	-	-
Chemical and Petrochem.	-	-	-	-	-	-	-	-	-	-	-
Non-Metallic Minerals	-	-	-	-	-	-	-	-	-	-	-
Non-specified	-	-	-	-	-	-	-	-	-	-	-
TRANSPORT SECTOR	**-**	**113**	**33886**	**100**	**5478**	**695**	**20024**	**5**	**-**	**-**	**-**
Air	-	-	-	100	5478	695	-	-	-	-	-
Road	-	113	33886	-	-	-	20024	-	-	-	-
Non-specified	-	-	-	-	-	-	-	-	-	-	-
OTHER SECTORS	**-**	**-**	**-**	**-**	**-**	**-**	**-**	**-**	**-**	**-**	**-**
Agriculture	-	-	-	-	-	-	-	-	-	-	-
Comm. and Publ. Services	-	-	-	-	-	-	-	-	-	-	-
Residential	-	-	-	-	-	-	-	-	-	-	-
Non-specified	-	-	-	-	-	-	-	-	-	-	-
NON-ENERGY USE	**-**	**-**	**-**	**-**	**-**	**-**	**-**	**-**	**-**	**-**	**8374**

Middle East / Moyen-Orient

SUPPLY AND CONSUMPTION 1997	Gas (TJ)				Comb. Renew. & Waste (TJ)				(GWh)	(TJ)
	Natural Gas	Gas Works	Coke Ovens	Blast Furnaces	Solid Biomass	Gas/Liquids from Biomass	Municipal Waste	Industrial Waste	Electricity	Heat
Production	6881974	-	-	22	42849	-	-	-	385234	-
Imports	34176	-	-	-	1365	-	-	-	608	-
Exports	-417677	-	-	-	-2	-	-	-	-1173	-
Intl. Marine Bunkers	-	-	-	-	-	-	-	-	-	-
Stock Changes	-	-	-	-	-	-	-	-	-	-
DOMESTIC SUPPLY	6498473	-	-	22	44212	-	-	-	384669	-
Transfers and Stat. Diff.	135097	-	-	-	-	-	-	-	2054	-
TRANSFORMATION	1968314	-	-	-	5782	-	-	-	-	-
Electricity and CHP Plants	1822118	-	-	-	-	-	-	-	-	-
Petroleum Refineries	-	-	-	-	-	-	-	-	-	-
Other Transform. Sector	146196	-	-	-	5782	-	-	-	-	-
ENERGY SECTOR	1288537	-	-	-	-	-	-	-	36009	-
DISTRIBUTION LOSSES	93031	-	-	-	-	-	-	-	31139	-
FINAL CONSUMPTION	3283688	-	-	22	38429	-	-	-	319575	-
INDUSTRY SECTOR	1881946	-	-	22	7789	-	-	-	64544	-
Iron and Steel	-	-	-	-	-	-	-	-	-	-
Chemical and Petrochem.	-	-	-	-	-	-	-	-	-	-
Non-Metallic Minerals	-	-	-	-	-	-	-	-	-	-
Non-specified	-	-	-	-	-	-	-	-	-	-
TRANSPORT SECTOR	-	-	-	-	-	-	-	-	-	-
Air	-	-	-	-	-	-	-	-	-	-
Road	-	-	-	-	-	-	-	-	-	-
Non-specified	-	-	-	-	-	-	-	-	-	-
OTHER SECTORS	-	-	-	-	-	-	-	-	-	-
Agriculture	-	-	-	-	-	-	-	-	-	-
Comm. and Publ. Services	-	-	-	-	-	-	-	-	-	-
Residential	-	-	-	-	-	-	-	-	-	-
Non-specified	-	-	-	-	-	-	-	-	-	-
NON-ENERGY USE	-	-	-	-	-	-	-	-	-	-

APPROVISIONNEMENT ET DEMANDE 1998	Gaz (TJ)				En. Re. Comb. & Déchets (TJ)				(GWh)	(TJ)
	Gaz naturel	Usines à gaz	Cokeries	Hauts fourneaux	Biomasse solide	Gaz/Liquides tirés de biomasse	Déchets urbains	Déchets industriels	Electricité	Chaleur
Production	7286161	-	-	22	42988	-	-	-	409638	-
Imports	91950	-	-	-	1371	-	-	-	654	-
Exports	-464706	-	-	-	-2	-	-	-	-1059	-
Intl. Marine Bunkers	-	-	-	-	-	-	-	-	-	-
Stock Changes	-	-	-	-	-	-	-	-	-	-
DOMESTIC SUPPLY	6913405	-	-	22	44357	-	-	-	409233	-
Transfers and Stat. Diff.	106142	-	-	-	-	-	-	-	2088	-
TRANSFORMATION	2193124	-	-	-	5878	-	-	-	-	-
Electricity and CHP Plants	2049176	-	-	-	-	-	-	-	-	-
Petroleum Refineries	-	-	-	-	-	-	-	-	-	-
Other Transform. Sector	143948	-	-	-	5878	-	-	-	-	-
ENERGY SECTOR	1335656	-	-	-	-	-	-	-	37846	-
DISTRIBUTION LOSSES	100255	-	-	-	-	-	-	-	35545	-
FINAL CONSUMPTION	3390512	-	-	22	38479	-	-	-	337930	-
INDUSTRY SECTOR	2030414	-	-	22	7789	-	-	-	67283	-
Iron and Steel	-	-	-	-	-	-	-	-	-	-
Chemical and Petrochem.	-	-	-	-	-	-	-	-	-	-
Non-Metallic Minerals	-	-	-	-	-	-	-	-	-	-
Non-specified	-	-	-	-	-	-	-	-	-	-
TRANSPORT SECTOR	-	-	-	-	-	-	-	-	-	-
Air	-	-	-	-	-	-	-	-	-	-
Road	-	-	-	-	-	-	-	-	-	-
Non-specified	-	-	-	-	-	-	-	-	-	-
OTHER SECTORS	-	-	-	-	-	-	-	-	-	-
Agriculture	-	-	-	-	-	-	-	-	-	-
Comm. and Publ. Services	-	-	-	-	-	-	-	-	-	-
Residential	-	-	-	-	-	-	-	-	-	-
Non-specified	-	-	-	-	-	-	-	-	-	-
NON-ENERGY USE	-	-	-	-	-	-	-	-	-	-

Albania / Albanie

SUPPLY AND CONSUMPTION 1997	Coal (1000 tonnes)							Oil (1000 tonnes)			
	Coking Coal	Other Bit. Coal	Sub-Bit. Coal	Lignite	Peat	Oven and Gas Coke	Pat. Fuel and BKB	Crude Oil	NGL	Feed-stocks	Additives
Production	-	-	-	70	-	-	-	360	1	-	-
Imports	-	-	-	-	-	-	-	6	-	-	-
Exports	-	-	-	-	-	-	-	-	-	-	-
Intl. Marine Bunkers	-	-	-	-	-	-	-	-	-	-	-
Stock Changes	-	-	-	-	-	-	-	-	-	-	-
DOMESTIC SUPPLY	-	-	-	70	-	-	-	366	1	-	-
Transfers and Stat. Diff.	-	-	-	-	-	-	-	-	-	-	-
TRANSFORMATION	-	-	-	-	-	-	-	366	1	-	-
Electricity and CHP Plants	-	-	-	-	-	-	-	-	-	-	-
Petroleum Refineries	-	-	-	-	-	-	-	366	1	-	-
Other Transform. Sector	-	-	-	-	-	-	-	-	-	-	-
ENERGY SECTOR	-	-	-	-	-	-	-	-	-	-	-
DISTRIBUTION LOSSES	-	-	-	-	-	-	-	-	-	-	-
FINAL CONSUMPTION	-	-	-	70	-	-	-	-	-	-	-
INDUSTRY SECTOR	-	-	-	-	-	-	-	-	-	-	-
Iron and Steel	-	-	-	-	-	-	-	-	-	-	-
Chemical and Petrochem.	-	-	-	-	-	-	-	-	-	-	-
Non-Metallic Minerals	-	-	-	-	-	-	-	-	-	-	-
Non-specified	-	-	-	-	-	-	-	-	-	-	-
TRANSPORT SECTOR	-	-	-	-	-	-	-	-	-	-	-
Air	-	-	-	-	-	-	-	-	-	-	-
Road	-	-	-	-	-	-	-	-	-	-	-
Non-specified	-	-	-	-	-	-	-	-	-	-	-
OTHER SECTORS	-	-	-	70	-	-	-	-	-	-	-
Agriculture	-	-	-	-	-	-	-	-	-	-	-
Comm. and Publ. Services	-	-	-	70	-	-	-	-	-	-	-
Residential	-	-	-	-	-	-	-	-	-	-	-
Non-specified	-	-	-	-	-	-	-	-	-	-	-
NON-ENERGY USE	-	-	-	-	-	-	-	-	-	-	-

APPROVISIONNEMENT ET DEMANDE 1998	Charbon (1000 tonnes)							Pétrole (1000 tonnes)			
	Charbon à coke	Autres charb. bit.	Charbon sous-bit.	Lignite	Tourbe	Coke de four/gaz	Agg./briq. de lignite	Pétrole brut	LGN	Produits d'aliment.	Additifs
Production	-	-	-	49	-	-	-	363	1	-	-
Imports	-	-	-	-	-	-	-	5	-	-	-
Exports	-	-	-	-	-	-	-	-	-	-	-
Intl. Marine Bunkers	-	-	-	-	-	-	-	-	-	-	-
Stock Changes	-	-	-	-	-	-	-	-	-	-	-
DOMESTIC SUPPLY	-	-	-	49	-	-	-	368	1	-	-
Transfers and Stat. Diff.	-	-	-	-	-	-	-	-	-	-	-
TRANSFORMATION	-	-	-	-	-	-	-	368	1	-	-
Electricity and CHP Plants	-	-	-	-	-	-	-	-	-	-	-
Petroleum Refineries	-	-	-	-	-	-	-	368	1	-	-
Other Transform. Sector	-	-	-	-	-	-	-	-	-	-	-
ENERGY SECTOR	-	-	-	-	-	-	-	-	-	-	-
DISTRIBUTION LOSSES	-	-	-	-	-	-	-	-	-	-	-
FINAL CONSUMPTION	-	-	-	49	-	-	-	-	-	-	-
INDUSTRY SECTOR	-	-	-	-	-	-	-	-	-	-	-
Iron and Steel	-	-	-	-	-	-	-	-	-	-	-
Chemical and Petrochem.	-	-	-	-	-	-	-	-	-	-	-
Non-Metallic Minerals	-	-	-	-	-	-	-	-	-	-	-
Non-specified	-	-	-	-	-	-	-	-	-	-	-
TRANSPORT SECTOR	-	-	-	-	-	-	-	-	-	-	-
Air	-	-	-	-	-	-	-	-	-	-	-
Road	-	-	-	-	-	-	-	-	-	-	-
Non-specified	-	-	-	-	-	-	-	-	-	-	-
OTHER SECTORS	-	-	-	49	-	-	-	-	-	-	-
Agriculture	-	-	-	-	-	-	-	-	-	-	-
Comm. and Publ. Services	-	-	-	49	-	-	-	-	-	-	-
Residential	-	-	-	-	-	-	-	-	-	-	-
Non-specified	-	-	-	-	-	-	-	-	-	-	-
NON-ENERGY USE	-	-	-	-	-	-	-	-	-	-	-

Albania / Albanie

SUPPLY AND CONSUMPTION 1997	Refinery Gas	LPG + Ethane	Motor Gasoline	Aviation Gasoline	Jet Fuel	Kerosene	Gas/ Diesel	Heavy Fuel Oil	Naphtha	Petrol. Coke	Other Prod.
Production	30	-	74	-	-	57	107	39	-	29	14
Imports	-	-	105	-	-	-	-	-	-	-	-
Exports	-	-	-	-	-	-	-	-	-	-	-
Intl. Marine Bunkers	-	-	-	-	-	-	-	-	-	-	-
Stock Changes	-	-	-	-	-	-	-	-	-	-	-
DOMESTIC SUPPLY	30	-	179	-	-	57	107	39	-	29	14
Transfers and Stat. Diff.	-	-	-	-	-	-	-	-	-	-	-
TRANSFORMATION	-	-	-	-	-	-	-	39	-	-	-
Electricity and CHP Plants	-	-	-	-	-	-	-	39	-	-	-
Petroleum Refineries	-	-	-	-	-	-	-	-	-	-	-
Other Transform. Sector	-	-	-	-	-	-	-	-	-	-	-
ENERGY SECTOR	30	-	-	-	-	-	-	-	-	-	-
DISTRIBUTION LOSSES	-	-	-	-	-	-	-	-	-	-	-
FINAL CONSUMPTION	-	-	179	-	-	57	107	-	-	29	14
INDUSTRY SECTOR	-	-	-	-	-	-	-	-	-	29	-
Iron and Steel	-	-	-	-	-	-	-	-	-	-	-
Chemical and Petrochem.	-	-	-	-	-	-	-	-	-	-	-
Non-Metallic Minerals	-	-	-	-	-	-	-	-	-	-	-
Non-specified	-	-	-	-	-	-	-	-	-	29	-
TRANSPORT SECTOR	-	-	179	-	-	-	107	-	-	-	-
Air	-	-	-	-	-	-	-	-	-	-	-
Road	-	-	179	-	-	-	107	-	-	-	-
Non-specified	-	-	-	-	-	-	-	-	-	-	-
OTHER SECTORS	-	-	-	-	-	57	-	-	-	-	-
Agriculture	-	-	-	-	-	-	-	-	-	-	-
Comm. and Publ. Services	-	-	-	-	-	-	-	-	-	-	-
Residential	-	-	-	-	-	57	-	-	-	-	-
Non-specified	-	-	-	-	-	-	-	-	-	-	-
NON-ENERGY USE	-	-	-	-	-	-	-	-	-	-	14

APPROVISIONNEMENT ET DEMANDE 1998	Gaz de raffinerie	GPL + éthane	Essence moteur	Essence aviation	Carbu- réacteurs	Kérosène	Gazole	Fioul lourd	Naphta	Coke de pétrole	Autres prod.
Production	30	-	74	-	-	57	107	39	-	29	14
Imports	-	-	105	-	-	-	-	-	-	-	-
Exports	-	-	-	-	-	-	-	-	-	-	-
Intl. Marine Bunkers	-	-	-	-	-	-	-	-	-	-	-
Stock Changes	-	-	-	-	-	-	-	-	-	-	-
DOMESTIC SUPPLY	30	-	179	-	-	57	107	39	-	29	14
Transfers and Stat. Diff.	-	-	-	-	-	-	-	-	-	-	-
TRANSFORMATION	-	-	-	-	-	-	-	39	-	-	-
Electricity and CHP Plants	-	-	-	-	-	-	-	39	-	-	-
Petroleum Refineries	-	-	-	-	-	-	-	-	-	-	-
Other Transform. Sector	-	-	-	-	-	-	-	-	-	-	-
ENERGY SECTOR	30	-	-	-	-	-	-	-	-	-	-
DISTRIBUTION LOSSES	-	-	-	-	-	-	-	-	-	-	-
FINAL CONSUMPTION	-	-	179	-	-	57	107	-	-	29	14
INDUSTRY SECTOR	-	-	-	-	-	-	-	-	-	29	-
Iron and Steel	-	-	-	-	-	-	-	-	-	-	-
Chemical and Petrochem.	-	-	-	-	-	-	-	-	-	-	-
Non-Metallic Minerals	-	-	-	-	-	-	-	-	-	-	-
Non-specified	-	-	-	-	-	-	-	-	-	29	-
TRANSPORT SECTOR	-	-	179	-	-	-	107	-	-	-	-
Air	-	-	-	-	-	-	-	-	-	-	-
Road	-	-	179	-	-	-	107	-	-	-	-
Non-specified	-	-	-	-	-	-	-	-	-	-	-
OTHER SECTORS	-	-	-	-	-	57	-	-	-	-	-
Agriculture	-	-	-	-	-	-	-	-	-	-	-
Comm. and Publ. Services	-	-	-	-	-	-	-	-	-	-	-
Residential	-	-	-	-	-	57	-	-	-	-	-
Non-specified	-	-	-	-	-	-	-	-	-	-	-
NON-ENERGY USE	-	-	-	-	-	-	-	-	-	-	14

Albania / Albanie

SUPPLY AND CONSUMPTION 1997	Gas (TJ) Natural Gas	Gas Works	Coke Ovens	Blast Furnaces	Comb. Renew. & Waste (TJ) Solid Biomass	Gas/Liquids from Biomass	Municipal Waste	Industrial Waste	(GWh) Electricity	(TJ) Heat
Production	700	-	-	-	2500	-	-	-	5184	491
Imports	-	-	-	-	-	-	-	-	80	-
Exports	-	-	-	-	-	-	-	-	-540	-
Intl. Marine Bunkers	-	-	-	-	-	-	-	-	-	-
Stock Changes	-	-	-	-	-	-	-	-	-	-
DOMESTIC SUPPLY	**700**	-	-	-	**2500**	-	-	-	**4724**	**491**
Transfers and Stat. Diff.	-	-	-	-	-	-	-	-	-	-
TRANSFORMATION	-	-	-	-	-	-	-	-	-	-
Electricity and CHP Plants	-	-	-	-	-	-	-	-	-	-
Petroleum Refineries	-	-	-	-	-	-	-	-	-	-
Other Transform. Sector	-	-	-	-	-	-	-	-	-	-
ENERGY SECTOR	-	-	-	-	-	-	-	-	**39**	-
DISTRIBUTION LOSSES	-	-	-	-	-	-	-	-	**2394**	-
FINAL CONSUMPTION	**700**	-	-	-	**2500**	-	-	-	**2291**	**491**
INDUSTRY SECTOR	**583**	-	-	-	-	-	-	-	**615**	-
Iron and Steel	-	-	-	-	-	-	-	-	-	-
Chemical and Petrochem.	233	-	-	-	-	-	-	-	-	-
Non-Metallic Minerals	39	-	-	-	-	-	-	-	-	-
Non-specified	311	-	-	-	-	-	-	-	615	-
TRANSPORT SECTOR	-	-	-	-	-	-	-	-	-	-
Air	-	-	-	-	-	-	-	-	-	-
Road	-	-	-	-	-	-	-	-	-	-
Non-specified	-	-	-	-	-	-	-	-	-	-
OTHER SECTORS	**117**	-	-	-	**2500**	-	-	-	**1676**	**491**
Agriculture	-	-	-	-	-	-	-	-	43	-
Comm. and Publ. Services	-	-	-	-	-	-	-	-	33	-
Residential	78	-	-	-	2500	-	-	-	974	491
Non-specified	39	-	-	-	-	-	-	-	626	-
NON-ENERGY USE	-	-	-	-	-	-	-	-	-	-

APPROVISIONNEMENT ET DEMANDE 1998	Gaz (TJ) Gaz naturel	Usines à gaz	Cokeries	Hauts fourneaux	En. Re. Comb. & Déchets (TJ) Biomasse solide	Gaz/Liquides tirés de biomasse	Déchets urbains	Déchets industriels	(GWh) Electricité	(TJ) Chaleur
Production	661	-	-	-	2500	-	-	-	5068	509
Imports	-	-	-	-	-	-	-	-	100	-
Exports	-	-	-	-	-	-	-	-	-500	-
Intl. Marine Bunkers	-	-	-	-	-	-	-	-	-	-
Stock Changes	-	-	-	-	-	-	-	-	-	-
DOMESTIC SUPPLY	**661**	-	-	-	**2500**	-	-	-	**4668**	**509**
Transfers and Stat. Diff.	-	-	-	-	-	-	-	-	-	-
TRANSFORMATION	-	-	-	-	-	-	-	-	-	-
Electricity and CHP Plants	-	-	-	-	-	-	-	-	-	-
Petroleum Refineries	-	-	-	-	-	-	-	-	-	-
Other Transform. Sector	-	-	-	-	-	-	-	-	-	-
ENERGY SECTOR	-	-	-	-	-	-	-	-	**38**	-
DISTRIBUTION LOSSES	-	-	-	-	-	-	-	-	**2366**	-
FINAL CONSUMPTION	**661**	-	-	-	**2500**	-	-	-	**2264**	**509**
INDUSTRY SECTOR	**544**	-	-	-	-	-	-	-	**607**	-
Iron and Steel	-	-	-	-	-	-	-	-	-	-
Chemical and Petrochem.	233	-	-	-	-	-	-	-	-	-
Non-Metallic Minerals	39	-	-	-	-	-	-	-	-	-
Non-specified	272	-	-	-	-	-	-	-	607	-
TRANSPORT SECTOR	-	-	-	-	-	-	-	-	-	-
Air	-	-	-	-	-	-	-	-	-	-
Road	-	-	-	-	-	-	-	-	-	-
Non-specified	-	-	-	-	-	-	-	-	-	-
OTHER SECTORS	**117**	-	-	-	**2500**	-	-	-	**1657**	**509**
Agriculture	-	-	-	-	-	-	-	-	43	-
Comm. and Publ. Services	-	-	-	-	-	-	-	-	33	-
Residential	78	-	-	-	2500	-	-	-	963	509
Non-specified	39	-	-	-	-	-	-	-	618	-
NON-ENERGY USE	-	-	-	-	-	-	-	-	-	-

Algeria / Algérie : 1997

SUPPLY AND CONSUMPTION APPROVISIONNEMENT ET DEMANDE	Coal / Charbon (1000 tonnes)							Oil / Pétrole (1000 tonnes)			
	Coking Coal Charbon à coke	Other Bit. Coal Autres charb. bit.	Sub-Bit. Coal Charbon sous-bit.	Lignite Lignite	Peat Tourbe	Oven and Gas Coke Coke de four/gaz	Pat. Fuel and BKB Agg./briq. de lignite	Crude Oil Pétrole brut	NGL LGN	Feed-stocks Produits d'aliment.	Additives Additifs
Production	-	-	-	-	-	258	-	38194	21462	-	-
From Other Sources	-	-	-	-	-	-	-	-	-	-	-
Imports	552	-	-	-	-	-	-	263	-	-	-
Exports	-	-	-	-	-	-	-	-17280	-15828	-	-
Intl. Marine Bunkers	-	-	-	-	-	-	-	-	-	-	-
Stock Changes	-116	-	-	-	-	91	-	31	-	-	-
DOMESTIC SUPPLY	**436**	**-**	**-**	**-**	**-**	**349**	**-**	**21208**	**5634**	**-**	**-**
Transfers	-	-	-	-	-	-	-	-	-5512	-	-
Statistical Differences	-	-	-	-	-	-	-	-394	1	-	-
TRANSFORMATION	**436**	**-**	**-**	**-**	**-**	**280**	**-**	**20156**	**-**	**-**	**-**
Electricity Plants	-	-	-	-	-	-	-	-	-	-	-
CHP Plants	-	-	-	-	-	-	-	-	-	-	-
Heat Plants	-	-	-	-	-	-	-	-	-	-	-
Blast Furnaces/Gas Works	-	-	-	-	-	280	-	-	-	-	-
Coke/Pat. Fuel/BKB Plants	436	-	-	-	-	-	-	-	-	-	-
Petroleum Refineries	-	-	-	-	-	-	-	20156	-	-	-
Petrochemical Industry	-	-	-	-	-	-	-	-	-	-	-
Liquefaction	-	-	-	-	-	-	-	-	-	-	-
Other Transform. Sector	-	-	-	-	-	-	-	-	-	-	-
ENERGY SECTOR	**-**	**-**	**-**	**-**	**-**	**-**	**-**	**450**	**-**	**-**	**-**
Coal Mines	-	-	-	-	-	-	-	-	-	-	-
Oil and Gas Extraction	-	-	-	-	-	-	-	31	-	-	-
Petroleum Refineries	-	-	-	-	-	-	-	419	-	-	-
Electr., CHP+Heat Plants	-	-	-	-	-	-	-	-	-	-	-
Pumped Storage (Elec.)	-	-	-	-	-	-	-	-	-	-	-
Other Energy Sector	-	-	-	-	-	-	-	-	-	-	-
Distribution Losses	-	-	-	-	-	-	-	180	123	-	-
FINAL CONSUMPTION	**-**	**-**	**-**	**-**	**-**	**69**	**-**	**28**	**-**	**-**	**-**
INDUSTRY SECTOR	**-**	**-**	**-**	**-**	**-**	**69**	**-**	**28**	**-**	**-**	**-**
Iron and Steel	-	-	-	-	-	69	-	-	-	-	-
Chemical and Petrochem.	-	-	-	-	-	-	-	-	-	-	-
of which: Feedstocks	-	-	-	-	-	-	-	-	-	-	-
Non-Ferrous Metals	-	-	-	-	-	-	-	-	-	-	-
Non-Metallic Minerals	-	-	-	-	-	-	-	-	-	-	-
Transport Equipment	-	-	-	-	-	-	-	-	-	-	-
Machinery	-	-	-	-	-	-	-	-	-	-	-
Mining and Quarrying	-	-	-	-	-	-	-	-	-	-	-
Food and Tobacco	-	-	-	-	-	-	-	-	-	-	-
Paper, Pulp and Print	-	-	-	-	-	-	-	-	-	-	-
Wood and Wood Products	-	-	-	-	-	-	-	-	-	-	-
Construction	-	-	-	-	-	-	-	-	-	-	-
Textile and Leather	-	-	-	-	-	-	-	-	-	-	-
Non-specified	-	-	-	-	-	-	-	28	-	-	-
TRANSPORT SECTOR	**-**	**-**	**-**	**-**	**-**	**-**	**-**	**-**	**-**	**-**	**-**
Air	-	-	-	-	-	-	-	-	-	-	-
Road	-	-	-	-	-	-	-	-	-	-	-
Rail	-	-	-	-	-	-	-	-	-	-	-
Pipeline Transport	-	-	-	-	-	-	-	-	-	-	-
Internal Navigation	-	-	-	-	-	-	-	-	-	-	-
Non-specified	-	-	-	-	-	-	-	-	-	-	-
OTHER SECTORS	**-**	**-**	**-**	**-**	**-**	**-**	**-**	**-**	**-**	**-**	**-**
Agriculture	-	-	-	-	-	-	-	-	-	-	-
Comm. and Publ. Services	-	-	-	-	-	-	-	-	-	-	-
Residential	-	-	-	-	-	-	-	-	-	-	-
Non-specified	-	-	-	-	-	-	-	-	-	-	-
NON-ENERGY USE	**-**	**-**	**-**	**-**	**-**	**-**	**-**	**-**	**-**	**-**	**-**
in Industry/Trans./Energy	-	-	-	-	-	-	-	-	-	-	-
in Transport	-	-	-	-	-	-	-	-	-	-	-
in Other Sectors	-	-	-	-	-	-	-	-	-	-	-

Algeria / Algérie : 1997

SUPPLY AND CONSUMPTION / APPROVISIONNEMENT ET DEMANDE	Refinery Gas / Gaz de raffinerie	LPG + Ethane / GPL + éthane	Motor Gasoline / Essence moteur	Aviation Gasoline / Essence aviation	Jet Fuel / Carbu-réacteurs	Kerosene / Kérosène	Gas/ Diesel / Gazole	Heavy Fuel Oil / Fioul lourd	Naphtha / Naphta	Petrol. Coke / Coke de pétrole	Other Prod. / Autres prod.
Production	-	477	1989	-	1197	10	6523	5556	4043	-	281
From Other Sources	-	862	-	-	-	-	-	-	-	-	-
Imports	-	-	-	-	-	-	-	83	-	-	1
Exports	-	-5163	-102	-	-813	-	-3515	-5637	-4304	-	-
Intl. Marine Bunkers	-	-	-	-	-	-	-85	-181	-	-	-
Stock Changes	-	-	76	-	-	-	43	195	-	-	-5
DOMESTIC SUPPLY	-	-3824	1963	-	384	10	2966	16	-261	-	277
Transfers	-	5512	-	-	-	-	-	-	-	-	-
Statistical Differences	-	-54	-	-	-54	-2	-	-	261	-	-
TRANSFORMATION	-	-	-	-	-	-	-	-	-	-	-
Electricity Plants	-	-	-	-	-	-	-	-	-	-	-
CHP Plants	-	-	-	-	-	-	-	-	-	-	-
Heat Plants	-	-	-	-	-	-	-	-	-	-	-
Blast Furnaces/Gas Works	-	-	-	-	-	-	-	-	-	-	-
Coke/Pat. Fuel/BKB Plants	-	-	-	-	-	-	-	-	-	-	-
Petroleum Refineries	-	-	-	-	-	-	-	-	-	-	-
Petrochemical Industry	-	-	-	-	-	-	-	-	-	-	-
Liquefaction	-	-	-	-	-	-	-	-	-	-	-
Other Transform. Sector	-	-	-	-	-	-	-	-	.	-	-
ENERGY SECTOR	-	-	-	-	-	-	-	-	-	-	-
Coal Mines	-	-	-	-	-	-	-	-	-	-	-
Oil and Gas Extraction	-	-	-	-	-	-	-	-	-	-	-
Petroleum Refineries	-	-	-	-	-	-	-	-	-	-	-
Electr., CHP+Heat Plants	-	-	-	-	-	-	-	-	-	-	-
Pumped Storage (Elec.)	-	-	-	-	-	-	-	-	-	-	-
Other Energy Sector	-	-	-	-	-	-	-	-	-	-	-
Distribution Losses	-	-	-	-	-	-	-	-	-	-	-
FINAL CONSUMPTION	-	1634	1963	-	330	8	2966	16	-	-	277
INDUSTRY SECTOR	-	206	-	-	-	-	-	16	-	-	-
Iron and Steel	-	-	-	-	-	-	-	-	-	-	-
Chemical and Petrochem.	-	188	-	-	-	-	-	-	-	-	-
of which: Feedstocks	-	188	-	-	-	-	-	-	-	-	-
Non-Ferrous Metals	-	-	-	-	-	-	-	-	-	-	-
Non-Metallic Minerals	-	-	-	-	-	-	-	-	-	-	-
Transport Equipment	-	-	-	-	-	-	-	-	-	-	-
Machinery	-	-	-	-	-	-	-	-	-	-	-
Mining and Quarrying	-	-	-	-	-	-	-	-	-	-	-
Food and Tobacco	-	-	-	-	-	-	-	-	-	-	-
Paper, Pulp and Print	-	-	-	-	-	-	-	-	-	-	-
Wood and Wood Products	-	-	-	-	-	-	-	-	-	-	-
Construction	-	-	-	-	-	-	-	-	-	-	-
Textile and Leather	-	-	-	-	-	-	-	-	-	-	-
Non-specified	-	18	-	-	-	-	-	16	-	-	-
TRANSPORT SECTOR	-	106	1963	-	330	-	-	-	-	-	-
Air	-	-	-	-	330	-	-	-	-	-	-
Road	-	106	1963	-	-	-	-	-	-	-	-
Rail	-	-	-	-	-	-	-	-	-	-	-
Pipeline Transport	-	-	-	-	-	-	-	-	-	-	-
Internal Navigation	-	-	-	-	-	-	-	-	-	-	-
Non-specified	-	-	-	-	-	-	-	-	-	-	-
OTHER SECTORS	-	1322	-	-	-	8	2966	-	-	-	-
Agriculture	-	-	-	-	-	-	-	-	-	-	-
Comm. and Publ. Services	-	-	-	-	-	-	-	-	-	-	-
Residential	-	1322	-	-	-	8	-	-	-	-	-
Non-specified	-	-	-	-	-	-	2966	-	-	-	-
NON-ENERGY USE	-	-	-	-	-	-	-	-	-	-	277
in Industry/Transf./Energy	-	-	-	-	-	-	-	-	-	-	277
in Transport	-	-	-	-	-	-	-	-	-	-	-
in Other Sectors	-	-	-	-	-	-	-	-	-	-	-

Header span: Oil cont. / *Pétrole cont.* (1000 tonnes)

Algeria / Algérie : 1997

SUPPLY AND CONSUMPTION	Gas / Gaz (TJ)				Comb. Renew. & Waste / En. Re. Comb. & Déchets (TJ)				(GWh)	(TJ)
	Natural Gas	Gas Works	Coke Ovens	Blast Furnaces	Solid Biomass	Gas/Liquids from Biomass	Municipal Waste	Industrial Waste	Electricity	Heat
APPROVISIONNEMENT ET DEMANDE	Gaz naturel	Usines à gaz	Cokeries	Hauts fourneaux	Biomasse solide	Gaz/Liquides tirés de biomasse	Déchets urbains	Déchets industriels	Electricité	Chaleur
Production	3007105	-	2076	3742	3014	-	-	-	21422	-
From Other Sources	-	-	-	-	-	-	-	-	-	-
Imports	-	-	-	-	-	-	-	-	312	-
Exports	-2162030	-	-	-	-	-	-	-	-313	-
Intl. Marine Bunkers	-	-	-	-	-	-	-	-	-	-
Stock Changes	-	-	-	-	-	-	-	-	-	-
DOMESTIC SUPPLY	845075	-	2076	3742	3014	-	-	-	21421	-
Transfers	-	-	-	-	-	-	-	-	-	-
Statistical Differences	-3043	-	-1	-	-	-	-	-	-11	-
TRANSFORMATION	332270	-	1126	-	-	-	-	-	-	-
Electricity Plants	274471	-	-	-	-	-	-	-	-	-
CHP Plants	-	-	-	-	-	-	-	-	-	-
Heat Plants	-	-	-	-	-	-	-	-	-	-
Blast Furnaces/Gas Works	-	-	1126	-	-	-	-	-	-	-
Coke/Pat. Fuel/BKB Plants	-	-	-	-	-	-	-	-	-	-
Petroleum Refineries	-	-	-	-	-	-	-	-	-	-
Petrochemical Industry	-	-	-	-	-	-	-	-	-	-
Liquefaction	57799	-	-	-	-	-	-	-	-	-
Other Transform. Sector	-	-	-	-	-	-	-	-	-	-
ENERGY SECTOR	267453	-	-	-	-	-	-	-	1926	-
Coal Mines	-	-	-	-	-	-	-	-	-	-
Oil and Gas Extraction	236613	-	-	-	-	-	-	-	297	-
Petroleum Refineries	23828	-	-	-	-	-	-	-	191	-
Electr., CHP+Heat Plants	-	-	-	-	-	-	-	-	1438	-
Pumped Storage (Elec.)	-	-	-	-	-	-	-	-	-	-
Other Energy Sector	7012	-	-	-	-	-	-	-	-	-
Distribution Losses	8128	-	668	2786	-	-	-	-	4374	-
FINAL CONSUMPTION	234181	-	281	956	3014	-	-	-	15110	-
INDUSTRY SECTOR	116419	-	281	956	-	-	-	-	6302	-
Iron and Steel	10670	-	281	956	-	-	-	-	529	-
Chemical and Petrochem.	53048	-	-	-	-	-	-	-	771	-
of which: Feedstocks	51748	-	-	-	-	-	-	-	-	-
Non-Ferrous Metals	-	-	-	-	-	-	-	-	-	-
Non-Metallic Minerals	-	-	-	-	-	-	-	-	-	-
Transport Equipment	-	-	-	-	-	-	-	-	-	-
Machinery	-	-	-	-	-	-	-	-	-	-
Mining and Quarrying	-	-	-	-	-	-	-	-	-	-
Food and Tobacco	-	-	-	-	-	-	-	-	-	-
Paper, Pulp and Print	-	-	-	-	-	-	-	-	-	-
Wood and Wood Products	-	-	-	-	-	-	-	-	-	-
Construction	43900	-	-	-	-	-	-	-	1383	-
Textile and Leather	-	-	-	-	-	-	-	-	-	-
Non-specified	8801	-	-	-	-	-	-	-	3619	-
TRANSPORT SECTOR	31357	-	-	-	-	-	-	-	267	-
Air	-	-	-	-	-	-	-	-	-	-
Road	-	-	-	-	-	-	-	-	-	-
Rail	-	-	-	-	-	-	-	-	182	-
Pipeline Transport	31357	-	-	-	-	-	-	-	85	-
Internal Navigation	-	-	-	-	-	-	-	-	-	-
Non-specified	-	-	-	-	-	-	-	-	-	-
OTHER SECTORS	86405	-	-	-	3014	-	-	-	8541	-
Agriculture	-	-	-	-	-	-	-	-	-	-
Comm. and Publ. Services	-	-	-	-	-	-	-	-	-	-
Residential	86405	-	-	-	3014	-	-	-	8541	-
Non-specified	-	-	-	-	-	-	-	-	-	-
NON-ENERGY USE	-	-	-	-	-	-	-	-	-	-
in Industry/Transf./Energy	-	-	-	-	-	-	-	-	-	-
in Transport	-	-	-	-	-	-	-	-	-	-
in Other Sectors	-	-	-	-	-	-	-	-	-	-

Algeria / Algérie : 1998

SUPPLY AND CONSUMPTION / APPROVISIONNEMENT ET DEMANDE	Coal / Charbon (1000 tonnes)							Oil / Pétrole (1000 tonnes)			
	Coking Coal / Charbon à coke	Other Bit. Coal / Autres charb. bit.	Sub-Bit. Coal / Charbon sous-bit.	Lignite / Lignite	Peat / Tourbe	Oven and Gas Coke / Coke de four/gaz	Pat. Fuel and BKB / Agg./briq. de lignite	Crude Oil / Pétrole brut	NGL / LGN	Feed-stocks / Produits d'aliment.	Additives / Additifs
Production	-	-	-	-	-	380	-	39103	22062	-	-
From Other Sources	-	-	-	-	-	-	-	-	-	-	-
Imports	628	-	-	-	-	-	-	232	-	-	-
Exports	-	-	-	-	-	-	-	-19294	-15341	-	-
Intl. Marine Bunkers	-	-	-	-	-	-	-	-	-	-	-
Stock Changes	14	-	-	-	-	105	-	-	-	-	-
DOMESTIC SUPPLY	**642**	**-**	**-**	**-**	**-**	**485**	**-**	**20041**	**6721**	**-**	**-**
Transfers	-	-	-	-	-	-	-	-	-6441	-	-
Statistical Differences	-	-	-	-	-	1	-	-381	-	-	-
TRANSFORMATION	**642**	**-**	**-**	**-**	**-**	**385**	**-**	**18946**	**-**	**-**	**-**
Electricity Plants	-	-	-	-	-	-	-	-	-	-	-
CHP Plants	-	-	-	-	-	-	-	-	-	-	-
Heat Plants	-	-	-	-	-	-	-	-	-	-	-
Blast Furnaces/Gas Works	-	-	-	-	-	385	-	-	-	-	-
Coke/Pat. Fuel/BKB Plants	642	-	-	-	-	-	-	-	-	-	-
Petroleum Refineries	-	-	-	-	-	-	-	18946	-	-	-
Petrochemical Industry	-	-	-	-	-	-	-	-	-	-	-
Liquefaction	-	-	-	-	-	-	-	-	-	-	-
Other Transform. Sector	-	-	-	-	-	-	-	-	-	-	-
ENERGY SECTOR	**-**	**-**	**-**	**-**	**-**	**-**	**-**	**265**	**-**	**-**	**-**
Coal Mines	-	-	-	-	-	-	-	-	-	-	-
Oil and Gas Extraction	-	-	-	-	-	-	-	26	-	-	-
Petroleum Refineries	-	-	-	-	-	-	-	239	-	-	-
Electr., CHP+Heat Plants	-	-	-	-	-	-	-	-	-	-	-
Pumped Storage (Elec.)	-	-	-	-	-	-	-	-	-	-	-
Other Energy Sector	-	-	-	-	-	-	-	-	-	-	-
Distribution Losses	-	-	-	-	-	-	-	421	280	-	-
FINAL CONSUMPTION	**-**	**-**	**-**	**-**	**-**	**101**	**-**	**28**	**-**	**-**	**-**
INDUSTRY SECTOR	**-**	**-**	**-**	**-**	**-**	**101**	**-**	**28**	**-**	**-**	**-**
Iron and Steel	-	-	-	-	-	101	-	-	-	-	-
Chemical and Petrochem.	-	-	-	-	-	-	-	-	-	-	-
of which: Feedstocks	-	-	-	-	-	-	-	-	-	-	-
Non-Ferrous Metals	-	-	-	-	-	-	-	-	-	-	-
Non-Metallic Minerals	-	-	-	-	-	-	-	-	-	-	-
Transport Equipment	-	-	-	-	-	-	-	-	-	-	-
Machinery	-	-	-	-	-	-	-	-	-	-	-
Mining and Quarrying	-	-	-	-	-	-	-	-	-	-	-
Food and Tobacco	-	-	-	-	-	-	-	-	-	-	-
Paper, Pulp and Print	-	-	-	-	-	-	-	-	-	-	-
Wood and Wood Products	-	-	-	-	-	-	-	-	-	-	-
Construction	-	-	-	-	-	-	-	-	-	-	-
Textile and Leather	-	-	-	-	-	-	-	-	-	-	-
Non-specified	-	-	-	-	-	-	-	28	-	-	-
TRANSPORT SECTOR	**-**	**-**	**-**	**-**	**-**	**-**	**-**	**-**	**-**	**-**	**-**
Air	-	-	-	-	-	-	-	-	-	-	-
Road	-	-	-	-	-	-	-	-	-	-	-
Rail	-	-	-	-	-	-	-	-	-	-	-
Pipeline Transport	-	-	-	-	-	-	-	-	-	-	-
Internal Navigation	-	-	-	-	-	-	-	-	-	-	-
Non-specified	-	-	-	-	-	-	-	-	-	-	-
OTHER SECTORS	**-**	**-**	**-**	**-**	**-**	**-**	**-**	**-**	**-**	**-**	**-**
Agriculture	-	-	-	-	-	-	-	-	-	-	-
Comm. and Publ. Services	-	-	-	-	-	-	-	-	-	-	-
Residential	-	-	-	-	-	-	-	-	-	-	-
Non-specified	-	-	-	-	-	-	-	-	-	-	-
NON-ENERGY USE	**-**	**-**	**-**	**-**	**-**	**-**	**-**	**-**	**-**	**-**	**-**
in Industry/Trans./Energy	-	-	-	-	-	-	-	-	-	-	-
in Transport	-	-	-	-	-	-	-	-	-	-	-
in Other Sectors	-	-	-	-	-	-	-	-	-	-	-

Algeria / Algérie : 1998

SUPPLY AND CONSUMPTION	Oil cont. / Pétrole cont. (1000 tonnes)										
	Refinery Gas	LPG + Ethane	Motor Gasoline	Aviation Gasoline	Jet Fuel	Kerosene	Gas/ Diesel	Heavy Fuel Oil	Naphtha	Petrol. Coke	Other Prod.
APPROVISIONNEMENT ET DEMANDE	Gaz de raffinerie	GPL + éthane	Essence moteur	Essence aviation	Carbu- réacteurs	Kérosène	Gazole	Fioul lourd	Naphta	Coke de pétrole	Autres prod.
Production	-	466	2037	-	970	9	6198	5061	3855	-	275
From Other Sources	-	730	-	-	-	-	-	-	-	-	-
Imports	-	-	-	-	-	-	-	100	-	-	19
Exports	-	-6025	-35	-	-706	-	-3006	-4859	-4170	-	-
Intl. Marine Bunkers	-	-	-	-	-	-	-76	-163	-	-	-
Stock Changes	-	-	-69	-	-	-	114	-122	-	-	-19
DOMESTIC SUPPLY	-	-4829	1933	-	264	9	3230	17	-315	-	275
Transfers	-	6441	-	-	-	-	-	-	-	-	-
Statistical Differences	-	-1	-	-	56	-	-	-	315	-	-
TRANSFORMATION	-	-	-	-	-	-	342	-	-	-	-
Electricity Plants	-	-	-	-	-	-	342	-	-	-	-
CHP Plants	-	-	-	-	-	-	-	-	-	-	-
Heat Plants	-	-	-	-	-	-	-	-	-	-	-
Blast Furnaces/Gas Works	-	-	-	-	-	-	-	-	-	-	-
Coke/Pat. Fuel/BKB Plants	-	-	-	-	-	-	-	-	-	-	-
Petroleum Refineries	-	-	-	-	-	-	-	-	-	-	-
Petrochemical Industry	-	-	-	-	-	-	-	-	-	-	-
Liquefaction	-	-	-	-	-	-	-	-	-	-	-
Other Transform. Sector	-	-	-	-	-	-	-	-	-	-	-
ENERGY SECTOR	-	-	-	-	-	-	-	-	-	-	-
Coal Mines	-	-	-	-	-	-	-	-	-	-	-
Oil and Gas Extraction	-	-	-	-	-	-	-	-	-	-	-
Petroleum Refineries	-	-	-	-	-	-	-	-	-	-	-
Electr., CHP+Heat Plants	-	-	-	-	-	-	-	-	-	-	-
Pumped Storage (Elec.)	-	-	-	-	-	-	-	-	-	-	-
Other Energy Sector	-	-	-	-	-	-	-	-	-	-	-
Distribution Losses	-	-	-	-	-	-	-	-	-	-	-
FINAL CONSUMPTION	-	1611	1933	-	320	9	2888	17	-	-	275
INDUSTRY SECTOR	-	101	-	-	-	-	-	17	-	-	-
Iron and Steel	-	-	-	-	-	-	-	-	-	-	-
Chemical and Petrochem.	-	82	-	-	-	-	-	-	-	-	-
of which: Feedstocks	-	82	-	-	-	-	-	-	-	-	-
Non-Ferrous Metals	-	-	-	-	-	-	-	-	-	-	-
Non-Metallic Minerals	-	-	-	-	-	-	-	-	-	-	-
Transport Equipment	-	-	-	-	-	-	-	-	-	-	-
Machinery	-	-	-	-	-	-	-	-	-	-	-
Mining and Quarrying	-	-	-	-	-	-	-	-	-	-	-
Food and Tobacco	-	-	-	-	-	-	-	-	-	-	-
Paper, Pulp and Print	-	-	-	-	-	-	-	-	-	-	-
Wood and Wood Products	-	-	-	-	-	-	-	-	-	-	-
Construction	-	-	-	-	-	-	-	-	-	-	-
Textile and Leather	-	-	-	-	-	-	-	-	-	-	-
Non-specified	-	19	-	-	-	-	-	17	-	-	-
TRANSPORT SECTOR	-	112	1933	-	320	-	-	-	-	-	-
Air	-	-	-	-	320	-	-	-	-	-	-
Road	-	112	1933	-	-	-	-	-	-	-	-
Rail	-	-	-	-	-	-	-	-	-	-	-
Pipeline Transport	-	-	-	-	-	-	-	-	-	-	-
Internal Navigation	-	-	-	-	-	-	-	-	-	-	-
Non-specified	-	-	-	-	-	-	-	-	-	-	-
OTHER SECTORS	-	1398	-	-	-	9	2888	-	-	-	-
Agriculture	-	-	-	-	-	-	-	-	-	-	-
Comm. and Publ. Services	-	-	-	-	-	-	-	-	-	-	-
Residential	-	1398	-	-	-	9	-	-	-	-	-
Non-specified	-	-	-	-	-	-	2888	-	-	-	-
NON-ENERGY USE	-	-	-	-	-	-	-	-	-	-	275
in Industry/Transf./Energy	-	-	-	-	-	-	-	-	-	-	275
in Transport	-	-	-	-	-	-	-	-	-	-	-
in Other Sectors	-	-	-	-	-	-	-	-	-	-	-

Algeria / Algérie : 1998

SUPPLY AND CONSUMPTION *APPROVISIONNEMENT ET DEMANDE*	Gas / *Gaz* (TJ)				Comb. Renew. & Waste / *En. Re. Comb. & Déchets* (TJ)				(GWh)	(TJ)
	Natural Gas *Gaz naturel*	Gas Works *Usines à gaz*	Coke Ovens *Cokeries*	Blast Furnaces *Hauts fourneaux*	Solid Biomass *Biomasse solide*	Gas/Liquids from Biomass *Gaz/Liquides tirés de biomasse*	Municipal Waste *Déchets urbains*	Industrial Waste *Déchets industriels*	Electricity *Electricité*	Heat *Chaleur*
Production	3207127	-	3448	5040	4271	-	-	-	23615	-
From Other Sources	-	-	-	-	-	-	-	-	-	-
Imports	-	-	-	-	-	-	-	-	255	-
Exports	-2342805	-	-	-	-	-	-	-	-262	-
Intl. Marine Bunkers	-	-	-	-	-	-	-	-	-	-
Stock Changes	-	-	-	-	-	-	-	-	-	-
DOMESTIC SUPPLY	**864322**	**-**	**3448**	**5040**	**4271**	**-**	**-**	**-**	**23608**	**-**
Transfers	-	-	-	-	-	-	-	-	-	-
Statistical Differences	-3445	-	-	-	-	-	-	-	-342	-
TRANSFORMATION	**356980**	**-**	**1323**	**-**	**-**	**-**	**-**	**-**	**-**	**-**
Electricity Plants	308020	-	-	-	-	-	-	-	-	-
CHP Plants	-	-	-	-	-	-	-	-	-	-
Heat Plants	-	-	-	-	-	-	-	-	-	-
Blast Furnaces/Gas Works	-	-	1323	-	-	-	-	-	-	-
Coke/Pat. Fuel/BKB Plants	-	-	-	-	-	-	-	-	-	-
Petroleum Refineries	-	-	-	-	-	-	-	-	-	-
Petrochemical Industry	-	-	-	-	-	-	-	-	-	-
Liquefaction	48960	-	-	-	-	-	-	-	-	-
Other Transform. Sector	-	-	-	-	-	-	-	-	-	-
ENERGY SECTOR	**265295**	**-**	**-**	**-**	**-**	**-**	**-**	**-**	**2092**	**-**
Coal Mines	-	-	-	-	-	-	-	-	-	-
Oil and Gas Extraction	233869	-	-	-	-	-	-	-	270	-
Petroleum Refineries	20912	-	-	-	-	-	-	-	178	-
Electr., CHP+Heat Plants	-	-	-	-	-	-	-	-	1644	-
Pumped Storage (Elec.)	-	-	-	-	-	-	-	-	-	-
Other Energy Sector	10514	-	-	-	-	-	-	-	-	-
Distribution Losses	9451	-	1794	3752	-	-	-	-	4562	-
FINAL CONSUMPTION	**229151**	**-**	**331**	**1288**	**4271**	**-**	**-**	**-**	**16612**	**-**
INDUSTRY SECTOR	**116938**	**-**	**331**	**1288**	**-**	**-**	**-**	**-**	**6800**	**-**
Iron and Steel	10334	-	331	1288	-	-	-	-	749	-
Chemical and Petrochem.	55762	-	-	-	-	-	-	-	886	-
of which: Feedstocks	*51330*	-	-	-	-	-	-	-	-	-
Non-Ferrous Metals	-	-	-	-	-	-	-	-	-	-
Non-Metallic Minerals	-	-	-	-	-	-	-	-	-	-
Transport Equipment	-	-	-	-	-	-	-	-	-	-
Machinery	-	-	-	-	-	-	-	-	-	-
Mining and Quarrying	-	-	-	-	-	-	-	-	-	-
Food and Tobacco	-	-	-	-	-	-	-	-	-	-
Paper, Pulp and Print	-	-	-	-	-	-	-	-	-	-
Wood and Wood Products	-	-	-	-	-	-	-	-	-	-
Construction	43586	-	-	-	-	-	-	-	1512	-
Textile and Leather	-	-	-	-	-	-	-	-	-	-
Non-specified	7256	-	-	-	-	-	-	-	3653	-
TRANSPORT SECTOR	**16356**	**-**	**-**	**-**	**-**	**-**	**-**	**-**	**261**	**-**
Air	-	-	-	-	-	-	-	-	-	-
Road	-	-	-	-	-	-	-	-	-	-
Rail	-	-	-	-	-	-	-	-	178	-
Pipeline Transport	16356	-	-	-	-	-	-	-	83	-
Internal Navigation	-	-	-	-	-	-	-	-	-	-
Non-specified	-	-	-	-	-	-	-	-	-	-
OTHER SECTORS	**95857**	**-**	**-**	**-**	**4271**	**-**	**-**	**-**	**9551**	**-**
Agriculture	-	-	-	-	-	-	-	-	-	-
Comm. and Publ. Services	-	-	-	-	-	-	-	-	-	-
Residential	95857	-	-	-	4271	-	-	-	9551	-
Non-specified	-	-	-	-	-	-	-	-	-	-
NON-ENERGY USE	**-**	**-**	**-**	**-**	**-**	**-**	**-**	**-**	**-**	**-**
in Industry/Transf./Energy	-	-	-	-	-	-	-	-	-	-
in Transport	-	-	-	-	-	-	-	-	-	-
in Other Sectors	-	-	-	-	-	-	-	-	-	-

Angola

SUPPLY AND CONSUMPTION 1997	Coal (1000 tonnes)							Oil (1000 tonnes)			
	Coking Coal	Other Bit. Coal	Sub-Bit. Coal	Lignite	Peat	Oven and Gas Coke	Pat. Fuel and BKB	Crude Oil	NGL	Feed-stocks	Additives
Production	-	-	-	-	-	-	-	35149	-	-	-
Imports	-	-	-	-	-	-	-	-	-	-	-
Exports	-	-	-	-	-	-	-	-32723	-	-	-
Intl. Marine Bunkers	-	-	-	-	-	-	-	-	-	-	-
Stock Changes	-	-	-	-	-	-	-	-448	-	-	-
DOMESTIC SUPPLY	-	-	-	-	-	-	-	1978	-	-	-
Transfers and Stat. Diff.	-	-	-	-	-	-	-	2	-	-	-
TRANSFORMATION	-	-	-	-	-	-	-	1918	-	-	-
Electricity and CHP Plants	-	-	-	-	-	-	-	-	-	-	-
Petroleum Refineries	-	-	-	-	-	-	-	1918	-	-	-
Other Transform. Sector	-	-	-	-	-	-	-	-	-	-	-
ENERGY SECTOR	-	-	-	-	-	-	-	62	-	-	-
DISTRIBUTION LOSSES	-	-	-	-	-	-	-	-	-	-	-
FINAL CONSUMPTION	-	-	-	-	-	-	-	-	-	-	-
INDUSTRY SECTOR	-	-	-	-	-	-	-	-	-	-	-
Iron and Steel	-	-	-	-	-	-	-	-	-	-	-
Chemical and Petrochem.	-	-	-	-	-	-	-	-	-	-	-
Non-Metallic Minerals	-	-	-	-	-	-	-	-	-	-	-
Non-specified	-	-	-	-	-	-	-	-	-	-	-
TRANSPORT SECTOR	-	-	-	-	-	-	-	-	-	-	-
Air	-	-	-	-	-	-	-	-	-	-	-
Road	-	-	-	-	-	-	-	-	-	-	-
Non-specified	-	-	-	-	-	-	-	-	-	-	-
OTHER SECTORS	-	-	-	-	-	-	-	-	-	-	-
Agriculture	-	-	-	-	-	-	-	-	-	-	-
Comm. and Publ. Services	-	-	-	-	-	-	-	-	-	-	-
Residential	-	-	-	-	-	-	-	-	-	-	-
Non-specified	-	-	-	-	-	-	-	-	-	-	-
NON-ENERGY USE	-	-	-	-	-	-	-	-	-	-	-

APPROVISIONNEMENT ET DEMANDE 1998	Charbon (1000 tonnes)							Pétrole (1000 tonnes)			
	Charbon à coke	Autres charb. bit.	Charbon sous-bit.	Lignite	Tourbe	Coke de four/gaz	Agg./briq. de lignite	Pétrole brut	LGN	Produits d'aliment.	Additifs
Production	-	-	-	-	-	-	-	36382	-	-	-
Imports	-	-	-	-	-	-	-	-	-	-	-
Exports	-	-	-	-	-	-	-	-34386	-	-	-
Intl. Marine Bunkers	-	-	-	-	-	-	-	-	-	-	-
Stock Changes	-	-	-	-	-	-	-	-153	-	-	-
DOMESTIC SUPPLY	-	-	-	-	-	-	-	1843	-	-	-
Transfers and Stat. Diff.	-	-	-	-	-	-	-	-1	-	-	-
TRANSFORMATION	-	-	-	-	-	-	-	1785	-	-	-
Electricity and CHP Plants	-	-	-	-	-	-	-	-	-	-	-
Petroleum Refineries	-	-	-	-	-	-	-	1785	-	-	-
Other Transform. Sector	-	-	-	-	-	-	-	-	-	-	-
ENERGY SECTOR	-	-	-	-	-	-	-	57	-	-	-
DISTRIBUTION LOSSES	-	-	-	-	-	-	-	-	-	-	-
FINAL CONSUMPTION	-	-	-	-	-	-	-	-	-	-	-
INDUSTRY SECTOR	-	-	-	-	-	-	-	-	-	-	-
Iron and Steel	-	-	-	-	-	-	-	-	-	-	-
Chemical and Petrochem.	-	-	-	-	-	-	-	-	-	-	-
Non-Metallic Minerals	-	-	-	-	-	-	-	-	-	-	-
Non-specified	-	-	-	-	-	-	-	-	-	-	-
TRANSPORT SECTOR	-	-	-	-	-	-	-	-	-	-	-
Air	-	-	-	-	-	-	-	-	-	-	-
Road	-	-	-	-	-	-	-	-	-	-	-
Non-specified	-	-	-	-	-	-	-	-	-	-	-
OTHER SECTORS	-	-	-	-	-	-	-	-	-	-	-
Agriculture	-	-	-	-	-	-	-	-	-	-	-
Comm. and Publ. Services	-	-	-	-	-	-	-	-	-	-	-
Residential	-	-	-	-	-	-	-	-	-	-	-
Non-specified	-	-	-	-	-	-	-	-	-	-	-
NON-ENERGY USE	-	-	-	-	-	-	-	-	-	-	-

Angola

SUPPLY AND CONSUMPTION 1997	Oil cont. (1000 tonnes)										
	Refinery Gas	LPG + Ethane	Motor Gasoline	Aviation Gasoline	Jet Fuel	Kerosene	Gas/ Diesel	Heavy Fuel Oil	Naphtha	Petrol. Coke	Other Prod.
Production	42	35	97	-	306	24	486	699	170	-	5
Imports	-	22	37	-	24	18	147	-	-	-	11
Exports	-	-	-	-	-	-	-	-632	-163	-	-
Intl. Marine Bunkers	-	-	-	-	-	-	-	-2	-	-	-
Stock Changes	-	-	-	-	-	-	-	-	-	-	-
DOMESTIC SUPPLY	42	57	134	-	330	42	633	65	7	-	16
Transfers and Stat. Diff.	-	-1	-	-	1	-1	-	-	-	-	-
TRANSFORMATION	-	-	-	-	-	-	12	51	-	-	-
Electricity and CHP Plants	-	-	-	-	-	-	12	51	-	-	-
Petroleum Refineries	-	-	-	-	-	-	-	-	-	-	-
Other Transform. Sector	-	-	-	-	-	-	-	-	-	-	-
ENERGY SECTOR	42	-	-	-	-	-	-	-	-	-	-
DISTRIBUTION LOSSES	-	-	-	-	-	-	-	-	-	-	-
FINAL CONSUMPTION	-	56	134	-	331	41	621	14	7	-	16
INDUSTRY SECTOR	-	2	-	-	-	41	187	14	7	-	-
Iron and Steel	-	-	-	-	-	-	-	-	-	-	-
Chemical and Petrochem.	-	-	-	-	-	-	-	-	7	-	-
Non-Metallic Minerals	-	-	-	-	-	-	-	-	-	-	-
Non-specified	-	2	-	-	-	41	187	14	-	-	-
TRANSPORT SECTOR	-	-	134	-	331	-	206	-	-	-	-
Air	-	-	-	-	331	-	-	-	-	-	-
Road	-	-	134	-	-	-	206	-	-	-	-
Non-specified	-	-	-	-	-	-	-	-	-	-	-
OTHER SECTORS	-	54	-	-	-	-	228	-	-	-	-
Agriculture	-	-	-	-	-	-	1	-	-	-	-
Comm. and Publ. Services	-	8	-	-	-	-	164	-	-	-	-
Residential	-	46	-	-	-	-	63	-	-	-	-
Non-specified	-	-	-	-	-	-	-	-	-	-	-
NON-ENERGY USE	-	-	-	-	-	-	-	-	-	-	16

APPROVISIONNEMENT ET DEMANDE 1998	Pétrole cont. (1000 tonnes)										
	Gaz de raffinerie	GPL + éthane	Essence moteur	Essence aviation	Carbu- réacteurs	Kérosène	Gazole	Fioul lourd	Naphta	Coke de pétrole	Autres prod.
Production	39	29	102	-	285	28	474	644	137	-	5
Imports	-	29	3	-	-	-	16	-	-	-	102
Exports	-	-	-	-	-71	-	-	-582	-130	-	-
Intl. Marine Bunkers	-	-	-	-	-	-	-	-	-	-	-
Stock Changes	-	-	-	-	-	-	-	-	-	-	-
DOMESTIC SUPPLY	39	58	105	-	214	28	490	62	7	-	107
Transfers and Stat. Diff.	-	-2	-	-	-1	1	-	-	-	-	-
TRANSFORMATION	-	-	-	-	-	-	11	51	-	-	-
Electricity and CHP Plants	-	-	-	-	-	-	11	51	-	-	-
Petroleum Refineries	-	-	-	-	-	-	-	-	-	-	-
Other Transform. Sector	-	-	-	-	-	-	-	-	-	-	-
ENERGY SECTOR	39	-	-	-	-	-	-	-	-	-	-
DISTRIBUTION LOSSES	-	-	-	-	-	-	-	-	-	-	-
FINAL CONSUMPTION	-	56	105	-	213	29	479	11	7	-	107
INDUSTRY SECTOR	-	2	-	-	-	29	53	11	7	-	-
Iron and Steel	-	-	-	-	-	-	-	-	-	-	-
Chemical and Petrochem.	-	-	-	-	-	-	-	-	7	-	-
Non-Metallic Minerals	-	-	-	-	-	-	-	-	-	-	-
Non-specified	-	2	-	-	-	29	53	11	-	-	-
TRANSPORT SECTOR	-	-	105	-	213	-	170	-	-	-	-
Air	-	-	-	-	213	-	-	-	-	-	-
Road	-	-	105	-	-	-	170	-	-	-	-
Non-specified	-	-	-	-	-	-	-	-	-	-	-
OTHER SECTORS	-	54	-	-	-	-	256	-	-	-	-
Agriculture	-	-	-	-	-	-	1	-	-	-	-
Comm. and Publ. Services	-	8	-	-	-	-	210	-	-	-	-
Residential	-	46	-	-	-	-	45	-	-	-	-
Non-specified	-	-	-	-	-	-	-	-	-	-	-
NON-ENERGY USE	-	-	-	-	-	-	-	-	-	-	107

Angola

SUPPLY AND CONSUMPTION 1997	Gas (TJ)				Comb. Renew. & Waste (TJ)				(GWh)	(TJ)
	Natural Gas	Gas Works	Coke Ovens	Blast Furnaces	Solid Biomass	Gas/Liquids from Biomass	Municipal Waste	Industrial Waste	Electricity	Heat
Production	21660	-	-	-	216825	-	-	-	1105	-
Imports	-	-	-	-	-	-	-	-	-	-
Exports	-	-	-	-	-	-	-	-	-	-
Intl. Marine Bunkers	-	-	-	-	-	-	-	-	-	-
Stock Changes	-	-	-	-	-	-	-	-	-	-
DOMESTIC SUPPLY	21660	-	-	-	216825	-	-	-	1105	-
Transfers and Stat. Diff.	-	-	-	-	-	-	-	-	-	-
TRANSFORMATION	-	-	-	-	57144	-	-	-	-	-
Electricity and CHP Plants	-	-	-	-	-	-	-	-	-	-
Petroleum Refineries	-	-	-	-	-	-	-	-	-	-
Other Transform. Sector	-	-	-	-	57144	-	-	-	-	-
ENERGY SECTOR	-	-	-	-	-	-	-	-	48	-
DISTRIBUTION LOSSES	-	-	-	-	-	-	-	-	314	-
FINAL CONSUMPTION	21660	-	-	-	159681	-	-	-	743	-
INDUSTRY SECTOR	21660	-	-	-	3738	-	-	-	232	-
Iron and Steel	-	-	-	-	-	-	-	-	-	-
Chemical and Petrochem.	-	-	-	-	-	-	-	-	-	-
Non-Metallic Minerals	-	-	-	-	-	-	-	-	-	-
Non-specified	21660	-	-	-	3738	-	-	-	232	-
TRANSPORT SECTOR	-	-	-	-	-	-	-	-	-	-
Air	-	-	-	-	-	-	-	-	-	-
Road	-	-	-	-	-	-	-	-	-	-
Non-specified	-	-	-	-	-	-	-	-	-	-
OTHER SECTORS	-	-	-	-	155943	-	-	-	511	-
Agriculture	-	-	-	-	-	-	-	-	-	-
Comm. and Publ. Services	-	-	-	-	-	-	-	-	-	-
Residential	-	-	-	-	155943	-	-	-	511	-
Non-specified	-	-	-	-	-	-	-	-	-	-
NON-ENERGY USE	-	-	-	-	-	-	-	-	-	-

APPROVISIONNEMENT ET DEMANDE 1998	Gaz (TJ)				En. Re. Comb. & Déchets (TJ)				(GWh)	(TJ)
	Gaz naturel	Usines à gaz	Cokeries	Hauts fourneaux	Biomasse solide	Gaz/Liquides tirés de biomasse	Déchets urbains	Déchets industriels	Electricité	Chaleur
Production	22040	-	-	-	223330	-	-	-	1063	-
Imports	-	-	-	-	-	-	-	-	-	-
Exports	-	-	-	-	-	-	-	-	-	-
Intl. Marine Bunkers	-	-	-	-	-	-	-	-	-	-
Stock Changes	-	-	-	-	-	-	-	-	-	-
DOMESTIC SUPPLY	22040	-	-	-	223330	-	-	-	1063	-
Transfers and Stat. Diff.	-	-	-	-	-	-	-	-	-	-
TRANSFORMATION	-	-	-	-	58858	-	-	-	-	-
Electricity and CHP Plants	-	-	-	-	-	-	-	-	-	-
Petroleum Refineries	-	-	-	-	-	-	-	-	-	-
Other Transform. Sector	-	-	-	-	58858	-	-	-	-	-
ENERGY SECTOR	-	-	-	-	-	-	-	-	46	-
DISTRIBUTION LOSSES	-	-	-	-	-	-	-	-	302	-
FINAL CONSUMPTION	22040	-	-	-	164471	-	-	-	715	-
INDUSTRY SECTOR	22040	-	-	-	3850	-	-	-	223	-
Iron and Steel	-	-	-	-	-	-	-	-	-	-
Chemical and Petrochem.	-	-	-	-	-	-	-	-	-	-
Non-Metallic Minerals	-	-	-	-	-	-	-	-	-	-
Non-specified	22040	-	-	-	3850	-	-	-	223	-
TRANSPORT SECTOR	-	-	-	-	-	-	-	-	-	-
Air	-	-	-	-	-	-	-	-	-	-
Road	-	-	-	-	-	-	-	-	-	-
Non-specified	-	-	-	-	-	-	-	-	-	-
OTHER SECTORS	-	-	-	-	160621	-	-	-	492	-
Agriculture	-	-	-	-	-	-	-	-	-	-
Comm. and Publ. Services	-	-	-	-	-	-	-	-	-	-
Residential	-	-	-	-	160621	-	-	-	492	-
Non-specified	-	-	-	-	-	-	-	-	-	-
NON-ENERGY USE	-	-	-	-	-	-	-	-	-	-

Argentina / Argentine : 1997

SUPPLY AND CONSUMPTION APPROVISIONNEMENT ET DEMANDE	Coal / *Charbon* (1000 tonnes)							Oil / *Pétrole* (1000 tonnes)			
	Coking Coal *Charbon à coke*	Other Bit. Coal *Autres charb. bit.*	Sub-Bit. Coal *Charbon sous-bit.*	Lignite *Lignite*	Peat *Tourbe*	Oven and Gas Coke *Coke de four/gaz*	Pat. Fuel and BKB *Agg./briq. de lignite*	Crude Oil *Pétrole brut*	NGL *LGN*	Feed-stocks *Produits d'aliment.*	Additives *Additifs*
Production	-	251	-	-	-	758	-	42837	1830	-	-
From Other Sources	-	-	-	-	-	-	-	-	-	-	-
Imports	881	-	-	-	-	-	-	935	-	-	-
Exports	-	-	-	-	-	-	-	-17104	-	-	-
Intl. Marine Bunkers	-	-	-	-	-	-	-	-	-	-	-
Stock Changes	-	66	-	-	-	-12	-	22	-	-	-
DOMESTIC SUPPLY	**881**	**317**	**-**	**-**	**-**	**746**	**-**	**26690**	**1830**	**-**	**-**
Transfers	-	-	-	-	-	-	-	-	-1830	-	-
Statistical Differences	-173	210	-	-	-	-	-	136	-	-	-
TRANSFORMATION	**708**	**503**	**-**	**-**	**-**	**731**	**-**	**26826**	**-**	**-**	**-**
Electricity Plants	-	503	-	-	-	-	-	-	-	-	-
CHP Plants	-	-	-	-	-	-	-	-	-	-	-
Heat Plants	-	-	-	-	-	-	-	-	-	-	-
Blast Furnaces/Gas Works	-	-	-	-	-	731	-	-	-	-	-
Coke/Pat. Fuel/BKB Plants	708	-	-	-	-	-	-	-	-	-	-
Petroleum Refineries	-	-	-	-	-	-	-	26826	-	-	-
Petrochemical Industry	-	-	-	-	-	-	-	-	-	-	-
Liquefaction	-	-	-	-	-	-	-	-	-	-	-
Other Transform. Sector	-	-	-	-	-	-	-	-	-	-	-
ENERGY SECTOR	**-**	**-**	**-**	**-**	**-**	**-**	**-**	**-**	**-**	**-**	**-**
Coal Mines	-	-	-	-	-	-	-	-	-	-	-
Oil and Gas Extraction	-	-	-	-	-	-	-	-	-	-	-
Petroleum Refineries	-	-	-	-	-	-	-	-	-	-	-
Electr., CHP+Heat Plants	-	-	-	-	-	-	-	-	-	-	-
Pumped Storage (Elec.)	-	-	-	-	-	-	-	-	-	-	-
Other Energy Sector	-	-	-	-	-	-	-	-	-	-	-
Distribution Losses	-	-	-	-	-	-	-	-	-	-	-
FINAL CONSUMPTION	**-**	**24**	**-**	**-**	**-**	**15**	**-**	**-**	**-**	**-**	**-**
INDUSTRY SECTOR	**-**	**24**	**-**	**-**	**-**	**15**	**-**	**-**	**-**	**-**	**-**
Iron and Steel	-	-	-	-	-	15	-	-	-	-	-
Chemical and Petrochem.	-	-	-	-	-	-	-	-	-	-	-
of which: Feedstocks	-	-	-	-	-	-	-	-	-	-	-
Non-Ferrous Metals	-	-	-	-	-	-	-	-	-	-	-
Non-Metallic Minerals	-	-	-	-	-	-	-	-	-	-	-
Transport Equipment	-	-	-	-	-	-	-	-	-	-	-
Machinery	-	-	-	-	-	-	-	-	-	-	-
Mining and Quarrying	-	-	-	-	-	-	-	-	-	-	-
Food and Tobacco	-	-	-	-	-	-	-	-	-	-	-
Paper, Pulp and Print	-	-	-	-	-	-	-	-	-	-	-
Wood and Wood Products	-	-	-	-	-	-	-	-	-	-	-
Construction	-	-	-	-	-	-	-	-	-	-	-
Textile and Leather	-	-	-	-	-	-	-	-	-	-	-
Non-specified	-	24	-	-	-	-	-	-	-	-	-
TRANSPORT SECTOR	**-**	**-**	**-**	**-**	**-**	**-**	**-**	**-**	**-**	**-**	**-**
Air	-	-	-	-	-	-	-	-	-	-	-
Road	-	-	-	-	-	-	-	-	-	-	-
Rail	-	-	-	-	-	-	-	-	-	-	-
Pipeline Transport	-	-	-	-	-	-	-	-	-	-	-
Internal Navigation	-	-	-	-	-	-	-	-	-	-	-
Non-specified	-	-	-	-	-	-	-	-	-	-	-
OTHER SECTORS	**-**	**-**	**-**	**-**	**-**	**-**	**-**	**-**	**-**	**-**	**-**
Agriculture	-	-	-	-	-	-	-	-	-	-	-
Comm. and Publ. Services	-	-	-	-	-	-	-	-	-	-	-
Residential	-	-	-	-	-	-	-	-	-	-	-
Non-specified	-	-	-	-	-	-	-	-	-	-	-
NON-ENERGY USE	**-**	**-**	**-**	**-**	**-**	**-**	**-**	**-**	**-**	**-**	**-**
in Industry/Trans./Energy	-	-	-	-	-	-	-	-	-	-	-
in Transport	-	-	-	-	-	-	-	-	-	-	-
in Other Sectors	-	-	-	-	-	-	-	-	-	-	-

Argentina / Argentine : 1997

SUPPLY AND CONSUMPTION / APPROVISIONNEMENT ET DEMANDE	Oil cont. / Pétrole cont. (1000 tonnes)										
	Refinery Gas / Gaz de raffinerie	LPG + Ethane / GPL + éthane	Motor Gasoline / Essence moteur	Aviation Gasoline / Essence aviation	Jet Fuel / Carbu-réacteurs	Kerosene / Kérosène	Gas/ Diesel / Gazole	Heavy Fuel Oil / Fioul lourd	Naphtha / Naphta	Petrol. Coke / Coke de pétrole	Other Prod. / Autres prod.
Production	373	832	5723	5	1181	231	10491	2554	1628	1475	738
From Other Sources	-	-	-	-	-	-	-	-	-	-	-
Imports	-	26	142	5	128	57	703	127	35	178	70
Exports	-	-609	-1261	-	-46	-	-989	-406	-726	-403	-77
Intl. Marine Bunkers	-	-	-	-	-	-	-278	-424	-	-	-
Stock Changes	-	-	-26	-	-12	-	-102	-135	-	150	-
DOMESTIC SUPPLY	**373**	**249**	**4578**	**10**	**1251**	**288**	**9825**	**1716**	**937**	**1400**	**731**
Transfers	-	1471	228	-	-	-	-	-	-	-	-
Statistical Differences	-2	-100	-398	-	-26	-	-106	181	-	-51	-
TRANSFORMATION	**30**	**-**	**-**	**-**	**-**	**-**	**127**	**515**	**-**	**424**	**-**
Electricity Plants	30	-	-	-	-	-	127	515	-	25	-
CHP Plants	-	-	-	-	-	-	-	-	-	-	-
Heat Plants	-	-	-	-	-	-	-	-	-	-	-
Blast Furnaces/Gas Works	-	-	-	-	-	-	-	-	-	13	-
Coke/Pat. Fuel/BKB Plants	-	-	-	-	-	-	-	-	-	386	-
Petroleum Refineries	-	-	-	-	-	-	-	-	-	-	-
Petrochemical Industry	-	-	-	-	-	-	-	-	-	-	-
Liquefaction	-	-	-	-	-	-	-	-	-	-	-
Other Transform. Sector	-	-	-	-	-	-	-	-	-	-	-
ENERGY SECTOR	**311**	**-**	**28**	**-**	**17**	**17**	**84**	**692**	**-**	**15**	**-**
Coal Mines	-	-	-	-	-	-	-	-	-	-	-
Oil and Gas Extraction	-	-	-	-	-	-	-	-	-	-	-
Petroleum Refineries	311	-	28	-	17	17	84	692	-	15	-
Electr., CHP+Heat Plants	-	-	-	-	-	-	-	-	-	-	-
Pumped Storage (Elec.)	-	-	-	-	-	-	-	-	-	-	-
Other Energy Sector	-	-	-	-	-	-	-	-	-	-	-
Distribution Losses	-	-	-	-	-	-	-	-	-	-	-
FINAL CONSUMPTION	**30**	**1620**	**4380**	**10**	**1208**	**271**	**9508**	**690**	**937**	**910**	**731**
INDUSTRY SECTOR	**30**	**801**	**-**	**-**	**-**	**-**	**60**	**557**	**937**	**910**	**-**
Iron and Steel	-	-	-	-	-	-	-	-	-	-	-
Chemical and Petrochem.	30	779	-	-	-	-	-	-	937	-	-
of which: Feedstocks	*30*	*779*	-	-	-	-	-	-	*937*	-	-
Non-Ferrous Metals	-	-	-	-	-	-	-	-	-	-	-
Non-Metallic Minerals	-	-	-	-	-	-	-	-	-	-	-
Transport Equipment	-	-	-	-	-	-	-	-	-	-	-
Machinery	-	-	-	-	-	-	-	-	-	-	-
Mining and Quarrying	-	-	-	-	-	-	-	-	-	-	-
Food and Tobacco	-	-	-	-	-	-	-	-	-	-	-
Paper, Pulp and Print	-	-	-	-	-	-	-	-	-	-	-
Wood and Wood Products	-	-	-	-	-	-	-	-	-	-	-
Construction	-	-	-	-	-	-	-	-	-	-	-
Textile and Leather	-	-	-	-	-	-	-	-	-	-	-
Non-specified	-	22	-	-	-	-	60	557	-	910	-
TRANSPORT SECTOR	**-**	**-**	**4380**	**10**	**1208**	**-**	**6829**	**76**	**-**	**-**	**-**
Air	-	-	-	10	1208	-	-	-	-	-	-
Road	-	-	4380	-	-	-	6829	-	-	-	-
Rail	-	-	-	-	-	-	-	-	-	-	-
Pipeline Transport	-	-	-	-	-	-	-	-	-	-	-
Internal Navigation	-	-	-	-	-	-	-	76	-	-	-
Non-specified	-	-	-	-	-	-	-	-	-	-	-
OTHER SECTORS	**-**	**819**	**-**	**-**	**-**	**271**	**2619**	**57**	**-**	**-**	**-**
Agriculture	-	-	-	-	-	-	2550	-	-	-	-
Comm. and Publ. Services	-	14	-	-	-	-	69	57	-	-	-
Residential	-	805	-	-	-	271	-	-	-	-	-
Non-specified	-	-	-	-	-	-	-	-	-	-	-
NON-ENERGY USE	**-**	**-**	**-**	**-**	**-**	**-**	**-**	**-**	**-**	**-**	**731**
in Industry/Transf./Energy	-	-	-	-	-	-	-	-	-	-	731
in Transport	-	-	-	-	-	-	-	-	-	-	-
in Other Sectors	-	-	-	-	-	-	-	-	-	-	-

Argentina / Argentine : 1997

SUPPLY AND CONSUMPTION / APPROVISIONNEMENT ET DEMANDE	Gas / Gaz (TJ)				Comb. Renew. & Waste / En. Re. Comb. & Déchets (TJ)				(GWh)	(TJ)
	Natural Gas / Gaz naturel	Gas Works / Usines à gaz	Coke Ovens / Cokeries	Blast Furnaces / Hauts fourneaux	Solid Biomass / Biomasse solide	Gas/Liquids from Biomass / Gaz/Liquides tirés de biomasse	Municipal Waste / Déchets urbains	Industrial Waste / Déchets industriels	Electricity / Electricité	Heat / Chaleur
Production	1221173	-	7725	11977	110875	-	-	-	72463	-
From Other Sources	-	-	-	-	-	-	-	-	-	-
Imports	65752	-	-	-	-	-	-	-	5466	-
Exports	-28571	-	-	-	-	-	-	-	-277	-
Intl. Marine Bunkers	-	-	-	-	-	-	-	-	-	-
Stock Changes	-	-	-	-	-	-	-	-	-	-
DOMESTIC SUPPLY	**1258354**	**-**	**7725**	**11977**	**110875**	**-**	**-**	**-**	**77652**	**-**
Transfers	-	-	-	-	-	-	-	-	-	-
Statistical Differences	-2	-	-1	-1	41	-	-	-	25	-
TRANSFORMATION	**373801**	**-**	**2187**	**5988**	**12267**	**-**	**-**	**-**	**-**	**-**
Electricity Plants	373801	-	2187	5988	4229	-	-	-	-	-
CHP Plants	-	-	-	-	-	-	-	-	-	-
Heat Plants	-	-	-	-	-	-	-	-	-	-
Blast Furnaces/Gas Works	-	-	-	-	1884	-	-	-	-	-
Coke/Pat. Fuel/BKB Plants	-	-	-	-	-	-	-	-	-	-
Petroleum Refineries	-	-	-	-	-	-	-	-	-	-
Petrochemical Industry	-	-	-	-	-	-	-	-	-	-
Liquefaction	-	-	-	-	-	-	-	-	-	-
Other Transform. Sector	-	-	-	-	6154	-	-	-	-	-
ENERGY SECTOR	**201861**	**-**	**-**	**-**	**-**	**-**	**-**	**-**	**2582**	**-**
Coal Mines	-	-	-	-	-	-	-	-	-	-
Oil and Gas Extraction	201861	-	-	-	-	-	-	-	-	-
Petroleum Refineries	-	-	-	-	-	-	-	-	-	-
Electr., CHP+Heat Plants	-	-	-	-	-	-	-	-	2306	-
Pumped Storage (Elec.)	-	-	-	-	-	-	-	-	276	-
Other Energy Sector	-	-	-	-	-	-	-	-	-	-
Distribution Losses	61052	-	-	377	-	-	-	-	11758	-
FINAL CONSUMPTION	**621638**	**-**	**5537**	**5611**	**98649**	**-**	**-**	**-**	**63337**	**-**
INDUSTRY SECTOR	**290600**	**-**	**5537**	**5611**	**82320**	**-**	**-**	**-**	**30075**	**-**
Iron and Steel	-	-	5537	-	-	-	-	-	-	-
Chemical and Petrochem.	8841	-	-	-	-	-	-	-	-	-
of which: Feedstocks	8841	-	-	-	-	-	-	-	-	-
Non-Ferrous Metals	-	-	-	-	-	-	-	-	-	-
Non-Metallic Minerals	-	-	-	-	-	-	-	-	-	-
Transport Equipment	-	-	-	-	-	-	-	-	-	-
Machinery	-	-	-	-	-	-	-	-	-	-
Mining and Quarrying	-	-	-	-	-	-	-	-	-	-
Food and Tobacco	-	-	-	-	-	-	-	-	-	-
Paper, Pulp and Print	-	-	-	-	-	-	-	-	-	-
Wood and Wood Products	-	-	-	-	-	-	-	-	-	-
Construction	-	-	-	-	-	-	-	-	-	-
Textile and Leather	-	-	-	-	-	-	-	-	-	-
Non-specified	281759	-	-	5611	82320	-	-	-	30075	-
TRANSPORT SECTOR	**48953**	**-**	**-**	**-**	**-**	**-**	**-**	**-**	**442**	**-**
Air	-	-	-	-	-	-	-	-	-	-
Road	48953	-	-	-	-	-	-	-	-	-
Rail	-	-	-	-	-	-	-	-	442	-
Pipeline Transport	-	-	-	-	-	-	-	-	-	-
Internal Navigation	-	-	-	-	-	-	-	-	-	-
Non-specified	-	-	-	-	-	-	-	-	-	-
OTHER SECTORS	**282085**	**-**	**-**	**-**	**16329**	**-**	**-**	**-**	**32820**	**-**
Agriculture	-	-	-	-	2554	-	-	-	535	-
Comm. and Publ. Services	53886	-	-	-	-	-	-	-	13770	-
Residential	228199	-	-	-	13775	-	-	-	18515	-
Non-specified	-	-	-	-	-	-	-	-	-	-
NON-ENERGY USE	**-**	**-**	**-**	**-**	**-**	**-**	**-**	**-**	**-**	**-**
in Industry/Transf./Energy	-	-	-	-	-	-	-	-	-	-
in Transport	-	-	-	-	-	-	-	-	-	-
in Other Sectors	-	-	-	-	-	-	-	-	-	-

Argentina / Argentine : 1998

SUPPLY AND CONSUMPTION	Coal / *Charbon* (1000 tonnes)							Oil / *Pétrole* (1000 tonnes)			
	Coking Coal	Other Bit. Coal	Sub-Bit. Coal	Lignite	Peat	Oven and Gas Coke	Pat. Fuel and BKB	Crude Oil	NGL	Feed-stocks	Additives
APPROVISIONNEMENT ET DEMANDE	*Charbon à coke*	*Autres charb. bit.*	*Charbon sous-bit.*	*Lignite*	*Tourbe*	*Coke de four/gaz*	*Agg./briq. de lignite*	*Pétrole brut*	*LGN*	*Produits d'aliment.*	*Additifs*
Production	-	290	-	-	-	1068	-	43513	1806	-	-
From Other Sources	-	-	-	-	-	-	-	-	-	-	-
Imports	990	-	-	-	-	79	-	1229	-	-	-
Exports	-	-	-	-	-	-177	-	-16979	-	-	-
Intl. Marine Bunkers	-	-	-	-	-	-	-	-	-	-	-
Stock Changes	-	-80	-	-	-	-34	-	270	-	-	-
DOMESTIC SUPPLY	990	210	-	-	-	936	-	28033	1806	-	-
Transfers	-	-	-	-	-	-	-	-	-1806	-	-
Statistical Differences	-250	304	-	-	-	-63	-	-132	-	-	-
TRANSFORMATION	740	482	-	-	-	799	-	27901	-	-	-
Electricity Plants	-	482	-	-	-	-	-	-	-	-	-
CHP Plants	-	-	-	-	-	-	-	-	-	-	-
Heat Plants	-	-	-	-	-	-	-	-	-	-	-
Blast Furnaces/Gas Works	-	-	-	-	-	799	-	-	-	-	-
Coke/Pat. Fuel/BKB Plants	740	-	-	-	-	-	-	-	-	-	-
Petroleum Refineries	-	-	-	-	-	-	-	27901	-	-	-
Petrochemical Industry	-	-	-	-	-	-	-	-	-	-	-
Liquefaction	-	-	-	-	-	-	-	-	-	-	-
Other Transform. Sector	-	-	-	-	-	-	-	-	-	-	-
ENERGY SECTOR	-	7	-	-	-	-	-	-	-	-	-
Coal Mines	-	7	-	-	-	-	-	-	-	-	-
Oil and Gas Extraction	-	-	-	-	-	-	-	-	-	-	-
Petroleum Refineries	-	-	-	-	-	-	-	-	-	-	-
Electr., CHP+Heat Plants	-	-	-	-	-	-	-	-	-	-	-
Pumped Storage (Elec.)	-	-	-	-	-	-	-	-	-	-	-
Other Energy Sector	-	-	-	-	-	-	-	-	-	-	-
Distribution Losses	-	-	-	-	-	-	-	-	-	-	-
FINAL CONSUMPTION	-	25	-	-	-	74	-	-	-	-	-
INDUSTRY SECTOR	-	25	-	-	-	74	-	-	-	-	-
Iron and Steel	-	-	-	-	-	74	-	-	-	-	-
Chemical and Petrochem.	-	-	-	-	-	-	-	-	-	-	-
of which: Feedstocks	-	-	-	-	-	-	-	-	-	-	-
Non-Ferrous Metals	-	-	-	-	-	-	-	-	-	-	-
Non-Metallic Minerals	-	-	-	-	-	-	-	-	-	-	-
Transport Equipment	-	-	-	-	-	-	-	-	-	-	-
Machinery	-	-	-	-	-	-	-	-	-	-	-
Mining and Quarrying	-	-	-	-	-	-	-	-	-	-	-
Food and Tobacco	-	-	-	-	-	-	-	-	-	-	-
Paper, Pulp and Print	-	-	-	-	-	-	-	-	-	-	-
Wood and Wood Products	-	-	-	-	-	-	-	-	-	-	-
Construction	-	-	-	-	-	-	-	-	-	-	-
Textile and Leather	-	-	-	-	-	-	-	-	-	-	-
Non-specified	-	25	-	-	-	-	-	-	-	-	-
TRANSPORT SECTOR	-	-	-	-	-	-	-	-	-	-	-
Air	-	-	-	-	-	-	-	-	-	-	-
Road	-	-	-	-	-	-	-	-	-	-	-
Rail	-	-	-	-	-	-	-	-	-	-	-
Pipeline Transport	-	-	-	-	-	-	-	-	-	-	-
Internal Navigation	-	-	-	-	-	-	-	-	-	-	-
Non-specified	-	-	-	-	-	-	-	-	-	-	-
OTHER SECTORS	-	-	-	-	-	-	-	-	-	-	-
Agriculture	-	-	-	-	-	-	-	-	-	-	-
Comm. and Publ. Services	-	-	-	-	-	-	-	-	-	-	-
Residential	-	-	-	-	-	-	-	-	-	-	-
Non-specified	-	-	-	-	-	-	-	-	-	-	-
NON-ENERGY USE	-	-	-	-	-	-	-	-	-	-	-
in Industry/Trans./Energy	-	-	-	-	-	-	-	-	-	-	-
in Transport	-	-	-	-	-	-	-	-	-	-	-
in Other Sectors	-	-	-	-	-	-	-	-	-	-	-

Argentina / Argentine : 1998

SUPPLY AND CONSUMPTION / *APPROVISIONNEMENT ET DEMANDE*	Oil cont. / *Pétrole cont.* (1000 tonnes)										
	Refinery Gas / *Gaz de raffinerie*	LPG + Ethane / *GPL + éthane*	Motor Gasoline / *Essence moteur*	Aviation Gasoline / *Essence aviation*	Jet Fuel / *Carbu- réacteurs*	Kerosene / *Kérosène*	Gas/ Diesel / *Gazole*	Heavy Fuel Oil / *Fioul lourd*	Naphtha / *Naphta*	Petrol. Coke / *Coke de pétrole*	Other Prod. / *Autres prod.*
Production	523	970	5387	5	1431	197	10742	2433	1837	1590	952
From Other Sources	-	-	-	-	-	-	-	-	-	-	-
Imports	-	55	85	5	112	49	1021	30	22	51	101
Exports	-	-731	-1488	-	-26	-	-1078	-233	-857	-271	-84
Intl. Marine Bunkers	-	-	-	-	-	-	-295	-253	-	-	-
Stock Changes	-	2	51	-	10	-	5	32	-	7	-
DOMESTIC SUPPLY	**523**	**296**	**4035**	**10**	**1527**	**246**	**10395**	**2009**	**1002**	**1377**	**969**
Transfers	-	1427	251	-	-	-	-	-	-	-	-
Statistical Differences	1	68	-48	-	-177	-	-248	-94	-	-8	-
TRANSFORMATION	**26**	**-**	**-**	**-**	**-**	**-**	**133**	**879**	**-**	**734**	**-**
Electricity Plants	26	-	-	-	-	-	133	879	-	33	-
CHP Plants	-	-	-	-	-	-	-	-	-	-	-
Heat Plants	-	-	-	-	-	-	-	-	-	-	-
Blast Furnaces/Gas Works	-	-	-	-	-	-	-	-	-	-	-
Coke/Pat. Fuel/BKB Plants	-	-	-	-	-	-	-	-	-	701	-
Petroleum Refineries	-	-	-	-	-	-	-	-	-	-	-
Petrochemical Industry	-	-	-	-	-	-	-	-	-	-	-
Liquefaction	-	-	-	-	-	-	-	-	-	-	-
Other Transform. Sector	-	-	-	-	-	-	-	-	-	-	-
ENERGY SECTOR	**481**	**-**	**2**	**-**	**-**	**-**	**41**	**514**	**-**	**97**	**-**
Coal Mines	-	-	-	-	-	-	-	-	-	-	-
Oil and Gas Extraction	-	-	-	-	-	-	-	-	-	-	-
Petroleum Refineries	481	-	2	-	-	-	41	514	-	97	-
Electr., CHP+Heat Plants	-	-	-	-	-	-	-	-	-	-	-
Pumped Storage (Elec.)	-	-	-	-	-	-	-	-	-	-	-
Other Energy Sector	-	-	-	-	-	-	-	-	-	-	-
Distribution Losses	-	-	-	-	-	-	-	-	-	-	-
FINAL CONSUMPTION	**17**	**1791**	**4236**	**10**	**1350**	**246**	**9973**	**522**	**1002**	**538**	**969**
INDUSTRY SECTOR	**17**	**856**	**-**	**-**	**-**	**-**	**76**	**371**	**1002**	**538**	**-**
Iron and Steel	-	-	-	-	-	-	-	-	-	-	-
Chemical and Petrochem.	17	830	-	-	-	-	-	-	1002	-	-
of which: Feedstocks	17	830	-	-	-	-	-	-	1002	-	-
Non-Ferrous Metals	-	-	-	-	-	-	-	-	-	-	-
Non-Metallic Minerals	-	-	-	-	-	-	-	-	-	-	-
Transport Equipment	-	-	-	-	-	-	-	-	-	-	-
Machinery	-	-	-	-	-	-	-	-	-	-	-
Mining and Quarrying	-	-	-	-	-	-	-	-	-	-	-
Food and Tobacco	-	-	-	-	-	-	-	-	-	-	-
Paper, Pulp and Print	-	-	-	-	-	-	-	-	-	-	-
Wood and Wood Products	-	-	-	-	-	-	-	-	-	-	-
Construction	-	-	-	-	-	-	-	-	-	-	-
Textile and Leather	-	-	-	-	-	-	-	-	-	-	-
Non-specified	-	26	-	-	-	-	76	371	-	538	-
TRANSPORT SECTOR	**-**	**-**	**4236**	**10**	**1350**	**-**	**7381**	**85**	**-**	**-**	**-**
Air	-	-	-	10	1350	-	-	-	-	-	-
Road	-	-	4236	-	-	-	7381	-	-	-	-
Rail	-	-	-	-	-	-	-	-	-	-	-
Pipeline Transport	-	-	-	-	-	-	-	-	-	-	-
Internal Navigation	-	-	-	-	-	-	-	85	-	-	-
Non-specified	-	-	-	-	-	-	-	-	-	-	-
OTHER SECTORS	**-**	**935**	**-**	**-**	**-**	**246**	**2516**	**66**	**-**	**-**	**-**
Agriculture	-	-	-	-	-	-	2431	-	-	-	-
Comm. and Publ. Services	-	22	-	-	-	-	85	66	-	-	-
Residential	-	913	-	-	-	246	-	-	-	-	-
Non-specified	-	-	-	-	-	-	-	-	-	-	-
NON-ENERGY USE	**-**	**-**	**-**	**-**	**-**	**-**	**-**	**-**	**-**	**-**	**969**
in Industry/Transf./Energy	-	-	-	-	-	-	-	-	-	-	969
in Transport	-	-	-	-	-	-	-	-	-	-	-
in Other Sectors	-	-	-	-	-	-	-	-	-	-	-

Argentina / Argentine : 1998

SUPPLY AND CONSUMPTION *APPROVISIONNEMENT ET DEMANDE*	Gas / *Gaz* (TJ)				Comb. Renew. & Waste / *En. Re. Comb. & Déchets* (TJ)				(GWh)	(TJ)
	Natural Gas *Gaz naturel*	Gas Works *Usines à gaz*	Coke Ovens *Cokeries*	Blast Furnaces *Hauts fourneaux*	Solid Biomass *Biomasse solide*	Gas/Liquids from Biomass *Gaz/Liquides tirés de biomasse*	Municipal Waste *Déchets urbains*	Industrial Waste *Déchets industriels*	Electricity *Electricité*	Heat *Chaleur*
Production	1298744	-	7492	12688	125520	-	-	-	74170	-
From Other Sources	-	-	-	-	-	-	-	-	-	-
Imports	67659	-	-	-	-	-	-	-	8000	-
Exports	-82922	-	-	-	-	-	-	-	-	-
Intl. Marine Bunkers	-	-	-	-	-	-	-	-	-	-
Stock Changes	-	-	-	-	-	-	-	-	-	-
DOMESTIC SUPPLY	**1283481**	**-**	**7492**	**12688**	**125520**	**-**	**-**	**-**	**82170**	**-**
Transfers	-	-	-	-	-	-	-	-	-	-
Statistical Differences	-92	-	-46	1	377	-	-	-	20	-
TRANSFORMATION	**390787**	**-**	**2932**	**5025**	**12769**	**-**	**-**	**-**	**-**	**-**
Electricity Plants	390787	-	2932	5025	4815	-	-	-	-	-
CHP Plants	-	-	-	-	-	-	-	-	-	-
Heat Plants	-	-	-	-	-	-	-	-	-	-
Blast Furnaces/Gas Works	-	-	-	-	1675	-	-	-	-	-
Coke/Pat. Fuel/BKB Plants	-	-	-	-	-	-	-	-	-	-
Petroleum Refineries	-	-	-	-	-	-	-	-	-	-
Petrochemical Industry	-	-	-	-	-	-	-	-	-	-
Liquefaction	-	-	-	-	-	-	-	-	-	-
Other Transform. Sector	-	-	-	-	6279	-	-	-	-	-
ENERGY SECTOR	**201210**	**-**	**-**	**-**	**-**	**-**	**-**	**-**	**2640**	**-**
Coal Mines	-	-	-	-	-	-	-	-	-	-
Oil and Gas Extraction	201210	-	-	-	-	-	-	-	-	-
Petroleum Refineries	-	-	-	-	-	-	-	-	-	-
Electr., CHP+Heat Plants	-	-	-	-	-	-	-	-	2379	-
Pumped Storage (Elec.)	-	-	-	-	-	-	-	-	261	-
Other Energy Sector	-	-	-	-	-	-	-	-	-	-
Distribution Losses	51745	-	-	838	-	-	-	-	11223	-
FINAL CONSUMPTION	**639647**	**-**	**4514**	**6826**	**113128**	**-**	**-**	**-**	**68327**	**-**
INDUSTRY SECTOR	**299675**	**-**	**4514**	**6826**	**95459**	**-**	**-**	**-**	**32483**	**-**
Iron and Steel	-	-	4514	-	-	-	-	-	-	-
Chemical and Petrochem.	9214	-	-	-	-	-	-	-	-	-
of which: Feedstocks	9214	-	-	-	-	-	-	-	-	-
Non-Ferrous Metals	-	-	-	-	-	-	-	-	-	-
Non-Metallic Minerals	-	-	-	-	-	-	-	-	-	-
Transport Equipment	-	-	-	-	-	-	-	-	-	-
Machinery	-	-	-	-	-	-	-	-	-	-
Mining and Quarrying	-	-	-	-	-	-	-	-	-	-
Food and Tobacco	-	-	-	-	-	-	-	-	-	-
Paper, Pulp and Print	-	-	-	-	-	-	-	-	-	-
Wood and Wood Products	-	-	-	-	-	-	-	-	-	-
Construction	-	-	-	-	-	-	-	-	-	-
Textile and Leather	-	-	-	-	-	-	-	-	-	-
Non-specified	290461	-	-	6826	95459	-	-	-	32483	-
TRANSPORT SECTOR	**54537**	**-**	**-**	**-**	**-**	**-**	**-**	**-**	**477**	**-**
Air	-	-	-	-	-	-	-	-	-	-
Road	54537	-	-	-	-	-	-	-	-	-
Rail	-	-	-	-	-	-	-	-	477	-
Pipeline Transport	-	-	-	-	-	-	-	-	-	-
Internal Navigation	-	-	-	-	-	-	-	-	-	-
Non-specified	-	-	-	-	-	-	-	-	-	-
OTHER SECTORS	**285435**	**-**	**-**	**-**	**17669**	**-**	**-**	**-**	**35367**	**-**
Agriculture	-	-	-	-	2554	-	-	-	582	-
Comm. and Publ. Services	53979	-	-	-	-	-	-	-	15607	-
Residential	231456	-	-	-	15115	-	-	-	19178	-
Non-specified	-	-	-	-	-	-	-	-	-	-
NON-ENERGY USE	**-**	**-**	**-**	**-**	**-**	**-**	**-**	**-**	**-**	**-**
in Industry/Transf./Energy	-	-	-	-	-	-	-	-	-	-
in Transport	-	-	-	-	-	-	-	-	-	-
in Other Sectors	-	-	-	-	-	-	-	-	-	-

Armenia / Arménie

SUPPLY AND CONSUMPTION 1997	Coal (1000 tonnes)							Oil (1000 tonnes)			
	Coking Coal	Other Bit. Coal	Sub-Bit. Coal	Lignite	Peat	Oven and Gas Coke	Pat. Fuel and BKB	Crude Oil	NGL	Feed-stocks	Additives
Production	-	-	-	-	-	-	-	-	-	-	-
Imports	-	5	-	-	-	-	-	-	-	-	-
Exports	-	-	-	-	-	-	-	-	-	-	-
Intl. Marine Bunkers	-	-	-	-	-	-	-	-	-	-	-
Stock Changes	-	-	-	-	-	-	-	-	-	-	-
DOMESTIC SUPPLY	-	5	-	-	-	-	-	-	-	-	-
Transfers and Stat. Diff.	-	-	-	-	-	-	-	-	-	-	-
TRANSFORMATION	-	-	-	-	-	-	-	-	-	-	-
Electricity and CHP Plants	-	-	-	-	-	-	-	-	-	-	-
Petroleum Refineries	-	-	-	-	-	-	-	-	-	-	-
Other Transform. Sector	-	-	-	-	-	-	-	-	-	-	-
ENERGY SECTOR	-	-	-	-	-	-	-	-	-	-	-
DISTRIBUTION LOSSES	-	-	-	-	-	-	-	-	-	-	-
FINAL CONSUMPTION	-	5	-	-	-	-	-	-	-	-	-
INDUSTRY SECTOR	-	-	-	-	-	-	-	-	-	-	-
Iron and Steel	-	-	-	-	-	-	-	-	-	-	-
Chemical and Petrochem.	-	-	-	-	-	-	-	-	-	-	-
Non-Metallic Minerals	-	-	-	-	-	-	-	-	-	-	-
Non-specified	-	-	-	-	-	-	-	-	-	-	-
TRANSPORT SECTOR	-	-	-	-	-	-	-	-	-	-	-
Air	-	-	-	-	-	-	-	-	-	-	-
Road	-	-	-	-	-	-	-	-	-	-	-
Non-specified	-	-	-	-	-	-	-	-	-	-	-
OTHER SECTORS	-	5	-	-	-	-	-	-	-	-	-
Agriculture	-	-	-	-	-	-	-	-	-	-	-
Comm. and Publ. Services	-	-	-	-	-	-	-	-	-	-	-
Residential	-	5	-	-	-	-	-	-	-	-	-
Non-specified	-	-	-	-	-	-	-	-	-	-	-
NON-ENERGY USE	-	-	-	-	-	-	-	-	-	-	-

APPROVISIONNEMENT ET DEMANDE 1998	Charbon (1000 tonnes)							Pétrole (1000 tonnes)			
	Charbon à coke	Autres charb. bit.	Charbon sous-bit.	Lignite	Tourbe	Coke de four/gaz	Agg./briq. de lignite	Pétrole brut	LGN	Produits d'aliment.	Additifs
Production	-	-	-	-	-	-	-	-	-	-	-
Imports	-	5	-	-	-	-	-	-	-	-	-
Exports	-	-	-	-	-	-	-	-	-	-	-
Intl. Marine Bunkers	-	-	-	-	-	-	-	-	-	-	-
Stock Changes	-	-	-	-	-	-	-	-	-	-	-
DOMESTIC SUPPLY	-	5	-	-	-	-	-	-	-	-	-
Transfers and Stat. Diff.	-	-	-	-	-	-	-	-	-	-	-
TRANSFORMATION	-	-	-	-	-	-	-	-	-	-	-
Electricity and CHP Plants	-	-	-	-	-	-	-	-	-	-	-
Petroleum Refineries	-	-	-	-	-	-	-	-	-	-	-
Other Transform. Sector	-	-	-	-	-	-	-	-	-	-	-
ENERGY SECTOR	-	-	-	-	-	-	-	-	-	-	-
DISTRIBUTION LOSSES	-	-	-	-	-	-	-	-	-	-	-
FINAL CONSUMPTION	-	5	-	-	-	-	-	-	-	-	-
INDUSTRY SECTOR	-	-	-	-	-	-	-	-	-	-	-
Iron and Steel	-	-	-	-	-	-	-	-	-	-	-
Chemical and Petrochem.	-	-	-	-	-	-	-	-	-	-	-
Non-Metallic Minerals	-	-	-	-	-	-	-	-	-	-	-
Non-specified	-	-	-	-	-	-	-	-	-	-	-
TRANSPORT SECTOR	-	-	-	-	-	-	-	-	-	-	-
Air	-	-	-	-	-	-	-	-	-	-	-
Road	-	-	-	-	-	-	-	-	-	-	-
Non-specified	-	-	-	-	-	-	-	-	-	-	-
OTHER SECTORS	-	5	-	-	-	-	-	-	-	-	-
Agriculture	-	-	-	-	-	-	-	-	-	-	-
Comm. and Publ. Services	-	-	-	-	-	-	-	-	-	-	-
Residential	-	5	-	-	-	-	-	-	-	-	-
Non-specified	-	-	-	-	-	-	-	-	-	-	-
NON-ENERGY USE	-	-	-	-	-	-	-	-	-	-	-

Armenia / Arménie

SUPPLY AND CONSUMPTION 1997	Oil cont. (1000 tonnes)										
	Refinery Gas	LPG + Ethane	Motor Gasoline	Aviation Gasoline	Jet Fuel	Kerosene	Gas/ Diesel	Heavy Fuel Oil	Naphtha	Petrol. Coke	Other Prod.
Production	-	-	-	-	-	-	-	-	-	-	-
Imports	-	1	21	-	20	2	36	68	-	-	16
Exports	-	-	-	-	-	-	-	-	-	-	-
Intl. Marine Bunkers	-	-	-	-	-	-	-	-	-	-	-
Stock Changes	-	-	-	-	-	-	-	-	-	-	-
DOMESTIC SUPPLY	-	1	21	-	20	2	36	68	-	-	16
Transfers and Stat. Diff.	-	-	-	-	-	-	-	-	-	-	-
TRANSFORMATION	-	-	-	-	-	-	-	3	-	-	-
Electricity and CHP Plants	-	-	-	-	-	-	-	3	-	-	-
Petroleum Refineries	-	-	-	-	-	-	-	-	-	-	-
Other Transform. Sector	-	-	-	-	-	-	-	-	-	-	-
ENERGY SECTOR	-	-	-	-	-	-	-	-	-	-	-
DISTRIBUTION LOSSES	-	-	-	-	-	-	-	-	-	-	-
FINAL CONSUMPTION	-	1	21	-	20	2	36	65	-	-	16
INDUSTRY SECTOR	-	-	-	-	-	-	-	65	-	-	-
Iron and Steel	-	-	-	-	-	-	-	-	-	-	-
Chemical and Petrochem.	-	-	-	-	-	-	-	-	-	-	-
Non-Metallic Minerals	-	-	-	-	-	-	-	-	-	-	-
Non-specified	-	-	-	-	-	-	-	65	-	-	-
TRANSPORT SECTOR	-	-	21	-	20	-	-	-	-	-	-
Air	-	-	-	-	20	-	-	-	-	-	-
Road	-	-	21	-	-	-	-	-	-	-	-
Non-specified	-	-	-	-	-	-	-	-	-	-	-
OTHER SECTORS	-	1	-	-	-	2	36	-	-	-	-
Agriculture	-	-	-	-	-	-	-	-	-	-	-
Comm. and Publ. Services	-	-	-	-	-	-	-	-	-	-	-
Residential	-	-	-	-	-	2	-	-	-	-	-
Non-specified	-	1	-	-	-	-	36	-	-	-	-
NON-ENERGY USE	-	-	-	-	-	-	-	-	-	-	16

APPROVISIONNEMENT ET DEMANDE 1998	Pétrole cont. (1000 tonnes)										
	Gaz de raffinerie	GPL + éthane	Essence moteur	Essence aviation	Carbu- réacteurs	Kérosène	Gazole	Fioul lourd	Naphta	Coke de pétrole	Autres prod.
Production	-	-	-	-	-	-	-	-	-	-	-
Imports	-	1	22	-	21	2	37	70	-	-	16
Exports	-	-	-	-	-	-	-	-	-	-	-
Intl. Marine Bunkers	-	-	-	-	-	-	-	-	-	-	-
Stock Changes	-	-	-	-	-	-	-	-	-	-	-
DOMESTIC SUPPLY	-	1	22	-	21	2	37	70	-	-	16
Transfers and Stat. Diff.	-	-	-	-	-	-	-	-	-	-	-
TRANSFORMATION	-	-	-	-	-	-	-	-	-	-	-
Electricity and CHP Plants	-	-	-	-	-	-	-	-	-	-	-
Petroleum Refineries	-	-	-	-	-	-	-	-	-	-	-
Other Transform. Sector	-	-	-	-	-	-	-	-	-	-	-
ENERGY SECTOR	-	-	-	-	-	-	-	-	-	-	-
DISTRIBUTION LOSSES	-	-	-	-	-	-	-	-	-	-	-
FINAL CONSUMPTION	-	1	22	-	21	2	37	70	-	-	16
INDUSTRY SECTOR	-	-	-	-	-	-	-	70	-	-	-
Iron and Steel	-	-	-	-	-	-	-	-	-	-	-
Chemical and Petrochem.	-	-	-	-	-	-	-	-	-	-	-
Non-Metallic Minerals	-	-	-	-	-	-	-	-	-	-	-
Non-specified	-	-	-	-	-	-	-	70	-	-	-
TRANSPORT SECTOR	-	-	22	-	21	-	-	-	-	-	-
Air	-	-	-	-	21	-	-	-	-	-	-
Road	-	-	22	-	-	-	-	-	-	-	-
Non-specified	-	-	-	-	-	-	-	-	-	-	-
OTHER SECTORS	-	1	-	-	-	2	37	-	-	-	-
Agriculture	-	-	-	-	-	-	-	-	-	-	-
Comm. and Publ. Services	-	-	-	-	-	-	-	-	-	-	-
Residential	-	-	-	-	-	2	-	-	-	-	-
Non-specified	-	1	-	-	-	-	37	-	-	-	-
NON-ENERGY USE	-	-	-	-	-	-	-	-	-	-	16

Armenia / Arménie

SUPPLY AND CONSUMPTION 1997	Gas (TJ)				Comb. Renew. & Waste (TJ)				(GWh)	(TJ)
	Natural Gas	Gas Works	Coke Ovens	Blast Furnaces	Solid Biomass	Gas/Liquids from Biomass	Municipal Waste	Industrial Waste	Electricity	Heat
Production	-	-	-	-	39	-	-	-	6022	2776
Imports	54477	-	-	-	-	-	-	-	-	-
Exports	-	-	-	-	-	-	-	-	-	-
Intl. Marine Bunkers	-	-	-	-	-	-	-	-	-	-
Stock Changes	-	-	-	-	-	-	-	-	-	-
DOMESTIC SUPPLY	54477	-	-	-	39	-	-	-	6022	2776
Transfers and Stat. Diff.	-	-	-	-	-	-	-	-	-	-
TRANSFORMATION	35493	-	-	-	-	-	-	-	-	-
Electricity and CHP Plants	35058	-	-	-	-	-	-	-	-	-
Petroleum Refineries	-	-	-	-	-	-	-	-	-	-
Other Transform. Sector	435	-	-	-	-	-	-	-	-	-
ENERGY SECTOR	-	-	-	-	-	-	-	-	891	-
DISTRIBUTION LOSSES	-	-	-	-	-	-	-	-	1252	555
FINAL CONSUMPTION	18984	-	-	-	39	-	-	-	3879	2221
INDUSTRY SECTOR	12877	-	-	-	-	-	-	-	722	1207
Iron and Steel	-	-	-	-	-	-	-	-	3	-
Chemical and Petrochem.	-	-	-	-	-	-	-	-	197	-
Non-Metallic Minerals	-	-	-	-	-	-	-	-	96	-
Non-specified	12877	-	-	-	-	-	-	-	426	1207
TRANSPORT SECTOR	-	-	-	-	-	-	-	-	151	-
Air	-	-	-	-	-	-	-	-	-	-
Road	-	-	-	-	-	-	-	-	-	-
Non-specified	-	-	-	-	-	-	-	-	151	-
OTHER SECTORS	6107	-	-	-	39	-	-	-	3006	1014
Agriculture	829	-	-	-	-	-	-	-	245	-
Comm. and Publ. Services	-	-	-	-	-	-	-	-	330	-
Residential	-	-	-	-	-	-	-	-	2431	1014
Non-specified	5278	-	-	-	39	-	-	-	-	-
NON-ENERGY USE	-	-	-	-	-	-	-	-	-	-

APPROVISIONNEMENT ET DEMANDE 1998	Gaz (TJ)				En. Re. Comb. & Déchets (TJ)				(GWh)	(TJ)
	Gaz naturel	Usines à gaz	Cokeries	Hauts fourneaux	Biomasse solide	Gaz/Liquides tirés de biomasse	Déchets urbains	Déchets industriels	Electricité	Chaleur
Production	-	-	-	-	42	-	-	-	6190	2478
Imports	56739	-	-	-	-	-	-	-	-	-
Exports	-	-	-	-	-	-	-	-	-	-
Intl. Marine Bunkers	-	-	-	-	-	-	-	-	-	-
Stock Changes	-	-	-	-	-	-	-	-	-	-
DOMESTIC SUPPLY	56739	-	-	-	42	-	-	-	6190	2478
Transfers and Stat. Diff.	-	-	-	-	-	-	-	-	-	-
TRANSFORMATION	35382	-	-	-	-	-	-	-	-	-
Electricity and CHP Plants	34974	-	-	-	-	-	-	-	-	-
Petroleum Refineries	-	-	-	-	-	-	-	-	-	-
Other Transform. Sector	408	-	-	-	-	-	-	-	-	-
ENERGY SECTOR	-	-	-	-	-	-	-	-	895	-
DISTRIBUTION LOSSES	-	-	-	-	-	-	-	-	1764	532
FINAL CONSUMPTION	21357	-	-	-	42	-	-	-	3531	1946
INDUSTRY SECTOR	14985	-	-	-	-	-	-	-	638	1057
Iron and Steel	-	-	-	-	-	-	-	-	-	-
Chemical and Petrochem.	-	-	-	-	-	-	-	-	160	-
Non-Metallic Minerals	-	-	-	-	-	-	-	-	97	-
Non-specified	14985	-	-	-	-	-	-	-	381	1057
TRANSPORT SECTOR	-	-	-	-	-	-	-	-	151	-
Air	-	-	-	-	-	-	-	-	-	-
Road	-	-	-	-	-	-	-	-	-	-
Non-specified	-	-	-	-	-	-	-	-	151	-
OTHER SECTORS	6372	-	-	-	42	-	-	-	2742	889
Agriculture	905	-	-	-	-	-	-	-	342	-
Comm. and Publ. Services	-	-	-	-	-	-	-	-	241	-
Residential	-	-	-	-	-	-	-	-	2159	889
Non-specified	5467	-	-	-	42	-	-	-	-	-
NON-ENERGY USE	-	-	-	-	-	-	-	-	-	-

Azerbaijan / Azerbaïdjan

SUPPLY AND CONSUMPTION 1997	Coal (1000 tonnes)							Oil (1000 tonnes)			
	Coking Coal	Other Bit. Coal	Sub-Bit. Coal	Lignite	Peat	Oven and Gas Coke	Pat. Fuel and BKB	Crude Oil	NGL	Feed-stocks	Additives
Production	-	-	-	-	-	-	-	8822	200	-	-
Imports	-	6	-	-	-	-	-	260	-	-	-
Exports	-	-	-	-	-	-	-	-300	-	-	-
Intl. Marine Bunkers	-	-	-	-	-	-	-	-	-	-	-
Stock Changes	-	-	-	-	-	-	-	-	-	-	-
DOMESTIC SUPPLY	-	6	-	-	-	-	-	8782	200	-	-
Transfers and Stat. Diff.	-	-	-	-	-	-	-	-	-	-	-
TRANSFORMATION	-	-	-	-	-	-	-	8782	200	-	-
Electricity and CHP Plants	-	-	-	-	-	-	-	-	-	-	-
Petroleum Refineries	-	-	-	-	-	-	-	8782	200	-	-
Other Transform. Sector	-	-	-	-	-	-	-	-	-	-	-
ENERGY SECTOR	-	-	-	-	-	-	-	-	-	-	-
DISTRIBUTION LOSSES	-	-	-	-	-	-	-	-	-	-	-
FINAL CONSUMPTION	-	6	-	-	-	-	-	-	-	-	-
INDUSTRY SECTOR	-	-	-	-	-	-	-	-	-	-	-
Iron and Steel	-	-	-	-	-	-	-	-	-	-	-
Chemical and Petrochem.	-	-	-	-	-	-	-	-	-	-	-
Non-Metallic Minerals	-	-	-	-	-	-	-	-	-	-	-
Non-specified	-	-	-	-	-	-	-	-	-	-	-
TRANSPORT SECTOR	-	6	-	-	-	-	-	-	-	-	-
Air	-	-	-	-	-	-	-	-	-	-	-
Road	-	-	-	-	-	-	-	-	-	-	-
Non-specified	-	6	-	-	-	-	-	-	-	-	-
OTHER SECTORS	-	-	-	-	-	-	-	-	-	-	-
Agriculture	-	-	-	-	-	-	-	-	-	-	-
Comm. and Publ. Services	-	-	-	-	-	-	-	-	-	-	-
Residential	-	-	-	-	-	-	-	-	-	-	-
Non-specified	-	-	-	-	-	-	-	-	-	-	-
NON-ENERGY USE	-	-	-	-	-	-	-	-	-	-	-

APPROVISIONNEMENT ET DEMANDE 1998	Charbon (1000 tonnes)							Pétrole (1000 tonnes)			
	Charbon à coke	Autres charb. bit.	Charbon sous-bit.	Lignite	Tourbe	Coke de four/gaz	Agg./briq. de lignite	Pétrole brut	LGN	Produits d'aliment.	Additifs
Production	-	-	-	-	-	-	-	11247	177	-	-
Imports	-	1	-	-	-	-	-	90	-	-	-
Exports	-	-	-	-	-	-	-	-2149	-	-	-
Intl. Marine Bunkers	-	-	-	-	-	-	-	-	-	-	-
Stock Changes	-	-	-	-	-	-	-	-	-	-	-
DOMESTIC SUPPLY	-	1	-	-	-	-	-	9188	177	-	-
Transfers and Stat. Diff.	-	-	-	-	-	-	-	-	-	-	-
TRANSFORMATION	-	-	-	-	-	-	-	9188	177	-	-
Electricity and CHP Plants	-	-	-	-	-	-	-	-	-	-	-
Petroleum Refineries	-	-	-	-	-	-	-	9188	177	-	-
Other Transform. Sector	-	-	-	-	-	-	-	-	-	-	-
ENERGY SECTOR	-	-	-	-	-	-	-	-	-	-	-
DISTRIBUTION LOSSES	-	-	-	-	-	-	-	-	-	-	-
FINAL CONSUMPTION	-	1	-	-	-	-	-	-	-	-	-
INDUSTRY SECTOR	-	-	-	-	-	-	-	-	-	-	-
Iron and Steel	-	-	-	-	-	-	-	-	-	-	-
Chemical and Petrochem.	-	-	-	-	-	-	-	-	-	-	-
Non-Metallic Minerals	-	-	-	-	-	-	-	-	-	-	-
Non-specified	-	-	-	-	-	-	-	-	-	-	-
TRANSPORT SECTOR	-	1	-	-	-	-	-	-	-	-	-
Air	-	-	-	-	-	-	-	-	-	-	-
Road	-	-	-	-	-	-	-	-	-	-	-
Non-specified	-	1	-	-	-	-	-	-	-	-	-
OTHER SECTORS	-	-	-	-	-	-	-	-	-	-	-
Agriculture	-	-	-	-	-	-	-	-	-	-	-
Comm. and Publ. Services	-	-	-	-	-	-	-	-	-	-	-
Residential	-	-	-	-	-	-	-	-	-	-	-
Non-specified	-	-	-	-	-	-	-	-	-	-	-
NON-ENERGY USE	-	-	-	-	-	-	-	-	-	-	-

Azerbaijan / Azerbaïdjan

SUPPLY AND CONSUMPTION 1997	Oil cont. (1000 tonnes)										
	Refinery Gas	LPG + Ethane	Motor Gasoline	Aviation Gasoline	Jet Fuel	Kerosene	Gas/ Diesel	Heavy Fuel Oil	Naphtha	Petrol. Coke	Other Prod.
Production	100	3	800	1	450	150	1900	3900	-	50	168
Imports	-	-	-	-	-	-	68	-	-	-	4
Exports	-	-	-308	-	-	-	-1629	-	-	-50	-139
Intl. Marine Bunkers	-	-	-	-	-	-	-	-	-	-	-
Stock Changes	-	-	-	-	-	-	-	-	-	-	-
DOMESTIC SUPPLY	100	3	492	1	450	150	339	3900	-	-	33
Transfers and Stat. Diff.	-	-	-	-	-	-	-	-	-	-	-
TRANSFORMATION	-	-	-	-	-	-	-	3683	-	-	-
Electricity and CHP Plants	-	-	-	-	-	-	-	3683	-	-	-
Petroleum Refineries	-	-	-	-	-	-	-	-	-	-	-
Other Transform. Sector	-	-	-	-	-	-	-	-	-	-	-
ENERGY SECTOR	100	-	-	-	-	-	-	-	-	-	-
DISTRIBUTION LOSSES	-	-	-	-	-	-	-	-	-	-	-
FINAL CONSUMPTION	-	3	492	1	450	150	339	217	-	-	33
INDUSTRY SECTOR	-	-	-	-	-	-	-	150	-	-	-
Iron and Steel	-	-	-	-	-	-	-	-	-	-	-
Chemical and Petrochem.	-	-	-	-	-	-	-	-	-	-	-
Non-Metallic Minerals	-	-	-	-	-	-	-	-	-	-	-
Non-specified	-	-	-	-	-	-	-	150	-	-	-
TRANSPORT SECTOR	-	-	492	1	450	-	-	10	-	-	-
Air	-	-	-	1	450	-	-	-	-	-	-
Road	-	-	492	-	-	-	-	-	-	-	-
Non-specified	-	-	-	-	-	-	-	10	-	-	-
OTHER SECTORS	-	3	-	-	-	150	339	57	-	-	-
Agriculture	-	-	-	-	-	-	-	35	-	-	-
Comm. and Publ. Services	-	-	-	-	-	-	-	12	-	-	-
Residential	-	-	-	-	-	-	-	10	-	-	-
Non-specified	-	3	-	-	-	150	339	-	-	-	-
NON-ENERGY USE	-	-	-	-	-	-	-	-	-	-	33

APPROVISIONNEMENT ET DEMANDE 1998	Pétrole cont. (1000 tonnes)										
	Gaz de raffinerie	GPL + éthane	Essence moteur	Essence aviation	Carbu- réacteurs	Kérosène	Gazole	Fioul lourd	Naphta	Coke de pétrole	Autres prod.
Production	100	56	630	-	526	178	2057	4028	-	30	107
Imports	-	2	7	-	-	21	49	140	-	-	10
Exports	-	-12	-191	-	-	-28	-1580	-29	-	-30	-58
Intl. Marine Bunkers	-	-	-	-	-	-	-	-	-	-	-
Stock Changes	-	-	-	-	-	-	-	-	-	-	-
DOMESTIC SUPPLY	100	46	446	-	526	171	526	4139	-	-	59
Transfers and Stat. Diff.	-	-	-	-	-	-	-	-	-	-	-
TRANSFORMATION	-	-	-	-	-	-	-	3908	-	-	-
Electricity and CHP Plants	-	-	-	-	-	-	-	3908	-	-	-
Petroleum Refineries	-	-	-	-	-	-	-	-	-	-	-
Other Transform. Sector	-	-	-	-	-	-	-	-	-	-	-
ENERGY SECTOR	100	-	-	-	-	-	-	-	-	-	-
DISTRIBUTION LOSSES	-	-	-	-	-	-	-	-	-	-	-
FINAL CONSUMPTION	-	46	446	-	526	171	526	231	-	-	59
INDUSTRY SECTOR	-	-	-	-	-	-	-	159	-	-	2
Iron and Steel	-	-	-	-	-	-	-	-	-	-	-
Chemical and Petrochem.	-	-	-	-	-	-	-	-	-	-	-
Non-Metallic Minerals	-	-	-	-	-	-	-	-	-	-	-
Non-specified	-	-	-	-	-	-	-	159	-	-	2
TRANSPORT SECTOR	-	-	446	-	526	-	-	11	-	-	-
Air	-	-	-	-	526	-	-	-	-	-	-
Road	-	-	446	-	-	-	-	-	-	-	-
Non-specified	-	-	-	-	-	-	-	11	-	-	-
OTHER SECTORS	-	46	-	-	-	171	526	61	-	-	-
Agriculture	-	-	-	-	-	-	-	37	-	-	-
Comm. and Publ. Services	-	-	-	-	-	-	-	13	-	-	-
Residential	-	-	-	-	-	-	-	11	-	-	-
Non-specified	-	46	-	-	-	171	526	-	-	-	-
NON-ENERGY USE	-	-	-	-	-	-	-	-	-	-	57

Azerbaijan / Azerbaïdjan

SUPPLY AND CONSUMPTION 1997	Gas (TJ)				Comb. Renew. & Waste (TJ)				(GWh)	(TJ)
	Natural Gas	Gas Works	Coke Ovens	Blast Furnaces	Solid Biomass	Gas/Liquids from Biomass	Municipal Waste	Industrial Waste	Electricity	Heat
Production	224839	-	-	-	68	-	-	-	16836	14715
Imports	-	-	-	-	117	-	-	-	715	-
Exports	-	-	-	-	-	-	-	-	-284	-
Intl. Marine Bunkers	-	-	-	-	-	-	-	-	-	-
Stock Changes	-	-	-	-	-	-	-	-	-	-
DOMESTIC SUPPLY	224839	-	-	-	185	-	-	-	17267	14715
Transfers and Stat. Diff.	-	-	-	-	-	-	-	-	-	-
TRANSFORMATION	77189	-	-	-	-	-	-	-	-	-
Electricity and CHP Plants	77189	-	-	-	-	-	-	-	-	-
Petroleum Refineries	-	-	-	-	-	-	-	-	-	-
Other Transform. Sector	-	-	-	-	-	-	-	-	-	-
ENERGY SECTOR	-	-	-	-	-	-	-	-	2885	-
DISTRIBUTION LOSSES	-	-	-	-	-	-	-	-	3108	-
FINAL CONSUMPTION	147650	-	-	-	185	-	-	-	11274	14715
INDUSTRY SECTOR	88590	-	-	-	-	-	-	-	1115	14715
Iron and Steel	-	-	-	-	-	-	-	-	52	-
Chemical and Petrochem.	-	-	-	-	-	-	-	-	496	-
Non-Metallic Minerals	-	-	-	-	-	-	-	-	76	-
Non-specified	88590	-	-	-	-	-	-	-	491	14715
TRANSPORT SECTOR	1450	-	-	-	-	-	-	-	496	-
Air	-	-	-	-	-	-	-	-	-	-
Road	-	-	-	-	-	-	-	-	-	-
Non-specified	1450	-	-	-	-	-	-	-	496	-
OTHER SECTORS	57610	-	-	-	185	-	-	-	9663	-
Agriculture	13320	-	-	-	-	-	-	-	1898	-
Comm. and Publ. Services	-	-	-	-	-	-	-	-	576	-
Residential	44290	-	-	-	-	-	-	-	360	-
Non-specified	-	-	-	-	185	-	-	-	6829	-
NON-ENERGY USE	-	-	-	-	-	-	-	-	-	-

APPROVISIONNEMENT ET DEMANDE 1998	Gaz (TJ)				En. Re. Comb. & Déchets (TJ)				(GWh)	(TJ)
	Gaz naturel	Usines à gaz	Cokeries	Hauts fourneaux	Biomasse solide	Gaz/Liquides tirés de biomasse	Déchets urbains	Déchets industriels	Electricité	Chaleur
Production	210724	-	-	-	68	-	-	-	17985	12659
Imports	-	-	-	-	117	-	-	-	903	-
Exports	-	-	-	-	-	-	-	-	-648	-
Intl. Marine Bunkers	-	-	-	-	-	-	-	-	-	-
Stock Changes	-	-	-	-	-	-	-	-	-	-
DOMESTIC SUPPLY	210724	-	-	-	185	-	-	-	18240	12659
Transfers and Stat. Diff.	-	-	-	-	-	-	-	-	-	-
TRANSFORMATION	96904	-	-	-	-	-	-	-	-	-
Electricity and CHP Plants	96904	-	-	-	-	-	-	-	-	-
Petroleum Refineries	-	-	-	-	-	-	-	-	-	-
Other Transform. Sector	-	-	-	-	-	-	-	-	-	-
ENERGY SECTOR	-	-	-	-	-	-	-	-	2792	-
DISTRIBUTION LOSSES	-	-	-	-	-	-	-	-	2921	-
FINAL CONSUMPTION	113820	-	-	-	185	-	-	-	12527	12659
INDUSTRY SECTOR	64880	-	-	-	-	-	-	-	911	12659
Iron and Steel	-	-	-	-	-	-	-	-	27	-
Chemical and Petrochem.	-	-	-	-	-	-	-	-	409	-
Non-Metallic Minerals	-	-	-	-	-	-	-	-	66	-
Non-specified	64880	-	-	-	-	-	-	-	409	12659
TRANSPORT SECTOR	1140	-	-	-	-	-	-	-	422	-
Air	-	-	-	-	-	-	-	-	-	-
Road	-	-	-	-	-	-	-	-	-	-
Non-specified	1140	-	-	-	-	-	-	-	422	-
OTHER SECTORS	47800	-	-	-	185	-	-	-	11194	-
Agriculture	10240	-	-	-	-	-	-	-	1659	-
Comm. and Publ. Services	-	-	-	-	-	-	-	-	844	-
Residential	37560	-	-	-	-	-	-	-	423	-
Non-specified	-	-	-	-	185	-	-	-	8268	-
NON-ENERGY USE	-	-	-	-	-	-	-	-	-	-

Bahrain / Bahrein

SUPPLY AND CONSUMPTION 1997	Coking Coal	Other Bit. Coal	Sub-Bit. Coal	Lignite	Peat	Oven and Gas Coke	Pat. Fuel and BKB	Crude Oil	NGL	Feed-stocks	Additives
	Coal (1000 tonnes)							Oil (1000 tonnes)			
Production	-	-	-	-	-	-	-	1934	413	-	-
Imports	-	-	-	-	-	-	-	10593	-	-	-
Exports	-	-	-	-	-	-	-	-	-	-	-
Intl. Marine Bunkers	-	-	-	-	-	-	-	-	-	-	-
Stock Changes	-	-	-	-	-	-	-	263	-	-	-
DOMESTIC SUPPLY	-	-	-	-	-	-	-	**12790**	**413**	-	-
Transfers and Stat. Diff.	-	-	-	-	-	-	-	-	-413	-	-
TRANSFORMATION	-	-	-	-	-	-	-	**12790**	-	-	-
Electricity and CHP Plants	-	-	-	-	-	-	-	-	-	-	-
Petroleum Refineries	-	-	-	-	-	-	-	12790	-	-	-
Other Transform. Sector	-	-	-	-	-	-	-	-	-	-	-
ENERGY SECTOR	-	-	-	-	-	-	-	-	-	-	-
DISTRIBUTION LOSSES	-	-	-	-	-	-	-	-	-	-	-
FINAL CONSUMPTION	-	-	-	-	-	-	-	-	-	-	-
INDUSTRY SECTOR	-	-	-	-	-	-	-	-	-	-	-
Iron and Steel	-	-	-	-	-	-	-	-	-	-	-
Chemical and Petrochem.	-	-	-	-	-	-	-	-	-	-	-
Non-Metallic Minerals	-	-	-	-	-	-	-	-	-	-	-
Non-specified	-	-	-	-	-	-	-	-	-	-	-
TRANSPORT SECTOR	-	-	-	-	-	-	-	-	-	-	-
Air	-	-	-	-	-	-	-	-	-	-	-
Road	-	-	-	-	-	-	-	-	-	-	-
Non-specified	-	-	-	-	-	-	-	-	-	-	-
OTHER SECTORS	-	-	-	-	-	-	-	-	-	-	-
Agriculture	-	-	-	-	-	-	-	-	-	-	-
Comm. and Publ. Services	-	-	-	-	-	-	-	-	-	-	-
Residential	-	-	-	-	-	-	-	-	-	-	-
Non-specified	-	-	-	-	-	-	-	-	-	-	-
NON-ENERGY USE	-	-	-	-	-	-	-	-	-	-	-

APPROVISIONNEMENT ET DEMANDE 1998	Charbon à coke	Autres charb. bit.	Charbon sous-bit.	Lignite	Tourbe	Coke de four/gaz	Agg./briq. de lignite	Pétrole brut	LGN	Produits d'aliment.	Additifs
	Charbon (1000 tonnes)							Pétrole (1000 tonnes)			
Production	-	-	-	-	-	-	-	1879	397	-	-
Imports	-	-	-	-	-	-	-	10526	-	-	-
Exports	-	-	-	-	-	-	-	-	-	-	-
Intl. Marine Bunkers	-	-	-	-	-	-	-	-	-	-	-
Stock Changes	-	-	-	-	-	-	-	252	-	-	-
DOMESTIC SUPPLY	-	-	-	-	-	-	-	**12657**	**397**	-	-
Transfers and Stat. Diff.	-	-	-	-	-	-	-	-1	-397	-	-
TRANSFORMATION	-	-	-	-	-	-	-	**12656**	-	-	-
Electricity and CHP Plants	-	-	-	-	-	-	-	-	-	-	-
Petroleum Refineries	-	-	-	-	-	-	-	12656	-	-	-
Other Transform. Sector	-	-	-	-	-	-	-	-	-	-	-
ENERGY SECTOR	-	-	-	-	-	-	-	-	-	-	-
DISTRIBUTION LOSSES	-	-	-	-	-	-	-	-	-	-	-
FINAL CONSUMPTION	-	-	-	-	-	-	-	-	-	-	-
INDUSTRY SECTOR	-	-	-	-	-	-	-	-	-	-	-
Iron and Steel	-	-	-	-	-	-	-	-	-	-	-
Chemical and Petrochem.	-	-	-	-	-	-	-	-	-	-	-
Non-Metallic Minerals	-	-	-	-	-	-	-	-	-	-	-
Non-specified	-	-	-	-	-	-	-	-	-	-	-
TRANSPORT SECTOR	-	-	-	-	-	-	-	-	-	-	-
Air	-	-	-	-	-	-	-	-	-	-	-
Road	-	-	-	-	-	-	-	-	-	-	-
Non-specified	-	-	-	-	-	-	-	-	-	-	-
OTHER SECTORS	-	-	-	-	-	-	-	-	-	-	-
Agriculture	-	-	-	-	-	-	-	-	-	-	-
Comm. and Publ. Services	-	-	-	-	-	-	-	-	-	-	-
Residential	-	-	-	-	-	-	-	-	-	-	-
Non-specified	-	-	-	-	-	-	-	-	-	-	-
NON-ENERGY USE	-	-	-	-	-	-	-	-	-	-	-

Bahrain / Bahrein

SUPPLY AND CONSUMPTION 1997	Oil cont. (1000 tonnes)										
	Refinery Gas	LPG + Ethane	Motor Gasoline	Aviation Gasoline	Jet Fuel	Kerosene	Gas/ Diesel	Heavy Fuel Oil	Naphtha	Petrol. Coke	Other Prod.
Production	325	31	865	-	1208	999	4299	2881	1511	-	365
Imports	-	-	-	-	-	-	-	-	-	-	-
Exports	-	-169	-567	-	-1235	-944	-4255	-2788	-1704	-	-313
Intl. Marine Bunkers	-	-	-	-	-	-	-	-	-	-	-
Stock Changes	-	-31	6	-	267	-	74	-93	-9	-	-28
DOMESTIC SUPPLY	325	-169	304	-	240	55	118	-	-202	-	24
Transfers and Stat. Diff.	-	195	-	-	-	-1	1	-	202	-	-
TRANSFORMATION	-	-	-	-	-	-	-	-	-	-	-
Electricity and CHP Plants	-	-	-	-	-	-	-	-	-	-	-
Petroleum Refineries	-	-	-	-	-	-	-	-	-	-	-
Other Transform. Sector	-	-	-	-	-	-	-	-	-	-	-
ENERGY SECTOR	325	-	-	-	-	-	-	-	-	-	-
DISTRIBUTION LOSSES	-	-	-	-	-	-	-	-	-	-	-
FINAL CONSUMPTION	-	26	304	-	240	54	119	-	-	-	24
INDUSTRY SECTOR	-	-	-	-	-	-	-	-	-	-	-
Iron and Steel	-	-	-	-	-	-	-	-	-	-	-
Chemical and Petrochem.	-	-	-	-	-	-	-	-	-	-	-
Non-Metallic Minerals	-	-	-	-	-	-	-	-	-	-	-
Non-specified	-	-	-	-	-	-	-	-	-	-	-
TRANSPORT SECTOR	-	-	304	-	240	-	119	-	-	-	-
Air	-	-	-	-	240	-	-	-	-	-	-
Road	-	-	304	-	-	-	119	-	-	-	-
Non-specified	-	-	-	-	-	-	-	-	-	-	-
OTHER SECTORS	-	26	-	-	-	54	-	-	-	-	-
Agriculture	-	-	-	-	-	-	-	-	-	-	-
Comm. and Publ. Services	-	-	-	-	-	-	-	-	-	-	-
Residential	-	26	-	-	-	54	-	-	-	-	-
Non-specified	-	-	-	-	-	-	-	-	-	-	-
NON-ENERGY USE	-	-	-	-	-	-	-	-	-	-	24

APPROVISIONNEMENT ET DEMANDE 1998	Pétrole cont. (1000 tonnes)										
	Gaz de raffinerie	GPL + éthane	Essence moteur	Essence aviation	Carbu- réacteurs	Kérosène	Gazole	Fioul lourd	Naphta	Coke de pétrole	Autres prod.
Production	313	26	744	-	1172	1050	4235	2961	1540	-	313
Imports	-	-	-	-	-	-	-	-	-	-	-
Exports	-	-192	-443	-	-1215	-973	-4074	-3063	-1721	-	-277
Intl. Marine Bunkers	-	-	-	-	-	-	-	-	-	-	-
Stock Changes	-	1	21	-	388	-	-36	102	-14	-	-12
DOMESTIC SUPPLY	313	-165	322	-	345	77	125	-	-195	-	24
Transfers and Stat. Diff.	-	191	-1	-	1	-1	-	-	195	-	-
TRANSFORMATION	-	-	-	-	-	-	-	-	-	-	-
Electricity and CHP Plants	-	-	-	-	-	-	-	-	-	-	-
Petroleum Refineries	-	-	-	-	-	-	-	-	-	-	-
Other Transform. Sector	-	-	-	-	-	-	-	-	-	-	-
ENERGY SECTOR	313	-	-	-	-	-	-	-	-	-	-
DISTRIBUTION LOSSES	-	-	-	-	-	-	-	-	-	-	-
FINAL CONSUMPTION	-	26	321	-	346	76	125	-	-	-	24
INDUSTRY SECTOR	-	-	-	-	-	-	-	-	-	-	-
Iron and Steel	-	-	-	-	-	-	-	-	-	-	-
Chemical and Petrochem.	-	-	-	-	-	-	-	-	-	-	-
Non-Metallic Minerals	-	-	-	-	-	-	-	-	-	-	-
Non-specified	-	-	-	-	-	-	-	-	-	-	-
TRANSPORT SECTOR	-	-	321	-	346	-	125	-	-	-	-
Air	-	-	-	-	346	-	-	-	-	-	-
Road	-	-	321	-	-	-	125	-	-	-	-
Non-specified	-	-	-	-	-	-	-	-	-	-	-
OTHER SECTORS	-	26	-	-	-	76	-	-	-	-	-
Agriculture	-	-	-	-	-	-	-	-	-	-	-
Comm. and Publ. Services	-	-	-	-	-	-	-	-	-	-	-
Residential	-	26	-	-	-	76	-	-	-	-	-
Non-specified	-	-	-	-	-	-	-	-	-	-	-
NON-ENERGY USE	-	-	-	-	-	-	-	-	-	-	24

Bahrain / Bahrein

SUPPLY AND CONSUMPTION 1997	Gas (TJ)				Comb. Renew. & Waste (TJ)				(GWh)	(TJ)
	Natural Gas	Gas Works	Coke Ovens	Blast Furnaces	Solid Biomass	Gas/Liquids from Biomass	Municipal Waste	Industrial Waste	Electricity	Heat
Production	203975	-	-	-	-	-	-	-	5040	-
Imports	-	-	-	-	-	-	-	-	-	-
Exports	-	-	-	-	-	-	-	-	-	-
Intl. Marine Bunkers	-	-	-	-	-	-	-	-	-	-
Stock Changes	-	-	-	-	-	-	-	-	-	-
DOMESTIC SUPPLY	203975	-	-	-	-	-	-	-	5040	-
Transfers and Stat. Diff.	59848	-	-	-	-	-	-	-	1	-
TRANSFORMATION	81025	-	-	-	-	-	-	-	-	-
Electricity and CHP Plants	81025	-	-	-	-	-	-	-	-	-
Petroleum Refineries	-	-	-	-	-	-	-	-	-	-
Other Transform. Sector	-	-	-	-	-	-	-	-	-	-
ENERGY SECTOR	39178	-	-	-	-	-	-	-	261	-
DISTRIBUTION LOSSES	205	-	-	-	-	-	-	-	85	-
FINAL CONSUMPTION	143415	-	-	-	-	-	-	-	4695	-
INDUSTRY SECTOR	143415	-	-	-	-	-	-	-	739	-
Iron and Steel	23007	-	-	-	-	-	-	-	-	-
Chemical and Petrochem.	119375	-	-	-	-	-	-	-	-	-
Non-Metallic Minerals	-	-	-	-	-	-	-	-	-	-
Non-specified	1033	-	-	-	-	-	-	-	739	-
TRANSPORT SECTOR	-	-	-	-	-	-	-	-	-	-
Air	-	-	-	-	-	-	-	-	-	-
Road	-	-	-	-	-	-	-	-	-	-
Non-specified	-	-	-	-	-	-	-	-	-	-
OTHER SECTORS	-	-	-	-	-	-	-	-	3956	-
Agriculture	-	-	-	-	-	-	-	-	31	-
Comm. and Publ. Services	-	-	-	-	-	-	-	-	1252	-
Residential	-	-	-	-	-	-	-	-	2673	-
Non-specified	-	-	-	-	-	-	-	-	-	-
NON-ENERGY USE	-	-	-	-	-	-	-	-	-	-

APPROVISIONNEMENT ET DEMANDE 1998	Gaz (TJ)				En. Re. Comb. & Déchets (TJ)				(GWh)	(TJ)
	Gaz naturel	Usines à gaz	Cokeries	Hauts fourneaux	Biomasse solide	Gaz/Liquids tirés de biomasse	Déchets urbains	Déchets industriels	Electricité	Chaleur
Production	220443	-	-	-	-	-	-	-	5773	-
Imports	-	-	-	-	-	-	-	-	-	-
Exports	-	-	-	-	-	-	-	-	-	-
Intl. Marine Bunkers	-	-	-	-	-	-	-	-	-	-
Stock Changes	-	-	-	-	-	-	-	-	-	-
DOMESTIC SUPPLY	220443	-	-	-	-	-	-	-	5773	-
Transfers and Stat. Diff.	36051	-	-	-	-	-	-	-	1	-
TRANSFORMATION	76838	-	-	-	-	-	-	-	-	-
Electricity and CHP Plants	76838	-	-	-	-	-	-	-	-	-
Petroleum Refineries	-	-	-	-	-	-	-	-	-	-
Other Transform. Sector	-	-	-	-	-	-	-	-	-	-
ENERGY SECTOR	37452	-	-	-	-	-	-	-	398	-
DISTRIBUTION LOSSES	199	-	-	-	-	-	-	-	460	-
FINAL CONSUMPTION	142005	-	-	-	-	-	-	-	4916	-
INDUSTRY SECTOR	142005	-	-	-	-	-	-	-	825	-
Iron and Steel	21028	-	-	-	-	-	-	-	-	-
Chemical and Petrochem.	119864	-	-	-	-	-	-	-	-	-
Non-Metallic Minerals	-	-	-	-	-	-	-	-	-	-
Non-specified	1113	-	-	-	-	-	-	-	825	-
TRANSPORT SECTOR	-	-	-	-	-	-	-	-	-	-
Air	-	-	-	-	-	-	-	-	-	-
Road	-	-	-	-	-	-	-	-	-	-
Non-specified	-	-	-	-	-	-	-	-	-	-
OTHER SECTORS	-	-	-	-	-	-	-	-	4091	-
Agriculture	-	-	-	-	-	-	-	-	38	-
Comm. and Publ. Services	-	-	-	-	-	-	-	-	1380	-
Residential	-	-	-	-	-	-	-	-	2673	-
Non-specified	-	-	-	-	-	-	-	-	-	-
NON-ENERGY USE	-	-	-	-	-	-	-	-	-	-

Bangladesh

SUPPLY AND CONSUMPTION 1997	Coal (1000 tonnes)							Oil (1000 tonnes)			
	Coking Coal	Other Bit. Coal	Sub-Bit. Coal	Lignite	Peat	Oven and Gas Coke	Pat. Fuel and BKB	Crude Oil	NGL	Feed-stocks	Additives
Production	-	-	-	-	-	-	-	1	-	-	-
Imports	-	635	-	-	-	-	-	1300	-	-	-
Exports	-	-	-	-	-	-	-	-	-	-	-
Intl. Marine Bunkers	-	-	-	-	-	-	-	-	-	-	-
Stock Changes	-	-	-	-	-	-	-	119	-	-	-
DOMESTIC SUPPLY	-	635	-	-	-	-	-	1420	-	-	-
Transfers and Stat. Diff.	-	-	-	-	-	-	-	-	-	-	-
TRANSFORMATION	-	-	-	-	-	-	-	1420	-	-	-
Electricity and CHP Plants	-	-	-	-	-	-	-	-	-	-	-
Petroleum Refineries	-	-	-	-	-	-	-	1420	-	-	-
Other Transform. Sector	-	-	-	-	-	-	-	-	-	-	-
ENERGY SECTOR	-	-	-	-	-	-	-	-	-	-	-
DISTRIBUTION LOSSES	-	-	-	-	-	-	-	-	-	-	-
FINAL CONSUMPTION	-	635	-	-	-	-	-	-	-	-	-
INDUSTRY SECTOR	-	635	-	-	-	-	-	-	-	-	-
Iron and Steel	-	-	-	-	-	-	-	-	-	-	-
Chemical and Petrochem.	-	-	-	-	-	-	-	-	-	-	-
Non-Metallic Minerals	-	635	-	-	-	-	-	-	-	-	-
Non-specified	-	-	-	-	-	-	-	-	-	-	-
TRANSPORT SECTOR	-	-	-	-	-	-	-	-	-	-	-
Air	-	-	-	-	-	-	-	-	-	-	-
Road	-	-	-	-	-	-	-	-	-	-	-
Non-specified	-	-	-	-	-	-	-	-	-	-	-
OTHER SECTORS	-	-	-	-	-	-	-	-	-	-	-
Agriculture	-	-	-	-	-	-	-	-	-	-	-
Comm. and Publ. Services	-	-	-	-	-	-	-	-	-	-	-
Residential	-	-	-	-	-	-	-	-	-	-	-
Non-specified	-	-	-	-	-	-	-	-	-	-	-
NON-ENERGY USE	-	-	-	-	-	-	-	-	-	-	-

APPROVISIONNEMENT ET DEMANDE 1998	Charbon (1000 tonnes)							Pétrole (1000 tonnes)			
	Charbon à coke	Autres charb. bit.	Charbon sous-bit.	Lignite	Tourbe	Coke de four/gaz	Agg./briq. de lignite	Pétrole brut	LGN	Produits d'aliment.	Additifs
Production	-	-	-	-	-	-	-	1	-	-	-
Imports	-	173	-	-	-	-	-	1119	-	-	-
Exports	-	-	-	-	-	-	-	-	-	-	-
Intl. Marine Bunkers	-	-	-	-	-	-	-	-	-	-	-
Stock Changes	-	-	-	-	-	-	-	-401	-	-	-
DOMESTIC SUPPLY	-	173	-	-	-	-	-	719	-	-	-
Transfers and Stat. Diff.	-	-	-	-	-	-	-	-1	-	-	-
TRANSFORMATION	-	-	-	-	-	-	-	718	-	-	-
Electricity and CHP Plants	-	-	-	-	-	-	-	-	-	-	-
Petroleum Refineries	-	-	-	-	-	-	-	718	-	-	-
Other Transform. Sector	-	-	-	-	-	-	-	-	-	-	-
ENERGY SECTOR	-	-	-	-	-	-	-	-	-	-	-
DISTRIBUTION LOSSES	-	-	-	-	-	-	-	-	-	-	-
FINAL CONSUMPTION	-	173	-	-	-	-	-	-	-	-	-
INDUSTRY SECTOR	-	173	-	-	-	-	-	-	-	-	-
Iron and Steel	-	-	-	-	-	-	-	-	-	-	-
Chemical and Petrochem.	-	-	-	-	-	-	-	-	-	-	-
Non-Metallic Minerals	-	173	-	-	-	-	-	-	-	-	-
Non-specified	-	-	-	-	-	-	-	-	-	-	-
TRANSPORT SECTOR	-	-	-	-	-	-	-	-	-	-	-
Air	-	-	-	-	-	-	-	-	-	-	-
Road	-	-	-	-	-	-	-	-	-	-	-
Non-specified	-	-	-	-	-	-	-	-	-	-	-
OTHER SECTORS	-	-	-	-	-	-	-	-	-	-	-
Agriculture	-	-	-	-	-	-	-	-	-	-	-
Comm. and Publ. Services	-	-	-	-	-	-	-	-	-	-	-
Residential	-	-	-	-	-	-	-	-	-	-	-
Non-specified	-	-	-	-	-	-	-	-	-	-	-
NON-ENERGY USE	-	-	-	-	-	-	-	-	-	-	-

Bangladesh

SUPPLY AND CONSUMPTION 1997	Oil cont. (1000 tonnes)										
	Refinery Gas	LPG + Ethane	Motor Gasoline	Aviation Gasoline	Jet Fuel	Kerosene	Gas/ Diesel	Heavy Fuel Oil	Naphtha	Petrol. Coke	Other Prod.
Production	45	16	173	-	5	304	272	127	13	-	397
Imports	-	-	106	-	96	221	1142	202	60	-	31
Exports	-	-	-	-	-	-	-	-	-	-	-
Intl. Marine Bunkers	-	-	-	-	-	-	-	-36	-	-	-
Stock Changes	-	-	-	-	7	14	68	-	-	-	-
DOMESTIC SUPPLY	45	16	279	-	108	539	1482	293	73	-	428
Transfers and Stat. Diff.	-	-	-	-	-	-	-	-	-	-	-2
TRANSFORMATION	-	-	-	-	-	-	256	118	-	-	-
Electricity and CHP Plants	-	-	-	-	-	-	256	118	-	-	-
Petroleum Refineries	-	-	-	-	-	-	-	-	-	-	-
Other Transform. Sector	-	-	-	-	-	-	-	-	-	-	-
ENERGY SECTOR	45	-	-	-	-	-	-	27	-	-	-
DISTRIBUTION LOSSES	-	-	-	-	-	-	-	-	-	-	-
FINAL CONSUMPTION	-	16	279	-	108	539	1226	148	73	-	426
INDUSTRY SECTOR	-	-	-	-	-	-	41	148	73	-	-
Iron and Steel	-	-	-	-	-	-	-	-	-	-	-
Chemical and Petrochem.	-	-	-	-	-	-	-	-	-	-	-
Non-Metallic Minerals	-	-	-	-	-	-	-	-	-	-	-
Non-specified	-	-	-	-	-	-	41	148	73	-	-
TRANSPORT SECTOR	-	-	279	-	108	-	764	-	-	-	-
Air	-	-	-	-	108	-	-	-	-	-	-
Road	-	-	279	-	-	-	489	-	-	-	-
Non-specified	-	-	-	-	-	-	275	-	-	-	-
OTHER SECTORS	-	16	-	-	-	539	421	-	-	-	-
Agriculture	-	-	-	-	-	-	421	-	-	-	-
Comm. and Publ. Services	-	-	-	-	-	-	-	-	-	-	-
Residential	-	16	-	-	-	539	-	-	-	-	-
Non-specified	-	-	-	-	-	-	-	-	-	-	-
NON-ENERGY USE	-	-	-	-	-	-	-	-	-	-	426

APPROVISIONNEMENT ET DEMANDE 1998	Pétrole cont. (1000 tonnes)										
	Gaz de raffinerie	GPL + éthane	Essence moteur	Essence aviation	Carbu- réacteurs	Kérosène	Gazole	Fioul lourd	Naphta	Coke de pétrole	Autres prod.
Production	23	9	80	-	6	187	100	64	11	-	204
Imports	-	-	204	-	52	305	1223	194	62	-	31
Exports	-	-	-	-	-	-	-	-	-	-	-
Intl. Marine Bunkers	-	-	-	-	-	-	-	-36	-	-	-
Stock Changes	-	-	-	-	-	94	219	-	-	-	-
DOMESTIC SUPPLY	23	9	284	-	58	586	1542	222	73	-	235
Transfers and Stat. Diff.	-	-	-	-	-	-	-1	-1	-	-	-
TRANSFORMATION	-	-	-	-	-	-	270	102	-	-	-
Electricity and CHP Plants	-	-	-	-	-	-	270	102	-	-	-
Petroleum Refineries	-	-	-	-	-	-	-	-	-	-	-
Other Transform. Sector	-	-	-	-	-	-	-	-	-	-	-
ENERGY SECTOR	23	-	-	-	-	-	-	20	-	-	-
DISTRIBUTION LOSSES	-	-	-	-	-	-	-	-	-	-	-
FINAL CONSUMPTION	-	9	284	-	58	586	1271	99	73	-	235
INDUSTRY SECTOR	-	-	-	-	-	-	42	99	73	-	-
Iron and Steel	-	-	-	-	-	-	-	-	-	-	-
Chemical and Petrochem.	-	-	-	-	-	-	-	-	-	-	-
Non-Metallic Minerals	-	-	-	-	-	-	-	-	-	-	-
Non-specified	-	-	-	-	-	-	42	99	73	-	-
TRANSPORT SECTOR	-	-	284	-	58	-	792	-	-	-	-
Air	-	-	-	-	58	-	-	-	-	-	-
Road	-	-	284	-	-	-	507	-	-	-	-
Non-specified	-	-	-	-	-	-	285	-	-	-	-
OTHER SECTORS	-	9	-	-	-	586	437	-	-	-	-
Agriculture	-	-	-	-	-	-	437	-	-	-	-
Comm. and Publ. Services	-	-	-	-	-	-	-	-	-	-	-
Residential	-	9	-	-	-	586	-	-	-	-	-
Non-specified	-	-	-	-	-	-	-	-	-	-	-
NON-ENERGY USE	-	-	-	-	-	-	-	-	-	-	235

Bangladesh

SUPPLY AND CONSUMPTION 1997	Gas (TJ) Natural Gas	Gas Works	Coke Ovens	Blast Furnaces	Comb. Renew. & Waste (TJ) Solid Biomass	Gas/Liquids from Biomass	Municipal Waste	Industrial Waste	(GWh) Electricity	(TJ) Heat
Production	269618	-	-	-	433920	-	-	-	11858	-
Imports	-	-	-	-	-	-	-	-	-	-
Exports	-	-	-	-	-	-	-	-	-	-
Intl. Marine Bunkers	-	-	-	-	-	-	-	-	-	-
Stock Changes	-	-	-	-	-	-	-	-	-	-
DOMESTIC SUPPLY	269618	-	-	-	433920	-	-	-	11858	-
Transfers and Stat. Diff.	-	-	-	-	-	-	-	-	1	-
TRANSFORMATION	114451	-	-	-	-	-	-	-	-	-
Electricity and CHP Plants	114451	-	-	-	-	-	-	-	-	-
Petroleum Refineries	-	-	-	-	-	-	-	-	-	-
Other Transform. Sector	-	-	-	-	-	-	-	-	-	-
ENERGY SECTOR	-	-	-	-	-	-	-	-	615	-
DISTRIBUTION LOSSES	17038	-	-	-	-	-	-	-	1796	-
FINAL CONSUMPTION	138129	-	-	-	433920	-	-	-	9448	-
INDUSTRY SECTOR	111167	-	-	-	-	-	-	-	7486	-
Iron and Steel	36	-	-	-	-	-	-	-	-	-
Chemical and Petrochem.	80448	-	-	-	-	-	-	-	-	-
Non-Metallic Minerals	-	-	-	-	-	-	-	-	-	-
Non-specified	30683	-	-	-	-	-	-	-	7486	-
TRANSPORT SECTOR	-	-	-	-	-	-	-	-	-	-
Air	-	-	-	-	-	-	-	-	-	-
Road	-	-	-	-	-	-	-	-	-	-
Non-specified	-	-	-	-	-	-	-	-	-	-
OTHER SECTORS	26962	-	-	-	433920	-	-	-	1962	-
Agriculture	-	-	-	-	-	-	-	-	154	-
Comm. and Publ. Services	3393	-	-	-	-	-	-	-	413	-
Residential	23569	-	-	-	433920	-	-	-	1319	-
Non-specified	-	-	-	-	-	-	-	-	76	-
NON-ENERGY USE	-	-	-	-	-	-	-	-	-	-

APPROVISIONNEMENT ET DEMANDE 1998	Gaz (TJ) Gaz naturel	Usines à gaz	Cokeries	Hauts fourneaux	En. Re. Comb. & Déchets (TJ) Biomasse solide	Gaz/Liquides tirés de biomasse	Déchets urbains	Déchets industriels	(GWh) Electricité	(TJ) Chaleur
Production	292557	-	-	-	433920	-	-	-	12882	-
Imports	-	-	-	-	-	-	-	-	-	-
Exports	-	-	-	-	-	-	-	-	-	-
Intl. Marine Bunkers	-	-	-	-	-	-	-	-	-	-
Stock Changes	-	-	-	-	-	-	-	-	-	-
DOMESTIC SUPPLY	292557	-	-	-	433920	-	-	-	12882	-
Transfers and Stat. Diff.	-	-	-	-	-	-	-	-	1	-
TRANSFORMATION	127512	-	-	-	-	-	-	-	-	-
Electricity and CHP Plants	127512	-	-	-	-	-	-	-	-	-
Petroleum Refineries	-	-	-	-	-	-	-	-	-	-
Other Transform. Sector	-	-	-	-	-	-	-	-	-	-
ENERGY SECTOR	-	-	-	-	-	-	-	-	688	-
DISTRIBUTION LOSSES	18488	-	-	-	-	-	-	-	2018	-
FINAL CONSUMPTION	146557	-	-	-	433920	-	-	-	10177	-
INDUSTRY SECTOR	117260	-	-	-	-	-	-	-	8167	-
Iron and Steel	-	-	-	-	-	-	-	-	-	-
Chemical and Petrochem.	82673	-	-	-	-	-	-	-	-	-
Non-Metallic Minerals	-	-	-	-	-	-	-	-	-	-
Non-specified	34587	-	-	-	-	-	-	-	8167	-
TRANSPORT SECTOR	-	-	-	-	-	-	-	-	-	-
Air	-	-	-	-	-	-	-	-	-	-
Road	-	-	-	-	-	-	-	-	-	-
Non-specified	-	-	-	-	-	-	-	-	-	-
OTHER SECTORS	29297	-	-	-	433920	-	-	-	2010	-
Agriculture	-	-	-	-	-	-	-	-	154	-
Comm. and Publ. Services	3539	-	-	-	-	-	-	-	430	-
Residential	25758	-	-	-	433920	-	-	-	1352	-
Non-specified	-	-	-	-	-	-	-	-	74	-
NON-ENERGY USE	-	-	-	-	-	-	-	-	-	-

Belarus / Bélarus : 1997

SUPPLY AND CONSUMPTION	Coal / Charbon (1000 tonnes)							Oil / Pétrole (1000 tonnes)			
	Coking Coal	Other Bit. Coal	Sub-Bit. Coal	Lignite	Peat	Oven and Gas Coke	Pat. Fuel and BKB	Crude Oil	NGL	Feed-stocks	Additives
APPROVISIONNEMENT ET DEMANDE	Charbon à coke	Autres charb. bit.	Charbon sous-bit.	Lignite	Tourbe	Coke de four/gaz	Agg./briq. de lignite	Pétrole brut	LGN	Produits d'aliment.	Additifs
Production	-	-	-	-	2763	-	1515	1822	-	-	-
From Other Sources	-	-	-	-	-	-	-	-	-	-	-
Imports	-	786	-	-	-	-	-	10461	-	-	-
Exports	-	-	-	-	-96	-	-79	-400	-	-	-
Intl. Marine Bunkers	-	-	-	-	-	-	-	-	-	-	-
Stock Changes	-	24	-	-	-1	-	-2	-95	-	-	-
DOMESTIC SUPPLY	-	810	-	-	2666	-	1434	11788	-	-	-
Transfers	-	-	-	-	-	-	-	-	-	-	-
Statistical Differences	-	-	-	-	-	-	-	72	-	-	-
TRANSFORMATION	-	252	-	-	2508	-	95	11199	-	-	-
Electricity Plants	-	-	-	-	-	-	-	-	-	-	-
CHP Plants	-	-	-	-	-	-	-	-	-	-	-
Heat Plants	-	252	-	-	260	-	95	-	-	-	-
Blast Furnaces/Gas Works	-	-	-	-	-	-	-	-	-	-	-
Coke/Pat. Fuel/BKB Plants	-	-	-	-	2248	-	-	-	-	-	-
Petroleum Refineries	-	-	-	-	-	-	-	11199	-	-	-
Petrochemical Industry	-	-	-	-	-	-	-	-	-	-	-
Liquefaction	-	-	-	-	-	-	-	-	-	-	-
Other Transform. Sector	-	-	-	-	-	-	-	-	-	-	-
ENERGY SECTOR	-	1	-	-	26	-	-	-	-	-	-
Coal Mines	-	-	-	-	-	-	-	-	-	-	-
Oil and Gas Extraction	-	-	-	-	-	-	-	-	-	-	-
Petroleum Refineries	-	-	-	-	-	-	-	-	-	-	-
Electr., CHP+Heat Plants	-	1	-	-	-	-	-	-	-	-	-
Pumped Storage (Elec.)	-	-	-	-	-	-	-	-	-	-	-
Other Energy Sector	-	-	-	-	26	-	-	-	-	-	-
Distribution Losses	-	-	-	-	-	-	-	135	-	-	-
FINAL CONSUMPTION	-	557	-	-	132	-	1339	526	-	-	-
INDUSTRY SECTOR	-	18	-	-	19	-	9	526	-	-	-
Iron and Steel	-	-	-	-	-	-	-	-	-	-	-
Chemical and Petrochem.	-	1	-	-	-	-	-	526	-	-	-
of which: Feedstocks	-	-	-	-	-	-	-	526	-	-	-
Non-Ferrous Metals	-	-	-	-	-	-	-	-	-	-	-
Non-Metallic Minerals	-	-	-	-	-	-	-	-	-	-	-
Transport Equipment	-	-	-	-	-	-	-	-	-	-	-
Machinery	-	-	-	-	2	-	1	-	-	-	-
Mining and Quarrying	-	-	-	-	-	-	-	-	-	-	-
Food and Tobacco	-	-	-	-	-	-	-	-	-	-	-
Paper, Pulp and Print	-	2	-	-	1	-	1	-	-	-	-
Wood and Wood Products	-	-	-	-	-	-	-	-	-	-	-
Construction	-	13	-	-	16	-	6	-	-	-	-
Textile and Leather	-	1	-	-	-	-	1	-	-	-	-
Non-specified	-	1	-	-	-	-	-	-	-	-	-
TRANSPORT SECTOR	-	12	-	-	-	-	-	-	-	-	-
Air	-	-	-	-	-	-	-	-	-	-	-
Road	-	-	-	-	-	-	-	-	-	-	-
Rail	-	10	-	-	-	-	-	-	-	-	-
Pipeline Transport	-	-	-	-	-	-	-	-	-	-	-
Internal Navigation	-	-	-	-	-	-	-	-	-	-	-
Non-specified	-	2	-	-	-	-	-	-	-	-	-
OTHER SECTORS	-	526	-	-	113	-	1330	-	-	-	-
Agriculture	-	6	-	-	3	-	5	-	-	-	-
Comm. and Publ. Services	-	1	-	-	-	-	1	-	-	-	-
Residential	-	66	-	-	3	-	1177	-	-	-	-
Non-specified	-	453	-	-	107	-	147	-	-	-	-
NON-ENERGY USE	-	1	-	-	-	-	-	-	-	-	-
in Industry/Trans./Energy	-	1	-	-	-	-	-	-	-	-	-
in Transport	-	-	-	-	-	-	-	-	-	-	-
in Other Sectors	-	-	-	-	-	-	-	-	-	-	-

Belarus / Bélarus : 1997

SUPPLY AND CONSUMPTION / APPROVISIONNEMENT ET DEMANDE	Oil cont. / Pétrole cont. (1000 tonnes)										
	Refinery Gas / Gaz de raffinerie	LPG + Ethane / GPL + éthane	Motor Gasoline / Essence moteur	Aviation Gasoline / Essence aviation	Jet Fuel / Carbu-réacteurs	Kerosene / Kérosène	Gas/ Diesel / Gazole	Heavy Fuel Oil / Fioul lourd	Naphtha / Naphta	Petrol. Coke / Coke de pétrole	Other Prod. / Autres prod.
Production	500	138	1954	-	-	86	3115	4524	-	-	755
From Other Sources	-	-	-	-	-	-	-	-	-	-	-
Imports	-	173	56	-	-	1	33	27	-	-	31
Exports	-	-	-666	-	-	-63	-990	-713	-	-	-224
Intl. Marine Bunkers	-	-	-	-	-	-	-	-	-	-	-
Stock Changes	-	4	-97	-	-	-	-242	-133	-	-	-
DOMESTIC SUPPLY	500	315	1247	-	-	24	1916	3705	-	-	562
Transfers	-	-	-	-	-	-	-	-	-	-	-
Statistical Differences	-	-	-	-	-	-	-	-	-	-	-
TRANSFORMATION	-	4	-	-	-	-	42	2759	-	-	-
Electricity Plants	-	-	-	-	-	-	-	314	-	-	-
CHP Plants	-	4	-	-	-	-	42	2445	-	-	-
Heat Plants	-	-	-	-	-	-	-	-	-	-	-
Blast Furnaces/Gas Works	-	-	-	-	-	-	-	-	-	-	-
Coke/Pat. Fuel/BKB Plants	-	-	-	-	-	-	-	-	-	-	-
Petroleum Refineries	-	-	-	-	-	-	-	-	-	-	-
Petrochemical Industry	-	-	-	-	-	-	-	-	-	-	-
Liquefaction	-	-	-	-	-	-	-	-	-	-	-
Other Transform. Sector	-	-	-	-	-	-	-	-	-	-	-
ENERGY SECTOR	500	-	-	-	-	-	-	-	-	-	-
Coal Mines	-	-	-	-	-	-	-	-	-	-	-
Oil and Gas Extraction	-	-	-	-	-	-	-	-	-	-	-
Petroleum Refineries	500	-	-	-	-	-	-	-	-	-	-
Electr., CHP+Heat Plants	-	-	-	-	-	-	-	-	-	-	-
Pumped Storage (Elec.)	-	-	-	-	-	-	-	-	-	-	-
Other Energy Sector	-	-	-	-	-	-	-	-	-	-	-
Distribution Losses	-	2	13	-	-	-	7	3	-	-	-
FINAL CONSUMPTION	-	309	1234	-	-	24	1867	943	-	-	562
INDUSTRY SECTOR	-	4	11	-	-	24	207	768	-	-	-
Iron and Steel	-	-	-	-	-	-	2	1	-	-	-
Chemical and Petrochem.	-	-	1	-	-	24	4	215	-	-	-
of which: Feedstocks	-	-	-	-	-	-	2	-	-	-	-
Non-Ferrous Metals	-	-	-	-	-	-	-	-	-	-	-
Non-Metallic Minerals	-	-	-	-	-	-	-	-	-	-	-
Transport Equipment	-	-	-	-	-	-	-	-	-	-	-
Machinery	-	1	1	-	-	-	7	14	-	-	-
Mining and Quarrying	-	-	-	-	-	-	-	-	-	-	-
Food and Tobacco	-	-	-	-	-	-	2	-	-	-	-
Paper, Pulp and Print	-	-	2	-	-	-	7	1	-	-	-
Wood and Wood Products	-	-	-	-	-	-	-	-	-	-	-
Construction	-	1	7	-	-	-	91	264	-	-	-
Textile and Leather	-	-	-	-	-	-	1	-	-	-	-
Non-specified	-	2	-	-	-	-	93	273	-	-	-
TRANSPORT SECTOR	-	20	1209	-	-	-	964	1	-	-	-
Air	-	-	-	-	-	-	-	-	-	-	-
Road	-	20	1206	-	-	-	593	-	-	-	-
Rail	-	-	-	-	-	-	334	1	-	-	-
Pipeline Transport	-	-	-	-	-	-	1	-	-	-	-
Internal Navigation	-	-	-	-	-	-	19	-	-	-	-
Non-specified	-	-	3	-	-	-	17	-	-	-	-
OTHER SECTORS	-	285	14	-	-	-	696	174	-	-	-
Agriculture	-	1	14	-	-	-	526	4	-	-	-
Comm. and Publ. Services	-	-	-	-	-	-	-	-	-	-	-
Residential	-	284	-	-	-	-	161	-	-	-	-
Non-specified	-	-	-	-	-	-	9	170	-	-	-
NON-ENERGY USE	-	-	-	-	-	-	-	-	-	-	562
in Industry/Transf./Energy	-	-	-	-	-	-	-	-	-	-	562
in Transport	-	-	-	-	-	-	-	-	-	-	-
in Other Sectors	-	-	-	-	-	-	-	-	-	-	-

Belarus / Bélarus : 1997

	Gas / Gaz (TJ)				Comb. Renew. & Waste / En. Re. Comb. & Déchets (TJ)				(GWh)	(TJ)
SUPPLY AND CONSUMPTION APPROVISIONNEMENT ET DEMANDE	Natural Gas Gaz naturel	Gas Works Usines à gaz	Coke Ovens Cokeries	Blast Furnaces Hauts fourneaux	Solid Biomass Biomasse solide	Gas/Liquids from Biomass Gaz/Liquides tirés de biomasse	Municipal Waste Déchets urbains	Industrial Waste Déchets industriels	Electricity Electricité	Heat Chaleur
Production	9501	-	-	-	28687	-	-	-	26057	334944
From Other Sources	-	-	-	-	-	-	-	-	-	-
Imports	627260	-	-	-	-	-	-	-	10308	-
Exports	-	-	-	-	-	-	-	-	-2688	-
Intl. Marine Bunkers	-	-	-	-	-	-	-	-	-	-
Stock Changes	4248	-	-	-	-	-	-	-	-	-
DOMESTIC SUPPLY	641009	-	-	-	28687	-	-	-	33677	334944
Transfers	-	-	-	-	-	-	-	-	-	-
Statistical Differences	-	-	-	-	-	-	-	-	-	-
TRANSFORMATION	494632	-	-	-	5831	-	-	-	-	-
Electricity Plants	119998	-	-	-	-	-	-	-	-	-
CHP Plants	214661	-	-	-	-	-	-	-	-	-
Heat Plants	159973	-	-	-	5831	-	-	-	-	-
Blast Furnaces/Gas Works	-	-	-	-	-	-	-	-	-	-
Coke/Pat. Fuel/BKB Plants	-	-	-	-	-	-	-	-	-	-
Petroleum Refineries	-	-	-	-	-	-	-	-	-	-
Petrochemical Industry	-	-	-	-	-	-	-	-	-	-
Liquefaction	-	-	-	-	-	-	-	-	-	-
Other Transform. Sector	-	-	-	-	-	-	-	-	-	-
ENERGY SECTOR	-	-	-	-	-	-	-	-	3112	4187
Coal Mines	-	-	-	-	-	-	-	-	-	-
Oil and Gas Extraction	-	-	-	-	-	-	-	-	158	-
Petroleum Refineries	-	-	-	-	-	-	-	-	643	-
Electr., CHP+Heat Plants	-	-	-	-	-	-	-	-	2198	-
Pumped Storage (Elec.)	-	-	-	-	-	-	-	-	-	-
Other Energy Sector	-	-	-	-	-	-	-	-	113	4187
Distribution Losses	5214	-	-	-	-	-	-	-	3801	20934
FINAL CONSUMPTION	141163	-	-	-	22856	-	-	-	26764	309823
INDUSTRY SECTOR	73150	-	-	-	469	-	-	-	12569	125604
Iron and Steel	3167	-	-	-	-	-	-	-	1100	-
Chemical and Petrochem.	46694	-	-	-	-	-	-	-	4635	46054
of which: Feedstocks	39394	-	-	-	-	-	-	-	-	-
Non-Ferrous Metals	-	-	-	-	-	-	-	-	8	-
Non-Metallic Minerals	13054	-	-	-	-	-	-	-	984	8374
Transport Equipment	-	-	-	-	-	-	-	-	31	-
Machinery	4364	-	-	-	-	-	-	-	2502	20933
Mining and Quarrying	-	-	-	-	-	-	-	-	-	-
Food and Tobacco	850	-	-	-	-	-	-	-	898	12560
Paper, Pulp and Print	-	-	-	-	29	-	-	-	239	8374
Wood and Wood Products	1468	-	-	-	-	-	-	-	355	8374
Construction	386	-	-	-	264	-	-	-	360	4187
Textile and Leather	425	-	-	-	117	-	-	-	580	8374
Non-specified	2742	-	-	-	59	-	-	-	877	8374
TRANSPORT SECTOR	3128	-	-	-	147	-	-	-	1715	-
Air	-	-	-	-	-	-	-	-	-	-
Road	1390	-	-	-	-	-	-	-	-	-
Rail	-	-	-	-	29	-	-	-	726	-
Pipeline Transport	1738	-	-	-	-	-	-	-	643	-
Internal Navigation	-	-	-	-	-	-	-	-	-	-
Non-specified	-	-	-	-	118	-	-	-	346	-
OTHER SECTORS	64885	-	-	-	22240	-	-	-	12480	184219
Agriculture	1777	-	-	-	176	-	-	-	2471	12560
Comm. and Publ. Services	-	-	-	-	-	-	-	-	2979	-
Residential	42909	-	-	-	18460	-	-	-	5347	163285
Non-specified	20199	-	-	-	3604	-	-	-	1683	8374
NON-ENERGY USE	-	-	-	-	-	-	-	-	-	-
in Industry/Transf./Energy	-	-	-	-	-	-	-	-	-	-
in Transport	-	-	-	-	-	-	-	-	-	-
in Other Sectors	-	-	-	-	-	-	-	-	-	-

Belarus / Bélarus : 1998

	Coal / *Charbon* (1000 tonnes)							Oil / *Pétrole* (1000 tonnes)			
SUPPLY AND CONSUMPTION *APPROVISIONNEMENT ET DEMANDE*	Coking Coal *Charbon à coke*	Other Bit. Coal *Autres charb. bit.*	Sub-Bit. Coal *Charbon sous-bit.*	Lignite *Lignite*	Peat *Tourbe*	Oven and Gas Coke *Coke de four/gaz*	Pat. Fuel and BKB *Agg./briq. de lignite*	Crude Oil *Pétrole brut*	NGL *LGN*	Feed-stocks *Produits d'aliment.*	Additives *Additifs*
Production	-	-	-	-	2993	-	1641	1817	-	-	-
From Other Sources	-	-	-	-	-	-	-	-	-	-	-
Imports	-	704	-	-	-	-	-	10430	-	-	-
Exports	-	-	-	-	-104	-	-86	-399	-	-	-
Intl. Marine Bunkers	-	-	-	-	-	-	-	-	-	-	-
Stock Changes	-	26	-	-	-1	-	-2	-	-	-	-
DOMESTIC SUPPLY	-	730	-	-	2888	-	1553	11848	-	-	-
Transfers	-	-	-	-	-	-	-	-	-	-	-
Statistical Differences	-	-	-	-	-	-	-	71	-	-	-
TRANSFORMATION	-	223	-	-	2717	-	103	11256	-	-	-
Electricity Plants	-	-	-	-	-	-	-	-	-	-	-
CHP Plants	-	-	-	-	-	-	-	-	-	-	-
Heat Plants	-	223	-	-	281	-	103	-	-	-	-
Blast Furnaces/Gas Works	-	-	-	-	-	-	-	-	-	-	-
Coke/Pat. Fuel/BKB Plants	-	-	-	-	2436	-	-	-	-	-	-
Petroleum Refineries	-	-	-	-	-	-	-	11256	-	-	-
Petrochemical Industry	-	-	-	-	-	-	-	-	-	-	-
Liquefaction	-	-	-	-	-	-	-	-	-	-	-
Other Transform. Sector	-	-	-	-	-	-	-	-	-	-	-
ENERGY SECTOR	-	1	-	-	28	-	-	-	-	-	-
Coal Mines	-	-	-	-	-	-	-	-	-	-	-
Oil and Gas Extraction	-	-	-	-	-	-	-	-	-	-	-
Petroleum Refineries	-	-	-	-	-	-	-	-	-	-	-
Electr., CHP+Heat Plants	-	1	-	-	-	-	-	-	-	-	-
Pumped Storage (Elec.)	-	-	-	-	-	-	-	-	-	-	-
Other Energy Sector	-	-	-	-	28	-	-	-	-	-	-
Distribution Losses	-	-	-	-	-	-	-	136	-	-	-
FINAL CONSUMPTION	-	506	-	-	143	-	1450	527	-	-	-
INDUSTRY SECTOR	-	16	-	-	21	-	10	527	-	-	-
Iron and Steel	-	-	-	-	-	-	-	-	-	-	-
Chemical and Petrochem.	-	1	-	-	-	-	-	527	-	-	-
of which: Feedstocks	-	-	-	-	-	-	-	*527*	-	-	-
Non-Ferrous Metals	-	-	-	-	-	-	-	-	-	-	-
Non-Metallic Minerals	-	-	-	-	-	-	-	-	-	-	-
Transport Equipment	-	-	-	-	-	-	-	-	-	-	-
Machinery	-	-	-	-	2	-	1	-	-	-	-
Mining and Quarrying	-	-	-	-	-	-	-	-	-	-	-
Food and Tobacco	-	-	-	-	-	-	-	-	-	-	-
Paper, Pulp and Print	-	1	-	-	1	-	1	-	-	-	-
Wood and Wood Products	-	-	-	-	-	-	-	-	-	-	-
Construction	-	12	-	-	18	-	7	-	-	-	-
Textile and Leather	-	1	-	-	-	-	1	-	-	-	-
Non-specified	-	1	-	-	-	-	-	-	-	-	-
TRANSPORT SECTOR	-	11	-	-	-	-	-	-	-	-	-
Air	-	-	-	-	-	-	-	-	-	-	-
Road	-	-	-	-	-	-	-	-	-	-	-
Rail	-	10	-	-	-	-	-	-	-	-	-
Pipeline Transport	-	-	-	-	-	-	-	-	-	-	-
Internal Navigation	-	-	-	-	-	-	-	-	-	-	-
Non-specified	-	1	-	-	-	-	-	-	-	-	-
OTHER SECTORS	-	478	-	-	122	-	1440	-	-	-	-
Agriculture	-	6	-	-	3	-	5	-	-	-	-
Comm. and Publ. Services	-	-	-	-	-	-	1	-	-	-	-
Residential	-	60	-	-	3	-	1275	-	-	-	-
Non-specified	-	412	-	-	116	-	159	-	-	-	-
NON-ENERGY USE	-	1	-	-	-	-	-	-	-	-	-
in Industry/Trans./Energy	-	1	-	-	-	-	-	-	-	-	-
in Transport	-	-	-	-	-	-	-	-	-	-	-
in Other Sectors	-	-	-	-	-	-	-	-	-	-	-

Belarus / Bélarus : 1998

SUPPLY AND CONSUMPTION *APPROVISIONNEMENT ET DEMANDE*	Oil cont. / *Pétrole cont.* (1000 tonnes)										
	Refinery Gas *Gaz de raffinerie*	LPG + Ethane *GPL + éthane*	Motor Gasoline *Essence moteur*	Aviation Gasoline *Essence aviation*	Jet Fuel *Carbu-réacteurs*	Kerosene *Kérosène*	Gas/ Diesel *Gazole*	Heavy Fuel Oil *Fioul lourd*	Naphtha *Naphta*	Petrol. Coke *Coke de pétrole*	Other Prod. *Autres prod.*
Production	500	139	1950	-	-	86	3109	4515	-	-	759
From Other Sources	-	-	-	-	-	-	-	-	-	-	-
Imports	-	174	65	-	-	-	41	30	-	-	31
Exports	-	-	-573	-	-	-50	-760	-594	-	-	-225
Intl. Marine Bunkers	-	-	-	-	-	-	-	-	-	-	-
Stock Changes	-	-	-	-	-	-	-	210	-	-	-
DOMESTIC SUPPLY	500	313	1442	-	-	36	2390	4161	-	-	565
Transfers	-	-	-	-	-	-	-	-	-	-	-
Statistical Differences	-	-	-	-	-	-	-	-	-	-	-
TRANSFORMATION	-	4	-	-	-	-	52	3099	-	-	-
Electricity Plants	-	-	-	-	-	-	-	353	-	-	-
CHP Plants	-	4	-	-	-	-	52	2746	-	-	-
Heat Plants	-	-	-	-	-	-	-	-	-	-	-
Blast Furnaces/Gas Works	-	-	-	-	-	-	-	-	-	-	-
Coke/Pat. Fuel/BKB Plants	-	-	-	-	-	-	-	-	-	-	-
Petroleum Refineries	-	-	-	-	-	-	-	-	-	-	-
Petrochemical Industry	-	-	-	-	-	-	-	-	-	-	-
Liquefaction	-	-	-	-	-	-	-	-	-	-	-
Other Transform. Sector	-	-	-	-	-	-	-	-	-	-	-
ENERGY SECTOR	500	-	-	-	-	-	-	-	-	-	-
Coal Mines	-	-	-	-	-	-	-	-	-	-	-
Oil and Gas Extraction	-	-	-	-	-	-	-	-	-	-	-
Petroleum Refineries	500	-	-	-	-	-	-	-	-	-	-
Electr., CHP+Heat Plants	-	-	-	-	-	-	-	-	-	-	-
Pumped Storage (Elec.)	-	-	-	-	-	-	-	-	-	-	-
Other Energy Sector	-	-	-	-	-	-	-	-	-	-	-
Distribution Losses	-	2	15	-	-	-	7	3	-	-	-
FINAL CONSUMPTION	-	307	1427	-	-	36	2331	1059	-	-	565
INDUSTRY SECTOR	-	4	13	-	-	36	258	862	-	-	-
Iron and Steel	-	-	-	-	-	-	3	1	-	-	-
Chemical and Petrochem.	-	-	1	-	-	36	5	241	-	-	-
of which: Feedstocks	-	-	-	-	-	-	3	-	-	-	-
Non-Ferrous Metals	-	-	-	-	-	-	-	-	-	-	-
Non-Metallic Minerals	-	-	-	-	-	-	-	-	-	-	-
Transport Equipment	-	-	-	-	-	-	-	-	-	-	-
Machinery	-	1	1	-	-	-	9	16	-	-	-
Mining and Quarrying	-	-	-	-	-	-	-	-	-	-	-
Food and Tobacco	-	-	-	-	-	-	3	-	-	-	-
Paper, Pulp and Print	-	-	2	-	-	-	9	1	-	-	-
Wood and Wood Products	-	-	-	-	-	-	-	-	-	-	-
Construction	-	1	9	-	-	-	113	296	-	-	-
Textile and Leather	-	-	-	-	-	-	1	-	-	-	-
Non-specified	-	2	-	-	-	-	115	307	-	-	-
TRANSPORT SECTOR	-	20	1398	-	-	-	1204	1	-	-	-
Air	-	-	-	-	-	-	-	-	-	-	-
Road	-	20	1395	-	-	-	741	-	-	-	-
Rail	-	-	-	-	-	-	417	1	-	-	-
Pipeline Transport	-	-	-	-	-	-	1	-	-	-	-
Internal Navigation	-	-	-	-	-	-	24	-	-	-	-
Non-specified	-	-	3	-	-	-	21	-	-	-	-
OTHER SECTORS	-	283	16	-	-	-	869	196	-	-	-
Agriculture	-	1	16	-	-	-	657	4	-	-	-
Comm. and Publ. Services	-	-	-	-	-	-	-	-	-	-	-
Residential	-	282	-	-	-	-	201	-	-	-	-
Non-specified	-	-	-	-	-	-	11	192	-	-	-
NON-ENERGY USE	-	-	-	-	-	-	-	-	-	-	565
in Industry/Transf./Energy	-	-	-	-	-	-	-	-	-	-	565
in Transport	-	-	-	-	-	-	-	-	-	-	-
in Other Sectors	-	-	-	-	-	-	-	-	-	-	-

Belarus / Bélarus : 1998

SUPPLY AND CONSUMPTION / APPROVISIONNEMENT ET DEMANDE	Gas / Gaz (TJ) Natural Gas / Gaz naturel	Gas Works / Usines à gaz	Coke Ovens / Cokeries	Blast Furnaces / Hauts fourneaux	Comb. Renew. & Waste / En. Re. Comb. & Déchets (TJ) Solid Biomass / Biomasse solide	Gas/Liquids from Biomass / Gaz/Liquides tirés de biomasse	Municipal Waste / Déchets urbains	Industrial Waste / Déchets industriels	(GWh) Electricity / Electricité	(TJ) Heat / Chaleur
Production	10562	-	-	-	31068	-	-	-	25300	327115
From Other Sources	-	-	-	-	-	-	-	-	-	-
Imports	629406	-	-	-	-	-	-	-	10009	-
Exports	-	-	-	-	-	-	-	-	-2610	-
Intl. Marine Bunkers	-	-	-	-	-	-	-	-	-	-
Stock Changes	4376	-	-	-	-	-	-	-	-	-
DOMESTIC SUPPLY	**644344**	**-**	**-**	**-**	**31068**	**-**	**-**	**-**	**32699**	**327115**
Transfers	-	-	-	-	-	-	-	-	-	-
Statistical Differences	-	-	-	-	-	-	-	-	-	-
TRANSFORMATION	**497205**	**-**	**-**	**-**	**6315**	**-**	**-**	**-**	**-**	**-**
Electricity Plants	120622	-	-	-	-	-	-	-	-	-
CHP Plants	215778	-	-	-	-	-	-	-	-	-
Heat Plants	160805	-	-	-	6315	-	-	-	-	-
Blast Furnaces/Gas Works	-	-	-	-	-	-	-	-	-	-
Coke/Pat. Fuel/BKB Plants	-	-	-	-	-	-	-	-	-	-
Petroleum Refineries	-	-	-	-	-	-	-	-	-	-
Petrochemical Industry	-	-	-	-	-	-	-	-	-	-
Liquefaction	-	-	-	-	-	-	-	-	-	-
Other Transform. Sector	-	-	-	-	-	-	-	-	-	-
ENERGY SECTOR	**-**	**-**	**-**	**-**	**-**	**-**	**-**	**-**	**3022**	**4089**
Coal Mines	-	-	-	-	-	-	-	-	-	-
Oil and Gas Extraction	-	-	-	-	-	-	-	-	153	-
Petroleum Refineries	-	-	-	-	-	-	-	-	624	-
Electr., CHP+Heat Plants	-	-	-	-	-	-	-	-	2135	-
Pumped Storage (Elec.)	-	-	-	-	-	-	-	-	-	-
Other Energy Sector	-	-	-	-	-	-	-	-	110	4089
Distribution Losses	5241	-	-	-	-	-	-	-	1717	20445
FINAL CONSUMPTION	**141898**	**-**	**-**	**-**	**24753**	**-**	**-**	**-**	**27960**	**302581**
INDUSTRY SECTOR	**73531**	**-**	**-**	**-**	**508**	**-**	**-**	**-**	**13492**	**122668**
Iron and Steel	3183	-	-	-	-	-	-	-	1181	-
Chemical and Petrochem.	46938	-	-	-	-	-	-	-	4975	44977
of which: Feedstocks	39559	-	-	-	-	-	-	-	-	-
Non-Ferrous Metals	-	-	-	-	-	-	-	-	9	-
Non-Metallic Minerals	13122	-	-	-	-	-	-	-	1056	8179
Transport Equipment	-	-	-	-	-	-	-	-	33	-
Machinery	4387	-	-	-	-	-	-	-	2686	20444
Mining and Quarrying	-	-	-	-	-	-	-	-	-	-
Food and Tobacco	854	-	-	-	-	-	-	-	964	12267
Paper, Pulp and Print	-	-	-	-	31	-	-	-	257	8178
Wood and Wood Products	1476	-	-	-	-	-	-	-	381	8178
Construction	388	-	-	-	286	-	-	-	386	4089
Textile and Leather	427	-	-	-	127	-	-	-	623	8178
Non-specified	2756	-	-	-	64	-	-	-	941	8178
TRANSPORT SECTOR	**3144**	**-**	**-**	**-**	**159**	**-**	**-**	**-**	**1977**	**-**
Air	-	-	-	-	-	-	-	-	-	-
Road	1397	-	-	-	-	-	-	-	-	-
Rail	-	-	-	-	31	-	-	-	837	-
Pipeline Transport	1747	-	-	-	-	-	-	-	741	-
Internal Navigation	-	-	-	-	-	-	-	-	-	-
Non-specified	-	-	-	-	128	-	-	-	399	-
OTHER SECTORS	**65223**	**-**	**-**	**-**	**24086**	**-**	**-**	**-**	**12491**	**179913**
Agriculture	1786	-	-	-	191	-	-	-	2473	12266
Comm. and Publ. Services	-	-	-	-	-	-	-	-	2982	-
Residential	43133	-	-	-	19992	-	-	-	5352	159468
Non-specified	20304	-	-	-	3903	-	-	-	1684	8179
NON-ENERGY USE	**-**	**-**	**-**	**-**	**-**	**-**	**-**	**-**	**-**	**-**
in Industry/Transf./Energy	-	-	-	-	-	-	-	-	-	-
in Transport	-	-	-	-	-	-	-	-	-	-
in Other Sectors	-	-	-	-	-	-	-	-	-	-

Benin / Bénin

SUPPLY AND CONSUMPTION 1997	Coal (1000 tonnes)							Oil (1000 tonnes)			
	Coking Coal	Other Bit. Coal	Sub-Bit. Coal	Lignite	Peat	Oven and Gas Coke	Pat. Fuel and BKB	Crude Oil	NGL	Feed-stocks	Additives
Production	-	-	-	-	-	-	-	66	-	-	-
Imports	-	-	-	-	-	-	-	-	-	-	-
Exports	-	-	-	-	-	-	-	-64	-	-	-
Intl. Marine Bunkers	-	-	-	-	-	-	-	-	-	-	-
Stock Changes	-	-	-	-	-	-	-	-2	-	-	-
DOMESTIC SUPPLY	-	-	-	-	-	-	-	-	-	-	-
Transfers and Stat. Diff.	-	-	-	-	-	-	-	-	-	-	-
TRANSFORMATION	-	-	-	-	-	-	-	-	-	-	-
Electricity and CHP Plants	-	-	-	-	-	-	-	-	-	-	-
Petroleum Refineries	-	-	-	-	-	-	-	-	-	-	-
Other Transform. Sector	-	-	-	-	-	-	-	-	-	-	-
ENERGY SECTOR	-	-	-	-	-	-	-	-	-	-	-
DISTRIBUTION LOSSES	-	-	-	-	-	-	-	-	-	-	-
FINAL CONSUMPTION	-	-	-	-	-	-	-	-	-	-	-
INDUSTRY SECTOR	-	-	-	-	-	-	-	-	-	-	-
Iron and Steel	-	-	-	-	-	-	-	-	-	-	-
Chemical and Petrochem.	-	-	-	-	-	-	-	-	-	-	-
Non-Metallic Minerals	-	-	-	-	-	-	-	-	-	-	-
Non-specified	-	-	-	-	-	-	-	-	-	-	-
TRANSPORT SECTOR	-	-	-	-	-	-	-	-	-	-	-
Air	-	-	-	-	-	-	-	-	-	-	-
Road	-	-	-	-	-	-	-	-	-	-	-
Non-specified	-	-	-	-	-	-	-	-	-	-	-
OTHER SECTORS	-	-	-	-	-	-	-	-	-	-	-
Agriculture	-	-	-	-	-	-	-	-	-	-	-
Comm. and Publ. Services	-	-	-	-	-	-	-	-	-	-	-
Residential	-	-	-	-	-	-	-	-	-	-	-
Non-specified	-	-	-	-	-	-	-	-	-	-	-
NON-ENERGY USE	-	-	-	-	-	-	-	-	-	-	-

APPROVISIONNEMENT ET DEMANDE 1998	Charbon (1000 tonnes)							Pétrole (1000 tonnes)			
	Charbon à coke	Autres charb. bit.	Charbon sous-bit.	Lignite	Tourbe	Coke de four/gaz	Agg./briq. de lignite	Pétrole brut	LGN	Produits d'aliment.	Additifs
Production	-	-	-	-	-	-	-	52	-	-	-
Imports	-	-	-	-	-	-	-	-	-	-	-
Exports	-	-	-	-	-	-	-	-52	-	-	-
Intl. Marine Bunkers	-	-	-	-	-	-	-	-	-	-	-
Stock Changes	-	-	-	-	-	-	-	-	-	-	-
DOMESTIC SUPPLY	-	-	-	-	-	-	-	-	-	-	-
Transfers and Stat. Diff.	-	-	-	-	-	-	-	-	-	-	-
TRANSFORMATION	-	-	-	-	-	-	-	-	-	-	-
Electricity and CHP Plants	-	-	-	-	-	-	-	-	-	-	-
Petroleum Refineries	-	-	-	-	-	-	-	-	-	-	-
Other Transform. Sector	-	-	-	-	-	-	-	-	-	-	-
ENERGY SECTOR	-	-	-	-	-	-	-	-	-	-	-
DISTRIBUTION LOSSES	-	-	-	-	-	-	-	-	-	-	-
FINAL CONSUMPTION	-	-	-	-	-	-	-	-	-	-	-
INDUSTRY SECTOR	-	-	-	-	-	-	-	-	-	-	-
Iron and Steel	-	-	-	-	-	-	-	-	-	-	-
Chemical and Petrochem.	-	-	-	-	-	-	-	-	-	-	-
Non-Metallic Minerals	-	-	-	-	-	-	-	-	-	-	-
Non-specified	-	-	-	-	-	-	-	-	-	-	-
TRANSPORT SECTOR	-	-	-	-	-	-	-	-	-	-	-
Air	-	-	-	-	-	-	-	-	-	-	-
Road	-	-	-	-	-	-	-	-	-	-	-
Non-specified	-	-	-	-	-	-	-	-	-	-	-
OTHER SECTORS	-	-	-	-	-	-	-	-	-	-	-
Agriculture	-	-	-	-	-	-	-	-	-	-	-
Comm. and Publ. Services	-	-	-	-	-	-	-	-	-	-	-
Residential	-	-	-	-	-	-	-	-	-	-	-
Non-specified	-	-	-	-	-	-	-	-	-	-	-
NON-ENERGY USE	-	-	-	-	-	-	-	-	-	-	-

Benin / Bénin

SUPPLY AND CONSUMPTION 1997	Oil cont. (1000 tonnes)										
	Refinery Gas	LPG + Ethane	Motor Gasoline	Aviation Gasoline	Jet Fuel	Kerosene	Gas/ Diesel	Heavy Fuel Oil	Naphtha	Petrol. Coke	Other Prod.
Production	-	-	-	-	-	-	-	-	-	-	-
Imports	-	1	131	-	49	34	71	18	-	-	-
Exports	-	-	-1	-	-	-	-4	-1	-	-	-
Intl. Marine Bunkers	-	-	-	-	-	-	-5	-	-	-	-
Stock Changes	-	-	-	-	-	-	25	-2	-	-	-
DOMESTIC SUPPLY	-	1	130	-	49	34	87	15	-	-	-
Transfers and Stat. Diff.	-	-	-4	-	-	-	-	10	-	-	-
TRANSFORMATION	-	-	-	-	-	-	8	6	-	-	-
Electricity and CHP Plants	-	-	-	-	-	-	8	6	-	-	-
Petroleum Refineries	-	-	-	-	-	-	-	-	-	-	-
Other Transform. Sector	-	-	-	-	-	-	-	-	-	-	-
ENERGY SECTOR	-	-	-	-	-	-	-	-	-	-	-
DISTRIBUTION LOSSES	-	-	-	-	-	-	-	-	-	-	-
FINAL CONSUMPTION	-	1	126	-	49	34	79	19	-	-	-
INDUSTRY SECTOR	-	-	-	-	-	-	-	19	-	-	-
Iron and Steel	-	-	-	-	-	-	-	-	-	-	-
Chemical and Petrochem.	-	-	-	-	-	-	-	-	-	-	-
Non-Metallic Minerals	-	-	-	-	-	-	-	-	-	-	-
Non-specified	-	-	-	-	-	-	-	19	-	-	-
TRANSPORT SECTOR	-	-	126	-	49	-	79	-	-	-	-
Air	-	-	-	-	49	-	-	-	-	-	-
Road	-	-	126	-	-	-	79	-	-	-	-
Non-specified	-	-	-	-	-	-	-	-	-	-	-
OTHER SECTORS	-	1	-	-	-	34	-	-	-	-	-
Agriculture	-	-	-	-	-	-	-	-	-	-	-
Comm. and Publ. Services	-	-	-	-	-	-	-	-	-	-	-
Residential	-	1	-	-	-	34	-	-	-	-	-
Non-specified	-	-	-	-	-	-	-	-	-	-	-
NON-ENERGY USE	-	-	-	-	-	-	-	-	-	-	-

APPROVISIONNEMENT ET DEMANDE 1998	Pétrole cont. (1000 tonnes)										
	Gaz de raffinerie	GPL + éthane	Essence moteur	Essence aviation	Carbu- réacteurs	Kérosène	Gazole	Fioul lourd	Naphta	Coke de pétrole	Autres prod.
Production	-	-	-	-	-	-	-	-	-	-	-
Imports	-	1	111	-	46	32	90	34	-	-	-
Exports	-	-	-1	-	-	-	-	-	-	-	-
Intl. Marine Bunkers	-	-	-	-	-	-	-5	-	-	-	-
Stock Changes	-	-	-	-	-	-	2	-	-	-	-
DOMESTIC SUPPLY	-	1	110	-	46	32	87	34	-	-	-
Transfers and Stat. Diff.	-	-	-	-	-	-	-	6	-	-	-
TRANSFORMATION	-	-	-	-	-	-	8	6	-	-	-
Electricity and CHP Plants	-	-	-	-	-	-	8	6	-	-	-
Petroleum Refineries	-	-	-	-	-	-	-	-	-	-	-
Other Transform. Sector	-	-	-	-	-	-	-	-	-	-	-
ENERGY SECTOR	-	-	-	-	-	-	-	-	-	-	-
DISTRIBUTION LOSSES	-	-	-	-	-	-	-	-	-	-	-
FINAL CONSUMPTION	-	1	110	-	46	32	79	34	-	-	-
INDUSTRY SECTOR	-	-	-	-	-	-	-	34	-	-	-
Iron and Steel	-	-	-	-	-	-	-	-	-	-	-
Chemical and Petrochem.	-	-	-	-	-	-	-	-	-	-	-
Non-Metallic Minerals	-	-	-	-	-	-	-	-	-	-	-
Non-specified	-	-	-	-	-	-	-	34	-	-	-
TRANSPORT SECTOR	-	-	110	-	46	-	79	-	-	-	-
Air	-	-	-	-	46	-	-	-	-	-	-
Road	-	-	110	-	-	-	79	-	-	-	-
Non-specified	-	-	-	-	-	-	-	-	-	-	-
OTHER SECTORS	-	1	-	-	-	32	-	-	-	-	-
Agriculture	-	-	-	-	-	-	-	-	-	-	-
Comm. and Publ. Services	-	-	-	-	-	-	-	-	-	-	-
Residential	-	1	-	-	-	32	-	-	-	-	-
Non-specified	-	-	-	-	-	-	-	-	-	-	-
NON-ENERGY USE	-	-	-	-	-	-	-	-	-	-	-

Benin / Bénin

SUPPLY AND CONSUMPTION 1997	Gas (TJ)				Comb. Renew. & Waste (TJ)				(GWh)	(TJ)
	Natural Gas	Gas Works	Coke Ovens	Blast Furnaces	Solid Biomass	Gas/Liquids from Biomass	Municipal Waste	Industrial Waste	Electricity	Heat
Production	-	-	-	-	77147	-	-	-	59	-
Imports	-	-	-	-	-	-	-	-	238	-
Exports	-	-	-	-	-	-	-	-	-	-
Intl. Marine Bunkers	-	-	-	-	-	-	-	-	-	-
Stock Changes	-	-	-	-	-	-	-	-	-	-
DOMESTIC SUPPLY	-	-	-	-	77147	-	-	-	297	-
Transfers and Stat. Diff.	-	-	-	-	-1	-	-	-	-1	-
TRANSFORMATION	-	-	-	-	1575	-	-	-	-	-
Electricity and CHP Plants	-	-	-	-	-	-	-	-	-	-
Petroleum Refineries	-	-	-	-	-	-	-	-	-	-
Other Transform. Sector	-	-	-	-	1575	-	-	-	-	-
ENERGY SECTOR	-	-	-	-	-	-	-	-	-	-
DISTRIBUTION LOSSES	-	-	-	-	-	-	-	-	38	-
FINAL CONSUMPTION	-	-	-	-	75571	-	-	-	258	-
INDUSTRY SECTOR	-	-	-	-	14169	-	-	-	129	-
Iron and Steel	-	-	-	-	-	-	-	-	-	-
Chemical and Petrochem.	-	-	-	-	-	-	-	-	-	-
Non-Metallic Minerals	-	-	-	-	-	-	-	-	-	-
Non-specified	-	-	-	-	14169	-	-	-	129	-
TRANSPORT SECTOR	-	-	-	-	-	-	-	-	-	-
Air	-	-	-	-	-	-	-	-	-	-
Road	-	-	-	-	-	-	-	-	-	-
Non-specified	-	-	-	-	-	-	-	-	-	-
OTHER SECTORS	-	-	-	-	61402	-	-	-	129	-
Agriculture	-	-	-	-	-	-	-	-	-	-
Comm. and Publ. Services	-	-	-	-	-	-	-	-	11	-
Residential	-	-	-	-	61402	-	-	-	118	-
Non-specified	-	-	-	-	-	-	-	-	-	-
NON-ENERGY USE	-	-	-	-	-	-	-	-	-	-

APPROVISIONNEMENT ET DEMANDE 1998	Gaz (TJ)				En. Re. Comb. & Déchets (TJ)				(GWh)	(TJ)
	Gaz naturel	Usines à gaz	Cokeries	Hauts fourneaux	Biomasse solide	Gaz/Liquides tirés de biomasse	Déchets urbains	Déchets industriels	Electricité	Chaleur
Production	-	-	-	-	79302	-	-	-	62	-
Imports	-	-	-	-	-	-	-	-	255	-
Exports	-	-	-	-	-	-	-	-	-	-
Intl. Marine Bunkers	-	-	-	-	-	-	-	-	-	-
Stock Changes	-	-	-	-	-	-	-	-	-	-
DOMESTIC SUPPLY	-	-	-	-	79302	-	-	-	317	-
Transfers and Stat. Diff.	-	-	-	-	-	-	-	-	-	-
TRANSFORMATION	-	-	-	-	1619	-	-	-	-	-
Electricity and CHP Plants	-	-	-	-	-	-	-	-	-	-
Petroleum Refineries	-	-	-	-	-	-	-	-	-	-
Other Transform. Sector	-	-	-	-	1619	-	-	-	-	-
ENERGY SECTOR	-	-	-	-	-	-	-	-	-	-
DISTRIBUTION LOSSES	-	-	-	-	-	-	-	-	44	-
FINAL CONSUMPTION	-	-	-	-	77683	-	-	-	273	-
INDUSTRY SECTOR	-	-	-	-	14565	-	-	-	136	-
Iron and Steel	-	-	-	-	-	-	-	-	-	-
Chemical and Petrochem.	-	-	-	-	-	-	-	-	-	-
Non-Metallic Minerals	-	-	-	-	-	-	-	-	-	-
Non-specified	-	-	-	-	14565	-	-	-	136	-
TRANSPORT SECTOR	-	-	-	-	-	-	-	-	-	-
Air	-	-	-	-	-	-	-	-	-	-
Road	-	-	-	-	-	-	-	-	-	-
Non-specified	-	-	-	-	-	-	-	-	-	-
OTHER SECTORS	-	-	-	-	63118	-	-	-	137	-
Agriculture	-	-	-	-	-	-	-	-	-	-
Comm. and Publ. Services	-	-	-	-	-	-	-	-	12	-
Residential	-	-	-	-	63118	-	-	-	125	-
Non-specified	-	-	-	-	-	-	-	-	-	-
NON-ENERGY USE	-	-	-	-	-	-	-	-	-	-

Bolivia / Bolivie

SUPPLY AND CONSUMPTION 1997	Coal (1000 tonnes)							Oil (1000 tonnes)			
	Coking Coal	Other Bit. Coal	Sub-Bit. Coal	Lignite	Peat	Oven and Gas Coke	Pat. Fuel and BKB	Crude Oil	NGL	Feed-stocks	Additives
Production	-	-	-	-	-	-	-	1628	352	-	-
Imports	-	-	-	-	-	-	-	-	-	-	-
Exports	-	-	-	-	-	-	-	-19	-	-	-
Intl. Marine Bunkers	-	-	-	-	-	-	-	-	-	-	-
Stock Changes	-	-	-	-	-	-	-	-	-	-	-
DOMESTIC SUPPLY	-	-	-	-	-	-	-	1609	352	-	-
Transfers and Stat. Diff.	-	-	-	-	-	-	-	43	-250	-	-
TRANSFORMATION	-	-	-	-	-	-	-	1652	102	-	-
Electricity and CHP Plants	-	-	-	-	-	-	-	-	-	-	-
Petroleum Refineries	-	-	-	-	-	-	-	1652	102	-	-
Other Transform. Sector	-	-	-	-	-	-	-	-	-	-	-
ENERGY SECTOR	-	-	-	-	-	-	-	-	-	-	-
DISTRIBUTION LOSSES	-	-	-	-	-	-	-	-	-	-	-
FINAL CONSUMPTION	-	-	-	-	-	-	-	-	-	-	-
INDUSTRY SECTOR	-	-	-	-	-	-	-	-	-	-	-
Iron and Steel	-	-	-	-	-	-	-	-	-	-	-
Chemical and Petrochem.	-	-	-	-	-	-	-	-	-	-	-
Non-Metallic Minerals	-	-	-	-	-	-	-	-	-	-	-
Non-specified	-	-	-	-	-	-	-	-	-	-	-
TRANSPORT SECTOR	-	-	-	-	-	-	-	-	-	-	-
Air	-	-	-	-	-	-	-	-	-	-	-
Road	-	-	-	-	-	-	-	-	-	-	-
Non-specified	-	-	-	-	-	-	-	-	-	-	-
OTHER SECTORS	-	-	-	-	-	-	-	-	-	-	-
Agriculture	-	-	-	-	-	-	-	-	-	-	-
Comm. and Publ. Services	-	-	-	-	-	-	-	-	-	-	-
Residential	-	-	-	-	-	-	-	-	-	-	-
Non-specified	-	-	-	-	-	-	-	-	-	-	-
NON-ENERGY USE	-	-	-	-	-	-	-	-	-	-	-

APPROVISIONNEMENT ET DEMANDE 1998	Charbon (1000 tonnes)							Pétrole (1000 tonnes)			
	Charbon à coke	Autres charb. bit.	Charbon sous-bit.	Lignite	Tourbe	Coke de four/gaz	Agg./briq. de lignite	Pétrole brut	LGN	Produits d'aliment.	Additifs
Production	-	-	-	-	-	-	-	1734	360	-	-
Imports	-	-	-	-	-	-	-	-	-	-	-
Exports	-	-	-	-	-	-	-	-20	-	-	-
Intl. Marine Bunkers	-	-	-	-	-	-	-	-	-	-	-
Stock Changes	-	-	-	-	-	-	-	-	-	-	-
DOMESTIC SUPPLY	-	-	-	-	-	-	-	1714	360	-	-
Transfers and Stat. Diff.	-	-	-	-	-	-	-	-4	-258	-	-
TRANSFORMATION	-	-	-	-	-	-	-	1710	102	-	-
Electricity and CHP Plants	-	-	-	-	-	-	-	-	-	-	-
Petroleum Refineries	-	-	-	-	-	-	-	1710	102	-	-
Other Transform. Sector	-	-	-	-	-	-	-	-	-	-	-
ENERGY SECTOR	-	-	-	-	-	-	-	-	-	-	-
DISTRIBUTION LOSSES	-	-	-	-	-	-	-	-	-	-	-
FINAL CONSUMPTION	-	-	-	-	-	-	-	-	-	-	-
INDUSTRY SECTOR	-	-	-	-	-	-	-	-	-	-	-
Iron and Steel	-	-	-	-	-	-	-	-	-	-	-
Chemical and Petrochem.	-	-	-	-	-	-	-	-	-	-	-
Non-Metallic Minerals	-	-	-	-	-	-	-	-	-	-	-
Non-specified	-	-	-	-	-	-	-	-	-	-	-
TRANSPORT SECTOR	-	-	-	-	-	-	-	-	-	-	-
Air	-	-	-	-	-	-	-	-	-	-	-
Road	-	-	-	-	-	-	-	-	-	-	-
Non-specified	-	-	-	-	-	-	-	-	-	-	-
OTHER SECTORS	-	-	-	-	-	-	-	-	-	-	-
Agriculture	-	-	-	-	-	-	-	-	-	-	-
Comm. and Publ. Services	-	-	-	-	-	-	-	-	-	-	-
Residential	-	-	-	-	-	-	-	-	-	-	-
Non-specified	-	-	-	-	-	-	-	-	-	-	-
NON-ENERGY USE	-	-	-	-	-	-	-	-	-	-	-

Bolivia / Bolivie

SUPPLY AND CONSUMPTION 1997	Oil cont. (1000 tonnes)										
	Refinery Gas	LPG + Ethane	Motor Gasoline	Aviation Gasoline	Jet Fuel	Kerosene	Gas/ Diesel	Heavy Fuel Oil	Naphtha	Petrol. Coke	Other Prod.
Production	84	57	433	4	138	25	381	2	-	-	497
Imports	-	-	-	-	-	-	235	-	-	-	-
Exports	-	-	-	-	-	-	-	-	-	-	-
Intl. Marine Bunkers	-	-	-	-	-	-	-	-	-	-	-
Stock Changes	-	-16	-	-	-6	-	-	-	-	-	-
DOMESTIC SUPPLY	84	41	433	4	132	25	616	2	-	-	497
Transfers and Stat. Diff.	-	250	24	-	-	-2	83	-1	-	-	-484
TRANSFORMATION	-	-	-	-	-	-	24	-	-	-	-
Electricity and CHP Plants	-	-	-	-	-	-	24	-	-	-	-
Petroleum Refineries	-	-	-	-	-	-	-	-	-	-	-
Other Transform. Sector	-	-	-	-	-	-	-	-	-	-	-
ENERGY SECTOR	84	-	11	-	-	-	80	-	-	-	-
DISTRIBUTION LOSSES	-	-	-	-	-	-	-	-	-	-	-
FINAL CONSUMPTION	-	291	446	4	132	23	595	1	-	-	13
INDUSTRY SECTOR	-	8	-	-	-	6	50	1	-	-	-
Iron and Steel	-	-	-	-	-	-	-	-	-	-	-
Chemical and Petrochem.	-	-	-	-	-	-	-	-	-	-	-
Non-Metallic Minerals	-	-	-	-	-	-	-	-	-	-	-
Non-specified	-	8	-	-	-	6	50	1	-	-	-
TRANSPORT SECTOR	-	-	446	4	132	-	531	-	-	-	-
Air	-	-	-	4	132	-	-	-	-	-	-
Road	-	-	446	-	-	-	479	-	-	-	-
Non-specified	-	-	-	-	-	-	52	-	-	-	-
OTHER SECTORS	-	283	-	-	-	17	14	-	-	-	-
Agriculture	-	-	-	-	-	-	14	-	-	-	-
Comm. and Publ. Services	-	-	-	-	-	-	-	-	-	-	-
Residential	-	283	-	-	-	17	-	-	-	-	-
Non-specified	-	-	-	-	-	-	-	-	-	-	-
NON-ENERGY USE	-	-	-	-	-	-	-	-	-	-	13

APPROVISIONNEMENT ET DEMANDE 1998	Pétrole cont. (1000 tonnes)										
	Gaz de raffinerie	GPL + éthane	Essence moteur	Essence aviation	Carbu- réacteurs	Kérosène	Gazole	Fioul lourd	Naphta	Coke de pétrole	Autres prod.
Production	84	59	455	4	138	26	399	2	-	-	503
Imports	-	-	-	-	-	-	290	-	-	-	-
Exports	-	-	-	-	-	-	-	-	-	-	-
Intl. Marine Bunkers	-	-	-	-	-	-	-	-	-	-	-
Stock Changes	-	-10	-	-	-	-	-	-	-	-	-
DOMESTIC SUPPLY	84	49	455	4	138	26	689	2	-	-	503
Transfers and Stat. Diff.	-	258	-1	-	-	-	27	-1	-	-	-490
TRANSFORMATION	-	-	-	-	-	-	24	-	-	-	-
Electricity and CHP Plants	-	-	-	-	-	-	24	-	-	-	-
Petroleum Refineries	-	-	-	-	-	-	-	-	-	-	-
Other Transform. Sector	-	-	-	-	-	-	-	-	-	-	-
ENERGY SECTOR	84	-	2	-	-	-	78	-	-	-	-
DISTRIBUTION LOSSES	-	-	-	-	-	-	-	-	-	-	-
FINAL CONSUMPTION	-	307	452	4	138	26	614	1	-	-	13
INDUSTRY SECTOR	-	8	-	-	-	9	54	1	-	-	-
Iron and Steel	-	-	-	-	-	-	-	-	-	-	-
Chemical and Petrochem.	-	-	-	-	-	-	-	-	-	-	-
Non-Metallic Minerals	-	-	-	-	-	-	-	-	-	-	-
Non-specified	-	8	-	-	-	9	54	1	-	-	-
TRANSPORT SECTOR	-	-	452	4	138	-	544	-	-	-	-
Air	-	-	-	4	138	-	-	-	-	-	-
Road	-	-	452	-	-	-	491	-	-	-	-
Non-specified	-	-	-	-	-	-	53	-	-	-	-
OTHER SECTORS	-	299	-	-	-	17	16	-	-	-	-
Agriculture	-	-	-	-	-	-	16	-	-	-	-
Comm. and Publ. Services	-	-	-	-	-	-	-	-	-	-	-
Residential	-	299	-	-	-	17	-	-	-	-	-
Non-specified	-	-	-	-	-	-	-	-	-	-	-
NON-ENERGY USE	-	-	-	-	-	-	-	-	-	-	13

Bolivia / Bolivie

SUPPLY AND CONSUMPTION 1997	Gas (TJ) Natural Gas	Gas Works	Coke Ovens	Blast Furnaces	Comb. Renew. & Waste (TJ) Solid Biomass	Gas/Liquids from Biomass	Municipal Waste	Industrial Waste	(GWh) Electricity	(TJ) Heat
Production	139720	-	-	-	36672	-	-	-	3472	-
Imports	-	-	-	-	-	-	-	-	8	-
Exports	-89240	-	-	-	-	-	-	-	-1	-
Intl. Marine Bunkers	-	-	-	-	-	-	-	-	-	-
Stock Changes	-	-	-	-	-	-	-	-	-	-
DOMESTIC SUPPLY	50480	-	-	-	36672	-	-	-	3479	-
Transfers and Stat. Diff.	-3336	-	-	-	-	-	-	-	-	-
TRANSFORMATION	20723	-	-	-	2792	-	-	-	-	-
Electricity and CHP Plants	20723	-	-	-	1133	-	-	-	-	-
Petroleum Refineries	-	-	-	-	-	-	-	-	-	-
Other Transform. Sector	-	-	-	-	1659	-	-	-	-	-
ENERGY SECTOR	8279	-	-	-	-	-	-	-	28	-
DISTRIBUTION LOSSES	2018	-	-	-	-	-	-	-	397	-
FINAL CONSUMPTION	16124	-	-	-	33880	-	-	-	3054	-
INDUSTRY SECTOR	16006	-	-	-	12766	-	-	-	1155	-
Iron and Steel	-	-	-	-	-	-	-	-	-	-
Chemical and Petrochem.	-	-	-	-	-	-	-	-	-	-
Non-Metallic Minerals	-	-	-	-	-	-	-	-	-	-
Non-specified	16006	-	-	-	12766	-	-	-	1155	-
TRANSPORT SECTOR	-	-	-	-	-	-	-	-	-	-
Air	-	-	-	-	-	-	-	-	-	-
Road	-	-	-	-	-	-	-	-	-	-
Non-specified	-	-	-	-	-	-	-	-	-	-
OTHER SECTORS	118	-	-	-	21114	-	-	-	1899	-
Agriculture	-	-	-	-	-	-	-	-	-	-
Comm. and Publ. Services	-	-	-	-	-	-	-	-	653	-
Residential	118	-	-	-	21114	-	-	-	1171	-
Non-specified	-	-	-	-	-	-	-	-	75	-
NON-ENERGY USE	-	-	-	-	-	-	-	-	-	-

APPROVISIONNEMENT ET DEMANDE 1998	Gaz (TJ) Gaz naturel	Usines à gaz	Cokeries	Hauts fourneaux	En. Re. Comb. & Déchets (TJ) Biomasse solide	Gaz/Liquides tirés de biomasse	Déchets urbains	Déchets industriels	(GWh) Electricité	(TJ) Chaleur
Production	123842	-	-	-	36736	-	-	-	3711	-
Imports	-	-	-	-	-	-	-	-	7	-
Exports	-69104	-	-	-	-	-	-	-	-1	-
Intl. Marine Bunkers	-	-	-	-	-	-	-	-	-	-
Stock Changes	-	-	-	-	-	-	-	-	-	-
DOMESTIC SUPPLY	54738	-	-	-	36736	-	-	-	3717	-
Transfers and Stat. Diff.	-3219	-	-	-	-	-	-	-	-	-
TRANSFORMATION	24458	-	-	-	2782	-	-	-	-	-
Electricity and CHP Plants	24458	-	-	-	1139	-	-	-	-	-
Petroleum Refineries	-	-	-	-	-	-	-	-	-	-
Other Transform. Sector	-	-	-	-	1643	-	-	-	-	-
ENERGY SECTOR	7759	-	-	-	-	-	-	-	29	-
DISTRIBUTION LOSSES	2018	-	-	-	-	-	-	-	437	-
FINAL CONSUMPTION	17284	-	-	-	33954	-	-	-	3251	-
INDUSTRY SECTOR	17088	-	-	-	12777	-	-	-	1227	-
Iron and Steel	-	-	-	-	-	-	-	-	-	-
Chemical and Petrochem.	-	-	-	-	-	-	-	-	-	-
Non-Metallic Minerals	-	-	-	-	-	-	-	-	-	-
Non-specified	17088	-	-	-	12777	-	-	-	1227	-
TRANSPORT SECTOR	-	-	-	-	-	-	-	-	-	-
Air	-	-	-	-	-	-	-	-	-	-
Road	-	-	-	-	-	-	-	-	-	-
Non-specified	-	-	-	-	-	-	-	-	-	-
OTHER SECTORS	196	-	-	-	21177	-	-	-	2024	-
Agriculture	-	-	-	-	-	-	-	-	-	-
Comm. and Publ. Services	-	-	-	-	-	-	-	-	720	-
Residential	196	-	-	-	21177	-	-	-	1251	-
Non-specified	-	-	-	-	-	-	-	-	53	-
NON-ENERGY USE	-	-	-	-	-	-	-	-	-	-

Bosnia-Herzegovina / Bosnie-Herzégovine

SUPPLY AND CONSUMPTION 1997	Coal (1000 tonnes)							Oil (1000 tonnes)			
	Coking Coal	Other Bit. Coal	Sub-Bit. Coal	Lignite	Peat	Oven and Gas Coke	Pat. Fuel and BKB	Crude Oil	NGL	Feed-stocks	Additives
Production	-	-	-	1741	-	-	-	-	-	-	-
Imports	-	-	-	-	-	-	-	-	-	-	-
Exports	-	-	-	-	-	-	-	-	-	-	-
Intl. Marine Bunkers	-	-	-	-	-	-	-	-	-	-	-
Stock Changes	-	-	-	-	-	-	-	-	-	-	-
DOMESTIC SUPPLY	-	-	-	1741	-	-	-	-	-	-	-
Transfers and Stat. Diff.	-	-	-	-	-	-	-	-	-	-	-
TRANSFORMATION	-	-	-	1433	-	-	-	-	-	-	-
Electricity and CHP Plants	-	-	-	1433	-	-	-	-	-	-	-
Petroleum Refineries	-	-	-	-	-	-	-	-	-	-	-
Other Transform. Sector	-	-	-	-	-	-	-	-	-	-	-
ENERGY SECTOR	-	-	-	-	-	-	-	-	-	-	-
DISTRIBUTION LOSSES	-	-	-	-	-	-	-	-	-	-	-
FINAL CONSUMPTION	-	-	-	308	-	-	-	-	-	-	-
INDUSTRY SECTOR	-	-	-	-	-	-	-	-	-	-	-
Iron and Steel	-	-	-	-	-	-	-	-	-	-	-
Chemical and Petrochem.	-	-	-	-	-	-	-	-	-	-	-
Non-Metallic Minerals	-	-	-	-	-	-	-	-	-	-	-
Non-specified	-	-	-	-	-	-	-	-	-	-	-
TRANSPORT SECTOR	-	-	-	308	-	-	-	-	-	-	-
Air	-	-	-	-	-	-	-	-	-	-	-
Road	-	-	-	-	-	-	-	-	-	-	-
Non-specified	-	-	-	308	-	-	-	-	-	-	-
OTHER SECTORS	-	-	-	-	-	-	-	-	-	-	-
Agriculture	-	-	-	-	-	-	-	-	-	-	-
Comm. and Publ. Services	-	-	-	-	-	-	-	-	-	-	-
Residential	-	-	-	-	-	-	-	-	-	-	-
Non-specified	-	-	-	-	-	-	-	-	-	-	-
NON-ENERGY USE	-	-	-	-	-	-	-	-	-	-	-

APPROVISIONNEMENT ET DEMANDE 1998	Charbon (1000 tonnes)							Pétrole (1000 tonnes)			
	Charbon à coke	Autres charb. bit.	Charbon sous-bit.	Lignite	Tourbe	Coke de four/gaz	Agg./briq. de lignite	Pétrole brut	LGN	Produits d'aliment.	Additifs
Production	-	-	-	1794	-	-	-	-	-	-	-
Imports	-	-	-	-	-	-	-	-	-	-	-
Exports	-	-	-	-	-	-	-	-	-	-	-
Intl. Marine Bunkers	-	-	-	-	-	-	-	-	-	-	-
Stock Changes	-	-	-	-	-	-	-	-	-	-	-
DOMESTIC SUPPLY	-	-	-	1794	-	-	-	-	-	-	-
Transfers and Stat. Diff.	-	-	-	-	-	-	-	-	-	-	-
TRANSFORMATION	-	-	-	1476	-	-	-	-	-	-	-
Electricity and CHP Plants	-	-	-	1476	-	-	-	-	-	-	-
Petroleum Refineries	-	-	-	-	-	-	-	-	-	-	-
Other Transform. Sector	-	-	-	-	-	-	-	-	-	-	-
ENERGY SECTOR	-	-	-	-	-	-	-	-	-	-	-
DISTRIBUTION LOSSES	-	-	-	-	-	-	-	-	-	-	-
FINAL CONSUMPTION	-	-	-	318	-	-	-	-	-	-	-
INDUSTRY SECTOR	-	-	-	-	-	-	-	-	-	-	-
Iron and Steel	-	-	-	-	-	-	-	-	-	-	-
Chemical and Petrochem.	-	-	-	-	-	-	-	-	-	-	-
Non-Metallic Minerals	-	-	-	-	-	-	-	-	-	-	-
Non-specified	-	-	-	-	-	-	-	-	-	-	-
TRANSPORT SECTOR	-	-	-	318	-	-	-	-	-	-	-
Air	-	-	-	-	-	-	-	-	-	-	-
Road	-	-	-	-	-	-	-	-	-	-	-
Non-specified	-	-	-	318	-	-	-	-	-	-	-
OTHER SECTORS	-	-	-	-	-	-	-	-	-	-	-
Agriculture	-	-	-	-	-	-	-	-	-	-	-
Comm. and Publ. Services	-	-	-	-	-	-	-	-	-	-	-
Residential	-	-	-	-	-	-	-	-	-	-	-
Non-specified	-	-	-	-	-	-	-	-	-	-	-
NON-ENERGY USE	-	-	-	-	-	-	-	-	-	-	-

Bosnia-Herzegovina / Bosnie-Herzégovine

SUPPLY AND CONSUMPTION 1997	Oil cont. (1000 tonnes)										
	Refinery Gas	LPG + Ethane	Motor Gasoline	Aviation Gasoline	Jet Fuel	Kerosene	Gas/ Diesel	Heavy Fuel Oil	Naphtha	Petrol. Coke	Other Prod.
Production	-	-	-	-	-	-	-	-	-	-	-
Imports	-	-	126	-	39	-	294	105	325	-	49
Exports	-	-	-	-	-	-	-	-	-	-	-
Intl. Marine Bunkers	-	-	-	-	-	-	-	-	-	-	-
Stock Changes	-	-	-	-	-	-	-	-	-	-	-
DOMESTIC SUPPLY	-	-	126	-	39	-	294	105	325	-	49
Transfers and Stat. Diff.	-	-	-	-	-	-	-	-	-	-	-
TRANSFORMATION	-	-	-	-	-	-	12	66	67	-	-
Electricity and CHP Plants	-	-	-	-	-	-	12	-	21	-	-
Petroleum Refineries	-	-	-	-	-	-	-	-	-	-	-
Other Transform. Sector	-	-	-	-	-	-	-	66	46	-	-
ENERGY SECTOR	-	-	-	-	-	-	-	-	-	-	-
DISTRIBUTION LOSSES	-	-	96	-	-	-	69	39	243	-	-
FINAL CONSUMPTION	-	-	30	-	39	-	213	-	15	-	49
INDUSTRY SECTOR	-	-	-	-	-	-	-	-	15	-	49
Iron and Steel	-	-	-	-	-	-	-	-	8	-	17
Chemical and Petrochem.	-	-	-	-	-	-	-	-	-	-	-
Non-Metallic Minerals	-	-	-	-	-	-	-	-	-	-	-
Non-specified	-	-	-	-	-	-	-	-	7	-	32
TRANSPORT SECTOR	-	-	30	-	39	-	213	-	-	-	-
Air	-	-	-	-	39	-	-	-	-	-	-
Road	-	-	30	-	-	-	209	-	-	-	-
Non-specified	-	-	-	-	-	-	4	-	-	-	-
OTHER SECTORS	-	-	-	-	-	-	-	-	-	-	-
Agriculture	-	-	-	-	-	-	-	-	-	-	-
Comm. and Publ. Services	-	-	-	-	-	-	-	-	-	-	-
Residential	-	-	-	-	-	-	-	-	-	-	-
Non-specified	-	-	-	-	-	-	-	-	-	-	-
NON-ENERGY USE	-	-	-	-	-	-	-	-	-	-	-

APPROVISIONNEMENT ET DEMANDE 1998	Pétrole cont. (1000 tonnes)										
	Gaz de raffinerie	GPL + éthane	Essence moteur	Essence aviation	Carbu- réacteurs	Kérosène	Gazole	Fioul lourd	Naphta	Coke de pétrole	Autres prod.
Production	-	-	-	-	-	-	-	-	-	-	-
Imports	-	-	130	-	40	-	303	108	335	-	51
Exports	-	-	-	-	-	-	-	-	-	-	-
Intl. Marine Bunkers	-	-	-	-	-	-	-	-	-	-	-
Stock Changes	-	-	-	-	-	-	-	-	-	-	-
DOMESTIC SUPPLY	-	-	130	-	40	-	303	108	335	-	51
Transfers and Stat. Diff.	-	-	-	-	-	-	-	-	-	-	-
TRANSFORMATION	-	-	-	-	-	-	12	68	69	-	-
Electricity and CHP Plants	-	-	-	-	-	-	12	-	23	-	-
Petroleum Refineries	-	-	-	-	-	-	-	-	-	-	-
Other Transform. Sector	-	-	-	-	-	-	-	68	46	-	-
ENERGY SECTOR	-	-	-	-	-	-	-	-	-	-	-
DISTRIBUTION LOSSES	-	-	99	-	-	-	71	40	251	-	-
FINAL CONSUMPTION	-	-	31	-	40	-	220	-	15	-	51
INDUSTRY SECTOR	-	-	-	-	-	-	-	-	15	-	51
Iron and Steel	-	-	-	-	-	-	-	-	8	-	18
Chemical and Petrochem.	-	-	-	-	-	-	-	-	-	-	-
Non-Metallic Minerals	-	-	-	-	-	-	-	-	-	-	-
Non-specified	-	-	-	-	-	-	-	-	7	-	33
TRANSPORT SECTOR	-	-	31	-	40	-	220	-	-	-	-
Air	-	-	-	-	40	-	-	-	-	-	-
Road	-	-	31	-	-	-	215	-	-	-	-
Non-specified	-	-	-	-	-	-	5	-	-	-	-
OTHER SECTORS	-	-	-	-	-	-	-	-	-	-	-
Agriculture	-	-	-	-	-	-	-	-	-	-	-
Comm. and Publ. Services	-	-	-	-	-	-	-	-	-	-	-
Residential	-	-	-	-	-	-	-	-	-	-	-
Non-specified	-	-	-	-	-	-	-	-	-	-	-
NON-ENERGY USE	-	-	-	-	-	-	-	-	-	-	-

Bosnia-Herzegovina / Bosnie-Herzégovine

SUPPLY AND CONSUMPTION 1997	Gas (TJ)				Comb. Renew. & Waste (TJ)				(GWh)	(TJ)
	Natural Gas	Gas Works	Coke Ovens	Blast Furnaces	Solid Biomass	Gas/Liquids from Biomass	Municipal Waste	Industrial Waste	Electricity	Heat
Production	-	-	-	-	6898	-	-	-	2461	17334
Imports	10752	-	-	-	-	-	-	-	411	-
Exports	-	-	-	-	-	-	-	-	-193	-
Intl. Marine Bunkers	-	-	-	-	-	-	-	-	-	-
Stock Changes	-	-	-	-	-	-	-	-	-	-
DOMESTIC SUPPLY	10752	-	-	-	6898	-	-	-	2679	17334
Transfers and Stat. Diff.	7199	-	-	-	-	-	-	-	-	-
TRANSFORMATION	17404	-	-	-	-	-	-	-	-	-
Electricity and CHP Plants	-	-	-	-	-	-	-	-	-	-
Petroleum Refineries	-	-	-	-	-	-	-	-	-	-
Other Transform. Sector	17404	-	-	-	-	-	-	-	-	-
ENERGY SECTOR	-	-	-	-	-	-	-	-	167	-
DISTRIBUTION LOSSES	547	-	-	-	-	-	-	-	543	-
FINAL CONSUMPTION	-	-	-	-	6898	-	-	-	1969	17334
INDUSTRY SECTOR	-	-	-	-	-	-	-	-	-	-
Iron and Steel	-	-	-	-	-	-	-	-	-	-
Chemical and Petrochem.	-	-	-	-	-	-	-	-	-	-
Non-Metallic Minerals	-	-	-	-	-	-	-	-	-	-
Non-specified	-	-	-	-	-	-	-	-	-	-
TRANSPORT SECTOR	-	-	-	-	-	-	-	-	-	-
Air	-	-	-	-	-	-	-	-	-	-
Road	-	-	-	-	-	-	-	-	-	-
Non-specified	-	-	-	-	-	-	-	-	-	-
OTHER SECTORS	-	-	-	-	6898	-	-	-	1969	17334
Agriculture	-	-	-	-	-	-	-	-	-	-
Comm. and Publ. Services	-	-	-	-	-	-	-	-	-	-
Residential	-	-	-	-	6898	-	-	-	-	-
Non-specified	-	-	-	-	-	-	-	-	1969	17334
NON-ENERGY USE	-	-	-	-	-	-	-	-	-	-

APPROVISIONNEMENT ET DEMANDE 1998	Gaz (TJ)				En. Re. Comb. & Déchets (TJ)				(GWh)	(TJ)
	Gaz naturel	Usines à gaz	Cokeries	Hauts fourneaux	Biomasse solide	Gaz/Liquides tirés de biomasse	Déchets urbains	Déchets industriels	Electricité	Chaleur
Production	-	-	-	-	7107	-	-	-	2538	17806
Imports	11076	-	-	-	-	-	-	-	423	-
Exports	-	-	-	-	-	-	-	-	-199	-
Intl. Marine Bunkers	-	-	-	-	-	-	-	-	-	-
Stock Changes	-	-	-	-	-	-	-	-	-	-
DOMESTIC SUPPLY	11076	-	-	-	7107	-	-	-	2762	17806
Transfers and Stat. Diff.	7416	-	-	-	-	-	-	-	-	-
TRANSFORMATION	17928	-	-	-	-	-	-	-	-	-
Electricity and CHP Plants	-	-	-	-	-	-	-	-	-	-
Petroleum Refineries	-	-	-	-	-	-	-	-	-	-
Other Transform. Sector	17928	-	-	-	-	-	-	-	-	-
ENERGY SECTOR	-	-	-	-	-	-	-	-	172	-
DISTRIBUTION LOSSES	564	-	-	-	-	-	-	-	560	-
FINAL CONSUMPTION	-	-	-	-	7107	-	-	-	2030	17806
INDUSTRY SECTOR	-	-	-	-	-	-	-	-	-	-
Iron and Steel	-	-	-	-	-	-	-	-	-	-
Chemical and Petrochem.	-	-	-	-	-	-	-	-	-	-
Non-Metallic Minerals	-	-	-	-	-	-	-	-	-	-
Non-specified	-	-	-	-	-	-	-	-	-	-
TRANSPORT SECTOR	-	-	-	-	-	-	-	-	-	-
Air	-	-	-	-	-	-	-	-	-	-
Road	-	-	-	-	-	-	-	-	-	-
Non-specified	-	-	-	-	-	-	-	-	-	-
OTHER SECTORS	-	-	-	-	7107	-	-	-	2030	17806
Agriculture	-	-	-	-	-	-	-	-	-	-
Comm. and Publ. Services	-	-	-	-	-	-	-	-	-	-
Residential	-	-	-	-	7107	-	-	-	-	-
Non-specified	-	-	-	-	-	-	-	-	2030	17806
NON-ENERGY USE	-	-	-	-	-	-	-	-	-	-

Brazil / Brésil : 1997

SUPPLY AND CONSUMPTION	Coal / *Charbon* (1000 tonnes)							Oil / *Pétrole* (1000 tonnes)			
	Coking Coal	Other Bit. Coal	Sub-Bit. Coal	Lignite	Peat	Oven and Gas Coke	Pat. Fuel and BKB	Crude Oil	NGL	Feed-stocks	Additives
APPROVISIONNEMENT ET DEMANDE	*Charbon à coke*	*Autres charb. bit.*	*Charbon sous-bit.*	*Lignite*	*Tourbe*	*Coke de four/gaz*	*Agg./briq. de lignite*	*Pétrole brut*	*LGN*	*Produits d'aliment.*	*Additifs*
Production	90	5557	-	-	-	8492	-	42777	2377	-	-
From Other Sources	-	-	-	-	-	-	-	-	-	-	-
Imports	12883	-	-	-	-	1708	-	29207	-	-	-
Exports	-	-	-	-	-	-	-	-130	-	-	-
Intl. Marine Bunkers	-	-	-	-	-	-	-	-	-	-	-
Stock Changes	-177	-278	-	-	-	-	-	-284	-	-	-
DOMESTIC SUPPLY	**12796**	**5279**	**-**	**-**	**-**	**10200**	**-**	**71570**	**2377**	**-**	**-**
Transfers	-	-	-	-	-	-	-	-	-	-	-
Statistical Differences	-	-2	-	-	-	-24	-	78	-7	-	-
TRANSFORMATION	**10562**	**4243**	**-**	**-**	**-**	**7736**	**-**	**71648**	**820**	**-**	**-**
Electricity Plants	-	4243	-	-	-	-	-	-	239	-	-
CHP Plants	-	-	-	-	-	-	-	-	-	-	-
Heat Plants	-	-	-	-	-	-	-	-	-	-	-
Blast Furnaces/Gas Works	-	-	-	-	-	7736	-	-	-	-	-
Coke/Pat. Fuel/BKB Plants	10562	-	-	-	-	-	-	-	-	-	-
Petroleum Refineries	-	-	-	-	-	-	-	71648	581	-	-
Petrochemical Industry	-	-	-	-	-	-	-	-	-	-	-
Liquefaction	-	-	-	-	-	-	-	-	-	-	-
Other Transform. Sector	-	-	-	-	-	-	-	-	-	-	-
ENERGY SECTOR	**-**	**-**	**-**	**-**	**-**	**-**	**-**	**-**	**1004**	**-**	**-**
Coal Mines	-	-	-	-	-	-	-	-	-	-	-
Oil and Gas Extraction	-	-	-	-	-	-	-	-	1004	-	-
Petroleum Refineries	-	-	-	-	-	-	-	-	-	-	-
Electr., CHP+Heat Plants	-	-	-	-	-	-	-	-	-	-	-
Pumped Storage (Elec.)	-	-	-	-	-	-	-	-	-	-	-
Other Energy Sector	-	-	-	-	-	-	-	-	-	-	-
Distribution Losses	-	-	-	-	-	97	-	-	-	-	-
FINAL CONSUMPTION	**2234**	**1034**	**-**	**-**	**-**	**2343**	**-**	**-**	**546**	**-**	**-**
INDUSTRY SECTOR	**2234**	**1034**	**-**	**-**	**-**	**2084**	**-**	**-**	**546**	**-**	**-**
Iron and Steel	1142	17	-	-	-	1934	-	-	-	-	-
Chemical and Petrochem.	-	315	-	-	-	-	-	-	536	-	-
of which: Feedstocks	-	-	-	-	-	-	-	-	*510*	-	-
Non-Ferrous Metals	131	24	-	-	-	117	-	-	-	-	-
Non-Metallic Minerals	590	326	-	-	-	33	-	-	-	-	-
Transport Equipment	-	-	-	-	-	-	-	-	-	-	-
Machinery	-	-	-	-	-	-	-	-	-	-	-
Mining and Quarrying	371	-	-	-	-	-	-	-	-	-	-
Food and Tobacco	-	154	-	-	-	-	-	-	9	-	-
Paper, Pulp and Print	-	189	-	-	-	-	-	-	-	-	-
Wood and Wood Products	-	-	-	-	-	-	-	-	-	-	-
Construction	-	-	-	-	-	-	-	-	-	-	-
Textile and Leather	-	4	-	-	-	-	-	-	-	-	-
Non-specified	-	5	-	-	-	-	-	-	1	-	-
TRANSPORT SECTOR	**-**	**-**	**-**	**-**	**-**	**-**	**-**	**-**	**-**	**-**	**-**
Air	-	-	-	-	-	-	-	-	-	-	-
Road	-	-	-	-	-	-	-	-	-	-	-
Rail	-	-	-	-	-	-	-	-	-	-	-
Pipeline Transport	-	-	-	-	-	-	-	-	-	-	-
Internal Navigation	-	-	-	-	-	-	-	-	-	-	-
Non-specified	-	-	-	-	-	-	-	-	-	-	-
OTHER SECTORS	**-**	**-**	**-**	**-**	**-**	**-**	**-**	**-**	**-**	**-**	**-**
Agriculture	-	-	-	-	-	-	-	-	-	-	-
Comm. and Publ. Services	-	-	-	-	-	-	-	-	-	-	-
Residential	-	-	-	-	-	-	-	-	-	-	-
Non-specified	-	-	-	-	-	-	-	-	-	-	-
NON-ENERGY USE	**-**	**-**	**-**	**-**	**-**	**259**	**-**	**-**	**-**	**-**	**-**
in Industry/Trans./Energy	-	-	-	-	-	259	-	-	-	-	-
in Transport	-	-	-	-	-	-	-	-	-	-	-
in Other Sectors	-	-	-	-	-	-	-	-	-	-	-

Brazil / Brésil : 1997

SUPPLY AND CONSUMPTION / APPROVISIONNEMENT ET DEMANDE	Oil cont. / Pétrole cont. (1000 tonnes)										
	Refinery Gas / Gaz de raffinerie	LPG + Ethane / GPL + éthane	Motor Gasoline / Essence moteur	Aviation Gasoline / Essence aviation	Jet Fuel / Carbu-réacteurs	Kerosene / Kérosène	Gas/Diesel / Gazole	Heavy Fuel Oil / Fioul lourd	Naphtha / Naphta	Petrol. Coke / Coke de pétrole	Other Prod. / Autres prod.
Production	3355	3238	13216	55	2716	66	23416	15973	4998	985	3956
From Other Sources	-	-	-	-	-	-	-	-	-	-	-
Imports	-	2577	291	-	681	5	4941	477	3407	5	608
Exports	-	-3	-465	-21	-838	-6	-154	-1140	-	-19	-323
Intl. Marine Bunkers	-	-	-	-	-	-	-348	-1372	-	-	-
Stock Changes	-	-56	20	12	62	-2	61	187	116	-28	81
DOMESTIC SUPPLY	3355	5756	13062	46	2621	63	27916	14125	8521	943	4322
Transfers	-	458	265	-	-	-	33	-	-1626	-	1353
Statistical Differences	-4	162	28	9	97	25	-104	-8	-248	-2	-247
TRANSFORMATION	144	-	-	-	-	-	1156	1240	47	-	129
Electricity Plants	144	-	-	-	-	-	1156	1240	-	-	129
CHP Plants	-	-	-	-	-	-	-	-	-	-	-
Heat Plants	-	-	-	-	-	-	-	-	-	-	-
Blast Furnaces/Gas Works	-	-	-	-	-	-	-	-	47	-	-
Coke/Pat. Fuel/BKB Plants	-	-	-	-	-	-	-	-	-	-	-
Petroleum Refineries	-	-	-	-	-	-	-	-	-	-	-
Petrochemical Industry	-	-	-	-	-	-	-	-	-	-	-
Liquefaction	-	-	-	-	-	-	-	-	-	-	-
Other Transform. Sector	-	-	-	-	-	-	-	-	-	-	-
ENERGY SECTOR	2961	15	-	-	-	1	177	1846	4	-	20
Coal Mines	-	-	-	-	-	-	-	-	-	-	-
Oil and Gas Extraction	-	-	-	-	-	-	-	-	-	-	-
Petroleum Refineries	2961	15	-	-	-	1	177	1846	-	-	20
Electr., CHP+Heat Plants	-	-	-	-	-	-	-	-	-	-	-
Pumped Storage (Elec.)	-	-	-	-	-	-	-	-	-	-	-
Other Energy Sector	-	-	-	-	-	-	-	-	4	-	-
Distribution Losses	63	1	-	-	-	-	133	64	35	-	29
FINAL CONSUMPTION	183	6360	13355	55	2718	87	26379	10967	6561	941	5250
INDUSTRY SECTOR	183	586	-	-	-	58	486	9541	6561	846	1223
Iron and Steel	-	62	-	-	-	6	32	333	-	106	-
Chemical and Petrochem.	183	15	-	-	-	38	94	2016	6561	31	1223
of which: Feedstocks	183	-	-	-	-	26	-	-	6561	-	-
Non-Ferrous Metals	-	42	-	-	-	-	-	750	-	512	-
Non-Metallic Minerals	-	271	-	-	-	2	25	2413	-	124	-
Transport Equipment	-	-	-	-	-	-	-	-	-	-	-
Machinery	-	-	-	-	-	-	-	-	-	-	-
Mining and Quarrying	-	4	-	-	-	2	143	553	-	42	-
Food and Tobacco	-	34	-	-	-	2	34	1067	-	-	-
Paper, Pulp and Print	-	14	-	-	-	2	31	936	-	-	-
Wood and Wood Products	-	-	-	-	-	-	-	-	-	-	-
Construction	-	-	-	-	-	-	-	-	-	-	-
Textile and Leather	-	3	-	-	-	-	3	342	-	-	-
Non-specified	-	141	-	-	-	6	124	1131	-	31	-
TRANSPORT SECTOR	-	-	13355	55	2718	-	21193	780	-	-	-
Air	-	-	-	55	2718	-	-	-	-	-	-
Road	-	-	13355	-	-	-	20635	-	-	-	-
Rail	-	-	-	-	-	-	317	-	-	-	-
Pipeline Transport	-	-	-	-	-	-	-	-	-	-	-
Internal Navigation	-	-	-	-	-	-	241	780	-	-	-
Non-specified	-	-	-	-	-	-	-	-	-	-	-
OTHER SECTORS	-	5774	-	-	-	29	4700	646	-	95	303
Agriculture	-	3	-	-	-	-	4488	83	-	-	-
Comm. and Publ. Services	-	236	-	-	-	-	212	563	-	-	-
Residential	-	5535	-	-	-	29	-	-	-	-	-
Non-specified	-	-	-	-	-	-	-	-	-	95	303
NON-ENERGY USE	-	-	-	-	-	-	-	-	-	-	3724
in Industry/Transf./Energy	-	-	-	-	-	-	-	-	-	-	3724
in Transport	-	-	-	-	-	-	-	-	-	-	-
in Other Sectors	-	-	-	-	-	-	-	-	-	-	-

Brazil / Brésil : 1997

SUPPLY AND CONSUMPTION / APPROVISIONNEMENT ET DEMANDE	Gas / Gaz (TJ)				Comb. Renew. & Waste / En. Re. Comb. & Déchets (TJ)				(GWh)	(TJ)
	Natural Gas / Gaz naturel	Gas Works / Usines à gaz	Coke Ovens / Cokeries	Blast Furnaces / Hauts fourneaux	Solid Biomass / Biomasse solide	Gas/Liquids from Biomass / Gaz/Liquides tirés de biomasse	Municipal Waste / Déchets urbains	Industrial Waste / Déchets industriels	Electricity / Electricité	Heat / Chaleur
Production	163189	5192	65094	85857	1396582	312790	-	-	307980	-
From Other Sources	-	-	-	-	-	-	-	-	-	-
Imports	-	-	-	-	324	19178	-	-	40478	-
Exports	-	-	-	-	-135	-3756	-	-	-8	-
Intl. Marine Bunkers	-	-	-	-	-	-	-	-	-	-
Stock Changes	-	-	-	-	-6410	-38908	-	-	-	-
DOMESTIC SUPPLY	**163189**	**5192**	**65094**	**85857**	**1390361**	**289304**	**-**	**-**	**348450**	**-**
Transfers	-	-	-	-	-	-	-	-	-	-
Statistical Differences	-350	1	-	-	24	-	-	-	-	-
TRANSFORMATION	**4921**	**-**	**3957**	**16723**	**231633**	**-**	**-**	**-**	**-**	**-**
Electricity Plants	1201	-	3957	16723	8148	-	-	-	-	-
CHP Plants	-	-	-	-	52761	-	-	-	-	-
Heat Plants	-	-	-	-	-	-	-	-	-	-
Blast Furnaces/Gas Works	3720	-	-	-	-	-	-	-	-	-
Coke/Pat. Fuel/BKB Plants	-	-	-	-	-	-	-	-	-	-
Petroleum Refineries	-	-	-	-	-	-	-	-	-	-
Petrochemical Industry	-	-	-	-	-	-	-	-	-	-
Liquefaction	-	-	-	-	-	-	-	-	-	-
Other Transform. Sector	-	-	-	-	170724	-	-	-	-	-
ENERGY SECTOR	**9533**	**474**	**15223**	**12176**	**-**	**-**	**-**	**-**	**9144**	**-**
Coal Mines	-	-	-	-	-	-	-	-	-	-
Oil and Gas Extraction	9533	-	-	-	-	-	-	-	-	-
Petroleum Refineries	-	-	-	-	-	-	-	-	-	-
Electr., CHP+Heat Plants	-	-	-	-	-	-	-	-	9144	-
Pumped Storage (Elec.)	-	-	-	-	-	-	-	-	-	-
Other Energy Sector	-	474	15223	12176	-	-	-	-	-	-
Distribution Losses	6976	555	546	18731	-	-	-	-	53761	-
FINAL CONSUMPTION	**141409**	**4164**	**45368**	**38227**	**1158752**	**289304**	**-**	**-**	**285545**	**-**
INDUSTRY SECTOR	**132884**	**82**	**45368**	**38227**	**804096**	**-**	**-**	**-**	**135521**	**-**
Iron and Steel	31157	33	45368	38227	153382	-	-	-	20612	-
Chemical and Petrochem.	51541	16	-	-	4657	-	-	-	15829	-
of which: Feedstocks	10463	-	-	-	-	-	-	-	-	-
Non-Ferrous Metals	1589	-	-	-	1147	-	-	-	27819	-
Non-Metallic Minerals	5929	-	-	-	81522	-	-	-	6892	-
Transport Equipment	-	-	-	-	-	-	-	-	-	-
Machinery	-	-	-	-	-	-	-	-	-	-
Mining and Quarrying	6782	-	-	-	-	-	-	-	6459	-
Food and Tobacco	6162	-	-	-	406093	-	-	-	14429	-
Paper, Pulp and Print	6278	-	-	-	127782	-	-	-	10611	-
Wood and Wood Products	-	-	-	-	-	-	-	-	-	-
Construction	-	-	-	-	-	-	-	-	-	-
Textile and Leather	3139	-	-	-	4246	-	-	-	6134	-
Non-specified	20307	33	-	-	25267	-	-	-	26736	-
TRANSPORT SECTOR	**1821**	**-**	**-**	**-**	**-**	**289304**	**-**	**-**	**1140**	**-**
Air	-	-	-	-	-	-	-	-	-	-
Road	1821	-	-	-	-	289304	-	-	-	-
Rail	-	-	-	-	-	-	-	-	1140	-
Pipeline Transport	-	-	-	-	-	-	-	-	-	-
Internal Navigation	-	-	-	-	-	-	-	-	-	-
Non-specified	-	-	-	-	-	-	-	-	-	-
OTHER SECTORS	**6704**	**4082**	**-**	**-**	**354656**	**-**	**-**	**-**	**148884**	**-**
Agriculture	-	-	-	-	77794	-	-	-	10799	-
Comm. and Publ. Services	3565	1012	-	-	6385	-	-	-	64014	-
Residential	3139	3070	-	-	270477	-	-	-	74071	-
Non-specified	-	-	-	-	-	-	-	-	-	-
NON-ENERGY USE	**-**	**-**	**-**	**-**	**-**	**-**	**-**	**-**	**-**	**-**
in Industry/Transf./Energy	-	-	-	-	-	-	-	-	-	-
in Transport	-	-	-	-	-	-	-	-	-	-
in Other Sectors	-	-	-	-	-	-	-	-	-	-

Brazil / Brésil : 1998

	Coal / *Charbon* (1000 tonnes)							Oil / *Pétrole* (1000 tonnes)			
SUPPLY AND CONSUMPTION	Coking Coal	Other Bit. Coal	Sub-Bit. Coal	Lignite	Peat	Oven and Gas Coke	Pat. Fuel and BKB	Crude Oil	NGL	Feed-stocks	Additives
APPROVISIONNEMENT ET DEMANDE	*Charbon à coke*	*Autres charb. bit.*	*Charbon sous-bit.*	*Lignite*	*Tourbe*	*Coke de four/gaz*	*Agg./briq. de lignite*	*Pétrole brut*	*LGN*	*Produits d'aliment.*	*Additifs*
Production	20	5496	-	-	-	8229	-	49570	1325	-	-
From Other Sources	-	-	-	-	-	-	-	-	-	-	-
Imports	12964	-	-	-	-	1516	-	27973	-	-	-
Exports	-	-	-	-	-	-	-	-	-	-	-
Intl. Marine Bunkers	-	-	-	-	-	-	-	-	-	-	-
Stock Changes	-10	-834	-	-	-	81	-	-230	-	-	-
DOMESTIC SUPPLY	12974	4662	-	-	-	9826	-	77313	1325	-	-
Transfers	-	-	-	-	-	-	-	-	-	-	-
Statistical Differences	1	2	-	-	-	-89	-	-139	-	-	-
TRANSFORMATION	10230	4011	-	-	-	7377	-	77174	897	-	-
Electricity Plants	-	4011	-	-	-	28	-	-	306	-	-
CHP Plants	-	-	-	-	-	-	-	-	-	-	-
Heat Plants	-	-	-	-	-	-	-	-	-	-	-
Blast Furnaces/Gas Works	-	-	-	-	-	7349	-	-	-	-	-
Coke/Pat. Fuel/BKB Plants	10230	-	-	-	-	-	-	-	-	-	-
Petroleum Refineries	-	-	-	-	-	-	-	77174	591	-	-
Petrochemical Industry	-	-	-	-	-	-	-	-	-	-	-
Liquefaction	-	-	-	-	-	-	-	-	-	-	-
Other Transform. Sector	-	-	-	-	-	-	-	-	-	-	-
ENERGY SECTOR	-	-	-	-	-	-	-	-	-	-	-
Coal Mines	-	-	-	-	-	-	-	-	-	-	-
Oil and Gas Extraction	-	-	-	-	-	-	-	-	-	-	-
Petroleum Refineries	-	-	-	-	-	-	-	-	-	-	-
Electr., CHP+Heat Plants	-	-	-	-	-	-	-	-	-	-	-
Pumped Storage (Elec.)	-	-	-	-	-	-	-	-	-	-	-
Other Energy Sector	-	-	-	-	-	-	-	-	-	-	-
Distribution Losses	39	-	-	-	-	94	-	-	-	-	-
FINAL CONSUMPTION	2706	653	-	-	-	2266	-	-	428	-	-
INDUSTRY SECTOR	2706	653	-	-	-	2044	-	-	428	-	-
Iron and Steel	1554	11	-	-	-	1837	-	-	-	-	-
Chemical and Petrochem.	-	263	-	-	-	-	-	-	428	-	-
of which: Feedstocks	-	-	-	-	-	-	-	-	428	-	-
Non-Ferrous Metals	145	-	-	-	-	109	-	-	-	-	-
Non-Metallic Minerals	653	127	-	-	-	98	-	-	-	-	-
Transport Equipment	-	-	-	-	-	-	-	-	-	-	-
Machinery	-	-	-	-	-	-	-	-	-	-	-
Mining and Quarrying	354	-	-	-	-	-	-	-	-	-	-
Food and Tobacco	-	107	-	-	-	-	-	-	-	-	-
Paper, Pulp and Print	-	138	-	-	-	-	-	-	-	-	-
Wood and Wood Products	-	-	-	-	-	-	-	-	-	-	-
Construction	-	-	-	-	-	-	-	-	-	-	-
Textile and Leather	-	-	-	-	-	-	-	-	-	-	-
Non-specified	-	7	-	-	-	-	-	-	-	-	-
TRANSPORT SECTOR	-	-	-	-	-	-	-	-	-	-	-
Air	-	-	-	-	-	-	-	-	-	-	-
Road	-	-	-	-	-	-	-	-	-	-	-
Rail	-	-	-	-	-	-	-	-	-	-	-
Pipeline Transport	-	-	-	-	-	-	-	-	-	-	-
Internal Navigation	-	-	-	-	-	-	-	-	-	-	-
Non-specified	-	-	-	-	-	-	-	-	-	-	-
OTHER SECTORS	-	-	-	-	-	-	-	-	-	-	-
Agriculture	-	-	-	-	-	-	-	-	-	-	-
Comm. and Publ. Services	-	-	-	-	-	-	-	-	-	-	-
Residential	-	-	-	-	-	-	-	-	-	-	-
Non-specified	-	-	-	-	-	-	-	-	-	-	-
NON-ENERGY USE	-	-	-	-	-	222	-	-	-	-	-
in Industry/Trans./Energy	-	-	-	-	-	222	-	-	-	-	-
in Transport	-	-	-	-	-	-	-	-	-	-	-
in Other Sectors	-	-	-	-	-	-	-	-	-	-	-

Brazil / Brésil : 1998

SUPPLY AND CONSUMPTION *APPROVISIONNEMENT ET DEMANDE*	Oil cont. / *Pétrole cont.* (1000 tonnes)										
	Refinery Gas *Gaz de raffinerie*	LPG + Ethane *GPL + éthane*	Motor Gasoline *Essence moteur*	Aviation Gasoline *Essence aviation*	Jet Fuel *Carbu- réacteurs*	Kerosene *Kérosène*	Gas/ Diesel *Gazole*	Heavy Fuel Oil *Fioul lourd*	Naphtha *Naphta*	Petrol. Coke *Coke de pétrole*	Other Prod. *Autres prod.*
Production	3425	3285	14618	79	2974	54	24603	18312	5040	919	4193
From Other Sources	-	-	-	-	-	-	-	-	-	-	-
Imports	-	2776	151	4	787	5	5206	59	3498	5	184
Exports	-	-3	-1274	-4	-838	-6	-34	-2767	-4	-20	-173
Intl. Marine Bunkers	-	-	-	-	-	-	-419	-1289	-	-	-
Stock Changes	-	55	173	-	-21	1	-57	-143	-25	14	163
DOMESTIC SUPPLY	**3425**	**6113**	**13668**	**79**	**2902**	**54**	**29299**	**14172**	**8509**	**918**	**4367**
Transfers	-	468	295	-	-	-	61	-	-1671	-	1359
Statistical Differences	-	-8	80	-20	104	25	-8	-149	-179	7	-55
TRANSFORMATION	**161**	**-**	**-**	**-**	**-**	**-**	**1449**	**1269**	**-**	**-**	**94**
Electricity Plants	161	-	-	-	-	-	1449	1269	-	-	94
CHP Plants	-	-	-	-	-	-	-	-	-	-	-
Heat Plants	-	-	-	-	-	-	-	-	-	-	-
Blast Furnaces/Gas Works	-	-	-	-	-	-	-	-	-	-	-
Coke/Pat. Fuel/BKB Plants	-	-	-	-	-	-	-	-	-	-	-
Petroleum Refineries	-	-	-	-	-	-	-	-	-	-	-
Petrochemical Industry	-	-	-	-	-	-	-	-	-	-	-
Liquefaction	-	-	-	-	-	-	-	-	-	-	-
Other Transform. Sector	-	-	-	-	-	-	-	-	-	-	-
ENERGY SECTOR	**2998**	**17**	**-**	**-**	**-**	**1**	**205**	**1669**	**4**	**-**	**-**
Coal Mines	-	-	-	-	-	-	-	-	-	-	-
Oil and Gas Extraction	-	-	-	-	-	-	-	-	-	-	-
Petroleum Refineries	2998	17	-	-	-	1	205	1669	-	-	-
Electr., CHP+Heat Plants	-	-	-	-	-	-	-	-	-	-	-
Pumped Storage (Elec.)	-	-	-	-	-	-	-	-	-	-	-
Other Energy Sector	-	-	-	-	-	-	-	-	4	-	-
Distribution Losses	69	33	-	-	-	-	139	64	84	-	-
FINAL CONSUMPTION	**197**	**6523**	**14043**	**59**	**3006**	**78**	**27559**	**11021**	**6571**	**925**	**5577**
INDUSTRY SECTOR	**197**	**666**	**-**	**-**	**-**	**50**	**512**	**9566**	**6571**	**925**	**1265**
Iron and Steel	-	80	-	-	-	4	27	283	-	112	-
Chemical and Petrochem.	197	17	-	-	-	33	79	1992	6571	42	1265
of which: Feedstocks	*197*	-	-	-	-	*24*	-	-	*6571*	-	-
Non-Ferrous Metals	-	53	-	-	-	-	-	683	-	465	-
Non-Metallic Minerals	-	260	-	-	-	2	27	2476	-	226	-
Transport Equipment	-	-	-	-	-	-	-	-	-	-	-
Machinery	-	-	-	-	-	-	-	-	-	-	-
Mining and Quarrying	-	9	-	-	-	3	159	628	-	48	-
Food and Tobacco	-	41	-	-	-	2	34	1120	-	-	-
Paper, Pulp and Print	-	18	-	-	-	1	23	969	-	-	-
Wood and Wood Products	-	-	-	-	-	-	-	-	-	-	-
Construction	-	-	-	-	-	-	-	-	-	-	-
Textile and Leather	-	7	-	-	-	-	6	335	-	-	-
Non-specified	-	181	-	-	-	5	157	1080	-	32	-
TRANSPORT SECTOR	**-**	**-**	**14043**	**59**	**3006**	**-**	**22473**	**810**	**-**	**-**	**-**
Air	-	-	-	59	3006	-	-	-	-	-	-
Road	-	-	14043	-	-	-	21843	-	-	-	-
Rail	-	-	-	-	-	-	341	-	-	-	-
Pipeline Transport	-	-	-	-	-	-	-	-	-	-	-
Internal Navigation	-	-	-	-	-	-	289	810	-	-	-
Non-specified	-	-	-	-	-	-	-	-	-	-	-
OTHER SECTORS	**-**	**5857**	**-**	**-**	**-**	**28**	**4574**	**645**	**-**	**-**	**62**
Agriculture	-	7	-	-	-	-	4354	48	-	-	-
Comm. and Publ. Services	-	320	-	-	-	-	220	597	-	-	-
Residential	-	5530	-	-	-	28	-	-	-	-	-
Non-specified	-	-	-	-	-	-	-	-	-	-	62
NON-ENERGY USE	**-**	**-**	**-**	**-**	**-**	**-**	**-**	**-**	**-**	**-**	**4250**
in Industry/Transf./Energy	-	-	-	-	-	-	-	-	-	-	4250
in Transport	-	-	-	-	-	-	-	-	-	-	-
in Other Sectors	-	-	-	-	-	-	-	-	-	-	-

Brazil / Brésil : 1998

SUPPLY AND CONSUMPTION APPROVISIONNEMENT ET DEMANDE	Gas / Gaz (TJ)				Comb. Renew. & Waste / En. Re. Comb. & Déchets (TJ)				(GWh)	(TJ)
	Natural Gas Gaz naturel	Gas Works Usines à gaz	Coke Ovens Cokeries	Blast Furnaces Hauts fourneaux	Solid Biomass Biomasse solide	Gas/Liquids from Biomass Gaz/Liquides tirés de biomasse	Municipal Waste Déchets urbains	Industrial Waste Déchets industriels	Electricity Electricité	Heat Chaleur
Production	220388	5405	63907	81638	1452711	278315	-	-	321588	-
From Other Sources	-	-	-	-	-	-	-	-	-	-
Imports	-	-	-	-	454	3474	-	-	39412	-
Exports	-	-	-	-	-270	-3009	-	-	-8	-
Intl. Marine Bunkers	-	-	-	-	-	-	-	-	-	-
Stock Changes	-	-	-	-	-6140	3725	-	-	-	-
DOMESTIC SUPPLY	**220388**	**5405**	**63907**	**81638**	**1446755**	**282505**	**-**	**-**	**360992**	**-**
Transfers	-	-	-	-	-	-	-	-	-	-
Statistical Differences	-154	16	-	-	-1	-	-	-	-	-
TRANSFORMATION	**7945**	**-**	**5313**	**16791**	**226100**	**-**	**-**	**-**	**-**	**-**
Electricity Plants	1938	-	5313	16791	8506	-	-	-	-	-
CHP Plants	-	-	-	-	53876	-	-	-	-	-
Heat Plants	-	-	-	-	-	-	-	-	-	-
Blast Furnaces/Gas Works	6007	-	-	-	-	-	-	-	-	-
Coke/Pat. Fuel/BKB Plants	-	-	-	-	-	-	-	-	-	-
Petroleum Refineries	-	-	-	-	-	-	-	-	-	-
Petrochemical Industry	-	-	-	-	-	-	-	-	-	-
Liquefaction	-	-	-	-	-	-	-	-	-	-
Other Transform. Sector	-	-	-	-	163718	-	-	-	-	-
ENERGY SECTOR	**57006**	**653**	**14149**	**11578**	**-**	**-**	**-**	**-**	**9627**	**-**
Coal Mines	-	-	-	-	-	-	-	-	-	-
Oil and Gas Extraction	57006	-	-	-	-	-	-	-	-	-
Petroleum Refineries	-	-	-	-	-	-	-	-	-	-
Electr., CHP+Heat Plants	-	-	-	-	-	-	-	-	9627	-
Pumped Storage (Elec.)	-	-	-	-	-	-	-	-	-	-
Other Energy Sector	-	653	14149	11578	-	-	-	-	-	-
Distribution Losses	6162	653	716	17811	-	-	-	-	54003	-
FINAL CONSUMPTION	**149121**	**4115**	**43729**	**35458**	**1220654**	**282505**	**-**	**-**	**297362**	**-**
INDUSTRY SECTOR	**137844**	**49**	**43729**	**35458**	**861938**	**-**	**-**	**-**	**136388**	**-**
Iron and Steel	30809	-	43729	35458	147296	-	-	-	19202	-
Chemical and Petrochem.	59486	16	-	-	4291	-	-	-	15924	-
of which: Feedstocks	16548	-	-	-	-	-	-	-	-	-
Non-Ferrous Metals	1085	-	-	-	998	-	-	-	27749	-
Non-Metallic Minerals	6743	-	-	-	80695	-	-	-	7197	-
Transport Equipment	-	-	-	-	-	-	-	-	-	-
Machinery	-	-	-	-	-	-	-	-	-	-
Mining and Quarrying	4689	-	-	-	-	-	-	-	7189	-
Food and Tobacco	6084	33	-	-	463554	-	-	-	15143	-
Paper, Pulp and Print	6394	-	-	-	136071	-	-	-	10927	-
Wood and Wood Products	-	-	-	-	-	-	-	-	-	-
Construction	-	-	-	-	-	-	-	-	-	-
Textile and Leather	2674	-	-	-	4116	-	-	-	6115	-
Non-specified	19880	-	-	-	24917	-	-	-	26942	-
TRANSPORT SECTOR	**3061**	**-**	**-**	**-**	**-**	**282505**	**-**	**-**	**1170**	**-**
Air	-	-	-	-	-	-	-	-	-	-
Road	3061	-	-	-	-	282505	-	-	-	-
Rail	-	-	-	-	-	-	-	-	1170	-
Pipeline Transport	-	-	-	-	-	-	-	-	-	-
Internal Navigation	-	-	-	-	-	-	-	-	-	-
Non-specified	-	-	-	-	-	-	-	-	-	-
OTHER SECTORS	**8216**	**4066**	**-**	**-**	**358716**	**-**	**-**	**-**	**159804**	**-**
Agriculture	-	-	-	-	74665	-	-	-	11614	-
Comm. and Publ. Services	4844	996	-	-	6136	-	-	-	68832	-
Residential	3372	3070	-	-	277915	-	-	-	79358	-
Non-specified	-	-	-	-	-	-	-	-	-	-
NON-ENERGY USE	**-**	**-**	**-**	**-**	**-**	**-**	**-**	**-**	**-**	**-**
in Industry/Transf./Energy	-	-	-	-	-	-	-	-	-	-
in Transport	-	-	-	-	-	-	-	-	-	-
in Other Sectors	-	-	-	-	-	-	-	-	-	-

Brunei

SUPPLY AND CONSUMPTION 1997	Coal (1000 tonnes)							Oil (1000 tonnes)			
	Coking Coal	Other Bit. Coal	Sub-Bit. Coal	Lignite	Peat	Oven and Gas Coke	Pat. Fuel and BKB	Crude Oil	NGL	Feed-stocks	Additives
Production	-	-	-	-	-	-	-	7825	715	-	-
Imports	-	-	-	-	-	-	-	-	-	-	-
Exports	-	-	-	-	-	-	-	-7300	-699	-	-
Intl. Marine Bunkers	-	-	-	-	-	-	-	-	-	-	-
Stock Changes	-	-	-	-	-	-	-	-174	-	-	-
DOMESTIC SUPPLY	-	-	-	-	-	-	-	351	16	-	-
Transfers and Stat. Diff.	-	-	-	-	-	-	-	173	-16	-	-
TRANSFORMATION	-	-	-	-	-	-	-	524	-	-	-
Electricity and CHP Plants	-	-	-	-	-	-	-	-	-	-	-
Petroleum Refineries	-	-	-	-	-	-	-	524	-	-	-
Other Transform. Sector	-	-	-	-	-	-	-	-	-	-	-
ENERGY SECTOR	-	-	-	-	-	-	-	-	-	-	-
DISTRIBUTION LOSSES	-	-	-	-	-	-	-	-	-	-	-
FINAL CONSUMPTION	-	-	-	-	-	-	-	-	-	-	-
INDUSTRY SECTOR	-	-	-	-	-	-	-	-	-	-	-
Iron and Steel	-	-	-	-	-	-	-	-	-	-	-
Chemical and Petrochem.	-	-	-	-	-	-	-	-	-	-	-
Non-Metallic Minerals	-	-	-	-	-	-	-	-	-	-	-
Non-specified	-	-	-	-	-	-	-	-	-	-	-
TRANSPORT SECTOR	-	-	-	-	-	-	-	-	-	-	-
Air	-	-	-	-	-	-	-	-	-	-	-
Road	-	-	-	-	-	-	-	-	-	-	-
Non-specified	-	-	-	-	-	-	-	-	-	-	-
OTHER SECTORS	-	-	-	-	-	-	-	-	-	-	-
Agriculture	-	-	-	-	-	-	-	-	-	-	-
Comm. and Publ. Services	-	-	-	-	-	-	-	-	-	-	-
Residential	-	-	-	-	-	-	-	-	-	-	-
Non-specified	-	-	-	-	-	-	-	-	-	-	-
NON-ENERGY USE	-	-	-	-	-	-	-	-	-	-	-

APPROVISIONNEMENT ET DEMANDE 1998	Charbon (1000 tonnes)							Pétrole (1000 tonnes)			
	Charbon à coke	Autres charb. bit.	Charbon sous-bit.	Lignite	Tourbe	Coke de four/gaz	Agg./briq. de lignite	Pétrole brut	LGN	Produits d'aliment.	Additifs
Production	-	-	-	-	-	-	-	7558	715	-	-
Imports	-	-	-	-	-	-	-	-	-	-	-
Exports	-	-	-	-	-	-	-	-7276	-699	-	-
Intl. Marine Bunkers	-	-	-	-	-	-	-	-	-	-	-
Stock Changes	-	-	-	-	-	-	-	72	-	-	-
DOMESTIC SUPPLY	-	-	-	-	-	-	-	354	16	-	-
Transfers and Stat. Diff.	-	-	-	-	-	-	-	170	-16	-	-
TRANSFORMATION	-	-	-	-	-	-	-	524	-	-	-
Electricity and CHP Plants	-	-	-	-	-	-	-	-	-	-	-
Petroleum Refineries	-	-	-	-	-	-	-	524	-	-	-
Other Transform. Sector	-	-	-	-	-	-	-	-	-	-	-
ENERGY SECTOR	-	-	-	-	-	-	-	-	-	-	-
DISTRIBUTION LOSSES	-	-	-	-	-	-	-	-	-	-	-
FINAL CONSUMPTION	-	-	-	-	-	-	-	-	-	-	-
INDUSTRY SECTOR	-	-	-	-	-	-	-	-	-	-	-
Iron and Steel	-	-	-	-	-	-	-	-	-	-	-
Chemical and Petrochem.	-	-	-	-	-	-	-	-	-	-	-
Non-Metallic Minerals	-	-	-	-	-	-	-	-	-	-	-
Non-specified	-	-	-	-	-	-	-	-	-	-	-
TRANSPORT SECTOR	-	-	-	-	-	-	-	-	-	-	-
Air	-	-	-	-	-	-	-	-	-	-	-
Road	-	-	-	-	-	-	-	-	-	-	-
Non-specified	-	-	-	-	-	-	-	-	-	-	-
OTHER SECTORS	-	-	-	-	-	-	-	-	-	-	-
Agriculture	-	-	-	-	-	-	-	-	-	-	-
Comm. and Publ. Services	-	-	-	-	-	-	-	-	-	-	-
Residential	-	-	-	-	-	-	-	-	-	-	-
Non-specified	-	-	-	-	-	-	-	-	-	-	-
NON-ENERGY USE	-	-	-	-	-	-	-	-	-	-	-

Brunei

SUPPLY AND CONSUMPTION 1997	Oil cont. (1000 tonnes)										
	Refinery Gas	LPG + Ethane	Motor Gasoline	Aviation Gasoline	Jet Fuel	Kerosene	Gas/ Diesel	Heavy Fuel Oil	Naphtha	Petrol. Coke	Other Prod.
Production	59	2	194	-	68	4	142	1	6	-	-
Imports	-	-	7	-	28	-	43	-	-	-	25
Exports	-	-	-	-	-	-	-	-	-	-	-1
Intl. Marine Bunkers	-	-	-	-	-	-	-	-	-	-	-
Stock Changes	-	-	-	-	-12	-	-2	-	-	-	-
DOMESTIC SUPPLY	59	2	201	-	84	4	183	1	6	-	24
Transfers and Stat. Diff.	-	16	-1	-	1	-1	-10	-	-	-	-
TRANSFORMATION	-	-	-	-	-	-	5	-	-	-	-
Electricity and CHP Plants	-	-	-	-	-	-	5	-	-	-	-
Petroleum Refineries	-	-	-	-	-	-	-	-	-	-	-
Other Transform. Sector	-	-	-	-	-	-	-	-	-	-	-
ENERGY SECTOR	59		-	-	-	-	-	-	-	-	-
DISTRIBUTION LOSSES	-	-	5	-	1	-	-	-	-	-	-
FINAL CONSUMPTION	-	18	195	-	84	3	168	1	6	-	24
INDUSTRY SECTOR	-	-	-	-	-	-	81	1	6	-	-
Iron and Steel	-	-	-	-	-	-	-	-	-	-	-
Chemical and Petrochem.	-	-	-	-	-	-	-	-	6	-	-
Non-Metallic Minerals	-	-	-	-	-	-	-	-	-	-	-
Non-specified	-	-	-	-	-	-	81	1	-	-	-
TRANSPORT SECTOR	-	-	195	-	84	2	84	-	-	-	-
Air	-	-	-	-	84	-	-	-	-	-	-
Road	-	-	195	-	-	-	84	-	-	-	-
Non-specified	-	-	-	-	-	2	-	-	-	-	-
OTHER SECTORS	-	18	-	-	-	1	3	-	-	-	-
Agriculture	-	-	-	-	-	-	1	-	-	-	-
Comm. and Publ. Services	-	-	-	-	-	-	-	-	-	-	-
Residential	-	18	-	-	-	1	2	-	-	-	-
Non-specified	-	-	-	-	-	-	-	-	-	-	-
NON-ENERGY USE	-	-	-	-	-	-	-	-	-	-	24

APPROVISIONNEMENT ET DEMANDE 1998	Pétrole cont. (1000 tonnes)										
	Gaz de raffinerie	GPL + éthane	Essence moteur	Essence aviation	Carbu- réacteurs	Kérosène	Gazole	Fioul lourd	Naphta	Coke de pétrole	Autres prod.
Production	59	1	184	-	74	4	147	1	6	-	-
Imports	-	-	4	-	5	-	5	-	-	-	20
Exports	-	-	-	-	-	-	-	-	-	-	-1
Intl. Marine Bunkers	-	-	-	-	-	-	-	-	-	-	-
Stock Changes	-	-	2	-	-6	-	3	-	-	-	-2
DOMESTIC SUPPLY	59	1	190	-	73	4	155	1	6	-	17
Transfers and Stat. Diff.	-	16	-	-	-	-1	-14	-	-	-	-
TRANSFORMATION	-	-	-	-	-	-	6	-	-	-	-
Electricity and CHP Plants	-	-	-	-	-	-	6	-	-	-	-
Petroleum Refineries	-	-	-	-	-	-	-	-	-	-	-
Other Transform. Sector	-	-	-	-	-	-	-	-	-	-	-
ENERGY SECTOR	59	-	-	-	-	-	-	-	-	-	-
DISTRIBUTION LOSSES	-	-	2	-	1	-	-	-	-	-	-
FINAL CONSUMPTION	-	17	188	-	72	3	135	1	6	-	17
INDUSTRY SECTOR	-	-	-	-	-	-	59	1	6	-	-
Iron and Steel	-	-	-	-	-	-	-	-	-	-	-
Chemical and Petrochem.	-	-	-	-	-	-	-	-	6	-	-
Non-Metallic Minerals	-	-	-	-	-	-	-	-	-	-	-
Non-specified	-	-	-	-	-	-	59	1	-	-	-
TRANSPORT SECTOR	-	-	188	-	72	2	74	-	-	-	-
Air	-	-	-	-	72	-	-	-	-	-	-
Road	-	-	188	-	-	-	74	-	-	-	-
Non-specified	-	-	-	-	-	2	-	-	-	-	-
OTHER SECTORS	-	17	-	-	-	1	2	-	-	-	-
Agriculture	-	-	-	-	-	-	-	-	-	-	-
Comm. and Publ. Services	-	-	-	-	-	-	-	-	-	-	-
Residential	-	17	-	-	-	1	2	-	-	-	-
Non-specified	-	-	-	-	-	-	-	-	-	-	-
NON-ENERGY USE	-	-	-	-	-	-	-	-	-	-	17

Brunei

SUPPLY AND CONSUMPTION 1997	Gas (TJ) Natural Gas	Gas Works	Coke Ovens	Blast Furnaces	Comb. Renew. & Waste (TJ) Solid Biomass	Gas/Liquids from Biomass	Municipal Waste	Industrial Waste	(GWh) Electricity	(TJ) Heat
Production	412149	-	-	-	772	-	-	-	2407	-
Imports	-	-	-	-	-	-	-	-	-	-
Exports	-336577	-	-	-	-	-	-	-	-	-
Intl. Marine Bunkers	-	-	-	-	-	-	-	-	-	-
Stock Changes	-	-	-	-	-	-	-	-	-	-
DOMESTIC SUPPLY	75572	-	-	-	772	-	-	-	2407	-
Transfers and Stat. Diff.	14996	-	-	-	-	-	-	-	1	-
TRANSFORMATION	38848	-	-	-	-	-	-	-	-	-
Electricity and CHP Plants	38848	-	-	-	-	-	-	-	-	-
Petroleum Refineries	-	-	-	-	-	-	-	-	-	-
Other Transform. Sector	-	-	-	-	-	-	-	-	-	-
ENERGY SECTOR	32711	-	-	-	-	-	-	-	58	-
DISTRIBUTION LOSSES	19009	-	-	-	-	-	-	-	29	-
FINAL CONSUMPTION	-	-	-	-	772	-	-	-	2321	-
INDUSTRY SECTOR	-	-	-	-	-	-	-	-	263	-
Iron and Steel	-	-	-	-	-	-	-	-	-	-
Chemical and Petrochem.	-	-	-	-	-	-	-	-	-	-
Non-Metallic Minerals	-	-	-	-	-	-	-	-	-	-
Non-specified	-	-	-	-	-	-	-	-	263	-
TRANSPORT SECTOR	-	-	-	-	-	-	-	-	-	-
Air	-	-	-	-	-	-	-	-	-	-
Road	-	-	-	-	-	-	-	-	-	-
Non-specified	-	-	-	-	-	-	-	-	-	-
OTHER SECTORS	-	-	-	-	772	-	-	-	2058	-
Agriculture	-	-	-	-	-	-	-	-	-	-
Comm. and Publ. Services	-	-	-	-	-	-	-	-	1522	-
Residential	-	-	-	-	-	-	-	-	536	-
Non-specified	-	-	-	-	772	-	-	-	-	-
NON-ENERGY USE	-	-	-	-	-	-	-	-	-	-

APPROVISIONNEMENT ET DEMANDE 1998	Gaz (TJ) Gaz naturel	Usines à gaz	Cokeries	Hauts fourneaux	En. Re. Comb. & Déchets (TJ) Biomasse solide	Gaz/Liquides tirés de biomasse	Déchets urbains	Déchets industriels	(GWh) Electricité	(TJ) Chaleur
Production	398823	-	-	-	772	-	-	-	2503	-
Imports	-	-	-	-	-	-	-	-	-	-
Exports	-321806	-	-	-	-	-	-	-	-	-
Intl. Marine Bunkers	-	-	-	-	-	-	-	-	-	-
Stock Changes	-	-	-	-	-	-	-	-	-	-
DOMESTIC SUPPLY	77017	-	-	-	772	-	-	-	2503	-
Transfers and Stat. Diff.	16082	-	-	-	-	-	-	-	1	-
TRANSFORMATION	40830	-	-	-	-	-	-	-	-	-
Electricity and CHP Plants	40830	-	-	-	-	-	-	-	-	-
Petroleum Refineries	-	-	-	-	-	-	-	-	-	-
Other Transform. Sector	-	-	-	-	-	-	-	-	-	-
ENERGY SECTOR	31408	-	-	-	-	-	-	-	58	-
DISTRIBUTION LOSSES	20861	-	-	-	-	-	-	-	29	-
FINAL CONSUMPTION	-	-	-	-	772	-	-	-	2417	-
INDUSTRY SECTOR	-	-	-	-	-	-	-	-	285	-
Iron and Steel	-	-	-	-	-	-	-	-	-	-
Chemical and Petrochem.	-	-	-	-	-	-	-	-	-	-
Non-Metallic Minerals	-	-	-	-	-	-	-	-	-	-
Non-specified	-	-	-	-	-	-	-	-	285	-
TRANSPORT SECTOR	-	-	-	-	-	-	-	-	-	-
Air	-	-	-	-	-	-	-	-	-	-
Road	-	-	-	-	-	-	-	-	-	-
Non-specified	-	-	-	-	-	-	-	-	-	-
OTHER SECTORS	-	-	-	-	772	-	-	-	2132	-
Agriculture	-	-	-	-	-	-	-	-	-	-
Comm. and Publ. Services	-	-	-	-	-	-	-	-	1617	-
Residential	-	-	-	-	-	-	-	-	515	-
Non-specified	-	-	-	-	772	-	-	-	-	-
NON-ENERGY USE	-	-	-	-	-	-	-	-	-	-

Bulgaria / Bulgarie : 1997

SUPPLY AND CONSUMPTION / APPROVISIONNEMENT ET DEMANDE	Coal / Charbon (1000 tonnes)							Oil / Pétrole (1000 tonnes)			
	Coking Coal / Charbon à coke	Other Bit. Coal / Autres charb. bit.	Sub-Bit. Coal / Charbon sous-bit.	Lignite / Lignite	Peat / Tourbe	Oven and Gas Coke / Coke de four/gaz	Pat. Fuel and BKB / Agg./briq. de lignite	Crude Oil / Pétrole brut	NGL / LGN	Feed-stocks / Produits d'aliment.	Additives / Additifs
Production	-	102	-	29606	-	1165	1082	28	-	-	-
From Other Sources	-	-	-	-	-	-	-	-	-	-	-
Imports	1677	2034	-	-	-	153	-	5888	-	-	-
Exports	-	-	-	-	-	-5	-	-	-	-	-
Intl. Marine Bunkers	-	-	-	-	-	-	-	-	-	-	-
Stock Changes	-21	293	-	322	-	-12	2	48	-	-	-
DOMESTIC SUPPLY	**1656**	**2429**	**-**	**29928**	**-**	**1301**	**1084**	**5964**	**-**	**-**	**-**
Transfers	-	-	-	-	-	-	-	-	-	-	-
Statistical Differences	-	33	-	327	-	9	4	1	-	-	-
TRANSFORMATION	**1656**	**2115**	**-**	**29650**	**-**	**981**	**135**	**5965**	**-**	**-**	**-**
Electricity Plants	-	1121	-	23241	-	-	5	-	-	-	-
CHP Plants	-	981	-	3960	-	-	130	-	-	-	-
Heat Plants	-	13	-	-	-	-	-	1	-	-	-
Blast Furnaces/Gas Works	-	-	-	-	-	981	-	-	-	-	-
Coke/Pat. Fuel/BKB Plants	1656	-	-	2449	-	-	-	-	-	-	-
Petroleum Refineries	-	-	-	-	-	-	-	5964	-	-	-
Petrochemical Industry	-	-	-	-	-	-	-	-	-	-	-
Liquefaction	-	-	-	-	-	-	-	-	-	-	-
Other Transform. Sector	-	-	-	-	-	-	-	-	-	-	-
ENERGY SECTOR	**-**	**4**	**-**	**71**	**-**	**-**	**3**	**-**	**-**	**-**	**-**
Coal Mines	-	4	-	66	-	-	3	-	-	-	-
Oil and Gas Extraction	-	-	-	-	-	-	-	-	-	-	-
Petroleum Refineries	-	-	-	-	-	-	-	-	-	-	-
Electr., CHP+Heat Plants	-	-	-	5	-	-	-	-	-	-	-
Pumped Storage (Elec.)	-	-	-	-	-	-	-	-	-	-	-
Other Energy Sector	-	-	-	-	-	-	-	-	-	-	-
Distribution Losses	-	-	-	-	-	-	-	-	-	-	-
FINAL CONSUMPTION	**-**	**343**	**-**	**534**	**-**	**329**	**950**	**-**	**-**	**-**	**-**
INDUSTRY SECTOR	**-**	**336**	**-**	**70**	**-**	**329**	**6**	**-**	**-**	**-**	**-**
Iron and Steel	-	-	-	-	-	170	-	-	-	-	-
Chemical and Petrochem.	-	153	-	5	-	85	-	-	-	-	-
of which: Feedstocks	-	-	-	-	-	-	-	-	-	-	-
Non-Ferrous Metals	-	-	-	-	-	70	-	-	-	-	-
Non-Metallic Minerals	-	45	-	37	-	-	1	-	-	-	-
Transport Equipment	-	-	-	-	-	-	-	-	-	-	-
Machinery	-	3	-	5	-	3	2	-	-	-	-
Mining and Quarrying	-	6	-	2	-	-	-	-	-	-	-
Food and Tobacco	-	44	-	7	-	1	1	-	-	-	-
Paper, Pulp and Print	-	-	-	-	-	-	-	-	-	-	-
Wood and Wood Products	-	-	-	1	-	-	-	-	-	-	-
Construction	-	-	-	6	-	-	1	-	-	-	-
Textile and Leather	-	8	-	2	-	-	1	-	-	-	-
Non-specified	-	77	-	5	-	-	-	-	-	-	-
TRANSPORT SECTOR	**-**	**-**	**-**	**-**	**-**	**-**	**-**	**-**	**-**	**-**	**-**
Air	-	-	-	-	-	-	-	-	-	-	-
Road	-	-	-	-	-	-	-	-	-	-	-
Rail	-	-	-	-	-	-	-	-	-	-	-
Pipeline Transport	-	-	-	-	-	-	-	-	-	-	-
Internal Navigation	-	-	-	-	-	-	-	-	-	-	-
Non-specified	-	-	-	-	-	-	-	-	-	-	-
OTHER SECTORS	**-**	**7**	**-**	**464**	**-**	**-**	**944**	**-**	**-**	**-**	**-**
Agriculture	-	3	-	1	-	-	2	-	-	-	-
Comm. and Publ. Services	-	3	-	4	-	-	1	-	-	-	-
Residential	-	-	-	428	-	-	939	-	-	-	-
Non-specified	-	1	-	31	-	-	2	-	-	-	-
NON-ENERGY USE	**-**	**-**	**-**	**-**	**-**	**-**	**-**	**-**	**-**	**-**	**-**
in Industry/Trans./Energy	-	-	-	-	-	-	-	-	-	-	-
in Transport	-	-	-	-	-	-	-	-	-	-	-
in Other Sectors	-	-	-	-	-	-	-	-	-	-	-

Bulgaria / Bulgarie : 1997

SUPPLY AND CONSUMPTION / APPROVISIONNEMENT ET DEMANDE	Oil cont. / *Pétrole cont.* (1000 tonnes)										
	Refinery Gas / *Gaz de raffinerie*	LPG + Ethane / *GPL + éthane*	Motor Gasoline / *Essence moteur*	Aviation Gasoline / *Essence aviation*	Jet Fuel / *Carbu- réacteurs*	Kerosene / *Kérosène*	Gas/ Diesel / *Gazole*	Heavy Fuel Oil / *Fioul lourd*	Naphtha / *Naphta*	Petrol. Coke / *Coke de pétrole*	Other Prod. / *Autres prod.*
Production	123	84	1112	-	152	-	2040	1477	615	-	217
From Other Sources	-	-	-	-	-	-	-	-	-	-	-
Imports	-	1	2	1	-	-	16	22	58	-	23
Exports	-	-	-453	-	-1	-	-1001	-30	-	-	-
Intl. Marine Bunkers	-	-	-	-	-	-	-9	-	-	-	-
Stock Changes	-	2	-20	-	18	-	-77	27	-	-	-
DOMESTIC SUPPLY	**123**	**87**	**641**	**1**	**169**	**-**	**969**	**1496**	**673**	**-**	**240**
Transfers	-	-	-	-	-	-	-	-	-	-	-
Statistical Differences	-	-	-32	-	-2	-	-2	53	-	-	-
TRANSFORMATION	**2**	**-**	**-**	**-**	**-**	**-**	**3**	**326**	**-**	**-**	**41**
Electricity Plants	-	-	-	-	-	-	-	23	-	-	-
CHP Plants	2	-	-	-	-	-	1	190	-	-	41
Heat Plants	-	-	-	-	-	-	2	113	-	-	-
Blast Furnaces/Gas Works	-	-	-	-	-	-	-	-	-	-	-
Coke/Pat. Fuel/BKB Plants	-	-	-	-	-	-	-	-	-	-	-
Petroleum Refineries	-	-	-	-	-	-	-	-	-	-	-
Petrochemical Industry	-	-	-	-	-	-	-	-	-	-	-
Liquefaction	-	-	-	-	-	-	-	-	-	-	-
Other Transform. Sector	-	-	-	-	-	-	-	-	-	-	-
ENERGY SECTOR	**94**	**-**	**-**	**-**	**-**	**-**	**2**	**86**	**-**	**-**	**-**
Coal Mines	-	-	-	-	-	-	-	-	-	-	-
Oil and Gas Extraction	-	-	-	-	-	-	-	2	-	-	-
Petroleum Refineries	94	-	-	-	-	-	-	84	-	-	-
Electr., CHP+Heat Plants	-	-	-	-	-	-	2	-	-	-	-
Pumped Storage (Elec.)	-	-	-	-	-	-	-	-	-	-	-
Other Energy Sector	-	-	-	-	-	-	-	-	-	-	-
Distribution Losses	-	2	-	-	-	-	-	-	-	-	-
FINAL CONSUMPTION	**27**	**85**	**609**	**1**	**167**	**-**	**962**	**1137**	**673**	**-**	**199**
INDUSTRY SECTOR	**27**	**12**	**-**	**-**	**-**	**-**	**36**	**1101**	**673**	**-**	**171**
Iron and Steel	-	-	-	-	-	-	-	34	-	-	-
Chemical and Petrochem.	27	-	-	-	-	-	1	375	673	-	171
of which: Feedstocks	-	-	-	-	-	-	-	-	673	-	-
Non-Ferrous Metals	-	1	-	-	-	-	2	59	-	-	-
Non-Metallic Minerals	-	6	-	-	-	-	4	172	-	-	-
Transport Equipment	-	-	-	-	-	-	2	5	-	-	-
Machinery	-	4	-	-	-	-	4	55	-	-	-
Mining and Quarrying	-	-	-	-	-	-	5	23	-	-	-
Food and Tobacco	-	1	-	-	-	-	8	160	-	-	-
Paper, Pulp and Print	-	-	-	-	-	-	1	61	-	-	-
Wood and Wood Products	-	-	-	-	-	-	1	36	-	-	-
Construction	-	-	-	-	-	-	6	8	-	-	-
Textile and Leather	-	-	-	-	-	-	1	65	-	-	-
Non-specified	-	-	-	-	-	-	1	48	-	-	-
TRANSPORT SECTOR	**-**	**66**	**609**	**1**	**167**	**-**	**674**	**-**	**-**	**-**	**-**
Air	-	-	-	1	167	-	-	-	-	-	-
Road	-	66	609	-	-	-	629	-	-	-	-
Rail	-	-	-	-	-	-	43	-	-	-	-
Pipeline Transport	-	-	-	-	-	-	-	-	-	-	-
Internal Navigation	-	-	-	-	-	-	2	-	-	-	-
Non-specified	-	-	-	-	-	-	-	-	-	-	-
OTHER SECTORS	**-**	**7**	**-**	**-**	**-**	**-**	**252**	**36**	**-**	**-**	**-**
Agriculture	-	-	-	-	-	-	248	23	-	-	-
Comm. and Publ. Services	-	-	-	-	-	-	-	-	-	-	-
Residential	-	7	-	-	-	-	-	-	-	-	-
Non-specified	-	-	-	-	-	-	4	13	-	-	-
NON-ENERGY USE	**-**	**-**	**-**	**-**	**-**	**-**	**-**	**-**	**-**	**-**	**28**
in Industry/Transf./Energy	-	-	-	-	-	-	-	-	-	-	28
in Transport	-	-	-	-	-	-	-	-	-	-	-
in Other Sectors	-	-	-	-	-	-	-	-	-	-	-

Bulgaria / Bulgarie : 1997

SUPPLY AND CONSUMPTION / APPROVISIONNEMENT ET DEMANDE	Gas / Gaz (TJ)				Comb. Renew. & Waste / En. Re. Comb. & Déchets (TJ)				(GWh)	(TJ)
	Natural Gas / Gaz naturel	Gas Works / Usines à gaz	Coke Ovens / Cokeries	Blast Furnaces / Hauts fourneaux	Solid Biomass / Biomasse solide	Gas/Liquids from Biomass / Gaz/Liquides tirés de biomasse	Municipal Waste / Déchets urbains	Industrial Waste / Déchets industriels	Electricity / Electricité	Heat / Chaleur
Production	1307	-	8663	13963	10491	-	-	859	42803	63273
From Other Sources	-	-	-	-	-	-	-	-	-	-
Imports	179192	-	-	-	-	-	-	-	785	-
Exports	-	-	-	-	-430	-	-	-	-4335	-
Intl. Marine Bunkers	-	-	-	-	-	-	-	-	-	-
Stock Changes	-8387	-	-	-	-24	-	-	-	-	-
DOMESTIC SUPPLY	**172112**	**-**	**8663**	**13963**	**10037**	**-**	**-**	**859**	**39253**	**63273**
Transfers	-	-	-	-	-	-	-	-	-	-
Statistical Differences	480	-	-	-	-191	-	-	-	-	-
TRANSFORMATION	**85192**	**-**	**1045**	**2272**	**-**	**-**	**-**	**-**	**115**	**-**
Electricity Plants	-	-	-	-	-	-	-	-	-	-
CHP Plants	54617	-	1045	2272	-	-	-	-	-	-
Heat Plants	30575	-	-	-	-	-	-	-	-	-
Blast Furnaces/Gas Works	-	-	-	-	-	-	-	-	-	-
Coke/Pat. Fuel/BKB Plants	-	-	-	-	-	-	-	-	-	-
Petroleum Refineries	-	-	-	-	-	-	-	-	-	-
Petrochemical Industry	-	-	-	-	-	-	-	-	-	-
Liquefaction	-	-	-	-	-	-	-	-	-	-
Other Transform. Sector	-	-	-	-	-	-	-	-	115	-
ENERGY SECTOR	**2990**	**-**	**4737**	**-**	**1**	**-**	**-**	**-**	**5969**	**9482**
Coal Mines	-	-	-	-	-	-	-	-	800	873
Oil and Gas Extraction	13	-	-	-	-	-	-	-	12	-
Petroleum Refineries	2973	-	-	-	-	-	-	-	309	-
Electr., CHP+Heat Plants	4	-	-	-	1	-	-	-	4390	2495
Pumped Storage (Elec.)	-	-	-	-	-	-	-	-	283	-
Other Energy Sector	-	-	4737	-	-	-	-	-	175	6114
Distribution Losses	2001	-	-	-	40	-	-	-	6253	5659
FINAL CONSUMPTION	**82409**	**-**	**2881**	**11691**	**9805**	**-**	**-**	**859**	**26916**	**48132**
INDUSTRY SECTOR	**81202**	**-**	**2881**	**11691**	**1099**	**-**	**-**	**859**	**11812**	**15219**
Iron and Steel	13676	-	2881	11691	2	-	-	-	2054	168
Chemical and Petrochem.	46105	-	-	-	2	-	-	54	3116	10363
of which: Feedstocks	25189	-	-	-	-	-	-	-	-	-
Non-Ferrous Metals	167	-	-	-	4	-	-	-	787	70
Non-Metallic Minerals	19879	-	-	-	9	-	-	-	856	153
Transport Equipment	-	-	-	-	7	-	-	-	159	38
Machinery	1134	-	-	-	39	-	-	-	1326	229
Mining and Quarrying	-	-	-	-	5	-	-	-	1008	3
Food and Tobacco	171	-	-	-	395	-	-	-	847	2060
Paper, Pulp and Print	59	-	-	-	1	-	-	805	323	300
Wood and Wood Products	-	-	-	-	524	-	-	-	123	2
Construction	8	-	-	-	27	-	-	-	303	99
Textile and Leather	1	-	-	-	16	-	-	-	548	1540
Non-specified	2	-	-	-	68	-	-	-	362	194
TRANSPORT SECTOR	**526**	**-**	**-**	**-**	**-**	**-**	**-**	**-**	**620**	**-**
Air	-	-	-	-	-	-	-	-	-	-
Road	-	-	-	-	-	-	-	-	-	-
Rail	-	-	-	-	-	-	-	-	555	-
Pipeline Transport	526	-	-	-	-	-	-	-	65	-
Internal Navigation	-	-	-	-	-	-	-	-	-	-
Non-specified	-	-	-	-	-	-	-	-	-	-
OTHER SECTORS	**681**	**-**	**-**	**-**	**8706**	**-**	**-**	**-**	**14484**	**32913**
Agriculture	-	-	-	-	224	-	-	-	355	-
Comm. and Publ. Services	656	-	-	-	-	-	-	-	1359	88
Residential	-	-	-	-	7521	-	-	-	9882	25987
Non-specified	25	-	-	-	961	-	-	-	2888	6838
NON-ENERGY USE	**-**	**-**	**-**	**-**	**-**	**-**	**-**	**-**	**-**	**-**
in Industry/Transf./Energy	-	-	-	-	-	-	-	-	-	-
in Transport	-	-	-	-	-	-	-	-	-	-
in Other Sectors	-	-	-	-	-	-	-	-	-	-

Bulgaria / Bulgarie : 1998

SUPPLY AND CONSUMPTION / APPROVISIONNEMENT ET DEMANDE	Coal / Charbon (1000 tonnes)							Oil / Pétrole (1000 tonnes)			
	Coking Coal / Charbon à coke	Other Bit. Coal / Autres charb. bit.	Sub-Bit. Coal / Charbon sous-bit.	Lignite / Lignite	Peat / Tourbe	Oven and Gas Coke / Coke de four/gaz	Pat. Fuel and BKB / Agg./briq. de lignite	Crude Oil / Pétrole brut	NGL / LGN	Feed-stocks / Produits d'aliment.	Additives / Additifs
Production	-	92	-	30019	-	868	1152	33	-	-	-
From Other Sources	-	-	-	-	-	-	-	-	-	2	-
Imports	1170	2379	-	-	-	196	-	5526	-	9	14
Exports	-	-	-	-	-	-12	-	-	-	-	-
Intl. Marine Bunkers	-	-	-	-	-	-	-	-	-	-	-
Stock Changes	19	-71	-	-388	-	11	-	18	-	-	-
DOMESTIC SUPPLY	**1189**	**2400**	**-**	**29631**	**-**	**1063**	**1152**	**5577**	**-**	**11**	**14**
Transfers	-	-	-	-	-	-	-	-	-	-	-
Statistical Differences	-	-23	-	109	-	39	11	-21	-	-	-
TRANSFORMATION	**1189**	**2120**	**-**	**29305**	**-**	**800**	**78**	**5556**	**-**	**11**	**14**
Electricity Plants	-	1058	-	22597	-	-	-	-	-	-	-
CHP Plants	-	1042	-	3776	-	-	77	-	-	-	-
Heat Plants	-	20	-	4	-	-	1	1	-	-	-
Blast Furnaces/Gas Works	-	-	-	-	-	800	-	-	-	-	-
Coke/Pat. Fuel/BKB Plants	1189	-	-	2928	-	-	-	-	-	-	-
Petroleum Refineries	-	-	-	-	-	-	-	5555	-	11	14
Petrochemical Industry	-	-	-	-	-	-	-	-	-	-	-
Liquefaction	-	-	-	-	-	-	-	-	-	-	-
Other Transform. Sector	-	-	-	-	-	-	-	-	-	-	-
ENERGY SECTOR	**-**	**4**	**-**	**50**	**-**	**-**	**3**	**-**	**-**	**-**	**-**
Coal Mines	-	4	-	50	-	-	3	-	-	-	-
Oil and Gas Extraction	-	-	-	-	-	-	-	-	-	-	-
Petroleum Refineries	-	-	-	-	-	-	-	-	-	-	-
Electr., CHP+Heat Plants	-	-	-	-	-	-	-	-	-	-	-
Pumped Storage (Elec.)	-	-	-	-	-	-	-	-	-	-	-
Other Energy Sector	-	-	-	-	-	-	-	-	-	-	-
Distribution Losses	-	-	-	-	-	-	-	-	-	-	-
FINAL CONSUMPTION	**-**	**253**	**-**	**385**	**-**	**302**	**1082**	**-**	**-**	**-**	**-**
INDUSTRY SECTOR	**-**	**250**	**-**	**56**	**-**	**302**	**7**	**-**	**-**	**-**	**-**
Iron and Steel	-	1	-	3	-	137	-	-	-	-	-
Chemical and Petrochem.	-	109	-	-	-	81	-	-	-	-	-
of which: Feedstocks	-	-	-	-	-	-	-	-	-	-	-
Non-Ferrous Metals	-	-	-	-	-	81	-	-	-	-	-
Non-Metallic Minerals	-	59	-	32	-	-	1	-	-	-	-
Transport Equipment	-	-	-	-	-	-	-	-	-	-	-
Machinery	-	2	-	2	-	2	3	-	-	-	-
Mining and Quarrying	-	9	-	2	-	-	-	-	-	-	-
Food and Tobacco	-	30	-	6	-	1	1	-	-	-	-
Paper, Pulp and Print	-	-	-	-	-	-	-	-	-	-	-
Wood and Wood Products	-	-	-	-	-	-	-	-	-	-	-
Construction	-	3	-	2	-	-	1	-	-	-	-
Textile and Leather	-	2	-	4	-	-	1	-	-	-	-
Non-specified	-	35	-	5	-	-	-	-	-	-	-
TRANSPORT SECTOR	**-**	**-**	**-**	**-**	**-**	**-**	**-**	**-**	**-**	**-**	**-**
Air	-	-	-	-	-	-	-	-	-	-	-
Road	-	-	-	-	-	-	-	-	-	-	-
Rail	-	-	-	-	-	-	-	-	-	-	-
Pipeline Transport	-	-	-	-	-	-	-	-	-	-	-
Internal Navigation	-	-	-	-	-	-	-	-	-	-	-
Non-specified	-	-	-	-	-	-	-	-	-	-	-
OTHER SECTORS	**-**	**3**	**-**	**329**	**-**	**-**	**1075**	**-**	**-**	**-**	**-**
Agriculture	-	1	-	1	-	-	2	-	-	-	-
Comm. and Publ. Services	-	-	-	-	-	-	3	-	-	-	-
Residential	-	-	-	319	-	-	1069	-	-	-	-
Non-specified	-	2	-	9	-	-	1	-	-	-	-
NON-ENERGY USE	**-**	**-**	**-**	**-**	**-**	**-**	**-**	**-**	**-**	**-**	**-**
in Industry/Trans./Energy	-	-	-	-	-	-	-	-	-	-	-
in Transport	-	-	-	-	-	-	-	-	-	-	-
in Other Sectors	-	-	-	-	-	-	-	-	-	-	-

Bulgaria / Bulgarie : 1998

SUPPLY AND CONSUMPTION / APPROVISIONNEMENT ET DEMANDE	Refinery Gas / Gaz de raffinerie	LPG + Ethane / GPL + éthane	Motor Gasoline / Essence moteur	Aviation Gasoline / Essence aviation	Jet Fuel / Carbu- réacteurs	Kerosene / Kérosène	Gas/ Diesel / Gazole	Heavy Fuel Oil / Fioul lourd	Naphtha / Naphta	Petrol. Coke / Coke de pétrole	Other Prod. / Autres prod.
Production	129	75	925	-	165	-	1957	1327	624	-	363
From Other Sources	-	-	-	-	-	-	-	-	-	-	-
Imports	-	-	74	1	-	1	313	31	30	-	41
Exports	-	-	-262	-	-2	-	-870	-125	-	-	-6
Intl. Marine Bunkers	-	-	-	-	-	-	-70	-	-	-	-
Stock Changes	-	-1	67	-	-16	-	-103	77	-4	-	-
DOMESTIC SUPPLY	129	74	804	1	147	1	1227	1310	650	-	398
Transfers	-	-	-	-	-	-	-	-	-	-	-
Statistical Differences	-	-	-4	-	-1	-	-8	-1	-	-	-
TRANSFORMATION	7	-	-	-	-	-	2	302	-	-	63
Electricity Plants	-	-	-	-	-	-	-	21	-	-	-
CHP Plants	7	-	-	-	-	-	-	202	-	-	63
Heat Plants	-	-	-	-	-	-	2	79	-	-	-
Blast Furnaces/Gas Works	-	-	-	-	-	-	-	-	-	-	-
Coke/Pat. Fuel/BKB Plants	-	-	-	-	-	-	-	-	-	-	-
Petroleum Refineries	-	-	-	-	-	-	-	-	-	-	-
Petrochemical Industry	-	-	-	-	-	-	-	-	-	-	-
Liquefaction	-	-	-	-	-	-	-	-	-	-	-
Other Transform. Sector	-	-	-	-	-	-	-	-	-	-	-
ENERGY SECTOR	81	-	-	-	-	-	1	73	-	-	-
Coal Mines	-	-	-	-	-	-	-	-	-	-	-
Oil and Gas Extraction	-	-	-	-	-	-	-	-	-	-	-
Petroleum Refineries	81	-	-	-	-	-	-	-	-	-	-
Electr., CHP+Heat Plants	-	-	-	-	-	-	1	3	-	-	-
Pumped Storage (Elec.)	-	-	-	-	-	-	-	-	-	-	-
Other Energy Sector	-	-	-	-	-	-	-	70	-	-	-
Distribution Losses	-	-	-	-	-	-	1	-	-	-	-
FINAL CONSUMPTION	41	74	800	1	146	1	1215	934	650	-	335
INDUSTRY SECTOR	41	7	-	-	-	1	90	899	650	-	196
Iron and Steel	-	-	-	-	-	-	1	34	-	-	-
Chemical and Petrochem.	41	-	-	-	-	1	5	244	650	-	196
of which: Feedstocks	-	-	-	-	-	1	-	-	650	-	-
Non-Ferrous Metals	-	1	-	-	-	-	1	56	-	-	-
Non-Metallic Minerals	-	3	-	-	-	-	2	178	-	-	-
Transport Equipment	-	-	-	-	-	-	1	4	-	-	-
Machinery	-	1	-	-	-	-	7	54	-	-	-
Mining and Quarrying	-	-	-	-	-	-	11	16	-	-	-
Food and Tobacco	-	2	-	-	-	-	33	128	-	-	-
Paper, Pulp and Print	-	-	-	-	-	-	1	49	-	-	-
Wood and Wood Products	-	-	-	-	-	-	-	30	-	-	-
Construction	-	-	-	-	-	-	9	12	-	-	-
Textile and Leather	-	-	-	-	-	-	16	58	-	-	-
Non-specified	-	-	-	-	-	-	3	36	-	-	-
TRANSPORT SECTOR	-	59	800	1	146	-	803	1	-	-	-
Air	-	-	-	1	146	-	-	-	-	-	-
Road	-	59	800	-	-	-	760	-	-	-	-
Rail	-	-	-	-	-	-	41	-	-	-	-
Pipeline Transport	-	-	-	-	-	-	-	-	-	-	-
Internal Navigation	-	-	-	-	-	-	2	1	-	-	-
Non-specified	-	-	-	-	-	-	-	-	-	-	-
OTHER SECTORS	-	8	-	-	-	-	322	34	-	-	-
Agriculture	-	-	-	-	-	-	247	15	-	-	-
Comm. and Publ. Services	-	-	-	-	-	-	26	5	-	-	-
Residential	-	7	-	-	-	-	6	-	-	-	-
Non-specified	-	1	-	-	-	-	43	14	-	-	-
NON-ENERGY USE	-	-	-	-	-	-	-	-	-	-	139
in Industry/Transf./Energy	-	-	-	-	-	-	-	-	-	-	139
in Transport	-	-	-	-	-	-	-	-	-	-	-
in Other Sectors	-	-	-	-	-	-	-	-	-	-	-

Oil cont. / Pétrole cont. (1000 tonnes)

Bulgaria / Bulgarie : 1998

SUPPLY AND CONSUMPTION / APPROVISIONNEMENT ET DEMANDE	Gas / Gaz (TJ)				Comb. Renew. & Waste / En. Re. Comb. & Déchets (TJ)				(GWh)	(TJ)
	Natural Gas / Gaz naturel	Gas Works / Usines à gaz	Coke Ovens / Cokeries	Blast Furnaces / Hauts fourneaux	Solid Biomass / Biomasse solide	Gas/Liquids from Biomass / Gaz/Liquides tirés de biomasse	Municipal Waste / Déchets urbains	Industrial Waste / Déchets industriels	Electricity / Electricité	Heat / Chaleur
Production	1088	-	6813	10561	17277	-	-	158	41711	58380
From Other Sources	-	-	-	-	-	-	-	-	-	-
Imports	145145	-	-	-	-	-	-	-	564	-
Exports	-	-	-	-	-49	-	-	-	-4211	-
Intl. Marine Bunkers	-	-	-	-	-	-	-	-	-	-
Stock Changes	-634	-	-	-	18	-	-	-4	-	-
DOMESTIC SUPPLY	**145599**	**-**	**6813**	**10561**	**17246**	**-**	**-**	**154**	**38064**	**58380**
Transfers	-	-	-	-	-	-	-	-	-	-
Statistical Differences	-1192	-	-	-	-28	-	-	-	-	-
TRANSFORMATION	**81799**	**-**	**868**	**2008**	**-**	**-**	**-**	**19**	**133**	**-**
Electricity Plants	-	-	-	-	-	-	-	-	-	-
CHP Plants	50097	-	868	2008	-	-	-	19	-	-
Heat Plants	31150	-	-	-	-	-	-	-	-	-
Blast Furnaces/Gas Works	-	-	-	-	-	-	-	-	-	-
Coke/Pat. Fuel/BKB Plants	-	-	-	-	-	-	-	-	-	-
Petroleum Refineries	-	-	-	-	-	-	-	-	-	-
Petrochemical Industry	-	-	-	-	-	-	-	-	-	-
Liquefaction	-	-	-	-	-	-	-	-	-	-
Other Transform. Sector	552	-	-	-	-	-	-	-	133	-
ENERGY SECTOR	**1843**	**-**	**3679**	**-**	**1**	**-**	**-**	**2**	**6195**	**8738**
Coal Mines	-	-	-	-	-	-	-	2	779	163
Oil and Gas Extraction	10	-	-	-	-	-	-	-	13	-
Petroleum Refineries	1833	-	-	-	-	-	-	-	286	-
Electr., CHP+Heat Plants	-	-	-	-	1	-	-	-	4387	2058
Pumped Storage (Elec.)	-	-	-	-	-	-	-	-	293	-
Other Energy Sector	-	-	3679	-	-	-	-	-	437	6517
Distribution Losses	1100	-	-	-	110	-	-	-	5595	5050
FINAL CONSUMPTION	**59665**	**-**	**2266**	**8553**	**17107**	**-**	**-**	**133**	**26141**	**44592**
INDUSTRY SECTOR	**58423**	**-**	**2266**	**8553**	**1054**	**-**	**-**	**133**	**10688**	**13489**
Iron and Steel	12526	-	2266	8553	-	-	-	-	1933	82
Chemical and Petrochem.	32011	-	-	-	5	-	-	20	2432	9754
of which: Feedstocks	15851	-	-	-	-	-	-	-	-	-
Non-Ferrous Metals	128	-	-	-	5	-	-	-	784	26
Non-Metallic Minerals	12499	-	-	-	8	-	-	-	736	181
Transport Equipment	-	-	-	-	9	-	-	-	145	47
Machinery	971	-	-	-	15	-	-	-	1154	292
Mining and Quarrying	-	-	-	-	8	-	-	-	954	3
Food and Tobacco	186	-	-	-	273	-	-	-	855	1399
Paper, Pulp and Print	80	-	-	-	2	-	-	113	315	176
Wood and Wood Products	-	-	-	-	571	-	-	-	111	1
Construction	19	-	-	-	46	-	-	-	265	92
Textile and Leather	1	-	-	-	14	-	-	-	541	1259
Non-specified	2	-	-	-	98	-	-	-	463	177
TRANSPORT SECTOR	**550**	**-**	**-**	**-**	**-**	**-**	**-**	**-**	**552**	**-**
Air	-	-	-	-	-	-	-	-	-	-
Road	-	-	-	-	-	-	-	-	-	-
Rail	-	-	-	-	-	-	-	-	499	-
Pipeline Transport	550	-	-	-	-	-	-	-	53	-
Internal Navigation	-	-	-	-	-	-	-	-	-	-
Non-specified	-	-	-	-	-	-	-	-	-	-
OTHER SECTORS	**692**	**-**	**-**	**-**	**16053**	**-**	**-**	**-**	**14901**	**31103**
Agriculture	-	-	-	-	42	-	-	-	221	-
Comm. and Publ. Services	656	-	-	-	-	-	-	-	1475	75
Residential	-	-	-	-	14955	-	-	-	10540	24752
Non-specified	36	-	-	-	1056	-	-	-	2665	6276
NON-ENERGY USE	**-**	**-**	**-**	**-**	**-**	**-**	**-**	**-**	**-**	**-**
in Industry/Transf./Energy	-	-	-	-	-	-	-	-	-	-
in Transport	-	-	-	-	-	-	-	-	-	-
in Other Sectors	-	-	-	-	-	-	-	-	-	-

Cameroon / Cameroun

SUPPLY AND CONSUMPTION 1997	Coal (1000 tonnes)							Oil (1000 tonnes)			
	Coking Coal	Other Bit. Coal	Sub-Bit. Coal	Lignite	Peat	Oven and Gas Coke	Pat. Fuel and BKB	Crude Oil	NGL	Feed-stocks	Additives
Production	-	-	-	-	-	-	-	7115	-	-	-
Imports	-	-	-	-	-	-	-	-	-	-	-
Exports	-	-	-	-	-	-	-	-5456	-	-	-
Intl. Marine Bunkers	-	-	-	-	-	-	-	-	-	-	-
Stock Changes	-	-	-	-	-	-	-	-	-	-	-
DOMESTIC SUPPLY	-	-	-	-	-	-	-	**1659**	-	-	-
Transfers and Stat. Diff.	-	-	-	-	-	-	-	-	-	-	-
TRANSFORMATION	-	-	-	-	-	-	-	**1659**	-	-	-
Electricity and CHP Plants	-	-	-	-	-	-	-	-	-	-	-
Petroleum Refineries	-	-	-	-	-	-	-	1659	-	-	-
Other Transform. Sector	-	-	-	-	-	-	-	-	-	-	-
ENERGY SECTOR	-	-	-	-	-	-	-	-	-	-	-
DISTRIBUTION LOSSES	-	-	-	-	-	-	-	-	-	-	-
FINAL CONSUMPTION	-	-	-	-	-	-	-	-	-	-	-
INDUSTRY SECTOR	-	-	-	-	-	-	-	-	-	-	-
Iron and Steel	-	-	-	-	-	-	-	-	-	-	-
Chemical and Petrochem.	-	-	-	-	-	-	-	-	-	-	-
Non-Metallic Minerals	-	-	-	-	-	-	-	-	-	-	-
Non-specified	-	-	-	-	-	-	-	-	-	-	-
TRANSPORT SECTOR	-	-	-	-	-	-	-	-	-	-	-
Air	-	-	-	-	-	-	-	-	-	-	-
Road	-	-	-	-	-	-	-	-	-	-	-
Non-specified	-	-	-	-	-	-	-	-	-	-	-
OTHER SECTORS	-	-	-	-	-	-	-	-	-	-	-
Agriculture	-	-	-	-	-	-	-	-	-	-	-
Comm. and Publ. Services	-	-	-	-	-	-	-	-	-	-	-
Residential	-	-	-	-	-	-	-	-	-	-	-
Non-specified	-	-	-	-	-	-	-	-	-	-	-
NON-ENERGY USE	-	-	-	-	-	-	-	-	-	-	-

APPROVISIONNEMENT ET DEMANDE 1998	Charbon (1000 tonnes)							Pétrole (1000 tonnes)			
	Charbon à coke	Autres charb. bit.	Charbon sous-bit.	Lignite	Tourbe	Coke de four/gaz	Agg./briq. de lignite	Pétrole brut	LGN	Produits d'aliment.	Additifs
Production	-	-	-	-	-	-	-	7822	-	-	-
Imports	-	-	-	-	-	-	-	-	-	-	-
Exports	-	-	-	-	-	-	-	-6114	-	-	-
Intl. Marine Bunkers	-	-	-	-	-	-	-	-	-	-	-
Stock Changes	-	-	-	-	-	-	-	-	-	-	-
DOMESTIC SUPPLY	-	-	-	-	-	-	-	**1708**	-	-	-
Transfers and Stat. Diff.	-	-	-	-	-	-	-	-	-	-	-
TRANSFORMATION	-	-	-	-	-	-	-	**1708**	-	-	-
Electricity and CHP Plants	-	-	-	-	-	-	-	-	-	-	-
Petroleum Refineries	-	-	-	-	-	-	-	1708	-	-	-
Other Transform. Sector	-	-	-	-	-	-	-	-	-	-	-
ENERGY SECTOR	-	-	-	-	-	-	-	-	-	-	-
DISTRIBUTION LOSSES	-	-	-	-	-	-	-	-	-	-	-
FINAL CONSUMPTION	-	-	-	-	-	-	-	-	-	-	-
INDUSTRY SECTOR	-	-	-	-	-	-	-	-	-	-	-
Iron and Steel	-	-	-	-	-	-	-	-	-	-	-
Chemical and Petrochem.	-	-	-	-	-	-	-	-	-	-	-
Non-Metallic Minerals	-	-	-	-	-	-	-	-	-	-	-
Non-specified	-	-	-	-	-	-	-	-	-	-	-
TRANSPORT SECTOR	-	-	-	-	-	-	-	-	-	-	-
Air	-	-	-	-	-	-	-	-	-	-	-
Road	-	-	-	-	-	-	-	-	-	-	-
Non-specified	-	-	-	-	-	-	-	-	-	-	-
OTHER SECTORS	-	-	-	-	-	-	-	-	-	-	-
Agriculture	-	-	-	-	-	-	-	-	-	-	-
Comm. and Publ. Services	-	-	-	-	-	-	-	-	-	-	-
Residential	-	-	-	-	-	-	-	-	-	-	-
Non-specified	-	-	-	-	-	-	-	-	-	-	-
NON-ENERGY USE	-	-	-	-	-	-	-	-	-	-	-

Cameroon / Cameroun

SUPPLY AND CONSUMPTION 1997	Refinery Gas	LPG + Ethane	Motor Gasoline	Aviation Gasoline	Jet Fuel	Kerosene	Gas/ Diesel	Heavy Fuel Oil	Naphtha	Petrol. Coke	Other Prod.
Production	-	28	329	-	51	280	398	468	-	-	31
Imports	-	-	2	-	-	9	9	-	-	-	-
Exports	-	-	-9	-	-	-8	-8	-418	-	-	-
Intl. Marine Bunkers	-	-	-	-	-	-	-53	-	-	-	-
Stock Changes	-	-4	-68	-	-	-	-15	-	-	-	-
DOMESTIC SUPPLY	-	24	254	-	51	281	331	50	-	-	31
Transfers and Stat. Diff.	-	-	-	-	-	-	-	-	-	-	-
TRANSFORMATION	-	-	-	-	-	-	9	-	-	-	-
Electricity and CHP Plants	-	-	-	-	-	-	9	-	-	-	-
Petroleum Refineries	-	-	-	-	-	-	-	-	-	-	-
Other Transform. Sector	-	-	-	-	-	-	-	-	-	-	-
ENERGY SECTOR	-	-	-	-	-	-	-	-	-	-	-
DISTRIBUTION LOSSES	-	-	-	-	-	-	-	-	-	-	-
FINAL CONSUMPTION	-	24	254	-	51	281	322	50	-	-	31
INDUSTRY SECTOR	-	-	-	-	-	-	-	50	-	-	-
Iron and Steel	-	-	-	-	-	-	-	-	-	-	-
Chemical and Petrochem.	-	-	-	-	-	-	-	-	-	-	-
Non-Metallic Minerals	-	-	-	-	-	-	-	-	-	-	-
Non-specified	-	-	-	-	-	-	-	50	-	-	-
TRANSPORT SECTOR	-	-	254	-	51	-	322	-	-	-	-
Air	-	-	-	-	51	-	-	-	-	-	-
Road	-	-	254	-	-	-	322	-	-	-	-
Non-specified	-	-	-	-	-	-	-	-	-	-	-
OTHER SECTORS	-	24	-	-	-	281	-	-	-	-	-
Agriculture	-	-	-	-	-	-	-	-	-	-	-
Comm. and Publ. Services	-	-	-	-	-	-	-	-	-	-	-
Residential	-	24	-	-	-	281	-	-	-	-	-
Non-specified	-	-	-	-	-	-	-	-	-	-	-
NON-ENERGY USE	-	-	-	-	-	-	-	-	-	-	31

APPROVISIONNEMENT ET DEMANDE 1998	Gaz de raffinerie	GPL + éthane	Essence moteur	Essence aviation	Carbu- réacteurs	Kérosène	Gazole	Fioul lourd	Naphta	Coke de pétrole	Autres prod.
Production	-	26	328	-	55	262	442	491	-	-	31
Imports	-	-	28	-	-	48	7	-	-	-	-
Exports	-	-	-29	-	-	-24	-25	-444	-	-	-
Intl. Marine Bunkers	-	-	-	-	-	-	-53	-	-	-	-
Stock Changes	-	-6	-40	-	-	-	-55	-	-	-	-
DOMESTIC SUPPLY	-	20	287	-	55	286	316	47	-	-	31
Transfers and Stat. Diff.	-	-	-	-	-	-	-	-	-	-	-
TRANSFORMATION	-	-	-	-	-	-	9	-	-	-	-
Electricity and CHP Plants	-	-	-	-	-	-	9	-	-	-	-
Petroleum Refineries	-	-	-	-	-	-	-	-	-	-	-
Other Transform. Sector	-	-	-	-	-	-	-	-	-	-	-
ENERGY SECTOR	-	-	-	-	-	-	-	-	-	-	-
DISTRIBUTION LOSSES	-	-	-	-	-	-	-	-	-	-	-
FINAL CONSUMPTION	-	20	287	-	55	286	307	47	-	-	31
INDUSTRY SECTOR	-	-	-	-	-	-	-	47	-	-	-
Iron and Steel	-	-	-	-	-	-	-	-	-	-	-
Chemical and Petrochem.	-	-	-	-	-	-	-	-	-	-	-
Non-Metallic Minerals	-	-	-	-	-	-	-	-	-	-	-
Non-specified	-	-	-	-	-	-	-	47	-	-	-
TRANSPORT SECTOR	-	-	287	-	55	-	307	-	-	-	-
Air	-	-	-	-	55	-	-	-	-	-	-
Road	-	-	287	-	-	-	307	-	-	-	-
Non-specified	-	-	-	-	-	-	-	-	-	-	-
OTHER SECTORS	-	20	-	-	-	286	-	-	-	-	-
Agriculture	-	-	-	-	-	-	-	-	-	-	-
Comm. and Publ. Services	-	-	-	-	-	-	-	-	-	-	-
Residential	-	20	-	-	-	286	-	-	-	-	-
Non-specified	-	-	-	-	-	-	-	-	-	-	-
NON-ENERGY USE	-	-	-	-	-	-	-	-	-	-	31

Cameroon / Cameroun

SUPPLY AND CONSUMPTION 1997	Gas (TJ)				Comb. Renew. & Waste (TJ)				(GWh)	(TJ)
	Natural Gas	Gas Works	Coke Ovens	Blast Furnaces	Solid Biomass	Gas/Liquids from Biomass	Municipal Waste	Industrial Waste	Electricity	Heat
Production	-	-	-	-	193768	-	-	-	3128	-
Imports	-	-	-	-	-	-	-	-	-	-
Exports	-	-	-	-	-	-	-	-	-	-
Intl. Marine Bunkers	-	-	-	-	-	-	-	-	-	-
Stock Changes	-	-	-	-	-	-	-	-	-	-
DOMESTIC SUPPLY	-	-	-	-	193768	-	-	-	3128	-
Transfers and Stat. Diff.	-	-	-	-	-1	-	-	-	-1	-
TRANSFORMATION	-	-	-	-	8788	-	-	-	-	-
Electricity and CHP Plants	-	-	-	-	-	-	-	-	-	-
Petroleum Refineries	-	-	-	-	-	-	-	-	-	-
Other Transform. Sector	-	-	-	-	8788	-	-	-	-	-
ENERGY SECTOR	-	-	-	-	-	-	-	-	-	-
DISTRIBUTION LOSSES	-	-	-	-	-	-	-	-	611	-
FINAL CONSUMPTION	-	-	-	-	184979	-	-	-	2516	-
INDUSTRY SECTOR	-	-	-	-	31765	-	-	-	1428	-
Iron and Steel	-	-	-	-	-	-	-	-	-	-
Chemical and Petrochem.	-	-	-	-	-	-	-	-	-	-
Non-Metallic Minerals	-	-	-	-	-	-	-	-	-	-
Non-specified	-	-	-	-	31765	-	-	-	1428	-
TRANSPORT SECTOR	-	-	-	-	-	-	-	-	-	-
Air	-	-	-	-	-	-	-	-	-	-
Road	-	-	-	-	-	-	-	-	-	-
Non-specified	-	-	-	-	-	-	-	-	-	-
OTHER SECTORS	-	-	-	-	153214	-	-	-	1088	-
Agriculture	-	-	-	-	-	-	-	-	-	-
Comm. and Publ. Services	-	-	-	-	-	-	-	-	298	-
Residential	-	-	-	-	153214	-	-	-	355	-
Non-specified	-	-	-	-	-	-	-	-	435	-
NON-ENERGY USE	-	-	-	-	-	-	-	-	-	-

APPROVISIONNEMENT ET DEMANDE 1998	Gaz (TJ)				En. Re. Comb. & Déchets (TJ)				(GWh)	(TJ)
	Gaz naturel	Usines à gaz	Cokeries	Hauts fourneaux	Biomasse solide	Gaz/Liquides tirés de biomasse	Déchets urbains	Déchets industriels	Electricité	Chaleur
Production	-	-	-	-	199071	-	-	-	3285	-
Imports	-	-	-	-	-	-	-	-	-	-
Exports	-	-	-	-	-	-	-	-	-	-
Intl. Marine Bunkers	-	-	-	-	-	-	-	-	-	-
Stock Changes	-	-	-	-	-	-	-	-	-	-
DOMESTIC SUPPLY	-	-	-	-	199071	-	-	-	3285	-
Transfers and Stat. Diff.	-	-	-	-	-1	-	-	-	-	-
TRANSFORMATION	-	-	-	-	9029	-	-	-	-	-
Electricity and CHP Plants	-	-	-	-	-	-	-	-	-	-
Petroleum Refineries	-	-	-	-	-	-	-	-	-	-
Other Transform. Sector	-	-	-	-	9029	-	-	-	-	-
ENERGY SECTOR	-	-	-	-	-	-	-	-	-	-
DISTRIBUTION LOSSES	-	-	-	-	-	-	-	-	642	-
FINAL CONSUMPTION	-	-	-	-	190041	-	-	-	2643	-
INDUSTRY SECTOR	-	-	-	-	32634	-	-	-	1501	-
Iron and Steel	-	-	-	-	-	-	-	-	-	-
Chemical and Petrochem.	-	-	-	-	-	-	-	-	-	-
Non-Metallic Minerals	-	-	-	-	-	-	-	-	-	-
Non-specified	-	-	-	-	32634	-	-	-	1501	-
TRANSPORT SECTOR	-	-	-	-	-	-	-	-	-	-
Air	-	-	-	-	-	-	-	-	-	-
Road	-	-	-	-	-	-	-	-	-	-
Non-specified	-	-	-	-	-	-	-	-	-	-
OTHER SECTORS	-	-	-	-	157407	-	-	-	1142	-
Agriculture	-	-	-	-	-	-	-	-	-	-
Comm. and Publ. Services	-	-	-	-	-	-	-	-	313	-
Residential	-	-	-	-	157407	-	-	-	372	-
Non-specified	-	-	-	-	-	-	-	-	457	-
NON-ENERGY USE	-	-	-	-	-	-	-	-	-	-

Chile / Chili : 1997

SUPPLY AND CONSUMPTION APPROVISIONNEMENT ET DEMANDE	Coal / Charbon (1000 tonnes)							Oil / Pétrole (1000 tonnes)			
	Coking Coal Charbon à coke	Other Bit. Coal Autres charb. bit.	Sub-Bit. Coal Charbon sous-bit.	Lignite Lignite	Peat Tourbe	Oven and Gas Coke Coke de four/gaz	Pat. Fuel and BKB Agg./briq. de lignite	Crude Oil Pétrole brut	NGL LGN	Feed-stocks Produits d'aliment.	Additives Additifs
Production	-	1044	-	-	-	490	-	270	156	-	-
From Other Sources	-	-	-	-	-	-	-	-	-	-	-
Imports	685	3940	-	-	-	64	-	8224	-	-	-
Exports	-	-	-	-	-	-18	-	-	-	-	-
Intl. Marine Bunkers	-	-	-	-	-	-	-	-	-	-	-
Stock Changes	-	448	-	-	-	81	-	-77	-	-	-
DOMESTIC SUPPLY	685	5432	-	-	-	617	-	8417	156	-	-
Transfers	-	-	-	-	-	-	-	-	-121	-	-
Statistical Differences	-	-	-	-	-	-	-	-	-1	-	-
TRANSFORMATION	685	3949	-	-	-	440	-	7773	34	-	-
Electricity Plants	-	3944	-	-	-	48	-	-	-	-	-
CHP Plants	-	-	-	-	-	-	-	-	-	-	-
Heat Plants	-	-	-	-	-	-	-	-	-	-	-
Blast Furnaces/Gas Works	-	5	-	-	-	392	-	-	-	-	-
Coke/Pat. Fuel/BKB Plants	685	-	-	-	-	-	-	-	-	-	-
Petroleum Refineries	-	-	-	-	-	-	-	8417	34	-	-
Petrochemical Industry	-	-	-	-	-	-	-	-	-	-	-
Liquefaction	-	-	-	-	-	-	-	-644	-	-	-
Other Transform. Sector	-	-	-	-	-	-	-	-	-	-	-
ENERGY SECTOR	-	-	-	-	-	-	-	-	-	-	-
Coal Mines	-	-	-	-	-	-	-	-	-	-	-
Oil and Gas Extraction	-	-	-	-	-	-	-	-	-	-	-
Petroleum Refineries	-	-	-	-	-	-	-	-	-	-	-
Electr., CHP+Heat Plants	-	-	-	-	-	-	-	-	-	-	-
Pumped Storage (Elec.)	-	-	-	-	-	-	-	-	-	-	-
Other Energy Sector	-	-	-	-	-	-	-	-	-	-	-
Distribution Losses	-	-	-	-	-	-	-	-	-	-	-
FINAL CONSUMPTION	-	1483	-	-	-	177	-	644	-	-	-
INDUSTRY SECTOR	-	1381	-	-	-	177	-	644	-	-	-
Iron and Steel	-	77	-	-	-	98	-	-	-	-	-
Chemical and Petrochem.	-	-	-	-	-	-	-	644	-	-	-
of which: Feedstocks	-	-	-	-	-	-	-	-	-	-	-
Non-Ferrous Metals	-	-	-	-	-	-	-	-	-	-	-
Non-Metallic Minerals	-	273	-	-	-	62	-	-	-	-	-
Transport Equipment	-	-	-	-	-	-	-	-	-	-	-
Machinery	-	-	-	-	-	-	-	-	-	-	-
Mining and Quarrying	-	-	-	-	-	-	-	-	-	-	-
Food and Tobacco	-	138	-	-	-	8	-	-	-	-	-
Paper, Pulp and Print	-	1	-	-	-	-	-	-	-	-	-
Wood and Wood Products	-	-	-	-	-	-	-	-	-	-	-
Construction	-	-	-	-	-	-	-	-	-	-	-
Textile and Leather	-	-	-	-	-	-	-	-	-	-	-
Non-specified	-	892	-	-	-	9	-	-	-	-	-
TRANSPORT SECTOR	-	-	-	-	-	-	-	-	-	-	-
Air	-	-	-	-	-	-	-	-	-	-	-
Road	-	-	-	-	-	-	-	-	-	-	-
Rail	-	-	-	-	-	-	-	-	-	-	-
Pipeline Transport	-	-	-	-	-	-	-	-	-	-	-
Internal Navigation	-	-	-	-	-	-	-	-	-	-	-
Non-specified	-	-	-	-	-	-	-	-	-	-	-
OTHER SECTORS	-	102	-	-	-	-	-	-	-	-	-
Agriculture	-	83	-	-	-	-	-	-	-	-	-
Comm. and Publ. Services	-	-	-	-	-	-	-	-	-	-	-
Residential	-	19	-	-	-	-	-	-	-	-	-
Non-specified	-	-	-	-	-	-	-	-	-	-	-
NON-ENERGY USE	-	-	-	-	-	-	-	-	-	-	-
in Industry/Trans./Energy	-	-	-	-	-	-	-	-	-	-	-
in Transport	-	-	-	-	-	-	-	-	-	-	-
in Other Sectors	-	-	-	-	-	-	-	-	-	-	-

Chile / Chili : 1997

SUPPLY AND CONSUMPTION / APPROVISIONNEMENT ET DEMANDE	Oil cont. / Pétrole cont. (1000 tonnes)										
	Refinery Gas / Gaz de raffinerie	LPG + Ethane / GPL + éthane	Motor Gasoline / Essence moteur	Aviation Gasoline / Essence aviation	Jet Fuel / Carbu-réacteurs	Kerosene / Kérosène	Gas/ Diesel / Gazole	Heavy Fuel Oil / Fioul lourd	Naphtha / Naphta	Petrol. Coke / Coke de pétrole	Other Prod. / Autres prod.
Production	438	348	1964	8	543	271	2937	1663	197	-	-
From Other Sources	-	-	-	-	-	-	-	-	-	-	-
Imports	-	510	296	-	97	-	945	404	-	-	-
Exports	-	-25	-55	-3	-5	-	-41	-	-	-	-
Intl. Marine Bunkers	-	-	-	-	-	-	-	-308	-	-	-
Stock Changes	-	-17	32	1	-85	47	139	20	-129	-	-
DOMESTIC SUPPLY	438	816	2237	6	550	318	3980	1779	68	-	-
Transfers	-	121	-	-	-	-	-	-	-	-	-
Statistical Differences	-	-	-	1	-	1	-	50	-	-	-
TRANSFORMATION	-	2	-	-	-	-	193	414	55	-	-
Electricity Plants	-	-	-	-	-	-	193	414	-	-	-
CHP Plants	-	-	-	-	-	-	-	-	-	-	-
Heat Plants	-	-	-	-	-	-	-	-	-	-	-
Blast Furnaces/Gas Works	-	2	-	-	-	-	-	-	55	-	-
Coke/Pat. Fuel/BKB Plants	-	-	-	-	-	-	-	-	-	-	-
Petroleum Refineries	-	-	-	-	-	-	-	-	-	-	-
Petrochemical Industry	-	-	-	-	-	-	-	-	-	-	-
Liquefaction	-	-	-	-	-	-	-	-	-	-	-
Other Transform. Sector	-	-	-	-	-	-	-	-	-	-	-
ENERGY SECTOR	438	2	-	-	-	-	12	77	-	-	-
Coal Mines	-	-	-	-	-	-	-	-	-	-	-
Oil and Gas Extraction	-	2	-	-	-	-	-	-	-	-	-
Petroleum Refineries	438	-	-	-	-	-	12	77	-	-	-
Electr., CHP+Heat Plants	-	-	-	-	-	-	-	-	-	-	-
Pumped Storage (Elec.)	-	-	-	-	-	-	-	-	-	-	-
Other Energy Sector	-	-	-	-	-	-	-	-	-	-	-
Distribution Losses	-	-	-	-	-	-	-	-	-	-	-
FINAL CONSUMPTION	-	933	2237	7	550	319	3775	1338	13	-	-
INDUSTRY SECTOR	-	124	-	-	-	137	1222	1196	13	-	-
Iron and Steel	-	-	-	-	-	-	29	40	-	-	-
Chemical and Petrochem.	-	-	-	-	-	-	-	4	-	-	-
of which: Feedstocks	-	-	-	-	-	-	-	-	-	-	-
Non-Ferrous Metals	-	-	-	-	-	-	292	-	-	-	-
Non-Metallic Minerals	-	-	-	-	-	-	4	55	-	-	-
Transport Equipment	-	-	-	-	-	-	-	-	-	-	-
Machinery	-	-	-	-	-	-	-	-	-	-	-
Mining and Quarrying	-	3	-	-	-	11	22	313	-	-	-
Food and Tobacco	-	-	-	-	-	-	-	12	-	-	-
Paper, Pulp and Print	-	1	-	-	-	-	3	148	-	-	-
Wood and Wood Products	-	-	-	-	-	-	-	-	-	-	-
Construction	-	-	-	-	-	-	-	-	-	-	-
Textile and Leather	-	-	-	-	-	-	-	-	-	-	-
Non-specified	-	120	-	-	-	126	872	624	13	-	-
TRANSPORT SECTOR	-	-	2237	7	550	-	2435	-	-	-	-
Air	-	-	-	7	550	-	-	-	-	-	-
Road	-	-	2237	-	-	-	1999	-	-	-	-
Rail	-	-	-	-	-	-	13	-	-	-	-
Pipeline Transport	-	-	-	-	-	-	-	-	-	-	-
Internal Navigation	-	-	-	-	-	-	423	-	-	-	-
Non-specified	-	-	-	-	-	-	-	-	-	-	-
OTHER SECTORS	-	809	-	-	-	182	118	142	-	-	-
Agriculture	-	-	-	-	-	-	23	121	-	-	-
Comm. and Publ. Services	-	106	-	-	-	-	-	-	-	-	-
Residential	-	703	-	-	-	182	-	-	-	-	-
Non-specified	-	-	-	-	-	-	95	21	-	-	-
NON-ENERGY USE	-	-	-	-	-	-	-	-	-	-	-
in Industry/Transf./Energy	-	-	-	-	-	-	-	-	-	-	-
in Transport	-	-	-	-	-	-	-	-	-	-	-
in Other Sectors	-	-	-	-	-	-	-	-	-	-	-

Chile / Chili : 1997

SUPPLY AND CONSUMPTION APPROVISIONNEMENT ET DEMANDE	Gas / Gaz (TJ)				Comb. Renew. & Waste / En. Re. Comb. & Déchets (TJ)				(GWh)	(TJ)
	Natural Gas Gaz naturel	Gas Works Usines à gaz	Coke Ovens Cokeries	Blast Furnaces Hauts fourneaux	Solid Biomass Biomasse solide	Gas/Liquids from Biomass Gaz/Liquides tirés de biomasse	Municipal Waste Déchets urbains	Industrial Waste Déchets industriels	Electricity Electricité	Heat Chaleur
Production	75155	2968	5217	5738	152775	1294	-	-	33292	-
From Other Sources	-	-	-	-	-	-	-	-	-	-
Imports	27524	-	-	-	-	-	-	-	-	-
Exports	-1825	-	-	-	-	-	-	-	-	-
Intl. Marine Bunkers	-	-	-	-	-	-	-	-	-	-
Stock Changes	-	-	-	-	-	-	-	-	-	-
DOMESTIC SUPPLY	**100854**	**2968**	**5217**	**5738**	**152775**	**1294**	**-**	**-**	**33292**	**-**
Transfers	-	-	-	-	-	-	-	-	-	-
Statistical Differences	-109	1	-	-	-	-	-	-	1	-
TRANSFORMATION	**61640**	**-**	**3560**	**-**	**11932**	**-**	**-**	**-**	**-**	**-**
Electricity Plants	7689	-	-	-	11932	-	-	-	-	-
CHP Plants	-	-	-	-	-	-	-	-	-	-
Heat Plants	-	-	-	-	-	-	-	-	-	-
Blast Furnaces/Gas Works	-	-	3560	-	-	-	-	-	-	-
Coke/Pat. Fuel/BKB Plants	-	-	-	-	-	-	-	-	-	-
Petroleum Refineries	-	-	-	-	-	-	-	-	-	-
Petrochemical Industry	-	-	-	-	-	-	-	-	-	-
Liquefaction	53951	-	-	-	-	-	-	-	-	-
Other Transform. Sector	-	-	-	-	-	-	-	-	-	-
ENERGY SECTOR	**17664**	**-**	**-**	**96**	**1294**	**1294**	**-**	**-**	**1362**	**-**
Coal Mines	-	-	-	-	-	-	-	-	43	-
Oil and Gas Extraction	17664	-	-	-	-	-	-	-	-	-
Petroleum Refineries	-	-	-	-	-	-	-	-	263	-
Electr., CHP+Heat Plants	-	-	-	-	-	-	-	-	946	-
Pumped Storage (Elec.)	-	-	-	-	-	-	-	-	-	-
Other Energy Sector	-	-	-	96	1294	1294	-	-	110	-
Distribution Losses	-	-	306	995	-	-	-	-	2520	-
FINAL CONSUMPTION	**21441**	**2969**	**1351**	**4647**	**139549**	**-**	**-**	**-**	**29411**	**-**
INDUSTRY SECTOR	**12121**	**-**	**1351**	**4647**	**26014**	**-**	**-**	**-**	**20276**	**-**
Iron and Steel	-	-	857	4647	-	-	-	-	317	-
Chemical and Petrochem.	-	-	-	-	-	-	-	-	571	-
of which: Feedstocks	-	-	-	-	-	-	-	-	-	-
Non-Ferrous Metals	-	-	-	-	-	-	-	-	8826	-
Non-Metallic Minerals	-	-	-	-	-	-	-	-	474	-
Transport Equipment	-	-	-	-	-	-	-	-	-	-
Machinery	-	-	-	-	-	-	-	-	-	-
Mining and Quarrying	-	-	-	-	-	-	-	-	236	-
Food and Tobacco	-	-	-	-	-	-	-	-	100	-
Paper, Pulp and Print	-	-	-	-	12422	-	-	-	2289	-
Wood and Wood Products	-	-	-	-	-	-	-	-	-	-
Construction	-	-	-	-	-	-	-	-	-	-
Textile and Leather	-	-	-	-	-	-	-	-	-	-
Non-specified	12121	-	494	-	13592	-	-	-	7463	-
TRANSPORT SECTOR	**243**	**-**	**-**	**-**	**-**	**-**	**-**	**-**	**211**	**-**
Air	-	-	-	-	-	-	-	-	-	-
Road	243	-	-	-	-	-	-	-	96	-
Rail	-	-	-	-	-	-	-	-	115	-
Pipeline Transport	-	-	-	-	-	-	-	-	-	-
Internal Navigation	-	-	-	-	-	-	-	-	-	-
Non-specified	-	-	-	-	-	-	-	-	-	-
OTHER SECTORS	**9077**	**2969**	**-**	**-**	**113535**	**-**	**-**	**-**	**8924**	**-**
Agriculture	-	-	-	-	-	-	-	-	173	-
Comm. and Publ. Services	1654	815	-	-	-	-	-	-	3527	-
Residential	7423	2154	-	-	113535	-	-	-	5224	-
Non-specified	-	-	-	-	-	-	-	-	-	-
NON-ENERGY USE	**-**	**-**	**-**	**-**	**-**	**-**	**-**	**-**	**-**	**-**
in Industry/Transf./Energy	-	-	-	-	-	-	-	-	-	-
in Transport	-	-	-	-	-	-	-	-	-	-
in Other Sectors	-	-	-	-	-	-	-	-	-	-

Chile / Chili : 1998

SUPPLY AND CONSUMPTION APPROVISIONNEMENT ET DEMANDE	Coal / Charbon (1000 tonnes)							Oil / Pétrole (1000 tonnes)			
	Coking Coal Charbon à coke	Other Bit. Coal Autres charb. bit.	Sub-Bit. Coal Charbon sous-bit.	Lignite Lignite	Peat Tourbe	Oven and Gas Coke Coke de four/gaz	Pat. Fuel and BKB Agg./briq. de lignite	Crude Oil Pétrole brut	NGL LGN	Feed-stocks Produits d'aliment.	Additives Additifs
Production	-	941	-	-	-	539	-	242	164	-	-
From Other Sources	-	-	-	-	-	-	-	-	-	-	-
Imports	712	3913	-	-	-	49	-	8924	-	-	-
Exports	-	-	-	-	-	-22	-	-	-	-	-
Intl. Marine Bunkers	-	-	-	-	-	-	-	-	-	-	-
Stock Changes	-	-192	-	-	-	118	-	-35	-	-	-
DOMESTIC SUPPLY	712	4662	-	-	-	684	-	9131	164	-	-
Transfers	-	-	-	-	-	-	-	-	-127	-	-
Statistical Differences	-	381	-	-	-	-	-	-	-	-	-
TRANSFORMATION	712	4162	-	-	-	472	-	8371	37	-	-
Electricity Plants	-	4162	-	-	-	66	-	-	-	-	-
CHP Plants	-	-	-	-	-	-	-	-	-	-	-
Heat Plants	-	-	-	-	-	-	-	-	-	-	-
Blast Furnaces/Gas Works	-	-	-	-	-	406	-	-	-	-	-
Coke/Pat. Fuel/BKB Plants	712	-	-	-	-	-	-	-	-	-	-
Petroleum Refineries	-	-	-	-	-	-	-	9131	37	-	-
Petrochemical Industry	-	-	-	-	-	-	-	-	-	-	-
Liquefaction	-	-	-	-	-	-	-	-760	-	-	-
Other Transform. Sector	-	-	-	-	-	-	-	-	-	-	-
ENERGY SECTOR	-	-	-	-	-	-	-	-	-	-	-
Coal Mines	-	-	-	-	-	-	-	-	-	-	-
Oil and Gas Extraction	-	-	-	-	-	-	-	-	-	-	-
Petroleum Refineries	-	-	-	-	-	-	-	-	-	-	-
Electr., CHP+Heat Plants	-	-	-	-	-	-	-	-	-	-	-
Pumped Storage (Elec.)	-	-	-	-	-	-	-	-	-	-	-
Other Energy Sector	-	-	-	-	-	-	-	-	-	-	-
Distribution Losses	-	-	-	-	-	-	-	-	-	-	-
FINAL CONSUMPTION	-	881	-	-	-	212	-	760	-	-	-
INDUSTRY SECTOR	-	834	-	-	-	212	-	760	-	-	-
Iron and Steel	-	86	-	-	-	101	-	-	-	-	-
Chemical and Petrochem.	-	-	-	-	-	-	-	760	-	-	-
of which: Feedstocks	-	-	-	-	-	-	-	-	-	-	-
Non-Ferrous Metals	-	-	-	-	-	-	-	-	-	-	-
Non-Metallic Minerals	-	177	-	-	-	60	-	-	-	-	-
Transport Equipment	-	-	-	-	-	-	-	-	-	-	-
Machinery	-	-	-	-	-	-	-	-	-	-	-
Mining and Quarrying	-	-	-	-	-	-	-	-	-	-	-
Food and Tobacco	-	162	-	-	-	9	-	-	-	-	-
Paper, Pulp and Print	-	1	-	-	-	-	-	-	-	-	-
Wood and Wood Products	-	-	-	-	-	-	-	-	-	-	-
Construction	-	-	-	-	-	-	-	-	-	-	-
Textile and Leather	-	-	-	-	-	-	-	-	-	-	-
Non-specified	-	408	-	-	-	42	-	-	-	-	-
TRANSPORT SECTOR	-	-	-	-	-	-	-	-	-	-	-
Air	-	-	-	-	-	-	-	-	-	-	-
Road	-	-	-	-	-	-	-	-	-	-	-
Rail	-	-	-	-	-	-	-	-	-	-	-
Pipeline Transport	-	-	-	-	-	-	-	-	-	-	-
Internal Navigation	-	-	-	-	-	-	-	-	-	-	-
Non-specified	-	-	-	-	-	-	-	-	-	-	-
OTHER SECTORS	-	47	-	-	-	-	-	-	-	-	-
Agriculture	-	40	-	-	-	-	-	-	-	-	-
Comm. and Publ. Services	-	-	-	-	-	-	-	-	-	-	-
Residential	-	7	-	-	-	-	-	-	-	-	-
Non-specified	-	-	-	-	-	-	-	-	-	-	-
NON-ENERGY USE	-	-	-	-	-	-	-	-	-	-	-
in Industry/Trans./Energy	-	-	-	-	-	-	-	-	-	-	-
in Transport	-	-	-	-	-	-	-	-	-	-	-
in Other Sectors	-	-	-	-	-	-	-	-	-	-	-

Chile / Chili : 1998

SUPPLY AND CONSUMPTION / *APPROVISIONNEMENT ET DEMANDE*	Oil cont. / *Pétrole cont.* (1000 tonnes)										
	Refinery Gas / *Gaz de raffinerie*	LPG + Ethane / *GPL + éthane*	Motor Gasoline / *Essence moteur*	Aviation Gasoline / *Essence aviation*	Jet Fuel / *Carbu-réacteurs*	Kerosene / *Kérosène*	Gas/ Diesel / *Gazole*	Heavy Fuel Oil / *Fioul lourd*	Naphtha / *Naphta*	Petrol. Coke / *Coke de pétrole*	Other Prod. / *Autres prod.*
Production	649	377	2040	12	502	288	3415	1576	242	-	-
From Other Sources	-	-	-	-	-	-	-	-	-	-	-
Imports	-	518	288	-	216	-	658	160	-	-	-
Exports	-	-28	-41	-2	-24	-	-39	-	-	-	-
Intl. Marine Bunkers	-	-	-	-	-	-	-	-439	-	-	-
Stock Changes	-	20	39	-2	-52	3	-116	188	-169	-	-
DOMESTIC SUPPLY	**649**	**887**	**2326**	**8**	**642**	**291**	**3918**	**1485**	**73**	**-**	**-**
Transfers	-	127	-	-	-	-	-	-	-	-	-
Statistical Differences	-	-1	1	-1	-	-	-2	1	-	-	-
TRANSFORMATION	**-**	**2**	**-**	**-**	**-**	**-**	**180**	**391**	**59**	**-**	**-**
Electricity Plants	-	-	-	-	-	-	180	391	-	-	-
CHP Plants	-	-	-	-	-	-	-	-	-	-	-
Heat Plants	-	-	-	-	-	-	-	-	-	-	-
Blast Furnaces/Gas Works	-	2	-	-	-	-	-	-	59	-	-
Coke/Pat. Fuel/BKB Plants	-	-	-	-	-	-	-	-	-	-	-
Petroleum Refineries	-	-	-	-	-	-	-	-	-	-	-
Petrochemical Industry	-	-	-	-	-	-	-	-	-	-	-
Liquefaction	-	-	-	-	-	-	-	-	-	-	-
Other Transform. Sector	-	-	-	-	-	-	-	-	-	-	-
ENERGY SECTOR	**649**	**2**	**-**	**-**	**-**	**-**	**9**	**140**	**-**	**-**	**-**
Coal Mines	-	-	-	-	-	-	-	-	-	-	-
Oil and Gas Extraction	-	2	-	-	-	-	-	-	-	-	-
Petroleum Refineries	649	-	-	-	-	-	9	140	-	-	-
Electr., CHP+Heat Plants	-	-	-	-	-	-	-	-	-	-	-
Pumped Storage (Elec.)	-	-	-	-	-	-	-	-	-	-	-
Other Energy Sector	-	-	-	-	-	-	-	-	-	-	-
Distribution Losses	-	-	-	-	-	-	-	-	-	-	-
FINAL CONSUMPTION	**-**	**1009**	**2327**	**7**	**642**	**291**	**3727**	**955**	**14**	**-**	**-**
INDUSTRY SECTOR	**-**	**202**	**-**	**-**	**-**	**137**	**1137**	**944**	**14**	**-**	**-**
Iron and Steel	-	-	-	-	-	-	27	37	-	-	-
Chemical and Petrochem.	-	-	-	-	-	-	-	2	-	-	-
of which: Feedstocks	-	-	-	-	-	-	-	-	-	-	-
Non-Ferrous Metals	-	-	-	-	-	-	372	-	-	-	-
Non-Metallic Minerals	-	-	-	-	-	-	9	70	-	-	-
Transport Equipment	-	-	-	-	-	-	-	-	-	-	-
Machinery	-	-	-	-	-	-	-	-	-	-	-
Mining and Quarrying	-	2	-	-	-	11	20	259	-	-	-
Food and Tobacco	-	-	-	-	-	-	-	10	-	-	-
Paper, Pulp and Print	-	2	-	-	-	-	3	141	-	-	-
Wood and Wood Products	-	-	-	-	-	-	-	-	-	-	-
Construction	-	-	-	-	-	-	-	-	-	-	-
Textile and Leather	-	-	-	-	-	-	-	-	-	-	-
Non-specified	-	198	-	-	-	126	706	425	14	-	-
TRANSPORT SECTOR	**-**	**-**	**2327**	**7**	**642**	**-**	**2540**	**-**	**-**	**-**	**-**
Air	-	-	-	7	642	-	-	-	-	-	-
Road	-	-	2327	-	-	-	2129	-	-	-	-
Rail	-	-	-	-	-	-	14	-	-	-	-
Pipeline Transport	-	-	-	-	-	-	-	-	-	-	-
Internal Navigation	-	-	-	-	-	-	397	-	-	-	-
Non-specified	-	-	-	-	-	-	-	-	-	-	-
OTHER SECTORS	**-**	**807**	**-**	**-**	**-**	**154**	**50**	**11**	**-**	**-**	**-**
Agriculture	-	-	-	-	-	-	10	9	-	-	-
Comm. and Publ. Services	-	97	-	-	-	-	-	-	-	-	-
Residential	-	710	-	-	-	154	-	-	-	-	-
Non-specified	-	-	-	-	-	-	40	2	-	-	-
NON-ENERGY USE	**-**	**-**	**-**	**-**	**-**	**-**	**-**	**-**	**-**	**-**	**-**
in Industry/Transf./Energy	-	-	-	-	-	-	-	-	-	-	-
in Transport	-	-	-	-	-	-	-	-	-	-	-
in Other Sectors	-	-	-	-	-	-	-	-	-	-	-

Chile / Chili : 1998

SUPPLY AND CONSUMPTION / APPROVISIONNEMENT ET DEMANDE	Gas / Gaz (TJ)				Comb. Renew. & Waste / En. Re. Comb. & Déchets (TJ)				(GWh)	(TJ)
	Natural Gas / Gaz naturel	Gas Works / Usines à gaz	Coke Ovens / Cokeries	Blast Furnaces / Hauts fourneaux	Solid Biomass / Biomasse solide	Gas/Liquids from Biomass / Gaz/Liquides tirés de biomasse	Municipal Waste / Déchets urbains	Industrial Waste / Déchets industriels	Electricity / Electricité	Heat / Chaleur
Production	67763	3155	4804	6000	164160	1030	-	-	35509	-
From Other Sources	-	-	-	-	-	-	-	-	-	-
Imports	78431	-	-	-	-	-	-	-	-	-
Exports	-582	-	-	-	-	-	-	-	-	-
Intl. Marine Bunkers	-	-	-	-	-	-	-	-	-	-
Stock Changes	-	-	-	-	-	-	-	-	-	-
DOMESTIC SUPPLY	**145612**	**3155**	**4804**	**6000**	**164160**	**1030**	**-**	**-**	**35509**	**-**
Transfers	-	-	-	-	-	-	-	-	-	-
Statistical Differences	-3605	-	-11	-	1	-	-	-	-	-
TRANSFORMATION	**101253**	**-**	**3629**	**-**	**16622**	**-**	**-**	**-**	**-**	**-**
Electricity Plants	37597	-	25	-	16622	-	-	-	-	-
CHP Plants	-	-	-	-	-	-	-	-	-	-
Heat Plants	-	-	-	-	-	-	-	-	-	-
Blast Furnaces/Gas Works	-	-	3604	-	-	-	-	-	-	-
Coke/Pat. Fuel/BKB Plants	-	-	-	-	-	-	-	-	-	-
Petroleum Refineries	-	-	-	-	-	-	-	-	-	-
Petrochemical Industry	-	-	-	-	-	-	-	-	-	-
Liquefaction	63656	-	-	-	-	-	-	-	-	-
Other Transform. Sector	-	-	-	-	-	-	-	-	-	-
ENERGY SECTOR	**16638**	**-**	**-**	**92**	**1030**	**1030**	**-**	**-**	**1742**	**-**
Coal Mines	-	-	-	-	-	-	-	-	31	-
Oil and Gas Extraction	16638	-	-	-	-	-	-	-	-	-
Petroleum Refineries	-	-	-	-	-	-	-	-	280	-
Electr., CHP+Heat Plants	-	-	-	-	-	-	-	-	1352	-
Pumped Storage (Elec.)	-	-	-	-	-	-	-	-	-	-
Other Energy Sector	-	-	-	92	1030	1030	-	-	79	-
Distribution Losses	-	-	209	1093	-	-	-	-	2903	-
FINAL CONSUMPTION	**24116**	**3155**	**955**	**4815**	**146509**	**-**	**-**	**-**	**30864**	**-**
INDUSTRY SECTOR	**15069**	**-**	**955**	**4815**	**32992**	**-**	**-**	**-**	**20943**	**-**
Iron and Steel	84	-	855	4815	-	-	-	-	309	-
Chemical and Petrochem.	1256	-	-	-	-	-	-	-	471	-
of which: Feedstocks	-	-	-	-	-	-	-	-	-	-
Non-Ferrous Metals	222	-	-	-	-	-	-	-	9825	-
Non-Metallic Minerals	-	-	-	-	-	-	-	-	467	-
Transport Equipment	-	-	-	-	-	-	-	-	-	-
Machinery	-	-	-	-	-	-	-	-	-	-
Mining and Quarrying	-	-	-	-	-	-	-	-	271	-
Food and Tobacco	-	-	-	-	-	-	-	-	115	-
Paper, Pulp and Print	1386	-	-	-	19343	-	-	-	2544	-
Wood and Wood Products	-	-	-	-	-	-	-	-	-	-
Construction	-	-	-	-	-	-	-	-	-	-
Textile and Leather	-	-	-	-	-	-	-	-	-	-
Non-specified	12121	-	100	-	13649	-	-	-	6941	-
TRANSPORT SECTOR	**255**	**-**	**-**	**-**	**-**	**-**	**-**	**-**	**208**	**-**
Air	-	-	-	-	-	-	-	-	-	-
Road	255	-	-	-	-	-	-	-	127	-
Rail	-	-	-	-	-	-	-	-	81	-
Pipeline Transport	-	-	-	-	-	-	-	-	-	-
Internal Navigation	-	-	-	-	-	-	-	-	-	-
Non-specified	-	-	-	-	-	-	-	-	-	-
OTHER SECTORS	**8792**	**3155**	**-**	**-**	**113517**	**-**	**-**	**-**	**9713**	**-**
Agriculture	-	-	-	-	-	-	-	-	111	-
Comm. and Publ. Services	1620	815	-	-	-	-	-	-	3870	-
Residential	7172	2340	-	-	113517	-	-	-	5732	-
Non-specified	-	-	-	-	-	-	-	-	-	-
NON-ENERGY USE	**-**	**-**	**-**	**-**	**-**	**-**	**-**	**-**	**-**	**-**
in Industry/Transf./Energy	-	-	-	-	-	-	-	-	-	-
in Transport	-	-	-	-	-	-	-	-	-	-
in Other Sectors	-	-	-	-	-	-	-	-	-	-

People's Republic of China / République populaire de Chine : 1997

SUPPLY AND CONSUMPTION / APPROVISIONNEMENT ET DEMANDE	Coal / Charbon (1000 tonnes)							Oil / Pétrole (1000 tonnes)			
	Coking Coal / Charbon à coke	Other Bit. Coal / Autres charb. bit.	Sub-Bit. Coal / Charbon sous-bit.	Lignite / Lignite	Peat / Tourbe	Oven and Gas Coke / Coke de four/gaz	Pat. Fuel and BKB / Agg./briq. de lignite	Crude Oil / Pétrole brut	NGL / LGN	Feed-stocks / Produits d'aliment.	Additives / Additifs
Production	192969	1179851	-	-	-	136531	10017	160741	-	-	-
From Other Sources	-	-	-	-	-	-	-	-	-	-	-
Imports	-	2013	-	-	-	1	-	35470	-	-	-
Exports	-	-35331	-	-	-	-10581	-	-19829	-	-	-
Intl. Marine Bunkers	-	-	-	-	-	-	-	-	-	-	-
Stock Changes	-	-13405	-	-	-	361	54	-1386	-	-	-
DOMESTIC SUPPLY	**192969**	**1133128**	**-**	**-**	**-**	**126312**	**10071**	**174996**	**-**	**-**	**-**
Transfers	-	-	-	-	-	-	-	-	-	-	-
Statistical Differences	-	-14327	-	-	-	-16110	-3	-1152	-	-	-
TRANSFORMATION	**192969**	**568448**	**-**	**-**	**-**	**63311**	**-**	**166865**	**-**	**-**	**-**
Electricity Plants	-	489792	-	-	-	-	-	643	-	-	-
CHP Plants	-	62447	-	-	-	-	-	149	-	-	-
Heat Plants	-	-	-	-	-	-	-	-	-	-	-
Blast Furnaces/Gas Works	-	7326	-	-	-	63311	-	-	-	-	-
Coke/Pat. Fuel/BKB Plants	192969	8883	-	-	-	-	-	-	-	-	-
Petroleum Refineries	-	-	-	-	-	-	-	166073	-	-	-
Petrochemical Industry	-	-	-	-	-	-	-	-	-	-	-
Liquefaction	-	-	-	-	-	-	-	-	-	-	-
Other Transform. Sector	-	-	-	-	-	-	-	-	-	-	-
ENERGY SECTOR	**-**	**69235**	**-**	**-**	**-**	**1549**	**-**	**3815**	**-**	**-**	**-**
Coal Mines	-	24592	-	-	-	525	-	13	-	-	-
Oil and Gas Extraction	-	3211	-	-	-	26	-	3165	-	-	-
Petroleum Refineries	-	-	-	-	-	-	-	636	-	-	-
Electr., CHP+Heat Plants	-	33344	-	-	-	73	-	1	-	-	-
Pumped Storage (Elec.)	-	-	-	-	-	-	-	-	-	-	-
Other Energy Sector	-	8088	-	-	-	925	-	-	-	-	-
Distribution Losses	-	-	-	-	-	-	-	1884	-	-	-
FINAL CONSUMPTION	**-**	**481118**	**-**	**-**	**-**	**45342**	**10068**	**1280**	**-**	**-**	**-**
INDUSTRY SECTOR	**-**	**328519**	**-**	**-**	**-**	**42304**	**-**	**926**	**-**	**-**	**-**
Iron and Steel	-	7668	-	-	-	17071	-	24	-	-	-
Chemical and Petrochem.	-	84548	-	-	-	13557	-	602	-	-	-
of which: Feedstocks	-	12774	-	-	-	8834	-	482	-	-	-
Non-Ferrous Metals	-	7671	-	-	-	2126	-	11	-	-	-
Non-Metallic Minerals	-	120116	-	-	-	2690	-	105	-	-	-
Transport Equipment	-	6201	-	-	-	367	-	1	-	-	-
Machinery	-	19530	-	-	-	4258	-	8	-	-	-
Mining and Quarrying	-	4851	-	-	-	1281	-	124	-	-	-
Food and Tobacco	-	28589	-	-	-	306	-	14	-	-	-
Paper, Pulp and Print	-	15356	-	-	-	79	-	1	-	-	-
Wood and Wood Products	-	4694	-	-	-	39	-	1	-	-	-
Construction	-	3721	-	-	-	125	-	30	-	-	-
Textile and Leather	-	20195	-	-	-	115	-	5	-	-	-
Non-specified	-	5379	-	-	-	290	-	-	-	-	-
TRANSPORT SECTOR	**-**	**14220**	**-**	**-**	**-**	**-**	**-**	**-**	**-**	**-**	**-**
Air	-	-	-	-	-	-	-	-	-	-	-
Road	-	-	-	-	-	-	-	-	-	-	-
Rail	-	14220	-	-	-	-	-	-	-	-	-
Pipeline Transport	-	-	-	-	-	-	-	-	-	-	-
Internal Navigation	-	-	-	-	-	-	-	-	-	-	-
Non-specified	-	-	-	-	-	-	-	-	-	-	-
OTHER SECTORS	**-**	**138379**	**-**	**-**	**-**	**3038**	**10027**	**354**	**-**	**-**	**-**
Agriculture	-	18794	-	-	-	1247	-	-	-	-	-
Comm. and Publ. Services	-	14412	-	-	-	544	79	170	-	-	-
Residential	-	105173	-	-	-	1247	9924	-	-	-	-
Non-specified	-	-	-	-	-	-	24	184	-	-	-
NON-ENERGY USE	**-**	**-**	**-**	**-**	**-**	**-**	**41**	**-**	**-**	**-**	**-**
in Industry/Trans./Energy	-	-	-	-	-	-	41	-	-	-	-
in Transport	-	-	-	-	-	-	-	-	-	-	-
in Other Sectors	-	-	-	-	-	-	-	-	-	-	-

People's Republic of China / République populaire de Chine : 1997

SUPPLY AND CONSUMPTION APPROVISIONNEMENT ET DEMANDE	Refinery Gas Gaz de raffinerie	LPG + Ethane GPL + éthane	Motor Gasoline Essence moteur	Aviation Gasoline Essence aviation	Jet Fuel Carbu-réacteurs	Kerosene Kérosène	Gas/Diesel Gazole	Heavy Fuel Oil Fioul lourd	Naphtha Naphta	Petrol. Coke Coke de pétrole	Other Prod. Autres prod.
Production	5341	6679	35102	77	4466	1304	49245	23112	17439	-	19488
From Other Sources	-	-	-	-	-	-	-	-	-	-	-
Imports	-	3582	84	-	-	1381	7428	13711	-	-	3582
Exports	-	-392	-1782	-	-	-723	-2321	-517	-	-	-1557
Intl. Marine Bunkers	-	-	-	-	-	-	-343	-335	-	-	-
Stock Changes	-	51	-564	-	-	-146	-1820	-57	-	-	-
DOMESTIC SUPPLY	**5341**	**9920**	**32840**	**77**	**4466**	**1816**	**52189**	**35914**	**17439**	**-**	**21513**
Transfers	-	-	-	-	-	-	-	-	-	-	-
Statistical Differences	728	161	279	-	-	802	2052	3943	-	-	-
TRANSFORMATION	**1039**	**12**	**8**	**-**	**-**	**-**	**6965**	**11828**	**-**	**-**	**-**
Electricity Plants	280	7	8	-	-	-	6638	8709	-	-	-
CHP Plants	759	5	-	-	-	-	327	2502	-	-	-
Heat Plants	-	-	-	-	-	-	-	-	-	-	-
Blast Furnaces/Gas Works	-	-	-	-	-	-	-	617	-	-	-
Coke/Pat. Fuel/BKB Plants	-	-	-	-	-	-	-	-	-	-	-
Petroleum Refineries	-	-	-	-	-	-	-	-	-	-	-
Petrochemical Industry	-	-	-	-	-	-	-	-	-	-	-
Liquefaction	-	-	-	-	-	-	-	-	-	-	-
Other Transform. Sector	-	-	-	-	-	-	-	-	-	-	-
ENERGY SECTOR	**3963**	**1395**	**-**	**-**	**-**	**155**	**3029**	**6790**	**-**	**-**	**-**
Coal Mines	-	-	-	-	-	34	285	20	-	-	-
Oil and Gas Extraction	94	73	-	-	-	7	1758	1378	-	-	-
Petroleum Refineries	3866	1280	-	-	-	106	301	3838	-	-	-
Electr., CHP+Heat Plants	3	5	-	-	-	7	662	1419	-	-	-
Pumped Storage (Elec.)	-	-	-	-	-	-	-	-	-	-	-
Other Energy Sector	-	37	-	-	-	1	23	135	-	-	-
Distribution Losses	-	18	-	-	-	-	-	-	-	-	-
FINAL CONSUMPTION	**1067**	**8656**	**33111**	**77**	**4466**	**2463**	**44247**	**21239**	**17439**	**-**	**21513**
INDUSTRY SECTOR	**1067**	**679**	**-**	**-**	**-**	**355**	**5867**	**13596**	**17439**	**-**	**-**
Iron and Steel	1	7	-	-	-	8	399	3295	-	-	-
Chemical and Petrochem.	949	487	-	-	-	74	1205	5134	17439	-	-
of which: Feedstocks	288	75	-	-	-	11	20	908	17439	-	-
Non-Ferrous Metals	-	4	-	-	-	8	164	592	-	-	-
Non-Metallic Minerals	8	90	-	-	-	20	1072	2844	-	-	-
Transport Equipment	4	7	-	-	-	35	207	135	-	-	-
Machinery	6	49	-	-	-	70	702	598	-	-	-
Mining and Quarrying	-	-	-	-	-	6	276	25	-	-	-
Food and Tobacco	-	18	-	-	-	11	307	267	-	-	-
Paper, Pulp and Print	94	1	-	-	-	47	128	90	-	-	-
Wood and Wood Products	-	-	-	-	-	7	110	22	-	-	-
Construction	-	4	-	-	-	42	870	132	-	-	-
Textile and Leather	-	10	-	-	-	24	280	367	-	-	-
Non-specified	5	2	-	-	-	3	147	95	-	-	-
TRANSPORT SECTOR	**-**	**16**	**33111**	**77**	**4466**	**-**	**18530**	**5752**	**-**	**-**	**-**
Air	-	-	-	77	4466	-	-	-	-	-	-
Road	-	16	33111	-	-	-	10624	197	-	-	-
Rail	-	-	-	-	-	-	3776	26	-	-	-
Pipeline Transport	-	-	-	-	-	-	10	151	-	-	-
Internal Navigation	-	-	-	-	-	-	4067	5378	-	-	-
Non-specified	-	-	-	-	-	-	53	-	-	-	-
OTHER SECTORS	**-**	**7961**	**-**	**-**	**-**	**2108**	**19850**	**1891**	**-**	**-**	**-**
Agriculture	-	-	-	-	-	14	10757	29	-	-	-
Comm. and Publ. Services	-	471	-	-	-	1467	8815	1862	-	-	-
Residential	-	7490	-	-	-	627	278	-	-	-	-
Non-specified	-	-	-	-	-	-	-	-	-	-	-
NON-ENERGY USE	**-**	**-**	**-**	**-**	**-**	**-**	**-**	**-**	**-**	**-**	**21513**
in Industry/Transf./Energy	-	-	-	-	-	-	-	-	-	-	21513
in Transport	-	-	-	-	-	-	-	-	-	-	-
in Other Sectors	-	-	-	-	-	-	-	-	-	-	-

People's Republic of China / République populaire de Chine : 1997

SUPPLY AND CONSUMPTION / APPROVISIONNEMENT ET DEMANDE	Gas / Gaz (TJ)				Comb. Renew. & Waste / En. Re. Comb. & Déchets (TJ)				(GWh)	(TJ)
	Natural Gas / Gaz naturel	Gas Works / Usines à gaz	Coke Ovens / Cokeries	Blast Furnaces / Hauts fourneaux	Solid Biomass / Biomasse solide	Gas/Liquids from Biomass / Gaz/Liquides tirés de biomasse	Municipal Waste / Déchets urbains	Industrial Waste / Déchets industriels	Electricity / Electricité	Heat / Chaleur
Production	983449	162207	424655	249475	8670677	50528	-	-	1134471	1175695
From Other Sources	-	-	-	-	-	-	-	-	-	-
Imports	-	-	-	-	-	-	-	-	89	-
Exports	-107462	-	-	-	-	-	-	-	-7204	-
Intl. Marine Bunkers	-	-	-	-	-	-	-	-	-	-
Stock Changes	-	-	-	-	-	-	-	-	-	-
DOMESTIC SUPPLY	**875987**	**162207**	**424655**	**249475**	**8670677**	**50528**	**-**	**-**	**1127356**	**1175695**
Transfers	-	-	-	-	-	-	-	-	-	-
Statistical Differences	5938	13744	4172	29851	-	-	-	-	-4304	1
TRANSFORMATION	**94910**	**-**	**34990**	**22910**	**-**	**-**	**-**	**-**	**-**	**-**
Electricity Plants	92181	-	22320	16125	-	-	-	-	-	-
CHP Plants	2729	-	12670	6785	-	-	-	-	-	-
Heat Plants	-	-	-	-	-	-	-	-	-	-
Blast Furnaces/Gas Works	-	-	-	-	-	-	-	-	-	-
Coke/Pat. Fuel/BKB Plants	-	-	-	-	-	-	-	-	-	-
Petroleum Refineries	-	-	-	-	-	-	-	-	-	-
Petrochemical Industry	-	-	-	-	-	-	-	-	-	-
Liquefaction	-	-	-	-	-	-	-	-	-	-
Other Transform. Sector	-	-	-	-	-	-	-	-	-	-
ENERGY SECTOR	**276860**	**23318**	**32277**	**-**	**-**	**-**	**-**	**-**	**185310**	**238198**
Coal Mines	-	-	252	-	-	-	-	-	38097	1522
Oil and Gas Extraction	227955	-	-	-	-	-	-	-	31468	10498
Petroleum Refineries	42971	-	-	-	-	-	-	-	18241	178857
Electr., CHP+Heat Plants	5501	11275	72	-	-	-	-	-	96182	43032
Pumped Storage (Elec.)	-	-	-	-	-	-	-	-	-	-
Other Energy Sector	433	12043	31953	-	-	-	-	-	1322	4289
Distribution Losses	24041	-	-	-	-	-	-	-	79838	16647
FINAL CONSUMPTION	**486114**	**152633**	**361560**	**256416**	**8670677**	**50528**	**-**	**-**	**857904**	**920851**
INDUSTRY SECTOR	**392937**	**56031**	**285435**	**256416**	**-**	**-**	**-**	**-**	**573092**	**711435**
Iron and Steel	11869	-	262862	256416	-	-	-	-	93416	119049
Chemical and Petrochem.	340393	6488	15078	-	-	-	-	-	150683	305810
of which: Feedstocks	-	-	-	-	-	-	-	-	-	-
Non-Ferrous Metals	2642	11697	288	-	-	-	-	-	51352	41932
Non-Metallic Minerals	13255	11381	3846	-	-	-	-	-	60808	15248
Transport Equipment	1949	4185	162	-	-	-	-	-	20123	18985
Machinery	15205	20579	2803	-	-	-	-	-	52566	32996
Mining and Quarrying	780	-	-	-	-	-	-	-	19055	4712
Food and Tobacco	2556	120	126	-	-	-	-	-	29884	48197
Paper, Pulp and Print	390	-	18	-	-	-	-	-	19546	41442
Wood and Wood Products	-	-	-	-	-	-	-	-	7025	6480
Construction	43	-	-	-	-	-	-	-	11741	2071
Textile and Leather	3855	241	252	-	-	-	-	-	39807	62876
Non-specified	-	1340	-	-	-	-	-	-	17086	11637
TRANSPORT SECTOR	**910**	**-**	**-**	**-**	**-**	**-**	**-**	**-**	**20200**	**-**
Air	-	-	-	-	-	-	-	-	-	-
Road	910	-	-	-	-	-	-	-	-	-
Rail	-	-	-	-	-	-	-	-	-	-
Pipeline Transport	-	-	-	-	-	-	-	-	-	-
Internal Navigation	-	-	-	-	-	-	-	-	-	-
Non-specified	-	-	-	-	-	-	-	-	20200	-
OTHER SECTORS	**92267**	**96602**	**76125**	**-**	**8670677**	**50528**	**-**	**-**	**264612**	**209416**
Agriculture	-	-	-	-	-	-	-	-	63977	533
Comm. and Publ. Services	6844	6609	10118	-	-	-	-	-	75320	38403
Residential	85423	89993	66007	-	8670677	50528	-	-	125315	164322
Non-specified	-	-	-	-	-	-	-	-	-	6158
NON-ENERGY USE	**-**	**-**	**-**	**-**	**-**	**-**	**-**	**-**	**-**	**-**
in Industry/Transf./Energy	-	-	-	-	-	-	-	-	-	-
in Transport	-	-	-	-	-	-	-	-	-	-
in Other Sectors	-	-	-	-	-	-	-	-	-	-

People's Republic of China / République populaire de Chine : 1998

SUPPLY AND CONSUMPTION APPROVISIONNEMENT ET DEMANDE	Coal / *Charbon* (1000 tonnes)							Oil / *Pétrole* (1000 tonnes)			
	Coking Coal *Charbon à coke*	Other Bit. Coal *Autres charb. bit.*	Sub-Bit. Coal *Charbon sous-bit.*	Lignite *Lignite*	Peat *Tourbe*	Oven and Gas Coke *Coke de four/gaz*	Pat. Fuel and BKB *Agg./briq. de lignite*	Crude Oil *Pétrole brut*	NGL *LGN*	Feed-stocks *Produits d'aliment.*	Additives *Additifs*
Production	156281	1057031	-	-	-	128991	9235	161000	-	-	-
From Other Sources	-	-	-	-	-	-	-	-	-	-	-
Imports	-	1586	-	-	-	-	-	27320	-	-	-
Exports	-	-32297	-	-	-	-11464	-	-15600	-	-	-
Intl. Marine Bunkers	-	-	-	-	-	-	-	-	-	-	-
Stock Changes	-	8813	-	-	-	-192	-3	449	-	-	-
DOMESTIC SUPPLY	**156281**	**1035133**	-	-	-	**117335**	**9232**	**173169**	-	-	-
Transfers	-	-	-	-	-	-	-	-	-	-	-
Statistical Differences	-	37212	-	-	-	-6226	-942	782	-	-	-
TRANSFORMATION	**156281**	**572811**	-	-	-	**64801**	-	**166807**	-	-	-
Electricity Plants	-	494893	-	-	-	-	-	744	-	-	-
CHP Plants	-	63199	-	-	-	-	-	245	-	-	-
Heat Plants	-	-	-	-	-	-	-	-	-	-	-
Blast Furnaces/Gas Works	-	6851	-	-	-	64801	-	-	-	-	-
Coke/Pat. Fuel/BKB Plants	156281	7868	-	-	-	-	-	-	-	-	-
Petroleum Refineries	-	-	-	-	-	-	-	165818	-	-	-
Petrochemical Industry	-	-	-	-	-	-	-	-	-	-	-
Liquefaction	-	-	-	-	-	-	-	-	-	-	-
Other Transform. Sector	-	-	-	-	-	-	-	-	-	-	-
ENERGY SECTOR	-	**72017**	-	-	-	**1300**	-	**4280**	-	-	-
Coal Mines	-	26928	-	-	-	450	-	13	-	-	-
Oil and Gas Extraction	-	1992	-	-	-	48	-	3400	-	-	-
Petroleum Refineries	-	-	-	-	-	-	-	855	-	-	-
Electr., CHP+Heat Plants	-	33287	-	-	-	89	-	12	-	-	-
Pumped Storage (Elec.)	-	-	-	-	-	-	-	-	-	-	-
Other Energy Sector	-	9810	-	-	-	713	-	-	-	-	-
Distribution Losses	-	-	-	-	-	-	-	1960	-	-	-
FINAL CONSUMPTION	-	**427517**	-	-	-	**45008**	**8290**	**904**	-	-	-
INDUSTRY SECTOR	-	**305000**	-	-	-	**41692**	-	**694**	-	-	-
Iron and Steel	-	7661	-	-	-	16806	-	68	-	-	-
Chemical and Petrochem.	-	78397	-	-	-	12537	-	430	-	-	-
of which: Feedstocks	-	*12485*	-	-	-	*6818*	-	*672*	-	-	-
Non-Ferrous Metals	-	8453	-	-	-	2084	-	9	-	-	-
Non-Metallic Minerals	-	109966	-	-	-	3023	-	119	-	-	-
Transport Equipment	-	5667	-	-	-	394	-	1	-	-	-
Machinery	-	16364	-	-	-	4573	-	15	-	-	-
Mining and Quarrying	-	4482	-	-	-	1273	-	-	-	-	-
Food and Tobacco	-	27481	-	-	-	319	-	9	-	-	-
Paper, Pulp and Print	-	14089	-	-	-	37	-	2	-	-	-
Wood and Wood Products	-	4687	-	-	-	31	-	-	-	-	-
Construction	-	6010	-	-	-	146	-	22	-	-	-
Textile and Leather	-	17815	-	-	-	182	-	19	-	-	-
Non-specified	-	3928	-	-	-	287	-	-	-	-	-
TRANSPORT SECTOR	-	**13754**	-	-	-	-	-	-	-	-	-
Air	-	-	-	-	-	-	-	-	-	-	-
Road	-	-	-	-	-	-	-	-	-	-	-
Rail	-	13754	-	-	-	-	-	-	-	-	-
Pipeline Transport	-	-	-	-	-	-	-	-	-	-	-
Internal Navigation	-	-	-	-	-	-	-	-	-	-	-
Non-specified	-	-	-	-	-	-	-	-	-	-	-
OTHER SECTORS	-	**108763**	-	-	-	**3316**	**8249**	**210**	-	-	-
Agriculture	-	18794	-	-	-	1400	-	-	-	-	-
Comm. and Publ. Services	-	15619	-	-	-	516	69	17	-	-	-
Residential	-	74350	-	-	-	1400	8156	-	-	-	-
Non-specified	-	-	-	-	-	-	24	193	-	-	-
NON-ENERGY USE	-	-	-	-	-	-	**41**	-	-	-	-
in Industry/Trans./Energy	-	-	-	-	-	-	41	-	-	-	-
in Transport	-	-	-	-	-	-	-	-	-	-	-
in Other Sectors	-	-	-	-	-	-	-	-	-	-	-

People's Republic of China / République populaire de Chine : 1998

SUPPLY AND CONSUMPTION APPROVISIONNEMENT ET DEMANDE	Oil cont. / *Pétrole cont.* (1000 tonnes)										
	Refinery Gas *Gaz de raffinerie*	LPG + Ethane *GPL + éthane*	Motor Gasoline *Essence moteur*	Aviation Gasoline *Essence aviation*	Jet Fuel *Carbu- réacteurs*	Kerosene *Kérosène*	Gas/ Diesel *Gazole*	Heavy Fuel Oil *Fioul lourd*	Naphtha *Naphta*	Petrol. Coke *Coke de pétrole*	Other Prod. *Autres prod.*
Production	5450	7474	34885	126	4264	1897	48977	21004	17412	-	21225
From Other Sources	-	-	-	-	-	-	-	-	-	-	-
Imports	-	4766	15	-	-	1291	3108	16272	-	-	1906
Exports	-	-502	-1820	-	-	-916	-985	-575	-	-	-2025
Intl. Marine Bunkers	-	-	-	-	-	-	-203	-154	-	-	-
Stock Changes	-	-81	-5	-	-	233	1192	-172	-	-	-
DOMESTIC SUPPLY	**5450**	**11657**	**33075**	**126**	**4264**	**2505**	**52089**	**36375**	**17412**	**-**	**21106**
Transfers	-	-	-	-	-	-	-	-	-	-	-
Statistical Differences	-2	203	212	-	-	305	748	1888	-	-	-
TRANSFORMATION	**1170**	**30**	**8**	**-**	**-**	**-**	**2039**	**13722**	**-**	**-**	**-**
Electricity Plants	290	30	8	-	-	-	2039	9938	-	-	-
CHP Plants	880	-	-	-	-	-	-	3430	-	-	-
Heat Plants	-	-	-	-	-	-	-	-	-	-	-
Blast Furnaces/Gas Works	-	-	-	-	-	-	-	354	-	-	-
Coke/Pat. Fuel/BKB Plants	-	-	-	-	-	-	-	-	-	-	-
Petroleum Refineries	-	-	-	-	-	-	-	-	-	-	-
Petrochemical Industry	-	-	-	-	-	-	-	-	-	-	-
Liquefaction	-	-	-	-	-	-	-	-	-	-	-
Other Transform. Sector	-	-	-	-	-	-	-	-	-	-	-
ENERGY SECTOR	**3493**	**978**	**-**	**-**	**-**	**146**	**2565**	**6078**	**-**	**-**	**-**
Coal Mines	-	-	-	-	-	30	390	49	-	-	-
Oil and Gas Extraction	438	147	-	-	-	6	1171	1537	-	-	-
Petroleum Refineries	3055	792	-	-	-	106	352	3404	-	-	-
Electr., CHP+Heat Plants	-	-	-	-	-	4	573	1063	-	-	-
Pumped Storage (Elec.)	-	-	-	-	-	-	-	-	-	-	-
Other Energy Sector	-	39	-	-	-	-	79	25	-	-	-
Distribution Losses	-	13	-	-	-	-	-	-	-	-	-
FINAL CONSUMPTION	**785**	**10839**	**33279**	**126**	**4264**	**2664**	**48233**	**18463**	**17412**	**-**	**21106**
INDUSTRY SECTOR	**785**	**1241**	**-**	**-**	**-**	**511**	**10385**	**12516**	**17412**	**-**	**-**
Iron and Steel	-	4	-	-	-	33	590	3073	-	-	-
Chemical and Petrochem.	736	515	-	-	-	82	1664	3690	17412	-	-
of which: Feedstocks	*343*	*78*	*-*	*-*	*-*	*1*	*70*	*914*	*17412*	*-*	*-*
Non-Ferrous Metals	-	4	-	-	-	5	490	566	-	-	-
Non-Metallic Minerals	38	424	-	-	-	19	2278	3118	-	-	-
Transport Equipment	-	8	-	-	-	43	386	86	-	-	-
Machinery	7	125	-	-	-	79	1258	492	-	-	-
Mining and Quarrying	-	-	-	-	-	5	431	5	-	-	-
Food and Tobacco	-	26	-	-	-	4	571	255	-	-	-
Paper, Pulp and Print	3	5	-	-	-	68	255	142	-	-	-
Wood and Wood Products	-	2	-	-	-	6	170	27	-	-	-
Construction	-	57	-	-	-	35	1535	166	-	-	-
Textile and Leather	1	55	-	-	-	32	511	606	-	-	-
Non-specified	-	16	-	-	-	100	246	290	-	-	-
TRANSPORT SECTOR	**-**	**5**	**33279**	**126**	**4264**	**-**	**19019**	**5656**	**-**	**-**	**-**
Air	-	-	-	126	4264	-	-	-	-	-	-
Road	-	5	33279	-	-	-	-	-	-	-	-
Rail	-	-	-	-	-	-	-	-	-	-	-
Pipeline Transport	-	-	-	-	-	-	-	-	-	-	-
Internal Navigation	-	-	-	-	-	-	-	-	-	-	-
Non-specified	-	-	-	-	-	-	19019	5656	-	-	-
OTHER SECTORS	**-**	**9593**	**-**	**-**	**-**	**2153**	**18829**	**291**	**-**	**-**	**-**
Agriculture	-	-	-	-	-	16	11202	3	-	-	-
Comm. and Publ. Services	-	1901	-	-	-	1506	7188	288	-	-	-
Residential	-	7692	-	-	-	631	439	-	-	-	-
Non-specified	-	-	-	-	-	-	-	-	-	-	-
NON-ENERGY USE	**-**	**-**	**-**	**-**	**-**	**-**	**-**	**-**	**-**	**-**	**21106**
in Industry/Transf./Energy	-	-	-	-	-	-	-	-	-	-	21106
in Transport	-	-	-	-	-	-	-	-	-	-	-
in Other Sectors	-	-	-	-	-	-	-	-	-	-	-

People's Republic of China / République populaire de Chine : 1998

SUPPLY AND CONSUMPTION *APPROVISIONNEMENT ET DEMANDE*	Gas / *Gaz* (TJ)				Comb. Renew. & Waste / *En. Re. Comb. & Déchets* (TJ)				(GWh)	(TJ)
	Natural Gas *Gaz naturel*	Gas Works *Usines à gaz*	Coke Ovens *Cokeries*	Blast Furnaces *Hauts fourneaux*	Solid Biomass *Biomasse solide*	Gas/Liquids from Biomass *Gaz/Liquides tirés de biomasse*	Municipal Waste *Déchets urbains*	Industrial Waste *Déchets industriels*	Electricity *Electricité*	Heat *Chaleur*
Production	1008400	155553	424891	255344	8717156	50799	-	-	1166200	1270677
From Other Sources	-	-	-	-	-	-	-	-	-	-
Imports	-	-	-	-	-	-	-	-	17	-
Exports	-100815	-	-	-	-	-	-	-	-7174	-
Intl. Marine Bunkers	-	-	-	-	-	-	-	-	-	-
Stock Changes	-	-	-	-	-	-	-	-	-	-
DOMESTIC SUPPLY	**907585**	**155553**	**424891**	**255344**	**8717156**	**50799**	**-**	**-**	**1159043**	**1270677**
Transfers	-	-	-	-	-	-	-	-	-	-
Statistical Differences	9581	30227	-36	-1693	-	-	-	-	-8355	1
TRANSFORMATION	**92440**	**-**	**35009**	**48575**	**-**	**-**	**-**	**-**	**-**	**-**
Electricity Plants	89525	-	22332	26334	-	-	-	-	-	-
CHP Plants	2915	-	12677	22241	-	-	-	-	-	-
Heat Plants	-	-	-	-	-	-	-	-	-	-
Blast Furnaces/Gas Works	-	-	-	-	-	-	-	-	-	-
Coke/Pat. Fuel/BKB Plants	-	-	-	-	-	-	-	-	-	-
Petroleum Refineries	-	-	-	-	-	-	-	-	-	-
Petrochemical Industry	-	-	-	-	-	-	-	-	-	-
Liquefaction	-	-	-	-	-	-	-	-	-	-
Other Transform. Sector	-	-	-	-	-	-	-	-	-	-
ENERGY SECTOR	**288795**	**18305**	**23555**	**-**	**-**	**-**	**-**	**-**	**179272**	**312073**
Coal Mines	-	-	701	-	-	-	-	-	42358	9003
Oil and Gas Extraction	233738	-	-	-	-	-	-	-	29498	32870
Petroleum Refineries	49079	-	-	-	-	-	-	-	19232	210853
Electr., CHP+Heat Plants	3292	9499	18	-	-	-	-	-	85553	55755
Pumped Storage (Elec.)	-	-	-	-	-	-	-	-	-	-
Other Energy Sector	2686	8806	22836	-	-	-	-	-	2631	3592
Distribution Losses	23478	-	-	-	-	-	-	-	79093	10723
FINAL CONSUMPTION	**512453**	**167475**	**366291**	**205076**	**8717156**	**50799**	**-**	**-**	**892323**	**947882**
INDUSTRY SECTOR	**410049**	**50340**	**290124**	**205076**	**-**	**-**	**-**	**-**	**588236**	**720904**
Iron and Steel	13255	-	261875	205076	-	-	-	-	93663	117629
Chemical and Petrochem.	350529	5690	21361	-	-	-	-	-	145941	291231
of which: Feedstocks	-	-	-	-	-	-	-	-	-	-
Non-Ferrous Metals	3076	11065	54	-	-	-	-	-	53860	52709
Non-Metallic Minerals	15248	11004	3902	-	-	-	-	-	64910	8109
Transport Equipment	6151	3809	126	-	-	-	-	-	20155	16483
Machinery	14858	17613	2356	-	-	-	-	-	56056	31522
Mining and Quarrying	520	-	-	-	-	-	-	-	19041	7563
Food and Tobacco	2123	241	234	-	-	-	-	-	35766	60172
Paper, Pulp and Print	520	30	18	-	-	-	-	-	20804	46127
Wood and Wood Products	-	105	-	-	-	-	-	-	4480	3740
Construction	520	-	18	-	-	-	-	-	18884	1609
Textile and Leather	3249	181	180	-	-	-	-	-	39509	73895
Non-specified	-	602	-	-	-	-	-	-	15167	10115
TRANSPORT SECTOR	**1126**	**-**	**-**	**-**	**-**	**-**	**-**	**-**	**20180**	**-**
Air	-	-	-	-	-	-	-	-	-	-
Road	1126	-	-	-	-	-	-	-	-	-
Rail	-	-	-	-	-	-	-	-	-	-
Pipeline Transport	-	-	-	-	-	-	-	-	-	-
Internal Navigation	-	-	-	-	-	-	-	-	-	-
Non-specified	-	-	-	-	-	-	-	-	20180	-
OTHER SECTORS	**101278**	**117135**	**76167**	**-**	**8717156**	**50799**	**-**	**-**	**283907**	**226978**
Agriculture	-	-	-	-	-	-	-	-	62349	533
Comm. and Publ. Services	13688	5615	10123	-	-	-	-	-	89108	33666
Residential	87590	111520	66044	-	8717156	50799	-	-	132450	187111
Non-specified	-	-	-	-	-	-	-	-	-	5668
NON-ENERGY USE	**-**	**-**	**-**	**-**	**-**	**-**	**-**	**-**	**-**	**-**
in Industry/Transf./Energy	-	-	-	-	-	-	-	-	-	-
in Transport	-	-	-	-	-	-	-	-	-	-
in Other Sectors	-	-	-	-	-	-	-	-	-	-

Chinese Taipei / Taipei chinois : 1997

SUPPLY AND CONSUMPTION / APPROVISIONNEMENT ET DEMANDE	Coal / Charbon (1000 tonnes)							Oil / Pétrole (1000 tonnes)			
	Coking Coal / Charbon à coke	Other Bit. Coal / Autres charb. bit.	Sub-Bit. Coal / Charbon sous-bit.	Lignite / Lignite	Peat / Tourbe	Oven and Gas Coke / Coke de four/gaz	Pat. Fuel and BKB / Agg./briq. de lignite	Crude Oil / Pétrole brut	NGL / LGN	Feed-stocks / Produits d'aliment.	Additives / Additifs
Production	-	99	-	-	-	3993	-	48	-	-	-
From Other Sources	-	-	-	-	-	-	-	-	-	-	-
Imports	6300	29952	-	-	-	-	-	32379	-	-	-
Exports	-	-	-	-	-	-	-	-	-	-	-
Intl. Marine Bunkers	-	-	-	-	-	-	-	-	-	-	-
Stock Changes	-612	-1954	-	-	-	90	-	224	-	-	-
DOMESTIC SUPPLY	5688	28097	-	-	-	4083	-	32651	-	-	-
Transfers	-	-	-	-	-	-	-	-	-	2382	-
Statistical Differences	1	-1	-	-	-	-	-	-1	-	-	-
TRANSFORMATION	5689	22429	-	-	-	3266	-	32650	-	2382	-
Electricity Plants	-	22201	-	-	-	-	-	-	-	-	-
CHP Plants	-	-	-	-	-	-	-	-	-	-	-
Heat Plants	-	-	-	-	-	-	-	-	-	-	-
Blast Furnaces/Gas Works	-	-	-	-	-	3266	-	-	-	-	-
Coke/Pat. Fuel/BKB Plants	5689	228	-	-	-	-	-	-	-	-	-
Petroleum Refineries	-	-	-	-	-	-	-	32650	-	2382	-
Petrochemical Industry	-	-	-	-	-	-	-	-	-	-	-
Liquefaction	-	-	-	-	-	-	-	-	-	-	-
Other Transform. Sector	-	-	-	-	-	-	-	-	-	-	-
ENERGY SECTOR	-	-	-	-	-	-	-	-	-	-	-
Coal Mines	-	-	-	-	-	-	-	-	-	-	-
Oil and Gas Extraction	-	-	-	-	-	-	-	-	-	-	-
Petroleum Refineries	-	-	-	-	-	-	-	-	-	-	-
Electr., CHP+Heat Plants	-	-	-	-	-	-	-	-	-	-	-
Pumped Storage (Elec.)	-	-	-	-	-	-	-	-	-	-	-
Other Energy Sector	-	-	-	-	-	-	-	-	-	-	-
Distribution Losses	-	-	-	-	-	-	-	-	-	-	-
FINAL CONSUMPTION	-	5667	-	-	-	817	-	-	-	-	-
INDUSTRY SECTOR	-	5667	-	-	-	817	-	-	-	-	-
Iron and Steel	-	659	-	-	-	817	-	-	-	-	-
Chemical and Petrochem.	-	1592	-	-	-	-	-	-	-	-	-
of which: Feedstocks	-	-	-	-	-	-	-	-	-	-	-
Non-Ferrous Metals	-	-	-	-	-	-	-	-	-	-	-
Non-Metallic Minerals	-	2849	-	-	-	-	-	-	-	-	-
Transport Equipment	-	-	-	-	-	-	-	-	-	-	-
Machinery	-	-	-	-	-	-	-	-	-	-	-
Mining and Quarrying	-	-	-	-	-	-	-	-	-	-	-
Food and Tobacco	-	-	-	-	-	-	-	-	-	-	-
Paper, Pulp and Print	-	490	-	-	-	-	-	-	-	-	-
Wood and Wood Products	-	-	-	-	-	-	-	-	-	-	-
Construction	-	-	-	-	-	-	-	-	-	-	-
Textile and Leather	-	77	-	-	-	-	-	-	-	-	-
Non-specified	-	-	-	-	-	-	-	-	-	-	-
TRANSPORT SECTOR	-	-	-	-	-	-	-	-	-	-	-
Air	-	-	-	-	-	-	-	-	-	-	-
Road	-	-	-	-	-	-	-	-	-	-	-
Rail	-	-	-	-	-	-	-	-	-	-	-
Pipeline Transport	-	-	-	-	-	-	-	-	-	-	-
Internal Navigation	-	-	-	-	-	-	-	-	-	-	-
Non-specified	-	-	-	-	-	-	-	-	-	-	-
OTHER SECTORS	-	-	-	-	-	-	-	-	-	-	-
Agriculture	-	-	-	-	-	-	-	-	-	-	-
Comm. and Publ. Services	-	-	-	-	-	-	-	-	-	-	-
Residential	-	-	-	-	-	-	-	-	-	-	-
Non-specified	-	-	-	-	-	-	-	-	-	-	-
NON-ENERGY USE	-	-	-	-	-	-	-	-	-	-	-
in Industry/Trans./Energy	-	-	-	-	-	-	-	-	-	-	-
in Transport	-	-	-	-	-	-	-	-	-	-	-
in Other Sectors	-	-	-	-	-	-	-	-	-	-	-

Chinese Taipei / Taipei chinois : 1997

SUPPLY AND CONSUMPTION / APPROVISIONNEMENT ET DEMANDE	Refinery Gas / Gaz de raffinerie	LPG + Ethane / GPL + éthane	Motor Gasoline / Essence moteur	Aviation Gasoline / Essence aviation	Jet Fuel / Carbu-réacteurs	Kerosene / Kérosène	Gas/ Diesel / Gazole	Heavy Fuel Oil / Fioul lourd	Naphtha / Naphta	Petrol. Coke / Coke de pétrole	Other Prod. / Autres prod.
Production	1043	817	4709	1	1501	465	5644	13800	3570	-	2295
From Other Sources	-	-	-	-	-	-	-	-	-	-	-
Imports	-	796	1164	-	416	-	-	3261	1348	-	801
Exports	-	-1	-	-	-	-	-1160	-1379	-63	-	-264
Intl. Marine Bunkers	-	-	-	-	-	-	-202	-2626	-	-	-2
Stock Changes	-	-44	535	-	23	-316	-103	479	-45	-	-170
DOMESTIC SUPPLY	1043	1568	6408	1	1940	149	4179	13535	4810	-	2660
Transfers	-	-	-19	-	-65	-123	-1	-596	-1284	-	-292
Statistical Differences	-	-1	-1	-	-1	3	-	1	-2	-	-
TRANSFORMATION	9	-	-	-	-	-	133	6354	-	-	-
Electricity Plants	9	-	-	-	-	-	133	6354	-	-	-
CHP Plants	-	-	-	-	-	-	-	-	-	-	-
Heat Plants	-	-	-	-	-	-	-	-	-	-	-
Blast Furnaces/Gas Works	-	-	-	-	-	-	-	-	-	-	-
Coke/Pat. Fuel/BKB Plants	-	-	-	-	-	-	-	-	-	-	-
Petroleum Refineries	-	-	-	-	-	-	-	-	-	-	-
Petrochemical Industry	-	-	-	-	-	-	-	-	-	-	-
Liquefaction	-	-	-	-	-	-	-	-	-	-	-
Other Transform. Sector	-	-	-	-	-	-	-	-	-	-	-
ENERGY SECTOR	1034	11	2	-	1	-	32	517	-	-	-
Coal Mines	-	-	-	-	-	-	-	-	-	-	-
Oil and Gas Extraction	-	-	-	-	-	-	-	-	-	-	-
Petroleum Refineries	1034	11	2	-	1	-	31	510	-	-	-
Electr., CHP+Heat Plants	-	-	-	-	-	-	-	-	-	-	-
Pumped Storage (Elec.)	-	-	-	-	-	-	-	-	-	-	-
Other Energy Sector	-	-	-	-	-	-	1	7	-	-	-
Distribution Losses	-	-	-	-	-	-	-	-	-	-	-
FINAL CONSUMPTION	-	1556	6386	1	1873	29	4013	6069	3524	-	2368
INDUSTRY SECTOR	-	427	4	-	-	11	169	5543	3524	-	1043
Iron and Steel	-	37	-	-	-	4	8	605	-	-	-
Chemical and Petrochem.	-	34	-	-	-	1	26	1791	3524	-	1043
of which: Feedstocks	-	-	-	-	-	-	-	-	3524	-	1043
Non-Ferrous Metals	-	19	-	-	-	2	3	104	-	-	-
Non-Metallic Minerals	-	82	-	-	-	-	32	462	-	-	-
Transport Equipment	-	16	-	-	-	-	11	-	-	-	-
Machinery	-	16	-	-	-	-	9	17	-	-	-
Mining and Quarrying	-	-	-	-	-	-	16	20	-	-	-
Food and Tobacco	-	9	-	-	-	1	12	413	-	-	-
Paper, Pulp and Print	-	13	-	-	-	-	2	400	-	-	-
Wood and Wood Products	-	-	-	-	-	-	-	-	-	-	-
Construction	-	-	-	-	-	-	41	121	-	-	-
Textile and Leather	-	6	-	-	-	-	3	1262	-	-	-
Non-specified	-	195	4	-	-	3	6	348	-	-	-
TRANSPORT SECTOR	-	15	6365	1	1873	-	2884	189	-	-	-
Air	-	-	-	1	1873	-	-	-	-	-	-
Road	-	15	6365	-	-	-	2708	1	-	-	-
Rail	-	-	-	-	-	-	39	-	-	-	-
Pipeline Transport	-	-	-	-	-	-	-	-	-	-	-
Internal Navigation	-	-	-	-	-	-	137	188	-	-	-
Non-specified	-	-	-	-	-	-	-	-	-	-	-
OTHER SECTORS	-	1114	17	-	-	18	960	337	-	-	-
Agriculture	-	-	-	-	-	-	661	92	-	-	-
Comm. and Publ. Services	-	-	17	-	-	16	212	218	-	-	-
Residential	-	1114	-	-	-	-	-	-	-	-	-
Non-specified	-	-	-	-	-	2	87	27	-	-	-
NON-ENERGY USE	-	-	-	-	-	-	-	-	-	-	1325
in Industry/Transf./Energy	-	-	-	-	-	-	-	-	-	-	1325
in Transport	-	-	-	-	-	-	-	-	-	-	-
in Other Sectors	-	-	-	-	-	-	-	-	-	-	-

Oil cont. / Pétrole cont. (1000 tonnes)

Chinese Taipei / Taipei chinois : 1997

SUPPLY AND CONSUMPTION / APPROVISIONNEMENT ET DEMANDE	Gas / Gaz (TJ)				Comb. Renew. & Waste / En. Re. Comb. & Déchets (TJ)				(GWh)	(TJ)
	Natural Gas / Gaz naturel	Gas Works / Usines à gaz	Coke Ovens / Cokeries	Blast Furnaces / Hauts fourneaux	Solid Biomass / Biomasse solide	Gas/Liquids from Biomass / Gaz/Liquides tirés de biomasse	Municipal Waste / Déchets urbains	Industrial Waste / Déchets industriels	Electricity / Electricité	Heat / Chaleur
Production	32036	-	43386	40718	537	-	-	-	153294	-
From Other Sources	-	-	-	-	-	-	-	-	-	-
Imports	174343	-	-	-	-	-	-	-	-	-
Exports	-	-	-	-	-	-	-	-	-	-
Intl. Marine Bunkers	-	-	-	-	-	-	-	-	-	-
Stock Changes	-7166	-	-	-	-	-	-	-	-	-
DOMESTIC SUPPLY	199213	-	43386	40718	537	-	-	-	153294	-
Transfers	-	-	-	-	-	-	-	-	-	-
Statistical Differences	-	-	-3	1	-	-	-	-	-	-
TRANSFORMATION	89104	-	6084	12367	-	-	-	-	-	-
Electricity Plants	88789	-	6084	12367	-	-	-	-	-	-
CHP Plants	-	-	-	-	-	-	-	-	-	-
Heat Plants	-	-	-	-	-	-	-	-	-	-
Blast Furnaces/Gas Works	-	-	-	-	-	-	-	-	-	-
Coke/Pat. Fuel/BKB Plants	-	-	-	-	-	-	-	-	-	-
Petroleum Refineries	315	-	-	-	-	-	-	-	-	-
Petrochemical Industry	-	-	-	-	-	-	-	-	-	-
Liquefaction	-	-	-	-	-	-	-	-	-	-
Other Transform. Sector	-	-	-	-	-	-	-	-	-	-
ENERGY SECTOR	28953	-	-	-	-	-	-	-	17774	-
Coal Mines	-	-	-	-	-	-	-	-	6	-
Oil and Gas Extraction	-	-	-	-	-	-	-	-	6	-
Petroleum Refineries	342	-	-	-	-	-	-	-	2462	-
Electr., CHP+Heat Plants	-	-	-	-	-	-	-	-	9125	-
Pumped Storage (Elec.)	-	-	-	-	-	-	-	-	5507	-
Other Energy Sector	28611	-	-	-	-	-	-	-	668	-
Distribution Losses	5508	-	-	397	-	-	-	-	6896	-
FINAL CONSUMPTION	75648	-	37299	27955	537	-	-	-	128624	-
INDUSTRY SECTOR	40423	-	37299	27955	-	-	-	-	67914	-
Iron and Steel	5397	-	37299	27955	-	-	-	-	9533	-
Chemical and Petrochem.	19538	-	-	-	-	-	-	-	19361	-
of which: Feedstocks	17673	-	-	-	-	-	-	-	-	-
Non-Ferrous Metals	-	-	-	-	-	-	-	-	387	-
Non-Metallic Minerals	11544	-	-	-	-	-	-	-	4552	-
Transport Equipment	-	-	-	-	-	-	-	-	1399	-
Machinery	2497	-	-	-	-	-	-	-	13932	-
Mining and Quarrying	-	-	-	-	-	-	-	-	178	-
Food and Tobacco	467	-	-	-	-	-	-	-	3082	-
Paper, Pulp and Print	-	-	-	-	-	-	-	-	3959	-
Wood and Wood Products	-	-	-	-	-	-	-	-	580	-
Construction	-	-	-	-	-	-	-	-	491	-
Textile and Leather	971	-	-	-	-	-	-	-	8435	-
Non-specified	9	-	-	-	-	-	-	-	2025	-
TRANSPORT SECTOR	-	-	-	-	-	-	-	-	370	-
Air	-	-	-	-	-	-	-	-	-	-
Road	-	-	-	-	-	-	-	-	-	-
Rail	-	-	-	-	-	-	-	-	370	-
Pipeline Transport	-	-	-	-	-	-	-	-	-	-
Internal Navigation	-	-	-	-	-	-	-	-	-	-
Non-specified	-	-	-	-	-	-	-	-	-	-
OTHER SECTORS	35225	-	-	-	537	-	-	-	60340	-
Agriculture	-	-	-	-	-	-	-	-	2264	-
Comm. and Publ. Services	6586	-	-	-	-	-	-	-	18258	-
Residential	27463	-	-	-	-	-	-	-	27841	-
Non-specified	1176	-	-	-	537	-	-	-	11977	-
NON-ENERGY USE	-	-	-	-	-	-	-	-	-	-
in Industry/Transf./Energy	-	-	-	-	-	-	-	-	-	-
in Transport	-	-	-	-	-	-	-	-	-	-
in Other Sectors	-	-	-	-	-	-	-	-	-	-

Chinese Taipei / Taipei chinois : 1998

SUPPLY AND CONSUMPTION *APPROVISIONNEMENT ET DEMANDE*	Coal / *Charbon* (1000 tonnes)							Oil / *Pétrole* (1000 tonnes)			
	Coking Coal *Charbon à coke*	Other Bit. Coal *Autres charb. bit.*	Sub-Bit. Coal *Charbon sous-bit.*	Lignite *Lignite*	Peat *Tourbe*	Oven and Gas Coke *Coke de four/gaz*	Pat. Fuel and BKB *Agg./briq. de lignite*	Crude Oil *Pétrole brut*	NGL *LGN*	Feed-stocks *Produits d'aliment.*	Additives *Additifs*
Production	-	79	-	-	-	4571	-	50	-	-	-
From Other Sources	-	-	-	-	-	-	-	-	-	-	-
Imports	5932	31120	-	-	-	-	-	34320	-	-	-
Exports	-	-	-	-	-	-	-	-	-	-	-
Intl. Marine Bunkers	-	-	-	-	-	-	-	-	-	-	-
Stock Changes	-425	108	-	-	-	-62	-	132	-	-	-
DOMESTIC SUPPLY	**5507**	**31307**	**-**	**-**	**-**	**4509**	**-**	**34502**	**-**	**-**	**-**
Transfers	-	-	-	-	-	-	-	-	-	1568	-
Statistical Differences	-	-2	-	-	-	-	-	-	-	-	-
TRANSFORMATION	**5507**	**24979**	**-**	**-**	**-**	**3607**	**-**	**34502**	**-**	**1568**	**-**
Electricity Plants	-	24476	-	-	-	-	-	-	-	-	-
CHP Plants	-	-	-	-	-	-	-	-	-	-	-
Heat Plants	-	-	-	-	-	-	-	-	-	-	-
Blast Furnaces/Gas Works	-	-	-	-	-	3607	-	-	-	-	-
Coke/Pat. Fuel/BKB Plants	5507	503	-	-	-	-	-	-	-	-	-
Petroleum Refineries	-	-	-	-	-	-	-	34502	-	1568	-
Petrochemical Industry	-	-	-	-	-	-	-	-	-	-	-
Liquefaction	-	-	-	-	-	-	-	-	-	-	-
Other Transform. Sector	-	-	-	-	-	-	-	-	-	-	-
ENERGY SECTOR	**-**	**-**	**-**	**-**	**-**	**-**	**-**	**-**	**-**	**-**	**-**
Coal Mines	-	-	-	-	-	-	-	-	-	-	-
Oil and Gas Extraction	-	-	-	-	-	-	-	-	-	-	-
Petroleum Refineries	-	-	-	-	-	-	-	-	-	-	-
Electr., CHP+Heat Plants	-	-	-	-	-	-	-	-	-	-	-
Pumped Storage (Elec.)	-	-	-	-	-	-	-	-	-	-	-
Other Energy Sector	-	-	-	-	-	-	-	-	-	-	-
Distribution Losses	-	-	-	-	-	-	-	-	-	-	-
FINAL CONSUMPTION	**-**	**6326**	**-**	**-**	**-**	**902**	**-**	**-**	**-**	**-**	**-**
INDUSTRY SECTOR	**-**	**6326**	**-**	**-**	**-**	**902**	**-**	**-**	**-**	**-**	**-**
Iron and Steel	-	923	-	-	-	902	-	-	-	-	-
Chemical and Petrochem.	-	2007	-	-	-	-	-	-	-	-	-
of which: Feedstocks	-	-	-	-	-	-	-	-	-	-	-
Non-Ferrous Metals	-	-	-	-	-	-	-	-	-	-	-
Non-Metallic Minerals	-	2857	-	-	-	-	-	-	-	-	-
Transport Equipment	-	-	-	-	-	-	-	-	-	-	-
Machinery	-	-	-	-	-	-	-	-	-	-	-
Mining and Quarrying	-	-	-	-	-	-	-	-	-	-	-
Food and Tobacco	-	-	-	-	-	-	-	-	-	-	-
Paper, Pulp and Print	-	453	-	-	-	-	-	-	-	-	-
Wood and Wood Products	-	-	-	-	-	-	-	-	-	-	-
Construction	-	-	-	-	-	-	-	-	-	-	-
Textile and Leather	-	86	-	-	-	-	-	-	-	-	-
Non-specified	-	-	-	-	-	-	-	-	-	-	-
TRANSPORT SECTOR	**-**	**-**	**-**	**-**	**-**	**-**	**-**	**-**	**-**	**-**	**-**
Air	-	-	-	-	-	-	-	-	-	-	-
Road	-	-	-	-	-	-	-	-	-	-	-
Rail	-	-	-	-	-	-	-	-	-	-	-
Pipeline Transport	-	-	-	-	-	-	-	-	-	-	-
Internal Navigation	-	-	-	-	-	-	-	-	-	-	-
Non-specified	-	-	-	-	-	-	-	-	-	-	-
OTHER SECTORS	**-**	**-**	**-**	**-**	**-**	**-**	**-**	**-**	**-**	**-**	**-**
Agriculture	-	-	-	-	-	-	-	-	-	-	-
Comm. and Publ. Services	-	-	-	-	-	-	-	-	-	-	-
Residential	-	-	-	-	-	-	-	-	-	-	-
Non-specified	-	-	-	-	-	-	-	-	-	-	-
NON-ENERGY USE	**-**	**-**	**-**	**-**	**-**	**-**	**-**	**-**	**-**	**-**	**-**
in Industry/Trans./Energy	-	-	-	-	-	-	-	-	-	-	-
in Transport	-	-	-	-	-	-	-	-	-	-	-
in Other Sectors	-	-	-	-	-	-	-	-	-	-	-

Chinese Taipei / Taipei chinois : 1998

SUPPLY AND CONSUMPTION APPROVISIONNEMENT ET DEMANDE	Oil cont. / *Pétrole cont.* (1000 tonnes)										
	Refinery Gas *Gaz de raffinerie*	LPG + Ethane *GPL + éthane*	Motor Gasoline *Essence moteur*	Aviation Gasoline *Essence aviation*	Jet Fuel *Carbu-réacteurs*	Kerosene *Kérosène*	Gas/ Diesel *Gazole*	Heavy Fuel Oil *Fioul lourd*	Naphtha *Naphta*	Petrol. Coke *Coke de pétrole*	Other Prod. *Autres prod.*
Production	1042	805	4755	1	1552	420	5973	14361	3301	-	2421
From Other Sources	-	-	-	-	-	-	-	-	-	-	-
Imports	-	766	1682	-	331	-	-	3215	740	-	1131
Exports	-	-8	-	-	-	-	-507	-850	-27	-	-234
Intl. Marine Bunkers	-	-	-	-	-	-	-218	-2911	-	-	-2
Stock Changes	-	-30	273	-	138	-279	-809	-270	-245	-	-195
DOMESTIC SUPPLY	**1042**	**1533**	**6710**	**1**	**2021**	**141**	**4439**	**13545**	**3769**	**-**	**3121**
Transfers	-	-	-1	-	-19	-111	-29	-662	-433	-	-313
Statistical Differences	-	2	1	-	-	1	-	-2	-1	-	-
TRANSFORMATION	**16**	**-**	**-**	**-**	**-**	**-**	**313**	**6311**	**-**	**-**	**-**
Electricity Plants	16	-	-	-	-	-	313	6311	-	-	-
CHP Plants	-	-	-	-	-	-	-	-	-	-	-
Heat Plants	-	-	-	-	-	-	-	-	-	-	-
Blast Furnaces/Gas Works	-	-	-	-	-	-	-	-	-	-	-
Coke/Pat. Fuel/BKB Plants	-	-	-	-	-	-	-	-	-	-	-
Petroleum Refineries	-	-	-	-	-	-	-	-	-	-	-
Petrochemical Industry	-	-	-	-	-	-	-	-	-	-	-
Liquefaction	-	-	-	-	-	-	-	-	-	-	-
Other Transform. Sector	-	-	-	-	-	-	-	-	-	-	-
ENERGY SECTOR	**1026**	**2**	**3**	**-**	**1**	**-**	**31**	**556**	**-**	**-**	**-**
Coal Mines	-	-	-	-	-	-	-	-	-	-	-
Oil and Gas Extraction	-	-	-	-	-	-	-	-	-	-	-
Petroleum Refineries	1026	2	2	-	1	-	30	496	-	-	-
Electr., CHP+Heat Plants	-	-	-	-	-	-	-	-	-	-	-
Pumped Storage (Elec.)	-	-	-	-	-	-	-	-	-	-	-
Other Energy Sector	-	-	1	-	-	-	1	60	-	-	-
Distribution Losses	-	-	-	-	-	-	-	-	-	-	-
FINAL CONSUMPTION	**-**	**1533**	**6707**	**1**	**2001**	**31**	**4066**	**6014**	**3335**	**-**	**2808**
INDUSTRY SECTOR	**-**	**375**	**5**	**-**	**-**	**9**	**160**	**5479**	**3335**	**-**	**1047**
Iron and Steel	-	37	-	-	-	5	8	610	-	-	-
Chemical and Petrochem.	-	27	-	-	-	-	25	1664	3335	-	1047
of which: Feedstocks	-	-	-	-	-	-	-	-	*3335*	-	*1047*
Non-Ferrous Metals	-	17	-	-	-	1	3	95	-	-	-
Non-Metallic Minerals	-	63	-	-	-	-	29	479	-	-	-
Transport Equipment	-	14	-	-	-	-	10	-	-	-	-
Machinery	-	11	-	-	-	-	11	13	-	-	-
Mining and Quarrying	-	-	-	-	-	-	15	19	-	-	-
Food and Tobacco	-	8	-	-	-	-	9	416	-	-	-
Paper, Pulp and Print	-	9	-	-	-	-	2	380	-	-	-
Wood and Wood Products	-	-	-	-	-	-	-	-	-	-	-
Construction	-	-	-	-	-	-	41	134	-	-	-
Textile and Leather	-	5	-	-	-	-	3	1323	-	-	-
Non-specified	-	184	5	-	-	3	4	346	-	-	-
TRANSPORT SECTOR	**-**	**14**	**6678**	**1**	**2001**	**-**	**3010**	**188**	**-**	**-**	**-**
Air	-	-	-	1	2001	-	-	-	-	-	-
Road	-	14	6678	-	-	-	2809	1	-	-	-
Rail	-	-	-	-	-	-	38	-	-	-	-
Pipeline Transport	-	-	-	-	-	-	-	-	-	-	-
Internal Navigation	-	-	-	-	-	-	163	187	-	-	-
Non-specified	-	-	-	-	-	-	-	-	-	-	-
OTHER SECTORS	**-**	**1144**	**24**	**-**	**-**	**22**	**896**	**347**	**-**	**-**	**-**
Agriculture	-	-	-	-	-	-	540	81	-	-	-
Comm. and Publ. Services	-	-	24	-	-	20	265	239	-	-	-
Residential	-	1144	-	-	-	-	-	-	-	-	-
Non-specified	-	-	-	-	-	2	91	27	-	-	-
NON-ENERGY USE	**-**	**-**	**-**	**-**	**-**	**-**	**-**	**-**	**-**	**-**	**1761**
in Industry/Transf./Energy	-	-	-	-	-	-	-	-	-	-	1761
in Transport	-	-	-	-	-	-	-	-	-	-	-
in Other Sectors	-	-	-	-	-	-	-	-	-	-	-

Chinese Taipei / Taipei chinois : 1998

SUPPLY AND CONSUMPTION *APPROVISIONNEMENT ET DEMANDE*	Gas / *Gaz* (TJ)				Comb. Renew. & Waste / *En. Re. Comb. & Déchets* (TJ)				(GWh)	(TJ)
	Natural Gas *Gaz naturel*	Gas Works *Usines à gaz*	Coke Ovens *Cokeries*	Blast Furnaces *Hauts fourneaux*	Solid Biomass *Biomasse solide*	Gas/Liquids from Biomass *Gaz/Liquides tirés de biomasse*	Municipal Waste *Déchets urbains*	Industrial Waste *Déchets industriels*	Electricity *Electricité*	Heat *Chaleur*
Production	32753	-	44252	42044	537	-	-	-	166179	-
From Other Sources	-	-	-	-	-	-	-	-	-	-
Imports	211727	-	-	-	-	-	-	-	-	-
Exports	-	-	-	-	-	-	-	-	-	-
Intl. Marine Bunkers	-	-	-	-	-	-	-	-	-	-
Stock Changes	1646	-	-	-	-	-	-	-	-	-
DOMESTIC SUPPLY	**246126**	**-**	**44252**	**42044**	**537**	**-**	**-**	**-**	**166179**	**-**
Transfers	-	-	-	-	-	-	-	-	-	-
Statistical Differences	-	-	-120	-	-	-	-	-	-	-
TRANSFORMATION	**134338**	**-**	**6727**	**12440**	**-**	**-**	**-**	**-**	**-**	**-**
Electricity Plants	134060	-	6727	12440	-	-	-	-	-	-
CHP Plants	-	-	-	-	-	-	-	-	-	-
Heat Plants	-	-	-	-	-	-	-	-	-	-
Blast Furnaces/Gas Works	-	-	-	-	-	-	-	-	-	-
Coke/Pat. Fuel/BKB Plants	-	-	-	-	-	-	-	-	-	-
Petroleum Refineries	278	-	-	-	-	-	-	-	-	-
Petrochemical Industry	-	-	-	-	-	-	-	-	-	-
Liquefaction	-	-	-	-	-	-	-	-	-	-
Other Transform. Sector	-	-	-	-	-	-	-	-	-	-
ENERGY SECTOR	**30302**	**-**	**-**	**-**	**-**	**-**	**-**	**-**	**18447**	**-**
Coal Mines	-	-	-	-	-	-	-	-	4	-
Oil and Gas Extraction	-	-	-	-	-	-	-	-	33	-
Petroleum Refineries	357	-	-	-	-	-	-	-	2357	-
Electr., CHP+Heat Plants	-	-	-	-	-	-	-	-	9806	-
Pumped Storage (Elec.)	-	-	-	-	-	-	-	-	5519	-
Other Energy Sector	29945	-	-	-	-	-	-	-	728	-
Distribution Losses	5764	-	-	557	-	-	-	-	10455	-
FINAL CONSUMPTION	**75722**	**-**	**37405**	**29047**	**537**	**-**	**-**	**-**	**137277**	**-**
INDUSTRY SECTOR	**40274**	**-**	**37405**	**29047**	**-**	**-**	**-**	**-**	**70929**	**-**
Iron and Steel	5887	-	37405	29047	-	-	-	-	11300	-
Chemical and Petrochem.	18945	-	-	-	-	-	-	-	20317	-
of which: Feedstocks	*17083*	-	-	-	-	-	-	-	-	-
Non-Ferrous Metals	-	-	-	-	-	-	-	-	738	-
Non-Metallic Minerals	10955	-	-	-	-	-	-	-	4437	-
Transport Equipment	-	-	-	-	-	-	-	-	1453	-
Machinery	2924	-	-	-	-	-	-	-	14051	-
Mining and Quarrying	-	-	-	-	-	-	-	-	316	-
Food and Tobacco	448	-	-	-	-	-	-	-	3058	-
Paper, Pulp and Print	-	-	-	-	-	-	-	-	3794	-
Wood and Wood Products	-	-	-	-	-	-	-	-	558	-
Construction	-	-	-	-	-	-	-	-	496	-
Textile and Leather	1108	-	-	-	-	-	-	-	8816	-
Non-specified	7	-	-	-	-	-	-	-	1595	-
TRANSPORT SECTOR	**-**	**-**	**-**	**-**	**-**	**-**	**-**	**-**	**522**	**-**
Air	-	-	-	-	-	-	-	-	-	-
Road	-	-	-	-	-	-	-	-	-	-
Rail	-	-	-	-	-	-	-	-	522	-
Pipeline Transport	-	-	-	-	-	-	-	-	-	-
Internal Navigation	-	-	-	-	-	-	-	-	-	-
Non-specified	-	-	-	-	-	-	-	-	-	-
OTHER SECTORS	**35448**	**-**	**-**	**-**	**537**	**-**	**-**	**-**	**65826**	**-**
Agriculture	-	-	-	-	-	-	-	-	2064	-
Comm. and Publ. Services	6864	-	-	-	-	-	-	-	21069	-
Residential	27470	-	-	-	-	-	-	-	31445	-
Non-specified	1114	-	-	-	537	-	-	-	11248	-
NON-ENERGY USE	**-**	**-**	**-**	**-**	**-**	**-**	**-**	**-**	**-**	**-**
in Industry/Transf./Energy	-	-	-	-	-	-	-	-	-	-
in Transport	-	-	-	-	-	-	-	-	-	-
in Other Sectors	-	-	-	-	-	-	-	-	-	-

Colombia / Colombie : 1997

SUPPLY AND CONSUMPTION *APPROVISIONNEMENT ET DEMANDE*	Coal / *Charbon* (1000 tonnes)							Oil / *Pétrole* (1000 tonnes)			
	Coking Coal *Charbon à coke*	Other Bit. Coal *Autres charb. bit.*	Sub-Bit. Coal *Charbon sous-bit.*	Lignite *Lignite*	Peat *Tourbe*	Oven and Gas Coke *Coke de four/gaz*	Pat. Fuel and BKB *Agg./briq. de lignite*	Crude Oil *Pétrole brut*	NGL *LGN*	Feed-stocks *Produits d'aliment.*	Additives *Additifs*
Production	1671	31064	-	-	-	610	-	33636	135	-	-
From Other Sources	-	-	-	-	-	-	-	-	-	-	-
Imports	-	-	-	-	-	-	-	-	-	-	-
Exports	-1101	-26479	-	-	-	-110	-	-17769	-	-	-
Intl. Marine Bunkers	-	-	-	-	-	-	-	-	-	-	-
Stock Changes	-	57	-	-	-	18	-	-123	-	-	-
DOMESTIC SUPPLY	570	4642	-	-	-	518	-	15744	135	-	-
Transfers	-	-	-	-	-	-	-	-	-135	86	-
Statistical Differences	-	-86	-	-	-	-	-	-14	-	-	-
TRANSFORMATION	570	1484	-	-	-	397	-	14707	-	86	-
Electricity Plants	-	1229	-	-	-	-	-	18	-	-	-
CHP Plants	-	-	-	-	-	-	-	-	-	-	-
Heat Plants	-	-	-	-	-	-	-	-	-	-	-
Blast Furnaces/Gas Works	-	255	-	-	-	397	-	-	-	-	-
Coke/Pat. Fuel/BKB Plants	570	-	-	-	-	-	-	-	-	-	-
Petroleum Refineries	-	-	-	-	-	-	-	14689	-	86	-
Petrochemical Industry	-	-	-	-	-	-	-	-	-	-	-
Liquefaction	-	-	-	-	-	-	-	-	-	-	-
Other Transform. Sector	-	-	-	-	-	-	-	-	-	-	-
ENERGY SECTOR	-	-	-	-	-	-	-	71	-	-	-
Coal Mines	-	-	-	-	-	-	-	-	-	-	-
Oil and Gas Extraction	-	-	-	-	-	-	-	-	-	-	-
Petroleum Refineries	-	-	-	-	-	-	-	71	-	-	-
Electr., CHP+Heat Plants	-	-	-	-	-	-	-	-	-	-	-
Pumped Storage (Elec.)	-	-	-	-	-	-	-	-	-	-	-
Other Energy Sector	-	-	-	-	-	-	-	-	-	-	-
Distribution Losses	-	51	-	-	-	22	-	35	-	-	-
FINAL CONSUMPTION	-	3021	-	-	-	99	-	917	-	-	-
INDUSTRY SECTOR	-	2811	-	-	-	99	-	909	-	-	-
Iron and Steel	-	510	-	-	-	99	-	56	-	-	-
Chemical and Petrochem.	-	159	-	-	-	-	-	89	-	-	-
of which: Feedstocks	-	-	-	-	-	-	-	-	-	-	-
Non-Ferrous Metals	-	-	-	-	-	-	-	-	-	-	-
Non-Metallic Minerals	-	1108	-	-	-	-	-	142	-	-	-
Transport Equipment	-	-	-	-	-	-	-	-	-	-	-
Machinery	-	-	-	-	-	-	-	79	-	-	-
Mining and Quarrying	-	-	-	-	-	-	-	-	-	-	-
Food and Tobacco	-	211	-	-	-	-	-	233	-	-	-
Paper, Pulp and Print	-	475	-	-	-	-	-	1	-	-	-
Wood and Wood Products	-	1	-	-	-	-	-	116	-	-	-
Construction	-	-	-	-	-	-	-	36	-	-	-
Textile and Leather	-	347	-	-	-	-	-	101	-	-	-
Non-specified	-	-	-	-	-	-	-	56	-	-	-
TRANSPORT SECTOR	-	-	-	-	-	-	-	-	-	-	-
Air	-	-	-	-	-	-	-	-	-	-	-
Road	-	-	-	-	-	-	-	-	-	-	-
Rail	-	-	-	-	-	-	-	-	-	-	-
Pipeline Transport	-	-	-	-	-	-	-	-	-	-	-
Internal Navigation	-	-	-	-	-	-	-	-	-	-	-
Non-specified	-	-	-	-	-	-	-	-	-	-	-
OTHER SECTORS	-	210	-	-	-	-	-	8	-	-	-
Agriculture	-	-	-	-	-	-	-	3	-	-	-
Comm. and Publ. Services	-	-	-	-	-	-	-	5	-	-	-
Residential	-	210	-	-	-	-	-	-	-	-	-
Non-specified	-	-	-	-	-	-	-	-	-	-	-
NON-ENERGY USE	-	-	-	-	-	-	-	-	-	-	-
in Industry/Trans./Energy	-	-	-	-	-	-	-	-	-	-	-
in Transport	-	-	-	-	-	-	-	-	-	-	-
in Other Sectors	-	-	-	-	-	-	-	-	-	-	-

Colombia / Colombie : 1997

SUPPLY AND CONSUMPTION APPROVISIONNEMENT ET DEMANDE	Oil cont. / *Pétrole cont.* (1000 tonnes)										
	Refinery Gas *Gaz de raffinerie*	LPG + Ethane *GPL + éthane*	Motor Gasoline *Essence moteur*	Aviation Gasoline *Essence aviation*	Jet Fuel *Carbu-réacteurs*	Kerosene *Kérosène*	Gas/ Diesel *Gazole*	Heavy Fuel Oil *Fioul lourd*	Naphtha *Naphta*	Petrol. Coke *Coke de pétrole*	Other Prod. *Autres prod.*
Production	392	686	4442	20	877	162	3301	2963	-	-	1664
From Other Sources	-	-	-	-	-	-	-	-	-	-	-
Imports	-	-	1312	-	7	-	1	6	-	-	-
Exports	-	-4	-	-	-47	-	-453	-2785	-	-	-
Intl. Marine Bunkers	-	-	-	-	-	-	-174	-31	-	-	-
Stock Changes	-	-	-113	-	32	-	-	45	-	-	236
DOMESTIC SUPPLY	**392**	**682**	**5641**	**20**	**869**	**162**	**2675**	**198**	**-**	**-**	**1900**
Transfers	-	-12	65	-	-	-	-	-	-	-	-
Statistical Differences	-	15	59	-	-160	-	92	69	-	-	-
TRANSFORMATION	**-**	**-**	**-**	**-**	**-**	**-**	**34**	**26**	**-**	**-**	**-**
Electricity Plants	-	-	-	-	-	-	34	26	-	-	-
CHP Plants	-	-	-	-	-	-	-	-	-	-	-
Heat Plants	-	-	-	-	-	-	-	-	-	-	-
Blast Furnaces/Gas Works	-	-	-	-	-	-	-	-	-	-	-
Coke/Pat. Fuel/BKB Plants	-	-	-	-	-	-	-	-	-	-	-
Petroleum Refineries	-	-	-	-	-	-	-	-	-	-	-
Petrochemical Industry	-	-	-	-	-	-	-	-	-	-	-
Liquefaction	-	-	-	-	-	-	-	-	-	-	-
Other Transform. Sector	-	-	-	-	-	-	-	-	-	-	-
ENERGY SECTOR	**392**	**5**	**7**	**-**	**-**	**1**	**14**	**48**	**-**	**-**	**-**
Coal Mines	-	-	-	-	-	-	-	-	-	-	-
Oil and Gas Extraction	-	-	-	-	-	-	-	-	-	-	-
Petroleum Refineries	392	5	7	-	-	1	14	48	-	-	-
Electr., CHP+Heat Plants	-	-	-	-	-	-	-	-	-	-	-
Pumped Storage (Elec.)	-	-	-	-	-	-	-	-	-	-	-
Other Energy Sector	-	-	-	-	-	-	-	-	-	-	-
Distribution Losses	-	-	-	-	-	-	-	-	-	-	-
FINAL CONSUMPTION	**-**	**680**	**5758**	**20**	**709**	**161**	**2719**	**193**	**-**	**-**	**1900**
INDUSTRY SECTOR	**-**	**70**	**59**	**-**	**-**	**93**	**463**	**176**	**-**	**-**	**-**
Iron and Steel	-	11	-	-	-	10	-	-	-	-	-
Chemical and Petrochem.	-	6	-	-	-	13	47	4	-	-	-
of which: Feedstocks	-	-	-	-	-	-	-	-	-	-	-
Non-Ferrous Metals	-	-	-	-	-	-	-	-	-	-	-
Non-Metallic Minerals	-	17	-	-	-	53	85	-	-	-	-
Transport Equipment	-	-	-	-	-	-	-	-	-	-	-
Machinery	-	2	-	-	-	-	17	125	-	-	-
Mining and Quarrying	-	5	-	-	-	-	-	-	-	-	-
Food and Tobacco	-	18	-	-	-	5	129	17	-	-	-
Paper, Pulp and Print	-	3	-	-	-	2	21	22	-	-	-
Wood and Wood Products	-	-	-	-	-	2	2	-	-	-	-
Construction	-	-	59	-	-	1	4	1	-	-	-
Textile and Leather	-	4	-	-	-	7	29	7	-	-	-
Non-specified	-	4	-	-	-	-	129	-	-	-	-
TRANSPORT SECTOR	**-**	**-**	**5627**	**20**	**709**	**-**	**1475**	**4**	**-**	**-**	**-**
Air	-	-	-	20	709	-	-	-	-	-	-
Road	-	-	5543	-	-	-	1382	-	-	-	-
Rail	-	-	-	-	-	-	14	4	-	-	-
Pipeline Transport	-	-	-	-	-	-	-	-	-	-	-
Internal Navigation	-	-	84	-	-	-	79	-	-	-	-
Non-specified	-	-	-	-	-	-	-	-	-	-	-
OTHER SECTORS	**-**	**610**	**72**	**-**	**-**	**68**	**781**	**13**	**-**	**-**	**-**
Agriculture	-	-	8	-	-	1	457	2	-	-	-
Comm. and Publ. Services	-	53	-	-	-	-	324	11	-	-	-
Residential	-	557	64	-	-	67	-	-	-	-	-
Non-specified	-	-	-	-	-	-	-	-	-	-	-
NON-ENERGY USE	**-**	**-**	**-**	**-**	**-**	**-**	**-**	**-**	**-**	**-**	**1900**
in Industry/Transf./Energy	-	-	-	-	-	-	-	-	-	-	1900
in Transport	-	-	-	-	-	-	-	-	-	-	-
in Other Sectors	-	-	-	-	-	-	-	-	-	-	-

Colombia / Colombie : 1997

SUPPLY AND CONSUMPTION / APPROVISIONNEMENT ET DEMANDE	Gas / Gaz (TJ)				Comb. Renew. & Waste / En. Re. Comb. & Déchets (TJ)				(GWh)	(TJ)
	Natural Gas / Gaz naturel	Gas Works / Usines à gaz	Coke Ovens / Cokeries	Blast Furnaces / Hauts fourneaux	Solid Biomass / Biomasse solide	Gas/Liquids from Biomass / Gaz/Liquides tirés de biomasse	Municipal Waste / Déchets urbains	Industrial Waste / Déchets industriels	Electricity / Electricité	Heat / Chaleur
Production	223603	-	2109	2544	220566	-	-	-	46115	-
From Other Sources	-									
Imports	-	-	-	-	-	-	-	-	199	-
Exports	-	-	-	-	-	-	-	-	-	-
Intl. Marine Bunkers	-	-	-	-						
Stock Changes	-	-	-	-	-	-	-	-	-	-
DOMESTIC SUPPLY	**223603**	**-**	**2109**	**2544**	**220566**	**-**	**-**	**-**	**46314**	**-**
Transfers	-	-	-		-					
Statistical Differences	-907	-	1	-	-3	-	-	-	23	-
TRANSFORMATION	**101307**	**-**	**-**	**1809**	**11880**	**-**	**-**	**-**	**-**	**-**
Electricity Plants	101307	-	-	1809	5419	-	-	-	-	-
CHP Plants	-	-	-	-	-	-	-	-	-	-
Heat Plants	-	-	-	-	-	-	-	-	-	-
Blast Furnaces/Gas Works	-	-	-	-	-	-	-	-	-	-
Coke/Pat. Fuel/BKB Plants	-	-	-	-	-	-	-	-	-	-
Petroleum Refineries	-	-	-	-	-	-	-	-	-	-
Petrochemical Industry	-	-	-	-	-	-	-	-	-	-
Liquefaction	-	-	-	-	-	-	-	-	-	-
Other Transform. Sector	-	-	-	-	6461	-	-	-	-	-
ENERGY SECTOR	**60038**	**-**	**2025**	**-**	**-**	**-**	**-**	**-**	**768**	**-**
Coal Mines	-	-	-	-	-	-	-	-	-	-
Oil and Gas Extraction	18866	-	-	-	-	-	-	-	-	-
Petroleum Refineries	41172	-	-	-	-	-	-	-	-	-
Electr., CHP+Heat Plants	-	-	-	-	-	-	-	-	768	-
Pumped Storage (Elec.)	-	-	-	-	-	-	-	-	-	-
Other Energy Sector	-	-	2025	-	-	-	-	-	-	-
Distribution Losses	-	-	-	251	-	-	-	-	10183	-
FINAL CONSUMPTION	**61351**	**-**	**85**	**484**	**208683**	**-**	**-**	**-**	**35386**	**-**
INDUSTRY SECTOR	**42298**	**-**	**85**	**484**	**72479**	**-**	**-**	**-**	**11349**	**-**
Iron and Steel	1317	-	85	484	-	-	-	-	2032	-
Chemical and Petrochem.	18246	-	-	-	1484	-	-	-	1885	-
of which: Feedstocks	-	-	-	-	-	-	-	-	-	-
Non-Ferrous Metals	-	-	-	-	6	-	-	-	-	-
Non-Metallic Minerals	16607	-	-	-	651	-	-	-	1479	-
Transport Equipment	-	-	-	-	-	-	-	-	-	-
Machinery	-	-	-	-	6	-	-	-	455	-
Mining and Quarrying	-	-	-	-	-	-	-	-	-	-
Food and Tobacco	2595	-	-	-	60748	-	-	-	2224	-
Paper, Pulp and Print	2156	-	-	-	9526	-	-	-	903	-
Wood and Wood Products	849	-	-	-	-	-	-	-	111	-
Construction	-	-	-	-	-	-	-	-	59	-
Textile and Leather	429	-	-	-	58	-	-	-	1741	-
Non-specified	99	-	-	-	-	-	-	-	460	-
TRANSPORT SECTOR	**2075**	**-**	**-**	**-**	**-**	**-**	**-**	**-**	**43**	**-**
Air	-	-	-	-	-	-	-	-	-	-
Road	-	-	-	-	-	-	-	-	43	-
Rail	-	-	-	-	-	-	-	-	-	-
Pipeline Transport	-	-	-	-	-	-	-	-	-	-
Internal Navigation	-	-	-	-	-	-	-	-	-	-
Non-specified	2075	-	-	-	-	-	-	-	-	-
OTHER SECTORS	**16978**	**-**	**-**	**-**	**136204**	**-**	**-**	**-**	**23994**	**-**
Agriculture	-	-	-	-	49326	-	-	-	1220	-
Comm. and Publ. Services	2547	-	-	-	-	-	-	-	6795	-
Residential	14431	-	-	-	86036	-	-	-	14675	-
Non-specified	-	-	-	-	842	-	-	-	1304	-
NON-ENERGY USE	**-**	**-**	**-**	**-**	**-**	**-**	**-**	**-**	**-**	**-**
in Industry/Transf./Energy	-	-	-	-	-	-	-	-	-	-
in Transport	-	-	-	-	-	-	-	-	-	-
in Other Sectors	-	-	-	-	-	-	-	-	-	-

Colombia / Colombie : 1998

SUPPLY AND CONSUMPTION / APPROVISIONNEMENT ET DEMANDE	Coal / *Charbon* (1000 tonnes)							Oil / *Pétrole* (1000 tonnes)			
	Coking Coal *Charbon à coke*	Other Bit. Coal *Autres charb. bit.*	Sub-Bit. Coal *Charbon sous-bit.*	Lignite *Lignite*	Peat *Tourbe*	Oven and Gas Coke *Coke de four/gaz*	Pat. Fuel and BKB *Agg./briq. de lignite*	Crude Oil *Pétrole brut*	NGL *LGN*	Feed-stocks *Produits d'aliment.*	Additives *Additifs*
Production	1458	32213	-	-	-	458	-	38506	356	-	-
From Other Sources	-	-	-	-	-	-	-	-	-	-	-
Imports	-	-	-	-	-	-	-	49	-	-	-
Exports	-660	-28741	-	-	-	-102	-	-22910	-	-	-
Intl. Marine Bunkers	-	-	-	-	-	-	-	-	-	-	-
Stock Changes	-172	97	-	-	-	-	-	429	-	-	-
DOMESTIC SUPPLY	**626**	**3569**	-	-	-	**356**	-	**16074**	**356**	-	-
Transfers	-	-	-	-	-	-	-	-	-356	86	-
Statistical Differences	-	1052	-	-	-	119	-	-695	-	-	-
TRANSFORMATION	**626**	**2222**	-	-	-	**366**	-	**14470**	-	**86**	-
Electricity Plants	-	1866	-	-	-	-	-	12	-	-	-
CHP Plants	-	-	-	-	-	-	-	-	-	-	-
Heat Plants	-	-	-	-	-	-	-	-	-	-	-
Blast Furnaces/Gas Works	-	356	-	-	-	366	-	-	-	-	-
Coke/Pat. Fuel/BKB Plants	626	-	-	-	-	-	-	-	-	-	-
Petroleum Refineries	-	-	-	-	-	-	-	14458	-	86	-
Petrochemical Industry	-	-	-	-	-	-	-	-	-	-	-
Liquefaction	-	-	-	-	-	-	-	-	-	-	-
Other Transform. Sector	-	-	-	-	-	-	-	-	-	-	-
ENERGY SECTOR	-	-	-	-	-	-	-	**129**	-	-	-
Coal Mines	-	-	-	-	-	-	-	-	-	-	-
Oil and Gas Extraction	-	-	-	-	-	-	-	-	-	-	-
Petroleum Refineries	-	-	-	-	-	-	-	129	-	-	-
Electr., CHP+Heat Plants	-	-	-	-	-	-	-	-	-	-	-
Pumped Storage (Elec.)	-	-	-	-	-	-	-	-	-	-	-
Other Energy Sector	-	-	-	-	-	-	-	-	-	-	-
Distribution Losses	-	74	-	-	-	17	-	9	-	-	-
FINAL CONSUMPTION	-	**2325**	-	-	-	**92**	-	**771**	-	-	-
INDUSTRY SECTOR	-	**2209**	-	-	-	**92**	-	**763**	-	-	-
Iron and Steel	-	401	-	-	-	92	-	47	-	-	-
Chemical and Petrochem.	-	125	-	-	-	-	-	72	-	-	-
of which: Feedstocks	-	-	-	-	-	-	-	-	-	-	-
Non-Ferrous Metals	-	-	-	-	-	-	-	-	-	-	-
Non-Metallic Minerals	-	871	-	-	-	-	-	126	-	-	-
Transport Equipment	-	-	-	-	-	-	-	-	-	-	-
Machinery	-	-	-	-	-	-	-	67	-	-	-
Mining and Quarrying	-	-	-	-	-	-	-	-	-	-	-
Food and Tobacco	-	166	-	-	-	-	-	198	-	-	-
Paper, Pulp and Print	-	373	-	-	-	-	-	30	-	-	-
Wood and Wood Products	-	1	-	-	-	-	-	65	-	-	-
Construction	-	-	-	-	-	-	-	30	-	-	-
Textile and Leather	-	272	-	-	-	-	-	87	-	-	-
Non-specified	-	-	-	-	-	-	-	41	-	-	-
TRANSPORT SECTOR	-	-	-	-	-	-	-	-	-	-	-
Air	-	-	-	-	-	-	-	-	-	-	-
Road	-	-	-	-	-	-	-	-	-	-	-
Rail	-	-	-	-	-	-	-	-	-	-	-
Pipeline Transport	-	-	-	-	-	-	-	-	-	-	-
Internal Navigation	-	-	-	-	-	-	-	-	-	-	-
Non-specified	-	-	-	-	-	-	-	-	-	-	-
OTHER SECTORS	-	**116**	-	-	-	-	-	**8**	-	-	-
Agriculture	-	-	-	-	-	-	-	3	-	-	-
Comm. and Publ. Services	-	-	-	-	-	-	-	5	-	-	-
Residential	-	116	-	-	-	-	-	-	-	-	-
Non-specified	-	-	-	-	-	-	-	-	-	-	-
NON-ENERGY USE	-	-	-	-	-	-	-	-	-	-	-
in Industry/Trans./Energy	-	-	-	-	-	-	-	-	-	-	-
in Transport	-	-	-	-	-	-	-	-	-	-	-
in Other Sectors	-	-	-	-	-	-	-	-	-	-	-

Colombia / Colombie : 1998

	Oil cont. / Pétrole cont. (1000 tonnes)										
SUPPLY AND CONSUMPTION *APPROVISIONNEMENT ET DEMANDE*	Refinery Gas *Gaz de raffinerie*	LPG + Ethane *GPL + éthane*	Motor Gasoline *Essence moteur*	Aviation Gasoline *Essence aviation*	Jet Fuel *Carbu-réacteurs*	Kerosene *Kérosène*	Gas/ Diesel *Gazole*	Heavy Fuel Oil *Fioul lourd*	Naphtha *Naphta*	Petrol. Coke *Coke de pétrole*	Other Prod. *Autres prod.*
Production	384	646	4578	20	907	120	3111	2817	-	-	1693
From Other Sources	-	-	-	-	-	-	-	-	-	-	-
Imports	-	2	1233	-	8	-	1	7	-	-	-
Exports	-	-3	-	-	-88	-	-368	-2449	-	-	-
Intl. Marine Bunkers	-	-	-	-	-	-	-176	-25	-	-	-
Stock Changes	-	-113	-131	-	32	-	107	-129	-	-	176
DOMESTIC SUPPLY	**384**	**532**	**5680**	**20**	**859**	**120**	**2675**	**221**	**-**	**-**	**1869**
Transfers	-	217	57	-	-	-	-	-	-	-	-
Statistical Differences	-	-10	-	-	-157	-	128	-2	-	-	-
TRANSFORMATION	**-**	**-**	**-**	**-**	**-**	**-**	**29**	**19**	**-**	**-**	**-**
Electricity Plants	-	-	-	-	-	-	29	19	-	-	-
CHP Plants	-	-	-	-	-	-	-	-	-	-	-
Heat Plants	-	-	-	-	-	-	-	-	-	-	-
Blast Furnaces/Gas Works	-	-	-	-	-	-	-	-	-	-	-
Coke/Pat. Fuel/BKB Plants	-	-	-	-	-	-	-	-	-	-	-
Petroleum Refineries	-	-	-	-	-	-	-	-	-	-	-
Petrochemical Industry	-	-	-	-	-	-	-	-	-	-	-
Liquefaction	-	-	-	-	-	-	-	-	-	-	-
Other Transform. Sector	-	-	-	-	-	-	-	-	-	-	-
ENERGY SECTOR	**384**	**5**	**7**	**-**	**-**	**1**	**15**	**49**	**-**	**-**	**-**
Coal Mines	-	-	-	-	-	-	-	-	-	-	-
Oil and Gas Extraction	-	-	-	-	-	-	-	-	-	-	-
Petroleum Refineries	384	5	7	-	-	1	15	49	-	-	-
Electr., CHP+Heat Plants	-	-	-	-	-	-	-	-	-	-	-
Pumped Storage (Elec.)	-	-	-	-	-	-	-	-	-	-	-
Other Energy Sector	-	-	-	-	-	-	-	-	-	-	-
Distribution Losses	-	-	-	-	-	-	-	-	-	-	-
FINAL CONSUMPTION	**-**	**734**	**5730**	**20**	**702**	**119**	**2759**	**151**	**-**	**-**	**1869**
INDUSTRY SECTOR	**-**	**77**	**52**	**-**	**-**	**68**	**465**	**143**	**-**	**-**	**-**
Iron and Steel	-	12	-	-	-	8	-	-	-	-	-
Chemical and Petrochem.	-	6	-	-	-	10	48	4	-	-	-
of which: Feedstocks	-	-	-	-	-	-	-	-	-	-	-
Non-Ferrous Metals	-	-	-	-	-	-	-	-	-	-	-
Non-Metallic Minerals	-	19	-	-	-	39	85	-	-	-	-
Transport Equipment	-	-	-	-	-	-	-	-	-	-	-
Machinery	-	2	-	-	-	-	17	101	-	-	-
Mining and Quarrying	-	5	-	-	-	-	-	-	-	-	-
Food and Tobacco	-	19	-	-	-	4	130	14	-	-	-
Paper, Pulp and Print	-	4	-	-	-	1	19	18	-	-	-
Wood and Wood Products	-	-	-	-	-	1	2	-	-	-	-
Construction	-	-	52	-	-	-	4	1	-	-	-
Textile and Leather	-	5	-	-	-	5	30	5	-	-	-
Non-specified	-	5	-	-	-	-	130	-	-	-	-
TRANSPORT SECTOR	**-**	**-**	**5435**	**20**	**702**	**-**	**1487**	**3**	**-**	**-**	**-**
Air	-	-	-	20	702	-	-	-	-	-	-
Road	-	-	5354	-	-	-	1393	-	-	-	-
Rail	-	-	-	-	-	-	14	3	-	-	-
Pipeline Transport	-	-	-	-	-	-	-	-	-	-	-
Internal Navigation	-	-	81	-	-	-	80	-	-	-	-
Non-specified	-	-	-	-	-	-	-	-	-	-	-
OTHER SECTORS	**-**	**657**	**243**	**-**	**-**	**51**	**807**	**5**	**-**	**-**	**-**
Agriculture	-	-	4	-	-	1	464	2	-	-	-
Comm. and Publ. Services	-	57	-	-	-	-	327	3	-	-	-
Residential	-	600	78	-	-	50	-	-	-	-	-
Non-specified	-	-	161	-	-	-	16	-	-	-	-
NON-ENERGY USE	**-**	**-**	**-**	**-**	**-**	**-**	**-**	**-**	**-**	**-**	**1869**
in Industry/Transf./Energy	-	-	-	-	-	-	-	-	-	-	1869
in Transport	-	-	-	-	-	-	-	-	-	-	-
in Other Sectors	-	-	-	-	-	-	-	-	-	-	-

Colombia / Colombie : 1998

SUPPLY AND CONSUMPTION / APPROVISIONNEMENT ET DEMANDE	Gas / Gaz (TJ)				Comb. Renew. & Waste / En. Re. Comb. & Déchets (TJ)				(GWh)	(TJ)
	Natural Gas / Gaz naturel	Gas Works / Usines à gaz	Coke Ovens / Cokeries	Blast Furnaces / Hauts fourneaux	Solid Biomass / Biomasse solide	Gas/Liquids from Biomass / Gaz/Liquides tirés de biomasse	Municipal Waste / Déchets urbains	Industrial Waste / Déchets industriels	Electricity / Electricité	Heat / Chaleur
Production	259886	-	2151	2691	213364	-	-	-	45946	-
From Other Sources	-	-	-	-	-	-	-	-	-	-
Imports	-	-	-	-	-	-	-	-	85	-
Exports	-	-	-	-	-	-	-	-	-64	-
Intl. Marine Bunkers	-	-	-	-	-	-	-	-	-	-
Stock Changes	-	-	-	-	-	-	-	-	-	-
DOMESTIC SUPPLY	**259886**	**-**	**2151**	**2691**	**213364**	**-**	**-**	**-**	**45967**	**-**
Transfers	-	-	-	-	-	-	-	-	-	-
Statistical Differences	-28375	-	-	-	1	-	-	-	70	-
TRANSFORMATION	**107410**	**-**	**-**	**1990**	**12165**	**-**	**-**	**-**	**-**	**-**
Electricity Plants	107410	-	-	1990	5553	-	-	-	-	-
CHP Plants	-	-	-	-	-	-	-	-	-	-
Heat Plants	-	-	-	-	-	-	-	-	-	-
Blast Furnaces/Gas Works	-	-	-	-	-	-	-	-	-	-
Coke/Pat. Fuel/BKB Plants	-	-	-	-	-	-	-	-	-	-
Petroleum Refineries	-	-	-	-	-	-	-	-	-	-
Petrochemical Industry	-	-	-	-	-	-	-	-	-	-
Liquefaction	-	-	-	-	-	-	-	-	-	-
Other Transform. Sector	-	-	-	-	6612	-	-	-	-	-
ENERGY SECTOR	**58654**	**-**	**1769**	**-**	**-**	**-**	**-**	**-**	**879**	**-**
Coal Mines	-	-	-	-	-	-	-	-	-	-
Oil and Gas Extraction	18017	-	-	-	-	-	-	-	-	-
Petroleum Refineries	40637	-	-	-	-	-	-	-	-	-
Electr., CHP+Heat Plants	-	-	-	-	-	-	-	-	879	-
Pumped Storage (Elec.)	-	-	-	-	-	-	-	-	-	-
Other Energy Sector	-	-	1769	-	-	-	-	-	-	-
Distribution Losses	-	-	-	282	-	-	-	-	9829	-
FINAL CONSUMPTION	**65447**	**-**	**382**	**419**	**201200**	**-**	**-**	**-**	**35329**	**-**
INDUSTRY SECTOR	**42376**	**-**	**382**	**419**	**73208**	**-**	**-**	**-**	**10267**	**-**
Iron and Steel	1224	-	382	419	1	-	-	-	1835	-
Chemical and Petrochem.	18067	-	-	-	1023	-	-	-	1661	-
of which: Feedstocks	-	-	-	-	-	-	-	-	-	-
Non-Ferrous Metals	-	-	-	-	-	-	-	-	-	-
Non-Metallic Minerals	16607	-	-	-	672	-	-	-	1416	-
Transport Equipment	-	-	-	-	-	-	-	-	-	-
Machinery	-	-	-	-	-	-	-	-	466	-
Mining and Quarrying	-	-	-	-	-	-	-	-	-	-
Food and Tobacco	2713	-	-	-	62965	-	-	-	1940	-
Paper, Pulp and Print	2075	-	-	-	8499	-	-	-	855	-
Wood and Wood Products	751	-	-	-	-	-	-	-	137	-
Construction	-	-	-	-	-	-	-	-	55	-
Textile and Leather	324	-	-	-	48	-	-	-	1566	-
Non-specified	615	-	-	-	-	-	-	-	336	-
TRANSPORT SECTOR	**2118**	**-**	**-**	**-**	**-**	**-**	**-**	**-**	**42**	**-**
Air	-	-	-	-	-	-	-	-	-	-
Road	-	-	-	-	-	-	-	-	42	-
Rail	-	-	-	-	-	-	-	-	-	-
Pipeline Transport	-	-	-	-	-	-	-	-	-	-
Internal Navigation	-	-	-	-	-	-	-	-	-	-
Non-specified	2118	-	-	-	-	-	-	-	-	-
OTHER SECTORS	**20953**	**-**	**-**	**-**	**127992**	**-**	**-**	**-**	**25020**	**-**
Agriculture	-	-	-	-	40882	-	-	-	1250	-
Comm. and Publ. Services	3143	-	-	-	-	-	-	-	6535	-
Residential	17810	-	-	-	86376	-	-	-	14692	-
Non-specified	-	-	-	-	734	-	-	-	2543	-
NON-ENERGY USE	**-**	**-**	**-**	**-**	**-**	**-**	**-**	**-**	**-**	**-**
in Industry/Transf./Energy	-	-	-	-	-	-	-	-	-	-
in Transport	-	-	-	-	-	-	-	-	-	-
in Other Sectors	-	-	-	-	-	-	-	-	-	-

Congo

SUPPLY AND CONSUMPTION 1997	Coal (1000 tonnes)							Oil (1000 tonnes)			
	Coking Coal	Other Bit. Coal	Sub-Bit. Coal	Lignite	Peat	Oven and Gas Coke	Pat. Fuel and BKB	Crude Oil	NGL	Feed-stocks	Additives
Production	-	-	-	-	-	-	-	12330	165	-	-
Imports	-	-	-	-	-	-	-	-	-	-	-
Exports	-	-	-	-	-	-	-	-11720	-	-	-
Intl. Marine Bunkers	-	-	-	-	-	-	-	-	-	-	-
Stock Changes	-	-	-	-	-	-	-	-	-	-	-
DOMESTIC SUPPLY	-	-	-	-	-	-	-	610	165	-	-
Transfers and Stat. Diff.	-	-	-	-	-	-	-	65	-165	-	-
TRANSFORMATION	-	-	-	-	-	-	-	675	-	-	-
Electricity and CHP Plants	-	-	-	-	-	-	-	-	-	-	-
Petroleum Refineries	-	-	-	-	-	-	-	675	-	-	-
Other Transform. Sector	-	-	-	-	-	-	-	-	-	-	-
ENERGY SECTOR	-	-	-	-	-	-	-	-	-	-	-
DISTRIBUTION LOSSES	-	-	-	-	-	-	-	-	-	-	-
FINAL CONSUMPTION	-	-	-	-	-	-	-	-	-	-	-
INDUSTRY SECTOR	-	-	-	-	-	-	-	-	-	-	-
Iron and Steel	-	-	-	-	-	-	-	-	-	-	-
Chemical and Petrochem.	-	-	-	-	-	-	-	-	-	-	-
Non-Metallic Minerals	-	-	-	-	-	-	-	-	-	-	-
Non-specified	-	-	-	-	-	-	-	-	-	-	-
TRANSPORT SECTOR	-	-	-	-	-	-	-	-	-	-	-
Air	-	-	-	-	-	-	-	-	-	-	-
Road	-	-	-	-	-	-	-	-	-	-	-
Non-specified	-	-	-	-	-	-	-	-	-	-	-
OTHER SECTORS	-	-	-	-	-	-	-	-	-	-	-
Agriculture	-	-	-	-	-	-	-	-	-	-	-
Comm. and Publ. Services	-	-	-	-	-	-	-	-	-	-	-
Residential	-	-	-	-	-	-	-	-	-	-	-
Non-specified	-	-	-	-	-	-	-	-	-	-	-
NON-ENERGY USE	-	-	-	-	-	-	-	-	-	-	-

APPROVISIONNEMENT ET DEMANDE 1998	Charbon (1000 tonnes)							Pétrole (1000 tonnes)			
	Charbon à coke	Autres charb. bit.	Charbon sous-bit.	Lignite	Tourbe	Coke de four/gaz	Agg./briq. de lignite	Pétrole brut	LGN	Produits d'aliment.	Additifs
Production	-	-	-	-	-	-	-	12915	165	-	-
Imports	-	-	-	-	-	-	-	-	-	-	-
Exports	-	-	-	-	-	-	-	-12498	-	-	-
Intl. Marine Bunkers	-	-	-	-	-	-	-	-	-	-	-
Stock Changes	-	-	-	-	-	-	-	-417	-	-	-
DOMESTIC SUPPLY	-	-	-	-	-	-	-	-	165	-	-
Transfers and Stat. Diff.	-	-	-	-	-	-	-	-	-165	-	-
TRANSFORMATION	-	-	-	-	-	-	-	-	-	-	-
Electricity and CHP Plants	-	-	-	-	-	-	-	-	-	-	-
Petroleum Refineries	-	-	-	-	-	-	-	-	-	-	-
Other Transform. Sector	-	-	-	-	-	-	-	-	-	-	-
ENERGY SECTOR	-	-	-	-	-	-	-	-	-	-	-
DISTRIBUTION LOSSES	-	-	-	-	-	-	-	-	-	-	-
FINAL CONSUMPTION	-	-	-	-	-	-	-	-	-	-	-
INDUSTRY SECTOR	-	-	-	-	-	-	-	-	-	-	-
Iron and Steel	-	-	-	-	-	-	-	-	-	-	-
Chemical and Petrochem.	-	-	-	-	-	-	-	-	-	-	-
Non-Metallic Minerals	-	-	-	-	-	-	-	-	-	-	-
Non-specified	-	-	-	-	-	-	-	-	-	-	-
TRANSPORT SECTOR	-	-	-	-	-	-	-	-	-	-	-
Air	-	-	-	-	-	-	-	-	-	-	-
Road	-	-	-	-	-	-	-	-	-	-	-
Non-specified	-	-	-	-	-	-	-	-	-	-	-
OTHER SECTORS	-	-	-	-	-	-	-	-	-	-	-
Agriculture	-	-	-	-	-	-	-	-	-	-	-
Comm. and Publ. Services	-	-	-	-	-	-	-	-	-	-	-
Residential	-	-	-	-	-	-	-	-	-	-	-
Non-specified	-	-	-	-	-	-	-	-	-	-	-
NON-ENERGY USE	-	-	-	-	-	-	-	-	-	-	-

Congo

SUPPLY AND CONSUMPTION 1997	Oil cont. (1000 tonnes)										
	Refinery Gas	LPG + Ethane	Motor Gasoline	Aviation Gasoline	Jet Fuel	Kerosene	Gas/ Diesel	Heavy Fuel Oil	Naphtha	Petrol. Coke	Other Prod.
Production	-	4	55	57	16	50	92	340	-	-	11
Imports	-	-	-	-	-	-	-	5	-	-	10
Exports	-	-	-3	-	-4	-	-10	-293	-	-	-
Intl. Marine Bunkers	-	-	-	-	-	-	-	-10	-	-	-
Stock Changes	-	-	-	-	-	-	-	-	-	-	-
DOMESTIC SUPPLY	-	4	52	57	12	50	82	42	-	-	21
Transfers and Stat. Diff.	-	-1	1	-	-	-2	-	-	-	-	-1
TRANSFORMATION	-	-	-	-	-	-	1	-	-	-	-
Electricity and CHP Plants	-	-	-	-	-	-	1	-	-	-	-
Petroleum Refineries	-	-	-	-	-	-	-	-	-	-	-
Other Transform. Sector	-	-	-	-	-	-	-	-	-	-	-
ENERGY SECTOR	-	-	-	-	-	-	-	38	-	-	-
DISTRIBUTION LOSSES	-	-	-	-	-	-	-	-	-	-	-
FINAL CONSUMPTION	-	3	53	57	12	48	81	4	-	-	20
INDUSTRY SECTOR	-	-	-	-	-	-	-	4	-	-	-
Iron and Steel	-	-	-	-	-	-	-	-	-	-	-
Chemical and Petrochem.	-	-	-	-	-	-	-	-	-	-	-
Non-Metallic Minerals	-	-	-	-	-	-	-	-	-	-	-
Non-specified	-	-	-	-	-	-	-	4	-	-	-
TRANSPORT SECTOR	-	-	53	57	12	-	81	-	-	-	-
Air	-	-	-	57	12	-	-	-	-	-	-
Road	-	-	53	-	-	-	81	-	-	-	-
Non-specified	-	-	-	-	-	-	-	-	-	-	-
OTHER SECTORS	-	3	-	-	-	48	-	-	-	-	-
Agriculture	-	-	-	-	-	-	-	-	-	-	-
Comm. and Publ. Services	-	-	-	-	-	-	-	-	-	-	-
Residential	-	3	-	-	-	48	-	-	-	-	-
Non-specified	-	-	-	-	-	-	-	-	-	-	-
NON-ENERGY USE	-	-	-	-	-	-	-	-	-	-	20

APPROVISIONNEMENT ET DEMANDE 1998	Pétrole cont. (1000 tonnes)										
	Gaz de raffinerie	GPL + éthane	Essence moteur	Essence aviation	Carbu- réacteurs	Kérosène	Gazole	Fioul lourd	Naphta	Coke de pétrole	Autres prod.
Production	-	-	-	-	-	-	-	-	-	-	-
Imports	-	4	52	57	12	50	61	50	-	-	20
Exports	-	-	-	-	-	-	-	-	-	-	-
Intl. Marine Bunkers	-	-	-	-	-	-	-	-46	-	-	-
Stock Changes	-	-	-	-	-	-	-	-	-	-	-
DOMESTIC SUPPLY	-	4	52	57	12	50	61	4	-	-	20
Transfers and Stat. Diff.	-	-1	1	-	-	-2	-	-	-	-	-
TRANSFORMATION	-	-	-	-	-	-	1	-	-	-	-
Electricity and CHP Plants	-	-	-	-	-	-	1	-	-	-	-
Petroleum Refineries	-	-	-	-	-	-	-	-	-	-	-
Other Transform. Sector	-	-	-	-	-	-	-	-	-	-	-
ENERGY SECTOR	-	-	-	-	-	-	-	-	-	-	-
DISTRIBUTION LOSSES	-	-	-	-	-	-	-	-	-	-	-
FINAL CONSUMPTION	-	3	53	57	12	48	60	4	-	-	20
INDUSTRY SECTOR	-	-	-	-	-	-	-	4	-	-	-
Iron and Steel	-	-	-	-	-	-	-	-	-	-	-
Chemical and Petrochem.	-	-	-	-	-	-	-	-	-	-	-
Non-Metallic Minerals	-	-	-	-	-	-	-	-	..	-	-
Non-specified	-	-	-	-	-	-	-	4	-	-	-
TRANSPORT SECTOR	-	-	53	57	12	-	60	-	-	-	-
Air	-	-	-	57	12	-	-	-	-	-	-
Road	-	-	53	-	-	-	60	-	-	-	-
Non-specified	-	-	-	-	-	-	-	-	-	-	-
OTHER SECTORS	-	3	-	-	-	48	-	-	-	-	-
Agriculture	-	-	-	-	-	-	-	-	-	-	-
Comm. and Publ. Services	-	-	-	-	-	-	-	-	-	-	-
Residential	-	3	-	-	-	48	-	-	-	-	-
Non-specified	-	-	-	-	-	-	-	-	-	-	-
NON-ENERGY USE	-	-	-	-	-	-	-	-	-	-	20

Congo

SUPPLY AND CONSUMPTION 1997	Gas (TJ)				Comb. Renew. & Waste (TJ)				(GWh)	(TJ)
	Natural Gas	Gas Works	Coke Ovens	Blast Furnaces	Solid Biomass	Gas/Liquids from Biomass	Municipal Waste	Industrial Waste	Electricity	Heat
Production	131	-	-	-	36284	-	-	-	453	-
Imports	-	-	-	-	-	-	-	-	38	-
Exports	-	-	-	-	-	-	-	-	-	-
Intl. Marine Bunkers	-	-	-	-	-	-	-	-	-	-
Stock Changes	-	-	-	-	-	-	-	-	-	-
DOMESTIC SUPPLY	131	-	-	-	36284	-	-	-	491	-
Transfers and Stat. Diff.	-	-	-	-	-1	-	-	-	-3	-
TRANSFORMATION	131	-	-	-	1420	-	-	-	-	-
Electricity and CHP Plants	131	-	-	-	-	-	-	-	-	-
Petroleum Refineries	-	-	-	-	-	-	-	-	-	-
Other Transform. Sector	-	-	-	-	1420	-	-	-	-	-
ENERGY SECTOR	-	-	-	-	-	-	-	-	92	-
DISTRIBUTION LOSSES	-	-	-	-	-	-	-	-	110	-
FINAL CONSUMPTION	-	-	-	-	34864	-	-	-	286	-
INDUSTRY SECTOR	-	-	-	-	5784	-	-	-	152	-
Iron and Steel	-	-	-	-	-	-	-	-	-	-
Chemical and Petrochem.	-	-	-	-	-	-	-	-	-	-
Non-Metallic Minerals	-	-	-	-	-	-	-	-	-	-
Non-specified	-	-	-	-	5784	-	-	-	152	-
TRANSPORT SECTOR	-	-	-	-	-	-	-	-	-	-
Air	-	-	-	-	-	-	-	-	-	-
Road	-	-	-	-	-	-	-	-	-	-
Non-specified	-	-	-	-	-	-	-	-	-	-
OTHER SECTORS	-	-	-	-	29080	-	-	-	134	-
Agriculture	-	-	-	-	-	-	-	-	-	-
Comm. and Publ. Services	-	-	-	-	-	-	-	-	-	-
Residential	-	-	-	-	29080	-	-	-	134	-
Non-specified	-	-	-	-	-	-	-	-	-	-
NON-ENERGY USE	-	-	-	-	-	-	-	-	-	-

APPROVISIONNEMENT ET DEMANDE 1998	Gaz (TJ)				En. Re. Comb. & Déchets (TJ)				(GWh)	(TJ)
	Gaz naturel	Usines à gaz	Cokeries	Hauts fourneaux	Biomasse solide	Gaz/Liquides tirés de biomasse	Déchets urbains	Déchets industriels	Electricité	Chaleur
Production	131	-	-	-	37288	-	-	-	344	-
Imports	-	-	-	-	-	-	-	-	126	-
Exports	-	-	-	-	-	-	-	-	-	-
Intl. Marine Bunkers	-	-	-	-	-	-	-	-	-	-
Stock Changes	-	-	-	-	-	-	-	-	-	-
DOMESTIC SUPPLY	131	-	-	-	37288	-	-	-	470	-
Transfers and Stat. Diff.	-	-	-	-	-2	-	-	-	-2	-
TRANSFORMATION	131	-	-	-	1459	-	-	-	-	-
Electricity and CHP Plants	131	-	-	-	-	-	-	-	-	-
Petroleum Refineries	-	-	-	-	-	-	-	-	-	-
Other Transform. Sector	-	-	-	-	1459	-	-	-	-	-
ENERGY SECTOR	-	-	-	-	-	-	-	-	108	-
DISTRIBUTION LOSSES	-	-	-	-	-	-	-	-	129	-
FINAL CONSUMPTION	-	-	-	-	35827	-	-	-	231	-
INDUSTRY SECTOR	-	-	-	-	5944	-	-	-	127	-
Iron and Steel	-	-	-	-	-	-	-	-	-	-
Chemical and Petrochem.	-	-	-	-	-	-	-	-	-	-
Non-Metallic Minerals	-	-	-	-	-	-	-	-	-	-
Non-specified	-	-	-	-	5944	-	-	-	127	-
TRANSPORT SECTOR	-	-	-	-	-	-	-	-	-	-
Air	-	-	-	-	-	-	-	-	-	-
Road	-	-	-	-	-	-	-	-	-	-
Non-specified	-	-	-	-	-	-	-	-	-	-
OTHER SECTORS	-	-	-	-	29883	-	-	-	104	-
Agriculture	-	-	-	-	-	-	-	-	-	-
Comm. and Publ. Services	-	-	-	-	-	-	-	-	-	-
Residential	-	-	-	-	29883	-	-	-	104	-
Non-specified	-	-	-	-	-	-	-	-	-	-
NON-ENERGY USE	-	-	-	-	-	-	-	-	-	-

Democratic Republic of Congo / République démocratique du Congo

SUPPLY AND CONSUMPTION 1997	Coking Coal	Other Bit. Coal	Sub-Bit. Coal	Lignite	Peat	Oven and Gas Coke	Pat. Fuel and BKB	Crude Oil	NGL	Feed-stocks	Additives
Production	-	93	-	-	-	-	-	1306	-	-	-
Imports	-	43	-	-	-	153	85	119	-	-	-
Exports	-	-	-	-	-	-	-	-1014	-	-	-
Intl. Marine Bunkers	-	-	-	-	-	-	-	-	-	-	-
Stock Changes	-	-	-	-	-	-	-	-	-	-	-
DOMESTIC SUPPLY	-	136	-	-	-	153	85	411	-	-	-
Transfers and Stat. Diff.	-	-	-	-	-	-	-	-47	-	-	-
TRANSFORMATION	-	-	-	-	-	122	-	364	-	-	-
Electricity and CHP Plants	-	-	-	-	-	-	-	-	-	-	-
Petroleum Refineries	-	-	-	-	-	-	-	364	-	-	-
Other Transform. Sector	-	-	-	-	-	122	-	-	-	-	-
ENERGY SECTOR	-	-	-	-	-	-	-	-	-	-	-
DISTRIBUTION LOSSES	-	-	-	-	-	-	-	-	-	-	-
FINAL CONSUMPTION	-	136	-	-	-	31	85	-	-	-	-
INDUSTRY SECTOR	-	136	-	-	-	31	-	-	-	-	-
Iron and Steel	-	-	-	-	-	31	-	-	-	-	-
Chemical and Petrochem.	-	-	-	-	-	-	-	-	-	-	-
Non-Metallic Minerals	-	-	-	-	-	-	-	-	-	-	-
Non-specified	-	136	-	-	-	-	-	-	-	-	-
TRANSPORT SECTOR	-	-	-	-	-	-	-	-	-	-	-
Air	-	-	-	-	-	-	-	-	-	-	-
Road	-	-	-	-	-	-	-	-	-	-	-
Non-specified	-	-	-	-	-	-	-	-	-	-	-
OTHER SECTORS	-	-	-	-	-	-	85	-	-	-	-
Agriculture	-	-	-	-	-	-	-	-	-	-	-
Comm. and Publ. Services	-	-	-	-	-	-	-	-	-	-	-
Residential	-	-	-	-	-	-	85	-	-	-	-
Non-specified	-	-	-	-	-	-	-	-	-	-	-
NON-ENERGY USE	-	-	-	-	-	-	-	-	-	-	-

APPROVISIONNEMENT ET DEMANDE 1998	Charbon à coke	Autres charb. bit.	Charbon sous-bit.	Lignite	Tourbe	Coke de four/gaz	Agg./briq. de lignite	Pétrole brut	LGN	Produits d'aliment.	Additifs
Production	-	88	-	-	-	-	-	1232	-	-	-
Imports	-	40	-	-	-	144	80	112	-	-	-
Exports	-	-	-	-	-	-	-	-956	-	-	-
Intl. Marine Bunkers	-	-	-	-	-	-	-	-	-	-	-
Stock Changes	-	-	-	-	-	-	-	-	-	-	-
DOMESTIC SUPPLY	-	128	-	-	-	144	80	388	-	-	-
Transfers and Stat. Diff.	-	-	-	-	-	-	-	-45	-	-	-
TRANSFORMATION	-	-	-	-	-	115	-	343	-	-	-
Electricity and CHP Plants	-	-	-	-	-	-	-	-	-	-	-
Petroleum Refineries	-	-	-	-	-	-	-	343	-	-	-
Other Transform. Sector	-	-	-	-	-	115	-	-	-	-	-
ENERGY SECTOR	-	-	-	-	-	-	-	-	-	-	-
DISTRIBUTION LOSSES	-	-	-	-	-	-	-	-	-	-	-
FINAL CONSUMPTION	-	128	-	-	-	29	80	-	-	-	-
INDUSTRY SECTOR	-	128	-	-	-	29	-	-	-	-	-
Iron and Steel	-	-	-	-	-	29	-	-	-	-	-
Chemical and Petrochem.	-	-	-	-	-	-	-	-	-	-	-
Non-Metallic Minerals	-	-	-	-	-	-	-	-	-	-	-
Non-specified	-	128	-	-	-	-	-	-	-	-	-
TRANSPORT SECTOR	-	-	-	-	-	-	-	-	-	-	-
Air	-	-	-	-	-	-	-	-	-	-	-
Road	-	-	-	-	-	-	-	-	-	-	-
Non-specified	-	-	-	-	-	-	-	-	-	-	-
OTHER SECTORS	-	-	-	-	-	-	80	-	-	-	-
Agriculture	-	-	-	-	-	-	-	-	-	-	-
Comm. and Publ. Services	-	-	-	-	-	-	-	-	-	-	-
Residential	-	-	-	-	-	-	80	-	-	-	-
Non-specified	-	-	-	-	-	-	-	-	-	-	-
NON-ENERGY USE	-	-	-	-	-	-	-	-	-	-	-

Democratic Republic of Congo / République démocratique du Congo

SUPPLY AND CONSUMPTION 1997	Oil cont. (1000 tonnes)										
	Refinery Gas	LPG + Ethane	Motor Gasoline	Aviation Gasoline	Jet Fuel	Kerosene	Gas/ Diesel	Heavy Fuel Oil	Naphtha	Petrol. Coke	Other Prod.
Production	7	-	54	-	23	47	65	134	-	-	12
Imports	-	12	181	21	124	66	272	212	-	-	47
Exports	-	-	-	-	-	-	-	-26	-	-	-
Intl. Marine Bunkers	-	-	-	-	-	-	-	-35	-	-	-
Stock Changes	-	1	-	-	-	-	-	-	-	-	-
DOMESTIC SUPPLY	7	13	235	21	147	113	337	285	-	-	59
Transfers and Stat. Diff.	-	-	-	-3	-6	-	4	4	-	-	-
TRANSFORMATION	-	-	-	-	-	-	30	16	-	-	-
Electricity and CHP Plants	-	-	-	-	-	-	30	16	-	-	-
Petroleum Refineries	-	-	-	-	-	-	-	-	-	-	-
Other Transform. Sector	-	-	-	-	-	-	-	-	-	-	-
ENERGY SECTOR	7	-	-	-	-	-	-	9	-	-	-
DISTRIBUTION LOSSES	-	-	-	-	-	-	-	-	-	-	-
FINAL CONSUMPTION	-	13	235	18	141	113	311	264	-	-	59
INDUSTRY SECTOR	-	2	-	-	-	-	-	56	-	-	-
Iron and Steel	-	-	-	-	-	-	-	-	-	-	-
Chemical and Petrochem.	-	-	-	-	-	-	-	-	-	-	-
Non-Metallic Minerals	-	-	-	-	-	-	-	-	-	-	-
Non-specified	-	2	-	-	-	-	-	56	-	-	-
TRANSPORT SECTOR	-	-	235	18	141	-	311	-	-	-	-
Air	-	-	-	18	141	-	-	-	-	-	-
Road	-	-	235	-	-	-	311	-	-	-	-
Non-specified	-	-	-	-	-	-	-	-	-	-	-
OTHER SECTORS	-	11	-	-	-	113	-	208	-	-	-
Agriculture	-	-	-	-	-	-	-	-	-	-	-
Comm. and Publ. Services	-	-	-	-	-	-	-	-	-	-	-
Residential	-	11	-	-	-	113	-	-	-	-	-
Non-specified	-	-	-	-	-	-	-	208	-	-	-
NON-ENERGY USE	-	-	-	-	-	-	-	-	-	-	59

APPROVISIONNEMENT ET DEMANDE 1998	Pétrole cont. (1000 tonnes)										
	Gaz de raffinerie	GPL + éthane	Essence moteur	Essence aviation	Carbu- réacteurs	Kérosène	Gazole	Fioul lourd	Naphta	Coke de pétrole	Autres prod.
Production	7	-	51	-	22	44	61	126	-	-	11
Imports	-	15	223	77	129	113	316	250	-	-	64
Exports	-	-4	-52	-57	-12	-50	-60	-75	-	-	-20
Intl. Marine Bunkers	-	-	-	-	-	-	-	-33	-	-	-
Stock Changes	-	1	-	-	-	-	-	-	-	-	-
DOMESTIC SUPPLY	7	12	222	20	139	107	317	268	-	-	55
Transfers and Stat. Diff.	-	-	-	-3	-6	-	4	4	-	-	-
TRANSFORMATION	-	-	-	-	-	-	28	15	-	-	-
Electricity and CHP Plants	-	-	-	-	-	-	28	15	-	-	-
Petroleum Refineries	-	-	-	-	-	-	-	-	-	-	-
Other Transform. Sector	-	-	-	-	-	-	-	-	-	-	-
ENERGY SECTOR	7	-	-	-	-	-	-	8	-	-	-
DISTRIBUTION LOSSES	-	-	-	-	-	-	-	-	-	-	-
FINAL CONSUMPTION	-	12	222	17	133	107	293	249	-	-	55
INDUSTRY SECTOR	-	2	-	-	-	-	-	53	-	-	-
Iron and Steel	-	-	-	-	-	-	-	-	-	-	-
Chemical and Petrochem.	-	-	-	-	-	-	-	-	-	-	-
Non-Metallic Minerals	-	-	-	-	-	-	-	-	-	-	-
Non-specified	-	2	-	-	-	-	-	53	-	-	-
TRANSPORT SECTOR	-	-	222	17	133	-	293	-	-	-	-
Air	-	-	-	17	133	-	-	-	-	-	-
Road	-	-	222	-	-	-	293	-	-	-	-
Non-specified	-	-	-	-	-	-	-	-	-	-	-
OTHER SECTORS	-	10	-	-	-	107	-	196	-	-	-
Agriculture	-	-	-	-	-	-	-	-	-	-	-
Comm. and Publ. Services	-	-	-	-	-	-	-	-	-	-	-
Residential	-	10	-	-	-	107	-	-	-	-	-
Non-specified	-	-	-	-	-	-	-	196	-	-	-
NON-ENERGY USE	-	-	-	-	-	-	-	-	-	-	55

Democratic Republic of Congo / République démocratique du Congo

SUPPLY AND CONSUMPTION 1997	Gas (TJ)				Comb. Renew. & Waste (TJ)				(GWh)	(TJ)
	Natural Gas	Gas Works	Coke Ovens	Blast Furnaces	Solid Biomass	Gas/Liquids from Biomass	Municipal Waste	Industrial Waste	Electricity	Heat
Production	-	-	-	1328	522927	-	-	-	6010	-
Imports	-	-	-	-	-	-	-	-	59	-
Exports	-	-	-	-	-	-	-	-	-199	-
Intl. Marine Bunkers	-	-	-	-	-	-	-	-	-	-
Stock Changes	-	-	-	-	-	-	-	-	-	-
DOMESTIC SUPPLY	-	-	-	1328	522927	-	-	-	5870	-
Transfers and Stat. Diff.	-	-	-	-	-	-	-	-	-24	-
TRANSFORMATION	-	-	-	-	23963	-	-	-	-	-
Electricity and CHP Plants	-	-	-	-	-	-	-	-	-	-
Petroleum Refineries	-	-	-	-	-	-	-	-	-	-
Other Transform. Sector	-	-	-	-	23963	-	-	-	-	-
ENERGY SECTOR	-	-	-	-	-	-	-	-	24	-
DISTRIBUTION LOSSES	-	-	-	-	-	-	-	-	197	-
FINAL CONSUMPTION	-	-	-	1328	498964	-	-	-	5625	-
INDUSTRY SECTOR	-	-	-	1328	105437	-	-	-	3650	-
Iron and Steel	-	-	-	1328	-	-	-	-	-	-
Chemical and Petrochem.	-	-	-	-	-	-	-	-	-	-
Non-Metallic Minerals	-	-	-	-	-	-	-	-	-	-
Non-specified	-	-	-	-	105437	-	-	-	3650	-
TRANSPORT SECTOR	-	-	-	-	-	-	-	-	-	-
Air	-	-	-	-	-	-	-	-	-	-
Road	-	-	-	-	-	-	-	-	-	-
Non-specified	-	-	-	-	-	-	-	-	-	-
OTHER SECTORS	-	-	-	-	393527	-	-	-	1975	-
Agriculture	-	-	-	-	-	-	-	-	-	-
Comm. and Publ. Services	-	-	-	-	-	-	-	-	-	-
Residential	-	-	-	-	393527	-	-	-	1873	-
Non-specified	-	-	-	-	-	-	-	-	102	-
NON-ENERGY USE	-	-	-	-	-	-	-	-	-	-

APPROVISIONNEMENT ET DEMANDE 1998	Gaz (TJ)				En. Re. Comb. & Déchets (TJ)				(GWh)	(TJ)
	Gaz naturel	Usines à gaz	Cokeries	Hauts fourneaux	Biomasse solide	Gaz/Liquides tirés de biomasse	Déchets urbains	Déchets industriels	Electricité	Chaleur
Production	-	-	-	1252	539786	-	-	-	5667	-
Imports	-	-	-	-	-	-	-	-	56	-
Exports	-	-	-	-	-	-	-	-	-188	-
Intl. Marine Bunkers	-	-	-	-	-	-	-	-	-	-
Stock Changes	-	-	-	-	-	-	-	-	-	-
DOMESTIC SUPPLY	-	-	-	1252	539786	-	-	-	5535	-
Transfers and Stat. Diff.	-	-	-	-	-	-	-	-	-22	-
TRANSFORMATION	-	-	-	-	24735	-	-	-	-	-
Electricity and CHP Plants	-	-	-	-	-	-	-	-	-	-
Petroleum Refineries	-	-	-	-	-	-	-	-	-	-
Other Transform. Sector	-	-	-	-	24735	-	-	-	-	-
ENERGY SECTOR	-	-	-	-	-	-	-	-	23	-
DISTRIBUTION LOSSES	-	-	-	-	-	-	-	-	186	-
FINAL CONSUMPTION	-	-	-	1252	515051	-	-	-	5304	-
INDUSTRY SECTOR	-	-	-	1252	108836	-	-	-	3442	-
Iron and Steel	-	-	-	1252	-	-	-	-	-	-
Chemical and Petrochem.	-	-	-	-	-	-	-	-	-	-
Non-Metallic Minerals	-	-	-	-	-	-	-	-	-	-
Non-specified	-	-	-	-	108836	-	-	-	3442	-
TRANSPORT SECTOR	-	-	-	-	-	-	-	-	-	-
Air	-	-	-	-	-	-	-	-	-	-
Road	-	-	-	-	-	-	-	-	-	-
Non-specified	-	-	-	-	-	-	-	-	-	-
OTHER SECTORS	-	-	-	-	406215	-	-	-	1862	-
Agriculture	-	-	-	-	-	-	-	-	-	-
Comm. and Publ. Services	-	-	-	-	-	-	-	-	-	-
Residential	-	-	-	-	406215	-	-	-	1766	-
Non-specified	-	-	-	-	-	-	-	-	96	-
NON-ENERGY USE	-	-	-	-	-	-	-	-	-	-

Costa Rica

SUPPLY AND CONSUMPTION 1997	Coal (1000 tonnes)							Oil (1000 tonnes)			
	Coking Coal	Other Bit. Coal	Sub-Bit. Coal	Lignite	Peat	Oven and Gas Coke	Pat. Fuel and BKB	Crude Oil	NGL	Feed-stocks	Additives
Production	-	-	-	-	-	-	-	-	-	-	-
Imports	-	-	-	-	-	-	-	626	-	-	-
Exports	-	-	-	-	-	-	-	-	-	-	-
Intl. Marine Bunkers	-	-	-	-	-	-	-	-	-	-	-
Stock Changes	-	-	-	-	-	-	-	-9	-	-	-
DOMESTIC SUPPLY	-	-	-	-	-	-	-	617	-	-	-
Transfers and Stat. Diff.	-	-	-	-	-	-	-	6	-	-	-
TRANSFORMATION	-	-	-	-	-	-	-	615	-	-	-
Electricity and CHP Plants	-	-	-	-	-	-	-	-	-	-	-
Petroleum Refineries	-	-	-	-	-	-	-	615	-	-	-
Other Transform. Sector	-	-	-	-	-	-	-	-	-	-	-
ENERGY SECTOR	-	-	-	-	-	-	-	-	-	-	-
DISTRIBUTION LOSSES	-	-	-	-	-	-	-	8	-	-	-
FINAL CONSUMPTION	-	-	-	-	-	-	-	-	-	-	-
INDUSTRY SECTOR	-	-	-	-	-	-	-	-	-	-	-
Iron and Steel	-	-	-	-	-	-	-	-	-	-	-
Chemical and Petrochem.	-	-	-	-	-	-	-	-	-	-	-
Non-Metallic Minerals	-	-	-	-	-	-	-	-	-	-	-
Non-specified	-	-	-	-	-	-	-	-	-	-	-
TRANSPORT SECTOR	-	-	-	-	-	-	-	-	-	-	-
Air	-	-	-	-	-	-	-	-	-	-	-
Road	-	-	-	-	-	-	-	-	-	-	-
Non-specified	-	-	-	-	-	-	-	-	-	-	-
OTHER SECTORS	-	-	-	-	-	-	-	-	-	-	-
Agriculture	-	-	-	-	-	-	-	-	-	-	-
Comm. and Publ. Services	-	-	-	-	-	-	-	-	-	-	-
Residential	-	-	-	-	-	-	-	-	-	-	-
Non-specified	-	-	-	-	-	-	-	-	-	-	-
NON-ENERGY USE	-	-	-	-	-	-	-	-	-	-	-

APPROVISIONNEMENT ET DEMANDE 1998	Charbon (1000 tonnes)							Pétrole (1000 tonnes)			
	Charbon à coke	Autres charb. bit.	Charbon sous-bit.	Lignite	Tourbe	Coke de four/gaz	Agg./briq. de lignite	Pétrole brut	LGN	Produits d'aliment.	Additifs
Production	-	-	-	-	-	-	-	-	-	-	-
Imports	-	-	-	-	-	-	-	39	-	-	-
Exports	-	-	-	-	-	-	-	-	-	-	-
Intl. Marine Bunkers	-	-	-	-	-	-	-	-	-	-	-
Stock Changes	-	-	-	-	-	-	-	10	-	-	-
DOMESTIC SUPPLY	-	-	-	-	-	-	-	49	-	-	-
Transfers and Stat. Diff.	-	-	-	-	-	-	-	8	-	-	-
TRANSFORMATION	-	-	-	-	-	-	-	49	-	-	-
Electricity and CHP Plants	-	-	-	-	-	-	-	-	-	-	-
Petroleum Refineries	-	-	-	-	-	-	-	49	-	-	-
Other Transform. Sector	-	-	-	-	-	-	-	-	-	-	-
ENERGY SECTOR	-	-	-	-	-	-	-	-	-	-	-
DISTRIBUTION LOSSES	-	-	-	-	-	-	-	8	-	-	-
FINAL CONSUMPTION	-	-	-	-	-	-	-	-	-	-	-
INDUSTRY SECTOR	-	-	-	-	-	-	-	-	-	-	-
Iron and Steel	-	-	-	-	-	-	-	-	-	-	-
Chemical and Petrochem.	-	-	-	-	-	-	-	-	-	-	-
Non-Metallic Minerals	-	-	-	-	-	-	-	-	-	-	-
Non-specified	-	-	-	-	-	-	-	-	-	-	-
TRANSPORT SECTOR	-	-	-	-	-	-	-	-	-	-	-
Air	-	-	-	-	-	-	-	-	-	-	-
Road	-	-	-	-	-	-	-	-	-	-	-
Non-specified	-	-	-	-	-	-	-	-	-	-	-
OTHER SECTORS	-	-	-	-	-	-	-	-	-	-	-
Agriculture	-	-	-	-	-	-	-	-	-	-	-
Comm. and Publ. Services	-	-	-	-	-	-	-	-	-	-	-
Residential	-	-	-	-	-	-	-	-	-	-	-
Non-specified	-	-	-	-	-	-	-	-	-	-	-
NON-ENERGY USE	-	-	-	-	-	-	-	-	-	-	-

Costa Rica

SUPPLY AND CONSUMPTION 1997	Oil cont. (1000 tonnes)										
	Refinery Gas	LPG + Ethane	Motor Gasoline	Aviation Gasoline	Jet Fuel	Kerosene	Gas/ Diesel	Heavy Fuel Oil	Naphtha	Petrol. Coke	Other Prod.
Production	-	2	80	-	29	6	171	292	7	-	26
Imports	-	52	398	3	72	-	383	32	-	-	-
Exports	-	-	-	-	-21	-	-10	-83	-	-	-
Intl. Marine Bunkers	-	-	-	-	-	-	-	-	-	-	-
Stock Changes	-	-2	-6	1	3	-1	17	-13	-	-	2
DOMESTIC SUPPLY	-	52	472	4	83	5	561	228	7	-	28
Transfers and Stat. Diff.	-	-2	-25	-	16	3	23	-7	-5	-	-5
TRANSFORMATION	-	-	-	-	-	-	45	15	-	-	-
Electricity and CHP Plants	-	-	-	-	-	-	45	15	-	-	-
Petroleum Refineries	-	-	-	-	-	-	-	-	-	-	-
Other Transform. Sector	-	-	-	-	-	-	-	-	-	-	-
ENERGY SECTOR	-	-	-	-	-	-	1	29	-	-	-
DISTRIBUTION LOSSES	-	-	-	-	-	-	-	-	-	-	-
FINAL CONSUMPTION	-	50	447	4	99	8	538	177	2	-	23
INDUSTRY SECTOR	-	13	-	-	-	2	33	174	2	-	-
Iron and Steel	-	-	-	-	-	-	-	-	-	-	-
Chemical and Petrochem.	-	-	-	-	-	1	11	15	2	-	-
Non-Metallic Minerals	-	-	-	-	-	-	-	-	-	-	-
Non-specified	-	13	-	-	-	1	22	159	-	-	-
TRANSPORT SECTOR	-	-	439	4	99	-	422	-	-	-	-
Air	-	-	-	4	99	-	-	-	-	-	-
Road	-	-	438	-	-	-	397	-	-	-	-
Non-specified	-	-	1	-	-	-	25	-	-	-	-
OTHER SECTORS	-	37	8	-	-	6	83	3	-	-	-
Agriculture	-	-	8	-	-	-	65	2	-	-	-
Comm. and Publ. Services	-	7	-	-	-	4	18	1	-	-	-
Residential	-	30	-	-	-	2	-	-	-	-	-
Non-specified	-	-	-	-	-	-	-	-	-	-	-
NON-ENERGY USE	-	-	-	-	-	-	-	-	-	-	23

APPROVISIONNEMENT ET DEMANDE 1998	Pétrole cont. (1000 tonnes)										
	Gaz de raffinerie	GPL + éthane	Essence moteur	Essence aviation	Carbu- réacteurs	Kérosène	Gazole	Fioul lourd	Naphta	Coke de pétrole	Autres prod.
Production	-	-	4	-	-	1	17	25	1	-	-
Imports	-	60	484	5	129	-	679	200	-	-	32
Exports	-	-	-	-	-21	-	-10	-	-	-	-
Intl. Marine Bunkers	-	-	-	-	-	-	-	-	-	-	-
Stock Changes	-	-1	8	-1	-4	1	5	5	2	-	-1
DOMESTIC SUPPLY	-	59	496	4	104	2	691	230	3	-	31
Transfers and Stat. Diff.	-	-1	6	-	14	6	100	-12	-1	-	-2
TRANSFORMATION	-	-	-	-	-	-	123	9	-	-	-
Electricity and CHP Plants	-	-	-	-	-	-	123	9	-	-	-
Petroleum Refineries	-	-	-	-	-	-	-	-	-	-	-
Other Transform. Sector	-	-	-	-	-	-	-	-	-	-	-
ENERGY SECTOR	-	-	-	-	-	-	1	11	-	-	-
DISTRIBUTION LOSSES	-	-	-	-	-	-	-	-	-	-	-
FINAL CONSUMPTION	-	58	502	4	118	8	667	198	2	-	29
INDUSTRY SECTOR	-	14	1	-	-	2	47	196	2	-	-
Iron and Steel	-	-	-	-	-	-	-	-	-	-	-
Chemical and Petrochem.	-	-	-	-	-	1	16	17	2	-	-
Non-Metallic Minerals	-	-	-	-	-	-	-	-	-	-	-
Non-specified	-	14	1	-	-	1	31	179	-	-	-
TRANSPORT SECTOR	-	-	496	4	118	-	521	-	-	-	-
Air	-	-	-	4	118	-	-	-	-	-	-
Road	-	-	496	-	-	-	490	-	-	-	-
Non-specified	-	-	-	-	-	-	31	-	-	-	-
OTHER SECTORS	-	44	5	-	-	6	99	2	-	-	-
Agriculture	-	-	5	-	-	-	77	2	-	-	-
Comm. and Publ. Services	-	9	-	-	-	4	22	-	-	-	-
Residential	-	35	-	-	-	2	-	-	-	-	-
Non-specified	-	-	-	-	-	-	-	-	-	-	-
NON-ENERGY USE	-	-	-	-	-	-	-	-	-	-	29

Costa Rica

SUPPLY AND CONSUMPTION 1997	Gas (TJ)				Comb. Renew. & Waste (TJ)				(GWh)	(TJ)
	Natural Gas	Gas Works	Coke Ovens	Blast Furnaces	Solid Biomass	Gas/Liquids from Biomass	Municipal Waste	Industrial Waste	Electricity	Heat
Production	-	-	-	-	10912	-	-	-	5635	-
Imports	-	-	-	-	-	-	-	-	107	-
Exports	-	-	-	-	-	-	-	-	-241	-
Intl. Marine Bunkers	-	-	-	-	-	-	-	-	-	-
Stock Changes	-	-	-	-	-	-	-	-	-	-
DOMESTIC SUPPLY	-	-	-	-	10912	-	-	-	5501	-
Transfers and Stat. Diff.	-	-	-	-	-	-	-	-	-377	-
TRANSFORMATION	-	-	-	-	424	-	-	-	-	-
Electricity and CHP Plants	-	-	-	-	172	-	-	-	-	-
Petroleum Refineries	-	-	-	-	-	-	-	-	-	-
Other Transform. Sector	-	-	-	-	252	-	-	-	-	-
ENERGY SECTOR	-	-	-	-	-	-	-	-	18	-
DISTRIBUTION LOSSES	-	-	-	-	-	-	-	-	420	-
FINAL CONSUMPTION	-	-	-	-	10488	-	-	-	4686	-
INDUSTRY SECTOR	-	-	-	-	7249	-	-	-	1096	-
Iron and Steel	-	-	-	-	-	-	-	-	-	-
Chemical and Petrochem.	-	-	-	-	-	-	-	-	219	-
Non-Metallic Minerals	-	-	-	-	-	-	-	-	-	-
Non-specified	-	-	-	-	7249	-	-	-	877	-
TRANSPORT SECTOR	-	-	-	-	-	-	-	-	-	-
Air	-	-	-	-	-	-	-	-	-	-
Road	-	-	-	-	-	-	-	-	-	-
Non-specified	-	-	-	-	-	-	-	-	-	-
OTHER SECTORS	-	-	-	-	3239	-	-	-	3590	-
Agriculture	-	-	-	-	-	-	-	-	7	-
Comm. and Publ. Services	-	-	-	-	5	-	-	-	1464	-
Residential	-	-	-	-	3234	-	-	-	2119	-
Non-specified	-	-	-	-	-	-	-	-	-	-
NON-ENERGY USE	-	-	-	-	-	-	-	-	-	-

APPROVISIONNEMENT ET DEMANDE 1998	Gaz (TJ)				En. Re. Comb. & Déchets (TJ)				(GWh)	(TJ)
	Gaz naturel	Usines à gaz	Cokeries	Hauts fourneaux	Biomasse solide	Gaz/Liquides tirés de biomasse	Déchets urbains	Déchets industriels	Electricité	Chaleur
Production	-	-	-	-	7560	-	-	-	5794	-
Imports	-	-	-	-	-	-	-	-	77	-
Exports	-	-	-	-	-	-	-	-	-148	-
Intl. Marine Bunkers	-	-	-	-	-	-	-	-	-	-
Stock Changes	-	-	-	-	-	-	-	-	-	-
DOMESTIC SUPPLY	-	-	-	-	7560	-	-	-	5723	-
Transfers and Stat. Diff.	-	-	-	-	27	-	-	-	-268	-
TRANSFORMATION	-	-	-	-	190	-	-	-	-	-
Electricity and CHP Plants	-	-	-	-	162	-	-	-	-	-
Petroleum Refineries	-	-	-	-	-	-	-	-	-	-
Other Transform. Sector	-	-	-	-	28	-	-	-	-	-
ENERGY SECTOR	-	-	-	-	-	-	-	-	36	-
DISTRIBUTION LOSSES	-	-	-	-	-	-	-	-	306	-
FINAL CONSUMPTION	-	-	-	-	7396	-	-	-	5113	-
INDUSTRY SECTOR	-	-	-	-	5153	-	-	-	1451	-
Iron and Steel	-	-	-	-	-	-	-	-	-	-
Chemical and Petrochem.	-	-	-	-	-	-	-	-	290	-
Non-Metallic Minerals	-	-	-	-	-	-	-	-	-	-
Non-specified	-	-	-	-	5153	-	-	-	1161	-
TRANSPORT SECTOR	-	-	-	-	-	-	-	-	-	-
Air	-	-	-	-	-	-	-	-	-	-
Road	-	-	-	-	-	-	-	-	-	-
Non-specified	-	-	-	-	-	-	-	-	-	-
OTHER SECTORS	-	-	-	-	2243	-	-	-	3662	-
Agriculture	-	-	-	-	-	-	-	-	22	-
Comm. and Publ. Services	-	-	-	-	4	-	-	-	1367	-
Residential	-	-	-	-	2239	-	-	-	2273	-
Non-specified	-	-	-	-	-	-	-	-	-	-
NON-ENERGY USE	-	-	-	-	-	-	-	-	-	-

Côte d'Ivoire

SUPPLY AND CONSUMPTION 1997	Coal (1000 tonnes)							Oil (1000 tonnes)			
	Coking Coal	Other Bit. Coal	Sub-Bit. Coal	Lignite	Peat	Oven and Gas Coke	Pat. Fuel and BKB	Crude Oil	NGL	Feed-stocks	Additives
Production	-	-	-	-	-	-	-	722	-	-	-
Imports	-	-	-	-	-	-	-	2959	-	-	-
Exports	-	-	-	-	-	-	-	-735	-	-	-
Intl. Marine Bunkers	-	-	-	-	-	-	-	-	-	-	-
Stock Changes	-	-	-	-	-	-	-	-	-	-	-
DOMESTIC SUPPLY	-	-	-	-	-	-	-	2946	-	-	-
Transfers and Stat. Diff.	-	-	-	-	-	-	-	792	-	-	-
TRANSFORMATION	-	-	-	-	-	-	-	3738	-	-	-
Electricity and CHP Plants	-	-	-	-	-	-	-	-	-	-	-
Petroleum Refineries	-	-	-	-	-	-	-	3738	-	-	-
Other Transform. Sector	-	-	-	-	-	-	-	-	-	-	-
ENERGY SECTOR	-	-	-	-	-	-	-	-	-	-	-
DISTRIBUTION LOSSES	-	-	-	-	-	-	-	-	-	-	-
FINAL CONSUMPTION	-	-	-	-	-	-	-	-	-	-	-
INDUSTRY SECTOR	-	-	-	-	-	-	-	-	-	-	-
Iron and Steel	-	-	-	-	-	-	-	-	-	-	-
Chemical and Petrochem.	-	-	-	-	-	-	-	-	-	-	-
Non-Metallic Minerals	-	-	-	-	-	-	-	-	-	-	-
Non-specified	-	-	-	-	-	-	-	-	-	-	-
TRANSPORT SECTOR	-	-	-	-	-	-	-	-	-	-	-
Air	-	-	-	-	-	-	-	-	-	-	-
Road	-	-	-	-	-	-	-	-	-	-	-
Non-specified	-	-	-	-	-	-	-	-	-	-	-
OTHER SECTORS	-	-	-	-	-	-	-	-	-	-	-
Agriculture	-	-	-	-	-	-	-	-	-	-	-
Comm. and Publ. Services	-	-	-	-	-	-	-	-	-	-	-
Residential	-	-	-	-	-	-	-	-	-	-	-
Non-specified	-	-	-	-	-	-	-	-	-	-	-
NON-ENERGY USE	-	-	-	-	-	-	-	-	-	-	-

APPROVISIONNEMENT ET DEMANDE 1998	Charbon (1000 tonnes)							Pétrole (1000 tonnes)			
	Charbon à coke	Autres charb. bit.	Charbon sous-bit.	Lignite	Tourbe	Coke de four/gaz	Agg./briq. de lignite	Pétrole brut	LGN	Produits d'aliment.	Additifs
Production	-	-	-	-	-	-	-	521	-	-	-
Imports	-	-	-	-	-	-	-	3612	-	-	-
Exports	-	-	-	-	-	-	-	-464	-	-	-
Intl. Marine Bunkers	-	-	-	-	-	-	-	-	-	-	-
Stock Changes	-	-	-	-	-	-	-	-	-	-	-
DOMESTIC SUPPLY	-	-	-	-	-	-	-	3669	-	-	-
Transfers and Stat. Diff.	-	-	-	-	-	-	-	363	-	-	-
TRANSFORMATION	-	-	-	-	-	-	-	4032	-	-	-
Electricity and CHP Plants	-	-	-	-	-	-	-	-	-	-	-
Petroleum Refineries	-	-	-	-	-	-	-	4032	-	-	-
Other Transform. Sector	-	-	-	-	-	-	-	-	-	-	-
ENERGY SECTOR	-	-	-	-	-	-	-	-	-	-	-
DISTRIBUTION LOSSES	-	-	-	-	-	-	-	-	-	-	-
FINAL CONSUMPTION	-	-	-	-	-	-	-	-	-	-	-
INDUSTRY SECTOR	-	-	-	-	-	-	-	-	-	-	-
Iron and Steel	-	-	-	-	-	-	-	-	-	-	-
Chemical and Petrochem.	-	-	-	-	-	-	-	-	-	-	-
Non-Metallic Minerals	-	-	-	-	-	-	-	-	-	-	-
Non-specified	-	-	-	-	-	-	-	-	-	-	-
TRANSPORT SECTOR	-	-	-	-	-	-	-	-	-	-	-
Air	-	-	-	-	-	-	-	-	-	-	-
Road	-	-	-	-	-	-	-	-	-	-	-
Non-specified	-	-	-	-	-	-	-	-	-	-	-
OTHER SECTORS	-	-	-	-	-	-	-	-	-	-	-
Agriculture	-	-	-	-	-	-	-	-	-	-	-
Comm. and Publ. Services	-	-	-	-	-	-	-	-	-	-	-
Residential	-	-	-	-	-	-	-	-	-	-	-
Non-specified	-	-	-	-	-	-	-	-	-	-	-
NON-ENERGY USE	-	-	-	-	-	-	-	-	-	-	-

Côte d'Ivoire

SUPPLY AND CONSUMPTION 1997	Oil cont. (1000 tonnes)										
	Refinery Gas	LPG + Ethane	Motor Gasoline	Aviation Gasoline	Jet Fuel	Kerosene	Gas/ Diesel	Heavy Fuel Oil	Naphtha	Petrol. Coke	Other Prod.
Production	54	42	617	-	108	744	1227	664	-	-	117
Imports	-	25	88	1	-	-	-	179	-	-	-
Exports	-	-	-541	-	-27	-685	-698	-714	-	-	-44
Intl. Marine Bunkers	-	-	-	-	-	-	-15	-74	-	-	-
Stock Changes	-	-	-	-	-	-	-	-	-	-	-
DOMESTIC SUPPLY	54	67	164	1	81	59	514	55	-	-	73
Transfers and Stat. Diff.	-	-8	-1	-	-	-	1	-	-	-	-
TRANSFORMATION	-	-	-	-	-	-	51	34	-	-	-
Electricity and CHP Plants	-	-	-	-	-	-	51	34	-	-	-
Petroleum Refineries	-	-	-	-	-	-	-	-	-	-	-
Other Transform. Sector	-	-	-	-	-	-	-	-	-	-	-
ENERGY SECTOR	54	-	-	-	-	-	-	13	-	-	-
DISTRIBUTION LOSSES	-	-	-	-	-	-	-	-	-	-	-
FINAL CONSUMPTION	-	59	163	1	81	59	464	8	-	-	73
INDUSTRY SECTOR	-	5	-	-	-	3	123	6	-	-	-
Iron and Steel	-	-	-	-	-	-	-	-	-	-	-
Chemical and Petrochem.	-	-	-	-	-	-	-	-	-	-	-
Non-Metallic Minerals	-	-	-	-	-	-	-	-	-	-	-
Non-specified	-	5	-	-	-	3	123	6	-	-	-
TRANSPORT SECTOR	-	-	163	1	81	-	265	2	-	-	-
Air	-	-	-	1	81	-	-	-	-	-	-
Road	-	-	163	-	-	-	212	-	-	-	-
Non-specified	-	-	-	-	-	-	53	2	-	-	-
OTHER SECTORS	-	54	-	-	-	56	76	-	-	-	-
Agriculture	-	-	-	-	-	-	48	-	-	-	-
Comm. and Publ. Services	-	-	-	-	-	-	28	-	-	-	-
Residential	-	54	-	-	-	56	-	-	-	-	-
Non-specified	-	-	-	-	-	-	-	-	-	-	-
NON-ENERGY USE	-	-	-	-	-	-	-	-	-	-	73

APPROVISIONNEMENT ET DEMANDE 1998	Pétrole cont. (1000 tonnes)										
	Gaz de raffinerie	GPL + éthane	Essence moteur	Essence aviation	Carbu- réacteurs	Kérosène	Gazole	Fioul lourd	Naphta	Coke de pétrole	Autres prod.
Production	59	49	657	-	128	880	1450	710	-	-	73
Imports	-	25	88	1	-	-	-	179	-	-	-
Exports	-	-	-583	-	-46	-821	-872	-753	-	-	-
Intl. Marine Bunkers	-	-	-	-	-	-	-15	-74	-	-	-
Stock Changes	-	-	-	-	-	-	-	-	-	-	-
DOMESTIC SUPPLY	59	74	162	1	82	59	563	62	-	-	73
Transfers and Stat. Diff.	-	1	1	-	-1	-	1	-	-	-	-
TRANSFORMATION	-	-	-	-	-	-	63	39	-	-	-
Electricity and CHP Plants	-	-	-	-	-	-	63	39	-	-	-
Petroleum Refineries	-	-	-	-	-	-	-	-	-	-	-
Other Transform. Sector	-	-	-	-	-	-	-	-	-	-	-
ENERGY SECTOR	59	-	-	-	-	-	-	14	-	-	-
DISTRIBUTION LOSSES	-	-	-	-	-	-	-	-	-	-	-
FINAL CONSUMPTION	-	75	163	1	81	59	501	9	-	-	73
INDUSTRY SECTOR	-	15	-	-	-	3	135	7	-	-	-
Iron and Steel	-	-	-	-	-	-	-	-	-	-	-
Chemical and Petrochem.	-	-	-	-	-	-	-	-	-	-	-
Non-Metallic Minerals	-	-	-	-	-	-	-	-	-	-	-
Non-specified	-	15	-	-	-	3	135	7	-	-	-
TRANSPORT SECTOR	-	-	163	1	81	-	283	2	-	-	-
Air	-	-	-	1	81	-	-	-	-	-	-
Road	-	-	163	-	-	-	225	-	-	-	-
Non-specified	-	-	-	-	-	-	58	2	-	-	-
OTHER SECTORS	-	60	-	-	-	56	83	-	-	-	-
Agriculture	-	-	-	-	-	-	53	-	-	-	-
Comm. and Publ. Services	-	-	-	-	-	-	30	-	-	-	-
Residential	-	60	-	-	-	56	-	-	-	-	-
Non-specified	-	-	-	-	-	-	-	-	-	-	-
NON-ENERGY USE	-	-	-	-	-	-	-	-	-	-	73

Côte d'Ivoire

SUPPLY AND CONSUMPTION 1997	Gas (TJ) Natural Gas	Gas Works	Coke Ovens	Blast Furnaces	Comb. Renew. & Waste (TJ) Solid Biomass	Gas/Liquids from Biomass	Municipal Waste	Industrial Waste	(GWh) Electricity	(TJ) Heat
Production	28554	-	-	-	164599	-	-	-	3982	-
Imports	-	-	-	-		-	-	-	-	-
Exports	-	-	-	-		-	-	-	-951	-
Intl. Marine Bunkers	-	-	-	-		-	-	-	-	-
Stock Changes	-	-	-	-		-	-	-	-	-
DOMESTIC SUPPLY	28554	-	-	-	164599	-	-	-	3031	-
Transfers and Stat. Diff.	-	-	-	-	-1	-	-	-	-	-
TRANSFORMATION	28554	-	-	-	61326	-	-	-	-	-
Electricity and CHP Plants	28554	-	-	-		-	-	-	-	-
Petroleum Refineries	-	-	-	-		-	-	-	-	-
Other Transform. Sector	-	-	-	-	61326	-	-	-	-	-
ENERGY SECTOR	-	-	-	-		-	-	-	30	-
DISTRIBUTION LOSSES	-	-	-	-		-	-	-	485	-
FINAL CONSUMPTION	-	-	-	-	103272	-	-	-	2516	-
INDUSTRY SECTOR	-	-	-	-		-	-	-	1251	-
Iron and Steel	-	-	-	-		-	-	-	-	-
Chemical and Petrochem.	-	-	-	-		-	-	-	225	-
Non-Metallic Minerals	-	-	-	-		-	-	-	-	-
Non-specified	-	-	-	-		-	-	-	1026	-
TRANSPORT SECTOR	-	-	-	-		-	-	-	-	-
Air	-	-	-	-		-	-	-	-	-
Road	-	-	-	-		-	-	-	-	-
Non-specified	-	-	-	-		-	-	-	-	-
OTHER SECTORS	-	-	-	-	103272	-	-	-	1265	-
Agriculture	-	-	-	-		-	-	-	187	-
Comm. and Publ. Services	-	-	-	-	13593	-	-	-	1078	-
Residential	-	-	-	-	89679	-	-	-	-	-
Non-specified	-	-	-	-		-	-	-	-	-
NON-ENERGY USE	-	-	-	-		-	-	-	-	-

APPROVISIONNEMENT ET DEMANDE 1998	Gaz (TJ) Gaz naturel	Usines à gaz	Cokeries	Hauts fourneaux	En. Re. Comb. & Déchets (TJ) Biomasse solide	Gaz/Liquides tirés de biomasse	Déchets urbains	Déchets industriels	(GWh) Electricité	(TJ) Chaleur
Production	35187	-	-	-	167853	-	-	-	3992	-
Imports	-	-	-	-	-	-	-	-	-	-
Exports	-	-	-	-	-	-	-	-	-593	-
Intl. Marine Bunkers	-	-	-	-	-	-	-	-	-	-
Stock Changes	-	-	-	-	-	-	-	-	-	-
DOMESTIC SUPPLY	35187	-	-	-	167853	-	-	-	3399	-
Transfers and Stat. Diff.	-	-	-	-	-1	-	-	-	1	-
TRANSFORMATION	35187	-	-	-	62539	-	-	-	-	-
Electricity and CHP Plants	35187	-	-	-		-	-	-	-	-
Petroleum Refineries	-	-	-	-	-	-	-	-	-	-
Other Transform. Sector	-	-	-	-	62539	-	-	-	-	-
ENERGY SECTOR	-	-	-	-	-	-	-	-	44	-
DISTRIBUTION LOSSES	-	-	-	-	-	-	-	-	544	-
FINAL CONSUMPTION	-	-	-	-	105314	-	-	-	2812	-
INDUSTRY SECTOR	-	-	-	-	-	-	-	-	1399	-
Iron and Steel	-	-	-	-	-	-	-	-	-	-
Chemical and Petrochem.	-	-	-	-	-	-	-	-	251	-
Non-Metallic Minerals	-	-	-	-	-	-	-	-	-	-
Non-specified	-	-	-	-	-	-	-	-	1148	-
TRANSPORT SECTOR	-	-	-	-	-	-	-	-	-	-
Air	-	-	-	-	-	-	-	-	-	-
Road	-	-	-	-	-	-	-	-	-	-
Non-specified	-	-	-	-	-	-	-	-	-	-
OTHER SECTORS	-	-	-	-	105314	-	-	-	1413	-
Agriculture	-	-	-	-	-	-	-	-	209	-
Comm. and Publ. Services	-	-	-	-	13862	-	-	-	1204	-
Residential	-	-	-	-	91452	-	-	-	-	-
Non-specified	-	-	-	-	-	-	-	-	-	-
NON-ENERGY USE	-	-	-	-	-	-	-	-	-	-

Croatia / Croatie : 1997

SUPPLY AND CONSUMPTION / APPROVISIONNEMENT ET DEMANDE	Coal / Charbon (1000 tonnes)							Oil / Pétrole (1000 tonnes)			
	Coking Coal / Charbon à coke	Other Bit. Coal / Autres charb. bit.	Sub-Bit. Coal / Charbon sous-bit.	Lignite / Lignite	Peat / Tourbe	Oven and Gas Coke / Coke de four/gaz	Pat. Fuel and BKB / Agg./briq. de lignite	Crude Oil / Pétrole brut	NGL / LGN	Feed-stocks / Produits d'aliment.	Additives / Additifs
Production	-	49	-	-	-	-	-	1533	252	-	-
From Other Sources	-	-	-	-	-	-	-	-	-	-	-
Imports	-	125	88	45	-	33	-	3699	-	-	-
Exports	-	-3	-	-	-	-	-	-40	-	-	-
Intl. Marine Bunkers	-	-	-	-	-	-	-	-	-	-	-
Stock Changes	-	115	-	-	-	-3	-	68	-	-	-
DOMESTIC SUPPLY	-	286	88	45	-	30	-	5260	252	-	-
Transfers	-	-	-	-	-	-	-	-	-252	-	-
Statistical Differences	-	-	-	-	-	-	-	-	-	-	-
TRANSFORMATION	-	230	12	5	-	-	-	5149	-	-	-
Electricity Plants	-	230	-	-	-	-	-	-	-	-	-
CHP Plants	-	-	12	5	-	-	-	-	-	-	-
Heat Plants	-	-	-	-	-	-	-	-	-	-	-
Blast Furnaces/Gas Works	-	-	-	-	-	-	-	-	-	-	-
Coke/Pat. Fuel/BKB Plants	-	-	-	-	-	-	-	-	-	-	-
Petroleum Refineries	-	-	-	-	-	-	-	5149	-	-	-
Petrochemical Industry	-	-	-	-	-	-	-	-	-	-	-
Liquefaction	-	-	-	-	-	-	-	-	-	-	-
Other Transform. Sector	-	-	-	-	-	-	-	-	-	-	-
ENERGY SECTOR	-	-	-	-	-	-	-	111	-	-	-
Coal Mines	-	-	-	-	-	-	-	-	-	-	-
Oil and Gas Extraction	-	-	-	-	-	-	-	111	-	-	-
Petroleum Refineries	-	-	-	-	-	-	-	-	-	-	-
Electr., CHP+Heat Plants	-	-	-	-	-	-	-	-	-	-	-
Pumped Storage (Elec.)	-	-	-	-	-	-	-	-	-	-	-
Other Energy Sector	-	-	-	-	-	-	-	-	-	-	-
Distribution Losses	-	-	-	-	-	-	-	-	-	-	-
FINAL CONSUMPTION	-	56	76	40	-	30	-	-	-	-	-
INDUSTRY SECTOR	-	56	55	28	-	30	-	-	-	-	-
Iron and Steel	-	-	-	-	-	17	-	-	-	-	-
Chemical and Petrochem.	-	-	4	2	-	-	-	-	-	-	-
of which: Feedstocks	-	-	-	-	-	-	-	-	-	-	-
Non-Ferrous Metals	-	-	-	-	-	-	-	-	-	-	-
Non-Metallic Minerals	-	53	3	2	-	6	-	-	-	-	-
Transport Equipment	-	-	-	-	-	-	-	-	-	-	-
Machinery	-	-	-	-	-	-	-	-	-	-	-
Mining and Quarrying	-	-	-	-	-	-	-	-	-	-	-
Food and Tobacco	-	3	48	24	-	7	-	-	-	-	-
Paper, Pulp and Print	-	-	-	-	-	-	-	-	-	-	-
Wood and Wood Products	-	-	-	-	-	-	-	-	-	-	-
Construction	-	-	-	-	-	-	-	-	-	-	-
Textile and Leather	-	-	-	-	-	-	-	-	-	-	-
Non-specified	-	-	-	-	-	-	-	-	-	-	-
TRANSPORT SECTOR	-	-	-	-	-	-	-	-	-	-	-
Air	-	-	-	-	-	-	-	-	-	-	-
Road	-	-	-	-	-	-	-	-	-	-	-
Rail	-	-	-	-	-	-	-	-	-	-	-
Pipeline Transport	-	-	-	-	-	-	-	-	-	-	-
Internal Navigation	-	-	-	-	-	-	-	-	-	-	-
Non-specified	-	-	-	-	-	-	-	-	-	-	-
OTHER SECTORS	-	-	21	12	-	-	-	-	-	-	-
Agriculture	-	-	-	-	-	-	-	-	-	-	-
Comm. and Publ. Services	-	-	11	2	-	-	-	-	-	-	-
Residential	-	-	10	10	-	-	-	-	-	-	-
Non-specified	-	-	-	-	-	-	-	-	-	-	-
NON-ENERGY USE	-	-	-	-	-	-	-	-	-	-	-
in Industry/Trans./Energy	-	-	-	-	-	-	-	-	-	-	-
in Transport	-	-	-	-	-	-	-	-	-	-	-
in Other Sectors	-	-	-	-	-	-	-	-	-	-	-

Croatia / Croatie : 1997

						Oil cont. / *Pétrole cont.* (1000 tonnes)					
SUPPLY AND CONSUMPTION / *APPROVISIONNEMENT ET DEMANDE*	Refinery Gas / *Gaz de raffinerie*	LPG + Ethane / *GPL + éthane*	Motor Gasoline / *Essence moteur*	Aviation Gasoline / *Essence aviation*	Jet Fuel / *Carbu-réacteurs*	Kerosene / *Kérosène*	Gas/ Diesel / *Gazole*	Heavy Fuel Oil / *Fioul lourd*	Naphtha / *Naphta*	Petrol. Coke / *Coke de pétrole*	Other Prod. / *Autres prod.*
Production	202	158	1089	-	95	5	1492	1484	238	28	323
From Other Sources	-	-	-	-	-	-	-	-	-	-	-
Imports	-	2	67	-	25	2	224	54	-	15	42
Exports	-	-146	-455	-	-34	-1	-503	-127	-182	-29	-185
Intl. Marine Bunkers	-	-	-	-	-	-	-7	-17	-	-	-
Stock Changes	-	1	-23	-	-4	-	-16	-74	2	1	21
DOMESTIC SUPPLY	202	15	678	-	82	6	1190	1320	58	15	201
Transfers	-	198	-	-	-	-	-	-	54	-	-
Statistical Differences	-	-	-	-	-	-	-	-	-	-	-
TRANSFORMATION	4	14	-	-	-	-	10	749	-	-	-
Electricity Plants	-	-	-	-	-	-	4	487	-	-	-
CHP Plants	4	-	-	-	-	-	-	225	-	-	-
Heat Plants	-	1	-	-	-	-	6	37	-	-	-
Blast Furnaces/Gas Works	-	13	-	-	-	-	-	-	-	-	-
Coke/Pat. Fuel/BKB Plants	-	-	-	-	-	-	-	-	-	-	-
Petroleum Refineries	-	-	-	-	-	-	-	-	-	-	-
Petrochemical Industry	-	-	-	-	-	-	-	-	-	-	-
Liquefaction	-	-	-	-	-	-	-	-	-	-	-
Other Transform. Sector	-	-	-	-	-	-	-	-	-	-	-
ENERGY SECTOR	198	-	-	-	-	-	4	281	-	-	-
Coal Mines	-	-	-	-	-	-	-	-	-	-	-
Oil and Gas Extraction	-	-	-	-	-	-	3	-	-	-	-
Petroleum Refineries	198	-	-	-	-	-	-	281	-	-	-
Electr., CHP+Heat Plants	-	-	-	-	-	-	1	-	-	-	-
Pumped Storage (Elec.)	-	-	-	-	-	-	-	-	-	-	-
Other Energy Sector	-	-	-	-	-	-	-	-	-	-	-
Distribution Losses	-	-	-	-	-	-	-	-	-	-	-
FINAL CONSUMPTION	-	199	678	-	82	6	1176	290	112	15	201
INDUSTRY SECTOR	-	121	11	-	-	-	122	251	112	15	-
Iron and Steel	-	1	-	-	-	-	3	5	-	-	-
Chemical and Petrochem.	-	102	-	-	-	-	2	72	112	-	-
of which: Feedstocks	-	*94*	-	-	-	-	-	-	*112*	-	-
Non-Ferrous Metals	-	-	-	-	-	-	3	1	-	-	-
Non-Metallic Minerals	-	3	-	-	-	-	6	87	-	15	-
Transport Equipment	-	2	-	-	-	-	-	2	-	-	-
Machinery	-	1	-	-	-	-	2	6	-	-	-
Mining and Quarrying	-	1	-	-	-	-	10	2	-	-	-
Food and Tobacco	-	4	-	-	-	-	14	50	-	-	-
Paper, Pulp and Print	-	-	-	-	-	-	1	3	-	-	-
Wood and Wood Products	-	2	-	-	-	-	1	1	-	-	-
Construction	-	1	11	-	-	-	78	-	-	-	-
Textile and Leather	-	1	-	-	-	-	1	18	-	-	-
Non-specified	-	3	-	-	-	-	1	4	-	-	-
TRANSPORT SECTOR	-	12	661	-	82	-	570	11	-	-	-
Air	-	-	-	-	82	-	-	-	-	-	-
Road	-	12	661	-	-	-	487	-	-	-	-
Rail	-	-	-	-	-	-	31	-	-	-	-
Pipeline Transport	-	-	-	-	-	-	-	-	-	-	-
Internal Navigation	-	-	-	-	-	-	26	11	-	-	-
Non-specified	-	-	-	-	-	-	26	-	-	-	-
OTHER SECTORS	-	66	6	-	-	6	484	28	-	-	-
Agriculture	-	2	6	-	-	-	166	2	-	-	-
Comm. and Publ. Services	-	3	-	-	-	1	109	7	-	-	-
Residential	-	61	-	-	-	5	209	19	-	-	-
Non-specified	-	-	-	-	-	-	-	-	-	-	-
NON-ENERGY USE	-	-	-	-	-	-	-	-	-	-	201
in Industry/Transf./Energy	-	-	-	-	-	-	-	-	-	-	171
in Transport	-	-	-	-	-	-	-	-	-	-	26
in Other Sectors	-	-	-	-	-	-	-	-	-	-	4

Croatia / Croatie : 1997

SUPPLY AND CONSUMPTION / APPROVISIONNEMENT ET DEMANDE	Gas / Gaz (TJ)				Comb. Renew. & Waste / En. Re. Comb. & Déchets (TJ)				(GWh)	(TJ)
	Natural Gas / Gaz naturel	Gas Works / Usines à gaz	Coke Ovens / Cokeries	Blast Furnaces / Hauts fourneaux	Solid Biomass / Biomasse solide	Gas/Liquids from Biomass / Gaz/Liquides tirés de biomasse	Municipal Waste / Déchets urbains	Industrial Waste / Déchets industriels	Electricity / Electricité	Heat / Chaleur
Production	65254	418	-	-	13557	-	-	10	9684	13327
From Other Sources	-	-	-	-	-	-	-	-	-	-
Imports	39729	-	-	-	-	-	-	-	4608	-
Exports	-	-	-	-	-	-	-	-	-660	-
Intl. Marine Bunkers	-	-	-	-	-	-	-	-	-	-
Stock Changes	-464	-	-	-	-	-	-	-	-	-
DOMESTIC SUPPLY	104519	418	-	-	13557	-	-	10	13632	13327
Transfers	-	-	-	-	-	-	-	-	-	-
Statistical Differences	-	-	-	-	-	-	-	-	-	-
TRANSFORMATION	20767	-	-	-	-	-	-	10	-	-
Electricity Plants	4856	-	-	-	-	-	-	-	-	-
CHP Plants	14433	-	-	-	-	-	-	10	-	-
Heat Plants	1478	-	-	-	-	-	-	-	-	-
Blast Furnaces/Gas Works	-	-	-	-	-	-	-	-	-	-
Coke/Pat. Fuel/BKB Plants	-	-	-	-	-	-	-	-	-	-
Petroleum Refineries	-	-	-	-	-	-	-	-	-	-
Petrochemical Industry	-	-	-	-	-	-	-	-	-	-
Liquefaction	-	-	-	-	-	-	-	-	-	-
Other Transform. Sector	-	-	-	-	-	-	-	-	-	-
ENERGY SECTOR	10431	-	-	-	-	-	-	-	785	1181
Coal Mines	-	-	-	-	-	-	-	-	12	-
Oil and Gas Extraction	10264	-	-	-	-	-	-	-	122	-
Petroleum Refineries	133	-	-	-	-	-	-	-	267	-
Electr., CHP+Heat Plants	34	-	-	-	-	-	-	-	367	1181
Pumped Storage (Elec.)	-	-	-	-	-	-	-	-	-	-
Other Energy Sector	-	-	-	-	-	-	-	-	17	-
Distribution Losses	2725	3	-	-	-	-	-	-	1809	1916
FINAL CONSUMPTION	70596	415	-	-	13557	-	-	-	11038	10230
INDUSTRY SECTOR	46930	145	-	-	-	-	-	-	3031	3188
Iron and Steel	1809	25	-	-	-	-	-	-	277	-
Chemical and Petrochem.	27413	-	-	-	-	-	-	-	547	958
of which: Feedstocks	22268	-	-	-	-	-	-	-	-	-
Non-Ferrous Metals	201	-	-	-	-	-	-	-	55	-
Non-Metallic Minerals	7615	43	-	-	-	-	-	-	448	125
Transport Equipment	19	45	-	-	-	-	-	-	116	28
Machinery	612	11	-	-	-	-	-	-	179	379
Mining and Quarrying	395	-	-	-	-	-	-	-	41	30
Food and Tobacco	4914	-	-	-	-	-	-	-	404	310
Paper, Pulp and Print	2212	-	-	-	-	-	-	-	241	270
Wood and Wood Products	315	-	-	-	-	-	-	-	135	395
Construction	-	-	-	-	-	-	-	-	235	-
Textile and Leather	1193	-	-	-	-	-	-	-	187	693
Non-specified	232	21	-	-	-	-	-	-	166	-
TRANSPORT SECTOR	-	-	-	-	-	-	-	-	214	-
Air	-	-	-	-	-	-	-	-	-	-
Road	-	-	-	-	-	-	-	-	-	-
Rail	-	-	-	-	-	-	-	-	140	-
Pipeline Transport	-	-	-	-	-	-	-	-	18	-
Internal Navigation	-	-	-	-	-	-	-	-	-	-
Non-specified	-	-	-	-	-	-	-	-	56	-
OTHER SECTORS	23666	270	-	-	13557	-	-	-	7793	7042
Agriculture	855	-	-	-	-	-	-	-	60	-
Comm. and Publ. Services	4313	68	-	-	-	-	-	-	2544	898
Residential	18498	202	-	-	13557	-	-	-	5189	6144
Non-specified	-	-	-	-	-	-	-	-	-	-
NON-ENERGY USE	-	-	-	-	-	-	-	-	-	-
in Industry/Transf./Energy	-	-	-	-	-	-	-	-	-	-
in Transport	-	-	-	-	-	-	-	-	-	-
in Other Sectors	-	-	-	-	-	-	-	-	-	-

Croatia / Croatie : 1998

SUPPLY AND CONSUMPTION APPROVISIONNEMENT ET DEMANDE	Coal / *Charbon* (1000 tonnes)							Oil / *Pétrole* (1000 tonnes)			
	Coking Coal *Charbon à coke*	Other Bit. Coal *Autres charb. bit.*	Sub-Bit. Coal *Charbon sous-bit.*	Lignite *Lignite*	Peat *Tourbe*	Oven and Gas Coke *Coke de four/gaz*	Pat. Fuel and BKB *Agg./briq. de lignite*	Crude Oil *Pétrole brut*	NGL *LGN*	Feed- stocks *Produits d'aliment.*	Additives *Additifs*
Production	-	51	-	-	-	-	-	1589	226	-	-
From Other Sources	-	-	-	-	-	-	-	-	-	-	-
Imports	-	230	62	39	-	20	-	3759	-	-	-
Exports	-	-	-	-	-	-1	-	-40	-	-	-
Intl. Marine Bunkers	-	-	-	-	-	-	-	-	-	-	-
Stock Changes	-	12	-	-	-	5	-	9	-	-	-
DOMESTIC SUPPLY	-	293	62	39	-	24	-	5317	226	-	-
Transfers	-	-	-	-	-	-	-	-	-226	-	-
Statistical Differences	-	-	-	-	-	-	-	-	-	-	-
TRANSFORMATION	-	229	5	2	-	-	-	5207	-	-	-
Electricity Plants	-	229	-	-	-	-	-	-	-	-	-
CHP Plants	-	-	5	2	-	-	-	-	-	-	-
Heat Plants	-	-	-	-	-	-	-	-	-	-	-
Blast Furnaces/Gas Works	-	-	-	-	-	-	-	-	-	-	-
Coke/Pat. Fuel/BKB Plants	-	-	-	-	-	-	-	-	-	-	-
Petroleum Refineries	-	-	-	-	-	-	-	5207	-	-	-
Petrochemical Industry	-	-	-	-	-	-	-	-	-	-	-
Liquefaction	-	-	-	-	-	-	-	-	-	-	-
Other Transform. Sector	-	-	-	-	-	-	-	-	-	-	-
ENERGY SECTOR	-	1	-	-	-	-	-	110	-	-	-
Coal Mines	-	1	-	-	-	-	-	-	-	-	-
Oil and Gas Extraction	-	-	-	-	-	-	-	110	-	-	-
Petroleum Refineries	-	-	-	-	-	-	-	-	-	-	-
Electr., CHP+Heat Plants	-	-	-	-	-	-	-	-	-	-	-
Pumped Storage (Elec.)	-	-	-	-	-	-	-	-	-	-	-
Other Energy Sector	-	-	-	-	-	-	-	-	-	-	-
Distribution Losses	-	-	-	-	-	-	-	-	-	-	-
FINAL CONSUMPTION	-	63	57	37	-	24	-	-	-	-	-
INDUSTRY SECTOR	-	63	29	19	-	24	-	-	-	-	-
Iron and Steel	-	-	-	-	-	11	-	-	-	-	-
Chemical and Petrochem.	-	-	-	-	-	-	-	-	-	-	-
of which: Feedstocks	-	-	-	-	-	-	-	-	-	-	-
Non-Ferrous Metals	-	-	-	-	-	-	-	-	-	-	-
Non-Metallic Minerals	-	60	2	1	-	6	-	-	-	-	-
Transport Equipment	-	-	-	-	-	-	-	-	-	-	-
Machinery	-	-	-	-	-	1	-	-	-	-	-
Mining and Quarrying	-	-	-	-	-	-	-	-	-	-	-
Food and Tobacco	-	3	27	18	-	6	-	-	-	-	-
Paper, Pulp and Print	-	-	-	-	-	-	-	-	-	-	-
Wood and Wood Products	-	-	-	-	-	-	-	-	-	-	-
Construction	-	-	-	-	-	-	-	-	-	-	-
Textile and Leather	-	-	-	-	-	-	-	-	-	-	-
Non-specified	-	-	-	-	-	-	-	-	-	-	-
TRANSPORT SECTOR	-	-	-	-	-	-	-	-	-	-	-
Air	-	-	-	-	-	-	-	-	-	-	-
Road	-	-	-	-	-	-	-	-	-	-	-
Rail	-	-	-	-	-	-	-	-	-	-	-
Pipeline Transport	-	-	-	-	-	-	-	-	-	-	-
Internal Navigation	-	-	-	-	-	-	-	-	-	-	-
Non-specified	-	-	-	-	-	-	-	-	-	-	-
OTHER SECTORS	-	-	28	18	-	-	-	-	-	-	-
Agriculture	-	-	-	-	-	-	-	-	-	-	-
Comm. and Publ. Services	-	-	17	1	-	-	-	-	-	-	-
Residential	-	-	11	17	-	-	-	-	-	-	-
Non-specified	-	-	-	-	-	-	-	-	-	-	-
NON-ENERGY USE	-	-	-	-	-	-	-	-	-	-	-
in Industry/Trans./Energy	-	-	-	-	-	-	-	-	-	-	-
in Transport	-	-	-	-	-	-	-	-	-	-	-
in Other Sectors	-	-	-	-	-	-	-	-	-	-	-

Croatia / Croatie : 1998

SUPPLY AND CONSUMPTION *APPROVISIONNEMENT ET DEMANDE*	Refinery Gas *Gaz de raffinerie*	LPG + Ethane *GPL + éthane*	Motor Gasoline *Essence moteur*	Aviation Gasoline *Essence aviation*	Jet Fuel *Carbu-réacteurs*	Kerosene *Kérosène*	Gas/ Diesel *Gazole*	Heavy Fuel Oil *Fioul lourd*	Naphtha *Naphta*	Petrol. Coke *Coke de pétrole*	Other Prod. *Autres prod.*
Production	238	196	1158	-	87	3	1522	1395	225	58	293
From Other Sources	-	-	-	-	-	-	-	-	-	-	-
Imports	-	3	70	-	30	-	208	226	-	32	127
Exports	-	-168	-456	-	-35	-1	-519	-201	-171	-52	-113
Intl. Marine Bunkers	-	-	-	-	-	-	-12	-14	-	-	-
Stock Changes	-	-7	-35	-	6	-	8	99	5	-1	35
DOMESTIC SUPPLY	**238**	**24**	**737**	**-**	**88**	**2**	**1207**	**1505**	**59**	**37**	**342**
Transfers	-	177	-	-	-	-	-	-	49	-	-
Statistical Differences	-	-	-	-	-	-	-	-	-	-	-
TRANSFORMATION	**3**	**16**	**-**	**-**	**-**	**-**	**7**	**964**	**-**	**-**	**-**
Electricity Plants	-	-	-	-	-	-	3	654	-	-	-
CHP Plants	3	1	-	-	-	-	-	274	-	-	-
Heat Plants	-	2	-	-	-	-	4	36	-	-	-
Blast Furnaces/Gas Works	-	13	-	-	-	-	-	-	-	-	-
Coke/Pat. Fuel/BKB Plants	-	-	-	-	-	-	-	-	-	-	-
Petroleum Refineries	-	-	-	-	-	-	-	-	-	-	-
Petrochemical Industry	-	-	-	-	-	-	-	-	-	-	-
Liquefaction	-	-	-	-	-	-	-	-	-	-	-
Other Transform. Sector	-	-	-	-	-	-	-	-	-	-	-
ENERGY SECTOR	**235**	**7**	**-**	**-**	**-**	**-**	**4**	**244**	**-**	**18**	**162**
Coal Mines	-	-	-	-	-	-	-	-	-	-	-
Oil and Gas Extraction	-	-	-	-	-	-	4	-	-	-	-
Petroleum Refineries	235	7	-	-	-	-	-	244	-	18	162
Electr., CHP+Heat Plants	-	-	-	-	-	-	-	-	-	-	-
Pumped Storage (Elec.)	-	-	-	-	-	-	-	-	-	-	-
Other Energy Sector	-	-	-	-	-	-	-	-	-	-	-
Distribution Losses	-	-	-	-	-	-	-	-	-	-	-
FINAL CONSUMPTION	**-**	**178**	**737**	**-**	**88**	**2**	**1196**	**297**	**108**	**19**	**180**
INDUSTRY SECTOR	**-**	**104**	**9**	**-**	**-**	**-**	**129**	**275**	**108**	**19**	**-**
Iron and Steel	-	1	-	-	-	-	4	3	-	-	-
Chemical and Petrochem.	-	90	-	-	-	-	3	107	108	-	-
of which: Feedstocks	-	83	-	-	-	-	-	-	108	-	-
Non-Ferrous Metals	-	1	-	-	-	-	2	1	-	-	-
Non-Metallic Minerals	-	5	-	-	-	-	6	79	-	19	-
Transport Equipment	-	1	-	-	-	-	1	1	-	-	-
Machinery	-	1	-	-	-	-	2	4	-	-	-
Mining and Quarrying	-	2	-	-	-	-	11	3	-	-	-
Food and Tobacco	-	1	-	-	-	-	18	51	-	-	-
Paper, Pulp and Print	-	-	-	-	-	-	1	4	-	-	-
Wood and Wood Products	-	-	-	-	-	-	1	3	-	-	-
Construction	-	1	9	-	-	-	76	-	-	-	-
Textile and Leather	-	-	-	-	-	-	3	15	-	-	-
Non-specified	-	1	-	-	-	-	1	4	-	-	-
TRANSPORT SECTOR	**-**	**11**	**721**	**-**	**88**	**-**	**571**	**3**	**-**	**-**	**-**
Air	-	-	-	-	88	-	-	-	-	-	-
Road	-	11	721	-	-	-	488	-	-	-	-
Rail	-	-	-	-	-	-	32	-	-	-	-
Pipeline Transport	-	-	-	-	-	-	-	-	-	-	-
Internal Navigation	-	-	-	-	-	-	25	3	-	-	-
Non-specified	-	-	-	-	-	-	26	-	-	-	-
OTHER SECTORS	**-**	**63**	**7**	**-**	**-**	**2**	**496**	**19**	**-**	**-**	**-**
Agriculture	-	2	7	-	-	-	184	3	-	-	-
Comm. and Publ. Services	-	3	-	-	-	-	107	5	-	-	-
Residential	-	58	-	-	-	2	205	11	-	-	-
Non-specified	-	-	-	-	-	-	-	-	-	-	-
NON-ENERGY USE	**-**	**-**	**-**	**-**	**-**	**-**	**-**	**-**	**-**	**-**	**180**
in Industry/Transf./Energy	-	-	-	-	-	-	-	-	-	-	152
in Transport	-	-	-	-	-	-	-	-	-	-	25
in Other Sectors	-	-	-	-	-	-	-	-	-	-	3

Oil cont. / *Pétrole cont.* (1000 tonnes)

Croatia / Croatie : 1998

	Gas / Gaz (TJ)				Comb. Renew. & Waste / En. Re. Comb. & Déchets (TJ)				(GWh)	(TJ)
SUPPLY AND CONSUMPTION	Natural Gas	Gas Works	Coke Ovens	Blast Furnaces	Solid Biomass	Gas/Liquids from Biomass	Municipal Waste	Industrial Waste	Electricity	Heat
APPROVISIONNEMENT ET DEMANDE	Gaz naturel	Usines à gaz	Cokeries	Hauts fourneaux	Biomasse solide	Gaz/Liquides tirés de biomasse	Déchets urbains	Déchets industriels	Electricité	Chaleur
Production	59660	418	-	-	12728	-	-	100	10898	12500
From Other Sources	-	-	-	-	-	-	-	-	-	-
Imports	41952	-	-	-	-	-	-	-	3782	-
Exports	-	-	-	-	-	-	-	-	-428	-
Intl. Marine Bunkers	-	-	-	-	-	-	-	-	-	-
Stock Changes	-1140	-	-	-	-	-	-	-	-	-
DOMESTIC SUPPLY	100472	418	-	-	12728	-	-	100	14252	12500
Transfers	-	-	-	-	-	-	-	-	-	-
Statistical Differences	-	-	-	-	-	-	-	-	-	-
TRANSFORMATION	21128	2	-	-	-	-	-	100	-	-
Electricity Plants	7714	-	-	-	-	-	-	-	-	-
CHP Plants	12578	-	-	-	-	-	-	100	-	-
Heat Plants	836	2	-	-	-	-	-	-	-	-
Blast Furnaces/Gas Works	-	-	-	-	-	-	-	-	-	-
Coke/Pat. Fuel/BKB Plants	-	-	-	-	-	-	-	-	-	-
Petroleum Refineries	-	-	-	-	-	-	-	-	-	-
Petrochemical Industry	-	-	-	-	-	-	-	-	-	-
Liquefaction	-	-	-	-	-	-	-	-	-	-
Other Transform. Sector	-	-	-	-	-	-	-	-	-	-
ENERGY SECTOR	8265	-	-	-	-	-	-	-	842	1336
Coal Mines	-	-	-	-	-	-	-	-	12	-
Oil and Gas Extraction	8170	-	-	-	-	-	-	-	109	-
Petroleum Refineries	57	-	-	-	-	-	-	-	267	-
Electr., CHP+Heat Plants	38	-	-	-	-	-	-	-	454	1336
Pumped Storage (Elec.)	-	-	-	-	-	-	-	-	-	-
Other Energy Sector	-	-	-	-	-	-	-	-	-	-
Distribution Losses	3914	3	-	-	-	-	-	-	2326	1632
FINAL CONSUMPTION	67165	413	-	-	12728	-	-	-	11084	9532
INDUSTRY SECTOR	41610	144	-	-	-	-	-	-	3054	3007
Iron and Steel	1330	25	-	-	-	-	-	-	233	-
Chemical and Petrochem.	22040	-	-	-	-	-	-	-	508	1193
of which: Feedstocks	17480	-	-	-	-	-	-	-	-	-
Non-Ferrous Metals	171	-	-	-	-	-	-	-	68	-
Non-Metallic Minerals	7981	43	-	-	-	-	-	-	519	109
Transport Equipment	11	44	-	-	-	-	-	-	127	85
Machinery	570	11	-	-	-	-	-	-	177	356
Mining and Quarrying	433	-	-	-	-	-	-	-	40	28
Food and Tobacco	5548	-	-	-	-	-	-	-	422	26
Paper, Pulp and Print	1710	-	-	-	-	-	-	-	242	123
Wood and Wood Products	646	-	-	-	-	-	-	-	88	180
Construction	-	-	-	-	-	-	-	-	286	-
Textile and Leather	1064	-	-	-	-	-	-	-	186	907
Non-specified	106	21	-	-	-	-	-	-	158	-
TRANSPORT SECTOR	-	-	-	-	-	-	-	-	225	-
Air	-	-	-	-	-	-	-	-	-	-
Road	-	-	-	-	-	-	-	-	-	-
Rail	-	-	-	-	-	-	-	-	151	-
Pipeline Transport	-	-	-	-	-	-	-	-	20	-
Internal Navigation	-	-	-	-	-	-	-	-	-	-
Non-specified	-	-	-	-	-	-	-	-	54	-
OTHER SECTORS	25555	269	-	-	12728	-	-	-	7805	6525
Agriculture	722	-	-	-	-	-	-	-	63	-
Comm. and Publ. Services	5548	68	-	-	-	-	-	-	2475	857
Residential	19285	201	-	-	12728	-	-	-	5267	5668
Non-specified	-	-	-	-	-	-	-	-	-	-
NON-ENERGY USE	-	-	-	-	-	-	-	-	-	-
in Industry/Transf./Energy	-	-	-	-	-	-	-	-	-	-
in Transport	-	-	-	-	-	-	-	-	-	-
in Other Sectors	-	-	-	-	-	-	-	-	-	-

Cuba

SUPPLY AND CONSUMPTION 1997	Coal (1000 tonnes)							Oil (1000 tonnes)			
	Coking Coal	Other Bit. Coal	Sub-Bit. Coal	Lignite	Peat	Oven and Gas Coke	Pat. Fuel and BKB	Crude Oil	NGL	Feed-stocks	Additives
Production	-	-	-	-	-	-	-	1462	-	-	-
Imports	-	16	-	-	-	19	-	1094	-	-	-
Exports	-	-	-	-	-	-	-	-	-	-	-
Intl. Marine Bunkers	-	-	-	-	-	-	-	-	-	-	-
Stock Changes	-	-4	-	-	-	-	-	-	-	-	-
DOMESTIC SUPPLY	-	12	-	-	-	19	-	2556	-	-	-
Transfers and Stat. Diff.	-	-1	-	-	-	18	-	414	-	-	-
TRANSFORMATION	-	-	-	-	-	31	-	1845	-	-	-
Electricity and CHP Plants	-	-	-	-	-	-	-	61	-	-	-
Petroleum Refineries	-	-	-	-	-	-	-	1784	-	-	-
Other Transform. Sector	-	-	-	-	-	31	-	-	-	-	-
ENERGY SECTOR	-	-	-	-	-	-	-	-	-	-	-
DISTRIBUTION LOSSES	-	-	-	-	-	-	-	-	-	-	-
FINAL CONSUMPTION	-	11	-	-	-	6	-	1125	-	-	-
INDUSTRY SECTOR	-	11	-	-	-	6	-	1125	-	-	-
Iron and Steel	-	11	-	-	-	6	-	-	-	-	-
Chemical and Petrochem.	-	-	-	-	-	-	-	-	-	-	-
Non-Metallic Minerals	-	-	-	-	-	-	-	-	-	-	-
Non-specified	-	-	-	-	-	-	-	1125	-	-	-
TRANSPORT SECTOR	-	-	-	-	-	-	-	-	-	-	-
Air	-	-	-	-	-	-	-	-	-	-	-
Road	-	-	-	-	-	-	-	-	-	-	-
Non-specified	-	-	-	-	-	-	-	-	-	-	-
OTHER SECTORS	-	-	-	-	-	-	-	-	-	-	-
Agriculture	-	-	-	-	-	-	-	-	-	-	-
Comm. and Publ. Services	-	-	-	-	-	-	-	-	-	-	-
Residential	-	-	-	-	-	-	-	-	-	-	-
Non-specified	-	-	-	-	-	-	-	-	-	-	-
NON-ENERGY USE	-	-	-	-	-	-	-	-	-	-	-

APPROVISIONNEMENT ET DEMANDE 1998	Charbon (1000 tonnes)							Pétrole (1000 tonnes)			
	Charbon à coke	Autres charb. bit.	Charbon sous-bit.	Lignite	Tourbe	Coke de four/gaz	Agg./briq. de lignite	Pétrole brut	LGN	Produits d'aliment.	Additifs
Production	-	-	-	-	-	-	-	1678	-	-	-
Imports	-	19	-	-	-	22	-	899	-	-	-
Exports	-	-	-	-	-	-	-	-	-	-	-
Intl. Marine Bunkers	-	-	-	-	-	-	-	-	-	-	-
Stock Changes	-	-5	-	-	-	-	-	-	-	-	-
DOMESTIC SUPPLY	-	14	-	-	-	22	-	2577	-	-	-
Transfers and Stat. Diff.	-	-	-	-	-	15	-	75	-	-	-
TRANSFORMATION	-	-	-	-	-	31	-	1498	-	-	-
Electricity and CHP Plants	-	-	-	-	-	-	-	61	-	-	-
Petroleum Refineries	-	-	-	-	-	-	-	1437	-	-	-
Other Transform. Sector	-	-	-	-	-	31	-	-	-	-	-
ENERGY SECTOR	-	-	-	-	-	-	-	-	-	-	-
DISTRIBUTION LOSSES	-	-	-	-	-	-	-	-	-	-	-
FINAL CONSUMPTION	-	14	-	-	-	6	-	1154	-	-	-
INDUSTRY SECTOR	-	14	-	-	-	6	-	1154	-	-	-
Iron and Steel	-	14	-	-	-	6	-	-	-	-	-
Chemical and Petrochem.	-	-	-	-	-	-	-	-	-	-	-
Non-Metallic Minerals	-	-	-	-	-	-	-	-	-	-	-
Non-specified	-	-	-	-	-	-	-	1154	-	-	-
TRANSPORT SECTOR	-	-	-	-	-	-	-	-	-	-	-
Air	-	-	-	-	-	-	-	-	-	-	-
Road	-	-	-	-	-	-	-	-	-	-	-
Non-specified	-	-	-	-	-	-	-	-	-	-	-
OTHER SECTORS	-	-	-	-	-	-	-	-	-	-	-
Agriculture	-	-	-	-	-	-	-	-	-	-	-
Comm. and Publ. Services	-	-	-	-	-	-	-	-	-	-	-
Residential	-	-	-	-	-	-	-	-	-	-	-
Non-specified	-	-	-	-	-	-	-	-	-	-	-
NON-ENERGY USE	-	-	-	-	-	-	-	-	-	-	-

Cuba

SUPPLY AND CONSUMPTION 1997	Oil cont. (1000 tonnes)										
	Refinery Gas	LPG + Ethane	Motor Gasoline	Aviation Gasoline	Jet Fuel	Kerosene	Gas/ Diesel	Heavy Fuel Oil	Naphtha	Petrol. Coke	Other Prod.
Production	26	67	321	-	-	168	156	689	56	19	159
Imports	-	30	114	9	346	11	1745	4565	-	-	72
Exports	-	-	-	-	-	-	-	-	-	-	-
Intl. Marine Bunkers	-	-	-	-	-	-	-15	-98	-	-	-
Stock Changes	-	-	-28	-	-52	16	-142	-	-	-	-
DOMESTIC SUPPLY	26	97	407	9	294	195	1744	5156	56	19	231
Transfers and Stat. Diff.	-	-9	-1	-	-	-2	-	-	324	-	-76
TRANSFORMATION	-	-	-	-	-	-	35	3581	380	-	-
Electricity and CHP Plants	-	-	-	-	-	-	35	3581	-	-	-
Petroleum Refineries	-	-	-	-	-	-	-	-	-	-	-
Other Transform. Sector	-	-	-	-	-	-	-	-	380	-	-
ENERGY SECTOR	26	-	-	-	-	-	15	-	-	-	-
DISTRIBUTION LOSSES	-	-	-	-	-	-	5	-	-	-	-
FINAL CONSUMPTION	-	88	406	9	294	193	1689	1575	-	19	155
INDUSTRY SECTOR	-	11	-	-	-	1	768	1572	-	19	-
Iron and Steel	-	-	-	-	-	-	-	-	-	-	-
Chemical and Petrochem.	-	-	-	-	-	-	-	-	-	-	-
Non-Metallic Minerals	-	-	-	-	-	-	-	-	-	-	-
Non-specified	-	11	-	-	-	1	768	1572	-	19	-
TRANSPORT SECTOR	-	-	406	9	294	-	292	-	-	-	-
Air	-	-	-	9	294	-	-	-	-	-	-
Road	-	-	406	-	-	-	292	-	-	-	-
Non-specified	-	-	-	-	-	-	-	-	-	-	-
OTHER SECTORS	-	77	-	-	-	192	629	3	-	-	-
Agriculture	-	-	-	-	-	-	256	3	-	-	-
Comm. and Publ. Services	-	8	-	-	-	3	-	-	-	-	-
Residential	-	65	-	-	-	189	-	-	-	-	-
Non-specified	-	4	-	-	-	-	373	-	-	-	-
NON-ENERGY USE	-	-	-	-	-	-	-	-	-	-	155

APPROVISIONNEMENT ET DEMANDE 1998	Pétrole cont. (1000 tonnes)										
	Gaz de raffinerie	GPL + éthane	Essence moteur	Essence aviation	Carbu- réacteurs	Kérosène	Gazole	Fioul lourd	Naphta	Coke de pétrole	Autres prod.
Production	28	48	274	-	-	75	203	511	45	15	170
Imports	-	55	172	10	460	10	1440	4683	-	-	60
Exports	-	-	-	-	-	-	-	-	-	-	-
Intl. Marine Bunkers	-	-	-	-	-	-	-15	-100	-	-	-
Stock Changes	-	-	-47	-	-193	67	-1	-	-	-	-
DOMESTIC SUPPLY	28	103	399	10	267	152	1627	5094	45	15	230
Transfers and Stat. Diff.	-	-7	1	-	-	1	-	-	328	-	-71
TRANSFORMATION	-	-	-	-	-	-	36	3617	373	-	-
Electricity and CHP Plants	-	-	-	-	-	-	36	3617	-	-	-
Petroleum Refineries	-	-	-	-	-	-	-	-	-	-	-
Other Transform. Sector	-	-	-	-	-	-	-	-	373	-	-
ENERGY SECTOR	28	-	-	-	-	-	16	-	-	-	-
DISTRIBUTION LOSSES	-	-	-	-	-	-	6	-	-	-	-
FINAL CONSUMPTION	-	96	400	10	267	153	1569	1477	-	15	159
INDUSTRY SECTOR	-	11	-	-	-	2	733	1474	-	15	-
Iron and Steel	-	-	-	-	-	-	-	-	-	-	-
Chemical and Petrochem.	-	-	-	-	-	-	-	-	-	-	-
Non-Metallic Minerals	-	-	-	-	-	-	-	-	-	-	-
Non-specified	-	11	-	-	-	2	733	1474	-	15	-
TRANSPORT SECTOR	-	-	400	10	267	-	253	-	-	-	-
Air	-	-	-	10	267	-	-	-	-	-	-
Road	-	-	400	-	-	-	253	-	-	-	-
Non-specified	-	-	-	-	-	-	-	-	-	-	-
OTHER SECTORS	-	85	-	-	-	151	583	3	-	-	-
Agriculture	-	-	-	-	-	1	233	3	-	-	-
Comm. and Publ. Services	-	9	-	-	-	2	-	-	-	-	-
Residential	-	72	-	-	-	148	-	-	-	-	-
Non-specified	-	4	-	-	-	-	350	-	-	-	-
NON-ENERGY USE	-	-	-	-	-	-	-	-	-	-	159

INTERNATIONAL ENERGY AGENCY

Cuba

SUPPLY AND CONSUMPTION 1997	Gas (TJ)				Comb. Renew. & Waste (TJ)				(GWh)	(TJ)
	Natural Gas	Gas Works	Coke Ovens	Blast Furnaces	Solid Biomass	Gas/Liquids from Biomass	Municipal Waste	Industrial Waste	Electricity	Heat
Production	1414	5886	-	272	128781	2439	-	-	14146	-
Imports	-	-	-	-	-	-	-	-	-	-
Exports	-	-	-	-	-	-	-	-	-	-
Intl. Marine Bunkers	-	-	-	-	-	-	-	-	-	-
Stock Changes	-	-	-	-	-	-	-	-	-	-
DOMESTIC SUPPLY	1414	5886	-	272	128781	2439	-	-	14146	-
Transfers and Stat. Diff.	-26	-1	-	-	-1	-	-	-	-	-
TRANSFORMATION	96	-	-	-	11182	-	-	-	-	-
Electricity and CHP Plants	96	-	-	-	6827	-	-	-	-	-
Petroleum Refineries	-	-	-	-	-	-	-	-	-	-
Other Transform. Sector	-	-	-	-	4355	-	-	-	-	-
ENERGY SECTOR	-	-	-	-	-	-	-	-	950	-
DISTRIBUTION LOSSES	-	-	-	-	-	-	-	-	2714	-
FINAL CONSUMPTION	1292	5885	-	272	117598	2439	-	-	10482	-
INDUSTRY SECTOR	1254	530	-	272	97865	-	-	-	4063	-
Iron and Steel	-	-	-	272	-	-	-	-	106	-
Chemical and Petrochem.	-	-	-	-	-	-	-	-	321	-
Non-Metallic Minerals	-	-	-	-	-	-	-	-	-	-
Non-specified	1254	530	-	-	97865	-	-	-	3636	-
TRANSPORT SECTOR	-	17	-	-	-	2439	-	-	66	-
Air	-	-	-	-	-	-	-	-	-	-
Road	-	-	-	-	-	2439	-	-	-	-
Non-specified	-	17	-	-	-	-	-	-	66	-
OTHER SECTORS	38	5338	-	-	19733	-	-	-	6353	-
Agriculture	38	3	-	-	1646	-	-	-	213	-
Comm. and Publ. Services	-	-	-	-	1048	-	-	-	2417	-
Residential	-	5335	-	-	2070	-	-	-	3723	-
Non-specified	-	-	-	-	14969	-	-	-	-	-
NON-ENERGY USE	-	-	-	-	-	-	-	-	-	-

APPROVISIONNEMENT ET DEMANDE 1998	Gaz (TJ)				En. Re. Comb. & Déchets (TJ)				(GWh)	(TJ)
	Gaz naturel	Usines à gaz	Cokeries	Hauts fourneaux	Biomasse solide	Gaz/Liquides tirés de biomasse	Déchets urbains	Déchets industriels	Electricité	Chaleur
Production	4720	5776	-	272	110592	2016	-	-	14145	-
Imports	-	-	-	-	-	-	-	-	-	-
Exports	-	-	-	-	-	-	-	-	-	-
Intl. Marine Bunkers	-	-	-	-	-	-	-	-	-	-
Stock Changes	-	-	-	-	-	-	-	-	-	-
DOMESTIC SUPPLY	4720	5776	-	272	110592	2016	-	-	14145	-
Transfers and Stat. Diff.	132	-	-	-	-	-	-	-	-1	-
TRANSFORMATION	185	-	-	-	10712	-	-	-	-	-
Electricity and CHP Plants	185	-	-	-	6903	-	-	-	-	-
Petroleum Refineries	-	-	-	-	-	-	-	-	-	-
Other Transform. Sector	-	-	-	-	3809	-	-	-	-	-
ENERGY SECTOR	-	-	-	-	-	-	-	-	958	-
DISTRIBUTION LOSSES	-	-	-	-	-	-	-	-	2578	-
FINAL CONSUMPTION	4667	5776	-	272	99880	2016	-	-	10608	-
INDUSTRY SECTOR	4659	520	-	272	82704	-	-	-	3850	-
Iron and Steel	-	-	-	272	-	-	-	-	100	-
Chemical and Petrochem.	-	-	-	-	-	-	-	-	304	-
Non-Metallic Minerals	-	-	-	-	-	-	-	-	-	-
Non-specified	4659	520	-	-	82704	-	-	-	3446	-
TRANSPORT SECTOR	-	17	-	-	-	2016	-	-	83	-
Air	-	-	-	-	-	-	-	-	-	-
Road	-	-	-	-	-	2016	-	-	-	-
Non-specified	-	17	-	-	-	-	-	-	83	-
OTHER SECTORS	8	5239	-	-	17176	-	-	-	6675	-
Agriculture	8	3	-	-	1524	-	-	-	194	-
Comm. and Publ. Services	-	-	-	-	898	-	-	-	2551	-
Residential	-	5236	-	-	2041	-	-	-	3930	-
Non-specified	-	-	-	-	12713	-	-	-	-	-
NON-ENERGY USE	-	-	-	-	-	-	-	-	-	-

Cyprus / Chypre

SUPPLY AND CONSUMPTION 1997	Coal (1000 tonnes)							Oil (1000 tonnes)			
	Coking Coal	Other Bit. Coal	Sub-Bit. Coal	Lignite	Peat	Oven and Gas Coke	Pat. Fuel and BKB	Crude Oil	NGL	Feed-stocks	Additives
Production	-	-	-	-	-	-	-	-	-	-	-
Imports	-	26	-	-	8	-	-	1039	-	-	-
Exports	-	-	-	-	-	-	-	-	-	-	-
Intl. Marine Bunkers	-	-	-	-	-	-	-	-	-	-	-
Stock Changes	-	-7	-	-	-	-	-	4	-	-	-
DOMESTIC SUPPLY	-	19	-	-	8	-	-	1043	-	-	-
Transfers and Stat. Diff.	-	-	-	-	-	-	-	-	-	-	-
TRANSFORMATION	-	-	-	-	-	-	-	1043	-	-	-
Electricity and CHP Plants	-	-	-	-	-	-	-	-	-	-	-
Petroleum Refineries	-	-	-	-	-	-	-	1043	-	-	-
Other Transform. Sector	-	-	-	-	-	-	-	-	-	-	-
ENERGY SECTOR	-	-	-	-	-	-	-	-	-	-	-
DISTRIBUTION LOSSES	-	-	-	-	-	-	-	-	-	-	-
FINAL CONSUMPTION	-	19	-	-	8	-	-	-	-	-	-
INDUSTRY SECTOR	-	19	-	-	8	-	-	-	-	-	-
Iron and Steel	-	-	-	-	-	-	-	-	-	-	-
Chemical and Petrochem.	-	-	-	-	-	-	-	-	-	-	-
Non-Metallic Minerals	-	19	-	-	-	-	-	-	-	-	-
Non-specified	-	-	-	-	8	-	-	-	-	-	-
TRANSPORT SECTOR	-	-	-	-	-	-	-	-	-	-	-
Air	-	-	-	-	-	-	-	-	-	-	-
Road	-	-	-	-	-	-	-	-	-	-	-
Non-specified	-	-	-	-	-	-	-	-	-	-	-
OTHER SECTORS	-	-	-	-	-	-	-	-	-	-	-
Agriculture	-	-	-	-	-	-	-	-	-	-	-
Comm. and Publ. Services	-	-	-	-	-	-	-	-	-	-	-
Residential	-	-	-	-	-	-	-	-	-	-	-
Non-specified	-	-	-	-	-	-	-	-	-	-	-
NON-ENERGY USE	-	-	-	-	-	-	-	-	-	-	-

APPROVISIONNEMENT ET DEMANDE 1998	Charbon (1000 tonnes)							Pétrole (1000 tonnes)			
	Charbon à coke	Autres charb. bit.	Charbon sous-bit.	Lignite	Tourbe	Coke de four/gaz	Agg./briq. de lignite	Pétrole brut	LGN	Produits d'aliment.	Additifs
Production	-	-	-	-	-	-	-	-	-	-	-
Imports	-	21	-	-	10	-	-	1075	-	-	-
Exports	-	-	-	-	-	-	-	-	-	-	-
Intl. Marine Bunkers	-	-	-	-	-	-	-	-	-	-	-
Stock Changes	-	5	-	-	-	-	-	7	-	-	-
DOMESTIC SUPPLY	-	26	-	-	10	-	-	1082	-	-	-
Transfers and Stat. Diff.	-	-	-	-	-	-	-	-	-	-	-
TRANSFORMATION	-	-	-	-	-	-	-	1082	-	-	-
Electricity and CHP Plants	-	-	-	-	-	-	-	-	-	-	-
Petroleum Refineries	-	-	-	-	-	-	-	1082	-	-	-
Other Transform. Sector	-	-	-	-	-	-	-	-	-	-	-
ENERGY SECTOR	-	-	-	-	-	-	-	-	-	-	-
DISTRIBUTION LOSSES	-	-	-	-	-	-	-	-	-	-	-
FINAL CONSUMPTION	-	26	-	-	10	-	-	-	-	-	-
INDUSTRY SECTOR	-	26	-	-	10	-	-	-	-	-	-
Iron and Steel	-	-	-	-	-	-	-	-	-	-	-
Chemical and Petrochem.	-	-	-	-	-	-	-	-	-	-	-
Non-Metallic Minerals	-	26	-	-	-	-	-	-	-	-	-
Non-specified	-	-	-	-	10	-	-	-	-	-	-
TRANSPORT SECTOR	-	-	-	-	-	-	-	-	-	-	-
Air	-	-	-	-	-	-	-	-	-	-	-
Road	-	-	-	-	-	-	-	-	-	-	-
Non-specified	-	-	-	-	-	-	-	-	-	-	-
OTHER SECTORS	-	-	-	-	-	-	-	-	-	-	-
Agriculture	-	-	-	-	-	-	-	-	-	-	-
Comm. and Publ. Services	-	-	-	-	-	-	-	-	-	-	-
Residential	-	-	-	-	-	-	-	-	-	-	-
Non-specified	-	-	-	-	-	-	-	-	-	-	-
NON-ENERGY USE	-	-	-	-	-	-	-	-	-	-	-

Cyprus / Chypre

SUPPLY AND CONSUMPTION 1997	Oil cont. (1000 tonnes)										
	Refinery Gas	LPG + Ethane	Motor Gasoline	Aviation Gasoline	Jet Fuel	Kerosene	Gas/ Diesel	Heavy Fuel Oil	Naphtha	Petrol. Coke	Other Prod.
Production	16	32	141	-	5	20	365	421	-	-	38
Imports	-	17	50	-	252	-	146	476	-	142	34
Exports	-	-	-	-	-	-	-	-	-	-	-
Intl. Marine Bunkers	-	-	-	-	-	-	-27	-71	-	-	-1
Stock Changes	-	-	3	-	-5	-	-8	-3	-	10	2
DOMESTIC SUPPLY	16	49	194	-	252	20	476	823	-	152	73
Transfers and Stat. Diff.	-	3	-3	-	-7	-	13	4	-	-	2
TRANSFORMATION	-	-	-	-	-	-	6	743	-	-	-
Electricity and CHP Plants	-	-	-	-	-	-	6	743	-	-	-
Petroleum Refineries	-	-	-	-	-	-	-	-	-	-	-
Other Transform. Sector	-	-	-	-	-	-	-	-	-	-	-
ENERGY SECTOR	16	-	-	-	-	-	-	14	-	-	-
DISTRIBUTION LOSSES	-	-	-	-	-	-	-	-	-	-	-
FINAL CONSUMPTION	-	52	191	-	245	20	483	70	-	152	75
INDUSTRY SECTOR	-	-	-	-	-	-	169	70	-	152	-
Iron and Steel	-	-	-	-	-	-	-	-	-	-	-
Chemical and Petrochem.	-	-	-	-	-	-	-	-	-	-	-
Non-Metallic Minerals	-	-	-	-	-	-	-	70	-	152	-
Non-specified	-	-	-	-	-	-	169	-	-	-	-
TRANSPORT SECTOR	-	-	191	-	245	-	314	-	-	-	-
Air	-	-	-	-	245	-	-	-	-	-	-
Road	-	-	191	-	-	-	314	-	-	-	-
Non-specified	-	-	-	-	-	-	-	-	-	-	-
OTHER SECTORS	-	52	-	-	-	20	-	-	-	-	-
Agriculture	-	-	-	-	-	-	-	-	-	-	-
Comm. and Publ. Services	-	-	-	-	-	-	-	-	-	-	-
Residential	-	52	-	-	-	20	-	-	-	-	-
Non-specified	-	-	-	-	-	-	-	-	-	-	-
NON-ENERGY USE	-	-	-	-	-	-	-	-	-	-	75

APPROVISIONNEMENT ET DEMANDE 1998	Pétrole cont. (1000 tonnes)										
	Gaz de raffinerie	GPL + éthane	Essence moteur	Essence aviation	Carbu- réacteurs	Kérosène	Gazole	Fioul lourd	Naphta	Coke de pétrole	Autres prod.
Production	16	30	142	-	4	23	382	442	-	-	37
Imports	-	20	51	-	260	-	175	515	-	133	44
Exports	-	-	-	-	-	-	-	-	-	-	-
Intl. Marine Bunkers	-	-	-	-	-	-	-35	-63	-	-	-1
Stock Changes	-	-	-2	-	4	-2	-5	10	-	17	2
DOMESTIC SUPPLY	16	50	191	-	268	21	517	904	-	150	82
Transfers and Stat. Diff.	-	-	4	-	-10	-	9	-10	-	-	-
TRANSFORMATION	-	-	-	-	-	-	12	811	-	-	-
Electricity and CHP Plants	-	-	-	-	-	-	12	811	-	-	-
Petroleum Refineries	-	-	-	-	-	-	-	-	-	-	-
Other Transform. Sector	-	-	-	-	-	-	-	-	-	-	-
ENERGY SECTOR	16	-	-	-	-	-	-	15	-	-	-
DISTRIBUTION LOSSES	-	-	-	-	-	-	-	-	-	-	-
FINAL CONSUMPTION	-	50	195	-	258	21	514	68	-	150	82
INDUSTRY SECTOR	-	-	-	-	-	-	180	68	-	150	-
Iron and Steel	-	-	-	-	-	-	-	-	-	-	-
Chemical and Petrochem.	-	-	-	-	-	-	-	-	-	-	-
Non-Metallic Minerals	-	-	-	-	-	-	-	68	-	150	-
Non-specified	-	-	-	-	-	-	180	-	-	-	-
TRANSPORT SECTOR	-	-	195	-	258	-	334	-	-	-	-
Air	-	-	-	-	258	-	-	-	-	-	-
Road	-	-	195	-	-	-	334	-	-	-	-
Non-specified	-	-	-	-	-	-	-	-	-	-	-
OTHER SECTORS	-	50	-	-	-	21	-	-	-	-	-
Agriculture	-	-	-	-	-	-	-	-	-	-	-
Comm. and Publ. Services	-	-	-	-	-	-	-	-	-	-	-
Residential	-	50	-	-	-	21	-	-	-	-	-
Non-specified	-	-	-	-	-	-	-	-	-	-	-
NON-ENERGY USE	-	-	-	-	-	-	-	-	-	-	82

Cyprus / Chypre

SUPPLY AND CONSUMPTION 1997	Gas (TJ)				Comb. Renew. & Waste (TJ)				(GWh)	(TJ)
	Natural Gas	Gas Works	Coke Ovens	Blast Furnaces	Solid Biomass	Gas/Liquids from Biomass	Municipal Waste	Industrial Waste	Electricity	Heat
Production	-	-	-	-	358	-	-	-	2711	-
Imports	-	-	-	-	119	-	-	-	-	-
Exports	-	-	-	-	-	-	-	-	-	-
Intl. Marine Bunkers	-	-	-	-	-	-	-	-	-	-
Stock Changes	-	-	-	-	-	-	-	-	-	-
DOMESTIC SUPPLY	-	-	-	-	477	-	-	-	2711	-
Transfers and Stat. Diff.	-	-	-	-	-	-	-	-	-	-
TRANSFORMATION	-	-	-	-	187	-	-	-	-	-
Electricity and CHP Plants	-	-	-	-	-	-	-	-	-	-
Petroleum Refineries	-	-	-	-	-	-	-	-	-	-
Other Transform. Sector	-	-	-	-	187	-	-	-	-	-
ENERGY SECTOR	-	-	-	-	-	-	-	-	156	-
DISTRIBUTION LOSSES	-	-	-	-	-	-	-	-	173	-
FINAL CONSUMPTION	-	-	-	-	290	-	-	-	2382	-
INDUSTRY SECTOR	-	-	-	-	-	-	-	-	395	-
Iron and Steel	-	-	-	-	-	-	-	-	-	-
Chemical and Petrochem.	-	-	-	-	-	-	-	-	11	-
Non-Metallic Minerals	-	-	-	-	-	-	-	-	165	-
Non-specified	-	-	-	-	-	-	-	-	219	-
TRANSPORT SECTOR	-	-	-	-	-	-	-	-	21	-
Air	-	-	-	-	-	-	-	-	-	-
Road	-	-	-	-	-	-	-	-	-	-
Non-specified	-	-	-	-	-	-	-	-	21	-
OTHER SECTORS	-	-	-	-	290	-	-	-	1966	-
Agriculture	-	-	-	-	-	-	-	-	76	-
Comm. and Publ. Services	-	-	-	-	-	-	-	-	1000	-
Residential	-	-	-	-	160	-	-	-	834	-
Non-specified	-	-	-	-	130	-	-	-	56	-
NON-ENERGY USE	-	-	-	-	-	-	-	-	-	-

APPROVISIONNEMENT ET DEMANDE 1998	Gaz (TJ)				En. Re. Comb. & Déchets (TJ)				(GWh)	(TJ)
	Gaz naturel	Usines à gaz	Cokeries	Hauts fourneaux	Biomasse solide	Gaz/Liquides tirés de biomasse	Déchets urbains	Déchets industriels	Electricité	Chaleur
Production	-	-	-	-	378	-	-	-	2954	-
Imports	-	-	-	-	99	-	-	-	-	-
Exports	-	-	-	-	-	-	-	-	-	-
Intl. Marine Bunkers	-	-	-	-	-	-	-	-	-	-
Stock Changes	-	-	-	-	-	-	-	-	-	-
DOMESTIC SUPPLY	-	-	-	-	477	-	-	-	2954	-
Transfers and Stat. Diff.	-	-	-	-	-	-	-	-	-	-
TRANSFORMATION	-	-	-	-	204	-	-	-	-	-
Electricity and CHP Plants	-	-	-	-	-	-	-	-	-	-
Petroleum Refineries	-	-	-	-	-	-	-	-	-	-
Other Transform. Sector	-	-	-	-	204	-	-	-	-	-
ENERGY SECTOR	-	-	-	-	-	-	-	-	163	-
DISTRIBUTION LOSSES	-	-	-	-	-	-	-	-	176	-
FINAL CONSUMPTION	-	-	-	-	273	-	-	-	2615	-
INDUSTRY SECTOR	-	-	-	-	-	-	-	-	421	-
Iron and Steel	-	-	-	-	-	-	-	-	-	-
Chemical and Petrochem.	-	-	-	-	-	-	-	-	12	-
Non-Metallic Minerals	-	-	-	-	-	-	-	-	181	-
Non-specified	-	-	-	-	-	-	-	-	228	-
TRANSPORT SECTOR	-	-	-	-	-	-	-	-	24	-
Air	-	-	-	-	-	-	-	-	-	-
Road	-	-	-	-	-	-	-	-	-	-
Non-specified	-	-	-	-	-	-	-	-	24	-
OTHER SECTORS	-	-	-	-	273	-	-	-	2170	-
Agriculture	-	-	-	-	-	-	-	-	85	-
Comm. and Publ. Services	-	-	-	-	-	-	-	-	1053	-
Residential	-	-	-	-	152	-	-	-	904	-
Non-specified	-	-	-	-	121	-	-	-	128	-
NON-ENERGY USE	-	-	-	-	-	-	-	-	-	-

Dominican Republic / République dominicaine

SUPPLY AND CONSUMPTION 1997	Coal (1000 tonnes)							Oil (1000 tonnes)			
	Coking Coal	Other Bit. Coal	Sub-Bit. Coal	Lignite	Peat	Oven and Gas Coke	Pat. Fuel and BKB	Crude Oil	NGL	Feed-stocks	Additives
Production	-	-	-	-	-	-	-	-	-	-	-
Imports	-	139	-	-	-	-	-	2221	-	-	-
Exports	-	-	-	-	-	-	-	-	-	-	-
Intl. Marine Bunkers	-	-	-	-	-	-	-	-	-	-	-
Stock Changes	-	-	-	-	-	-	-	1	-	-	-
DOMESTIC SUPPLY	-	139	-	-	-	-	-	2222	-	-	-
Transfers and Stat. Diff.	-	-	-	-	-	-	-	-	-	-	-
TRANSFORMATION	-	139	-	-	-	-	-	2222	-	-	-
Electricity and CHP Plants	-	139	-	-	-	-	-	-	-	-	-
Petroleum Refineries	-	-	-	-	-	-	-	2222	-	-	-
Other Transform. Sector	-	-	-	-	-	-	-	-	-	-	-
ENERGY SECTOR	-	-	-	-	-	-	-	-	-	-	-
DISTRIBUTION LOSSES	-	-	-	-	-	-	-	-	-	-	-
FINAL CONSUMPTION	-	-	-	-	-	-	-	-	-	-	-
INDUSTRY SECTOR	-	-	-	-	-	-	-	-	-	-	-
Iron and Steel	-	-	-	-	-	-	-	-	-	-	-
Chemical and Petrochem.	-	-	-	-	-	-	-	-	-	-	-
Non-Metallic Minerals	-	-	-	-	-	-	-	-	-	-	-
Non-specified	-	-	-	-	-	-	-	-	-	-	-
TRANSPORT SECTOR	-	-	-	-	-	-	-	-	-	-	-
Air	-	-	-	-	-	-	-	-	-	-	-
Road	-	-	-	-	-	-	-	-	-	-	-
Non-specified	-	-	-	-	-	-	-	-	-	-	-
OTHER SECTORS	-	-	-	-	-	-	-	-	-	-	-
Agriculture	-	-	-	-	-	-	-	-	-	-	-
Comm. and Publ. Services	-	-	-	-	-	-	-	-	-	-	-
Residential	-	-	-	-	-	-	-	-	-	-	-
Non-specified	-	-	-	-	-	-	-	-	-	-	-
NON-ENERGY USE	-	-	-	-	-	-	-	-	-	-	-

APPROVISIONNEMENT ET DEMANDE 1998	Charbon (1000 tonnes)							Pétrole (1000 tonnes)			
	Charbon à coke	Autres charb. bit.	Charbon sous-bit.	Lignite	Tourbe	Coke de four/gaz	Agg./briq. de lignite	Pétrole brut	LGN	Produits d'aliment.	Additifs
Production	-	-	-	-	-	-	-	-	-	-	-
Imports	-	147	-	-	-	-	-	2283	-	-	-
Exports	-	-	-	-	-	-	-	-	-	-	-
Intl. Marine Bunkers	-	-	-	-	-	-	-	-	-	-	-
Stock Changes	-	-	-	-	-	-	-	-	-	-	-
DOMESTIC SUPPLY	-	147	-	-	-	-	-	2283	-	-	-
Transfers and Stat. Diff.	-	-	-	-	-	-	-	-	-	-	-
TRANSFORMATION	-	147	-	-	-	-	-	2283	-	-	-
Electricity and CHP Plants	-	147	-	-	-	-	-	-	-	-	-
Petroleum Refineries	-	-	-	-	-	-	-	2283	-	-	-
Other Transform. Sector	-	-	-	-	-	-	-	-	-	-	-
ENERGY SECTOR	-	-	-	-	-	-	-	-	-	-	-
DISTRIBUTION LOSSES	-	-	-	-	-	-	-	-	-	-	-
FINAL CONSUMPTION	-	-	-	-	-	-	-	-	-	-	-
INDUSTRY SECTOR	-	-	-	-	-	-	-	-	-	-	-
Iron and Steel	-	-	-	-	-	-	-	-	-	-	-
Chemical and Petrochem.	-	-	-	-	-	-	-	-	-	-	-
Non-Metallic Minerals	-	-	-	-	-	-	-	-	-	-	-
Non-specified	-	-	-	-	-	-	-	-	-	-	-
TRANSPORT SECTOR	-	-	-	-	-	-	-	-	-	-	-
Air	-	-	-	-	-	-	-	-	-	-	-
Road	-	-	-	-	-	-	-	-	-	-	-
Non-specified	-	-	-	-	-	-	-	-	-	-	-
OTHER SECTORS	-	-	-	-	-	-	-	-	-	-	-
Agriculture	-	-	-	-	-	-	-	-	-	-	-
Comm. and Publ. Services	-	-	-	-	-	-	-	-	-	-	-
Residential	-	-	-	-	-	-	-	-	-	-	-
Non-specified	-	-	-	-	-	-	-	-	-	-	-
NON-ENERGY USE	-	-	-	-	-	-	-	-	-	-	-

Dominican Republic / République dominicaine

SUPPLY AND CONSUMPTION 1997	Oil cont. (1000 tonnes)										
	Refinery Gas	LPG + Ethane	Motor Gasoline	Aviation Gasoline	Jet Fuel	Kerosene	Gas/ Diesel	Heavy Fuel Oil	Naphtha	Petrol. Coke	Other Prod.
Production	-	36	369	-	47	200	449	1100	-	-	-
Imports	-	389	252	-	15	63	538	373	-	-	-
Exports	-	-	-	-	-	-	-	-	-	-	-
Intl. Marine Bunkers	-	-	-	-	-	-	-	-	-	-	-
Stock Changes	-	-	-	-	-	-	7	3	-	-	-
DOMESTIC SUPPLY	-	425	621	-	62	263	994	1476	-	-	-
Transfers and Stat. Diff.	-	-	-	-	-	-	1	1	-	-	-
TRANSFORMATION	-	-	30	-	-	-	338	1293	-	-	-
Electricity and CHP Plants	-	-	30	-	-	-	338	1293	-	-	-
Petroleum Refineries	-	-	-	-	-	-	-	-	-	-	-
Other Transform. Sector	-	-	-	-	-	-	-	-	-	-	-
ENERGY SECTOR	-	-	-	-	-	-	-	-	-	-	-
DISTRIBUTION LOSSES	-	-	-	-	-	-	-	-	-	-	-
FINAL CONSUMPTION	-	425	591	-	62	263	657	184	-	-	-
INDUSTRY SECTOR	-	38	-	-	-	-	159	176	-	-	-
Iron and Steel	-	-	-	-	-	-	-	-	-	-	-
Chemical and Petrochem.	-	-	-	-	-	-	-	-	-	-	-
Non-Metallic Minerals	-	-	-	-	-	-	-	-	-	-	-
Non-specified	-	38	-	-	-	-	159	176	-	-	-
TRANSPORT SECTOR	-	90	591	-	62	-	418	8	-	-	-
Air	-	-	-	-	62	-	-	-	-	-	-
Road	-	-	591	-	-	-	-	-	-	-	-
Non-specified	-	90	-	-	-	-	418	8	-	-	-
OTHER SECTORS	-	297	-	-	-	263	80	-	-	-	-
Agriculture	-	-	-	-	-	-	36	-	-	-	-
Comm. and Publ. Services	-	-	-	-	-	-	-	-	-	-	-
Residential	-	297	-	-	-	263	44	-	-	-	-
Non-specified	-	-	-	-	-	-	-	-	-	-	-
NON-ENERGY USE	-	-	-	-	-	-	-	-	-	-	-

APPROVISIONNEMENT ET DEMANDE 1998	Pétrole cont. (1000 tonnes)										
	Gaz de raffinerie	GPL + éthane	Essence moteur	Essence aviation	Carbu- réacteurs	Kérosène	Gazole	Fioul lourd	Naphta	Coke de pétrole	Autres prod.
Production	-	38	375	-	51	217	464	1117	-	-	-
Imports	-	398	257	-	16	67	544	395	-	-	-
Exports	-	-	-	-	-	-	-	-	-	-	-
Intl. Marine Bunkers	-	-	-	-	-	-	-	-	-	-	-
Stock Changes	-	-	-	-	-	-	5	2	-	-	-
DOMESTIC SUPPLY	-	436	632	-	67	284	1013	1514	-	-	-
Transfers and Stat. Diff.	-	-1	-	-	-1	-1	2	-	-	-	-
TRANSFORMATION	-	-	30	-	-	-	350	1314	-	-	-
Electricity and CHP Plants	-	-	30	-	-	-	350	1314	-	-	-
Petroleum Refineries	-	-	-	-	-	-	-	-	-	-	-
Other Transform. Sector	-	-	-	-	-	-	-	-	-	-	-
ENERGY SECTOR	-	-	-	-	-	-	-	-	-	-	-
DISTRIBUTION LOSSES	-	-	-	-	-	-	-	-	-	-	-
FINAL CONSUMPTION	-	435	602	-	66	283	665	200	-	-	-
INDUSTRY SECTOR	-	39	-	-	-	-	161	190	-	-	-
Iron and Steel	-	-	-	-	-	-	-	-	-	-	-
Chemical and Petrochem.	-	-	-	-	-	-	-	-	-	-	-
Non-Metallic Minerals	-	-	-	-	-	-	-	-	-	-	-
Non-specified	-	39	-	-	-	-	161	190	-	-	-
TRANSPORT SECTOR	-	92	602	-	66	-	423	10	-	-	-
Air	-	-	-	-	66	-	-	-	-	-	-
Road	-	-	602	-	-	-	-	-	-	-	-
Non-specified	-	92	-	-	-	-	423	10	-	-	-
OTHER SECTORS	-	304	-	-	-	283	81	-	-	-	-
Agriculture	-	-	-	-	-	-	37	-	-	-	-
Comm. and Publ. Services	-	-	-	-	-	-	-	-	-	-	-
Residential	-	304	-	-	-	283	44	-	-	-	-
Non-specified	-	-	-	-	-	-	-	-	-	-	-
NON-ENERGY USE	-	-	-	-	-	-	-	-	-	-	-

Dominican Republic / République dominicaine

SUPPLY AND CONSUMPTION 1997	Gas (TJ) Natural Gas	Gas Works	Coke Ovens	Blast Furnaces	Comb. Renew. & Waste (TJ) Solid Biomass	Gas/Liquids from Biomass	Municipal Waste	Industrial Waste	(GWh) Electricity	(TJ) Heat
Production	-	-	-	-	54785	-	-	-	7335	-
Imports	-	-	-	-	-	-	-	-	-	-
Exports	-	-	-	-	-	-	-	-	-	-
Intl. Marine Bunkers	-	-	-	-	-	-	-	-	-	-
Stock Changes	-	-	-	-	-	-	-	-	-	-
DOMESTIC SUPPLY	-	-	-	-	54785	-	-	-	7335	-
Transfers and Stat. Diff.	-	-	-	-	-	-	-	-	-	-
TRANSFORMATION	-	-	-	-	8809	-	-	-	-	-
Electricity and CHP Plants	-	-	-	-	854	-	-	-	-	-
Petroleum Refineries	-	-	-	-	-	-	-	-	-	-
Other Transform. Sector	-	-	-	-	7955	-	-	-	-	-
ENERGY SECTOR	-	-	-	-	-	-	-	-	276	-
DISTRIBUTION LOSSES	-	-	-	-	-	-	-	-	2036	-
FINAL CONSUMPTION	-	-	-	-	45976	-	-	-	5023	-
INDUSTRY SECTOR	-	-	-	-	12116	-	-	-	1392	-
Iron and Steel	-	-	-	-	-	-	-	-	-	-
Chemical and Petrochem.	-	-	-	-	-	-	-	-	-	-
Non-Metallic Minerals	-	-	-	-	-	-	-	-	-	-
Non-specified	-	-	-	-	12116	-	-	-	1392	-
TRANSPORT SECTOR	-	-	-	-	-	-	-	-	-	-
Air	-	-	-	-	-	-	-	-	-	-
Road	-	-	-	-	-	-	-	-	-	-
Non-specified	-	-	-	-	-	-	-	-	-	-
OTHER SECTORS	-	-	-	-	33860	-	-	-	3631	-
Agriculture	-	-	-	-	-	-	-	-	-	-
Comm. and Publ. Services	-	-	-	-	-	-	-	-	-	-
Residential	-	-	-	-	33860	-	-	-	3631	-
Non-specified	-	-	-	-	-	-	-	-	-	-
NON-ENERGY USE	-	-	-	-	-	-	-	-	-	-

APPROVISIONNEMENT ET DEMANDE 1998	Gaz (TJ) Gaz naturel	Usines à gaz	Cokeries	Hauts fourneaux	En. Re. Comb. & Déchets (TJ) Biomasse solide	Gaz/Liquides tirés de biomasse	Déchets urbains	Déchets industriels	(GWh) Electricité	(TJ) Chaleur
Production	-	-	-	-	54898	-	-	-	7555	-
Imports	-	-	-	-	-	-	-	-	-	-
Exports	-	-	-	-	-	-	-	-	-	-
Intl. Marine Bunkers	-	-	-	-	-	-	-	-	-	-
Stock Changes	-	-	-	-	-	-	-	-	-	-
DOMESTIC SUPPLY	-	-	-	-	54898	-	-	-	7555	-
Transfers and Stat. Diff.	-	-	-	-	-	-	-	-	-	-
TRANSFORMATION	-	-	-	-	8695	-	-	-	-	-
Electricity and CHP Plants	-	-	-	-	854	-	-	-	-	-
Petroleum Refineries	-	-	-	-	-	-	-	-	-	-
Other Transform. Sector	-	-	-	-	7841	-	-	-	-	-
ENERGY SECTOR	-	-	-	-	-	-	-	-	282	-
DISTRIBUTION LOSSES	-	-	-	-	-	-	-	-	2095	-
FINAL CONSUMPTION	-	-	-	-	46203	-	-	-	5178	-
INDUSTRY SECTOR	-	-	-	-	12168	-	-	-	1432	-
Iron and Steel	-	-	-	-	-	-	-	-	-	-
Chemical and Petrochem.	-	-	-	-	-	-	-	-	-	-
Non-Metallic Minerals	-	-	-	-	-	-	-	-	-	-
Non-specified	-	-	-	-	12168	-	-	-	1432	-
TRANSPORT SECTOR	-	-	-	-	-	-	-	-	-	-
Air	-	-	-	-	-	-	-	-	-	-
Road	-	-	-	-	-	-	-	-	-	-
Non-specified	-	-	-	-	-	-	-	-	-	-
OTHER SECTORS	-	-	-	-	34035	-	-	-	3746	-
Agriculture	-	-	-	-	-	-	-	-	-	-
Comm. and Publ. Services	-	-	-	-	-	-	-	-	-	-
Residential	-	-	-	-	34035	-	-	-	3746	-
Non-specified	-	-	-	-	-	-	-	-	-	-
NON-ENERGY USE	-	-	-	-	-	-	-	-	-	-

Ecuador / Equateur

SUPPLY AND CONSUMPTION 1997	Coal (1000 tonnes)							Oil (1000 tonnes)			
	Coking Coal	Other Bit. Coal	Sub-Bit. Coal	Lignite	Peat	Oven and Gas Coke	Pat. Fuel and BKB	Crude Oil	NGL	Feed-stocks	Additives
Production	-	-	-	-	-	-	-	21029	305	-	-
Imports	-	-	-	-	-	-	-	-	-	-	-
Exports	-	-	-	-	-	-	-	-13376	-	-	-
Intl. Marine Bunkers	-	-	-	-	-	-	-	-	-	-	-
Stock Changes	-	-	-	-	-	-	-	-25	-	-	-
DOMESTIC SUPPLY	-	-	-	-	-	-	-	7628	305	-	-
Transfers and Stat. Diff.	-	-	-	-	-	-	-	1	-305	282	-
TRANSFORMATION	-	-	-	-	-	-	-	7481	-	282	-
Electricity and CHP Plants	-	-	-	-	-	-	-	-	-	-	-
Petroleum Refineries	-	-	-	-	-	-	-	7481	-	282	-
Other Transform. Sector	-	-	-	-	-	-	-	-	-	-	-
ENERGY SECTOR	-	-	-	-	-	-	-	82	-	-	-
DISTRIBUTION LOSSES	-	-	-	-	-	-	-	66	-	-	-
FINAL CONSUMPTION	-	-	-	-	-	-	-	-	-	-	-
INDUSTRY SECTOR	-	-	-	-	-	-	-	-	-	-	-
Iron and Steel	-	-	-	-	-	-	-	-	-	-	-
Chemical and Petrochem.	-	-	-	-	-	-	-	-	-	-	-
Non-Metallic Minerals	-	-	-	-	-	-	-	-	-	-	-
Non-specified	-	-	-	-	-	-	-	-	-	-	-
TRANSPORT SECTOR	-	-	-	-	-	-	-	-	-	-	-
Air	-	-	-	-	-	-	-	-	-	-	-
Road	-	-	-	-	-	-	-	-	-	-	-
Non-specified	-	-	-	-	-	-	-	-	-	-	-
OTHER SECTORS	-	-	-	-	-	-	-	-	-	-	-
Agriculture	-	-	-	-	-	-	-	-	-	-	-
Comm. and Publ. Services	-	-	-	-	-	-	-	-	-	-	-
Residential	-	-	-	-	-	-	-	-	-	-	-
Non-specified	-	-	-	-	-	-	-	-	-	-	-
NON-ENERGY USE	-	-	-	-	-	-	-	-	-	-	-

APPROVISIONNEMENT ET DEMANDE 1998	Charbon (1000 tonnes)							Pétrole (1000 tonnes)			
	Charbon à coke	Autres charb. bit.	Charbon sous-bit.	Lignite	Tourbe	Coke de four/gaz	Agg./briq. de lignite	Pétrole brut	LGN	Produits d'aliment.	Additifs
Production	-	-	-	-	-	-	-	20496	305	-	-
Imports	-	-	-	-	-	-	-	-	-	-	-
Exports	-	-	-	-	-	-	-	-12659	-	-	-
Intl. Marine Bunkers	-	-	-	-	-	-	-	-	-	-	-
Stock Changes	-	-	-	-	-	-	-	-271	-	-	-
DOMESTIC SUPPLY	-	-	-	-	-	-	-	7566	305	-	-
Transfers and Stat. Diff.	-	-	-	-	-	-	-	-	-305	303	-
TRANSFORMATION	-	-	-	-	-	-	-	7412	-	303	-
Electricity and CHP Plants	-	-	-	-	-	-	-	-	-	-	-
Petroleum Refineries	-	-	-	-	-	-	-	7412	-	303	-
Other Transform. Sector	-	-	-	-	-	-	-	-	-	-	-
ENERGY SECTOR	-	-	-	-	-	-	-	82	-	-	-
DISTRIBUTION LOSSES	-	-	-	-	-	-	-	72	-	-	-
FINAL CONSUMPTION	-	-	-	-	-	-	-	-	-	-	-
INDUSTRY SECTOR	-	-	-	-	-	-	-	-	-	-	-
Iron and Steel	-	-	-	-	-	-	-	-	-	-	-
Chemical and Petrochem.	-	-	-	-	-	-	-	-	-	-	-
Non-Metallic Minerals	-	-	-	-	-	-	-	-	-	-	-
Non-specified	-	-	-	-	-	-	-	-	-	-	-
TRANSPORT SECTOR	-	-	-	-	-	-	-	-	-	-	-
Air	-	-	-	-	-	-	-	-	-	-	-
Road	-	-	-	-	-	-	-	-	-	-	-
Non-specified	-	-	-	-	-	-	-	-	-	-	-
OTHER SECTORS	-	-	-	-	-	-	-	-	-	-	-
Agriculture	-	-	-	-	-	-	-	-	-	-	-
Comm. and Publ. Services	-	-	-	-	-	-	-	-	-	-	-
Residential	-	-	-	-	-	-	-	-	-	-	-
Non-specified	-	-	-	-	-	-	-	-	-	-	-
NON-ENERGY USE	-	-	-	-	-	-	-	-	-	-	-

Ecuador / Equateur

SUPPLY AND CONSUMPTION 1997	Oil cont. (1000 tonnes)										
	Refinery Gas	LPG + Ethane	Motor Gasoline	Aviation Gasoline	Jet Fuel	Kerosene	Gas/ Diesel	Heavy Fuel Oil	Naphtha	Petrol. Coke	Other Prod.
Production	-	112	1319	7	247	50	1853	3706	-	-	113
Imports	-	410	-	-	5	-	420	-	-	-	-
Exports	-	-	-	-	-	-	-21	-2184	-	-	-
Intl. Marine Bunkers	-	-	-	-	-	-	-	-445	-	-	-
Stock Changes	-	-	-	-	-	-3	31	-14	-	-	-
DOMESTIC SUPPLY	-	522	1319	7	252	47	2283	1063	-	-	113
Transfers and Stat. Diff.	-	111	11	-	-40	-17	-181	6	-	-	-
TRANSFORMATION	-	-	-	-	-	-	282	488	-	-	-
Electricity and CHP Plants	-	-	-	-	-	-	282	488	-	-	-
Petroleum Refineries	-	-	-	-	-	-	-	-	-	-	-
Other Transform. Sector	-	-	-	-	-	-	-	-	-	-	-
ENERGY SECTOR	-	-	4	-	-	-	27	2	-	-	-
DISTRIBUTION LOSSES	-	-	-	-	-	-	-	-	-	-	-
FINAL CONSUMPTION	-	633	1326	7	212	30	1793	579	-	-	113
INDUSTRY SECTOR	-	31	185	-	-	7	243	267	-	-	-
Iron and Steel	-	-	-	-	-	-	-	-	-	-	-
Chemical and Petrochem.	-	-	-	-	-	-	-	-	-	-	-
Non-Metallic Minerals	-	-	-	-	-	-	-	-	-	-	-
Non-specified	-	31	185	-	-	7	243	267	-	-	-
TRANSPORT SECTOR	-	-	1028	7	212	-	1022	301	-	-	-
Air	-	-	-	7	212	-	-	-	-	-	-
Road	-	-	1028	-	-	-	824	-	-	-	-
Non-specified	-	-	-	-	-	-	198	301	-	-	-
OTHER SECTORS	-	602	113	-	-	23	528	11	-	-	-
Agriculture	-	19	30	-	-	3	310	-	-	-	-
Comm. and Publ. Services	-	-	43	-	-	1	218	11	-	-	-
Residential	-	558	40	-	-	10	-	-	-	-	-
Non-specified	-	25	-	-	-	9	-	-	-	-	-
NON-ENERGY USE	-	-	-	-	-	-	-	-	-	-	113

APPROVISIONNEMENT ET DEMANDE 1998	Pétrole cont. (1000 tonnes)										
	Gaz de raffinerie	GPL + éthane	Essence moteur	Essence aviation	Carbu- réacteurs	Kérosène	Gazole	Fioul lourd	Naphta	Coke de pétrole	Autres prod.
Production	-	119	1335	7	256	50	1303	3301	-	-	148
Imports	-	404	43	-	5	-	964	-	-	-	-
Exports	-	-	-	-	-74	-	-14	-1793	-	-	-
Intl. Marine Bunkers	-	-	-	-	-	-	-	-365	-	-	-
Stock Changes	-	2	17	-	-	-3	30	-	-	-	-
DOMESTIC SUPPLY	-	525	1395	7	187	47	2283	1143	-	-	148
Transfers and Stat. Diff.	-	111	-36	-	-36	-17	-91	66	-	-	-
TRANSFORMATION	-	-	-	-	-	-	397	675	-	-	-
Electricity and CHP Plants	-	-	-	-	-	-	397	675	-	-	-
Petroleum Refineries	-	-	-	-	-	-	-	-	-	-	-
Other Transform. Sector	-	-	-	-	-	-	-	-	-	-	-
ENERGY SECTOR	-	-	4	-	-	-	31	-	-	-	-
DISTRIBUTION LOSSES	-	-	-	-	-	-	-	-	-	-	-
FINAL CONSUMPTION	-	636	1355	7	151	30	1764	534	-	-	148
INDUSTRY SECTOR	-	31	189	-	-	7	243	246	-	-	-
Iron and Steel	-	-	-	-	-	-	-	-	-	-	-
Chemical and Petrochem.	-	-	-	-	-	-	-	-	-	-	-
Non-Metallic Minerals	-	-	-	-	-	-	-	-	-	-	-
Non-specified	-	31	189	-	-	7	243	246	-	-	-
TRANSPORT SECTOR	-	-	1050	7	151	-	1001	277	-	-	-
Air	-	-	-	7	151	-	-	-	-	-	-
Road	-	-	1050	-	-	-	807	-	-	-	-
Non-specified	-	-	-	-	-	-	194	277	-	-	-
OTHER SECTORS	-	605	116	-	-	23	520	11	-	-	-
Agriculture	-	19	31	-	-	3	305	-	-	-	-
Comm. and Publ. Services	-	-	45	-	-	1	215	11	-	-	-
Residential	-	561	40	-	-	10	-	-	-	-	-
Non-specified	-	25	-	-	-	9	-	-	-	-	-
NON-ENERGY USE	-	-	-	-	-	-	-	-	-	-	148

Ecuador / Equateur

SUPPLY AND CONSUMPTION 1997	Gas (TJ)				Comb. Renew. & Waste (TJ)				(GWh)	(TJ)
	Natural Gas	Gas Works	Coke Ovens	Blast Furnaces	Solid Biomass	Gas/Liquids from Biomass	Municipal Waste	Industrial Waste	Electricity	Heat
Production	-	-	-	-	47608	-	-	-	9595	-
Imports	-	-	-	-	-	-	-	-	-	-
Exports	-	-	-	-	-	-	-	-	-	-
Intl. Marine Bunkers	-	-	-	-	-	-	-	-	-	-
Stock Changes	-	-	-	-	-	-	-	-	-	-
DOMESTIC SUPPLY	-	-	-	-	47608	-	-	-	9595	-
Transfers and Stat. Diff.	-	-	-	-	-	-	-	-	-12	-
TRANSFORMATION	-	-	-	-	-	-	-	-	-	-
Electricity and CHP Plants	-	-	-	-	-	-	-	-	-	-
Petroleum Refineries	-	-	-	-	-	-	-	-	-	-
Other Transform. Sector	-	-	-	-	-	-	-	-	-	-
ENERGY SECTOR	-	-	-	-	-	-	-	-	128	-
DISTRIBUTION LOSSES	-	-	-	-	-	-	-	-	2163	-
FINAL CONSUMPTION	-	-	-	-	47608	-	-	-	7292	-
INDUSTRY SECTOR	-	-	-	-	13343	-	-	-	2361	-
Iron and Steel	-	-	-	-	-	-	-	-	-	-
Chemical and Petrochem.	-	-	-	-	-	-	-	-	-	-
Non-Metallic Minerals	-	-	-	-	-	-	-	-	-	-
Non-specified	-	-	-	-	13343	-	-	-	2361	-
TRANSPORT SECTOR	-	-	-	-	-	-	-	-	-	-
Air	-	-	-	-	-	-	-	-	-	-
Road	-	-	-	-	-	-	-	-	-	-
Non-specified	-	-	-	-	-	-	-	-	-	-
OTHER SECTORS	-	-	-	-	34265	-	-	-	4931	-
Agriculture	-	-	-	-	-	-	-	-	-	-
Comm. and Publ. Services	-	-	-	-	-	-	-	-	2151	-
Residential	-	-	-	-	34265	-	-	-	2733	-
Non-specified	-	-	-	-	-	-	-	-	47	-
NON-ENERGY USE	-	-	-	-	-	-	-	-	-	-

APPROVISIONNEMENT ET DEMANDE 1998	Gaz (TJ)				En. Re. Comb. & Déchets (TJ)				(GWh)	(TJ)
	Gaz naturel	Usines à gaz	Cokeries	Hauts fourneaux	Biomasse solide	Gaz/Liquides tirés de biomasse	Déchets urbains	Déchets industriels	Electricité	Chaleur
Production	-	-	-	-	48023	-	-	-	9897	-
Imports	-	-	-	-	-	-	-	-	-	-
Exports	-	-	-	-	-	-	-	-	-	-
Intl. Marine Bunkers	-	-	-	-	-	-	-	-	-	-
Stock Changes	-	-	-	-	-	-	-	-	-	-
DOMESTIC SUPPLY	-	-	-	-	48023	-	-	-	9897	-
Transfers and Stat. Diff.	-	-	-	-	-	-	-	-	-11	-
TRANSFORMATION	-	-	-	-	-	-	-	-	-	-
Electricity and CHP Plants	-	-	-	-	-	-	-	-	-	-
Petroleum Refineries	-	-	-	-	-	-	-	-	-	-
Other Transform. Sector	-	-	-	-	-	-	-	-	-	-
ENERGY SECTOR	-	-	-	-	-	-	-	-	163	-
DISTRIBUTION LOSSES	-	-	-	-	-	-	-	-	2117	-
FINAL CONSUMPTION	-	-	-	-	48024	-	-	-	7606	-
INDUSTRY SECTOR	-	-	-	-	13721	-	-	-	2457	-
Iron and Steel	-	-	-	-	-	-	-	-	-	-
Chemical and Petrochem.	-	-	-	-	-	-	-	-	-	-
Non-Metallic Minerals	-	-	-	-	-	-	-	-	-	-
Non-specified	-	-	-	-	13721	-	-	-	2457	-
TRANSPORT SECTOR	-	-	-	-	-	-	-	-	-	-
Air	-	-	-	-	-	-	-	-	-	-
Road	-	-	-	-	-	-	-	-	-	-
Non-specified	-	-	-	-	-	-	-	-	-	-
OTHER SECTORS	-	-	-	-	34303	-	-	-	5149	-
Agriculture	-	-	-	-	-	-	-	-	-	-
Comm. and Publ. Services	-	-	-	-	-	-	-	-	2248	-
Residential	-	-	-	-	34303	-	-	-	2853	-
Non-specified	-	-	-	-	-	-	-	-	48	-
NON-ENERGY USE	-	-	-	-	-	-	-	-	-	-

Egypt / Egypte

SUPPLY AND CONSUMPTION 1997	Coal (1000 tonnes)							Oil (1000 tonnes)			
	Coking Coal	Other Bit. Coal	Sub-Bit. Coal	Lignite	Peat	Oven and Gas Coke	Pat. Fuel and BKB	Crude Oil	NGL	Feed-stocks	Additives
Production	-	-	-	-	-	1938	-	41167	2629	-	-
Imports	1876	-	-	-	-	3	-	-	-	-	-
Exports	-	-	-	-	-	-394	-	-6143	-	-	-
Intl. Marine Bunkers	-	-	-	-	-	-	-	-	-	-	-
Stock Changes	-	-	-	-	-	-	-	-6375	-	-	-
DOMESTIC SUPPLY	1876	-	-	-	-	1547	-	28649	2629	-	-
Transfers and Stat. Diff.	351	-	-	-	-	-	-	-	-986	-	-
TRANSFORMATION	2227	-	-	-	-	1179	-	28649	1643	-	-
Electricity and CHP Plants	-	-	-	-	-	-	-	-	-	-	-
Petroleum Refineries	-	-	-	-	-	-	-	28649	1643	-	-
Other Transform. Sector	2227	-	-	-	-	1179	-	-	-	-	-
ENERGY SECTOR	-	-	-	-	-	-	-	-	-	-	-
DISTRIBUTION LOSSES	-	-	-	-	-	-	-	-	-	-	-
FINAL CONSUMPTION	-	-	-	-	-	368	-	-	-	-	-
INDUSTRY SECTOR	-	-	-	-	-	368	-	-	-	-	-
Iron and Steel	-	-	-	-	-	368	-	-	-	-	-
Chemical and Petrochem.	-	-	-	-	-	-	-	-	-	-	-
Non-Metallic Minerals	-	-	-	-	-	-	-	-	-	-	-
Non-specified	-	-	-	-	-	-	-	-	-	-	-
TRANSPORT SECTOR	-	-	-	-	-	-	-	-	-	-	-
Air	-	-	-	-	-	-	-	-	-	-	-
Road	-	-	-	-	-	-	-	-	-	-	-
Non-specified	-	-	-	-	-	-	-	-	-	-	-
OTHER SECTORS	-	-	-	-	-	-	-	-	-	-	-
Agriculture	-	-	-	-	-	-	-	-	-	-	-
Comm. and Publ. Services	-	-	-	-	-	-	-	-	-	-	-
Residential	-	-	-	-	-	-	-	-	-	-	-
Non-specified	-	-	-	-	-	-	-	-	-	-	-
NON-ENERGY USE	-	-	-	-	-	-	-	-	-	-	-

APPROVISIONNEMENT ET DEMANDE 1998	Charbon (1000 tonnes)							Pétrole (1000 tonnes)			
	Charbon à coke	Autres charb. bit.	Charbon sous-bit.	Lignite	Tourbe	Coke de four/gaz	Agg./briq. de lignite	Pétrole brut	LGN	Produits d'aliment.	Additifs
Production	-	-	-	-	-	1938	-	40286	2676	-	-
Imports	1876	-	-	-	-	3	-	-	-	-	-
Exports	-	-	-	-	-	-394	-	-2791	-	-	-
Intl. Marine Bunkers	-	-	-	-	-	-	-	-	-	-	-
Stock Changes	-	-	-	-	-	-	-	-8195	-	-	-
DOMESTIC SUPPLY	1876	-	-	-	-	1547	-	29300	2676	-	-
Transfers and Stat. Diff.	351	-	-	-	-	-	-	-	-997	-	-
TRANSFORMATION	2227	-	-	-	-	1179	-	29300	1679	-	-
Electricity and CHP Plants	-	-	-	-	-	-	-	-	-	-	-
Petroleum Refineries	-	-	-	-	-	-	-	29300	1679	-	-
Other Transform. Sector	2227	-	-	-	-	1179	-	-	-	-	-
ENERGY SECTOR	-	-	-	-	-	-	-	-	-	-	-
DISTRIBUTION LOSSES	-	-	-	-	-	-	-	-	-	-	-
FINAL CONSUMPTION	-	-	-	-	-	368	-	-	-	-	-
INDUSTRY SECTOR	-	-	-	-	-	368	-	-	-	-	-
Iron and Steel	-	-	-	-	-	368	-	-	-	-	-
Chemical and Petrochem.	-	-	-	-	-	-	-	-	-	-	-
Non-Metallic Minerals	-	-	-	-	-	-	-	-	-	-	-
Non-specified	-	-	-	-	-	-	-	-	-	-	-
TRANSPORT SECTOR	-	-	-	-	-	-	-	-	-	-	-
Air	-	-	-	-	-	-	-	-	-	-	-
Road	-	-	-	-	-	-	-	-	-	-	-
Non-specified	-	-	-	-	-	-	-	-	-	-	-
OTHER SECTORS	-	-	-	-	-	-	-	-	-	-	-
Agriculture	-	-	-	-	-	-	-	-	-	-	-
Comm. and Publ. Services	-	-	-	-	-	-	-	-	-	-	-
Residential	-	-	-	-	-	-	-	-	-	-	-
Non-specified	-	-	-	-	-	-	-	-	-	-	-
NON-ENERGY USE	-	-	-	-	-	-	-	-	-	-	-

Egypt / Egypte

SUPPLY AND CONSUMPTION 1997	Oil cont. (1000 tonnes)										
	Refinery Gas	LPG + Ethane	Motor Gasoline	Aviation Gasoline	Jet Fuel	Kerosene	Gas/ Diesel	Heavy Fuel Oil	Naphtha	Petrol. Coke	Other Prod.
Production	423	459	1747	-	883	1197	5867	12914	3071	122	1238
Imports	-	419	158	-	-	-	855	-	-	-	58
Exports	-	-	-	-	-50	-	-	-2001	-3071	-	-20
Intl. Marine Bunkers	-	-	-	-	-	-	-269	-2763	-	-	-
Stock Changes	-	-	175	-	-401	-	-	-	-	-	39
DOMESTIC SUPPLY	423	878	2080	-	432	1197	6453	8150	-	122	1315
Transfers and Stat. Diff.	-	986	-	-	-	-	-237	-2639	-	-	-26
TRANSFORMATION	-	-	-	-	-	-	6	2926	-	-	-
Electricity and CHP Plants	-	-	-	-	-	-	6	2926	-	-	-
Petroleum Refineries	-	-	-	-	-	-	-	-	-	-	-
Other Transform. Sector	-	-	-	-	-	-	-	-	-	-	-
ENERGY SECTOR	423	-	-	-	-	-	182	485	-	-	-
DISTRIBUTION LOSSES	-	106	-	-	-	-	-	-	-	-	-
FINAL CONSUMPTION	-	1758	2080	-	432	1197	6028	2100	-	122	1289
INDUSTRY SECTOR	-	126	-	-	-	115	3441	2100	-	-	-
Iron and Steel	-	-	-	-	-	-	-	-	-	-	-
Chemical and Petrochem.	-	-	-	-	-	-	-	-	-	-	-
Non-Metallic Minerals	-	-	-	-	-	-	-	-	-	-	-
Non-specified	-	126	-	-	-	115	3441	2100	-	-	-
TRANSPORT SECTOR	-	-	2080	-	432	-	2587	-	-	-	-
Air	-	-	-	-	432	-	-	-	-	-	-
Road	-	-	2080	-	-	-	2587	-	-	-	-
Non-specified	-	-	-	-	-	-	-	-	-	-	-
OTHER SECTORS	-	1632	-	-	-	1082	-	-	-	-	-
Agriculture	-	-	-	-	-	-	-	-	-	-	-
Comm. and Publ. Services	-	-	-	-	-	-	-	-	-	-	-
Residential	-	1632	-	-	-	1082	-	-	-	-	-
Non-specified	-	-	-	-	-	-	-	-	-	-	-
NON-ENERGY USE	-	-	-	-	-	-	-	-	-	122	1289

APPROVISIONNEMENT ET DEMANDE 1998	Pétrole cont. (1000 tonnes)										
	Gaz de raffinerie	GPL + éthane	Essence moteur	Essence aviation	Carbu- réacteurs	Kérosène	Gazole	Fioul lourd	Naphta	Coke de pétrole	Autres prod.
Production	409	435	2099	-	865	1114	6078	13131	3145	181	1333
Imports	-	584	102	-	-	-	1237	-	-	-	78
Exports	-	-	-	-	-105	-	-	-1473	-3145	-	-45
Intl. Marine Bunkers	-	-	-	-	-	-	-247	-1995	-	-	-
Stock Changes	-	-	-51	-	-365	-	-	-	-	-	57
DOMESTIC SUPPLY	409	1019	2150	-	395	1114	7068	9663	-	181	1423
Transfers and Stat. Diff.	-	997	-	-	-	-	-343	-1928	-	-	-42
TRANSFORMATION	-	-	-	-	-	-	18	4142	-	-	-
Electricity and CHP Plants	-	-	-	-	-	-	18	4142	-	-	-
Petroleum Refineries	-	-	-	-	-	-	-	-	-	-	-
Other Transform. Sector	-	-	-	-	-	-	-	-	-	-	-
ENERGY SECTOR	409	-	-	-	-	-	182	485	-	-	-
DISTRIBUTION LOSSES	-	53	-	-	-	-	-	-	-	-	-
FINAL CONSUMPTION	-	1963	2150	-	395	1114	6525	3108	-	181	1381
INDUSTRY SECTOR	-	141	-	-	-	107	3725	3108	-	-	-
Iron and Steel	-	-	-	-	-	-	-	-	-	-	-
Chemical and Petrochem.	-	-	-	-	-	-	-	-	-	-	-
Non-Metallic Minerals	-	-	-	-	-	-	-	-	-	-	-
Non-specified	-	141	-	-	-	107	3725	3108	-	-	-
TRANSPORT SECTOR	-	-	2150	-	395	-	2800	-	-	-	-
Air	-	-	-	-	395	-	-	-	-	-	-
Road	-	-	2150	-	-	-	2800	-	-	-	-
Non-specified	-	-	-	-	-	-	-	-	-	-	-
OTHER SECTORS	-	1822	-	-	-	1007	-	-	-	-	-
Agriculture	-	-	-	-	-	-	-	-	-	-	-
Comm. and Publ. Services	-	-	-	-	-	-	-	-	-	-	-
Residential	-	1822	-	-	-	1007	-	-	-	-	-
Non-specified	-	-	-	-	-	-	-	-	-	-	-
NON-ENERGY USE	-	-	-	-	-	-	-	-	-	181	1381

Egypt / Egypte

SUPPLY AND CONSUMPTION 1997	Gas (TJ)				Comb. Renew. & Waste (TJ)				(GWh)	(TJ)
	Natural Gas	Gas Works	Coke Ovens	Blast Furnaces	Solid Biomass	Gas/Liquids from Biomass	Municipal Waste	Industrial Waste	Electricity	Heat
Production	509490	380	-	12837	52577	-	-	-	58428	-
Imports	-	-	-	-	31	-	-	-	-	-
Exports	-	-	-	-	-782	-	-	-	-	-
Intl. Marine Bunkers	-	-	-	-	-	-	-	-	-	-
Stock Changes	-	-	-	-	-	-	-	-	-	-
DOMESTIC SUPPLY	509490	380	-	12837	51825	-	-	-	58428	-
Transfers and Stat. Diff.	-8080	-	-	-	-	-	-	-	240	-
TRANSFORMATION	315970	-	-	-	-	-	-	-	-	-
Electricity and CHP Plants	315970	-	-	-	-	-	-	-	-	-
Petroleum Refineries	-	-	-	-	-	-	-	-	-	-
Other Transform. Sector	-	-	-	-	-	-	-	-	-	-
ENERGY SECTOR	31312	-	-	-	-	-	-	-	2019	-
DISTRIBUTION LOSSES	-	-	-	-	-	-	-	-	7112	-
FINAL CONSUMPTION	154128	380	-	12837	51825	-	-	-	49537	-
INDUSTRY SECTOR	111492	-	-	12837	26454	-	-	-	21356	-
Iron and Steel	-	-	-	12837	-	-	-	-	-	-
Chemical and Petrochem.	63536	-	-	-	-	-	-	-	-	-
Non-Metallic Minerals	-	-	-	-	-	-	-	-	-	-
Non-specified	47956	-	-	-	26454	-	-	-	21356	-
TRANSPORT SECTOR	-	-	-	-	-	-	-	-	-	-
Air	-	-	-	-	-	-	-	-	-	-
Road	-	-	-	-	-	-	-	-	-	-
Non-specified	-	-	-	-	-	-	-	-	-	-
OTHER SECTORS	42636	380	-	-	25371	-	-	-	28181	-
Agriculture	-	-	-	-	-	-	-	-	1977	-
Comm. and Publ. Services	-	-	-	-	-	-	-	-	-	-
Residential	32376	380	-	-	25371	-	-	-	17695	-
Non-specified	10260	-	-	-	-	-	-	-	8509	-
NON-ENERGY USE	-	-	-	-	-	-	-	-	-	-

APPROVISIONNEMENT ET DEMANDE 1998	Gaz (TJ)				En. Re. Comb. & Déchets (TJ)				(GWh)	(TJ)
	Gaz naturel	Usines à gaz	Cokeries	Hauts fourneaux	Biomasse solide	Gaz/Liquides tirés de biomasse	Déchets urbains	Déchets industriels	Electricité	Chaleur
Production	534483	380	-	12837	53494	-	-	-	62966	-
Imports	-	-	-	-	32	-	-	-	-	-
Exports	-	-	-	-	-796	-	-	-	-	-
Intl. Marine Bunkers	-	-	-	-	-	-	-	-	-	-
Stock Changes	-	-	-	-	-	-	-	-	-	-
DOMESTIC SUPPLY	534483	380	-	12837	52730	-	-	-	62966	-
Transfers and Stat. Diff.	-15479	-	-	-	-	-	-	-	-246	-
TRANSFORMATION	309852	-	-	-	-	-	-	-	-	-
Electricity and CHP Plants	309852	-	-	-	-	-	-	-	-	-
Petroleum Refineries	-	-	-	-	-	-	-	-	-	-
Other Transform. Sector	-	-	-	-	-	-	-	-	-	-
ENERGY SECTOR	37696	-	-	-	-	-	-	-	2078	-
DISTRIBUTION LOSSES	-	-	-	-	-	-	-	-	7664	-
FINAL CONSUMPTION	171456	380	-	12837	52730	-	-	-	52978	-
INDUSTRY SECTOR	131784	-	-	12837	26916	-	-	-	22093	-
Iron and Steel	-	-	-	12837	-	-	-	-	-	-
Chemical and Petrochem.	70148	-	-	-	-	-	-	-	-	-
Non-Metallic Minerals	-	-	-	-	-	-	-	-	-	-
Non-specified	61636	-	-	-	26916	-	-	-	22093	-
TRANSPORT SECTOR	-	-	-	-	-	-	-	-	-	-
Air	-	-	-	-	-	-	-	-	-	-
Road	-	-	-	-	-	-	-	-	-	-
Non-specified	-	-	-	-	-	-	-	-	-	-
OTHER SECTORS	39672	380	-	-	25814	-	-	-	30885	-
Agriculture	-	-	-	-	-	-	-	-	2131	-
Comm. and Publ. Services	-	-	-	-	-	-	-	-	8338	-
Residential	39672	380	-	-	25814	-	-	-	18533	-
Non-specified	-	-	-	-	-	-	-	-	1883	-
NON-ENERGY USE	-	-	-	-	-	-	-	-	-	-

El Salvador

SUPPLY AND CONSUMPTION 1997	Coal (1000 tonnes)							Oil (1000 tonnes)			
	Coking Coal	Other Bit. Coal	Sub-Bit. Coal	Lignite	Peat	Oven and Gas Coke	Pat. Fuel and BKB	Crude Oil	NGL	Feed-stocks	Additives
Production	-	-	-	-	-	-	-	-	-	-	-
Imports	-	-	-	-	-	1	-	834	-	-	-
Exports	-	-	-	-	-	-	-	-	-	-	-
Intl. Marine Bunkers	-	-	-	-	-	-	-	-	-	-	-
Stock Changes	-	-	-	-	-	-	-	-15	-	-	-
DOMESTIC SUPPLY	-	-	-	-	-	1	-	819	-	-	-
Transfers and Stat. Diff.	-	-	-	-	-	-	-	-	-	-	-
TRANSFORMATION	-	-	-	-	-	1	-	819	-	-	-
Electricity and CHP Plants	-	-	-	-	-	-	-	-	-	-	-
Petroleum Refineries	-	-	-	-	-	-	-	819	-	-	-
Other Transform. Sector	-	-	-	-	-	1	-	-	-	-	-
ENERGY SECTOR	-	-	-	-	-	-	-	-	-	-	-
DISTRIBUTION LOSSES	-	-	-	-	-	-	-	-	-	-	-
FINAL CONSUMPTION	-	-	-	-	-	-	-	-	-	-	-
INDUSTRY SECTOR	-	-	-	-	-	-	-	-	-	-	-
Iron and Steel	-	-	-	-	-	-	-	-	-	-	-
Chemical and Petrochem.	-	-	-	-	-	-	-	-	-	-	-
Non-Metallic Minerals	-	-	-	-	-	-	-	-	-	-	-
Non-specified	-	-	-	-	-	-	-	-	-	-	-
TRANSPORT SECTOR	-	-	-	-	-	-	-	-	-	-	-
Air	-	-	-	-	-	-	-	-	-	-	-
Road	-	-	-	-	-	-	-	-	-	-	-
Non-specified	-	-	-	-	-	-	-	-	-	-	-
OTHER SECTORS	-	-	-	-	-	-	-	-	-	-	-
Agriculture	-	-	-	-	-	-	-	-	-	-	-
Comm. and Publ. Services	-	-	-	-	-	-	-	-	-	-	-
Residential	-	-	-	-	-	-	-	-	-	-	-
Non-specified	-	-	-	-	-	-	-	-	-	-	-
NON-ENERGY USE	-	-	-	-	-	-	-	-	-	-	-

APPROVISIONNEMENT ET DEMANDE 1998	Charbon (1000 tonnes)							Pétrole (1000 tonnes)			
	Charbon à coke	Autres charb. bit.	Charbon sous-bit.	Lignite	Tourbe	Coke de four/gaz	Agg./briq. de lignite	Pétrole brut	LGN	Produits d'aliment.	Additifs
Production	-	-	-	-	-	-	-	-	-	-	-
Imports	-	-	-	-	-	1	-	948	-	-	-
Exports	-	-	-	-	-	-	-	-	-	-	-
Intl. Marine Bunkers	-	-	-	-	-	-	-	-	-	-	-
Stock Changes	-	-	-	-	-	-	-	-	-	-	-
DOMESTIC SUPPLY	-	-	-	-	-	1	-	948	-	-	-
Transfers and Stat. Diff.	-	-	-	-	-	-	-	-	-	-	-
TRANSFORMATION	-	-	-	-	-	1	-	948	-	-	-
Electricity and CHP Plants	-	-	-	-	-	-	-	-	-	-	-
Petroleum Refineries	-	-	-	-	-	-	-	948	-	-	-
Other Transform. Sector	-	-	-	-	-	1	-	-	-	-	-
ENERGY SECTOR	-	-	-	-	-	-	-	-	-	-	-
DISTRIBUTION LOSSES	-	-	-	-	-	-	-	-	-	-	-
FINAL CONSUMPTION	-	-	-	-	-	-	-	-	-	-	-
INDUSTRY SECTOR	-	-	-	-	-	-	-	-	-	-	-
Iron and Steel	-	-	-	-	-	-	-	-	-	-	-
Chemical and Petrochem.	-	-	-	-	-	-	-	-	-	-	-
Non-Metallic Minerals	-	-	-	-	-	-	-	-	-	-	-
Non-specified	-	-	-	-	-	-	-	-	-	-	-
TRANSPORT SECTOR	-	-	-	-	-	-	-	-	-	-	-
Air	-	-	-	-	-	-	-	-	-	-	-
Road	-	-	-	-	-	-	-	-	-	-	-
Non-specified	-	-	-	-	-	-	-	-	-	-	-
OTHER SECTORS	-	-	-	-	-	-	-	-	-	-	-
Agriculture	-	-	-	-	-	-	-	-	-	-	-
Comm. and Publ. Services	-	-	-	-	-	-	-	-	-	-	-
Residential	-	-	-	-	-	-	-	-	-	-	-
Non-specified	-	-	-	-	-	-	-	-	-	-	-
NON-ENERGY USE	-	-	-	-	-	-	-	-	-	-	-

El Salvador

SUPPLY AND CONSUMPTION 1997	Oil cont. (1000 tonnes)										
	Refinery Gas	LPG + Ethane	Motor Gasoline	Aviation Gasoline	Jet Fuel	Kerosene	Gas/ Diesel	Heavy Fuel Oil	Naphtha	Petrol. Coke	Other Prod.
Production	16	16	149	-	36	14	121	382	-	-	32
Imports	-	93	152	-	11	4	465	347	-	-	8
Exports	-	1	-1	-	-	-	-	-223	-	-	-10
Intl. Marine Bunkers	-	-	-	-	-	-	-	-	-	-	-
Stock Changes	-	-	7	-	-	-	25	-2	-	-	7
DOMESTIC SUPPLY	16	110	307	-	47	18	611	504	-	-	37
Transfers and Stat. Diff.	-	-5	-	-	-	-	-2	-	-	-	-5
TRANSFORMATION	-	-	-	-	-	-	125	314	-	-	-
Electricity and CHP Plants	-	-	-	-	-	-	125	314	-	-	-
Petroleum Refineries	-	-	-	-	-	-	-	-	-	-	-
Other Transform. Sector	-	-	-	-	-	-	-	-	-	-	-
ENERGY SECTOR	16	-	-	-	-	-	-	-	-	-	-
DISTRIBUTION LOSSES	-	-	-	-	-	-	-	-	-	-	-
FINAL CONSUMPTION	-	105	307	-	47	18	484	190	-	-	32
INDUSTRY SECTOR	-	16	-	-	-	3	86	187	-	-	-
Iron and Steel	-	-	-	-	-	-	-	-	-	-	-
Chemical and Petrochem.	-	-	-	-	-	-	-	-	-	-	-
Non-Metallic Minerals	-	-	-	-	-	-	-	-	-	-	-
Non-specified	-	16	-	-	-	3	86	187	-	-	-
TRANSPORT SECTOR	-	-	307	-	47	-	368	-	-	-	-
Air	-	-	-	-	47	-	-	-	-	-	-
Road	-	-	307	-	-	-	368	-	-	-	-
Non-specified	-	-	-	-	-	-	-	-	-	-	-
OTHER SECTORS	-	89	-	-	-	15	30	3	-	-	-
Agriculture	-	-	-	-	-	-	-	-	-	-	-
Comm. and Publ. Services	-	-	-	-	-	-	-	-	-	-	-
Residential	-	89	-	-	-	15	30	3	-	-	-
Non-specified	-	-	-	-	-	-	-	-	-	-	-
NON-ENERGY USE	-	-	-	-	-	-	-	-	-	-	32

APPROVISIONNEMENT ET DEMANDE 1998	Pétrole cont. (1000 tonnes)										
	Gaz de raffinerie	GPL + éthane	Essence moteur	Essence aviation	Carbu- réacteurs	Kérosène	Gazole	Fioul lourd	Naphta	Coke de pétrole	Autres prod.
Production	16	18	134	-	41	16	146	497	-	-	32
Imports	-	98	200	-	23	9	511	249	-	-	14
Exports	-	1	-	-	-	-	-	-223	-	-	-8
Intl. Marine Bunkers	-	-	-	-	-	-	-	-	-	-	-
Stock Changes	-	-	7	-	-4	-2	-11	1	-	-	-
DOMESTIC SUPPLY	16	117	341	-	60	23	646	524	-	-	38
Transfers and Stat. Diff.	-	1	-	-	-	1	2	-	-	-	-6
TRANSFORMATION	-	-	-	-	-	-	120	336	-	-	-
Electricity and CHP Plants	-	-	-	-	-	-	120	336	-	-	-
Petroleum Refineries	-	-	-	-	-	-	-	-	-	-	-
Other Transform. Sector	-	-	-	-	-	-	-	-	-	-	-
ENERGY SECTOR	16	-	-	-	-	-	-	-	-	-	-
DISTRIBUTION LOSSES	-	-	-	-	-	-	-	-	-	-	-
FINAL CONSUMPTION	-	118	341	-	60	24	528	188	-	-	32
INDUSTRY SECTOR	-	20	-	-	-	4	112	184	-	-	-
Iron and Steel	-	-	-	-	-	-	-	-	-	-	-
Chemical and Petrochem.	-	-	-	-	-	-	-	-	-	-	-
Non-Metallic Minerals	-	-	-	-	-	-	-	-	-	-	-
Non-specified	-	20	-	-	-	4	112	184	-	-	-
TRANSPORT SECTOR	-	-	341	-	60	-	386	-	-	-	-
Air	-	-	-	-	60	-	-	-	-	-	-
Road	-	-	341	-	-	-	386	-	-	-	-
Non-specified	-	-	-	-	-	-	-	-	-	-	-
OTHER SECTORS	-	98	-	-	-	20	30	4	-	-	-
Agriculture	-	-	-	-	-	-	-	-	-	-	-
Comm. and Publ. Services	-	-	-	-	-	-	-	-	-	-	-
Residential	-	98	-	-	-	20	30	4	-	-	-
Non-specified	-	-	-	-	-	-	-	-	-	-	-
NON-ENERGY USE	-	-	-	-	-	-	-	-	-	-	32

El Salvador

SUPPLY AND CONSUMPTION 1997	Gas (TJ)				Comb. Renew. & Waste (TJ)				(GWh)	(TJ)
	Natural Gas	Gas Works	Coke Ovens	Blast Furnaces	Solid Biomass	Gas/Liquids from Biomass	Municipal Waste	Industrial Waste	Electricity	Heat
Production	-	-	-	11	60780	-	-	-	3643	-
Imports	-	-	-	-	-	-	-	-	106	-
Exports	-	-	-	-	-	-	-	-	18	-
Intl. Marine Bunkers	-	-	-	-	-	-	-	-	-	-
Stock Changes	-	-	-	-	-	-	-	-	-	-
DOMESTIC SUPPLY	-	-	-	11	60780	-	-	-	3767	-
Transfers and Stat. Diff.	-	-	-	-	-1790	-	-	-	-37	-
TRANSFORMATION	-	-	-	-	4089	-	-	-	-	-
Electricity and CHP Plants	-	-	-	-	3521	-	-	-	-	-
Petroleum Refineries	-	-	-	-	-	-	-	-	-	-
Other Transform. Sector	-	-	-	-	568	-	-	-	-	-
ENERGY SECTOR	-	-	-	-	-	-	-	-	66	-
DISTRIBUTION LOSSES	-	-	-	-	-	-	-	-	480	-
FINAL CONSUMPTION	-	-	-	11	53122	-	-	-	3184	-
INDUSTRY SECTOR	-	-	-	11	9634	-	-	-	905	-
Iron and Steel	-	-	-	11	-	-	-	-	-	-
Chemical and Petrochem.	-	-	-	-	-	-	-	-	-	-
Non-Metallic Minerals	-	-	-	-	-	-	-	-	-	-
Non-specified	-	-	-	-	9634	-	-	-	905	-
TRANSPORT SECTOR	-	-	-	-	-	-	-	-	-	-
Air	-	-	-	-	-	-	-	-	-	-
Road	-	-	-	-	-	-	-	-	-	-
Non-specified	-	-	-	-	-	-	-	-	-	-
OTHER SECTORS	-	-	-	-	43488	-	-	-	2279	-
Agriculture	-	-	-	-	-	-	-	-	-	-
Comm. and Publ. Services	-	-	-	-	-	-	-	-	1130	-
Residential	-	-	-	-	43488	-	-	-	1149	-
Non-specified	-	-	-	-	-	-	-	-	-	-
NON-ENERGY USE	-	-	-	-	-	-	-	-	-	-

APPROVISIONNEMENT ET DEMANDE 1998	Gaz (TJ)				En. Re. Comb. & Déchets (TJ)				(GWh)	(TJ)
	Gaz naturel	Usines à gaz	Cokeries	Hauts fourneaux	Biomasse solide	Gaz/Liquides tirés de biomasse	Déchets urbains	Déchets industriels	Electricité	Chaleur
Production	-	-	-	11	61344	-	-	-	3837	-
Imports	-	-	-	-	-	-	-	-	61	-
Exports	-	-	-	-	-	-	-	-	23	-
Intl. Marine Bunkers	-	-	-	-	-	-	-	-	-	-
Stock Changes	-	-	-	-	-	-	-	-	-	-
DOMESTIC SUPPLY	-	-	-	11	61344	-	-	-	3921	-
Transfers and Stat. Diff.	-	-	-	-	-1865	-	-	-	30	-
TRANSFORMATION	-	-	-	-	4452	-	-	-	-	-
Electricity and CHP Plants	-	-	-	-	3893	-	-	-	-	-
Petroleum Refineries	-	-	-	-	-	-	-	-	-	-
Other Transform. Sector	-	-	-	-	559	-	-	-	-	-
ENERGY SECTOR	-	-	-	-	-	-	-	-	69	-
DISTRIBUTION LOSSES	-	-	-	-	-	-	-	-	506	-
FINAL CONSUMPTION	-	-	-	11	53233	-	-	-	3376	-
INDUSTRY SECTOR	-	-	-	11	9646	-	-	-	1028	-
Iron and Steel	-	-	-	11	-	-	-	-	-	-
Chemical and Petrochem.	-	-	-	-	-	-	-	-	-	-
Non-Metallic Minerals	-	-	-	-	-	-	-	-	-	-
Non-specified	-	-	-	-	9646	-	-	-	1028	-
TRANSPORT SECTOR	-	-	-	-	-	-	-	-	-	-
Air	-	-	-	-	-	-	-	-	-	-
Road	-	-	-	-	-	-	-	-	-	-
Non-specified	-	-	-	-	-	-	-	-	-	-
OTHER SECTORS	-	-	-	-	43587	-	-	-	2348	-
Agriculture	-	-	-	-	-	-	-	-	-	-
Comm. and Publ. Services	-	-	-	-	-	-	-	-	1149	-
Residential	-	-	-	-	43587	-	-	-	1199	-
Non-specified	-	-	-	-	-	-	-	-	-	-
NON-ENERGY USE	-	-	-	-	-	-	-	-	-	-

Eritrea / Erythrée

SUPPLY AND CONSUMPTION 1997	Coal (1000 tonnes)							Oil (1000 tonnes)			
	Coking Coal	Other Bit. Coal	Sub-Bit. Coal	Lignite	Peat	Oven and Gas Coke	Pat. Fuel and BKB	Crude Oil	NGL	Feed-stocks	Additives
Production	-	-	-	-	-	-	-	-	-	-	-
Imports	-	-	-	-	-	-	-	408	-	-	-
Exports	-	-	-	-	-	-	-	-	-	-	-
Intl. Marine Bunkers	-	-	-	-	-	-	-	-	-	-	-
Stock Changes	-	-	-	-	-	-	-	-	-	-	-
DOMESTIC SUPPLY	-	-	-	-	-	-	-	408	-	-	-
Transfers and Stat. Diff.	-	-	-	-	-	-	-	-	-	-	-
TRANSFORMATION	-	-	-	-	-	-	-	403	-	-	-
Electricity and CHP Plants	-	-	-	-	-	-	-	-	-	-	-
Petroleum Refineries	-	-	-	-	-	-	-	403	-	-	-
Other Transform. Sector	-	-	-	-	-	-	-	-	-	-	-
ENERGY SECTOR	-	-	-	-	-	-	-	-	-	-	-
DISTRIBUTION LOSSES	-	-	-	-	-	-	-	5	-	-	-
FINAL CONSUMPTION	-	-	-	-	-	-	-	-	-	-	-
INDUSTRY SECTOR	-	-	-	-	-	-	-	-	-	-	-
Iron and Steel	-	-	-	-	-	-	-	-	-	-	-
Chemical and Petrochem.	-	-	-	-	-	-	-	-	-	-	-
Non-Metallic Minerals	-	-	-	-	-	-	-	-	-	-	-
Non-specified	-	-	-	-	-	-	-	-	-	-	-
TRANSPORT SECTOR	-	-	-	-	-	-	-	-	-	-	-
Air	-	-	-	-	-	-	-	-	-	-	-
Road	-	-	-	-	-	-	-	-	-	-	-
Non-specified	-	-	-	-	-	-	-	-	-	-	-
OTHER SECTORS	-	-	-	-	-	-	-	-	-	-	-
Agriculture	-	-	-	-	-	-	-	-	-	-	-
Comm. and Publ. Services	-	-	-	-	-	-	-	-	-	-	-
Residential	-	-	-	-	-	-	-	-	-	-	-
Non-specified	-	-	-	-	-	-	-	-	-	-	-
NON-ENERGY USE	-	-	-	-	-	-	-	-	-	-	-

APPROVISIONNEMENT ET DEMANDE 1998	Charbon (1000 tonnes)							Pétrole (1000 tonnes)			
	Charbon à coke	Autres charb. bit.	Charbon sous-bit.	Lignite	Tourbe	Coke de four/gaz	Agg./briq. de lignite	Pétrole brut	LGN	Produits d'aliment.	Additifs
Production	-	-	-	-	-	-	-	-	-	-	-
Imports	-	-	-	-	-	-	-	-	-	-	-
Exports	-	-	-	-	-	-	-	-	-	-	-
Intl. Marine Bunkers	-	-	-	-	-	-	-	-	-	-	-
Stock Changes	-	-	-	-	-	-	-	-	-	-	-
DOMESTIC SUPPLY	-	-	-	-	-	-	-	-	-	-	-
Transfers and Stat. Diff.	-	-	-	-	-	-	-	-	-	-	-
TRANSFORMATION	-	-	-	-	-	-	-	-	-	-	-
Electricity and CHP Plants	-	-	-	-	-	-	-	-	-	-	-
Petroleum Refineries	-	-	-	-	-	-	-	-	-	-	-
Other Transform. Sector	-	-	-	-	-	-	-	-	-	-	-
ENERGY SECTOR	-	-	-	-	-	-	-	-	-	-	-
DISTRIBUTION LOSSES	-	-	-	-	-	-	-	-	-	-	-
FINAL CONSUMPTION	-	-	-	-	-	-	-	-	-	-	-
INDUSTRY SECTOR	-	-	-	-	-	-	-	-	-	-	-
Iron and Steel	-	-	-	-	-	-	-	-	-	-	-
Chemical and Petrochem.	-	-	-	-	-	-	-	-	-	-	-
Non-Metallic Minerals	-	-	-	-	-	-	-	-	-	-	-
Non-specified	-	-	-	-	-	-	-	-	-	-	-
TRANSPORT SECTOR	-	-	-	-	-	-	-	-	-	-	-
Air	-	-	-	-	-	-	-	-	-	-	-
Road	-	-	-	-	-	-	-	-	-	-	-
Non-specified	-	-	-	-	-	-	-	-	-	-	-
OTHER SECTORS	-	-	-	-	-	-	-	-	-	-	-
Agriculture	-	-	-	-	-	-	-	-	-	-	-
Comm. and Publ. Services	-	-	-	-	-	-	-	-	-	-	-
Residential	-	-	-	-	-	-	-	-	-	-	-
Non-specified	-	-	-	-	-	-	-	-	-	-	-
NON-ENERGY USE	-	-	-	-	-	-	-	-	-	-	-

Eritrea / Erythrée

SUPPLY AND CONSUMPTION 1997	Oil cont. (1000 tonnes)										
	Refinery Gas	LPG + Ethane	Motor Gasoline	Aviation Gasoline	Jet Fuel	Kerosene	Gas/ Diesel	Heavy Fuel Oil	Naphtha	Petrol. Coke	Other Prod.
Production	4	4	50	-	9	30	104	173	-	-	28
Imports	-	-	-	-	1	-	33	-	-	-	14
Exports	-	-2	-35	-	-	-9	-	-56	-	-	-28
Intl. Marine Bunkers	-	-	-	-	-	-	-	-74	-	-	-
Stock Changes	-	-	-	-	-	-	-	-	-	-	-
DOMESTIC SUPPLY	4	2	15	-	10	21	137	43	-	-	14
Transfers and Stat. Diff.	-	-	-	-	-	1	1	54	-	-	-
TRANSFORMATION	-	-	-	-	-	-	15	54	-	-	-
Electricity and CHP Plants	-	-	-	-	-	-	15	54	-	-	-
Petroleum Refineries	-	-	-	-	-	-	-	-	-	-	-
Other Transform. Sector	-	-	-	-	-	-	-	-	-	-	-
ENERGY SECTOR	4	-	-	-	-	-	-	-	-	-	-
DISTRIBUTION LOSSES	-	-	-	-	-	-	-	-	-	-	-
FINAL CONSUMPTION	-	2	15	-	10	22	123	43	-	-	14
INDUSTRY SECTOR	-	-	-	-	-	-	1	15	-	-	-
Iron and Steel	-	-	-	-	-	-	-	-	-	-	-
Chemical and Petrochem.	-	-	-	-	-	-	-	-	-	-	-
Non-Metallic Minerals	-	-	-	-	-	-	-	-	-	-	-
Non-specified	-	-	-	-	-	-	1	15	-	-	-
TRANSPORT SECTOR	-	-	15	-	10	-	98	-	-	-	-
Air	-	-	-	-	10	-	-	-	-	-	-
Road	-	-	15	-	-	-	98	-	-	-	-
Non-specified	-	-	-	-	-	-	-	-	-	-	-
OTHER SECTORS	-	2	-	-	-	22	24	28	-	-	-
Agriculture	-	-	-	-	-	-	-	-	-	-	-
Comm. and Publ. Services	-	-	-	-	-	1	24	28	-	-	-
Residential	-	2	-	-	-	21	-	-	-	-	-
Non-specified	-	-	-	-	-	-	-	-	-	-	-
NON-ENERGY USE	-	-	-	-	-	-	-	-	-	-	14

APPROVISIONNEMENT ET DEMANDE 1998	Pétrole cont. (1000 tonnes)										
	Gaz de raffinerie	GPL + éthane	Essence moteur	Essence aviation	Carbu- réacteurs	Kérosène	Gazole	Fioul lourd	Naphta	Coke de pétrole	Autres prod.
Production	-	-	-	-	-	-	-	-	-	-	-
Imports	-	1	16	-	7	21	104	43	-	-	6
Exports	-	-	-	-	-	-	-	-	-	-	-
Intl. Marine Bunkers	-	-	-	-	-	-	-	-	-	-	-
Stock Changes	-	-	-	-	-	-	-	-	-	-	-
DOMESTIC SUPPLY	-	1	16	-	7	21	104	43	-	-	6
Transfers and Stat. Diff.	-	-	-	-	-	-	-	27	-	-	-
TRANSFORMATION	-	-	-	-	-	-	14	28	-	-	-
Electricity and CHP Plants	-	-	-	-	-	-	14	28	-	-	-
Petroleum Refineries	-	-	-	-	-	-	-	-	-	-	-
Other Transform. Sector	-	-	-	-	-	-	-	-	-	-	-
ENERGY SECTOR	-	-	-	-	-	-	-	-	-	-	-
DISTRIBUTION LOSSES	-	-	-	-	-	-	-	-	-	-	-
FINAL CONSUMPTION	-	1	16	-	7	21	90	42	-	-	6
INDUSTRY SECTOR	-	-	-	-	-	-	1	11	-	-	-
Iron and Steel	-	-	-	-	-	-	-	-	-	-	-
Chemical and Petrochem.	-	-	-	-	-	-	-	-	-	-	-
Non-Metallic Minerals	-	-	-	-	-	-	-	-	-	-	-
Non-specified	-	-	-	-	-	-	1	11	-	-	-
TRANSPORT SECTOR	-	-	16	-	7	-	56	-	-	-	-
Air	-	-	-	-	7	-	-	-	-	-	-
Road	-	-	16	-	-	-	56	-	-	-	-
Non-specified	-	-	-	-	-	-	-	-	-	-	-
OTHER SECTORS	-	1	-	-	-	21	33	31	-	-	-
Agriculture	-	-	-	-	-	-	-	-	-	-	-
Comm. and Publ. Services	-	-	-	-	-	2	33	31	-	-	-
Residential	-	1	-	-	-	19	-	-	-	-	-
Non-specified	-	-	-	-	-	-	-	-	-	-	-
NON-ENERGY USE	-	-	-	-	-	-	-	-	-	-	6

Eritrea / Erythrée

SUPPLY AND CONSUMPTION 1997	Gas (TJ)				Comb. Renew. & Waste (TJ)				(GWh)	(TJ)
	Natural Gas	Gas Works	Coke Ovens	Blast Furnaces	Solid Biomass	Gas/Liquids from Biomass	Municipal Waste	Industrial Waste	Electricity	Heat
Production	-	-	-	-	30004	-	-	-	212	-
Imports	-	-	-	-	-	-	-	-	-	-
Exports	-	-	-	-	-	-	-	-	-	-
Intl. Marine Bunkers	-	-	-	-	-	-	-	-	-	-
Stock Changes	-	-	-	-	-	-	-	-	-	-
DOMESTIC SUPPLY	-	-	-	-	30004	-	-	-	212	-
Transfers and Stat. Diff.	-	-	-	-	-32	-	-	-	-6	-
TRANSFORMATION	-	-	-	-	1854	-	-	-	-	-
Electricity and CHP Plants	-	-	-	-	-	-	-	-	-	-
Petroleum Refineries	-	-	-	-	-	-	-	-	-	-
Other Transform. Sector	-	-	-	-	1854	-	-	-	-	-
ENERGY SECTOR	-	-	-	-	-	-	-	-	7	-
DISTRIBUTION LOSSES	-	-	-	-	-	-	-	-	29	-
FINAL CONSUMPTION	-	-	-	-	28118	-	-	-	170	-
INDUSTRY SECTOR	-	-	-	-	-	-	-	-	88	-
Iron and Steel	-	-	-	-	-	-	-	-	-	-
Chemical and Petrochem.	-	-	-	-	-	-	-	-	-	-
Non-Metallic Minerals	-	-	-	-	-	-	-	-	-	-
Non-specified	-	-	-	-	-	-	-	-	88	-
TRANSPORT SECTOR	-	-	-	-	-	-	-	-	-	-
Air	-	-	-	-	-	-	-	-	-	-
Road	-	-	-	-	-	-	-	-	-	-
Non-specified	-	-	-	-	-	-	-	-	-	-
OTHER SECTORS	-	-	-	-	28118	-	-	-	82	-
Agriculture	-	-	-	-	-	-	-	-	5	-
Comm. and Publ. Services	-	-	-	-	826	-	-	-	28	-
Residential	-	-	-	-	27292	-	-	-	49	-
Non-specified	-	-	-	-	-	-	-	-	-	-
NON-ENERGY USE	-	-	-	-	-	-	-	-	-	-

APPROVISIONNEMENT ET DEMANDE 1998	Gaz (TJ)				En. Re. Comb. & Déchets (TJ)				(GWh)	(TJ)
	Gaz naturel	Usines à gaz	Cokeries	Hauts fourneaux	Biomasse solide	Gaz/Liquides tirés de biomasse	Déchets urbains	Déchets industriels	Electricité	Chaleur
Production	-	-	-	-	17299	-	-	-	194	-
Imports	-	-	-	-	-	-	-	-	-	-
Exports	-	-	-	-	-	-	-	-	-	-
Intl. Marine Bunkers	-	-	-	-	-	-	-	-	-	-
Stock Changes	-	-	-	-	-	-	-	-	-	-
DOMESTIC SUPPLY	-	-	-	-	17299	-	-	-	194	-
Transfers and Stat. Diff.	-	-	-	-	7	-	-	-	4	-
TRANSFORMATION	-	-	-	-	969	-	-	-	-	-
Electricity and CHP Plants	-	-	-	-	-	-	-	-	-	-
Petroleum Refineries	-	-	-	-	-	-	-	-	-	-
Other Transform. Sector	-	-	-	-	969	-	-	-	-	-
ENERGY SECTOR	-	-	-	-	-	-	-	-	8	-
DISTRIBUTION LOSSES	-	-	-	-	-	-	-	-	32	-
FINAL CONSUMPTION	-	-	-	-	16337	-	-	-	158	-
INDUSTRY SECTOR	-	-	-	-	-	-	-	-	64	-
Iron and Steel	-	-	-	-	-	-	-	-	-	-
Chemical and Petrochem.	-	-	-	-	-	-	-	-	-	-
Non-Metallic Minerals	-	-	-	-	-	-	-	-	-	-
Non-specified	-	-	-	-	-	-	-	-	64	-
TRANSPORT SECTOR	-	-	-	-	-	-	-	-	-	-
Air	-	-	-	-	-	-	-	-	-	-
Road	-	-	-	-	-	-	-	-	-	-
Non-specified	-	-	-	-	-	-	-	-	-	-
OTHER SECTORS	-	-	-	-	16337	-	-	-	94	-
Agriculture	-	-	-	-	-	-	-	-	6	-
Comm. and Publ. Services	-	-	-	-	427	-	-	-	31	-
Residential	-	-	-	-	15910	-	-	-	57	-
Non-specified	-	-	-	-	-	-	-	-	-	-
NON-ENERGY USE	-	-	-	-	-	-	-	-	-	-

Estonia / Estonie : 1997

SUPPLY AND CONSUMPTION APPROVISIONNEMENT ET DEMANDE	Coal / *Charbon* (1000 tonnes)							Oil / *Pétrole* (1000 tonnes)			
	Coking Coal *Charbon à coke*	Other Bit. Coal *Autres charb. bit.*	Sub-Bit. Coal *Charbon sous-bit.*	Lignite *Lignite*	Peat *Tourbe*	Oven and Gas Coke *Coke de four/gaz*	Pat. Fuel and BKB *Agg./briq. de lignite*	Crude Oil *Pétrole brut*	NGL *LGN*	Feed-stocks *Produits d'aliment.*	Additives *Additifs*
Production	-	-	-	11895	571	42	129	-	-	-	-
From Other Sources	-	-	-	-	-	-	-	404	-	-	-
Imports	-	48	-	1534	-	1	-	-	-	-	-
Exports	-	-	-	-16	-4	-35	-50	-	-	-	-
Intl. Marine Bunkers	-	-	-	-	-	-	-	-	-	-	-
Stock Changes	-	50	-	50	67	-7	-12	-	-	-	-
DOMESTIC SUPPLY	-	98	-	13463	634	1	67	404	-	-	-
Transfers	-	-	-	-	-	-	-	-	-	-	-
Statistical Differences	-	-	-	-	-	-	-	-	-	-	-
TRANSFORMATION	-	27	-	13029	453	-	3	404	-	-	-
Electricity Plants	-	-	-	9763	-	-	-	-	-	-	-
CHP Plants	-	-	-	2050	14	-	-	-	-	-	-
Heat Plants	-	27	-	111	118	-	3	-	-	-	-
Blast Furnaces/Gas Works	-	-	-	938	-	-	-	-	-	-	-
Coke/Pat. Fuel/BKB Plants	-	-	-	167	321	-	-	-	-	-	-
Petroleum Refineries	-	-	-	-	-	-	-	404	-	-	-
Petrochemical Industry	-	-	-	-	-	-	-	-	-	-	-
Liquefaction	-	-	-	-	-	-	-	-	-	-	-
Other Transform. Sector	-	-	-	-	-	-	-	-	-	-	-
ENERGY SECTOR	-	-	-	-	-	-	1	-	-	-	-
Coal Mines	-	-	-	-	-	-	-	-	-	-	-
Oil and Gas Extraction	-	-	-	-	-	-	-	-	-	-	-
Petroleum Refineries	-	-	-	-	-	-	-	-	-	-	-
Electr., CHP+Heat Plants	-	-	-	-	-	-	-	-	-	-	-
Pumped Storage (Elec.)	-	-	-	-	-	-	-	-	-	-	-
Other Energy Sector	-	-	-	-	-	-	1	-	-	-	-
Distribution Losses	-	-	-	-	163	-	-	-	-	-	-
FINAL CONSUMPTION	-	71	-	434	18	1	63	-	-	-	-
INDUSTRY SECTOR	-	10	-	342	7	1	3	-	-	-	-
Iron and Steel	-	-	-	-	-	-	-	-	-	-	-
Chemical and Petrochem.	-	-	-	10	-	-	-	-	-	-	-
of which: Feedstocks	-	-	-	-	-	-	-	-	-	-	-
Non-Ferrous Metals	-	-	-	-	-	-	-	-	-	-	-
Non-Metallic Minerals	-	6	-	326	-	-	3	-	-	-	-
Transport Equipment	-	-	-	-	-	-	-	-	-	-	-
Machinery	-	-	-	-	-	-	-	-	-	-	-
Mining and Quarrying	-	-	-	-	-	1	-	-	-	-	-
Food and Tobacco	-	2	-	-	5	-	-	-	-	-	-
Paper, Pulp and Print	-	-	-	-	-	-	-	-	-	-	-
Wood and Wood Products	-	-	-	5	-	-	-	-	-	-	-
Construction	-	1	-	-	1	-	-	-	-	-	-
Textile and Leather	-	1	-	1	-	-	-	-	-	-	-
Non-specified	-	-	-	-	1	-	-	-	-	-	-
TRANSPORT SECTOR	-	2	-	-	-	-	-	-	-	-	-
Air	-	-	-	-	-	-	-	-	-	-	-
Road	-	-	-	-	-	-	-	-	-	-	-
Rail	-	2	-	-	-	-	-	-	-	-	-
Pipeline Transport	-	-	-	-	-	-	-	-	-	-	-
Internal Navigation	-	-	-	-	-	-	-	-	-	-	-
Non-specified	-	-	-	-	-	-	-	-	-	-	-
OTHER SECTORS	-	59	-	1	11	-	60	-	-	-	-
Agriculture	-	2	-	1	6	-	-	-	-	-	-
Comm. and Publ. Services	-	11	-	-	2	-	-	-	-	-	-
Residential	-	46	-	-	3	-	60	-	-	-	-
Non-specified	-	-	-	-	-	-	-	-	-	-	-
NON-ENERGY USE	-	-	-	91	-	-	-	-	-	-	-
in Industry/Trans./Energy	-	-	-	91	-	-	-	-	-	-	-
in Transport	-	-	-	-	-	-	-	-	-	-	-
in Other Sectors	-	-	-	-	-	-	-	-	-	-	-

Estonia / Estonie : 1997

SUPPLY AND CONSUMPTION / APPROVISIONNEMENT ET DEMANDE	Oil cont. / Pétrole cont. (1000 tonnes)										
	Refinery Gas / Gaz de raffinerie	LPG + Ethane / GPL + éthane	Motor Gasoline / Essence moteur	Aviation Gasoline / Essence aviation	Jet Fuel / Carbu-réacteurs	Kerosene / Kérosène	Gas/ Diesel / Gazole	Heavy Fuel Oil / Fioul lourd	Naphtha / Naphta	Petrol. Coke / Coke de pétrole	Other Prod. / Autres prod.
Production	-	-	6	-	-	-	-	368	-	-	7
From Other Sources	-	-	-	-	-	-	-	-	-	-	-
Imports	-	9	423	2	27	-	584	514	-	-	100
Exports	-	-	-136	-	-9	-	-176	-370	-	-	-61
Intl. Marine Bunkers	-	-	-	-	-	-	-31	-71	-	-	-
Stock Changes	-	-	12	-	2	-	1	43	-	-	-4
DOMESTIC SUPPLY	-	9	305	2	20	-	378	484	-	-	42
Transfers	-	-	-	-	-	-	-	-	-	-	-
Statistical Differences	-	-	-	-	-	-	-	-	-	-	-
TRANSFORMATION	-	-	-	-	-	-	12	295	-	-	-
Electricity Plants	-	-	-	-	-	-	-	9	-	-	-
CHP Plants	-	-	-	-	-	-	-	47	-	-	-
Heat Plants	-	-	-	-	-	-	12	239	-	-	-
Blast Furnaces/Gas Works	-	-	-	-	-	-	-	-	-	-	-
Coke/Pat. Fuel/BKB Plants	-	-	-	-	-	-	-	-	-	-	-
Petroleum Refineries	-	-	-	-	-	-	-	-	-	-	-
Petrochemical Industry	-	-	-	-	-	-	-	-	-	-	-
Liquefaction	-	-	-	-	-	-	-	-	-	-	-
Other Transform. Sector	-	-	-	-	-	-	-	-	-	-	-
ENERGY SECTOR	-	-	-	-	-	-	12	2	-	-	-
Coal Mines	-	-	-	-	-	-	6	2	-	-	-
Oil and Gas Extraction	-	-	-	-	-	-	-	-	-	-	-
Petroleum Refineries	-	-	-	-	-	-	-	-	-	-	-
Electr., CHP+Heat Plants	-	-	-	-	-	-	-	-	-	-	-
Pumped Storage (Elec.)	-	-	-	-	-	-	-	-	-	-	-
Other Energy Sector	-	-	-	-	-	-	6	-	-	-	-
Distribution Losses	-	1	-	-	-	-	-	-	-	-	-
FINAL CONSUMPTION	-	8	305	2	20	-	354	187	-	-	42
INDUSTRY SECTOR	-	1	1	-	-	-	47	154	-	-	-
Iron and Steel	-	-	-	-	-	-	-	-	-	-	-
Chemical and Petrochem.	-	-	-	-	-	-	-	42	-	-	-
of which: Feedstocks	-	-	-	-	-	-	-	-	-	-	-
Non-Ferrous Metals	-	-	-	-	-	-	-	-	-	-	-
Non-Metallic Minerals	-	-	-	-	-	-	2	2	-	-	-
Transport Equipment	-	-	-	-	-	-	-	1	-	-	-
Machinery	-	-	-	-	-	-	1	1	-	-	-
Mining and Quarrying	-	-	-	-	-	-	-	-	-	-	-
Food and Tobacco	-	-	-	-	-	-	31	61	-	-	-
Paper, Pulp and Print	-	-	-	-	-	-	-	6	-	-	-
Wood and Wood Products	-	-	-	-	-	-	1	17	-	-	-
Construction	-	-	1	-	-	-	11	4	-	-	-
Textile and Leather	-	-	-	-	-	-	-	6	-	-	-
Non-specified	-	1	-	-	-	-	1	14	-	-	-
TRANSPORT SECTOR	-	1	302	2	20	-	201	5	-	-	-
Air	-	-	-	2	20	-	-	-	-	-	-
Road	-	1	302	-	-	-	162	2	-	-	-
Rail	-	-	-	-	-	-	33	3	-	-	-
Pipeline Transport	-	-	-	-	-	-	-	-	-	-	-
Internal Navigation	-	-	-	-	-	-	6	-	-	-	-
Non-specified	-	-	-	-	-	-	-	-	-	-	-
OTHER SECTORS	-	6	2	-	-	-	106	28	-	-	-
Agriculture	-	-	2	-	-	-	37	9	-	-	-
Comm. and Publ. Services	-	-	-	-	-	-	12	19	-	-	-
Residential	-	6	-	-	-	-	57	-	-	-	-
Non-specified	-	-	-	-	-	-	-	-	-	-	-
NON-ENERGY USE	-	-	-	-	-	-	-	-	-	-	42
in Industry/Transf./Energy	-	-	-	-	-	-	-	-	-	-	37
in Transport	-	-	-	-	-	-	-	-	-	-	3
in Other Sectors	-	-	-	-	-	-	-	-	-	-	2

Estonia / Estonie : 1997

SUPPLY AND CONSUMPTION / APPROVISIONNEMENT ET DEMANDE	Gas / Gaz (TJ)				Comb. Renew. & Waste / En. Re. Comb. & Déchets (TJ)				(GWh)	(TJ)
	Natural Gas / Gaz naturel	Gas Works / Usines à gaz	Coke Ovens / Cokeries	Blast Furnaces / Hauts fourneaux	Solid Biomass / Biomasse solide	Gas/Liquids from Biomass / Gaz/Liquides tirés de biomasse	Municipal Waste / Déchets urbains	Industrial Waste / Déchets industriels	Electricity / Electricité	Heat / Chaleur
Production	-	5313	-	-	26118	58	-	-	9218	32593
From Other Sources	-	-	-	-	-	-	-	-	-	-
Imports	29038	-	-	-	-	-	-	-	210	-
Exports	-	-	-	-	-1540	-	-	-	-1184	-
Intl. Marine Bunkers	-	-	-	-	-	-	-	-	-	-
Stock Changes	-	-	-	-	157	-	-	-	-	-
DOMESTIC SUPPLY	29038	5313	-	-	24735	58	-	-	8244	32593
Transfers	-	-	-	-	-	-	-	-	-	-
Statistical Differences	-	-	-	-	-	-	-	-	-	-
TRANSFORMATION	14775	1627	-	-	3213	58	-	-	105	-
Electricity Plants	-	-	-	-	19	-	-	-	-	-
CHP Plants	3404	1455	-	-	55	-	-	-	-	-
Heat Plants	11371	172	-	-	3139	58	-	-	-	-
Blast Furnaces/Gas Works	-	-	-	-	-	-	-	-	-	-
Coke/Pat. Fuel/BKB Plants	-	-	-	-	-	-	-	-	-	-
Petroleum Refineries	-	-	-	-	-	-	-	-	-	-
Petrochemical Industry	-	-	-	-	-	-	-	-	-	-
Liquefaction	-	-	-	-	-	-	-	-	-	-
Other Transform. Sector	-	-	-	-	-	-	-	-	105	-
ENERGY SECTOR	26	-	-	-	9	-	-	-	1483	-
Coal Mines	-	-	-	-	9	-	-	-	275	-
Oil and Gas Extraction	26	-	-	-	-	-	-	-	-	-
Petroleum Refineries	-	-	-	-	-	-	-	-	-	-
Electr., CHP+Heat Plants	-	-	-	-	-	-	-	-	1060	-
Pumped Storage (Elec.)	-	-	-	-	-	-	-	-	-	-
Other Energy Sector	-	-	-	-	-	-	-	-	148	-
Distribution Losses	28	136	-	-	15	-	-	-	1510	5589
FINAL CONSUMPTION	14209	3550	-	-	21498	-	-	-	5146	27004
INDUSTRY SECTOR	12063	3550	-	-	2540	-	-	-	2208	2662
Iron and Steel	-	-	-	-	-	-	-	-	7	10
Chemical and Petrochem.	8415	3550	-	-	-	-	-	-	613	700
of which: Feedstocks	7611	-	-	-	-	-	-	-	-	-
Non-Ferrous Metals	-	-	-	-	-	-	-	-	-	-
Non-Metallic Minerals	927	-	-	-	23	-	-	-	259	35
Transport Equipment	20	-	-	-	7	-	-	-	63	24
Machinery	109	-	-	-	30	-	-	-	150	167
Mining and Quarrying	-	-	-	-	1	-	-	-	24	10
Food and Tobacco	1251	-	-	-	113	-	-	-	349	269
Paper, Pulp and Print	712	-	-	-	370	-	-	-	85	174
Wood and Wood Products	113	-	-	-	642	-	-	-	130	218
Construction	20	-	-	-	36	-	-	-	104	51
Textile and Leather	179	-	-	-	16	-	-	-	206	698
Non-specified	317	-	-	-	1302	-	-	-	218	306
TRANSPORT SECTOR	-	-	-	-	1	-	-	-	108	-
Air	-	-	-	-	-	-	-	-	-	-
Road	-	-	-	-	-	-	-	-	-	-
Rail	-	-	-	-	1	-	-	-	14	-
Pipeline Transport	-	-	-	-	-	-	-	-	-	-
Internal Navigation	-	-	-	-	-	-	-	-	-	-
Non-specified	-	-	-	-	-	-	-	-	94	-
OTHER SECTORS	2146	-	-	-	18957	-	-	-	2830	24342
Agriculture	144	-	-	-	234	-	-	-	229	108
Comm. and Publ. Services	303	-	-	-	410	-	-	-	1396	3095
Residential	1699	-	-	-	18313	-	-	-	1205	21139
Non-specified	-	-	-	-	-	-	-	-	-	-
NON-ENERGY USE	-	-	-	-	-	-	-	-	-	-
in Industry/Transf./Energy	-	-	-	-	-	-	-	-	-	-
in Transport	-	-	-	-	-	-	-	-	-	-
in Other Sectors	-	-	-	-	-	-	-	-	-	-

Estonia / Estonie : 1998

SUPPLY AND CONSUMPTION / APPROVISIONNEMENT ET DEMANDE	Coal / Charbon (1000 tonnes)							Oil / Pétrole (1000 tonnes)			
	Coking Coal / Charbon à coke	Other Bit. Coal / Autres charb. bit.	Sub-Bit. Coal / Charbon sous-bit.	Lignite / Lignite	Peat / Tourbe	Oven and Gas Coke / Coke de four/gaz	Pat. Fuel and BKB / Agg./briq. de lignite	Crude Oil / Pétrole brut	NGL / LGN	Feed-stocks / Produits d'aliment.	Additives / Additifs
Production	-	-	-	10774	157	27	101	-	-	-	-
From Other Sources	-	-	-	-	-	-	-	239	-	-	-
Imports	-	80	-	1318	-	2	-	-	-	-	-
Exports	-	-	-	-6	-8	-27	-80	-	-	-	-
Intl. Marine Bunkers	-	-	-	-	-	-	-	-	-	-	-
Stock Changes	-	-5	-	112	268	-1	13	-	-	-	-
DOMESTIC SUPPLY	-	75	-	12198	417	1	34	239	-	-	-
Transfers	-	-	-	-	-	-	-	-	-	-	-
Statistical Differences	-	-	-	-	-	-	-	-	-	-	-
TRANSFORMATION	-	18	-	11665	358	-	2	239	-	-	-
Electricity Plants	-	-	-	8893	-	-	-	-	-	-	-
CHP Plants	-	-	-	2016	21	-	-	-	-	-	-
Heat Plants	-	18	-	46	88	-	2	-	-	-	-
Blast Furnaces/Gas Works	-	-	-	603	-	-	-	-	-	-	-
Coke/Pat. Fuel/BKB Plants	-	-	-	107	249	-	-	-	-	-	-
Petroleum Refineries	-	-	-	-	-	-	-	239	-	-	-
Petrochemical Industry	-	-	-	-	-	-	-	-	-	-	-
Liquefaction	-	-	-	-	-	-	-	-	-	-	-
Other Transform. Sector	-	-	-	-	-	-	-	-	-	-	-
ENERGY SECTOR	-	-	-	42	-	-	1	-	-	-	-
Coal Mines	-	-	-	42	-	-	-	-	-	-	-
Oil and Gas Extraction	-	-	-	-	-	-	-	-	-	-	-
Petroleum Refineries	-	-	-	-	-	-	-	-	-	-	-
Electr., CHP+Heat Plants	-	-	-	-	-	-	-	-	-	-	-
Pumped Storage (Elec.)	-	-	-	-	-	-	-	-	-	-	-
Other Energy Sector	-	-	-	-	-	-	1	-	-	-	-
Distribution Losses	-	-	-	1	38	-	-	-	-	-	-
FINAL CONSUMPTION	-	57	-	490	21	1	31	-	-	-	-
INDUSTRY SECTOR	-	11	-	366	11	1	-	-	-	-	-
Iron and Steel	-	-	-	-	-	-	-	-	-	-	-
Chemical and Petrochem.	-	-	-	27	-	-	-	-	-	-	-
of which: Feedstocks	-	-	-	-	-	-	-	-	-	-	-
Non-Ferrous Metals	-	-	-	-	-	-	-	-	-	-	-
Non-Metallic Minerals	-	7	-	334	6	-	-	-	-	-	-
Transport Equipment	-	-	-	-	-	-	-	-	-	-	-
Machinery	-	1	-	-	-	1	-	-	-	-	-
Mining and Quarrying	-	-	-	-	-	-	-	-	-	-	-
Food and Tobacco	-	1	-	-	4	-	-	-	-	-	-
Paper, Pulp and Print	-	-	-	-	-	-	-	-	-	-	-
Wood and Wood Products	-	-	-	5	-	-	-	-	-	-	-
Construction	-	1	-	-	-	-	-	-	-	-	-
Textile and Leather	-	1	-	-	-	-	-	-	-	-	-
Non-specified	-	-	-	-	1	-	-	-	-	-	-
TRANSPORT SECTOR	-	1	-	-	-	-	-	-	-	-	-
Air	-	-	-	-	-	-	-	-	-	-	-
Road	-	-	-	-	-	-	-	-	-	-	-
Rail	-	1	-	-	-	-	-	-	-	-	-
Pipeline Transport	-	-	-	-	-	-	-	-	-	-	-
Internal Navigation	-	-	-	-	-	-	-	-	-	-	-
Non-specified	-	-	-	-	-	-	-	-	-	-	-
OTHER SECTORS	-	45	-	-	10	-	31	-	-	-	-
Agriculture	-	1	-	-	6	-	-	-	-	-	-
Comm. and Publ. Services	-	10	-	-	3	-	1	-	-	-	-
Residential	-	34	-	-	1	-	30	-	-	-	-
Non-specified	-	-	-	-	-	-	-	-	-	-	-
NON-ENERGY USE	-	-	-	124	-	-	-	-	-	-	-
in Industry/Trans./Energy	-	-	-	124	-	-	-	-	-	-	-
in Transport	-	-	-	-	-	-	-	-	-	-	-
in Other Sectors	-	-	-	-	-	-	-	-	-	-	-

Estonia / Estonie : 1998

SUPPLY AND CONSUMPTION	Oil cont. / *Pétrole cont.* (1000 tonnes)										
	Refinery Gas	LPG + Ethane	Motor Gasoline	Aviation Gasoline	Jet Fuel	Kerosene	Gas/ Diesel	Heavy Fuel Oil	Naphtha	Petrol. Coke	Other Prod.
APPROVISIONNEMENT ET DEMANDE	*Gaz de raffinerie*	*GPL + éthane*	*Essence moteur*	*Essence aviation*	*Carbu- réacteurs*	*Kérosène*	*Gazole*	*Fioul lourd*	*Naphta*	*Coke de pétrole*	*Autres prod.*
Production	-	-	-	-	-	-	-	220	-	-	5
From Other Sources	-	-	-	-	-	-	-	-	-	-	-
Imports	-	9	336	3	18	-	611	662	-	-	108
Exports	-	-	-54	-	-5	-	-162	-320	-	-	-67
Intl. Marine Bunkers	-	-	-	-	-	-	-30	-78	-	-	-
Stock Changes	-	-	13	-	-	-	-11	-35	-	-	1
DOMESTIC SUPPLY	**-**	**9**	**295**	**3**	**13**	**-**	**408**	**449**	**-**	**-**	**47**
Transfers	-	-	-	-	-	-	-	-	-	-	-
Statistical Differences	-	-	-	-	-	-	-	-	-	-	-
TRANSFORMATION	**-**	**-**	**-**	**-**	**-**	**-**	**13**	**308**	**-**	**-**	**-**
Electricity Plants	-	-	-	-	-	-	-	9	-	-	-
CHP Plants	-	-	-	-	-	-	-	80	-	-	-
Heat Plants	-	-	-	-	-	-	13	219	-	-	-
Blast Furnaces/Gas Works	-	-	-	-	-	-	-	-	-	-	-
Coke/Pat. Fuel/BKB Plants	-	-	-	-	-	-	-	-	-	-	-
Petroleum Refineries	-	-	-	-	-	-	-	-	-	-	-
Petrochemical Industry	-	-	-	-	-	-	-	-	-	-	-
Liquefaction	-	-	-	-	-	-	-	-	-	-	-
Other Transform. Sector	-	-	-	-	-	-	-	-	-	-	-
ENERGY SECTOR	**-**	**-**	**-**	**-**	**-**	**-**	**11**	**3**	**-**	**-**	**-**
Coal Mines	-	-	-	-	-	-	8	3	-	-	-
Oil and Gas Extraction	-	-	-	-	-	-	1	-	-	-	-
Petroleum Refineries	-	-	-	-	-	-	-	-	-	-	-
Electr., CHP+Heat Plants	-	-	-	-	-	-	-	-	-	-	-
Pumped Storage (Elec.)	-	-	-	-	-	-	-	-	-	-	-
Other Energy Sector	-	-	-	-	-	-	2	-	-	-	-
Distribution Losses	-	-	-	-	-	-	-	-	-	-	-
FINAL CONSUMPTION	**-**	**9**	**295**	**3**	**13**	**-**	**384**	**138**	**-**	**-**	**47**
INDUSTRY SECTOR	**-**	**2**	**2**	**-**	**-**	**-**	**43**	**105**	**-**	**-**	**-**
Iron and Steel	-	-	-	-	-	-	-	-	-	-	-
Chemical and Petrochem.	-	-	-	-	-	-	-	1	-	-	-
of which: Feedstocks	-	-	-	-	-	-	-	-	-	-	-
Non-Ferrous Metals	-	-	-	-	-	-	-	-	-	-	-
Non-Metallic Minerals	-	-	-	-	-	-	3	3	-	-	-
Transport Equipment	-	-	-	-	-	-	1	1	-	-	-
Machinery	-	-	-	-	-	-	1	1	-	-	-
Mining and Quarrying	-	-	-	-	-	-	1	-	-	-	-
Food and Tobacco	-	-	-	-	-	-	19	49	-	-	-
Paper, Pulp and Print	-	-	-	-	-	-	-	22	-	-	-
Wood and Wood Products	-	-	-	-	-	-	2	18	-	-	-
Construction	-	1	2	-	-	-	15	2	-	-	-
Textile and Leather	-	-	-	-	-	-	-	4	-	-	-
Non-specified	-	1	-	-	-	-	1	4	-	-	-
TRANSPORT SECTOR	**-**	**-**	**290**	**3**	**13**	**-**	**242**	**3**	**-**	**-**	**-**
Air	-	-	-	3	13	-	-	-	-	-	-
Road	-	-	290	-	-	-	194	3	-	-	-
Rail	-	-	-	-	-	-	42	-	-	-	-
Pipeline Transport	-	-	-	-	-	-	-	-	-	-	-
Internal Navigation	-	-	-	-	-	-	6	-	-	-	-
Non-specified	-	-	-	-	-	-	-	-	-	-	-
OTHER SECTORS	**-**	**7**	**3**	**-**	**-**	**-**	**99**	**30**	**-**	**-**	**-**
Agriculture	-	-	2	-	-	-	39	8	-	-	-
Comm. and Publ. Services	-	-	1	-	-	-	13	22	-	-	-
Residential	-	7	-	-	-	-	47	-	-	-	-
Non-specified	-	-	-	-	-	-	-	-	-	-	-
NON-ENERGY USE	**-**	**-**	**-**	**-**	**-**	**-**	**-**	**-**	**-**	**-**	**47**
in Industry/Transf./Energy	-	-	-	-	-	-	-	-	-	-	39
in Transport	-	-	-	-	-	-	-	-	-	-	5
in Other Sectors	-	-	-	-	-	-	-	-	-	-	3

Estonia / Estonie : 1998

SUPPLY AND CONSUMPTION APPROVISIONNEMENT ET DEMANDE	Gas / Gaz (TJ)				Comb. Renew. & Waste / En. Re. Comb. & Déchets (TJ)				(GWh)	(TJ)
	Natural Gas Gaz naturel	Gas Works Usines à gaz	Coke Ovens Cokeries	Blast Furnaces Hauts fourneaux	Solid Biomass Biomasse solide	Gas/Liquids from Biomass Gaz/Liquides tirés de biomasse	Municipal Waste Déchets urbains	Industrial Waste Déchets industriels	Electricity Electricité	Heat Chaleur
Production	-	3416	-	-	21341	70	-	-	8521	29113
From Other Sources	-	-	-	-	-	-	-	-	-	-
Imports	27560	-	-	-	-	-	-	-	138	-
Exports	-	-	-	-	-	-	-	-	-528	-
Intl. Marine Bunkers	-	-	-	-	-	-	-	-	-	-
Stock Changes	-	-	-	-	-194	-	-	-	-	-
DOMESTIC SUPPLY	27560	3416	-	-	21147	70	-	-	8131	29113
Transfers	-	-	-	-	-	-	-	-	-	-
Statistical Differences	-	-	-	-	-	-	-	-	-	-
TRANSFORMATION	12649	1321	-	-	3699	70	-	-	37	-
Electricity Plants	-	-	-	-	33	-	-	-	-	-
CHP Plants	3440	1135	-	-	106	-	-	-	-	-
Heat Plants	9209	186	-	-	3560	70	-	-	-	-
Blast Furnaces/Gas Works	-	-	-	-	-	-	-	-	-	-
Coke/Pat. Fuel/BKB Plants	-	-	-	-	-	-	-	-	-	-
Petroleum Refineries	-	-	-	-	-	-	-	-	-	-
Petrochemical Industry	-	-	-	-	-	-	-	-	-	-
Liquefaction	-	-	-	-	-	-	-	-	-	-
Other Transform. Sector	-	-	-	-	-	-	-	-	37	-
ENERGY SECTOR	24	-	-	-	16	-	-	-	1407	90
Coal Mines	-	-	-	-	1	-	-	-	257	89
Oil and Gas Extraction	24	-	-	-	-	-	-	-	-	-
Petroleum Refineries	-	-	-	-	-	-	-	-	-	-
Electr., CHP+Heat Plants	-	-	-	-	7	-	-	-	983	-
Pumped Storage (Elec.)	-	-	-	-	-	-	-	-	-	-
Other Energy Sector	-	-	-	-	8	-	-	-	167	1
Distribution Losses	31	60	-	-	7	-	-	-	1568	5270
FINAL CONSUMPTION	14856	2035	-	-	17425	-	-	-	5119	23753
INDUSTRY SECTOR	12315	2035	-	-	2647	-	-	-	1919	2135
Iron and Steel	-	-	-	-	-	-	-	-	4	4
Chemical and Petrochem.	8973	2035	-	-	-	-	-	-	418	515
of which: Feedstocks	7680	-	-	-	-	-	-	-	-	-
Non-Ferrous Metals	1	-	-	-	-	-	-	-	-	-
Non-Metallic Minerals	919	-	-	-	12	-	-	-	171	58
Transport Equipment	79	-	-	-	1	-	-	-	76	32
Machinery	154	-	-	-	23	-	-	-	152	158
Mining and Quarrying	-	-	-	-	-	-	-	-	5	4
Food and Tobacco	1169	-	-	-	101	-	-	-	302	115
Paper, Pulp and Print	304	-	-	-	335	-	-	-	96	76
Wood and Wood Products	91	-	-	-	967	-	-	-	195	61
Construction	67	-	-	-	30	-	-	-	99	104
Textile and Leather	254	-	-	-	9	-	-	-	239	893
Non-specified	304	-	-	-	1169	-	-	-	162	115
TRANSPORT SECTOR	-	-	-	-	1	-	-	-	108	-
Air	-	-	-	-	-	-	-	-	-	-
Road	-	-	-	-	-	-	-	-	-	-
Rail	-	-	-	-	1	-	-	-	21	-
Pipeline Transport	-	-	-	-	-	-	-	-	-	-
Internal Navigation	-	-	-	-	-	-	-	-	-	-
Non-specified	-	-	-	-	-	70	-	-	87	-
OTHER SECTORS	2541	-	-	-	14777	-	-	-	3092	21618
Agriculture	222	-	-	-	232	-	-	-	252	101
Comm. and Publ. Services	299	-	-	-	307	-	-	-	1491	2513
Residential	2020	-	-	-	14238	-	-	-	1349	19004
Non-specified	-	-	-	-	-	-	-	-	-	-
NON-ENERGY USE	-	-	-	-	-	-	-	-	-	-
in Industry/Transf./Energy	-	-	-	-	-	-	-	-	-	-
in Transport	-	-	-	-	-	-	-	-	-	-
in Other Sectors	-	-	-	-	-	-	-	-	-	-

Ethiopia / Ethiopie

SUPPLY AND CONSUMPTION 1997	Coal (1000 tonnes)							Oil (1000 tonnes)			
	Coking Coal	Other Bit. Coal	Sub-Bit. Coal	Lignite	Peat	Oven and Gas Coke	Pat. Fuel and BKB	Crude Oil	NGL	Feed-stocks	Additives
Production	-	-	-	-	-	-	-	-	-	-	-
Imports	-	-	-	-	-	-	-	362	-	-	-
Exports	-	-	-	-	-	-	-	-	-	-	-
Intl. Marine Bunkers	-	-	-	-	-	-	-	-	-	-	-
Stock Changes	-	-	-	-	-	-	-	-	-	-	-
DOMESTIC SUPPLY	-	-	-	-	-	-	-	362	-	-	-
Transfers and Stat. Diff.	-	-	-	-	-	-	-	-	-	-	-
TRANSFORMATION	-	-	-	-	-	-	-	362	-	-	-
Electricity and CHP Plants	-	-	-	-	-	-	-	-	-	-	-
Petroleum Refineries	-	-	-	-	-	-	-	362	-	-	-
Other Transform. Sector	-	-	-	-	-	-	-	-	-	-	-
ENERGY SECTOR	-	-	-	-	-	-	-	-	-	-	-
DISTRIBUTION LOSSES	-	-	-	-	-	-	-	-	-	-	-
FINAL CONSUMPTION	-	-	-	-	-	-	-	-	-	-	-
INDUSTRY SECTOR	-	-	-	-	-	-	-	-	-	-	-
Iron and Steel	-	-	-	-	-	-	-	-	-	-	-
Chemical and Petrochem.	-	-	-	-	-	-	-	-	-	-	-
Non-Metallic Minerals	-	-	-	-	-	-	-	-	-	-	-
Non-specified	-	-	-	-	-	-	-	-	-	-	-
TRANSPORT SECTOR	-	-	-	-	-	-	-	-	-	-	-
Air	-	-	-	-	-	-	-	-	-	-	-
Road	-	-	-	-	-	-	-	-	-	-	-
Non-specified	-	-	-	-	-	-	-	-	-	-	-
OTHER SECTORS	-	-	-	-	-	-	-	-	-	-	-
Agriculture	-	-	-	-	-	-	-	-	-	-	-
Comm. and Publ. Services	-	-	-	-	-	-	-	-	-	-	-
Residential	-	-	-	-	-	-	-	-	-	-	-
Non-specified	-	-	-	-	-	-	-	-	-	-	-
NON-ENERGY USE	-	-	-	-	-	-	-	-	-	-	-

APPROVISIONNEMENT ET DEMANDE 1998	Charbon (1000 tonnes)							Pétrole (1000 tonnes)			
	Charbon à coke	Autres charb. bit.	Charbon sous-bit.	Lignite	Tourbe	Coke de four/gaz	Agg./briq. de lignite	Pétrole brut	LGN	Produits d'aliment.	Additifs
Production	-	-	-	-	-	-	-	-	-	-	-
Imports	-	-	-	-	-	-	-	-	-	-	-
Exports	-	-	-	-	-	-	-	-	-	-	-
Intl. Marine Bunkers	-	-	-	-	-	-	-	-	-	-	-
Stock Changes	-	-	-	-	-	-	-	-	-	-	-
DOMESTIC SUPPLY	-	-	-	-	-	-	-	-	-	-	-
Transfers and Stat. Diff.	-	-	-	-	-	-	-	-	-	-	-
TRANSFORMATION	-	-	-	-	-	-	-	-	-	-	-
Electricity and CHP Plants	-	-	-	-	-	-	-	-	-	-	-
Petroleum Refineries	-	-	-	-	-	-	-	-	-	-	-
Other Transform. Sector	-	-	-	-	-	-	-	-	-	-	-
ENERGY SECTOR	-	-	-	-	-	-	-	-	-	-	-
DISTRIBUTION LOSSES	-	-	-	-	-	-	-	-	-	-	-
FINAL CONSUMPTION	-	-	-	-	-	-	-	-	-	-	-
INDUSTRY SECTOR	-	-	-	-	-	-	-	-	-	-	-
Iron and Steel	-	-	-	-	-	-	-	-	-	-	-
Chemical and Petrochem.	-	-	-	-	-	-	-	-	-	-	-
Non-Metallic Minerals	-	-	-	-	-	-	-	-	-	-	-
Non-specified	-	-	-	-	-	-	-	-	-	-	-
TRANSPORT SECTOR	-	-	-	-	-	-	-	-	-	-	-
Air	-	-	-	-	-	-	-	-	-	-	-
Road	-	-	-	-	-	-	-	-	-	-	-
Non-specified	-	-	-	-	-	-	-	-	-	-	-
OTHER SECTORS	-	-	-	-	-	-	-	-	-	-	-
Agriculture	-	-	-	-	-	-	-	-	-	-	-
Comm. and Publ. Services	-	-	-	-	-	-	-	-	-	-	-
Residential	-	-	-	-	-	-	-	-	-	-	-
Non-specified	-	-	-	-	-	-	-	-	-	-	-
NON-ENERGY USE	-	-	-	-	-	-	-	-	-	-	-

INTERNATIONAL ENERGY AGENCY

Ethiopia / Ethiopie

SUPPLY AND CONSUMPTION 1997	Oil cont. (1000 tonnes)										
	Refinery Gas	LPG + Ethane	Motor Gasoline	Aviation Gasoline	Jet Fuel	Kerosene	Gas/ Diesel	Heavy Fuel Oil	Naphtha	Petrol. Coke	Other Prod.
Production	2	3	51	-	31	-	100	153	-	-	-
Imports	-	2	95	6	18	211	489	69	-	-	40
Exports	-	-3	-23	-	-	-41	-135	-52	-	-	-
Intl. Marine Bunkers	-	-	-	-	-	-	-	-70	-	-	-
Stock Changes	-	-	7	-	-	-2	-1	4	-	-	-
DOMESTIC SUPPLY	2	2	130	6	49	168	453	104	-	-	40
Transfers and Stat. Diff.	-	-	-	-	-	1	-	1	-	-	-
TRANSFORMATION	-	-	-	-	-	-	13	-	-	-	-
Electricity and CHP Plants	-	-	-	-	-	-	13	-	-	-	-
Petroleum Refineries	-	-	-	-	-	-	-	-	-	-	-
Other Transform. Sector	-	-	-	-	-	-	-	-	-	-	-
ENERGY SECTOR	2	-	-	-	-	-	-	-	-	-	-
DISTRIBUTION LOSSES	-	-	-	-	-	-	-	-	-	-	-
FINAL CONSUMPTION	-	2	130	6	49	169	440	105	-	-	40
INDUSTRY SECTOR	-	-	-	-	-	-	110	105	-	-	-
Iron and Steel	-	-	-	-	-	-	-	-	-	-	-
Chemical and Petrochem.	-	-	-	-	-	-	-	-	-	-	-
Non-Metallic Minerals	-	-	-	-	-	-	-	-	-	-	-
Non-specified	-	-	-	-	-	-	110	105	-	-	-
TRANSPORT SECTOR	-	-	130	6	49	-	330	-	-	-	-
Air	-	-	-	6	49	-	-	-	-	-	-
Road	-	-	130	-	-	-	330	-	-	-	-
Non-specified	-	-	-	-	-	-	-	-	-	-	-
OTHER SECTORS	-	2	-	-	-	169	-	-	-	-	-
Agriculture	-	-	-	-	-	-	-	-	-	-	-
Comm. and Publ. Services	-	-	-	-	-	-	-	-	-	-	-
Residential	-	2	-	-	-	169	-	-	-	-	-
Non-specified	-	-	-	-	-	-	-	-	-	-	-
NON-ENERGY USE	-	-	-	-	-	-	-	-	-	-	40

APPROVISIONNEMENT ET DEMANDE 1998	Pétrole cont. (1000 tonnes)										
	Gaz de raffinerie	GPL + éthane	Essence moteur	Essence aviation	Carbu- réacteurs	Kérosène	Gazole	Fioul lourd	Naphta	Coke de pétrole	Autres prod.
Production	-	-	-	-	-	-	-	-	-	-	-
Imports	-	2	129	-	63	181	549	117	-	-	40
Exports	-	-	-	-	-	-	-	-	-	-	-
Intl. Marine Bunkers	-	-	-	-	-	-	-	-	-	-	-
Stock Changes	-	1	5	-	-	-22	-42	-6	-	-	-
DOMESTIC SUPPLY	-	3	134	-	63	159	507	111	-	-	40
Transfers and Stat. Diff.	-	-	-	-	-	-	-1	-1	-	-	-
TRANSFORMATION	-	-	-	-	-	-	13	-	-	-	-
Electricity and CHP Plants	-	-	-	-	-	-	13	-	-	-	-
Petroleum Refineries	-	-	-	-	-	-	-	-	-	-	-
Other Transform. Sector	-	-	-	-	-	-	-	-	-	-	-
ENERGY SECTOR	-	-	-	-	-	-	-	-	-	-	-
DISTRIBUTION LOSSES	-	-	-	-	-	-	-	-	-	-	-
FINAL CONSUMPTION	-	3	134	-	63	159	493	110	-	-	40
INDUSTRY SECTOR	-	-	-	-	-	-	123	110	-	-	-
Iron and Steel	-	-	-	-	-	-	-	-	-	-	-
Chemical and Petrochem.	-	-	-	-	-	-	-	-	-	-	-
Non-Metallic Minerals	-	-	-	-	-	-	-	-	-	-	-
Non-specified	-	-	-	-	-	-	123	110	-	-	-
TRANSPORT SECTOR	-	-	134	-	63	-	370	-	-	-	-
Air	-	-	-	-	63	-	-	-	-	-	-
Road	-	-	134	-	-	-	370	-	-	-	-
Non-specified	-	-	-	-	-	-	-	-	-	-	-
OTHER SECTORS	-	3	-	-	-	159	-	-	-	-	-
Agriculture	-	-	-	-	-	-	-	-	-	-	-
Comm. and Publ. Services	-	-	-	-	-	-	-	-	-	-	-
Residential	-	3	-	-	-	159	-	-	-	-	-
Non-specified	-	-	-	-	-	-	-	-	-	-	-
NON-ENERGY USE	-	-	-	-	-	-	-	-	-	-	40

Ethiopia / Ethiopie

SUPPLY AND CONSUMPTION 1997	Natural Gas	Gas Works	Coke Ovens	Blast Furnaces	Solid Biomass	Gas/Liquids from Biomass	Municipal Waste	Industrial Waste	Electricity	Heat
	Gas (TJ)				Comb. Renew. & Waste (TJ)				(GWh)	(TJ)
Production	-	-	-	-	648057	-	-	-	1614	-
Imports	-	-	-	-	-	-	-	-	-	-
Exports	-	-	-	-	-	-	-	-	-	-
Intl. Marine Bunkers	-	-	-	-	-	-	-	-	-	-
Stock Changes	-	-	-	-	-	-	-	-	-	-
DOMESTIC SUPPLY	-	-	-	-	648057	-	-	-	1614	-
Transfers and Stat. Diff.	-	-	-	-	1	-	-	-	-	-
TRANSFORMATION	-	-	-	-	23543	-	-	-	-	-
Electricity and CHP Plants	-	-	-	-	-	-	-	-	-	-
Petroleum Refineries	-	-	-	-	-	-	-	-	-	-
Other Transform. Sector	-	-	-	-	23543	-	-	-	-	-
ENERGY SECTOR	-	-	-	-	-	-	-	-	128	-
DISTRIBUTION LOSSES	-	-	-	-	-	-	-	-	161	-
FINAL CONSUMPTION	-	-	-	-	624515	-	-	-	1325	-
INDUSTRY SECTOR	-	-	-	-	-	-	-	-	563	-
Iron and Steel	-	-	-	-	-	-	-	-	-	-
Chemical and Petrochem.	-	-	-	-	-	-	-	-	-	-
Non-Metallic Minerals	-	-	-	-	-	-	-	-	-	-
Non-specified	-	-	-	-	-	-	-	-	563	-
TRANSPORT SECTOR	-	-	-	-	-	-	-	-	-	-
Air	-	-	-	-	-	-	-	-	-	-
Road	-	-	-	-	-	-	-	-	-	-
Non-specified	-	-	-	-	-	-	-	-	-	-
OTHER SECTORS	-	-	-	-	624515	-	-	-	762	-
Agriculture	-	-	-	-	-	-	-	-	-	-
Comm. and Publ. Services	-	-	-	-	-	-	-	-	185	-
Residential	-	-	-	-	3914	-	-	-	569	-
Non-specified	-	-	-	-	620601	-	-	-	8	-
NON-ENERGY USE	-	-	-	-	-	-	-	-	-	-

APPROVISIONNEMENT ET DEMANDE 1998	Gaz naturel	Usines à gaz	Cokeries	Hauts fourneaux	Biomasse solide	Gaz/Liquides tirés de biomasse	Déchets urbains	Déchets industriels	Electricité	Chaleur
	Gaz (TJ)				En. Re. Comb. & Déchets (TJ)				(GWh)	(TJ)
Production	-	-	-	-	680220	-	-	-	1627	-
Imports	-	-	-	-	-	-	-	-	-	-
Exports	-	-	-	-	-	-	-	-	-	-
Intl. Marine Bunkers	-	-	-	-	-	-	-	-	-	-
Stock Changes	-	-	-	-	-	-	-	-	-	-
DOMESTIC SUPPLY	-	-	-	-	680220	-	-	-	1627	-
Transfers and Stat. Diff.	-	-	-	-	-	-	-	-	-	-
TRANSFORMATION	-	-	-	-	24055	-	-	-	-	-
Electricity and CHP Plants	-	-	-	-	-	-	-	-	-	-
Petroleum Refineries	-	-	-	-	-	-	-	-	-	-
Other Transform. Sector	-	-	-	-	24055	-	-	-	-	-
ENERGY SECTOR	-	-	-	-	-	-	-	-	106	-
DISTRIBUTION LOSSES	-	-	-	-	-	-	-	-	163	-
FINAL CONSUMPTION	-	-	-	-	636167	-	-	-	1358	-
INDUSTRY SECTOR	-	-	-	-	-	-	-	-	559	-
Iron and Steel	-	-	-	-	-	-	-	-	-	-
Chemical and Petrochem.	-	-	-	-	-	-	-	-	-	-
Non-Metallic Minerals	-	-	-	-	-	-	-	-	-	-
Non-specified	-	-	-	-	-	-	-	-	559	-
TRANSPORT SECTOR	-	-	-	-	-	-	-	-	-	-
Air	-	-	-	-	-	-	-	-	-	-
Road	-	-	-	-	-	-	-	-	-	-
Non-specified	-	-	-	-	-	-	-	-	-	-
OTHER SECTORS	-	-	-	-	636167	-	-	-	799	-
Agriculture	-	-	-	-	-	-	-	-	-	-
Comm. and Publ. Services	-	-	-	-	-	-	-	-	274	-
Residential	-	-	-	-	5737	-	-	-	516	-
Non-specified	-	-	-	-	630430	-	-	-	9	-
NON-ENERGY USE	-	-	-	-	-	-	-	-	-	-

Gabon

SUPPLY AND CONSUMPTION 1997	Coal (1000 tonnes)							Oil (1000 tonnes)			
	Coking Coal	Other Bit. Coal	Sub-Bit. Coal	Lignite	Peat	Oven and Gas Coke	Pat. Fuel and BKB	Crude Oil	NGL	Feed-stocks	Additives
Production	-	-	-	-	-	-	-	18462	-	-	-
Imports	-	-	-	-	-	-	-	-	-	-	-
Exports	-	-	-	-	-	-	-	-17539	-	-	-
Intl. Marine Bunkers	-	-	-	-	-	-	-	-	-	-	-
Stock Changes	-	-	-	-	-	-	-	-198	-	-	-
DOMESTIC SUPPLY	-	-	-	-	-	-	-	725	-	-	-
Transfers and Stat. Diff.	-	-	-	-	-	-	-	-	-	-	-
TRANSFORMATION	-	-	-	-	-	-	-	725	-	-	-
Electricity and CHP Plants	-	-	-	-	-	-	-	-	-	-	-
Petroleum Refineries	-	-	-	-	-	-	-	725	-	-	-
Other Transform. Sector	-	-	-	-	-	-	-	-	-	-	-
ENERGY SECTOR	-	-	-	-	-	-	-	-	-	-	-
DISTRIBUTION LOSSES	-	-	-	-	-	-	-	-	-	-	-
FINAL CONSUMPTION	-	-	-	-	-	-	-	-	-	-	-
INDUSTRY SECTOR	-	-	-	-	-	-	-	-	-	-	-
Iron and Steel	-	-	-	-	-	-	-	-	-	-	-
Chemical and Petrochem.	-	-	-	-	-	-	-	-	-	-	-
Non-Metallic Minerals	-	-	-	-	-	-	-	-	-	-	-
Non-specified	-	-	-	-	-	-	-	-	-	-	-
TRANSPORT SECTOR	-	-	-	-	-	-	-	-	-	-	-
Air	-	-	-	-	-	-	-	-	-	-	-
Road	-	-	-	-	-	-	-	-	-	-	-
Non-specified	-	-	-	-	-	-	-	-	-	-	-
OTHER SECTORS	-	-	-	-	-	-	-	-	-	-	-
Agriculture	-	-	-	-	-	-	-	-	-	-	-
Comm. and Publ. Services	-	-	-	-	-	-	-	-	-	-	-
Residential	-	-	-	-	-	-	-	-	-	-	-
Non-specified	-	-	-	-	-	-	-	-	-	-	-
NON-ENERGY USE	-	-	-	-	-	-	-	-	-	-	-

APPROVISIONNEMENT ET DEMANDE 1998	Charbon (1000 tonnes)							Pétrole (1000 tonnes)			
	Charbon à coke	Autres charb. bit.	Charbon sous-bit.	Lignite	Tourbe	Coke de four/gaz	Agg./briq. de lignite	Pétrole brut	LGN	Produits d'aliment.	Additifs
Production	-	-	-	-	-	-	-	17564	-	-	-
Imports	-	-	-	-	-	-	-	-	-	-	-
Exports	-	-	-	-	-	-	-	-16781	-	-	-
Intl. Marine Bunkers	-	-	-	-	-	-	-	-	-	-	-
Stock Changes	-	-	-	-	-	-	-	-	-	-	-
DOMESTIC SUPPLY	-	-	-	-	-	-	-	783	-	-	-
Transfers and Stat. Diff.	-	-	-	-	-	-	-	-	-	-	-
TRANSFORMATION	-	-	-	-	-	-	-	783	-	-	-
Electricity and CHP Plants	-	-	-	-	-	-	-	-	-	-	-
Petroleum Refineries	-	-	-	-	-	-	-	783	-	-	-
Other Transform. Sector	-	-	-	-	-	-	-	-	-	-	-
ENERGY SECTOR	-	-	-	-	-	-	-	-	-	-	-
DISTRIBUTION LOSSES	-	-	-	-	-	-	-	-	-	-	-
FINAL CONSUMPTION	-	-	-	-	-	-	-	-	-	-	-
INDUSTRY SECTOR	-	-	-	-	-	-	-	-	-	-	-
Iron and Steel	-	-	-	-	-	-	-	-	-	-	-
Chemical and Petrochem.	-	-	-	-	-	-	-	-	-	-	-
Non-Metallic Minerals	-	-	-	-	-	-	-	-	-	-	-
Non-specified	-	-	-	-	-	-	-	-	-	-	-
TRANSPORT SECTOR	-	-	-	-	-	-	-	-	-	-	-
Air	-	-	-	-	-	-	-	-	-	-	-
Road	-	-	-	-	-	-	-	-	-	-	-
Non-specified	-	-	-	-	-	-	-	-	-	-	-
OTHER SECTORS	-	-	-	-	-	-	-	-	-	-	-
Agriculture	-	-	-	-	-	-	-	-	-	-	-
Comm. and Publ. Services	-	-	-	-	-	-	-	-	-	-	-
Residential	-	-	-	-	-	-	-	-	-	-	-
Non-specified	-	-	-	-	-	-	-	-	-	-	-
NON-ENERGY USE	-	-	-	-	-	-	-	-	-	-	-

Gabon

SUPPLY AND CONSUMPTION 1997	Oil cont. (1000 tonnes)										
	Refinery Gas	LPG + Ethane	Motor Gasoline	Aviation Gasoline	Jet Fuel	Kerosene	Gas/ Diesel	Heavy Fuel Oil	Naphtha	Petrol. Coke	Other Prod.
Production	30	10	55	-	35	27	192	222	71	-	45
Imports	-	6	2	13	31	24	61	-	-	-	16
Exports	-	-	-15	-	-	-	-	-186	-61	-	-
Intl. Marine Bunkers	-	-	-	-	-	-	-	-	-	-	-
Stock Changes	-	-	-1	-	15	11	-2	-4	-	-	-2
DOMESTIC SUPPLY	30	16	41	13	81	62	251	32	10	-	59
Transfers and Stat. Diff.	-	-	-	-	-	-	-	-	-	-	-
TRANSFORMATION	-	-	-	-	-	-	48	-	-	-	-
Electricity and CHP Plants	-	-	-	-	-	-	48	-	-	-	-
Petroleum Refineries	-	-	-	-	-	-	-	-	-	-	-
Other Transform. Sector	-	-	-	-	-	-	-	-	-	-	-
ENERGY SECTOR	30	-	-	-	-	-	-	-	-	-	-
DISTRIBUTION LOSSES	-	-	-	-	-	-	-	-	-	-	-
FINAL CONSUMPTION	-	16	41	13	81	62	203	32	10	-	59
INDUSTRY SECTOR	-	-	-	-	-	-	72	32	10	-	-
Iron and Steel	-	-	-	-	-	-	-	-	-	-	-
Chemical and Petrochem.	-	-	-	-	-	-	-	-	10	-	-
Non-Metallic Minerals	-	-	-	-	-	-	-	-	-	-	-
Non-specified	-	-	-	-	-	-	72	32	-	-	-
TRANSPORT SECTOR	-	-	41	13	81	-	119	-	-	-	-
Air	-	-	-	13	81	-	-	-	-	-	-
Road	-	-	41	-	-	-	96	-	-	-	-
Non-specified	-	-	-	-	-	-	23	-	-	-	-
OTHER SECTORS	-	16	-	-	-	62	12	-	-	-	-
Agriculture	-	-	-	-	-	-	-	-	-	-	-
Comm. and Publ. Services	-	-	-	-	-	-	-	-	-	-	-
Residential	-	16	-	-	-	62	-	-	-	-	-
Non-specified	-	-	-	-	-	-	12	-	-	-	-
NON-ENERGY USE	-	-	-	-	-	-	-	-	-	-	59

APPROVISIONNEMENT ET DEMANDE 1998	Pétrole cont. (1000 tonnes)										
	Gaz de raffinerie	GPL + éthane	Essence moteur	Essence aviation	Carbu- réacteurs	Kérosène	Gazole	Fioul lourd	Naphta	Coke de pétrole	Autres prod.
Production	32	10	60	-	58	19	213	242	78	-	52
Imports	-	6	-	13	20	7	74	-	-	-	16
Exports	-	-	-13	-	-	-	-	-211	-68	-	-
Intl. Marine Bunkers	-	-	-	-	-	-	-	-	-	-	-
Stock Changes	-	1	-2	-	1	-	7	1	-	-	-1
DOMESTIC SUPPLY	32	17	45	13	79	26	294	32	10	-	67
Transfers and Stat. Diff.	-	-	-	-	-	-	-	-	-	-	-
TRANSFORMATION	-	-	-	-	-	-	56	-	-	-	-
Electricity and CHP Plants	-	-	-	-	-	-	56	-	-	-	-
Petroleum Refineries	-	-	-	-	-	-	-	-	-	-	-
Other Transform. Sector	-	-	-	-	-	-	-	-	-	-	-
ENERGY SECTOR	32	-	-	-	-	-	-	-	-	-	-
DISTRIBUTION LOSSES	-	-	-	-	-	-	-	-	-	-	-
FINAL CONSUMPTION	-	17	45	13	79	26	238	32	10	-	67
INDUSTRY SECTOR	-	-	-	-	-	-	84	32	10	-	-
Iron and Steel	-	-	-	-	-	-	-	-	-	-	-
Chemical and Petrochem.	-	-	-	-	-	-	-	-	10	-	-
Non-Metallic Minerals	-	-	-	-	-	-	-	-	-	-	-
Non-specified	-	-	-	-	-	-	84	32	-	-	-
TRANSPORT SECTOR	-	-	45	13	79	-	140	-	-	-	-
Air	-	-	-	13	79	-	-	-	-	-	-
Road	-	-	45	-	-	-	113	-	-	-	-
Non-specified	-	-	-	-	-	-	27	-	-	-	-
OTHER SECTORS	-	17	-	-	-	26	14	-	-	-	-
Agriculture	-	-	-	-	-	-	-	-	-	-	-
Comm. and Publ. Services	-	-	-	-	-	-	-	-	-	-	-
Residential	-	17	-	-	-	26	-	-	-	-	-
Non-specified	-	-	-	-	-	-	14	-	-	-	-
NON-ENERGY USE	-	-	-	-	-	-	-	-	-	-	67

Gabon

SUPPLY AND CONSUMPTION 1997	Gas (TJ)				Comb. Renew. & Waste (TJ)				(GWh)	(TJ)
	Natural Gas	Gas Works	Coke Ovens	Blast Furnaces	Solid Biomass	Gas/Liquids from Biomass	Municipal Waste	Industrial Waste	Electricity	Heat
Production	3162	-	-	-	36022	-	-	-	1007	-
Imports	-	-	-	-	-	-	-	-	-	-
Exports	-	-	-	-	-	-	-	-	-	-
Intl. Marine Bunkers	-	-	-	-	-	-	-	-	-	-
Stock Changes	-	-	-	-	-	-	-	-	-	-
DOMESTIC SUPPLY	3162	-	-	-	36022	-	-	-	1007	-
Transfers and Stat. Diff.	-	-	-	-	-	-	-	-	-	-
TRANSFORMATION	1897	-	-	-	-	-	-	-	-	-
Electricity and CHP Plants	1897	-	-	-	-	-	-	-	-	-
Petroleum Refineries	-	-	-	-	-	-	-	-	-	-
Other Transform. Sector	-	-	-	-	-	-	-	-	-	-
ENERGY SECTOR	1237	-	-	-	-	-	-	-	39	-
DISTRIBUTION LOSSES	-	-	-	-	-	-	-	-	101	-
FINAL CONSUMPTION	28	-	-	-	36022	-	-	-	867	-
INDUSTRY SECTOR	28	-	-	-	7204	-	-	-	451	-
Iron and Steel	-	-	-	-	-	-	-	-	-	-
Chemical and Petrochem.	-	-	-	-	-	-	-	-	-	-
Non-Metallic Minerals	-	-	-	-	-	-	-	-	-	-
Non-specified	28	-	-	-	7204	-	-	-	451	-
TRANSPORT SECTOR	-	-	-	-	-	-	-	-	-	-
Air	-	-	-	-	-	-	-	-	-	-
Road	-	-	-	-	-	-	-	-	-	-
Non-specified	-	-	-	-	-	-	-	-	-	-
OTHER SECTORS	-	-	-	-	28818	-	-	-	416	-
Agriculture	-	-	-	-	-	-	-	-	-	-
Comm. and Publ. Services	-	-	-	-	-	-	-	-	161	-
Residential	-	-	-	-	28818	-	-	-	255	-
Non-specified	-	-	-	-	-	-	-	-	-	-
NON-ENERGY USE	-	-	-	-	-	-	-	-	-	-

APPROVISIONNEMENT ET DEMANDE 1998	Gaz (TJ)				En. Re. Comb. & Déchets (TJ)				(GWh)	(TJ)
	Gaz naturel	Usines à gaz	Cokeries	Hauts fourneaux	Biomasse solide	Gaz/Liquides tirés de biomasse	Déchets urbains	Déchets industriels	Electricité	Chaleur
Production	3162	-	-	-	36891	-	-	-	1027	-
Imports	-	-	-	-	-	-	-	-	-	-
Exports	-	-	-	-	-	-	-	-	-	-
Intl. Marine Bunkers	-	-	-	-	-	-	-	-	-	-
Stock Changes	-	-	-	-	-	-	-	-	-	-
DOMESTIC SUPPLY	3162	-	-	-	36891	-	-	-	1027	-
Transfers and Stat. Diff.	-	-	-	-	-	-	-	-	-	-
TRANSFORMATION	1897	-	-	-	-	-	-	-	-	-
Electricity and CHP Plants	1897	-	-	-	-	-	-	-	-	-
Petroleum Refineries	-	-	-	-	-	-	-	-	-	-
Other Transform. Sector	-	-	-	-	-	-	-	-	-	-
ENERGY SECTOR	1237	-	-	-	-	-	-	-	40	-
DISTRIBUTION LOSSES	-	-	-	-	-	-	-	-	103	-
FINAL CONSUMPTION	28	-	-	-	36891	-	-	-	884	-
INDUSTRY SECTOR	28	-	-	-	7378	-	-	-	460	-
Iron and Steel	-	-	-	-	-	-	-	-	-	-
Chemical and Petrochem.	-	-	-	-	-	-	-	-	-	-
Non-Metallic Minerals	-	-	-	-	-	-	-	-	-	-
Non-specified	28	-	-	-	7378	-	-	-	460	-
TRANSPORT SECTOR	-	-	-	-	-	-	-	-	-	-
Air	-	-	-	-	-	-	-	-	-	-
Road	-	-	-	-	-	-	-	-	-	-
Non-specified	-	-	-	-	-	-	-	-	-	-
OTHER SECTORS	-	-	-	-	29513	-	-	-	424	-
Agriculture	-	-	-	-	-	-	-	-	-	-
Comm. and Publ. Services	-	-	-	-	-	-	-	-	164	-
Residential	-	-	-	-	29513	-	-	-	260	-
Non-specified	-	-	-	-	-	-	-	-	-	-
NON-ENERGY USE	-	-	-	-	-	-	-	-	-	-

Georgia / Géorgie

SUPPLY AND CONSUMPTION 1997	Coking Coal	Other Bit. Coal	Sub-Bit. Coal	Lignite	Peat	Oven and Gas Coke	Pat. Fuel and BKB	Crude Oil	NGL	Feed-stocks	Additives
	Coal (1000 tonnes)							Oil (1000 tonnes)			
Production	-	5	-	-	-	-	-	134	-	-	-
Imports	-	12	-	-	-	-	-	27	-	-	-
Exports	-	-2	-	-	-	-	-	-131	-	-	-
Intl. Marine Bunkers	-	-	-	-	-	-	-		-	-	-
Stock Changes	-	-2	-	-	-	-	-	-	-	-	-
DOMESTIC SUPPLY	-	13	-	-	-	-	-	30	-	-	-
Transfers and Stat. Diff.	-	-	-	-	-	-	-	-			-
TRANSFORMATION	-	-	-	-	-	-	-	30	-	-	-
Electricity and CHP Plants	-	-	-	-	-	-	-	-	-	-	-
Petroleum Refineries	-	-	-	-	-	-	-	30	-	-	-
Other Transform. Sector	-	-	-	-	-	-	-	-	-	-	-
ENERGY SECTOR	-	-	-	-	-	-	-	-	-	-	-
DISTRIBUTION LOSSES	-	-	-	-	-	-	-	-	-	-	-
FINAL CONSUMPTION	-	13	-	-	-	-	-	-	-	-	-
INDUSTRY SECTOR	-	-	-	-	-	-	-	-	-	-	-
Iron and Steel	-	-	-	-	-	-	-	-	-	-	-
Chemical and Petrochem.	-	-	-	-	-	-	-	-	-	-	-
Non-Metallic Minerals	-	-	-	-	-	-	-	-	-	-	-
Non-specified	-	-	-	-	-	-	-	-	-	-	-
TRANSPORT SECTOR	-	-	-	-	-	-	-	-	-	-	-
Air	-	-	-	-	-	-	-	-	-	-	-
Road	-	-	-	-	-	-	-	-	-	-	-
Non-specified	-	-	-	-	-	-	-	-	-	-	-
OTHER SECTORS	-	13	-	-	-	-	-	-	-	-	-
Agriculture	-	-	-	-	-	-	-	-	-	-	-
Comm. and Publ. Services	-	-	-	-	-	-	-	-	-	-	-
Residential	-	-	-	-	-	-	-	-	-	-	-
Non-specified	-	13	-	-	-	-	-	-	-	-	-
NON-ENERGY USE	-	-	-	-	-	-	-	-	-	-	-

APPROVISIONNEMENT ET DEMANDE 1998	Charbon à coke	Autres charb. bit.	Charbon sous-bit.	Lignite	Tourbe	Coke de four/gaz	Agg./briq. de lignite	Pétrole brut	LGN	Produits d'aliment.	Additifs
	Charbon (1000 tonnes)							Pétrole (1000 tonnes)			
Production	-	14	-	-	-	-	-	119	-	-	-
Imports	-	4	-	-	-	-	-		-	-	-
Exports	-	-1	-	-	-	-	-	-54	-	-	-
Intl. Marine Bunkers	-	-	-	-	-	-	-		-	-	-
Stock Changes	-	2	-	-	-	-	-	-9	-	-	-
DOMESTIC SUPPLY	-	19	-	-	-	-	-	56	-	-	-
Transfers and Stat. Diff.	-	-	-	-	-	-	-	-	-	-	-
TRANSFORMATION	-	-	-	-	-	-	-	56	-	-	-
Electricity and CHP Plants	-	-	-	-	-	-	-	-	-	-	-
Petroleum Refineries	-	-	-	-	-	-	-	56	-	-	-
Other Transform. Sector	-	-	-	-	-	-	-	-	-	-	-
ENERGY SECTOR	-	-	-	-	-	-	-	-	-	-	-
DISTRIBUTION LOSSES	-	-	-	-	-	-	-	-	-	-	-
FINAL CONSUMPTION	-	19	-	-	-	-	-	-	-	-	-
INDUSTRY SECTOR	-	-	-	-	-	-	-	-	-	-	-
Iron and Steel	-	-	-	-	-	-	-	-	-	-	-
Chemical and Petrochem.	-	-	-	-	-	-	-	-	-	-	-
Non-Metallic Minerals	-	-	-	-	-	-	-	-	-	-	-
Non-specified	-	-	-	-	-	-	-	-	-	-	-
TRANSPORT SECTOR	-	-	-	-	-	-	-	-	-	-	-
Air	-	-	-	-	-	-	-	-	-	-	-
Road	-	-	-	-	-	-	-	-	-	-	-
Non-specified	-	-	-	-	-	-	-	-	-	-	-
OTHER SECTORS	-	19	-	-	-	-	-	-	-	-	-
Agriculture	-	-	-	-	-	-	-	-	-	-	-
Comm. and Publ. Services	-	-	-	-	-	-	-	-	-	-	-
Residential	-	-	-	-	-	-	-	-	-	-	-
Non-specified	-	19	-	-	-	-	-	-	-	-	-
NON-ENERGY USE	-	-	-	-	-	-	-	-	-	-	-

Georgia / Géorgie

SUPPLY AND CONSUMPTION 1997	Oil cont. (1000 tonnes)										
	Refinery Gas	LPG + Ethane	Motor Gasoline	Aviation Gasoline	Jet Fuel	Kerosene	Gas/ Diesel	Heavy Fuel Oil	Naphtha	Petrol. Coke	Other Prod.
Production	1	-	5	-	-	-	7	10	-	-	1
Imports	-	42	456	-	48	28	236	75	-	-	50
Exports	-	-	-2	-	-	-	-	-3	-	-	-
Intl. Marine Bunkers	-	-	-	-	-	-	-	-	-	-	-
Stock Changes	-	-3	13	-	-	3	-77	3	-	-	-
DOMESTIC SUPPLY	1	39	472	-	48	31	166	85	-	-	51
Transfers and Stat. Diff.	-	-	-	-	-	-	-	-	-	-	-
TRANSFORMATION	-	-	-	-	-	-	-	85	-	-	-
Electricity and CHP Plants	-	-	-	-	-	-	-	85	-	-	-
Petroleum Refineries	-	-	-	-	-	-	-	-	-	-	-
Other Transform. Sector	-	-	-	-	-	-	-	-	-	-	-
ENERGY SECTOR	1	-	-	-	-	-	-	-	-	-	-
DISTRIBUTION LOSSES	-	-	-	-	-	-	-	-	-	-	-
FINAL CONSUMPTION	-	39	472	-	48	31	166	-	-	-	51
INDUSTRY SECTOR	-	-	28	-	-	-	33	-	-	-	-
Iron and Steel	-	-	-	-	-	-	-	-	-	-	-
Chemical and Petrochem.	-	-	-	-	-	-	-	-	-	-	-
Non-Metallic Minerals	-	-	-	-	-	-	-	-	-	-	-
Non-specified	-	-	28	-	-	-	33	-	-	-	-
TRANSPORT SECTOR	-	-	363	-	48	-	62	-	-	-	-
Air	-	-	-	-	48	-	-	-	-	-	-
Road	-	-	363	-	-	-	62	-	-	-	-
Non-specified	-	-	-	-	-	-	-	-	-	-	-
OTHER SECTORS	-	39	81	-	-	31	71	-	-	-	36
Agriculture	-	-	16	-	-	-	58	-	-	-	-
Comm. and Publ. Services	-	-	65	-	-	-	9	-	-	-	-
Residential	-	39	-	-	-	31	-	-	-	-	-
Non-specified	-	-	-	-	-	-	4	-	-	-	36
NON-ENERGY USE	-	-	-	-	-	-	-	-	-	-	15

APPROVISIONNEMENT ET DEMANDE 1998	Pétrole cont. (1000 tonnes)										
	Gaz de raffinerie	GPL + éthane	Essence moteur	Essence aviation	Carbu- réacteurs	Kérosène	Gazole	Fioul lourd	Naphta	Coke de pétrole	Autres prod.
Production	1	-	4	-	-	1	15	18	-	-	-
Imports	-	101	421	-	103	110	256	157	-	-	22
Exports	-	-	-12	-	-	-	-	-16	-	-	-
Intl. Marine Bunkers	-	-	-	-	-	-	-	-	-	-	-
Stock Changes	-	-5	5	-	-	4	-30	9	-	-	-
DOMESTIC SUPPLY	1	96	418	-	103	115	241	168	-	-	22
Transfers and Stat. Diff.	-	-	-	-	-	-	-	-	-	-	-
TRANSFORMATION	-	-	-	-	-	-	-	168	-	-	-
Electricity and CHP Plants	-	-	-	-	-	-	-	168	-	-	-
Petroleum Refineries	-	-	-	-	-	-	-	-	-	-	-
Other Transform. Sector	-	-	-	-	-	-	-	-	-	-	-
ENERGY SECTOR	1	-	-	-	-	-	-	-	-	-	-
DISTRIBUTION LOSSES	-	-	-	-	-	-	-	-	-	-	-
FINAL CONSUMPTION	-	96	418	-	103	115	241	-	-	-	22
INDUSTRY SECTOR	-	-	24	-	-	-	48	-	-	-	-
Iron and Steel	-	-	-	-	-	-	-	-	-	-	-
Chemical and Petrochem.	-	-	-	-	-	-	-	-	-	-	-
Non-Metallic Minerals	-	-	-	-	-	-	-	-	-	-	-
Non-specified	-	-	24	-	-	-	48	-	-	-	-
TRANSPORT SECTOR	-	-	322	-	5	-	92	-	-	-	-
Air	-	-	-	-	5	-	-	-	-	-	-
Road	-	-	322	-	-	-	92	-	-	-	-
Non-specified	-	-	-	-	-	-	-	-	-	-	-
OTHER SECTORS	-	96	72	-	98	115	101	-	-	-	-
Agriculture	-	-	15	-	-	1	86	-	-	-	-
Comm. and Publ. Services	-	-	57	-	-	16	11	-	-	-	-
Residential	-	96	-	-	-	96	-	-	-	-	-
Non-specified	-	-	-	-	98	2	4	-	-	-	-
NON-ENERGY USE	-	-	-	-	-	-	-	-	-	-	22

Georgia / Géorgie

SUPPLY AND CONSUMPTION 1997	Gas (TJ)				Comb. Renew. & Waste (TJ)				(GWh)	(TJ)
	Natural Gas	Gas Works	Coke Ovens	Blast Furnaces	Solid Biomass	Gas/Liquids from Biomass	Municipal Waste	Industrial Waste	Electricity	Heat
Production	-	-	-	-	2554	-	-	-	7172	19950
Imports	35589	-	-	-	-	-	-	-	653	-
Exports	-	-	-	-	-	-	-	-	-462	-
Intl. Marine Bunkers	-	-	-	-	-	-	-	-	-	-
Stock Changes	-	-	-	-	-	-	-	-	-	-
DOMESTIC SUPPLY	35589	-	-	-	2554	-	-	-	7363	19950
Transfers and Stat. Diff.	113	-	-	-	-	-	-	-	-	-
TRANSFORMATION	14100	-	-	-	-	-	-	-	-	-
Electricity and CHP Plants	14100	-	-	-	-	-	-	-	-	-
Petroleum Refineries	-	-	-	-	-	-	-	-	-	-
Other Transform. Sector	-	-	-	-	-	-	-	-	-	-
ENERGY SECTOR	-	-	-	-	-	-	-	-	-	-
DISTRIBUTION LOSSES	-	-	-	-	-	-	-	-	1163	-
FINAL CONSUMPTION	21602	-	-	-	2554	-	-	-	6200	19950
INDUSTRY SECTOR	16324	-	-	-	-	-	-	-	828	-
Iron and Steel	2752	-	-	-	-	-	-	-	285	-
Chemical and Petrochem.	7766	-	-	-	-	-	-	-	-	-
Non-Metallic Minerals	-	-	-	-	-	-	-	-	-	-
Non-specified	5806	-	-	-	-	-	-	-	543	-
TRANSPORT SECTOR	-	-	-	-	-	-	-	-	231	-
Air	-	-	-	-	-	-	-	-	-	-
Road	-	-	-	-	-	-	-	-	-	-
Non-specified	-	-	-	-	-	-	-	-	231	-
OTHER SECTORS	5278	-	-	-	2554	-	-	-	5141	19950
Agriculture	-	-	-	-	-	-	-	-	14	-
Comm. and Publ. Services	-	-	-	-	-	-	-	-	2534	-
Residential	5278	-	-	-	2554	-	-	-	2593	-
Non-specified	-	-	-	-	-	-	-	-	-	19950
NON-ENERGY USE	-	-	-	-	-	-	-	-	-	-

APPROVISIONNEMENT ET DEMANDE 1998	Gaz (TJ)				En. Re. Comb. & Déchets (TJ)				(GWh)	(TJ)
	Gaz naturel	Usines à gaz	Cokeries	Hauts fourneaux	Biomasse solide	Gaz/Liquids tirés de biomasse	Déchets urbains	Déchets industriels	Electricité	Chaleur
Production	-	-	-	-	2303	-	-	-	8069	19950
Imports	32000	-	-	-	-	-	-	-	698	-
Exports	-	-	-	-	-	-	-	-	-796	-
Intl. Marine Bunkers	-	-	-	-	-	-	-	-	-	-
Stock Changes	-	-	-	-	-	-	-	-	-	-
DOMESTIC SUPPLY	32000	-	-	-	2303	-	-	-	7971	19950
Transfers and Stat. Diff.	-	-	-	-	-	-	-	-	-	-
TRANSFORMATION	17200	-	-	-	-	-	-	-	-	-
Electricity and CHP Plants	17200	-	-	-	-	-	-	-	-	-
Petroleum Refineries	-	-	-	-	-	-	-	-	-	-
Other Transform. Sector	-	-	-	-	-	-	-	-	-	-
ENERGY SECTOR	-	-	-	-	-	-	-	-	-	-
DISTRIBUTION LOSSES	-	-	-	-	-	-	-	-	1132	-
FINAL CONSUMPTION	14800	-	-	-	2303	-	-	-	6839	19950
INDUSTRY SECTOR	10000	-	-	-	-	-	-	-	801	-
Iron and Steel	1330	-	-	-	-	-	-	-	276	-
Chemical and Petrochem.	8000	-	-	-	-	-	-	-	-	-
Non-Metallic Minerals	-	-	-	-	-	-	-	-	-	-
Non-specified	670	-	-	-	-	-	-	-	525	-
TRANSPORT SECTOR	-	-	-	-	-	-	-	-	315	-
Air	-	-	-	-	-	-	-	-	-	-
Road	-	-	-	-	-	-	-	-	-	-
Non-specified	-	-	-	-	-	-	-	-	315	-
OTHER SECTORS	4800	-	-	-	2303	-	-	-	5723	19950
Agriculture	-	-	-	-	-	-	-	-	14	-
Comm. and Publ. Services	-	-	-	-	-	-	-	-	2759	-
Residential	4800	-	-	-	2303	-	-	-	2950	-
Non-specified	-	-	-	-	-	-	-	-	-	19950
NON-ENERGY USE	-	-	-	-	-	-	-	-	-	-

Ghana

SUPPLY AND CONSUMPTION 1997	Coal (1000 tonnes)							Oil (1000 tonnes)			
	Coking Coal	Other Bit. Coal	Sub-Bit. Coal	Lignite	Peat	Oven and Gas Coke	Pat. Fuel and BKB	Crude Oil	NGL	Feed-stocks	Additives
Production	-	-	-	-	-	-	-	-	-	-	-
Imports	-	3	-	-	-	-	-	178	-	-	-
Exports	-	-	-	-	-	-	-	-	-	-	-
Intl. Marine Bunkers	-	-	-	-	-	-	-	-	-	-	-
Stock Changes	-	-	-	-	-	-	-	-	-	-	-
DOMESTIC SUPPLY	-	3	-	-	-	-	-	178	-	-	-
Transfers and Stat. Diff.	-	-	-	-	-	-	-	-148	-	-	-
TRANSFORMATION	-	-	-	-	-	-	-	30	-	-	-
Electricity and CHP Plants	-	-	-	-	-	-	-	-	-	-	-
Petroleum Refineries	-	-	-	-	-	-	-	30	-	-	-
Other Transform. Sector	-	-	-	-	-	-	-	-	-	-	-
ENERGY SECTOR	-	-	-	-	-	-	-	-	-	-	-
DISTRIBUTION LOSSES	-	-	-	-	-	-	-	-	-	-	-
FINAL CONSUMPTION	-	3	-	-	-	-	-	-	-	-	-
INDUSTRY SECTOR	-	-	-	-	-	-	-	-	-	-	-
Iron and Steel	-	-	-	-	-	-	-	-	-	-	-
Chemical and Petrochem.	-	-	-	-	-	-	-	-	-	-	-
Non-Metallic Minerals	-	-	-	-	-	-	-	-	-	-	-
Non-specified	-	-	-	-	-	-	-	-	-	-	-
TRANSPORT SECTOR	-	-	-	-	-	-	-	-	-	-	-
Air	-	-	-	-	-	-	-	-	-	-	-
Road	-	-	-	-	-	-	-	-	-	-	-
Non-specified	-	-	-	-	-	-	-	-	-	-	-
OTHER SECTORS	-	3	-	-	-	-	-	-	-	-	-
Agriculture	-	-	-	-	-	-	-	-	-	-	-
Comm. and Publ. Services	-	-	-	-	-	-	-	-	-	-	-
Residential	-	3	-	-	-	-	-	-	-	-	-
Non-specified	-	-	-	-	-	-	-	-	-	-	-
NON-ENERGY USE	-	-	-	-	-	-	-	-	-	-	-

APPROVISIONNEMENT ET DEMANDE 1998	Charbon (1000 tonnes)							Pétrole (1000 tonnes)			
	Charbon à coke	Autres charb. bit.	Charbon sous-bit.	Lignite	Tourbe	Coke de four/gaz	Agg./briq. de lignite	Pétrole brut	LGN	Produits d'aliment.	Additifs
Production	-	-	-	-	-	-	-	-	-	-	-
Imports	-	-	-	-	-	-	-	750	-	-	-
Exports	-	-	-	-	-	-	-	-	-	-	-
Intl. Marine Bunkers	-	-	-	-	-	-	-	-	-	-	-
Stock Changes	-	-	-	-	-	-	-	-	-	-	-
DOMESTIC SUPPLY	-	-	-	-	-	-	-	750	-	-	-
Transfers and Stat. Diff.	-	-	-	-	-	-	-	-	-	-	-
TRANSFORMATION	-	-	-	-	-	-	-	750	-	-	-
Electricity and CHP Plants	-	-	-	-	-	-	-	-	-	-	-
Petroleum Refineries	-	-	-	-	-	-	-	750	-	-	-
Other Transform. Sector	-	-	-	-	-	-	-	-	-	-	-
ENERGY SECTOR	-	-	-	-	-	-	-	-	-	-	-
DISTRIBUTION LOSSES	-	-	-	-	-	-	-	-	-	-	-
FINAL CONSUMPTION	-	-	-	-	-	-	-	-	-	-	-
INDUSTRY SECTOR	-	-	-	-	-	-	-	-	-	-	-
Iron and Steel	-	-	-	-	-	-	-	-	-	-	-
Chemical and Petrochem.	-	-	-	-	-	-	-	-	-	-	-
Non-Metallic Minerals	-	-	-	-	-	-	-	-	-	-	-
Non-specified	-	-	-	-	-	-	-	-	-	-	-
TRANSPORT SECTOR	-	-	-	-	-	-	-	-	-	-	-
Air	-	-	-	-	-	-	-	-	-	-	-
Road	-	-	-	-	-	-	-	-	-	-	-
Non-specified	-	-	-	-	-	-	-	-	-	-	-
OTHER SECTORS	-	-	-	-	-	-	-	-	-	-	-
Agriculture	-	-	-	-	-	-	-	-	-	-	-
Comm. and Publ. Services	-	-	-	-	-	-	-	-	-	-	-
Residential	-	-	-	-	-	-	-	-	-	-	-
Non-specified	-	-	-	-	-	-	-	-	-	-	-
NON-ENERGY USE	-	-	-	-	-	-	-	-	-	-	-

Ghana

SUPPLY AND CONSUMPTION 1997	Oil cont. (1000 tonnes)										
	Refinery Gas	LPG + Ethane	Motor Gasoline	Aviation Gasoline	Jet Fuel	Kerosene	Gas/ Diesel	Heavy Fuel Oil	Naphtha	Petrol. Coke	Other Prod.
Production	-	-	-	-	-	-	5	5	-	-	-
Imports	-	72	400	5	52	132	550	77	-	-	60
Exports	-	-2	-	-	-	-	-1	-28	-	-	-
Intl. Marine Bunkers	-	-	-	-	-	-	-25	-	-	-	-
Stock Changes	-	-	-	-	-	-	-	-	-	-	-
DOMESTIC SUPPLY	-	70	400	5	52	132	529	54	-	-	60
Transfers and Stat. Diff.	-	-	-11	-	-	-	1	-	-	-	-
TRANSFORMATION	-	-	-	-	-	-	7	-	-	-	-
Electricity and CHP Plants	-	-	-	-	-	-	7	-	-	-	-
Petroleum Refineries	-	-	-	-	-	-	-	-	-	-	-
Other Transform. Sector	-	-	-	-	-	-	-	-	-	-	-
ENERGY SECTOR	-	-	-	-	-	-	-	-	-	-	-
DISTRIBUTION LOSSES	-	-	-	-	-	-	-	-	-	-	-
FINAL CONSUMPTION	-	70	389	5	52	132	523	54	-	-	60
INDUSTRY SECTOR	-	5	-	-	-	-	84	54	-	-	-
Iron and Steel	-	-	-	-	-	-	-	-	-	-	-
Chemical and Petrochem.	-	-	-	-	-	-	-	-	-	-	-
Non-Metallic Minerals	-	-	-	-	-	-	-	-	-	-	-
Non-specified	-	5	-	-	-	-	84	54	-	-	-
TRANSPORT SECTOR	-	-	389	5	52	-	348	-	-	-	-
Air	-	-	-	5	52	-	-	-	-	-	-
Road	-	-	389	-	-	-	299	-	-	-	-
Non-specified	-	-	-	-	-	-	49	-	-	-	-
OTHER SECTORS	-	65	-	-	-	132	91	-	-	-	-
Agriculture	-	-	-	-	-	-	69	-	-	-	-
Comm. and Publ. Services	-	-	-	-	-	-	22	-	-	-	-
Residential	-	65	-	-	-	132	-	-	-	-	-
Non-specified	-	-	-	-	-	-	-	-	-	-	-
NON-ENERGY USE	-	-	-	-	-	-	-	-	-	-	60

APPROVISIONNEMENT ET DEMANDE 1998	Pétrole cont. (1000 tonnes)										
	Gaz de raffinerie	GPL + éthane	Essence moteur	Essence aviation	Carbu- réacteurs	Kérosène	Gazole	Fioul lourd	Naphta	Coke de pétrole	Autres prod.
Production	45	4	162	-	31	85	213	199	-	-	-
Imports	-	68	350	5	21	50	428	-	-	-	60
Exports	-	-3	-95	-	-	-	-	-160	-	-	-
Intl. Marine Bunkers	-	-	-	-	-	-	-25	-	-	-	-
Stock Changes	-	-	-	-	-	-	-	12	-	-	-
DOMESTIC SUPPLY	45	69	417	5	52	135	616	51	-	-	60
Transfers and Stat. Diff.	-	-	-1	-	-	-	-	-1	-	-	-
TRANSFORMATION	-	-	-	-	-	-	7	-	-	-	-
Electricity and CHP Plants	-	-	-	-	-	-	7	-	-	-	-
Petroleum Refineries	-	-	-	-	-	-	-	-	-	-	-
Other Transform. Sector	-	-	-	-	-	-	-	-	-	-	-
ENERGY SECTOR	45	-	-	-	-	-	-	-	-	-	-
DISTRIBUTION LOSSES	-	-	-	-	-	-	-	-	-	-	-
FINAL CONSUMPTION	-	69	416	5	52	135	609	50	-	-	60
INDUSTRY SECTOR	-	5	-	-	-	-	98	50	-	-	-
Iron and Steel	-	-	-	-	-	-	-	-	-	-	-
Chemical and Petrochem.	-	-	-	-	-	-	-	-	-	-	-
Non-Metallic Minerals	-	-	-	-	-	-	-	-	-	-	-
Non-specified	-	5	-	-	-	-	98	50	-	-	-
TRANSPORT SECTOR	-	-	416	5	52	-	405	-	-	-	-
Air	-	-	-	5	52	-	-	-	-	-	-
Road	-	-	416	-	-	-	348	-	-	-	-
Non-specified	-	-	-	-	-	-	57	-	-	-	-
OTHER SECTORS	-	64	-	-	-	135	106	-	-	-	-
Agriculture	-	-	-	-	-	-	80	-	-	-	-
Comm. and Publ. Services	-	-	-	-	-	-	26	-	-	-	-
Residential	-	64	-	-	-	135	-	-	-	-	-
Non-specified	-	-	-	-	-	-	-	-	-	-	-
NON-ENERGY USE	-	-	-	-	-	-	-	-	-	-	60

Ghana

SUPPLY AND CONSUMPTION 1997	Gas (TJ)				Comb. Renew. & Waste (TJ)				(GWh)	(TJ)
	Natural Gas	Gas Works	Coke Ovens	Blast Furnaces	Solid Biomass	Gas/Liquids from Biomass	Municipal Waste	Industrial Waste	Electricity	Heat
Production	-	-	-	-	207415	-	-	-	6914	-
Imports	-	-	-	-	-	-	-	-	912	-
Exports	-	-	-	-	-	-	-	-	-422	-
Intl. Marine Bunkers	-	-	-	-	-	-	-	-	-	-
Stock Changes	-	-	-	-	-	-	-	-	-	-
DOMESTIC SUPPLY	-	-	-	-	207415	-	-	-	7404	-
Transfers and Stat. Diff.	-	-	-	-	-	-	-	-	-	-
TRANSFORMATION	-	-	-	-	50702	-	-	-	-	-
Electricity and CHP Plants	-	-	-	-	-	-	-	-	-	-
Petroleum Refineries	-	-	-	-	-	-	-	-	-	-
Other Transform. Sector	-	-	-	-	50702	-	-	-	-	-
ENERGY SECTOR	-	-	-	-	-	-	-	-	2280	-
DISTRIBUTION LOSSES	-	-	-	-	-	-	-	-	49	-
FINAL CONSUMPTION	-	-	-	-	156713	-	-	-	5075	-
INDUSTRY SECTOR	-	-	-	-	14246	-	-	-	3590	-
Iron and Steel	-	-	-	-	-	-	-	-	-	-
Chemical and Petrochem.	-	-	-	-	-	-	-	-	-	-
Non-Metallic Minerals	-	-	-	-	-	-	-	-	-	-
Non-specified	-	-	-	-	14246	-	-	-	3590	-
TRANSPORT SECTOR	-	-	-	-	-	-	-	-	-	-
Air	-	-	-	-	-	-	-	-	-	-
Road	-	-	-	-	-	-	-	-	-	-
Non-specified	-	-	-	-	-	-	-	-	-	-
OTHER SECTORS	-	-	-	-	142467	-	-	-	1485	-
Agriculture	-	-	-	-	-	-	-	-	-	-
Comm. and Publ. Services	-	-	-	-	-	-	-	-	-	-
Residential	-	-	-	-	142467	-	-	-	1478	-
Non-specified	-	-	-	-	-	-	-	-	7	-
NON-ENERGY USE	-	-	-	-	-	-	-	-	-	-

APPROVISIONNEMENT ET DEMANDE 1998	Gaz (TJ)				En. Re. Comb. & Déchets (TJ)				(GWh)	(TJ)
	Gaz naturel	Usines à gaz	Cokeries	Hauts fourneaux	Biomasse solide	Gaz/Liquides tirés de biomasse	Déchets urbains	Déchets industriels	Electricité	Chaleur
Production	-	-	-	-	212893	-	-	-	7251	-
Imports	-	-	-	-	-	-	-	-	912	-
Exports	-	-	-	-	-	-	-	-	-422	-
Intl. Marine Bunkers	-	-	-	-	-	-	-	-	-	-
Stock Changes	-	-	-	-	-	-	-	-	-	-
DOMESTIC SUPPLY	-	-	-	-	212893	-	-	-	7741	-
Transfers and Stat. Diff.	-	-	-	-	-1	-	-	-	-	-
TRANSFORMATION	-	-	-	-	52041	-	-	-	-	-
Electricity and CHP Plants	-	-	-	-	-	-	-	-	-	-
Petroleum Refineries	-	-	-	-	-	-	-	-	-	-
Other Transform. Sector	-	-	-	-	52041	-	-	-	-	-
ENERGY SECTOR	-	-	-	-	-	-	-	-	2384	-
DISTRIBUTION LOSSES	-	-	-	-	-	-	-	-	51	-
FINAL CONSUMPTION	-	-	-	-	160851	-	-	-	5306	-
INDUSTRY SECTOR	-	-	-	-	14622	-	-	-	3754	-
Iron and Steel	-	-	-	-	-	-	-	-	-	-
Chemical and Petrochem.	-	-	-	-	-	-	-	-	-	-
Non-Metallic Minerals	-	-	-	-	-	-	-	-	-	-
Non-specified	-	-	-	-	14622	-	-	-	3754	-
TRANSPORT SECTOR	-	-	-	-	-	-	-	-	-	-
Air	-	-	-	-	-	-	-	-	-	-
Road	-	-	-	-	-	-	-	-	-	-
Non-specified	-	-	-	-	-	-	-	-	-	-
OTHER SECTORS	-	-	-	-	146229	-	-	-	1552	-
Agriculture	-	-	-	-	-	-	-	-	-	-
Comm. and Publ. Services	-	-	-	-	-	-	-	-	-	-
Residential	-	-	-	-	146229	-	-	-	1545	-
Non-specified	-	-	-	-	-	-	-	-	7	-
NON-ENERGY USE	-	-	-	-	-	-	-	-	-	-

Gibraltar

SUPPLY AND CONSUMPTION 1997	Coal (1000 tonnes)							Oil (1000 tonnes)			
	Coking Coal	Other Bit. Coal	Sub-Bit. Coal	Lignite	Peat	Oven and Gas Coke	Pat. Fuel and BKB	Crude Oil	NGL	Feed-stocks	Additives
Production	-	-	-	-	-	-	-	-	-	-	-
Imports	-	-	-	-	-	-	-	-	-	-	-
Exports	-	-	-	-	-	-	-	-	-	-	-
Intl. Marine Bunkers	-	-	-	-	-	-	-	-	-	-	-
Stock Changes	-	-	-	-	-	-	-	-	-	-	-
DOMESTIC SUPPLY	-	-	-	-	-	-	-	-	-	-	-
Transfers and Stat. Diff.	-	-	-	-	-	-	-	-	-	-	-
TRANSFORMATION	-	-	-	-	-	-	-	-	-	-	-
Electricity and CHP Plants	-	-	-	-	-	-	-	-	-	-	-
Petroleum Refineries	-	-	-	-	-	-	-	-	-	-	-
Other Transform. Sector	-	-	-	-	-	-	-	-	-	-	-
ENERGY SECTOR	-	-	-	-	-	-	-	-	-	-	-
DISTRIBUTION LOSSES	-	-	-	-	-	-	-	-	-	-	-
FINAL CONSUMPTION	-	-	-	-	-	-	-	-	-	-	-
INDUSTRY SECTOR	-	-	-	-	-	-	-	-	-	-	-
Iron and Steel	-	-	-	-	-	-	-	-	-	-	-
Chemical and Petrochem.	-	-	-	-	-	-	-	-	-	-	-
Non-Metallic Minerals	-	-	-	-	-	-	-	-	-	-	-
Non-specified	-	-	-	-	-	-	-	-	-	-	-
TRANSPORT SECTOR	-	-	-	-	-	-	-	-	-	-	-
Air	-	-	-	-	-	-	-	-	-	-	-
Road	-	-	-	-	-	-	-	-	-	-	-
Non-specified	-	-	-	-	-	-	-	-	-	-	-
OTHER SECTORS	-	-	-	-	-	-	-	-	-	-	-
Agriculture	-	-	-	-	-	-	-	-	-	-	-
Comm. and Publ. Services	-	-	-	-	-	-	-	-	-	-	-
Residential	-	-	-	-	-	-	-	-	-	-	-
Non-specified	-	-	-	-	-	-	-	-	-	-	-
NON-ENERGY USE	-	-	-	-	-	-	-	-	-	-	-

APPROVISIONNEMENT ET DEMANDE 1998	Charbon (1000 tonnes)							Pétrole (1000 tonnes)			
	Charbon à coke	Autres charb. bit.	Charbon sous-bit.	Lignite	Tourbe	Coke de four/gaz	Agg./briq. de lignite	Pétrole brut	LGN	Produits d'aliment.	Additifs
Production	-	-	-	-	-	-	-	-	-	-	-
Imports	-	-	-	-	-	-	-	-	-	-	-
Exports	-	-	-	-	-	-	-	-	-	-	-
Intl. Marine Bunkers	-	-	-	-	-	-	-	-	-	-	-
Stock Changes	-	-	-	-	-	-	-	-	-	-	-
DOMESTIC SUPPLY	-	-	-	-	-	-	-	-	-	-	-
Transfers and Stat. Diff.	-	-	-	-	-	-	-	-	-	-	-
TRANSFORMATION	-	-	-	-	-	-	-	-	-	-	-
Electricity and CHP Plants	-	-	-	-	-	-	-	-	-	-	-
Petroleum Refineries	-	-	-	-	-	-	-	-	-	-	-
Other Transform. Sector	-	-	-	-	-	-	-	-	-	-	-
ENERGY SECTOR	-	-	-	-	-	-	-	-	-	-	-
DISTRIBUTION LOSSES	-	-	-	-	-	-	-	-	-	-	-
FINAL CONSUMPTION	-	-	-	-	-	-	-	-	-	-	-
INDUSTRY SECTOR	-	-	-	-	-	-	-	-	-	-	-
Iron and Steel	-	-	-	-	-	-	-	-	-	-	-
Chemical and Petrochem.	-	-	-	-	-	-	-	-	-	-	-
Non-Metallic Minerals	-	-	-	-	-	-	-	-	-	-	-
Non-specified	-	-	-	-	-	-	-	-	-	-	-
TRANSPORT SECTOR	-	-	-	-	-	-	-	-	-	-	-
Air	-	-	-	-	-	-	-	-	-	-	-
Road	-	-	-	-	-	-	-	-	-	-	-
Non-specified	-	-	-	-	-	-	-	-	-	-	-
OTHER SECTORS	-	-	-	-	-	-	-	-	-	-	-
Agriculture	-	-	-	-	-	-	-	-	-	-	-
Comm. and Publ. Services	-	-	-	-	-	-	-	-	-	-	-
Residential	-	-	-	-	-	-	-	-	-	-	-
Non-specified	-	-	-	-	-	-	-	-	-	-	-
NON-ENERGY USE	-	-	-	-	-	-	-	-	-	-	-

Gibraltar

SUPPLY AND CONSUMPTION 1997	Oil cont. (1000 tonnes)										
	Refinery Gas	LPG + Ethane	Motor Gasoline	Aviation Gasoline	Jet Fuel	Kerosene	Gas/ Diesel	Heavy Fuel Oil	Naphtha	Petrol. Coke	Other Prod.
Production	-	-	-	-	-	-	-	-	-	-	-
Imports	-	-	16	-	4	-	152	800	-	-	15
Exports	-	-	-	-	-	-	-	-	-	-	-
Intl. Marine Bunkers	-	-	-	-	-	-	-104	-750	-	-	-
Stock Changes	-	-	-	-	-	-	-	-	-	-	-
DOMESTIC SUPPLY	-	-	16	-	4	-	48	50	-	-	15
Transfers and Stat. Diff.	-	-	-	-	-	-	-	-	-	-	-
TRANSFORMATION	-	-	-	-	-	-	-	50	-	-	-
Electricity and CHP Plants	-	-	-	-	-	-	-	50	-	-	-
Petroleum Refineries	-	-	-	-	-	-	-	-	-	-	-
Other Transform. Sector	-	-	-	-	-	-	-	-	-	-	-
ENERGY SECTOR	-	-	-	-	-	-	-	-	-	-	-
DISTRIBUTION LOSSES	-	-	-	-	-	-	-	-	-	-	-
FINAL CONSUMPTION	-	-	16	-	4	-	48	-	-	-	15
INDUSTRY SECTOR	-	-	-	-	-	-	-	-	-	-	-
Iron and Steel	-	-	-	-	-	-	-	-	-	-	-
Chemical and Petrochem.	-	-	-	-	-	-	-	-	-	-	-
Non-Metallic Minerals	-	-	-	-	-	-	-	-	-	-	-
Non-specified	-	-	-	-	-	-	-	-	-	-	-
TRANSPORT SECTOR	-	-	16	-	4	-	48	-	-	-	-
Air	-	-	-	-	4	-	-	-	-	-	-
Road	-	-	16	-	-	-	48	-	-	-	-
Non-specified	-	-	-	-	-	-	-	-	-	-	-
OTHER SECTORS	-	-	-	-	-	-	-	-	-	-	-
Agriculture	-	-	-	-	-	-	-	-	-	-	-
Comm. and Publ. Services	-	-	-	-	-	-	-	-	-	-	-
Residential	-	-	-	-	-	-	-	-	-	-	-
Non-specified	-	-	-	-	-	-	-	-	-	-	-
NON-ENERGY USE	-	-	-	-	-	-	-	-	-	-	15

APPROVISIONNEMENT ET DEMANDE 1998	Pétrole cont. (1000 tonnes)										
	Gaz de raffinerie	GPL + éthane	Essence moteur	Essence aviation	Carbu- réacteurs	Kérosène	Gazole	Fioul lourd	Naphta	Coke de pétrole	Autres prod.
Production	-	-	-	-	-	-	-	-	-	-	-
Imports	-	-	16	-	4	-	152	800	-	-	15
Exports	-	-	-	-	-	-	-	-	-	-	-
Intl. Marine Bunkers	-	-	-	-	-	-	-104	-750	-	-	-
Stock Changes	-	-	-	-	-	-	-	-	-	-	-
DOMESTIC SUPPLY	-	-	16	-	4	-	48	50	-	-	15
Transfers and Stat. Diff.	-	-	-	-	-	-	-	-	-	-	-
TRANSFORMATION	-	-	-	-	-	-	-	50	-	-	-
Electricity and CHP Plants	-	-	-	-	-	-	-	50	-	-	-
Petroleum Refineries	-	-	-	-	-	-	-	-	-	-	-
Other Transform. Sector	-	-	-	-	-	-	-	-	-	-	-
ENERGY SECTOR	-	-	-	-	-	-	-	-	-	-	-
DISTRIBUTION LOSSES	-	-	-	-	-	-	-	-	-	-	-
FINAL CONSUMPTION	-	-	16	-	4	-	48	-	-	-	15
INDUSTRY SECTOR	-	-	-	-	-	-	-	-	-	-	-
Iron and Steel	-	-	-	-	-	-	-	-	-	-	-
Chemical and Petrochem.	-	-	-	-	-	-	-	-	-	-	-
Non-Metallic Minerals	-	-	-	-	-	-	-	-	-	-	-
Non-specified	-	-	-	-	-	-	-	-	-	-	-
TRANSPORT SECTOR	-	-	16	-	4	-	48	-	-	-	-
Air	-	-	-	-	4	-	-	-	-	-	-
Road	-	-	16	-	-	-	48	-	-	-	-
Non-specified	-	-	-	-	-	-	-	-	-	-	-
OTHER SECTORS	-	-	-	-	-	-	-	-	-	-	-
Agriculture	-	-	-	-	-	-	-	-	-	-	-
Comm. and Publ. Services	-	-	-	-	-	-	-	-	-	-	-
Residential	-	-	-	-	-	-	-	-	-	-	-
Non-specified	-	-	-	-	-	-	-	-	-	-	-
NON-ENERGY USE	-	-	-	-	-	-	-	-	-	-	15

Gibraltar

SUPPLY AND CONSUMPTION 1997	Gas (TJ)				Comb. Renew. & Waste (TJ)				(GWh) Electricity	(TJ) Heat
	Natural Gas	Gas Works	Coke Ovens	Blast Furnaces	Solid Biomass	Gas/Liquids from Biomass	Municipal Waste	Industrial Waste		
Production	-	-	-	-	-	-	-	-	93	-
Imports	-	-	-	-	-	-	-	-	-	-
Exports	-	-	-	-	-	-	-	-	-	-
Intl. Marine Bunkers	-	-	-	-	-	-	-	-	-	-
Stock Changes	-	-	-	-	-	-	-	-	-	-
DOMESTIC SUPPLY	-	-	-	-	-	-	-	-	93	-
Transfers and Stat. Diff.	-	-	-	-	-	-	-	-	-	-
TRANSFORMATION	-	-	-	-	-	-	-	-	-	-
Electricity and CHP Plants	-	-	-	-	-	-	-	-	-	-
Petroleum Refineries	-	-	-	-	-	-	-	-	-	-
Other Transform. Sector	-	-	-	-	-	-	-	-	-	-
ENERGY SECTOR	-	-	-	-	-	-	-	-	5	-
DISTRIBUTION LOSSES	-	-	-	-	-	-	-	-	-	-
FINAL CONSUMPTION	-	-	-	-	-	-	-	-	88	-
INDUSTRY SECTOR	-	-	-	-	-	-	-	-	-	-
Iron and Steel	-	-	-	-	-	-	-	-	-	-
Chemical and Petrochem.	-	-	-	-	-	-	-	-	-	-
Non-Metallic Minerals	-	-	-	-	-	-	-	-	-	-
Non-specified	-	-	-	-	-	-	-	-	-	-
TRANSPORT SECTOR	-	-	-	-	-	-	-	-	-	-
Air	-	-	-	-	-	-	-	-	-	-
Road	-	-	-	-	-	-	-	-	-	-
Non-specified	-	-	-	-	-	-	-	-	-	-
OTHER SECTORS	-	-	-	-	-	-	-	-	88	-
Agriculture	-	-	-	-	-	-	-	-	-	-
Comm. and Publ. Services	-	-	-	-	-	-	-	-	-	-
Residential	-	-	-	-	-	-	-	-	-	-
Non-specified	-	-	-	-	-	-	-	-	88	-
NON-ENERGY USE	-	-	-	-	-	-	-	-	-	-

APPROVISIONNEMENT ET DEMANDE 1998	Gaz (TJ)				En. Re. Comb. & Déchets (TJ)				(GWh) Electricité	(TJ) Chaleur
	Gaz naturel	Usines à gaz	Cokeries	Hauts fourneaux	Biomasse solide	Gaz/Liquides tirés de biomasse	Déchets urbains	Déchets industriels		
Production	-	-	-	-	-	-	-	-	93	-
Imports	-	-	-	-	-	-	-	-	-	-
Exports	-	-	-	-	-	-	-	-	-	-
Intl. Marine Bunkers	-	-	-	-	-	-	-	-	-	-
Stock Changes	-	-	-	-	-	-	-	-	-	-
DOMESTIC SUPPLY	-	-	-	-	-	-	-	-	93	-
Transfers and Stat. Diff.	-	-	-	-	-	-	-	-	-	-
TRANSFORMATION	-	-	-	-	-	-	-	-	-	-
Electricity and CHP Plants	-	-	-	-	-	-	-	-	-	-
Petroleum Refineries	-	-	-	-	-	-	-	-	-	-
Other Transform. Sector	-	-	-	-	-	-	-	-	-	-
ENERGY SECTOR	-	-	-	-	-	-	-	-	5	-
DISTRIBUTION LOSSES	-	-	-	-	-	-	-	-	-	-
FINAL CONSUMPTION	-	-	-	-	-	-	-	-	88	-
INDUSTRY SECTOR	-	-	-	-	-	-	-	-	-	-
Iron and Steel	-	-	-	-	-	-	-	-	-	-
Chemical and Petrochem.	-	-	-	-	-	-	-	-	-	-
Non-Metallic Minerals	-	-	-	-	-	-	-	-	-	-
Non-specified	-	-	-	-	-	-	-	-	-	-
TRANSPORT SECTOR	-	-	-	-	-	-	-	-	-	-
Air	-	-	-	-	-	-	-	-	-	-
Road	-	-	-	-	-	-	-	-	-	-
Non-specified	-	-	-	-	-	-	-	-	-	-
OTHER SECTORS	-	-	-	-	-	-	-	-	88	-
Agriculture	-	-	-	-	-	-	-	-	-	-
Comm. and Publ. Services	-	-	-	-	-	-	-	-	-	-
Residential	-	-	-	-	-	-	-	-	-	-
Non-specified	-	-	-	-	-	-	-	-	88	-
NON-ENERGY USE	-	-	-	-	-	-	-	-	-	-

Guatemala

SUPPLY AND CONSUMPTION 1997	Coal (1000 tonnes)							Oil (1000 tonnes)			
	Coking Coal	Other Bit. Coal	Sub-Bit. Coal	Lignite	Peat	Oven and Gas Coke	Pat. Fuel and BKB	Crude Oil	NGL	Feed-stocks	Additives
Production	-	-	-	-	-	-	-	1065	-	-	-
Imports	-	-	-	-	-	-	-	839	-	-	-
Exports	-	-	-	-	-	-	-	-974	-	-	-
Intl. Marine Bunkers	-	-	-	-	-	-	-			-	-
Stock Changes	-	-	-	-	-	-	-	-	-	-	-
DOMESTIC SUPPLY	-	-	-	-	-	-	-	930	-	-	-
Transfers and Stat. Diff.	-	-	-	-	-	-	-	-83	-	-	-
TRANSFORMATION	-	-	-	-	-	-	-	847	-	-	-
Electricity and CHP Plants	-	-	-	-	-	-	-	-	-	-	-
Petroleum Refineries	-	-	-	-	-	-	-	847	-	-	-
Other Transform. Sector	-	-	-	-	-	-	-	-	-	-	-
ENERGY SECTOR	-	-	-	-	-	-	-	-	-	-	-
DISTRIBUTION LOSSES	-	-	-	-	-	-	-	-	-	-	-
FINAL CONSUMPTION	-	-	-	-	-	-	-	-	-	-	-
INDUSTRY SECTOR	-	-	-	-	-	-	-	-	-	-	-
Iron and Steel	-	-	-	-	-	-	-	-	-	-	-
Chemical and Petrochem.	-	-	-	-	-	-	-	-	-	-	-
Non-Metallic Minerals	-	-	-	-	-	-	-	-	-	-	-
Non-specified	-	-	-	-	-	-	-	-	-	-	-
TRANSPORT SECTOR	-	-	-	-	-	-	-	-	-	-	-
Air	-	-	-	-	-	-	-	-	-	-	-
Road	-	-	-	-	-	-	-	-	-	-	-
Non-specified	-	-	-	-	-	-	-	-	-	-	-
OTHER SECTORS	-	-	-	-	-	-	-	-	-	-	-
Agriculture	-	-	-	-	-	-	-	-	-	-	-
Comm. and Publ. Services	-	-	-	-	-	-	-	-	-	-	-
Residential	-	-	-	-	-	-	-	-	-	-	-
Non-specified	-	-	-	-	-	-	-	-	-	-	-
NON-ENERGY USE	-	-	-	-	-	-	-	-	-	-	-

APPROVISIONNEMENT ET DEMANDE 1998	Charbon (1000 tonnes)							Pétrole (1000 tonnes)			
	Charbon à coke	Autres charb. bit.	Charbon sous-bit.	Lignite	Tourbe	Coke de four/gaz	Agg./briq. de lignite	Pétrole brut	LGN	Produits d'aliment.	Additifs
Production	-	-	-	-	-	-	-	1389	-	-	-
Imports	-	-	-	-	-	-	-	836	-	-	-
Exports	-	-	-	-	-	-	-	-1204	-	-	-
Intl. Marine Bunkers	-	-	-	-	-	-	-	-	-	-	-
Stock Changes	-	-	-	-	-	-	-	-	-	-	-
DOMESTIC SUPPLY	-	-	-	-	-	-	-	1021	-	-	-
Transfers and Stat. Diff.	-	-	-	-	-	-	-	-177	-	-	-
TRANSFORMATION	-	-	-	-	-	-	-	844	-	-	-
Electricity and CHP Plants	-	-	-	-	-	-	-	-	-	-	-
Petroleum Refineries	-	-	-	-	-	-	-	844	-	-	-
Other Transform. Sector	-	-	-	-	-	-	-	-	-	-	-
ENERGY SECTOR	-	-	-	-	-	-	-	-	-	-	-
DISTRIBUTION LOSSES	-	-	-	-	-	-	-	-	-	-	-
FINAL CONSUMPTION	-	-	-	-	-	-	-	-	-	-	-
INDUSTRY SECTOR	-	-	-	-	-	-	-	-	-	-	-
Iron and Steel	-	-	-	-	-	-	-	-	-	-	-
Chemical and Petrochem.	-	-	-	-	-	-	-	-	-	-	-
Non-Metallic Minerals	-	-	-	-	-	-	-	-	-	-	-
Non-specified	-	-	-	-	-	-	-	-	-	-	-
TRANSPORT SECTOR	-	-	-	-	-	-	-	-	-	-	-
Air	-	-	-	-	-	-	-	-	-	-	-
Road	-	-	-	-	-	-	-	-	-	-	-
Non-specified	-	-	-	-	-	-	-	-	-	-	-
OTHER SECTORS	-	-	-	-	-	-	-	-	-	-	-
Agriculture	-	-	-	-	-	-	-	-	-	-	-
Comm. and Publ. Services	-	-	-	-	-	-	-	-	-	-	-
Residential	-	-	-	-	-	-	-	-	-	-	-
Non-specified	-	-	-	-	-	-	-	-	-	-	-
NON-ENERGY USE	-	-	-	-	-	-	-	-	-	-	-

Guatemala

SUPPLY AND CONSUMPTION 1997	Oil cont. (1000 tonnes)										
	Refinery Gas	LPG + Ethane	Motor Gasoline	Aviation Gasoline	Jet Fuel	Kerosene	Gas/ Diesel	Heavy Fuel Oil	Naphtha	Petrol. Coke	Other Prod.
Production	21	7	139	-	17	16	266	283	-	-	-
Imports	-	148	430	-	31	31	542	203	-	-	27
Exports	-	-	-	-	-	-	-	-	-	-	-
Intl. Marine Bunkers	-	-	-	-	-	-	-120	-	-	-	-
Stock Changes	-	2	-19	-	3	3	-5	29	-	-	-
DOMESTIC SUPPLY	**21**	**157**	**550**	**-**	**51**	**50**	**683**	**515**	**-**	**-**	**27**
Transfers and Stat. Diff.	-	2	-	-	-1	-1	1	-	-	-	-
TRANSFORMATION	**-**	**-**	**-**	**-**	**-**	**-**	**97**	**221**	**-**	**-**	**-**
Electricity and CHP Plants	-	-	-	-	-	-	97	221	-	-	-
Petroleum Refineries	-	-	-	-	-	-	-	-	-	-	-
Other Transform. Sector	-	-	-	-	-	-	-	-	-	-	-
ENERGY SECTOR	**21**	**-**	**-**	**-**	**-**	**-**	**-**	**-**	**-**	**-**	**-**
DISTRIBUTION LOSSES	**-**	**-**	**-**	**-**	**-**	**-**	**-**	**-**	**-**	**-**	**-**
FINAL CONSUMPTION	**-**	**159**	**550**	**-**	**50**	**49**	**587**	**294**	**-**	**-**	**27**
INDUSTRY SECTOR	**-**	**21**	**-**	**-**	**-**	**5**	**86**	**290**	**-**	**-**	**-**
Iron and Steel	-	-	-	-	-	-	-	-	-	-	-
Chemical and Petrochem.	-	-	-	-	-	-	-	-	-	-	-
Non-Metallic Minerals	-	-	-	-	-	-	-	-	-	-	-
Non-specified	-	21	-	-	-	5	86	290	-	-	-
TRANSPORT SECTOR	**-**	**1**	**550**	**-**	**50**	**-**	**418**	**-**	**-**	**-**	**-**
Air	-	-	-	-	50	-	-	-	-	-	-
Road	-	-	550	-	-	-	418	-	-	-	-
Non-specified	-	1	-	-	-	-	-	-	-	-	-
OTHER SECTORS	**-**	**137**	**-**	**-**	**-**	**44**	**83**	**4**	**-**	**-**	**-**
Agriculture	-	2	-	-	-	-	34	-	-	-	-
Comm. and Publ. Services	-	23	-	-	-	1	49	4	-	-	-
Residential	-	112	-	-	-	43	-	-	-	-	-
Non-specified	-	-	-	-	-	-	-	-	-	-	-
NON-ENERGY USE	**-**	**-**	**-**	**-**	**-**	**-**	**-**	**-**	**-**	**-**	**27**

APPROVISIONNEMENT ET DEMANDE 1998	Pétrole cont. (1000 tonnes)										
	Gaz de raffinerie	GPL + éthane	Essence moteur	Essence aviation	Carbu- réacteurs	Kérosène	Gazole	Fioul lourd	Naphta	Coke de pétrole	Autres prod.
Production	11	5	123	-	16	16	328	248	-	-	-
Imports	-	165	561	-	36	36	633	561	-	-	27
Exports	-	-	-1	-	-	-	-	-	-	-	-
Intl. Marine Bunkers	-	-	-	-	-	-	-120	-	-	-	-
Stock Changes	-	-3	-26	-	-1	-1	-39	13	-	-	-
DOMESTIC SUPPLY	**11**	**167**	**657**	**-**	**51**	**51**	**802**	**822**	**-**	**-**	**27**
Transfers and Stat. Diff.	-	1	-	-	-1	-2	-	1	-	-	-
TRANSFORMATION	**-**	**-**	**-**	**-**	**-**	**-**	**116**	**531**	**-**	**-**	**-**
Electricity and CHP Plants	-	-	-	-	-	-	116	531	-	-	-
Petroleum Refineries	-	-	-	-	-	-	-	-	-	-	-
Other Transform. Sector	-	-	-	-	-	-	-	-	-	-	-
ENERGY SECTOR	**11**	**-**	**-**	**-**	**-**	**-**	**-**	**-**	**-**	**-**	**-**
DISTRIBUTION LOSSES	**-**	**-**	**-**	**-**	**-**	**-**	**-**	**-**	**-**	**-**	**-**
FINAL CONSUMPTION	**-**	**168**	**657**	**-**	**50**	**49**	**686**	**292**	**-**	**-**	**27**
INDUSTRY SECTOR	**-**	**22**	**-**	**-**	**-**	**5**	**98**	**289**	**-**	**-**	**-**
Iron and Steel	-	-	-	-	-	-	-	-	-	-	-
Chemical and Petrochem.	-	-	-	-	-	-	-	-	-	-	-
Non-Metallic Minerals	-	-	-	-	-	-	-	-	-	-	-
Non-specified	-	22	-	-	-	5	98	289	-	-	-
TRANSPORT SECTOR	**-**	**1**	**657**	**-**	**50**	**-**	**494**	**-**	**-**	**-**	**-**
Air	-	-	-	-	50	-	-	-	-	-	-
Road	-	-	657	-	-	-	494	-	-	-	-
Non-specified	-	1	-	-	-	-	-	-	-	-	-
OTHER SECTORS	**-**	**145**	**-**	**-**	**-**	**44**	**94**	**3**	**-**	**-**	**-**
Agriculture	-	2	-	-	-	-	39	-	-	-	-
Comm. and Publ. Services	-	24	-	-	-	1	55	3	-	-	-
Residential	-	119	-	-	-	43	-	-	-	-	-
Non-specified	-	-	-	-	-	-	-	-	-	-	-
NON-ENERGY USE	**-**	**-**	**-**	**-**	**-**	**-**	**-**	**-**	**-**	**-**	**27**

Guatemala

SUPPLY AND CONSUMPTION 1997	Gas (TJ)				Comb. Renew. & Waste (TJ)				(GWh)	(TJ)
	Natural Gas	Gas Works	Coke Ovens	Blast Furnaces	Solid Biomass	Gas/Liquids from Biomass	Municipal Waste	Industrial Waste	Electricity	Heat
Production	-	-	-	-	126878	-	-	-	4132	-
Imports	-	-	-	-	-	-	-	-	-	-
Exports	-	-	-	-	-	-	-	-	-88	-
Intl. Marine Bunkers	-	-	-	-	-	-	-	-	-	-
Stock Changes	-	-	-	-	-	-	-	-	-	-
DOMESTIC SUPPLY	-	-	-	-	126878	-	-	-	4044	-
Transfers and Stat. Diff.	-	-	-	-	-	-	-	-	-	-
TRANSFORMATION	-	-	-	-	6403	-	-	-	-	-
Electricity and CHP Plants	-	-	-	-	4032	-	-	-	-	-
Petroleum Refineries	-	-	-	-	105	-	-	-	-	-
Other Transform. Sector	-	-	-	-	2266	-	-	-	-	-
ENERGY SECTOR	-	-	-	-	-	-	-	-	23	-
DISTRIBUTION LOSSES	-	-	-	-	-	-	-	-	764	-
FINAL CONSUMPTION	-	-	-	-	120475	-	-	-	3257	-
INDUSTRY SECTOR	-	-	-	-	7679	-	-	-	1064	-
Iron and Steel	-	-	-	-	-	-	-	-	-	-
Chemical and Petrochem.	-	-	-	-	-	-	-	-	-	-
Non-Metallic Minerals	-	-	-	-	-	-	-	-	-	-
Non-specified	-	-	-	-	7679	-	-	-	1064	-
TRANSPORT SECTOR	-	-	-	-	-	-	-	-	-	-
Air	-	-	-	-	-	-	-	-	-	-
Road	-	-	-	-	-	-	-	-	-	-
Non-specified	-	-	-	-	-	-	-	-	-	-
OTHER SECTORS	-	-	-	-	112796	-	-	-	2193	-
Agriculture	-	-	-	-	-	-	-	-	16	-
Comm. and Publ. Services	-	-	-	-	-	-	-	-	1122	-
Residential	-	-	-	-	112796	-	-	-	1055	-
Non-specified	-	-	-	-	-	-	-	-	-	-
NON-ENERGY USE	-	-	-	-	-	-	-	-	-	-

APPROVISIONNEMENT ET DEMANDE 1998	Gaz (TJ)				En. Re. Comb. & Déchets (TJ)				(GWh)	(TJ)
	Gaz naturel	Usines à gaz	Cokeries	Hauts fourneaux	Biomasse solide	Gaz/Liquides tirés de biomasse	Déchets urbains	Déchets industriels	Electricité	Chaleur
Production	-	-	-	-	127082	-	-	-	4456	-
Imports	-	-	-	-	-	-	-	-	-	-
Exports	-	-	-	-	-	-	-	-	-38	-
Intl. Marine Bunkers	-	-	-	-	-	-	-	-	-	-
Stock Changes	-	-	-	-	-	-	-	-	-	-
DOMESTIC SUPPLY	-	-	-	-	127082	-	-	-	4418	-
Transfers and Stat. Diff.	-	-	-	-	-	-	-	-	-	-
TRANSFORMATION	-	-	-	-	6426	-	-	-	-	-
Electricity and CHP Plants	-	-	-	-	4056	-	-	-	-	-
Petroleum Refineries	-	-	-	-	116	-	-	-	-	-
Other Transform. Sector	-	-	-	-	2254	-	-	-	-	-
ENERGY SECTOR	-	-	-	-	-	-	-	-	25	-
DISTRIBUTION LOSSES	-	-	-	-	-	-	-	-	915	-
FINAL CONSUMPTION	-	-	-	-	120656	-	-	-	3478	-
INDUSTRY SECTOR	-	-	-	-	7758	-	-	-	1136	-
Iron and Steel	-	-	-	-	-	-	-	-	-	-
Chemical and Petrochem.	-	-	-	-	-	-	-	-	-	-
Non-Metallic Minerals	-	-	-	-	-	-	-	-	-	-
Non-specified	-	-	-	-	7758	-	-	-	1136	-
TRANSPORT SECTOR	-	-	-	-	-	-	-	-	-	-
Air	-	-	-	-	-	-	-	-	-	-
Road	-	-	-	-	-	-	-	-	-	-
Non-specified	-	-	-	-	-	-	-	-	-	-
OTHER SECTORS	-	-	-	-	112898	-	-	-	2342	-
Agriculture	-	-	-	-	-	-	-	-	17	-
Comm. and Publ. Services	-	-	-	-	-	-	-	-	1198	-
Residential	-	-	-	-	112898	-	-	-	1127	-
Non-specified	-	-	-	-	-	-	-	-	-	-
NON-ENERGY USE	-	-	-	-	-	-	-	-	-	-

Haiti

SUPPLY AND CONSUMPTION 1997	Coal (1000 tonnes)							Oil (1000 tonnes)			
	Coking Coal	Other Bit. Coal	Sub-Bit. Coal	Lignite	Peat	Oven and Gas Coke	Pat. Fuel and BKB	Crude Oil	NGL	Feed-stocks	Additives
Production	-	-	-	-	-	-	-	-	-	-	-
Imports	-	-	-	-	-	-	-	-	-	-	-
Exports	-	-	-	-	-	-	-	-	-	-	-
Intl. Marine Bunkers	-	-	-	-	-	-	-	-	-	-	-
Stock Changes	-	-	-	-	-	-	-	-	-	-	-
DOMESTIC SUPPLY	-	-	-	-	-	-	-	-	-	-	-
Transfers and Stat. Diff.	-	-	-	-	-	-	-	-	-	-	-
TRANSFORMATION	-	-	-	-	-	-	-	-	-	-	-
Electricity and CHP Plants	-	-	-	-	-	-	-	-	-	-	-
Petroleum Refineries	-	-	-	-	-	-	-	-	-	-	-
Other Transform. Sector	-	-	-	-	-	-	-	-	-	-	-
ENERGY SECTOR	-	-	-	-	-	-	-	-	-	-	-
DISTRIBUTION LOSSES	-	-	-	-	-	-	-	-	-	-	-
FINAL CONSUMPTION	-	-	-	-	-	-	-	-	-	-	-
INDUSTRY SECTOR	-	-	-	-	-	-	-	-	-	-	-
Iron and Steel	-	-	-	-	-	-	-	-	-	-	-
Chemical and Petrochem.	-	-	-	-	-	-	-	-	-	-	-
Non-Metallic Minerals	-	-	-	-	-	-	-	-	-	-	-
Non-specified	-	-	-	-	-	-	-	-	-	-	-
TRANSPORT SECTOR	-	-	-	-	-	-	-	-	-	-	-
Air	-	-	-	-	-	-	-	-	-	-	-
Road	-	-	-	-	-	-	-	-	-	-	-
Non-specified	-	-	-	-	-	-	-	-	-	-	-
OTHER SECTORS	-	-	-	-	-	-	-	-	-	-	-
Agriculture	-	-	-	-	-	-	-	-	-	-	-
Comm. and Publ. Services	-	-	-	-	-	-	-	-	-	-	-
Residential	-	-	-	-	-	-	-	-	-	-	-
Non-specified	-	-	-	-	-	-	-	-	-	-	-
NON-ENERGY USE	-	-	-	-	-	-	-	-	-	-	-

APPROVISIONNEMENT ET DEMANDE 1998	Charbon (1000 tonnes)							Pétrole (1000 tonnes)			
	Charbon à coke	Autres charb. bit.	Charbon sous-bit.	Lignite	Tourbe	Coke de four/gaz	Agg./briq. de lignite	Pétrole brut	LGN	Produits d'aliment.	Additifs
Production	-	-	-	-	-	-	-	-	-	-	-
Imports	-	-	-	-	-	-	-	-	-	-	-
Exports	-	-	-	-	-	-	-	-	-	-	-
Intl. Marine Bunkers	-	-	-	-	-	-	-	-	-	-	-
Stock Changes	-	-	-	-	-	-	-	-	-	-	-
DOMESTIC SUPPLY	-	-	-	-	-	-	-	-	-	-	-
Transfers and Stat. Diff.	-	-	-	-	-	-	-	-	-	-	-
TRANSFORMATION	-	-	-	-	-	-	-	-	-	-	-
Electricity and CHP Plants	-	-	-	-	-	-	-	-	-	-	-
Petroleum Refineries	-	-	-	-	-	-	-	-	-	-	-
Other Transform. Sector	-	-	-	-	-	-	-	-	-	-	-
ENERGY SECTOR	-	-	-	-	-	-	-	-	-	-	-
DISTRIBUTION LOSSES	-	-	-	-	-	-	-	-	-	-	-
FINAL CONSUMPTION	-	-	-	-	-	-	-	-	-	-	-
INDUSTRY SECTOR	-	-	-	-	-	-	-	-	-	-	-
Iron and Steel	-	-	-	-	-	-	-	-	-	-	-
Chemical and Petrochem.	-	-	-	-	-	-	-	-	-	-	-
Non-Metallic Minerals	-	-	-	-	-	-	-	-	-	-	-
Non-specified	-	-	-	-	-	-	-	-	-	-	-
TRANSPORT SECTOR	-	-	-	-	-	-	-	-	-	-	-
Air	-	-	-	-	-	-	-	-	-	-	-
Road	-	-	-	-	-	-	-	-	-	-	-
Non-specified	-	-	-	-	-	-	-	-	-	-	-
OTHER SECTORS	-	-	-	-	-	-	-	-	-	-	-
Agriculture	-	-	-	-	-	-	-	-	-	-	-
Comm. and Publ. Services	-	-	-	-	-	-	-	-	-	-	-
Residential	-	-	-	-	-	-	-	-	-	-	-
Non-specified	-	-	-	-	-	-	-	-	-	-	-
NON-ENERGY USE	-	-	-	-	-	-	-	-	-	-	-

Haiti

SUPPLY AND CONSUMPTION 1997	Oil cont. (1000 tonnes)										
	Refinery Gas	LPG + Ethane	Motor Gasoline	Aviation Gasoline	Jet Fuel	Kerosene	Gas/ Diesel	Heavy Fuel Oil	Naphtha	Petrol. Coke	Other Prod.
Production	-	-	-	-	-	-	-	-	-	-	-
Imports	-	6	85	-	19	34	209	106	-	-	9
Exports	-	-	-	-	-	-	-	-	-	-	-
Intl. Marine Bunkers	-	-	-	-	-	-	-	-	-	-	-
Stock Changes	-	-	-	-	-	-	-	-	-	-	-
DOMESTIC SUPPLY	-	6	85	-	19	34	209	106	-	-	9
Transfers and Stat. Diff.	-	-	-	-	-	1	-	-38	-	-	-
TRANSFORMATION	-	-	-	-	-	-	50	60	-	-	-
Electricity and CHP Plants	-	-	-	-	-	-	50	60	-	-	-
Petroleum Refineries	-	-	-	-	-	-	-	-	-	-	-
Other Transform. Sector	-	-	-	-	-	-	-	-	-	-	-
ENERGY SECTOR	-	-	-	-	-	-	-	-	-	-	-
DISTRIBUTION LOSSES	-	-	-	-	-	-	-	-	-	-	-
FINAL CONSUMPTION	-	6	85	-	19	35	159	8	-	-	9
INDUSTRY SECTOR	-	-	-	-	-	8	61	8	-	-	-
Iron and Steel	-	-	-	-	-	-	-	-	-	-	-
Chemical and Petrochem.	-	-	-	-	-	-	-	-	-	-	-
Non-Metallic Minerals	-	-	-	-	-	-	-	-	-	-	-
Non-specified	-	-	-	-	-	8	61	8	-	-	-
TRANSPORT SECTOR	-	-	85	-	19	-	98	-	-	-	-
Air	-	-	-	-	19	-	-	-	-	-	-
Road	-	-	85	-	-	-	-	-	-	-	-
Non-specified	-	-	-	-	-	-	98	-	-	-	-
OTHER SECTORS	-	6	-	-	-	27	-	-	-	-	-
Agriculture	-	-	-	-	-	-	-	-	-	-	-
Comm. and Publ. Services	-	-	-	-	-	-	-	-	-	-	-
Residential	-	6	-	-	-	27	-	-	-	-	-
Non-specified	-	-	-	-	-	-	-	-	-	-	-
NON-ENERGY USE	-	-	-	-	-	-	-	-	-	-	9

APPROVISIONNEMENT ET DEMANDE 1998	Pétrole cont. (1000 tonnes)										
	Gaz de raffinerie	GPL + éthane	Essence moteur	Essence aviation	Carbu- réacteurs	Kérosène	Gazole	Fioul lourd	Naphta	Coke de pétrole	Autres prod.
Production	-	-	-	-	-	-	-	-	-	-	-
Imports	-	7	86	-	21	34	215	45	-	-	8
Exports	-	-	-	-	-	-	-	-	-	-	-
Intl. Marine Bunkers	-	-	-	-	-	-	-	-	-	-	-
Stock Changes	-	-	7	-	-	-	6	1	-	-	-
DOMESTIC SUPPLY	-	7	93	-	21	34	221	46	-	-	8
Transfers and Stat. Diff.	-	-	-	-	-	1	-	-	-	-	-
TRANSFORMATION	-	-	-	-	-	-	44	26	-	-	-
Electricity and CHP Plants	-	-	-	-	-	-	44	26	-	-	-
Petroleum Refineries	-	-	-	-	-	-	-	-	-	-	-
Other Transform. Sector	-	-	-	-	-	-	-	-	-	-	-
ENERGY SECTOR	-	-	-	-	-	-	-	-	-	-	-
DISTRIBUTION LOSSES	-	-	-	-	-	-	-	-	-	-	-
FINAL CONSUMPTION	-	7	93	-	21	35	177	20	-	-	8
INDUSTRY SECTOR	-	-	-	-	-	8	68	20	-	-	-
Iron and Steel	-	-	-	-	-	-	-	-	-	-	-
Chemical and Petrochem.	-	-	-	-	-	-	-	-	-	-	-
Non-Metallic Minerals	-	-	-	-	-	-	-	-	-	-	-
Non-specified	-	-	-	-	-	8	68	20	-	-	-
TRANSPORT SECTOR	-	-	93	-	21	-	109	-	-	-	-
Air	-	-	-	-	21	-	-	-	-	-	-
Road	-	-	93	-	-	-	-	-	-	-	-
Non-specified	-	-	-	-	-	-	109	-	-	-	-
OTHER SECTORS	-	7	-	-	-	27	-	-	-	-	-
Agriculture	-	-	-	-	-	-	-	-	-	-	-
Comm. and Publ. Services	-	-	-	-	-	-	-	-	-	-	-
Residential	-	7	-	-	-	27	-	-	-	-	-
Non-specified	-	-	-	-	-	-	-	-	-	-	-
NON-ENERGY USE	-	-	-	-	-	-	-	-	-	-	8

Haiti

SUPPLY AND CONSUMPTION 1997	Gas (TJ)				Comb. Renew. & Waste (TJ)				(GWh)	(TJ)
	Natural Gas	Gas Works	Coke Ovens	Blast Furnaces	Solid Biomass	Gas/Liquids from Biomass	Municipal Waste	Industrial Waste	Electricity	Heat
Production	-	-	-	-	65785	-	-	-	605	-
Imports	-	-	-	-	-	-	-	-	-	-
Exports	-	-	-	-	-	-	-	-	-	-
Intl. Marine Bunkers	-	-	-	-	-	-	-	-	-	-
Stock Changes	-	-	-	-	-	-	-	-	-	-
DOMESTIC SUPPLY	-	-	-	-	65785	-	-	-	605	-
Transfers and Stat. Diff.	-	-	-	-	-	-	-	-	-35	-
TRANSFORMATION	-	-	-	-	9957	-	-	-	-	-
Electricity and CHP Plants	-	-	-	-	-	-	-	-	-	-
Petroleum Refineries	-	-	-	-	-	-	-	-	-	-
Other Transform. Sector	-	-	-	-	9957	-	-	-	-	-
ENERGY SECTOR	-	-	-	-	-	-	-	-	12	-
DISTRIBUTION LOSSES	-	-	-	-	-	-	-	-	267	-
FINAL CONSUMPTION	-	-	-	-	55828	-	-	-	291	-
INDUSTRY SECTOR	-	-	-	-	5057	-	-	-	105	-
Iron and Steel	-	-	-	-	-	-	-	-	-	-
Chemical and Petrochem.	-	-	-	-	-	-	-	-	-	-
Non-Metallic Minerals	-	-	-	-	-	-	-	-	-	-
Non-specified	-	-	-	-	5057	-	-	-	105	-
TRANSPORT SECTOR	-	-	-	-	-	-	-	-	-	-
Air	-	-	-	-	-	-	-	-	-	-
Road	-	-	-	-	-	-	-	-	-	-
Non-specified	-	-	-	-	-	-	-	-	-	-
OTHER SECTORS	-	-	-	-	50771	-	-	-	186	-
Agriculture	-	-	-	-	-	-	-	-	-	-
Comm. and Publ. Services	-	-	-	-	1647	-	-	-	105	-
Residential	-	-	-	-	49124	-	-	-	81	-
Non-specified	-	-	-	-	-	-	-	-	-	-
NON-ENERGY USE	-	-	-	-	-	-	-	-	-	-

APPROVISIONNEMENT ET DEMANDE 1998	Gaz (TJ)				En. Re. Comb. & Déchets (TJ)				(GWh)	(TJ)
	Gaz naturel	Usines à gaz	Cokeries	Hauts fourneaux	Biomasse solide	Gaz/Liquides tirés de biomasse	Déchets urbains	Déchets industriels	Electricité	Chaleur
Production	-	-	-	-	66994	-	-	-	663	-
Imports	-	-	-	-	-	-	-	-	-	-
Exports	-	-	-	-	-	-	-	-	-	-
Intl. Marine Bunkers	-	-	-	-	-	-	-	-	-	-
Stock Changes	-	-	-	-	-	-	-	-	-	-
DOMESTIC SUPPLY	-	-	-	-	66994	-	-	-	663	-
Transfers and Stat. Diff.	-	-	-	-	-	-	-	-	-47	-
TRANSFORMATION	-	-	-	-	10133	-	-	-	-	-
Electricity and CHP Plants	-	-	-	-	-	-	-	-	-	-
Petroleum Refineries	-	-	-	-	-	-	-	-	-	-
Other Transform. Sector	-	-	-	-	10133	-	-	-	-	-
ENERGY SECTOR	-	-	-	-	-	-	-	-	-	-
DISTRIBUTION LOSSES	-	-	-	-	-	-	-	-	361	-
FINAL CONSUMPTION	-	-	-	-	56861	-	-	-	255	-
INDUSTRY SECTOR	-	-	-	-	5117	-	-	-	116	-
Iron and Steel	-	-	-	-	-	-	-	-	-	-
Chemical and Petrochem.	-	-	-	-	-	-	-	-	-	-
Non-Metallic Minerals	-	-	-	-	-	-	-	-	-	-
Non-specified	-	-	-	-	5117	-	-	-	116	-
TRANSPORT SECTOR	-	-	-	-	-	-	-	-	-	-
Air	-	-	-	-	-	-	-	-	-	-
Road	-	-	-	-	-	-	-	-	-	-
Non-specified	-	-	-	-	-	-	-	-	-	-
OTHER SECTORS	-	-	-	-	51744	-	-	-	139	-
Agriculture	-	-	-	-	-	-	-	-	-	-
Comm. and Publ. Services	-	-	-	-	1676	-	-	-	23	-
Residential	-	-	-	-	50068	-	-	-	116	-
Non-specified	-	-	-	-	-	-	-	-	-	-
NON-ENERGY USE	-	-	-	-	-	-	-	-	-	-

Honduras

SUPPLY AND CONSUMPTION 1997	Coal (1000 tonnes)							Oil (1000 tonnes)			
	Coking Coal	Other Bit. Coal	Sub-Bit. Coal	Lignite	Peat	Oven and Gas Coke	Pat. Fuel and BKB	Crude Oil	NGL	Feed-stocks	Additives
Production	-	-	-	-	-	-	-	-	-	-	-
Imports	-	-	-	-	-	1	-	-	-	-	-
Exports	-	-	-	-	-	-	-	-	-	-	-
Intl. Marine Bunkers	-	-	-	-	-	-	-	-	-	-	-
Stock Changes	-	-	-	-	-	-	-	-	-	-	-
DOMESTIC SUPPLY	-	-	-	-	-	1	-	-	-	-	-
Transfers and Stat. Diff.	-	-	-	-	-	-	-	-	-	-	-
TRANSFORMATION	-	-	-	-	-	1	-	-	-	-	-
Electricity and CHP Plants	-	-	-	-	-	-	-	-	-	-	-
Petroleum Refineries	-	-	-	-	-	-	-	-	-	-	-
Other Transform. Sector	-	-	-	-	-	1	-	-	-	-	-
ENERGY SECTOR	-	-	-	-	-	-	-	-	-	-	-
DISTRIBUTION LOSSES	-	-	-	-	-	-	-	-	-	-	-
FINAL CONSUMPTION	-	-	-	-	-	-	-	-	-	-	-
INDUSTRY SECTOR	-	-	-	-	-	-	-	-	-	-	-
Iron and Steel	-	-	-	-	-	-	-	-	-	-	-
Chemical and Petrochem.	-	-	-	-	-	-	-	-	-	-	-
Non-Metallic Minerals	-	-	-	-	-	-	-	-	-	-	-
Non-specified	-	-	-	-	-	-	-	-	-	-	-
TRANSPORT SECTOR	-	-	-	-	-	-	-	-	-	-	-
Air	-	-	-	-	-	-	-	-	-	-	-
Road	-	-	-	-	-	-	-	-	-	-	-
Non-specified	-	-	-	-	-	-	-	-	-	-	-
OTHER SECTORS	-	-	-	-	-	-	-	-	-	-	-
Agriculture	-	-	-	-	-	-	-	-	-	-	-
Comm. and Publ. Services	-	-	-	-	-	-	-	-	-	-	-
Residential	-	-	-	-	-	-	-	-	-	-	-
Non-specified	-	-	-	-	-	-	-	-	-	-	-
NON-ENERGY USE	-	-	-	-	-	-	-	-	-	-	-

APPROVISIONNEMENT ET DEMANDE 1998	Charbon (1000 tonnes)							Pétrole (1000 tonnes)			
	Charbon à coke	Autres charb. bit.	Charbon sous-bit.	Lignite	Tourbe	Coke de four/gaz	Agg./briq. de lignite	Pétrole brut	LGN	Produits d'aliment.	Additifs
Production	-	-	-	-	-	-	-	-	-	-	-
Imports	-	-	-	-	-	1	-	-	-	-	-
Exports	-	-	-	-	-	-	-	-	-	-	-
Intl. Marine Bunkers	-	-	-	-	-	-	-	-	-	-	-
Stock Changes	-	-	-	-	-	-	-	-	-	-	-
DOMESTIC SUPPLY	-	-	-	-	-	1	-	-	-	-	-
Transfers and Stat. Diff.	-	-	-	-	-	-	-	-	-	-	-
TRANSFORMATION	-	-	-	-	-	1	-	-	-	-	-
Electricity and CHP Plants	-	-	-	-	-	-	-	-	-	-	-
Petroleum Refineries	-	-	-	-	-	-	-	-	-	-	-
Other Transform. Sector	-	-	-	-	-	1	-	-	-	-	-
ENERGY SECTOR	-	-	-	-	-	-	-	-	-	-	-
DISTRIBUTION LOSSES	-	-	-	-	-	-	-	-	-	-	-
FINAL CONSUMPTION	-	-	-	-	-	-	-	-	-	-	-
INDUSTRY SECTOR	-	-	-	-	-	-	-	-	-	-	-
Iron and Steel	-	-	-	-	-	-	-	-	-	-	-
Chemical and Petrochem.	-	-	-	-	-	-	-	-	-	-	-
Non-Metallic Minerals	-	-	-	-	-	-	-	-	-	-	-
Non-specified	-	-	-	-	-	-	-	-	-	-	-
TRANSPORT SECTOR	-	-	-	-	-	-	-	-	-	-	-
Air	-	-	-	-	-	-	-	-	-	-	-
Road	-	-	-	-	-	-	-	-	-	-	-
Non-specified	-	-	-	-	-	-	-	-	-	-	-
OTHER SECTORS	-	-	-	-	-	-	-	-	-	-	-
Agriculture	-	-	-	-	-	-	-	-	-	-	-
Comm. and Publ. Services	-	-	-	-	-	-	-	-	-	-	-
Residential	-	-	-	-	-	-	-	-	-	-	-
Non-specified	-	-	-	-	-	-	-	-	-	-	-
NON-ENERGY USE	-	-	-	-	-	-	-	-	-	-	-

Honduras

SUPPLY AND CONSUMPTION 1997	Oil cont. (1000 tonnes)										
	Refinery Gas	LPG + Ethane	Motor Gasoline	Aviation Gasoline	Jet Fuel	Kerosene	Gas/ Diesel	Heavy Fuel Oil	Naphtha	Petrol. Coke	Other Prod.
Production	-	-	-	-	-	-	-	-	-	-	-
Imports	-	41	250	-	21	49	569	336	-	-	-
Exports	-	-	-	-	-	-	-	-	-	-	-
Intl. Marine Bunkers	-	-	-	-	-	-	-	-	-	-	-
Stock Changes	-	2	-2	-	-	1	-71	-18	-	-	-
DOMESTIC SUPPLY	-	43	248	-	21	50	498	318	-	-	-
Transfers and Stat. Diff.	-	-	-	-	-	-	-	-	-	-	-
TRANSFORMATION	-	-	-	-	-	-	49	229	-	-	-
Electricity and CHP Plants	-	-	-	-	-	-	49	229	-	-	-
Petroleum Refineries	-	-	-	-	-	-	-	-	-	-	-
Other Transform. Sector	-	-	-	-	-	-	-	-	-	-	-
ENERGY SECTOR	-	-	-	-	-	-	-	1	-	-	-
DISTRIBUTION LOSSES	-	-	-	-	-	-	-	-	-	-	-
FINAL CONSUMPTION	-	43	248	-	21	50	449	88	-	-	-
INDUSTRY SECTOR	-	3	16	-	-	-	148	85	-	-	-
Iron and Steel	-	-	-	-	-	-	-	-	-	-	-
Chemical and Petrochem.	-	-	-	-	-	-	-	-	-	-	-
Non-Metallic Minerals	-	-	-	-	-	-	-	-	-	-	-
Non-specified	-	3	16	-	-	-	148	85	-	-	-
TRANSPORT SECTOR	-	-	217	-	21	-	269	-	-	-	-
Air	-	-	-	-	21	-	-	-	-	-	-
Road	-	-	217	-	-	-	269	-	-	-	-
Non-specified	-	-	-	-	-	-	-	-	-	-	-
OTHER SECTORS	-	40	15	-	-	50	32	3	-	-	-
Agriculture	-	-	-	-	-	-	2	1	-	-	-
Comm. and Publ. Services	-	-	15	-	-	30	30	2	-	-	-
Residential	-	40	-	-	-	20	-	-	-	-	-
Non-specified	-	-	-	-	-	-	-	-	-	-	-
NON-ENERGY USE	-	-	-	-	-	-	-	-	-	-	-

APPROVISIONNEMENT ET DEMANDE 1998	Pétrole cont. (1000 tonnes)										
	Gaz de raffinerie	GPL + éthane	Essence moteur	Essence aviation	Carbu- réacteurs	Kérosène	Gazole	Fioul lourd	Naphta	Coke de pétrole	Autres prod.
Production	-	-	-	-	-	-	-	-	-	-	-
Imports	-	43	291	-	24	45	807	299	-	-	-
Exports	-	-	-	-	-	-	-	-1	-	-	-
Intl. Marine Bunkers	-	-	-	-	-	-	-	-	-	-	-
Stock Changes	-	-11	-9	-	-	13	-131	22	-	-	-
DOMESTIC SUPPLY	-	32	282	-	24	58	676	320	-	-	-
Transfers and Stat. Diff.	-	-	-	-	-	-	-	-	-	-	-
TRANSFORMATION	-	-	-	-	-	-	192	198	-	-	-
Electricity and CHP Plants	-	-	-	-	-	-	192	198	-	-	-
Petroleum Refineries	-	-	-	-	-	-	-	-	-	-	-
Other Transform. Sector	-	-	-	-	-	-	-	-	-	-	-
ENERGY SECTOR	-	-	-	-	-	-	-	2	-	-	-
DISTRIBUTION LOSSES	-	-	-	-	-	-	-	-	-	-	-
FINAL CONSUMPTION	-	32	282	-	24	58	484	120	-	-	-
INDUSTRY SECTOR	-	3	18	-	-	-	160	116	-	-	-
Iron and Steel	-	-	-	-	-	-	-	-	-	-	-
Chemical and Petrochem.	-	-	-	-	-	-	-	-	-	-	-
Non-Metallic Minerals	-	-	-	-	-	-	-	-	-	-	-
Non-specified	-	3	18	-	-	-	160	116	-	-	-
TRANSPORT SECTOR	-	-	247	-	24	-	290	-	-	-	-
Air	-	-	-	-	24	-	-	-	-	-	-
Road	-	-	247	-	-	-	290	-	-	-	-
Non-specified	-	-	-	-	-	-	-	-	-	-	-
OTHER SECTORS	-	29	17	-	-	58	34	4	-	-	-
Agriculture	-	-	-	-	-	-	2	1	-	-	-
Comm. and Publ. Services	-	-	17	-	-	35	32	3	-	-	-
Residential	-	29	-	-	-	23	-	-	-	-	-
Non-specified	-	-	-	-	-	-	-	-	-	-	-
NON-ENERGY USE	-	-	-	-	-	-	-	-	-	-	-

Honduras

SUPPLY AND CONSUMPTION 1997	Gas (TJ)				Comb. Renew. & Waste (TJ)				(GWh)	(TJ)
	Natural Gas	Gas Works	Coke Ovens	Blast Furnaces	Solid Biomass	Gas/Liquids from Biomass	Municipal Waste	Industrial Waste	Electricity	Heat
Production	-	-	-	11	72204	-	-	-	3303	-
Imports	-	-	-	-	-	-	-	-	161	-
Exports	-	-	-	-	-	-	-	-	-6	-
Intl. Marine Bunkers	-	-	-	-	-	-	-	-	-	-
Stock Changes	-	-	-	-	-	-	-	-	-	-
DOMESTIC SUPPLY	-	-	-	11	72204	-	-	-	3458	-
Transfers and Stat. Diff.	-	-	-	-	-	-	-	-	-155	-
TRANSFORMATION	-	-	-	-	651	-	-	-	-	-
Electricity and CHP Plants	-	-	-	-	-	-	-	-	-	-
Petroleum Refineries	-	-	-	-	-	-	-	-	-	-
Other Transform. Sector	-	-	-	-	651	-	-	-	-	-
ENERGY SECTOR	-	-	-	-	-	-	-	-	12	-
DISTRIBUTION LOSSES	-	-	-	-	-	-	-	-	800	-
FINAL CONSUMPTION	-	-	-	11	71553	-	-	-	2491	-
INDUSTRY SECTOR	-	-	-	11	12533	-	-	-	732	-
Iron and Steel	-	-	-	11	-	-	-	-	-	-
Chemical and Petrochem.	-	-	-	-	-	-	-	-	-	-
Non-Metallic Minerals	-	-	-	-	-	-	-	-	-	-
Non-specified	-	-	-	-	12533	-	-	-	732	-
TRANSPORT SECTOR	-	-	-	-	-	-	-	-	-	-
Air	-	-	-	-	-	-	-	-	-	-
Road	-	-	-	-	-	-	-	-	-	-
Non-specified	-	-	-	-	-	-	-	-	-	-
OTHER SECTORS	-	-	-	-	59020	-	-	-	1759	-
Agriculture	-	-	-	-	-	-	-	-	-	-
Comm. and Publ. Services	-	-	-	-	-	-	-	-	775	-
Residential	-	-	-	-	59020	-	-	-	984	-
Non-specified	-	-	-	-	-	-	-	-	-	-
NON-ENERGY USE	-	-	-	-	-	-	-	-	-	-

APPROVISIONNEMENT ET DEMANDE 1998	Gaz (TJ)				En. Re. Comb. & Déchets (TJ)				(GWh)	(TJ)
	Gaz naturel	Usines à gaz	Cokeries	Hauts fourneaux	Biomasse solide	Gaz/Liquides tirés de biomasse	Déchets urbains	Déchets industriels	Electricité	Chaleur
Production	-	-	-	11	72523	-	-	-	3471	-
Imports	-	-	-	-	-	-	-	-	60	-
Exports	-	-	-	-	-	-	-	-	-16	-
Intl. Marine Bunkers	-	-	-	-	-	-	-	-	-	-
Stock Changes	-	-	-	-	-	-	-	-	-	-
DOMESTIC SUPPLY	-	-	-	11	72523	-	-	-	3515	-
Transfers and Stat. Diff.	-	-	-	-	-	-	-	-	-46	-
TRANSFORMATION	-	-	-	-	695	-	-	-	-	-
Electricity and CHP Plants	-	-	-	-	-	-	-	-	-	-
Petroleum Refineries	-	-	-	-	-	-	-	-	-	-
Other Transform. Sector	-	-	-	-	695	-	-	-	-	-
ENERGY SECTOR	-	-	-	-	-	-	-	-	13	-
DISTRIBUTION LOSSES	-	-	-	-	-	-	-	-	713	-
FINAL CONSUMPTION	-	-	-	11	71828	-	-	-	2743	-
INDUSTRY SECTOR	-	-	-	11	12535	-	-	-	779	-
Iron and Steel	-	-	-	11	-	-	-	-	-	-
Chemical and Petrochem.	-	-	-	-	-	-	-	-	-	-
Non-Metallic Minerals	-	-	-	-	-	-	-	-	-	-
Non-specified	-	-	-	-	12535	-	-	-	779	-
TRANSPORT SECTOR	-	-	-	-	-	-	-	-	-	-
Air	-	-	-	-	-	-	-	-	-	-
Road	-	-	-	-	-	-	-	-	-	-
Non-specified	-	-	-	-	-	-	-	-	-	-
OTHER SECTORS	-	-	-	-	59293	-	-	-	1964	-
Agriculture	-	-	-	-	-	-	-	-	-	-
Comm. and Publ. Services	-	-	-	-	-	-	-	-	856	-
Residential	-	-	-	-	59293	-	-	-	1108	-
Non-specified	-	-	-	-	-	-	-	-	-	-
NON-ENERGY USE	-	-	-	-	-	-	-	-	-	-

Hong Kong, China / Hong Kong, Chine : 1997

SUPPLY AND CONSUMPTION / APPROVISIONNEMENT ET DEMANDE	Coal / Charbon (1000 tonnes)							Oil / Pétrole (1000 tonnes)			
	Coking Coal / Charbon à coke	Other Bit. Coal / Autres charb. bit.	Sub-Bit. Coal / Charbon sous-bit.	Lignite / Lignite	Peat / Tourbe	Oven and Gas Coke / Coke de four/gaz	Pat. Fuel and BKB / Agg./briq. de lignite	Crude Oil / Pétrole brut	NGL / LGN	Feed-stocks / Produits d'aliment.	Additives / Additifs
Production	-	-	-	-	-	-	-	-	-	-	-
From Other Sources	-	-	-	-	-	-	-	-	-	-	-
Imports	-	5711	-	-	-	-	-	-	-	-	-
Exports	-	-	-	-	-	-	-	-	-	-	-
Intl. Marine Bunkers	-	-	-	-	-	-	-	-	-	-	-
Stock Changes	-	-	-	-	-	-	-	-	-	-	-
DOMESTIC SUPPLY	-	**5711**	-	-	-	-	-	-	-	-	-
Transfers	-	-	-	-	-	-	-	-	-	-	-
Statistical Differences	-	376	-	-	-	-	-	-	-	-	-
TRANSFORMATION	-	**6087**	-	-	-	-	-	-	-	-	-
Electricity Plants	-	6087	-	-	-	-	-	-	-	-	-
CHP Plants	-	-	-	-	-	-	-	-	-	-	-
Heat Plants	-	-	-	-	-	-	-	-	-	-	-
Blast Furnaces/Gas Works	-	-	-	-	-	-	-	-	-	-	-
Coke/Pat. Fuel/BKB Plants	-	-	-	-	-	-	-	-	-	-	-
Petroleum Refineries	-	-	-	-	-	-	-	-	-	-	-
Petrochemical Industry	-	-	-	-	-	-	-	-	-	-	-
Liquefaction	-	-	-	-	-	-	-	-	-	-	-
Other Transform. Sector	-	-	-	-	-	-	-	-	-	-	-
ENERGY SECTOR	-	-	-	-	-	-	-	-	-	-	-
Coal Mines	-	-	-	-	-	-	-	-	-	-	-
Oil and Gas Extraction	-	-	-	-	-	-	-	-	-	-	-
Petroleum Refineries	-	-	-	-	-	-	-	-	-	-	-
Electr., CHP+Heat Plants	-	-	-	-	-	-	-	-	-	-	-
Pumped Storage (Elec.)	-	-	-	-	-	-	-	-	-	-	-
Other Energy Sector	-	-	-	-	-	-	-	-	-	-	-
Distribution Losses	-	-	-	-	-	-	-	-	-	-	-
FINAL CONSUMPTION	-	-	-	-	-	-	-	-	-	-	-
INDUSTRY SECTOR	-	-	-	-	-	-	-	-	-	-	-
Iron and Steel	-	-	-	-	-	-	-	-	-	-	-
Chemical and Petrochem.	-	-	-	-	-	-	-	-	-	-	-
of which: Feedstocks	-	-	-	-	-	-	-	-	-	-	-
Non-Ferrous Metals	-	-	-	-	-	-	-	-	-	-	-
Non-Metallic Minerals	-	-	-	-	-	-	-	-	-	-	-
Transport Equipment	-	-	-	-	-	-	-	-	-	-	-
Machinery	-	-	-	-	-	-	-	-	-	-	-
Mining and Quarrying	-	-	-	-	-	-	-	-	-	-	-
Food and Tobacco	-	-	-	-	-	-	-	-	-	-	-
Paper, Pulp and Print	-	-	-	-	-	-	-	-	-	-	-
Wood and Wood Products	-	-	-	-	-	-	-	-	-	-	-
Construction	-	-	-	-	-	-	-	-	-	-	-
Textile and Leather	-	-	-	-	-	-	-	-	-	-	-
Non-specified	-	-	-	-	-	-	-	-	-	-	-
TRANSPORT SECTOR	-	-	-	-	-	-	-	-	-	-	-
Air	-	-	-	-	-	-	-	-	-	-	-
Road	-	-	-	-	-	-	-	-	-	-	-
Rail	-	-	-	-	-	-	-	-	-	-	-
Pipeline Transport	-	-	-	-	-	-	-	-	-	-	-
Internal Navigation	-	-	-	-	-	-	-	-	-	-	-
Non-specified	-	-	-	-	-	-	-	-	-	-	-
OTHER SECTORS	-	-	-	-	-	-	-	-	-	-	-
Agriculture	-	-	-	-	-	-	-	-	-	-	-
Comm. and Publ. Services	-	-	-	-	-	-	-	-	-	-	-
Residential	-	-	-	-	-	-	-	-	-	-	-
Non-specified	-	-	-	-	-	-	-	-	-	-	-
NON-ENERGY USE	-	-	-	-	-	-	-	-	-	-	-
in Industry/Trans./Energy	-	-	-	-	-	-	-	-	-	-	-
in Transport	-	-	-	-	-	-	-	-	-	-	-
in Other Sectors	-	-	-	-	-	-	-	-	-	-	-

Hong Kong, China / Hong Kong, Chine : 1997

SUPPLY AND CONSUMPTION / APPROVISIONNEMENT ET DEMANDE	Oil cont. / Pétrole cont. (1000 tonnes)										
	Refinery Gas / Gaz de raffinerie	LPG + Ethane / GPL + éthane	Motor Gasoline / Essence moteur	Aviation Gasoline / Essence aviation	Jet Fuel / Carbu- réacteurs	Kerosene / Kérosène	Gas/ Diesel / Gazole	Heavy Fuel Oil / Fioul lourd	Naphtha / Naphta	Petrol. Coke / Coke de pétrole	Other Prod. / Autres prod.
Production	-	-	-	-	-	-	-	-	-	-	-
From Other Sources	-	-	-	-	-	-	-	-	-	-	-
Imports	-	166	470	-	3438	69	10554	4074	531	-	437
Exports	-	-18	-123	-	-265	-17	-7342	-2321	-	-	-123
Intl. Marine Bunkers	-	-	-	-	-	-	-509	-1640	-	-	-
Stock Changes	-	-2	-4	-	28	1	-49	-28	-	-	-
DOMESTIC SUPPLY	-	146	343	-	3201	53	2654	85	531	-	314
Transfers	-	-	-	-	-	-	-	-	-	-	-
Statistical Differences	-	-	1	-	1	-	-	-	-	-	-1
TRANSFORMATION	-	-	-	-	-	-	16	68	531	-	-
Electricity Plants	-	-	-	-	-	-	16	68	-	-	-
CHP Plants	-	-	-	-	-	-	-	-	-	-	-
Heat Plants	-	-	-	-	-	-	-	-	-	-	-
Blast Furnaces/Gas Works	-	-	-	-	-	-	-	-	531	-	-
Coke/Pat. Fuel/BKB Plants	-	-	-	-	-	-	-	-	-	-	-
Petroleum Refineries	-	-	-	-	-	-	-	-	-	-	-
Petrochemical Industry	-	-	-	-	-	-	-	-	-	-	-
Liquefaction	-	-	-	-	-	-	-	-	-	-	-
Other Transform. Sector	-	-	-	-	-	-	-	-	-	-	-
ENERGY SECTOR	-	-	-	-	-	-	-	-	-	-	-
Coal Mines	-	-	-	-	-	-	-	-	-	-	-
Oil and Gas Extraction	-	-	-	-	-	-	-	-	-	-	-
Petroleum Refineries	-	-	-	-	-	-	-	-	-	-	-
Electr., CHP+Heat Plants	-	-	-	-	-	-	-	-	-	-	-
Pumped Storage (Elec.)	-	-	-	-	-	-	-	-	-	-	-
Other Energy Sector	-	-	-	-	-	-	-	-	-	-	-
Distribution Losses	-	-	-	-	-	-	-	-	-	-	-
FINAL CONSUMPTION	-	146	344	-	3202	53	2638	17	-	-	313
INDUSTRY SECTOR	-	13	-	-	-	8	755	17	-	-	-
Iron and Steel	-	-	-	-	-	-	-	-	-	-	-
Chemical and Petrochem.	-	-	-	-	-	-	-	-	-	-	-
of which: Feedstocks	-	-	-	-	-	-	-	-	-	-	-
Non-Ferrous Metals	-	-	-	-	-	-	-	-	-	-	-
Non-Metallic Minerals	-	-	-	-	-	-	-	-	-	-	-
Transport Equipment	-	-	-	-	-	-	-	-	-	-	-
Machinery	-	-	-	-	-	-	-	-	-	-	-
Mining and Quarrying	-	-	-	-	-	-	-	-	-	-	-
Food and Tobacco	-	-	-	-	-	-	-	-	-	-	-
Paper, Pulp and Print	-	-	-	-	-	-	-	-	-	-	-
Wood and Wood Products	-	-	-	-	-	-	-	-	-	-	-
Construction	-	-	-	-	-	-	-	-	-	-	-
Textile and Leather	-	-	-	-	-	-	-	-	-	-	-
Non-specified	-	13	-	-	-	8	755	17	-	-	-
TRANSPORT SECTOR	-	-	344	-	3202	-	1883	-	-	-	-
Air	-	-	-	-	3202	-	-	-	-	-	-
Road	-	-	344	-	-	-	1883	-	-	-	-
Rail	-	-	-	-	-	-	-	-	-	-	-
Pipeline Transport	-	-	-	-	-	-	-	-	-	-	-
Internal Navigation	-	-	-	-	-	-	-	-	-	-	-
Non-specified	-	-	-	-	-	-	-	-	-	-	-
OTHER SECTORS	-	133	-	-	-	45	-	-	-	-	-
Agriculture	-	-	-	-	-	-	-	-	-	-	-
Comm. and Publ. Services	-	133	-	-	-	-	-	-	-	-	-
Residential	-	-	-	-	-	45	-	-	-	-	-
Non-specified	-	-	-	-	-	-	-	-	-	-	-
NON-ENERGY USE	-	-	-	-	-	-	-	-	-	-	313
in Industry/Transf./Energy	-	-	-	-	-	-	-	-	-	-	313
in Transport	-	-	-	-	-	-	-	-	-	-	-
in Other Sectors	-	-	-	-	-	-	-	-	-	-	-

Hong Kong, China / Hong Kong, Chine : 1997

SUPPLY AND CONSUMPTION *APPROVISIONNEMENT ET DEMANDE*	Gas / *Gaz* (TJ)				Comb. Renew. & Waste / *En. Re. Comb. & Déchets* (TJ)				(GWh)	(TJ)
	Natural Gas *Gaz naturel*	Gas Works *Usines à gaz*	Coke Ovens *Cokeries*	Blast Furnaces *Hauts fourneaux*	Solid Biomass *Biomasse solide*	Gas/Liquids from Biomass *Gaz/Liquides tirés de biomasse*	Municipal Waste *Déchets urbains*	Industrial Waste *Déchets industriels*	Electricity *Electricité*	Heat *Chaleur*
Production	-	23905	-	-	2012	-	-	-	28945	-
From Other Sources	-	-	-	-	-	-	-	-	-	-
Imports	107462	-	-	-	378	-	-	-	7876	-
Exports	-	-	-	-	-60	-	-	-	-559	-
Intl. Marine Bunkers	-	-	-	-	-	-	-	-	-	-
Stock Changes	-	-	-	-	-	-	-	-	-	-
DOMESTIC SUPPLY	**107462**	**23905**	**-**	**-**	**2330**	**-**	**-**	**-**	**36262**	**-**
Transfers	-	-	-	-	-	-	-	-	-	-
Statistical Differences	-	-	-	-	-	-	-	-	-	-
TRANSFORMATION	**107462**	**-**	**-**	**-**	**-**	**-**	**-**	**-**	**-**	**-**
Electricity Plants	107462	-	-	-	-	-	-	-	-	-
CHP Plants	-	-	-	-	-	-	-	-	-	-
Heat Plants	-	-	-	-	-	-	-	-	-	-
Blast Furnaces/Gas Works	-	-	-	-	-	-	-	-	-	-
Coke/Pat. Fuel/BKB Plants	-	-	-	-	-	-	-	-	-	-
Petroleum Refineries	-	-	-	-	-	-	-	-	-	-
Petrochemical Industry	-	-	-	-	-	-	-	-	-	-
Liquefaction	-	-	-	-	-	-	-	-	-	-
Other Transform. Sector	-	-	-	-	-	-	-	-	-	-
ENERGY SECTOR	**-**	**-**	**-**	**-**	**-**	**-**	**-**	**-**	**-**	**-**
Coal Mines	-	-	-	-	-	-	-	-	-	-
Oil and Gas Extraction	-	-	-	-	-	-	-	-	-	-
Petroleum Refineries	-	-	-	-	-	-	-	-	-	-
Electr., CHP+Heat Plants	-	-	-	-	-	-	-	-	-	-
Pumped Storage (Elec.)	-	-	-	-	-	-	-	-	-	-
Other Energy Sector	-	-	-	-	-	-	-	-	-	-
Distribution Losses	-	-	-	-	-	-	-	-	4017	-
FINAL CONSUMPTION	**-**	**23905**	**-**	**-**	**2330**	**-**	**-**	**-**	**32245**	**-**
INDUSTRY SECTOR	**-**	**911**	**-**	**-**	**-**	**-**	**-**	**-**	**5268**	**-**
Iron and Steel	-	-	-	-	-	-	-	-	-	-
Chemical and Petrochem.	-	-	-	-	-	-	-	-	-	-
of which: Feedstocks	-	-	-	-	-	-	-	-	-	-
Non-Ferrous Metals	-	-	-	-	-	-	-	-	-	-
Non-Metallic Minerals	-	-	-	-	-	-	-	-	-	-
Transport Equipment	-	-	-	-	-	-	-	-	-	-
Machinery	-	-	-	-	-	-	-	-	-	-
Mining and Quarrying	-	-	-	-	-	-	-	-	-	-
Food and Tobacco	-	-	-	-	-	-	-	-	-	-
Paper, Pulp and Print	-	-	-	-	-	-	-	-	-	-
Wood and Wood Products	-	-	-	-	-	-	-	-	-	-
Construction	-	-	-	-	-	-	-	-	-	-
Textile and Leather	-	-	-	-	-	-	-	-	-	-
Non-specified	-	911	-	-	-	-	-	-	5268	-
TRANSPORT SECTOR	**-**	**-**	**-**	**-**	**-**	**-**	**-**	**-**	**-**	**-**
Air	-	-	-	-	-	-	-	-	-	-
Road	-	-	-	-	-	-	-	-	-	-
Rail	-	-	-	-	-	-	-	-	-	-
Pipeline Transport	-	-	-	-	-	-	-	-	-	-
Internal Navigation	-	-	-	-	-	-	-	-	-	-
Non-specified	-	-	-	-	-	-	-	-	-	-
OTHER SECTORS	**-**	**22994**	**-**	**-**	**2330**	**-**	**-**	**-**	**26977**	**-**
Agriculture	-	-	-	-	-	-	-	-	-	-
Comm. and Publ. Services	-	10529	-	-	-	-	-	-	18938	-
Residential	-	12465	-	-	2012	-	-	-	8039	-
Non-specified	-	-	-	-	318	-	-	-	-	-
NON-ENERGY USE	**-**	**-**	**-**	**-**	**-**	**-**	**-**	**-**	**-**	**-**
in Industry/Transf./Energy	-	-	-	-	-	-	-	-	-	-
in Transport	-	-	-	-	-	-	-	-	-	-
in Other Sectors	-	-	-	-	-	-	-	-	-	-

Hong Kong, China / Hong Kong, Chine : 1998

SUPPLY AND CONSUMPTION / APPROVISIONNEMENT ET DEMANDE	Coal / Charbon (1000 tonnes)							Oil / Pétrole (1000 tonnes)			
	Coking Coal / Charbon à coke	Other Bit. Coal / Autres charb. bit.	Sub-Bit. Coal / Charbon sous-bit.	Lignite / Lignite	Peat / Tourbe	Oven and Gas Coke / Coke de four/gaz	Pat. Fuel and BKB / Agg./briq. de lignite	Crude Oil / Pétrole brut	NGL / LGN	Feed-stocks / Produits d'aliment.	Additives / Additifs
Production	-	-	-	-	-	-	-	-	-	-	-
From Other Sources	-	-	-	-	-	-	-	-	-	-	-
Imports	-	7102	-	-	-	-	-	-	-	-	-
Exports	-	-	-	-	-	-	-	-	-	-	-
Intl. Marine Bunkers	-	-	-	-	-	-	-	-	-	-	-
Stock Changes	-	-	-	-	-	-	-	-	-	-	-
DOMESTIC SUPPLY	-	**7102**	-	-	-	-	-	-	-	-	-
Transfers	-	-	-	-	-	-	-	-	-	-	-
Statistical Differences	-	121	-	-	-	-	-	-	-	-	-
TRANSFORMATION	-	**7223**	-	-	-	-	-	-	-	-	-
Electricity Plants	-	7223	-	-	-	-	-	-	-	-	-
CHP Plants	-	-	-	-	-	-	-	-	-	-	-
Heat Plants	-	-	-	-	-	-	-	-	-	-	-
Blast Furnaces/Gas Works	-	-	-	-	-	-	-	-	-	-	-
Coke/Pat. Fuel/BKB Plants	-	-	-	-	-	-	-	-	-	-	-
Petroleum Refineries	-	-	-	-	-	-	-	-	-	-	-
Petrochemical Industry	-	-	-	-	-	-	-	-	-	-	-
Liquefaction	-	-	-	-	-	-	-	-	-	-	-
Other Transform. Sector	-	-	-	-	-	-	-	-	-	-	-
ENERGY SECTOR	-	-	-	-	-	-	-	-	-	-	-
Coal Mines	-	-	-	-	-	-	-	-	-	-	-
Oil and Gas Extraction	-	-	-	-	-	-	-	-	-	-	-
Petroleum Refineries	-	-	-	-	-	-	-	-	-	-	-
Electr., CHP+Heat Plants	-	-	-	-	-	-	-	-	-	-	-
Pumped Storage (Elec.)	-	-	-	-	-	-	-	-	-	-	-
Other Energy Sector	-	-	-	-	-	-	-	-	-	-	-
Distribution Losses	-	-	-	-	-	-	-	-	-	-	-
FINAL CONSUMPTION	-	-	-	-	-	-	-	-	-	-	-
INDUSTRY SECTOR	-	-	-	-	-	-	-	-	-	-	-
Iron and Steel	-	-	-	-	-	-	-	-	-	-	-
Chemical and Petrochem.	-	-	-	-	-	-	-	-	-	-	-
of which: Feedstocks	-	-	-	-	-	-	-	-	-	-	-
Non-Ferrous Metals	-	-	-	-	-	-	-	-	-	-	-
Non-Metallic Minerals	-	-	-	-	-	-	-	-	-	-	-
Transport Equipment	-	-	-	-	-	-	-	-	-	-	-
Machinery	-	-	-	-	-	-	-	-	-	-	-
Mining and Quarrying	-	-	-	-	-	-	-	-	-	-	-
Food and Tobacco	-	-	-	-	-	-	-	-	-	-	-
Paper, Pulp and Print	-	-	-	-	-	-	-	-	-	-	-
Wood and Wood Products	-	-	-	-	-	-	-	-	-	-	-
Construction	-	-	-	-	-	-	-	-	-	-	-
Textile and Leather	-	-	-	-	-	-	-	-	-	-	-
Non-specified	-	-	-	-	-	-	-	-	-	-	-
TRANSPORT SECTOR	-	-	-	-	-	-	-	-	-	-	-
Air	-	-	-	-	-	-	-	-	-	-	-
Road	-	-	-	-	-	-	-	-	-	-	-
Rail	-	-	-	-	-	-	-	-	-	-	-
Pipeline Transport	-	-	-	-	-	-	-	-	-	-	-
Internal Navigation	-	-	-	-	-	-	-	-	-	-	-
Non-specified	-	-	-	-	-	-	-	-	-	-	-
OTHER SECTORS	-	-	-	-	-	-	-	-	-	-	-
Agriculture	-	-	-	-	-	-	-	-	-	-	-
Comm. and Publ. Services	-	-	-	-	-	-	-	-	-	-	-
Residential	-	-	-	-	-	-	-	-	-	-	-
Non-specified	-	-	-	-	-	-	-	-	-	-	-
NON-ENERGY USE	-	-	-	-	-	-	-	-	-	-	-
in Industry/Trans./Energy	-	-	-	-	-	-	-	-	-	-	-
in Transport	-	-	-	-	-	-	-	-	-	-	-
in Other Sectors	-	-	-	-	-	-	-	-	-	-	-

Hong Kong, China / Hong Kong, Chine : 1998

SUPPLY AND CONSUMPTION / APPROVISIONNEMENT ET DEMANDE	Oil cont. / Pétrole cont. (1000 tonnes)										
	Refinery Gas / Gaz de raffinerie	LPG + Ethane / GPL + éthane	Motor Gasoline / Essence moteur	Aviation Gasoline / Essence aviation	Jet Fuel / Carbu- réacteurs	Kerosene / Kérosène	Gas/ Diesel / Gazole	Heavy Fuel Oil / Fioul lourd	Naphtha / Naphta	Petrol. Coke / Coke de pétrole	Other Prod. / Autres prod.
Production	-	-	-	-	-	-	-	-	-	-	-
From Other Sources	-	-	-	-	-	-	-	-	-	-	-
Imports	-	144	413	-	3238	49	10952	4269	552	-	472
Exports	-	-9	-63	-	-255	-6	-5144	-2650	-	-	-306
Intl. Marine Bunkers	-	-	-	-	-	-	-1281	-1562	-	-	-
Stock Changes	-	5	6	-	27	-2	84	23	-	-	-
DOMESTIC SUPPLY	-	140	356	-	3010	41	4611	80	552	-	166
Transfers	-	-	-	-	-	-	-	-	-	-	-
Statistical Differences	-	-	-	-	-	-	-	-	-	-	-
TRANSFORMATION	-	-	-	-	-	-	5	65	552	-	-
Electricity Plants	-	-	-	-	-	-	5	65	-	-	-
CHP Plants	-	-	-	-	-	-	-	-	-	-	-
Heat Plants	-	-	-	-	-	-	-	-	-	-	-
Blast Furnaces/Gas Works	-	-	-	-	-	-	-	-	552	-	-
Coke/Pat. Fuel/BKB Plants	-	-	-	-	-	-	-	-	-	-	-
Petroleum Refineries	-	-	-	-	-	-	-	-	-	-	-
Petrochemical Industry	-	-	-	-	-	-	-	-	-	-	-
Liquefaction	-	-	-	-	-	-	-	-	-	-	-
Other Transform. Sector	-	-	-	-	-	-	-	-	-	-	-
ENERGY SECTOR	-	-	-	-	-	-	-	-	-	-	-
Coal Mines	-	-	-	-	-	-	-	-	-	-	-
Oil and Gas Extraction	-	-	-	-	-	-	-	-	-	-	-
Petroleum Refineries	-	-	-	-	-	-	-	-	-	-	-
Electr., CHP+Heat Plants	-	-	-	-	-	-	-	-	-	-	-
Pumped Storage (Elec.)	-	-	-	-	-	-	-	-	-	-	-
Other Energy Sector	-	-	-	-	-	-	-	-	-	-	-
Distribution Losses	-	-	-	-	-	-	-	-	-	-	-
FINAL CONSUMPTION	-	140	356	-	3010	41	4606	15	-	-	166
INDUSTRY SECTOR	-	13	-	-	-	6	1319	15	-	-	-
Iron and Steel	-	-	-	-	-	-	-	-	-	-	-
Chemical and Petrochem.	-	-	-	-	-	-	-	-	-	-	-
of which: Feedstocks	-	-	-	-	-	-	-	-	-	-	-
Non-Ferrous Metals	-	-	-	-	-	-	-	-	-	-	-
Non-Metallic Minerals	-	-	-	-	-	-	-	-	-	-	-
Transport Equipment	-	-	-	-	-	-	-	-	-	-	-
Machinery	-	-	-	-	-	-	-	-	-	-	-
Mining and Quarrying	-	-	-	-	-	-	-	-	-	-	-
Food and Tobacco	-	-	-	-	-	-	-	-	-	-	-
Paper, Pulp and Print	-	-	-	-	-	-	-	-	-	-	-
Wood and Wood Products	-	-	-	-	-	-	-	-	-	-	-
Construction	-	-	-	-	-	-	-	-	-	-	-
Textile and Leather	-	-	-	-	-	-	-	-	-	-	-
Non-specified	-	13	-	-	-	6	1319	15	-	-	-
TRANSPORT SECTOR	-	-	356	-	3010	-	3287	-	-	-	-
Air	-	-	-	-	3010	-	-	-	-	-	-
Road	-	-	356	-	-	-	3287	-	-	-	-
Rail	-	-	-	-	-	-	-	-	-	-	-
Pipeline Transport	-	-	-	-	-	-	-	-	-	-	-
Internal Navigation	-	-	-	-	-	-	-	-	-	-	-
Non-specified	-	-	-	-	-	-	-	-	-	-	-
OTHER SECTORS	-	127	-	-	-	35	-	-	-	-	-
Agriculture	-	-	-	-	-	-	-	-	-	-	-
Comm. and Publ. Services	-	127	-	-	-	-	-	-	-	-	-
Residential	-	-	-	-	-	35	-	-	-	-	-
Non-specified	-	-	-	-	-	-	-	-	-	-	-
NON-ENERGY USE	-	-	-	-	-	-	-	-	-	-	166
in Industry/Transf./Energy	-	-	-	-	-	-	-	-	-	-	166
in Transport	-	-	-	-	-	-	-	-	-	-	-
in Other Sectors	-	-	-	-	-	-	-	-	-	-	-

Hong Kong, China / Hong Kong, Chine : 1998

SUPPLY AND CONSUMPTION APPROVISIONNEMENT ET DEMANDE	Gas / Gaz (TJ)				Comb. Renew. & Waste / En. Re. Comb. & Déchets (TJ)				(GWh)	(TJ)
	Natural Gas Gaz naturel	Gas Works Usines à gaz	Coke Ovens Cokeries	Blast Furnaces Hauts fourneaux	Solid Biomass Biomasse solide	Gas/Liquids from Biomass Gaz/Liquides tirés de biomasse	Municipal Waste Déchets urbains	Industrial Waste Déchets industriels	Electricity Electricité	Heat Chaleur
Production	-	23943	-	-	2012	-	-	-	31417	-
From Other Sources	-	-	-	-	-	-	-	-	-	-
Imports	100815	-	-	-	274	-	-	-	7760	-
Exports	-	-	-	-	-7	-	-	-	-610	-
Intl. Marine Bunkers	-	-	-	-	-	-	-	-	-	-
Stock Changes	-	-	-	-	-	-	-	-	-	-
DOMESTIC SUPPLY	100815	23943	-	-	2279	-	-	-	38567	-
Transfers	-	-	-	-	-	-	-	-	-	-
Statistical Differences	-	-	-	-	-	-	-	-	-	-
TRANSFORMATION	100815	-	-	-	-	-	-	-	-	-
Electricity Plants	100815	-	-	-	-	-	-	-	-	-
CHP Plants	-	-	-	-	-	-	-	-	-	-
Heat Plants	-	-	-	-	-	-	-	-	-	-
Blast Furnaces/Gas Works	-	-	-	-	-	-	-	-	-	-
Coke/Pat. Fuel/BKB Plants	-	-	-	-	-	-	-	-	-	-
Petroleum Refineries	-	-	-	-	-	-	-	-	-	-
Petrochemical Industry	-	-	-	-	-	-	-	-	-	-
Liquefaction	-	-	-	-	-	-	-	-	-	-
Other Transform. Sector	-	-	-	-	-	-	-	-	-	-
ENERGY SECTOR	-	-	-	-	-	-	-	-	-	-
Coal Mines	-	-	-	-	-	-	-	-	-	-
Oil and Gas Extraction	-	-	-	-	-	-	-	-	-	-
Petroleum Refineries	-	-	-	-	-	-	-	-	-	-
Electr., CHP+Heat Plants	-	-	-	-	-	-	-	-	-	-
Pumped Storage (Elec.)	-	-	-	-	-	-	-	-	-	-
Other Energy Sector	-	-	-	-	-	-	-	-	-	-
Distribution Losses	-	-	-	-	-	-	-	-	3718	-
FINAL CONSUMPTION	-	23943	-	-	2279	-	-	-	34849	-
INDUSTRY SECTOR	-	888	-	-	-	-	-	-	5136	-
Iron and Steel	-	-	-	-	-	-	-	-	-	-
Chemical and Petrochem.	-	-	-	-	-	-	-	-	-	-
of which: Feedstocks	-	-	-	-	-	-	-	-	-	-
Non-Ferrous Metals	-	-	-	-	-	-	-	-	-	-
Non-Metallic Minerals	-	-	-	-	-	-	-	-	-	-
Transport Equipment	-	-	-	-	-	-	-	-	-	-
Machinery	-	-	-	-	-	-	-	-	-	-
Mining and Quarrying	-	-	-	-	-	-	-	-	-	-
Food and Tobacco	-	-	-	-	-	-	-	-	-	-
Paper, Pulp and Print	-	-	-	-	-	-	-	-	-	-
Wood and Wood Products	-	-	-	-	-	-	-	-	-	-
Construction	-	-	-	-	-	-	-	-	-	-
Textile and Leather	-	-	-	-	-	-	-	-	-	-
Non-specified	-	888	-	-	-	-	-	-	5136	-
TRANSPORT SECTOR	-	-	-	-	-	-	-	-	-	-
Air	-	-	-	-	-	-	-	-	-	-
Road	-	-	-	-	-	-	-	-	-	-
Rail	-	-	-	-	-	-	-	-	-	-
Pipeline Transport	-	-	-	-	-	-	-	-	-	-
Internal Navigation	-	-	-	-	-	-	-	-	-	-
Non-specified	-	-	-	-	-	-	-	-	-	-
OTHER SECTORS	-	23055	-	-	2279	-	-	-	29713	-
Agriculture	-	-	-	-	-	-	-	-	-	-
Comm. and Publ. Services	-	10536	-	-	-	-	-	-	20603	-
Residential	-	12519	-	-	2012	-	-	-	9110	-
Non-specified	-	-	-	-	267	-	-	-	-	-
NON-ENERGY USE	-	-	-	-	-	-	-	-	-	-
in Industry/Transf./Energy	-	-	-	-	-	-	-	-	-	-
in Transport	-	-	-	-	-	-	-	-	-	-
in Other Sectors	-	-	-	-	-	-	-	-	-	-

India / Inde : 1997

SUPPLY AND CONSUMPTION *APPROVISIONNEMENT ET DEMANDE*	Coal / *Charbon* (1000 tonnes)							Oil / *Pétrole* (1000 tonnes)			
	Coking Coal *Charbon à coke*	Other Bit. Coal *Autres charb. bit.*	Sub-Bit. Coal *Charbon sous-bit.*	Lignite *Lignite*	Peat *Tourbe*	Oven and Gas Coke *Coke de four/gaz*	Pat. Fuel and BKB *Agg./briq. de lignite*	Crude Oil *Pétrole brut*	NGL *LGN*	Feed-stocks *Produits d'aliment.*	Additives *Additifs*
Production	44760	251040	-	23500	-	13584	262	33830	3535	-	-
From Other Sources	-	-	-	-	-	-	-	-	-	-	-
Imports	10650	6560	-	-	-	-	-	33573	-	-	-
Exports	-	-60	-	-	-	-	-	-	-985	-	-
Intl. Marine Bunkers	-	-	-	-	-	-	-	-	-	-	-
Stock Changes	279	601	-	-	-	-	-	-3968	-	-	-
DOMESTIC SUPPLY	55689	258141	-	23500	-	13584	262	63435	2550	-	-
Transfers	-	-	-	-	-	-	-	-	-1782	-	-
Statistical Differences	-	-	-	-	-	-	-	-	-	-	-
TRANSFORMATION	33960	193520	-	22800	-	10867	-	63435	-	-	-
Electricity Plants	-	193520	-	22400	-	-	-	-	-	-	-
CHP Plants	-	-	-	-	-	-	-	-	-	-	-
Heat Plants	-	-	-	-	-	-	-	-	-	-	-
Blast Furnaces/Gas Works	-	-	-	-	-	10867	-	-	-	-	-
Coke/Pat. Fuel/BKB Plants	33960	-	-	400	-	-	-	-	-	-	-
Petroleum Refineries	-	-	-	-	-	-	-	63435	-	-	-
Petrochemical Industry	-	-	-	-	-	-	-	-	-	-	-
Liquefaction	-	-	-	-	-	-	-	-	-	-	-
Other Transform. Sector	-	-	-	-	-	-	-	-	-	-	-
ENERGY SECTOR	-	3060	-	-	-	-	-	-	-	-	-
Coal Mines	-	3060	-	-	-	-	-	-	-	-	-
Oil and Gas Extraction	-	-	-	-	-	-	-	-	-	-	-
Petroleum Refineries	-	-	-	-	-	-	-	-	-	-	-
Electr., CHP+Heat Plants	-	-	-	-	-	-	-	-	-	-	-
Pumped Storage (Elec.)	-	-	-	-	-	-	-	-	-	-	-
Other Energy Sector	-	-	-	-	-	-	-	-	-	-	-
Distribution Losses	-	-	-	-	-	-	-	-	-	-	-
FINAL CONSUMPTION	21729	61561	-	700	-	2717	262	-	768	-	-
INDUSTRY SECTOR	21729	61471	-	650	-	2717	-	-	768	-	-
Iron and Steel	19320	-	-	-	-	2717	-	-	-	-	-
Chemical and Petrochem.	-	4640	-	650	-	-	-	-	768	-	-
of which: Feedstocks	-	-	-	-	-	-	-	-	768	-	-
Non-Ferrous Metals	-	-	-	-	-	-	-	-	-	-	-
Non-Metallic Minerals	-	14680	-	-	-	-	-	-	-	-	-
Transport Equipment	-	-	-	-	-	-	-	-	-	-	-
Machinery	-	-	-	-	-	-	-	-	-	-	-
Mining and Quarrying	-	-	-	-	-	-	-	-	-	-	-
Food and Tobacco	-	-	-	-	-	-	-	-	-	-	-
Paper, Pulp and Print	-	3430	-	-	-	-	-	-	-	-	-
Wood and Wood Products	-	-	-	-	-	-	-	-	-	-	-
Construction	-	-	-	-	-	-	-	-	-	-	-
Textile and Leather	-	1360	-	-	-	-	-	-	-	-	-
Non-specified	2409	37361	-	-	-	-	-	-	-	-	-
TRANSPORT SECTOR	-	60	-	-	-	-	-	-	-	-	-
Air	-	-	-	-	-	-	-	-	-	-	-
Road	-	-	-	-	-	-	-	-	-	-	-
Rail	-	60	-	-	-	-	-	-	-	-	-
Pipeline Transport	-	-	-	-	-	-	-	-	-	-	-
Internal Navigation	-	-	-	-	-	-	-	-	-	-	-
Non-specified	-	-	-	-	-	-	-	-	-	-	-
OTHER SECTORS	-	30	-	50	-	-	262	-	-	-	-
Agriculture	-	-	-	-	-	-	-	-	-	-	-
Comm. and Publ. Services	-	-	-	-	-	-	-	-	-	-	-
Residential	-	30	-	50	-	-	262	-	-	-	-
Non-specified	-	-	-	-	-	-	-	-	-	-	-
NON-ENERGY USE	-	-	-	-	-	-	-	-	-	-	-
in Industry/Trans./Energy	-	-	-	-	-	-	-	-	-	-	-
in Transport	-	-	-	-	-	-	-	-	-	-	-
in Other Sectors	-	-	-	-	-	-	-	-	-	-	-

India / Inde : 1997

SUPPLY AND CONSUMPTION *APPROVISIONNEMENT ET DEMANDE*	Oil cont. / *Pétrole cont.* (1000 tonnes)										
	Refinery Gas *Gaz de raffinerie*	LPG + Ethane *GPL + éthane*	Motor Gasoline *Essence moteur*	Aviation Gasoline *Essence aviation*	Jet Fuel *Carbu- réacteurs*	Kerosene *Kérosène*	Gas/ Diesel *Gazole*	Heavy Fuel Oil *Fioul lourd*	Naphtha *Naphta*	Petrol. Coke *Coke de pétrole*	Other Prod. *Autres prod.*
Production	1148	1615	4876	-	2147	6701	23304	13943	6015	266	3232
From Other Sources	-	-	-	-	-	-	-	-	-	-	-
Imports	-	1615	32	26	55	3812	14016	358	-	-	-
Exports	-	-21	-	-	-	-	-220	-358	-1590	-	-
Intl. Marine Bunkers	-	-	-	-	-	-	-471	-612	-	-	-
Stock Changes	-	-702	97	-	-	-	-	-	-	-	-
DOMESTIC SUPPLY	**1148**	**2507**	**5005**	**26**	**2202**	**10513**	**36629**	**13331**	**4425**	**266**	**3232**
Transfers	-	1782	-	-	-	-	-	-	-	-	-
Statistical Differences	-	-	-	-	-	-644	-	-1	-	-	-
TRANSFORMATION	**-**	**-**	**-**	**-**	**-**	**-**	**270**	**1111**	**-**	**-**	**-**
Electricity Plants	-	-	-	-	-	-	270	1111	-	-	-
CHP Plants	-	-	-	-	-	-	-	-	-	-	-
Heat Plants	-	-	-	-	-	-	-	-	-	-	-
Blast Furnaces/Gas Works	-	-	-	-	-	-	-	-	-	-	-
Coke/Pat. Fuel/BKB Plants	-	-	-	-	-	-	-	-	-	-	-
Petroleum Refineries	-	-	-	-	-	-	-	-	-	-	-
Petrochemical Industry	-	-	-	-	-	-	-	-	-	-	-
Liquefaction	-	-	-	-	-	-	-	-	-	-	-
Other Transform. Sector	-	-	-	-	-	-	-	-	-	-	-
ENERGY SECTOR	**1148**	**-**	**-**	**-**	**-**	**-**	**713**	**2330**	**-**	**-**	**-**
Coal Mines	-	-	-	-	-	-	713	-	-	-	-
Oil and Gas Extraction	-	-	-	-	-	-	-	-	-	-	-
Petroleum Refineries	1148	-	-	-	-	-	-	2330	-	-	-
Electr., CHP+Heat Plants	-	-	-	-	-	-	-	-	-	-	-
Pumped Storage (Elec.)	-	-	-	-	-	-	-	-	-	-	-
Other Energy Sector	-	-	-	-	-	-	-	-	-	-	-
Distribution Losses	-	-	-	-	-	-	-	-	-	-	-
FINAL CONSUMPTION	**-**	**4289**	**5005**	**26**	**2202**	**9869**	**35646**	**9889**	**4425**	**266**	**3232**
INDUSTRY SECTOR	**-**	**769**	**-**	**-**	**-**	**-**	**2011**	**7823**	**4425**	**-**	**-**
Iron and Steel	-	-	-	-	-	-	118	671	-	-	-
Chemical and Petrochem.	-	-	-	-	-	-	208	4665	4425	-	-
of which: Feedstocks	-	-	-	-	-	-	-	-	4425	-	-
Non-Ferrous Metals	-	-	-	-	-	-	7	177	-	-	-
Non-Metallic Minerals	-	-	-	-	-	-	193	382	-	-	-
Transport Equipment	-	-	-	-	-	-	-	-	-	-	-
Machinery	-	-	-	-	-	-	193	232	-	-	-
Mining and Quarrying	-	-	-	-	-	-	387	84	-	-	-
Food and Tobacco	-	-	-	-	-	-	428	282	-	-	-
Paper, Pulp and Print	-	-	-	-	-	-	-	-	-	-	-
Wood and Wood Products	-	-	-	-	-	-	-	-	-	-	-
Construction	-	-	-	-	-	-	-	-	-	-	-
Textile and Leather	-	-	-	-	-	-	173	854	-	-	-
Non-specified	-	769	-	-	-	-	304	476	-	-	-
TRANSPORT SECTOR	**-**	**-**	**5005**	**26**	**2202**	**-**	**31690**	**581**	**-**	**-**	**-**
Air	-	-	-	26	2202	-	-	-	-	-	-
Road	-	-	5005	-	-	-	29551	7	-	-	-
Rail	-	-	-	-	-	-	1787	90	-	-	-
Pipeline Transport	-	-	-	-	-	-	-	-	-	-	-
Internal Navigation	-	-	-	-	-	-	343	484	-	-	-
Non-specified	-	-	-	-	-	-	9	-	-	-	-
OTHER SECTORS	**-**	**3520**	**-**	**-**	**-**	**9869**	**1945**	**1485**	**-**	**-**	**-**
Agriculture	-	-	-	-	-	-	270	518	-	-	-
Comm. and Publ. Services	-	-	-	-	-	-	-	-	-	-	-
Residential	-	3520	-	-	-	9869	1675	698	-	-	-
Non-specified	-	-	-	-	-	-	-	269	-	-	-
NON-ENERGY USE	**-**	**-**	**-**	**-**	**-**	**-**	**-**	**-**	**-**	**266**	**3232**
in Industry/Transf./Energy	-	-	-	-	-	-	-	-	-	266	3232
in Transport	-	-	-	-	-	-	-	-	-	-	-
in Other Sectors	-	-	-	-	-	-	-	-	-	-	-

India / Inde : 1997

SUPPLY AND CONSUMPTION APPROVISIONNEMENT ET DEMANDE	Gas / Gaz (TJ)				Comb. Renew. & Waste / En. Re. Comb. & Déchets (TJ)				(GWh)	(TJ)
	Natural Gas Gaz naturel	Gas Works Usines à gaz	Coke Ovens Cokeries	Blast Furnaces Hauts fourneaux	Solid Biomass Biomasse solide	Gas/Liquids from Biomass Gaz/Liquides tirés de biomasse	Municipal Waste Déchets urbains	Industrial Waste Déchets industriels	Electricity Electricité	Heat Chaleur
Production	866983	-	-	118297	8073745	-	-	-	463401	-
From Other Sources	-	-	-	-	-	-	-	-	-	-
Imports	-	-	-	-	-	-	-	-	1675	-
Exports	-	-	-	-	-	-	-	-	-130	-
Intl. Marine Bunkers	-	-	-	-	-	-	-	-	-	-
Stock Changes	-	-	-	-	-	-	-	-	-	-
DOMESTIC SUPPLY	**866983**	-	-	**118297**	**8073745**	-	-	-	**464946**	-
Transfers	-	-	-	-	-	-	-	-	-	-
Statistical Differences	1	-	-	-	-	-	-	-	-46	-
TRANSFORMATION	**261207**	-	-	-	-	-	-	-	-	-
Electricity Plants	261207	-	-	-	-	-	-	-	-	-
CHP Plants	-	-	-	-	-	-	-	-	-	-
Heat Plants	-	-	-	-	-	-	-	-	-	-
Blast Furnaces/Gas Works	-	-	-	-	-	-	-	-	-	-
Coke/Pat. Fuel/BKB Plants	-	-	-	-	-	-	-	-	-	-
Petroleum Refineries	-	-	-	-	-	-	-	-	-	-
Petrochemical Industry	-	-	-	-	-	-	-	-	-	-
Liquefaction	-	-	-	-	-	-	-	-	-	-
Other Transform. Sector	-	-	-	-	-	-	-	-	-	-
ENERGY SECTOR	**118151**	-	-	-	-	-	-	-	**29561**	-
Coal Mines	-	-	-	-	-	-	-	-	-	-
Oil and Gas Extraction	-	-	-	-	-	-	-	-	-	-
Petroleum Refineries	118151	-	-	-	-	-	-	-	-	-
Electr., CHP+Heat Plants	-	-	-	-	-	-	-	-	29561	-
Pumped Storage (Elec.)	-	-	-	-	-	-	-	-	-	-
Other Energy Sector	-	-	-	-	-	-	-	-	-	-
Distribution Losses	-	-	-	-	-	-	-	-	82770	-
FINAL CONSUMPTION	**487626**	-	-	**118297**	**8073745**	-	-	-	**352569**	-
INDUSTRY SECTOR	**467761**	-	-	**118297**	**919794**	-	-	-	**158749**	-
Iron and Steel	-	-	-	118297	-	-	-	-	28853	-
Chemical and Petrochem.	457153	-	-	-	-	-	-	-	35824	-
of which: Feedstocks	31883	-	-	-	-	-	-	-	-	-
Non-Ferrous Metals	-	-	-	-	-	-	-	-	26011	-
Non-Metallic Minerals	-	-	-	-	-	-	-	-	17055	-
Transport Equipment	-	-	-	-	-	-	-	-	5987	-
Machinery	-	-	-	-	-	-	-	-	-	-
Mining and Quarrying	-	-	-	-	-	-	-	-	1096	-
Food and Tobacco	-	-	-	-	-	-	-	-	5336	-
Paper, Pulp and Print	-	-	-	-	-	-	-	-	6669	-
Wood and Wood Products	-	-	-	-	-	-	-	-	-	-
Construction	-	-	-	-	-	-	-	-	-	-
Textile and Leather	-	-	-	-	-	-	-	-	19913	-
Non-specified	10608	-	-	-	919794	-	-	-	12005	-
TRANSPORT SECTOR	-	-	-	-	-	-	-	-	**7073**	-
Air	-	-	-	-	-	-	-	-	-	-
Road	-	-	-	-	-	-	-	-	-	-
Rail	-	-	-	-	-	-	-	-	7073	-
Pipeline Transport	-	-	-	-	-	-	-	-	-	-
Internal Navigation	-	-	-	-	-	-	-	-	-	-
Non-specified	-	-	-	-	-	-	-	-	-	-
OTHER SECTORS	**19865**	-	-	-	**7153951**	-	-	-	**186747**	-
Agriculture	8215	-	-	-	-	-	-	-	95295	-
Comm. and Publ. Services	-	-	-	-	-	-	-	-	27737	-
Residential	11650	-	-	-	7153951	-	-	-	57579	-
Non-specified	-	-	-	-	-	-	-	-	6136	-
NON-ENERGY USE	-	-	-	-	-	-	-	-	-	-
in Industry/Transf./Energy	-	-	-	-	-	-	-	-	-	-
in Transport	-	-	-	-	-	-	-	-	-	-
in Other Sectors	-	-	-	-	-	-	-	-	-	-

India / Inde : 1998

SUPPLY AND CONSUMPTION *APPROVISIONNEMENT ET DEMANDE*	Coal / *Charbon* (1000 tonnes)							Oil / *Pétrole* (1000 tonnes)			
	Coking Coal *Charbon à coke*	Other Bit. Coal *Autres charb. bit.*	Sub-Bit. Coal *Charbon sous-bit.*	Lignite *Lignite*	Peat *Tourbe*	Oven and Gas Coke *Coke de four/gaz*	Pat. Fuel and BKB *Agg./briq. de lignite*	Crude Oil *Pétrole brut*	NGL *LGN*	Feed-stocks *Produits d'aliment.*	Additives *Additifs*
Production	44810	253090	-	23170	-	12768	262	32972	3901	-	-
From Other Sources	-	-		-		-		-	-		
Imports	9640	6000	-	-	-	-	-	38686	-	-	-
Exports	-	-40	-	-	-	-	-	-	-1640	-	-
Intl. Marine Bunkers	-	-		-		-		-	-	-	-
Stock Changes	159	341	-	-	-	-	-	-3160	-	-	-
DOMESTIC SUPPLY	54609	259391	-	23170	-	12768	262	68498	2261	-	-
Transfers	-	-	-	-	-	-	-	-	-1948	-	-
Statistical Differences	-	-	-	-	-	-	-	-	-1	-	-
TRANSFORMATION	31920	207680	-	22470	-	10214	-	68498	-	-	-
Electricity Plants	-	207680	-	22070	-	-	-	-	-	-	-
CHP Plants	-	-	-	-	-	-	-	-	-	-	-
Heat Plants	-	-	-	-	-	-	-	-	-	-	-
Blast Furnaces/Gas Works	-	-	-	-	-	10214	-	-	-	-	-
Coke/Pat. Fuel/BKB Plants	31920	-	-	400	-	-	-	-	-	-	-
Petroleum Refineries	-	-	-	-	-	-	-	68498	-	-	-
Petrochemical Industry	-	-	-	-	-	-	-	-	-	-	-
Liquefaction	-	-	-	-	-	-	-	-	-	-	-
Other Transform. Sector	-	-	-	-	-	-	-	-	-	-	-
ENERGY SECTOR	-	2950	-	-	-	-	-	-	-	-	-
Coal Mines	-	2950	-	-	-	-	-	-	-	-	-
Oil and Gas Extraction	-	-	-	-	-	-	-	-	-	-	-
Petroleum Refineries	-	-	-	-	-	-	-	-	-	-	-
Electr., CHP+Heat Plants	-	-	-	-	-	-	-	-	-	-	-
Pumped Storage (Elec.)	-	-	-	-	-	-	-	-	-	-	-
Other Energy Sector	-	-	-	-	-	-	-	-	-	-	-
Distribution Losses	-	-	-	-	-	-	-	-	-	-	-
FINAL CONSUMPTION	22689	48761	-	700	-	2554	262	-	312	-	-
INDUSTRY SECTOR	22689	48721	-	650	-	2554	-	-	312	-	-
Iron and Steel	17620	-	-	-	-	2554	-	-	-	-	-
Chemical and Petrochem.	-	4110	-	650	-	-	-	-	312	-	-
of which: Feedstocks	-	-	-	-	-	-	-	-	*312*	-	-
Non-Ferrous Metals	-	-	-	-	-	-	-	-	-	-	-
Non-Metallic Minerals	-	13360	-	-	-	-	-	-	-	-	-
Transport Equipment	-	-	-	-	-	-	-	-	-	-	-
Machinery	-	-	-	-	-	-	-	-	-	-	-
Mining and Quarrying	-	-	-	-	-	-	-	-	-	-	-
Food and Tobacco	-	-	-	-	-	-	-	-	-	-	-
Paper, Pulp and Print	-	3500	-	-	-	-	-	-	-	-	-
Wood and Wood Products	-	-	-	-	-	-	-	-	-	-	-
Construction	-	-	-	-	-	-	-	-	-	-	-
Textile and Leather	-	1400	-	-	-	-	-	-	-	-	-
Non-specified	5069	26351	-	-	-	-	-	-	-	-	-
TRANSPORT SECTOR	-	30	-	-	-	-	-	-	-	-	-
Air	-	-	-	-	-	-	-	-	-	-	-
Road	-	-	-	-	-	-	-	-	-	-	-
Rail	-	30	-	-	-	-	-	-	-	-	-
Pipeline Transport	-	-	-	-	-	-	-	-	-	-	-
Internal Navigation	-	-	-	-	-	-	-	-	-	-	-
Non-specified	-	-	-	-	-	-	-	-	-	-	-
OTHER SECTORS	-	10	-	50	-	-	262	-	-	-	-
Agriculture	-	-	-	-	-	-	-	-	-	-	-
Comm. and Publ. Services	-	-	-	-	-	-	-	-	-	-	-
Residential	-	10	-	50	-	-	262	-	-	-	-
Non-specified	-	-	-	-	-	-	-	-	-	-	-
NON-ENERGY USE	-	-	-	-	-	-	-	-	-	-	-
in Industry/Trans./Energy	-	-	-	-	-	-	-	-	-	-	-
in Transport	-	-	-	-	-	-	-	-	-	-	-
in Other Sectors	-	-	-	-	-	-	-	-	-	-	-

India / Inde : 1998

SUPPLY AND CONSUMPTION *APPROVISIONNEMENT ET DEMANDE*	Oil cont. / *Pétrole cont.* (1000 tonnes)										
	Refinery Gas *Gaz de raffinerie*	LPG + Ethane *GPL + éthane*	Motor Gasoline *Essence moteur*	Aviation Gasoline *Essence aviation*	Jet Fuel *Carbu-réacteurs*	Kerosene *Kérosène*	Gas/ Diesel *Gazole*	Heavy Fuel Oil *Fioul lourd*	Naphtha *Naphta*	Petrol. Coke *Coke de pétrole*	Other Prod. *Autres prod.*
Production	1196	2011	5247	-	2027	6328	24984	15002	6006	287	3415
From Other Sources	-	-	-	-	-	-	-	-	-	-	-
Imports	-	1971	36	26	79	5479	11782	771	40	-	-
Exports	-	-22	-	-	-	-	-261	-223	-874	-	-
Intl. Marine Bunkers	-	-	-	-	-	-	-559	-381	-	-	-
Stock Changes	-	-702	228	-	-	-	-	-	-	-	-
DOMESTIC SUPPLY	**1196**	**3258**	**5511**	**26**	**2106**	**11807**	**35946**	**15169**	**5172**	**287**	**3415**
Transfers	-	1948	-	-	-	-	-	-	-	-	-
Statistical Differences	-	1	-	-	-	-	-1	-	-	-	-
TRANSFORMATION	**-**	**-**	**-**	**-**	**-**	**-**	**265**	**1264**	**-**	**-**	**-**
Electricity Plants	-	-	-	-	-	-	265	1264	-	-	-
CHP Plants	-	-	-	-	-	-	-	-	-	-	-
Heat Plants	-	-	-	-	-	-	-	-	-	-	-
Blast Furnaces/Gas Works	-	-	-	-	-	-	-	-	-	-	-
Coke/Pat. Fuel/BKB Plants	-	-	-	-	-	-	-	-	-	-	-
Petroleum Refineries	-	-	-	-	-	-	-	-	-	-	-
Petrochemical Industry	-	-	-	-	-	-	-	-	-	-	-
Liquefaction	-	-	-	-	-	-	-	-	-	-	-
Other Transform. Sector	-	-	-	-	-	-	-	-	-	-	-
ENERGY SECTOR	**1196**	**-**	**-**	**-**	**-**	**-**	**764**	**2427**	**-**	**-**	**-**
Coal Mines	-	-	-	-	-	-	764	-	-	-	-
Oil and Gas Extraction	-	-	-	-	-	-	-	-	-	-	-
Petroleum Refineries	1196	-	-	-	-	-	-	2427	-	-	-
Electr., CHP+Heat Plants	-	-	-	-	-	-	-	-	-	-	-
Pumped Storage (Elec.)	-	-	-	-	-	-	-	-	-	-	-
Other Energy Sector	-	-	-	-	-	-	-	-	-	-	-
Distribution Losses	-	-	-	-	-	-	-	-	-	-	-
FINAL CONSUMPTION	**-**	**5207**	**5511**	**26**	**2106**	**11807**	**34916**	**11478**	**5172**	**287**	**3415**
INDUSTRY SECTOR	**-**	**934**	**-**	**-**	**-**	**-**	**1907**	**9127**	**5172**	**-**	**-**
Iron and Steel	-	-	-	-	-	-	112	782	-	-	-
Chemical and Petrochem.	-	-	-	-	-	-	197	5441	5172	-	-
of which: Feedstocks	-	-	-	-	-	-	-	-	*5172*	-	-
Non-Ferrous Metals	-	-	-	-	-	-	6	207	-	-	-
Non-Metallic Minerals	-	-	-	-	-	-	183	446	-	-	-
Transport Equipment	-	-	-	-	-	-	-	-	-	-	-
Machinery	-	-	-	-	-	-	183	271	-	-	-
Mining and Quarrying	-	-	-	-	-	-	367	98	-	-	-
Food and Tobacco	-	-	-	-	-	-	406	329	-	-	-
Paper, Pulp and Print	-	-	-	-	-	-	-	-	-	-	-
Wood and Wood Products	-	-	-	-	-	-	-	-	-	-	-
Construction	-	-	-	-	-	-	-	-	-	-	-
Textile and Leather	-	-	-	-	-	-	164	997	-	-	-
Non-specified	-	934	-	-	-	-	289	556	-	-	-
TRANSPORT SECTOR	**-**	**-**	**5511**	**26**	**2106**	**-**	**31100**	**662**	**-**	**-**	**-**
Air	-	-	-	26	2106	-	-	-	-	-	-
Road	-	-	5511	-	-	-	29000	8	-	-	-
Rail	-	-	-	-	-	-	1754	103	-	-	-
Pipeline Transport	-	-	-	-	-	-	-	-	-	-	-
Internal Navigation	-	-	-	-	-	-	337	551	-	-	-
Non-specified	-	-	-	-	-	-	9	-	-	-	-
OTHER SECTORS	**-**	**4273**	**-**	**-**	**-**	**11807**	**1909**	**1689**	**-**	**-**	**-**
Agriculture	-	-	-	-	-	-	265	589	-	-	-
Comm. and Publ. Services	-	-	-	-	-	-	-	-	-	-	-
Residential	-	4273	-	-	-	11807	1644	794	-	-	-
Non-specified	-	-	-	-	-	-	-	306	-	-	-
NON-ENERGY USE	**-**	**-**	**-**	**-**	**-**	**-**	**-**	**-**	**-**	**287**	**3415**
in Industry/Transf./Energy	-	-	-	-	-	-	-	-	-	287	3415
in Transport	-	-	-	-	-	-	-	-	-	-	-
in Other Sectors	-	-	-	-	-	-	-	-	-	-	-

India / Inde : 1998

SUPPLY AND CONSUMPTION / APPROVISIONNEMENT ET DEMANDE	Gas / Gaz (TJ)				Comb. Renew. & Waste / En. Re. Comb. & Déchets (TJ)				(GWh)	(TJ)
	Natural Gas / Gaz naturel	Gas Works / Usines à gaz	Coke Ovens / Cokeries	Blast Furnaces / Hauts fourneaux	Solid Biomass / Biomasse solide	Gas/Liquids from Biomass / Gaz/Liquides tirés de biomasse	Municipal Waste / Déchets urbains	Industrial Waste / Déchets industriels	Electricity / Electricité	Heat / Chaleur
Production	907994	-	-	111191	8177729	-	-	-	493968	-
From Other Sources	-	-	-	-	-	-	-	-	-	-
Imports	-	-	-	-	-	-	-	-	1675	-
Exports	-	-	-	-	-	-	-	-	-130	-
Intl. Marine Bunkers	-	-	-	-	-	-	-	-	-	-
Stock Changes	-	-	-	-	-	-	-	-	-	-
DOMESTIC SUPPLY	907994	-	-	111191	8177729	-	-	-	495513	-
Transfers	-	-	-	-	-	-	-	-	-	-
Statistical Differences	-1	-	-	-	-	-	-	-	-	-
TRANSFORMATION	273562	-	-	-	-	-	-	-	-	-
Electricity Plants	273562	-	-	-	-	-	-	-	-	-
CHP Plants	-	-	-	-	-	-	-	-	-	-
Heat Plants	-	-	-	-	-	-	-	-	-	-
Blast Furnaces/Gas Works	-	-	-	-	-	-	-	-	-	-
Coke/Pat. Fuel/BKB Plants	-	-	-	-	-	-	-	-	-	-
Petroleum Refineries	-	-	-	-	-	-	-	-	-	-
Petrochemical Industry	-	-	-	-	-	-	-	-	-	-
Liquefaction	-	-	-	-	-	-	-	-	-	-
Other Transform. Sector	-	-	-	-	-	-	-	-	-	-
ENERGY SECTOR	123740	-	-	-	-	-	-	-	31504	-
Coal Mines	-	-	-	-	-	-	-	-	-	-
Oil and Gas Extraction	-	-	-	-	-	-	-	-	-	-
Petroleum Refineries	123740	-	-	-	-	-	-	-	-	-
Electr., CHP+Heat Plants	-	-	-	-	-	-	-	-	31504	-
Pumped Storage (Elec.)	-	-	-	-	-	-	-	-	-	-
Other Energy Sector	-	-	-	-	-	-	-	-	-	-
Distribution Losses	-	-	-	-	-	-	-	-	88211	-
FINAL CONSUMPTION	510691	-	-	111191	8177729	-	-	-	375798	-
INDUSTRY SECTOR	489886	-	-	111191	931640	-	-	-	169235	-
Iron and Steel	-	-	-	111191	-	-	-	-	30750	-
Chemical and Petrochem.	478777	-	-	-	-	-	-	-	38179	-
of which: Feedstocks	33391	-	-	-	-	-	-	-	-	-
Non-Ferrous Metals	-	-	-	-	-	-	-	-	27721	-
Non-Metallic Minerals	-	-	-	-	-	-	-	-	18176	-
Transport Equipment	-	-	-	-	-	-	-	-	6380	-
Machinery	-	-	-	-	-	-	-	-	-	-
Mining and Quarrying	-	-	-	-	-	-	-	-	1168	-
Food and Tobacco	-	-	-	-	-	-	-	-	5686	-
Paper, Pulp and Print	-	-	-	-	-	-	-	-	7108	-
Wood and Wood Products	-	-	-	-	-	-	-	-	-	-
Construction	-	-	-	-	-	-	-	-	-	-
Textile and Leather	-	-	-	-	-	-	-	-	21222	-
Non-specified	11109	-	-	-	931640	-	-	-	12845	-
TRANSPORT SECTOR	-	-	-	-	-	-	-	-	7538	-
Air	-	-	-	-	-	-	-	-	-	-
Road	-	-	-	-	-	-	-	-	-	-
Rail	-	-	-	-	-	-	-	-	7538	-
Pipeline Transport	-	-	-	-	-	-	-	-	-	-
Internal Navigation	-	-	-	-	-	-	-	-	-	-
Non-specified	-	-	-	-	-	-	-	-	-	-
OTHER SECTORS	20805	-	-	-	7246089	-	-	-	199025	-
Agriculture	8604	-	-	-	-	-	-	-	101560	-
Comm. and Publ. Services	-	-	-	-	-	-	-	-	29561	-
Residential	12201	-	-	-	7246089	-	-	-	61365	-
Non-specified	-	-	-	-	-	-	-	-	6539	-
NON-ENERGY USE	-	-	-	-	-	-	-	-	-	-
in Industry/Transf./Energy	-	-	-	-	-	-	-	-	-	-
in Transport	-	-	-	-	-	-	-	-	-	-
in Other Sectors	-	-	-	-	-	-	-	-	-	-

Indonesia / Indonésie : 1997

SUPPLY AND CONSUMPTION / APPROVISIONNEMENT ET DEMANDE	Coal / *Charbon* (1000 tonnes)							Oil / *Pétrole* (1000 tonnes)			
	Coking Coal / *Charbon à coke*	Other Bit. Coal / *Autres charb. bit.*	Sub-Bit. Coal / *Charbon sous-bit.*	Lignite / *Lignite*	Peat / *Tourbe*	Oven and Gas Coke / *Coke de four/gaz*	Pat. Fuel and BKB / *Agg./briq. de lignite*	Crude Oil / *Pétrole brut*	NGL / *LGN*	Feed-stocks / *Produits d'aliment.*	Additives / *Additifs*
Production	-	54877	-	-	-	-	5	65807	2358	-	-
From Other Sources	-	-	-	-	-	-	-	-	-	-	-
Imports	-	385	-	-	-	-	6	8544	-	-	-
Exports	-	-41727	-	-	-	-	-	-35597	-	-	-
Intl. Marine Bunkers	-	-	-	-	-	-	-	-	-	-	-
Stock Changes	-	-136	-	-	-	-	-	-	-	-	-
DOMESTIC SUPPLY	-	13399	-	-	-	-	11	38754	2358	-	-
Transfers	-	-	-	-	-	-	-	-	-2358	-	-
Statistical Differences	-	-	-	-	-	-	-	7256	-	-	-
TRANSFORMATION	-	10037	-	-	-	-	-	46010	-	-	-
Electricity Plants	-	10010	-	-	-	-	-	-	-	-	-
CHP Plants	-	-	-	-	-	-	-	-	-	-	-
Heat Plants	-	-	-	-	-	-	-	-	-	-	-
Blast Furnaces/Gas Works	-	-	-	-	-	-	-	-	-	-	-
Coke/Pat. Fuel/BKB Plants	-	27	-	-	-	-	-	-	-	-	-
Petroleum Refineries	-	-	-	-	-	-	-	46010	-	-	-
Petrochemical Industry	-	-	-	-	-	-	-	-	-	-	-
Liquefaction	-	-	-	-	-	-	-	-	-	-	-
Other Transform. Sector	-	-	-	-	-	-	-	-	-	-	-
ENERGY SECTOR	-	-	-	-	-	-	-	-	-	-	-
Coal Mines	-	-	-	-	-	-	-	-	-	-	-
Oil and Gas Extraction	-	-	-	-	-	-	-	-	-	-	-
Petroleum Refineries	-	-	-	-	-	-	-	-	-	-	-
Electr., CHP+Heat Plants	-	-	-	-	-	-	-	-	-	-	-
Pumped Storage (Elec.)	-	-	-	-	-	-	-	-	-	-	-
Other Energy Sector	-	-	-	-	-	-	-	-	-	-	-
Distribution Losses	-	-	-	-	-	-	-	-	-	-	-
FINAL CONSUMPTION	-	3362	-	-	-	-	11	-	-	-	-
INDUSTRY SECTOR	-	2049	-	-	-	-	11	-	-	-	-
Iron and Steel	-	131	-	-	-	-	-	-	-	-	-
Chemical and Petrochem.	-	-	-	-	-	-	-	-	-	-	-
of which: Feedstocks	-	-	-	-	-	-	-	-	-	-	-
Non-Ferrous Metals	-	-	-	-	-	-	-	-	-	-	-
Non-Metallic Minerals	-	1333	-	-	-	-	-	-	-	-	-
Transport Equipment	-	-	-	-	-	-	-	-	-	-	-
Machinery	-	-	-	-	-	-	-	-	-	-	-
Mining and Quarrying	-	-	-	-	-	-	-	-	-	-	-
Food and Tobacco	-	-	-	-	-	-	-	-	-	-	-
Paper, Pulp and Print	-	-	-	-	-	-	-	-	-	-	-
Wood and Wood Products	-	-	-	-	-	-	-	-	-	-	-
Construction	-	-	-	-	-	-	-	-	-	-	-
Textile and Leather	-	-	-	-	-	-	-	-	-	-	-
Non-specified	-	585	-	-	-	-	11	-	-	-	-
TRANSPORT SECTOR	-	-	-	-	-	-	-	-	-	-	-
Air	-	-	-	-	-	-	-	-	-	-	-
Road	-	-	-	-	-	-	-	-	-	-	-
Rail	-	-	-	-	-	-	-	-	-	-	-
Pipeline Transport	-	-	-	-	-	-	-	-	-	-	-
Internal Navigation	-	-	-	-	-	-	-	-	-	-	-
Non-specified	-	-	-	-	-	-	-	-	-	-	-
OTHER SECTORS	-	1313	-	-	-	-	-	-	-	-	-
Agriculture	-	-	-	-	-	-	-	-	-	-	-
Comm. and Publ. Services	-	-	-	-	-	-	-	-	-	-	-
Residential	-	1313	-	-	-	-	-	-	-	-	-
Non-specified	-	-	-	-	-	-	-	-	-	-	-
NON-ENERGY USE	-	-	-	-	-	-	-	-	-	-	-
in Industry/Trans./Energy	-	-	-	-	-	-	-	-	-	-	-
in Transport	-	-	-	-	-	-	-	-	-	-	-
in Other Sectors	-	-	-	-	-	-	-	-	-	-	-

Indonesia / Indonésie : 1997

SUPPLY AND CONSUMPTION / APPROVISIONNEMENT ET DEMANDE	Oil cont. / *Pétrole cont.* (1000 tonnes)										
	Refinery Gas / *Gaz de raffinerie*	LPG + Ethane / *GPL + éthane*	Motor Gasoline / *Essence moteur*	Aviation Gasoline / *Essence aviation*	Jet Fuel / *Carbu- réacteurs*	Kerosene / *Kérosène*	Gas/ Diesel / *Gazole*	Heavy Fuel Oil / *Fioul lourd*	Naphtha / *Naphta*	Petrol. Coke / *Coke de pétrole*	Other Prod. / *Autres prod.*
Production	581	685	8277	7	969	6179	12275	11170	955	-	1341
From Other Sources	-	-	-	-	-	-	-	-	-	-	-
Imports	-	-	565	-	1138	1335	7515	2023	-	-	-
Exports	-	-2133	-314	-	-731	-	-15	-8806	-772	-	-271
Intl. Marine Bunkers	-	-	-	-	-	-	-19	-296	-	-	-
Stock Changes	-	-	-245	-2	-	-	-	-	-	-	-
DOMESTIC SUPPLY	**581**	**-1448**	**8283**	**5**	**1376**	**7514**	**19756**	**4091**	**183**	**-**	**1070**
Transfers	-	2358	-	-	-	-	-	-	-	-	-
Statistical Differences	-	-81	-	-	-448	599	-66	1961	-	-	-18
TRANSFORMATION	**-**	**-**	**-**	**-**	**-**	**-**	**3433**	**1735**	**-**	**-**	**-**
Electricity Plants	-	-	-	-	-	-	3433	1735	-	-	-
CHP Plants	-	-	-	-	-	-	-	-	-	-	-
Heat Plants	-	-	-	-	-	-	-	-	-	-	-
Blast Furnaces/Gas Works	-	-	-	-	-	-	-	-	-	-	-
Coke/Pat. Fuel/BKB Plants	-	-	-	-	-	-	-	-	-	-	-
Petroleum Refineries	-	-	-	-	-	-	-	-	-	-	-
Petrochemical Industry	-	-	-	-	-	-	-	-	-	-	-
Liquefaction	-	-	-	-	-	-	-	-	-	-	-
Other Transform. Sector	-	-	-	-	-	-	-	-	-	-	-
ENERGY SECTOR	**581**	**-**	**-**	**-**	**-**	**-**	**55**	**1222**	**-**	**-**	**-**
Coal Mines	-	-	-	-	-	-	-	-	-	-	-
Oil and Gas Extraction	-	-	-	-	-	-	-	-	-	-	-
Petroleum Refineries	581	-	-	-	-	-	55	1222	-	-	-
Electr., CHP+Heat Plants	-	-	-	-	-	-	-	-	-	-	-
Pumped Storage (Elec.)	-	-	-	-	-	-	-	-	-	-	-
Other Energy Sector	-	-	-	-	-	-	-	-	-	-	-
Distribution Losses	-	-	-	-	-	-	-	-	-	-	-
FINAL CONSUMPTION	**-**	**829**	**8283**	**5**	**928**	**8113**	**16202**	**3095**	**183**	**-**	**1052**
INDUSTRY SECTOR	**-**	**360**	**-**	**-**	**-**	**73**	**4689**	**2779**	**183**	**-**	**-**
Iron and Steel	-	-	-	-	-	-	208	767	-	-	-
Chemical and Petrochem.	-	-	-	-	-	-	434	474	183	-	-
of which: Feedstocks	-	-	-	-	-	-	-	-	-	-	-
Non-Ferrous Metals	-	-	-	-	-	-	-	-	-	-	-
Non-Metallic Minerals	-	-	-	-	-	-	633	594	-	-	-
Transport Equipment	-	-	-	-	-	-	-	-	-	-	-
Machinery	-	-	-	-	-	-	69	-	-	-	-
Mining and Quarrying	-	-	-	-	-	-	948	134	-	-	-
Food and Tobacco	-	-	-	-	-	-	545	186	-	-	-
Paper, Pulp and Print	-	-	-	-	-	-	-	-	-	-	-
Wood and Wood Products	-	-	-	-	-	-	-	-	-	-	-
Construction	-	-	-	-	-	-	338	-	-	-	-
Textile and Leather	-	-	-	-	-	-	1049	193	-	-	-
Non-specified	-	360	-	-	-	73	465	431	-	-	-
TRANSPORT SECTOR	**-**	**-**	**8283**	**5**	**928**	**-**	**9739**	**219**	**-**	**-**	**-**
Air	-	-	-	5	928	-	-	-	-	-	-
Road	-	-	8283	-	-	-	8288	-	-	-	-
Rail	-	-	-	-	-	-	-	-	-	-	-
Pipeline Transport	-	-	-	-	-	-	-	-	-	-	-
Internal Navigation	-	-	-	-	-	-	1451	219	-	-	-
Non-specified	-	-	-	-	-	-	-	-	-	-	-
OTHER SECTORS	**-**	**469**	**-**	**-**	**-**	**8040**	**1774**	**97**	**-**	**-**	**-**
Agriculture	-	-	-	-	-	-	1481	97	-	-	-
Comm. and Publ. Services	-	-	-	-	-	-	293	-	-	-	-
Residential	-	469	-	-	-	8040	-	-	-	-	-
Non-specified	-	-	-	-	-	-	-	-	-	-	-
NON-ENERGY USE	**-**	**-**	**-**	**-**	**-**	**-**	**-**	**-**	**-**	**-**	**1052**
in Industry/Transf./Energy	-	-	-	-	-	-	-	-	-	-	1052
in Transport	-	-	-	-	-	-	-	-	-	-	-
in Other Sectors	-	-	-	-	-	-	-	-	-	-	-

Indonesia / Indonésie : 1997

SUPPLY AND CONSUMPTION / APPROVISIONNEMENT ET DEMANDE	Gas / Gaz (TJ)				Comb. Renew. & Waste / En. Re. Comb. & Déchets (TJ)				(GWh)	(TJ)
	Natural Gas / Gaz naturel	Gas Works / Usines à gaz	Coke Ovens / Cokeries	Blast Furnaces / Hauts fourneaux	Solid Biomass / Biomasse solide	Gas/Liquids from Biomass / Gaz/Liquides tirés de biomasse	Municipal Waste / Déchets urbains	Industrial Waste / Déchets industriels	Electricity / Electricité	Heat / Chaleur
Production	2664687	22000	-	-	1905645	-	-	-	76620	-
From Other Sources	-									
Imports	-	-	-	-	-	-	-	-	-	-
Exports	-1464003	-	-	-	-4365	-	-	-	-	-
Intl. Marine Bunkers	-	-	-	-	-					
Stock Changes	-	-	-	-	-					
DOMESTIC SUPPLY	**1200684**	**22000**	**-**	**-**	**1901280**	**-**	**-**	**-**	**76620**	**-**
Transfers	-	-	-	-	-	-	-	-	-	-
Statistical Differences	-117	-	-	-	-	-	-	-	-3168	-
TRANSFORMATION	**276523**	**-**	**-**	**-**	**3534**	**-**	**-**	**-**	**-**	**-**
Electricity Plants	276523	-	-	-	-	-	-	-	-	-
CHP Plants	-	-	-	-	-	-	-	-	-	-
Heat Plants	-	-	-	-	-	-	-	-	-	-
Blast Furnaces/Gas Works	-	-	-	-	-	-	-	-	-	-
Coke/Pat. Fuel/BKB Plants	-	-	-	-	-	-	-	-	-	-
Petroleum Refineries	-	-	-	-	-	-	-	-	-	-
Petrochemical Industry	-	-	-	-	-	-	-	-	-	-
Liquefaction	-	-	-	-	-	-	-	-	-	-
Other Transform. Sector	-	-	-	-	3534	-	-	-	-	-
ENERGY SECTOR	**550248**	**-**	**-**	**-**	**-**	**-**	**-**	**-**	**246**	**-**
Coal Mines	-	-	-	-	-	-	-	-	-	-
Oil and Gas Extraction	504989	-	-	-	-	-	-	-	-	-
Petroleum Refineries	45259	-	-	-	-	-	-	-	-	-
Electr., CHP+Heat Plants	-	-	-	-	-	-	-	-	246	-
Pumped Storage (Elec.)	-	-	-	-	-	-	-	-	-	-
Other Energy Sector	-	-	-	-	-	-	-	-	-	-
Distribution Losses	-	-	-	-	-	-	-	-	8894	-
FINAL CONSUMPTION	**373796**	**22000**	**-**	**-**	**1897746**	**-**	**-**	**-**	**64312**	**-**
INDUSTRY SECTOR	**313760**	**-**	**-**	**-**	**-**	**-**	**-**	**-**	**30769**	**-**
Iron and Steel	41402	-	-	-	-	-	-	-	-	-
Chemical and Petrochem.	260046	-	-	-	-	-	-	-	-	-
of which: Feedstocks	260046	-	-	-	-	-	-	-	-	-
Non-Ferrous Metals	-	-	-	-	-	-	-	-	-	-
Non-Metallic Minerals	5229	-	-	-	-	-	-	-	-	-
Transport Equipment	-	-	-	-	-	-	-	-	-	-
Machinery	-	-	-	-	-	-	-	-	-	-
Mining and Quarrying	-	-	-	-	-	-	-	-	-	-
Food and Tobacco	-	-	-	-	-	-	-	-	-	-
Paper, Pulp and Print	-	-	-	-	-	-	-	-	-	-
Wood and Wood Products	-	-	-	-	-	-	-	-	-	-
Construction	-	-	-	-	-	-	-	-	-	-
Textile and Leather	-	-	-	-	-	-	-	-	-	-
Non-specified	7083	-	-	-	-	-	-	-	30769	-
TRANSPORT SECTOR	**-**	**-**	**-**	**-**	**-**	**-**	**-**	**-**	**-**	**-**
Air	-	-	-	-	-	-	-	-	-	-
Road	-	-	-	-	-	-	-	-	-	-
Rail	-	-	-	-	-	-	-	-	-	-
Pipeline Transport	-	-	-	-	-	-	-	-	-	-
Internal Navigation	-	-	-	-	-	-	-	-	-	-
Non-specified	-	-	-	-	-	-	-	-	-	-
OTHER SECTORS	**60036**	**22000**	**-**	**-**	**1897746**	**-**	**-**	**-**	**33543**	**-**
Agriculture	-	-	-	-	-	-	-	-	-	-
Comm. and Publ. Services	-	11000	-	-	-	-	-	-	7250	-
Residential	60036	11000	-	-	1897746	-	-	-	22739	-
Non-specified	-	-	-	-	-	-	-	-	3554	-
NON-ENERGY USE	**-**	**-**	**-**	**-**	**-**	**-**	**-**	**-**	**-**	**-**
in Industry/Transf./Energy	-	-	-	-	-	-	-	-	-	-
in Transport	-	-	-	-	-	-	-	-	-	-
in Other Sectors	-	-	-	-	-	-	-	-	-	-

Indonesia / Indonésie : 1998

SUPPLY AND CONSUMPTION / APPROVISIONNEMENT ET DEMANDE	Coal / Charbon (1000 tonnes)							Oil / Pétrole (1000 tonnes)			
	Coking Coal / Charbon à coke	Other Bit. Coal / Autres charb. bit.	Sub-Bit. Coal / Charbon sous-bit.	Lignite / Lignite	Peat / Tourbe	Oven and Gas Coke / Coke de four/gaz	Pat. Fuel and BKB / Agg./briq. de lignite	Crude Oil / Pétrole brut	NGL / LGN	Feed-stocks / Produits d'aliment.	Additives / Additifs
Production	-	61206	-	-	-	-	4	65232	1969	-	-
From Other Sources	-	-	-	-	-	-	-	-	-	-	-
Imports	-	385	-	-	-	-	11	9073	-	-	-
Exports	-	-46933	-	-	-	-	-	-33593	-	-	-
Intl. Marine Bunkers	-	-	-	-	-	-	-	-	-	-	-
Stock Changes	-	706	-	-	-	-	-	-	-	-	-
DOMESTIC SUPPLY	-	15364	-	-	-	-	15	40712	1969	-	-
Transfers	-	-	-	-	-	-	-	-	-1969	-	-
Statistical Differences	-	-	-	-	-	-	-	4158	-	-	-
TRANSFORMATION	-	10650	-	-	-	-	-	44870	-	-	-
Electricity Plants	-	10623	-	-	-	-	-	-	-	-	-
CHP Plants	-	-	-	-	-	-	-	-	-	-	-
Heat Plants	-	-	-	-	-	-	-	-	-	-	-
Blast Furnaces/Gas Works	-	-	-	-	-	-	-	-	-	-	-
Coke/Pat. Fuel/BKB Plants	-	27	-	-	-	-	-	-	-	-	-
Petroleum Refineries	-	-	-	-	-	-	-	44870	-	-	-
Petrochemical Industry	-	-	-	-	-	-	-	-	-	-	-
Liquefaction	-	-	-	-	-	-	-	-	-	-	-
Other Transform. Sector	-	-	-	-	-	-	-	-	-	-	-
ENERGY SECTOR	-	-	-	-	-	-	-	-	-	-	-
Coal Mines	-	-	-	-	-	-	-	-	-	-	-
Oil and Gas Extraction	-	-	-	-	-	-	-	-	-	-	-
Petroleum Refineries	-	-	-	-	-	-	-	-	-	-	-
Electr., CHP+Heat Plants	-	-	-	-	-	-	-	-	-	-	-
Pumped Storage (Elec.)	-	-	-	-	-	-	-	-	-	-	-
Other Energy Sector	-	-	-	-	-	-	-	-	-	-	-
Distribution Losses	-	-	-	-	-	-	-	-	-	-	-
FINAL CONSUMPTION	-	4714	-	-	-	-	15	-	-	-	-
INDUSTRY SECTOR	-	2113	-	-	-	-	15	-	-	-	-
Iron and Steel	-	145	-	-	-	-	-	-	-	-	-
Chemical and Petrochem.	-	-	-	-	-	-	-	-	-	-	-
of which: Feedstocks	-	-	-	-	-	-	-	-	-	-	-
Non-Ferrous Metals	-	-	-	-	-	-	-	-	-	-	-
Non-Metallic Minerals	-	1265	-	-	-	-	-	-	-	-	-
Transport Equipment	-	-	-	-	-	-	-	-	-	-	-
Machinery	-	-	-	-	-	-	-	-	-	-	-
Mining and Quarrying	-	-	-	-	-	-	-	-	-	-	-
Food and Tobacco	-	-	-	-	-	-	-	-	-	-	-
Paper, Pulp and Print	-	-	-	-	-	-	-	-	-	-	-
Wood and Wood Products	-	-	-	-	-	-	-	-	-	-	-
Construction	-	-	-	-	-	-	-	-	-	-	-
Textile and Leather	-	-	-	-	-	-	-	-	-	-	-
Non-specified	-	703	-	-	-	-	15	-	-	-	-
TRANSPORT SECTOR	-	-	-	-	-	-	-	-	-	-	-
Air	-	-	-	-	-	-	-	-	-	-	-
Road	-	-	-	-	-	-	-	-	-	-	-
Rail	-	-	-	-	-	-	-	-	-	-	-
Pipeline Transport	-	-	-	-	-	-	-	-	-	-	-
Internal Navigation	-	-	-	-	-	-	-	-	-	-	-
Non-specified	-	-	-	-	-	-	-	-	-	-	-
OTHER SECTORS	-	2601	-	-	-	-	-	-	-	-	-
Agriculture	-	-	-	-	-	-	-	-	-	-	-
Comm. and Publ. Services	-	-	-	-	-	-	-	-	-	-	-
Residential	-	2601	-	-	-	-	-	-	-	-	-
Non-specified	-	-	-	-	-	-	-	-	-	-	-
NON-ENERGY USE	-	-	-	-	-	-	-	-	-	-	-
in Industry/Trans./Energy	-	-	-	-	-	-	-	-	-	-	-
in Transport	-	-	-	-	-	-	-	-	-	-	-
in Other Sectors	-	-	-	-	-	-	-	-	-	-	-

Indonesia / Indonésie : 1998

SUPPLY AND CONSUMPTION / APPROVISIONNEMENT ET DEMANDE	Oil cont. / Pétrole cont. (1000 tonnes)										
	Refinery Gas / Gaz de raffinerie	LPG + Ethane / GPL + éthane	Motor Gasoline / Essence moteur	Aviation Gasoline / Essence aviation	Jet Fuel / Carbu-réacteurs	Kerosene / Kérosène	Gas/ Diesel / Gazole	Heavy Fuel Oil / Fioul lourd	Naphtha / Naphta	Petrol. Coke / Coke de pétrole	Other Prod. / Autres prod.
Production	697	616	8029	3	907	6885	13423	12290	792	-	1087
From Other Sources	-	-	-	-	-	-	-	-	-	-	-
Imports	-	-	431	-	203	865	4391	1392	-	-	-
Exports	-	-1756	-153	-	-458	-	-53	-7295	-704	-	-404
Intl. Marine Bunkers	-	-	-	-	-	-	-17	-325	-	-	-
Stock Changes	-	-	182	1	-	-	-	-	-8	-	-
DOMESTIC SUPPLY	**697**	**-1140**	**8489**	**4**	**652**	**7750**	**17744**	**6062**	**80**	**-**	**683**
Transfers	-	1969	-	-	-	-	-	-	-	-	-
Statistical Differences	-	-	-	-	-19	507	-	505	-	-	-
TRANSFORMATION	**-**	**-**	**-**	**-**	**-**	**-**	**2566**	**1267**	**-**	**-**	**-**
Electricity Plants	-	-	-	-	-	-	2566	1267	-	-	-
CHP Plants	-	-	-	-	-	-	-	-	-	-	-
Heat Plants	-	-	-	-	-	-	-	-	-	-	-
Blast Furnaces/Gas Works	-	-	-	-	-	-	-	-	-	-	-
Coke/Pat. Fuel/BKB Plants	-	-	-	-	-	-	-	-	-	-	-
Petroleum Refineries	-	-	-	-	-	-	-	-	-	-	-
Petrochemical Industry	-	-	-	-	-	-	-	-	-	-	-
Liquefaction	-	-	-	-	-	-	-	-	-	-	-
Other Transform. Sector	-	-	-	-	-	-	-	-	-	-	-
ENERGY SECTOR	**697**	**-**	**-**	**-**	**-**	**-**	**55**	**1620**	**-**	**-**	**-**
Coal Mines	-	-	-	-	-	-	-	-	-	-	-
Oil and Gas Extraction	-	-	-	-	-	-	-	-	-	-	-
Petroleum Refineries	697	-	-	-	-	-	55	1620	-	-	-
Electr., CHP+Heat Plants	-	-	-	-	-	-	-	-	-	-	-
Pumped Storage (Elec.)	-	-	-	-	-	-	-	-	-	-	-
Other Energy Sector	-	-	-	-	-	-	-	-	-	-	-
Distribution Losses	-	-	-	-	-	-	-	-	-	-	-
FINAL CONSUMPTION	**-**	**829**	**8489**	**4**	**633**	**8257**	**15123**	**3680**	**80**	**-**	**683**
INDUSTRY SECTOR	**-**	**360**	**-**	**-**	**-**	**73**	**3965**	**3189**	**80**	**-**	**-**
Iron and Steel	-	-	-	-	-	-	176	880	-	-	-
Chemical and Petrochem.	-	-	-	-	-	-	367	544	80	-	-
of which: Feedstocks	-	-	-	-	-	-	-	-	-	-	-
Non-Ferrous Metals	-	-	-	-	-	-	-	-	-	-	-
Non-Metallic Minerals	-	-	-	-	-	-	535	681	-	-	-
Transport Equipment	-	-	-	-	-	-	-	-	-	-	-
Machinery	-	-	-	-	-	-	59	-	-	-	-
Mining and Quarrying	-	-	-	-	-	-	802	154	-	-	-
Food and Tobacco	-	-	-	-	-	-	461	213	-	-	-
Paper, Pulp and Print	-	-	-	-	-	-	-	-	-	-	-
Wood and Wood Products	-	-	-	-	-	-	-	-	-	-	-
Construction	-	-	-	-	-	-	286	-	-	-	-
Textile and Leather	-	-	-	-	-	-	887	221	-	-	-
Non-specified	-	360	-	-	-	73	392	496	-	-	-
TRANSPORT SECTOR	**-**	**-**	**8489**	**4**	**633**	**-**	**9218**	**384**	**-**	**-**	**-**
Air	-	-	-	4	633	-	-	-	-	-	-
Road	-	-	8489	-	-	-	7845	-	-	-	-
Rail	-	-	-	-	-	-	-	-	-	-	-
Pipeline Transport	-	-	-	-	-	-	-	-	-	-	-
Internal Navigation	-	-	-	-	-	-	1373	384	-	-	-
Non-specified	-	-	-	-	-	-	-	-	-	-	-
OTHER SECTORS	**-**	**469**	**-**	**-**	**-**	**8184**	**1940**	**107**	**-**	**-**	**-**
Agriculture	-	-	-	-	-	-	1620	107	-	-	-
Comm. and Publ. Services	-	-	-	-	-	-	320	-	-	-	-
Residential	-	469	-	-	-	8184	-	-	-	-	-
Non-specified	-	-	-	-	-	-	-	-	-	-	-
NON-ENERGY USE	**-**	**-**	**-**	**-**	**-**	**-**	**-**	**-**	**-**	**-**	**683**
in Industry/Transf./Energy	-	-	-	-	-	-	-	-	-	-	683
in Transport	-	-	-	-	-	-	-	-	-	-	-
in Other Sectors	-	-	-	-	-	-	-	-	-	-	-

Indonesia / Indonésie : 1998

SUPPLY AND CONSUMPTION APPROVISIONNEMENT ET DEMANDE	Gas / Gaz (TJ)				Comb. Renew. & Waste / En. Re. Comb. & Déchets (TJ)				(GWh)	(TJ)
	Natural Gas Gaz naturel	Gas Works Usines à gaz	Coke Ovens Cokeries	Blast Furnaces Hauts fourneaux	Solid Biomass Biomasse solide	Gas/Liquids from Biomass Gaz/Liquides tirés de biomasse	Municipal Waste Déchets urbains	Industrial Waste Déchets industriels	Electricity Electricité	Heat Chaleur
Production	2643585	22000	-	-	1905645	-	-	-	77903	-
From Other Sources	-	-	-	-	-	-	-	-	-	-
Imports	-	-	-	-	-	-	-	-	-	-
Exports	-1472856	-	-	-	-4365	-	-	-	-	-
Intl. Marine Bunkers	-	-	-	-	-	-	-	-	-	-
Stock Changes	-	-	-	-	-	-	-	-	-	-
DOMESTIC SUPPLY	**1170729**	**22000**	-	-	**1901280**	-	-	-	**77903**	-
Transfers	-	-	-	-	-	-	-	-	-	-
Statistical Differences	1150	-	-	-	-	-	-	-	-642	-
TRANSFORMATION	**259293**	-	-	-	**2830**	-	-	-	-	-
Electricity Plants	259293	-	-	-	-	-	-	-	-	-
CHP Plants	-	-	-	-	-	-	-	-	-	-
Heat Plants	-	-	-	-	-	-	-	-	-	-
Blast Furnaces/Gas Works	-	-	-	-	-	-	-	-	-	-
Coke/Pat. Fuel/BKB Plants	-	-	-	-	-	-	-	-	-	-
Petroleum Refineries	-	-	-	-	-	-	-	-	-	-
Petrochemical Industry	-	-	-	-	-	-	-	-	-	-
Liquefaction	-	-	-	-	-	-	-	-	-	-
Other Transform. Sector	-	-	-	-	2830	-	-	-	-	-
ENERGY SECTOR	**557232**	-	-	-	-	-	-	-	**2783**	-
Coal Mines	-	-	-	-	-	-	-	-	-	-
Oil and Gas Extraction	501344	-	-	-	-	-	-	-	-	-
Petroleum Refineries	55888	-	-	-	-	-	-	-	-	-
Electr., CHP+Heat Plants	-	-	-	-	-	-	-	-	2783	-
Pumped Storage (Elec.)	-	-	-	-	-	-	-	-	-	-
Other Energy Sector	-	-	-	-	-	-	-	-	-	-
Distribution Losses	-	-	-	-	-	-	-	-	9217	-
FINAL CONSUMPTION	**355354**	**22000**	-	-	**1898450**	-	-	-	**65261**	-
INDUSTRY SECTOR	**312264**	-	-	-	-	-	-	-	**27985**	-
Iron and Steel	37736	-	-	-	-	-	-	-	-	-
Chemical and Petrochem.	254323	-	-	-	-	-	-	-	-	-
of which: Feedstocks	254323	-	-	-	-	-	-	-	-	-
Non-Ferrous Metals	-	-	-	-	-	-	-	-	-	-
Non-Metallic Minerals	1709	-	-	-	-	-	-	-	-	-
Transport Equipment	-	-	-	-	-	-	-	-	-	-
Machinery	-	-	-	-	-	-	-	-	-	-
Mining and Quarrying	-	-	-	-	-	-	-	-	-	-
Food and Tobacco	-	-	-	-	-	-	-	-	-	-
Paper, Pulp and Print	-	-	-	-	-	-	-	-	-	-
Wood and Wood Products	-	-	-	-	-	-	-	-	-	-
Construction	-	-	-	-	-	-	-	-	-	-
Textile and Leather	-	-	-	-	-	-	-	-	-	-
Non-specified	18496	-	-	-	-	-	-	-	27985	-
TRANSPORT SECTOR	-	-	-	-	-	-	-	-	-	-
Air	-	-	-	-	-	-	-	-	-	-
Road	-	-	-	-	-	-	-	-	-	-
Rail	-	-	-	-	-	-	-	-	-	-
Pipeline Transport	-	-	-	-	-	-	-	-	-	-
Internal Navigation	-	-	-	-	-	-	-	-	-	-
Non-specified	-	-	-	-	-	-	-	-	-	-
OTHER SECTORS	**43090**	**22000**	-	-	**1898450**	-	-	-	**37276**	-
Agriculture	-	-	-	-	-	-	-	-	-	-
Comm. and Publ. Services	-	11000	-	-	-	-	-	-	8667	-
Residential	43090	11000	-	-	1898450	-	-	-	24866	-
Non-specified	-	-	-	-	-	-	-	-	3743	-
NON-ENERGY USE	-	-	-	-	-	-	-	-	-	-
in Industry/Transf./Energy	-	-	-	-	-	-	-	-	-	-
in Transport	-	-	-	-	-	-	-	-	-	-
in Other Sectors	-	-	-	-	-	-	-	-	-	-

Islamic Republic of Iran / République Islamique d'Iran

SUPPLY AND CONSUMPTION 1997	Coal (1000 tonnes)							Oil (1000 tonnes)			
	Coking Coal	Other Bit. Coal	Sub-Bit. Coal	Lignite	Peat	Oven and Gas Coke	Pat. Fuel and BKB	Crude Oil	NGL	Feed-stocks	Additives
Production	922	165	-	-	-	1215	-	178726	2390	-	-
Imports	598	-	-	-	-	-	-	-	-	-	-
Exports	-	-13	-	-	-	-	-	-105185	-	-	-
Intl. Marine Bunkers	-	-	-	-	-	-	-	-	-	-	-
Stock Changes	-	-	-	-	-	-	-	-14446	-	-	-
DOMESTIC SUPPLY	1520	152	-	-	-	1215	-	59095	2390	-	-
Transfers and Stat. Diff.	-1	-	-	-	-	-	-	-	-	-	-
TRANSFORMATION	1519	-	-	-	-	972	-	59095	2390	-	-
Electricity and CHP Plants	-	-	-	-	-	-	-	-	-	-	-
Petroleum Refineries	-	-	-	-	-	-	-	59095	2390	-	-
Other Transform. Sector	1519	-	-	-	-	972	-	-	-	-	-
ENERGY SECTOR	-	-	-	-	-	-	-	-	-	-	-
DISTRIBUTION LOSSES	-	-	-	-	-	-	-	-	-	-	-
FINAL CONSUMPTION	-	152	-	-	-	243	-	-	-	-	-
INDUSTRY SECTOR	-	152	-	-	-	243	-	-	-	-	-
Iron and Steel	-	-	-	-	-	243	-	-	-	-	-
Chemical and Petrochem.	-	-	-	-	-	-	-	-	-	-	-
Non-Metallic Minerals	-	-	-	-	-	-	-	-	-	-	-
Non-specified	-	152	-	-	-	-	-	-	-	-	-
TRANSPORT SECTOR	-	-	-	-	-	-	-	-	-	-	-
Air	-	-	-	-	-	-	-	-	-	-	-
Road	-	-	-	-	-	-	-	-	-	-	-
Non-specified	-	-	-	-	-	-	-	-	-	-	-
OTHER SECTORS	-	-	-	-	-	-	-	-	-	-	-
Agriculture	-	-	-	-	-	-	-	-	-	-	-
Comm. and Publ. Services	-	-	-	-	-	-	-	-	-	-	-
Residential	-	-	-	-	-	-	-	-	-	-	-
Non-specified	-	-	-	-	-	-	-	-	-	-	-
NON-ENERGY USE	-	-	-	-	-	-	-	-	-	-	-

APPROVISIONNEMENT ET DEMANDE 1998	Charbon (1000 tonnes)							Pétrole (1000 tonnes)			
	Charbon à coke	Autres charb. bit.	Charbon sous-bit.	Lignite	Tourbe	Coke de four/gaz	Agg./briq. de lignite	Pétrole brut	LGN	Produits d'aliment.	Additifs
Production	1000	169	-	-	-	1363	-	180265	2490	-	-
Imports	704	-	-	-	-	-	-	-	-	-	-
Exports	-	-11	-	-	-	-	-	-102410	-	-	-
Intl. Marine Bunkers	-	-	-	-	-	-	-	-	-	-	-
Stock Changes	-	-	-	-	-	-	-	-15655	-	-	-
DOMESTIC SUPPLY	1704	158	-	-	-	1363	-	62200	2490	-	-
Transfers and Stat. Diff.	-1	-	-	-	-	-	-	1	-	-	-
TRANSFORMATION	1703	-	-	-	-	1090	-	62201	2490	-	-
Electricity and CHP Plants	-	-	-	-	-	-	-	-	-	-	-
Petroleum Refineries	-	-	-	-	-	-	-	62201	2490	-	-
Other Transform. Sector	1703	-	-	-	-	1090	-	-	-	-	-
ENERGY SECTOR	-	-	-	-	-	-	-	-	-	-	-
DISTRIBUTION LOSSES	-	-	-	-	-	-	-	-	-	-	-
FINAL CONSUMPTION	-	158	-	-	-	273	-	-	-	-	-
INDUSTRY SECTOR	-	158	-	-	-	273	-	-	-	-	-
Iron and Steel	-	-	-	-	-	273	-	-	-	-	-
Chemical and Petrochem.	-	-	-	-	-	-	-	-	-	-	-
Non-Metallic Minerals	-	-	-	-	-	-	-	-	-	-	-
Non-specified	-	158	-	-	-	-	-	-	-	-	-
TRANSPORT SECTOR	-	-	-	-	-	-	-	-	-	-	-
Air	-	-	-	-	-	-	-	-	-	-	-
Road	-	-	-	-	-	-	-	-	-	-	-
Non-specified	-	-	-	-	-	-	-	-	-	-	-
OTHER SECTORS	-	-	-	-	-	-	-	-	-	-	-
Agriculture	-	-	-	-	-	-	-	-	-	-	-
Comm. and Publ. Services	-	-	-	-	-	-	-	-	-	-	-
Residential	-	-	-	-	-	-	-	-	-	-	-
Non-specified	-	-	-	-	-	-	-	-	-	-	-
NON-ENERGY USE	-	-	-	-	-	-	-	-	-	-	-

Islamic Republic of Iran / République Islamique d'Iran

SUPPLY AND CONSUMPTION 1997	Oil cont. (1000 tonnes)										
	Refinery Gas	LPG + Ethane	Motor Gasoline	Aviation Gasoline	Jet Fuel	Kerosene	Gas/ Diesel	Heavy Fuel Oil	Naphtha	Petrol. Coke	Other Prod.
Production	-	1540	7893	99	141	1201	18992	25834	679	-	3366
Imports	-	-	1572	-	-	174	-	-	-	-	-
Exports	-	-9	-	-	-	-	-	-12042	-141	-	-142
Intl. Marine Bunkers	-	-	-	-	-	-	-	-345	-	-	-
Stock Changes	-	391	-50	-4	2	-3	-336	3316	-	-	-
DOMESTIC SUPPLY	-	1922	9415	95	143	1372	18656	16763	538	-	3224
Transfers and Stat. Diff.	-	-	-1	-	1	-	-	1	1	-	-
TRANSFORMATION	-	-	-	-	-	-	1322	6646	-	-	-
Electricity and CHP Plants	-	-	-	-	-	-	1322	6646	-	-	-
Petroleum Refineries	-	-	-	-	-	-	-	-	-	-	-
Other Transform. Sector	-	-	-	-	-	-	-	-	-	-	-
ENERGY SECTOR	-	85	-	-	-	-	421	1850	-	-	-
DISTRIBUTION LOSSES	-	-	-	-	-	-	-	-	-	-	-
FINAL CONSUMPTION	-	1837	9414	95	144	1372	16913	8268	539	-	3224
INDUSTRY SECTOR	-	11	-	-	-	19	39	5812	539	-	-
Iron and Steel	-	-	-	-	-	-	-	-	-	-	-
Chemical and Petrochem.	-	-	-	-	-	-	-	-	539	-	-
Non-Metallic Minerals	-	-	-	-	-	-	-	-	-	-	-
Non-specified	-	11	-	-	-	19	39	5812	-	-	-
TRANSPORT SECTOR	-	54	9414	95	144	-	10572	-	-	-	-
Air	-	-	-	95	144	-	-	-	-	-	-
Road	-	54	9414	-	-	-	10572	-	-	-	-
Non-specified	-	-	-	-	-	-	-	-	-	-	-
OTHER SECTORS	-	1772	-	-	-	1353	6302	2456	-	-	-
Agriculture	-	-	-	-	-	21	3322	-	-	-	-
Comm. and Publ. Services	-	73	-	-	-	86	1091	2456	-	-	-
Residential	-	1699	-	-	-	1246	1889	-	-	-	-
Non-specified	-	-	-	-	-	-	-	-	-	-	-
NON-ENERGY USE	-	-	-	-	-	-	-	-	-	-	3224

APPROVISIONNEMENT ET DEMANDE 1998	Pétrole cont. (1000 tonnes)										
	Gaz de raffinerie	GPL + éthane	Essence moteur	Essence aviation	Carbu- réacteurs	Kérosène	Gazole	Fioul lourd	Naphta	Coke de pétrole	Autres prod.
Production	-	1598	9025	96	148	1351	19918	27251	1871	-	3323
Imports	-	-	1017	-	-	-	-	-	-	-	-
Exports	-	-7	-	-	-	-	-	-14925	-107	-	-151
Intl. Marine Bunkers	-	-	-	-	-	-	-	-608	-	-	-
Stock Changes	-	375	53	-4	-21	-75	-550	2359	-6	-	-
DOMESTIC SUPPLY	-	1966	10095	92	127	1276	19368	14077	1758	-	3172
Transfers and Stat. Diff.	-	1	1	-	-	-	-1	1	1	-	-
TRANSFORMATION	-	-	-	-	-	-	946	4599	-	-	-
Electricity and CHP Plants	-	-	-	-	-	-	946	4599	-	-	-
Petroleum Refineries	-	-	-	-	-	-	-	-	-	-	-
Other Transform. Sector	-	-	-	-	-	-	-	-	-	-	-
ENERGY SECTOR	-	26	-	-	-	-	443	1850	-	-	-
DISTRIBUTION LOSSES	-	-	-	-	-	-	-	-	-	-	-
FINAL CONSUMPTION	-	1941	10096	92	127	1276	17978	7629	1759	-	3172
INDUSTRY SECTOR	-	13	-	-	-	10	44	5803	1759	-	-
Iron and Steel	-	-	-	-	-	-	-	-	-	-	-
Chemical and Petrochem.	-	-	-	-	-	-	-	-	1759	-	-
Non-Metallic Minerals	-	-	-	-	-	-	-	-	-	-	-
Non-specified	-	13	-	-	-	10	44	5803	-	-	-
TRANSPORT SECTOR	-	113	10096	92	127	-	11383	-	-	-	-
Air	-	-	-	92	127	-	-	-	-	-	-
Road	-	113	10096	-	-	-	11383	-	-	-	-
Non-specified	-	-	-	-	-	-	-	-	-	-	-
OTHER SECTORS	-	1815	-	-	-	1266	6551	1826	-	-	-
Agriculture	-	-	-	-	-	23	3537	-	-	-	-
Comm. and Publ. Services	-	76	-	-	-	86	969	1826	-	-	-
Residential	-	1739	-	-	-	1157	2045	-	-	-	-
Non-specified	-	-	-	-	-	-	-	-	-	-	-
NON-ENERGY USE	-	-	-	-	-	-	-	-	-	-	3172

Islamic Republic of Iran / République Islamique d'Iran

SUPPLY AND CONSUMPTION 1997	Gas (TJ)				Comb. Renew. & Waste (TJ)				(GWh)	(TJ)
	Natural Gas	Gas Works	Coke Ovens	Blast Furnaces	Solid Biomass	Gas/Liquids from Biomass	Municipal Waste	Industrial Waste	Electricity	Heat
Production	1924194	-	-	-	32914	-	-	-	97744	-
Imports	15546	-	-	-	-	-	-	-	-	-
Exports	-	-	-	-	-	-	-	-	-	-
Intl. Marine Bunkers	-	-	-	-	-	-	-	-	-	-
Stock Changes	-	-	-	-	-	-	-	-	-	-
DOMESTIC SUPPLY	**1939740**	**-**	**-**	**-**	**32914**	**-**	**-**	**-**	**97744**	**-**
Transfers and Stat. Diff.	75249	-	-	-	-	-	-	-	-	-
TRANSFORMATION	**652326**	**-**	**-**	**-**	**2959**	**-**	**-**	**-**	**-**	**-**
Electricity and CHP Plants	652326	-	-	-	-	-	-	-	-	-
Petroleum Refineries	-	-	-	-	-	-	-	-	-	-
Other Transform. Sector	-	-	-	-	2959	-	-	-	-	-
ENERGY SECTOR	**46322**	**-**	**-**	**-**	**-**	**-**	**-**	**-**	**4592**	**-**
DISTRIBUTION LOSSES	**89338**	**-**	**-**	**-**	**-**	**-**	**-**	**-**	**14360**	**-**
FINAL CONSUMPTION	**1227003**	**-**	**-**	**-**	**29955**	**-**	**-**	**-**	**78792**	**-**
INDUSTRY SECTOR	**510999**	**-**	**-**	**-**	**7789**	**-**	**-**	**-**	**29095**	**-**
Iron and Steel	-	-	-	-	-	-	-	-	-	-
Chemical and Petrochem.	182494	-	-	-	-	-	-	-	-	-
Non-Metallic Minerals	-	-	-	-	-	-	-	-	-	-
Non-specified	328505	-	-	-	7789	-	-	-	29095	-
TRANSPORT SECTOR	**-**	**-**	**-**	**-**	**-**	**-**	**-**	**-**	**-**	**-**
Air	-	-	-	-	-	-	-	-	-	-
Road	-	-	-	-	-	-	-	-	-	-
Non-specified	-	-	-	-	-	-	-	-	-	-
OTHER SECTORS	**716004**	**-**	**-**	**-**	**22166**	**-**	**-**	**-**	**49697**	**-**
Agriculture	-	-	-	-	-	-	-	-	6009	-
Comm. and Publ. Services	75249	-	-	-	-	-	-	-	14887	-
Residential	640755	-	-	-	-	-	-	-	26523	-
Non-specified	-	-	-	-	22166	-	-	-	2278	-
NON-ENERGY USE	**-**	**-**	**-**	**-**	**-**	**-**	**-**	**-**	**-**	**-**

APPROVISIONNEMENT ET DEMANDE 1998	Gaz (TJ)				En. Re. Comb. & Déchets (TJ)				(GWh)	(TJ)
	Gaz naturel	Usines à gaz	Cokeries	Hauts fourneaux	Biomasse solide	Gaz/Liquides tirés de biomasse	Déchets urbains	Déchets industriels	Electricité	Chaleur
Production	2054580	-	-	-	32914	-	-	-	103413	-
Imports	73320	-	-	-	-	-	-	-	-	-
Exports	-	-	-	-	-	-	-	-	-	-
Intl. Marine Bunkers	-	-	-	-	-	-	-	-	-	-
Stock Changes	-	-	-	-	-	-	-	-	-	-
DOMESTIC SUPPLY	**2127900**	**-**	**-**	**-**	**32914**	**-**	**-**	**-**	**103413**	**-**
Transfers and Stat. Diff.	70094	-	-	-	-	-	-	-	-	-
TRANSFORMATION	**800383**	**-**	**-**	**-**	**3061**	**-**	**-**	**-**	**-**	**-**
Electricity and CHP Plants	800383	-	-	-	-	-	-	-	-	-
Petroleum Refineries	-	-	-	-	-	-	-	-	-	-
Other Transform. Sector	-	-	-	-	3061	-	-	-	-	-
ENERGY SECTOR	**78712**	**-**	**-**	**-**	**-**	**-**	**-**	**-**	**4431**	**-**
DISTRIBUTION LOSSES	**95596**	**-**	**-**	**-**	**-**	**-**	**-**	**-**	**15760**	**-**
FINAL CONSUMPTION	**1223303**	**-**	**-**	**-**	**29853**	**-**	**-**	**-**	**83222**	**-**
INDUSTRY SECTOR	**576684**	**-**	**-**	**-**	**7789**	**-**	**-**	**-**	**29716**	**-**
Iron and Steel	-	-	-	-	-	-	-	-	-	-
Chemical and Petrochem.	212247	-	-	-	-	-	-	-	-	-
Non-Metallic Minerals	-	-	-	-	-	-	-	-	-	-
Non-specified	364437	-	-	-	7789	-	-	-	29716	-
TRANSPORT SECTOR	**-**	**-**	**-**	**-**	**-**	**-**	**-**	**-**	**-**	**-**
Air	-	-	-	-	-	-	-	-	-	-
Road	-	-	-	-	-	-	-	-	-	-
Non-specified	-	-	-	-	-	-	-	-	-	-
OTHER SECTORS	**646619**	**-**	**-**	**-**	**22064**	**-**	**-**	**-**	**53506**	**-**
Agriculture	-	-	-	-	-	-	-	-	6782	-
Comm. and Publ. Services	70132	-	-	-	-	-	-	-	15561	-
Residential	576487	-	-	-	-	-	-	-	28686	-
Non-specified	-	-	-	-	22064	-	-	-	2477	-
NON-ENERGY USE	**-**	**-**	**-**	**-**	**-**	**-**	**-**	**-**	**-**	**-**

Iraq / Irak

SUPPLY AND CONSUMPTION 1997	Coal (1000 tonnes)							Oil (1000 tonnes)			
	Coking Coal	Other Bit. Coal	Sub-Bit. Coal	Lignite	Peat	Oven and Gas Coke	Pat. Fuel and BKB	Crude Oil	NGL	Feed-stocks	Additives
Production	-	-	-	-	-	-	-	56502	498	-	-
Imports	-	-	-	-	-	-	-	-	-	-	-
Exports	-	-	-	-	-	-	-	-35063	-	-	-
Intl. Marine Bunkers	-	-	-	-	-	-	-	-	-	-	-
Stock Changes	-	-	-	-	-	-	-	3196	-	-	-
DOMESTIC SUPPLY	-	-	-	-	-	-	-	24635	498	-	-
Transfers and Stat. Diff.	-	-	-	-	-	-	-	-	-498	-	-
TRANSFORMATION	-	-	-	-	-	-	-	24635	-	-	-
Electricity and CHP Plants	-	-	-	-	-	-	-	1800	-	-	-
Petroleum Refineries	-	-	-	-	-	-	-	22835	-	-	-
Other Transform. Sector	-	-	-	-	-	-	-	-	-	-	-
ENERGY SECTOR	-	-	-	-	-	-	-	-	-	-	-
DISTRIBUTION LOSSES	-	-	-	-	-	-	-	-	-	-	-
FINAL CONSUMPTION	-	-	-	-	-	-	-	-	-	-	-
INDUSTRY SECTOR	-	-	-	-	-	-	-	-	-	-	-
Iron and Steel	-	-	-	-	-	-	-	-	-	-	-
Chemical and Petrochem.	-	-	-	-	-	-	-	-	-	-	-
Non-Metallic Minerals	-	-	-	-	-	-	-	-	-	-	-
Non-specified	-	-	-	-	-	-	-	-	-	-	-
TRANSPORT SECTOR	-	-	-	-	-	-	-	-	-	-	-
Air	-	-	-	-	-	-	-	-	-	-	-
Road	-	-	-	-	-	-	-	-	-	-	-
Non-specified	-	-	-	-	-	-	-	-	-	-	-
OTHER SECTORS	-	-	-	-	-	-	-	-	-	-	-
Agriculture	-	-	-	-	-	-	-	-	-	-	-
Comm. and Publ. Services	-	-	-	-	-	-	-	-	-	-	-
Residential	-	-	-	-	-	-	-	-	-	-	-
Non-specified	-	-	-	-	-	-	-	-	-	-	-
NON-ENERGY USE	-	-	-	-	-	-	-	-	-	-	-

APPROVISIONNEMENT ET DEMANDE 1998	Charbon (1000 tonnes)							Pétrole (1000 tonnes)			
	Charbon à coke	Autres charb. bit.	Charbon sous-bit.	Lignite	Tourbe	Coke de four/gaz	Agg./briq. de lignite	Pétrole brut	LGN	Produits d'aliment.	Additifs
Production	-	-	-	-	-	-	-	104038	498	-	-
Imports	-	-	-	-	-	-	-	-	-	-	-
Exports	-	-	-	-	-	-	-	-75219	-	-	-
Intl. Marine Bunkers	-	-	-	-	-	-	-	-	-	-	-
Stock Changes	-	-	-	-	-	-	-	-3436	-	-	-
DOMESTIC SUPPLY	-	-	-	-	-	-	-	25383	498	-	-
Transfers and Stat. Diff.	-	-	-	-	-	-	-	-	-498	-	-
TRANSFORMATION	-	-	-	-	-	-	-	25383	-	-	-
Electricity and CHP Plants	-	-	-	-	-	-	-	1800	-	-	-
Petroleum Refineries	-	-	-	-	-	-	-	23583	-	-	-
Other Transform. Sector	-	-	-	-	-	-	-	-	-	-	-
ENERGY SECTOR	-	-	-	-	-	-	-	-	-	-	-
DISTRIBUTION LOSSES	-	-	-	-	-	-	-	-	-	-	-
FINAL CONSUMPTION	-	-	-	-	-	-	-	-	-	-	-
INDUSTRY SECTOR	-	-	-	-	-	-	-	-	-	-	-
Iron and Steel	-	-	-	-	-	-	-	-	-	-	-
Chemical and Petrochem.	-	-	-	-	-	-	-	-	-	-	-
Non-Metallic Minerals	-	-	-	-	-	-	-	-	-	-	-
Non-specified	-	-	-	-	-	-	-	-	-	-	-
TRANSPORT SECTOR	-	-	-	-	-	-	-	-	-	-	-
Air	-	-	-	-	-	-	-	-	-	-	-
Road	-	-	-	-	-	-	-	-	-	-	-
Non-specified	-	-	-	-	-	-	-	-	-	-	-
OTHER SECTORS	-	-	-	-	-	-	-	-	-	-	-
Agriculture	-	-	-	-	-	-	-	-	-	-	-
Comm. and Publ. Services	-	-	-	-	-	-	-	-	-	-	-
Residential	-	-	-	-	-	-	-	-	-	-	-
Non-specified	-	-	-	-	-	-	-	-	-	-	-
NON-ENERGY USE	-	-	-	-	-	-	-	-	-	-	-

Iraq / Irak

SUPPLY AND CONSUMPTION 1997	Oil cont. (1000 tonnes)										
	Refinery Gas	LPG + Ethane	Motor Gasoline	Aviation Gasoline	Jet Fuel	Kerosene	Gas/ Diesel	Heavy Fuel Oil	Naphtha	Petrol. Coke	Other Prod.
Production	456	993	2968	-	549	991	6701	7601	489	-	767
Imports	-	-	-	-	-	-	-	-	-	-	-
Exports	-	-	-	-	-	-	-	-	-	-	-
Intl. Marine Bunkers	-	-	-	-	-	-	-	-	-	-	-
Stock Changes	-	-347	-	-	-	-	-	-	-	-	-
DOMESTIC SUPPLY	456	646	2968	-	549	991	6701	7601	489	-	767
Transfers and Stat. Diff.	-	498	-82	-	-130	-27	-1326	-1291	-6	-	-8
TRANSFORMATION	-	-	-	-	-	-	-	3411	-	-	-
Electricity and CHP Plants	-	-	-	-	-	-	-	3411	-	-	-
Petroleum Refineries	-	-	-	-	-	-	-	-	-	-	-
Other Transform. Sector	-	-	-	-	-	-	-	-	-	-	-
ENERGY SECTOR	456	-	-	-	-	-	-	1153	-	-	-
DISTRIBUTION LOSSES	-	-	-	-	-	-	-	-	-	-	-
FINAL CONSUMPTION	-	1144	2886	-	419	964	5375	1746	483	-	759
INDUSTRY SECTOR	-	-	-	-	-	-	-	1746	483	-	-
Iron and Steel	-	-	-	-	-	-	-	-	-	-	-
Chemical and Petrochem.	-	-	-	-	-	-	-	-	483	-	-
Non-Metallic Minerals	-	-	-	-	-	-	-	-	-	-	-
Non-specified	-	-	-	-	-	-	-	1746	-	-	-
TRANSPORT SECTOR	-	-	2886	-	419	-	5375	-	-	-	-
Air	-	-	-	-	419	-	-	-	-	-	-
Road	-	-	2886	-	-	-	5375	-	-	-	-
Non-specified	-	-	-	-	-	-	-	-	-	-	-
OTHER SECTORS	-	1144	-	-	-	964	-	-	-	-	-
Agriculture	-	-	-	-	-	-	-	-	-	-	-
Comm. and Publ. Services	-	-	-	-	-	-	-	-	-	-	-
Residential	-	1144	-	-	-	964	-	-	-	-	-
Non-specified	-	-	-	-	-	-	-	-	-	-	-
NON-ENERGY USE	-	-	-	-	-	-	-	-	-	-	759

APPROVISIONNEMENT ET DEMANDE 1998	Pétrole cont. (1000 tonnes)										
	Gaz de raffinerie	GPL + éthane	Essence moteur	Essence aviation	Carbu- réacteurs	Kérosène	Gazole	Fioul lourd	Naphta	Coke de pétrole	Autres prod.
Production	471	1025	3066	-	567	1023	6921	7850	505	-	792
Imports	-	-	-	-	-	-	-	-	-	-	-
Exports	-	-	-	-	-	-	-	-	-	-	-
Intl. Marine Bunkers	-	-	-	-	-	-	-	-	-	-	-
Stock Changes	-	-343	-	-	-	-	-	-	-	-	-
DOMESTIC SUPPLY	471	682	3066	-	567	1023	6921	7850	505	-	792
Transfers and Stat. Diff.	-	498	-78	-	-136	-31	-1493	-1280	-7	-	-8
TRANSFORMATION	-	-	-	-	-	-	-	3552	-	-	-
Electricity and CHP Plants	-	-	-	-	-	-	-	3552	-	-	-
Petroleum Refineries	-	-	-	-	-	-	-	-	-	-	-
Other Transform. Sector	-	-	-	-	-	-	-	-	-	-	-
ENERGY SECTOR	471	-	-	-	-	-	-	1200	-	-	-
DISTRIBUTION LOSSES	-	-	-	-	-	-	-	-	-	-	-
FINAL CONSUMPTION	-	1180	2988	-	431	992	5428	1818	498	-	784
INDUSTRY SECTOR	-	-	-	-	-	-	-	1818	498	-	-
Iron and Steel	-	-	-	-	-	-	-	-	-	-	-
Chemical and Petrochem.	-	-	-	-	-	-	-	-	498	-	-
Non-Metallic Minerals	-	-	-	-	-	-	-	-	-	-	-
Non-specified	-	-	-	-	-	-	-	1818	-	-	-
TRANSPORT SECTOR	-	-	2988	-	431	-	5428	-	-	-	-
Air	-	-	-	-	431	-	-	-	-	-	-
Road	-	-	2988	-	-	-	5428	-	-	-	-
Non-specified	-	-	-	-	-	-	-	-	-	-	-
OTHER SECTORS	-	1180	-	-	-	992	-	-	-	-	-
Agriculture	-	-	-	-	-	-	-	-	-	-	-
Comm. and Publ. Services	-	-	-	-	-	-	-	-	-	-	-
Residential	-	1180	-	-	-	992	-	-	-	-	-
Non-specified	-	-	-	-	-	-	-	-	-	-	-
NON-ENERGY USE	-	-	-	-	-	-	-	-	-	-	784

Iraq / Irak

SUPPLY AND CONSUMPTION 1997	Gas (TJ) Natural Gas	Gas Works	Coke Ovens	Blast Furnaces	Comb. Renew. & Waste (TJ) Solid Biomass	Gas/Liquids from Biomass	Municipal Waste	Industrial Waste	(GWh) Electricity	(TJ) Heat
Production	168299	-	-	-	1100	-	-	-	29561	-
Imports	-	-	-	-	-	-	-	-	-	-
Exports	-	-	-	-	-	-	-	-	-	-
Intl. Marine Bunkers	-	-	-	-	-	-	-	-	-	-
Stock Changes	-	-	-	-	-	-	-	-	-	-
DOMESTIC SUPPLY	168299	-	-	-	1100	-	-	-	29561	-
Transfers and Stat. Diff.	-	-	-	-	-	-	-	-	-	-
TRANSFORMATION	-	-	-	-	476	-	-	-	-	-
Electricity and CHP Plants	-	-	-	-	-	-	-	-	-	-
Petroleum Refineries	-	-	-	-	-	-	-	-	-	-
Other Transform. Sector	-	-	-	-	476	-	-	-	-	-
ENERGY SECTOR	-	-	-	-	-	-	-	-	-	-
DISTRIBUTION LOSSES	-	-	-	-	-	-	-	-	-	-
FINAL CONSUMPTION	168299	-	-	-	624	-	-	-	29561	-
INDUSTRY SECTOR	168299	-	-	-	-	-	-	-	-	-
Iron and Steel	-	-	-	-	-	-	-	-	-	-
Chemical and Petrochem.	-	-	-	-	-	-	-	-	-	-
Non-Metallic Minerals	-	-	-	-	-	-	-	-	-	-
Non-specified	168299	-	-	-	-	-	-	-	-	-
TRANSPORT SECTOR	-	-	-	-	-	-	-	-	-	-
Air	-	-	-	-	-	-	-	-	-	-
Road	-	-	-	-	-	-	-	-	-	-
Non-specified	-	-	-	-	-	-	-	-	-	-
OTHER SECTORS	-	-	-	-	624	-	-	-	29561	-
Agriculture	-	-	-	-	-	-	-	-	-	-
Comm. and Publ. Services	-	-	-	-	-	-	-	-	-	-
Residential	-	-	-	-	-	-	-	-	-	-
Non-specified	-	-	-	-	624	-	-	-	29561	-
NON-ENERGY USE	-	-	-	-	-	-	-	-	-	-

APPROVISIONNEMENT ET DEMANDE 1998	Gaz (TJ) Gaz naturel	Usines à gaz	Cokeries	Hauts fourneaux	En. Re. Comb. & Déchets (TJ) Biomasse solide	Gaz/Liquides tirés de biomasse	Déchets urbains	Déchets industriels	(GWh) Electricité	(TJ) Chaleur
Production	177156	-	-	-	1100	-	-	-	30346	-
Imports	-	-	-	-	-	-	-	-	-	-
Exports	-	-	-	-	-	-	-	-	-	-
Intl. Marine Bunkers	-	-	-	-	-	-	-	-	-	-
Stock Changes	-	-	-	-	-	-	-	-	-	-
DOMESTIC SUPPLY	177156	-	-	-	1100	-	-	-	30346	-
Transfers and Stat. Diff.	-	-	-	-	-	-	-	-	-	-
TRANSFORMATION	-	-	-	-	476	-	-	-	-	-
Electricity and CHP Plants	-	-	-	-	-	-	-	-	-	-
Petroleum Refineries	-	-	-	-	-	-	-	-	-	-
Other Transform. Sector	-	-	-	-	476	-	-	-	-	-
ENERGY SECTOR	-	-	-	-	-	-	-	-	-	-
DISTRIBUTION LOSSES	-	-	-	-	-	-	-	-	-	-
FINAL CONSUMPTION	177156	-	-	-	624	-	-	-	30346	-
INDUSTRY SECTOR	177156	-	-	-	-	-	-	-	-	-
Iron and Steel	-	-	-	-	-	-	-	-	-	-
Chemical and Petrochem.	-	-	-	-	-	-	-	-	-	-
Non-Metallic Minerals	-	-	-	-	-	-	-	-	-	-
Non-specified	177156	-	-	-	-	-	-	-	-	-
TRANSPORT SECTOR	-	-	-	-	-	-	-	-	-	-
Air	-	-	-	-	-	-	-	-	-	-
Road	-	-	-	-	-	-	-	-	-	-
Non-specified	-	-	-	-	-	-	-	-	-	-
OTHER SECTORS	-	-	-	-	624	-	-	-	30346	-
Agriculture	-	-	-	-	-	-	-	-	-	-
Comm. and Publ. Services	-	-	-	-	-	-	-	-	-	-
Residential	-	-	-	-	-	-	-	-	-	-
Non-specified	-	-	-	-	624	-	-	-	30346	-
NON-ENERGY USE	-	-	-	-	-	-	-	-	-	-

Israel / Israël : 1997

SUPPLY AND CONSUMPTION / APPROVISIONNEMENT ET DEMANDE	Coal / Charbon (1000 tonnes)							Oil / Pétrole (1000 tonnes)			
	Coking Coal / Charbon à coke	Other Bit. Coal / Autres charb. bit.	Sub-Bit. Coal / Charbon sous-bit.	Lignite / Lignite	Peat / Tourbe	Oven and Gas Coke / Coke de four/gaz	Pat. Fuel and BKB / Agg./briq. de lignite	Crude Oil / Pétrole brut	NGL / LGN	Feedstocks / Produits d'aliment.	Additives / Additifs
Production	-	-	-	472	-	-	-	-	21	-	-
From Other Sources	-	-	-	-	-	-	-	-	-	-	-
Imports	-	8804	-	-	-	-	-	11265	-	-	-
Exports	-	-	-	-	-	-	-	-	-	-	-
Intl. Marine Bunkers	-	-	-	-	-	-	-	-	-	-	-
Stock Changes	-	-164	-	-	-	-	-	-179	-	-	-
DOMESTIC SUPPLY	-	8640	-	472	-	-	-	11086	21	-	-
Transfers	-	-	-	-	-	-	-	-	-	-	-
Statistical Differences	-	-	-	-	-	-	-	-	-	-	-
TRANSFORMATION	-	8639	-	472	-	-	-	11086	-	-	-
Electricity Plants	-	8639	-	472	-	-	-	-	-	-	-
CHP Plants	-	-	-	-	-	-	-	-	-	-	-
Heat Plants	-	-	-	-	-	-	-	-	-	-	-
Blast Furnaces/Gas Works	-	-	-	-	-	-	-	-	-	-	-
Coke/Pat. Fuel/BKB Plants	-	-	-	-	-	-	-	-	-	-	-
Petroleum Refineries	-	-	-	-	-	-	-	11086	-	-	-
Petrochemical Industry	-	-	-	-	-	-	-	-	-	-	-
Liquefaction	-	-	-	-	-	-	-	-	-	-	-
Other Transform. Sector	-	-	-	-	-	-	-	-	-	-	-
ENERGY SECTOR	-	-	-	-	-	-	-	-	-	-	-
Coal Mines	-	-	-	-	-	-	-	-	-	-	-
Oil and Gas Extraction	-	-	-	-	-	-	-	-	-	-	-
Petroleum Refineries	-	-	-	-	-	-	-	-	-	-	-
Electr., CHP+Heat Plants	-	-	-	-	-	-	-	-	-	-	-
Pumped Storage (Elec.)	-	-	-	-	-	-	-	-	-	-	-
Other Energy Sector	-	-	-	-	-	-	-	-	-	-	-
Distribution Losses	-	-	-	-	-	-	-	-	-	-	-
FINAL CONSUMPTION	-	1	-	-	-	-	-	-	21	-	-
INDUSTRY SECTOR	-	1	-	-	-	-	-	-	21	-	-
Iron and Steel	-	-	-	-	-	-	-	-	-	-	-
Chemical and Petrochem.	-	-	-	-	-	-	-	-	21	-	-
of which: Feedstocks	-	-	-	-	-	-	-	-	21	-	-
Non-Ferrous Metals	-	-	-	-	-	-	-	-	-	-	-
Non-Metallic Minerals	-	-	-	-	-	-	-	-	-	-	-
Transport Equipment	-	-	-	-	-	-	-	-	-	-	-
Machinery	-	-	-	-	-	-	-	-	-	-	-
Mining and Quarrying	-	-	-	-	-	-	-	-	-	-	-
Food and Tobacco	-	-	-	-	-	-	-	-	-	-	-
Paper, Pulp and Print	-	-	-	-	-	-	-	-	-	-	-
Wood and Wood Products	-	-	-	-	-	-	-	-	-	-	-
Construction	-	-	-	-	-	-	-	-	-	-	-
Textile and Leather	-	-	-	-	-	-	-	-	-	-	-
Non-specified	-	1	-	-	-	-	-	-	-	-	-
TRANSPORT SECTOR	-	-	-	-	-	-	-	-	-	-	-
Air	-	-	-	-	-	-	-	-	-	-	-
Road	-	-	-	-	-	-	-	-	-	-	-
Rail	-	-	-	-	-	-	-	-	-	-	-
Pipeline Transport	-	-	-	-	-	-	-	-	-	-	-
Internal Navigation	-	-	-	-	-	-	-	-	-	-	-
Non-specified	-	-	-	-	-	-	-	-	-	-	-
OTHER SECTORS	-	-	-	-	-	-	-	-	-	-	-
Agriculture	-	-	-	-	-	-	-	-	-	-	-
Comm. and Publ. Services	-	-	-	-	-	-	-	-	-	-	-
Residential	-	-	-	-	-	-	-	-	-	-	-
Non-specified	-	-	-	-	-	-	-	-	-	-	-
NON-ENERGY USE	-	-	-	-	-	-	-	-	-	-	-
in Industry/Trans./Energy	-	-	-	-	-	-	-	-	-	-	-
in Transport	-	-	-	-	-	-	-	-	-	-	-
in Other Sectors	-	-	-	-	-	-	-	-	-	-	-

Israel / Israël : 1997

SUPPLY AND CONSUMPTION / *APPROVISIONNEMENT ET DEMANDE*	Oil cont. / *Pétrole cont.* (1000 tonnes)										
	Refinery Gas / *Gaz de raffinerie*	LPG + Ethane / *GPL + éthane*	Motor Gasoline / *Essence moteur*	Aviation Gasoline / *Essence aviation*	Jet Fuel / *Carbu-réacteurs*	Kerosene / *Kérosène*	Gas/Diesel / *Gazole*	Heavy Fuel Oil / *Fioul lourd*	Naphtha / *Naphta*	Petrol. Coke / *Coke de pétrole*	Other Prod. / *Autres prod.*
Production	-	493	2112	-	-	1207	2796	2956	743	-	310
From Other Sources	-	-	-	-	-	-	-	-	-	-	-
Imports	-	-	2	-	-	283	57	2035	66	-	256
Exports	-	-83	-301	-	-	-567	-753	-734	-50	-	-38
Intl. Marine Bunkers	-	-	-	-	-	-	-69	-111	-	-	-
Stock Changes	-	-	192	-	-	89	238	250	16	-	-87
DOMESTIC SUPPLY	-	410	2005	-	-	1012	2269	4396	775	-	441
Transfers	-	-	-	-	-	-	-	-	-	-	-
Statistical Differences	-	-19	-	-	-	-	-	-	-	-	-
TRANSFORMATION	-	-	-	-	-	-	119	2326	-	-	-
Electricity Plants	-	-	-	-	-	-	119	2326	-	-	-
CHP Plants	-	-	-	-	-	-	-	-	-	-	-
Heat Plants	-	-	-	-	-	-	-	-	-	-	-
Blast Furnaces/Gas Works	-	-	-	-	-	-	-	-	-	-	-
Coke/Pat. Fuel/BKB Plants	-	-	-	-	-	-	-	-	-	-	-
Petroleum Refineries	-	-	-	-	-	-	-	-	-	-	-
Petrochemical Industry	-	-	-	-	-	-	-	-	-	-	-
Liquefaction	-	-	-	-	-	-	-	-	-	-	-
Other Transform. Sector	-	-	-	-	-	-	-	-	-	-	-
ENERGY SECTOR	-	-	-	-	-	-	-	586	-	-	-
Coal Mines	-	-	-	-	-	-	-	-	-	-	-
Oil and Gas Extraction	-	-	-	-	-	-	-	-	-	-	-
Petroleum Refineries	-	-	-	-	-	-	-	586	-	-	-
Electr., CHP+Heat Plants	-	-	-	-	-	-	-	-	-	-	-
Pumped Storage (Elec.)	-	-	-	-	-	-	-	-	-	-	-
Other Energy Sector	-	-	-	-	-	-	-	-	-	-	-
Distribution Losses	-	-	-	-	-	-	-	-	-	-	-
FINAL CONSUMPTION	-	391	2005	-	-	1012	2150	1484	775	-	441
INDUSTRY SECTOR	-	124	-	-	-	-	145	975	775	-	-
Iron and Steel	-	-	-	-	-	-	-	-	-	-	-
Chemical and Petrochem.	-	-	-	-	-	-	-	-	775	-	-
of which: Feedstocks	-	-	-	-	-	-	-	-	775	-	-
Non-Ferrous Metals	-	-	-	-	-	-	-	-	-	-	-
Non-Metallic Minerals	-	-	-	-	-	-	-	-	-	-	-
Transport Equipment	-	-	-	-	-	-	-	-	-	-	-
Machinery	-	-	-	-	-	-	-	-	-	-	-
Mining and Quarrying	-	-	-	-	-	-	-	-	-	-	-
Food and Tobacco	-	-	-	-	-	-	-	-	-	-	-
Paper, Pulp and Print	-	-	-	-	-	-	-	-	-	-	-
Wood and Wood Products	-	-	-	-	-	-	-	-	-	-	-
Construction	-	-	-	-	-	-	-	-	-	-	-
Textile and Leather	-	-	-	-	-	-	-	-	-	-	-
Non-specified	-	124	-	-	-	-	145	975	-	-	-
TRANSPORT SECTOR	-	-	2005	-	-	642	963	-	-	-	-
Air	-	-	-	-	-	642	-	-	-	-	-
Road	-	-	2005	-	-	-	963	-	-	-	-
Rail	-	-	-	-	-	-	-	-	-	-	-
Pipeline Transport	-	-	-	-	-	-	-	-	-	-	-
Internal Navigation	-	-	-	-	-	-	-	-	-	-	-
Non-specified	-	-	-	-	-	-	-	-	-	-	-
OTHER SECTORS	-	267	-	-	-	370	1042	509	-	-	-
Agriculture	-	-	-	-	-	-	-	-	-	-	-
Comm. and Publ. Services	-	-	-	-	-	-	-	-	-	-	-
Residential	-	267	-	-	-	370	-	-	-	-	-
Non-specified	-	-	-	-	-	-	1042	509	-	-	-
NON-ENERGY USE	-	-	-	-	-	-	-	-	-	-	441
in Industry/Transf./Energy	-	-	-	-	-	-	-	-	-	-	441
in Transport	-	-	-	-	-	-	-	-	-	-	-
in Other Sectors	-	-	-	-	-	-	-	-	-	-	-

Israel / Israël : 1997

SUPPLY AND CONSUMPTION / APPROVISIONNEMENT ET DEMANDE	Gas / Gaz (TJ)				Comb. Renew. & Waste / En. Re. Comb. & Déchets (TJ)				(GWh)	(TJ)
	Natural Gas / Gaz naturel	Gas Works / Usines à gaz	Coke Ovens / Cokeries	Blast Furnaces / Hauts fourneaux	Solid Biomass / Biomasse solide	Gas/Liquids from Biomass / Gaz/Liquides tirés de biomasse	Municipal Waste / Déchets urbains	Industrial Waste / Déchets industriels	Electricity / Electricité	Heat / Chaleur
Production	605	-	-	-	152	-	-	-	35098	-
From Other Sources	-	-	-	-	-	-	-	-	-	-
Imports	-	-	-	-	150	-	-	-	-	-
Exports	-	-	-	-	-	-	-	-	-1170	-
Intl. Marine Bunkers	-	-	-	-	-	-	-	-	-	-
Stock Changes	-	-	-	-	-	-	-	-	-	-
DOMESTIC SUPPLY	605	-	-	-	302	-	-	-	33928	-
Transfers	-	-	-	-	-	-	-	-	-	-
Statistical Differences	-	-	-	-	-	-	-	-	1140	-
TRANSFORMATION	209	-	-	-	-	-	-	-	-	-
Electricity Plants	209	-	-	-	-	-	-	-	-	-
CHP Plants	-	-	-	-	-	-	-	-	-	-
Heat Plants	-	-	-	-	-	-	-	-	-	-
Blast Furnaces/Gas Works	-	-	-	-	-	-	-	-	-	-
Coke/Pat. Fuel/BKB Plants	-	-	-	-	-	-	-	-	-	-
Petroleum Refineries	-	-	-	-	-	-	-	-	-	-
Petrochemical Industry	-	-	-	-	-	-	-	-	-	-
Liquefaction	-	-	-	-	-	-	-	-	-	-
Other Transform. Sector	-	-	-	-	-	-	-	-	-	-
ENERGY SECTOR	-	-	-	-	-	-	-	-	2765	-
Coal Mines	-	-	-	-	-	-	-	-	-	-
Oil and Gas Extraction	-	-	-	-	-	-	-	-	-	-
Petroleum Refineries	-	-	-	-	-	-	-	-	-	-
Electr., CHP+Heat Plants	-	-	-	-	-	-	-	-	2765	-
Pumped Storage (Elec.)	-	-	-	-	-	-	-	-	-	-
Other Energy Sector	-	-	-	-	-	-	-	-	-	-
Distribution Losses	-	-	-	-	-	-	-	-	1899	-
FINAL CONSUMPTION	396	-	-	-	302	-	-	-	30404	-
INDUSTRY SECTOR	396	-	-	-	-	-	-	-	9496	-
Iron and Steel	-	-	-	-	-	-	-	-	957	-
Chemical and Petrochem.	396	-	-	-	-	-	-	-	2098	-
of which: Feedstocks	396	-	-	-	-	-	-	-	-	-
Non-Ferrous Metals	-	-	-	-	-	-	-	-	-	-
Non-Metallic Minerals	-	-	-	-	-	-	-	-	776	-
Transport Equipment	-	-	-	-	-	-	-	-	264	-
Machinery	-	-	-	-	-	-	-	-	702	-
Mining and Quarrying	-	-	-	-	-	-	-	-	1167	-
Food and Tobacco	-	-	-	-	-	-	-	-	977	-
Paper, Pulp and Print	-	-	-	-	-	-	-	-	374	-
Wood and Wood Products	-	-	-	-	-	-	-	-	118	-
Construction	-	-	-	-	-	-	-	-	154	-
Textile and Leather	-	-	-	-	-	-	-	-	569	-
Non-specified	-	-	-	-	-	-	-	-	1340	-
TRANSPORT SECTOR	-	-	-	-	-	-	-	-	-	-
Air	-	-	-	-	-	-	-	-	-	-
Road	-	-	-	-	-	-	-	-	-	-
Rail	-	-	-	-	-	-	-	-	-	-
Pipeline Transport	-	-	-	-	-	-	-	-	-	-
Internal Navigation	-	-	-	-	-	-	-	-	-	-
Non-specified	-	-	-	-	-	-	-	-	-	-
OTHER SECTORS	-	-	-	-	302	-	-	-	20908	-
Agriculture	-	-	-	-	-	-	-	-	1397	-
Comm. and Publ. Services	-	-	-	-	-	-	-	-	7679	-
Residential	-	-	-	-	152	-	-	-	9476	-
Non-specified	-	-	-	-	150	-	-	-	2356	-
NON-ENERGY USE	-	-	-	-	-	-	-	-	-	-
in Industry/Transf./Energy	-	-	-	-	-	-	-	-	-	-
in Transport	-	-	-	-	-	-	-	-	-	-
in Other Sectors	-	-	-	-	-	-	-	-	-	-

Israel / Israël : 1998

| SUPPLY AND CONSUMPTION | Coal / *Charbon* (1000 tonnes) | | | | | | | Oil / *Pétrole* (1000 tonnes) | | | |
| | Coking Coal | Other Bit. Coal | Sub-Bit. Coal | Lignite | Peat | Oven and Gas Coke | Pat. Fuel and BKB | Crude Oil | NGL | Feed-stocks | Additives |
APPROVISIONNEMENT ET DEMANDE	Charbon à coke	Autres charb. bit.	Charbon sous-bit.	Lignite	Tourbe	Coke de four/gaz	Agg./briq. de lignite	Pétrole brut	LGN	Produits d'aliment.	Additifs
Production	-	-	-	444	-	-	-	-	21	-	-
From Other Sources	-	-	-	-	-	-	-	-	-	-	-
Imports	-	9636	-	-	-	-	-	11730	-	-	-
Exports	-	-	-	-	-	-	-	-	-	-	-
Intl. Marine Bunkers	-	-	-	-	-	-	-	-	-	-	-
Stock Changes	-	-352	-	-	-	-	-	-53	-	-	-
DOMESTIC SUPPLY	**-**	**9284**	**-**	**444**	**-**	**-**	**-**	**11677**	**21**	**-**	**-**
Transfers	-	-	-	-	-	-	-	-	-	-	-
Statistical Differences	-	-	-	-	-	-	-	-	-	-	-
TRANSFORMATION	**-**	**9283**	**-**	**444**	**-**	**-**	**-**	**11677**	**-**	**-**	**-**
Electricity Plants	-	9283	-	444	-	-	-	-	-	-	-
CHP Plants	-	-	-	-	-	-	-	-	-	-	-
Heat Plants	-	-	-	-	-	-	-	-	-	-	-
Blast Furnaces/Gas Works	-	-	-	-	-	-	-	-	-	-	-
Coke/Pat. Fuel/BKB Plants	-	-	-	-	-	-	-	-	-	-	-
Petroleum Refineries	-	-	-	-	-	-	-	11677	-	-	-
Petrochemical Industry	-	-	-	-	-	-	-	-	-	-	-
Liquefaction	-	-	-	-	-	-	-	-	-	-	-
Other Transform. Sector	-	-	-	-	-	-	-	-	-	-	-
ENERGY SECTOR	**-**	**-**	**-**	**-**	**-**	**-**	**-**	**-**	**-**	**-**	**-**
Coal Mines	-	-	-	-	-	-	-	-	-	-	-
Oil and Gas Extraction	-	-	-	-	-	-	-	-	-	-	-
Petroleum Refineries	-	-	-	-	-	-	-	-	-	-	-
Electr., CHP+Heat Plants	-	-	-	-	-	-	-	-	-	-	-
Pumped Storage (Elec.)	-	-	-	-	-	-	-	-	-	-	-
Other Energy Sector	-	-	-	-	-	-	-	-	-	-	-
Distribution Losses	-	-	-	-	-	-	-	-	-	-	-
FINAL CONSUMPTION	**-**	**1**	**-**	**-**	**-**	**-**	**-**	**-**	**21**	**-**	**-**
INDUSTRY SECTOR	**-**	**1**	**-**	**-**	**-**	**-**	**-**	**-**	**21**	**-**	**-**
Iron and Steel	-	-	-	-	-	-	-	-	-	-	-
Chemical and Petrochem.	-	-	-	-	-	-	-	-	21	-	-
of which: Feedstocks	-	-	-	-	-	-	-	-	21	-	-
Non-Ferrous Metals	-	-	-	-	-	-	-	-	-	-	-
Non-Metallic Minerals	-	-	-	-	-	-	-	-	-	-	-
Transport Equipment	-	-	-	-	-	-	-	-	-	-	-
Machinery	-	-	-	-	-	-	-	-	-	-	-
Mining and Quarrying	-	-	-	-	-	-	-	-	-	-	-
Food and Tobacco	-	-	-	-	-	-	-	-	-	-	-
Paper, Pulp and Print	-	-	-	-	-	-	-	-	-	-	-
Wood and Wood Products	-	-	-	-	-	-	-	-	-	-	-
Construction	-	-	-	-	-	-	-	-	-	-	-
Textile and Leather	-	-	-	-	-	-	-	-	-	-	-
Non-specified	-	1	-	-	-	-	-	-	-	-	-
TRANSPORT SECTOR	**-**	**-**	**-**	**-**	**-**	**-**	**-**	**-**	**-**	**-**	**-**
Air	-	-	-	-	-	-	-	-	-	-	-
Road	-	-	-	-	-	-	-	-	-	-	-
Rail	-	-	-	-	-	-	-	-	-	-	-
Pipeline Transport	-	-	-	-	-	-	-	-	-	-	-
Internal Navigation	-	-	-	-	-	-	-	-	-	-	-
Non-specified	-	-	-	-	-	-	-	-	-	-	-
OTHER SECTORS	**-**	**-**	**-**	**-**	**-**	**-**	**-**	**-**	**-**	**-**	**-**
Agriculture	-	-	-	-	-	-	-	-	-	-	-
Comm. and Publ. Services	-	-	-	-	-	-	-	-	-	-	-
Residential	-	-	-	-	-	-	-	-	-	-	-
Non-specified	-	-	-	-	-	-	-	-	-	-	-
NON-ENERGY USE	**-**	**-**	**-**	**-**	**-**	**-**	**-**	**-**	**-**	**-**	**-**
in Industry/Trans./Energy	-	-	-	-	-	-	-	-	-	-	-
in Transport	-	-	-	-	-	-	-	-	-	-	-
in Other Sectors	-	-	-	-	-	-	-	-	-	-	-

Israel / Israël : 1998

SUPPLY AND CONSUMPTION *APPROVISIONNEMENT ET DEMANDE*	Refinery Gas *Gaz de raffinerie*	LPG + Ethane *GPL + éthane*	Motor Gasoline *Essence moteur*	Aviation Gasoline *Essence aviation*	Jet Fuel *Carbu- réacteurs*	Oil cont. / *Pétrole cont.* (1000 tonnes) Kerosene *Kérosène*	Gas/ Diesel *Gazole*	Heavy Fuel Oil *Fioul lourd*	Naphtha *Naphta*	Petrol. Coke *Coke de pétrole*	Other Prod. *Autres prod.*
Production	-	498	2109	-	-	1330	2958	3166	887	-	331
From Other Sources	-	-	-	-	-	-	-	-	-	-	-
Imports	-	-	5	-	-	279	34	2213	60	-	258
Exports	-	-89	-286	-	-	-462	-786	-1124	-25	-	-33
Intl. Marine Bunkers	-	-	-	-	-	-	-64	-89	-	-	-
Stock Changes	-	-	174	-	-	-77	444	362	-35	-	11
DOMESTIC SUPPLY	-	409	2002	-	-	1070	2586	4528	887	-	567
Transfers	-	-	-	-	-	-	-	-	-	-	-
Statistical Differences	-	-17	-	-	-	-	-	-	-	-	-
TRANSFORMATION	-	-	-	-	-	-	225	2527	-	-	-
Electricity Plants	-	-	-	-	-	-	225	2527	-	-	-
CHP Plants	-	-	-	-	-	-	-	-	-	-	-
Heat Plants	-	-	-	-	-	-	-	-	-	-	-
Blast Furnaces/Gas Works	-	-	-	-	-	-	-	-	-	-	-
Coke/Pat. Fuel/BKB Plants	-	-	-	-	-	-	-	-	-	-	-
Petroleum Refineries	-	-	-	-	-	-	-	-	-	-	-
Petrochemical Industry	-	-	-	-	-	-	-	-	-	-	-
Liquefaction	-	-	-	-	-	-	-	-	-	-	-
Other Transform. Sector	-	-	-	-	-	-	-	-	-	-	-
ENERGY SECTOR	-	-	-	-	-	-	-	508	-	-	-
Coal Mines	-	-	-	-	-	-	-	-	-	-	-
Oil and Gas Extraction	-	-	-	-	-	-	-	-	-	-	-
Petroleum Refineries	-	-	-	-	-	-	-	508	-	-	-
Electr., CHP+Heat Plants	-	-	-	-	-	-	-	-	-	-	-
Pumped Storage (Elec.)	-	-	-	-	-	-	-	-	-	-	-
Other Energy Sector	-	-	-	-	-	-	-	-	-	-	-
Distribution Losses	-	-	-	-	-	-	-	-	-	-	-
FINAL CONSUMPTION	-	392	2002	-	-	1070	2361	1493	887	-	567
INDUSTRY SECTOR	-	123	-	-	-	-	159	981	887	-	-
Iron and Steel	-	-	-	-	-	-	-	-	-	-	-
Chemical and Petrochem.	-	-	-	-	-	-	-	-	887	-	-
of which: Feedstocks	-	-	-	-	-	-	-	-	887	-	-
Non-Ferrous Metals	-	-	-	-	-	-	-	-	-	-	-
Non-Metallic Minerals	-	-	-	-	-	-	-	-	-	-	-
Transport Equipment	-	-	-	-	-	-	-	-	-	-	-
Machinery	-	-	-	-	-	-	-	-	-	-	-
Mining and Quarrying	-	-	-	-	-	-	-	-	-	-	-
Food and Tobacco	-	-	-	-	-	-	-	-	-	-	-
Paper, Pulp and Print	-	-	-	-	-	-	-	-	-	-	-
Wood and Wood Products	-	-	-	-	-	-	-	-	-	-	-
Construction	-	-	-	-	-	-	-	-	-	-	-
Textile and Leather	-	-	-	-	-	-	-	-	-	-	-
Non-specified	-	123	-	-	-	-	159	981	-	-	-
TRANSPORT SECTOR	-	-	2002	-	-	695	1058	-	-	-	-
Air	-	-	-	-	-	695	-	-	-	-	-
Road	-	-	2002	-	-	-	1058	-	-	-	-
Rail	-	-	-	-	-	-	-	-	-	-	-
Pipeline Transport	-	-	-	-	-	-	-	-	-	-	-
Internal Navigation	-	-	-	-	-	-	-	-	-	-	-
Non-specified	-	-	-	-	-	-	-	-	-	-	-
OTHER SECTORS	-	269	-	-	-	375	1144	512	-	-	-
Agriculture	-	-	-	-	-	-	-	-	-	-	-
Comm. and Publ. Services	-	-	-	-	-	-	-	-	-	-	-
Residential	-	269	-	-	-	375	-	-	-	-	-
Non-specified	-	-	-	-	-	-	1144	512	-	-	-
NON-ENERGY USE	-	-	-	-	-	-	-	-	-	-	567
in Industry/Transf./Energy	-	-	-	-	-	-	-	-	-	-	567
in Transport	-	-	-	-	-	-	-	-	-	-	-
in Other Sectors	-	-	-	-	-	-	-	-	-	-	-

Israel / Israël : 1998

SUPPLY AND CONSUMPTION APPROVISIONNEMENT ET DEMANDE	Gas / Gaz (TJ)				Comb. Renew. & Waste / En. Re. Comb. & Déchets (TJ)				(GWh)	(TJ)
	Natural Gas Gaz naturel	Gas Works Usines à gaz	Coke Ovens Cokeries	Blast Furnaces Hauts fourneaux	Solid Biomass Biomasse solide	Gas/Liquids from Biomass Gaz/Liquides tirés de biomasse	Municipal Waste Déchets urbains	Industrial Waste Déchets industriels	Electricity Electricité	Heat Chaleur
Production	479	-	-	-	183	-	-	-	37964	-
From Other Sources	-	-	-	-	-	-	-	-	-	
Imports	-	-	-	-	195	-	-	-	-	-
Exports	-	-	-	-	-	-	-	-	-1055	-
Intl. Marine Bunkers	-	-	-	-		-	-	-		-
Stock Changes	-	-	-	-	-	-	-	-	-	-
DOMESTIC SUPPLY	479	-	-	-	378	-	-	-	36909	-
Transfers	-	-	-	-		-	-	-	-	
Statistical Differences	-	-	-	-		-	-	-	1024	-
TRANSFORMATION	200	-	-	-	-	-	-	-		-
Electricity Plants	200	-	-	-	-	-	-	-		-
CHP Plants	-	-	-	-	-	-	-	-	-	-
Heat Plants	-	-	-	-	-	-	-	-	-	-
Blast Furnaces/Gas Works	-	-	-	-	-	-	-	-	-	-
Coke/Pat. Fuel/BKB Plants	-	-	-	-	-	-	-	-	-	-
Petroleum Refineries	-	-	-	-	-	-	-	-	-	-
Petrochemical Industry	-	-	-	-	-	-	-	-	-	-
Liquefaction	-	-	-	-	-	-	-	-	-	-
Other Transform. Sector	-	-	-	-	-	-	-	-	-	-
ENERGY SECTOR	-	-	-	-	-	-	-	-	3133	-
Coal Mines	-	-	-	-	-	-	-	-	-	-
Oil and Gas Extraction	-	-	-	-	-	-	-	-	-	-
Petroleum Refineries	-	-	-	-	-	-	-	-	-	-
Electr., CHP+Heat Plants	-	-	-	-	-	-	-	-	3133	-
Pumped Storage (Elec.)	-	-	-	-	-	-	-	-	-	-
Other Energy Sector	-	-	-	-	-	-	-	-		-
Distribution Losses	-	-	-	-	-	-	-	-	2152	-
FINAL CONSUMPTION	279	-	-	-	378	-	-	-	32648	-
INDUSTRY SECTOR	279	-	-	-	-	-	-	-	9921	-
Iron and Steel	-	-	-	-	-	-	-	-	960	-
Chemical and Petrochem.	279	-	-	-	-	-	-	-	2051	-
of which: Feedstocks	279	-	-	-	-	-	-	-		-
Non-Ferrous Metals	-	-	-	-	-	-	-	-	-	-
Non-Metallic Minerals	-	-	-	-	-	-	-	-	874	-
Transport Equipment	-	-	-	-	-	-	-	-	281	-
Machinery	-	-	-	-	-	-	-	-	803	-
Mining and Quarrying	-	-	-	-	-	-	-	-	1267	-
Food and Tobacco	-	-	-	-	-	-	-	-	1036	-
Paper, Pulp and Print	-	-	-	-	-	-	-	-	408	-
Wood and Wood Products	-	-	-	-	-	-	-	-	110	-
Construction	-	-	-	-	-	-	-	-	161	-
Textile and Leather	-	-	-	-	-	-	-	-	548	-
Non-specified	-	-	-	-	-	-	-	-	1422	-
TRANSPORT SECTOR	-	-	-	-	-	-	-	-	-	-
Air	-	-	-	-	-	-	-	-	-	-
Road	-	-	-	-	-	-	-	-	-	-
Rail	-	-	-	-	-	-	-	-	-	-
Pipeline Transport	-	-	-	-	-	-	-	-	-	-
Internal Navigation	-	-	-	-	-	-	-	-	-	-
Non-specified	-	-	-	-	-	-	-	-	-	-
OTHER SECTORS	-	-	-	-	378	-	-	-	22727	-
Agriculture	-	-	-	-	-	-	-	-	1487	-
Comm. and Publ. Services	-	-	-	-	-	-	-	-	8583	-
Residential	-	-	-	-	183	-	-	-	10147	-
Non-specified	-	-	-	-	195	-	-	-	2510	-
NON-ENERGY USE	-	-	-	-	-	-	-	-	-	-
in Industry/Transf./Energy	-	-	-	-	-	-	-	-	-	-
in Transport	-	-	-	-	-	-	-	-	-	-
in Other Sectors	-	-	-	-	-	-	-	-	-	-

Jamaica / Jamaïque

SUPPLY AND CONSUMPTION 1997	Coal (1000 tonnes)							Oil (1000 tonnes)			
	Coking Coal	Other Bit. Coal	Sub-Bit. Coal	Lignite	Peat	Oven and Gas Coke	Pat. Fuel and BKB	Crude Oil	NGL	Feed-stocks	Additives
Production	-		-	-	-	-	-	-	-	-	-
Imports	56	-	-	-	-	-	-	1125	-	-	-
Exports	-	-	-	-	-	-	-	-	-	-	-
Intl. Marine Bunkers	-	-	-	-	-	-	-	-	-	-	-
Stock Changes	-	-	-	-	-	-	-	44	-	-	-
DOMESTIC SUPPLY	56	-	-	-	-	-	-	1169	-	-	-
Transfers and Stat. Diff.	-	-	-	-	-	-	-	-	-	-	-
TRANSFORMATION	-	-	-	-	-	-	-	1169	-	-	-
Electricity and CHP Plants	-	-	-	-	-	-	-	-	-	-	-
Petroleum Refineries	-	-	-	-	-	-	-	1169	-	-	-
Other Transform. Sector	-	-	-	-	-	-	-	-	-	-	-
ENERGY SECTOR	-	-	-	-	-	-	-	-	-	-	-
DISTRIBUTION LOSSES	-	-	-	-	-	-	-	-	-	-	-
FINAL CONSUMPTION	56	-	-	-	-	-	-	-	-	-	-
INDUSTRY SECTOR	56	-	-	-	-	-	-	-	-	-	-
Iron and Steel	-	-	-	-	-	-	-	-	-	-	-
Chemical and Petrochem.	-	-	-	-	-	-	-	-	-	-	-
Non-Metallic Minerals	56	-	-	-	-	-	-	-	-	-	-
Non-specified	-	-	-	-	-	-	-	-	-	-	-
TRANSPORT SECTOR	-	-	-	-	-	-	-	-	-	-	-
Air	-	-	-	-	-	-	-	-	-	-	-
Road	-	-	-	-	-	-	-	-	-	-	-
Non-specified	-	-	-	-	-	-	-	-	-	-	-
OTHER SECTORS	-	-	-	-	-	-	-	-	-	-	-
Agriculture	-	-	-	-	-	-	-	-	-	-	-
Comm. and Publ. Services	-	-	-	-	-	-	-	-	-	-	-
Residential	-	-	-	-	-	-	-	-	-	-	-
Non-specified	-	-	-	-	-	-	-	-	-	-	-
NON-ENERGY USE	-	-	-	-	-	-	-	-	-	-	-

APPROVISIONNEMENT ET DEMANDE 1998	Charbon (1000 tonnes)							Pétrole (1000 tonnes)			
	Charbon à coke	Autres charb. bit.	Charbon sous-bit.	Lignite	Tourbe	Coke de four/gaz	Agg./briq. de lignite	Pétrole brut	LGN	Produits d'aliment.	Additifs
Production	-	-	-	-	-	-	-	-	-	-	-
Imports	27	-	-	-	-	-	-	1198	-	-	-
Exports	-	-	-	-	-	-	-	-	-	-	-
Intl. Marine Bunkers	-	-	-	-	-	-	-	-	-	-	-
Stock Changes	-	-	-	-	-	-	-	8	-	-	-
DOMESTIC SUPPLY	27	-	-	-	-	-	-	1206	-	-	-
Transfers and Stat. Diff.	-	-	-	-	-	-	-	-	-	-	-
TRANSFORMATION	-	-	-	-	-	-	-	1206	-	-	-
Electricity and CHP Plants	-	-	-	-	-	-	-	-	-	-	-
Petroleum Refineries	-	-	-	-	-	-	-	1206	-	-	-
Other Transform. Sector	-	-	-	-	-	-	-	-	-	-	-
ENERGY SECTOR	-	-	-	-	-	-	-	-	-	-	-
DISTRIBUTION LOSSES	-	-	-	-	-	-	-	-	-	-	-
FINAL CONSUMPTION	27	-	-	-	-	-	-	-	-	-	-
INDUSTRY SECTOR	27	-	-	-	-	-	-	-	-	-	-
Iron and Steel	-	-	-	-	-	-	-	-	-	-	-
Chemical and Petrochem.	-	-	-	-	-	-	-	-	-	-	-
Non-Metallic Minerals	27	-	-	-	-	-	-	-	-	-	-
Non-specified	-	-	-	-	-	-	-	-	-	-	-
TRANSPORT SECTOR	-	-	-	-	-	-	-	-	-	-	-
Air	-	-	-	-	-	-	-	-	-	-	-
Road	-	-	-	-	-	-	-	-	-	-	-
Non-specified	-	-	-	-	-	-	-	-	-	-	-
OTHER SECTORS	-	-	-	-	-	-	-	-	-	-	-
Agriculture	-	-	-	-	-	-	-	-	-	-	-
Comm. and Publ. Services	-	-	-	-	-	-	-	-	-	-	-
Residential	-	-	-	-	-	-	-	-	-	-	-
Non-specified	-	-	-	-	-	-	-	-	-	-	-
NON-ENERGY USE	-	-	-	-	-	-	-	-	-	-	-

Jamaica / Jamaïque

SUPPLY AND CONSUMPTION 1997	Oil cont. (1000 tonnes)										
	Refinery Gas	LPG + Ethane	Motor Gasoline	Aviation Gasoline	Jet Fuel	Kerosene	Gas/ Diesel	Heavy Fuel Oil	Naphtha	Petrol. Coke	Other Prod.
Production	-	7	179	-	41	16	273	506	-	-	16
Imports	-	62	220	1	153	8	271	1431	-	-	17
Exports	-	-	-	-	-	-	-20	-38	-	-	-
Intl. Marine Bunkers	-	-	-	-	-	-	-37	-	-	-	-
Stock Changes	-	-	8	-	2	1	25	40	-	-	-5
DOMESTIC SUPPLY	-	69	407	1	196	25	512	1939	-	-	28
Transfers and Stat. Diff.	-	-7	30	-	-10	-10	-	-2	-	-	-6
TRANSFORMATION	-	-	-	-	-	-	245	1566	-	-	-
Electricity and CHP Plants	-	-	-	-	-	-	245	1566	-	-	-
Petroleum Refineries	-	-	-	-	-	-	-	-	-	-	-
Other Transform. Sector	-	-	-	-	-	-	-	-	-	-	-
ENERGY SECTOR	-	-	-	-	-	-	-	2	-	-	-
DISTRIBUTION LOSSES	-	-	-	-	-	-	-	-	-	-	-
FINAL CONSUMPTION	-	62	437	1	186	15	267	369	-	-	22
INDUSTRY SECTOR	-	-	-	-	-	-	95	100	-	-	-
Iron and Steel	-	-	-	-	-	-	-	-	-	-	-
Chemical and Petrochem.	-	-	-	-	-	-	-	-	-	-	-
Non-Metallic Minerals	-	-	-	-	-	-	-	-	-	-	-
Non-specified	-	-	-	-	-	-	95	100	-	-	-
TRANSPORT SECTOR	-	-	437	1	186	-	162	12	-	-	-
Air	-	-	-	1	186	-	-	-	-	-	-
Road	-	-	437	-	-	-	-	-	-	-	-
Non-specified	-	-	-	-	-	-	162	12	-	-	-
OTHER SECTORS	-	62	-	-	-	15	10	257	-	-	-
Agriculture	-	-	-	-	-	-	6	249	-	-	-
Comm. and Publ. Services	-	1	-	-	-	2	4	8	-	-	-
Residential	-	61	-	-	-	13	-	-	-	-	-
Non-specified	-	-	-	-	-	-	-	-	-	-	-
NON-ENERGY USE	-	-	-	-	-	-	-	-	-	-	22

APPROVISIONNEMENT ET DEMANDE 1998	Pétrole cont. (1000 tonnes)										
	Gaz de raffinerie	GPL + éthane	Essence moteur	Essence aviation	Carbu- réacteurs	Kérosène	Gazole	Fioul lourd	Naphta	Coke de pétrole	Autres prod.
Production	-	7	194	-	43	2	280	520	-	-	4
Imports	-	68	240	-	163	6	278	1453	-	-	15
Exports	-	-	-	-	-	-	-19	-39	-	-	-
Intl. Marine Bunkers	-	-	-	-	-	-	-37	-	-	-	-
Stock Changes	-	-	7	-	2	-	24	26	-	-	-3
DOMESTIC SUPPLY	-	75	441	-	208	8	526	1960	-	-	16
Transfers and Stat. Diff.	-	-9	25	-	-16	3	-	-	-	-	-
TRANSFORMATION	-	-	-	-	-	-	252	1584	-	-	-
Electricity and CHP Plants	-	-	-	-	-	-	252	1584	-	-	-
Petroleum Refineries	-	-	-	-	-	-	-	-	-	-	-
Other Transform. Sector	-	-	-	-	-	-	-	-	-	-	-
ENERGY SECTOR	-	-	-	-	-	-	-	2	-	-	-
DISTRIBUTION LOSSES	-	-	-	-	-	-	-	-	-	-	-
FINAL CONSUMPTION	-	66	466	-	192	11	274	374	-	-	16
INDUSTRY SECTOR	-	-	-	-	-	-	97	101	-	-	-
Iron and Steel	-	-	-	-	-	-	-	-	-	-	-
Chemical and Petrochem.	-	-	-	-	-	-	-	-	-	-	-
Non-Metallic Minerals	-	-	-	-	-	-	-	-	-	-	-
Non-specified	-	-	-	-	-	-	97	101	-	-	-
TRANSPORT SECTOR	-	-	466	-	192	-	167	12	-	-	-
Air	-	-	-	-	192	-	-	-	-	-	-
Road	-	-	466	-	-	-	-	-	-	-	-
Non-specified	-	-	-	-	-	-	167	12	-	-	-
OTHER SECTORS	-	66	-	-	-	11	10	261	-	-	-
Agriculture	-	-	-	-	-	-	6	252	-	-	-
Comm. and Publ. Services	-	1	-	-	-	1	4	9	-	-	-
Residential	-	65	-	-	-	10	-	-	-	-	-
Non-specified	-	-	-	-	-	-	-	-	-	-	-
NON-ENERGY USE	-	-	-	-	-	-	-	-	-	-	16

Jamaica / Jamaïque

SUPPLY AND CONSUMPTION 1997	Gas (TJ)				Comb. Renew. & Waste (TJ)				(GWh)	(TJ)
	Natural Gas	Gas Works	Coke Ovens	Blast Furnaces	Solid Biomass	Gas/Liquids from Biomass	Municipal Waste	Industrial Waste	Electricity	Heat
Production	-	-	-	-	24501	-	-	-	6255	-
Imports	-	-	-	-	-	-	-	-	-	-
Exports	-	-	-	-	-	-	-	-	-	-
Intl. Marine Bunkers	-	-	-	-	-	-	-	-	-	-
Stock Changes	-	-	-	-	-	-	-	-	-	-
DOMESTIC SUPPLY	-	-	-	-	24501	-	-	-	6255	-
Transfers and Stat. Diff.	-	-	-	-	-	-	-	-	-19	-
TRANSFORMATION	-	-	-	-	21556	-	-	-	-	-
Electricity and CHP Plants	-	-	-	-	13272	-	-	-	-	-
Petroleum Refineries	-	-	-	-	-	-	-	-	-	-
Other Transform. Sector	-	-	-	-	8284	-	-	-	-	-
ENERGY SECTOR	-	-	-	-	-	-	-	-	16	-
DISTRIBUTION LOSSES	-	-	-	-	-	-	-	-	676	-
FINAL CONSUMPTION	-	-	-	-	2945	-	-	-	5544	-
INDUSTRY SECTOR	-	-	-	-	-	-	-	-	3548	-
Iron and Steel	-	-	-	-	-	-	-	-	-	-
Chemical and Petrochem.	-	-	-	-	-	-	-	-	-	-
Non-Metallic Minerals	-	-	-	-	-	-	-	-	-	-
Non-specified	-	-	-	-	-	-	-	-	3548	-
TRANSPORT SECTOR	-	-	-	-	-	-	-	-	-	-
Air	-	-	-	-	-	-	-	-	-	-
Road	-	-	-	-	-	-	-	-	-	-
Non-specified	-	-	-	-	-	-	-	-	-	-
OTHER SECTORS	-	-	-	-	2945	-	-	-	1996	-
Agriculture	-	-	-	-	-	-	-	-	-	-
Comm. and Publ. Services	-	-	-	-	-	-	-	-	1162	-
Residential	-	-	-	-	2945	-	-	-	829	-
Non-specified	-	-	-	-	-	-	-	-	5	-
NON-ENERGY USE	-	-	-	-	-	-	-	-	-	-

APPROVISIONNEMENT ET DEMANDE 1998	Gaz (TJ)				En. Re. Comb. & Déchets (TJ)				(GWh)	(TJ)
	Gaz naturel	Usines à gaz	Cokeries	Hauts fourneaux	Biomasse solide	Gaz/Liquides tirés de biomasse	Déchets urbains	Déchets industriels	Electricité	Chaleur
Production	-	-	-	-	27105	-	-	-	6480	-
Imports	-	-	-	-	-	-	-	-	-	-
Exports	-	-	-	-	-	-	-	-	-	-
Intl. Marine Bunkers	-	-	-	-	-	-	-	-	-	-
Stock Changes	-	-	-	-	-	-	-	-	-	-
DOMESTIC SUPPLY	-	-	-	-	27105	-	-	-	6480	-
Transfers and Stat. Diff.	-	-	-	-	-	-	-	-	-19	-
TRANSFORMATION	-	-	-	-	24057	-	-	-	-	-
Electricity and CHP Plants	-	-	-	-	15544	-	-	-	-	-
Petroleum Refineries	-	-	-	-	-	-	-	-	-	-
Other Transform. Sector	-	-	-	-	8513	-	-	-	-	-
ENERGY SECTOR	-	-	-	-	-	-	-	-	17	-
DISTRIBUTION LOSSES	-	-	-	-	-	-	-	-	642	-
FINAL CONSUMPTION	-	-	-	-	3047	-	-	-	5802	-
INDUSTRY SECTOR	-	-	-	-	-	-	-	-	3713	-
Iron and Steel	-	-	-	-	-	-	-	-	-	-
Chemical and Petrochem.	-	-	-	-	-	-	-	-	-	-
Non-Metallic Minerals	-	-	-	-	-	-	-	-	-	-
Non-specified	-	-	-	-	-	-	-	-	3713	-
TRANSPORT SECTOR	-	-	-	-	-	-	-	-	-	-
Air	-	-	-	-	-	-	-	-	-	-
Road	-	-	-	-	-	-	-	-	-	-
Non-specified	-	-	-	-	-	-	-	-	-	-
OTHER SECTORS	-	-	-	-	3047	-	-	-	2089	-
Agriculture	-	-	-	-	-	-	-	-	-	-
Comm. and Publ. Services	-	-	-	-	-	-	-	-	1158	-
Residential	-	-	-	-	3047	-	-	-	926	-
Non-specified	-	-	-	-	-	-	-	-	5	-
NON-ENERGY USE	-	-	-	-	-	-	-	-	-	-

Jordan / Jordanie

SUPPLY AND CONSUMPTION 1997	Coal (1000 tonnes)							Oil (1000 tonnes)			
	Coking Coal	Other Bit. Coal	Sub-Bit. Coal	Lignite	Peat	Oven and Gas Coke	Pat. Fuel and BKB	Crude Oil	NGL	Feed-stocks	Additives
Production	-	-	-	-	-	-	-	2	-	-	-
Imports	-	-	-	-	-	-	-	3441	-	-	-
Exports	-	-	-	-	-	-	-	-	-	-	-
Intl. Marine Bunkers	-	-	-	-	-	-	-	-	-	-	-
Stock Changes	-	-	-	-	-	-	-	32	-	-	-
DOMESTIC SUPPLY	-	-	-	-	-	-	-	3475	-	-	-
Transfers and Stat. Diff.	-	-	-	-	-	-	-	-	-	-	-
TRANSFORMATION	-	-	-	-	-	-	-	3475	-	-	-
Electricity and CHP Plants	-	-	-	-	-	-	-	-	-	-	-
Petroleum Refineries	-	-	-	-	-	-	-	3475	-	-	-
Other Transform. Sector	-	-	-	-	-	-	-	-	-	-	-
ENERGY SECTOR	-	-	-	-	-	-	-	-	-	-	-
DISTRIBUTION LOSSES	-	-	-	-	-	-	-	-	-	-	-
FINAL CONSUMPTION	-	-	-	-	-	-	-	-	-	-	-
INDUSTRY SECTOR	-	-	-	-	-	-	-	-	-	-	-
Iron and Steel	-	-	-	-	-	-	-	-	-	-	-
Chemical and Petrochem.	-	-	-	-	-	-	-	-	-	-	-
Non-Metallic Minerals	-	-	-	-	-	-	-	-	-	-	-
Non-specified	-	-	-	-	-	-	-	-	-	-	-
TRANSPORT SECTOR	-	-	-	-	-	-	-	-	-	-	-
Air	-	-	-	-	-	-	-	-	-	-	-
Road	-	-	-	-	-	-	-	-	-	-	-
Non-specified	-	-	-	-	-	-	-	-	-	-	-
OTHER SECTORS	-	-	-	-	-	-	-	-	-	-	-
Agriculture	-	-	-	-	-	-	-	-	-	-	-
Comm. and Publ. Services	-	-	-	-	-	-	-	-	-	-	-
Residential	-	-	-	-	-	-	-	-	-	-	-
Non-specified	-	-	-	-	-	-	-	-	-	-	-
NON-ENERGY USE	-	-	-	-	-	-	-	-	-	-	-

APPROVISIONNEMENT ET DEMANDE 1998	Charbon (1000 tonnes)							Pétrole (1000 tonnes)			
	Charbon à coke	Autres charb. bit.	Charbon sous-bit.	Lignite	Tourbe	Coke de four/gaz	Agg./briq. de lignite	Pétrole brut	LGN	Produits d'aliment.	Additifs
Production	-	-	-	-	-	-	-	2	-	-	-
Imports	-	-	-	-	-	-	-	3556	-	-	-
Exports	-	-	-	-	-	-	-	-	-	-	-
Intl. Marine Bunkers	-	-	-	-	-	-	-	-	-	-	-
Stock Changes	-	-	-	-	-	-	-	-73	-	-	-
DOMESTIC SUPPLY	-	-	-	-	-	-	-	3485	-	-	-
Transfers and Stat. Diff.	-	-	-	-	-	-	-	-	-	-	-
TRANSFORMATION	-	-	-	-	-	-	-	3485	-	-	-
Electricity and CHP Plants	-	-	-	-	-	-	-	-	-	-	-
Petroleum Refineries	-	-	-	-	-	-	-	3485	-	-	-
Other Transform. Sector	-	-	-	-	-	-	-	-	-	-	-
ENERGY SECTOR	-	-	-	-	-	-	-	-	-	-	-
DISTRIBUTION LOSSES	-	-	-	-	-	-	-	-	-	-	-
FINAL CONSUMPTION	-	-	-	-	-	-	-	-	-	-	-
INDUSTRY SECTOR	-	-	-	-	-	-	-	-	-	-	-
Iron and Steel	-	-	-	-	-	-	-	-	-	-	-
Chemical and Petrochem.	-	-	-	-	-	-	-	-	-	-	-
Non-Metallic Minerals	-	-	-	-	-	-	-	-	-	-	-
Non-specified	-	-	-	-	-	-	-	-	-	-	-
TRANSPORT SECTOR	-	-	-	-	-	-	-	-	-	-	-
Air	-	-	-	-	-	-	-	-	-	-	-
Road	-	-	-	-	-	-	-	-	-	-	-
Non-specified	-	-	-	-	-	-	-	-	-	-	-
OTHER SECTORS	-	-	-	-	-	-	-	-	-	-	-
Agriculture	-	-	-	-	-	-	-	-	-	-	-
Comm. and Publ. Services	-	-	-	-	-	-	-	-	-	-	-
Residential	-	-	-	-	-	-	-	-	-	-	-
Non-specified	-	-	-	-	-	-	-	-	-	-	-
NON-ENERGY USE	-	-	-	-	-	-	-	-	-	-	-

Jordan / Jordanie

SUPPLY AND CONSUMPTION 1997	Oil cont. (1000 tonnes)										
	Refinery Gas	LPG + Ethane	Motor Gasoline	Aviation Gasoline	Jet Fuel	Kerosene	Gas/ Diesel	Heavy Fuel Oil	Naphtha	Petrol. Coke	Other Prod.
Production	10	143	531	6	267	196	923	1199	-	-	194
Imports	-	106	-	-	-	-	215	611	-	-	-
Exports	-	-	-	-	-	-	-	-	-	-	-
Intl. Marine Bunkers	-	-	-	-	-	-	-2	-4	-	-	-
Stock Changes	-	-	-2	-	4	11	26	-6	-	-	-1
DOMESTIC SUPPLY	10	249	529	6	271	207	1162	1800	-	-	193
Transfers and Stat. Diff.	-	-	4	-	-	1	75	1	-	-	-
TRANSFORMATION	-	-	-	-	-	-	252	1194	-	-	-
Electricity and CHP Plants	-	-	-	-	-	-	252	1194	-	-	-
Petroleum Refineries	-	-	-	-	-	-	-	-	-	-	-
Other Transform. Sector	-	-	-	-	-	-	-	-	-	-	-
ENERGY SECTOR	10	-	-	-	-	-	1	157	-	-	64
DISTRIBUTION LOSSES	-	-	-	-	-	-	-	-	-	-	-
FINAL CONSUMPTION	-	249	533	6	271	208	984	450	-	-	129
INDUSTRY SECTOR	-	3	-	-	-	2	139	445	-	-	-
Iron and Steel	-	-	-	-	-	-	-	-	-	-	-
Chemical and Petrochem.	-	-	-	-	-	-	-	-	-	-	-
Non-Metallic Minerals	-	-	-	-	-	-	-	-	-	-	-
Non-specified	-	3	-	-	-	2	139	445	-	-	-
TRANSPORT SECTOR	-	-	533	6	271	-	495	5	-	-	-
Air	-	-	-	6	271	-	-	-	-	-	-
Road	-	-	533	-	-	-	495	-	-	-	-
Non-specified	-	-	-	-	-	-	-	5	-	-	-
OTHER SECTORS	-	246	-	-	-	206	350	-	-	-	-
Agriculture	-	-	-	-	-	-	-	-	-	-	-
Comm. and Publ. Services	-	15	-	-	-	11	110	-	-	-	-
Residential	-	225	-	-	-	195	152	-	-	-	-
Non-specified	-	6	-	-	-	-	88	-	-	-	-
NON-ENERGY USE	-	-	-	-	-	-	-	-	-	-	129

APPROVISIONNEMENT ET DEMANDE 1998	Pétrole cont. (1000 tonnes)										
	Gaz de raffinerie	GPL + éthane	Essence moteur	Essence aviation	Carbu- réacteurs	Kérosène	Gazole	Fioul lourd	Naphta	Coke de pétrole	Autres prod.
Production	10	133	524	6	221	195	928	1253	-	-	211
Imports	-	104	7	-	-	-	246	796	-	-	-
Exports	-	-	-	-	-	-	-	-	-	-	-
Intl. Marine Bunkers	-	-	-	-	-	-	-1	-2	-	-	-
Stock Changes	-	7	-5	-	-2	-13	-33	-56	-	-	-
DOMESTIC SUPPLY	10	244	526	6	219	182	1140	1991	-	-	211
Transfers and Stat. Diff.	-	-1	4	-	-	-	93	1	-	-	-
TRANSFORMATION	-	-	-	-	-	-	238	1345	-	-	-
Electricity and CHP Plants	-	-	-	-	-	-	238	1345	-	-	-
Petroleum Refineries	-	-	-	-	-	-	-	-	-	-	-
Other Transform. Sector	-	-	-	-	-	-	-	-	-	-	-
ENERGY SECTOR	10	-	-	-	-	-	1	163	-	-	68
DISTRIBUTION LOSSES	-	-	-	-	-	-	-	-	-	-	-
FINAL CONSUMPTION	-	243	530	6	219	182	994	484	-	-	143
INDUSTRY SECTOR	-	3	-	-	-	3	150	479	-	-	-
Iron and Steel	-	-	-	-	-	-	-	-	-	-	-
Chemical and Petrochem.	-	-	-	-	-	-	-	-	-	-	-
Non-Metallic Minerals	-	-	-	-	-	-	-	-	-	-	-
Non-specified	-	3	-	-	-	3	150	479	-	-	-
TRANSPORT SECTOR	-	-	530	6	219	-	510	5	-	-	-
Air	-	-	-	6	219	-	-	-	-	-	-
Road	-	-	530	-	-	-	510	-	-	-	-
Non-specified	-	-	-	-	-	-	-	5	-	-	-
OTHER SECTORS	-	240	-	-	-	179	334	-	-	-	-
Agriculture	-	-	-	-	-	-	-	-	-	-	-
Comm. and Publ. Services	-	14	-	-	-	9	115	-	-	-	-
Residential	-	220	-	-	-	170	140	-	-	-	-
Non-specified	-	6	-	-	-	-	79	-	-	-	-
NON-ENERGY USE	-	-	-	-	-	-	-	-	-	-	143

INTERNATIONAL ENERGY AGENCY

Jordan / Jordanie

SUPPLY AND CONSUMPTION 1997	Gas (TJ)				Comb. Renew. & Waste (TJ)				(GWh)	(TJ)
	Natural Gas	Gas Works	Coke Ovens	Blast Furnaces	Solid Biomass	Gas/Liquids from Biomass	Municipal Waste	Industrial Waste	Electricity	Heat
Production	10377	-	-	-	91	-	-	-	6264	-
Imports	-	-	-	-	20	-	-	-	-	-
Exports	-	-	-	-	-	-	-	-	-3	-
Intl. Marine Bunkers	-	-	-	-	-	-	-	-	-	-
Stock Changes	-	-	-	-	-	-	-	-	-	-
DOMESTIC SUPPLY	10377	-	-	-	111	-	-	-	6261	-
Transfers and Stat. Diff.	-	-	-	-	-	-	-	-	-16	-
TRANSFORMATION	10377	-	-	-	-	-	-	-	-	-
Electricity and CHP Plants	10377	-	-	-	-	-	-	-	-	-
Petroleum Refineries	-	-	-	-	-	-	-	-	-	-
Other Transform. Sector	-	-	-	-	-	-	-	-	-	-
ENERGY SECTOR	-	-	-	-	-	-	-	-	414	-
DISTRIBUTION LOSSES	-	-	-	-	-	-	-	-	637	-
FINAL CONSUMPTION	-	-	-	-	111	-	-	-	5194	-
INDUSTRY SECTOR	-	-	-	-	-	-	-	-	1712	-
Iron and Steel	-	-	-	-	-	-	-	-	13	-
Chemical and Petrochem.	-	-	-	-	-	-	-	-	166	-
Non-Metallic Minerals	-	-	-	-	-	-	-	-	366	-
Non-specified	-	-	-	-	-	-	-	-	1167	-
TRANSPORT SECTOR	-	-	-	-	-	-	-	-	-	-
Air	-	-	-	-	-	-	-	-	-	-
Road	-	-	-	-	-	-	-	-	-	-
Non-specified	-	-	-	-	-	-	-	-	-	-
OTHER SECTORS	-	-	-	-	111	-	-	-	3482	-
Agriculture	-	-	-	-	-	-	-	-	936	-
Comm. and Publ. Services	-	-	-	-	-	-	-	-	918	-
Residential	-	-	-	-	91	-	-	-	1628	-
Non-specified	-	-	-	-	20	-	-	-	-	-
NON-ENERGY USE	-	-	-	-	-	-	-	-	-	-

APPROVISIONNEMENT ET DEMANDE 1998	Gaz (TJ)				En. Re. Comb. & Déchets (TJ)				(GWh)	(TJ)
	Gaz naturel	Usines à gaz	Cokeries	Hauts fourneaux	Biomasse solide	Gaz/Liquides tirés de biomasse	Déchets urbains	Déchets industriels	Electricité	Chaleur
Production	10517	-	-	-	106	-	-	-	6745	-
Imports	-	-	-	-	20	-	-	-	-	-
Exports	-	-	-	-	-	-	-	-	-4	-
Intl. Marine Bunkers	-	-	-	-	-	-	-	-	-	-
Stock Changes	-	-	-	-	-	-	-	-	-	-
DOMESTIC SUPPLY	10517	-	-	-	126	-	-	-	6741	-
Transfers and Stat. Diff.	-	-	-	-	-	-	-	-	-11	-
TRANSFORMATION	10517	-	-	-	-	-	-	-	-	-
Electricity and CHP Plants	10517	-	-	-	-	-	-	-	-	-
Petroleum Refineries	-	-	-	-	-	-	-	-	-	-
Other Transform. Sector	-	-	-	-	-	-	-	-	-	-
ENERGY SECTOR	-	-	-	-	-	-	-	-	494	-
DISTRIBUTION LOSSES	-	-	-	-	-	-	-	-	695	-
FINAL CONSUMPTION	-	-	-	-	126	-	-	-	5541	-
INDUSTRY SECTOR	-	-	-	-	-	-	-	-	1809	-
Iron and Steel	-	-	-	-	-	-	-	-	12	-
Chemical and Petrochem.	-	-	-	-	-	-	-	-	171	-
Non-Metallic Minerals	-	-	-	-	-	-	-	-	307	-
Non-specified	-	-	-	-	-	-	-	-	1319	-
TRANSPORT SECTOR	-	-	-	-	-	-	-	-	-	-
Air	-	-	-	-	-	-	-	-	-	-
Road	-	-	-	-	-	-	-	-	-	-
Non-specified	-	-	-	-	-	-	-	-	-	-
OTHER SECTORS	-	-	-	-	126	-	-	-	3732	-
Agriculture	-	-	-	-	-	-	-	-	945	-
Comm. and Publ. Services	-	-	-	-	-	-	-	-	1007	-
Residential	-	-	-	-	106	-	-	-	1780	-
Non-specified	-	-	-	-	20	-	-	-	-	-
NON-ENERGY USE	-	-	-	-	-	-	-	-	-	-

Kazakhstan

SUPPLY AND CONSUMPTION 1997	Coking Coal	Other Bit. Coal	Sub-Bit. Coal	Lignite	Peat	Oven and Gas Coke	Pat. Fuel and BKB	Crude Oil	NGL	Feed-stocks	Additives
	Coal (1000 tonnes)							**Oil (1000 tonnes)**			
Production	10526	59648	-	2473	-	-	-	23409	2500	-	-
Imports	44	931	-	59	5	656	-	1723	15	-	-
Exports	-1371	-23486	-	-223	-	-12	-	-14605	-	-	-
Intl. Marine Bunkers	-	-	-	-	-	-	-	-	-	-	-
Stock Changes	-	-	-	-	-	-	-	-	-	-	-
DOMESTIC SUPPLY	**9199**	**37093**	**-**	**2309**	**5**	**644**	**-**	**10527**	**2515**	**-**	**-**
Transfers and Stat. Diff.	-	-	-	-	-	-	-	-	-	-	-
TRANSFORMATION	**-**	**29485**	**-**	**-**	**-**	**-**	**-**	**10527**	**2515**	**-**	**-**
Electricity and CHP Plants	-	29485						-	-	-	-
Petroleum Refineries	-	-						10527	2515	-	-
Other Transform. Sector	-	-						-	-	-	-
ENERGY SECTOR	**-**	**-**	**-**	**-**	**-**	**-**	**-**	**-**	**-**	**-**	**-**
DISTRIBUTION LOSSES	**-**	**-**	**-**	**-**	**-**	**-**	**-**	**-**	**-**	**-**	**-**
FINAL CONSUMPTION	**9199**	**7608**	**-**	**2309**	**5**	**644**	**-**	**-**	**-**	**-**	**-**
INDUSTRY SECTOR	**9199**	**7608**	**-**	**2309**	**-**	**644**	**-**	**-**	**-**	**-**	**-**
Iron and Steel	-	-	-	-	-	644	-	-	-	-	-
Chemical and Petrochem.	-	-	-	-	-	-	-	-	-	-	-
Non-Metallic Minerals	-	-	-	-	-	-	-	-	-	-	-
Non-specified	9199	7608	-	2309	-	-	-	-	-	-	-
TRANSPORT SECTOR	**-**	**-**	**-**	**-**	**-**	**-**	**-**	**-**	**-**	**-**	**-**
Air	-	-	-	-	-	-	-	-	-	-	-
Road	-	-	-	-	-	-	-	-	-	-	-
Non-specified	-	-	-	-	-	-	-	-	-	-	-
OTHER SECTORS	**-**	**-**	**-**	**-**	**5**	**-**	**-**	**-**	**-**	**-**	**-**
Agriculture	-	-	-	-	-	-	-	-	-	-	-
Comm. and Publ. Services	-	-	-	-	-	-	-	-	-	-	-
Residential	-	-	-	-	5	-	-	-	-	-	-
Non-specified	-	-	-	-	-	-	-	-	-	-	-
NON-ENERGY USE	**-**	**-**	**-**	**-**	**-**	**-**	**-**	**-**	**-**	**-**	**-**

APPROVISIONNEMENT ET DEMANDE 1998	Charbon à coke	Autres charb. bit.	Charbon sous-bit.	Lignite	Tourbe	Coke de four/gaz	Agg./briq. de lignite	Pétrole brut	LGN	Produits d'aliment.	Additifs
	Charbon (1000 tonnes)							**Pétrole (1000 tonnes)**			
Production	8718	59340	-	1715	-	848	-	23819	2300	-	-
Imports	-	1211	-	16	-	649	-	2074	100	-	-
Exports	-271	-23308	-	-111	-	-15	-	-18260	-50	-	-
Intl. Marine Bunkers	-	-	-	-	-	-	-	-	-	-	-
Stock Changes	-	-	-	-	-	-	-	-	-	-	-
DOMESTIC SUPPLY	**8447**	**37243**	**-**	**1620**	**-**	**1482**	**-**	**7633**	**2350**	**-**	**-**
Transfers and Stat. Diff.	-	-	-	-	-	-	-	-	-	-	-
TRANSFORMATION	**2166**	**29129**	**-**	**-**	**-**	**-**	**-**	**7633**	**2350**	**-**	**-**
Electricity and CHP Plants	-	29129						-	-	-	-
Petroleum Refineries	-	-						7633	2350	-	-
Other Transform. Sector	2166	-						-	-	-	-
ENERGY SECTOR	**-**	**-**	**-**	**-**	**-**	**-**	**-**	**-**	**-**	**-**	**-**
DISTRIBUTION LOSSES	**-**	**-**	**-**	**-**	**-**	**-**	**-**	**-**	**-**	**-**	**-**
FINAL CONSUMPTION	**6281**	**8114**	**-**	**1620**	**-**	**1482**	**-**	**-**	**-**	**-**	**-**
INDUSTRY SECTOR	**6281**	**8114**	**-**	**1620**	**-**	**1482**	**-**	**-**	**-**	**-**	**-**
Iron and Steel	-	-	-	-	-	1482	-	-	-	-	-
Chemical and Petrochem.	-	-	-	-	-	-	-	-	-	-	-
Non-Metallic Minerals	-	-	-	-	-	-	-	-	-	-	-
Non-specified	6281	8114	-	1620	-	-	-	-	-	-	-
TRANSPORT SECTOR	**-**	**-**	**-**	**-**	**-**	**-**	**-**	**-**	**-**	**-**	**-**
Air	-	-	-	-	-	-	-	-	-	-	-
Road	-	-	-	-	-	-	-	-	-	-	-
Non-specified	-	-	-	-	-	-	-	-	-	-	-
OTHER SECTORS	**-**	**-**	**-**	**-**	**-**	**-**	**-**	**-**	**-**	**-**	**-**
Agriculture	-	-	-	-	-	-	-	-	-	-	-
Comm. and Publ. Services	-	-	-	-	-	-	-	-	-	-	-
Residential	-	-	-	-	-	-	-	-	-	-	-
Non-specified	-	-	-	-	-	-	-	-	-	-	-
NON-ENERGY USE	**-**	**-**	**-**	**-**	**-**	**-**	**-**	**-**	**-**	**-**	**-**

Kazakhstan

SUPPLY AND CONSUMPTION 1997	Refinery Gas	LPG + Ethane	Motor Gasoline	Aviation Gasoline	Jet Fuel	Kerosene	Gas/ Diesel	Heavy Fuel Oil	Naphtha	Petrol. Coke	Other Prod.
						Oil cont. (1000 tonnes)					
Production	95	60	1787	-	200	231	2839	3000	-	-	650
Imports	-	21	195	1	28	13	77	171	-	-	133
Exports	-	-6	-76	-	-8	-7	-719	-591	-	-	-126
Intl. Marine Bunkers	-	-	-	-	-	-	-	-	-	-	-
Stock Changes	-	-	-	-	-	-	-	-	-	-	-
DOMESTIC SUPPLY	95	75	1906	1	220	237	2197	2580	-	-	657
Transfers and Stat. Diff.	-	-	-	-	-	-	-	-	-	-	-
TRANSFORMATION	-	-	-	-	-	-	-	1229	-	-	-
Electricity and CHP Plants	-	-	-	-	-	-	-	1229	-	-	-
Petroleum Refineries	-	-	-	-	-	-	-	-	-	-	-
Other Transform. Sector	-	-	-	-	-	-	-	-	-	-	-
ENERGY SECTOR	95	-	-	-	-	-	-	272	-	-	-
DISTRIBUTION LOSSES	-	-	-	-	-	-	-	-	-	-	-
FINAL CONSUMPTION	-	75	1906	1	220	237	2197	1079	-	-	657
INDUSTRY SECTOR	-	-	-	-	-	-	-	-	-	-	-
Iron and Steel	-	-	-	-	-	-	-	-	-	-	-
Chemical and Petrochem.	-	-	-	-	-	-	-	-	-	-	-
Non-Metallic Minerals	-	-	-	-	-	-	-	-	-	-	-
Non-specified	-	-	-	-	-	-	-	-	-	-	-
TRANSPORT SECTOR	-	-	1906	1	220	-	-	-	-	-	-
Air	-	-	-	1	220	-	-	-	-	-	-
Road	-	-	1906	-	-	-	-	-	-	-	-
Non-specified	-	-	-	-	-	-	-	-	-	-	-
OTHER SECTORS	-	75	-	-	-	237	2197	1079	-	-	525
Agriculture	-	-	-	-	-	-	-	-	-	-	-
Comm. and Publ. Services	-	-	-	-	-	-	-	-	-	-	-
Residential	-	-	-	-	-	-	-	-	-	-	-
Non-specified	-	75	-	-	-	237	2197	1079	-	-	525
NON-ENERGY USE	-	-	-	-	-	-	-	-	-	-	132

APPROVISIONNEMENT ET DEMANDE 1998	Gaz de raffinerie	GPL + éthane	Essence moteur	Essence aviation	Carbu- réacteurs	Kérosène	Gazole	Fioul lourd	Naphta	Coke de pétrole	Autres prod.
						Pétrole cont. (1000 tonnes)					
Production	95	60	1732	-	200	229	2495	3052	-	-	880
Imports	-	3	216	1	81	42	172	128	-	-	156
Exports	-	-1	-29	-	-4	-3	-201	-799	-	-	-2
Intl. Marine Bunkers	-	-	-	-	-	-	-	-	-	-	-
Stock Changes	-	-	-	-	-	-	-	-	-	-	-
DOMESTIC SUPPLY	95	62	1919	1	277	268	2466	2381	-	-	1034
Transfers and Stat. Diff.	-	-	-	-	-	-	-	-	-	-	-
TRANSFORMATION	-	-	-	-	-	-	-	995	-	-	-
Electricity and CHP Plants	-	-	-	-	-	-	-	995	-	-	-
Petroleum Refineries	-	-	-	-	-	-	-	-	-	-	-
Other Transform. Sector	-	-	-	-	-	-	-	-	-	-	-
ENERGY SECTOR	95	-	-	-	-	-	-	248	-	-	-
DISTRIBUTION LOSSES	-	-	-	-	-	-	-	-	-	-	-
FINAL CONSUMPTION	-	62	1919	1	277	268	2466	1138	-	-	1034
INDUSTRY SECTOR	-	-	-	-	-	-	-	-	-	-	-
Iron and Steel	-	-	-	-	-	-	-	-	-	-	-
Chemical and Petrochem.	-	-	-	-	-	-	-	-	-	-	-
Non-Metallic Minerals	-	-	-	-	-	-	-	-	-	-	-
Non-specified	-	-	-	-	-	-	-	-	-	-	-
TRANSPORT SECTOR	-	-	1919	1	277	-	-	-	-	-	-
Air	-	-	-	1	277	-	-	-	-	-	-
Road	-	-	1919	-	-	-	-	-	-	-	-
Non-specified	-	-	-	-	-	-	-	-	-	-	-
OTHER SECTORS	-	62	-	-	-	268	2466	1138	-	-	880
Agriculture	-	-	-	-	-	-	-	-	-	-	-
Comm. and Publ. Services	-	-	-	-	-	-	-	-	-	-	-
Residential	-	-	-	-	-	-	-	-	-	-	-
Non-specified	-	62	-	-	-	268	2466	1138	-	-	880
NON-ENERGY USE	-	-	-	-	-	-	-	-	-	-	154

Kazakhstan

SUPPLY AND CONSUMPTION 1997	Gas (TJ)				Comb. Renew. & Waste (TJ)				(GWh)	(TJ)
	Natural Gas	Gas Works	Coke Ovens	Blast Furnaces	Solid Biomass	Gas/Liquids from Biomass	Municipal Waste	Industrial Waste	Electricity	Heat
Production	305902	-	-	-	3100	-	-	-	52006	256
Imports	113239	-	-	-	-	-	-	-	4704	-
Exports	-91679	-	-	-	-	-	-	-	-1	-
Intl. Marine Bunkers	-	-	-	-	-	-	-	-	-	-
Stock Changes	-	-	-	-	-	-	-	-	-	-
DOMESTIC SUPPLY	**327462**	-	-	-	**3100**	-	-	-	**56709**	**256**
Transfers and Stat. Diff.	-	-	-	-	-	-	-	-	-	-
TRANSFORMATION	**100000**	-	-	-	-	-	-	-	-	-
Electricity and CHP Plants	100000	-	-	-	-	-	-	-	-	-
Petroleum Refineries	-	-	-	-	-	-	-	-	-	-
Other Transform. Sector	-	-	-	-	-	-	-	-	-	-
ENERGY SECTOR	**50000**	-	-	-	-	-	-	-	**9784**	-
DISTRIBUTION LOSSES	**1700**	-	-	-	-	-	-	-	**8043**	-
FINAL CONSUMPTION	**175762**	-	-	-	**3100**	-	-	-	**38882**	**256**
INDUSTRY SECTOR	**15000**	-	-	-	-	-	-	-	**18686**	-
Iron and Steel	-	-	-	-	-	-	-	-	7168	-
Chemical and Petrochem.	-	-	-	-	-	-	-	-	2815	-
Non-Metallic Minerals	-	-	-	-	-	-	-	-	-	-
Non-specified	15000	-	-	-	-	-	-	-	8703	-
TRANSPORT SECTOR	-	-	-	-	-	-	-	-	**3075**	-
Air	-	-	-	-	-	-	-	-	-	-
Road	-	-	-	-	-	-	-	-	-	-
Non-specified	-	-	-	-	-	-	-	-	3075	-
OTHER SECTORS	**160762**	-	-	-	**3100**	-	-	-	**17121**	**256**
Agriculture	-	-	-	-	-	-	-	-	7909	-
Comm. and Publ. Services	-	-	-	-	-	-	-	-	-	-
Residential	-	-	-	-	-	-	-	-	5431	-
Non-specified	160762	-	-	-	3100	-	-	-	3781	256
NON-ENERGY USE	-	-	-	-	-	-	-	-	-	-

APPROVISIONNEMENT ET DEMANDE 1998	Gaz (TJ)				En. Re. Comb. & Déchets (TJ)				(GWh)	(TJ)
	Gaz naturel	Usines à gaz	Cokeries	Hauts fourneaux	Biomasse solide	Gaz/Liquides tirés de biomasse	Déchets urbains	Déchets industriels	Electricité	Chaleur
Production	299632	-	-	-	3069	-	-	-	49144	280
Imports	115053	746	-	-	-	-	-	-	7252	-
Exports	-86925	-	-	-	-	-	-	-	-3441	-
Intl. Marine Bunkers	-	-	-	-	-	-	-	-	-	-
Stock Changes	-	-	-	-	-	-	-	-	-	-
DOMESTIC SUPPLY	**327760**	**746**	-	-	**3069**	-	-	-	**52955**	**280**
Transfers and Stat. Diff.	-	-	-	-	-	-	-	-	-	-
TRANSFORMATION	**98435**	-	-	-	-	-	-	-	-	-
Electricity and CHP Plants	98435	-	-	-	-	-	-	-	-	-
Petroleum Refineries	-	-	-	-	-	-	-	-	-	-
Other Transform. Sector	-	-	-	-	-	-	-	-	-	-
ENERGY SECTOR	**52780**	-	-	-	-	-	-	-	**8279**	-
DISTRIBUTION LOSSES	**1700**	-	-	-	-	-	-	-	**8513**	-
FINAL CONSUMPTION	**174845**	**746**	-	-	**3069**	-	-	-	**36163**	**280**
INDUSTRY SECTOR	**15080**	**746**	-	-	-	-	-	-	**10104**	-
Iron and Steel	-	746	-	-	-	-	-	-	2847	-
Chemical and Petrochem.	-	-	-	-	-	-	-	-	675	-
Non-Metallic Minerals	-	-	-	-	-	-	-	-	362	-
Non-specified	15080	-	-	-	-	-	-	-	6220	-
TRANSPORT SECTOR	-	-	-	-	-	-	-	-	**3269**	-
Air	-	-	-	-	-	-	-	-	-	-
Road	-	-	-	-	-	-	-	-	-	-
Non-specified	-	-	-	-	-	-	-	-	3269	-
OTHER SECTORS	**159765**	-	-	-	**3069**	-	-	-	**22790**	**280**
Agriculture	-	-	-	-	-	-	-	-	7647	-
Comm. and Publ. Services	-	-	-	-	-	-	-	-	-	-
Residential	-	-	-	-	-	-	-	-	5409	-
Non-specified	159765	-	-	-	3069	-	-	-	9734	280
NON-ENERGY USE	-	-	-	-	-	-	-	-	-	-

Kenya

SUPPLY AND CONSUMPTION 1997	Coal (1000 tonnes)							Oil (1000 tonnes)			
	Coking Coal	Other Bit. Coal	Sub-Bit. Coal	Lignite	Peat	Oven and Gas Coke	Pat. Fuel and BKB	Crude Oil	NGL	Feed-stocks	Additives
Production	-	-	-	-	-	-	-	-	-	-	-
Imports	-	148	-	-	-	1	-	1834	-	-	-
Exports	-	-	-	-	-	-	-	-	-	-	-
Intl. Marine Bunkers	-	-	-	-	-	-	-	-	-	-	-
Stock Changes	-	-	-	-	-	-	-	-	-	-	-
DOMESTIC SUPPLY	-	148	-	-	-	1	-	1834	-	-	-
Transfers and Stat. Diff.	-	-	-	-	-	-	-	-94	-	-	-
TRANSFORMATION	-	-	-	-	-	-	-	1647	-	-	-
Electricity and CHP Plants	-	-	-	-	-	-	-	-	-	-	-
Petroleum Refineries	-	-	-	-	-	-	-	1647	-	-	-
Other Transform. Sector	-	-	-	-	-	-	-	-	-	-	-
ENERGY SECTOR	-	-	-	-	-	-	-	93	-	-	-
DISTRIBUTION LOSSES	-	-	-	-	-	-	-	-	-	-	-
FINAL CONSUMPTION	-	148	-	-	-	1	-	-	-	-	-
INDUSTRY SECTOR	-	148	-	-	-	-	-	-	-	-	-
Iron and Steel	-	-	-	-	-	-	-	-	-	-	-
Chemical and Petrochem.	-	-	-	-	-	-	-	-	-	-	-
Non-Metallic Minerals	-	148	-	-	-	-	-	-	-	-	-
Non-specified	-	-	-	-	-	-	-	-	-	-	-
TRANSPORT SECTOR	-	-	-	-	-	-	-	-	-	-	-
Air	-	-	-	-	-	-	-	-	-	-	-
Road	-	-	-	-	-	-	-	-	-	-	-
Non-specified	-	-	-	-	-	-	-	-	-	-	-
OTHER SECTORS	-	-	-	-	-	1	-	-	-	-	-
Agriculture	-	-	-	-	-	-	-	-	-	-	-
Comm. and Publ. Services	-	-	-	-	-	-	-	-	-	-	-
Residential	-	-	-	-	-	1	-	-	-	-	-
Non-specified	-	-	-	-	-	-	-	-	-	-	-
NON-ENERGY USE	-	-	-	-	-	-	-	-	-	-	-

APPROVISIONNEMENT ET DEMANDE 1998	Charbon (1000 tonnes)							Pétrole (1000 tonnes)			
	Charbon à coke	Autres charb. bit.	Charbon sous-bit.	Lignite	Tourbe	Coke de four/gaz	Agg./briq. de lignite	Pétrole brut	LGN	Produits d'aliment.	Additifs
Production	-	-	-	-	-	-	-	-	-	-	-
Imports	-	122	-	-	-	1	-	2158	-	-	-
Exports	-	-	-	-	-	-	-	-	-	-	-
Intl. Marine Bunkers	-	-	-	-	-	-	-	-	-	-	-
Stock Changes	-	-	-	-	-	-	-	-	-	-	-
DOMESTIC SUPPLY	-	122	-	-	-	1	-	2158	-	-	-
Transfers and Stat. Diff.	-	-	-	-	-	-	-	-342	-	-	-
TRANSFORMATION	-	-	-	-	-	-	-	1722	-	-	-
Electricity and CHP Plants	-	-	-	-	-	-	-	-	-	-	-
Petroleum Refineries	-	-	-	-	-	-	-	1722	-	-	-
Other Transform. Sector	-	-	-	-	-	-	-	-	-	-	-
ENERGY SECTOR	-	-	-	-	-	-	-	94	-	-	-
DISTRIBUTION LOSSES	-	-	-	-	-	-	-	-	-	-	-
FINAL CONSUMPTION	-	122	-	-	-	1	-	-	-	-	-
INDUSTRY SECTOR	-	122	-	-	-	-	-	-	-	-	-
Iron and Steel	-	-	-	-	-	-	-	-	-	-	-
Chemical and Petrochem.	-	-	-	-	-	-	-	-	-	-	-
Non-Metallic Minerals	-	122	-	-	-	-	-	-	-	-	-
Non-specified	-	-	-	-	-	-	-	-	-	-	-
TRANSPORT SECTOR	-	-	-	-	-	-	-	-	-	-	-
Air	-	-	-	-	-	-	-	-	-	-	-
Road	-	-	-	-	-	-	-	-	-	-	-
Non-specified	-	-	-	-	-	-	-	-	-	-	-
OTHER SECTORS	-	-	-	-	-	1	-	-	-	-	-
Agriculture	-	-	-	-	-	-	-	-	-	-	-
Comm. and Publ. Services	-	-	-	-	-	-	-	-	-	-	-
Residential	-	-	-	-	-	1	-	-	-	-	-
Non-specified	-	-	-	-	-	-	-	-	-	-	-
NON-ENERGY USE	-	-	-	-	-	-	-	-	-	-	-

Kenya

SUPPLY AND CONSUMPTION 1997	Oil cont. (1000 tonnes)										
	Refinery Gas	LPG + Ethane	Motor Gasoline	Aviation Gasoline	Jet Fuel	Kerosene	Gas/ Diesel	Heavy Fuel Oil	Naphtha	Petrol. Coke	Other Prod.
Production	61	24	273	-	234	101	412	499	-	-	19
Imports	-	-	159	4	194	135	308	91	-	-	25
Exports	-	-	-60	-	-109	-	-184	-299	-	-	-1
Intl. Marine Bunkers	-	-	-	-	-	-	-11	-49	-	-	-
Stock Changes	-	7	19	-	113	32	139	145	-	-	-
DOMESTIC SUPPLY	61	31	391	4	432	268	664	387	-	-	43
Transfers and Stat. Diff.	-	-	-	-	-	-	-	-	-	-	-
TRANSFORMATION	-	-	-	-	-	-	126	66	-	-	-
Electricity and CHP Plants	-	-	-	-	-	-	126	66	-	-	-
Petroleum Refineries	-	-	-	-	-	-	-	-	-	-	-
Other Transform. Sector	-	-	-	-	-	-	-	-	-	-	-
ENERGY SECTOR	61	-	-	-	-	-	-	30	-	-	-
DISTRIBUTION LOSSES	-	-	-	-	-	-	-	-	-	-	-
FINAL CONSUMPTION	-	31	391	4	432	268	538	291	-	-	43
INDUSTRY SECTOR	-	9	14	-	-	27	96	247	-	-	-
Iron and Steel	-	-	-	-	-	-	-	-	-	-	-
Chemical and Petrochem.	-	-	-	-	-	-	-	-	-	-	-
Non-Metallic Minerals	-	-	-	-	-	-	-	-	-	-	-
Non-specified	-	9	14	-	-	27	96	247	-	-	-
TRANSPORT SECTOR	-	-	361	4	432	-	394	6	-	-	-
Air	-	-	-	4	432	-	-	-	-	-	-
Road	-	-	361	-	-	-	365	-	-	-	-
Non-specified	-	-	-	-	-	-	29	6	-	-	-
OTHER SECTORS	-	22	16	-	-	241	48	38	-	-	-
Agriculture	-	-	4	-	-	-	28	36	-	-	-
Comm. and Publ. Services	-	-	-	-	-	-	-	-	-	-	-
Residential	-	16	-	-	-	241	-	-	-	-	-
Non-specified	-	6	12	-	-	-	20	2	-	-	-
NON-ENERGY USE	-	-	-	-	-	-	-	-	-	-	43

APPROVISIONNEMENT ET DEMANDE 1998	Pétrole cont. (1000 tonnes)										
	Gaz de raffinerie	GPL + éthane	Essence moteur	Essence aviation	Carbu- réacteurs	Kérosène	Gazole	Fioul lourd	Naphta	Coke de pétrole	Autres prod.
Production	61	29	294	-	248	107	429	506	-	-	22
Imports	-	-	246	3	301	209	481	141	-	-	39
Exports	-	-	-163	-	-107	-	-217	-153	-	-	-1
Intl. Marine Bunkers	-	-	-	-	-	-	-11	-50	-	-	-
Stock Changes	-	2	19	-	-23	2	-48	-47	-	-	-
DOMESTIC SUPPLY	61	31	396	3	419	318	634	397	-	-	60
Transfers and Stat. Diff.	-	-	-	-	-	-	-	-	-	-	-
TRANSFORMATION	-	-	-	-	-	-	128	66	-	-	-
Electricity and CHP Plants	-	-	-	-	-	-	128	66	-	-	-
Petroleum Refineries	-	-	-	-	-	-	-	-	-	-	-
Other Transform. Sector	-	-	-	-	-	-	-	-	-	-	-
ENERGY SECTOR	61	-	-	-	-	-	-	30	-	-	-
DISTRIBUTION LOSSES	-	-	-	-	-	-	-	-	-	-	-
FINAL CONSUMPTION	-	31	396	3	419	318	506	301	-	-	60
INDUSTRY SECTOR	-	9	14	-	-	32	90	255	-	-	-
Iron and Steel	-	-	-	-	-	-	-	-	-	-	-
Chemical and Petrochem.	-	-	-	-	-	-	-	-	-	-	-
Non-Metallic Minerals	-	-	-	-	-	-	-	-	-	-	-
Non-specified	-	9	14	-	-	32	90	255	-	-	-
TRANSPORT SECTOR	-	-	366	3	419	-	370	6	-	-	-
Air	-	-	-	3	419	-	-	-	-	-	-
Road	-	-	366	-	-	-	343	-	-	-	-
Non-specified	-	-	-	-	-	-	27	6	-	-	-
OTHER SECTORS	-	22	16	-	-	286	46	40	-	-	-
Agriculture	-	-	4	-	-	-	26	37	-	-	-
Comm. and Publ. Services	-	-	-	-	-	-	-	-	-	-	-
Residential	-	16	-	-	-	286	-	-	-	-	-
Non-specified	-	6	12	-	-	-	20	3	-	-	-
NON-ENERGY USE	-	-	-	-	-	-	-	-	-	-	60

Kenya

SUPPLY AND CONSUMPTION 1997	Gas (TJ)				Comb. Renew. & Waste (TJ)				(GWh)	(TJ)
	Natural Gas	Gas Works	Coke Ovens	Blast Furnaces	Solid Biomass	Gas/Liquids from Biomass	Municipal Waste	Industrial Waste	Electricity	Heat
Production	-	-	-	-	461164	-	-	-	4223	-
Imports	-	-	-	-	-	-	-	-	144	-
Exports	-	-	-	-	-	-	-	-	-	-
Intl. Marine Bunkers	-	-	-	-	-	-	-	-	-	-
Stock Changes	-	-	-	-	-	-	-	-	-	-
DOMESTIC SUPPLY	-	-	-	-	461164	-	-	-	4367	-
Transfers and Stat. Diff.	-	-	-	-	-2	-	-	-	-	-
TRANSFORMATION	-	-	-	-	130576	-	-	-	-	-
Electricity and CHP Plants	-	-	-	-	-	-	-	-	-	-
Petroleum Refineries	-	-	-	-	-	-	-	-	-	-
Other Transform. Sector	-	-	-	-	130576	-	-	-	-	-
ENERGY SECTOR	-	-	-	-	-	-	-	-	29	-
DISTRIBUTION LOSSES	-	-	-	-	-	-	-	-	713	-
FINAL CONSUMPTION	-	-	-	-	330586	-	-	-	3625	-
INDUSTRY SECTOR	-	-	-	-	22902	-	-	-	2263	-
Iron and Steel	-	-	-	-	-	-	-	-	-	-
Chemical and Petrochem.	-	-	-	-	-	-	-	-	-	-
Non-Metallic Minerals	-	-	-	-	-	-	-	-	-	-
Non-specified	-	-	-	-	22902	-	-	-	2263	-
TRANSPORT SECTOR	-	-	-	-	-	-	-	-	-	-
Air	-	-	-	-	-	-	-	-	-	-
Road	-	-	-	-	-	-	-	-	-	-
Non-specified	-	-	-	-	-	-	-	-	-	-
OTHER SECTORS	-	-	-	-	307684	-	-	-	1362	-
Agriculture	-	-	-	-	24872	-	-	-	-	-
Comm. and Publ. Services	-	-	-	-	1969	-	-	-	246	-
Residential	-	-	-	-	280843	-	-	-	1116	-
Non-specified	-	-	-	-	-	-	-	-	-	-
NON-ENERGY USE	-	-	-	-	-	-	-	-	-	-

APPROVISIONNEMENT ET DEMANDE 1998	Gaz (TJ)				En. Re. Comb. & Déchets (TJ)				(GWh)	(TJ)
	Gaz naturel	Usines à gaz	Cokeries	Hauts fourneaux	Biomasse solide	Gaz/Liquides tirés de biomasse	Déchets urbains	Déchets industriels	Electricité	Chaleur
Production	-	-	-	-	472172	-	-	-	4804	-
Imports	-	-	-	-	-	-	-	-	146	-
Exports	-	-	-	-	-	-	-	-	-	-
Intl. Marine Bunkers	-	-	-	-	-	-	-	-	-	-
Stock Changes	-	-	-	-	-	-	-	-	-	-
DOMESTIC SUPPLY	-	-	-	-	472172	-	-	-	4950	-
Transfers and Stat. Diff.	-	-	-	-	-2	-	-	-	-	-
TRANSFORMATION	-	-	-	-	133693	-	-	-	-	-
Electricity and CHP Plants	-	-	-	-	-	-	-	-	-	-
Petroleum Refineries	-	-	-	-	-	-	-	-	-	-
Other Transform. Sector	-	-	-	-	133693	-	-	-	-	-
ENERGY SECTOR	-	-	-	-	-	-	-	-	29	-
DISTRIBUTION LOSSES	-	-	-	-	-	-	-	-	1200	-
FINAL CONSUMPTION	-	-	-	-	338477	-	-	-	3721	-
INDUSTRY SECTOR	-	-	-	-	23448	-	-	-	2261	-
Iron and Steel	-	-	-	-	-	-	-	-	-	-
Chemical and Petrochem.	-	-	-	-	-	-	-	-	-	-
Non-Metallic Minerals	-	-	-	-	-	-	-	-	-	-
Non-specified	-	-	-	-	23448	-	-	-	2261	-
TRANSPORT SECTOR	-	-	-	-	-	-	-	-	-	-
Air	-	-	-	-	-	-	-	-	-	-
Road	-	-	-	-	-	-	-	-	-	-
Non-specified	-	-	-	-	-	-	-	-	-	-
OTHER SECTORS	-	-	-	-	315029	-	-	-	1460	-
Agriculture	-	-	-	-	25466	-	-	-	-	-
Comm. and Publ. Services	-	-	-	-	2016	-	-	-	253	-
Residential	-	-	-	-	287547	-	-	-	1207	-
Non-specified	-	-	-	-	-	-	-	-	-	-
NON-ENERGY USE	-	-	-	-	-	-	-	-	-	-

Dem. People's Rep. of Korea / Rép. pop. dém. de Corée

SUPPLY AND CONSUMPTION 1997	Coal (1000 tonnes)							Oil (1000 tonnes)			
	Coking Coal	Other Bit. Coal	Sub-Bit. Coal	Lignite	Peat	Oven and Gas Coke	Pat. Fuel and BKB	Crude Oil	NGL	Feed-stocks	Additives
Production	8287	57326	23766	-	-	3215	-	-	-	-	-
Imports	2120	-	-	-	-	252	-	2004	-	-	-
Exports	-	-359	-	-	-	-	-	-	-	-	-
Intl. Marine Bunkers	-	-	-	-	-	-	-	-	-	-	-
Stock Changes	-	-	-	-	-	-	-	-	-	-	-
DOMESTIC SUPPLY	10407	56967	23766	-	-	3467	-	2004	-	-	-
Transfers and Stat. Diff.	-	-	-	-	-	-	-	606	-	-	-
TRANSFORMATION	10407	13194	-	-	-	2774	-	2610	-	-	-
Electricity and CHP Plants	-	13194	-	-	-	-	-	-	-	-	-
Petroleum Refineries	-	-	-	-	-	-	-	2610	-	-	-
Other Transform. Sector	10407	-	-	-	-	2774	-	-	-	-	-
ENERGY SECTOR	-	-	-	-	-	-	-	-	-	-	-
DISTRIBUTION LOSSES	-	-	-	-	-	-	-	-	-	-	-
FINAL CONSUMPTION	-	43773	23766	-	-	693	-	-	-	-	-
INDUSTRY SECTOR	-	43773	23766	-	-	693	-	-	-	-	-
Iron and Steel	-	-	-	-	-	693	-	-	-	-	-
Chemical and Petrochem.	-	-	-	-	-	-	-	-	-	-	-
Non-Metallic Minerals	-	-	-	-	-	-	-	-	-	-	-
Non-specified	-	43773	23766	-	-	-	-	-	-	-	-
TRANSPORT SECTOR	-	-	-	-	-	-	-	-	-	-	-
Air	-	-	-	-	-	-	-	-	-	-	-
Road	-	-	-	-	-	-	-	-	-	-	-
Non-specified	-	-	-	-	-	-	-	-	-	-	-
OTHER SECTORS	-	-	-	-	-	-	-	-	-	-	-
Agriculture	-	-	-	-	-	-	-	-	-	-	-
Comm. and Publ. Services	-	-	-	-	-	-	-	-	-	-	-
Residential	-	-	-	-	-	-	-	-	-	-	-
Non-specified	-	-	-	-	-	-	-	-	-	-	-
NON-ENERGY USE	-	-	-	-	-	-	-	-	-	-	-

APPROVISIONNEMENT ET DEMANDE 1998	Charbon (1000 tonnes)							Pétrole (1000 tonnes)			
	Charbon à coke	Autres charb. bit.	Charbon sous-bit.	Lignite	Tourbe	Coke de four/gaz	Agg./briq. de lignite	Pétrole brut	LGN	Produits d'aliment.	Additifs
Production	7873	54459	22578	-	-	3054	-	-	-	-	-
Imports	2014	-	-	-	-	239	-	1904	-	-	-
Exports	-	-341	-	-	-	-	-	-	-	-	-
Intl. Marine Bunkers	-	-	-	-	-	-	-	-	-	-	-
Stock Changes	-	-	-	-	-	-	-	-	-	-	-
DOMESTIC SUPPLY	9887	54118	22578	-	-	3293	-	1904	-	-	-
Transfers and Stat. Diff.	-	-	-	-	-	-	-	576	-	-	-
TRANSFORMATION	9887	12534	-	-	-	2635	-	2480	-	-	-
Electricity and CHP Plants	-	12534	-	-	-	-	-	-	-	-	-
Petroleum Refineries	-	-	-	-	-	-	-	2480	-	-	-
Other Transform. Sector	9887	-	-	-	-	2635	-	-	-	-	-
ENERGY SECTOR	-	-	-	-	-	-	-	-	-	-	-
DISTRIBUTION LOSSES	-	-	-	-	-	-	-	-	-	-	-
FINAL CONSUMPTION	-	41584	22578	-	-	658	-	-	-	-	-
INDUSTRY SECTOR	-	41584	22578	-	-	658	-	-	-	-	-
Iron and Steel	-	-	-	-	-	658	-	-	-	-	-
Chemical and Petrochem.	-	-	-	-	-	-	-	-	-	-	-
Non-Metallic Minerals	-	-	-	-	-	-	-	-	-	-	-
Non-specified	-	41584	22578	-	-	-	-	-	-	-	-
TRANSPORT SECTOR	-	-	-	-	-	-	-	-	-	-	-
Air	-	-	-	-	-	-	-	-	-	-	-
Road	-	-	-	-	-	-	-	-	-	-	-
Non-specified	-	-	-	-	-	-	-	-	-	-	-
OTHER SECTORS	-	-	-	-	-	-	-	-	-	-	-
Agriculture	-	-	-	-	-	-	-	-	-	-	-
Comm. and Publ. Services	-	-	-	-	-	-	-	-	-	-	-
Residential	-	-	-	-	-	-	-	-	-	-	-
Non-specified	-	-	-	-	-	-	-	-	-	-	-
NON-ENERGY USE	-	-	-	-	-	-	-	-	-	-	-

Dem. People's Rep. of Korea / Rép. pop. dém. de Corée

SUPPLY AND CONSUMPTION 1997	Oil cont. (1000 tonnes)										
	Refinery Gas	LPG + Ethane	Motor Gasoline	Aviation Gasoline	Jet Fuel	Kerosene	Gas/ Diesel	Heavy Fuel Oil	Naphtha	Petrol. Coke	Other Prod.
Production	-	-	885	-	-	186	965	559	-	-	-
Imports	-	-	778	-	-	44	256	303	-	-	-
Exports	-	-	-	-	-	-	-	-8	-	-	-
Intl. Marine Bunkers	-	-	-	-	-	-	-	-	-	-	-
Stock Changes	-	-	-	-	-	-	-	-	-	-	-
DOMESTIC SUPPLY	-	-	1663	-	-	230	1221	854	-	-	-
Transfers and Stat. Diff.	-	-	-	-	-	-	-	-	-	-	-
TRANSFORMATION	-	-	-	-	-	-	-	-	-	-	-
Electricity and CHP Plants	-	-	-	-	-	-	-	-	-	-	-
Petroleum Refineries	-	-	-	-	-	-	-	-	-	-	-
Other Transform. Sector	-	-	-	-	-	-	-	-	-	-	-
ENERGY SECTOR	-	-	-	-	-	-	-	102	-	-	-
DISTRIBUTION LOSSES	-	-	-	-	-	-	-	-	-	-	-
FINAL CONSUMPTION	-	-	1663	-	-	230	1221	752	-	-	-
INDUSTRY SECTOR	-	-	-	-	-	-	-	752	-	-	-
Iron and Steel	-	-	-	-	-	-	-	-	-	-	-
Chemical and Petrochem.	-	-	-	-	-	-	-	-	-	-	-
Non-Metallic Minerals	-	-	-	-	-	-	-	-	-	-	-
Non-specified	-	-	-	-	-	-	-	752	-	-	-
TRANSPORT SECTOR	-	-	1663	-	-	-	1221	-	-	-	-
Air	-	-	-	-	-	-	-	-	-	-	-
Road	-	-	1663	-	-	-	1221	-	-	-	-
Non-specified	-	-	-	-	-	-	-	-	-	-	-
OTHER SECTORS	-	-	-	-	-	230	-	-	-	-	-
Agriculture	-	-	-	-	-	-	-	-	-	-	-
Comm. and Publ. Services	-	-	-	-	-	-	-	-	-	-	-
Residential	-	-	-	-	-	230	-	-	-	-	-
Non-specified	-	-	-	-	-	-	-	-	-	-	-
NON-ENERGY USE	-	-	-	-	-	-	-	-	-	-	-

APPROVISIONNEMENT ET DEMANDE 1998	Pétrole cont. (1000 tonnes)										
	Gaz de raffinerie	GPL + éthane	Essence moteur	Essence aviation	Carbu- réacteurs	Kérosène	Gazole	Fioul lourd	Naphta	Coke de pétrole	Autres prod.
Production	-	-	841	-	-	177	917	531	-	-	-
Imports	-	-	739	-	-	42	243	288	-	-	-
Exports	-	-	-	-	-	-	-	-8	-	-	-
Intl. Marine Bunkers	-	-	-	-	-	-	-	-	-	-	-
Stock Changes	-	-	-	-	-	-	-	-	-	-	-
DOMESTIC SUPPLY	-	-	1580	-	-	219	1160	811	-	-	-
Transfers and Stat. Diff.	-	-	-	-	-	-	-	-	-	-	-
TRANSFORMATION	-	-	-	-	-	-	-	-	-	-	-
Electricity and CHP Plants	-	-	-	-	-	-	-	-	-	-	-
Petroleum Refineries	-	-	-	-	-	-	-	-	-	-	-
Other Transform. Sector	-	-	-	-	-	-	-	-	-	-	-
ENERGY SECTOR	-	-	-	-	-	-	-	97	-	-	-
DISTRIBUTION LOSSES	-	-	-	-	-	-	-	-	-	-	-
FINAL CONSUMPTION	-	-	1580	-	-	219	1160	714	-	-	-
INDUSTRY SECTOR	-	-	-	-	-	-	-	714	-	-	-
Iron and Steel	-	-	-	-	-	-	-	-	-	-	-
Chemical and Petrochem.	-	-	-	-	-	-	-	-	-	-	-
Non-Metallic Minerals	-	-	-	-	-	-	-	-	-	-	-
Non-specified	-	-	-	-	-	-	-	714	-	-	-
TRANSPORT SECTOR	-	-	1580	-	-	-	1160	-	-	-	-
Air	-	-	-	-	-	-	-	-	-	-	-
Road	-	-	1580	-	-	-	1160	-	-	-	-
Non-specified	-	-	-	-	-	-	-	-	-	-	-
OTHER SECTORS	-	-	-	-	-	219	-	-	-	-	-
Agriculture	-	-	-	-	-	-	-	-	-	-	-
Comm. and Publ. Services	-	-	-	-	-	-	-	-	-	-	-
Residential	-	-	-	-	-	219	-	-	-	-	-
Non-specified	-	-	-	-	-	-	-	-	-	-	-
NON-ENERGY USE	-	-	-	-	-	-	-	-	-	-	-

Dem. People's Rep. of Korea / Rép. pop. dém. de Corée

SUPPLY AND CONSUMPTION 1997	Gas (TJ)				Comb. Renew. & Waste (TJ)				(GWh)	(TJ)
	Natural Gas	Gas Works	Coke Ovens	Blast Furnaces	Solid Biomass	Gas/Liquids from Biomass	Municipal Waste	Industrial Waste	Electricity	Heat
Production	-	-	-	30203	43223	-	-	-	32620	-
Imports	-	-	-	-	-	-	-	-	-	-
Exports	-	-	-	-	-	-	-	-	-	-
Intl. Marine Bunkers	-	-	-	-	-	-	-	-	-	-
Stock Changes	-	-	-	-	-	-	-	-	-	-
DOMESTIC SUPPLY	-	-	-	30203	43223	-	-	-	32620	-
Transfers and Stat. Diff.	-	-	-	-	-	-	-	-	-	-
TRANSFORMATION	-	-	-	-	-	-	-	-	-	-
Electricity and CHP Plants	-	-	-	-	-	-	-	-	-	-
Petroleum Refineries	-	-	-	-	-	-	-	-	-	-
Other Transform. Sector	-	-	-	-	-	-	-	-	-	-
ENERGY SECTOR	-	-	-	-	-	-	-	-	-	-
DISTRIBUTION LOSSES	-	-	-	-	-	-	-	-	27401	-
FINAL CONSUMPTION	-	-	-	30203	43223	-	-	-	5219	-
INDUSTRY SECTOR	-	-	-	30203	-	-	-	-	-	-
Iron and Steel	-	-	-	30203	-	-	-	-	-	-
Chemical and Petrochem.	-	-	-	-	-	-	-	-	-	-
Non-Metallic Minerals	-	-	-	-	-	-	-	-	-	-
Non-specified	-	-	-	-	-	-	-	-	-	-
TRANSPORT SECTOR	-	-	-	-	-	-	-	-	-	-
Air	-	-	-	-	-	-	-	-	-	-
Road	-	-	-	-	-	-	-	-	-	-
Non-specified	-	-	-	-	-	-	-	-	-	-
OTHER SECTORS	-	-	-	-	43223	-	-	-	5219	-
Agriculture	-	-	-	-	-	-	-	-	-	-
Comm. and Publ. Services	-	-	-	-	-	-	-	-	-	-
Residential	-	-	-	-	-	-	-	-	-	-
Non-specified	-	-	-	-	43223	-	-	-	5219	-
NON-ENERGY USE	-	-	-	-	-	-	-	-	-	-

APPROVISIONNEMENT ET DEMANDE 1998	Gaz (TJ)				En. Re. Comb. & Déchets (TJ)				(GWh)	(TJ)
	Gaz naturel	Usines à gaz	Cokeries	Hauts fourneaux	Biomasse solide	Gaz/Liquides tirés de biomasse	Déchets urbains	Déchets industriels	Electricité	Chaleur
Production	-	-	-	28689	43742	-	-	-	30989	-
Imports	-	-	-	-	-	-	-	-	-	-
Exports	-	-	-	-	-	-	-	-	-	-
Intl. Marine Bunkers	-	-	-	-	-	-	-	-	-	-
Stock Changes	-	-	-	-	-	-	-	-	-	-
DOMESTIC SUPPLY	-	-	-	28689	43742	-	-	-	30989	-
Transfers and Stat. Diff.	-	-	-	-	-	-	-	-	-	-
TRANSFORMATION	-	-	-	-	-	-	-	-	-	-
Electricity and CHP Plants	-	-	-	-	-	-	-	-	-	-
Petroleum Refineries	-	-	-	-	-	-	-	-	-	-
Other Transform. Sector	-	-	-	-	-	-	-	-	-	-
ENERGY SECTOR	-	-	-	-	-	-	-	-	-	-
DISTRIBUTION LOSSES	-	-	-	-	-	-	-	-	26031	-
FINAL CONSUMPTION	-	-	-	28689	43742	-	-	-	4958	-
INDUSTRY SECTOR	-	-	-	28689	-	-	-	-	-	-
Iron and Steel	-	-	-	28689	-	-	-	-	-	-
Chemical and Petrochem.	-	-	-	-	-	-	-	-	-	-
Non-Metallic Minerals	-	-	-	-	-	-	-	-	-	-
Non-specified	-	-	-	-	-	-	-	-	-	-
TRANSPORT SECTOR	-	-	-	-	-	-	-	-	-	-
Air	-	-	-	-	-	-	-	-	-	-
Road	-	-	-	-	-	-	-	-	-	-
Non-specified	-	-	-	-	-	-	-	-	-	-
OTHER SECTORS	-	-	-	-	43742	-	-	-	4958	-
Agriculture	-	-	-	-	-	-	-	-	-	-
Comm. and Publ. Services	-	-	-	-	-	-	-	-	-	-
Residential	-	-	-	-	-	-	-	-	-	-
Non-specified	-	-	-	-	43742	-	-	-	4958	-
NON-ENERGY USE	-	-	-	-	-	-	-	-	-	-

Kuwait / Koweit

SUPPLY AND CONSUMPTION 1997	Coal (1000 tonnes)							Oil (1000 tonnes)			
	Coking Coal	Other Bit. Coal	Sub-Bit. Coal	Lignite	Peat	Oven and Gas Coke	Pat. Fuel and BKB	Crude Oil	NGL	Feed-stocks	Additives
Production	-	-	-	-	-	-	-	101687	740	-	-
Imports	-	-	-	-	-	-	-	-	-	-	-
Exports	-	-	-	-	-	-	-	-57449	-	-	-
Intl. Marine Bunkers	-	-	-	-	-	-	-	-	-	-	-
Stock Changes	-	-	-	-	-	-	-	-	-	-	-
DOMESTIC SUPPLY	-	-	-	-	-	-	-	44238	740	-	-
Transfers and Stat. Diff.	-	-	-	-	-	-	-	7154	-740	-	-
TRANSFORMATION	-	-	-	-	-	-	-	51392	-	-	-
Electricity and CHP Plants	-	-	-	-	-	-	-	4249	-	-	-
Petroleum Refineries	-	-	-	-	-	-	-	47143	-	-	-
Other Transform. Sector	-	-	-	-	-	-	-	-	-	-	-
ENERGY SECTOR	-	-	-	-	-	-	-	-	-	-	-
DISTRIBUTION LOSSES	-	-	-	-	-	-	-	-	-	-	-
FINAL CONSUMPTION	-	-	-	-	-	-	-	-	-	-	-
INDUSTRY SECTOR	-	-	-	-	-	-	-	-	-	-	-
Iron and Steel	-	-	-	-	-	-	-	-	-	-	-
Chemical and Petrochem.	-	-	-	-	-	-	-	-	-	-	-
Non-Metallic Minerals	-	-	-	-	-	-	-	-	-	-	-
Non-specified	-	-	-	-	-	-	-	-	-	-	-
TRANSPORT SECTOR	-	-	-	-	-	-	-	-	-	-	-
Air	-	-	-	-	-	-	-	-	-	-	-
Road	-	-	-	-	-	-	-	-	-	-	-
Non-specified	-	-	-	-	-	-	-	-	-	-	-
OTHER SECTORS	-	-	-	-	-	-	-	-	-	-	-
Agriculture	-	-	-	-	-	-	-	-	-	-	-
Comm. and Publ. Services	-	-	-	-	-	-	-	-	-	-	-
Residential	-	-	-	-	-	-	-	-	-	-	-
Non-specified	-	-	-	-	-	-	-	-	-	-	-
NON-ENERGY USE	-	-	-	-	-	-	-	-	-	-	-

APPROVISIONNEMENT ET DEMANDE 1998	Charbon (1000 tonnes)							Pétrole (1000 tonnes)			
	Charbon à coke	Autres charb. bit.	Charbon sous-bit.	Lignite	Tourbe	Coke de four/gaz	Agg./briq. de lignite	Pétrole brut	LGN	Produits d'aliment.	Additifs
Production	-	-	-	-	-	-	-	103939	755	-	-
Imports	-	-	-	-	-	-	-	-	-	-	-
Exports	-	-	-	-	-	-	-	-60282	-	-	-
Intl. Marine Bunkers	-	-	-	-	-	-	-	-	-	-	-
Stock Changes	-	-	-	-	-	-	-	-	-	-	-
DOMESTIC SUPPLY	-	-	-	-	-	-	-	43657	755	-	-
Transfers and Stat. Diff.	-	-	-	-	-	-	-	8391	-755	-	-
TRANSFORMATION	-	-	-	-	-	-	-	52048	-	-	-
Electricity and CHP Plants	-	-	-	-	-	-	-	4981	-	-	-
Petroleum Refineries	-	-	-	-	-	-	-	47067	-	-	-
Other Transform. Sector	-	-	-	-	-	-	-	-	-	-	-
ENERGY SECTOR	-	-	-	-	-	-	-	-	-	-	-
DISTRIBUTION LOSSES	-	-	-	-	-	-	-	-	-	-	-
FINAL CONSUMPTION	-	-	-	-	-	-	-	-	-	-	-
INDUSTRY SECTOR	-	-	-	-	-	-	-	-	-	-	-
Iron and Steel	-	-	-	-	-	-	-	-	-	-	-
Chemical and Petrochem.	-	-	-	-	-	-	-	-	-	-	-
Non-Metallic Minerals	-	-	-	-	-	-	-	-	-	-	-
Non-specified	-	-	-	-	-	-	-	-	-	-	-
TRANSPORT SECTOR	-	-	-	-	-	-	-	-	-	-	-
Air	-	-	-	-	-	-	-	-	-	-	-
Road	-	-	-	-	-	-	-	-	-	-	-
Non-specified	-	-	-	-	-	-	-	-	-	-	-
OTHER SECTORS	-	-	-	-	-	-	-	-	-	-	-
Agriculture	-	-	-	-	-	-	-	-	-	-	-
Comm. and Publ. Services	-	-	-	-	-	-	-	-	-	-	-
Residential	-	-	-	-	-	-	-	-	-	-	-
Non-specified	-	-	-	-	-	-	-	-	-	-	-
NON-ENERGY USE	-	-	-	-	-	-	-	-	-	-	-

Kuwait / Koweit

SUPPLY AND CONSUMPTION 1997	Oil cont. (1000 tonnes)										
	Refinery Gas	LPG + Ethane	Motor Gasoline	Aviation Gasoline	Jet Fuel	Kerosene	Gas/ Diesel	Heavy Fuel Oil	Naphtha	Petrol. Coke	Other Prod.
Production	333	2665	1822	-	1595	6641	12704	11555	7936	-	1759
Imports	-	-	-	1	-	-	-	-	-	-	-
Exports	-	-2573	-128	-	-1187	-6615	-12073	-7243	-7936	-	-5
Intl. Marine Bunkers	-	-	-	-	-	-	-140	-525	-	-	-
Stock Changes	-	-	-	-	-	-	-	-	-	-	6
DOMESTIC SUPPLY	333	92	1694	1	408	26	491	3787	-	-	1760
Transfers and Stat. Diff.	-	925	-1	-	-	-	2	-	-	-	-
TRANSFORMATION	-	-	-	-	-	-	123	1337	-	-	-
Electricity and CHP Plants	-	-	-	-	-	-	123	1337	-	-	-
Petroleum Refineries	-	-	-	-	-	-	-	-	-	-	-
Other Transform. Sector	-	-	-	-	-	-	-	-	-	-	-
ENERGY SECTOR	333	-	-	-	-	-	-	2450	-	-	-
DISTRIBUTION LOSSES	-	-	-	-	-	-	-	-	-	-	-
FINAL CONSUMPTION	-	1017	1693	1	408	26	370	-	-	-	1760
INDUSTRY SECTOR	-	-	-	-	-	-	148	-	-	-	-
Iron and Steel	-	-	-	-	-	-	-	-	-	-	-
Chemical and Petrochem.	-	-	-	-	-	-	-	-	-	-	-
Non-Metallic Minerals	-	-	-	-	-	-	-	-	-	-	-
Non-specified	-	-	-	-	-	-	148	-	-	-	-
TRANSPORT SECTOR	-	-	1693	1	408	-	222	-	-	-	-
Air	-	-	-	1	408	-	-	-	-	-	-
Road	-	-	1693	-	-	-	222	-	-	-	-
Non-specified	-	-	-	-	-	-	-	-	-	-	-
OTHER SECTORS	-	1017	-	-	-	26	-	-	-	-	-
Agriculture	-	-	-	-	-	-	-	-	-	-	-
Comm. and Publ. Services	-	-	-	-	-	-	-	-	-	-	-
Residential	-	1017	-	-	-	26	-	-	-	-	-
Non-specified	-	-	-	-	-	-	-	-	-	-	-
NON-ENERGY USE	-	-	-	-	-	-	-	-	-	-	1760

APPROVISIONNEMENT ET DEMANDE 1998	Pétrole cont. (1000 tonnes)										
	Gaz de raffinerie	GPL + éthane	Essence moteur	Essence aviation	Carbu- réacteurs	Kérosène	Gazole	Fioul lourd	Naphta	Coke de pétrole	Autres prod.
Production	341	3124	2057	-	1880	6655	12302	10742	7872	-	1968
Imports	-	-	-	1	-	-	-	-	-	-	-
Exports	-	-3016	-275	-	-1399	-6625	-11640	-5379	-7872	-	-4
Intl. Marine Bunkers	-	-	-	-	-	-	-135	-488	-	-	-
Stock Changes	-	-	-	-	-	-	-	-	-	-	16
DOMESTIC SUPPLY	341	108	1782	1	481	30	527	4875	-	-	1980
Transfers and Stat. Diff.	-	1201	-	-	-1	-	-	-	-	-	-
TRANSFORMATION	-	-	-	-	-	-	144	1567	-	-	-
Electricity and CHP Plants	-	-	-	-	-	-	144	1567	-	-	-
Petroleum Refineries	-	-	-	-	-	-	-	-	-	-	-
Other Transform. Sector	-	-	-	-	-	-	-	-	-	-	-
ENERGY SECTOR	341	-	-	-	-	-	-	3308	-	-	-
DISTRIBUTION LOSSES	-	-	-	-	-	-	-	-	-	-	-
FINAL CONSUMPTION	-	1309	1782	1	480	30	383	-	-	-	1980
INDUSTRY SECTOR	-	-	-	-	-	-	153	-	-	-	-
Iron and Steel	-	-	-	-	-	-	-	-	-	-	-
Chemical and Petrochem.	-	-	-	-	-	-	-	-	-	-	-
Non-Metallic Minerals	-	-	-	-	-	-	-	-	-	-	-
Non-specified	-	-	-	-	-	-	153	-	-	-	-
TRANSPORT SECTOR	-	-	1782	1	480	-	230	-	-	-	-
Air	-	-	-	1	480	-	-	-	-	-	-
Road	-	-	1782	-	-	-	230	-	-	-	-
Non-specified	-	-	-	-	-	-	-	-	-	-	-
OTHER SECTORS	-	1309	-	-	-	30	-	-	-	-	-
Agriculture	-	-	-	-	-	-	-	-	-	-	-
Comm. and Publ. Services	-	-	-	-	-	-	-	-	-	-	-
Residential	-	1309	-	-	-	30	-	-	-	-	-
Non-specified	-	-	-	-	-	-	-	-	-	-	-
NON-ENERGY USE	-	-	-	-	-	-	-	-	-	-	1980

Kuwait / Koweit

SUPPLY AND CONSUMPTION 1997	Gas (TJ)				Comb. Renew. & Waste (TJ)				(GWh)	(TJ)
	Natural Gas	Gas Works	Coke Ovens	Blast Furnaces	Solid Biomass	Gas/Liquids from Biomass	Municipal Waste	Industrial Waste	Electricity	Heat
Production	358525	-	-	-	-	-	-	-	26724	-
Imports	-	-	-	-	165	-	-	-	-	-
Exports	-	-	-	-	-	-	-	-	-	-
Intl. Marine Bunkers	-	-	-	-	-	-	-	-	-	-
Stock Changes	-	-	-	-	-	-	-	-	-	-
DOMESTIC SUPPLY	358525	-	-	-	165	-	-	-	26724	-
Transfers and Stat. Diff.	-	-	-	-	-	-	-	-	-	-
TRANSFORMATION	79166	-	-	-	-	-	-	-	-	-
Electricity and CHP Plants	79166	-	-	-	-	-	-	-	-	-
Petroleum Refineries	-	-	-	-	-	-	-	-	-	-
Other Transform. Sector	-	-	-	-	-	-	-	-	-	-
ENERGY SECTOR	-	-	-	-	-	-	-	-	3864	-
DISTRIBUTION LOSSES	-	-	-	-	-	-	-	-	-	-
FINAL CONSUMPTION	279359	-	-	-	165	-	-	-	22860	-
INDUSTRY SECTOR	212078	-	-	-	-	-	-	-	-	-
Iron and Steel	-	-	-	-	-	-	-	-	-	-
Chemical and Petrochem.	-	-	-	-	-	-	-	-	-	-
Non-Metallic Minerals	-	-	-	-	-	-	-	-	-	-
Non-specified	212078	-	-	-	-	-	-	-	-	-
TRANSPORT SECTOR	-	-	-	-	-	-	-	-	-	-
Air	-	-	-	-	-	-	-	-	-	-
Road	-	-	-	-	-	-	-	-	-	-
Non-specified	-	-	-	-	-	-	-	-	-	-
OTHER SECTORS	67281	-	-	-	165	-	-	-	22860	-
Agriculture	-	-	-	-	-	-	-	-	-	-
Comm. and Publ. Services	-	-	-	-	-	-	-	-	388	-
Residential	-	-	-	-	165	-	-	-	18217	-
Non-specified	67281	-	-	-	-	-	-	-	4255	-
NON-ENERGY USE	-	-	-	-	-	-	-	-	-	-

APPROVISIONNEMENT ET DEMANDE 1998	Gaz (TJ)				En. Re. Comb. & Déchets (TJ)				(GWh)	(TJ)
	Gaz naturel	Usines à gaz	Cokeries	Hauts fourneaux	Biomasse solide	Gaz/Liquides tirés de biomasse	Déchets urbains	Déchets industriels	Electricité	Chaleur
Production	365497	-	-	-	-	-	-	-	29984	-
Imports	-	-	-	-	165	-	-	-	-	-
Exports	-	-	-	-	-	-	-	-	-	-
Intl. Marine Bunkers	-	-	-	-	-	-	-	-	-	-
Stock Changes	-	-	-	-	-	-	-	-	-	-
DOMESTIC SUPPLY	365497	-	-	-	165	-	-	-	29984	-
Transfers and Stat. Diff.	-	-	-	-	-	-	-	-	-	-
TRANSFORMATION	80705	-	-	-	-	-	-	-	-	-
Electricity and CHP Plants	80705	-	-	-	-	-	-	-	-	-
Petroleum Refineries	-	-	-	-	-	-	-	-	-	-
Other Transform. Sector	-	-	-	-	-	-	-	-	-	-
ENERGY SECTOR	-	-	-	-	-	-	-	-	4231	-
DISTRIBUTION LOSSES	-	-	-	-	-	-	-	-	-	-
FINAL CONSUMPTION	284792	-	-	-	165	-	-	-	25753	-
INDUSTRY SECTOR	216202	-	-	-	-	-	-	-	-	-
Iron and Steel	-	-	-	-	-	-	-	-	-	-
Chemical and Petrochem.	-	-	-	-	-	-	-	-	-	-
Non-Metallic Minerals	-	-	-	-	-	-	-	-	-	-
Non-specified	216202	-	-	-	-	-	-	-	-	-
TRANSPORT SECTOR	-	-	-	-	-	-	-	-	-	-
Air	-	-	-	-	-	-	-	-	-	-
Road	-	-	-	-	-	-	-	-	-	-
Non-specified	-	-	-	-	-	-	-	-	-	-
OTHER SECTORS	68590	-	-	-	165	-	-	-	25753	-
Agriculture	-	-	-	-	-	-	-	-	-	-
Comm. and Publ. Services	-	-	-	-	-	-	-	-	435	-
Residential	-	-	-	-	165	-	-	-	20439	-
Non-specified	68590	-	-	-	-	-	-	-	4879	-
NON-ENERGY USE	-	-	-	-	-	-	-	-	-	-

Kyrgyzstan / Kirghizistan

SUPPLY AND CONSUMPTION 1997	Coal (1000 tonnes)							Oil (1000 tonnes)			
	Coking Coal	Other Bit. Coal	Sub-Bit. Coal	Lignite	Peat	Oven and Gas Coke	Pat. Fuel and BKB	Crude Oil	NGL	Feed-stocks	Additives
Production	-	500	-	280	-	-	-	85	-	-	-
Imports	-	1500	-	50	-	-	-	65	-	-	-
Exports	-	-100	-	-	-	-	-	-40	-	-	-
Intl. Marine Bunkers	-	-	-	-	-	-	-	-	-	-	-
Stock Changes	-	-	-	-	-	-	-	-	-	-	-
DOMESTIC SUPPLY	-	1900	-	330	-	-	-	110	-	-	-
Transfers and Stat. Diff.	-	-	-	-	-	-	-	-	-	-	-
TRANSFORMATION	-	510	-	-	-	-	-	110	-	-	-
Electricity and CHP Plants	-	510	-	-	-	-	-	-	-	-	-
Petroleum Refineries	-	-	-	-	-	-	-	110	-	-	-
Other Transform. Sector	-	-	-	-	-	-	-	-	-	-	-
ENERGY SECTOR	-	-	-	-	-	-	-	-	-	-	-
DISTRIBUTION LOSSES	-	-	-	-	-	-	-	-	-	-	-
FINAL CONSUMPTION	-	1390	-	330	-	-	-	-	-	-	-
INDUSTRY SECTOR	-	1390	-	330	-	-	-	-	-	-	-
Iron and Steel	-	-	-	-	-	-	-	-	-	-	-
Chemical and Petrochem.	-	-	-	-	-	-	-	-	-	-	-
Non-Metallic Minerals	-	-	-	-	-	-	-	-	-	-	-
Non-specified	-	1390	-	330	-	-	-	-	-	-	-
TRANSPORT SECTOR	-	-	-	-	-	-	-	-	-	-	-
Air	-	-	-	-	-	-	-	-	-	-	-
Road	-	-	-	-	-	-	-	-	-	-	-
Non-specified	-	-	-	-	-	-	-	-	-	-	-
OTHER SECTORS	-	-	-	-	-	-	-	-	-	-	-
Agriculture	-	-	-	-	-	-	-	-	-	-	-
Comm. and Publ. Services	-	-	-	-	-	-	-	-	-	-	-
Residential	-	-	-	-	-	-	-	-	-	-	-
Non-specified	-	-	-	-	-	-	-	-	-	-	-
NON-ENERGY USE	-	-	-	-	-	-	-	-	-	-	-

APPROVISIONNEMENT ET DEMANDE 1998	Charbon (1000 tonnes)							Pétrole (1000 tonnes)			
	Charbon à coke	Autres charb. bit.	Charbon sous-bit.	Lignite	Tourbe	Coke de four/gaz	Agg./briq. de lignite	Pétrole brut	LGN	Produits d'aliment.	Additifs
Production	-	400	-	280	-	-	-	77	-	-	-
Imports	-	711	-	91	-	-	-	62	-	-	-
Exports	-	-16	-	-7	-	-	-	-	-	-	-
Intl. Marine Bunkers	-	-	-	-	-	-	-	-	-	-	-
Stock Changes	-	-	-	-	-	-	-	-	-	-	-
DOMESTIC SUPPLY	-	1095	-	364	-	-	-	139	-	-	-
Transfers and Stat. Diff.	-	-	-	-	-	-	-	-	-	-	-
TRANSFORMATION	-	446	-	-	-	-	-	139	-	-	-
Electricity and CHP Plants	-	446	-	-	-	-	-	-	-	-	-
Petroleum Refineries	-	-	-	-	-	-	-	139	-	-	-
Other Transform. Sector	-	-	-	-	-	-	-	-	-	-	-
ENERGY SECTOR	-	-	-	-	-	-	-	-	-	-	-
DISTRIBUTION LOSSES	-	-	-	-	-	-	-	-	-	-	-
FINAL CONSUMPTION	-	649	-	364	-	-	-	-	-	-	-
INDUSTRY SECTOR	-	649	-	364	-	-	-	-	-	-	-
Iron and Steel	-	-	-	-	-	-	-	-	-	-	-
Chemical and Petrochem.	-	-	-	-	-	-	-	-	-	-	-
Non-Metallic Minerals	-	-	-	-	-	-	-	-	-	-	-
Non-specified	-	649	-	364	-	-	-	-	-	-	-
TRANSPORT SECTOR	-	-	-	-	-	-	-	-	-	-	-
Air	-	-	-	-	-	-	-	-	-	-	-
Road	-	-	-	-	-	-	-	-	-	-	-
Non-specified	-	-	-	-	-	-	-	-	-	-	-
OTHER SECTORS	-	-	-	-	-	-	-	-	-	-	-
Agriculture	-	-	-	-	-	-	-	-	-	-	-
Comm. and Publ. Services	-	-	-	-	-	-	-	-	-	-	-
Residential	-	-	-	-	-	-	-	-	-	-	-
Non-specified	-	-	-	-	-	-	-	-	-	-	-
NON-ENERGY USE	-	-	-	-	-	-	-	-	-	-	-

Kyrgyzstan / Kirghizistan

SUPPLY AND CONSUMPTION 1997	Oil cont. (1000 tonnes)										
	Refinery Gas	LPG + Ethane	Motor Gasoline	Aviation Gasoline	Jet Fuel	Kerosene	Gas/ Diesel	Heavy Fuel Oil	Naphtha	Petrol. Coke	Other Prod.
Production	-	-	-	-	-	-	-	-	-	-	110
Imports	-	-	115	-	45	-	100	88	-	-	12
Exports	-	-	-	-	-	-	-	-	-	-	-
Intl. Marine Bunkers	-	-	-	-	-	-	-	-	-	-	-
Stock Changes	-	-	-	-	-	-	-	-	-	-	-
DOMESTIC SUPPLY	-	-	115	-	45	-	100	88	-	-	122
Transfers and Stat. Diff.	-	-	-	-	-	-	-	-	-	-	-
TRANSFORMATION	-	-	-	-	-	-	-	-	-	-	-
Electricity and CHP Plants	-	-	-	-	-	-	-	-	-	-	-
Petroleum Refineries	-	-	-	-	-	-	-	-	-	-	-
Other Transform. Sector	-	-	-	-	-	-	-	-	-	-	-
ENERGY SECTOR	-	-	-	-	-	-	-	-	-	-	-
DISTRIBUTION LOSSES	-	-	-	-	-	-	-	-	-	-	-
FINAL CONSUMPTION	-	-	115	-	45	-	100	88	-	-	122
INDUSTRY SECTOR	-	-	-	-	-	-	-	-	-	-	-
Iron and Steel	-	-	-	-	-	-	-	-	-	-	-
Chemical and Petrochem.	-	-	-	-	-	-	-	-	-	-	-
Non-Metallic Minerals	-	-	-	-	-	-	-	-	-	-	-
Non-specified	-	-	-	-	-	-	-	-	-	-	-
TRANSPORT SECTOR	-	-	115	-	45	-	-	-	-	-	-
Air	-	-	-	-	45	-	-	-	-	-	-
Road	-	-	115	-	-	-	-	-	-	-	-
Non-specified	-	-	-	-	-	-	-	-	-	-	-
OTHER SECTORS	-	-	-	-	-	-	100	88	-	-	110
Agriculture	-	-	-	-	-	-	-	-	-	-	-
Comm. and Publ. Services	-	-	-	-	-	-	-	-	-	-	-
Residential	-	-	-	-	-	-	-	-	-	-	-
Non-specified	-	-	-	-	-	-	100	88	-	-	110
NON-ENERGY USE	-	-	-	-	-	-	-	-	-	-	12

APPROVISIONNEMENT ET DEMANDE 1998	Pétrole cont. (1000 tonnes)										
	Gaz de raffinerie	GPL + éthane	Essence moteur	Essence aviation	Carbu- réacteurs	Kérosène	Gazole	Fioul lourd	Naphta	Coke de pétrole	Autres prod.
Production	-	-	-	-	-	-	-	-	-	-	139
Imports	-	-	232	-	57	-	82	108	-	-	34
Exports	-	-	-5	-	-4	-	-1	-	-	-	-1
Intl. Marine Bunkers	-	-	-	-	-	-	-	-	-	-	-
Stock Changes	-	-	-	-	-	-	-	-	-	-	-
DOMESTIC SUPPLY	-	-	227	-	53	-	81	108	-	-	172
Transfers and Stat. Diff.	-	-	-	-	-	-	-	-	-	-	-
TRANSFORMATION	-	-	-	-	-	-	-	-	-	-	-
Electricity and CHP Plants	-	-	-	-	-	-	-	-	-	-	-
Petroleum Refineries	-	-	-	-	-	-	-	-	-	-	-
Other Transform. Sector	-	-	-	-	-	-	-	-	-	-	-
ENERGY SECTOR	-	-	-	-	-	-	-	-	-	-	-
DISTRIBUTION LOSSES	-	-	-	-	-	-	-	-	-	-	-
FINAL CONSUMPTION	-	-	227	-	53	-	81	108	-	-	172
INDUSTRY SECTOR	-	-	-	-	-	-	-	-	-	-	-
Iron and Steel	-	-	-	-	-	-	-	-	-	-	-
Chemical and Petrochem.	-	-	-	-	-	-	-	-	-	-	-
Non-Metallic Minerals	-	-	-	-	-	-	-	-	-	-	-
Non-specified	-	-	-	-	-	-	-	-	-	-	-
TRANSPORT SECTOR	-	-	227	-	53	-	-	-	-	-	-
Air	-	-	-	-	53	-	-	-	-	-	-
Road	-	-	227	-	-	-	-	-	-	-	-
Non-specified	-	-	-	-	-	-	-	-	-	-	-
OTHER SECTORS	-	-	-	-	-	-	81	108	-	-	139
Agriculture	-	-	-	-	-	-	-	-	-	-	-
Comm. and Publ. Services	-	-	-	-	-	-	-	-	-	-	-
Residential	-	-	-	-	-	-	-	-	-	-	-
Non-specified	-	-	-	-	-	-	81	108	-	-	139
NON-ENERGY USE	-	-	-	-	-	-	-	-	-	-	33

Kyrgyzstan / Kirghizistan

SUPPLY AND CONSUMPTION 1997	Gas (TJ)				Comb. Renew. & Waste (TJ)				(GWh)	(TJ)
	Natural Gas	Gas Works	Coke Ovens	Blast Furnaces	Solid Biomass	Gas/Liquids from Biomass	Municipal Waste	Industrial Waste	Electricity	Heat
Production	1561	-	-	-	150	-	-	-	12637	16546
Imports	22241	-	-	-	-	-	-	-	5882	-
Exports	-	-	-	-	-	-	-	-	-7666	-
Intl. Marine Bunkers	-	-	-	-	-	-	-	-	-	-
Stock Changes	-	-	-	-	-	-	-	-	-	-
DOMESTIC SUPPLY	23802	-	-	-	150	-	-	-	10853	16546
Transfers and Stat. Diff.	-	-	-	-	-	-	-	-	-	-
TRANSFORMATION	14437	-	-	-	-	-	-	-	-	-
Electricity and CHP Plants	14437	-	-	-	-	-	-	-	-	-
Petroleum Refineries	-	-	-	-	-	-	-	-	-	-
Other Transform. Sector	-	-	-	-	-	-	-	-	-	-
ENERGY SECTOR	-	-	-	-	-	-	-	-	316	-
DISTRIBUTION LOSSES	-	-	-	-	-	-	-	-	4034	-
FINAL CONSUMPTION	9365	-	-	-	150	-	-	-	6503	16546
INDUSTRY SECTOR	-	-	-	-	-	-	-	-	1899	-
Iron and Steel	-	-	-	-	-	-	-	-	-	-
Chemical and Petrochem.	-	-	-	-	-	-	-	-	3	-
Non-Metallic Minerals	-	-	-	-	-	-	-	-	152	-
Non-specified	-	-	-	-	-	-	-	-	1744	-
TRANSPORT SECTOR	-	-	-	-	-	-	-	-	111	-
Air	-	-	-	-	-	-	-	-	-	-
Road	-	-	-	-	-	-	-	-	-	-
Non-specified	-	-	-	-	-	-	-	-	111	-
OTHER SECTORS	9365	-	-	-	150	-	-	-	4493	16546
Agriculture	-	-	-	-	-	-	-	-	2258	-
Comm. and Publ. Services	-	-	-	-	-	-	-	-	-	-
Residential	-	-	-	-	-	-	-	-	1585	-
Non-specified	9365	-	-	-	150	-	-	-	650	16546
NON-ENERGY USE	-	-	-	-	-	-	-	-	-	-

APPROVISIONNEMENT ET DEMANDE 1998	Gaz (TJ)				En. Re. Comb. & Déchets (TJ)				(GWh)	(TJ)
	Gaz naturel	Usines à gaz	Cokeries	Hauts fourneaux	Biomasse solide	Gaz/Liquides tirés de biomasse	Déchets urbains	Déchets industriels	Electricité	Chaleur
Production	702	-	-	-	150	-	-	-	11615	15815
Imports	38992	-	-	-	-	-	-	-	6399	-
Exports	-	-	-	-	-	-	-	-	-7125	-
Intl. Marine Bunkers	-	-	-	-	-	-	-	-	-	-
Stock Changes	-	-	-	-	-	-	-	-	-	-
DOMESTIC SUPPLY	39694	-	-	-	150	-	-	-	10889	15815
Transfers and Stat. Diff.	-	-	-	-	-	-	-	-	-	-
TRANSFORMATION	22824	-	-	-	-	-	-	-	-	-
Electricity and CHP Plants	22824	-	-	-	-	-	-	-	-	-
Petroleum Refineries	-	-	-	-	-	-	-	-	-	-
Other Transform. Sector	-	-	-	-	-	-	-	-	-	-
ENERGY SECTOR	-	-	-	-	-	-	-	-	329	-
DISTRIBUTION LOSSES	-	-	-	-	-	-	-	-	3698	-
FINAL CONSUMPTION	16870	-	-	-	150	-	-	-	6862	15815
INDUSTRY SECTOR	-	-	-	-	-	-	-	-	1774	-
Iron and Steel	-	-	-	-	-	-	-	-	-	-
Chemical and Petrochem.	-	-	-	-	-	-	-	-	2	-
Non-Metallic Minerals	-	-	-	-	-	-	-	-	173	-
Non-specified	-	-	-	-	-	-	-	-	1599	-
TRANSPORT SECTOR	-	-	-	-	-	-	-	-	117	-
Air	-	-	-	-	-	-	-	-	-	-
Road	-	-	-	-	-	-	-	-	-	-
Non-specified	-	-	-	-	-	-	-	-	117	-
OTHER SECTORS	16870	-	-	-	150	-	-	-	4971	15815
Agriculture	-	-	-	-	-	-	-	-	2605	-
Comm. and Publ. Services	-	-	-	-	-	-	-	-	-	-
Residential	-	-	-	-	-	-	-	-	1608	-
Non-specified	16870	-	-	-	150	-	-	-	758	15815
NON-ENERGY USE	-	-	-	-	-	-	-	-	-	-

Latvia / Lettonie : 1997

SUPPLY AND CONSUMPTION APPROVISIONNEMENT ET DEMANDE	Coal / *Charbon* (1000 tonnes)							Oil / *Pétrole* (1000 tonnes)			
	Coking Coal *Charbon à coke*	Other Bit. Coal *Autres charb. bit.*	Sub-Bit. Coal *Charbon sous-bit.*	Lignite *Lignite*	Peat *Tourbe*	Oven and Gas Coke *Coke de four/gaz*	Pat. Fuel and BKB *Agg./briq. de lignite*	Crude Oil *Pétrole brut*	NGL *LGN*	Feed-stocks *Produits d'aliment.*	Additives *Additifs*
Production	-	-	-	-	391	-	22	-	-	-	-
From Other Sources	-	-	-	-	-	-	-	-	-	-	-
Imports	-	218	-	-	-	12	-	-	-	-	-
Exports	-	-2	-	-	-	-	-	-	-	-	-
Intl. Marine Bunkers	-	-	-	-	-	-	-	-	-	-	-
Stock Changes	-	-20	-	-	-46	-	-	-	-	-	-
DOMESTIC SUPPLY	**-**	**196**	**-**	**-**	**345**	**12**	**22**	**-**	**-**	**-**	**-**
Transfers	-	-	-	-	-	-	-	-	-	-	-
Statistical Differences	-	-	-	-	-	-	-	-	-	-	-
TRANSFORMATION	**-**	**40**	**-**	**-**	**307**	**-**	**7**	**-**	**-**	**-**	**-**
Electricity Plants	-	-	-	-	-	-	-	-	-	-	-
CHP Plants	-	-	-	-	228	-	-	-	-	-	-
Heat Plants	-	40	-	-	43	-	7	-	-	-	-
Blast Furnaces/Gas Works	-	-	-	-	-	-	-	-	-	-	-
Coke/Pat. Fuel/BKB Plants	-	-	-	-	36	-	-	-	-	-	-
Petroleum Refineries	-	-	-	-	-	-	-	-	-	-	-
Petrochemical Industry	-	-	-	-	-	-	-	-	-	-	-
Liquefaction	-	-	-	-	-	-	-	-	-	-	-
Other Transform. Sector	-	-	-	-	-	-	-	-	-	-	-
ENERGY SECTOR	**-**	**-**	**-**	**-**	**32**	**-**	**-**	**-**	**-**	**-**	**-**
Coal Mines	-	-	-	-	-	-	-	-	-	-	-
Oil and Gas Extraction	-	-	-	-	-	-	-	-	-	-	-
Petroleum Refineries	-	-	-	-	-	-	-	-	-	-	-
Electr., CHP+Heat Plants	-	-	-	-	-	-	-	-	-	-	-
Pumped Storage (Elec.)	-	-	-	-	-	-	-	-	-	-	-
Other Energy Sector	-	-	-	-	32	-	-	-	-	-	-
Distribution Losses	-	-	-	-	-	-	-	-	-	-	-
FINAL CONSUMPTION	**-**	**156**	**-**	**-**	**6**	**12**	**15**	**-**	**-**	**-**	**-**
INDUSTRY SECTOR	**-**	**12**	**-**	**-**	**1**	**2**	**1**	**-**	**-**	**-**	**-**
Iron and Steel	-	-	-	-	-	-	-	-	-	-	-
Chemical and Petrochem.	-	-	-	-	-	-	-	-	-	-	-
of which: Feedstocks	-	-	-	-	-	-	-	-	-	-	-
Non-Ferrous Metals	-	-	-	-	-	-	-	-	-	-	-
Non-Metallic Minerals	-	3	-	-	1	-	-	-	-	-	-
Transport Equipment	-	-	-	-	-	-	-	-	-	-	-
Machinery	-	-	-	-	-	-	-	-	-	-	-
Mining and Quarrying	-	-	-	-	-	-	-	-	-	-	-
Food and Tobacco	-	5	-	-	-	2	1	-	-	-	-
Paper, Pulp and Print	-	2	-	-	-	-	-	-	-	-	-
Wood and Wood Products	-	-	-	-	-	-	-	-	-	-	-
Construction	-	1	-	-	-	-	-	-	-	-	-
Textile and Leather	-	1	-	-	-	-	-	-	-	-	-
Non-specified	-	-	-	-	-	-	-	-	-	-	-
TRANSPORT SECTOR	**-**	**-**	**-**	**-**	**-**	**-**	**-**	**-**	**-**	**-**	**-**
Air	-	-	-	-	-	-	-	-	-	-	-
Road	-	-	-	-	-	-	-	-	-	-	-
Rail	-	-	-	-	-	-	-	-	-	-	-
Pipeline Transport	-	-	-	-	-	-	-	-	-	-	-
Internal Navigation	-	-	-	-	-	-	-	-	-	-	-
Non-specified	-	-	-	-	-	-	-	-	-	-	-
OTHER SECTORS	**-**	**144**	**-**	**-**	**5**	**-**	**14**	**-**	**-**	**-**	**-**
Agriculture	-	5	-	-	-	-	1	-	-	-	-
Comm. and Publ. Services	-	79	-	-	1	-	4	-	-	-	-
Residential	-	60	-	-	4	-	9	-	-	-	-
Non-specified	-	-	-	-	-	-	-	-	-	-	-
NON-ENERGY USE	**-**	**-**	**-**	**-**	**-**	**10**	**-**	**-**	**-**	**-**	**-**
in Industry/Trans./Energy	-	-	-	-	-	10	-	-	-	-	-
in Transport	-	-	-	-	-	-	-	-	-	-	-
in Other Sectors	-	-	-	-	-	-	-	-	-	-	-

Latvia / Lettonie : 1997

SUPPLY AND CONSUMPTION / APPROVISIONNEMENT ET DEMANDE	Oil cont. / Pétrole cont. (1000 tonnes)										
	Refinery Gas / Gaz de raffinerie	LPG + Ethane / GPL + éthane	Motor Gasoline / Essence moteur	Aviation Gasoline / Essence aviation	Jet Fuel / Carbu-réacteurs	Kerosene / Kérosène	Gas/ Diesel / Gazole	Heavy Fuel Oil / Fioul lourd	Naphtha / Naphta	Petrol. Coke / Coke de pétrole	Other Prod. / Autres prod.
Production	-	-	-	-	-	-	-	-	-	-	-
From Other Sources	-	-	-	-	-	-	-	-	-	-	-
Imports	-	39	365	-	46	4	221	898	-	-	47
Exports	-	-	-5	-	-5	-	-31	-20	-	-	-3
Intl. Marine Bunkers	-	-	-	-	-	-	-	-	-	-	-
Stock Changes	-	-1	-6	-	7	-	206	-87	-	-	-3
DOMESTIC SUPPLY	-	38	354	-	48	4	396	791	-	-	41
Transfers	-	-	-	-	-	-	-	-	-	-	-
Statistical Differences	-	-	-	-	-	-	-	-	-	-	-
TRANSFORMATION	-	-	-	-	-	1	7	493	-	-	1
Electricity Plants	-	-	-	-	-	-	-	-	-	-	-
CHP Plants	-	-	-	-	-	-	-	152	-	-	1
Heat Plants	-	-	-	-	-	1	7	341	-	-	-
Blast Furnaces/Gas Works	-	-	-	-	-	-	-	-	-	-	-
Coke/Pat. Fuel/BKB Plants	-	-	-	-	-	-	-	-	-	-	-
Petroleum Refineries	-	-	-	-	-	-	-	-	-	-	-
Petrochemical Industry	-	-	-	-	-	-	-	-	-	-	-
Liquefaction	-	-	-	-	-	-	-	-	-	-	-
Other Transform. Sector	-	-	-	-	-	-	-	-	-	-	-
ENERGY SECTOR	-	1	-	-	-	-	3	4	-	-	-
Coal Mines	-	-	-	-	-	-	-	-	-	-	-
Oil and Gas Extraction	-	-	-	-	-	-	-	-	-	-	-
Petroleum Refineries	-	-	-	-	-	-	-	-	-	-	-
Electr., CHP+Heat Plants	-	-	-	-	-	-	1	-	-	-	-
Pumped Storage (Elec.)	-	-	-	-	-	-	-	-	-	-	-
Other Energy Sector	-	1	-	-	-	-	2	4	-	-	-
Distribution Losses	-	1	-	-	-	-	-	-	-	-	-
FINAL CONSUMPTION	-	36	354	-	48	3	386	294	-	-	40
INDUSTRY SECTOR	-	1	2	-	2	2	41	216	-	-	19
Iron and Steel	-	-	-	-	-	-	-	8	-	-	19
Chemical and Petrochem.	-	-	-	-	-	-	-	47	-	-	-
of which: Feedstocks	-	-	-	-	-	-	-	-	-	-	-
Non-Ferrous Metals	-	-	-	-	-	-	-	-	-	-	-
Non-Metallic Minerals	-	-	-	-	-	1	1	54	-	-	-
Transport Equipment	-	-	-	-	-	-	2	3	-	-	-
Machinery	-	-	-	-	-	-	1	5	-	-	-
Mining and Quarrying	-	-	-	-	-	-	-	-	-	-	-
Food and Tobacco	-	1	-	-	-	1	19	61	-	-	-
Paper, Pulp and Print	-	-	-	-	-	-	-	-	-	-	-
Wood and Wood Products	-	-	1	-	-	-	3	21	-	-	-
Construction	-	-	1	-	-	-	11	3	-	-	-
Textile and Leather	-	-	-	-	-	-	4	13	-	-	-
Non-specified	-	-	-	-	2	-	-	1	-	-	-
TRANSPORT SECTOR	-	2	348	-	42	-	277	52	-	-	21
Air	-	-	-	-	42	-	-	-	-	-	-
Road	-	2	348	-	-	-	177	-	-	-	21
Rail	-	-	-	-	-	-	80	-	-	-	-
Pipeline Transport	-	-	-	-	-	-	-	-	-	-	-
Internal Navigation	-	-	-	-	-	-	20	52	-	-	-
Non-specified	-	-	-	-	-	-	-	-	-	-	-
OTHER SECTORS	-	33	4	-	4	1	68	26	-	-	-
Agriculture	-	-	2	-	1	-	55	6	-	-	-
Comm. and Publ. Services	-	4	2	-	1	1	13	20	-	-	-
Residential	-	29	-	-	2	-	-	-	-	-	-
Non-specified	-	-	-	-	-	-	-	-	-	-	-
NON-ENERGY USE	-	-	-	-	-	-	-	-	-	-	-
in Industry/Transf./Energy	-	-	-	-	-	-	-	-	-	-	-
in Transport	-	-	-	-	-	-	-	-	-	-	-
in Other Sectors	-	-	-	-	-	-	-	-	-	-	-

Latvia / Lettonie : 1997

SUPPLY AND CONSUMPTION / APPROVISIONNEMENT ET DEMANDE	Gas / Gaz (TJ)				Comb. Renew. & Waste / En. Re. Comb. & Déchets (TJ)				(GWh)	(TJ)
	Natural Gas / Gaz naturel	Gas Works / Usines à gaz	Coke Ovens / Cokeries	Blast Furnaces / Hauts fourneaux	Solid Biomass / Biomasse solide	Gas/Liquids from Biomass / Gaz/Liquides tirés de biomasse	Municipal Waste / Déchets urbains	Industrial Waste / Déchets industriels	Electricity / Electricité	Heat / Chaleur
Production	-	-	-	-	54585	-	-	-	4456	46540
From Other Sources	-	-	-	-	-	-	-	-	-	-
Imports	49249	-	-	-	-	-	-	-	1824	-
Exports	-	-	-	-	-9869	-	-	-	-1	-
Intl. Marine Bunkers	-	-	-	-	-	-	-	-	-	-
Stock Changes	299	-	-	-	556	-	-	-	-	-
DOMESTIC SUPPLY	49548	-	-	-	45272	-	-	-	6279	46540
Transfers	-	-	-	-	-	-	-	-	-	-
Statistical Differences	-	-	-	-	-	-	-	-	-	-
TRANSFORMATION	31949	-	-	-	9936	-	-	-	6	-
Electricity Plants	-	-	-	-	-	-	-	-	-	-
CHP Plants	22932	-	-	-	-	-	-	-	-	-
Heat Plants	9017	-	-	-	9936	-	-	-	-	-
Blast Furnaces/Gas Works	-	-	-	-	-	-	-	-	-	-
Coke/Pat. Fuel/BKB Plants	-	-	-	-	-	-	-	-	-	-
Petroleum Refineries	-	-	-	-	-	-	-	-	-	-
Petrochemical Industry	-	-	-	-	-	-	-	-	-	-
Liquefaction	-	-	-	-	-	-	-	-	-	-
Other Transform. Sector	-	-	-	-	-	-	-	-	6	-
ENERGY SECTOR	638	-	-	-	26	-	-	-	478	1565
Coal Mines	-	-	-	-	-	-	-	-	-	-
Oil and Gas Extraction	-	-	-	-	-	-	-	-	-	-
Petroleum Refineries	-	-	-	-	-	-	-	-	-	-
Electr., CHP+Heat Plants	-	-	-	-	26	-	-	-	174	1197
Pumped Storage (Elec.)	-	-	-	-	-	-	-	-	-	-
Other Energy Sector	638	-	-	-	-	-	-	-	304	368
Distribution Losses	1160	-	-	-	-	-	-	-	1279	6752
FINAL CONSUMPTION	15801	-	-	-	35310	-	-	-	4516	38223
INDUSTRY SECTOR	9965	-	-	-	4167	-	-	-	1584	11384
Iron and Steel	4399	-	-	-	-	-	-	-	102	725
Chemical and Petrochem.	212	-	-	-	110	-	-	-	202	1111
of which: Feedstocks	-	-	-	-	-	-	-	-	-	-
Non-Ferrous Metals	-	-	-	-	-	-	-	-	-	-
Non-Metallic Minerals	711	-	-	-	20	-	-	-	98	485
Transport Equipment	124	-	-	-	2	-	-	-	74	393
Machinery	82	-	-	-	197	-	-	-	173	481
Mining and Quarrying	3	-	-	-	5	-	-	-	6	10
Food and Tobacco	3127	-	-	-	610	-	-	-	325	4273
Paper, Pulp and Print	117	-	-	-	22	-	-	-	24	211
Wood and Wood Products	569	-	-	-	2492	-	-	-	229	1353
Construction	22	-	-	-	355	-	-	-	21	996
Textile and Leather	519	-	-	-	34	-	-	-	215	986
Non-specified	80	-	-	-	320	-	-	-	115	360
TRANSPORT SECTOR	29	-	-	-	-	-	-	-	176	-
Air	-	-	-	-	-	-	-	-	-	-
Road	29	-	-	-	-	-	-	-	-	-
Rail	-	-	-	-	-	-	-	-	54	-
Pipeline Transport	-	-	-	-	-	-	-	-	37	-
Internal Navigation	-	-	-	-	-	-	-	-	-	-
Non-specified	-	-	-	-	-	-	-	-	85	-
OTHER SECTORS	5807	-	-	-	31143	-	-	-	2756	26839
Agriculture	647	-	-	-	529	-	-	-	187	15
Comm. and Publ. Services	1722	-	-	-	3700	-	-	-	1487	2263
Residential	3438	-	-	-	26914	-	-	-	1082	24561
Non-specified	-	-	-	-	-	-	-	-	-	-
NON-ENERGY USE	-	-	-	-	-	-	-	-	-	-
in Industry/Transf./Energy	-	-	-	-	-	-	-	-	-	-
in Transport	-	-	-	-	-	-	-	-	-	-
in Other Sectors	-	-	-	-	-	-	-	-	-	-

Latvia / Lettonie : 1998

SUPPLY AND CONSUMPTION	Coal / Charbon (1000 tonnes)							Oil / Pétrole (1000 tonnes)			
	Coking Coal	Other Bit. Coal	Sub-Bit. Coal	Lignite	Peat	Oven and Gas Coke	Pat. Fuel and BKB	Crude Oil	NGL	Feed-stocks	Additives
APPROVISIONNEMENT ET DEMANDE	Charbon à coke	Autres charb. bit.	Charbon sous-bit.	Lignite	Tourbe	Coke de four/gaz	Agg./briq. de lignite	Pétrole brut	LGN	Produits d'aliment.	Additifs
Production	-	-	-	-	59	-	11	-	-	-	-
From Other Sources	-	-	-	-	-	-	-	-	-	-	-
Imports	-	142	-	-	-	11	-	-	-	-	-
Exports	-	-	-	-	-	-	-	-	-	-	-
Intl. Marine Bunkers	-	-	-	-	-	-	-	-	-	-	-
Stock Changes	-	4	-	-	198	-	4	-	-	-	-
DOMESTIC SUPPLY	-	146	-	-	257	11	15	-	-	-	-
Transfers	-	-	-	-	-	-	-	-	-	-	-
Statistical Differences	-	-	-	-	-	-	-	-	-	-	-
TRANSFORMATION	-	21	-	-	215	-	1	-	-	-	-
Electricity Plants	-	-	-	-	-	-	-	-	-	-	-
CHP Plants	-	-	-	-	172	-	-	-	-	-	-
Heat Plants	-	21	-	-	26	-	1	-	-	-	-
Blast Furnaces/Gas Works	-	-	-	-	-	-	-	-	-	-	-
Coke/Pat. Fuel/BKB Plants	-	-	-	-	17	-	-	-	-	-	-
Petroleum Refineries	-	-	-	-	-	-	-	-	-	-	-
Petrochemical Industry	-	-	-	-	-	-	-	-	-	-	-
Liquefaction	-	-	-	-	-	-	-	-	-	-	-
Other Transform. Sector	-	-	-	-	-	-	-	-	-	-	-
ENERGY SECTOR	-	1	-	-	32	-	-	-	-	-	-
Coal Mines	-	-	-	-	-	-	-	-	-	-	-
Oil and Gas Extraction	-	-	-	-	-	-	-	-	-	-	-
Petroleum Refineries	-	-	-	-	-	-	-	-	-	-	-
Electr., CHP+Heat Plants	-	1	-	-	-	-	-	-	-	-	-
Pumped Storage (Elec.)	-	-	-	-	-	-	-	-	-	-	-
Other Energy Sector	-	-	-	-	32	-	-	-	-	-	-
Distribution Losses	-	-	-	-	-	-	-	-	-	-	-
FINAL CONSUMPTION	-	124	-	-	10	11	14	-	-	-	-
INDUSTRY SECTOR	-	13	-	-	1	1	1	-	-	-	-
Iron and Steel	-	-	-	-	-	-	-	-	-	-	-
Chemical and Petrochem.	-	-	-	-	-	-	-	-	-	-	-
of which: Feedstocks	-	-	-	-	-	-	-	-	-	-	-
Non-Ferrous Metals	-	-	-	-	-	-	-	-	-	-	-
Non-Metallic Minerals	-	1	-	-	1	-	-	-	-	-	-
Transport Equipment	-	-	-	-	-	-	-	-	-	-	-
Machinery	-	1	-	-	-	-	-	-	-	-	-
Mining and Quarrying	-	-	-	-	-	-	-	-	-	-	-
Food and Tobacco	-	5	-	-	-	1	1	-	-	-	-
Paper, Pulp and Print	-	1	-	-	-	-	-	-	-	-	-
Wood and Wood Products	-	-	-	-	-	-	-	-	-	-	-
Construction	-	1	-	-	-	-	-	-	-	-	-
Textile and Leather	-	1	-	-	-	-	-	-	-	-	-
Non-specified	-	3	-	-	-	-	-	-	-	-	-
TRANSPORT SECTOR	-	-	-	-	-	-	-	-	-	-	-
Air	-	-	-	-	-	-	-	-	-	-	-
Road	-	-	-	-	-	-	-	-	-	-	-
Rail	-	-	-	-	-	-	-	-	-	-	-
Pipeline Transport	-	-	-	-	-	-	-	-	-	-	-
Internal Navigation	-	-	-	-	-	-	-	-	-	-	-
Non-specified	-	-	-	-	-	-	-	-	-	-	-
OTHER SECTORS	-	111	-	-	9	-	13	-	-	-	-
Agriculture	-	4	-	-	-	-	-	-	-	-	-
Comm. and Publ. Services	-	79	-	-	1	-	1	-	-	-	-
Residential	-	28	-	-	8	-	12	-	-	-	-
Non-specified	-	-	-	-	-	-	-	-	-	-	-
NON-ENERGY USE	-	-	-	-	-	10	-	-	-	-	-
in Industry/Trans./Energy	-	-	-	-	-	10	-	-	-	-	-
in Transport	-	-	-	-	-	-	-	-	-	-	-
in Other Sectors	-	-	-	-	-	-	-	-	-	-	-

Latvia / Lettonie : 1998

SUPPLY AND CONSUMPTION / APPROVISIONNEMENT ET DEMANDE	Oil cont. / Pétrole cont. (1000 tonnes)										
	Refinery Gas / Gaz de raffinerie	LPG + Ethane / GPL + éthane	Motor Gasoline / Essence moteur	Aviation Gasoline / Essence aviation	Jet Fuel / Carbu- réacteurs	Kerosene / Kérosène	Gas/ Diesel / Gazole	Heavy Fuel Oil / Fioul lourd	Naphtha / Naphta	Petrol. Coke / Coke de pétrole	Other Prod. / Autres prod.
Production	-	-	-	-	-	-	-	-	-	-	-
From Other Sources	-	-	-	-	-	-	-	-	-	-	-
Imports	-	45	434	-	38	2	413	785	-	-	129
Exports	-	-	-89	-	-3	-	-13	-15	-	-	-3
Intl. Marine Bunkers	-	-	-	-	-	-	-	-	-	-	-
Stock Changes	-	-1	5	-	1	-	-9	-37	-	-	-51
DOMESTIC SUPPLY	-	44	350	-	36	2	391	733	-	-	75
Transfers	-	-	-	-	-	-	-	-	-	-	-
Statistical Differences	-	-	-	-	-	-	-	-	-	-	-
TRANSFORMATION	-	-	-	-	-	-	2	550	-	-	3
Electricity Plants	-	-	-	-	-	-	-	-	-	-	-
CHP Plants	-	-	-	-	-	-	-	197	-	-	3
Heat Plants	-	-	-	-	-	-	2	353	-	-	-
Blast Furnaces/Gas Works	-	-	-	-	-	-	-	-	-	-	-
Coke/Pat. Fuel/BKB Plants	-	-	-	-	-	-	-	-	-	-	-
Petroleum Refineries	-	-	-	-	-	-	-	-	-	-	-
Petrochemical Industry	-	-	-	-	-	-	-	-	-	-	-
Liquefaction	-	-	-	-	-	-	-	-	-	-	-
Other Transform. Sector	-	-	-	-	-	-	-	-	-	-	-
ENERGY SECTOR	-	-	-	-	-	-	3	12	-	-	-
Coal Mines	-	-	-	-	-	-	-	-	-	-	-
Oil and Gas Extraction	-	-	-	-	-	-	-	-	-	-	-
Petroleum Refineries	-	-	-	-	-	-	-	-	-	-	-
Electr., CHP+Heat Plants	-	-	-	-	-	-	1	-	-	-	-
Pumped Storage (Elec.)	-	-	-	-	-	-	-	-	-	-	-
Other Energy Sector	-	-	-	-	-	-	2	12	-	-	-
Distribution Losses	-	-	-	-	-	-	-	-	-	-	-
FINAL CONSUMPTION	-	44	350	-	36	2	386	171	-	-	72
INDUSTRY SECTOR	-	1	2	-	1	1	39	144	-	-	48
Iron and Steel	-	-	-	-	-	-	-	-	-	-	26
Chemical and Petrochem.	-	-	-	-	-	-	-	2	-	-	-
of which: Feedstocks	-	-	-	-	-	-	-	-	-	-	-
Non-Ferrous Metals	-	-	-	-	-	-	-	-	-	-	-
Non-Metallic Minerals	-	-	-	-	-	-	2	45	-	-	-
Transport Equipment	-	-	-	-	-	-	1	3	-	-	-
Machinery	-	-	-	-	-	-	-	6	-	-	-
Mining and Quarrying	-	-	-	-	-	-	1	-	-	-	-
Food and Tobacco	-	1	-	-	-	1	17	39	-	-	-
Paper, Pulp and Print	-	-	-	-	-	-	-	-	-	-	-
Wood and Wood Products	-	-	1	-	-	-	4	24	-	-	-
Construction	-	-	1	-	-	-	9	4	-	-	22
Textile and Leather	-	-	-	-	-	-	4	20	-	-	-
Non-specified	-	-	-	-	1	-	1	1	-	-	-
TRANSPORT SECTOR	-	3	346	-	34	-	291	2	-	-	24
Air	-	-	-	-	34	-	-	-	-	-	-
Road	-	3	346	-	-	-	205	-	-	-	24
Rail	-	-	-	-	-	-	73	-	-	-	-
Pipeline Transport	-	-	-	-	-	-	-	-	-	-	-
Internal Navigation	-	-	-	-	-	-	13	2	-	-	-
Non-specified	-	-	-	-	-	-	-	-	-	-	-
OTHER SECTORS	-	40	2	-	1	1	56	25	-	-	-
Agriculture	-	-	1	-	-	-	36	5	-	-	-
Comm. and Publ. Services	-	9	1	-	1	1	20	20	-	-	-
Residential	-	31	-	-	-	-	-	-	-	-	-
Non-specified	-	-	-	-	-	-	-	-	-	-	-
NON-ENERGY USE	-	-	-	-	-	-	-	-	-	-	-
in Industry/Transf./Energy	-	-	-	-	-	-	-	-	-	-	-
in Transport	-	-	-	-	-	-	-	-	-	-	-
in Other Sectors	-	-	-	-	-	-	-	-	-	-	-

Latvia / Lettonie : 1998

SUPPLY AND CONSUMPTION / APPROVISIONNEMENT ET DEMANDE	Gas / Gaz (TJ)				Comb. Renew. & Waste / En. Re. Comb. & Déchets (TJ)				(GWh)	(TJ)
	Natural Gas / Gaz naturel	Gas Works / Usines à gaz	Coke Ovens / Cokeries	Blast Furnaces / Hauts fourneaux	Solid Biomass / Biomasse solide	Gas/Liquids from Biomass / Gaz/Liquides tirés de biomasse	Municipal Waste / Déchets urbains	Industrial Waste / Déchets industriels	Electricity / Electricité	Heat / Chaleur
Production	-	-	-	-	58243	-	-	-	5738	44057
From Other Sources	-	-	-	-	-	-	-	-	-	-
Imports	51641	-	-	-	-	-	-	-	914	-
Exports	-	-	-	-	-14405	-	-	-	-385	-
Intl. Marine Bunkers	-	-	-	-	-	-	-	-	-	-
Stock Changes	-3101	-	-	-	-33	-	-	-	-	-
DOMESTIC SUPPLY	48540	-	-	-	43805	-	-	-	6267	44057
Transfers	-	-	-	-	-	-	-	-	-	-
Statistical Differences	-	-	-	-	-	-	-	-	-	-
TRANSFORMATION	30342	-	-	-	9306	-	-	-	6	-
Electricity Plants	-	-	-	-	-	-	-	-	-	-
CHP Plants	17973	-	-	-	-	-	-	-	-	-
Heat Plants	12369	-	-	-	9306	-	-	-	-	-
Blast Furnaces/Gas Works	-	-	-	-	-	-	-	-	-	-
Coke/Pat. Fuel/BKB Plants	-	-	-	-	-	-	-	-	-	-
Petroleum Refineries	-	-	-	-	-	-	-	-	-	-
Petrochemical Industry	-	-	-	-	-	-	-	-	-	-
Liquefaction	-	-	-	-	-	-	-	-	-	-
Other Transform. Sector	-	-	-	-	-	-	-	-	6	-
ENERGY SECTOR	822	-	-	-	34	-	-	-	501	1508
Coal Mines	-	-	-	-	-	-	-	-	-	-
Oil and Gas Extraction	-	-	-	-	-	-	-	-	-	-
Petroleum Refineries	-	-	-	-	-	-	-	-	-	-
Electr., CHP+Heat Plants	-	-	-	-	34	-	-	-	183	1152
Pumped Storage (Elec.)	-	-	-	-	-	-	-	-	-	-
Other Energy Sector	822	-	-	-	-	-	-	-	318	356
Distribution Losses	1161	-	-	-	-	-	-	-	1159	7462
FINAL CONSUMPTION	16215	-	-	-	34465	-	-	-	4601	35087
INDUSTRY SECTOR	10352	-	-	-	5052	-	-	-	1503	12374
Iron and Steel	4544	-	-	-	-	-	-	-	112	1390
Chemical and Petrochem.	237	-	-	-	22	-	-	-	165	1418
of which: Feedstocks	-	-	-	-	-	-	-	-	-	-
Non-Ferrous Metals	-	-	-	-	-	-	-	-	-	-
Non-Metallic Minerals	1201	-	-	-	53	-	-	-	111	463
Transport Equipment	91	-	-	-	-	-	-	-	74	358
Machinery	112	-	-	-	491	-	-	-	176	540
Mining and Quarrying	8	-	-	-	-	-	-	-	9	9
Food and Tobacco	2903	-	-	-	578	-	-	-	332	3946
Paper, Pulp and Print	101	-	-	-	21	-	-	-	25	132
Wood and Wood Products	572	-	-	-	3096	-	-	-	221	1705
Construction	149	-	-	-	462	-	-	-	21	1090
Textile and Leather	348	-	-	-	27	-	-	-	207	1058
Non-specified	86	-	-	-	302	-	-	-	50	265
TRANSPORT SECTOR	33	-	-	-	-	-	-	-	170	-
Air	-	-	-	-	-	-	-	-	-	-
Road	33	-	-	-	-	-	-	-	-	-
Rail	-	-	-	-	-	-	-	-	43	-
Pipeline Transport	-	-	-	-	-	-	-	-	44	-
Internal Navigation	-	-	-	-	-	-	-	-	-	-
Non-specified	-	-	-	-	-	-	-	-	83	-
OTHER SECTORS	5830	-	-	-	29413	-	-	-	2928	22713
Agriculture	748	-	-	-	690	-	-	-	178	10
Comm. and Publ. Services	1831	-	-	-	4489	-	-	-	1639	1417
Residential	3251	-	-	-	24234	-	-	-	1111	21286
Non-specified	-	-	-	-	-	-	-	-	-	-
NON-ENERGY USE	-	-	-	-	-	-	-	-	-	-
in Industry/Transf./Energy	-	-	-	-	-	-	-	-	-	-
in Transport	-	-	-	-	-	-	-	-	-	-
in Other Sectors	-	-	-	-	-	-	-	-	-	-

Lebanon / Liban

SUPPLY AND CONSUMPTION 1997	Coal (1000 tonnes)							Oil (1000 tonnes)			
	Coking Coal	Other Bit. Coal	Sub-Bit. Coal	Lignite	Peat	Oven and Gas Coke	Pat. Fuel and BKB	Crude Oil	NGL	Feed-stocks	Additives
Production	-	-	-	-	-	-	-	-	-	-	-
Imports	-	200	-	-	-	-	-	-	-	-	-
Exports	-	-	-	-	-	-	-	-	-	-	-
Intl. Marine Bunkers	-	-	-	-	-	-	-	-	-	-	-
Stock Changes	-	-	-	-	-	-	-	-	-	-	-
DOMESTIC SUPPLY	-	200	-	-	-	-	-	-	-	-	-
Transfers and Stat. Diff.	-	-	-	-	-	-	-	-	-	-	-
TRANSFORMATION	-	-	-	-	-	-	-	-	-	-	-
Electricity and CHP Plants	-	-	-	-	-	-	-	-	-	-	-
Petroleum Refineries	-	-	-	-	-	-	-	-	-	-	-
Other Transform. Sector	-	-	-	-	-	-	-	-	-	-	-
ENERGY SECTOR	-	-	-	-	-	-	-	-	-	-	-
DISTRIBUTION LOSSES	-	-	-	-	-	-	-	-	-	-	-
FINAL CONSUMPTION	-	200	-	-	-	-	-	-	-	-	-
INDUSTRY SECTOR	-	200	-	-	-	-	-	-	-	-	-
Iron and Steel	-	-	-	-	-	-	-	-	-	-	-
Chemical and Petrochem.	-	-	-	-	-	-	-	-	-	-	-
Non-Metallic Minerals	-	200	-	-	-	-	-	-	-	-	-
Non-specified	-	-	-	-	-	-	-	-	-	-	-
TRANSPORT SECTOR	-	-	-	-	-	-	-	-	-	-	-
Air	-	-	-	-	-	-	-	-	-	-	-
Road	-	-	-	-	-	-	-	-	-	-	-
Non-specified	-	-	-	-	-	-	-	-	-	-	-
OTHER SECTORS	-	-	-	-	-	-	-	-	-	-	-
Agriculture	-	-	-	-	-	-	-	-	-	-	-
Comm. and Publ. Services	-	-	-	-	-	-	-	-	-	-	-
Residential	-	-	-	-	-	-	-	-	-	-	-
Non-specified	-	-	-	-	-	-	-	-	-	-	-
NON-ENERGY USE	-	-	-	-	-	-	-	-	-	-	-

APPROVISIONNEMENT ET DEMANDE 1998	Charbon (1000 tonnes)							Pétrole (1000 tonnes)			
	Charbon à coke	Autres charb. bit.	Charbon sous-bit.	Lignite	Tourbe	Coke de four/gaz	Agg./briq. de lignite	Pétrole brut	LGN	Produits d'aliment.	Additifs
Production	-	-	-	-	-	-	-	-	-	-	-
Imports	-	200	-	-	-	-	-	-	-	-	-
Exports	-	-	-	-	-	-	-	-	-	-	-
Intl. Marine Bunkers	-	-	-	-	-	-	-	-	-	-	-
Stock Changes	-	-	-	-	-	-	-	-	-	-	-
DOMESTIC SUPPLY	-	200	-	-	-	-	-	-	-	-	-
Transfers and Stat. Diff.	-	-	-	-	-	-	-	-	-	-	-
TRANSFORMATION	-	-	-	-	-	-	-	-	-	-	-
Electricity and CHP Plants	-	-	-	-	-	-	-	-	-	-	-
Petroleum Refineries	-	-	-	-	-	-	-	-	-	-	-
Other Transform. Sector	-	-	-	-	-	-	-	-	-	-	-
ENERGY SECTOR	-	-	-	-	-	-	-	-	-	-	-
DISTRIBUTION LOSSES	-	-	-	-	-	-	-	-	-	-	-
FINAL CONSUMPTION	-	200	-	-	-	-	-	-	-	-	-
INDUSTRY SECTOR	-	200	-	-	-	-	-	-	-	-	-
Iron and Steel	-	-	-	-	-	-	-	-	-	-	-
Chemical and Petrochem.	-	-	-	-	-	-	-	-	-	-	-
Non-Metallic Minerals	-	200	-	-	-	-	-	-	-	-	-
Non-specified	-	-	-	-	-	-	-	-	-	-	-
TRANSPORT SECTOR	-	-	-	-	-	-	-	-	-	-	-
Air	-	-	-	-	-	-	-	-	-	-	-
Road	-	-	-	-	-	-	-	-	-	-	-
Non-specified	-	-	-	-	-	-	-	-	-	-	-
OTHER SECTORS	-	-	-	-	-	-	-	-	-	-	-
Agriculture	-	-	-	-	-	-	-	-	-	-	-
Comm. and Publ. Services	-	-	-	-	-	-	-	-	-	-	-
Residential	-	-	-	-	-	-	-	-	-	-	-
Non-specified	-	-	-	-	-	-	-	-	-	-	-
NON-ENERGY USE	-	-	-	-	-	-	-	-	-	-	-

Lebanon / Liban

SUPPLY AND CONSUMPTION 1997	Oil cont. (1000 tonnes)										
	Refinery Gas	LPG + Ethane	Motor Gasoline	Aviation Gasoline	Jet Fuel	Kerosene	Gas/ Diesel	Heavy Fuel Oil	Naphtha	Petrol. Coke	Other Prod.
Production	-	-	-	-	-	-	-	-	-	-	-
Imports	-	141	1310	-	109	4	1376	1805	-	-	88
Exports	-	-	-	-	-	-	-	-	-	-	-
Intl. Marine Bunkers	-	-	-	-	-	-	-	-	-	-	-
Stock Changes	-	-	-	-	-	-	11	65	-	-	-
DOMESTIC SUPPLY	-	141	1310	-	109	4	1387	1870	-	-	88
Transfers and Stat. Diff.	-	-	-	-	-	-	-	-	-	-	-
TRANSFORMATION	-	-	-	-	-	-	349	1532	-	-	-
Electricity and CHP Plants	-	-	-	-	-	-	349	1532	-	-	-
Petroleum Refineries	-	-	-	-	-	-	-	-	-	-	-
Other Transform. Sector	-	-	-	-	-	-	-	-	-	-	-
ENERGY SECTOR	-	-	-	-	-	-	-	-	-	-	-
DISTRIBUTION LOSSES	-	-	-	-	-	-	-	-	-	-	-
FINAL CONSUMPTION	-	141	1310	-	109	4	1038	338	-	-	88
INDUSTRY SECTOR	-	-	-	-	-	-	489	338	-	-	-
Iron and Steel	-	-	-	-	-	-	-	-	-	-	-
Chemical and Petrochem.	-	-	-	-	-	-	-	-	-	-	-
Non-Metallic Minerals	-	-	-	-	-	-	-	-	-	-	-
Non-specified	-	-	-	-	-	-	489	338	-	-	-
TRANSPORT SECTOR	-	-	1310	-	109	-	21	-	-	-	-
Air	-	-	-	-	109	-	-	-	-	-	-
Road	-	-	1310	-	-	-	21	-	-	-	-
Non-specified	-	-	-	-	-	-	-	-	-	-	-
OTHER SECTORS	-	141	-	-	-	4	528	-	-	-	-
Agriculture	-	-	-	-	-	-	-	-	-	-	-
Comm. and Publ. Services	-	-	-	-	-	-	-	-	-	-	-
Residential	-	141	-	-	-	4	528	-	-	-	-
Non-specified	-	-	-	-	-	-	-	-	-	-	-
NON-ENERGY USE	-	-	-	-	-	-	-	-	-	-	88

APPROVISIONNEMENT ET DEMANDE 1998	Pétrole cont. (1000 tonnes)										
	Gaz de raffinerie	GPL + éthane	Essence moteur	Essence aviation	Carbu- réacteurs	Kérosène	Gazole	Fioul lourd	Naphta	Coke de pétrole	Autres prod.
Production	-	-	-	-	-	-	-	-	-	-	-
Imports	-	100	1412	-	107	4	1425	1588	-	-	110
Exports	-	-	-	-	-	-	-	-	-	-	-
Intl. Marine Bunkers	-	-	-	-	-	-	-	-	-	-	-
Stock Changes	-	-	-	-	-	-	90	-114	-	-	-
DOMESTIC SUPPLY	-	100	1412	-	107	4	1515	1474	-	-	110
Transfers and Stat. Diff.	-	-	-	-	-	-	-	-	-	-	-
TRANSFORMATION	-	-	-	-	-	-	634	1325	-	-	-
Electricity and CHP Plants	-	-	-	-	-	-	634	1325	-	-	-
Petroleum Refineries	-	-	-	-	-	-	-	-	-	-	-
Other Transform. Sector	-	-	-	-	-	-	-	-	-	-	-
ENERGY SECTOR	-	-	-	-	-	-	-	-	-	-	-
DISTRIBUTION LOSSES	-	-	-	-	-	-	-	-	-	-	-
FINAL CONSUMPTION	-	100	1412	-	107	4	881	149	-	-	110
INDUSTRY SECTOR	-	-	-	-	-	-	415	149	-	-	-
Iron and Steel	-	-	-	-	-	-	-	-	-	-	-
Chemical and Petrochem.	-	-	-	-	-	-	-	-	-	-	-
Non-Metallic Minerals	-	-	-	-	-	-	-	-	-	-	-
Non-specified	-	-	-	-	-	-	415	149	-	-	-
TRANSPORT SECTOR	-	-	1412	-	107	-	18	-	-	-	-
Air	-	-	-	-	107	-	-	-	-	-	-
Road	-	-	1412	-	-	-	18	-	-	-	-
Non-specified	-	-	-	-	-	-	-	-	-	-	-
OTHER SECTORS	-	100	-	-	-	4	448	-	-	-	-
Agriculture	-	-	-	-	-	-	-	-	-	-	-
Comm. and Publ. Services	-	-	-	-	-	-	-	-	-	-	-
Residential	-	100	-	-	-	4	448	-	-	-	-
Non-specified	-	-	-	-	-	-	-	-	-	-	-
NON-ENERGY USE	-	-	-	-	-	-	-	-	-	-	110

INTERNATIONAL ENERGY AGENCY

Lebanon / Liban

SUPPLY AND CONSUMPTION 1997	Gas (TJ) Natural Gas	Gas Works	Coke Ovens	Blast Furnaces	Comb. Renew. & Waste (TJ) Solid Biomass	Gas/Liquids from Biomass	Municipal Waste	Industrial Waste	(GWh) Electricity	(TJ) Heat
Production	-	-	-	-	5152	-	-	-	8515	-
Imports	-	-	-	-	64	-	-	-	608	-
Exports	-	-	-	-	-	-	-	-	-	-
Intl. Marine Bunkers	-	-	-	-	-	-	-	-	-	-
Stock Changes	-	-	-	-	-	-	-	-	-	-
DOMESTIC SUPPLY	-	-	-	-	**5216**	-	-	-	**9123**	-
Transfers and Stat. Diff.	-	-	-	-	-	-	-	-	-	-
TRANSFORMATION	-	-	-	-	**647**	-	-	-	-	-
Electricity and CHP Plants	-	-	-	-	-	-	-	-	-	-
Petroleum Refineries	-	-	-	-	-	-	-	-	-	-
Other Transform. Sector	-	-	-	-	647	-	-	-	-	-
ENERGY SECTOR	-	-	-	-	-	-	-	-	-	-
DISTRIBUTION LOSSES	-	-	-	-	-	-	-	-	1123	-
FINAL CONSUMPTION	-	-	-	-	**4568**	-	-	-	**8000**	-
INDUSTRY SECTOR	-	-	-	-	-	-	-	-	**2100**	-
Iron and Steel	-	-	-	-	-	-	-	-	-	-
Chemical and Petrochem.	-	-	-	-	-	-	-	-	-	-
Non-Metallic Minerals	-	-	-	-	-	-	-	-	-	-
Non-specified	-	-	-	-	-	-	-	-	2100	-
TRANSPORT SECTOR	-	-	-	-	-	-	-	-	-	-
Air	-	-	-	-	-	-	-	-	-	-
Road	-	-	-	-	-	-	-	-	-	-
Non-specified	-	-	-	-	-	-	-	-	-	-
OTHER SECTORS	-	-	-	-	**4568**	-	-	-	**5900**	-
Agriculture	-	-	-	-	-	-	-	-	-	-
Comm. and Publ. Services	-	-	-	-	-	-	-	-	1340	-
Residential	-	-	-	-	-	-	-	-	3050	-
Non-specified	-	-	-	-	4568	-	-	-	1510	-
NON-ENERGY USE	-	-	-	-	-	-	-	-	-	-

APPROVISIONNEMENT ET DEMANDE 1998	Gaz (TJ) Gaz naturel	Usines à gaz	Cokeries	Hauts fourneaux	En. Re. Comb. & Déchets (TJ) Biomasse solide	Gaz/Liquides tirés de biomasse	Déchets urbains	Déchets industriels	(GWh) Electricité	(TJ) Chaleur
Production	-	-	-	-	5245	-	-	-	8357	-
Imports	-	-	-	-	67	-	-	-	654	-
Exports	-	-	-	-	-	-	-	-	-	-
Intl. Marine Bunkers	-	-	-	-	-	-	-	-	-	-
Stock Changes	-	-	-	-	-	-	-	-	-	-
DOMESTIC SUPPLY	-	-	-	-	**5312**	-	-	-	**9011**	-
Transfers and Stat. Diff.	-	-	-	-	-	-	-	-	-	-
TRANSFORMATION	-	-	-	-	**641**	-	-	-	-	-
Electricity and CHP Plants	-	-	-	-	-	-	-	-	-	-
Petroleum Refineries	-	-	-	-	-	-	-	-	-	-
Other Transform. Sector	-	-	-	-	641	-	-	-	-	-
ENERGY SECTOR	-	-	-	-	-	-	-	-	-	-
DISTRIBUTION LOSSES	-	-	-	-	-	-	-	-	1350	-
FINAL CONSUMPTION	-	-	-	-	**4671**	-	-	-	**7661**	-
INDUSTRY SECTOR	-	-	-	-	-	-	-	-	**2011**	-
Iron and Steel	-	-	-	-	-	-	-	-	-	-
Chemical and Petrochem.	-	-	-	-	-	-	-	-	-	-
Non-Metallic Minerals	-	-	-	-	-	-	-	-	-	-
Non-specified	-	-	-	-	-	-	-	-	2011	-
TRANSPORT SECTOR	-	-	-	-	-	-	-	-	-	-
Air	-	-	-	-	-	-	-	-	-	-
Road	-	-	-	-	-	-	-	-	-	-
Non-specified	-	-	-	-	-	-	-	-	-	-
OTHER SECTORS	-	-	-	-	**4671**	-	-	-	**5650**	-
Agriculture	-	-	-	-	-	-	-	-	-	-
Comm. and Publ. Services	-	-	-	-	-	-	-	-	1283	-
Residential	-	-	-	-	-	-	-	-	2921	-
Non-specified	-	-	-	-	4671	-	-	-	1446	-
NON-ENERGY USE	-	-	-	-	-	-	-	-	-	-

Libya / Libye

SUPPLY AND CONSUMPTION 1997	Coal (1000 tonnes)							Oil (1000 tonnes)			
	Coking Coal	Other Bit. Coal	Sub-Bit. Coal	Lignite	Peat	Oven and Gas Coke	Pat. Fuel and BKB	Crude Oil	NGL	Feed-stocks	Additives
Production	-	-	-	-	-	-	-	69347	1738	-	-
Imports	-	-	-	-	-	-	-	-	-	-	-
Exports	-	-	-	-	-	-	-	-52223	-	-	-
Intl. Marine Bunkers	-	-	-	-	-	-	-	-	-	-	-
Stock Changes	-	-	-	-	-	-	-	-	-	-	-
DOMESTIC SUPPLY	-	-	-	-	-	-	-	17124	1738	-	-
Transfers and Stat. Diff.	-	-	-	-	-	-	-	-233	-1738	-	-
TRANSFORMATION	-	-	-	-	-	-	-	16891	-	-	-
Electricity and CHP Plants	-	-	-	-	-	-	-	-	-	-	-
Petroleum Refineries	-	-	-	-	-	-	-	16891	-	-	-
Other Transform. Sector	-	-	-	-	-	-	-	-	-	-	-
ENERGY SECTOR	-	-	-	-	-	-	-	-	-	-	-
DISTRIBUTION LOSSES	-	-	-	-	-	-	-	-	-	-	-
FINAL CONSUMPTION	-	-	-	-	-	-	-	-	-	-	-
INDUSTRY SECTOR	-	-	-	-	-	-	-	-	-	-	-
Iron and Steel	-	-	-	-	-	-	-	-	-	-	-
Chemical and Petrochem.	-	-	-	-	-	-	-	-	-	-	-
Non-Metallic Minerals	-	-	-	-	-	-	-	-	-	-	-
Non-specified	-	-	-	-	-	-	-	-	-	-	-
TRANSPORT SECTOR	-	-	-	-	-	-	-	-	-	-	-
Air	-	-	-	-	-	-	-	-	-	-	-
Road	-	-	-	-	-	-	-	-	-	-	-
Non-specified	-	-	-	-	-	-	-	-	-	-	-
OTHER SECTORS	-	-	-	-	-	-	-	-	-	-	-
Agriculture	-	-	-	-	-	-	-	-	-	-	-
Comm. and Publ. Services	-	-	-	-	-	-	-	-	-	-	-
Residential	-	-	-	-	-	-	-	-	-	-	-
Non-specified	-	-	-	-	-	-	-	-	-	-	-
NON-ENERGY USE	-	-	-	-	-	-	-	-	-	-	-

APPROVISIONNEMENT ET DEMANDE 1998	Charbon (1000 tonnes)							Pétrole (1000 tonnes)			
	Charbon à coke	Autres charb. bit.	Charbon sous-bit.	Lignite	Tourbe	Coke de four/gaz	Agg./briq. de lignite	Pétrole brut	LGN	Produits d'aliment.	Additifs
Production	-	-	-	-	-	-	-	67461	1872	-	-
Imports	-	-	-	-	-	-	-	-	-	-	-
Exports	-	-	-	-	-	-	-	-54386	-	-	-
Intl. Marine Bunkers	-	-	-	-	-	-	-	-	-	-	-
Stock Changes	-	-	-	-	-	-	-	-	-	-	-
DOMESTIC SUPPLY	-	-	-	-	-	-	-	13075	1872	-	-
Transfers and Stat. Diff.	-	-	-	-	-	-	-	3342	-1872	-	-
TRANSFORMATION	-	-	-	-	-	-	-	16417	-	-	-
Electricity and CHP Plants	-	-	-	-	-	-	-	-	-	-	-
Petroleum Refineries	-	-	-	-	-	-	-	16417	-	-	-
Other Transform. Sector	-	-	-	-	-	-	-	-	-	-	-
ENERGY SECTOR	-	-	-	-	-	-	-	-	-	-	-
DISTRIBUTION LOSSES	-	-	-	-	-	-	-	-	-	-	-
FINAL CONSUMPTION	-	-	-	-	-	-	-	-	-	-	-
INDUSTRY SECTOR	-	-	-	-	-	-	-	-	-	-	-
Iron and Steel	-	-	-	-	-	-	-	-	-	-	-
Chemical and Petrochem.	-	-	-	-	-	-	-	-	-	-	-
Non-Metallic Minerals	-	-	-	-	-	-	-	-	-	-	-
Non-specified	-	-	-	-	-	-	-	-	-	-	-
TRANSPORT SECTOR	-	-	-	-	-	-	-	-	-	-	-
Air	-	-	-	-	-	-	-	-	-	-	-
Road	-	-	-	-	-	-	-	-	-	-	-
Non-specified	-	-	-	-	-	-	-	-	-	-	-
OTHER SECTORS	-	-	-	-	-	-	-	-	-	-	-
Agriculture	-	-	-	-	-	-	-	-	-	-	-
Comm. and Publ. Services	-	-	-	-	-	-	-	-	-	-	-
Residential	-	-	-	-	-	-	-	-	-	-	-
Non-specified	-	-	-	-	-	-	-	-	-	-	-
NON-ENERGY USE	-	-	-	-	-	-	-	-	-	-	-

Libya / Libye

SUPPLY AND CONSUMPTION 1997	Refinery Gas	LPG + Ethane	Motor Gasoline	Aviation Gasoline	Jet Fuel	Kerosene	Gas/ Diesel	Heavy Fuel Oil	Naphtha	Petrol. Coke	Other Prod.
Oil cont. (1000 tonnes)											
Production	555	263	1970	3	1339	274	4272	4834	1143	-	122
Imports	-	-	-	-	-	-	-	-	-	-	30
Exports	-	-261	-314	-	-1232	-70	-2164	-1517	-1615	-	-
Intl. Marine Bunkers	-	-	-	-	-	-	-	-90	-	-	-
Stock Changes	-	-	-	-	-	-	-	-	-	-	-
DOMESTIC SUPPLY	555	2	1656	3	107	204	2108	3227	-472	-	152
Transfers and Stat. Diff.	-	420	60	-	189	67	223	198	1408	-	73
TRANSFORMATION	-	-	-	-	-	-	1054	2807	-	-	-
Electricity and CHP Plants	-	-	-	-	-	-	1054	2807	-	-	-
Petroleum Refineries	-	-	-	-	-	-	-	-	-	-	-
Other Transform. Sector	-	-	-	-	-	-	-	-	-	-	-
ENERGY SECTOR	555	-	-	-	-	-	-	173	-	-	-
DISTRIBUTION LOSSES	-	-	-	-	-	-	-	-	-	-	-
FINAL CONSUMPTION	-	422	1716	3	296	271	1277	445	936	-	225
INDUSTRY SECTOR	-	-	-	-	-	-	-	445	936	-	-
Iron and Steel	-	-	-	-	-	-	-	-	-	-	-
Chemical and Petrochem.	-	-	-	-	-	-	-	-	936	-	-
Non-Metallic Minerals	-	-	-	-	-	-	-	-	-	-	-
Non-specified	-	-	-	-	-	-	-	445	-	-	-
TRANSPORT SECTOR	-	-	1716	3	296	-	1277	-	-	-	-
Air	-	-	-	3	296	-	-	-	-	-	-
Road	-	-	1716	-	-	-	1277	-	-	-	-
Non-specified	-	-	-	-	-	-	-	-	-	-	-
OTHER SECTORS	-	422	-	-	-	271	-	-	-	-	-
Agriculture	-	-	-	-	-	-	-	-	-	-	-
Comm. and Publ. Services	-	-	-	-	-	-	-	-	-	-	-
Residential	-	422	-	-	-	271	-	-	-	-	-
Non-specified	-	-	-	-	-	-	-	-	-	-	-
NON-ENERGY USE	-	-	-	-	-	-	-	-	-	-	225

APPROVISIONNEMENT ET DEMANDE 1998	Gaz de raffinerie	GPL + éthane	Essence moteur	Essence aviation	Carbu- réacteurs	Kérosène	Gazole	Fioul lourd	Naphta	Coke de pétrole	Autres prod.
Pétrole cont. (1000 tonnes)											
Production	555	263	1970	3	1339	274	4272	4834	1143	-	122
Imports	-	-	-	-	-	-	-	-	-	-	30
Exports	-	-261	-314	-	-1232	-70	-2164	-1517	-1615	-	-
Intl. Marine Bunkers	-	-	-	-	-	-	-	-90	-	-	-
Stock Changes	-	-	-	-	-	-	-	-	-	-	-
DOMESTIC SUPPLY	555	2	1656	3	107	204	2108	3227	-472	-	152
Transfers and Stat. Diff.	-	437	132	-	189	67	291	293	1408	-	73
TRANSFORMATION	-	-	-	-	-	-	1083	2884	-	-	-
Electricity and CHP Plants	-	-	-	-	-	-	1083	2884	-	-	-
Petroleum Refineries	-	-	-	-	-	-	-	-	-	-	-
Other Transform. Sector	-	-	-	-	-	-	-	-	-	-	-
ENERGY SECTOR	555	-	-	-	-	-	-	173	-	-	-
DISTRIBUTION LOSSES	-	-	-	-	-	-	-	-	-	-	-
FINAL CONSUMPTION	-	439	1788	3	296	271	1316	463	936	-	225
INDUSTRY SECTOR	-	-	-	-	-	-	-	463	936	-	-
Iron and Steel	-	-	-	-	-	-	-	-	-	-	-
Chemical and Petrochem.	-	-	-	-	-	-	-	-	936	-	-
Non-Metallic Minerals	-	-	-	-	-	-	-	-	-	-	-
Non-specified	-	-	-	-	-	-	-	463	-	-	-
TRANSPORT SECTOR	-	-	1788	3	296	-	1316	-	-	-	-
Air	-	-	-	3	296	-	-	-	-	-	-
Road	-	-	1788	-	-	-	1316	-	-	-	-
Non-specified	-	-	-	-	-	-	-	-	-	-	-
OTHER SECTORS	-	439	-	-	-	271	-	-	-	-	-
Agriculture	-	-	-	-	-	-	-	-	-	-	-
Comm. and Publ. Services	-	-	-	-	-	-	-	-	-	-	-
Residential	-	439	-	-	-	271	-	-	-	-	-
Non-specified	-	-	-	-	-	-	-	-	-	-	-
NON-ENERGY USE	-	-	-	-	-	-	-	-	-	-	225

Libya / Libye

SUPPLY AND CONSUMPTION 1997	Gas (TJ)				Comb. Renew. & Waste (TJ)				(GWh)	(TJ)
	Natural Gas	Gas Works	Coke Ovens	Blast Furnaces	Solid Biomass	Gas/Liquids from Biomass	Municipal Waste	Industrial Waste	Electricity	Heat
Production	249660	-	-	-	5469	-	-	-	18974	-
Imports	-	-	-	-	-	-	-	-	-	-
Exports	-41800	-	-	-	-	-	-	-	-	-
Intl. Marine Bunkers	-	-	-	-	-	-	-	-	-	-
Stock Changes	-	-	-	-	-	-	-	-	-	-
DOMESTIC SUPPLY	207860	-	-	-	5469	-	-	-	18974	-
Transfers and Stat. Diff.	-	-	-	-	-	-	-	-	-	-
TRANSFORMATION	-	-	-	-	-	-	-	-	-	-
Electricity and CHP Plants	-	-	-	-	-	-	-	-	-	-
Petroleum Refineries	-	-	-	-	-	-	-	-	-	-
Other Transform. Sector	-	-	-	-	-	-	-	-	-	-
ENERGY SECTOR	121506	-	-	-	-	-	-	-	-	-
DISTRIBUTION LOSSES	-	-	-	-	-	-	-	-	-	-
FINAL CONSUMPTION	86354	-	-	-	5469	-	-	-	18974	-
INDUSTRY SECTOR	86354	-	-	-	-	-	-	-	-	-
Iron and Steel	-	-	-	-	-	-	-	-	-	-
Chemical and Petrochem.	86354	-	-	-	-	-	-	-	-	-
Non-Metallic Minerals	-	-	-	-	-	-	-	-	-	-
Non-specified	-	-	-	-	-	-	-	-	-	-
TRANSPORT SECTOR	-	-	-	-	-	-	-	-	-	-
Air	-	-	-	-	-	-	-	-	-	-
Road	-	-	-	-	-	-	-	-	-	-
Non-specified	-	-	-	-	-	-	-	-	-	-
OTHER SECTORS	-	-	-	-	5469	-	-	-	18974	-
Agriculture	-	-	-	-	-	-	-	-	-	-
Comm. and Publ. Services	-	-	-	-	-	-	-	-	-	-
Residential	-	-	-	-	5469	-	-	-	-	-
Non-specified	-	-	-	-	-	-	-	-	18974	-
NON-ENERGY USE	-	-	-	-	-	-	-	-	-	-

APPROVISIONNEMENT ET DEMANDE 1998	Gaz (TJ)				En. Re. Comb. & Déchets (TJ)				(GWh)	(TJ)
	Gaz naturel	Usines à gaz	Cokeries	Hauts fourneaux	Biomasse solide	Gaz/Liquides tirés de biomasse	Déchets urbains	Déchets industriels	Electricité	Chaleur
Production	241680	-	-	-	5590	-	-	-	19496	-
Imports	-	-	-	-	-	-	-	-	-	-
Exports	-34580	-	-	-	-	-	-	-	-	-
Intl. Marine Bunkers	-	-	-	-	-	-	-	-	-	-
Stock Changes	-	-	-	-	-	-	-	-	-	-
DOMESTIC SUPPLY	207100	-	-	-	5590	-	-	-	19496	-
Transfers and Stat. Diff.	-	-	-	-	-	-	-	-	-	-
TRANSFORMATION	-	-	-	-	-	-	-	-	-	-
Electricity and CHP Plants	-	-	-	-	-	-	-	-	-	-
Petroleum Refineries	-	-	-	-	-	-	-	-	-	-
Other Transform. Sector	-	-	-	-	-	-	-	-	-	-
ENERGY SECTOR	121062	-	-	-	-	-	-	-	-	-
DISTRIBUTION LOSSES	-	-	-	-	-	-	-	-	-	-
FINAL CONSUMPTION	86038	-	-	-	5590	-	-	-	19496	-
INDUSTRY SECTOR	86038	-	-	-	-	-	-	-	-	-
Iron and Steel	-	-	-	-	-	-	-	-	-	-
Chemical and Petrochem.	86038	-	-	-	-	-	-	-	-	-
Non-Metallic Minerals	-	-	-	-	-	-	-	-	-	-
Non-specified	-	-	-	-	-	-	-	-	-	-
TRANSPORT SECTOR	-	-	-	-	-	-	-	-	-	-
Air	-	-	-	-	-	-	-	-	-	-
Road	-	-	-	-	-	-	-	-	-	-
Non-specified	-	-	-	-	-	-	-	-	-	-
OTHER SECTORS	-	-	-	-	5590	-	-	-	19496	-
Agriculture	-	-	-	-	-	-	-	-	-	-
Comm. and Publ. Services	-	-	-	-	-	-	-	-	-	-
Residential	-	-	-	-	5590	-	-	-	-	-
Non-specified	-	-	-	-	-	-	-	-	19496	-
NON-ENERGY USE	-	-	-	-	-	-	-	-	-	-

Lithuania / Lituanie : 1997

SUPPLY AND CONSUMPTION / APPROVISIONNEMENT ET DEMANDE	Coal / Charbon (1000 tonnes)							Oil / Pétrole (1000 tonnes)			
	Coking Coal / Charbon à coke	Other Bit. Coal / Autres charb. bit.	Sub-Bit. Coal / Charbon sous-bit.	Lignite / Lignite	Peat / Tourbe	Oven and Gas Coke / Coke de four/gaz	Pat. Fuel and BKB / Agg./briq. de lignite	Crude Oil / Pétrole brut	NGL / LGN	Feed-stocks / Produits d'aliment.	Additives / Additifs
Production	-	-	-	-	89	-	21	212	-	-	12
From Other Sources	-	-	-	-	-	-	-	-	-	-	-
Imports	-	259	-	3	-	10	1	5074	-	701	2
Exports	-	-48	-	-	-	-	-	-193	-	-	-
Intl. Marine Bunkers	-	-	-	-	-	-	-	-	-	-	-
Stock Changes	-	45	-	3	-10	1	1	-26	-	-74	-
DOMESTIC SUPPLY	-	256	-	6	79	11	23	5067	-	627	14
Transfers	-	-	-	-	-	-	-	-49	-	-	-
Statistical Differences	-	-	-	-	-	-	-	-	-	-	-
TRANSFORMATION	-	25	-	-	46	-	-	5018	-	627	14
Electricity Plants	-	-	-	-	-	-	-	-	-	-	-
CHP Plants	-	-	-	-	-	-	-	-	-	-	-
Heat Plants	-	25	-	-	14	-	-	4	-	-	-
Blast Furnaces/Gas Works	-	-	-	-	-	-	-	-	-	-	-
Coke/Pat. Fuel/BKB Plants	-	-	-	-	32	-	-	-	-	-	-
Petroleum Refineries	-	-	-	-	-	-	-	5014	-	627	14
Petrochemical Industry	-	-	-	-	-	-	-	-	-	-	-
Liquefaction	-	-	-	-	-	-	-	-	-	-	-
Other Transform. Sector	-	-	-	-	-	-	-	-	-	-	-
ENERGY SECTOR	-	-	-	-	-	-	-	-	-	-	-
Coal Mines	-	-	-	-	-	-	-	-	-	-	-
Oil and Gas Extraction	-	-	-	-	-	-	-	-	-	-	-
Petroleum Refineries	-	-	-	-	-	-	-	-	-	-	-
Electr., CHP+Heat Plants	-	-	-	-	-	-	-	-	-	-	-
Pumped Storage (Elec.)	-	-	-	-	-	-	-	-	-	-	-
Other Energy Sector	-	-	-	-	-	-	-	-	-	-	-
Distribution Losses	-	1	-	-	2	-	-	-	-	-	-
FINAL CONSUMPTION	-	230	-	6	31	11	23	-	-	-	-
INDUSTRY SECTOR	-	7	-	-	9	11	-	-	-	-	-
Iron and Steel	-	-	-	-	-	2	-	-	-	-	-
Chemical and Petrochem.	-	-	-	-	-	-	-	-	-	-	-
of which: Feedstocks	-	-	-	-	-	-	-	-	-	-	-
Non-Ferrous Metals	-	-	-	-	-	-	-	-	-	-	-
Non-Metallic Minerals	-	1	-	-	9	4	-	-	-	-	-
Transport Equipment	-	-	-	-	-	-	-	-	-	-	-
Machinery	-	-	-	-	-	-	-	-	-	-	-
Mining and Quarrying	-	-	-	-	-	-	-	-	-	-	-
Food and Tobacco	-	-	-	-	-	-	-	-	-	-	-
Paper, Pulp and Print	-	2	-	-	-	5	-	-	-	-	-
Wood and Wood Products	-	-	-	-	-	-	-	-	-	-	-
Construction	-	1	-	-	-	-	-	-	-	-	-
Textile and Leather	-	-	-	-	-	-	-	-	-	-	-
Non-specified	-	3	-	-	-	-	-	-	-	-	-
TRANSPORT SECTOR	-	-	-	-	-	-	-	-	-	-	-
Air	-	-	-	-	-	-	-	-	-	-	-
Road	-	-	-	-	-	-	-	-	-	-	-
Rail	-	-	-	-	-	-	-	-	-	-	-
Pipeline Transport	-	-	-	-	-	-	-	-	-	-	-
Internal Navigation	-	-	-	-	-	-	-	-	-	-	-
Non-specified	-	-	-	-	-	-	-	-	-	-	-
OTHER SECTORS	-	223	-	6	22	-	23	-	-	-	-
Agriculture	-	2	-	-	-	-	-	-	-	-	-
Comm. and Publ. Services	-	136	-	6	3	-	2	-	-	-	-
Residential	-	85	-	-	19	-	21	-	-	-	-
Non-specified	-	-	-	-	-	-	-	-	-	-	-
NON-ENERGY USE	-	-	-	-	-	-	-	-	-	-	-
in Industry/Trans./Energy	-	-	-	-	-	-	-	-	-	-	-
in Transport	-	-	-	-	-	-	-	-	-	-	-
in Other Sectors	-	-	-	-	-	-	-	-	-	-	-

Lithuania / Lituanie : 1997

SUPPLY AND CONSUMPTION *APPROVISIONNEMENT ET DEMANDE*	Refinery Gas *Gaz de raffinerie*	LPG + Ethane *GPL + éthane*	Motor Gasoline *Essence moteur*	Aviation Gasoline *Essence aviation*	Jet Fuel *Carbu-réacteurs*	Kerosene *Kérosène*	Gas/ Diesel *Gazole*	Heavy Fuel Oil *Fioul lourd*	Naphtha *Naphta*	Petrol. Coke *Coke de pétrole*	Other Prod. *Autres prod.*
Production	176	256	1664	-	225	-	1632	1374	-	84	103
From Other Sources	-	-	-	-	-	-	-	-	-	-	-
Imports	-	28	142	-	-	-	128	421	-	-	89
Exports	-	-189	-1157	-	-198	-	-1098	-443	-	-	-47
Intl. Marine Bunkers	-	-	-	-	-	-	-7	-55	-	-	-
Stock Changes	-	-1	13	-	4	-	18	13	-	-	-9
DOMESTIC SUPPLY	**176**	**94**	**662**	**-**	**31**	**-**	**673**	**1310**	**-**	**84**	**136**
Transfers	-	-	-	-	-	-	-	-	-	-	49
Statistical Differences	-	-	-	-	-	-	-	-	-	-	-
TRANSFORMATION	**-**	**1**	**-**	**-**	**-**	**-**	**29**	**1015**	**-**	**-**	**7**
Electricity Plants	-	-	-	-	-	-	-	-	-	-	-
CHP Plants	-	-	-	-	-	-	-	470	-	-	7
Heat Plants	-	1	-	-	-	-	29	545	-	-	-
Blast Furnaces/Gas Works	-	-	-	-	-	-	-	-	-	-	-
Coke/Pat. Fuel/BKB Plants	-	-	-	-	-	-	-	-	-	-	-
Petroleum Refineries	-	-	-	-	-	-	-	-	-	-	-
Petrochemical Industry	-	-	-	-	-	-	-	-	-	-	-
Liquefaction	-	-	-	-	-	-	-	-	-	-	-
Other Transform. Sector	-	-	-	-	-	-	-	-	-	-	-
ENERGY SECTOR	**174**	**-**	**-**	**-**	**-**	**-**	**2**	**134**	**-**	**84**	**-**
Coal Mines	-	-	-	-	-	-	-	-	-	-	-
Oil and Gas Extraction	-	-	-	-	-	-	-	-	-	-	-
Petroleum Refineries	174	-	-	-	-	-	-	134	-	84	-
Electr., CHP+Heat Plants	-	-	-	-	-	-	-	-	-	-	-
Pumped Storage (Elec.)	-	-	-	-	-	-	-	-	-	-	-
Other Energy Sector	-	-	-	-	-	-	2	-	-	-	-
Distribution Losses	2	3	4	-	-	-	1	-	-	-	-
FINAL CONSUMPTION	**-**	**90**	**658**	**-**	**31**	**-**	**641**	**161**	**-**	**-**	**178**
INDUSTRY SECTOR	**-**	**2**	**5**	**-**	**-**	**-**	**42**	**135**	**-**	**-**	**2**
Iron and Steel	-	-	-	-	-	-	-	-	-	-	-
Chemical and Petrochem.	-	-	-	-	-	-	1	1	-	-	2
of which: Feedstocks	-	-	-	-	-	-	-	-	-	-	-
Non-Ferrous Metals	-	-	-	-	-	-	-	-	-	-	-
Non-Metallic Minerals	-	-	-	-	-	-	1	125	-	-	-
Transport Equipment	-	-	-	-	-	-	-	-	-	-	-
Machinery	-	-	1	-	-	-	-	-	-	-	-
Mining and Quarrying	-	-	-	-	-	-	-	1	-	-	-
Food and Tobacco	-	1	-	-	-	-	11	6	-	-	-
Paper, Pulp and Print	-	-	-	-	-	-	-	-	-	-	-
Wood and Wood Products	-	-	-	-	-	-	4	-	-	-	-
Construction	-	1	3	-	-	-	20	2	-	-	-
Textile and Leather	-	-	-	-	-	-	-	-	-	-	-
Non-specified	-	-	1	-	-	-	5	-	-	-	-
TRANSPORT SECTOR	**-**	**11**	**640**	**-**	**31**	**-**	**500**	**5**	**-**	**-**	**-**
Air	-	-	-	-	31	-	-	-	-	-	-
Road	-	11	639	-	-	-	408	-	-	-	-
Rail	-	-	-	-	-	-	78	2	-	-	-
Pipeline Transport	-	-	-	-	-	-	-	-	-	-	-
Internal Navigation	-	-	-	-	-	-	1	-	-	-	-
Non-specified	-	-	1	-	-	-	13	3	-	-	-
OTHER SECTORS	**-**	**77**	**13**	**-**	**-**	**-**	**99**	**21**	**-**	**-**	**-**
Agriculture	-	-	12	-	-	-	77	2	-	-	-
Comm. and Publ. Services	-	1	1	-	-	-	22	18	-	-	-
Residential	-	76	-	-	-	-	-	1	-	-	-
Non-specified	-	-	-	-	-	-	-	-	-	-	-
NON-ENERGY USE	**-**	**-**	**-**	**-**	**-**	**-**	**-**	**-**	**-**	**-**	**176**
in Industry/Transf./Energy	-	-	-	-	-	-	-	-	-	-	146
in Transport	-	-	-	-	-	-	-	-	-	-	30
in Other Sectors	-	-	-	-	-	-	-	-	-	-	-

Oil cont. / *Pétrole cont.* (1000 tonnes)

Lithuania / Lituanie : 1997

SUPPLY AND CONSUMPTION / APPROVISIONNEMENT ET DEMANDE	Gas / Gaz (TJ)				Comb. Renew. & Waste / En. Re. Comb. & Déchets (TJ)				(GWh)	(TJ)
	Natural Gas / Gaz naturel	Gas Works / Usines à gaz	Coke Ovens / Cokeries	Blast Furnaces / Hauts fourneaux	Solid Biomass / Biomasse solide	Gas/Liquids from Biomass / Gaz/Liquides tirés de biomasse	Municipal Waste / Déchets urbains	Industrial Waste / Déchets industriels	Electricity / Electricité	Heat / Chaleur
Production	-	-	-	-	21614	-	-	98	14861	70342
From Other Sources	-	-	-	-	-	-	-	-	-	-
Imports	93134	-	-	-	5	-	-	-	232	-
Exports	-	-	-	-	-48	-	-	-	-3757	-
Intl. Marine Bunkers	-	-	-	-	-	-	-	-	-	-
Stock Changes	-	-	-	-	102	-	-	3	-	-
DOMESTIC SUPPLY	93134	-	-	-	21673	-	-	101	11336	70342
Transfers	-	-	-	-	-	-	-	-	-	-
Statistical Differences	-	-	-	-	-	-	-	-	-	-
TRANSFORMATION	53686	-	-	-	1617	-	-	46	76	-
Electricity Plants	-	-	-	-	-	-	-	-	-	-
CHP Plants	23815	-	-	-	-	-	-	-	-	-
Heat Plants	29871	-	-	-	1617	-	-	46	-	-
Blast Furnaces/Gas Works	-	-	-	-	-	-	-	-	-	-
Coke/Pat. Fuel/BKB Plants	-	-	-	-	-	-	-	-	-	-
Petroleum Refineries	-	-	-	-	-	-	-	-	-	-
Petrochemical Industry	-	-	-	-	-	-	-	-	-	-
Liquefaction	-	-	-	-	-	-	-	-	-	-
Other Transform. Sector	-	-	-	-	-	-	-	-	76	-
ENERGY SECTOR	-	-	-	-	-	-	-	-	2716	4877
Coal Mines	-	-	-	-	-	-	-	-	-	-
Oil and Gas Extraction	-	-	-	-	-	-	-	-	-	-
Petroleum Refineries	-	-	-	-	-	-	-	-	481	4877
Electr., CHP+Heat Plants	-	-	-	-	-	-	-	-	1569	-
Pumped Storage (Elec.)	-	-	-	-	-	-	-	-	663	-
Other Energy Sector	-	-	-	-	-	-	-	-	3	-
Distribution Losses	1909	-	-	-	18	-	-	-	1563	8285
FINAL CONSUMPTION	37539	-	-	-	20038	-	-	55	6981	57180
INDUSTRY SECTOR	28649	-	-	-	333	-	-	-	2777	17099
Iron and Steel	19	-	-	-	-	-	-	-	62	18
Chemical and Petrochem.	24447	-	-	-	-	-	-	-	588	5174
of which: Feedstocks	21921	-	-	-	-	-	-	-	-	-
Non-Ferrous Metals	-	-	-	-	-	-	-	-	-	-
Non-Metallic Minerals	1816	-	-	-	117	-	-	-	267	619
Transport Equipment	33	-	-	-	-	-	-	-	59	184
Machinery	997	-	-	-	22	-	-	-	435	972
Mining and Quarrying	178	-	-	-	-	-	-	-	30	4
Food and Tobacco	662	-	-	-	82	-	-	-	503	5666
Paper, Pulp and Print	26	-	-	-	-	-	-	-	118	1326
Wood and Wood Products	398	-	-	-	33	-	-	-	145	409
Construction	62	-	-	-	10	-	-	-	65	51
Textile and Leather	11	-	-	-	20	-	-	-	395	2141
Non-specified	-	-	-	-	49	-	-	-	110	535
TRANSPORT SECTOR	25	-	-	-	-	-	-	-	101	-
Air	-	-	-	-	-	-	-	-	-	-
Road	25	-	-	-	-	-	-	-	-	-
Rail	-	-	-	-	-	-	-	-	23	-
Pipeline Transport	-	-	-	-	-	-	-	-	-	-
Internal Navigation	-	-	-	-	-	-	-	-	-	-
Non-specified	-	-	-	-	-	-	-	-	78	-
OTHER SECTORS	8865	-	-	-	19705	-	-	55	4103	40081
Agriculture	342	-	-	-	41	-	-	-	426	1561
Comm. and Publ. Services	1641	-	-	-	1367	-	-	55	1712	10102
Residential	6882	-	-	-	18297	-	-	-	1720	26662
Non-specified	-	-	-	-	-	-	-	-	245	1756
NON-ENERGY USE	-	-	-	-	-	-	-	-	-	-
in Industry/Transf./Energy	-	-	-	-	-	-	-	-	-	-
in Transport	-	-	-	-	-	-	-	-	-	-
in Other Sectors	-	-	-	-	-	-	-	-	-	-

Lithuania / Lituanie : 1998

SUPPLY AND CONSUMPTION *APPROVISIONNEMENT ET DEMANDE*	Coal / *Charbon* (1000 tonnes)							Oil / *Pétrole* (1000 tonnes)			
	Coking Coal *Charbon à coke*	Other Bit. Coal *Autres charb. bit.*	Sub-Bit. Coal *Charbon sous-bit.*	Lignite *Lignite*	Peat *Tourbe*	Oven and Gas Coke *Coke de four/gaz*	Pat. Fuel and BKB *Agg./briq. de lignite*	Crude Oil *Pétrole brut*	NGL *LGN*	Feed-stocks *Produits d'aliment.*	Additives *Additifs*
Production	-	-	-	-	62	-	14	277	-	-	15
From Other Sources	-	-	-	-	-	-	-	-	-	-	-
Imports	-	232	-	2	-	14	-	6321	94	358	2
Exports	-	-8	-	-	-	-	-	-265	-	-	-
Intl. Marine Bunkers	-	-	-	-	-	-	-	-	-	-	-
Stock Changes	-	-5	-	1	-1	-1	1	91	-20	95	-
DOMESTIC SUPPLY	-	**219**	-	**3**	**61**	**13**	**15**	**6424**	**74**	**453**	**17**
Transfers	-	-	-	-	-	-	-	-60	-	-	-
Statistical Differences	-	-	-	-	-	-	-	-	-	-	-
TRANSFORMATION	-	**26**	-	-	**30**	-	-	**6364**	**74**	**453**	**17**
Electricity Plants	-	-	-	-	-	-	-	-	-	-	-
CHP Plants	-	-	-	-	-	-	-	-	-	-	-
Heat Plants	-	26	-	-	12	-	-	1	-	-	-
Blast Furnaces/Gas Works	-	-	-	-	-	-	-	-	-	-	-
Coke/Pat. Fuel/BKB Plants	-	-	-	-	18	-	-	-	-	-	-
Petroleum Refineries	-	-	-	-	-	-	-	6363	74	453	17
Petrochemical Industry	-	-	-	-	-	-	-	-	-	-	-
Liquefaction	-	-	-	-	-	-	-	-	-	-	-
Other Transform. Sector	-	-	-	-	-	-	-	-	-	-	-
ENERGY SECTOR	-	-	-	-	**3**	-	-	-	-	-	-
Coal Mines	-	-	-	-	-	-	-	-	-	-	-
Oil and Gas Extraction	-	-	-	-	-	-	-	-	-	-	-
Petroleum Refineries	-	-	-	-	-	-	-	-	-	-	-
Electr., CHP+Heat Plants	-	-	-	-	2	-	-	-	-	-	-
Pumped Storage (Elec.)	-	-	-	-	-	-	-	-	-	-	-
Other Energy Sector	-	-	-	-	1	-	-	-	-	-	-
Distribution Losses	-	2	-	-	1	-	-	-	-	-	-
FINAL CONSUMPTION	-	**191**	-	**3**	**27**	**13**	**15**	-	-	-	-
INDUSTRY SECTOR	-	**6**	-	-	**5**	**13**	-	-	-	-	-
Iron and Steel	-	-	-	-	-	1	-	-	-	-	-
Chemical and Petrochem.	-	-	-	-	-	-	-	-	-	-	-
of which: Feedstocks	-	-	-	-	-	-	-	-	-	-	-
Non-Ferrous Metals	-	-	-	-	-	-	-	-	-	-	-
Non-Metallic Minerals	-	1	-	-	5	6	-	-	-	-	-
Transport Equipment	-	-	-	-	-	-	-	-	-	-	-
Machinery	-	-	-	-	-	-	-	-	-	-	-
Mining and Quarrying	-	-	-	-	-	-	-	-	-	-	-
Food and Tobacco	-	1	-	-	-	-	-	-	-	-	-
Paper, Pulp and Print	-	2	-	-	-	6	-	-	-	-	-
Wood and Wood Products	-	-	-	-	-	-	-	-	-	-	-
Construction	-	1	-	-	-	-	-	-	-	-	-
Textile and Leather	-	1	-	-	-	-	-	-	-	-	-
Non-specified	-	-	-	-	-	-	-	-	-	-	-
TRANSPORT SECTOR	-	-	-	-	-	-	-	-	-	-	-
Air	-	-	-	-	-	-	-	-	-	-	-
Road	-	-	-	-	-	-	-	-	-	-	-
Rail	-	-	-	-	-	-	-	-	-	-	-
Pipeline Transport	-	-	-	-	-	-	-	-	-	-	-
Internal Navigation	-	-	-	-	-	-	-	-	-	-	-
Non-specified	-	-	-	-	-	-	-	-	-	-	-
OTHER SECTORS	-	**185**	-	**3**	**22**	-	**15**	-	-	-	-
Agriculture	-	2	-	-	-	-	-	-	-	-	-
Comm. and Publ. Services	-	141	-	3	-	-	-	-	-	-	-
Residential	-	42	-	-	22	-	15	-	-	-	-
Non-specified	-	-	-	-	-	-	-	-	-	-	-
NON-ENERGY USE	-	-	-	-	-	-	-	-	-	-	-
in Industry/Trans./Energy	-	-	-	-	-	-	-	-	-	-	-
in Transport	-	-	-	-	-	-	-	-	-	-	-
in Other Sectors	-	-	-	-	-	-	-	-	-	-	-

Lithuania / Lituanie : 1998

SUPPLY AND CONSUMPTION / APPROVISIONNEMENT ET DEMANDE	Refinery Gas / Gaz de raffinerie	LPG + Ethane / GPL + éthane	Motor Gasoline / Essence moteur	Aviation Gasoline / Essence aviation	Jet Fuel / Carbu- réacteurs	Kerosene / Kérosène	Gas/ Diesel / Gazole	Heavy Fuel Oil / Fioul lourd	Naphtha / Naphta	Petrol. Coke / Coke de pétrole	Other Prod. / Autres prod.
Production	241	213	2112	-	373	-	2055	1515	-	101	151
From Other Sources	-	-	-	-	-	-	-	-	-	-	-
Imports	-	39	108	-	4	-	85	637	-	-	99
Exports	-	-143	-1596	-	-350	-	-1390	-463	-	-	-55
Intl. Marine Bunkers	-	-	-	-	-	-	-22	-29	-	-	-
Stock Changes	-	-1	10	-	1	-	-12	5	-	-	1
DOMESTIC SUPPLY	241	108	634	-	28	-	716	1665	-	101	196
Transfers	-	-	-	-	-	-	-	-	-	-	60
Statistical Differences	-	-	-	-	-	-	-	-	-	-	-
TRANSFORMATION	14	1	-	-	-	-	12	1385	-	-	10
Electricity Plants	-	-	-	-	-	-	-	-	-	-	-
CHP Plants	-	-	-	-	-	-	-	849	-	-	10
Heat Plants	14	1	-	-	-	-	12	536	-	-	-
Blast Furnaces/Gas Works	-	-	-	-	-	-	-	-	-	-	-
Coke/Pat. Fuel/BKB Plants	-	-	-	-	-	-	-	-	-	-	-
Petroleum Refineries	-	-	-	-	-	-	-	-	-	-	-
Petrochemical Industry	-	-	-	-	-	-	-	-	-	-	-
Liquefaction	-	-	-	-	-	-	-	-	-	-	-
Other Transform. Sector	-	-	-	-	-	-	-	-	-	-	-
ENERGY SECTOR	225	-	-	-	-	-	3	141	-	101	-
Coal Mines	-	-	-	-	-	-	1	-	-	-	-
Oil and Gas Extraction	-	-	-	-	-	-	-	-	-	-	-
Petroleum Refineries	225	-	-	-	-	-	-	141	-	101	-
Electr., CHP+Heat Plants	-	-	-	-	-	-	1	-	-	-	-
Pumped Storage (Elec.)	-	-	-	-	-	-	-	-	-	-	-
Other Energy Sector	-	-	-	-	-	-	1	-	-	-	-
Distribution Losses	2	4	3	-	-	-	1	-	-	-	-
FINAL CONSUMPTION	-	103	631	-	28	-	700	139	-	-	246
INDUSTRY SECTOR	-	4	4	-	-	-	45	125	-	-	-
Iron and Steel	-	-	-	-	-	-	-	-	-	-	-
Chemical and Petrochem.	-	-	-	-	-	-	1	2	-	-	-
of which: Feedstocks	-	-	-	-	-	-	-	-	-	-	-
Non-Ferrous Metals	-	-	-	-	-	-	-	-	-	-	-
Non-Metallic Minerals	-	-	-	-	-	-	1	111	-	-	-
Transport Equipment	-	-	-	-	-	-	1	-	-	-	-
Machinery	-	-	1	-	-	-	-	-	-	-	-
Mining and Quarrying	-	-	-	-	-	-	-	1	-	-	-
Food and Tobacco	-	2	-	-	-	-	13	5	-	-	-
Paper, Pulp and Print	-	-	-	-	-	-	-	-	-	-	-
Wood and Wood Products	-	-	-	-	-	-	4	-	-	-	-
Construction	-	2	3	-	-	-	24	2	-	-	-
Textile and Leather	-	-	-	-	-	-	-	-	-	-	-
Non-specified	-	-	-	-	-	-	1	4	-	-	-
TRANSPORT SECTOR	-	15	617	-	28	-	579	1	-	-	-
Air	-	-	-	-	28	-	-	-	-	-	-
Road	-	15	616	-	-	-	490	-	-	-	-
Rail	-	-	-	-	-	-	75	-	-	-	-
Pipeline Transport	-	-	-	-	-	-	-	-	-	-	-
Internal Navigation	-	-	-	-	-	-	4	1	-	-	-
Non-specified	-	-	1	-	-	-	10	-	-	-	-
OTHER SECTORS	-	84	10	-	-	-	76	13	-	-	-
Agriculture	-	-	8	-	-	-	63	1	-	-	-
Comm. and Publ. Services	-	2	2	-	-	-	13	12	-	-	-
Residential	-	82	-	-	-	-	-	-	-	-	-
Non-specified	-	-	-	-	-	-	-	-	-	-	-
NON-ENERGY USE	-	-	-	-	-	-	-	-	-	-	246
in Industry/Transf./Energy	-	-	-	-	-	-	-	-	-	-	194
in Transport	-	-	-	-	-	-	-	-	-	-	46
in Other Sectors	-	-	-	-	-	-	-	-	-	-	6

Oil cont. / Pétrole cont. (1000 tonnes)

Lithuania / Lituanie : 1998

SUPPLY AND CONSUMPTION / APPROVISIONNEMENT ET DEMANDE	Gas / Gaz (TJ)				Comb. Renew. & Waste / En. Re. Comb. & Déchets (TJ)				(GWh)	(TJ)
	Natural Gas / Gaz naturel	Gas Works / Usines à gaz	Coke Ovens / Cokeries	Blast Furnaces / Hauts fourneaux	Solid Biomass / Biomasse solide	Gas/Liquids from Biomass / Gaz/Liquides tirés de biomasse	Municipal Waste / Déchets urbains	Industrial Waste / Déchets industriels	Electricity / Electricité	Heat / Chaleur
Production	-	-	-	-	24104	-	-	32	17631	67005
From Other Sources	-	-	-	-	-	-	-	-	-	-
Imports	81578	-	-	-	12	-	-	-	384	-
Exports	-	-	-	-	-117	-	-	-	-6466	-
Intl. Marine Bunkers	-	-	-	-	-	-	-	-	-	-
Stock Changes	-	-	-	-	-108	-	-	-	-	-
DOMESTIC SUPPLY	81578	-	-	-	23891	-	-	32	11549	67005
Transfers	-	-	-	-	-	-	-	-	-	-
Statistical Differences	-	-	-	-	-	-	-	-	-	-
TRANSFORMATION	42482	-	-	-	1815	-	-	32	61	-
Electricity Plants	-	-	-	-	-	-	-	-	-	-
CHP Plants	20462	-	-	-	-	-	-	-	-	-
Heat Plants	22020	-	-	-	1815	-	-	32	-	-
Blast Furnaces/Gas Works	-	-	-	-	-	-	-	-	-	-
Coke/Pat. Fuel/BKB Plants	-	-	-	-	-	-	-	-	-	-
Petroleum Refineries	-	-	-	-	-	-	-	-	-	-
Petrochemical Industry	-	-	-	-	-	-	-	-	-	-
Liquefaction	-	-	-	-	-	-	-	-	-	-
Other Transform. Sector	-	-	-	-	-	-	-	-	61	-
ENERGY SECTOR	-	-	-	-	11	-	-	-	2866	5474
Coal Mines	-	-	-	-	-	-	-	-	-	-
Oil and Gas Extraction	-	-	-	-	-	-	-	-	6	-
Petroleum Refineries	-	-	-	-	-	-	-	-	554	5468
Electr., CHP+Heat Plants	-	-	-	-	11	-	-	-	1649	-
Pumped Storage (Elec.)	-	-	-	-	-	-	-	-	654	-
Other Energy Sector	-	-	-	-	-	-	-	-	3	6
Distribution Losses	2114	-	-	-	17	-	-	-	1554	7315
FINAL CONSUMPTION	36982	-	-	-	22048	-	-	-	7068	54216
INDUSTRY SECTOR	29929	-	-	-	716	-	-	-	2620	17110
Iron and Steel	11	-	-	-	1	-	-	-	54	8
Chemical and Petrochem.	25482	-	-	-	-	-	-	-	512	5332
of which: Feedstocks	22743	-	-	-	-	-	-	-	-	-
Non-Ferrous Metals	-	-	-	-	-	-	-	-	-	-
Non-Metallic Minerals	1697	-	-	-	103	-	-	-	248	1012
Transport Equipment	71	-	-	-	-	-	-	-	75	183
Machinery	826	-	-	-	65	-	-	-	313	700
Mining and Quarrying	198	-	-	-	15	-	-	-	34	1
Food and Tobacco	874	-	-	-	89	-	-	-	491	5309
Paper, Pulp and Print	33	-	-	-	2	-	-	-	113	1293
Wood and Wood Products	383	-	-	-	270	-	-	-	126	432
Construction	321	-	-	-	107	-	-	-	146	164
Textile and Leather	11	-	-	-	28	-	-	-	395	2144
Non-specified	22	-	-	-	36	-	-	-	113	532
TRANSPORT SECTOR	17	-	-	-	-	-	-	-	103	-
Air	-	-	-	-	-	-	-	-	-	-
Road	17	-	-	-	-	-	-	-	-	-
Rail	-	-	-	-	-	-	-	-	32	-
Pipeline Transport	-	-	-	-	-	-	-	-	-	-
Internal Navigation	-	-	-	-	-	-	-	-	-	-
Non-specified	-	-	-	-	-	-	-	-	71	-
OTHER SECTORS	7036	-	-	-	21332	-	-	-	4345	37106
Agriculture	323	-	-	-	277	-	-	-	414	1501
Comm. and Publ. Services	967	-	-	-	1179	-	-	-	1873	8550
Residential	5746	-	-	-	19876	-	-	-	1743	25283
Non-specified	-	-	-	-	-	-	-	-	315	1772
NON-ENERGY USE	-	-	-	-	-	-	-	-	-	-
in Industry/Transf./Energy	-	-	-	-	-	-	-	-	-	-
in Transport	-	-	-	-	-	-	-	-	-	-
in Other Sectors	-	-	-	-	-	-	-	-	-	-

FYR of Macedonia / ex-République yougoslave de Macédoine

SUPPLY AND CONSUMPTION 1997	Coal (1000 tonnes)							Oil (1000 tonnes)			
	Coking Coal	Other Bit. Coal	Sub-Bit. Coal	Lignite	Peat	Oven and Gas Coke	Pat. Fuel and BKB	Crude Oil	NGL	Feed-stocks	Additives
Production	-	-	-	6700	-	-	-	-	-	-	-
Imports	49	51	-	-	-	-	-	400	-	-	-
Exports	-	-	-	-	-	-	-	-	-	-	-
Intl. Marine Bunkers	-	-	-	-	-	-	-	-	-	-	-
Stock Changes	-	-	-	-	-	-	-	-	-	-	-
DOMESTIC SUPPLY	49	51	-	6700	-	-	-	400	-	-	-
Transfers and Stat. Diff.	-	-	-	-	-	-	-	-	-	-	-
TRANSFORMATION	-	-	-	6550	-	-	-	400	-	-	-
Electricity and CHP Plants	-	-	-	6550	-	-	-	-	-	-	-
Petroleum Refineries	-	-	-	-	-	-	-	400	-	-	-
Other Transform. Sector	-	-	-	-	-	-	-	-	-	-	-
ENERGY SECTOR	-	-	-	-	-	-	-	-	-	-	-
DISTRIBUTION LOSSES	-	-	-	-	-	-	-	-	-	-	-
FINAL CONSUMPTION	49	51	-	150	-	-	-	-	-	-	-
INDUSTRY SECTOR	49	51	-	105	-	-	-	-	-	-	-
Iron and Steel	11	22	-	98	-	-	-	-	-	-	-
Chemical and Petrochem.	-	-	-	-	-	-	-	-	-	-	-
Non-Metallic Minerals	-	-	-	-	-	-	-	-	-	-	-
Non-specified	38	29	-	7	-	-	-	-	-	-	-
TRANSPORT SECTOR	-	-	-	-	-	-	-	-	-	-	-
Air	-	-	-	-	-	-	-	-	-	-	-
Road	-	-	-	-	-	-	-	-	-	-	-
Non-specified	-	-	-	-	-	-	-	-	-	-	-
OTHER SECTORS	-	-	-	45	-	-	-	-	-	-	-
Agriculture	-	-	-	-	-	-	-	-	-	-	-
Comm. and Publ. Services	-	-	-	20	-	-	-	-	-	-	-
Residential	-	-	-	25	-	-	-	-	-	-	-
Non-specified	-	-	-	-	-	-	-	-	-	-	-
NON-ENERGY USE	-	-	-	-	-	-	-	-	-	-	-

APPROVISIONNEMENT ET DEMANDE 1998	Charbon (1000 tonnes)							Pétrole (1000 tonnes)			
	Charbon à coke	Autres charb. bit.	Charbon sous-bit.	Lignite	Tourbe	Coke de four/gaz	Agg./briq. de lignite	Pétrole brut	LGN	Produits d'aliment.	Additifs
Production	-	-	-	8176	-	-	-	-	-	-	-
Imports	122	101	-	196	-	-	-	787	-	-	-
Exports	-	-	-	-	-	-	-	-	-	-	-
Intl. Marine Bunkers	-	-	-	-	-	-	-	-	-	-	-
Stock Changes	-8	-6	-	-80	-	-	-	-29	-	-3	-
DOMESTIC SUPPLY	114	95	-	8292	-	-	-	758	-	-3	-
Transfers and Stat. Diff.	-	-	-	-	-	-	-	-	-	215	-
TRANSFORMATION	1	-	-	8124	-	-	-	758	-	212	-
Electricity and CHP Plants	-	-	-	8036	-	-	-	-	-	-	-
Petroleum Refineries	-	-	-	-	-	-	-	758	-	212	-
Other Transform. Sector	1	-	-	88	-	-	-	-	-	-	-
ENERGY SECTOR	-	-	-	-	-	-	-	-	-	-	-
DISTRIBUTION LOSSES	-	-	-	21	-	-	-	-	-	-	-
FINAL CONSUMPTION	113	95	-	147	-	-	-	-	-	-	-
INDUSTRY SECTOR	113	95	-	123	-	-	-	-	-	-	-
Iron and Steel	90	29	-	120	-	-	-	-	-	-	-
Chemical and Petrochem.	-	-	-	-	-	-	-	-	-	-	-
Non-Metallic Minerals	-	-	-	-	-	-	-	-	-	-	-
Non-specified	23	66	-	3	-	-	-	-	-	-	-
TRANSPORT SECTOR	-	-	-	-	-	-	-	-	-	-	-
Air	-	-	-	-	-	-	-	-	-	-	-
Road	-	-	-	-	-	-	-	-	-	-	-
Non-specified	-	-	-	-	-	-	-	-	-	-	-
OTHER SECTORS	-	-	-	24	-	-	-	-	-	-	-
Agriculture	-	-	-	-	-	-	-	-	-	-	-
Comm. and Publ. Services	-	-	-	1	-	-	-	-	-	-	-
Residential	-	-	-	23	-	-	-	-	-	-	-
Non-specified	-	-	-	-	-	-	-	-	-	-	-
NON-ENERGY USE	-	-	-	-	-	-	-	-	-	-	-

FYR of Macedonia / ex-République yougoslave de Macédoine

SUPPLY AND CONSUMPTION 1997	Oil cont. (1000 tonnes)										
	Refinery Gas	LPG + Ethane	Motor Gasoline	Aviation Gasoline	Jet Fuel	Kerosene	Gas/ Diesel	Heavy Fuel Oil	Naphtha	Petrol. Coke	Other Prod.
Production	-	3	35	-	-	-	173	174	-	-	-
Imports	-	13	184	-	30	-	193	186	-	-	41
Exports	-	-	-5	-	-	-	-	-	-	-	-
Intl. Marine Bunkers	-	-	-	-	-	-	-	-	-	-	-
Stock Changes	-	-	-	-	-	-	-	-	-	-	-
DOMESTIC SUPPLY	-	16	214	-	30	-	366	360	-	-	41
Transfers and Stat. Diff.	-	-	-	-	-	-	-	-	-	-	-
TRANSFORMATION	-	-	-	-	-	-	-	200	-	-	-
Electricity and CHP Plants	-	-	-	-	-	-	-	20	-	-	-
Petroleum Refineries	-	-	-	-	-	-	-	-	-	-	-
Other Transform. Sector	-	-	-	-	-	-	-	180	-	-	-
ENERGY SECTOR	-	-	-	-	-	-	-	-	-	-	-
DISTRIBUTION LOSSES	-	-	-	-	-	-	-	-	-	-	-
FINAL CONSUMPTION	-	16	214	-	30	-	366	160	-	-	41
INDUSTRY SECTOR	-	15	-	-	-	-	52	120	-	-	-
Iron and Steel	-	2	-	-	-	-	2	20	-	-	-
Chemical and Petrochem.	-	-	-	-	-	-	-	39	-	-	-
Non-Metallic Minerals	-	10	-	-	-	-	3	8	-	-	-
Non-specified	-	3	-	-	-	-	47	53	-	-	-
TRANSPORT SECTOR	-	-	211	-	30	-	264	-	-	-	-
Air	-	-	-	-	30	-	-	-	-	-	-
Road	-	-	211	-	-	-	230	-	-	-	-
Non-specified	-	-	-	-	-	-	34	-	-	-	-
OTHER SECTORS	-	1	3	-	-	-	50	40	-	-	-
Agriculture	-	-	3	-	-	-	50	10	-	-	-
Comm. and Publ. Services	-	-	-	-	-	-	-	-	-	-	-
Residential	-	1	-	-	-	-	-	30	-	-	-
Non-specified	-	-	-	-	-	-	-	-	-	-	-
NON-ENERGY USE	-	-	-	-	-	-	-	-	-	-	41

APPROVISIONNEMENT ET DEMANDE 1998	Pétrole cont. (1000 tonnes)										
	Gaz de raffinerie	GPL + éthane	Essence moteur	Essence aviation	Carbu- réacteurs	Kérosène	Gazole	Fioul lourd	Naphta	Coke de pétrole	Autres prod.
Production	-	9	153	-	-	-	274	290	188	-	-
Imports	-	19	31	-	20	-	54	76	-	8	34
Exports	-	-	-25	-2	-4	-	-4	-	-	-	-
Intl. Marine Bunkers	-	-	-	-	-	-	-	-	-	-	-
Stock Changes	-	-2	11	2	-	-	-20	-7	-10	1	1
DOMESTIC SUPPLY	-	26	170	-	16	-	304	359	178	9	35
Transfers and Stat. Diff.	-	-	-	-	-	-	-	-37	-178	-	-
TRANSFORMATION	-	1	-	-	-	-	15	180	-	-	-
Electricity and CHP Plants	-	-	-	-	-	-	-	47	-	-	-
Petroleum Refineries	-	-	-	-	-	-	-	-	-	-	-
Other Transform. Sector	-	1	-	-	-	-	15	133	-	-	-
ENERGY SECTOR	-	-	1	-	-	-	4	9	-	-	-
DISTRIBUTION LOSSES	-	-	-	-	-	-	-	-	-	-	-
FINAL CONSUMPTION	-	25	169	-	16	-	285	133	-	9	35
INDUSTRY SECTOR	-	15	-	-	-	-	29	107	-	9	1
Iron and Steel	-	8	-	-	-	-	-	36	-	-	-
Chemical and Petrochem.	-	-	-	-	-	-	-	2	-	-	1
Non-Metallic Minerals	-	6	-	-	-	-	1	38	-	9	-
Non-specified	-	1	-	-	-	-	28	31	-	-	-
TRANSPORT SECTOR	-	1	167	-	16	-	158	-	-	-	-
Air	-	-	-	-	16	-	-	-	-	-	-
Road	-	1	167	-	-	-	154	-	-	-	-
Non-specified	-	-	-	-	-	-	4	-	-	-	-
OTHER SECTORS	-	9	2	-	-	-	98	26	-	-	-
Agriculture	-	-	2	-	-	-	17	-	-	-	-
Comm. and Publ. Services	-	4	-	-	-	-	58	-	-	-	-
Residential	-	5	-	-	-	-	23	26	-	-	-
Non-specified	-	-	-	-	-	-	-	-	-	-	-
NON-ENERGY USE	-	-	-	-	-	-	-	-	-	-	34

FYR of Macedonia / ex-République yougoslave de Macédoine

SUPPLY AND CONSUMPTION 1997	Gas (TJ) Natural Gas	Gas Works	Coke Ovens	Blast Furnaces	Comb. Renew. & Waste (TJ) Solid Biomass	Gas/Liquids from Biomass	Municipal Waste	Industrial Waste	(GWh) Electricity	(TJ) Heat
Production	-	-	-	-	7814	-	-	-	6719	6208
Imports	-	-	-	-	150	-	-	-	-	-
Exports	-	-	-	-	-	-	-	-	-	-
Intl. Marine Bunkers	-	-	-	-	-	-	-	-	-	-
Stock Changes	-	-	-	-	-	-	-	-	-	-
DOMESTIC SUPPLY	-	-	-	-	7964	-	-	-	6719	6208
Transfers and Stat. Diff.	-	-	-	-	-	-	-	-	-	-
TRANSFORMATION	-	-	-	-	-	-	-	-	-	-
Electricity and CHP Plants	-	-	-	-	-	-	-	-	-	-
Petroleum Refineries	-	-	-	-	-	-	-	-	-	-
Other Transform. Sector	-	-	-	-	-	-	-	-	-	-
ENERGY SECTOR	-	-	-	-	-	-	-	-	603	123
DISTRIBUTION LOSSES	-	-	-	-	-	-	-	-	839	898
FINAL CONSUMPTION	-	-	-	-	7964	-	-	-	5277	5187
INDUSTRY SECTOR	-	-	-	-	-	-	-	-	2087	2811
Iron and Steel	-	-	-	-	-	-	-	-	1130	132
Chemical and Petrochem.	-	-	-	-	-	-	-	-	145	624
Non-Metallic Minerals	-	-	-	-	-	-	-	-	28	16
Non-specified	-	-	-	-	-	-	-	-	784	2039
TRANSPORT SECTOR	-	-	-	-	-	-	-	-	14	-
Air	-	-	-	-	-	-	-	-	-	-
Road	-	-	-	-	-	-	-	-	-	-
Non-specified	-	-	-	-	-	-	-	-	14	-
OTHER SECTORS	-	-	-	-	7964	-	-	-	3176	2376
Agriculture	-	-	-	-	-	-	-	-	25	440
Comm. and Publ. Services	-	-	-	-	-	-	-	-	533	774
Residential	-	-	-	-	7814	-	-	-	2618	1162
Non-specified	-	-	-	-	150	-	-	-	-	-
NON-ENERGY USE	-	-	-	-	-	-	-	-	-	-

APPROVISIONNEMENT ET DEMANDE 1998	Gaz (TJ) Gaz naturel	Usines à gaz	Cokeries	Hauts fourneaux	En. Re. Comb. & Déchets (TJ) Biomasse solide	Gaz/Liquides tirés de biomasse	Déchets urbains	Déchets industriels	(GWh) Electricité	(TJ) Chaleur
Production	-	-	-	-	6388	-	-	-	7048	8488
Imports	867	-	-	-	151	-	-	-	-	-
Exports	-38	-	-	-	-	-	-	-	-2	-
Intl. Marine Bunkers	-	-	-	-	-	-	-	-	-	-
Stock Changes	-	-	-	-	-12	-	-	-	-	-
DOMESTIC SUPPLY	829	-	-	-	6527	-	-	-	7046	8488
Transfers and Stat. Diff.	-	-	-	-	-	-	-	-	-	-
TRANSFORMATION	717	-	-	-	129	-	-	-	-	-
Electricity and CHP Plants	103	-	-	-	-	-	-	-	-	-
Petroleum Refineries	-	-	-	-	-	-	-	-	-	-
Other Transform. Sector	614	-	-	-	129	-	-	-	-	-
ENERGY SECTOR	79	-	-	-	-	-	-	-	719	633
DISTRIBUTION LOSSES	33	-	-	-	-	-	-	-	866	1058
FINAL CONSUMPTION	-	-	-	-	6398	-	-	-	5461	6797
INDUSTRY SECTOR	-	-	-	-	84	-	-	-	2118	3880
Iron and Steel	-	-	-	-	-	-	-	-	1286	484
Chemical and Petrochem.	-	-	-	-	1	-	-	-	70	104
Non-Metallic Minerals	-	-	-	-	16	-	-	-	110	150
Non-specified	-	-	-	-	67	-	-	-	652	3142
TRANSPORT SECTOR	-	-	-	-	-	-	-	-	24	-
Air	-	-	-	-	-	-	-	-	-	-
Road	-	-	-	-	-	-	-	-	-	-
Non-specified	-	-	-	-	-	-	-	-	24	-
OTHER SECTORS	-	-	-	-	6314	-	-	-	3319	2917
Agriculture	-	-	-	-	-	-	-	-	40	623
Comm. and Publ. Services	-	-	-	-	-	-	-	-	722	1085
Residential	-	-	-	-	6164	-	-	-	2555	1209
Non-specified	-	-	-	-	150	-	-	-	2	-
NON-ENERGY USE	-	-	-	-	-	-	-	-	-	-

Malaysia / Malaisie : 1997

SUPPLY AND CONSUMPTION	Coal / *Charbon* (1000 tonnes)							Oil / *Pétrole* (1000 tonnes)			
	Coking Coal	Other Bit. Coal	Sub-Bit. Coal	Lignite	Peat	Oven and Gas Coke	Pat. Fuel and BKB	Crude Oil	NGL	Feed-stocks	Additives
APPROVISIONNEMENT ET DEMANDE	*Charbon à coke*	*Autres charb. bit.*	*Charbon sous-bit.*	*Lignite*	*Tourbe*	*Coke de four/gaz*	*Agg./briq. de lignite*	*Pétrole brut*	*LGN*	*Produits d'aliment.*	*Additifs*
Production	-	-	218	-	-	-	-	36842	1694	-	-
From Other Sources	-	-		-	-	-	-	-	-		-
Imports	-	1023	1041	-	-	-	-	1324	-	1715	-
Exports	-	-13	-	-	-	-	-	-17926	-		-
Intl. Marine Bunkers	-	-		-	-	-	-	-	-		-
Stock Changes	-	-30	-	-	-	-	-	83	-		-
DOMESTIC SUPPLY	-	980	1259	-	-	-	-	20323	1694	1715	-
Transfers	-	-	-	-	-	-	-	-	-1156	-	-
Statistical Differences	-	76	-	-	-	-	-	-1782	-	-	-
TRANSFORMATION	-	-	1259	-	-	-	-	18541	-	1715	-
Electricity Plants	-	-	1259	-	-	-	-		-		-
CHP Plants	-	-	-	-	-	-	-	-	-	-	-
Heat Plants	-	-	-	-	-	-	-	-	-	-	-
Blast Furnaces/Gas Works	-	-	-	-	-	-	-	-	-	-	-
Coke/Pat. Fuel/BKB Plants	-	-	-	-	-	-	-	-	-	-	-
Petroleum Refineries	-	-	-	-	-	-	-	18541	-	1715	-
Petrochemical Industry	-	-	-	-	-	-	-	-	-	-	-
Liquefaction	-	-	-	-	-	-	-	-	-	-	-
Other Transform. Sector	-	-	-	-	-	-	-	-	-	-	-
ENERGY SECTOR	-	-	-	-	-	-	-	-	-	-	-
Coal Mines	-	-	-	-	-	-	-	-	-	-	-
Oil and Gas Extraction	-	-	-	-	-	-	-	-	-	-	-
Petroleum Refineries	-	-	-	-	-	-	-	-	-	-	-
Electr., CHP+Heat Plants	-	-	-	-	-	-	-	-	-	-	-
Pumped Storage (Elec.)	-	-	-	-	-	-	-	-	-	-	-
Other Energy Sector	-	-	-	-	-	-	-	-	-	-	-
Distribution Losses	-	-	-	-	-	-	-	-	538	-	-
FINAL CONSUMPTION	-	1056	-	-	-	-	-	-	-	-	-
INDUSTRY SECTOR	-	1056	-	-	-	-	-	-	-	-	-
Iron and Steel	-	-	-	-	-	-	-	-	-	-	-
Chemical and Petrochem.	-	-	-	-	-	-	-	-	-	-	-
of which: Feedstocks	-	-	-	-	-	-	-	-	-	-	-
Non-Ferrous Metals	-	-	-	-	-	-	-	-	-	-	-
Non-Metallic Minerals	-	-	-	-	-	-	-	-	-	-	-
Transport Equipment	-	-	-	-	-	-	-	-	-	-	-
Machinery	-	-	-	-	-	-	-	-	-	-	-
Mining and Quarrying	-	-	-	-	-	-	-	-	-	-	-
Food and Tobacco	-	-	-	-	-	-	-	-	-	-	-
Paper, Pulp and Print	-	-	-	-	-	-	-	-	-	-	-
Wood and Wood Products	-	-	-	-	-	-	-	-	-	-	-
Construction	-	-	-	-	-	-	-	-	-	-	-
Textile and Leather	-	-	-	-	-	-	-	-	-	-	-
Non-specified	-	1056	-	-	-	-	-	-	-	-	-
TRANSPORT SECTOR	-	-	-	-	-	-	-	-	-	-	-
Air	-	-	-	-	-	-	-	-	-	-	-
Road	-	-	-	-	-	-	-	-	-	-	-
Rail	-	-	-	-	-	-	-	-	-	-	-
Pipeline Transport	-	-	-	-	-	-	-	-	-	-	-
Internal Navigation	-	-	-	-	-	-	-	-	-	-	-
Non-specified	-	-	-	-	-	-	-	-	-	-	-
OTHER SECTORS	-	-	-	-	-	-	-	-	-	-	-
Agriculture	-	-	-	-	-	-	-	-	-	-	-
Comm. and Publ. Services	-	-	-	-	-	-	-	-	-	-	-
Residential	-	-	-	-	-	-	-	-	-	-	-
Non-specified	-	-	-	-	-	-	-	-	-	-	-
NON-ENERGY USE	-	-	-	-	-	-	-	-	-	-	-
in Industry/Trans./Energy	-	-	-	-	-	-	-	-	-	-	-
in Transport	-	-	-	-	-	-	-	-	-	-	-
in Other Sectors	-	-	-	-	-	-	-	-	-	-	-

Malaysia / Malaisie : 1997

SUPPLY AND CONSUMPTION / APPROVISIONNEMENT ET DEMANDE	Oil cont. / Pétrole cont. (1000 tonnes)										
	Refinery Gas / Gaz de raffinerie	LPG + Ethane / GPL + éthane	Motor Gasoline / Essence moteur	Aviation Gasoline / Essence aviation	Jet Fuel / Carbu- réacteurs	Kerosene / Kérosène	Gas/ Diesel / Gazole	Heavy Fuel Oil / Fioul lourd	Naphtha / Naphta	Petrol. Coke / Coke de pétrole	Other Prod. / Autres prod.
Production	177	341	2371	-	1937	257	6640	2738	-	-	1785
From Other Sources	-	-	-	-	-	35	96	-	-	-	255
Imports	-	673	2925	2	42	6	1682	3813	-	-	985
Exports	-	-684	-168	-	-449	-226	-741	-2767	-	-	-2728
Intl. Marine Bunkers	-	-	-	-	-	-	-19	-150	-	-	-
Stock Changes	-	1	-21	-	3	-31	-105	-46	-	-	-8
DOMESTIC SUPPLY	177	331	5107	2	1533	41	7553	3588	-	-	289
Transfers	-	1156	-	-	-	-	-	-	-	-	-
Statistical Differences	4	-229	210	-	-139	122	-170	908	-	-	555
TRANSFORMATION	-	-	-	-	-	-	182	2502	-	-	-
Electricity Plants	-	-	-	-	-	-	182	2502	-	-	-
CHP Plants	-	-	-	-	-	-	-	-	-	-	-
Heat Plants	-	-	-	-	-	-	-	-	-	-	-
Blast Furnaces/Gas Works	-	-	-	-	-	-	-	-	-	-	-
Coke/Pat. Fuel/BKB Plants	-	-	-	-	-	-	-	-	-	-	-
Petroleum Refineries	-	-	-	-	-	-	-	-	-	-	-
Petrochemical Industry	-	-	-	-	-	-	-	-	-	-	-
Liquefaction	-	-	-	-	-	-	-	-	-	-	-
Other Transform. Sector	-	-	-	-	-	-	-	-	-	-	-
ENERGY SECTOR	178	-	-	-	-	-	-	-	-	-	-
Coal Mines	-	-	-	-	-	-	-	-	-	-	-
Oil and Gas Extraction	-	-	-	-	-	-	-	-	-	-	-
Petroleum Refineries	178	-	-	-	-	-	-	-	-	-	-
Electr., CHP+Heat Plants	-	-	-	-	-	-	-	-	-	-	-
Pumped Storage (Elec.)	-	-	-	-	-	-	-	-	-	-	-
Other Energy Sector	-	-	-	-	-	-	-	-	-	-	-
Distribution Losses	-	114	-	-	-	-	-	-	-	-	-
FINAL CONSUMPTION	3	1144	5317	2	1394	163	7201	1994	-	-	844
INDUSTRY SECTOR	3	270	9	-	-	15	3686	1897	-	-	-
Iron and Steel	-	-	-	-	-	-	-	-	-	-	-
Chemical and Petrochem.	-	71	-	-	-	-	-	-	-	-	-
of which: Feedstocks	-	71	-	-	-	-	-	-	-	-	-
Non-Ferrous Metals	-	-	-	-	-	-	-	-	-	-	-
Non-Metallic Minerals	-	-	-	-	-	-	-	-	-	-	-
Transport Equipment	-	-	-	-	-	-	-	-	-	-	-
Machinery	-	-	-	-	-	-	-	-	-	-	-
Mining and Quarrying	-	-	-	-	-	-	-	-	-	-	-
Food and Tobacco	-	-	-	-	-	-	-	-	-	-	-
Paper, Pulp and Print	-	-	-	-	-	-	-	-	-	-	-
Wood and Wood Products	-	-	-	-	-	-	-	-	-	-	-
Construction	-	-	-	-	-	-	-	-	-	-	-
Textile and Leather	-	-	-	-	-	-	-	-	-	-	-
Non-specified	3	199	9	-	-	15	3686	1897	-	-	-
TRANSPORT SECTOR	-	-	5305	2	1394	-	3058	76	-	-	-
Air	-	-	-	2	1394	-	-	-	-	-	-
Road	-	-	5305	-	-	-	3058	76	-	-	-
Rail	-	-	-	-	-	-	-	-	-	-	-
Pipeline Transport	-	-	-	-	-	-	-	-	-	-	-
Internal Navigation	-	-	-	-	-	-	-	-	-	-	-
Non-specified	-	-	-	-	-	-	-	-	-	-	-
OTHER SECTORS	-	874	3	-	-	148	457	21	-	-	-
Agriculture	-	20	3	-	-	-	457	-	-	-	-
Comm. and Publ. Services	-	342	-	-	-	-	-	21	-	-	-
Residential	-	512	-	-	-	148	-	-	-	-	-
Non-specified	-	-	-	-	-	-	-	-	-	-	-
NON-ENERGY USE	-	-	-	-	-	-	-	-	-	-	844
in Industry/Transf./Energy	-	-	-	-	-	-	-	-	-	-	844
in Transport	-	-	-	-	-	-	-	-	-	-	-
in Other Sectors	-	-	-	-	-	-	-	-	-	-	-

Malaysia / Malaisie : 1997

SUPPLY AND CONSUMPTION / APPROVISIONNEMENT ET DEMANDE	Gas / Gaz (TJ)				Comb. Renew. & Waste / En. Re. Comb. & Déchets (TJ)				(GWh)	(TJ)
	Natural Gas / Gaz naturel	Gas Works / Usines à gaz	Coke Ovens / Cokeries	Blast Furnaces / Hauts fourneaux	Solid Biomass / Biomasse solide	Gas/Liquids from Biomass / Gaz/Liquides tirés de biomasse	Municipal Waste / Déchets urbains	Industrial Waste / Déchets industriels	Electricity / Electricité	Heat / Chaleur
Production	1462517	-	-	-	100794	-	-	-	57844	-
From Other Sources									-	-
Imports	-	-	-	-	185	-	-	-	81	-
Exports	-742074	-	-	-	-878	-	-	-	-70	-
Intl. Marine Bunkers	-	-	-	-	-	-	-	-	-	-
Stock Changes	-	-	-	-	-	-	-	-	-	-
DOMESTIC SUPPLY	**720443**	**-**	**-**	**-**	**100101**	**-**	**-**	**-**	**57855**	**-**
Transfers	-	-	-	-	-	-	-	-	-	-
Statistical Differences	-19623	-	-	-	-	-	-	-	-859	-
TRANSFORMATION	**390158**	**-**	**-**	**-**	**39941**	**-**	**-**	**-**	**-**	**-**
Electricity Plants	315097	-	-	-	-	-	-	-	-	-
CHP Plants	-	-	-	-	-	-	-	-	-	-
Heat Plants	-	-	-	-	-	-	-	-	-	-
Blast Furnaces/Gas Works	-	-	-	-	-	-	-	-	-	-
Coke/Pat. Fuel/BKB Plants	-	-	-	-	-	-	-	-	-	-
Petroleum Refineries	-	-	-	-	-	-	-	-	-	-
Petrochemical Industry	-	-	-	-	-	-	-	-	-	-
Liquefaction	75061	-	-	-	-	-	-	-	-	-
Other Transform. Sector	-	-	-	-	39941	-	-	-	-	-
ENERGY SECTOR	**206397**	**-**	**-**	**-**	**-**	**-**	**-**	**-**	**207**	**-**
Coal Mines	-	-	-	-	-	-	-	-	-	-
Oil and Gas Extraction	206397	-	-	-	-	-	-	-	-	-
Petroleum Refineries	-	-	-	-	-	-	-	-	-	-
Electr., CHP+Heat Plants	-	-	-	-	-	-	-	-	207	-
Pumped Storage (Elec.)	-	-	-	-	-	-	-	-	-	-
Other Energy Sector	-	-	-	-	-	-	-	-	-	-
Distribution Losses	1130	-	-	-	-	-	-	-	4628	-
FINAL CONSUMPTION	**103135**	**-**	**-**	**-**	**60160**	**-**	**-**	**-**	**52161**	**-**
INDUSTRY SECTOR	**102592**	**-**	**-**	**-**	**2772**	**-**	**-**	**-**	**28149**	**-**
Iron and Steel	-	-	-	-	-	-	-	-	-	-
Chemical and Petrochem.	57656	-	-	-	-	-	-	-	-	-
of which: Feedstocks	57656	-	-	-	-	-	-	-	-	-
Non-Ferrous Metals	-	-	-	-	-	-	-	-	-	-
Non-Metallic Minerals	-	-	-	-	-	-	-	-	-	-
Transport Equipment	-	-	-	-	-	-	-	-	-	-
Machinery	-	-	-	-	-	-	-	-	-	-
Mining and Quarrying	-	-	-	-	-	-	-	-	-	-
Food and Tobacco	-	-	-	-	-	-	-	-	-	-
Paper, Pulp and Print	-	-	-	-	-	-	-	-	-	-
Wood and Wood Products	-	-	-	-	-	-	-	-	-	-
Construction	-	-	-	-	-	-	-	-	-	-
Textile and Leather	-	-	-	-	-	-	-	-	-	-
Non-specified	44936	-	-	-	2772	-	-	-	28149	-
TRANSPORT SECTOR	**209**	**-**	**-**	**-**	**-**	**-**	**-**	**-**	**12**	**-**
Air	-	-	-	-	-	-	-	-	-	-
Road	-	-	-	-	-	-	-	-	-	-
Rail	-	-	-	-	-	-	-	-	-	-
Pipeline Transport	-	-	-	-	-	-	-	-	-	-
Internal Navigation	-	-	-	-	-	-	-	-	-	-
Non-specified	209	-	-	-	-	-	-	-	12	-
OTHER SECTORS	**334**	**-**	**-**	**-**	**57388**	**-**	**-**	**-**	**24000**	**-**
Agriculture	-	-	-	-	-	-	-	-	-	-
Comm. and Publ. Services	167	-	-	-	-	-	-	-	13842	-
Residential	167	-	-	-	57388	-	-	-	10158	-
Non-specified	-	-	-	-	-	-	-	-	-	-
NON-ENERGY USE	**-**	**-**	**-**	**-**	**-**	**-**	**-**	**-**	**-**	**-**
in Industry/Transf./Energy	-	-	-	-	-	-	-	-	-	-
in Transport	-	-	-	-	-	-	-	-	-	-
in Other Sectors	-	-	-	-	-	-	-	-	-	-

Malaysia / Malaisie : 1998

SUPPLY AND CONSUMPTION / APPROVISIONNEMENT ET DEMANDE	Coal / Charbon (1000 tonnes)							Oil / Pétrole (1000 tonnes)			
	Coking Coal / Charbon à coke	Other Bit. Coal / Autres charb. bit.	Sub-Bit. Coal / Charbon sous-bit.	Lignite / Lignite	Peat / Tourbe	Oven and Gas Coke / Coke de four/gaz	Pat. Fuel and BKB / Agg./briq. de lignite	Crude Oil / Pétrole brut	NGL / LGN	Feed-stocks / Produits d'aliment.	Additives / Additifs
Production	-	-	315	-	-	-	-	37033	1792	-	-
From Other Sources	-	-	-	-	-	-	-	-	-	-	-
Imports	-	1121	1061	-	-	-	-	2051	-	170	-
Exports	-	-10	-	-	-	-	-	-19291	-	-	-
Intl. Marine Bunkers	-	-	-	-	-	-	-	-	-	-	-
Stock Changes	-	-118	-	-	-	-	-	-364	-	-	-
DOMESTIC SUPPLY	-	993	1376	-	-	-	-	19429	1792	170	-
Transfers	-	-	-	-	-	-	-	-	-1402	-	-
Statistical Differences	-	102	-	-	-	-	-	-1699	-	-	-
TRANSFORMATION	-	-	1376	-	-	-	-	17730	-	170	-
Electricity Plants	-	-	1376	-	-	-	-	-	-	-	-
CHP Plants	-	-	-	-	-	-	-	-	-	-	-
Heat Plants	-	-	-	-	-	-	-	-	-	-	-
Blast Furnaces/Gas Works	-	-	-	-	-	-	-	-	-	-	-
Coke/Pat. Fuel/BKB Plants	-	-	-	-	-	-	-	-	-	-	-
Petroleum Refineries	-	-	-	-	-	-	-	17730	-	170	-
Petrochemical Industry	-	-	-	-	-	-	-	-	-	-	-
Liquefaction	-	-	-	-	-	-	-	-	-	-	-
Other Transform. Sector	-	-	-	-	-	-	-	-	-	-	-
ENERGY SECTOR	-	-	-	-	-	-	-	-	-	-	-
Coal Mines	-	-	-	-	-	-	-	-	-	-	-
Oil and Gas Extraction	-	-	-	-	-	-	-	-	-	-	-
Petroleum Refineries	-	-	-	-	-	-	-	-	-	-	-
Electr., CHP+Heat Plants	-	-	-	-	-	-	-	-	-	-	-
Pumped Storage (Elec.)	-	-	-	-	-	-	-	-	-	-	-
Other Energy Sector	-	-	-	-	-	-	-	-	-	-	-
Distribution Losses	-	-	-	-	-	-	-	-	390	-	-
FINAL CONSUMPTION	-	1095	-	-	-	-	-	-	-	-	-
INDUSTRY SECTOR	-	1095	-	-	-	-	-	-	-	-	-
Iron and Steel	-	-	-	-	-	-	-	-	-	-	-
Chemical and Petrochem.	-	-	-	-	-	-	-	-	-	-	-
of which: Feedstocks	-	-	-	-	-	-	-	-	-	-	-
Non-Ferrous Metals	-	-	-	-	-	-	-	-	-	-	-
Non-Metallic Minerals	-	-	-	-	-	-	-	-	-	-	-
Transport Equipment	-	-	-	-	-	-	-	-	-	-	-
Machinery	-	-	-	-	-	-	-	-	-	-	-
Mining and Quarrying	-	-	-	-	-	-	-	-	-	-	-
Food and Tobacco	-	-	-	-	-	-	-	-	-	-	-
Paper, Pulp and Print	-	-	-	-	-	-	-	-	-	-	-
Wood and Wood Products	-	-	-	-	-	-	-	-	-	-	-
Construction	-	-	-	-	-	-	-	-	-	-	-
Textile and Leather	-	-	-	-	-	-	-	-	-	-	-
Non-specified	-	1095	-	-	-	-	-	-	-	-	-
TRANSPORT SECTOR	-	-	-	-	-	-	-	-	-	-	-
Air	-	-	-	-	-	-	-	-	-	-	-
Road	-	-	-	-	-	-	-	-	-	-	-
Rail	-	-	-	-	-	-	-	-	-	-	-
Pipeline Transport	-	-	-	-	-	-	-	-	-	-	-
Internal Navigation	-	-	-	-	-	-	-	-	-	-	-
Non-specified	-	-	-	-	-	-	-	-	-	-	-
OTHER SECTORS	-	-	-	-	-	-	-	-	-	-	-
Agriculture	-	-	-	-	-	-	-	-	-	-	-
Comm. and Publ. Services	-	-	-	-	-	-	-	-	-	-	-
Residential	-	-	-	-	-	-	-	-	-	-	-
Non-specified	-	-	-	-	-	-	-	-	-	-	-
NON-ENERGY USE	-	-	-	-	-	-	-	-	-	-	-
in Industry/Trans./Energy	-	-	-	-	-	-	-	-	-	-	-
in Transport	-	-	-	-	-	-	-	-	-	-	-
in Other Sectors	-	-	-	-	-	-	-	-	-	-	-

Malaysia / Malaisie : 1998

SUPPLY AND CONSUMPTION / APPROVISIONNEMENT ET DEMANDE	Oil cont. / Pétrole cont. (1000 tonnes)										
	Refinery Gas / Gaz de raffinerie	LPG + Ethane / GPL + éthane	Motor Gasoline / Essence moteur	Aviation Gasoline / Essence aviation	Jet Fuel / Carbu- réacteurs	Kerosene / Kérosène	Gas/ Diesel / Gazole	Heavy Fuel Oil / Fioul lourd	Naphtha / Naphta	Petrol. Coke / Coke de pétrole	Other Prod. / Autres prod.
Production	167	413	2422	-	1923	276	5835	3259	-	-	2119
From Other Sources	-	-	-	-	-	-	-	-	-	-	-
Imports	-	257	3049	2	102	9	1483	4179	-	-	109
Exports	-	-723	-114	-	-319	-228	-1074	-2880	-	-	-1678
Intl. Marine Bunkers	-	-	-	-	-	-	-38	-410	-	-	-
Stock Changes	-	3	32	-	-26	-52	40	29	-	-	18
DOMESTIC SUPPLY	167	-50	5389	2	1680	5	6246	4177	-	-	568
Transfers	-	1402	-	-	-	-	-	-	-	-	-
Statistical Differences	-	-156	183	-	-112	155	356	-337	-	-	47
TRANSFORMATION	-	-	-	-	-	-	447	2148	-	-	-
Electricity Plants	-	-	-	-	-	-	447	2148	-	-	-
CHP Plants	-	-	-	-	-	-	-	-	-	-	-
Heat Plants	-	-	-	-	-	-	-	-	-	-	-
Blast Furnaces/Gas Works	-	-	-	-	-	-	-	-	-	-	-
Coke/Pat. Fuel/BKB Plants	-	-	-	-	-	-	-	-	-	-	-
Petroleum Refineries	-	-	-	-	-	-	-	-	-	-	-
Petrochemical Industry	-	-	-	-	-	-	-	-	-	-	-
Liquefaction	-	-	-	-	-	-	-	-	-	-	-
Other Transform. Sector	-	-	-	-	-	-	-	-	-	-	-
ENERGY SECTOR	163	-	-	-	-	-	-	-	-	-	-
Coal Mines	-	-	-	-	-	-	-	-	-	-	-
Oil and Gas Extraction	-	-	-	-	-	-	-	-	-	-	-
Petroleum Refineries	163	-	-	-	-	-	-	-	-	-	-
Electr., CHP+Heat Plants	-	-	-	-	-	-	-	-	-	-	-
Pumped Storage (Elec.)	-	-	-	-	-	-	-	-	-	-	-
Other Energy Sector	-	-	-	-	-	-	-	-	-	-	-
Distribution Losses	-	1	-	-	-	-	-	-	-	-	-
FINAL CONSUMPTION	4	1195	5572	2	1568	160	6155	1692	-	-	615
INDUSTRY SECTOR	4	323	2	-	-	19	3581	1639	-	-	-
Iron and Steel	-	-	-	-	-	-	-	-	-	-	-
Chemical and Petrochem.	-	116	-	-	-	-	-	-	-	-	-
of which: Feedstocks	-	116	-	-	-	-	-	-	-	-	-
Non-Ferrous Metals	-	-	-	-	-	-	-	-	-	-	-
Non-Metallic Minerals	-	-	-	-	-	-	-	-	-	-	-
Transport Equipment	-	-	-	-	-	-	-	-	-	-	-
Machinery	-	-	-	-	-	-	-	-	-	-	-
Mining and Quarrying	-	-	-	-	-	-	-	-	-	-	-
Food and Tobacco	-	-	-	-	-	-	-	-	-	-	-
Paper, Pulp and Print	-	-	-	-	-	-	-	-	-	-	-
Wood and Wood Products	-	-	-	-	-	-	-	-	-	-	-
Construction	-	-	-	-	-	-	-	-	-	-	-
Textile and Leather	-	-	-	-	-	-	-	-	-	-	-
Non-specified	4	207	2	-	-	19	3581	1639	-	-	-
TRANSPORT SECTOR	-	-	5567	2	1568	-	2275	9	-	-	-
Air	-	-	-	2	1568	-	-	-	-	-	-
Road	-	-	5567	-	-	-	2275	9	-	-	-
Rail	-	-	-	-	-	-	-	-	-	-	-
Pipeline Transport	-	-	-	-	-	-	-	-	-	-	-
Internal Navigation	-	-	-	-	-	-	-	-	-	-	-
Non-specified	-	-	-	-	-	-	-	-	-	-	-
OTHER SECTORS	-	872	3	-	-	141	299	44	-	-	-
Agriculture	-	-	3	-	-	-	299	-	-	-	-
Comm. and Publ. Services	-	349	-	-	-	-	-	44	-	-	-
Residential	-	523	-	-	-	141	-	-	-	-	-
Non-specified	-	-	-	-	-	-	-	-	-	-	-
NON-ENERGY USE	-	-	-	-	-	-	-	-	-	-	615
in Industry/Transf./Energy	-	-	-	-	-	-	-	-	-	-	615
in Transport	-	-	-	-	-	-	-	-	-	-	-
in Other Sectors	-	-	-	-	-	-	-	-	-	-	-

Malaysia / Malaisie : 1998

SUPPLY AND CONSUMPTION / APPROVISIONNEMENT ET DEMANDE	Gas / Gaz (TJ)				Comb. Renew. & Waste / En. Re. Comb. & Déchets (TJ)				(GWh)	(TJ)
	Natural Gas / Gaz naturel	Gas Works / Usines à gaz	Coke Ovens / Cokeries	Blast Furnaces / Hauts fourneaux	Solid Biomass / Biomasse solide	Gas/Liquids from Biomass / Gaz/Liquides tirés de biomasse	Municipal Waste / Déchets urbains	Industrial Waste / Déchets industriels	Electricity / Electricité	Heat / Chaleur
Production	1478427	-	-	-	100990	-	-	-	60671	-
From Other Sources	-				-				-	-
Imports	-	-	-	-	185	-	-	-	1	-
Exports	-747797	-	-	-	-880	-	-	-	-8	-
Intl. Marine Bunkers	-	-	-	-	-	-	-	-	-	-
Stock Changes	-	-	-	-	-	-	-	-	-	-
DOMESTIC SUPPLY	730630	-	-	-	100295	-	-	-	60664	-
Transfers	-	-	-	-	-	-	-	-	-	-
Statistical Differences	-31099	-	-	-	-	-	-	-	632	-
TRANSFORMATION	383158	-	-	-	40019	-	-	-	-	-
Electricity Plants	383158	-	-	-	-	-	-	-	-	-
CHP Plants	-	-	-	-	-	-	-	-	-	-
Heat Plants	-	-	-	-	-	-	-	-	-	-
Blast Furnaces/Gas Works	-	-	-	-	-	-	-	-	-	-
Coke/Pat. Fuel/BKB Plants	-	-	-	-	-	-	-	-	-	-
Petroleum Refineries	-	-	-	-	-	-	-	-	-	-
Petrochemical Industry	-	-	-	-	-	-	-	-	-	-
Liquefaction	-	-	-	-	-	-	-	-	-	-
Other Transform. Sector	-	-	-	-	40019	-	-	-	-	-
ENERGY SECTOR	201310	-	-	-	-	-	-	-	402	-
Coal Mines	-	-	-	-	-	-	-	-	-	-
Oil and Gas Extraction	201310	-	-	-	-	-	-	-	-	-
Petroleum Refineries	-	-	-	-	-	-	-	-	-	-
Electr., CHP+Heat Plants	-	-	-	-	-	-	-	-	402	-
Pumped Storage (Elec.)	-	-	-	-	-	-	-	-	-	-
Other Energy Sector	-	-	-	-	-	-	-	-	-	-
Distribution Losses	1018	-	-	-	-	-	-	-	4247	-
FINAL CONSUMPTION	114045	-	-	-	60276	-	-	-	56647	-
INDUSTRY SECTOR	113450	-	-	-	2777	-	-	-	28021	-
Iron and Steel	-	-	-	-	-	-	-	-	-	-
Chemical and Petrochem.	53637	-	-	-	-	-	-	-	-	-
of which: Feedstocks	53637	-	-	-	-	-	-	-	-	-
Non-Ferrous Metals	-	-	-	-	-	-	-	-	-	-
Non-Metallic Minerals	-	-	-	-	-	-	-	-	-	-
Transport Equipment	-	-	-	-	-	-	-	-	-	-
Machinery	-	-	-	-	-	-	-	-	-	-
Mining and Quarrying	-	-	-	-	-	-	-	-	-	-
Food and Tobacco	-	-	-	-	-	-	-	-	-	-
Paper, Pulp and Print	-	-	-	-	-	-	-	-	-	-
Wood and Wood Products	-	-	-	-	-	-	-	-	-	-
Construction	-	-	-	-	-	-	-	-	-	-
Textile and Leather	-	-	-	-	-	-	-	-	-	-
Non-specified	59813	-	-	-	2777	-	-	-	28021	-
TRANSPORT SECTOR	164	-	-	-	-	-	-	-	12	-
Air	-	-	-	-	-	-	-	-	-	-
Road	-	-	-	-	-	-	-	-	-	-
Rail	-	-	-	-	-	-	-	-	-	-
Pipeline Transport	-	-	-	-	-	-	-	-	-	-
Internal Navigation	-	-	-	-	-	-	-	-	-	-
Non-specified	164	-	-	-	-	-	-	-	12	-
OTHER SECTORS	431	-	-	-	57499	-	-	-	28614	-
Agriculture	-	-	-	-	-	-	-	-	-	-
Comm. and Publ. Services	254	-	-	-	-	-	-	-	18456	-
Residential	177	-	-	-	57499	-	-	-	10158	-
Non-specified	-	-	-	-	-	-	-	-	-	-
NON-ENERGY USE	-	-	-	-	-	-	-	-	-	-
in Industry/Transf./Energy	-	-	-	-	-	-	-	-	-	-
in Transport	-	-	-	-	-	-	-	-	-	-
in Other Sectors	-	-	-	-	-	-	-	-	-	-

Malta / Malte

SUPPLY AND CONSUMPTION 1997	Coal (1000 tonnes)							Oil (1000 tonnes)			
	Coking Coal	Other Bit. Coal	Sub-Bit. Coal	Lignite	Peat	Oven and Gas Coke	Pat. Fuel and BKB	Crude Oil	NGL	Feed-stocks	Additives
Production	-	-	-	-	-	-	-	-	-	-	-
Imports	-	-	-	-	-	-	-	-	-	-	-
Exports	-	-	-	-	-	-	-	-	-	-	-
Intl. Marine Bunkers	-	-	-	-	-	-	-	-	-	-	-
Stock Changes	-	-	-	-	-	-	-	-	-	-	-
DOMESTIC SUPPLY	-	-	-	-	-	-	-	-	-	-	-
Transfers and Stat. Diff.	-	-	-	-	-	-	-	-	-	-	-
TRANSFORMATION	-	-	-	-	-	-	-	-	-	-	-
Electricity and CHP Plants	-	-	-	-	-	-	-	-	-	-	-
Petroleum Refineries	-	-	-	-	-	-	-	-	-	-	-
Other Transform. Sector	-	-	-	-	-	-	-	-	-	-	-
ENERGY SECTOR	-	-	-	-	-	-	-	-	-	-	-
DISTRIBUTION LOSSES	-	-	-	-	-	-	-	-	-	-	-
FINAL CONSUMPTION	-	-	-	-	-	-	-	-	-	-	-
INDUSTRY SECTOR	-	-	-	-	-	-	-	-	-	-	-
Iron and Steel	-	-	-	-	-	-	-	-	-	-	-
Chemical and Petrochem.	-	-	-	-	-	-	-	-	-	-	-
Non-Metallic Minerals	-	-	-	-	-	-	-	-	-	-	-
Non-specified	-	-	-	-	-	-	-	-	-	-	-
TRANSPORT SECTOR	-	-	-	-	-	-	-	-	-	-	-
Air	-	-	-	-	-	-	-	-	-	-	-
Road	-	-	-	-	-	-	-	-	-	-	-
Non-specified	-	-	-	-	-	-	-	-	-	-	-
OTHER SECTORS	-	-	-	-	-	-	-	-	-	-	-
Agriculture	-	-	-	-	-	-	-	-	-	-	-
Comm. and Publ. Services	-	-	-	-	-	-	-	-	-	-	-
Residential	-	-	-	-	-	-	-	-	-	-	-
Non-specified	-	-	-	-	-	-	-	-	-	-	-
NON-ENERGY USE	-	-	-	-	-	-	-	-	-	-	-

APPROVISIONNEMENT ET DEMANDE 1998	Charbon (1000 tonnes)							Pétrole (1000 tonnes)			
	Charbon à coke	Autres charb. bit.	Charbon sous-bit.	Lignite	Tourbe	Coke de four/gaz	Agg./briq. de lignite	Pétrole brut	LGN	Produits d'aliment.	Additifs
Production	-	-	-	-	-	-	-	-	-	-	-
Imports	-	-	-	-	-	-	-	-	-	-	-
Exports	-	-	-	-	-	-	-	-	-	-	-
Intl. Marine Bunkers	-	-	-	-	-	-	-	-	-	-	-
Stock Changes	-	-	-	-	-	-	-	-	-	-	-
DOMESTIC SUPPLY	-	-	-	-	-	-	-	-	-	-	-
Transfers and Stat. Diff.	-	-	-	-	-	-	-	-	-	-	-
TRANSFORMATION	-	-	-	-	-	-	-	-	-	-	-
Electricity and CHP Plants	-	-	-	-	-	-	-	-	-	-	-
Petroleum Refineries	-	-	-	-	-	-	-	-	-	-	-
Other Transform. Sector	-	-	-	-	-	-	-	-	-	-	-
ENERGY SECTOR	-	-	-	-	-	-	-	-	-	-	-
DISTRIBUTION LOSSES	-	-	-	-	-	-	-	-	-	-	-
FINAL CONSUMPTION	-	-	-	-	-	-	-	-	-	-	-
INDUSTRY SECTOR	-	-	-	-	-	-	-	-	-	-	-
Iron and Steel	-	-	-	-	-	-	-	-	-	-	-
Chemical and Petrochem.	-	-	-	-	-	-	-	-	-	-	-
Non-Metallic Minerals	-	-	-	-	-	-	-	-	-	-	-
Non-specified	-	-	-	-	-	-	-	-	-	-	-
TRANSPORT SECTOR	-	-	-	-	-	-	-	-	-	-	-
Air	-	-	-	-	-	-	-	-	-	-	-
Road	-	-	-	-	-	-	-	-	-	-	-
Non-specified	-	-	-	-	-	-	-	-	-	-	-
OTHER SECTORS	-	-	-	-	-	-	-	-	-	-	-
Agriculture	-	-	-	-	-	-	-	-	-	-	-
Comm. and Publ. Services	-	-	-	-	-	-	-	-	-	-	-
Residential	-	-	-	-	-	-	-	-	-	-	-
Non-specified	-	-	-	-	-	-	-	-	-	-	-
NON-ENERGY USE	-	-	-	-	-	-	-	-	-	-	-

Malta / Malte

SUPPLY AND CONSUMPTION 1997	Refinery Gas	LPG + Ethane	Motor Gasoline	Aviation Gasoline	Jet Fuel	Kerosene	Gas/ Diesel	Heavy Fuel Oil	Naphtha	Petrol. Coke	Other Prod.
	Oil cont. (1000 tonnes)										
Production	-	-	-	-	-	-	-	-	-	-	-
Imports	-	18	77	-	112	55	190	557	-	-	11
Exports	-	-	-	-	-	-	-	-	-	-	-
Intl. Marine Bunkers	-	-	-	-	-	-	-20	-60	-	-	-
Stock Changes	-	-	-	-	-	-	-	-	-	-	-
DOMESTIC SUPPLY	-	18	77	-	112	55	170	497	-	-	11
Transfers and Stat. Diff.	-	-	-	-	-	-	-	-	-	-	-
TRANSFORMATION	-	-	-	-	-	-	21	497	-	-	-
Electricity and CHP Plants	-	-	-	-	-	-	21	497	-	-	-
Petroleum Refineries	-	-	-	-	-	-	-	-	-	-	-
Other Transform. Sector	-	-	-	-	-	-	-	-	-	-	-
ENERGY SECTOR	-	-	-	-	-	-	-	-	-	-	-
DISTRIBUTION LOSSES	-	-	-	-	-	-	-	-	-	-	-
FINAL CONSUMPTION	-	18	77	-	112	55	149	-	-	-	11
INDUSTRY SECTOR	-	-	-	-	-	-	-	-	-	-	-
Iron and Steel	-	-	-	-	-	-	-	-	-	-	-
Chemical and Petrochem.	-	-	-	-	-	-	-	-	-	-	-
Non-Metallic Minerals	-	-	-	-	-	-	-	-	-	-	-
Non-specified	-	-	-	-	-	-	-	-	-	-	-
TRANSPORT SECTOR	-	-	77	-	112	-	149	-	-	-	-
Air	-	-	-	-	112	-	-	-	-	-	-
Road	-	-	77	-	-	-	149	-	-	-	-
Non-specified	-	-	-	-	-	-	-	-	-	-	-
OTHER SECTORS	-	18	-	-	-	55	-	-	-	-	11
Agriculture	-	-	-	-	-	-	-	-	-	-	-
Comm. and Publ. Services	-	-	-	-	-	-	-	-	-	-	-
Residential	-	18	-	-	-	55	-	-	-	-	-
Non-specified	-	-	-	-	-	-	-	-	-	-	11
NON-ENERGY USE	-	-	-	-	-	-	-	-	-	-	-

APPROVISIONNEMENT ET DEMANDE 1998	Gaz de raffinerie	GPL + éthane	Essence moteur	Essence aviation	Carbu- réacteurs	Kérosène	Gazole	Fioul lourd	Naphta	Coke de pétrole	Autres prod.
	Pétrole cont. (1000 tonnes)										
Production	-	-	-	-	-	-	-	-	-	-	-
Imports	-	19	81	-	205	57	157	549	-	3	15
Exports	-	-	-	-	-42	-	-51	-	-	-	-
Intl. Marine Bunkers	-	-	-	-	-	-	-20	-26	-	-	-
Stock Changes	-	-	-	-	-	-	-	-	-	-	-
DOMESTIC SUPPLY	-	19	81	-	163	57	86	523	-	3	15
Transfers and Stat. Diff.	-	-	-	-	-	-	-	-	-	-	-
TRANSFORMATION	-	-	-	-	-	-	18	523	-	-	-
Electricity and CHP Plants	-	-	-	-	-	-	18	523	-	-	-
Petroleum Refineries	-	-	-	-	-	-	-	-	-	-	-
Other Transform. Sector	-	-	-	-	-	-	-	-	-	-	-
ENERGY SECTOR	-	-	-	-	-	-	-	-	-	-	-
DISTRIBUTION LOSSES	-	-	-	-	-	-	-	-	-	-	-
FINAL CONSUMPTION	-	19	81	-	163	57	68	-	-	3	15
INDUSTRY SECTOR	-	-	-	-	-	-	-	-	-	3	-
Iron and Steel	-	-	-	-	-	-	-	-	-	-	-
Chemical and Petrochem.	-	-	-	-	-	-	-	-	-	-	-
Non-Metallic Minerals	-	-	-	-	-	-	-	-	-	-	-
Non-specified	-	-	-	-	-	-	-	-	-	3	-
TRANSPORT SECTOR	-	-	81	-	163	-	68	-	-	-	-
Air	-	-	-	-	163	-	-	-	-	-	-
Road	-	-	81	-	-	-	68	-	-	-	-
Non-specified	-	-	-	-	-	-	-	-	-	-	-
OTHER SECTORS	-	19	-	-	-	57	-	-	-	-	11
Agriculture	-	-	-	-	-	-	-	-	-	-	-
Comm. and Publ. Services	-	-	-	-	-	-	-	-	-	-	-
Residential	-	19	-	-	-	57	-	-	-	-	-
Non-specified	-	-	-	-	-	-	-	-	-	-	11
NON-ENERGY USE	-	-	-	-	-	-	-	-	-	-	4

Malta / Malte

SUPPLY AND CONSUMPTION 1997	Gas (TJ)				Comb. Renew. & Waste (TJ)				(GWh)	(TJ)
	Natural Gas	Gas Works	Coke Ovens	Blast Furnaces	Solid Biomass	Gas/Liquids from Biomass	Municipal Waste	Industrial Waste	Electricity	Heat
Production	-	-	-	-	-	-	-	-	1686	-
Imports	-	-	-	-	-	-	-	-	-	-
Exports	-	-	-	-	-	-	-	-	-	-
Intl. Marine Bunkers	-	-	-	-	-	-	-	-	-	-
Stock Changes	-	-	-	-	-	-	-	-	-	-
DOMESTIC SUPPLY	-	-	-	-	-	-	-	-	1686	-
Transfers and Stat. Diff.	-	-	-	-	-	-	-	-	-	-
TRANSFORMATION	-	-	-	-	-	-	-	-	-	-
Electricity and CHP Plants	-	-	-	-	-	-	-	-	-	-
Petroleum Refineries	-	-	-	-	-	-	-	-	-	-
Other Transform. Sector	-	-	-	-	-	-	-	-	-	-
ENERGY SECTOR	-	-	-	-	-	-	-	-	105	-
DISTRIBUTION LOSSES	-	-	-	-	-	-	-	-	221	-
FINAL CONSUMPTION	-	-	-	-	-	-	-	-	1360	-
INDUSTRY SECTOR	-	-	-	-	-	-	-	-	453	-
Iron and Steel	-	-	-	-	-	-	-	-	-	-
Chemical and Petrochem.	-	-	-	-	-	-	-	-	-	-
Non-Metallic Minerals	-	-	-	-	-	-	-	-	-	-
Non-specified	-	-	-	-	-	-	-	-	453	-
TRANSPORT SECTOR	-	-	-	-	-	-	-	-	-	-
Air	-	-	-	-	-	-	-	-	-	-
Road	-	-	-	-	-	-	-	-	-	-
Non-specified	-	-	-	-	-	-	-	-	-	-
OTHER SECTORS	-	-	-	-	-	-	-	-	907	-
Agriculture	-	-	-	-	-	-	-	-	-	-
Comm. and Publ. Services	-	-	-	-	-	-	-	-	445	-
Residential	-	-	-	-	-	-	-	-	462	-
Non-specified	-	-	-	-	-	-	-	-	-	-
NON-ENERGY USE	-	-	-	-	-	-	-	-	-	-

APPROVISIONNEMENT ET DEMANDE 1998	Gaz (TJ)				En. Re. Comb. & Déchets (TJ)				(GWh)	(TJ)
	Gaz naturel	Usines à gaz	Cokeries	Hauts fourneaux	Biomasse solide	Gaz/Liquides tirés de biomasse	Déchets urbains	Déchets industriels	Electricité	Chaleur
Production	-	-	-	-	-	-	-	-	1721	-
Imports	-	-	-	-	-	-	-	-	-	-
Exports	-	-	-	-	-	-	-	-	-	-
Intl. Marine Bunkers	-	-	-	-	-	-	-	-	-	-
Stock Changes	-	-	-	-	-	-	-	-	-	-
DOMESTIC SUPPLY	-	-	-	-	-	-	-	-	1721	-
Transfers and Stat. Diff.	-	-	-	-	-	-	-	-	-	-
TRANSFORMATION	-	-	-	-	-	-	-	-	-	-
Electricity and CHP Plants	-	-	-	-	-	-	-	-	-	-
Petroleum Refineries	-	-	-	-	-	-	-	-	-	-
Other Transform. Sector	-	-	-	-	-	-	-	-	-	-
ENERGY SECTOR	-	-	-	-	-	-	-	-	109	-
DISTRIBUTION LOSSES	-	-	-	-	-	-	-	-	210	-
FINAL CONSUMPTION	-	-	-	-	-	-	-	-	1402	-
INDUSTRY SECTOR	-	-	-	-	-	-	-	-	451	-
Iron and Steel	-	-	-	-	-	-	-	-	-	-
Chemical and Petrochem.	-	-	-	-	-	-	-	-	-	-
Non-Metallic Minerals	-	-	-	-	-	-	-	-	-	-
Non-specified	-	-	-	-	-	-	-	-	451	-
TRANSPORT SECTOR	-	-	-	-	-	-	-	-	-	-
Air	-	-	-	-	-	-	-	-	-	-
Road	-	-	-	-	-	-	-	-	-	-
Non-specified	-	-	-	-	-	-	-	-	-	-
OTHER SECTORS	-	-	-	-	-	-	-	-	951	-
Agriculture	-	-	-	-	-	-	-	-	-	-
Comm. and Publ. Services	-	-	-	-	-	-	-	-	451	-
Residential	-	-	-	-	-	-	-	-	500	-
Non-specified	-	-	-	-	-	-	-	-	-	-
NON-ENERGY USE	-	-	-	-	-	-	-	-	-	-

Republic of Moldova / République de Moldovie : 1997

SUPPLY AND CONSUMPTION / APPROVISIONNEMENT ET DEMANDE	Coal / Charbon (1000 tonnes)							Oil / Pétrole (1000 tonnes)			
	Coking Coal / Charbon à coke	Other Bit. Coal / Autres charb. bit.	Sub-Bit. Coal / Charbon sous-bit.	Lignite / Lignite	Peat / Tourbe	Oven and Gas Coke / Coke de four/gaz	Pat. Fuel and BKB / Agg./briq. de lignite	Crude Oil / Pétrole brut	NGL / LGN	Feed-stocks / Produits d'aliment.	Additives / Additifs
Production	-	-	-	-	-	-	-	-	-	-	-
From Other Sources	-	-	-	-	-	-	-	-	-	-	-
Imports	-	653	-	-	-	8	-	-	-	-	-
Exports	-	-	-	-	-	-	-	-	-	-	-
Intl. Marine Bunkers	-	-	-	-	-	-	-	-	-	-	-
Stock Changes	-	-50	-	-	-	5	-	-	-	-	-
DOMESTIC SUPPLY	-	603	-	-	-	13	-	-	-	-	-
Transfers	-	-	-	-	-	-	-	-	-	-	-
Statistical Differences	-	-	-	-	-	-	-	-	-	-	-
TRANSFORMATION	-	359	-	-	-	1	-	-	-	-	-
Electricity Plants	-	281	-	-	-	-	-	-	-	-	-
CHP Plants	-	78	-	-	-	-	-	-	-	-	-
Heat Plants	-	-	-	-	-	-	-	-	-	-	-
Blast Furnaces/Gas Works	-	-	-	-	-	-	-	-	-	-	-
Coke/Pat. Fuel/BKB Plants	-	-	-	-	-	1	-	-	-	-	-
Petroleum Refineries	-	-	-	-	-	-	-	-	-	-	-
Petrochemical Industry	-	-	-	-	-	-	-	-	-	-	-
Liquefaction	-	-	-	-	-	-	-	-	-	-	-
Other Transform. Sector	-	-	-	-	-	-	-	-	-	-	-
ENERGY SECTOR	-	-	-	-	-	-	-	-	-	-	-
Coal Mines	-	-	-	-	-	-	-	-	-	-	-
Oil and Gas Extraction	-	-	-	-	-	-	-	-	-	-	-
Petroleum Refineries	-	-	-	-	-	-	-	-	-	-	-
Electr., CHP+Heat Plants	-	-	-	-	-	-	-	-	-	-	-
Pumped Storage (Elec.)	-	-	-	-	-	-	-	-	-	-	-
Other Energy Sector	-	-	-	-	-	-	-	-	-	-	-
Distribution Losses	-	-	-	-	-	-	-	-	-	-	-
FINAL CONSUMPTION	-	244	-	-	-	12	-	-	-	-	-
INDUSTRY SECTOR	-	3	-	-	-	12	-	-	-	-	-
Iron and Steel	-	-	-	-	-	-	-	-	-	-	-
Chemical and Petrochem.	-	-	-	-	-	-	-	-	-	-	-
of which: Feedstocks	-	-	-	-	-	-	-	-	-	-	-
Non-Ferrous Metals	-	-	-	-	-	-	-	-	-	-	-
Non-Metallic Minerals	-	3	-	-	-	1	-	-	-	-	-
Transport Equipment	-	-	-	-	-	-	-	-	-	-	-
Machinery	-	-	-	-	-	-	-	-	-	-	-
Mining and Quarrying	-	-	-	-	-	-	-	-	-	-	-
Food and Tobacco	-	-	-	-	-	11	-	-	-	-	-
Paper, Pulp and Print	-	-	-	-	-	-	-	-	-	-	-
Wood and Wood Products	-	-	-	-	-	-	-	-	-	-	-
Construction	-	-	-	-	-	-	-	-	-	-	-
Textile and Leather	-	-	-	-	-	-	-	-	-	-	-
Non-specified	-	-	-	-	-	-	-	-	-	-	-
TRANSPORT SECTOR	-	5	-	-	-	-	-	-	-	-	-
Air	-	-	-	-	-	-	-	-	-	-	-
Road	-	-	-	-	-	-	-	-	-	-	-
Rail	-	5	-	-	-	-	-	-	-	-	-
Pipeline Transport	-	-	-	-	-	-	-	-	-	-	-
Internal Navigation	-	-	-	-	-	-	-	-	-	-	-
Non-specified	-	-	-	-	-	-	-	-	-	-	-
OTHER SECTORS	-	236	-	-	-	-	-	-	-	-	-
Agriculture	-	2	-	-	-	-	-	-	-	-	-
Comm. and Publ. Services	-	117	-	-	-	-	-	-	-	-	-
Residential	-	111	-	-	-	-	-	-	-	-	-
Non-specified	-	6	-	-	-	-	-	-	-	-	-
NON-ENERGY USE	-	-	-	-	-	-	-	-	-	-	-
in Industry/Trans./Energy	-	-	-	-	-	-	-	-	-	-	-
in Transport	-	-	-	-	-	-	-	-	-	-	-
in Other Sectors	-	-	-	-	-	-	-	-	-	-	-

Republic of Moldova / République de Moldovie : 1997

SUPPLY AND CONSUMPTION *APPROVISIONNEMENT ET DEMANDE*	Oil cont. / *Pétrole cont.* (1000 tonnes)										
	Refinery Gas *Gaz de raffinerie*	LPG + Ethane *GPL + éthane*	Motor Gasoline *Essence moteur*	Aviation Gasoline *Essence aviation*	Jet Fuel *Carbu- réacteurs*	Kerosene *Kérosène*	Gas/ Diesel *Gazole*	Heavy Fuel Oil *Fioul lourd*	Naphtha *Naphta*	Petrol. Coke *Coke de pétrole*	Other Prod. *Autres prod.*
Production	-	-	-	-	-	-	-	-	-	-	-
From Other Sources	-	-	-	-	-	-	-	-	-	-	-
Imports	-	28	262	-	25	10	328	275	-	-	16
Exports	-	-	-	-	-	-	-	-	-	-	-
Intl. Marine Bunkers	-	-	-	-	-	-	-	-	-	-	-
Stock Changes	-	-1	-13	-	-2	-3	-16	-33	-	-	-
DOMESTIC SUPPLY	-	27	249	-	23	7	312	242	-	-	16
Transfers	-	-	-	-	-	-	-	-	-	-	-
Statistical Differences	-	-	-	-	-	-	-	-	-	-	-
TRANSFORMATION	-	-	-	-	-	1	4	225	-	-	-
Electricity Plants	-	-	-	-	-	-	3	47	-	-	-
CHP Plants	-	-	-	-	-	-	-	178	-	-	-
Heat Plants	-	-	-	-	-	1	1	-	-	-	-
Blast Furnaces/Gas Works	-	-	-	-	-	-	-	-	-	-	-
Coke/Pat. Fuel/BKB Plants	-	-	-	-	-	-	-	-	-	-	-
Petroleum Refineries	-	-	-	-	-	-	-	-	-	-	-
Petrochemical Industry	-	-	-	-	-	-	-	-	-	-	-
Liquefaction	-	-	-	-	-	-	-	-	-	-	-
Other Transform. Sector	-	-	-	-	-	-	-	-	-	-	-
ENERGY SECTOR	-	-	-	-	-	-	-	-	-	-	-
Coal Mines	-	-	-	-	-	-	-	-	-	-	-
Oil and Gas Extraction	-	-	-	-	-	-	-	-	-	-	-
Petroleum Refineries	-	-	-	-	-	-	-	-	-	-	-
Electr., CHP+Heat Plants	-	-	-	-	-	-	-	-	-	-	-
Pumped Storage (Elec.)	-	-	-	-	-	-	-	-	-	-	-
Other Energy Sector	-	-	-	-	-	-	-	-	-	-	-
Distribution Losses	-	1	5	-	1	1	7	-	-	-	-
FINAL CONSUMPTION	-	26	244	-	22	5	301	17	-	-	16
INDUSTRY SECTOR	-	-	-	-	-	3	13	7	-	-	-
Iron and Steel	-	-	-	-	-	-	-	-	-	-	-
Chemical and Petrochem.	-	-	-	-	-	-	-	-	-	-	-
of which: Feedstocks	-	-	-	-	-	-	-	-	-	-	-
Non-Ferrous Metals	-	-	-	-	-	-	-	-	-	-	-
Non-Metallic Minerals	-	-	-	-	-	-	-	1	-	-	-
Transport Equipment	-	-	-	-	-	-	-	-	-	-	-
Machinery	-	-	-	-	-	-	-	-	-	-	-
Mining and Quarrying	-	-	-	-	-	-	1	-	-	-	-
Food and Tobacco	-	-	-	-	-	3	7	4	-	-	-
Paper, Pulp and Print	-	-	-	-	-	-	-	-	-	-	-
Wood and Wood Products	-	-	-	-	-	-	-	-	-	-	-
Construction	-	-	-	-	-	-	5	2	-	-	-
Textile and Leather	-	-	-	-	-	-	-	-	-	-	-
Non-specified	-	-	-	-	-	-	-	-	-	-	-
TRANSPORT SECTOR	-	5	112	-	21	-	110	1	-	-	-
Air	-	-	-	-	21	-	-	-	-	-	-
Road	-	5	111	-	-	-	88	-	-	-	-
Rail	-	-	-	-	-	-	21	1	-	-	-
Pipeline Transport	-	-	-	-	-	-	-	-	-	-	-
Internal Navigation	-	-	-	-	-	-	-	-	-	-	-
Non-specified	-	-	1	-	-	-	1	-	-	-	-
OTHER SECTORS	-	21	132	-	1	2	178	9	-	-	-
Agriculture	-	-	4	-	-	2	146	1	-	-	-
Comm. and Publ. Services	-	-	-	-	-	-	2	1	-	-	-
Residential	-	20	127	-	-	-	28	-	-	-	-
Non-specified	-	1	1	-	1	-	2	7	-	-	-
NON-ENERGY USE	-	-	-	-	-	-	-	-	-	-	16
in Industry/Transf./Energy	-	-	-	-	-	-	-	-	-	-	-
in Transport	-	-	-	-	-	-	-	-	-	-	7
in Other Sectors	-	-	-	-	-	-	-	-	-	-	9

Republic of Moldova / République de Moldovie : 1997

SUPPLY AND CONSUMPTION / APPROVISIONNEMENT ET DEMANDE	Gas / Gaz (TJ)				Comb. Renew. & Waste / En. Re. Comb. & Déchets (TJ)				(GWh)	(TJ)
	Natural Gas / Gaz naturel	Gas Works / Usines à gaz	Coke Ovens / Cokeries	Blast Furnaces / Hauts fourneaux	Solid Biomass / Biomasse solide	Gas/Liquids from Biomass / Gaz/Liquides tirés de biomasse	Municipal Waste / Déchets urbains	Industrial Waste / Déchets industriels	Electricity / Electricité	Heat / Chaleur
Production	-	-	-	-	2817	-	-	-	5273	13819
From Other Sources	-	-	-	-	-	-	-	-	-	-
Imports	145726	-	-	-	-	-	-	-	1979	-
Exports	-	-	-	-	-	-	-	-	-27	-
Intl. Marine Bunkers	-	-	-	-	-	-	-	-	-	-
Stock Changes	-222	-	-	-	-264	-	-	-	-	-
DOMESTIC SUPPLY	**145504**	**-**	**-**	**-**	**2553**	**-**	**-**	**-**	**7225**	**13819**
Transfers	-	-	-	-	-	-	-	-	-	-
Statistical Differences	-	-	-	-	-	-	-	-	-	-
TRANSFORMATION	**66329**	**-**	**-**	**-**	**-**	**-**	**-**	**-**	**27**	**-**
Electricity Plants	38458	-	-	-	-	-	-	-	-	-
CHP Plants	25215	-	-	-	-	-	-	-	-	-
Heat Plants	2656	-	-	-	-	-	-	-	-	-
Blast Furnaces/Gas Works	-	-	-	-	-	-	-	-	-	-
Coke/Pat. Fuel/BKB Plants	-	-	-	-	-	-	-	-	-	-
Petroleum Refineries	-	-	-	-	-	-	-	-	-	-
Petrochemical Industry	-	-	-	-	-	-	-	-	-	-
Liquefaction	-	-	-	-	-	-	-	-	-	-
Other Transform. Sector	-	-	-	-	-	-	-	-	27	-
ENERGY SECTOR	**-**	**-**	**-**	**-**	**-**	**-**	**-**	**-**	**524**	**-**
Coal Mines	-	-	-	-	-	-	-	-	-	-
Oil and Gas Extraction	-	-	-	-	-	-	-	-	-	-
Petroleum Refineries	-	-	-	-	-	-	-	-	-	-
Electr., CHP+Heat Plants	-	-	-	-	-	-	-	-	524	-
Pumped Storage (Elec.)	-	-	-	-	-	-	-	-	-	-
Other Energy Sector	-	-	-	-	-	-	-	-	-	-
Distribution Losses	6027	-	-	-	-	-	-	-	1425	1380
FINAL CONSUMPTION	**73148**	**-**	**-**	**-**	**2553**	**-**	**-**	**-**	**5249**	**12439**
INDUSTRY SECTOR	**21171**	**-**	**-**	**-**	**-**	**-**	**-**	**-**	**1802**	**-**
Iron and Steel	-	-	-	-	-	-	-	-	-	-
Chemical and Petrochem.	-	-	-	-	-	-	-	-	-	-
of which: Feedstocks	-	-	-	-	-	-	-	-	-	-
Non-Ferrous Metals	-	-	-	-	-	-	-	-	-	-
Non-Metallic Minerals	6343	-	-	-	-	-	-	-	-	-
Transport Equipment	-	-	-	-	-	-	-	-	-	-
Machinery	833	-	-	-	-	-	-	-	-	-
Mining and Quarrying	-	-	-	-	-	-	-	-	-	-
Food and Tobacco	436	-	-	-	-	-	-	-	-	-
Paper, Pulp and Print	-	-	-	-	-	-	-	-	-	-
Wood and Wood Products	198	-	-	-	-	-	-	-	-	-
Construction	-	-	-	-	-	-	-	-	-	-
Textile and Leather	6581	-	-	-	-	-	-	-	-	-
Non-specified	6780	-	-	-	-	-	-	-	1802	-
TRANSPORT SECTOR	**3965**	**-**	**-**	**-**	**-**	**-**	**-**	**-**	**118**	**-**
Air	-	-	-	-	-	-	-	-	-	-
Road	515	-	-	-	-	-	-	-	-	-
Rail	-	-	-	-	-	-	-	-	46	-
Pipeline Transport	3450	-	-	-	-	-	-	-	4	-
Internal Navigation	-	-	-	-	-	-	-	-	-	-
Non-specified	-	-	-	-	-	-	-	-	68	-
OTHER SECTORS	**48012**	**-**	**-**	**-**	**2553**	**-**	**-**	**-**	**3329**	**12439**
Agriculture	4004	-	-	-	-	-	-	-	526	-
Comm. and Publ. Services	23550	-	-	-	-	-	-	-	1036	-
Residential	16176	-	-	-	-	-	-	-	1767	12439
Non-specified	4282	-	-	-	2553	-	-	-	-	-
NON-ENERGY USE	**-**	**-**	**-**	**-**	**-**	**-**	**-**	**-**	**-**	**-**
in Industry/Transf./Energy	-	-	-	-	-	-	-	-	-	-
in Transport	-	-	-	-	-	-	-	-	-	-
in Other Sectors	-	-	-	-	-	-	-	-	-	-

Republic of Moldova / République de Moldovie : 1998

SUPPLY AND CONSUMPTION / APPROVISIONNEMENT ET DEMANDE	Coal / Charbon (1000 tonnes)							Oil / Pétrole (1000 tonnes)			
	Coking Coal / Charbon à coke	Other Bit. Coal / Autres charb. bit.	Sub-Bit. Coal / Charbon sous-bit.	Lignite / Lignite	Peat / Tourbe	Oven and Gas Coke / Coke de four/gaz	Pat. Fuel and BKB / Agg./briq. de lignite	Crude Oil / Pétrole brut	NGL / LGN	Feed-stocks / Produits d'aliment.	Additives / Additifs
Production	-	-	-	-	-	-	-	-	-	-	-
From Other Sources	-	-	-	-	-	-	-	-	-	-	-
Imports	-	504	-	-	-	7	-	-	-	-	-
Exports	-	-	-	-	-	-	-	-	-	-	-
Intl. Marine Bunkers	-	-	-	-	-	-	-	-	-	-	-
Stock Changes	-	42	-	-	-	3	-	-	-	-	-
DOMESTIC SUPPLY	-	546	-	-	-	10	-	-	-	-	-
Transfers	-	-	-	-	-	-	-	-	-	-	-
Statistical Differences	-	-	-	-	-	-	-	-	-	-	-
TRANSFORMATION	-	368	-	-	-	-	-	-	-	-	-
Electricity Plants	-	288	-	-	-	-	-	-	-	-	-
CHP Plants	-	80	-	-	-	-	-	-	-	-	-
Heat Plants	-	-	-	-	-	-	-	-	-	-	-
Blast Furnaces/Gas Works	-	-	-	-	-	-	-	-	-	-	-
Coke/Pat. Fuel/BKB Plants	-	-	-	-	-	-	-	-	-	-	-
Petroleum Refineries	-	-	-	-	-	-	-	-	-	-	-
Petrochemical Industry	-	-	-	-	-	-	-	-	-	-	-
Liquefaction	-	-	-	-	-	-	-	-	-	-	-
Other Transform. Sector	-	-	-	-	-	-	-	-	-	-	-
ENERGY SECTOR	-	-	-	-	-	-	-	-	-	-	-
Coal Mines	-	-	-	-	-	-	-	-	-	-	-
Oil and Gas Extraction	-	-	-	-	-	-	-	-	-	-	-
Petroleum Refineries	-	-	-	-	-	-	-	-	-	-	-
Electr., CHP+Heat Plants	-	-	-	-	-	-	-	-	-	-	-
Pumped Storage (Elec.)	-	-	-	-	-	-	-	-	-	-	-
Other Energy Sector	-	-	-	-	-	-	-	-	-	-	-
Distribution Losses	-	1	-	-	-	-	-	-	-	-	-
FINAL CONSUMPTION	-	177	-	-	-	10	-	-	-	-	-
INDUSTRY SECTOR	-	5	-	-	-	9	-	-	-	-	-
Iron and Steel	-	-	-	-	-	-	-	-	-	-	-
Chemical and Petrochem.	-	-	-	-	-	-	-	-	-	-	-
of which: Feedstocks	-	-	-	-	-	-	-	-	-	-	-
Non-Ferrous Metals	-	-	-	-	-	-	-	-	-	-	-
Non-Metallic Minerals	-	2	-	-	-	-	-	-	-	-	-
Transport Equipment	-	-	-	-	-	-	-	-	-	-	-
Machinery	-	-	-	-	-	-	-	-	-	-	-
Mining and Quarrying	-	-	-	-	-	-	-	-	-	-	-
Food and Tobacco	-	2	-	-	-	9	-	-	-	-	-
Paper, Pulp and Print	-	-	-	-	-	-	-	-	-	-	-
Wood and Wood Products	-	-	-	-	-	-	-	-	-	-	-
Construction	-	-	-	-	-	-	-	-	-	-	-
Textile and Leather	-	1	-	-	-	-	-	-	-	-	-
Non-specified	-	-	-	-	-	-	-	-	-	-	-
TRANSPORT SECTOR	-	-	-	-	-	-	-	-	-	-	-
Air	-	-	-	-	-	-	-	-	-	-	-
Road	-	-	-	-	-	-	-	-	-	-	-
Rail	-	-	-	-	-	-	-	-	-	-	-
Pipeline Transport	-	-	-	-	-	-	-	-	-	-	-
Internal Navigation	-	-	-	-	-	-	-	-	-	-	-
Non-specified	-	-	-	-	-	-	-	-	-	-	-
OTHER SECTORS	-	172	-	-	-	1	-	-	-	-	-
Agriculture	-	2	-	-	-	-	-	-	-	-	-
Comm. and Publ. Services	-	123	-	-	-	-	-	-	-	-	-
Residential	-	41	-	-	-	-	-	-	-	-	-
Non-specified	-	6	-	-	-	1	-	-	-	-	-
NON-ENERGY USE	-	-	-	-	-	-	-	-	-	-	-
in Industry/Trans./Energy	-	-	-	-	-	-	-	-	-	-	-
in Transport	-	-	-	-	-	-	-	-	-	-	-
in Other Sectors	-	-	-	-	-	-	-	-	-	-	-

Republic of Moldova / République de Moldovie : 1998

SUPPLY AND CONSUMPTION / APPROVISIONNEMENT ET DEMANDE	Oil cont. / Pétrole cont. (1000 tonnes)										
	Refinery Gas / Gaz de raffinerie	LPG + Ethane / GPL + éthane	Motor Gasoline / Essence moteur	Aviation Gasoline / Essence aviation	Jet Fuel / Carbu- réacteurs	Kerosene / Kérosène	Gas/ Diesel / Gazole	Heavy Fuel Oil / Fioul lourd	Naphtha / Naphta	Petrol. Coke / Coke de pétrole	Other Prod. / Autres prod.
Production	-	-	-	-	-	-	-	-	-	-	-
From Other Sources	-	-	-	-	-	-	-	-	-	-	-
Imports	-	24	192	-	21	9	235	139	-	-	9
Exports	-	-	-	-	-	-	-	-	-	-	-
Intl. Marine Bunkers	-	-	-	-	-	-	-	-	-	-	-
Stock Changes	-	4	19	-	-	2	13	56	-	-	2
DOMESTIC SUPPLY	-	28	211	-	21	11	248	195	-	-	11
Transfers	-	-	-	-	-	-	-	-	-	-	-
Statistical Differences	-	-	-	-	-	-	-	-	-	-	-
TRANSFORMATION	-	-	-	-	-	-	1	184	-	-	-
Electricity Plants	-	-	-	-	-	-	1	43	-	-	-
CHP Plants	-	-	-	-	-	-	-	141	-	-	-
Heat Plants	-	-	-	-	-	-	-	-	-	-	-
Blast Furnaces/Gas Works	-	-	-	-	-	-	-	-	-	-	-
Coke/Pat. Fuel/BKB Plants	-	-	-	-	-	-	-	-	-	-	-
Petroleum Refineries	-	-	-	-	-	-	-	-	-	-	-
Petrochemical Industry	-	-	-	-	-	-	-	-	-	-	-
Liquefaction	-	-	-	-	-	-	-	-	-	-	-
Other Transform. Sector	-	-	-	-	-	-	-	-	-	-	-
ENERGY SECTOR	-	-	-	-	-	-	-	-	-	-	-
Coal Mines	-	-	-	-	-	-	-	-	-	-	-
Oil and Gas Extraction	-	-	-	-	-	-	-	-	-	-	-
Petroleum Refineries	-	-	-	-	-	-	-	-	-	-	-
Electr., CHP+Heat Plants	-	-	-	-	-	-	-	-	-	-	-
Pumped Storage (Elec.)	-	-	-	-	-	-	-	-	-	-	-
Other Energy Sector	-	-	-	-	-	-	-	-	-	-	-
Distribution Losses	-	2	4	-	-	-	1	-	-	-	-
FINAL CONSUMPTION	-	26	207	-	21	11	246	11	-	-	11
INDUSTRY SECTOR	-	-	-	-	-	1	7	7	-	-	-
Iron and Steel	-	-	-	-	-	-	-	-	-	-	-
Chemical and Petrochem.	-	-	-	-	-	-	-	-	-	-	-
of which: Feedstocks	-	-	-	-	-	-	-	-	-	-	-
Non-Ferrous Metals	-	-	-	-	-	-	-	-	-	-	-
Non-Metallic Minerals	-	-	-	-	-	-	-	1	-	-	-
Transport Equipment	-	-	-	-	-	-	-	-	-	-	-
Machinery	-	-	-	-	-	-	-	-	-	-	-
Mining and Quarrying	-	-	-	-	-	-	1	-	-	-	-
Food and Tobacco	-	-	-	-	-	1	3	5	-	-	-
Paper, Pulp and Print	-	-	-	-	-	-	-	-	-	-	-
Wood and Wood Products	-	-	-	-	-	-	-	-	-	-	-
Construction	-	-	-	-	-	-	3	1	-	-	-
Textile and Leather	-	-	-	-	-	-	-	-	-	-	-
Non-specified	-	-	-	-	-	-	-	-	-	-	-
TRANSPORT SECTOR	-	4	93	-	17	-	93	1	-	-	-
Air	-	-	-	-	17	-	-	-	-	-	-
Road	-	4	92	-	-	-	76	-	-	-	-
Rail	-	-	-	-	-	-	16	1	-	-	-
Pipeline Transport	-	-	-	-	-	-	-	-	-	-	-
Internal Navigation	-	-	-	-	-	-	-	-	-	-	-
Non-specified	-	-	1	-	-	-	1	-	-	-	-
OTHER SECTORS	-	22	114	-	4	10	146	3	-	-	-
Agriculture	-	-	2	-	-	6	107	-	-	-	-
Comm. and Publ. Services	-	-	-	-	-	1	33	1	-	-	-
Residential	-	20	110	-	-	-	-	-	-	-	-
Non-specified	-	2	2	-	4	3	6	2	-	-	-
NON-ENERGY USE	-	-	-	-	-	-	-	-	-	-	11
in Industry/Transf./Energy	-	-	-	-	-	-	-	-	-	-	-
in Transport	-	-	-	-	-	-	-	-	-	-	6
in Other Sectors	-	-	-	-	-	-	-	-	-	-	5

Republic of Moldova / République de Moldovie : 1998

	Gas / *Gaz* (TJ)				Comb. Renew. & Waste / *En. Re. Comb. & Déchets* (TJ)				(GWh)	(TJ)
SUPPLY AND CONSUMPTION *APPROVISIONNEMENT ET DEMANDE*	Natural Gas *Gaz naturel*	Gas Works *Usines à gaz*	Coke Ovens *Cokeries*	Blast Furnaces *Hauts fourneaux*	Solid Biomass *Biomasse solide*	Gas/Liquids from Biomass *Gaz/Liquides tirés de biomasse*	Municipal Waste *Déchets urbains*	Industrial Waste *Déchets industriels*	Electricity *Electricité*	Heat *Chaleur*
Production	-	-	-	-	2318	-	-	-	4584	15033
From Other Sources	-	-	-	-	-	-	-	-	-	-
Imports	131861	-	-	-	-	-	-	-	1916	-
Exports	-	-	-	-	-	-	-	-	-	-
Intl. Marine Bunkers	-	-	-	-	-	-	-	-	-	-
Stock Changes	-255	-	-	-	146	-	-	-	-	-
DOMESTIC SUPPLY	**131606**	-	-	-	**2464**	-	-	-	**6500**	**15033**
Transfers	-	-	-	-	-	-	-	-	-	-
Statistical Differences	-	-	-	-	-	-	-	-	-	-
TRANSFORMATION	**59994**	-	-	-	-	-	-	-	-	-
Electricity Plants	34784	-	-	-	-	-	-	-	-	-
CHP Plants	22807	-	-	-	-	-	-	-	-	-
Heat Plants	2403	-	-	-	-	-	-	-	-	-
Blast Furnaces/Gas Works	-	-	-	-	-	-	-	-	-	-
Coke/Pat. Fuel/BKB Plants	-	-	-	-	-	-	-	-	-	-
Petroleum Refineries	-	-	-	-	-	-	-	-	-	-
Petrochemical Industry	-	-	-	-	-	-	-	-	-	-
Liquefaction	-	-	-	-	-	-	-	-	-	-
Other Transform. Sector	-	-	-	-	-	-	-	-	-	-
ENERGY SECTOR	-	-	-	-	-	-	-	-	**2336**	-
Coal Mines	-	-	-	-	-	-	-	-	-	-
Oil and Gas Extraction	-	-	-	-	-	-	-	-	-	-
Petroleum Refineries	-	-	-	-	-	-	-	-	-	-
Electr., CHP+Heat Plants	-	-	-	-	-	-	-	-	2082	-
Pumped Storage (Elec.)	-	-	-	-	-	-	-	-	-	-
Other Energy Sector	-	-	-	-	-	-	-	-	254	-
Distribution Losses	5450	-	-	-	-	-	-	-	1205	1203
FINAL CONSUMPTION	**66162**	-	-	-	**2464**	-	-	-	**2959**	**13830**
INDUSTRY SECTOR	**19149**	-	-	-	-	-	-	-	**566**	-
Iron and Steel	-	-	-	-	-	-	-	-	4	-
Chemical and Petrochem.	-	-	-	-	-	-	-	-	11	-
of which: Feedstocks	-	-	-	-	-	-	-	-	-	-
Non-Ferrous Metals	-	-	-	-	-	-	-	-	-	-
Non-Metallic Minerals	5738	-	-	-	-	-	-	-	95	-
Transport Equipment	-	-	-	-	-	-	-	-	1	-
Machinery	753	-	-	-	-	-	-	-	49	-
Mining and Quarrying	-	-	-	-	-	-	-	-	17	-
Food and Tobacco	394	-	-	-	-	-	-	-	308	-
Paper, Pulp and Print	-	-	-	-	-	-	-	-	23	-
Wood and Wood Products	179	-	-	-	-	-	-	-	11	-
Construction	-	-	-	-	-	-	-	-	17	-
Textile and Leather	5953	-	-	-	-	-	-	-	29	-
Non-specified	6132	-	-	-	-	-	-	-	1	-
TRANSPORT SECTOR	**3586**	-	-	-	-	-	-	-	**116**	-
Air	-	-	-	-	-	-	-	-	-	-
Road	466	-	-	-	-	-	-	-	-	-
Rail	-	-	-	-	-	-	-	-	-	-
Pipeline Transport	3120	-	-	-	-	-	-	-	1	-
Internal Navigation	-	-	-	-	-	-	-	-	-	-
Non-specified	-	-	-	-	-	-	-	-	115	-
OTHER SECTORS	**43427**	-	-	-	**2464**	-	-	-	**2277**	**13830**
Agriculture	3622	-	-	-	-	-	-	-	213	-
Comm. and Publ. Services	21301	-	-	-	-	-	-	-	960	-
Residential	14631	-	-	-	-	-	-	-	1104	13830
Non-specified	3873	-	-	-	2464	-	-	-	-	-
NON-ENERGY USE	-	-	-	-	-	-	-	-	-	-
in Industry/Transf./Energy	-	-	-	-	-	-	-	-	-	-
in Transport	-	-	-	-	-	-	-	-	-	-
in Other Sectors	-	-	-	-	-	-	-	-	-	-

Morocco / Maroc : 1997

SUPPLY AND CONSUMPTION / APPROVISIONNEMENT ET DEMANDE	Coal / Charbon (1000 tonnes)							Oil / Pétrole (1000 tonnes)			
	Coking Coal / Charbon à coke	Other Bit. Coal / Autres charb. bit.	Sub-Bit. Coal / Charbon sous-bit.	Lignite / Lignite	Peat / Tourbe	Oven and Gas Coke / Coke de four/gaz	Pat. Fuel and BKB / Agg./briq. de lignite	Crude Oil / Pétrole brut	NGL / LGN	Feed-stocks / Produits d'aliment.	Additives / Additifs
Production	-	376	-	-	-	-	-	12	-	-	-
From Other Sources	-	-	-	-	-	-	-	-	-	-	-
Imports	-	2831	-	-	-	-	-	6017	-	-	-
Exports	-	-	-	-	-	-	-	-	-	-	-
Intl. Marine Bunkers	-	-	-	-	-	-	-	-	-	-	-
Stock Changes	-	-92	-	-	-	-	-	-108	-	-	-
DOMESTIC SUPPLY	-	3115	-	-	-	-	-	5921	-	-	-
Transfers	-	-	-	-	-	-	-	-	-	-	-
Statistical Differences	-	-	-	-	-	-	-	-	-	-	-
TRANSFORMATION	-	2443	-	-	-	-	-	5921	-	-	-
Electricity Plants	-	2443	-	-	-	-	-	-	-	-	-
CHP Plants	-	-	-	-	-	-	-	-	-	-	-
Heat Plants	-	-	-	-	-	-	-	-	-	-	-
Blast Furnaces/Gas Works	-	-	-	-	-	-	-	-	-	-	-
Coke/Pat. Fuel/BKB Plants	-	-	-	-	-	-	-	-	-	-	-
Petroleum Refineries	-	-	-	-	-	-	-	5921	-	-	-
Petrochemical Industry	-	-	-	-	-	-	-	-	-	-	-
Liquefaction	-	-	-	-	-	-	-	-	-	-	-
Other Transform. Sector	-	-	-	-	-	-	-	-	-	-	-
ENERGY SECTOR	-	-	-	-	-	-	-	-	-	-	-
Coal Mines	-	-	-	-	-	-	-	-	-	-	-
Oil and Gas Extraction	-	-	-	-	-	-	-	-	-	-	-
Petroleum Refineries	-	-	-	-	-	-	-	-	-	-	-
Electr., CHP+Heat Plants	-	-	-	-	-	-	-	-	-	-	-
Pumped Storage (Elec.)	-	-	-	-	-	-	-	-	-	-	-
Other Energy Sector	-	-	-	-	-	-	-	-	-	-	-
Distribution Losses	-	-	-	-	-	-	-	-	-	-	-
FINAL CONSUMPTION	-	672	-	-	-	-	-	-	-	-	-
INDUSTRY SECTOR	-	672	-	-	-	-	-	-	-	-	-
Iron and Steel	-	-	-	-	-	-	-	-	-	-	-
Chemical and Petrochem.	-	-	-	-	-	-	-	-	-	-	-
of which: Feedstocks	-	-	-	-	-	-	-	-	-	-	-
Non-Ferrous Metals	-	-	-	-	-	-	-	-	-	-	-
Non-Metallic Minerals	-	672	-	-	-	-	-	-	-	-	-
Transport Equipment	-	-	-	-	-	-	-	-	-	-	-
Machinery	-	-	-	-	-	-	-	-	-	-	-
Mining and Quarrying	-	-	-	-	-	-	-	-	-	-	-
Food and Tobacco	-	-	-	-	-	-	-	-	-	-	-
Paper, Pulp and Print	-	-	-	-	-	-	-	-	-	-	-
Wood and Wood Products	-	-	-	-	-	-	-	-	-	-	-
Construction	-	-	-	-	-	-	-	-	-	-	-
Textile and Leather	-	-	-	-	-	-	-	-	-	-	-
Non-specified	-	-	-	-	-	-	-	-	-	-	-
TRANSPORT SECTOR	-	-	-	-	-	-	-	-	-	-	-
Air	-	-	-	-	-	-	-	-	-	-	-
Road	-	-	-	-	-	-	-	-	-	-	-
Rail	-	-	-	-	-	-	-	-	-	-	-
Pipeline Transport	-	-	-	-	-	-	-	-	-	-	-
Internal Navigation	-	-	-	-	-	-	-	-	-	-	-
Non-specified	-	-	-	-	-	-	-	-	-	-	-
OTHER SECTORS	-	-	-	-	-	-	-	-	-	-	-
Agriculture	-	-	-	-	-	-	-	-	-	-	-
Comm. and Publ. Services	-	-	-	-	-	-	-	-	-	-	-
Residential	-	-	-	-	-	-	-	-	-	-	-
Non-specified	-	-	-	-	-	-	-	-	-	-	-
NON-ENERGY USE	-	-	-	-	-	-	-	-	-	-	-
in Industry/Trans./Energy	-	-	-	-	-	-	-	-	-	-	-
in Transport	-	-	-	-	-	-	-	-	-	-	-
in Other Sectors	-	-	-	-	-	-	-	-	-	-	-

Morocco / Maroc : 1997

	Oil cont. / *Pétrole cont.* (1000 tonnes)										
SUPPLY AND CONSUMPTION	Refinery Gas	LPG + Ethane	Motor Gasoline	Aviation Gasoline	Jet Fuel	Kerosene	Gas/ Diesel	Heavy Fuel Oil	Naphtha	Petrol. Coke	Other Prod.
APPROVISIONNEMENT ET DEMANDE	*Gaz de raffinerie*	*GPL + éthane*	*Essence moteur*	*Essence aviation*	*Carbu- réacteurs*	*Kérosène*	*Gazole*	*Fioul lourd*	*Naphta*	*Coke de pétrole*	*Autres prod.*
Production	112	239	409	-	241	62	2394	1711	303	-	259
From Other Sources	-	-	-	-	-	-	-	-	-	-	-
Imports	-	745	-	-	-	-	423	-	-	-	22
Exports	-	-	-	-	-	-	-	-	-199	-	-61
Intl. Marine Bunkers	-	-	-	-	-	-	-13	-	-	-	-
Stock Changes	-	-2	-29	-	-3	-1	-94	82	-12	-	3
DOMESTIC SUPPLY	**112**	**982**	**380**	**-**	**238**	**61**	**2710**	**1793**	**92**	**-**	**223**
Transfers	-	-	-	-	-	-	-	-	-	-	-
Statistical Differences	-	25	-	-	9	-4	-89	96	-92	-	6
TRANSFORMATION	**-**	**-**	**-**	**-**	**-**	**-**	**156**	**1187**	**-**	**-**	**-**
Electricity Plants	-	-	-	-	-	-	156	1187	-	-	-
CHP Plants	-	-	-	-	-	-	-	-	-	-	-
Heat Plants	-	-	-	-	-	-	-	-	-	-	-
Blast Furnaces/Gas Works	-	-	-	-	-	-	-	-	-	-	-
Coke/Pat. Fuel/BKB Plants	-	-	-	-	-	-	-	-	-	-	-
Petroleum Refineries	-	-	-	-	-	-	-	-	-	-	-
Petrochemical Industry	-	-	-	-	-	-	-	-	-	-	-
Liquefaction	-	-	-	-	-	-	-	-	-	-	-
Other Transform. Sector	-	-	-	-	-	-	-	-	-	-	-
ENERGY SECTOR	**112**	**-**	**-**	**-**	**-**	**-**	**-**	**-**	**-**	**-**	**-**
Coal Mines	-	-	-	-	-	-	-	-	-	-	-
Oil and Gas Extraction	-	-	-	-	-	-	-	-	-	-	-
Petroleum Refineries	112	-	-	-	-	-	-	-	-	-	-
Electr., CHP+Heat Plants	-	-	-	-	-	-	-	-	-	-	-
Pumped Storage (Elec.)	-	-	-	-	-	-	-	-	-	-	-
Other Energy Sector	-	-	-	-	-	-	-	-	-	-	-
Distribution Losses	-	-	-	-	-	-	-	-	-	-	-
FINAL CONSUMPTION	**-**	**1007**	**380**	**-**	**247**	**57**	**2465**	**702**	**-**	**-**	**229**
INDUSTRY SECTOR	**-**	**92**	**-**	**-**	**-**	**-**	**-**	**702**	**-**	**-**	**-**
Iron and Steel	-	-	-	-	-	-	-	-	-	-	-
Chemical and Petrochem.	-	-	-	-	-	-	-	-	-	-	-
of which: Feedstocks	-	-	-	-	-	-	-	-	-	-	-
Non-Ferrous Metals	-	-	-	-	-	-	-	-	-	-	-
Non-Metallic Minerals	-	-	-	-	-	-	-	-	-	-	-
Transport Equipment	-	-	-	-	-	-	-	-	-	-	-
Machinery	-	-	-	-	-	-	-	-	-	-	-
Mining and Quarrying	-	-	-	-	-	-	-	-	-	-	-
Food and Tobacco	-	-	-	-	-	-	-	-	-	-	-
Paper, Pulp and Print	-	-	-	-	-	-	-	-	-	-	-
Wood and Wood Products	-	-	-	-	-	-	-	-	-	-	-
Construction	-	-	-	-	-	-	-	-	-	-	-
Textile and Leather	-	-	-	-	-	-	-	-	-	-	-
Non-specified	-	92	-	-	-	-	-	702	-	-	-
TRANSPORT SECTOR	**-**	**-**	**380**	**-**	**247**	**-**	**148**	**-**	**-**	**-**	**-**
Air	-	-	-	-	247	-	-	-	-	-	-
Road	-	-	380	-	-	-	-	-	-	-	-
Rail	-	-	-	-	-	-	-	-	-	-	-
Pipeline Transport	-	-	-	-	-	-	-	-	-	-	-
Internal Navigation	-	-	-	-	-	-	148	-	-	-	-
Non-specified	-	-	-	-	-	-	-	-	-	-	-
OTHER SECTORS	**-**	**915**	**-**	**-**	**-**	**57**	**2317**	**-**	**-**	**-**	**-**
Agriculture	-	-	-	-	-	-	-	-	-	-	-
Comm. and Publ. Services	-	-	-	-	-	-	-	-	-	-	-
Residential	-	915	-	-	-	57	-	-	-	-	-
Non-specified	-	-	-	-	-	-	2317	-	-	-	-
NON-ENERGY USE	**-**	**-**	**-**	**-**	**-**	**-**	**-**	**-**	**-**	**-**	**229**
in Industry/Transf./Energy	-	-	-	-	-	-	-	-	-	-	229
in Transport	-	-	-	-	-	-	-	-	-	-	-
in Other Sectors	-	-	-	-	-	-	-	-	-	-	-

Morocco / Maroc : 1997

SUPPLY AND CONSUMPTION APPROVISIONNEMENT ET DEMANDE	Gas / Gaz (TJ)				Comb. Renew. & Waste / En. Re. Comb. & Déchets (TJ)				(GWh)	(TJ)
	Natural Gas Gaz naturel	Gas Works Usines à gaz	Coke Ovens Cokeries	Blast Furnaces Hauts fourneaux	Solid Biomass Biomasse solide	Gas/Liquids from Biomass Gaz/Liquides tirés de biomasse	Municipal Waste Déchets urbains	Industrial Waste Déchets industriels	Electricity Electricité	Heat Chaleur
Production	1200	-	-	-	17373	-	-	-	13871	-
From Other Sources	-	-	-	-	-	-	-	-	-	-
Imports	-	-	-	-	-	-	-	-	124	-
Exports	-	-	-	-	-	-	-	-	-	-
Intl. Marine Bunkers	-	-	-	-	-	-	-	-	-	-
Stock Changes	-	-	-	-	-	-	-	-	-	-
DOMESTIC SUPPLY	**1200**	**-**	**-**	**-**	**17373**	**-**	**-**	**-**	**13995**	**-**
Transfers	-	-	-	-	-	-	-	-	-	-
Statistical Differences	-	-	-	-	-	-	-	-	-	-
TRANSFORMATION	**-**	**-**	**-**	**-**	**-**	**-**	**-**	**-**	**-**	**-**
Electricity Plants	-	-	-	-	-	-	-	-	-	-
CHP Plants	-	-	-	-	-	-	-	-	-	-
Heat Plants	-	-	-	-	-	-	-	-	-	-
Blast Furnaces/Gas Works	-	-	-	-	-	-	-	-	-	-
Coke/Pat. Fuel/BKB Plants	-	-	-	-	-	-	-	-	-	-
Petroleum Refineries	-	-	-	-	-	-	-	-	-	-
Petrochemical Industry	-	-	-	-	-	-	-	-	-	-
Liquefaction	-	-	-	-	-	-	-	-	-	-
Other Transform. Sector	-	-	-	-	-	-	-	-	-	-
ENERGY SECTOR	**-**	**-**	**-**	**-**	**-**	**-**	**-**	**-**	**1892**	**-**
Coal Mines	-	-	-	-	-	-	-	-	75	-
Oil and Gas Extraction	-	-	-	-	-	-	-	-	26	-
Petroleum Refineries	-	-	-	-	-	-	-	-	194	-
Electr., CHP+Heat Plants	-	-	-	-	-	-	-	-	1597	-
Pumped Storage (Elec.)	-	-	-	-	-	-	-	-	-	-
Other Energy Sector	-	-	-	-	-	-	-	-	-	-
Distribution Losses	-	-	-	-	-	-	-	-	560	-
FINAL CONSUMPTION	**1200**	**-**	**-**	**-**	**17373**	**-**	**-**	**-**	**11543**	**-**
INDUSTRY SECTOR	**1200**	**-**	**-**	**-**	**2628**	**-**	**-**	**-**	**5233**	**-**
Iron and Steel	-	-	-	-	-	-	-	-	-	-
Chemical and Petrochem.	-	-	-	-	-	-	-	-	393	-
of which: Feedstocks	-	-	-	-	-	-	-	-	-	-
Non-Ferrous Metals	-	-	-	-	-	-	-	-	-	-
Non-Metallic Minerals	-	-	-	-	-	-	-	-	682	-
Transport Equipment	-	-	-	-	-	-	-	-	-	-
Machinery	-	-	-	-	-	-	-	-	263	-
Mining and Quarrying	1200	-	-	-	-	-	-	-	751	-
Food and Tobacco	-	-	-	-	-	-	-	-	527	-
Paper, Pulp and Print	-	-	-	-	-	-	-	-	396	-
Wood and Wood Products	-	-	-	-	-	-	-	-	-	-
Construction	-	-	-	-	-	-	-	-	218	-
Textile and Leather	-	-	-	-	-	-	-	-	623	-
Non-specified	-	-	-	-	2628	-	-	-	1380	-
TRANSPORT SECTOR	**-**	**-**	**-**	**-**	**-**	**-**	**-**	**-**	**214**	**-**
Air	-	-	-	-	-	-	-	-	-	-
Road	-	-	-	-	-	-	-	-	-	-
Rail	-	-	-	-	-	-	-	-	214	-
Pipeline Transport	-	-	-	-	-	-	-	-	-	-
Internal Navigation	-	-	-	-	-	-	-	-	-	-
Non-specified	-	-	-	-	-	-	-	-	-	-
OTHER SECTORS	**-**	**-**	**-**	**-**	**14745**	**-**	**-**	**-**	**6096**	**-**
Agriculture	-	-	-	-	-	-	-	-	542	-
Comm. and Publ. Services	-	-	-	-	-	-	-	-	1947	-
Residential	-	-	-	-	14745	-	-	-	3607	-
Non-specified	-	-	-	-	-	-	-	-	-	-
NON-ENERGY USE	**-**	**-**	**-**	**-**	**-**	**-**	**-**	**-**	**-**	**-**
in Industry/Transf./Energy	-	-	-	-	-	-	-	-	-	-
in Transport	-	-	-	-	-	-	-	-	-	-
in Other Sectors	-	-	-	-	-	-	-	-	-	-

Morocco / Maroc : 1998

SUPPLY AND CONSUMPTION	Coal / *Charbon* (1000 tonnes)							Oil / *Pétrole* (1000 tonnes)			
	Coking Coal	Other Bit. Coal	Sub-Bit. Coal	Lignite	Peat	Oven and Gas Coke	Pat. Fuel and BKB	Crude Oil	NGL	Feed-stocks	Additives
APPROVISIONNEMENT ET DEMANDE	*Charbon à coke*	*Autres charb. bit.*	*Charbon sous-bit.*	*Lignite*	*Tourbe*	*Coke de four/gaz*	*Agg./briq. de lignite*	*Pétrole brut*	*LGN*	*Produits d'aliment.*	*Additifs*
Production	-	269	-	-	-	-	-	12	-	-	-
From Other Sources	-	-	-	-	-	-	-	-	-	-	-
Imports	-	3205	-	-	-	-	-	5957	-	-	-
Exports	-	-	-	-	-	-	-	-	-	-	-
Intl. Marine Bunkers	-	-	-	-	-	-	-	-	-	-	-
Stock Changes	-	413	-	-	-	-	-	143	-	-	-
DOMESTIC SUPPLY	-	**3887**	-	-	-	-	-	**6112**	-	-	-
Transfers	-	-	-	-	-	-	-	-	-	-	-
Statistical Differences	-	-	-	-	-	-	-	-	-	-	-
TRANSFORMATION	-	**3165**	-	-	-	-	-	**6112**	-	-	-
Electricity Plants	-	3165	-	-	-	-	-	-	-	-	-
CHP Plants	-	-	-	-	-	-	-	-	-	-	-
Heat Plants	-	-	-	-	-	-	-	-	-	-	-
Blast Furnaces/Gas Works	-	-	-	-	-	-	-	-	-	-	-
Coke/Pat. Fuel/BKB Plants	-	-	-	-	-	-	-	-	-	-	-
Petroleum Refineries	-	-	-	-	-	-	-	6112	-	-	-
Petrochemical Industry	-	-	-	-	-	-	-	-	-	-	-
Liquefaction	-	-	-	-	-	-	-	-	-	-	-
Other Transform. Sector	-	-	-	-	-	-	-	-	-	-	-
ENERGY SECTOR	-	-	-	-	-	-	-	-	-	-	-
Coal Mines	-	-	-	-	-	-	-	-	-	-	-
Oil and Gas Extraction	-	-	-	-	-	-	-	-	-	-	-
Petroleum Refineries	-	-	-	-	-	-	-	-	-	-	-
Electr., CHP+Heat Plants	-	-	-	-	-	-	-	-	-	-	-
Pumped Storage (Elec.)	-	-	-	-	-	-	-	-	-	-	-
Other Energy Sector	-	-	-	-	-	-	-	-	-	-	-
Distribution Losses	-	-	-	-	-	-	-	-	-	-	-
FINAL CONSUMPTION	-	**722**	-	-	-	-	-	-	-	-	-
INDUSTRY SECTOR	-	**722**	-	-	-	-	-	-	-	-	-
Iron and Steel	-	-	-	-	-	-	-	-	-	-	-
Chemical and Petrochem.	-	-	-	-	-	-	-	-	-	-	-
of which: Feedstocks	-	-	-	-	-	-	-	-	-	-	-
Non-Ferrous Metals	-	-	-	-	-	-	-	-	-	-	-
Non-Metallic Minerals	-	722	-	-	-	-	-	-	-	-	-
Transport Equipment	-	-	-	-	-	-	-	-	-	-	-
Machinery	-	-	-	-	-	-	-	-	-	-	-
Mining and Quarrying	-	-	-	-	-	-	-	-	-	-	-
Food and Tobacco	-	-	-	-	-	-	-	-	-	-	-
Paper, Pulp and Print	-	-	-	-	-	-	-	-	-	-	-
Wood and Wood Products	-	-	-	-	-	-	-	-	-	-	-
Construction	-	-	-	-	-	-	-	-	-	-	-
Textile and Leather	-	-	-	-	-	-	-	-	-	-	-
Non-specified	-	-	-	-	-	-	-	-	-	-	-
TRANSPORT SECTOR	-	-	-	-	-	-	-	-	-	-	-
Air	-	-	-	-	-	-	-	-	-	-	-
Road	-	-	-	-	-	-	-	-	-	-	-
Rail	-	-	-	-	-	-	-	-	-	-	-
Pipeline Transport	-	-	-	-	-	-	-	-	-	-	-
Internal Navigation	-	-	-	-	-	-	-	-	-	-	-
Non-specified	-	-	-	-	-	-	-	-	-	-	-
OTHER SECTORS	-	-	-	-	-	-	-	-	-	-	-
Agriculture	-	-	-	-	-	-	-	-	-	-	-
Comm. and Publ. Services	-	-	-	-	-	-	-	-	-	-	-
Residential	-	-	-	-	-	-	-	-	-	-	-
Non-specified	-	-	-	-	-	-	-	-	-	-	-
NON-ENERGY USE	-	-	-	-	-	-	-	-	-	-	-
in Industry/Trans./Energy	-	-	-	-	-	-	-	-	-	-	-
in Transport	-	-	-	-	-	-	-	-	-	-	-
in Other Sectors	-	-	-	-	-	-	-	-	-	-	-

Morocco / Maroc : 1998

	Oil cont. / *Pétrole cont.* (1000 tonnes)										
SUPPLY AND CONSUMPTION *APPROVISIONNEMENT ET DEMANDE*	Refinery Gas *Gaz de raffinerie*	LPG + Ethane *GPL + éthane*	Motor Gasoline *Essence moteur*	Aviation Gasoline *Essence aviation*	Jet Fuel *Carbu- réacteurs*	Kerosene *Kérosène*	Gas/ Diesel *Gazole*	Heavy Fuel Oil *Fioul lourd*	Naphtha *Naphta*	Petrol. Coke *Coke de pétrole*	Other Prod. *Autres prod.*
Production	115	252	354	-	272	79	2236	1918	385	-	259
From Other Sources	-	-	-	-	-	-	-	-	-	-	-
Imports	-	796	-	-	-	-	-	-	-	-	22
Exports	-	-	-	-	-	-	-	-	-330	-	-43
Intl. Marine Bunkers	-	-	-	-	-	-	-13	-	-	-	-
Stock Changes	-	-3	21	-	-	1	95	-87	-40	-	-5
DOMESTIC SUPPLY	115	1045	375	-	272	80	2318	1831	15	-	233
Transfers	-	-	-	-	-	-	-	-	-	-	-
Statistical Differences	-	-9	10	-	8	-	344	21	-15	-	3
TRANSFORMATION	-	-	-	-	-	-	167	1063	-	-	-
Electricity Plants	-	-	-	-	-	-	167	1063	-	-	-
CHP Plants	-	-	-	-	-	-	-	-	-	-	-
Heat Plants	-	-	-	-	-	-	-	-	-	-	-
Blast Furnaces/Gas Works	-	-	-	-	-	-	-	-	-	-	-
Coke/Pat. Fuel/BKB Plants	-	-	-	-	-	-	-	-	-	-	-
Petroleum Refineries	-	-	-	-	-	-	-	-	-	-	-
Petrochemical Industry	-	-	-	-	-	-	-	-	-	-	-
Liquefaction	-	-	-	-	-	-	-	-	-	-	-
Other Transform. Sector	-	-	-	-	-	-	-	-	-	-	-
ENERGY SECTOR	115	-	-	-	-	-	-	-	-	-	-
Coal Mines	-	-	-	-	-	-	-	-	-	-	-
Oil and Gas Extraction	-	-	-	-	-	-	-	-	-	-	-
Petroleum Refineries	115	-	-	-	-	-	-	-	-	-	-
Electr., CHP+Heat Plants	-	-	-	-	-	-	-	-	-	-	-
Pumped Storage (Elec.)	-	-	-	-	-	-	-	-	-	-	-
Other Energy Sector	-	-	-	-	-	-	-	-	-	-	-
Distribution Losses	-	-	-	-	-	-	-	-	-	-	-
FINAL CONSUMPTION	-	1036	385	-	280	80	2495	789	-	-	236
INDUSTRY SECTOR	-	95	-	-	-	-	-	789	-	-	-
Iron and Steel	-	-	-	-	-	-	-	-	-	-	-
Chemical and Petrochem.	-	-	-	-	-	-	-	-	-	-	-
of which: Feedstocks	-	-	-	-	-	-	-	-	-	-	-
Non-Ferrous Metals	-	-	-	-	-	-	-	-	-	-	-
Non-Metallic Minerals	-	-	-	-	-	-	-	-	-	-	-
Transport Equipment	-	-	-	-	-	-	-	-	-	-	-
Machinery	-	-	-	-	-	-	-	-	-	-	-
Mining and Quarrying	-	-	-	-	-	-	-	-	-	-	-
Food and Tobacco	-	-	-	-	-	-	-	-	-	-	-
Paper, Pulp and Print	-	-	-	-	-	-	-	-	-	-	-
Wood and Wood Products	-	-	-	-	-	-	-	-	-	-	-
Construction	-	-	-	-	-	-	-	-	-	-	-
Textile and Leather	-	-	-	-	-	-	-	-	-	-	-
Non-specified	-	95	-	-	-	-	-	789	-	-	-
TRANSPORT SECTOR	-	-	385	-	280	-	150	-	-	-	-
Air	-	-	-	-	280	-	-	-	-	-	-
Road	-	-	385	-	-	-	-	-	-	-	-
Rail	-	-	-	-	-	-	-	-	-	-	-
Pipeline Transport	-	-	-	-	-	-	-	-	-	-	-
Internal Navigation	-	-	-	-	-	-	150	-	-	-	-
Non-specified	-	-	-	-	-	-	-	-	-	-	-
OTHER SECTORS	-	941	-	-	-	80	2345	-	-	-	-
Agriculture	-	-	-	-	-	-	-	-	-	-	-
Comm. and Publ. Services	-	-	-	-	-	-	-	-	-	-	-
Residential	-	941	-	-	-	80	-	-	-	-	-
Non-specified	-	-	-	-	-	-	2345	-	-	-	-
NON-ENERGY USE	-	-	-	-	-	-	-	-	-	-	236
in Industry/Transf./Energy	-	-	-	-	-	-	-	-	-	-	236
in Transport	-	-	-	-	-	-	-	-	-	-	-
in Other Sectors	-	-	-	-	-	-	-	-	-	-	-

Morocco / Maroc : 1998

SUPPLY AND CONSUMPTION *APPROVISIONNEMENT ET DEMANDE*	Gas / *Gaz* (TJ)				Comb. Renew. & Waste / *En. Re. Comb. & Déchets* (TJ)				(GWh)	(TJ)
	Natural Gas *Gaz naturel*	Gas Works *Usines à gaz*	Coke Ovens *Cokeries*	Blast Furnaces *Hauts fourneaux*	Solid Biomass *Biomasse solide*	Gas/Liquids from Biomass *Gaz/Liquides tirés de biomasse*	Municipal Waste *Déchets urbains*	Industrial Waste *Déchets industriels*	Electricity *Electricité*	Heat *Chaleur*
Production	1317	-	-	-	17669	-	-	-	14136	-
From Other Sources	-	-	-	-	-	-	-	-	-	-
Imports	-	-	-	-	-	-	-	-	705	-
Exports	-	-	-	-	-	-	-	-	-	-
Intl. Marine Bunkers	-	-	-	-	-	-	-	-	-	-
Stock Changes	-	-	-	-	-	-	-	-	-	-
DOMESTIC SUPPLY	**1317**	**-**	**-**	**-**	**17669**	**-**	**-**	**-**	**14841**	**-**
Transfers	-	-	-	-	-	-	-	-	-	-
Statistical Differences	-	-	-	-	-	-	-	-	-	-
TRANSFORMATION	**-**	**-**	**-**	**-**	**-**	**-**	**-**	**-**	**-**	**-**
Electricity Plants	-	-	-	-	-	-	-	-	-	-
CHP Plants	-	-	-	-	-	-	-	-	-	-
Heat Plants	-	-	-	-	-	-	-	-	-	-
Blast Furnaces/Gas Works	-	-	-	-	-	-	-	-	-	-
Coke/Pat. Fuel/BKB Plants	-	-	-	-	-	-	-	-	-	-
Petroleum Refineries	-	-	-	-	-	-	-	-	-	-
Petrochemical Industry	-	-	-	-	-	-	-	-	-	-
Liquefaction	-	-	-	-	-	-	-	-	-	-
Other Transform. Sector	-	-	-	-	-	-	-	-	-	-
ENERGY SECTOR	**-**	**-**	**-**	**-**	**-**	**-**	**-**	**-**	**1967**	**-**
Coal Mines	-	-	-	-	-	-	-	-	80	-
Oil and Gas Extraction	-	-	-	-	-	-	-	-	28	-
Petroleum Refineries	-	-	-	-	-	-	-	-	206	-
Electr., CHP+Heat Plants	-	-	-	-	-	-	-	-	1653	-
Pumped Storage (Elec.)	-	-	-	-	-	-	-	-	-	-
Other Energy Sector	-	-	-	-	-	-	-	-	-	-
Distribution Losses	-	-	-	-	-	-	-	-	571	-
FINAL CONSUMPTION	**1317**	**-**	**-**	**-**	**17669**	**-**	**-**	**-**	**12303**	**-**
INDUSTRY SECTOR	**1317**	**-**	**-**	**-**	**2673**	**-**	**-**	**-**	**5573**	**-**
Iron and Steel	-	-	-	-	-	-	-	-	-	-
Chemical and Petrochem.	-	-	-	-	-	-	-	-	419	-
of which: Feedstocks	-	-	-	-	-	-	-	-	-	-
Non-Ferrous Metals	-	-	-	-	-	-	-	-	-	-
Non-Metallic Minerals	-	-	-	-	-	-	-	-	727	-
Transport Equipment	-	-	-	-	-	-	-	-	-	-
Machinery	-	-	-	-	-	-	-	-	281	-
Mining and Quarrying	1317	-	-	-	-	-	-	-	801	-
Food and Tobacco	-	-	-	-	-	-	-	-	562	-
Paper, Pulp and Print	-	-	-	-	-	-	-	-	422	-
Wood and Wood Products	-	-	-	-	-	-	-	-	-	-
Construction	-	-	-	-	-	-	-	-	233	-
Textile and Leather	-	-	-	-	-	-	-	-	665	-
Non-specified	-	-	-	-	2673	-	-	-	1463	-
TRANSPORT SECTOR	**-**	**-**	**-**	**-**	**-**	**-**	**-**	**-**	**228**	**-**
Air	-	-	-	-	-	-	-	-	-	-
Road	-	-	-	-	-	-	-	-	-	-
Rail	-	-	-	-	-	-	-	-	228	-
Pipeline Transport	-	-	-	-	-	-	-	-	-	-
Internal Navigation	-	-	-	-	-	-	-	-	-	-
Non-specified	-	-	-	-	-	-	-	-	-	-
OTHER SECTORS	**-**	**-**	**-**	**-**	**14996**	**-**	**-**	**-**	**6502**	**-**
Agriculture	-	-	-	-	-	-	-	-	578	-
Comm. and Publ. Services	-	-	-	-	-	-	-	-	2077	-
Residential	-	-	-	-	14996	-	-	-	3847	-
Non-specified	-	-	-	-	-	-	-	-	-	-
NON-ENERGY USE	**-**	**-**	**-**	**-**	**-**	**-**	**-**	**-**	**-**	**-**
in Industry/Transf./Energy	-	-	-	-	-	-	-	-	-	-
in Transport	-	-	-	-	-	-	-	-	-	-
in Other Sectors	-	-	-	-	-	-	-	-	-	-

Mozambique

SUPPLY AND CONSUMPTION 1997	Coking Coal	Other Bit. Coal	Sub-Bit. Coal	Lignite	Peat	Oven and Gas Coke	Pat. Fuel and BKB	Crude Oil	NGL	Feed-stocks	Additives
	Coal (1000 tonnes)							Oil (1000 tonnes)			
Production	-	18	-	-	-	-	-	-	-	-	-
Imports	-	-	-	-	-	-	-	-	-	-	-
Exports	-	-18	-	-	-	-	-	-	-	-	-
Intl. Marine Bunkers	-	-	-	-	-	-	-	-	-	-	-
Stock Changes	-	-	-	-	-	-	-	-	-	-	-
DOMESTIC SUPPLY	-	-	-	-	-	-	-	-	-	-	-
Transfers and Stat. Diff.	-	-	-	-	-	-	-	-	-	-	-
TRANSFORMATION	-	-	-	-	-	-	-	-	-	-	-
Electricity and CHP Plants	-	-	-	-	-	-	-	-	-	-	-
Petroleum Refineries	-	-	-	-	-	-	-	-	-	-	-
Other Transform. Sector	-	-	-	-	-	-	-	-	-	-	-
ENERGY SECTOR	-	-	-	-	-	-	-	-	-	-	-
DISTRIBUTION LOSSES	-	-	-	-	-	-	-	-	-	-	-
FINAL CONSUMPTION	-	-	-	-	-	-	-	-	-	-	-
INDUSTRY SECTOR	-	-	-	-	-	-	-	-	-	-	-
Iron and Steel	-	-	-	-	-	-	-	-	-	-	-
Chemical and Petrochem.	-	-	-	-	-	-	-	-	-	-	-
Non-Metallic Minerals	-	-	-	-	-	-	-	-	-	-	-
Non-specified	-	-	-	-	-	-	-	-	-	-	-
TRANSPORT SECTOR	-	-	-	-	-	-	-	-	-	-	-
Air	-	-	-	-	-	-	-	-	-	-	-
Road	-	-	-	-	-	-	-	-	-	-	-
Non-specified	-	-	-	-	-	-	-	-	-	-	-
OTHER SECTORS	-	-	-	-	-	-	-	-	-	-	-
Agriculture	-	-	-	-	-	-	-	-	-	-	-
Comm. and Publ. Services	-	-	-	-	-	-	-	-	-	-	-
Residential	-	-	-	-	-	-	-	-	-	-	-
Non-specified	-	-	-	-	-	-	-	-	-	-	-
NON-ENERGY USE	-	-	-	-	-	-	-	-	-	-	-

APPROVISIONNEMENT ET DEMANDE 1998	Charbon à coke	Autres charb. bit.	Charbon sous-bit.	Lignite	Tourbe	Coke de four/gaz	Agg./briq. de lignite	Pétrole brut	LGN	Produits d'aliment.	Additifs
	Charbon (1000 tonnes)							Pétrole (1000 tonnes)			
Production	-	18	-	-	-	-	-	-	-	-	-
Imports	-	-	-	-	-	-	-	-	-	-	-
Exports	-	-18	-	-	-	-	-	-	-	-	-
Intl. Marine Bunkers	-	-	-	-	-	-	-	-	-	-	-
Stock Changes	-	-	-	-	-	-	-	-	-	-	-
DOMESTIC SUPPLY	-	-	-	-	-	-	-	-	-	-	-
Transfers and Stat. Diff.	-	-	-	-	-	-	-	-	-	-	-
TRANSFORMATION	-	-	-	-	-	-	-	-	-	-	-
Electricity and CHP Plants	-	-	-	-	-	-	-	-	-	-	-
Petroleum Refineries	-	-	-	-	-	-	-	-	-	-	-
Other Transform. Sector	-	-	-	-	-	-	-	-	-	-	-
ENERGY SECTOR	-	-	-	-	-	-	-	-	-	-	-
DISTRIBUTION LOSSES	-	-	-	-	-	-	-	-	-	-	-
FINAL CONSUMPTION	-	-	-	-	-	-	-	-	-	-	-
INDUSTRY SECTOR	-	-	-	-	-	-	-	-	-	-	-
Iron and Steel	-	-	-	-	-	-	-	-	-	-	-
Chemical and Petrochem.	-	-	-	-	-	-	-	-	-	-	-
Non-Metallic Minerals	-	-	-	-	-	-	-	-	-	-	-
Non-specified	-	-	-	-	-	-	-	-	-	-	-
TRANSPORT SECTOR	-	-	-	-	-	-	-	-	-	-	-
Air	-	-	-	-	-	-	-	-	-	-	-
Road	-	-	-	-	-	-	-	-	-	-	-
Non-specified	-	-	-	-	-	-	-	-	-	-	-
OTHER SECTORS	-	-	-	-	-	-	-	-	-	-	-
Agriculture	-	-	-	-	-	-	-	-	-	-	-
Comm. and Publ. Services	-	-	-	-	-	-	-	-	-	-	-
Residential	-	-	-	-	-	-	-	-	-	-	-
Non-specified	-	-	-	-	-	-	-	-	-	-	-
NON-ENERGY USE	-	-	-	-	-	-	-	-	-	-	-

Mozambique

SUPPLY AND CONSUMPTION 1997	\multicolumn{11}{c}{Oil cont. (1000 tonnes)}
	Refinery Gas	LPG + Ethane	Motor Gasoline	Aviation Gasoline	Jet Fuel	Kerosene	Gas/ Diesel	Heavy Fuel Oil	Naphtha	Petrol. Coke	Other Prod.
Production	-	-	-	-	-	-	-	-	-	-	-
Imports	-	5	44	-	28	24	264	15	-	-	10
Exports	-	-	-	-	-2	-	-	-	-	-	-
Intl. Marine Bunkers	-	-	-	-	-	-	-2	-1	-	-	-
Stock Changes	-	-	-2	-	-2	1	-	-	-	-	2
DOMESTIC SUPPLY	-	5	42	-	24	25	262	14	-	-	12
Transfers and Stat. Diff.	-	-	1	-	1	-1	-29	1	-	-	-5
TRANSFORMATION	-	-	-	-	-	-	3	-	-	-	-
Electricity and CHP Plants	-	-	-	-	-	-	3	-	-	-	-
Petroleum Refineries	-	-	-	-	-	-	-	-	-	-	-
Other Transform. Sector	-	-	-	-	-	-	-	-	-	-	-
ENERGY SECTOR	-	-	-	-	-	-	-	-	-	-	-
DISTRIBUTION LOSSES	-	-	-	-	-	-	-	-	-	-	-
FINAL CONSUMPTION	-	5	43	-	25	24	230	15	-	-	7
INDUSTRY SECTOR	-	-	-	-	-	-	6	10	-	-	-
Iron and Steel	-	-	-	-	-	-	-	-	-	-	-
Chemical and Petrochem.	-	-	-	-	-	-	-	-	-	-	-
Non-Metallic Minerals	-	-	-	-	-	-	-	-	-	-	-
Non-specified	-	-	-	-	-	-	6	10	-	-	-
TRANSPORT SECTOR	-	-	43	-	25	-	215	-	-	-	-
Air	-	-	-	-	25	-	-	-	-	-	-
Road	-	-	43	-	-	-	192	-	-	-	-
Non-specified	-	-	-	-	-	-	23	-	-	-	-
OTHER SECTORS	-	5	-	-	-	24	9	5	-	-	-
Agriculture	-	-	-	-	-	-	2	-	-	-	-
Comm. and Publ. Services	-	-	-	-	-	-	7	-	-	-	-
Residential	-	5	-	-	-	24	-	-	-	-	-
Non-specified	-	-	-	-	-	-	-	5	-	-	-
NON-ENERGY USE	-	-	-	-	-	-	-	-	-	-	7

APPROVISIONNEMENT ET DEMANDE 1998	\multicolumn{11}{c}{Pétrole cont. (1000 tonnes)}
	Gaz de raffinerie	GPL + éthane	Essence moteur	Essence aviation	Carbu- réacteurs	Kérosène	Gazole	Fioul lourd	Naphta	Coke de pétrole	Autres prod.
Production	-	-	-	-	-	-	-	-	-	-	-
Imports	-	6	52	-	26	30	235	11	-	-	10
Exports	-	-	-	-	-2	-	-	-	-	-	-
Intl. Marine Bunkers	-	-	-	-	-	-	-1	-	-	-	-
Stock Changes	-	-	1	-	2	-	-1	1	-	-	2
DOMESTIC SUPPLY	-	6	53	-	26	30	233	12	-	-	12
Transfers and Stat. Diff.	-	-	-4	-	1	-	30	-	-	-	-5
TRANSFORMATION	-	-	-	-	-	-	3	-	-	-	-
Electricity and CHP Plants	-	-	-	-	-	-	3	-	-	-	-
Petroleum Refineries	-	-	-	-	-	-	-	-	-	-	-
Other Transform. Sector	-	-	-	-	-	-	-	-	-	-	-
ENERGY SECTOR	-	-	-	-	-	-	-	-	-	-	-
DISTRIBUTION LOSSES	-	-	-	-	-	-	-	-	-	-	-
FINAL CONSUMPTION	-	6	49	-	27	30	260	12	-	-	7
INDUSTRY SECTOR	-	1	-	-	-	-	31	7	-	-	-
Iron and Steel	-	-	-	-	-	-	-	-	-	-	-
Chemical and Petrochem.	-	-	-	-	-	-	-	-	-	-	-
Non-Metallic Minerals	-	-	-	-	-	-	-	-	-	-	-
Non-specified	-	1	-	-	-	-	31	7	-	-	-
TRANSPORT SECTOR	-	-	49	-	27	-	215	-	-	-	-
Air	-	-	-	-	27	-	-	-	-	-	-
Road	-	-	49	-	-	-	192	-	-	-	-
Non-specified	-	-	-	-	-	-	23	-	-	-	-
OTHER SECTORS	-	5	-	-	-	30	14	5	-	-	-
Agriculture	-	-	-	-	-	-	5	-	-	-	-
Comm. and Publ. Services	-	-	-	-	-	-	9	-	-	-	-
Residential	-	5	-	-	-	30	-	-	-	-	-
Non-specified	-	-	-	-	-	-	-	5	-	-	-
NON-ENERGY USE	-	-	-	-	-	-	-	-	-	-	7

Mozambique

SUPPLY AND CONSUMPTION 1997	Natural Gas	Gas Works	Coke Ovens	Blast Furnaces	Solid Biomass	Gas/Liquids from Biomass	Municipal Waste	Industrial Waste	Electricity	Heat
	Gas (TJ)				Comb. Renew. & Waste (TJ)				(GWh)	(TJ)
Production	15	-	-	-	265800	-	-	-	1005	-
Imports	-	-	-	-	-	-	-	-	686	-
Exports	-	-	-	-	-	-	-	-	-483	-
Intl. Marine Bunkers	-	-	-	-	-	-	-	-	-	-
Stock Changes	-	-	-	-	-	-	-	-	-	-
DOMESTIC SUPPLY	15	-	-	-	265800	-	-	-	1208	-
Transfers and Stat. Diff.	-	-	-	-	-	-	-	-	-42	-
TRANSFORMATION	15	-	-	-	-	-	-	-	-	-
Electricity and CHP Plants	15	-	-	-	-	-	-	-	-	-
Petroleum Refineries	-	-	-	-	-	-	-	-	-	-
Other Transform. Sector	-	-	-	-	-	-	-	-	-	-
ENERGY SECTOR	-	-	-	-	-	-	-	-	72	-
DISTRIBUTION LOSSES	-	-	-	-	-	-	-	-	315	-
FINAL CONSUMPTION	-	-	-	-	265800	-	-	-	779	-
INDUSTRY SECTOR	-	-	-	-	63947	-	-	-	291	-
Iron and Steel	-	-	-	-	-	-	-	-	-	-
Chemical and Petrochem.	-	-	-	-	-	-	-	-	-	-
Non-Metallic Minerals	-	-	-	-	-	-	-	-	-	-
Non-specified	-	-	-	-	63947	-	-	-	291	-
TRANSPORT SECTOR	-	-	-	-	-	-	-	-	-	-
Air	-	-	-	-	-	-	-	-	-	-
Road	-	-	-	-	-	-	-	-	-	-
Non-specified	-	-	-	-	-	-	-	-	-	-
OTHER SECTORS	-	-	-	-	201853	-	-	-	488	-
Agriculture	-	-	-	-	-	-	-	-	-	-
Comm. and Publ. Services	-	-	-	-	-	-	-	-	184	-
Residential	-	-	-	-	201853	-	-	-	304	-
Non-specified	-	-	-	-	-	-	-	-	-	-
NON-ENERGY USE	-	-	-	-	-	-	-	-	-	-

APPROVISIONNEMENT ET DEMANDE 1998	Gaz naturel	Usines à gaz	Cokeries	Hauts fourneaux	Biomasse solide	Gaz/Liquides tirés de biomasse	Déchets urbains	Déchets industriels	Electricité	Chaleur
	Gaz (TJ)				En. Re. Comb. & Déchets (TJ)				(GWh)	(TJ)
Production	21	-	-	-	265800	-	-	-	6864	-
Imports	-	-	-	-	-	-	-	-	350	-
Exports	-	-	-	-	-	-	-	-	-5677	-
Intl. Marine Bunkers	-	-	-	-	-	-	-	-	-	-
Stock Changes	-	-	-	-	-	-	-	-	-	-
DOMESTIC SUPPLY	21	-	-	-	265800	-	-	-	1537	-
Transfers and Stat. Diff.	-	-	-	-	-	-	-	-	178	-
TRANSFORMATION	21	-	-	-	-	-	-	-	-	-
Electricity and CHP Plants	21	-	-	-	-	-	-	-	-	-
Petroleum Refineries	-	-	-	-	-	-	-	-	-	-
Other Transform. Sector	-	-	-	-	-	-	-	-	-	-
ENERGY SECTOR	-	-	-	-	-	-	-	-	86	-
DISTRIBUTION LOSSES	-	-	-	-	-	-	-	-	712	-
FINAL CONSUMPTION	-	-	-	-	265800	-	-	-	917	-
INDUSTRY SECTOR	-	-	-	-	63947	-	-	-	376	-
Iron and Steel	-	-	-	-	-	-	-	-	-	-
Chemical and Petrochem.	-	-	-	-	-	-	-	-	-	-
Non-Metallic Minerals	-	-	-	-	-	-	-	-	-	-
Non-specified	-	-	-	-	63947	-	-	-	376	-
TRANSPORT SECTOR	-	-	-	-	-	-	-	-	-	-
Air	-	-	-	-	-	-	-	-	-	-
Road	-	-	-	-	-	-	-	-	-	-
Non-specified	-	-	-	-	-	-	-	-	-	-
OTHER SECTORS	-	-	-	-	201853	-	-	-	541	-
Agriculture	-	-	-	-	-	-	-	-	-	-
Comm. and Publ. Services	-	-	-	-	-	-	-	-	204	-
Residential	-	-	-	-	201853	-	-	-	337	-
Non-specified	-	-	-	-	-	-	-	-	-	-
NON-ENERGY USE	-	-	-	-	-	-	-	-	-	-

Myanmar

SUPPLY AND CONSUMPTION 1997	Coal (1000 tonnes)							Oil (1000 tonnes)			
	Coking Coal	Other Bit. Coal	Sub-Bit. Coal	Lignite	Peat	Oven and Gas Coke	Pat. Fuel and BKB	Crude Oil	NGL	Feed-stocks	Additives
Production	-	10	-	20	-	-	-	395	1	-	-
Imports	-	1	-	-	-	-	-	647	-	-	-
Exports	-	-	-	-	-	-	-	-	-	-	-
Intl. Marine Bunkers	-	-	-	-	-	-	-	-	-	-	-
Stock Changes	-	-	-	-	-	-	-	-1	-	-	-
DOMESTIC SUPPLY	-	11	-	20	-	-	-	1041	1	-	-
Transfers and Stat. Diff.	-	-	-	-	-	-	-	-7	-	-	-
TRANSFORMATION	-	-	-	-	-	-	-	1034	1	-	-
Electricity and CHP Plants	-	-	-	-	-	-	-	-	-	-	-
Petroleum Refineries	-	-	-	-	-	-	-	1034	1	-	-
Other Transform. Sector	-	-	-	-	-	-	-	-	-	-	-
ENERGY SECTOR	-	-	-	-	-	-	-	-	-	-	-
DISTRIBUTION LOSSES	-	-	-	-	-	-	-	-	-	-	-
FINAL CONSUMPTION	-	11	-	20	-	-	-	-	-	-	-
INDUSTRY SECTOR	-	11	-	20	-	-	-	-	-	-	-
Iron and Steel	-	-	-	-	-	-	-	-	-	-	-
Chemical and Petrochem.	-	-	-	-	-	-	-	-	-	-	-
Non-Metallic Minerals	-	-	-	-	-	-	-	-	-	-	-
Non-specified	-	11	-	20	-	-	-	-	-	-	-
TRANSPORT SECTOR	-	-	-	-	-	-	-	-	-	-	-
Air	-	-	-	-	-	-	-	-	-	-	-
Road	-	-	-	-	-	-	-	-	-	-	-
Non-specified	-	-	-	-	-	-	-	-	-	-	-
OTHER SECTORS	-	-	-	-	-	-	-	-	-	-	-
Agriculture	-	-	-	-	-	-	-	-	-	-	-
Comm. and Publ. Services	-	-	-	-	-	-	-	-	-	-	-
Residential	-	-	-	-	-	-	-	-	-	-	-
Non-specified	-	-	-	-	-	-	-	-	-	-	-
NON-ENERGY USE	-	-	-	-	-	-	-	-	-	-	-

APPROVISIONNEMENT ET DEMANDE 1998	Charbon (1000 tonnes)							Pétrole (1000 tonnes)			
	Charbon à coke	Autres charb. bit.	Charbon sous-bit.	Lignite	Tourbe	Coke de four/gaz	Agg./briq. de lignite	Pétrole brut	LGN	Produits d'aliment.	Additifs
Production	-	10	-	21	-	-	-	382	1	-	-
Imports	-	2	-	-	-	-	-	685	-	-	-
Exports	-	-	-	-	-	-	-	-	-	-	-
Intl. Marine Bunkers	-	-	-	-	-	-	-	-	-	-	-
Stock Changes	-	-	-	-	-	-	-	-32	-	-	-
DOMESTIC SUPPLY	-	12	-	21	-	-	-	1035	1	-	-
Transfers and Stat. Diff.	-	-	-	-	-	-	-	-6	-	-	-
TRANSFORMATION	-	-	-	-	-	-	-	1029	1	-	-
Electricity and CHP Plants	-	-	-	-	-	-	-	-	-	-	-
Petroleum Refineries	-	-	-	-	-	-	-	1029	1	-	-
Other Transform. Sector	-	-	-	-	-	-	-	-	-	-	-
ENERGY SECTOR	-	-	-	-	-	-	-	-	-	-	-
DISTRIBUTION LOSSES	-	-	-	-	-	-	-	-	-	-	-
FINAL CONSUMPTION	-	12	-	21	-	-	-	-	-	-	-
INDUSTRY SECTOR	-	12	-	21	-	-	-	-	-	-	-
Iron and Steel	-	-	-	-	-	-	-	-	-	-	-
Chemical and Petrochem.	-	-	-	-	-	-	-	-	-	-	-
Non-Metallic Minerals	-	-	-	-	-	-	-	-	-	-	-
Non-specified	-	12	-	21	-	-	-	-	-	-	-
TRANSPORT SECTOR	-	-	-	-	-	-	-	-	-	-	-
Air	-	-	-	-	-	-	-	-	-	-	-
Road	-	-	-	-	-	-	-	-	-	-	-
Non-specified	-	-	-	-	-	-	-	-	-	-	-
OTHER SECTORS	-	-	-	-	-	-	-	-	-	-	-
Agriculture	-	-	-	-	-	-	-	-	-	-	-
Comm. and Publ. Services	-	-	-	-	-	-	-	-	-	-	-
Residential	-	-	-	-	-	-	-	-	-	-	-
Non-specified	-	-	-	-	-	-	-	-	-	-	-
NON-ENERGY USE	-	-	-	-	-	-	-	-	-	-	-

Myanmar

SUPPLY AND CONSUMPTION 1997	Refinery Gas	LPG + Ethane	Motor Gasoline	Aviation Gasoline	Jet Fuel	Kerosene	Gas/ Diesel	Heavy Fuel Oil	Naphtha	Petrol. Coke	Other Prod.
Production	38	11	255	-	56	1	465	103	-	26	28
Imports	-	-	-	-	-	-	289	-	-	-	-
Exports	-	-1	-	-	-	-	-	-	-	-9	-
Intl. Marine Bunkers	-	-	-	-	-	-	-4	-	-	-	-
Stock Changes	-	1	-4	-	-2	1	285	11	-	9	-1
DOMESTIC SUPPLY	38	11	251	-	54	2	1035	114	-	26	27
Transfers and Stat. Diff.	-	1	-3	-	-2	1	-4	-7	-	2	1
TRANSFORMATION	-	-	-	-	-	-	13	20	-	-	-
Electricity and CHP Plants	-	-	-	-	-	-	13	20	-	-	-
Petroleum Refineries	-	-	-	-	-	-	-	-	-	-	-
Other Transform. Sector	-	-	-	-	-	-	-	-	-	-	-
ENERGY SECTOR	38	-	-	-	-	-	22	1	-	-	12
DISTRIBUTION LOSSES	-	-	-	-	-	1	-	-	-	-	-
FINAL CONSUMPTION	-	12	248	-	52	2	996	86	-	28	16
INDUSTRY SECTOR	-	1	-	-	-	-	385	66	-	28	-
Iron and Steel	-	-	-	-	-	-	-	-	-	-	-
Chemical and Petrochem.	-	-	-	-	-	-	-	-	-	-	-
Non-Metallic Minerals	-	-	-	-	-	-	-	-	-	-	-
Non-specified	-	1	-	-	-	-	385	66	-	28	-
TRANSPORT SECTOR	-	2	248	-	52	-	457	7	-	-	-
Air	-	-	-	-	52	-	-	-	-	-	-
Road	-	2	248	-	-	-	457	-	-	-	-
Non-specified	-	-	-	-	-	-	-	7	-	-	-
OTHER SECTORS	-	9	-	-	-	2	154	13	-	-	-
Agriculture	-	-	-	-	-	-	-	9	-	-	-
Comm. and Publ. Services	-	-	-	-	-	-	-	-	-	-	-
Residential	-	5	-	-	-	2	148	-	-	-	-
Non-specified	-	4	-	-	-	-	6	4	-	-	-
NON-ENERGY USE	-	-	-	-	-	-	-	-	-	-	16

APPROVISIONNEMENT ET DEMANDE 1998	Gaz de raffinerie	GPL + éthane	Essence moteur	Essence aviation	Carbu- réacteurs	Kérosène	Gazole	Fioul lourd	Naphta	Coke de pétrole	Autres prod.
Production	38	17	271	-	50	-	450	101	-	29	22
Imports	-	-	39	-	-	-	495	-	-	-	-
Exports	-	-	-	-	-	-	-	-	-	-3	-
Intl. Marine Bunkers	-	-	-	-	-	-	-2	-	-	-	-
Stock Changes	-	-4	-3	-	-1	2	7	14	-	3	-
DOMESTIC SUPPLY	38	13	307	-	49	2	950	115	-	29	22
Transfers and Stat. Diff.	-	-	-8	-	-	-	-4	4	-	-	-
TRANSFORMATION	-	-	-	-	-	-	71	21	-	-	-
Electricity and CHP Plants	-	-	-	-	-	-	71	21	-	-	-
Petroleum Refineries	-	-	-	-	-	-	-	-	-	-	-
Other Transform. Sector	-	-	-	-	-	-	-	-	-	-	-
ENERGY SECTOR	38	-	-	-	-	-	-	14	-	-	10
DISTRIBUTION LOSSES	-	-	-	-	-	-	-	-	-	-	-
FINAL CONSUMPTION	-	13	299	-	49	2	875	84	-	29	12
INDUSTRY SECTOR	-	1	-	-	-	-	72	60	-	29	-
Iron and Steel	-	-	-	-	-	-	-	-	-	-	-
Chemical and Petrochem.	-	-	-	-	-	-	-	-	-	-	-
Non-Metallic Minerals	-	-	-	-	-	-	-	-	-	-	-
Non-specified	-	1	-	-	-	-	72	60	-	29	-
TRANSPORT SECTOR	-	2	299	-	49	-	580	8	-	-	-
Air	-	-	-	-	49	-	-	-	-	-	-
Road	-	2	299	-	-	-	580	-	-	-	-
Non-specified	-	-	-	-	-	-	-	8	-	-	-
OTHER SECTORS	-	10	-	-	-	2	223	16	-	-	-
Agriculture	-	-	-	-	-	-	-	10	-	-	-
Comm. and Publ. Services	-	-	-	-	-	-	-	-	-	-	-
Residential	-	6	-	-	-	2	197	-	-	-	-
Non-specified	-	4	-	-	-	-	26	6	-	-	-
NON-ENERGY USE	-	-	-	-	-	-	-	-	-	-	12

Myanmar

SUPPLY AND CONSUMPTION 1997	Gas (TJ) Natural Gas	Gas Works	Coke Ovens	Blast Furnaces	Comb. Renew. & Waste (TJ) Solid Biomass	Gas/Liquids from Biomass	Municipal Waste	Industrial Waste	(GWh) Electricity	(TJ) Heat
Production	66444	-	-	-	430070	-	-	-	4445	-
Imports	-	-	-	-	-	-	-	-	-	-
Exports	-	-	-	-	-	-	-	-	-	-
Intl. Marine Bunkers	-	-	-	-	-	-	-	-	-	-
Stock Changes	-	-	-	-	-	-	-	-	-	-
DOMESTIC SUPPLY	**66444**	-	-	-	**430070**	-	-	-	**4445**	-
Transfers and Stat. Diff.	3	-	-	-		-	-	-	2	-
TRANSFORMATION	**44363**	-	-	-	**58787**	-	-	-	-	-
Electricity and CHP Plants	44363	-	-	-		-	-	-	-	-
Petroleum Refineries	-	-	-	-		-	-	-	-	-
Other Transform. Sector	-	-	-	-	58787	-	-	-	-	-
ENERGY SECTOR	**6739**	-	-	-	-	-	-	-	**226**	-
DISTRIBUTION LOSSES	-	-	-	-	-	-	-	-	**1544**	-
FINAL CONSUMPTION	**15345**	-	-	-	**371283**	-	-	-	**2677**	-
INDUSTRY SECTOR	**15177**	-	-	-	**12376**	-	-	-	**914**	-
Iron and Steel	-	-	-	-	-	-	-	-	-	-
Chemical and Petrochem.	3039	-	-	-	-	-	-	-	-	-
Non-Metallic Minerals	-	-	-	-	-	-	-	-	-	-
Non-specified	12138	-	-	-	12376	-	-	-	914	-
TRANSPORT SECTOR	**78**	-	-	-	-	-	-	-	-	-
Air	-	-	-	-	-	-	-	-	-	-
Road	-	-	-	-	-	-	-	-	-	-
Non-specified	78	-	-	-	-	-	-	-	-	-
OTHER SECTORS	**90**	-	-	-	**358907**	-	-	-	**1763**	-
Agriculture	-	-	-	-	-	-	-	-	-	-
Comm. and Publ. Services	-	-	-	-	-	-	-	-	556	-
Residential	-	-	-	-	358907	-	-	-	1207	-
Non-specified	90	-	-	-	-	-	-	-	-	-
NON-ENERGY USE	-	-	-	-	-	-	-	-	-	-

APPROVISIONNEMENT ET DEMANDE 1998	Gaz (TJ) Gaz naturel	Usines à gaz	Cokeries	Hauts fourneaux	En. Re. Comb. & Déchets (TJ) Biomasse solide	Gaz/Liquides tirés de biomasse	Déchets urbains	Déchets industriels	(GWh) Electricité	(TJ) Chaleur
Production	64045	-	-	-	439213	-	-	-	4579	-
Imports	-	-	-	-	-	-	-	-	-	-
Exports	-	-	-	-	-	-	-	-	-	-
Intl. Marine Bunkers	-	-	-	-	-	-	-	-	-	-
Stock Changes	-	-	-	-	-	-	-	-	-	-
DOMESTIC SUPPLY	**64045**	-	-	-	**439213**	-	-	-	**4579**	-
Transfers and Stat. Diff.	-513	-	-	-		-	-	-	1	-
TRANSFORMATION	**43712**	-	-	-	**57770**	-	-	-	-	-
Electricity and CHP Plants	43712	-	-	-		-	-	-	-	-
Petroleum Refineries	-	-	-	-		-	-	-	-	-
Other Transform. Sector	-	-	-	-	57770	-	-	-	-	-
ENERGY SECTOR	**5871**	-	-	-	-	-	-	-	**231**	-
DISTRIBUTION LOSSES	-	-	-	-	-	-	-	-	**1501**	-
FINAL CONSUMPTION	**13949**	-	-	-	**381443**	-	-	-	**2848**	-
INDUSTRY SECTOR	**13858**	-	-	-	**12730**	-	-	-	**963**	-
Iron and Steel	-	-	-	-	-	-	-	-	-	-
Chemical and Petrochem.	3588	-	-	-	-	-	-	-	-	-
Non-Metallic Minerals	-	-	-	-	-	-	-	-	-	-
Non-specified	10270	-	-	-	12730	-	-	-	963	-
TRANSPORT SECTOR	**67**	-	-	-	-	-	-	-	-	-
Air	-	-	-	-	-	-	-	-	-	-
Road	-	-	-	-	-	-	-	-	-	-
Non-specified	67	-	-	-	-	-	-	-	-	-
OTHER SECTORS	**24**	-	-	-	**368713**	-	-	-	**1885**	-
Agriculture	-	-	-	-	-	-	-	-	-	-
Comm. and Publ. Services	-	-	-	-	-	-	-	-	608	-
Residential	-	-	-	-	368713	-	-	-	1277	-
Non-specified	24	-	-	-	-	-	-	-	-	-
NON-ENERGY USE	-	-	-	-	-	-	-	-	-	-

Nepal / Népal

SUPPLY AND CONSUMPTION 1997	Coal (1000 tonnes)							Oil (1000 tonnes)			
	Coking Coal	Other Bit. Coal	Sub-Bit. Coal	Lignite	Peat	Oven and Gas Coke	Pat. Fuel and BKB	Crude Oil	NGL	Feed-stocks	Additives
Production	-	-	-	-	-	-	-	-	-	-	-
Imports	-	297	-	-	-	-	-	-	-	-	-
Exports	-	-	-	-	-	-	-	-	-	-	-
Intl. Marine Bunkers	-	-	-	-	-	-	-	-	-	-	-
Stock Changes	-	-	-	-	-	-	-	-	-	-	-
DOMESTIC SUPPLY	-	297	-	-	-	-	-	-	-	-	-
Transfers and Stat. Diff.	-	-	-	-	-	-	-	-	-	-	-
TRANSFORMATION	-	-	-	-	-	-	-	-	-	-	-
Electricity and CHP Plants	-	-	-	-	-	-	-	-	-	-	-
Petroleum Refineries	-	-	-	-	-	-	-	-	-	-	-
Other Transform. Sector	-	-	-	-	-	-	-	-	-	-	-
ENERGY SECTOR	-	-	-	-	-	-	-	-	-	-	-
DISTRIBUTION LOSSES	-	-	-	-	-	-	-	-	-	-	-
FINAL CONSUMPTION	-	297	-	-	-	-	-	-	-	-	-
INDUSTRY SECTOR	-	249	-	-	-	-	-	-	-	-	-
Iron and Steel	-	-	-	-	-	-	-	-	-	-	-
Chemical and Petrochem.	-	-	-	-	-	-	-	-	-	-	-
Non-Metallic Minerals	-	-	-	-	-	-	-	-	-	-	-
Non-specified	-	249	-	-	-	-	-	-	-	-	-
TRANSPORT SECTOR	-	10	-	-	-	-	-	-	-	-	-
Air	-	-	-	-	-	-	-	-	-	-	-
Road	-	-	-	-	-	-	-	-	-	-	-
Non-specified	-	10	-	-	-	-	-	-	-	-	-
OTHER SECTORS	-	38	-	-	-	-	-	-	-	-	-
Agriculture	-	-	-	-	-	-	-	-	-	-	-
Comm. and Publ. Services	-	36	-	-	-	-	-	-	-	-	-
Residential	-	2	-	-	-	-	-	-	-	-	-
Non-specified	-	-	-	-	-	-	-	-	-	-	-
NON-ENERGY USE	-	-	-	-	-	-	-	-	-	-	-

APPROVISIONNEMENT ET DEMANDE 1998	Charbon (1000 tonnes)							Pétrole (1000 tonnes)			
	Charbon à coke	Autres charb. bit.	Charbon sous-bit.	Lignite	Tourbe	Coke de four/gaz	Agg./briq. de lignite	Pétrole brut	LGN	Produits d'aliment.	Additifs
Production	-	-	-	-	-	-	-	-	-	-	-
Imports	-	323	-	-	-	-	-	-	-	-	-
Exports	-	-	-	-	-	-	-	-	-	-	-
Intl. Marine Bunkers	-	-	-	-	-	-	-	-	-	-	-
Stock Changes	-	-	-	-	-	-	-	-	-	-	-
DOMESTIC SUPPLY	-	323	-	-	-	-	-	-	-	-	-
Transfers and Stat. Diff.	-	-	-	-	-	-	-	-	-	-	-
TRANSFORMATION	-	-	-	-	-	-	-	-	-	-	-
Electricity and CHP Plants	-	-	-	-	-	-	-	-	-	-	-
Petroleum Refineries	-	-	-	-	-	-	-	-	-	-	-
Other Transform. Sector	-	-	-	-	-	-	-	-	-	-	-
ENERGY SECTOR	-	-	-	-	-	-	-	-	-	-	-
DISTRIBUTION LOSSES	-	-	-	-	-	-	-	-	-	-	-
FINAL CONSUMPTION	-	323	-	-	-	-	-	-	-	· -	-
INDUSTRY SECTOR	-	273	-	-	-	-	-	-	-	-	-
Iron and Steel	-	-	-	-	-	-	-	-	-	-	-
Chemical and Petrochem.	-	-	-	-	-	-	-	-	-	-	-
Non-Metallic Minerals	-	-	-	-	-	-	-	-	-	-	-
Non-specified	-	273	-	-	-	-	-	-	-	-	-
TRANSPORT SECTOR	-	10	-	-	-	-	-	-	-	-	-
Air	-	-	-	-	-	-	-	-	-	-	-
Road	-	-	-	-	-	-	-	-	-	-	-
Non-specified	-	10	-	-	-	-	-	-	-	-	-
OTHER SECTORS	-	40	-	-	-	-	-	-	-	-	-
Agriculture	-	-	-	-	-	-	-	-	-	-	-
Comm. and Publ. Services	-	38	-	-	-	-	-	-	-	-	-
Residential	-	2	-	-	-	-	-	-	-	-	-
Non-specified	-	-	-	-	-	-	-	-	-	-	-
NON-ENERGY USE	-	-	-	-	-	-	-	-	-	-	-

Nepal / Népal

SUPPLY AND CONSUMPTION 1997	Oil cont. (1000 tonnes)										
	Refinery Gas	LPG + Ethane	Motor Gasoline	Aviation Gasoline	Jet Fuel	Kerosene	Gas/ Diesel	Heavy Fuel Oil	Naphtha	Petrol. Coke	Other Prod.
Production	-	-	-	-	-	-	-	-	-	-	-
Imports	-	29	34	-	38	222	253	26	-	-	10
Exports	-	-	-	-	-	-	-	-	-	-	-
Intl. Marine Bunkers	-	-	-	-	-	-	-	-	-	-	-
Stock Changes	-	-	1	-	1	1	1	-	-	-	-
DOMESTIC SUPPLY	-	29	35	-	39	223	254	26	-	-	10
Transfers and Stat. Diff.	-	-	-	-	-	-	1	-	-	-	-
TRANSFORMATION	-	-	-	-	-	-	-	26	-	-	-
Electricity and CHP Plants	-	-	-	-	-	-	-	26	-	-	-
Petroleum Refineries	-	-	-	-	-	-	-	-	-	-	-
Other Transform. Sector	-	-	-	-	-	-	-	-	-	-	-
ENERGY SECTOR	-	-	-	-	-	-	-	-	-	-	-
DISTRIBUTION LOSSES	-	-	-	-	-	-	-	-	-	-	-
FINAL CONSUMPTION	-	29	35	-	39	223	255	-	-	-	10
INDUSTRY SECTOR	-	-	-	-	-	11	90	-	-	-	-
Iron and Steel	-	-	-	-	-	-	-	-	-	-	-
Chemical and Petrochem.	-	-	-	-	-	-	-	-	-	-	-
Non-Metallic Minerals	-	-	-	-	-	-	-	-	-	-	-
Non-specified	-	-	-	-	-	11	90	-	-	-	-
TRANSPORT SECTOR	-	-	35	-	39	-	154	-	-	-	-
Air	-	-	-	-	39	-	-	-	-	-	-
Road	-	-	35	-	-	-	154	-	-	-	-
Non-specified	-	-	-	-	-	-	-	-	-	-	-
OTHER SECTORS	-	29	-	-	-	212	11	-	-	-	-
Agriculture	-	-	-	-	-	-	11	-	-	-	-
Comm. and Publ. Services	-	12	-	-	-	33	-	-	-	-	-
Residential	-	17	-	-	-	179	-	-	-	-	-
Non-specified	-	-	-	-	-	-	-	-	-	-	-
NON-ENERGY USE	-	-	-	-	-	-	-	-	-	-	10

APPROVISIONNEMENT ET DEMANDE 1998	Pétrole cont. (1000 tonnes)										
	Gaz de raffinerie	GPL + éthane	Essence moteur	Essence aviation	Carbu- réacteurs	Kérosène	Gazole	Fioul lourd	Naphta	Coke de pétrole	Autres prod.
Production	-	-	-	-	-	-	-	-	-	-	-
Imports	-	39	37	-	47	231	271	29	-	-	11
Exports	-	-	-	-	-	-	-	-	-	-	-
Intl. Marine Bunkers	-	-	-	-	-	-	-	-	-	-	-
Stock Changes	-	-	-	-	-	-	-	-	-	-	-
DOMESTIC SUPPLY	-	39	37	-	47	231	271	29	-	-	11
Transfers and Stat. Diff.	-	-	-	-	-	-	-	-	-	-	-
TRANSFORMATION	-	-	-	-	-	-	-	29	-	-	-
Electricity and CHP Plants	-	-	-	-	-	-	-	29	-	-	-
Petroleum Refineries	-	-	-	-	-	-	-	-	-	-	-
Other Transform. Sector	-	-	-	-	-	-	-	-	-	-	-
ENERGY SECTOR	-	-	-	-	-	-	-	-	-	-	-
DISTRIBUTION LOSSES	-	-	-	-	-	-	-	-	-	-	-
FINAL CONSUMPTION	-	39	37	-	47	231	271	-	-	-	11
INDUSTRY SECTOR	-	-	-	-	-	12	96	-	-	-	-
Iron and Steel	-	-	-	-	-	-	-	-	-	-	-
Chemical and Petrochem.	-	-	-	-	-	-	-	-	-	-	-
Non-Metallic Minerals	-	-	-	-	-	-	-	-	-	-	-
Non-specified	-	-	-	-	-	12	96	-	-	-	-
TRANSPORT SECTOR	-	-	37	-	47	-	163	-	-	-	-
Air	-	-	-	-	47	-	-	-	-	-	-
Road	-	-	37	-	-	-	163	-	-	-	-
Non-specified	-	-	-	-	-	-	-	-	-	-	-
OTHER SECTORS	-	39	-	-	-	219	12	-	-	-	-
Agriculture	-	-	-	-	-	-	12	-	-	-	-
Comm. and Publ. Services	-	16	-	-	-	34	-	-	-	-	-
Residential	-	23	-	-	-	185	-	-	-	-	-
Non-specified	-	-	-	-	-	-	-	-	-	-	-
NON-ENERGY USE	-	-	-	-	-	-	-	-	-	-	11

Nepal / Népal

SUPPLY AND CONSUMPTION 1997	Gas (TJ)				Comb. Renew. & Waste (TJ)				(GWh)	(TJ)
	Natural Gas	Gas Works	Coke Ovens	Blast Furnaces	Solid Biomass	Gas/Liquids from Biomass	Municipal Waste	Industrial Waste	Electricity	Heat
Production	-	-	-	-	278224	-	-	-	1173	-
Imports	-	-	-	-	-	-	-	-	215	-
Exports	-	-	-	-	-	-	-	-	-72	-
Intl. Marine Bunkers	-	-	-	-	-	-	-	-	-	-
Stock Changes	-	-	-	-	-	-	-	-	-	-
DOMESTIC SUPPLY	-	-	-	-	278224	-	-	-	1316	-
Transfers and Stat. Diff.	-	-	-	-	-	-	-	-	-	-
TRANSFORMATION	-	-	-	-	-	-	-	-	-	-
Electricity and CHP Plants	-	-	-	-	-	-	-	-	-	-
Petroleum Refineries	-	-	-	-	-	-	-	-	-	-
Other Transform. Sector	-	-	-	-	-	-	-	-	-	-
ENERGY SECTOR	-	-	-	-	-	-	-	-	60	-
DISTRIBUTION LOSSES	-	-	-	-	-	-	-	-	263	-
FINAL CONSUMPTION	-	-	-	-	278224	-	-	-	993	-
INDUSTRY SECTOR	-	-	-	-	4049	-	-	-	419	-
Iron and Steel	-	-	-	-	-	-	-	-	-	-
Chemical and Petrochem.	-	-	-	-	-	-	-	-	-	-
Non-Metallic Minerals	-	-	-	-	-	-	-	-	-	-
Non-specified	-	-	-	-	4049	-	-	-	419	-
TRANSPORT SECTOR	-	-	-	-	-	-	-	-	-	-
Air	-	-	-	-	-	-	-	-	-	-
Road	-	-	-	-	-	-	-	-	-	-
Non-specified	-	-	-	-	-	-	-	-	-	-
OTHER SECTORS	-	-	-	-	274175	-	-	-	574	-
Agriculture	-	-	-	-	-	-	-	-	12	-
Comm. and Publ. Services	-	-	-	-	991	-	-	-	144	-
Residential	-	-	-	-	273184	-	-	-	394	-
Non-specified	-	-	-	-	-	-	-	-	24	-
NON-ENERGY USE	-	-	-	-	-	-	-	-	-	-

APPROVISIONNEMENT ET DEMANDE 1998	Gaz (TJ)				En. Re. Comb. & Déchets (TJ)				(GWh)	(TJ)
	Gaz naturel	Usines à gaz	Cokeries	Hauts fourneaux	Biomasse solide	Gaz/Liquides tirés de biomasse	Déchets urbains	Déchets industriels	Electricité	Chaleur
Production	-	-	-	-	284256	-	-	-	1257	-
Imports	-	-	-	-	-	-	-	-	239	-
Exports	-	-	-	-	-	-	-	-	-60	-
Intl. Marine Bunkers	-	-	-	-	-	-	-	-	-	-
Stock Changes	-	-	-	-	-	-	-	-	-	-
DOMESTIC SUPPLY	-	-	-	-	284256	-	-	-	1436	-
Transfers and Stat. Diff.	-	-	-	-	-	-	-	-	-	-
TRANSFORMATION	-	-	-	-	-	-	-	-	-	-
Electricity and CHP Plants	-	-	-	-	-	-	-	-	-	-
Petroleum Refineries	-	-	-	-	-	-	-	-	-	-
Other Transform. Sector	-	-	-	-	-	-	-	-	-	-
ENERGY SECTOR	-	-	-	-	-	-	-	-	84	-
DISTRIBUTION LOSSES	-	-	-	-	-	-	-	-	287	-
FINAL CONSUMPTION	-	-	-	-	284256	-	-	-	1065	-
INDUSTRY SECTOR	-	-	-	-	4222	-	-	-	443	-
Iron and Steel	-	-	-	-	-	-	-	-	-	-
Chemical and Petrochem.	-	-	-	-	-	-	-	-	-	-
Non-Metallic Minerals	-	-	-	-	-	-	-	-	-	-
Non-specified	-	-	-	-	4222	-	-	-	443	-
TRANSPORT SECTOR	-	-	-	-	-	-	-	-	-	-
Air	-	-	-	-	-	-	-	-	-	-
Road	-	-	-	-	-	-	-	-	-	-
Non-specified	-	-	-	-	-	-	-	-	-	-
OTHER SECTORS	-	-	-	-	280034	-	-	-	622	-
Agriculture	-	-	-	-	-	-	-	-	12	-
Comm. and Publ. Services	-	-	-	-	991	-	-	-	156	-
Residential	-	-	-	-	279043	-	-	-	418	-
Non-specified	-	-	-	-	-	-	-	-	36	-
NON-ENERGY USE	-	-	-	-	-	-	-	-	-	-

Netherlands Antilles / Antilles néerlandaises

SUPPLY AND CONSUMPTION 1997	Coal (1000 tonnes)							Oil (1000 tonnes)			
	Coking Coal	Other Bit. Coal	Sub-Bit. Coal	Lignite	Peat	Oven and Gas Coke	Pat. Fuel and BKB	Crude Oil	NGL	Feed-stocks	Additives
Production	-	-	-	-	-	-	-	-	-	-	-
Imports	-	-	-	-	-	-	-	11778	-	664	-
Exports	-	-	-	-	-	-	-	-	-	-	-
Intl. Marine Bunkers	-	-	-	-	-	-	-	-	-	-	-
Stock Changes	-	-	-	-	-	-	-	-112	-	-	-
DOMESTIC SUPPLY	-	-	-	-	-	-	-	11666	-	664	-
Transfers and Stat. Diff.	-	-	-	-	-	-	-	929	-	-	-
TRANSFORMATION	-	-	-	-	-	-	-	12595	-	664	-
Electricity and CHP Plants	-	-	-	-	-	-	-	-	-	-	-
Petroleum Refineries	-	-	-	-	-	-	-	12595	-	664	-
Other Transform. Sector	-	-	-	-	-	-	-	-	-	-	-
ENERGY SECTOR	-	-	-	-	-	-	-	-	-	-	-
DISTRIBUTION LOSSES	-	-	-	-	-	-	-	-	-	-	-
FINAL CONSUMPTION	-	-	-	-	-	-	-	-	-	-	-
INDUSTRY SECTOR	-	-	-	-	-	-	-	-	-	-	-
Iron and Steel	-	-	-	-	-	-	-	-	-	-	-
Chemical and Petrochem.	-	-	-	-	-	-	-	-	-	-	-
Non-Metallic Minerals	-	-	-	-	-	-	-	-	-	-	-
Non-specified	-	-	-	-	-	-	-	-	-	-	-
TRANSPORT SECTOR	-	-	-	-	-	-	-	-	-	-	-
Air	-	-	-	-	-	-	-	-	-	-	-
Road	-	-	-	-	-	-	-	-	-	-	-
Non-specified	-	-	-	-	-	-	-	-	-	-	-
OTHER SECTORS	-	-	-	-	-	-	-	-	-	-	-
Agriculture	-	-	-	-	-	-	-	-	-	-	-
Comm. and Publ. Services	-	-	-	-	-	-	-	-	-	-	-
Residential	-	-	-	-	-	-	-	-	-	-	-
Non-specified	-	-	-	-	-	-	-	-	-	-	-
NON-ENERGY USE	-	-	-	-	-	-	-	-	-	-	-

APPROVISIONNEMENT ET DEMANDE 1998	Charbon (1000 tonnes)							Pétrole (1000 tonnes)			
	Charbon à coke	Autres charb. bit.	Charbon sous-bit.	Lignite	Tourbe	Coke de four/gaz	Agg./briq. de lignite	Pétrole brut	LGN	Produits d'aliment.	Additifs
Production	-	-	-	-	-	-	-	-	-	-	-
Imports	-	-	-	-	-	-	-	12440	-	698	-
Exports	-	-	-	-	-	-	-	-	-	-	-
Intl. Marine Bunkers	-	-	-	-	-	-	-	-	-	-	-
Stock Changes	-	-	-	-	-	-	-	-112	-	-	-
DOMESTIC SUPPLY	-	-	-	-	-	-	-	12328	-	698	-
Transfers and Stat. Diff.	-	-	-	-	-	-	-	982	-	-	-
TRANSFORMATION	-	-	-	-	-	-	-	13310	-	698	-
Electricity and CHP Plants	-	-	-	-	-	-	-	-	-	-	-
Petroleum Refineries	-	-	-	-	-	-	-	13310	-	698	-
Other Transform. Sector	-	-	-	-	-	-	-	-	-	-	-
ENERGY SECTOR	-	-	-	-	-	-	-	-	-	-	-
DISTRIBUTION LOSSES	-	-	-	-	-	-	-	-	-	-	-
FINAL CONSUMPTION	-	-	-	-	-	-	-	-	-	-	-
INDUSTRY SECTOR	-	-	-	-	-	-	-	-	-	-	-
Iron and Steel	-	-	-	-	-	-	-	-	-	-	-
Chemical and Petrochem.	-	-	-	-	-	-	-	-	-	-	-
Non-Metallic Minerals	-	-	-	-	-	-	-	-	-	-	-
Non-specified	-	-	-	-	-	-	-	-	-	-	-
TRANSPORT SECTOR	-	-	-	-	-	-	-	-	-	-	-
Air	-	-	-	-	-	-	-	-	-	-	-
Road	-	-	-	-	-	-	-	-	-	-	-
Non-specified	-	-	-	-	-	-	-	-	-	-	-
OTHER SECTORS	-	-	-	-	-	-	-	-	-	-	-
Agriculture	-	-	-	-	-	-	-	-	-	-	-
Comm. and Publ. Services	-	-	-	-	-	-	-	-	-	-	-
Residential	-	-	-	-	-	-	-	-	-	-	-
Non-specified	-	-	-	-	-	-	-	-	-	-	-
NON-ENERGY USE	-	-	-	-	-	-	-	-	-	-	-

Netherlands Antilles / Antilles néerlandaises

SUPPLY AND CONSUMPTION 1997	Oil cont. (1000 tonnes)										
	Refinery Gas	LPG + Ethane	Motor Gasoline	Aviation Gasoline	Jet Fuel	Kerosene	Gas/ Diesel	Heavy Fuel Oil	Naphtha	Petrol. Coke	Other Prod.
Production	-	97	1715	14	881	111	2415	5383	-	-	1818
Imports	-	24	306	-	58	-	317	715	-	-	246
Exports	-	-66	-1952	-12	-879	-	-2227	-4193	-	-	-1675
Intl. Marine Bunkers	-	-	-	-	-	-	-212	-1512	-	-	-
Stock Changes	-	-	-	-	-	-	-	-	-	-	-
DOMESTIC SUPPLY	-	55	69	2	60	111	293	393	-	-	389
Transfers and Stat. Diff.	-	-	-	-	-	-	-	-	-	-	-324
TRANSFORMATION	-	-	-	-	-	-	-	246	-	-	-
Electricity and CHP Plants	-	-	-	-	-	-	-	246	-	-	-
Petroleum Refineries	-	-	-	-	-	-	-	-	-	-	-
Other Transform. Sector	-	-	-	-	-	-	-	-	-	-	-
ENERGY SECTOR	-	-	-	-	-	-	-	10	-	-	-
DISTRIBUTION LOSSES	-	-	-	-	-	-	-	-	-	-	-
FINAL CONSUMPTION	-	55	69	2	60	111	293	137	-	-	65
INDUSTRY SECTOR	-	-	-	-	-	-	-	137	-	-	-
Iron and Steel	-	-	-	-	-	-	-	-	-	-	-
Chemical and Petrochem.	-	-	-	-	-	-	-	-	-	-	-
Non-Metallic Minerals	-	-	-	-	-	-	-	-	-	-	-
Non-specified	-	-	-	-	-	-	-	137	-	-	-
TRANSPORT SECTOR	-	-	69	2	60	-	293	-	-	-	-
Air	-	-	-	2	60	-	-	-	-	-	-
Road	-	-	69	-	-	-	293	-	-	-	-
Non-specified	-	-	-	-	-	-	-	-	-	-	-
OTHER SECTORS	-	55	-	-	-	111	-	-	-	-	-
Agriculture	-	-	-	-	-	-	-	-	-	-	-
Comm. and Publ. Services	-	-	-	-	-	-	-	-	-	-	-
Residential	-	55	-	-	-	111	-	-	-	-	-
Non-specified	-	-	-	-	-	-	-	-	-	-	-
NON-ENERGY USE	-	-	-	-	-	-	-	-	-	-	65

APPROVISIONNEMENT ET DEMANDE 1998	Pétrole cont. (1000 tonnes)										
	Gaz de raffinerie	GPL + éthane	Essence moteur	Essence aviation	Carbu- réacteurs	Kérosène	Gazole	Fioul lourd	Naphta	Coke de pétrole	Autres prod.
Production	-	90	1448	7	935	111	2768	5239	-	-	2239
Imports	-	24	306	-	58	-	317	715	-	-	246
Exports	-	-59	-1685	-5	-933	-	-2580	-4035	-	-	-2096
Intl. Marine Bunkers	-	-	-	-	-	-	-212	-1512	-	-	-
Stock Changes	-	-	-	-	-	-	-	-	-	-	-
DOMESTIC SUPPLY	-	55	69	2	60	111	293	407	-	-	389
Transfers and Stat. Diff.	-	-	-	-	-	-	-	-	-	-	-324
TRANSFORMATION	-	-	-	-	-	-	-	260	-	-	-
Electricity and CHP Plants	-	-	-	-	-	-	-	260	-	-	-
Petroleum Refineries	-	-	-	-	-	-	-	-	-	-	-
Other Transform. Sector	-	-	-	-	-	-	-	-	-	-	-
ENERGY SECTOR	-	-	-	-	-	-	-	10	-	-	-
DISTRIBUTION LOSSES	-	-	-	-	-	-	-	-	-	-	-
FINAL CONSUMPTION	-	55	69	2	60	111	293	137	-	-	65
INDUSTRY SECTOR	-	-	-	-	-	-	-	137	-	-	-
Iron and Steel	-	-	-	-	-	-	-	-	-	-	-
Chemical and Petrochem.	-	-	-	-	-	-	-	-	-	-	-
Non-Metallic Minerals	-	-	-	-	-	-	-	-	-	-	-
Non-specified	-	-	-	-	-	-	-	137	-	-	-
TRANSPORT SECTOR	-	-	69	2	60	-	293	-	-	-	-
Air	-	-	-	2	60	-	-	-	-	-	-
Road	-	-	69	-	-	-	293	-	-	-	-
Non-specified	-	-	-	-	-	-	-	-	-	-	-
OTHER SECTORS	-	55	-	-	-	111	-	-	-	-	-
Agriculture	-	-	-	-	-	-	-	-	-	-	-
Comm. and Publ. Services	-	-	-	-	-	-	-	-	-	-	-
Residential	-	55	-	-	-	111	-	-	-	-	-
Non-specified	-	-	-	-	-	-	-	-	-	-	-
NON-ENERGY USE	-	-	-	-	-	-	-	-	-	-	65

Netherlands Antilles / Antilles néerlandaises

SUPPLY AND CONSUMPTION 1997	Gas (TJ)				Comb. Renew. & Waste (TJ)				(GWh)	(TJ)
	Natural Gas	Gas Works	Coke Ovens	Blast Furnaces	Solid Biomass	Gas/Liquids from Biomass	Municipal Waste	Industrial Waste	Electricity	Heat
Production	-	-	-	-	-	-	-	-	1054	-
Imports	-	-	-	-	-	-	-	-	-	-
Exports	-	-	-	-	-	-	-	-	-	-
Intl. Marine Bunkers	-	-	-	-	-	-	-	-	-	-
Stock Changes	-	-	-	-	-	-	-	-	-	-
DOMESTIC SUPPLY	-	-	-	-	-	-	-	-	1054	-
Transfers and Stat. Diff.	-	-	-	-	-	-	-	-	-	-
TRANSFORMATION	-	-	-	-	-	-	-	-	-	-
Electricity and CHP Plants	-	-	-	-	-	-	-	-	-	-
Petroleum Refineries	-	-	-	-	-	-	-	-	-	-
Other Transform. Sector	-	-	-	-	-	-	-	-	-	-
ENERGY SECTOR	-	-	-	-	-	-	-	-	98	-
DISTRIBUTION LOSSES	-	-	-	-	-	-	-	-	130	-
FINAL CONSUMPTION	-	-	-	-	-	-	-	-	826	-
INDUSTRY SECTOR	-	-	-	-	-	-	-	-	455	-
Iron and Steel	-	-	-	-	-	-	-	-	-	-
Chemical and Petrochem.	-	-	-	-	-	-	-	-	-	-
Non-Metallic Minerals	-	-	-	-	-	-	-	-	-	-
Non-specified	-	-	-	-	-	-	-	-	455	-
TRANSPORT SECTOR	-	-	-	-	-	-	-	-	-	-
Air	-	-	-	-	-	-	-	-	-	-
Road	-	-	-	-	-	-	-	-	-	-
Non-specified	-	-	-	-	-	-	-	-	-	-
OTHER SECTORS	-	-	-	-	-	-	-	-	371	-
Agriculture	-	-	-	-	-	-	-	-	-	-
Comm. and Publ. Services	-	-	-	-	-	-	-	-	-	-
Residential	-	-	-	-	-	-	-	-	-	-
Non-specified	-	-	-	-	-	-	-	-	371	-
NON-ENERGY USE	-	-	-	-	-	-	-	-	-	-

APPROVISIONNEMENT ET DEMANDE 1998	Gaz (TJ)				En. Re. Comb. & Déchets (TJ)				(GWh)	(TJ)
	Gaz naturel	Usines à gaz	Cokeries	Hauts fourneaux	Biomasse solide	Gaz/Liquides tirés de biomasse	Déchets urbains	Déchets industriels	Electricité	Chaleur
Production	-	-	-	-	-	-	-	-	1116	-
Imports	-	-	-	-	-	-	-	-	-	-
Exports	-	-	-	-	-	-	-	-	-	-
Intl. Marine Bunkers	-	-	-	-	-	-	-	-	-	-
Stock Changes	-	-	-	-	-	-	-	-	-	-
DOMESTIC SUPPLY	-	-	-	-	-	-	-	-	1116	-
Transfers and Stat. Diff.	-	-	-	-	-	-	-	-	-	-
TRANSFORMATION	-	-	-	-	-	-	-	-	-	-
Electricity and CHP Plants	-	-	-	-	-	-	-	-	-	-
Petroleum Refineries	-	-	-	-	-	-	-	-	-	-
Other Transform. Sector	-	-	-	-	-	-	-	-	-	-
ENERGY SECTOR	-	-	-	-	-	-	-	-	104	-
DISTRIBUTION LOSSES	-	-	-	-	-	-	-	-	137	-
FINAL CONSUMPTION	-	-	-	-	-	-	-	-	875	-
INDUSTRY SECTOR	-	-	-	-	-	-	-	-	482	-
Iron and Steel	-	-	-	-	-	-	-	-	-	-
Chemical and Petrochem.	-	-	-	-	-	-	-	-	-	-
Non-Metallic Minerals	-	-	-	-	-	-	-	-	-	-
Non-specified	-	-	-	-	-	-	-	-	482	-
TRANSPORT SECTOR	-	-	-	-	-	-	-	-	-	-
Air	-	-	-	-	-	-	-	-	-	-
Road	-	-	-	-	-	-	-	-	-	-
Non-specified	-	-	-	-	-	-	-	-	-	-
OTHER SECTORS	-	-	-	-	-	-	-	-	393	-
Agriculture	-	-	-	-	-	-	-	-	-	-
Comm. and Publ. Services	-	-	-	-	-	-	-	-	-	-
Residential	-	-	-	-	-	-	-	-	-	-
Non-specified	-	-	-	-	-	-	-	-	393	-
NON-ENERGY USE	-	-	-	-	-	-	-	-	-	-

Nicaragua

SUPPLY AND CONSUMPTION 1997	Coal (1000 tonnes)							Oil (1000 tonnes)			
	Coking Coal	Other Bit. Coal	Sub-Bit. Coal	Lignite	Peat	Oven and Gas Coke	Pat. Fuel and BKB	Crude Oil	NGL	Feed- stocks	Additives
Production	-	-	-	-	-	-	-	-	-	-	-
Imports	-	-	-	-	-	-	-	729	-	-	-
Exports	-	-	-	-	-	-	-	-	-	-	-
Intl. Marine Bunkers	-	-	-	-	-	-	-	-	-	-	-
Stock Changes	-	-	-	-	-	-	-	101	-	-	-
DOMESTIC SUPPLY	-	-	-	-	-	-	-	830	-	-	-
Transfers and Stat. Diff.	-	-	-	-	-	-	-	-	-	-	-
TRANSFORMATION	-	-	-	-	-	-	-	822	-	-	-
Electricity and CHP Plants	-	-	-	-	-	-	-	-	-	-	-
Petroleum Refineries	-	-	-	-	-	-	-	822	-	-	-
Other Transform. Sector	-	-	-	-	-	-	-	-	-	-	-
ENERGY SECTOR	-	-	-	-	-	-	-	8	-	-	-
DISTRIBUTION LOSSES	-	-	-	-	-	-	-	-	-	-	-
FINAL CONSUMPTION	-	-	-	-	-	-	-	-	-	-	-
INDUSTRY SECTOR	-	-	-	-	-	-	-	-	-	-	-
Iron and Steel	-	-	-	-	-	-	-	-	-	-	-
Chemical and Petrochem.	-	-	-	-	-	-	-	-	-	-	-
Non-Metallic Minerals	-	-	-	-	-	-	-	-	-	-	-
Non-specified	-	-	-	-	-	-	-	-	-	-	-
TRANSPORT SECTOR	-	-	-	-	-	-	-	-	-	-	-
Air	-	-	-	-	-	-	-	-	-	-	-
Road	-	-	-	-	-	-	-	-	-	-	-
Non-specified	-	-	-	-	-	-	-	-	-	-	-
OTHER SECTORS	-	-	-	-	-	-	-	-	-	-	-
Agriculture	-	-	-	-	-	-	-	-	-	-	-
Comm. and Publ. Services	-	-	-	-	-	-	-	-	-	-	-
Residential	-	-	-	-	-	-	-	-	-	-	-
Non-specified	-	-	-	-	-	-	-	-	-	-	-
NON-ENERGY USE	-	-	-	-	-	-	-	-	-	-	-

APPROVISIONNEMENT ET DEMANDE 1998	Charbon (1000 tonnes)							Pétrole (1000 tonnes)			
	Charbon à coke	Autres charb. bit.	Charbon sous-bit.	Lignite	Tourbe	Coke de four/gaz	Agg./briq. de lignite	Pétrole brut	LGN	Produits d'aliment.	Additifs
Production	-	-	-	-	-	-	-	-	-	-	-
Imports	-	-	-	-	-	-	-	862	-	-	-
Exports	-	-	-	-	-	-	-	-	-	-	-
Intl. Marine Bunkers	-	-	-	-	-	-	-	-	-	-	-
Stock Changes	-	-	-	-	-	-	-	86	-	-	-
DOMESTIC SUPPLY	-	-	-	-	-	-	-	948	-	-	-
Transfers and Stat. Diff.	-	-	-	-	-	-	-	-	-	-	-
TRANSFORMATION	-	-	-	-	-	-	-	939	-	-	-
Electricity and CHP Plants	-	-	-	-	-	-	-	-	-	-	-
Petroleum Refineries	-	-	-	-	-	-	-	939	-	-	-
Other Transform. Sector	-	-	-	-	-	-	-	-	-	-	-
ENERGY SECTOR	-	-	-	-	-	-	-	9	-	-	-
DISTRIBUTION LOSSES	-	-	-	-	-	-	-	-	-	-	-
FINAL CONSUMPTION	-	-	-	-	-	-	-	-	-	-	-
INDUSTRY SECTOR	-	-	-	-	-	-	-	-	-	-	-
Iron and Steel	-	-	-	-	-	-	-	-	-	-	-
Chemical and Petrochem.	-	-	-	-	-	-	-	-	-	-	-
Non-Metallic Minerals	-	-	-	-	-	-	-	-	-	-	-
Non-specified	-	-	-	-	-	-	-	-	-	-	-
TRANSPORT SECTOR	-	-	-	-	-	-	-	-	-	-	-
Air	-	-	-	-	-	-	-	-	-	-	-
Road	-	-	-	-	-	-	-	-	-	-	-
Non-specified	-	-	-	-	-	-	-	-	-	-	-
OTHER SECTORS	-	-	-	-	-	-	-	-	-	-	-
Agriculture	-	-	-	-	-	-	-	-	-	-	-
Comm. and Publ. Services	-	-	-	-	-	-	-	-	-	-	-
Residential	-	-	-	-	-	-	-	-	-	-	-
Non-specified	-	-	-	-	-	-	-	-	-	-	-
NON-ENERGY USE	-	-	-	-	-	-	-	-	-	-	-

Nicaragua

SUPPLY AND CONSUMPTION 1997	Oil cont. (1000 tonnes)										
	Refinery Gas	LPG + Ethane	Motor Gasoline	Aviation Gasoline	Jet Fuel	Kerosene	Gas/ Diesel	Heavy Fuel Oil	Naphtha	Petrol. Coke	Other Prod.
Production	8	13	94	-	25	12	224	353	-	-	66
Imports	-	19	16	1	-	-	168	20	-	-	7
Exports	-	-	-	-	-	-	-	-	-	-	-37
Intl. Marine Bunkers	-	-	-	-	-	-	-	-	-	-	-
Stock Changes	-	-3	6	-	-4	-	-35	6	-	-	20
DOMESTIC SUPPLY	8	29	116	1	21	12	357	379	-	-	56
Transfers and Stat. Diff.	-	-	-	-	-	-	-	-	-	-	-
TRANSFORMATION	-	-	-	-	-	-	37	295	-	-	-
Electricity and CHP Plants	-	-	-	-	-	-	37	295	-	-	-
Petroleum Refineries	-	-	-	-	-	-	-	-	-	-	-
Other Transform. Sector	-	-	-	-	-	-	-	-	-	-	-
ENERGY SECTOR	8	-	-	-	-	-	-	-	-	-	-
DISTRIBUTION LOSSES	-	-	-	-	-	-	-	-	-	-	-
FINAL CONSUMPTION	-	29	116	1	21	12	320	84	-	-	56
INDUSTRY SECTOR	-	2	-	-	-	-	23	35	-	-	-
Iron and Steel	-	-	-	-	-	-	-	-	-	-	-
Chemical and Petrochem.	-	-	-	-	-	-	-	-	-	-	-
Non-Metallic Minerals	-	-	-	-	-	-	-	-	-	-	-
Non-specified	-	2	-	-	-	-	23	35	-	-	-
TRANSPORT SECTOR	-	-	116	1	21	-	277	-	-	-	-
Air	-	-	-	1	21	-	-	-	-	-	-
Road	-	-	116	-	-	-	248	-	-	-	-
Non-specified	-	-	-	-	-	-	29	-	-	-	-
OTHER SECTORS	-	27	-	-	-	12	20	49	-	-	-
Agriculture	-	-	-	-	-	-	4	2	-	-	-
Comm. and Publ. Services	-	-	-	-	-	-	16	47	-	-	-
Residential	-	27	-	-	-	12	-	-	-	-	-
Non-specified	-	-	-	-	-	-	-	-	-	-	-
NON-ENERGY USE	-	-	-	-	-	-	-	-	-	-	56

APPROVISIONNEMENT ET DEMANDE 1998	Pétrole cont. (1000 tonnes)										
	Gaz de raffinerie	GPL + éthane	Essence moteur	Essence aviation	Carbu- réacteurs	Kérosène	Gazole	Fioul lourd	Naphta	Coke de pétrole	Autres prod.
Production	9	11	102	-	37	12	200	449	-	-	75
Imports	-	24	32	1	-	-	200	-	-	-	9
Exports	-	-	-	-	-	-	-	-	-	-	-41
Intl. Marine Bunkers	-	-	-	-	-	-	-	-	-	-	-
Stock Changes	-	-	-1	-	-10	-	29	-22	-	-	3
DOMESTIC SUPPLY	9	35	133	1	27	12	429	427	-	-	46
Transfers and Stat. Diff.	-	-	-	-	-	-	-	-	-	-	-
TRANSFORMATION	-	-	-	-	-	-	88	352	-	-	-
Electricity and CHP Plants	-	-	-	-	-	-	88	352	-	-	-
Petroleum Refineries	-	-	-	-	-	-	-	-	-	-	-
Other Transform. Sector	-	-	-	-	-	-	-	-	-	-	-
ENERGY SECTOR	9	-	-	-	-	-	-	-	-	-	-
DISTRIBUTION LOSSES	-	-	-	-	-	-	-	-	-	-	-
FINAL CONSUMPTION	-	35	133	1	27	12	341	75	-	-	46
INDUSTRY SECTOR	-	7	-	-	-	-	29	73	-	-	-
Iron and Steel	-	-	-	-	-	-	-	-	-	-	-
Chemical and Petrochem.	-	-	-	-	-	-	-	-	-	-	-
Non-Metallic Minerals	-	-	-	-	-	-	-	-	-	-	-
Non-specified	-	7	-	-	-	-	29	73	-	-	-
TRANSPORT SECTOR	-	-	133	1	27	-	295	-	-	-	-
Air	-	-	-	1	27	-	-	-	-	-	-
Road	-	-	133	-	-	-	268	-	-	-	-
Non-specified	-	-	-	-	-	-	27	-	-	-	-
OTHER SECTORS	-	28	-	-	-	12	17	2	-	-	-
Agriculture	-	-	-	-	-	-	5	1	-	-	-
Comm. and Publ. Services	-	6	-	-	-	-	12	1	-	-	-
Residential	-	22	-	-	-	12	-	-	-	-	-
Non-specified	-	-	-	-	-	-	-	-	-	-	-
NON-ENERGY USE	-	-	-	-	-	-	-	-	-	-	46

Nicaragua

SUPPLY AND CONSUMPTION 1997	Gas (TJ)				Comb. Renew. & Waste (TJ)				(GWh)	(TJ)
	Natural Gas	Gas Works	Coke Ovens	Blast Furnaces	Solid Biomass	Gas/Liquids from Biomass	Municipal Waste	Industrial Waste	Electricity	Heat
Production	-	-	-	-	57788	-	-	-	1928	-
Imports	-	-	-	-	-	-	-	-	14	-
Exports	-	-	-	-	-	-	-	-	-10	-
Intl. Marine Bunkers	-	-	-	-	-	-	-	-	-	-
Stock Changes	-	-	-	-	-	-	-	-	-	-
DOMESTIC SUPPLY	-	-	-	-	57788	-	-	-	1932	-
Transfers and Stat. Diff.	-	-	-	-	-	-	-	-	-	-
TRANSFORMATION	-	-	-	-	8468	-	-	-	-	-
Electricity and CHP Plants	-	-	-	-	6894	-	-	-	-	-
Petroleum Refineries	-	-	-	-	-	-	-	-	-	-
Other Transform. Sector	-	-	-	-	1574	-	-	-	-	-
ENERGY SECTOR	-	-	-	-	-	-	-	-	229	-
DISTRIBUTION LOSSES	-	-	-	-	-	-	-	-	466	-
FINAL CONSUMPTION	-	-	-	-	49320	-	-	-	1237	-
INDUSTRY SECTOR	-	-	-	-	5862	-	-	-	289	-
Iron and Steel	-	-	-	-	-	-	-	-	-	-
Chemical and Petrochem.	-	-	-	-	-	-	-	-	-	-
Non-Metallic Minerals	-	-	-	-	-	-	-	-	-	-
Non-specified	-	-	-	-	5862	-	-	-	289	-
TRANSPORT SECTOR	-	-	-	-	-	-	-	-	-	-
Air	-	-	-	-	-	-	-	-	-	-
Road	-	-	-	-	-	-	-	-	-	-
Non-specified	-	-	-	-	-	-	-	-	-	-
OTHER SECTORS	-	-	-	-	43458	-	-	-	948	-
Agriculture	-	-	-	-	188	-	-	-	120	-
Comm. and Publ. Services	-	-	-	-	528	-	-	-	369	-
Residential	-	-	-	-	42742	-	-	-	459	-
Non-specified	-	-	-	-	-	-	-	-	-	-
NON-ENERGY USE	-	-	-	-	-	-	-	-	-	-

APPROVISIONNEMENT ET DEMANDE 1998	Gaz (TJ)				En. Re. Comb. & Déchets (TJ)				(GWh)	(TJ)
	Gaz naturel	Usines à gaz	Cokeries	Hauts fourneaux	Biomasse solide	Gaz/Liquides tirés de biomasse	Déchets urbains	Déchets industriels	Electricité	Chaleur
Production	-	-	-	-	55617	-	-	-	2266	-
Imports	-	-	-	-	-	-	-	-	7	-
Exports	-	-	-	-	-	-	-	-	-11	-
Intl. Marine Bunkers	-	-	-	-	-	-	-	-	-	-
Stock Changes	-	-	-	-	-	-	-	-	-	-
DOMESTIC SUPPLY	-	-	-	-	55617	-	-	-	2262	-
Transfers and Stat. Diff.	-	-	-	-	-	-	-	-	-7	-
TRANSFORMATION	-	-	-	-	6567	-	-	-	-	-
Electricity and CHP Plants	-	-	-	-	4824	-	-	-	-	-
Petroleum Refineries	-	-	-	-	-	-	-	-	-	-
Other Transform. Sector	-	-	-	-	1743	-	-	-	-	-
ENERGY SECTOR	-	-	-	-	-	-	-	-	249	-
DISTRIBUTION LOSSES	-	-	-	-	-	-	-	-	657	-
FINAL CONSUMPTION	-	-	-	-	49050	-	-	-	1349	-
INDUSTRY SECTOR	-	-	-	-	4259	-	-	-	410	-
Iron and Steel	-	-	-	-	-	-	-	-	-	-
Chemical and Petrochem.	-	-	-	-	-	-	-	-	-	-
Non-Metallic Minerals	-	-	-	-	-	-	-	-	-	-
Non-specified	-	-	-	-	4259	-	-	-	410	-
TRANSPORT SECTOR	-	-	-	-	-	-	-	-	-	-
Air	-	-	-	-	-	-	-	-	-	-
Road	-	-	-	-	-	-	-	-	-	-
Non-specified	-	-	-	-	-	-	-	-	-	-
OTHER SECTORS	-	-	-	-	44791	-	-	-	939	-
Agriculture	-	-	-	-	331	-	-	-	99	-
Comm. and Publ. Services	-	-	-	-	528	-	-	-	390	-
Residential	-	-	-	-	43932	-	-	-	450	-
Non-specified	-	-	-	-	-	-	-	-	-	-
NON-ENERGY USE	-	-	-	-	-	-	-	-	-	-

Nigeria / Nigéria

SUPPLY AND	Coal (1000 tonnes)							Oil (1000 tonnes)			
CONSUMPTION 1997	Coking Coal	Other Bit. Coal	Sub-Bit. Coal	Lignite	Peat	Oven and Gas Coke	Pat. Fuel and BKB	Crude Oil	NGL	Feed-stocks	Additives
Production	-	140	-	-	-	-	-	111057	3821	-	-
Imports	-	-	-	-	-	-	-	-	-	-	-
Exports	-	-	-	-	-	-	-	-96892	-3821	-	-
Intl. Marine Bunkers	-	-	-	-	-	-	-	-	-	-	-
Stock Changes	-	-	-	-	-	-	-	-	-	-	-
DOMESTIC SUPPLY	-	140	-	-	-	-	-	14165	-	-	-
Transfers and Stat. Diff.	-	-130	-	-	-	-	-	-	-	-	-
TRANSFORMATION	-	-	-	-	-	-	-	14165	-	-	-
Electricity and CHP Plants	-	-	-	-	-	-	-	-	-	-	-
Petroleum Refineries	-	-	-	-	-	-	-	14165	-	-	-
Other Transform. Sector	-	-	-	-	-	-	-	-	-	-	-
ENERGY SECTOR	-	-	-	-	-	-	-	-	-	-	-
DISTRIBUTION LOSSES	-	-	-	-	-	-	-	-	-	-	-
FINAL CONSUMPTION	-	10	-	-	-	-	-	-	-	-	-
INDUSTRY SECTOR	-	10	-	-	-	-	-	-	-	-	-
Iron and Steel	-	-	-	-	-	-	-	-	-	-	-
Chemical and Petrochem.	-	-	-	-	-	-	-	-	-	-	-
Non-Metallic Minerals	-	10	-	-	-	-	-	-	-	-	-
Non-specified	-	-	-	-	-	-	-	-	-	-	-
TRANSPORT SECTOR	-	-	-	-	-	-	-	-	-	-	-
Air	-	-	-	-	-	-	-	-	-	-	-
Road	-	-	-	-	-	-	-	-	-	-	-
Non-specified	-	-	-	-	-	-	-	-	-	-	-
OTHER SECTORS	-	-	-	-	-	-	-	-	-	-	-
Agriculture	-	-	-	-	-	-	-	-	-	-	-
Comm. and Publ. Services	-	-	-	-	-	-	-	-	-	-	-
Residential	-	-	-	-	-	-	-	-	-	-	-
Non-specified	-	-	-	-	-	-	-	-	-	-	-
NON-ENERGY USE	-	-	-	-	-	-	-	-	-	-	-

APPROVISIONNEMENT	Charbon (1000 tonnes)							Pétrole (1000 tonnes)			
ET DEMANDE 1998	Charbon à coke	Autres charb. bit.	Charbon sous-bit.	Lignite	Tourbe	Coke de four/gaz	Agg./briq. de lignite	Pétrole brut	LGN	Produits d'aliment.	Additifs
Production	-	59	-	-	-	-	-	102930	3541	-	-
Imports	-	-	-	-	-	-	-	-	-	-	-
Exports	-	-	-	-	-	-	-	-94615	-3541	-	-
Intl. Marine Bunkers	-	-	-	-	-	-	-	-	-	-	-
Stock Changes	-	-	-	-	-	-	-	-	-	-	-
DOMESTIC SUPPLY	-	59	-	-	-	-	-	8315	-	-	-
Transfers and Stat. Diff.	-	-49	-	-	-	-	-	-	-	-	-
TRANSFORMATION	-	-	-	-	-	-	-	8315	-	-	-
Electricity and CHP Plants	-	-	-	-	-	-	-	-	-	-	-
Petroleum Refineries	-	-	-	-	-	-	-	8315	-	-	-
Other Transform. Sector	-	-	-	-	-	-	-	-	-	-	-
ENERGY SECTOR	-	-	-	-	-	-	-	-	-	-	-
DISTRIBUTION LOSSES	-	-	-	-	-	-	-	-	-	-	-
FINAL CONSUMPTION	-	10	-	-	-	-	-	-	-	-	-
INDUSTRY SECTOR	-	10	-	-	-	-	-	-	-	-	-
Iron and Steel	-	-	-	-	-	-	-	-	-	-	-
Chemical and Petrochem.	-	-	-	-	-	-	-	-	-	-	-
Non-Metallic Minerals	-	10	-	-	-	-	-	-	-	-	-
Non-specified	-	-	-	-	-	-	-	-	-	-	-
TRANSPORT SECTOR	-	-	-	-	-	-	-	-	-	-	-
Air	-	-	-	-	-	-	-	-	-	-	-
Road	-	-	-	-	-	-	-	-	-	-	-
Non-specified	-	-	-	-	-	-	-	-	-	-	-
OTHER SECTORS	-	-	-	-	-	-	-	-	-	-	-
Agriculture	-	-	-	-	-	-	-	-	-	-	-
Comm. and Publ. Services	-	-	-	-	-	-	-	-	-	-	-
Residential	-	-	-	-	-	-	-	-	-	-	-
Non-specified	-	-	-	-	-	-	-	-	-	-	-
NON-ENERGY USE	-	-	-	-	-	-	-	-	-	-	-

Nigeria / Nigéria

SUPPLY AND CONSUMPTION 1997	Oil cont. (1000 tonnes)										
	Refinery Gas	LPG + Ethane	Motor Gasoline	Aviation Gasoline	Jet Fuel	Kerosene	Gas/ Diesel	Heavy Fuel Oil	Naphtha	Petrol. Coke	Other Prod.
Production	345	119	3600	-	473	1327	4122	2697	-	-	604
Imports	-	13	-	-	-	-	40	-	-	-	-
Exports	-	-121	-292	-	-	-274	-1331	-817	-	-	-
Intl. Marine Bunkers	-	-	-	-	-	-	-90	-254	-	-	-
Stock Changes	-	-	-	-	25	-	-	-	-	-	-
DOMESTIC SUPPLY	345	11	3308	-	498	1053	2741	1626	-	-	604
Transfers and Stat. Diff.	-	-	-	-	-	-	276	-1	-	-	1147
TRANSFORMATION	-	-	-	-	-	-	205	392	-	-	-
Electricity and CHP Plants	-	-	-	-	-	-	205	392	-	-	-
Petroleum Refineries	-	-	-	-	-	-	-	-	-	-	-
Other Transform. Sector	-	-	-	-	-	-	-	-	-	-	-
ENERGY SECTOR	345	-	-	-	-	-	-	490	-	-	-
DISTRIBUTION LOSSES	-	-	-	-	-	-	-	-	-	-	-
FINAL CONSUMPTION	-	11	3308	-	498	1053	2812	743	-	-	1751
INDUSTRY SECTOR	-	-	-	-	-	-	103	743	-	-	-
Iron and Steel	-	-	-	-	-	-	-	-	-	-	-
Chemical and Petrochem.	-	-	-	-	-	-	-	-	-	-	-
Non-Metallic Minerals	-	-	-	-	-	-	-	-	-	-	-
Non-specified	-	-	-	-	-	-	103	743	-	-	-
TRANSPORT SECTOR	-	-	3308	-	498	-	2709	-	-	-	-
Air	-	-	-	-	498	-	-	-	-	-	-
Road	-	-	3308	-	-	-	2603	-	-	-	-
Non-specified	-	-	-	-	-	-	106	-	-	-	-
OTHER SECTORS	-	11	-	-	-	1053	-	-	-	-	-
Agriculture	-	-	-	-	-	-	-	-	-	-	-
Comm. and Publ. Services	-	-	-	-	-	-	-	-	-	-	-
Residential	-	11	-	-	-	1053	-	-	-	-	-
Non-specified	-	-	-	-	-	-	-	-	-	-	-
NON-ENERGY USE	-	-	-	-	-	-	-	-	-	-	1751

APPROVISIONNEMENT ET DEMANDE 1998	Pétrole cont. (1000 tonnes)										
	Gaz de raffinerie	GPL + éthane	Essence moteur	Essence aviation	Carbu- réacteurs	Kérosène	Gazole	Fioul lourd	Naphta	Coke de pétrole	Autres prod.
Production	204	70	2124	-	279	783	2433	1592	-	-	356
Imports	-	13	1141	-	-	256	358	259	-	-	-
Exports	-	-73	-	-	-	-	-	-	-	-	-
Intl. Marine Bunkers	-	-	-	-	-	-	-90	-254	-	-	-
Stock Changes	-	-	-	-	147	-	-	-	-	-	-
DOMESTIC SUPPLY	204	10	3265	-	426	1039	2701	1597	-	-	356
Transfers and Stat. Diff.	-	-	-	-	-	-	275	-	-	-	1236
TRANSFORMATION	-	-	-	-	-	-	205	385	-	-	-
Electricity and CHP Plants	-	-	-	-	-	-	205	385	-	-	-
Petroleum Refineries	-	-	-	-	-	-	-	-	-	-	-
Other Transform. Sector	-	-	-	-	-	-	-	-	-	-	-
ENERGY SECTOR	204	-	-	-	-	-	-	482	-	-	-
DISTRIBUTION LOSSES	-	-	-	-	-	-	-	-	-	-	-
FINAL CONSUMPTION	-	10	3265	-	426	1039	2771	730	-	-	1592
INDUSTRY SECTOR	-	-	-	-	-	-	102	730	-	-	-
Iron and Steel	-	-	-	-	-	-	-	-	-	-	-
Chemical and Petrochem.	-	-	-	-	-	-	-	-	-	-	-
Non-Metallic Minerals	-	-	-	-	-	-	-	-	-	-	-
Non-specified	-	-	-	-	-	-	102	730	-	-	-
TRANSPORT SECTOR	-	-	3265	-	426	-	2669	-	-	-	-
Air	-	-	-	-	426	-	-	-	-	-	-
Road	-	-	3265	-	-	-	2565	-	-	-	-
Non-specified	-	-	-	-	-	-	104	-	-	-	-
OTHER SECTORS	-	10	-	-	-	1039	-	-	-	-	-
Agriculture	-	-	-	-	-	-	-	-	-	-	-
Comm. and Publ. Services	-	-	-	-	-	-	-	-	-	-	-
Residential	-	10	-	-	-	1039	-	-	-	-	-
Non-specified	-	-	-	-	-	-	-	-	-	-	-
NON-ENERGY USE	-	-	-	-	-	-	-	-	-	-	1592

Nigeria / Nigéria

SUPPLY AND CONSUMPTION 1997	Gas (TJ) Natural Gas	Gas Works	Coke Ovens	Blast Furnaces	Comb. Renew. & Waste (TJ) Solid Biomass	Gas/Liquids from Biomass	Municipal Waste	Industrial Waste	(GWh) Electricity	(TJ) Heat
Production	216833	-	-	-	2880624	-	-	-	15338	-
Imports	-	-	-	-	-	-	-	-	-	-
Exports	-	-	-	-	-	-	-	-	-	-
Intl. Marine Bunkers	-	-	-	-	-	-	-	-	-	-
Stock Changes	-	-	-	-	-	-	-	-	-	-
DOMESTIC SUPPLY	216833	-	-	-	2880624	-	-	-	15338	-
Transfers and Stat. Diff.	1	-	-	-	-	-	-	-	-	-
TRANSFORMATION	77683	-	-	-	76533	-	-	-	-	-
Electricity and CHP Plants	77683	-	-	-	-	-	-	-	-	-
Petroleum Refineries	-	-	-	-	-	-	-	-	-	-
Other Transform. Sector	-	-	-	-	76533	-	-	-	-	-
ENERGY SECTOR	80201	-	-	-	-	-	-	-	461	-
DISTRIBUTION LOSSES	22460	-	-	-	-	-	-	-	4829	-
FINAL CONSUMPTION	36490	-	-	-	2804091	-	-	-	10048	-
INDUSTRY SECTOR	36490	-	-	-	283811	-	-	-	2134	-
Iron and Steel	2742	-	-	-	-	-	-	-	-	-
Chemical and Petrochem.	14764	-	-	-	-	-	-	-	-	-
Non-Metallic Minerals	-	-	-	-	-	-	-	-	-	-
Non-specified	18984	-	-	-	283811	-	-	-	2134	-
TRANSPORT SECTOR	-	-	-	-	-	-	-	-	-	-
Air	-	-	-	-	-	-	-	-	-	-
Road	-	-	-	-	-	-	-	-	-	-
Non-specified	-	-	-	-	-	-	-	-	-	-
OTHER SECTORS	-	-	-	-	2520280	-	-	-	7914	-
Agriculture	-	-	-	-	-	-	-	-	-	-
Comm. and Publ. Services	-	-	-	-	-	-	-	-	2567	-
Residential	-	-	-	-	2520280	-	-	-	5347	-
Non-specified	-	-	-	-	-	-	-	-	-	-
NON-ENERGY USE	-	-	-	-	-	-	-	-	-	-

APPROVISIONNEMENT ET DEMANDE 1998	Gaz (TJ) Gaz naturel	Usines à gaz	Cokeries	Hauts fourneaux	En. Re. Comb. & Déchets (TJ) Biomasse solide	Gaz/Liquides tirés de biomasse	Déchets urbains	Déchets industriels	(GWh) Electricité	(TJ) Chaleur
Production	232602	-	-	-	2957479	-	-	-	15716	-
Imports	-	-	-	-	-	-	-	-	-	-
Exports	-	-	-	-	-	-	-	-	-	-
Intl. Marine Bunkers	-	-	-	-	-	-	-	-	-	-
Stock Changes	-	-	-	-	-	-	-	-	-	-
DOMESTIC SUPPLY	232602	-	-	-	2957479	-	-	-	15716	-
Transfers and Stat. Diff.	-	-	-	-	-	-	-	-	-	-
TRANSFORMATION	83332	-	-	-	78575	-	-	-	-	-
Electricity and CHP Plants	83332	-	-	-	-	-	-	-	-	-
Petroleum Refineries	-	-	-	-	-	-	-	-	-	-
Other Transform. Sector	-	-	-	-	78575	-	-	-	-	-
ENERGY SECTOR	86033	-	-	-	-	-	-	-	477	-
DISTRIBUTION LOSSES	24093	-	-	-	-	-	-	-	5000	-
FINAL CONSUMPTION	39144	-	-	-	2878904	-	-	-	10239	-
INDUSTRY SECTOR	39144	-	-	-	291383	-	-	-	2209	-
Iron and Steel	2941	-	-	-	-	-	-	-	-	-
Chemical and Petrochem.	15838	-	-	-	-	-	-	-	-	-
Non-Metallic Minerals	-	-	-	-	-	-	-	-	-	-
Non-specified	20365	-	-	-	291383	-	-	-	2209	-
TRANSPORT SECTOR	-	-	-	-	-	-	-	-	-	-
Air	-	-	-	-	-	-	-	-	-	-
Road	-	-	-	-	-	-	-	-	-	-
Non-specified	-	-	-	-	-	-	-	-	-	-
OTHER SECTORS	-	-	-	-	2587521	-	-	-	8030	-
Agriculture	-	-	-	-	-	-	-	-	-	-
Comm. and Publ. Services	-	-	-	-	-	-	-	-	2658	-
Residential	-	-	-	-	2587521	-	-	-	5372	-
Non-specified	-	-	-	-	-	-	-	-	-	-
NON-ENERGY USE	-	-	-	-	-	-	-	-	-	-

Oman

SUPPLY AND CONSUMPTION 1997	Coal (1000 tonnes)							Oil (1000 tonnes)			
	Coking Coal	Other Bit. Coal	Sub-Bit. Coal	Lignite	Peat	Oven and Gas Coke	Pat. Fuel and BKB	Crude Oil	NGL	Feed-stocks	Additives
Production	-	-	-	-	-	-	-	45916	332	-	-
Imports	-	-	-	-	-	-	-	-	-	-	-
Exports	-	-	-	-	-	-	-	-42448	-332	-	-
Intl. Marine Bunkers	-	-	-	-	-	-	-	-	-	-	-
Stock Changes	-	-	-	-	-	-	-	-159	-	-	-
DOMESTIC SUPPLY	-	-	-	-	-	-	-	3309	-	-	-
Transfers and Stat. Diff.	-	-	-	-	-	-	-	-	-	-	-
TRANSFORMATION	-	-	-	-	-	-	-	3309	-	-	-
Electricity and CHP Plants	-	-	-	-	-	-	-	-	-	-	-
Petroleum Refineries	-	-	-	-	-	-	-	3309	-	-	-
Other Transform. Sector	-	-	-	-	-	-	-	-	-	-	-
ENERGY SECTOR	-	-	-	-	-	-	-	-	-	-	-
DISTRIBUTION LOSSES	-	-	-	-	-	-	-	-	-	-	-
FINAL CONSUMPTION	-	-	-	-	-	-	-	-	-	-	-
INDUSTRY SECTOR	-	-	-	-	-	-	-	-	-	-	-
Iron and Steel	-	-	-	-	-	-	-	-	-	-	-
Chemical and Petrochem.	-	-	-	-	-	-	-	-	-	-	-
Non-Metallic Minerals	-	-	-	-	-	-	-	-	-	-	-
Non-specified	-	-	-	-	-	-	-	-	-	-	-
TRANSPORT SECTOR	-	-	-	-	-	-	-	-	-	-	-
Air	-	-	-	-	-	-	-	-	-	-	-
Road	-	-	-	-	-	-	-	-	-	-	-
Non-specified	-	-	-	-	-	-	-	-	-	-	-
OTHER SECTORS	-	-	-	-	-	-	-	-	-	-	-
Agriculture	-	-	-	-	-	-	-	-	-	-	-
Comm. and Publ. Services	-	-	-	-	-	-	-	-	-	-	-
Residential	-	-	-	-	-	-	-	-	-	-	-
Non-specified	-	-	-	-	-	-	-	-	-	-	-
NON-ENERGY USE	-	-	-	-	-	-	-	-	-	-	-

APPROVISIONNEMENT ET DEMANDE 1998	Charbon (1000 tonnes)							Pétrole (1000 tonnes)			
	Charbon à coke	Autres charb. bit.	Charbon sous-bit.	Lignite	Tourbe	Coke de four/gaz	Agg./briq. de lignite	Pétrole brut	LGN	Produits d'aliment.	Additifs
Production	-	-	-	-	-	-	-	45735	332	-	-
Imports	-	-	-	-	-	-	-	-	-	-	-
Exports	-	-	-	-	-	-	-	-41960	-332	-	-
Intl. Marine Bunkers	-	-	-	-	-	-	-	-	-	-	-
Stock Changes	-	-	-	-	-	-	-	220	-	-	-
DOMESTIC SUPPLY	-	-	-	-	-	-	-	3995	-	-	-
Transfers and Stat. Diff.	-	-	-	-	-	-	-	-	-	-	-
TRANSFORMATION	-	-	-	-	-	-	-	3995	-	-	-
Electricity and CHP Plants	-	-	-	-	-	-	-	-	-	-	-
Petroleum Refineries	-	-	-	-	-	-	-	3995	-	-	-
Other Transform. Sector	-	-	-	-	-	-	-	-	-	-	-
ENERGY SECTOR	-	-	-	-	-	-	-	-	-	-	-
DISTRIBUTION LOSSES	-	-	-	-	-	-	-	-	-	-	-
FINAL CONSUMPTION	-	-	-	-	-	-	-	-	-	-	-
INDUSTRY SECTOR	-	-	-	-	-	-	-	-	-	-	-
Iron and Steel	-	-	-	-	-	-	-	-	-	-	-
Chemical and Petrochem.	-	-	-	-	-	-	-	-	-	-	-
Non-Metallic Minerals	-	-	-	-	-	-	-	-	-	-	-
Non-specified	-	-	-	-	-	-	-	-	-	-	-
TRANSPORT SECTOR	-	-	-	-	-	-	-	-	-	-	-
Air	-	-	-	-	-	-	-	-	-	-	-
Road	-	-	-	-	-	-	-	-	-	-	-
Non-specified	-	-	-	-	-	-	-	-	-	-	-
OTHER SECTORS	-	-	-	-	-	-	-	-	-	-	-
Agriculture	-	-	-	-	-	-	-	-	-	-	-
Comm. and Publ. Services	-	-	-	-	-	-	-	-	-	-	-
Residential	-	-	-	-	-	-	-	-	-	-	-
Non-specified	-	-	-	-	-	-	-	-	-	-	-
NON-ENERGY USE	-	-	-	-	-	-	-	-	-	-	-

Oman

SUPPLY AND CONSUMPTION 1997	Oil cont. (1000 tonnes)										
	Refinery Gas	LPG + Ethane	Motor Gasoline	Aviation Gasoline	Jet Fuel	Kerosene	Gas/ Diesel	Heavy Fuel Oil	Naphtha	Petrol. Coke	Other Prod.
Production	-	81	479	-	134	4	651	1902	-	-	25
Imports	-	19	170	-	39	-	77	-	-	-	16
Exports	-	-	-	-	-	-	-	-1293	-	-	-
Intl. Marine Bunkers	-	-	-	-	-	-	-	-30	-	-	-
Stock Changes	-	-48	29	-	-1	-	28	-	-	-	-
DOMESTIC SUPPLY	-	52	678	-	172	4	756	579	-	-	41
Transfers and Stat. Diff.	-	-	-	-	-	-	-78	-	-	-	-
TRANSFORMATION	-	-	-	-	-	-	400	-	-	-	-
Electricity and CHP Plants	-	-	-	-	-	-	400	-	-	-	-
Petroleum Refineries	-	-	-	-	-	-	-	-	-	-	-
Other Transform. Sector	-	-	-	-	-	-	-	-	-	-	-
ENERGY SECTOR	-	-	-	-	-	-	-	137	-	-	-
DISTRIBUTION LOSSES	-	-	-	-	-	-	-	-	-	-	-
FINAL CONSUMPTION	-	52	678	-	172	4	278	442	-	-	41
INDUSTRY SECTOR	-	-	-	-	-	4	-	442	-	-	-
Iron and Steel	-	-	-	-	-	-	-	-	-	-	-
Chemical and Petrochem.	-	-	-	-	-	-	-	-	-	-	-
Non-Metallic Minerals	-	-	-	-	-	-	-	-	-	-	-
Non-specified	-	-	-	-	-	4	-	442	-	-	-
TRANSPORT SECTOR	-	-	678	-	172	-	96	-	-	-	-
Air	-	-	-	-	172	-	-	-	-	-	-
Road	-	-	678	-	-	-	96	-	-	-	-
Non-specified	-	-	-	-	-	-	-	-	-	-	-
OTHER SECTORS	-	52	-	-	-	-	182	-	-	-	-
Agriculture	-	-	-	-	-	-	-	-	-	-	-
Comm. and Publ. Services	-	-	-	-	-	-	-	-	-	-	-
Residential	-	52	-	-	-	-	-	-	-	-	-
Non-specified	-	-	-	-	-	-	182	-	-	-	-
NON-ENERGY USE	-	-	-	-	-	-	-	-	-	-	41

APPROVISIONNEMENT ET DEMANDE 1998	Pétrole cont. (1000 tonnes)										
	Gaz de raffinerie	GPL + éthane	Essence moteur	Essence aviation	Carbu- réacteurs	Kérosène	Gazole	Fioul lourd	Naphta	Coke de pétrole	Autres prod.
Production	-	102	613	-	146	3	809	2245	-	-	25
Imports	-	11	53	-	46	-	12	-	-	-	16
Exports	-	-	-	-	-	-	-43	-1764	-	-	-
Intl. Marine Bunkers	-	-	-	-	-	-	-	-39	-	-	-
Stock Changes	-	-47	34	-	-11	-	19	-	-	-	-
DOMESTIC SUPPLY	-	66	700	-	181	3	797	442	-	-	41
Transfers and Stat. Diff.	-	-	-	-	-	-	-52	-	-	-	-
TRANSFORMATION	-	-	-	-	-	-	452	-	-	-	-
Electricity and CHP Plants	-	-	-	-	-	-	452	-	-	-	-
Petroleum Refineries	-	-	-	-	-	-	-	-	-	-	-
Other Transform. Sector	-	-	-	-	-	-	-	-	-	-	-
ENERGY SECTOR	-	-	-	-	-	-	-	161	-	-	-
DISTRIBUTION LOSSES	-	-	-	-	-	-	-	-	-	-	-
FINAL CONSUMPTION	-	66	700	-	181	3	293	281	-	-	41
INDUSTRY SECTOR	-	-	-	-	-	3	-	281	-	-	-
Iron and Steel	-	-	-	-	-	-	-	-	-	-	-
Chemical and Petrochem.	-	-	-	-	-	-	-	-	-	-	-
Non-Metallic Minerals	-	-	-	-	-	-	-	-	-	-	-
Non-specified	-	-	-	-	-	3	-	281	-	-	-
TRANSPORT SECTOR	-	-	700	-	181	-	101	-	-	-	-
Air	-	-	-	-	181	-	-	-	-	-	-
Road	-	-	700	-	-	-	101	-	-	-	-
Non-specified	-	-	-	-	-	-	-	-	-	-	-
OTHER SECTORS	-	66	-	-	-	-	192	-	-	-	-
Agriculture	-	-	-	-	-	-	-	-	-	-	-
Comm. and Publ. Services	-	-	-	-	-	-	-	-	-	-	-
Residential	-	66	-	-	-	-	-	-	-	-	-
Non-specified	-	-	-	-	-	-	192	-	-	-	-
NON-ENERGY USE	-	-	-	-	-	-	-	-	-	-	41

Oman

SUPPLY AND CONSUMPTION 1997	Gas (TJ) Natural Gas	Gas Works	Coke Ovens	Blast Furnaces	Comb. Renew. & Waste (TJ) Solid Biomass	Gas/Liquids from Biomass	Municipal Waste	Industrial Waste	(GWh) Electricity	(TJ) Heat
Production	218657	-	-	-	-	-	-	-	7318	-
Imports	-	-	-	-	-	-	-	-	-	-
Exports	-18850	-	-	-	-	-	-	-	-	-
Intl. Marine Bunkers	-	-	-	-	-	-	-	-	-	-
Stock Changes	-	-	-	-	-	-	-	-	-	-
DOMESTIC SUPPLY	199807	-	-	-	-	-	-	-	7318	-
Transfers and Stat. Diff.	-2	-	-	-	-	-	-	-	-	-
TRANSFORMATION	85011	-	-	-	-	-	-	-	-	-
Electricity and CHP Plants	85011	-	-	-	-	-	-	-	-	-
Petroleum Refineries	-	-	-	-	-	-	-	-	-	-
Other Transform. Sector	-	-	-	-	-	-	-	-	-	-
ENERGY SECTOR	62732	-	-	-	-	-	-	-	369	-
DISTRIBUTION LOSSES	3488	-	-	-	-	-	-	-	1053	-
FINAL CONSUMPTION	48574	-	-	-	-	-	-	-	5896	-
INDUSTRY SECTOR	30634	-	-	-	-	-	-	-	455	-
Iron and Steel	-	-	-	-	-	-	-	-	-	-
Chemical and Petrochem.	-	-	-	-	-	-	-	-	-	-
Non-Metallic Minerals	2202	-	-	-	-	-	-	-	-	-
Non-specified	28432	-	-	-	-	-	-	-	455	-
TRANSPORT SECTOR	-	-	-	-	-	-	-	-	-	-
Air	-	-	-	-	-	-	-	-	-	-
Road	-	-	-	-	-	-	-	-	-	-
Non-specified	-	-	-	-	-	-	-	-	-	-
OTHER SECTORS	17940	-	-	-	-	-	-	-	5441	-
Agriculture	-	-	-	-	-	-	-	-	-	-
Comm. and Publ. Services	-	-	-	-	-	-	-	-	2099	-
Residential	-	-	-	-	-	-	-	-	3246	-
Non-specified	17940	-	-	-	-	-	-	-	96	-
NON-ENERGY USE	-	-	-	-	-	-	-	-	-	-

APPROVISIONNEMENT ET DEMANDE 1998	Gaz (TJ) Gaz naturel	Usines à gaz	Cokeries	Hauts fourneaux	En. Re. Comb. & Déchets (TJ) Biomasse solide	Gaz/Liquides tirés de biomasse	Déchets urbains	Déchets industriels	(GWh) Electricité	(TJ) Chaleur
Production	242591	-	-	-	-	-	-	-	8198	-
Imports	-	-	-	-	-	-	-	-	-	-
Exports	-16965	-	-	-	-	-	-	-	-	-
Intl. Marine Bunkers	-	-	-	-	-	-	-	-	-	-
Stock Changes	-	-	-	-	-	-	-	-	-	-
DOMESTIC SUPPLY	225626	-	-	-	-	-	-	-	8198	-
Transfers and Stat. Diff.	-3	-	-	-	-	-	-	-	-	-
TRANSFORMATION	94007	-	-	-	-	-	-	-	-	-
Electricity and CHP Plants	94007	-	-	-	-	-	-	-	-	-
Petroleum Refineries	-	-	-	-	-	-	-	-	-	-
Other Transform. Sector	-	-	-	-	-	-	-	-	-	-
ENERGY SECTOR	54005	-	-	-	-	-	-	-	389	-
DISTRIBUTION LOSSES	4460	-	-	-	-	-	-	-	1300	-
FINAL CONSUMPTION	73151	-	-	-	-	-	-	-	6509	-
INDUSTRY SECTOR	53482	-	-	-	-	-	-	-	467	-
Iron and Steel	-	-	-	-	-	-	-	-	-	-
Chemical and Petrochem.	-	-	-	-	-	-	-	-	-	-
Non-Metallic Minerals	4043	-	-	-	-	-	-	-	-	-
Non-specified	49439	-	-	-	-	-	-	-	467	-
TRANSPORT SECTOR	-	-	-	-	-	-	-	-	-	-
Air	-	-	-	-	-	-	-	-	-	-
Road	-	-	-	-	-	-	-	-	-	-
Non-specified	-	-	-	-	-	-	-	-	-	-
OTHER SECTORS	19669	-	-	-	-	-	-	-	6042	-
Agriculture	-	-	-	-	-	-	-	-	-	-
Comm. and Publ. Services	-	-	-	-	-	-	-	-	2247	-
Residential	-	-	-	-	-	-	-	-	3678	-
Non-specified	19669	-	-	-	-	-	-	-	117	-
NON-ENERGY USE	-	-	-	-	-	-	-	-	-	-

Pakistan : 1997

SUPPLY AND CONSUMPTION *APPROVISIONNEMENT ET DEMANDE*	Coal / *Charbon* (1000 tonnes)							Oil / *Pétrole* (1000 tonnes)			
	Coking Coal *Charbon à coke*	Other Bit. Coal *Autres charb. bit.*	Sub-Bit. Coal *Charbon sous-bit.*	Lignite *Lignite*	Peat *Tourbe*	Oven and Gas Coke *Coke de four/gaz*	Pat. Fuel and BKB *Agg./briq. de lignite*	Crude Oil *Pétrole brut*	NGL *LGN*	Feed-stocks *Produits d'aliment.*	Additives *Additifs*
Production	-	3553	-	-	-	588	-	2850	114	-	67
From Other Sources	-	-	-	-	-	-	-	-	-	-	-
Imports	840	-	-	-	-	-	-	3835	-	-	96
Exports	-	-	-	-	-	-	-	-472	-	-	-
Intl. Marine Bunkers	-	-	-	-	-	-	-	-	-	-	-
Stock Changes	-	-	-	-	-	-	-	-45	-	-	-1
DOMESTIC SUPPLY	**840**	**3553**	**-**	**-**	**-**	**588**	**-**	**6168**	**114**	**-**	**162**
Transfers	-	-	-	-	-	-	-	-	-114	-	-
Statistical Differences	-	-	-	-	-	-	-	-	-	-	-
TRANSFORMATION	**840**	**352**	**-**	**-**	**-**	**470**	**-**	**6168**	**-**	**-**	**162**
Electricity Plants	-	352	-	-	-	-	-	-	-	-	-
CHP Plants	-	-	-	-	-	-	-	-	-	-	-
Heat Plants	-	-	-	-	-	-	-	-	-	-	-
Blast Furnaces/Gas Works	-	-	-	-	-	470	-	-	-	-	-
Coke/Pat. Fuel/BKB Plants	840	-	-	-	-	-	-	-	-	-	-
Petroleum Refineries	-	-	-	-	-	-	-	6168	-	-	162
Petrochemical Industry	-	-	-	-	-	-	-	-	-	-	-
Liquefaction	-	-	-	-	-	-	-	-	-	-	-
Other Transform. Sector	-	-	-	-	-	-	-	-	-	-	-
ENERGY SECTOR	**-**	**-**	**-**	**-**	**-**	**-**	**-**	**-**	**-**	**-**	**-**
Coal Mines	-	-	-	-	-	-	-	-	-	-	-
Oil and Gas Extraction	-	-	-	-	-	-	-	-	-	-	-
Petroleum Refineries	-	-	-	-	-	-	-	-	-	-	-
Electr., CHP+Heat Plants	-	-	-	-	-	-	-	-	-	-	-
Pumped Storage (Elec.)	-	-	-	-	-	-	-	-	-	-	-
Other Energy Sector	-	-	-	-	-	-	-	-	-	-	-
Distribution Losses	-	-	-	-	-	-	-	-	-	-	-
FINAL CONSUMPTION	**-**	**3201**	**-**	**-**	**-**	**118**	**-**	**-**	**-**	**-**	**-**
INDUSTRY SECTOR	**-**	**3191**	**-**	**-**	**-**	**118**	**-**	**-**	**-**	**-**	**-**
Iron and Steel	-	-	-	-	-	118	-	-	-	-	-
Chemical and Petrochem.	-	-	-	-	-	-	-	-	-	-	-
of which: Feedstocks	-	-	-	-	-	-	-	-	-	-	-
Non-Ferrous Metals	-	-	-	-	-	-	-	-	-	-	-
Non-Metallic Minerals	-	3191	-	-	-	-	-	-	-	-	-
Transport Equipment	-	-	-	-	-	-	-	-	-	-	-
Machinery	-	-	-	-	-	-	-	-	-	-	-
Mining and Quarrying	-	-	-	-	-	-	-	-	-	-	-
Food and Tobacco	-	-	-	-	-	-	-	-	-	-	-
Paper, Pulp and Print	-	-	-	-	-	-	-	-	-	-	-
Wood and Wood Products	-	-	-	-	-	-	-	-	-	-	-
Construction	-	-	-	-	-	-	-	-	-	-	-
Textile and Leather	-	-	-	-	-	-	-	-	-	-	-
Non-specified	-	-	-	-	-	-	-	-	-	-	-
TRANSPORT SECTOR	**-**	**-**	**-**	**-**	**-**	**-**	**-**	**-**	**-**	**-**	**-**
Air	-	-	-	-	-	-	-	-	-	-	-
Road	-	-	-	-	-	-	-	-	-	-	-
Rail	-	-	-	-	-	-	-	-	-	-	-
Pipeline Transport	-	-	-	-	-	-	-	-	-	-	-
Internal Navigation	-	-	-	-	-	-	-	-	-	-	-
Non-specified	-	-	-	-	-	-	-	-	-	-	-
OTHER SECTORS	**-**	**10**	**-**	**-**	**-**	**-**	**-**	**-**	**-**	**-**	**-**
Agriculture	-	-	-	-	-	-	-	-	-	-	-
Comm. and Publ. Services	-	-	-	-	-	-	-	-	-	-	-
Residential	-	10	-	-	-	-	-	-	-	-	-
Non-specified	-	-	-	-	-	-	-	-	-	-	-
NON-ENERGY USE	**-**	**-**	**-**	**-**	**-**	**-**	**-**	**-**	**-**	**-**	**-**
in Industry/Trans./Energy	-	-	-	-	-	-	-	-	-	-	-
in Transport	-	-	-	-	-	-	-	-	-	-	-
in Other Sectors	-	-	-	-	-	-	-	-	-	-	-

Pakistan : 1997

SUPPLY AND CONSUMPTION *APPROVISIONNEMENT ET DEMANDE*	Oil cont. / *Pétrole cont.* (1000 tonnes)										
	Refinery Gas *Gaz de raffinerie*	LPG + Ethane *GPL + éthane*	Motor Gasoline *Essence moteur*	Aviation Gasoline *Essence aviation*	Jet Fuel *Carbu-réacteurs*	Kerosene *Kérosène*	Gas/ Diesel *Gazole*	Heavy Fuel Oil *Fioul lourd*	Naphtha *Naphta*	Petrol. Coke *Coke de pétrole*	Other Prod. *Autres prod.*
Production	105	41	909	-	594	413	1596	1794	81	-	435
From Other Sources	-	-	-	-	-	-	-	-	-	-	-
Imports	-	32	210	-	-	103	4870	5117	-	-	-
Exports	-	-	-	-	-	-	-	-	-84	-	-
Intl. Marine Bunkers	-	-	-	-	-	-	-3	-15	-	-	-
Stock Changes	-	-	78	-	-12	6	-58	63	3	-	-
DOMESTIC SUPPLY	**105**	**73**	**1197**	**-**	**582**	**522**	**6405**	**6959**	**-**	**-**	**435**
Transfers	-	114	-	-	-	-	-	-	-	-	-
Statistical Differences	-	-	-	-	-	-	-	100	-	-	-93
TRANSFORMATION	**-**	**-**	**-**	**-**	**-**	**-**	**143**	**4973**	**-**	**-**	**-**
Electricity Plants	-	-	-	-	-	-	143	4973	-	-	-
CHP Plants	-	-	-	-	-	-	-	-	-	-	-
Heat Plants	-	-	-	-	-	-	-	-	-	-	-
Blast Furnaces/Gas Works	-	-	-	-	-	-	-	-	-	-	-
Coke/Pat. Fuel/BKB Plants	-	-	-	-	-	-	-	-	-	-	-
Petroleum Refineries	-	-	-	-	-	-	-	-	-	-	-
Petrochemical Industry	-	-	-	-	-	-	-	-	-	-	-
Liquefaction	-	-	-	-	-	-	-	-	-	-	-
Other Transform. Sector	-	-	-	-	-	-	-	-	-	-	-
ENERGY SECTOR	**105**	**-**	**-**	**-**	**-**	**-**	**-**	**100**	**-**	**-**	**-**
Coal Mines	-	-	-	-	-	-	-	-	-	-	-
Oil and Gas Extraction	-	-	-	-	-	-	-	-	-	-	-
Petroleum Refineries	105	-	-	-	-	-	-	100	-	-	-
Electr., CHP+Heat Plants	-	-	-	-	-	-	-	-	-	-	-
Pumped Storage (Elec.)	-	-	-	-	-	-	-	-	-	-	-
Other Energy Sector	-	-	-	-	-	-	-	-	-	-	-
Distribution Losses	-	-	-	-	-	-	-	-	-	-	-
FINAL CONSUMPTION	**-**	**187**	**1197**	**-**	**582**	**522**	**6262**	**1986**	**-**	**-**	**342**
INDUSTRY SECTOR	**-**	**-**	**-**	**-**	**-**	**-**	**219**	**1922**	**-**	**-**	**-**
Iron and Steel	-	-	-	-	-	-	-	-	-	-	-
Chemical and Petrochem.	-	-	-	-	-	-	219	-	-	-	-
of which: Feedstocks	-	-	-	-	-	-	-	-	-	-	-
Non-Ferrous Metals	-	-	-	-	-	-	-	-	-	-	-
Non-Metallic Minerals	-	-	-	-	-	-	-	1378	-	-	-
Transport Equipment	-	-	-	-	-	-	-	-	-	-	-
Machinery	-	-	-	-	-	-	-	-	-	-	-
Mining and Quarrying	-	-	-	-	-	-	-	-	-	-	-
Food and Tobacco	-	-	-	-	-	-	-	-	-	-	-
Paper, Pulp and Print	-	-	-	-	-	-	-	-	-	-	-
Wood and Wood Products	-	-	-	-	-	-	-	-	-	-	-
Construction	-	-	-	-	-	-	-	-	-	-	-
Textile and Leather	-	-	-	-	-	-	-	-	-	-	-
Non-specified	-	-	-	-	-	-	-	544	-	-	-
TRANSPORT SECTOR	**-**	**-**	**1197**	**-**	**582**	**-**	**5662**	**19**	**-**	**-**	**-**
Air	-	-	-	-	582	-	-	-	-	-	-
Road	-	-	1197	-	-	-	5462	-	-	-	-
Rail	-	-	-	-	-	-	200	19	-	-	-
Pipeline Transport	-	-	-	-	-	-	-	-	-	-	-
Internal Navigation	-	-	-	-	-	-	-	-	-	-	-
Non-specified	-	-	-	-	-	-	-	-	-	-	-
OTHER SECTORS	**-**	**187**	**-**	**-**	**-**	**522**	**381**	**45**	**-**	**-**	**-**
Agriculture	-	-	-	-	-	-	269	-	-	-	-
Comm. and Publ. Services	-	-	-	-	-	14	110	45	-	-	-
Residential	-	187	-	-	-	508	2	-	-	-	-
Non-specified	-	-	-	-	-	-	-	-	-	-	-
NON-ENERGY USE	**-**	**-**	**-**	**-**	**-**	**-**	**-**	**-**	**-**	**-**	**342**
in Industry/Transf./Energy	-	-	-	-	-	-	-	-	-	-	342
in Transport	-	-	-	-	-	-	-	-	-	-	-
in Other Sectors	-	-	-	-	-	-	-	-	-	-	-

Pakistan : 1997

SUPPLY AND CONSUMPTION / APPROVISIONNEMENT ET DEMANDE	Gas / Gaz (TJ)				Comb. Renew. & Waste / En. Re. Comb. & Déchets (TJ)				(GWh)	(TJ)
	Natural Gas / Gaz naturel	Gas Works / Usines à gaz	Coke Ovens / Cokeries	Blast Furnaces / Hauts fourneaux	Solid Biomass / Biomasse solide	Gas/Liquids from Biomass / Gaz/Liquides tirés de biomasse	Municipal Waste / Déchets urbains	Industrial Waste / Déchets industriels	Electricity / Electricité	Heat / Chaleur
Production	637183	-	-	5117	937950	-	-	-	59125	-
From Other Sources	-	-	-	-	-	-	-	-	-	-
Imports	-	-	-	-	-	-	-	-	-	-
Exports	-	-	-	-	-	-	-	-	-	-
Intl. Marine Bunkers	-	-	-	-	-	-	-	-	-	-
Stock Changes	131	-	-	-	-	-	-	-	-	-
DOMESTIC SUPPLY	637314	-	-	5117	937950	-	-	-	59125	-
Transfers	-	-	-	-	-	-	-	-	-	-
Statistical Differences	1	-	-	-	-	-	-	-	-	-
TRANSFORMATION	187303	-	-	-	18738	-	-	-	-	-
Electricity Plants	187303	-	-	-	-	-	-	-	-	-
CHP Plants	-	-	-	-	-	-	-	-	-	-
Heat Plants	-	-	-	-	-	-	-	-	-	-
Blast Furnaces/Gas Works	-	-	-	-	-	-	-	-	-	-
Coke/Pat. Fuel/BKB Plants	-	-	-	-	-	-	-	-	-	-
Petroleum Refineries	-	-	-	-	-	-	-	-	-	-
Petrochemical Industry	-	-	-	-	-	-	-	-	-	-
Liquefaction	-	-	-	-	-	-	-	-	-	-
Other Transform. Sector	-	-	-	-	18738	-	-	-	-	-
ENERGY SECTOR	11207	-	-	-	-	-	-	-	2357	-
Coal Mines	-	-	-	-	-	-	-	-	-	-
Oil and Gas Extraction	11207	-	-	-	-	-	-	-	-	-
Petroleum Refineries	-	-	-	-	-	-	-	-	-	-
Electr., CHP+Heat Plants	-	-	-	-	-	-	-	-	2357	-
Pumped Storage (Elec.)	-	-	-	-	-	-	-	-	-	-
Other Energy Sector	-	-	-	-	-	-	-	-	-	-
Distribution Losses	49459	-	-	-	-	-	-	-	14053	-
FINAL CONSUMPTION	389346	-	-	5117	919212	-	-	-	42715	-
INDUSTRY SECTOR	250483	-	-	5117	102019	-	-	-	11982	-
Iron and Steel	-	-	-	5117	-	-	-	-	-	-
Chemical and Petrochem.	127309	-	-	-	-	-	-	-	-	-
of which: Feedstocks	94889	-	-	-	-	-	-	-	-	-
Non-Ferrous Metals	-	-	-	-	-	-	-	-	-	-
Non-Metallic Minerals	9017	-	-	-	-	-	-	-	-	-
Transport Equipment	-	-	-	-	-	-	-	-	-	-
Machinery	-	-	-	-	-	-	-	-	-	-
Mining and Quarrying	-	-	-	-	-	-	-	-	-	-
Food and Tobacco	-	-	-	-	-	-	-	-	-	-
Paper, Pulp and Print	-	-	-	-	-	-	-	-	-	-
Wood and Wood Products	-	-	-	-	-	-	-	-	-	-
Construction	-	-	-	-	-	-	-	-	-	-
Textile and Leather	-	-	-	-	-	-	-	-	-	-
Non-specified	114157	-	-	-	102019	-	-	-	11982	-
TRANSPORT SECTOR	371	-	-	-	-	-	-	-	19	-
Air	-	-	-	-	-	-	-	-	-	-
Road	-	-	-	-	-	-	-	-	-	-
Rail	-	-	-	-	-	-	-	-	19	-
Pipeline Transport	371	-	-	-	-	-	-	-	-	-
Internal Navigation	-	-	-	-	-	-	-	-	-	-
Non-specified	-	-	-	-	-	-	-	-	-	-
OTHER SECTORS	138492	-	-	-	817193	-	-	-	30714	-
Agriculture	-	-	-	-	-	-	-	-	7086	-
Comm. and Publ. Services	19036	-	-	-	-	-	-	-	5944	-
Residential	119456	-	-	-	817193	-	-	-	17684	-
Non-specified	-	-	-	-	-	-	-	-	-	-
NON-ENERGY USE	-	-	-	-	-	-	-	-	-	-
in Industry/Transf./Energy	-	-	-	-	-	-	-	-	-	-
in Transport	-	-	-	-	-	-	-	-	-	-
in Other Sectors	-	-	-	-	-	-	-	-	-	-

Pakistan : 1998

| | Coal / *Charbon* (1000 tonnes) | | | | | | | Oil / *Pétrole* (1000 tonnes) | | | |
SUPPLY AND CONSUMPTION *APPROVISIONNEMENT ET DEMANDE*	Coking Coal *Charbon à coke*	Other Bit. Coal *Autres charb. bit.*	Sub-Bit. Coal *Charbon sous-bit.*	Lignite *Lignite*	Peat *Tourbe*	Oven and Gas Coke *Coke de four/gaz*	Pat. Fuel and BKB *Agg./briq. de lignite*	Crude Oil *Pétrole brut*	NGL *LGN*	Feed-stocks *Produits d'aliment.*	Additives *Additifs*
Production	-	3159	-	-	-	672	-	2753	121	-	65
From Other Sources	-	-	-	-	-	-	-	-	-	-	-
Imports	960	-	-	-	-	-	-	4063	-	-	100
Exports	-	-	-	-	-	-	-	-227	-	-	-
Intl. Marine Bunkers	-	-	-	-	-	-	-	-	-	-	-
Stock Changes	-	-	-	-	-	-	-	-144	-	-	-7
DOMESTIC SUPPLY	960	3159	-	-	-	672	-	6445	121	-	158
Transfers	-	-	-	-	-	-	-	-	-121	-	-
Statistical Differences	-	-	-	-	-	-	-	-	-	-	-
TRANSFORMATION	960	347	-	-	-	538	-	6445	-	-	158
Electricity Plants	-	347	-	-	-	-	-	-	-	-	-
CHP Plants	-	-	-	-	-	-	-	-	-	-	-
Heat Plants	-	-	-	-	-	-	-	-	-	-	-
Blast Furnaces/Gas Works	-	-	-	-	-	538	-	-	-	-	-
Coke/Pat. Fuel/BKB Plants	960	-	-	-	-	-	-	-	-	-	-
Petroleum Refineries	-	-	-	-	-	-	-	6445	-	-	158
Petrochemical Industry	-	-	-	-	-	-	-	-	-	-	-
Liquefaction	-	-	-	-	-	-	-	-	-	-	-
Other Transform. Sector	-	-	-	-	-	-	-	-	-	-	-
ENERGY SECTOR	-	-	-	-	-	-	-	-	-	-	-
Coal Mines	-	-	-	-	-	-	-	-	-	-	-
Oil and Gas Extraction	-	-	-	-	-	-	-	-	-	-	-
Petroleum Refineries	-	-	-	-	-	-	-	-	-	-	-
Electr., CHP+Heat Plants	-	-	-	-	-	-	-	-	-	-	-
Pumped Storage (Elec.)	-	-	-	-	-	-	-	-	-	-	-
Other Energy Sector	-	-	-	-	-	-	-	-	-	-	-
Distribution Losses	-	-	-	-	-	-	-	-	-	-	-
FINAL CONSUMPTION	-	2812	-	-	-	134	-	-	-	-	-
INDUSTRY SECTOR	-	2810	-	-	-	134	-	-	-	-	-
Iron and Steel	-	-	-	-	-	134	-	-	-	-	-
Chemical and Petrochem.	-	-	-	-	-	-	-	-	-	-	-
of which: Feedstocks	-	-	-	-	-	-	-	-	-	-	-
Non-Ferrous Metals	-	-	-	-	-	-	-	-	-	-	-
Non-Metallic Minerals	-	2810	-	-	-	-	-	-	-	-	-
Transport Equipment	-	-	-	-	-	-	-	-	-	-	-
Machinery	-	-	-	-	-	-	-	-	-	-	-
Mining and Quarrying	-	-	-	-	-	-	-	-	-	-	-
Food and Tobacco	-	-	-	-	-	-	-	-	-	-	-
Paper, Pulp and Print	-	-	-	-	-	-	-	-	-	-	-
Wood and Wood Products	-	-	-	-	-	-	-	-	-	-	-
Construction	-	-	-	-	-	-	-	-	-	-	-
Textile and Leather	-	-	-	-	-	-	-	-	-	-	-
Non-specified	-	-	-	-	-	-	-	-	-	-	-
TRANSPORT SECTOR	-	-	-	-	-	-	-	-	-	-	-
Air	-	-	-	-	-	-	-	-	-	-	-
Road	-	-	-	-	-	-	-	-	-	-	-
Rail	-	-	-	-	-	-	-	-	-	-	-
Pipeline Transport	-	-	-	-	-	-	-	-	-	-	-
Internal Navigation	-	-	-	-	-	-	-	-	-	-	-
Non-specified	-	-	-	-	-	-	-	-	-	-	-
OTHER SECTORS	-	2	-	-	-	-	-	-	-	-	-
Agriculture	-	-	-	-	-	-	-	-	-	-	-
Comm. and Publ. Services	-	-	-	-	-	-	-	-	-	-	-
Residential	-	2	-	-	-	-	-	-	-	-	-
Non-specified	-	-	-	-	-	-	-	-	-	-	-
NON-ENERGY USE	-	-	-	-	-	-	-	-	-	-	-
in Industry/Trans./Energy	-	-	-	-	-	-	-	-	-	-	-
in Transport	-	-	-	-	-	-	-	-	-	-	-
in Other Sectors	-	-	-	-	-	-	-	-	-	-	-

Pakistan : 1998

SUPPLY AND CONSUMPTION / APPROVISIONNEMENT ET DEMANDE	Refinery Gas / Gaz de raffinerie	LPG + Ethane / GPL + éthane	Motor Gasoline / Essence moteur	Aviation Gasoline / Essence aviation	Jet Fuel / Carbu- réacteurs	Kerosene / Kérosène	Gas/ Diesel / Gazole	Heavy Fuel Oil / Fioul lourd	Naphtha / Naphta	Petrol. Coke / Coke de pétrole	Other Prod. / Autres prod.
Production	110	55	989	-	583	478	1602	2014	71	-	400
From Other Sources	-	-	-	-	-	-	-	-	-	-	-
Imports	-	30	148	-	-	43	4925	5846	-	-	-
Exports	-	-	-	-	-	-	-	-	-73	-	-
Intl. Marine Bunkers	-	-	-	-	-	-	-3	-12	-	-	-
Stock Changes	-	-	86	-	14	-7	-15	-29	2	-	-
DOMESTIC SUPPLY	110	85	1223	-	597	514	6509	7819	-	-	400
Transfers	-	121	-	-	-	-	-	-	-	-	-
Statistical Differences	-	-	-	-	-	-	-	100	-	-	-30
TRANSFORMATION	-	-	-	-	-	-	124	5910	-	-	-
Electricity Plants	-	-	-	-	-	-	124	5910	-	-	-
CHP Plants	-	-	-	-	-	-	-	-	-	-	-
Heat Plants	-	-	-	-	-	-	-	-	-	-	-
Blast Furnaces/Gas Works	-	-	-	-	-	-	-	-	-	-	-
Coke/Pat. Fuel/BKB Plants	-	-	-	-	-	-	-	-	-	-	-
Petroleum Refineries	-	-	-	-	-	-	-	-	-	-	-
Petrochemical Industry	-	-	-	-	-	-	-	-	-	-	-
Liquefaction	-	-	-	-	-	-	-	-	-	-	-
Other Transform. Sector	-	-	-	-	-	-	-	-	-	-	-
ENERGY SECTOR	110	-	-	-	-	-	-	100	-	-	-
Coal Mines	-	-	-	-	-	-	-	-	-	-	-
Oil and Gas Extraction	-	-	-	-	-	-	-	-	-	-	-
Petroleum Refineries	110	-	-	-	-	-	-	100	-	-	-
Electr., CHP+Heat Plants	-	-	-	-	-	-	-	-	-	-	-
Pumped Storage (Elec.)	-	-	-	-	-	-	-	-	-	-	-
Other Energy Sector	-	-	-	-	-	-	-	-	-	-	-
Distribution Losses	-	-	-	-	-	-	-	-	-	-	-
FINAL CONSUMPTION	-	206	1223	-	597	514	6385	1909	-	-	370
INDUSTRY SECTOR	-	-	-	-	-	-	219	1862	-	-	-
Iron and Steel	-	-	-	-	-	-	-	-	-	-	-
Chemical and Petrochem.	-	-	-	-	-	-	219	-	-	-	-
of which: Feedstocks	-	-	-	-	-	-	-	-	-	-	-
Non-Ferrous Metals	-	-	-	-	-	-	-	-	-	-	-
Non-Metallic Minerals	-	-	-	-	-	-	-	1335	-	-	-
Transport Equipment	-	-	-	-	-	-	-	-	-	-	-
Machinery	-	-	-	-	-	-	-	-	-	-	-
Mining and Quarrying	-	-	-	-	-	-	-	-	-	-	-
Food and Tobacco	-	-	-	-	-	-	-	-	-	-	-
Paper, Pulp and Print	-	-	-	-	-	-	-	-	-	-	-
Wood and Wood Products	-	-	-	-	-	-	-	-	-	-	-
Construction	-	-	-	-	-	-	-	-	-	-	-
Textile and Leather	-	-	-	-	-	-	-	-	-	-	-
Non-specified	-	-	-	-	-	-	-	527	-	-	-
TRANSPORT SECTOR	-	-	1223	-	597	-	5815	8	-	-	-
Air	-	-	-	-	597	-	-	-	-	-	-
Road	-	-	1223	-	-	-	5610	-	-	-	-
Rail	-	-	-	-	-	-	205	8	-	-	-
Pipeline Transport	-	-	-	-	-	-	-	-	-	-	-
Internal Navigation	-	-	-	-	-	-	-	-	-	-	-
Non-specified	-	-	-	-	-	-	-	-	-	-	-
OTHER SECTORS	-	206	-	-	-	514	351	39	-	-	-
Agriculture	-	-	-	-	-	-	245	-	-	-	-
Comm. and Publ. Services	-	-	-	-	-	18	103	39	-	-	-
Residential	-	206	-	-	-	496	3	-	-	-	-
Non-specified	-	-	-	-	-	-	-	-	-	-	-
NON-ENERGY USE	-	-	-	-	-	-	-	-	-	-	370
in Industry/Transf./Energy	-	-	-	-	-	-	-	-	-	-	370
in Transport	-	-	-	-	-	-	-	-	-	-	-
in Other Sectors	-	-	-	-	-	-	-	-	-	-	-

Oil cont. / Pétrole cont. (1000 tonnes)

Pakistan : 1998

SUPPLY AND CONSUMPTION *APPROVISIONNEMENT ET DEMANDE*	Gas / *Gaz* (TJ)				Comb. Renew. & Waste / *En. Re. Comb. & Déchets* (TJ)				(GWh)	(TJ)
	Natural Gas *Gaz naturel*	Gas Works *Usines à gaz*	Coke Ovens *Cokeries*	Blast Furnaces *Hauts fourneaux*	Solid Biomass *Biomasse solide*	Gas/Liquids from Biomass *Gaz/Liquides tirés de biomasse*	Municipal Waste *Déchets urbains*	Industrial Waste *Déchets industriels*	Electricity *Electricité*	Heat *Chaleur*
Production	633062	-	-	5858	958525	-	-	-	62154	-
From Other Sources	-	-	-	-	-	-	-	-	-	-
Imports	-	-	-	-	-	-	-	-	-	-
Exports	-	-	-	-	-	-	-	-	-	-
Intl. Marine Bunkers	-	-	-	-	-	-	-	-	-	-
Stock Changes	131	-	-	-	-	-	-	-	-	-
DOMESTIC SUPPLY	**633193**	**-**	**-**	**5858**	**958525**	**-**	**-**	**-**	**62154**	**-**
Transfers	-	-	-	-	-	-	-	-	-	-
Statistical Differences	2	-	-	-	-	-	-	-	-	-
TRANSFORMATION	**171985**	**-**	**-**	**-**	**19149**	**-**	**-**	**-**	**-**	**-**
Electricity Plants	171985	-	-	-	-	-	-	-	-	-
CHP Plants	-	-	-	-	-	-	-	-	-	-
Heat Plants	-	-	-	-	-	-	-	-	-	-
Blast Furnaces/Gas Works	-	-	-	-	-	-	-	-	-	-
Coke/Pat. Fuel/BKB Plants	-	-	-	-	-	-	-	-	-	-
Petroleum Refineries	-	-	-	-	-	-	-	-	-	-
Petrochemical Industry	-	-	-	-	-	-	-	-	-	-
Liquefaction	-	-	-	-	-	-	-	-	-	-
Other Transform. Sector	-	-	-	-	19149	-	-	-	-	-
ENERGY SECTOR	**9974**	**-**	**-**	**-**	**-**	**-**	**-**	**-**	**2285**	**-**
Coal Mines	-	-	-	-	-	-	-	-	-	-
Oil and Gas Extraction	9974	-	-	-	-	-	-	-	-	-
Petroleum Refineries	-	-	-	-	-	-	-	-	-	-
Electr., CHP+Heat Plants	-	-	-	-	-	-	-	-	2285	-
Pumped Storage (Elec.)	-	-	-	-	-	-	-	-	-	-
Other Energy Sector	-	-	-	-	-	-	-	-	-	-
Distribution Losses	36372	-	-	-	-	-	-	-	15502	-
FINAL CONSUMPTION	**414864**	**-**	**-**	**5858**	**939376**	**-**	**-**	**-**	**44367**	**-**
INDUSTRY SECTOR	**255826**	**-**	**-**	**5858**	**104257**	**-**	**-**	**-**	**12297**	**-**
Iron and Steel	-	-	-	5858	-	-	-	-	-	-
Chemical and Petrochem.	124108	-	-	-	-	-	-	-	-	-
of which: Feedstocks	*92327*	-	-	-	-	-	-	-	-	-
Non-Ferrous Metals	-	-	-	-	-	-	-	-	-	-
Non-Metallic Minerals	12508	-	-	-	-	-	-	-	-	-
Transport Equipment	-	-	-	-	-	-	-	-	-	-
Machinery	-	-	-	-	-	-	-	-	-	-
Mining and Quarrying	-	-	-	-	-	-	-	-	-	-
Food and Tobacco	-	-	-	-	-	-	-	-	-	-
Paper, Pulp and Print	-	-	-	-	-	-	-	-	-	-
Wood and Wood Products	-	-	-	-	-	-	-	-	-	-
Construction	-	-	-	-	-	-	-	-	-	-
Textile and Leather	-	-	-	-	-	-	-	-	-	-
Non-specified	119210	-	-	-	104257	-	-	-	12297	-
TRANSPORT SECTOR	**507**	**-**	**-**	**-**	**-**	**-**	**-**	**-**	**16**	**-**
Air	-	-	-	-	-	-	-	-	-	-
Road	-	-	-	-	-	-	-	-	-	-
Rail	-	-	-	-	-	-	-	-	16	-
Pipeline Transport	507	-	-	-	-	-	-	-	-	-
Internal Navigation	-	-	-	-	-	-	-	-	-	-
Non-specified	-	-	-	-	-	-	-	-	-	-
OTHER SECTORS	**158531**	**-**	**-**	**-**	**835119**	**-**	**-**	**-**	**32054**	**-**
Agriculture	-	-	-	-	-	-	-	-	6937	-
Comm. and Publ. Services	19409	-	-	-	-	-	-	-	6448	-
Residential	139122	-	-	-	835119	-	-	-	18669	-
Non-specified	-	-	-	-	-	-	-	-	-	-
NON-ENERGY USE	**-**	**-**	**-**	**-**	**-**	**-**	**-**	**-**	**-**	**-**
in Industry/Transf./Energy	-	-	-	-	-	-	-	-	-	-
in Transport	-	-	-	-	-	-	-	-	-	-
in Other Sectors	-	-	-	-	-	-	-	-	-	-

Panama

SUPPLY AND CONSUMPTION 1997	Coal (1000 tonnes)							Oil (1000 tonnes)			
	Coking Coal	Other Bit. Coal	Sub-Bit. Coal	Lignite	Peat	Oven and Gas Coke	Pat. Fuel and BKB	Crude Oil	NGL	Feed-stocks	Additives
Production	-	-	-	-	-	-	-	-	-	-	-
Imports	-	57	-	-	-	-	-	2017	-	-	-
Exports	-	-	-	-	-	-	-	-	-	-	-
Intl. Marine Bunkers	-	-	-	-	-	-	-	-	-	-	-
Stock Changes	-	-	-	-	-	-	-	16	-	-	-
DOMESTIC SUPPLY	-	57	-	-	-	-	-	2033	-	-	-
Transfers and Stat. Diff.	-	-	-	-	-	-	-	-	-	-	-
TRANSFORMATION	-	-	-	-	-	-	-	2033	-	-	-
Electricity and CHP Plants	-	-	-	-	-	-	-	-	-	-	-
Petroleum Refineries	-	-	-	-	-	-	-	2033	-	-	-
Other Transform. Sector	-	-	-	-	-	-	-	-	-	-	-
ENERGY SECTOR	-	-	-	-	-	-	-	-	-	-	-
DISTRIBUTION LOSSES	-	-	-	-	-	-	-	-	-	-	-
FINAL CONSUMPTION	-	57	-	-	-	-	-	-	-	-	-
INDUSTRY SECTOR	-	57	-	-	-	-	-	-	-	-	-
Iron and Steel	-	-	-	-	-	-	-	-	-	-	-
Chemical and Petrochem.	-	-	-	-	-	-	-	-	-	-	-
Non-Metallic Minerals	-	-	-	-	-	-	-	-	-	-	-
Non-specified	-	57	-	-	-	-	-	-	-	-	-
TRANSPORT SECTOR	-	-	-	-	-	-	-	-	-	-	-
Air	-	-	-	-	-	-	-	-	-	-	-
Road	-	-	-	-	-	-	-	-	-	-	-
Non-specified	-	-	-	-	-	-	-	-	-	-	-
OTHER SECTORS	-	-	-	-	-	-	-	-	-	-	-
Agriculture	-	-	-	-	-	-	-	-	-	-	-
Comm. and Publ. Services	-	-	-	-	-	-	-	-	-	-	-
Residential	-	-	-	-	-	-	-	-	-	-	-
Non-specified	-	-	-	-	-	-	-	-	-	-	-
NON-ENERGY USE	-	-	-	-	-	-	-	-	-	-	-

APPROVISIONNEMENT ET DEMANDE 1998	Charbon (1000 tonnes)							Pétrole (1000 tonnes)			
	Charbon à coke	Autres charb. bit.	Charbon sous-bit.	Lignite	Tourbe	Coke de four/gaz	Agg./briq. de lignite	Pétrole brut	LGN	Produits d'aliment.	Additifs
Production	-	-	-	-	-	-	-	-	-	-	-
Imports	-	59	-	-	-	-	-	2274	-	-	-
Exports	-	-	-	-	-	-	-	-	-	-	-
Intl. Marine Bunkers	-	-	-	-	-	-	-	-	-	-	-
Stock Changes	-	-	-	-	-	-	-	84	-	-	-
DOMESTIC SUPPLY	-	59	-	-	-	-	-	2358	-	-	-
Transfers and Stat. Diff.	-	-	-	-	-	-	-	-	-	-	-
TRANSFORMATION	-	-	-	-	-	-	-	2358	-	-	-
Electricity and CHP Plants	-	-	-	-	-	-	-	-	-	-	-
Petroleum Refineries	-	-	-	-	-	-	-	2358	-	-	-
Other Transform. Sector	-	-	-	-	-	-	-	-	-	-	-
ENERGY SECTOR	-	-	-	-	-	-	-	-	-	-	-
DISTRIBUTION LOSSES	-	-	-	-	-	-	-	-	-	-	-
FINAL CONSUMPTION	-	59	-	-	-	-	-	-	-	-	-
INDUSTRY SECTOR	-	59	-	-	-	-	-	-	-	-	-
Iron and Steel	-	-	-	-	-	-	-	-	-	-	-
Chemical and Petrochem.	-	-	-	-	-	-	-	-	-	-	-
Non-Metallic Minerals	-	-	-	-	-	-	-	-	-	-	-
Non-specified	-	59	-	-	-	-	-	-	-	-	-
TRANSPORT SECTOR	-	-	-	-	-	-	-	-	-	-	-
Air	-	-	-	-	-	-	-	-	-	-	-
Road	-	-	-	-	-	-	-	-	-	-	-
Non-specified	-	-	-	-	-	-	-	-	-	-	-
OTHER SECTORS	-	-	-	-	-	-	-	-	-	-	-
Agriculture	-	-	-	-	-	-	-	-	-	-	-
Comm. and Publ. Services	-	-	-	-	-	-	-	-	-	-	-
Residential	-	-	-	-	-	-	-	-	-	-	-
Non-specified	-	-	-	-	-	-	-	-	-	-	-
NON-ENERGY USE	-	-	-	-	-	-	-	-	-	-	-

Panama

SUPPLY AND CONSUMPTION 1997	Oil cont. (1000 tonnes)										
	Refinery Gas	LPG + Ethane	Motor Gasoline	Aviation Gasoline	Jet Fuel	Kerosene	Gas/ Diesel	Heavy Fuel Oil	Naphtha	Petrol. Coke	Other Prod.
Production	42	15	261	-	-	-	569	1096	-	-	47
Imports	-	76	77	-	4	14	135	719	-	-	3
Exports	-	-	-	-	-	-	-69	-507	-	-	-
Intl. Marine Bunkers	-	-	-	-	-	-	-60	-1000	-	-	-
Stock Changes	-	-5	2	-	-1	-4	-23	-56	-	-	-12
DOMESTIC SUPPLY	42	86	340	-	3	10	552	252	-	-	38
Transfers and Stat. Diff.	-	-1	-	-	-	-1	-1	-1	-	-	9
TRANSFORMATION	-	-	-	-	-	-	165	198	-	-	-
Electricity and CHP Plants	-	-	-	-	-	-	165	198	-	-	-
Petroleum Refineries	-	-	-	-	-	-	-	-	-	-	-
Other Transform. Sector	-	-	-	-	-	-	-	-	-	-	-
ENERGY SECTOR	42	-	-	-	-	-	2	-	-	-	-
DISTRIBUTION LOSSES	-	-	-	-	-	-	-	-	-	-	-
FINAL CONSUMPTION	-	85	340	-	3	9	384	53	-	-	47
INDUSTRY SECTOR	-	2	-	-	-	-	153	52	-	-	-
Iron and Steel	-	-	-	-	-	-	-	-	-	-	-
Chemical and Petrochem.	-	-	-	-	-	-	-	-	-	-	-
Non-Metallic Minerals	-	-	-	-	-	-	-	-	-	-	-
Non-specified	-	2	-	-	-	-	153	52	-	-	-
TRANSPORT SECTOR	-	-	340	-	3	-	220	-	-	-	-
Air	-	-	-	-	3	-	-	-	-	-	-
Road	-	-	340	-	-	-	220	-	-	-	-
Non-specified	-	-	-	-	-	-	-	-	-	-	-
OTHER SECTORS	-	83	-	-	-	9	11	1	-	-	-
Agriculture	-	-	-	-	-	-	-	-	-	-	-
Comm. and Publ. Services	-	19	-	-	-	-	11	1	-	-	-
Residential	-	64	-	-	-	9	-	-	-	-	-
Non-specified	-	-	-	-	-	-	-	-	-	-	-
NON-ENERGY USE	-	-	-	-	-	-	-	-	-	-	47

APPROVISIONNEMENT ET DEMANDE 1998	Pétrole cont. (1000 tonnes)										
	Gaz de raffinerie	GPL + éthane	Essence moteur	Essence aviation	Carbu- réacteurs	Kérosène	Gazole	Fioul lourd	Naphta	Coke de pétrole	Autres prod.
Production	42	27	312	-	25	77	612	1208	-	-	47
Imports	-	70	104	-	-	-	286	731	-	-	-
Exports	-	-	-2	-	-21	-65	-115	-592	-	-	-
Intl. Marine Bunkers	-	-	-	-	-	-	-60	-1000	-	-	-
Stock Changes	-	-4	-32	-	-	-	-3	-18	-	-	-
DOMESTIC SUPPLY	42	93	382	-	4	12	720	329	-	-	47
Transfers and Stat. Diff.	-	-3	-1	-	-	-1	-	-	-	-	-
TRANSFORMATION	-	-	-	-	-	-	286	275	-	-	-
Electricity and CHP Plants	-	-	-	-	-	-	286	275	-	-	-
Petroleum Refineries	-	-	-	-	-	-	-	-	-	-	-
Other Transform. Sector	-	-	-	-	-	-	-	-	-	-	-
ENERGY SECTOR	42	-	-	-	-	-	2	-	-	-	-
DISTRIBUTION LOSSES	-	-	-	-	-	-	-	-	-	-	-
FINAL CONSUMPTION	-	90	381	-	4	11	432	54	-	-	47
INDUSTRY SECTOR	-	2	-	-	-	-	170	53	-	-	-
Iron and Steel	-	-	-	-	-	-	-	-	-	-	-
Chemical and Petrochem.	-	-	-	-	-	-	-	-	-	-	-
Non-Metallic Minerals	-	-	-	-	-	-	-	-	-	-	-
Non-specified	-	2	-	-	-	-	170	53	-	-	-
TRANSPORT SECTOR	-	-	381	-	4	-	250	-	-	-	-
Air	-	-	-	-	4	-	-	-	-	-	-
Road	-	-	381	-	-	-	250	-	-	-	-
Non-specified	-	-	-	-	-	-	-	-	-	-	-
OTHER SECTORS	-	88	-	-	-	11	12	1	-	-	-
Agriculture	-	-	-	-	-	-	-	-	-	-	-
Comm. and Publ. Services	-	20	-	-	-	-	12	1	-	-	-
Residential	-	68	-	-	-	11	-	-	-	-	-
Non-specified	-	-	-	-	-	-	-	-	-	-	-
NON-ENERGY USE	-	-	-	-	-	-	-	-	-	-	47

Panama

SUPPLY AND CONSUMPTION 1997	Gas (TJ)				Comb. Renew. & Waste (TJ)				(GWh)	(TJ)
	Natural Gas	Gas Works	Coke Ovens	Blast Furnaces	Solid Biomass	Gas/Liquids from Biomass	Municipal Waste	Industrial Waste	Electricity	Heat
Production	-	-	-	-	19039	-	-	-	4151	-
Imports	-	-	-	-	-	-	-	-	65	-
Exports	-	-	-	-		-	-	-	-118	-
Intl. Marine Bunkers	-	-	-	-	-	-	-	-	-	-
Stock Changes	-	-	-	-	-	-	-	-	-	-
DOMESTIC SUPPLY	-	-	-	-	19039	-	-	-	4098	-
Transfers and Stat. Diff.	-	-	-	-	-1	-	-	-	-	-
TRANSFORMATION	-	-	-	-	787	-	-	-	-	-
Electricity and CHP Plants	-	-	-	-	535	-	-	-	-	-
Petroleum Refineries	-	-	-	-	-	-	-	-	-	-
Other Transform. Sector	-	-	-	-	252	-	-	-	-	-
ENERGY SECTOR	-	-	-	-	-	-	-	-	52	-
DISTRIBUTION LOSSES	-	-	-	-	-	-	-	-	914	-
FINAL CONSUMPTION	-	-	-	-	18251	-	-	-	3132	-
INDUSTRY SECTOR	-	-	-	-	2984	-	-	-	498	-
Iron and Steel	-	-	-	-	-	-	-	-	-	-
Chemical and Petrochem.	-	-	-	-	-	-	-	-	-	-
Non-Metallic Minerals	-	-	-	-	-	-	-	-	-	-
Non-specified	-	-	-	-	2984	-	-	-	498	-
TRANSPORT SECTOR	-	-	-	-	-	-	-	-	93	-
Air	-	-	-	-	-	-	-	-	-	-
Road	-	-	-	-	-	-	-	-	-	-
Non-specified	-	-	-	-	-	-	-	-	93	-
OTHER SECTORS	-	-	-	-	15267	-	-	-	2541	-
Agriculture	-	-	-	-	-	-	-	-	-	-
Comm. and Publ. Services	-	-	-	-	44	-	-	-	-	-
Residential	-	-	-	-	15223	-	-	-	2541	-
Non-specified	-	-	-	-	-	-	-	-	-	-
NON-ENERGY USE	-	-	-	-	-	-	-	-	-	-

APPROVISIONNEMENT ET DEMANDE 1998	Gaz (TJ)				En. Re. Comb. & Déchets (TJ)				(GWh)	(TJ)
	Gaz naturel	Usines à gaz	Cokeries	Hauts fourneaux	Biomasse solide	Gaz/Liquides tirés de biomasse	Déchets urbains	Déchets industriels	Electricité	Chaleur
Production	-	-	-	-	19238	-	-	-	4353	-
Imports	-	-	-	-	-	-	-	-	57	-
Exports	-	-	-	-	-	-	-	-	-23	-
Intl. Marine Bunkers	-	-	-	-	-	-	-	-	-	-
Stock Changes	-	-	-	-	-	-	-	-	-	-
DOMESTIC SUPPLY	-	-	-	-	19238	-	-	-	4387	-
Transfers and Stat. Diff.	-	-	-	-	-	-	-	-	-	-
TRANSFORMATION	-	-	-	-	815	-	-	-	-	-
Electricity and CHP Plants	-	-	-	-	552	-	-	-	-	-
Petroleum Refineries	-	-	-	-	-	-	-	-	-	-
Other Transform. Sector	-	-	-	-	263	-	-	-	-	-
ENERGY SECTOR	-	-	-	-	-	-	-	-	50	-
DISTRIBUTION LOSSES	-	-	-	-	-	-	-	-	991	-
FINAL CONSUMPTION	-	-	-	-	18423	-	-	-	3346	-
INDUSTRY SECTOR	-	-	-	-	3026	-	-	-	532	-
Iron and Steel	-	-	-	-	-	-	-	-	-	-
Chemical and Petrochem.	-	-	-	-	-	-	-	-	-	-
Non-Metallic Minerals	-	-	-	-	-	-	-	-	-	-
Non-specified	-	-	-	-	3026	-	-	-	532	-
TRANSPORT SECTOR	-	-	-	-	-	-	-	-	99	-
Air	-	-	-	-	-	-	-	-	-	-
Road	-	-	-	-	-	-	-	-	-	-
Non-specified	-	-	-	-	-	-	-	-	99	-
OTHER SECTORS	-	-	-	-	15397	-	-	-	2715	-
Agriculture	-	-	-	-	-	-	-	-	-	-
Comm. and Publ. Services	-	-	-	-	46	-	-	-	-	-
Residential	-	-	-	-	15351	-	-	-	2715	-
Non-specified	-	-	-	-	-	-	-	-	-	-
NON-ENERGY USE	-	-	-	-	-	-	-	-	-	-

Paraguay

SUPPLY AND CONSUMPTION 1997	Coal (1000 tonnes)							Oil (1000 tonnes)			
	Coking Coal	Other Bit. Coal	Sub-Bit. Coal	Lignite	Peat	Oven and Gas Coke	Pat. Fuel and BKB	Crude Oil	NGL	Feed-stocks	Additives
Production	-	-	-	-	-	-	-	-	-	-	-
Imports	-	-	-	-	-	-	-	155	-	-	-
Exports	-	-	-	-	-	-	-	-	-	-	-
Intl. Marine Bunkers	-	-	-	-	-	-	-	-	-	-	-
Stock Changes	-	-	-	-	-	-	-	3	-	-	-
DOMESTIC SUPPLY	-	-	-	-	-	-	-	158	-	-	-
Transfers and Stat. Diff.	-	-	-	-	-	-	-	-	-	-	-
TRANSFORMATION	-	-	-	-	-	-	-	158	-	-	-
Electricity and CHP Plants	-	-	-	-	-	-	-	-	-	-	-
Petroleum Refineries	-	-	-	-	-	-	-	158	-	-	-
Other Transform. Sector	-	-	-	-	-	-	-	-	-	-	-
ENERGY SECTOR	-	-	-	-	-	-	-	-	-	-	-
DISTRIBUTION LOSSES	-	-	-	-	-	-	-	-	-	-	-
FINAL CONSUMPTION	-	-	-	-	-	-	-	-	-	-	-
INDUSTRY SECTOR	-	-	-	-	-	-	-	-	-	-	-
Iron and Steel	-	-	-	-	-	-	-	-	-	-	-
Chemical and Petrochem.	-	-	-	-	-	-	-	-	-	-	-
Non-Metallic Minerals	-	-	-	-	-	-	-	-	-	-	-
Non-specified	-	-	-	-	-	-	-	-	-	-	-
TRANSPORT SECTOR	-	-	-	-	-	-	-	-	-	-	-
Air	-	-	-	-	-	-	-	-	-	-	-
Road	-	-	-	-	-	-	-	-	-	-	-
Non-specified	-	-	-	-	-	-	-	-	-	-	-
OTHER SECTORS	-	-	-	-	-	-	-	-	-	-	-
Agriculture	-	-	-	-	-	-	-	-	-	-	-
Comm. and Publ. Services	-	-	-	-	-	-	-	-	-	-	-
Residential	-	-	-	-	-	-	-	-	-	-	-
Non-specified	-	-	-	-	-	-	-	-	-	-	-
NON-ENERGY USE	-	-	-	-	-	-	-	-	-	-	-

APPROVISIONNEMENT ET DEMANDE 1998	Charbon (1000 tonnes)							Pétrole (1000 tonnes)			
	Charbon à coke	Autres charb. bit.	Charbon sous-bit.	Lignite	Tourbe	Coke de four/gaz	Agg./briq. de lignite	Pétrole brut	LGN	Produits d'aliment.	Additifs
Production	-	-	-	-	-	-	-	-	-	-	-
Imports	-	-	-	-	-	-	-	138	-	-	-
Exports	-	-	-	-	-	-	-	-	-	-	-
Intl. Marine Bunkers	-	-	-	-	-	-	-	-	-	-	-
Stock Changes	-	-	-	-	-	-	-	-	-	-	-
DOMESTIC SUPPLY	-	-	-	-	-	-	-	138	-	-	-
Transfers and Stat. Diff.	-	-	-	-	-	-	-	-	-	-	-
TRANSFORMATION	-	-	-	-	-	-	-	138	-	-	-
Electricity and CHP Plants	-	-	-	-	-	-	-	-	-	-	-
Petroleum Refineries	-	-	-	-	-	-	-	138	-	-	-
Other Transform. Sector	-	-	-	-	-	-	-	-	-	-	-
ENERGY SECTOR	-	-	-	-	-	-	-	-	-	-	-
DISTRIBUTION LOSSES	-	-	-	-	-	-	-	-	-	-	-
FINAL CONSUMPTION	-	-	-	-	-	-	-	-	-	-	-
INDUSTRY SECTOR	-	-	-	-	-	-	-	-	-	-	-
Iron and Steel	-	-	-	-	-	-	-	-	-	-	-
Chemical and Petrochem.	-	-	-	-	-	-	-	-	-	-	-
Non-Metallic Minerals	-	-	-	-	-	-	-	-	-	-	-
Non-specified	-	-	-	-	-	-	-	-	-	-	-
TRANSPORT SECTOR	-	-	-	-	-	-	-	-	-	-	-
Air	-	-	-	-	-	-	-	-	-	-	-
Road	-	-	-	-	-	-	-	-	-	-	-
Non-specified	-	-	-	-	-	-	-	-	-	-	-
OTHER SECTORS	-	-	-	-	-	-	-	-	-	-	-
Agriculture	-	-	-	-	-	-	-	-	-	-	-
Comm. and Publ. Services	-	-	-	-	-	-	-	-	-	-	-
Residential	-	-	-	-	-	-	-	-	-	-	-
Non-specified	-	-	-	-	-	-	-	-	-	-	-
NON-ENERGY USE	-	-	-	-	-	-	-	-	-	-	-

Paraguay

SUPPLY AND CONSUMPTION 1997	Oil cont. (1000 tonnes)										
	Refinery Gas	LPG + Ethane	Motor Gasoline	Aviation Gasoline	Jet Fuel	Kerosene	Gas/ Diesel	Heavy Fuel Oil	Naphtha	Petrol. Coke	Other Prod.
Production	-	1	27	-	-	5	56	56	-	-	-
Imports	-	64	144	3	-	2	693	-	-	-	12
Exports	-	-	-	-	-	-	-	-19	-	-	-
Intl. Marine Bunkers	-	-	-	-	-	-	-	-	-	-	-
Stock Changes	-	1	-1	-	-	-1	58	-	-	-	-
DOMESTIC SUPPLY	-	66	170	3	-	6	807	37	-	-	12
Transfers and Stat. Diff.	-	-	-1	-	-	1	-	1	-	-	-
TRANSFORMATION	-	-	-	-	-	-	34	-	-	-	-
Electricity and CHP Plants	-	-	-	-	-	-	34	-	-	-	-
Petroleum Refineries	-	-	-	-	-	-	-	-	-	-	-
Other Transform. Sector	-	-	-	-	-	-	-	-	-	-	-
ENERGY SECTOR	-	-	-	-	-	-	-	-	-	-	-
DISTRIBUTION LOSSES	-	-	-	-	-	-	-	-	-	-	-
FINAL CONSUMPTION	-	66	169	3	-	7	773	38	-	-	12
INDUSTRY SECTOR	-	-	-	-	-	-	2	38	-	-	-
Iron and Steel	-	-	-	-	-	-	-	-	-	-	-
Chemical and Petrochem.	-	-	-	-	-	-	-	-	-	-	-
Non-Metallic Minerals	-	-	-	-	-	-	-	-	-	-	-
Non-specified	-	-	-	-	-	-	2	38	-	-	-
TRANSPORT SECTOR	-	10	169	3	-	-	771	-	-	-	-
Air	-	-	-	3	-	-	-	-	-	-	-
Road	-	-	169	-	-	-	771	-	-	-	-
Non-specified	-	10	-	-	-	-	-	-	-	-	-
OTHER SECTORS	-	56	-	-	-	7	-	-	-	-	-
Agriculture	-	-	-	-	-	-	-	-	-	-	-
Comm. and Publ. Services	-	-	-	-	-	-	-	-	-	-	-
Residential	-	56	-	-	-	7	-	-	-	-	-
Non-specified	-	-	-	-	-	-	-	-	-	-	-
NON-ENERGY USE	-	-	-	-	-	-	-	-	-	-	12

APPROVISIONNEMENT ET DEMANDE 1998	Pétrole cont. (1000 tonnes)										
	Gaz de raffinerie	GPL + éthane	Essence moteur	Essence aviation	Carbu- réacteurs	Kérosène	Gazole	Fioul lourd	Naphta	Coke de pétrole	Autres prod.
Production	-	-	19	-	-	1	60	43	-	-	-
Imports	-	78	182	3	-	7	685	6	-	-	12
Exports	-	-	-	-	-	-	-	-2	-	-	-
Intl. Marine Bunkers	-	-	-	-	-	-	-	-	-	-	-
Stock Changes	-	-	19	-	-	-	139	-	-	-	-
DOMESTIC SUPPLY	-	78	220	3	-	8	884	47	-	-	12
Transfers and Stat. Diff.	-	-	-	-	-	-	-	-9	-	-	-
TRANSFORMATION	-	-	-	-	-	-	6	-	-	-	-
Electricity and CHP Plants	-	-	-	-	-	-	6	-	-	-	-
Petroleum Refineries	-	-	-	-	-	-	-	-	-	-	-
Other Transform. Sector	-	-	-	-	-	-	-	-	-	-	-
ENERGY SECTOR	-	-	-	-	-	-	-	-	-	-	-
DISTRIBUTION LOSSES	-	-	-	-	-	-	-	-	-	-	-
FINAL CONSUMPTION	-	78	220	3	-	8	878	38	-	-	12
INDUSTRY SECTOR	-	-	-	-	-	-	-	38	-	-	-
Iron and Steel	-	-	-	-	-	-	-	-	-	-	-
Chemical and Petrochem.	-	-	-	-	-	-	-	-	-	-	-
Non-Metallic Minerals	-	-	-	-	-	-	-	-	-	-	-
Non-specified	-	-	-	-	-	-	-	38	-	-	-
TRANSPORT SECTOR	-	10	220	3	-	-	878	-	-	-	-
Air	-	-	-	3	-	-	-	-	-	-	-
Road	-	-	220	-	-	-	878	-	-	-	-
Non-specified	-	10	-	-	-	-	-	-	-	-	-
OTHER SECTORS	-	68	-	-	-	8	-	-	-	-	-
Agriculture	-	-	-	-	-	-	-	-	-	-	-
Comm. and Publ. Services	-	-	-	-	-	-	-	-	-	-	-
Residential	-	68	-	-	-	8	-	-	-	-	-
Non-specified	-	-	-	-	-	-	-	-	-	-	-
NON-ENERGY USE	-	-	-	-	-	-	-	-	-	-	12

Paraguay

SUPPLY AND CONSUMPTION 1997	Gas (TJ)				Comb. Renew. & Waste (TJ)				(GWh)	(TJ)
	Natural Gas	Gas Works	Coke Ovens	Blast Furnaces	Solid Biomass	Gas/Liquids from Biomass	Municipal Waste	Industrial Waste	Electricity	Heat
Production	-	-	-	-	110149	477	-	-	50619	-
Imports	-	-	-	-	-	-	-	-	-	-
Exports	-	-	-	-	-87	-	-	-	-45673	-
Intl. Marine Bunkers	-	-	-	-	-	-	-	-	-	-
Stock Changes	-	-	-	-	-	-	-	-	-	-
DOMESTIC SUPPLY	-	-	-	-	110062	477	-	-	4946	-
Transfers and Stat. Diff.	-	-	-	-	-	-	-	-	-	-
TRANSFORMATION	-	-	-	-	4433	-	-	-	-	-
Electricity and CHP Plants	-	-	-	-	716	-	-	-	-	-
Petroleum Refineries	-	-	-	-	-	-	-	-	-	-
Other Transform. Sector	-	-	-	-	3717	-	-	-	-	-
ENERGY SECTOR	-	-	-	-	-	-	-	-	158	-
DISTRIBUTION LOSSES	-	-	-	-	-	-	-	-	927	-
FINAL CONSUMPTION	-	-	-	-	105629	477	-	-	3861	-
INDUSTRY SECTOR	-	-	-	-	56729	-	-	-	1386	-
Iron and Steel	-	-	-	-	-	-	-	-	-	-
Chemical and Petrochem.	-	-	-	-	-	-	-	-	-	-
Non-Metallic Minerals	-	-	-	-	-	-	-	-	-	-
Non-specified	-	-	-	-	56729	-	-	-	1386	-
TRANSPORT SECTOR	-	-	-	-	347	477	-	-	-	-
Air	-	-	-	-	-	-	-	-	-	-
Road	-	-	-	-	-	477	-	-	-	-
Non-specified	-	-	-	-	347	-	-	-	-	-
OTHER SECTORS	-	-	-	-	48553	-	-	-	2475	-
Agriculture	-	-	-	-	-	-	-	-	-	-
Comm. and Publ. Services	-	-	-	-	181	-	-	-	753	-
Residential	-	-	-	-	48372	-	-	-	1722	-
Non-specified	-	-	-	-	-	-	-	-	-	-
NON-ENERGY USE	-	-	-	-	-	-	-	-	-	-

APPROVISIONNEMENT ET DEMANDE 1998	Gaz (TJ)				En. Re. Comb. & Déchets (TJ)				(GWh)	(TJ)
	Gaz naturel	Usines à gaz	Cokeries	Hauts fourneaux	Biomasse solide	Gaz/Liquides tirés de biomasse	Déchets urbains	Déchets industriels	Electricité	Chaleur
Production	-	-	-	-	104443	94	-	-	50893	-
Imports	-	-	-	-	-	-	-	-	-	-
Exports	-	-	-	-	-	-	-	-	-45357	-
Intl. Marine Bunkers	-	-	-	-	-	-	-	-	-	-
Stock Changes	-	-	-	-	-	-	-	-	-	-
DOMESTIC SUPPLY	-	-	-	-	104443	94	-	-	5536	-
Transfers and Stat. Diff.	-	-	-	-	1	-	-	-	-	-
TRANSFORMATION	-	-	-	-	4922	-	-	-	-	-
Electricity and CHP Plants	-	-	-	-	1832	-	-	-	-	-
Petroleum Refineries	-	-	-	-	-	-	-	-	-	-
Other Transform. Sector	-	-	-	-	3090	-	-	-	-	-
ENERGY SECTOR	-	-	-	-	-	-	-	-	201	-
DISTRIBUTION LOSSES	-	-	-	-	-	-	-	-	1391	-
FINAL CONSUMPTION	-	-	-	-	99522	94	-	-	3944	-
INDUSTRY SECTOR	-	-	-	-	56591	-	-	-	1382	-
Iron and Steel	-	-	-	-	-	-	-	-	-	-
Chemical and Petrochem.	-	-	-	-	-	-	-	-	-	-
Non-Metallic Minerals	-	-	-	-	-	-	-	-	-	-
Non-specified	-	-	-	-	56591	-	-	-	1382	-
TRANSPORT SECTOR	-	-	-	-	302	94	-	-	-	-
Air	-	-	-	-	-	-	-	-	-	-
Road	-	-	-	-	-	94	-	-	-	-
Non-specified	-	-	-	-	302	-	-	-	-	-
OTHER SECTORS	-	-	-	-	42629	-	-	-	2562	-
Agriculture	-	-	-	-	-	-	-	-	-	-
Comm. and Publ. Services	-	-	-	-	151	-	-	-	782	-
Residential	-	-	-	-	42478	-	-	-	1780	-
Non-specified	-	-	-	-	-	-	-	-	-	-
NON-ENERGY USE	-	-	-	-	-	-	-	-	-	-

Peru / Pérou

SUPPLY AND CONSUMPTION 1997	Coal (1000 tonnes)							Oil (1000 tonnes)			
	Coking Coal	Other Bit. Coal	Sub-Bit. Coal	Lignite	Peat	Oven and Gas Coke	Pat. Fuel and BKB	Crude Oil	NGL	Feed-stocks	Additives
Production	-	22	-	-	-	26	-	5824	71	-	-
Imports	41	282	-	-	-	150	-	3938	-	-	-
Exports	-	-	-	-	-	-	-	-2246	-	-	-
Intl. Marine Bunkers	-	-	-	-	-	-	-	-	-	-	-
Stock Changes	2	33	-	-	-	-19	-	602	-	-	-
DOMESTIC SUPPLY	43	337	-	-	-	157	-	8118	71	-	-
Transfers and Stat. Diff.	-	40	-	-	-	5	-	-	-71	-	-
TRANSFORMATION	43	-	-	-	-	139	-	8118	-	-	-
Electricity and CHP Plants	-	-	-	-	-	-	-	-	-	-	-
Petroleum Refineries	-	-	-	-	-	-	-	8118	-	-	-
Other Transform. Sector	43	-	-	-	-	139	-	-	-	-	-
ENERGY SECTOR	-	-	-	-	-	-	-	-	-	-	-
DISTRIBUTION LOSSES	-	-	-	-	-	-	-	-	-	-	-
FINAL CONSUMPTION	-	377	-	-	-	23	-	-	-	-	-
INDUSTRY SECTOR	-	368	-	-	-	23	-	-	-	-	-
Iron and Steel	-	53	-	-	-	23	-	-	-	-	-
Chemical and Petrochem.	-	-	-	-	-	-	-	-	-	-	-
Non-Metallic Minerals	-	-	-	-	-	-	-	-	-	-	-
Non-specified	-	315	-	-	-	-	-	-	-	-	-
TRANSPORT SECTOR	-	-	-	-	-	-	-	-	-	-	-
Air	-	-	-	-	-	-	-	-	-	-	-
Road	-	-	-	-	-	-	-	-	-	-	-
Non-specified	-	-	-	-	-	-	-	-	-	-	-
OTHER SECTORS	-	9	-	-	-	-	-	-	-	-	-
Agriculture	-	-	-	-	-	-	-	-	-	-	-
Comm. and Publ. Services	-	-	-	-	-	-	-	-	-	-	-
Residential	-	9	-	-	-	-	-	-	-	-	-
Non-specified	-	-	-	-	-	-	-	-	-	-	-
NON-ENERGY USE	-	-	-	-	-	-	-	-	-	-	-

APPROVISIONNEMENT ET DEMANDE 1998	Charbon (1000 tonnes)							Pétrole (1000 tonnes)			
	Charbon à coke	Autres charb. bit.	Charbon sous-bit.	Lignite	Tourbe	Coke de four/gaz	Agg./briq. de lignite	Pétrole brut	LGN	Produits d'aliment.	Additifs
Production	-	21	-	-	-	39	-	5694	102	-	-
Imports	38	399	-	-	-	166	-	4665	-	-	-
Exports	-	-	-	-	-	-	-	-2117	-	-	-
Intl. Marine Bunkers	-	-	-	-	-	-	-	-	-	-	-
Stock Changes	-	-9	-	-	-	19	-	114	-	-	-
DOMESTIC SUPPLY	38	411	-	-	-	224	-	8356	102	-	-
Transfers and Stat. Diff.	-	23	-	-	-	-	-	-	-102	-	-
TRANSFORMATION	38	8	-	-	-	184	-	8356	-	-	-
Electricity and CHP Plants	-	-	-	-	-	-	-	-	-	-	-
Petroleum Refineries	-	-	-	-	-	-	-	8356	-	-	-
Other Transform. Sector	38	8	-	-	-	184	-	-	-	-	-
ENERGY SECTOR	-	-	-	-	-	-	-	-	-	-	-
DISTRIBUTION LOSSES	-	-	-	-	-	-	-	-	-	-	-
FINAL CONSUMPTION	-	426	-	-	-	40	-	-	-	-	-
INDUSTRY SECTOR	-	411	-	-	-	40	-	-	-	-	-
Iron and Steel	-	59	-	-	-	40	-	-	-	-	-
Chemical and Petrochem.	-	-	-	-	-	-	-	-	-	-	-
Non-Metallic Minerals	-	-	-	-	-	-	-	-	-	-	-
Non-specified	-	352	-	-	-	-	-	-	-	-	-
TRANSPORT SECTOR	-	-	-	-	-	-	-	-	-	-	-
Air	-	-	-	-	-	-	-	-	-	-	-
Road	-	-	-	-	-	-	-	-	-	-	-
Non-specified	-	-	-	-	-	-	-	-	-	-	-
OTHER SECTORS	-	15	-	-	-	-	-	-	-	-	-
Agriculture	-	6	-	-	-	-	-	-	-	-	-
Comm. and Publ. Services	-	-	-	-	-	-	-	-	-	-	-
Residential	-	9	-	-	-	-	-	-	-	-	-
Non-specified	-	-	-	-	-	-	-	-	-	-	-
NON-ENERGY USE	-	-	-	-	-	-	-	-	-	-	-

Peru / Pérou

SUPPLY AND CONSUMPTION 1997	Oil cont. (1000 tonnes)										
	Refinery Gas	LPG + Ethane	Motor Gasoline	Aviation Gasoline	Jet Fuel	Kerosene	Gas/ Diesel	Heavy Fuel Oil	Naphtha	Petrol. Coke	Other Prod.
Production	76	201	1312	-	361	656	2023	2847	-	-	173
Imports	-	107	38	31	125	-	1075	-	-	-	6
Exports	-	-	-157	-	-6	-	-	-1006	-	-	-20
Intl. Marine Bunkers	-	-	-	-	-	-	-21	-	-	-	-
Stock Changes	-	-28	23	-20	-52	56	24	55	-	-	-
DOMESTIC SUPPLY	76	280	1216	11	428	712	3101	1896	-	-	159
Transfers and Stat. Diff.	-	14	7	-	1	1	-2	-2	-	-	-7
TRANSFORMATION	-	-	-	-	-	5	652	471	-	-	-
Electricity and CHP Plants	-	-	-	-	-	5	652	471	-	-	-
Petroleum Refineries	-	-	-	-	-	-	-	-	-	-	-
Other Transform. Sector	-	-	-	-	-	-	-	-	-	-	-
ENERGY SECTOR	76	-	3	-	-	24	104	135	-	-	-
DISTRIBUTION LOSSES	-	-	-	-	-	-	-	-	-	-	-
FINAL CONSUMPTION	-	294	1220	11	429	684	2343	1288	-	-	152
INDUSTRY SECTOR	-	2	6	-	-	47	318	877	-	-	-
Iron and Steel	-	-	1	-	-	18	163	-	-	-	-
Chemical and Petrochem.	-	-	-	-	-	-	-	-	-	-	-
Non-Metallic Minerals	-	2	-	-	-	-	-	-	-	-	-
Non-specified	-	-	5	-	-	29	155	877	-	-	-
TRANSPORT SECTOR	-	-	972	11	429	-	1735	47	-	-	-
Air	-	-	-	11	429	-	-	-	-	-	-
Road	-	-	972	-	-	-	1735	-	-	-	-
Non-specified	-	-	-	-	-	-	-	47	-	-	-
OTHER SECTORS	-	292	242	-	-	637	290	364	-	-	-
Agriculture	-	-	5	-	-	6	202	356	-	-	-
Comm. and Publ. Services	-	-	138	-	-	60	88	8	-	-	-
Residential	-	292	-	-	-	571	-	-	-	-	-
Non-specified	-	-	99	-	-	-	-	-	-	-	-
NON-ENERGY USE	-	-	-	-	-	-	-	-	-	-	152

APPROVISIONNEMENT ET DEMANDE 1998	Pétrole cont. (1000 tonnes)										
	Gaz de raffinerie	GPL + éthane	Essence moteur	Essence aviation	Carbu- réacteurs	Kérosène	Gazole	Fioul lourd	Naphta	Coke de pétrole	Autres prod.
Production	40	240	1478	-	431	644	2015	3295	-	-	102
Imports	-	140	47	2	-	-	930	-	-	-	96
Exports	-	-	-150	-	-6	-	-	-1468	-	-	-1
Intl. Marine Bunkers	-	-	-	-	-	-	-45	-	-	-	-
Stock Changes	-	-65	-253	-	-8	51	-46	36	-	-	-33
DOMESTIC SUPPLY	40	315	1122	2	417	695	2854	1863	-	-	164
Transfers and Stat. Diff.	-	32	69	-	-1	-2	-	-2	-	-	5
TRANSFORMATION	-	-	-	-	-	6	451	537	-	-	-
Electricity and CHP Plants	-	-	-	-	-	6	451	537	-	-	-
Petroleum Refineries	-	-	-	-	-	-	-	-	-	-	-
Other Transform. Sector	-	-	-	-	-	-	-	-	-	-	-
ENERGY SECTOR	40	-	3	-	-	24	104	135	-	-	-
DISTRIBUTION LOSSES	-	-	-	-	-	-	-	-	-	-	-
FINAL CONSUMPTION	-	347	1188	2	416	663	2299	1189	-	-	169
INDUSTRY SECTOR	-	6	8	-	-	45	356	960	-	-	-
Iron and Steel	-	-	2	-	-	19	170	357	-	-	-
Chemical and Petrochem.	-	-	-	-	-	-	-	-	-	-	-
Non-Metallic Minerals	-	2	-	-	-	-	-	-	-	-	-
Non-specified	-	4	6	-	-	26	186	603	-	-	-
TRANSPORT SECTOR	-	-	1030	2	416	-	1743	49	-	-	-
Air	-	-	-	2	416	-	-	-	-	-	-
Road	-	-	1030	-	-	-	1743	-	-	-	-
Non-specified	-	-	-	-	-	-	-	49	-	-	-
OTHER SECTORS	-	341	150	-	-	618	200	180	-	-	-
Agriculture	-	-	4	-	-	6	107	171	-	-	-
Comm. and Publ. Services	-	-	144	-	-	63	93	9	-	-	-
Residential	-	341	-	-	-	549	-	-	-	-	-
Non-specified	-	-	2	-	-	-	-	-	-	-	-
NON-ENERGY USE	-	-	-	-	-	-	-	-	-	-	169

Peru / Pérou

SUPPLY AND CONSUMPTION 1997	Gas (TJ) Natural Gas	Gas Works	Coke Ovens	Blast Furnaces	Comb. Renew. & Waste (TJ) Solid Biomass	Gas/Liquids from Biomass	Municipal Waste	Industrial Waste	(GWh) Electricity	(TJ) Heat
Production	16010	14696	-	1424	181079	-	-	-	17954	-
Imports	-	-	-	-	-	-	-	-	2	-
Exports	-	-	-	-	-	-	-	-	-	-
Intl. Marine Bunkers	-	-	-	-	-	-	-	-	-	-
Stock Changes	-	-	-	-	-	-	-	-	-	-
DOMESTIC SUPPLY	16010	14696	-	1424	181079	-	-	-	17956	-
Transfers and Stat. Diff.	-	-	-	-	-	-	-	-	2	-
TRANSFORMATION	16010	4354	-	-	10635	-	-	-	-	-
Electricity and CHP Plants	-	4354	-	-	2219	-	-	-	-	-
Petroleum Refineries	-	-	-	-	-	-	-	-	-	-
Other Transform. Sector	16010	-	-	-	8416	-	-	-	-	-
ENERGY SECTOR	-	10216	-	-	-	-	-	-	280	-
DISTRIBUTION LOSSES	-	-	-	-	-	-	-	-	2885	-
FINAL CONSUMPTION	-	126	-	1424	170444	-	-	-	14793	-
INDUSTRY SECTOR	-	-	-	1424	20473	-	-	-	8536	-
Iron and Steel	-	-	-	1424	-	-	-	-	3756	-
Chemical and Petrochem.	-	-	-	-	-	-	-	-	-	-
Non-Metallic Minerals	-	-	-	-	-	-	-	-	-	-
Non-specified	-	-	-	-	20473	-	-	-	4780	-
TRANSPORT SECTOR	-	-	-	-	-	-	-	-	-	-
Air	-	-	-	-	-	-	-	-	-	-
Road	-	-	-	-	-	-	-	-	-	-
Non-specified	-	-	-	-	-	-	-	-	-	-
OTHER SECTORS	-	126	-	-	149971	-	-	-	6257	-
Agriculture	-	-	-	-	5987	-	-	-	582	-
Comm. and Publ. Services	-	-	-	-	-	-	-	-	717	-
Residential	-	126	-	-	143984	-	-	-	4958	-
Non-specified	-	-	-	-	-	-	-	-	-	-
NON-ENERGY USE	-	-	-	-	-	-	-	-	-	-

APPROVISIONNEMENT ET DEMANDE 1998	Gaz (TJ) Gaz naturel	Usines à gaz	Cokeries	Hauts fourneaux	En. Re. Comb. & Déchets (TJ) Biomasse solide	Gaz/Liquides tirés de biomasse	Déchets urbains	Déchets industriels	(GWh) Electricité	(TJ) Chaleur
Production	22207	17446	-	1583	182869	-	-	-	18583	-
Imports	-	-	-	-	-	-	-	-	1	-
Exports	-	-	-	-	-	-	-	-	-	-
Intl. Marine Bunkers	-	-	-	-	-	-	-	-	-	-
Stock Changes	-	-	-	-	-	-	-	-	-	-
DOMESTIC SUPPLY	22207	17446	-	1583	182869	-	-	-	18584	-
Transfers and Stat. Diff.	-	-	-	-	1	-	-	-	-1	-
TRANSFORMATION	22207	9819	-	-	10338	-	-	-	-	-
Electricity and CHP Plants	-	9819	-	-	1815	-	-	-	-	-
Petroleum Refineries	-	-	-	-	-	-	-	-	-	-
Other Transform. Sector	22207	-	-	-	8523	-	-	-	-	-
ENERGY SECTOR	-	7448	-	-	-	-	-	-	284	-
DISTRIBUTION LOSSES	-	-	-	-	-	-	-	-	2387	-
FINAL CONSUMPTION	-	179	-	1583	172532	-	-	-	15912	-
INDUSTRY SECTOR	-	-	-	1583	20747	-	-	-	9180	-
Iron and Steel	-	-	-	1583	-	-	-	-	4035	-
Chemical and Petrochem.	-	-	-	-	-	-	-	-	-	-
Non-Metallic Minerals	-	-	-	-	-	-	-	-	-	-
Non-specified	-	-	-	-	20747	-	-	-	5145	-
TRANSPORT SECTOR	-	-	-	-	-	-	-	-	-	-
Air	-	-	-	-	-	-	-	-	-	-
Road	-	-	-	-	-	-	-	-	-	-
Non-specified	-	-	-	-	-	-	-	-	-	-
OTHER SECTORS	-	179	-	-	151785	-	-	-	6732	-
Agriculture	-	-	-	-	4911	-	-	-	626	-
Comm. and Publ. Services	-	-	-	-	-	-	-	-	771	-
Residential	-	179	-	-	146874	-	-	-	5335	-
Non-specified	-	-	-	-	-	-	-	-	-	-
NON-ENERGY USE	-	-	-	-	-	-	-	-	-	-

Philippines : 1997

SUPPLY AND CONSUMPTION *APPROVISIONNEMENT ET DEMANDE*	Coal / *Charbon* (1000 tonnes)							Oil / *Pétrole* (1000 tonnes)			
	Coking Coal *Charbon à coke*	Other Bit. Coal *Autres charb. bit.*	Sub-Bit. Coal *Charbon sous-bit.*	Lignite *Lignite*	Peat *Tourbe*	Oven and Gas Coke *Coke de four/gaz*	Pat. Fuel and BKB *Agg./briq. de lignite*	Crude Oil *Pétrole brut*	NGL *LGN*	Feed-stocks *Produits d'aliment.*	Additives *Additifs*
Production	-	930	-	3	-	-	-	24	-	-	-
From Other Sources	-	-	-	-	-	-	-	-	-	-	-
Imports	-	4248	-	-	-	190	-	17687	-	-	-
Exports	-	-	-	-	-	-	-	-	-	-	-
Intl. Marine Bunkers	-	-	-	-	-	-	-	-	-	-	-
Stock Changes	-	-	-	-	-	-	-	-	-	-	-
DOMESTIC SUPPLY	-	5178	-	3	-	190	-	17711	-	-	-
Transfers	-	-	-	-	-	-	-	-	-	-	-
Statistical Differences	-	2	-	-	-	-	-	1750	-	-	-
TRANSFORMATION	-	2160	-	-	-	152	-	19461	-	-	-
Electricity Plants	-	2160	-	-	-	-	-	-	-	-	-
CHP Plants	-	-	-	-	-	-	-	-	-	-	-
Heat Plants	-	-	-	-	-	-	-	-	-	-	-
Blast Furnaces/Gas Works	-	-	-	-	-	152	-	-	-	-	-
Coke/Pat. Fuel/BKB Plants	-	-	-	-	-	-	-	-	-	-	-
Petroleum Refineries	-	-	-	-	-	-	-	19461	-	-	-
Petrochemical Industry	-	-	-	-	-	-	-	-	-	-	-
Liquefaction	-	-	-	-	-	-	-	-	-	-	-
Other Transform. Sector	-	-	-	-	-	-	-	-	-	-	-
ENERGY SECTOR	-	836	-	-	-	-	-	-	-	-	-
Coal Mines	-	836	-	-	-	-	-	-	-	-	-
Oil and Gas Extraction	-	-	-	-	-	-	-	-	-	-	-
Petroleum Refineries	-	-	-	-	-	-	-	-	-	-	-
Electr., CHP+Heat Plants	-	-	-	-	-	-	-	-	-	-	-
Pumped Storage (Elec.)	-	-	-	-	-	-	-	-	-	-	-
Other Energy Sector	-	-	-	-	-	-	-	-	-	-	-
Distribution Losses	-	-	-	-	-	-	-	-	-	-	-
FINAL CONSUMPTION	-	2184	-	3	-	38	-	-	-	-	-
INDUSTRY SECTOR	-	2184	-	3	-	38	-	-	-	-	-
Iron and Steel	-	37	-	-	-	38	-	-	-	-	-
Chemical and Petrochem.	-	63	-	-	-	-	-	-	-	-	-
of which: Feedstocks	-	-	-	-	-	-	-	-	-	-	-
Non-Ferrous Metals	-	-	-	-	-	-	-	-	-	-	-
Non-Metallic Minerals	-	1961	-	-	-	-	-	-	-	-	-
Transport Equipment	-	-	-	-	-	-	-	-	-	-	-
Machinery	-	-	-	-	-	-	-	-	-	-	-
Mining and Quarrying	-	-	-	-	-	-	-	-	-	-	-
Food and Tobacco	-	123	-	-	-	-	-	-	-	-	-
Paper, Pulp and Print	-	-	-	-	-	-	-	-	-	-	-
Wood and Wood Products	-	-	-	-	-	-	-	-	-	-	-
Construction	-	-	-	-	-	-	-	-	-	-	-
Textile and Leather	-	-	-	-	-	-	-	-	-	-	-
Non-specified	-	-	-	3	-	-	-	-	-	-	-
TRANSPORT SECTOR	-	-	-	-	-	-	-	-	-	-	-
Air	-	-	-	-	-	-	-	-	-	-	-
Road	-	-	-	-	-	-	-	-	-	-	-
Rail	-	-	-	-	-	-	-	-	-	-	-
Pipeline Transport	-	-	-	-	-	-	-	-	-	-	-
Internal Navigation	-	-	-	-	-	-	-	-	-	-	-
Non-specified	-	-	-	-	-	-	-	-	-	-	-
OTHER SECTORS	-	-	-	-	-	-	-	-	-	-	-
Agriculture	-	-	-	-	-	-	-	-	-	-	-
Comm. and Publ. Services	-	-	-	-	-	-	-	-	-	-	-
Residential	-	-	-	-	-	-	-	-	-	-	-
Non-specified	-	-	-	-	-	-	-	-	-	-	-
NON-ENERGY USE	-	-	-	-	-	-	-	-	-	-	-
in Industry/Trans./Energy	-	-	-	-	-	-	-	-	-	-	-
in Transport	-	-	-	-	-	-	-	-	-	-	-
in Other Sectors	-	-	-	-	-	-	-	-	-	-	-

Philippines : 1997

SUPPLY AND CONSUMPTION / APPROVISIONNEMENT ET DEMANDE	Oil cont. / Pétrole cont. (1000 tonnes)										
	Refinery Gas / Gaz de raffinerie	LPG + Ethane / GPL + éthane	Motor Gasoline / Essence moteur	Aviation Gasoline / Essence aviation	Jet Fuel / Carbu-réacteurs	Kerosene / Kérosène	Gas/ Diesel / Gazole	Heavy Fuel Oil / Fioul lourd	Naphtha / Naphta	Petrol. Coke / Coke de pétrole	Other Prod. / Autres prod.
Production	491	456	2199	-	809	582	5385	6849	667	-	253
From Other Sources	-	-	-	-	-	-	-	-	-	-	-
Imports	-	443	337	4	80	99	537	643	4	-	108
Exports	-	-	-12	-	-	-17	-55	-610	-615	-	-9
Intl. Marine Bunkers	-	-	-	-	-	-	-23	-50	-	-	-
Stock Changes	-	-10	2	-	81	-11	75	205	-15	-	1
DOMESTIC SUPPLY	491	889	2526	4	970	653	5919	7037	41	-	353
Transfers	-	-	-	-	-	-	-	-	-	-	-
Statistical Differences	-	-1	-	-	-	-	487	-1054	-	-	-
TRANSFORMATION	-	-	-	-	-	-	2512	2582	-	-	-
Electricity Plants	-	-	-	-	-	-	2512	2576	-	-	-
CHP Plants	-	-	-	-	-	-	-	-	-	-	-
Heat Plants	-	-	-	-	-	-	-	-	-	-	-
Blast Furnaces/Gas Works	-	-	-	-	-	-	-	6	-	-	-
Coke/Pat. Fuel/BKB Plants	-	-	-	-	-	-	-	-	-	-	-
Petroleum Refineries	-	-	-	-	-	-	-	-	-	-	-
Petrochemical Industry	-	-	-	-	-	-	-	-	-	-	-
Liquefaction	-	-	-	-	-	-	-	-	-	-	-
Other Transform. Sector	-	-	-	-	-	-	-	-	-	-	-
ENERGY SECTOR	491	-	-	-	-	29	293	411	41	-	-
Coal Mines	-	-	-	-	-	29	293	246	-	-	-
Oil and Gas Extraction	-	-	-	-	-	-	-	-	-	-	-
Petroleum Refineries	491	-	-	-	-	-	-	165	41	-	-
Electr., CHP+Heat Plants	-	-	-	-	-	-	-	-	-	-	-
Pumped Storage (Elec.)	-	-	-	-	-	-	-	-	-	-	-
Other Energy Sector	-	-	-	-	-	-	-	-	-	-	-
Distribution Losses	-	-	-	-	-	-	-	-	-	-	-
FINAL CONSUMPTION	-	888	2526	4	970	624	3601	2990	-	-	353
INDUSTRY SECTOR	-	333	-	-	-	49	738	2333	-	-	-
Iron and Steel	-	-	-	-	-	23	32	352	-	-	-
Chemical and Petrochem.	-	-	-	-	-	1	20	64	-	-	-
of which: Feedstocks	-	-	-	-	-	-	-	-	-	-	-
Non-Ferrous Metals	-	-	-	-	-	-	-	-	-	-	-
Non-Metallic Minerals	-	-	-	-	-	-	-	352	-	-	-
Transport Equipment	-	-	-	-	-	-	-	-	-	-	-
Machinery	-	-	-	-	-	-	-	-	-	-	-
Mining and Quarrying	-	-	-	-	-	-	-	-	-	-	-
Food and Tobacco	-	-	-	-	-	-	284	165	-	-	-
Paper, Pulp and Print	-	-	-	-	-	-	31	265	-	-	-
Wood and Wood Products	-	-	-	-	-	-	161	7	-	-	-
Construction	-	-	-	-	-	-	122	22	-	-	-
Textile and Leather	-	-	-	-	-	-	25	116	-	-	-
Non-specified	-	333	-	-	-	25	63	990	-	-	-
TRANSPORT SECTOR	-	-	2526	4	970	6	787	54	-	-	-
Air	-	-	-	4	970	-	-	-	-	-	-
Road	-	-	2526	-	-	-	336	-	-	-	-
Rail	-	-	-	-	-	-	-	-	-	-	-
Pipeline Transport	-	-	-	-	-	-	-	-	-	-	-
Internal Navigation	-	-	-	-	-	6	451	54	-	-	-
Non-specified	-	-	-	-	-	-	-	-	-	-	-
OTHER SECTORS	-	555	-	-	-	569	2076	603	-	-	-
Agriculture	-	-	-	-	-	9	1576	384	-	-	-
Comm. and Publ. Services	-	-	-	-	-	-	-	219	-	-	-
Residential	-	555	-	-	-	557	500	-	-	-	-
Non-specified	-	-	-	-	-	3	-	-	-	-	-
NON-ENERGY USE	-	-	-	-	-	-	-	-	-	-	353
in Industry/Transf./Energy	-	-	-	-	-	-	-	-	-	-	353
in Transport	-	-	-	-	-	-	-	-	-	-	-
in Other Sectors	-	-	-	-	-	-	-	-	-	-	-

Philippines : 1997

SUPPLY AND CONSUMPTION / APPROVISIONNEMENT ET DEMANDE	Gas / Gaz (TJ)				Comb. Renew. & Waste / En. Re. Comb. & Déchets (TJ)				(GWh)	(TJ)
	Natural Gas / Gaz naturel	Gas Works / Usines à gaz	Coke Ovens / Cokeries	Blast Furnaces / Hauts fourneaux	Solid Biomass / Biomasse solide	Gas/Liquids from Biomass / Gaz/Liquides tirés de biomasse	Municipal Waste / Déchets urbains	Industrial Waste / Déchets industriels	Electricity / Electricité	Heat / Chaleur
Production	217	205	-	1655	395958	-	-	-	40053	-
From Other Sources	-	-	-	-	-	-	-	-	-	-
Imports	-	-	-	-	-	-	-	-	-	-
Exports	-	-	-	-	-	-	-	-	-	-
Intl. Marine Bunkers	-	-	-	-	-	-	-	-	-	-
Stock Changes	-	-	-	-	-	-	-	-	-	-
DOMESTIC SUPPLY	217	205	-	1655	395958	-	-	-	40053	-
Transfers	-	-	-	-	-	-	-	-	-	-
Statistical Differences	-	-	-	-	-	-	-	-	-	-
TRANSFORMATION	217	-	-	-	87725	-	-	-	-	-
Electricity Plants	217	-	-	-	-	-	-	-	-	-
CHP Plants	-	-	-	-	-	-	-	-	-	-
Heat Plants	-	-	-	-	-	-	-	-	-	-
Blast Furnaces/Gas Works	-	-	-	-	-	-	-	-	-	-
Coke/Pat. Fuel/BKB Plants	-	-	-	-	-	-	-	-	-	-
Petroleum Refineries	-	-	-	-	-	-	-	-	-	-
Petrochemical Industry	-	-	-	-	-	-	-	-	-	-
Liquefaction	-	-	-	-	-	-	-	-	-	-
Other Transform. Sector	-	-	-	-	87725	-	-	-	-	-
ENERGY SECTOR	-	-	-	-	-	-	-	-	1438	-
Coal Mines	-	-	-	-	-	-	-	-	-	-
Oil and Gas Extraction	-	-	-	-	-	-	-	-	-	-
Petroleum Refineries	-	-	-	-	-	-	-	-	-	-
Electr., CHP+Heat Plants	-	-	-	-	-	-	-	-	1438	-
Pumped Storage (Elec.)	-	-	-	-	-	-	-	-	-	-
Other Energy Sector	-	-	-	-	-	-	-	-	-	-
Distribution Losses	-	63	-	-	-	-	-	-	6867	-
FINAL CONSUMPTION	-	142	-	1655	308233	-	-	-	31748	-
INDUSTRY SECTOR	-	1	-	1655	151962	-	-	-	12531	-
Iron and Steel	-	-	-	1655	-	-	-	-	934	-
Chemical and Petrochem.	-	-	-	-	-	-	-	-	1452	-
of which: Feedstocks	-	-	-	-	-	-	-	-	-	-
Non-Ferrous Metals	-	-	-	-	-	-	-	-	-	-
Non-Metallic Minerals	-	-	-	-	-	-	-	-	-	-
Transport Equipment	-	-	-	-	-	-	-	-	-	-
Machinery	-	-	-	-	-	-	-	-	-	-
Mining and Quarrying	-	-	-	-	-	-	-	-	-	-
Food and Tobacco	-	-	-	-	-	-	-	-	-	-
Paper, Pulp and Print	-	-	-	-	-	-	-	-	-	-
Wood and Wood Products	-	-	-	-	-	-	-	-	-	-
Construction	-	-	-	-	-	-	-	-	-	-
Textile and Leather	-	-	-	-	-	-	-	-	-	-
Non-specified	-	1	-	-	151962	-	-	-	10145	-
TRANSPORT SECTOR	-	-	-	-	-	-	-	-	-	-
Air	-	-	-	-	-	-	-	-	-	-
Road	-	-	-	-	-	-	-	-	-	-
Rail	-	-	-	-	-	-	-	-	-	-
Pipeline Transport	-	-	-	-	-	-	-	-	-	-
Internal Navigation	-	-	-	-	-	-	-	-	-	-
Non-specified	-	-	-	-	-	-	-	-	-	-
OTHER SECTORS	-	141	-	-	156271	-	-	-	19217	-
Agriculture	-	-	-	-	-	-	-	-	-	-
Comm. and Publ. Services	-	-	-	-	-	-	-	-	7723	-
Residential	-	38	-	-	125251	-	-	-	10121	-
Non-specified	-	103	-	-	31020	-	-	-	1373	-
NON-ENERGY USE	-	-	-	-	-	-	-	-	-	-
in Industry/Transf./Energy	-	-	-	-	-	-	-	-	-	-
in Transport	-	-	-	-	-	-	-	-	-	-
in Other Sectors	-	-	-	-	-	-	-	-	-	-

Philippines : 1998

SUPPLY AND CONSUMPTION / APPROVISIONNEMENT ET DEMANDE	Coal / *Charbon* (1000 tonnes)							Oil / *Pétrole* (1000 tonnes)			
	Coking Coal / *Charbon à coke*	Other Bit. Coal / *Autres charb. bit.*	Sub-Bit. Coal / *Charbon sous-bit.*	Lignite / *Lignite*	Peat / *Tourbe*	Oven and Gas Coke / *Coke de four/gaz*	Pat. Fuel and BKB / *Agg./briq. de lignite*	Crude Oil / *Pétrole brut*	NGL / *LGN*	Feed-stocks / *Produits d'aliment.*	Additives / *Additifs*
Production	-	998	-	3	-	-	-	41	-	-	-
From Other Sources	-	-	-	-	-	-	-	-	-	-	-
Imports	-	3713	-	-	-	190	-	16053	-	-	-
Exports	-	-	-	-	-	-	-	-	-	-	-
Intl. Marine Bunkers	-	-	-	-	-	-	-	-	-	-	-
Stock Changes	-	-	-	-	-	-	-	-	-	-	-
DOMESTIC SUPPLY	-	**4711**	-	**3**	-	**190**	-	**16094**	-	-	-
Transfers	-	-	-	-	-	-	-	-	-	-	-
Statistical Differences	-	2	-	-	-	-	-	2817	-	-	-
TRANSFORMATION	-	**1965**	-	-	-	**152**	-	**18911**	-	-	-
Electricity Plants	-	1965	-	-	-	-	-	-	-	-	-
CHP Plants	-	-	-	-	-	-	-	-	-	-	-
Heat Plants	-	-	-	-	-	-	-	-	-	-	-
Blast Furnaces/Gas Works	-	-	-	-	-	152	-	-	-	-	-
Coke/Pat. Fuel/BKB Plants	-	-	-	-	-	-	-	-	-	-	-
Petroleum Refineries	-	-	-	-	-	-	-	18911	-	-	-
Petrochemical Industry	-	-	-	-	-	-	-	-	-	-	-
Liquefaction	-	-	-	-	-	-	-	-	-	-	-
Other Transform. Sector	-	-	-	-	-	-	-	-	-	-	-
ENERGY SECTOR	-	**761**	-	-	-	-	-	-	-	-	-
Coal Mines	-	761	-	-	-	-	-	-	-	-	-
Oil and Gas Extraction	-	-	-	-	-	-	-	-	-	-	-
Petroleum Refineries	-	-	-	-	-	-	-	-	-	-	-
Electr., CHP+Heat Plants	-	-	-	-	-	-	-	-	-	-	-
Pumped Storage (Elec.)	-	-	-	-	-	-	-	-	-	-	-
Other Energy Sector	-	-	-	-	-	-	-	-	-	-	-
Distribution Losses	-	-	-	-	-	-	-	-	-	-	-
FINAL CONSUMPTION	-	**1987**	-	**3**	-	**38**	-	-	-	-	-
INDUSTRY SECTOR	-	**1987**	-	**3**	-	**38**	-	-	-	-	-
Iron and Steel	-	34	-	-	-	38	-	-	-	-	-
Chemical and Petrochem.	-	57	-	-	-	-	-	-	-	-	-
of which: Feedstocks	-	-	-	-	-	-	-	-	-	-	-
Non-Ferrous Metals	-	-	-	-	-	-	-	-	-	-	-
Non-Metallic Minerals	-	1784	-	-	-	-	-	-	-	-	-
Transport Equipment	-	-	-	-	-	-	-	-	-	-	-
Machinery	-	-	-	-	-	-	-	-	-	-	-
Mining and Quarrying	-	-	-	-	-	-	-	-	-	-	-
Food and Tobacco	-	112	-	-	-	-	-	-	-	-	-
Paper, Pulp and Print	-	-	-	-	-	-	-	-	-	-	-
Wood and Wood Products	-	-	-	-	-	-	-	-	-	-	-
Construction	-	-	-	-	-	-	-	-	-	-	-
Textile and Leather	-	-	-	-	-	-	-	-	-	-	-
Non-specified	-	-	-	3	-	-	-	-	-	-	-
TRANSPORT SECTOR	-	-	-	-	-	-	-	-	-	-	-
Air	-	-	-	-	-	-	-	-	-	-	-
Road	-	-	-	-	-	-	-	-	-	-	-
Rail	-	-	-	-	-	-	-	-	-	-	-
Pipeline Transport	-	-	-	-	-	-	-	-	-	-	-
Internal Navigation	-	-	-	-	-	-	-	-	-	-	-
Non-specified	-	-	-	-	-	-	-	-	-	-	-
OTHER SECTORS	-	-	-	-	-	-	-	-	-	-	-
Agriculture	-	-	-	-	-	-	-	-	-	-	-
Comm. and Publ. Services	-	-	-	-	-	-	-	-	-	-	-
Residential	-	-	-	-	-	-	-	-	-	-	-
Non-specified	-	-	-	-	-	-	-	-	-	-	-
NON-ENERGY USE	-	-	-	-	-	-	-	-	-	-	-
in Industry/Trans./Energy	-	-	-	-	-	-	-	-	-	-	-
in Transport	-	-	-	-	-	-	-	-	-	-	-
in Other Sectors	-	-	-	-	-	-	-	-	-	-	-

Philippines : 1998

SUPPLY AND CONSUMPTION / APPROVISIONNEMENT ET DEMANDE	Oil cont. / *Pétrole cont.* (1000 tonnes)										
	Refinery Gas / *Gaz de raffinerie*	LPG + Ethane / *GPL + éthane*	Motor Gasoline / *Essence moteur*	Aviation Gasoline / *Essence aviation*	Jet Fuel / *Carbu-réacteurs*	Kerosene / *Kérosène*	Gas/ Diesel / *Gazole*	Heavy Fuel Oil / *Fioul lourd*	Naphtha / *Naphta*	Petrol. Coke / *Coke de pétrole*	Other Prod. / *Autres prod.*
Production	437	422	2248	-	562	551	4937	6132	602	-	253
From Other Sources	-	-	-	-	-	-	-	-	-	-	-
Imports	-	533	445	1	89	95	953	1018	-	-	104
Exports	-	-	-3	-	-	-	-31	-187	-590	-	-78
Intl. Marine Bunkers	-	-	-	-	-	-	-23	-50	-	-	-
Stock Changes	-	-9	-36	3	24	1	-74	-70	-12	-	63
DOMESTIC SUPPLY	437	946	2654	4	675	647	5762	6843	-	-	342
Transfers	-	-	-	-	-	-	-	-	-	-	-
Statistical Differences	-	-2	-	-	-	-	474	-1026	-	-	-
TRANSFORMATION	-	-	-	-	-	-	2445	2511	-	-	-
Electricity Plants	-	-	-	-	-	-	2445	2505	-	-	-
CHP Plants	-	-	-	-	-	-	-	-	-	-	-
Heat Plants	-	-	-	-	-	-	-	-	-	-	-
Blast Furnaces/Gas Works	-	-	-	-	-	-	-	6	-	-	-
Coke/Pat. Fuel/BKB Plants	-	-	-	-	-	-	-	-	-	-	-
Petroleum Refineries	-	-	-	-	-	-	-	-	-	-	-
Petrochemical Industry	-	-	-	-	-	-	-	-	-	-	-
Liquefaction	-	-	-	-	-	-	-	-	-	-	-
Other Transform. Sector	-	-	-	-	-	-	-	-	-	-	-
ENERGY SECTOR	437	-	-	-	-	29	285	399	-	-	-
Coal Mines	-	-	-	-	-	29	285	239	-	-	-
Oil and Gas Extraction	-	-	-	-	-	-	-	-	-	-	-
Petroleum Refineries	437	-	-	-	-	-	-	160	-	-	-
Electr., CHP+Heat Plants	-	-	-	-	-	-	-	-	-	-	-
Pumped Storage (Elec.)	-	-	-	-	-	-	-	-	-	-	-
Other Energy Sector	-	-	-	-	-	-	-	-	-	-	-
Distribution Losses	-	-	-	-	-	-	-	-	-	-	-
FINAL CONSUMPTION	-	944	2654	4	675	618	3506	2907	-	-	342
INDUSTRY SECTOR	-	354	-	-	-	49	718	2268	-	-	-
Iron and Steel	-	-	-	-	-	23	31	342	-	-	-
Chemical and Petrochem.	-	-	-	-	-	1	19	62	-	-	-
of which: Feedstocks	-	-	-	-	-	-	-	-	-	-	-
Non-Ferrous Metals	-	-	-	-	-	-	-	-	-	-	-
Non-Metallic Minerals	-	-	-	-	-	-	-	342	-	-	-
Transport Equipment	-	-	-	-	-	-	-	-	-	-	-
Machinery	-	-	-	-	-	-	-	-	-	-	-
Mining and Quarrying	-	-	-	-	-	-	-	-	-	-	-
Food and Tobacco	-	-	-	-	-	-	277	160	-	-	-
Paper, Pulp and Print	-	-	-	-	-	-	30	258	-	-	-
Wood and Wood Products	-	-	-	-	-	-	157	7	-	-	-
Construction	-	-	-	-	-	-	119	21	-	-	-
Textile and Leather	-	-	-	-	-	-	24	113	-	-	-
Non-specified	-	354	-	-	-	25	61	963	-	-	-
TRANSPORT SECTOR	-	-	2654	4	675	6	766	53	-	-	-
Air	-	-	-	4	675	-	-	-	-	-	-
Road	-	-	2654	-	-	-	327	-	-	-	-
Rail	-	-	-	-	-	-	-	-	-	-	-
Pipeline Transport	-	-	-	-	-	-	-	-	-	-	-
Internal Navigation	-	-	-	-	-	6	439	53	-	-	-
Non-specified	-	-	-	-	-	-	-	-	-	-	-
OTHER SECTORS	-	590	-	-	-	563	2022	586	-	-	-
Agriculture	-	-	-	-	-	9	1535	373	-	-	-
Comm. and Publ. Services	-	-	-	-	-	-	-	213	-	-	-
Residential	-	590	-	-	-	551	487	-	-	-	-
Non-specified	-	-	-	-	-	3	-	-	-	-	-
NON-ENERGY USE	-	-	-	-	-	-	-	-	-	-	342
in Industry/Transf./Energy	-	-	-	-	-	-	-	-	-	-	342
in Transport	-	-	-	-	-	-	-	-	-	-	-
in Other Sectors	-	-	-	-	-	-	-	-	-	-	-

Philippines : 1998

SUPPLY AND CONSUMPTION *APPROVISIONNEMENT ET DEMANDE*	Gas / *Gaz* (TJ)				Comb. Renew. & Waste / *En. Re. Comb. & Déchets* (TJ)				(GWh)	(TJ)
	Natural Gas *Gaz naturel*	Gas Works *Usines à gaz*	Coke Ovens *Cokeries*	Blast Furnaces *Hauts fourneaux*	Solid Biomass *Biomasse solide*	Gas/Liquids from Biomass *Gaz/Liquides tirés de biomasse*	Municipal Waste *Déchets urbains*	Industrial Waste *Déchets industriels*	Electricity *Electricité*	Heat *Chaleur*
Production	351	205	-	1655	384794	-	-	-	41192	-
From Other Sources	-	-	-	-	-	-	-	-	-	-
Imports	-	-	-	-	-	-	-	-	-	-
Exports	-	-	-	-	-	-	-	-	-	-
Intl. Marine Bunkers	-	-	-	-	-	-	-	-	-	-
Stock Changes	-	-	-	-	-	-	-	-	-	-
DOMESTIC SUPPLY	**351**	**205**	**-**	**1655**	**384794**	**-**	**-**	**-**	**41192**	**-**
Transfers	-	-	-	-	-	-	-	-	-	-
Statistical Differences	-	-	-	-	-	-	-	-	-	-
TRANSFORMATION	**351**	**-**	**-**	**-**	**80261**	**-**	**-**	**-**	**-**	**-**
Electricity Plants	351	-	-	-	-	-	-	-	-	-
CHP Plants	-	-	-	-	-	-	-	-	-	-
Heat Plants	-	-	-	-	-	-	-	-	-	-
Blast Furnaces/Gas Works	-	-	-	-	-	-	-	-	-	-
Coke/Pat. Fuel/BKB Plants	-	-	-	-	-	-	-	-	-	-
Petroleum Refineries	-	-	-	-	-	-	-	-	-	-
Petrochemical Industry	-	-	-	-	-	-	-	-	-	-
Liquefaction	-	-	-	-	-	-	-	-	-	-
Other Transform. Sector	-	-	-	-	80261	-	-	-	-	-
ENERGY SECTOR	**-**	**-**	**-**	**-**	**-**	**-**	**-**	**-**	**1599**	**-**
Coal Mines	-	-	-	-	-	-	-	-	-	-
Oil and Gas Extraction	-	-	-	-	-	-	-	-	-	-
Petroleum Refineries	-	-	-	-	-	-	-	-	-	-
Electr., CHP+Heat Plants	-	-	-	-	-	-	-	-	1599	-
Pumped Storage (Elec.)	-	-	-	-	-	-	-	-	-	-
Other Energy Sector	-	-	-	-	-	-	-	-	-	-
Distribution Losses	-	63	-	-	-	-	-	-	6736	-
FINAL CONSUMPTION	**-**	**142**	**-**	**1655**	**304533**	**-**	**-**	**-**	**32857**	**-**
INDUSTRY SECTOR	**-**	**1**	**-**	**1655**	**151479**	**-**	**-**	**-**	**11388**	**-**
Iron and Steel	-	-	-	1655	-	-	-	-	849	-
Chemical and Petrochem.	-	-	-	-	-	-	-	-	1320	-
of which: Feedstocks	-	-	-	-	-	-	-	-	-	-
Non-Ferrous Metals	-	-	-	-	-	-	-	-	-	-
Non-Metallic Minerals	-	-	-	-	-	-	-	-	-	-
Transport Equipment	-	-	-	-	-	-	-	-	-	-
Machinery	-	-	-	-	-	-	-	-	-	-
Mining and Quarrying	-	-	-	-	-	-	-	-	-	-
Food and Tobacco	-	-	-	-	-	-	-	-	-	..
Paper, Pulp and Print	-	-	-	-	-	-	-	-	-	-
Wood and Wood Products	-	-	-	-	-	-	-	-	-	-
Construction	-	-	-	-	-	-	-	-	-	-
Textile and Leather	-	-	-	-	-	-	-	-	-	-
Non-specified	-	1	-	-	151479	-	-	-	9219	-
TRANSPORT SECTOR	**-**	**-**	**-**	**-**	**-**	**-**	**-**	**-**	**-**	**-**
Air	-	-	-	-	-	-	-	-	-	-
Road	-	-	-	-	-	-	-	-	-	-
Rail	-	-	-	-	-	-	-	-	-	-
Pipeline Transport	-	-	-	-	-	-	-	-	-	-
Internal Navigation	-	-	-	-	-	-	-	-	-	-
Non-specified	-	-	-	-	-	-	-	-	-	-
OTHER SECTORS	**-**	**141**	**-**	**-**	**153054**	**-**	**-**	**-**	**21469**	**-**
Agriculture	-	-	-	-	-	-	-	-	-	-
Comm. and Publ. Services	-	-	-	-	-	-	-	-	8522	-
Residential	-	38	-	-	124448	-	-	-	11532	-
Non-specified	-	103	-	-	28606	-	-	-	1415	-
NON-ENERGY USE	**-**	**-**	**-**	**-**	**-**	**-**	**-**	**-**	**-**	**-**
in Industry/Transf./Energy	-	-	-	-	-	-	-	-	-	-
in Transport	-	-	-	-	-	-	-	-	-	-
in Other Sectors	-	-	-	-	-	-	-	-	-	-

Qatar

SUPPLY AND CONSUMPTION 1997	Coal (1000 tonnes)							Oil (1000 tonnes)			
	Coking Coal	Other Bit. Coal	Sub-Bit. Coal	Lignite	Peat	Oven and Gas Coke	Pat. Fuel and BKB	Crude Oil	NGL	Feed-stocks	Additives
Production	-	-	-	-	-	-	-	26168	2599	-	-
Imports	-	-	-	-	-	-	-	-	-	-	-
Exports	-	-	-	-	-	-	-	-22010	-1372	-	-
Intl. Marine Bunkers	-	-	-	-	-	-	-	-	-	-	-
Stock Changes	-	-	-	-	-	-	-	-730	-	-	-
DOMESTIC SUPPLY	-	-	-	-	-	-	-	**3428**	**1227**	-	-
Transfers and Stat. Diff.	-	-	-	-	-	-	-	-600	-1227	-	-
TRANSFORMATION	-	-	-	-	-	-	-	**2828**	-	-	-
Electricity and CHP Plants	-	-	-	-	-	-	-	-	-	-	-
Petroleum Refineries	-	-	-	-	-	-	-	2828	-	-	-
Other Transform. Sector	-	-	-	-	-	-	-	-	-	-	-
ENERGY SECTOR	-	-	-	-	-	-	-	-	-	-	-
DISTRIBUTION LOSSES	-	-	-	-	-	-	-	-	-	-	-
FINAL CONSUMPTION	-	-	-	-	-	-	-	-	-	-	-
INDUSTRY SECTOR	-	-	-	-	-	-	-	-	-	-	-
Iron and Steel	-	-	-	-	-	-	-	-	-	-	-
Chemical and Petrochem.	-	-	-	-	-	-	-	-	-	-	-
Non-Metallic Minerals	-	-	-	-	-	-	-	-	-	-	-
Non-specified	-	-	-	-	-	-	-	-	-	-	-
TRANSPORT SECTOR	-	-	-	-	-	-	-	-	-	-	-
Air	-	-	-	-	-	-	-	-	-	-	-
Road	-	-	-	-	-	-	-	-	-	-	-
Non-specified	-	-	-	-	-	-	-	-	-	-	-
OTHER SECTORS	-	-	-	-	-	-	-	-	-	-	-
Agriculture	-	-	-	-	-	-	-	-	-	-	-
Comm. and Publ. Services	-	-	-	-	-	-	-	-	-	-	-
Residential	-	-	-	-	-	-	-	-	-	-	-
Non-specified	-	-	-	-	-	-	-	-	-	-	-
NON-ENERGY USE	-	-	-	-	-	-	-	-	-	-	-

APPROVISIONNEMENT ET DEMANDE 1998	Charbon (1000 tonnes)							Pétrole (1000 tonnes)			
	Charbon à coke	Autres charb. bit.	Charbon sous-bit.	Lignite	Tourbe	Coke de four/gaz	Agg./briq. de lignite	Pétrole brut	LGN	Produits d'aliment.	Additifs
Production	-	-	-	-	-	-	-	29009	2783	-	-
Imports	-	-	-	-	-	-	-	-	-	-	-
Exports	-	-	-	-	-	-	-	-27094	-1498	-	-
Intl. Marine Bunkers	-	-	-	-	-	-	-	-	-	-	-
Stock Changes	-	-	-	-	-	-	-	1767	-	-	-
DOMESTIC SUPPLY	-	-	-	-	-	-	-	**3682**	**1285**	-	-
Transfers and Stat. Diff.	-	-	-	-	-	-	-	-600	-1285	-	-
TRANSFORMATION	-	-	-	-	-	-	-	**3082**	-	-	-
Electricity and CHP Plants	-	-	-	-	-	-	-	-	-	-	-
Petroleum Refineries	-	-	-	-	-	-	-	3082	-	-	-
Other Transform. Sector	-	-	-	-	-	-	-	-	-	-	-
ENERGY SECTOR	-	-	-	-	-	-	-	-	-	-	-
DISTRIBUTION LOSSES	-	-	-	-	-	-	-	-	-	-	-
FINAL CONSUMPTION	-	-	-	-	-	-	-	-	-	-	-
INDUSTRY SECTOR	-	-	-	-	-	-	-	-	-	-	-
Iron and Steel	-	-	-	-	-	-	-	-	-	-	-
Chemical and Petrochem.	-	-	-	-	-	-	-	-	-	-	-
Non-Metallic Minerals	-	-	-	-	-	-	-	-	-	-	-
Non-specified	-	-	-	-	-	-	-	-	-	-	-
TRANSPORT SECTOR	-	-	-	-	-	-	-	-	-	-	-
Air	-	-	-	-	-	-	-	-	-	-	-
Road	-	-	-	-	-	-	-	-	-	-	-
Non-specified	-	-	-	-	-	-	-	-	-	-	-
OTHER SECTORS	-	-	-	-	-	-	-	-	-	-	-
Agriculture	-	-	-	-	-	-	-	-	-	-	-
Comm. and Publ. Services	-	-	-	-	-	-	-	-	-	-	-
Residential	-	-	-	-	-	-	-	-	-	-	-
Non-specified	-	-	-	-	-	-	-	-	-	-	-
NON-ENERGY USE	-	-	-	-	-	-	-	-	-	-	-

Qatar

SUPPLY AND CONSUMPTION 1997	Oil cont. (1000 tonnes)										
	Refinery Gas	LPG + Ethane	Motor Gasoline	Aviation Gasoline	Jet Fuel	Kerosene	Gas/ Diesel	Heavy Fuel Oil	Naphtha	Petrol. Coke	Other Prod.
Production	58	69	525	-	413	21	646	899	-	-	-
Imports	-	-	-	-	-	-	-	-	-	-	-
Exports	-	-167	-111	-	-216	-	-315	-898	-	-	-
Intl. Marine Bunkers	-	-	-	-	-	-	-	-	-	-	-
Stock Changes	-	-	-	-	-	-	-	-	-	-	-
DOMESTIC SUPPLY	58	-98	414	-	197	21	331	1	-	-	-
Transfers and Stat. Diff.	-	1227	-	-	-	-	-	-	-	-	-
TRANSFORMATION	-	-	-	-	-	-	-	-	-	-	-
Electricity and CHP Plants	-	-	-	-	-	-	-	-	-	-	-
Petroleum Refineries	-	-	-	-	-	-	-	-	-	-	-
Other Transform. Sector	-	-	-	-	-	-	-	-	-	-	-
ENERGY SECTOR	58	-	-	-	-	-	-	1	-	-	-
DISTRIBUTION LOSSES	-	-	-	-	-	-	-	-	-	-	-
FINAL CONSUMPTION	-	1129	414	-	197	21	331	-	-	-	-
INDUSTRY SECTOR	-	1129	-	-	-	-	-	-	-	-	-
Iron and Steel	-	-	-	-	-	-	-	-	-	-	-
Chemical and Petrochem.	-	1100	-	-	-	-	-	-	-	-	-
Non-Metallic Minerals	-	-	-	-	-	-	-	-	-	-	-
Non-specified	-	29	-	-	-	-	-	-	-	-	-
TRANSPORT SECTOR	-	-	414	-	197	-	331	-	-	-	-
Air	-	-	-	-	197	-	-	-	-	-	-
Road	-	-	414	-	-	-	331	-	-	-	-
Non-specified	-	-	-	-	-	-	-	-	-	-	-
OTHER SECTORS	-	-	-	-	-	21	-	-	-	-	-
Agriculture	-	-	-	-	-	-	-	-	-	-	-
Comm. and Publ. Services	-	-	-	-	-	-	-	-	-	-	-
Residential	-	-	-	-	-	21	-	-	-	-	-
Non-specified	-	-	-	-	-	-	-	-	-	-	-
NON-ENERGY USE	-	-	-	-	-	-	-	-	-	-	-

APPROVISIONNEMENT ET DEMANDE 1998	Pétrole cont. (1000 tonnes)										
	Gaz de raffinerie	GPL + éthane	Essence moteur	Essence aviation	Carbu- réacteurs	Kérosène	Gazole	Fioul lourd	Naphta	Coke de pétrole	Autres prod.
Production	62	78	603	-	437	21	668	983	-	-	-
Imports	-	-	-	-	-	-	-	-	-	-	-
Exports	-	-179	-164	-	-258	-	-324	-982	-	-	-
Intl. Marine Bunkers	-	-	-	-	-	-	-	-	-	-	-
Stock Changes	-	-	-	-	-	-	-	-	-	-	-
DOMESTIC SUPPLY	62	-101	439	-	179	21	344	1	-	-	-
Transfers and Stat. Diff.	-	1285	-	-	-	-	-	-	-	-	-
TRANSFORMATION	-	-	-	-	-	-	-	-	-	-	-
Electricity and CHP Plants	-	-	-	-	-	-	-	-	-	-	-
Petroleum Refineries	-	-	-	-	-	-	-	-	-	-	-
Other Transform. Sector	-	-	-	-	-	-	-	-	-	-	-
ENERGY SECTOR	62	-	-	-	-	-	-	1	-	-	-
DISTRIBUTION LOSSES	-	-	-	-	-	-	-	-	-	-	-
FINAL CONSUMPTION	-	1184	439	-	179	21	344	-	-	-	-
INDUSTRY SECTOR	-	1184	-	-	-	-	-	-	-	-	-
Iron and Steel	-	-	-	-	-	-	-	-	-	-	-
Chemical and Petrochem.	-	1152	-	-	-	-	-	-	-	-	-
Non-Metallic Minerals	-	-	-	-	-	-	-	-	-	-	-
Non-specified	-	32	-	-	-	-	-	-	-	-	-
TRANSPORT SECTOR	-	-	439	-	179	-	344	-	-	-	-
Air	-	-	-	-	179	-	-	-	-	-	-
Road	-	-	439	-	-	-	344	-	-	-	-
Non-specified	-	-	-	-	-	-	-	-	-	-	-
OTHER SECTORS	-	-	-	-	-	21	-	-	-	-	-
Agriculture	-	-	-	-	-	-	-	-	-	-	-
Comm. and Publ. Services	-	-	-	-	-	-	-	-	-	-	-
Residential	-	-	-	-	-	21	-	-	-	-	-
Non-specified	-	-	-	-	-	-	-	-	-	-	-
NON-ENERGY USE	-	-	-	-	-	-	-	-	-	-	-

Qatar

SUPPLY AND CONSUMPTION 1997	Gas (TJ)				Comb. Renew. & Waste (TJ)				(GWh)	(TJ)
	Natural Gas	Gas Works	Coke Ovens	Blast Furnaces	Solid Biomass	Gas/Liquids from Biomass	Municipal Waste	Industrial Waste	Electricity	Heat
Production	655980	-	-	-	-	-	-	-	6868	-
Imports	-	-	-	-	63	-	-	-	-	-
Exports	-107822	-	-	-	-	-	-	-	-	-
Intl. Marine Bunkers	-	-	-	-	-	-	-	-	-	-
Stock Changes	-	-	-	-	-	-	-	-	-	-
DOMESTIC SUPPLY	548158	-	-	-	63	-	-	-	6868	-
Transfers and Stat. Diff.	1	-	-	-	-	-	-	-	1	-
TRANSFORMATION	82416	-	-	-	-	-	-	-	-	-
Electricity and CHP Plants	82416	-	-	-	-	-	-	-	-	-
Petroleum Refineries	-	-	-	-	-	-	-	-	-	-
Other Transform. Sector	-	-	-	-	-	-	-	-	-	-
ENERGY SECTOR	230197	-	-	-	-	-	-	-	-	-
DISTRIBUTION LOSSES	-	-	-	-	-	-	-	-	412	-
FINAL CONSUMPTION	235546	-	-	-	63	-	-	-	6457	-
INDUSTRY SECTOR	235546	-	-	-	-	-	-	-	1485	-
Iron and Steel	17622	-	-	-	-	-	-	-	1485	-
Chemical and Petrochem.	197913	-	-	-	-	-	-	-	-	-
Non-Metallic Minerals	6217	-	-	-	-	-	-	-	-	-
Non-specified	13794	-	-	-	-	-	-	-	-	-
TRANSPORT SECTOR	-	-	-	-	-	-	-	-	-	-
Air	-	-	-	-	-	-	-	-	-	-
Road	-	-	-	-	-	-	-	-	-	-
Non-specified	-	-	-	-	-	-	-	-	-	-
OTHER SECTORS	-	-	-	-	63	-	-	-	4972	-
Agriculture	-	-	-	-	-	-	-	-	-	-
Comm. and Publ. Services	-	-	-	-	-	-	-	-	646	-
Residential	-	-	-	-	63	-	-	-	-	-
Non-specified	-	-	-	-	-	-	-	-	4326	-
NON-ENERGY USE	-	-	-	-	-	-	-	-	-	-

APPROVISIONNEMENT ET DEMANDE 1998	Gaz (TJ)				En. Re. Comb. & Déchets (TJ)				(GWh)	(TJ)
	Gaz naturel	Usines à gaz	Cokeries	Hauts fourneaux	Biomasse solide	Gaz/Liquides tirés de biomasse	Déchets urbains	Déchets industriels	Electricité	Chaleur
Production	738166	-	-	-	-	-	-	-	8125	-
Imports	-	-	-	-	40	-	-	-	-	-
Exports	-180583	-	-	-	-	-	-	-	-	-
Intl. Marine Bunkers	-	-	-	-	-	-	-	-	-	-
Stock Changes	-	-	-	-	-	-	-	-	-	-
DOMESTIC SUPPLY	557583	-	-	-	40	-	-	-	8125	-
Transfers and Stat. Diff.	-	-	-	-	-	-	-	-	-	-
TRANSFORMATION	97500	-	-	-	-	-	-	-	-	-
Electricity and CHP Plants	97500	-	-	-	-	-	-	-	-	-
Petroleum Refineries	-	-	-	-	-	-	-	-	-	-
Other Transform. Sector	-	-	-	-	-	-	-	-	-	-
ENERGY SECTOR	227400	-	-	-	-	-	-	-	-	-
DISTRIBUTION LOSSES	-	-	-	-	-	-	-	-	529	-
FINAL CONSUMPTION	232683	-	-	-	40	-	-	-	7596	-
INDUSTRY SECTOR	232683	-	-	-	-	-	-	-	1747	-
Iron and Steel	17408	-	-	-	-	-	-	-	1747	-
Chemical and Petrochem.	195508	-	-	-	-	-	-	-	-	-
Non-Metallic Minerals	6141	-	-	-	-	-	-	-	-	-
Non-specified	13626	-	-	-	-	-	-	-	-	-
TRANSPORT SECTOR	-	-	-	-	-	-	-	-	-	-
Air	-	-	-	-	-	-	-	-	-	-
Road	-	-	-	-	-	-	-	-	-	-
Non-specified	-	-	-	-	-	-	-	-	-	-
OTHER SECTORS	-	-	-	-	40	-	-	-	5849	-
Agriculture	-	-	-	-	-	-	-	-	-	-
Comm. and Publ. Services	-	-	-	-	-	-	-	-	760	-
Residential	-	-	-	-	40	-	-	-	-	-
Non-specified	-	-	-	-	-	-	-	-	5089	-
NON-ENERGY USE	-	-	-	-	-	-	-	-	-	-

Romania / Roumanie : 1997

SUPPLY AND CONSUMPTION / APPROVISIONNEMENT ET DEMANDE	Coal / Charbon (1000 tonnes)							Oil / Pétrole (1000 tonnes)			
	Coking Coal / Charbon à coke	Other Bit. Coal / Autres charb. bit.	Sub-Bit. Coal / Charbon sous-bit.	Lignite / Lignite	Peat / Tourbe	Oven and Gas Coke / Coke de four/gaz	Pat. Fuel and BKB / Agg./briq. de lignite	Crude Oil / Pétrole brut	NGL / LGN	Feed-stocks / Produits d'aliment.	Additives / Additifs
Production	324	10	1418	32055	-	3316	-	6517	234	-	104
From Other Sources	-	-	-	-	-	-	-	-	-	-	-
Imports	5014	7	228	-	-	189	-	6245	-	-	1
Exports	-	-	-	-	-	-146	-	-	-	-	-
Intl. Marine Bunkers	-	-	-	-	-	-	-	-	-	-	-
Stock Changes	-364	-	15	-551	-	-38	-	19	-5	-	-1
DOMESTIC SUPPLY	4974	17	1661	31504	-	3321	-	12781	229	-	104
Transfers	-	-	-	-	-	-	-	-	-	-	-
Statistical Differences	-97	3	-283	-61	-	63	-	-91	-65	-	-
TRANSFORMATION	4833	-	1252	30468	-	2703	-	12431	127	-	104
Electricity Plants	-	-	-	8251	-	-	-	-	-	-	-
CHP Plants	-	-	1173	21984	-	-	-	-	-	-	-
Heat Plants	-	-	79	233	-	-	-	-	-	-	-
Blast Furnaces/Gas Works	-	-	-	-	-	2703	-	-	-	-	-
Coke/Pat. Fuel/BKB Plants	4833	-	-	-	-	-	-	-	-	-	-
Petroleum Refineries	-	-	-	-	-	-	-	12431	127	-	104
Petrochemical Industry	-	-	-	-	-	-	-	-	-	-	-
Liquefaction	-	-	-	-	-	-	-	-	-	-	-
Other Transform. Sector	-	-	-	-	-	-	-	-	-	-	-
ENERGY SECTOR	42	-	22	119	-	-	-	148	37	-	-
Coal Mines	13	-	21	108	-	-	-	-	-	-	-
Oil and Gas Extraction	-	-	-	-	-	-	-	148	37	-	-
Petroleum Refineries	-	-	-	-	-	-	-	-	-	-	-
Electr., CHP+Heat Plants	-	-	1	11	-	-	-	-	-	-	-
Pumped Storage (Elec.)	-	-	-	-	-	-	-	-	-	-	-
Other Energy Sector	29	-	-	-	-	-	-	-	-	-	-
Distribution Losses	-	-	4	293	-	-	-	109	-	-	-
FINAL CONSUMPTION	2	20	100	563	-	681	-	2	-	-	-
INDUSTRY SECTOR	1	6	10	116	-	681	-	2	-	-	-
Iron and Steel	-	-	-	-	-	584	-	-	-	-	-
Chemical and Petrochem.	-	6	-	15	-	62	-	2	-	-	-
of which: Feedstocks	-	-	-	-	-	-	-	-	-	-	-
Non-Ferrous Metals	-	-	-	-	-	-	-	-	-	-	-
Non-Metallic Minerals	1	-	-	54	-	3	-	-	-	-	-
Transport Equipment	-	-	-	-	-	-	-	-	-	-	-
Machinery	-	-	1	11	-	17	-	-	-	-	-
Mining and Quarrying	-	-	5	1	-	-	-	-	-	-	-
Food and Tobacco	-	-	1	27	-	15	-	-	-	-	-
Paper, Pulp and Print	-	-	-	2	-	-	-	-	-	-	-
Wood and Wood Products	-	-	1	-	-	-	-	-	-	-	-
Construction	-	-	-	2	-	-	-	-	-	-	-
Textile and Leather	-	-	1	3	-	-	-	-	-	-	-
Non-specified	-	-	1	1	-	-	-	-	-	-	-
TRANSPORT SECTOR	1	-	3	4	-	-	-	-	-	-	-
Air	-	-	-	-	-	-	-	-	-	-	-
Road	-	-	-	-	-	-	-	-	-	-	-
Rail	1	-	3	4	-	-	-	-	-	-	-
Pipeline Transport	-	-	-	-	-	-	-	-	-	-	-
Internal Navigation	-	-	-	-	-	-	-	-	-	-	-
Non-specified	-	-	-	-	-	-	-	-	-	-	-
OTHER SECTORS	-	-	87	443	-	-	-	-	-	-	-
Agriculture	-	-	-	3	-	-	-	-	-	-	-
Comm. and Publ. Services	-	-	-	-	-	-	-	-	-	-	-
Residential	-	-	84	438	-	-	-	-	-	-	-
Non-specified	-	-	3	2	-	-	-	-	-	-	-
NON-ENERGY USE	-	14	-	-	-	-	-	-	-	-	-
in Industry/Trans./Energy	-	14	-	-	-	-	-	-	-	-	-
in Transport	-	-	-	-	-	-	-	-	-	-	-
in Other Sectors	-	-	-	-	-	-	-	-	-	-	-

Romania / Roumanie : 1997

SUPPLY AND CONSUMPTION APPROVISIONNEMENT ET DEMANDE	Oil cont. / Pétrole cont. (1000 tonnes)										
	Refinery Gas Gaz de raffinerie	LPG + Ethane GPL + éthane	Motor Gasoline Essence moteur	Aviation Gasoline Essence aviation	Jet Fuel Carbu- réacteurs	Kerosene Kérosène	Gas/ Diesel Gazole	Heavy Fuel Oil Fioul lourd	Naphtha Naphta	Petrol. Coke Coke de pétrole	Other Prod. Autres prod.
Production	884	242	3205	3	88	66	3953	2079	433	442	907
From Other Sources	-	-	-	-	-	-	-	-	-	-	-
Imports	-	6	191	-	79	-	110	3567	2	-	47
Exports	-	-	-1575	-	-	-4	-1031	-1	-	-217	-46
Intl. Marine Bunkers	-	-	-	-	-	-	-	-	-	-	-
Stock Changes	3	1	-294	-	45	-5	-81	-472	-10	-47	3
DOMESTIC SUPPLY	**887**	**249**	**1527**	**3**	**212**	**57**	**2951**	**5173**	**425**	**178**	**911**
Transfers	-	-	-	-	-	-	-17	-47	-179	-	250
Statistical Differences	-57	-	34	-	-92	9	208	127	-204	102	552
TRANSFORMATION	**59**	**-**	**-**	**-**	**-**	**-**	**7**	**3719**	**-**	**-**	**184**
Electricity Plants	1	-	-	-	-	-	3	127	-	-	-
CHP Plants	58	-	-	-	-	-	-	3300	-	-	-
Heat Plants	-	-	-	-	-	-	4	292	-	-	184
Blast Furnaces/Gas Works	-	-	-	-	-	-	-	-	-	-	-
Coke/Pat. Fuel/BKB Plants	-	-	-	-	-	-	-	-	-	-	-
Petroleum Refineries	-	-	-	-	-	-	-	-	-	-	-
Petrochemical Industry	-	-	-	-	-	-	-	-	-	-	-
Liquefaction	-	-	-	-	-	-	-	-	-	-	-
Other Transform. Sector	-	-	-	-	-	-	-	-	-	-	-
ENERGY SECTOR	**535**	**-**	**6**	**-**	**-**	**24**	**70**	**385**	**-**	**-**	**396**
Coal Mines	-	-	-	-	-	-	25	3	-	-	-
Oil and Gas Extraction	-	-	2	-	-	-	36	-	-	-	-
Petroleum Refineries	535	-	-	-	-	-	-	-	-	-	-
Electr., CHP+Heat Plants	-	-	3	-	-	-	9	19	-	-	-
Pumped Storage (Elec.)	-	-	-	-	-	-	-	-	-	-	-
Other Energy Sector	-	-	1	-	-	24	-	363	-	-	396
Distribution Losses	-	-	-	-	-	4	-	-	-	-	140
FINAL CONSUMPTION	**236**	**249**	**1555**	**3**	**120**	**38**	**3065**	**1149**	**42**	**280**	**993**
INDUSTRY SECTOR	**236**	**3**	**36**	**-**	**-**	**8**	**299**	**892**	**42**	**-**	**273**
Iron and Steel	-	-	1	-	-	-	8	148	-	-	4
Chemical and Petrochem.	236	-	-	-	-	2	1	239	42	-	34
of which: Feedstocks	-	-	-	-	-	-	-	-	2	-	-
Non-Ferrous Metals	-	-	-	-	-	-	-	-	-	-	-
Non-Metallic Minerals	-	-	-	-	-	-	8	201	-	-	7
Transport Equipment	-	-	-	-	-	-	-	-	-	-	-
Machinery	-	2	4	-	-	5	19	57	-	-	28
Mining and Quarrying	-	-	-	-	-	-	12	14	-	-	6
Food and Tobacco	-	1	20	-	-	-	36	129	-	-	101
Paper, Pulp and Print	-	-	-	-	-	-	1	46	-	-	2
Wood and Wood Products	-	-	1	-	-	-	6	21	-	-	5
Construction	-	-	1	-	-	1	176	1	-	-	34
Textile and Leather	-	-	9	-	-	-	27	20	-	-	39
Non-specified	-	-	-	-	-	-	5	16	-	-	13
TRANSPORT SECTOR	**-**	**-**	**1429**	**3**	**120**	**8**	**2056**	**239**	**-**	**-**	**11**
Air	-	-	-	3	120	-	-	-	-	-	-
Road	-	-	1427	-	-	7	1664	-	-	-	11
Rail	-	-	1	-	-	1	285	-	-	-	-
Pipeline Transport	-	-	-	-	-	-	1	-	-	-	-
Internal Navigation	-	-	1	-	-	-	106	239	-	-	-
Non-specified	-	-	-	-	-	-	-	-	-	-	-
OTHER SECTORS	**-**	**246**	**90**	**-**	**-**	**22**	**710**	**18**	**-**	**-**	**180**
Agriculture	-	-	7	-	-	-	497	1	-	-	15
Comm. and Publ. Services	-	-	-	-	-	-	-	-	-	-	-
Residential	-	246	-	-	-	19	-	13	-	-	116
Non-specified	-	-	83	-	-	3	213	4	-	-	49
NON-ENERGY USE	**-**	**-**	**-**	**-**	**-**	**-**	**-**	**-**	**-**	**280**	**529**
in Industry/Transf./Energy	-	-	-	-	-	-	-	-	-	280	400
in Transport	-	-	-	-	-	-	-	-	-	-	73
in Other Sectors	-	-	-	-	-	-	-	-	-	-	56

Romania / Roumanie : 1997

	Gas / Gaz (TJ)				Comb. Renew. & Waste / En. Re. Comb. & Déchets (TJ)				(GWh)	(TJ)
SUPPLY AND CONSUMPTION	Natural Gas	Gas Works	Coke Ovens	Blast Furnaces	Solid Biomass	Gas/Liquids from Biomass	Municipal Waste	Industrial Waste	Electricity	Heat
APPROVISIONNEMENT ET DEMANDE	Gaz naturel	Usines à gaz	Cokeries	Hauts fourneaux	Biomasse solide	Gaz/Liquides tirés de biomasse	Déchets urbains	Déchets industriels	Electricité	Chaleur
Production	553958	-	19883	27164	140672	-	-	1181	57148	284612
From Other Sources	-	-	-	-	-	-	-	-	-	-
Imports	187503	-	-	-	-	-	-	-	777	-
Exports	-	-	-	-	-347	-	-	-	-556	-
Intl. Marine Bunkers	-	-	-	-	-	-	-	-	-	-
Stock Changes	-	-	-	-	-221	-	-	-	-	-
DOMESTIC SUPPLY	741461	-	19883	27164	140104	-	-	1181	57369	284612
Transfers	-	-	-	-	-	-	-	-	-	-
Statistical Differences	33273	-	603	-119	-1130	-	-	-	-285	-1693
TRANSFORMATION	270523	-	1623	5059	-870	-	-	1181	-	-
Electricity Plants	26489	-	-	1827	125	-	-	-	-	-
CHP Plants	170472	-	1623	3232	157	-	-	727	-	-
Heat Plants	73562	-	-	-	572	-	-	454	-	-
Blast Furnaces/Gas Works	-	-	-	-	-	-	-	-	-	-
Coke/Pat. Fuel/BKB Plants	-	-	-	-	38	-	-	-	-	-
Petroleum Refineries	-	-	-	-	-	-	-	-	-	-
Petrochemical Industry	-	-	-	-	-	-	-	-	-	-
Liquefaction	-	-	-	-	-	-	-	-	-	-
Other Transform. Sector	-	-	-	-	-1762	-	-	-	-	-
ENERGY SECTOR	84171	-	5137	3232	26	-	-	-	12055	39754
Coal Mines	102	-	-	-	-	-	-	-	1589	615
Oil and Gas Extraction	70996	-	-	-	-	-	-	-	1975	-
Petroleum Refineries	10250	-	-	-	-	-	-	-	1361	12673
Electr., CHP+Heat Plants	2823	-	-	-	26	-	-	-	3745	25372
Pumped Storage (Elec.)	-	-	-	-	-	-	-	-	-	-
Other Energy Sector	-	-	5137	3232	-	-	-	-	3385	1094
Distribution Losses	6140	-	92	1429	-	-	-	-	6600	38627
FINAL CONSUMPTION	413900	-	13634	17325	139818	-	-	-	38429	204538
INDUSTRY SECTOR	302123	-	13634	17325	12653	-	-	-	25125	41459
Iron and Steel	50381	-	13634	17325	88	-	-	-	7912	3305
Chemical and Petrochem.	124054	-	-	-	813	-	-	-	4815	14572
of which: Feedstocks	58018	-	-	-	-	-	-	-	-	-
Non-Ferrous Metals	-	-	-	-	-	-	-	-	-	-
Non-Metallic Minerals	35719	-	-	-	45	-	-	-	1719	887
Transport Equipment	-	-	-	-	-	-	-	-	-	-
Machinery	38533	-	-	-	70	-	-	-	3688	8024
Mining and Quarrying	2809	-	-	-	27	-	-	-	1071	20
Food and Tobacco	26368	-	-	-	914	-	-	-	1449	5445
Paper, Pulp and Print	6519	-	-	-	2534	-	-	-	776	1619
Wood and Wood Products	1404	-	-	-	4815	-	-	-	344	1572
Construction	1541	-	-	-	240	-	-	-	674	1073
Textile and Leather	11012	-	-	-	144	-	-	-	905	3198
Non-specified	3783	-	-	-	2963	-	-	-	1772	1744
TRANSPORT SECTOR	102	-	-	-	412	-	-	-	2230	-
Air	-	-	-	-	-	-	-	-	-	-
Road	-	-	-	-	-	-	-	-	-	-
Rail	-	-	-	-	189	-	-	-	1517	-
Pipeline Transport	-	-	-	-	-	-	-	-	59	-
Internal Navigation	-	-	-	-	-	-	-	-	-	-
Non-specified	102	-	-	-	223	-	-	-	654	-
OTHER SECTORS	111675	-	-	-	126753	-	-	-	11074	163079
Agriculture	3324	-	-	-	1291	-	-	-	1791	5270
Comm. and Publ. Services	12631	-	-	-	-	-	-	-	1337	-
Residential	95720	-	-	-	121839	-	-	-	7946	144803
Non-specified	-	-	-	-	3623	-	-	-	-	13006
NON-ENERGY USE	-	-	-	-	-	-	-	-	-	-
in Industry/Transf./Energy	-	-	-	-	-	-	-	-	-	-
in Transport	-	-	-	-	-	-	-	-	-	-
in Other Sectors	-	-	-	-	-	-	-	-	-	-

Romania / Roumanie : 1998

SUPPLY AND CONSUMPTION / APPROVISIONNEMENT ET DEMANDE	Coal / Charbon (1000 tonnes)							Oil / Pétrole (1000 tonnes)			
	Coking Coal / Charbon à coke	Other Bit. Coal / Autres charb. bit.	Sub-Bit. Coal / Charbon sous-bit.	Lignite / Lignite	Peat / Tourbe	Oven and Gas Coke / Coke de four/gaz	Pat. Fuel and BKB / Agg./briq. de lignite	Crude Oil / Pétrole brut	NGL / LGN	Feed-stocks / Produits d'aliment.	Additives / Additifs
Production	192	2	1212	24825	4	3132	-	6309	244	-	171
From Other Sources	-	-	-	-	-	-	-	-	-	419	-
Imports	3711	10	-	146	-	106	-	5975	-	-	2
Exports	-	-	-	-	-	-143	-	-2	-	-	-
Intl. Marine Bunkers	-	-	-	-	-	-	-	-	-	-	-
Stock Changes	396	4	-80	886	-	-54	-	-2	-21	-15	2
DOMESTIC SUPPLY	4299	16	1132	25857	4	3041	-	12280	223	404	175
Transfers	-	-	-	-	-	-	-	-	-	-	-
Statistical Differences	335	-2	-35	-90	9	-99	-	449	-10	-	-2
TRANSFORMATION	4579	-	1042	24944	-	2459	-	12520	136	404	173
Electricity Plants	-	-	-	8193	-	-	-	-	-	-	-
CHP Plants	-	-	977	16432	-	-	-	-	-	-	-
Heat Plants	-	-	65	319	-	-	-	-	-	-	-
Blast Furnaces/Gas Works	-	-	-	-	-	2459	-	-	-	-	-
Coke/Pat. Fuel/BKB Plants	4579	-	-	-	-	-	-	-	-	-	-
Petroleum Refineries	-	-	-	-	-	-	-	12520	136	404	173
Petrochemical Industry	-	-	-	-	-	-	-	-	-	-	-
Liquefaction	-	-	-	-	-	-	-	-	-	-	-
Other Transform. Sector	-	-	-	-	-	-	-	-	-	-	-
ENERGY SECTOR	54	-	22	109	-	-	-	145	77	-	-
Coal Mines	1	-	22	105	-	-	-	-	-	-	-
Oil and Gas Extraction	-	-	-	-	-	-	-	145	77	-	-
Petroleum Refineries	-	-	-	-	-	-	-	-	-	-	-
Electr., CHP+Heat Plants	-	-	-	4	-	-	-	-	-	-	-
Pumped Storage (Elec.)	-	-	-	-	-	-	-	-	-	-	-
Other Energy Sector	53	-	-	-	-	-	-	-	-	-	-
Distribution Losses	-	-	13	223	-	-	-	64	-	-	-
FINAL CONSUMPTION	1	14	20	491	13	483	-	-	-	-	-
INDUSTRY SECTOR	1	14	5	422	-	483	-	-	-	-	-
Iron and Steel	-	2	-	-	-	412	-	-	-	-	-
Chemical and Petrochem.	-	6	-	375	-	45	-	-	-	-	-
of which: Feedstocks	-	-	-	-	-	-	-	-	-	-	-
Non-Ferrous Metals	-	-	-	-	-	-	-	-	-	-	-
Non-Metallic Minerals	1	6	-	36	-	5	-	-	-	-	-
Transport Equipment	-	-	-	-	-	-	-	-	-	-	-
Machinery	-	-	2	2	-	13	-	-	-	-	-
Mining and Quarrying	-	-	2	1	-	-	-	-	-	-	-
Food and Tobacco	-	-	-	2	-	8	-	-	-	-	-
Paper, Pulp and Print	-	-	-	1	-	-	-	-	-	-	-
Wood and Wood Products	-	-	-	-	-	-	-	-	-	-	-
Construction	-	-	-	1	-	-	-	-	-	-	-
Textile and Leather	-	-	-	2	-	-	-	-	-	-	-
Non-specified	-	-	1	2	-	-	-	-	-	-	-
TRANSPORT SECTOR	-	-	2	-	-	-	-	-	-	-	-
Air	-	-	-	-	-	-	-	-	-	-	-
Road	-	-	-	-	-	-	-	-	-	-	-
Rail	-	-	2	-	-	-	-	-	-	-	-
Pipeline Transport	-	-	-	-	-	-	-	-	-	-	-
Internal Navigation	-	-	-	-	-	-	-	-	-	-	-
Non-specified	-	-	-	-	-	-	-	-	-	-	-
OTHER SECTORS	-	-	13	69	13	-	-	-	-	-	-
Agriculture	-	-	-	2	6	-	-	-	-	-	-
Comm. and Publ. Services	-	-	-	-	-	-	-	-	-	-	-
Residential	-	-	11	58	7	-	-	-	-	-	-
Non-specified	-	-	2	9	-	-	-	-	-	-	-
NON-ENERGY USE	-	-	-	-	-	-	-	-	-	-	-
in Industry/Trans./Energy	-	-	-	-	-	-	-	-	-	-	-
in Transport	-	-	-	-	-	-	-	-	-	-	-
in Other Sectors	-	-	-	-	-	-	-	-	-	-	-

Romania / Roumanie : 1998

	Oil cont. / *Pétrole cont.* (1000 tonnes)										
SUPPLY AND CONSUMPTION *APPROVISIONNEMENT ET DEMANDE*	Refinery Gas *Gaz de raffinerie*	LPG + Ethane *GPL + éthane*	Motor Gasoline *Essence moteur*	Aviation Gasoline *Essence aviation*	Jet Fuel *Carbu- réacteurs*	Kerosene *Kérosène*	Gas/ Diesel *Gazole*	Heavy Fuel Oil *Fioul lourd*	Naphtha *Naphta*	Petrol. Coke *Coke de pétrole*	Other Prod. *Autres prod.*
Production	882	304	3197	6	84	80	4035	1941	424	684	1388
From Other Sources	-	-	-	-	-	-	-	-	-	-	-
Imports	-	6	107	3	18	5	137	2368	1	-	100
Exports	-	-8	-1502	-	-9	-	-1441	-9	-2	-163	-98
Intl. Marine Bunkers	-	-	-	-	-	-	-32	-76	-	-	-
Stock Changes	4	-9	51	-	-2	-2	31	75	18	-64	-25
DOMESTIC SUPPLY	886	293	1853	9	91	83	2730	4299	441	457	1365
Transfers	-	-	-	-	-	-	-	-	-	-	-
Statistical Differences	-36	-35	-305	-1	11	-28	233	-649	-150	-47	82
TRANSFORMATION	26	-	-	-	-	-	6	2277	-	-	245
Electricity Plants	-	-	-	-	-	-	1	54	-	-	-
CHP Plants	25	-	-	-	-	-	-	1907	-	-	-
Heat Plants	1	-	-	-	-	-	5	316	-	-	245
Blast Furnaces/Gas Works	-	-	-	-	-	-	-	-	-	-	-
Coke/Pat. Fuel/BKB Plants	-	-	-	-	-	-	-	-	-	-	-
Petroleum Refineries	-	-	-	-	-	-	-	-	-	-	-
Petrochemical Industry	-	-	-	-	-	-	-	-	-	-	-
Liquefaction	-	-	-	-	-	-	-	-	-	-	-
Other Transform. Sector	-	-	-	-	-	-	-	-	-	-	-
ENERGY SECTOR	647	1	4	-	-	2	46	398	-	-	345
Coal Mines	-	-	-	-	-	-	8	3	-	-	-
Oil and Gas Extraction	-	-	-	-	-	-	17	-	-	-	-
Petroleum Refineries	647	-	-	-	-	-	-	-	-	-	-
Electr., CHP+Heat Plants	-	-	4	-	-	-	20	18	-	-	-
Pumped Storage (Elec.)	-	-	-	-	-	-	-	-	-	-	-
Other Energy Sector	-	1	-	-	-	2	1	377	-	-	345
Distribution Losses	-	-	-	-	-	-	-	-	-	-	-
FINAL CONSUMPTION	177	257	1544	8	102	53	2911	975	291	410	857
INDUSTRY SECTOR	177	3	61	-	-	3	322	784	291	-	213
Iron and Steel	-	-	-	-	-	-	3	201	-	-	7
Chemical and Petrochem.	177	-	-	-	-	-	3	129	291	-	12
of which: Feedstocks	-	-	-	-	-	-	-	46	275	-	-
Non-Ferrous Metals	-	-	-	-	-	-	-	-	-	-	-
Non-Metallic Minerals	-	-	1	-	-	-	5	250	-	-	5
Transport Equipment	-	-	-	-	-	-	-	-	-	-	-
Machinery	-	2	4	-	-	3	14	45	-	-	26
Mining and Quarrying	-	-	-	-	-	-	29	13	-	-	6
Food and Tobacco	-	1	10	-	-	-	26	37	-	-	37
Paper, Pulp and Print	-	-	-	-	-	-	-	53	-	-	1
Wood and Wood Products	-	-	6	-	-	-	7	8	-	-	3
Construction	-	-	20	-	-	-	188	-	-	-	24
Textile and Leather	-	-	19	-	-	-	43	12	-	-	68
Non-specified	-	-	1	-	-	-	4	36	-	-	24
TRANSPORT SECTOR	-	-	1416	8	102	2	1932	164	-	-	6
Air	-	-	-	8	102	-	-	-	-	-	-
Road	-	-	1415	-	-	2	1626	1	-	-	6
Rail	-	-	1	-	-	-	246	-	-	-	-
Pipeline Transport	-	-	-	-	-	-	1	-	-	-	-
Internal Navigation	-	-	-	-	-	-	59	163	-	-	-
Non-specified	-	-	-	-	-	-	-	-	-	-	-
OTHER SECTORS	-	254	67	-	-	48	657	27	-	-	143
Agriculture	-	3	12	-	-	1	439	-	-	-	11
Comm. and Publ. Services	-	4	-	-	-	-	-	17	-	-	15
Residential	-	247	-	-	-	46	-	10	-	-	117
Non-specified	-	-	55	-	-	1	218	-	-	-	-
NON-ENERGY USE	-	-	-	-	-	-	-	-	-	410	495
in Industry/Transf./Energy	-	-	-	-	-	-	-	-	-	410	377
in Transport	-	-	-	-	-	-	-	-	-	-	82
in Other Sectors	-	-	-	-	-	-	-	-	-	-	36

Romania / Roumanie : 1998

SUPPLY AND CONSUMPTION / APPROVISIONNEMENT ET DEMANDE	Gas / Gaz (TJ)				Comb. Renew. & Waste / En. Re. Comb. & Déchets (TJ)				(GWh)	(TJ)
	Natural Gas / Gaz naturel	Gas Works / Usines à gaz	Coke Ovens / Cokeries	Blast Furnaces / Hauts fourneaux	Solid Biomass / Biomasse solide	Gas/Liquids from Biomass / Gaz/Liquides tirés de biomasse	Municipal Waste / Déchets urbains	Industrial Waste / Déchets industriels	Electricity / Electricité	Heat / Chaleur
Production	518676	-	18881	24715	126312	-	-	-	53496	267463
From Other Sources	-	-	-	-	-	-	-	-	-	-
Imports	175546	-	-	-	-	-	-	-	724	-
Exports	-	-	-	-	-347	-	-	-	-337	-
Intl. Marine Bunkers	-	-	-	-	-	-	-	-	-	-
Stock Changes	29	-	-	-	187	-	-	-	-	-
DOMESTIC SUPPLY	694251	-	18881	24715	126152	-	-	-	53883	267463
Transfers	-	-	-	-	-	-	-	-	-	-
Statistical Differences	1761	-	140	715	-1511	-	-	-	-	-
TRANSFORMATION	288736	-	1531	5497	-932	-	-	-	-	-
Electricity Plants	21678	-	-	1521	31	-	-	-	-	-
CHP Plants	180243	-	1525	3959	118	-	-	-	-	-
Heat Plants	86815	-	6	17	681	-	-	-	-	-
Blast Furnaces/Gas Works	-	-	-	-	-	-	-	-	-	-
Coke/Pat. Fuel/BKB Plants	-	-	-	-	-	-	-	-	-	-
Petroleum Refineries	-	-	-	-	-	-	-	-	-	-
Petrochemical Industry	-	-	-	-	-	-	-	-	-	-
Liquefaction	-	-	-	-	-	-	-	-	-	-
Other Transform. Sector	-	-	-	-	-1762	-	-	-	-	-
ENERGY SECTOR	72711	-	3609	2221	3	-	-	-	10838	25894
Coal Mines	92	-	-	-	-	-	-	-	1221	116
Oil and Gas Extraction	45958	-	-	-	-	-	-	-	1921	-
Petroleum Refineries	17109	-	-	-	-	-	-	-	1471	7379
Electr., CHP+Heat Plants	9552	-	2	-	3	-	-	-	3404	17277
Pumped Storage (Elec.)	-	-	-	-	-	-	-	-	-	-
Other Energy Sector	-	-	3607	2221	-	-	-	-	2821	1122
Distribution Losses	13903	-	72	929	4	-	-	-	6450	31651
FINAL CONSUMPTION	320662	-	13809	16783	125566	-	-	-	36595	209918
INDUSTRY SECTOR	192187	-	13800	16783	9250	-	-	-	22686	50361
Iron and Steel	50177	-	12772	16783	2	-	-	-	7558	3556
Chemical and Petrochem.	59101	-	1028	-	336	-	-	-	4329	10645
of which: Feedstocks	24987	-	-	-	-	-	-	-	-	-
Non-Ferrous Metals	-	-	-	-	-	-	-	-	-	-
Non-Metallic Minerals	26660	-	-	-	119	-	-	-	1806	1552
Transport Equipment	-	-	-	-	-	-	-	-	-	-
Machinery	23149	-	-	-	23	-	-	-	4109	6355
Mining and Quarrying	1162	-	-	-	13	-	-	-	672	27
Food and Tobacco	12019	-	-	-	531	-	-	-	1032	14397
Paper, Pulp and Print	4586	-	-	-	2563	-	-	-	553	593
Wood and Wood Products	953	-	-	-	3381	-	-	-	202	757
Construction	1010	-	-	-	193	-	-	-	413	879
Textile and Leather	6128	-	-	-	21	-	-	-	546	7022
Non-specified	7242	-	-	-	2068	-	-	-	1466	4578
TRANSPORT SECTOR	1487	-	-	-	227	-	-	-	1996	-
Air	-	-	-	-	-	-	-	-	-	-
Road	-	-	-	-	-	-	-	-	-	-
Rail	-	-	-	-	-	-	-	-	1375	-
Pipeline Transport	1443	-	-	-	-	-	-	-	45	-
Internal Navigation	-	-	-	-	-	-	-	-	-	-
Non-specified	44	-	-	-	227	-	-	-	576	-
OTHER SECTORS	126988	-	9	-	116089	-	-	-	11913	159557
Agriculture	2156	-	-	-	1253	-	-	-	1311	4467
Comm. and Publ. Services	21071	-	-	-	-	-	-	-	2685	-
Residential	103761	-	-	-	112184	-	-	-	7917	144066
Non-specified	-	-	9	-	2652	-	-	-	-	11024
NON-ENERGY USE	-	-	-	-	-	-	-	-	-	-
in Industry/Transf./Energy	-	-	-	-	-	-	-	-	-	-
in Transport	-	-	-	-	-	-	-	-	-	-
in Other Sectors	-	-	-	-	-	-	-	-	-	-

Russia / Russie : 1997

SUPPLY AND CONSUMPTION / APPROVISIONNEMENT ET DEMANDE	Coal / Charbon (1000 tonnes)							Oil / Pétrole (1000 tonnes)			
	Coking Coal / Charbon à coke	Other Bit. Coal / Autres charb. bit.	Sub-Bit. Coal / Charbon sous-bit.	Lignite / Lignite	Peat / Tourbe	Oven and Gas Coke / Coke de four/gaz	Pat. Fuel and BKB / Agg./briq. de lignite	Crude Oil / Pétrole brut	NGL / LGN	Feed-stocks / Produits d'aliment.	Additives / Additifs
Production	52368	107439	-	85224	3363	21879	443	303868	-	-	-
From Other Sources	-	-	-	-	-	-	-	-	-	-	-
Imports	1274	19441	-	89	-	70	-	4004	-	-	-
Exports	-8450	-15043	-	-883	-	-1458	-	-126889	-	-	-
Intl. Marine Bunkers	-	-	-	-	-	-	-	-	-	-	-
Stock Changes	158	4775	-	401	180	-95	23	-77	-	-	-
DOMESTIC SUPPLY	**45350**	**116612**	**-**	**84831**	**3543**	**20396**	**466**	**180906**	**-**	**-**	**-**
Transfers	-	-	-	-	-	-	-	-	-	-	-
Statistical Differences	-5284	-7821	-	-2522	-236	-	-	-2513	-	-	-
TRANSFORMATION	**40066**	**79129**	**-**	**73119**	**2505**	**15192**	**52**	**175302**	**-**	**-**	**-**
Electricity Plants	-	-	-	-	-	-	-	-	-	-	-
CHP Plants	-	56973	-	56216	1787	-	52	995	-	-	-
Heat Plants	-	22156	-	16251	450	-	-	-	-	-	-
Blast Furnaces/Gas Works	-	-	-	-	-	15192	-	-	-	-	-
Coke/Pat. Fuel/BKB Plants	40066	-	-	652	268	-	-	-	-	-	-
Petroleum Refineries	-	-	-	-	-	-	-	174307	-	-	-
Petrochemical Industry	-	-	-	-	-	-	-	-	-	-	-
Liquefaction	-	-	-	-	-	-	-	-	-	-	-
Other Transform. Sector	-	-	-	-	-	-	-	-	-	-	-
ENERGY SECTOR	**-**	**392**	**-**	**184**	**96**	**36**	**44**	**1428**	**-**	**-**	**-**
Coal Mines	-	309	-	160	96	-	44	-	-	-	-
Oil and Gas Extraction	-	-	-	-	-	-	-	1428	-	-	-
Petroleum Refineries	-	-	-	-	-	-	-	-	-	-	-
Electr., CHP+Heat Plants	-	83	-	24	-	-	-	-	-	-	-
Pumped Storage (Elec.)	-	-	-	-	-	-	-	-	-	-	-
Other Energy Sector	-	-	-	-	-	36	-	-	-	-	-
Distribution Losses	-	5511	-	3774	60	-	-	1275	-	-	-
FINAL CONSUMPTION	**-**	**23759**	**-**	**5232**	**646**	**5168**	**370**	**388**	**-**	**-**	**-**
INDUSTRY SECTOR	**-**	**2703**	**-**	**270**	**19**	**4689**	**18**	**165**	**-**	**-**	**-**
Iron and Steel	-	467	-	34	-	3831	2	-	-	-	-
Chemical and Petrochem.	-	81	-	8	-	159	1	-	-	-	-
of which: Feedstocks	-	-	-	-	-	-	-	-	-	-	-
Non-Ferrous Metals	-	214	-	30	1	417	-	9	-	-	-
Non-Metallic Minerals	-	545	-	99	-	64	-	-	-	-	-
Transport Equipment	-	-	-	-	-	-	-	-	-	-	-
Machinery	-	394	-	9	6	169	-	9	-	-	-
Mining and Quarrying	-	-	-	-	-	-	-	-	-	-	-
Food and Tobacco	-	154	-	40	-	26	4	9	-	-	-
Paper, Pulp and Print	-	-	-	-	-	-	-	-	-	-	-
Wood and Wood Products	-	39	-	6	3	1	-	2	-	-	-
Construction	-	725	-	25	3	1	1	98	-	-	-
Textile and Leather	-	48	-	6	4	-	-	-	-	-	-
Non-specified	-	36	-	13	2	21	10	38	-	-	-
TRANSPORT SECTOR	**-**	**590**	**-**	**118**	**-**	**-**	**22**	**16**	**-**	**-**	**-**
Air	-	-	-	-	-	-	-	-	-	-	-
Road	-	-	-	-	-	-	-	-	-	-	-
Rail	-	545	-	106	-	-	22	-	-	-	-
Pipeline Transport	-	-	-	-	-	-	-	14	-	-	-
Internal Navigation	-	3	-	-	-	-	-	-	-	-	-
Non-specified	-	42	-	12	-	-	-	2	-	-	-
OTHER SECTORS	**-**	**20085**	**-**	**4662**	**578**	**5**	**330**	**207**	**-**	**-**	**-**
Agriculture	-	2032	-	921	2	-	4	-	-	-	-
Comm. and Publ. Services	-	-	-	-	-	-	-	-	-	-	-
Residential	-	18053	-	3652	558	3	326	207	-	-	-
Non-specified	-	-	-	89	18	2	-	-	-	-	-
NON-ENERGY USE	**-**	**381**	**-**	**182**	**49**	**474**	**-**	**-**	**-**	**-**	**-**
in Industry/Trans./Energy	-	381	-	182	49	474	-	-	-	-	-
in Transport	-	-	-	-	-	-	-	-	-	-	-
in Other Sectors	-	-	-	-	-	-	-	-	-	-	-

Russia / Russie : 1997

SUPPLY AND CONSUMPTION / APPROVISIONNEMENT ET DEMANDE	Oil cont. / *Pétrole cont.* (1000 tonnes)										
	Refinery Gas / *Gaz de raffinerie*	LPG + Ethane / *GPL + éthane*	Motor Gasoline / *Essence moteur*	Aviation Gasoline / *Essence aviation*	Jet Fuel / *Carbu- réacteurs*	Kerosene / *Kérosène*	Gas/ Diesel / *Gazole*	Heavy Fuel Oil / *Fioul lourd*	Naphtha / *Naphta*	Petrol. Coke / *Coke de pétrole*	Other Prod. / *Autres prod.*
Production	4678	5041	27178	39	9189	107	47226	63887	-	794	13655
From Other Sources	-	-	-	-	-	-	-	-	-	-	-
Imports	-	35	2298	3	-	-	2036	734	-	599	-
Exports	-	-861	-4966	-2	-	-	-23818	-27822	-	-72	-103
Intl. Marine Bunkers	-	-	-	-	-	-	-	-	-	-	-
Stock Changes	-	-8	231	-	-	-	41	-26	-	-	3
DOMESTIC SUPPLY	4678	4207	24741	40	9189	107	25485	36773	-	1321	13555
Transfers	-	-	-	-	-	-	-	-	-	-	-
Statistical Differences	-	-	-	-	-	-	-	-	-	-	-
TRANSFORMATION	2176	312	-	-	-	9	1001	28710	-	-	-
Electricity Plants	-	-	-	-	-	-	-	-	-	-	-
CHP Plants	2176	312	-	-	-	9	-	-	-	-	-
Heat Plants	-	-	-	-	-	-	1001	28710	-	-	-
Blast Furnaces/Gas Works	-	-	-	-	-	-	-	-	-	-	-
Coke/Pat. Fuel/BKB Plants	-	-	-	-	-	-	-	-	-	-	-
Petroleum Refineries	-	-	-	-	-	-	-	-	-	-	-
Petrochemical Industry	-	-	-	-	-	-	-	-	-	-	-
Liquefaction	-	-	-	-	-	-	-	-	-	-	-
Other Transform. Sector	-	-	-	-	-	-	-	-	-	-	-
ENERGY SECTOR	1839	-	909	-	-	3	1678	2465	-	-	-
Coal Mines	-	-	-	-	-	-	-	-	-	-	-
Oil and Gas Extraction	-	-	909	-	-	-	1355	-	-	-	-
Petroleum Refineries	-	-	-	-	-	-	-	-	-	-	-
Electr., CHP+Heat Plants	-	-	-	-	-	-	323	-	-	-	-
Pumped Storage (Elec.)	-	-	-	-	-	-	-	-	-	-	-
Other Energy Sector	1839	-	-	-	-	3	-	2465	-	-	-
Distribution Losses	-	-	-	-	-	-	-	-	-	-	-
FINAL CONSUMPTION	663	3895	23832	40	9189	95	22806	5598	-	1321	13555
INDUSTRY SECTOR	661	1406	2972	-	-	37	4558	3798	-	1321	-
Iron and Steel	-	-	110	-	-	2	298	921	-	-	-
Chemical and Petrochem.	659	1390	228	-	-	17	153	223	-	-	-
of which: Feedstocks	-	*1390*	-	-	-	*11*	*3*	*11*	-	-	-
Non-Ferrous Metals	-	-	116	-	-	1	418	808	-	-	-
Non-Metallic Minerals	1	-	150	-	-	-	209	535	-	-	-
Transport Equipment	-	-	-	-	-	-	-	-	-	-	-
Machinery	1	-	416	-	-	2	250	309	-	-	-
Mining and Quarrying	-	-	-	-	-	-	-	-	-	-	-
Food and Tobacco	-	-	203	-	-	2	119	277	-	-	-
Paper, Pulp and Print	-	-	-	-	-	-	-	-	-	-	-
Wood and Wood Products	-	-	212	-	-	-	451	206	-	-	-
Construction	-	16	1084	-	-	6	1799	166	-	-	-
Textile and Leather	-	-	62	-	-	-	22	4	-	-	-
Non-specified	-	-	391	-	-	7	839	349	-	1321	-
TRANSPORT SECTOR	-	90	17412	40	9189	24	7175	304	-	-	-
Air	-	-	-	40	9189	-	-	-	-	-	-
Road	-	90	16600	-	-	-	4477	15	-	-	-
Rail	-	-	360	-	-	-	1729	75	-	-	-
Pipeline Transport	-	-	-	-	-	-	-	-	-	-	-
Internal Navigation	-	-	26	-	-	-	855	193	-	-	-
Non-specified	-	-	426	-	-	24	114	21	-	-	-
OTHER SECTORS	2	2399	3448	-	-	34	11073	1496	-	-	8245
Agriculture	-	158	3448	-	-	-	6172	540	-	-	-
Comm. and Publ. Services	-	-	-	-	-	-	-	-	-	-	-
Residential	-	2095	-	-	-	-	1868	956	-	-	-
Non-specified	2	146	-	-	-	34	3033	-	-	-	8245
NON-ENERGY USE	-	-	-	-	-	-	-	-	-	-	5310
in Industry/Transf./Energy	-	-	-	-	-	-	-	-	-	-	5299
in Transport	-	-	-	-	-	-	-	-	-	-	-
in Other Sectors	-	-	-	-	-	-	-	-	-	-	11

Russia / Russie : 1997

	Gas / *Gaz* (TJ)				Comb. Renew. & Waste / *En. Re. Comb. & Déchets* (TJ)				(GWh)	(TJ)
SUPPLY AND CONSUMPTION *APPROVISIONNEMENT ET DEMANDE*	Natural Gas *Gaz naturel*	Gas Works *Usines à gaz*	Coke Ovens *Cokeries*	Blast Furnaces *Hauts fourneaux*	Solid Biomass *Biomasse solide*	Gas/Liquids from Biomass *Gaz/Liquides tirés de biomasse*	Municipal Waste *Déchets urbains*	Industrial Waste *Déchets industriels*	Electricity *Electricité*	Heat *Chaleur*
Production	21440210	-	165277	190854	202033	-	-	363171	834132	6400200
From Other Sources	-	-	-	-	-	-	-	-	-	-
Imports	168335	-	-	-	-	-	-	-	7151	-
Exports	-7548043	-	-	-	-13760	-	-	-	-26840	-
Intl. Marine Bunkers	-	-	-	-	-	-	-	-	-	-
Stock Changes	433511	-	-	-	3813	-	-	-	-	-
DOMESTIC SUPPLY	**14494013**	-	**165277**	**190854**	**192086**	-	-	**363171**	**814443**	**6400200**
Transfers	-	-	-	-	-	-	-	-	-	-
Statistical Differences	-	-	-66103	-	-1820	-	-	-17746	-	-
TRANSFORMATION	**8297222**	-	**34194**	**35060**	**58427**	-	-	**67018**	-	-
Electricity Plants	-	-	-	-	8337	-	-	-	-	-
CHP Plants	5735778	-	-	24366	432	-	-	67018	-	-
Heat Plants	2109889	-	32382	10694	47350	-	-	-	-	-
Blast Furnaces/Gas Works	-	-	1812	-	-	-	-	-	-	-
Coke/Pat. Fuel/BKB Plants	-	-	-	-	-	-	-	-	-	-
Petroleum Refineries	-	-	-	-	-	-	-	-	-	-
Petrochemical Industry	-	-	-	-	-	-	-	-	-	-
Liquefaction	-	-	-	-	-	-	-	-	-	-
Other Transform. Sector	451555	-	-	-	2308	-	-	-	-	-
ENERGY SECTOR	**350111**	-	**1978**	-	**251**	-	-	**8071**	**140176**	**480300**
Coal Mines	-	-	-	-	-	-	-	-	8955	65400
Oil and Gas Extraction	349333	-	-	-	-	-	-	-	41203	159300
Petroleum Refineries	-	-	-	-	-	-	-	-	11415	198600
Electr., CHP+Heat Plants	778	-	-	-	-	-	-	-	60787	-
Pumped Storage (Elec.)	-	-	-	-	-	-	-	-	-	-
Other Energy Sector	-	-	1978	-	251	-	-	8071	17816	57000
Distribution Losses	250124	-	-	-	-	-	-	20726	84389	204800
FINAL CONSUMPTION	**5596556**	-	**63002**	**155794**	**131588**	-	-	**249610**	**589878**	**5715100**
INDUSTRY SECTOR	**2530666**	-	**63002**	**155794**	**8979**	-	-	**242092**	**291573**	**2183400**
Iron and Steel	817556	-	63002	155794	124	-	-	150934	56112	248800
Chemical and Petrochem.	810000	-	-	-	68	-	-	35922	41734	538400
of which: Feedstocks	*619000*	-	-	-	-	-	-	-	-	-
Non-Ferrous Metals	157444	-	-	-	59	-	-	18	93990	176500
Non-Metallic Minerals	234222	-	-	-	608	-	-	9	11223	101200
Transport Equipment	-	-	-	-	-	-	-	-	-	-
Machinery	254000	-	-	-	126	-	-	67	43488	478300
Mining and Quarrying	-	-	-	-	-	-	-	-	-	-
Food and Tobacco	94667	-	-	-	228	-	-	158	11223	214500
Paper, Pulp and Print	-	-	-	-	4209	-	-	-	11779	200200
Wood and Wood Products	108778	-	-	-	1822	-	-	38836	3757	-
Construction	34222	-	-	-	1703	-	-	-	10305	91000
Textile and Leather	1333	-	-	-	16	-	-	26	4208	61400
Non-specified	18444	-	-	-	16	-	-	16122	3754	73100
TRANSPORT SECTOR	**519445**	-	-	-	**2181**	-	-	**595**	**63464**	-
Air	-	-	-	-	-	-	-	-	-	-
Road	9556	-	-	-	-	-	-	-	-	-
Rail	-	-	-	-	927	-	-	595	30456	-
Pipeline Transport	474667	-	-	-	-	-	-	-	17231	-
Internal Navigation	-	-	-	-	780	-	-	-	-	-
Non-specified	35222	-	-	-	474	-	-	-	15777	-
OTHER SECTORS	**2546445**	-	-	-	**120428**	-	-	**6923**	**234841**	**3531700**
Agriculture	363778	-	-	-	13102	-	-	-	42094	223100
Comm. and Publ. Services	-	-	-	-	-	-	-	-	-	149700
Residential	1920889	-	-	-	105948	-	-	6923	132966	2762000
Non-specified	261778	-	-	-	1378	-	-	-	59781	396900
NON-ENERGY USE	-	-	-	-	-	-	-	-	-	-
in Industry/Transf./Energy	-	-	-	-	-	-	-	-	-	-
in Transport	-	-	-	-	-	-	-	-	-	-
in Other Sectors	-	-	-	-	-	-	-	-	-	-

Russia / Russie : 1998

SUPPLY AND CONSUMPTION	Coal / Charbon (1000 tonnes)							Oil / Pétrole (1000 tonnes)			
	Coking Coal	Other Bit. Coal	Sub-Bit. Coal	Lignite	Peat	Oven and Gas Coke	Pat. Fuel and BKB	Crude Oil	NGL	Feed-stocks	Additives
APPROVISIONNEMENT ET DEMANDE	Charbon à coke	Autres charb. bit.	Charbon sous-bit.	Lignite	Tourbe	Coke de four/gaz	Agg./briq. de lignite	Pétrole brut	LGN	Produits d'aliment.	Additifs
Production	51996	101088	-	78835	1767	19772	276	301406	-	-	-
From Other Sources	-	-	-	-	-	-	-	-	-	-	-
Imports	292	21508	-	22	-	83	-	5575	-	-	-
Exports	-6440	-17602	-	-321	-	-1337	-	-137235	-	-	-
Intl. Marine Bunkers	-	-	-	-	-	-	-	-	-	-	-
Stock Changes	499	1988	-	234	592	-5	14	-256	-	-	-
DOMESTIC SUPPLY	46347	106982	-	78770	2359	18513	290	169490	-	-	-
Transfers	-	-	-	-	-	-	-	-	-	-	-
Statistical Differences	-14273	1720	-	-2365	-134	-	-	-2230	-	-	-
TRANSFORMATION	32074	80817	-	67915	2021	13891	89	163488	-	-	-
Electricity Plants	-	-	-	-	-	-	-	-	-	-	-
CHP Plants	-	62979	-	55304	1372	-	18	2006	-	-	-
Heat Plants	-	17838	-	12280	410	-	71	-	-	-	-
Blast Furnaces/Gas Works	-	-	-	-	-	13891	-	-	-	-	-
Coke/Pat. Fuel/BKB Plants	32074	-	-	331	239	-	-	-	-	-	-
Petroleum Refineries	-	-	-	-	-	-	-	161482	-	-	-
Petrochemical Industry	-	-	-	-	-	-	-	-	-	-	-
Liquefaction	-	-	-	-	-	-	-	-	-	-	-
Other Transform. Sector	-	-	-	-	-	-	-	-	-	-	-
ENERGY SECTOR	-	673	-	173	37	60	7	1819	-	-	-
Coal Mines	-	530	-	151	37	-	7	-	-	-	-
Oil and Gas Extraction	-	-	-	-	-	-	-	1819	-	-	-
Petroleum Refineries	-	-	-	-	-	-	-	-	-	-	-
Electr., CHP+Heat Plants	-	143	-	22	-	-	-	-	-	-	-
Pumped Storage (Elec.)	-	-	-	-	-	-	-	-	-	-	-
Other Energy Sector	-	-	-	-	-	60	-	-	-	-	-
Distribution Losses	-	5183	-	3562	31	-	-	1615	-	-	-
FINAL CONSUMPTION	-	22029	-	4755	136	4562	194	338	-	-	-
INDUSTRY SECTOR	-	2351	-	290	3	4175	6	53	-	-	-
Iron and Steel	-	244	-	11	-	3467	-	18	-	-	-
Chemical and Petrochem.	-	19	-	7	-	41	-	-	-	-	-
of which: Feedstocks	-	-	-	-	-	-	-	-	-	-	-
Non-Ferrous Metals	-	488	-	7	-	337	-	-	-	-	-
Non-Metallic Minerals	-	526	-	69	-	81	-	2	-	-	-
Transport Equipment	-	-	-	-	-	-	-	-	-	-	-
Machinery	-	121	-	18	1	210	2	18	-	-	-
Mining and Quarrying	-	-	-	-	-	-	-	-	-	-	-
Food and Tobacco	-	233	-	33	-	37	-	11	-	-	-
Paper, Pulp and Print	-	-	-	-	-	-	-	-	-	-	-
Wood and Wood Products	-	22	-	4	1	-	1	-	-	-	-
Construction	-	533	-	122	1	2	2	4	-	-	-
Textile and Leather	-	100	-	2	-	-	-	-	-	-	-
Non-specified	-	65	-	17	-	-	1	-	-	-	-
TRANSPORT SECTOR	-	526	-	105	-	-	13	11	-	-	-
Air	-	-	-	-	-	-	-	-	-	-	-
Road	-	-	-	-	-	-	-	-	-	-	-
Rail	-	465	-	92	-	-	13	-	-	-	-
Pipeline Transport	-	-	-	-	-	-	-	11	-	-	-
Internal Navigation	-	3	-	-	-	-	-	-	-	-	-
Non-specified	-	58	-	13	-	-	-	-	-	-	-
OTHER SECTORS	-	18998	-	4192	126	2	175	274	-	-	-
Agriculture	-	1845	-	741	3	-	4	13	-	-	-
Comm. and Publ. Services	-	-	-	-	-	-	-	-	-	-	-
Residential	-	17153	-	3451	123	2	171	261	-	-	-
Non-specified	-	-	-	-	-	-	-	-	-	-	-
NON-ENERGY USE	-	154	-	168	7	385	-	-	-	-	-
in Industry/Trans./Energy	-	154	-	168	7	385	-	-	-	-	-
in Transport	-	-	-	-	-	-	-	-	-	-	-
in Other Sectors	-	-	-	-	-	-	-	-	-	-	-

Russia / Russie : 1998

SUPPLY AND CONSUMPTION / APPROVISIONNEMENT ET DEMANDE	Oil cont. / Pétrole cont. (1000 tonnes)										
	Refinery Gas / Gaz de raffinerie	LPG + Ethane / GPL + éthane	Motor Gasoline / Essence moteur	Aviation Gasoline / Essence aviation	Jet Fuel / Carbu- réacteurs	Kerosene / Kérosène	Gas/ Diesel / Gazole	Heavy Fuel Oil / Fioul lourd	Naphtha / Naphta	Petrol. Coke / Coke de pétrole	Other Prod. / Autres prod.
Production	5016	4808	25923	31	9189	51	45102	56701	-	997	11434
From Other Sources	-	-	-	-	-	-	-	-	-	-	-
Imports	-	30	1570	-	-	-	1365	773	-	-	19
Exports	-	-1619	-2797	-	-	-	-24413	-21839	-	-	-122
Intl. Marine Bunkers	-	-	-	-	-	-	-	-	-	-	-
Stock Changes	-	-6	-292	-	-	-	1134	261	-	-	-21
DOMESTIC SUPPLY	5016	3213	24404	31	9189	51	23188	35896	-	997	11310
Transfers	-	-	-	-	-	-	-	-	-	-	-
Statistical Differences	-	-	-	-	-	-	-	-	-	-	-
TRANSFORMATION	2333	669	-	-	-	4	955	28499	-	-	-
Electricity Plants	-	-	-	-	-	-	-	-	-	-	-
CHP Plants	2333	669	-	-	-	4	-	-	-	-	-
Heat Plants	-	-	-	-	-	-	955	28499	-	-	-
Blast Furnaces/Gas Works	-	-	-	-	-	-	-	-	-	-	-
Coke/Pat. Fuel/BKB Plants	-	-	-	-	-	-	-	-	-	-	-
Petroleum Refineries	-	-	-	-	-	-	-	-	-	-	-
Petrochemical Industry	-	-	-	-	-	-	-	-	-	-	-
Liquefaction	-	-	-	-	-	-	-	-	-	-	-
Other Transform. Sector	-	-	-	-	-	-	-	-	-	-	-
ENERGY SECTOR	1975	-	1070	-	-	1	1576	2541	-	-	-
Coal Mines	-	-	-	-	-	-	-	-	-	-	-
Oil and Gas Extraction	-	-	1070	-	-	-	1271	-	-	-	-
Petroleum Refineries	-	-	-	-	-	-	-	-	-	-	-
Electr., CHP+Heat Plants	-	-	-	-	-	-	305	-	-	-	-
Pumped Storage (Elec.)	-	-	-	-	-	-	-	-	-	-	-
Other Energy Sector	1975	-	-	-	-	1	-	2541	-	-	-
Distribution Losses	-	-	-	-	-	-	-	-	-	-	-
FINAL CONSUMPTION	708	2544	23334	31	9189	46	20657	4856	-	997	11310
INDUSTRY SECTOR	706	1134	3142	-	-	18	4299	3291	-	997	-
Iron and Steel	-	-	101	-	-	1	280	807	-	-	-
Chemical and Petrochem.	704	1122	126	-	-	8	144	247	-	-	-
of which: Feedstocks	-	1122	-	-	-	8	3	10	-	-	-
Non-Ferrous Metals	-	-	115	-	-	-	393	850	-	-	-
Non-Metallic Minerals	1	-	119	-	-	-	196	467	-	-	-
Transport Equipment	-	-	-	-	-	-	-	-	-	-	-
Machinery	1	-	403	-	-	1	235	256	-	-	-
Mining and Quarrying	-	-	-	-	-	-	-	-	-	-	-
Food and Tobacco	-	-	374	-	-	1	112	271	-	-	-
Paper, Pulp and Print	-	-	-	-	-	-	-	-	-	-	-
Wood and Wood Products	-	-	228	-	-	-	424	202	-	-	-
Construction	-	12	1184	-	-	3	1705	164	-	-	-
Textile and Leather	-	-	57	-	-	-	21	5	-	-	-
Non-specified	-	-	435	-	-	4	789	22	-	997	-
TRANSPORT SECTOR	-	24	17374	31	9189	12	6744	224	-	-	-
Air	-	-	-	31	9189	-	-	-	-	-	-
Road	-	24	16846	-	-	-	4207	52	-	-	-
Rail	-	-	187	-	-	-	1626	50	-	-	-
Pipeline Transport	-	-	-	-	-	-	-	-	-	-	-
Internal Navigation	-	-	21	-	-	-	804	103	-	-	-
Non-specified	-	-	320	-	-	12	107	19	-	-	-
OTHER SECTORS	2	1386	2818	-	-	16	9614	1341	-	-	4937
Agriculture	-	23	2818	-	-	-	6129	402	-	-	-
Comm. and Publ. Services	-	-	-	-	-	-	-	-	-	-	-
Residential	-	1363	-	-	-	-	1779	919	-	-	-
Non-specified	2	-	-	-	-	16	1706	20	-	-	4937
NON-ENERGY USE	-	-	-	-	-	-	-	-	-	-	6373
in Industry/Transf./Energy	-	-	-	-	-	-	-	-	-	-	6357
in Transport	-	-	-	-	-	-	-	-	-	-	-
in Other Sectors	-	-	-	-	-	-	-	-	-	-	16

Russia / Russie : 1998

SUPPLY AND CONSUMPTION	Gas / Gaz (TJ)				Comb. Renew. & Waste / En. Re. Comb. & Déchets (TJ)				(GWh)	(TJ)
	Natural Gas	Gas Works	Coke Ovens	Blast Furnaces	Solid Biomass	Gas/Liquids from Biomass	Municipal Waste	Industrial Waste	Electricity	Heat
APPROVISIONNEMENT ET DEMANDE	Gaz naturel	Usines à gaz	Cokeries	Hauts fourneaux	Biomasse solide	Gaz/Liquides tirés de biomasse	Déchets urbains	Déchets industriels	Electricité	Chaleur
Production	22197314	-	151454	174510	167191	-	-	341539	827138	6262200
From Other Sources	-	-	-	-	-	-	-	-	-	-
Imports	112889	-	-	-	-	-	-	-	8261	-
Exports	-7643869	-	-	-	-11430	-	-	-	-26275	-
Intl. Marine Bunkers	-	-	-	-	-	-	-	-	-	-
Stock Changes	-200221	-	-	-	3061	-	-	-	-	-
DOMESTIC SUPPLY	**14466113**	**-**	**151454**	**174510**	**158822**	**-**	**-**	**341539**	**809124**	**6262200**
Transfers	-	-	-	-	-	-	-	-	-	-
Statistical Differences	-	-	-60574	-	-452	-	-	-16732	-	-
TRANSFORMATION	**8233889**	**-**	**31356**	**33522**	**50086**	**-**	**-**	**61713**	**-**	**-**
Electricity Plants	-	-	-	-	7125	-	-	-	-	-
CHP Plants	5583000	-	-	23297	369	-	-	61713	-	-
Heat Plants	2122000	-	29694	10225	40469	-	-	-	-	-
Blast Furnaces/Gas Works	-	-	1662	-	-	-	-	-	-	-
Coke/Pat. Fuel/BKB Plants	-	-	-	-	-	-	-	-	-	-
Petroleum Refineries	-	-	-	-	-	-	-	-	-	-
Petrochemical Industry	-	-	-	-	-	-	-	-	-	-
Liquefaction	-	-	-	-	-	-	-	-	-	-
Other Transform. Sector	528889	-	-	-	2123	-	-	-	-	-
ENERGY SECTOR	**421667**	**-**	**1976**	**-**	**18**	**-**	**-**	**15753**	**137927**	**509400**
Coal Mines	-	-	-	-	-	-	-	-	8152	58500
Oil and Gas Extraction	421667	-	-	-	-	-	-	-	40854	153000
Petroleum Refineries	-	-	-	-	-	-	-	-	10433	200000
Electr., CHP+Heat Plants	-	-	-	-	-	-	-	-	65376	-
Pumped Storage (Elec.)	-	-	-	-	-	-	-	-	-	-
Other Energy Sector	-	-	1976	-	18	-	-	15753	13112	97900
Distribution Losses	259556	-	-	-	-	-	-	19489	93216	274700
FINAL CONSUMPTION	**5551001**	**-**	**57548**	**140988**	**108266**	**-**	**-**	**227852**	**577981**	**5478100**
INDUSTRY SECTOR	**2399890**	**-**	**57548**	**140988**	**9165**	**-**	**-**	**218778**	**283142**	**2020100**
Iron and Steel	795444	-	57548	140988	9	-	-	144947	54338	243700
Chemical and Petrochem.	814668	-	-	-	12	-	-	28607	39734	484400
of which: Feedstocks	626222	-	-	-	-	-	-	-	-	-
Non-Ferrous Metals	132556	-	-	-	75	-	-	21	92714	174200
Non-Metallic Minerals	205667	-	-	-	391	-	-	9	10572	93100
Transport Equipment	-	-	-	-	-	-	-	-	-	-
Machinery	238333	-	-	-	250	-	-	1729	41090	430500
Mining and Quarrying	-	-	-	-	-	-	-	-	-	-
Food and Tobacco	81556	-	-	-	507	-	-	97	10484	206700
Paper, Pulp and Print	-	-	-	-	3220	-	-	-	13399	194300
Wood and Wood Products	82778	-	-	-	3616	-	-	29025	3582	-
Construction	26444	-	-	-	1015	-	-	-	9078	73000
Textile and Leather	3222	-	-	-	70	-	-	3	3736	52300
Non-specified	19222	-	-	-	-	-	-	14340	4415	67900
TRANSPORT SECTOR	**659889**	**-**	**-**	**-**	**1097**	**-**	**-**	**3159**	**60034**	**-**
Air	-	-	-	-	-	-	-	-	-	-
Road	13889	-	-	-	-	-	-	-	-	-
Rail	-	-	-	-	573	-	-	3159	27227	-
Pipeline Transport	609111	-	-	-	-	-	-	-	17865	-
Internal Navigation	-	-	-	-	86	-	-	-	-	-
Non-specified	36889	-	-	-	438	-	-	-	14942	-
OTHER SECTORS	**2491222**	**-**	**-**	**-**	**98004**	**-**	**-**	**5915**	**234805**	**3458000**
Agriculture	339333	-	-	-	11210	-	-	-	38364	203000
Comm. and Publ. Services	-	-	-	-	-	-	-	-	-	155400
Residential	1933889	-	-	-	86794	-	-	5915	134574	2747000
Non-specified	218000	-	-	-	-	-	-	-	61867	352600
NON-ENERGY USE	**-**	**-**	**-**	**-**	**-**	**-**	**-**	**-**	**-**	**-**
in Industry/Transf./Energy	-	-	-	-	-	-	-	-	-	-
in Transport	-	-	-	-	-	-	-	-	-	-
in Other Sectors	-	-	-	-	-	-	-	-	-	-

Saudi Arabia / Arabie saoudite

SUPPLY AND CONSUMPTION 1997	Coal (1000 tonnes)							Oil (1000 tonnes)			
	Coking Coal	Other Bit. Coal	Sub-Bit. Coal	Lignite	Peat	Oven and Gas Coke	Pat. Fuel and BKB	Crude Oil	NGL	Feed-stocks	Additives
Production	-	-	-	-	-	-	-	417326	25057	-	-
Imports	-	-	-	-	-	-	-	-	-	-	-
Exports	-	-	-	-	-	-	-	-325499	-13313	-	-
Intl. Marine Bunkers	-	-	-	-	-	-	-	-	-	-	-
Stock Changes	-	-	-	-	-	-	-	-	-	-	-
DOMESTIC SUPPLY	-	-	-	-	-	-	-	**91827**	**11744**	-	-
Transfers and Stat. Diff.	-	-	-	-	-	-	-	-	-5794	-	-
TRANSFORMATION	-	-	-	-	-	-	-	**91827**	-	-	-
Electricity and CHP Plants	-	-	-	-	-	-	-	9293	-	-	-
Petroleum Refineries	-	-	-	-	-	-	-	82534	-	-	-
Other Transform. Sector	-	-	-	-	-	-	-	-	-	-	-
ENERGY SECTOR	-	-	-	-	-	-	-	-	-	-	-
DISTRIBUTION LOSSES	-	-	-	-	-	-	-	-	-	-	-
FINAL CONSUMPTION	-	-	-	-	-	-	-	-	5950	-	-
INDUSTRY SECTOR	-	-	-	-	-	-	-	-	5950	-	-
Iron and Steel	-	-	-	-	-	-	-	-	-	-	-
Chemical and Petrochem.	-	-	-	-	-	-	-	-	5950	-	-
Non-Metallic Minerals	-	-	-	-	-	-	-	-	-	-	-
Non-specified	-	-	-	-	-	-	-	-	-	-	-
TRANSPORT SECTOR	-	-	-	-	-	-	-	-	-	-	-
Air	-	-	-	-	-	-	-	-	-	-	-
Road	-	-	-	-	-	-	-	-	-	-	-
Non-specified	-	-	-	-	-	-	-	-	-	-	-
OTHER SECTORS	-	-	-	-	-	-	-	-	-	-	-
Agriculture	-	-	-	-	-	-	-	-	-	-	-
Comm. and Publ. Services	-	-	-	-	-	-	-	-	-	-	-
Residential	-	-	-	-	-	-	-	-	-	-	-
Non-specified	-	-	-	-	-	-	-	-	-	-	-
NON-ENERGY USE	-	-	-	-	-	-	-	-	-	-	-

APPROVISIONNEMENT ET DEMANDE 1998	Charbon (1000 tonnes)							Pétrole (1000 tonnes)			
	Charbon à coke	Autres charb. bit.	Charbon sous-bit.	Lignite	Tourbe	Coke de four/gaz	Agg./briq. de lignite	Pétrole brut	LGN	Produits d'aliment.	Additifs
Production	-	-	-	-	-	-	-	431312	25896	-	-
Imports	-	-	-	-	-	-	-	-	-	-	-
Exports	-	-	-	-	-	-	-	-339077	-13756	-	-
Intl. Marine Bunkers	-	-	-	-	-	-	-	-	-	-	-
Stock Changes	-	-	-	-	-	-	-	-	-	-	-
DOMESTIC SUPPLY	-	-	-	-	-	-	-	**92235**	**12140**	-	-
Transfers and Stat. Diff.	-	-	-	-	-	-	-	1204	-4490	-	-
TRANSFORMATION	-	-	-	-	-	-	-	**93439**	-	-	-
Electricity and CHP Plants	-	-	-	-	-	-	-	8609	-	-	-
Petroleum Refineries	-	-	-	-	-	-	-	84830	-	-	-
Other Transform. Sector	-	-	-	-	-	-	-	-	-	-	-
ENERGY SECTOR	-	-	-	-	-	-	-	-	-	-	-
DISTRIBUTION LOSSES	-	-	-	-	-	-	-	-	-	-	-
FINAL CONSUMPTION	-	-	-	-	-	-	-	-	7650	-	-
INDUSTRY SECTOR	-	-	-	-	-	-	-	-	7650	-	-
Iron and Steel	-	-	-	-	-	-	-	-	-	-	-
Chemical and Petrochem.	-	-	-	-	-	-	-	-	7650	-	-
Non-Metallic Minerals	-	-	-	-	-	-	-	-	-	-	-
Non-specified	-	-	-	-	-	-	-	-	-	-	-
TRANSPORT SECTOR	-	-	-	-	-	-	-	-	-	-	-
Air	-	-	-	-	-	-	-	-	-	-	-
Road	-	-	-	-	-	-	-	-	-	-	-
Non-specified	-	-	-	-	-	-	-	-	-	-	-
OTHER SECTORS	-	-	-	-	-	-	-	-	-	-	-
Agriculture	-	-	-	-	-	-	-	-	-	-	-
Comm. and Publ. Services	-	-	-	-	-	-	-	-	-	-	-
Residential	-	-	-	-	-	-	-	-	-	-	-
Non-specified	-	-	-	-	-	-	-	-	-	-	-
NON-ENERGY USE	-	-	-	-	-	-	-	-	-	-	-

Saudi Arabia / Arabie saoudite

SUPPLY AND CONSUMPTION 1997	Oil cont. (1000 tonnes)										
	Refinery Gas	LPG + Ethane	Motor Gasoline	Aviation Gasoline	Jet Fuel	Kerosene	Gas/ Diesel	Heavy Fuel Oil	Naphtha	Petrol. Coke	Other Prod.
Production	1787	1240	11309	-	4092	3666	24514	24834	4948	-	731
Imports	-	-	-	-	-	-	-	-	-	-	-
Exports	-	-299	-7460	-	-1283	-3451	-6798	-17588	-4948	-	-
Intl. Marine Bunkers	-	-	-	-	-	-	-	-1936	-	-	-
Stock Changes	-	-	-	-	-	-	-	-	-	-	-
DOMESTIC SUPPLY	1787	941	3849	-	2809	215	17716	5310	-	-	731
Transfers and Stat. Diff.	-	-83	5794	-	-361	-	-1170	-70	-	-	299
TRANSFORMATION	-	-	-	-	-	-	5604	78	-	-	-
Electricity and CHP Plants	-	-	-	-	-	-	5604	78	-	-	-
Petroleum Refineries	-	-	-	-	-	-	-	-	-	-	-
Other Transform. Sector	-	-	-	-	-	-	-	-	-	-	-
ENERGY SECTOR	1787	-	-	-	-	-	368	1599	-	-	-
DISTRIBUTION LOSSES	-	-	-	-	-	-	-	-	-	-	-
FINAL CONSUMPTION	-	858	9643	-	2448	215	10574	3563	-	-	1030
INDUSTRY SECTOR	-	-	-	-	-	-	-	3563	-	-	-
Iron and Steel	-	-	-	-	-	-	-	-	-	-	-
Chemical and Petrochem.	-	-	-	-	-	-	-	-	-	-	-
Non-Metallic Minerals	-	-	-	-	-	-	-	-	-	-	-
Non-specified	-	-	-	-	-	-	-	3563	-	-	-
TRANSPORT SECTOR	-	-	9643	-	2448	-	-	-	-	-	-
Air	-	-	-	-	2448	-	-	-	-	-	-
Road	-	-	9643	-	-	-	-	-	-	-	-
Non-specified	-	-	-	-	-	-	-	-	-	-	-
OTHER SECTORS	-	858	-	-	-	215	10574	-	-	-	-
Agriculture	-	-	-	-	-	-	-	-	-	-	-
Comm. and Publ. Services	-	-	-	-	-	-	-	-	-	-	-
Residential	-	858	-	-	-	215	-	-	-	-	-
Non-specified	-	-	-	-	-	-	10574	-	-	-	-
NON-ENERGY USE	-	-	-	-	-	-	-	-	-	-	1030

APPROVISIONNEMENT ET DEMANDE 1998	Pétrole cont. (1000 tonnes)										
	Gaz de raffinerie	GPL + éthane	Essence moteur	Essence aviation	Carbu- réacteurs	Kérosène	Gazole	Fioul lourd	Naphta	Coke de pétrole	Autres prod.
Production	1837	1249	12242	-	3582	3209	23352	26500	5356	-	663
Imports	-	-	-	-	-	-	-	-	-	-	-
Exports	-	-281	-7004	-	-1205	-2997	-6383	-17646	-5356	-	-
Intl. Marine Bunkers	-	-	-	-	-	-	-	-1936	-	-	-
Stock Changes	-	-	-	-	-	-	-	-	-	-	-
DOMESTIC SUPPLY	1837	968	5238	-	2377	212	16969	6918	-	-	663
Transfers and Stat. Diff.	-	-95	4490	-	37	-	-492	-91	-	-	401
TRANSFORMATION	-	-	-	-	-	-	5368	102	-	-	-
Electricity and CHP Plants	-	-	-	-	-	-	5368	102	-	-	-
Petroleum Refineries	-	-	-	-	-	-	-	-	-	-	-
Other Transform. Sector	-	-	-	-	-	-	-	-	-	-	-
ENERGY SECTOR	1837	-	-	-	-	-	353	2083	-	-	-
DISTRIBUTION LOSSES	-	-	-	-	-	-	-	-	-	-	-
FINAL CONSUMPTION	-	873	9728	-	2414	212	10756	4642	-	-	1064
INDUSTRY SECTOR	-	-	-	-	-	-	-	4642	-	-	-
Iron and Steel	-	-	-	-	-	-	-	-	-	-	-
Chemical and Petrochem.	-	-	-	-	-	-	-	-	-	-	-
Non-Metallic Minerals	-	-	-	-	-	-	-	-	-	-	-
Non-specified	-	-	-	-	-	-	-	4642	-	-	-
TRANSPORT SECTOR	-	-	9728	-	2414	-	-	-	-	-	-
Air	-	-	-	-	2414	-	-	-	-	-	-
Road	-	-	9728	-	-	-	-	-	-	-	-
Non-specified	-	-	-	-	-	-	-	-	-	-	-
OTHER SECTORS	-	873	-	-	-	212	10756	-	-	-	-
Agriculture	-	-	-	-	-	-	-	-	-	-	-
Comm. and Publ. Services	-	-	-	-	-	-	-	-	-	-	-
Residential	-	873	-	-	-	212	-	-	-	-	-
Non-specified	-	-	-	-	-	-	10756	-	-	-	-
NON-ENERGY USE	-	-	-	-	-	-	-	-	-	-	1064

Saudi Arabia / Arabie saoudite

SUPPLY AND CONSUMPTION 1997	Gas (TJ)				Comb. Renew. & Waste (TJ)				(GWh)	(TJ)
	Natural Gas	Gas Works	Coke Ovens	Blast Furnaces	Solid Biomass	Gas/Liquids from Biomass	Municipal Waste	Industrial Waste	Electricity	Heat
Production	1827092	-	-	-	-	-	-	-	113125	-
Imports	-	-	-	-	173	-	-	-	-	-
Exports	-	-	-	-	-	-	-	-	-	-
Intl. Marine Bunkers	-	-	-	-	-	-	-	-	-	-
Stock Changes	-	-	-	-	-	-	-	-	-	-
DOMESTIC SUPPLY	1827092	-	-	-	173	-	-	-	113125	-
Transfers and Stat. Diff.	1	-	-	-	-	-	-	-	-	-
TRANSFORMATION	361252	-	-	-	-	-	-	-	-	-
Electricity and CHP Plants	361252	-	-	-	-	-	-	-	-	-
Petroleum Refineries	-	-	-	-	-	-	-	-	-	-
Other Transform. Sector	-	-	-	-	-	-	-	-	-	-
ENERGY SECTOR	874857	-	-	-	-	-	-	-	14078	-
DISTRIBUTION LOSSES	-	-	-	-	-	-	-	-	8351	-
FINAL CONSUMPTION	590984	-	-	-	173	-	-	-	90696	-
INDUSTRY SECTOR	-	-	-	-	-	-	-	-	12263	-
Iron and Steel	-	-	-	-	-	-	-	-	-	-
Chemical and Petrochem.	-	-	-	-	-	-	-	-	-	-
Non-Metallic Minerals	-	-	-	-	-	-	-	-	-	-
Non-specified	-	-	-	-	-	-	-	-	12263	-
TRANSPORT SECTOR	-	-	-	-	-	-	-	-	-	-
Air	-	-	-	-	-	-	-	-	-	-
Road	-	-	-	-	-	-	-	-	-	-
Non-specified	-	-	-	-	-	-	-	-	-	-
OTHER SECTORS	590984	-	-	-	173	-	-	-	78433	-
Agriculture	-	-	-	-	-	-	-	-	1934	-
Comm. and Publ. Services	-	-	-	-	-	-	-	-	28807	-
Residential	-	-	-	-	173	-	-	-	47692	-
Non-specified	590984	-	-	-	-	-	-	-	-	-
NON-ENERGY USE	-	-	-	-	-	-	-	-	-	-

APPROVISIONNEMENT ET DEMANDE 1998	Gaz (TJ)				En. Re. Comb. & Déchets (TJ)				(GWh)	(TJ)
	Gaz naturel	Usines à gaz	Cokeries	Hauts fourneaux	Biomasse solide	Gaz/Liquides tirés de biomasse	Déchets urbains	Déchets industriels	Electricité	Chaleur
Production	1886733	-	-	-	-	-	-	-	116519	-
Imports	-	-	-	-	174	-	-	-	-	-
Exports	-	-	-	-	-	-	-	-	-	-
Intl. Marine Bunkers	-	-	-	-	-	-	-	-	-	-
Stock Changes	-	-	-	-	-	-	-	-	-	-
DOMESTIC SUPPLY	1886733	-	-	-	174	-	-	-	116519	-
Transfers and Stat. Diff.	-	-	-	-	-	-	-	-	-	-
TRANSFORMATION	373044	-	-	-	-	-	-	-	-	-
Electricity and CHP Plants	373044	-	-	-	-	-	-	-	-	-
Petroleum Refineries	-	-	-	-	-	-	-	-	-	-
Other Transform. Sector	-	-	-	-	-	-	-	-	-	-
ENERGY SECTOR	903414	-	-	-	-	-	-	-	14500	-
DISTRIBUTION LOSSES	-	-	-	-	-	-	-	-	9650	-
FINAL CONSUMPTION	610275	-	-	-	174	-	-	-	92369	-
INDUSTRY SECTOR	-	-	-	-	-	-	-	-	12489	-
Iron and Steel	-	-	-	-	-	-	-	-	-	-
Chemical and Petrochem.	-	-	-	-	-	-	-	-	-	-
Non-Metallic Minerals	-	-	-	-	-	-	-	-	-	-
Non-specified	-	-	-	-	-	-	-	-	12489	-
TRANSPORT SECTOR	-	-	-	-	-	-	-	-	-	-
Air	-	-	-	-	-	-	-	-	-	-
Road	-	-	-	-	-	-	-	-	-	-
Non-specified	-	-	-	-	-	-	-	-	-	-
OTHER SECTORS	610275	-	-	-	174	-	-	-	79880	-
Agriculture	-	-	-	-	-	-	-	-	1970	-
Comm. and Publ. Services	-	-	-	-	-	-	-	-	29338	-
Residential	-	-	-	-	174	-	-	-	48572	-
Non-specified	610275	-	-	-	-	-	-	-	-	-
NON-ENERGY USE	-	-	-	-	-	-	-	-	-	-

Senegal / Sénégal

SUPPLY AND CONSUMPTION 1997	Coal (1000 tonnes)							Oil (1000 tonnes)			
	Coking Coal	Other Bit. Coal	Sub-Bit. Coal	Lignite	Peat	Oven and Gas Coke	Pat. Fuel and BKB	Crude Oil	NGL	Feed-stocks	Additives
Production	-	-	-	-	-	-	-	1	-	-	-
Imports	-	-	-	-	-	-	-	771	-	-	-
Exports	-	-	-	-	-	-	-	-	-	-	-
Intl. Marine Bunkers	-	-	-	-	-	-	-	-	-	-	-
Stock Changes	-	-	-	-	-	-	-	66	-	-	-
DOMESTIC SUPPLY	-	-	-	-	-	-	-	838	-	-	-
Transfers and Stat. Diff.	-	-	-	-	-	-	-	-1	-	-	-
TRANSFORMATION	-	-	-	-	-	-	-	837	-	-	-
Electricity and CHP Plants	-	-	-	-	-	-	-	-	-	-	-
Petroleum Refineries	-	-	-	-	-	-	-	837	-	-	-
Other Transform. Sector	-	-	-	-	-	-	-	-	-	-	-
ENERGY SECTOR	-	-	-	-	-	-	-	-	-	-	-
DISTRIBUTION LOSSES	-	-	-	-	-	-	-	-	-	-	-
FINAL CONSUMPTION	-	-	-	-	-	-	-	-	-	-	-
INDUSTRY SECTOR	-	-	-	-	-	-	-	-	-	-	-
Iron and Steel	-	-	-	-	-	-	-	-	-	-	-
Chemical and Petrochem.	-	-	-	-	-	-	-	-	-	-	-
Non-Metallic Minerals	-	-	-	-	-	-	-	-	-	-	-
Non-specified	-	-	-	-	-	-	-	-	-	-	-
TRANSPORT SECTOR	-	-	-	-	-	-	-	-	-	-	-
Air	-	-	-	-	-	-	-	-	-	-	-
Road	-	-	-	-	-	-	-	-	-	-	-
Non-specified	-	-	-	-	-	-	-	-	-	-	-
OTHER SECTORS	-	-	-	-	-	-	-	-	-	-	-
Agriculture	-	-	-	-	-	-	-	-	-	-	-
Comm. and Publ. Services	-	-	-	-	-	-	-	-	-	-	-
Residential	-	-	-	-	-	-	-	-	-	-	-
Non-specified	-	-	-	-	-	-	-	-	-	-	-
NON-ENERGY USE	-	-	-	-	-	-	-	-	-	-	-

APPROVISIONNEMENT ET DEMANDE 1998	Charbon (1000 tonnes)							Pétrole (1000 tonnes)			
	Charbon à coke	Autres charb. bit.	Charbon sous-bit.	Lignite	Tourbe	Coke de four/gaz	Agg./briq. de lignite	Pétrole brut	LGN	Produits d'aliment.	Additifs
Production	-	-	-	-	-	-	-	2	-	-	-
Imports	-	-	-	-	-	-	-	862	-	-	-
Exports	-	-	-	-	-	-	-	-	-	-	-
Intl. Marine Bunkers	-	-	-	-	-	-	-	-	-	-	-
Stock Changes	-	-	-	-	-	-	-	-	-	-	-
DOMESTIC SUPPLY	-	-	-	-	-	-	-	864	-	-	-
Transfers and Stat. Diff.	-	-	-	-	-	-	-	-	-	-	-
TRANSFORMATION	-	-	-	-	-	-	-	864	-	-	-
Electricity and CHP Plants	-	-	-	-	-	-	-	-	-	-	-
Petroleum Refineries	-	-	-	-	-	-	-	864	-	-	-
Other Transform. Sector	-	-	-	-	-	-	-	-	-	-	-
ENERGY SECTOR	-	-	-	-	-	-	-	-	-	-	-
DISTRIBUTION LOSSES	-	-	-	-	-	-	-	-	-	-	-
FINAL CONSUMPTION	-	-	-	-	-	-	-	-	-	-	-
INDUSTRY SECTOR	-	-	-	-	-	-	-	-	-	-	-
Iron and Steel	-	-	-	-	-	-	-	-	-	-	-
Chemical and Petrochem.	-	-	-	-	-	-	-	-	-	-	-
Non-Metallic Minerals	-	-	-	-	-	-	-	-	-	-	-
Non-specified	-	-	-	-	-	-	-	-	-	-	-
TRANSPORT SECTOR	-	-	-	-	-	-	-	-	-	-	-
Air	-	-	-	-	-	-	-	-	-	-	-
Road	-	-	-	-	-	-	-	-	-	-	-
Non-specified	-	-	-	-	-	-	-	-	-	-	-
OTHER SECTORS	-	-	-	-	-	-	-	-	-	-	-
Agriculture	-	-	-	-	-	-	-	-	-	-	-
Comm. and Publ. Services	-	-	-	-	-	-	-	-	-	-	-
Residential	-	-	-	-	-	-	-	-	-	-	-
Non-specified	-	-	-	-	-	-	-	-	-	-	-
NON-ENERGY USE	-	-	-	-	-	-	-	-	-	-	-

Senegal / Sénégal

SUPPLY AND CONSUMPTION 1997	Oil cont. (1000 tonnes)										
	Refinery Gas	LPG + Ethane	Motor Gasoline	Aviation Gasoline	Jet Fuel	Kerosene	Gas/ Diesel	Heavy Fuel Oil	Naphtha	Petrol. Coke	Other Prod.
Production	6	10	118	-	125	15	267	251	-	-	5
Imports	-	68	-	-	45	-	104	169	-	-	13
Exports	-	-2	-27	-	-5	-11	-32	-20	-	-	-1
Intl. Marine Bunkers	-	-	-	-	-	-	-	-	-	-	-
Stock Changes	-	-	-	-	-5	8	-	-	-	-	2
DOMESTIC SUPPLY	6	76	91	-	160	12	339	400	-	-	19
Transfers and Stat. Diff.	-	-	-13	-	-	-	-17	2	-	-	-
TRANSFORMATION	-	-	-	-	-	-	40	308	-	-	-
Electricity and CHP Plants	-	-	-	-	-	-	40	308	-	-	-
Petroleum Refineries	-	-	-	-	-	-	-	-	-	-	-
Other Transform. Sector	-	-	-	-	-	-	-	-	-	-	-
ENERGY SECTOR	6	-	-	-	-	-	-	-	-	-	-
DISTRIBUTION LOSSES	-	-	-	-	-	-	-	-	-	-	-
FINAL CONSUMPTION	-	76	78	-	160	12	282	94	-	-	19
INDUSTRY SECTOR	-	-	-	-	-	1	49	73	-	-	-
Iron and Steel	-	-	-	-	-	-	-	-	-	-	-
Chemical and Petrochem.	-	-	-	-	-	-	-	-	-	-	-
Non-Metallic Minerals	-	-	-	-	-	-	-	-	-	-	-
Non-specified	-	-	-	-	-	1	49	73	-	-	-
TRANSPORT SECTOR	-	-	78	-	160	-	212	-	-	-	-
Air	-	-	-	-	160	-	-	-	-	-	-
Road	-	-	50	-	-	-	212	-	-	-	-
Non-specified	-	-	28	-	-	-	-	-	-	-	-
OTHER SECTORS	-	76	-	-	-	11	21	21	-	-	-
Agriculture	-	-	-	-	-	-	21	21	-	-	-
Comm. and Publ. Services	-	-	-	-	-	-	-	-	-	-	-
Residential	-	76	-	-	-	11	-	-	-	-	-
Non-specified	-	-	-	-	-	-	-	-	-	-	-
NON-ENERGY USE	-	-	-	-	-	-	-	-	-	-	19

APPROVISIONNEMENT ET DEMANDE 1998	Pétrole cont. (1000 tonnes)										
	Gaz de raffinerie	GPL + éthane	Essence moteur	Essence aviation	Carbu- réacteurs	Kérosène	Gazole	Fioul lourd	Naphta	Coke de pétrole	Autres prod.
Production	6	5	123	-	113	17	303	244	-	-	5
Imports	-	84	-	-	70	-	112	148	-	-	13
Exports	-	-2	-42	-	-5	-3	-10	-13	-	-	-1
Intl. Marine Bunkers	-	-	-	-	-	-	-70	-	-	-	-
Stock Changes	-	-	-	-	1	-	-	-	-	-	2
DOMESTIC SUPPLY	6	87	81	-	179	14	335	379	-	-	19
Transfers and Stat. Diff.	-	-	3	-	-	-	56	39	-	-	-
TRANSFORMATION	-	-	-	-	-	-	62	306	-	-	-
Electricity and CHP Plants	-	-	-	-	-	-	62	306	-	-	-
Petroleum Refineries	-	-	-	-	-	-	-	-	-	-	-
Other Transform. Sector	-	-	-	-	-	-	-	-	-	-	-
ENERGY SECTOR	6	-	-	-	-	-	-	-	-	-	-
DISTRIBUTION LOSSES	-	-	-	-	-	-	-	-	-	-	-
FINAL CONSUMPTION	-	87	84	-	179	14	329	112	-	-	19
INDUSTRY SECTOR	-	-	-	-	-	1	65	87	-	-	-
Iron and Steel	-	-	-	-	-	-	-	-	-	-	-
Chemical and Petrochem.	-	-	-	-	-	-	-	-	-	-	-
Non-Metallic Minerals	-	-	-	-	-	-	-	-	-	-	-
Non-specified	-	-	-	-	-	1	65	87	-	-	-
TRANSPORT SECTOR	-	-	84	-	179	-	236	-	-	-	-
Air	-	-	-	-	179	-	-	-	-	-	-
Road	-	-	50	-	-	-	236	-	-	-	-
Non-specified	-	-	34	-	-	-	-	-	-	-	-
OTHER SECTORS	-	87	-	-	-	13	28	25	-	-	-
Agriculture	-	-	-	-	-	-	28	25	-	-	-
Comm. and Publ. Services	-	-	-	-	-	-	-	-	-	-	-
Residential	-	87	-	-	-	13	-	-	-	-	-
Non-specified	-	-	-	-	-	-	-	-	-	-	-
NON-ENERGY USE	-	-	-	-	-	-	-	-	-	-	19

Senegal / Sénégal

SUPPLY AND CONSUMPTION 1997	Gas (TJ) Natural Gas	Gas Works	Coke Ovens	Blast Furnaces	Comb. Renew. & Waste (TJ) Solid Biomass	Gas/Liquids from Biomass	Municipal Waste	Industrial Waste	(GWh) Electricity	(TJ) Heat
Production	928	-	-	-	68426	-	-	-	1226	-
Imports	-	-	-	-	-	-	-	-	-	-
Exports	-	-	-	-	-	-	-	-	-	-
Intl. Marine Bunkers	-	-	-	-	-	-	-	-	-	-
Stock Changes	-	-	-	-	-	-	-	-	-	-
DOMESTIC SUPPLY	928	-	-	-	68426	-	-	-	1226	-
Transfers and Stat. Diff.	-	-	-	-	1	-	-	-	-60	-
TRANSFORMATION	928	-	-	-	16251	-	-	-	-	-
Electricity and CHP Plants	928	-	-	-	-	-	-	-	-	-
Petroleum Refineries	-	-	-	-	-	-	-	-	-	-
Other Transform. Sector	-	-	-	-	16251	-	-	-	-	-
ENERGY SECTOR	-	-	-	-	-	-	-	-	67	-
DISTRIBUTION LOSSES	-	-	-	-	-	-	-	-	154	-
FINAL CONSUMPTION	-	-	-	-	52176	-	-	-	945	-
INDUSTRY SECTOR	-	-	-	-	6950	-	-	-	547	-
Iron and Steel	-	-	-	-	-	-	-	-	-	-
Chemical and Petrochem.	-	-	-	-	-	-	-	-	-	-
Non-Metallic Minerals	-	-	-	-	-	-	-	-	-	-
Non-specified	-	-	-	-	6950	-	-	-	547	-
TRANSPORT SECTOR	-	-	-	-	-	-	-	-	-	-
Air	-	-	-	-	-	-	-	-	-	-
Road	-	-	-	-	-	-	-	-	-	-
Non-specified	-	-	-	-	-	-	-	-	-	-
OTHER SECTORS	-	-	-	-	45226	-	-	-	398	-
Agriculture	-	-	-	-	-	-	-	-	48	-
Comm. and Publ. Services	-	-	-	-	-	-	-	-	143	-
Residential	-	-	-	-	45226	-	-	-	207	-
Non-specified	-	-	-	-	-	-	-	-	-	-
NON-ENERGY USE	-	-	-	-	-	-	-	-	-	-

APPROVISIONNEMENT ET DEMANDE 1998	Gaz (TJ) Gaz naturel	Usines à gaz	Cokeries	Hauts fourneaux	En. Re. Comb. & Déchets (TJ) Biomasse solide	Gaz/Liquides tirés de biomasse	Déchets urbains	Déchets industriels	(GWh) Electricité	(TJ) Chaleur
Production	793	-	-	-	68426	-	-	-	1286	-
Imports	-	-	-	-	-	-	-	-	-	-
Exports	-	-	-	-	-	-	-	-	-	-
Intl. Marine Bunkers	-	-	-	-	-	-	-	-	-	-
Stock Changes	-	-	-	-	-	-	-	-	-	-
DOMESTIC SUPPLY	793	-	-	-	68426	-	-	-	1286	-
Transfers and Stat. Diff.	-	-	-	-	1	-	-	-	-69	-
TRANSFORMATION	793	-	-	-	16251	-	-	-	-	-
Electricity and CHP Plants	793	-	-	-	-	-	-	-	-	-
Petroleum Refineries	-	-	-	-	-	-	-	-	-	-
Other Transform. Sector	-	-	-	-	16251	-	-	-	-	-
ENERGY SECTOR	-	-	-	-	-	-	-	-	67	-
DISTRIBUTION LOSSES	-	-	-	-	-	-	-	-	144	-
FINAL CONSUMPTION	-	-	-	-	52176	-	-	-	1006	-
INDUSTRY SECTOR	-	-	-	-	6950	-	-	-	547	-
Iron and Steel	-	-	-	-	-	-	-	-	-	-
Chemical and Petrochem.	-	-	-	-	-	-	-	-	-	-
Non-Metallic Minerals	-	-	-	-	-	-	-	-	-	-
Non-specified	-	-	-	-	6950	-	-	-	547	-
TRANSPORT SECTOR	-	-	-	-	-	-	-	-	-	-
Air	-	-	-	-	-	-	-	-	-	-
Road	-	-	-	-	-	-	-	-	-	-
Non-specified	-	-	-	-	-	-	-	-	-	-
OTHER SECTORS	-	-	-	-	45226	-	-	-	459	-
Agriculture	-	-	-	-	-	-	-	-	55	-
Comm. and Publ. Services	-	-	-	-	-	-	-	-	165	-
Residential	-	-	-	-	45226	-	-	-	239	-
Non-specified	-	-	-	-	-	-	-	-	-	-
NON-ENERGY USE	-	-	-	-	-	-	-	-	-	-

Singapore / Singapour

SUPPLY AND CONSUMPTION 1997	Coal (1000 tonnes)							Oil (1000 tonnes)			
	Coking Coal	Other Bit. Coal	Sub-Bit. Coal	Lignite	Peat	Oven and Gas Coke	Pat. Fuel and BKB	Crude Oil	NGL	Feed-stocks	Additives
Production	-	-	-	-	-	-	-	-	-	-	-
Imports	-	-	-	1	-	9	-	55507	-	-	-
Exports	-	-	-	-	-	-9	-	-14	-	-	-
Intl. Marine Bunkers	-	-	-	-	-	-	-	-	-	-	-
Stock Changes	-	-	-	-	-	-	-	-	-	-	-
DOMESTIC SUPPLY	-	-	-	1	-	-	-	**55493**	-	-	-
Transfers and Stat. Diff.	-	-	-	-1	-	-	-	-	-	-	-
TRANSFORMATION	-	-	-	-	-	-	-	**55493**	-	-	-
Electricity and CHP Plants	-	-	-	-	-	-	-	-	-	-	-
Petroleum Refineries	-	-	-	-	-	-	-	55493	-	-	-
Other Transform. Sector	-	-	-	-	-	-	-	-	-	-	-
ENERGY SECTOR	-	-	-	-	-	-	-	-	-	-	-
DISTRIBUTION LOSSES	-	-	-	-	-	-	-	-	-	-	-
FINAL CONSUMPTION	-	-	-	-	-	-	-	-	-	-	-
INDUSTRY SECTOR	-	-	-	-	-	-	-	-	-	-	-
Iron and Steel	-	-	-	-	-	-	-	-	-	-	-
Chemical and Petrochem.	-	-	-	-	-	-	-	-	-	-	-
Non-Metallic Minerals	-	-	-	-	-	-	-	-	-	-	-
Non-specified	-	-	-	-	-	-	-	-	-	-	-
TRANSPORT SECTOR	-	-	-	-	-	-	-	-	-	-	-
Air	-	-	-	-	-	-	-	-	-	-	-
Road	-	-	-	-	-	-	-	-	-	-	-
Non-specified	-	-	-	-	-	-	-	-	-	-	-
OTHER SECTORS	-	-	-	-	-	-	-	-	-	-	-
Agriculture	-	-	-	-	-	-	-	-	-	-	-
Comm. and Publ. Services	-	-	-	-	-	-	-	-	-	-	-
Residential	-	-	-	-	-	-	-	-	-	-	-
Non-specified	-	-	-	-	-	-	-	-	-	-	-
NON-ENERGY USE	-	-	-	-	-	-	-	-	-	-	-

APPROVISIONNEMENT ET DEMANDE 1998	Charbon (1000 tonnes)							Pétrole (1000 tonnes)			
	Charbon à coke	Autres charb. bit.	Charbon sous-bit.	Lignite	Tourbe	Coke de four/gaz	Agg./briq. de lignite	Pétrole brut	LGN	Produits d'aliment.	Additifs
Production	-	-	-	-	-	-	-	-	-	-	-
Imports	-	-	-	1	-	9	-	51908	-	-	-
Exports	-	-	-	-	-	-9	-	-	-	-	-
Intl. Marine Bunkers	-	-	-	-	-	-	-	-	-	-	-
Stock Changes	-	-	-	-	-	-	-	-	-	-	-
DOMESTIC SUPPLY	-	-	-	1	-	-	-	**51908**	-	-	-
Transfers and Stat. Diff.	-	-	-	-1	-	-	-	-	-	-	-
TRANSFORMATION	-	-	-	-	-	-	-	**51908**	-	-	-
Electricity and CHP Plants	-	-	-	-	-	-	-	-	-	-	-
Petroleum Refineries	-	-	-	-	-	-	-	51908	-	-	-
Other Transform. Sector	-	-	-	-	-	-	-	-	-	-	-
ENERGY SECTOR	-	-	-	-	-	-	-	-	-	-	-
DISTRIBUTION LOSSES	-	-	-	-	-	-	-	-	-	-	-
FINAL CONSUMPTION	-	-	-	-	-	-	-	-	-	-	-
INDUSTRY SECTOR	-	-	-	-	-	-	-	-	-	-	-
Iron and Steel	-	-	-	-	-	-	-	-	-	-	-
Chemical and Petrochem.	-	-	-	-	-	-	-	-	-	-	-
Non-Metallic Minerals	-	-	-	-	-	-	-	-	-	-	-
Non-specified	-	-	-	-	-	-	-	-	-	-	-
TRANSPORT SECTOR	-	-	-	-	-	-	-	-	-	-	-
Air	-	-	-	-	-	-	-	-	-	-	-
Road	-	-	-	-	-	-	-	-	-	-	-
Non-specified	-	-	-	-	-	-	-	-	-	-	-
OTHER SECTORS	-	-	-	-	-	-	-	-	-	-	-
Agriculture	-	-	-	-	-	-	-	-	-	-	-
Comm. and Publ. Services	-	-	-	-	-	-	-	-	-	-	-
Residential	-	-	-	-	-	-	-	-	-	-	-
Non-specified	-	-	-	-	-	-	-	-	-	-	-
NON-ENERGY USE	-	-	-	-	-	-	-	-	-	-	-

Singapore / Singapour

SUPPLY AND CONSUMPTION 1997	Oil cont. (1000 tonnes)										
	Refinery Gas	LPG + Ethane	Motor Gasoline	Aviation Gasoline	Jet Fuel	Kerosene	Gas/ Diesel	Heavy Fuel Oil	Naphtha	Petrol. Coke	Other Prod.
Production	582	965	5032	-	7995	278	15238	11699	3765	-	2557
Imports	-	9	1528	1	1394	513	5124	21591	2140	10	416
Exports	-	-760	-5870	-	-6585	-471	-16262	-10563	-3487	-	-2437
Intl. Marine Bunkers	-	-	-	-	-	-	-2887	-13304	-	-	-
Stock Changes	-	-	-	-	-	-	-	-	-	-	-
DOMESTIC SUPPLY	582	214	690	1	2804	320	1213	9423	2418	10	536
Transfers and Stat. Diff.	-	-	-	-	1	-277	-	-2483	-	5	22
TRANSFORMATION	-	-	-	-	-	-	18	4234	-	-	-
Electricity and CHP Plants	-	-	-	-	-	-	18	4234	-	-	-
Petroleum Refineries	-	-	-	-	-	-	-	-	-	-	-
Other Transform. Sector	-	-	-	-	-	-	-	-	-	-	-
ENERGY SECTOR	582	-	-	-	-	-	-	2706	-	15	-
DISTRIBUTION LOSSES	-	-	-	-	-	-	-	-	-	-	-
FINAL CONSUMPTION	-	214	690	1	2805	43	1195	-	2418	-	558
INDUSTRY SECTOR	-	214	-	-	-	43	-	-	2418	-	-
Iron and Steel	-	-	-	-	-	-	-	-	-	-	-
Chemical and Petrochem.	-	214	-	-	-	-	-	-	2418	-	-
Non-Metallic Minerals	-	-	-	-	-	-	-	-	-	-	-
Non-specified	-	-	-	-	-	43	-	-	-	-	-
TRANSPORT SECTOR	-	-	690	1	2805	-	1195	-	-	-	-
Air	-	-	-	1	2805	-	-	-	-	-	-
Road	-	-	690	-	-	-	1195	-	-	-	-
Non-specified	-	-	-	-	-	-	-	-	-	-	-
OTHER SECTORS	-	-	-	-	-	-	-	-	-	-	-
Agriculture	-	-	-	-	-	-	-	-	-	-	-
Comm. and Publ. Services	-	-	-	-	-	-	-	-	-	-	-
Residential	-	-	-	-	-	-	-	-	-	-	-
Non-specified	-	-	-	-	-	-	-	-	-	-	-
NON-ENERGY USE	-	-	-	-	-	-	-	-	-	-	558

APPROVISIONNEMENT ET DEMANDE 1998	Pétrole cont. (1000 tonnes)										
	Gaz de raffinerie	GPL + éthane	Essence moteur	Essence aviation	Carbu- réacteurs	Kérosène	Gazole	Fioul lourd	Naphta	Coke de pétrole	Autres prod.
Production	582	965	4920	-	6463	332	14805	12136	4934	-	2557
Imports	-	9	1726	1	1342	182	5124	23970	1093	10	416
Exports	-	-760	-5928	-	-5538	-471	-15518	-12182	-2820	-	-2437
Intl. Marine Bunkers	-	-	-	-	-	-	-3186	-14012	-	-	-
Stock Changes	-	-	-	-	-	-	-	-	-	-	-
DOMESTIC SUPPLY	582	214	718	1	2267	43	1225	9912	3207	10	536
Transfers and Stat. Diff.	-	-	-	-	-	-	-	-2778	-	5	22
TRANSFORMATION	-	-	-	-	-	-	30	4428	-	-	-
Electricity and CHP Plants	-	-	-	-	-	-	30	4428	-	-	-
Petroleum Refineries	-	-	-	-	-	-	-	-	-	-	-
Other Transform. Sector	-	-	-	-	-	-	-	-	-	-	-
ENERGY SECTOR	582	-	-	-	-	-	-	2706	-	15	-
DISTRIBUTION LOSSES	-	-	-	-	-	-	-	-	-	-	-
FINAL CONSUMPTION	-	214	718	1	2267	43	1195	-	3207	-	558
INDUSTRY SECTOR	-	214	-	-	-	43	-	-	3207	-	-
Iron and Steel	-	-	-	-	-	-	-	-	-	-	-
Chemical and Petrochem.	-	214	-	-	-	-	-	-	3207	-	-
Non-Metallic Minerals	-	-	-	-	-	-	-	-	-	-	-
Non-specified	-	-	-	-	-	43	-	-	-	-	-
TRANSPORT SECTOR	-	-	718	1	2267	-	1195	-	-	-	-
Air	-	-	-	1	2267	-	-	-	-	-	-
Road	-	-	718	-	-	-	1195	-	-	-	-
Non-specified	-	-	-	-	-	-	-	-	-	-	-
OTHER SECTORS	-	-	-	-	-	-	-	-	-	-	-
Agriculture	-	-	-	-	-	-	-	-	-	-	-
Comm. and Publ. Services	-	-	-	-	-	-	-	-	-	-	-
Residential	-	-	-	-	-	-	-	-	-	-	-
Non-specified	-	-	-	-	-	-	-	-	-	-	-
NON-ENERGY USE	-	-	-	-	-	-	-	-	-	-	558

Singapore / Singapour

SUPPLY AND CONSUMPTION 1997	Gas (TJ)				Comb. Renew. & Waste (TJ)				(GWh)	(TJ)
	Natural Gas	Gas Works	Coke Ovens	Blast Furnaces	Solid Biomass	Gas/Liquids from Biomass	Municipal Waste	Industrial Waste	Electricity	Heat
Production	-	4394	-	-	-	-	-	-	26898	-
Imports	60465	-	-	-	-	-	-	-	-	-
Exports	-	-	-	-	-	-	-	-	-	-
Intl. Marine Bunkers	-	-	-	-	-	-	-	-	-	-
Stock Changes	-	-	-	-	-	-	-	-	-	-
DOMESTIC SUPPLY	60465	4394	-	-	-	-	-	-	26898	-
Transfers and Stat. Diff.	-	189	-	-	-	-	-	-	2711	-
TRANSFORMATION	60465	-	-	-	-	-	-	-	-	-
Electricity and CHP Plants	60465	-	-	-	-	-	-	-	-	-
Petroleum Refineries	-	-	-	-	-	-	-	-	-	-
Other Transform. Sector	-	-	-	-	-	-	-	-	-	-
ENERGY SECTOR	-	-	-	-	-	-	-	-	3836	-
DISTRIBUTION LOSSES	-	425	-	-	-	-	-	-	1118	-
FINAL CONSUMPTION	-	4158	-	-	-	-	-	-	24655	-
INDUSTRY SECTOR	-	2097	-	-	-	-	-	-	11077	-
Iron and Steel	-	-	-	-	-	-	-	-	972	-
Chemical and Petrochem.	-	-	-	-	-	-	-	-	583	-
Non-Metallic Minerals	-	-	-	-	-	-	-	-	-	-
Non-specified	-	2097	-	-	-	-	-	-	9522	-
TRANSPORT SECTOR	-	-	-	-	-	-	-	-	248	-
Air	-	-	-	-	-	-	-	-	-	-
Road	-	-	-	-	-	-	-	-	-	-
Non-specified	-	-	-	-	-	-	-	-	248	-
OTHER SECTORS	-	2061	-	-	-	-	-	-	13330	-
Agriculture	-	-	-	-	-	-	-	-	25	-
Comm. and Publ. Services	-	-	-	-	-	-	-	-	8102	-
Residential	-	2061	-	-	-	-	-	-	5203	-
Non-specified	-	-	-	-	-	-	-	-	-	-
NON-ENERGY USE	-	-	-	-	-	-	-	-	-	-

APPROVISIONNEMENT ET DEMANDE 1998	Gaz (TJ)				En. Re. Comb. & Déchets (TJ)				(GWh)	(TJ)
	Gaz naturel	Usines à gaz	Cokeries	Hauts fourneaux	Biomasse solide	Gaz/Liquides tirés de biomasse	Déchets urbains	Déchets industriels	Electricité	Chaleur
Production	-	4581	-	-	-	-	-	-	28557	-
Imports	66213	-	-	-	-	-	-	-	-	-
Exports	-	-	-	-	-	-	-	-	-	-
Intl. Marine Bunkers	-	-	-	-	-	-	-	-	-	-
Stock Changes	-	-	-	-	-	-	-	-	-	-
DOMESTIC SUPPLY	66213	4581	-	-	-	-	-	-	28557	-
Transfers and Stat. Diff.	-	-128	-	-	-	-	-	-	2878	-
TRANSFORMATION	66213	-	-	-	-	-	-	-	-	-
Electricity and CHP Plants	66213	-	-	-	-	-	-	-	-	-
Petroleum Refineries	-	-	-	-	-	-	-	-	-	-
Other Transform. Sector	-	-	-	-	-	-	-	-	-	-
ENERGY SECTOR	-	-	-	-	-	-	-	-	4072	-
DISTRIBUTION LOSSES	-	170	-	-	-	-	-	-	1187	-
FINAL CONSUMPTION	-	4283	-	-	-	-	-	-	26176	-
INDUSTRY SECTOR	-	2097	-	-	-	-	-	-	11760	-
Iron and Steel	-	-	-	-	-	-	-	-	1032	-
Chemical and Petrochem.	-	-	-	-	-	-	-	-	619	-
Non-Metallic Minerals	-	-	-	-	-	-	-	-	-	-
Non-specified	-	2097	-	-	-	-	-	-	10109	-
TRANSPORT SECTOR	-	-	-	-	-	-	-	-	263	-
Air	-	-	-	-	-	-	-	-	-	-
Road	-	-	-	-	-	-	-	-	-	-
Non-specified	-	-	-	-	-	-	-	-	263	-
OTHER SECTORS	-	2186	-	-	-	-	-	-	14153	-
Agriculture	-	-	-	-	-	-	-	-	27	-
Comm. and Publ. Services	-	-	-	-	-	-	-	-	8602	-
Residential	-	2186	-	-	-	-	-	-	5524	-
Non-specified	-	-	-	-	-	-	-	-	-	-
NON-ENERGY USE	-	-	-	-	-	-	-	-	-	-

Slovak Republic / République slovaque : 1997

SUPPLY AND CONSUMPTION APPROVISIONNEMENT ET DEMANDE	Coal / *Charbon* (1000 tonnes)							Oil / *Pétrole* (1000 tonnes)			
	Coking Coal *Charbon à coke*	Other Bit. Coal *Autres charb. bit.*	Sub-Bit. Coal *Charbon sous-bit.*	Lignite *Lignite*	Peat *Tourbe*	Oven and Gas Coke *Coke de four/gaz*	Pat. Fuel and BKB *Agg./briq. de lignite*	Crude Oil *Pétrole brut*	NGL *LGN*	Feed-stocks *Produits d'aliment.*	Additives *Additifs*
Production	-	-	-	3915	-	1730	-	64	1	-	-
From Other Sources	-	-	-	-	-	-	-	-	-	-	-
Imports	2568	2668	-	2196	-	146	5	5270	-	-	-
Exports	-	-	-	-8	-	-21	-	-	-	-	-
Intl. Marine Bunkers	-	-	-	-	-	-	-	-	-	-	-
Stock Changes	-2	-359	-	93	-	38	3	-8	-	-	-
DOMESTIC SUPPLY	**2566**	**2309**	**-**	**6196**	**-**	**1893**	**8**	**5326**	**1**	**-**	**-**
Transfers	-	-	-	-	-	-	-	-	-1	-	-
Statistical Differences	-	-1	-	-1	-	-	-	73	-	-	-
TRANSFORMATION	**2271**	**1467**	**-**	**3815**	**-**	**1313**	**-**	**5399**	**-**	**-**	**-**
Electricity Plants	-	1467	-	3755	-	-	-	-	-	-	-
CHP Plants	-	-	-	-	-	-	-	-	-	-	-
Heat Plants	-	-	-	60	-	6	-	-	-	-	-
Blast Furnaces/Gas Works	-	-	-	-	-	1307	-	-	-	-	-
Coke/Pat. Fuel/BKB Plants	2271	-	-	-	-	-	-	-	-	-	-
Petroleum Refineries	-	-	-	-	-	-	-	5399	-	-	-
Petrochemical Industry	-	-	-	-	-	-	-	-	-	-	-
Liquefaction	-	-	-	-	-	-	-	-	-	-	-
Other Transform. Sector	-	-	-	-	-	-	-	-	-	-	-
ENERGY SECTOR	**-**	**-**	**-**	**-**	**-**	**-**	**-**	**-**	**-**	**-**	**-**
Coal Mines	-	-	-	-	-	-	-	-	-	-	-
Oil and Gas Extraction	-	-	-	-	-	-	-	-	-	-	-
Petroleum Refineries	-	-	-	-	-	-	-	-	-	-	-
Electr., CHP+Heat Plants	-	-	-	-	-	-	-	-	-	-	-
Pumped Storage (Elec.)	-	-	-	-	-	-	-	-	-	-	-
Other Energy Sector	-	-	-	-	-	-	-	-	-	-	-
Distribution Losses	-	1	-	7	-	-	-	-	-	-	-
FINAL CONSUMPTION	**295**	**840**	**-**	**2373**	**-**	**580**	**8**	**-**	**-**	**-**	**-**
INDUSTRY SECTOR	**293**	**706**	**-**	**1318**	**-**	**480**	**1**	**-**	**-**	**-**	**-**
Iron and Steel	289	452	-	81	-	327	-	-	-	-	-
Chemical and Petrochem.	-	13	-	398	-	51	-	-	-	-	-
of which: Feedstocks	-	-	-	-	-	-	-	-	-	-	-
Non-Ferrous Metals	-	-	-	-	-	-	-	-	-	-	-
Non-Metallic Minerals	3	172	-	30	-	84	-	-	-	-	-
Transport Equipment	-	-	-	110	-	-	-	-	-	-	-
Machinery	-	20	-	163	-	3	1	-	-	-	-
Mining and Quarrying	-	-	-	8	-	4	-	-	-	-	-
Food and Tobacco	-	23	-	93	-	9	-	-	-	-	-
Paper, Pulp and Print	-	24	-	228	-	-	-	-	-	-	-
Wood and Wood Products	-	-	-	37	-	-	-	-	-	-	-
Construction	1	1	-	21	-	-	-	-	-	-	-
Textile and Leather	-	-	-	124	-	1	-	-	-	-	-
Non-specified	-	1	-	25	-	1	-	-	-	-	-
TRANSPORT SECTOR	**-**	**-**	**-**	**-**	**-**	**-**	**-**	**-**	**-**	**-**	**-**
Air	-	-	-	-	-	-	-	-	-	-	-
Road	-	-	-	-	-	-	-	-	-	-	-
Rail	-	-	-	-	-	-	-	-	-	-	-
Pipeline Transport	-	-	-	-	-	-	-	-	-	-	-
Internal Navigation	-	-	-	-	-	-	-	-	-	-	-
Non-specified	-	-	-	-	-	-	-	-	-	-	-
OTHER SECTORS	**2**	**134**	**-**	**1055**	**-**	**100**	**7**	**-**	**-**	**-**	**-**
Agriculture	1	3	-	56	-	8	3	-	-	-	-
Comm. and Publ. Services	-	127	-	550	-	74	-	-	-	-	-
Residential	1	4	-	449	-	18	4	-	-	-	-
Non-specified	-	-	-	-	-	-	-	-	-	-	-
NON-ENERGY USE	**-**	**-**	**-**	**-**	**-**	**-**	**-**	**-**	**-**	**-**	**-**
in Industry/Trans./Energy	-	-	-	-	-	-	-	-	-	-	-
in Transport	-	-	-	-	-	-	-	-	-	-	-
in Other Sectors	-	-	-	-	-	-	-	-	-	-	-

Slovak Republic / République slovaque : 1997

SUPPLY AND CONSUMPTION *APPROVISIONNEMENT ET DEMANDE*	Oil cont. / *Pétrole cont.* (1000 tonnes)										
	Refinery Gas *Gaz de raffinerie*	LPG + Ethane *GPL + éthane*	Motor Gasoline *Essence moteur*	Aviation Gasoline *Essence aviation*	Jet Fuel *Carbu-réacteurs*	Kerosene *Kérosène*	Gas/ Diesel *Gazole*	Heavy Fuel Oil *Fioul lourd*	Naphtha *Naphta*	Petrol. Coke *Coke de pétrole*	Other Prod. *Autres prod.*
Production	75	53	832	-	62	26	1782	1146	670	-	741
From Other Sources	-	-	-	-	-	-	-	-	-	-	-
Imports	-	15	92	-	-	-	57	41	-	-	-
Exports	-	-34	-397	-	-27	-18	-1052	-610	-	-	-20
Intl. Marine Bunkers	-	-	-	-	-	-	-	-	-	-	-
Stock Changes	-	-	21	-	-	-7	-44	-12	-	-	-
DOMESTIC SUPPLY	75	34	548	-	35	1	743	565	670	-	721
Transfers	-	1	-	-	-	-	-	-	-	-	-
Statistical Differences	-	-	-	-	-	-	-	-	-	-	-
TRANSFORMATION	-	-	-	-	-	-	-	172	-	-	-
Electricity Plants	-	-	-	-	-	-	-	-	-	-	-
CHP Plants	-	-	-	-	-	-	-	152	-	-	-
Heat Plants	-	-	-	-	-	-	-	20	-	-	-
Blast Furnaces/Gas Works	-	-	-	-	-	-	-	-	-	-	-
Coke/Pat. Fuel/BKB Plants	-	-	-	-	-	-	-	-	-	-	-
Petroleum Refineries	-	-	-	-	-	-	-	-	-	-	-
Petrochemical Industry	-	-	-	-	-	-	-	-	-	-	-
Liquefaction	-	-	-	-	-	-	-	-	-	-	-
Other Transform. Sector	-	-	-	-	-	-	-	-	-	-	-
ENERGY SECTOR	-	-	15	-	-	-	1	27	-	-	-
Coal Mines	-	-	-	-	-	-	-	-	-	-	-
Oil and Gas Extraction	-	-	-	-	-	-	-	-	-	-	-
Petroleum Refineries	-	-	-	-	-	-	-	27	-	-	-
Electr., CHP+Heat Plants	-	-	-	-	-	-	-	-	-	-	-
Pumped Storage (Elec.)	-	-	-	-	-	-	-	-	-	-	-
Other Energy Sector	-	-	15	-	-	-	1	-	-	-	-
Distribution Losses	-	-	-	-	-	-	-	-	-	-	-
FINAL CONSUMPTION	75	35	533	-	35	1	742	366	670	-	721
INDUSTRY SECTOR	75	5	-	-	-	1	-	296	670	-	-
Iron and Steel	-	-	-	-	-	-	-	38	-	-	-
Chemical and Petrochem.	-	-	-	-	-	-	-	17	670	-	-
of which: Feedstocks	-	-	-	-	-	-	-	-	670	-	-
Non-Ferrous Metals	-	-	-	-	-	-	-	-	-	-	-
Non-Metallic Minerals	-	-	-	-	-	-	-	22	-	-	-
Transport Equipment	-	-	-	-	-	1	-	6	-	-	-
Machinery	-	-	-	-	-	-	-	11	-	-	-
Mining and Quarrying	-	-	-	-	-	-	-	1	-	-	-
Food and Tobacco	-	-	-	-	-	-	-	33	-	-	-
Paper, Pulp and Print	-	-	-	-	-	-	-	77	-	-	-
Wood and Wood Products	-	-	-	-	-	-	-	3	-	-	-
Construction	-	-	-	-	-	-	-	5	-	-	-
Textile and Leather	-	-	-	-	-	-	-	60	-	-	-
Non-specified	75	5	-	-	-	-	-	23	-	-	-
TRANSPORT SECTOR	-	-	523	-	35	-	566	-	-	-	-
Air	-	-	-	-	35	-	-	-	-	-	-
Road	-	-	523	-	-	-	566	-	-	-	-
Rail	-	-	-	-	-	-	-	-	-	-	-
Pipeline Transport	-	-	-	-	-	-	-	-	-	-	-
Internal Navigation	-	-	-	-	-	-	-	-	-	-	-
Non-specified	-	-	-	-	-	-	-	-	-	-	-
OTHER SECTORS	-	30	10	-	-	-	176	70	-	-	-
Agriculture	-	1	10	-	-	-	176	8	-	-	-
Comm. and Publ. Services	-	6	-	-	-	-	-	62	-	-	-
Residential	-	23	-	-	-	-	-	-	-	-	-
Non-specified	-	-	-	-	-	-	-	-	-	-	-
NON-ENERGY USE	-	-	-	-	-	-	-	-	-	-	721
in Industry/Transf./Energy	-	-	-	-	-	-	-	-	-	-	721
in Transport	-	-	-	-	-	-	-	-	-	-	-
in Other Sectors	-	-	-	-	-	-	-	-	-	-	-

Slovak Republic / République slovaque : 1997

SUPPLY AND CONSUMPTION / APPROVISIONNEMENT ET DEMANDE	Gas / Gaz (TJ)				Comb. Renew. & Waste / En. Re. Comb. & Déchets (TJ)				(GWh)	(TJ)
	Natural Gas / Gaz naturel	Gas Works / Usines à gaz	Coke Ovens / Cokeries	Blast Furnaces / Hauts fourneaux	Solid Biomass / Biomasse solide	Gas/Liquids from Biomass / Gaz/Liquides tirés de biomasse	Municipal Waste / Déchets urbains	Industrial Waste / Déchets industriels	Electricity / Electricité	Heat / Chaleur
Production	10451	-	12919	16256	3454	-	-	-	24547	36339
From Other Sources	-	-			-	-	-	-		
Imports	242593	-	-	-	-	-	-	-	6825	-
Exports	-	-	-	-	-	-	-	-	-2743	-
Intl. Marine Bunkers	-	-	-	-	-	-	-	-	-	-
Stock Changes	9139	-	-	-	24	-	-	-	-	-
DOMESTIC SUPPLY	262183	-	12919	16256	3478	-	-	-	28629	36339
Transfers	-	-	-	-	-	-	-	-	-	-
Statistical Differences	-162	-	-	-	-	-	-	-	-	-
TRANSFORMATION	45411	-	1167	1458	3283	-	-	-	1688	-
Electricity Plants	-	-	-	-	-	-	-	-	-	-
CHP Plants	45411	-	1167	1458	-	-	-	-	-	-
Heat Plants	-	-	-	-	-	-	-	-	-	-
Blast Furnaces/Gas Works	-	-	-	-	-	-	-	-	-	-
Coke/Pat. Fuel/BKB Plants	-	-	-	-	-	-	-	-	-	-
Petroleum Refineries	-	-	-	-	-	-	-	-	-	-
Petrochemical Industry	-	-	-	-	-	-	-	-	-	-
Liquefaction	-	-	-	-	-	-	-	-	-	-
Other Transform. Sector	-	-	-	-	3283	-	-	-	1688	-
ENERGY SECTOR	-	-	1170	4006	-	-	-	-	2016	5000
Coal Mines	-	-	-	-	-	-	-	-	-	-
Oil and Gas Extraction	-	-	-	-	-	-	-	-	-	-
Petroleum Refineries	-	-	-	-	-	-	-	-	-	-
Electr., CHP+Heat Plants	-	-	-	-	-	-	-	-	1721	-
Pumped Storage (Elec.)	-	-	-	-	-	-	-	-	295	-
Other Energy Sector	-	-	1170	4006	-	-	-	-	-	5000
Distribution Losses	4144	-	230	184	-	-	-	-	2085	2000
FINAL CONSUMPTION	212466	-	10352	10608	195	-	-	-	22840	29339
INDUSTRY SECTOR	113621	-	10352	10608	3	-	-	-	10053	1251
Iron and Steel	11535	-	10302	10608	-	-	-	-	3353	-
Chemical and Petrochem.	54590	-	-	-	-	-	-	-	1490	-
of which: Feedstocks	27222	-	-	-	-	-	-	-	-	-
Non-Ferrous Metals	-	-	-	-	-	-	-	-	-	-
Non-Metallic Minerals	12513	-	-	-	-	-	-	-	1047	-
Transport Equipment	2341	-	-	-	-	-	-	-	256	305
Machinery	6767	-	50	-	-	-	-	-	864	285
Mining and Quarrying	3234	-	-	-	-	-	-	-	136	-
Food and Tobacco	10257	-	-	-	-	-	-	-	1173	102
Paper, Pulp and Print	2809	-	-	-	-	-	-	-	704	-
Wood and Wood Products	1064	-	-	-	-	-	-	-	213	203
Construction	42	-	-	-	3	-	-	-	183	51
Textile and Leather	3107	-	-	-	-	-	-	-	337	102
Non-specified	5362	-	-	-	-	-	-	-	297	203
TRANSPORT SECTOR	-	-	-	-	-	-	-	-	1010	-
Air	-	-	-	-	-	-	-	-	-	-
Road	-	-	-	-	-	-	-	-	-	-
Rail	-	-	-	-	-	-	-	-	1010	-
Pipeline Transport	-	-	-	-	-	-	-	-	-	-
Internal Navigation	-	-	-	-	-	-	-	-	-	-
Non-specified	-	-	-	-	-	-	-	-	-	-
OTHER SECTORS	98845	-	-	-	192	-	-	-	11777	28088
Agriculture	2363	-	-	-	129	-	-	-	1136	934
Comm. and Publ. Services	32586	-	-	-	49	-	-	-	5134	11334
Residential	63896	-	-	-	14	-	-	-	5507	15820
Non-specified	-	-	-	-	-	-	-	-	-	-
NON-ENERGY USE	-	-	-	-	-	-	-	-	-	-
in Industry/Transf./Energy	-	-	-	-	-	-	-	-	-	-
in Transport	-	-	-	-	-	-	-	-	-	-
in Other Sectors	-	-	-	-	-	-	-	-	-	-

Slovak Republic / République slovaque : 1998

SUPPLY AND CONSUMPTION APPROVISIONNEMENT ET DEMANDE	Coal / *Charbon* (1000 tonnes)							Oil / *Pétrole* (1000 tonnes)			
	Coking Coal *Charbon à coke*	Other Bit. Coal *Autres charb. bit.*	Sub-Bit. Coal *Charbon sous-bit.*	Lignite *Lignite*	Peat *Tourbe*	Oven and Gas Coke *Coke de four/gaz*	Pat. Fuel and BKB *Agg./briq. de lignite*	Crude Oil *Pétrole brut*	NGL *LGN*	Feed-stocks *Produits d'aliment.*	Additives *Additifs*
Production	-	-	-	3951	-	1515	-	60	1	-	-
From Other Sources	-	-	-	-	-	-	-	-	-	-	-
Imports	2287	2244	-	1415	-	162	2	5421	-	-	-
Exports	-	-	-	-11	-	-27	-	-	-	-	-
Intl. Marine Bunkers	-	-	-	-	-	-	-	-	-	-	-
Stock Changes	-4	230	-	211	-	1	-	-31	-	-	-
DOMESTIC SUPPLY	2283	2474	-	5566	-	1651	2	5450	1	-	-
Transfers	-	-	-	-	-	-	-	-	-1	-	-
Statistical Differences	-	-	-	-	-	-	-	-	-	-	-
TRANSFORMATION	1971	1879	-	3821	-	1145	-	5450	-	-	-
Electricity Plants	-	1879	-	3758	-	-	-	-	-	-	-
CHP Plants	-	-	-	-	-	-	-	-	-	-	-
Heat Plants	-	-	-	63	-	6	-	-	-	-	-
Blast Furnaces/Gas Works	-	-	-	-	-	1139	-	-	-	-	-
Coke/Pat. Fuel/BKB Plants	1971	-	-	-	-	-	-	-	-	-	-
Petroleum Refineries	-	-	-	-	-	-	-	5450	-	-	-
Petrochemical Industry	-	-	-	-	-	-	-	-	-	-	-
Liquefaction	-	-	-	-	-	-	-	-	-	-	-
Other Transform. Sector	-	-	-	-	-	-	-	-	-	-	-
ENERGY SECTOR	-	-	-	-	-	-	-	-	-	-	-
Coal Mines	-	-	-	-	-	-	-	-	-	-	-
Oil and Gas Extraction	-	-	-	-	-	-	-	-	-	-	-
Petroleum Refineries	-	-	-	-	-	-	-	-	-	-	-
Electr., CHP+Heat Plants	-	-	-	-	-	-	-	-	-	-	-
Pumped Storage (Elec.)	-	-	-	-	-	-	-	-	-	-	-
Other Energy Sector	-	-	-	-	-	-	-	-	-	-	-
Distribution Losses	-	1	-	5	-	-	-	-	-	-	-
FINAL CONSUMPTION	312	594	-	1740	-	506	2	-	-	-	-
INDUSTRY SECTOR	312	238	-	1039	-	451	1	-	-	-	-
Iron and Steel	307	152	-	64	-	307	-	-	-	-	-
Chemical and Petrochem.	-	4	-	314	-	48	-	-	-	-	-
of which: Feedstocks	-	-	-	-	-	-	-	-	-	-	-
Non-Ferrous Metals	-	-	-	-	-	-	-	-	-	-	-
Non-Metallic Minerals	4	57	-	24	-	79	-	-	-	-	-
Transport Equipment	-	-	-	87	-	-	-	-	-	-	-
Machinery	-	7	-	128	-	3	1	-	-	-	-
Mining and Quarrying	-	-	-	6	-	4	-	-	-	-	-
Food and Tobacco	-	8	-	73	-	8	-	-	-	-	-
Paper, Pulp and Print	-	8	-	180	-	-	-	-	-	-	-
Wood and Wood Products	-	-	-	29	-	-	-	-	-	-	-
Construction	1	1	-	17	-	-	-	-	-	-	-
Textile and Leather	-	-	-	97	-	1	-	-	-	-	-
Non-specified	-	1	-	20	-	1	-	-	-	-	-
TRANSPORT SECTOR	-	2	-	-	-	-	-	-	-	-	-
Air	-	-	-	-	-	-	-	-	-	-	-
Road	-	-	-	-	-	-	-	-	-	-	-
Rail	-	2	-	-	-	-	-	-	-	-	-
Pipeline Transport	-	-	-	-	-	-	-	-	-	-	-
Internal Navigation	-	-	-	-	-	-	-	-	-	-	-
Non-specified	-	-	-	-	-	-	-	-	-	-	-
OTHER SECTORS	-	354	-	701	-	55	1	-	-	-	-
Agriculture	-	3	-	42	-	3	-	-	-	-	-
Comm. and Publ. Services	-	345	-	242	-	42	-	-	-	-	-
Residential	-	6	-	417	-	10	1	-	-	-	-
Non-specified	-	-	-	-	-	-	-	-	-	-	-
NON-ENERGY USE	-	-	-	-	-	-	-	-	-	-	-
in Industry/Trans./Energy	-	-	-	-	-	-	-	-	-	-	-
in Transport	-	-	-	-	-	-	-	-	-	-	-
in Other Sectors	-	-	-	-	-	-	-	-	-	-	-

Slovak Republic / République slovaque : 1998

SUPPLY AND CONSUMPTION / APPROVISIONNEMENT ET DEMANDE	Refinery Gas / Gaz de raffinerie	LPG + Ethane / GPL + éthane	Motor Gasoline / Essence moteur	Aviation Gasoline / Essence aviation	Jet Fuel / Carbu- réacteurs	Kerosene / Kérosène	Gas/ Diesel / Gazole	Heavy Fuel Oil / Fioul lourd	Naphtha / Naphta	Petrol. Coke / Coke de pétrole	Other Prod. / Autres prod.
Production	75	31	912	-	58	26	1898	1217	670	-	563
From Other Sources	-	-	-	-	-	-	-	-	-	-	-
Imports	-	15	116	-	-	-	103	47	-	-	-
Exports	-	-15	-408	-	-29	-21	-1130	-752	-	-	-20
Intl. Marine Bunkers	-	-	-	-	-	-	-	-	-	-	-
Stock Changes	-	-	-50	-	-	-3	-51	14	-	-	-
DOMESTIC SUPPLY	75	31	570	-	29	2	820	526	670	-	543
Transfers	-	1	-	-	-	-	-	-	-	-	-
Statistical Differences	-	-	-	-	-	-	-	-17	-	-	-
TRANSFORMATION	-	-	-	-	-	-	-	260	-	-	-
Electricity Plants	-	-	-	-	-	-	-	-	-	-	-
CHP Plants	-	-	-	-	-	-	-	239	-	-	-
Heat Plants	-	-	-	-	-	-	-	21	-	-	-
Blast Furnaces/Gas Works	-	-	-	-	-	-	-	-	-	-	-
Coke/Pat. Fuel/BKB Plants	-	-	-	-	-	-	-	-	-	-	-
Petroleum Refineries	-	-	-	-	-	-	-	-	-	-	-
Petrochemical Industry	-	-	-	-	-	-	-	-	-	-	-
Liquefaction	-	-	-	-	-	-	-	-	-	-	-
Other Transform. Sector	-	-	-	-	-	-	-	-	-	-	-
ENERGY SECTOR	-	-	55	-	-	-	-	-	-	-	-
Coal Mines	-	-	-	-	-	-	-	-	-	-	-
Oil and Gas Extraction	-	-	-	-	-	-	-	-	-	-	-
Petroleum Refineries	-	-	-	-	-	-	-	-	-	-	-
Electr., CHP+Heat Plants	-	-	-	-	-	-	-	-	-	-	-
Pumped Storage (Elec.)	-	-	-	-	-	-	-	-	-	-	-
Other Energy Sector	-	-	55	-	-	-	-	-	-	-	-
Distribution Losses	-	-	-	-	-	-	-	-	-	-	-
FINAL CONSUMPTION	75	32	515	-	29	2	820	249	670	-	543
INDUSTRY SECTOR	75	1	-	-	-	2	-	219	670	-	-
Iron and Steel	-	-	-	-	-	-	-	28	-	-	-
Chemical and Petrochem.	-	-	-	-	-	-	-	13	670	-	-
of which: Feedstocks	-	-	-	-	-	-	-	-	670	-	-
Non-Ferrous Metals	-	-	-	-	-	-	-	-	-	-	-
Non-Metallic Minerals	-	-	-	-	-	-	-	16	-	-	-
Transport Equipment	-	-	-	-	-	2	-	4	-	-	-
Machinery	-	-	-	-	-	-	-	8	-	-	-
Mining and Quarrying	-	-	-	-	-	-	-	1	-	-	-
Food and Tobacco	-	-	-	-	-	-	-	24	-	-	-
Paper, Pulp and Print	-	-	-	-	-	-	-	57	-	-	-
Wood and Wood Products	-	-	-	-	-	-	-	2	-	-	-
Construction	-	-	-	-	-	-	-	4	-	-	-
Textile and Leather	-	-	-	-	-	-	-	44	-	-	-
Non-specified	75	1	-	-	-	-	-	18	-	-	-
TRANSPORT SECTOR	-	-	506	-	29	-	659	-	-	-	-
Air	-	-	-	-	29	-	-	-	-	-	-
Road	-	-	506	-	-	-	659	-	-	-	-
Rail	-	-	-	-	-	-	-	-	-	-	-
Pipeline Transport	-	-	-	-	-	-	-	-	-	-	-
Internal Navigation	-	-	-	-	-	-	-	-	-	-	-
Non-specified	-	-	-	-	-	-	-	-	-	-	-
OTHER SECTORS	-	31	9	-	-	-	161	30	-	-	-
Agriculture	-	4	9	-	-	-	161	5	-	-	-
Comm. and Publ. Services	-	3	-	-	-	-	-	25	-	-	-
Residential	-	24	-	-	-	-	-	-	-	-	-
Non-specified	-	-	-	-	-	-	-	-	-	-	-
NON-ENERGY USE	-	-	-	-	-	-	-	-	-	-	543
in Industry/Transf./Energy	-	-	-	-	-	-	-	-	-	-	543
in Transport	-	-	-	-	-	-	-	-	-	-	-
in Other Sectors	-	-	-	-	-	-	-	-	-	-	-

Oil cont. / Pétrole cont. (1000 tonnes)

Slovak Republic / République slovaque : 1998

SUPPLY AND CONSUMPTION APPROVISIONNEMENT ET DEMANDE	Gas / Gaz (TJ)				Comb. Renew. & Waste / En. Re. Comb. & Déchets (TJ)				(GWh)	(TJ)
	Natural Gas Gaz naturel	Gas Works Usines à gaz	Coke Ovens Cokeries	Blast Furnaces Hauts fourneaux	Solid Biomass Biomasse solide	Gas/Liquids from Biomass Gaz/Liquides tirés de biomasse	Municipal Waste Déchets urbains	Industrial Waste Déchets industriels	Electricity Electricité	Heat Chaleur
Production	9393	-	10659	14565	3086	-	-	-	25465	38664
From Other Sources	-	-	-	-	-	-	-	-	-	-
Imports	251657	-	-	-	-	-	-	-	1447	-
Exports	-	-	-	-	-	-	-	-	-157	-
Intl. Marine Bunkers	-	-	-	-	-	-	-	-	-	-
Stock Changes	5361	-	-	-	-67	-	-	-	-	-
DOMESTIC SUPPLY	**266411**	**-**	**10659**	**14565**	**3019**	**-**	**-**	**-**	**26755**	**38664**
Transfers	-	-	-	-	-	-	-	-	-	-
Statistical Differences	2	-	-	-	-	-	-	-	-	-
TRANSFORMATION	**45959**	**-**	**1345**	**1223**	**2787**	**-**	**-**	**-**	**1798**	**-**
Electricity Plants	-	-	-	-	-	-	-	-	-	-
CHP Plants	45959	-	1345	1223	-	-	-	-	-	-
Heat Plants	-	-	-	-	-	-	-	-	-	-
Blast Furnaces/Gas Works	-	-	-	-	-	-	-	-	-	-
Coke/Pat. Fuel/BKB Plants	-	-	-	-	-	-	-	-	-	-
Petroleum Refineries	-	-	-	-	-	-	-	-	-	-
Petrochemical Industry	-	-	-	-	-	-	-	-	-	-
Liquefaction	-	-	-	-	-	-	-	-	-	-
Other Transform. Sector	-	-	-	-	2787	-	-	-	1798	-
ENERGY SECTOR	**-**	**-**	**960**	**3645**	**-**	**-**	**-**	**-**	**1898**	**5200**
Coal Mines	-	-	-	-	-	-	-	-	-	-
Oil and Gas Extraction	-	-	-	-	-	-	-	-	-	-
Petroleum Refineries	-	-	-	-	-	-	-	-	-	-
Electr., CHP+Heat Plants	-	-	-	-	-	-	-	-	1386	-
Pumped Storage (Elec.)	-	-	-	-	-	-	-	-	512	-
Other Energy Sector	-	-	960	3645	-	-	-	-	-	5200
Distribution Losses	5424	-	218	172	-	-	-	-	2039	2200
FINAL CONSUMPTION	**215030**	**-**	**8136**	**9525**	**232**	**-**	**-**	**-**	**21020**	**31264**
INDUSTRY SECTOR	**114992**	**-**	**8136**	**9525**	**2**	**-**	**-**	**-**	**9441**	**799**
Iron and Steel	11674	-	8136	9525	-	-	-	-	3149	-
Chemical and Petrochem.	55248	-	-	-	-	-	-	-	1401	-
of which: Feedstocks	27671	-	-	-	-	-	-	-	-	-
Non-Ferrous Metals	-	-	-	-	-	-	-	-	-	-
Non-Metallic Minerals	12664	-	-	-	-	-	-	-	983	-
Transport Equipment	2369	-	-	-	-	-	-	-	241	195
Machinery	6849	-	-	-	-	-	-	-	812	182
Mining and Quarrying	3273	-	-	-	-	-	-	-	128	-
Food and Tobacco	10381	-	-	-	-	-	-	-	1085	65
Paper, Pulp and Print	2843	-	-	-	-	-	-	-	661	-
Wood and Wood Products	1077	-	-	-	-	-	-	-	200	130
Construction	43	-	-	-	2	-	-	-	176	33
Textile and Leather	3144	-	-	-	-	-	-	-	316	65
Non-specified	5427	-	-	-	-	-	-	-	289	129
TRANSPORT SECTOR	**-**	**-**	**-**	**-**	**-**	**-**	**-**	**-**	**915**	**-**
Air	-	-	-	-	-	-	-	-	-	-
Road	-	-	-	-	-	-	-	-	-	-
Rail	-	-	-	-	-	-	-	-	915	-
Pipeline Transport	-	-	-	-	-	-	-	-	-	-
Internal Navigation	-	-	-	-	-	-	-	-	-	-
Non-specified	-	-	-	-	-	-	-	-	-	-
OTHER SECTORS	**100038**	**-**	**-**	**-**	**230**	**-**	**-**	**-**	**10664**	**30465**
Agriculture	2392	-	-	-	160	-	-	-	941	443
Comm. and Publ. Services	32979	-	-	-	59	-	-	-	4104	13996
Residential	64667	-	-	-	11	-	-	-	5619	16026
Non-specified	-	-	-	-	-	-	-	-	-	-
NON-ENERGY USE	**-**	**-**	**-**	**-**	**-**	**-**	**-**	**-**	**-**	**-**
in Industry/Transf./Energy	-	-	-	-	-	-	-	-	-	-
in Transport	-	-	-	-	-	-	-	-	-	-
in Other Sectors	-	-	-	-	-	-	-	-	-	-

Slovenia / Slovénie : 1997

SUPPLY AND CONSUMPTION / APPROVISIONNEMENT ET DEMANDE	Coal / Charbon (1000 tonnes)							Oil / Pétrole (1000 tonnes)			
	Coking Coal / Charbon à coke	Other Bit. Coal / Autres charb. bit.	Sub-Bit. Coal / Charbon sous-bit.	Lignite / Lignite	Peat / Tourbe	Oven and Gas Coke / Coke de four/gaz	Pat. Fuel and BKB / Agg./briq. de lignite	Crude Oil / Pétrole brut	NGL / LGN	Feed-stocks / Produits d'aliment.	Additives / Additifs
Production	-	-	648	4305	-	-	-	1	-	-	-
From Other Sources	-	-	-	-	-	-	-	-	-	-	-
Imports	-	19	300	7	-	72	-	545	-	77	-
Exports	-	-	-1	-1	-	-	-	-	-	-	-
Intl. Marine Bunkers	-	-	-	-	-	-	-	-	-	-	-
Stock Changes	-	-	51	58	-	-1	-	-12	-	-4	-
DOMESTIC SUPPLY	-	19	998	4369	-	71	-	534	-	73	-
Transfers	-	-	-	-	-	-	-	-	-	-	-
Statistical Differences	-	-	-	-	-	-	-	-	-	-	-
TRANSFORMATION	-	-	965	4304	-	-	-	534	-	73	-
Electricity Plants	-	-	625	-	-	-	-	-	-	-	-
CHP Plants	-	-	340	4304	-	-	-	-	-	-	-
Heat Plants	-	-	-	-	-	-	-	-	-	-	-
Blast Furnaces/Gas Works	-	-	-	-	-	-	-	-	-	-	-
Coke/Pat. Fuel/BKB Plants	-	-	-	-	-	-	-	-	-	-	-
Petroleum Refineries	-	-	-	-	-	-	-	534	-	73	-
Petrochemical Industry	-	-	-	-	-	-	-	-	-	-	-
Liquefaction	-	-	-	-	-	-	-	-	-	-	-
Other Transform. Sector	-	-	-	-	-	-	-	-	-	-	-
ENERGY SECTOR	-	-	3	-	-	-	-	-	-	-	-
Coal Mines	-	-	3	-	-	-	-	-	-	-	-
Oil and Gas Extraction	-	-	-	-	-	-	-	-	-	-	-
Petroleum Refineries	-	-	-	-	-	-	-	-	-	-	-
Electr., CHP+Heat Plants	-	-	-	-	-	-	-	-	-	-	-
Pumped Storage (Elec.)	-	-	-	-	-	-	-	-	-	-	-
Other Energy Sector	-	-	-	-	-	-	-	-	-	-	-
Distribution Losses	-	-	-	-	-	-	-	-	-	-	-
FINAL CONSUMPTION	-	19	30	65	-	71	-	-	-	-	-
INDUSTRY SECTOR	-	19	-	7	-	71	-	-	-	-	-
Iron and Steel	-	-	-	-	-	25	-	-	-	-	-
Chemical and Petrochem.	-	-	-	-	-	-	-	-	-	-	-
of which: Feedstocks	-	-	-	-	-	-	-	-	-	-	-
Non-Ferrous Metals	-	2	-	-	-	-	-	-	-	-	-
Non-Metallic Minerals	-	16	-	-	-	44	-	-	-	-	-
Transport Equipment	-	-	-	1	-	-	-	-	-	-	-
Machinery	-	-	-	-	-	1	-	-	-	-	-
Mining and Quarrying	-	-	-	-	-	-	-	-	-	-	-
Food and Tobacco	-	-	-	1	-	1	-	-	-	-	-
Paper, Pulp and Print	-	-	-	-	-	-	-	-	-	-	-
Wood and Wood Products	-	-	-	-	-	-	-	-	-	-	-
Construction	-	-	-	-	-	-	-	-	-	-	-
Textile and Leather	-	-	-	3	-	-	-	-	-	-	-
Non-specified	-	1	-	2	-	-	-	-	-	-	-
TRANSPORT SECTOR	-	-	-	-	-	-	-	-	-	-	-
Air	-	-	-	-	-	-	-	-	-	-	-
Road	-	-	-	-	-	-	-	-	-	-	-
Rail	-	-	-	-	-	-	-	-	-	-	-
Pipeline Transport	-	-	-	-	-	-	-	-	-	-	-
Internal Navigation	-	-	-	-	-	-	-	-	-	-	-
Non-specified	-	-	-	-	-	-	-	-	-	-	-
OTHER SECTORS	-	-	30	58	-	-	-	-	-	-	-
Agriculture	-	-	-	-	-	-	-	-	-	-	-
Comm. and Publ. Services	-	-	12	20	-	-	-	-	-	-	-
Residential	-	-	18	38	-	-	-	-	-	-	-
Non-specified	-	-	-	-	-	-	-	-	-	-	-
NON-ENERGY USE	-	-	-	-	-	-	-	-	-	-	-
in Industry/Trans./Energy	-	-	-	-	-	-	-	-	-	-	-
in Transport	-	-	-	-	-	-	-	-	-	-	-
in Other Sectors	-	-	-	-	-	-	-	-	-	-	-

Slovenia / Slovénie : 1997

SUPPLY AND CONSUMPTION / APPROVISIONNEMENT ET DEMANDE	Oil cont. / Pétrole cont. (1000 tonnes)										
	Refinery Gas / Gaz de raffinerie	LPG + Ethane / GPL + éthane	Motor Gasoline / Essence moteur	Aviation Gasoline / Essence aviation	Jet Fuel / Carbu- réacteurs	Kerosene / Kérosène	Gas/ Diesel / Gazole	Heavy Fuel Oil / Fioul lourd	Naphtha / Naphta	Petrol. Coke / Coke de pétrole	Other Prod. / Autres prod.
Production	-	-	89	-	-	-	200	173	74	-	5
From Other Sources	-	-	-	-	-	-	-	-	-	-	-
Imports	-	74	853	1	21	-	1134	56	-	-	-
Exports	-	-	-	-	-3	-	-	-3	-63	-	-1
Intl. Marine Bunkers	-	-	-	-	-	-	-	-	-	-	-
Stock Changes	-	-	-28	-	-	-	-31	-2	-	-	-
DOMESTIC SUPPLY	-	74	914	1	18	-	1303	224	11	-	4
Transfers	-	-	-	-	-	-	-	-	-	-	-
Statistical Differences	-	-	-	-	-	-	-	1	-	-	-
TRANSFORMATION	-	-	-	-	-	-	7	56	-	-	-
Electricity Plants	-	-	-	-	-	-	3	-	-	-	-
CHP Plants	-	-	-	-	-	-	1	51	-	-	-
Heat Plants	-	-	-	-	-	-	3	5	-	-	-
Blast Furnaces/Gas Works	-	-	-	-	-	-	-	-	-	-	-
Coke/Pat. Fuel/BKB Plants	-	-	-	-	-	-	-	-	-	-	-
Petroleum Refineries	-	-	-	-	-	-	-	-	-	-	-
Petrochemical Industry	-	-	-	-	-	-	-	-	-	-	-
Liquefaction	-	-	-	-	-	-	-	-	-	-	-
Other Transform. Sector	-	-	-	-	-	-	-	-	-	-	-
ENERGY SECTOR	-	-	1	-	-	-	-	37	-	-	-
Coal Mines	-	-	-	-	-	-	-	-	-	-	-
Oil and Gas Extraction	-	-	-	-	-	-	-	22	-	-	-
Petroleum Refineries	-	-	1	-	-	-	-	15	-	-	-
Electr., CHP+Heat Plants	-	-	-	-	-	-	-	-	-	-	-
Pumped Storage (Elec.)	-	-	-	-	-	-	-	-	-	-	-
Other Energy Sector	-	-	-	-	-	-	-	-	-	-	-
Distribution Losses	-	-	-	-	-	-	-	-	-	-	-
FINAL CONSUMPTION	-	74	913	1	18	-	1296	132	11	-	4
INDUSTRY SECTOR	-	11	-	-	-	-	36	103	11	-	-
Iron and Steel	-	2	-	-	-	-	-	-	-	-	-
Chemical and Petrochem.	-	-	-	-	-	-	2	12	11	-	-
of which: Feedstocks	-	-	-	-	-	-	-	-	11	-	-
Non-Ferrous Metals	-	1	-	-	-	-	1	3	-	-	-
Non-Metallic Minerals	-	5	-	-	-	-	5	29	-	-	-
Transport Equipment	-	-	-	-	-	-	2	2	-	-	-
Machinery	-	2	-	-	-	-	7	7	-	-	-
Mining and Quarrying	-	-	-	-	-	-	1	-	-	-	-
Food and Tobacco	-	1	-	-	-	-	10	24	-	-	-
Paper, Pulp and Print	-	-	-	-	-	-	1	-	-	-	-
Wood and Wood Products	-	-	-	-	-	-	-	8	-	-	-
Construction	-	-	-	-	-	-	-	-	-	-	-
Textile and Leather	-	-	-	-	-	-	4	12	-	-	-
Non-specified	-	-	-	-	-	-	3	6	-	-	-
TRANSPORT SECTOR	-	-	913	1	18	-	563	-	-	-	-
Air	-	-	-	1	18	-	-	-	-	-	-
Road	-	-	913	-	-	-	551	-	-	-	-
Rail	-	-	-	-	-	-	12	-	-	-	-
Pipeline Transport	-	-	-	-	-	-	-	-	-	-	-
Internal Navigation	-	-	-	-	-	-	-	-	-	-	-
Non-specified	-	-	-	-	-	-	-	-	-	-	-
OTHER SECTORS	-	63	-	-	-	-	697	29	-	-	-
Agriculture	-	-	-	-	-	-	-	-	-	-	-
Comm. and Publ. Services	-	13	-	-	-	-	246	29	-	-	-
Residential	-	50	-	-	-	-	451	-	-	-	-
Non-specified	-	-	-	-	-	-	-	-	-	-	-
NON-ENERGY USE	-	-	-	-	-	-	-	-	-	-	4
in Industry/Transf./Energy	-	-	-	-	-	-	-	-	-	-	4
in Transport	-	-	-	-	-	-	-	-	-	-	-
in Other Sectors	-	-	-	-	-	-	-	-	-	-	-

Slovenia / Slovénie : 1997

SUPPLY AND CONSUMPTION *APPROVISIONNEMENT ET DEMANDE*	Gas / *Gaz* (TJ)				Comb. Renew. & Waste / *En. Re. Comb. & Déchets* (TJ)				(GWh)	(TJ)
	Natural Gas *Gaz naturel*	Gas Works *Usines à gaz*	Coke Ovens *Cokeries*	Blast Furnaces *Hauts fourneaux*	Solid Biomass *Biomasse solide*	Gas/Liquids from Biomass *Gaz/Liquides tirés de biomasse*	Municipal Waste *Déchets urbains*	Industrial Waste *Déchets industriels*	Electricity *Electricité*	Heat *Chaleur*
Production	432	-	-	-	9792	-	-	-	13176	9078
From Other Sources	-	-	-	-	-	-	-	-	-	-
Imports	32847	-	-	-	1236	-	-	-	824	-
Exports	-	-	-	-	-	-	-	-	-2520	-
Intl. Marine Bunkers	-	-	-	-	-	-	-	-	-	-
Stock Changes	-	-	-	-	-33	-	-	-	-	-
DOMESTIC SUPPLY	**33279**	**-**	**-**	**-**	**10995**	**-**	**-**	**-**	**11480**	**9078**
Transfers	-	-	-	-	-	-	-	-	-	-
Statistical Differences	147	-	-	-	-	-	-	-	-	-
TRANSFORMATION	**3630**	**-**	**-**	**-**	**94**	**-**	**-**	**-**	**-**	**-**
Electricity Plants	98	-	-	-	-	-	-	-	-	-
CHP Plants	612	-	-	-	-	-	-	-	-	-
Heat Plants	2920	-	-	-	94	-	-	-	-	-
Blast Furnaces/Gas Works	-	-	-	-	-	-	-	-	-	-
Coke/Pat. Fuel/BKB Plants	-	-	-	-	-	-	-	-	-	-
Petroleum Refineries	-	-	-	-	-	-	-	-	-	-
Petrochemical Industry	-	-	-	-	-	-	-	-	-	-
Liquefaction	-	-	-	-	-	-	-	-	-	-
Other Transform. Sector	-	-	-	-	-	-	-	-	-	-
ENERGY SECTOR	**40**	**-**	**-**	**-**	**-**	**-**	**-**	**-**	**943**	**945**
Coal Mines	-	-	-	-	-	-	-	-	94	-
Oil and Gas Extraction	-	-	-	-	-	-	-	-	-	-
Petroleum Refineries	40	-	-	-	-	-	-	-	30	-
Electr., CHP+Heat Plants	-	-	-	-	-	-	-	-	819	945
Pumped Storage (Elec.)	-	-	-	-	-	-	-	-	-	-
Other Energy Sector	-	-	-	-	-	-	-	-	-	-
Distribution Losses	-	-	-	-	-	-	-	-	687	-
FINAL CONSUMPTION	**29756**	**-**	**-**	**-**	**10901**	**-**	**-**	**-**	**9850**	**8133**
INDUSTRY SECTOR	**25590**	**-**	**-**	**-**	**2556**	**-**	**-**	**-**	**4884**	**1116**
Iron and Steel	2686	-	-	-	-	-	-	-	748	106
Chemical and Petrochem.	8384	-	-	-	-	-	-	-	406	155
of which: Feedstocks	*4449*	-	-	-	-	-	-	-	-	-
Non-Ferrous Metals	749	-	-	-	-	-	-	-	1219	-
Non-Metallic Minerals	4054	-	-	-	-	-	-	-	431	31
Transport Equipment	830	-	-	-	-	-	-	-	98	-
Machinery	1648	-	-	-	-	-	-	-	504	-
Mining and Quarrying	28	-	-	-	-	-	-	-	26	-
Food and Tobacco	1378	-	-	-	-	-	-	-	220	-
Paper, Pulp and Print	2873	-	-	-	-	-	-	-	525	-
Wood and Wood Products	344	-	-	-	-	-	-	-	160	-
Construction	-	-	-	-	-	-	-	-	-	-
Textile and Leather	1420	-	-	-	-	-	-	-	286	216
Non-specified	1196	-	-	-	2556	-	-	-	261	608
TRANSPORT SECTOR	**-**	**-**	**-**	**-**	**-**	**-**	**-**	**-**	**164**	**-**
Air	-	-	-	-	-	-	-	-	-	-
Road	-	-	-	-	-	-	-	-	-	-
Rail	-	-	-	-	-	-	-	-	164	-
Pipeline Transport	-	-	-	-	-	-	-	-	-	-
Internal Navigation	-	-	-	-	-	-	-	-	-	-
Non-specified	-	-	-	-	-	-	-	-	-	-
OTHER SECTORS	**4166**	**-**	**-**	**-**	**8345**	**-**	**-**	**-**	**4802**	**7017**
Agriculture	-	-	-	-	-	-	-	-	-	-
Comm. and Publ. Services	2261	-	-	-	2141	-	-	-	2118	2124
Residential	1905	-	-	-	6204	-	-	-	2684	4893
Non-specified	-	-	-	-	-	-	-	-	-	-
NON-ENERGY USE	**-**	**-**	**-**	**-**	**-**	**-**	**-**	**-**	**-**	**-**
in Industry/Transf./Energy	-	-	-	-	-	-	-	-	-	-
in Transport	-	-	-	-	-	-	-	-	-	-
in Other Sectors	-	-	-	-	-	-	-	-	-	-

Slovenia / Slovénie : 1998

SUPPLY AND CONSUMPTION / APPROVISIONNEMENT ET DEMANDE	Coal / Charbon (1000 tonnes)							Oil / Pétrole (1000 tonnes)			
	Coking Coal / Charbon à coke	Other Bit. Coal / Autres charb. bit.	Sub-Bit. Coal / Charbon sous-bit.	Lignite / Lignite	Peat / Tourbe	Oven and Gas Coke / Coke de four/gaz	Pat. Fuel and BKB / Agg./briq. de lignite	Crude Oil / Pétrole brut	NGL / LGN	Feed-stocks / Produits d'aliment.	Additives / Additifs
Production	-	-	698	4194	-	-	-	1	-	-	-
From Other Sources	-	-	-	-	-	-	-	-	-	-	-
Imports	-	17	420	-	-	74	-	594	-	10	-
Exports	-	-	-1	-1	-	-	-	-	-	-	-
Intl. Marine Bunkers	-	-	-	-	-	-	-	-	-	-	-
Stock Changes	-	-3	28	-14	-	-1	-	12	-	4	-
DOMESTIC SUPPLY	-	14	1145	4179	-	73	-	607	-	14	-
Transfers	-	-	-	-	-	-	-	-	-	-	-
Statistical Differences	-	-	-	151	-	-	-	-	-	-	-
TRANSFORMATION	-	-	1119	4286	-	-	-	607	-	14	-
Electricity Plants	-	-	682	-	-	-	-	-	-	-	-
CHP Plants	-	-	437	4286	-	-	-	-	-	-	-
Heat Plants	-	-	-	-	-	-	-	-	-	-	-
Blast Furnaces/Gas Works	-	-	-	-	-	-	-	-	-	-	-
Coke/Pat. Fuel/BKB Plants	-	-	-	-	-	-	-	-	-	-	-
Petroleum Refineries	-	-	-	-	-	-	-	607	-	14	-
Petrochemical Industry	-	-	-	-	-	-	-	-	-	-	-
Liquefaction	-	-	-	-	-	-	-	-	-	-	-
Other Transform. Sector	-	-	-	-	-	-	-	-	-	-	-
ENERGY SECTOR	-	-	3	-	-	-	-	-	-	-	-
Coal Mines	-	-	3	-	-	-	-	-	-	-	-
Oil and Gas Extraction	-	-	-	-	-	-	-	-	-	-	-
Petroleum Refineries	-	-	-	-	-	-	-	-	-	-	-
Electr., CHP+Heat Plants	-	-	-	-	-	-	-	-	-	-	-
Pumped Storage (Elec.)	-	-	-	-	-	-	-	-	-	-	-
Other Energy Sector	-	-	-	-	-	-	-	-	-	-	-
Distribution Losses	-	-	-	-	-	-	-	-	-	-	-
FINAL CONSUMPTION	-	14	23	44	-	73	-	-	-	-	-
INDUSTRY SECTOR	-	14	-	6	-	73	-	-	-	-	-
Iron and Steel	-	-	-	-	-	29	-	-	-	-	-
Chemical and Petrochem.	-	-	-	-	-	-	-	-	-	-	-
of which: Feedstocks	-	-	-	-	-	-	-	-	-	-	-
Non-Ferrous Metals	-	2	-	-	-	-	-	-	-	-	-
Non-Metallic Minerals	-	11	-	-	-	43	-	-	-	-	-
Transport Equipment	-	-	-	1	-	-	-	-	-	-	-
Machinery	-	-	-	-	-	-	-	-	-	-	-
Mining and Quarrying	-	-	-	-	-	-	-	-	-	-	-
Food and Tobacco	-	-	-	-	-	1	-	-	-	-	-
Paper, Pulp and Print	-	-	-	-	-	-	-	-	-	-	-
Wood and Wood Products	-	-	-	-	-	-	-	-	-	-	-
Construction	-	-	-	-	-	-	-	-	-	-	-
Textile and Leather	-	-	-	3	-	-	-	-	-	-	-
Non-specified	-	1	-	2	-	-	-	-	-	-	-
TRANSPORT SECTOR	-	-	-	-	-	-	-	-	-	-	-
Air	-	-	-	-	-	-	-	-	-	-	-
Road	-	-	-	-	-	-	-	-	-	-	-
Rail	-	-	-	-	-	-	-	-	-	-	-
Pipeline Transport	-	-	-	-	-	-	-	-	-	-	-
Internal Navigation	-	-	-	-	-	-	-	-	-	-	-
Non-specified	-	-	-	-	-	-	-	-	-	-	-
OTHER SECTORS	-	-	23	38	-	-	-	-	-	-	-
Agriculture	-	-	-	-	-	-	-	-	-	-	-
Comm. and Publ. Services	-	-	9	13	-	-	-	-	-	-	-
Residential	-	-	14	25	-	-	-	-	-	-	-
Non-specified	-	-	-	-	-	-	-	-	-	-	-
NON-ENERGY USE	-	-	-	-	-	-	-	-	-	-	-
in Industry/Trans./Energy	-	-	-	-	-	-	-	-	-	-	-
in Transport	-	-	-	-	-	-	-	-	-	-	-
in Other Sectors	-	-	-	-	-	-	-	-	-	-	-

Slovenia / Slovénie : 1998

	Oil cont. / Pétrole cont. (1000 tonnes)										
SUPPLY AND CONSUMPTION	Refinery Gas	LPG + Ethane	Motor Gasoline	Aviation Gasoline	Jet Fuel	Kerosene	Gas/ Diesel	Heavy Fuel Oil	Naphtha	Petrol. Coke	Other Prod.
APPROVISIONNEMENT ET DEMANDE	Gaz de raffinerie	GPL + éthane	Essence moteur	Essence aviation	Carbu- réacteurs	Kérosène	Gazole	Fioul lourd	Naphta	Coke de pétrole	Autres prod.
Production	-	-	14	-	-	1	396	78	53	-	11
From Other Sources	-	-	-	-	-	-	-	-	-	-	-
Imports	-	74	841	1	19	-	1262	99	-	-	110
Exports	-	-	-43	-	-2	-	-132	-1	-33	-	-9
Intl. Marine Bunkers	-	-	-	-	-	-	-	-	-	-	-
Stock Changes	-	1	-14	-	-	-	36	-	-2	-	-1
DOMESTIC SUPPLY	-	75	798	1	17	1	1562	176	18	-	111
Transfers	-	-	-	-	-	-	-	-	-	-	-
Statistical Differences	-	-	-	-	-	-	-	-2	-	-	1
TRANSFORMATION	-	-	-	-	-	-	6	48	-	-	-
Electricity Plants	-	-	-	-	-	-	3	-	-	-	-
CHP Plants	-	-	-	-	-	-	1	47	-	-	-
Heat Plants	-	-	-	-	-	-	2	1	-	-	-
Blast Furnaces/Gas Works	-	-	-	-	-	-	-	-	-	-	-
Coke/Pat. Fuel/BKB Plants	-	-	-	-	-	-	-	-	-	-	-
Petroleum Refineries	-	-	-	-	-	-	-	-	-	-	-
Petrochemical Industry	-	-	-	-	-	-	-	-	-	-	-
Liquefaction	-	-	-	-	-	-	-	-	-	-	-
Other Transform. Sector	-	-	-	-	-	-	-	-	-	-	-
ENERGY SECTOR	-	-	-	-	-	-	-	19	-	-	-
Coal Mines	-	-	-	-	-	-	-	-	-	-	-
Oil and Gas Extraction	-	-	-	-	-	-	-	3	-	-	-
Petroleum Refineries	-	-	-	-	-	-	-	16	-	-	-
Electr., CHP+Heat Plants	-	-	-	-	-	-	-	-	-	-	-
Pumped Storage (Elec.)	-	-	-	-	-	-	-	-	-	-	-
Other Energy Sector	-	-	-	-	-	-	-	-	-	-	-
Distribution Losses	-	-	-	-	-	-	-	-	-	-	-
FINAL CONSUMPTION	-	75	798	1	17	1	1556	107	18	-	112
INDUSTRY SECTOR	-	11	-	-	-	-	36	94	18	-	-
Iron and Steel	-	2	-	-	-	-	-	-	-	-	-
Chemical and Petrochem.	-	-	-	-	-	-	2	8	18	-	-
of which: Feedstocks	-	-	-	-	-	-	-	-	18	-	-
Non-Ferrous Metals	-	-	-	-	-	-	1	3	-	-	-
Non-Metallic Minerals	-	4	-	-	-	-	5	27	-	-	-
Transport Equipment	-	-	-	-	-	-	2	1	-	-	-
Machinery	-	3	-	-	-	-	8	6	-	-	-
Mining and Quarrying	-	-	-	-	-	-	1	-	-	-	-
Food and Tobacco	-	2	-	-	-	-	10	23	-	-	-
Paper, Pulp and Print	-	-	-	-	-	-	1	-	-	-	-
Wood and Wood Products	-	-	-	-	-	-	-	9	-	-	-
Construction	-	-	-	-	-	-	-	-	-	-	-
Textile and Leather	-	-	-	-	-	-	4	9	-	-	-
Non-specified	-	-	-	-	-	-	2	8	-	-	-
TRANSPORT SECTOR	-	-	798	1	17	-	802	-	-	-	-
Air	-	-	-	1	17	-	-	-	-	-	-
Road	-	-	798	-	-	-	790	-	-	-	-
Rail	-	-	-	-	-	-	12	-	-	-	-
Pipeline Transport	-	-	-	-	-	-	-	-	-	-	-
Internal Navigation	-	-	-	-	-	-	-	-	-	-	-
Non-specified	-	-	-	-	-	-	-	-	-	-	-
OTHER SECTORS	-	64	-	-	-	1	718	13	-	-	-
Agriculture	-	-	-	-	-	-	-	-	-	-	-
Comm. and Publ. Services	-	30	-	-	-	-	287	13	-	-	-
Residential	-	34	-	-	-	-	431	-	-	-	-
Non-specified	-	-	-	-	-	1	-	-	-	-	-
NON-ENERGY USE	-	-	-	-	-	-	-	-	-	-	112
in Industry/Transf./Energy	-	-	-	-	-	-	-	-	-	-	87
in Transport	-	-	-	-	-	-	-	-	-	-	-
in Other Sectors	-	-	-	-	-	-	-	-	-	-	25

Slovenia / Slovénie : 1998

SUPPLY AND CONSUMPTION APPROVISIONNEMENT ET DEMANDE	Gas / Gaz (TJ)				Comb. Renew. & Waste / En. Re. Comb. & Déchets (TJ)				(GWh)	(TJ)
	Natural Gas Gaz naturel	Gas Works Usines à gaz	Coke Ovens Cokeries	Blast Furnaces Hauts fourneaux	Solid Biomass Biomasse solide	Gas/Liquids from Biomass Gaz/Liquides tirés de biomasse	Municipal Waste Déchets urbains	Industrial Waste Déchets industriels	Electricity Electricité	Heat Chaleur
Production	286	-	-	-	9792	-	-	-	13744	9194
From Other Sources	-	-	-	-	-	-	-	-	-	-
Imports	33932	-	-	-	1236	-	-	-	706	-
Exports	-	-	-	-	-	-	-	-	-2630	-
Intl. Marine Bunkers	-	-	-	-	-	-	-	-	-	-
Stock Changes	-	-	-	-	-24	-	-	-	-	-
DOMESTIC SUPPLY	**34218**	**-**	**-**	**-**	**11004**	**-**	**-**	**-**	**11820**	**9194**
Transfers	-	-	-	-	-	-	-	-	-	-
Statistical Differences	-	-	-	-	-	-	-	-	-	-
TRANSFORMATION	**4241**	**-**	**-**	**-**	**103**	**-**	**-**	**-**	**-**	**-**
Electricity Plants	1027	-	-	-	-	-	-	-	-	-
CHP Plants	1027	-	-	-	-	-	-	-	-	-
Heat Plants	2187	-	-	-	103	-	-	-	-	-
Blast Furnaces/Gas Works	-	-	-	-	-	-	-	-	-	-
Coke/Pat. Fuel/BKB Plants	-	-	-	-	-	-	-	-	-	-
Petroleum Refineries	-	-	-	-	-	-	-	-	-	-
Petrochemical Industry	-	-	-	-	-	-	-	-	-	-
Liquefaction	-	-	-	-	-	-	-	-	-	-
Other Transform. Sector	-	-	-	-	-	-	-	-	-	-
ENERGY SECTOR	**14**	**-**	**-**	**-**	**-**	**-**	**-**	**-**	**958**	**972**
Coal Mines	-	-	-	-	-	-	-	-	93	-
Oil and Gas Extraction	-	-	-	-	-	-	-	-	-	-
Petroleum Refineries	14	-	-	-	-	-	-	-	27	-
Electr., CHP+Heat Plants	-	-	-	-	-	-	-	-	838	972
Pumped Storage (Elec.)	-	-	-	-	-	-	-	-	-	-
Other Energy Sector	-	-	-	-	-	-	-	-	-	-
Distribution Losses	-	-	-	-	-	-	-	-	759	-
FINAL CONSUMPTION	**29963**	**-**	**-**	**-**	**10901**	**-**	**-**	**-**	**10103**	**8222**
INDUSTRY SECTOR	**25412**	**-**	**-**	**-**	**2556**	**-**	**-**	**-**	**5041**	**1082**
Iron and Steel	2758	-	-	-	-	-	-	-	771	77
Chemical and Petrochem.	8029	-	-	-	-	-	-	-	417	73
of which: Feedstocks	4390	-	-	-	-	-	-	-	-	-
Non-Ferrous Metals	439	-	-	-	-	-	-	-	1167	13
Non-Metallic Minerals	4400	-	-	-	-	-	-	-	443	104
Transport Equipment	427	-	-	-	-	-	-	-	111	12
Machinery	1765	-	-	-	-	-	-	-	592	42
Mining and Quarrying	32	-	-	-	-	-	-	-	28	-
Food and Tobacco	1400	-	-	-	-	-	-	-	239	21
Paper, Pulp and Print	3168	-	-	-	-	-	-	-	535	79
Wood and Wood Products	370	-	-	-	-	-	-	-	177	10
Construction	-	-	-	-	-	-	-	-	-	7
Textile and Leather	1451	-	-	-	-	-	-	-	279	48
Non-specified	1173	-	-	-	2556	-	-	-	282	596
TRANSPORT SECTOR	**-**	**-**	**-**	**-**	**-**	**-**	**-**	**-**	**160**	**-**
Air	-	-	-	-	-	-	-	-	-	-
Road	-	-	-	-	-	-	-	-	-	-
Rail	-	-	-	-	-	-	-	-	160	-
Pipeline Transport	-	-	-	-	-	-	-	-	-	-
Internal Navigation	-	-	-	-	-	-	-	-	-	-
Non-specified	-	-	-	-	-	-	-	-	-	-
OTHER SECTORS	**4551**	**-**	**-**	**-**	**8345**	**-**	**-**	**-**	**4902**	**7140**
Agriculture	-	-	-	-	-	-	-	-	-	-
Comm. and Publ. Services	2470	-	-	-	2141	-	-	-	2211	2244
Residential	2081	-	-	-	6204	-	-	-	2691	4896
Non-specified	-	-	-	-	-	-	-	-	-	-
NON-ENERGY USE	**-**	**-**	**-**	**-**	**-**	**-**	**-**	**-**	**-**	**-**
in Industry/Transf./Energy	-	-	-	-	-	-	-	-	-	-
in Transport	-	-	-	-	-	-	-	-	-	-
in Other Sectors	-	-	-	-	-	-	-	-	-	-

South Africa / Afrique du Sud : 1997

SUPPLY AND CONSUMPTION / APPROVISIONNEMENT ET DEMANDE	Coal / Charbon (1000 tonnes)							Oil / Pétrole (1000 tonnes)			
	Coking Coal / Charbon à coke	Other Bit. Coal / Autres charb. bit.	Sub-Bit. Coal / Charbon sous-bit.	Lignite / Lignite	Peat / Tourbe	Oven and Gas Coke / Coke de four/gaz	Pat. Fuel and BKB / Agg./briq. de lignite	Crude Oil / Pétrole brut	NGL / LGN	Feed-stocks / Produits d'aliment.	Additives / Additifs
Production	3647	216426	-	-	-	2839	-	-	393	-	-
From Other Sources	-	-	-	-	-	-	-	-	-	-	-
Imports	425	-	-	-	-	-	-	12616	-	-	-
Exports	-	-64200	-	-	-	-	-	-	-	-	-
Intl. Marine Bunkers	-	-	-	-	-	-	-	-	-	-	-
Stock Changes	-	-2445	-	-	-	-	-	-	-	-	-
DOMESTIC SUPPLY	**4072**	**149781**	**-**	**-**	**-**	**2839**	**-**	**12616**	**393**	**-**	**-**
Transfers	-	-	-	-	-	-	-	-	-	-	-
Statistical Differences	-	9944	-	-	-	-	-	-	-	-	-
TRANSFORMATION	**4072**	**128780**	**-**	**-**	**-**	**2271**	**-**	**11337**	**393**	**-**	**-**
Electricity Plants	-	95651	-	-	-	-	-	-	-	-	-
CHP Plants	-	-	-	-	-	-	-	-	-	-	-
Heat Plants	-	-	-	-	-	-	-	-	-	-	-
Blast Furnaces/Gas Works	-	5363	-	-	-	2271	-	-	-	-	-
Coke/Pat. Fuel/BKB Plants	4072	-	-	-	-	-	-	-	-	-	-
Petroleum Refineries	-	-	-	-	-	-	-	18588	393	-	-
Petrochemical Industry	-	-	-	-	-	-	-	-	-	-	-
Liquefaction	-	27766	-	-	-	-	-	-7251	-	-	-
Other Transform. Sector	-	-	-	-	-	-	-	-	-	-	-
ENERGY SECTOR	**-**	**-**	**-**	**-**	**-**	**-**	**-**	**1279**	**-**	**-**	**-**
Coal Mines	-	-	-	-	-	-	-	-	-	-	-
Oil and Gas Extraction	-	-	-	-	-	-	-	-	-	-	-
Petroleum Refineries	-	-	-	-	-	-	-	1279	-	-	-
Electr., CHP+Heat Plants	-	-	-	-	-	-	-	-	-	-	-
Pumped Storage (Elec.)	-	-	-	-	-	-	-	-	-	-	-
Other Energy Sector	-	-	-	-	-	-	-	-	-	-	-
Distribution Losses	-	-	-	-	-	-	-	-	-	-	-
FINAL CONSUMPTION	**-**	**30945**	**-**	**-**	**-**	**568**	**-**	**-**	**-**	**-**	**-**
INDUSTRY SECTOR	**-**	**19479**	**-**	**-**	**-**	**568**	**-**	**-**	**-**	**-**	**-**
Iron and Steel	-	1758	-	-	-	568	-	-	-	-	-
Chemical and Petrochem.	-	8973	-	-	-	-	-	-	-	-	-
of which: Feedstocks	-	-	-	-	-	-	-	-	-	-	-
Non-Ferrous Metals	-	-	-	-	-	-	-	-	-	-	-
Non-Metallic Minerals	-	1072	-	-	-	-	-	-	-	-	-
Transport Equipment	-	-	-	-	-	-	-	-	-	-	-
Machinery	-	-	-	-	-	-	-	-	-	-	-
Mining and Quarrying	-	1248	-	-	-	-	-	-	-	-	-
Food and Tobacco	-	-	-	-	-	-	-	-	-	-	-
Paper, Pulp and Print	-	-	-	-	-	-	-	-	-	-	-
Wood and Wood Products	-	-	-	-	-	-	-	-	-	-	-
Construction	-	-	-	-	-	-	-	-	-	-	-
Textile and Leather	-	-	-	-	-	-	-	-	-	-	-
Non-specified	-	6428	-	-	-	-	-	-	-	-	-
TRANSPORT SECTOR	**-**	**2**	**-**	**-**	**-**	**-**	**-**	**-**	**-**	**-**	**-**
Air	-	-	-	-	-	-	-	-	-	-	-
Road	-	-	-	-	-	-	-	-	-	-	-
Rail	-	2	-	-	-	-	-	-	-	-	-
Pipeline Transport	-	-	-	-	-	-	-	-	-	-	-
Internal Navigation	-	-	-	-	-	-	-	-	-	-	-
Non-specified	-	-	-	-	-	-	-	-	-	-	-
OTHER SECTORS	**-**	**3847**	**-**	**-**	**-**	**-**	**-**	**-**	**-**	**-**	**-**
Agriculture	-	241	-	-	-	-	-	-	-	-	-
Comm. and Publ. Services	-	1356	-	-	-	-	-	-	-	-	-
Residential	-	2250	-	-	-	-	-	-	-	-	-
Non-specified	-	-	-	-	-	-	-	-	-	-	-
NON-ENERGY USE	**-**	**7617**	**-**	**-**	**-**	**-**	**-**	**-**	**-**	**-**	**-**
in Industry/Trans./Energy	-	7617	-	-	-	-	-	-	-	-	-
in Transport	-	-	-	-	-	-	-	-	-	-	-
in Other Sectors	-	-	-	-	-	-	-	-	-	-	-

South Africa / Afrique du Sud : 1997

SUPPLY AND CONSUMPTION / APPROVISIONNEMENT ET DEMANDE	Refinery Gas / Gaz de raffinerie	LPG + Ethane / GPL + éthane	Motor Gasoline / Essence moteur	Aviation Gasoline / Essence aviation	Jet Fuel / Carbu- réacteurs	Kerosene / Kérosène	Gas/ Diesel / Gazole	Heavy Fuel Oil / Fioul lourd	Naphtha / Naphta	Petrol. Coke / Coke de pétrole	Other Prod. / Autres prod.
Production	64	286	7898	20	1409	780	5147	2337	-	-	716
From Other Sources	-	-	-	-	-	-	-	-	-	-	-
Imports	-	-	-	-	-	-	-	-	-	-	-
Exports	-	-	-	-	-	-	-	-	-	-	-
Intl. Marine Bunkers	-	-	-	-	-	-	-343	-1802	-	-	-
Stock Changes	-	-	-	-	-	-	-	-	-	-	-
DOMESTIC SUPPLY	64	286	7898	20	1409	780	4804	535	-	-	716
Transfers	-	-	-	-	-	-	-	-	-	-	-
Statistical Differences	-	-1	-	-	-	-1	-	-1	-	-	-
TRANSFORMATION	-	-	-	-	-	-	-	-	-	-	-
Electricity Plants	-	-	-	-	-	-	-	-	-	-	-
CHP Plants	-	-	-	-	-	-	-	-	-	-	-
Heat Plants	-	-	-	-	-	-	-	-	-	-	-
Blast Furnaces/Gas Works	-	-	-	-	-	-	-	-	-	-	-
Coke/Pat. Fuel/BKB Plants	-	-	-	-	-	-	-	-	-	-	-
Petroleum Refineries	-	-	-	-	-	-	-	-	-	-	-
Petrochemical Industry	-	-	-	-	-	-	-	-	-	-	-
Liquefaction	-	-	-	-	-	-	-	-	-	-	-
Other Transform. Sector	-	-	-	-	-	-	-	-	-	-	-
ENERGY SECTOR	-	-	-	-	-	-	-	-	-	-	-
Coal Mines	-	-	-	-	-	-	-	-	-	-	-
Oil and Gas Extraction	-	-	-	-	-	-	-	-	-	-	-
Petroleum Refineries	-	-	-	-	-	-	-	-	-	-	-
Electr., CHP+Heat Plants	-	-	-	-	-	-	-	-	-	-	-
Pumped Storage (Elec.)	-	-	-	-	-	-	-	-	-	-	-
Other Energy Sector	-	-	-	-	-	-	-	-	-	-	-
Distribution Losses	-	-	-	-	-	-	-	-	-	-	-
FINAL CONSUMPTION	64	285	7898	20	1409	779	4804	534	-	-	716
INDUSTRY SECTOR	64	128	7	-	-	163	696	427	-	-	-
Iron and Steel	-	-	-	-	-	-	-	179	-	-	-
Chemical and Petrochem.	64	-	-	-	-	-	-	14	-	-	-
of which: Feedstocks	44	-	-	-	-	-	-	-	-	-	-
Non-Ferrous Metals	-	-	-	-	-	-	-	-	-	-	-
Non-Metallic Minerals	-	-	-	-	-	-	-	78	-	-	-
Transport Equipment	-	-	-	-	-	-	-	-	-	-	-
Machinery	-	-	-	-	-	-	-	12	-	-	-
Mining and Quarrying	-	2	-	-	-	10	411	16	-	-	-
Food and Tobacco	-	-	-	-	-	-	-	29	-	-	-
Paper, Pulp and Print	-	-	-	-	-	-	-	-	-	-	-
Wood and Wood Products	-	-	-	-	-	-	-	4	-	-	-
Construction	-	-	5	-	-	7	283	1	-	-	-
Textile and Leather	-	-	-	-	-	-	-	8	-	-	-
Non-specified	-	126	2	-	-	146	2	86	-	-	-
TRANSPORT SECTOR	-	-	7830	20	1409	11	3015	-	-	-	-
Air	-	-	-	20	1409	-	-	-	-	-	-
Road	-	-	7830	-	-	-	2855	-	-	-	-
Rail	-	-	-	-	-	-	160	-	-	-	-
Pipeline Transport	-	-	-	-	-	-	-	-	-	-	-
Internal Navigation	-	-	-	-	-	-	-	-	-	-	-
Non-specified	-	-	-	-	-	11	-	-	-	-	-
OTHER SECTORS	-	157	61	-	-	605	1093	107	-	-	-
Agriculture	-	6	61	-	-	60	1093	2	-	-	-
Comm. and Publ. Services	-	57	-	-	-	2	-	-	-	-	-
Residential	-	94	-	-	-	542	-	-	-	-	-
Non-specified	-	-	-	-	-	1	-	105	-	-	-
NON-ENERGY USE	-	-	-	-	-	-	-	-	-	-	716
in Industry/Transf./Energy	-	-	-	-	-	-	-	-	-	-	716
in Transport	-	-	-	-	-	-	-	-	-	-	-
in Other Sectors	-	-	-	-	-	-	-	-	-	-	-

South Africa / Afrique du Sud : 1997

SUPPLY AND CONSUMPTION APPROVISIONNEMENT ET DEMANDE	Gas / Gaz (TJ)				Comb. Renew. & Waste / En. Re. Comb. & Déchets (TJ)				(GWh)	(TJ)
	Natural Gas Gaz naturel	Gas Works Usines à gaz	Coke Ovens Cokeries	Blast Furnaces Hauts fourneaux	Solid Biomass Biomasse solide	Gas/Liquids from Biomass Gaz/Liquides tirés de biomasse	Municipal Waste Déchets urbains	Industrial Waste Déchets industriels	Electricity Electricité	Heat Chaleur
Production	64120	29902	22576	20968	504237	-	-	-	210052	-
From Other Sources									-	-
Imports	-	-	-	-	-	-	-	-	5	-
Exports	-	-	-	-	-9299	-	-	-	-6617	-
Intl. Marine Bunkers	-	-	-	-	-	-	-	-	-	-
Stock Changes	-	-	-	-	-	-	-	-	-	-
DOMESTIC SUPPLY	64120	29902	22576	20968	494938	-	-	-	203440	-
Transfers	-	-	-	-	-	-	-	-	-	-
Statistical Differences	-	1	-4	-	-1	-	-	-	-772	-
TRANSFORMATION	64120	-	-	-	129159	-	-	-	-	-
Electricity Plants	-	-	-	-	-	-	-	-	-	-
CHP Plants	-	-	-	-	-	-	-	-	-	-
Heat Plants	-	-	-	-	-	-	-	-	-	-
Blast Furnaces/Gas Works	-	-	-	-	-	-	-	-	-	-
Coke/Pat. Fuel/BKB Plants	-	-	-	-	-	-	-	-	-	-
Petroleum Refineries	-	-	-	-	-	-	-	-	-	-
Petrochemical Industry	-	-	-	-	-	-	-	-	-	-
Liquefaction	64120	-	-	-	-	-	-	-	-	-
Other Transform. Sector	-	-	-	-	129159	-	-	-	-	-
ENERGY SECTOR	-	7	-	-	-	-	-	-	32020	-
Coal Mines	-	-	-	-	-	-	-	-	2848	-
Oil and Gas Extraction	-	-	-	-	-	-	-	-	-	-
Petroleum Refineries	-	7	-	-	-	-	-	-	12908	-
Electr., CHP+Heat Plants	-	-	-	-	-	-	-	-	12691	-
Pumped Storage (Elec.)	-	-	-	-	-	-	-	-	3573	-
Other Energy Sector	-	-	-	-	-	-	-	-	-	-
Distribution Losses	-	-	-	-	-	-	-	-	16094	-
FINAL CONSUMPTION	-	29896	22572	20968	365778	-	-	-	154554	-
INDUSTRY SECTOR	-	28469	22572	20968	66129	-	-	-	91460	-
Iron and Steel	-	8372	22572	20968	-	-	-	-	17875	-
Chemical and Petrochem.	-	2948	-	-	-	-	-	-	2432	-
of which: Feedstocks	-	-	-	-	-	-	-	-	-	-
Non-Ferrous Metals	-	1230	-	-	-	-	-	-	14584	-
Non-Metallic Minerals	-	5918	-	-	-	-	-	-	1188	-
Transport Equipment	-	174	-	-	-	-	-	-	11	-
Machinery	-	5414	-	-	-	-	-	-	127	-
Mining and Quarrying	-	549	-	-	-	-	-	-	30390	-
Food and Tobacco	-	1181	-	-	-	-	-	-	539	-
Paper, Pulp and Print	-	393	-	-	-	-	-	-	1029	-
Wood and Wood Products	-	879	-	-	-	-	-	-	596	-
Construction	-	-	-	-	-	-	-	-	17	-
Textile and Leather	-	20	-	-	-	-	-	-	514	-
Non-specified	-	1391	-	-	66129	-	-	-	22158	-
TRANSPORT SECTOR	-	-	-	-	-	-	-	-	4548	-
Air	-	-	-	-	-	-	-	-	-	-
Road	-	-	-	-	-	-	-	-	8	-
Rail	-	-	-	-	-	-	-	-	3385	-
Pipeline Transport	-	-	-	-	-	-	-	-	66	-
Internal Navigation	-	-	-	-	-	-	-	-	-	-
Non-specified	-	-	-	-	-	-	-	-	1089	-
OTHER SECTORS	-	1427	-	-	299649	-	-	-	58546	-
Agriculture	-	-	-	-	-	-	-	-	5640	-
Comm. and Publ. Services	-	915	-	-	-	-	-	-	22184	-
Residential	-	512	-	-	299649	-	-	-	30722	-
Non-specified	-	-	-	-	-	-	-	-	-	-
NON-ENERGY USE	-	-	-	-	-	-	-	-	-	-
in Industry/Transf./Energy	-	-	-	-	-	-	-	-	-	-
in Transport	-	-	-	-	-	-	-	-	-	-
in Other Sectors	-	-	-	-	-	-	-	-	-	-

South Africa / Afrique du Sud : 1998

SUPPLY AND CONSUMPTION *APPROVISIONNEMENT ET DEMANDE*	Coal / *Charbon* (1000 tonnes)							Oil / *Pétrole* (1000 tonnes)			
	Coking Coal *Charbon à coke*	Other Bit. Coal *Autres charb. bit.*	Sub-Bit. Coal *Charbon sous-bit.*	Lignite *Lignite*	Peat *Tourbe*	Oven and Gas Coke *Coke de four/gaz*	Pat. Fuel and BKB *Agg./briq. de lignite*	Crude Oil *Pétrole brut*	NGL *LGN*	Feed-stocks *Produits d'aliment.*	Additives *Additifs*
Production	1858	222942	-	-	-	2141	-	-	290	-	-
From Other Sources	-	-	-	-	-	-	-	-	-	-	-
Imports	1213	-	-	-	-	-	-	15542	-	-	-
Exports	-	-66100	-	-	-	-	-	-501	-	-	-
Intl. Marine Bunkers	-	-	-	-	-	-	-	-	-	-	-
Stock Changes	-	-2756	-	-	-	-	-	-	-	-	-
DOMESTIC SUPPLY	**3071**	**154086**	**-**	**-**	**-**	**2141**	**-**	**15041**	**290**	**-**	**-**
Transfers	-	-	-	-	-	-	-	-	-	-	-
Statistical Differences	-	11436	-	-	-	-	-	-	-	-	-
TRANSFORMATION	**3071**	**131971**	**-**	**-**	**-**	**1713**	**-**	**13516**	**290**	**-**	**-**
Electricity Plants	-	99631	-	-	-	-	-	-	-	-	-
CHP Plants	-	-	-	-	-	-	-	-	-	-	-
Heat Plants	-	-	-	-	-	-	-	-	-	-	-
Blast Furnaces/Gas Works	-	4799	-	-	-	1713	-	-	-	-	-
Coke/Pat. Fuel/BKB Plants	3071	-	-	-	-	-	-	-	-	-	-
Petroleum Refineries	-	-	-	-	-	-	-	20709	290	-	-
Petrochemical Industry	-	-	-	-	-	-	-	-	-	-	-
Liquefaction	-	27541	-	-	-	-	-	-7193	-	-	-
Other Transform. Sector	-	-	-	-	-	-	-	-	-	-	-
ENERGY SECTOR	**-**	**-**	**-**	**-**	**-**	**-**	**-**	**1525**	**-**	**-**	**-**
Coal Mines	-	-	-	-	-	-	-	-	-	-	-
Oil and Gas Extraction	-	-	-	-	-	-	-	-	-	-	-
Petroleum Refineries	-	-	-	-	-	-	-	1525	-	-	-
Electr., CHP+Heat Plants	-	-	-	-	-	-	-	-	-	-	-
Pumped Storage (Elec.)	-	-	-	-	-	-	-	-	-	-	-
Other Energy Sector	-	-	-	-	-	-	-	-	-	-	-
Distribution Losses	-	-	-	-	-	-	-	-	-	-	-
FINAL CONSUMPTION	**-**	**33551**	**-**	**-**	**-**	**428**	**-**	**-**	**-**	**-**	**-**
INDUSTRY SECTOR	**-**	**23873**	**-**	**-**	**-**	**428**	**-**	**-**	**-**	**-**	**-**
Iron and Steel	-	5919	-	-	-	428	-	-	-	-	-
Chemical and Petrochem.	-	9702	-	-	-	-	-	-	-	-	-
of which: Feedstocks	-	-	-	-	-	-	-	-	-	-	-
Non-Ferrous Metals	-	-	-	-	-	-	-	-	-	-	-
Non-Metallic Minerals	-	1259	-	-	-	-	-	-	-	-	-
Transport Equipment	-	-	-	-	-	-	-	-	-	-	-
Machinery	-	-	-	-	-	-	-	-	-	-	-
Mining and Quarrying	-	1517	-	-	-	-	-	-	-	-	-
Food and Tobacco	-	-	-	-	-	-	-	-	-	-	-
Paper, Pulp and Print	-	-	-	-	-	-	-	-	-	-	-
Wood and Wood Products	-	-	-	-	-	-	-	-	-	-	-
Construction	-	-	-	-	-	-	-	-	-	-	-
Textile and Leather	-	-	-	-	-	-	-	-	-	-	-
Non-specified	-	5476	-	-	-	-	-	-	-	-	-
TRANSPORT SECTOR	**-**	**23**	**-**	**-**	**-**	**-**	**-**	**-**	**-**	**-**	**-**
Air	-	-	-	-	-	-	-	-	-	-	-
Road	-	-	-	-	-	-	-	-	-	-	-
Rail	-	23	-	-	-	-	-	-	-	-	-
Pipeline Transport	-	-	-	-	-	-	-	-	-	-	-
Internal Navigation	-	-	-	-	-	-	-	-	-	-	-
Non-specified	-	-	-	-	-	-	-	-	-	-	-
OTHER SECTORS	**-**	**2100**	**-**	**-**	**-**	**-**	**-**	**-**	**-**	**-**	**-**
Agriculture	-	182	-	-	-	-	-	-	-	-	-
Comm. and Publ. Services	-	1731	-	-	-	-	-	-	-	-	-
Residential	-	187	-	-	-	-	-	-	-	-	-
Non-specified	-	-	-	-	-	-	-	-	-	-	-
NON-ENERGY USE	**-**	**7555**	**-**	**-**	**-**	**-**	**-**	**-**	**-**	**-**	**-**
in Industry/Trans./Energy	-	7555	-	-	-	-	-	-	-	-	-
in Transport	-	-	-	-	-	-	-	-	-	-	-
in Other Sectors	-	-	-	-	-	-	-	-	-	-	-

South Africa / Afrique du Sud : 1998

SUPPLY AND CONSUMPTION APPROVISIONNEMENT ET DEMANDE	Refinery Gas Gaz de raffinerie	LPG + Ethane GPL + éthane	Motor Gasoline Essence moteur	Aviation Gasoline Essence aviation	Jet Fuel Carbu- réacteurs	Kerosene Kérosène	Gas/ Diesel Gazole	Heavy Fuel Oil Fioul lourd	Naphtha Naphta	Petrol. Coke Coke de pétrole	Other Prod. Autres prod.
Production	103	284	8024	19	1488	842	5030	2819	-	-	657
From Other Sources	-	-	-	-	-	-	-	-	-	-	-
Imports	-	-	-	-	-	-	-	-	-	-	-
Exports	-	-	-	-	-	-	-	-	-	-	-
Intl. Marine Bunkers	-	-	-	-	-	-	-188	-2300	-	-	-
Stock Changes	-	-	-	-	-	-	-	-	-	-	-
DOMESTIC SUPPLY	**103**	**284**	**8024**	**19**	**1488**	**842**	**4842**	**519**	**-**	**-**	**657**
Transfers	-	-	-	-	-	-	-	-	-	-	-
Statistical Differences	-	-	-	-	-	-	-	-	-	-	-
TRANSFORMATION	**-**	**-**	**-**	**-**	**-**	**-**	**-**	**-**	**-**	**-**	**-**
Electricity Plants	-	-	-	-	-	-	-	-	-	-	-
CHP Plants	-	-	-	-	-	-	-	-	-	-	-
Heat Plants	-	-	-	-	-	-	-	-	-	-	-
Blast Furnaces/Gas Works	-	-	-	-	-	-	-	-	-	-	-
Coke/Pat. Fuel/BKB Plants	-	-	-	-	-	-	-	-	-	-	-
Petroleum Refineries	-	-	-	-	-	-	-	-	-	-	-
Petrochemical Industry	-	-	-	-	-	-	-	-	-	-	-
Liquefaction	-	-	-	-	-	-	-	-	-	-	-
Other Transform. Sector	-	-	-	-	-	-	-	-	-	-	-
ENERGY SECTOR	**-**	**-**	**-**	**-**	**-**	**-**	**-**	**-**	**-**	**-**	**-**
Coal Mines	-	-	-	-	-	-	-	-	-	-	-
Oil and Gas Extraction	-	-	-	-	-	-	-	-	-	-	-
Petroleum Refineries	-	-	-	-	-	-	-	-	-	-	-
Electr., CHP+Heat Plants	-	-	-	-	-	-	-	-	-	-	-
Pumped Storage (Elec.)	-	-	-	-	-	-	-	-	-	-	-
Other Energy Sector	-	-	-	-	-	-	-	-	-	-	-
Distribution Losses	-	-	-	-	-	-	-	-	-	-	-
FINAL CONSUMPTION	**103**	**284**	**8024**	**19**	**1488**	**842**	**4842**	**519**	**-**	**-**	**657**
INDUSTRY SECTOR	**103**	**150**	**4**	**-**	**-**	**250**	**664**	**415**	**-**	**-**	**-**
Iron and Steel	-	-	-	-	-	-	-	174	-	-	-
Chemical and Petrochem.	103	-	-	-	-	-	-	14	-	-	-
of which: Feedstocks	71	-	-	-	-	-	-	-	-	-	-
Non-Ferrous Metals	-	-	-	-	-	-	-	-	-	-	-
Non-Metallic Minerals	-	-	-	-	-	-	-	76	-	-	-
Transport Equipment	-	-	-	-	-	-	-	-	-	-	-
Machinery	-	-	-	-	-	-	-	12	-	-	-
Mining and Quarrying	-	1	-	-	-	16	417	15	-	-	-
Food and Tobacco	-	-	-	-	-	-	-	28	-	-	-
Paper, Pulp and Print	-	-	-	-	-	-	-	-	-	-	-
Wood and Wood Products	-	-	-	-	-	-	-	4	-	-	-
Construction	-	1	4	-	-	4	247	1	-	-	-
Textile and Leather	-	-	-	-	-	-	-	8	-	-	-
Non-specified	-	148	-	-	-	230	-	83	-	-	-
TRANSPORT SECTOR	**-**	**-**	**7950**	**19**	**1488**	**14**	**3150**	**-**	**-**	**-**	**-**
Air	-	-	-	19	1488	-	-	-	-	-	-
Road	-	-	7950	-	-	-	2981	-	-	-	-
Rail	-	-	-	-	-	-	169	-	-	-	-
Pipeline Transport	-	-	-	-	-	-	-	-	-	-	-
Internal Navigation	-	-	-	-	-	-	-	-	-	-	-
Non-specified	-	-	-	-	-	14	-	-	-	-	-
OTHER SECTORS	**-**	**134**	**70**	**-**	**-**	**578**	**1028**	**104**	**-**	**-**	**-**
Agriculture	-	4	61	-	-	66	1026	2	-	-	-
Comm. and Publ. Services	-	2	-	-	-	6	-	-	-	-	-
Residential	-	69	-	-	-	506	-	-	-	-	-
Non-specified	-	59	9	-	-	-	2	102	-	-	-
NON-ENERGY USE	**-**	**-**	**-**	**-**	**-**	**-**	**-**	**-**	**-**	**-**	**657**
in Industry/Transf./Energy	-	-	-	-	-	-	-	-	-	-	657
in Transport	-	-	-	-	-	-	-	-	-	-	-
in Other Sectors	-	-	-	-	-	-	-	-	-	-	-

South Africa / Afrique du Sud : 1998

SUPPLY AND CONSUMPTION / APPROVISIONNEMENT ET DEMANDE	Gas / Gaz (TJ)				Comb. Renew. & Waste / En. Re. Comb. & Déchets (TJ)				(GWh)	(TJ)
	Natural Gas / Gaz naturel	Gas Works / Usines à gaz	Coke Ovens / Cokeries	Blast Furnaces / Hauts fourneaux	Solid Biomass / Biomasse solide	Gas/Liquids from Biomass / Gaz/Liquides tirés de biomasse	Municipal Waste / Déchets urbains	Industrial Waste / Déchets industriels	Electricity / Electricité	Heat / Chaleur
Production	53983	31378	17024	15811	513313	-	-	-	205374	-
From Other Sources	-	-	-	-	-	-	-	-	-	-
Imports	-	-	-	-	-	-	-	-	2375	-
Exports	-	-	-	-	-9466	-	-	-	-4532	-
Intl. Marine Bunkers	-	-	-	-	-	-	-	-	-	-
Stock Changes	-	-	-	-	-	-	-	-	-	-
DOMESTIC SUPPLY	53983	31378	17024	15811	503847	-	-	-	203217	-
Transfers	-	-	-	-	-	-	-	-	-	-
Statistical Differences	-	-	-	-	-2	-	-	-	-14630	-
TRANSFORMATION	53983	-	-	-	131483	-	-	-	-	-
Electricity Plants	-	-	-	-	-	-	-	-	-	-
CHP Plants	-	-	-	-	-	-	-	-	-	-
Heat Plants	-	-	-	-	-	-	-	-	-	-
Blast Furnaces/Gas Works	-	-	-	-	-	-	-	-	-	-
Coke/Pat. Fuel/BKB Plants	-	-	-	-	-	-	-	-	-	-
Petroleum Refineries	-	-	-	-	-	-	-	-	-	-
Petrochemical Industry	-	-	-	-	-	-	-	-	-	-
Liquefaction	53983	-	-	-	-	-	-	-	-	-
Other Transform. Sector	-	-	-	-	131483	-	-	-	-	-
ENERGY SECTOR	-	187	-	-	-	-	-	-	13410	-
Coal Mines	-	-	-	-	-	-	-	-	2836	-
Oil and Gas Extraction	-	-	-	-	-	-	-	-	-	-
Petroleum Refineries	-	187	-	-	-	-	-	-	6698	-
Electr., CHP+Heat Plants	-	-	-	-	-	-	-	-	303	-
Pumped Storage (Elec.)	-	-	-	-	-	-	-	-	3573	-
Other Energy Sector	-	-	-	-	-	-	-	-	-	-
Distribution Losses	-	-	-	-	-	-	-	-	16534	-
FINAL CONSUMPTION	-	31191	17024	15811	372362	-	-	-	158643	-
INDUSTRY SECTOR	-	28711	17024	15811	67319	-	-	-	96159	-
Iron and Steel	-	15329	17024	15811	-	-	-	-	18865	-
Chemical and Petrochem.	-	3051	-	-	-	-	-	-	2630	-
of which: Feedstocks	-	-	-	-	-	-	-	-	-	-
Non-Ferrous Metals	-	957	-	-	-	-	-	-	14769	-
Non-Metallic Minerals	-	6314	-	-	-	-	-	-	1155	-
Transport Equipment	-	106	-	-	-	-	-	-	15	-
Machinery	-	604	-	-	-	-	-	-	37	-
Mining and Quarrying	-	498	-	-	-	-	-	-	29204	-
Food and Tobacco	-	1154	-	-	-	-	-	-	578	-
Paper, Pulp and Print	-	371	-	-	-	-	-	-	1039	-
Wood and Wood Products	-	170	-	-	-	-	-	-	623	-
Construction	-	-	-	-	-	-	-	-	20	-
Textile and Leather	-	18	-	-	-	-	-	-	373	-
Non-specified	-	139	-	-	67319	-	-	-	26851	-
TRANSPORT SECTOR	-	-	-	-	-	-	-	-	4546	-
Air	-	-	-	-	-	-	-	-	-	-
Road	-	-	-	-	-	-	-	-	23	-
Rail	-	-	-	-	-	-	-	-	3572	-
Pipeline Transport	-	-	-	-	-	-	-	-	67	-
Internal Navigation	-	-	-	-	-	-	-	-	-	-
Non-specified	-	-	-	-	-	-	-	-	884	-
OTHER SECTORS	-	2480	-	-	305043	-	-	-	57938	-
Agriculture	-	-	-	-	-	-	-	-	5627	-
Comm. and Publ. Services	-	960	-	-	-	-	-	-	13999	-
Residential	-	1520	-	-	305043	-	-	-	30163	-
Non-specified	-	-	-	-	-	-	-	-	8149	-
NON-ENERGY USE	-	-	-	-	-	-	-	-	-	-
in Industry/Transf./Energy	-	-	-	-	-	-	-	-	-	-
in Transport	-	-	-	-	-	-	-	-	-	-
in Other Sectors	-	-	-	-	-	-	-	-	-	-

Sri Lanka

SUPPLY AND CONSUMPTION 1997	Coal (1000 tonnes)							Oil (1000 tonnes)			
	Coking Coal	Other Bit. Coal	Sub-Bit. Coal	Lignite	Peat	Oven and Gas Coke	Pat. Fuel and BKB	Crude Oil	NGL	Feed-stocks	Additives
Production	-	-	-	-	-	-	-	-	-	-	-
Imports	-	1	-	-	-	-	-	1820	-	-	-
Exports	-	-	-	-	-	-	-	-	-	-	-
Intl. Marine Bunkers	-	-	-	-	-	-	-	-	-	-	-
Stock Changes	-	-	-	-	-	-	-	-	-	-	-
DOMESTIC SUPPLY	-	1	-	-	-	-	-	1820	-	-	-
Transfers and Stat. Diff.	-	-	-	-	-	-	-	-	-	-	-
TRANSFORMATION	-	-	-	-	-	-	-	1820	-	-	-
Electricity and CHP Plants	-	-	-	-	-	-	-	-	-	-	-
Petroleum Refineries	-	-	-	-	-	-	-	1820	-	-	-
Other Transform. Sector	-	-	-	-	-	-	-	-	-	-	-
ENERGY SECTOR	-	-	-	-	-	-	-	-	-	-	-
DISTRIBUTION LOSSES	-	-	-	-	-	-	-	-	-	-	-
FINAL CONSUMPTION	-	1	-	-	-	-	-	-	-	-	-
INDUSTRY SECTOR	-	-	-	-	-	-	-	-	-	-	-
Iron and Steel	-	-	-	-	-	-	-	-	-	-	-
Chemical and Petrochem.	-	-	-	-	-	-	-	-	-	-	-
Non-Metallic Minerals	-	-	-	-	-	-	-	-	-	-	-
Non-specified	-	-	-	-	-	-	-	-	-	-	-
TRANSPORT SECTOR	-	1	-	-	-	-	-	-	-	-	-
Air	-	-	-	-	-	-	-	-	-	-	-
Road	-	-	-	-	-	-	-	-	-	-	-
Non-specified	-	1	-	-	-	-	-	-	-	-	-
OTHER SECTORS	-	-	-	-	-	-	-	-	-	-	-
Agriculture	-	-	-	-	-	-	-	-	-	-	-
Comm. and Publ. Services	-	-	-	-	-	-	-	-	-	-	-
Residential	-	-	-	-	-	-	-	-	-	-	-
Non-specified	-	-	-	-	-	-	-	-	-	-	-
NON-ENERGY USE	-	-	-	-	-	-	-	-	-	-	-

APPROVISIONNEMENT ET DEMANDE 1998	Charbon (1000 tonnes)							Pétrole (1000 tonnes)			
	Charbon à coke	Autres charb. bit.	Charbon sous-bit.	Lignite	Tourbe	Coke de four/gaz	Agg./briq. de lignite	Pétrole brut	LGN	Produits d'aliment.	Additifs
Production	-	-	-	-	-	-	-	-	-	-	-
Imports	-	1	-	-	-	-	-	2155	-	-	-
Exports	-	-	-	-	-	-	-	-	-	-	-
Intl. Marine Bunkers	-	-	-	-	-	-	-	-	-	-	-
Stock Changes	-	-	-	-	-	-	-	-	-	-	-
DOMESTIC SUPPLY	-	1	-	-	-	-	-	2155	-	-	-
Transfers and Stat. Diff.	-	-	-	-	-	-	-	-	-	-	-
TRANSFORMATION	-	-	-	-	-	-	-	2155	-	-	-
Electricity and CHP Plants	-	-	-	-	-	-	-	-	-	-	-
Petroleum Refineries	-	-	-	-	-	-	-	2155	-	-	-
Other Transform. Sector	-	-	-	-	-	-	-	-	-	-	-
ENERGY SECTOR	-	-	-	-	-	-	-	-	-	-	-
DISTRIBUTION LOSSES	-	-	-	-	-	-	-	-	-	-	-
FINAL CONSUMPTION	-	1	-	-	-	-	-	-	-	-	-
INDUSTRY SECTOR	-	-	-	-	-	-	-	-	-	-	-
Iron and Steel	-	-	-	-	-	-	-	-	-	-	-
Chemical and Petrochem.	-	-	-	-	-	-	-	-	-	-	-
Non-Metallic Minerals	-	-	-	-	-	-	-	-	-	-	-
Non-specified	-	-	-	-	-	-	-	-	-	-	-
TRANSPORT SECTOR	-	1	-	-	-	-	-	-	-	-	-
Air	-	-	-	-	-	-	-	-	-	-	-
Road	-	-	-	-	-	-	-	-	-	-	-
Non-specified	-	1	-	-	-	-	-	-	-	-	-
OTHER SECTORS	-	-	-	-	-	-	-	-	-	-	-
Agriculture	-	-	-	-	-	-	-	-	-	-	-
Comm. and Publ. Services	-	-	-	-	-	-	-	-	-	-	-
Residential	-	-	-	-	-	-	-	-	-	-	-
Non-specified	-	-	-	-	-	-	-	-	-	-	-
NON-ENERGY USE	-	-	-	-	-	-	-	-	-	-	-

Sri Lanka

SUPPLY AND CONSUMPTION 1997	Oil cont. (1000 tonnes)										
	Refinery Gas	LPG + Ethane	Motor Gasoline	Aviation Gasoline	Jet Fuel	Kerosene	Gas/ Diesel	Heavy Fuel Oil	Naphtha	Petrol. Coke	Other Prod.
Production	14	53	161	-	74	147	532	697	99	-	38
Imports	-	87	23	-	207	10	848	-	-	-	4
Exports	-	-	-	-	-	-	-	-80	-93	-	-
Intl. Marine Bunkers	-	-	-	-	-	-	-37	-221	-	-	-
Stock Changes	-	-2	1	-	-5	3	-19	-	-6	-	-
DOMESTIC SUPPLY	14	138	185	-	276	160	1324	396	-	-	42
Transfers and Stat. Diff.	-	-4	10	-	-71	65	369	127	-	-	13
TRANSFORMATION	-	-	-	-	-	-	361	121	-	-	-
Electricity and CHP Plants	-	-	-	-	-	-	361	121	-	-	-
Petroleum Refineries	-	-	-	-	-	-	-	-	-	-	-
Other Transform. Sector	-	-	-	-	-	-	-	-	-	-	-
ENERGY SECTOR	14	35	-	-	-	-	3	52	-	-	-
DISTRIBUTION LOSSES	-	-	-	-	-	-	-	1	-	-	-
FINAL CONSUMPTION	-	99	195	-	205	225	1329	349	-	-	55
INDUSTRY SECTOR	-	24	-	-	-	4	383	330	-	-	-
Iron and Steel	-	-	-	-	-	-	-	-	-	-	-
Chemical and Petrochem.	-	-	-	-	-	-	-	-	-	-	-
Non-Metallic Minerals	-	-	-	-	-	-	-	-	-	-	-
Non-specified	-	24	-	-	-	4	383	330	-	-	-
TRANSPORT SECTOR	-	-	195	-	205	-	946	19	-	-	-
Air	-	-	-	-	205	-	-	-	-	-	-
Road	-	-	195	-	-	-	922	-	-	-	-
Non-specified	-	-	-	-	-	-	24	19	-	-	-
OTHER SECTORS	-	75	-	-	-	221	-	-	-	-	-
Agriculture	-	-	-	-	-	-	-	-	-	-	-
Comm. and Publ. Services	-	13	-	-	-	-	-	-	-	-	-
Residential	-	62	-	-	-	-	-	-	-	-	-
Non-specified	-	-	-	-	-	221	-	-	-	-	-
NON-ENERGY USE	-	-	-	-	-	-	-	-	-	-	55

APPROVISIONNEMENT ET DEMANDE 1998	Pétrole cont. (1000 tonnes)										
	Gaz de raffinerie	GPL + éthane	Essence moteur	Essence aviation	Carbu- réacteurs	Kérosène	Gazole	Fioul lourd	Naphta	Coke de pétrole	Autres prod.
Production	17	62	187	-	54	204	680	750	119	-	61
Imports	-	98	5	-	167	5	586	-	-	-	2
Exports	-	-	-	-	-	-	-	-20	-130	-	-
Intl. Marine Bunkers	-	-	-	-	-	-	-	-	-	-	-
Stock Changes	-	1	-3	-	-	1	10	-6	11	-	-
DOMESTIC SUPPLY	17	161	189	-	221	210	1276	724	-	-	63
Transfers and Stat. Diff.	-	-5	15	-	-21	26	153	-52	-	-	-8
TRANSFORMATION	-	-	-	-	-	-	127	223	-	-	-
Electricity and CHP Plants	-	-	-	-	-	-	127	223	-	-	-
Petroleum Refineries	-	-	-	-	-	-	-	-	-	-	-
Other Transform. Sector	-	-	-	-	-	-	-	-	-	-	-
ENERGY SECTOR	17	43	-	-	-	-	5	34	-	-	-
DISTRIBUTION LOSSES	-	-	-	-	-	-	-	1	-	-	-
FINAL CONSUMPTION	-	113	204	-	200	236	1297	414	-	-	55
INDUSTRY SECTOR	-	27	-	-	-	4	36	414	-	-	-
Iron and Steel	-	-	-	-	-	-	-	-	-	-	-
Chemical and Petrochem.	-	-	-	-	-	-	-	-	-	-	-
Non-Metallic Minerals	-	-	-	-	-	-	-	-	-	-	-
Non-specified	-	27	-	-	-	4	36	414	-	-	-
TRANSPORT SECTOR	-	-	204	-	200	-	1261	-	-	-	-
Air	-	-	-	-	200	-	-	-	-	-	-
Road	-	-	204	-	-	-	1237	-	-	-	-
Non-specified	-	-	-	-	-	-	24	-	-	-	-
OTHER SECTORS	-	86	-	-	-	232	-	-	-	-	-
Agriculture	-	-	-	-	-	-	-	-	-	-	-
Comm. and Publ. Services	-	15	-	-	-	-	-	-	-	-	-
Residential	-	71	-	-	-	-	-	-	-	-	-
Non-specified	-	-	-	-	-	232	-	-	-	-	-
NON-ENERGY USE	-	-	-	-	-	-	-	-	-	-	55

INTERNATIONAL ENERGY AGENCY

Sri Lanka

SUPPLY AND CONSUMPTION 1997	Gas (TJ)				Comb. Renew. & Waste (TJ)				(GWh)	(TJ)
	Natural Gas	Gas Works	Coke Ovens	Blast Furnaces	Solid Biomass	Gas/Liquids from Biomass	Municipal Waste	Industrial Waste	Electricity	Heat
Production	-	-	-	-	169591	-	-	-	5145	-
Imports	-	-	-	-	-	-	-	-	-	-
Exports	-	-	-	-	-	-	-	-	-	-
Intl. Marine Bunkers	-	-	-	-	-	-	-	-	-	-
Stock Changes	-	-	-	-	-	-	-	-	-	-
DOMESTIC SUPPLY	-	-	-	-	169591	-	-	-	5145	-
Transfers and Stat. Diff.	-	-	-	-	-	-	-	-	1	-
TRANSFORMATION	-	-	-	-	1	-	-	-	-	-
Electricity and CHP Plants	-	-	-	-	-	-	-	-	-	-
Petroleum Refineries	-	-	-	-	-	-	-	-	-	-
Other Transform. Sector	-	-	-	-	1	-	-	-	-	-
ENERGY SECTOR	-	-	-	-	-	-	-	-	39	-
DISTRIBUTION LOSSES	-	-	-	-	-	-	-	-	900	-
FINAL CONSUMPTION	-	-	-	-	169590	-	-	-	4207	-
INDUSTRY SECTOR	-	-	-	-	30700	-	-	-	1819	-
Iron and Steel	-	-	-	-	-	-	-	-	-	-
Chemical and Petrochem.	-	-	-	-	-	-	-	-	-	-
Non-Metallic Minerals	-	-	-	-	-	-	-	-	-	-
Non-specified	-	-	-	-	30700	-	-	-	1819	-
TRANSPORT SECTOR	-	-	-	-	-	-	-	-	-	-
Air	-	-	-	-	-	-	-	-	-	-
Road	-	-	-	-	-	-	-	-	-	-
Non-specified	-	-	-	-	-	-	-	-	-	-
OTHER SECTORS	-	-	-	-	138890	-	-	-	2388	-
Agriculture	-	-	-	-	-	-	-	-	-	-
Comm. and Publ. Services	-	-	-	-	5454	-	-	-	903	-
Residential	-	-	-	-	133436	-	-	-	1485	-
Non-specified	-	-	-	-	-	-	-	-	-	-
NON-ENERGY USE	-	-	-	-	-	-	-	-	-	-

APPROVISIONNEMENT ET DEMANDE 1998	Gaz (TJ)				En. Re. Comb. & Déchets (TJ)				(GWh)	(TJ)
	Gaz naturel	Usines à gaz	Cokeries	Hauts fourneaux	Biomasse solide	Gaz/Liquides tirés de biomasse	Déchets urbains	Déchets industriels	Electricité	Chaleur
Production	-	-	-	-	166749	-	-	-	5683	-
Imports	-	-	-	-	-	-	-	-	-	-
Exports	-	-	-	-	-	-	-	-	-	-
Intl. Marine Bunkers	-	-	-	-	-	-	-	-	-	-
Stock Changes	-	-	-	-	-	-	-	-	-	-
DOMESTIC SUPPLY	-	-	-	-	166749	-	-	-	5683	-
Transfers and Stat. Diff.	-	-	-	-	34	-	-	-	-1	-
TRANSFORMATION	-	-	-	-	14	-	-	-	-	-
Electricity and CHP Plants	-	-	-	-	-	-	-	-	-	-
Petroleum Refineries	-	-	-	-	-	-	-	-	-	-
Other Transform. Sector	-	-	-	-	14	-	-	-	-	-
ENERGY SECTOR	-	-	-	-	-	-	-	-	51	-
DISTRIBUTION LOSSES	-	-	-	-	-	-	-	-	1057	-
FINAL CONSUMPTION	-	-	-	-	166769	-	-	-	4574	-
INDUSTRY SECTOR	-	-	-	-	29818	-	-	-	1901	-
Iron and Steel	-	-	-	-	-	-	-	-	-	-
Chemical and Petrochem.	-	-	-	-	-	-	-	-	-	-
Non-Metallic Minerals	-	-	-	-	-	-	-	-	-	-
Non-specified	-	-	-	-	29818	-	-	-	1901	-
TRANSPORT SECTOR	-	-	-	-	-	-	-	-	-	-
Air	-	-	-	-	-	-	-	-	-	-
Road	-	-	-	-	-	-	-	-	-	-
Non-specified	-	-	-	-	-	-	-	-	-	-
OTHER SECTORS	-	-	-	-	136951	-	-	-	2673	-
Agriculture	-	-	-	-	-	-	-	-	-	-
Comm. and Publ. Services	-	-	-	-	5378	-	-	-	1011	-
Residential	-	-	-	-	131573	-	-	-	1662	-
Non-specified	-	-	-	-	-	-	-	-	-	-
NON-ENERGY USE	-	-	-	-	-	-	-	-	-	-

Sudan / Soudan

SUPPLY AND CONSUMPTION 1997	Coal (1000 tonnes)							Oil (1000 tonnes)			
	Coking Coal	Other Bit. Coal	Sub-Bit. Coal	Lignite	Peat	Oven and Gas Coke	Pat. Fuel and BKB	Crude Oil	NGL	Feed-stocks	Additives
Production	-	-	-	-	-	-	-	254	-	-	-
Imports	-	-	-	-	-	-	-	630	-	-	-
Exports	-	-	-	-	-	-	-	-	-	-	-
Intl. Marine Bunkers	-	-	-	-	-	-	-	-	-	-	-
Stock Changes	-	-	-	-	-	-	-	88	-	-	-
DOMESTIC SUPPLY	-	-	-	-	-	-	-	972	-	-	-
Transfers and Stat. Diff.	-	-	-	-	-	-	-	-	-	-	-
TRANSFORMATION	-	-	-	-	-	-	-	972	-	-	-
Electricity and CHP Plants	-	-	-	-	-	-	-	-	-	-	-
Petroleum Refineries	-	-	-	-	-	-	-	972	-	-	-
Other Transform. Sector	-	-	-	-	-	-	-	-	-	-	-
ENERGY SECTOR	-	-	-	-	-	-	-	-	-	-	-
DISTRIBUTION LOSSES	-	-	-	-	-	-	-	-	-	-	-
FINAL CONSUMPTION	-	-	-	-	-	-	-	-	-	-	-
INDUSTRY SECTOR	-	-	-	-	-	-	-	-	-	-	-
Iron and Steel	-	-	-	-	-	-	-	-	-	-	-
Chemical and Petrochem.	-	-	-	-	-	-	-	-	-	-	-
Non-Metallic Minerals	-	-	-	-	-	-	-	-	-	-	-
Non-specified	-	-	-	-	-	-	-	-	-	-	-
TRANSPORT SECTOR	-	-	-	-	-	-	-	-	-	-	-
Air	-	-	-	-	-	-	-	-	-	-	-
Road	-	-	-	-	-	-	-	-	-	-	-
Non-specified	-	-	-	-	-	-	-	-	-	-	-
OTHER SECTORS	-	-	-	-	-	-	-	-	-	-	-
Agriculture	-	-	-	-	-	-	-	-	-	-	-
Comm. and Publ. Services	-	-	-	-	-	-	-	-	-	-	-
Residential	-	-	-	-	-	-	-	-	-	-	-
Non-specified	-	-	-	-	-	-	-	-	-	-	-
NON-ENERGY USE	-	-	-	-	-	-	-	-	-	-	-

APPROVISIONNEMENT ET DEMANDE 1998	Charbon (1000 tonnes)							Pétrole (1000 tonnes)			
	Charbon à coke	Autres charb. bit.	Charbon sous-bit.	Lignite	Tourbe	Coke de four/gaz	Agg./briq. de lignite	Pétrole brut	LGN	Produits d'aliment.	Additifs
Production	-	-	-	-	-	-	-	328	-	-	-
Imports	-	-	-	-	-	-	-	258	-	-	-
Exports	-	-	-	-	-	-	-	-	-	-	-
Intl. Marine Bunkers	-	-	-	-	-	-	-	-	-	-	-
Stock Changes	-	-	-	-	-	-	-	-	-	-	-
DOMESTIC SUPPLY	-	-	-	-	-	-	-	586	-	-	-
Transfers and Stat. Diff.	-	-	-	-	-	-	-	1	-	-	-
TRANSFORMATION	-	-	-	-	-	-	-	587	-	-	-
Electricity and CHP Plants	-	-	-	-	-	-	-	-	-	-	-
Petroleum Refineries	-	-	-	-	-	-	-	587	-	-	-
Other Transform. Sector	-	-	-	-	-	-	-	-	-	-	-
ENERGY SECTOR	-	-	-	-	-	-	-	-	-	-	-
DISTRIBUTION LOSSES	-	-	-	-	-	-	-	-	-	-	-
FINAL CONSUMPTION	-	-	-	-	-	-	-	-	-	-	-
INDUSTRY SECTOR	-	-	-	-	-	-	-	-	-	-	-
Iron and Steel	-	-	-	-	-	-	-	-	-	-	-
Chemical and Petrochem.	-	-	-	-	-	-	-	-	-	-	-
Non-Metallic Minerals	-	-	-	-	-	-	-	-	-	-	-
Non-specified	-	-	-	-	-	-	-	-	-	-	-
TRANSPORT SECTOR	-	-	-	-	-	-	-	-	-	-	-
Air	-	-	-	-	-	-	-	-	-	-	-
Road	-	-	-	-	-	-	-	-	-	-	-
Non-specified	-	-	-	-	-	-	-	-	-	-	-
OTHER SECTORS	-	-	-	-	-	-	-	-	-	-	-
Agriculture	-	-	-	-	-	-	-	-	-	-	-
Comm. and Publ. Services	-	-	-	-	-	-	-	-	-	-	-
Residential	-	-	-	-	-	-	-	-	-	-	-
Non-specified	-	-	-	-	-	-	-	-	-	-	-
NON-ENERGY USE	-	-	-	-	-	-	-	-	-	-	-

Sudan / Soudan

SUPPLY AND CONSUMPTION 1997	Oil cont. (1000 tonnes)										
	Refinery Gas	LPG + Ethane	Motor Gasoline	Aviation Gasoline	Jet Fuel	Kerosene	Gas/ Diesel	Heavy Fuel Oil	Naphtha	Petrol. Coke	Other Prod.
Production	5	5	121	-	21	40	242	462	-	-	-
Imports	-	20	121	20	8	16	579	-	24	-	106
Exports	-	-	-	-	-	-	-	-	-	-	-
Intl. Marine Bunkers	-	-	-	-	-	-	-8	-	-	-	-
Stock Changes	-	-	-	-	3	6	-	-	-	-	-
DOMESTIC SUPPLY	5	25	242	20	32	62	813	462	24	-	106
Transfers and Stat. Diff.	-	-	-	-	-	-	-1	-	-	-	-2
TRANSFORMATION	-	-	-	-	-	-	65	290	-	-	-
Electricity and CHP Plants	-	-	-	-	-	-	65	290	-	-	-
Petroleum Refineries	-	-	-	-	-	-	-	-	-	-	-
Other Transform. Sector	-	-	-	-	-	-	-	-	-	-	-
ENERGY SECTOR	5	-	-	-	-	-	-	-	-	-	-
DISTRIBUTION LOSSES	-	-	-	-	-	-	-	-	-	-	-
FINAL CONSUMPTION	-	25	242	20	32	62	747	172	24	-	104
INDUSTRY SECTOR	-	-	3	-	-	-	-	172	24	-	-
Iron and Steel	-	-	-	-	-	-	-	-	-	-	-
Chemical and Petrochem.	-	-	-	-	-	-	-	-	24	-	-
Non-Metallic Minerals	-	-	-	-	-	-	-	-	-	-	-
Non-specified	-	-	3	-	-	-	-	172	-	-	-
TRANSPORT SECTOR	-	-	236	20	32	-	747	-	-	-	-
Air	-	-	-	20	32	-	-	-	-	-	-
Road	-	-	236	-	-	-	747	-	-	-	-
Non-specified	-	-	-	-	-	-	-	-	-	-	-
OTHER SECTORS	-	25	3	-	-	62	-	-	-	-	-
Agriculture	-	-	3	-	-	-	-	-	-	-	-
Comm. and Publ. Services	-	-	-	-	-	-	-	-	-	-	-
Residential	-	25	-	-	-	62	-	-	-	-	-
Non-specified	-	-	-	-	-	-	-	-	-	-	-
NON-ENERGY USE	-	-	-	-	-	-	-	-	-	-	104

APPROVISIONNEMENT ET DEMANDE 1998	Pétrole cont. (1000 tonnes)										
	Gaz de raffinerie	GPL + éthane	Essence moteur	Essence aviation	Carbu- réacteurs	Kérosène	Gazole	Fioul lourd	Naphta	Coke de pétrole	Autres prod.
Production	3	4	64	-	9	18	151	331	-	-	-
Imports	-	29	159	20	26	50	652	21	24	-	106
Exports	-	-	-	-	-	-	-	-	-	-	-
Intl. Marine Bunkers	-	-	-	-	-	-	-8	-	-	-	-
Stock Changes	-	-	-	-	-3	-6	-	-	-	-	-
DOMESTIC SUPPLY	3	33	223	20	32	62	795	352	24	-	106
Transfers and Stat. Diff.	-	-	-	-	-	-	-	-	-	-	-2
TRANSFORMATION	-	-	-	-	-	-	74	221	-	-	-
Electricity and CHP Plants	-	-	-	-	-	-	74	221	-	-	-
Petroleum Refineries	-	-	-	-	-	-	-	-	-	-	-
Other Transform. Sector	-	-	-	-	-	-	-	-	-	-	-
ENERGY SECTOR	3	-	-	-	-	-	-	-	-	-	-
DISTRIBUTION LOSSES	-	-	-	-	-	-	-	-	-	-	-
FINAL CONSUMPTION	-	33	223	20	32	62	721	131	24	-	104
INDUSTRY SECTOR	-	-	3	-	-	-	-	131	24	-	-
Iron and Steel	-	-	-	-	-	-	-	-	-	-	-
Chemical and Petrochem.	-	-	-	-	-	-	-	-	24	-	-
Non-Metallic Minerals	-	-	-	-	-	-	-	-	-	-	-
Non-specified	-	-	3	-	-	-	-	131	-	-	-
TRANSPORT SECTOR	-	-	217	20	32	-	721	-	-	-	-
Air	-	-	-	20	32	-	-	-	-	-	-
Road	-	-	217	-	-	-	721	-	-	-	-
Non-specified	-	-	-	-	-	-	-	-	-	-	-
OTHER SECTORS	-	33	3	-	-	62	-	-	-	-	-
Agriculture	-	-	3	-	-	-	-	-	-	-	-
Comm. and Publ. Services	-	-	-	-	-	-	-	-	-	-	-
Residential	-	33	-	-	-	62	-	-	-	-	-
Non-specified	-	-	-	-	-	-	-	-	-	-	-
NON-ENERGY USE	-	-	-	-	-	-	-	-	-	-	104

Sudan / Soudan

SUPPLY AND CONSUMPTION 1997	Gas (TJ) Natural Gas	Gas Works	Coke Ovens	Blast Furnaces	Comb. Renew. & Waste (TJ) Solid Biomass	Gas/Liquids from Biomass	Municipal Waste	Industrial Waste	(GWh) Electricity	(TJ) Heat
Production	-	-	-	-	527347	-	-	-	2150	-
Imports	-	-	-	-	-	-	-	-	-	-
Exports	-	-	-	-	-	-	-	-	-	-
Intl. Marine Bunkers	-	-	-	-	-	-	-	-	-	-
Stock Changes	-	-	-	-	-	-	-	-	-	-
DOMESTIC SUPPLY	-	-	-	-	527347	-	-	-	2150	-
Transfers and Stat. Diff.	-	-	-	-	-2	-	-	-	-	-
TRANSFORMATION	-	-	-	-	270212	-	-	-	-	-
Electricity and CHP Plants	-	-	-	-	-	-	-	-	-	-
Petroleum Refineries	-	-	-	-	-	-	-	-	-	-
Other Transform. Sector	-	-	-	-	270212	-	-	-	-	-
ENERGY SECTOR	-	-	-	-	-	-	-	-	16	-
DISTRIBUTION LOSSES	-	-	-	-	-	-	-	-	790	-
FINAL CONSUMPTION	-	-	-	-	257133	-	-	-	1344	-
INDUSTRY SECTOR	-	-	-	-	10995	-	-	-	433	-
Iron and Steel	-	-	-	-	-	-	-	-	-	-
Chemical and Petrochem.	-	-	-	-	-	-	-	-	-	-
Non-Metallic Minerals	-	-	-	-	-	-	-	-	-	-
Non-specified	-	-	-	-	10995	-	-	-	433	-
TRANSPORT SECTOR	-	-	-	-	-	-	-	-	-	-
Air	-	-	-	-	-	-	-	-	-	-
Road	-	-	-	-	-	-	-	-	-	-
Non-specified	-	-	-	-	-	-	-	-	-	-
OTHER SECTORS	-	-	-	-	246138	-	-	-	911	-
Agriculture	-	-	-	-	-	-	-	-	36	-
Comm. and Publ. Services	-	-	-	-	6391	-	-	-	211	-
Residential	-	-	-	-	239747	-	-	-	664	-
Non-specified	-	-	-	-	-	-	-	-	-	-
NON-ENERGY USE	-	-	-	-	-	-	-	-	-	-

APPROVISIONNEMENT ET DEMANDE 1998	Gaz (TJ) Gaz naturel	Usines à gaz	Cokeries	Hauts fourneaux	En. Re. Comb. & Déchets (TJ) Biomasse solide	Gaz/Liquides tirés de biomasse	Déchets urbains	Déchets industriels	(GWh) Electricité	(TJ) Chaleur
Production	-	-	-	-	538943	-	-	-	1966	-
Imports	-	-	-	-	-	-	-	-	-	-
Exports	-	-	-	-	-	-	-	-	-	-
Intl. Marine Bunkers	-	-	-	-	-	-	-	-	-	-
Stock Changes	-	-	-	-	-	-	-	-	-	-
DOMESTIC SUPPLY	-	-	-	-	538943	-	-	-	1966	-
Transfers and Stat. Diff.	-	-	-	-	-1	-	-	-	1	-
TRANSFORMATION	-	-	-	-	276154	-	-	-	-	-
Electricity and CHP Plants	-	-	-	-	-	-	-	-	-	-
Petroleum Refineries	-	-	-	-	-	-	-	-	-	-
Other Transform. Sector	-	-	-	-	276154	-	-	-	-	-
ENERGY SECTOR	-	-	-	-	-	-	-	-	18	-
DISTRIBUTION LOSSES	-	-	-	-	-	-	-	-	611	-
FINAL CONSUMPTION	-	-	-	-	262788	-	-	-	1338	-
INDUSTRY SECTOR	-	-	-	-	11237	-	-	-	438	-
Iron and Steel	-	-	-	-	-	-	-	-	-	-
Chemical and Petrochem.	-	-	-	-	-	-	-	-	-	-
Non-Metallic Minerals	-	-	-	-	-	-	-	-	-	-
Non-specified	-	-	-	-	11237	-	-	-	438	-
TRANSPORT SECTOR	-	-	-	-	-	-	-	-	-	-
Air	-	-	-	-	-	-	-	-	-	-
Road	-	-	-	-	-	-	-	-	-	-
Non-specified	-	-	-	-	-	-	-	-	-	-
OTHER SECTORS	-	-	-	-	251551	-	-	-	900	-
Agriculture	-	-	-	-	-	-	-	-	40	-
Comm. and Publ. Services	-	-	-	-	6532	-	-	-	280	-
Residential	-	-	-	-	245019	-	-	-	580	-
Non-specified	-	-	-	-	-	-	-	-	-	-
NON-ENERGY USE	-	-	-	-	-	-	-	-	-	-

Syria / Syrie

SUPPLY AND CONSUMPTION 1997	Coal (1000 tonnes)							Oil (1000 tonnes)			
	Coking Coal	Other Bit. Coal	Sub-Bit. Coal	Lignite	Peat	Oven and Gas Coke	Pat. Fuel and BKB	Crude Oil	NGL	Feed-stocks	Additives
Production	-	-	-	-	-	-	-	30186	-	-	-
Imports	-	-	-	-	-	2	-	-	-	-	-
Exports	-	-	-	-	-	-	-	-17467	-	-	-
Intl. Marine Bunkers	-	-	-	-	-	-	-	-	-	-	-
Stock Changes	-	-	-	-	-	-	-	-	-	-	-
DOMESTIC SUPPLY	-	-	-	-	-	2	-	12719	-	-	-
Transfers and Stat. Diff.	-	-	-	-	-	-	-	-	-	-	-
TRANSFORMATION	-	-	-	-	-	2	-	12719	-	-	-
Electricity and CHP Plants	-	-	-	-	-	-	-	-	-	-	-
Petroleum Refineries	-	-	-	-	-	-	-	12719	-	-	-
Other Transform. Sector	-	-	-	-	-	2	-	-	-	-	-
ENERGY SECTOR	-	-	-	-	-	-	-	-	-	-	-
DISTRIBUTION LOSSES	-	-	-	-	-	-	-	-	-	-	-
FINAL CONSUMPTION	-	-	-	-	-	-	-	-	-	-	-
INDUSTRY SECTOR	-	-	-	-	-	-	-	-	-	-	-
Iron and Steel	-	-	-	-	-	-	-	-	-	-	-
Chemical and Petrochem.	-	-	-	-	-	-	-	-	-	-	-
Non-Metallic Minerals	-	-	-	-	-	-	-	-	-	-	-
Non-specified	-	-	-	-	-	-	-	-	-	-	-
TRANSPORT SECTOR	-	-	-	-	-	-	-	-	-	-	-
Air	-	-	-	-	-	-	-	-	-	-	-
Road	-	-	-	-	-	-	-	-	-	-	-
Non-specified	-	-	-	-	-	-	-	-	-	-	-
OTHER SECTORS	-	-	-	-	-	-	-	-	-	-	-
Agriculture	-	-	-	-	-	-	-	-	-	-	-
Comm. and Publ. Services	-	-	-	-	-	-	-	-	-	-	-
Residential	-	-	-	-	-	-	-	-	-	-	-
Non-specified	-	-	-	-	-	-	-	-	-	-	-
NON-ENERGY USE	-	-	-	-	-	-	-	-	-	-	-

APPROVISIONNEMENT ET DEMANDE 1998	Charbon (1000 tonnes)							Pétrole (1000 tonnes)			
	Charbon à coke	Autres charb. bit.	Charbon sous-bit.	Lignite	Tourbe	Coke de four/gaz	Agg./briq. de lignite	Pétrole brut	LGN	Produits d'aliment.	Additifs
Production	-	-	-	-	-	-	-	30152	-	-	-
Imports	-	-	-	-	-	2	-	-	-	-	-
Exports	-	-	-	-	-	-	-	-17433	-	-	-
Intl. Marine Bunkers	-	-	-	-	-	-	-	-	-	-	-
Stock Changes	-	-	-	-	-	-	-	-	-	-	-
DOMESTIC SUPPLY	-	-	-	-	-	2	-	12719	-	-	-
Transfers and Stat. Diff.	-	-	-	-	-	-	-	-	-	-	-
TRANSFORMATION	-	-	-	-	-	2	-	12719	-	-	-
Electricity and CHP Plants	-	-	-	-	-	-	-	-	-	-	-
Petroleum Refineries	-	-	-	-	-	-	-	12719	-	-	-
Other Transform. Sector	-	-	-	-	-	2	-	-	-	-	-
ENERGY SECTOR	-	-	-	-	-	-	-	-	-	-	-
DISTRIBUTION LOSSES	-	-	-	-	-	-	-	-	-	-	-
FINAL CONSUMPTION	-	-	-	-	-	-	-	-	-	-	-
INDUSTRY SECTOR	-	-	-	-	-	-	-	-	-	-	-
Iron and Steel	-	-	-	-	-	-	-	-	-	-	-
Chemical and Petrochem.	-	-	-	-	-	-	-	-	-	-	-
Non-Metallic Minerals	-	-	-	-	-	-	-	-	-	-	-
Non-specified	-	-	-	-	-	-	-	-	-	-	-
TRANSPORT SECTOR	-	-	-	-	-	-	-	-	-	-	-
Air	-	-	-	-	-	-	-	-	-	-	-
Road	-	-	-	-	-	-	-	-	-	-	-
Non-specified	-	-	-	-	-	-	-	-	-	-	-
OTHER SECTORS	-	-	-	-	-	-	-	-	-	-	-
Agriculture	-	-	-	-	-	-	-	-	-	-	-
Comm. and Publ. Services	-	-	-	-	-	-	-	-	-	-	-
Residential	-	-	-	-	-	-	-	-	-	-	-
Non-specified	-	-	-	-	-	-	-	-	-	-	-
NON-ENERGY USE	-	-	-	-	-	-	-	-	-	-	-

Syria / Syrie

SUPPLY AND CONSUMPTION 1997	Refinery Gas	LPG + Ethane	Motor Gasoline	Aviation Gasoline	Jet Fuel	Kerosene	Gas/ Diesel	Heavy Fuel Oil	Naphtha	Petrol. Coke	Other Prod.
Production	-	286	1591	-	188	169	4088	5216	244	-	389
Imports	-	106	-	-	47	115	604	-	-	-	-
Exports	-	-	-449	-	-	-	-	-1650	-133	-	-
Intl. Marine Bunkers	-	-	-	-	-	-	-	-	-	-	-
Stock Changes	-	-	-	-	-	-	-	-	-	-	-
DOMESTIC SUPPLY	-	392	1142	-	235	284	4692	3566	111	-	389
Transfers and Stat. Diff.	-	-	-	-	-	-	359	871	-	-	-
TRANSFORMATION	-	-	-	-	-	-	390	1906	-	-	-
Electricity and CHP Plants	-	-	-	-	-	-	390	1906	-	-	-
Petroleum Refineries	-	-	-	-	-	-	-	-	-	-	-
Other Transform. Sector	-	-	-	-	-	-	-	-	-	-	-
ENERGY SECTOR	-	-	-	-	-	-	-	391	-	-	-
DISTRIBUTION LOSSES	-	-	-	-	-	-	-	-	-	-	-
FINAL CONSUMPTION	-	392	1142	-	235	284	4661	2140	111	-	389
INDUSTRY SECTOR	-	-	-	-	-	-	-	828	111	-	-
Iron and Steel	-	-	-	-	-	-	-	-	-	-	-
Chemical and Petrochem.	-	-	-	-	-	-	-	-	111	-	-
Non-Metallic Minerals	-	-	-	-	-	-	-	-	-	-	-
Non-specified	-	-	-	-	-	-	-	828	-	-	-
TRANSPORT SECTOR	-	-	1142	-	235	-	-	-	-	-	-
Air	-	-	-	-	235	-	-	-	-	-	-
Road	-	-	1142	-	-	-	-	-	-	-	-
Non-specified	-	-	-	-	-	-	-	-	-	-	-
OTHER SECTORS	-	392	-	-	-	284	4661	1312	-	-	-
Agriculture	-	-	-	-	-	-	-	-	-	-	-
Comm. and Publ. Services	-	-	-	-	-	-	-	-	-	-	-
Residential	-	392	-	-	-	284	-	-	-	-	-
Non-specified	-	-	-	-	-	-	4661	1312	-	-	-
NON-ENERGY USE	-	-	-	-	-	-	-	-	-	-	389

APPROVISIONNEMENT ET DEMANDE 1998	Gaz de raffinerie	GPL + éthane	Essence moteur	Essence aviation	Carbu- réacteurs	Kérosène	Gazole	Fioul lourd	Naphta	Coke de pétrole	Autres prod.
Production	-	286	1591	-	188	169	4088	5216	244	-	389
Imports	-	134	-	-	58	103	747	-	-	-	-
Exports	-	-	-415	-	-	-	-	-1120	-129	-	-
Intl. Marine Bunkers	-	-	-	-	-	-	-	-	-	-	-
Stock Changes	-	-	-	-	-	-	-	-	-	-	-
DOMESTIC SUPPLY	-	420	1176	-	246	272	4835	4096	115	-	389
Transfers and Stat. Diff.	-	-	-	-	-	-	368	1002	-	-	-
TRANSFORMATION	-	-	-	-	-	-	401	2190	-	-	-
Electricity and CHP Plants	-	-	-	-	-	-	401	2190	-	-	-
Petroleum Refineries	-	-	-	-	-	-	-	-	-	-	-
Other Transform. Sector	-	-	-	-	-	-	-	-	-	-	-
ENERGY SECTOR	-	-	-	-	-	-	-	450	-	-	-
DISTRIBUTION LOSSES	-	-	-	-	-	-	-	-	-	-	-
FINAL CONSUMPTION	-	420	1176	-	246	272	4802	2458	115	-	389
INDUSTRY SECTOR	-	-	-	-	-	-	-	951	115	-	-
Iron and Steel	-	-	-	-	-	-	-	-	-	-	-
Chemical and Petrochem.	-	-	-	-	-	-	-	-	115	-	-
Non-Metallic Minerals	-	-	-	-	-	-	-	-	-	-	-
Non-specified	-	-	-	-	-	-	-	951	-	-	-
TRANSPORT SECTOR	-	-	1176	-	246	-	-	-	-	-	-
Air	-	-	-	-	246	-	-	-	-	-	-
Road	-	-	1176	-	-	-	-	-	-	-	-
Non-specified	-	-	-	-	-	-	-	-	-	-	-
OTHER SECTORS	-	420	-	-	-	272	4802	1507	-	-	-
Agriculture	-	-	-	-	-	-	-	-	-	-	-
Comm. and Publ. Services	-	-	-	-	-	-	-	-	-	-	-
Residential	-	420	-	-	-	272	-	-	-	-	-
Non-specified	-	-	-	-	-	-	4802	1507	-	-	-
NON-ENERGY USE	-	-	-	-	-	-	-	-	-	-	389

Syria / Syrie

SUPPLY AND CONSUMPTION 1997	Gas (TJ) Natural Gas	Gas Works	Coke Ovens	Blast Furnaces	Comb. Renew. & Waste (TJ) Solid Biomass	Gas/Liquids from Biomass	Municipal Waste	Industrial Waste	(GWh) Electricity	(TJ) Heat
Production	161340	-	-	22	200	-	-	-	17956	-
Imports	-	-	-	-	-	-	-	-	-	-
Exports	-	-	-	-	-2	-	-	-	-	-
Intl. Marine Bunkers	-	-	-	-	-	-	-	-	-	-
Stock Changes	-	-	-	-	-	-	-	-	-	-
DOMESTIC SUPPLY	161340	-	-	22	198	-	-	-	17956	-
Transfers and Stat. Diff.	-	-	-	-	-	-	-	-	928	-
TRANSFORMATION	69390	-	-	-	20	-	-	-	-	-
Electricity and CHP Plants	69390	-	-	-	-	-	-	-	-	-
Petroleum Refineries	-	-	-	-	-	-	-	-	-	-
Other Transform. Sector	-	-	-	-	20	-	-	-	-	-
ENERGY SECTOR	6150	-	-	-	-	-	-	-	6431	-
DISTRIBUTION LOSSES	-	-	-	-	-	-	-	-	-	-
FINAL CONSUMPTION	85800	-	-	22	178	-	-	-	12453	-
INDUSTRY SECTOR	76267	-	-	22	-	-	-	-	5766	-
Iron and Steel	-	-	-	22	-	-	-	-	-	-
Chemical and Petrochem.	-	-	-	-	-	-	-	-	-	-
Non-Metallic Minerals	-	-	-	-	-	-	-	-	-	-
Non-specified	76267	-	-	-	-	-	-	-	5766	-
TRANSPORT SECTOR	-	-	-	-	-	-	-	-	-	-
Air	-	-	-	-	-	-	-	-	-	-
Road	-	-	-	-	-	-	-	-	-	-
Non-specified	-	-	-	-	-	-	-	-	-	-
OTHER SECTORS	9533	-	-	-	178	-	-	-	6687	-
Agriculture	-	-	-	-	-	-	-	-	-	-
Comm. and Publ. Services	-	-	-	-	-	-	-	-	-	-
Residential	-	-	-	-	-	-	-	-	6687	-
Non-specified	9533	-	-	-	178	-	-	-	-	-
NON-ENERGY USE	-	-	-	-	-	-	-	-	-	-

APPROVISIONNEMENT ET DEMANDE 1998	Gaz (TJ) Gaz naturel	Usines à gaz	Cokeries	Hauts fourneaux	En. Re. Comb. & Déchets (TJ) Biomasse solide	Gaz/Liquides tirés de biomasse	Déchets urbains	Déchets industriels	(GWh) Electricité	(TJ) Chaleur
Production	208751	-	-	22	200	-	-	-	18315	-
Imports	-	-	-	-	-	-	-	-	-	-
Exports	-	-	-	-	-2	-	-	-	-	-
Intl. Marine Bunkers	-	-	-	-	-	-	-	-	-	-
Stock Changes	-	-	-	-	-	-	-	-	-	-
DOMESTIC SUPPLY	208751	-	-	22	198	-	-	-	18315	-
Transfers and Stat. Diff.	-	-	-	-	-	-	-	-	1074	-
TRANSFORMATION	66287	-	-	-	20	-	-	-	-	-
Electricity and CHP Plants	66287	-	-	-	-	-	-	-	-	-
Petroleum Refineries	-	-	-	-	-	-	-	-	-	-
Other Transform. Sector	-	-	-	-	20	-	-	-	-	-
ENERGY SECTOR	7957	-	-	-	-	-	-	-	6560	-
DISTRIBUTION LOSSES	-	-	-	-	-	-	-	-	-	-
FINAL CONSUMPTION	134507	-	-	22	178	-	-	-	12829	-
INDUSTRY SECTOR	119562	-	-	22	-	-	-	-	5940	-
Iron and Steel	-	-	-	22	-	-	-	-	-	-
Chemical and Petrochem.	-	-	-	-	-	-	-	-	-	-
Non-Metallic Minerals	-	-	-	-	-	-	-	-	-	-
Non-specified	119562	-	-	-	-	-	-	-	5940	-
TRANSPORT SECTOR	-	-	-	-	-	-	-	-	-	-
Air	-	-	-	-	-	-	-	-	-	-
Road	-	-	-	-	-	-	-	-	-	-
Non-specified	-	-	-	-	-	-	-	-	-	-
OTHER SECTORS	14945	-	-	-	178	-	-	-	6889	-
Agriculture	-	-	-	-	-	-	-	-	-	-
Comm. and Publ. Services	-	-	-	-	-	-	-	-	-	-
Residential	-	-	-	-	-	-	-	-	6889	-
Non-specified	14945	-	-	-	178	-	-	-	-	-
NON-ENERGY USE	-	-	-	-	-	-	-	-	-	-

Tajikistan / Tadjikistan

SUPPLY AND CONSUMPTION 1997	Coal (1000 tonnes)							Oil (1000 tonnes)			
	Coking Coal	Other Bit. Coal	Sub-Bit. Coal	Lignite	Peat	Oven and Gas Coke	Pat. Fuel and BKB	Crude Oil	NGL	Feed-stocks	Additives
Production	-	-	20	-	-	4	-	25	1	-	-
Imports	7	100	-	-	-	-	-	-	-	-	-
Exports	-	-	-	-	-	-	-	-	-	-	-
Intl. Marine Bunkers	-	-	-	-	-	-	-	-	-	-	-
Stock Changes	-	-	-	-	-	-	-	-	-	-	-
DOMESTIC SUPPLY	**7**	**100**	**20**	**-**	**-**	**4**	**-**	**25**	**1**	**-**	**-**
Transfers and Stat. Diff.	-	-	-	-	-	-	-	-	-	-	-
TRANSFORMATION	**-**	**-**	**-**	**-**	**-**	**-**	**-**	**25**	**1**	**-**	**-**
Electricity and CHP Plants	-	-	-	-	-	-	-	-	-	-	-
Petroleum Refineries	-	-	-	-	-	-	-	25	1	-	-
Other Transform. Sector	-	-	-	-	-	-	-	-	-	-	-
ENERGY SECTOR	**-**	**-**	**-**	**-**	**-**	**-**	**-**	**-**	**-**	**-**	**-**
DISTRIBUTION LOSSES	**-**	**-**	**-**	**-**	**-**	**-**	**-**	**-**	**-**	**-**	**-**
FINAL CONSUMPTION	**7**	**100**	**20**	**-**	**-**	**4**	**-**	**-**	**-**	**-**	**-**
INDUSTRY SECTOR	**-**	**-**	**-**	**-**	**-**	**-**	**-**	**-**	**-**	**-**	**-**
Iron and Steel	-	-	-	-	-	-	-	-	-	-	-
Chemical and Petrochem.	-	-	-	-	-	-	-	-	-	-	-
Non-Metallic Minerals	-	-	-	-	-	-	-	-	-	-	-
Non-specified	-	-	-	-	-	-	-	-	-	-	-
TRANSPORT SECTOR	**-**	**-**	**-**	**-**	**-**	**-**	**-**	**-**	**-**	**-**	**-**
Air	-	-	-	-	-	-	-	-	-	-	-
Road	-	-	-	-	-	-	-	-	-	-	-
Non-specified	-	-	-	-	-	-	-	-	-	-	-
OTHER SECTORS	**7**	**100**	**20**	**-**	**-**	**4**	**-**	**-**	**-**	**-**	**-**
Agriculture	-	-	-	-	-	-	-	-	-	-	-
Comm. and Publ. Services	-	-	-	-	-	-	-	-	-	-	-
Residential	-	-	-	-	-	-	-	-	-	-	-
Non-specified	7	100	20	-	-	4	-	-	-	-	-
NON-ENERGY USE	**-**	**-**	**-**	**-**	**-**	**-**	**-**	**-**	**-**	**-**	**-**

APPROVISIONNEMENT ET DEMANDE 1998	Charbon (1000 tonnes)							Pétrole (1000 tonnes)			
	Charbon à coke	Autres charb. bit.	Charbon sous-bit.	Lignite	Tourbe	Coke de four/gaz	Agg./briq. de lignite	Pétrole brut	LGN	Produits d'aliment.	Additifs
Production	-	-	20	-	-	1	-	19	1	-	-
Imports	6	100	-	-	-	-	-	-	-	-	-
Exports	-	-	-	-	-	-	-	-5	-	-	-
Intl. Marine Bunkers	-	-	-	-	-	-	-	-	-	-	-
Stock Changes	-	-	-	-	-	-	-	-	-	-	-
DOMESTIC SUPPLY	**6**	**100**	**20**	**-**	**-**	**1**	**-**	**14**	**1**	**-**	**-**
Transfers and Stat. Diff.	-	-	-	-	-	-	-	-	-	-	-
TRANSFORMATION	**-**	**-**	**-**	**-**	**-**	**-**	**-**	**14**	**1**	**-**	**-**
Electricity and CHP Plants	-	-	-	-	-	-	-	-	-	-	-
Petroleum Refineries	-	-	-	-	-	-	-	14	1	-	-
Other Transform. Sector	-	-	-	-	-	-	-	-	-	-	-
ENERGY SECTOR	**-**	**-**	**-**	**-**	**-**	**-**	**-**	**-**	**-**	**-**	**-**
DISTRIBUTION LOSSES	**-**	**-**	**-**	**-**	**-**	**-**	**-**	**-**	**-**	**-**	**-**
FINAL CONSUMPTION	**6**	**100**	**20**	**-**	**-**	**1**	**-**	**-**	**-**	**-**	**-**
INDUSTRY SECTOR	**-**	**-**	**-**	**-**	**-**	**-**	**-**	**-**	**-**	**-**	**-**
Iron and Steel	-	-	-	-	-	-	-	-	-	-	-
Chemical and Petrochem.	-	-	-	-	-	-	-	-	-	-	-
Non-Metallic Minerals	-	-	-	-	-	-	-	-	-	-	-
Non-specified	-	-	-	-	-	-	-	-	-	-	-
TRANSPORT SECTOR	**-**	**-**	**-**	**-**	**-**	**-**	**-**	**-**	**-**	**-**	**-**
Air	-	-	-	-	-	-	-	-	-	-	-
Road	-	-	-	-	-	-	-	-	-	-	-
Non-specified	-	-	-	-	-	-	-	-	-	-	-
OTHER SECTORS	**6**	**100**	**20**	**-**	**-**	**1**	**-**	**-**	**-**	**-**	**-**
Agriculture	-	-	-	-	-	-	-	-	-	-	-
Comm. and Publ. Services	-	-	-	-	-	-	-	-	-	-	-
Residential	-	-	-	-	-	-	-	-	-	-	-
Non-specified	6	100	20	-	-	1	-	-	-	-	-
NON-ENERGY USE	**-**	**-**	**-**	**-**	**-**	**-**	**-**	**-**	**-**	**-**	**-**

Tajikistan / Tadjikistan

SUPPLY AND CONSUMPTION 1997	Oil cont. (1000 tonnes)										
	Refinery Gas	LPG + Ethane	Motor Gasoline	Aviation Gasoline	Jet Fuel	Kerosene	Gas/ Diesel	Heavy Fuel Oil	Naphtha	Petrol. Coke	Other Prod.
Production	-	-	-	-	-	-	-	-	23	-	-
Imports	-	7	996	-	5	-	75	33	-	-	43
Exports	-	-	-	-	-	-	-	-	-23	-	-
Intl. Marine Bunkers	-	-	-	-	-	-	-	-	-	-	-
Stock Changes	-	-	-	-	-	-	-	-	-	-	-
DOMESTIC SUPPLY	-	7	996	-	5	-	75	33	-	-	43
Transfers and Stat. Diff.	-	-	-	-	-	-	-	-	-	-	-
TRANSFORMATION	-	-	-	-	-	-	-	-	-	-	-
Electricity and CHP Plants	-	-	-	-	-	-	-	-	-	-	-
Petroleum Refineries	-	-	-	-	-	-	-	-	-	-	-
Other Transform. Sector	-	-	-	-	-	-	-	-	-	-	-
ENERGY SECTOR	-	-	-	-	-	-	-	-	-	-	-
DISTRIBUTION LOSSES	-	-	-	-	-	-	-	-	-	-	-
FINAL CONSUMPTION	-	7	996	-	5	-	75	33	-	-	43
INDUSTRY SECTOR	-	-	-	-	-	-	-	-	-	-	-
Iron and Steel	-	-	-	-	-	-	-	-	-	-	-
Chemical and Petrochem.	-	-	-	-	-	-	-	-	-	-	-
Non-Metallic Minerals	-	-	-	-	-	-	-	-	-	-	-
Non-specified	-	-	-	-	-	-	-	-	-	-	-
TRANSPORT SECTOR	-	-	996	-	5	-	-	-	-	-	-
Air	-	-	-	-	5	-	-	-	-	-	-
Road	-	-	996	-	-	-	-	-	-	-	-
Non-specified	-	-	-	-	-	-	-	-	-	-	-
OTHER SECTORS	-	7	-	-	-	-	75	33	-	-	43
Agriculture	-	-	-	-	-	-	-	-	-	-	-
Comm. and Publ. Services	-	-	-	-	-	-	-	-	-	-	-
Residential	-	-	-	-	-	-	-	-	-	-	-
Non-specified	-	7	-	-	-	-	75	33	-	-	43
NON-ENERGY USE	-	-	-	-	-	-	-	-	-	-	-

APPROVISIONNEMENT ET DEMANDE 1998	Pétrole cont. (1000 tonnes)										
	Gaz de raffinerie	GPL + éthane	Essence moteur	Essence aviation	Carbu- réacteurs	Kérosène	Gazole	Fioul lourd	Naphta	Coke de pétrole	Autres prod.
Production	-	-	-	-	-	-	-	-	12	-	-
Imports	-	7	996	-	5	-	75	33	-	117	44
Exports	-	-	-	-	-	-	-	-	-12	-	-
Intl. Marine Bunkers	-	-	-	-	-	-	-	-	-	-	-
Stock Changes	-	-	-	-	-	-	-	-	-	-	-
DOMESTIC SUPPLY	-	7	996	-	5	-	75	33	-	117	44
Transfers and Stat. Diff.	-	-	-	-	-	-	-	-	-	-	-
TRANSFORMATION	-	-	-	-	-	-	-	-	-	-	-
Electricity and CHP Plants	-	-	-	-	-	-	-	-	-	-	-
Petroleum Refineries	-	-	-	-	-	-	-	-	-	-	-
Other Transform. Sector	-	-	-	-	-	-	-	-	-	-	-
ENERGY SECTOR	-	-	-	-	-	-	-	-	-	-	-
DISTRIBUTION LOSSES	-	-	-	-	-	-	-	-	-	-	-
FINAL CONSUMPTION	-	7	996	-	5	-	75	33	-	117	44
INDUSTRY SECTOR	-	-	-	-	-	-	-	-	-	-	-
Iron and Steel	-	-	-	-	-	-	-	-	-	-	-
Chemical and Petrochem.	-	-	-	-	-	-	-	-	-	-	-
Non-Metallic Minerals	-	-	-	-	-	-	-	-	-	-	-
Non-specified	-	-	-	-	-	-	-	-	-	-	-
TRANSPORT SECTOR	-	-	996	-	5	-	-	-	-	-	-
Air	-	-	-	-	5	-	-	-	-	-	-
Road	-	-	996	-	-	-	-	-	-	-	-
Non-specified	-	-	-	-	-	-	-	-	-	-	-
OTHER SECTORS	-	7	-	-	-	-	75	33	-	117	43
Agriculture	-	-	-	-	-	-	-	-	-	-	-
Comm. and Publ. Services	-	-	-	-	-	-	-	-	-	-	-
Residential	-	-	-	-	-	-	-	-	-	-	-
Non-specified	-	7	-	-	-	-	75	33	-	117	43
NON-ENERGY USE	-	-	-	-	-	-	-	-	-	-	1

Tajikistan / Tadjikistan

SUPPLY AND CONSUMPTION 1997	Gas (TJ) Natural Gas	Gas Works	Coke Ovens	Blast Furnaces	Comb. Renew. & Waste (TJ) Solid Biomass	Gas/Liquids from Biomass	Municipal Waste	Industrial Waste	(GWh) Electricity	(TJ) Heat
Production	1444	-	-	-	-	-	-	-	14005	3777
Imports	27854	-	-	-	-	-	-	-	4345	-
Exports	-	-	-	-	-	-	-	-	-4247	-
Intl. Marine Bunkers	-	-	-	-	-	-	-	-	-	-
Stock Changes	-	-	-	-	-	-	-	-	-	-
DOMESTIC SUPPLY	29298	-	-	-	-	-	-	-	14103	3777
Transfers and Stat. Diff.	-	-	-	-	-	-	-	-	-	-
TRANSFORMATION	13795	-	-	-	-	-	-	-	-	-
Electricity and CHP Plants	13795	-	-	-	-	-	-	-	-	-
Petroleum Refineries	-	-	-	-	-	-	-	-	-	-
Other Transform. Sector	-	-	-	-	-	-	-	-	-	-
ENERGY SECTOR	-	-	-	-	-	-	-	-	159	-
DISTRIBUTION LOSSES	-	-	-	-	-	-	-	-	1951	-
FINAL CONSUMPTION	15503	-	-	-	-	-	-	-	11993	3777
INDUSTRY SECTOR	-	-	-	-	-	-	-	-	4962	-
Iron and Steel	-	-	-	-	-	-	-	-	-	-
Chemical and Petrochem.	-	-	-	-	-	-	-	-	142	-
Non-Metallic Minerals	-	-	-	-	-	-	-	-	-	-
Non-specified	-	-	-	-	-	-	-	-	4820	-
TRANSPORT SECTOR	-	-	-	-	-	-	-	-	71	-
Air	-	-	-	-	-	-	-	-	-	-
Road	-	-	-	-	-	-	-	-	-	-
Non-specified	-	-	-	-	-	-	-	-	71	-
OTHER SECTORS	15503	-	-	-	-	-	-	-	6960	3777
Agriculture	-	-	-	-	-	-	-	-	4354	-
Comm. and Publ. Services	-	-	-	-	-	-	-	-	232	-
Residential	-	-	-	-	-	-	-	-	2374	-
Non-specified	15503	-	-	-	-	-	-	-	-	3777
NON-ENERGY USE	-	-	-	-	-	-	-	-	-	-

APPROVISIONNEMENT ET DEMANDE 1998	Gaz (TJ) Gaz naturel	Usines à gaz	Cokeries	Hauts fourneaux	En. Re. Comb. & Déchets (TJ) Biomasse solide	Gaz/Liquides tirés de biomasse	Déchets urbains	Déchets industriels	(GWh) Electricité	(TJ) Chaleur
Production	1140	-	-	-	-	-	-	-	14422	3952
Imports	28804	-	-	-	-	-	-	-	3969	-
Exports	-	-	-	-	-	-	-	-	-3724	-
Intl. Marine Bunkers	-	-	-	-	-	-	-	-	-	-
Stock Changes	-	-	-	-	-	-	-	-	-	-
DOMESTIC SUPPLY	29944	-	-	-	-	-	-	-	14667	3952
Transfers and Stat. Diff.	-	-	-	-	-	-	-	-	-	-
TRANSFORMATION	14099	-	-	-	-	-	-	-	-	-
Electricity and CHP Plants	14099	-	-	-	-	-	-	-	-	-
Petroleum Refineries	-	-	-	-	-	-	-	-	-	-
Other Transform. Sector	-	-	-	-	-	-	-	-	-	-
ENERGY SECTOR	-	-	-	-	-	-	-	-	168	-
DISTRIBUTION LOSSES	-	-	-	-	-	-	-	-	1991	-
FINAL CONSUMPTION	15845	-	-	-	-	-	-	-	12508	3952
INDUSTRY SECTOR	-	-	-	-	-	-	-	-	5087	-
Iron and Steel	-	-	-	-	-	-	-	-	-	-
Chemical and Petrochem.	-	-	-	-	-	-	-	-	201	-
Non-Metallic Minerals	-	-	-	-	-	-	-	-	-	-
Non-specified	-	-	-	-	-	-	-	-	4886	-
TRANSPORT SECTOR	-	-	-	-	-	-	-	-	67	-
Air	-	-	-	-	-	-	-	-	-	-
Road	-	-	-	-	-	-	-	-	-	-
Non-specified	-	-	-	-	-	-	-	-	67	-
OTHER SECTORS	15845	-	-	-	-	-	-	-	7354	3952
Agriculture	-	-	-	-	-	-	-	-	4471	-
Comm. and Publ. Services	-	-	-	-	-	-	-	-	247	-
Residential	-	-	-	-	-	-	-	-	2636	-
Non-specified	15845	-	-	-	-	-	-	-	-	3952
NON-ENERGY USE	-	-	-	-	-	-	-	-	-	-

United Republic of Tanzania / République-Unie de Tanzanie

SUPPLY AND CONSUMPTION 1997	Coal (1000 tonnes)							Oil (1000 tonnes)			
	Coking Coal	Other Bit. Coal	Sub-Bit. Coal	Lignite	Peat	Oven and Gas Coke	Pat. Fuel and BKB	Crude Oil	NGL	Feed-stocks	Additives
Production	1	4	-	-	-	-	-	-	-	-	-
Imports	-	-	-	-	-	-	-	589	-	-	-
Exports	-	-	-	-	-	-	-	-	-	-	-
Intl. Marine Bunkers	-	-	-	-	-	-	-	-	-	-	-
Stock Changes	-	-	-	-	-	-	-	-	-	-	-
DOMESTIC SUPPLY	1	4	-	-	-	-	-	589	-	-	-
Transfers and Stat. Diff.	-	-	-	-	-	-	-	22	-	-	-
TRANSFORMATION	1	-	-	-	-	-	-	611	-	-	-
Electricity and CHP Plants	-	-	-	-	-	-	-	-	-	-	-
Petroleum Refineries	-	-	-	-	-	-	-	611	-	-	-
Other Transform. Sector	1	-	-	-	-	-	-	-	-	-	-
ENERGY SECTOR	-	-	-	-	-	-	-	-	-	-	-
DISTRIBUTION LOSSES	-	-	-	-	-	-	-	-	-	-	-
FINAL CONSUMPTION	-	4	-	-	-	-	-	-	-	-	-
INDUSTRY SECTOR	-	4	-	-	-	-	-	-	-	-	-
Iron and Steel	-	-	-	-	-	-	-	-	-	-	-
Chemical and Petrochem.	-	-	-	-	-	-	-	-	-	-	-
Non-Metallic Minerals	-	1	-	-	-	-	-	-	-	-	-
Non-specified	-	3	-	-	-	-	-	-	-	-	-
TRANSPORT SECTOR	-	-	-	-	-	-	-	-	-	-	-
Air	-	-	-	-	-	-	-	-	-	-	-
Road	-	-	-	-	-	-	-	-	-	-	-
Non-specified	-	-	-	-	-	-	-	-	-	-	-
OTHER SECTORS	-	-	-	-	-	-	-	-	-	-	-
Agriculture	-	-	-	-	-	-	-	-	-	-	-
Comm. and Publ. Services	-	-	-	-	-	-	-	-	-	-	-
Residential	-	-	-	-	-	-	-	-	-	-	-
Non-specified	-	-	-	-	-	-	-	-	-	-	-
NON-ENERGY USE	-	-	-	-	-	-	-	-	-	-	-

APPROVISIONNEMENT ET DEMANDE 1998	Charbon (1000 tonnes)							Pétrole (1000 tonnes)			
	Charbon à coke	Autres charb. bit.	Charbon sous-bit.	Lignite	Tourbe	Coke de four/gaz	Agg./briq. de lignite	Pétrole brut	LGN	Produits d'aliment.	Additifs
Production	1	4	-	-	-	-	-	-	-	-	-
Imports	-	-	-	-	-	-	-	589	-	-	-
Exports	-	-	-	-	-	-	-	-	-	-	-
Intl. Marine Bunkers	-	-	-	-	-	-	-	-	-	-	-
Stock Changes	-	-	-	-	-	-	-	-	-	-	-
DOMESTIC SUPPLY	1	4	-	-	-	-	-	589	-	-	-
Transfers and Stat. Diff.	-	-	-	-	-	-	-	22	-	-	-
TRANSFORMATION	1	-	-	-	-	-	-	611	-	-	-
Electricity and CHP Plants	-	-	-	-	-	-	-	-	-	-	-
Petroleum Refineries	-	-	-	-	-	-	-	611	-	-	-
Other Transform. Sector	1	-	-	-	-	-	-	-	-	-	-
ENERGY SECTOR	-	-	-	-	-	-	-	-	-	-	-
DISTRIBUTION LOSSES	-	-	-	-	-	-	-	-	-	-	-
FINAL CONSUMPTION	-	4	-	-	-	-	-	-	-	-	-
INDUSTRY SECTOR	-	4	-	-	-	-	-	-	-	-	-
Iron and Steel	-	-	-	-	-	-	-	-	-	-	-
Chemical and Petrochem.	-	-	-	-	-	-	-	-	-	-	-
Non-Metallic Minerals	-	1	-	-	-	-	-	-	-	-	-
Non-specified	-	3	-	-	-	-	-	-	-	-	-
TRANSPORT SECTOR	-	-	-	-	-	-	-	-	-	-	-
Air	-	-	-	-	-	-	-	-	-	-	-
Road	-	-	-	-	-	-	-	-	-	-	-
Non-specified	-	-	-	-	-	-	-	-	-	-	-
OTHER SECTORS	-	-	-	-	-	-	-	-	-	-	-
Agriculture	-	-	-	-	-	-	-	-	-	-	-
Comm. and Publ. Services	-	-	-	-	-	-	-	-	-	-	-
Residential	-	-	-	-	-	-	-	-	-	-	-
Non-specified	-	-	-	-	-	-	-	-	-	-	-
NON-ENERGY USE	-	-	-	-	-	-	-	-	-	-	-

United Republic of Tanzania / République-Unie de Tanzanie

SUPPLY AND CONSUMPTION 1997	Oil cont. (1000 tonnes)										
	Refinery Gas	LPG + Ethane	Motor Gasoline	Aviation Gasoline	Jet Fuel	Kerosene	Gas/ Diesel	Heavy Fuel Oil	Naphtha	Petrol. Coke	Other Prod.
Production	14	6	105	-	33	45	148	246	-	-	2
Imports	-	-	11	8	9	27	102	-	-	-	14
Exports	-	-	-4	-	-	-	-12	-13	-	-	-
Intl. Marine Bunkers	-	-	-	-	-	-	-1	-22	-	-	-
Stock Changes	-	-	-	-	-	-	-	-	-	-	-
DOMESTIC SUPPLY	14	6	112	8	42	72	237	211	-	-	16
Transfers and Stat. Diff.	-	-	-14	-	-14	-	44	-106	-	-	-
TRANSFORMATION	-	-	-	-	-	-	160	-	-	-	-
Electricity and CHP Plants	-	-	-	-	-	-	160	-	-	-	-
Petroleum Refineries	-	-	-	-	-	-	-	-	-	-	-
Other Transform. Sector	-	-	-	-	-	-	-	-	-	-	-
ENERGY SECTOR	14	-	-	-	-	-	-	-	-	-	-
DISTRIBUTION LOSSES	-	-	-	-	-	-	-	-	-	-	-
FINAL CONSUMPTION	-	6	98	8	28	72	121	105	-	-	16
INDUSTRY SECTOR	-	-	-	-	-	-	-	105	-	-	-
Iron and Steel	-	-	-	-	-	-	-	-	-	-	-
Chemical and Petrochem.	-	-	-	-	-	-	-	-	-	-	-
Non-Metallic Minerals	-	-	-	-	-	-	-	-	-	-	-
Non-specified	-	-	-	-	-	-	-	105	-	-	-
TRANSPORT SECTOR	-	-	98	8	28	-	121	-	-	-	-
Air	-	-	-	8	28	-	-	-	-	-	-
Road	-	-	98	-	-	-	121	-	-	-	-
Non-specified	-	-	-	-	-	-	-	-	-	-	-
OTHER SECTORS	-	6	-	-	-	72	-	-	-	-	-
Agriculture	-	-	-	-	-	-	-	-	-	-	-
Comm. and Publ. Services	-	-	-	-	-	-	-	-	-	-	-
Residential	-	6	-	-	-	72	-	-	-	-	-
Non-specified	-	-	-	-	-	-	-	-	-	-	-
NON-ENERGY USE	-	-	-	-	-	-	-	-	-	-	16

APPROVISIONNEMENT ET DEMANDE 1998	Pétrole cont. (1000 tonnes)										
	Gaz de raffinerie	GPL + éthane	Essence moteur	Essence aviation	Carbu- réacteurs	Kérosène	Gazole	Fioul lourd	Naphta	Coke de pétrole	Autres prod.
Production	14	6	105	-	33	45	148	246	-	-	2
Imports	-	-	11	8	9	27	102	-	-	-	14
Exports	-	-	-4	-	-	-	-12	-13	-	-	-
Intl. Marine Bunkers	-	-	-	-	-	-	-1	-22	-	-	-
Stock Changes	-	-	-	-	-	-	-	-	-	-	-
DOMESTIC SUPPLY	14	6	112	8	42	72	237	211	-	-	16
Transfers and Stat. Diff.	-	-	-14	-	-14	-	-82	-106	-	-	-
TRANSFORMATION	-	-	-	-	-	-	25	-	-	-	-
Electricity and CHP Plants	-	-	-	-	-	-	25	-	-	-	-
Petroleum Refineries	-	-	-	-	-	-	-	-	-	-	-
Other Transform. Sector	-	-	-	-	-	-	-	-	-	-	-
ENERGY SECTOR	14	-	-	-	-	-	-	-	-	-	-
DISTRIBUTION LOSSES	-	-	-	-	-	-	-	-	-	-	-
FINAL CONSUMPTION	-	6	98	8	28	72	130	105	-	-	16
INDUSTRY SECTOR	-	-	-	-	-	-	-	105	-	-	-
Iron and Steel	-	-	-	-	-	-	-	-	-	-	-
Chemical and Petrochem.	-	-	-	-	-	-	-	-	-	-	-
Non-Metallic Minerals	-	-	-	-	-	-	-	-	-	-	-
Non-specified	-	-	-	-	-	-	-	105	-	-	-
TRANSPORT SECTOR	-	-	98	8	28	-	130	-	-	-	-
Air	-	-	-	8	28	-	-	-	-	-	-
Road	-	-	98	-	-	-	130	-	-	-	-
Non-specified	-	-	-	-	-	-	-	-	-	-	-
OTHER SECTORS	-	6	-	-	-	72	-	-	-	-	-
Agriculture	-	-	-	-	-	-	-	-	-	-	-
Comm. and Publ. Services	-	-	-	-	-	-	-	-	-	-	-
Residential	-	6	-	-	-	72	-	-	-	-	-
Non-specified	-	-	-	-	-	-	-	-	-	-	-
NON-ENERGY USE	-	-	-	-	-	-	-	-	-	-	16

United Republic of Tanzania / République-Unie de Tanzanie

SUPPLY AND CONSUMPTION 1997	Gas (TJ)				Comb. Renew. & Waste (TJ)				(GWh)	(TJ)
	Natural Gas	Gas Works	Coke Ovens	Blast Furnaces	Solid Biomass	Gas/Liquids from Biomass	Municipal Waste	Industrial Waste	Electricity	Heat
Production	-	-	-	-	561202	-	-	-	1934	-
Imports	-	-	-	-	-	-	-	-	44	-
Exports	-	-	-	-	-	-	-	-	-	-
Intl. Marine Bunkers	-	-	-	-	-	-	-	-	-	-
Stock Changes	-	-	-	-	-	-	-	-	-	-
DOMESTIC SUPPLY	-	-	-	-	561202	-	-	-	1978	-
Transfers and Stat. Diff.	-	-	-	-	-	-	-	-	1	-
TRANSFORMATION	-	-	-	-	45644	-	-	-	-	-
Electricity and CHP Plants	-	-	-	-	-	-	-	-	-	-
Petroleum Refineries	-	-	-	-	-	-	-	-	-	-
Other Transform. Sector	-	-	-	-	45644	-	-	-	-	-
ENERGY SECTOR	-	-	-	-	-	-	-	-	-	-
DISTRIBUTION LOSSES	-	-	-	-	-	-	-	-	278	-
FINAL CONSUMPTION	-	-	-	-	515559	-	-	-	1701	-
INDUSTRY SECTOR	-	-	-	-	57054	-	-	-	460	-
Iron and Steel	-	-	-	-	-	-	-	-	-	-
Chemical and Petrochem.	-	-	-	-	-	-	-	-	-	-
Non-Metallic Minerals	-	-	-	-	-	-	-	-	-	-
Non-specified	-	-	-	-	57054	-	-	-	460	-
TRANSPORT SECTOR	-	-	-	-	-	-	-	-	-	-
Air	-	-	-	-	-	-	-	-	-	-
Road	-	-	-	-	-	-	-	-	-	-
Non-specified	-	-	-	-	-	-	-	-	-	-
OTHER SECTORS	-	-	-	-	458505	-	-	-	1241	-
Agriculture	-	-	-	-	17116	-	-	-	77	-
Comm. and Publ. Services	-	-	-	-	-	-	-	-	444	-
Residential	-	-	-	-	420642	-	-	-	651	-
Non-specified	-	-	-	-	20747	-	-	-	69	-
NON-ENERGY USE	-	-	-	-	-	-	-	-	-	-

APPROVISIONNEMENT ET DEMANDE 1998	Gaz (TJ)				En. Re. Comb. & Déchets (TJ)				(GWh)	(TJ)
	Gaz naturel	Usines à gaz	Cokeries	Hauts fourneaux	Biomasse solide	Gaz/Liquides tirés de biomasse	Déchets urbains	Déchets industriels	Electricité	Chaleur
Production	-	-	-	-	575754	-	-	-	2157	-
Imports	-	-	-	-	-	-	-	-	43	-
Exports	-	-	-	-	-	-	-	-	-	-
Intl. Marine Bunkers	-	-	-	-	-	-	-	-	-	-
Stock Changes	-	-	-	-	-	-	-	-	-	-
DOMESTIC SUPPLY	-	-	-	-	575754	-	-	-	2200	-
Transfers and Stat. Diff.	-	-	-	-	2	-	-	-	-1	-
TRANSFORMATION	-	-	-	-	46828	-	-	-	-	-
Electricity and CHP Plants	-	-	-	-	-	-	-	-	-	-
Petroleum Refineries	-	-	-	-	-	-	-	-	-	-
Other Transform. Sector	-	-	-	-	46828	-	-	-	-	-
ENERGY SECTOR	-	-	-	-	-	-	-	-	-	-
DISTRIBUTION LOSSES	-	-	-	-	-	-	-	-	481	-
FINAL CONSUMPTION	-	-	-	-	528928	-	-	-	1718	-
INDUSTRY SECTOR	-	-	-	-	58533	-	-	-	465	-
Iron and Steel	-	-	-	-	-	-	-	-	-	-
Chemical and Petrochem.	-	-	-	-	-	-	-	-	-	-
Non-Metallic Minerals	-	-	-	-	-	-	-	-	-	-
Non-specified	-	-	-	-	58533	-	-	-	465	-
TRANSPORT SECTOR	-	-	-	-	-	-	-	-	-	-
Air	-	-	-	-	-	-	-	-	-	-
Road	-	-	-	-	-	-	-	-	-	-
Non-specified	-	-	-	-	-	-	-	-	-	-
OTHER SECTORS	-	-	-	-	470395	-	-	-	1253	-
Agriculture	-	-	-	-	17561	-	-	-	78	-
Comm. and Publ. Services	-	-	-	-	-	-	-	-	448	-
Residential	-	-	-	-	431549	-	-	-	658	-
Non-specified	-	-	-	-	21285	-	-	-	69	-
NON-ENERGY USE	-	-	-	-	-	-	-	-	-	-

Thailand / Thaïlande : 1997

	Coal / *Charbon* (1000 tonnes)							Oil / *Pétrole* (1000 tonnes)			
SUPPLY AND CONSUMPTION	Coking Coal	Other Bit. Coal	Sub-Bit. Coal	Lignite	Peat	Oven and Gas Coke	Pat. Fuel and BKB	Crude Oil	NGL	Feed-stocks	Additives
APPROVISIONNEMENT ET DEMANDE	*Charbon à coke*	*Autres charb. bit.*	*Charbon sous-bit.*	*Lignite*	*Tourbe*	*Coke de four/gaz*	*Agg./briq. de lignite*	*Pétrole brut*	*LGN*	*Produits d'aliment.*	*Additifs*
Production	-	-	-	23393	-	-	-	3306	2094	-	-
From Other Sources	-	-	-	-	-	-	-	-	-	-	-
Imports	-	3074	-	-	-	83	126	34947	-	-	-
Exports	-	-	-	-	-	-	-	-1043	-85	-	-
Intl. Marine Bunkers	-	·	-	-	-	-	-	-	-	-	-
Stock Changes	-	4	-	-779	-	-	-	-142	-	-	-
DOMESTIC SUPPLY	-	**3078**	-	**22614**	-	**83**	**126**	**37068**	**2009**	-	-
Transfers	-	-	-	-	-	-	-	-	-1895	-	-
Statistical Differences	-	-	-	-	-	-	-	-	-	-	-
TRANSFORMATION	-	**42**	-	**18144**	-	**66**	**13**	**36723**	**13**	-	-
Electricity Plants	-	42	-	18144	-	-	13	-	-	-	-
CHP Plants	-	-	-	-	-	-	-	-	-	-	-
Heat Plants	-	-	-	-	-	-	-	-	-	-	-
Blast Furnaces/Gas Works	-	-	-	-	-	66	-	-	-	-	-
Coke/Pat. Fuel/BKB Plants	-	-	-	-	-	-	-	-	-	-	-
Petroleum Refineries	-	-	-	-	-	-	-	36723	13	-	-
Petrochemical Industry	-	-	-	-	-	-	-	-	-	-	-
Liquefaction	-	-	-	-	-	-	-	-	-	-	-
Other Transform. Sector	-	-	-	-	-	-	-	-	-	-	-
ENERGY SECTOR	-	-	-	-	-	-	-	-	-	-	-
Coal Mines	-	-	-	-	-	-	-	-	-	-	-
Oil and Gas Extraction	-	-	-	-	-	-	-	-	-	-	-
Petroleum Refineries	-	-	-	-	-	-	-	-	-	-	-
Electr., CHP+Heat Plants	-	-	-	-	-	-	-	-	-	-	-
Pumped Storage (Elec.)	-	-	-	-	-	-	-	-	-	-	-
Other Energy Sector	-	-	-	-	-	-	-	-	-	-	-
Distribution Losses	-	-	-	-	-	-	-	67	14	-	-
FINAL CONSUMPTION	-	**3036**	-	**4470**	-	**17**	**113**	**278**	**87**	-	-
INDUSTRY SECTOR	-	**3036**	-	**4468**	-	**17**	**113**	**278**	**87**	-	-
Iron and Steel	-	-	-	-	-	17	-	-	-	-	-
Chemical and Petrochem.	-	-	-	-	-	-	-	278	87	-	-
of which: Feedstocks	-	-	-	-	-	-	-	*278*	*87*	-	-
Non-Ferrous Metals	-	-	-	-	-	-	-	-	-	-	-
Non-Metallic Minerals	-	2568	-	2998	-	-	-	-	-	-	-
Transport Equipment	-	-	-	-	-	-	-	-	-	-	-
Machinery	-	-	-	-	-	-	-	-	-	-	-
Mining and Quarrying	-	-	-	-	-	-	-	-	-	-	-
Food and Tobacco	-	161	-	205	-	-	-	-	-	-	-
Paper, Pulp and Print	-	-	-	-	-	-	-	-	-	-	-
Wood and Wood Products	-	-	-	-	-	-	-	-	-	-	-
Construction	-	-	-	-	-	-	-	-	-	-	-
Textile and Leather	-	-	-	-	-	-	-	-	-	-	-
Non-specified	-	307	-	1265	-	-	113	-	-	-	-
TRANSPORT SECTOR	-	-	-	-	-	-	-	-	-	-	-
Air	-	-	-	-	-	-	-	-	-	-	-
Road	-	-	-	-	-	-	-	-	-	-	-
Rail	-	-	-	-	-	-	-	-	-	-	-
Pipeline Transport	-	-	-	-	-	-	-	-	-	-	-
Internal Navigation	-	-	-	-	-	-	-	-	-	-	-
Non-specified	-	-	-	-	-	-	-	-	-	-	-
OTHER SECTORS	-	-	-	-	-	-	-	-	-	-	-
Agriculture	-	-	-	-	-	-	-	-	-	-	-
Comm. and Publ. Services	-	-	-	-	-	-	-	-	-	-	-
Residential	-	-	-	-	-	-	-	-	-	-	-
Non-specified	-	-	-	-	-	-	-	-	-	-	-
NON-ENERGY USE	-	-	-	**2**	-	-	-	-	-	-	-
in Industry/Trans./Energy	-	-	-	2	-	-	-	-	-	-	-
in Transport	-	-	-	-	-	-	-	-	-	-	-
in Other Sectors	-	-	-	-	-	-	-	-	-	-	-

Thailand / Thaïlande : 1997

SUPPLY AND CONSUMPTION / APPROVISIONNEMENT ET DEMANDE	Oil cont. / Pétrole cont. (1000 tonnes)										
	Refinery Gas / Gaz de raffinerie	LPG + Ethane / GPL + éthane	Motor Gasoline / Essence moteur	Aviation Gasoline / Essence aviation	Jet Fuel / Carbu- réacteurs	Kerosene / Kérosène	Gas/ Diesel / Gazole	Heavy Fuel Oil / Fioul lourd	Naphtha / Naphta	Petrol. Coke / Coke de pétrole	Other Prod. / Autres prod.
Production	367	1002	6866	-	2851	103	14794	8217	-	-	616
From Other Sources	-	-	-	-	-	-	-	-	-	-	-
Imports	-	2	32	10	17	29	1725	889	-	-	112
Exports	-	-471	-1414	-	-208	-11	-1680	-672	-	-	-
Intl. Marine Bunkers	-	-	-	-	-	-	-102	-707	-	-	-
Stock Changes	-	-61	-14	-	-4	-53	255	127	-	-	-9
DOMESTIC SUPPLY	367	472	5470	10	2656	68	14992	7854	-	-	719
Transfers	-	1895	-	-	-	-	-	-	-	-	-
Statistical Differences	-	-1	-86	-	1	2	1	-1	-	-	-
TRANSFORMATION	-	-	-	-	-	-	632	4413	-	-	-
Electricity Plants	-	-	-	-	-	-	632	4413	-	-	-
CHP Plants	-	-	-	-	-	-	-	-	-	-	-
Heat Plants	-	-	-	-	-	-	-	-	-	-	-
Blast Furnaces/Gas Works	-	-	-	-	-	-	-	-	-	-	-
Coke/Pat. Fuel/BKB Plants	-	-	-	-	-	-	-	-	-	-	-
Petroleum Refineries	-	-	-	-	-	-	-	-	-	-	-
Petrochemical Industry	-	-	-	-	-	-	-	-	-	-	-
Liquefaction	-	-	-	-	-	-	-	-	-	-	-
Other Transform. Sector	-	-	-	-	-	-	-	-	-	-	-
ENERGY SECTOR	367	-	-	-	-	-	-	-	-	-	-
Coal Mines	-	-	-	-	-	-	-	-	-	-	-
Oil and Gas Extraction	-	-	-	-	-	-	-	-	-	-	-
Petroleum Refineries	367	-	-	-	-	-	-	-	-	-	-
Electr., CHP+Heat Plants	-	-	-	-	-	-	-	-	-	-	-
Pumped Storage (Elec.)	-	-	-	-	-	-	-	-	-	-	-
Other Energy Sector	-	-	-	-	-	-	-	-	-	-	-
Distribution Losses	-	85	-	-	-	-	-	-	-	-	-
FINAL CONSUMPTION	-	2281	5384	10	2657	70	14361	3440	-	-	719
INDUSTRY SECTOR	-	943	30	-	-	42	815	3411	-	-	-
Iron and Steel	-	68	-	-	-	4	36	228	-	-	-
Chemical and Petrochem.	-	680	3	-	-	24	128	349	-	-	-
of which: Feedstocks	-	527	-	-	-	-	-	-	-	-	-
Non-Ferrous Metals	-	-	-	-	-	-	-	-	-	-	-
Non-Metallic Minerals	-	82	6	-	-	3	60	571	-	-	-
Transport Equipment	-	-	-	-	-	-	-	-	-	-	-
Machinery	-	52	5	-	-	4	28	46	-	-	-
Mining and Quarrying	-	-	-	-	-	-	27	18	-	-	-
Food and Tobacco	-	17	5	-	-	-	163	553	-	-	-
Paper, Pulp and Print	-	2	1	-	-	1	16	185	-	-	-
Wood and Wood Products	-	-	1	-	-	-	10	18	-	-	-
Construction	-	-	-	-	-	-	266	101	-	-	-
Textile and Leather	-	11	-	-	-	2	8	527	-	-	-
Non-specified	-	31	9	-	-	4	73	815	-	-	-
TRANSPORT SECTOR	-	112	5307	10	2657	-	12124	-	-	-	-
Air	-	-	-	10	2657	-	-	-	-	-	-
Road	-	112	5303	-	-	-	11913	-	-	-	-
Rail	-	-	-	-	-	-	114	-	-	-	-
Pipeline Transport	-	-	-	-	-	-	-	-	-	-	-
Internal Navigation	-	-	4	-	-	-	97	-	-	-	-
Non-specified	-	-	-	-	-	-	-	-	-	-	-
OTHER SECTORS	-	1226	47	-	-	28	1422	29	-	-	-
Agriculture	-	2	47	-	-	1	1416	8	-	-	-
Comm. and Publ. Services	-	-	-	-	-	-	-	-	-	-	-
Residential	-	1224	-	-	-	27	6	21	-	-	-
Non-specified	-	-	-	-	-	-	-	-	-	-	-
NON-ENERGY USE	-	-	-	-	-	-	-	-	-	-	719
in Industry/Transf./Energy	-	-	-	-	-	-	-	-	-	-	719
in Transport	-	-	-	-	-	-	-	-	-	-	-
in Other Sectors	-	-	-	-	-	-	-	-	-	-	-

Thailand / Thaïlande : 1997

SUPPLY AND CONSUMPTION / APPROVISIONNEMENT ET DEMANDE	Gas / Gaz (TJ)				Comb. Renew. & Waste / En. Re. Comb. & Déchets (TJ)				(GWh)	(TJ)
	Natural Gas / Gaz naturel	Gas Works / Usines à gaz	Coke Ovens / Cokeries	Blast Furnaces / Hauts fourneaux	Solid Biomass / Biomasse solide	Gas/Liquids from Biomass / Gaz/Liquides tirés de biomasse	Municipal Waste / Déchets urbains	Industrial Waste / Déchets industriels	Electricity / Electricité	Heat / Chaleur
Production	588108	-	-	719	584733	-	-	-	93253	-
From Other Sources	-	-	-	-	-	-	-	-	-	-
Imports	-	-	-	-	415	-	-	-	746	-
Exports	-	-	-	-	-124	-	-	-	-104	-
Intl. Marine Bunkers	- ·	-	-	-	-	-	-	-	-	-
Stock Changes	-	-	-	-	-	-	-	-	-	-
DOMESTIC SUPPLY	588108	-	-	719	585024	-	-	-	93895	-
Transfers	-	-	-	-	-	-	-	-	-	-
Statistical Differences	-	-	-	-	-	-	-	-	-	-
TRANSFORMATION	461984	-	-	-	204503	-	-	-	-	-
Electricity Plants	461984	-	-	-	9083	-	-	-	-	-
CHP Plants	-	-	-	-	-	-	-	-	-	-
Heat Plants	-	-	-	-	-	-	-	-	-	-
Blast Furnaces/Gas Works	-	-	-	-	-	-	-	-	-	-
Coke/Pat. Fuel/BKB Plants	-	-	-	-	-	-	-	-	-	-
Petroleum Refineries	-	-	-	-	-	-	-	-	-	-
Petrochemical Industry	-	-	-	-	-	-	-	-	-	-
Liquefaction	-	-	-	-	-	-	-	-	-	-
Other Transform. Sector	-	-	-	-	195420	-	-	-	-	-
ENERGY SECTOR	85596	-	-	-	-	-	-	-	3378	-
Coal Mines	-	-	-	-	-	-	-	-	-	-
Oil and Gas Extraction	-	-	-	-	-	-	-	-	-	-
Petroleum Refineries	-	-	-	-	-	-	-	-	-	-
Electr., CHP+Heat Plants	-	-	-	-	-	-	-	-	3378	-
Pumped Storage (Elec.)	-	-	-	-	-	-	-	-	-	-
Other Energy Sector	85596	-	-	-	-	-	-	-	-	-
Distribution Losses	-	-	-	-	-	-	-	-	8088	-
FINAL CONSUMPTION	40528	-	-	719	380521	-	-	-	82429	-
INDUSTRY SECTOR	40292	-	-	719	187732	-	-	-	34541	-
Iron and Steel	-	-	-	719	-	-	-	-	3404	-
Chemical and Petrochem.	6120	-	-	-	4501	-	-	-	5433	-
of which: Feedstocks	6120	-	-	-	-	-	-	-	-	-
Non-Ferrous Metals	-	-	-	-	-	-	-	-	-	-
Non-Metallic Minerals	315	-	-	-	7728	-	-	-	5689	-
Transport Equipment	-	-	-	-	-	-	-	-	-	-
Machinery	-	-	-	-	-	-	-	-	5736	-
Mining and Quarrying	-	-	-	-	-	-	-	-	-	-
Food and Tobacco	-	-	-	-	175078	-	-	-	5436	-
Paper, Pulp and Print	-	-	- ·	-	-	-	-	-	1211	-
Wood and Wood Products	-	-	-	-	425	-	-	-	630	-
Construction	-	-	-	-	-	-	-	-	-	-
Textile and Leather	-	-	-	-	-	-	-	-	5811	-
Non-specified	33857	- ·	-	-	-	-	-	-	1191	-
TRANSPORT SECTOR	236	-	-	-	-	-	-	-	-	-
Air	-	-	-	-	-	-	-	-	-	-
Road	236	-	-	-	-	-	-	-	-	-
Rail	-	-	-	-	-	-	-	-	-	-
Pipeline Transport	-	-	-	-	-	-	-	-	-	-
Internal Navigation	-	-	-	-	-	-	-	-	-	-
Non-specified	-	-	-	-	-	-	-	-	-	-
OTHER SECTORS	-	-	-	-	192789	-	-	-	47888	-
Agriculture	-	-	-	-	-	-	-	-	165	-
Comm. and Publ. Services	-	-	-	-	-	-	-	-	29204	-
Residential	-	-	-	-	192789	-	-	-	17667	-
Non-specified	-	-	-	-	-	-	-	-	852	-
NON-ENERGY USE	-	-	-	-	-	-	-	-	-	-
in Industry/Transf./Energy	-	-	-	-	-	-	-	-	-	-
in Transport	-	-	-	-	-	-	-	-	-	-
in Other Sectors	-	-	-	-	-	-	-	-	-	-

Thailand / Thaïlande : 1998

SUPPLY AND CONSUMPTION / APPROVISIONNEMENT ET DEMANDE	Coal / Charbon (1000 tonnes)							Oil / Pétrole (1000 tonnes)			
	Coking Coal / Charbon à coke	Other Bit. Coal / Autres charb. bit.	Sub-Bit. Coal / Charbon sous-bit.	Lignite / Lignite	Peat / Tourbe	Oven and Gas Coke / Coke de four/gaz	Pat. Fuel and BKB / Agg./briq. de lignite	Crude Oil / Pétrole brut	NGL / LGN	Feed-stocks / Produits d'aliment.	Additives / Additifs
Production	-	-	-	20163	-	-	-	3503	2139	-	-
From Other Sources	-	-	-	-	-	-	-	-	-	-	-
Imports	-	1380	-	-	-	80	164	32388	-	-	-
Exports	-	-	-	-	-	-	-	-801	-95	-	-
Intl. Marine Bunkers	-	-	-	-	-	-	-	-	-	-	-
Stock Changes	-	8	-	417	-	-	-	-301	-	-	-
DOMESTIC SUPPLY	-	1388	-	20580	-	80	164	34789	2044	-	-
Transfers	-	-	-	-	-	-	-	-	-1929	-	-
Statistical Differences	-	-	-	-	-	-	-	-	-1	-	-
TRANSFORMATION	-	54	-	15424	-	64	-	34616	10	-	-
Electricity Plants	-	54	-	15424	-	-	-	-	-	-	-
CHP Plants	-	-	-	-	-	-	-	-	-	-	-
Heat Plants	-	-	-	-	-	-	-	-	-	-	-
Blast Furnaces/Gas Works	-	-	-	-	-	64	-	-	-	-	-
Coke/Pat. Fuel/BKB Plants	-	-	-	-	-	-	-	-	-	-	-
Petroleum Refineries	-	-	-	-	-	-	-	34616	10	-	-
Petrochemical Industry	-	-	-	-	-	-	-	-	-	-	-
Liquefaction	-	-	-	-	-	-	-	-	-	-	-
Other Transform. Sector	-	-	-	-	-	-	-	-	-	-	-
ENERGY SECTOR	-	-	-	-	-	-	-	-	-	-	-
Coal Mines	-	-	-	-	-	-	-	-	-	-	-
Oil and Gas Extraction	-	-	-	-	-	-	-	-	-	-	-
Petroleum Refineries	-	-	-	-	-	-	-	-	-	-	-
Electr., CHP+Heat Plants	-	-	-	-	-	-	-	-	-	-	-
Pumped Storage (Elec.)	-	-	-	-	-	-	-	-	-	-	-
Other Energy Sector	-	-	-	-	-	-	-	-	-	-	-
Distribution Losses	-	-	-	-	-	-	-	134	10	-	-
FINAL CONSUMPTION	-	1334	-	5156	-	16	164	39	94	-	-
INDUSTRY SECTOR	-	1334	-	5155	-	16	164	39	94	-	-
Iron and Steel	-	-	-	-	-	16	-	-	-	-	-
Chemical and Petrochem.	-	-	-	-	-	-	-	39	94	-	-
of which: Feedstocks	-	-	-	-	-	-	-	39	94	-	-
Non-Ferrous Metals	-	-	-	-	-	-	-	-	-	-	-
Non-Metallic Minerals	-	1128	-	3458	-	-	-	-	-	-	-
Transport Equipment	-	-	-	-	-	-	-	-	-	-	-
Machinery	-	-	-	-	-	-	-	-	-	-	-
Mining and Quarrying	-	-	-	-	-	-	-	-	-	-	-
Food and Tobacco	-	71	-	237	-	-	-	-	-	-	-
Paper, Pulp and Print	-	-	-	-	-	-	-	-	-	-	-
Wood and Wood Products	-	-	-	-	-	-	-	-	-	-	-
Construction	-	-	-	-	-	-	-	-	-	-	-
Textile and Leather	-	-	-	-	-	-	-	-	-	-	-
Non-specified	-	135	-	1460	-	-	164	-	-	-	-
TRANSPORT SECTOR	-	-	-	-	-	-	-	-	-	-	-
Air	-	-	-	-	-	-	-	-	-	-	-
Road	-	-	-	-	-	-	-	-	-	-	-
Rail	-	-	-	-	-	-	-	-	-	-	-
Pipeline Transport	-	-	-	-	-	-	-	-	-	-	-
Internal Navigation	-	-	-	-	-	-	-	-	-	-	-
Non-specified	-	-	-	-	-	-	-	-	-	-	-
OTHER SECTORS	-	-	-	-	-	-	-	-	-	-	-
Agriculture	-	-	-	-	-	-	-	-	-	-	-
Comm. and Publ. Services	-	-	-	-	-	-	-	-	-	-	-
Residential	-	-	-	-	-	-	-	-	-	-	-
Non-specified	-	-	-	-	-	-	-	-	-	-	-
NON-ENERGY USE	-	-	-	1	-	-	-	-	-	-	-
in Industry/Trans./Energy	-	-	-	1	-	-	-	-	-	-	-
in Transport	-	-	-	-	-	-	-	-	-	-	-
in Other Sectors	-	-	-	-	-	-	-	-	-	-	-

Thailand / Thaïlande : 1998

SUPPLY AND CONSUMPTION / APPROVISIONNEMENT ET DEMANDE	Refinery Gas / Gaz de raffinerie	LPG + Ethane / GPL + éthane	Motor Gasoline / Essence moteur	Aviation Gasoline / Essence aviation	Jet Fuel / Carbu-réacteurs	Kerosene / Kérosène	Gas/ Diesel / Gazole	Heavy Fuel Oil / Fioul lourd	Naphtha / Naphta	Petrol. Coke / Coke de pétrole	Other Prod. / Autres prod.
Production	346	924	6418	-	2659	96	14125	7317	-	-	697
From Other Sources	-	-	-	-	-	-	-	-	-	-	-
Imports	-	2	21	10	1	-	616	622	-	-	1
Exports	-	-472	-1251	-	-178	-21	-1895	-187	-	-	-79
Intl. Marine Bunkers	-	-	-	-	-	-	-79	-489	-	-	-
Stock Changes	-	8	95	-	3	-32	311	-272	-	-	29
DOMESTIC SUPPLY	346	462	5283	10	2485	43	13078	6991	-	-	648
Transfers	-	1929	-	-	-	-	-	-	-	-	-
Statistical Differences	-	-1	-45	-	-	-	-1	1	-	-	1
TRANSFORMATION	-	-	-	-	-	-	270	4024	-	-	-
Electricity Plants	-	-	-	-	-	-	270	4024	-	-	-
CHP Plants	-	-	-	-	-	-	-	-	-	-	-
Heat Plants	-	-	-	-	-	-	-	-	-	-	-
Blast Furnaces/Gas Works	-	-	-	-	-	-	-	-	-	-	-
Coke/Pat. Fuel/BKB Plants	-	-	-	-	-	-	-	-	-	-	-
Petroleum Refineries	-	-	-	-	-	-	-	-	-	-	-
Petrochemical Industry	-	-	-	-	-	-	-	-	-	-	-
Liquefaction	-	-	-	-	-	-	-	-	-	-	-
Other Transform. Sector	-	-	-	-	-	-	-	-	-	-	-
ENERGY SECTOR	346	-	-	-	-	-	-	-	-	-	-
Coal Mines	-	-	-	-	-	-	-	-	-	-	-
Oil and Gas Extraction	-	-	-	-	-	-	-	-	-	-	-
Petroleum Refineries	346	-	-	-	-	-	-	-	-	-	-
Electr., CHP+Heat Plants	-	-	-	-	-	-	-	-	-	-	-
Pumped Storage (Elec.)	-	-	-	-	-	-	-	-	-	-	-
Other Energy Sector	-	-	-	-	-	-	-	-	-	-	-
Distribution Losses	-	52	-	-	-	-	-	-	-	-	-
FINAL CONSUMPTION	-	2338	5238	10	2485	43	12807	2968	-	-	649
INDUSTRY SECTOR	-	1115	51	-	-	25	801	2948	-	-	-
Iron and Steel	-	53	-	-	-	1	28	132	-	-	-
Chemical and Petrochem.	-	866	2	-	-	15	33	305	-	-	-
of which: Feedstocks	-	607	-	-	-	-	-	-	-	-	-
Non-Ferrous Metals	-	-	-	-	-	-	-	-	-	-	-
Non-Metallic Minerals	-	56	1	-	-	2	33	261	-	-	-
Transport Equipment	-	-	-	-	-	-	-	-	-	-	-
Machinery	-	43	2	-	-	3	17	54	-	-	-
Mining and Quarrying	-	-	-	-	-	-	20	12	-	-	-
Food and Tobacco	-	13	2	-	-	-	156	532	-	-	-
Paper, Pulp and Print	-	1	1	-	-	-	14	155	-	-	-
Wood and Wood Products	-	-	-	-	-	-	9	19	-	-	-
Construction	-	-	-	-	-	-	205	59	-	-	-
Textile and Leather	-	11	-	-	-	1	9	475	-	-	-
Non-specified	-	72	43	-	-	3	277	944	-	-	-
TRANSPORT SECTOR	-	93	5139	10	2485	-	10200	-	-	-	-
Air	-	-	-	10	2485	-	-	-	-	-	-
Road	-	93	5138	-	-	-	9998	-	-	-	-
Rail	-	-	-	-	-	-	96	-	-	-	-
Pipeline Transport	-	-	-	-	-	-	-	-	-	-	-
Internal Navigation	-	-	1	-	-	-	106	-	-	-	-
Non-specified	-	-	-	-	-	-	-	-	-	-	-
OTHER SECTORS	-	1130	48	-	-	18	1806	20	-	-	-
Agriculture	-	1	48	-	-	1	1801	3	-	-	-
Comm. and Publ. Services	-	-	-	-	-	-	-	-	-	-	-
Residential	-	1129	-	-	-	17	5	17	-	-	-
Non-specified	-	-	-	-	-	-	-	-	-	-	-
NON-ENERGY USE	-	-	-	-	-	-	-	-	-	-	649
in Industry/Transf./Energy	-	-	-	-	-	-	-	-	-	-	649
in Transport	-	-	-	-	-	-	-	-	-	-	-
in Other Sectors	-	-	-	-	-	-	-	-	-	-	-

Thailand / Thaïlande : 1998

SUPPLY AND CONSUMPTION *APPROVISIONNEMENT ET DEMANDE*	Gas / *Gaz* (TJ)				Comb. Renew. & Waste / *En. Re. Comb. & Déchets* (TJ)				(GWh)	(TJ)
	Natural Gas *Gaz naturel*	Gas Works *Usines à gaz*	Coke Ovens *Cokeries*	Blast Furnaces *Hauts fourneaux*	Solid Biomass *Biomasse solide*	Gas/Liquids from Biomass *Gaz/Liquides tirés de biomasse*	Municipal Waste *Déchets urbains*	Industrial Waste *Déchets industriels*	Electricity *Electricité*	Heat *Chaleur*
Production	640711	-	-	697	558019	-	-	-	90069	-
From Other Sources	-	-	-	-	-	-	-	-	-	-
Imports	813	-	-	-	207	-	-	-	1623	-
Exports	-	-	-	-	-166	-	-	-	-153	-
Intl. Marine Bunkers	-	-	-	-	-	-	-	-	-	-
Stock Changes	-	-	-	-	-	-	-	-	-	-
DOMESTIC SUPPLY	**641524**	**-**	**-**	**697**	**558060**	**-**	**-**	**-**	**91539**	**-**
Transfers	-	-	-	-	-	-	-	-	-	-
Statistical Differences	-1	-	-	-	-	-	-	-	-1	-
TRANSFORMATION	**481704**	**-**	**-**	**-**	**207163**	**-**	**-**	**-**	**-**	**-**
Electricity Plants	481704	-	-	-	8545	-	-	-	-	-
CHP Plants	-	-	-	-	-	-	-	-	-	-
Heat Plants	-	-	-	-	-	-	-	-	-	-
Blast Furnaces/Gas Works	-	-	-	-	-	-	-	-	-	-
Coke/Pat. Fuel/BKB Plants	-	-	-	-	-	-	-	-	-	-
Petroleum Refineries	-	-	-	-	-	-	-	-	-	-
Petrochemical Industry	-	-	-	-	-	-	-	-	-	-
Liquefaction	-	-	-	-	-	-	-	-	-	-
Other Transform. Sector	-	-	-	-	198618	-	-	-	-	-
ENERGY SECTOR	**122236**	**-**	**-**	**-**	**-**	**-**	**-**	**-**	**3267**	**-**
Coal Mines	-	-	-	-	-	-	-	-	-	-
Oil and Gas Extraction	-	-	-	-	-	-	-	-	-	-
Petroleum Refineries	-	-	-	-	-	-	-	-	-	-
Electr., CHP+Heat Plants	-	-	-	-	-	-	-	-	3267	-
Pumped Storage (Elec.)	-	-	-	-	-	-	-	-	-	-
Other Energy Sector	122236	-	-	-	-	-	-	-	-	-
Distribution Losses	-	-	-	-	-	-	-	-	7838	-
FINAL CONSUMPTION	**37583**	**-**	**-**	**697**	**350897**	**-**	**-**	**-**	**80433**	**-**
INDUSTRY SECTOR	**37350**	**-**	**-**	**697**	**157425**	**-**	**-**	**-**	**30834**	**-**
Iron and Steel	-	-	-	697	-	-	-	-	2864	-
Chemical and Petrochem.	5790	-	-	-	4457	-	-	-	5371	-
of which: Feedstocks	*5790*	-	-	-	-	-	-	-	-	-
Non-Ferrous Metals	-	-	-	-	-	-	-	-	-	-
Non-Metallic Minerals	-	-	-	-	7652	-	-	-	3721	-
Transport Equipment	-	-	-	-	-	-	-	-	-	-
Machinery	-	-	-	-	-	-	-	-	5130	-
Mining and Quarrying	-	-	-	-	-	-	-	-	-	-
Food and Tobacco	-	-	-	-	144896	-	-	-	5409	-
Paper, Pulp and Print	-	-	-	-	-	-	-	-	1161	-
Wood and Wood Products	-	-	-	-	420	-	-	-	611	-
Construction	-	-	-	-	-	-	-	-	-	-
Textile and Leather	-	-	-	-	-	-	-	-	5534	-
Non-specified	31560	-	-	-	-	-	-	-	1033	-
TRANSPORT SECTOR	**233**	**-**	**-**	**-**	**-**	**-**	**-**	**-**	**-**	**-**
Air	-	-	-	-	-	-	-	-	-	-
Road	233	-	-	-	-	-	-	-	-	-
Rail	-	-	-	-	-	-	-	-	-	-
Pipeline Transport	-	-	-	-	-	-	-	-	-	-
Internal Navigation	-	-	-	-	-	-	-	-	-	-
Non-specified	-	-	-	-	-	-	-	-	-	-
OTHER SECTORS	**-**	**-**	**-**	**-**	**193472**	**-**	**-**	**-**	**49599**	**-**
Agriculture	-	-	-	-	-	-	-	-	211	-
Comm. and Publ. Services	-	-	-	-	-	-	-	-	29921	-
Residential	-	-	-	-	193472	-	-	-	18868	-
Non-specified	-	-	-	-	-	-	-	-	599	-
NON-ENERGY USE	**-**	**-**	**-**	**-**	**-**	**-**	**-**	**-**	**-**	**-**
in Industry/Transf./Energy	-	-	-	-	-	-	-	-	-	-
in Transport	-	-	-	-	-	-	-	-	-	-
in Other Sectors	-	-	-	-	-	-	-	-	-	-

Togo

SUPPLY AND CONSUMPTION 1997	Coal (1000 tonnes)							Oil (1000 tonnes)			
	Coking Coal	Other Bit. Coal	Sub-Bit. Coal	Lignite	Peat	Oven and Gas Coke	Pat. Fuel and BKB	Crude Oil	NGL	Feed-stocks	Additives
Production	-	-	-	-	-	-	-	-	-	-	-
Imports	-	-	-	-	-	-	-	-	-	-	-
Exports	-	-	-	-	-	-	-	-	-	-	-
Intl. Marine Bunkers	-	-	-	-	-	-	-	-	-	-	-
Stock Changes	-	-	-	-	-	-	-	-	-	-	-
DOMESTIC SUPPLY	-	-	-	-	-	-	-	-	-	-	-
Transfers and Stat. Diff.	-	-	-	-	-	-	-	-	-	-	-
TRANSFORMATION	-	-	-	-	-	-	-	-	-	-	-
Electricity and CHP Plants	-	-	-	-	-	-	-	-	-	-	-
Petroleum Refineries	-	-	-	-	-	-	-	-	-	-	-
Other Transform. Sector	-	-	-	-	-	-	-	-	-	-	-
ENERGY SECTOR	-	-	-	-	-	-	-	-	-	-	-
DISTRIBUTION LOSSES	-	-	-	-	-	-	-	-	-	-	-
FINAL CONSUMPTION	-	-	-	-	-	-	-	-	-	-	-
INDUSTRY SECTOR	-	-	-	-	-	-	-	-	-	-	-
Iron and Steel	-	-	-	-	-	-	-	-	-	-	-
Chemical and Petrochem.	-	-	-	-	-	-	-	-	-	-	-
Non-Metallic Minerals	-	-	-	-	-	-	-	-	-	-	-
Non-specified	-	-	-	-	-	-	-	-	-	-	-
TRANSPORT SECTOR	-	-	-	-	-	-	-	-	-	-	-
Air	-	-	-	-	-	-	-	-	-	-	-
Road	-	-	-	-	-	-	-	-	-	-	-
Non-specified	-	-	-	-	-	-	-	-	-	-	-
OTHER SECTORS	-	-	-	-	-	-	-	-	-	-	-
Agriculture	-	-	-	-	-	-	-	-	-	-	-
Comm. and Publ. Services	-	-	-	-	-	-	-	-	-	-	-
Residential	-	-	-	-	-	-	-	-	-	-	-
Non-specified	-	-	-	-	-	-	-	-	-	-	-
NON-ENERGY USE	-	-	-	-	-	-	-	-	-	-	-

APPROVISIONNEMENT ET DEMANDE 1998	Charbon (1000 tonnes)							Pétrole (1000 tonnes)			
	Charbon à coke	Autres charb. bit.	Charbon sous-bit.	Lignite	Tourbe	Coke de four/gaz	Agg./briq. de lignite	Pétrole brut	LGN	Produits d'aliment.	Additifs
Production	-	-	-	-	-	-	-	-	-	-	-
Imports	-	-	-	-	-	-	-	-	-	-	-
Exports	-	-	-	-	-	-	-	-	-	-	-
Intl. Marine Bunkers	-	-	-	-	-	-	-	-	-	-	-
Stock Changes	-	-	-	-	-	-	-	-	-	-	-
DOMESTIC SUPPLY	-	-	-	-	-	-	-	-	-	-	-
Transfers and Stat. Diff.	-	-	-	-	-	-	-	-	-	-	-
TRANSFORMATION	-	-	-	-	-	-	-	-	-	-	-
Electricity and CHP Plants	-	-	-	-	-	-	-	-	-	-	-
Petroleum Refineries	-	-	-	-	-	-	-	-	-	-	-
Other Transform. Sector	-	-	-	-	-	-	-	-	-	-	-
ENERGY SECTOR	-	-	-	-	-	-	-	-	-	-	-
DISTRIBUTION LOSSES	-	-	-	-	-	-	-	-	-	-	-
FINAL CONSUMPTION	-	-	-	-	-	-	-	-	-	-	-
INDUSTRY SECTOR	-	-	-	-	-	-	-	-	-	-	-
Iron and Steel	-	-	-	-	-	-	-	-	-	-	-
Chemical and Petrochem.	-	-	-	-	-	-	-	-	-	-	-
Non-Metallic Minerals	-	-	-	-	-	-	-	-	-	-	-
Non-specified	-	-	-	-	-	-	-	-	-	-	-
TRANSPORT SECTOR	-	-	-	-	-	-	-	-	-	-	-
Air	-	-	-	-	-	-	-	-	-	-	-
Road	-	-	-	-	-	-	-	-	-	-	-
Non-specified	-	-	-	-	-	-	-	-	-	-	-
OTHER SECTORS	-	-	-	-	-	-	-	-	-	-	-
Agriculture	-	-	-	-	-	-	-	-	-	-	-
Comm. and Publ. Services	-	-	-	-	-	-	-	-	-	-	-
Residential	-	-	-	-	-	-	-	-	-	-	-
Non-specified	-	-	-	-	-	-	-	-	-	-	-
NON-ENERGY USE	-	-	-	-	-	-	-	-	-	-	-

Togo

SUPPLY AND CONSUMPTION 1997	Oil cont. (1000 tonnes)										
	Refinery Gas	LPG + Ethane	Motor Gasoline	Aviation Gasoline	Jet Fuel	Kerosene	Gas/ Diesel	Heavy Fuel Oil	Naphtha	Petrol. Coke	Other Prod.
Production	-	-	-	-	-	-	-	-	-	-	-
Imports	-	-	88	-	14	28	69	-	-	-	86
Exports	-	-	-	-	-	-	-	-	-	-	-
Intl. Marine Bunkers	-	-	-	-	-	-	-	-	-	-	-
Stock Changes	-	-	-1	-	-	3	-2	-	-	-	-23
DOMESTIC SUPPLY	-	-	87	-	14	31	67	-	-	-	63
Transfers and Stat. Diff.	-	-	-	-	-	-	-	-	-	-	1
TRANSFORMATION	-	-	-	-	-	-	21	-	-	-	-
Electricity and CHP Plants	-	-	-	-	-	-	21	-	-	-	-
Petroleum Refineries	-	-	-	-	-	-	-	-	-	-	-
Other Transform. Sector	-	-	-	-	-	-	-	-	-	-	-
ENERGY SECTOR	-	-	-	-	-	-	-	-	-	-	-
DISTRIBUTION LOSSES	-	-	-	-	-	-	-	-	-	-	-
FINAL CONSUMPTION	-	-	87	-	14	31	46	-	-	-	64
INDUSTRY SECTOR	-	-	-	-	-	-	46	-	-	-	-
Iron and Steel	-	-	-	-	-	-	-	-	-	-	-
Chemical and Petrochem.	-	-	-	-	-	-	-	-	-	-	-
Non-Metallic Minerals	-	-	-	-	-	-	-	-	-	-	-
Non-specified	-	-	-	-	-	-	46	-	-	-	-
TRANSPORT SECTOR	-	-	87	-	14	-	-	-	-	-	-
Air	-	-	-	-	14	-	-	-	-	-	-
Road	-	-	87	-	-	-	-	-	-	-	-
Non-specified	-	-	-	-	-	-	-	-	-	-	-
OTHER SECTORS	-	-	-	-	-	31	-	-	-	-	-
Agriculture	-	-	-	-	-	-	-	-	-	-	-
Comm. and Publ. Services	-	-	-	-	-	-	-	-	-	-	-
Residential	-	-	-	-	-	31	-	-	-	-	-
Non-specified	-	-	-	-	-	-	-	-	-	-	-
NON-ENERGY USE	-	-	-	-	-	-	-	-	-	-	64

APPROVISIONNEMENT ET DEMANDE 1998	Pétrole cont. (1000 tonnes)										
	Gaz de raffinerie	GPL + éthane	Essence moteur	Essence aviation	Carbu- réacteurs	Kérosène	Gazole	Fioul lourd	Naphta	Coke de pétrole	Autres prod.
Production	-	-	-	-	-	-	-	-	-	-	-
Imports	-	-	94	-	19	38	79	-	-	-	68
Exports	-	-	-	-	-	-	-	-	-	-	-
Intl. Marine Bunkers	-	-	-	-	-	-	-	-	-	-	-
Stock Changes	-	-	4	-	-	-1	4	-	-	-	1
DOMESTIC SUPPLY	-	-	98	-	19	37	83	-	-	-	69
Transfers and Stat. Diff.	-	-	-	-	-	-	-1	-	-	-	-
TRANSFORMATION	-	-	-	-	-	-	39	-	-	-	-
Electricity and CHP Plants	-	-	-	-	-	-	39	-	-	-	-
Petroleum Refineries	-	-	-	-	-	-	-	-	-	-	-
Other Transform. Sector	-	-	-	-	-	-	-	-	-	-	-
ENERGY SECTOR	-	-	-	-	-	-	-	-	-	-	-
DISTRIBUTION LOSSES	-	-	-	-	-	-	-	-	-	-	-
FINAL CONSUMPTION	-	-	98	-	19	37	43	-	-	-	69
INDUSTRY SECTOR	-	-	-	-	-	-	43	-	-	-	-
Iron and Steel	-	-	-	-	-	-	-	-	-	-	-
Chemical and Petrochem.	-	-	-	-	-	-	-	-	-	-	-
Non-Metallic Minerals	-	-	-	-	-	-	-	-	-	-	-
Non-specified	-	-	-	-	-	-	43	-	-	-	-
TRANSPORT SECTOR	-	-	98	-	19	-	-	-	-	-	-
Air	-	-	-	-	19	-	-	-	-	-	-
Road	-	-	98	-	-	-	-	-	-	-	-
Non-specified	-	-	-	-	-	-	-	-	-	-	-
OTHER SECTORS	-	-	-	-	-	37	-	-	-	-	-
Agriculture	-	-	-	-	-	-	-	-	-	-	-
Comm. and Publ. Services	-	-	-	-	-	-	-	-	-	-	-
Residential	-	-	-	-	-	37	-	-	-	-	-
Non-specified	-	-	-	-	-	-	-	-	-	-	-
NON-ENERGY USE	-	-	-	-	-	-	-	-	-	-	69

Togo

SUPPLY AND CONSUMPTION 1997	Gas (TJ) Natural Gas	Gas Works	Coke Ovens	Blast Furnaces	Comb. Renew. & Waste (TJ) Solid Biomass	Gas/Liquids from Biomass	Municipal Waste	Industrial Waste	(GWh) Electricity	(TJ) Heat
Production	-	-	-	-	40219	-	-	-	53	-
Imports	-	-	-	-	-	-	-	-	452	-
Exports	-	-	-	-	-	-	-	-	-	-
Intl. Marine Bunkers	-	-	-	-	-	-	-	-	-	-
Stock Changes	-	-	-	-	-	-	-	-	-	-
DOMESTIC SUPPLY	-	-	-	-	40219	-	-	-	505	-
Transfers and Stat. Diff.	-·	-	-	-	-	-	-	-	-	-
TRANSFORMATION	-	-	-	-	32881	-	-	-	-	-
Electricity and CHP Plants	-	-	-	-	-	-	-	-	-	-
Petroleum Refineries	-	-	-	-	-	-	-	-	-	-
Other Transform. Sector	-	-	-	-	32881	-	-	-	-	-
ENERGY SECTOR	-	-	-	-	-	-	-	-	-	-
DISTRIBUTION LOSSES	-	-	-	-	-	-	-	-	17	-
FINAL CONSUMPTION	-	-	-	-	7338	-	-	-	488	-
INDUSTRY SECTOR	-	-	-	-	-	-	-	-	203	-
Iron and Steel	-	-	-	-	-	-	-	-	-	-
Chemical and Petrochem.	-	-	-	-	-	-	-	-	75	-
Non-Metallic Minerals	-	-	-	-	-	-	-	-	-	-
Non-specified	-	-	-	-	-	-	-	-	128	-
TRANSPORT SECTOR	-	-	-	-	-	-	-	-	-	-
Air	-	-	-	-	-	-	-	-	-	-
Road	-	-	-	-	-	-	-	-	-	-
Non-specified	-	-	-	-	-	-	-	-	-	-
OTHER SECTORS	-	-	-	-	7338	-	-	-	285	-
Agriculture	-	-	-	-	-	-	-	-	-	-
Comm. and Publ. Services	-	-	-	-	-	-	-	-	123	-
Residential	-	-	-	-	7338	-	-	-	162	-
Non-specified	-	-	-	-	-	-	-	-	-	-
NON-ENERGY USE	-	-	-	-	-	-	-	-	-	-

APPROVISIONNEMENT ET DEMANDE 1998	Gaz (TJ) Gaz naturel	Usines à gaz	Cokeries	Hauts fourneaux	En. Re. Comb. & Déchets (TJ) Biomasse solide	Gaz/Liquides tirés de biomasse	Déchets urbains	Déchets industriels	(GWh) Electricité	(TJ) Chaleur
Production	-	-	-	-	41479	-	-	-	96	-
Imports	-	-	-	-	-	-	-	-	426	-
Exports	-	-	-	-	-	-	-	-	-	-
Intl. Marine Bunkers	-	-	-	-	-	-	-	-	-	-
Stock Changes	-	-	-	-	-	-	-	-	-	-
DOMESTIC SUPPLY	-	-	-·	-	41479	-	-	-	522	-
Transfers and Stat. Diff.	-	-	-	-	-	-	-	-	-	-
TRANSFORMATION	-	-	-	-	33910	-	-	-	-	-
Electricity and CHP Plants	-	-	-	-	-	-	-	-	-	-
Petroleum Refineries	-	-	-	-	-	-	-	-	-	-
Other Transform. Sector	-	-	-	-	33910	-	-	-	-	-
ENERGY SECTOR	-	-	-	-	-	-	-	-	-	-
DISTRIBUTION LOSSES	-	-	-	-	-	-	-	-	51	-
FINAL CONSUMPTION	-	-	-	-	7569	-	-	-	471	-
INDUSTRY SECTOR	-	-	-	-	-	-	-	-	204	-
Iron and Steel	-	-	-	-	-	-	-	-	-	-
Chemical and Petrochem.	-	-	-	-	-	-	-	-	55	-
Non-Metallic Minerals	-	-	-	-	-	-	-	-	-	-
Non-specified	-	-	-	-	-	-	-	-	149	-
TRANSPORT SECTOR	-	-	-	-	-	-	-	-	-	-
Air	-	-	-	-	-	-	-	-	-	-
Road	-	-	-	-	-	-	-	-	-	-
Non-specified	-	-	-	-	-	-	-	-	-	-
OTHER SECTORS	-	-	-	-	7569	-	-	-	267	-
Agriculture	-	-	-	-	-	-	-	-	-	-
Comm. and Publ. Services	-	-	-	-	-	-	-	-	109	-
Residential	-	-	-	-	7569	-	-	-	158	-
Non-specified	-	-	-	-	-	-	-	-	-	-
NON-ENERGY USE	-	-	-	-	-	-	-	-	-	-

Trinidad-&-Tobago / Trinité-et-Tobago

SUPPLY AND CONSUMPTION 1997	Coal (1000 tonnes)							Oil (1000 tonnes)			
	Coking Coal	Other Bit. Coal	Sub-Bit. Coal	Lignite	Peat	Oven and Gas Coke	Pat. Fuel and BKB	Crude Oil	NGL	Feed-stocks	Additives
Production	-	-	-	-	-	-	-	6384	374	-	1520
Imports	-	-	-	-	-	-	-	2020	-	-	-
Exports	-	-	-	-	-	-	-	-2867	-123	-	-1546
Intl. Marine Bunkers	-	-	-	-	-	-	-				
Stock Changes	-	-	-	-	-	-	-	-395	3	-	36
DOMESTIC SUPPLY	-	-	-	-	-	-	-	5142	254	-	10
Transfers and Stat. Diff.	-	-	-	-	-	-	-	-	-254	-	-3
TRANSFORMATION	-	-	-	-	-	-	-	5142	-	-	-
Electricity and CHP Plants	-	-	-	-	-	-	-	-	-	-	-
Petroleum Refineries	-	-	-	-	-	-	-	5142	-	-	-
Other Transform. Sector	-	-	-	-	-	-	-	-	-	-	-
ENERGY SECTOR	-	-	-	-	-	-	-	-	-	-	-
DISTRIBUTION LOSSES	-	-	-	-	-	-	-	-	-	-	-
FINAL CONSUMPTION	-	-	-	-	-	-	-	-	-	-	7
INDUSTRY SECTOR	-	-	-	-	-	-	-	-	-	-	7
Iron and Steel	-	-	-	-	-	-	-	-	-	-	-
Chemical and Petrochem.	-	-	-	-	-	-	-	-	-	-	7
Non-Metallic Minerals	-	-	-	-	-	-	-	-	-	-	-
Non-specified	-	-	-	-	-	-	-	-	-	-	-
TRANSPORT SECTOR	-	-	-	-	-	-	-	-	-	-	-
Air	-	-	-	-	-	-	-	-	-	-	-
Road	-	-	-	-	-	-	-	-	-	-	-
Non-specified	-	-	-	-	-	-	-	-	-	-	-
OTHER SECTORS	-	-	-	-	-	-	-	-	-	-	-
Agriculture	-	-	-	-	-	-	-	-	-	-	-
Comm. and Publ. Services	-	-	-	-	-	-	-	-	-	-	-
Residential	-	-	-	-	-	-	-	-	-	-	-
Non-specified	-	-	-	-	-	-	-	-	-	-	-
NON-ENERGY USE	-	-	-	-	-	-	-	-	-	-	-

APPROVISIONNEMENT ET DEMANDE 1998	Charbon (1000 tonnes)							Pétrole (1000 tonnes)			
	Charbon à coke	Autres charb. bit.	Charbon sous-bit.	Lignite	Tourbe	Coke de four/gaz	Agg./briq. de lignite	Pétrole brut	LGN	Produits d'aliment.	Additifs
Production	-	-	-	-	-	-	-	6322	377	-	1882
Imports	-	-	-	-	-	-	-	2377	-	-	-
Exports	-	-	-	-	-	-	-	-3573	-113	-	-1890
Intl. Marine Bunkers	-	-	-	-	-	-	-	-	-	-	-
Stock Changes	-	-	-	-	-	-	-	187	-2	-	32
DOMESTIC SUPPLY	-	-	-	-	-	-	-	5313	262	-	24
Transfers and Stat. Diff.	-	-	-	-	-	-	-	-	-262	-	-14
TRANSFORMATION	-	-	-	-	-	-	-	5313	-	-	-
Electricity and CHP Plants	-	-	-	-	-	-	-	-	-	-	-
Petroleum Refineries	-	-	-	-	-	-	-	5313	-	-	-
Other Transform. Sector	-	-	-	-	-	-	-	-	-	-	-
ENERGY SECTOR	-	-	-	-	-	-	-	-	-	-	-
DISTRIBUTION LOSSES	-	-	-	-	-	-	-	-	-	-	-
FINAL CONSUMPTION	-	-	-	-	-	-	-	-	-	-	10
INDUSTRY SECTOR	-	-	-	-	-	-	-	-	-	-	10
Iron and Steel	-	-	-	-	-	-	-	-	-	-	-
Chemical and Petrochem.	-	-	-	-	-	-	-	-	-	-	10
Non-Metallic Minerals	-	-	-	-	-	-	-	-	-	-	-
Non-specified	-	-	-	-	-	-	-	-	-	-	-
TRANSPORT SECTOR	-	-	-	-	-	-	-	-	-	-	-
Air	-	-	-	-	-	-	-	-	-	-	-
Road	-	-	-	-	-	-	-	-	-	-	-
Non-specified	-	-	-	-	-	-	-	-	-	-	-
OTHER SECTORS	-	-	-	-	-	-	-	-	-	-	-
Agriculture	-	-	-	-	-	-	-	-	-	-	-
Comm. and Publ. Services	-	-	-	-	-	-	-	-	-	-	-
Residential	-	-	-	-	-	-	-	-	-	-	-
Non-specified	-	-	-	-	-	-	-	-	-	-	-
NON-ENERGY USE	-	-	-	-	-	-	-	-	-	-	-

Trinidad-&-Tobago / Trinité-et-Tobago

SUPPLY AND CONSUMPTION 1997	Oil cont. (1000 tonnes)										
	Refinery Gas	LPG + Ethane	Motor Gasoline	Aviation Gasoline	Jet Fuel	Kerosene	Gas/ Diesel	Heavy Fuel Oil	Naphtha	Petrol. Coke	Other Prod.
Production	120	78	872	-	340	42	1114	2497	-	-	40
Imports	-	-	-	-	21	-	70	-	-	-	8
Exports	-	-268	-473	-	-261	-	-930	-2447	-	-	-38
Intl. Marine Bunkers	-	-	-	-	-	-	-12	-5	-	-	-
Stock Changes	-	-	-90	-	-47	-25	-9	-42	-	-	-
DOMESTIC SUPPLY	120	-190	309	-	53	17	233	3	-	-	10
Transfers and Stat. Diff.	-	254	-	-	-1	12	12	-	-	-	-3
TRANSFORMATION	-	-	-	-	-	-	-	-	-	-	-
Electricity and CHP Plants	-	-	-	-	-	-	-	-	-	-	-
Petroleum Refineries	-	-	-	-	-	-	-	-	-	-	-
Other Transform. Sector	-	-	-	-	-	-	-	-	-	-	-
ENERGY SECTOR	120	12	1	-	-	-	4	-	-	-	-
DISTRIBUTION LOSSES	-	-	-	-	-	-	-	-	-	-	-
FINAL CONSUMPTION	-	52	308	-	52	29	241	3	-	-	7
INDUSTRY SECTOR	-	10	-	-	-	23	60	3	-	-	-
Iron and Steel	-	-	-	-	-	-	-	-	-	-	-
Chemical and Petrochem.	-	-	-	-	-	-	-	-	-	-	-
Non-Metallic Minerals	-	-	-	-	-	-	-	-	-	-	-
Non-specified	-	10	-	-	-	23	60	3	-	-	-
TRANSPORT SECTOR	-	-	308	-	52	-	171	-	-	-	-
Air	-	-	-	-	52	-	-	-	-	-	-
Road	-	-	308	-	-	-	171	-	-	-	-
Non-specified	-	-	-	-	-	-	-	-	-	-	-
OTHER SECTORS	-	42	-	-	-	6	10	-	-	-	-
Agriculture	-	-	-	-	-	-	-	-	-	-	-
Comm. and Publ. Services	-	23	-	-	-	-	10	-	-	-	-
Residential	-	19	-	-	-	6	-	-	-	-	-
Non-specified	-	-	-	-	-	-	-	-	-	-	-
NON-ENERGY USE	-	-	-	-	-	-	-	-	-	-	7

APPROVISIONNEMENT ET DEMANDE 1998	Pétrole cont. (1000 tonnes)										
	Gaz de raffinerie	GPL + éthane	Essence moteur	Essence aviation	Carbu- réacteurs	Kérosène	Gazole	Fioul lourd	Naphta	Coke de pétrole	Autres prod.
Production	122	83	884	-	347	21	1118	2520	-	-	44
Imports	-	-	-	-	17	-	69	-	-	-	11
Exports	-	-274	-568	-	-308	-	-937	-2512	-	-	-43
Intl. Marine Bunkers	-	-	-	-	-	-	-12	-5	-	-	-
Stock Changes	-	-	-	-	-	-	-	-	-	-	3
DOMESTIC SUPPLY	122	-191	316	-	56	21	238	3	-	-	15
Transfers and Stat. Diff.	-	259	1	-	-	9	13	-	-	-	-6
TRANSFORMATION	-	-	-	-	-	-	-	-	-	-	-
Electricity and CHP Plants	-	-	-	-	-	-	-	-	-	-	-
Petroleum Refineries	-	-	-	-	-	-	-	-	-	-	-
Other Transform. Sector	-	-	-	-	-	-	-	-	-	-	-
ENERGY SECTOR	122	13	2	-	-	-	5	-	-	-	-
DISTRIBUTION LOSSES	-	-	-	-	-	-	-	-	-	-	-
FINAL CONSUMPTION	-	55	315	-	56	30	246	3	-	-	9
INDUSTRY SECTOR	-	10	-	-	-	24	62	3	-	-	-
Iron and Steel	-	-	-	-	-	-	-	-	-	-	-
Chemical and Petrochem.	-	-	-	-	-	-	-	-	-	-	-
Non-Metallic Minerals	-	-	-	-	-	-	-	-	-	-	-
Non-specified	-	10	-	-	-	24	62	3	-	-	-
TRANSPORT SECTOR	-	-	315	-	56	-	174	-	-	-	-
Air	-	-	-	-	56	-	-	-	-	-	-
Road	-	-	315	-	-	-	174	-	-	-	-
Non-specified	-	-	-	-	-	-	-	-	-	-	-
OTHER SECTORS	-	45	-	-	-	6	10	-	-	-	-
Agriculture	-	-	-	-	-	-	-	-	-	-	-
Comm. and Publ. Services	-	24	-	-	-	-	10	-	-	-	-
Residential	-	21	-	-	-	6	-	-	-	-	-
Non-specified	-	-	-	-	-	-	-	-	-	-	-
NON-ENERGY USE	-	-	-	-	-	-	-	-	-	-	9

Trinidad-&-Tobago / Trinité-et-Tobago

SUPPLY AND CONSUMPTION 1997	Gas (TJ)				Comb. Renew. & Waste (TJ)				(GWh)	(TJ)
	Natural Gas	Gas Works	Coke Ovens	Blast Furnaces	Solid Biomass	Gas/Liquids from Biomass	Municipal Waste	Industrial Waste	Electricity	Heat
Production	303781	-	-	-	1238	-	-	-	4988	-
Imports	-	-	-	-	-	-	-	-	-	-
Exports	-	-	-	-	-	-	-	-	-	-
Intl. Marine Bunkers	-	-	-	-	-	-	-	-	-	-
Stock Changes	-	-	-	-	-	-	-	-	-	-
DOMESTIC SUPPLY	303781	-	-	-	1238	-	-	-	4988	-
Transfers and Stat. Diff.	-119	-	-	-	-	-	-	-	-147	-
TRANSFORMATION	116537	-	-	-	1238	-	-	-	-	-
Electricity and CHP Plants	68486	-	-	-	1238	-	-	-	-	-
Petroleum Refineries	-	-	-	-	-	-	-	-	-	-
Other Transform. Sector	48051	-	-	-	-	-	-	-	-	-
ENERGY SECTOR	38894	-	-	-	-	-	-	-	137	-
DISTRIBUTION LOSSES	-	-	-	-	-	-	-	-	401	-
FINAL CONSUMPTION	148231	-	-	-	-	-	-	-	4303	-
INDUSTRY SECTOR	148231	-	-	-	-	-	-	-	2882	-
Iron and Steel	19216	-	-	-	-	-	-	-	-	-
Chemical and Petrochem.	119254	-	-	-	-	-	-	-	-	-
Non-Metallic Minerals	5042	-	-	-	-	-	-	-	-	-
Non-specified	4719	-	-	-	-	-	-	-	2882	-
TRANSPORT SECTOR	-	-	-	-	-	-	-	-	-	-
Air	-	-	-	-	-	-	-	-	-	-
Road	-	-	-	-	-	-	-	-	-	-
Non-specified	-	-	-	-	-	-	-	-	-	-
OTHER SECTORS	-	-	-	-	-	-	-	-	1421	-
Agriculture	-	-	-	-	-	-	-	-	-	-
Comm. and Publ. Services	-	-	-	-	-	-	-	-	413	-
Residential	-	-	-	-	-	-	-	-	1008	-
Non-specified	-	-	-	-	-	-	-	-	-	-
NON-ENERGY USE	-	-	-	-	-	-	-	-	-	-

APPROVISIONNEMENT ET DEMANDE 1998	Gaz (TJ)				En. Re. Comb. & Déchets (TJ)				(GWh)	(TJ)
	Gaz naturel	Usinès à gaz	Cokeries	Hauts fourneaux	Biomasse solide	Gaz/Liquides tirés de biomasse	Déchets urbains	Déchets industriels	Electricité	Chaleur
Production	366002	-	-	-	1238	-	-	-	5169	-
Imports	-	-	-	-	-	-	-	-	-	-
Exports	-	-	-	-	-	-	-	-	-	-
Intl. Marine Bunkers	-	-	-	-	-	-	-	-	-	-
Stock Changes	-	-	-	-	-	-	-	-	-	-
DOMESTIC SUPPLY	366002	-	-	-	1238	-	-	-	5169	-
Transfers and Stat. Diff.	612	-	-	-	-	-	-	-	-139	-
TRANSFORMATION	145586	-	-	-	1238	-	-	-	-	-
Electricity and CHP Plants	73710	-	-	-	1238	-	-	-	-	-
Petroleum Refineries	-	-	-	-	-	-	-	-	-	-
Other Transform. Sector	71876	-	-	-	-	-	-	-	-	-
ENERGY SECTOR	48694	-	-	-	-	-	-	-	142	-
DISTRIBUTION LOSSES	-	-	-	-	-	-	-	-	418	-
FINAL CONSUMPTION	172334	-	-	-	-	-	-	-	4470	-
INDUSTRY SECTOR	172334	-	-	-	-	-	-	-	2938	-
Iron and Steel	19100	-	-	-	-	-	-	-	-	-
Chemical and Petrochem.	143680	-	-	-	-	-	-	-	-	-
Non-Metallic Minerals	4835	-	-	-	-	-	-	-	-	-
Non-specified	4719	-	-	-	-	-	-	-	2938	-
TRANSPORT SECTOR	-	-	-	-	-	-	-	-	-	-
Air	-	-	-	-	-	-	-	-	-	-
Road	-	-	-	-	-	-	-	-	-	-
Non-specified	-	-	-	-	-	-	-	-	-	-
OTHER SECTORS	-	-	-	-	-	-	-	-	1532	-
Agriculture	-	-	-	-	-	-	-	-	-	-
Comm. and Publ. Services	-	-	-	-	-	-	-	-	434	-
Residential	-	-	-	-	-	-	-	-	1098	-
Non-specified	-	-	-	-	-	-	-	-	-	-
NON-ENERGY USE	-	-	-	-	-	-	-	-	-	-

Tunisia / Tunisie : 1997

SUPPLY AND CONSUMPTION	Coal / *Charbon* (1000 tonnes)							Oil / *Pétrole* (1000 tonnes)			
	Coking Coal	Other Bit. Coal	Sub-Bit. Coal	Lignite	Peat	Oven and Gas Coke	Pat. Fuel and BKB	Crude Oil	NGL	Feed-stocks	Additives
APPROVISIONNEMENT ET DEMANDE	*Charbon à coke*	*Autres charb. bit.*	*Charbon sous-bit.*	*Lignite*	*Tourbe*	*Coke de four/gaz*	*Agg./briq. de lignite*	*Pétrole brut*	*LGN*	*Produits d'aliment.*	*Additifs*
Production	-	-	-	-	-	-	-	3789	107	-	-
From Other Sources	-	-	-	-	-	-	-	-	-	-	-
Imports	-	-	-	-	-	106	-	915	-	-	-
Exports	-	-	-	-	-	-	-	-2780	-	-	-
Intl. Marine Bunkers	-	-	-	-	-	-	-	-	-	-	-
Stock Changes	-	-	-	-	-	-	-	-116	-	-	-
DOMESTIC SUPPLY	-	-	-	-	-	106	-	1808	107	-	-
Transfers	-	-	-	-	-	-	-	-	-107	-	-
Statistical Differences	-	-	-	-	-	-	-	201	-	-	-
TRANSFORMATION	-	-	-	-	-	85	-	2009	-	-	-
Electricity Plants	-	-	-	-	-	-	-	-	-	-	-
CHP Plants	-	-	-	-	-	-	-	-	-	-	-
Heat Plants	-	-	-	-	-	-	-	-	-	-	-
Blast Furnaces/Gas Works	-	-	-	-	-	85	-	-	-	-	-
Coke/Pat. Fuel/BKB Plants	-	-	-	-	-	-	-	-	-	-	-
Petroleum Refineries	-	-	-	-	-	-	-	2009	-	-	-
Petrochemical Industry	-	-	-	-	-	-	-	-	-	-	-
Liquefaction	-	-	-	-	-	-	-	-	-	-	-
Other Transform. Sector	-	-	-	-	-	-	-	-	-	-	-
ENERGY SECTOR	-	-	-	-	-	-	-	-	-	-	-
Coal Mines	-	-	-	-	-	-	-	-	-	-	-
Oil and Gas Extraction	-	-	-	-	-	-	-	-	-	-	-
Petroleum Refineries	-	-	-	-	-	-	-	-	-	-	-
Electr., CHP+Heat Plants	-	-	-	-	-	-	-	-	-	-	-
Pumped Storage (Elec.)	-	-	-	-	-	-	-	-	-	-	-
Other Energy Sector	-	-	-	-	-	-	-	-	-	-	-
Distribution Losses	-	-	-	-	-	-	-	-	-	-	-
FINAL CONSUMPTION	-	-	-	-	-	21	-	-	-	-	-
INDUSTRY SECTOR	-	-	-	-	-	21	-	-	-	-	-
Iron and Steel	-	-	-	-	-	21	-	-	-	-	-
Chemical and Petrochem.	-	-	-	-	-	-	-	-	-	-	-
of which: Feedstocks	-	-	-	-	-	-	-	-	-	-	-
Non-Ferrous Metals	-	-	-	-	-	-	-	-	-	-	-
Non-Metallic Minerals	-	-	-	-	-	-	-	-	-	-	-
Transport Equipment	-	-	-	-	-	-	-	-	-	-	-
Machinery	-	-	-	-	-	-	-	-	-	-	-
Mining and Quarrying	-	-	-	-	-	-	-	-	-	-	-
Food and Tobacco	-	-	-	-	-	-	-	-	-	-	-
Paper, Pulp and Print	-	-	-	-	-	-	-	-	-	-	-
Wood and Wood Products	-	-	-	-	-	-	-	-	-	-	-
Construction	-	-	-	-	-	-	-	-	-	-	-
Textile and Leather	-	-	-	-	-	-	-	-	-	-	-
Non-specified	-	-	-	-	-	-	-	-	-	-	-
TRANSPORT SECTOR	-	-	-	-	-	-	-	-	-	-	-
Air	-	-	-	-	-	-	-	-	-	-	-
Road	-	-	-	-	-	-	-	-	-	-	-
Rail	-	-	-	-	-	-	-	-	-	-	-
Pipeline Transport	-	-	-	-	-	-	-	-	-	-	-
Internal Navigation	-	-	-	-	-	-	-	-	-	-	-
Non-specified	-	-	-	-	-	-	-	-	-	-	-
OTHER SECTORS	-	-	-	-	-	-	-	-	-	-	-
Agriculture	-	-	-	-	-	-	-	-	-	-	-
Comm. and Publ. Services	-	-	-	-	-	-	-	-	-	-	-
Residential	-	-	-	-	-	-	-	-	-	-	-
Non-specified	-	-	-	-	-	-	-	-	-	-	-
NON-ENERGY USE	-	-	-	-	-	-	-	-	-	-	-
in Industry/Trans./Energy	-	-	-	-	-	-	-	-	-	-	-
in Transport	-	-	-	-	-	-	-	-	-	-	-
in Other Sectors	-	-	-	-	-	-	-	-	-	-	-

Tunisia / Tunisie : 1997

SUPPLY AND CONSUMPTION APPROVISIONNEMENT ET DEMANDE	Oil cont. / *Pétrole cont.* (1000 tonnes)										
	Refinery Gas *Gaz de raffinerie*	LPG + Ethane *GPL + éthane*	Motor Gasoline *Essence moteur*	Aviation Gasoline *Essence aviation*	Jet Fuel *Carbu-réacteurs*	Kerosene *Kérosène*	Gas/ Diesel *Gazole*	Heavy Fuel Oil *Fioul lourd*	Naphtha *Naphta*	Petrol. Coke *Coke de pétrole*	Other Prod. *Autres prod.*
Production	34	13	339	-	-	107	595	646	234	-	41
From Other Sources	-	-	-	-	-	-	-	-	-	-	-
Imports	-	226	201	-	292	64	939	684	-	-	133
Exports	-	-4	-	-	-	-	-	-565	-223	-	-36
Intl. Marine Bunkers	-	-	-	-	-	-	-	-18	-	-	-
Stock Changes	-	-	-200	-	-9	8	-100	61	-11	-	8
DOMESTIC SUPPLY	**34**	**235**	**340**	**-**	**283**	**179**	**1434**	**808**	**-**	**-**	**146**
Transfers	-	107	-	-	-	-	-	-	-	-	-
Statistical Differences	-	-	-	-	-	-1	-	-1	-	-	2
TRANSFORMATION	**-**	**-**	**-**	**-**	**-**	**-**	**11**	**360**	**-**	**-**	**-**
Electricity Plants	-	-	-	-	-	-	11	360	-	-	-
CHP Plants	-	-	-	-	-	-	-	-	-	-	-
Heat Plants	-	-	-	-	-	-	-	-	-	-	-
Blast Furnaces/Gas Works	-	-	-	-	-	-	-	-	-	-	-
Coke/Pat. Fuel/BKB Plants	-	-	-	-	-	-	-	-	-	-	-
Petroleum Refineries	-	-	-	-	-	-	-	-	-	-	-
Petrochemical Industry	-	-	-	-	-	-	-	-	-	-	-
Liquefaction	-	-	-	-	-	-	-	-	-	-	-
Other Transform. Sector	-	-	-	-	-	-	-	-	-	-	-
ENERGY SECTOR	**34**	**-**	**-**	**-**	**-**	**-**	**-**	**34**	**-**	**-**	**-**
Coal Mines	-	-	-	-	-	-	-	-	-	-	-
Oil and Gas Extraction	-	-	-	-	-	-	-	-	-	-	-
Petroleum Refineries	34	-	-	-	-	-	-	34	-	-	-
Electr., CHP+Heat Plants	-	-	-	-	-	-	-	-	-	-	-
Pumped Storage (Elec.)	-	-	-	-	-	-	-	-	-	-	-
Other Energy Sector	-	-	-	-	-	-	-	-	-	-	-
Distribution Losses	-	-	-	-	-	-	-	-	-	-	-
FINAL CONSUMPTION	**-**	**342**	**340**	**-**	**283**	**178**	**1423**	**413**	**-**	**-**	**148**
INDUSTRY SECTOR	**-**	**11**	**-**	**-**	**-**	**6**	**181**	**383**	**-**	**-**	**-**
Iron and Steel	-	-	-	-	-	-	-	-	-	-	-
Chemical and Petrochem.	-	-	-	-	-	-	-	-	-	-	-
of which: Feedstocks	-	-	-	-	-	-	-	-	-	-	-
Non-Ferrous Metals	-	-	-	-	-	-	-	-	-	-	-
Non-Metallic Minerals	-	-	-	-	-	-	-	-	-	-	-
Transport Equipment	-	-	-	-	-	-	-	-	-	-	-
Machinery	-	-	-	-	-	-	-	-	-	-	-
Mining and Quarrying	-	-	-	-	-	-	-	-	-	-	-
Food and Tobacco	-	-	-	-	-	-	-	-	-	-	-
Paper, Pulp and Print	-	-	-	-	-	-	-	-	-	-	-
Wood and Wood Products	-	-	-	-	-	-	-	-	-	-	-
Construction	-	-	-	-	-	-	-	-	-	-	-
Textile and Leather	-	-	-	-	-	-	-	-	-	-	-
Non-specified	-	11	-	-	-	6	181	383	-	-	-
TRANSPORT SECTOR	**-**	**6**	**340**	**-**	**283**	**-**	**766**	**6**	**-**	**-**	**-**
Air	-	-	-	-	283	-	-	-	-	-	-
Road	-	6	340	-	-	-	766	-	-	-	-
Rail	-	-	-	-	-	-	-	-	-	-	-
Pipeline Transport	-	-	-	-	-	-	-	-	-	-	-
Internal Navigation	-	-	-	-	-	-	-	-	-	-	-
Non-specified	-	-	-	-	-	-	-	6	-	-	-
OTHER SECTORS	**-**	**325**	**-**	**-**	**-**	**172**	**476**	**24**	**-**	**-**	**-**
Agriculture	-	-	-	-	-	-	256	23	-	-	-
Comm. and Publ. Services	-	49	-	-	-	9	195	1	-	-	-
Residential	-	276	-	-	-	163	25	-	-	-	-
Non-specified	-	-	-	-	-	-	-	-	-	-	-
NON-ENERGY USE	**-**	**-**	**-**	**-**	**-**	**-**	**-**	**-**	**-**	**-**	**148**
in Industry/Transf./Energy	-	-	-	-	-	-	-	-	-	-	148
in Transport	-	-	-	-	-	-	-	-	-	-	-
in Other Sectors	-	-	-	-	-	-	-	-	-	-	-

Tunisia / Tunisie : 1997

SUPPLY AND CONSUMPTION APPROVISIONNEMENT ET DEMANDE	Gas / Gaz (TJ)				Comb. Renew. & Waste / En. Re. Comb. & Déchets (TJ)				(GWh)	(TJ)
	Natural Gas Gaz naturel	Gas Works Usines à gaz	Coke Ovens Cokeries	Blast Furnaces Hauts fourneaux	Solid Biomass Biomasse solide	Gas/Liquids from Biomass Gaz/Liquides tirés de biomasse	Municipal Waste Déchets urbains	Industrial Waste Déchets industriels	Electricity Electricité	Heat Chaleur
Production	73139	-	-	925	49085	-	-	-	8485	-
From Other Sources	-	-	-	-	-	-	-	-	-	-
Imports	29836	-	-	-	-	-	-	-	7	-
Exports	-	-	-	-	-	-	-	-	-	-
Intl. Marine Bunkers	-	-	-	-	-	-	-	-	-	-
Stock Changes	-	-	-	-	-	-	-	-	-	-
DOMESTIC SUPPLY	**102975**	**-**	**-**	**925**	**49085**	**-**	**-**	**-**	**8492**	**-**
Transfers	-	-	-	-	-	-	-	-	-	-
Statistical Differences	209	-	-	-	-1	-	-	-	4	-
TRANSFORMATION	**79523**	**-**	**-**	**-**	**7229**	**-**	**-**	**-**	**-**	**-**
Electricity Plants	79523	-	-	-	-	-	-	-	-	-
CHP Plants	-	-	-	-	-	-	-	-	-	-
Heat Plants	-	-	-	-	-	-	-	-	-	-
Blast Furnaces/Gas Works	-	-	-	-	-	-	-	-	-	-
Coke/Pat. Fuel/BKB Plants	-	-	-	-	-	-	-	-	-	-
Petroleum Refineries	-	-	-	-	-	-	-	-	-	-
Petrochemical Industry	-	-	-	-	-	-	-	-	-	-
Liquefaction	-	-	-	-	-	-	-	-	-	-
Other Transform. Sector	-	-	-	-	7229	-	-	-	-	-
ENERGY SECTOR	**-**	**-**	**-**	**-**	**-**	**-**	**-**	**-**	**404**	**-**
Coal Mines	-	-	-	-	-	-	-	-	-	-
Oil and Gas Extraction	-	-	-	-	-	-	-	-	-	-
Petroleum Refineries	-	-	-	-	-	-	-	-	-	-
Electr., CHP+Heat Plants	-	-	-	-	-	-	-	-	404	-
Pumped Storage (Elec.)	-	-	-	-	-	-	-	-	-	-
Other Energy Sector	-	-	-	-	-	-	-	-	-	-
Distribution Losses	**-**	**-**	**-**	**-**	**-**	**-**	**-**	**-**	**877**	**-**
FINAL CONSUMPTION	**23661**	**-**	**-**	**925**	**41855**	**-**	**-**	**-**	**7215**	**-**
INDUSTRY SECTOR	**17896**	**-**	**-**	**925**	**-**	**-**	**-**	**-**	**3548**	**-**
Iron and Steel	-	-	-	925	-	-	-	-	177	-
Chemical and Petrochem.	-	-	-	-	-	-	-	-	164	-
of which: Feedstocks	-	-	-	-	-	-	-	-	-	-
Non-Ferrous Metals	-	-	-	-	-	-	-	-	-	-
Non-Metallic Minerals	-	-	-	-	-	-	-	-	893	-
Transport Equipment	-	-	-	-	-	-	-	-	-	-
Machinery	-	-	-	-	-	-	-	-	-	-
Mining and Quarrying	-	-	-	-	-	-	-	-	237	-
Food and Tobacco	-	-	-	-	-	-	-	-	329	-
Paper, Pulp and Print	-	-	-	-	-	-	-	-	94	-
Wood and Wood Products	-	-	-	-	-	-	-	-	-	-
Construction	-	-	-	-	-	-	-	-	-	-
Textile and Leather	-	-	-	-	-	-	-	-	348	-
Non-specified	17896	-	-	-	-	-	-	-	1306	-
TRANSPORT SECTOR	**-**	**-**	**-**	**-**	**-**	**-**	**-**	**-**	**158**	**-**
Air	-	-	-	-	-	-	-	-	-	-
Road	-	-	-	-	-	-	-	-	-	-
Rail	-	-	-	-	-	-	-	-	-	-
Pipeline Transport	-	-	-	-	-	-	-	-	-	-
Internal Navigation	-	-	-	-	-	-	-	-	-	-
Non-specified	-	-	-	-	-	-	-	-	158	-
OTHER SECTORS	**5765**	**-**	**-**	**-**	**41855**	**-**	**-**	**-**	**3509**	**-**
Agriculture	-	-	-	-	-	-	-	-	309	-
Comm. and Publ. Services	1701	-	-	-	-	-	-	-	1188	-
Residential	4064	-	-	-	41855	-	-	-	2012	-
Non-specified	-	-	-	-	-	-	-	-	-	-
NON-ENERGY USE	**-**	**-**	**-**	**-**	**-**	**-**	**-**	**-**	**-**	**-**
in Industry/Transf./Energy	-	-	-	-	-	-	-	-	-	-
in Transport	-	-	-	-	-	-	-	-	-	-
in Other Sectors	-	-	-	-	-	-	-	-	-	-

Tunisia / Tunisie : 1998

SUPPLY AND CONSUMPTION / APPROVISIONNEMENT ET DEMANDE	Coal / Charbon (1000 tonnes)							Oil / Pétrole (1000 tonnes)			
	Coking Coal / Charbon à coke	Other Bit. Coal / Autres charb. bit.	Sub-Bit. Coal / Charbon sous-bit.	Lignite / Lignite	Peat / Tourbe	Oven and Gas Coke / Coke de four/gaz	Pat. Fuel and BKB / Agg./briq. de lignite	Crude Oil / Pétrole brut	NGL / LGN	Feed-stocks / Produits d'aliment.	Additives / Additifs
Production	-	-	-	-	-	-	-	3898	113	-	-
From Other Sources	-	-	-	-	-	-	-	-	-	-	-
Imports	-	-	-	-	-	87	-	953	-	-	-
Exports	-	-	-	-	-	-	-	-2842	-	-	-
Intl. Marine Bunkers	-	-	-	-	-	-	-	-	-	-	-
Stock Changes	-	-	-	-	-	-	-	-113	-	-	-
DOMESTIC SUPPLY	-	-	-	-	-	87	-	1896	113	-	-
Transfers	-	-	-	-	-	-	-	-	-113	-	-
Statistical Differences	-	-	-	-	-	-	-	-88	-	-	-
TRANSFORMATION	-	-	-	-	-	70	-	1808	-	-	-
Electricity Plants	-	-	-	-	-	-	-	-	-	-	-
CHP Plants	-	-	-	-	-	-	-	-	-	-	-
Heat Plants	-	-	-	-	-	-	-	-	-	-	-
Blast Furnaces/Gas Works	-	-	-	-	-	70	-	-	-	-	-
Coke/Pat. Fuel/BKB Plants	-	-	-	-	-	-	-	-	-	-	-
Petroleum Refineries	-	-	-	-	-	-	-	1808	-	-	-
Petrochemical Industry	-	-	-	-	-	-	-	-	-	-	-
Liquefaction	-	-	-	-	-	-	-	-	-	-	-
Other Transform. Sector	-	-	-	-	-	-	-	-	-	-	-
ENERGY SECTOR	-	-	-	-	-	-	-	-	-	-	-
Coal Mines	-	-	-	-	-	-	-	-	-	-	-
Oil and Gas Extraction	-	-	-	-	-	-	-	-	-	-	-
Petroleum Refineries	-	-	-	-	-	-	-	-	-	-	-
Electr., CHP+Heat Plants	-	-	-	-	-	-	-	-	-	-	-
Pumped Storage (Elec.)	-	-	-	-	-	-	-	-	-	-	-
Other Energy Sector	-	-	-	-	-	-	-	-	-	-	-
Distribution Losses	-	-	-	-	-	-	-	-	-	-	-
FINAL CONSUMPTION	-	-	-	-	-	17	-	-	-	-	-
INDUSTRY SECTOR	-	-	-	-	-	17	-	-	-	-	-
Iron and Steel	-	-	-	-	-	17	-	-	-	-	-
Chemical and Petrochem.	-	-	-	-	-	-	-	-	-	-	-
of which: Feedstocks	-	-	-	-	-	-	-	-	-	-	-
Non-Ferrous Metals	-	-	-	-	-	-	-	-	-	-	-
Non-Metallic Minerals	-	-	-	-	-	-	-	-	-	-	-
Transport Equipment	-	-	-	-	-	-	-	-	-	-	-
Machinery	-	-	-	-	-	-	-	-	-	-	-
Mining and Quarrying	-	-	-	-	-	-	-	-	-	-	-
Food and Tobacco	-	-	-	-	-	-	-	-	-	-	-
Paper, Pulp and Print	-	-	-	-	-	-	-	-	-	-	-
Wood and Wood Products	-	-	-	-	-	-	-	-	-	-	-
Construction	-	-	-	-	-	-	-	-	-	-	-
Textile and Leather	-	-	-	-	-	-	-	-	-	-	-
Non-specified	-	-	-	-	-	-	-	-	-	-	-
TRANSPORT SECTOR	-	-	-	-	-	-	-	-	-	-	-
Air	-	-	-	-	-	-	-	-	-	-	-
Road	-	-	-	-	-	-	-	-	-	-	-
Rail	-	-	-	-	-	-	-	-	-	-	-
Pipeline Transport	-	-	-	-	-	-	-	-	-	-	-
Internal Navigation	-	-	-	-	-	-	-	-	-	-	-
Non-specified	-	-	-	-	-	-	-	-	-	-	-
OTHER SECTORS	-	-	-	-	-	-	-	-	-	-	-
Agriculture	-	-	-	-	-	-	-	-	-	-	-
Comm. and Publ. Services	-	-	-	-	-	-	-	-	-	-	-
Residential	-	-	-	-	-	-	-	-	-	-	-
Non-specified	-	-	-	-	-	-	-	-	-	-	-
NON-ENERGY USE	-	-	-	-	-	-	-	-	-	-	-
in Industry/Trans./Energy	-	-	-	-	-	-	-	-	-	-	-
in Transport	-	-	-	-	-	-	-	-	-	-	-
in Other Sectors	-	-	-	-	-	-	-	-	-	-	-

Tunisia / Tunisie : 1998

SUPPLY AND CONSUMPTION *APPROVISIONNEMENT ET DEMANDE*	Oil cont. / *Pétrole cont.* (1000 tonnes)										
	Refinery Gas *Gaz de raffinerie*	LPG + Ethane *GPL + éthane*	Motor Gasoline *Essence moteur*	Aviation Gasoline *Essence aviation*	Jet Fuel *Carbu- réacteurs*	Kerosene *Kérosène*	Gas/ Diesel *Gazole*	Heavy Fuel Oil *Fioul lourd*	Naphtha *Naphta*	Petrol. Coke *Coke de pétrole*	Other Prod. *Autres prod.*
Production	34	14	344	-	-	127	560	631	58	-	40
From Other Sources	-	-	-	-	-	-	-	-	-	-	-
Imports	-	222	79	-	297	41	889	846	-	-	145
Exports	-	-1	-	-	-	-	-33	-549	-78	-	-34
Intl. Marine Bunkers	-	-	-	-	-	-	-	-18	-	-	-
Stock Changes	-	5	-77	-	4	14	73	-99	20	-	14
DOMESTIC SUPPLY	**34**	**240**	**346**	**-**	**301**	**182**	**1489**	**811**	**-**	**-**	**165**
Transfers	-	113	-	-	-	-	-	-	-	-	-
Statistical Differences	-	-1	-	-	-	-	-1	1	-	-	1
TRANSFORMATION	**-**	**-**	**-**	**-**	**-**	**-**	**5**	**305**	**-**	**-**	**-**
Electricity Plants	-	-	-	-	-	-	5	305	-	-	-
CHP Plants	-	-	-	-	-	-	-	-	-	-	-
Heat Plants	-	-	-	-	-	-	-	-	-	-	-
Blast Furnaces/Gas Works	-	-	-	-	-	-	-	-	-	-	-
Coke/Pat. Fuel/BKB Plants	-	-	-	-	-	-	-	-	-	-	-
Petroleum Refineries	-	-	-	-	-	-	-	-	-	-	-
Petrochemical Industry	-	-	-	-	-	-	-	-	-	-	-
Liquefaction	-	-	-	-	-	-	-	-	-	-	-
Other Transform. Sector	-	-	-	-	-	-	-	-	-	-	-
ENERGY SECTOR	**34**	**-**	**-**	**-**	**-**	**-**	**-**	**34**	**-**	**-**	**-**
Coal Mines	-	-	-	-	-	-	-	-	-	-	-
Oil and Gas Extraction	-	-	-	-	-	-	-	-	-	-	-
Petroleum Refineries	34	-	-	-	-	-	-	34	-	-	-
Electr., CHP+Heat Plants	-	-	-	-	-	-	-	-	-	-	-
Pumped Storage (Elec.)	-	-	-	-	-	-	-	-	-	-	-
Other Energy Sector	-	-	-	-	-	-	-	-	-	-	-
Distribution Losses	-	-	-	-	-	-	-	-	-	-	-
FINAL CONSUMPTION	**-**	**352**	**346**	**-**	**301**	**182**	**1483**	**473**	**-**	**-**	**166**
INDUSTRY SECTOR	**-**	**43**	**-**	**-**	**-**	**7**	**196**	**442**	**-**	**-**	**-**
Iron and Steel	-	-	-	-	-	-	-	-	-	-	-
Chemical and Petrochem.	-	-	-	-	-	-	-	-	-	-	-
of which: Feedstocks	-	-	-	-	-	-	-	-	-	-	-
Non-Ferrous Metals	-	-	-	-	-	-	-	-	-	-	-
Non-Metallic Minerals	-	-	-	-	-	-	-	-	-	-	-
Transport Equipment	-	-	-	-	-	-	-	-	-	-	-
Machinery	-	-	-	-	-	-	-	-	-	-	-
Mining and Quarrying	-	-	-	-	-	-	-	-	-	-	-
Food and Tobacco	-	-	-	-	-	-	-	-	-	-	-
Paper, Pulp and Print	-	-	-	-	-	-	-	-	-	-	-
Wood and Wood Products	-	-	-	-	-	-	-	-	-	-	-
Construction	-	-	-	-	-	-	-	-	-	-	-
Textile and Leather	-	-	-	-	-	-	-	-	-	-	-
Non-specified	-	43	-	-	-	7	196	442	-	-	-
TRANSPORT SECTOR	**-**	**7**	**346**	**-**	**301**	**-**	**790**	**4**	**-**	**-**	**-**
Air	-	-	-	-	301	-	-	-	-	-	-
Road	-	7	346	-	-	-	790	-	-	-	-
Rail	-	-	-	-	-	-	-	-	-	-	-
Pipeline Transport	-	-	-	-	-	-	-	-	-	-	-
Internal Navigation	-	-	-	-	-	-	-	-	-	-	-
Non-specified	-	-	-	-	-	-	-	4	-	-	-
OTHER SECTORS	**-**	**302**	**-**	**-**	**-**	**175**	**497**	**27**	**-**	**-**	**-**
Agriculture	-	-	-	-	-	-	269	26	-	-	-
Comm. and Publ. Services	-	16	-	-	-	9	202	1	-	-	-
Residential	-	286	-	-	-	166	26	-	-	-	-
Non-specified	-	-	-	-	-	-	-	-	-	-	-
NON-ENERGY USE	**-**	**-**	**-**	**-**	**-**	**-**	**-**	**-**	**-**	**-**	**166**
in Industry/Transf./Energy	-	-	-	-	-	-	-	-	-	-	166
in Transport	-	-	-	-	-	-	-	-	-	-	-
in Other Sectors	-	-	-	-	-	-	-	-	-	-	-

Tunisia / Tunisie : 1998

SUPPLY AND CONSUMPTION	Gas / Gaz (TJ)				Comb. Renew. & Waste / En. Re. Comb. & Déchets (TJ)				(GWh)	(TJ)
APPROVISIONNEMENT ET DEMANDE	Natural Gas	Gas Works	Coke Ovens	Blast Furnaces	Solid Biomass	Gas/Liquids from Biomass	Municipal Waste	Industrial Waste	Electricity	Heat
	Gaz naturel	Usines à gaz	Cokeries	Hauts fourneaux	Biomasse solide	Gaz/Liquides tirés de biomasse	Déchets urbains	Déchets industriels	Electricité	Chaleur
Production	83887	-	-	762	50582	-	-	-	9106	-
From Other Sources	-	-	-	-	-					-
Imports	30678	-	-	-	-		-	-	-	-
Exports	-	-	-	-	-			-	-	-
Intl. Marine Bunkers	-	-	-	-	-			-	-19	-
Stock Changes	-	-	-	-	-				-	-
DOMESTIC SUPPLY	114565	-	-	762	50582	-	-	-	9087	-
Transfers	-	-	-	-	-		-	-		-
Statistical Differences	-	-	-	-	-1		-	-	1	-
TRANSFORMATION	89532	-	-	-	7449	-	-	-	-	-
Electricity Plants	89532	-	-	-	-		-	-	-	
CHP Plants	-	-	-	-	-		-	-	-	
Heat Plants	-	-	-	-	-		-	-	-	
Blast Furnaces/Gas Works	-	-	-	-	-		-	-	-	
Coke/Pat. Fuel/BKB Plants	-	-	-	-	-		-	-	-	
Petroleum Refineries	-	-	-	-	-		-	-	-	
Petrochemical Industry	-	-	-	-	-		-	-		
Liquefaction	-	-	-	-	-		-	-	-	
Other Transform. Sector	-	-	-	-	7449		-	-	-	
ENERGY SECTOR	-	-	-	-	-	-	-	-	434	-
Coal Mines	-	-	-	-	-		-	-	-	
Oil and Gas Extraction	-	-	-	-	-		-	-	-	
Petroleum Refineries	-	-	-	-	-		-	-	-	
Electr., CHP+Heat Plants	-	-	-	-	-		-	-	434	
Pumped Storage (Elec.)	-	-	-	-	-		-	-	-	
Other Energy Sector	-	-	-	-	-		-	-	-	
Distribution Losses	-	-	-	-	-	-	-	-	963	-
FINAL CONSUMPTION	25033	-	-	762	43132	-	-	-	7691	-
INDUSTRY SECTOR	18603	-	-	762	-	-	-	-	3753	-
Iron and Steel	-	-	-	762	-		-	-	176	-
Chemical and Petrochem.	-	-	-	-	-		-	-	151	-
of which: Feedstocks	-	-	-	-	-		-	-	-	-
Non-Ferrous Metals	-	-	-	-	-		-	-	-	
Non-Metallic Minerals	-	-	-	-	-		-	-	921	
Transport Equipment	-	-	-	-	-		-	-	-	-
Machinery	-	-	-	-	-		-	-	-	-
Mining and Quarrying	-	-	-	-	-		-	-	267	
Food and Tobacco	-	-	-	-	-		-	-	361	
Paper, Pulp and Print	-	-	-	-	-		-	-	102	
Wood and Wood Products	-	-	-	-	-		-	-	-	
Construction	-	-	-	-	-		-	-	-	
Textile and Leather	-	-	-	-	-		-	-	365	
Non-specified	18603	-	-	-	-		-	-	1410	-
TRANSPORT SECTOR	-	-	-	-	-	-	-	-	170	-
Air	-	-	-	-	-		-	-	-	-
Road	-	-	-	-	-		-	-		
Rail	-	-	-	-	-		-	-		
Pipeline Transport	-	-	-	-	-		-	-	-	
Internal Navigation	-	-	-	-	-		-	-	-	
Non-specified	-	-	-	-	-		-	-	170	-
OTHER SECTORS	6430	-	-	-	43132	-	-	-	3768	-
Agriculture	-	-	-	-	-		-	-	352	-
Comm. and Publ. Services	1816	-	-	-	-		-	-	1356	-
Residential	4614	-	-	-	43132		-	-	2060	-
Non-specified	-	-	-	-	-		-	-	-	-
NON-ENERGY USE	-	-	-	-	-	-	-	-	-	-
in Industry/Transf./Energy	-	-	-	-	-		-	-	-	-
in Transport	-	-	-	-	-		-	-	-	-
in Other Sectors	-	-	-	-	-		-	-	-	-

Turkmenistan / Turkménistan

SUPPLY AND CONSUMPTION 1997	Coal (1000 tonnes)							Oil (1000 tonnes)			
	Coking Coal	Other Bit. Coal	Sub-Bit. Coal	Lignite	Peat	Oven and Gas Coke	Pat. Fuel and BKB	Crude Oil	NGL	Feed-stocks	Additives
Production	-	-	-	-	-	-	-	5149	219	-	-
Imports	-	-	-	-	-	-	-	500	-	-	-
Exports	-	-	-	-	-	-	-	-1400	-	-	-
Intl. Marine Bunkers	-	-	-	-	-	-	-	-	-	-	-
Stock Changes	-	-	-	-	-	-	-	-	-	-	-
DOMESTIC SUPPLY	-	-	-	-	-	-	-	**4249**	**219**	-	-
Transfers and Stat. Diff.	-	-	-	-	-	-	-	-	-	-	-
TRANSFORMATION	-	-	-	-	-	-	-	**4249**	**219**	-	-
Electricity and CHP Plants	-	-	-	-	-	-	-	-	-	-	-
Petroleum Refineries	-	-	-	-	-	-	-	4249	-	-	-
Other Transform. Sector	-	-	-	-	-	-	-	-	219	-	-
ENERGY SECTOR	-	-	-	-	-	-	-	-	-	-	-
DISTRIBUTION LOSSES	-	-	-	-	-	-	-	-	-	-	-
FINAL CONSUMPTION	-	-	-	-	-	-	-	-	-	-	-
INDUSTRY SECTOR	-	-	-	-	-	-	-	-	-	-	-
Iron and Steel	-	-	-	-	-	-	-	-	-	-	-
Chemical and Petrochem.	-	-	-	-	-	-	-	-	-	-	-
Non-Metallic Minerals	-	-	-	-	-	-	-	-	-	-	-
Non-specified	-	-	-	-	-	-	-	-	-	-	-
TRANSPORT SECTOR	-	-	-	-	-	-	-	-	-	-	-
Air	-	-	-	-	-	-	-	-	-	-	-
Road	-	-	-	-	-	-	-	-	-	-	-
Non-specified	-	-	-	-	-	-	-	-	-	-	-
OTHER SECTORS	-	-	-	-	-	-	-	-	-	-	-
Agriculture	-	-	-	-	-	-	-	-	-	-	-
Comm. and Publ. Services	-	-	-	-	-	-	-	-	-	-	-
Residential	-	-	-	-	-	-	-	-	-	-	-
Non-specified	-	-	-	-	-	-	-	-	-	-	-
NON-ENERGY USE	-	-	-	-	-	-	-	-	-	-	-

APPROVISIONNEMENT ET DEMANDE 1998	Charbon (1000 tonnes)							Pétrole (1000 tonnes)			
	Charbon à coke	Autres charb. bit.	Charbon sous-bit.	Lignite	Tourbe	Coke de four/gaz	Agg./briq. de lignite	Pétrole brut	LGN	Produits d'aliment.	Additifs
Production	-	-	-	-	-	-	-	6375	263	-	-
Imports	-	-	-	-	-	-	-	500	-	-	-
Exports	-	-	-	-	-	-	-	-1400	-	-	-
Intl. Marine Bunkers	-	-	-	-	-	-	-	-	-	-	-
Stock Changes	-	-	-	-	-	-	-	-	-	-	-
DOMESTIC SUPPLY	-	-	-	-	-	-	-	**5475**	**263**	-	-
Transfers and Stat. Diff.	-	-	-	-	-	-	-	-	-	-	-
TRANSFORMATION	-	-	-	-	-	-	-	**5475**	**263**	-	-
Electricity and CHP Plants	-	-	-	-	-	-	-	-	-	-	-
Petroleum Refineries	-	-	-	-	-	-	-	5475	-	-	-
Other Transform. Sector	-	-	-	-	-	-	-	-	263	-	-
ENERGY SECTOR	-	-	-	-	-	-	-	-	-	-	-
DISTRIBUTION LOSSES	-	-	-	-	-	-	-	-	-	-	-
FINAL CONSUMPTION	-	-	-	-	-	-	-	-	-	-	-
INDUSTRY SECTOR	-	-	-	-	-	-	-	-	-	-	-
Iron and Steel	-	-	-	-	-	-	-	-	-	-	-
Chemical and Petrochem.	-	-	-	-	-	-	-	-	-	-	-
Non-Metallic Minerals	-	-	-	-	-	-	-	-	-	-	-
Non-specified	-	-	-	-	-	-	-	-	-	-	-
TRANSPORT SECTOR	-	-	-	-	-	-	-	-	-	-	-
Air	-	-	-	-	-	-	-	-	-	-	-
Road	-	-	-	-	-	-	-	-	-	-	-
Non-specified	-	-	-	-	-	-	-	-	-	-	-
OTHER SECTORS	-	-	-	-	-	-	-	-	-	-	-
Agriculture	-	-	-	-	-	-	-	-	-	-	-
Comm. and Publ. Services	-	-	-	-	-	-	-	-	-	-	-
Residential	-	-	-	-	-	-	-	-	-	-	-
Non-specified	-	-	-	-	-	-	-	-	-	-	-
NON-ENERGY USE	-	-	-	-	-	-	-	-	-	-	-

Turkmenistan / Turkménistan

SUPPLY AND CONSUMPTION 1997	Oil cont. (1000 tonnes)										
	Refinery Gas	LPG + Ethane	Motor Gasoline	Aviation Gasoline	Jet Fuel	Kerosene	Gas/ Diesel	Heavy Fuel Oil	Naphtha	Petrol. Coke	Other Prod.
Production	250	-	766	-	-	-	1520	1600	-	-	-
Imports	-	53	-	-	-	-	-	-	-	-	-
Exports	-	-	-419	-	-	-	-1063	-231	-	-	-
Intl. Marine Bunkers	-	-	-	-	-	-	-	-	-	-	-
Stock Changes	-	-	-	-	-	-	-	-	-	-	-
DOMESTIC SUPPLY	250	53	347	-	-	-	457	1369	-	-	-
Transfers and Stat. Diff.	-	-	-	-	-	-	-	-	-	-	-
TRANSFORMATION	-	-	-	-	-	-	-	-	-	-	-
Electricity and CHP Plants	-	-	-	-	-	-	-	-	-	-	-
Petroleum Refineries	-	-	-	-	-	-	-	-	-	-	-
Other Transform. Sector	-	-	-	-	-	-	-	-	-	-	-
ENERGY SECTOR	250	-	-	-	-	-	-	-	-	-	-
DISTRIBUTION LOSSES	-	-	-	-	-	-	-	-	-	-	-
FINAL CONSUMPTION	-	53	347	-	-	-	457	1369	-	-	-
INDUSTRY SECTOR	-	-	-	-	-	-	-	-	-	-	-
Iron and Steel	-	-	-	-	-	-	-	-	-	-	-
Chemical and Petrochem.	-	-	-	-	-	-	-	-	-	-	-
Non-Metallic Minerals	-	-	-	-	-	-	-	-	-	-	-
Non-specified	-	-	-	-	-	-	-	-	-	-	-
TRANSPORT SECTOR	-	-	347	-	-	-	-	-	-	-	-
Air	-	-	-	-	-	-	-	-	-	-	-
Road	-	-	347	-	-	-	-	-	-	-	-
Non-specified	-	-	-	-	-	-	-	-	-	-	-
OTHER SECTORS	-	53	-	-	-	-	457	1369	-	-	-
Agriculture	-	-	-	-	-	-	-	-	-	-	-
Comm. and Publ. Services	-	-	-	-	-	-	-	-	-	-	-
Residential	-	-	-	-	-	-	-	-	-	-	-
Non-specified	-	53	-	-	-	-	457	1369	-	-	-
NON-ENERGY USE	-	-	-	-	-	-	-	-	-	-	-

APPROVISIONNEMENT ET DEMANDE 1998	Pétrole cont. (1000 tonnes)										
	Gaz de raffinerie	GPL + éthane	Essence moteur	Essence aviation	Carbu- réacteurs	Kérosène	Gazole	Fioul lourd	Naphta	Coke de pétrole	Autres prod.
Production	287	-	693	-	-	-	1575	2214	-	-	-
Imports	-	61	-	-	-	-	-	-	-	-	-
Exports	-	-	-380	-	-	-	-1218	-1117	-	-	-
Intl. Marine Bunkers	-	-	-	-	-	-	-	-	-	-	-
Stock Changes	-	-	-	-	-	-	-	-	-	-	-
DOMESTIC SUPPLY	287	61	313	-	-	-	357	1097	-	-	-
Transfers and Stat. Diff.	-	-	-	-	-	-	-	-	-	-	-
TRANSFORMATION	-	-	-	-	-	-	-	-	-	-	-
Electricity and CHP Plants	-	-	-	-	-	-	-	-	-	-	-
Petroleum Refineries	-	-	-	-	-	-	-	-	-	-	-
Other Transform. Sector	-	-	-	-	-	-	-	-	-	-	-
ENERGY SECTOR	287	-	-	-	-	-	-	-	-	-	-
DISTRIBUTION LOSSES	-	-	-	-	-	-	-	-	-	-	-
FINAL CONSUMPTION	-	61	313	-	-	-	357	1097	-	-	-
INDUSTRY SECTOR	-	-	-	-	-	-	-	-	-	-	-
Iron and Steel	-	-	-	-	-	-	-	-	-	-	-
Chemical and Petrochem.	-	-	-	-	-	-	-	-	-	-	-
Non-Metallic Minerals	-	-	-	-	-	-	-	-	-	-	-
Non-specified	-	-	-	-	-	-	-	-	-	-	-
TRANSPORT SECTOR	-	-	313	-	-	-	-	-	-	-	-
Air	-	-	-	-	-	-	-	-	-	-	-
Road	-	-	313	-	-	-	-	-	-	-	-
Non-specified	-	-	-	-	-	-	-	-	-	-	-
OTHER SECTORS	-	61	-	-	-	-	357	1097	-	-	-
Agriculture	-	-	-	-	-	-	-	-	-	-	-
Comm. and Publ. Services	-	-	-	-	-	-	-	-	-	-	-
Residential	-	-	-	-	-	-	-	-	-	-	-
Non-specified	-	61	-	-	-	-	357	1097	-	-	-
NON-ENERGY USE	-	-	-	-	-	-	-	-	-	-	-

Turkmenistan / Turkménistan

SUPPLY AND CONSUMPTION 1997	Gas (TJ)				Comb. Renew. & Waste (TJ)				(GWh)	(TJ)
	Natural Gas	Gas Works	Coke Ovens	Blast Furnaces	Solid Biomass	Gas/Liquids from Biomass	Municipal Waste	Industrial Waste	Electricity	Heat
Production	652889	8260	-	-	-	-	-	-	9498	-
Imports	-	-	-	-	-	-	-	-	950	-
Exports	-222654	-	-	-	-	-	-	-	-4411	-
Intl. Marine Bunkers	-	-	-	-	-	-	-	-	-	-
Stock Changes	-	-	-	-	-	-	-	-	-	-
DOMESTIC SUPPLY	**430235**	**8260**	-	-	-	-	-	-	**6037**	-
Transfers and Stat. Diff.	-	-	-	-	-	-	-	-	-	-
TRANSFORMATION	**149786**	-	-	-	-	-	-	-	-	-
Electricity and CHP Plants	149786	-	-	-	-	-	-	-	-	-
Petroleum Refineries	-	-	-	-	-	-	-	-	-	-
Other Transform. Sector	-	-	-	-	-	-	-	-	-	-
ENERGY SECTOR	**41430**	-	-	-	-	-	-	-	**1207**	-
DISTRIBUTION LOSSES	-	-	-	-	-	-	-	-	**935**	-
FINAL CONSUMPTION	**239019**	**8260**	-	-	-	-	-	-	**3895**	-
INDUSTRY SECTOR	-	-	-	-	-	-	-	-	**1406**	-
Iron and Steel	-	-	-	-	-	-	-	-	1	-
Chemical and Petrochem.	-	-	-	-	-	-	-	-	456	-
Non-Metallic Minerals	-	-	-	-	-	-	-	-	-	-
Non-specified	-	-	-	-	-	-	-	-	949	-
TRANSPORT SECTOR	-	-	-	-	-	-	-	-	**98**	-
Air	-	-	-	-	-	-	-	-	-	-
Road	-	-	-	-	-	-	-	-	-	-
Non-specified	-	-	-	-	-	-	-	-	98	-
OTHER SECTORS	**239019**	**8260**	-	-	-	-	-	-	**2391**	-
Agriculture	-	-	-	-	-	-	-	-	1240	-
Comm. and Publ. Services	-	-	-	-	-	-	-	-	-	-
Residential	-	-	-	-	-	-	-	-	819	-
Non-specified	239019	8260	-	-	-	-	-	-	332	-
NON-ENERGY USE	-	-	-	-	-	-	-	-	-	-

APPROVISIONNEMENT ET DEMANDE 1998	Gaz (TJ)				En. Re. Comb. & Déchets (TJ)				(GWh)	(TJ)
	Gaz naturel	Usines à gaz	Cokeries	Hauts fourneaux	Biomasse solide	Gaz/Liquides tirés de biomasse	Déchets urbains	Déchets industriels	Electricité	Chaleur
Production	499789	11022	-	-	-	-	-	-	9416	-
Imports	-	-	-	-	-	-	-	-	950	-
Exports	-113839	-	-	-	-	-	-	-	-4059	-
Intl. Marine Bunkers	-	-	-	-	-	-	-	-	-	-
Stock Changes	-	-	-	-	-	-	-	-	-	-
DOMESTIC SUPPLY	**385950**	**11022**	-	-	-	-	-	-	**6307**	-
Transfers and Stat. Diff.	-	-	-	-	-	-	-	-	-	-
TRANSFORMATION	**134471**	-	-	-	-	-	-	-	-	-
Electricity and CHP Plants	134471	-	-	-	-	-	-	-	-	-
Petroleum Refineries	-	-	-	-	-	-	-	-	-	-
Other Transform. Sector	-	-	-	-	-	-	-	-	-	-
ENERGY SECTOR	**37191**	-	-	-	-	-	-	-	**1277**	-
DISTRIBUTION LOSSES	-	-	-	-	-	-	-	-	**978**	-
FINAL CONSUMPTION	**214288**	**11022**	-	-	-	-	-	-	**4052**	-
INDUSTRY SECTOR	-	-	-	-	-	-	-	-	**1462**	-
Iron and Steel	-	-	-	-	-	-	-	-	1	-
Chemical and Petrochem.	-	-	-	-	-	-	-	-	474	-
Non-Metallic Minerals	-	-	-	-	-	-	-	-	-	-
Non-specified	-	-	-	-	-	-	-	-	987	-
TRANSPORT SECTOR	-	-	-	-	-	-	-	-	**102**	-
Air	-	-	-	-	-	-	-	-	-	-
Road	-	-	-	-	-	-	-	-	-	-
Non-specified	-	-	-	-	-	-	-	-	102	-
OTHER SECTORS	**214288**	**11022**	-	-	-	-	-	-	**2488**	-
Agriculture	-	-	-	-	-	-	-	-	1290	-
Comm. and Publ. Services	-	-	-	-	-	-	-	-	-	-
Residential	-	-	-	-	-	-	-	-	852	-
Non-specified	214288	11022	-	-	-	-	-	-	346	-
NON-ENERGY USE	-	-	-	-	-	-	-	-	-	-

Ukraine : 1997

SUPPLY AND CONSUMPTION	Coal / Charbon (1000 tonnes)							Oil / Pétrole (1000 tonnes)			
	Coking Coal	Other Bit. Coal	Sub-Bit. Coal	Lignite	Peat	Oven and Gas Coke	Pat. Fuel and BKB	Crude Oil	NGL	Feed-stocks	Additives
APPROVISIONNEMENT ET DEMANDE	Charbon à coke	Autres charb. bit.	Charbon sous-bit.	Lignite	Tourbe	Coke de four/gaz	Agg./briq. de lignite	Pétrole brut	LGN	Produits d'aliment.	Additifs
Production	31838	43676	-	1436	770	15000	2882	4134	439	-	-
From Other Sources	-	-	-	-	-	-	-	-	-	-	-
Imports	3562	4807	-	482	1	181	-	8957	5	-	-
Exports	-44	-2330	-	-	-2	-359	-	-5	-	-	-
Intl. Marine Bunkers	-	-	-	-	-	-	-	-	-	-	-
Stock Changes	-	-	-	-	-	-	-	-	-	-	-
DOMESTIC SUPPLY	35356	46153	-	1918	769	14822	2882	13086	444	-	-
Transfers	-	-	-	-	-	-	-	-	-	-	-
Statistical Differences	-	-	-	-	-	-	-	-	-	-	-
TRANSFORMATION	35356	32973	-	918	-	11858	-	13086	444	-	-
Electricity Plants	-	-	-	918	-	-	-	-	-	-	-
CHP Plants	-	26039	-	-	-	-	-	-	-	-	-
Heat Plants	-	2955	-	-	-	-	-	-	-	-	-
Blast Furnaces/Gas Works	-	-	-	-	-	11858	-	-	-	-	-
Coke/Pat. Fuel/BKB Plants	35356	3979	-	-	-	-	-	-	-	-	-
Petroleum Refineries	-	-	-	-	-	-	-	13086	444	-	-
Petrochemical Industry	-	-	-	-	-	-	-	-	-	-	-
Liquefaction	-	-	-	-	-	-	-	-	-	-	-
Other Transform. Sector	-	-	-	-	-	-	-	-	-	-	-
ENERGY SECTOR	-	198	-	-	-	-	-	-	-	-	-
Coal Mines	-	198	-	-	-	-	-	-	-	-	-
Oil and Gas Extraction	-	-	-	-	-	-	-	-	-	-	-
Petroleum Refineries	-	-	-	-	-	-	-	-	-	-	-
Electr., CHP+Heat Plants	-	-	-	-	-	-	-	-	-	-	-
Pumped Storage (Elec.)	-	-	-	-	-	-	-	-	-	-	-
Other Energy Sector	-	-	-	-	-	-	-	-	-	-	-
Distribution Losses	-	-	-	-	-	-	-	-	-	-	-
FINAL CONSUMPTION	-	12982	-	1000	769	2964	2882	-	-	-	-
INDUSTRY SECTOR	-	8634	-	-	-	2964	-	-	-	-	-
Iron and Steel	-	-	-	-	-	2964	-	-	-	-	-
Chemical and Petrochem.	-	-	-	-	-	-	-	-	-	-	-
of which: Feedstocks	-	-	-	-	-	-	-	-	-	-	-
Non-Ferrous Metals	-	-	-	-	-	-	-	-	-	-	-
Non-Metallic Minerals	-	-	-	-	-	-	-	-	-	-	-
Transport Equipment	-	-	-	-	-	-	-	-	-	-	-
Machinery	-	-	-	-	-	-	-	-	-	-	-
Mining and Quarrying	-	-	-	-	-	-	-	-	-	-	-
Food and Tobacco	-	-	-	-	-	-	-	-	-	-	-
Paper, Pulp and Print	-	-	-	-	-	-	-	-	-	-	-
Wood and Wood Products	-	-	-	-	-	-	-	-	-	-	-
Construction	-	-	-	-	-	-	-	-	-	-	-
Textile and Leather	-	-	-	-	-	-	-	-	-	-	-
Non-specified	-	8634	-	-	-	-	-	-	-	-	-
TRANSPORT SECTOR	-	-	-	-	-	-	-	-	-	-	-
Air	-	-	-	-	-	-	-	-	-	-	-
Road	-	-	-	-	-	-	-	-	-	-	-
Rail	-	-	-	-	-	-	-	-	-	-	-
Pipeline Transport	-	-	-	-	-	-	-	-	-	-	-
Internal Navigation	-	-	-	-	-	-	-	-	-	-	-
Non-specified	-	-	-	-	-	-	-	-	-	-	-
OTHER SECTORS	-	4348	-	1000	769	-	2882	-	-	-	-
Agriculture	-	-	-	-	-	-	-	-	-	-	-
Comm. and Publ. Services	-	-	-	-	-	-	-	-	-	-	-
Residential	-	4348	-	1000	769	-	2882	-	-	-	-
Non-specified	-	-	-	-	-	-	-	-	-	-	-
NON-ENERGY USE	-	-	-	-	-	-	-	-	-	-	-
in Industry/Trans./Energy	-	-	-	-	-	-	-	-	-	-	-
in Transport	-	-	-	-	-	-	-	-	-	-	-
in Other Sectors	-	-	-	-	-	-	-	-	-	-	-

Ukraine : 1997

SUPPLY AND CONSUMPTION / APPROVISIONNEMENT ET DEMANDE	Oil cont. / Pétrole cont. (1000 tonnes)										
	Refinery Gas / Gaz de raffinerie	LPG + Ethane / GPL + éthane	Motor Gasoline / Essence moteur	Aviation Gasoline / Essence aviation	Jet Fuel / Carbu-réacteurs	Kerosene / Kérosène	Gas/ Diesel / Gazole	Heavy Fuel Oil / Fioul lourd	Naphtha / Naphta	Petrol. Coke / Coke de pétrole	Other Prod. / Autres prod.
Production	181	147	2803	-	620	11	3779	4326	-	-	981
From Other Sources	-	-	-	-	-	-	-	-	-	-	-
Imports	-	1437	437	36	23	4	1146	2501	-	-	461
Exports	-	-	-40	.	-5	-5	-130	-1165	-	-	-42
Intl. Marine Bunkers	-	-	-	-	-	-	-	-	-	-	-
Stock Changes	-	-	-	-	-	.	-	-	-	-	-
DOMESTIC SUPPLY	181	1584	3200	36	638	10	4795	5662	-	-	1400
Transfers	-	-	-	-	-	-	-	-	-	-	-
Statistical Differences	-	-	-	-	-	-	-	-	-	-	-
TRANSFORMATION	-	-	-	-	-	-	71	2507	-	-	222
Electricity Plants	-	-	-	-	-	-	71	1689	-	-	222
CHP Plants	-	-	-	-	-	-	-	-	-	-	-
Heat Plants	-	-	-	-	-	-	-	818	-	-	-
Blast Furnaces/Gas Works	-	-	-	-	-	-	-	-	-	-	-
Coke/Pat. Fuel/BKB Plants	-	-	-	-	-	-	-	-	-	-	-
Petroleum Refineries	-	-	-	-	-	-	-	-	-	-	-
Petrochemical Industry	-	-	-	-	-	-	-	-	-	-	-
Liquefaction	-	-	-	-	-	-	-	-	-	-	-
Other Transform. Sector	-	-	-	-	-	-	-	-	-	-	-
ENERGY SECTOR	181	-	-	-	-	-	-	203	-	-	-
Coal Mines	-	-	-	-	-	-	-	-	-	-	-
Oil and Gas Extraction	-	-	-	-	-	-	-	-	-	-	-
Petroleum Refineries	181	-	-	-	-	-	-	203	-	-	-
Electr., CHP+Heat Plants	-	-	-	-	-	-	-	-	-	-	-
Pumped Storage (Elec.)	-	-	-	-	-	-	-	-	-	-	-
Other Energy Sector	-	-	-	-	-	-	-	-	-	-	-
Distribution Losses	-	-	-	-	-	-	-	-	-	-	-
FINAL CONSUMPTION	-	1584	3200	36	638	10	4724	2952	-	-	1178
INDUSTRY SECTOR	-	-	-	-	-	-	896	2952	-	-	-
Iron and Steel	-	-	-	-	-	-	-	-	-	-	-
Chemical and Petrochem.	-	-	-	-	-	-	-	-	-	-	-
of which: Feedstocks	-	-	-	-	-	-	-	-	-	-	-
Non-Ferrous Metals	-	-	-	-	-	-	-	-	-	-	-
Non-Metallic Minerals	-	-	-	-	-	-	-	-	-	-	-
Transport Equipment	-	-	-	-	-	-	-	-	-	-	-
Machinery	-	-	-	-	-	-	-	-	-	-	-
Mining and Quarrying	-	-	-	-	-	-	-	-	-	-	-
Food and Tobacco	-	-	-	-	-	-	-	-	-	-	-
Paper, Pulp and Print	-	-	-	-	-	-	-	-	-	-	-
Wood and Wood Products	-	-	-	-	-	-	-	-	-	-	-
Construction	-	-	-	-	-	-	-	-	-	-	-
Textile and Leather	-	-	-	-	-	-	-	-	-	-	-
Non-specified	-	-	-	-	-	-	896	2952	-	-	-
TRANSPORT SECTOR	-	-	3200	36	638	-	1699	-	-	-	-
Air	-	-	-	36	638	-	-	-	-	-	-
Road	-	-	3200	-	-	-	-	-	-	-	-
Rail	-	-	-	-	-	-	1699	-	-	-	-
Pipeline Transport	-	-	-	-	-	-	-	-	-	-	-
Internal Navigation	-	-	-	-	-	-	-	-	-	-	-
Non-specified	-	-	-	-	-	-	-	-	-	-	-
OTHER SECTORS	-	1584	-	-	-	10	2129	-	-	-	-
Agriculture	-	-	-	-	-	-	2129	-	-	-	-
Comm. and Publ. Services	-	-	-	-	-	-	-	-	-	-	-
Residential	-	1584	-	-	-	10	-	-	-	-	-
Non-specified	-	-	-	-	-	-	-	-	-	-	-
NON-ENERGY USE	-	-	-	-	-	-	-	-	-	-	1178
in Industry/Transf./Energy	-	-	-	-	-	-	-	-	-	-	1178
in Transport	-	-	-	-	-	-	-	-	-	-	-
in Other Sectors	-	-	-	-	-	-	-	-	-	-	-

Ukraine : 1997

SUPPLY AND CONSUMPTION / APPROVISIONNEMENT ET DEMANDE	Gas / Gaz (TJ) Natural Gas / Gaz naturel	Gas Works / Usines à gaz	Coke Ovens / Cokeries	Blast Furnaces / Hauts fourneaux	Comb. Renew. & Waste / En. Re. Comb. & Déchets (TJ) Solid Biomass / Biomasse solide	Gas/Liquids from Biomass / Gaz/Liquides tirés de biomasse	Municipal Waste / Déchets urbains	Industrial Waste / Déchets industriels	(GWh) Electricity / Electricité	(TJ) Heat / Chaleur
Production	707472	-	-	119176	10994	-	-	-	178002	631001
From Other Sources	-	-	-	-	-	-	-	-	-	-
Imports	2433127	-	-	-	-	-	-	-	9719	-
Exports	-54784	-	-	-	-	-	-	-	-9873	-
Intl. Marine Bunkers	-	-	-	-	-	-	-	-	-	-
Stock Changes	-	-	-	-	-	-	-	-	-	-
DOMESTIC SUPPLY	**3085815**	**-**	**-**	**119176**	**10994**	**-**	**-**	**-**	**177848**	**631001**
Transfers	-	-	-	-	-	-	-	-	-	-
Statistical Differences	-	-	-	-	-	-	-	-	-	-
TRANSFORMATION	**1048000**	**-**	**-**	**-**	**-**	**-**	**-**	**-**	**-**	**-**
Electricity Plants	618000	-	-	-	-	-	-	-	-	-
CHP Plants	160000	-	-	-	-	-	-	-	-	-
Heat Plants	270000	-	-	-	-	-	-	-	-	-
Blast Furnaces/Gas Works	-	-	-	-	-	-	-	-	-	-
Coke/Pat. Fuel/BKB Plants	-	-	-	-	-	-	-	-	-	-
Petroleum Refineries	-	-	-	-	-	-	-	-	-	-
Petrochemical Industry	-	-	-	-	-	-	-	-	-	-
Liquefaction	-	-	-	-	-	-	-	-	-	-
Other Transform. Sector	-	-	-	-	-	-	-	-	-	-
ENERGY SECTOR	**60000**	**-**	**-**	**-**	**-**	**-**	**-**	**-**	**25290**	**5000**
Coal Mines	-	-	-	-	-	-	-	-	9036	-
Oil and Gas Extraction	57000	-	-	-	-	-	-	-	381	-
Petroleum Refineries	3000	-	-	-	-	-	-	-	1032	-
Electr., CHP+Heat Plants	-	-	-	-	-	-	-	-	13166	5000
Pumped Storage (Elec.)	-	-	-	-	-	-	-	-	-	-
Other Energy Sector	-	-	-	-	-	-	-	-	1675	-
Distribution Losses	60815	-	-	-	-	-	-	-	28407	188927
FINAL CONSUMPTION	**1917000**	**-**	**-**	**119176**	**10994**	**-**	**-**	**-**	**124151**	**437074**
INDUSTRY SECTOR	**700000**	**-**	**-**	**119176**	**-**	**-**	**-**	**-**	**64822**	**282244**
Iron and Steel	-	-	-	119176	-	-	-	-	22347	-
Chemical and Petrochem.	-	-	-	-	-	-	-	-	7569	-
of which: Feedstocks	-	-	-	-	-	-	-	-	-	-
Non-Ferrous Metals	-	-	-	-	-	-	-	-	3390	-
Non-Metallic Minerals	-	-	-	-	-	-	-	-	2717	-
Transport Equipment	-	-	-	-	-	-	-	-	887	-
Machinery	-	-	-	-	-	-	-	-	6049	-
Mining and Quarrying	-	-	-	-	-	-	-	-	8724	-
Food and Tobacco	-	-	-	-	-	-	-	-	3896	-
Paper, Pulp and Print	-	-	-	-	-	-	-	-	536	-
Wood and Wood Products	-	-	-	-	-	-	-	-	309	-
Construction	-	-	-	-	-	-	-	-	1450	-
Textile and Leather	-	-	-	-	-	-	-	-	573	-
Non-specified	700000	-	-	-	-	-	-	-	6375	282244
TRANSPORT SECTOR	**-**	**-**	**-**	**-**	**-**	**-**	**-**	**-**	**9545**	**-**
Air	-	-	-	-	-	-	-	-	-	-
Road	-	-	-	-	-	-	-	-	-	-
Rail	-	-	-	-	-	-	-	-	4499	-
Pipeline Transport	-	-	-	-	-	-	-	-	1745	-
Internal Navigation	-	-	-	-	-	-	-	-	-	-
Non-specified	-	-	-	-	-	-	-	-	3301	-
OTHER SECTORS	**1217000**	**-**	**-**	**-**	**10994**	**-**	**-**	**-**	**49784**	**154830**
Agriculture	20000	-	-	-	-	-	-	-	9184	17759
Comm. and Publ. Services	397000	-	-	-	-	-	-	-	9277	-
Residential	800000	-	-	-	-	-	-	-	-	137071
Non-specified	-	-	-	-	10994	-	-	-	31323	-
NON-ENERGY USE	**-**	**-**	**-**	**-**	**-**	**-**	**-**	**-**	**-**	**-**
in Industry/Transf./Energy	-	-	-	-	-	-	-	-	-	-
in Transport	-	-	-	-	-	-	-	-	-	-
in Other Sectors	-	-	-	-	-	-	-	-	-	-

Ukraine : 1998

SUPPLY AND CONSUMPTION / APPROVISIONNEMENT ET DEMANDE	Coal / Charbon (1000 tonnes)							Oil / Pétrole (1000 tonnes)			
	Coking Coal / Charbon à coke	Other Bit. Coal / Autres charb. bit.	Sub-Bit. Coal / Charbon sous-bit.	Lignite / Lignite	Peat / Tourbe	Oven and Gas Coke / Coke de four/gaz	Pat. Fuel and BKB / Agg./briq. de lignite	Crude Oil / Pétrole brut	NGL / LGN	Feed-stocks / Produits d'aliment.	Additives / Additifs
Production	32656	43111	-	1409	687	15150	2764	3900	495	-	-
From Other Sources	-	-	-	-	-	-	-	-	-	-	-
Imports	3089	5301	-	21	-	71	-	9884	-	9	-
Exports	-12	-1869	-	-	-2	-324	-	-88	-	-4	-
Intl. Marine Bunkers	-	-	-	-	-	-	-	-	-	-	-
Stock Changes	-	-	-	-	-	-	-	-	-	-	-
DOMESTIC SUPPLY	35733	46543	-	1430	685	14897	2764	13696	495	5	-
Transfers	-	-	-	-	-	-	-	-	-	-	-
Statistical Differences	-	-	-	-	-	-	-	-	-	-	-
TRANSFORMATION	35733	33252	-	430	-	11918	-	13696	495	5	-
Electricity Plants	-	-	-	430	-	-	-	-	-	-	-
CHP Plants	-	26259	-	-	-	-	-	-	-	-	-
Heat Plants	-	2980	-	-	-	-	-	-	-	-	-
Blast Furnaces/Gas Works	-	-	-	-	-	11918	-	-	-	-	-
Coke/Pat. Fuel/BKB Plants	35733	4013	-	-	-	-	-	-	-	-	-
Petroleum Refineries	-	-	-	-	-	-	-	13696	495	5	-
Petrochemical Industry	-	-	-	-	-	-	-	-	-	-	-
Liquefaction	-	-	-	-	-	-	-	-	-	-	-
Other Transform. Sector	-	-	-	-	-	-	-	-	-	-	-
ENERGY SECTOR	-	200	-	-	-	-	-	-	-	-	-
Coal Mines	-	200	-	-	-	-	-	-	-	-	-
Oil and Gas Extraction	-	-	-	-	-	-	-	-	-	-	-
Petroleum Refineries	-	-	-	-	-	-	-	-	-	-	-
Electr., CHP+Heat Plants	-	-	-	-	-	-	-	-	-	-	-
Pumped Storage (Elec.)	-	-	-	-	-	-	-	-	-	-	-
Other Energy Sector	-	-	-	-	-	-	-	-	-	-	-
Distribution Losses	-	-	-	-	-	-	-	-	-	-	-
FINAL CONSUMPTION	-	13091	-	1000	685	2979	2764	-	-	-	-
INDUSTRY SECTOR	-	8707	-	-	-	2979	-	-	-	-	-
Iron and Steel	-	-	-	-	-	2979	-	-	-	-	-
Chemical and Petrochem.	-	-	-	-	-	-	-	-	-	-	-
of which: Feedstocks	-	-	-	-	-	-	-	-	-	-	-
Non-Ferrous Metals	-	-	-	-	-	-	-	-	-	-	-
Non-Metallic Minerals	-	-	-	-	-	-	-	-	-	-	-
Transport Equipment	-	-	-	-	-	-	-	-	-	-	-
Machinery	-	-	-	-	-	-	-	-	-	-	-
Mining and Quarrying	-	-	-	-	-	-	-	-	-	-	-
Food and Tobacco	-	-	-	-	-	-	-	-	-	-	-
Paper, Pulp and Print	-	-	-	-	-	-	-	-	-	-	-
Wood and Wood Products	-	-	-	-	-	-	-	-	-	-	-
Construction	-	-	-	-	-	-	-	-	-	-	-
Textile and Leather	-	-	-	-	-	-	-	-	-	-	-
Non-specified	-	8707	-	-	-	-	-	-	-	-	-
TRANSPORT SECTOR	-	-	-	-	-	-	-	-	-	-	-
Air	-	-	-	-	-	-	-	-	-	-	-
Road	-	-	-	-	-	-	-	-	-	-	-
Rail	-	-	-	-	-	-	-	-	-	-	-
Pipeline Transport	-	-	-	-	-	-	-	-	-	-	-
Internal Navigation	-	-	-	-	-	-	-	-	-	-	-
Non-specified	-	-	-	-	-	-	-	-	-	-	-
OTHER SECTORS	-	4384	-	1000	685	-	2764	-	-	-	-
Agriculture	-	-	-	-	-	-	-	-	-	-	-
Comm. and Publ. Services	-	-	-	-	-	-	-	-	-	-	-
Residential	-	4384	-	1000	685	-	2764	-	-	-	-
Non-specified	-	-	-	-	-	-	-	-	-	-	-
NON-ENERGY USE	-	-	-	-	-	-	-	-	-	-	-
in Industry/Trans./Energy	-	-	-	-	-	-	-	-	-	-	-
in Transport	-	-	-	-	-	-	-	-	-	-	-
in Other Sectors	-	-	-	-	-	-	-	-	-	-	-

Ukraine : 1998

SUPPLY AND CONSUMPTION APPROVISIONNEMENT ET DEMANDE	Oil cont. / *Pétrole cont.* (1000 tonnes)										
	Refinery Gas *Gaz de raffinerie*	LPG + Ethane *GPL + éthane*	Motor Gasoline *Essence moteur*	Aviation Gasoline *Essence aviation*	Jet Fuel *Carbu- réacteurs*	Kerosene *Kérosène*	Gas/ Diesel *Gazole*	Heavy Fuel Oil *Fioul lourd*	Naphtha *Naphta*	Petrol. Coke *Coke de pétrole*	Other Prod. *Autres prod.*
Production	190	154	2943	-	651	11	3968	4542	-	-	1030
From Other Sources	-	-	-	-	-	-	-	-	-	-	-
Imports	-	1331	419	36	15	-	1211	2672	-	-	234
Exports	-	-	-51	-	-3	-3	-270	-1435	-	-	-14
Intl. Marine Bunkers	-	-	-	-	-	-	-	-	-	-	-
Stock Changes	-	-	-	-	-	-	-	-	-	-	-
DOMESTIC SUPPLY	**190**	**1485**	**3311**	**36**	**663**	**8**	**4909**	**5779**	**-**	**-**	**1250**
Transfers	-	-	-	-	-	-	-	-	-	-	-
Statistical Differences	-	-	-	-	-	-	-	-	-	-	-
TRANSFORMATION	**-**	**-**	**-**	**-**	**-**	**-**	**73**	**2559**	**-**	**-**	**217**
Electricity Plants	-	-	-	-	-	-	73	1724	-	-	217
CHP Plants	-	-	-	-	-	-	-	-	-	-	-
Heat Plants	-	-	-	-	-	-	-	835	-	-	-
Blast Furnaces/Gas Works	-	-	-	-	-	-	-	-	-	-	-
Coke/Pat. Fuel/BKB Plants	-	-	-	-	-	-	-	-	-	-	-
Petroleum Refineries	-	-	-	-	-	-	-	-	-	-	-
Petrochemical Industry	-	-	-	-	-	-	-	-	-	-	-
Liquefaction	-	-	-	-	-	-	-	-	-	-	-
Other Transform. Sector	-	-	-	-	-	-	-	-	-	-	-
ENERGY SECTOR	**190**	**-**	**-**	**-**	**-**	**-**	**-**	**207**	**-**	**-**	**-**
Coal Mines	-	-	-	-	-	-	-	-	-	-	-
Oil and Gas Extraction	-	-	-	-	-	-	-	-	-	-	-
Petroleum Refineries	190	-	-	-	-	-	-	207	-	-	-
Electr., CHP+Heat Plants	-	-	-	-	-	-	-	-	-	-	-
Pumped Storage (Elec.)	-	-	-	-	-	-	-	-	-	-	-
Other Energy Sector	-	-	-	-	-	-	-	-	-	-	-
Distribution Losses	-	-	-	-	-	-	-	-	-	-	-
FINAL CONSUMPTION	**-**	**1485**	**3311**	**36**	**663**	**8**	**4836**	**3013**	**-**	**-**	**1033**
INDUSTRY SECTOR	**-**	**-**	**-**	**-**	**-**	**-**	**917**	**3013**	**-**	**-**	**-**
Iron and Steel	-	-	-	-	-	-	-	-	-	-	-
Chemical and Petrochem.	-	-	-	-	-	-	-	-	-	-	-
of which: Feedstocks	-	-	-	-	-	-	-	-	-	-	-
Non-Ferrous Metals	-	-	-	-	-	-	-	-	-	-	-
Non-Metallic Minerals	-	-	-	-	-	-	-	-	-	-	-
Transport Equipment	-	-	-	-	-	-	-	-	-	-	-
Machinery	-	-	-	-	-	-	-	-	-	-	-
Mining and Quarrying	-	-	-	-	-	-	-	-	-	-	-
Food and Tobacco	-	-	-	-	-	-	-	-	-	-	-
Paper, Pulp and Print	-	-	-	-	-	-	-	-	-	-	-
Wood and Wood Products	-	-	-	-	-	-	-	-	-	-	-
Construction	-	-	-	-	-	-	-	-	-	-	-
Textile and Leather	-	-	-	-	-	-	-	-	-	-	-
Non-specified	-	-	-	-	-	-	917	3013	-	-	-
TRANSPORT SECTOR	**-**	**-**	**3311**	**36**	**663**	**-**	**1739**	**-**	**-**	**-**	**-**
Air	-	-	-	36	663	-	-	-	-	-	-
Road	-	-	3311	-	-	-	-	-	-	-	-
Rail	-	-	-	-	-	-	1739	-	-	-	-
Pipeline Transport	-	-	-	-	-	-	-	-	-	-	-
Internal Navigation	-	-	-	-	-	-	-	-	-	-	-
Non-specified	-	-	-	-	-	-	-	-	-	-	-
OTHER SECTORS	**-**	**1485**	**-**	**-**	**-**	**8**	**2180**	**-**	**-**	**-**	**-**
Agriculture	-	-	-	-	-	-	2180	-	-	-	-
Comm. and Publ. Services	-	-	-	-	-	-	-	-	-	-	-
Residential	-	1485	-	-	-	8	-	-	-	-	-
Non-specified	-	-	-	-	-	-	-	-	-	-	-
NON-ENERGY USE	**-**	**-**	**-**	**-**	**-**	**-**	**-**	**-**	**-**	**-**	**1033**
in Industry/Transf./Energy	-	-	-	-	-	-	-	-	-	-	1033
in Transport	-	-	-	-	-	-	-	-	-	-	-
in Other Sectors	-	-	-	-	-	-	-	-	-	-	-

Ukraine : 1998

SUPPLY AND CONSUMPTION / APPROVISIONNEMENT ET DEMANDE	Gas / Gaz (TJ)				Comb. Renew. & Waste / En. Re. Comb. & Déchets (TJ)				(GWh)	(TJ)
	Natural Gas / Gaz naturel	Gas Works / Usines à gaz	Coke Ovens / Cokeries	Blast Furnaces / Hauts fourneaux	Solid Biomass / Biomasse solide	Gas/Liquids from Biomass / Gaz/Liquides tirés de biomasse	Municipal Waste / Déchets urbains	Industrial Waste / Déchets industriels	Electricity / Electricité	Heat / Chaleur
Production	701072	-	-	119779	10952	-	-	-	172822	594593
From Other Sources	-	-	-	-	-	-	-	-	-	-
Imports	2089490	-	-	-	-	-	-	-	10056	-
Exports	-22983	-	-	-	-	-	-	-	-10728	-
Intl. Marine Bunkers	-	-	-	-	-	-	-	-	-	-
Stock Changes	-	-	-	-	-	-	-	-	-	-
DOMESTIC SUPPLY	**2767579**	**-**	**-**	**119779**	**10952**	**-**	**-**	**-**	**172150**	**594593**
Transfers	-	-	-	-	-	-	-	-	-	-
Statistical Differences	-	-	-	-	-	-	-	-	-	-
TRANSFORMATION	**766000**	**-**	**-**	**-**	**-**	**-**	**-**	**-**	**-**	**-**
Electricity Plants	436000	-	-	-	-	-	-	-	-	-
CHP Plants	120000	-	-	-	-	-	-	-	-	-
Heat Plants	210000	-	-	-	-	-	-	-	-	-
Blast Furnaces/Gas Works	-	-	-	-	-	-	-	-	-	-
Coke/Pat. Fuel/BKB Plants	-	-	-	-	-	-	-	-	-	-
Petroleum Refineries	-	-	-	-	-	-	-	-	-	-
Petrochemical Industry	-	-	-	-	-	-	-	-	-	-
Liquefaction	-	-	-	-	-	-	-	-	-	-
Other Transform. Sector	-	-	-	-	-	-	-	-	-	-
ENERGY SECTOR	**60000**	**-**	**-**	**-**	**-**	**-**	**-**	**-**	**23934**	**5000**
Coal Mines	-	-	-	-	-	-	-	-	8625	-
Oil and Gas Extraction	57000	-	-	-	-	-	-	-	401	-
Petroleum Refineries	3000	-	-	-	-	-	-	-	963	-
Electr., CHP+Heat Plants	-	-	-	-	-	-	-	-	12334	5000
Pumped Storage (Elec.)	-	-	-	-	-	-	-	-	-	-
Other Energy Sector	-	-	-	-	-	-	-	-	1611	-
Distribution Losses	61579	-	-	-	-	-	-	-	30013	177939
FINAL CONSUMPTION	**1880000**	**-**	**-**	**119779**	**10952**	**-**	**-**	**-**	**118203**	**411654**
INDUSTRY SECTOR	**680000**	**-**	**-**	**119779**	**-**	**-**	**-**	**-**	**60192**	**265828**
Iron and Steel	-	-	-	119779	-	-	-	-	20558	-
Chemical and Petrochem.	-	-	-	-	-	-	-	-	7208	-
of which: Feedstocks	-	-	-	-	-	-	-	-	-	-
Non-Ferrous Metals	-	-	-	-	-	-	-	-	3208	-
Non-Metallic Minerals	-	-	-	-	-	-	-	-	2831	-
Transport Equipment	-	-	-	-	-	-	-	-	900	-
Machinery	-	-	-	-	-	-	-	-	5828	-
Mining and Quarrying	-	-	-	-	-	-	-	-	8087	-
Food and Tobacco	-	-	-	-	-	-	-	-	3619	-
Paper, Pulp and Print	-	-	-	-	-	-	-	-	518	-
Wood and Wood Products	-	-	-	-	-	-	-	-	282	-
Construction	-	-	-	-	-	-	-	-	1414	-
Textile and Leather	-	-	-	-	-	-	-	-	537	-
Non-specified	680000	-	-	-	-	-	-	-	5202	265828
TRANSPORT SECTOR	**-**	**-**	**-**	**-**	**-**	**-**	**-**	**-**	**9703**	**-**
Air	-	-	-	-	-	-	-	-	-	-
Road	-	-	-	-	-	-	-	-	-	-
Rail	-	-	-	-	-	-	-	-	4403	-
Pipeline Transport	-	-	-	-	-	-	-	-	2001	-
Internal Navigation	-	-	-	-	-	-	-	-	-	-
Non-specified	-	-	-	-	-	-	-	-	3299	-
OTHER SECTORS	**1200000**	**-**	**-**	**-**	**10952**	**-**	**-**	**-**	**48308**	**145826**
Agriculture	20000	-	-	-	-	-	-	-	8148	16727
Comm. and Publ. Services	410000	-	-	-	-	-	-	-	9192	-
Residential	770000	-	-	-	-	-	-	-	-	129099
Non-specified	-	-	-	-	10952	-	-	-	30968	-
NON-ENERGY USE	**-**	**-**	**-**	**-**	**-**	**-**	**-**	**-**	**-**	**-**
in Industry/Transf./Energy	-	-	-	-	-	-	-	-	-	-
in Transport	-	-	-	-	-	-	-	-	-	-
in Other Sectors	-	-	-	-	-	-	-	-	-	-

United Arab Emirates / Emirats arabes unis

SUPPLY AND CONSUMPTION 1997	Coal (1000 tonnes)							Oil (1000 tonnes)			
	Coking Coal	Other Bit. Coal	Sub-Bit. Coal	Lignite	Peat	Oven and Gas Coke	Pat. Fuel and BKB	Crude Oil	NGL	Feed-stocks	Additives
Production	-	-	-	-	-	-	-	105892	3808	-	-
Imports	-	-	-	-	-	-	-	-	-	-	-
Exports	-	-	-	-	-	-	-	-93558	-	-	-
Intl. Marine Bunkers	-	-	-	-	-	-	-	-	-	-	-
Stock Changes	-	-	-	-	-	-	-	-	-	-	-
DOMESTIC SUPPLY	-	-	-	-	-	-	-	12334	3808	-	-
Transfers and Stat. Diff.	-	-	-	-	-	-	-	-	-3808	-	-
TRANSFORMATION	-	-	-	-	-	-	-	12334	-	-	-
Electricity and CHP Plants	-	-	-	-	-	-	-	31	-	-	-
Petroleum Refineries	-	-	-	-	-	-	-	12303	-	-	-
Other Transform. Sector	-	-	-	-	-	-	-	-	-	-	-
ENERGY SECTOR	-	-	-	-	-	-	-	-	-	-	-
DISTRIBUTION LOSSES	-	-	-	-	-	-	-	-	-	-	-
FINAL CONSUMPTION	-	-	-	-	-	-	-	-	-	-	-
INDUSTRY SECTOR	-	-	-	-	-	-	-	-	-	-	-
Iron and Steel	-	-	-	-	-	-	-	-	-	-	-
Chemical and Petrochem.	-	-	-	-	-	-	-	-	-	-	-
Non-Metallic Minerals	-	-	-	-	-	-	-	-	-	-	-
Non-specified	-	-	-	-	-	-	-	-	-	-	-
TRANSPORT SECTOR	-	-	-	-	-	-	-	-	-	-	-
Air	-	-	-	-	-	-	-	-	-	-	-
Road	-	-	-	-	-	-	-	-	-	-	-
Non-specified	-	-	-	-	-	-	-	-	-	-	-
OTHER SECTORS	-	-	-	-	-	-	-	-	-	-	-
Agriculture	-	-	-	-	-	-	-	-	-	-	-
Comm. and Publ. Services	-	-	-	-	-	-	-	-	-	-	-
Residential	-	-	-	-	-	-	-	-	-	-	-
Non-specified	-	-	-	-	-	-	-	-	-	-	-
NON-ENERGY USE	-	-	-	-	-	-	-	-	-	-	-

APPROVISIONNEMENT ET DEMANDE 1998	Charbon (1000 tonnes)							Pétrole (1000 tonnes)			
	Charbon à coke	Autres charb. bit.	Charbon sous-bit.	Lignite	Tourbe	Coke de four/gaz	Agg./briq. de lignite	Pétrole brut	LGN	Produits d'aliment.	Additifs
Production	-	-	-	-	-	-	-	108010	5204	-	-
Imports	-	-	-	-	-	-	-	-	-	-	-
Exports	-	-	-	-	-	-	-	-96085	-	-	-
Intl. Marine Bunkers	-	-	-	-	-	-	-	-	-	-	-
Stock Changes	-	-	-	-	-	-	-	-	-	-	-
DOMESTIC SUPPLY	-	-	-	-	-	-	-	11925	5204	-	-
Transfers and Stat. Diff.	-	-	-	-	-	-	-	-	-5204	-	-
TRANSFORMATION	-	-	-	-	-	-	-	11925	-	-	-
Electricity and CHP Plants	-	-	-	-	-	-	-	30	-	-	-
Petroleum Refineries	-	-	-	-	-	-	-	11895	-	-	-
Other Transform. Sector	-	-	-	-	-	-	-	-	-	-	-
ENERGY SECTOR	-	-	-	-	-	-	-	-	-	-	-
DISTRIBUTION LOSSES	-	-	-	-	-	-	-	-	-	-	-
FINAL CONSUMPTION	-	-	-	-	-	-	-	-	-	-	-
INDUSTRY SECTOR	-	-	-	-	-	-	-	-	-	-	-
Iron and Steel	-	-	-	-	-	-	-	-	-	-	-
Chemical and Petrochem.	-	-	-	-	-	-	-	-	-	-	-
Non-Metallic Minerals	-	-	-	-	-	-	-	-	-	-	-
Non-specified	-	-	-	-	-	-	-	-	-	-	-
TRANSPORT SECTOR	-	-	-	-	-	-	-	-	-	-	-
Air	-	-	-	-	-	-	-	-	-	-	-
Road	-	-	-	-	-	-	-	-	-	-	-
Non-specified	-	-	-	-	-	-	-	-	-	-	-
OTHER SECTORS	-	-	-	-	-	-	-	-	-	-	-
Agriculture	-	-	-	-	-	-	-	-	-	-	-
Comm. and Publ. Services	-	-	-	-	-	-	-	-	-	-	-
Residential	-	-	-	-	-	-	-	-	-	-	-
Non-specified	-	-	-	-	-	-	-	-	-	-	-
NON-ENERGY USE	-	-	-	-	-	-	-	-	-	-	-

United Arab Emirates / Emirats arabes unis

SUPPLY AND CONSUMPTION 1997	Oil cont. (1000 tonnes)										
	Refinery Gas	LPG + Ethane	Motor Gasoline	Aviation Gasoline	Jet Fuel	Kerosene	Gas/ Diesel	Heavy Fuel Oil	Naphtha	Petrol. Coke	Other Prod.
Production	369	2613	1507	-	2739	-	3309	2113	1250	-	18
Imports	-	-	-	-	-	-	-	-	-	-	-
Exports	-	-6334	-	-	-2070	-	-2308	-	-1250	-	-
Intl. Marine Bunkers	-	-	-	-	-	-	-	-117	-	-	-
Stock Changes	-	-	44	-	-	-	-	-469	-	-	22
DOMESTIC SUPPLY	369	-3721	1551	-	669	-	1001	1527	-	-	40
Transfers and Stat. Diff.	-	3808	-	-	-	-	-	-	-	-	-
TRANSFORMATION	-	-	-	-	-	-	268	55	-	-	-
Electricity and CHP Plants	-	-	-	-	-	-	268	55	-	-	-
Petroleum Refineries	-	-	-	-	-	-	-	-	-	-	-
Other Transform. Sector	-	-	-	-	-	-	-	-	-	-	-
ENERGY SECTOR	369	-	-	-	-	-	-	-	-	-	-
DISTRIBUTION LOSSES	-	-	-	-	-	-	-	-	-	-	-
FINAL CONSUMPTION	-	87	1551	-	669	-	733	1472	-	-	40
INDUSTRY SECTOR	-	-	-	-	-	-	-	1472	-	-	-
Iron and Steel	-	-	-	-	-	-	-	-	-	-	-
Chemical and Petrochem.	-	-	-	-	-	-	-	-	-	-	-
Non-Metallic Minerals	-	-	-	-	-	-	-	-	-	-	-
Non-specified	-	-	-	-	-	-	-	1472	-	-	-
TRANSPORT SECTOR	-	-	1551	-	669	-	733	-	-	-	-
Air	-	-	-	-	669	-	-	-	-	-	-
Road	-	-	1551	-	-	-	733	-	-	-	-
Non-specified	-	-	-	-	-	-	-	-	-	-	-
OTHER SECTORS	-	87	-	-	-	-	-	-	-	-	-
Agriculture	-	-	-	-	-	-	-	-	-	-	-
Comm. and Publ. Services	-	-	-	-	-	-	-	-	-	-	-
Residential	-	87	-	-	-	-	-	-	-	-	-
Non-specified	-	-	-	-	-	-	-	-	-	-	-
NON-ENERGY USE	-	-	-	-	-	-	-	-	-	-	40

APPROVISIONNEMENT ET DEMANDE 1998	Pétrole cont. (1000 tonnes)										
	Gaz de raffinerie	GPL + éthane	Essence moteur	Essence aviation	Carbu- réacteurs	Kérosène	Gazole	Fioul lourd	Naphta	Coke de pétrole	Autres prod.
Production	357	2577	1539	-	2659	-	3258	1845	1236	-	28
Imports	-	-	-	-	-	-	-	-	-	-	-
Exports	-	-7692	-	-	-1998	-	-2086	-	-1236	-	-
Intl. Marine Bunkers	-	-	-	-	-	-	-	-	-	-	-
Stock Changes	-	-	141	-	-	-	-	-329	-	-	13
DOMESTIC SUPPLY	357	-5115	1680	-	661	-	1172	1516	-	-	41
Transfers and Stat. Diff.	-	5204	-	-	-	-	-	-	-	-	-
TRANSFORMATION	-	-	-	-	-	-	791	83	-	-	-
Electricity and CHP Plants	-	-	-	-	-	-	791	83	-	-	-
Petroleum Refineries	-	-	-	-	-	-	-	-	-	-	-
Other Transform. Sector	-	-	-	-	-	-	-	-	-	-	-
ENERGY SECTOR	357	-	-	-	-	-	-	-	-	-	-
DISTRIBUTION LOSSES	-	-	-	-	-	-	-	-	-	-	-
FINAL CONSUMPTION	-	89	1680	-	661	-	381	1433	-	-	41
INDUSTRY SECTOR	-	-	-	-	-	-	-	1433	-	-	-
Iron and Steel	-	-	-	-	-	-	-	-	-	-	-
Chemical and Petrochem.	-	-	-	-	-	-	-	-	-	-	-
Non-Metallic Minerals	-	-	-	-	-	-	-	-	-	-	-
Non-specified	-	-	-	-	-	-	-	1433	-	-	-
TRANSPORT SECTOR	-	-	1680	-	661	-	381	-	-	-	-
Air	-	-	-	-	661	-	-	-	-	-	-
Road	-	-	1680	-	-	-	381	-	-	-	-
Non-specified	-	-	-	-	-	-	-	-	-	-	-
OTHER SECTORS	-	89	-	-	-	-	-	-	-	-	-
Agriculture	-	-	-	-	-	-	-	-	-	-	-
Comm. and Publ. Services	-	-	-	-	-	-	-	-	-	-	-
Residential	-	89	-	-	-	-	-	-	-	-	-
Non-specified	-	-	-	-	-	-	-	-	-	-	-
NON-ENERGY USE	-	-	-	-	-	-	-	-	-	-	41

United Arab Emirates / Emirats arabes unis

SUPPLY AND CONSUMPTION 1997	Gas (TJ) Natural Gas	Gas Works	Coke Ovens	Blast Furnaces	Comb. Renew. & Waste (TJ) Solid Biomass	Gas/Liquids from Biomass	Municipal Waste	Industrial Waste	(GWh) Electricity	(TJ) Heat
Production	1352930	-	-	-	-	-	-	-	28464	-
Imports	18630	-	-	-	730	-	-	-	-	-
Exports	-291005	-	-	-	-	-	-	-	-	-
Intl. Marine Bunkers	-	-	-	-	-	-	-	-	-	-
Stock Changes	-	-	-	-	-	-	-	-	-	-
DOMESTIC SUPPLY	1080555	-	-	-	730	-	-	-	28464	-
Transfers and Stat. Diff.	-	-	-	-	-	-	-	-	-	-
TRANSFORMATION	547142	-	-	-	-	-	-	-	-	-
Electricity and CHP Plants	400946	-	-	-	-	-	-	-	-	-
Petroleum Refineries	-	-	-	-	-	-	-	-	-	-
Other Transform. Sector	146196	-	-	-	-	-	-	-	-	-
ENERGY SECTOR	29101	-	-	-	-	-	-	-	2960	-
DISTRIBUTION LOSSES	-	-	-	-	-	-	-	-	2562	-
FINAL CONSUMPTION	504312	-	-	-	730	-	-	-	22942	-
INDUSTRY SECTOR	504312	-	-	-	-	-	-	-	1433	-
Iron and Steel	-	-	-	-	-	-	-	-	-	-
Chemical and Petrochem.	504312	-	-	-	-	-	-	-	-	-
Non-Metallic Minerals	-	-	-	-	-	-	-	-	-	-
Non-specified	-	-	-	-	-	-	-	-	1433	-
TRANSPORT SECTOR	-	-	-	-	-	-	-	-	-	-
Air	-	-	-	-	-	-	-	-	-	-
Road	-	-	-	-	-	-	-	-	-	-
Non-specified	-	-	-	-	-	-	-	-	-	-
OTHER SECTORS	-	-	-	-	730	-	-	-	21509	-
Agriculture	-	-	-	-	-	-	-	-	657	-
Comm. and Publ. Services	-	-	-	-	-	-	-	-	9354	-
Residential	-	-	-	-	-	-	-	-	11498	-
Non-specified	-	-	-	-	730	-	-	-	-	-
NON-ENERGY USE	-	-	-	-	-	-	-	-	-	-

APPROVISIONNEMENT ET DEMANDE 1998	Gaz (TJ) Gaz naturel	Usines à gaz	Cokeries	Hauts fourneaux	En. Re. Comb. & Déchets (TJ) Biomasse solide	Gaz/Liquides tirés de biomasse	Déchets urbains	Déchets industriels	(GWh) Electricité	(TJ) Chaleur
Production	1381248	-	-	-	-	-	-	-	33392	-
Imports	18630	-	-	-	710	-	-	-	-	-
Exports	-267158	-	-	-	-	-	-	-	-	-
Intl. Marine Bunkers	-	-	-	-	-	-	-	-	-	-
Stock Changes	-	-	-	-	-	-	-	-	-	-
DOMESTIC SUPPLY	1132720	-	-	-	710	-	-	-	33392	-
Transfers and Stat. Diff.	-	-	-	-	-	-	-	-	-	-
TRANSFORMATION	593643	-	-	-	-	-	-	-	-	-
Electricity and CHP Plants	449695	-	-	-	-	-	-	-	-	-
Petroleum Refineries	-	-	-	-	-	-	-	-	-	-
Other Transform. Sector	143948	-	-	-	-	-	-	-	-	-
ENERGY SECTOR	26716	-	-	-	-	-	-	-	3440	-
DISTRIBUTION LOSSES	-	-	-	-	-	-	-	-	3005	-
FINAL CONSUMPTION	512361	-	-	-	710	-	-	-	26947	-
INDUSTRY SECTOR	512361	-	-	-	-	-	-	-	2358	-
Iron and Steel	-	-	-	-	-	-	-	-	-	-
Chemical and Petrochem.	512361	-	-	-	-	-	-	-	-	-
Non-Metallic Minerals	-	-	-	-	-	-	-	-	-	-
Non-specified	-	-	-	-	-	-	-	-	2358	-
TRANSPORT SECTOR	-	-	-	-	-	-	-	-	-	-
Air	-	-	-	-	-	-	-	-	-	-
Road	-	-	-	-	-	-	-	-	-	-
Non-specified	-	-	-	-	-	-	-	-	-	-
OTHER SECTORS	-	-	-	-	710	-	-	-	24589	-
Agriculture	-	-	-	-	-	-	-	-	802	-
Comm. and Publ. Services	-	-	-	-	-	-	-	-	11623	-
Residential	-	-	-	-	-	-	-	-	12164	-
Non-specified	-	-	-	-	710	-	-	-	-	-
NON-ENERGY USE	-	-	-	-	-	-	-	-	-	-

Uruguay

SUPPLY AND CONSUMPTION 1997	Coal (1000 tonnes)							Oil (1000 tonnes)			
	Coking Coal	Other Bit. Coal	Sub-Bit. Coal	Lignite	Peat	Oven and Gas Coke	Pat. Fuel and BKB	Crude Oil	NGL	Feed-stocks	Additives
Production	-	-	-	-	-	-	-	-	-	-	-
Imports	-	-	-	-	-	-	-	1465	-	-	-
Exports	-	-	-	-	-	-	-	-	-	-	-
Intl. Marine Bunkers	-	-	-	-	-	-	-	-	-	-	-
Stock Changes	-	-	-	-	-	-	-	-71	-	-	-
DOMESTIC SUPPLY	-	-	-	-	-	-	-	1394	-	-	-
Transfers and Stat. Diff.	-	-	-	-	-	-	-	-	-	-	-
TRANSFORMATION	-	-	-	-	-	-	-	1394	-	-	-
Electricity and CHP Plants	-	-	-	-	-	-	-	-	-	-	-
Petroleum Refineries	-	-	-	-	-	-	-	1394	-	-	-
Other Transform. Sector	-	-	-	-	-	-	-	-	-	-	-
ENERGY SECTOR	-	-	-	-	-	-	-	-	-	-	-
DISTRIBUTION LOSSES	-	-	-	-	-	-	-	-	-	-	-
FINAL CONSUMPTION	-	-	-	-	-	-	-	-	-	-	-
INDUSTRY SECTOR	-	-	-	-	-	-	-	-	-	-	-
Iron and Steel	-	-	-	-	-	-	-	-	-	-	-
Chemical and Petrochem.	-	-	-	-	-	-	-	-	-	-	-
Non-Metallic Minerals	-	-	-	-	-	-	-	-	-	-	-
Non-specified	-	-	-	-	-	-	-	-	-	-	-
TRANSPORT SECTOR	-	-	-	-	-	-	-	-	-	-	-
Air	-	-	-	-	-	-	-	-	-	-	-
Road	-	-	-	-	-	-	-	-	-	-	-
Non-specified	-	-	-	-	-	-	-	-	-	-	-
OTHER SECTORS	-	-	-	-	-	-	-	-	-	-	-
Agriculture	-	-	-	-	-	-	-	-	-	-	-
Comm. and Publ. Services	-	-	-	-	-	-	-	-	-	-	-
Residential	-	-	-	-	-	-	-	-	-	-	-
Non-specified	-	-	-	-	-	-	-	-	-	-	-
NON-ENERGY USE	-	-	-	-	-	-	-	-	-	-	-

APPROVISIONNEMENT ET DEMANDE 1998	Charbon (1000 tonnes)							Pétrole (1000 tonnes)			
	Charbon à coke	Autres charb. bit.	Charbon sous-bit.	Lignite	Tourbe	Coke de four/gaz	Agg./briq. de lignite	Pétrole brut	LGN	Produits d'aliment.	Additifs
Production	-	-	-	-	-	-	-	-	-	-	-
Imports	-	-	-	-	-	-	-	1780	-	-	-
Exports	-	-	-	-	-	-	-	-	-	-	-
Intl. Marine Bunkers	-	-	-	-	-	-	-	-	-	-	-
Stock Changes	-	-	-	-	-	-	-	78	-	-	-
DOMESTIC SUPPLY	-	-	-	-	-	-	-	1858	-	-	-
Transfers and Stat. Diff.	-	-	-	-	-	-	-	-	-	-	-
TRANSFORMATION	-	-	-	-	-	-	-	1858	-	-	-
Electricity and CHP Plants	-	-	-	-	-	-	-	-	-	-	-
Petroleum Refineries	-	-	-	-	-	-	-	1858	-	-	-
Other Transform. Sector	-	-	-	-	-	-	-	-	-	-	-
ENERGY SECTOR	-	-	-	-	-	-	-	-	-	-	-
DISTRIBUTION LOSSES	-	-	-	-	-	-	-	-	-	-	-
FINAL CONSUMPTION	-	-	-	-	-	-	-	-	-	-	-
INDUSTRY SECTOR	-	-	-	-	-	-	-	-	-	-	-
Iron and Steel	-	-	-	-	-	-	-	-	-	-	-
Chemical and Petrochem.	-	-	-	-	-	-	-	-	-	-	-
Non-Metallic Minerals	-	-	-	-	-	-	-	-	-	-	-
Non-specified	-	-	-	-	-	-	-	-	-	-	-
TRANSPORT SECTOR	-	-	-	-	-	-	-	-	-	-	-
Air	-	-	-	-	-	-	-	-	-	-	-
Road	-	-	-	-	-	-	-	-	-	-	-
Non-specified	-	-	-	-	-	-	-	-	-	-	-
OTHER SECTORS	-	-	-	-	-	-	-	-	-	-	-
Agriculture	-	-	-	-	-	-	-	-	-	-	-
Comm. and Publ. Services	-	-	-	-	-	-	-	-	-	-	-
Residential	-	-	-	-	-	-	-	-	-	-	-
Non-specified	-	-	-	-	-	-	-	-	-	-	-
NON-ENERGY USE	-	-	-	-	-	-	-	-	-	-	-

Uruguay

SUPPLY AND CONSUMPTION 1997	Refinery Gas	LPG + Ethane	Motor Gasoline	Aviation Gasoline	Jet Fuel	Kerosene	Gas/ Diesel	Heavy Fuel Oil	Naphtha	Petrol. Coke	Other Prod.
Production	14	62	268	-	37	21	431	429	11	-	89
Imports	-	40	83	3	4	-	384	119	-	-	18
Exports	-	-	-12	-	-39	-	-	-	-2	-	-1
Intl. Marine Bunkers	-	-	-	-	-	-	-128	-179	-	-	-
Stock Changes	-	-2	-15	-	3	6	46	44	1	-	-4
DOMESTIC SUPPLY	14	100	324	3	5	27	733	413	10	-	102
Transfers and Stat. Diff.	-	-	-2	-	-1	-2	-1	1	1	-	-1
TRANSFORMATION	-	4	-	-	-	-	31	124	11	-	-
Electricity and CHP Plants	-	-	-	-	-	-	31	124	-	-	-
Petroleum Refineries	-	-	-	-	-	-	-	-	-	-	-
Other Transform. Sector	-	4	-	-	-	-	-	-	11	-	-
ENERGY SECTOR	14	3	-	-	-	-	-	64	-	-	17
DISTRIBUTION LOSSES	-	1	2	-	-	-	1	1	-	-	-
FINAL CONSUMPTION	-	92	320	3	4	25	700	225	-	-	84
INDUSTRY SECTOR	-	3	-	-	-	1	10	195	-	-	-
Iron and Steel	-	-	-	-	-	-	-	-	-	-	-
Chemical and Petrochem.	-	-	-	-	-	-	-	-	-	-	-
Non-Metallic Minerals	-	-	-	-	-	-	-	-	-	-	-
Non-specified	-	3	-	-	-	1	10	195	-	-	-
TRANSPORT SECTOR	-	-	308	3	4	-	477	-	-	-	-
Air	-	-	-	3	4	-	-	-	-	-	-
Road	-	-	308	-	-	-	477	-	-	-	-
Non-specified	-	-	-	-	-	-	-	-	-	-	-
OTHER SECTORS	-	89	12	-	-	24	213	30	-	-	-
Agriculture	-	-	10	-	-	-	168	-	-	-	-
Comm. and Publ. Services	-	1	-	-	-	-	36	6	-	-	-
Residential	-	88	-	-	-	24	9	24	-	-	-
Non-specified	-	-	2	-	-	-	-	-	-	-	-
NON-ENERGY USE	-	-	-	-	-	-	-	-	-	-	84

APPROVISIONNEMENT ET DEMANDE 1998	Gaz de raffinerie	GPL + éthane	Essence moteur	Essence aviation	Carbu- réacteurs	Kérosène	Gazole	Fioul lourd	Naphta	Coke de pétrole	Autres prod.
Production	21	84	348	-	58	27	592	575	13	-	104
Imports	-	21	-	3	-	-	299	49	-	-	17
Exports	-	-	-	-	-52	-1	-	-	-1	-	-1
Intl. Marine Bunkers	-	-	-	-	-	-	-103	-177	-	-	-
Stock Changes	-	-	3	-	-2	-2	-14	-34	-1	-	3
DOMESTIC SUPPLY	21	105	351	3	4	24	774	413	11	-	123
Transfers and Stat. Diff.	-	-	-1	-	-	-	-6	-1	-	-	-23
TRANSFORMATION	-	5	-	-	-	-	29	68	11	-	-
Electricity and CHP Plants	-	-	-	-	-	-	29	68	-	-	-
Petroleum Refineries	-	-	-	-	-	-	-	-	-	-	-
Other Transform. Sector	-	5	-	-	-	-	-	-	11	-	-
ENERGY SECTOR	21	1	-	-	-	-	-	40	-	-	9
DISTRIBUTION LOSSES	-	-	3	-	-	-	1	-	-	-	1
FINAL CONSUMPTION	-	99	347	3	4	24	738	304	-	-	90
INDUSTRY SECTOR	-	8	-	-	-	1	8	270	-	-	-
Iron and Steel	-	-	-	-	-	-	-	-	-	-	-
Chemical and Petrochem.	-	-	-	-	-	-	-	-	-	-	-
Non-Metallic Minerals	-	-	-	-	-	-	-	-	-	-	-
Non-specified	-	8	-	-	-	1	8	270	-	-	-
TRANSPORT SECTOR	-	-	333	3	4	-	508	-	-	-	-
Air	-	-	-	3	4	-	-	-	-	-	-
Road	-	-	333	-	-	-	508	-	-	-	-
Non-specified	-	-	-	-	-	-	-	-	-	-	-
OTHER SECTORS	-	91	14	-	-	23	222	34	-	-	-
Agriculture	-	-	11	-	-	-	175	-	-	-	-
Comm. and Publ. Services	-	1	-	-	-	-	41	5	-	-	-
Residential	-	90	-	-	-	23	6	29	-	-	-
Non-specified	-	-	3	-	-	-	-	-	-	-	-
NON-ENERGY USE	-	-	-	-	-	-	-	-	-	-	90

Uruguay

SUPPLY AND CONSUMPTION 1997	Natural Gas	Gas Works	Coke Ovens	Blast Furnaces	Solid Biomasse	Gas/Liquids from Biomass	Municipal Waste	Industrial Waste	Electricity	Heat
Production	-	515	-	-	21397	-	-	-	7148	-
Imports	-	-	-	-	21	-	-	-	271	-
Exports	-	-	-	-	-	-	-	-	-415	-
Intl. Marine Bunkers	-	-	-	-	-	-	-	-	-	-
Stock Changes	-	-	-	-	-	-	-	-	-	-
DOMESTIC SUPPLY	-	515	-	-	21418	-	-	-	7004	-
Transfers and Stat. Diff.	-	-1	-	-	-1	-	-	-	26	-
TRANSFORMATION	-	-	-	-	246	-	-	-	-	-
Electricity and CHP Plants	-	-	-	-	200	-	-	-	-	-
Petroleum Refineries	-	-	-	-	-	-	-	-	-	-
Other Transform. Sector	-	-	-	-	46	-	-	-	-	-
ENERGY SECTOR	-	-	-	-	-	-	-	-	107	-
DISTRIBUTION LOSSES	-	75	-	-	-	-	-	-	1340	-
FINAL CONSUMPTION	-	439	-	-	21163	-	-	-	5583	-
INDUSTRY SECTOR	-	71	-	-	8347	-	-	-	1300	-
Iron and Steel	-	-	-	-	-	-	-	-	-	-
Chemical and Petrochem.	-	-	-	-	-	-	-	-	-	-
Non-Metallic Minerals	-	-	-	-	-	-	-	-	-	-
Non-specified	-	71	-	-	8347	-	-	-	1300	-
TRANSPORT SECTOR	-	-	-	-	-	-	-	-	-	-
Air	-	-	-	-	-	-	-	-	-	-
Road	-	-	-	-	-	-	-	-	-	-
Non-specified	-	-	-	-	-	-	-	-	-	-
OTHER SECTORS	-	368	-	-	12816	-	-	-	4283	-
Agriculture	-	-	-	-	-	-	-	-	190	-
Comm. and Publ. Services	-	142	-	-	130	-	-	-	1632	-
Residential	-	226	-	-	12686	-	-	-	2461	-
Non-specified	-	-	-	-	-	-	-	-	-	-
NON-ENERGY USE	-	-	-	-	-	-	-	-	-	-

APPROVISIONNEMENT ET DEMANDE 1998	Gaz naturel	Usines à gaz	Cokeries	Hauts fourneaux	Biomasse solide	Gaz/Liquides tirés de biomasse	Déchets urbains	Déchets industriels	Electricité	Chaleur
Production	-	548	-	-	19865	-	-	-	9568	-
Imports	92	-	-	-	29	-	-	-	78	-
Exports	-	-	-	-	-	-	-	-	-2234	-
Intl. Marine Bunkers	-	-	-	-	-	-	-	-	-	-
Stock Changes	-	-	-	-	-	-	-	-	-	-
DOMESTIC SUPPLY	92	548	-	-	19894	-	-	-	7412	-
Transfers and Stat. Diff.	-	1	-	-	-	-	-	-	37	-
TRANSFORMATION	-	-	-	-	324	-	-	-	-	-
Electricity and CHP Plants	-	-	-	-	283	-	-	-	-	-
Petroleum Refineries	-	-	-	-	-	-	-	-	-	-
Other Transform. Sector	-	-	-	-	41	-	-	-	-	-
ENERGY SECTOR	-	-	-	-	-	-	-	-	104	-
DISTRIBUTION LOSSES	-	46	-	-	-	-	-	-	1464	-
FINAL CONSUMPTION	92	503	-	-	19570	-	-	-	5881	-
INDUSTRY SECTOR	92	84	-	-	6742	-	-	-	1405	-
Iron and Steel	-	-	-	-	-	-	-	-	-	-
Chemical and Petrochem.	-	-	-	-	-	-	-	-	-	-
Non-Metallic Minerals	-	-	-	-	-	-	-	-	-	-
Non-specified	92	84	-	-	6742	-	-	-	1405	-
TRANSPORT SECTOR	-	-	-	-	-	-	-	-	-	-
Air	-	-	-	-	-	-	-	-	-	-
Road	-	-	-	-	-	-	-	-	-	-
Non-specified	-	-	-	-	-	-	-	-	-	-
OTHER SECTORS	-	419	-	-	12828	-	-	-	4476	-
Agriculture	-	-	-	-	-	-	-	-	184	-
Comm. and Publ. Services	-	159	-	-	130	-	-	-	1699	-
Residential	-	260	-	-	12686	-	-	-	2593	-
Non-specified	-	-	-	-	12	-	-	-	-	-
NON-ENERGY USE	-	-	-	-	-	-	-	-	-	-

Uzbekistan / Ouzbékistan : 1997

SUPPLY AND CONSUMPTION APPROVISIONNEMENT ET DEMANDE	Coal / Charbon (1000 tonnes)							Oil / Pétrole (1000 tonnes)			
	Coking Coal Charbon à coke	Other Bit. Coal Autres charb. bit.	Sub-Bit. Coal Charbon sous-bit.	Lignite Lignite	Peat Tourbe	Oven and Gas Coke Coke de four/gaz	Pat. Fuel and BKB Agg./briq. de lignite	Crude Oil Pétrole brut	NGL LGN	Feed-stocks Produits d'aliment.	Additives Additifs
Production	-	59	-	2888	-	-	18	5456	2668	-	-
From Other Sources	-	-	-	-	-	-	-	-	-	-	-
Imports	-	-	-	27	-	-	-	-	-	-	7
Exports	-	-	-	-30	-	-	-	-324	-589	-	-
Intl. Marine Bunkers	-	-	-	-	-	-	-	-	-	-	-
Stock Changes	-	16	-	-168	-	-	-	17	24	-	-
DOMESTIC SUPPLY	-	75	-	2717	-	-	18	5149	2103	-	7
Transfers	-	-	-	-	-	-	-	-	-32	-	-
Statistical Differences	-	-	-	-	-	-	-	-13	-10	-	-
TRANSFORMATION	-	42	-	2223	-	-	-	4929	2046	-	7
Electricity Plants	-	-	-	1191	-	-	-	-	-	-	-
CHP Plants	-	-	-	1028	-	-	-	-	-	-	-
Heat Plants	-	-	-	4	-	-	-	-	-	-	-
Blast Furnaces/Gas Works	-	-	-	-	-	-	-	-	-	-	-
Coke/Pat. Fuel/BKB Plants	-	42	-	-	-	-	-	-	-	-	-
Petroleum Refineries	-	-	-	-	-	-	-	4929	2046	-	7
Petrochemical Industry	-	-	-	-	-	-	-	-	-	-	-
Liquefaction	-	-	-	-	-	-	-	-	-	-	-
Other Transform. Sector	-	-	-	-	-	-	-	-	-	-	-
ENERGY SECTOR	-	-	-	6	-	-	-	14	1	-	-
Coal Mines	-	-	-	-	-	-	-	-	-	-	-
Oil and Gas Extraction	-	-	-	-	-	-	-	14	1	-	-
Petroleum Refineries	-	-	-	-	-	-	-	-	-	-	-
Electr., CHP+Heat Plants	-	-	-	6	-	-	-	-	-	-	-
Pumped Storage (Elec.)	-	-	-	-	-	-	-	-	-	-	-
Other Energy Sector	-	-	-	-	-	-	-	-	-	-	-
Distribution Losses	-	8	-	28	-	-	-	74	14	-	-
FINAL CONSUMPTION	-	25	-	460	-	-	18	119	-	-	-
INDUSTRY SECTOR	-	17	-	131	-	-	-	119	-	-	-
Iron and Steel	-	-	-	-	-	-	-	-	-	-	-
Chemical and Petrochem.	-	-	-	-	-	-	-	119	-	-	-
of which: Feedstocks	-	-	-	-	-	-	-	119	-	-	-
Non-Ferrous Metals	-	-	-	-	-	-	-	-	-	-	-
Non-Metallic Minerals	-	-	-	-	-	-	-	-	-	-	-
Transport Equipment	-	-	-	-	-	-	-	-	-	-	-
Machinery	-	-	-	-	-	-	-	-	-	-	-
Mining and Quarrying	-	-	-	-	-	-	-	-	-	-	-
Food and Tobacco	-	-	-	-	-	-	-	-	-	-	-
Paper, Pulp and Print	-	-	-	-	-	-	-	-	-	-	-
Wood and Wood Products	-	-	-	-	-	-	-	-	-	-	-
Construction	-	-	-	-	-	-	-	-	-	-	-
Textile and Leather	-	-	-	-	-	-	-	-	-	-	-
Non-specified	-	17	-	131	-	-	-	-	-	-	-
TRANSPORT SECTOR	-	-	-	-	-	-	-	-	-	-	-
Air	-	-	-	-	-	-	-	-	-	-	-
Road	-	-	-	-	-	-	-	-	-	-	-
Rail	-	-	-	-	-	-	-	-	-	-	-
Pipeline Transport	-	-	-	-	-	-	-	-	-	-	-
Internal Navigation	-	-	-	-	-	-	-	-	-	-	-
Non-specified	-	-	-	-	-	-	-	-	-	-	-
OTHER SECTORS	-	8	-	329	-	-	18	-	-	-	-
Agriculture	-	-	-	9	-	-	-	-	-	-	-
Comm. and Publ. Services	-	-	-	-	-	-	-	-	-	-	-
Residential	-	-	-	34	-	-	-	-	-	-	-
Non-specified	-	8	-	286	-	-	18	-	-	-	-
NON-ENERGY USE	-	-	-	-	-	-	-	-	-	-	-
in Industry/Trans./Energy	-	-	-	-	-	-	-	-	-	-	-
in Transport	-	-	-	-	-	-	-	-	-	-	-
in Other Sectors	-	-	-	-	-	-	-	-	-	-	-

Uzbekistan / Ouzbékistan : 1997

SUPPLY AND CONSUMPTION / APPROVISIONNEMENT ET DEMANDE	Refinery Gas / Gaz de raffinerie	LPG + Ethane / GPL + éthane	Motor Gasoline / Essence moteur	Aviation Gasoline / Essence aviation	Jet Fuel / Carbu-réacteurs	Kerosene / Kérosène	Gas/ Diesel / Gazole	Heavy Fuel Oil / Fioul lourd	Naphtha / Naphta	Petrol. Coke / Coke de pétrole	Other Prod. / Autres prod.
Production	221	10	1344	3	272	62	2035	1979	-	46	795
From Other Sources	-	-	-	-	-	-	-	-	-	-	-
Imports	-	-	-	-	33	-	-	-	-	-	1
Exports	-	-	-29	-	-5	-	-245	-4	-	-	-83
Intl. Marine Bunkers	-	-	-	-	-	-	-	-	-	-	-
Stock Changes	-	-	92	-	-26	7	221	73	-	-46	24
DOMESTIC SUPPLY	221	10	1407	3	274	69	2011	2048	-	-	737
Transfers	-	32	-	-	-	-	-	-	-	-	-
Statistical Differences	-	-	-	-	-	-	-	-	-	-	-
TRANSFORMATION	-	-	-	-	-	6	6	2013	-	-	8
Electricity Plants	-	-	-	-	-	-	6	737	-	-	-
CHP Plants	-	-	-	-	-	-	-	954	-	-	-
Heat Plants	-	-	-	-	-	6	-	322	-	-	8
Blast Furnaces/Gas Works	-	-	-	-	-	-	-	-	-	-	-
Coke/Pat. Fuel/BKB Plants	-	-	-	-	-	-	-	-	-	-	-
Petroleum Refineries	-	-	-	-	-	-	-	-	-	-	-
Petrochemical Industry	-	-	-	-	-	-	-	-	-	-	-
Liquefaction	-	-	-	-	-	-	-	-	-	-	-
Other Transform. Sector	-	-	-	-	-	-	-	-	-	-	-
ENERGY SECTOR	221	-	-	-	-	-	-	27	-	-	-
Coal Mines	-	-	-	-	-	-	-	-	-	-	-
Oil and Gas Extraction	-	-	-	-	-	-	-	-	-	-	-
Petroleum Refineries	221	-	-	-	-	-	-	-	-	-	-
Electr., CHP+Heat Plants	-	-	-	-	-	-	-	27	-	-	-
Pumped Storage (Elec.)	-	-	-	-	-	-	-	-	-	-	-
Other Energy Sector	-	-	-	-	-	-	-	-	-	-	-
Distribution Losses	-	-	-	-	-	-	-	-	-	-	-
FINAL CONSUMPTION	-	42	1407	3	274	63	2005	8	-	-	729
INDUSTRY SECTOR	-	-	63	-	-	12	196	5	-	-	3
Iron and Steel	-	-	-	-	-	-	-	-	-	-	-
Chemical and Petrochem.	-	-	-	-	-	-	-	-	-	-	-
of which: Feedstocks	-	-	-	-	-	-	-	-	-	-	-
Non-Ferrous Metals	-	-	-	-	-	-	-	-	-	-	-
Non-Metallic Minerals	-	-	-	-	-	-	-	5	-	-	-
Transport Equipment	-	-	-	-	-	-	-	-	-	-	-
Machinery	-	-	4	-	-	-	-	-	-	-	-
Mining and Quarrying	-	-	-	-	-	-	-	-	-	-	-
Food and Tobacco	-	-	-	-	-	-	-	-	-	-	-
Paper, Pulp and Print	-	-	-	-	-	-	-	-	-	-	-
Wood and Wood Products	-	-	-	-	-	-	-	-	-	-	-
Construction	-	-	58	-	-	-	196	-	-	-	-
Textile and Leather	-	-	-	-	-	12	-	-	-	-	3
Non-specified	-	-	1	-	-	-	-	-	-	-	-
TRANSPORT SECTOR	-	14	1309	3	274	-	539	-	-	-	-
Air	-	-	-	3	274	-	-	-	-	-	-
Road	-	14	1309	-	-	-	418	-	-	-	-
Rail	-	-	-	-	-	-	121	-	-	-	-
Pipeline Transport	-	-	-	-	-	-	-	-	-	-	-
Internal Navigation	-	-	-	-	-	-	-	-	-	-	-
Non-specified	-	-	-	-	-	-	-	-	-	-	-
OTHER SECTORS	-	28	35	-	-	51	1270	3	-	-	14
Agriculture	-	-	30	-	-	11	1061	1	-	-	1
Comm. and Publ. Services	-	-	-	-	-	-	-	-	-	-	-
Residential	-	14	1	-	-	1	6	-	-	-	1
Non-specified	-	14	4	-	-	39	203	2	-	-	12
NON-ENERGY USE	-	-	-	-	-	-	-	-	-	-	712
in Industry/Transf./Energy	-	-	-	-	-	-	-	-	-	-	537
in Transport	-	-	-	-	-	-	-	-	-	-	-
in Other Sectors	-	-	-	-	-	-	-	-	-	-	175

Uzbekistan / Ouzbékistan : 1997

SUPPLY AND CONSUMPTION *APPROVISIONNEMENT ET DEMANDE*	Gas / *Gaz* (TJ)				Comb. Renew. & Waste / *En. Re. Comb. & Déchets* (TJ)				(GWh)	(TJ)
	Natural Gas *Gaz naturel*	Gas Works *Usines à gaz*	Coke Ovens *Cokeries*	Blast Furnaces *Hauts fourneaux*	Solid Biomass *Biomasse solide*	Gas/Liquids from Biomass *Gaz/Liquides tirés de biomasse*	Municipal Waste *Déchets urbains*	Industrial Waste *Déchets industriels*	Electricity *Electricité*	Heat *Chaleur*
Production	1848372	-	-	-	10	-	-	-	46054	100100
From Other Sources	-	-	-	-	-	-	-	-		-
Imports	104573	-	-	-	-	-	-	-	12418	-
Exports	-374986	-	-	-	-	-	-	-	-11489	-
Intl. Marine Bunkers	-	-	-	-	-	-	-	-		-
Stock Changes	5987	-	-	-	-	-	-	-		-
DOMESTIC SUPPLY	**1583946**	**-**	**-**	**-**	**10**	**-**	**-**	**-**	**46983**	**100100**
Transfers	-	-	-	-	-	-	-	-		-
Statistical Differences	-	-	-	-	-	-	-	-		-
TRANSFORMATION	**478523**	**-**	**-**	**-**	**-**	**-**	**-**	**-**	**-**	**-**
Electricity Plants	207343	-	-	-	-	-	-	-	-	-
CHP Plants	211978	-	-	-	-	-	-	-	-	-
Heat Plants	59202	-	-	-	-	-	-	-	-	-
Blast Furnaces/Gas Works	-	-	-	-	-	-	-	-	-	-
Coke/Pat. Fuel/BKB Plants	-	-	-	-	-	-	-	-	-	-
Petroleum Refineries	-	-	-	-	-	-	-	-	-	-
Petrochemical Industry	-	-	-	-	-	-	-	-	-	-
Liquefaction	-	-	-	-	-	-	-	-	-	-
Other Transform. Sector	-	-	-	-	-	-	-	-	-	-
ENERGY SECTOR	**75285**	**-**	**-**	**-**	**-**	**-**	**-**	**-**	**4035**	**-**
Coal Mines	-	-	-	-	-	-	-	-	-	-
Oil and Gas Extraction	15724	-	-	-	-	-	-	-	-	-
Petroleum Refineries	-	-	-	-	-	-	-	-	-	-
Electr., CHP+Heat Plants	8676	-	-	-	-	-	-	-	2682	-
Pumped Storage (Elec.)	-	-	-	-	-	-	-	-	-	-
Other Energy Sector	50885	-	-	-	-	-	-	-	1353	-
Distribution Losses	65851	-	-	-	-	-	-	-	4008	-
FINAL CONSUMPTION	**964287**	**-**	**-**	**-**	**10**	**-**	**-**	**-**	**38940**	**100100**
INDUSTRY SECTOR	**190102**	**-**	**-**	**-**	**-**	**-**	**-**	**-**	**14921**	**-**
Iron and Steel	-	-	-	-	-	-	-	-	-	-
Chemical and Petrochem.	59963	-	-	-	-	-	-	-	-	-
of which: Feedstocks	59963	-	-	-	-	-	-	-	-	-
Non-Ferrous Metals	-	-	-	-	-	-	-	-	-	-
Non-Metallic Minerals	-	-	-	-	-	-	-	-	-	-
Transport Equipment	-	-	-	-	-	-	-	-	-	-
Machinery	-	-	-	-	-	-	-	-	-	-
Mining and Quarrying	-	-	-	-	-	-	-	-	-	-
Food and Tobacco	-	-	-	-	-	-	-	-	-	-
Paper, Pulp and Print	-	-	-	-	-	-	-	-	-	-
Wood and Wood Products	-	-	-	-	-	-	-	-	-	-
Construction	-	-	-	-	-	-	-	-	-	-
Textile and Leather	-	-	-	-	-	-	-	-	-	-
Non-specified	130139	-	-	-	-	-	-	-	14921	-
TRANSPORT SECTOR	**57478**	**-**	**-**	**-**	**-**	**-**	**-**	**-**	**1282**	**-**
Air	-	-	-	-	-	-	-	-	-	-
Road	2539	-	-	-	-	-	-	-	-	-
Rail	-	-	-	-	-	-	-	-	156	-
Pipeline Transport	54939	-	-	-	-	-	-	-	820	-
Internal Navigation	-	-	-	-	-	-	-	-	-	-
Non-specified	-	-	-	-	-	-	-	-	306	-
OTHER SECTORS	**716707**	**-**	**-**	**-**	**10**	**-**	**-**	**-**	**22737**	**100100**
Agriculture	6328	-	-	-	-	-	-	-	12647	-
Comm. and Publ. Services	118403	-	-	-	-	-	-	-	3019	-
Residential	591976	-	-	-	-	-	-	-	7071	-
Non-specified	-	-	-	-	10	-	-	-	-	100100
NON-ENERGY USE	**-**	**-**	**-**	**-**	**-**	**-**	**-**	**-**	**-**	**-**
in Industry/Transf./Energy	-	-	-	-	-	-	-	-	-	-
in Transport	-	-	-	-	-	-	-	-	-	-
in Other Sectors	-	-	-	-	-	-	-	-	-	-

Uzbekistan / Ouzbékistan : 1998

SUPPLY AND CONSUMPTION *APPROVISIONNEMENT ET DEMANDE*	Coal / *Charbon* (1000 tonnes)							Oil / *Pétrole* (1000 tonnes)			
	Coking Coal *Charbon à coke*	Other Bit. Coal *Autres charb. bit.*	Sub-Bit. Coal *Charbon sous-bit.*	Lignite *Lignite*	Peat *Tourbe*	Oven and Gas Coke *Coke de four/gaz*	Pat. Fuel and BKB *Agg./briq. de lignite*	Crude Oil *Pétrole brut*	NGL *LGN*	Feed-stocks *Produits d'aliment.*	Additives *Additifs*
Production	-	68	-	2855	-	-	15	5700	2739	-	-
From Other Sources	-	-	-	-	-	-	-	-	-	-	-
Imports	-	-	-	28	-	-	-	-	-	-	7
Exports	-	-	-	-31	-	-	-	-399	-725	-	-
Intl. Marine Bunkers	-	-	-	-	-	-	-	-	-	-	-
Stock Changes	-	-	-	-	-	-	-	-	-	-	-
DOMESTIC SUPPLY	-	68	-	2852	-	-	15	5301	2014	-	7
Transfers	-	-	-	-	-	-	-	-	-32	-	-
Statistical Differences	-	-	-	-	-	-	-	-13	-10	-	-
TRANSFORMATION	-	38	-	2333	-	-	-	5079	1958	-	7
Electricity Plants	-	-	-	1250	-	-	-	-	-	-	-
CHP Plants	-	-	-	1079	-	-	-	-	-	-	-
Heat Plants	-	-	-	4	-	-	-	-	-	-	-
Blast Furnaces/Gas Works	-	-	-	-	-	-	-	-	-	-	-
Coke/Pat. Fuel/BKB Plants	-	38	-	-	-	-	-	-	-	-	-
Petroleum Refineries	-	-	-	-	-	-	-	5079	1958	-	7
Petrochemical Industry	-	-	-	-	-	-	-	-	-	-	-
Liquefaction	-	-	-	-	-	-	-	-	-	-	-
Other Transform. Sector	-	-	-	-	-	-	-	-	-	-	-
ENERGY SECTOR	-	-	-	6	-	-	-	14	1	-	-
Coal Mines	-	-	-	-	-	-	-	-	-	-	-
Oil and Gas Extraction	-	-	-	-	-	-	-	14	1	-	-
Petroleum Refineries	-	-	-	-	-	-	-	-	-	-	-
Electr., CHP+Heat Plants	-	-	-	6	-	-	-	-	-	-	-
Pumped Storage (Elec.)	-	-	-	-	-	-	-	-	-	-	-
Other Energy Sector	-	-	-	-	-	-	-	-	-	-	-
Distribution Losses	-	7	-	29	-	-	-	75	13	-	-
FINAL CONSUMPTION	-	23	-	484	-	-	15	120	-	-	-
INDUSTRY SECTOR	-	16	-	138	-	-	-	120	-	-	-
Iron and Steel	-	-	-	-	-	-	-	-	-	-	-
Chemical and Petrochem.	-	-	-	-	-	-	-	120	-	-	-
of which: Feedstocks	-	-	-	-	-	-	-	*120*	-	-	-
Non-Ferrous Metals	-	-	-	-	-	-	-	-	-	-	-
Non-Metallic Minerals	-	-	-	-	-	-	-	-	-	-	-
Transport Equipment	-	-	-	-	-	-	-	-	-	-	-
Machinery	-	-	-	-	-	-	-	-	-	-	-
Mining and Quarrying	-	-	-	-	-	-	-	-	-	-	-
Food and Tobacco	-	-	-	-	-	-	-	-	-	-	-
Paper, Pulp and Print	-	-	-	-	-	-	-	-	-	-	-
Wood and Wood Products	-	-	-	-	-	-	-	-	-	-	-
Construction	-	-	-	-	-	-	-	-	-	-	-
Textile and Leather	-	-	-	-	-	-	-	-	-	-	-
Non-specified	-	16	-	138	-	-	-	-	-	-	-
TRANSPORT SECTOR	-	-	-	-	-	-	-	-	-	-	-
Air	-	-	-	-	-	-	-	-	-	-	-
Road	-	-	-	-	-	-	-	-	-	-	-
Rail	-	-	-	-	-	-	-	-	-	-	-
Pipeline Transport	-	-	-	-	-	-	-	-	-	-	-
Internal Navigation	-	-	-	-	-	-	-	-	-	-	-
Non-specified	-	-	-	-	-	-	-	-	-	-	-
OTHER SECTORS	-	7	-	346	-	-	15	-	-	-	-
Agriculture	-	-	-	9	-	-	-	-	-	-	-
Comm. and Publ. Services	-	-	-	-	-	-	-	-	-	-	-
Residential	-	-	-	36	-	-	-	-	-	-	-
Non-specified	-	7	-	301	-	-	15	-	-	-	-
NON-ENERGY USE	-	-	-	-	-	-	-	-	-	-	-
in Industry/Trans./Energy	-	-	-	-	-	-	-	-	-	-	-
in Transport	-	-	-	-	-	-	-	-	-	-	-
in Other Sectors	-	-	-	-	-	-	-	-	-	-	-

Uzbekistan / Ouzbékistan : 1998

SUPPLY AND CONSUMPTION / APPROVISIONNEMENT ET DEMANDE	Oil cont. / Pétrole cont. (1000 tonnes)										
	Refinery Gas / Gaz de raffinerie	LPG + Ethane / GPL + éthane	Motor Gasoline / Essence moteur	Aviation Gasoline / Essence aviation	Jet Fuel / Carbu- réacteurs	Kerosene / Kérosène	Gas/ Diesel / Gazole	Heavy Fuel Oil / Fioul lourd	Naphtha / Naphta	Petrol. Coke / Coke de pétrole	Other Prod. / Autres prod.
Production	219	10	1406	3	270	100	2318	1801	-	46	789
From Other Sources	-	-	-	-	-	-	-	-	-	-	-
Imports	-	-	-	-	27	-	-	-	-	-	1
Exports	-	-	-36	-	-6	-	-301	-5	-	-	-102
Intl. Marine Bunkers	-	-	-	-	-	-	-	-	-	-	-
Stock Changes	-	-	-	-	-	-	-	-	-	-46	-
DOMESTIC SUPPLY	219	10	1370	3	291	100	2017	1796	-	-	688
Transfers	-	32	-	-	-	-	-	-	-	-	-
Statistical Differences	-	-	-	-	-	-	-	-	-	-	-
TRANSFORMATION	-	-	-	-	-	9	6	1765	-	-	8
Electricity Plants	-	-	-	-	-	-	6	646	-	-	-
CHP Plants	-	-	-	-	-	-	-	836	-	-	-
Heat Plants	-	-	-	-	-	9	-	283	-	-	8
Blast Furnaces/Gas Works	-	-	-	-	-	-	-	-	-	-	-
Coke/Pat. Fuel/BKB Plants	-	-	-	-	-	-	-	-	-	-	-
Petroleum Refineries	-	-	-	-	-	-	-	-	-	-	-
Petrochemical Industry	-	-	-	-	-	-	-	-	-	-	-
Liquefaction	-	-	-	-	-	-	-	-	-	-	-
Other Transform. Sector	-	-	-	-	-	-	-	-	-	-	-
ENERGY SECTOR	219	-	-	-	-	-	-	24	-	-	-
Coal Mines	-	-	-	-	-	-	-	-	-	-	-
Oil and Gas Extraction	-	-	-	-	-	-	-	-	-	-	-
Petroleum Refineries	219	-	-	-	-	-	-	-	-	-	-
Electr., CHP+Heat Plants	-	-	-	-	-	-	-	24	-	-	-
Pumped Storage (Elec.)	-	-	-	-	-	-	-	-	-	-	-
Other Energy Sector	-	-	-	-	-	-	-	-	-	-	-
Distribution Losses	-	-	-	-	-	-	-	-	-	-	-
FINAL CONSUMPTION	-	42	1370	3	291	91	2011	7	-	-	680
INDUSTRY SECTOR	-	-	61	-	-	17	197	4	-	-	3
Iron and Steel	-	-	-	-	-	-	-	-	-	-	-
Chemical and Petrochem.	-	-	-	-	-	-	-	-	-	-	-
of which: Feedstocks	-	-	-	-	-	-	-	-	-	-	-
Non-Ferrous Metals	-	-	-	-	-	-	-	-	-	-	-
Non-Metallic Minerals	-	-	-	-	-	-	-	4	-	-	-
Transport Equipment	-	-	-	-	-	-	-	-	-	-	-
Machinery	-	-	4	-	-	-	-	-	-	-	-
Mining and Quarrying	-	-	-	-	-	-	-	-	-	-	-
Food and Tobacco	-	-	-	-	-	-	-	-	-	-	-
Paper, Pulp and Print	-	-	-	-	-	-	-	-	-	-	-
Wood and Wood Products	-	-	-	-	-	-	-	-	-	-	-
Construction	-	-	56	-	-	-	197	-	-	-	-
Textile and Leather	-	-	-	-	-	17	-	-	-	-	3
Non-specified	-	-	1	-	-	-	-	-	-	-	-
TRANSPORT SECTOR	-	14	1275	3	291	-	541	-	-	-	-
Air	-	-	-	3	291	-	-	-	-	-	-
Road	-	14	1275	-	-	-	420	-	-	-	-
Rail	-	-	-	-	-	-	121	-	-	-	-
Pipeline Transport	-	-	-	-	-	-	-	-	-	-	-
Internal Navigation	-	-	-	-	-	-	-	-	-	-	-
Non-specified	-	-	-	-	-	-	-	-	-	-	-
OTHER SECTORS	-	28	34	-	-	74	1273	3	-	-	14
Agriculture	-	-	29	-	-	16	1064	1	-	-	1
Comm. and Publ. Services	-	-	-	-	-	-	-	-	-	-	-
Residential	-	14	1	-	-	1	6	-	-	-	1
Non-specified	-	14	4	-	-	57	203	2	-	-	12
NON-ENERGY USE	-	-	-	-	-	-	-	-	-	-	663
in Industry/Transf./Energy	-	-	-	-	-	-	-	-	-	-	533
in Transport	-	-	-	-	-	-	-	-	-	-	-
in Other Sectors	-	-	-	-	-	-	-	-	-	-	130

Uzbekistan / Ouzbékistan : 1998

SUPPLY AND CONSUMPTION / APPROVISIONNEMENT ET DEMANDE	Gas / Gaz (TJ)				Comb. Renew. & Waste / En. Re. Comb. & Déchets (TJ)				(GWh)	(TJ)
	Natural Gas / Gaz naturel	Gas Works / Usines à gaz	Coke Ovens / Cokeries	Blast Furnaces / Hauts fourneaux	Solid Biomass / Biomasse solide	Gas/Liquids from Biomass / Gaz/Liquides tirés de biomasse	Municipal Waste / Déchets urbains	Industrial Waste / Déchets industriels	Electricity / Electricité	Heat / Chaleur
Production	1877194	-	-	-	10	-	-	-	45900	106763
From Other Sources	-				-				-	-
Imports	184600	-	-	-	-			-	6000	-
Exports	-303112	-	-	-	-			-	-5100	-
Intl. Marine Bunkers	-									-
Stock Changes	-	-	-	-	-			-		-
DOMESTIC SUPPLY	1758682	-	-	-	10	-	-	-	46800	106763
Transfers									-	-
Statistical Differences	-	-	-	-	-			-	-	-
TRANSFORMATION	531312	-	-	-	-	-	-	-	-	-
Electricity Plants	230216	-	-	-	-	-	-	-	-	-
CHP Plants	235363	-	-	-	-	-	-	-	-	-
Heat Plants	65733	-	-	-	-	-	-	-	-	-
Blast Furnaces/Gas Works	-	-	-	-	-	-	-	-	-	-
Coke/Pat. Fuel/BKB Plants	-	-	-	-	-	-	-	-	-	-
Petroleum Refineries	-	-	-	-	-	-	-	-	-	-
Petrochemical Industry	-	-	-	-	-	-	-	-	-	-
Liquefaction	-	-	-	-	-	-	-	-	-	-
Other Transform. Sector	-	-	-	-	-	-	-	-	-	-
ENERGY SECTOR	83590	-	-	-	-	-	-	-	4019	-
Coal Mines	-	-	-	-	-	-	-	-	-	-
Oil and Gas Extraction	17459	-	-	-	-	-	-	-	-	-
Petroleum Refineries	-	-	-	-	-	-	-	-	-	-
Electr., CHP+Heat Plants	9633	-	-	-	-	-	-	-	2671	-
Pumped Storage (Elec.)	-	-	-	-	-	-	-	-	-	-
Other Energy Sector	56498	-	-	-	-	-	-	-	1348	-
Distribution Losses	73115	-	-	-	-	-	-	-	3992	-
FINAL CONSUMPTION	1070665	-	-	-	10	-	-	-	38789	106763
INDUSTRY SECTOR	211074	-	-	-	-	-	-	-	14863	-
Iron and Steel	-	-	-	-	-	-	-	-	-	-
Chemical and Petrochem.	66578	-	-	-	-	-	-	-	-	-
of which: Feedstocks	66578	-	-	-	-	-	-	-	-	-
Non-Ferrous Metals	-	-	-	-	-	-	-	-	-	-
Non-Metallic Minerals	-	-	-	-	-	-	-	-	-	-
Transport Equipment	-	-	-	-	-	-	-	-	-	-
Machinery	-	-	-	-	-	-	-	-	-	-
Mining and Quarrying	-	-	-	-	-	-	-	-	-	-
Food and Tobacco	-	-	-	-	-	-	-	-	-	-
Paper, Pulp and Print	-	-	-	-	-	-	-	-	-	-
Wood and Wood Products	-	-	-	-	-	-	-	-	-	-
Construction	-	-	-	-	-	-	-	-	-	-
Textile and Leather	-	-	-	-	-	-	-	-	-	-
Non-specified	144496	-	-	-	-	-	-	-	14863	-
TRANSPORT SECTOR	63819	-	-	-	-	-	-	-	1277	-
Air	-	-	-	-	-	-	-	-	-	-
Road	2819	-	-	-	-	-	-	-	-	-
Rail	-	-	-	-	-	-	-	-	155	-
Pipeline Transport	61000	-	-	-	-	-	-	-	817	-
Internal Navigation	-	-	-	-	-	-	-	-	-	-
Non-specified	-	-	-	-	-	-	-	-	305	-
OTHER SECTORS	795772	-	-	-	10	-	-	-	22649	106763
Agriculture	7026	-	-	-	-	-	-	-	12598	-
Comm. and Publ. Services	131465	-	-	-	-	-	-	-	3007	-
Residential	657281	-	-	-	-	-	-	-	7044	-
Non-specified	-	-	-	-	10	-	-	-	-	106763
NON-ENERGY USE	-	-	-	-	-	-	-	-	-	-
in Industry/Transf./Energy	-	-	-	-	-	-	-	-	-	-
in Transport	-	-	-	-	-	-	-	-	-	-
in Other Sectors	-	-	-	-	-	-	-	-	-	-

Venezuela / Vénézuela : 1997

SUPPLY AND CONSUMPTION	Coal / *Charbon* (1000 tonnes)							Oil / *Pétrole* (1000 tonnes)			
	Coking Coal	Other Bit. Coal	Sub-Bit. Coal	Lignite	Peat	Oven and Gas Coke	Pat. Fuel and BKB	Crude Oil	NGL	Feed-stocks	Additives
APPROVISIONNEMENT ET DEMANDE	*Charbon à coke*	*Autres charb. bit.*	*Charbon sous-bit.*	*Lignite*	*Tourbe*	*Coke de four/gaz*	*Agg./briq. de lignite*	*Pétrole brut*	*LGN*	*Produits d'aliment.*	*Additifs*
Production	-	5146	-	-	-	-	-	181715	3497	-	-
From Other Sources	-	-	-	-	-	-	-	-	-	-	-
Imports	-	-	-	-	-	-	-	387	-	-	-
Exports	-	-5105	-	-	-	-	-	-128822	-	-	-
Intl. Marine Bunkers	-	-	-	-	-	-	-	-	-	-	-
Stock Changes	-	6	-	-	-	-	-	297	-	-	-
DOMESTIC SUPPLY	-	47	-	-	-	-	-	53577	3497	-	-
Transfers	-	-	-	-	-	-	-	-3320	-3497	-	-
Statistical Differences	-	-	-	-	-	-	-	6917	-	-	-
TRANSFORMATION	-	-	-	-	-	-	-	53707	-	-	-
Electricity Plants	-	-	-	-	-	-	-	-	-	-	-
CHP Plants	-	-	-	-	-	-	-	-	-	-	-
Heat Plants	-	-	-	-	-	-	-	-	-	-	-
Blast Furnaces/Gas Works	-	-	-	-	-	-	-	-	-	-	-
Coke/Pat. Fuel/BKB Plants	-	-	-	-	-	-	-	-	-	-	-
Petroleum Refineries	-	-	-	-	-	-	-	53707	-	-	-
Petrochemical Industry	-	-	-	-	-	-	-	-	-	-	-
Liquefaction	-	-	-	-	-	-	-	-	-	-	-
Other Transform. Sector	-	-	-	-	-	-	-	-	-	-	-
ENERGY SECTOR	-	-	-	-	-	-	-	-	-	-	-
Coal Mines	-	-	-	-	-	-	-	-	-	-	-
Oil and Gas Extraction	-	-	-	-	-	-	-	-	-	-	-
Petroleum Refineries	-	-	-	-	-	-	-	-	-	-	-
Electr., CHP+Heat Plants	-	-	-	-	-	-	-	-	-	-	-
Pumped Storage (Elec.)	-	-	-	-	-	-	-	-	-	-	-
Other Energy Sector	-	-	-	-	-	-	-	-	-	-	-
Distribution Losses	-	-	-	-	-	-	-	3467	-	-	-
FINAL CONSUMPTION	-	47	-	-	-	-	-	-	-	-	-
INDUSTRY SECTOR	-	47	-	-	-	-	-	-	-	-	-
Iron and Steel	-	-	-	-	-	-	-	-	-	-	-
Chemical and Petrochem.	-	-	-	-	-	-	-	-	-	-	-
of which: Feedstocks	-	-	-	-	-	-	-	-	-	-	-
Non-Ferrous Metals	-	-	-	-	-	-	-	-	-	-	-
Non-Metallic Minerals	-	47	-	-	-	-	-	-	-	-	-
Transport Equipment	-	-	-	-	-	-	-	-	-	-	-
Machinery	-	-	-	-	-	-	-	-	-	-	-
Mining and Quarrying	-	-	-	-	-	-	-	-	-	-	-
Food and Tobacco	-	-	-	-	-	-	-	-	-	-	-
Paper, Pulp and Print	-	-	-	-	-	-	-	-	-	-	-
Wood and Wood Products	-	-	-	-	-	-	-	-	-	-	-
Construction	-	-	-	-	-	-	-	-	-	-	-
Textile and Leather	-	-	-	-	-	-	-	-	-	-	-
Non-specified	-	-	-	-	-	-	-	-	-	-	-
TRANSPORT SECTOR	-	-	-	-	-	-	-	-	-	-	-
Air	-	-	-	-	-	-	-	-	-	-	-
Road	-	-	-	-	-	-	-	-	-	-	-
Rail	-	-	-	-	-	-	-	-	-	-	-
Pipeline Transport	-	-	-	-	-	-	-	-	-	-	-
Internal Navigation	-	-	-	-	-	-	-	-	-	-	-
Non-specified	-	-	-	-	-	-	-	-	-	-	-
OTHER SECTORS	-	-	-	-	-	-	-	-	-	-	-
Agriculture	-	-	-	-	-	-	-	-	-	-	-
Comm. and Publ. Services	-	-	-	-	-	-	-	-	-	-	-
Residential	-	-	-	-	-	-	-	-	-	-	-
Non-specified	-	-	-	-	-	-	-	-	-	-	-
NON-ENERGY USE	-	-	-	-	-	-	-	-	-	-	-
in Industry/Trans./Energy	-	-	-	-	-	-	-	-	-	-	-
in Transport	-	-	-	-	-	-	-	-	-	-	-
in Other Sectors	-	-	-	-	-	-	-	-	-	-	-

Venezuela / Vénézuela : 1997

SUPPLY AND CONSUMPTION / APPROVISIONNEMENT ET DEMANDE	Oil cont. / *Pétrole cont.* (1000 tonnes)										
	Refinery Gas / *Gaz de raffinerie*	LPG + Ethane / *GPL + éthane*	Motor Gasoline / *Essence moteur*	Aviation Gasoline / *Essence aviation*	Jet Fuel / *Carbu- réacteurs*	Kerosene / *Kérosène*	Gas/ Diesel / *Gazole*	Heavy Fuel Oil / *Fioul lourd*	Naphtha / *Naphta*	Petrol. Coke / *Coke de pétrole*	Other Prod. / *Autres prod.*
Production	931	653	17476	30	4014	124	15866	9974	-	1980	2689
From Other Sources	-	-		-		-			-		
Imports	-	-	-	-	-	-	-	-	-	-	-
Exports	-	-1929	-9648	-8	-3720	-33	-11512	-11046	-	-788	-1069
Intl. Marine Bunkers	- -	-	-	-		-	-77	-533	-	-	
Stock Changes	-	3	-32	-	-26	2	452	-389	-	-109	-149
DOMESTIC SUPPLY	**931**	**-1273**	**7796**	**22**	**268**	**93**	**4729**	**-1994**	**-**	**1083**	**1471**
Transfers	-	3497	797	-	-	-	-	3472	-	-	-
Statistical Differences	-	-2	-34	-1	-1	-	27	-55	-	249	99
TRANSFORMATION	**-**	**-**	**-**	**-**	**-**	**-**	**513**	**558**	**-**	**-**	**-**
Electricity Plants	-	-	-	-	-	-	513	558	-	-	-
CHP Plants	-	-	-	-	-	-	-	-	-	-	-
Heat Plants	-	-	-	-	-	-	-	-	-	-	-
Blast Furnaces/Gas Works	-	-	-	-	-	-	-	-	-	-	-
Coke/Pat. Fuel/BKB Plants	-	-	-	-	-	-	-	-	-	-	-
Petroleum Refineries	-	-	-	-	-	-	-	-	-	-	-
Petrochemical Industry	-	-	-	-	-	-	-	-	-	-	-
Liquefaction	-	-	-	-	-	-	-	-	-	-	-
Other Transform. Sector	-	-	-	-	-	-	-	-	-	-	-
ENERGY SECTOR	**931**	**169**	**-**	**-**	**-**	**-**	**702**	**518**	**-**	**856**	**235**
Coal Mines	-	-	-	-	-	-	-	-	-	-	-
Oil and Gas Extraction	-	44	-	-	-	-	697	501	-	-	-
Petroleum Refineries	931	78	-	-	-	-	5	17	-	856	235
Electr., CHP+Heat Plants	-	-	-	-	-	-	-	-	-	-	-
Pumped Storage (Elec.)	-	-	-	-	-	-	-	-	-	-	-
Other Energy Sector	-	47	-	-	-	-	-	-	-	-	-
Distribution Losses	-	-	-	-	-	-	-	-	-	-	-
FINAL CONSUMPTION	**-**	**2053**	**8559**	**21**	**267**	**93**	**3541**	**347**	**-**	**476**	**1335**
INDUSTRY SECTOR	**-**	**1113**	**444**	**-**	**-**	**93**	**1712**	**347**	**-**	**-**	**-**
Iron and Steel	-	-	-	-	-	-	-	-	-	-	-
Chemical and Petrochem.	-	997	350	-	-	-	-	-	-	-	-
of which: Feedstocks	-	-	-	-	-	-	-	-	-	-	-
Non-Ferrous Metals	-	-	94	-	-	-	-	-	-	-	-
Non-Metallic Minerals	-	-	-	-	-	-	-	149	-	-	-
Transport Equipment	-	-	-	-	-	-	-	-	-	-	-
Machinery	-	-	-	-	-	-	-	-	-	-	-
Mining and Quarrying	-	-	-	-	-	-	-	-	-	-	-
Food and Tobacco	-	-	-	-	-	-	-	-	-	-	-
Paper, Pulp and Print	-	-	-	-	-	-	-	-	-	-	-
Wood and Wood Products	-	-	-	-	-	-	-	-	-	-	-
Construction	-	-	-	-	-	-	-	-	-	-	-
Textile and Leather	-	-	-	-	-	-	-	-	-	-	-
Non-specified	-	116	-	-	-	93	1712	198	-	-	-
TRANSPORT SECTOR	**-**	**9**	**8115**	**21**	**267**	**-**	**1741**	**-**	**-**	**-**	**-**
Air	-	-	-	21	267	-	-	-	-	-	-
Road	-	9	8115	-	-	-	1741	-	-	-	-
Rail	-	-	-	-	-	-	-	-	-	-	-
Pipeline Transport	-	-	-	-	-	-	-	-	-	-	-
Internal Navigation	-	-	-	-	-	-	-	-	-	-	-
Non-specified	-	-	-	-	-	-	-	-	-	-	-
OTHER SECTORS	**-**	**931**	**-**	**-**	**-**	**-**	**88**	**-**	**-**	**-**	**-**
Agriculture	-	-	-	-	-	-	60	-	-	-	-
Comm. and Publ. Services	-	-	-	-	-	-	28	-	-	-	-
Residential	-	931	-	-	-	-	-	-	-	-	-
Non-specified	-	-	-	-	-	-	-	-	-	-	-
NON-ENERGY USE	**-**	**-**	**-**	**-**	**-**	**-**	**-**	**-**	**-**	**476**	**1335**
in Industry/Transf./Energy	-	-	-	-	-	-	-	-	-	476	1335
in Transport	-	-	-	-	-	-	-	-	-	-	-
in Other Sectors	-	-	-	-	-	-	-	-	-	-	-

Venezuela / Vénézuela : 1997

SUPPLY AND CONSUMPTION *APPROVISIONNEMENT ET DEMANDE*	Gas / *Gaz* (TJ)				Comb. Renew. & Waste / *En. Re. Comb. & Déchets* (TJ)				(GWh)	(TJ)
	Natural Gas *Gaz naturel*	Gas Works *Usines à gaz*	Coke Ovens *Cokeries*	Blast Furnaces *Hauts fourneaux*	Solid Biomass *Biomasse solide*	Gas/Liquids from Biomass *Gaz/Liquides tirés de biomasse*	Municipal Waste *Déchets urbains*	Industrial Waste *Déchets industriels*	Electricity *Electricité*	Heat *Chaleur*
Production	1334909	-	-	-	22647	-	-	-	78066	-
From Other Sources	-	-	-	-	-	-	-	-	-	-
Imports	-	-	-	-	-	-	-	-	-	-
Exports	-	-	-	-	-	-	-	-	-	-
Intl. Marine Bunkers	-	-	-	-	-	-	-	-	-	-
Stock Changes	-	-	-	-	-	-	-	-	-	-
DOMESTIC SUPPLY	**1334909**	**-**	**-**	**-**	**22647**	**-**	**-**	**-**	**78066**	**-**
Transfers	-	-	-	-	-	-	-	-	-	-
Statistical Differences	886	-	-	-	-	-	-	-	-	-
TRANSFORMATION	**274600**	**-**	**-**	**-**	**600**	**-**	**-**	**-**	**-**	**-**
Electricity Plants	274600	-	-	-	-	-	-	-	-	-
CHP Plants	-	-	-	-	-	-	-	-	-	-
Heat Plants	-	-	-	-	-	-	-	-	-	-
Blast Furnaces/Gas Works	-	-	-	-	-	-	-	-	-	-
Coke/Pat. Fuel/BKB Plants	-	-	-	-	-	-	-	-	-	-
Petroleum Refineries	-	-	-	-	-	-	-	-	-	-
Petrochemical Industry	-	-	-	-	-	-	-	-	-	-
Liquefaction	-	-	-	-	-	-	-	-	-	-
Other Transform. Sector	-	-	-	-	600	-	-	-	-	-
ENERGY SECTOR	**532705**	**-**	**-**	**-**	**-**	**-**	**-**	**-**	**3183**	**-**
Coal Mines	-	-	-	-	-	-	-	-	-	-
Oil and Gas Extraction	497982	-	-	-	-	-	-	-	909	-
Petroleum Refineries	34723	-	-	-	-	-	-	-	1069	-
Electr., CHP+Heat Plants	-	-	-	-	-	-	-	-	1064	-
Pumped Storage (Elec.)	-	-	-	-	-	-	-	-	-	-
Other Energy Sector	-	-	-	-	-	-	-	-	141	-
Distribution Losses	-	-	-	-	-	-	-	-	16397	-
FINAL CONSUMPTION	**528490**	**-**	**-**	**-**	**22047**	**-**	**-**	**-**	**58486**	**-**
INDUSTRY SECTOR	**489824**	**-**	**-**	**-**	**13846**	**-**	**-**	**-**	**28595**	**-**
Iron and Steel	127261	-	-	-	-	-	-	-	6715	-
Chemical and Petrochem.	191711	-	-	-	-	-	-	-	1287	-
of which: Feedstocks	-	-	-	-	-	-	-	-	-	-
Non-Ferrous Metals	24608	-	-	-	-	-	-	-	11249	-
Non-Metallic Minerals	42655	-	-	-	-	-	-	-	377	-
Transport Equipment	-	-	-	-	-	-	-	-	-	-
Machinery	-	-	-	-	-	-	-	-	-	-
Mining and Quarrying	-	-	-	-	-	-	-	-	-	-
Food and Tobacco	-	-	-	-	13846	-	-	-	-	-
Paper, Pulp and Print	-	-	-	-	-	-	-	-	-	-
Wood and Wood Products	-	-	-	-	-	-	-	-	-	-
Construction	-	-	-	-	-	-	-	-	-	-
Textile and Leather	-	-	-	-	-	-	-	-	-	-
Non-specified	103589	-	-	-	-	-	-	-	8967	-
TRANSPORT SECTOR	**4690**	**-**	**-**	**-**	**-**	**-**	**-**	**-**	**154**	**-**
Air	-	-	-	-	-	-	-	-	-	-
Road	-	-	-	-	-	-	-	-	-	-
Rail	-	-	-	-	-	-	-	-	154	-
Pipeline Transport	-	-	-	-	-	-	-	-	-	-
Internal Navigation	-	-	-	-	-	-	-	-	-	-
Non-specified	4690	-	-	-	-	-	-	-	-	-
OTHER SECTORS	**33976**	**-**	**-**	**-**	**8201**	**-**	**-**	**-**	**29737**	**-**
Agriculture	-	-	-	-	-	-	-	-	-	-
Comm. and Publ. Services	23950	-	-	-	-	-	-	-	16078	-
Residential	10026	-	-	-	8201	-	-	-	13659	-
Non-specified	-	-	-	-	-	-	-	-	-	-
NON-ENERGY USE	**-**	**-**	**-**	**-**	**-**	**-**	**-**	**-**	**-**	**-**
in Industry/Transf./Energy	-	-	-	-	-	-	-	-	-	-
in Transport	-	-	-	-	-	-	-	-	-	-
in Other Sectors	-	-	-	-	-	-	-	-	-	-

Venezuela / Vénézuela : 1998

SUPPLY AND CONSUMPTION / APPROVISIONNEMENT ET DEMANDE	Coal / Charbon (1000 tonnes)							Oil / Pétrole (1000 tonnes)			
	Coking Coal / Charbon à coke	Other Bit. Coal / Autres charb. bit.	Sub-Bit. Coal / Charbon sous-bit.	Lignite / Lignite	Peat / Tourbe	Oven and Gas Coke / Coke de four/gaz	Pat. Fuel and BKB / Agg./briq. de lignite	Crude Oil / Pétrole brut	NGL / LGN	Feed-stocks / Produits d'aliment.	Additives / Additifs
Production	-	7456	-	-	-	-	-	176572	3221	-	-
From Other Sources	-	-	-	-	-	-	-	-	-	-	-
Imports	-	-	-	-	-	-	-	1068	-	-	-
Exports	-	-5908	-	-	-	-	-	-122920	-	-	-
Intl. Marine Bunkers	-	-	-	-	-	-	-	-	-	-	-
Stock Changes	-	-221	-	-	-	-	-	214	-	-	-
DOMESTIC SUPPLY	-	1327	-	-	-	-	-	54934	3221	-	-
Transfers	-	-	-	-	-	-	-	-	-3221	-	-
Statistical Differences	-	-	-	-	-	-	-	1	-	-	-
TRANSFORMATION	-	-	-	-	-	-	-	54930	-	-	-
Electricity Plants	-	-	-	-	-	-	-	-	-	-	-
CHP Plants	-	-	-	-	-	-	-	-	-	-	-
Heat Plants	-	-	-	-	-	-	-	-	-	-	-
Blast Furnaces/Gas Works	-	-	-	-	-	-	-	-	-	-	-
Coke/Pat. Fuel/BKB Plants	-	-	-	-	-	-	-	-	-	-	-
Petroleum Refineries	-	-	-	-	-	-	-	54930	-	-	-
Petrochemical Industry	-	-	-	-	-	-	-	-	-	-	-
Liquefaction	-	-	-	-	-	-	-	-	-	-	-
Other Transform. Sector	-	-	-	-	-	-	-	-	-	-	-
ENERGY SECTOR	-	-	-	-	-	-	-	-	-	-	-
Coal Mines	-	-	-	-	-	-	-	-	-	-	-
Oil and Gas Extraction	-	-	-	-	-	-	-	-	-	-	-
Petroleum Refineries	-	-	-	-	-	-	-	-	-	-	-
Electr., CHP+Heat Plants	-	-	-	-	-	-	-	-	-	-	-
Pumped Storage (Elec.)	-	-	-	-	-	-	-	-	-	-	-
Other Energy Sector	-	-	-	-	-	-	-	-	-	-	-
Distribution Losses	-	-	-	-	-	-	-	5	-	-	-
FINAL CONSUMPTION	-	1327	-	-	-	-	-	-	-	-	-
INDUSTRY SECTOR	-	1327	-	-	-	-	-	-	-	-	-
Iron and Steel	-	-	-	-	-	-	-	-	-	-	-
Chemical and Petrochem.	-	-	-	-	-	-	-	-	-	-	-
of which: Feedstocks	-	-	-	-	-	-	-	-	-	-	-
Non-Ferrous Metals	-	-	-	-	-	-	-	-	-	-	-
Non-Metallic Minerals	-	1327	-	-	-	-	-	-	-	-	-
Transport Equipment	-	-	-	-	-	-	-	-	-	-	-
Machinery	-	-	-	-	-	-	-	-	-	-	-
Mining and Quarrying	-	-	-	-	-	-	-	-	-	-	-
Food and Tobacco	-	-	-	-	-	-	-	-	-	-	-
Paper, Pulp and Print	-	-	-	-	-	-	-	-	-	-	-
Wood and Wood Products	-	-	-	-	-	-	-	-	-	-	-
Construction	-	-	-	-	-	-	-	-	-	-	-
Textile and Leather	-	-	-	-	-	-	-	-	-	-	-
Non-specified	-	-	-	-	-	-	-	-	-	-	-
TRANSPORT SECTOR	-	-	-	-	-	-	-	-	-	-	-
Air	-	-	-	-	-	-	-	-	-	-	-
Road	-	-	-	-	-	-	-	-	-	-	-
Rail	-	-	-	-	-	-	-	-	-	-	-
Pipeline Transport	-	-	-	-	-	-	-	-	-	-	-
Internal Navigation	-	-	-	-	-	-	-	-	-	-	-
Non-specified	-	-	-	-	-	-	-	-	-	-	-
OTHER SECTORS	-	-	-	-	-	-	-	-	-	-	-
Agriculture	-	-	-	-	-	-	-	-	-	-	-
Comm. and Publ. Services	-	-	-	-	-	-	-	-	-	-	-
Residential	-	-	-	-	-	-	-	-	-	-	-
Non-specified	-	-	-	-	-	-	-	-	-	-	-
NON-ENERGY USE	-	-	-	-	-	-	-	-	-	-	-
in Industry/Trans./Energy	-	-	-	-	-	-	-	-	-	-	-
in Transport	-	-	-	-	-	-	-	-	-	-	-
in Other Sectors	-	-	-	-	-	-	-	-	-	-	-

Venezuela / Vénézuela : 1998

	Oil cont. / *Pétrole cont.* (1000 tonnes)										
SUPPLY AND CONSUMPTION *APPROVISIONNEMENT ET DEMANDE*	Refinery Gas *Gaz de raffinerie*	LPG + Ethane *GPL + éthane*	Motor Gasoline *Essence moteur*	Aviation Gasoline *Essence aviation*	Jet Fuel *Carbu- réacteurs*	Kerosene *Kérosène*	Gas/ Diesel *Gazole*	Heavy Fuel Oil *Fioul lourd*	Naphtha *Naphta*	Petrol. Coke *Coke de pétrole*	Other Prod. *Autres prod.*
Production	1975	771	19438	17	3722	119	13495	10893	-	1522	2729
From Other Sources	-	-	-	-	-	-	-	-	-	-	-
Imports	-	-	-	-	-	-	-	-	-	-	-
Exports	-	-1880	-12691	-	-3478	-35	-8893	-9253	-	-568	-1018
Intl. Marine Bunkers	-	-	-	-	-	-	-60	-447	-	-	-
Stock Changes	-	-	-	-	-	-	-1		-	-	-
DOMESTIC SUPPLY	1975	-1109	6747	17	244	84	4541	1193	-	954	1711
Transfers	-	3221	1606	-	-	-	-	-	-	-	-
Statistical Differences	-	-1	-1	-	-	-	-	-	-	406	241
TRANSFORMATION	-	-	-	-	-	-	541	780	-	-	-
Electricity Plants	-	-	-	-	-	-	541	780	-	-	-
CHP Plants	-	-	-	-	-	-	-	-	-	-	-
Heat Plants	-	-	-	-	-	-	-	-	-	-	-
Blast Furnaces/Gas Works	-	-	-	-	-	-	-	-	-	-	-
Coke/Pat. Fuel/BKB Plants	-	-	-	-	-	-	-	-	-	-	-
Petroleum Refineries	-	-	-	-	-	-	-	-	-	-	-
Petrochemical Industry	-	-	-	-	-	-	-	-	-	-	-
Liquefaction	-	-	-	-	-	-	-	-	-	-	-
Other Transform. Sector	-	-	-	-	-	-	-	-	-	-	-
ENERGY SECTOR	1975	66	2	-	-	-	601	-	-	989	459
Coal Mines	-	-	2	-	-	-	38	-	-	1	-
Oil and Gas Extraction	-	-	-	-	-	-	563	-	-	-	-
Petroleum Refineries	1975	66	-	-	-	-	-	-	-	988	459
Electr., CHP+Heat Plants	-	-	-	-	-	-	-	-	-	-	-
Pumped Storage (Elec.)	-	-	-	-	-	-	-	-	-	-	-
Other Energy Sector	-	-	-	-	-	-	-	-	-	-	-
Distribution Losses	-	-	-	-	-	-	-	-	-	-	-
FINAL CONSUMPTION	-	2045	8350	17	244	84	3399	413	-	371	1493
INDUSTRY SECTOR	-	1131	99	-	-	70	1501	413	-	-	-
Iron and Steel	-	-	-	-	-	-	-	-	-	-	-
Chemical and Petrochem.	-	1018	-	-	-	-	-	-	-	-	-
of which: Feedstocks	-	-	-	-	-	-	-	-	-	-	-
Non-Ferrous Metals	-	-	99	-	-	-	-	-	-	-	-
Non-Metallic Minerals	-	-	-	-	-	-	--	176	-	-	-
Transport Equipment	-	-	-	-	-	-	-	-	-	-	-
Machinery	-	-	-	-	-	-	-	-	-	-	-
Mining and Quarrying	-	-	-	-	-	-	-	-	-	-	-
Food and Tobacco	-	-	-	-	-	-	-	-	-	-	-
Paper, Pulp and Print	-	-	-	-	-	-	-	-	-	-	-
Wood and Wood Products	-	-	-	-	-	-	-	-	-	-	-
Construction	-	-	-	-	-	-	-	-	-	-	-
Textile and Leather	-	-	-	-	-	-	-	-	-	-	-
Non-specified	-	113	-	-	-	70	1501	237	-	-	-
TRANSPORT SECTOR	-	9	8251	17	244	-	1812	-	-	-	-
Air	-	-	-	17	244	-	-	-	-	-	-
Road	-	9	8251	-	-	-	1812	-	-	-	-
Rail	-	-	-	-	-	-	-	-	-	-	-
Pipeline Transport	-	-	-	-	-	-	-	-	-	-	-
Internal Navigation	-	-	-	-	-	-	-	-	-	-	-
Non-specified	-	-	-	-	-	-	-	-	-	-	-
OTHER SECTORS	-	905	-	-	-	14	86	-	-	-	-
Agriculture	-	-	-	-	-	-	57	-	-	-	-
Comm. and Publ. Services	-	-	-	-	-	-	29	-	-	-	-
Residential	-	905	-	-	-	14	-	-	-	-	-
Non-specified	-	-	-	-	-	-	-	-	-	-	-
NON-ENERGY USE	-	-	-	-	-	-	-	-	-	371	1493
in Industry/Transf./Energy	-	-	-	-	-	-	-	-	-	371	1493
in Transport	-	-	-	-	-	-	-	-	-	-	-
in Other Sectors	-	-	-	-	-	-	-	-	-	-	-

Venezuela / Vénézuela : 1998

SUPPLY AND CONSUMPTION APPROVISIONNEMENT ET DEMANDE	Gas / Gaz (TJ)				Comb. Renew. & Waste / En. Re. Comb. & Déchets (TJ)				(GWh)	(TJ)
	Natural Gas Gaz naturel	Gas Works Usines à gaz	Coke Ovens Cokeries	Blast Furnaces Hauts fourneaux	Solid Biomass Biomasse solide	Gas/Liquids from Biomass Gaz/Liquides tirés de biomasse	Municipal Waste Déchets urbains	Industrial Waste Déchets industriels	Electricity Electricité	Heat Chaleur
Production	1377020	-	-	-	22647	-	-	-	80904	-
From Other Sources	-	-	-	-	-	-	-	-	-	-
Imports	-	-	-	-	-	-	-	-	-	-
Exports	-	-	-	-	-	-	-	-	-	-
Intl. Marine Bunkers	-	-	-	-	-	-	-	-	-	-
Stock Changes	-	-	-	-	-	-	-	-	-	-
DOMESTIC SUPPLY	**1377020**	**-**	**-**	**-**	**22647**	**-**	**-**	**-**	**80904**	**-**
Transfers	-	-	-	-	-	-	-	-	-	-
Statistical Differences	-447	-	-	-	-	-	-	-	1088	-
TRANSFORMATION	**295227**	**-**	**-**	**-**	**600**	**-**	**-**	**-**	**-**	**-**
Electricity Plants	295227	-	-	-	-	-	-	-	-	-
CHP Plants	-	-	-	-	-	-	-	-	-	-
Heat Plants	-	-	-	-	-	-	-	-	-	-
Blast Furnaces/Gas Works	-	-	-	-	-	-	-	-	-	-
Coke/Pat. Fuel/BKB Plants	-	-	-	-	-	-	-	-	-	-
Petroleum Refineries	-	-	-	-	-	-	-	-	-	-
Petrochemical Industry	-	-	-	-	-	-	-	-	-	-
Liquefaction	-	-	-	-	-	-	-	-	-	-
Other Transform. Sector	-	-	-	-	600	-	-	-	-	-
ENERGY SECTOR	**522942**	**-**	**-**	**-**	**-**	**-**	**-**	**-**	**3573**	**-**
Coal Mines	-	-	-	-	-	-	-	-	21	-
Oil and Gas Extraction	495084	-	-	-	-	-	-	-	942	-
Petroleum Refineries	27858	-	-	-	-	-	-	-	1317	-
Electr., CHP+Heat Plants	-	-	-	-	-	-	-	-	1147	-
Pumped Storage (Elec.)	-	-	-	-	-	-	-	-	-	-
Other Energy Sector	-	-	-	-	-	-	-	-	146	-
Distribution Losses	-	-	-	-	-	-	-	-	18789	-
FINAL CONSUMPTION	**558404**	**-**	**-**	**-**	**22047**	**-**	**-**	**-**	**59630**	**-**
INDUSTRY SECTOR	**516949**	**-**	**-**	**-**	**13846**	**-**	**-**	**-**	**27689**	**-**
Iron and Steel	137588	-	-	-	-	-	-	-	6078	-
Chemical and Petrochem.	182920	-	-	-	-	-	-	-	1038	-
of which: Feedstocks	-	-	-	-	-	-	-	-	-	-
Non-Ferrous Metals	24920	-	-	-	-	-	-	-	10593	-
Non-Metallic Minerals	50900	-	-	-	-	-	-	-	403	-
Transport Equipment	-	-	-	-	-	-	-	-	-	-
Machinery	-	-	-	-	-	-	-	-	-	-
Mining and Quarrying	-	-	-	-	-	-	-	-	-	-
Food and Tobacco	-	-	-	-	13846	-	-	-	-	-
Paper, Pulp and Print	-	-	-	-	-	-	-	-	-	-
Wood and Wood Products	-	-	-	-	-	-	-	-	-	-
Construction	-	-	-	-	-	-	-	-	-	-
Textile and Leather	-	-	-	-	-	-	-	-	-	-
Non-specified	120621	-	-	-	-	-	-	-	9577	-
TRANSPORT SECTOR	**6893**	**-**	**-**	**-**	**-**	**-**	**-**	**-**	**161**	**-**
Air	-	-	-	-	-	-	-	-	-	-
Road	-	-	-	-	-	-	-	-	-	-
Rail	-	-	-	-	-	-	-	-	161	-
Pipeline Transport	-	-	-	-	-	-	-	-	-	-
Internal Navigation	-	-	-	-	-	-	-	-	-	-
Non-specified	6893	-	-	-	-	-	-	-	-	-
OTHER SECTORS	**34562**	**-**	**-**	**-**	**8201**	**-**	**-**	**-**	**31780**	**-**
Agriculture	-	-	-	-	-	-	-	-	-	-
Comm. and Publ. Services	24495	-	-	-	-	-	-	-	17258	-
Residential	10067	-	-	-	8201	-	-	-	14522	-
Non-specified	-	-	-	-	-	-	-	-	-	-
NON-ENERGY USE	**-**	**-**	**-**	**-**	**-**	**-**	**-**	**-**	**-**	**-**
in Industry/Transf./Energy	-	-	-	-	-	-	-	-	-	-
in Transport	-	-	-	-	-	-	-	-	-	-
in Other Sectors	-	-	-	-	-	-	-	-	-	-

Vietnam / Viêt-Nam

SUPPLY AND CONSUMPTION 1997	Coal (1000 tonnes)							Oil (1000 tonnes)			
	Coking Coal	Other Bit. Coal	Sub-Bit. Coal	Lignite	Peat	Oven and Gas Coke	Pat. Fuel and BKB	Crude Oil	NGL	Feed-stocks	Additives
Production	-	11344	-	-	-	-	-	10909	-	-	-
Imports	-	-	-	-	-	-	-	-	-	-	-
Exports	-	-3454	-	-	-	-	-	-10909	-	-	-
Intl. Marine Bunkers	-	-	-	-	-	-	-	-	-	-	-
Stock Changes	-	-1760	-	-	-	-	-	-	-	-	-
DOMESTIC SUPPLY	-	6130	-	-	-	-	-	-	-	-	-
Transfers and Stat. Diff.	-	-	-	-	-	-	-	-	-	-	-
TRANSFORMATION	-	2172	-	-	-	-	-	-	-	-	-
Electricity and CHP Plants	-	2172	-	-	-	-	-	-	-	-	-
Petroleum Refineries	-	-	-	-	-	-	-	-	-	-	-
Other Transform. Sector	-	-	-	-	-	-	-	-	-	-	-
ENERGY SECTOR	-	-	-	-	-	-	-	-	-	-	-
DISTRIBUTION LOSSES	-	-	-	-	-	-	-	-	-	-	-
FINAL CONSUMPTION	-	3958	-	-	-	-	-	-	-	-	-
INDUSTRY SECTOR	-	2758	-	-	-	-	-	-	-	-	-
Iron and Steel	-	-	-	-	-	-	-	-	-	-	-
Chemical and Petrochem.	-	-	-	-	-	-	-	-	-	-	-
Non-Metallic Minerals	-	-	-	-	-	-	-	-	-	-	-
Non-specified	-	2758	-	-	-	-	-	-	-	-	-
TRANSPORT SECTOR	-	7	-	-	-	-	-	-	-	-	-
Air	-	-	-	-	-	-	-	-	-	-	-
Road	-	-	-	-	-	-	-	-	-	-	-
Non-specified	-	7	-	-	-	-	-	-	-	-	-
OTHER SECTORS	-	1193	-	-	-	-	-	-	-	-	-
Agriculture	-	75	-	-	-	-	-	-	-	-	-
Comm. and Publ. Services	-	226	-	-	-	-	-	-	-	-	-
Residential	-	892	-	-	-	-	-	-	-	-	-
Non-specified	-	-	-	-	-	-	-	-	-	-	-
NON-ENERGY USE	-	-	-	-	-	-	-	-	-	-	-

APPROVISIONNEMENT ET DEMANDE 1998	Charbon (1000 tonnes)							Pétrole (1000 tonnes)			
	Charbon à coke	Autres charb. bit.	Charbon sous-bit.	Lignite	Tourbe	Coke de four/gaz	Agg./briq. de lignite	Pétrole brut	LGN	Produits d'aliment.	Additifs
Production	-	10772	-	-	-	-	-	12400	-	-	-
Imports	-	-	-	-	-	-	-	-	-	-	-
Exports	-	-2900	-	-	-	-	-	-12400	-	-	-
Intl. Marine Bunkers	-	-	-	-	-	-	-	-	-	-	-
Stock Changes	-	-2348	-	-	-	-	-	-	-	-	-
DOMESTIC SUPPLY	-	5524	-	-	-	-	-	-	-	-	-
Transfers and Stat. Diff.	-	-	-	-	-	-	-	-	-	-	-
TRANSFORMATION	-	2434	-	-	-	-	-	-	-	-	-
Electricity and CHP Plants	-	2434	-	-	-	-	-	-	-	-	-
Petroleum Refineries	-	-	-	-	-	-	-	-	-	-	-
Other Transform. Sector	-	-	-	-	-	-	-	-	-	-	-
ENERGY SECTOR	-	-	-	-	-	-	-	-	-	-	-
DISTRIBUTION LOSSES	-	-	-	-	-	-	-	-	-	-	-
FINAL CONSUMPTION	-	3090	-	-	-	-	-	-	-	-	-
INDUSTRY SECTOR	-	2143	-	-	-	-	-	-	-	-	-
Iron and Steel	-	-	-	-	-	-	-	-	-	-	-
Chemical and Petrochem.	-	-	-	-	-	-	-	-	-	-	-
Non-Metallic Minerals	-	-	-	-	-	-	-	-	-	-	-
Non-specified	-	2143	-	-	-	-	-	-	-	-	-
TRANSPORT SECTOR	-	6	-	-	-	-	-	-	-	-	-
Air	-	-	-	-	-	-	-	-	-	-	-
Road	-	-	-	-	-	-	-	-	-	-	-
Non-specified	-	6	-	-	-	-	-	-	-	-	-
OTHER SECTORS	-	941	-	-	-	-	-	-	-	-	-
Agriculture	-	92	-	-	-	-	-	-	-	-	-
Comm. and Publ. Services	-	135	-	-	-	-	-	-	-	-	-
Residential	-	714	-	-	-	-	-	-	-	-	-
Non-specified	-	-	-	-	-	-	-	-	-	-	-
NON-ENERGY USE	-	-	-	-	-	-	-	-	-	-	-

Vietnam / Viêt-Nam

SUPPLY AND CONSUMPTION 1997	Refinery Gas	LPG + Ethane	Motor Gasoline	Aviation Gasoline	Jet Fuel	Kerosene	Gas/ Diesel	Heavy Fuel Oil	Naphtha	Petrol. Coke	Other Prod.
Production	-	-	-	-	-	-	-	-	-	-	-
Imports	-	125	1469	-	288	265	2665	1032	-	-	113
Exports	-	-	-	-	-	-	-	-	-	-	-
Intl. Marine Bunkers	-	-	-	-	-	-	-	-	-	-	-
Stock Changes	-	-	-	-	-	-	80	185	-	-	-
DOMESTIC SUPPLY	-	125	1469	-	288	265	2745	1217	-	-	113
Transfers and Stat. Diff.	-	-	-	-	-	-	-	-	-	-	-
TRANSFORMATION	-	-	-	-	-	-	306	279	-	-	-
Electricity and CHP Plants	-	-	-	-	-	-	306	279	-	-	-
Petroleum Refineries	-	-	-	-	-	-	-	-	-	-	-
Other Transform. Sector	-	-	-	-	-	-	-	-	-	-	-
ENERGY SECTOR	-	-	-	-	-	-	-	-	-	-	-
DISTRIBUTION LOSSES	-	-	-	-	-	-	-	-	-	-	-
FINAL CONSUMPTION	-	125	1469	-	288	265	2439	938	-	-	113
INDUSTRY SECTOR	-	8	-	-	-	6	54	700	-	-	18
Iron and Steel	-	-	-	-	-	-	-	-	-	-	-
Chemical and Petrochem.	-	-	-	-	-	-	-	-	-	-	-
Non-Metallic Minerals	-	-	-	-	-	-	-	-	-	-	-
Non-specified	-	8	-	-	-	6	54	700	-	-	18
TRANSPORT SECTOR	-	-	1377	-	288	-	1946	112	-	-	-
Air	-	-	-	-	288	-	-	-	-	-	-
Road	-	-	1377	-	-	-	1946	-	-	-	-
Non-specified	-	-	-	-	-	-	-	112	-	-	-
OTHER SECTORS	-	117	92	-	-	259	439	126	-	-	95
Agriculture	-	-	92	-	-	-	220	9	-	-	-
Comm. and Publ. Services	-	86	-	-	-	182	193	103	-	-	95
Residential	-	31	-	-	-	77	26	14	-	-	-
Non-specified	-	-	-	-	-	-	-	-	-	-	-
NON-ENERGY USE	-	-	-	-	-	-	-	-	-	-	-

APPROVISIONNEMENT ET DEMANDE 1998	Gaz de raffinerie	GPL + éthane	Essence moteur	Essence aviation	Carbu- réacteurs	Kérosène	Gazole	Fioul lourd	Naphta	Coke de pétrole	Autres prod.
Production	-	-	-	-	-	-	-	-	-	-	-
Imports	-	245	1174	-	230	264	3067	1420	-	-	118
Exports	-	-	-	-	-	-	-	-	-	-	-
Intl. Marine Bunkers	-	-	-	-	-	-	-	-	-	-	-
Stock Changes	-	-50	-	-	-	-	-	-	-	-	-
DOMESTIC SUPPLY	-	195	1174	-	230	264	3067	1420	-	-	118
Transfers and Stat. Diff.	-	-	-	-	-	-	-	-	-	-	-
TRANSFORMATION	-	-	-	-	-	-	407	371	-	-	-
Electricity and CHP Plants	-	-	-	-	-	-	407	371	-	-	-
Petroleum Refineries	-	-	-	-	-	-	-	-	-	-	-
Other Transform. Sector	-	-	-	-	-	-	-	-	-	-	-
ENERGY SECTOR	-	-	-	-	-	-	-	-	-	-	-
DISTRIBUTION LOSSES	-	-	-	-	-	-	-	-	-	-	-
FINAL CONSUMPTION	-	195	1174	-	230	264	2660	1049	-	-	118
INDUSTRY SECTOR	-	32	-	-	-	5	84	787	-	-	16
Iron and Steel	-	-	-	-	-	-	-	-	-	-	-
Chemical and Petrochem.	-	-	-	-	-	-	-	-	-	-	-
Non-Metallic Minerals	-	-	-	-	-	-	-	-	-	-	-
Non-specified	-	32	-	-	-	5	84	787	-	-	16
TRANSPORT SECTOR	-	-	1059	-	230	-	2098	133	-	-	-
Air	-	-	-	-	230	-	-	-	-	-	-
Road	-	-	1059	-	-	-	2098	-	-	-	-
Non-specified	-	-	-	-	-	-	-	133	-	-	-
OTHER SECTORS	-	163	115	-	-	259	478	129	-	-	102
Agriculture	-	-	115	-	-	-	252	7	-	-	-
Comm. and Publ. Services	-	100	-	-	-	157	193	109	-	-	102
Residential	-	63	-	-	-	102	33	13	-	-	-
Non-specified	-	-	-	-	-	-	-	-	-	-	-
NON-ENERGY USE	-	-	-	-	-	-	-	-	-	-	-

Vietnam / Viêt-Nam

SUPPLY AND CONSUMPTION 1997	Natural Gas	Gas Works	Coke Ovens	Blast Furnaces	Solid Biomass	Gas/Liquids from Biomass	Municipal Waste	Industrial Waste	Electricity (GWh)	Heat (TJ)
Production	23151	-	-	-	937950	-	-	-	19151	-
Imports	-	-	-	-	-	-	-	-	-	-
Exports	-	-	-	-	-	-	-	-	-	-
Intl. Marine Bunkers	-	-	-	-	-	-	-	-	-	-
Stock Changes	-	-	-	-	-	-	-	-	-	-
DOMESTIC SUPPLY	23151	-	-	-	937950	-	-	-	19151	-
Transfers and Stat. Diff.	-	-	-	-	1	-	-	-	1	-
TRANSFORMATION	23151	-	-	-	29912	-	-	-	-	-
Electricity and CHP Plants	23151	-	-	-	-	-	-	-	-	-
Petroleum Refineries	-	-	-	-	-	-	-	-	-	-
Other Transform. Sector	-	-	-	-	29912	-	-	-	-	-
ENERGY SECTOR	-	-	-	-	-	-	-	-	465	-
DISTRIBUTION LOSSES	-	-	-	-	-	-	-	-	3384	-
FINAL CONSUMPTION	-	-	-	-	908039	-	-	-	15303	-
INDUSTRY SECTOR	-	-	-	-	-	-	-	-	6163	-
Iron and Steel	-	-	-	-	-	-	-	-	-	-
Chemical and Petrochem.	-	-	-	-	-	-	-	-	-	-
Non-Metallic Minerals	-	-	-	-	-	-	-	-	-	-
Non-specified	-	-	-	-	-	-	-	-	6163	-
TRANSPORT SECTOR	-	-	-	-	-	-	-	-	128	-
Air	-	-	-	-	-	-	-	-	-	-
Road	-	-	-	-	-	-	-	-	-	-
Non-specified	-	-	-	-	-	-	-	-	128	-
OTHER SECTORS	-	-	-	-	908039	-	-	-	9012	-
Agriculture	-	-	-	-	-	-	-	-	691	-
Comm. and Publ. Services	-	-	-	-	-	-	-	-	1100	-
Residential	-	-	-	-	908039	-	-	-	7221	-
Non-specified	-	-	-	-	-	-	-	-	-	-
NON-ENERGY USE	-	-	-	-	-	-	-	-	-	-

APPROVISIONNEMENT ET DEMANDE 1998	Gaz naturel	Usines à gaz	Cokeries	Hauts fourneaux	Biomasse solide	Gaz/Liquides tirés de biomasse	Déchets urbains	Déchets industriels	Electricité (GWh)	Chaleur (TJ)
Production	34252	-	-	-	958525	-	-	-	21665	-
Imports	-	-	-	-	-	-	-	-	-	-
Exports	-	-	-	-	-	-	-	-	-	-
Intl. Marine Bunkers	-	-	-	-	-	-	-	-	-	-
Stock Changes	-	-	-	-	-	-	-	-	-	-
DOMESTIC SUPPLY	34252	-	-	-	958525	-	-	-	21665	-
Transfers and Stat. Diff.	-	-	-	-	-	-	-	-	-	-
TRANSFORMATION	34252	-	-	-	30568	-	-	-	-	-
Electricity and CHP Plants	34252	-	-	-	-	-	-	-	-	-
Petroleum Refineries	-	-	-	-	-	-	-	-	-	-
Other Transform. Sector	-	-	-	-	30568	-	-	-	-	-
ENERGY SECTOR	-	-	-	-	-	-	-	-	560	-
DISTRIBUTION LOSSES	-	-	-	-	-	-	-	-	3380	-
FINAL CONSUMPTION	-	-	-	-	927957	-	-	-	17725	-
INDUSTRY SECTOR	-	-	-	-	-	-	-	-	6781	-
Iron and Steel	-	-	-	-	-	-	-	-	-	-
Chemical and Petrochem.	-	-	-	-	-	-	-	-	-	-
Non-Metallic Minerals	-	-	-	-	-	-	-	-	-	-
Non-specified	-	-	-	-	-	-	-	-	6781	-
TRANSPORT SECTOR	-	-	-	-	-	-	-	-	158	-
Air	-	-	-	-	-	-	-	-	-	-
Road	-	-	-	-	-	-	-	-	-	-
Non-specified	-	-	-	-	-	-	-	-	158	-
OTHER SECTORS	-	-	-	-	927957	-	-	-	10786	-
Agriculture	-	-	-	-	-	-	-	-	715	-
Comm. and Publ. Services	-	-	-	-	-	-	-	-	1222	-
Residential	-	-	-	-	927957	-	-	-	8849	-
Non-specified	-	-	-	-	-	-	-	-	-	-
NON-ENERGY USE	-	-	-	-	-	-	-	-	-	-

Yemen / Yémen

SUPPLY AND CONSUMPTION 1997	Coal (1000 tonnes)							Oil (1000 tonnes)			
	Coking Coal	Other Bit. Coal	Sub-Bit. Coal	Lignite	Peat	Oven and Gas Coke	Pat. Fuel and BKB	Crude Oil	NGL	Feed-stocks	Additives
Production	-	-	-	-	-	-	-	18124	408	-	-
Imports	-	-	-	-	-	-	-	-	-	-	-
Exports	-	-	-	-	-	-	-	-13613	-	-	-
Intl. Marine Bunkers	-	-	-	-	-	-	-	-	-	-	-
Stock Changes	-	-	-	-	-	-	-	-	-	-	-
DOMESTIC SUPPLY	-	-	-	-	-	-	-	4511	408	-	-
Transfers and Stat. Diff.	-	-	-	-	-	-	-	-	-408	-	-
TRANSFORMATION	-	-	-	-	-	-	-	4511	-	-	-
Electricity and CHP Plants	-	-	-	-	-	-	-	-	-	-	-
Petroleum Refineries	-	-	-	-	-	-	-	4511	-	-	-
Other Transform. Sector	-	-	-	-	-	-	-	-	-	-	-
ENERGY SECTOR	-	-	-	-	-	-	-	-	-	-	-
DISTRIBUTION LOSSES	-	-	-	-	-	-	-	-	-	-	-
FINAL CONSUMPTION	-	-	-	-	-	-	-	-	-	-	-
INDUSTRY SECTOR	-	-	-	-	-	-	-	-	-	-	-
Iron and Steel	-	-	-	-	-	-	-	-	-	-	-
Chemical and Petrochem.	-	-	-	-	-	-	-	-	-	-	-
Non-Metallic Minerals	-	-	-	-	-	-	-	-	-	-	-
Non-specified	-	-	-	-	-	-	-	-	-	-	-
TRANSPORT SECTOR	-	-	-	-	-	-	-	-	-	-	-
Air	-	-	-	-	-	-	-	-	-	-	-
Road	-	-	-	-	-	-	-	-	-	-	-
Non-specified	-	-	-	-	-	-	-	-	-	-	-
OTHER SECTORS	-	-	-	-	-	-	-	-	-	-	-
Agriculture	-	-	-	-	-	-	-	-	-	-	-
Comm. and Publ. Services	-	-	-	-	-	-	-	-	-	-	-
Residential	-	-	-	-	-	-	-	-	-	-	-
Non-specified	-	-	-	-	-	-	-	-	-	-	-
NON-ENERGY USE	-	-	-	-	-	-	-	-	-	-	-

APPROVISIONNEMENT ET DEMANDE 1998	Charbon (1000 tonnes)							Pétrole (1000 tonnes)			
	Charbon à coke	Autres charb. bit.	Charbon sous-bit.	Lignite	Tourbe	Coke de four/gaz	Agg./briq. de lignite	Pétrole brut	LGN	Produits .d'aliment.	Additifs
Production	-	-	-	-	-	-	-	18562	418	-	-
Imports	-	-	-	-	-	-	-	-	-	-	-
Exports	-	-	-	-	-	-	-	-14051	-	-	-
Intl. Marine Bunkers	-	-	-	-	-	-	-	-	-	-	-
Stock Changes	-	-	-	-	-	-	-	-	-	-	-
DOMESTIC SUPPLY	-	-	-	-	-	-	-	4511	418	-	-
Transfers and Stat. Diff.	-	-	-	-	-	-	-	-	-418	-	-
TRANSFORMATION	-	-	-	-	-	-	-	4511	-	-	-
Electricity and CHP Plants	-	-	-	-	-	-	-	-	-	-	-
Petroleum Refineries	-	-	-	-	-	-	-	4511	-	-	-
Other Transform. Sector	-	-	-	-	-	-	-	-	-	-	-
ENERGY SECTOR	-	-	-	-	-	-	-	-	-	-	-
DISTRIBUTION LOSSES	-	-	-	-	-	-	-	-	-	-	-
FINAL CONSUMPTION	-	-	-	-	-	-	-	-	-	-	-
INDUSTRY SECTOR	-	-	-	-	-	-	-	-	-	-	-
Iron and Steel	-	-	-	-	-	-	-	-	-	-	-
Chemical and Petrochem.	-	-	-	-	-	-	-	-	-	-	-
Non-Metallic Minerals	-	-	-	-	-	-	-	-	-	-	-
Non-specified	-	-	-	-	-	-	-	-	-	-	-
TRANSPORT SECTOR	-	-	-	-	-	-	-	-	-	-	-
Air	-	-	-	-	-	-	-	-	-	-	-
Road	-	-	-	-	-	-	-	-	-	-	-
Non-specified	-	-	-	-	-	-	-	-	-	-	-
OTHER SECTORS	-	-	-	-	-	-	-	-	-	-	-
Agriculture	-	-	-	-	-	-	-	-	-	-	-
Comm. and Publ. Services	-	-	-	-	-	-	-	-	-	-	-
Residential	-	-	-	-	-	-	-	-	-	-	-
Non-specified	-	-	-	-	-	-	-	-	-	-	-
NON-ENERGY USE	-	-	-	-	-	-	-	-	-	-	-

Yemen / Yémen

SUPPLY AND CONSUMPTION 1997	Refinery Gas	LPG + Ethane	Motor Gasoline	Aviation Gasoline	Jet Fuel	Kerosene	Gas/ Diesel	Heavy Fuel Oil	Naphtha	Petrol. Coke	Other Prod.
						Oil cont. (1000 tonnes)					
Production	-	20	1032	-	337	118	784	1531	415	-	59
Imports	-	-	-	1	-	-	-	-	-	-	-
Exports	-	-	-	-	-250	-	-34	-996	-415	-	-
Intl. Marine Bunkers	-	-	-	-	-	-	-40	-60	-	-	-
Stock Changes	-	-	-	-	-	-	-	-	-	-	-
DOMESTIC SUPPLY	-	20	1032	1	87	118	710	475	-	-	59
Transfers and Stat. Diff.	-	408	-	-	-	-	-1	-	-	-	-
TRANSFORMATION	-	-	-	-	-	-	263	181	-	-	-
Electricity and CHP Plants	-	-	-	-	-	-	263	181	-	-	-
Petroleum Refineries	-	-	-	-	-	-	-	-	-	-	-
Other Transform. Sector	-	-	-	-	-	-	-	-	-	-	-
ENERGY SECTOR	-	-	-	-	-	-	-	130	-	-	-
DISTRIBUTION LOSSES	-	-	-	-	-	-	-	-	-	-	-
FINAL CONSUMPTION	-	428	1032	1	87	118	446	164	-	-	59
INDUSTRY SECTOR	-	-	-	-	-	-	-	164	-	-	-
Iron and Steel	-	-	-	-	-	-	-	-	-	-	-
Chemical and Petrochem.	-	-	-	-	-	-	-	-	-	-	-
Non-Metallic Minerals	-	-	-	-	-	-	-	-	-	-	-
Non-specified	-	-	-	-	-	-	-	164	-	-	-
TRANSPORT SECTOR	-	-	1032	1	87	-	446	-	-	-	-
Air	-	-	-	1	87	-	-	-	-	-	-
Road	-	-	1032	-	-	-	446	-	-	-	-
Non-specified	-	-	-	-	-	-	-	-	-	-	-
OTHER SECTORS	-	428	-	-	-	118	-	-	-	-	-
Agriculture	-	-	-	-	-	-	-	-	-	-	-
Comm. and Publ. Services	-	-	-	-	-	-	-	-	-	-	-
Residential	-	428	-	-	-	118	-	-	-	-	-
Non-specified	-	-	-	-	-	-	-	-	-	-	-
NON-ENERGY USE	-	-	-	-	-	-	-	-	-	-	59

APPROVISIONNEMENT ET DEMANDE 1998	Gaz de raffinerie	GPL + éthane	Essence moteur	Essence aviation	Carbu- réacteurs	Kérosène	Gazole	Fioul lourd	Naphta	Coke de pétrole	Autres prod.
						Pétrole cont. (1000 tonnes)					
Production	-	20	1032	-	337	118	784	1531	415	-	59
Imports	-	-	-	1	-	-	-	-	-	-	-
Exports	-	-	-	-	-250	-	-34	-996	-415	-	-
Intl. Marine Bunkers	-	-	-	-	-	-	-40	-60	-	-	-
Stock Changes	-	-	-	-	-	-	-	-	-	-	-
DOMESTIC SUPPLY	-	20	1032	1	87	118	710	475	-	-	59
Transfers and Stat. Diff.	-	418	-	-	-	-	-1	-	-	-	-
TRANSFORMATION	-	-	-	-	-	-	263	181	-	-	-
Electricity and CHP Plants	-	-	-	-	-	-	263	181	-	-	-
Petroleum Refineries	-	-	-	-	-	-	-	-	-	-	-
Other Transform. Sector	-	-	-	-	-	-	-	-	-	-	-
ENERGY SECTOR	-	-	-	-	-	-	-	130	-	-	-
DISTRIBUTION LOSSES	-	-	-	-	-	-	-	-	-	-	-
FINAL CONSUMPTION	-	438	1032	1	87	118	446	164	-	-	59
INDUSTRY SECTOR	-	-	-	-	-	-	-	164	-	-	-
Iron and Steel	-	-	-	-	-	-	-	-	-	-	-
Chemical and Petrochem.	-	-	-	-	-	-	-	-	-	-	-
Non-Metallic Minerals	-	-	-	-	-	-	-	-	-	-	-
Non-specified	-	-	-	-	-	-	-	164	-	-	-
TRANSPORT SECTOR	-	-	1032	1	87	-	446	-	-	-	-
Air	-	-	-	1	87	-	-	-	-	-	-
Road	-	-	1032	-	-	-	446	-	-	-	-
Non-specified	-	-	-	-	-	-	-	-	-	-	-
OTHER SECTORS	-	438	-	-	-	118	-	-	-	-	-
Agriculture	-	-	-	-	-	-	-	-	-	-	-
Comm. and Publ. Services	-	-	-	-	-	-	-	-	-	-	-
Residential	-	438	-	-	-	118	-	-	-	-	-
Non-specified	-	-	-	-	-	-	-	-	-	-	-
NON-ENERGY USE	-	-	-	-	-	-	-	-	-	-	59

Yemen / Yémen

SUPPLY AND CONSUMPTION 1997	Gas (TJ)				Comb. Renew. & Waste (TJ)				(GWh)	(TJ)
	Natural Gas	Gas Works	Coke Ovens	Blast Furnaces	Solid Biomass	Gas/Liquids from Biomass	Municipal Waste	Industrial Waste	Electricity	Heat
Production	-	-	-	-	3240	-	-	-	2557	-
Imports	-	-	-	-	-	-	-	-	-	-
Exports	-	-	-	-	-	-	-	-	-	-
Intl. Marine Bunkers	-	-	-	-	-	-	-	-	-	-
Stock Changes	-	-	-	-	-	-	-	-	-	-
DOMESTIC SUPPLY	-	-	-	-	3240	-	-	-	2557	-
Transfers and Stat. Diff.	-	-	-	-	-	-	-	-	-	-
TRANSFORMATION	-	-	-	-	1680	-	-	-	-	-
Electricity and CHP Plants	-	-	-	-	-	-	-	-	-	-
Petroleum Refineries	-	-	-	-	-	-	-	-	-	-
Other Transform. Sector	-	-	-	-	1680	-	-	-	-	-
ENERGY SECTOR	-	-	-	-	-	-	-	-	275	-
DISTRIBUTION LOSSES	-	-	-	-	-	-	-	-	657	-
FINAL CONSUMPTION	-	-	-	-	1560	-	-	-	1625	-
INDUSTRY SECTOR	-	-	-	-	-	-	-	-	-	-
Iron and Steel	-	-	-	-	-	-	-	-	-	-
Chemical and Petrochem.	-	-	-	-	-	-	-	-	-	-
Non-Metallic Minerals	-	-	-	-	-	-	-	-	-	-
Non-specified	-	-	-	-	-	-	-	-	-	-
TRANSPORT SECTOR	-	-	-	-	-	-	-	-	-	-
Air	-	-	-	-	-	-	-	-	-	-
Road	-	-	-	-	-	-	-	-	-	-
Non-specified	-	-	-	-	-	-	-	-	-	-
OTHER SECTORS	-	-	-	-	1560	-	-	-	1625	-
Agriculture	-	-	-	-	-	-	-	-	-	-
Comm. and Publ. Services	-	-	-	-	-	-	-	-	-	-
Residential	-	-	-	-	-	-	-	-	-	-
Non-specified	-	-	-	-	1560	-	-	-	1625	-
NON-ENERGY USE	-	-	-	-	-	-	-	-	-	-

APPROVISIONNEMENT ET DEMANDE 1998	Gaz (TJ)				En. Re. Comb. & Déchets (TJ)				(GWh)	(TJ)
	Gaz naturel	Usines à gaz	Cokeries	Hauts fourneaux	Biomasse solide	Gaz/Liquides tirés de biomasse	Déchets urbains	Déchets industriels	Electricité	Chaleur
Production	-	-	-	-	3240	-	-	-	2507	-
Imports	-	-	-	-	-	-	-	-	-	-
Exports	-	-	-	-	-	-	-	-	-	-
Intl. Marine Bunkers	-	-	-	-	-	-	-	-	-	-
Stock Changes	-	-	-	-	-	-	-	-	-	-
DOMESTIC SUPPLY	-	-	-	-	3240	-	-	-	2507	-
Transfers and Stat. Diff.	-	-	-	-	-	-	-	-	-	-
TRANSFORMATION	-	-	-	-	1680	-	-	-	-	-
Electricity and CHP Plants	-	-	-	-	-	-	-	-	-	-
Petroleum Refineries	-	-	-	-	-	-	-	-	-	-
Other Transform. Sector	-	-	-	-	1680	-	-	-	-	-
ENERGY SECTOR	-	-	-	-	-	-	-	-	270	-
DISTRIBUTION LOSSES	-	-	-	-	-	-	-	-	644	-
FINAL CONSUMPTION	-	-	-	-	1560	-	-	-	1593	-
INDUSTRY SECTOR	-	-	-	-	-	-	-	-	-	-
Iron and Steel	-	-	-	-	-	-	-	-	-	-
Chemical and Petrochem.	-	-	-	-	-	-	-	-	-	-
Non-Metallic Minerals	-	-	-	-	-	-	-	-	-	-
Non-specified	-	-	-	-	-	-	-	-	-	-
TRANSPORT SECTOR	-	-	-	-	-	-	-	-	-	-
Air	-	-	-	-	-	-	-	-	-	-
Road	-	-	-	-	-	-	-	-	-	-
Non-specified	-	-	-	-	-	-	-	-	-	-
OTHER SECTORS	-	-	-	-	1560	-	-	-	1593	-
Agriculture	-	-	-	-	-	-	-	-	-	-
Comm. and Publ. Services	-	-	-	-	-	-	-	-	-	-
Residential	-	-	-	-	-	-	-	-	-	-
Non-specified	-	-	-	-	1560	-	-	-	1593	-
NON-ENERGY USE	-	-	-	-	-	-	-	-	-	-

Federal Republic of Yugoslavia / République fédérative de Yougoslavie

SUPPLY AND CONSUMPTION 1997	Coal (1000 tonnes)							Oil (1000 tonnes)			
	Coking Coal	Other Bit. Coal	Sub-Bit. Coal	Lignite	Peat	Oven and Gas Coke	Pat. Fuel and BKB	Crude Oil	NGL	Feed-stocks	Additives
Production	-	93	-	40563	-	-	-	979	-	-	-
Imports	-	52	-	-	-	-	-	2292	-	-	-
Exports	-	-	-	-	-	-	-	-	-	-	-
Intl. Marine Bunkers	-	-	-	-	-	-	-	-	-	-	-
Stock Changes	-	-	-	-	-	-	-	-	-	-	-
DOMESTIC SUPPLY	-	145	-	40563	-	-	-	3271	-	-	-
Transfers and Stat. Diff.	-	-	-	-	-	-	-	-	-	-	'
TRANSFORMATION	-	15	-	37000	-	-	-	3271	-	-	-
Electricity and CHP Plants	-	15	-	37000	-	-	-	-	-	-	-
Petroleum Refineries	-	-	-	-	-	-	-	3271	-	-	-
Other Transform. Sector	-	-	-	-	-	-	-	-	-	-	-
ENERGY SECTOR	-	-	-	-	-	-	-	-	-	-	-
DISTRIBUTION LOSSES	-	-	-	-	-	-	-	-	-	-	-
FINAL CONSUMPTION	-	130	-	3563	-	-	-	-	-	-	-
INDUSTRY SECTOR	-	-	-	2400	-	-	-	-	-	-	-
Iron and Steel	-	-	-	-	-	-	-	-	-	-	-
Chemical and Petrochem.	-	-	-	-	-	-	-	-	-	-	-
Non-Metallic Minerals	-	-	-	-	-	-	-	-	-	-	-
Non-specified	-	-	-	2400	-	-	-	-	-	-	-
TRANSPORT SECTOR	-	-	-	-	-	-	-	-	-	-	-
Air	-	-	-	-	-	-	-	-	-	-	-
Road	-	-	-	-	-	-	-	-	-	-	-
Non-specified	-	-	-	-	-	-	-	-	-	-	-
OTHER SECTORS	-	130	-	1163	-	-	-	-	-	-	-
Agriculture	-	-	-	-	-	-	-	-	-	-	-
Comm. and Publ. Services	-	-	-	-	-	-	-	-	-	-	-
Residential	-	-	-	-	-	-	-	-	-	-	-
Non-specified	-	130	-	1163	-	-	-	-	-	-	-
NON-ENERGY USE	-	-	-	-	-	-	-	-	-	-	-

APPROVISIONNEMENT ET DEMANDE 1998	Charbon (1000 tonnes)							Pétrole (1000 tonnes)			
	Charbon à coke	Autres charb. bit.	Charbon sous-bit.	Lignite	Tourbe	Coke de four/gaz	Agg./briq. de lignite	Pétrole brut	LGN	Produits d'aliment.	Additifs
Production	-	105	-	43967	-	-	-	913	-	-	-
Imports	-	52	-	-	-	-	-	1975	-	-	-
Exports	-	-	-	-	-	-	-	-	-	-	-
Intl. Marine Bunkers	-	-	-	-	-	-	-	-	-	-	-
Stock Changes	-	-	-	-	-	-	-	-	-	-	-
DOMESTIC SUPPLY	-	157	-	43967	-	-	-	2888	-	-	-
Transfers and Stat. Diff.	-	-	-	-	-	-	-	-	-	-	-
TRANSFORMATION	-	16	-	40105	-	-	-	2888	-	-	-
Electricity and CHP Plants	-	16	-	40105	-	-	-	-	-	-	-
Petroleum Refineries	-	-	-	-	-	-	-	2888	-	-	-
Other Transform. Sector	-	-	-	-	-	-	-	-	-	-	-
ENERGY SECTOR	-	-	-	-	-	-	-	-	-	-	-
DISTRIBUTION LOSSES	-	-	-	-	-	-	-	-	-	-	-
FINAL CONSUMPTION	-	141	-	3862	-	-	-	-	-	-	-
INDUSTRY SECTOR	-	-	-	2601	-	-	-	-	-	-	-
Iron and Steel	-	-	-	-	-	-	-	-	-	-	-
Chemical and Petrochem.	-	-	-	-	-	-	-	-	-	-	-
Non-Metallic Minerals	-	-	-	-	-	-	-	-	-	-	-
Non-specified	-	-	-	2601	-	-	-	-	-	-	-
TRANSPORT SECTOR	-	-	-	-	-	-	-	-	-	-	-
Air	-	-	-	-	-	-	-	-	-	-	-
Road	-	-	-	-	-	-	-	-	-	-	-
Non-specified	-	-	-	-	-	-	-	-	-	-	-
OTHER SECTORS	-	141	-	1261	-	-	-	-	-	-	-
Agriculture	-	-	-	-	-	-	-	-	-	-	-
Comm. and Publ. Services	-	-	-	-	-	-	-	-	-	-	-
Residential	-	-	-	-	-	-	-	-	-	-	-
Non-specified	-	141	-	1261	-	-	-	-	-	-	-
NON-ENERGY USE	-	-	-	-	-	-	-	-	-	-	-

Federal Republic of Yugoslavia / République fédérative de Yougoslavie

SUPPLY AND CONSUMPTION 1997	Oil cont. (1000 tonnes)										
	Refinery Gas	LPG + Ethane	Motor Gasoline	Aviation Gasoline	Jet Fuel	Kerosene	Gas/ Diesel	Heavy Fuel Oil	Naphtha	Petrol. Coke	Other Prod.
Production	-	47	590	-	95	54	784	959	385	-	292
Imports	-	-	200	-	-	-	-	100	-	-	-
Exports	-	-	-	-	-	-	-	-	-	-	-
Intl. Marine Bunkers	-	-	-	-	-	-	-	-	-	-	-
Stock Changes	-	-	-	-	-	-	-	-	-	-	-
DOMESTIC SUPPLY	-	47	790	-	95	54	784	1059	385	-	292
Transfers and Stat. Diff.	-	-	-	-	-	-	-	-	-	-	-
TRANSFORMATION	-	-	-	-	-	-	-	414	-	-	-
Electricity and CHP Plants	-	-	-	-	-	-	-	414	-	-	-
Petroleum Refineries	-	-	-	-	-	-	-	-	-	-	-
Other Transform. Sector	-	-	-	-	-	-	-	-	-	-	-
ENERGY SECTOR	-	-	-	-	-	-	-	-	-	-	-
DISTRIBUTION LOSSES	-	-	-	-	-	-	-	-	-	-	-
FINAL CONSUMPTION	-	47	790	-	95	54	784	645	385	-	292
INDUSTRY SECTOR	-	47	-	-	-	-	131	485	385	-	-
Iron and Steel	-	-	-	-	-	-	-	-	-	-	-
Chemical and Petrochem.	-	-	-	-	-	-	-	-	385	-	-
Non-Metallic Minerals	-	-	-	-	-	-	-	-	-	-	-
Non-specified	-	47	-	-	-	-	131	485	-	-	-
TRANSPORT SECTOR	-	-	790	-	95	-	653	-	-	-	-
Air	-	-	-	-	95	-	-	-	-	-	-
Road	-	-	790	-	-	-	653	-	-	-	-
Non-specified	-	-	-	-	-	-	-	-	-	-	-
OTHER SECTORS	-	-	-	-	-	54	-	160	-	-	-
Agriculture	-	-	-	-	-	-	-	-	-	-	-
Comm. and Publ. Services	-	-	-	-	-	-	-	-	-	-	-
Residential	-	-	-	-	-	54	-	-	-	-	-
Non-specified	-	-	-	-	-	-	-	160	-	-	-
NON-ENERGY USE	-	-	-	-	-	-	-	-	-	-	292

APPROVISIONNEMENT ET DEMANDE 1998	Pétrole cont. (1000 tonnes)										
	Gaz de raffinerie	GPL + éthane	Essence moteur	Essence aviation	Carbu- réacteurs	Kérosène	Gazole	Fioul lourd	Naphta	Coke de pétrole	Autres prod.
Production	-	45	570	-	95	52	600	584	385	-	288
Imports	-	-	200	-	-	-	-	100	-	-	-
Exports	-	-	-	-	-	-	-	-	-	-	-
Intl. Marine Bunkers	-	-	-	-	-	-	-	-	-	-	-
Stock Changes	-	-	-	-	-	-	-	-	-	-	-
DOMESTIC SUPPLY	-	45	770	-	95	52	600	684	385	-	288
Transfers and Stat. Diff.	-	-	-	-	-	-	-	-	-	-	-
TRANSFORMATION	-	-	-	-	-	-	-	267	-	-	-
Electricity and CHP Plants	-	-	-	-	-	-	-	267	-	-	-
Petroleum Refineries	-	-	-	-	-	-	-	-	-	-	-
Other Transform. Sector	-	-	-	-	-	-	-	-	-	-	-
ENERGY SECTOR	-	-	-	-	-	-	-	-	-	-	-
DISTRIBUTION LOSSES	-	-	-	-	-	-	-	-	-	-	-
FINAL CONSUMPTION	-	45	770	-	95	52	600	417	385	-	288
INDUSTRY SECTOR	-	45	-	-	-	-	100	314	385	-	-
Iron and Steel	-	-	-	-	-	-	-	-	-	-	-
Chemical and Petrochem.	-	-	-	-	-	-	-	-	385	-	-
Non-Metallic Minerals	-	-	-	-	-	-	-	-	-	-	-
Non-specified	-	45	-	-	-	-	100	314	-	-	-
TRANSPORT SECTOR	-	-	770	-	95	-	500	-	-	-	-
Air	-	-	-	-	95	-	-	-	-	-	-
Road	-	-	770	-	-	-	500	-	-	-	-
Non-specified	-	-	-	-	-	-	-	-	-	-	-
OTHER SECTORS	-	-	-	-	-	52	-	103	-	-	-
Agriculture	-	-	-	-	-	-	-	-	-	-	-
Comm. and Publ. Services	-	-	-	-	-	-	-	-	-	-	-
Residential	-	-	-	-	-	52	-	-	-	-	-
Non-specified	-	-	-	-	-	-	-	103	-	-	-
NON-ENERGY USE	-	-	-	-	-	-	-	-	-	-	288

Federal Republic of Yugoslavia / République fédérative de Yougoslavie

SUPPLY AND CONSUMPTION 1997	Gas (TJ)				Comb. Renew. & Waste (TJ)				(GWh)	(TJ)
	Natural Gas	Gas Works	Coke Ovens	Blast Furnaces	Solid Biomass	Gas/Liquids from Biomass	Municipal Waste	Industrial Waste	Electricity	Heat
Production	25937	-	-	-	8791	-	-	-	40312	19000
Imports	77813	-	-	-	-	-	-	-	-	-
Exports	-	-	-	-	-	-	-	-	-9	-
Intl. Marine Bunkers	-	-	-	-	-	-	-	-	-	-
Stock Changes	-	-	-	-	-	-	-	-	-	-
DOMESTIC SUPPLY	103750	-	-	-	8791	-	-	-	40303	19000
Transfers and Stat. Diff.	-	-	-	-	-	-	-	-	-	-
TRANSFORMATION	11988	-	-	-	-	-	-	-	-	-
Electricity and CHP Plants	11988	-	-	-	-	-	-	-	-	-
Petroleum Refineries	-	-	-	-	-	-	-	-	-	-
Other Transform. Sector	-	-	-	-	-	-	-	-	-	-
ENERGY SECTOR	1056	-	-	-	-	-	-	-	3169	-
DISTRIBUTION LOSSES	-	-	-	-	-	-	-	-	3745	-
FINAL CONSUMPTION	90706	-	-	-	8791	-	-	-	33389	19000
INDUSTRY SECTOR	29858	-	-	-	-	-	-	-	6615	-
Iron and Steel	-	-	-	-	-	-	-	-	360	-
Chemical and Petrochem.	-	-	-	-	-	-	-	-	429	-
Non-Metallic Minerals	-	-	-	-	-	-	-	-	-	-
Non-specified	29858	-	-	-	-	-	-	-	5826	-
TRANSPORT SECTOR	-	-	-	-	-	-	-	-	287	-
Air	-	-	-	-	-	-	-	-	-	-
Road	-	-	-	-	-	-	-	-	57	-
Non-specified	-	-	-	-	-	-	-	-	230	-
OTHER SECTORS	60848	-	-	-	8791	-	-	-	26487	19000
Agriculture	-	-	-	-	-	-	-	-	194	-
Comm. and Publ. Services	-	-	-	-	-	-	-	-	310	-
Residential	-	-	-	-	-	-	-	-	17469	-
Non-specified	60848	-	-	-	8791	-	-	-	8514	19000
NON-ENERGY USE	-	-	-	-	-	-	-	-	-	-

APPROVISIONNEMENT ET DEMANDE 1998	Gaz (TJ)				En. Re. Comb. & Déchets (TJ)				(GWh)	(TJ)
	Gaz naturel	Usines à gaz	Cokeries	Hauts fourneaux	Biomasse solide	Gaz/Liquides tirés de biomasse	Déchets urbains	Déchets industriels	Electricité	Chaleur
Production	27559	-	-	-	8791	-	-	-	40651	19000
Imports	70687	-	-	-	-	-	-	-	-	-
Exports	-	-	-	-	-	-	-	-	-268	-
Intl. Marine Bunkers	-	-	-	-	-	-	-	-	-	-
Stock Changes	-	-	-	-	-	-	-	-	-	-
DOMESTIC SUPPLY	98246	-	-	-	8791	-	-	-	40383	19000
Transfers and Stat. Diff.	-	-	-	-	-	-	-	-	-	-
TRANSFORMATION	11348	-	-	-	-	-	-	-	-	-
Electricity and CHP Plants	11348	-	-	-	-	-	-	-	-	-
Petroleum Refineries	-	-	-	-	-	-	-	-	-	-
Other Transform. Sector	-	-	-	-	-	-	-	-	-	-
ENERGY SECTOR	1018	-	-	-	-	-	-	-	3175	-
DISTRIBUTION LOSSES	-	-	-	-	-	-	-	-	3752	-
FINAL CONSUMPTION	85880	-	-	-	8791	-	-	-	33456	19000
INDUSTRY SECTOR	28275	-	-	-	-	-	-	-	6628	-
Iron and Steel	-	-	-	-	-	-	-	-	361	-
Chemical and Petrochem.	-	-	-	-	-	-	-	-	430	-
Non-Metallic Minerals	-	-	-	-	-	-	-	-	-	-
Non-specified	28275	-	-	-	-	-	-	-	5837	-
TRANSPORT SECTOR	-	-	-	-	-	-	-	-	288	-
Air	-	-	-	-	-	-	-	-	-	-
Road	-	-	-	-	-	-	-	-	58	-
Non-specified	-	-	-	-	-	-	-	-	230	-
OTHER SECTORS	57605	-	-	-	8791	-	-	-	26540	19000
Agriculture	-	-	-	-	-	-	-	-	194	-
Comm. and Publ. Services	-	-	-	-	-	-	-	-	311	-
Residential	-	-	-	-	-	-	-	-	17504	-
Non-specified	57605	-	-	-	8791	-	-	-	8531	19000
NON-ENERGY USE	-	-	-	-	-	-	-	-	-	-

Former Yugoslavia / ex-Yougoslavie

SUPPLY AND CONSUMPTION 1997	Coal (1000 tonnes)							Oil (1000 tonnes)			
	Coking Coal	Other Bit. Coal	Sub-Bit. Coal	Lignite	Peat	Oven and Gas Coke	Pat. Fuel and BKB	Crude Oil	NGL	Feed-stocks	Additives
Production	-	142	648	53309	-	-	-	2513	252	-	-
Imports	49	247	388	52	-	105	-	6936	-	77	-
Exports	-	-3	-1	-1	-	-	-	-40	-	-	-
Intl. Marine Bunkers	-	-	-	-	-	-	-	-	-	-	-
Stock Changes	-	115	51	58	-	-4	-	56	-	-4	-
DOMESTIC SUPPLY	49	501	1086	53418	-	101	-	9465	252	73	-
Transfers and Stat. Diff.	-	-	-	-	-	-	-	-	-252	-	-
TRANSFORMATION	-	245	977	49292	-	-	-	9354	-	73	-
Electricity and CHP Plants	-	245	977	49292	-	-	-	-	-	-	-
Petroleum Refineries	-	-	-	-	-	-	-	9354	-	73	-
Other Transform. Sector	-	-	-	-	-	-	-	-	-	-	-
ENERGY SECTOR	-	-	3	-	-	-	-	111	-	-	-
DISTRIBUTION LOSSES	-	-	-	-	-	-	-	-	-	-	-
FINAL CONSUMPTION	49	256	106	4126	-	101	-	-	-	-	-
INDUSTRY SECTOR	49	126	55	2540	-	101	-	-	-	-	-
Iron and Steel	11	22	-	98	-	42	-	-	-	-	-
Chemical and Petrochem.	-	-	4	2	-	-	-	-	-	-	-
Non-Metallic Minerals	-	69	3	2	-	50	-	-	-	-	-
Non-specified	38	35	48	2438	-	9	-	-	-	-	-
TRANSPORT SECTOR	-	-	-	308	-	-	-	-	-	-	-
Air	-	-	-	-	-	-	-	-	-	-	-
Road	-	-	-	-	-	-	-	-	-	-	-
Non-specified	-	-	-	308	-	-	-	-	-	-	-
OTHER SECTORS	-	130	51	1278	-	-	-	-	-	-	-
Agriculture	-	-	-	-	-	-	-	-	-	-	-
Comm. and Publ. Services	-	-	23	42	-	-	-	-	-	-	-
Residential	-	-	28	73	-	-	-	-	-	-	-
Non-specified	-	130	-	1163	-	-	-	-	-	-	-
NON-ENERGY USE	-	-	-	-	-	-	-	-	-	-	-

APPROVISIONNEMENT ET DEMANDE 1998	Charbon (1000 tonnes)							Pétrole (1000 tonnes)			
	Charbon à coke	Autres charb. bit.	Charbon sous-bit.	Lignite	Tourbe	Coke de four/gaz	Agg./briq. de lignite	Pétrole brut	LGN	Produits d'aliment.	Additifs
Production	-	156	698	58131	-	-	-	2503	226	-	-
Imports	122	400	482	235	-	94	-	7115	-	10	-
Exports	-	-	-1	-1	-	-1	-	-40	-	-	-
Intl. Marine Bunkers	-	-	-	-	-	-	-	-	-	-	-
Stock Changes	-8	3	28	-94	-	4	-	-8	-	1	-
DOMESTIC SUPPLY	114	559	1207	58271	-	97	-	9570	226	11	-
Transfers and Stat. Diff.	-	-	-	151	-	-	-	-	-226	215	-
TRANSFORMATION	1	245	1124	53993	-	-	-	9460	-	226	-
Electricity and CHP Plants	-	245	1124	53905	-	-	-	-	-	-	-
Petroleum Refineries	-	-	-	-	-	-	-	9460	-	226	-
Other Transform. Sector	1	-	-	88	-	-	-	-	-	-	-
ENERGY SECTOR	-	1	3	-	-	-	-	110	-	-	-
DISTRIBUTION LOSSES	-	-	-	21	-	-	-	-	-	-	-
FINAL CONSUMPTION	113	313	80	4408	-	97	-	-	-	-	-
INDUSTRY SECTOR	113	172	29	2749	-	97	-	-	-	-	-
Iron and Steel	90	29	-	120	-	40	-	-	-	-	-
Chemical and Petrochem.	-	-	-	-	-	-	-	-	-	-	-
Non-Metallic Minerals	-	71	2	1	-	49	-	-	-	-	-
Non-specified	23	72	27	2628	-	8	-	-	-	-	-
TRANSPORT SECTOR	-	-	-	318	-	-	-	-	-	-	-
Air	-	-	-	-	-	-	-	-	-	-	-
Road	-	-	-	-	-	-	-	-	-	-	-
Non-specified	-	-	-	318	-	-	-	-	-	-	-
OTHER SECTORS	-	141	51	1341	-	-	-	-	-	-	-
Agriculture	-	-	-	-	-	-	-	-	-	-	-
Comm. and Publ. Services	-	-	26	15	-	-	-	-	-	-	-
Residential	-	-	25	65	-	-	-	-	-	-	-
Non-specified	-	141	-	1261	-	-	-	-	-	-	-
NON-ENERGY USE	-	-	-	-	-	-	-	-	-	-	-

Former Yugoslavia / ex-Yougoslavie

SUPPLY AND CONSUMPTION 1997	Refinery Gas	LPG + Ethane	Motor Gasoline	Aviation Gasoline	Jet Fuel	Kerosene	Gas/ Diesel	Heavy Fuel Oil	Naphtha	Petrol. Coke	Other Prod.
Production	202	208	1803	-	190	59	2649	2790	697	28	620
Imports	-	89	1430	1	115	2	1845	501	325	15	132
Exports	-	-146	-460	-	-37	-1	-503	-130	-245	-29	-186
Intl. Marine Bunkers	-	-	-	-	-	-	-7	-17	-	-	-
Stock Changes	-	1	-51	-	-4	-	-47	-76	2	1	21
DOMESTIC SUPPLY	202	152	2722	1	264	60	3937	3068	779	15	587
Transfers and Stat. Diff.	-	198	-	-	-	-	-	1	54	-	-
TRANSFORMATION	4	14	-	-	-	-	29	1485	67	-	-
Electricity and CHP Plants	4	-	-	-	-	-	20	1197	21	-	-
Petroleum Refineries	-	-	-	-	-	-	-	-	-	-	-
Other Transform. Sector	-	14	-	-	-	-	9	288	46	-	-
ENERGY SECTOR	198	-	1	-	-	-	4	318	-	-	-
DISTRIBUTION LOSSES	-	-	96	-	-	-	69	39	243	-	-
FINAL CONSUMPTION	-	336	2625	1	264	60	3835	1227	523	15	587
INDUSTRY SECTOR	-	194	11	-	-	-	341	959	523	15	49
Iron and Steel	-	5	-	-	-	-	5	25	8	-	17
Chemical and Petrochem.	-	102	-	-	-	-	4	123	508	-	-
Non-Metallic Minerals	-	18	-	-	-	-	14	124	-	15	-
Non-specified	-	69	11	-	-	-	318	687	7	-	32
TRANSPORT SECTOR	-	12	2605	1	264	-	2263	11	-	-	-
Air	-	-	-	1	264	-	-	-	-	-	-
Road	-	12	2605	-	-	-	2130	-	-	-	-
Non-specified	-	-	-	-	-	-	133	11	-	-	-
OTHER SECTORS	-	130	9	-	-	60	1231	257	-	-	-
Agriculture	-	2	9	-	-	-	216	12	-	-	-
Comm. and Publ. Services	-	16	-	-	-	1	355	36	-	-	-
Residential	-	112	-	-	-	59	660	49	-	-	-
Non-specified	-	-	-	-	-	-	-	160	-	-	-
NON-ENERGY USE	-	-	-	-	-	-	-	-	-	-	538

APPROVISIONNEMENT ET DEMANDE 1998	Gaz de raffinerie	GPL + éthane	Essence moteur	Essence aviation	Carbu- réacteurs	Kérosène	Gazole	Fioul lourd	Naphta	Coke de pétrole	Autres prod.
Production	238	250	1895	-	182	56	2792	2347	851	58	592
Imports	-	96	1272	1	109	-	1827	609	335	40	322
Exports	-	-168	-524	-2	-41	-1	-655	-202	-204	-52	-122
Intl. Marine Bunkers	-	-	-	-	-	-	-12	-14	-	-	-
Stock Changes	-	-8	-38	2	6	-	24	92	-7	-	35
DOMESTIC SUPPLY	238	170	2605	1	256	55	3976	2832	975	46	827
Transfers and Stat. Diff.	-	177	-	-	-	-	-	-39	-129	-	1
TRANSFORMATION	3	17	-	-	-	-	40	1527	69	-	-
Electricity and CHP Plants	3	1	-	-	-	-	19	1289	23	-	-
Petroleum Refineries	-	-	-	-	-	-	-	-	-	-	-
Other Transform. Sector	-	16	-	-	-	-	21	238	46	-	-
ENERGY SECTOR	235	7	1	-	-	-	8	272	-	18	162
DISTRIBUTION LOSSES	-	-	99	-	-	-	71	40	251	-	-
FINAL CONSUMPTION	-	323	2505	1	256	55	3857	954	526	28	666
INDUSTRY SECTOR	-	175	9	-	-	-	294	790	526	28	52
Iron and Steel	-	11	-	-	-	-	4	39	8	-	18
Chemical and Petrochem.	-	90	-	-	-	-	5	117	511	-	1
Non-Metallic Minerals	-	15	-	-	-	-	12	144	-	28	-
Non-specified	-	59	9	-	-	-	273	490	7	-	33
TRANSPORT SECTOR	-	12	2487	1	256	-	2251	3	-	-	-
Air	-	-	-	1	256	-	-	-	-	-	-
Road	-	12	2487	-	-	-	2147	-	-	-	-
Non-specified	-	-	-	-	-	-	104	3	-	-	-
OTHER SECTORS	-	136	9	-	-	55	1312	161	-	-	-
Agriculture	-	2	9	-	-	-	201	3	-	-	-
Comm. and Publ. Services	-	37	-	-	-	-	452	18	-	-	-
Residential	-	97	-	-	-	54	659	37	-	-	-
Non-specified	-	-	-	-	-	1	-	103	-	-	-
NON-ENERGY USE	-	-	-	-	-	-	-	-	-	-	614

Former Yugoslavia / ex-Yougoslavie

SUPPLY AND CONSUMPTION 1997	Gas (TJ)				Comb. Renew. & Waste (TJ)				(GWh)	(TJ)
	Natural Gas	Gas Works	Coke Ovens	Blast Furnaces	Solid Biomass	Gas/Liquids from Biomass	Municipal Waste	Industrial Waste	Electricity	Heat
Production	91623	418	-	-	46852	-	-	10	72352	64947
Imports	161141	-	-	-	1386	-	-	-	5843	-
Exports	-	-	-	-	-	-	-	-	-3382	-
Intl. Marine Bunkers	-	-	-	-	-	-	-	-		
Stock Changes	-464	-	-	-	-33	-	-	-	-	-
DOMESTIC SUPPLY	252300	418	-	-	48205	-	-	10	74813	64947
Transfers and Stat. Diff.	7346	-	-	-	-	-	-	-	-	-
TRANSFORMATION	53789	-	-	-	94	-	-	10	-	-
Electricity and CHP Plants	31987	-	-	-	-	-	-	10	-	-
Petroleum Refineries	-	-	-	-	-	-	-	-	-	-
Other Transform. Sector	21802	-	-	-	94	-	-	-	-	-
ENERGY SECTOR	11527	-	-	-	-	-	-	-	5667	2249
DISTRIBUTION LOSSES	3272	3	-	-	-	-	-	-	7623	2814
FINAL CONSUMPTION	191058	415	-	-	48111	-	-	-	61523	59884
INDUSTRY SECTOR	102378	145	-	-	2556	-	-	-	16617	7115
Iron and Steel	4495	25	-	-	-	-	-	-	2515	238
Chemical and Petrochem.	35797	-	-	-	-	-	-	-	1527	1737
Non-Metallic Minerals	11669	43	-	-	-	-	-	-	907	172
Non-specified	50417	77	-	-	2556	-	-	-	11668	4968
TRANSPORT SECTOR	-	-	-	-	-	-	-	-	679	-
Air	-	-	-	-	-	-	-	-	-	-
Road	-	-	-	-	-	-	-	-	57	-
Non-specified	-	-	-	-	-	-	-	-	622	-
OTHER SECTORS	88680	270	-	-	45555	-	-	-	44227	52769
Agriculture	855	-	-	-	-	-	-	-	279	440
Comm. and Publ. Services	6574	68	-	-	2141	-	-	-	5505	3796
Residential	20403	202	-	-	34473	-	-	-	27960	12199
Non-specified	60848	-	-	-	8941	-	-	-	10483	36334
NON-ENERGY USE	-	-	-	-	-	-	-	-	-	-

APPROVISIONNEMENT ET DEMANDE 1998	Gaz (TJ)				En. Re. Comb. & Déchets (TJ)				(GWh)	(TJ)
	Gaz naturel	Usines à gaz	Cokeries	Hauts fourneaux	Biomasse solide	Gaz/Liquides tirés de biomasse	Déchets urbains	Déchets industriels	Electricité	Chaleur
Production	87505	418	-	-	44806	-	-	100	74879	66988
Imports	158514	-	-	-	1387	-	-	-	4911	-
Exports	-38	-	-	-	-	-	-	-	-3527	-
Intl. Marine Bunkers	-	-	-	-	-	-	-	-	-	-
Stock Changes	-1140	-	-	-	-36	-	-	-	-	-
DOMESTIC SUPPLY	244841	418	-	-	46157	-	-	100	76263	66988
Transfers and Stat. Diff.	7416	-	-	-	-	-	-	-	-	-
TRANSFORMATION	55362	2	-	-	232	-	-	100	-	-
Electricity and CHP Plants	33797	-	-	-	-	-	-	100	-	-
Petroleum Refineries	-	-	-	-	-	-	-	-	-	-
Other Transform. Sector	21565	2	-	-	232	-	-	-	-	-
ENERGY SECTOR	9376	-	-	-	-	-	-	-	5866	2941
DISTRIBUTION LOSSES	4511	3	-	-	-	-	-	-	8263	2690
FINAL CONSUMPTION	183008	413	-	-	45925	-	-	-	62134	61357
INDUSTRY SECTOR	95297	144	-	-	2640	-	-	-	16841	7969
Iron and Steel	4088	25	-	-	-	-	-	-	2651	561
Chemical and Petrochem.	30069	-	-	-	1	-	-	-	1425	1370
Non-Metallic Minerals	12381	43	-	-	16	-	-	-	1072	363
Non-specified	48759	76	-	-	2623	-	-	-	11693	5675
TRANSPORT SECTOR	-	-	-	-	-	-	-	-	697	-
Air	-	-	-	-	-	-	-	-	-	-
Road	-	-	-	-	-	-	-	-	58	-
Non-specified	-	-	-	-	-	-	-	-	639	-
OTHER SECTORS	87711	269	-	-	43285	-	-	-	44596	53388
Agriculture	722	-	-	-	-	-	-	-	297	623
Comm. and Publ. Services	8018	68	-	-	2141	-	-	-	5719	4186
Residential	21366	201	-	-	32203	-	-	-	28017	11773
Non-specified	57605	-	-	-	8941	-	-	-	10563	36806
NON-ENERGY USE	-	-	-	-	-	-	-	-	-	-

Zambia / Zambie

SUPPLY AND CONSUMPTION 1997	Coal (1000 tonnes)							Oil (1000 tonnes)			
	Coking Coal	Other Bit. Coal	Sub-Bit. Coal	Lignite	Peat	Oven and Gas Coke	Pat. Fuel and BKB	Crude Oil	NGL	Feed-stocks	Additives
Production	24	155	-	-	-	32	-	-	-	-	-
Imports	-	-	-	-	-	-	-	588	-	-	-
Exports	-	-7	-	-	-	-	-	-	-	-	-
Intl. Marine Bunkers	-	-	-	-	-	-	-	-	-	-	-
Stock Changes	-	-	-	-	-	-	-	-	-	-	-
DOMESTIC SUPPLY	24	148	-	-	-	32	-	588	-	-	-
Transfers and Stat. Diff.	-	-	-	-	-	-	-	-	-	-	-
TRANSFORMATION	24	30	-	-	-	26	-	588	-	-	-
Electricity and CHP Plants	-	15	-	-	-	-	-	-	-	-	-
Petroleum Refineries	-	-	-	-	-	-	-	588	-	-	-
Other Transform. Sector	24	15	-	-	-	26	-	-	-	-	-
ENERGY SECTOR	-	-	-	-	-	-	-	-	-	-	-
DISTRIBUTION LOSSES	-	-	-	-	-	-	-	-	-	-	-
FINAL CONSUMPTION	-	118	-	-	-	6	-	-	-	-	-
INDUSTRY SECTOR	-	118	-	-	-	6	-	-	-	-	-
Iron and Steel	-	-	-	-	-	6	-	-	-	-	-
Chemical and Petrochem.	-	-	-	-	-	-	-	-	-	-	-
Non-Metallic Minerals	-	-	-	-	-	-	-	-	-	-	-
Non-specified	-	118	-	-	-	-	-	-	-	-	-
TRANSPORT SECTOR	-	-	-	-	-	-	-	-	-	-	-
Air	-	-	-	-	-	-	-	-	-	-	-
Road	-	-	-	-	-	-	-	-	-	-	-
Non-specified	-	-	-	-	-	-	-	-	-	-	-
OTHER SECTORS	-	-	-	-	-	-	-	-	-	-	-
Agriculture	-	-	-	-	-	-	-	-	-	-	-
Comm. and Publ. Services	-	-	-	-	-	-	-	-	-	-	-
Residential	-	-	-	-	-	-	-	-	-	-	-
Non-specified	-	-	-	-	-	-	-	-	-	-	-
NON-ENERGY USE	-	-	-	-	-	-	-	-	-	-	-

APPROVISIONNEMENT ET DEMANDE 1998	Charbon (1000 tonnes)							Pétrole (1000 tonnes)			
	Charbon à coke	Autres charb. bit.	Charbon sous-bit.	Lignite	Tourbe	Coke de four/gaz	Agg./briq. de lignite	Pétrole brut	LGN	Produits d'aliment.	Additifs
Production	24	155	-	-	-	32	-	-	-	-	-
Imports	-	-	-	-	-	-	-	588	-	-	-
Exports	-	-7	-	-	-	-	-	-	-	-	-
Intl. Marine Bunkers	-	-	-	-	-	-	-	-	-	-	-
Stock Changes	-	-	-	-	-	-	-	-	-	-	-
DOMESTIC SUPPLY	24	148	-	-	-	32	-	588	-	-	-
Transfers and Stat. Diff.	-	-	-	-	-	-	-	-	-	-	-
TRANSFORMATION	24	30	-	-	-	26	-	588	-	-	-
Electricity and CHP Plants	-	15	-	-	-	-	-	-	-	-	-
Petroleum Refineries	-	-	-	-	-	-	-	588	-	-	-
Other Transform. Sector	24	15	-	-	-	26	-	-	-	-	-
ENERGY SECTOR	-	-	-	-	-	-	-	-	-	-	-
DISTRIBUTION LOSSES	-	-	-	-	-	-	-	-	-	-	-
FINAL CONSUMPTION	-	118	-	-	-	6	-	-	-	-	-
INDUSTRY SECTOR	-	118	-	-	-	6	-	-	-	-	-
Iron and Steel	-	-	-	-	-	6	-	-	-	-	-
Chemical and Petrochem.	-	-	-	-	-	-	-	-	-	-	-
Non-Metallic Minerals	-	-	-	-	-	-	-	-	-	-	-
Non-specified	-	118	-	-	-	-	-	-	-	-	-
TRANSPORT SECTOR	-	-	-	-	-	-	-	-	-	-	-
Air	-	-	-	-	-	-	-	-	-	-	-
Road	-	-	-	-	-	-	-	-	-	-	-
Non-specified	-	-	-	-	-	-	-	-	-	-	-
OTHER SECTORS	-	-	-	-	-	-	-	-	-	-	-
Agriculture	-	-	-	-	-	-	-	-	-	-	-
Comm. and Publ. Services	-	-	-	-	-	-	-	-	-	-	-
Residential	-	-	-	-	-	-	-	-	-	-	-
Non-specified	-	-	-	-	-	-	-	-	-	-	-
NON-ENERGY USE	-	-	-	-	-	-	-	-	-	-	-

Zambia / Zambie

SUPPLY AND CONSUMPTION 1997	Oil cont. (1000 tonnes)										
	Refinery Gas	LPG + Ethane	Motor Gasoline	Aviation Gasoline	Jet Fuel	Kerosene	Gas/ Diesel	Heavy Fuel Oil	Naphtha	Petrol. Coke	Other Prod.
Production	15	10	120	-	40	31	219	95	-	-	32
Imports	-	-	-	3	-	-	10	-	-	-	-
Exports	-	-	-10	-	-	-	-30	-	-	-	-
Intl. Marine Bunkers	-	-	-	-	-	-	-	-	-	-	-
Stock Changes	-	-	-	-	-	-	-	-	-	-	-
DOMESTIC SUPPLY	15	10	110	3	40	31	199	95	-	-	32
Transfers and Stat. Diff.	-	-	-	-	-5	-	-2	-	-	-	-
TRANSFORMATION	-	-	-	-	-	-	1	-	-	-	-
Electricity and CHP Plants	-	-	-	-	-	-	1	-	-	-	-
Petroleum Refineries	-	-	-	-	-	-	-	-	-	-	-
Other Transform. Sector	-	-	-	-	-	-	-	-	-	-	-
ENERGY SECTOR	15	-	-	-	-	-	-	21	-	-	-
DISTRIBUTION LOSSES	-	-	-	-	-	-	-	-	-	-	-
FINAL CONSUMPTION	-	10	110	3	35	31	196	74	-	-	32
INDUSTRY SECTOR	-	10	-	-	-	-	36	74	-	-	-
Iron and Steel	-	-	-	-	-	-	-	-	-	-	-
Chemical and Petrochem.	-	-	-	-	-	-	-	-	-	-	-
Non-Metallic Minerals	-	-	-	-	-	-	-	-	-	-	-
Non-specified	-	10	-	-	-	-	36	74	-	-	-
TRANSPORT SECTOR	-	-	110	3	35	-	70	-	-	-	-
Air	-	-	-	3	35	-	-	-	-	-	-
Road	-	-	110	-	-	-	70	-	-	-	-
Non-specified	-	-	-	-	-	-	-	-	-	-	-
OTHER SECTORS	-	-	-	-	-	31	90	-	-	-	-
Agriculture	-	-	-	-	-	-	15	-	-	-	-
Comm. and Publ. Services	-	-	-	-	-	-	56	-	-	-	-
Residential	-	-	-	-	-	-	-	-	-	-	-
Non-specified	-	-	-	-	-	31	19	-	-	-	-
NON-ENERGY USE	-	-	-	-	-	-	-	-	-	-	32

APPROVISIONNEMENT ET DEMANDE 1998	Pétrole cont. (1000 tonnes)										
	Gaz de raffinerie	GPL + éthane	Essence moteur	Essence aviation	Carbu- réacteurs	Kérosène	Gazole	Fioul lourd	Naphta	Coke de pétrole	Autres prod.
Production	15	10	120	-	40	31	219	95	-	-	32
Imports	-	-	-	3	-	-	10	-	-	-	-
Exports	-	-	-10	-	-	-	-30	-	-	-	-
Intl. Marine Bunkers	-	-	-	-	-	-	-	-	-	-	-
Stock Changes	-	-	-	-	-	-	-	-	-	-	-
DOMESTIC SUPPLY	15	10	110	3	40	31	199	95	-	-	32
Transfers and Stat. Diff.	-	-	-	-	-5	-	-2	-	-	-	-
TRANSFORMATION	-	-	-	-	-	-	1	-	-	-	-
Electricity and CHP Plants	-	-	-	-	-	-	1	-	-	-	-
Petroleum Refineries	-	-	-	-	-	-	-	-	-	-	-
Other Transform. Sector	-	-	-	-	-	-	-	-	-	-	-
ENERGY SECTOR	15	-	-	-	-	-	-	21	-	-	-
DISTRIBUTION LOSSES	-	-	-	-	-	-	-	-	-	-	-
FINAL CONSUMPTION	-	10	110	3	35	31	196	74	-	-	32
INDUSTRY SECTOR	-	10	-	-	-	-	36	74	-	-	-
Iron and Steel	-	-	-	-	-	-	-	-	-	-	-
Chemical and Petrochem.	-	-	-	-	-	-	-	-	-	-	-
Non-Metallic Minerals	-	-	-	-	-	-	-	-	-	-	-
Non-specified	-	10	-	-	-	-	36	74	-	-	-
TRANSPORT SECTOR	-	-	110	3	35	-	70	-	-	-	-
Air	-	-	-	3	35	-	-	-	-	-	-
Road	-	-	110	-	-	-	70	-	-	-	-
Non-specified	-	-	-	-	-	-	-	-	-	-	-
OTHER SECTORS	-	-	-	-	-	31	90	-	-	-	-
Agriculture	-	-	-	-	-	-	15	-	-	-	-
Comm. and Publ. Services	-	-	-	-	-	-	56	-	-	-	-
Residential	-	-	-	-	-	-	-	-	-	-	-
Non-specified	-	-	-	-	-	31	19	-	-	-	-
NON-ENERGY USE	-	-	-	-	-	-	-	-	-	-	32

Zambia / Zambie

SUPPLY AND CONSUMPTION 1997	Natural Gas	Gas Works	Coke Ovens	Blast Furnaces	Solid Biomass	Gas/Liquids from Biomass	Municipal Waste	Industrial Waste	Electricity (GWh)	Heat (TJ)
	Gas (TJ)				Comb. Renew. & Waste (TJ)				(GWh)	(TJ)
Production	-	-	-	283	199560	-	-	-	8006	-
Imports	-	-	-	-	-	-	-	-	-	-
Exports	-	-	-	-	-	-	-	-	-1500	-
Intl. Marine Bunkers	-	-	-	-	-	-	-	-	-	-
Stock Changes	-	-	-	-	-	-	-	-	-	-
DOMESTIC SUPPLY	-	-	-	283	199560	-	-	-	6506	-
Transfers and Stat. Diff.	-	-	-	-	-	-	-	-	-2	-
TRANSFORMATION	-	-	-	-	53788	-	-	-	-	-
Electricity and CHP Plants	-	-	-	-	-	-	-	-	-	-
Petroleum Refineries	-	-	-	-	-	-	-	-	-	-
Other Transform. Sector	-	-	-	-	53788	-	-	-	-	-
ENERGY SECTOR	-	-	-	-	-	-	-	-	278	-
DISTRIBUTION LOSSES	-	-	-	-	-	-	-	-	910	-
FINAL CONSUMPTION	-	-	-	283	145772	-	-	-	5316	-
INDUSTRY SECTOR	-	-	-	283	21827	-	-	-	4118	-
Iron and Steel	-	-	-	283	-	-	-	-	-	-
Chemical and Petrochem.	-	-	-	-	-	-	-	-	-	-
Non-Metallic Minerals	-	-	-	-	-	-	-	-	-	-
Non-specified	-	-	-	-	21827	-	-	-	4118	-
TRANSPORT SECTOR	-	-	-	-	-	-	-	-	-	-
Air	-	-	-	-	-	-	-	-	-	-
Road	-	-	-	-	-	-	-	-	-	-
Non-specified	-	-	-	-	-	-	-	-	-	-
OTHER SECTORS	-	-	-	-	123945	-	-	-	1198	-
Agriculture	-	-	-	-	-	-	-	-	206	-
Comm. and Publ. Services	-	-	-	-	-	-	-	-	402	-
Residential	-	-	-	-	123945	-	-	-	590	-
Non-specified	-	-	-	-	-	-	-	-	-	-
NON-ENERGY USE	-	-	-	-	-	-	-	-	-	-

APPROVISIONNEMENT ET DEMANDE 1998	Gaz naturel	Usines à gaz	Cokeries	Hauts fourneaux	Biomasse solide	Gaz/Liquides tirés de biomasse	Déchets urbains	Déchets industriels	Electricité (GWh)	Chaleur (TJ)
	Gaz (TJ)				En. Re. Comb. & Déchets (TJ)				(GWh)	(TJ)
Production	-	-	-	283	204262	-	-	-	7876	-
Imports	-	-	-	-	-	-	-	-	-	-
Exports	-	-	-	-	-	-	-	-	-1500	-
Intl. Marine Bunkers	-	-	-	-	-	-	-	-	-	-
Stock Changes	-	-	-	-	-	-	-	-	-	-
DOMESTIC SUPPLY	-	-	-	283	204262	-	-	-	6376	-
Transfers and Stat. Diff.	-	-	-	-	-	-	-	-	-	-
TRANSFORMATION	-	-	-	-	55056	-	-	-	-	-
Electricity and CHP Plants	-	-	-	-	-	-	-	-	-	-
Petroleum Refineries	-	-	-	-	-	-	-	-	-	-
Other Transform. Sector	-	-	-	-	55056	-	-	-	-	-
ENERGY SECTOR	-	-	-	-	-	-	-	-	273	-
DISTRIBUTION LOSSES	-	-	-	-	-	-	-	-	892	-
FINAL CONSUMPTION	-	-	-	283	149206	-	-	-	5211	-
INDUSTRY SECTOR	-	-	-	283	22341	-	-	-	4037	-
Iron and Steel	-	-	-	283	-	-	-	-	-	-
Chemical and Petrochem.	-	-	-	-	-	-	-	-	-	-
Non-Metallic Minerals	-	-	-	-	-	-	-	-	-	-
Non-specified	-	-	-	-	22341	-	-	-	4037	-
TRANSPORT SECTOR	-	-	-	-	-	-	-	-	-	-
Air	-	-	-	-	-	-	-	-	-	-
Road	-	-	-	-	-	-	-	-	-	-
Non-specified	-	-	-	-	-	-	-	-	-	-
OTHER SECTORS	-	-	-	-	126865	-	-	-	1174	-
Agriculture	-	-	-	-	-	-	-	-	202	-
Comm. and Publ. Services	-	-	-	-	-	-	-	-	394	-
Residential	-	-	-	-	126865	-	-	-	578	-
Non-specified	-	-	-	-	-	-	-	-	-	-
NON-ENERGY USE	-	-	-	-	-	-	-	-	-	-

Zimbabwe

SUPPLY AND CONSUMPTION 1997	Coal (1000 tonnes)							Oil (1000 tonnes)			
	Coking Coal	Other Bit. Coal	Sub-Bit. Coal	Lignite	Peat	Oven and Gas Coke	Pat. Fuel and BKB	Crude Oil	NGL	Feed-stocks	Additives
Production	571	3434	-	-	-	540	-	-	-	-	-
Imports	-	-	-	-	-	20	-	-	-	-	-
Exports	-	-112	-	-	-	-30	-	-	-	-	-
Intl. Marine Bunkers	-	-	-	-	-	-	-	-	-	-	-
Stock Changes	-	-	-	-	-	-	-	-	-	-	-
DOMESTIC SUPPLY	571	3322	-	-	-	530	-	-	-	-	-
Transfers and Stat. Diff.	-	-	-	-	-	-	-	-	-	-	-
TRANSFORMATION	571	2270	-	-	-	348	-	-	-	-	-
Electricity and CHP Plants	-	2270	-	-	-	-	-	-	-	-	-
Petroleum Refineries	-	-	-	-	-	-	-	-	-	-	-
Other Transform. Sector	571	-	-	-	-	348	-	-	-	-	-
ENERGY SECTOR	-	-	-	-	-	41	-	-	-	-	-
DISTRIBUTION LOSSES	-	-	-	-	-	-	-	-	-	-	-
FINAL CONSUMPTION	-	1052	-	-	-	141	-	-	-	-	-
INDUSTRY SECTOR	-	285	-	-	-	140	-	-	-	-	-
Iron and Steel	-	91	-	-	-	87	-	-	-	-	-
Chemical and Petrochem.	-	-	-	-	-	-	-	-	-	-	-
Non-Metallic Minerals	-	-	-	-	-	-	-	-	-	-	-
Non-specified	-	194	-	-	-	53	-	-	-	-	-
TRANSPORT SECTOR	-	31	-	-	-	1	-	-	-	-	-
Air	-	-	-	-	-	-	-	-	-	-	-
Road	-	-	-	-	-	-	-	-	-	-	-
Non-specified	-	31	-	-	-	1	-	-	-	-	-
OTHER SECTORS	-	736	-	-	-	-	-	-	-	-	-
Agriculture	-	489	-	-	-	-	-	-	-	-	-
Comm. and Publ. Services	-	236	-	-	-	-	-	-	-	-	-
Residential	-	11	-	-	-	-	-	-	-	-	-
Non-specified	-	-	-	-	-	-	-	-	-	-	-
NON-ENERGY USE	-	-	-	-	-	-	-	-	-	-	-

APPROVISIONNEMENT ET DEMANDE 1998	Charbon (1000 tonnes)							Pétrole (1000 tonnes)			
	Charbon à coke	Autres charb. bit.	Charbon sous-bit.	Lignite	Tourbe	Coke de four/gaz	Agg./briq. de lignite	Pétrole brut	LGN	Produits d'aliment.	Additifs
Production	571	3591	-	-	-	540	-	-	-	-	-
Imports	-	-	-	-	-	20	-	-	-	-	-
Exports	-	-172	-	-	-	-30	-	-	-	-	-
Intl. Marine Bunkers	-	-	-	-	-	-	-	-	-	-	-
Stock Changes	-	-	-	-	-	-	-	-	-	-	-
DOMESTIC SUPPLY	571	3419	-	-	-	530	-	-	-	-	-
Transfers and Stat. Diff.	-	1	-	-	-	-	-	-	-	-	-
TRANSFORMATION	571	2364	-	-	-	348	-	-	-	-	-
Electricity and CHP Plants	-	2364	-	-	-	-	-	-	-	-	-
Petroleum Refineries	-	-	-	-	-	-	-	-	-	-	-
Other Transform. Sector	571	-	-	-	-	348	-	-	-	-	-
ENERGY SECTOR	-	-	-	-	-	41	-	-	-	-	-
DISTRIBUTION LOSSES	-	-	-	-	-	-	-	-	-	-	-
FINAL CONSUMPTION	-	1056	-	-	-	141	-	-	-	-	-
INDUSTRY SECTOR	-	328	-	-	-	140	-	-	-	-	-
Iron and Steel	-	171	-	-	-	87	-	-	-	-	-
Chemical and Petrochem.	-	-	-	-	-	-	-	-	-	-	-
Non-Metallic Minerals	-	-	-	-	-	-	-	-	-	-	-
Non-specified	-	157	-	-	-	53	-	-	-	-	-
TRANSPORT SECTOR	-	25	-	-	-	1	-	-	-	-	-
Air	-	-	-	-	-	-	-	-	-	-	-
Road	-	-	-	-	-	-	-	-	-	-	-
Non-specified	-	25	-	-	-	1	-	-	-	-	-
OTHER SECTORS	-	703	-	-	-	-	-	-	-	-	-
Agriculture	-	503	-	-	-	-	-	-	-	-	-
Comm. and Publ. Services	-	191	-	-	-	-	-	-	-	-	-
Residential	-	9	-	-	-	-	-	-	-	-	-
Non-specified	-	-	-	-	-	-	-	-	-	-	-
NON-ENERGY USE	-	-	-	-	-	-	-	-	-	-	-

Zimbabwe

SUPPLY AND CONSUMPTION 1997	Oil cont. (1000 tonnes)										
	Refinery Gas	LPG + Ethane	Motor Gasoline	Aviation Gasoline	Jet Fuel	Kerosene	Gas/ Diesel	Heavy Fuel Oil	Naphtha	Petrol. Coke	Other Prod.
Production	-	-	-	-	-	-	-	-	-	-	-
Imports	-	10	450	5	180	42	750	-	-	-	26
Exports	-	-	-	-	-	-	-	-	-	-	-
Intl. Marine Bunkers	-	-	-	-	-	-	-	-	-	-	-
Stock Changes	-	-	-30	-	-	-	-	-	-	-	1
DOMESTIC SUPPLY	-	10	420	5	180	42	750	-	-	-	27
Transfers and Stat. Diff.	-	2	15	-	-67	-	-116	-	-	-	-
TRANSFORMATION	-	-	-	-	-	-	-	-	-	-	-
Electricity and CHP Plants	-	-	-	-	-	-	-	-	-	-	-
Petroleum Refineries	-	-	-	-	-	-	-	-	-	-	-
Other Transform. Sector	-	-	-	-	-	-	-	-	-	-	-
ENERGY SECTOR	-	-	-	-	-	-	20	-	-	-	-
DISTRIBUTION LOSSES	-	-	-	-	-	-	-	-	-	-	-
FINAL CONSUMPTION	-	12	435	5	113	42	614	-	-	-	27
INDUSTRY SECTOR	-	8	15	-	-	4	73	-	-	-	-
Iron and Steel	-	-	-	-	-	-	-	-	-	-	-
Chemical and Petrochem.	-	-	-	-	-	-	-	-	-	-	-
Non-Metallic Minerals	-	-	-	-	-	-	-	-	-	-	-
Non-specified	-	8	15	-	-	4	73	-	-	-	-
TRANSPORT SECTOR	-	-	350	5	113	-	302	-	-	-	-
Air	-	-	-	5	113	-	-	-	-	-	-
Road	-	-	350	-	-	-	302	-	-	-	-
Non-specified	-	-	-	-	-	-	-	-	-	-	-
OTHER SECTORS	-	4	70	-	-	38	239	-	-	-	-
Agriculture	-	-	10	-	-	2	85	-	-	-	-
Comm. and Publ. Services	-	-	-	-	-	-	-	-	-	-	-
Residential	-	2	-	-	-	25	-	-	-	-	-
Non-specified	-	2	60	-	-	11	154	-	-	-	-
NON-ENERGY USE	-	-	-	-	-	-	-	-	-	-	27

APPROVISIONNEMENT ET DEMANDE 1998	Pétrole cont. (1000 tonnes)										
	Gaz de raffinerie	GPL + éthane	Essence moteur	Essence aviation	Carbu- réacteurs	Kérosène	Gazole	Fioul lourd	Naphta	Coke de pétrole	Autres prod.
Production	-	-	-	-	-	-	-	-	-	-	-
Imports	-	10	450	5	180	42	750	-	-	-	26
Exports	-	-	-	-	-	-	-	-	-	-	-
Intl. Marine Bunkers	-	-	-	-	-	-	-	-	-	-	-
Stock Changes	-	-	-30	-	-	-	-	-	-	-	1
DOMESTIC SUPPLY	-	10	420	5	180	42	750	-	-	-	27
Transfers and Stat. Diff.	-	2	15	-	-67	-	-116	-	-	-	-
TRANSFORMATION	-	-	-	-	-	-	-	-	-	-	-
Electricity and CHP Plants	-	-	-	-	-	-	-	-	-	-	-
Petroleum Refineries	-	-	-	-	-	-	-	-	-	-	-
Other Transform. Sector	-	-	-	-	-	-	-	-	-	-	-
ENERGY SECTOR	-	-	-	-	-	-	20	-	-	-	-
DISTRIBUTION LOSSES	-	-	-	-	-	-	-	-	-	-	-
FINAL CONSUMPTION	-	12	435	5	113	42	614	-	-	-	27
INDUSTRY SECTOR	-	8	15	-	-	4	73	-	-	-	-
Iron and Steel	-	-	-	-	-	-	-	-	-	-	-
Chemical and Petrochem.	-	-	-	-	-	-	-	-	-	-	-
Non-Metallic Minerals	-	-	-	-	-	-	-	-	-	-	-
Non-specified	-	8	15	-	-	4	73	-	-	-	-
TRANSPORT SECTOR	-	-	350	5	113	-	302	-	-	-	-
Air	-	-	-	5	113	-	-	-	-	-	-
Road	-	-	350	-	-	-	302	-	-	-	-
Non-specified	-	-	-	-	-	-	-	-	-	-	-
OTHER SECTORS	-	4	70	-	-	38	239	-	-	-	-
Agriculture	-	-	10	-	-	2	85	-	-	-	-
Comm. and Publ. Services	-	-	-	-	-	-	-	-	-	-	-
Residential	-	2	-	-	-	25	-	-	-	-	-
Non-specified	-	2	60	-	-	11	154	-	-	-	-
NON-ENERGY USE	-	-	-	-	-	-	-	-	-	-	27

Zimbabwe

SUPPLY AND CONSUMPTION 1997	Gas (TJ)				Comb. Renew. & Waste (TJ)				(GWh)	(TJ)
	Natural Gas	Gas Works	Coke Ovens	Blast Furnaces	Solid Biomass	Gas/Liquids from Biomass	Municipal Waste	Industrial Waste	Electricity	Heat
Production	-	-	2500	3789	221434	-	-	-	7297	-
Imports	-	-	-	-	-	-	-	-	4013	-
Exports	-	-	-	-	-	-	-	-	-	-
Intl. Marine Bunkers	-	-	-	-	-	-	-	-	-	-
Stock Changes	-	-	-	-	-	-	-	-	-	-
DOMESTIC SUPPLY	-	-	**2500**	**3789**	**221434**	-	-	-	**11310**	-
Transfers and Stat. Diff.	-	-	-	-	-	-	-	-	415	-
TRANSFORMATION	-	-	-	-	-	-	-	-	-	-
Electricity and CHP Plants	-	-	-	-	-	-	-	-	-	-
Petroleum Refineries	-	-	-	-	-	-	-	-	-	-
Other Transform. Sector	-	-	-	-	-	-	-	-	-	-
ENERGY SECTOR	-	-	-	-	-	-	-	-	**210**	-
DISTRIBUTION LOSSES	-	-	-	-	-	-	-	-	**978**	-
FINAL CONSUMPTION	-	-	**2500**	**3789**	**221434**	-	-	-	**10537**	-
INDUSTRY SECTOR	-	-	**2500**	**3789**	**4101**	-	-	-	**5670**	-
Iron and Steel	-	-	2500	3789	-	-	-	-	-	-
Chemical and Petrochem.	-	-	-	-	-	-	-	-	-	-
Non-Metallic Minerals	-	-	-	-	-	-	-	-	-	-
Non-specified	-	-	-	-	4101	-	-	-	5670	-
TRANSPORT SECTOR	-	-	-	-	-	-	-	-	-	-
Air	-	-	-	-	-	-	-	-	-	-
Road	-	-	-	-	-	-	-	-	-	-
Non-specified	-	-	-	-	-	-	-	-	-	-
OTHER SECTORS	-	-	-	-	**217333**	-	-	-	**4867**	-
Agriculture	-	-	-	-	11277	-	-	-	1055	-
Comm. and Publ. Services	-	-	-	-	-	-	-	-	1821	-
Residential	-	-	-	-	206056	-	-	-	1991	-
Non-specified	-	-	-	-	-	-	-	-	-	-
NON-ENERGY USE	-	-	-	-	-	-	-	-	-	-

APPROVISIONNEMENT ET DEMANDE 1998	Gaz (TJ)				En. Re. Comb. & Déchets (TJ)				(GWh)	(TJ)
	Gaz naturel	Usines à gaz	Cokeries	Hauts fourneaux	Biomasse solide	Gaz/Liquides tirés de biomasse	Déchets urbains	Déchets industriels	Electricité	Chaleur
Production	-	-	2500	3789	225701	-	-	-	6608	-
Imports	-	-	-	-	-	-	-	-	5149	-
Exports	-	-	-	-	-	-	-	-	-	-
Intl. Marine Bunkers	-	-	-	-	-	-	-	-	-	-
Stock Changes	-	-	-	-	-	-	-	-	-	-
DOMESTIC SUPPLY	-	-	**2500**	**3789**	**225701**	-	-	-	**11757**	-
Transfers and Stat. Diff.	-	-	-	-	-	-	-	-	-15	-
TRANSFORMATION	-	-	-	-	-	-	-	-	-	-
Electricity and CHP Plants	-	-	-	-	-	-	-	-	-	-
Petroleum Refineries	-	-	-	-	-	-	-	-	-	-
Other Transform. Sector	-	-	-	-	-	-	-	-	-	-
ENERGY SECTOR	-	-	-	-	-	-	-	-	**144**	-
DISTRIBUTION LOSSES	-	-	-	-	-	-	-	-	**1124**	-
FINAL CONSUMPTION	-	-	**2500**	**3789**	**225701**	-	-	-	**10474**	-
INDUSTRY SECTOR	-	-	**2500**	**3789**	**4180**	-	-	-	**5384**	-
Iron and Steel	-	-	2500	3789	-	-	-	-	-	-
Chemical and Petrochem.	-	-	-	-	-	-	-	-	-	-
Non-Metallic Minerals	-	-	-	-	-	-	-	-	-	-
Non-specified	-	-	-	-	4180	-	-	-	5384	-
TRANSPORT SECTOR	-	-	-	-	-	-	-	-	-	-
Air	-	-	-	-	-	-	-	-	-	-
Road	-	-	-	-	-	-	-	-	-	-
Non-specified	-	-	-	-	-	-	-	-	-	-
OTHER SECTORS	-	-	-	-	**221521**	-	-	-	**5090**	-
Agriculture	-	-	-	-	11494	-	-	-	1165	-
Comm. and Publ. Services	-	-	-	-	-	-	-	-	1884	-
Residential	-	-	-	-	210027	-	-	-	2041	-
Non-specified	-	-	-	-	-	-	-	-	-	-
NON-ENERGY USE	-	-	-	-	-	-	-	-	-	-

SUMMARY TABLES

TABLEAUX RECAPITULATIFS

Production of Coking Coal (1000 tonnes)
Production de charbon à coke (1000 tonnes)
Erzeugung von Kokskohle (1000 Tonnen)
Produzione di carbone siderurgico (1000 tonnellate)
原料炭の生産量（千トン）
Producción de carbón coquizable (1000 toneladas)
Производство коксующихся углей (тыс. m)

	1971	1973	1978	1984	1985	1986	1987	1988	1989
Algérie	-	-	-	23	23	11	15	8	7
Afrique du Sud	-	-	9 718	10 803	11 142	12 115	10 830	11 187	9 799
Rép. Unie de Tanzanie	-	-	-	-	-	-	-	-	1
Zambie	-	-	115	56	50	50	50	65	65
Zimbabwe	-	-	946	245	270	300	373	237	200
Afrique	-	-	**10 779**	**11 127**	**11 485**	**12 476**	**11 268**	**11 497**	**10 072**
Brésil	-	-	1 317	1 303	1 407	1 330	991	1 229	1 052
Colombie	-	-	682	785	684	891	963	1 085	905
Amérique latine	-	-	**1 999**	**2 088**	**2 091**	**2 221**	**1 954**	**2 314**	**1 957**
Inde	-	-	22 520	36 500	35 650	39 530	41 010	42 720	44 430
RPD de Corée	-	-	4 958	5 179	5 558	6 063	6 947	7 831	8 210
Asie	-	-	**27 478**	**41 679**	**41 208**	**45 593**	**47 957**	**50 551**	**52 640**
Rép. populaire de Chine	-	-	52 604	72 034	75 338	82 769	91 060	92 943	99 511
Chine	-	-	**52 604**	**72 034**	**75 338**	**82 769**	**91 060**	**92 943**	**99 511**
Bulgarie	-	-	-	-	-	-	-	-	-
Roumanie	-	-	2 134	3 984	3 825	4 110	3 300	3 500	3 685
Europe non-OCDE	-	-	**2 134**	**3 984**	**3 825**	**4 110**	**3 300**	**3 500**	**3 685**
Kazakhstan	-	-	-	-	-	-	-	-	-
Russie	-	-	-	-	-	-	-	-	-
Tadjikistan	-	-	-	-	-	-	-	-	-
Ukraine	-	-	-	-	-	-	-	-	-
Ex-URSS	-	-	**139 250**	**131 985**	**134 985**	**137 551**	**142 985**	**144 985**	**143 985**
République Islamique d'Iran	-	-	377	320	324	280	230	230	150
Moyen-Orient	-	-	**377**	**320**	**324**	**280**	**230**	**230**	**150**
Total non-OCDE	-	-	**234 621**	**263 217**	**269 256**	**285 000**	**298 754**	**306 020**	**312 000**
OCDE Amérique du N.	-	-	109 066	118 641	117 547	107 828	107 094	124 185	127 413
OCDE Pacifique	-	-	48 943	54 670	60 173	58 549	61 203	62 170	64 582
OCDE Europe	-	-	100 223	79 988	81 266	81 857	75 122	72 136	69 680
Total OCDE	-	-	**258 232**	**253 299**	**258 986**	**248 234**	**243 419**	**258 491**	**261 675**
Monde	-	-	**492 853**	**516 516**	**528 242**	**533 234**	**542 173**	**564 511**	**573 675**

Production of Coking Coal (1000 tonnes)
Production de charbon à coke (1000 tonnes)
Erzeugung von Kokskohle (1000 Tonnen)
Produzione di carbone siderurgico (1000 tonnellate)
原料炭の生産量（千トン）
Producción de carbón coquizable (1000 toneladas)
Производство коксующихся углей (тыс. т)

1990	1991	1992	1993	1994	1995	1996	1997	1998	
-	-	-	-	-	-	-	-	-	Algeria
9 308	9 187	9 857	4 702	4 283	3 860	3 533	3 647	1 858	South Africa
1	1	1	1	1	1	1	1	1	United Rep. of Tanzania
60	47	54	40	24	25	24	24	24	Zambia
224	200	220	552	571	571	571	571	571	Zimbabwe
9 593	**9 435**	**10 132**	**5 295**	**4 879**	**4 457**	**4 129**	**4 243**	**2 454**	**Africa**
499	229	126	58	119	106	133	90	20	Brazil
657	835	1 036	1 703	1 701	1 743	1 776	1 671	1 458	Colombia
1 156	**1 064**	**1 162**	**1 761**	**1 820**	**1 849**	**1 909**	**1 761**	**1 478**	**Latin America**
45 300	46 280	45 360	45 060	41 970	40 100	40 540	44 760	44 810	India
8 589	8 842	8 968	9 094	9 031	8 968	8 892	8 287	7 873	DPR of Korea
53 889	**55 122**	**54 328**	**54 154**	**51 001**	**49 068**	**49 432**	**53 047**	**52 683**	**Asia**
110 076	112 208	114 162	120 929	139 479	183 964	184 558	192 969	156 281	People's Rep. of China
110 076	**112 208**	**114 162**	**120 929**	**139 479**	**183 964**	**184 558**	**192 969**	**156 281**	**China**
-	61	11	-	-	-	-	-	-	Bulgaria
1 482	523	1 027	539	444	349	312	324	192	Romania
1 482	**584**	**1 038**	**539**	**444**	**349**	**312**	**324**	**192**	**Non-OECD Europe**
-	-	20 855	19 222	15 491	11 107	10 986	10 526	8 718	Kazakhstan
-	-	70 900	62 419	56 582	60 616	55 276	52 368	51 996	Russia
-	-	-	-	3	7	-	-	-	Tajikistan
-	-	53 726	42 546	36 390	29 595	28 951	31 838	32 656	Ukraine
141 062	**100 000**	**145 481**	**124 187**	**108 466**	**101 325**	**95 213**	**94 732**	**93 370**	**Former USSR**
50	50	-	850	485	1 042	922	922	1 000	Islamic Republic of Iran
50	**50**	**-**	**850**	**485**	**1 042**	**922**	**922**	**1 000**	**Middle East**
317 308	**278 463**	**326 303**	**307 715**	**306 574**	**342 054**	**336 475**	**347 998**	**307 458**	**Non-OECD Total**
123 882	120 458	106 760	100 785	101 339	107 449	108 248	106 968	99 094	OECD North America
65 959	66 146	71 594	79 021	78 192	81 058	83 439	85 126	92 706	OECD Pacific
64 653	57 634	55 109	75 676	70 213	73 163	66 663	65 329	56 230	OECD Europe
254 494	**244 238**	**233 463**	**255 482**	**249 744**	**261 670**	**258 350**	**257 423**	**248 030**	**OECD Total**
571 802	**522 701**	**559 766**	**563 197**	**556 318**	**603 724**	**594 825**	**605 421**	**555 488**	**World**

Production of Other Bituminous Coal (1000 tonnes)
Production d'autres charbons bitumineux (1000 tonnes)
Erzeugung von sonstiger bituminöser Kohle (1000 Tonnen)
Produzione di altri carboni bituminosi (1000 tonnellate)
その他歴青炭の生産量（チトン）
Producción de otro carbón bituminoso (1000 toneladas)
Производство других битуминозных углей (тыс. т)

	1971	1973	1978	1984	1985	1986	1987	1988	1989
RD du Congo	-	-	107	121	121	119	122	123	125
Maroc	-	-	720	838	775	775	634	636	504
Mozambique	-	-	149	40	35	4	43	45	50
Nigéria	-	-	219	76	140	144	145	82	84
Afrique du Sud	-	-	80 640	152 097	162 358	164 585	165 770	170 213	166 501
Rép. Unie de Tanzanie	-	-	1	10	15	4	3	3	2
Zambie	-	-	500	455	461	507	413	559	332
Zimbabwe	-	-	1 567	2 473	2 835	3 247	4 490	4 328	4 391
Autre Afrique	-	-	480	668	754	795	908	929	969
Afrique	**-**	**-**	**84 383**	**156 778**	**167 494**	**170 180**	**172 528**	**176 918**	**172 958**
Argentine	-	-	434	509	400	364	373	512	514
Brésil	-	-	3 265	6 216	6 305	6 061	5 893	6 102	5 619
Chili	-	-	1 129	1 184	1 291	1 633	1 562	1 926	1 949
Colombie	-	-	3 984	5 847	8 082	9 658	12 401	13 885	17 330
Pérou	-	-	26	103	127	147	107	130	100
Vénézuela	-	-	84	51	40	57	62	1 090	2 125
Amérique latine	**-**	**-**	**8 922**	**13 910**	**16 245**	**17 920**	**20 398**	**23 645**	**27 637**
Inde	-	-	79 430	110 910	118 590	126 240	138 710	151 880	156 460
Indonésie	-	-	264	1 085	2 000	2 559	3 027	4 613	8 812
RPD de Corée	-	-	30 042	35 821	38 442	41 937	48 053	54 169	56 790
Myanmar	-	-	28	44	43	51	39	30	29
Pakistan	-	-	1 251	1 869	2 238	2 202	2 261	2 750	2 536
Philippines	-	-	255	1 216	1 261	1 239	1 169	1 336	1 328
Taipei chinois	-	-	2 884	2 011	1 858	1 725	1 499	1 225	784
Thailande	-	-	-	-	-	-	-	-	-
Viêt-Nam	-	-	6 000	5 000	5 594	6 122	6 332	6 059	5 136
Autre Asie	-	-	218	148	151	160	167	138	129
Asie	**-**	**-**	**120 372**	**158 104**	**170 177**	**182 235**	**201 257**	**222 200**	**232 004**
Rép. populaire de Chine	-	-	565 396	717 196	796 946	811 270	836 905	886 933	954 632
Chine	**-**	**-**	**565 396**	**717 196**	**796 946**	**811 270**	**836 905**	**886 933**	**954 632**
Bulgarie	-	-	273	223	223	207	198	196	193
Roumanie	-	-	5 284	4 474	4 832	4 586	5 799	5 642	4 615
Croatie	-	-							
RF de Yougoslavie	-	-							
Ex-Yougoslavie	-	-	471	388	400	407	379	363	293
Europe non-OCDE	**-**	**-**	**6 028**	**5 085**	**5 455**	**5 200**	**6 376**	**6 201**	**5 101**
Géorgie	-	-							
Kazakhstan	-	-							
Kirghizistan	-	-							
Russie	-	-							
Ukraine	-	-							
Ouzbékistan	-	-							
Ex-URSS	**-**	**-**	**417 850**	**424 015**	**434 015**	**450 449**	**452 015**	**454 015**	**433 015**
République Islamique d'Iran	-	-	523	921	928	982	1 010	1 030	1 050
Moyen-Orient	**-**	**-**	**523**	**921**	**928**	**982**	**1 010**	**1 030**	**1 050**
Total non-OCDE	**-**	**-**	**1 203 474**	**1 476 009**	**1 591 260**	**1 638 236**	**1 690 489**	**1 770 942**	**1 826 397**
OCDE Amérique du N.	-	-	400 185	510 223	484 544	492 266	509 679	498 964	515 764
OCDE Pacifique	-	-	52 198	78 922	87 342	103 503	109 538	94 534	100 710
OCDE Europe	-	-	375 676	315 468	358 467	368 086	364 202	360 078	340 276
Total OCDE	**-**	**-**	**828 059**	**904 613**	**930 353**	**963 855**	**983 419**	**953 576**	**956 750**
Monde	**-**	**-**	**2 031 533**	**2 380 622**	**2 521 613**	**2 602 091**	**2 673 908**	**2 724 518**	**2 783 147**

Production of Other Bituminous Coal (1000 tonnes)
Production d'autres charbons bitumineux (1000 tonnes)
Erzeugung von sonstiger bituminöser Kohle (1000 Tonnen)
Produzione di altri carboni bituminosi (1000 tonnellate)
その他歴青炭の生産量（チトン）
Producción de otro carbón bituminoso (1000 toneladas)
Производство других битуминозных углей (тыс. т)

1990	1991	1992	1993	1994	1995	1996	1997	1998	
126	128	128	92	93	93	93	93	88	DR of Congo
526	551	576	604	650	650	506	376	269	Morocco
40	42	40	40	40	38	20	18	18	Mozambique
90	138	100	120	130	140	140	140	59	Nigeria
165 492	169 013	164 543	183 512	191 522	202 351	202 829	216 426	222 942	South Africa
3	3	3	3	3	4	4	4	4	United Rep. of Tanzania
377	312	349	268	161	159	155	155	155	Zambia
4 733	5 400	5 328	4 733	4 334	4 072	3 840	3 434	3 591	Zimbabwe
1 126	1 095	1 173	1 112	1 254	1 160	1 067	1 067	1 067	Other Africa
172 513	**176 682**	**172 240**	**190 484**	**198 187**	**208 667**	**208 654**	**221 713**	**228 193**	**Africa**
276	292	202	168	347	305	310	251	290	Argentina
4 096	4 959	4 605	4 537	5 015	5 093	4 672	5 557	5 496	Brazil
2 184	2 208	1 626	1 355	1 182	1 038	1 004	1 044	941	Chile
20 718	19 059	20 781	19 436	20 898	23 908	28 223	31 064	32 213	Colombia
102	60	90	98	74	143	58	22	21	Peru
2 189	2 175	2 450	3 815	4 278	4 064	3 639	5 146	7 456	Venezuela
29 565	**28 753**	**29 754**	**29 409**	**31 794**	**34 551**	**37 906**	**43 084**	**46 417**	**Latin America**
166 430	183 000	192 900	200 980	211 760	230 030	245 090	251 040	253 090	India
10 486	13 715	22 357	29 328	32 275	41 145	50 157	54 877	61 206	Indonesia
59 411	61 158	62 032	62 906	62 469	62 032	61 508	57 326	54 459	DPR of Korea
40	44	30	9	35	12	9	10	10	Myanmar
2 745	3 054	3 099	3 267	3 534	3 043	3 638	3 553	3 159	Pakistan
1 243	1 262	1 661	1 675	1 231	1 120	976	930	998	Philippines
472	403	335	328	285	235	147	99	79	Chinese Taipei
-	-	22	16	12	5	3	-	-	Thailand
5 130	5 204	5 232	5 575	6 157	8 350	9 774	11 344	10 772	Vietnam
107	96	10	9	8	7	6	6	6	Other Asia
246 064	**267 936**	**287 678**	**304 093**	**317 766**	**345 979**	**371 308**	**379 185**	**383 779**	**Asia**
969 807	975 198	1 002 218	1 029 742	1 100 423	1 176 767	1 212 141	1 179 851	1 057 031	People's Rep. of China
969 807	**975 198**	**1 002 218**	**1 029 742**	**1 100 423**	**1 176 767**	**1 212 141**	**1 179 851**	**1 057 031**	**China**
143	67	237	263	173	194	138	102	92	Bulgaria
2 964	3 313	3 071	685	921	10	10	10	2	Romania
-	-	120	105	96	75	64	49	51	*Croatia*
-	-	102	73	82	71	78	93	105	*FR of Yugoslavia*
292	260	222	178	178	146	142	142	156	Former Yugoslavia
3 399	**3 640**	**3 530**	**1 126**	**1 272**	**350**	**290**	**254**	**250**	**Non-OECD Europe**
-	-	200	82	44	43	23	5	14	Georgia
-	-	101 529	87 989	84 320	68 508	62 254	59 648	59 340	Kazakhstan
-	-	1 040	736	746	463	370	500	400	Kyrgyzstan
-	-	139 532	130 717	120 172	116 302	111 240	107 439	101 088	Russia
-	-	74 148	69 054	54 910	53 933	45 181	43 676	43 111	Ukraine
-	-	200	152	200	74	74	59	68	Uzbekistan
401 938	**397 968**	**316 649**	**288 730**	**260 392**	**239 323**	**219 142**	**211 327**	**204 021**	**Former USSR**
1 250	1 350	1 500	673	415	156	160	165	169	Islamic Republic of Iran
1 250	**1 350**	**1 500**	**673**	**415**	**156**	**160**	**165**	**169**	**Middle East**
1 824 536	**1 851 527**	**1 813 569**	**1 844 257**	**1 910 249**	**2 005 793**	**2 049 601**	**2 035 579**	**1 919 860**	**Non-OECD Total**
548 800	515 205	521 913	463 300	522 457	493 888	510 017	528 297	523 785	OECD North America
102 238	105 026	106 044	97 775	93 721	103 564	102 240	110 720	112 613	OECD Pacific
302 374	295 158	278 660	229 996	211 618	215 236	212 350	208 494	180 046	OECD Europe
953 412	**915 389**	**906 617**	**791 071**	**827 796**	**812 688**	**824 607**	**847 511**	**816 444**	**OECD Total**
2 777 948	**2 766 916**	**2 720 186**	**2 635 328**	**2 738 045**	**2 818 481**	**2 874 208**	**2 883 090**	**2 736 304**	**World**

Production of Sub-Bituminous Coal (1000 tonnes)
Production de charbons sous-bitumineux (1000 tonnes)
Erzeugung von subbituminöser Kohle (1000 Tonnen)
Produzione di carbone sub-bituminoso (1000 tonnellate)
亜歴青炭の生産量（チトン）
Producción de carbón sub-bituminoso (1000 toneladas)
Производство полубитуминозных углей (тыс. т)

	1971	1973	1978	1984	1985	1986	1987	1988	1989
RPD de Corée	-	-	9 000	12 000	13 000	14 000	15 000	18 000	20 000
Malaisie	-	-	-	-	-	-	-	21	108
Asie	**-**	**-**	**9 000**	**12 000**	**13 000**	**14 000**	**15 000**	**18 021**	**20 108**
Bulgarie	-	-	-	-	-	-	-	-	-
Roumanie	-	-	-	-	-	-	-	-	-
Slovénie	-	-	-	-	-	-	-	-	-
Ex-Yougoslavie	-	-	-	-	-	-	-	-	-
Europe non-OCDE	**-**	**-**	**-**	**-**	**-**	**-**	**-**	**-**	**-**
Tadjikistan	-	-	-	-	-	-	-	-	-
Ex-URSS	**-**	**-**	**-**	**-**	**-**	**-**	**-**	**-**	**-**
Total non-OCDE	**-**	**-**	**9 000**	**12 000**	**13 000**	**14 000**	**15 000**	**18 021**	**20 108**
OCDE Amérique du N.	-	-	96 054	179 232	193 127	192 711	202 971	225 760	233 793
OCDE Pacifique	-	-	7 387	11 316	11 192	13 773	16 630	15 851	16 044
OCDE Europe	-	-	92 684	104 496	101 667	102 033	100 572	97 604	92 382
Total OCDE	**-**	**-**	**196 125**	**295 044**	**305 986**	**308 517**	**320 173**	**339 215**	**342 219**
Monde	**-**	**-**	**205 125**	**307 044**	**318 986**	**322 517**	**335 173**	**357 236**	**362 327**

Production of Sub-Bituminous Coal (1000 tonnes)
Production de charbons sous-bitumineux (1000 tonnes)
Erzeugung von subbituminöser Kohle (1000 Tonnen)
Produzione di carbone sub-bituminoso (1000 tonnellate)
亜歴青炭の生産量（千トン）
Producción de carbón sub-bituminoso (1000 toneladas)
Производство полубитуминозных углей (тыс. т)

1990	1991	1992	1993	1994	1995	1996	1997	1998	
22 000	23 000	24 000	27 000	26 500	26 000	25 500	23 766	22 578	DPR of Korea
100	180	76	121	127	121	218	218	315	Malaysia
22 100	**23 180**	**24 076**	**27 121**	**26 627**	**26 121**	**25 718**	**23 984**	**22 893**	**Asia**
-	3 092	3 352	3 419	3 155	3 187	3 060	-	-	Bulgaria
-	-	-	-	-	789	1 001	1 418	1 212	Romania
-	-	1 323	1 201	1 079	967	830	648	698	*Slovenia*
-	-	1 323	1 201	1 079	967	830	648	698	Former Yugoslavia
-	**3 092**	**4 675**	**4 620**	**4 234**	**4 943**	**4 891**	**2 066**	**1 910**	**Non-OECD Europe**
-	-	-	-	103	27	20	20	20	Tajikistan
-	-	-	-	**103**	**27**	**20**	**20**	**20**	**Former USSR**
22 100	**26 272**	**28 751**	**31 741**	**30 964**	**31 091**	**30 629**	**26 070**	**24 823**	**Non-OECD Total**
245 931	257 043	256 273	277 952	304 914	330 852	342 286	353 003	386 721	OECD North America
18 544	19 086	19 828	19 745	21 854	21 668	22 523	22 894	24 701	OECD Pacific
84 242	80 804	72 212	70 966	63 972	61 339	63 781	61 706	55 526	OECD Europe
348 717	**356 933**	**348 313**	**368 663**	**390 740**	**413 859**	**428 590**	**437 603**	**466 948**	**OECD Total**
370 817	**383 205**	**377 064**	**400 404**	**421 704**	**444 950**	**459 219**	**463 673**	**491 771**	**World**

Production of Lignite (1000 tonnes)
Production de lignite (1000 tonnes)
Erzeugung von Braunkohle (1000 Tonnen)

Produzione di lignite (1000 tonnellate)

亜炭の生産量（千トン）

Producción de lignito (1000 toneladas)

Производство лигнита (тыс. т)

	1971	1973	1978	1984	1985	1986	1987	1988	1989
Chili	-	-	30	40	35	35	35	35	38
Amérique latine	-	-	**30**	**40**	**35**	**35**	**35**	**35**	**38**
Inde	-	-	3 270	7 110	7 776	9 600	11 270	12 590	12 360
Myanmar	-	-	3	43	43	47	35	35	35
Philippines	-	-	-	3	4	4	3	3	3
Thaïlande	-	-	684	2 493	5 188	5 476	6 901	7 259	8 901
Asie	-	-	**3 957**	**9 649**	**13 011**	**15 127**	**18 209**	**19 887**	**21 299**
Albanie	-	-	1 200	2 010	2 150	2 167	2 134	2 184	2 193
Bulgarie	-	-	25 531	33 433	30 657	35 016	36 621	33 951	34 105
Roumanie	-	-	21 845	35 822	37 924	38 824	42 425	49 612	53 043
République slovaque	-	-	5 804	5 785	5 751	5 374	5 589	5 639	5 269
Bosnie-Herzegovine	-	-	-	-	-	-	-	-	-
Croatie	-	-	-	-	-	-	-	-	-
Ex-RYM	-	-	-	-	-	-	-	-	-
Slovénie	-	-	-	-	-	-	-	-	-
RF de Yougoslavie	-	-	-	-	-	-	-	-	-
Ex-Yougoslavie	-	-	39 021	64 684	68 072	68 381	70 754	70 498	70 861
Europe non-OCDE	-	-	**93 401**	**141 734**	**144 554**	**149 762**	**157 523**	**161 884**	**165 471**
Estonie	-	-	-	-	-	-	-	-	-
Kazakhstan	-	-	-	-	-	-	-	-	-
Kirghizistan	-	-	-	-	-	-	-	-	-
Russie	-	-	-	-	-	-	-	-	-
Tadjikistan	-	-	-	-	-	-	-	-	-
Ukraine	-	-	-	-	-	-	-	-	-
Ouzbékistan	-	-	-	-	-	-	-	-	-
Ex-URSS	-	-	**166 500**	**156 000**	**157 000**	**163 000**	**165 000**	**172 000**	**164 000**
Israël	-	-	-	-	-	-	-	-	-
Moyen-Orient	-	-	**-**	**-**	**-**	**-**	**-**	**-**	**-**
Total non-OCDE	-	-	**263 888**	**307 423**	**314 600**	**327 924**	**340 767**	**353 806**	**350 808**
OCDE Amérique du N.	-	-	36 227	67 134	72 451	77 548	81 166	89 350	89 222
OCDE Pacifique	-	-	30 661	33 425	38 627	36 270	41 890	43 572	48 448
OCDE Europe	-	-	492 686	582 376	611 153	620 046	623 456	615 766	626 827
Total OCDE	-	-	**559 574**	**682 935**	**722 231**	**733 864**	**746 512**	**748 688**	**764 497**
Monde	-	-	**823 462**	**990 358**	**1 036 831**	**1 061 788**	**1 087 279**	**1 102 494**	**1 115 305**

Production of Lignite (1000 tonnes)
Production de lignite (1000 tonnes)
Erzeugung von Braunkohle (1000 Tonnen)
Produzione di lignite (1000 tonnellate)
亜炭の生産量（チトン）
Producción de lignito (1000 toneladas)
Производство лигнита (тыс. т)

1990	1991	1992	1993	1994	1995	1996	1997	1998	
40	14	-	-	-	-	-	-	-	Chile
40	**14**	**-**	**-**	**-**	**-**	**-**	**-**	**-**	**Latin America**
14 100	15 970	16 620	18 100	19 340	22 140	22 540	23 500	23 170	India
38	39	39	25	22	23	22	20	21	Myanmar
3	3	3	3	3	3	3	3	3	Philippines
12 421	14 689	15 335	15 530	17 083	18 416	21 474	23 393	20 163	Thailand
26 562	**30 701**	**31 997**	**33 658**	**36 448**	**40 582**	**44 039**	**46 916**	**43 357**	**Asia**
2 071	1 087	800	600	169	163	101	70	49	Albania
31 544	25 231	26 735	25 350	25 429	27 449	28 104	29 606	30 019	Bulgaria
33 737	28 578	34 272	38 527	39 182	39 973	40 546	32 055	24 825	Romania
4 766	4 119	3 497	3 547	3 634	3 759	3 829	3 915	3 951	Slovak Republic
-	-	15 000	13 000	1 400	1 640	1 690	1 741	1 794	*Bosnia-Herzegovina*
-	-	-	11	7	7	2	-	-	*Croatia*
-	-	6 978	6 917	6 860	7 249	7 145	6 700	8 176	*FYROM*
-	-	4 251	3 920	3 775	3 917	3 937	4 305	4 194	*Slovenia*
-	-	40 003	37 360	38 269	39 939	38 367	40 563	43 967	*FR of Yugoslavia*
75 556	71 066	66 232	61 208	50 311	52 752	51 141	53 309	58 131	Former Yugoslavia
147 674	**130 081**	**131 536**	**129 232**	**118 725**	**124 096**	**123 721**	**118 955**	**116 975**	**Non-OECD Europe**
-	-	16 976	12 498	12 192	11 011	12 297	11 895	10 774	Estonia
-	-	4 152	4 669	4 814	3 740	3 591	2 473	1 715	Kazakhstan
-	-	1 111	985	550	280	280	280	280	Kyrgyzstan
-	-	126 832	112 778	95 292	85 984	90 179	85 224	78 835	Russia
-	-	200	200	-	-	-	-	-	Tajikistan
-	-	5 779	4 100	3 100	2 296	1 587	1 436	1 409	Ukraine
-	-	4 481	3 655	3 600	2 980	2 763	2 888	2 855	Uzbekistan
160 000	**149 500**	**159 531**	**138 885**	**119 548**	**106 291**	**110 697**	**104 196**	**95 868**	**Former USSR**
303	336	355	399	381	470	421	472	444	Israel
303	**336**	**355**	**399**	**381**	**470**	**421**	**472**	**444**	**Middle East**
334 579	**310 632**	**323 419**	**302 174**	**275 102**	**271 439**	**278 878**	**270 539**	**256 644**	**Non-OECD Total**
89 321	87 465	91 730	91 282	90 591	89 211	90 722	90 054	90 383	OECD North America
46 151	49 559	50 902	47 824	49 004	50 945	53 926	58 387	65 883	OECD Pacific
561 121	482 837	446 901	422 666	410 565	394 885	391 610	382 881	381 248	OECD Europe
696 593	**619 861**	**589 533**	**561 772**	**550 160**	**535 041**	**536 258**	**531 322**	**537 514**	**OECD Total**
1 031 172	**930 493**	**912 952**	**863 946**	**825 262**	**806 480**	**815 136**	**801 861**	**794 158**	**World**

Production of Peat (1000 tonnes)
Production de tourbe (1000 tonnes)
Erzeugung von Torf (1000 Tonnen)

Produzione di torba (1000 tonnellate)

泥炭の生産量（千トン）

Producción de turba (1000 toneladas)

Производство торфа (тыс. т)

	1971	1973	1978	1984	1985	1986	1987	1988	1989
Autre Afrique	1	1	2	14	10	12	18	12	14
Afrique	**1**	**1**	**2**	**14**	**10**	**12**	**18**	**12**	**14**
Roumanie	-	-	-	-	-	-	-	-	-
Europe non-OCDE	**-**	**-**	**-**	**-**	**-**	**-**	**-**	**-**	**-**
Bélarus	-	-	-	-	-	-	-	-	-
Estonie	-	-	-	-	-	-	-	-	-
Lettonie	-	-	-	-	-	-	-	-	-
Lituanie	-	-	-	-	-	-	-	-	-
Russie	-	-	-	-	-	-	-	-	-
Ukraine	-	-	-	-	-	-	-	-	-
Ex-URSS	**61 808**	**66 549**	**31 184**	**22 193**	**22 762**	**27 314**	**17 754**	**25 038**	**24 128**
Total non-OCDE	**61 809**	**66 550**	**31 186**	**22 207**	**22 772**	**27 326**	**17 772**	**25 050**	**24 142**
OCDE Amérique du N.	-	-	-	-	-	-	-	-	-
OCDE Europe	7 635	6 337	8 050	12 103	8 276	15 558	12 334	12 150	18 061
Total OCDE	**7 635**	**6 337**	**8 050**	**12 103**	**8 276**	**15 558**	**12 334**	**12 150**	**18 061**
Monde	**69 444**	**72 887**	**39 236**	**34 310**	**31 048**	**42 884**	**30 106**	**37 200**	**42 203**

Production of Peat (1000 tonnes)
Production de tourbe (1000 tonnes)
Erzeugung von Torf (1000 Tonnen)
Produzione di torba (1000 tonnellate)
泥炭の生産量（千トン）
Producción de turba (1000 toneladas)
Производство торфа (тыс. т)

1990	1991	1992	1993	1994	1995	1996	1997	1998	
11	10	12	12	12	12	12	12	12	Other Africa
11	**10**	**12**	**12**	**12**	**12**	**12**	**12**	**12**	**Africa**
-	-	-	26	19	6	2	-	4	Romania
-	**-**	**-**	**26**	**19**	**6**	**2**	**-**	**4**	**Non-OECD Europe**
-	-	6 262	2 909	3 477	3 133	2 846	2 763	2 993	Belarus
-	-	797	465	645	583	700	571	157	Estonia
-	-	616	417	637	346	389	391	59	Latvia
-	-	109	52	91	62	77	89	62	Lithuania
-	-	5 533	7 651	2 928	4 401	4 103	3 363	1 767	Russia
-	-	2 990	2 990	2 096	1 577	1 026	770	687	Ukraine
17 641	**16 719**	**16 307**	**14 484**	**9 874**	**10 102**	**9 141**	**7 947**	**5 725**	**Former USSR**
17 652	**16 729**	**16 319**	**14 522**	**9 905**	**10 120**	**9 155**	**7 959**	**5 741**	**Non-OECD Total**
-	-	11	24	-	-	-	-	-	OECD North America
14 924	10 712	12 607	10 611	14 771	17 151	15 470	14 634	6 459	OECD Europe
14 924	**10 712**	**12 618**	**10 635**	**14 771**	**17 151**	**15 470**	**14 634**	**6 459**	**OECD Total**
32 576	**27 441**	**28 937**	**25 157**	**24 676**	**27 271**	**24 625**	**22 593**	**12 200**	**World**

Production of Crude Oil, NGL and Additives (1000 tonnes)
Production de pétrole brut, LGN et additifs (1000 tonnes)
Erzeugung von Rohöl, Kondensaten und Additiven (1000 Tonnen)
Produzione petrolio grezzo, LNG e additivi (1000 tonnellate)
原油、ＮＧＬ等随伴物の生産量（チトン）
Producción de petróleo crudo, liquidos de gas natural y aditivos (1000 toneladas)
Производство сырой нефти, газовых конденсатов и присадок (тыс. т)

	1971	1973	1978	1984	1985	1986	1987	1988	1989
Algérie	38 000	51 118	58 021	49 050	50 421	51 851	52 432	53 043	54 101
Angola	5 721	8 154	6 469	10 078	11 430	13 894	17 705	22 270	22 297
Bénin	-	-	-	5	325	365	312	238	191
Cameroun	-	-	505	9 340	8 795	8 651	8 716	8 064	8 536
Congo	14	2 091	2 570	6 025	5 937	5 951	6 317	7 038	7 969
RD du Congo	-	-	1 145	1 375	1 672	1 621	1 561	1 495	1 497
Côte d'Ivoire	-	-	-	1 003	1 008	945	793	523	104
Egypte	14 962	8 504	25 193	41 611	44 825	40 915	46 002	44 000	44 287
Gabon	5 785	7 598	10 355	7 810	8 520	8 172	7 666	7 790	10 137
Ghana	-	-	-	33	14	-	-	-	-
Libye	132 535	106 172	97 061	53 499	51 461	50 827	49 629	52 659	53 980
Maroc	23	42	24	16	22	24	18	20	13
Nigéria	75 524	101 412	93 550	68 741	74 031	72 435	66 288	73 237	86 658
Sénégal	-	-	-	-	-	-	-	3	2
Afrique du Sud	-	-	-	-	-	-	-	-	-
Soudan	-	-	-	-	-	-	-	-	-
Tunisie	4 109	3 878	4 944	5 480	5 408	5 249	5 038	4 997	5 020
Autre Afrique	-	-	-	-	-	-	-	-	-
Afrique	**276 673**	**288 969**	**299 837**	**254 066**	**263 869**	**260 900**	**262 477**	**275 377**	**294 792**
Argentine	21 918	21 938	23 549	25 850	24 805	23 686	23 450	24 541	25 074
Bolivie	1 728	2 216	1 555	1 249	1 188	1 162	1 186	1 142	1 203
Brésil	8 313	8 296	8 002	23 216	27 493	28 784	28 463	28 567	30 728
Chili	1 723	1 748	1 017	2 042	1 884	1 776	1 712	1 342	1 219
Colombie	11 376	9 758	6 977	8 944	9 392	15 886	20 328	19 534	20 999
Cuba	137	157	330	770	876	947	903	724	726
Equateur	196	10 631	10 487	13 358	14 401	14 820	8 901	15 310	14 392
Guatemala	-	-	3	235	146	247	183	184	182
Pérou	3 140	3 567	7 640	9 432	9 547	8 997	8 156	7 071	6 486
Trinité-et-Tobago	6 468	8 291	11 446	8 749	9 086	8 713	7 960	7 798	7 699
Vénézuela	186 632	178 351	114 843	96 856	89 442	96 058	97 603	101 477	101 124
Autre Amérique latine	-	2	67	218	250	276	284	344	368
Amérique latine	**241 631**	**244 955**	**185 916**	**190 919**	**188 510**	**201 352**	**199 129**	**208 034**	**210 200**
Bangladesh	-	5	6	-	80	92	110	115	119
Brunei	6 403	11 369	11 796	9 513	8 620	8 453	7 474	7 224	7 603
Inde	7 185	7 198	11 271	28 194	30 302	31 795	30 920	32 672	34 967
Indonésie	44 071	66 112	79 272	73 650	65 565	70 531	70 086	67 695	72 135
Malaisie	3 276	4 340	11 301	22 982	22 961	25 851	25 605	27 862	29 978
Myanmar	872	981	1 353	1 586	1 452	1 431	980	733	732
Pakistan	415	419	479	660	1 284	1 931	2 018	2 152	2 251
Philippines	-	-	-	540	368	433	267	290	262
Taipei chinois	111	148	218	244	227	200	226	256	140
Thailande	13	6	11	1 111	2 171	2 221	2 072	2 348	2 169
Viêt-Nam	-	-	-	-	-	-	-	681	1 490
Autre Asie	1	2	1	7	8	7	6	3	4
Asie	**62 347**	**90 580**	**115 708**	**138 487**	**133 038**	**142 945**	**139 764**	**142 031**	**151 850**
Rép. populaire de Chine	39 410	53 610	104 050	114 613	124 895	130 688	134 140	137 046	137 654
Chine	**39 410**	**53 610**	**104 050**	**114 613**	**124 895**	**130 688**	**134 140**	**137 046**	**137 654**

Production of Crude Oil, NGL and Additives (1000 tonnes)
Production de pétrole brut, LGN et additifs (1000 tonnes)
Erzeugung von Rohöl, Kondensaten und Additiven (1000 Tonnen)
Produzione petrolio grezzo, LNG e additivi (1000 tonnellate)
原油、ＮＧＬ等随伴物の生産量（千トン）
Producción de petróleo crudo, liquidos de gas natural y aditivos (1000 toneladas)
Производство сырой нефти, газовых конденсатов и присадок (тыс. т)

1990	1991	1992	1993	1994	1995	1996	1997	1998	
56 711	56 005	55 900	55 994	55 391	55 747	58 613	59 656	61 165	Algeria
23 337	24 479	27 122	24 849	27 116	30 371	33 548	35 149	36 382	Angola
206	197	136	154	129	95	72	66	52	Benin
7 930	7 285	7 090	7 148	6 659	6 208	6 843	7 115	7 822	Cameroon
8 029	8 054	8 654	9 537	9 562	9 268	10 370	12 495	13 080	Congo
1 410	1 372	1 238	1 280	1 345	1 371	1 371	1 306	1 232	DR of Congo
91	60	293	324	335	345	797	722	521	Cote d'Ivoire
45 499	45 418	45 869	47 527	46 497	46 588	45 080	43 796	42 962	Egypt
13 352	14 648	14 489	15 481	14 732	18 131	18 277	18 462	17 564	Gabon
-	-	-	12	-	-	-	-	-	Ghana
66 198	74 207	70 427	66 976	67 780	69 025	68 917	71 085	69 333	Libya
15	12	11	10	8	5	5	12	12	Morocco
88 322	92 144	95 575	96 509	96 216	97 541	108 031	114 878	106 471	Nigeria
1	1	1	1	1	1	1	1	2	Senegal
-	-	-	393	393	393	393	393	290	South Africa
-	-	-	-	-	-	102	254	328	Sudan
4 571	5 269	5 279	5 020	4 450	4 352	4 291	3 896	4 011	Tunisia
-	-	-	221	242	339	857	2 167	2 167	Other Africa
315 672	**329 151**	**332 084**	**331 436**	**330 856**	**339 780**	**357 568**	**371 453**	**363 394**	**Africa**
26 181	26 689	30 053	32 157	35 934	38 702	42 080	44 667	45 319	Argentina
1 311	1 375	1 263	1 312	1 510	1 693	1 736	1 980	2 094	Bolivia
32 662	32 043	32 575	33 766	34 155	35 626	40 520	45 154	50 895	Brazil
1 076	988	843	826	822	698	543	426	406	Chile
22 790	22 049	22 887	23 593	23 731	30 433	32 641	33 771	38 862	Colombia
671	527	882	1 108	1 299	1 471	1 476	1 462	1 678	Cuba
14 895	15 674	16 789	18 104	19 974	20 160	20 096	21 334	20 801	Ecuador
197	185	281	337	366	468	795	1 065	1 389	Guatemala
6 393	5 724	5 772	6 295	6 345	6 072	6 000	5 895	5 796	Peru
7 797	7 385	7 397	6 720	7 098	7 136	7 070	6 758	6 699	Trinidad and Tobago
113 372	128 898	129 307	135 814	141 121	154 664	175 316	185 212	179 793	Venezuela
375	425	431	432	432	436	393	393	393	Other Latin America
227 720	**241 962**	**248 480**	**260 464**	**272 787**	**297 559**	**328 666**	**348 117**	**354 125**	**Latin America**
93	65	67	72	16	10	1	1	1	Bangladesh
7 712	8 269	9 258	8 965	9 242	9 076	8 588	8 540	8 273	Brunei
34 792	33 120	30 195	28 951	33 213	37 669	36 309	37 365	36 873	India
73 387	80 318	76 163	76 129	76 268	75 517	77 043	68 165	67 201	Indonesia
31 698	32 954	33 940	33 841	35 434	38 259	38 498	38 536	38 825	Malaysia
722	760	723	675	681	474	402	396	383	Myanmar
2 721	3 359	3 219	3 084	2 883	2 813	3 009	3 031	2 939	Pakistan
237	160	404	471	262	146	45	24	41	Philippines
178	109	66	61	63	57	55	48	50	Chinese Taipei
2 618	3 028	3 401	3 496	4 051	3 913	4 962	5 400	5 642	Thailand
2 701	3 921	5 501	6 312	6 901	7 652	8 705	10 909	12 400	Vietnam
4 504	5 003	5 300	5 400	5 500	5 000	4 000	4 000	4 000	Other Asia
161 363	**171 066**	**168 237**	**167 457**	**174 514**	**180 586**	**181 617**	**176 415**	**176 628**	**Asia**
138 306	140 992	141 747	145 174	146 082	150 044	157 334	160 741	161 000	People's Rep. of China
138 306	**140 992**	**141 747**	**145 174**	**146 082**	**150 044**	**157 334**	**160 741**	**161 000**	**China**

Production of Crude Oil, NGL and Additives (1000 tonnes)
Production de pétrole brut, LGN et additifs (1000 tonnes)
Erzeugung von Rohöl, Kondensaten und Additiven (1000 Tonnen)

Produzione petrolio grezzo, LNG e additivi (1000 tonnellate)

原油、ＮＧＬ等随伴物の生産量（チトン）

Producción de petróleo crudo, liquidos de gas natural y aditivos (1000 toneladas)

Производство сырой нефти, газовых конденсатов и присадок (тыс. т)

	1971	1973	1978	1984	1985	1986	1987	1988	1989
Albanie	1 657	2 107	2 200	1 388	1 188	1 205	1 182	1 167	1 226
Bulgarie	305	190	246	150	200	100	90	80	80
Roumanie	13 793	14 287	13 724	11 453	10 718	10 125	9 504	9 389	9 173
République slovaque	163	130	64	35	66	87	96	98	99
Croatie	-	-	-	-	-	-	-	-	-
Slovénie	-	-	-	-	-	-	-	-	-
RF de Yougoslavie	-	-	-	-	-	-	-	-	-
Ex-Yougoslavie	2 961	3 332	4 076	4 059	4 152	4 152	3 864	3 684	3 396
Europe non-OCDE	**18 879**	**20 046**	**20 310**	**17 085**	**16 324**	**15 669**	**14 736**	**14 418**	**13 974**
Azerbaïdjan	-	-	-	-	-	-	-	-	-
Bélarus	-	-	-	-	-	-	-	-	-
Géorgie	-	-	-	-	-	-	-	-	-
Kazakhstan	-	-	-	-	-	-	-	-	-
Kirghizistan	-	-	-	-	-	-	-	-	-
Lituanie	-	-	-	-	-	-	-	-	-
Russie	-	-	-	-	-	-	-	-	-
Tadjikistan	-	-	-	-	-	-	-	-	-
Turkménistan	-	-	-	-	-	-	-	-	-
Ukraine	-	-	-	-	-	-	-	-	-
Ouzbékistan	-	-	-	-	-	-	-	-	-
Ex-URSS	**377 100**	**429 100**	**567 500**	**612 700**	**595 291**	**614 753**	**624 177**	**624 324**	**607 254**
Bahrein	3 739	3 411	2 758	2 375	2 355	2 395	2 368	2 456	2 442
République Islamique d'Iran	226 946	293 159	263 734	109 100	111 750	91 500	116 110	112 244	143 110
Irak	83 266	99 542	127 080	59 246	70 656	83 976	103 694	129 356	139 399
Israël	5 738	6 005	24	9	9	12	13	18	16
Jordanie	-	-	-	-	-	15	21	16	
Koweit	152 396	150 565	107 890	59 396	54 166	57 050	61 857	73 210	87 536
Oman	14 598	14 598	16 021	21 175	25 333	28 580	29 732	31 698	32 696
Qatar	20 645	27 502	23 382	20 526	15 623	16 399	14 919	16 331	19 322
Arabie saoudite	238 703	380 179	424 553	244 700	182 100	260 600	222 400	268 500	267 400
Syrie	5 289	5 543	9 924	9 417	9 236	9 267	12 364	14 164	17 500
Emirats arabes unis	50 892	73 617	87 996	53 190	50 381	58 449	63 876	66 973	81 582
Yémen	-	-	-	-	-	-	948	8 094	9 136
Moyen-Orient	**802 212**	**1 054 121**	**1 063 362**	**579 134**	**521 609**	**608 243**	**628 302**	**723 060**	**800 139**
Total non-OCDE	**1 818 252**	**2 181 381**	**2 356 683**	**1 907 004**	**1 843 536**	**1 974 550**	**2 002 725**	**2 124 290**	**2 215 863**
OCDE Amérique du N.	623 660	634 606	627 383	730 920	732 110	708 021	705 132	699 651	668 717
OCDE Pacifique	15 103	20 060	23 031	24 376	28 548	29 465	29 193	29 379	26 949
OCDE Europe	22 205	22 848	90 324	188 100	195 017	200 191	205 359	204 233	199 556
Total OCDE	**660 968**	**677 514**	**740 738**	**943 396**	**955 675**	**937 677**	**939 684**	**933 263**	**895 222**
Monde	**2 479 220**	**2 858 895**	**3 097 421**	**2 850 400**	**2 799 211**	**2 912 227**	**2 942 409**	**3 057 553**	**3 111 085**
Pour mémoire: OPEP	1 249 610	1 527 729	1 477 382	887 954	815 596	909 676	918 894	1 014 725	1 106 347

Production of Crude Oil, NGL and Additives (1000 tonnes)
Production de pétrole brut, LGN et additifs (1000 tonnes)
Erzeugung von Rohöl, Kondensaten und Additiven (1000 Tonnen)

Produzione petrolio grezzo, LNG e additivi (1000 tonnellate)

原油、ＮＧＬ等随伴物の生産量（千トン）

Producción de petróleo crudo, liquidos de gas natural y aditivos (1000 toneladas)

Производство сырой нефти, газовых конденсатов и присадок (тыс. т)

1990	1991	1992	1993	1994	1995	1996	1997	1998	
1 162	884	592	582	545	522	489	361	364	Albania
60	58	53	43	36	43	32	28	33	Bulgaria
7 929	6 791	6 827	6 931	6 974	7 040	6 965	6 855	6 724	Romania
75	120	92	69	70	76	72	65	61	Slovak Republic
-	-	2 115	2 139	1 823	1 761	1 696	1 785	1 815	*Croatia*
-	-	2	2	2	2	1	1	1	*Slovenia*
-	-	1 165	1 148	1 077	1 066	1 030	979	913	*FR of Yugoslavia*
3 145	3 069	3 282	3 289	2 902	2 829	2 727	2 765	2 729	Former Yugoslavia
12 371	**10 922**	**10 846**	**10 914**	**10 527**	**10 510**	**10 285**	**10 074**	**9 911**	**Non-OECD Europe**
-	-	11 566	10 619	9 563	9 162	9 100	9 022	11 424	Azerbaijan
-	-	2 000	2 005	2 000	1 932	1 860	1 822	1 817	Belarus
-	-	100	100	74	47	128	134	119	Georgia
-	-	26 084	23 622	21 294	20 723	23 850	25 909	26 119	Kazakhstan
-	-	112	88	88	89	100	85	77	Kyrgyzstan
-	-	64	73	93	128	155	224	292	Lithuania
-	-	399 337	351 496	315 767	305 107	299 498	303 868	301 406	Russia
-	-	60	51	33	25	21	26	20	Tajikistan
-	-	4 950	4 154	4 060	4 427	4 332	5 368	6 638	Turkmenistan
-	-	4 474	4 248	4 200	4 058	4 482	4 573	4 395	Ukraine
-	-	3 293	3 944	5 500	7 530	7 717	8 124	8 439	Uzbekistan
570 753	**512 332**	**452 040**	**400 400**	**362 672**	**353 228**	**351 243**	**359 155**	**360 746**	**Former USSR**
2 403	2 425	2 473	2 495	2 453	2 422	2 366	2 347	2 276	Bahrain
155 900	167 619	173 936	182 870	180 949	183 018	183 812	181 116	182 755	Islamic Republic of Iran
99 156	14 590	21 266	23 648	26 006	27 185	28 540	57 000	104 536	Iraq
13	11	9	28	25	28	25	21	21	Israel
-	-	3	1	1	1	2	2	2	Jordan
44 462	9 201	54 175	95 985	102 404	102 406	102 633	102 427	104 694	Kuwait
35 173	36 277	37 924	39 822	41 546	43 936	45 552	46 248	46 067	Oman
20 514	20 103	22 189	21 398	20 744	20 856	21 336	28 767	31 792	Qatar
337 950	430 750	439 967	433 794	428 721	430 844	432 231	442 383	457 208	Saudi Arabia
20 630	24 600	25 750	26 767	29 000	31 329	30 745	30 186	30 152	Syria
90 736	104 402	112 318	108 548	108 017	107 213	107 970	109 700	113 214	United Arab Emirates
9 459	9 921	8 996	11 048	17 138	17 475	17 701	18 532	18 980	Yemen
816 396	**819 899**	**899 006**	**946 404**	**957 004**	**966 713**	**972 913**	**1 018 729**	**1 091 697**	**Middle East**
2 242 581	**2 226 324**	**2 252 440**	**2 262 249**	**2 254 442**	**2 298 420**	**2 359 626**	**2 444 684**	**2 517 501**	**Non-OECD Total**
656 008	662 861	661 608	655 520	650 496	647 177	656 472	668 627	663 126	OECD North America
29 916	30 246	29 673	29 456	27 767	29 268	29 257	30 720	32 788	OECD Pacific
206 541	218 485	234 685	247 489	289 938	302 476	320 253	318 966	315 902	OECD Europe
892 465	**911 592**	**925 966**	**932 465**	**968 201**	**978 921**	**1 005 982**	**1 018 313**	**1 011 816**	**OECD Total**
3 135 046	**3 137 916**	**3 178 406**	**3 194 714**	**3 222 643**	**3 277 341**	**3 365 608**	**3 462 997**	**3 529 317**	**World**
1 146 708	1 178 237	1 251 223	1 297 665	1 303 617	1 324 016	1 364 442	1 420 389	1 478 162	*Memo: OPEC*

Production of Natural Gas (TJ)
Production de gaz naturel (TJ)

Erzeugung von Erdgas (TJ)

Produzione di gas naturale (TJ)

天然ガスの生産量 (TJ)

Producción de gas natural (TJ)

Производство природного газа (ТДж)

	1971	1973	1978	1984	1985	1986	1987	1988	1989
Algérie	111 757	188 055	500 460	1 317 721	1 435 166	1 527 577	1 724 031	1 802 324	1 943 326
Angola	1 678	2 536	2 758	4 499	4 499	4 997	5 998	6 098	6 500
Congo	2 842	607	88	80	126	80	84	88	80
Côte d'Ivoire	-	-	-	-	-	-	-	-	-
Egypte	3 269	3 390	26 611	139 030	173 602	301 860	313 188	345 040	301 860
Gabon	3 618	18 465	1 438	1 536	1 972	3 512	4 486	6 824	3 300
Libye	60 648	159 045	199 724	174 800	197 600	212 800	190 000	209 000	258 400
Maroc	2 009	2 721	3 177	3 244	3 666	3 549	2 884	2 947	2 233
Mozambique	-	-	-	-	21	6	-	-	6
Nigéria	8 011	16 491	36 138	116 842	137 921	124 129	139 373	138 122	161 500
Sénégal	-	-	-	-	-	-	11	314	299
Afrique du Sud	-	-	-	-	-	-	-	-	-
Tunisie	42	5 240	13 172	19 630	18 789	17 466	14 930	13 812	13 830
Autre Afrique	37	37	37	16	15	10	4	7	5
Afrique	**193 911**	**396 587**	**783 603**	**1 777 398**	**1 973 377**	**2 195 986**	**2 394 989**	**2 524 576**	**2 691 339**
Argentine	262 985	267 382	296 230	535 920	556 005	597 394	610 982	731 550	781 666
Bolivie	2 744	78 727	85 105	115 753	118 381	117 183	116 565	121 628	121 666
Brésil	5 038	7 673	30 150	84 753	101 572	121 723	134 977	114 128	116 298
Chili	29 746	24 457	56 254	36 777	37 483	34 276	33 611	44 718	66 112
Colombie	52 018	65 840	91 233	145 363	148 270	144 506	152 277	146 843	148 410
Cuba	67	565	427	126	266	228	912	836	1 292
Pérou	20 664	19 433	27 822	29 630	23 131	24 500	24 392	22 997	19 487
Trinité-et-Tobago	74 540	73 992	97 041	164 704	168 608	172 538	182 774	210 377	210 779
Vénézuela	414 958	500 067	551 594	772 143	766 383	775 598	774 199	762 395	826 479
Autre Amérique latine	117	117	374	760	993	1 028	865	1 025	1 104
Amérique latine	**862 877**	**1 038 253**	**1 236 230**	**1 885 929**	**1 921 092**	**1 988 974**	**2 031 554**	**2 156 497**	**2 293 293**
Bangladesh	17 596	24 322	34 041	91 565	101 823	116 016	137 327	153 195	171 366
Brunei	1 934	61 811	281 935	340 611	334 509	322 410	340 348	334 908	340 416
Inde	28 168	29 197	65 810	146 626	178 137	247 677	285 320	339 597	413 553
Indonésie	11 261	15 456	414 000	1 134 139	1 350 057	1 441 376	1 537 346	1 634 248	1 769 481
Malaisie	3 315	4 642	85 103	364 636	402 877	541 828	585 844	604 797	654 587
Myanmar	2 578	3 976	10 644	26 971	36 708	41 044	42 233	42 649	41 263
Pakistan	109 874	133 215	189 134	282 724	300 616	310 648	401 750	413 627	470 264
Philippines	-	-	-	-	-	-	-	-	-
Taipei chinois	42 244	56 701	73 882	55 767	49 996	45 554	46 074	52 553	52 943
Thaïlande	-	-	-	88 071	136 244	184 017	184 017	217 991	217 741
Viêt-Nam	-	-	-	2 367	1 506	1 530	1 527	1 147	307
Autre Asie	96 601	102 111	90 223	105 940	111 240	111 444	70 950	58 938	12 754
Asie	**313 571**	**431 431**	**1 244 772**	**2 639 417**	**3 003 713**	**3 363 544**	**3 632 736**	**3 853 650**	**4 144 675**
Rép. populaire de Chine	145 756	233 032	534 158	492 186	504 000	536 353	541 420	555 842	586 471
Chine	**145 756**	**233 032**	**534 158**	**492 186**	**504 000**	**536 353**	**541 420**	**555 842**	**586 471**
Albanie	4 914	7 413	13 804	14 997	14 997	14 997	14 997	15 997	12 997
Bulgarie	11 506	7 810	1 134	1 836	800	670	521	391	312
Roumanie	1 044 291	1 130 968	1 474 669	1 525 552	1 458 152	1 475 663	1 402 488	1 379 495	1 239 397
République slovaque	23 047	18 006	22 285	14 955	14 473	15 827	17 889	24 044	21 789
Croatie	-	-	-	-	-	-	-	-	-
Slovénie	-	-	-	-	-	-	-	-	-
RF de Yougoslavie	-	-	-	-	-	-	-	-	-
Ex-Yougoslavie	46 534	61 724	82 019	72 870	93 374	93 542	99 401	114 868	110 267
Europe non-OCDE	**1 130 292**	**1 225 921**	**1 593 911**	**1 630 210**	**1 581 796**	**1 600 699**	**1 535 296**	**1 534 795**	**1 384 762**

Production of Natural Gas (TJ)
Production de gaz naturel (TJ)

Erzeugung von Erdgas (TJ)

Produzione di gas naturale (TJ)

天然ガスの生産量 (TJ)

Producción de gas natural (TJ)

Производство природного газа (ТДж)

1990	1991	1992	1993	1994	1995	1996	1997	1998	
2 063 244	2 227 827	2 317 602	2 349 091	2 161 621	2 458 328	2 610 705	3 007 105	3 207 127	Algeria
20 520	22 040	21 660	21 280	19 760	21 280	21 280	21 660	22 040	Angola
90	83	115	113	200	130	131	131	131	Congo
-	-	-	-	-	1 934	17 406	28 554	35 187	Cote d'Ivoire
313 188	345 040	371 600	438 683	461 889	479 075	503 495	509 490	534 483	Egypt
3 200	3 200	3 100	3 200	2 853	3 109	3 317	3 162	3 162	Gabon
235 600	248 520	257 260	241 680	242 820	240 920	243 960	249 660	241 680	Libya
2 014	1 377	847	828	809	512	651	1 200	1 317	Morocco
-	-	6	6	6	13	13	15	21	Mozambique
152 000	185 350	195 000	213 000	208 740	204 611	207 371	216 833	232 602	Nigeria
219	170	105	498	761	1 843	1 685	928	793	Senegal
70 000	71 000	71 500	79 793	79 793	79 793	71 814	64 120	53 983	South Africa
15 402	12 099	10 979	8 234	8 096	6 188	35 636	73 139	83 887	Tunisia
6	5	6	7	7	7	7	7	7	Other Africa
2 875 483	**3 116 711**	**3 249 780**	**3 356 413**	**3 187 355**	**3 497 743**	**3 717 471**	**4 176 004**	**4 416 420**	**Africa**
737 040	793 020	820 195	865 100	907 120	997 813	1 113 541	1 221 173	1 298 744	Argentina
124 178	124 217	127 425	129 082	136 568	123 389	116 023	139 720	123 842	Bolivia
111 841	118 778	134 589	127 614	176 249	181 790	193 339	163 189	220 388	Brazil
69 244	61 266	69 753	67 131	72 803	69 990	69 170	75 155	67 763	Chile
156 984	155 446	159 236	155 895	160 818	168 596	181 874	223 603	259 886	Colombia
1 281	1 125	794	874	752	657	733	1 414	4 720	Cuba
18 647	17 514	15 796	16 611	17 343	16 810	17 015	16 010	22 207	Peru
218 535	220 473	225 817	223 997	237 819	234 080	268 267	303 781	366 002	Trinidad and Tobago
965 461	991 189	994 801	1 031 747	1 088 395	1 162 046	1 308 453	1 334 909	1 377 020	Venezuela
1 132	1 242	1 746	1 707	1 668	1 630	1 513	1 513	1 319	Other Latin America
2 404 343	**2 484 270**	**2 550 152**	**2 619 758**	**2 799 535**	**2 956 801**	**3 269 928**	**3 480 467**	**3 741 891**	**Latin America**
172 789	177 604	205 991	220 770	237 517	264 857	274 361	269 618	292 557	Bangladesh
352 914	361 890	367 021	368 914	379 424	415 463	412 233	412 149	398 823	Brunei
471 217	526 309	608 797	602 923	647 768	725 500	805 113	866 983	907 994	India
2 045 547	2 176 978	2 332 980	2 397 568	2 449 874	2 468 487	2 592 460	2 664 687	2 643 585	Indonesia
647 976	769 438	797 512	867 887	899 058	1 030 017	1 284 321	1 462 517	1 478 427	Malaysia
35 359	33 355	32 820	40 550	50 632	56 717	61 495	66 444	64 045	Myanmar
469 925	468 924	495 193	536 845	571 400	575 891	612 080	637 183	633 062	Pakistan
-	-	-	-	249	272	406	217	351	Philippines
49 148	36 762	31 870	30 843	33 818	35 088	33 574	32 036	32 753	Chinese Taipei
237 169	293 903	314 468	352 018	390 142	412 227	482 841	588 108	640 711	Thailand
121	100	-	-	-	7 778	12 851	23 151	34 252	Vietnam
11 167	10 920	10 373	10 230	9 930	9 750	9 500	9 788	9 788	Other Asia
4 493 332	**4 856 183**	**5 197 025**	**5 428 548**	**5 669 812**	**6 002 047**	**6 581 235**	**7 032 881**	**7 136 348**	**Asia**
595 644	625 621	634 012	726 278	760 630	777 436	871 307	983 449	1 008 400	People's Rep. of China
595 644	**625 621**	**634 012**	**726 278**	**760 630**	**777 436**	**871 307**	**983 449**	**1 008 400**	**China**
9 482	5 593	4 051	3 500	1 976	1 089	894	700	661	Albania
453	377	1 405	2 537	2 113	1 841	1 534	1 307	1 088	Bulgaria
1 065 811	933 074	819 293	779 681	689 594	672 012	640 302	553 958	518 676	Romania
16 031	11 322	15 000	9 186	10 588	12 412	11 360	10 451	9 393	Slovak Republic
-	-	68 514	77 862	68 096	74 723	67 853	65 254	59 660	*Croatia*
-	-	600	479	451	652	466	432	286	*Slovenia*
-	-	31 894	36 267	31 065	34 156	25 297	25 937	27 559	*FR of Yugoslavia*
100 976	99 035	101 008	114 608	99 612	109 531	93 616	91 623	87 505	Former Yugoslavia
1 192 753	**1 049 401**	**940 757**	**909 512**	**803 883**	**796 885**	**747 706**	**658 039**	**617 323**	**Non-OECD Europe**

Production of Natural Gas (TJ)
Production de gaz naturel (TJ)
Erzeugung von Erdgas (TJ)
Produzione di gas naturale (TJ)
天然ガスの生産量 (TJ)
Producción de gas natural (TJ)
Производство природного газа (ТДж)

	1971	1973	1978	1984	1985	1986	1987	1988	1989
Azerbaïdjan	-	-	-	-	-	-	-	-	-
Bélarus	-	-	-	-	-	-	-	-	-
Géorgie	-	-	-	-	-	-	-	-	-
Kazakhstan	-	-	-	-	-	-	-	-	-
Kirghizistan	-	-	-	-	-	-	-	-	-
Russie	-	-	-	-	-	-	-	-	-
Tadjikistan	-	-	-	-	-	-	-	-	-
Turkménistan	-	-	-	-	-	-	-	-	-
Ukraine	-	-	-	-	-	-	-	-	-
Ouzbékistan	-	-	-	-	-	-	-	-	-
Ex-URSS	**8 175 896**	**9 092 601**	**14 322 941**	**22 148 172**	**24 199 704**	**25 818 600**	**27 358 412**	**28 977 308**	**29 958 880**
Bahrein	35 326	62 494	101 084	140 292	171 261	144 365	142 457	150 944	161 225
République Islamique d'Iran	577 600	725 800	708 700	513 000	554 800	577 600	608 000	788 500	843 600
Irak	35 340	45 980	64 600	114 300	121 250	172 996	291 021	375 958	434 481
Israël	4 880	2 097	2 210	2 051	1 837	1 444	1 624	1 461	1 464
Jordanie	-	-	-	-	-	-	-	-	2 378
Koweit	197 540	230 608	237 873	196 743	192 344	223 470	186 420	169 650	212 340
Oman	-	-	5 990	37 240	42 375	52 141	57 433	60 156	61 999
Qatar	38 380	60 040	56 240	225 340	209 040	219 640	220 020	237 120	224 960
Arabie saoudite	66 441	90 302	235 912	733 074	733 074	982 630	1 045 437	1 134 271	1 160 000
Syrie	-	-	1 348	5 050	6 000	15 171	14 937	35 334	57 837
Emirats arabes unis	40 750	48 782	212 758	409 866	492 957	567 105	629 331	646 843	759 370
Moyen-Orient	**996 257**	**1 266 103**	**1 626 715**	**2 376 956**	**2 524 898**	**2 956 562**	**3 196 680**	**3 600 237**	**3 919 654**
Total non-OCDE	**11 818 560**	**13 683 928**	**21 342 330**	**32 950 268**	**35 708 580**	**38 460 718**	**40 691 087**	**43 202 905**	**44 979 074**
OCDE Amérique du N.	26 301 914	26 734 297	24 071 675	23 140 565	22 334 868	21 529 971	22 382 518	23 434 304	23 893 591
OCDE Pacifique	188 142	276 933	452 396	687 248	754 228	825 982	840 559	872 960	891 305
OCDE Europe	4 079 755	5 807 276	7 605 197	7 482 135	7 746 253	7 594 282	7 867 079	7 422 071	7 657 097
Total OCDE	**30 569 811**	**32 818 506**	**32 129 268**	**31 309 948**	**30 835 349**	**29 950 235**	**31 090 156**	**31 729 335**	**32 441 993**
Monde	**42 388 371**	**46 502 434**	**53 471 598**	**64 260 216**	**66 543 929**	**68 410 953**	**71 781 243**	**74 932 240**	**77 421 067**

Production of Natural Gas (TJ)
Production de gaz naturel (TJ)
Erzeugung von Erdgas (TJ)
Produzione di gas naturale (TJ)
天然ガスの生産量 (TJ)
Producción de gas natural (TJ)
Производство природного газа (ТДж)

1990	1991	1992	1993	1994	1995	1996	1997	1998	
-	-	296 774	256 549	240 488	250 479	237 698	224 839	210 724	Azerbaijan
-	-	11 272	11 239	11 354	10 273	9 617	9 501	10 562	Belarus
-	-	1 430	1 885	188	377	113	-	-	Georgia
-	-	305 860	252 025	169 198	223 033	245 955	305 902	299 632	Kazakhstan
-	-	2 714	1 583	1 470	1 405	1 015	1 561	702	Kyrgyzstan
-	-	24 128 000	23 158 851	22 375 556	22 153 111	22 339 444	21 440 210	22 197 314	Russia
-	-	2 444	1 664	1 254	1 482	1 862	1 444	1 140	Tajikistan
-	-	2 040 403	2 477 600	1 360 514	1 352 800	1 326 361	652 889	499 789	Turkmenistan
-	-	787 251	723 840	689 910	708 642	718 280	707 472	701 072	Ukraine
-	-	1 626 514	1 711 330	1 745 504	1 774 223	1 783 203	1 848 372	1 877 194	Uzbekistan
30 540 161	**30 681 816**	**29 202 662**	**28 596 566**	**26 595 436**	**26 475 825**	**26 663 548**	**25 192 190**	**25 798 129**	**Former USSR**
165 279	139 889	162 223	182 214	184 364	184 450	194 994	203 975	220 443	Bahrain
919 600	978 500	950 000	1 028 850	1 208 900	1 593 721	1 716 630	1 924 194	2 054 580	Islamic Republic of Iran
234 135	65 021	108 680	122 048	151 706	150 796	154 126	168 299	177 156	Iraq
1 316	967	953	1 005	903	903	572	605	479	Israel
4 724	4 571	4 975	5 853	9 772	10 098	9 819	10 377	10 517	Jordan
155 631	32 204	189 627	335 974	358 445	358 450	359 246	358 525	365 497	Kuwait
113 322	116 437	142 267	172 852	180 063	178 646	186 000	218 657	242 591	Oman
237 510	287 651	475 774	508 950	508 950	508 950	516 490	655 980	738 166	Qatar
1 180 000	1 248 000	1 292 000	1 364 200	1 432 410	1 532 920	1 665 902	1 827 092	1 886 733	Saudi Arabia
63 687	74 334	76 888	76 100	82 492	104 022	104 022	161 340	208 751	Syria
749 309	887 173	826 066	856 620	1 000 818	1 167 000	1 350 695	1 352 930	1 381 248	United Arab Emirates
3 824 513	**3 834 747**	**4 229 453**	**4 654 666**	**5 118 823**	**5 789 956**	**6 258 496**	**6 881 974**	**7 286 161**	**Middle East**
45 926 229	**46 648 749**	**46 003 841**	**46 291 741**	**44 935 474**	**46 296 693**	**48 109 691**	**48 405 004**	**50 004 672**	**Non-OECD Total**
24 686 438	24 766 401	25 325 335	26 119 756	27 461 512	27 466 263	28 061 677	28 306 871	28 502 728	OECD North America
1 061 382	1 112 672	1 198 996	1 263 356	1 327 863	1 431 581	1 489 547	1 501 432	1 525 409	OECD Pacific
7 604 588	8 207 913	8 362 046	8 880 873	9 045 275	9 422 941	10 826 899	10 703 186	10 684 132	OECD Europe
33 352 408	**34 086 986**	**34 886 377**	**36 263 985**	**37 834 650**	**38 320 785**	**40 378 123**	**40 511 489**	**40 712 269**	**OECD Total**
79 278 637	**80 735 735**	**80 890 218**	**82 555 726**	**82 770 124**	**84 617 478**	**88 487 814**	**88 916 493**	**90 716 941**	**World**

Production of Combustible Renewables and Waste (TJ)
Production d'énergies renouvelables combustibles et déchets (TJ)
Erzeugung von erneuerbaren Brennstoffen und Abfällen (TJ)
Produzione di energia da combustibili rinnovabili e da rifiuti (TJ)
可燃性再生可能エネルギー及び廃棄物の生産量 *(TJ)*
Producción de combustibles renovables y desechos (TJ)
Производство возобновляемых видов топлива и отходов (ТДж)

	1971	1973	1978	1984	1985	1986	1987	1988	1989
Algérie	370	392	444	535	547	565	581	597	613
Angola	133 542	135 265	145 360	161 231	164 009	166 807	169 829	173 147	176 833
Bénin	41 970	43 662	48 406	56 015	57 506	59 022	60 594	62 196	63 839
Cameroun	98 931	103 013	117 596	137 397	140 711	144 174	147 825	151 676	155 726
Congo	20 235	21 205	23 482	26 484	27 100	27 769	28 459	29 183	29 928
RD du Congo	233 072	246 090	284 325	343 279	354 740	366 661	379 096	391 957	405 176
Côte d'Ivoire	68 142	73 019	86 900	107 423	111 312	115 304	119 388	123 586	128 229
Egypte	27 257	28 468	30 494	36 368	39 517	40 924	42 427	43 522	42 418
Ethiopie	353 660	370 917	419 286	489 328	502 344	516 793	532 586	549 822	568 611
Erythrée	-	-	-	-	-	-	-	-	-
Gabon	20 368	21 263	23 755	27 006	27 624	28 147	28 666	29 214	29 792
Ghana	87 400	96 000	113 302	133 100	138 000	143 100	148 500	154 000	159 800
Kenya	265 434	276 972	310 692	361 879	370 642	378 821	386 787	394 545	402 102
Libye	4 063	4 415	5 236	5 236	5 236	5 236	5 236	5 236	5 236
Maroc	5 275	5 558	8 664	12 298	12 503	12 708	12 884	13 089	13 294
Mozambique	304 855	301 187	301 349	302 418	300 765	297 413	293 845	290 308	287 107
Nigéria	1 421 913	1 493 666	1 701 819	2 017 829	2 074 773	2 132 032	2 189 941	2 248 623	2 308 276
Sénégal	34 549	36 513	41 659	48 349	49 619	50 944	52 323	53 751	55 240
Afrique du Sud	196 800	215 000	245 000	344 000	349 000	355 000	382 000	404 000	434 000
Soudan	236 454	246 015	279 234	326 372	333 729	340 306	346 576	352 587	358 378
Rép. Unie de Tanzanie	374 052	376 166	387 435	430 029	439 450	448 535	457 920	467 584	477 564
Togo	19 144	20 143	22 367	26 496	27 309	28 219	29 715	30 620	31 577
Tunisie	28 245	29 090	32 449	37 667	38 808	40 023	41 019	41 912	42 420
Zambie	103 368	107 507	120 140	148 438	152 919	155 817	158 770	161 806	164 942
Zimbabwe	127 788	132 901	146 442	170 710	175 426	180 008	184 591	189 147	193 620
Autre Afrique	996 802	1 032 342	1 148 761	1 319 885	1 351 351	1 384 003	1 417 370	1 452 699	1 489 299
Afrique	**5 203 689**	**5 416 769**	**6 044 597**	**7 069 772**	**7 244 940**	**7 418 331**	**7 616 928**	**7 814 807**	**8 024 020**
Argentine	94 738	87 793	87 690	87 151	85 134	74 185	79 546	76 369	65 443
Bolivie	8 150	9 348	12 064	37 850	38 529	39 134	27 458	30 580	29 104
Brésil	1 486 640	1 525 756	1 504 286	1 858 665	1 887 274	1 853 366	1 933 014	1 884 433	1 882 430
Chili	58 422	55 478	69 501	92 327	93 951	97 487	101 873	105 813	106 631
Colombie	185 183	142 485	196 856	206 756	216 492	222 059	220 127	227 234	231 446
Costa Rica	24 066	24 618	24 723	30 785	32 085	30 933	31 550	31 273	29 648
Cuba	150 464	142 026	176 451	180 456	168 269	177 012	179 058	193 196	215 412
République dominicaine	48 856	48 691	54 141	76 362	52 513	52 873	53 595	43 332	41 484
Equateur	46 295	43 985	41 157	45 971	44 464	42 329	43 124	46 558	45 134
El Salvador	50 670	53 985	63 683	64 289	65 200	52 016	56 239	50 172	48 948
Guatemala	80 073	84 716	110 377	103 622	106 050	108 625	112 321	114 652	119 561
Haïti	57 608	61 138	71 971	67 016	67 580	50 053	50 912	51 877	52 209
Honduras	41 515	43 054	48 361	55 674	56 770	56 744	59 093	60 116	61 125
Jamaïque	11 147	10 152	10 182	10 870	10 023	10 155	10 273	10 531	12 509
Nicaragua	30 108	31 988	36 108	40 369	42 593	40 821	41 802	44 497	44 452
Panama	14 118	14 119	17 492	19 454	18 257	17 734	17 281	17 597	17 065
Paraguay	48 486	52 339	58 541	69 208	70 689	73 784	86 868	90 563	92 864
Pérou	149 413	152 110	150 394	156 050	159 596	160 507	161 264	157 576	159 102
Trinité-et-Tobago	866	790	744	535	1 069	1 778	2 109	1 720	1 586
Uruguay	16 443	16 860	18 834	22 713	24 360	26 604	26 569	24 359	24 549
Vénézuela	17 147	16 589	15 431	16 106	18 918	21 416	22 723	20 063	21 418
Autre Amérique latine	19 827	20 123	24 418	29 016	28 476	28 853	25 989	23 648	23 479
Amérique latine	**2 640 235**	**2 638 143**	**2 793 405**	**3 271 245**	**3 288 292**	**3 238 468**	**3 342 788**	**3 306 159**	**3 325 599**

Production of Combustible Renewables and Waste (TJ)
Production d'énergies renouvelables combustibles et déchets (TJ)
Erzeugung von erneuerbaren Brennstoffen und Abfällen (TJ)
Produzione di energia da combustibili rinnovabili e da rifiuti (TJ)
可燃性再生可能エネルギー及び廃棄物の生産量 *(TJ)*
Producción de combustibles renovables y desechos (TJ)
Производство возобновляемых видов топлива и отходов (ТДж)

1990	1991	1992	1993	1994	1995	1996	1997	1998	
628	1 214	1 340	2 261	2 261	2 261	2 261	3 014	4 271	Algeria
180 978	185 199	189 515	193 925	198 000	203 000	210 714	216 825	223 330	Angola
65 497	67 134	68 758	70 395	72 000	73 500	75 044	77 147	79 302	Benin
159 987	164 362	168 863	173 490	178 000	183 000	188 307	193 768	199 071	Cameroon
30 693	31 476	32 252	33 009	34 000	34 500	35 294	36 284	37 288	Congo
418 675	432 476	446 590	461 018	476 000	491 000	506 712	522 927	539 786	DR of Congo
133 005	137 693	142 291	146 768	151 100	155 000	159 960	164 599	167 853	Cote d'Ivoire
44 272	46 938	47 599	48 770	51 805	50 585	51 597	52 577	53 494	Egypt
589 075	608 403	603 669	585 490	602 278	618 479	633 703	648 057	680 220	Ethiopia
-	-	24 722	24 722	24 722	25 521	26 397	27 225	15 624	Eritrea
31 097	32 244	33 216	33 984	34 000	35 000	35 525	36 022	36 891	Gabon
163 300	172 200	178 700	185 600	192 800	198 000	202 752	207 415	212 893	Ghana
409 498	416 700	423 923	431 140	438 000	445 000	452 565	461 164	472 172	Kenya
5 236	5 236	5 236	5 236	5 236	5 236	5 351	5 469	5 590	Libya
13 255	15 982	15 997	16 152	16 251	16 698	17 032	17 373	17 669	Morocco
284 652	284 166	285 636	289 037	269 498	265 800	265 800	265 800	265 800	Mozambique
2 369 215	2 432 000	2 497 578	2 565 617	2 636 000	2 710 000	2 794 010	2 880 624	2 957 479	Nigeria
56 783	58 249	59 684	61 081	62 439	64 000	66 176	68 426	68 426	Senegal
443 000	452 000	460 000	469 000	478 000	488 000	496 296	504 237	513 313	South Africa
364 000	369 067	374 164	379 305	384 000	389 000	516 394	527 347	538 943	Sudan
487 819	497 955	508 421	519 145	530 000	541 000	550 738	561 202	575 754	United Rep. of Tanzania
32 571	33 529	34 524	35 519	36 641	37 800	38 984	40 219	41 479	Togo
43 433	44 412	45 313	46 157	47 000	47 645	48 360	49 085	50 582	Tunisia
168 224	172 653	177 244	182 016	187 000	192 000	195 264	199 560	204 262	Zambia
197 970	201 950	205 720	209 264	213 000	216 000	218 592	221 434	225 701	Zimbabwe
1 527 308	1 556 437	1 597 140	1 619 842	1 663 359	1 707 200	1 747 896	1 812 776	1 867 107	Other Africa
8 220 171	**8 419 675**	**8 628 095**	**8 787 943**	**8 983 390**	**9 195 225**	**9 541 724**	**9 800 576**	**10 054 300**	**Africa**
72 093	75 408	77 713	82 438	98 011	110 742	112 676	110 875	125 520	Argentina
31 580	33 016	32 165	32 631	33 283	34 132	32 413	36 672	36 736	Bolivia
1 697 602	1 669 277	1 639 884	1 623 666	1 709 846	1 646 756	1 648 445	1 709 372	1 731 026	Brazil
112 854	126 143	139 964	130 000	136 507	146 453	156 555	154 069	165 190	Chile
243 676	238 368	246 934	252 833	262 598	292 309	288 427	220 566	213 364	Colombia
31 030	31 051	28 739	14 795	14 785	15 629	15 902	10 912	7 560	Costa Rica
233 520	205 530	205 226	147 473	144 208	118 048	139 196	131 220	112 608	Cuba
42 592	42 547	54 348	54 292	54 235	54 360	54 544	54 785	54 898	Dominican Republic
44 840	46 180	47 060	49 823	49 530	51 213	50 913	47 608	48 023	Ecuador
51 110	54 363	55 768	57 906	58 952	58 986	58 823	60 780	61 344	El Salvador
122 305	124 880	126 639	123 148	120 931	122 876	124 866	126 878	127 082	Guatemala
50 805	51 898	55 254	55 088	56 671	57 394	65 873	65 785	66 994	Haiti
62 732	62 939	63 083	63 233	63 383	63 532	63 744	72 204	72 523	Honduras
18 840	15 179	15 486	17 115	22 188	20 512	22 452	24 501	27 105	Jamaica
47 239	46 836	45 970	46 823	47 900	49 965	53 249	57 788	55 617	Nicaragua
16 954	17 130	17 252	19 494	18 659	18 152	18 565	19 039	19 238	Panama
93 925	97 874	91 650	88 854	94 301	101 571	107 513	110 626	104 537	Paraguay
165 238	166 922	169 592	169 812	172 662	174 715	176 778	181 079	182 869	Peru
2 034	1 737	1 743	1 615	1 860	1 070	1 145	1 238	1 238	Trinidad and Tobago
23 778	24 550	25 301	24 755	23 579	22 362	21 220	21 397	19 865	Uruguay
22 378	23 196	23 501	23 050	23 603	22 647	22 647	22 647	22 647	Venezuela
24 213	24 336	24 093	24 108	24 640	26 121	26 408	26 566	27 203	Other Latin America
3 211 338	**3 179 360**	**3 187 365**	**3 102 952**	**3 232 332**	**3 209 545**	**3 262 354**	**3 266 607**	**3 283 187**	**Latin America**

Production of Combustible Renewables and Waste (TJ)
Production d'énergies renouvelables combustibles et déchets (TJ)
Erzeugung von erneuerbaren Brennstoffen und Abfällen (TJ)
Produzione di energia da combustibili rinnovabili e da rifiuti (TJ)
可燃性再生可能エネルギー及び廃棄物の生産量 *(TJ)*
Producción de combustibles renovables y desechos (TJ)
Производство возобновляемых видов топлива и отходов (ТДж)

	1971	1973	1978	1984	1985	1986	1987	1988	1989
Bangladesh	268 720	288 693	325 322	364 754	372 333	389 121	352 403	384 557	396 113
Brunei	664	694	781	772	772	772	772	772	772
Inde	5 056 264	5 290 474	5 928 712	6 682 171	6 797 788	6 928 534	7 059 289	7 153 440	7 253 657
Indonésie	1 151 476	1 209 156	1 352 941	1 533 381	1 562 905	1 593 576	1 623 026	1 651 256	1 679 485
RPD de Corée	28 523	30 281	34 676	37 998	38 515	38 359	38 740	39 140	39 551
Malaisie	53 079	56 334	64 335	74 537	75 334	77 043	81 013	83 142	85 535
Myanmar	266 124	279 829	306 851	340 498	347 350	355 583	364 908	376 938	383 419
Népal	104 712	109 295	123 631	210 327	214 893	219 633	224 372	229 241	232 429
Pakistan	445 097	474 496	553 261	658 930	678 064	698 542	719 397	740 471	762 712
Philippines	256 417	265 836	318 839	372 252	387 363	392 348	394 390	396 297	402 263
Sri Lanka	114 668	116 707	126 421	148 489	149 461	154 428	158 961	162 406	165 071
Taipei chinois	3 390	4 122	918	664	664	606	586	567	547
Thailande	318 287	331 320	436 307	398 205	450 081	485 208	511 962	535 074	595 062
Viêt-Nam	522 498	547 592	614 642	689 399	703 452	719 954	737 890	755 970	774 026
Autre Asie	79 307	83 399	96 489	99 300	98 876	101 228	102 093	103 802	101 943
Asie	**8 669 226**	**9 088 228**	**10 284 126**	**11 611 677**	**11 877 793**	**12 154 915**	**12 369 783**	**12 613 053**	**12 872 585**
Rép. populaire de Chine	6 458 822	6 772 393	7 342 365	7 836 222	7 894 774	7 983 778	8 072 335	8 163 445	8 279 229
Hong-Kong, Chine	1 299	1 368	1 495	1 690	1 758	1 758	1 758	1 817	1 817
Chine	**6 460 121**	**6 773 761**	**7 343 860**	**7 837 912**	**7 896 532**	**7 985 536**	**8 074 093**	**8 165 262**	**8 281 046**
Albanie	15 707	15 707	15 707	15 707	15 707	15 707	15 707	15 707	15 707
Bulgarie	11 145	10 012	8 996	16 869	17 280	17 113	8 850	7 970	15 131
Chypre	382	382	342	277	282	282	272	252	252
Roumanie	58 061	57 319	50 247	40 557	49 974	45 265	39 913	37 529	27 273
République slovaque	8 498	7 619	7 756	5 099	5 402	5 402	5 763	5 988	5 372
Bosnie-Herzegovine	-	-	-	-	-	-	-	-	-
Croatie	-	-	-	-	-	-	-	-	-
Ex-RYM	-	-	-	-	-	-	-	-	-
Slovénie	-	-	-	-	-	-	-	-	-
RF de Yougoslavie	-	-	-	-	-	-	-	-	-
Ex-Yougoslavie	69 949	37 372	32 547	41 260	40 840	42 315	37 021	37 714	39 512
Europe non-OCDE	**163 742**	**128 411**	**115 595**	**119 769**	**129 485**	**126 084**	**107 526**	**105 160**	**103 247**
Arménie	-	-	-	-	-	-	-	-	-
Azerbaïdjan	-	-	-	-	-	-	-	-	-
Bélarus	-	-	-	-	-	-	-	-	-
Estonie	-	-	-	-	-	-	-	-	-
Géorgie	-	-	-	-	-	-	-	-	-
Kazakhstan	-	-	-	-	-	-	-	-	-
Kirghizistan	-	-	-	-	-	-	-	-	-
Lettonie	-	-	-	-	-	-	-	-	-
Lituanie	-	-	-	-	-	-	-	-	-
République de Moldavie	-	-	-	-	-	-	-	-	-
Russie	-	-	-	-	-	-	-	-	-
Ukraine	-	-	-	-	-	-	-	-	-
Ouzbékistan	-	-	-	-	-	-	-	-	-
Ex-URSS	**843 869**	**815 545**	**759 873**	**832 148**	**847 502**	**788 207**	**842 960**	**791 889**	**728 110**

Production of Combustible Renewables and Waste (TJ)
Production d'énergies renouvelables combustibles et déchets (TJ)
Erzeugung von erneuerbaren Brennstoffen und Abfällen (TJ)
Produzione di energia da combustibili rinnovabili e da rifiuti (TJ)
可燃性再生可能エネルギー及び廃棄物の生産量 (TJ)
Producción de combustibles renovables y desechos (TJ)
Производство возобновляемых видов топлива и отходов (ТДж)

1990	1991	1992	1993	1994	1995	1996	1997	1998	
413 949	414 660	421 694	423 411	434 129	435 301	419 893	433 920	433 920	Bangladesh
772	772	772	772	772	772	772	772	772	Brunei
7 362 496	7 526 756	7 624 816	7 737 372	7 820 000	7 900 000	7 976 282	8 073 745	8 177 729	India
1 707 715	1 736 018	1 763 026	1 790 035	1 817 044	1 845 273	1 867 465	1 905 645	1 905 645	Indonesia
39 971	40 410	40 860	41 273	41 731	42 190	42 639	43 223	43 742	DPR of Korea
87 939	89 957	92 155	94 294	96 564	98 684	100 794	100 794	100 990	Malaysia
390 002	397 953	401 861	407 239	412 000	417 000	423 549	430 070	439 213	Myanmar
239 321	244 492	249 834	255 262	260 863	266 463	272 323	278 224	284 256	Nepal
785 890	808 477	829 040	854 286	878 000	901 000	919 378	937 950	958 525	Pakistan
412 161	415 820	424 461	422 257	431 016	430 100	388 894	395 958	384 794	Philippines
164 184	164 639	170 780	164 700	152 654	152 385	164 334	169 591	166 749	Sri Lanka
547	537	537	537	537	537	537	537	537	Chinese Taipei
613 273	645 381	654 050	588 550	567 808	585 065	577 681	584 733	558 019	Thailand
791 461	808 883	826 270	843 597	861 000	878 000	919 378	937 950	958 525	Vietnam
103 353	105 515	107 991	111 097	114 668	118 466	120 361	120 602	116 261	Other Asia
13 113 034	**13 400 270**	**13 608 147**	**13 734 682**	**13 888 786**	**14 071 236**	**14 194 280**	**14 413 714**	**14 529 677**	**Asia**
8 392 235	8 473 543	8 521 378	8 565 497	8 605 000	8 630 000	8 674 758	8 721 205	8 767 955	People's Rep. of China
1 817	1 817	1 827	1 885	1 885	1 954	2 012	2 012	2 012	Hong Kong, China
8 394 052	**8 475 360**	**8 523 205**	**8 567 382**	**8 606 885**	**8 631 954**	**8 676 770**	**8 723 217**	**8 769 967**	**China**
15 199	15 199	15 199	5 255	2 500	2 500	2 500	2 500	2 500	Albania
15 000	5 900	6 945	6 639	7 175	9 160	10 471	11 350	17 435	Bulgaria
257	232	230	229	490	479	464	358	378	Cyprus
25 212	29 949	44 553	48 520	49 772	72 285	206 102	141 853	126 312	Romania
6 965	5 939	4 933	4 884	7 157	3 196	3 184	3 454	3 086	Slovak Republic
-	-	6 838	6 838	6 838	6 500	6 696	6 898	7 107	*Bosnia-Herzegovina*
-	-	10 770	10 190	10 860	11 170	13 780	13 567	12 828	*Croatia*
-	-	7 844	8 088	7 814	7 814	7 814	7 814	6 388	*FYROM*
-	-	9 912	9 792	9 792	9 792	9 792	9 792	9 792	*Slovenia*
-	-	8 996	8 996	8 791	8 791	8 791	8 791	8 791	*FR of Yugoslavia*
32 322	32 020	44 360	43 904	44 095	44 067	46 873	46 862	44 906	Former Yugoslavia
94 955	**89 239**	**116 220**	**109 431**	**111 189**	**131 687**	**269 594**	**206 377**	**194 617**	**Non-OECD Europe**
-	-	39	39	39	39	39	39	42	Armenia
-	-	68	68	68	68	68	68	68	Azerbaijan
-	-	25 000	24 000	23 000	22 000	21 596	28 687	31 068	Belarus
-	-	7 709	7 539	12 989	14 521	24 745	26 176	21 411	Estonia
-	-	21 981	33 704	9 462	2 219	3 098	2 554	2 303	Georgia
-	-	4 800	3 300	3 200	3 300	3 100	3 100	3 069	Kazakhstan
-	-	186	137	166	156	150	150	150	Kyrgyzstan
-	-	20 093	20 108	23 287	16 844	31 814	54 585	58 243	Latvia
-	-	7 035	10 455	9 795	9 953	11 312	21 712	24 136	Lithuania
-	-	1 500	1 300	1 200	1 100	2 494	2 817	2 318	Republic of Moldova
-	-	723 724	776 442	591 027	668 541	505 298	565 204	508 730	Russia
-	-	12 454	11 398	11 120	11 078	11 036	10 994	10 952	Ukraine
-	-	10	10	10	10	10	10	10	Uzbekistan
800 000	**800 000**	**824 599**	**888 500**	**685 363**	**749 829**	**614 760**	**716 096**	**662 500**	**Former USSR**

Production of Combustible Renewables and Waste (TJ)
Production d'énergies renouvelables combustibles et déchets (TJ)
Erzeugung von erneuerbaren Brennstoffen und Abfällen (TJ)
Produzione di energia da combustibili rinnovabili e da rifiuti (TJ)
可燃性再生可能エネルギー及び廃棄物の生産量 (TJ)
Producción de combustibles renovables y desechos (TJ)
Производство возобновляемых видов топлива и отходов (ТДж)

	1971	1973	1978	1984	1985	1986	1987	1988	1989
République Islamique d'Iran	13 873	14 645	24 257	27 092	28 527	28 703	27 206	29 629	27 658
Irak	932	868	1 029	791	850	850	908	908	908
Israël	156	127	107	107	107	107	107	107	68
Jordanie	39	39	29	49	49	49	49	59	59
Koweit	46	40	18	21	21	21	21	21	21
Liban	4 034	4 239	4 513	4 366	4 366	4 376	4 386	4 396	4 484
Syrie	30	142	75	147	147	147	147	137	156
Yémen	2 040	2 100	2 400	2 760	2 880	2 940	3 060	3 120	3 240
Moyen-Orient	**21 150**	**22 200**	**32 428**	**35 333**	**36 947**	**37 193**	**35 884**	**38 377**	**36 594**
Total non-OCDE	24 002 032	24 883 057	27 373 884	30 777 856	31 321 491	31 748 734	32 389 962	32 834 707	33 371 201
OCDE Amérique du N.	2 129 642	2 240 726	2 739 805	3 420 154	3 389 486	3 477 112	3 761 835	3 721 931	3 564 489
OCDE Pacifique	148 975	147 710	174 926	405 758	419 551	408 688	422 026	445 049	466 195
OCDE Europe	894 403	943 151	1 102 218	1 438 525	1 460 052	1 462 220	1 469 236	1 557 532	1 947 012
Total OCDE	**3 173 020**	**3 331 587**	**4 016 949**	**5 264 437**	**5 269 089**	**5 348 020**	**5 653 097**	**5 724 512**	**5 977 696**
Monde	27 175 052	28 214 644	31 390 833	36 042 293	36 590 580	37 096 754	38 043 059	38 559 219	39 348 897

Avant 1978 les données pour les pays de l'OCDE sont incomplètes.

Production of Combustible Renewables and Waste (TJ)
Production d'énergies renouvelables combustibles et déchets (TJ)
Erzeugung von erneuerbaren Brennstoffen und Abfällen (TJ)
Produzione di energia da combustibili rinnovabili e da rifiuti (TJ)
可燃性再生可能エネルギー及び廃棄物の生産量 *(TJ)*
Producción de combustibles renovables y desechos (TJ)
Производство возобновляемых видов топлива и отходов (ТДж)

1990	1991	1992	1993	1994	1995	1996	1997	1998	
28 591	29 021	30 143	30 391	30 700	30 100	32 914	32 914	32 914	Islamic Republic of Iran
967	967	1 026	1 026	1 050	1 100	1 100	1 100	1 100	Iraq
127	127	127	127	127	127	127	152	183	Israel
59	68	68	68	78	78	78	91	106	Jordan
21	21	-	-	-	-	-	-	-	Kuwait
4 181	4 308	4 405	4 601	4 800	4 960	5 060	5 152	5 245	Lebanon
127	225	244	200	200	200	200	200	200	Syria
3 240	3 240	3 240	3 240	3 240	3 240	3 240	3 240	3 240	Yemen
37 313	**37 977**	**39 253**	**39 653**	**40 195**	**39 805**	**42 719**	**42 849**	**42 988**	**Middle East**
33 870 863	34 401 881	34 926 884	35 230 543	35 548 140	36 029 281	36 602 201	37 169 436	37 537 236	**Non-OECD Total**
3 260 627	3 347 236	3 733 661	3 466 163	3 583 040	3 689 103	3 753 521	3 716 113	3 937 966	OECD North America
473 247	478 067	459 259	484 211	512 429	541 059	579 603	615 117	624 539	OECD Pacific
1 909 124	1 941 984	1 972 616	2 152 697	2 193 219	2 257 356	2 336 978	2 407 019	2 480 549	OECD Europe
5 642 998	**5 767 287**	**6 165 536**	**6 103 071**	**6 288 688**	**6 487 518**	**6 670 102**	**6 738 249**	**7 043 054**	**OECD Total**
39 513 861	**40 169 168**	**41 092 420**	**41 333 614**	**41 836 828**	**42 516 799**	**43 272 303**	**43 907 685**	**44 580 290**	**World**

Prior to 1978 data for OECD countries are incomplete.

Production of Charcoal (TJ)
Production de charbon de bois (TJ)
Erzeugung von Holzkohle (TJ)

Produzione di carbone di legna (TJ)

木炭の生産量 （TJ）

Producción de carbón vegetal (TJ)

Производство древесного угля (ТДж)

	1992	1993	1994	1995	1996	1997	1998	
Angola	-	-	22 000	22 500	23 355	24 032	24 753	Angola
Bénin	-	-	500	500	510	524	539	Benin
Cameroun	-	-	2 700	2 700	2 778	2 859	2 937	Cameroon
Congo	-	-	450	450	460	473	486	Congo
RD du Congo	-	-	7 250	7 500	7 740	7 988	8 246	DR of Congo
Côte d'Ivoire	-	-	18 800	19 250	19 866	20 442	20 846	Cote d'Ivoire
Ethiopie	-	-	3 805	3 896	3 856	3 914	5 737	Ethiopia
Erythrée	-	-	968	999	1 063	1 098	574	Eritrea
Ghana	-	-	15 700	16 100	16 486	16 865	17 310	Ghana
Kenya	-	-	41 500	42 000	42 714	43 526	44 565	Kenya
Mozambique	-	-	9 543	9 425	9 425	9 425	9 425	Mozambique
Nigéria	-	-	23 300	24 000	24 744	25 511	26 192	Nigeria
Sénégal	-	-	3 732	3 800	3 929	4 063	4 063	Senegal
Afrique du Sud	-	-	41 000	41 000	41 697	42 364	43 127	South Africa
Soudan	-	-	71 500	72 500	88 635	90 515	92 505	Sudan
Rép.-Unie de Tanzanie	-	-	14 500	14 500	14 761	15 041	15 431	United Rep. of Tanzania
Togo	-	-	6 702	6 905	7 136	7 338	7 569	Togo
Tunisie	-	-	4 600	4 677	4 747	4 818	4 965	Tunisia
Zambie	-	-	17 000	17 250	17 543	17 929	18 351	Zambia
Autre Afrique	-	-	26 931	27 431	28 271	29 597	30 712	Other Africa
Afrique	**-**	**-**	**332 481**	**337 383**	**359 716**	**368 322**	**378 333**	**Africa**
Argentine	-	-	9 253	10 178	10 383	7 997	8 165	Argentina
Bolivie	-	-	60	60	96	96	126	Bolivia
Brésil	-	-	232 413	213 913	197 225	189 679	181 889	Brazil
Colombie	-	-	1 613	1 643	1 658	1 829	1 872	Colombia
Costa Rica	-	-	136	148	148	154	157	Costa Rica
Cuba	-	-	1 239	1 321	1 474	1 546	1 592	Cuba
République dominicaine	-	-	8 843	9 448	10 081	10 629	10 773	Dominican Republic
El Salvador	-	-	490	490	517	517	571	El Salvador
Guatemala	-	-	918	975	946	975	1 032	Guatemala
Haiti	-	-	9 764	9 764	10 082	8 147	8 291	Haiti
Honduras	-	-	356	377	419	419	481	Honduras
Jamaïque	-	-	2 051	2 138	2 224	2 282	2 369	Jamaica
Nicaragua	-	-	880	880	733	762	850	Nicaragua
Panama	-	-	103	107	107	107	112	Panama
Paraguay	-	-	6 875	6 500	7 078	7 280	6 353	Paraguay
Pérou	-	-	5 434	3 478	5 543	5 610	5 681	Peru
Uruguay	-	-	71	63	54	63	59	Uruguay
Vénézuela	-	-	200	200	200	200	200	Venezuela
Autre Amérique latine	-	-	87	87	88	89	91	Other Latin America
Amérique latine	**-**	**-**	**280 786**	**261 770**	**249 056**	**238 381**	**230 664**	**Latin America**
Indonésie	-	-	5 338	5 419	5 501	5 581	6 285	Indonesia
Malaisie	-	-	12 740	13 028	13 317	13 317	13 343	Malaysia
Myanmar	-	-	9 900	10 000	10 157	10 313	10 134	Myanmar
Pakistan	-	-	5 500	6 000	6 122	6 246	6 383	Pakistan
Philippines	-	-	-	12 700	12 720	29 089	26 614	Philippines
Sri Lanka	-	-	368	4	3	1	3	Sri Lanka
Thaïlande	-	-	115 652	103 871	94 828	93 999	93 459	Thailand
Viêt-Nam	-	-	16 300	17 000	17 801	18 161	18 559	Vietnam
Asie	**-**	**-**	**165 798**	**168 022**	**160 449**	**176 707**	**174 780**	**Asia**

Production of Charcoal (TJ)
Production de charbon de bois (TJ)
Erzeugung von Holzkohle (TJ)
Produzione di carbone di legna (TJ)
木炭の生産量（TJ）
Producción de carbón vegetal (TJ)
Производство древесного угля (ТДж)

	1992	1993	1994	1995	1996	1997	1998	
Chypre	-	-	142	136	115	101	110	Cyprus
Roumanie	-	-	1 618	1 733	1 762	1 762	1 762	Romania
Europe non-OCDE	-	-	**1 760**	**1 869**	**1 877**	**1 863**	**1 872**	**Non-OECD Europe**
Russie	-	-	175	175	175	175	-	Russia
Ex-URSS	-	-	**175**	**175**	**175**	**175**	-	**Former USSR**
Rép. Islamique d'Iran	-	-	2 687	2 744	2 773	2 666	2 564	Islamic Republic of Iran
Irak	-	-	433	462	462	462	462	Iraq
Liban	-	-	607	636	636	665	695	Lebanon
Syrie	-	-	20	20	20	20	20	Syria
Yémen	-	-	1 560	1 560	1 560	1 560	1 560	Yemen
Moyen-Orient	-	-	**5 307**	**5 422**	**5 451**	**5 373**	**5 301**	**Middle East**
Total non-OCDE	-	-	**786 307**	**774 641**	**776 724**	**790 821**	**790 950**	**Non-OECD Total**
Monde	-	-	**786 307**	**774 641**	**776 724**	**790 821**	**790 950**	**World**

Production of Electricity from Fossil Fuels (GWh)
Production d'électricité à partir de combustibles fossiles (GWh)
Elektrizitätserzeugung aus fossilen Energieträgern (GWh)

Produzione di energia elettrica da combustibili fossili (GWh)

化石燃料発電量 (GWh)

Producción de electricidad a partir de combustibles fósiles (GWh)

Производство электроэнергии из ископаемых видов топлива (ГВт.ч)

	1971	1973	1978	1984	1985	1986	1987	1988	1989
Algérie	1 899	2 054	5 230	10 730	11 628	12 731	12 223	13 783	15 098
Angola	137	170	57	69	105	105	109	114	109
Bénin	-	9	5	61	24	26	26	20	20
Cameroun	29	51	78	37	53	47	57	51	42
Congo	33	46	6	4	4	4	6	6	6
RD du Congo	108	80	80	140	144	147	149	144	22
Côte d'Ivoire	449	628	1 251	1 316	484	668	1 197	1 069	607
Egypte	2 957	2 950	5 830	19 416	22 795	24 183	29 187	30 580	31 675
Ethiopie	289	258	161	165	156	139	110	102	94
Erythrée	-	-	-	-	-	-	-	-	-
Gabon	114	160	140	185	194	195	197	201	201
Ghana	35	38	52	20	31	30	5	34	29
Kenya	288	385	309	225	139	259	384	126	109
Libye	508	1 127	3 363	9 200	11 844	13 280	15 600	16 551	16 700
Maroc	770	1 683	2 972	6 555	6 859	7 121	7 176	8 048	7 867
Mozambique	450	450	72	154	175	137	156	171	188
Nigéria	242	749	2 324	6 367	7 340	7 135	7 981	7 646	8 672
Sénégal	344	404	580	765	761	759	822	848	878
Afrique du Sud	54 535	63 405	82 589	130 775	135 445	136 030	142 815	143 083	148 462
Soudan	250	183	320	298	492	466	480	486	470
Rép. Unie de Tanzanie	182	286	175	140	130	112	121	126	90
Togo	77	63	65	110	35	31	35	43	33
Tunisie	874	1 106	2 053	3 814	4 106	4 439	4 658	5 154	5 380
Zambie	308	271	139	89	77	87	73	76	95
Zimbabwe	1 177	1 685	707	1 080	1 928	2 833	5 237	5 366	5 240
Autre Afrique	1 705	1 980	2 755	3 335	3 377	3 695	4 093	4 341	4 469
Afrique	**67 760**	**80 221**	**111 313**	**195 050**	**208 326**	**214 659**	**232 897**	**238 169**	**246 556**
Argentine	22 023	23 610	22 725	20 350	18 737	22 176	23 625	31 294	32 388
Bolivie	111	158	384	481	518	567	611	659	796
Brésil	7 744	6 057	8 621	8 141	8 819	16 410	13 197	11 927	11 975
Chili	3 646	3 117	3 016	3 320	2 774	2 568	2 522	4 222	6 945
Colombie	2 837	4 657	5 046	8 731	8 520	7 976	8 155	8 101	7 466
Costa Rica	110	209	439	53	50	48	123	137	82
Cuba	4 132	4 955	7 486	11 134	11 051	11 962	12 385	13 195	13 918
République dominicaine	478	1 585	1 974	2 364	2 865	2 750	2 921	2 175	2 701
Equateur	610	821	1 814	1 303	1 279	1 000	814	815	815
El Salvador	205	398	192	69	120	73	331	253	166
Guatemala	393	594	1 190	1 024	1 086	73	254	248	207
Haiti	43	26	43	210	210	105	158	127	153
Honduras	149	127	112	320	90	45	50	54	49
Jamaïque	1 447	1 883	1 343	1 439	1 436	1 522	1 659	1 641	1 583
Antilles néerlandaises	710	775	825	818	760	637	692	700	737
Nicaragua	435	330	805	338	349	469	535	505	354
Panama	862	1 195	874	822	572	539	692	407	473
Paraguay	19	34	242	9	18	22	18	27	4
Pérou	1 372	1 588	2 289	2 881	2 500	2 691	2 951	2 957	3 048
Trinité-et-Tobago	956	1 076	1 528	2 993	3 006	3 265	3 450	3 458	3 402
Uruguay	901	963	1 404	109	102	80	260	1 527	1 797
Vénézuela	8 199	10 220	13 252	24 897	24 642	24 953	23 289	24 042	22 954
Autre Amérique latine	1 996	2 614	3 226	3 427	3 595	3 949	4 307	4 607	4 474
Amérique latine	**59 378**	**66 992**	**78 830**	**95 233**	**93 099**	**103 880**	**102 999**	**113 078**	**116 487**

Production of Electricity from Fossil Fuels (GWh)
Production d'électricité à partir de combustibles fossiles (GWh)
Elektrizitätserzeugung aus fossilen Energieträgern (GWh)
Produzione di energia elettrica da combustibili fossili (GWh)
化石燃料発電量 (GWh)
Producción de electricidad a partir de combustibles fósiles (GWh)
Производство электроэнергии из ископаемых видов топлива (ГВт.ч)

1990	1991	1992	1993	1994	1995	1996	1997	1998	
15 642	16 365	17 634	18 262	19 329	19 042	20 202	21 183	22 866	Algeria
116	162	107	60	60	60	103	111	106	Angola
21	23	25	26	52	33	47	59	60	Benin
41	39	35	30	32	32	34	36	36	Cameroon
3	2	4	3	2	2	2	2	2	Congo
14	9	7	105	118	115	130	130	122	DR of Congo
557	606	668	1 102	1 314	1 272	1 450	2 114	2 626	Cote d'Ivoire
32 324	34 396	35 978	37 311	38 333	40 587	43 585	46 441	50 744	Egypt
74	60	27	45	35	42	44	48	48	Ethiopia
-	-	146	151	161	172	193	212	194	Eritrea
210	207	207	212	215	215	236	267	295	Gabon
11	7	1	22	28	19	28	28	28	Ghana
231	160	147	131	210	334	467	457	1 161	Kenya
16 800	16 800	16 950	17 000	17 800	18 000	18 300	18 974	19 496	Libya
8 408	7 979	8 755	9 467	10 730	12 193	11 093	11 809	12 377	Morocco
170	148	91	48	29	29	30	40	39	Mozambique
8 177	8 236	8 775	8 933	9 969	8 983	9 491	9 745	10 123	Nigeria
901	925	1 007	991	1 025	1 086	1 155	1 226	1 286	Senegal
155 926	157 192	156 443	165 835	170 164	174 721	184 952	192 705	187 758	South Africa
557	623	545	595	737	892	989	1 096	924	Sudan
79	96	166	177	211	302	243	485	76	United Rep. of Tanzania
62	91	39	59	99	53	48	50	93	Togo
5 765	5 921	6 422	6 561	6 932	7 629	7 859	8 438	9 030	Tunisia
39	41	40	42	42	42	42	42	42	Zambia
5 571	5 361	5 327	5 683	5 890	5 973	5 160	5 174	4 728	Zimbabwe
4 586	4 822	5 057	5 344	5 343	5 460	5 418	5 451	5 591	Other Africa
256 285	**260 271**	**264 603**	**278 195**	**288 860**	**297 288**	**311 301**	**326 323**	**329 851**	**Africa**
25 476	29 546	29 314	29 852	29 600	33 000	39 176	36 181	39 932	Argentina
905	950	1 116	1 114	1 470	1 712	1 760	1 861	2 117	Bolivia
10 497	11 487	11 731	11 513	11 890	13 583	16 327	18 461	19 450	Brazil
7 945	5 228	3 493	4 013	5 778	7 518	11 057	13 520	17 681	Chile
8 354	9 177	10 781	9 994	8 897	10 934	8 444	13 751	14 470	Colombia
86	198	585	496	826	825	466	190	396	Costa Rica
13 774	12 130	10 387	10 212	11 198	11 833	12 404	13 257	13 281	Cuba
3 277	3 311	3 620	4 294	4 503	4 722	5 552	5 967	6 106	Dominican Republic
1 360	1 896	2 210	1 594	1 605	3 297	3 011	3 174	3 363	Ecuador
153	610	575	896	1 310	1 445	1 084	1 706	1 796	El Salvador
194	253	983	608	701	717	590	678	1 021	Guatemala
123	119	102	90	48	242	333	405	357	Haiti
40	34	151	304	518	1 055	1 028	1 218	1 548	Honduras
1 786	1 854	2 011	3 505	4 430	5 450	5 649	5 820	6 017	Jamaica
790	801	853	909	971	1 009	1 056	1 054	1 116	Netherlands Antilles
565	631	827	773	922	1 033	1 155	1 219	1 672	Nicaragua
572	909	1 022	1 108	1 270	1 170	1 080	1 444	2 208	Panama
7	5	8	9	4	66	69	6	12	Paraguay
2 857	2 537	3 210	2 893	1 862	3 828	3 842	4 584	4 647	Peru
3 546	3 695	3 946	3 789	4 037	4 274	4 524	4 970	5 151	Trinidad and Tobago
395	856	917	622	95	414	867	638	380	Uruguay
22 338	18 795	20 169	21 904	19 937	21 996	21 744	20 798	22 948	Venezuela
4 816	5 048	5 240	5 422	5 487	5 667	5 706	5 952	6 095	Other Latin America
109 856	**110 070**	**113 251**	**115 914**	**117 359**	**135 790**	**146 924**	**156 854**	**171 764**	**Latin America**

Production of Electricity from Fossil Fuels (GWh)
Production d'électricité à partir de combustibles fossiles (GWh)
Elektrizitätserzeugung aus fossilen Energieträgern (GWh)
Produzione di energia elettrica da combustibili fossili (GWh)
化石燃料発電量 (GWh)
Producción de electricidad a partir de combustibles fósiles (GWh)
Производство электроэнергии из ископаемых видов топлива (ГВт.ч)

	1971	1973	1978	1984	1985	1986	1987	1988	1989
Bangladesh	855	1 073	1 713	3 069	3 789	4 350	5 070	5 866	6 195
Brunei	262	258	262	750	766	776	1 043	1 098	1 131
Inde	37 161	41 418	60 188	111 164	127 369	142 398	166 486	177 607	201 959
Indonésie	1 620	2 021	4 235	12 442	13 681	14 288	17 123	19 376	21 923
RPD de Corée	5 310	7 500	12 500	18 000	20 000	21 000	21 100	21 500	21 750
Malaisie	2 749	3 665	7 357	10 320	11 204	12 041	12 494	13 667	16 129
Myanmar	214	245	507	879	1 116	1 203	1 296	1 291	1 254
Népal	19	23	21	-	-	-	12	24	-
Pakistan	3 789	3 718	4 702	8 723	10 416	11 355	12 951	16 147	17 601
Philippines	7 212	11 311	12 733	11 370	12 284	11 204	12 863	13 431	13 772
Singapour	2 585	3 719	5 898	9 490	9 960	10 640	11 909	13 113	14 136
Sri Lanka	66	323	19	170	69	7	530	202	56
Taipei chinois	12 720	17 336	28 210	23 195	19 901	27 969	28 930	38 925	49 097
Thaïlande	3 035	5 091	10 527	16 943	19 383	19 163	24 577	28 686	31 836
Viêt-Nam	1 686	1 930	2 800	3 179	3 592	4 120	4 667	4 994	3 946
Autre Asie	1 345	2 635	2 497	2 252	2 434	2 739	2 959	2 683	2 755
Asie	**80 628**	**102 266**	**154 169**	**231 946**	**255 964**	**283 253**	**324 010**	**358 610**	**403 540**
Rép. populaire de Chine	108 400	128 800	211 952	290 610	318 320	355 000	397 260	436 060	466 410
Hong-Kong, Chine	5 574	6 799	10 370	17 918	19 230	21 406	23 746	25 501	27 361
Chine	**113 974**	**135 599**	**222 322**	**308 528**	**337 550**	**376 406**	**421 006**	**461 561**	**493 771**
Albanie	525	575	700	442	203	1 706	1 045	394	533
Bulgarie	18 846	19 386	22 673	28 677	26 265	27 424	28 499	26 395	27 071
Chypre	665	830	919	1 250	1 319	1 423	1 501	1 647	1 831
Gibraltar	47	49	52	61	63	65	70	74	81
Malte	305	365	459	700	784	850	944	1 030	1 100
Roumanie	34 959	39 232	53 641	60 341	58 218	63 590	61 655	60 546	62 173
République slovaque	9 389	10 745	13 154	11 153	10 446	10 338	9 677	9 263	9 281
Bosnie-Herzegovine	-	-	-	-	-	-	-	-	-
Croatie	-	-	-	-	-	-	-	-	-
Ex-RYM	-	-	-	-	-	-	-	-	-
Slovénie	-	-	-	-	-	-	-	-	-
RF de Yougoslavie	-	-	-	-	-	-	-	-	-
Ex-Yougoslavie	13 865	18 668	26 051	42 673	46 479	46 393	50 044	53 645	55 289
Europe non-OCDE	**78 601**	**89 850**	**117 649**	**145 297**	**143 777**	**151 789**	**153 435**	**152 994**	**157 359**
Arménie	-	-	-	-	-	-	-	-	-
Azerbaïdjan	-	-	-	-	-	-	-	-	-
Bélarus	-	-	-	-	-	-	-	-	-
Estonie	-	-	-	-	-	-	-	-	-
Géorgie	-	-	-	-	-	-	-	-	-
Kazakhstan	-	-	-	-	-	-	-	-	-
Lettonie	-	-	-	-	-	-	-	-	-
Lituanie	-	-	-	-	-	-	-	-	-
République de Moldavie	-	-	-	-	-	-	-	-	-
Russie	-	-	-	-	-	-	-	-	-
Tadjikistan	-	-	-	-	-	-	-	-	-
Turkménistan	-	-	-	-	-	-	-	-	-
Ukraine	-	-	-	-	-	-	-	-	-
Ouzbékistan	-	-	-	-	-	-	-	-	-
Ex-URSS	**645 600**	**757 500**	**956 200**	**1 122 900**	**1 139 497**	**1 200 462**	**1 234 875**	**1 236 320**	**1 264 605**

Production of Electricity from Fossil Fuels (GWh)
Production d'électricité à partir de combustibles fossiles (GWh)
Elektrizitätserzeugung aus fossilen Energieträgern (GWh)
Produzione di energia elettrica da combustibili fossili (GWh)
化石燃料発電量 (GWh)
Producción de electricidad a partir de combustibles fósiles (GWh)
Производство электроэнергии из ископаемых видов топлива (ГВт.ч)

1990	1991	1992	1993	1994	1995	1996	1997	1998	
6 848	7 432	8 098	8 598	8 937	10 434	10 735	11 139	12 017	Bangladesh
1 172	1 269	1 408	1 547	1 663	1 966	2 123	2 407	2 503	Brunei
211 610	237 294	256 050	280 402	297 125	337 462	357 546	378 469	399 391	India
29 337	30 897	32 489	37 695	42 933	49 577	56 311	68 866	65 637	Indonesia
21 750	21 750	14 000	14 000	13 500	13 000	12 500	11 650	11 067	DPR of Korea
19 013	22 129	24 942	29 846	32 554	39 213	46 198	53 974	55 819	Malaysia
1 285	1 437	1 478	1 680	1 980	2 431	2 294	2 790	3 121	Myanmar
-	36	48	60	84	36	36	108	120	Nepal
20 455	22 388	26 400	27 116	30 713	30 186	33 267	37 921	39 719	Pakistan
13 717	14 752	15 917	16 115	18 222	21 198	23 055	26 747	27 195	Philippines
15 714	16 921	17 674	18 962	20 849	22 244	23 458	26 188	28 283	Singapore
5	261	640	183	298	350	1 278	1 698	1 768	Sri Lanka
49 425	58 889	64 361	75 398	82 172	91 082	98 098	107 458	119 047	Chinese Taipei
39 199	45 599	52 859	59 703	66 662	73 083	79 860	83 893	81 696	Thailand
3 337	2 973	2 598	2 797	3 368	4 083	4 936	7 474	10 573	Vietnam
2 803	2 921	2 824	2 841	2 951	3 244	3 108	3 112	3 112	Other Asia
435 670	**486 948**	**521 786**	**576 943**	**624 011**	**699 589**	**754 803**	**823 894**	**861 068**	**Asia**
494 480	552 460	622 965	683 877	745 927	804 316	877 713	924 070	944 100	People's Rep. of China
28 938	31 807	34 907	35 950	26 743	27 916	28 442	28 945	31 417	Hong Kong, China
523 418	**584 267**	**657 872**	**719 827**	**772 670**	**832 232**	**906 155**	**953 015**	**975 517**	**China**
341	168	131	170	133	210	242	212	207	Albania
25 598	23 292	21 995	22 082	21 330	22 214	21 715	22 116	21 485	Bulgaria
1 974	2 077	2 404	2 581	2 681	2 473	2 592	2 711	2 954	Cyprus
82	82	88	88	88	88	93	93	93	Gibraltar
1 100	1 419	1 490	1 500	1 541	1 632	1 658	1 686	1 721	Malta
52 898	42 215	42 438	42 639	42 090	42 570	44 209	34 228	29 299	Romania
9 516	9 163	8 965	8 530	8 051	9 643	9 496	9 392	9 505	Slovak Republic
-	-	9 200	9 700	789	901	930	954	985	*Bosnia-Herzegovina*
-	-	4 547	4 995	3 339	3 593	3 310	4 384	5 427	*Croatia*
-	-	5 217	4 658	4 816	5 313	5 639	5 819	5 965	*FYROM*
-	-	4 702	4 714	4 623	4 634	4 543	5 065	5 253	*Slovenia*
-	-	25 145	23 442	23 401	25 956	26 592	28 141	28 398	*FR of Yugoslavia*
58 484	51 304	48 811	47 509	36 968	40 397	41 014	44 363	46 028	Former Yugoslavia
149 993	**129 720**	**126 322**	**125 099**	**112 882**	**119 227**	**121 019**	**114 801**	**111 292**	**Non-OECD Europe**
-	-	5 960	2 002	2 144	3 338	2 318	3 032	3 064	Armenia
-	-	17 926	16 700	15 742	15 488	15 550	15 124	16 034	Azerbaijan
-	-	33 000	33 350	31 378	24 898	23 712	26 036	25 280	Belarus
-	-	11 830	9 117	9 148	8 685	9 096	9 207	8 504	Estonia
-	-	5 005	3 116	2 090	1 590	1 214	1 125	1 697	Georgia
-	-	75 835	69 815	57 218	58 333	51 707	45 507	43 000	Kazakhstan
-	-	1 314	1 049	1 135	1 042	1 263	1 503	1 420	Latvia
-	-	1 591	1 579	1 633	1 369	1 973	2 068	3 182	Lithuania
-	-	10 989	9 890	7 950	5 744	5 756	4 891	4 500	Republic of Moldova
-	-	711 201	632 112	574 853	580 805	582 803	567 212	563 923	Russia
-	-	894	623	291	172	175	292	273	Tajikistan
-	-	13 179	12 632	10 492	9 796	10 095	9 493	9 411	Turkmenistan
-	-	170 723	143 426	121 747	113 345	94 576	88 537	81 665	Ukraine
-	-	44 630	41 791	40 644	41 265	38 894	40 277	40 142	Uzbekistan
1 261 500	**1 212 758**	**1 104 077**	**977 202**	**876 465**	**865 870**	**839 132**	**814 304**	**802 095**	**Former USSR**

Production of Electricity from Fossil Fuels (GWh)
Production d'électricité à partir de combustibles fossiles (GWh)

Elektrizitätserzeugung aus fossilen Energieträgern (GWh)

Produzione di energia elettrica da combustibili fossili (GWh)

化石燃料発電量 (GWh)

Producción de electricidad a partir de combustibles fósiles (GWh)

Производство электроэнергии из ископаемых видов топлива (ГВт.ч)

	1971	1973	1978	1984	1985	1986	1987	1988	1989
Bahrein	430	500	1 212	2 187	2 637	2 892	2 996	3 162	3 293
République Islamique d'Iran	5 411	9 230	13 757	30 844	33 670	34 054	37 807	40 289	45 190
Irak	2 600	3 229	7 127	18 849	20 363	21 687	19 910	20 850	21 260
Israël	7 639	8 720	11 874	14 663	15 430	16 028	17 667	19 342	20 450
Jordanie	230	315	704	2 265	2 495	2 952	3 467	3 236	3 416
Koweit	2 623	3 651	6 983	13 894	15 417	16 934	18 092	19 599	21 085
Liban	536	1 313	1 500	3 215	3 276	3 610	3 990	3 400	2 000
Oman	13	47	468	2 016	2 498	3 180	3 392	3 773	3 927
Qatar	316	419	1 349	3 563	3 949	4 303	4 371	4 501	4 624
Arabie saoudite	2 082	2 949	10 551	40 069	44 311	47 646	50 649	57 229	61 568
Syrie	1 294	1 406	776	4 433	5 038	6 309	5 866	4 807	5 728
Emirats arabes unis	203	720	3 764	11 000	12 000	12 814	13 657	14 840	15 612
Yémen	209	206	360	856	895	1 117	1 165	1 318	1 345
Moyen-Orient	**23 586**	**32 705**	**60 425**	**147 854**	**161 979**	**173 526**	**183 029**	**196 346**	**209 498**
Total non-OCDE	**1 069 527**	**1 265 133**	**1 700 908**	**2 246 808**	**2 340 192**	**2 503 975**	**2 652 251**	**2 757 078**	**2 891 816**
OCDE Amérique du N.	1 470 253	1 689 133	1 852 193	2 041 151	2 078 047	2 053 320	2 166 448	2 275 747	2 452 509
OCDE Pacifique	342 226	457 808	531 183	569 187	560 870	555 519	587 223	632 548	682 392
OCDE Europe	1 009 654	1 181 174	1 319 834	1 255 997	1 259 733	1 276 547	1 297 648	1 275 508	1 358 715
Total OCDE	**2 822 133**	**3 328 115**	**3 703 210**	**3 866 335**	**3 898 650**	**3 885 386**	**4 051 319**	**4 183 803**	**4 493 616**
Monde	**3 891 660**	**4 593 248**	**5 404 118**	**6 113 143**	**6 238 842**	**6 389 361**	**6 703 570**	**6 940 881**	**7 385 432**

Production of Electricity from Fossil Fuels (GWh)
Production d'électricité à partir de combustibles fossiles (GWh)

Elektrizitätserzeugung aus fossilen Energieträgern (GWh)

Produzione di energia elettrica da combustibili fossili (GWh)

化石燃料発電量 (GWh)

Producción de electricidad a partir de combustibles fósiles (GWh)

Производство электроэнергии из ископаемых видов топлива (ГВт.ч)

1990	1991	1992	1993	1994	1995	1996	1997	1998	
3 482	3 320	3 896	4 245	4 550	4 612	5 016	5 040	5 773	Bahrain
53 019	57 070	59 089	66 191	74 574	77 694	83 475	90 836	96 398	Islamic Republic of Iran
21 400	19 910	24 600	25 700	27 440	28 430	28 430	28 980	29 765	Iraq
20 895	21 508	24 657	25 974	28 293	30 400	32 533	35 073	37 939	Israel
3 626	3 716	4 406	4 738	5 061	5 597	6 035	6 244	6 729	Jordan
18 477	10 780	16 885	20 178	22 802	23 724	25 475	26 724	29 984	Kuwait
1 000	2 440	3 051	3 776	4 367	4 564	6 167	7 614	7 570	Lebanon
4 501	4 625	5 113	5 833	6 197	6 460	6 802	7 318	8 198	Oman
4 818	4 653	5 153	5 525	5 814	5 976	6 575	6 868	8 125	Qatar
64 899	69 212	74 009	82 183	91 019	100 748	106 606	113 125	116 519	Saudi Arabia
5 963	5 930	5 176	5 929	8 382	8 356	9 941	10 572	10 783	Syria
17 080	17 351	18 689	21 730	23 736	24 982	26 572	28 464	33 392	United Arab Emirates
1 663	1 802	1 953	2 051	2 159	2 369	2 334	2 557	2 507	Yemen
220 823	**222 317**	**246 677**	**274 053**	**304 394**	**323 912**	**345 961**	**369 415**	**393 682**	**Middle East**
2 957 545	**3 006 351**	**3 034 588**	**3 067 233**	**3 096 641**	**3 273 908**	**3 425 295**	**3 558 606**	**3 645 269**	**Non-OECD Total**
2 411 492	2 448 081	2 487 835	2 587 799	2 655 102	2 680 015	2 736 616	2 856 746	3 005 983	OECD North America
734 925	753 990	782 743	764 421	873 734	875 242	907 267	929 826	913 105	OECD Pacific
1 373 393	1 398 043	1 369 649	1 330 061	1 370 291	1 421 225	1 471 426	1 474 338	1 530 193	OECD Europe
4 519 810	**4 600 114**	**4 640 227**	**4 682 281**	**4 899 127**	**4 976 482**	**5 115 309**	**5 260 910**	**5 449 281**	**OECD Total**
7 477 355	**7 606 465**	**7 674 815**	**7 749 514**	**7 995 768**	**8 250 390**	**8 540 604**	**8 819 516**	**9 094 550**	**World**

Production of Nuclear Electricity (GWh)
Production d'électricité d'origine nucléaire (GWh)
Elektrizitätserzeugung in Kernkraftwerken (GWh)
Produzione di energia nucleotermoelettrica (GWh)
原子力発電量 (GWh)
Producción de electricidad nuclear (GWh)
Производство атомной электроэнергии (ГВт.ч)

	1971	1973	1978	1984	1985	1986	1987	1988	1989
Afrique du Sud	-	-	-	3 925	5 315	8 803	6 167	10 493	11 099
Afrique	-	-	-	**3 925**	**5 315**	**8 803**	**6 167**	**10 493**	**11 099**
Argentine	-	-	2 896	4 641	5 766	5 711	6 465	5 798	5 039
Brésil	-	-	-	1 643	3 381	144	973	608	1 830
Amérique latine	-	-	**2 896**	**6 284**	**9 147**	**5 855**	**7 438**	**6 406**	**6 869**
Inde	1 189	2 396	2 770	4 075	4 982	5 022	5 035	5 817	4 625
Pakistan	104	304	231	324	346	430	502	254	30
Taipei chinois	-	-	2 670	24 589	28 727	26 941	33 128	30 651	28 276
Asie	**1 293**	**2 700**	**5 671**	**28 988**	**34 055**	**32 393**	**38 665**	**36 722**	**32 931**
Rép. populaire de Chine	-	-	-	-	-	-	-	-	-
Chine	**-**	**-**	**-**	**-**	**-**	**-**	**-**	**-**	**-**
Bulgarie	-	-	5 910	12 735	13 131	12 070	12 436	16 030	14 566
Roumanie	-	-	-	-	-	-	-	-	-
République slovaque	-	232	19	7 239	9 382	11 716	11 513	11 474	12 157
Slovénie	-	-	-	-	-	-	-	-	-
Ex-Yougoslavie	-	-	-	4 420	4 053	4 019	4 495	4 135	4 688
Europe non-OCDE	-	**232**	**5 929**	**24 394**	**26 566**	**27 805**	**28 444**	**31 639**	**31 411**
Arménie	-	-	-	-	-	-	-	-	-
Lituanie	-	-	-	-	-	-	-	-	-
Russie	-	-	-	-	-	-	-	-	-
Ukraine	-	-	-	-	-	-	-	-	-
Ex-URSS	**6 100**	**12 000**	**50 000**	**142 000**	**167 000**	**161 000**	**189 000**	**215 700**	**212 600**
Total non-OCDE	**7 393**	**14 932**	**64 496**	**205 591**	**242 083**	**235 856**	**269 714**	**300 960**	**294 910**
OCDE Amérique du N.	44 819	104 421	324 189	399 502	467 233	510 147	559 847	641 458	641 409
OCDE Pacifique	8 000	9 707	61 637	146 056	176 323	196 615	227 072	218 760	230 234
OCDE Europe	50 874	74 136	175 670	504 195	606 459	658 713	681 172	730 071	772 536
Total OCDE	**103 693**	**188 264**	**561 496**	**1 049 753**	**1 250 015**	**1 365 475**	**1 468 091**	**1 590 289**	**1 644 179**
Monde	**111 086**	**203 196**	**625 992**	**1 255 344**	**1 492 098**	**1 601 331**	**1 737 805**	**1 891 249**	**1 939 089**

Production of Nuclear Electricity (GWh)
Production d'électricité d'origine nucléaire (GWh)
Elektrizitätserzeugung in Kernkraftwerken (GWh)
Produzione di energia nucleotermoelettrica (GWh)
原子力発電量 (GWh)
Producción de electricidad nuclear (GWh)
Производство атомной электроэнергии (ГВт.ч)

1990	1991	1992	1993	1994	1995	1996	1997	1998	
8 449	9 144	9 288	7 255	9 697	11 301	11 775	12 647	13 601	South Africa
8 449	**9 144**	**9 288**	**7 255**	**9 697**	**11 301**	**11 775**	**12 647**	**13 601**	**Africa**
7 281	7 756	7 081	7 750	8 235	7 066	7 459	7 961	7 453	Argentina
2 237	1 442	1 759	442	55	2 519	2 429	3 169	3 265	Brazil
9 518	**9 198**	**8 840**	**8 192**	**8 290**	**9 585**	**9 888**	**11 130**	**10 718**	**Latin America**
6 141	5 525	6 726	5 398	5 648	7 600	8 400	10 100	11 500	India
293	385	418	582	497	511	483	346	375	Pakistan
32 866	35 290	33 845	34 354	34 871	35 316	37 788	36 269	36 524	Chinese Taipei
39 300	**41 200**	**40 989**	**40 334**	**41 016**	**43 427**	**46 671**	**46 715**	**48 399**	**Asia**
-	-	-	1 604	13 906	12 833	14 339	14 418	14 100	People's Rep. of China
-	**-**	**-**	**1 604**	**13 906**	**12 833**	**14 339**	**14 418**	**14 100**	**China**
14 665	13 184	11 552	13 973	15 335	17 261	18 082	17 751	16 899	Bulgaria
-	-	-	-	-	-	1 386	5 400	5 307	Romania
12 036	11 689	11 050	11 022	12 135	11 437	11 261	10 797	11 394	Slovak Republic
-	-	3 971	3 956	4 609	4 779	4 562	5 019	5 042	*Slovenia*
4 622	4 952	3 971	3 956	4 609	4 779	4 562	5 019	5 042	Former Yugoslavia
31 323	**29 825**	**26 573**	**28 951**	**32 079**	**33 477**	**35 291**	**38 967**	**38 642**	**Non-OECD Europe**
-	-	-	-	-	304	2 324	1 600	1 589	Armenia
-	-	14 638	12 260	7 706	11 822	13 942	12 024	13 554	Lithuania
-	-	119 626	119 186	97 820	99 532	109 026	108 499	103 719	Russia
-	-	73 732	75 243	68 848	70 523	79 577	79 433	75 239	Ukraine
211 500	**212 655**	**207 996**	**206 689**	**174 374**	**182 181**	**204 869**	**201 556**	**194 101**	**Former USSR**
300 090	**302 022**	**293 686**	**293 025**	**279 362**	**292 804**	**322 833**	**325 433**	**319 561**	**Non-OECD Total**
687 493	738 571	740 471	745 855	790 993	820 093	815 857	759 347	794 900	OECD North America
255 159	269 771	279 789	307 394	327 777	358 283	376 124	396 263	422 032	OECD Pacific
770 141	795 963	809 560	843 858	842 842	860 417	901 775	911 764	907 139	OECD Europe
1 712 793	**1 804 305**	**1 829 820**	**1 897 107**	**1 961 612**	**2 038 793**	**2 093 756**	**2 067 374**	**2 124 071**	**OECD Total**
2 012 883	**2 106 327**	**2 123 506**	**2 190 132**	**2 240 974**	**2 331 597**	**2 416 589**	**2 392 807**	**2 443 632**	**World**

Production of Hydro Electricity (GWh)
Production d'électricité d'origine hydraulique (GWh)
Elektrizitätserzeugung in Wasserkraftwerken (GWh)

Produzione di energia idroelettrica (GWh)

水力発電量 (GWh)

Producción de electricidad hidráulica (GWh)

Производство гидроэлектроэнергии (ГВт.ч)

	1971	1973	1978	1984	1985	1986	1987	1988	1989
Algérie	330	752	250	452	646	250	499	183	226
Angola	605	814	528	601	701	701	701	701	710
Bénin	-	-	-	-	-	-	-	-	-
Cameroun	1 046	1 068	1 259	2 178	2 388	2 431	2 451	2 572	2 664
Congo	55	50	49	231	310	266	275	286	391
RD du Congo	3 437	3 768	4 000	4 695	5 027	5 259	5 242	5 248	6 915
Côte d'Ivoire	139	168	167	374	1 358	1 418	877	1 240	1 593
Egypte	5 041	5 156	9 320	9 633	8 663	9 281	8 658	9 000	9 974
Ethiopie	304	333	406	690	675	717	721	712	721
Gabon	-	5	296	610	667	693	698	709	675
Ghana	2 909	3 872	3 721	1 811	2 986	4 372	4 861	4 808	5 231
Kenya	339	408	1 073	1 724	2 016	1 680	1 911	2 323	2 469
Maroc	1 520	1 192	1 416	366	486	643	825	936	1 157
Mozambique	224	191	804	244	375	173	180	166	284
Nigéria	1 574	1 858	2 324	2 668	3 091	3 630	3 284	4 008	4 140
Afrique du Sud	112	985	1 907	2 554	2 731	3 408	3 391	4 565	3 798
Soudan	245	427	545	695	737	872	899	892	895
Rép. Unie de Tanzanie	309	296	520	782	885	1 034	1 151	1 252	1 419
Togo	2	1	6	1	2	2	2	2	3
Tunisie	52	73	29	66	114	54	118	49	36
Zambie	908	3 097	7 953	9 972	10 247	10 014	8 612	8 409	6 687
Zimbabwe	2 425	3 487	5 545	3 040	2 857	2 371	1 108	2 755	3 329
Autre Afrique	806	1 057	1 269	2 986	3 181	3 284	3 464	3 290	3 485
Afrique	**22 382**	**29 058**	**43 387**	**46 373**	**50 143**	**52 553**	**49 928**	**54 106**	**56 802**
Argentine	1 544	2 994	7 752	19 880	20 646	21 028	21 917	15 287	13 317
Bolivie	898	972	1 259	1 164	1 142	1 114	1 061	1 163	1 167
Brésil	43 199	57 890	102 746	166 593	178 375	182 419	185 600	199 093	204 690
Chili	4 812	5 598	7 253	9 976	11 071	11 896	12 844	12 366	10 545
Colombie	6 663	7 939	12 070	17 051	18 343	21 220	23 133	24 377	26 659
Costa Rica	1 028	1 131	1 478	3 004	2 756	2 891	3 000	3 046	3 328
Cuba	110	62	83	70	54	59	44	73	82
République dominicaine	590	583	818	1 268	1 258	1 525	1 689	1 436	521
Equateur	440	435	791	3 233	3 279	4 012	4 570	4 826	4 954
El Salvador	522	492	666	869	1 171	1 231	1 134	1 302	1 425
Guatemala	260	233	283	520	522	1 707	1 656	1 825	2 012
Haiti	27	84	190	200	216	318	320	392	406
Honduras	242	359	647	874	1 305	1 428	1 747	1 900	1 994
Jamaïque	126	99	115	150	150	150	125	101	127
Nicaragua	203	371	325	320	368	406	488	468	572
Panama	83	102	719	1 492	1 929	2 096	2 032	2 199	2 181
Paraguay	154	280	249	1 024	4 029	11 842	18 524	19 934	24 307
Pérou	4 283	4 769	6 199	8 704	9 396	10 058	10 881	10 425	10 518
Uruguay	1 469	1 556	1 631	7 108	6 452	7 294	7 263	5 421	3 903
Vénézuela	5 390	6 225	12 232	20 243	22 648	25 159	30 843	34 203	34 666
Autre Amérique latine	1 093	1 017	893	898	1 017	810	376	546	1 413
Amérique latine	**73 136**	**93 191**	**158 399**	**264 641**	**286 127**	**308 663**	**329 247**	**340 383**	**348 787**

Production of Hydro Electricity (GWh)
Production d'électricité d'origine hydraulique (GWh)
Elektrizitätserzeugung in Wasserkraftwerken (GWh)
Produzione di energia idroelettrica (GWh)
水力発電量 *(GWh)*
Producción de electricidad hidráulica (GWh)
Производство гидроэлектроэнергии (ГВт.ч)

1990	1991	1992	1993	1994	1995	1996	1997	1998	
463	980	652	1 153	554	672	448	239	749	Algeria
725	772	840	890	895	900	925	994	957	Angola
-	-	-	-	-	-	-	-	2	Benin
2 656	2 671	2 702	2 802	2 688	2 753	2 868	3 092	3 249	Cameroon
490	479	425	359	322	352	456	451	342	Congo
5 636	5 272	6 066	5 780	5 427	6 065	6 131	5 880	5 545	DR of Congo
1 464	1 288	116	1 048	971	1 727	1 777	1 868	1 366	Cote d'Ivoire
9 932	9 900	9 700	10 485	10 971	11 413	11 555	11 987	12 222	Egypt
1 062	1 082	1 151	1 263	1 354	1 428	1 510	1 566	1 579	Ethiopia
705	707	712	710	718	725	734	740	732	Gabon
5 720	6 109	6 602	6 313	6 103	6 117	6 627	6 886	7 223	Ghana
2 477	2 770	2 796	2 993	3 068	3 123	3 183	3 373	3 277	Kenya
1 220	1 226	964	443	840	605	1 938	2 062	1 759	Morocco
284	323	325	344	366	379	446	965	6 825	Mozambique
4 387	5 931	6 059	5 572	5 562	5 500	5 500	5 593	5 593	Nigeria
2 851	3 784	2 085	1 491	2 591	1 803	3 539	4 700	4 015	South Africa
958	1 038	1 089	1 091	1 121	972	1 074	1 054	1 042	Sudan
1 549	1 726	1 650	1 698	1 492	1 539	1 748	1 449	2 081	United Rep. of Tanzania
3	4	3	3	3	2	4	3	3	Togo
46	110	68	67	42	41	71	47	76	Tunisia
6 291	7 793	7 740	7 743	7 743	7 748	7 754	7 964	7 834	Zambia
3 790	3 563	2 909	1 784	1 644	1 832	2 163	2 123	1 880	Zimbabwe
3 653	3 623	3 713	3 772	3 813	3 939	3 960	3 996	4 098	Other Africa
56 362	**61 151**	**58 367**	**57 804**	**58 288**	**59 635**	**64 411**	**67 032**	**72 449**	**Africa**
18 141	16 453	19 610	24 159	27 667	26 986	22 985	28 181	26 592	Argentina
1 180	1 276	1 240	1 478	1 295	1 247	1 422	1 547	1 529	Bolivia
206 708	217 782	223 343	235 065	242 705	253 905	265 736	278 972	291 371	Brazil
10 156	14 442	18 378	19 466	18 925	19 814	18 806	18 943	16 674	Chile
27 497	27 726	22 281	27 857	32 059	32 160	35 491	31 686	30 782	Colombia
3 382	3 656	3 584	3 932	3 943	3 555	3 805	4 813	4 730	Costa Rica
91	105	81	82	49	74	95	130	97	Cuba
395	558	1 934	1 552	1 650	1 754	1 265	1 338	1 419	Dominican Republic
5 013	5 117	4 989	5 838	6 652	5 229	6 249	6 421	6 534	Ecuador
1 648	1 268	1 416	1 518	1 448	1 471	1 883	1 429	1 566	El Salvador
2 036	2 167	1 773	2 423	2 453	2 696	3 106	3 454	3 435	Guatemala
457	332	321	274	233	253	272	200	306	Haiti
2 279	2 312	2 192	2 258	1 817	1 673	2 039	2 085	1 923	Honduras
110	120	109	90	111	100	110	115	88	Jamaica
403	337	257	483	383	407	431	407	296	Nicaragua
2 213	1 906	2 314	2 274	2 072	2 334	2 816	2 685	2 122	Panama
27 158	29 305	27 116	31 408	36 394	42 092	48 035	50 613	50 834	Paraguay
10 170	11 231	9 690	11 676	12 816	12 938	13 324	13 215	13 809	Peru
7 009	6 113	7 922	7 299	7 468	5 856	5 768	6 486	9 154	Uruguay
36 983	44 542	47 271	47 478	51 284	51 450	53 844	57 268	57 956	Venezuela
1 228	1 284	1 307	1 313	1 317	1 324	1 390	1 449	1 482	Other Latin America
364 257	**388 032**	**397 128**	**427 923**	**452 741**	**467 318**	**488 872**	**511 437**	**522 699**	**Latin America**

Production of Hydro Electricity (GWh)
Production d'électricité d'origine hydraulique (GWh)
Elektrizitätserzeugung in Wasserkraftwerken (GWh)
Produzione di energia idroelettrica (GWh)
水力発電量 *(GWh)*
Producción de electricidad hidráulica (GWh)
Производство гидроэлектроэнергии (ГВт.ч)

	1971	1973	1978	1984	1985	1986	1987	1988	1989
Bangladesh	175	331	506	897	739	450	517	675	920
Inde	28 034	28 982	47 172	53 966	51 039	53 859	47 462	57 884	62 074
Indonésie	1 425	1 603	1 299	2 118	2 990	4 935	5 183	5 892	7 450
RPD de Corée	11 600	12 500	19 500	27 000	28 000	29 000	29 100	31 500	31 750
Malaisie	1 046	1 118	895	3 417	3 731	4 079	4 916	5 672	4 405
Myanmar	477	576	736	1 011	1 003	1 042	1 024	935	1 240
Népal	67	81	153	383	431	539	562	539	718
Pakistan	3 679	4 355	7 442	12 826	12 241	13 804	15 250	16 690	16 970
Philippines	1 933	1 875	2 806	5 278	5 553	6 017	5 247	6 264	6 485
Sri Lanka	834	708	1 366	2 091	2 395	2 645	2 177	2 597	2 802
Taipei chinois	3 091	3 399	4 966	4 429	6 926	7 419	7 118	6 682	6 682
Thailande	2 048	1 880	2 110	4 081	3 691	5 554	4 075	3 778	5 570
Viêt-Nam	614	420	800	1 599	1 477	1 407	1 384	1 791	3 838
Autre Asie	911	865	1 254	1 911	2 032	2 305	3 019	3 692	3 722
Asie	**55 934**	**58 693**	**91 005**	**121 007**	**122 248**	**133 055**	**127 034**	**144 591**	**154 626**
Rép. populaire de Chine	30 000	38 000	44 600	86 780	92 370	94 530	100 010	109 150	118 400
Chine	**30 000**	**38 000**	**44 600**	**86 780**	**92 370**	**94 530**	**100 010**	**109 150**	**118 400**
Albanie	700	1 127	2 400	3 275	2 944	3 400	3 350	3 590	3 590
Bulgarie	2 170	2 570	2 909	3 260	2 236	2 326	2 538	2 596	2 691
Roumanie	4 495	7 547	10 614	11 326	12 713	10 810	11 209	13 622	12 629
République slovaque	1 476	1 322	2 259	1 970	2 678	2 115	2 439	2 303	2 631
Bosnie-Herzegovine	-	-	-	-	-	-	-	-	-
Croatie	-	-	-	-	-	-	-	-	-
Ex-RYM	-	-	-	-	-	-	-	-	-
Slovénie	-	-	-	-	-	-	-	-	-
RF de Yougoslavie	-	-	-	-	-	-	-	-	-
Ex-Yougoslavie	15 644	16 394	25 199	25 915	24 270	27 504	26 253	25 871	23 491
Europe non-OCDE	**24 485**	**28 960**	**43 381**	**45 746**	**44 841**	**46 155**	**45 789**	**47 982**	**45 032**
Arménie	-	-	-	-	-	-	-	-	-
Azerbaïdjan	-	-	-	-	-	-	-	-	-
Bélarus	-	-	-	-	-	-	-	-	-
Estonie	-	-	-	-	-	-	-	-	-
Géorgie	-	-	-	-	-	-	-	-	-
Kazakhstan	-	-	-	-	-	-	-	-	-
Kirghizistan	-	-	-	-	-	-	-	-	-
Lettonie	-	-	-	-	-	-	-	-	-
Lituanie	-	-	-	-	-	-	-	-	-
République de Moldavie	-	-	-	-	-	-	-	-	-
Russie	-	-	-	-	-	-	-	-	-
Tadjikistan	-	-	-	-	-	-	-	-	-
Turkménistan	-	-	-	-	-	-	-	-	-
Ukraine	-	-	-	-	-	-	-	-	-
Ouzbékistan	-	-	-	-	-	-	-	-	-
Ex-URSS	**126 000**	**122 300**	**169 700**	**203 000**	**214 403**	**215 738**	**219 825**	**231 880**	**223 900**

Production of Hydro Electricity (GWh)
Production d'électricité d'origine hydraulique (GWh)
Elektrizitätserzeugung in Wasserkraftwerken (GWh)
Produzione di energia idroelettrica (GWh)
水力発電量 (GWh)
Producción de electricidad hidráulica (GWh)
Производство гидроэлектроэнергии (ГВт.ч)

1990	1991	1992	1993	1994	1995	1996	1997	1998	
884	838	796	608	847	372	739	719	865	Bangladesh
71 656	72 774	69 885	70 478	82 727	72 717	69 072	74 775	83 020	India
6 489	7 396	9 836	8 845	7 043	7 509	8 142	5 149	9 649	Indonesia
31 750	31 750	24 000	24 000	23 500	23 000	22 500	20 970	19 922	DPR of Korea
3 987	4 405	4 358	4 870	6 520	6 218	5 184	3 870	4 852	Malaysia
1 193	1 240	1 518	1 705	1 614	1 624	1 651	1 655	1 458	Myanmar
886	910	838	874	933	1 161	1 185	1 065	1 137	Nepal
16 925	18 303	18 647	21 112	19 436	22 858	23 206	20 858	22 060	Pakistan
6 062	5 145	4 252	4 994	5 946	6 199	7 074	6 069	5 087	Philippines
3 145	3 116	2 900	3 796	4 089	4 452	3 252	3 447	3 915	Sri Lanka
8 188	5 508	8 351	6 719	8 887	8 879	9 044	9 567	10 608	Chinese Taipei
4 975	4 586	4 238	3 702	4 514	6 712	7 340	7 200	5 177	Thailand
5 385	6 327	7 093	7 862	8 902	10 582	12 008	11 677	11 092	Vietnam
3 717	3 546	3 409	3 410	3 519	3 625	3 681	3 846	3 846	Other Asia
165 242	**165 844**	**160 121**	**162 975**	**178 477**	**175 908**	**174 078**	**170 867**	**182 688**	**Asia**
126 720	125 090	131 230	152 775	168 250	190 577	187 966	195 983	208 000	People's Rep. of China
126 720	**125 090**	**131 230**	**152 775**	**168 250**	**190 577**	**187 966**	**195 983**	**208 000**	**China**
2 848	3 518	3 226	3 314	3 771	4 204	5 684	4 972	4 861	Albania
1 878	2 441	2 063	1 942	1 468	2 314	2 919	2 936	3 325	Bulgaria
10 982	14 234	11 700	12 768	13 046	16 693	15 755	17 509	18 879	Romania
2 515	1 894	2 332	3 865	4 554	5 226	4 533	4 358	4 566	Slovak Republic
-	-	3 000	2 000	1 250	1 420	1 463	1 507	1 553	*Bosnia-Herzegovina*
-	-	4 341	4 345	4 930	5 265	7 228	5 298	5 466	*Croatia*
-	-	848	522	695	801	850	900	1 083	*FYROM*
-	-	3 413	3 022	3 399	3 241	3 673	3 092	3 449	*Slovenia*
-	-	11 343	10 015	11 127	11 220	11 501	12 171	12 253	*FR of Yugoslavia*
19 799	25 629	22 945	19 904	21 401	21 947	24 715	22 968	23 804	Former Yugoslavia
38 022	**47 716**	**42 266**	**41 793**	**44 240**	**50 384**	**53 606**	**52 743**	**55 435**	**Non-OECD Europe**
-	-	3 044	4 293	3 514	1 919	1 572	1 390	1 537	Armenia
-	-	1 747	2 400	1 829	1 556	1 538	1 712	1 951	Azerbaijan
-	-	17	19	19	20	16	21	20	Belarus
-	-	1	1	3	2	2	3	4	Estonia
-	-	6 515	7 034	4 713	5 310	6 012	6 047	6 372	Georgia
-	-	6 866	7 629	9 179	8 331	7 331	6 499	6 141	Kazakhstan
-	-	9 200	9 085	11 724	11 118	12 255	10 934	9 943	Kyrgyzstan
-	-	2 520	2 875	3 305	2 937	1 860	2 952	4 316	Latvia
-	-	311	393	718	751	874	769	895	Lithuania
-	-	259	375	278	324	366	382	84	Republic of Moldova
-	-	172 594	174 627	176 959	177 256	155 326	158 392	159 466	Russia
-	-	15 928	17 118	16 691	14 596	14 825	13 713	14 149	Tajikistan
-	-	4	5	4	4	5	5	5	Turkmenistan
-	-	8 069	11 237	12 327	10 150	8 833	10 032	15 915	Ukraine
-	-	6 281	7 358	7 156	6 188	6 525	5 777	5 758	Uzbekistan
233 000	**234 670**	**233 356**	**244 449**	**248 419**	**240 462**	**217 340**	**218 628**	**226 556**	**Former USSR**

Production of Hydro Electricity (GWh)
Production d'électricité d'origine hydraulique (GWh)
Elektrizitätserzeugung in Wasserkraftwerken (GWh)
Produzione di energia idroelettrica (GWh)
水力発電量 *(GWh)*
Producción de electricidad hidráulica (GWh)
Производство гидроэлектроэнергии (ГВт.ч)

	1971	1973	1978	1984	1985	1986	1987	1988	1989
République Islamique d'Iran	2 679	2 842	6 249	5 750	5 550	7 517	8 390	7 311	7 522
Irak	200	290	708	610	610	600	2 600	2 600	2 600
Israël	-	-	-	2	2	6	10	10	3
Jordanie	-	-	-	-	-	3	19	27	17
Liban	839	478	800	585	585	560	610	600	500
Syrie	51	17	2 095	2 877	2 860	1 633	2 123	4 807	4 727
Moyen-Orient	**3 769**	**3 627**	**9 852**	**9 824**	**9 607**	**10 319**	**13 752**	**15 355**	**15 369**
Total non-OCDE	**335 706**	**373 829**	**560 324**	**777 371**	**819 739**	**861 013**	**885 585**	**943 447**	**962 916**
OCDE Amérique du N.	440 469	476 344	534 849	634 151	613 917	624 465	586 819	553 570	589 417
OCDE Pacifique	113 175	99 078	107 123	110 325	126 216	127 393	123 027	137 575	139 737
OCDE Europe	329 708	348 813	420 904	458 946	457 170	436 950	478 732	511 967	444 900
Total OCDE	**883 352**	**924 235**	**1 062 876**	**1 203 422**	**1 197 303**	**1 188 808**	**1 188 578**	**1 203 112**	**1 174 054**
Monde	**1 219 058**	**1 298 064**	**1 623 200**	**1 980 793**	**2 017 042**	**2 049 821**	**2 074 163**	**2 146 559**	**2 136 970**

Production of Hydro Electricity (GWh)
Production d'électricité d'origine hydraulique (GWh)
Elektrizitätserzeugung in Wasserkraftwerken (GWh)
Produzione di energia idroelettrica (GWh)
水力発電量 *(GWh)*
Producción de electricidad hidráulica (GWh)
Производство гидроэлектроэнергии (ГВт.ч)

1990	1991	1992	1993	1994	1995	1996	1997	1998	
6 083	7 056	9 530	9 823	7 445	7 275	7 376	6 908	7 015	Islamic Republic of Iran
2 600	900	700	600	560	570	570	581	581	Iraq
3	6	29	26	23	25	24	25	25	Israel
11	7	15	22	14	18	22	17	13	Jordan
500	560	720	708	817	717	798	901	787	Lebanon
5 648	6 249	7 386	6 709	6 800	6 944	6 944	7 384	7 532	Syria
14 845	**14 778**	**18 380**	**17 888**	**15 659**	**15 549**	**15 734**	**15 816**	**15 953**	**Middle East**
998 448	**1 037 281**	**1 040 848**	**1 105 607**	**1 166 074**	**1 199 833**	**1 202 007**	**1 232 506**	**1 283 780**	**Non-OECD Total**
609 286	639 384	617 542	652 962	634 366	701 418	763 902	735 925	678 696	OECD North America
140 416	149 872	130 877	151 908	122 238	140 192	136 704	146 844	149 327	OECD Pacific
460 743	460 702	491 360	506 124	506 562	514 743	496 760	512 942	531 326	OECD Europe
1 210 445	**1 249 958**	**1 239 779**	**1 310 994**	**1 263 166**	**1 356 353**	**1 397 366**	**1 395 711**	**1 359 349**	**OECD Total**
2 208 893	**2 287 239**	**2 280 627**	**2 416 601**	**2 429 240**	**2 556 186**	**2 599 373**	**2 628 217**	**2 643 129**	**World**

Production of Geothermal Electricity (GWh)
Production d'électricité d'origine géothermique (GWh)
Geothermische Elektrizitätserzeugung (GWh)

Produzione di energia geotermoelettrica (GWh)

地熱発電量 (GWh)

Producción de electricidad geotérmica (GWh)

Производство геотермальной электроэнергии (ГВт.ч)

	1971	1973	1978	1984	1985	1986	1987	1988	1989
Ethiopie	-	-	-	55	57	59	61	63	64
Kenya	-	-	-	-	-	368	359	323	323
Afrique	**-**	**-**	**-**	**55**	**57**	**427**	**420**	**386**	**387**
Costa Rica	-	-	-	-	-	-	-	-	-
El Salvador	-	-	600	680	422	373	435	430	441
Nicaragua	-	-	-	272	301	300	234	190	381
Amérique latine	**-**	**-**	**600**	**952**	**723**	**673**	**669**	**620**	**822**
Indonésie	-	-	-	217	224	232	719	1 012	1 007
Philippines	-	-	3	4 532	4 929	4 576	4 532	4 844	5 316
Asie	**-**	**-**	**3**	**4 749**	**5 153**	**4 808**	**5 251**	**5 856**	**6 323**
Russie	-	-	-	-	-	-	-	-	-
Ex-URSS	**-**	**-**	**-**	**-**	**-**	**-**	**-**	**-**	**-**
Total non-OCDE	**-**	**-**	**603**	**5 756**	**5 933**	**5 908**	**6 340**	**6 862**	**7 532**
OCDE Amérique du N.	586	2 612	3 740	9 266	11 526	14 320	15 840	15 551	19 623
OCDE Pacifique	1 256	1 512	1 816	2 594	2 633	2 558	2 593	2 595	3 293
OCDE Europe	2 677	2 506	2 514	3 059	2 883	3 031	3 293	3 413	3 510
Total OCDE	**4 519**	**6 630**	**8 070**	**14 919**	**17 042**	**19 909**	**21 726**	**21 559**	**26 426**
Monde	**4 519**	**6 630**	**8 673**	**20 675**	**22 975**	**25 817**	**28 066**	**28 421**	**33 958**

Production of Geothermal Electricity (GWh)
Production d'électricité d'origine géothermique (GWh)
Geothermische Elektrizitätserzeugung (GWh)
Produzione di energia geotermoelettrica (GWh)
地熱発電量 *(GWh)*
Producción de electricidad geotérmica (GWh)
Производство геотермальной электроэнергии (ГВт.ч)

1990	1991	1992	1993	1994	1995	1996	1997	1998	
66	67	63	75	62	57	49	-	-	Ethiopia
326	297	272	272	261	290	390	393	366	Kenya
392	**364**	**335**	**347**	**323**	**347**	**439**	**393**	**366**	**Africa**
-	-	-	-	-	468	510	544	592	Costa Rica
419	425	391	380	407	443	431	486	451	El Salvador
386	458	468	406	360	310	277	209	121	Nicaragua
805	**883**	**859**	**786**	**767**	**1 221**	**1 218**	**1 239**	**1 164**	**Latin America**
1 125	1 049	1 084	1 090	1 602	2 196	2 253	2 605	2 617	Indonesia
5 466	5 757	5 700	5 668	6 297	6 134	6 534	7 237	8 910	Philippines
6 591	**6 806**	**6 784**	**6 758**	**7 899**	**8 330**	**8 787**	**9 842**	**11 527**	**Asia**
-	-	29	28	28	30	28	29	30	Russia
-	**-**	**29**	**28**	**28**	**30**	**28**	**29**	**30**	**Former USSR**
7 788	**8 053**	**8 007**	**7 919**	**9 017**	**9 928**	**10 472**	**11 503**	**13 087**	**Non-OECD Total**
21 136	21 702	22 972	23 651	23 077	20 610	21 475	20 373	21 026	OECD North America
3 951	3 979	4 059	3 936	4 275	5 222	5 814	6 014	6 009	OECD Pacific
3 626	3 571	3 770	4 005	3 789	3 854	4 241	4 414	5 012	OECD Europe
28 713	**29 252**	**30 801**	**31 592**	**31 141**	**29 686**	**31 530**	**30 801**	**32 047**	**OECD Total**
36 501	**37 305**	**38 808**	**39 511**	**40 158**	**39 614**	**42 002**	**42 304**	**45 134**	**World**

Production of Solar, Wind, Tide and Wave Electricity (GWh)
Production d'électricité d'origine solaire, éolienne, marémotrice et houlomotrice (GWh)

Elektrizitätserzeugung aus Sonnenenergie, Wind, Gezeiten und Wellen (GWh)

Produzione di energia elettrica solare, eolica e energia da moto ondoso (GWh)

太陽光、風力、潮力、波力発電量 *(GWh)*

Producción de electricidad solar, eólica, maremotriz y de oleaje (GWh)

Производство электроэнергии солнца, ветра, приливов и волн (ГВт.ч)

	1971	1973	1978	1984	1985	1986	1987	1988	1989
Argentine	-	-	-	-	-	-	-	-	-
Costa Rica	-	-	-	-	-	-	-	-	-
Pérou	-	-	-	-	-	-	-	-	-
Amérique latine	-	-	-	-	-	-	-	-	-
Inde	-	-	-	-	-	2	3	6	6
Asie	-	-	-	-	-	**2**	**3**	**6**	**6**
Estonie	-	-	-	-	-	-	-	-	-
Lettonie	-	-	-	-	-	-	-	-	-
Ukraine	-	-	-	-	-	-	-	-	-
Ex-URSS	-	-	-	-	-	-	-	-	-
Jordanie	-	-	-	-	-	-	-	-	-
Moyen-Orient	-	-	-	-	-	-	-	-	-
Total non-OCDE	-	-	-	-	-	**2**	**3**	**6**	**6**
OCDE Amérique du N.	-	-	-	20	42	53	50	33	2 720
OCDE Pacifique	-	-	-	-	-	1	2	1	1
OCDE Europe	501	559	473	642	661	721	761	897	1 082
Total OCDE	**501**	**559**	**473**	**662**	**703**	**775**	**813**	**931**	**3 803**
Monde	**501**	**559**	**473**	**662**	**703**	**777**	**816**	**937**	**3 809**

Production of Solar, Wind, Tide and Wave Electricity (GWh)
Production d'électricité d'origine solaire, éolienne, marémotrice et houlomotrice (GWh)
Elektrizitätserzeugung aus Sonnenenergie, Wind, Gezeiten und Wellen (GWh)
Produzione di energia elettrica solare, eolica e energia da moto ondoso (GWh)
太陽光、風力、潮力、波力発電量 (GWh)
Producción de electricidad solar, eólica, maremotriz y de oleaje (GWh)
Производство электроэнергии солнца, ветра, приливов и волн (ГВт.ч)

1990	1991	1992	1993	1994	1995	1996	1997	1998	
-	-	-	-	-	-	10	15	33	Argentina
-	-	-	-	-	-	23	76	65	Costa Rica
-	-	-	-	-	-	-	1	1	Peru
-	**-**	**-**	**-**	**-**	**-**	**33**	**92**	**99**	**Latin America**
32	38	52	57	57	57	57	57	57	India
32	**38**	**52**	**57**	**57**	**57**	**57**	**57**	**57**	**Asia**
-	-	-	-	-	-	-	-	1	Estonia
-	-	-	-	-	-	1	1	2	Latvia
-	-	-	-	-	-	-	-	3	Ukraine
-	**-**	**-**	**-**	**-**	**-**	**1**	**1**	**6**	**Former USSR**
1	1	1	1	1	1	1	3	3	Jordan
1	**1**	**1**	**1**	**1**	**1**	**1**	**3**	**3**	**Middle East**
33	**39**	**53**	**58**	**58**	**58**	**92**	**153**	**165**	**Non-OECD Total**
2 988	3 465	3 761	4 049	4 410	4 125	4 418	4 336	3 918	OECD North America
1	1	1	1	35	59	74	90	138	OECD Pacific
1 329	1 626	2 079	2 870	3 584	4 442	5 409	7 957	11 888	OECD Europe
4 318	**5 092**	**5 841**	**6 920**	**8 029**	**8 626**	**9 901**	**12 383**	**15 944**	**OECD Total**
4 351	**5 131**	**5 894**	**6 978**	**8 087**	**8 684**	**9 993**	**12 536**	**16 109**	**World**

Production of Electricity from Combustible Renewables and Waste (GWh)
Production d'électricité à partir d'énergies renouv. combustibles et de déchets (GWh)
Elektrizitätserzeugung aus erneuerbaren Brennstoffen und Abfällen (GWh)
Produzione di energia elettrica da combustibili rinnovabili e da rifiuti (GWh)
可燃性再生可能エネルギー及び廃棄物からの発電量 *(GWh)*
Producción de Electricidad a partir de combustibles renovables y desechos (GWh)
Производство электроэнергии из возобновляемых видов топлива и отходов (ГВт.ч)

	1971	1973	1978	1984	1985	1986	1987	1988	1989
Argentine	57	57	61	95	116	107	108	121	119
Bolivie	15	21	32	41	37	39	49	50	47
Brésil	649	780	1 362	3 011	3 107	3 155	3 561	3 324	3 243
Chili	66	51	91	201	195	280	271	327	321
Colombie	-	-	247	512	264	434	309	534	545
Costa Rica	10	7	8	10	10	10	10	10	-
Cuba	779	686	912	1 088	1 094	1 146	1 165	1 274	1 240
République dominicaine	-	78	82	137	52	52	53	29	33
El Salvador	16	22	31	25	25	27	18	22	24
Guatemala	36	47	123	61	61	61	68	80	94
Haiti	12	12	12	12	15	20	16	16	16
Jamaïque	103	205	59	56	50	50	71	89	93
Nicaragua	19	16	33	44	42	36	43	24	45
Panama	5	5	29	152	102	111	101	135	72
Paraguay	63	65	78	57	42	48	56	27	22
Pérou	294	298	277	185	219	200	178	178	171
Trinité-et-Tobago	35	29	29	13	17	26	31	27	26
Uruguay	11	11	12	27	46	56	56	52	50
Autre Amérique latine	108	110	140	123	121	131	110	86	83
Amérique latine	**2 278**	**2 500**	**3 618**	**5 850**	**5 615**	**5 989**	**6 274**	**6 405**	**6 244**
Thailande	-	-	-	-	-	-	-	-	-
Asie	**-**	**-**	**-**	**-**	**-**	**-**	**-**	**-**	**-**
Bulgarie	-	-	-	-	-	-	-	-	-
Roumanie	-	-	-	-	-	-	-	-	-
Croatie	-	-	-	-	-	-	-	-	-
Ex-Yougoslavie	-	-	-	-	-	-	-	-	-
Europe non-OCDE	**-**	**-**	**-**	**-**	**-**	**-**	**-**	**-**	**-**
Estonie	-	-	-	-	-	-	-	-	-
Russie	-	-	-	-	-	-	-	-	-
Ex-URSS	**22 700**	**22 800**	**26 000**	**24 100**	**23 100**	**21 800**	**21 200**	**21 100**	**20 895**
Total non-OCDE	**24 978**	**25 300**	**29 618**	**29 950**	**28 715**	**27 789**	**27 474**	**27 505**	**27 139**
OCDE Amérique du N.	258	297	1 488	2 696	3 138	3 095	3 862	4 258	56 958
OCDE Pacifique	263	338	752	12 202	12 774	14 101	15 151	16 760	17 930
OCDE Europe	5 870	6 517	8 912	10 865	11 492	12 470	14 375	14 504	15 157
Total OCDE	**6 391**	**7 152**	**11 152**	**25 763**	**27 404**	**29 666**	**33 388**	**35 522**	**90 045**
Monde	**31 369**	**32 452**	**40 770**	**55 713**	**56 119**	**57 455**	**60 862**	**63 027**	**117 184**

Production of Electricity from Combustible Renewables and Waste (GWh)
Production d'électricité à partir d'énergies renouv. combustibles et de déchets (GWh)
Elektrizitätserzeugung aus erneuerbaren Brennstoffen und Abfällen (GWh)
Produzione di energia elettrica da combustibili rinnovabili e da rifiuti (GWh)
可燃性再生可能エネルギー及び廃棄物からの発電量 (GWh)
Producción de Electricidad a partir de combustibles renovables y desechos (GWh)
Производство электроэнергии из возобновляемых видов топлива и отходов (ГВт.ч)

1990	1991	1992	1993	1994	1995	1996	1997	1998	
107	100	102	107	120	117	129	125	160	Argentina
48	51	57	58	60	61	57	64	65	Bolivia
3 560	3 655	4 898	4 953	5 391	5 594	6 732	7 378	7 502	Brazil
271	291	491	525	573	695	927	829	1 154	Chile
306	287	382	368	403	522	670	678	694	Colombia
-	10	4	3	3	17	11	12	11	Costa Rica
1 160	1 012	1 070	710	717	552	737	759	767	Cuba
26	26	27	28	29	30	30	30	30	Dominican Republic
22	17	30	16	5	6	20	22	24	El Salvador
88	10	11	-	-	-	-	-	-	Guatemala
17	17	-	-	-	-	-	-	-	Haiti
120	80	79	196	234	279	279	320	375	Jamaica
45	45	31	19	21	45	57	93	177	Nicaragua
86	82	96	62	19	15	19	22	23	Panama
20	18	17	32	17	78	96	-	47	Paraguay
136	133	144	109	108	114	114	154	126	Peru
31	25	30	28	32	33	17	18	18	Trinidad and Tobago
40	50	60	57	56	37	35	24	34	Uruguay
87	88	89	89	90	90	90	90	94	Other Latin America
6 170	**5 997**	**7 618**	**7 360**	**7 878**	**8 285**	**10 020**	**10 618**	**11 301**	**Latin America**
-	-	-	-	-	265	265	2 158	3 194	Thailand
-	**-**	**-**	**-**	**-**	**265**	**265**	**2 158**	**3 194**	**Asia**
-	-	-	-	-	-	-	-	2	Bulgaria
-	-	57	69	-	3	-	11	11	Romania
-	-	6	19	6	5	10	2	5	Croatia
-	-	6	19	6	5	10	2	5	Former Yugoslavia
-	**-**	**63**	**88**	**6**	**8**	**10**	**13**	**18**	**Non-OECD Europe**
-	-	-	-	-	6	5	8	12	Estonia
-	-	5 000	30 634	26 254	2 404	-	-	-	Russia
21 000	**21 000**	**5 000**	**30 634**	**26 254**	**2 410**	**5**	**8**	**12**	**Former USSR**
27 170	**26 997**	**12 681**	**38 082**	**34 138**	**10 968**	**10 300**	**12 797**	**14 525**	**Non-OECD Total**
69 697	57 388	69 800	65 164	68 530	68 382	70 190	70 163	72 210	OECD North America
17 844	18 761	20 859	21 104	22 129	23 198	23 934	25 641	25 311	OECD Pacific
15 357	15 977	23 579	28 049	31 397	34 424	35 390	40 868	45 015	OECD Europe
102 898	**92 126**	**114 238**	**114 317**	**122 056**	**126 004**	**129 514**	**136 672**	**142 536**	**OECD Total**
130 068	**119 123**	**126 919**	**152 399**	**156 194**	**136 972**	**139 814**	**149 469**	**157 061**	**World**

Total Production of Electricity (GWh)
Production totale d'électricité (GWh)
Gesamterzeugung von Elektrizität (GWh)
Produzione totale di energia elettrica (GWh)
総発電量 (GWh)
Producción total de electricidad (GWh)
Общее производство электроэнергии (ГВт.ч)

	1971	1973	1978	1984	1985	1986	1987	1988	1989
Algérie	2 229	2 806	5 480	11 182	12 274	12 981	12 722	13 966	15 324
Angola	742	984	585	670	806	806	810	815	819
Bénin	-	9	5	61	24	26	26	20	20
Cameroun	1 075	1 119	1 337	2 215	2 441	2 478	2 508	2 623	2 706
Congo	88	96	55	235	314	270	281	292	397
RD du Congo	3 545	3 848	4 080	4 835	5 171	5 406	5 391	5 392	6 937
Côte d'Ivoire	588	796	1 418	1 690	1 842	2 086	2 074	2 309	2 200
Egypte	7 998	8 106	15 150	29 049	31 458	33 464	37 845	39 580	41 649
Ethiopie	593	591	567	910	888	915	892	877	879
Erythrée	-	-	-	-	-	-	-	-	-
Gabon	114	165	436	795	861	888	895	910	876
Ghana	2 944	3 910	3 773	1 831	3 017	4 402	4 866	4 842	5 260
Kenya	627	793	1 382	1 949	2 155	2 307	2 654	2 772	2 901
Libye	508	1 127	3 363	9 200	11 844	13 280	15 600	16 551	16 700
Maroc	2 290	2 875	4 388	6 921	7 345	7 764	8 001	8 984	9 024
Mozambique	674	641	876	398	550	310	336	337	472
Nigéria	1 816	2 607	4 648	9 035	10 431	10 765	11 265	11 654	12 812
Sénégal	344	404	580	765	761	759	822	848	878
Afrique du Sud	54 647	64 390	84 496	137 254	143 491	148 241	152 373	158 141	163 359
Soudan	495	610	865	993	1 229	1 338	1 379	1 378	1 365
Rép. Unie de Tanzanie	491	582	695	922	1 015	1 146	1 272	1 378	1 509
Togo	79	64	71	111	37	33	37	45	36
Tunisie	926	1 179	2 082	3 880	4 220	4 493	4 776	5 203	5 416
Zambie	1 216	3 368	8 092	10 061	10 324	10 101	8 685	8 485	6 782
Zimbabwe	3 602	5 172	6 252	4 120	4 785	5 204	6 345	8 121	8 569
Autre Afrique	2 511	3 037	4 024	6 321	6 558	6 979	7 557	7 631	7 954
Afrique	**90 142**	**109 279**	**154 700**	**245 403**	**263 841**	**276 442**	**289 412**	**303 154**	**314 844**
Argentine	23 624	26 661	33 434	44 966	45 265	49 022	52 115	52 500	50 863
Bolivie	1 024	1 151	1 675	1 686	1 697	1 720	1 721	1 872	2 010
Brésil	51 592	64 727	112 729	179 388	193 682	202 128	203 331	214 952	221 738
Chili	8 524	8 766	10 360	13 497	14 040	14 744	15 637	16 915	17 811
Colombie	9 500	12 596	17 363	26 294	27 127	29 630	31 597	33 012	34 670
Costa Rica	1 148	1 347	1 925	3 067	2 816	2 949	3 133	3 193	3 410
Cuba	5 021	5 703	8 481	12 292	12 199	13 167	13 594	14 542	15 240
République dominicaine	1 068	2 246	2 874	3 769	4 175	4 327	4 663	3 640	3 255
Equateur	1 050	1 256	2 605	4 536	4 558	5 012	5 384	5 641	5 769
El Salvador	743	912	1 489	1 643	1 738	1 704	1 918	2 007	2 056
Guatemala	689	874	1 596	1 605	1 669	1 841	1 978	2 153	2 313
Haiti	82	122	245	422	441	443	494	535	575
Honduras	391	486	759	1 194	1 395	1 473	1 797	1 954	2 043
Jamaïque	1 676	2 187	1 517	1 645	1 636	1 722	1 855	1 831	1 803
Antilles néerlandaises	710	775	825	818	760	637	692	700	737
Nicaragua	657	717	1 163	974	1 060	1 211	1 300	1 187	1 352
Panama	950	1 302	1 622	2 466	2 603	2 746	2 825	2 741	2 726
Paraguay	236	379	569	1 090	4 089	11 912	18 598	19 988	24 333
Pérou	5 949	6 655	8 765	11 770	12 115	12 949	14 010	13 560	13 737
Trinité-et-Tobago	991	1 105	1 557	3 006	3 023	3 291	3 481	3 485	3 428
Uruguay	2 381	2 530	3 047	7 244	6 600	7 430	7 579	7 000	5 750
Vénézuela	13 589	16 445	25 484	45 140	47 290	50 112	54 132	58 245	57 620
Autre Amérique latine	3 197	3 741	4 259	4 448	4 733	4 890	4 793	5 239	5 970
Amérique latine	**134 792**	**162 683**	**244 343**	**372 960**	**394 711**	**425 060**	**446 627**	**466 892**	**479 209**

La production totale d'électricité peut être supérieure à la somme des productions par source d'énergie.

Total Production of Electricity (GWh)
Production totale d'électricité (GWh)
Gesamterzeugung von Elektrizität (GWh)

Produzione totale di energia elettrica (GWh)

総発電量 (GWh)

Producción total de electricidad (GWh)

Общее производство электроэнергии (ГВт.ч)

1990	1991	1992	1993	1994	1995	1996	1997	1998	
16 105	17 345	18 286	19 415	19 883	19 714	20 650	21 422	23 615	Algeria
841	934	947	950	955	960	1 028	1 105	1 063	Angola
21	23	25	26	52	33	47	59	62	Benin
2 697	2 710	2 737	2 832	2 720	2 785	2 902	3 128	3 285	Cameroon
493	481	429	362	324	354	458	453	344	Congo
5 650	5 281	6 073	5 885	5 545	6 180	6 261	6 010	5 667	DR of Congo
2 021	1 894	784	2 150	2 285	2 999	3 227	3 982	3 992	Cote d'Ivoire
42 256	44 296	45 678	47 796	49 304	52 000	55 140	58 428	62 966	Egypt
1 202	1 209	1 241	1 383	1 451	1 527	1 603	1 614	1 627	Ethiopia
-	-	146	151	161	172	193	212	194	Eritrea
915	914	919	922	933	940	970	1 007	1 027	Gabon
5 731	6 116	6 603	6 335	6 131	6 136	6 655	6 914	7 251	Ghana
3 034	3 227	3 215	3 396	3 539	3 747	4 040	4 223	4 804	Kenya
16 800	16 800	16 950	17 000	17 800	18 000	18 300	18 974	19 496	Libya
9 628	9 205	9 719	9 910	11 570	12 798	13 031	13 871	14 136	Morocco
454	471	416	392	395	408	476	1 005	6 864	Mozambique
12 564	14 167	14 834	14 505	15 531	14 483	14 991	15 338	15 716	Nigeria
901	925	1 007	991	1 025	1 086	1 155	1 226	1 286	Senegal
167 226	170 120	167 816	174 581	182 452	187 825	200 266	210 052	205 374	South Africa
1 515	1 661	1 634	1 686	1 858	1 864	2 063	2 150	1 966	Sudan
1 628	1 822	1 816	1 875	1 703	1 841	1 991	1 934	2 157	United Rep. of Tanzania
65	95	42	62	102	55	52	53	96	Togo
5 811	6 031	6 490	6 628	6 974	7 670	7 930	8 485	9 106	Tunisia
6 330	7 834	7 780	7 785	7 785	7 790	7 796	8 006	7 876	Zambia
9 361	8 924	8 236	7 467	7 534	7 805	7 323	7 297	6 608	Zimbabwe
8 239	8 445	8 770	9 116	9 156	9 399	9 378	9 447	9 689	Other Africa
321 488	**330 930**	**332 593**	**343 601**	**357 168**	**368 571**	**387 926**	**406 395**	**416 267**	**Africa**
51 005	53 855	56 107	61 868	65 622	67 169	69 759	72 463	74 170	Argentina
2 133	2 277	2 413	2 650	2 825	3 020	3 239	3 472	3 711	Bolivia
223 002	234 366	241 731	251 973	260 041	275 601	291 224	307 980	321 588	Brazil
18 372	19 961	22 362	24 004	25 276	28 027	30 790	33 292	35 509	Chile
36 157	37 190	33 444	38 219	41 359	43 616	44 605	46 115	45 946	Colombia
3 468	3 864	4 173	4 431	4 772	4 865	4 815	5 635	5 794	Costa Rica
15 025	13 247	11 538	11 004	11 964	12 459	13 236	14 146	14 145	Cuba
3 698	3 895	5 581	5 874	6 182	6 506	6 847	7 335	7 555	Dominican Republic
6 373	7 013	7 199	7 432	8 257	8 526	9 260	9 595	9 897	Ecuador
2 242	2 320	2 412	2 810	3 170	3 365	3 418	3 643	3 837	El Salvador
2 318	2 430	2 767	3 031	3 154	3 413	3 696	4 132	4 456	Guatemala
597	468	423	364	281	495	605	605	663	Haiti
2 319	2 346	2 343	2 562	2 335	2 728	3 067	3 303	3 471	Honduras
2 016	2 054	2 199	3 791	4 775	5 829	6 038	6 255	6 480	Jamaica
790	801	853	909	971	1 009	1 056	1 054	1 116	Netherlands Antilles
1 399	1 471	1 583	1 681	1 686	1 795	1 920	1 928	2 266	Nicaragua
2 871	2 897	3 432	3 444	3 361	3 519	3 915	4 151	4 353	Panama
27 185	29 328	27 141	31 449	36 415	42 236	48 200	50 619	50 893	Paraguay
13 163	13 901	13 044	14 678	14 786	16 880	17 280	17 954	18 583	Peru
3 577	3 720	3 976	3 817	4 069	4 307	4 541	4 988	5 169	Trinidad and Tobago
7 444	7 019	8 899	7 978	7 619	6 307	6 670	7 148	9 568	Uruguay
59 321	63 337	67 440	69 382	71 221	73 446	75 588	78 066	80 904	Venezuela
6 131	6 420	6 636	6 824	6 894	7 081	7 186	7 491	7 671	Other Latin America
490 606	**514 180**	**527 696**	**560 175**	**587 035**	**622 199**	**656 955**	**691 370**	**717 745**	**Latin America**

Total electricity production may be greater than the sum of production by energy source.

Total Production of Electricity (GWh)
Production totale d'électricité (GWh)
Gesamterzeugung von Elektrizität (GWh)

Produzione totale di energia elettrica (GWh)

総発電量 (GWh)

Producción total de electricidad (GWh)

Общее производство электроэнергии (ГВт.ч)

	1971	1973	1978	1984	1985	1986	1987	1988	1989
Bangladesh	1 030	1 404	2 219	3 966	4 528	4 800	5 587	6 541	7 115
Brunei	262	258	262	750	766	776	1 043	1 098	1 131
Inde	66 384	72 796	110 130	169 205	183 390	201 281	218 986	241 314	268 664
Indonésie	3 045	3 624	5 534	14 777	16 895	19 455	23 025	26 280	30 380
RPD de Corée	16 910	20 000	32 000	45 000	48 000	50 000	50 200	53 000	53 500
Malaisie	3 795	4 783	8 252	13 737	14 935	16 120	17 410	19 339	20 534
Myanmar	691	821	1 243	1 890	2 119	2 245	2 320	2 226	2 494
Népal	86	104	174	383	431	539	574	563	718
Pakistan	7 572	8 377	12 375	21 873	23 003	25 589	28 703	33 091	34 601
Philippines	9 145	13 186	15 542	21 180	22 766	21 797	22 642	24 539	25 573
Singapour	2 585	3 719	5 898	9 490	9 960	10 640	11 909	13 113	14 136
Sri Lanka	900	1 031	1 385	2 261	2 464	2 652	2 707	2 799	2 858
Taipei chinois	15 811	20 735	35 846	52 213	55 554	62 329	69 176	76 258	84 055
Thaïlande	5 083	6 971	12 637	21 024	23 074	24 717	28 652	32 464	37 406
Viêt-Nam	2 300	2 350	3 600	4 778	5 069	5 527	6 051	6 785	7 784
Autre Asie	2 256	3 500	3 751	4 163	4 466	5 044	5 978	6 375	6 477
Asie	**137 855**	**163 659**	**250 848**	**386 690**	**417 420**	**453 511**	**494 963**	**545 785**	**597 426**
Rép. populaire de Chine	138 400	166 800	256 552	377 390	410 690	449 530	497 270	545 210	584 810
Hong-Kong, Chine	5 574	6 799	10 370	17 918	19 230	21 406	23 746	25 501	27 361
Chine	**143 974**	**173 599**	**266 922**	**395 308**	**429 920**	**470 936**	**521 016**	**570 711**	**612 171**
Albanie	1 225	1 702	3 100	3 717	3 147	5 106	4 395	3 984	4 123
Bulgarie	21 016	21 956	31 492	44 672	41 632	41 820	43 473	45 021	44 328
Chypre	665	830	919	1 250	1 319	1 423	1 501	1 647	1 831
Gibraltar	47	49	52	61	63	65	70	74	81
Malte	305	365	459	700	784	850	944	1 030	1 100
Roumanie	39 454	46 779	64 255	71 667	71 818	75 478	74 077	75 322	75 851
République slovaque	10 865	12 299	15 432	20 362	22 506	24 169	23 629	23 040	24 069
Bosnie-Herzegovine	-	-	-	-	-	-	-	-	-
Croatie	-	-	-	-	-	-	-	-	-
Ex-RYM	-	-	-	-	-	-	-	-	-
Slovénie	-	-	-	-	-	-	-	-	-
RF de Yougoslavie	-	-	-	-	-	-	-	-	-
Ex-Yougoslavie	29 509	35 062	51 250	73 008	74 802	77 916	80 792	83 651	83 468
Europe non-OCDE	**103 086**	**119 042**	**166 959**	**215 437**	**216 071**	**226 827**	**228 881**	**233 769**	**234 851**
Arménie	-	-	-	-	-	-	-	-	-
Azerbaïdjan	-	-	-	-	-	-	-	-	-
Bélarus	-	-	-	-	-	-	-	-	-
Estonie	-	-	-	-	-	-	-	-	-
Géorgie	-	-	-	-	-	-	-	-	-
Kazakhstan	-	-	-	-	-	-	-	-	-
Kirghizistan	-	-	-	-	-	-	-	-	-
Lettonie	-	-	-	-	-	-	-	-	-
Lituanie	-	-	-	-	-	-	-	-	-
République de Moldavie	-	-	-	-	-	-	-	-	-
Russie	-	-	-	-	-	-	-	-	-
Tadjikistan	-	-	-	-	-	-	-	-	-
Turkménistan	-	-	-	-	-	-	-	-	-
Ukraine	-	-	-	-	-	-	-	-	-
Ouzbékistan	-	-	-	-	-	-	-	-	-
Ex-URSS	**800 400**	**914 600**	**1 201 900**	**1 492 000**	**1 544 000**	**1 599 000**	**1 664 900**	**1 705 000**	**1 722 000**

La production totale d'électricité peut être supérieure à la somme des productions par source d'énergie.

Total Production of Electricity (GWh)
Production totale d'électricité (GWh)
Gesamterzeugung von Elektrizität (GWh)
Produzione totale di energia elettrica (GWh)
総発電量 (GWh)
Producción total de electricidad (GWh)
Общее производство электроэнергии (ГВт.ч)

1990	1991	1992	1993	1994	1995	1996	1997	1998	
7 732	8 270	8 894	9 206	9 784	10 806	11 474	11 858	12 882	Bangladesh
1 172	1 269	1 408	1 547	1 663	1 966	2 123	2 407	2 503	Brunei
289 439	315 631	332 713	356 335	385 557	417 836	435 075	463 401	493 968	India
36 951	39 342	43 409	47 630	51 578	59 282	66 706	76 620	77 903	Indonesia
53 500	53 500	38 000	38 000	37 000	36 000	35 000	32 620	30 989	DPR of Korea
23 000	26 534	29 300	34 716	39 074	45 431	51 382	57 844	60 671	Malaysia
2 478	2 677	2 996	3 385	3 594	4 055	3 945	4 445	4 579	Myanmar
886	946	886	934	1 017	1 197	1 221	1 173	1 257	Nepal
37 673	41 076	45 465	48 810	50 646	53 555	56 956	59 125	62 154	Pakistan
25 245	25 654	25 869	26 777	30 465	33 531	36 663	40 053	41 192	Philippines
15 714	16 921	17 674	18 962	20 849	22 244	24 100	26 898	28 557	Singapore
3 150	3 377	3 540	3 979	4 387	4 802	4 530	5 145	5 683	Sri Lanka
90 479	99 687	106 557	116 471	125 930	135 277	144 930	153 294	166 179	Chinese Taipei
44 175	50 186	57 098	63 406	71 177	80 061	87 467	93 253	90 069	Thailand
8 722	9 300	9 691	10 659	12 270	14 665	16 944	19 151	21 665	Vietnam
6 520	6 467	6 233	6 251	6 470	6 869	6 789	6 958	6 958	Other Asia
646 836	**700 837**	**729 733**	**787 068**	**851 461**	**927 577**	**985 305**	**1 054 245**	**1 107 209**	**Asia**
621 200	677 550	754 195	838 256	928 083	1 007 726	1 080 018	1 134 471	1 166 200	People's Rep. of China
28 938	31 807	34 907	35 950	26 743	27 916	28 442	28 945	31 417	Hong Kong, China
650 138	**709 357**	**789 102**	**874 206**	**954 826**	**1 035 642**	**1 108 460**	**1 163 416**	**1 197 617**	**China**
3 189	3 686	3 357	3 484	3 904	4 414	5 926	5 184	5 068	Albania
42 141	38 917	35 610	37 997	38 133	41 789	42 716	42 803	41 711	Bulgaria
1 974	2 077	2 404	2 581	2 681	2 473	2 592	2 711	2 954	Cyprus
82	82	88	88	88	88	93	93	93	Gibraltar
1 100	1 419	1 490	1 500	1 541	1 632	1 658	1 686	1 721	Malta
64 309	56 803	54 195	55 476	55 136	59 266	61 350	57 148	53 496	Romania
24 067	22 746	22 347	23 417	24 740	26 306	25 290	24 547	25 465	Slovak Republic
-	-	12 200	11 700	2 039	2 321	2 393	2 461	2 538	*Bosnia-Herzegovina*
-	-	8 894	9 359	8 275	8 863	10 548	9 684	10 898	*Croatia*
-	-	6 065	5 180	5 511	6 114	6 489	6 719	7 048	*FYROM*
-	-	12 086	11 692	12 631	12 654	12 778	13 176	13 744	*Slovenia*
-	-	36 488	33 457	34 528	37 176	38 093	40 312	40 651	*FR of Yugoslavia*
82 905	81 885	75 733	71 388	62 984	67 128	70 301	72 352	74 879	Former Yugoslavia
219 767	**207 615**	**195 224**	**195 931**	**189 207**	**203 096**	**209 926**	**206 524**	**205 387**	**Non-OECD Europe**
-	-	9 004	6 295	5 658	5 561	6 214	6 022	6 190	Armenia
-	-	19 673	19 100	17 571	17 044	17 088	16 836	17 985	Azerbaijan
-	-	37 595	33 369	31 397	24 918	23 728	26 057	25 300	Belarus
-	-	11 831	9 118	9 151	8 693	9 103	9 218	8 521	Estonia
-	-	11 520	10 150	6 803	6 900	7 226	7 172	8 069	Georgia
-	-	82 701	77 444	66 397	66 664	59 038	52 006	49 144	Kazakhstan
-	-	11 892	11 273	12 932	12 349	13 758	12 637	11 615	Kyrgyzstan
-	-	3 834	3 924	4 440	3 979	3 124	4 456	5 738	Latvia
-	-	16 540	14 232	10 057	13 942	16 789	14 861	17 631	Lithuania
-	-	11 248	10 265	8 228	6 068	6 122	5 273	4 584	Republic of Moldova
-	-	1 008 450	956 587	875 914	860 027	847 183	834 132	827 138	Russia
-	-	16 822	17 741	16 982	14 768	15 000	14 005	14 422	Tajikistan
-	-	13 183	12 637	10 496	9 800	10 100	9 498	9 416	Turkmenistan
-	-	252 524	229 906	202 922	194·018	182 986	178 002	172 822	Ukraine
-	-	50 911	49 149	47 800	47 453	45 419	46 054	45 900	Uzbekistan
1 727 000	**1 681 083**	**1 557 728**	**1 461 190**	**1 326 748**	**1 292 184**	**1 262 878**	**1 236 229**	**1 224 475**	**Former USSR**

Total electricity production may be greater than the sum of production by energy source.

Total Production of Electricity (GWh)
Production totale d'électricité (GWh)
Gesamterzeugung von Elektrizität (GWh)
Produzione totale di energia elettrica (GWh)
総発電量 *(GWh)*
Producción total de electricidad (GWh)
Общее производство электроэнергии (ГВт.ч)

	1971	1973	1978	1984	1985	1986	1987	1988	1989
Bahrein	430	500	1 212	2 187	2 637	2 892	2 996	3 162	3 293
République Islamique d'Iran	8 090	12 072	20 006	36 594	39 220	41 571	46 197	47 600	52 712
Irak	2 800	3 519	7 835	19 459	20 973	22 287	22 510	23 450	23 860
Israël	7 639	8 720	11 874	14 665	15 432	16 034	17 677	19 352	20 453
Jordanie	230	315	704	2 265	2 495	2 955	3 486	3 263	3 433
Koweit	2 623	3 651	6 983	13 894	15 417	16 934	18 092	19 599	21 085
Liban	1 375	1 791	2 300	3 800	3 861	4 170	4 600	4 000	2 500
Oman	13	47	468	2 016	2 498	3 180	3 392	3 773	3 927
Qatar	316	419	1 349	3 563	3 949	4 303	4 371	4 501	4 624
Arabie saoudite	2 082	2 949	10 551	40 069	44 311	47 646	50 649	57 229	61 568
Syrie	1 345	1 423	2 871	7 310	7 898	7 942	7 989	9 614	10 455
Emirats arabes unis	203	720	3 764	11 000	12 000	12 814	13 657	14 840	15 612
Yémen	209	206	360	856	895	1 117	1 165	1 318	1 345
Moyen-Orient	**27 355**	**36 332**	**70 277**	**157 678**	**171 586**	**183 845**	**196 781**	**211 701**	**224 867**
Total non-OCDE	**1 437 604**	**1 679 194**	**2 355 949**	**3 265 476**	**3 437 549**	**3 635 621**	**3 842 580**	**4 037 012**	**4 185 368**
OCDE Amérique du N.	1 956 385	2 272 807	2 716 459	3 086 786	3 173 903	3 205 400	3 332 866	3 490 617	3 762 636
OCDE Pacifique	464 920	568 443	702 511	840 364	878 816	896 187	955 068	1 008 239	1 073 587
OCDE Europe	1 399 284	1 613 705	1 928 307	2 233 704	2 338 398	2 388 432	2 475 981	2 536 360	2 595 900
Total OCDE	**3 820 589**	**4 454 955**	**5 347 277**	**6 160 854**	**6 391 117**	**6 490 019**	**6 763 915**	**7 035 216**	**7 432 123**
Monde	**5 258 193**	**6 134 149**	**7 703 226**	**9 426 330**	**9 828 666**	**10 125 640**	**10 606 495**	**11 072 228**	**11 617 491**

La production totale d'électricité peut être supérieure à la somme des productions par source d'énergie.

Total Production of Electricity (GWh)
Production totale d'électricité (GWh)
Gesamterzeugung von Elektrizität (GWh)

Produzione totale di energia elettrica (GWh)

総発電量 (GWh)

Producción total de electricidad (GWh)

Общее производство электроэнергии (ГВт.ч)

1990	1991	1992	1993	1994	1995	1996	1997	1998	
3 482	3 320	3 896	4 245	4 550	4 612	5 016	5 040	5 773	Bahrain
59 102	64 126	68 619	76 014	82 019	84 969	90 851	97 744	103 413	Islamic Republic of Iran
24 000	20 810	25 300	26 300	28 000	29 000	29 000	29 561	30 346	Iraq
20 898	21 514	24 686	26 000	28 316	30 425	32 557	35 098	37 964	Israel
3 638	3 724	4 422	4 761	5 076	5 616	6 058	6 264	6 745	Jordan
18 477	10 780	16 885	20 178	22 802	23 724	25 475	26 724	29 984	Kuwait
1 500	3 000	3 771	4 484	5 184	5 281	6 965	8 515	8 357	Lebanon
4 501	4 625	5 113	5 833	6 197	6 460	6 802	7 318	8 198	Oman
4 818	4 653	5 153	5 525	5 814	5 976	6 575	6 868	8 125	Qatar
64 899	69 212	74 009	82 183	91 019	100 748	106 606	113 125	116 519	Saudi Arabia
11 611	12 179	12 562	12 638	15 182	15 300	16 885	17 956	18 315	Syria
17 080	17 351	18 689	21 730	23 736	24 982	26 572	28 464	33 392	United Arab Emirates
1 663	1 802	1 953	2 051	2 159	2 369	2 334	2 557	2 507	Yemen
235 669	**237 096**	**265 058**	**291 942**	**320 054**	**339 462**	**361 696**	**385 234**	**409 638**	**Middle East**
4 291 504	**4 381 098**	**4 397 134**	**4 514 113**	**4 586 499**	**4 788 731**	**4 973 146**	**5 143 413**	**5 278 338**	**Non-OECD Total**
3 802 092	3 908 591	3 942 381	4 079 480	4 176 478	4 294 643	4 412 458	4 446 890	4 576 733	OECD North America
1 152 296	1 196 376	1 218 345	1 248 784	1 350 202	1 402 225	1 449 946	1 504 692	1 515 934	OECD Pacific
2 624 589	2 675 882	2 700 310	2 715 323	2 758 772	2 839 715	2 915 884	2 953 661	3 032 041	OECD Europe
7 578 977	**7 780 849**	**7 861 036**	**8 043 587**	**8 285 452**	**8 536 583**	**8 778 288**	**8 905 243**	**9 124 708**	**OECD Total**
11 870 481	**12 161 947**	**12 258 170**	**12 557 700**	**12 871 951**	**13 325 314**	**13 751 434**	**14 048 656**	**14 403 046**	**World**

Total electricity production may be greater than the sum of production by energy source.

Production of Heat (TJ)
Production de chaleur (TJ)
Erzeugung von Wärme (TJ)

Produzione di calore (TJ)

総発熱量 (TJ)

Producción de calor (TJ)

Производство теплоэнергии (ТДж)

	1971	1973	1978	1984	1985	1986	1987	1988	1989
Rép. populaire de Chine	-	-	-	354 478	382 277	413 204	463 720	477 142	529 298
Chine	**-**	**-**	**-**	**354 478**	**382 277**	**413 204**	**463 720**	**477 142**	**529 298**
Albanie	-	-	-	-	-	-	-	-	-
Bulgarie	28 043	33 065	93 415	73 399	64 500	69 100	91 100	63 300	200 000
Roumanie	150 000	150 000	170 000	200 000	213 916	228 201	238 066	263 002	269 994
République slovaque	30 623	31 899	35 300	28 151	30 069	30 088	31 343	31 757	32 023
Bosnie-Herzegovine	-	-	-	-	-	-	-	-	-
Croatie	-	-	-	-	-	-	-	-	-
Ex-RYM	-	-	-	-	-	-	-	-	-
Slovénie	-	-	-	-	-	-	-	-	-
RF de Yougoslavie	-	-	-	-	-	-	-	-	-
Ex-Yougoslavie	-	-	-	101 490	88 502	93 777	88 460	78 312	73 743
Europe non-OCDE	**208 666**	**214 964**	**298 715**	**403 040**	**396 987**	**421 166**	**448 969**	**436 371**	**575 760**
Arménie	-	-	-	-	-	-	-	-	-
Azerbaïdjan	-	-	-	-	-	-	-	-	-
Bélarus	-	-	-	-	-	-	-	-	-
Estonie	-	-	-	-	-	-	-	-	-
Géorgie	-	-	-	-	-	-	-	-	-
Kazakhstan	-	-	-	-	-	-	-	-	-
Kirghizistan	-	-	-	-	-	-	-	-	-
Lettonie	-	-	-	-	-	-	-	-	-
Lituanie	-	-	-	-	-	-	-	-	-
République de Moldavie	-	-	-	-	-	-	-	-	-
Russie	-	-	-	-	-	-	-	-	-
Tadjikistan	-	-	-	-	-	-	-	-	-
Ukraine	-	-	-	-	-	-	-	-	-
Ouzbékistan	-	-	-	-	-	-	-	-	-
Ex-URSS	**2 310 396**	**2 595 010**	**3 335 844**	**4 546 865**	**4 848 314**	**4 978 105**	**5 112 083**	**5 250 247**	**5 283 742**
Total non-OCDE	**2 519 062**	**2 809 974**	**3 634 559**	**5 304 383**	**5 627 578**	**5 812 475**	**6 024 772**	**6 163 760**	**6 388 800**
OCDE Amérique du N.	-	4 003	17 011	132 339	117 844	114 003	116 937	95 866	118 824
OCDE Pacifique	-	1 160	8 264	8 217	9 222	9 275	9 641	9 629	10 759
OCDE Europe	849 069	955 525	1 397 362	1 646 721	1 783 245	1 765 100	1 820 267	1 724 097	1 707 623
Total OCDE	**849 069**	**960 688**	**1 422 637**	**1 787 277**	**1 910 311**	**1 888 378**	**1 946 845**	**1 829 592**	**1 837 206**
Monde	**3 368 131**	**3 770 662**	**5 057 196**	**7 091 660**	**7 537 889**	**7 700 853**	**7 971 617**	**7 993 352**	**8 226 006**

Production of Heat (TJ)
Production de chaleur (TJ)
Erzeugung von Wärme (TJ)

Produzione di calore (TJ)

総発熱量 *(TJ)*

Producción de calor (TJ)

Производство теплоэнергии (ТДж)

1990	1991	1992	1993	1994	1995	1996	1997	1998	
626 364	690 157	724 508	945 442	1 119 551	1 072 002	1 148 301	1 175 695	1 270 677	People's Rep. of China
626 364	**690 157**	**724 508**	**945 442**	**1 119 551**	**1 072 002**	**1 148 301**	**1 175 695**	**1 270 677**	**China**
4 758	4 115	2 874	2 734	647	627	608	491	509	Albania
210 056	183 725	147 733	136 317	126 671	133 463	140 120	63 273	58 380	Bulgaria
258 112	202 056	501 336	509 537	279 432	286 999	296 264	284 612	267 463	Romania
35 880	38 111	36 997	32 142	31 500	35 676	41 198	36 339	38 664	Slovak Republic
-	-	3 215	27 125	15 294	16 314	16 800	17 334	17 806	*Bosnia-Herzegovina*
-	-	12 916	12 691	12 067	13 219	13 737	13 327	12 500	*Croatia*
-	-	4 293	8 636	6 623	6 208	6 208	6 208	8 488	*FYROM*
-	-	8 274	8 455	8 118	8 917	9 702	9 078	9 194	*Slovenia*
-	-	20 000	20 000	19 000	19 000	19 000	19 000	19 000	*FR of Yugoslavia*
75 000	65 000	48 698	76 907	61 102	63 658	65 447	64 947	66 988	Former Yugoslavia
583 806	**493 007**	**737 638**	**757 637**	**499 352**	**520 423**	**543 637**	**449 662**	**432 004**	**Non-OECD Europe**
-	-	11 179	3 926	2 883	3 285	3 762	2 776	2 478	Armenia
-	-	-	15 200	50 125	40 866	31 067	14 715	12 659	Azerbaijan
-	-	400 000	393 559	326 570	305 636	326 570	334 944	327 115	Belarus
-	-	63 256	48 352	46 758	30 625	33 132	32 593	29 113	Estonia
-	-	42 705	29 740	19 950	19 950	19 950	19 950	19 950	Georgia
-	-	300	300	284	256	256	256	280	Kazakhstan
-	-	20 000	6 378	4 454	18 556	19 162	16 546	15 815	Kyrgyzstan
-	-	60 074	56 812	45 882	43 106	54 925	46 540	44 057	Latvia
-	-	94 953	80 846	85 444	78 418	78 269	70 342	67 005	Lithuania
-	-	25 653	20 360	15 792	14 881	14 598	13 819	15 033	Republic of Moldova
-	-	8 700 000	9 466 604	8 631 400	8 052 800	6 708 300	6 400 200	6 262 200	Russia
-	-	8 508	6 671	3 933	4 981	6 564	3 777	3 952	Tajikistan
-	-	1 014 527	858 882	795 495	663 814	626 043	631 001	594 593	Ukraine
-	-	115 974	111 655	112 033	108 692	107 874	100 100	106 763	Uzbekistan
5 191 632	4 867 551	10 557 129	11 099 285	10 141 003	9 385 866	8 030 472	7 687 559	7 501 013	**Former USSR**
6 401 802	**6 050 715**	**12 019 275**	**12 802 364**	**11 759 906**	**10 978 291**	**9 722 410**	**9 312 916**	**9 203 694**	**Non-OECD Total**
107 893	326 990	400 070	425 158	414 970	439 406	443 335	443 619	442 224	OECD North America
10 727	11 808	13 680	14 498	50 714	61 789	73 163	81 996	82 988	OECD Pacific
1 715 755	1 732 337	1 700 897	1 739 582	1 727 266	1 664 912	1 785 495	1 710 949	1 676 125	OECD Europe
1 834 375	**2 071 135**	**2 114 647**	**2 179 238**	**2 192 950**	**2 166 107**	**2 301 993**	**2 236 564**	**2 201 337**	**OECD Total**
8 236 177	8 121 850	14 133 922	14 981 602	13 952 856	13 144 398	12 024 403	11 549 480	11 405 031	**World**

Refinery Output of Petroleum Products (1000 tonnes)
Production de produits pétroliers en raffineries (1000 tonnes)
Raffinerieausstoß von Ölprodukten (1000 Tonnen)
Produzione di prodotti petroliferi nelle raffinerie (1000 tonnellate)
石油製品の生産量（チトン）
Producción de productos petrolíferos en refinerías (1000 toneladas)
Производство нефтепродуктов нефтеперерабатующими заводами (тыс. т)

	1971	1973	1978	1984	1985	1986	1987	1988	1989
Algérie	2 416	6 127	5 441	20 976	20 316	21 750	20 845	22 040	21 187
Angola	661	725	989	1 344	1 412	1 411	1 517	1 407	1 546
Cameroun	-	-	-	1 527	1 542	1 551	1 548	1 297	1 214
Congo	-	-	-	513	467	513	509	501	615
RD du Congo	669	719	200	416	320	189	305	351	334
Côte d'Ivoire	808	1 154	1 618	1 864	1 941	2 001	1 679	2 330	2 364
Egypte	5 003	6 851	11 753	19 597	20 119	20 921	22 050	22 238	22 482
Ethiopie	664	616	586	715	711	737	759	658	662
Erythrée	-	-	-	-	-	-	-	-	-
Gabon	1 002	1 058	1 688	924	741	733	584	628	799
Ghana	870	992	1 170	768	965	911	784	894	957
Kenya	2 522	2 638	2 562	2 047	1 964	1 931	2 118	2 031	2 148
Libye	435	1 608	5 217	6 558	10 379	10 948	12 341	13 723	13 258
Maroc	1 485	2 224	2 969	4 625	4 752	4 487	4 515	5 051	5 430
Mozambique	785	739	599	120					
Nigéria	1 949	2 762	2 633	8 401	8 140	7 234	9 106	10 612	11 471
Sénégal	553	670	762	349	303	477	524	711	586
Afrique du Sud	12 668	13 106	13 147	14 912	14 469	14 160	14 786	16 052	16 264
Soudan	682	1 145	980	587	604	778	462	772	539
Rép. Unie de Tanzanie	755	785	547	534	569	565	606	606	570
Togo	-	-	161	-	-	-	-	-	-
Tunisie	1 117	1 036	1 181	1 588	1 563	1 577	1 657	1 691	1 699
Zambie	-	403	777	624	580	548	573	579	568
Autre Afrique	880	854	611	494	719	645	1 752	1 764	1 510
Afrique	**35 924**	**46 212**	**55 591**	**89 483**	**92 576**	**94 067**	**99 020**	**105 936**	**106 203**
Argentine	23 096	23 491	24 215	23 275	22 652	21 323	20 737	21 176	22 134
Bolivie	647	729	1 121	988	1 006	937	988	1 034	1 098
Brésil	26 309	36 920	52 448	53 894	54 237	57 729	59 453	59 053	59 455
Chili	4 655	4 628	4 669	3 768	3 739	4 025	4 220	4 775	5 761
Colombie	7 679	8 121	7 420	8 706	8 695	9 143	10 352	10 501	11 223
Costa Rica	400	382	410	400	398	614	606	596	641
Cuba	4 402	5 436	6 565	6 639	6 829	6 765	7 067	7 862	8 300
République dominicaine	-	1 162	1 483	1 703	1 435	1 710	1 573	1 559	1 781
Equateur	1 350	1 570	4 094	4 407	4 326	4 870	4 159	5 856	5 646
El Salvador	442	584	695	596	634	632	691	670	623
Guatemala	788	919	812	679	667	485	599	595	587
Honduras	576	616	417	391	352	192	304	407	451
Jamaïque	1 617	1 761	983	566	879	706	596	713	849
Antilles néerlandaises	40 927	43 644	26 931	19 470	8 420	8 065	9 566	9 872	9 649
Nicaragua	470	603	598	411	480	499	512	512	560
Panama	3 701	3 313	2 314	1 435	1 216	918	1 345	945	920
Paraguay	176	226	293	147	193	207	232	274	268
Pérou	4 188	4 656	5 873	8 290	8 259	8 156	8 308	8 122	7 355
Trinité-et-Tobago	20 224	19 583	11 765	4 048	4 256	4 162	4 916	4 365	3 976
Uruguay	1 675	1 654	1 821	1 172	1 012	989	1 205	1 130	1 109
Vénézuela	66 482	70 102	51 872	42 783	44 781	43 869	40 391	47 837	46 561
Autre Amérique latine	9 908	12 948	8 556	6 520	1 562	747	777	775	851
Amérique latine	**219 712**	**243 048**	**215 355**	**190 288**	**176 028**	**176 743**	**178 597**	**188 629**	**189 798**

Refinery Output of Petroleum Products (1000 tonnes)
Production de produits pétroliers en raffineries (1000 tonnes)
Raffinerieausstoß von Ölprodukten (1000 Tonnen)
Produzione di prodotti petroliferi nelle raffinerie (1000 tonnellate)
石油製品の生産量（チトン）
Producción de productos petrolíferos en refinerías (1000 toneladas)
Производство нефтепродуктов нефтеперерабатующими заводами (тыс. т)

1990	1991	1992	1993	1994	1995	1996	1997	1998	
20 584	20 597	20 680	20 614	19 246	19 861	18 830	20 076	18 871	Algeria
1 551	1 614	1 704	1 557	1 743	1 785	1 817	1 864	1 743	Angola
839	880	1 040	1 098	1 403	1 220	1 495	1 585	1 635	Cameroon
628	645	605	650	634	614	625	625	-	Congo
349	360	333	364	372	342	342	342	322	DR of Congo
2 035	2 267	3 199	3 150	3 188	3 220	3 602	3 573	4 006	Cote d'Ivoire
23 648	23 997	23 600	24 901	25 919	26 715	27 615	27 921	28 790	Egypt
655	680	431	611	641	622	619	340	-	Ethiopia
-	-	646	586	718	604	656	402	-	Eritrea
419	363	625	655	666	681	721	687	764	Gabon
783	979	1 028	787	1 093	951	1 016	10	739	Ghana
2 198	2 053	2 217	2 083	2 058	1 809	1 733	1 623	1 696	Kenya
13 249	13 396	13 463	12 996	14 230	14 522	14 730	14 775	14 775	Libya
5 559	5 417	5 976	6 031	6 410	6 028	5 417	5 730	5 870	Morocco
-	-	-	-	-	-	-	-	-	Mozambique
12 849	13 902	13 140	12 842	8 219	11 200	13 415	13 287	7 841	Nigeria
675	529	601	548	171	679	662	797	816	Senegal
17 551	17 691	17 838	18 857	20 970	22 942	21 196	18 657	19 266	South Africa
809	957	724	312	617	710	689	896	580	Sudan
584	590	589	589	592	598	599	599	599	United Rep. of Tanzania
-	-	-	-	-	-	-	-	-	Togo
1 788	1 759	1 670	1 695	1 669	1 800	1 815	2 009	1 808	Tunisia
533	546	510	547	543	554	562	562	562	Zambia
1 152	1 207	1 297	1 313	1 322	1 333	1 344	1 373	1 373	Other Africa
108 438	**110 429**	**111 916**	**112 786**	**112 424**	**118 790**	**119 500**	**117 733**	**112 056**	**Africa**
22 336	22 909	23 797	23 949	22 746	21 508	23 111	25 231	26 067	Argentina
1 159	1 251	1 245	1 252	1 363	1 474	1 551	1 621	1 670	Bolivia
58 969	57 057	59 133	59 560	63 308	62 089	66 574	71 974	77 502	Brazil
5 978	6 003	6 249	6 744	7 078	7 677	7 987	8 369	9 101	Chile
11 026	11 598	11 710	12 587	12 354	13 460	14 465	14 507	14 276	Colombia
442	351	530	538	568	738	624	613	48	Costa Rica
6 846	4 749	1 715	1 819	1 562	1 611	1 977	1 661	1 369	Cuba
1 509	1 800	1 938	1 884	1 971	2 038	2 106	2 201	2 262	Dominican Republic
6 044	6 327	5 995	6 006	6 437	6 585	7 176	7 407	6 519	Ecuador
652	745	803	901	832	732	751	766	900	El Salvador
428	543	744	704	752	764	717	749	747	Guatemala
407	391	288	-	729	1 012	989	1 038	1 050	Honduras
1 005	861	1 186	702	729	1 012	989	1 038	1 050	Jamaica
10 004	9 782	11 592	12 269	12 354	12 380	12 404	12 434	12 837	Netherlands Antilles
618	625	664	638	678	589	620	795	895	Nicaragua
1 139	1 090	1 701	1 720	1 070	1 176	2 036	2 030	2 350	Panama
300	268	299	248	254	200	155	145	123	Paraguay
7 302	7 471	7 510	7 547	7 174	6 751	7 148	7 649	8 245	Peru
4 260	5 203	5 143	4 777	5 131	4 773	4 999	5 103	5 139	Trinidad and Tobago
1 142	1 194	1 111	367	-	1 275	1 580	1 362	1 822	Uruguay
49 008	51 598	46 392	48 250	49 113	50 553	51 771	53 737	54 681	Venezuela
957	965	947	985	975	1 002	998	1 022	1 022	Other Latin America
191 531	**192 781**	**190 692**	**193 447**	**196 449**	**198 387**	**209 739**	**220 414**	**228 625**	**Latin America**

Refinery Output of Petroleum Products (1000 tonnes)
Production de produits pétroliers en raffineries (1000 tonnes)
Raffinerieausstoß von Ölprodukten (1000 Tonnen)
Produzione di prodotti petroliferi nelle raffinerie (1000 tonnellate)
石油製品の生産量（千トン）
Producción de productos petrolíferos en refinerías (1000 toneladas)
Производство нефтепродуктов нефтеперерабатующими заводами (тыс. т)

	1971	1973	1978	1984	1985	1986	1987	1988	1989
Bangladesh	732	609	1 037	1 130	1 026	984	1 050	1 063	1 135
Brunei	-	-	-	252	241	240	243	258	273
Inde	19 290	20 215	25 683	34 968	40 558	44 506	47 033	46 799	50 858
Indonésie	8 989	10 130	13 923	21 461	22 956	25 888	30 686	32 328	33 085
RPD de Corée	-	-	1 190	2 330	2 515	2 690	2 690	2 795	2 907
Malaisie	3 746	3 858	5 650	7 459	7 372	7 682	7 893	8 258	9 039
Myanmar	1 262	974	1 202	1 214	1 173	1 200	796	717	707
Pakistan	3 291	3 360	3 965	4 824	5 201	5 452	5 646	5 925	5 686
Philippines	8 406	8 645	9 443	7 861	6 449	7 186	8 249	9 284	9 557
Singapour	16 489	22 799	27 101	34 213	30 241	32 847	33 824	32 014	33 781
Sri Lanka	1 495	1 711	1 439	1 741	1 591	1 632	1 670	1 728	1 139
Taipei chinois	5 477	9 004	17 103	19 784	19 935	19 894	21 623	22 986	24 526
Thaïlande	4 889	6 956	7 906	7 620	7 711	8 265	8 331	8 741	10 705
Viêt-Nam	-	-	-	-	-	-	1	42	42
Autre Asie	-	-	-	-	-	-	-	-	-
Asie	**74 066**	**88 261**	**115 642**	**144 857**	**146 969**	**158 466**	**169 735**	**172 938**	**183 440**
Rép. populaire de Chine	32 300	41 600	67 161	82 315	85 685	92 316	97 495	101 645	105 193
Chine	**32 300**	**41 600**	**67 161**	**82 315**	**85 685**	**92 316**	**97 495**	**101 645**	**105 193**
Albanie	1 477	1 646	2 145	1 260	1 111	1 134	1 130	1 074	1 067
Bulgarie	7 742	9 359	11 851	11 159	11 464	11 303	11 492	11 950	12 001
Chypre	-	663	431	552	442	539	612	717	654
Roumanie	16 178	17 986	25 927	23 546	23 724	25 524	28 639	28 810	28 423
République slovaque	5 361	6 043	7 411	6 637	6 614	6 661	6 825	6 585	8 138
Croatie	-	-	-	-	-	-	-	-	-
Ex-RYM	-	-	-	-	-	-	-	-	-
Slovénie	-	-	-	-	-	-	-	-	-
RF de Yougoslavie	-	-	-	-	-	-	-	-	-
Ex-Yougoslavie	8 007	9 029	14 301	12 927	12 714	14 666	15 266	16 890	16 047
Europe non-OCDE	**38 765**	**44 726**	**62 066**	**56 081**	**56 069**	**59 827**	**63 964**	**66 026**	**66 330**
Azerbaïdjan	-	-	-	-	-	-	-	-	-
Bélarus	-	-	-	-	-	-	-	-	-
Estonie	-	-	-	-	-	-	-	-	-
Géorgie	-	-	-	-	-	-	-	-	-
Kazakhstan	-	-	-	-	-	-	-	-	-
Kirghizistan	-	-	-	-	-	-	-	-	-
Lituanie	-	-	-	-	-	-	-	-	-
Russie	-	-	-	-	-	-	-	-	-
Tadjikistan	-	-	-	-	-	-	-	-	-
Turkménistan	-	-	-	-	-	-	-	-	-
Ukraine	-	-	-	-	-	-	-	-	-
Ouzbékistan	-	-	-	-	-	-	-	-	-
Ex-URSS	**289 600**	**332 100**	**420 690**	**451 300**	**449 000**	**446 000**	**461 300**	**461 500**	**456 000**

Refinery Output of Petroleum Products (1000 tonnes)
Production de produits pétroliers en raffineries (1000 tonnes)
Raffinerieausstoß von Ölprodukten (1000 Tonnen)
Produzione di prodotti petroliferi nelle raffinerie (1000 tonnellate)
石油製品の生産量（千トン）
Producción de productos petrolíferos en refinerías (1000 toneladas)
Производство нефтепродуктов нефтеперерабатующими заводами (тыс. т)

1990	1991	1992	1993	1994	1995	1996	1997	1998	
1 044	1 134	1 026	1 347	1 207	1 388	1 169	1 352	684	Bangladesh
285	315	378	389	403	460	450	476	476	Brunei
50 301	49 594	53 007	52 604	55 356	57 184	61 652	63 247	66 503	India
36 603	37 496	39 575	39 778	39 225	39 756	43 192	42 439	44 729	Indonesia
2 942	2 885	2 885	2 860	2 835	2 810	2 785	2 595	2 466	DPR of Korea
10 219	10 373	9 309	10 281	13 674	13 574	18 095	16 246	16 414	Malaysia
713	671	745	774	846	929	761	983	978	Myanmar
5 711	6 416	6 330	6 219	6 309	5 931	6 386	5 968	6 302	Pakistan
10 853	10 410	11 670	11 440	11 809	15 977	17 472	17 691	16 144	Philippines
40 264	42 538	44 063	47 737	47 612	46 629	50 343	48 111	47 694	Singapore
1 689	1 543	1 533	1 642	1 907	1 856	2 043	1 815	2 134	Sri Lanka
23 434	23 710	23 747	26 226	27 047	31 045	32 773	33 845	34 631	Chinese Taipei
11 259	11 674	14 431	16 638	19 291	22 064	30 078	34 816	32 582	Thailand
42	42	42	1	1	-	2	-	-	Vietnam
-	50	50	50	50	50	50	50	50	Other Asia
195 359	**198 851**	**208 791**	**217 986**	**227 572**	**239 653**	**267 251**	**269 634**	**271 787**	**Asia**
107 204	113 530	119 693	127 546	128 381	139 334	148 629	162 253	162 714	People's Rep. of China
107 204	**113 530**	**119 693**	**127 546**	**128 381**	**139 334**	**148 629**	**162 253**	**162 714**	**China**
1 055	1 007	753	631	554	456	434	350	350	Albania
8 003	4 263	2 420	5 388	6 595	7 255	6 722	5 820	5 565	Bulgaria
626	747	724	778	902	826	758	1 038	1 076	Cyprus
22 618	14 736	12 955	13 217	14 752	14 559	13 246	12 302	13 025	Romania
6 434	5 184	4 557	4 576	4 950	4 954	5 214	5 387	5 450	Slovak Republic
-	-	3 960	4 975	4 925	5 294	4 806	5 114	5 175	*Croatia*
-	-	884	902	164	111	675	385	914	*FYROM*
-	-	593	558	373	593	521	541	553	*Slovenia*
-	-	2 074	1 265	1 178	1 297	2 057	3 206	2 619	*FR of Yugoslavia*
14 859	12 259	7 511	7 700	6 640	7 295	8 059	9 246	9 261	Former Yugoslavia
53 595	**38 196**	**28 920**	**32 290**	**34 393**	**35 345**	**34 433**	**34 143**	**34 727**	**Non-OECD Europe**
-	-	11 279	10 409	9 161	8 888	7 956	7 522	7 712	Azerbaijan
-	-	20 120	13 735	11 847	12 262	11 200	11 072	11 058	Belarus
-	-	273	178	301	313	351	381	225	Estonia
-	-	716	300	201	41	13	24	39	Georgia
-	-	17 458	14 692	11 603	10 589	10 483	8 862	8 743	Kazakhstan
-	-	-	-	-	70	91	110	139	Kyrgyzstan
-	-	4 116	5 185	4 104	3 471	4 111	5 514	6 761	Lithuania
-	-	248 531	223 624	187 129	179 459	173 132	171 794	159 252	Russia
-	-	56	39	30	22	18	23	12	Tajikistan
-	-	4 770	3 560	3 499	3 200	3 424	4 136	4 769	Turkmenistan
-	-	36 667	22 282	19 362	16 550	13 143	12 848	13 489	Ukraine
-	-	6 465	6 302	5 898	6 607	6 571	6 767	6 962	Uzbekistan
430 230	**418 260**	**350 451**	**300 306**	**253 135**	**241 472**	**230 493**	**229 053**	**219 161**	**Former USSR**

Refinery Output of Petroleum Products (1000 tonnes)
Production de produits pétroliers en raffineries (1000 tonnes)
Raffinerieausstoß von Ölprodukten (1000 Tonnen)
Produzione di prodotti petroliferi nelle raffinerie (1000 tonnellate)
石油製品の生産量（千トン）
Producción de productos petrolíferos en refinerías (1000 toneladas)
Производство нефтепродуктов нефтеперерабатующими заводами (тыс. т)

	1971	1973	1978	1984	1985	1986	1987	1988	1989
Bahrein	12 640	12 042	12 540	10 098	9 141	12 233	12 061	11 858	12 070
République Islamique d'Iran	27 437	29 111	36 567	33 198	35 368	35 714	34 908	35 903	39 804
Irak	3 578	4 035	9 015	17 516	18 634	19 473	20 169	21 491	22 406
Israël	5 207	6 113	7 392	6 712	6 603	6 823	7 363	7 016	7 212
Jordanie	554	671	1 370	2 456	2 410	2 236	2 386	2 321	2 330
Koweit	18 241	19 252	19 353	27 320	30 876	32 240	36 147	41 589	49 052
Liban	2 034	2 420	1 829	735	615	1 030	1 150	540	92
Oman	-	-	-	1 999	2 369	2 767	2 334	2 472	2 606
Qatar	21	20	304	732	1 249	1 580	1 744	2 084	2 437
Arabie saoudite	27 514	28 244	30 804	42 475	50 400	55 121	66 121	70 154	69 052
Syrie	2 112	1 963	4 090	10 159	10 041	10 090	11 132	11 511	11 179
Emirats arabes unis	-	-	-	6 196	7 371	8 668	9 263	8 893	9 131
Yémen	3 505	2 852	1 845	4 362	3 716	3 296	4 996	5 015	3 624
Moyen-Orient	**102 843**	**106 723**	**125 109**	**163 958**	**178 793**	**191 271**	**209 774**	**220 847**	**230 995**
Total non-OCDE	**793 210**	**902 670**	**1 061 614**	**1 178 282**	**1 185 120**	**1 218 690**	**1 279 885**	**1 317 521**	**1 337 959**
OCDE Amérique du N.	689 448	773 666	916 952	788 677	784 786	817 234	832 181	854 246	867 803
OCDE Pacifique	215 293	265 838	271 128	227 127	213 149	209 480	206 069	220 541	234 273
OCDE Europe	667 556	776 260	723 302	601 030	586 588	624 915	621 731	646 211	648 789
Total OCDE	**1 572 297**	**1 815 764**	**1 911 382**	**1 616 834**	**1 584 523**	**1 651 629**	**1 659 981**	**1 720 998**	**1 750 865**
Monde	**2 365 507**	**2 718 434**	**2 972 996**	**2 795 116**	**2 769 643**	**2 870 319**	**2 939 866**	**3 038 519**	**3 088 824**

Refinery Output of Petroleum Products (1000 tonnes)
Production de produits pétroliers en raffineries (1000 tonnes)
Raffinerieausstoß von Ölprodukten (1000 Tonnen)

Produzione di prodotti petroliferi nelle raffinerie (1000 tonnellate)

石油製品の生産量（千トン）

Producción de productos petrolíferos en refinerías (1000 toneladas)

Производство нефтепродуктов нефтеперерабатующими заводами (тыс. m)

1990	1991	1992	1993	1994	1995	1996	1997	1998	
12 335	12 438	12 851	12 307	12 067	12 539	12 936	12 484	12 354	Bahrain
43 168	44 822	46 143	52 576	54 564	53 793	55 961	59 745	64 581	Islamic Republic of Iran
19 865	13 365	17 520	20 101	22 120	21 673	21 216	21 515	22 220	Iraq
8 060	8 278	10 063	11 257	11 330	10 901	9 872	10 617	11 279	Israel
2 618	2 332	2 874	2 847	2 988	3 147	3 239	3 469	3 481	Jordan
11 768	2 889	17 750	24 629	40 121	43 996	41 188	47 010	46 941	Kuwait
92	530	429	-	-	-	-	-	-	Lebanon
3 360	3 189	2 923	2 405	3 632	3 638	3 647	3 276	3 943	Oman
2 893	2 574	2 690	2 424	2 819	2 646	2 919	2 631	2 852	Qatar
78 477	69 512	74 178	74 192	75 143	72 080	81 823	77 121	77 990	Saudi Arabia
11 317	11 676	11 587	11 416	11 630	11 704	12 114	12 171	12 171	Syria
9 296	10 469	10 845	10 328	11 361	11 361	11 222	11 577	11 194	United Arab Emirates
4 655	4 510	5 180	3 497	3 504	3 457	4 051	4 296	4 296	Yemen
207 904	**186 584**	**215 033**	**227 979**	**251 279**	**250 935**	**260 188**	**265 912**	**273 302**	**Middle East**
1 294 261	1 258 631	1 225 496	1 212 340	1 203 633	1 223 916	1 270 233	1 299 142	1 302 372	**Non-OECD Total**
880 663	873 743	880 031	901 351	910 046	910 900	928 264	948 324	966 033	OECD North America
251 622	278 350	302 564	314 545	328 470	339 795	350 341	377 878	368 448	OECD Pacific
655 382	664 024	682 868	694 917	702 448	699 780	720 264	733 788	752 675	OECD Europe
1 787 667	**1 816 117**	**1 865 463**	**1 910 813**	**1 940 964**	**1 950 475**	**1 998 869**	**2 059 990**	**2 087 156**	**OECD Total**
3 081 928	3 074 748	3 090 959	3 123 153	3 144 597	3 174 391	3 269 102	3 359 132	3 389 528	World

Net Imports of Coking Coal (1000 tonnes)
Importations nettes de charbon à coke (1000 tonnes)
Nettoimporte von Kokskohle (1000 Tonnen)
Importazioni nette di carbone siderurgico (1000 tonnellate)
原料炭の純輸入量 (チトン)
Importaciones netas de carbón coquizable (1000 toneladas)
Чистый импорт коксующихся углей (тыс. m)

	1971	1973	1978	1984	1985	1986	1987	1988	1989
Algérie	-	-	7	1 320	1 379	1 427	1 705	1 596	1 350
Egypte	-	-	857	1 100	1 200	1 200	1 250	1 400	1 448
Afrique du Sud	-	-	- 2 700	- 4 803	- 5 142	- 5 867	- 4 755	- 5 016	- 4 161
Afrique	**-**	**-**	**- 1 836**	**- 2 383**	**- 2 563**	**- 3 240**	**- 1 800**	**- 2 020**	**- 1 363**
Argentine	-	-	814	539	764	1 157	1 103	1 149	1 249
Brésil	-	-	3 547	7 576	8 049	8 442	9 660	9 282	9 552
Chili	-	-	185	342	372	424	430	359	434
Colombie	-	-	-	- 46	-	- 140	- 207	- 322	- 136
Jamaïque	-	-	-	-	-	-	-	8	56
Pérou	-	-	41	47	13	27	36	20	60
Amérique latine	**-**	**-**	**4 587**	**8 458**	**9 198**	**9 910**	**11 022**	**10 496**	**11 215**
Inde	-	-	220	350	2 030	2 100	2 670	3 500	4 210
RPD de Corée	-	-	442	2 200	2 500	2 500	2 600	2 600	2 600
Pakistan	-	-	16	491	716	852	918	853	895
Taipei chinois	-	-	1 386	2 460	2 562	2 771	2 984	4 328	4 921
Asie	**-**	**-**	**2 064**	**5 501**	**7 808**	**8 223**	**9 172**	**11 281**	**12 626**
Rép. populaire de Chine	-	-	- 300	- 2 400	- 2 300	- 2 200	- 3 400	- 4 150	- 3 200
Chine	**-**	**-**	**- 300**	**- 2 400**	**- 2 300**	**- 2 200**	**- 3 400**	**- 4 150**	**- 3 200**
Albanie	-	-	25	32	33	35	35	37	37
Bulgarie	-	-	1 921	1 694	1 553	1 651	1 651	1 400	1 208
Roumanie	-	-	3 600	4 000	4 000	3 800	4 700	4 500	4 400
République slovaque	-	-	2 180	2 512	2 530	2 548	2 621	2 793	3 101
Croatie	-	-	-	-	-	-	-	-	-
Ex-RYM	-	-	-	-	-	-	-	-	-
Ex-Yougoslavie	-	-	1 447	4 854	4 689	4 837	3 980	4 269	3 501
Europe non-OCDE	**-**	**-**	**9 173**	**13 092**	**12 805**	**12 871**	**12 987**	**12 999**	**12 247**
Ex-URSS	**-**	**-**	**- 10 000**	**- 8 000**	**- 11 000**	**- 13 000**	**- 19 000**	**- 21 000**	**- 20 000**
République Islamique d'Iran	-	-	100	50	100	150	200	200	300
Moyen-Orient	**-**	**-**	**100**	**50**	**100**	**150**	**200**	**200**	**300**
Total non-OCDE	**-**	**-**	**3 788**	**14 318**	**14 048**	**12 714**	**9 181**	**7 806**	**11 825**
OCDE Amérique du N.	-	-	- 34 426	- 65 918	- 70 420	- 65 427	- 63 770	- 77 560	- 81 857
OCDE Pacifique	-	-	19 386	30 684	25 847	24 856	23 044	25 769	23 345
OCDE Europe	-	-	17 359	31 242	37 151	35 038	37 631	41 328	40 605
Total OCDE	**-**	**-**	**2 319**	**- 3 992**	**- 7 422**	**- 5 533**	**- 3 095**	**- 10 463**	**- 17 907**
Monde	**-**	**-**	**6 107**	**10 326**	**6 626**	**7 181**	**6 086**	**- 2 657**	**- 6 082**

Un chiffre négatif correspond à des exportations nettes.
La ligne Monde montre la divergence entre le total des exportations et des importations mondiales.
Les divergences dans les échanges de charbon s'expliquent principalement par des différences de classification.

Net Imports of Coking Coal (1000 tonnes)
Importations nettes de charbon à coke (1000 tonnes)
Nettoimporte von Kokskohle (1000 Tonnen)
Importazioni nette di carbone siderurgico (1000 tonnellate)
原料炭の純輸入量（チトン）
Importaciones netas de carbón coquizable (1000 toneladas)
Чистый импорт коксующихся углей (тыс. т)

1990	1991	1992	1993	1994	1995	1996	1997	1998	
749	868	1 035	805	817	480	318	552	628	Algeria
1 339	1 206	1 154	1 500	1 852	1 540	1 876	1 876	1 876	Egypt
- 3 633	- 3 523	- 5 088	-	-	360	425	425	1 213	South Africa
- 1 545	**- 1 449**	**- 2 899**	**2 305**	**2 669**	**2 380**	**2 619**	**2 853**	**3 717**	**Africa**
1 233	882	992	967	1 279	1 154	1 049	881	990	Argentina
10 146	10 758	10 399	10 975	11 212	11 790	12 847	12 883	12 964	Brazil
477	602	594	665	659	704	713	685	712	Chile
-	- 121	- 280	- 1 067	- 1 114	- 1 062	- 1 118	- 1 101	- 660	Colombia
62	40	40	49	53	55	64	56	27	Jamaica
45	51	52	53	47	51	51	41	38	Peru
11 963	**12 212**	**11 797**	**11 642**	**12 136**	**12 692**	**13 606**	**13 445**	**14 071**	**Latin America**
5 000	5 270	6 320	6 820	10 150	9 370	9 780	10 650	9 640	India
2 600	2 500	2 500	2 400	2 350	2 300	2 275	2 120	2 014	DPR of Korea
900	971	985	994	1 094	1 096	1 085	840	960	Pakistan
4 237	4 350	3 998	4 125	4 161	4 580	4 112	6 300	5 932	Chinese Taipei
12 737	**13 091**	**13 803**	**14 339**	**17 755**	**17 346**	**17 252**	**19 910**	**18 546**	**Asia**
- 3 100	- 3 600	- 3 300	-	-	-	-	-	-	People's Rep. of China
- 3 100	**- 3 600**	**- 3 300**	**-**	**-**	**-**	**-**	**-**	**-**	**China**
62	60	-	-	-	-	-	-	-	Albania
1 100	933	1 143	1 258	1 527	1 734	1 438	1 677	1 170	Bulgaria
3 600	2 100	4 563	2 615	4 005	4 675	3 996	5 014	3 711	Romania
3 132	2 720	2 517	2 612	2 727	2 746	2 850	2 568	2 287	Slovak Republic
-	-	613	490	317	-	-	-	-	*Croatia*
-	-	73	44	47	54	49	49	122	*FYROM*
3 297	2 200	686	534	364	54	49	49	122	Former Yugoslavia
11 191	**8 013**	**8 909**	**7 019**	**8 623**	**9 209**	**8 333**	**9 308**	**7 290**	**Non-OECD Europe**
- 17 000	**-**	**- 32 222**	**- 14 145**	**- 22 105**	**- 2 522**	**- 759**	**- 4 978**	**- 3 336**	**Former USSR**
400	400	500	-	400	552	598	598	704	Islamic Republic of Iran
400	**400**	**500**	**-**	**400**	**552**	**598**	**598**	**704**	**Middle East**
14 646	**28 667**	**- 3 412**	**21 160**	**19 478**	**39 657**	**41 649**	**41 136**	**40 992**	**Non-OECD Total**
- 79 706	- 82 663	- 72 186	- 64 286	- 65 540	- 71 690	- 71 925	- 73 105	- 65 388	OECD North America
21 940	23 515	13 349	9 091	7 659	7 437	5 906	3 340	- 5 299	OECD Pacific
43 366	44 944	42 653	30 494	32 107	30 926	36 617	40 556	45 389	OECD Europe
- 14 400	**- 14 204**	**- 16 184**	**- 24 701**	**- 25 774**	**- 33 327**	**- 29 402**	**- 29 209**	**- 25 298**	**OECD Total**
246	**14 463**	**- 19 596**	**- 3 541**	**- 6 296**	**6 330**	**12 247**	**11 927**	**15 694**	**World**

A negative number shows net exports.
The row World shows the discrepancy between total world exports and imports.
Discrepancies in coal trade are mainly due to inconsistent classifications.

Net Imports of Other Bituminous Coal (1000 tonnes)
Importations nettes d'autres charbons bitumineux (1000 tonnes)
Nettoimporte von sonstiger bituminöser Kohle (1000 Tonnen)

Importazioni nette di altri carboni bituminosi (1000 tonnellate)

その他歴青炭の純輸入量（チトン）

Importaciones netas de otro carbón bituminoso (1000 toneladas)

Чистый импорт других битуминозных углей (тыс. т)

	1971	1973	1978	1984	1985	1986	1987	1988	1989
RD du Congo	-	-	60	36	35	41	43	43	45
Egypte	-	-	-	-	-	-	-	-	2
Ghana	-	-	2	2	2	2	3	3	3
Kenya	-	-	50	83	90	85	92	113	131
Maroc	-	-	- 41	129	355	742	990	1 043	1 279
Mozambique	-	-	98	85	71	62	20	20	22
Nigéria	-	-	- 4	- 7	- 7	- 40	- 45	- 40	- 40
Afrique du Sud	-	-	- 12 689	- 34 054	- 42 488	- 40 318	- 36 945	- 39 154	- 42 770
Tunisie	-	-	27	24	21	12	16	29	30
Zambie	-	-	-	-	-	- 10	-	-	-
Zimbabwe	-	-	- 197	- 138	- 78	- 13	- 25	- 34	- 30
Autre Afrique	-	-	62	80	71	100	76	74	141
Afrique	**-**	**-**	**- 12 632**	**- 33 760**	**- 41 928**	**- 39 337**	**- 35 775**	**- 37 903**	**- 41 187**
Argentine	-	-	-	-	-	-	-	-	-
Brésil	-	-	-	-	-	- 107	-	-	-
Chili	-	-	-	-	-	35	3	-	1 049
Colombie	-	-	- 150	- 767	- 3 166	- 4 944	- 8 731	- 9 799	- 12 351
Costa Rica	-	-	1	1	1	1	1	1	1
Cuba	-	-	91	92	126	119	85	95	214
République dominicaine	-	-	-	-	224	28	232	97	10
Haiti	-	-	-	60	61	18	18	36	6
Panama	-	-	-	9	60	39	39	16	18
Pérou	-	-	-	-	-	- 21	-	- 30	-
Uruguay	-	-	2	-	-	-	1	-	-
Vénézuela	-	-	- 21	-	-	-	-	- 1 007	- 1 688
Amérique latine	**-**	**-**	**- 77**	**- 605**	**- 2 694**	**- 4 832**	**- 8 352**	**- 10 591**	**- 12 741**
Bangladesh	-	-	319	62	98	148	233	240	250
Inde	-	-	- 270	- 120	- 210	- 160	130	-	40
Indonésie	-	-	- 29	- 882	- 1 066	330	708	324	- 1 461
RPD de Corée	-	-	- 27	- 150	- 300	- 500	- 2 000	- 500	- 500
Malaisie	-	-	33	385	517	383	467	270	793
Myanmar	-	-	210	175	180	180	40	40	40
Népal	-	-	16	145	17	92	85	77	12
Philippines	-	-	20	513	1 240	956	615	1 320	987
Sri Lanka	-	-	-	-	1	-	-	-	2
Taipei chinois	-	-	-	5 053	7 530	7 914	11 038	13 148	11 859
Thailande	-	-	52	227	212	183	223	304	392
Viêt-Nam	-	-	- 1 430	- 470	- 604	- 620	- 200	- 314	- 527
Autre Asie	-	-	68	125	194	191	176	195	219
Asie	**-**	**-**	**- 1 038**	**5 063**	**7 809**	**9 097**	**11 515**	**15 104**	**12 106**
Rép. populaire de Chine	-	-	- 380	- 2 065	- 3 163	- 2 687	- 4 834	- 9 803	- 9 848
Hong-Kong, Chine	-	-	8	4 462	5 523	6 392	8 009	9 267	9 927
Chine	**-**	**-**	**- 372**	**2 397**	**2 360**	**3 705**	**3 175**	**- 536**	**79**

Un chiffre négatif correspond à des exportations nettes.

Net Imports of Other Bituminous Coal (1000 tonnes)
Importations nettes d'autres charbons bitumineux (1000 tonnes)
Nettoimporte von sonstiger bituminöser Kohle (1000 Tonnen)

Importazioni nette di altri carboni bituminosi (1000 tonnellate)

その他歴青炭の純輸入量（チトン）

Importaciones netas de otro carbón bituminoso (1000 toneladas)

Чистый импорт других битуминозных углей (тыс. т)

1990	1991	1992	1993	1994	1995	1996	1997	1998	
45	46	42	42	43	43	43	43	40	DR of Congo
1	1	1	-	-	-	-	-	-	Egypt
3	3	3	3	3	3	3	3	-	Ghana
151	151	158	131	123	156	144	148	122	Kenya
1 201	1 373	1 189	1 010	1 297	1 931	2 859	2 831	3 205	Morocco
18	20	20	20	20	18	- 20	- 18	- 18	Mozambique
- 35	- 35	- 35	- 20	-	-	-	-	-	Nigeria
- 46 267	- 43 834	- 46 971	- 51 711	- 54 838	- 59 676	- 60 224	- 64 200	- 66 100	South Africa
15	12	15	14	-	-	-	-	-	Tunisia
- 5	- 18	- 15	- 19	- 15	- 7	- 7	- 7	- 7	Zambia
10	- 25	- 11	-	- 225	- 199	- 151	- 112	- 172	Zimbabwe
138	113	112	108	88	113	114	119	119	Other Africa
- 44 725	**- 42 193**	**- 45 492**	**- 50 422**	**- 53 504**	**- 57 618**	**- 57 239**	**- 61 193**	**- 62 811**	**Africa**
- 32	- 15	- 7	-	- 41	- 37	-	-	-	Argentina
-	-	-	-	107	-	-	-	-	Brazil
1 183	888	285	423	1 271	1 504	2 882	3 940	3 913	Chile
- 13 505	- 16 258	- 14 334	- 16 549	- 17 323	- 17 212	- 23 663	- 26 479	- 28 741	Colombia
1	-	-	-	-	-	-	-	-	Costa Rica
153	106	47	70	87	77	1	16	19	Cuba
9	57	139	120	104	111	128	139	147	Dominican Republic
12	31	26	-	-	-	-	-	-	Haiti
32	46	50	75	61	57	108	57	59	Panama
-	148	138	311	321	256	204	282	399	Peru
1	-	-	-	-	-	-	-	-	Uruguay
- 1 834	- 2 196	- 2 309	- 3 825	- 4 135	- 4 242	- 3 617	- 5 105	- 5 908	Venezuela
- 13 980	**- 17 193**	**- 15 965**	**- 19 375**	**- 19 548**	**- 19 486**	**- 23 957**	**- 27 150**	**- 30 112**	**Latin America**
563	180	169	63	59	59	345	635	173	Bangladesh
-	550	290	480	1 130	3 050	4 400	6 500	5 960	India
- 4 110	- 6 933	- 15 604	- 17 932	- 20 997	- 30 684	- 36 032	- 41 342	- 46 548	Indonesia
- 500	- 475	- 450	- 450	- 425	- 400	- 385	- 359	- 341	DPR of Korea
932	720	728	951	678	951	1 608	1 010	1 111	Malaysia
40	40	-	-	- 29	3	-	1	2	Myanmar
82	92	111	104	115	123	274	297	323	Nepal
1 364	1 356	676	725	1 049	1 032	2 984	4 248	3 713	Philippines
8	2	2	2	2	1	1	1	1	Sri Lanka
14 231	14 032	18 094	21 169	22 507	24 101	26 976	29 952	31 120	Chinese Taipei
250	436	471	968	1 417	2 308	3 750	3 074	1 380	Thailand
- 745	- 945	- 1 306	- 1 736	- 2 060	- 2 821	- 3 647	- 3 454	- 2 900	Vietnam
200	211	207	208	204	206	212	222	222	Other Asia
12 315	**9 266**	**3 388**	**4 552**	**3 650**	**- 2 071**	**486**	**785**	**- 5 784**	**Asia**
- 12 177	- 15 033	- 15 133	- 18 387	- 22 985	- 26 982	- 33 267	- 33 318	- 30 711	People's Rep. of China
8 928	9 635	10 229	11 830	8 451	9 109	6 769	5 711	7 102	Hong Kong, China
- 3 249	**- 5 398**	**- 4 904**	**- 6 557**	**- 14 534**	**- 17 873**	**- 26 498**	**- 27 607**	**- 23 609**	**China**

A negative number shows net exports.

Net Imports of Other Bituminous Coal (1000 tonnes)
Importations nettes d'autres charbons bitumineux (1000 tonnes)
Nettoimporte von sonstiger bituminöser Kohle (1000 Tonnen)

Importazioni nette di altri carboni bituminosi (1000 tonnellate)

その他瀝青炭の純輸入量（千トン）

Importaciones netas de otro carbón bituminoso (1000 toneladas)

Чистый импорт других битуминозных углей (тыс. т)

	1971	1973	1978	1984	1985	1986	1987	1988	1989
Albanie	-	-	143	178	187	195	195	203	203
Bulgarie	-	-	4 280	5 219	6 501	5 317	5 536	5 074	5 032
Chypre	-	-	-	52	74	55	151	91	102
Malte	-	-	-	94	212	146	182	252	303
Roumanie	-	-	1 069	1 300	2 048	2 524	2 382	3 585	1 408
République slovaque	-	-	3 971	3 423	3 507	3 827	3 660	3 643	3 629
Croatie	-	-	-	-	-	-	-	-	-
Ex-RYM	-	-	-	-	-	-	-	-	-
Slovénie	-	-	-	-	-	-	-	-	-
RF de Yougoslavie	-	-	-	-	-	-	-	-	-
Ex-Yougoslavie	-	-	-	-	-	-	-	-	-
Europe non-OCDE	-	-	**9 463**	**10 266**	**12 529**	**12 064**	**12 106**	**12 848**	**10 677**
Ex-URSS	-	-	**- 8 100**	**- 6 300**	**- 8 800**	**- 9 900**	**- 7 800**	**- 11 700**	**- 7 700**
République Islamique d'Iran	-	-	-	-	-	-	-	-	-
Israël	-	-	-	2 669	3 063	3 084	3 518	3 536	3 695
Liban	-	-	1	-	-	-	-	-	-
Syrie	-	-	1	2	-	-	-	-	-
Moyen-Orient	-	-	**2**	**2 671**	**3 063**	**3 084**	**3 518**	**3 536**	**3 695**
Total non-OCDE	-	-	**- 12 754**	**- 20 268**	**- 27 661**	**- 26 119**	**- 21 613**	**- 29 242**	**- 35 071**
OCDE Amérique du N.	-	-	667	- 13 354	- 24 066	- 22 459	- 19 267	- 20 968	- 25 322
OCDE Pacifique	-	-	- 1 677	2 648	2 872	- 4 278	- 7 525	- 3 509	2 804
OCDE Europe	-	-	- 176	26 430	42 877	44 553	45 603	40 696	50 180
Total OCDE	-	-	**- 1 186**	**15 724**	**21 683**	**17 816**	**18 811**	**16 219**	**27 662**
Monde	-	-	**- 13 940**	**- 4 544**	**- 5 978**	**- 8 303**	**- 2 802**	**- 13 023**	**- 7 409**

Un chiffre négatif correspond à des exportations nettes.

La ligne Monde montre la divergence entre le total des exportations et des importations mondiales.

Les divergences dans les échanges de charbon s'expliquent principalement par des différences de classification.

Net Imports of Other Bituminous Coal (1000 tonnes)
Importations nettes d'autres charbons bitumineux (1000 tonnes)
Nettoimporte von sonstiger bituminöser Kohle (1000 Tonnen)

Importazioni nette di altri carboni bituminosi (1000 tonnellate)

その他歴青炭の純輸入量（千トン）

Importaciones netas de otro carbón bituminoso (1000 toneladas)

Чистый импорт других битуминозных углей (тыс. т)

1990	1991	1992	1993	1994	1995	1996	1997	1998	
240	140	50	-	-	-	-	-	-	Albania
4 690	3 595	2 531	2 977	1 834	1 719	2 275	2 034	2 379	Bulgaria
97	97	26	33	30	26	17	26	21	Cyprus
300	243	205	300	210	37	-	-	-	Malta
1 381	871	1 223	51	30	6	9	7	10	Romania
2 734	2 440	2 428	2 512	2 441	2 182	2 650	2 668	2 244	Slovak Republic
-	-	146	143	60	106	51	122	230	*Croatia*
-	-	12	65	30	44	51	51	101	*FYROM*
-	-	14	20	23	20	22	19	17	*Slovenia*
-	-	1 000	60	50	60	60	52	52	*FR of Yugoslavia*
-	-	1 172	288	163	230	184	244	400	*Former Yugoslavia*
9 442	**7 386**	**7 635**	**6 161**	**4 708**	**4 200**	**5 135**	**4 979**	**5 054**	**Non-OECD Europe**
- 9 400	**- 22 088**	**5 736**	**- 3 511**	**9 513**	**- 4 182**	**- 12 016**	**- 12 245**	**- 12 301**	**Former USSR**
-	-	-	-	-	- 50	- 13	- 13	- 11	Islamic Republic of Iran
3 998	3 921	5 432	5 381	6 121	7 090	7 156	8 804	9 636	Israel
-	-	-	111	112	180	200	200	200	Lebanon
-	-	-	-	-	-	-	-	-	Syria
3 998	**3 921**	**5 432**	**5 492**	**6 233**	**7 220**	**7 343**	**8 991**	**9 825**	**Middle East**
- 45 599	**- 66 299**	**- 44 170**	**- 63 660**	**- 63 482**	**- 89 810**	**- 106 746**	**- 113 440**	**- 119 738**	**Non-OECD Total**
- 30 366	- 34 583	- 31 847	- 14 484	- 12 186	- 22 460	- 22 151	- 16 643	- 11 407	OECD North America
1 102	1 113	1 921	9 104	18 413	21 429	26 988	25 980	19 937	OECD Pacific
65 345	85 843	86 076	76 938	71 771	78 919	82 611	85 075	83 428	OECD Europe
36 081	**52 373**	**56 150**	**71 558**	**77 998**	**77 888**	**87 448**	**94 412**	**91 958**	**OECD Total**
- 9 518	**- 13 926**	**11 980**	**7 898**	**14 516**	**- 11 922**	**- 19 298**	**- 19 028**	**- 27 780**	**World**

A negative number shows net exports.

The row World shows the discrepancy between total world exports and imports.

Discrepancies in coal trade are mainly due to inconsistent classifications.

Net Imports of Sub-Bituminous Coal (1000 tonnes)
Importations nettes de charbons sous-bitumineux (1000 tonnes)
Nettoimporte von subbituminöser Kohle (1000 Tonnen)
Importazioni nette di carbone sub-bituminoso (1000 tonnellate)
亜歴青炭の純輸入量（チトン）
Importaciones netas de otro carbón sub-bituminoso (1000 toneladas)
Чистый импорт полубитуминозных углей (тыс. m)

	1971	1973	1978	1984	1985	1986	1987	1988	1989
Malaisie	-	-	-	-	-	-	-	80	751
Asie	-	-	-	-	-	-	-	80	751
Roumanie	-	-	-	-	-	-	-	-	-
Croatie	-	-	-	-	-	-	-	-	-
Slovénie	-	-	-	-	-	-	-	-	-
Ex-Yougoslavie	-	-	-	-	-	-	-	-	-
Europe non-OCDE	-	-	-	-	-	-	-	-	-
Total non-OCDE	-	-	-	-	-	-	-	80	751
OCDE Amérique du N.	-	-	-	-	-	-	-	-	-
OCDE Pacifique	-	-	-	-	-	-	-	-	-
OCDE Europe	-	-	- 6 778	- 9 414	- 9 345	- 10 315	- 10 410	- 9 705	- 9 051
Total OCDE	-	-	- 6 778	- 9 414	- 9 345	- 10 315	- 10 410	- 9 705	- 9 051
Monde	-	-	- 6 778	- 9 414	- 9 345	- 10 315	- 10 410	- 9 625	- 8 300

Un chiffre négatif correspond à des exportations nettes.
La ligne Monde montre la divergence entre le total des exportations et des importations mondiales.
Les divergences dans les échanges de charbon s'expliquent principalement par des différences de classification.

Net Imports of Sub-Bituminous Coal (1000 tonnes)
Importations nettes de charbons sous-bitumineux (1000 tonnes)
Nettoimporte von subbituminöser Kohle (1000 Tonnen)

Importazioni nette di carbone sub-bituminoso (1000 tonnellate)

亜歴青炭の純輸入量（千トン）

Importaciones netas de otro carbón sub-bituminoso (1000 toneladas)

Чистый импорт полубитуминозных углей (тыс. т)

1990	1991	1992	1993	1994	1995	1996	1997	1998	
1 061	1 195	1 306	1 245	1 194	1 245	1 138	1 041	1 061	Malaysia
1 061	**1 195**	**1 306**	**1 245**	**1 194**	**1 245**	**1 138**	**1 041**	**1 061**	**Asia**
-	-	-	-	-	24	-	228	-	Romania
-	-	-	-	-	-	-	88	62	*Croatia*
-	-	118	297	228	329	385	299	419	*Slovenia*
-	-	118	297	228	329	385	387	481	Former Yugoslavia
-	**-**	**118**	**297**	**228**	**353**	**385**	**615**	**481**	**Non-OECD Europe**
1 061	1 195	1 424	1 542	1 422	1 598	1 523	1 656	1 542	**Non-OECD Total**
-	-	-	- 1 931	- 2 170	- 3 086	- 2 856	- 133	- 349	OECD North America
-	-	-	- 7	-	-	-	2 305	2 527	OECD Pacific
- 8 405	- 7 900	- 7 923	- 8 061	- 5 380	- 5 565	- 5 058	- 4 333	- 3 471	OECD Europe
- 8 405	**- 7 900**	**- 7 923**	**- 9 999**	**- 7 550**	**- 8 651**	**- 7 914**	**- 2 161**	**- 1 293**	**OECD Total**
- 7 344	**- 6 705**	**- 6 499**	**- 8 457**	**- 6 128**	**- 7 053**	**- 6 391**	**- 505**	**249**	**World**

A negative number shows net exports.

The row World shows the discrepancy between total world exports and imports.

Discrepancies in coal trade are mainly due to inconsistent classifications.

Net Imports of Lignite (1000 tonnes)
Importations nettes de lignite (1000 tonnes)

Nettoimporte von Braunkohle (1000 Tonnen)

Importazioni nette di lignite (1000 tonnellate)

亜炭の純輸入量（千トン）

Importaciones netas de lignito (1000 toneladas)

Чистый импорт лигнита (тыс. т)

	1971	1973	1978	1984	1985	1986	1987	1988	1989
Singapour	-	-	2	1	2	1	1	1	1
Asie	-	-	2	1	2	1	1	1	1
Albanie	-	-	-	-	-	-	-	-	-
Bulgarie	-	-	-	-	-	-	21	-	-
Roumanie	-	-	430	210	480	1 002	1 586	1 953	2 539
République slovaque	-	-	7 560	7 455	8 072	8 831	9 209	9 037	8 223
Croatie	-	-	-	-	-	-	-	-	-
Ex-RYM	-	-	-	-	-	-	-	-	-
Slovénie	-	-	-	-	-	-	-	-	-
Ex-Yougoslavie	-	-	- 376	- 123	- 390	- 351	- 33	- 47	93
Europe non-OCDE	-	-	7 614	7 542	8 162	9 482	10 783	10 943	10 855
Ex-URSS	-	-	-	15	15	15	15	-	-
Total non-OCDE	-	-	7 616	7 558	8 179	9 498	10 799	10 944	10 856
OCDE Amérique du N.	-	-	- 83	- 8	- 30	- 37	- 32	- 7	- 10
OCDE Europe	-	-	1 559	3 053	3 445	3 424	2 741	2 258	2 416
Total OCDE	-	-	1 476	3 045	3 415	3 387	2 709	2 251	2 406
Monde	-	-	9 092	10 603	11 594	12 885	13 508	13 195	13 262

Un chiffre négatif correspond à des exportations nettes.

La ligne Monde montre la divergence entre le total des exportations et des importations mondiales.

Les divergences dans les échanges de charbon s'expliquent principalement par des différences de classification.

Net Imports of Lignite (1000 tonnes)
Importations nettes de lignite (1000 tonnes)
Nettoimporte von Braunkohle (1000 Tonnen)
Importazioni nette di lignite (1000 tonnellate)
亜炭の純輸入量（チトン）
Importaciones netas de lignito (1000 toneladas)
Чистый импорт лигнита (тыс. т)

1990	1991	1992	1993	1994	1995	1996	1997	1998	
2	2	1	1	1	1	1	1	1	Singapore
2	**2**	**1**	**1**	**1**	**1**	**1**	**1**	**1**	**Asia**
- 228	- 100	- 100	- 8	-	-	-	-	-	Albania
-	-	-	-	-	-	-	-	-	Bulgaria
3 451	2 181	1 295	541	197	29	729	-	146	Romania
6 753	6 767	5 791	5 558	3 656	3 413	3 365	2 188	1 404	Slovak Republic
-	-	204	208	173	181	147	45	39	*Croatia*
-	-	130	17	-	19	19	-	196	*FYROM*
-	-	- 36	- 8	- 6	9	19	6	- 1	*Slovenia*
- 37	-	298	217	167	209	185	51	234	Former Yugoslavia
9 939	**8 848**	**7 284**	**6 308**	**4 020**	**3 651**	**4 279**	**2 239**	**1 784**	**Non-OECD Europe**
-	**-**	**- 8 375**	**442**	**1 288**	**- 114**	**18**	**1 092**	**1 022**	**Former USSR**
9 941	**8 850**	**- 1 090**	**6 751**	**5 309**	**3 538**	**4 298**	**3 332**	**2 807**	**Non-OECD Total**
- 79	- 55	- 54	- 27	- 50	2	- 9	-	-	OECD North America
2 606	3 926	3 656	3 269	2 858	2 524	2 854	3 053	2 587	OECD Europe
2 527	**3 871**	**3 602**	**3 242**	**2 808**	**2 526**	**2 845**	**3 053**	**2 587**	**OECD Total**
12 468	**12 721**	**2 512**	**9 993**	**8 117**	**6 064**	**7 143**	**6 385**	**5 394**	**World**

A negative number shows net exports.

The row World shows the discrepancy between total world exports and imports.

Discrepancies in coal trade are mainly due to inconsistent classifications.

Net Imports of Crude Oil, NGL, Refinery Feedstocks and Additives (1000 tonnes)
Import. nettes de pétr. brut, LGN, produits d'aliment. des raffin. et additifs (1000 tonnes)
Nettoimporte von Rohöl, Kondensaten, Raffinerie-Feedstocks und Additiven (1000 Tonnen)
Importazioni nette di petrolio grezzo, GNL, prodotti intermedi e additivi (1000 tonnellate)
原油、ＮＧＬ等随伴物の純輸入量（千トン）
Import. netas de petróleo crudo, líquidos de gas natural, prod. de alim. de refinerias y aditivos (1000 toneladas)
Чистый имп. сырой нефти, газ. конденсатов, нефтезаводского сырья и присадок (тыс. т)

	1971	1973	1978	1984	1985	1986	1987	1988	1989
Algérie	- 34 300	- 44 856	- 49 088	- 26 610	- 27 769	- 26 641	- 27 405	- 26 742	- 28 959
Angola	- 4 747	- 7 323	- 5 500	- 8 659	- 9 744	- 12 458	- 15 686	- 20 797	- 20 580
Bénin	-	-	-	- 5	- 325	- 365	- 312	- 238	- 191
Cameroun	-	-	- 505	- 7 357	- 7 206	- 7 050	- 7 122	- 6 725	- 7 286
Congo	- 20	- 1 461	- 2 000	- 5 407	- 5 470	- 5 513	- 5 416	- 5 749	- 6 701
RD du Congo	679	732	- 930	- 955	- 1 352	- 1 430	- 1 268	- 1 115	- 1 136
Côte d'Ivoire	751	1 255	1 551	455	767	1 156	970	1 923	2 379
Egypte	- 9 915	- 1 610	- 13 251	- 22 413	- 24 734	- 20 828	- 24 046	- 20 632	- 19 942
Ethiopie	665	624	612	731	670	815	812	732	735
Erythrée	-	-	-	-	-	-	-	-	-
Gabon	- 4 842	- 6 500	- 8 609	- 6 907	- 8 113	- 7 448	- 6 867	- 7 225	- 8 584
Ghana	900	1 069	1 146	809	888	841	883	836	914
Kenya	2 541	2 696	2 369	1 874	1 981	2 006	2 131	2 042	2 101
Libye	- 132 369	- 106 278	- 91 617	- 45 458	- 39 331	- 38 834	- 35 276	- 37 946	- 46 000
Maroc	1 651	2 094	2 815	4 644	4 805	4 507	4 833	5 050	5 472
Mozambique	836	792	549	130	-	-	-	-	-
Nigéria	- 73 317	- 97 975	- 90 183	- 57 185	- 69 616	- 64 847	- 56 419	- 60 978	- 74 387
Sénégal	563	666	758	384	133	648	529	757	583
Afrique du Sud	13 220	13 665	14 965	13 423	13 750	13 250	12 756	11 223	10 000
Soudan	970	1 100	1 058	617	639	811	480	798	557
Rép. Unie de Tanzanie	944	879	566	565	610	570	550	553	553
Togo	-	-	169	-	-	-	-	-	-
Tunisie	- 2 622	- 2 644	- 3 567	- 3 821	- 3 746	- 3 575	- 3 641	- 3 270	- 3 200
Zambie	-	658	780	615	594	588	594	619	609
Autre Afrique	778	648	983	706	791	1 145	1 715	1 868	1 477
Afrique	**- 237 634**	**- 241 769**	**- 236 929**	**- 159 824**	**- 171 778**	**- 162 652**	**- 157 205**	**- 165 016**	**- 191 586**
Argentine	2 221	2 978	2 192	-	- 461	- 101	- 102	- 410	- 610
Bolivie	- 1 062	- 1 489	- 364	- 39	-	- 40	-	-	-
Brésil	19 187	33 510	44 715	32 689	27 422	30 234	31 110	31 924	29 666
Chili	3 308	3 141	3 686	1 968	2 057	2 596	2 675	4 074	4 508
Colombie	- 3 512	- 1 340	1 247	1 389	953	- 4 526	- 7 503	- 7 466	- 8 464
Costa Rica	417	420	463	387	417	635	629	629	660
Cuba	5 438	5 993	7 269	5 937	6 217	5 975	6 741	7 505	7 505
République dominicaine	-	1 227	1 571	1 778	1 710	1 804	1 896	1 803	1 888
Equateur	1 163	- 8 932	- 6 316	- 8 520	- 9 576	- 9 916	- 4 776	- 9 583	- 9 105
El Salvador	480	677	793	641	679	708	748	693	663
Guatemala	837	946	794	557	609	269	435	473	462
Honduras	557	607	423	449	326	241	334	440	452
Jamaïque	1 528	1 830	965	632	855	784	625	705	686
Antilles néerlandaises	39 868	43 441	27 525	19 800	8 350	8 050	9 315	9 858	9 424
Nicaragua	480	594	622	433	485	532	505	517	569
Panama	4 010	3 717	2 382	1 500	1 257	975	1 360	958	962
Paraguay	180	240	320	179	157	205	275	268	317
Pérou	1 381	1 729	- 1 388	- 1 000	- 1 264	- 650	121	936	880
Trinité-et-Tobago	13 623	10 913	383	- 4 508	- 4 583	- 4 296	- 3 619	- 3 817	- 3 989
Uruguay	1 773	1 747	2 132	1 224	1 065	1 337	1 126	1 118	1 077
Vénézuela	- 120 811	- 111 350	- 65 136	- 53 287	- 43 640	- 49 015	- 54 174	- 52 436	- 50 227
Autre Amérique latine	12 494	15 530	9 854	6 209	1 616	634	810	740	743
Amérique latine	**- 16 440**	**6 129**	**34 132**	**8 418**	**- 5 349**	**- 13 565**	**- 11 469**	**- 11 071**	**- 11 933**

Un chiffre négatif correspond à des exportations nettes.

Net Imports of Crude Oil, NGL, Refinery Feedstocks and Additives (1000 tonnes)
Import. nettes de pétr. brut, LGN, produits d'aliment. des raffin. et additifs (1000 tonnes)

Nettoimporte von Rohöl, Kondensaten, Raffinerie-Feedstocks und Additiven (1000 Tonnen)

Importazioni nette di petrolio grezzo, GNL, prodotti intermedi e additivi (1000 tonnellate)

原油、NGL 等随伴物の純輸入量（千トン）

Import. netas de petróleo crudo, líquidos de gas natural, prod. de alim. de refinerias y aditivos (1000 toneladas)

Чистый имп. сырой нефти, газ. конденсатов, нефтезаводского сырья и присадок (тыс. т)

1990	1991	1992	1993	1994	1995	1996	1997	1998	
- 30 906	- 30 307	- 29 499	- 30 525	- 30 643	- 30 365	- 33 923	- 32 845	- 34 403	Algeria
- 21 741	- 22 777	- 25 177	- 23 014	- 24 794	- 28 065	- 30 794	- 32 723	- 34 386	Angola
- 206	- 197	- 136	- 154	- 129	- 95	- 75	- 64	- 52	Benin
- 7 067	- 6 367	- 6 013	- 6 007	- 5 205	- 4 945	- 5 280	- 5 456	- 6 114	Cameroon
- 7 425	- 7 315	- 7 938	- 9 002	- 8 621	- 7 806	- 8 996	- 11 720	- 12 498	Congo
- 1 035	- 997	- 900	- 902	- 965	- 960	- 960	- 895	- 844	DR of Congo
2 046	2 314	3 055	2 975	2 999	3 019	2 331	2 224	3 148	Cote d'Ivoire
- 19 615	- 18 170	- 20 091	- 19 610	- 8 346	- 8 369	- 6 728	- 6 143	- 2 791	Egypt
700	740	474	667	712	707	718	362	-	Ethiopia
-	-	692	612	749	633	679	408	-	Eritrea
- 11 909	- 13 819	- 13 546	- 14 757	- 15 432	- 16 770	- 17 363	- 17 539	- 16 781	Gabon
817	988	942	690	1 089	830	994	178	750	Ghana
2 178	2 059	2 235	2 274	2 173	1 680	1 413	1 834	2 158	Kenya
- 54 300	- 58 815	- 56 926	- 53 450	- 54 198	- 53 138	- 53 909	- 52 223	- 54 386	Libya
5 681	5 139	6 455	6 352	6 625	6 377	5 336	6 017	5 957	Morocco
-	-	-	-	-	-	-	-	-	Mozambique
- 75 556	- 78 519	- 81 299	- 90 298	- 87 422	- 85 521	- 93 766	- 100 713	- 98 156	Nigeria
642	506	643	546	303	659	643	771	862	Senegal
12 378	10 828	11 813	11 705	13 820	15 604	13 972	12 616	15 041	South Africa
837	1 013	750	327	641	761	712	630	258	Sudan
557	574	502	578	580	588	589	589	589	United Rep. of Tanzania
-	-	-	-	-	-	-	-	-	Togo
- 2 839	- 3 275	- 3 648	- 3 016	- 2 512	- 2 607	- 2 376	- 1 865	- 1 889	Tunisia
557	557	563	568	568	578	588	588	588	Zambia
1 190	1 270	1 325	1 140	1 130	1 060	533	- 782	- 782	Other Africa
- 205 016	**- 214 570**	**- 215 724**	**- 222 301**	**- 206 878**	**- 206 145**	**- 225 662**	**- 236 751**	**- 233 731**	**Africa**
- 579	- 1 054	- 2 703	- 4 315	- 10 029	- 13 627	- 15 940	- 16 169	- 15 750	Argentina
-	-	- 17	- 62	- 67	-	-	- 19	- 20	Bolivia
28 617	26 361	26 566	25 477	27 701	24 563	28 490	29 077	27 973	Brazil
5 260	5 408	5 867	6 133	6 831	7 368	7 597	8 224	8 924	Chile
- 9 902	- 8 753	- 9 238	- 10 150	- 9 718	- 16 113	- 16 606	- 17 769	- 22 861	Colombia
445	349	552	566	548	720	622	626	39	Costa Rica
6 308	4 420	1 411	1 630	1 386	1 199	1 636	1 094	899	Cuba
1 585	1 873	1 913	1 954	1 995	2 080	2 110	2 221	2 283	Dominican Republic
- 8 851	- 8 894	- 10 626	- 11 386	- 12 401	- 12 667	- 12 054	- 13 376	- 12 659	Ecuador
697	811	865	991	817	801	785	834	948	El Salvador
385	515	543	399	601	360	76	- 135	- 368	Guatemala
425	408	419	-	-	-	-	-	-	Honduras
1 095	908	1 191	740	800	945	1 030	1 125	1 198	Jamaica
10 495	11 100	12 450	13 715	13 693	13 680	13 678	12 442	13 138	Netherlands Antilles
609	616	679	652	558	574	617	729	862	Nicaragua
1 219	1 168	1 769	1 749	1 170	1 203	2 078	2 017	2 274	Panama
307	288	285	213	236	195	158	155	138	Paraguay
723	1 624	1 389	1 148	1 068	923	934	1 692	2 548	Peru
- 3 370	- 2 047	- 1 643	- 1 430	- 2 709	- 2 752	- 2 356	- 2 516	- 3 199	Trinidad and Tobago
1 122	1 349	1 322	144	181	1 429	1 497	1 465	1 780	Uruguay
- 64 358	- 73 260	- 66 852	- 87 049	- 88 177	- 99 535	- 113 812	- 128 435	- 121 852	Venezuela
932	885	871	933	858	905	906	906	906	Other Latin America
- 26 836	**- 35 925**	**- 32 987**	**- 57 948**	**- 64 658**	**- 87 749**	**- 98 554**	**- 115 812**	**- 112 799**	**Latin America**

A negative number shows net exports.

Net Imports of Crude Oil, NGL, Refinery Feedstocks and Additives (1000 tonnes)
Import. nettes de pétr. brut, LGN, produits d'aliment. des raffin. et additifs (1000 tonnes)
Nettoimporte von Rohöl, Kondensaten, Raffinerie-Feedstocks und Additiven (1000 Tonnen)
Importazioni nette di petrolio grezzo, GNL, prodotti intermedi e additivi (1000 tonnellate)
原油、NGL等随伴物の純輸入量（千トン）
Import. netas de petróleo crudo, líquidos de gas natural, prod. de alim. de refinerias y aditivos (1000 toneladas)
Чистый имп. сырой нефти, газ. конденсатов, нефтезаводского сырья и присадок (тыс. т)

	1971	1973	1978	1984	1985	1986	1987	1988	1989
Bangladesh	850	719	1 064	1 104	992	983	1 000	1 212	1 253
Brunei	- 5 889	- 11 597	- 11 967	- 9 251	- 8 410	- 8 087	- 7 274	- 7 081	- 7 402
Inde	12 688	13 425	14 892	7 860	12 767	14 481	17 959	17 690	18 609
Indonésie	- 32 416	- 49 237	- 67 346	- 52 117	- 41 248	- 40 720	- 35 557	- 33 348	- 35 763
RPD de Corée	-	-	1 250	2 400	2 600	2 700	2 800	2 900	2 900
Malaisie	824	- 333	- 5 420	- 14 929	- 15 598	- 18 219	- 17 514	- 19 265	- 20 927
Myanmar	266	-	-	-	-	-	-	-	-
Pakistan	3 057	3 044	3 461	4 294	4 017	3 797	3 713	3 801	3 467
Philippines	9 027	9 266	10 022	9 821	6 725	6 922	8 301	9 769	9 777
Singapour	16 309	22 320	27 900	35 866	29 248	31 916	34 500	32 839	36 591
Sri Lanka	1 536	1 747	1 449	1 772	1 657	1 638	1 704	1 848	1 284
Taipei chinois	5 805	9 075	17 417	17 468	16 114	16 116	17 308	18 250	21 502
Thailande	5 413	7 643	8 203	6 685	6 348	6 685	7 423	6 833	9 013
Viêt-Nam	-	-	-	-	-	-	-	- 641	- 1 450
Autre Asie	-	-	-	-	-	-	-	-	-
Asie	**17 470**	**6 072**	**925**	**10 973**	**15 212**	**18 212**	**34 363**	**34 807**	**38 854**
Rép. populaire de Chine	- 213	- 1 834	- 10 623	- 22 047	- 29 780	- 28 042	- 26 725	- 25 190	- 21 128
Chine	**- 213**	**- 1 834**	**- 10 623**	**- 22 047**	**- 29 780**	**- 28 042**	**- 26 725**	**- 25 190**	**- 21 128**
Albanie	- 143	- 412	-	-	-	-	-	-	-
Bulgarie	7 547	9 652	12 644	12 650	12 400	13 300	12 700	13 100	13 600
Chypre	-	656	451	594	466	559	616	707	680
Roumanie	2 858	4 143	12 937	13 534	14 626	17 047	21 366	20 957	21 809
République slovaque	5 470	6 684	8 843	7 924	7 942	8 075	8 159	7 923	7 926
Croatie	-	-	-	-	-	-	-	-	-
Ex-RYM	-	-	-	-	-	-	-	-	-
Slovénie	-	-	-	-	-	-	-	-	-
RF de Yougoslavie	-	-	-	-	-	-	-	-	-
Ex-Yougoslavie	4 747	8 226	10 380	9 693	8 661	10 436	11 870	13 230	13 100
Europe non-OCDE	**20 479**	**28 949**	**45 255**	**44 395**	**44 095**	**49 417**	**54 711**	**55 917**	**57 115**
Azerbaïdjan	-	-	-	-	-	-	-	-	-
Bélarus	-	-	-	-	-	-	-	-	-
Géorgie	-	-	-	-	-	-	-	-	-
Kazakhstan	-	-	-	-	-	-	-	-	-
Kirghizistan	-	-	-	-	-	-	-	-	-
Lituanie	-	-	-	-	-	-	-	-	-
Russie	-	-	-	-	-	-	-	-	-
Tadjikistan	-	-	-	-	-	-	-	-	-
Turkménistan	-	-	-	-	-	-	-	-	-
Ukraine	-	-	-	-	-	-	-	-	-
Ouzbékistan	-	-	-	-	-	-	-	-	-
Ex-URSS	**- 69 700**	**- 72 100**	**- 112 920**	**- 120 500**	**- 104 600**	**- 118 100**	**- 122 600**	**- 124 500**	**- 113 700**

Un chiffre négatif correspond à des exportations nettes.

Net Imports of Crude Oil, NGL, Refinery Feedstocks and Additives (1000 tonnes)
Import. nettes de pétr. brut, LGN, produits d'aliment. des raffin. et additifs (1000 tonnes)
Nettoimporte von Rohöl, Kondensaten, Raffinerie-Feedstocks und Additiven (1000 Tonnen)
Importazioni nette di petrolio grezzo, GNL, prodotti intermedi e additivi (1000 tonnellate)
原油、NGL 等随伴物の純輸入量（チトン）
Import. netas de petróleo crudo, líquidos de gas natural, prod. de alim. de refinerias y aditivos (1000 toneladas)
Чистый имп. сырой нефти, газ. конденсатов, нефтезаводского сырья и присадок (тыс. т)

1990	1991	1992	1993	1994	1995	1996	1997	1998	
948	1 182	1 018	1 129	1 239	1 364	1 141	1 300	1 119	Bangladesh
- 7 578	- 8 221	- 8 953	- 8 739	- 9 010	- 8 728	- 8 374	- 7 999	- 7 975	Brunei
20 154	21 141	29 115	29 445	27 590	26 354	31 982	32 588	37 046	India
- 33 748	- 36 121	- 32 109	- 30 030	- 34 258	- 31 448	- 29 107	- 27 053	- 24 520	Indonesia
2 800	2 700	2 500	2 300	2 250	2 200	2 150	2 004	1 904	DPR of Korea
- 21 468	- 22 182	- 22 756	- 19 206	- 18 236	- 18 372	- 16 338	- 14 887	- 17 070	Malaysia
-	72	108	215	329	479	365	647	685	Myanmar
3 412	3 752	3 788	3 642	4 013	3 645	4 043	3 459	3 936	Pakistan
10 981	10 466	11 827	11 074	11 211	15 371	17 561	17 687	16 053	Philippines
42 661	45 025	46 314	52 627	55 776	51 442	55 192	55 493	51 908	Singapore
1 758	1 636	1 296	1 799	1 913	1 915	2 040	1 820	2 155	Sri Lanka
21 996	21 683	23 629	25 520	24 642	29 890	32 535	32 379	34 320	Chinese Taipei
9 641	9 647	12 416	14 847	16 979	21 247	29 311	33 819	31 492	Thailand
- 2 661	- 3 881	- 5 461	- 6 312	- 6 901	- 7 652	- 8 705	- 10 909	- 12 400	Vietnam
- 4 500	- 4 900	- 5 220	- 5 320	- 5 420	- 4 945	- 3 945	- 3 945	- 3 945	Other Asia
44 396	**41 999**	**57 512**	**72 991**	**72 117**	**82 762**	**109 851**	**116 403**	**114 708**	**Asia**
- 21 067	- 16 625	- 10 147	- 3 763	- 6 145	- 1 137	2 215	15 641	11 720	People's Rep. of China
- 21 067	**- 16 625**	**- 10 147**	**- 3 763**	**- 6 145**	**- 1 137**	**2 215**	**15 641**	**11 720**	**China**
-	-	-	-	-	-	5	6	5	Albania
8 166	4 430	2 215	5 744	6 944	7 973	6 996	5 888	5 549	Bulgaria
624	763	749	789	916	797	804	1 039	1 075	Cyprus
16 058	8 398	6 572	7 581	8 122	8 657	7 157	6 246	5 975	Romania
6 169	4 924	4 299	4 495	4 762	5 390	5 341	5 270	5 421	Slovak Republic
-	-	2 159	3 247	3 580	3 959	3 513	3 659	3 719	*Croatia*
-	-	1 000	933	112	159	656	400	787	*FYROM*
-	-	587	570	395	595	524	622	604	*Slovenia*
-	-	1 000	200	200	300	1 321	2 292	1 975	*FR of Yugoslavia*
11 995	9 506	4 746	4 950	4 287	5 013	6 014	6 973	7 085	Former Yugoslavia
43 012	**28 021**	**18 581**	**23 559**	**25 031**	**27 830**	**26 317**	**25 422**	**25 110**	**Non-OECD Europe**
-	-	- 1 945	- 812	853	62	-	- 40	- 2 059	Azerbaijan
-	-	18 689	12 094	11 050	11 355	10 345	10 061	10 031	Belarus
-	-	672	207	200	-	- 113	- 104	- 54	Georgia
-	-	- 7 800	- 8 477	- 6 128	- 8 990	- 13 228	- 12 867	- 16 136	Kazakhstan
-	-	- 112	- 88	- 88	- 22	- 9	25	62	Kyrgyzstan
-	-	4 083	5 161	3 952	3 554	4 048	5 584	6 510	Lithuania
-	-	- 117 387	- 117 256	- 121 897	- 113 832	- 119 713	- 122 885	- 131 660	Russia
-	-	-	-	- 10	-	-	-	- 5	Tajikistan
-	-	1 119	- 69	419	-	- 100	- 900	- 900	Turkmenistan
-	-	34 360	19 679	15 800	13 300	9 315	8 957	9 801	Ukraine
-	-	3 106	3 999	1 503	- 368	- 771	- 906	- 1 117	Uzbekistan
- 108 000	**- 60 375**	**- 65 215**	**- 85 562**	**- 94 346**	**- 94 941**	**- 110 226**	**- 113 075**	**- 125 527**	**Former USSR**

A negative number shows net exports.

Net Imports of Crude Oil, NGL, Refinery Feedstocks and Additives (1000 tonnes)
Import. nettes de pétr. brut, LGN, produits d'aliment. des raffin. et additifs (1000 tonnes)
Nettoimporte von Rohöl, Kondensaten, Raffinerie-Feedstocks und Additiven (1000 Tonnen)

Importazioni nette di petrolio grezzo, GNL, prodotti intermedi e additivi (1000 tonnellate)

原油、NGL 等随伴物の純輸入量（千トン）

Import. netas de petróleo crudo, líquidos de gas natural, prod. de alim. de refinerias y aditivos (1000 toneladas)

Чистый имп. сырой нефти, газ. конденсатов, нефтезаводского сырья и присадок (тыс. т)

	1971	1973	1978	1984	1985	1986	1987	1988	1989
Bahrein	9 022	8 812	9 513	7 922	7 067	9 804	9 584	9 957	10 021
République Islamique d'Iran	- 197 994	- 262 650	- 186 456	- 74 823	- 75 582	- 55 625	- 80 681	- 75 944	- 99 600
Irak	- 79 572	- 94 991	- 116 949	- 40 706	- 50 691	- 63 551	- 82 189	- 106 375	- 115 700
Israël	500	2 700	7 975	6 993	6 292	7 100	6 900	8 400	8 400
Jordanie	592	706	1 332	2 616	2 486	2 245	2 495	2 412	2 452
Koweit	- 137 924	- 131 524	- 88 656	- 33 677	- 24 112	- 25 285	- 27 490	- 32 604	- 38 971
Liban	2 060	2 413	1 829	760	625	1 120	1 250	600	100
Oman	- 14 461	- 14 598	- 16 021	- 18 807	- 22 954	- 26 234	- 27 579	- 29 608	- 30 214
Qatar	- 20 545	- 27 485	- 23 395	- 17 493	- 13 686	- 15 333	- 12 533	- 14 262	- 16 952
Arabie saoudite	- 210 396	- 351 025	- 392 501	- 193 897	- 122 046	- 195 688	- 143 333	- 181 976	- 184 870
Syrie	- 2 750	- 3 288	- 5 869	1 096	622	1 833	- 1 493	- 2 246	- 6 060
Emirats arabes unis	- 50 892	- 73 617	- 87 922	- 44 769	- 40 814	- 46 740	- 51 292	- 55 467	- 69 182
Yémen	3 748	3 124	1 930	4 410	3 750	3 353	4 052	- 3 094	- 5 136
Moyen-Orient	**- 698 612**	**- 941 423**	**- 895 190**	**- 400 375**	**- 329 043**	**- 403 001**	**- 402 309**	**- 480 207**	**- 545 712**
Total non-OCDE	**- 984 650**	**-1 215 976**	**-1 175 350**	**- 638 960**	**- 581 243**	**- 657 731**	**- 631 234**	**- 715 260**	**- 788 090**
OCDE Amérique du N.	87 221	161 233	344 553	112 039	103 253	162 225	183 293	201 326	246 495
OCDE Pacifique	213 202	270 931	265 812	217 926	197 970	198 690	191 201	205 615	227 724
OCDE Europe	655 931	756 882	624 348	376 235	361 347	391 658	383 378	407 358	414 300
Total OCDE	**956 354**	**1 189 046**	**1 234 713**	**706 200**	**662 570**	**752 573**	**757 872**	**814 299**	**888 519**
Monde	**- 28 296**	**- 26 930**	**59 363**	**67 240**	**81 327**	**94 842**	**126 638**	**99 039**	**100 429**
Pour mémoire: OPEP	**-1 090 536**	**-1 350 988**	**-1 259 249**	**- 640 022**	**- 548 535**	**- 622 279**	**- 606 349**	**- 678 078**	**- 760 611**

Un chiffre négatif correspond à des exportations nettes.

La ligne Monde montre la divergence entre le total des exportations et des importations mondiales.

Net Imports of Crude Oil, NGL, Refinery Feedstocks and Additives (1000 tonnes)
Import. nettes de pétr. brut, LGN, produits d'aliment. des raffin. et additifs (1000 tonnes)
Nettoimporte von Rohöl, Kondensaten, Raffinerie-Feedstocks und Additiven (1000 Tonnen)
Importazioni nette di petrolio grezzo, GNL, prodotti intermedi e additivi (1000 tonnellate)
原油、NGL 等随伴物の純輸入量（チトン）
Import. netas de petróleo crudo, líquidos de gas natural, prod. de alim. de refinerias y aditivos (1000 toneladas)
Чистый имп. сырой нефти, газ. конденсатов, нефтезаводского сырья и присадок (тыс. т)

1990	1991	1992	1993	1994	1995	1996	1997	1998	
10 406	10 705	10 921	10 307	10 344	10 538	11 078	10 593	10 526	Bahrain
- 110 000	- 120 177	- 126 095	- 129 116	- 120 376	- 121 021	- 119 631	- 105 185	- 102 410	Islamic Republic of Iran
- 78 821	- 2 127	- 2 973	- 2 899	- 2 938	- 3 119	- 4 327	- 35 063	- 75 219	Iraq
8 300	8 355	10 982	12 092	11 199	11 123	10 090	11 265	11 730	Israel
2 690	2 344	2 975	2 900	2 977	3 163	3 272	3 441	3 556	Jordan
- 31 780	- 6 278	- 37 580	- 73 022	- 63 532	- 60 416	- 62 185	- 57 449	- 60 282	Kuwait
100	540	436	-	-	-	-	-	-	Lebanon
- 32 284	- 32 964	- 35 379	- 37 654	- 38 008	- 40 244	- 41 791	- 42 780	- 42 292	Oman
- 16 440	- 16 355	- 18 896	- 17 803	- 16 999	- 17 034	- 18 870	- 23 382	- 28 592	Qatar
- 249 675	- 341 490	- 343 026	- 336 481	- 335 073	- 336 766	- 327 049	- 338 812	- 352 833	Saudi Arabia
- 9 240	- 12 780	- 13 053	- 14 275	- 17 012	- 18 109	- 17 651	- 17 467	- 17 433	Syria
- 77 537	- 89 464	- 96 106	- 92 750	- 91 999	- 91 147	- 91 893	- 93 558	- 96 085	United Arab Emirates
- 4 459	- 5 139	- 3 458	- 7 144	- 12 605	- 14 026	- 13 079	- 13 613	- 14 051	Yemen
- 588 740	**- 604 830**	**- 651 252**	**- 685 845**	**- 674 022**	**- 677 058**	**- 672 036**	**- 702 010**	**- 763 385**	**Middle East**
- 862 251	**- 862 305**	**- 899 232**	**- 958 869**	**- 948 901**	**- 956 438**	**- 968 095**	**-1 010 182**	**-1 083 904**	**Non-OECD Total**
259 704	241 621	256 623	297 653	313 390	308 181	319 595	344 122	356 101	OECD North America
243 104	266 349	290 631	303 894	322 449	325 997	338 174	364 456	341 532	OECD Pacific
417 428	413 319	424 197	411 636	382 003	366 960	371 542	378 823	405 810	OECD Europe
920 236	**921 289**	**971 451**	**1 013 183**	**1 017 842**	**1 001 138**	**1 029 311**	**1 087 401**	**1 103 443**	**OECD Total**
57 985	**58 984**	**72 219**	**54 314**	**68 941**	**44 700**	**61 216**	**77 219**	**19 539**	**World**
- 823 121	**- 852 913**	**- 891 361**	**- 943 423**	**- 925 615**	**- 929 510**	**- 948 472**	**- 994 718**	**-1 048 738**	***Memo: OPEC***

A negative number shows net exports.
The row World shows the discrepancy between total world exports and imports.

Net Imports of Petroleum Products (1000 tonnes)
Importations nettes de produits pétroliers (1000 tonnes)
Nettoimporte von Ölprodukten (1000 Tonnen)

Importazioni nette di prodotti petroliferi (1000 tonnellate)

石油製品の純輸入量（千トン）

Importaciones netas de productos petrolíferos (1000 toneladas)

Чистый импорт нефтепродуктов (тыс. т)

	1971	1973	1978	1984	1985	1986	1987	1988	1989
Algérie	- 118	- 2 933	- 1 120	- 14 460	- 14 079	- 16 858	- 16 557	- 18 001	- 16 459
Angola	172	183	31	- 524	- 229	- 251	- 331	- 103	- 226
Bénin	101	134	132	138	169	149	152	132	109
Cameroun	286	314	531	- 560	- 645	- 682	- 628	- 497	- 351
Congo	158	164	215	- 238	- 207	- 230	- 248	- 266	- 293
RD du Congo	127	139	624	647	672	692	709	725	727
Côte d'Ivoire	58	- 191	- 249	- 443	- 590	- 665	- 314	- 1 015	- 1 094
Egypte	1 197	- 237	- 691	394	- 359	- 1 047	- 939	- 1 560	- 1 729
Ethiopie	- 96	- 75	- 119	- 130	- 117	- 39	57	177	183
Erythrée	-	-	-	-	-	-	-	-	-
Gabon	- 535	- 512	- 811	- 22	- 11	- 4	58	53	- 237
Ghana	- 171	- 197	- 262	- 91	- 245	- 154	18	24	8
Kenya	- 884	- 964	- 471	- 277	- 265	- 231	- 161	78	67
Libye	189	- 262	- 1 849	- 1 316	- 4 734	- 5 585	- 6 340	- 7 930	- 7 456
Maroc	485	267	799	- 119	- 68	11	- 103	- 119	- 35
Mozambique	- 1	- 11	- 46	385	458	482	490	491	473
Nigéria	- 134	- 304	3 937	236	418	1 248	- 570	- 1 641	- 1 820
Sénégal	886	891	499	606	647	427	378	118	286
Afrique du Sud	93	523	525	-	-	-	-	-	-
Soudan	471	464	58	638	846	633	641	809	893
Rép. Unie de Tanzanie	- 98	91	203	128	130	156	156	166	172
Togo	97	90	- 53	115	98	146	156	154	157
Tunisie	81	400	1 157	1 160	1 049	1 218	803	1 177	1 491
Zambie	482	278	28	15	17	- 24	- 25	- 17	- 27
Zimbabwe	539	679	666	725	752	779	731	898	833
Autre Afrique	2 430	2 559	3 205	4 067	3 911	4 121	4 287	4 387	4 654
Afrique	**5 815**	**1 490**	**6 939**	**- 8 926**	**- 12 382**	**- 15 708**	**- 17 580**	**- 21 760**	**- 19 674**
Argentine	918	788	- 198	- 1 399	- 3 366	- 1 158	1 250	201	- 2 542
Bolivie	- 23	3	- 12	- 22	- 15	- 13	- 10	- 6	- 1
Brésil	1 337	- 775	- 822	- 7 614	- 5 833	- 3 891	- 4 522	- 3 398	- 2 684
Chili	447	275	22	546	382	625	244	203	356
Colombie	- 1 460	- 1 544	- 683	- 1 838	- 1 745	- 2 027	- 3 053	- 2 293	- 2 502
Costa Rica	38	160	401	228	237	81	127	185	169
Cuba	2 038	2 837	3 162	3 739	3 968	3 854	3 697	3 508	3 444
République dominicaine	1 139	425	434	637	475	632	862	1 073	1 007
Equateur	- 29	- 95	- 998	- 508	- 202	- 671	312	- 1 220	- 1 107
El Salvador	19	10	-	- 74	- 55	- 42	8	32	92
Guatemala	76	86	575	464	522	467	472	530	546
Haiti	130	128	225	216	220	250	271	287	316
Honduras	- 200	- 217	95	194	223	314	306	294	348
Jamaïque	388	1 126	1 364	1 239	771	860	1 078	1 051	1 296
Antilles néerlandaises	- 33 699	- 36 246	- 21 627	- 15 246	- 4 919	- 4 801	- 6 620	- 6 804	- 6 535
Nicaragua	41	- 1	195	164	179	175	262	204	41
Panama	3 165	3 276	935	168	332	629	474	537	637
Paraguay	28	31	152	347	290	336	357	341	404
Pérou	498	122	- 198	- 2 289	- 2 678	- 2 349	- 1 995	- 1 401	- 1 672
Trinité-et-Tobago	- 17 408	- 16 153	- 9 865	- 2 910	- 3 163	- 3 310	- 3 831	- 3 541	- 3 147
Uruguay	81	92	96	37	36	122	112	451	462
Vénézuela	- 53 766	- 59 536	- 36 731	- 25 046	- 26 112	- 27 127	- 22 808	- 30 517	- 31 771
Autre Amérique latine	- 5 412	- 7 629	- 3 568	- 2 753	1 781	2 655	3 125	3 129	3 175
Amérique latine	**- 101 654**	**- 112 837**	**- 67 046**	**- 51 720**	**- 38 672**	**- 34 389**	**- 29 882**	**- 37 154**	**- 39 668**

Un chiffre négatif correspond à des exportations nettes.

Net Imports of Petroleum Products (1000 tonnes)
Importations nettes de produits pétroliers (1000 tonnes)
Nettoimporte von Ölprodukten (1000 Tonnen)

Importazioni nette di prodotti petroliferi (1000 tonnellate)

石油製品の純輸入量（チトン）

Importaciones netas de productos petrolíferos (1000 toneladas)

Чистый импорт нефтепродуктов (тыс. m)

1990	1991	1992	1993	1994	1995	1996	1997	1998	
- 17 287	- 16 679	- 17 240	- 16 635	- 16 361	- 16 746	- 16 512	- 19 450	- 18 682	Algeria
- 276	- 382	- 535	- 246	- 323	- 508	- 508	- 536	- 633	Angola
97	78	80	85	88	89	296	298	313	Benin
- 43	- 105	- 228	- 253	- 339	- 298	- 339	- 423	- 439	Cameroon
- 307	- 316	- 277	- 318	- 302	- 293	- 295	- 295	306	Congo
797	813	754	849	861	909	909	909	857	DR of Congo
- 1 013	- 1 163	- 2 138	- 2 152	- 2 083	- 2 045	- 2 503	- 2 416	- 2 782	Cote d'Ivoire
- 1 557	- 2 978	- 3 081	- 4 191	- 6 098	- 5 155	- 4 156	- 3 652	- 2 767	Egypt
268	262	223	227	300	391	436	676	1 081	Ethiopia
-	-	- 301	- 260	- 321	- 257	- 265	- 82	198	Eritrea
- 77	27	- 26	- 15	- 92	- 143	- 133	- 109	- 156	Gabon
133	- 74	- 46	275	81	300	230	1 317	724	Ghana
- 56	- 44	- 193	- 19	- 104	369	536	263	779	Kenya
- 7 547	- 7 699	- 7 771	- 6 844	- 6 860	- 6 134	- 5 678	- 7 143	- 7 143	Libya
- 70	317	15	335	527	411	896	930	445	Morocco
331	289	313	339	320	327	331	388	368	Mozambique
- 2 919	- 2 860	- 1 424	- 1 484	- 147	- 1 847	- 3 366	- 2 782	1 954	Nigeria
187	297	321	322	733	254	354	301	351	Senegal
-	-	-	- 2 315	- 2 918	- 2 867	- 2 200	-	-	South Africa
1 046	705	867	833	1 044	836	775	894	1 087	Sudan
195	174	168	137	139	139	142	142	142	United Rep. of Tanzania
185	169	133	63	159	249	260	285	298	Togo
1 498	1 851	1 921	2 053	1 642	1 312	1 582	1 711	1 824	Tunisia
- 26	- 27	- 24	- 36	- 27	- 29	- 27	- 27	- 27	Zambia
772	935	1 026	1 101	1 287	1 504	1 463	1 463	1 463	Zimbabwe
4 627	4 723	4 816	4 969	5 080	5 192	5 240	5 450	5 450	Other Africa
- 21 042	- 21 687	- 22 647	- 23 180	- 23 714	- 24 040	- 22 532	- 21 888	- 14 989	**Africa**
- 3 572	- 3 179	- 3 288	- 2 805	- 2 022	- 978	- 2 136	- 3 046	- 3 237	Argentina
- 10	- 15	47	124	95	192	133	235	290	Bolivia
- 1 608	383	990	6 987	4 639	7 844	9 621	10 023	7 552	Brazil
305	553	835	874	1 380	1 538	2 324	2 123	1 706	Chile
- 2 521	- 2 734	- 1 287	- 1 393	- 1 469	- 1 260	- 2 313	- 1 963	- 1 657	Colombia
469	614	796	819	927	776	781	826	1 558	Costa Rica
3 871	3 788	5 550	4 737	5 240	5 886	6 151	6 892	6 890	Cuba
1 308	989	1 328	1 283	1 387	1 440	1 494	1 630	1 677	Dominican Republic
- 1 154	- 1 256	- 834	- 1 284	- 1 248	- 932	- 1 308	- 1 370	- 465	Ecuador
79	193	237	308	549	817	637	847	874	El Salvador
731	791	860	1 003	1 041	1 315	1 336	1 412	2 018	Guatemala
309	284	289	216	60	311	353	468	416	Haiti
327	396	590	907	1 085	1 295	1 249	1 266	1 508	Honduras
1 452	1 557	1 415	1 912	1 894	1 996	2 040	2 105	2 165	Jamaica
- 6 793	- 7 043	- 8 680	- 9 233	- 9 258	- 9 271	- 9 301	- 9 338	- 9 727	Netherlands Antilles
21	-	59	69	82	276	298	194	225	Nicaragua
754	954	623	431	1 214	1 041	294	452	396	Panama
354	351	456	568	796	875	734	899	971	Paraguay
- 1 513	- 1 933	- 1 984	- 1 909	- 1 356	253	353	193	- 410	Peru
- 3 404	- 4 329	- 4 487	- 4 354	- 4 586	- 4 376	- 4 097	- 4 318	- 4 545	Trinidad and Tobago
210	295	575	1 358	1 418	522	593	597	334	Uruguay
- 30 002	- 35 062	- 30 013	- 30 063	- 32 769	- 33 823	- 35 989	- 39 753	- 37 816	Venezuela
3 247	3 216	3 031	3 039	3 081	3 076	3 101	3 191	3 191	Other Latin America
- 37 140	- 41 187	- 32 892	- 26 406	- 27 820	- 21 187	- 23 652	- 26 435	- 26 086	**Latin America**

A negative number shows net exports.

Net Imports of Petroleum Products (1000 tonnes)
Importations nettes de produits pétroliers (1000 tonnes)
Nettoimporte von Ölprodukten (1000 Tonnen)
Importazioni nette di prodotti petroliferi (1000 tonnellate)
石油製品の純輸入量（千トン）
Importaciones netas de productos petrolíferos (1000 toneladas)
Чистый импорт нефтепродуктов (тыс. т)

	1971	1973	1978	1984	1985	1986	1987	1988	1989
Bangladesh	97	169	288	365	530	709	631	685	774
Brunei	62	67	98	10	14	12	7	10	13
Inde	1 797	3 579	3 868	5 043	2 561	847	184	3 256	4 261
Indonésie	- 1 063	- 834	1 642	- 160	- 1 092	- 3 916	- 7 600	- 7 558	- 7 255
RPD de Corée	683	765	580	490	490	1 040	2 110	1 715	1 720
Malaisie	560	527	1 205	1 716	2 077	1 881	1 805	1 781	1 476
Myanmar	11	48	- 12	- 17	- 18	- 13	- 6	- 6	- 6
Népal	54	70	88	152	145	168	157	171	204
Pakistan	- 238	39	434	2 033	2 182	2 300	2 991	3 579	4 185
Philippines	- 465	- 174	999	343	414	475	1 056	268	1 114
Singapour	- 9 877	- 10 265	- 16 356	- 24 201	- 18 792	- 17 040	- 13 777	- 10 423	- 15 194
Sri Lanka	- 228	- 108	- 8	- 91	- 86	- 74	134	- 87	228
Taipei chinois	1 443	1 201	1 008	439	- 49	2 581	1 708	4 318	3 957
Thailande	840	506	2 287	3 620	2 323	2 061	3 151	4 256	5 148
Viêt-Nam	5 660	5 654	933	1 839	1 872	2 099	2 432	2 473	2 277
Autre Asie	1 609	2 053	1 923	2 048	2 281	2 177	2 530	2 413	2 489
Asie	**945**	**3 297**	**- 1 023**	**- 6 371**	**- 5 148**	**- 4 693**	**- 2 487**	**6 851**	**5 391**
Rép. populaire de Chine	- 40	16	- 1 505	- 5 314	- 5 729	- 2 894	- 2 228	- 1 435	37
Hong-Kong, Chine	4 044	4 861	6 187	4 918	5 056	5 462	4 981	5 521	6 476
Chine	**4 004**	**4 877**	**4 682**	**- 396**	**- 673**	**2 568**	**2 753**	**4 086**	**6 513**
Albanie	- 666	- 946	- 1 054	- 166	- 208	- 205	- 176	- 117	- 126
Bulgarie	2 147	1 490	1 464	- 852	- 970	- 728	- 997	- 1 169	- 1 360
Chypre	629	192	401	388	449	465	630	561	844
Gibraltar	217	233	173	152	328	411	448	580	496
Malte	330	343	385	461	298	597	600	616	616
Roumanie	- 4 946	- 4 593	- 7 217	- 10 240	- 9 418	- 10 164	- 11 600	- 12 916	- 13 291
République slovaque	- 1 260	- 1 386	- 1 798	- 1 988	- 2 017	- 2 195	- 2 399	- 2 302	- 2 456
Bosnie-Herzegovine	-	-	-	-	-	-	-	-	-
Croatie	-	-	-	-	-	-	-	-	-
Ex-RYM	-	-	-	-	-	-	-	-	-
Slovénie	-	-	-	-	-	-	-	-	-
RF de Yougoslavie	-	-	-	-	-	-	-	-	-
Ex-Yougoslavie	507	882	898	327	842	323	70	45	- 143
Europe non-OCDE	**- 3 042**	**- 3 785**	**- 6 748**	**- 11 918**	**- 10 696**	**- 11 496**	**- 13 424**	**- 14 702**	**- 15 420**
Arménie	-	-	-	-	-	-	-	-	-
Azerbaïdjan	-	-	-	-	-	-	-	-	-
Bélarus	-	-	-	-	-	-	-	-	-
Estonie	-	-	-	-	-	-	-	-	-
Géorgie	-	-	-	-	-	-	-	-	-
Kazakhstan	-	-	-	-	-	-	-	-	-
Kirghizistan	-	-	-	-	-	-	-	-	-
Lettonie	-	-	-	-	-	-	-	-	-
Lituanie	-	-	-	-	-	-	-	-	-
République de Moldavie	-	-	-	-	-	-	-	-	-
Russie	-	-	-	-	-	-	-	-	-
Tadjikistan	-	-	-	-	-	-	-	-	-
Turkménistan	-	-	-	-	-	-	-	-	-
Ukraine	-	-	-	-	-	-	-	-	-
Ouzbékistan	-	-	-	-	-	-	-	-	-
Ex-URSS	**- 28 700**	**- 31 500**	**- 43 400**	**- 49 800**	**- 47 350**	**- 54 350**	**- 57 575**	**- 59 425**	**- 56 508**

Un chiffre négatif correspond à des exportations nettes.

Net Imports of Petroleum Products (1000 tonnes)
Importations nettes de produits pétroliers (1000 tonnes)
Nettoimporte von Ölprodukten (1000 Tonnen)

Importazioni nette di prodotti petroliferi (1000 tonnellate)

石油製品の純輸入量（チトン）

Importaciones netas de productos petrolíferos (1000 toneladas)

Чистый импорт нефтепродуктов (тыс. т)

1990	1991	1992	1993	1994	1995	1996	1997	1998	
879	521	735	732	916	1 318	1 653	1 858	2 071	Bangladesh
10	7	16	20	22	39	88	102	33	Brunei
5 359	7 799	7 140	8 776	10 941	15 898	18 524	17 725	18 804	India
- 7 008	- 7 051	- 7 125	- 3 622	- 4 955	- 3 273	- 3 737	- 466	- 3 541	Indonesia
1 590	1 540	1 540	1 530	1 511	1 489	1 473	1 373	1 304	DPR of Korea
2 539	3 365	3 900	4 220	2 303	2 849	- 311	2 365	2 174	Malaysia
- 29	- 23	57	53	55	162	291	279	531	Myanmar
208	343	333	370	408	481	519	612	665	Nepal
5 056	4 119	5 125	6 412	7 732	8 428	9 952	10 248	10 919	Pakistan
601	1 052	1 609	2 134	3 094	718	- 98	937	2 349	Philippines
- 17 808	- 17 950	- 16 935	- 19 587	- 20 923	- 19 935	- 17 409	- 13 709	- 11 781	Singapore
- 109	- 10	386	270	140	364	663	1 006	713	Sri Lanka
6 313	5 652	5 996	5 579	8 704	6 916	4 787	4 919	6 239	Chinese Taipei
7 543	7 882	7 891	8 678	8 261	8 894	4 398	- 1 640	- 2 810	Thailand
2 764	2 483	3 178	3 900	4 309	5 003	5 898	5 957	6 518	Vietnam
2 591	2 497	2 277	2 251	2 153	2 180	2 186	2 262	2 262	Other Asia
10 499	**12 226**	**16 123**	**21 716**	**24 671**	**31 531**	**28 877**	**33 828**	**36 450**	**Asia**
- 2 204	- 1 041	2 829	14 856	11 377	11 723	13 926	22 476	20 535	People's Rep. of China
6 387	6 452	8 088	8 561	8 853	9 229	9 455	9 530	11 656	Hong Kong, China
4 183	**5 411**	**10 917**	**23 417**	**20 230**	**20 952**	**23 381**	**32 006**	**32 191**	**China**
14	- 72	- 55	14	61	90	99	105	105	Albania
315	1 619	3 077	939	- 987	- 1 461	- 1 143	- 1 362	- 774	Bulgaria
947	944	1 154	1 258	1 247	1 258	1 403	1 117	1 198	Cyprus
515	934	944	980	985	987	987	987	987	Gibraltar
617	707	762	785	764	890	971	1 020	993	Malta
- 4 779	- 726	- 577	- 939	- 3 277	- 1 749	- 505	1 128	- 487	Romania
- 1 424	- 1 138	- 809	- 1 286	- 1 663	- 1 691	- 1 834	- 1 953	- 2 094	Slovak Republic
-	-	536	716	859	884	911	938	967	Bosnia-Herzegovina
-	-	- 812	- 1 830	- 1 722	- 1 742	- 1 301	- 1 231	- 1 020	Croatia
-	-	79	220	706	808	562	642	207	FYROM
-	-	979	1 324	1 679	1 606	2 082	2 069	2 186	Slovenia
-	-	455	160	160	160	353	300	300	FR of Yugoslavia
560	937	1 237	590	1 682	1 716	2 607	2 718	2 640	Former Yugoslavia
- 3 235	**3 205**	**5 733**	**2 341**	**- 1 188**	**40**	**2 585**	**3 760**	**2 568**	**Non-OECD Europe**
-	-	2 427	1 211	391	278	154	164	169	Armenia
-	-	- 2 952	- 2 129	- 1 686	- 2 198	- 2 159	- 2 054	- 1 699	Azerbaijan
-	-	404	133	- 1 404	- 2 240	- 3 030	- 2 335	- 1 861	Belarus
-	-	1 323	1 530	1 325	1 025	1 038	907	1 139	Estonia
-	-	594	431	83	87	865	930	1 142	Georgia
-	-	3 915	1 698	599	- 3	- 1 272	- 894	- 240	Kazakhstan
-	-	1 909	1 165	417	529	555	360	502	Kyrgyzstan
-	-	2 662	2 491	2 454	2 059	2 309	1 556	1 723	Latvia
-	-	- 99	- 1 362	- 691	134	- 933	- 2 324	- 3 025	Lithuania
-	-	2 872	1 866	1 132	1 076	820	944	629	Republic of Moldova
-	-	- 41 509	- 40 523	- 43 288	- 43 423	- 50 362	- 51 939	- 47 033	Russia
-	-	5 533	3 446	1 129	1 137	1 141	1 136	1 265	Tajikistan
-	-	- 122	- 1 000	- 925	- 901	- 936	- 1 660	- 2 654	Turkmenistan
-	-	1 751	4 986	3 442	7 421	6 079	4 658	4 142	Ukraine
-	-	1 840	1 278	912	- 130	- 325	- 332	- 422	Uzbekistan
- 49 494	**- 43 387**	**- 19 452**	**- 24 779**	**- 36 110**	**- 35 149**	**- 46 056**	**- 50 883**	**- 46 223**	**Former USSR**

A negative number shows net exports.

Net Imports of Petroleum Products (1000 tonnes)
Importations nettes de produits pétroliers (1000 tonnes)

Nettoimporte von Ölprodukten (1000 Tonnen)

Importazioni nette di prodotti petroliferi (1000 tonnellate)

石油製品の純輸入量（千トン）

Importaciones netas de productos petrolíferos (1000 toneladas)

Чистый импорт нефтепродуктов (тыс. т)

	1971	1973	1978	1984	1985	1986	1987	1988	1989
Bahrein	- 11 044	- 10 431	- 11 850	- 9 138	- 8 030	- 11 306	- 11 646	- 11 262	- 11 661
République Islamique d'Iran	- 10 911	- 7 750	- 7 329	5 806	4 708	6 878	8 823	8 144	6 579
Irak	- 63	54	- 1 929	- 4 616	- 4 866	- 5 709	- 5 697	- 6 504	- 6 440
Israël	- 101	- 284	- 535	- 559	- 107	85	89	1 781	1 199
Jordanie	16	- 38	- 13	144	404	665	733	628	625
Koweit	- 11 924	- 12 176	- 13 015	- 18 215	- 20 827	- 22 859	- 26 492	- 30 196	- 37 711
Liban	- 42	- 39	226	1 434	1 937	1 593	1 515	1 495	2 023
Oman	1 335	1 326	1 151	- 1 040	- 1 005	- 896	- 615	- 1 035	- 1 034
Qatar	58	114	66	- 600	- 836	- 1 073	- 1 225	- 1 606	- 1 904
Arabie saoudite	- 11 152	- 9 749	- 9 647	- 12 882	- 15 571	- 21 945	- 34 435	- 39 336	- 42 595
Syrie	211	180	394	- 1 283	- 2 468	- 2 054	- 1 675	- 2 643	- 2 863
Emirats arabes unis	136	280	1 381	- 5 528	- 6 664	- 8 741	- 8 641	- 7 602	- 8 312
Yémen	- 2 730	- 1 905	- 235	- 2 219	- 1 157	- 778	- 2 197	- 2 120	- 836
Moyen-Orient	**- 46 211**	**- 40 418**	**- 41 335**	**- 48 696**	**- 54 482**	**- 66 140**	**- 81 463**	**- 90 256**	**- 102 930**
Total non-OCDE	**- 168 843**	**- 178 876**	**- 147 931**	**- 177 827**	**- 169 403**	**- 184 208**	**- 199 658**	**- 212 360**	**- 222 296**
OCDE Amérique du N.	103 321	129 112	64 484	34 647	22 735	29 100	30 037	38 452	34 130
OCDE Pacifique	28 506	24 047	29 188	36 377	39 683	46 316	57 911	61 622	65 508
OCDE Europe	20 534	12 000	41 141	52 321	52 516	49 778	51 292	33 696	37 989
Total OCDE	**152 361**	**165 159**	**134 813**	**123 345**	**114 934**	**125 194**	**139 240**	**133 770**	**137 627**
Monde	**- 16 482**	**- 13 717**	**- 13 118**	**- 54 482**	**- 54 469**	**- 59 014**	**- 60 418**	**- 78 590**	**- 84 669**
Pour mémoire: OPEP	*- 88 748*	*- 93 096*	*- 64 594*	*- 76 781*	*- 89 655*	*- 105 687*	*- 121 542*	*- 142 747*	*- 155 144*

Un chiffre négatif correspond à des exportations nettes.
La ligne Monde montre la divergence entre le total des exportations et des importations mondiales.

Net Imports of Petroleum Products (1000 tonnes)
Importations nettes de produits pétroliers (1000 tonnes)
Nettoimporte von Ölprodukten (1000 Tonnen)
Importazioni nette di prodotti petroliferi (1000 tonnellate)
石油製品の純輸入量（チトン）
Importaciones netas de productos petrolíferos (1000 toneladas)
Чистый импорт нефтепродуктов (тыс. m)

1990	1991	1992	1993	1994	1995	1996	1997	1998	
- 11 914	- 12 059	- 12 369	- 12 070	- 11 859	- 12 067	- 12 529	- 11 975	- 11 958	Bahrain
6 415	7 603	6 482	2 458	1 943	- 5 826	- 6 653	- 10 588	- 14 173	Islamic Republic of Iran
- 3 540	-	- 50	- 70	- 80	-	-	-	-	Iraq
638	330	- 491	- 1 478	- 1 570	406	497	173	44	Israel
781	776	842	902	922	923	1 190	932	1 153	Jordan
- 9 340	- 354	- 11 759	- 18 799	- 32 566	- 35 444	- 32 910	- 37 759	- 36 209	Kuwait
2 023	2 210	2 248	3 291	3 640	4 176	4 276	4 833	4 746	Lebanon
- 1 317	- 87	- 41	568	- 1 321	- 861	- 1 022	- 972	- 1 669	Oman
- 2 293	- 1 975	- 2 360	- 1 999	- 2 233	- 2 049	- 2 217	- 1 707	- 1 907	Qatar
- 49 427	- 40 508	- 44 438	- 42 520	- 42 272	- 42 246	- 47 710	- 41 827	- 40 872	Saudi Arabia
- 1 936	- 1 479	- 1 688	- 1 403	- 1 313	- 1 064	- 1 091	- 1 360	- 622	Syria
- 8 976	- 10 239	- 11 267	- 11 368	- 10 818	- 11 469	- 11 442	- 11 962	- 13 012	United Arab Emirates
- 2 161	- 1 633	- 1 906	- 1 195	- 1 093	- 831	- 1 478	- 1 694	- 1 694	Yemen
- 81 047	**- 57 415**	**- 76 797**	**- 83 683**	**- 98 620**	**- 106 352**	**- 111 089**	**- 113 906**	**- 116 173**	**Middle East**
- 177 276	**- 142 834**	**- 119 015**	**- 110 574**	**- 142 551**	**- 134 205**	**- 148 486**	**- 143 518**	**- 132 262**	**Non-OECD Total**
18 519	4 644	7 145	- 2 136	7 026	- 6 391	5 103	7 188	19 379	OECD North America
67 355	50 257	46 897	44 169	51 240	50 704	54 017	33 125	16 944	OECD Pacific
33 010	33 804	21 022	14 746	8 968	18 076	18 355	14 190	15 876	OECD Europe
118 884	**88 705**	**75 064**	**56 779**	**67 234**	**62 389**	**77 475**	**54 503**	**52 199**	**OECD Total**
- 58 392	**- 54 129**	**- 43 951**	**- 53 795**	**- 75 317**	**- 71 816**	**- 71 011**	**- 89 015**	**- 80 063**	**World**
- 131 924	**- 114 824**	**- 126 965**	**- 130 946**	**- 147 118**	**- 158 857**	**- 166 214**	**- 173 437**	**- 171 401**	***Memo: OPEC***

A negative number shows net exports.
The row World shows the discrepancy between total world exports and imports.

Net Imports of Natural Gas (TJ)
Importations nettes de gaz naturel (TJ)
Nettoimporte von Erdgas (TJ)

Importazioni nette di gas naturale (TJ)

天然ガスの純輸入量 (TJ)

Importaciones netas de gas natural (TJ)

Чистый импорт природного газа (ТДж)

	1971	1973	1978	1984	1985	1986	1987	1988	1989
Algérie	- 58 484	- 107 986	- 288 381	- 832 077	- 948 434	- 933 052	-1 140 856	-1 164 675	-1 314 423
Libye	- 18 793	- 119 789	- 153 407	- 36 860	- 46 740	- 32 680	- 30 400	- 40 280	- 54 340
Tunisie	-	-	-	13 963	24 226	13 276	36 238	28 043	30 000
Afrique	**- 77 277**	**- 227 775**	**- 441 788**	**- 854 974**	**- 970 948**	**- 952 456**	**-1 135 018**	**-1 176 912**	**-1 338 763**
Argentine	-	67 511	93 814	92 613	91 805	90 740	88 506	92 787	92 508
Bolivie	-	- 74 747	- 75 520	- 93 994	- 94 226	- 94 148	- 90 168	- 94 767	- 95 733
Chili	-	-	- 26 984	-	-	-	-	-	-
Uruguay	-	-	-	-	-	-	-	-	-
Amérique latine	**-**	**- 7 236**	**- 8 690**	**- 1 381**	**- 2 421**	**- 3 408**	**- 1 662**	**- 1 980**	**- 3 225**
Brunei	-	- 59 070	- 277 900	- 290 153	- 279 602	- 278 546	- 284 877	- 292 263	- 291 208
Indonésie	-	-	- 199 347	- 778 813	- 818 705	- 822 846	- 909 047	-1 003 855	-1 013 113
Malaisie	-	-	- 544	- 192 590	- 236 731	- 284 010	- 328 653	- 342 419	- 354 134
Singapour	-	-	-	-	-	-	-	-	-
Taipei chinois	-	-	-	-	-	-	-	-	-
Thailande	-	-	-	-	-	-	-	-	-
Autre Asie	- 87 634	- 95 365	- 84 129	- 87 000	- 87 000	- 87 000	- 63 000	- 50 000	- 3 000
Asie	**- 87 634**	**- 154 435**	**- 561 920**	**-1 348 556**	**-1 422 038**	**-1 472 402**	**-1 585 577**	**-1 688 537**	**-1 661 455**
Rép. populaire de Chine	-	-	-	-	-	-	-	-	-
Hong-Kong, Chine	-	-	-	-	-	-	-	-	-
Chine	**-**	**-**	**-**	**-**	**-**	**-**	**-**	**-**	**-**
Bulgarie	-	-	106 467	194 944	213 078	221 840	237 171	236 162	228 802
Roumanie	- 7 555	- 7 555	30 416	71 480	70 811	110 445	119 404	146 128	277 701
République slovaque	37 848	54 618	120 235	184 524	182 259	182 779	183 076	180 032	194 889
Bosnie-Herzegovine	-	-	-	-	-	-	-	-	-
Croatie	-	-	-	-	-	-	-	-	-
Ex-RYM	-	-	-	-	-	-	-	-	-
Slovénie	-	-	-	-	-	-	-	-	-
RF de Yougoslavie	-	-	-	-	-	-	-	-	-
Ex-Yougoslavie	-	-	-	130 596	137 075	146 032	164 247	153 849	157 992
Europe non-OCDE	**30 293**	**47 063**	**257 118**	**581 544**	**603 223**	**661 096**	**703 898**	**716 171**	**859 384**
Arménie	-	-	-	-	-	-	-	-	-
Azerbaïdjan	-	-	-	-	-	-	-	-	-
Bélarus	-	-	-	-	-	-	-	-	-
Estonie	-	-	-	-	-	-	-	-	-
Géorgie	-	-	-	-	-	-	-	-	-
Kazakhstan	-	-	-	-	-	-	-	-	-
Kirghizistan	-	-	-	-	-	-	-	-	-
Lettonie	-	-	-	-	-	-	-	-	-
Lituanie	-	-	-	-	-	-	-	-	-
République de Moldavie	-	-	-	-	-	-	-	-	-
Russie	-	-	-	-	-	-	-	-	-
Tadjikistan	-	-	-	-	-	-	-	-	-
Turkménistan	-	-	-	-	-	-	-	-	-
Ukraine	-	-	-	-	-	-	-	-	-
Ouzbékistan	-	-	-	-	-	-	-	-	-
Ex-URSS	**134 947**	**176 827**	**-1 023 732**	**-2 480 251**	**-2 550 050**	**-2 964 193**	**-3 182 894**	**-3 270 222**	**-3 798 277**

Un chiffre négatif correspond à des exportations nettes.

Net Imports of Natural Gas (TJ)
Importations nettes de gaz naturel (TJ)

Nettoimporte von Erdgas (TJ)
Importazioni nette di gas naturale (TJ)
天然ガスの純輸入量 *(TJ)*
Importaciones netas de gas natural (TJ)
Чистый импорт природного газа (ТДж)

1990	1991	1992	1993	1994	1995	1996	1997	1998	
-1 379 394	-1 504 156	-1 582 796	-1 567 050	-1 412 872	-1 665 932	-1 825 297	-2 162 030	-2 342 805	Algeria
- 47 120	- 60 040	- 69 920	- 60 800	- 56 240	- 56 620	- 45 600	- 41 800	- 34 580	Libya
42 020	27 734	45 775	46 566	69 374	83 333	59 982	29 836	30 678	Tunisia
-1 384 494	**-1 536 462**	**-1 606 941**	**-1 581 284**	**-1 399 738**	**-1 639 219**	**-1 810 915**	**-2 173 994**	**-2 346 707**	**Africa**
84 691	84 132	82 131	80 782	87 157	82 085	81 899	37 181	- 15 263	Argentina
- 95 462	- 94 380	- 92 100	- 90 617	- 93 925	- 88 947	- 82 293	- 89 240	- 69 104	Bolivia
-	-	-	-	- 3 329	- 2 675	- 1 557	25 699	77 849	Chile
-	-	-	-	-	-	-	-	92	Uruguay
- 10 771	**- 10 248**	**- 9 969**	**- 9 835**	**- 10 097**	**- 9 537**	**- 1 951**	**- 26 360**	**- 6 426**	**Latin America**
- 291 208	- 290 153	- 293 318	- 301 759	- 314 420	- 340 797	- 338 687	- 336 577	- 321 806	Brunei
-1 121 462	-1 225 212	-1 296 124	-1 324 766	-1 434 447	-1 357 846	-1 450 383	-1 464 003	-1 472 856	Indonesia
- 363 422	- 346 352	- 345 724	- 414 718	- 440 450	- 513 126	- 699 774	- 742 074	- 747 797	Malaysia
-	-	18 621	55 427	66 031	68 766	61 332	60 465	66 213	Singapore
35 465	85 819	89 637	96 291	120 545	137 393	142 400	174 343	211 727	Chinese Taipei
-	-	-	-	-	-	-	-	813	Thailand
-	-	-	-	-	-	-	-	-	Other Asia
-1 740 627	**-1 775 898**	**-1 826 908**	**-1 889 525**	**-2 002 741**	**-2 005 610**	**-2 285 112**	**-2 307 846**	**-2 263 706**	**Asia**
-	-	-	-	-	- 1 086	- 68 823	- 107 462	- 100 815	People's Rep. of China
-	-	-	-	-	1 086	68 823	107 462	100 815	Hong Kong, China
-	**-**	**-**	**-**	**-**	**-**	**-**	**-**	**-**	**China**
227 353	209 228	188 931	175 690	173 913	212 258	220 040	179 192	145 145	Bulgaria
275 754	173 989	166 688	168 206	173 770	223 038	263 046	187 503	175 546	Romania
260 714	223 987	221 560	199 640	201 238	212 780	237 880	242 593	251 657	Slovak Republic
-	-	16 000	5 473	9 834	10 131	10 437	10 752	11 076	*Bosnia-Herzegovina*
-	-	27 398	30 020	28 234	10 408	33 402	39 729	41 952	*Croatia*
-	-	10 000	10 500	-	-	-	-	829	*FYROM*
-	-	25 164	24 943	26 491	31 412	30 039	32 847	33 932	*Slovenia*
-	-	73 176	33 628	-	7 502	79 170	77 813	70 687	*FR of Yugoslavia*
180 151	184 421	151 738	104 564	64 559	59 453	153 048	161 141	158 476	Former Yugoslavia
943 972	**791 625**	**728 917**	**648 100**	**613 480**	**707 529**	**874 014**	**770 429**	**730 824**	**Non-OECD Europe**
-	-	70 680	30 628	32 870	53 200	41 470	54 477	56 739	Armenia
-	-	162 637	94 778	98 435	20 019	905	-	-	Azerbaijan
-	-	677 771	629 735	552 182	522 594	554 032	627 260	629 406	Belarus
-	-	33 381	16 546	23 763	27 098	29 884	29 038	27 560	Estonia
-	-	182 717	138 284	92 591	33 930	29 368	35 589	32 000	Georgia
-	-	388 762	238 339	208 745	247 312	118 868	21 560	28 128	Kazakhstan
-	-	68 426	51 385	32 271	33 050	40 074	22 241	38 992	Kyrgyzstan
-	-	99 298	35 309	37 798	46 491	40 692	49 249	51 641	Latvia
-	-	134 073	72 511	80 518	94 502	100 749	93 134	81 578	Lithuania
-	-	136 163	126 333	118 325	119 118	136 679	145 726	131 861	Republic of Moldova
-	-	-6 940 834	-6 279 921	-6 924 444	-7 062 222	-7 252 444	-7 379 708	-7 530 980	Russia
-	-	63 032	52 394	28 956	32 528	41 800	27 854	28 804	Tajikistan
-	-	-1 783 948	-2 113 256	- 865 640	- 831 440	- 904 800	- 222 654	- 113 839	Turkmenistan
-	-	3 364 763	2 880 770	2 494 496	2 477 224	2 747 125	2 378 343	2 066 507	Ukraine
-	-	- 37 555	- 53 580	- 156 788	- 159 436	- 184 747	- 270 413	- 118 512	Uzbekistan
-4 042 188	**-3 826 289**	**-3 380 634**	**-4 079 745**	**-4 145 922**	**-4 346 032**	**-4 460 345**	**-4 388 304**	**-4 590 115**	**Former USSR**

A negative number shows net exports.

Net Imports of Natural Gas (TJ)
Importations nettes de gaz naturel (TJ)
Nettoimporte von Erdgas (TJ)
Importazioni nette di gas naturale (TJ)
天然ガスの純輸入量 *(TJ)*
Importaciones netas de gas natural (TJ)
Чистый импорт природного газа (ТДж)

	1971	1973	1978	1984	1985	1986	1987	1988	1989
République Islamique d'Iran	- 196 049	- 317 010	- 275 500	-	-	-	-	-	-
Irak	-	-	-	-	-	- 24 700	- 104 500	- 117 800	- 136 800
Koweit	-	-	-	-	-	50 000	107 250	120 900	140 400
Oman	-	-	-	-	-	-	-	-	-
Qatar	-	-	-	-	-	-	-	-	-
Emirats arabes unis	-	-	- 63 343	- 105 075	- 115 880	- 109 173	- 106 938	- 118 489	- 115 880
Moyen-Orient	**- 196 049**	**- 317 010**	**- 338 843**	**- 105 075**	**- 115 880**	**- 83 873**	**- 104 188**	**- 115 389**	**- 112 280**
Total non-OCDE	**- 195 720**	**- 482 566**	**-2 117 855**	**-4 208 693**	**-4 458 114**	**-4 815 236**	**-5 305 441**	**-5 536 869**	**-6 054 616**
OCDE Amérique du N.	- 31 451	- 33 113	50 896	- 35 624	- 46 389	- 61 733	- 66 063	- 55 432	- 36 662
OCDE Pacifique	54 387	129 581	625 219	1 433 158	1 534 416	1 580 057	1 679 138	1 786 552	1 916 515
OCDE Europe	210 720	415 674	1 419 544	2 477 774	2 699 589	3 106 693	3 437 829	3 557 428	3 944 690
Total OCDE	**233 656**	**512 142**	**2 095 659**	**3 875 308**	**4 187 616**	**4 625 017**	**5 050 904**	**5 288 548**	**5 824 543**
Monde	**37 936**	**29 576**	**- 22 196**	**- 333 385**	**- 270 498**	**- 190 219**	**- 254 537**	**- 248 321**	**- 230 073**

Un chiffre négatif correspond à des exportations nettes.
La ligne Monde montre la divergence entre le total des exportations et des importations mondiales.

Net Imports of Natural Gas (TJ)
Importations nettes de gaz naturel (TJ)
Nettoimporte von Erdgas (TJ)

Importazioni nette di gas naturale (TJ)

天然ガスの純輸入量 (TJ)

Importaciones netas de gas natural (TJ)

Чистый импорт природного газа (ТДж)

1990	1991	1992	1993	1994	1995	1996	1997	1998	
- 57 000	- 114 380	-	- 19 000	- 30 400	-	-	15 546	73 320	Islamic Republic of Iran
- 76 000	-	-	-	-	-	-	-	-	Iraq
80 000	-	-	-	-	-	-	-	-	Kuwait
-	-	-	-	-	- 18 850	- 18 850	- 18 850	- 16 965	Oman
-	-	-	-	-	-	-	- 107 822	- 180 583	Qatar
- 119 234	- 128 549	- 127 431	- 124 823	- 154 631	- 243 312	- 248 118	- 272 375	- 248 528	United Arab Emirates
- 172 234	**- 242 929**	**- 127 431**	**- 143 823**	**- 185 031**	**- 262 162**	**- 266 968**	**- 383 501**	**- 372 756**	**Middle East**
-6 406 342	**-6 600 201**	**-6 222 966**	**-7 056 112**	**-7 130 049**	**-7 555 031**	**-7 951 277**	**-8 509 576**	**-8 848 886**	**Non-OECD Total**
48 372	30 891	18 688	51 635	- 15 168	- 31 147	15 773	21 942	- 78 174	OECD North America
1 954 208	2 037 803	2 075 322	2 100 560	2 296 558	2 327 094	2 636 172	2 805 686	2 846 967	OECD Pacific
4 246 773	4 344 317	4 360 303	4 405 149	4 481 100	5 118 818	5 346 569	5 502 467	5 879 892	OECD Europe
6 249 353	**6 413 011**	**6 454 313**	**6 557 344**	**6 762 490**	**7 414 765**	**7 998 514**	**8 330 095**	**8 648 685**	**OECD Total**
- 156 989	**- 187 190**	**231 347**	**- 498 768**	**- 367 559**	**- 140 266**	**47 237**	**- 179 481**	**- 200 201**	**World**

A negative number shows net exports.

The row World shows the discrepancy between total world exports and imports.

Net Imports of Electricity (GWh)
Importations nettes d'électricité (GWh)
Nettoimporte von Elektrizität (GWh)
Importazioni nette di energia elettrica (GWh)
電力の純輸入量 (GWh)
Importaciones netas de electricidad (GWh)
Чистый импорт электроэнергии (ГВт.ч)

	1971	1973	1978	1984	1985	1986	1987	1988	1989
Algérie	1	- 1	1	71	36	84	- 50	- 52	- 15
Bénin	33	50	164	102	149	141	153	193	183
Congo	-	-	71	53	55	129	148	149	61
RD du Congo	- 26	- 26	- 55	- 128	- 147	- 184	- 187	- 109	- 533
Côte d'Ivoire	-	-	-	383	217	-	-	-	-
Ghana	-	- 100	- 217	- 622	- 611	- 664	- 493	- 211	- 521
Kenya	293	303	217	217	215	220	176	110	112
Maroc	-	-	-	-	-	-	-	-	-
Mozambique	2	153	77	37	- 12	222	273	183	116
Nigéria	-	-	- 45	- 135	- 140	-	-	-	-
Afrique du Sud	- 14	- 194	6 609	- 405	- 314	- 1 223	- 1 187	- 997	- 1 120
Rép. Unie de Tanzanie	-	-	-	-	-	-	-	-	-
Togo	-	59	157	175	253	271	304	306	338
Tunisie	-	-	-	- 73	35	14	- 1	- 31	22
Zambie	3 430	2 029	- 1 731	- 2 901	- 2 903	- 3 080	- 1 500	- 1 500	- 1 500
Zimbabwe	176	94	53	3 433	3 996	3 308	2 536	855	835
Autre Afrique	21	42	69	- 70	- 66	- 71	- 17	55	107
Afrique	**3 916**	**2 409**	**5 370**	**137**	**763**	**- 833**	**155**	**- 1 049**	**- 1 915**
Argentine	- 10	50	74	- 5	- 6	- 5	174	820	213
Bolivie	-	-	-	-	-	-	-	-	8
Brésil	- 18	- 17	- 127	- 86	1 913	10 292	16 803	17 943	22 106
Chili	2	3	1	-	-	-	-	-	-
Colombie	-	-	-	-	-	-	73	164	191
Costa Rica	-	-	-	- 430	- 59	78	170	190	143
El Salvador	-	-	-	-	-	90	28	43	8
Guatemala	-	-	-	-	-	- 88	- 11	-	-
Honduras	-	-	-	- 3	- 134	- 148	- 313	- 302	- 229
Nicaragua	-	-	- 1	265	187	69	80	76	12
Panama	- 46	- 50	- 13	58	22	- 18	119	19	72
Paraguay	-	- 77	63	54	- 2 821	- 10 462	- 16 813	- 17 960	- 20 525
Pérou	-	-	-	-	-	-	-	-	-
Uruguay	35	27	29	- 3 332	- 2 678	- 3 152	- 2 955	- 2 091	- 1 154
Autre Amérique latine	-	-	-	-	-	-	-	-	-
Amérique latine	**- 37**	**- 64**	**26**	**- 3 479**	**- 3 576**	**- 3 344**	**- 2 645**	**- 1 098**	**845**
Inde	-	- 11	- 44	- 105	- 91	103	855	1 200	1 297
Malaisie	-	-	12	81	58	-	- 23	- 70	- 23
Népal	1	-	27	24	36	12	60	108	36
Singapour	-	-	- 48	- 73	- 50	-	-	-	-
Thailande	- 41	156	216	688	703	741	398	410	620
Autre Asie	-	-	4	9	31	- 292	- 1 092	- 1 392	- 1 392
Asie	**- 40**	**145**	**167**	**624**	**687**	**564**	**198**	**256**	**538**
Rép. populaire de Chine	-	-	-	770	1 070	1 168	1 250	1 470	1 720
Hong-Kong, Chine	-	-	-	- 740	- 1 050	- 1 208	- 1 354	- 1 434	- 1 771
Chine	**-**	**-**	**-**	**30**	**20**	**- 40**	**- 104**	**36**	**- 51**

Un chiffre négatif correspond à des exportations nettes.

Net Imports of Electricity (GWh)
Importations nettes d'électricité (GWh)
Nettoimporte von Elektrizität (GWh)
Importazioni nette di energia elettrica (GWh)
電力の純輸入量 *(GWh)*
Importaciones netas de electricidad (GWh)
Чистый импорт электроэнергии (ГВт.ч)

1990	1991	1992	1993	1994	1995	1996	1997	1998	
- 60	- 661	- 928	- 1 241	- 1 124	- 273	- 142	- 1	- 7	Algeria
196	215	214	233	211	256	264	238	255	Benin
14	58	106	162	184	166	96	38	126	Congo
- 5	- 42	- 143	- 142	- 140	- 138	- 140	- 140	- 132	DR of Congo
294	367	484	91	34	34	- 436	- 951	- 593	Cote d'Ivoire
- 761	- 805	- 893	- 765	- 387	35	- 120	490	490	Ghana
181	134	240	273	264	172	149	144	146	Kenya
103	641	932	1 027	793	243	129	124	705	Morocco
166	301	391	422	465	501	599	203	- 5 327	Mozambique
-	- 92	- 92	- 104	- 138	-	-	-	-	Nigeria
- 1 263	- 1 627	- 1 481	- 2 489	- 2 625	- 2 851	- 5 550	- 6 612	- 2 157	South Africa
-	-	-	-	15	35	34	44	43	United Rep. of Tanzania
337	323	368	274	301	413	434	452	426	Togo
- 4	16	- 4	215	403	31	15	7	- 19	Tunisia
- 1 480	- 1 480	- 1 480	- 1 471	- 1 480	- 1 480	- 1 481	- 1 500	- 1 500	Zambia
332	1 144	2 027	1 214	2 009	2 312	3 172	4 013	5 149	Zimbabwe
135	245	194	326	418	356	693	734	734	Other Africa
- 1 815	**- 1 263**	**- 65**	**- 1 975**	**- 797**	**- 188**	**- 2 284**	**- 2 717**	**- 1 661**	**Africa**
821	875	2 579	1 461	992	2 122	3 363	5 189	8 000	Argentina
9	9	12	13	16	10	- 3	7	6	Bolivia
26 538	27 080	24 014	27 550	31 767	35 352	36 558	40 470	39 404	Brazil
-	-	-	-	-	-	-	-	-	Chile
200	212	348	303	280	370	161	199	21	Colombia
163	20	- 64	- 3	- 6	29	124	- 134	- 71	Costa Rica
20	9	144	101	75	95	63	124	84	El Salvador
- 2	- 5	- 54	- 79	11	35	- 23	- 88	- 38	Guatemala
- 334	- 214	4	51	56	- 18	- 64	155	44	Honduras
66	91	31	- 47	- 21	-	4	4	- 4	Nicaragua
114	136	143	86	83	87	- 62	- 53	34	Panama
- 24 970	- 26 975	- 24 618	- 28 120	- 32 600	- 37 827	- 42 644	- 45 673	- 45 357	Paraguay
-	-	-	-	3	-	-	2	1	Peru
- 2 538	- 1 792	- 3 385	- 2 241	- 1 661	- 45	- 35	- 144	- 2 156	Uruguay
-	3	17	12	12	16	16	17	17	Other Latin America
87	**- 551**	**- 829**	**- 913**	**- 993**	**226**	**- 2 542**	**75**	**- 15**	**Latin America**
1 376	1 453	1 206	1 240	1 423	1 545	1 545	1 545	1 545	India
- 58	- 23	- 23	23	47	- 23	- 11	11	- 7	Malaysia
- 48	-	36	60	60	- 12	60	143	179	Nepal
-	-	-	- 52	- 91	-	-	-	-	Singapore
621	555	440	596	826	620	717	642	1 470	Thailand
- 1 392	- 1 397	- 1 311	- 1 312	- 1 324	- 1 351	- 1 366	- 1 433	- 1 433	Other Asia
499	**588**	**348**	**555**	**941**	**779**	**945**	**908**	**1 754**	**Asia**
1 840	2 850	2 850	4 389	- 2 046	- 5 386	- 3 588	- 7 115	- 7 157	People's Rep. of China
- 1 799	- 3 061	- 4 962	- 4 054	6 495	6 063	7 247	7 317	7 150	Hong Kong, China
41	**- 211**	**- 2 112**	**335**	**4 449**	**677**	**3 659**	**202**	**- 7**	**China**

A negative number shows net exports.

Net Imports of Electricity (GWh)
Importations nettes d'électricité (GWh)
Nettoimporte von Elektrizität (GWh)
Importazioni nette di energia elettrica (GWh)
電力の純輸入量 (GWh)
Importaciones netas de electricidad (GWh)
Чистый импорт электроэнергии (ГВт.ч)

	1971	1973	1978	1984	1985	1986	1987	1988	1989
Albanie	-	- 267	- 133	- 600	- 625	- 650	- 650	- 650	- 607
Bulgarie	218	3 256	3 813	2 392	4 304	3 972	4 349	4 146	4 389
Gibraltar	-	-	-	-	-	-	-	-	- 1
Roumanie	- 3 155	- 3 548	- 1 812	1 956	3 259	4 430	5 150	7 199	7 811
République slovaque	2 381	2 840	2 849	5 054	4 185	3 256	4 733	5 833	5 586
Bosnie-Herzegovine	-	-	-	-	-	-	-	-	-
Croatie	-	-	-	-	-	-	-	-	-
Ex-RYM	-	-	-	-	-	-	-	-	-
Slovénie	-	-	-	-	-	-	-	-	-
RF de Yougoslavie	-	-	-	-	-	-	-	-	-
Ex-Yougoslavie	- 162	- 49	- 652	- 1 130	627	466	375	- 1 412	- 408
Europe non-OCDE	**- 718**	**2 232**	**4 065**	**7 672**	**11 750**	**11 474**	**13 957**	**15 116**	**16 770**
Arménie	-	-	-	-	-	-	-	-	-
Azerbaïdjan	-	-	-	-	-	-	-	-	-
Bélarus	-	-	-	-	-	-	-	-	-
Estonie	-	-	-	-	-	-	-	-	-
Géorgie	-	-	-	-	-	-	-	-	-
Kazakhstan	-	-	-	-	-	-	-	-	-
Kirghizistan	-	-	-	-	-	-	-	-	-
Lettonie	-	-	-	-	-	-	-	-	-
Lituanie	-	-	-	-	-	-	-	-	-
République de Moldavie	-	-	-	-	-	-	-	-	-
Russie	-	-	-	-	-	-	-	-	-
Tadjikistan	-	-	-	-	-	-	-	-	-
Turkménistan	-	-	-	-	-	-	-	-	-
Ukraine	-	-	-	-	-	-	-	-	-
Ouzbékistan	-	-	-	-	-	-	-	-	-
Ex-URSS	**- 6 700**	**- 9 700**	**- 12 200**	**- 24 700**	**- 28 900**	**- 29 000**	**- 34 600**	**- 38 900**	**- 39 300**
Israël	- 33	- 57	- 163	- 254	- 300	- 369	- 359	- 535	- 413
Jordanie	33	57	-	-	- 21	- 215	- 334	-	-
Liban	-	-	50	30	40	30	40	-	-
Syrie	- 71	-	-	- 135	- 140	- 130	- 130	-	-
Moyen-Orient	**- 71**	-	**- 113**	**- 359**	**- 421**	**- 684**	**- 783**	**- 535**	**- 413**
Total non-OCDE	**- 3 650**	**- 4 978**	**- 2 685**	**- 20 075**	**- 19 677**	**- 21 863**	**- 23 822**	**- 26 174**	**- 23 526**
OCDE Amérique du N.	110	558	71	484	505	554	458	2 096	295
OCDE Europe	5 003	2 905	2 384	15 566	16 551	17 404	20 101	22 662	20 284
Total OCDE	**5 113**	**3 463**	**2 455**	**16 050**	**17 056**	**17 958**	**20 559**	**24 758**	**20 579**
Monde	**1 463**	**- 1 515**	**- 230**	**- 4 025**	**- 2 621**	**- 3 905**	**- 3 263**	**- 1 416**	**- 2 947**

Un chiffre négatif correspond à des exportations nettes.
La ligne Monde montre la divergence entre le total des exportations et des importations mondiales.

Net Imports of Electricity (GWh)
Importations nettes d'électricité (GWh)
Nettoimporte von Elektrizität (GWh)

Importazioni nette di energia elettrica (GWh)

電力の純輸入量 (GWh)

Importaciones netas de electricidad (GWh)

Чистый импорт электроэнергии (ГВт.ч)

1990	1991	1992	1993	1994	1995	1996	1997	1998	
206	- 1 173	- 510	- 141	- 185	- 74	- 230	- 460	- 400	Albania
3 790	2 124	2 705	110	- 72	- 160	- 449	- 3 550	- 3 647	Bulgaria
- 1	- 1	-	- 5	-	-	-	-	-	Gibraltar
9 476	7 047	4 203	1 873	725	299	807	221	387	Romania
5 196	4 338	3 468	1 113	438	1 383	3 592	4 082	1 290	Slovak Republic
-	-	-	-	160	205	211	218	224	*Bosnia-Herzegovina*
-	-	2 787	2 326	3 565	3 496	2 330	3 948	3 354	*Croatia*
-	-	274	607	167	-	-	-	- 2	*FYROM*
-	-	- 1 813	- 1 418	- 1 934	- 1 652	- 1 661	- 1 696	- 1 924	*Slovenia*
-	-	- 400	-	-	-	1	- 9	- 268	*FR of Yugoslavia*
- 359	- 300	848	1 515	1 958	2 049	881	2 461	1 384	Former Yugoslavia
18 308	**12 035**	**10 714**	**4 465**	**2 864**	**3 497**	**4 601**	**2 754**	**- 986**	**Non-OECD Europe**
-	-	283	114	16	13	-	-	-	Armenia
-	-	- 513	100	260	399	442	431	255	Azerbaijan
-	-	6 504	6 005	3 820	7 159	8 543	7 620	7 399	Belarus
-	-	- 3 238	- 1 596	- 1 191	- 760	- 860	- 974	- 390	Estonia
-	-	1 016	711	800	670	86	191	- 98	Georgia
-	-	14 173	6 000	13 031	7 395	6 589	4 703	3 811	Kazakhstan
-	-	- 2 088	- 1 023	- 2 505	- 1 368	- 2 080	- 1 784	- 726	Kyrgyzstan
-	-	4 077	2 502	1 818	2 256	3 227	1 823	529	Latvia
-	-	- 5 303	- 2 731	1 144	- 2 678	- 5 159	- 3 525	- 6 082	Lithuania
-	-	- 768	83	615	1 870	1 606	1 952	1 916	Republic of Moldova
-	-	- 16 242	- 18 732	- 20 496	- 19 605	- 19 490	- 19 689	- 18 014	Russia
-	-	832	- 1 172	- 501	662	320	98	245	Tajikistan
-	-	- 4 295	- 3 186	- 2 650	- 2 020	- 2 800	- 3 461	- 3 109	Turkmenistan
-	-	- 5 085	- 1 543	- 1 042	- 2 952	- 2 017	- 154	- 672	Ukraine
-	-	- 490	- 407	- 400	- 1 290	1 092	929	900	Uzbekistan
- 35 000	**- 21 629**	**- 11 137**	**- 14 875**	**- 7 281**	**- 10 249**	**- 10 501**	**- 11 840**	**- 14 036**	**Former USSR**
- 456	- 744	- 615	- 630	- 735	- 914	- 979	- 1 170	- 1 055	Israel
-	-	- 67	- 46	-	-	- 2	- 3	- 4	Jordan
-	-	-	-	-	292	683	608	654	Lebanon
-	-	-	-	-	-	-	-	-	Syria
- 456	**- 744**	**- 682**	**- 676**	**- 735**	**- 622**	**- 298**	**- 565**	**- 405**	**Middle East**
- 18 336	**- 11 775**	**- 3 763**	**- 13 084**	**- 1 552**	**- 5 880**	**- 6 420**	**- 11 183**	**- 15 356**	**Non-OECD Total**
262	2 443	2 247	- 96	- 106	812	452	- 212	850	OECD North America
17 751	11 205	2 042	4 473	3 313	2 792	5 444	5 844	8 597	OECD Europe
18 013	**13 648**	**4 289**	**4 377**	**3 207**	**3 604**	**5 896**	**5 632**	**9 447**	**OECD Total**
- 323	**1 873**	**526**	**- 8 707**	**1 655**	**- 2 276**	**- 524**	**- 5 551**	**- 5 909**	**World**

A negative number shows net exports.

The row World shows the discrepancy between total world exports and imports.

Former USSR: data for individual republics may not add to the total.

Final Consumption of Coking Coal (1000 tonnes)
Consommation finale de charbon à coke (1000 tonnes)
Endverbrauch von Kokskohle (1000 Tonnen)
Consumo finale di carbone siderurgico (1000 tonnellate)
原料炭の最終消費量（千トン）
Consumo final de carbón coquizable (1000 toneladas)
Конечное потребление коксующихся углей (тыс. т)

	1971	1973	1978	1984	1985	1986	1987	1988	1989
Brésil	-	-	-	-	-	-	-	-	-
Jamaïque	-	-	-	-	-	-	-	8	56
Amérique latine	-	-	-	-	-	-	-	**8**	**56**
Inde	-	-	8 921	13 769	15 215	16 009	17 838	17 714	19 742
Asie	-	-	**8 921**	**13 769**	**15 215**	**16 009**	**17 838**	**17 714**	**19 742**
Roumanie	-	-	-	-	-	-	-	-	-
République slovaque	-	-	27	31	32	31	32	34	38
Ex-RYM	-	-	-	-	-	-	-	-	-
Ex-Yougoslavie	-	-	-	-	-	-	-	-	-
Europe non-OCDE	-	-	**27**	**31**	**32**	**31**	**32**	**34**	**38**
Kazakhstan	-	-	-	-	-	-	-	-	-
Tadjikistan	-	-	-	-	-	-	-	-	-
Ex-URSS	-	-	-	-	-	-	-	-	-
Total non-OCDE	-	-	**8 948**	**13 800**	**15 247**	**16 040**	**17 870**	**17 756**	**19 836**
OCDE Pacifique	-	-	121	38	264	749	742	580	914
OCDE Europe	-	-	801	808	2 748	1 681	1 502	1 755	1 462
Total OCDE	-	-	**922**	**846**	**3 012**	**2 430**	**2 244**	**2 335**	**2 376**
Monde	-	-	**9 870**	**14 646**	**18 259**	**18 470**	**20 114**	**20 091**	**22 212**

Ex-URSS: les séries antérieures à 1990-1992 ne sont pas comparables aux années récentes; 1991 a été estimé.

Final Consumption of Coking Coal (1000 tonnes)
Consommation finale de charbon à coke (1000 tonnes)
Endverbrauch von Kokskohle (1000 Tonnen)
Consumo finale di carbone siderurgico (1000 tonnellate)
原料炭の最終消費量（千トン）
Consumo final de carbón coquizable (1000 toneladas)
Конечное потребление коксующихся углей (тыс. т)

1990	1991	1992	1993	1994	1995	1996	1997	1998	
-	-	-	235	354	834	1 596	2 234	2 706	Brazil
62	9	65	71	53	55	64	56	27	Jamaica
62	**9**	**65**	**306**	**407**	**889**	**1 660**	**2 290**	**2 733**	**Latin America**
18 700	20 110	20 150	21 740	22 650	18 320	17 643	21 729	22 689	India
18 700	**20 110**	**20 150**	**21 740**	**22 650**	**18 320**	**17 643**	**21 729**	**22 689**	**Asia**
-	-	915	37	6	121	30	2	1	Romania
6	28	-	166	323	314	628	295	312	Slovak Republic
-	-	73	74	66	70	49	49	113	*FYROM*
-	77	73	74	66	70	49	49	113	Former Yugoslavia
6	**105**	**988**	**277**	**395**	**505**	**707**	**346**	**426**	**Non-OECD Europe**
-	-	20 855	15 206	12 116	9 204	9 503	9 199	6 281	Kazakhstan
-	-	-	-	3	7	-	7	6	Tajikistan
-	**-**	**20 855**	**15 206**	**12 119**	**9 211**	**9 503**	**9 206**	**6 287**	**Former USSR**
18 768	20 224	42 058	37 529	35 571	28 925	29 513	33 571	32 135	**Non-OECD Total**
1 298	1 699	1 338	1 584	1 860	1 922	1 913	2 190	2 533	OECD Pacific
2 597	3 084	2 978	2 882	2 716	1 606	2 018	2 848	2 486	OECD Europe
3 895	**4 783**	**4 316**	**4 466**	**4 576**	**3 528**	**3 931**	**5 038**	**5 019**	**OECD Total**
22 663	25 007	46 374	41 995	40 147	32 453	33 444	38 609	37 154	**World**

Former USSR: series up to 1990-1992 are not comparable with recent years; 1991 is estimated.

Final Consumption of Other Bituminous Coal (1000 tonnes)
Consommation finale d'autres charbons bitumineux (1000 tonnes)
Endverbrauch von sonstiger bituminöser Kohle (1000 Tonnen)
Consumo finale di altri carboni bituminosi (1000 tonnellate)
その他歴青炭の最終消費量（チトン）
Consumo final de otro carbón bituminoso (1000 toneladas)
Конечное потребление других битуминозных углей (тыс. т)

	1971	1973	1978	1984	1985	1986	1987	1988	1989
RD du Congo	-	-	167	157	156	160	165	166	170
Ghana	-	-	2	2	2	2	3	3	3
Kenya	-	-	50	83	90	85	92	113	131
Maroc	-	-	28	282	523	495	439	447	772
Mozambique	-	-	247	94	65	39	32	28	34
Nigéria	-	-	159	64	91	100	96	34	36
Afrique du Sud	-	-	24 428	20 836	21 288	20 111	20 631	21 596	22 385
Rép. Unie de Tanzanie	-	-	1	10	15	4	3	3	2
Tunisie	-	-	27	24	21	12	16	29	30
Zambie	-	-	450	380	401	444	500	541	312
Zimbabwe	-	-	1 110	1 638	1 706	1 765	1 786	1 733	1 641
Autre Afrique	-	-	261	298	338	426	431	435	436
Afrique	**-**	**-**	**26 930**	**23 868**	**24 696**	**23 643**	**24 194**	**25 128**	**25 952**
Argentine	-	-	34	131	14	17	7	29	17
Brésil	-	-	479	3 163	3 427	3 726	3 911	3 380	2 705
Chili	-	-	527	559	632	724	667	732	802
Colombie	-	-	1 994	1 836	2 043	2 139	2 342	2 120	2 415
Costa Rica	-	-	1	1	1	1	1	1	1
Cuba	-	-	91	81	126	119	85	95	167
Haiti	-	-	-	54	61	18	18	31	12
Panama	-	-	-	-	38	38	38	18	19
Pérou	-	-	26	89	114	107	100	80	126
Uruguay	-	-	2	-	-	-	-	-	-
Vénézuela	-	-	63	51	42	57	62	71	413
Amérique latine	**-**	**-**	**3 217**	**5 965**	**6 498**	**6 946**	**7 231**	**6 557**	**6 677**
Bangladesh	-	-	250	62	98	148	233	240	250
Inde	-	-	42 867	36 224	37 542	39 632	42 386	46 585	48 668
Indonésie	-	-	137	302	331	290	300	760	1 517
RPD de Corée	-	-	23 616	27 408	29 307	31 839	35 386	41 238	43 252
Malaisie	-	-	33	385	517	383	467	270	849
Myanmar	-	-	218	204	203	206	59	52	51
Népal	-	-	16	145	17	92	85	77	12
Pakistan	-	-	1 241	1 845	2 206	2 176	2 242	2 730	2 602
Philippines	-	-	171	813	954	743	958	1 155	724
Sri Lanka	-	-	-	-	-	-	-	-	1
Taipei chinois	-	-	1 956	3 122	3 220	3 441	3 643	4 385	4 170
Thailande	-	-	52	227	212	183	223	304	392
Viêt-Nam	-	-	3 172	3 244	3 208	3 518	3 882	2 884	2 642
Autre Asie	-	-	286	273	345	351	343	334	348
Asie	**-**	**-**	**74 015**	**74 254**	**78 160**	**83 002**	**90 207**	**101 014**	**105 478**
Rép. populaire de Chine	-	-	424 824	504 779	537 425	566 657	591 973	613 526	616 531
Hong-Kong, Chine	-	-	6	-	-	-	-	-	-
Chine	**-**	**-**	**424 830**	**504 779**	**537 425**	**566 657**	**591 973**	**613 526**	**616 531**

Final Consumption of Other Bituminous Coal (1000 tonnes)
Consommation finale d'autres charbons bitumineux (1000 tonnes)
Endverbrauch von sonstiger bituminöser Kohle (1000 Tonnen)
Consumo finale di altri carboni bituminosi (1000 tonnellate)
その他歴青炭の最終消費量（千トン）
Consumo final de otro carbón bituminoso (1000 toneladas)
Конечное потребление других битуминозных углей (тыс. т)

1990	1991	1992	1993	1994	1995	1996	1997	1998	
169	174	170	134	136	136	136	136	128	DR of Congo
3	3	3	3	3	3	3	3	-	Ghana
151	151	158	131	123	156	144	148	122	Kenya
574	798	605	505	526	289	874	672	722	Morocco
34	32	41	54	60	56	-	-	-	Mozambique
46	46	55	60	70	10	10	10	10	Nigeria
21 690	20 333	19 038	23 691	23 904	30 977	29 432	30 945	33 551	South Africa
3	3	3	3	3	4	4	4	4	United Rep. of Tanzania
15	12	15	14	-	-	-	-	-	Tunisia
297	234	267	200	119	124	118	118	118	Zambia
2 056	2 040	1 988	1 918	943	1 033	1 142	1 052	1 056	Zimbabwe
400	223	364	360	149	133	134	135	135	Other Africa
25 438	**24 049**	**22 707**	**27 073**	**26 036**	**32 921**	**31 997**	**33 223**	**35 846**	**Africa**
10	25	92	75	32	51	-	24	25	Argentina
2·189	2 851	2 111	1 802	1 893	1 463	1 277	1 034	653	Brazil
733	788	900	790	628	584	763	1 483	881	Chile
2 364	2 437	2 482	3 154	3 189	2 784	2 943	3 021	2 325	Colombia
1	-	-	-	-	-	-	-	-	Costa Rica
153	117	107	86	70	63	12	11	14	Cuba
12	25	26	-	-	-	-	-	-	Haiti
32	46	58	63	52	51	108	57	59	Panama
107	229	236	401	421	376	246	377	426	Peru
1	-	-	-	-	-	-	-	-	Uruguay
332	-	5	39	76	7	22	47	1 327	Venezuela
5 934	**6 518**	**6 017**	**6 410**	**6 361**	**5 379**	**5 371**	**6 054**	**5 710**	**Latin America**
563	180	169	63	59	59	345	635	173	Bangladesh
51 409	56 210	54 440	51 800	61 350	78 320	67 847	61 561	48 761	India
1 804	1 404	1 555	1 737	2 209	1 868	5 122	3 362	4 714	Indonesia
45 266	46 628	47 319	47 991	47 674	47 357	46 966	43 773	41 584	DPR of Korea
732	855	959	1 016	854	1 016	1 038	1 056	1 095	Malaysia
60	61	31	6	4	15	9	11	12	Myanmar
82	92	111	104	115	123	274	297	323	Nepal
3 103	3 030	3 059	3 220	3 490	3 002	3 239	3 201	2 812	Pakistan
795	1 243	1 004	1 093	961	907	1 670	2 184	1 987	Philippines
8	1	1	1	1	5	1	1	1	Sri Lanka
4 308	4 601	5 325	5 768	5 981	5 795	6 145	5 667	6 326	Chinese Taipei
250	436	489	986	1 424	2 305	3 723	3 036	1 334	Thailand
2 834	3 295	3 227	3 236	3 196	3 706	4 133	3 958	3 090	Vietnam
307	307	217	217	212	213	218	228	228	Other Asia
111 521	**118 343**	**117 906**	**117 238**	**127 530**	**144 691**	**140 730**	**128 970**	**112 440**	**Asia**
597 762	588 675	567 033	582 441	588 573	601 182	551 609	481 118	427 517	People's Rep. of China
-	-	-	-	-	-	-	-	-	Hong Kong, China
597 762	**588 675**	**567 033**	**582 441**	**588 573**	**601 182**	**551 609**	**481 118**	**427 517**	**China**

Final Consumption of Other Bituminous Coal (1000 tonnes)
Consommation finale d'autres charbons bitumineux (1000 tonnes)
Endverbrauch von sonstiger bituminöser Kohle (1000 Tonnen)

Consumo finale di altri carboni bituminosi (1000 tonnellate)

その他歴青炭の最終消費量（チトン）

Consumo final de otro carbón bituminoso (1000 toneladas)

Конечное потребление других битуминозных углей (тыс. т)

	1971	1973	1978	1984	1985	1986	1987	1988	1989
Albanie	-	-	143	178	187	195	195	203	203
Bulgarie	-	-	2 253	2 722	4 004	2 804	3 014	70	31
Chypre	-	-	-	52	74	55	151	91	102
Roumanie	-	-	1 572	282	2 759	3 134	4 213	5 174	700
République slovaque	-	-	1 312	1 189	1 181	1 209	1 212	1 194	1 260
Croatie	-	-	-	-	-	-	-	-	-
Ex-RYM	-	-	-	-	-	-	-	-	-
Slovénie	-	-	-	-	-	-	-	-	-
RF de Yougoslavie	-	-	-	-	-	-	-	-	-
Ex-Yougoslavie	-	-	572	288	89	259	236	75	247
Europe non-OCDE	-	-	**5 852**	**4 711**	**8 294**	**7 656**	**9 021**	**6 807**	**2 543**
Arménie	-	-	-	-	-	-	-	-	-
Azerbaïdjan	-	-	-	-	-	-	-	-	-
Bélarus	-	-	-	-	-	-	-	-	-
Estonie	-	-	-	-	-	-	-	-	-
Géorgie	-	-	-	-	-	-	-	-	-
Kazakhstan	-	-	-	-	-	-	-	-	-
Kirghizistan	-	-	-	-	-	-	-	-	-
Lettonie	-	-	-	-	-	-	-	-	-
Lituanie	-	-	-	-	-	-	-	-	-
République de Moldavie	-	-	-	-	-	-	-	-	-
Russie	-	-	-	-	-	-	-	-	-
Tadjikistan	-	-	-	-	-	-	-	-	-
Turkménistan	-	-	-	-	-	-	-	-	-
Ukraine	-	-	-	-	-	-	-	-	-
Ouzbékistan	-	-	-	-	-	-	-	-	-
Ex-URSS	-	-	**243 465**	**216 687**	**206 551**	**196 377**	**206 945**	**212 228**	**209 517**
République Islamique d'Iran	-	-	523	921	928	982	1 010	1 030	1 050
Israël	-	-	-	-	-	-	162	137	160
Liban	-	-	1	-	-	-	-	-	-
Syrie	-	-	1	2	-	-	-	-	-
Moyen-Orient	-	-	**525**	**923**	**928**	**982**	**1 172**	**1 167**	**1 210**
Total non-OCDE	-	-	**778 834**	**831 187**	**862 552**	**885 263**	**930 743**	**966 427**	**967 908**
OCDE Amérique du N.	-	-	62 375	67 152	64 955	64 207	62 667	63 732	61 402
OCDE Pacifique	-	-	23 219	40 677	44 024	45 145	43 080	44 573	41 814
OCDE Europe	-	-	90 885	94 980	103 441	101 766	104 163	100 326	92 111
Total OCDE	-	-	**176 479**	**202 809**	**212 420**	**211 118**	**209 910**	**208 631**	**195 327**
Monde	-	-	**955 313**	**1 033 996**	**1 074 972**	**1 096 381**	**1 140 653**	**1 175 058**	**1 163 235**

Ex-URSS: les séries antérieures à 1990-1992 ne sont pas comparables aux années récentes; 1991 a été estimé.

Final Consumption of Other Bituminous Coal (1000 tonnes)
Consommation finale d'autres charbons bitumineux (1000 tonnes)
Endverbrauch von sonstiger bituminöser Kohle (1000 Tonnen)
Consumo finale di altri carboni bituminosi (1000 tonnellate)
その他歴青炭の最終消費量（千トン）
Consumo final de otro carbón bituminoso (1000 toneladas)
Конечное потребление других битуминозных углей (тыс. m)

1990	1991	1992	1993	1994	1995	1996	1997	1998	
240	160	50	-	-	-	-	-	-	Albania
60	123	272	178	209	135	125	343	253	Bulgaria
97	97	26	31	27	20	17	19	26	Cyprus
648	1 434	606	108	105	18	15	20	14	Romania
1 077	1 196	1 106	1 046	988	1 033	684	840	594	Slovak Republic
-	-	83	68	60	46	61	56	63	*Croatia*
-	-	12	68	61	72	51	51	95	*FYROM*
-	-	13	17	22	20	26	19	14	*Slovenia*
-	-	1 070	121	120	119	126	130	141	*FR of Yugoslavia*
212	110	1 178	274	263	257	264	256	313	Former Yugoslavia
2 334	**3 120**	**3 238**	**1 637**	**1 592**	**1 463**	**1 105**	**1 478**	**1 200**	**Non-OECD Europe**
-	-	141	3	36	3	5	5	5	Armenia
-	-	27	7	8	6	6	6	1	Azerbaijan
-	-	1 112	1 182	894	808	778	557	506	Belarus
-	-	113	33	9	54	88	71	57	Estonia
-	-	406	288	152	71	72	13	19	Georgia
-	-	7 827	9 450	7 921	12 600	9 906	7 608	8 114	Kazakhstan
-	-	990	893	1 520	395	581	1 390	649	Kyrgyzstan
-	-	233	294	312	169	163	156	124	Latvia
-	-	563	502	414	318	303	230	191	Lithuania
-	-	640	487	435	397	348	244	177	Republic of Moldova
-	-	54 573	60 494	46 277	39 340	32 825	23 759	22 029	Russia
-	-	413	176	-	-	100	100	100	Tajikistan
-	-	600	135	-	-	100	-	-	Turkmenistan
-	-	28 877	33 879	22 955	18 274	13 541	12 982	13 091	Ukraine
-	-	1 381	688	788	42	36	25	23	Uzbekistan
194 312	**149 208**	**97 896**	**108 511**	**81 721**	**72 477**	**58 852**	**47 146**	**45 086**	**Former USSR**
1 250	1 350	1 500	607	632	106	148	152	158	Islamic Republic of Iran
18	11	11	10	8	12	26	1	1	Israel
-	-	-	111	112	180	200	200	200	Lebanon
-	-	-	-	-	-	-	-	-	Syria
1 268	**1 361**	**1 511**	**728**	**752**	**298**	**374**	**353**	**359**	**Middle East**
938 569	**891 274**	**816 308**	**844 038**	**832 565**	**858 411**	**790 038**	**698 342**	**628 158**	**Non-OECD Total**
61 975	60 340	29 356	31 717	28 726	27 035	26 918	31 685	30 763	OECD North America
40 563	37 686	34 090	30 959	27 560	26 779	26 020	25 510	23 167	OECD Pacific
74 818	78 103	72 459	74 486	66 907	70 642	69 123	64 602	55 280	OECD Europe
177 356	**176 129**	**135 905**	**137 162**	**123 193**	**124 456**	**122 061**	**121 797**	**109 210**	**OECD Total**
1 115 925	**1 067 403**	**952 213**	**981 200**	**955 758**	**982 867**	**912 099**	**820 139**	**737 368**	**World**

Former USSR: series up to 1990-1992 are not comparable with recent years; 1991 is estimated.

Final Consumption of Sub-Bituminous Coal (1000 tonnes)
Consommation finale de charbons sous-bitumineux (1000 tonnes)
Endverbrauch von subbituminöser Kohle (1000 Tonnen)
Consumo finale di carbone sub-bituminoso (1000 tonnellate)
亜歴青炭の最終消費量 （チトン）
Consumo final de carbón sub-bituminoso (1000 toneladas)
Конечное потребление полубитуминозных углей (тыс. m)

	1971	1973	1978	1984	1985	1986	1987	1988	1989
RPD de Corée	-	-	9 000	12 000	13 000	14 000	15 000	18 000	20 000
Asie	-	-	9 000	12 000	13 000	14 000	15 000	18 000	20 000
Bulgarie	-	-	-	-	-	-	-	-	-
Roumanie	-	-	-	-	-	-	-	-	-
Croatie	-	-	-	-	-	-	-	-	-
Slovénie	-	-	-	-	-	-	-	-	-
Ex-Yougoslavie	-	-	-	-	-	-	-	-	-
Europe non-OCDE	-	-	-	-	-	-	-	-	-
Tadjikistan	-	-	-	-	-	-	-	-	-
Ex-URSS	-	-	-	-	-	-	-	-	-
Total non-OCDE	-	-	9 000	12 000	13 000	14 000	15 000	18 000	20 000
OCDE Amérique du N.	-	-	2 810	5 137	5 240	5 218	5 313	5 663	6 253
OCDE Pacifique	-	-	2 182	2 861	3 111	3 223	3 265	3 208	3 593
OCDE Europe	-	-	44 454	42 236	41 099	42 220	45 288	42 605	35 531
Total OCDE	-	-	49 446	50 234	49 450	50 661	53 866	51 476	45 377
Monde	-	-	58 446	62 234	62 450	64 661	68 866	69 476	65 377

Ex-URSS: les séries antérieures à 1990-1992 ne sont pas comparables aux années récentes; 1991 a été estimé.

Final Consumption of Sub-Bituminous Coal (1000 tonnes)
Consommation finale de charbons sous-bitumineux (1000 tonnes)
Endverbrauch von subbituminöser Kohle (1000 Tonnen)
Consumo finale di carbone sub-bituminoso (1000 tonnellate)
亜歴青炭の最終消費量 （チトン）
Consumo final de carbón sub-bituminoso (1000 toneladas)
Конечное потребление полубитуминозных углей (тыс. т)

1990	1991	1992	1993	1994	1995	1996	1997	1998	
22 000	23 000	24 000	27 000	26 500	26 000	25 500	23 766	22 578	DPR of Korea
22 000	**23 000**	**24 000**	**27 000**	**26 500**	**26 000**	**25 500**	**23 766**	**22 578**	**Asia**
-	510	548	527	420	388	588	-	-	Bulgaria
-	-	-	-	-	86	95	100	20	Romania
-	-	-	-	-	-	-	76	57	*Croatia*
-	-	242	199	128	69	48	30	23	*Slovenia*
-	-	242	199	128	69	48	106	80	Former Yugoslavia
-	**510**	**790**	**726**	**548**	**543**	**731**	**206**	**100**	**Non-OECD Europe**
-	-	-	-	103	27	20	20	20	Tajikistan
-	**-**	**-**	**-**	**103**	**27**	**20**	**20**	**20**	**Former USSR**
22 000	**23 510**	**24 790**	**27 726**	**27 151**	**26 570**	**26 251**	**23 992**	**22 698**	**Non-OECD Total**
6 713	7 102	3 988	4 122	3 493	3 724	3 727	6 474	6 774	OECD North America
3 972	4 065	4 350	4 454	4 286	4 235	4 242	4 046	3 993	OECD Pacific
30 352	23 948	17 831	8 407	6 830	7 716	6 507	5 561	4 243	OECD Europe
41 037	**35 115**	**26 169**	**16 983**	**14 609**	**15 675**	**14 476**	**16 081**	**15 010**	**OECD Total**
63 037	**58 625**	**50 959**	**44 709**	**41 760**	**42 245**	**40 727**	**40 073**	**37 708**	**World**

Former USSR: series up to 1990-1992 are not comparable with recent years; 1991 is estimated.

Final Consumption of Lignite (1000 tonnes)
Consommation finale de lignite (1000 tonnes)
Endverbrauch von Braunkohle (1000 Tonnen)

Consumo finale di lignite (1000 tonnellate)

亜炭の最終消費量（千トン）

Consumo final de lignito (1000 toneladas)

Конечное потребление лигнита (тыс. т)

	1971	1973	1978	1984	1985	1986	1987	1988	1989
Inde	-	-	601	504	1 024	639	1 754	1 853	3 287
Myanmar	-	-	3	43	43	47	35	35	35
Philippines	-	-	-	3	4	4	3	3	3
Thailande	-	-	132	361	535	741	1 096	1 303	1 794
Asie	-	-	**736**	**911**	**1 606**	**1 431**	**2 888**	**3 194**	**5 119**
Albanie	-	-	1 014	1 719	1 859	1 811	1 911	1 910	1 910
Bulgarie	-	-	3 655	4 499	4 292	4 902	5 000	2 800	2 505
Roumanie	-	-	4 015	7 990	5 426	2 883	5 696	2 632	5 079
République slovaque	-	-	6 672	6 547	6 926	7 232	7 256	7 076	8 506
Bosnie-Herzegovine	-	-	-	-	-	-	-	-	-
Croatie	-	-	-	-	-	-	-	-	-
Ex-RYM	-	-	-	-	-	-	-	-	-
Slovénie	-	-	-	-	-	-	-	-	-
RF de Yougoslavie	-	-	-	-	-	-	-	-	-
Ex-Yougoslavie	-	-	9 084	8 297	8 292	6 635	6 865	3 514	4 907
Europe non-OCDE	-	-	**24 440**	**29 052**	**26 795**	**23 463**	**26 728**	**17 932**	**22 907**
Estonie	-	-	-	-	-	-	-	-	-
Kazakhstan	-	-	-	-	-	-	-	-	-
Kirghizistan	-	-	-	-	-	-	-	-	-
Lituanie	-	-	-	-	-	-	-	-	-
Russie	-	-	-	-	-	-	-	-	-
Tadjikistan	-	-	-	-	-	-	-	-	-
Ukraine	-	-	-	-	-	-	-	-	-
Ouzbékistan	-	-	-	-	-	-	-	-	-
Ex-URSS	-	-	-	-	-	-	-	-	-
Total non-OCDE	-	-	**25 176**	**29 963**	**28 401**	**24 894**	**29 616**	**21 126**	**28 026**
OCDE Amérique du N.	-	-	2 665	2 912	2 308	2 552	2 590	2 709	2 767
OCDE Pacifique	-	-	477	509	532	473	340	520	327
OCDE Europe	-	-	40 229	60 273	67 183	64 516	68 361	64 780	61 928
Total OCDE	-	-	**43 371**	**63 694**	**70 023**	**67 541**	**71 291**	**68 009**	**65 022**
Monde	-	-	**68 547**	**93 657**	**98 424**	**92 435**	**100 907**	**89 135**	**93 048**

Ex-URSS: les séries antérieures à 1990-1992 ne sont pas comparables aux années récentes; 1991 a été estimé.

Final Consumption of Lignite (1000 tonnes)
Consommation finale de lignite (1000 tonnes)
Endverbrauch von Braunkohle (1000 Tonnen)

Consumo finale di lignite (1000 tonnellate)

亜炭の最終消費量（千トン）

Consumo final de lignito (1000 toneladas)

Конечное потребление лигнита (тыс. т)

1990	1991	1992	1993	1994	1995	1996	1997	1998	
700	700	700	700	700	700	700	700	700	India
38	39	39	25	22	23	22	20	21	Myanmar
3	3	3	3	3	3	3	3	3	Philippines
2 582	2 818	3 147	4 263	4 916	4 915	4 551	4 470	5 156	Thailand
3 323	**3 560**	**3 889**	**4 991**	**5 641**	**5 641**	**5 276**	**5 193**	**5 880**	**Asia**
1 395	631	350	242	169	163	101	70	49	Albania
963	200	218	215	141	123	184	534	385	Bulgaria
2 267	1 531	1 398	1 034	477	297	735	563	491	Romania
7 983	7 451	6 239	5 364	3 855	3 352	3 354	2 373	1 740	Slovak Republic
-	-	3 000	3 000	337	290	299	308	318	*Bosnia-Herzegovina*
-	-	195	211	172	177	143	40	37	*Croatia*
-	-	162	201	213	160	158	150	147	*FYROM*
-	-	366	245	180	170	166	65	44	*Slovenia*
-	-	5 829	3 314	3 389	3 539	3 367	3 563	3 862	*FR of Yugoslavia*
5·163	4 626	9 552	6 971	4 291	4 336	4 133	4 126	4 408	Former Yugoslavia
17 771	**14 439**	**17 757**	**13 826**	**8 933**	**8 271**	**8 507**	**7 666**	**7 073**	**Non-OECD Europe**
-	-	337	264	439	593	596	434	490	Estonia
-	-	4 152	4 669	4 559	3 630	3 401	2 309	1 620	Kazakhstan
-	-	1 111	985	550	287	332	330	364	Kyrgyzstan
-	-	-	-	-	-	-	6	3	Lithuania
-	-	15 139	13 162	9 327	7 933	8 454	5 232	4 755	Russia
-	-	200	200	-	-	-	-	-	Tajikistan
-	-	4 779	3 121	2 269	2 000	2 000	1 000	1 000	Ukraine
-	-	575	795	704	846	841	460	484	Uzbekistan
-	**-**	**26 293**	**23 196**	**17 848**	**15 289**	**15 624**	**9 771**	**8 716**	**Former USSR**
21 094	**17 999**	**47 939**	**42 013**	**32 422**	**29 201**	**29 407**	**22 630**	**21 669**	**Non-OECD Total**
2 452	2 415	300	531	314	303	335	671	567	OECD North America
248	263	271	281	302	304	305	328	325	OECD Pacific
49 774	37 422	29 011	23 808	18 385	18 346	17 475	17 869	15 550	OECD Europe
52 474	**40 100**	**29 582**	**24 620**	**19 001**	**18 953**	**18 115**	**18 868**	**16 442**	**OECD Total**
73 568	**58 099**	**77 521**	**66 633**	**51 423**	**48 154**	**47 522**	**41 498**	**38 111**	**World**

Former USSR: series up to 1990-1992 are not comparable with recent years; 1991 is estimated.

Final Consumption of Oil (1000 tonnes)
Consommation finale de pétrole (1000 tonnes)
Endverbrauch von Öl (1000 Tonnen)
Consumo finale di petrolio (1000 tonnellate)
石油の最終消費量（千トン）
Consumo final de petróleo (1000 toneladas)
Конечное потребление нефти и нефтепродуктов (тыс. т)

	1971	1973	1978	1984	1985	1986	1987	1988	1989
Algérie	1 607	2 283	3 701	6 698	6 975	7 170	7 293	7 216	7 482
Angola	523	683	654	752	1 066	1 042	1 081	1 168	1 187
Bénin	101	126	127	126	160	140	125	122	102
Cameroun	276	301	511	780	841	818	815	866	948
Congo	158	164	212	237	246	243	221	200	264
RD du Congo	628	700	749	972	907	794	925	987	975
Côte d'Ivoire	659	694	937	837	819	878	915	896	921
Egypte	5 054	5 286	7 657	13 954	12 536	12 045	12 842	12 764	13 253
Ethiopie	361	379	390	462	474	576	698	720	763
Erythrée	-	-	-	-	-	-	-	-	-
Gabon	135	170	566	546	561	506	434	478	361
Ghana	642	675	825	624	676	699	717	884	935
Kenya	1 037	1 192	1 498	1 582	1 643	1 743	1 805	1 787	1 870
Libye	438	891	1 897	3 407	3 334	3 154	3 279	3 154	3 593
Maroc	1 742	2 134	3 058	3 105	3 121	3 088	3 059	3 224	3 571
Mozambique	349	359	331	279	310	331	334	337	322
Nigéria	1 496	2 353	6 243	7 674	7 655	7 102	7 239	7 624	7 878
Sénégal	369	437	484	567	563	537	508	490	536
Afrique du Sud	8 653	10 213	11 691	12 992	12 589	12 380	13 006	14 099	14 461
Soudan	1 070	1 529	959	1 143	1 282	1 254	970	1 373	1 283
Rép. Unie de Tanzanie	433	507	496	408	429	450	450	488	496
Togo	65	63	81	83	83	133	142	136	143
Tunisie	892	1 062	1 624	2 149	2 196	2 105	1 958	2 201	2 455
Zambie	462	637	687	571	540	477	491	501	493
Zimbabwe	539	680	658	707	758	797	807	839	785
Autre Afrique	2 263	2 535	2 634	2 641	2 750	2 816	4 113	4 181	4 409
Afrique	**29 952**	**36 053**	**48 670**	**63 296**	**62 514**	**61 278**	**64 227**	**66 735**	**69 486**
Argentine	16 212	16 756	17 523	17 566	16 201	15 250	16 629	15 643	14 892
Bolivie	551	653	976	854	831	848	912	923	979
Brésil	23 976	32 414	46 592	40 630	43 714	46 531	48 261	49 293	50 766
Chili	3 795	3 820	3 816	3 855	3 765	4 010	4 176	4 804	5 479
Colombie	5 409	5 836	6 395	7 186	7 268	7 269	7 779	8 301	8 675
Costa Rica	354	435	660	565	611	629	665	692	821
Cuba	4 235	5 605	6 168	6 435	6 929	6 766	6 974	7 202	7 578
République dominicaine	750	877	1 080	1 177	982	1 121	1 328	1 396	1 599
Equateur	1 000	1 133	2 372	3 266	3 614	3 783	3 687	4 060	3 974
El Salvador	348	434	614	482	506	506	545	616	629
Guatemala	643	711	891	770	790	790	871	931	977
Haiti	107	121	204	166	169	185	193	205	229
Honduras	297	354	461	534	526	541	599	664	741
Jamaïque	1 408	2 068	1 952	1 353	1 202	1 143	1 171	1 188	1 565
Antilles néerlandaises	2 159	2 248	1 234	933	860	773	752	693	624
Nicaragua	390	497	545	483	480	522	551	488	455
Panama	478	523	618	568	582	618	613	533	554
Paraguay	170	218	366	464	490	508	535	554	624
Pérou	4 073	4 280	4 855	4 685	4 656	5 032	5 565	5 461	4 874
Trinité-et-Tobago	407	515	758	868	820	755	699	678	612
Uruguay	1 325	1 330	1 354	976	908	922	992	1 010	1 026
Vénézuela	5 921	7 015	10 739	11 855	12 266	12 673	13 102	13 666	13 049
Autre Amérique latine	1 685	2 113	1 904	1 822	1 718	1 862	2 224	2 150	2 466
Amérique latine	**75 693**	**89 956**	**112 077**	**107 493**	**109 888**	**113 037**	**118 823**	**121 151**	**123 188**

Final Consumption of Oil (1000 tonnes)
Consommation finale de pétrole (1000 tonnes)
Endverbrauch von Öl (1000 Tonnen)
Consumo finale di petrolio (1000 tonnellate)
石油の最終消費量（千トン）
Consumo final de petróleo (1000 toneladas)
Конечное потребление нефти и нефтепродуктов (тыс. т)

1990	1991	1992	1993	1994	1995	1996	1997	1998	
7 648	8 008	7 903	8 000	7 401	7 141	6 990	7 222	7 081	Algeria
1 140	1 103	1 069	1 229	1 318	1 172	1 198	1 220	1 007	Angola
89	70	71	75	77	81	316	308	302	Benin
921	850	798	845	954	914	968	1 013	1 033	Cameroon
254	264	249	257	251	274	278	278	257	Congo
1 057	1 083	1 000	1 122	1 140	1 154	1 154	1 154	1 088	DR of Congo
790	813	764	665	707	800	866	908	962	Cote d'Ivoire
13 780	13 433	13 277	12 456	12 173	14 000	14 346	15 006	16 817	Egypt
796	817	499	635	743	832	890	941	1 002	Ethiopia
-	-	86	144	159	176	215	229	183	Eritrea
290	334	346	384	375	411	444	517	527	Gabon
897	855	919	986	1 067	1 133	1 216	1 285	1 396	Ghana
1 822	1 728	1 798	1 825	1 920	1 968	2 086	1 998	2 034	Kenya
3 592	3 757	3 742	4 212	4 883	5 359	5 421	5 591	5 737	Libya
3 717	3 855	4 336	4 331	4 720	4 818	4 825	5 087	5 301	Morocco
329	264	317	384	333	335	322	349	391	Mozambique
6 534	7 965	11 508	9 027	7 376	8 293	9 247	10 176	9 833	Nigeria
558	548	577	513	564	626	671	721	824	Senegal
14 874	14 823	14 670	14 098	14 837	16 139	16 237	16 509	16 778	South Africa
1 659	1 470	1 393	864	1 417	1 256	1 155	1 428	1 350	Sudan
521	514	508	467	471	464	454	454	463	United Rep. of Tanzania
159	131	117	38	143	188	267	242	266	Togo
2 497	2 412	2 560	2 834	2 834	2 881	3 043	3 127	3 303	Tunisia
473	485	452	466	481	485	491	491	491	Zambia
917	987	962	1 004	1 112	1 254	1 248	1 248	1 248	Zimbabwe
4 235	4 434	4 589	4 761	4 755	4 837	4 891	5 073	5 073	Other Africa
69 549	**71 003**	**74 510**	**71 622**	**72 211**	**76 991**	**79 239**	**82 575**	**84 747**	**Africa**
15 161	15 601	16 279	17 090	18 162	18 646	19 421	20 295	20 654	Argentina
1 012	1 038	1 046	1 079	1 134	1 255	1 356	1 505	1 555	Bolivia
51 829	52 601	53 758	55 751	58 818	63 243	67 802	73 402	75 987	Brazil
5 717	5 893	6 534	7 101	7 573	8 213	8 785	9 816	9 732	Chile
8 908	8 958	9 700	10 522	11 196	12 076	12 429	13 057	12 855	Colombia
816	838	1 110	1 162	1 274	1 280	1 278	1 348	1 586	Costa Rica
6 688	4 878	4 604	4 471	4 734	5 035	5 456	5 553	5 300	Cuba
1 579	1 532	1 909	1 769	1 885	1 958	2 032	2 182	2 251	Dominican Republic
3 991	4 029	3 827	3 789	4 040	4 489	4 425	4 693	4 625	Ecuador
673	716	875	933	994	1 105	1 072	1 183	1 291	El Salvador
1 045	1 089	1 169	1 278	1 346	1 623	1 618	1 716	1 929	Guatemala
239	222	217	200	52	260	279	321	361	Haiti
696	697	782	808	869	980	913	899	1 000	Honduras
1 778	1 841	1 857	955	1 330	1 257	1 328	1 359	1 399	Jamaica
596	639	674	712	800	804	792	792	792	Netherlands Antilles
456	435	479	485	546	541	575	639	670	Nicaragua
582	620	670	681	757	817	837	921	1 019	Panama
625	614	750	839	997	1 094	990	1 068	1 237	Paraguay
4 787	4 751	4 632	4 957	4 966	5 971	6 405	6 421	6 273	Peru
646	717	692	603	613	630	669	699	724	Trinidad and Tobago
1 017	1 073	1 131	1 189	1 248	1 254	1 362	1 453	1 609	Uruguay
13 266	13 886	14 246	15 047	15 028	16 530	16 803	16 692	16 416	Venezuela
2 449	2 384	2 182	2 212	2 295	2 323	2 385	2 462	2 462	Other Latin America
124 556	**125 052**	**129 123**	**133 633**	**140 657**	**151 384**	**159 012**	**168 476**	**171 727**	**Latin America**

Final Consumption of Oil (1000 tonnes)
Consommation finale de pétrole (1000 tonnes)

Endverbrauch von Öl (1000 Tonnen)

Consumo finale di petrolio (1000 tonnellate)

石油の最終消費量（千トン）

Consumo final de petróleo (1000 toneladas)

Конечное потребление нефти и нефтепродуктов (тыс. т)

	1971	1973	1978	1984	1985	1986	1987	1988	1989
Bangladesh	569	604	993	1 140	1 243	1 239	1 319	1 342	1 531
Brunei	61	67	97	205	220	225	237	261	272
Inde	17 865	20 350	24 700	34 498	37 310	39 684	42 418	45 740	49 652
Indonésie	6 142	7 413	13 373	17 266	17 870	17 919	18 571	20 337	23 009
RPD de Corée	683	765	1 732	2 748	2 927	3 627	4 690	4 392	4 507
Malaisie	3 471	3 320	4 343	6 424	6 463	6 675	7 049	7 576	8 402
Myanmar	1 177	947	1 064	1 043	1 000	1 029	639	616	606
Népal	45	61	78	148	151	181	175	187	225
Pakistan	2 960	3 109	4 031	5 825	6 201	6 626	7 114	7 553	7 854
Philippines	5 227	5 870	6 927	5 086	4 576	4 850	5 548	6 023	6 843
Singapour	1 023	1 381	2 294	3 131	3 777	3 834	4 239	4 263	4 583
Sri Lanka	797	1 013	1 034	1 141	1 079	1 085	1 085	1 126	1 106
Taipei chinois	3 714	5 751	10 508	13 304	14 095	15 374	16 488	17 505	18 347
Thaïlande	4 873	6 100	7 467	9 164	9 204	9 748	10 863	12 213	14 329
Viêt-Nam	5 660	5 654	933	1 540	1 524	1 719	2 005	2 059	1 978
Autre Asie	1 391	1 622	1 285	1 198	1 376	1 185	1 411	1 351	1 529
Asie	**55 658**	**64 027**	**80 859**	**103 861**	**109 016**	**115 000**	**123 851**	**132 544**	**144 773**
Rép. populaire de Chine	34 402	41 215	71 580	62 260	65 710	71 174	77 040	81 175	84 805
Hong-Kong, Chine	1 991	1 938	3 127	3 202	2 955	3 454	3 538	4 209	4 371
Chine	**36 393**	**43 153**	**74 707**	**65 462**	**68 665**	**74 628**	**80 578**	**85 384**	**89 176**
Albanie	583	468	866	834	635	656	678	712	669
Bulgarie	8 758	9 317	7 971	6 471	5 922	5 964	5 752	7 565	7 591
Chypre	423	549	520	647	629	719	769	842	873
Gibraltar	27	27	20	21	26	32	38	48	44
Malte	156	182	162	213	119	237	254	251	251
Roumanie	9 218	10 472	12 235	8 087	10 001	9 928	11 519	11 031	9 898
République slovaque	2 870	3 255	3 318	4 113	4 048	3 924	3 887	3 786	5 171
Bosnie-Herzegovine	-	-	-	-	-	-	-	-	-
Croatie	-	-	-	-	-	-	-	-	-
Ex-RYM	-	-	-	-	-	-	-	-	-
Slovénie	-	-	-	-	-	-	-	-	-
RF de Yougoslavie	-	-	-	-	-	-	-	-	-
Ex-Yougoslavie	7 718	8 782	12 050	9 190	9 700	10 815	10 796	11 364	10 652
Europe non-OCDE	**29 753**	**33 052**	**37 142**	**29 576**	**31 080**	**32 275**	**33 693**	**35 599**	**35 149**
Arménie	-	-	-	-	-	-	-	-	-
Azerbaïdjan	-	-	-	-	-	-	-	-	-
Bélarus	-	-	-	-	-	-	-	-	-
Estonie	-	-	-	-	-	-	-	-	-
Géorgie	-	-	-	-	-	-	-	-	-
Kazakhstan	-	-	-	-	-	-	-	-	-
Kirghizistan	-	-	-	-	-	-	-	-	-
Lettonie	-	-	-	-	-	-	-	-	-
Lituanie	-	-	-	-	-	-	-	-	-
République de Moldavie	-	-	-	-	-	-	-	-	-
Russie	-	-	-	-	-	-	-	-	-
Tadjikistan	-	-	-	-	-	-	-	-	-
Turkménistan	-	-	-	-	-	-	-	-	-
Ukraine	-	-	-	-	-	-	-	-	-
Ouzbékistan	-	-	-	-	-	-	-	-	-
Ex-URSS	**198 700**	**225 100**	**279 500**	**299 200**	**296 416**	**300 912**	**304 617**	**312 548**	**313 118**

Ex-URSS: les séries antérieures à 1990-1992 ne sont pas comparables aux années récentes; 1991 a été estimé.

Final Consumption of Oil (1000 tonnes)
Consommation finale de pétrole (1000 tonnes)
Endverbrauch von Öl (1000 Tonnen)
Consumo finale di petrolio (1000 tonnellate)
石油の最終消費量（千トン）
Consumo final de petróleo (1000 toneladas)
Конечное потребление нефти и нефтепродуктов (тыс. m)

1990	1991	1992	1993	1994	1995	1996	1997	1998	
1 527	1 575	1 652	1 798	1 857	2 443	2 529	2 815	2 615	Bangladesh
281	298	347	370	392	426	464	499	439	Brunei
50 309	53 722	56 408	57 913	61 875	67 607	73 594	75 617	80 237	India
24 524	25 629	26 981	29 391	31 853	34 132	37 100	38 690	37 778	Indonesia
4 416	4 312	4 312	4 278	4 235	4 189	4 149	3 866	3 673	DPR of Korea
9 556	10 619	11 608	12 706	13 513	15 531	16 715	18 062	16 963	Malaysia
591	552	629	750	870	1 136	1 242	1 440	1 363	Myanmar
231	356	345	362	386	490	512	591	636	Nepal
7 937	7 903	8 230	9 325	9 724	10 216	11 328	11 078	11 204	Pakistan
7 117	6 927	7 799	8 840	9 519	10 438	11 083	11 956	11 650	Philippines
5 472	5 298	5 068	5 578	6 194	6 643	7 070	7 924	8 203	Singapore
1 139	1 145	1 344	1 434	1 588	1 644	2 036	2 457	2 519	Sri Lanka
19 318	19 675	20 958	21 879	23 772	24 420	25 607	25 819	26 496	Chinese Taipei
16 107	16 660	18 406	20 312	22 825	25 829	28 840	29 287	26 671	Thailand
2 433	2 118	2 765	3 311	3 716	4 748	5 209	5 637	5 690	Vietnam
1 650	1 644	1 423	1 405	1 346	1 445	1 426	1 464	1 464	Other Asia
152 608	**158 433**	**168 275**	**179 652**	**193 665**	**211 337**	**228 904**	**237 202**	**237 601**	**Asia**
84 459	91 282	97 811	105 219	109 799	122 620	137 994	155 558	158 075	People's Rep. of China
4 553	4 357	5 634	5 891	6 598	6 537	6 395	6 713	8 334	Hong Kong, China
89 012	**95 639**	**103 445**	**111 110**	**116 397**	**129 157**	**144 389**	**162 271**	**166 409**	**China**
834	682	523	468	590	453	447	386	386	Albania
6 142	3 244	3 330	3 730	3 564	3 476	3 437	3 860	4 197	Bulgaria
878	1 067	1 147	1 141	1 176	1 237	1 291	1 288	1 338	Cyprus
46	62	72	72	85	83	83	83	83	Gibraltar
252	281	288	310	310	317	383	422	406	Malta
8 936	8 374	7 601	5 354	6 114	6 192	7 343	7 732	7 585	Romania
4 540	3 765	3 401	2 799	3 084	3 027	3 113	3 178	2 935	Slovak Republic
-	-	436	418	318	326	337	346	357	*Bosnia-Herzegovina*
-	-	2 083	2 154	2 363	2 401	2 293	2 759	2 805	*Croatia*
-	-	765	759	687	643	998	827	672	*FYROM*
-	-	1 424	1 734	1 869	2 078	2 450	2 449	2 685	*Slovenia*
-	-	2 073	1 204	1 163	1 275	2 104	3 092	2 652	*FR of Yugoslavia*
10 457	9 634	6 781	6 269	6 400	6 723	8 182	9 473	9 171	Former Yugoslavia
32 085	**27 109**	**23 143**	**20 143**	**21 323**	**21 508**	**24 279**	**26 422**	**26 101**	**Non-OECD Europe**
-	-	1 227	1 134	312	229	104	161	169	Armenia
-	-	4 935	3 453	3 448	2 737	2 000	1 685	2 005	Azerbaijan
-	-	11 808	8 107	6 051	5 548	5 318	5 465	6 252	Belarus
-	-	726	733	815	876	941	918	889	Estonia
-	-	936	539	205	95	739	807	995	Georgia
-	-	18 193	13 450	9 453	8 509	7 152	6 372	7 165	Kazakhstan
-	-	1 909	1 165	417	599	646	470	641	Kyrgyzstan
-	-	1 832	1 640	1 518	1 272	1 057	1 161	1 061	Latvia
-	-	2 790	1 933	1 760	1 984	1 769	1 759	1 847	Lithuania
-	-	1 200	865	678	685	624	631	533	Republic of Moldova
-	-	136 762	117 206	76 461	85 151	82 268	81 382	74 010	Russia
-	-	5 595	3 483	1 159	1 159	1 159	1 159	1 277	Tajikistan
-	-	4 418	2 360	2 377	2 119	2 295	2 226	1 828	Turkmenistan
-	-	30 125	20 754	18 115	19 323	15 710	14 322	14 385	Ukraine
-	-	6 090	5 531	5 195	4 674	4 301	4 650	4 615	Uzbekistan
303 755	**300 115**	**228 546**	**182 353**	**127 964**	**134 960**	**126 083**	**123 168**	**117 672**	**Former USSR**

Former USSR: series up to 1990-1992 are not comparable with recent years; 1991 is estimated.

Final Consumption of Oil (1000 tonnes)
Consommation finale de pétrole (1000 tonnes)
Endverbrauch von Öl (1000 Tonnen)
Consumo finale di petrolio (1000 tonnellate)
石油の最終消費量（千トン）
Consumo final de petróleo (1000 toneladas)
Конечное потребление нефти и нефтепродуктов (тыс. т)

	1971	1973	1978	1984	1985	1986	1987	1988	1989
Bahrein	208	289	639	724	725	724	721	764	798
République Islamique d'Iran	7 599	10 823	21 030	29 587	29 376	31 166	34 528	34 451	35 824
Irak	1 939	2 253	5 056	9 214	9 714	10 295	10 298	11 183	12 344
Israël	3 154	3 640	3 966	4 203	4 165	4 633	4 958	5 064	5 065
Jordanie	415	494	1 001	1 735	1 799	1 906	1 902	1 909	1 941
Koweit	938	976	1 635	2 922	2 793	3 092	3 342	3 526	3 847
Liban	1 477	1 772	1 444	1 427	1 618	1 547	1 632	1 106	1 216
Oman	81	82	446	911	980	934	837	882	941
Qatar	78	133	354	833	831	951	1 012	974	1 137
Arabie saoudite	2 203	3 287	11 404	23 480	24 914	25 826	24 705	26 289	25 795
Syrie	1 861	1 735	4 186	6 668	6 312	6 417	6 999	6 834	6 819
Emirats arabes unis	136	280	1 477	3 380	3 598	3 520	3 659	4 036	4 149
Yémen	292	484	790	1 426	1 691	1 762	1 883	1 977	1 938
Moyen-Orient	**20 381**	**26 248**	**53 428**	**86 510**	**88 516**	**92 773**	**96 476**	**98 995**	**101 814**
Total non-OCDE	**446 530**	**517 589**	**686 383**	**755 398**	**766 095**	**789 903**	**822 265**	**852 956**	**876 704**
OCDE Amérique du N.	697 407	766 782	842 083	740 378	742 672	756 886	778 994	801 046	800 880
OCDE Pacifique	174 580	205 201	212 564	202 971	201 582	208 727	217 236	232 146	242 972
OCDE Europe	515 276	584 632	580 067	491 414	492 950	510 898	515 554	526 418	520 378
Total OCDE	**1 387 263**	**1 556 615**	**1 634 714**	**1 434 763**	**1 437 204**	**1 476 511**	**1 511 784**	**1 559 610**	**1 564 230**
Monde	**1 833 793**	**2 074 204**	**2 321 097**	**2 190 161**	**2 203 299**	**2 266 414**	**2 334 049**	**2 412 566**	**2 440 934**

Final Consumption of Oil (1000 tonnes)
Consommation finale de pétrole (1000 tonnes)
Endverbrauch von Öl (1000 Tonnen)

Consumo finale di petrolio (1000 tonnellate)

石油の最終消費量（千トン）

Consumo final de petróleo (1000 toneladas)

Конечное потребление нефти и нефтепродуктов (тыс. m)

1990	1991	1992	1993	1994	1995	1996	1997	1998	
849	737	751	750	810	836	822	767	918	Bahrain
38 109	40 938	42 433	44 344	45 719	37 138	39 002	41 806	44 070	Islamic Republic of Iran
12 436	10 035	10 791	13 032	14 192	13 965	13 638	13 776	14 119	Iraq
5 349	5 458	5 906	6 469	6 820	7 690	8 054	8 279	8 793	Israel
2 194	2 023	2 355	2 386	2 525	2 646	2 719	2 830	2 801	Jordan
2 236	1 537	3 099	3 711	4 331	4 987	4 998	5 275	5 965	Kuwait
1 216	1 630	1 807	2 266	2 461	2 920	2 739	3 028	2 763	Lebanon
1 533	2 721	2 452	2 080	1 647	2 079	1 984	1 667	1 565	Oman
1 215	1 237	1 339	1 425	1 533	1 684	1 835	2 092	2 167	Qatar
28 325	29 805	30 539	32 753	32 647	30 873	32 993	34 281	37 339	Saudi Arabia
7 351	7 925	8 029	8 094	8 495	8 663	8 968	9 354	9 878	Syria
4 251	4 416	4 635	4 826	4 836	4 454	4 427	4 552	4 285	United Arab Emirates
1 664	2 211	2 490	1 914	2 041	2 284	2 279	2 335	2 345	Yemen
106 728	**110 673**	**116 626**	**124 050**	**128 057**	**120 219**	**124 458**	**130 042**	**137 008**	**Middle East**
878 293	**888 024**	**843 668**	**822 563**	**800 274**	**845 556**	**886 364**	**930 156**	**941 265**	**Non-OECD Total**
786 070	769 811	791 249	800 876	827 158	829 594	854 632	862 536	870 882	OECD North America
256 879	267 395	281 909	289 232	302 207	315 597	325 015	330 662	317 379	OECD Pacific
524 481	538 117	544 017	544 924	548 812	558 877	571 909	577 040	587 184	OECD Europe
1 567 430	**1 575 323**	**1 617 175**	**1 635 032**	**1 678 177**	**1 704 068**	**1 751 556**	**1 770 238**	**1 775 445**	**OECD Total**
2 445 723	**2 463 347**	**2 460 843**	**2 457 595**	**2 478 451**	**2 549 624**	**2 637 920**	**2 700 394**	**2 716 710**	**World**

Final Consumption of Natural Gas (TJ)
Consommation finale de gaz naturel (TJ)

Endverbrauch von Erdgas (TJ)

Consumo finale di gas naturale (TJ)

天然ガスの最終消費量 (TJ)

Consumo final de gas natural (TJ)

Конечное потребление природного газа (ТДж)

	1971	1973	1978	1984	1985	1986	1987	1988	1989
Algérie	11 455	13 900	42 645	71 764	88 397	109 100	118 500	119 800	109 500
Angola	1 678	2 536	2 758	4 499	4 499	4 997	5 998	6 098	6 500
Congo	2 256	-	-	-	-	-	-	-	-
Egypte	-	-	15 101	66 801	68 070	120 396	112 408	114 200	120 396
Gabon	-	-	1 546	1 984	3 512	4 486	6 824	30
Libye	-	-	30 408	57 306	62 674	74 830	66 305	70 094	84 776
Maroc	2 009	2 721	3 177	3 244	3 666	3 549	2 884	2 947	2 177
Nigéria	1 411	1 352	1 398	14 343	17 188	18 149	21 296	26 957	31 825
Tunisie	38	331	2 888	5 085	5 902	7 969	13 268	7 124	9 830
Autre Afrique	37	37	37	16	15	10	4	7	5
Afrique	**18 884**	**20 877**	**98 412**	**224 604**	**252 395**	**342 512**	**345 149**	**354 051**	**365 039**
Argentine	134 479	171 102	221 648	360 389	355 784	401 024	421 265	425 360	438 111
Bolivie	-	155	2 203	2 087	2 783	3 208	3 672	3 710	6 996
Brésil	1 279	3 875	24 298	54 100	63 091	73 941	86 806	84 522	90 258
Chili	155	1 838	4 693	6 066	5 987	6 578	6 301	6 824	7 026
Colombie	4 905	9 832	22 925	30 864	32 812	35 823	36 779	39 476	40 719
Cuba	67	565	427	126	266	228	874	722	988
Trinité-et-Tobago	21 363	24 192	39 896	76 235	76 094	82 363	83 502	111 476	113 602
Uruguay	-	-	-	-	-	-	-	-	-
Vénézuela	98 756	182 049	251 784	359 134	378 983	331 222	370 524	334 214	342 037
Autre Amérique latine	117	117	374	760	993	675	592	479	636
Amérique latine	**261 121**	**393 725**	**568 248**	**889 761**	**916 793**	**935 062**	**1 010 315**	**1 006 783**	**1 040 373**
Bangladesh	9 049	15 403	21 075	51 917	55 365	64 185	71 460	79 972	85 747
Inde	9 218	8 617	34 299	73 570	112 884	149 155	156 615	238 889	292 008
Indonésie	4 637	5 630	64 584	185 804	251 208	256 651	265 396	262 699	260 932
Malaisie	293	419	1 297	5 606	21 547	44 183	47 363	44 266	44 769
Myanmar	762	1 175	3 145	7 969	10 845	12 126	12 478	12 602	12 190
Pakistan	69 684	83 917	119 992	201 729	208 309	212 680	246 773	256 774	278 550
Taipei chinois	30 295	43 741	67 810	45 889	41 088	37 485	38 929	41 913	42 170
Thaïlande	-	-	-	8 268	7 591	3 706	1 684	2 564	4 869
Autre Asie	8 967	6 746	6 094	18 940	24 240	24 444	7 950	8 938	9 754
Asie	**132 905**	**165 648**	**318 296**	**599 692**	**733 077**	**804 615**	**848 648**	**948 617**	**1 030 989**
Rép. populaire de Chine	65 591	104 864	293 789	290 160	334 020	327 419	347 295	357 709	384 348
Chine	**65 591**	**104 864**	**293 789**	**290 160**	**334 020**	**327 419**	**347 295**	**357 709**	**384 348**
Albanie	4 914	7 413	13 804	14 997	14 997	14 997	14 997	15 997	12 597
Bulgarie	11 506	7 810	107 601	196 780	171 632	177 181	187 692	99 898	103 693
Roumanie	605 483	646 572	1 170 245	1 290 428	1 015 452	1 035 853	1 029 288	1 046 281	1 050 489
République slovaque	54 763	65 312	128 169	179 391	176 922	178 607	180 728	183 525	193 498
Bosnie-Herzegovine	-	-	-	-	-	-	-	-	-
Croatie	-	-	-	-	-	-	-	-	-
Ex-RYM	-	-	-	-	-	-	-	-	-
Slovénie	-	-	-	-	-	-	-	-	-
RF de Yougoslavie	-	-	-	-	-	-	-	-	-
Ex-Yougoslavie	37 152	45 773	68 010	126 176	137 380	144 085	192 123	147 526	134 074
Europe non-OCDE	**713 818**	**772 880**	**1 487 829**	**1 807 772**	**1 516 383**	**1 550 723**	**1 604 828**	**1 493 227**	**1 494 351**

Final Consumption of Natural Gas (TJ)
Consommation finale de gaz naturel (TJ)
Endverbrauch von Erdgas (TJ)
Consumo finale di gas naturale (TJ)
天然ガスの最終消費量 (TJ)
Consumo final de gas natural (TJ)
Конечное потребление природного газа (ТДж)

1990	1991	1992	1993	1994	1995	1996	1997	1998	
173 580	198 071	218 404	231 293	222 079	223 165	225 902	234 181	229 151	Algeria
20 520	22 040	21 660	21 280	19 760	21 280	21 280	21 660	22 040	Angola
-	-	-	-	-	-	-	-	-	Congo
112 408	114 200	119 776	144 086	132 958	143 811	146 452	154 128	171 456	Egypt
30	30	30	30	25	27	29	28	28	Gabon
78 303	78 303	77 829	75 145	77 513	76 566	82 407	86 354	86 038	Libya
2 014	1 289	847	828	809	512	651	1 200	1 317	Morocco
33 326	30 326	32 500	34 600	37 714	34 433	34 898	36 490	39 144	Nigeria
14 611	15 775	17 537	18 654	19 588	21 775	23 120	23 661	25 033	Tunisia
6	5	6	7	7	7	7	7	7	Other Africa
434 798	**460 039**	**488 589**	**525 923**	**510 453**	**521 576**	**534 746**	**557 709**	**574 214**	**Africa**
445 371	451 420	475 384	537 412	568 450	593 718	597 163	621 638	639 647	Argentina
7 382	8 194	9 740	11 001	12 307	13 495	11 898	16 124	17 284	Bolivia
87 890	92 893	110 640	117 888	125 485	133 544	150 942	141 409	149 121	Brazil
7 394	7 904	8 122	8 281	8 630	8 914	9 199	21 441	24 116	Chile
42 517	45 838	47 201	46 716	52 871	55 854	59 312	61 351	65 447	Colombia
1 250	1 033	566	810	639	543	608	1 292	4 667	Cuba
116 681	111 138	112 701	107 727	118 670	123 502	135 932	148 231	172 334	Trinidad and Tobago
-	-	-	-	-	-	-	-	92	Uruguay
387 950	443 915	427 809	446 641	449 217	490 758	559 259	528 490	558 404	Venezuela
742	504	931	854	1 048	893	970	1 126	854	Other Latin America
1 097 177	**1 162 839**	**1 193 094**	**1 277 330**	**1 337 317**	**1 421 221**	**1 525 283**	**1 541 102**	**1 631 966**	**Latin America**
83 384	81 241	99 385	106 328	121 497	139 161	143 095	138 129	146 557	Bangladesh
279 961	288 792	333 680	320 862	340 969	381 885	423 792	487 626	510 691	India
324 005	338 804	377 878	333 368	315 431	355 163	370 303	373 796	355 354	Indonesia
45 732	47 070	57 237	71 797	77 738	80 751	103 345	103 135	114 045	Malaysia
10 447	8 451	8 371	10 709	13 347	13 341	13 100	15 345	13 949	Myanmar
278 603	279 628	284 417	315 350	338 117	351 782	381 435	389 346	414 864	Pakistan
43 767	52 061	58 505	59 733	63 434	71 352	74 823	75 648	75 722	Chinese Taipei
11 258	15 334	18 802	20 993	25 017	33 516	40 045	40 528	37 583	Thailand
11 167	10 920	10 373	10 230	9 930	9 750	9 500	9 788	9 788	Other Asia
1 088 324	**1 122 301**	**1 248 648**	**1 249 370**	**1 305 480**	**1 436 701**	**1 559 438**	**1 633 341**	**1 678 553**	**Asia**
404 103	434 761	447 356	447 782	455 840	469 003	489 710	486 114	512 453	People's Rep. of China
404 103	**434 761**	**447 356**	**447 782**	**455 840**	**469 003**	**489 710**	**486 114**	**512 453**	**China**
9 059	5 126	3 584	3 080	1 976	1 089	894	700	661	Albania
106 508	90 855	75 096	73 858	83 265	97 520	98 539	82 409	59 665	Bulgaria
923 608	696 575	342 355	332 571	430 410	471 352	464 141	413 900	320 662	Romania
214 116	190 334	193 586	187 090	174 262	189 428	211 143	212 466	215 030	Slovak Republic
-	-	16 000	-	-	-	-	-	-	*Bosnia-Herzegovina*
-	-	58 938	59 063	60 397	62 822	66 192	70 596	67 165	*Croatia*
-	-	10 000	10 500	-	-	-	-	-	*FYROM*
-	-	18 546	18 369	21 825	23 187	25 397	29 756	29 963	*Slovenia*
-	-	93 478	58 782	21 374	30 839	91 359	90 706	85 880	*FR of Yugoslavia*
273 295	215 337	196 962	146 714	103 596	116 848	182 948	191 058	183 008	Former Yugoslavia
1 526 586	**1 198 227**	**811 583**	**743 313**	**793 509**	**876 237**	**957 665**	**900 533**	**779 026**	**Non-OECD Europe**

Final Consumption of Natural Gas (TJ)
Consommation finale de gaz naturel (TJ)
Endverbrauch von Erdgas (TJ)
Consumo finale di gas naturale (TJ)
天然ガスの最終消費量 (TJ)
Consumo final de gas natural (TJ)
Конечное потребление природного газа (ТДж)

	1971	1973	1978	1984	1985	1986	1987	1988	1989
Arménie	-	-	-	-	-	-	-	-	-
Azerbaïdjan	-	-	-	-	-	-	-	-	-
Bélarus	-	-	-	-	-	-	-	-	-
Estonie	-	-	-	-	-	-	-	-	-
Géorgie	-	-	-	-	-	-	-	-	-
Kazakhstan	-	-	-	-	-	-	-	-	-
Kirghizistan	-	-	-	-	-	-	-	-	-
Lettonie	-	-	-	-	-	-	-	-	-
Lituanie	-	-	-	-	-	-	-	-	-
République de Moldavie	-	-	-	-	-	-	-	-	-
Russie	-	-	-	-	-	-	-	-	-
Tadjikistan	-	-	-	-	-	-	-	-	-
Turkménistan	-	-	-	-	-	-	-	-	-
Ukraine	-	-	-	-	-	-	-	-	-
Ouzbékistan	-	-	-	-	-	-	-	-	-
Ex-URSS	**4 695 208**	**5 407 165**	**7 226 618**	**10 077 430**	**10 603 194**	**11 063 827**	**11 589 600**	**12 096 708**	**12 374 594**
Bahrein	8 371	27 369	43 408	56 591	60 683	58 610	53 670	57 546	120 677
République Islamique d'Iran	376 386	383 782	389 252	403 403	464 824	477 629	395 460	503 020	507 828
Irak	35 340	45 980	64 600	114 300	121 250	148 296	186 521	258 158	297 681
Israël	4 232	2 097	2 210	2 051	1 838	1 444	1 623	1 461	1 464
Koweit	154 134	179 936	178 647	151 901	149 873	213 086	228 825	226 394	274 852
Oman	-	-	-	11 950	17 111	6 986	9 090	13 194	13 846
Qatar	17 674	28 065	21 185	92 680	81 922	85 198	84 976	92 824	85 927
Arabie saoudite	14 783	14 783	14 783	75 946	110 510	314 093	430 534	430 039	397 475
Syrie	-	-	-	-	-	8 463	7 985	23 336	35 270
Emirats arabes unis	37 782	38 254	85 753	126 782	133 260	207 023	262 394	248 221	353 988
Moyen-Orient	**648 702**	**720 266**	**799 838**	**1 035 604**	**1 141 271**	**1 520 828**	**1 661 078**	**1 854 193**	**2 089 008**
Total non-OCDE	**6 536 229**	**7 585 425**	**10 793 030**	**14 925 023**	**15 497 133**	**16 544 986**	**17 406 913**	**18 111 288**	**18 778 702**
OCDE Amérique du N.	17 908 538	18 518 244	16 970 340	16 917 149	16 427 363	15 597 639	16 129 494	17 471 245	17 257 914
OCDE Pacifique	165 396	248 571	463 030	736 618	780 304	835 137	875 275	905 439	982 559
OCDE Europe	2 829 662	4 360 087	6 855 630	7 774 469	8 181 967	8 321 869	8 820 632	8 722 716	8 963 765
Total OCDE	**20 903 596**	**23 126 902**	**24 289 000**	**25 428 236**	**25 389 634**	**24 754 645**	**25 825 401**	**27 099 400**	**27 204 238**
Monde	**27 439 825**	**30 712 327**	**35 082 030**	**40 353 259**	**40 886 767**	**41 299 631**	**43 232 314**	**45 210 688**	**45 982 940**

Ex-URSS: les séries antérieures à 1990-1992 ne sont pas comparables aux années récentes; 1991 a été estimé.

Final Consumption of Natural Gas (TJ)
Consommation finale de gaz naturel (TJ)
Endverbrauch von Erdgas (TJ)

Consumo finale di gas naturale (TJ)

天然ガスの最終消費量 (TJ)

Consumo final de gas natural (TJ)

Конечное потребление природного газа (ТДж)

1990	1991	1992	1993	1994	1995	1996	1997	1998	
-	-	46 951	18 806	21 452	28 628	12 687	18 984	21 357	Armenia
-	-	287 310	232 497	231 214	188 880	162 150	147 650	113 820	Azerbaijan
-	-	174 542	165 998	122 934	145 489	142 283	141 163	141 898	Belarus
-	-	14 054	9 043	12 452	13 846	14 247	14 209	14 856	Estonia
-	-	128 636	103 072	60 320	22 469	16 173	21 602	14 800	Georgia
-	-	413 003	281 883	217 378	270 498	209 800	175 762	174 845	Kazakhstan
-	-	35 740	23 034	14 363	14 673	17 481	9 365	16 870	Kyrgyzstan
-	-	28 183	17 165	14 722	13 603	21 635	15 801	16 215	Latvia
-	-	51 534	29 840	32 783	39 390	42 769	37 539	36 982	Lithuania
-	-	39 442	60 055	57 834	57 834	68 469	73 148	66 162	Republic of Moldova
-	-	7 132 727	5 969 417	5 501 111	5 330 889	4 911 322	5 596 556	5 551 001	Russia
-	-	34 511	28 590	15 980	18 010	23 109	15 503	15 845	Tajikistan
-	-	142 395	202 294	274 764	289 470	234 060	239 019	214 288	Turkmenistan
-	-	2 239 000	2 125 000	1 759 000	2 000 000	1 950 000	1 917 000	1 880 000	Ukraine
-	-	884 902	950 018	961 540	948 953	949 410	964 287	1 070 665	Uzbekistan
12 890 966	12 237 812	11 652 930	10 216 712	9 297 847	9 382 632	8 775 595	9 387 588	9 349 604	**Former USSR**
65 897	58 181	85 508	108 305	110 614	117 439	121 187	143 415	142 005	Bahrain
377 568	653 667	811 188	847 564	870 995	1 005 091	1 153 948	1 227 003	1 223 303	Islamic Republic of Iran
158 135	65 021	108 680	122 048	151 706	150 796	154 126	168 299	177 156	Iraq
1 316	967	953	1 005	899	754	377	396	279	Israel
150 758	25 125	147 756	261 788	279 296	229 338	280 268	279 359	284 792	Kuwait
30 054	31 999	38 518	53 829	52 902	31 597	28 510	48 574	73 151	Oman
90 879	117 239	209 345	223 867	222 111	221 129	221 310	235 546	232 683	Qatar
417 321	408 636	412 606	437 712	445 800	495 832	538 846	590 984	610 275	Saudi Arabia
34 908	27 202	29 004	31 500	34 146	43 058	43 058	85 800	134 507	Syria
317 877	436 979	358 831	320 261	420 207	430 586	577 164	504 312	512 361	United Arab Emirates
1 644 713	1 825 016	2 202 389	2 407 879	2 588 676	2 725 620	3 118 794	3 283 688	3 390 512	**Middle East**
19 086 667	18 440 995	18 044 589	16 868 309	16 289 122	16 832 990	16 961 231	17 790 075	17 916 328	**Non-OECD Total**
16 772 941	16 728 867	16 944 459	17 707 516	17 894 642	18 324 006	19 089 098	18 542 336	17 265 219	OECD North America
1 040 572	1 113 430	1 174 275	1 259 255	1 356 587	1 469 852	1 674 003	1 748 634	1 811 480	OECD Pacific
9 139 357	9 850 799	9 811 997	10 170 046	10 181 722	10 810 449	11 783 042	11 435 581	11 762 719	OECD Europe
26 952 870	27 693 096	27 930 731	29 136 817	29 432 951	30 604 307	32 546 143	31 726 551	30 839 418	**OECD Total**
46 039 537	46 134 091	45 975 320	46 005 126	45 722 073	47 437 297	49 507 374	49 516 626	48 755 746	**World**

Former USSR: series up to 1990-1992 are not comparable with recent years; 1991 is estimated.

Consumption of Electricity (GWh)
Consommation d'électricité (GWh)
Verbrauch von Elektrizität (GWh)
Consumo di energia elettrica (GWh)
電力消費量 (GWh)
Consumo de electricidad (GWh)
Потребление электроэнергии (ГВт.ч)

	1971	1973	1978	1984	1985	1986	1987	1988	1989
Algérie	1 991	2 494	5 050	9 604	10 464	11 211	10 827	11 779	13 109
Angola	556	738	439	502	604	604	607	611	614
Bénin	32	58	168	135	143	137	140	167	162
Cameroun	1 019	1 080	1 265	1 996	2 253	2 167	2 228	2 323	2 402
Congo	73	80	103	238	305	330	355	365	379
RD du Congo	3 337	3 630	3 703	4 116	4 427	4 623	4 598	4 733	5 672
Côte d'Ivoire	517	697	1 226	2 029	1 853	1 752	1 742	1 940	1 848
Egypte	7 216	7 282	13 309	25 024	24 487	29 867	32 255	33 914	36 015
Ethiopie	552	551	535	842	809	835	881	864	866
Erythrée	-	-	-	-	-	-	-	-	-
Gabon	112	162	432	787	852	878	883	897	771
Ghana	2 809	3 728	3 542	1 198	2 382	3 727	4 341	4 601	4 706
Kenya	808	959	1 396	1 888	2 060	2 192	2 463	2 407	2 558
Libye	508	1 127	3 363	9 200	11 844	13 280	15 600	16 551	16 700
Maroc	2 047	2 594	3 928	6 226	6 627	7 060	7 297	8 149	8 174
Mozambique	476	559	671	372	402	473	540	441	507
Nigéria	1 534	2 184	4 086	5 967	6 471	7 793	7 862	8 264	9 304
Sénégal	323	374	517	690	673	678	711	741	735
Afrique du Sud	50 768	59 870	84 337	128 480	134 565	140 874	143 197	149 360	153 469
Soudan	373	491	744	961	1 055	1 198	1 242	1 119	1 075
Rép. Unie de Tanzanie	424	503	601	727	787	909	946	1 061	1 151
Togo	67	105	195	244	247	259	291	299	319
Tunisie	803	1 036	1 833	3 290	3 708	3 977	4 257	4 607	4 865
Zambie	4 411	5 174	5 838	6 660	6 565	6 183	6 335	6 135	4 632
Zimbabwe	3 560	4 962	5 941	7 108	8 309	8 019	8 373	8 452	8 898
Autre Afrique	2 524	3 068	4 052	5 990	6 170	6 612	7 161	7 129	7 560
Afrique	**86 840**	**103 506**	**147 274**	**224 274**	**238 062**	**255 638**	**265 132**	**276 909**	**286 491**
Argentine	21 214	24 111	29 589	38 682	38 945	40 841	44 090	45 993	42 447
Bolivie	919	1 035	1 512	1 477	1 488	1 499	1 488	1 628	1 757
Brésil	44 846	56 717	99 039	160 000	173 564	187 069	192 755	203 903	212 381
Chili	7 569	7 780	9 249	11 678	12 112	12 796	13 331	14 409	15 785
Colombie	8 386	10 932	14 621	22 129	22 919	24 943	24 592	25 605	27 337
Costa Rica	1 148	1 347	1 925	2 637	2 757	3 027	3 303	3 383	3 272
Cuba	4 519	5 133	7 554	10 794	10 621	11 433	11 749	12 436	12 955
République dominicaine	1 068	1 955	2 269	2 815	3 082	3 199	3 222	2 617	2 348
Equateur	903	1 080	2 221	3 641	3 744	3 884	4 303	4 408	4 455
El Salvador	644	802	1 315	1 423	1 495	1 541	1 662	1 721	1 744
Guatemala	666	851	1 446	1 400	1 456	1 511	1 679	1 753	1 903
Haiti	59	99	187	309	323	351	350	367	391
Honduras	354	426	628	996	1 085	1 095	1 184	1 297	1 389
Jamaïque	1 552	1 969	1 317	1 369	1 356	1 432	1 527	1 467	1 393
Antilles néerlandaises	603	658	701	695	646	525	580	588	625
Nicaragua	603	624	1 023	1 111	1 098	1 095	1 177	1 071	1 104
Panama	819	1 096	1 407	2 114	2 156	2 197	2 384	2 230	2 146
Paraguay	209	279	567	1 046	1 081	1 270	1 498	1 728	1 837
Pérou	5 399	6 018	7 768	10 593	10 702	11 396	12 154	11 636	11 838
Trinité-et-Tobago	991	1 105	1 557	3 006	3 023	3 291	3 481	3 157	3 056
Uruguay	2 033	1 969	2 604	3 132	3 220	3 367	3 561	3 875	3 753
Vénézuela	12 343	14 469	22 159	39 754	40 472	41 507	44 670	48 872	47 856
Autre Amérique latine	3 134	3 646	4 100	4 234	4 537	4 712	4 565	5 000	5 695
Amérique latine	**119 981**	**144 101**	**214 758**	**325 035**	**341 882**	**363 981**	**379 305**	**399 144**	**407 467**

Consumption of Electricity (GWh)
Consommation d'électricité (GWh)
Verbrauch von Elektrizität (GWh)
Consumo di energia elettrica (GWh)
電力消費量 (GWh)
Consumo de electricidad (GWh)
Потребление электроэнергии (ГВт.ч)

1990	1991	1992	1993	1994	1995	1996	1997	1998	
13 845	14 484	14 688	15 145	15 545	16 102	16 694	17 047	19 046	Algeria
630	667	677	680	684	687	736	791	761	Angola
172	189	210	216	225	246	270	259	273	Benin
2 345	2 334	2 360	2 457	2 213	2 177	2 315	2 517	2 643	Cameroon
419	458	450	420	409	420	443	381	341	Congo
4 922	4 944	5 700	5 204	5 205	5 840	5 916	5 673	5 349	DR of Congo
1 945	1 899	1 065	1 882	1 948	2 548	2 345	2 546	2 855	Cote d'Ivoire
37 237	38 976	40 255	42 089	43 307	45 670	48 428	51 316	55 302	Egypt
1 082	1 088	1 117	1 245	1 306	1 374	1 443	1 453	1 464	Ethiopia
-	-	123	128	141	148	169	183	162	Eritrea
810	814	814	816	847	846	873	906	924	Gabon
4 927	5 263	5 657	5 521	5 695	6 122	6 486	7 355	7 690	Ghana
2 729	2 854	2 945	3 103	3 234	3 318	3 507	3 654	3 750	Kenya
16 800	16 800	16 950	17 000	17 800	18 000	18 300	18 974	19 496	Libya
8 910	9 385	10 339	10 517	11 906	12 569	12 670	13 435	14 270	Morocco
548	742	746	728	662	714	732	893	825	Mozambique
7 833	9 063	9 693	8 695	11 065	9 876	10 222	10 509	10 716	Nigeria
780	797	874	859	931	950	981	1 072	1 142	Senegal
155 988	157 518	154 302	159 333	165 796	173 442	179 422	187 346	186 683	South Africa
1 282	1 264	1 295	1 271	1 345	1 362	1 413	1 360	1 355	Sudan
1 303	1 428	1 441	1 433	1 456	1 623	1 793	1 700	1 719	United Rep. of Tanzania
343	328	372	282	321	426	447	488	471	Togo
5 206	5 422	5 837	6 152	6 663	6 942	7 147	7 615	8 124	Tunisia
4 164	5 484	5 410	5 423	5 423	5 427	5 431	5 596	5 484	Zambia
9 133	9 436	9 813	8 242	8 997	9 523	9 743	10 332	10 633	Zimbabwe
7 882	8 226	8 441	8 986	9 115	9 284	9 555	9 614	9 838	Other Africa
291 235	**299 863**	**301 574**	**307 827**	**322 239**	**335 636**	**347 481**	**363 015**	**371 316**	**Africa**
42 534	44 577	47 289	50 850	55 054	57 428	60 910	65 894	70 947	Argentina
1 827	1 967	2 077	2 286	2 484	2 680	2 877	3 082	3 280	Bolivia
217 839	225 372	230 472	241 167	249 793	264 805	277 665	294 689	306 989	Brazil
16 428	17 732	19 990	21 124	22 506	25 100	28 102	30 772	32 606	Chile
28 652	29 610	27 305	30 288	32 501	34 425	35 023	36 131	36 138	Colombia
3 344	3 592	3 791	4 084	4 389	4 526	4 553	5 081	5 417	Costa Rica
12 848	11 195	9 711	9 067	9 737	10 010	10 660	11 432	11 567	Cuba
2 791	2 918	4 035	4 380	4 610	4 852	5 106	5 299	5 460	Dominican Republic
4 896	5 362	5 617	5 908	6 629	6 844	7 345	7 432	7 780	Ecuador
1 899	2 008	2 207	2 452	2 743	3 030	3 019	3 287	3 415	El Salvador
1 989	2 102	2 371	2 571	2 769	2 962	3 085	3 280	3 503	Guatemala
414	323	295	213	142	234	267	338	302	Haiti
1 516	1 607	1 755	1 932	1 779	2 003	2 294	2 658	2 802	Honduras
1 658	1 671	1 760	3 791	4 257	5 197	5 385	5 579	5 838	Jamaica
678	689	743	799	851	884	925	924	979	Netherlands Antilles
1 220	1 243	1 267	1 247	1 211	1 298	1 392	1 466	1 605	Nicaragua
2 282	2 324	2 443	2 649	2 749	2 940	3 129	3 184	3 396	Panama
2 131	2 200	2 392	2 854	3 269	3 743	4 713	4 019	4 145	Paraguay
10 819	11 426	10 721	12 064	12 156	13 874	14 612	15 071	16 197	Peru
3 278	3 257	3 522	3 502	3 611	3 892	4 091	4 587	4 751	Trinidad and Tobago
3 870	4 212	4 309	4 623	4 759	5 089	5 313	5 664	5 948	Uruguay
48 643	51 236	55 286	56 697	57 246	58 152	59 834	61 669	62 115	Venezuela
5 849	6 094	6 311	6 484	6 541	6 684	6 792	7 081	7 251	Other Latin America
417 405	**432 717**	**445 669**	**471 032**	**491 786**	**520 652**	**547 092**	**578 619**	**602 431**	**Latin America**

Consumption of Electricity (GWh)
Consommation d'électricité (GWh)
Verbrauch von Elektrizität (GWh)
Consumo di energia elettrica (GWh)
電力消費量 (GWh)
Consumo de electricidad (GWh)
Потребление электроэнергии (ГВт.ч)

	1971	1973	1978	1984	1985	1986	1987	1988	1989
Bangladesh	710	1 042	1 651	2 866	3 040	3 534	3 771	4 173	5 093
Brunei	250	246	250	716	731	740	995	1 048	1 079
Inde	55 522	59 854	90 727	137 886	149 105	163 600	177 610	196 482	216 701
Indonésie	2 530	3 010	4 404	11 735	13 561	15 614	18 847	21 954	25 704
RPD de Corée	15 320	18 120	28 992	40 750	43 470	45 000	45 200	48 000	48 500
Malaisie	3 464	4 347	7 521	12 582	13 649	14 669	15 820	17 528	18 663
Myanmar	553	666	1 020	1 313	1 595	1 626	1 665	1 503	1 638
Népal	74	74	151	326	374	441	507	537	603
Pakistan	5 584	6 389	8 913	16 384	18 336	20 398	22 454	25 919	27 652
Philippines	8 688	12 562	14 740	18 047	19 020	18 199	18 304	20 191	21 776
Singapour	2 440	3 506	5 574	8 896	9 422	10 104	11 440	12 616	13 596
Sri Lanka	735	881	1 178	1 886	2 073	2 243	2 285	2 383	2 367
Taipei chinois	14 520	18 913	33 345	49 019	52 681	59 392	65 474	72 373	79 243
Thaïlande	4 568	6 485	11 886	19 456	21 117	23 047	26 085	29 560	34 297
Viêt-Nam	1 794	1 833	2 808	3 866	4 141	4 423	4 949	5 515	6 099
Autre Asie	2 170	3 386	3 559	3 829	4 145	4 441	4 562	4 670	4 764
Asie	**118 922**	**141 314**	**216 719**	**329 557**	**356 460**	**387 471**	**419 968**	**464 452**	**507 775**
Rép. populaire de Chine	138 400	166 800	256 552	348 800	381 330	417 488	462 400	508 730	545 200
Hong-Kong, Chine	5 250	6 009	9 107	15 041	15 923	17 660	21 430	22 918	22 385
Chine	**143 650**	**172 809**	**265 659**	**363 841**	**397 253**	**435 148**	**483 830**	**531 648**	**567 585**
Albanie	1 164	1 363	2 819	2 957	2 361	4 294	3 583	3 161	3 297
Bulgarie	19 571	23 131	32 157	42 763	41 920	41 622	43 409	44 373	44 058
Chypre	615	774	865	1 175	1 243	1 334	1 411	1 540	1 724
Gibraltar	45	47	50	59	61	63	62	66	76
Malte	305	365	459	635	710	770	857	935	1 000
Roumanie	33 395	39 732	58 495	69 732	71 222	75 795	75 175	78 226	79 027
République slovaque	12 290	14 046	16 962	23 708	24 990	25 666	26 559	27 043	27 784
Bosnie-Herzegovine	-	-	-	-	-	-	-	-	-
Croatie	-	-	-	-	-	-	-	-	-
Ex-RYM	-	-	-	-	-	-	-	-	-
Slovénie	-	-	-	-	-	-	-	-	-
RF de Yougoslavie	-	-	-	-	-	-	-	-	-
Ex-Yougoslavie	26 127	31 260	45 608	65 990	67 899	70 645	73 736	75 956	75 711
Europe non-OCDE	**93 512**	**110 718**	**157 415**	**207 019**	**210 406**	**220 189**	**224 792**	**231 300**	**232 677**
Arménie	-	-	-	-	-	-	-	-	-
Azerbaïdjan	-	-	-	-	-	-	-	-	-
Bélarus	-	-	-	-	-	-	-	-	-
Estonie	-	-	-	-	-	-	-	-	-
Géorgie	-	-	-	-	-	-	-	-	-
Kazakhstan	-	-	-	-	-	-	-	-	-
Kirghizistan	-	-	-	-	-	-	-	-	-
Lettonie	-	-	-	-	-	-	-	-	-
Lituanie	-	-	-	-	-	-	-	-	-
République de Moldavie	-	-	-	-	-	-	-	-	-
Russie	-	-	-	-	-	-	-	-	-
Tadjikistan	-	-	-	-	-	-	-	-	-
Turkménistan	-	-	-	-	-	-	-	-	-
Ukraine	-	-	-	-	-	-	-	-	-
Ouzbékistan	-	-	-	-	-	-	-	-	-
Ex-URSS	**730 500**	**832 300**	**1 091 900**	**1 341 200**	**1 381 400**	**1 432 700**	**1 487 700**	**1 526 200**	**1 541 100**

Consumption of Electricity (GWh)
Consommation d'électricité (GWh)
Verbrauch von Elektrizität (GWh)
Consumo di energia elettrica (GWh)
電力消費量 *(GWh)*
Consumo de electricidad (GWh)
Потребление электроэнергии *(ГВт.ч)*

1990	1991	1992	1993	1994	1995	1996	1997	1998	
5 135	5 324	6 522	7 411	8 010	9 011	9 637	10 062	10 864	Bangladesh
1 118	1 192	1 320	1 483	1 615	1 897	2 080	2 378	2 474	Brunei
238 238	260 917	277 706	297 795	317 411	344 723	358 893	382 176	407 302	India
31 497	33 653	38 366	41 533	44 835	51 532	57 986	67 726	68 686	Indonesia
48 500	28 500	6 080	6 080	5 920	5 760	5 600	5 219	4 958	DPR of Korea
20 872	24 123	26 640	31 615	35 604	41 319	46 747	53 227	56 417	Malaysia
1 823	1 712	1 875	2 173	2 342	2 510	2 562	2 901	3 078	Myanmar
670	757	738	795	862	948	1 025	1 053	1 149	Nepal
29 865	32 922	35 381	37 979	39 099	41 192	43 777	45 072	46 652	Pakistan
22 020	22 451	21 778	22 862	25 725	28 109	30 535	33 186	34 456	Philippines
15 184	16 136	16 849	18 035	19 838	21 258	23 016	25 780	27 370	Singapore
2 624	2 765	2 946	3 272	3 613	3 936	3 756	4 245	4 626	Sri Lanka
85 124	93 931	100 638	110 556	119 495	128 198	137 274	146 398	155 724	Chinese Taipei
40 130	45 334	51 647	58 834	65 136	74 178	80 581	85 807	83 701	Thailand
6 610	6 970	7 358	8 159	9 586	11 522	13 741	15 767	18 285	Vietnam
4 820	4 735	4 592	4 609	4 809	5 261	5 161	5 251	5 251	Other Asia
554 230	**581 422**	**600 436**	**653 191**	**703 900**	**771 354**	**822 371**	**886 248**	**930 993**	**Asia**
579 580	631 650	701 582	780 735	866 430	927 888	999 485	1 047 518	1 079 950	People's Rep. of China
23 833	25 316	26 148	27 727	29 184	29 856	31 635	32 245	34 849	Hong Kong, China
603 413	**656 966**	**727 730**	**808 462**	**895 614**	**957 744**	**1 031 120**	**1 079 763**	**1 114 799**	**China**
2 828	2 063	2 397	2 866	1 876	2 094	2 809	2 330	2 302	Albania
41 488	35 702	33 367	33 333	33 316	36 203	36 690	33 000	32 469	Bulgaria
1 863	1 943	2 246	2 451	2 556	2 358	2 443	2 538	2 778	Cyprus
76	76	85	83	88	88	93	93	93	Gibraltar
1 000	1 309	1 370	1 380	1 400	1 480	1 440	1 465	1 511	Malta
67 856	57 986	52 873	51 721	50 796	52 827	54 971	50 769	47 433	Romania
27 436	25 261	23 752	22 663	23 198	25 974	26 857	26 544	24 716	Slovak Republic
-	-	12 200	11 700	1 793	2 014	2 077	2 136	2 202	*Bosnia-Herzegovina*
-	-	10 194	10 205	10 256	10 699	10 970	11 823	11 926	*Croatia*
-	-	5 900	5 186	4 999	5 351	5 679	5 880	6 180	*FYROM*
-	-	9 611	9 607	10 161	10 313	10 391	10 793	11 061	*Slovenia*
-	-	32 685	30 150	31 108	33 722	34 555	36 558	36 631	*FR of Yugoslavia*
74 794	75 542	70 590	66 848	58 317	62 099	63 672	67 190	68 000	Former Yugoslavia
217 341	**199 882**	**186 680**	**181 345**	**171 547**	**183 123**	**188 975**	**183 929**	**179 302**	**Non-OECD Europe**
-	-	6 763	4 019	3 433	3 379	3 865	4 770	4 426	Armenia
-	-	16 374	15 400	15 526	14 667	14 809	14 159	15 319	Azerbaijan
-	-	39 974	35 327	31 380	28 441	28 514	29 876	30 982	Belarus
-	-	7 564	6 052	6 368	6 098	6 450	6 734	6 563	Estonia
-	-	9 736	8 331	5 903	5 870	6 238	6 200	6 839	Georgia
-	-	88 023	74 170	69 865	63 913	56 651	48 666	44 442	Kazakhstan
-	-	8 616	8 466	8 190	7 524	7 127	6 819	7 191	Kyrgyzstan
-	-	6 898	5 250	4 974	4 963	4 895	5 000	5 108	Latvia
-	-	9 541	8 874	8 700	8 632	9 838	9 773	9 995	Lithuania
-	-	9 281	8 901	7 316	6 510	6 336	5 800	5 295	Republic of Moldova
-	-	908 115	850 142	769 972	756 947	743 237	730 054	715 908	Russia
-	-	16 454	14 373	14 326	13 811	13 559	12 152	12 676	Tajikistan
-	-	7 793	8 274	6 868	6 574	6 169	5 102	5 329	Turkmenistan
-	-	224 662	206 002	180 141	172 235	155 973	149 441	142 137	Ukraine
-	-	45 651	44 284	43 064	42 020	42 450	42 975	42 808	Uzbekistan
1 550 000	**1 518 528**	**1 405 445**	**1 297 865**	**1 176 026**	**1 141 584**	**1 106 111**	**1 077 521**	**1 055 018**	**Former USSR**

Consumption of Electricity (GWh)
Consommation d'électricité (GWh)
Verbrauch von Elektrizität (GWh)

Consumo di energia elettrica (GWh)

電力消費量 *(GWh)*

Consumo de electricidad (GWh)

Потребление электроэнергии (ГВт.ч)

	1971	1973	1978	1984	1985	1986	1987	1988	1989
Bahrein	430	500	1 212	2 093	2 429	2 887	2 708	2 954	3 092
République Islamique d'Iran	7 366	11 027	17 600	32 119	34 840	36 889	40 540	42 092	46 518
Irak	2 660	3 343	7 443	18 486	19 924	21 173	21 332	22 250	22 660
Israël	7 024	8 187	11 114	13 972	14 753	14 647	16 412	17 884	19 104
Jordanie	237	338	574	1 961	2 160	2 343	2 673	2 780	2 930
Koweit	2 623	3 651	6 983	13 894	15 417	16 934	18 092	19 599	21 085
Liban	1 237	1 612	2 115	3 447	3 511	3 780	4 176	3 600	2 250
Oman	11	40	399	1 720	2 132	2 714	2 895	3 220	3 351
Qatar	316	419	1 213	3 263	3 734	4 088	4 151	4 271	4 394
Arabie saoudite	1 927	2 716	9 381	36 986	41 904	45 866	48 906	51 531	55 201
Syrie	1 158	1 283	2 406	6 068	6 583	6 674	6 709	7 360	8 179
Emirats arabes unis	189	670	3 501	10 230	11 160	11 917	12 701	13 801	14 207
Yémen	209	206	360	776	823	972	1 041	1 185	1 212
Moyen-Orient	**25 387**	**33 992**	**64 301**	**145 015**	**159 370**	**170 884**	**182 336**	**192 527**	**204 183**
Total non-OCDE	**1 318 792**	**1 538 740**	**2 158 026**	**2 935 941**	**3 084 833**	**3 266 011**	**3 443 063**	**3 622 180**	**3 747 278**
OCDE Amérique du N.	1 789 307	2 079 898	2 473 291	2 869 343	2 943 602	2 989 260	3 145 986	3 296 548	3 396 906
OCDE Pacifique	432 675	528 815	662 578	801 277	833 158	851 119	906 767	960 575	1 021 573
OCDE Europe	1 302 703	1 501 890	1 796 937	2 091 135	2 186 840	2 241 102	2 324 634	2 386 677	2 444 405
Total OCDE	**3 524 685**	**4 110 603**	**4 932 806**	**5 761 755**	**5 963 600**	**6 081 481**	**6 377 387**	**6 643 800**	**6 862 884**
Monde	**4 843 477**	**5 649 343**	**7 090 832**	**8 697 696**	**9 048 433**	**9 347 492**	**9 820 450**	**10 265 980**	**10 610 162**

Consumption of Electricity (GWh)
Consommation d'électricité (GWh)
Verbrauch von Elektrizität (GWh)
Consumo di energia elettrica (GWh)
電力消費量 *(GWh)*
Consumo de electricidad (GWh)
Потребление электроэнергии (ГВт.ч)

1990	1991	1992	1993	1994	1995	1996	1997	1998	
3 240	3 163	3 700	4 044	4 386	4 550	4 800	4 955	5 313	Bahrain
52 227	56 559	60 351	66 199	72 174	74 689	79 265	83 384	87 653	Islamic Republic of Iran
22 800	19 610	25 300	26 300	28 000	29 000	29 000	29 561	30 346	Iraq
19 461	19 703	22 946	24 285	26 532	28 434	30 423	32 029	34 757	Israel
3 330	3 387	3 969	4 291	4 640	5 088	5 462	5 624	6 046	Jordan
18 477	10 780	16 885	20 178	22 802	23 724	25 475	26 724	29 984	Kuwait
1 400	2 800	3 371	3 964	4 355	4 681	6 730	8 000	7 661	Lebanon
3 963	4 080	4 453	5 067	5 366	5 650	5 862	6 265	6 898	Oman
4 568	4 381	4 873	5 199	5 455	5 314	6 184	6 456	7 596	Qatar
58 973	63 632	67 492	74 171	82 182	92 507	98 398	104 774	106 869	Saudi Arabia
8 573	8 978	9 262	12 638	15 182	15 300	16 885	17 956	18 315	Syria
15 543	15 789	17 007	19 774	21 600	22 734	24 181	25 902	30 387	United Arab Emirates
1 471	1 586	1 738	1 826	1 664	1 760	1 734	1 900	1 863	Yemen
214 026	**214 448**	**241 347**	**267 936**	**294 338**	**313 431**	**334 399**	**353 530**	**373 688**	**Middle East**
3 847 650	**3 903 826**	**3 908 881**	**3 987 658**	**4 055 450**	**4 223 524**	**4 377 549**	**4 522 625**	**4 627 547**	**Non-OECD Total**
3 476 484	3 632 527	3 653 401	3 771 535	3 878 494	3 985 277	4 093 463	4 164 249	4 255 271	OECD North America
1 099 032	1 140 788	1 161 274	1 190 803	1 290 618	1 339 897	1 388 805	1 441 492	1 448 379	OECD Pacific
2 470 523	2 514 905	2 534 039	2 537 825	2 569 353	2 638 631	2 710 909	2 759 224	2 826 803	OECD Europe
7 046 039	**7 288 220**	**7 348 714**	**7 500 163**	**7 738 465**	**7 963 805**	**8 193 177**	**8 364 965**	**8 530 453**	**OECD Total**
10 893 689	11 192 046	11 257 595	11 487 821	11 793 915	12 187 329	12 570 726	12 887 590	13 158 000	**World**

Industry Consumption of Coking Coal (1000 tonnes)
Consommation industrielle de charbon à coke (1000 tonnes)
Industrieverbrauch von Kokskohle (1000 Tonnen)
Consumo di carbone siderurgico nell'industria (1000 tonnellate)
原料炭の産業用消費量（千トン）
Consumo industrial de carbón coquizable (1000 toneladas)
Потребление коксующихся углей промышленным сектором (тыс. m)

	1971	1973	1978	1984	1985	1986	1987	1988	1989
Brésil	-	-	-	-	-	-	-	-	-
Jamaïque	-	-	-	-	-	-	-	8	56
Amérique latine	-	-	-	-	-	-	-	8	56
Inde	-	-	8 921	13 769	15 215	16 009	17 838	17 714	19 742
Asie	-	-	**8 921**	**13 769**	**15 215**	**16 009**	**17 838**	**17 714**	**19 742**
Roumanie	-	-	-	-	-	-	-	-	-
République slovaque	-	-	27	31	32	31	32	34	38
Ex-RYM	-	-	-	-	-	-	-	-	-
Ex-Yougoslavie	-	-	-	-	-	-	-	-	-
Europe non-OCDE	-	-	**27**	**31**	**32**	**31**	**32**	**34**	**38**
Kazakhstan	-	-	-	-	-	-	-	-	-
Ex-URSS	-	-	-	-	-	-	-	-	-
Total non-OCDE	-	-	**8 948**	**13 800**	**15 247**	**16 040**	**17 870**	**17 756**	**19 836**
OCDE Pacifique	-	-	116	38	264	749	738	558	865
OCDE Europe	-	-	116	412	2 194	1 433	1 319	1 403	1 292
Total OCDE	-	-	**232**	**450**	**2 458**	**2 182**	**2 057**	**1 961**	**2 157**
Monde	-	-	**9 180**	**14 250**	**17 705**	**18 222**	**19 927**	**19 717**	**21 993**

Ex-URSS: les séries antérieures à 1990-1992 ne sont pas comparables aux années récentes; 1991 a été estimé.

Industry Consumption of Coking Coal (1000 tonnes)
Consommation industrielle de charbon à coke (1000 tonnes)
Industrieverbrauch von Kokskohle (1000 Tonnen)

Consumo di carbone siderurgico nell'industria (1000 tonnellate)

原料炭の産業用消費量（千トン）

Consumo industrial de carbón coquizable (1000 toneladas)

Потребление коксующихся углей промышленным сектором (тыс. т)

1990	1991	1992	1993	1994	1995	1996	1997	1998	
-	-	-	235	354	834	1 596	2 234	2 706	Brazil
62	9	65	71	53	55	64	56	27	Jamaica
62	**9**	**65**	**306**	**407**	**889**	**1 660**	**2 290**	**2 733**	**Latin America**
18 700	20 110	20 150	21 740	22 650	18 320	17 643	21 729	22 689	India
18 700	**20 110**	**20 150**	**21 740**	**22 650**	**18 320**	**17 643**	**21 729**	**22 689**	**Asia**
-	-	5	12	6	94	29	1	1	Romania
6	28	-	158	296	314	394	293	312	Slovak Republic
-	-	71	74	66	70	49	49	113	*FYROM*
-	75	71	74	66	70	49	49	113	Former Yugoslavia
6	**103**	**76**	**244**	**368**	**478**	**472**	**343**	**426**	**Non-OECD Europe**
-	-	20 855	15 206	12 116	9 204	9 503	9 199	6 281	Kazakhstan
-	**-**	**20 855**	**15 206**	**12 116**	**9 204**	**9 503**	**9 199**	**6 281**	**Former USSR**
18 768	**20 222**	**41 146**	**37 496**	**35 541**	**28 891**	**29 278**	**33 561**	**32 129**	**Non-OECD Total**
1 238	1 699	1 338	1 584	1 860	1 922	1 913	2 190	2 533	OECD Pacific
2 448	2 011	2 231	2 357	2 316	1 199	1 578	2 439	2 100	OECD Europe
3 686	**3 710**	**3 569**	**3 941**	**4 176**	**3 121**	**3 491**	**4 629**	**4 633**	**OECD Total**
22 454	**23 932**	**44 715**	**41 437**	**39 717**	**32 012**	**32 769**	**38 190**	**36 762**	**World**

Former USSR: series up to 1990-1992 are not comparable with recent years; 1991 is estimated.

Industry Consumption of Other Bituminous Coal (1000 tonnes)
Consommation industrielle d'autres charbons bitumineux (1000 tonnes)
Industrieverbrauch von sonstiger bituminöser Kohle (1000 Tonnen)
Consumo di altri carboni bituminosi nell'industria (1000 tonnellate)
その他歴青炭の産業用消費量（千トン）
Consumo industrial de otro carbón bituminoso (1000 toneladas)
Потребление других битуминозных углей промышленным сектором (тыс. m)

	1971	1973	1978	1984	1985	1986	1987	1988	1989
RD du Congo	-	-	167	157	156	160	165	166	170
Kenya	-	-	10	83	90	80	89	113	131
Maroc	-	-	28	282	523	495	439	447	772
Mozambique	-	-	247	94	65	39	32	28	34
Nigéria	-	-	124	53	86	75	80	31	33
Afrique du Sud	-	-	16 837	16 360	16 984	16 581	17 223	18 245	18 812
Rép. Unie de Tanzanie	-	-	1	10	15	4	3	3	2
Tunisie	-	-	27	24	21	12	16	29	30
Zambie	-	-	435	365	386	425	500	541	312
Zimbabwe	-	-	514	1 008	981	931	966	933	845
Autre Afrique	-	-	225	265	291	247	251	280	264
Afrique	**-**	**-**	**18 615**	**18 701**	**19 598**	**19 049**	**19 764**	**20 816**	**21 405**
Argentine	-	-	31	131	14	17	7	29	17
Brésil	-	-	440	3 115	3 401	3 712	3 898	3 365	2 691
Chili	-	-	378	540	614	714	643	702	771
Colombie	-	-	1 740	1 648	1 836	1 920	2 124	1 920	2 196
Costa Rica	-	-	1	1	1	1	1	1	1
Cuba	-	-	91	79	126	119	85	95	167
Haiti	-	-		54	61	18	18	31	12
Panama	-	-	-	-	38	38	38	18	19
Pérou	-	-	26	89	114	107	100	80	91
Uruguay	-	-	2	-	-	-	-	-	-
Vénézuela	-	-	63	51	42	57	62	71	413
Amérique latine	**-**	**-**	**2 772**	**5 708**	**6 247**	**6 703**	**6 976**	**6 312**	**6 378**
Bangladesh	-	-	250	62	98	148	233	240	250
Inde	-	-	27 757	24 454	25 772	30 052	32 946	38 073	41 483
Indonésie	-	-	100	291	283	290	300	760	1 407
RPD de Corée	-	-	23 616	27 408	29 307	31 839	35 386	41 238	43 252
Malaisie	-	-	33	385	517	383	467	270	849
Myanmar	-	-	218	204	203	206	59	52	51
Népal	-	-	14	135	17	87	80	72	12
Pakistan	-	-	1 187	1 811	2 174	2 148	2 221	2 705	2 586
Philippines	-	-	165	813	954	743	958	1 155	724
Taipei chinois	-	-	1 658	2 968	3 118	3 366	3 580	4 353	4 170
Thailande	-	-	52	227	212	183	223	304	392
Viêt-Nam	-	-	-	2 246	2 576	2 850	3 146	2 321	2 163
Autre Asie	-	-	-	273	345	351	343	333	348
Asie	**-**	**-**	**55 050**	**61 277**	**65 576**	**72 646**	**79 942**	**91 876**	**97 687**
Rép. populaire de Chine	-	-	-	295 575	312 848	338 003	356 266	363 511	372 430
Hong-Kong, Chine	-	-	6	-	-	-	-	-	-
Chine	**-**	**-**	**6**	**295 575**	**312 848**	**338 003**	**356 266**	**363 511**	**372 430**
Albanie	-	-	143	178	187	195	195	203	203
Bulgarie	-	-	1 910	2 472	3 754	2 554	2 733	70	31
Chypre	-	-	-	52	74	55	151	91	102
Roumanie	-	-	-	-	1 460	1 705	1 812	1 747	232
République slovaque	-	-	1 237	1 121	1 114	1 140	1 143	1 126	1 058
Croatie	-	-	-	-	-	-	-	-	-
Ex-RYM	-	-	-	-	-	-	-	-	-
Slovénie	-	-	-	-	-	-	-	-	-
Ex-Yougoslavie	-	-	172	288	89	259	236	75	247
Europe non-OCDE	**-**	**-**	**3 462**	**4 111**	**6 678**	**5 908**	**6 270**	**3 312**	**1 873**

Rép. populaire de Chine: jusqu'en 1978 la ventilation de la consommation finale par secteur est incomplète

Industry Consumption of Other Bituminous Coal (1000 tonnes)
Consommation industrielle d'autres charbons bitumineux (1000 tonnes)
Industrieverbrauch von sonstiger bituminöser Kohle (1000 Tonnen)

Consumo di altri carboni bituminosi nell'industria (1000 tonnellate)

その他歴青炭の産業用消費量（千トン）

Consumo industrial de otro carbón bituminoso (1000 toneladas)

Потребление других битуминозных углей промышленным сектором (тыс. m)

1990	1991	1992	1993	1994	1995	1996	1997	1998	
169	174	170	134	136	136	136	136	128	DR of Congo
151	151	158	131	123	156	144	148	122	Kenya
574	798	605	505	526	289	874	672	722	Morocco
34	32	41	54	60	56	-	-	-	Mozambique
43	43	53	60	70	10	10	10	10	Nigeria
17 865	16 351	14 987	13 170	13 677	16 354	18 040	19 479	23 873	South Africa
3	3	3	3	3	4	4	4	4	United Rep. of Tanzania
15	12	15	14	-	-	-	-	-	Tunisia
297	234	267	200	119	124	118	118	118	Zambia
1 190	1 150	1 150	1 096	275	333	322	285	328	Zimbabwe
226	48	209	205	14	9	10	11	11	Other Africa
20 567	**18 996**	**17 658**	**15 572**	**15 003**	**17 471**	**19 658**	**20 863**	**25 316**	**Africa**
10	25	92	75	32	51	-	24	25	Argentina
2 178	2 845	2 111	1 802	1 893	1 463	1 277	1 034	653	Brazil
647	640	755	692	536	526	635	1 381	834	Chile
2 173	2 264	2 278	2 892	2 899	2 579	2 726	2 811	2 209	Colombia
1	-	-	-	-	-	-	-	-	Costa Rica
153	117	107	86	70	63	12	11	14	Cuba
12	25	26	-	-	-	-	-	-	Haiti
32	46	58	63	52	51	108	57	59	Panama
70	206	207	388	408	364	238	368	411	Peru
1	-	-	-	-	-	-	-	-	Uruguay
332	-	5	39	76	7	22	47	1 327	Venezuela
5 609	**6 168**	**5 639**	**6 037**	**5 966**	**5 104**	**5 018**	**5 733**	**5 532**	**Latin America**
563	180	169	63	59	59	345	635	173	Bangladesh
44 969	50 120	49 130	49 360	60 350	77 760	67 397	61 471	48 721	India
1 804	1 404	1 555	1 737	2 209	1 868	5 122	2 049	2 113	Indonesia
45 266	46 628	47 319	47 991	47 674	47 357	46 966	43 773	41 584	DPR of Korea
732	855	959	1 016	854	1 016	1 038	1 056	1 095	Malaysia
60	61	31	6	4	15	9	11	12	Myanmar
73	82	97	94	103	105	231	249	273	Nepal
3 096	3 026	3 052	3 217	3 487	2 999	3 236	3 191	2 810	Pakistan
795	1 243	1 004	1 093	961	907	1 670	2 184	1 987	Philippines
4 308	4 601	5 325	5 768	5 981	5 795	6 145	5 667	6 326	Chinese Taipei
250	436	489	986	1 424	2 305	3 723	3 036	1 334	Thailand
2 357	2 847	2 793	2 503	2 429	2 693	2 912	2 758	2 143	Vietnam
307	307	217	216	211	213	218	228	228	Other Asia
104 580	**111 790**	**112 140**	**114 050**	**125 746**	**143 092**	**139 012**	**126 308**	**108 799**	**Asia**
357 817	352 265	353 341	372 151	394 340	404 639	372 526	328 519	305 000	People's Rep. of China
-	-	-	-	-	-	-	-	-	Hong Kong, China
357 817	**352 265**	**353 341**	**372 151**	**394 340**	**404 639**	**372 526**	**328 519**	**305 000**	**China**
240	160	50	-	-	-	-	-	-	Albania
60	71	57	79	132	131	120	336	250	Bulgaria
97	97	26	31	27	20	17	19	26	Cyprus
136	37	30	8	10	1	-	6	14	Romania
860	1 083	985	789	699	781	623	706	238	Slovak Republic
-	-	83	68	60	46	61	56	63	*Croatia*
-	-	12	68	61	72	51	51	95	*FYROM*
-	-	13	17	22	20	26	19	14	*Slovenia*
212	110	108	153	143	138	138	126	172	Former Yugoslavia
1 605	**1 558**	**1 256**	**1 060**	**1 011**	**1 071**	**898**	**1 193**	**700**	**Non-OECD Europe**

People's Rep. of China: up to 1978 the breakdown of final consumption by sector is incomplete

Industry Consumption of Other Bituminous Coal (1000 tonnes)
Consommation industrielle d'autres charbons bitumineux (1000 tonnes)
Industrieverbrauch von sonstiger bituminöser Kohle (1000 Tonnen)
Consumo di altri carboni bituminosi nell'industria (1000 tonnellate)
その他歴青炭の産業用消費量（千トン）
Consumo industrial de otro carbón bituminoso (1000 toneladas)
Потребление других битуминозных углей промышленным сектором (тыс. т)

	1971	1973	1978	1984	1985	1986	1987	1988	1989
Azerbaïdjan	-	-	-	-	-	-	-	-	-
Bélarus	-	-	-	-	-	-	-	-	-
Estonie	-	-	-	-	-	-	-	-	-
Kazakhstan	-	-	-	-	-	-	-	-	-
Kirghizistan	-	-	-	-	-	-	-	-	-
Lettonie	-	-	-	-	-	-	-	-	-
Lituanie	-	-	-	-	-	-	-	-	-
République de Moldavie	-	-	-	-	-	-	-	-	-
Russie	-	-	-	-	-	-	-	-	-
Ukraine	-	-	-	-	-	-	-	-	-
Ouzbékistan	-	-	-	-	-	-	-	-	-
Ex-URSS	-	-	92 144	65 964	59 422	52 884	65 742	70 640	68 872
République Islamique d'Iran	-	-	523	921	928	982	1 010	1 030	1 050
Israël	-	-	-	-	-	-	162	137	160
Liban	-	-	1	-	-	-	-	-	-
Syrie	-	-	1	2	-	-	-	-	-
Moyen-Orient	-	-	525	923	928	982	1 172	1 167	1 210
Total non-OCDE	-	-	172 574	452 259	471 297	496 175	536 132	557 634	569 855
OCDE Amérique du N.	-	-	43 377	51 127	51 829	51 446	50 635	51 067	49 256
OCDE Pacifique	-	-	5 575	18 545	20 256	20 303	18 974	21 198	21 363
OCDE Europe	-	-	36 839	47 266	50 148	47 737	48 565	48 154	46 026
Total OCDE	-	-	85 791	116 938	122 233	119 486	118 174	120 419	116 645
Monde	-	-	258 365	569 197	593 530	615 661	654 306	678 053	686 500

Ex-URSS: les séries antérieures à 1990-1992 ne sont pas comparables aux années récentes; 1991 a été estimé.

Industry Consumption of Other Bituminous Coal (1000 tonnes)
Consommation industrielle d'autres charbons bitumineux (1000 tonnes)
Industrieverbrauch von sonstiger bituminöser Kohle (1000 Tonnen)

Consumo di altri carboni bituminosi nell'industria (1000 tonnellate)

その他歴青炭の産業用消費量（千トン）

Consumo industrial de otro carbón bituminoso (1000 toneladas)

Потребление других битуминозных углей промышленным сектором (тыс. т)

1990	1991	1992	1993	1994	1995	1996	1997	1998	
-	-	-	1	-	-	-	-	-	Azerbaijan
-	-	64	43	37	29	21	18	16	Belarus
-	-	56	2	1	44	37	10	11	Estonia
-	-	7 827	9 450	7 921	12 600	9 906	7 608	8 114	Kazakhstan
-	-	990	893	1 520	395	581	1 390	649	Kyrgyzstan
-	-	25	21	30	13	11	12	13	Latvia
-	-	103	68	54	21	16	7	6	Lithuania
-	-	57	18	10	6	3	3	5	Republic of Moldova
-	-	5 879	12 243	8 062	5 181	3 077	2 703	2 351	Russia
-	-	14 983	21 373	13 015	12 154	9 006	8 634	8 707	Ukraine
-	-	-	-	-	30	24	17	16	Uzbekistan
66 484	51 051	29 984	44 112	30 650	30 473	22 682	20 402	19 888	**Former USSR**
1 250	1 350	1 500	607	632	106	148	152	158	Islamic Republic of Iran
18	11	11	10	8	12	26	1	1	Israel
-	-	-	111	112	180	200	200	200	Lebanon
-	-	-	-	-	-	-	-	-	Syria
1 268	1 361	1 511	728	752	298	374	353	359	**Middle East**
557 930	543 189	521 529	553 710	573 468	602 148	560 168	503 371	465 594	**Non-OECD Total**
47 873	46 514	25 592	28 563	25 688	24 178	24 053	26 798	26 787	OECD North America
21 480	22 378	22 841	22 968	22 721	23 629	23 940	24 012	21 850	OECD Pacific
42 399	39 961	38 252	38 685	37 527	43 282	43 016	40 547	36 710	OECD Europe
111 752	108 853	86 685	90 216	85 936	91 089	91 009	91 357	85 347	**OECD Total**
669 682	652 042	608 214	643 926	659 404	693 237	651 177	594 728	550 941	**World**

Former USSR: series up to 1990-1992 are not comparable with recent years; 1991 is estimated.

Industry Consumption of Sub-Bituminous Coal (1000 tonnes)
Consommation industrielle de charbons sous-bitumineux (1000 tonnes)
Industrieverbrauch von subbituminöser Kohle (1000 Tonnen)
Consumo di carbone sub-bituminoso nell'industria (1000 tonnellate)
亜歴青炭の産業用消費量（千トン）
Consumo industrial de carbón sub-bituminoso (1000 toneladas)
Потребление полубитуминозных углей промышленным сектором (тыс. т)

	1971	1973	1978	1984	1985	1986	1987	1988	1989
RPD de Corée	-	-	9 000	12 000	13 000	14 000	15 000	18 000	20 000
Asie	-	-	**9 000**	**12 000**	**13 000**	**14 000**	**15 000**	**18 000**	**20 000**
Bulgarie	-	-	-	-	-	-	-	-	-
Roumanie	-	-	-	-	-	-	-	-	-
Croatie	-	-	-	-	-	-	-	-	-
Slovénie	-	-	-	-	-	-	-	-	-
Ex-Yougoslavie	-	-	-	-	-	-	-	-	-
Europe non-OCDE	-	-	-	-	-	-	-	-	-
Total non-OCDE	-	-	**9 000**	**12 000**	**13 000**	**14 000**	**15 000**	**18 000**	**20 000**
OCDE Amérique du N.	-	-	2 674	5 062	5 155	5 150	5 265	5 614	6 203
OCDE Pacifique	-	-	1 791	2 378	2 715	2 873	2 897	2 961	3 282
OCDE Europe	-	-	21 011	19 433	18 404	18 875	20 374	19 253	16 261
Total OCDE	-	-	**25 476**	**26 873**	**26 274**	**26 898**	**28 536**	**27 828**	**25 746**
Monde	-	-	**34 476**	**38 873**	**39 274**	**40 898**	**43 536**	**45 828**	**45 746**

Industry Consumption of Sub-Bituminous Coal (1000 tonnes)
Consommation industrielle de charbons sous-bitumineux (1000 tonnes)
Industrieverbrauch von subbituminöser Kohle (1000 Tonnen)
Consumo di carbone sub-bituminoso nell'industria (1000 tonnellate)
亜歴青炭の産業用消費量（千トン）
Consumo industrial de carbón sub-bituminoso (1000 toneladas)
Потребление полубитуминозных углей промышленным сектором (тыс. т)

1990	1991	1992	1993	1994	1995	1996	1997	1998	
22 000	23 000	24 000	27 000	26 500	26 000	25 500	23 766	22 578	DPR of Korea
22 000	**23 000**	**24 000**	**27 000**	**26 500**	**26 000**	**25 500**	**23 766**	**22 578**	**Asia**
-	128	108	73	77	80	72	-	-	Bulgaria
-	-	-	-	-	6	14	10	5	Romania
-	-	-	-	-	-	-	55	29	*Croatia*
-	-	64	56	25	14	10	-	-	*Slovenia*
-	128	64	56	25	14	10	55	29	Former Yugoslavia
-	**128**	**172**	**129**	**102**	**100**	**96**	**65**	**34**	**Non-OECD Europe**
22 000	**23 128**	**24 172**	**27 129**	**26 602**	**26 100**	**25 596**	**23 831**	**22 612**	**Non-OECD Total**
6 663	7 053	3 950	4 079	3 453	3 685	3 685	4 413	4 614	OECD North America
3 663	3 723	4 051	3 984	3 851	3 774	3 804	3 653	3 636	OECD Pacific
13 580	10 165	8 860	4 350	3 131	3 633	4 456	3 994	3 459	OECD Europe
23 906	**20 941**	**16 861**	**12 413**	**10 435**	**11 092**	**11 945**	**12 060**	**11 709**	**OECD Total**
45 906	**44 069**	**41 033**	**39 542**	**37 037**	**37 192**	**37 541**	**35 891**	**34 321**	**World**

INTERNATIONAL ENERGY AGENCY

Industry Consumption of Lignite (1000 tonnes)
Consommation industrielle de lignite (1000 tonnes)
Industrieverbrauch von Braunkohle (1000 Tonnen)

Consumo di lignite nell'industria (1000 tonnellate)

亜炭の産業用消費量（チトン）

Consumo industrial de lignito (1000 toneladas)

Потребление лигнита промышленным сектором (тыс. m)

	1971	1973	1978	1984	1985	1986	1987	1988	1989
Inde	-	-	601	504	1 024	639	1 754	1 853	3 287
Myanmar	-	-	3	43	43	47	35	35	35
Philippines	-	-	-	3	4	4	3	3	3
Thailande	-	-	132	361	535	741	1 096	1 303	1 794
Asie	**-**	**-**	**736**	**911**	**1 606**	**1 431**	**2 888**	**3 194**	**5 119**
Albanie	-	-	600	1 005	1 075	950	1 050	1 006	1 006
Bulgarie	-	-	2 753	3 535	3 372	3 852	4 000	2 000	1 805
Roumanie	-	-	-	-	-	-	-	-	2 887
République slovaque	-	-	3 820	3 748	3 965	4 140	4 154	4 051	3 641
Croatie	-	-	-	-	-	-	-	-	-
Ex-RYM	-	-	-	-	-	-	-	-	-
Slovénie	-	-	-	-	-	-	-	-	-
RF de Yougoslavie	-	-	-	-	-	-	-	-	-
Ex-Yougoslavie	-	-	3 300	1 424	1 242	1 361	1 336	1 300	1 300
Europe non-OCDE	**-**	**-**	**10 473**	**9 712**	**9 654**	**10 303**	**10 540**	**8 357**	**10 639**
Estonie	-	-	-	-	-	-	-	-	-
Kazakhstan	-	-	-	-	-	-	-	-	-
Kirghizistan	-	-	-	-	-	-	-	-	-
Russie	-	-	-	-	-	-	-	-	-
Ouzbékistan	-	-	-	-	-	-	-	-	-
Ex-URSS	**-**	**-**	**-**	**-**	**-**	**-**	**-**	**-**	**-**
Total non-OCDE	**-**	**-**	**11 209**	**10 623**	**11 260**	**11 734**	**13 428**	**11 551**	**15 758**
OCDE Amérique du N.	-	-	2 464	2 620	2 049	2 300	2 411	2 512	2 599
OCDE Pacifique	-	-	424	471	438	390	257	427	263
OCDE Europe	-	-	27 628	36 017	39 071	36 372	37 431	37 783	37 620
Total OCDE	**-**	**-**	**30 516**	**39 108**	**41 558**	**39 062**	**40 099**	**40 722**	**40 482**
Monde	**-**	**-**	**41 725**	**49 731**	**52 818**	**50 796**	**53 527**	**52 273**	**56 240**

Ex-URSS: les séries antérieures à 1990-1992 ne sont pas comparables aux années récentes; 1991 a été estimé.

Industry Consumption of Lignite (1000 tonnes)
Consommation industrielle de lignite (1000 tonnes)
Industrieverbrauch von Braunkohle (1000 Tonnen)
Consumo di lignite nell'industria (1000 tonnellate)
亜炭の産業用消費量（チトン）
Consumo industrial de lignito (1000 toneladas)
Потребление лигнита промышленным сектором (тыс. m)

1990	1991	1992	1993	1994	1995	1996	1997	1998	
650	650	650	650	650	650	650	650	650	India
38	39	39	25	22	23	22	20	21	Myanmar
3	3	3	3	3	3	3	3	3	Philippines
2 582	2 818	3 114	4 248	4 916	4 915	4 544	4 468	5 155	Thailand
3 273	**3 510**	**3 806**	**4 926**	**5 591**	**5 591**	**5 219**	**5 141**	**5 829**	**Asia**
843	321	-	-	-	-	-	-	-	Albania
271	19	8	8	7	9	5	70	56	Bulgaria
184	185	119	123	261	166	178	116	422	Romania
3 347	3 284	3 107	2 166	1 879	1 693	1 614	1 318	1 039	Slovak Republic
-	-	137	129	128	141	106	28	19	*Croatia*
-	-	122	169	166	115	113	105	123	*FYROM*
-	-	94	66	42	58	51	7	6	*Slovenia*
-	-	386	2 255	2 300	2 400	2 300	2 400	2 601	*FR of Yugoslavia*
1 666	764	739	2 619	2 636	2 714	2 570	2 540	2 749	Former Yugoslavia
6·311	**4 573**	**3 973**	**4 916**	**4 783**	**4 582**	**4 367**	**4 044**	**4 266**	**Non-OECD Europe**
-	-	336	260	368	500	498	342	366	Estonia
-	-	4 152	4 669	4 559	3 630	3 401	2 309	1 620	Kazakhstan
-	-	1 111	985	550	287	332	330	364	Kyrgyzstan
-	-	2 025	6 148	2 739	2 193	2 714	270	290	Russia
-	-	-	-	-	196	177	131	138	Uzbekistan
-	**-**	**7 624**	**12 062**	**8 216**	**6 806**	**7 122**	**3 382**	**2 778**	**Former USSR**
9 584	**8 083**	**15 403**	**21 904**	**18 590**	**16 979**	**16 708**	**12 567**	**12 873**	**Non-OECD Total**
2 245	2 314	198	384	200	154	170	232	205	OECD North America
174	181	164	170	192	206	214	226	227	OECD Pacific
30 788	22 344	16 761	12 770	9 337	9 494	9 020	9 191	8 402	OECD Europe
33 207	**24 839**	**17 123**	**13 324**	**9 729**	**9 854**	**9 404**	**9 649**	**8 834**	**OECD Total**
42 791	**32 922**	**32 526**	**35 228**	**28 319**	**26 833**	**26 112**	**22 216**	**21 707**	**World**

Former USSR: series up to 1990-1992 are not comparable with recent years; 1991 is estimated.

Industry Consumption of Oil (1000 tonnes)
Consommation industrielle de pétrole (1000 tonnes)
Industrieverbrauch von Öl (1000 Tonnen)
Consumo di petrolio nell'industria (1000 tonnellate)
石油の産業用消費量 （千トン）
Consumo industrial de petróleo (1000 toneladas)
Потребление нефти и нефтепродуктов промышленным сектором (тыс. т)

	1971	1973	1978	1984	1985	1986	1987	1988	1989
Algérie	102	316	318	474	507	412	417	392	315
Angola	56	74	101	151	200	160	169	157	167
Bénin	4	3	5	5	9	9	7	8	10
Cameroun	26	23	52	60	68	68	56	54	58
Congo	32	27	8	18	18	18	-	-	1
RD du Congo	1	2	1	1	1	46	48	49	56
Côte d'Ivoire	225	235	204	64	126	130	135	140	134
Egypte	2 135	2 147	3 241	5 251	3 819	3 060	4 618	4 783	5 003
Ethiopie	87	91	72	67	74	103	102	131	154
Erythrée	-	-	-	-	-	-	-	-	-
Gabon	127	135	344	317	299	198	184	200	90
Ghana	130	138	135	89	76	92	78	95	117
Kenya	189	195	286	306	304	311	377	344	319
Libye	-	63	168	619	685	648	560	480	782
Maroc	620	812	1 232	1 152	1 101	1 000	870	604	829
Mozambique	-	-	-	-	-	-	-	-	-
Nigéria	212	375	833	948	1 029	733	539	794	800
Sénégal	103	111	116	139	136	128	138	123	66
Afrique du Sud	1 900	2 252	2 320	3 227	3 106	2 973	2 966	3 186	3 194
Soudan	188	525	190	210	215	244	142	179	200
Rép. Unie de Tanzanie	76	106	90	86	100	99	96	97	102
Togo	21	18	22	-	17	56	61	52	56
Tunisie	227	259	352	579	608	602	395	585	745
Zambie	149	259	313	196	152	136	135	120	120
Zimbabwe	49	60	49	85	89	103	87	91	99
Autre Afrique	16	18	104	207	201	212	217	265	271
Afrique	**6 675**	**8 244**	**10 556**	**14 251**	**12 940**	**11 541**	**12 397**	**12 929**	**13 688**
Argentine	3 749	3 088	3 109	3 005	2 660	2 071	2 770	2 547	2 250
Bolivie	134	144	137	75	54	31	21	43	51
Brésil	6 799	9 868	15 596	10 506	11 202	11 929	13 462	13 685	14 043
Chili	1 171	1 221	1 295	1 096	1 060	1 104	1 150	1 416	1 828
Colombie	1 432	1 346	1 315	1 150	1 045	1 053	1 028	1 107	1 120
Costa Rica	106	131	173	150	177	161	165	135	201
Cuba	1 810	2 211	2 602	2 505	2 296	2 171	2 313	2 376	2 610
République dominicaine	211	285	391	355	151	232	250	397	290
Equateur	151	133	346	651	783	770	789	699	718
El Salvador	109	139	178	123	112	119	133	161	159
Guatemala	276	221	271	141	142	165	184	204	217
Haiti	48	51	87	40	40	44	46	45	54
Honduras	95	116	161	201	170	160	190	223	263
Jamaïque	625	1 197	1 136	812	624	527	530	495	898
Antilles néerlandaises	1 491	1 637	531	359	344	327	313	257	209
Nicaragua	78	109	95	116	113	121	118	112	95
Panama	26	36	71	123	181	199	199	131	141
Paraguay	11	39	6	11	19	20	27	40	40
Pérou	687	950	1 268	1 078	1 162	1 201	990	844	688
Trinité-et-Tobago	28	25	37	21	6	1	11	86	90
Uruguay	422	417	425	200	160	178	183	182	173
Vénézuela	1 016	983	1 052	1 406	1 388	1 669	1 909	1 983	2 142
Autre Amérique latine	16	15	22	30	34	54	174	170	171
Amérique latine	**20 491**	**24 362**	**30 304**	**24 154**	**23 923**	**24 307**	**26 955**	**27 338**	**28 451**

Industry Consumption of Oil (1000 tonnes)
Consommation industrielle de pétrole (1000 tonnes)
Industrieverbrauch von Öl (1000 Tonnen)

Consumo di petrolio nell'industria (1000 tonnellate)

石油の産業用消費量（チトン）

Consumo industrial de petróleo (1000 toneladas)

Потребление нефти и нефтепродуктов промышленным сектором (тыс. m)

1990	1991	1992	1993	1994	1995	1996	1997	1998	
347	325	280	261	275	256	251	250	146	Algeria
234	161	198	357	178	229	258	251	102	Angola
10	6	8	9	7	9	19	19	34	Benin
54	48	44	48	45	50	49	50	47	Cameroon
2	2	5	4	4	4	4	4	4	Congo
60	61	56	62	62	58	58	58	55	DR of Congo
126	133	129	124	138	149	134	137	160	Cote d'Ivoire
5 350	4 665	5 100	4 485	4 123	5 405	5 216	5 782	7 081	Egypt
174	178	89	129	170	186	199	215	233	Ethiopia
-	-	7	14	16	18	15	16	12	Eritrea
47	49	89	99	92	98	114	114	126	Gabon
93	100	104	112	114	123	127	143	153	Ghana
335	332	349	398	395	369	430	393	400	Kenya
850	970	880	960	1 062	1 238	1 243	1 381	1 399	Libya
770	734	829	542	675	650	602	794	884	Morocco
16	17	15	20	17	19	7	16	39	Mozambique
695	647	1 795	1 418	874	503	746	846	832	Nigeria
76	79	88	71	98	113	110	123	153	Senegal
3 370	3 319	3 050	1 283	1 532	1 363	1 468	1 485	1 586	South Africa
238	295	226	24	270	164	77	199	158	Sudan
105	109	109	105	105	105	105	105	105	United Rep. of Tanzania
44	21	23	3	15	33	70	46	43	Togo
782	694	760	641	586	559	612	581	688	Tunisia
115	120	109	115	116	118	120	120	120	Zambia
93	211	68	73	72	99	100	100	100	Zimbabwe
279	269	260	254	264	237	240	246	246	Other Africa
14 265	**13 545**	**14 670**	**11 611**	**11 305**	**12 155**	**12 374**	**13 474**	**14 906**	**Africa**
2 346	2 276	2 422	2 414	2 406	2 367	2 618	3 295	2 860	Argentina
70	73	70	66	69	78	82	65	72	Bolivia
13 972	13 800	14 343	14 906	15 869	16 684	17 501	20 030	20 180	Brazil
1 840	1 720	1 978	2 158	2 200	2 469	2 654	3 336	3 194	Chile
1 320	1 129	1 496	1 147	1 416	1 592	1 628	1 770	1 568	Colombia
217	208	185	186	209	215	198	224	262	Costa Rica
2 999	2 088	2 555	2 835	3 126	3 367	3 541	3 496	3 389	Cuba
251	238	267	308	326	338	349	373	390	Dominican Republic
883	878	653	586	619	710	667	733	716	Ecuador
162	177	234	234	249	253	223	292	320	El Salvador
229	243	269	331	347	409	389	402	414	Guatemala
49	49	51	41	6	44	49	77	96	Haiti
247	249	262	256	263	329	319	252	297	Honduras
146	161	159	135	218	186	190	195	198	Jamaica
174	174	174	137	137	136	137	137	137	Netherlands Antilles
103	78	81	76	98	89	91	60	109	Nicaragua
125	167	176	164	197	217	150	207	225	Panama
51	46	78	64	104	82	34	40	38	Paraguay
687	634	516	537	600	945	1 279	1 250	1 375	Peru
97	106	91	71	75	83	109	103	109	Trinidad and Tobago
167	171	173	158	145	148	171	209	287	Uruguay
2 126	2 546	2 085	2 736	2 720	2 969	2 901	3 709	3 214	Venezuela
149	148	134	134	138	84	85	89	89	Other Latin America
28 410	**27 359**	**28 452**	**29 680**	**31 537**	**33 794**	**35 365**	**40 344**	**39 539**	**Latin America**

Industry Consumption of Oil (1000 tonnes)
Consommation industrielle de pétrole (1000 tonnes)
Industrieverbrauch von Öl (1000 Tonnen)

Consumo di petrolio nell'industria (1000 tonnellate)

石油の産業用消費量（千トン）

Consumo industrial de petróleo (1000 toneladas)

Потребление нефти и нефтепродуктов промышленным сектором (тыс. m)

	1971	1973	1978	1984	1985	1986	1987	1988	1989
Bangladesh	179	237	272	229	151	98	60	39	30
Brunei	22	22	17	32	36	32	33	38	42
Inde	4 544	5 806	6 819	8 685	9 193	9 411	9 386	10 072	11 603
Indonésie	1 398	1 639	3 604	4 645	5 098	4 621	4 527	4 917	4 835
RPD de Corée	90	95	326	528	572	752	805	867	885
Malaisie	1 917	1 496	1 910	2 460	2 374	2 214	2 405	2 506	2 961
Myanmar	363	256	384	355	349	310	200	174	167
Népal	-	1	1	7	17	20	16	15	2
Pakistan	153	247	223	689	815	946	1 227	1 225	1 291
Philippines	1 574	2 012	2 483	1 607	1 408	1 455	1 734	1 898	2 109
Singapour	75	101	235	606	1 175	1 111	1 383	1 376	1 460
Sri Lanka	95	175	179	197	100	92	87	109	132
Taipei chinois	1 167	2 004	5 440	6 528	6 993	7 819	8 270	8 552	8 192
Thailande	1 324	1 570	1 858	1 622	1 621	1 637	1 865	1 918	2 316
Viêt-Nam	-	-	-	642	383	402	413	372	374
Autre Asie	40	49	57	119	120	109	106	100	100
Asie	**12 941**	**15 710**	**23 808**	**28 951**	**30 405**	**31 029**	**32 517**	**34 178**	**36 499**
Rép. populaire de Chine	6 697	8 776	22 719	21 086	24 018	24 452	27 361	28 973	30 803
Hong-Kong, Chine	772	654	1 461	1 063	909	1 010	1 088	1 307	1 274
Chine	**7 469**	**9 430**	**24 180**	**22 149**	**24 927**	**25 462**	**28 449**	**30 280**	**32 077**
Albanie	213	50	-	189	117	157	187	112	152
Bulgarie	-	-	-	-	-	-	-	2 349	2 287
Chypre	125	164	210	179	149	212	209	267	233
Malte	-	-	-	-	-	-	-	-	-
Roumanie	724	754	1 061	-	1 171	1 328	2 206	2 302	2 865
République slovaque	804	882	995	2 234	2 226	2 108	2 044	1 914	2 936
Bosnie-Herzegovine	-	-	-	-	-	-	-	-	-
Croatie	-	-	-	-	-	-	-	-	-
Ex-RYM	-	-	-	-	-	-	-	-	-
Slovénie	-	-	-	-	-	-	-	-	-
RF de Yougoslavie	-	-	-	-	-	-	-	-	-
Ex-Yougoslavie	2 320	3 205	4 175	2 936	2 808	3 055	3 014	3 251	3 022
Europe non-OCDE	**4 186**	**5 055**	**6 441**	**5 538**	**6 471**	**6 860**	**7 660**	**10 195**	**11 495**
Arménie	-	-	-	-	-	-	-	-	-
Azerbaïdjan	-	-	-	-	-	-	-	-	-
Bélarus	-	-	-	-	-	-	-	-	-
Estonie	-	-	-	-	-	-	-	-	-
Géorgie	-	-	-	-	-	-	-	-	-
Lettonie	-	-	-	-	-	-	-	-	-
Lituanie	-	-	-	-	-	-	-	-	-
République de Moldavie	-	-	-	-	-	-	-	-	-
Russie	-	-	-	-	-	-	-	-	-
Ukraine	-	-	-	-	-	-	-	-	-
Ouzbékistan	-	-	-	-	-	-	-	-	-
Ex-URSS	**57 000**	**67 100**	**82 900**	**81 000**	**76 816**	**76 812**	**76 117**	**76 648**	**74 451**

Ex-URSS: les séries antérieures à 1990-1992 ne sont pas comparables aux années récentes; 1991 a été estimé.

Rép. populaire de Chine: jusqu'en 1978 la ventilation de la consommation finale par secteur est incomplète.

Industry Consumption of Oil (1000 tonnes)
Consommation industrielle de pétrole (1000 tonnes)
Industrieverbrauch von Öl (1000 Tonnen)

Consumo di petrolio nell'industria (1000 tonnellate)

石油の産業用消費量（千トン）

Consumo industrial de petróleo (1000 toneladas)

Потребление нефти и нефтепродуктов промышленным сектором (тыс. т)

1990	1991	1992	1993	1994	1995	1996	1997	1998	
25	150	141	164	197	230	248	262	214	Bangladesh
44	45	56	58	62	67	76	88	66	Brunei
12 173	12 297	12 608	12 477	13 376	14 139	15 404	15 796	17 452	India
5 243	5 295	5 586	6 332	6 824	7 245	7 650	8 084	7 667	Indonesia
859	837	837	833	826	816	807	752	714	DPR of Korea
3 440	3 741	3 920	4 853	4 578	5 997	5 553	5 880	5 568	Malaysia
130	131	215	104	107	83	139	480	162	Myanmar
22	29	29	28	28	92	90	101	108	Nepal
1 297	1 147	1 163	1 479	1 654	1 869	2 416	2 141	2 081	Pakistan
1 972	1 903	2 009	2 232	2 640	3 323	3 354	3 453	3 389	Philippines
1 702	1 570	1 383	1 521	1 519	1 881	2 174	2 675	3 464	Singapore
123	117	191	207	198	172	435	741	481	Sri Lanka
8 526	8 475	8 687	8 937	10 014	10 271	10 579	10 721	10 410	Chinese Taipei
2 730	2 943	3 560	3 887	4 790	5 224	6 246	5 606	5 073	Thailand
569	478	1 192	1 171	1 315	679	747	786	924	Vietnam
103	103	103	96	98	20	20	20	20	Other Asia
38 958	**39 261**	**41 680**	**44 379**	**48 226**	**52 108**	**55 938**	**57 586**	**57 793**	**Asia**
29 706	33 215	35 585	33 012	36 585	37 350	40 169	39 929	43 544	People's Rep. of China
946	674	1 167	922	1 015	895	808	793	1 353	Hong Kong, China
30 652	**33 889**	**36 752**	**33 934**	**37 600**	**38 245**	**40 977**	**40 722**	**44 897**	**China**
303	195	111	138	209	69	61	29	29	Albania
2 067	757	1 205	1 182	1 077	1 198	1 368	2 020	1 884	Bulgaria
135	326	335	351	363	375	419	391	398	Cyprus
-	-	-	-	-	-	-	-	3	Malta
2 082	2 486	1 487	900	1 709	1 701	1 915	1 791	1 854	Romania
2 122	1 836	1 814	1 191	1 382	1 146	1 091	1 047	967	Slovak Republic
-	-	80	179	59	60	63	64	66	Bosnia-Herzegovina
-	-	562	580	628	614	535	632	644	Croatia
-	-	339	208	141	134	263	187	161	FYROM
-	-	181	174	202	210	210	161	159	Slovenia
-	-	836	474	421	455	683	1 048	844	FR of Yugoslavia
2 879	2 437	1 998	1 615	1 451	1 473	1 754	2 092	1 874	Former Yugoslavia
9 588	**8 037**	**6 950**	**5 377**	**6 191**	**5 962**	**6 608**	**7 370**	**7 009**	**Non-OECD Europe**
-	-	200	523	84	66	13	65	70	Armenia
-	-	-	302	196	163	157	150	161	Azerbaijan
-	-	1 507	2 024	1 707	1 585	1 444	1 540	1 700	Belarus
-	-	170	116	156	267	255	203	152	Estonia
-	-	-	-	-	-	58	61	72	Georgia
-	-	199	195	117	141	64	283	236	Latvia
-	-	996	387	438	542	272	186	178	Lithuania
-	-	40	88	32	30	27	23	15	Republic of Moldova
-	-	25 624	28 501	17 316	14 379	12 562	14 918	13 640	Russia
-	-	9 164	6 244	5 172	5 637	4 205	3 848	3 930	Ukraine
-	-	273	-	-	148	413	398	402	Uzbekistan
72 542	**74 258**	**38 173**	**38 380**	**25 218**	**22 958**	**19 470**	**21 675**	**20 556**	**Former USSR**

Former USSR: series up to 1990-1992 are not comparable with recent years; 1991 is estimated.

People's Rep. of China: up to 1978 the breakdown of final consumption by sector is incomplete.

Industry Consumption of Oil (1000 tonnes)
Consommation industrielle de pétrole (1000 tonnes)
Industrieverbrauch von Öl (1000 Tonnen)

Consumo di petrolio nell'industria (1000 tonnellate)

石油の産業用消費量（千トン）

Consumo industrial de petróleo (1000 toneladas)

Потребление нефти и нефтепродуктов промышленным сектором (тыс. т)

	1971	1973	1978	1984	1985	1986	1987	1988	1989
République Islamique d'Iran	1 286	1 748	3 495	5 175	5 040	7 009	8 065	8 035	7 533
Irak	65	32	90	310	244	379	-	100	800
Israël	833	944	1 215	1 319	1 318	1 502	1 511	1 560	1 447
Jordanie	47	74	116	327	386	431	420	382	383
Koweit	-	-	170	226	229	228	229	237	142
Liban	464	494	431	36	51	50	97	67	105
Oman	6	9	13	11	9	10	10	5	10
Qatar	-	-	6	335	343	450	504	454	622
Arabie saoudite	632	1 179	3 618	5 844	7 012	7 612	7 433	6 632	7 093
Syrie	511	249	1 572	2 015	2 084	2 020	2 045	1 758	670
Emirats arabes unis	-	-	534	1 631	1 751	1 793	1 797	1 916	1 942
Yémen	-	4	107	224	379	362	317	304	310
Moyen-Orient	**3 844**	**4 733**	**11 367**	**17 453**	**18 846**	**21 846**	**22 428**	**21 450**	**21 057**
Total non-OCDE	**112 606**	**134 634**	**189 556**	**193 496**	**194 328**	**197 857**	**206 523**	**213 018**	**217 718**
OCDE Amérique du N.	112 634	131 228	142 872	122 669	117 317	119 779	121 296	119 994	117 829
OCDE Pacifique	85 855	98 157	86 803	67 833	66 080	66 692	69 588	73 148	76 658
OCDE Europe	172 897	191 180	169 791	117 348	112 988	113 907	113 886	114 503	110 261
Total OCDE	**371 386**	**420 565**	**399 466**	**307 850**	**296 385**	**300 378**	**304 770**	**307 645**	**304 748**
Monde	**483 992**	**555 199**	**589 022**	**501 346**	**490 713**	**498 235**	**511 293**	**520 663**	**522 466**

Industry Consumption of Oil (1000 tonnes)
Consommation industrielle de pétrole (1000 tonnes)
Industrieverbrauch von Öl (1000 Tonnen)
Consumo di petrolio nell'industria (1000 tonnellate)
石油の産業用消費量（チトン）
Consumo industrial de petróleo (1000 toneladas)
Потребление нефти и нефтепродуктов промышленным сектором (тыс. т)

1990	1991	1992	1993	1994	1995	1996	1997	1998	
130	140	140	5 078	5 126	5 208	5 358	6 420	7 629	Islamic Republic of Iran
900	2 090	1 310	1 900	2 409	2 310	2 217	2 229	2 316	Iraq
1 431	1 480	1 337	1 407	1 532	1 821	1 923	2 040	2 171	Israel
416	367	415	413	527	554	556	589	635	Jordan
43	236	166	135	134	162	128	148	153	Kuwait
105	100	70	411	547	726	627	827	564	Lebanon
511	1 666	1 432	970	525	958	828	446	284	Oman
627	674	726	781	815	904	974	1 129	1 184	Qatar
7 450	7 560	7 677	8 024	7 648	8 028	8 855	9 513	12 292	Saudi Arabia
650	750	810	832	948	1 047	1 084	939	1 066	Syria
1 967	2 021	2 118	2 255	2 235	1 879	1 463	1 472	1 433	United Arab Emirates
108	328	403	136	145	145	163	164	164	Yemen
14 338	**17 412**	**16 604**	**22 342**	**22 591**	**23 742**	**24 176**	**25 916**	**29 891**	**Middle East**
208 753	**213 761**	**183 281**	**185 703**	**182 668**	**188 964**	**194 908**	**207 087**	**214 591**	**Non-OECD Total**
109 361	105 417	111 046	110 310	114 379	110 910	115 758	113 999	112 725	OECD North America
80 596	83 609	88 498	91 428	94 964	97 817	99 485	102 538	100 348	OECD Pacific
104 779	105 435	103 261	100 422	103 168	107 659	105 872	108 090	108 485	OECD Europe
294 736	**294 461**	**302 805**	**302 160**	**312 511**	**316 386**	**321 115**	**324 627**	**321 558**	**OECD Total**
503 489	**508 222**	**486 086**	**487 863**	**495 179**	**505 350**	**516 023**	**531 714**	**536 149**	**World**

Industry Consumption of Natural Gas (TJ)
Consommation industrielle de gaz naturel (TJ)
Industrieverbrauch von Erdgas (TJ)
Consumo di gas naturale nell'industria (TJ)
天然ガスの産業用消費量 (TJ)
Consumo industrial de gas natural (TJ)
Потребление природного газа промышленным сектором (ТДж)

	1971	1973	1978	1984	1985	1986	1987	1988	1989
Algérie	8 379	9 531	29 055	42 353	54 951	64 800	70 200	68 700	55 200
Angola	1 678	2 536	2 758	4 499	4 499	4 997	5 998	6 098	6 500
Congo	2 256	-	-	-	-	-	-	-	-
Egypte	-	-	15 101	65 763	66 722	117 396	109 408	110 750	117 396
Gabon	-	-	-	1 546	1 972	2 263	2 224	3 125	30
Libye	-	-	30 408	57 306	62 674	74 830	66 305	70 094	84 776
Maroc	2 009	2 721	3 177	3 244	3 666	3 549	2 884	2 947	2 177
Nigéria	1 411	1 352	1 398	14 343	17 188	18 149	21 296	26 957	31 825
Tunisie	38	331	2 888	5 085	5 902	7 425	12 364	6 203	8 830
Afrique	**15 771**	**16 471**	**84 785**	**194 139**	**217 574**	**293 409**	**290 679**	**294 874**	**306 734**
Argentine	79 484	107 298	130 036	174 499	174 996	195 347	206 282	178 595	203 583
Bolivie	-	155	2 203	2 087	2 783	3 208	3 556	3 401	6 648
Brésil	1 279	3 875	24 298	54 100	63 091	73 941	86 728	84 444	90 102
Chili	38	80	389	301	297	331	402	406	398
Colombie	4 905	9 832	22 857	29 974	32 508	33 901	34 184	35 970	36 578
Cuba	67	565	427	126	266	228	874	722	988
Trinité-et-Tobago	21 363	24 192	39 896	76 235	76 094	82 363	83 502	111 476	113 602
Uruguay	-	-	-	-	-	-	-	-	-
Vénézuela	79 483	158 745	223 163	320 234	341 191	303 753	337 520	301 582	311 985
Autre Amérique latine	-	-	135	200	393	39	39	39	39
Amérique latine	**186 619**	**304 742**	**443 404**	**657 756**	**691 619**	**693 111**	**753 087**	**716 635**	**763 923**
Bangladesh	8 978	15 248	19 325	43 585	46 238	54 764	61 244	68 375	72 642
Inde	8 403	7 674	31 898	70 140	108 640	144 096	150 913	233 358	286 906
Indonésie	4 637	4 968	52 330	182 602	95 087	97 134	102 809	95 180	70 758
Malaisie	-	-	42	3 640	19 581	41 547	44 267	41 756	42 886
Myanmar	755	1 164	3 116	7 895	10 745	12 014	12 363	12 485	12 078
Pakistan	65 540	78 450	104 363	161 193	162 048	161 299	183 924	188 370	200 933
Taipei chinois	28 881	40 670	57 355	27 180	21 195	16 650	17 052	18 420	16 805
Thailande	-	-	-	8 268	7 591	3 706	1 684	2 563	4 866
Asie	**117 194**	**148 174**	**268 429**	**504 503**	**471 125**	**531 210**	**574 256**	**660 507**	**707 874**
Rép. populaire de Chine	65 591	104 864	293 789	269 104	314 921	296 237	311 053	291 834	312 161
Chine	**65 591**	**104 864**	**293 789**	**269 104**	**314 921**	**296 237**	**311 053**	**291 834**	**312 161**
Albanie	-	-	-	-	-	-	-	-	10 000
Bulgarie	-	-	-	-	171 632	177 181	187 692	99 898	103 693
Roumanie	605 483	532 592	1 170 245	1 290 428	873 008	897 612	887 519	919 990	909 942
République slovaque	32 162	38 358	75 274	105 357	103 907	104 896	106 142	107 785	113 642
Croatie	-	-	-	-	-	-	-	-	-
Ex-RYM	-	-	-	-	-	-	-	-	-
Slovénie	-	-	-	-	-	-	-	-	-
RF de Yougoslavie	-	-	-	-	-	-	-	-	-
Ex-Yougoslavie	32 786	40 013	56 964	111 824	119 692	126 657	140 524	104 914	105 480
Europe non-OCDE	**670 431**	**610 963**	**1 302 483**	**1 507 609**	**1 268 239**	**1 306 346**	**1 321 877**	**1 232 587**	**1 242 757**

Rép. populaire de Chine: jusqu'en 1978 la ventilation de la consommation finale par secteur est incomplète.

Industry Consumption of Natural Gas (TJ)
Consommation industrielle de gaz naturel (TJ)
Industrieverbrauch von Erdgas (TJ)
Consumo di gas naturale nell'industria (TJ)
天然ガスの産業用消費量 (TJ)
Consumo industrial de gas natural (TJ)
Потребление природного газа промышленным сектором (ТДж)

1990	1991	1992	1993	1994	1995	1996	1997	1998	
104 345	111 194	124 821	127 875	113 929	111 869	106 982	116 419	116 938	Algeria
20 520	22 040	21 660	21 280	19 760	21 280	21 280	21 660	22 040	Angola
-	-	-	-	-	-	-	-	-	Congo
109 408	110 750	116 060	140 086	101 251	105 948	106 058	111 492	131 784	Egypt
30	30	30	30	25	27	29	28	28	Gabon
78 303	78 303	77 829	75 145	77 513	76 566	82 407	86 354	86 038	Libya
2 014	1 289	847	828	809	512	651	1 200	1 317	Morocco
33 326	30 326	32 500	34 600	37 714	34 433	34 898	36 490	39 144	Nigeria
11 866	12 517	13 816	14 514	15 308	16 843	17 720	17 896	18 603	Tunisia
359 812	**366 449**	**387 563**	**414 358**	**366 309**	**367 478**	**370 025**	**391 539**	**415 892**	**Africa**
200 279	204 049	215 868	218 707	249 465	273 429	274 454	290 600	299 675	Argentina
7 382	8 155	9 701	10 962	12 229	13 417	11 820	16 006	17 088	Bolivia
87 502	92 427	110 291	115 640	121 841	128 235	144 858	132 884	137 844	Brazil
134	230	293	230	465	431	507	12 121	15 069	Chile
36 936	38 240	37 463	37 093	40 517	42 210	43 661	42 298	42 376	Colombia
1 250	1 033	566	810	616	448	547	1 254	4 659	Cuba
116 681	111 138	112 701	107 727	118 670	123 502	135 932	148 231	172 334	Trinidad and Tobago
-	-	-	-	-	-	-	-	92	Uruguay
359 783	412 067	394 716	412 811	413 824	454 118	523 841	489 824	516 949	Venezuela
39	310	737	621	815	660	737	854	621	Other Latin America
809 986	**867 649**	**882 336**	**904 601**	**958 442**	**1 036 450**	**1 136 357**	**1 134 072**	**1 206 707**	**Latin America**
69 819	67 173	83 936	89 371	100 521	116 301	117 765	111 167	117 260	Bangladesh
274 001	281 075	321 161	307 571	327 078	366 327	406 527	467 761	489 886	India
309 258	316 986	352 711	306 247	286 032	311 084	317 171	313 760	312 264	Indonesia
44 393	45 647	55 898	70 835	68 910	68 659	86 483	102 592	113 450	Malaysia
10 351	8 317	8 200	10 544	13 202	13 188	12 899	15 177	13 858	Myanmar
200 971	197 779	197 714	222 112	237 016	234 739	249 849	250 483	255 826	Pakistan
18 057	24 772	29 179	30 533	32 127	38 071	39 959	40 423	40 274	Chinese Taipei
11 256	15 332	18 801	20 965	24 811	33 375	39 826	40 292	37 350	Thailand
938 106	**957 081**	**1 067 600**	**1 058 178**	**1 089 697**	**1 181 744**	**1 270 479**	**1 341 655**	**1 380 168**	**Asia**
319 623	353 003	356 154	327 313	356 164	372 058	389 689	392 937	410 049	People's Rep. of China
319 623	**353 003**	**356 154**	**327 313**	**356 164**	**372 058**	**389 689**	**392 937**	**410 049**	**China**
7 639	4 349	2 807	2 380	1 623	895	739	583	544	Albania
105 743	90 623	74 774	73 547	82 923	96 585	96 901	81 202	58 423	Bulgaria
779 979	532 075	252 697	242 600	336 689	366 965	364 926	302 123	192 187	Romania
133 541	105 800	104 791	99 238	76 064	100 891	115 405	113 621	114 992	Slovak Republic
-	-	45 562	42 081	44 004	42 705	42 871	46 930	41 610	*Croatia*
-	-	6 000	6 250	-	-	-	-	-	*FYROM*
-	-	16 908	16 663	20 291	20 900	22 677	25 590	25 412	*Slovenia*
-	-	30 851	19 400	7 125	10 280	30 100	29 858	28 275	*FR of Yugoslavia*
30 795	92 916	99 321	84 394	71 420	73 885	95 648	102 378	95 297	Former Yugoslavia
1 057 697	**825 763**	**534 390**	**502 159**	**568 719**	**639 221**	**673 619**	**599 907**	**461 443**	**Non-OECD Europe**

People's Rep. of China: up to 1978 the breakdown of final consumption by sector is incomplete.

Industry Consumption of Natural Gas (TJ)
Consommation industrielle de gaz naturel (TJ)
Industrieverbrauch von Erdgas (TJ)
Consumo di gas naturale nell'industria (TJ)
天然ガスの産業用消費量 (TJ)
Consumo industrial de gas natural (TJ)
Потребление природного газа промышленным сектором (ТДж)

	1971	1973	1978	1984	1985	1986	1987	1988	1989
Arménie	-	-	-	-	-	-	-	-	-
Azerbaïdjan	-	-	-	-	-	-	-	-	-
Bélarus	-	-	-	-	-	-	-	-	-
Estonie	-	-	-	-	-	-	-	-	-
Géorgie	-	-	-	-	-	-	-	-	-
Kazakhstan	-	-	-	-	-	-	-	-	-
Lettonie	-	-	-	-	-	-	-	-	-
Lituanie	-	-	-	-	-	-	-	-	-
République de Moldavie	-	-	-	-	-	-	-	-	-
Russie	-	-	-	-	-	-	-	-	-
Ukraine	-	-	-	-	-	-	-	-	-
Ouzbékistan	-	-	-	-	-	-	-	-	-
Ex-URSS	**3 592 369**	**4 132 154**	**5 444 394**	**6 765 103**	**7 039 643**	**7 295 568**	**7 565 454**	**7 839 943**	**7 978 454**
Bahrein	8 371	27 369	43 408	56 591	60 683	58 610	53 670	57 546	120 677
République Islamique d'Iran	376 386	383 782	389 252	403 403	464 824	477 629	395 460	286 612	289 351
Irak	35 340	45 980	64 600	114 300	121 250	148 296	186 521	258 158	297 681
Israël	4 232	2 097	2 185	2 034	1 821	1 427	1 615	1 453	1 456
Koweit	89 845	104 885	90 926	85 485	86 967	123 648	132 781	131 370	159 489
Oman	-	-	-	9 909	14 792	6 982	8 898	9 846	10 628
Qatar	17 674	28 065	21 185	92 680	81 922	85 198	84 976	92 824	85 927
Arabie saoudite	14 783	14 783	14 783	75 946	110 510	130 031	-	-	-
Syrie	-	-	-	-	-	-	-	-	-
Emirats arabes unis	37 782	38 254	85 753	126 782	133 260	207 023	262 394	248 221	353 988
Moyen-Orient	**584 413**	**645 215**	**712 092**	**967 130**	**1 076 029**	**1 238 844**	**1 126 315**	**1 086 030**	**1 319 197**
Total non-OCDE	**5 232 388**	**5 962 583**	**8 549 376**	**10 865 344**	**11 079 150**	**11 654 725**	**11 942 721**	**12 122 410**	**12 631 100**
OCDE Amérique du N.	8 463 968	9 118 124	7 571 515	7 658 377	7 338 862	6 847 240	7 297 490	7 823 356	7 292 871
OCDE Pacifique	114 847	147 115	243 796	366 502	406 397	423 965	445 157	456 871	486 872
OCDE Europe	1 832 274	2 603 588	3 554 951	3 658 792	3 668 798	3 619 136	3 876 636	3 940 426	4 145 743
Total OCDE	**10 411 089**	**11 868 827**	**11 370 262**	**11 683 671**	**11 414 057**	**10 890 341**	**11 619 283**	**12 220 653**	**11 925 486**
Monde	**15 643 477**	**17 831 410**	**19 919 638**	**22 549 015**	**22 493 207**	**22 545 066**	**23 562 004**	**24 343 063**	**24 556 586**

Ex-URSS: les séries antérieures à 1990-1992 ne sont pas comparables aux années récentes; 1991 a été estimé.

Industry Consumption of Natural Gas (TJ)
Consommation industrielle de gaz naturel (TJ)
Industrieverbrauch von Erdgas (TJ)

Consumo di gas naturale nell'industria (TJ)

天然ガスの産業用消費量 (TJ)

Consumo industrial de gas natural (TJ)

Потребление природного газа промышленным сектором (ТДж)

1990	1991	1992	1993	1994	1995	1996	1997	1998	
-	-	15 011	7 535	7 454	5 962	8 048	12 877	14 985	Armenia
-	-	-	162 563	158 755	129 640	105 380	88 590	64 880	Azerbaijan
-	-	98 472	93 659	54 495	78 943	76 008	73 150	73 531	Belarus
-	-	11 049	5 924	9 190	11 488	12 055	12 063	12 315	Estonia
-	-	-	74 872	29 406	18 209	11 574	16 324	10 000	Georgia
-	-	-	-	-	-	-	15 000	15 080	Kazakhstan
-	-	15 231	7 215	11 053	7 696	10 089	9 965	10 352	Latvia
-	-	28 735	14 713	21 071	28 741	32 078	28 649	29 929	Lithuania
-	-	19 344	17 759	16 733	16 728	19 803	21 171	19 149	Republic of Moldova
-	-	3 861 800	2 243 294	1 825 556	1 896 555	1 604 445	2 530 666	2 399 890	Russia
-	-	1 289 000	1 025 000	659 000	850 000	700 000	700 000	680 000	Ukraine
-	-	-	-	-	312 722	243 919	190 102	211 074	Uzbekistan
8 332 006	7 909 844	5 338 642	3 652 534	2 792 713	3 356 684	2 823 399	3 698 557	3 541 185	**Former USSR**
65 897	58 181	85 508	108 305	110 614	117 439	121 187	143 415	142 005	Bahrain
264 924	402 602	541 188	558 893	536 458	509 779	590 364	510 999	576 684	Islamic Republic of Iran
158 135	65 021	108 680	122 048	151 706	150 796	154 126	168 299	177 156	Iraq
1 308	959	949	1 001	894	749	372	396	279	Israel
106 539	19 081	112 170	198 738	212 030	162 071	212 851	212 078	216 202	Kuwait
26 817	28 760	34 952	49 814	48 383	27 212	20 733	30 634	53 482	Oman
90 879	117 239	209 345	223 867	222 111	221 129	221 310	235 546	232 683	Qatar
-	-	-	-	-	-	-	-	-	Saudi Arabia
-	17 000	27 000	28 000	30 352	38 274	38 274	76 267	119 562	Syria
317 877	436 979	358 831	320 261	420 207	430 586	577 164	504 312	512 361	United Arab Emirates
1 032 376	1 145 822	1 478 623	1 610 927	1 732 755	1 658 035	1 936 381	1 881 946	2 030 414	**Middle East**
12 849 606	12 425 611	10 045 308	8 470 070	7 864 799	8 611 670	8 599 949	9 440 613	9 445 858	**Non-OECD Total**
7 322 189	7 001 817	6 923 015	7 242 672	7 421 869	7 661 959	7 755 298	7 430 093	7 090 820	OECD North America
513 028	547 405	558 394	595 510	648 626	689 359	764 181	805 672	839 683	OECD Pacific
4 141 712	4 057 653	4 094 372	4 136 087	4 211 087	4 449 068	4 615 542	4 709 403	4 832 420	OECD Europe
11 976 929	11 606 875	11 575 781	11 974 269	12 281 582	12 800 386	13 135 021	12 945 168	12 762 923	**OECD Total**
24 826 535	24 032 486	21 621 089	20 444 339	20 146 381	21 412 056	21 734 970	22 385 781	22 208 781	**World**

Former USSR: series up to 1990-1992 are not comparable with recent years; 1991 is estimated.

Industry Consumption of Electricity (GWh)
Consommation industrielle d'électricité (GWh)
Industrieverbrauch von Elektrizität (GWh)
Consumo di energia elettrica nell'industria (GWh)
電力の産業用消費量 (GWh)
Consumo industrial de electricidad (GWh)
Потребление электроэнергии промышленным сектором (ГВт.ч)

	1971	1973	1978	1984	1985	1986	1987	1988	1989
Algérie	793	944	1 975	3 939	4 291	5 117	4 330	4 650	5 265
Angola	129	172	102	117	141	141	141	142	143
Bénin	16	21	30	68	72	64	59	83	76
Cameroun	884	837	825	1 183	1 289	1 228	1 200	1 296	1 364
Congo	42	46	60	141	142	161	173	178	185
RD du Congo	-	-	2 450	2 690	2 567	2 516	2 520	2 525	3 020
Côte d'Ivoire	132	183	386	900	585	852	838	942	919
Egypte	4 113	4 203	8 141	11 708	12 708	13 798	14 710	15 390	16 282
Ethiopie	365	359	316	468	400	424	518	463	470
Erythrée	-	-	-	-	-	-	-	-	-
Gabon	-	32	179	349	377	350	383	385	380
Ghana	2 439	3 238	2 826	548	1 448	2 745	3 148	3 458	3 527
Kenya	360	434	711	1 206	1 354	1 476	1 522	1 573	1 626
Maroc	975	1 218	1 849	2 844	2 919	3 201	3 163	3 785	3 629
Mozambique	201	236	283	154	167	186	252	164	185
Nigéria	590	850	1 312	1 728	1 902	2 457	2 576	2 377	2 676
Sénégal	181	252	336	435	409	412	427	459	437
Afrique du Sud	32 529	35 557	51 297	68 123	71 359	74 205	76 512	79 499	81 352
Soudan	206	234	360	182	186	170	230	239	190
Rép. Unie de Tanzanie	151	179	214	259	281	324	338	345	384
Togo	22	35	64	80	82	85	95	98	104
Tunisie	459	579	963	1 528	2 040	2 207	2 350	2 576	2 706
Zambie	3 619	4 202	4 661	5 316	5 225	4 911	5 123	4 773	3 465
Zimbabwe	2 206	3 407	3 983	4 736	5 101	5 318	5 382	5 586	5 765
Autre Afrique	229	283	946	1 390	1 857	1 744	1 904	2 052	1 999
Afrique	**50 641**	**57 501**	**84 269**	**110 092**	**116 902**	**124 092**	**127 894**	**133 038**	**136 149**
Argentine	10 370	12 313	14 465	19 969	19 255	20 387	21 399	23 307	21 516
Bolivie	152	174	337	837	802	709	593	639	701
Brésil	22 302	29 514	54 473	87 189	96 233	104 361	104 911	111 452	114 543
Chili	4 908	4 757	5 759	7 259	7 486	7 841	8 137	9 030	9 868
Colombie	2 994	3 449	4 725	6 676	6 162	6 677	6 807	7 145	8 522
Costa Rica	309	365	568	739	765	821	850	847	711
Cuba	2 008	2 281	3 372	5 190	5 023	5 491	5 679	5 975	6 079
République dominicaine	298	675	826	1 302	1 383	1 256	1 526	744	721
Equateur	290	320	628	1 360	1 395	1 407	1 360	1 384	1 407
El Salvador	274	352	589	469	480	472	515	527	509
Guatemala	349	360	583	442	483	507	499	639	598
Haiti	11	23	81	166	173	174	157	165	169
Honduras	184	229	307	467	584	569	739	776	745
Jamaïque	1 205	1 507	1 513	657	771	727	872	855	748
Antilles néerlandaises	319	337	360	337	313	247	277	282	302
Nicaragua	272	304	398	347	319	306	340	299	325
Panama	234	255	273	468	331	332	359	276	321
Paraguay	100	131	281	371	225	244	450	568	559
Pérou	3 668	4 151	5 253	6 494	6 391	6 808	7 124	6 777	7 038
Trinité-et-Tobago	340	514	696	1 526	1 525	1 742	1 887	1 782	1 734
Uruguay	663	649	978	1 103	1 140	1 235	1 273	1 362	1 430
Vénézuela	5 132	5 621	8 460	18 275	19 455	20 171	21 813	23 713	23 446
Autre Amérique latine	113	144	205	291	307	326	351	371	360
Amérique latine	**56 495**	**68 425**	**105 130**	**161 934**	**171 001**	**182 810**	**187 918**	**198 915**	**202 352**

Industry Consumption of Electricity (GWh)
Consommation industrielle d'électricité (GWh)
Industrieverbrauch von Elektrizität (GWh)
Consumo di energia elettrica nell'industria (GWh)
電力の産業用消費量 (GWh)
Consumo industrial de electricidad (GWh)
Потребление электроэнергии промышленным сектором (ГВт.ч)

1990	1991	1992	1993	1994	1995	1996	1997	1998	
5 455	5 625	5 584	6 220	6 089	6 216	6 259	6 302	6 800	Algeria
157	191	200	200	201	202	216	232	223	Angola
87	102	121	111	109	121	130	129	136	Benin
1 359	1 342	1 304	1 376	1 295	1 279	1 315	1 428	1 501	Cameroon
205	218	215	190	185	187	185	152	127	Congo
3 031	3 300	3 400	4 234	4 321	3 757	3 806	3 650	3 442	DR of Congo
914	892	501	885	916	1 198	1 148	1 251	1 399	Cote d'Ivoire
16 844	17 425	18 026	18 648	19 291	19 956	20 644	21 356	22 093	Egypt
487	441	441	499	521	546	564	563	559	Ethiopia
-	-	62	67	77	79	85	88	64	Eritrea
390	395	395	398	420	421	434	451	460	Gabon
3 558	3 714	3 865	3 969	3 590	3 682	3 846	3 590	3 754	Ghana
1 754	1 832	1 835	1 915	1 955	1 995	2 179	2 263	2 261	Kenya
4 016	4 180	4 513	4 411	4 736	4 815	4 969	5 233	5 573	Morocco
177	311	294	266	244	260	247	291	376	Mozambique
1 905	2 235	2 183	2 067	2 042	2 037	2 108	2 134	2 209	Nigeria
450	461	507	464	505	508	524	547	547	Senegal
82 341	82 430	72 804	75 706	75 681	80 657	89 904	91 460	96 159	South Africa
204	221	231	231	416	378	388	433	438	Sudan
399	465	391	389	395	439	485	460	465	United Rep. of Tanzania
113	88	137	89	90	156	185	203	204	Togo
2 840	2 923	3 058	3 074	3 281	3 387	3 438	3 548	3 753	Tunisia
2 793	4 018	3 990	3 992	3 992	3 995	3 998	4 118	4 037	Zambia
5 660	4 955	5 329	4 517	4 958	5 375	5 699	5 670	5 384	Zimbabwe
2 076	1 949	1 842	1 481	1 518	1 096	739	720	744	Other Africa
137 215	**139 713**	**131 228**	**135 399**	**136 828**	**142 742**	**153 495**	**156 272**	**162 708**	**Africa**
21 388	21 376	21 934	23 504	24 621	25 667	27 342	30 075	32 483	Argentina
725	760	779	906	983	1 032	1 106	1 155	1 227	Bolivia
112 339	115 041	116 586	122 462	126 177	127 171	129 755	135 521	136 388	Brazil
10 094	11 333	13 192	13 747	14 438	16 485	18 349	20 276	20 943	Chile
7 958	8 411	7 696	8 956	9 846	10 207	10 417	11 349	10 267	Colombia
740	759	816	998	1 110	1 005	1 021	1 096	1 451	Costa Rica
5 674	4 521	3 916	3 331	3 533	3 541	3 976	4 063	3 850	Cuba
616	674	826	1 066	1 122	1 181	1 243	1 392	1 432	Dominican Republic
1 535	1 686	1 640	1 814	2 059	2 083	2 262	2 361	2 457	Ecuador
569	588	628	731	774	830	842	905	1 028	El Salvador
646	683	770	836	901	962	1 002	1 064	1 136	Guatemala
179	133	104	66	41	80	93	105	116	Haiti
852	767	596	637	570	581	778	732	779	Honduras
208	289	316	2 841	2 710	3 308	3 425	3 548	3 713	Jamaica
331	337	365	397	420	436	456	455	482	Netherlands Antilles
355	333	288	249	255	299	332	289	410	Nicaragua
374	419	451	529	562	603	642	498	532	Panama
624	518	637	725	774	928	1 174	1 386	1 382	Paraguay
4 535	4 789	4 494	4 511	4 988	5 466	5 965	8 536	9 180	Peru
1 916	1 932	2 089	2 062	2 209	2 263	2 566	2 882	2 938	Trinidad and Tobago
1 480	1 582	1 533	1 546	1 547	1 310	1 285	1 300	1 405	Uruguay
24 612	25 571	26 896	26 114	25 884	27 148	27 442	28 595	27 689	Venezuela
382	399	402	414	422	425	433	451	462	Other Latin America
198 132	**202 901**	**206 954**	**218 442**	**225 946**	**233 011**	**241 906**	**258 034**	**261 750**	**Latin America**

Industry Consumption of Electricity (GWh)
Consommation industrielle d'électricité (GWh)

Industrieverbrauch von Elektrizität (GWh)

Consumo di energia elettrica nell'industria (GWh)

電力の産業用消費量 (GWh)

Consumo industrial de electricidad (GWh)

Потребление электроэнергии промышленным сектором (ГВт.ч)

	1971	1973	1978	1984	1985	1986	1987	1988	1989
Bangladesh	475	781	1 227	1 682	1 678	1 843	1 907	2 040	2 564
Brunei	53	52	53	152	155	157	211	223	229
Inde	34 713	37 210	50 929	67 643	72 277	76 329	75 718	83 289	100 373
Indonésie	511	638	1 155	4 011	4 874	6 183	7 402	9 087	11 418
RPD de Corée	7 155	8 460	13 536	19 125	20 385	21 100	21 245	22 500	23 000
Malaisie	1 814	2 378	3 754	5 288	5 521	5 892	6 485	7 322	8 500
Myanmar	302	359	546	645	942	919	904	737	805
Népal	12	7	34	96	108	144	168	180	180
Pakistan	2 857	3 662	4 589	5 862	6 249	7 288	8 012	8 973	9 455
Philippines	3 397	4 743	6 918	8 554	9 008	5 843	7 750	8 566	9 763
Singapour	920	1 349	2 607	3 270	3 398	3 513	4 021	4 592	5 099
Sri Lanka	379	497	535	857	872	926	867	905	909
Taipei chinois	9 029	11 688	20 586	27 622	28 227	32 236	34 807	37 946	41 063
Thailande	2 998	3 939	5 091	8 537	9 110	9 956	11 099	12 952	15 432
Viêt-Nam	-	-	-	2 027	2 113	2 196	2 384	2 590	2 623
Autre Asie	-	-	1 898	2 152	2 366	2 322	2 300	2 425	2 523
Asie	**64 615**	**75 763**	**113 458**	**157 523**	**167 283**	**176 847**	**185 280**	**204 327**	**233 936**
Rép. populaire de Chine	-	-	-	240 860	271 770	300 120	314 320	346 530	365 610
Hong-Kong, Chine	2 030	2 419	3 549	5 149	5 226	5 941	6 399	6 647	6 994
Chine	**2 030**	**2 419**	**3 549**	**246 009**	**276 996**	**306 061**	**320 719**	**353 177**	**372 604**
Albanie	-	-	-	-	-	1 226	1 055	884	1 500
Bulgarie	12 399	14 036	17 431	21 647	18 894	21 372	20 485	21 047	19 507
Chypre	177	231	237	296	206	244	273	294	318
Malte	-	-	-	-	-	-	-	-	-
Roumanie	22 440	26 710	40 318	54 731	44 356	47 480	47 012	48 766	49 044
République slovaque	7 281	8 321	10 049	13 745	14 488	14 849	15 288	15 604	16 005
Croatie	-	-	-	-	-	-	-	-	-
Ex-RYM	-	-	-	-	-	-	-	-	-
Slovénie	-	-	-	-	-	-	-	-	-
RF de Yougoslavie	-	-	-	-	-	-	-	-	-
Ex-Yougoslavie	13 535	15 749	23 986	35 348	36 714	37 941	37 116	34 734	37 052
Europe non-OCDE	**55 832**	**65 047**	**92 021**	**125 767**	**114 658**	**123 112**	**121 229**	**121 329**	**123 426**
Arménie	-	-	-	-	-	-	-	-	-
Azerbaïdjan	-	-	-	-	-	-	-	-	-
Bélarus	-	-	-	-	-	-	-	-	-
Estonie	-	-	-	-	-	-	-	-	-
Géorgie	-	-	-	-	-	-	-	-	-
Kazakhstan	-	-	-	-	-	-	-	-	-
Kirghizistan	-	-	-	-	-	-	-	-	-
Lettonie	-	-	-	-	-	-	-	-	-
Lituanie	-	-	-	-	-	-	-	-	-
République de Moldavie	-	-	-	-	-	-	-	-	-
Russie	-	-	-	-	-	-	-	-	-
Tadjikistan	-	-	-	-	-	-	-	-	-
Turkménistan	-	-	-	-	-	-	-	-	-
Ukraine	-	-	-	-	-	-	-	-	-
Ouzbékistan	-	-	-	-	-	-	-	-	-
Ex-URSS	**451 400**	**508 900**	**639 100**	**730 000**	**742 100**	**756 600**	**783 300**	**805 100**	**799 500**

Ex-URSS: les séries antérieures à 1990-1992 ne sont pas comparables aux années récentes; 1991 a été estimé.

Rép. populaire de Chine: jusqu'en 1978 la ventilation de la consommation finale par secteur est incomplète.

Industry Consumption of Electricity (GWh)
Consommation industrielle d'électricité (GWh)
Industrieverbrauch von Elektrizität (GWh)
Consumo di energia elettrica nell'industria (GWh)
電力の産業用消費量 (GWh)
Consumo industrial de electricidad (GWh)
Потребление электроэнергии промышленным сектором (ГВт.ч)

1990	1991	1992	1993	1994	1995	1996	1997	1998	
2 716	2 616	4 201	5 140	5 623	6 440	6 982	7 486	8 167	Bangladesh
238	248	257	280	283	291	281	263	285	Brunei
109 319	115 891	121 520	126 144	133 930	143 701	149 608	158 749	169 235	India
14 166	16 026	17 755	21 881	21 622	24 722	27 949	30 769	27 985	Indonesia
23 000	-	-	-	-	-	-	-	-	DPR of Korea
9 646	10 890	13 214	15 144	18 212	21 222	23 581	28 149	28 021	Malaysia
862	713	766	829	843	918	890	914	963	Myanmar
203	251	275	311	335	359	383	419	443	Nepal
10 337	11 263	12 598	13 044	12 635	12 528	12 183	11 982	12 297	Pakistan
8 982	9 339	8 646	9 426	10 684	11 228	11 851	12 531	11 388	Philippines
5 638	7 234	7 446	7 930	8 725	9 349	9 787	11 077	11 760	Singapore
910	1 049	1 057	1 223	1 560	1 674	1 650	1 819	1 901	Sri Lanka
42 783	46 270	49 155	51 805	55 160	59 139	62 002	67 914	70 929	Chinese Taipei
17 929	19 814	20 407	22 374	28 921	32 860	34 645	34 541	30 834	Thailand
2 850	3 107	3 192	3 476	3 840	4 614	5 503	6 163	6 781	Vietnam
2 487	2 380	2 375	2 374	2 404	1 440	1 360	1 360	1 360	Other Asia
252 066	**247 091**	**262 864**	**281 381**	**304 777**	**330 485**	**348 655**	**374 136**	**382 349**	**Asia**
381 860	416 490	459 311	449 968	495 608	534 609	569 406	573 092	588 236	People's Rep. of China
6 926	6 958	6 721	6 454	5 955	5 618	5 538	5 268	5 136	Hong Kong, China
388 786	**423 448**	**466 032**	**456 422**	**501 563**	**540 227**	**574 944**	**578 360**	**593 372**	**China**
1 374	1 056	1 242	1 504	527	506	741	615	607	Albania
18 552	13 998	11 952	10 914	11 461	12 167	12 258	11 812	10 688	Bulgaria
334	334	363	381	394	397	403	395	421	Cyprus
-	236	257	260	488	489	510	453	451	Malta
38 553	30 881	25 489	23 559	21 878	23 343	24 512	25 125	22 686	Romania
15 008	11 082	11 235	9 953	10 128	9 146	10 501	10 053	9 441	Slovak Republic
-	-	3 420	3 064	3 054	2 747	2 651	3 031	3 054	*Croatia*
-	-	2 540	2 263	2 130	1 899	2 015	2 087	2 118	*FYROM*
-	-	4 660	4 553	4 956	4 943	4 785	4 884	5 041	*Slovenia*
-	-	10 930	5 841	6 060	6 102	6 253	6 615	6 628	*FR of Yugoslavia*
38 650	22 075	21 550	15 721	16 200	15 691	15 704	16 617	16 841	Former Yugoslavia
112 471	**79 662**	**72 088**	**62 292**	**61 076**	**61 739**	**64 629**	**65 070**	**61 135**	**Non-OECD Europe**
-	-	1 807	828	668	981	815	722	638	Armenia
-	-	4 461	4 400	2 017	1 591	1 241	1 115	911	Azerbaijan
-	-	19 952	14 976	11 925	10 617	10 785	12 569	13 492	Belarus
-	-	2 229	1 384	1 710	1 751	1 907	2 208	1 919	Estonia
-	-	-	-	-	-	876	828	801	Georgia
-	-	57 560	44 927	24 158	23 336	20 854	18 686	10 104	Kazakhstan
-	-	3 460	2 506	1 958	1 479	1 589	1 899	1 774	Kyrgyzstan
-	-	2 613	1 439	1 492	1 427	1 697	1 584	1 503	Latvia
-	-	4 525	3 116	2 795	2 806	2 520	2 777	2 620	Lithuania
-	-	2 971	3 000	2 579	1 847	1 738	1 802	566	Republic of Moldova
-	-	469 801	383 544	325 763	293 252	294 110	291 573	283 142	Russia
-	-	10 500	8 168	7 277	6 876	6 750	4 962	5 087	Tajikistan
-	-	1 929	2 191	1 821	1 806	1 695	1 406	1 462	Turkmenistan
-	-	108 637	95 077	76 656	71 208	65 888	64 822	60 192	Ukraine
-	-	19 122	18 163	17 664	14 400	14 070	14 921	14 863	Uzbekistan
792 700	**793 254**	**709 567**	**583 719**	**478 483**	**433 377**	**426 535**	**421 874**	**399 074**	**Former USSR**

Former USSR: series up to 1990-1992 are not comparable with recent years; 1991 is estimated.
People's Rep. of China: up to 1978 the breakdown of final consumption by sector is incomplete.

Industry Consumption of Electricity (GWh)
Consommation industrielle d'électricité (GWh)
Industrieverbrauch von Elektrizität (GWh)

Consumo di energia elettrica nell'industria (GWh)

電力の産業用消費量 (GWh)

Consumo industrial de electricidad (GWh)

Потребление электроэнергии промышленным сектором (ГВт.ч)

	1971	1973	1978	1984	1985	1986	1987	1988	1989
Bahrein	172	170	220	193	232	255	264	278	290
République Islamique d'Iran	4 635	6 796	8 441	11 131	11 334	11 229	11 491	11 677	12 453
Irak	904	1 137	3 275	6 563	7 073	7 516	7 942	9 050	9 000
Israël	1 983	2 308	3 237	4 130	4 203	4 390	4 840	4 834	5 068
Jordanie	80	110	210	851	903	906	1 061	1 040	1 097
Liban	-	-	-	-	-	-	-	-	-
Oman	-	3	33	141	175	222	238	264	274
Qatar	-	-	-	473	499	471	461	485	502
Arabie saoudite	1 281	1 745	1 481	2 922	6 200	5 269	5 585	5 935	7 185
Syrie	774	876	1 442	3 213	3 532	3 710	3 728	3 785	3 894
Emirats arabes unis	13	44	232	678	739	789	841	914	940
Yémen	24	32	118	25	31	37	66	70	-
Moyen-Orient	**9 866**	**13 221**	**18 689**	**30 320**	**34 921**	**34 794**	**36 517**	**38 332**	**40 703**
Total non-OCDE	**690 879**	**791 276**	**1 056 216**	**1 561 645**	**1 623 861**	**1 704 316**	**1 762 857**	**1 854 218**	**1 908 670**
OCDE Amérique du N.	684 057	769 752	897 398	962 463	970 104	945 573	986 316	1 042 324	1 065 163
OCDE Pacifique	272 660	329 041	376 315	382 322	393 787	396 058	419 061	448 397	475 358
OCDE Europe	632 540	729 003	820 794	875 088	892 111	904 344	932 402	970 137	1 000 097
Total OCDE	**1 589 257**	**1 827 796**	**2 094 507**	**2 219 873**	**2 256 002**	**2 245 975**	**2 337 779**	**2 460 858**	**2 540 618**
Monde	**2 280 136**	**2 619 072**	**3 150 723**	**3 781 518**	**3 879 863**	**3 950 291**	**4 100 636**	**4 315 076**	**4 449 288**

Industry Consumption of Electricity (GWh)
Consommation industrielle d'électricité (GWh)
Industrieverbrauch von Elektrizität (GWh)
Consumo di energia elettrica nell'industria (GWh)
電力の産業用消費量 (GWh)
Consumo industrial de electricidad (GWh)
Потребление электроэнергии промышленным сектором (ГВт.ч)

1990	1991	1992	1993	1994	1995	1996	1997	1998	
336	479	563	613	668	678	737	739	825	Bahrain
14 426	15 053	17 899	20 251	25 403	26 315	27 951	29 095	29 716	Islamic Republic of Iran
-	-	-	-	-	-	-	-	-	Iraq
5 289	5 348	5 687	5 990	7 640	8 318	8 860	9 496	9 921	Israel
1 189	1 181	1 342	1 366	1 434	1 592	1 685	1 712	1 809	Jordan
			1 189	1 350	1 224	1 759	2 100	2 011	Lebanon
315	339	360	410	402	372	391	455	467	Oman
600	696	799	755	800	751	1 422	1 485	1 747	Qatar
7 964	8 177	8 394	8 745	9 581	10 813	11 511	12 263	12 489	Saudi Arabia
4 201	4 236	3 840	3 828	4 529	4 564	5 037	5 766	5 940	Syria
1 004	1 020	1 099	1 278	1 396	1 506	383	1 433	2 358	United Arab Emirates
-	-	-	-	-	-	-	-	-	Yemen
35 324	**36 529**	**39 983**	**44 425**	**53 203**	**56 133**	**59 736**	**64 544**	**67 283**	**Middle East**
1 916 694	**1 922 598**	**1 888 716**	**1 782 080**	**1 761 876**	**1 797 714**	**1 869 900**	**1 918 290**	**1 927 671**	**Non-OECD Total**
1 087 880	1 179 190	1 205 090	1 220 846	1 271 693	1 294 629	1 322 757	1 340 468	1 359 606	OECD North America
505 297	518 340	518 299	522 036	567 814	585 700	607 453	628 804	612 238	OECD Pacific
990 914	975 896	969 676	960 537	971 954	1 001 481	1 009 190	1 046 715	1 070 126	OECD Europe
2 584 091	**2 673 426**	**2 693 065**	**2 703 419**	**2 811 461**	**2 881 810**	**2 939 400**	**3 015 987**	**3 041 970**	**OECD Total**
4 500 785	**4 596 024**	**4 581 781**	**4 485 499**	**4 573 337**	**4 679 524**	**4 809 300**	**4 934 277**	**4 969 641**	**World**

Consumption of Oil in Transport (1000 tonnes)
Consommation de pétrole dans les transports (1000 tonnes)
Ölverbrauch im Verkehrssektor (1000 Tonnen)
Consumo di petrolio nel settore dei trasporti (1000 tonnellate)
運輸部門における石油の消費量（チトン）
Consumo de petróleo en el transporte (1000 toneladas)
Потребление нефти и нефтепродуктов в транспорте (тыс. т)

	1971	1973	1978	1984	1985	1986	1987	1988	1989
Algérie	853	1 074	1 892	3 593	3 956	4 236	4 293	4 280	4 405
Angola	394	529	503	400	647	658	687	734	749
Bénin	84	104	95	108	133	122	107	102	81
Cameroun	232	264	385	549	599	571	577	640	630
Congo	116	125	173	196	193	193	209	189	214
RD du Congo	519	550	606	674	628	529	611	625	636
Côte d'Ivoire	321	340	526	463	468	451	436	458	494
Egypte	1 312	1 468	2 102	3 855	4 171	4 110	4 091	3 913	4 043
Ethiopie	218	224	238	298	313	355	429	447	462
Erythrée	-	-	-	-	-	-	-	-	-
Gabon	8	22	195	102	129	194	146	162	217
Ghana	373	377	467	365	430	426	409	545	554
Kenya	617	741	869	947	1 000	1 084	1 057	1 045	1 114
Libye	297	632	1 360	2 265	2 198	2 116	2 239	2 190	2 253
Maroc	663	790	1 002	1 020	1 044	1 078	1 121	586	594
Mozambique	137	135	108	89	70	73	76	78	78
Nigéria	918	1 328	3 975	5 005	4 763	4 460	4 368	4 764	4 797
Sénégal	233	275	319	375	371	359	317	303	365
Afrique du Sud	5 164	6 367	7 647	8 149	7 898	7 835	8 417	9 096	9 422
Soudan	751	892	704	783	931	876	694	1 061	948
Rép. Unie de Tanzanie	252	283	263	240	243	239	244	272	272
Togo	24	31	45	66	55	61	67	68	68
Tunisie	400	474	727	794	806	783	833	882	919
Zambie	204	251	253	232	236	226	228	245	240
Zimbabwe	343	443	447	520	562	578	533	617	601
Autre Afrique	628	769	1 131	1 484	1 521	1 672	2 178	2 244	2 362
Afrique	**15 061**	**18 488**	**26 032**	**32 572**	**33 365**	**33 285**	**34 367**	**35 546**	**36 518**
Argentine	7 644	8 340	8 855	9 525	8 589	9 304	9 497	8 918	8 895
Bolivie	294	356	624	526	527	587	651	654	691
Brésil	13 561	18 024	23 614	21 278	22 112	24 317	23 908	24 024	24 981
Chili	1 862	1 717	1 795	2 145	2 114	2 209	2 364	2 668	2 864
Colombie	2 411	2 757	4 009	4 854	4 997	5 043	5 222	5 673	5 768
Costa Rica	221	275	433	379	390	434	463	526	564
Cuba	1 460	1 832	2 096	2 054	2 745	2 722	2 813	2 929	3 050
République dominicaine	477	527	608	710	683	756	934	842	1 022
Equateur	677	786	1 443	1 862	1 979	2 053	1 988	2 378	2 298
El Salvador	191	228	359	305	336	324	353	391	402
Guatemala	265	365	453	429	452	434	508	546	553
Haiti	52	64	104	108	110	120	128	141	151
Honduras	140	166	198	247	266	287	308	328	359
Jamaïque	647	672	387	360	385	373	429	467	461
Antilles néerlandaises	520	484	574	446	391	323	338	350	323
Nicaragua	241	278	348	282	275	299	325	282	248
Panama	344	374	456	362	334	349	341	333	342
Paraguay	132	147	327	405	409	429	449	465	523
Pérou	1 828	2 166	2 136	2 312	2 168	2 301	2 601	2 613	2 309
Trinité-et-Tobago	298	395	441	607	605	634	598	477	457
Uruguay	548	573	558	430	418	422	442	453	474
Vénézuela	3 759	4 564	7 870	8 665	8 649	8 741	8 951	9 285	8 801
Autre Amérique latine	408	296	486	642	624	583	750	804	830
Amérique latine	**37 980**	**45 386**	**58 174**	**58 933**	**59 558**	**63 044**	**64 361**	**65 547**	**66 366**

Consumption of Oil in Transport (1000 tonnes)
Consommation de pétrole dans les transports (1000 tonnes)
Ölverbrauch im Verkehrssektor (1000 Tonnen)
Consumo di petrolio nel settore dei trasporti (1000 tonnellate)
運輸部門における石油の消費量（千トン）
Consumo de petróleo en el transporte (1000 toneladas)
Потребление нефти и нефтепродуктов в транспорте (тыс. m)

1990	1991	1992	1993	1994	1995	1996	1997	1998	
4 483	4 593	2 608	2 677	2 547	2 409	2 397	2 399	2 365	Algeria
656	683	639	675	907	720	652	671	488	Angola
68	56	51	56	58	59	257	254	235	Benin
603	543	543	573	615	593	603	627	649	Cameroon
203	208	185	193	186	198	203	203	182	Congo
650	659	603	677	690	705	705	705	665	DR of Congo
461	458	428	351	355	416	494	512	530	Cote d'Ivoire
4 226	4 735	4 191	4 243	4 389	4 710	5 199	5 099	5 345	Egypt
460	469	300	364	414	455	494	515	567	Ethiopia
-	-	40	70	94	108	130	123	79	Eritrea
179	178	173	179	181	195	211	254	277	Gabon
550	491	547	607	642	687	741	794	878	Ghana
1 115	1 101	1 116	1 096	1 205	1 204	1 247	1 197	1 164	Kenya
2 162	2 187	2 230	2 592	3 087	3 306	3 337	3 292	3 403	Libya
630	568	613	771	758	758	763	775	815	Morocco
105	99	242	301	261	247	273	283	291	Mozambique
4 050	5 243	6 324	5 486	4 180	5 142	5 857	6 515	6 360	Nigeria
374	363	376	340	360	391	429	450	499	Senegal
9 702	9 744	9 965	10 211	10 606	11 924	11 877	12 285	12 621	South Africa
1 277	1 023	1 022	689	985	927	884	1 035	990	Sudan
291	277	283	271	274	265	255	255	264	United Rep. of Tanzania
85	72	73	27	73	91	125	101	117	Togo
963	939	1 021	1 146	1 212	1 261	1 307	1 401	1 448	Tunisia
219	223	203	205	214	214	218	218	218	Zambia
577	481	625	562	625	772	770	770	770	Zimbabwe
2 365	2 251	2 338	2 562	2 501	2 320	2 330	2 427	2 427	Other Africa
36 454	**37 644**	**36 739**	**36 924**	**37 419**	**40 077**	**41 758**	**43 160**	**43 647**	**Africa**
9 092	9 471	10 060	10 628	11 244	11 720	12 119	12 503	13 062	Argentina
719	737	729	759	791	876	954	1 113	1 138	Bolivia
25 806	27 094	27 297	28 377	29 682	32 863	35 904	38 101	40 391	Brazil
2 975	3 186	3 412	3 736	4 122	4 497	4 912	5 229	5 516	Chile
5 845	6 077	6 442	6 514	6 206	7 322	7 472	7 835	7 647	Colombia
543	562	834	868	930	935	934	964	1 139	Costa Rica
1 738	1 271	892	739	688	740	871	1 001	930	Cuba
1 036	970	1 208	1 018	1 057	1 088	1 122	1 169	1 193	Dominican Republic
2 186	2 160	2 161	2 158	2 246	2 474	2 463	2 570	2 486	Ecuador
434	461	547	595	632	716	711	722	787	El Salvador
598	622	659	723	767	952	957	1 019	1 202	Guatemala
158	146	141	131	36	181	197	202	223	Haiti
334	333	384	429	479	513	464	507	561	Honduras
471	464	451	509	602	682	761	798	837	Jamaica
319	362	387	324	416	419	424	424	424	Netherlands Antilles
258	277	308	313	337	371	390	415	456	Nicaragua
377	371	400	415	452	486	563	563	635	Panama
514	506	606	706	822	938	880	953	1 111	Paraguay
2 491	2 539	2 621	2 807	2 619	3 093	3 183	3 194	3 240	Peru
493	516	510	442	457	479	488	531	545	Trinidad and Tobago
476	506	544	614	692	686	729	792	848	Uruguay
9 039	9 530	9 393	10 108	10 193	10 549	10 936	10 153	10 333	Venezuela
827	825	797	791	862	875	870	881	881	Other Latin America
66 729	**68 986**	**70 783**	**73 704**	**76 332**	**83 455**	**88 304**	**91 639**	**95 585**	**Latin America**

Consumption of Oil in Transport (1000 tonnes)
Consommation de pétrole dans les transports (1000 tonnes)
Ölverbrauch im Verkehrssektor (1000 Tonnen)
Consumo di petrolio nel settore dei trasporti (1000 tonnellate)
運輸部門における石油の消費量（千トン）
Consumo de petróleo en el transporte (1000 toneladas)
Потребление нефти и нефтепродуктов в транспорте (тыс. m)

	1971	1973	1978	1984	1985	1986	1987	1988	1989
Bangladesh	98	92	243	402	492	491	516	552	593
Brunei	27	33	71	152	160	172	183	198	201
Inde	6 554	7 297	10 284	15 149	16 583	17 991	18 810	19 819	21 949
Indonésie	2 579	2 974	4 868	6 684	6 865	7 283	8 012	8 685	9 448
RPD de Corée	563	635	1 305	2 000	2 125	2 630	3 630	3 260	3 340
Malaisie	1 425	1 651	1 869	3 184	3 255	3 594	3 787	4 128	4 522
Myanmar	495	421	535	647	607	671	397	402	392
Népal	23	25	41	84	67	78	85	80	115
Pakistan	1 359	1 366	2 058	3 364	3 522	3 723	4 198	4 481	4 637
Philippines	2 264	2 349	2 254	1 713	1 641	1 735	1 963	2 144	2 422
Singapour	704	880	1 595	2 141	2 086	2 232	2 348	2 401	2 614
Sri Lanka	426	515	577	732	764	768	787	795	755
Taipei chinois	1 137	1 938	2 815	4 052	4 267	4 611	5 109	5 706	6 508
Thailande	2 264	3 065	3 685	5 394	5 584	5 956	6 846	7 955	9 443
Viêt-Nam	3 169	2 920	185	497	870	990	1 138	1 246	1 169
Autre Asie	555	544	687	942	955	864	984	931	905
Asie	**23 642**	**26 705**	**33 072**	**47 137**	**49 843**	**53 789**	**58 793**	**62 783**	**69 013**
Rép. populaire de Chine	4 465	6 110	9 562	20 018	22 548	25 199	27 453	29 832	31 062
Hong-Kong, Chine	852	1 006	1 235	1 768	1 687	2 049	2 051	2 525	2 687
Chine	**5 317**	**7 116**	**10 797**	**21 786**	**24 235**	**27 248**	**29 504**	**32 357**	**33 749**
Albanie	233	252	360	425	346	324	335	370	306
Bulgarie	1 552	1 525	1 620	1 594	1 579	1 572	1 588	2 634	2 796
Chypre	243	313	246	381	400	420	464	479	541
Gibraltar	16	15	11	13	16	23	29	36	32
Malte	123	142	121	177	93	212	217	215	215
Roumanie	2 030	2 291	2 596	1 656	1 405	1 308	1 821	1 700	3 644
République slovaque	1 367	1 576	1 490	875	867	890	918	965	954
Bosnie-Herzegovine	-	-	-	-	-	-	-	-	-
Croatie	-	-	-	-	-	-	-	-	-
Ex-RYM	-	-	-	-	-	-	-	-	-
Slovénie	-	-	-	-	-	-	-	-	-
RF de Yougoslavie	-	-	-	-	-	-	-	-	-
Ex-Yougoslavie	3 563	3 634	4 798	3 984	4 504	5 549	4 962	5 467	5 146
Europe non-OCDE	**9 127**	**9 748**	**11 242**	**9 105**	**9 210**	**10 298**	**10 334**	**11 866**	**13 634**
Arménie	-	-	-	-	-	-	-	-	-
Azerbaïdjan	-	-	-	-	-	-	-	-	-
Bélarus	-	-	-	-	-	-	-	-	-
Estonie	-	-	-	-	-	-	-	-	-
Géorgie	-	-	-	-	-	-	-	-	-
Kazakhstan	-	-	-	-	-	-	-	-	-
Kirghizistan	-	-	-	-	-	-	-	-	-
Lettonie	-	-	-	-	-	-	-	-	-
Lituanie	-	-	-	-	-	-	-	-	-
République de Moldavie	-	-	-	-	-	-	-	-	-
Russie	-	-	-	-	-	-	-	-	-
Tadjikistan	-	-	-	-	-	-	-	-	-
Turkménistan	-	-	-	-	-	-	-	-	-
Ukraine	-	-	-	-	-	-	-	-	-
Ouzbékistan	-	-	-	-	-	-	-	-	-
Ex-URSS	**73 100**	**81 200**	**100 000**	**112 600**	**113 600**	**117 800**	**120 100**	**121 600**	**123 900**

Ex-URSS: les séries antérieures à 1990-1992 ne sont pas comparables aux années récentes; 1991 a été estimé.

Rép. populaire de Chine: jusqu'en 1978 la ventilation de la consommation finale par secteur est incomplète

Consumption of Oil in Transport (1000 tonnes)
Consommation de pétrole dans les transports (1000 tonnes)
Ölverbrauch im Verkehrssektor (1000 Tonnen)
Consumo di petrolio nel settore dei trasporti (1000 tonnellate)
運輸部門における石油の消費量（千トン）
Consumo de petróleo en el transporte (1000 toneladas)
Потребление нефти и нефтепродуктов в транспорте (тыс. т)

1990	1991	1992	1993	1994	1995	1996	1997	1998	
609	653	736	756	784	931	1 018	1 151	1 134	Bangladesh
212	230	255	270	289	313	341	365	336	Brunei
22 641	25 756	27 263	28 907	31 078	34 753	38 291	39 504	39 405	India
10 812	11 609	12 306	13 140	14 735	16 137	17 953	19 174	18 728	Indonesia
3 275	3 205	3 205	3 180	3 150	3 120	3 095	2 884	2 740	DPR of Korea
5 202	5 605	6 013	6 328	6 999	7 538	8 619	9 835	9 421	Malaysia
418	382	333	505	583	860	930	766	938	Myanmar
100	157	152	149	165	198	197	228	247	Nepal
4 719	4 924	5 518	6 255	6 563	6 800	7 346	7 460	7 643	Pakistan
2 502	2 303	2 654	2 983	3 148	3 451	3 882	4 347	4 158	Philippines
3 066	3 049	3 111	3 561	4 152	4 255	4 338	4 691	4 181	Singapore
784	789	878	939	1 080	1 146	1 236	1 365	1 665	Sri Lanka
7 091	7 507	8 721	9 334	10 075	10 669	11 074	11 327	11 892	Chinese Taipei
10 581	10 844	11 790	13 522	15 078	17 470	18 969	20 210	17 927	Thailand
1 384	1 171	1 134	1 552	1 669	3 078	3 393	3 723	3 520	Vietnam
935	897	838	826	742	646	659	663	663	Other Asia
74 331	**79 081**	**84 907**	**92 207**	**100 290**	**111 365**	**121 341**	**127 693**	**124 598**	**Asia**
31 714	33 358	35 993	46 623	42 483	49 845	54 469	61 952	62 349	People's Rep. of China
3 196	3 309	4 061	4 587	5 169	5 246	5 209	5 429	6 653	Hong Kong, China
34 910	**36 667**	**40 054**	**51 210**	**47 652**	**55 091**	**59 678**	**67 381**	**69 002**	**China**
265	226	171	204	240	270	281	286	286	Albania
2 377	1 354	1 376	1 666	1 481	1 578	1 450	1 517	1 810	Bulgaria
609	652	690	655	678	728	733	750	787	Cyprus
34	51	57	57	70	68	68	68	68	Gibraltar
215	243	248	270	270	283	333	338	312	Malta
4 012	3 481	3 320	2 881	3 004	2 797	3 762	3 866	3 630	Romania
933	784	887	848	1 045	1 193	1 115	1 124	1 194	Slovak Republic
-	-	356	239	259	266	274	282	291	*Bosnia-Herzegovina*
-	-	915	998	1 080	1 135	1 191	1 336	1 394	*Croatia*
-	-	257	474	322	316	563	505	342	*FYROM*
-	-	840	1 017	1 135	1 264	1 428	1 495	1 618	*Slovenia*
-	-	787	405	449	498	947	1 538	1 365	*FR of Yugoslavia*
5 163	4 617	3 155	3 133	3 245	3 479	4 403	5 156	5 010	Former Yugoslavia
13 608	**11 408**	**9 904**	**9 714**	**10 033**	**10 396**	**12 145**	**13 105**	**13 097**	**Non-OECD Europe**
-	-	750	455	100	71	39	41	43	Armenia
-	-	1 470	1 551	1 699	1 464	1 274	953	983	Azerbaijan
-	-	4 562	3 526	2 536	2 187	2 194	2 194	2 623	Belarus
-	-	353	417	461	464	496	531	551	Estonia
-	-	401	244	71	46	501	473	419	Georgia
-	-	4 815	3 736	2 896	2 613	2 577	2 127	2 197	Kazakhstan
-	-	472	289	115	293	298	160	280	Kyrgyzstan
-	-	959	858	772	791	786	742	700	Latvia
-	-	1 054	1 070	967	1 125	1 193	1 187	1 240	Lithuania
-	-	738	461	322	299	282	249	208	Republic of Moldova
-	-	69 040	47 701	30 497	33 755	35 561	34 250	33 609	Russia
-	-	4 778	3 005	1 001	1 001	1 001	1 001	1 001	Tajikistan
-	-	713	527	493	451	483	347	313	Turkmenistan
-	-	10 204	8 389	7 279	7 275	6 160	5 573	5 749	Ukraine
-	-	2 185	1 995	1 978	1 542	2 015	2 139	2 124	Uzbekistan
116 085	**140 512**	**102 494**	**74 224**	**51 187**	**53 377**	**54 860**	**51 967**	**52 040**	**Former USSR**

Former USSR: series up to 1990-1992 are not comparable with recent years; 1991 is estimated.
People's Rep. of China: up to 1978 the breakdown of final consumption by sector is incomplete

Consumption of Oil in Transport (1000 tonnes)
Consommation de pétrole dans les transports (1000 tonnes)
Ölverbrauch im Verkehrssektor (1000 Tonnen)

Consumo di petrolio nel settore dei trasporti (1000 tonnellate)

運輸部門における石油の消費量（千トン）

Consumo de petróleo en el transporte (1000 toneladas)

Потребление нефти и нефтепродуктов в транспорте (тыс. m)

	1971	1973	1978	1984	1985	1986	1987	1988	1989
Bahrein	193	277	597	663	661	647	657	688	710
République Islamique d'Iran	1 610	2 694	4 958	5 313	5 761	10 469	11 319	11 098	12 000
Irak	1 037	1 282	3 237	6 459	6 899	7 244	7 460	7 675	8 066
Israël	1 593	1 896	2 030	2 125	2 237	2 194	2 402	2 414	2 482
Jordanie	218	289	610	936	950	1 060	1 075	1 093	1 146
Koweit	816	872	1 273	1 765	1 765	1 831	1 825	1 931	1 936
Liban	921	1 182	924	1 276	1 407	1 326	1 380	914	660
Oman	34	41	269	532	579	610	529	544	578
Qatar	74	127	343	488	473	486	493	505	500
Arabie saoudite	723	1 016	3 613	11 114	11 855	11 936	11 719	8 105	7 455
Syrie	775	956	1 921	2 580	2 597	2 642	2 874	2 860	1 436
Emirats arabes unis	136	280	943	1 731	1 814	1 689	1 823	2 043	2 128
Yémen	237	426	580	989	1 053	1 142	1 225	1 301	1 261
Moyen-Orient	**8 367**	**11 338**	**21 298**	**35 971**	**38 051**	**43 276**	**44 781**	**41 171**	**40 358**
Total non-OCDE	**172 594**	**199 981**	**260 615**	**318 104**	**327 862**	**348 740**	**362 240**	**370 870**	**383 538**
OCDE Amérique du N.	378 385	424 261	479 623	466 997	471 295	482 528	498 487	519 743	526 820
OCDE Pacifique	48 489	56 408	71 475	78 992	80 848	84 577	88 865	94 520	101 616
OCDE Europe	148 626	170 329	201 272	216 826	220 717	232 242	242 361	256 089	266 034
Total OCDE	**575 500**	**650 998**	**752 370**	**762 815**	**772 860**	**799 347**	**829 713**	**870 352**	**894 470**
Monde	**748 094**	**850 979**	**1 012 985**	**1 080 919**	**1 100 722**	**1 148 087**	**1 191 953**	**1 241 222**	**1 278 008**

Consumption of Oil in Transport (1000 tonnes)
Consommation de pétrole dans les transports (1000 tonnes)
Ölverbrauch im Verkehrssektor (1000 Tonnen)

Consumo di petrolio nel settore dei trasporti (1000 tonnellate)

運輸部門における石油の消費量（千トン）

Consumo de petróleo en el transporte (1000 toneladas)

Потребление нефти и нефтепродуктов в транспорте (тыс. m)

1990	1991	1992	1993	1994	1995	1996	1997	1998	
772	656	663	674	725	759	739	663	792	Bahrain
6 637	7 390	7 655	18 738	19 395	18 144	19 163	20 279	21 811	Islamic Republic of Iran
8 416	6 550	7 618	8 375	8 797	8 737	8 592	8 680	8 847	Iraq
2 609	2 666	2 739	2 953	3 131	3 414	3 580	3 610	3 755	Israel
1 096	972	1 064	1 103	1 117	1 215	1 285	1 310	1 270	Jordan
1 013	1 085	1 683	1 778	1 889	2 068	2 125	2 324	2 493	Kuwait
660	825	1 148	1 333	1 285	1 553	1 501	1 440	1 537	Lebanon
836	835	790	860	842	851	881	946	982	Oman
578	559	603	631	698	760	840	942	962	Qatar
9 245	10 280	9 666	10 612	11 503	11 313	12 015	12 091	12 142	Saudi Arabia
1 610	1 582	1 390	1 324	1 327	1 219	1 263	1 377	1 422	Syria
2 204	2 316	2 439	2 493	2 524	2 474	2 838	2 953	2 722	United Arab Emirates
1 262	1 545	1 675	1 319	1 392	1 609	1 555	1 566	1 566	Yemen
36 938	**37 261**	**39 133**	**52 193**	**54 625**	**54 116**	**56 377**	**58 181**	**60 301**	**Middle East**
379 055	**411 559**	**384 014**	**390 176**	**377 538**	**407 877**	**434 463**	**453 126**	**458 270**	**Non-OECD Total**
525 619	518 776	530 824	539 491	558 297	569 081	582 319	593 037	609 930	OECD North America
107 388	111 371	116 518	119 851	126 862	133 673	139 844	142 724	139 345	OECD Pacific
273 805	277 183	285 924	291 699	294 913	299 527	308 589	313 205	322 326	OECD Europe
906 812	**907 330**	**933 266**	**951 041**	**980 072**	**1 002 281**	**1 030 752**	**1 048 966**	**1 071 601**	**OECD Total**
1 285 867	**1 318 889**	**1 317 280**	**1 341 217**	**1 357 610**	**1 410 158**	**1 465 215**	**1 502 092**	**1 529 871**	**World**

Consumption of Electricity in Transport (GWh)
Consommation d'électricité dans les transports (GWh)
Elektrizitätsverbrauch im Verkehrssektor (GWh)
Consumo di energia elettrica nel settore dei trasporti (GWh)
運輸部門における電力の消費量 (GWh)
Consumo de electricidad en el transporte (GWh)
Потребление электроэнергии в транспорте (ГВт.ч)

	1971	1973	1978	1984	1985	1986	1987	1988	1989
Algérie	19	28	32	35	39	180	198	163	213
Maroc	76	85	103	115	150	153	165	183	178
Afrique du Sud	3 278	2 896	3 517	4 595	4 587	4 501	4 049	4 120	4 229
Tunisie	-	-	-	74	80	84	87	88	92
Afrique	**3 373**	**3 009**	**3 652**	**4 819**	**4 856**	**4 918**	**4 499**	**4 554**	**4 712**
Argentine	290	292	257	269	267	267	337	337	326
Brésil	619	601	675	1 112	1 146	1 158	1 181	1 200	1 293
Chili	216	193	211	225	226	223	233	219	200
Colombie	-	-	-	-	-	-	-	-	-
Costa Rica	12	11	9	10	8	8	7	7	15
Cuba	-	-	-	-	6	5	6	5	5
Panama	-	-	-	88	75	-	-	-	60
Paraguay	-	-	-	-	1	1	1	1	1
Vénézuela	-	-	-	97	104	111	175	186	219
Amérique latine	**1 137**	**1 097**	**1 152**	**1 801**	**1 833**	**1 773**	**1 940**	**1 955**	**2 119**
Inde	1 643	1 531	2 307	3 064	3 315	3 543	3 985	4 067	4 097
Malaisie	-	-	-	-	-	-	-	-	-
Pakistan	28	28	31	38	37	36	38	40	35
Singapour	-	-	-	-	-	-	29	107	126
Taipei chinois	4	6	55	257	255	280	263	255	260
Viêt-Nam	-	-	-	-	-	-	-	38	44
Asie	**1 675**	**1 565**	**2 393**	**3 359**	**3 607**	**3 859**	**4 315**	**4 507**	**4 562**
Rép. populaire de Chine	-	-	-	4 140	6 340	6 690	7 670	8 950	9 870
Hong-Kong, Chine	-	-	-	-	-	-	603	593	-
Chine	**-**	**-**	**-**	**4 140**	**6 340**	**6 690**	**8 273**	**9 543**	**9 870**
Bulgarie	-	-	-	1 347	1 319	1 340	1 441	1 488	1 317
Chypre	34	38	4	8	28	30	36	43	47
Roumanie	-	-	-	-	2 430	2 580	2 583	2 824	2 921
République slovaque	554	633	764	1 095	1 080	1 092	1 103	1 163	1 182
Croatie	-	-	-	-	-	-	-	-	-
Ex-RYM	-	-	-	-	-	-	-	-	-
Slovénie	-	-	-	-	-	-	-	-	-
RF de Yougoslavie	-	-	-	-	-	-	-	-	-
Ex-Yougoslavie	533	717	830	1 224	1 237	1 493	1 231	1 298	1 412
Europe non-OCDE	**1 121**	**1 388**	**1 598**	**3 674**	**6 094**	**6 535**	**6 394**	**6 816**	**6 879**

Rép. populaire de Chine: jusqu'en 1978 la ventilation de la consommation finale par secteur est incomplète.

Consumption of Electricity in Transport (GWh)
Consommation d'électricité dans les transports (GWh)
Elektrizitätsverbrauch im Verkehrssektor (GWh)

Consumo di energia elettrica nel settore dei trasporti (GWh)

運輸部門における電力の消費量 (GWh)

Consumo de electricidad en el transporte (GWh)

Потребление электроэнергии в транспорте (ГВт.ч)

1990	1991	1992	1993	1994	1995	1996	1997	1998	
185	200	192	314	298	323	338	267	261	Algeria
201	216	269	184	189	199	204	214	228	Morocco
3 958	3 685	4 629	4 017	4 388	4 290	4 262	4 548	4 546	South Africa
103	107	123	126	132	138	148	158	170	Tunisia
4 447	**4 208**	**5 213**	**4 641**	**5 007**	**4 950**	**4 952**	**5 187**	**5 205**	**Africa**
314	244	279	279	302	349	419	442	477	Argentina
1 194	1 081	1 192	1 200	1 176	1 211	1 150	1 140	1 170	Brazil
212	223	226	250	218	200	201	211	208	Chile
-	-	-	-	-	-	-	43	42	Colombia
15	14	5	5	6	-	-	-	-	Costa Rica
89	77	55	53	52	64	69	66	83	Cuba
63	56	22	11	67	72	77	93	99	Panama
1	1	1	1	1	1	-	-	-	Paraguay
277	196	194	145	124	128	156	154	161	Venezuela
2 165	**1 892**	**1 974**	**1 944**	**1 946**	**2 025**	**2 072**	**2 149**	**2 240**	**Latin America**
4 112	4 519	5 068	5 620	5 886	6 315	6 575	7 073	7 538	India
-	-	-	-	-	-	-	12	12	Malaysia
38	33	29	27	27	22	20	19	16	Pakistan
186	186	186	186	198	198	236	248	263	Singapore
305	292	290	295	293	295	329	370	522	Chinese Taipei
51	38	40	45	87	100	114	128	158	Vietnam
4 692	**5 068**	**5 613**	**6 173**	**6 491**	**6 930**	**7 274**	**7 850**	**8 509**	**Asia**
10 590	9 290	11 847	14 623	16 404	10 913	15 621	20 200	20 180	People's Rep. of China
-	-	-	-	-	-	-	-	-	Hong Kong, China
10 590	**9 290**	**11 847**	**14 623**	**16 404**	**10 913**	**15 621**	**20 200**	**20 180**	**China**
1 305	1 206	1 040	726	636	803	811	620	552	Bulgaria
51	53	25	27	31	33	37	21	24	Cyprus
2 614	1 786	2 827	2 207	1 880	2 173	2 326	2 230	1 996	Romania
1 164	1 439	950	1 125	1 467	1 379	984	1 010	915	Slovak Republic
-	-	219	229	238	240	242	214	225	*Croatia*
-	-	29	21	-	13	14	14	24	*FYROM*
-	-	147	146	147	170	160	164	160	*Slovenia*
-	-	342	255	264	265	272	287	288	*FR of Yugoslavia*
1 059	740	737	651	649	688	688	679	697	Former Yugoslavia
6 193	**5 224**	**5 579**	**4 736**	**4 663**	**5 076**	**4 846**	**4 560**	**4 184**	**Non-OECD Europe**

People's Rep. of China: up to 1978 the breakdown of final consumption by sector is incomplete.

Consumption of Electricity in Transport (GWh)
Consommation d'électricité dans les transports (GWh)
Elektrizitätsverbrauch im Verkehrssektor (GWh)
Consumo di energia elettrica nel settore dei trasporti (GWh)
運輸部門における電力の消費量 *(GWh)*
Consumo de electricidad en el transporte (GWh)
Потребление электроэнергии в транспорте (ГВт.ч)

	1971	1973	1978	1984	1985	1986	1987	1988	1989
Arménie	-	-	-	-	-	-	-	-	-
Azerbaïdjan	-	-	-	-	-	-	-	-	-
Bélarus	-	-	-	-	-	-	-	-	-
Estonie	-	-	-	-	-	-	-	-	-
Géorgie	-	-	-	-	-	-	-	-	-
Kazakhstan	-	-	-	-	-	-	-	-	-
Kirghizistan	-	-	-	-	-	-	-	-	-
Lettonie	-	-	-	-	-	-	-	-	-
Lituanie	-	-	-	-	-	-	-	-	-
République de Moldavie	-	-	-	-	-	-	-	-	-
Russie	-	-	-	-	-	-	-	-	-
Tadjikistan	-	-	-	-	-	-	-	-	-
Turkménistan	-	-	-	-	-	-	-	-	-
Ukraine	-	-	-	-	-	-	-	-	-
Ouzbékistan	-	-	-	-	-	-	-	-	-
Ex-URSS	**48 800**	**53 800**	**70 100**	**80 800**	**82 000**	**83 200**	**84 400**	**85 600**	**86 500**
Total non-OCDE	**56 106**	**60 859**	**78 895**	**98 593**	**104 730**	**106 975**	**109 821**	**112 975**	**114 642**
OCDE Amérique du N.	6 715	7 878	4 979	6 430	7 344	7 514	7 718	8 156	8 128
OCDE Pacifique	12 297	14 058	16 166	17 582	18 233	18 762	19 304	20 543	21 701
OCDE Europe	36 576	38 905	43 979	50 308	52 175	54 057	55 238	56 866	57 825
Total OCDE	**55 588**	**60 841**	**65 124**	**74 320**	**77 752**	**80 333**	**82 260**	**85 565**	**87 654**
Monde	**111 694**	**121 700**	**144 019**	**172 913**	**182 482**	**187 308**	**192 081**	**198 540**	**202 296**

Ex-URSS: les séries antérieures à 1990-1992 ne sont pas comparables aux années récentes; 1991 a été estimé.

Consumption of Electricity in Transport (GWh)
Consommation d'électricité dans les transports (GWh)
Elektrizitätsverbrauch im Verkehrssektor (GWh)
Consumo di energia elettrica nel settore dei trasporti (GWh)
運輸部門における電力の消費量 (GWh)
Consumo de electricidad en el transporte (GWh)
Потребление электроэнергии в транспорте (ГВт.ч)

1990	1991	1992	1993	1994	1995	1996	1997	1998	
-	-	312	187	178	182	173	151	151	Armenia
-	-	854	700	551	481	455	496	422	Azerbaijan
-	-	965	2 029	2 173	1 832	1 824	1 715	1 977	Belarus
-	-	330	147	120	116	105	108	108	Estonia
-	-	619	608	405	430	250	231	315	Georgia
-	-	6 020	4 969	4 375	3 841	3 432	3 075	3 269	Kazakhstan
-	-	90	142	128	133	127	111	117	Kyrgyzstan
-	-	292	221	230	203	178	176	170	Latvia
-	-	141	93	102	96	103	101	103	Lithuania
-	-	211	92	83	135	125	118	116	Republic of Moldova
-	-	86 774	76 721	68 396	65 160	64 932	63 464	60 034	Russia
-	-	-	-	98	80	79	71	67	Tajikistan
-	-	155	154	130	129	121	98	102	Turkmenistan
-	-	12 698	11 941	10 873	10 777	9 754	9 545	9 703	Ukraine
-	-	1 256	1 090	1 111	1 366	1 398	1 282	1 277	Uzbekistan
87 000	87 061	110 717	99 094	88 953	84 961	83 056	80 742	77 931	**Former USSR**
115 087	112 743	140 943	131 211	123 464	114 855	117 821	120 688	118 249	**Non-OECD Total**
8 200	8 003	8 288	8 382	8 588	8 707	8 770	9 407	10 222	OECD North America
22 815	23 543	23 888	24 155	24 761	25 177	25 227	25 898	26 029	OECD Pacific
60 603	62 059	63 145	65 534	66 805	67 641	69 719	69 680	69 956	OECD Europe
91 618	93 605	95 321	98 071	100 154	101 525	103 716	104 985	106 207	**OECD Total**
206 705	206 348	236 264	229 282	223 618	216 380	221 537	225 673	224 456	**World**

Former USSR: series up to 1990-1992 are not comparable with recent years; 1991 is estimated.

OIL DEMAND BY PRODUCT

DEMANDE DE PETROLE PAR PRODUIT

Oil Demand by Main Product Groups[1]
Demande de produits raffinés par groupes principaux

World

	1971	1973	1978	1984	1985	1986	1987	1988	1989
Thousand tonnes									
NGL/LPG/Ethane	83 438	100 088	111 173	145 391	147 886	151 726	163 359	168 199	173 026
Naphtha	76 415	97 912	91 651	81 485	82 646	85 265	90 642	98 436	103 829
Motor Gasoline	494 543	559 137	659 585	658 323	663 433	685 192	707 699	730 393	745 244
Aviation Fuels	104 753	114 799	122 985	139 033	143 328	150 524	157 139	164 278	171 006
Kerosene	70 973	75 243	78 372	74 742	73 543	72 869	74 557	77 175	75 756
Gas Diesel	499 525	591 962	689 186	672 122	684 341	705 356	723 542	751 062	760 486
Heavy Fuel Oil	646 369	743 807	811 689	599 701	554 525	555 738	541 523	543 910	533 939
Other	159 054	196 199	250 570	231 456	230 451	242 649	250 317	262 955	261 550
Refinery Fuel	125 438	153 860	158 742	146 322	148 055	156 751	161 148	166 039	173 402
Sub-Total	2 260 508	2 633 007	2 973 953	2 748 575	2 728 208	2 806 070	2 869 926	2 962 447	2 998 238
Bunkers	120 725	132 805	114 138	86 669	92 092	96 984	95 032	98 787	99 100
Total	2 381 233	2 765 812	3 088 091	2 835 244	2 820 300	2 903 054	2 964 958	3 061 234	3 097 338
Thousand barrels/day									
NGL/LPG/Ethane	2 651.6	3 176.6	3 509.3	4 649.3	4 758.7	4 888.7	5 251.6	5 365.8	5 534.7
Naphtha	1 867.9	2 406.0	2 430.6	2 231.9	2 233.2	2 354.7	2 458.8	2 613.3	2 742.5
Motor Gasoline	11 535.4	13 043.6	15 385.4	15 311.4	15 468.1	15 976.3	16 500.5	16 983.9	17 378.6
Aviation Fuels	2 287.0	2 501.8	2 685.3	3 025.6	3 125.2	3 284.0	3 423.4	3 568.5	3 726.9
Kerosene	1 506.2	1 602.7	1 664.2	1 592.8	1 567.3	1 554.9	1 591.0	1 645.0	1 617.2
Gas Diesel	10 211.0	12 103.2	14 109.0	13 729.7	14 016.9	14 467.5	14 814.7	15 333.6	15 562.2
Heavy Fuel Oil	11 689.0	13 452.1	14 693.1	10 850.0	10 051.1	10 071.9	9 812.5	9 826.4	9 674.7
Other	2 943.0	3 649.4	4 795.4	4 362.4	4 344.9	4 575.9	4 701.2	4 923.1	4 894.2
Refinery Fuel	2 442.6	2 980.0	3 095.1	2 840.3	2 896.9	3 072.4	3 155.9	3 248.2	3 410.1
Sub-Total	47 133.6	54 915.4	62 367.4	58 593.3	58 462.2	60 246.3	61 709.7	63 507.7	64 541.1
Bunkers	2 212.5	2 433.3	2 095.6	1 593.4	1 693.8	1 784.8	1 751.6	1 814.1	1 824.2
Total	49 346.2	57 348.6	64 463.0	60 186.7	60 156.0	62 031.2	63 461.3	65 321.9	66 365.3

	1990	1991	1992	1993	1994	1995	1996	1997	1998
Thousand tonnes									
NGL/LPG/Ethane	169 740	178 796	171 512	174 501	183 341	191 881	200 433	205 055	205 029
Naphtha	102 652	102 043	114 449	117 029	120 351	128 696	133 973	145 251	151 038
Motor Gasoline	750 062	752 392	753 233	763 818	769 891	787 545	805 212	817 752	832 268
Aviation Fuels	171 839	167 815	166 113	168 223	175 573	179 930	188 302	193 550	196 322
Kerosene	74 756	76 167	74 875	77 300	77 893	75 936	80 082	80 116	80 010
Gas Diesel	765 697	782 238	804 589	818 855	820 289	847 569	882 974	905 736	904 948
Heavy Fuel Oil	515 454	514 742	515 037	502 347	479 925	459 263	453 135	445 619	446 139
Other	270 924	260 800	263 162	255 166	267 759	263 508	275 890	300 140	290 879
Refinery Fuel	170 385	168 964	173 465	165 085	166 417	170 120	171 551	174 588	178 491
Sub-Total	2 991 509	3 003 957	3 036 435	3 042 324	3 061 439	3 104 448	3 191 552	3 267 807	3 285 124
Bunkers	113 371	116 114	118 029	116 747	117 348	119 787	121 346	124 654	127 931
Total	3 104 880	3 120 071	3 154 464	3 159 071	3 178 787	3 224 235	3 312 898	3 392 461	3 413 055
Thousand barrels/day									
NGL/LPG/Ethane	5 566.4	5 878.3	5 725.4	5 848.6	6 131.7	6 409.0	6 689.3	6 859.6	6 848.6
Naphtha	2 728.1	2 728.9	3 023.8	3 093.5	3 209.0	3 384.0	3 503.1	3 873.0	3 998.6
Motor Gasoline	17 486.2	17 539.7	17 512.0	17 805.2	17 950.9	18 364.6	18 721.5	19 070.9	19 406.3
Aviation Fuels	3 747.1	3 651.5	3 607.6	3 659.5	3 817.4	3 911.6	4 082.6	4 204.9	4 267.2
Kerosene	1 593.4	1 622.7	1 586.5	1 646.7	1 660.7	1 619.8	1 702.3	1 712.3	1 710.4
Gas Diesel	15 660.9	16 002.4	16 411.5	16 751.4	16 776.0	17 334.6	18 003.4	18 518.2	18 503.4
Heavy Fuel Oil	9 340.7	9 329.3	9 304.5	9 100.0	8 694.1	8 336.8	8 200.1	8 083.2	8 096.1
Other	5 050.0	4 878.9	4 848.8	4 738.1	4 964.9	4 850.3	5 101.1	5 604.2	5 379.0
Refinery Fuel	3 347.4	3 311.1	3 426.7	3 238.2	3 259.8	3 334.7	3 336.9	3 411.5	3 476.5
Sub-Total	64 520.1	64 942.8	65 446.7	65 880.9	66 464.6	67 545.5	69 340.3	71 337.8	71 686.1
Bunkers	2 095.0	2 145.5	2 175.3	2 156.8	2 170.9	2 218.0	2 243.5	2 308.0	2 374.5
Total	66 615.1	67 088.3	67 622.0	68 037.7	68 635.5	69 763.4	71 583.7	73 645.8	74 060.6

[1] See Notes in Part I.4.

Oil Demand by Main Product Groups[1]
Demande de produits raffinés par groupes principaux

OECD Total

	1971	1973	1978	1984	1985	1986	1987	1988	1989
Thousand tonnes									
NGL/LPG/Ethane	70 519	82 980	86 557	107 518	108 109	105 943	113 978	115 361	117 656
Naphtha	72 337	89 527	78 419	62 674	60 467	63 827	66 750	71 513	73 321
Motor Gasoline	393 521	443 111	504 374	481 291	482 428	496 146	511 440	527 655	535 864
Aviation Fuels	70 449	76 947	82 048	91 551	95 384	101 284	105 762	111 441	116 429
Kerosene	37 979	41 660	40 379	33 584	32 280	32 533	32 466	34 639	32 906
Gas Diesel	377 311	442 365	486 106	417 206	425 074	436 895	443 695	460 978	460 642
Heavy Fuel Oil	462 907	519 507	492 413	278 463	243 068	246 560	233 986	239 233	245 243
Other	108 067	135 523	156 345	139 185	137 524	145 706	147 871	155 328	152 510
Refinery Fuel	96 373	120 791	112 363	94 796	94 953	100 465	103 015	106 914	110 244
Sub-Total	1 689 463	1 952 411	2 039 004	1 706 268	1 679 287	1 729 359	1 758 963	1 823 062	1 844 815
Bunkers	65 190	73 780	74 828	55 254	56 834	60 214	59 278	61 221	62 900
Total	1 754 653	2 026 191	2 113 832	1 761 522	1 736 121	1 789 573	1 818 241	1 884 283	1 907 715
Thousand barrels/day									
NGL/LPG/Ethane	2 262.0	2 663.0	2 768.0	3 507.0	3 552.0	3 504.0	3 759.0	3 774.0	3 857.0
Naphtha	1 773.0	2 211.0	2 119.0	1 792.0	1 713.0	1 851.0	1 897.0	1 982.0	2 027.0
Motor Gasoline	9 176.0	10 334.0	11 762.0	11 196.0	11 251.0	11 575.0	11 931.0	12 277.0	12 505.0
Aviation Fuels	1 537.0	1 675.0	1 793.0	1 994.0	2 081.0	2 211.0	2 304.0	2 420.0	2 537.0
Kerosene	806.0	890.0	858.0	722.0	692.0	699.0	698.0	745.0	708.0
Gas Diesel	7 712.0	9 044.0	9 955.0	8 534.0	8 718.0	8 986.0	9 101.0	9 427.0	9 440.0
Heavy Fuel Oil	8 338.0	9 357.0	8 866.0	5 006.0	4 376.0	4 440.0	4 211.0	4 294.0	4 416.0
Other	1 964.0	2 482.0	2 969.0	2 598.0	2 569.0	2 720.0	2 743.0	2 875.0	2 818.0
Refinery Fuel	1 873.0	2 335.0	2 171.0	1 811.0	1 830.0	1 948.0	1 992.0	2 066.0	2 144.0
Sub-Total	35 441.0	40 991.0	43 261.0	37 160.0	36 782.0	37 934.0	38 636.0	39 860.0	40 452.0
Bunkers	1 190.0	1 349.0	1 369.0	1 015.0	1 044.0	1 107.0	1 091.0	1 122.0	1 157.0
Total	36 631.0	42 340.0	44 630.0	38 175.0	37 826.0	39 041.0	39 727.0	40 982.0	41 609.0

	1990	1991	1992	1993	1994	1995	1996	1997	1998
Thousand tonnes									
NGL/LPG/Ethane	112 037	120 712	125 591	125 784	131 315	133 529	138 404	139 029	135 168
Naphtha	72 853	71 945	80 288	78 778	84 686	93 793	96 550	103 496	106 380
Motor Gasoline	538 383	538 684	553 113	561 777	571 447	579 632	587 217	594 686	606 039
Aviation Fuels	120 174	117 499	118 608	121 386	127 550	130 530	136 604	140 005	143 676
Kerosene	31 847	32 537	33 879	36 079	36 715	39 942	43 540	43 017	40 689
Gas Diesel	466 091	476 327	486 642	497 870	505 841	518 661	542 779	544 356	544 517
Heavy Fuel Oil	228 764	222 489	224 684	216 966	221 389	202 396	196 046	192 169	193 129
Other	160 721	160 034	160 241	155 144	167 681	162 079	165 100	175 535	169 299
Refinery Fuel	112 029	112 111	115 636	116 045	118 159	118 248	122 173	121 555	123 429
Sub-Total	1 842 899	1 852 338	1 898 682	1 909 829	1 964 783	1 978 810	2 028 413	2 053 848	2 062 326
Bunkers	75 008	77 581	80 096	78 573	77 385	80 151	79 416	80 733	83 121
Total	1 917 907	1 929 919	1 978 778	1 988 402	2 042 168	2 058 961	2 107 829	2 134 581	2 145 447
Thousand barrels/day									
NGL/LPG/Ethane	3 816.0	4 117.0	4 280.0	4 312.0	4 488.0	4 565.0	4 727.0	4 776.0	4 647.0
Naphtha	2 028.0	2 022.0	2 224.0	2 196.0	2 371.0	2 564.0	2 627.0	2 892.0	2 950.0
Motor Gasoline	12 558.0	12 565.0	12 867.0	13 103.0	13 334.0	13 527.0	13 663.0	13 880.0	14 141.0
Aviation Fuels	2 619.0	2 554.0	2 573.0	2 637.0	2 769.0	2 833.0	2 957.0	3 036.0	3 118.0
Kerosene	683.0	697.0	719.0	772.0	787.0	856.0	929.0	925.0	876.0
Gas Diesel	9 542.0	9 755.0	9 936.0	10 196.0	10 355.0	10 619.0	11 077.0	11 140.0	11 144.0
Heavy Fuel Oil	4 122.0	4 009.0	4 034.0	3 907.0	3 991.0	3 665.0	3 539.0	3 476.0	3 497.0
Other	2 961.0	2 966.0	2 928.0	2 840.0	3 085.0	2 959.0	3 010.0	3 244.0	3 080.0
Refinery Fuel	2 176.0	2 172.0	2 235.0	2 244.0	2 284.0	2 286.0	2 336.0	2 336.0	2 368.0
Sub-Total	40 505.0	40 857.0	41 796.0	42 207.0	43 464.0	43 874.0	44 865.0	45 705.0	45 821.0
Bunkers	1 388.0	1 436.0	1 478.0	1 450.0	1 430.0	1 484.0	1 468.0	1 493.0	1 541.0
Total	41 893.0	42 293.0	43 274.0	43 657.0	44 894.0	45 358.0	46 333.0	47 198.0	47 362.0

[1] See Notes in Part I.4.

Oil Demand by Main Product Groups[1]
Demande de produits raffinés par groupes principaux

OECD Europe

	1971	1973	1978	1984	1985	1986	1987	1988	1989
Thousand tonnes									
NGL/LPG/Ethane	12 092	14 709	16 817	20 290	20 701	21 006	23 083	22 808	22 484
Naphtha	29 662	38 280	30 803	29 286	29 545	30 674	32 041	35 036	34 900
Motor Gasoline	83 035	95 282	110 021	114 057	113 325	118 368	121 692	126 371	129 006
Aviation Fuels	16 368	18 928	21 028	22 135	23 054	24 399	25 812	27 802	29 383
Kerosene	7 920	8 959	5 990	3 600	3 908	3 944	4 167	3 898	3 743
Gas Diesel	194 298	226 102	237 324	201 641	209 267	218 663	217 868	218 874	213 846
Heavy Fuel Oil	218 505	243 267	214 072	127 599	109 899	104 903	99 341	95 401	96 455
Other	39 389	44 996	42 450	42 015	40 491	42 944	44 849	45 757	45 681
Refinery Fuel	31 002	35 123	37 488	29 669	29 485	31 125	31 297	33 142	34 239
Sub-Total	632 271	725 646	715 993	590 292	579 675	596 026	600 150	609 089	609 737
Bunkers	39 856	44 078	36 964	26 988	29 261	34 343	33 562	34 211	34 379
Total	672 127	769 724	752 957	617 280	608 936	630 369	633 712	643 300	644 116
Thousand barrels/day									
NGL/LPG/Ethane	394.0	479.0	550.0	706.0	718.0	740.0	800.0	766.0	759.0
Naphtha	779.0	1 017.0	953.0	966.0	937.0	1 025.0	1 029.0	1 066.0	1 059.0
Motor Gasoline	1 922.0	2 206.0	2 548.0	2 639.0	2 628.0	2 750.0	2 826.0	2 929.0	3 000.0
Aviation Fuels	356.0	408.0	453.0	477.0	498.0	527.0	556.0	597.0	635.0
Kerosene	169.0	196.0	129.0	87.0	89.0	92.0	99.0	95.0	91.0
Gas Diesel	3 973.0	4 625.0	4 871.0	4 142.0	4 308.0	4 528.0	4 487.0	4 494.0	4 398.0
Heavy Fuel Oil	3 881.0	4 320.0	3 796.0	2 273.0	1 959.0	1 868.0	1 768.0	1 692.0	1 718.0
Other	729.0	833.0	792.0	771.0	747.0	792.0	822.0	833.0	827.0
Refinery Fuel	608.0	684.0	731.0	577.0	578.0	616.0	616.0	656.0	677.0
Sub-Total	12 811.0	14 768.0	14 823.0	12 638.0	12 462.0	12 938.0	13 003.0	13 128.0	13 164.0
Bunkers	722.0	799.0	672.0	495.0	536.0	631.0	614.0	625.0	630.0
Total	13 533.0	15 567.0	15 495.0	13 133.0	12 998.0	13 569.0	13 617.0	13 753.0	13 794.0

	1990	1991	1992	1993	1994	1995	1996	1997	1998
Thousand tonnes									
NGL/LPG/Ethane	22 896	24 866	24 392	24 648	26 131	26 616	27 417	27 603	28 109
Naphtha	33 957	33 116	33 534	31 354	32 857	37 821	37 393	39 199	40 467
Motor Gasoline	132 556	134 271	137 908	137 556	136 214	135 598	136 993	136 649	136 525
Aviation Fuels	30 237	29 536	31 164	32 595	34 168	35 851	37 654	39 584	42 264
Kerosene	3 892	4 259	4 247	4 352	4 534	4 627	5 391	5 519	5 817
Gas Diesel	218 008	228 951	230 711	234 684	232 179	237 428	250 365	248 523	254 600
Heavy Fuel Oil	91 555	93 334	94 082	88 237	87 025	86 982	81 713	75 714	75 764
Other	45 267	46 644	48 252	46 609	50 407	51 145	50 713	55 737	55 858
Refinery Fuel	34 104	33 797	36 065	37 293	38 045	39 187	40 396	40 040	40 617
Sub-Total	612 472	628 774	640 355	637 328	641 560	655 255	668 035	668 568	680 021
Bunkers	36 154	35 737	36 204	37 055	35 713	36 997	39 240	42 800	44 855
Total	648 626	664 511	676 559	674 383	677 273	692 252	707 275	711 368	724 876
Thousand barrels/day									
NGL/LPG/Ethane	769.0	833.0	823.0	835.0	885.0	892.0	914.0	925.0	942.0
Naphtha	1 043.0	1 012.0	1 032.0	987.0	1 037.0	1 115.0	1 103.0	1 161.0	1 192.0
Motor Gasoline	3 076.0	3 115.0	3 193.0	3 192.0	3 165.0	3 153.0	3 174.0	3 178.0	3 173.0
Aviation Fuels	655.0	634.0	672.0	703.0	736.0	774.0	811.0	853.0	914.0
Kerosene	90.0	98.0	93.0	98.0	105.0	107.0	124.0	128.0	136.0
Gas Diesel	4 480.0	4 707.0	4 728.0	4 826.0	4 771.0	4 880.0	5 127.0	5 103.0	5 229.0
Heavy Fuel Oil	1 635.0	1 669.0	1 673.0	1 573.0	1 553.0	1 572.0	1 469.0	1 362.0	1 371.0
Other	816.0	846.0	877.0	851.0	923.0	933.0	920.0	1 034.0	1 032.0
Refinery Fuel	680.0	669.0	714.0	740.0	753.0	781.0	798.0	793.0	804.0
Sub-Total	13 244.0	13 583.0	13 805.0	13 805.0	13 928.0	14 207.0	14 440.0	14 537.0	14 793.0
Bunkers	660.0	654.0	660.0	675.0	651.0	678.0	717.0	782.0	822.0
Total	13 904.0	14 237.0	14 465.0	14 480.0	14 579.0	14 885.0	15 157.0	15 319.0	15 615.0

[1] See Notes in Part I.4.

Oil Demand by Main Product Groups[1]
Demande de produits raffinés par groupes principaux

OECD North America

	1971	1973	1978	1984	1985	1986	1987	1988	1989
Thousand tonnes									
NGL/LPG/Ethane	48 958	56 188	52 398	64 706	65 612	62 785	67 989	70 001	70 445
Naphtha	20 255	23 435	19 670	14 471	12 016	12 710	13 128	13 202	14 292
Motor Gasoline	284 388	317 088	356 735	326 940	328 222	335 772	346 855	356 721	359 336
Aviation Fuels	50 515	53 158	54 989	61 775	64 381	68 187	70 543	73 419	75 621
Kerosene	15 673	13 951	11 776	7 524	6 994	6 271	6 149	6 287	5 359
Gas Diesel	154 100	177 892	200 980	162 720	162 106	161 859	165 780	175 662	175 259
Heavy Fuel Oil	140 576	162 544	173 834	84 562	77 301	89 974	85 072	89 915	92 412
Other	46 935	55 470	81 097	67 648	70 110	73 400	73 968	75 107	71 878
Refinery Fuel	56 701	74 076	60 466	54 157	54 483	58 697	60 530	61 989	63 668
Sub-Total	818 101	933 802	1 011 945	844 503	841 225	869 655	890 014	922 303	928 270
Bunkers	7 620	9 629	23 343	19 503	18 737	18 301	18 337	19 781	20 732
Total	825 721	943 431	1 035 288	864 006	859 962	887 956	908 351	942 084	949 002
Thousand barrels/day									
NGL/LPG/Ethane	1 567.0	1 800.0	1 685.0	2 102.0	2 147.0	2 063.0	2 233.0	2 303.0	2 325.0
Naphtha	472.0	547.0	458.0	338.0	282.0	298.0	307.0	308.0	335.0
Motor Gasoline	6 644.0	7 409.0	8 334.0	7 617.0	7 667.0	7 844.0	8 103.0	8 310.0	8 394.0
Aviation Fuels	1 103.0	1 160.0	1 208.0	1 351.0	1 411.0	1 495.0	1 544.0	1 602.0	1 655.0
Kerosene	332.0	296.0	250.0	160.0	149.0	133.0	130.0	133.0	113.0
Gas Diesel	3 149.0	3 634.0	4 106.0	3 315.0	3 312.0	3 306.0	3 387.0	3 579.0	3 580.0
Heavy Fuel Oil	2 562.0	2 962.0	3 163.0	1 527.0	1 397.0	1 629.0	1 538.0	1 622.0	1 670.0
Other	817.0	970.0	1 544.0	1 273.0	1 318.0	1 380.0	1 378.0	1 396.0	1 336.0
Refinery Fuel	1 102.0	1 434.0	1 151.0	1 023.0	1 040.0	1 123.0	1 156.0	1 178.0	1 223.0
Sub-Total	17 748.0	20 212.0	21 899.0	18 706.0	18 723.0	19 271.0	19 776.0	20 431.0	20 631.0
Bunkers	141.0	178.0	430.0	359.0	345.0	337.0	340.0	364.0	383.0
Total	17 889.0	20 390.0	22 329.0	19 065.0	19 068.0	19 608.0	20 116.0	20 795.0	21 014.0

	1990	1991	1992	1993	1994	1995	1996	1997	1998
Thousand tonnes									
NGL/LPG/Ethane	61 995	67 031	71 319	71 119	75 640	76 878	80 099	80 956	78 027
Naphtha	13 314	11 298	13 102	12 212	13 716	13 645	15 877	16 603	18 003
Motor Gasoline	355 442	353 043	361 942	369 468	377 347	384 384	388 267	394 538	406 259
Aviation Fuels	77 735	75 279	74 374	75 158	78 624	78 012	81 695	82 453	84 014
Kerosene	3 084	2 814	2 597	3 127	2 990	3 298	3 591	3 807	4 252
Gas Diesel	171 102	165 882	171 179	176 084	182 242	185 694	193 724	198 469	200 400
Heavy Fuel Oil	76 286	69 069	67 356	68 952	69 090	53 217	54 153	58 323	68 698
Other	78 673	76 099	74 725	74 423	78 383	75 206	78 633	84 497	82 953
Refinery Fuel	65 054	64 253	64 902	63 547	64 274	62 799	64 960	63 620	64 560
Sub-Total	902 685	884 768	901 496	914 090	942 306	933 133	960 999	983 266	1 007 166
Bunkers	30 934	32 654	33 969	29 889	29 321	30 936	29 225	25 578	25 317
Total	933 619	917 422	935 465	943 979	971 627	964 069	990 224	1 008 844	1 032 483
Thousand barrels/day									
NGL/LPG/Ethane	2 203.0	2 384.0	2 524.0	2 530.0	2 667.0	2 721.0	2 837.0	2 875.0	2 771.0
Naphtha	313.0	265.0	307.0	286.0	321.0	320.0	371.0	389.0	421.0
Motor Gasoline	8 305.0	8 248.0	8 433.0	8 632.0	8 816.0	8 980.0	9 046.0	9 218.0	9 491.0
Aviation Fuels	1 698.0	1 644.0	1 618.0	1 638.0	1 712.0	1 697.0	1 772.0	1 792.0	1 826.0
Kerosene	66.0	60.0	54.0	67.0	63.0	69.0	75.0	81.0	90.0
Gas Diesel	3 488.0	3 381.0	3 480.0	3 589.0	3 715.0	3 785.0	3 938.0	4 046.0	4 085.0
Heavy Fuel Oil	1 375.0	1 244.0	1 210.0	1 243.0	1 243.0	955.0	970.0	1 047.0	1 234.0
Other	1 451.0	1 417.0	1 355.0	1 358.0	1 432.0	1 359.0	1 425.0	1 555.0	1 485.0
Refinery Fuel	1 244.0	1 228.0	1 237.0	1 210.0	1 225.0	1 190.0	1 217.0	1 195.0	1 211.0
Sub-Total	20 143.0	19 871.0	20 218.0	20 553.0	21 194.0	21 076.0	21 651.0	22 198.0	22 614.0
Bunkers	580.0	611.0	634.0	560.0	551.0	580.0	550.0	482.0	479.0
Total	20 723.0	20 482.0	20 852.0	21 113.0	21 745.0	21 656.0	22 201.0	22 680.0	23 093.0

[1] See Notes in Part I.4.

Oil Demand by Main Product Groups[1]
Demande de produits raffinés par groupes principaux

OECD Pacific

	1971	1973	1978	1984	1985	1986	1987	1988	1989
Thousand tonnes									
NGL/LPG/Ethane	9 469	12 083	17 342	22 522	21 796	22 152	22 906	22 552	24 727
Naphtha	22 420	27 812	27 946	18 917	18 906	20 443	21 581	23 275	24 129
Motor Gasoline	26 098	30 741	37 618	40 294	40 881	42 006	42 893	44 563	47 522
Aviation Fuels	3 566	4 861	6 031	7 641	7 949	8 698	9 407	10 220	11 425
Kerosene	14 386	18 750	22 613	22 460	21 378	22 318	22 150	24 454	23 804
Gas Diesel	28 913	38 371	47 802	52 845	53 701	56 373	60 047	66 442	71 537
Heavy Fuel Oil	103 826	113 696	104 507	66 302	55 868	51 683	49 573	53 917	56 376
Other	21 743	35 057	32 798	29 522	26 923	29 362	29 054	34 464	34 951
Refinery Fuel	8 670	11 592	14 409	10 970	10 985	10 643	11 188	11 783	12 337
Sub-Total	239 091	292 963	311 066	271 473	258 387	263 678	268 799	291 670	306 808
Bunkers	17 714	20 073	14 521	8 763	8 836	7 570	7 379	7 229	7 789
Total	256 805	313 036	325 587	280 236	267 223	271 248	276 178	298 899	314 597
Thousand barrels/day									
NGL/LPG/Ethane	301.0	384.0	533.0	699.0	687.0	701.0	726.0	705.0	773.0
Naphtha	522.0	647.0	708.0	488.0	494.0	528.0	561.0	608.0	633.0
Motor Gasoline	610.0	719.0	880.0	940.0	956.0	981.0	1 002.0	1 038.0	1 111.0
Aviation Fuels	78.0	107.0	132.0	166.0	172.0	189.0	204.0	221.0	247.0
Kerosene	305.0	398.0	479.0	475.0	454.0	474.0	469.0	517.0	504.0
Gas Diesel	590.0	785.0	978.0	1 077.0	1 098.0	1 152.0	1 227.0	1 354.0	1 462.0
Heavy Fuel Oil	1 895.0	2 075.0	1 907.0	1 206.0	1 020.0	943.0	905.0	980.0	1 028.0
Other	418.0	679.0	633.0	554.0	504.0	548.0	543.0	646.0	655.0
Refinery Fuel	163.0	217.0	289.0	211.0	212.0	209.0	220.0	232.0	244.0
Sub-Total	4 882.0	6 011.0	6 539.0	5 816.0	5 597.0	5 725.0	5 857.0	6 301.0	6 657.0
Bunkers	327.0	372.0	267.0	161.0	163.0	139.0	137.0	133.0	144.0
Total	5 209.0	6 383.0	6 806.0	5 977.0	5 760.0	5 864.0	5 994.0	6 434.0	6 801.0

	1990	1991	1992	1993	1994	1995	1996	1997	1998
Thousand tonnes									
NGL/LPG/Ethane	27 146	28 815	29 880	30 017	29 544	30 035	30 888	30 470	29 032
Naphtha	25 582	27 531	33 652	35 212	38 113	42 327	43 280	47 694	47 910
Motor Gasoline	50 385	51 370	53 263	54 753	57 886	59 650	61 957	63 499	63 255
Aviation Fuels	12 202	12 684	13 070	13 633	14 758	16 667	17 255	17 968	17 398
Kerosene	24 871	25 464	27 035	28 600	29 191	32 017	34 558	33 691	30 620
Gas Diesel	76 981	81 494	84 752	87 102	91 420	95 539	98 690	97 364	89 517
Heavy Fuel Oil	60 923	60 086	63 246	59 777	65 274	62 197	60 180	58 132	48 667
Other	36 781	37 291	37 264	34 112	38 891	35 728	35 754	35 301	30 488
Refinery Fuel	12 871	14 061	14 669	15 205	15 840	16 262	16 817	17 895	18 252
Sub-Total	327 742	338 796	356 831	358 411	380 917	390 422	399 379	402 014	375 139
Bunkers	7 920	9 190	9 923	11 629	12 351	12 218	10 951	12 355	12 949
Total	335 662	347 986	366 754	370 040	393 268	402 640	410 330	414 369	388 088
Thousand barrels/day									
NGL/LPG/Ethane	844.0	900.0	933.0	947.0	936.0	952.0	976.0	976.0	934.0
Naphtha	672.0	745.0	885.0	923.0	1 013.0	1 129.0	1 153.0	1 342.0	1 337.0
Motor Gasoline	1 177.0	1 202.0	1 241.0	1 279.0	1 353.0	1 394.0	1 443.0	1 484.0	1 477.0
Aviation Fuels	266.0	276.0	283.0	296.0	321.0	362.0	374.0	391.0	378.0
Kerosene	527.0	539.0	572.0	607.0	619.0	680.0	730.0	716.0	650.0
Gas Diesel	1 574.0	1 667.0	1 728.0	1 781.0	1 869.0	1 954.0	2 012.0	1 991.0	1 830.0
Heavy Fuel Oil	1 112.0	1 096.0	1 151.0	1 091.0	1 195.0	1 138.0	1 100.0	1 067.0	892.0
Other	694.0	703.0	696.0	631.0	730.0	667.0	665.0	655.0	563.0
Refinery Fuel	252.0	275.0	284.0	294.0	306.0	315.0	321.0	348.0	353.0
Sub-Total	7 118.0	7 403.0	7 773.0	7 849.0	8 342.0	8 591.0	8 774.0	8 970.0	8 414.0
Bunkers	148.0	171.0	184.0	215.0	228.0	226.0	201.0	229.0	240.0
Total	7 266.0	7 574.0	7 957.0	8 064.0	8 570.0	8 817.0	8 975.0	9 199.0	8 654.0

[1] See Notes in Part I.4.

Oil Demand by Main Product Groups[1]
Demande de produits raffinés par groupes principaux

Non-OECD Total

	1971	1973	1978	1984	1985	1986	1987	1988	1989
Thousand tonnes									
NGL/LPG/Ethane	12 919	17 108	24 616	37 873	39 777	45 783	49 381	52 838	55 370
Naphtha	4 078	8 385	13 232	18 811	22 179	21 438	23 892	26 923	30 508
Motor Gasoline	101 022	116 026	155 211	177 032	181 005	189 046	196 259	202 738	209 380
Aviation Fuels	34 304	37 852	40 937	47 482	47 944	49 240	51 377	52 837	54 577
Kerosene	32 994	33 583	37 993	41 158	41 263	. 40 336	42 091	42 536	42 850
Gas Diesel	122 214	149 597	203 080	254 916	259 267	268 461	279 847	290 084	299 844
Heavy Fuel Oil	183 462	224 300	319 276	321 238	311 457	309 178	307 537	304 677	288 696
Other	50 987	60 676	94 225	92 271	92 927	96 943	102 446	107 627	109 040
Refinery Fuel	29 065	33 069	46 379	51 526	53 102	56 286	58 133	59 125	63 158
Sub-Total	**571 045**	**680 596**	**934 949**	**1 042 307**	**1 048 921**	**1 076 711**	**1 110 963**	**1 139 385**	**1 153 423**
Bunkers	55 535	59 025	39 310	31 415	35 258	36 770	35 754	37 566	36 200
Total	**626 580**	**739 621**	**974 259**	**1 073 722**	**1 084 179**	**1 113 481**	**1 146 717**	**1 176 951**	**1 189 623**
Thousand barrels/day									
NGL/LPG/Ethane	389.6	513.6	741.3	1 142.3	1 206.7	1 384.7	1 492.6	1 591.8	1 677.7
Naphtha	94.9	195.0	311.6	439.9	520.2	503.7	561.8	631.3	715.5
Motor Gasoline	2 359.4	2 709.6	3 623.4	4 115.4	4 217.1	4 401.3	4 569.5	4 706.9	4 873.6
Aviation Fuels	750.0	826.8	892.3	1 031.6	1 044.2	1 073.0	1 119.4	1 148.5	1 189.9
Kerosene	700.2	712.7	806.2	870.8	875.3	855.9	893.0	900.0	909.2
Gas Diesel	2 499.0	3 059.2	4 154.0	5 195.7	5 298.9	5 481.5	5 713.7	5 906.6	6 122.2
Heavy Fuel Oil	3 351.0	4 095.1	5 827.1	5 844.0	5 675.1	5 631.9	5 601.5	5 532.4	5 258.7
Other	979.0	1 167.4	1 826.4	1 764.4	1 775.9	1 855.9	1 958.2	2 048.1	2 076.2
Refinery Fuel	569.6	645.0	924.1	1 029.3	1 066.9	1 124.4	1 163.9	1 182.2	1 266.1
Sub-Total	**11 692.6**	**13 924.4**	**19 106.4**	**21 433.3**	**21 680.2**	**22 312.3**	**23 073.7**	**23 647.7**	**24 089.1**
Bunkers	1 022.5	1 084.3	726.6	578.4	649.8	677.8	660.6	692.1	667.2
Total	**12 715.2**	**15 008.6**	**19 833.0**	**22 011.7**	**22 330.0**	**22 990.2**	**23 734.3**	**24 339.9**	**24 756.3**

	1990	1991	1992	1993	1994	1995	1996	1997	1998
Thousand tonnes									
NGL/LPG/Ethane	57 703	58 084	45 921	48 717	52 026	58 352	62 029	66 026	69 861
Naphtha	29 799	30 098	34 161	38 251	35 665	34 903	37 423	41 755	44 658
Motor Gasoline	211 679	213 708	200 120	202 041	198 444	207 913	217 995	223 066	226 229
Aviation Fuels	51 665	50 316	47 505	46 837	48 023	49 400	51 698	53 545	52 646
Kerosene	42 909	43 630	40 996	41 221	41 178	35 994	36 542	37 099	39 321
Gas Diesel	299 606	305 911	317 947	320 985	314 448	328 908	340 195	361 380	360 431
Heavy Fuel Oil	286 690	292 253	290 353	285 381	258 536	256 867	257 089	253 450	253 010
Other	110 203	100 766	102 921	100 022	100 078	101 429	110 790	124 605	121 580
Refinery Fuel	58 356	56 853	57 829	49 040	48 258	51 872	49 378	53 033	55 062
Sub-Total	**1 148 610**	**1 151 619**	**1 137 753**	**1 132 495**	**1 096 656**	**1 125 638**	**1 163 139**	**1 213 959**	**1 222 798**
Bunkers	38 363	38 533	37 933	38 174	39 963	39 636	41 930	43 921	44 810
Total	**1 186 973**	**1 190 152**	**1 175 686**	**1 170 669**	**1 136 619**	**1 165 274**	**1 205 069**	**1 257 880**	**1 267 608**
Thousand barrels/day									
NGL/LPG/Ethane	1 750.4	1 761.3	1 445.4	1 536.6	1 643.7	1 844.0	1 962.3	2 083.6	2 201.6
Naphtha	700.1	706.9	799.8	897.5	838.0	820.0	876.1	981.0	1 048.6
Motor Gasoline	4 928.2	4 974.7	4 645.0	4 702.2	4 616.9	4 837.6	5 058.5	5 190.9	5 265.3
Aviation Fuels	1 128.1	1 097.5	1 034.6	1 022.5	1 048.4	1 078.6	1 125.6	1 168.9	1 149.2
Kerosene	910.4	925.7	867.5	874.7	873.7	763.8	773.3	787.3	834.4
Gas Diesel	6 118.9	6 247.4	6 475.5	6 555.4	6 421.0	6 715.6	6 926.4	7 378.2	7 359.4
Heavy Fuel Oil	5 218.7	5 320.3	5 270.5	5 193.0	4 703.1	4 671.8	4 661.1	4 607.2	4 599.1
Other	2 089.0	1 912.9	1 920.8	1 898.1	1 879.9	1 891.3	2 091.1	2 360.2	2 299.0
Refinery Fuel	1 171.4	1 139.1	1 191.7	994.2	975.8	1 048.7	1 000.9	1 075.5	1 108.5
Sub-Total	**24 015.1**	**24 085.8**	**23 650.7**	**23 673.9**	**23 000.6**	**23 671.5**	**24 475.3**	**25 632.8**	**25 865.1**
Bunkers	707.0	709.5	697.3	706.8	740.9	734.0	775.5	815.0	833.5
Total	**24 722.1**	**24 795.3**	**24 348.0**	**24 380.7**	**23 741.5**	**24 405.4**	**25 250.7**	**26 447.8**	**26 698.6**

[1] See Notes in Part I.4.

Oil Demand by Main Product Groups[1]
Demande de produits raffinés par groupes principaux

Africa

	1971	1973	1978	1984	1985	1986	1987	1988	1989
Thousand tonnes									
NGL/LPG/Ethane	623	847	1 534	2 805	2 840	3 011	3 405	3 486	3 866
Naphtha	90	75	87	488	622	647	571	523	719
Motor Gasoline	7 745	9 139	12 554	15 816	16 049	16 437	17 592	18 589	19 192
Aviation Fuels	1 970	2 440	3 449	4 188	4 350	4 091	4 467	4 491	4 858
Kerosene	2 773	3 007	3 816	5 149	5 339	5 703	6 039	6 018	6 154
Gas Diesel	9 884	12 302	17 809	23 543	23 418	22 686	23 444	24 325	24 557
Heavy Fuel Oil	7 825	9 205	12 347	17 349	15 789	15 181	15 580	16 637	16 514
Other	1 920	2 226	3 518	5 383	5 108	5 241	5 460	5 536	5 791
Refinery Fuel	1 276	1 455	2 059	2 963	3 016	3 300	3 427	3 489	3 516
Sub-Total	34 106	40 696	57 173	77 684	76 531	76 297	79 985	83 094	85 167
Bunkers	7 028	6 114	4 095	4 790	4 449	4 528	4 372	4 779	4 951
Total	41 134	46 810	61 268	82 474	80 980	80 825	84 357	87 873	90 118
Thousand barrels/day									
NGL/LPG/Ethane	19.8	26.9	48.8	88.9	90.3	95.7	108.2	110.5	122.9
Naphtha	2.1	1.7	2.0	11.3	14.5	15.1	13.3	12.1	16.7
Motor Gasoline	181.0	213.6	293.4	368.6	375.1	384.1	411.1	433.2	448.5
Aviation Fuels	43.2	53.4	75.3	91.1	94.9	89.3	97.4	97.7	106.3
Kerosene	58.8	63.8	80.9	108.9	113.2	120.9	128.1	127.3	130.5
Gas Diesel	202.0	251.4	364.0	479.9	478.6	463.7	479.2	495.8	501.9
Heavy Fuel Oil	142.8	168.0	225.3	315.7	288.1	277.0	284.3	302.7	301.3
Other	36.2	41.9	67.0	99.8	94.7	96.3	100.6	101.9	107.0
Refinery Fuel	24.5	28.1	40.0	58.6	59.9	65.7	68.3	69.7	70.5
Sub-Total	710.4	848.8	1 196.6	1 622.8	1 609.2	1 607.7	1 690.4	1 751.1	1 805.6
Bunkers	130.9	114.3	77.2	89.8	83.7	85.0	82.0	89.4	92.7
Total	841.3	963.1	1 273.8	1 712.5	1 692.9	1 692.8	1 772.5	1 840.4	1 898.3

	1990	1991	1992	1993	1994	1995	1996	1997	1998
Thousand tonnes									
NGL/LPG/Ethane	4 533	4 481	4 857	4 925	5 442	5 726	5 769	6 230	6 602
Naphtha	786	909	806	857	868	973	982	985	985
Motor Gasoline	18 568	19 347	19 956	19 848	19 662	21 445	21 739	22 033	22 258
Aviation Fuels	4 793	4 554	4 669	4 999	5 391	5 702	6 107	6 021	5 923
Kerosene	6 029	5 749	5 590	5 358	4 999	5 169	5 430	5 560	5 560
Gas Diesel	24 771	25 476	25 826	26 497	27 204	28 580	29 846	31 586	32 492
Heavy Fuel Oil	16 501	16 647	17 661	15 896	15 077	15 265	14 579	15 872	18 038
Other	6 205	6 584	8 327	5 373	6 175	6 514	6 215	6 239	6 587
Refinery Fuel	3 555	3 516	3 603	4 534	4 836	5 058	5 137	4 851	4 760
Sub-Total	85 741	87 263	91 295	88 287	89 654	94 432	95 804	99 377	103 205
Bunkers	5 413	5 696	6 313	6 105	7 520	8 404	7 910	6 996	6 553
Total	91 154	92 959	97 608	94 392	97 174	102 836	103 714	106 373	109 758
Thousand barrels/day									
NGL/LPG/Ethane	147.0	144.8	155.9	158.4	175.6	184.7	185.5	200.7	211.0
Naphtha	18.3	21.2	18.7	20.0	20.2	22.7	22.8	22.9	22.9
Motor Gasoline	433.9	452.1	465.1	463.8	459.5	501.2	506.6	514.9	520.2
Aviation Fuels	105.0	99.8	102.0	109.5	118.0	124.8	133.2	131.8	129.6
Kerosene	127.8	121.9	118.2	113.6	106.0	109.6	114.8	117.9	117.9
Gas Diesel	506.3	520.7	526.4	541.6	556.0	584.1	608.3	645.6	664.1
Heavy Fuel Oil	301.1	303.8	321.4	290.0	275.1	278.5	265.3	289.6	329.1
Other	116.5	124.1	158.3	100.0	116.3	122.4	116.1	116.2	123.3
Refinery Fuel	71.5	70.8	72.6	97.6	104.2	109.0	110.3	103.3	101.8
Sub-Total	1 827.4	1 859.2	1 938.7	1 894.5	1 930.9	2 036.9	2 063.1	2 142.9	2 219.9
Bunkers	101.0	106.1	117.1	113.9	140.1	156.5	146.8	130.3	122.0
Total	1 928.4	1 965.3	2 055.8	2 008.3	2 071.1	2 193.4	2 209.9	2 273.2	2 341.9

[1] See Notes in Part I.4.

Oil Demand by Main Product Groups[1]
Demande de produits raffinés par groupes principaux

Latin America

	1971	1973	1978	1984	1985	1986	1987	1988	1989
Thousand tonnes									
NGL/LPG/Ethane	3 545	4 071	5 337	7 697	8 381	8 552	9 040	10 165	11 048
Naphtha	556	1 603	2 638	4 593	5 260	5 170	5 608	5 551	5 985
Motor Gasoline	23 562	27 922	32 591	34 291	34 977	38 447	38 699	39 626	41 117
Aviation Fuels	2 859	3 129	4 564	4 995	4 752	4 947	5 268	5 292	5 294
Kerosene	4 379	4 687	4 797	4 024	3 815	4 003	4 113	4 113	4 087
Gas Diesel	20 135	24 623	33 534	37 549	37 843	40 380	42 965	43 795	44 478
Heavy Fuel Oil	31 391	35 477	41 046	28 623	27 287	28 466	29 489	30 517	29 542
Other	5 224	5 523	7 613	9 141	9 516	9 652	11 536	10 487	9 630
Refinery Fuel	9 397	10 735	10 323	9 514	8 593	9 893	9 786	9 943	10 363
Sub-Total	101 048	117 770	142 443	140 427	140 424	149 510	156 504	159 489	161 544
Bunkers	15 049	14 620	8 471	5 736	5 806	4 946	5 190	5 189	5 690
Total	116 097	132 390	150 914	146 163	146 230	154 456	161 694	164 678	167 234
Thousand barrels/day									
NGL/LPG/Ethane	112.6	129.2	169.4	241.0	263.2	273.1	290.0	323.2	351.8
Naphtha	12.9	37.1	64.9	109.7	126.2	124.9	136.0	134.9	144.4
Motor Gasoline	548.8	650.2	757.2	788.1	803.9	881.2	886.7	904.7	940.6
Aviation Fuels	63.3	69.1	100.3	108.8	103.6	108.2	115.1	115.6	115.8
Kerosene	93.1	99.7	102.1	85.3	81.0	85.3	87.6	87.4	87.0
Gas Diesel	413.5	506.0	690.0	766.9	775.1	821.7	874.3	889.0	905.8
Heavy Fuel Oil	575.9	649.4	749.8	518.9	489.5	509.4	527.8	543.2	529.6
Other	97.6	102.8	144.6	171.3	174.0	180.6	212.6	189.6	169.8
Refinery Fuel	188.6	211.9	205.8	191.2	174.2	196.5	194.8	198.4	208.6
Sub-Total	2 106.2	2 455.4	2 984.1	2 981.1	2 990.6	3 180.9	3 324.9	3 386.0	3 453.3
Bunkers	278.2	267.4	155.7	104.6	105.9	90.3	94.8	94.5	104.1
Total	2 384.4	2 722.8	3 139.8	3 085.7	3 096.5	3 271.2	3 419.8	3 480.4	3 557.4

	1990	1991	1992	1993	1994	1995	1996	1997	1998
Thousand tonnes									
NGL/LPG/Ethane	11 525	11 569	12 387	13 318	12 560	13 662	14 817	16 242	15 730
Naphtha	6 208	5 840	6 327	6 426	6 730	6 666	6 662	8 045	8 120
Motor Gasoline	41 162	42 376	43 219	44 649	46 569	49 831	52 713	53 333	53 870
Aviation Fuels	5 438	5 386	5 443	5 657	5 947	6 648	7 107	7 766	8 212
Kerosene	3 781	3 605	3 429	3 260	2 997	2 994	2 905	2 771	2 634
Gas Diesel	44 734	46 099	47 413	49 556	53 369	57 944	60 341	64 543	66 946
Heavy Fuel Oil	27 587	26 327	26 319	26 843	26 254	28 059	30 602	31 711	31 913
Other	9 685	10 522	11 236	11 080	16 694	15 364	16 158	20 007	17 034
Refinery Fuel	9 940	9 699	10 914	10 303	9 820	9 727	8 691	10 197	11 639
Sub-Total	160 060	161 423	166 687	171 092	180 940	190 895	199 996	214 615	216 098
Bunkers	5 987	6 382	6 753	6 797	6 963	7 134	7 130	7 621	7 398
Total	166 047	167 805	173 440	177 889	187 903	198 029	207 126	222 236	223 496
Thousand barrels/day									
NGL/LPG/Ethane	365.7	367.9	393.5	423.2	400.4	435.5	470.6	514.2	499.8
Naphtha	150.7	142.0	153.4	156.3	164.1	162.4	161.7	196.0	197.8
Motor Gasoline	943.7	971.1	988.7	1 024.2	1 067.8	1 143.4	1 206.5	1 224.3	1 237.4
Aviation Fuels	118.6	117.4	118.3	123.3	129.6	144.8	154.3	169.0	178.6
Kerosene	80.4	76.7	72.7	69.3	63.7	63.6	61.6	58.9	56.0
Gas Diesel	911.0	938.6	962.7	1 008.9	1 086.3	1 179.6	1 224.9	1 314.0	1 362.9
Heavy Fuel Oil	491.7	468.9	466.5	476.9	466.0	498.1	541.1	562.7	566.2
Other	170.0	188.8	196.4	196.8	300.1	273.1	285.3	354.3	294.2
Refinery Fuel	199.6	197.3	219.7	210.4	200.2	197.3	177.7	204.5	235.3
Sub-Total	3 431.6	3 468.7	3 572.0	3 689.4	3 878.3	4 097.7	4 283.7	4 597.7	4 628.2
Bunkers	109.6	116.6	123.2	124.3	127.4	130.5	129.8	138.9	135.1
Total	3 541.2	3 585.4	3 695.3	3 813.7	4 005.7	4 228.2	4 413.6	4 736.6	4 763.3

[1] See Notes in Part I.4.

Oil Demand by Main Product Groups[1]
Demande de produits raffinés par groupes principaux

Asia (excluding China)

	1971	1973	1978	1984	1985	1986	1987	1988	1989
Thousand tonnes									
NGL/LPG/Ethane	631	930	1 614	2 952	3 889	4 394	4 789	5 429	6 165
Naphtha	1 146	1 586	3 678	5 676	6 562	6 962	6 879	7 337	7 695
Motor Gasoline	9 070	10 083	11 965	14 602	15 168	16 642	19 243	20 382	22 297
Aviation Fuels	5 334	6 063	5 130	6 182	6 300	6 887	7 417	7 886	8 469
Kerosene	8 200	8 632	12 087	14 352	14 441	14 902	15 503	16 286	17 243
Gas Diesel	18 386	21 136	28 899	42 749	44 431	45 755	49 845	54 007	59 938
Heavy Fuel Oil	19 519	25 266	35 091	33 502	31 485	32 101	33 717	37 312	40 070
Other	4 066	5 532	5 070	5 079	5 907	6 237	6 676	7 217	8 813
Refinery Fuel	2 582	3 122	4 517	7 085	7 687	8 367	8 757	9 069	8 538
Sub-Total	68 934	82 350	108 051	132 179	135 870	142 247	152 826	164 925	179 228
Bunkers	4 648	6 628	6 850	7 041	6 513	10 406	10 709	12 534	12 816
Total	73 582	88 978	114 901	139 220	142 383	152 653	163 535	177 459	192 044
Thousand barrels/day									
NGL/LPG/Ethane	20.1	29.6	51.4	93.2	123.0	139.0	151.4	171.3	195.0
Naphtha	26.7	36.9	85.7	131.8	152.8	162.1	160.2	170.4	179.2
Motor Gasoline	212.3	236.1	280.2	340.9	355.1	389.6	450.4	475.7	521.8
Aviation Fuels	118.8	134.4	112.6	135.4	138.3	151.3	162.9	172.8	186.2
Kerosene	174.1	183.3	256.5	303.7	306.3	316.1	328.9	344.5	365.8
Gas Diesel	375.0	431.0	589.4	869.7	906.3	933.3	1 016.7	1 098.4	1 222.2
Heavy Fuel Oil	356.5	461.3	640.9	610.1	574.9	586.2	615.5	679.3	731.5
Other	74.5	103.0	92.4	91.6	107.1	113.1	121.2	130.1	160.6
Refinery Fuel	49.7	59.8	88.0	139.9	152.3	164.8	172.9	179.1	169.4
Sub-Total	1 407.8	1 675.3	2 197.0	2 716.3	2 816.3	2 955.4	3 180.0	3 421.6	3 731.8
Bunkers	86.3	123.4	128.5	130.3	120.8	192.3	199.0	232.0	235.9
Total	1 494.0	1 798.7	2 325.5	2 846.5	2 937.1	3 147.8	3 379.0	3 653.6	3 967.7

	1990	1991	1992	1993	1994	1995	1996	1997	1998
Thousand tonnes									
NGL/LPG/Ethane	6 747	7 135	7 787	8 385	10 239	11 694	12 725	13 329	13 747
Naphtha	7 591	6 983	6 871	7 123	8 211	8 828	9 443	10 629	11 873
Motor Gasoline	24 062	24 700	26 410	28 441	30 997	33 533	36 546	39 216	40 220
Aviation Fuels	9 366	8 951	10 069	10 964	12 428	13 198	14 285	14 465	13 265
Kerosene	18 135	17 919	18 004	18 417	18 934	19 746	20 532	21 021	23 114
Gas Diesel	62 704	68 153	73 821	81 703	86 700	93 021	101 811	106 288	101 115
Heavy Fuel Oil	45 898	46 728	49 625	52 670	53 814	58 815	59 365	61 073	62 532
Other	8 562	8 673	9 023	9 352	9 577	9 578	10 682	10 793	10 427
Refinery Fuel	10 117	9 954	10 274	10 569	11 261	11 543	12 038	12 085	12 531
Sub-Total	193 182	199 196	211 884	227 624	242 161	259 956	277 427	288 899	288 824
Bunkers	14 634	13 633	16 766	15 427	15 996	15 865	19 131	21 868	22 835
Total	207 816	212 829	228 650	243 051	258 157	275 821	296 558	310 767	311 659
Thousand barrels/day									
NGL/LPG/Ethane	213.4	225.7	245.5	266.0	328.9	375.2	412.1	427.7	442.7
Naphtha	176.8	162.6	159.6	165.9	191.2	205.6	219.3	247.5	276.5
Motor Gasoline	562.0	576.9	615.2	664.5	724.3	783.7	851.8	916.7	940.1
Aviation Fuels	205.9	196.8	221.0	241.0	273.1	289.9	312.7	317.7	291.4
Kerosene	384.7	380.2	380.9	390.8	401.8	419.0	434.5	446.1	490.5
Gas Diesel	1 280.0	1 391.5	1 503.1	1 668.5	1 770.5	1 899.0	2 072.2	2 169.4	2 064.5
Heavy Fuel Oil	836.6	851.7	902.3	959.6	980.6	1 071.7	1 078.7	1 112.7	1 139.4
Other	155.4	156.3	162.2	169.1	173.0	172.1	191.4	193.8	187.0
Refinery Fuel	199.1	195.3	200.6	206.3	220.3	227.0	236.6	238.5	246.6
Sub-Total	4 014.1	4 137.0	4 390.3	4 731.8	5 063.6	5 443.1	5 809.6	6 070.3	6 078.7
Bunkers	269.8	251.4	307.7	284.5	295.8	293.3	354.6	407.1	425.5
Total	4 283.8	4 388.4	4 698.0	5 016.3	5 359.4	5 736.5	6 164.1	6 477.4	6 504.2

[1] See Notes in Part I.4.

Oil Demand by Main Product Groups[1]
Demande de produits raffinés par groupes principaux

China

	1971	1973	1978	1984	1985	1986	1987	1988	1989
Thousand tonnes									
NGL/LPG/Ethane	46	72	897	1 676	1 748	2 194	2 327	2 502	2 704
Naphtha	26	32	1 103	3 429	5 124	4 084	5 845	7 324	8 266
Motor Gasoline	4 579	6 242	9 727	12 391	14 187	15 191	16 437	18 194	18 813
Aviation Fuels	447	523	633	1 396	1 441	1 530	1 729	1 977	2 215
Kerosene	3 210	3 161	3 509	3 441	3 396	3 447	3 264	3 005	2 982
Gas Diesel	10 111	13 385	18 415	18 585	20 289	22 426	24 200	27 071	28 776
Heavy Fuel Oil	12 915	16 987	27 783	28 561	27 486	28 548	29 170	30 168	30 303
Other	9 461	12 866	30 972	18 693	17 155	18 865	18 785	18 229	17 800
Refinery Fuel	1 600	1 800	4 500	3 251	4 707	5 295	5 541	5 787	7 166
Sub-Total	42 395	55 068	97 539	91 423	95 533	101 580	107 298	114 257	119 025
Bunkers	642	1 179	551	317	1 003	838	911	1 075	1 367
Total	43 037	56 247	98 090	91 740	96 536	102 418	108 209	115 332	120 392
Thousand barrels/day									
NGL/LPG/Ethane	1.5	2.3	28.5	53.1	55.6	69.7	74.0	79.3	85.9
Naphtha	0.6	0.7	25.7	79.6	119.3	95.1	136.1	170.1	192.5
Motor Gasoline	107.0	145.9	227.3	288.8	331.5	355.0	384.1	424.0	439.7
Aviation Fuels	9.7	11.4	13.8	30.4	31.5	33.4	37.8	43.0	48.3
Kerosene	68.1	67.0	74.4	72.8	72.0	73.1	69.2	63.5	63.2
Gas Diesel	206.7	273.6	376.4	378.8	414.7	458.4	494.6	551.8	588.1
Heavy Fuel Oil	235.7	310.0	506.9	519.7	501.5	520.9	532.3	549.0	552.9
Other	182.9	250.5	611.3	357.7	328.9	361.5	359.7	348.3	340.8
Refinery Fuel	31.4	35.8	85.8	62.9	93.9	105.5	110.4	114.8	140.5
Sub-Total	843.4	1 097.1	1 950.0	1 843.8	1 948.9	2 072.6	2 198.1	2 343.8	2 452.0
Bunkers	11.7	21.5	10.1	6.1	18.6	15.6	17.1	20.0	25.7
Total	855.1	1 118.7	1 960.1	1 849.8	1 967.5	2 088.3	2 215.2	2 363.8	2 477.7

	1990	1991	1992	1993	1994	1995	1996	1997	1998
Thousand tonnes									
NGL/LPG/Ethane	2 745	3 211	3 670	4 430	4 995	6 959	8 479	8 947	11 208
Naphtha	8 824	10 063	10 311	11 942	13 184	14 959	16 400	17 970	17 964
Motor Gasoline	19 078	22 365	24 752	30 983	27 086	29 195	32 159	33 463	33 643
Aviation Fuels	2 792	2 983	3 581	4 134	4 730	5 670	6 373	7 745	7 400
Kerosene	2 660	2 587	2 589	2 548	2 582	2 755	2 601	2 565	2 745
Gas Diesel	28 446	31 835	34 227	40 693	39 518	45 675	49 198	56 594	57 096
Heavy Fuel Oil	29 870	29 746	31 861	34 513	31 714	33 346	35 614	36 104	34 939
Other	16 019	14 830	15 388	13 387	20 256	17 451	22 953	31 164	30 943
Refinery Fuel	7 152	7 134	7 880	7 045	6 322	7 763	8 143	10 027	8 564
Sub-Total	117 586	124 754	134 259	149 675	150 387	163 773	181 920	204 579	204 502
Bunkers	1 457	1 244	1 527	4 206	4 337	3 100	3 069	2 827	3 200
Total	119 043	125 998	135 786	153 881	154 724	166 873	184 989	207 406	207 702
Thousand barrels/day									
NGL/LPG/Ethane	87.2	102.0	116.3	140.8	158.7	221.2	268.7	284.3	356.2
Naphtha	205.5	234.3	239.5	278.1	307.0	348.4	380.9	418.5	418.3
Motor Gasoline	445.9	522.7	576.9	724.1	633.0	682.3	749.5	782.0	786.2
Aviation Fuels	60.9	65.0	77.8	89.8	102.8	123.4	138.3	168.5	161.1
Kerosene	56.4	54.9	54.8	54.0	54.8	58.4	55.0	54.4	58.2
Gas Diesel	581.4	650.7	697.6	831.7	807.7	933.5	1 002.8	1 156.7	1 166.9
Heavy Fuel Oil	545.0	542.8	579.8	629.7	578.7	608.5	648.1	658.8	637.5
Other	305.3	277.8	287.0	265.6	383.8	324.9	449.3	609.8	606.2
Refinery Fuel	140.8	140.4	155.0	150.5	136.0	162.2	174.1	216.6	180.8
Sub-Total	2 428.4	2 590.6	2 784.7	3 164.4	3 162.5	3 462.8	3 866.7	4 349.6	4 371.6
Bunkers	27.3	23.4	28.4	80.0	82.5	58.8	57.9	53.5	61.6
Total	2 455.7	2 613.9	2 813.1	3 244.4	3 245.0	3 521.6	3 924.6	4 403.0	4 433.2

[1] See Notes in Part I.4.

Oil Demand by Main Product Groups[1]
Demande de produits raffinés par groupes principaux

Non-OECD Europe

	1971	1973	1978	1984	1985	1986	1987	1988	1989
Thousand tonnes									
NGL/LPG/Ethane	319	519	683	848	768	812	888	891	848
Naphtha	420	880	1 317	668	474	647	828	1 049	2 078
Motor Gasoline	5 402	5 866	6 989	5 349	4 996	5 433	5 763	6 221	6 353
Aviation Fuels	543	600	709	888	886	936	985	1 055	1 371
Kerosene	1 188	1 262	1 299	907	791	877	883	943	673
Gas Diesel	9 547	10 850	13 537	11 882	12 138	12 612	12 830	12 797	11 968
Heavy Fuel Oil	12 308	14 487	20 587	16 028	17 177	18 905	19 327	19 789	19 282
Other	3 035	3 089	4 230	3 681	3 878	3 968	4 209	4 084	5 001
Refinery Fuel	1 777	2 081	3 115	3 545	3 444	3 688	3 814	3 838	3 983
Sub-Total	34 539	39 634	52 466	43 796	44 552	47 878	49 527	50 667	51 557
Bunkers	242	228	185	432	568	716	752	859	861
Total	34 781	39 862	52 651	44 228	45 120	48 594	50 279	51 526	52 418
Thousand barrels/day									
NGL/LPG/Ethane	10.1	16.5	21.7	26.9	24.4	25.8	28.2	28.2	27.0
Naphtha	9.8	20.5	30.7	15.5	11.0	15.1	19.3	24.4	48.4
Motor Gasoline	126.2	137.1	163.3	124.7	116.8	127.0	134.7	145.0	148.5
Aviation Fuels	11.9	13.1	15.5	19.3	19.3	20.4	21.4	22.9	29.8
Kerosene	25.2	26.8	27.5	19.2	16.8	18.6	18.7	19.9	14.3
Gas Diesel	195.1	221.8	276.7	242.2	248.1	257.8	262.2	260.8	244.6
Heavy Fuel Oil	224.6	264.3	375.6	291.7	313.4	345.0	352.7	360.1	351.8
Other	56.9	57.4	78.4	67.5	71.5	72.3	77.5	74.5	92.8
Refinery Fuel	34.4	40.5	61.4	69.8	68.0	72.9	75.6	75.9	79.3
Sub-Total	694.3	798.0	1 050.8	876.6	889.2	954.7	990.2	1 011.7	1 036.4
Bunkers	4.6	4.4	3.5	8.4	10.9	13.6	14.3	16.2	16.3
Total	698.9	802.3	1 054.4	884.9	900.1	968.3	1 004.5	1 027.9	1 052.8

	1990	1991	1992	1993	1994	1995	1996	1997	1998
Thousand tonnes									
NGL/LPG/Ethane	888	1 016	641	626	608	680	619	828	850
Naphtha	1 307	1 285	1 542	2 521	1 847	1 986	2 276	2 218	2 457
Motor Gasoline	6 523	5 779	5 435	4 531	4 954	4 962	5 681	5 902	5 994
Aviation Fuels	1 209	970	1 024	1 128	1 041	1 134	919	952	968
Kerosene	595	468	517	216	146	171	186	259	248
Gas Diesel	11 200	8 930	9 182	8 321	7 740	8 039	9 743	9 603	9 745
Heavy Fuel Oil	19 818	15 986	12 429	12 014	10 620	10 453	11 527	11 428	9 492
Other	5 127	4 129	2 941	3 094	3 600	3 896	4 012	4 626	4 449
Refinery Fuel	3 689	2 770	1 418	1 503	1 795	1 545	1 380	1 295	1 471
Sub-Total	50 356	41 333	35 129	33 954	32 351	32 866	36 343	37 111	35 674
Bunkers	857	1 253	1 236	1 201	1 252	1 276	1 292	1 066	1 203
Total	51 213	42 586	36 365	35 155	33 603	34 142	37 635	38 177	36 877
Thousand barrels/day									
NGL/LPG/Ethane	28.2	31.4	21.1	21.1	20.6	22.9	20.8	27.6	28.1
Naphtha	30.4	29.9	35.8	58.7	43.0	46.2	52.9	51.7	57.2
Motor Gasoline	152.4	135.1	126.7	105.9	115.8	116.0	132.4	137.9	140.1
Aviation Fuels	26.3	21.1	22.2	24.5	22.6	24.7	19.9	20.7	21.1
Kerosene	12.6	9.9	10.9	4.6	3.1	3.6	3.9	5.5	5.3
Gas Diesel	228.9	182.5	187.2	170.1	158.2	164.3	198.6	196.3	199.2
Heavy Fuel Oil	361.6	291.7	226.2	219.2	193.8	190.7	209.8	208.5	173.2
Other	96.6	76.7	55.1	58.1	67.7	72.7	73.8	86.0	81.4
Refinery Fuel	74.8	55.6	28.8	30.8	36.8	32.3	28.7	26.8	30.7
Sub-Total	1 011.9	833.8	713.9	692.9	661.5	673.5	740.8	761.0	736.2
Bunkers	16.3	23.4	23.0	22.3	23.3	23.8	24.0	19.8	22.6
Total	1 028.2	857.2	736.8	715.2	684.8	697.3	764.7	780.9	758.7

[1] See Notes in Part I.4.

Oil Demand by Main Product Groups[1]
Demande de produits raffinés par groupes principaux

Former USSR

	1971	1973	1978	1984	1985	1986	1987	1988	1989
Thousand tonnes									
NGL/LPG/Ethane	7 000	9 800	13 200	18 500	16 916	20 012	21 817	22 748	22 351
Naphtha	1 400	3 800	3 800	3 500	3 700	3 400	3 600	4 700	4 600
Motor Gasoline	46 300	51 200	70 500	76 300	75 900	77 400	78 500	79 500	80 500
Aviation Fuels	21 900	22 900	22 200	24 900	25 200	25 700	26 300	27 000	27 300
Kerosene	9 700	8 500	5 400	4 600	4 500	4 400	4 300	4 200	4 200
Gas Diesel	47 500	58 800	70 800	81 400	82 200	84 800	85 600	85 400	86 400
Heavy Fuel Oil	92 000	113 298	165 800	171 400	165 400	156 300	149 900	138 600	121 600
Other	25 300	29 000	38 900	40 754	41 255	41 554	43 553	49 553	49 722
Refinery Fuel	5 600	6 300	15 340	15 800	15 600	14 900	15 100	14 900	17 200
Sub-Total	256 700	303 598	405 940	437 154	430 671	428 466	428 670	426 601	413 873
Bunkers	4 300	4 700	4 900	4 700	4 500	4 700	4 700	4 800	4 700
Total	261 000	308 298	410 840	441 854	435 171	433 166	433 370	431 401	418 573
Thousand barrels/day									
NGL/LPG/Ethane	201.6	281.4	378.6	531.6	490.1	575.0	624.5	648.2	639.1
Naphtha	32.6	88.5	88.5	81.3	86.2	79.2	83.8	109.2	107.1
Motor Gasoline	1 082.0	1 196.5	1 647.6	1 778.2	1 773.8	1 808.8	1 834.5	1 852.8	1 881.3
Aviation Fuels	475.8	497.5	482.3	539.5	547.5	558.4	571.4	585.0	593.1
Kerosene	205.7	180.2	114.5	97.3	95.4	93.3	91.2	88.8	89.1
Gas Diesel	970.8	1 201.8	1 447.0	1 659.1	1 680.0	1 733.2	1 749.5	1 740.7	1 765.9
Heavy Fuel Oil	1 678.7	2 067.3	3 025.3	3 118.9	3 018.0	2 851.9	2 735.2	2 522.1	2 218.8
Other	492.9	565.0	757.8	792.0	803.5	809.6	848.0	960.5	966.0
Refinery Fuel	112.8	127.4	320.8	330.3	327.6	313.0	317.7	313.2	361.2
Sub-Total	5 252.9	6 205.7	8 262.4	8 928.3	8 822.0	8 822.4	8 855.9	8 820.5	8 621.5
Bunkers	78.5	85.8	89.4	85.5	82.1	85.8	85.8	87.3	85.8
Total	5 331.4	6 291.4	8 351.8	9 013.8	8 904.1	8 908.1	8 941.6	8 907.8	8 707.3

	1990	1991	1992	1993	1994	1995	1996	1997	1998
Thousand tonnes									
NGL/LPG/Ethane	22 942	23 080	7 857	6 761	6 918	8 129	7 611	6 728	5 792
Naphtha	3 900	3 900	7 032	8 240	3 577	-	-	-	-
Motor Gasoline	79 500	76 000	54 668	45 544	39 933	38 920	37 490	36 514	36 358
Aviation Fuels	21 585	21 200	17 664	14 691	12 589	11 639	11 462	11 094	11 300
Kerosene	4 200	5 000	2 471	1 615	1 098	1 029	655	641	764
Gas Diesel	85 000	80 199	79 475	64 049	48 255	47 133	39 066	39 336	38 050
Heavy Fuel Oil	116 500	123 459	118 703	108 698	84 711	73 316	65 752	58 529	58 074
Other	51 428	40 777	38 082	37 070	22 342	28 080	28 218	28 419	28 346
Refinery Fuel	14 500	13 700	12 380	4 024	2 291	3 863	1 975	2 215	2 314
Sub-Total	399 555	387 315	338 332	290,692	221 714	212 109	192 229	183 476	180 998
Bunkers	4 600	4 600	157	158	101	90	93	164	159
Total	404 155	391 915	338 489	290 850	221 815	212 199	192 322	183 640	181 157
Thousand barrels/day									
NGL/LPG/Ethane	655.3	659.1	247.7	213.8	216.8	254.3	239.8	212.8	182.9
Naphtha	90.8	90.8	163.3	191.9	83.3	-	-	-	-
Motor Gasoline	1 857.9	1 776.1	1 274.1	1 064.4	933.2	909.6	873.7	853.3	849.7
Aviation Fuels	470.0	460.6	383.2	319.6	273.8	253.2	248.6	241.3	245.7
Kerosene	89.1	106.0	52.3	34.2	23.3	21.8	13.9	13.6	16.2
Gas Diesel	1 737.3	1 639.1	1 619.9	1 309.1	986.3	963.3	796.3	804.0	777.7
Heavy Fuel Oil	2 125.7	2 252.7	2 160.0	1 983.3	1 545.6	1 337.8	1 196.5	1 068.0	1 059.7
Other	988.5	789.7	710.6	702.5	418.5	528.7	540.0	547.5	546.1
Refinery Fuel	303.5	286.0	298.0	84.0	47.1	82.0	40.7	45.7	47.8
Sub-Total	8 318.1	8 060.1	6 909.2	5 902.7	4 527.8	4 350.6	3 949.4	3 786.1	3 725.7
Bunkers	83.9	83.9	3.0	3.1	1.9	1.7	1.8	3.1	3.0
Total	8 402.0	8 144.0	6 912.1	5 905.8	4 529.7	4 352.3	3 951.2	3 789.2	3 728.8

[1] See Notes in Part I.4.

Oil Demand by Main Product Groups[1]
Demande de produits raffinés par groupes principaux

Middle East

	1971	1973	1978	1984	1985	1986	1987	1988	1989
Thousand tonnes									
NGL/LPG/Ethane	755	869	1 351	3 395	5 235	6 808	7 115	7 617	8 388
Naphtha	440	409	609	457	437	528	561	439	1 165
Motor Gasoline	4 364	5 574	10 885	18 283	19 728	19 496	20 025	20 226	21 108
Aviation Fuels	1 251	2 197	4 252	4 933	5 015	5 149	5 211	5 136	5 070
Kerosene	3 544	4 334	7 085	8 685	8 981	7 004	7 989	7 971	7 511
Gas Diesel	6 651	8 501	20 086	39 208	38 948	39 802	40 963	42 689	43 727
Heavy Fuel Oil	7 504	9 580	16 622	25 775	26 833	29 677	30 354	31 654	31 385
Other	1 981	2 440	3 922	9 540	10 108	11 426	12 227	12 521	12 283
Refinery Fuel	6 833	7 576	6 525	9 368	10 055	10 843	11 708	12 099	12 392
Sub-Total	33 323	41 480	71 337	119 644	125 340	130 733	136 153	140 352	143 029
Bunkers	23 626	25 556	14 258	8 399	12 419	10 636	9 120	8 330	5 815
Total	56 949	67 036	85 595	128 043	137 759	141 369	145 273	148 682	148 844
Thousand barrels/day									
NGL/LPG/Ethane	24.0	27.6	42.9	107.6	160.3	206.6	216.3	231.0	256.1
Naphtha	10.2	9.5	14.2	10.6	10.2	12.3	13.1	10.2	27.1
Motor Gasoline	102.0	130.3	254.4	426.1	461.0	455.6	468.0	471.4	493.3
Aviation Fuels	27.3	47.8	92.6	107.1	109.2	112.1	113.5	111.5	110.4
Kerosene	75.2	91.9	150.2	183.7	190.4	148.5	169.4	168.6	159.3
Gas Diesel	135.9	173.7	410.5	799.2	796.0	813.5	837.2	870.1	893.7
Heavy Fuel Oil	136.9	174.8	303.3	469.0	489.6	541.5	553.9	576.0	572.7
Other	38.1	46.9	74.9	184.6	196.1	222.5	238.6	243.2	239.2
Refinery Fuel	128.1	141.6	122.4	176.7	191.0	205.9	224.2	231.1	236.7
Sub-Total	677.7	844.1	1 465.4	2 464.6	2 603.9	2 718.5	2 834.1	2 913.1	2 988.5
Bunkers	432.3	467.5	262.2	153.8	227.8	195.2	167.6	152.7	106.7
Total	1 110.0	1 311.7	1 727.6	2 618.4	2 831.7	2 913.7	3 001.7	3 065.8	3 095.1

	1990	1991	1992	1993	1994	1995	1996	1997	1998
Thousand tonnes									
NGL/LPG/Ethane	8 323	7 592	8 722	10 272	11 264	11 502	12 009	13 722	15 932
Naphtha	1 183	1 118	1 272	1 142	1 248	1 491	1 660	1 908	3 259
Motor Gasoline	22 786	23 141	25 680	28 045	29 243	30 027	31 667	32 605	33 886
Aviation Fuels	6 482	6 272	5 055	5 264	5 897	5 409	5 445	5 502	5 578
Kerosene	7 509	8 302	8 396	9 807	10 422	4 130	4 233	4 282	4 256
Gas Diesel	42 751	45 219	48 003	50 166	51 662	48 516	50 190	53 430	54 987
Heavy Fuel Oil	30 516	33 360	33 755	34 747	36 346	37 613	39 650	38 733	38 022
Other	13 177	15 251	17 924	20 666	21 434	20 546	22 552	23 357	23 794
Refinery Fuel	9 403	10 080	11 360	11 062	11 933	12 373	12 014	12 363	13 783
Sub-Total	142 130	150 335	160 167	171 171	179 449	171 607	179 420	185 902	193 497
Bunkers	5 415	5 725	5 181	4 280	3 794	3 767	3 305	3 379	3 462
Total	147 545	156 060	165 348	175 451	183 243	175 374	182 725	189 281	196 959
Thousand barrels/day									
NGL/LPG/Ethane	253.5	230.3	265.4	313.3	342.7	350.2	364.8	416.2	480.9
Naphtha	27.5	26.0	29.5	26.6	29.1	34.7	38.6	44.4	75.9
Motor Gasoline	532.4	540.7	598.3	655.3	683.2	701.6	737.8	761.8	791.7
Aviation Fuels	141.4	136.7	110.0	114.8	128.6	118.0	118.5	120.1	121.7
Kerosene	159.3	176.1	177.7	208.1	221.1	87.7	89.6	90.9	90.4
Gas Diesel	873.9	924.4	978.6	1 025.5	1 056.1	991.8	1 023.2	1 092.3	1 124.1
Heavy Fuel Oil	556.9	608.8	614.4	634.2	663.4	686.5	721.7	706.9	694.0
Other	256.6	299.5	351.1	406.0	420.7	397.5	435.1	452.6	460.8
Refinery Fuel	182.0	193.8	216.9	214.6	231.2	238.8	232.7	240.0	265.3
Sub-Total	2 983.6	3 136.3	3 342.1	3 598.3	3 776.0	3 606.8	3 762.1	3 925.2	4 104.7
Bunkers	99.2	104.7	94.8	78.7	69.9	69.3	60.6	62.2	63.7
Total	3 082.8	3 241.0	3 436.9	3 677.0	3 845.9	3 676.1	3 822.7	3 987.4	4 168.4

[1] See Notes in Part I.4.

Oil Demand by Main Product Groups[1]
Demande de produits raffinés par groupes principaux

Albania

	1971	1973	1978	1984	1985	1986	1987	1988	1989
Thousand tonnes									
NGL/LPG/Ethane	-	-	-	-	-	-	-	-	-
Naphtha	-	-	-	-	-	-	-	-	-
Motor Gasoline	88	78	140	200	121	99	110	145	81
Aviation Fuels	-	-	-	-	-	-	-	-	-
Kerosene	15	35	60	80	55	55	56	60	61
Gas Diesel	145	174	220	225	225	225	225	225	225
Heavy Fuel Oil	290	315	400	365	300	345	375	300	340
Other	239	64	215	140	117	120	100	170	150
Refinery Fuel	34	34	56	84	85	85	84	73	84
Sub-Total	811	700	1 091	1 094	903	929	950	973	941
Bunkers	-	-	-	-	-	-	-	-	-
Total	811	700	1 091	1 094	903	929	950	973	941
Thousand barrels/day									
NGL/LPG/Ethane	-	-	-	-	-	-	-	-	-
Naphtha	-	-	-	-	-	-	-	-	-
Motor Gasoline	2.1	1.8	3.3	4.7	2.8	2.3	2.6	3.4	1.9
Aviation Fuels	-	-	-	-	-	-	-	-	-
Kerosene	0.3	0.7	1.3	1.7	1.2	1.2	1.2	1.3	1.3
Gas Diesel	3.0	3.6	4.5	4.6	4.6	4.6	4.6	4.6	4.6
Heavy Fuel Oil	5.3	5.7	7.3	6.6	5.5	6.3	6.8	5.5	6.2
Other	4.6	1.2	4.1	2.7	2.2	2.3	1.9	3.3	2.9
Refinery Fuel	0.7	0.7	1.1	1.7	1.8	1.8	1.7	1.5	1.7
Sub-Total	15.9	13.7	21.5	22.0	18.1	18.4	18.9	19.5	18.6
Bunkers	-	-	-	-	-	-	-	-	-
Total	15.9	13.7	21.5	22.0	18.1	18.4	18.9	19.5	18.6

	1990	1991	1992	1993	1994	1995	1996	1997	1998
Thousand tonnes									
NGL/LPG/Ethane	-	-	-	-	-	-	-	-	-
Naphtha	-	-	-	-	-	-	-	-	-
Motor Gasoline	100	101	90	120	155	164	173	179	179
Aviation Fuels	-	-	-	-	-	-	-	-	-
Kerosene	56	61	41	29	29	74	69	57	57
Gas Diesel	238	180	116	120	123	106	108	107	107
Heavy Fuel Oil	412	320	176	202	206	43	46	39	39
Other	210	200	200	97	112	109	97	43	43
Refinery Fuel	73	73	75	75	75	50	40	30	30
Sub-Total	1 089	935	698	643	700	546	533	455	455
Bunkers	-	-	-	-	-	-	-	-	-
Total	1 089	935	698	643	700	546	533	455	455
Thousand barrels/day									
NGL/LPG/Ethane	-	-	-	-	-	-	-	-	-
Naphtha	-	-	-	-	-	-	-	-	-
Motor Gasoline	2.3	2.4	2.1	2.8	3.6	3.8	4.0	4.2	4.2
Aviation Fuels	-	-	-	-	-	-	-	-	-
Kerosene	1.2	1.3	0.9	0.6	0.6	1.6	1.5	1.2	1.2
Gas Diesel	4.9	3.7	2.4	2.5	2.5	2.2	2.2	2.2	2.2
Heavy Fuel Oil	7.5	5.8	3.2	3.7	3.8	0.8	0.8	0.7	0.7
Other	4.0	3.8	3.8	1.8	2.1	1.7	1.5	0.7	0.7
Refinery Fuel	1.5	1.5	1.6	1.6	1.6	1.1	0.9	0.7	0.7
Sub-Total	21.5	18.5	13.9	12.9	14.1	11.2	10.9	9.6	9.6
Bunkers	-	-	-	-	-	-	-	-	-
Total	21.5	18.5	13.9	12.9	14.1	11.2	10.9	9.6	9.6

[1] See Notes in Part I.4.

Oil Demand by Main Product Groups[1]
Demande de produits raffinés par groupes principaux

Algeria

	1971	1973	1978	1984	1985	1986	1987	1988	1989
Thousand tonnes									
NGL/LPG/Ethane	173	300	551	1 161	1 087	1 165	1 239	1 199	1 363
Naphtha	26	8	-	-	-	-	-	-	-
Motor Gasoline	449	538	958	1 658	1 785	1 897	1 963	1 987	2 109
Aviation Fuels	92	128	243	394	415	383	355	347	354
Kerosene	84	83	58	33	33	30	29	27	15
Gas Diesel	771	1 000	1 739	3 025	3 267	3 320	3 332	3 303	3 288
Heavy Fuel Oil	221	225	199	191	184	190	155	137	135
Other	169	199	435	597	556	553	558	556	558
Refinery Fuel	35	66	63	273	267	292	279	302	287
Sub-Total	2 020	2 547	4 246	7 332	7 594	7 830	7 910	7 858	8 109
Bunkers	196	158	275	428	372	248	248	310	444
Total	2 216	2 705	4 521	7 760	7 966	8 078	8 158	8 168	8 553
Thousand barrels/day									
NGL/LPG/Ethane	5.5	9.5	17.5	36.8	34.5	37.0	39.4	38.0	43.3
Naphtha	0.6	0.2	-	-	-	-	-	-	-
Motor Gasoline	10.5	12.6	22.4	38.6	41.7	44.3	45.9	46.3	49.3
Aviation Fuels	2.0	2.8	5.3	8.5	9.0	8.3	7.7	7.5	7.7
Kerosene	1.8	1.8	1.2	0.7	0.7	0.6	0.6	0.6	0.3
Gas Diesel	15.8	20.4	35.5	61.7	66.8	67.9	68.1	67.3	67.2
Heavy Fuel Oil	4.0	4.1	3.6	3.5	3.4	3.5	2.8	2.5	2.5
Other	3.3	3.9	8.7	11.4	10.7	9.6	9.7	9.6	9.7
Refinery Fuel	0.8	1.4	1.4	6.0	5.9	6.4	6.1	6.6	6.3
Sub-Total	44.3	56.7	95.7	167.2	172.6	177.7	180.3	178.5	186.3
Bunkers	3.7	3.0	5.1	8.0	7.0	4.7	4.7	5.9	8.4
Total	48.0	59.8	100.8	175.2	179.7	182.4	185.0	184.3	194.6

	1990	1991	1992	1993	1994	1995	1996	1997	1998
Thousand tonnes									
NGL/LPG/Ethane	2 024	1 721	1 885	1 790	1 967	1 910	1 670	1 757	1 891
Naphtha	-	-	-	-	-	-	-	-	-
Motor Gasoline	2 182	2 268	2 269	2 328	2 210	2 060	2 023	1 963	1 933
Aviation Fuels	345	303	306	309	302	304	302	330	320
Kerosene	12	10	10	11	12	16	10	8	9
Gas Diesel	3 256	3 536	3 438	3 365	3 039	3 032	2 937	2 966	3 230
Heavy Fuel Oil	124	101	84	78	36	29	23	16	17
Other	1 021	1 060	1 436	891	998	965	789	516	750
Refinery Fuel	302	312	327	412	376	336	419	419	239
Sub-Total	9 266	9 311	9 755	9 184	8 940	8 652	8 173	7 975	8 389
Bunkers	438	381	375	334	358	375	334	266	239
Total	9 704	9 692	10 130	9 518	9 298	9 027	8 507	8 241	8 628
Thousand barrels/day									
NGL/LPG/Ethane	67.2	57.1	61.7	58.8	65.2	63.4	55.6	58.5	61.3
Naphtha	-	-	-	-	-	-	-	-	-
Motor Gasoline	51.0	53.0	52.9	54.4	51.6	48.1	47.1	45.9	45.2
Aviation Fuels	7.5	6.6	6.6	6.7	6.6	6.6	6.5	7.2	7.0
Kerosene	0.3	0.2	0.2	0.2	0.3	0.3	0.2	0.2	0.2
Gas Diesel	66.5	72.3	70.1	68.8	62.1	62.0	59.9	60.6	66.0
Heavy Fuel Oil	2.3	1.8	1.5	1.4	0.7	0.5	0.4	0.3	0.3
Other	20.5	21.5	29.9	17.7	20.5	19.7	15.9	10.2	15.4
Refinery Fuel	6.7	6.9	7.3	9.2	8.4	7.5	9.3	9.3	5.3
Sub-Total	222.0	219.5	230.2	217.2	215.3	208.2	195.0	192.2	200.6
Bunkers	8.2	7.2	7.0	6.3	6.8	7.1	6.3	5.0	4.5
Total	230.2	226.7	237.2	223.4	222.0	215.3	201.3	197.2	205.2

[1] See Notes in Part I.4.

Oil Demand by Main Product Groups[1]
Demande de produits raffinés par groupes principaux

Angola

	1971	1973	1978	1984	1985	1986	1987	1988	1989
Thousand tonnes									
NGL/LPG/Ethane	16	21	13	25	28	29	33	42	46
Naphtha	-	-	-	5	7	5	6	7	7
Motor Gasoline	67	119	100	130	129	138	139	152	152
Aviation Fuels	72	74	93	106	231	251	300	311	321
Kerosene	23	25	21	81	99	92	67	57	35
Gas Diesel	262	344	322	365	521	462	477	525	537
Heavy Fuel Oil	93	120	200	48	72	97	81	112	120
Other	34	34	16	29	30	33	33	45	47
Refinery Fuel	44	51	75	74	79	79	85	79	87
Sub-Total	611	788	840	863	1 196	1 186	1 221	1 330	1 352
Bunkers	249	151	232	1	34	19	14	17	18
Total	860	939	1 072	864	1 230	1 205	1 235	1 347	1 370
Thousand barrels/day									
NGL/LPG/Ethane	0.5	0.7	0.4	0.8	0.9	0.9	1.0	1.3	1.5
Naphtha	-	-	-	0.1	0.2	0.1	0.1	0.2	0.2
Motor Gasoline	1.6	2.8	2.3	3.0	3.0	3.2	3.2	3.5	3.6
Aviation Fuels	1.6	1.6	2.0	2.3	5.0	5.5	6.5	6.7	7.0
Kerosene	0.5	0.5	0.4	1.7	2.1	2.0	1.4	1.2	0.7
Gas Diesel	5.4	7.0	6.6	7.4	10.6	9.4	9.7	10.7	11.0
Heavy Fuel Oil	1.7	2.2	3.6	0.9	1.3	1.8	1.5	2.0	2.2
Other	0.6	0.6	0.3	0.5	0.6	0.6	0.6	0.8	0.9
Refinery Fuel	0.9	1.1	1.6	1.5	1.7	1.7	1.8	1.7	1.8
Sub-Total	12.7	16.5	17.3	18.3	25.4	25.1	26.0	28.2	28.8
Bunkers	4.6	2.8	4.4	0.0	0.6	0.3	0.3	0.3	0.3
Total	17.3	19.3	21.7	18.4	26.0	25.5	26.2	28.5	29.1

	1990	1991	1992	1993	1994	1995	1996	1997	1998
Thousand tonnes									
NGL/LPG/Ethane	46	51	47	50	59	58	57	56	56
Naphtha	7	7	7	7	7	6	7	7	7
Motor Gasoline	138	142	136	139	153	143	142	134	105
Aviation Fuels	326	272	254	286	415	370	324	331	213
Kerosene	97	20	61	134	47	24	34	41	29
Gas Diesel	447	543	519	569	598	548	621	633	490
Heavy Fuel Oil	128	101	57	46	60	58	63	65	62
Other	44	45	43	38	35	20	13	16	107
Refinery Fuel	88	91	94	87	98	99	101	104	96
Sub-Total	1 321	1 272	1 218	1 356	1 472	1 326	1 362	1 387	1 165
Bunkers	5	14	8	8	5	9	7	2	-
Total	1 326	1 286	1 226	1 364	1 477	1 335	1 369	1 389	1 165
Thousand barrels/day									
NGL/LPG/Ethane	1.5	1.6	1.5	1.6	1.9	1.8	1.8	1.8	1.8
Naphtha	0.2	0.2	0.2	0.2	0.2	0.1	0.2	0.2	0.2
Motor Gasoline	3.2	3.3	3.2	3.2	3.6	3.3	3.3	3.1	2.5
Aviation Fuels	7.1	5.9	5.5	6.2	9.0	8.0	7.0	7.2	4.6
Kerosene	2.1	0.4	1.3	2.8	1.0	0.5	0.7	0.9	0.6
Gas Diesel	9.1	11.1	10.6	11.6	12.2	11.2	12.7	12.9	10.0
Heavy Fuel Oil	2.3	1.8	1.0	0.8	1.1	1.1	1.1	1.2	1.1
Other	0.8	0.8	0.8	0.7	0.6	0.4	0.2	0.3	2.0
Refinery Fuel	1.8	1.9	2.0	1.8	2.1	2.1	2.1	2.2	2.0
Sub-Total	28.1	27.1	26.0	29.0	31.6	28.6	29.2	29.7	24.8
Bunkers	0.1	0.3	0.1	0.1	0.1	0.2	0.1	0.0	-
Total	28.2	27.4	26.1	29.2	31.7	28.7	29.3	29.8	24.8

[1] See Notes in Part I.4.

Oil Demand by Main Product Groups[1]
Demande de produits raffinés par groupes principaux

Argentina

	1971	1973	1978	1984	1985	1986	1987	1988	1989
Thousand tonnes									
NGL/LPG/Ethane	1 012	969	1 107	1 352	1 352	1 386	1 454	1 412	1 400
Naphtha	148	210	495	707	634	571	571	525	610
Motor Gasoline	4 022	4 524	4 339	4 784	4 312	4 571	4 617	4 109	4 123
Aviation Fuels	371	385	634	736	645	652	686	686	672
Kerosene	795	737	668	609	443	470	485	446	445
Gas Diesel	5 849	5 894	6 759	6 827	6 820	6 667	7 073	6 981	6 490
Heavy Fuel Oil	8 814	8 860	8 090	4 582	3 078	2 840	3 602	3 911	2 925
Other	1 637	1 686	1 415	1 605	1 548	1 661	1 745	1 813	1 536
Refinery Fuel	689	819	1 002	763	909	2 005	2 018	1 911	1 770
Sub-Total	23 337	24 084	24 509	21 965	19 741	20 823	22 251	21 794	19 971
Bunkers	208	140	355	699	645	452	637	562	850
Total	23 545	24 224	24 864	22 664	20 386	21 275	22 888	22 356	20 821
Thousand barrels/day									
NGL/LPG/Ethane	32.2	30.8	35.2	42.4	42.5	48.5	51.0	49.4	49.3
Naphtha	3.4	4.9	11.5	16.4	14.8	13.5	13.5	12.3	14.4
Motor Gasoline	94.0	105.7	101.4	111.5	100.8	107.3	108.4	96.2	96.8
Aviation Fuels	8.2	8.5	13.8	16.0	14.1	14.0	14.7	14.7	14.4
Kerosene	16.9	15.6	14.2	12.9	9.4	10.0	10.4	9.5	9.5
Gas Diesel	119.5	120.5	138.1	139.2	139.4	130.8	138.7	136.6	127.3
Heavy Fuel Oil	160.8	161.7	147.6	83.4	56.2	51.9	65.8	71.3	53.5
Other	27.6	28.1	24.5	26.6	25.5	27.4	29.1	30.1	25.4
Refinery Fuel	14.2	16.8	20.1	15.2	18.0	37.5	37.9	36.1	33.8
Sub-Total	476.8	492.5	506.4	463.5	420.6	440.8	469.4	456.1	424.3
Bunkers	4.1	2.6	6.7	13.1	12.2	8.6	12.0	10.6	16.0
Total	480.9	495.1	513.1	476.6	432.8	449.4	481.4	466.8	440.3

	1990	1991	1992	1993	1994	1995	1996	1997	1998
Thousand tonnes									
NGL/LPG/Ethane	1 435	1 416	1 534	1 532	1 520	1 627	1 670	1 620	1 791
Naphtha	650	662	732	874	552	572	785	937	1 002
Motor Gasoline	4 322	4 283	4 649	4 751	4 867	4 867	4 703	4 380	4 236
Aviation Fuels	666	628	729	740	918	1 008	1 095	1 218	1 360
Kerosene	479	415	404	375	339	326	336	271	246
Gas Diesel	6 442	7 000	7 168	7 493	8 272	8 816	9 212	9 635	10 106
Heavy Fuel Oil	2 166	2 617	2 300	2 259	1 660	996	1 386	1 205	1 401
Other	1 305	1 382	1 330	1 098	1 733	1 639	1 637	2 125	2 284
Refinery Fuel	1 608	1 993	2 053	1 991	1 233	1 191	965	1 164	1 135
Sub-Total	19 073	20 396	20 899	21 113	21 094	21 042	21 789	22 555	23 561
Bunkers	708	479	485	382	435	563	579	702	548
Total	19 781	20 875	21 384	21 495	21 529	21 605	22 368	23 257	24 109
Thousand barrels/day									
NGL/LPG/Ethane	49.9	49.4	53.7	53.6	53.0	57.0	58.5	56.3	61.7
Naphtha	15.3	15.6	17.2	20.6	13.0	13.5	18.4	22.1	23.6
Motor Gasoline	101.5	100.6	108.9	111.6	114.3	114.3	110.1	102.8	99.5
Aviation Fuels	14.3	13.5	15.6	15.8	19.7	21.6	23.4	26.1	29.1
Kerosene	10.2	8.9	8.6	8.0	7.2	7.0	7.2	5.8	5.3
Gas Diesel	126.4	137.3	140.2	147.0	162.3	172.9	180.2	189.0	198.2
Heavy Fuel Oil	39.6	47.8	41.9	41.3	30.3	18.2	25.3	22.0	25.6
Other	21.4	22.7	21.8	19.0	28.9	27.2	27.3	34.5	37.5
Refinery Fuel	30.8	38.0	38.9	38.5	24.5	23.1	19.2	22.7	22.2
Sub-Total	409.3	433.8	446.8	455.4	453.1	454.7	469.6	481.3	502.7
Bunkers	13.3	9.1	9.2	7.2	8.2	10.6	10.8	13.2	10.4
Total	422.6	442.9	456.0	462.6	461.3	465.3	480.4	494.5	513.1

[1] See Notes in Part I.4.

Oil Demand by Main Product Groups[1]
Demande de produits raffinés par groupes principaux

Armenia

	1971	1973	1978	1984	1985	1986	1987	1988	1989
Thousand tonnes									
NGL/LPG/Ethane	-	-	-	-	-	-	-	-	-
Naphtha	-	-	-	-	-	-	-	-	-
Motor Gasoline	-	-	-	-	-	-	-	-	-
Aviation Fuels	-	-	-	-	-	-	-	-	-
Kerosene	-	-	-	-	-	-	-	-	-
Gas Diesel	-	-	-	-	-	-	-	-	-
Heavy Fuel Oil	-	-	-	-	-	-	-	-	-
Other	-	-	-	-	-	-	-	-	-
Refinery Fuel	-	-	-	-	-	-	-	-	-
Sub-Total	-	-	-	-	-	-	-	-	-
Bunkers	-	-	-	-	-	-	-	-	-
Total	-	-	-	-	-	-	-	-	-
Thousand barrels/day									
NGL/LPG/Ethane	-	-	-	-	-	-	-	-	-
Naphtha	-	-	-	-	-	-	-	-	-
Motor Gasoline	-	-	-	-	-	-	-	-	-
Aviation Fuels	-	-	-	-	-	-	-	-	-
Kerosene	-	-	-	-	-	-	-	-	-
Gas Diesel	-	-	-	-	-	-	-	-	-
Heavy Fuel Oil	-	-	-	-	-	-	-	-	-
Other	-	-	-	-	-	-	-	-	-
Refinery Fuel	-	-	-	-	-	-	-	-	-
Sub-Total	-	-	-	-	-	-	-	-	-
Bunkers	-	-	-	-	-	-	-	-	-
Total	-	-	-	-	-	-	-	-	-

	1990	1991	1992	1993	1994	1995	1996	1997	1998
Thousand tonnes									
NGL/LPG/Ethane	-	-	10	4	1	1	1	1	1
Naphtha	-	-	-	-	-	-	-	-	-
Motor Gasoline	-	-	600	390	52	37	20	21	22
Aviation Fuels	-	-	150	65	48	34	19	20	21
Kerosene	-	-	7	5	5	4	2	2	2
Gas Diesel	-	-	213	100	86	61	34	36	37
Heavy Fuel Oil	-	-	1 400	600	163	115	63	68	70
Other	-	-	47	47	36	26	15	16	16
Refinery Fuel	-	-	-	-	-	-	-	-	-
Sub-Total	-	-	2 427	1 211	391	278	154	164	169
Bunkers	-	-	-	-	-	-	-	-	-
Total	-	-	2 427	1 211	391	278	154	164	169
Thousand barrels/day									
NGL/LPG/Ethane	-	-	0.3	0.1	0.0	0.0	0.0	0.0	0.0
Naphtha	-	-	-	-	-	-	-	-	-
Motor Gasoline	-	-	14.0	9.1	1.2	0.9	0.5	0.5	0.5
Aviation Fuels	-	-	3.3	1.4	1.0	0.7	0.4	0.4	0.5
Kerosene	-	-	0.1	0.1	0.1	0.1	0.0	0.0	0.0
Gas Diesel	-	-	4.3	2.0	1.8	1.2	0.7	0.7	0.8
Heavy Fuel Oil	-	-	25.5	10.9	3.0	2.1	1.1	1.2	1.3
Other	-	-	0.9	0.9	0.7	0.5	0.3	0.3	0.3
Refinery Fuel	-	-	-	-	-	-	-	-	-
Sub-Total	-	-	48.4	24.7	7.8	5.6	3.1	3.3	3.4
Bunkers	-	-	-	-	-	-	-	-	-
Total	-	-	48.4	24.7	7.8	5.6	3.1	3.3	3.4

[1] See Notes in Part I.4.

Oil Demand by Main Product Groups[1]
Demande de produits raffinés par groupes principaux

Azerbaijan

	1971	1973	1978	1984	1985	1986	1987	1988	1989
Thousand tonnes									
NGL/LPG/Ethane	-	-	-	-	-	-	-	-	-
Naphtha	-	-	-	-	-	-	-	-	-
Motor Gasoline	-	-	-	-	-	-	-	-	-
Aviation Fuels	-	-	-	-	-	-	-	-	-
Kerosene	-	-	-	-	-	-	-	-	-
Gas Diesel	-	-	-	-	-	-	-	-	-
Heavy Fuel Oil	-	-	-	-	-	-	-	-	-
Other	-	-	-	-	-	-	-	-	-
Refinery Fuel	-	-	-	-	-	-	-	-	-
Sub-Total	-	-	-	-	-	-	-	-	-
Bunkers	-	-	-	-	-	-	-	-	-
Total	-	-	-	-	-	-	-	-	-
Thousand barrels/day									
NGL/LPG/Ethane	-	-	-	-	-	-	-	-	-
Naphtha	-	-	-	-	-	-	-	-	-
Motor Gasoline	-	-	-	-	-	-	-	-	-
Aviation Fuels	-	-	-	-	-	-	-	-	-
Kerosene	-	-	-	-	-	-	-	-	-
Gas Diesel	-	-	-	-	-	-	-	-	-
Heavy Fuel Oil	-	-	-	-	-	-	-	-	-
Other	-	-	-	-	-	-	-	-	-
Refinery Fuel	-	-	-	-	-	-	-	-	-
Sub-Total	-	-	-	-	-	-	-	-	-
Bunkers	-	-	-	-	-	-	-	-	-
Total	-	-	-	-	-	-	-	-	-

	1990	1991	1992	1993	1994	1995	1996	1997	1998
Thousand tonnes									
NGL/LPG/Ethane	-	-	-	2	5	1	4	3	46
Naphtha	-	-	-	-	-	-	-	-	-
Motor Gasoline	-	-	1 024	1 066	1 269	950	737	492	446
Aviation Fuels	-	-	446	470	419	505	528	451	526
Kerosene	-	-	346	359	434	405	116	150	171
Gas Diesel	-	-	1 854	952	824	588	302	339	526
Heavy Fuel Oil	-	-	3 049	4 996	4 081	4 093	3 927	3 900	4 139
Other	-	-	1 265	165	223	48	83	33	59
Refinery Fuel	-	-	343	270	220	100	100	100	100
Sub-Total	-	-	8 327	8 280	7 475	6 690	5 797	5 468	6 013
Bunkers	-	-	-	-	-	-	-	-	-
Total	-	-	8 327	8 280	7 475	6 690	5 797	5 468	6 013
Thousand barrels/day									
NGL/LPG/Ethane	-	-	-	0.1	0.2	0.0	0.1	0.1	1.5
Naphtha	-	-	-	-	-	-	-	-	-
Motor Gasoline	-	-	23.9	24.9	29.7	22.2	17.2	11.5	10.4
Aviation Fuels	-	-	9.7	10.2	9.1	11.0	11.4	9.8	11.4
Kerosene	-	-	7.3	7.6	9.2	8.6	2.5	3.2	3.6
Gas Diesel	-	-	37.8	19.5	16.8	12.0	6.2	6.9	10.8
Heavy Fuel Oil	-	-	55.5	91.1	74.4	74.7	71.5	71.2	75.5
Other	-	-	24.1	2.9	4.0	0.8	1.3	0.6	1.1
Refinery Fuel	-	-	7.5	5.9	4.8	2.2	2.2	2.2	2.2
Sub-Total	-	-	165.8	162.1	148.2	131.5	112.3	105.4	116.5
Bunkers	-	-	-	-	-	-	-	-	-
Total	-	-	165.8	162.1	148.2	131.5	112.3	105.4	116.5

[1] See Notes in Part I.4.

Oil Demand by Main Product Groups[1]
Demande de produits raffinés par groupes principaux

Bahrain

	1971	1973	1978	1984	1985	1986	1987	1988	1989
Thousand tonnes									
NGL/LPG/Ethane	6	4	10	17	18	18	18	20	21
Naphtha	-	-	-	-	-	-	-	-	-
Motor Gasoline	27	37	87	161	173	179	186	198	211
Aviation Fuels	136	204	386	392	383	380	390	405	412
Kerosene	9	8	8	10	10	12	16	16	17
Gas Diesel	30	36	124	110	105	88	81	85	87
Heavy Fuel Oil	-	-	-	-	-	-	-	-	-
Other	-	-	24	34	36	47	30	40	50
Refinery Fuel	319	305	307	256	232	307	308	296	302
Sub-Total	527	594	946	980	957	1 031	1 029	1 060	1 100
Bunkers	1 060	1 016	406	191	206	206	100		
Total	1 587	1 610	1 352	1 171	1 163	1 237	1 129	1 060	1 100
Thousand barrels/day									
NGL/LPG/Ethane	0.2	0.1	0.3	0.5	0.6	0.6	0.6	0.6	0.7
Naphtha	-	-	-	-	-	-	-	-	-
Motor Gasoline	0.6	0.9	2.0	3.8	4.0	4.2	4.3	4.6	4.9
Aviation Fuels	3.0	4.4	8.4	8.5	8.3	8.3	8.5	8.8	9.0
Kerosene	0.2	0.2	0.2	0.2	0.2	0.3	0.3	0.3	0.4
Gas Diesel	0.6	0.7	2.5	2.2	2.1	1.8	1.7	1.7	1.8
Heavy Fuel Oil	-	-	-	-	-	-	-	-	-
Other	-	-	0.4	0.7	0.7	0.9	0.6	0.7	0.9
Refinery Fuel	7.0	6.7	6.7	5.6	5.1	6.7	6.8	6.5	6.6
Sub-Total	11.6	13.0	20.6	21.5	21.1	22.7	22.7	23.3	24.2
Bunkers	19.6	18.7	7.6	3.7	4.0	4.0	2.0		
Total	31.1	31.7	28.1	25.2	25.0	26.7	24.8	23.3	24.2

	1990	1991	1992	1993	1994	1995	1996	1997	1998
Thousand tonnes									
NGL/LPG/Ethane	20	22	23	23	24	25	25	26	26
Naphtha	-	-	-	-	-	-	-	-	-
Motor Gasoline	224	230	250	265	274	280	289	304	321
Aviation Fuels	453	333	304	302	346	364	336	240	346
Kerosene	21	23	25	25	27	26	32	54	76
Gas Diesel	95	93	109	107	105	115	114	119	125
Heavy Fuel Oil	-	-	-	-	-	-	-	-	-
Other	36	36	40	28	34	26	26	24	24
Refinery Fuel	304	312	320	325	308	309	313	325	313
Sub-Total	1 153	1 049	1 071	1 075	1 118	1 145	1 135	1 092	1 231
Bunkers	-	-	-	-	-	-	-	-	-
Total	1 153	1 049	1 071	1 075	1 118	1 145	1 135	1 092	1 231
Thousand barrels/day									
NGL/LPG/Ethane	0.6	0.7	0.7	0.7	0.8	0.8	0.8	0.8	0.8
Naphtha	-	-	-	-	-	-	-	-	-
Motor Gasoline	5.2	5.4	5.8	6.2	6.4	6.5	6.7	7.1	7.5
Aviation Fuels	9.8	7.2	6.6	6.6	7.5	7.9	7.3	5.2	7.5
Kerosene	0.4	0.5	0.5	0.5	0.6	0.6	0.7	1.1	1.6
Gas Diesel	1.9	1.9	2.2	2.2	2.1	2.4	2.3	2.4	2.6
Heavy Fuel Oil	-	-	-	-	-	-	-	-	-
Other	0.6	0.6	0.7	0.5	0.6	0.5	0.5	0.4	0.4
Refinery Fuel	6.7	6.8	7.0	7.1	6.8	6.8	6.8	7.1	6.9
Sub-Total	25.4	23.2	23.6	23.8	24.8	25.4	25.1	24.3	27.3
Bunkers	-	-	-	-	-	-	-	-	-
Total	25.4	23.2	23.6	23.8	24.8	25.4	25.1	24.3	27.3

[1] See Notes in Part I.4.

Oil Demand by Main Product Groups[1]
Demande de produits raffinés par groupes principaux

Bangladesh

	1971	1973	1978	1984	1985	1986	1987	1988	1989
Thousand tonnes									
NGL/LPG/Ethane	-	-	-	6	7	9	10	9	9
Naphtha	36	46	15	-	-	-	-	-	-
Motor Gasoline	43	31	53	48	49	62	62	72	93
Aviation Fuels	19	18	33	65	70	77	72	81	78
Kerosene	265	219	382	322	351	385	369	384	420
Gas Diesel	134	197	316	506	582	659	685	696	856
Heavy Fuel Oil	179	249	359	395	324	346	296	284	207
Other	17	44	38	78	92	79	121	156	153
Refinery Fuel	43	41	58	61	56	56	55	55	55
Sub-Total	736	845	1 254	1 481	1 531	1 673	1 670	1 737	1 871
Bunkers	21	31	32	34	22	12	10	10	10
Total	757	876	1 286	1 515	1 553	1 685	1 680	1 747	1 881
Thousand barrels/day									
NGL/LPG/Ethane	-	-	-	0.2	0.2	0.3	0.3	0.3	0.3
Naphtha	0.8	1.1	0.3	-	-	-	-	-	-
Motor Gasoline	1.0	0.7	1.2	1.1	1.1	1.4	1.4	1.7	2.2
Aviation Fuels	0.4	0.4	0.7	1.4	1.5	1.7	1.6	1.8	1.7
Kerosene	5.6	4.6	8.1	6.8	7.4	8.2	7.8	8.1	8.9
Gas Diesel	2.7	4.0	6.5	10.3	11.9	13.5	14.0	14.2	17.5
Heavy Fuel Oil	3.3	4.5	6.6	7.2	5.9	6.3	5.4	5.2	3.8
Other	0.3	0.9	0.7	1.5	1.8	1.5	2.3	3.0	2.9
Refinery Fuel	0.9	0.8	1.2	1.2	1.1	1.1	1.1	1.1	1.1
Sub-Total	15.1	17.1	25.3	29.7	31.0	34.0	34.0	35.3	38.4
Bunkers	0.4	0.6	0.6	0.6	0.4	0.2	0.2	0.2	0.2
Total	15.5	17.7	25.9	30.4	31.4	34.2	34.2	35.5	38.6

	1990	1991	1992	1993	1994	1995	1996	1997	1998
Thousand tonnes									
NGL/LPG/Ethane	9	8	8	9	13	15	13	16	9
Naphtha	-	6	16	29	73	69	72	73	73
Motor Gasoline	98	107	120	138	153	187	209	279	284
Aviation Fuels	86	78	74	77	90	95	97	108	58
Kerosene	475	388	358	375	392	434	452	539	586
Gas Diesel	833	807	920	948	964	1 225	1 312	1 482	1 541
Heavy Fuel Oil	177	136	161	226	211	238	209	266	201
Other	166	118	110	195	173	475	405	426	235
Refinery Fuel	50	44	46	55	57	62	60	72	43
Sub-Total	1 894	1 692	1 813	2 052	2 126	2 800	2 829	3 261	3 030
Bunkers	15	22	15	26	27	36	36	36	36
Total	1 909	1 714	1 828	2 078	2 153	2 836	2 865	3 297	3 066
Thousand barrels/day									
NGL/LPG/Ethane	0.3	0.3	0.3	0.3	0.4	0.5	0.4	0.5	0.3
Naphtha	-	0.1	0.4	0.7	1.7	1.6	1.7	1.7	1.7
Motor Gasoline	2.3	2.5	2.8	3.2	3.6	4.4	4.9	6.5	6.6
Aviation Fuels	1.9	1.7	1.6	1.7	2.0	2.1	2.1	2.3	1.3
Kerosene	10.1	8.2	7.6	8.0	8.3	9.2	9.6	11.4	12.4
Gas Diesel	17.0	16.5	18.8	19.4	19.7	25.0	26.7	30.3	31.5
Heavy Fuel Oil	3.2	2.5	2.9	4.1	3.9	4.3	3.8	4.9	3.7
Other	3.2	2.3	2.1	3.8	3.3	9.1	7.8	8.2	4.5
Refinery Fuel	1.0	0.9	0.9	1.1	1.2	1.3	1.2	1.5	0.9
Sub-Total	39.0	35.0	37.3	42.2	44.0	57.5	58.2	67.3	62.9
Bunkers	0.3	0.4	0.3	0.5	0.5	0.7	0.7	0.7	0.7
Total	39.3	35.4	37.6	42.7	44.5	58.2	58.8	68.0	63.5

[1] See Notes in Part I.4.

Oil Demand by Main Product Groups[1]
Demande de produits raffinés par groupes principaux

Belarus

	1971	1973	1978	1984	1985	1986	1987	1988	1989
Thousand tonnes									
NGL/LPG/Ethane	-	-	-	-	-	-	-	-	-
Naphtha	-	-	-	-	-	-	-	-	-
Motor Gasoline	-	-	-	-	-	-	-	-	-
Aviation Fuels	-	-	-	-	-	-	-	-	-
Kerosene	-	-	-	-	-	-	-	-	-
Gas Diesel	-	-	-	-	-	-	-	-	-
Heavy Fuel Oil	-	-	-	-	-	-	-	-	-
Other	-	-	-	-	-	-	-	-	-
Refinery Fuel	-	-	-	-	-	-	-	-	-
Sub-Total	-	-	-	-	-	-	-	-	-
Bunkers	-	-	-	-	-	-	-	-	-
Total	-	-	-	-	-	-	-	-	-
Thousand barrels/day									
NGL/LPG/Ethane	-	-	-	-	-	-	-	-	-
Naphtha	-	-	-	-	-	-	-	-	-
Motor Gasoline	-	-	-	-	-	-	-	-	-
Aviation Fuels	-	-	-	-	-	-	-	-	-
Kerosene	-	-	-	-	-	-	-	-	-
Gas Diesel	-	-	-	-	-	-	-	-	-
Heavy Fuel Oil	-	-	-	-	-	-	-	-	-
Other	-	-	-	-	-	-	-	-	-
Refinery Fuel	-	-	-	-	-	-	-	-	-
Sub-Total	-	-	-	-	-	-	-	-	-
Bunkers	-	-	-	-	-	-	-	-	-
Total	-	-	-	-	-	-	-	-	-

	1990	1991	1992	1993	1994	1995	1996	1997	1998
Thousand tonnes									
NGL/LPG/Ethane	-	-	338	336	359	343	316	315	313
Naphtha	-	-	380	-	-	-	-	-	-
Motor Gasoline	-	-	2 235	1 558	1 094	1 269	1 305	1 247	1 442
Aviation Fuels	-	-	900	444	444	-	-	-	-
Kerosene	-	-	46	21	30	15	16	24	36
Gas Diesel	-	-	2 893	3 039	1 966	1 805	1 900	1 916	2 390
Heavy Fuel Oil	-	-	9 027	7 063	6 135	5 178	4 711	3 705	4 161
Other	-	-	4 264	1 514	1 287	1 364	1 189	1 223	1 228
Refinery Fuel	-	-	800	548	492	500	500	500	500
Sub-Total	-	-	20 883	14 523	11 807	10 474	9 937	8 930	10 070
Bunkers	-	-	-	-	-	-	-	-	-
Total	-	-	20 883	14 523	11 807	10 474	9 937	8 930	10 070
Thousand barrels/day									
NGL/LPG/Ethane	-	-	10.7	10.7	11.4	10.9	10.0	10.0	9.9
Naphtha	-	-	8.8	-	-	-	-	-	-
Motor Gasoline	-	-	52.1	36.4	25.6	29.7	30.4	29.1	33.7
Aviation Fuels	-	-	19.5	9.6	9.6	-	-	-	-
Kerosene	-	-	1.0	0.4	0.6	0.3	0.3	0.5	0.8
Gas Diesel	-	-	59.0	62.1	40.2	36.9	38.7	39.2	48.8
Heavy Fuel Oil	-	-	164.3	128.9	111.9	94.5	85.7	67.6	75.9
Other	-	-	79.5	28.4	24.3	25.6	22.3	22.9	23.0
Refinery Fuel	-	-	17.5	12.0	10.8	11.0	10.9	11.0	11.0
Sub-Total	-	-	412.3	288.6	234.4	208.8	198.5	180.2	203.1
Bunkers	-	-	-	-	-	-	-	-	-
Total	-	-	412.3	288.6	234.4	208.8	198.5	180.2	203.1

[1] See Notes in Part I.4.

Oil Demand by Main Product Groups[1]
Demande de produits raffinés par groupes principaux

Benin

	1971	1973	1978	1984	1985	1986	1987	1988	1989
Thousand tonnes									
NGL/LPG/Ethane	-	-	-	-	-	-	-	-	-
Naphtha	-	-	-	-	-	-	-	-	-
Motor Gasoline	21	27	37	41	60	58	41	45	23
Aviation Fuels	5	4	10	18	19	19	19	16	20
Kerosene	13	19	27	13	18	9	11	12	11
Gas Diesel	58	79	51	54	60	51	52	46	43
Heavy Fuel Oil	4	4	6	12	12	12	10	10	12
Other	-	-	-	-	-	-	-	-	-
Refinery Fuel	-	-	-	-	-	-	-	-	-
Sub-Total	101	133	131	138	169	149	133	129	109
Bunkers	-	-	-	-	-	-	-	-	-
Total	101	133	131	138	169	149	133	129	109
Thousand barrels/day									
NGL/LPG/Ethane	-	-	-	-	-	-	-	-	-
Naphtha	-	-	-	-	-	-	-	-	-
Motor Gasoline	0.5	0.6	0.9	1.0	1.4	1.4	1.0	1.0	0.5
Aviation Fuels	0.1	0.1	0.2	0.4	0.4	0.4	0.4	0.3	0.4
Kerosene	0.3	0.4	0.6	0.3	0.4	0.2	0.2	0.3	0.2
Gas Diesel	1.2	1.6	1.0	1.1	1.2	1.0	1.1	0.9	0.9
Heavy Fuel Oil	0.1	0.1	0.1	0.2	0.2	0.2	0.2	0.2	0.2
Other	-	-	-	-	-	-	-	-	-
Refinery Fuel	-	-	-	-	-	-	-	-	-
Sub-Total	2.1	2.8	2.8	2.9	3.6	3.2	2.9	2.8	2.3
Bunkers	-	-	-	-	-	-	-	-	-
Total	2.1	2.8	2.8	2.9	3.6	3.2	2.9	2.8	2.3

	1990	1991	1992	1993	1994	1995	1996	1997	1998
Thousand tonnes									
NGL/LPG/Ethane	-	-	-	-	-	-	1	1	1
Naphtha	-	-	-	-	-	-	-	-	-
Motor Gasoline	16	10	10	10	10	10	105	126	110
Aviation Fuels	16	15	14	19	20	21	56	49	46
Kerosene	11	8	12	10	12	13	39	34	32
Gas Diesel	42	37	33	34	33	34	102	87	87
Heavy Fuel Oil	12	8	11	12	13	13	24	25	40
Other	-	-	-	-	-	-	-	-	-
Refinery Fuel	-	-	-	-	-	-	-	-	-
Sub-Total	97	78	80	85	88	91	327	322	316
Bunkers	-	-	-	-	-	-	5	5	5
Total	97	78	80	85	88	91	332	327	321
Thousand barrels/day									
NGL/LPG/Ethane	-	-	-	-	-	-	0.0	0.0	0.0
Naphtha	-	-	-	-	-	-	-	-	-
Motor Gasoline	0.4	0.2	0.2	0.2	0.2	0.2	2.4	2.9	2.6
Aviation Fuels	0.3	0.3	0.3	0.4	0.4	0.5	1.2	1.1	1.0
Kerosene	0.2	0.2	0.3	0.2	0.3	0.3	0.8	0.7	0.7
Gas Diesel	0.9	0.8	0.7	0.7	0.7	0.7	2.1	1.8	1.8
Heavy Fuel Oil	0.2	0.1	0.2	0.2	0.2	0.2	0.4	0.5	0.7
Other	-	-	-	-	-	-	-	-	-
Refinery Fuel	-	-	-	-	-	-	-	-	-
Sub-Total	2.0	1.6	1.7	1.8	1.8	1.9	7.0	7.0	6.8
Bunkers	-	-	-	-	-	-	0.1	0.1	0.1
Total	2.0	1.6	1.7	1.8	1.8	1.9	7.1	7.1	6.9

[1] See Notes in Part I.4.

Oil Demand by Main Product Groups[1]
Demande de produits raffinés par groupes principaux

Bolivia

	1971	1973	1978	1984	1985	1986	1987	1988	1989
Thousand tonnes									
NGL/LPG/Ethane	8	14	60	147	154	169	181	172	167
Naphtha	-	-	-	-	-	-	-	-	-
Motor Gasoline	225	261	319	321	326	340	380	379	392
Aviation Fuels	27	32	95	85	82	81	79	78	86
Kerosene	109	127	141	84	73	43	40	33	37
Gas Diesel	92	118	267	220	219	272	263	255	288
Heavy Fuel Oil	147	157	151	75	54	20	9	31	27
Other	6	12	14	7	8	8	9	11	19
Refinery Fuel	10	11	23	76	76	80	85	77	76
Sub-Total	624	732	1 070	1 015	992	1 013	1 046	1 036	1 092
Bunkers	-	-	-	-	-	-	-	-	-
Total	624	732	1 070	1 015	992	1 013	1 046	1 036	1 092
Thousand barrels/day									
NGL/LPG/Ethane	0.3	0.4	1.9	4.7	4.9	5.4	5.8	5.5	5.3
Naphtha	-	-	-	-	-	-	-	-	-
Motor Gasoline	5.3	6.1	7.5	7.5	7.6	7.9	8.9	8.8	9.2
Aviation Fuels	0.6	0.7	2.1	1.9	1.8	1.8	1.7	1.7	1.9
Kerosene	2.3	2.7	3.0	1.8	1.5	0.9	0.8	0.7	0.8
Gas Diesel	1.9	2.4	5.5	4.5	4.5	5.6	5.4	5.2	5.9
Heavy Fuel Oil	2.7	2.9	2.8	1.4	1.0	0.4	0.2	0.6	0.5
Other	0.1	0.2	0.3	0.1	0.1	0.2	0.2	0.2	0.4
Refinery Fuel	0.2	0.2	0.5	1.6	1.6	1.6	1.7	1.7	1.6
Sub-Total	13.3	15.7	23.4	23.3	23.0	23.7	24.7	24.3	25.5
Bunkers	-	-	-	-	-	-	-	-	-
Total	13.3	15.7	23.4	23.3	23.0	23.7	24.7	24.3	25.5

	1990	1991	1992	1993	1994	1995	1996	1997	1998
Thousand tonnes									
NGL/LPG/Ethane	175	180	190	200	217	235	249	291	307
Naphtha	-	-	-	-	-	-	-	-	-
Motor Gasoline	390	370	360	361	376	401	428	446	452
Aviation Fuels	89	85	87	93	95	108	128	136	142
Kerosene	30	29	23	21	21	24	25	23	26
Gas Diesel	326	376	393	438	462	530	538	619	638
Heavy Fuel Oil	25	23	15	4	3	2	1	1	1
Other	14	11	14	7	9	11	11	13	13
Refinery Fuel	82	81	81	87	96	109	105	175	164
Sub-Total	1 131	1 155	1 163	1 211	1 279	1 420	1 485	1 704	1 743
Bunkers	-	-	-	-	-	-	-	-	-
Total	1 131	1 155	1 163	1 211	1 279	1 420	1 485	1 704	1 743
Thousand barrels/day									
NGL/LPG/Ethane	5.6	5.7	6.0	6.4	6.9	7.5	7.9	9.2	9.8
Naphtha	-	-	-	-	-	-	-	-	-
Motor Gasoline	9.1	8.6	8.4	8.4	8.8	9.4	10.0	10.4	10.6
Aviation Fuels	2.0	1.9	1.9	2.0	2.1	2.4	2.8	3.0	3.1
Kerosene	0.6	0.6	0.5	0.4	0.4	0.5	0.5	0.5	0.6
Gas Diesel	6.7	7.7	8.0	9.0	9.4	10.8	11.0	12.7	13.0
Heavy Fuel Oil	0.5	0.4	0.3	0.1	0.1	0.0	0.0	0.0	0.0
Other	0.3	0.2	0.3	0.1	0.2	0.2	0.2	0.2	0.2
Refinery Fuel	1.8	1.8	1.8	1.9	2.1	2.4	2.3	3.7	3.5
Sub-Total	26.4	26.9	27.1	28.3	30.0	33.1	34.6	39.8	40.8
Bunkers	-	-	-	-	-	-	-	-	-
Total	26.4	26.9	27.1	28.3	30.0	33.1	34.6	39.8	40.8

[1] See Notes in Part I.4.

Oil Demand by Main Product Groups[1]
Demande de produits raffinés par groupes principaux

Bosnia-Herzegovina

	1971	1973	1978	1984	1985	1986	1987	1988	1989
Thousand tonnes									
NGL/LPG/Ethane	-	-	-	-	-	-	-	-	-
Naphtha	-	-	-	-	-	-	-	-	-
Motor Gasoline	-	-	-	-	-	-	-	-	-
Aviation Fuels	-	-	-	-	-	-	-	-	-
Kerosene	-	-	-	-	-	-	-	-	-
Gas Diesel	-	-	-	-	-	-	-	-	-
Heavy Fuel Oil	-	-	-	-	-	-	-	-	-
Other	-	-	-	-	-	-	-	-	-
Refinery Fuel	-	-	-	-	-	-	-	-	-
Sub-Total	-	-	-	-	-	-	-	-	-
Bunkers	-	-	-	-	-	-	-	-	-
Total	-	-	-	-	-	-	-	-	-
Thousand barrels/day									
NGL/LPG/Ethane	-	-	-	-	-	-	-	-	-
Naphtha	-	-	-	-	-	-	-	-	-
Motor Gasoline	-	-	-	-	-	-	-	-	-
Aviation Fuels	-	-	-	-	-	-	-	-	-
Kerosene	-	-	-	-	-	-	-	-	-
Gas Diesel	-	-	-	-	-	-	-	-	-
Heavy Fuel Oil	-	-	-	-	-	-	-	-	-
Other	-	-	-	-	-	-	-	-	-
Refinery Fuel	-	-	-	-	-	-	-	-	-
Sub-Total	-	-	-	-	-	-	-	-	-
Bunkers	-	-	-	-	-	-	-	-	-
Total	-	-	-	-	-	-	-	-	-

	1990	1991	1992	1993	1994	1995	1996	1997	1998
Thousand tonnes									
NGL/LPG/Ethane	-	-	-	-	-	-	-	-	-
Naphtha	-	-	-	964	298	307	316	325	335
Motor Gasoline	-	-	100	96	115	118	122	126	130
Aviation Fuels	-	-	16	30	36	37	38	39	40
Kerosene	-	-	-	-	-	-	-	-	-
Gas Diesel	-	-	320	224	269	277	285	294	303
Heavy Fuel Oil	-	-	100	80	96	99	102	105	108
Other	-	-	-	38	45	46	48	49	51
Refinery Fuel	-	-	-	-	-	-	-	-	-
Sub-Total	-	-	536	1 432	859	884	911	938	967
Bunkers	-	-	-	-	-	-	-	-	-
Total	-	-	536	1 432	859	884	911	938	967
Thousand barrels/day									
NGL/LPG/Ethane	-	-	-	-	-	-	-	-	-
Naphtha	-	-	-	22.4	6.9	7.1	7.3	7.6	7.8
Motor Gasoline	-	-	2.3	2.2	2.7	2.8	2.8	2.9	3.0
Aviation Fuels	-	-	0.3	0.7	0.8	0.8	0.8	0.8	0.9
Kerosene	-	-	-	-	-	-	-	-	-
Gas Diesel	-	-	6.5	4.6	5.5	5.7	5.8	6.0	6.2
Heavy Fuel Oil	-	-	1.8	1.5	1.8	1.8	1.9	1.9	2.0
Other	-	-	-	0.7	0.9	0.9	0.9	0.9	1.0
Refinery Fuel	-	-	-	-	-	-	-	-	-
Sub-Total	-	-	11.0	32.1	18.5	19.1	19.6	20.2	20.9
Bunkers	-	-	-	-	-	-	-	-	-
Total	-	-	11.0	32.1	18.5	19.1	19.6	20.2	20.9

[1] See Notes in Part I.4.

Oil Demand by Main Product Groups[1]
Demande de produits raffinés par groupes principaux

Brazil

	1971	1973	1978	1984	1985	1986	1987	1988	1989
Thousand tonnes									
NGL/LPG/Ethane	1 306	1 600	2 282	3 381	3 669	3 937	4 330	5 294	5 618
Naphtha	109	1 079	1 820	3 214	3 968	3 938	4 420	4 399	4 699
Motor Gasoline	7 815	10 250	11 164	11 134	12 279	15 116	14 442	14 891	16 323
Aviation Fuels	784	1 047	1 470	1 671	1 769	1 909	1 925	1 864	1 970
Kerosene	476	556	597	349	316	328	345	333	317
Gas Diesel	5 878	7 936	13 237	15 580	16 579	18 407	19 526	20 093	20 586
Heavy Fuel Oil	8 253	10 398	15 488	7 986	8 345	9 923	9 911	9 937	9 340
Other	1 204	1 295	3 410	3 784	4 641	4 605	4 697	4 713	4 406
Refinery Fuel	1 552	2 098	3 231	3 549	3 403	3 560	3 911	4 001	4 055
Sub-Total	27 377	36 259	52 699	50 648	54 969	61 723	63 507	65 525	67 314
Bunkers	319	345	420	531	553	574	597	622	646
Total	27 696	36 604	53 119	51 179	55 522	62 297	64 104	66 147	67 960
Thousand barrels/day									
NGL/LPG/Ethane	41.4	50.7	72.3	104.7	113.9	122.0	135.6	164.2	174.4
Naphtha	2.5	24.9	45.8	77.6	96.1	96.0	108.1	108.1	114.3
Motor Gasoline	181.0	237.4	256.8	248.8	273.7	335.8	319.7	328.3	361.1
Aviation Fuels	17.4	23.3	32.5	36.3	38.3	42.0	42.4	41.2	43.3
Kerosene	10.4	12.1	13.0	7.6	6.8	7.3	7.6	7.3	7.0
Gas Diesel	122.3	165.1	275.3	319.2	340.6	378.2	401.2	411.7	422.9
Heavy Fuel Oil	153.8	193.7	284.3	144.0	144.5	171.5	171.1	169.1	161.5
Other	24.7	26.6	68.3	75.6	87.4	91.2	89.1	89.5	84.2
Refinery Fuel	31.0	41.5	64.8	71.8	67.7	70.8	77.8	79.2	81.3
Sub-Total	584.3	775.2	1 113.1	1 085.6	1 169.1	1 314.9	1 352.5	1 398.5	1 450.0
Bunkers	6.1	6.6	7.9	9.9	10.0	10.3	10.7	11.1	11.6
Total	590.4	781.8	1 121.0	1 095.5	1 179.1	1 325.2	1 363.2	1 409.6	1 461.6

	1990	1991	1992	1993	1994	1995	1996	1997	1998
Thousand tonnes									
NGL/LPG/Ethane	5 907	5 717	5 965	6 427	5 564	6 088	6 682	8 150	7 290
Naphtha	4 711	4 581	4 947	5 111	5 750	5 584	5 399	6 647	6 659
Motor Gasoline	16 162	17 138	16 900	17 689	19 090	21 164	23 385	24 150	24 584
Aviation Fuels	1 864	1 951	1 834	1 937	1 988	2 309	2 465	2 773	3 065
Kerosene	255	247	200	178	142	129	112	87	78
Gas Diesel	20 212	21 023	21 546	22 323	23 402	25 051	25 936	27 668	29 147
Heavy Fuel Oil	9 123	8 966	9 669	10 071	10 136	10 750	12 315	12 271	12 354
Other	4 728	4 402	4 540	4 529	5 182	5 352	5 788	6 739	7 023
Refinery Fuel	3 924	3 625	3 752	3 974	4 250	4 067	4 420	5 020	4 890
Sub-Total	66 886	67 650	69 353	72 239	75 504	80 494	86 502	93 505	95 090
Bunkers	554	784	846	1 060	1 228	1 176	1 352	1 720	1 708
Total	67 440	68 434	70 199	73 299	76 732	81 670	87 854	95 225	96 798
Thousand barrels/day									
NGL/LPG/Ethane	182.9	177.5	184.9	199.3	173.4	189.5	207.2	252.2	226.5
Naphtha	115.6	112.5	121.1	125.5	141.2	137.1	132.2	163.2	163.5
Motor Gasoline	359.4	381.4	375.3	393.9	425.4	473.1	522.7	542.8	553.4
Aviation Fuels	40.8	42.6	40.0	42.3	43.5	50.5	53.7	60.6	67.0
Kerosene	5.6	5.4	4.4	3.9	3.1	2.8	2.4	1.9	1.7
Gas Diesel	415.3	431.9	441.5	458.6	480.8	514.7	531.4	568.4	598.8
Heavy Fuel Oil	155.2	152.5	164.1	171.3	172.5	182.9	209.0	208.8	210.2
Other	89.4	83.5	85.4	85.8	97.7	101.0	107.5	127.1	131.5
Refinery Fuel	77.3	73.8	76.2	80.1	84.9	80.9	89.3	101.3	99.4
Sub-Total	1 441.6	1 461.2	1 492.6	1 560.8	1 622.4	1 732.4	1 855.4	2 026.3	2 052.0
Bunkers	9.9	13.6	14.7	18.5	21.8	21.0	23.9	30.5	30.5
Total	1 451.5	1 474.9	1 507.3	1 579.2	1 644.2	1 753.4	1 879.3	2 056.8	2 082.6

[1] See Notes in Part I.4.

Oil Demand by Main Product Groups[1]
Demande de produits raffinés par groupes principaux

Brunei

	1971	1973	1978	1984	1985	1986	1987	1988	1989
Thousand tonnes									
NGL/LPG/Ethane	5	5	5	8	9	9	9	10	10
Naphtha	1	5	4	4	5	4	4	4	5
Motor Gasoline	26	27	54	102	113	120	128	132	134
Aviation Fuels	1	6	17	18	14	23	25	30	29
Kerosene	1	1	1	5	3	3	3	3	4
Gas Diesel	22	17	14	60	65	59	61	72	78
Heavy Fuel Oil	-	-	-	-	1	1	1	1	1
Other	6	6	3	9	11	9	9	11	14
Refinery Fuel	-	-	-	1	1	1	1	1	1
Sub-Total	62	67	98	207	222	229	241	264	276
Bunkers	-	-	-	-	-	-	-	-	-
Total	62	67	98	207	222	229	241	264	276
Thousand barrels/day									
NGL/LPG/Ethane	0.2	0.2	0.2	0.3	0.3	0.3	0.3	0.3	0.3
Naphtha	0.0	0.1	0.1	0.1	0.1	0.1	0.1	0.1	0.1
Motor Gasoline	0.6	0.6	1.3	2.4	2.6	2.8	3.0	3.1	3.1
Aviation Fuels	0.0	0.1	0.4	0.4	0.3	0.5	0.5	0.7	0.6
Kerosene	0.0	0.0	0.0	0.1	0.1	0.1	0.1	0.1	0.1
Gas Diesel	0.5	0.3	0.3	1.2	1.3	1.2	1.2	1.5	1.6
Heavy Fuel Oil	-	-	-	-	0.0	0.0	0.0	0.0	0.0
Other	0.1	0.1	0.1	0.2	0.2	0.2	0.2	0.2	0.2
Refinery Fuel	-	-	-	0.0	0.0	0.0	0.0	0.0	0.0
Sub-Total	1.4	1.5	2.2	4.6	5.0	5.2	5.4	5.9	6.2
Bunkers	-	-	-	-	-	-	-	-	-
Total	1.4	1.5	2.2	4.6	5.0	5.2	5.4	5.9	6.2

	1990	1991	1992	1993	1994	1995	1996	1997	1998
Thousand tonnes									
NGL/LPG/Ethane	10	11	13	13	14	16	17	18	17
Naphtha	6	4	4	5	6	6	6	6	6
Motor Gasoline	140	151	157	164	176	187	195	200	190
Aviation Fuels	33	37	45	50	56	63	72	85	73
Kerosene	3	3	3	5	5	5	5	3	3
Gas Diesel	80	87	110	114	120	130	150	173	141
Heavy Fuel Oil	1	1	1	1	1	1	1	1	1
Other	11	8	18	22	20	27	27	24	17
Refinery Fuel	1	8	25	24	27	48	55	59	59
Sub-Total	285	310	376	398	425	483	528	569	507
Bunkers	-	-	-	-	-	-	-	-	-
Total	285	310	376	398	425	483	528	569	507
Thousand barrels/day									
NGL/LPG/Ethane	0.3	0.4	0.4	0.4	0.4	0.5	0.5	0.6	0.5
Naphtha	0.1	0.1	0.1	0.1	0.1	0.1	0.1	0.1	0.1
Motor Gasoline	3.3	3.5	3.7	3.8	4.1	4.4	4.5	4.7	4.4
Aviation Fuels	0.7	0.8	1.0	1.1	1.2	1.4	1.6	1.8	1.6
Kerosene	0.1	0.1	0.1	0.1	0.1	0.1	0.1	0.1	0.1
Gas Diesel	1.6	1.8	2.2	2.3	2.5	2.7	3.1	3.5	2.9
Heavy Fuel Oil	0.0	0.0	0.0	0.0	0.0	0.0	0.0	0.0	0.0
Other	0.2	0.1	0.3	0.4	0.3	0.5	0.5	0.4	0.3
Refinery Fuel	0.0	0.2	0.5	0.5	0.6	1.1	1.2	1.3	1.3
Sub-Total	6.4	6.9	8.3	8.8	9.4	10.7	11.6	12.5	11.3
Bunkers	-	-	-	-	-	-	-	-	-
Total	6.4	6.9	8.3	8.8	9.4	10.7	11.6	12.5	11.3

[1] See Notes in Part I.4.

Oil Demand by Main Product Groups[1]
Demande de produits raffinés par groupes principaux

Bulgaria

	1971	1973	1978	1984	1985	1986	1987	1988	1989
Thousand tonnes									
NGL/LPG/Ethane	7	23	29	156	150	155	173	168	170
Naphtha	-	-	-	-	-	-	-	-	-
Motor Gasoline	1 352	1 325	1 320	1 257	1 213	1 210	1 218	1 383	1 533
Aviation Fuels	200	200	300	337	366	362	370	416	444
Kerosene	143	155	200	220	220	220	220	220	165
Gas Diesel	2 527	3 000	4 000	2 751	2 572	2 881	2 643	2 801	2 697
Heavy Fuel Oil	4 983	5 290	6 175	4 187	4 245	4 383	4 176	4 465	4 178
Other	282	366	666	763	732	789	757	774	748
Refinery Fuel	395	490	625	635	635	635	635	600	600
Sub-Total	9 889	10 849	13 315	10 306	10 133	10 635	10 192	10 827	10 535
Bunkers	-	-	-	261	229	286	299	303	310
Total	9 889	10 849	13 315	10 567	10 362	10 921	10 491	11 130	10 845
Thousand barrels/day									
NGL/LPG/Ethane	0.2	0.7	0.9	4.9	4.8	4.9	5.5	5.3	5.4
Naphtha	-	-	-	-	-	-	-	-	-
Motor Gasoline	31.6	31.0	30.8	29.3	28.3	28.3	28.5	32.2	35.8
Aviation Fuels	4.3	4.3	6.5	7.3	8.0	7.9	8.0	9.0	9.6
Kerosene	3.0	3.3	4.2	4.7	4.7	4.7	4.7	4.7	3.5
Gas Diesel	51.6	61.3	81.8	56.1	52.6	58.9	54.0	57.1	55.1
Heavy Fuel Oil	90.9	96.5	112.7	76.2	77.5	80.0	76.2	81.2	76.2
Other	5.1	6.4	11.4	13.3	12.8	13.8	13.2	13.5	13.0
Refinery Fuel	7.2	8.9	11.4	11.6	11.6	11.6	11.6	10.9	10.9
Sub-Total	194.0	212.5	259.8	203.3	200.1	210.0	201.7	214.0	209.7
Bunkers	-	-	-	5.2	4.6	5.6	5.9	5.9	6.0
Total	194.0	212.5	259.8	208.4	204.7	215.6	207.5	219.9	215.7

	1990	1991	1992	1993	1994	1995	1996	1997	1998
Thousand tonnes									
NGL/LPG/Ethane	174	148	32	51	68	70	67	87	74
Naphtha	-	-	520	584	484	599	678	673	650
Motor Gasoline	1 332	673	830	978	984	1 083	932	609	800
Aviation Fuels	293	187	308	398	329	324	220	168	147
Kerosene	145	75	2	2	2	1	-	-	1
Gas Diesel	2 063	1 281	1 394	1 432	1 180	1 087	1 169	967	1 219
Heavy Fuel Oil	3 347	2 576	2 319	2 099	2 120	1 762	1 614	1 465	1 309
Other	678	311	131	156	129	112	374	270	447
Refinery Fuel	468	333	156	216	219	268	250	178	81
Sub-Total	8 500	5 584	5 692	5 916	5 515	5 306	5 304	4 417	4 728
Bunkers	326	302	272	263	265	275	239	9	70
Total	8 826	5 886	5 964	6 179	5 780	5 581	5 543	4 426	4 798
Thousand barrels/day									
NGL/LPG/Ethane	5.5	4.7	1.0	1.6	2.2	2.2	2.1	2.8	2.4
Naphtha	-	-	12.1	13.6	11.3	13.9	15.7	15.7	15.1
Motor Gasoline	31.1	15.7	19.3	22.9	23.0	25.3	21.7	14.2	18.7
Aviation Fuels	6.4	4.1	6.7	8.7	7.2	7.0	4.8	3.7	3.2
Kerosene	3.1	1.6	0.0	0.0	0.0	0.0	-	-	0.0
Gas Diesel	42.2	26.2	28.4	29.3	24.1	22.2	23.8	19.8	24.9
Heavy Fuel Oil	61.1	47.0	42.2	38.3	38.7	32.2	29.4	26.7	23.9
Other	12.6	5.8	2.7	3.2	2.6	2.3	7.3	5.3	8.5
Refinery Fuel	9.2	6.6	3.1	4.4	4.5	5.4	4.9	3.6	1.8
Sub-Total	171.0	111.7	115.5	121.9	113.5	110.6	109.7	91.7	98.5
Bunkers	6.3	5.7	5.1	4.9	5.0	5.2	4.5	0.2	1.4
Total	177.4	117.4	120.6	126.8	118.5	115.8	114.2	91.9	99.9

[1] See Notes in Part I.4.

Oil Demand by Main Product Groups[1]
Demande de produits raffinés par groupes principaux

Cameroon

	1971	1973	1978	1984	1985	1986	1987	1988	1989
Thousand tonnes									
NGL/LPG/Ethane	2	2	4	13	16	21	21	22	23
Naphtha	-	-	-	-	-	-	-	-	-
Motor Gasoline	93	101	142	230	267	270	299	272	242
Aviation Fuels	53	61	59	48	48	48	48	50	51
Kerosene	16	12	51	87	106	105	104	111	192
Gas Diesel	96	115	204	289	298	266	245	332	349
Heavy Fuel Oil	26	23	52	60	68	68	56	54	58
Other	-	-	19	71	52	53	57	39	45
Refinery Fuel	-	-	-	-	-	-	-	-	-
Sub-Total	286	314	531	798	855	831	830	880	960
Bunkers	-	-	-	7	9	5	3	11	10
Total	286	314	531	805	864	836	833	891	970
Thousand barrels/day									
NGL/LPG/Ethane	0.1	0.1	0.1	0.4	0.5	0.7	0.7	0.7	0.7
Naphtha	-	-	-	-	-	-	-	-	-
Motor Gasoline	2.2	2.4	3.3	5.4	6.2	6.3	7.0	6.3	5.7
Aviation Fuels	1.2	1.3	1.3	1.0	1.0	1.0	1.0	1.1	1.1
Kerosene	0.3	0.3	1.1	1.8	2.2	2.2	2.2	2.3	4.1
Gas Diesel	2.0	2.4	4.2	5.9	6.1	5.4	5.0	6.8	7.1
Heavy Fuel Oil	0.5	0.4	0.9	1.1	1.2	1.2	1.0	1.0	1.1
Other	-	-	0.4	1.4	1.0	1.0	1.1	0.7	0.8
Refinery Fuel	-	-	-	-	-	-	-	-	-
Sub-Total	6.2	6.8	11.3	17.0	18.4	17.9	18.0	18.9	20.6
Bunkers	-	-	-	0.1	0.2	0.1	0.1	0.2	0.2
Total	6.2	6.8	11.3	17.1	18.5	18.0	18.1	19.2	20.8

	1990	1991	1992	1993	1994	1995	1996	1997	1998
Thousand tonnes									
NGL/LPG/Ethane	25	21	24	23	21	22	23	24	20
Naphtha	-	-	-	-	-	-	-	-	-
Motor Gasoline	218	189	206	241	249	231	238	254	287
Aviation Fuels	49	48	47	39	58	53	56	51	55
Kerosene	194	196	147	156	233	204	262	281	286
Gas Diesel	347	317	300	301	317	318	318	331	316
Heavy Fuel Oil	54	48	44	48	45	50	49	50	47
Other	45	42	40	45	40	45	31	31	31
Refinery Fuel	-	-	-	-	-	-	-	-	-
Sub-Total	932	861	808	853	963	923	977	1 022	1 042
Bunkers	13	23	19	15	22	28	44	53	53
Total	945	884	827	868	985	951	1 021	1 075	1 095
Thousand barrels/day									
NGL/LPG/Ethane	0.8	0.7	0.8	0.7	0.7	0.7	0.7	0.8	0.6
Naphtha	-	-	-	-	-	-	-	-	-
Motor Gasoline	5.1	4.4	4.8	5.6	5.8	5.4	5.5	5.9	6.7
Aviation Fuels	1.1	1.0	1.0	0.8	1.3	1.2	1.2	1.1	1.2
Kerosene	4.1	4.2	3.1	3.3	4.9	4.3	5.5	6.0	6.1
Gas Diesel	7.1	6.5	6.1	6.2	6.5	6.5	6.5	6.8	6.5
Heavy Fuel Oil	1.0	0.9	0.8	0.9	0.8	0.9	0.9	0.9	0.9
Other	0.8	0.8	0.7	0.8	0.8	0.8	0.6	0.6	0.6
Refinery Fuel	-	-	-	-	-	-	-	-	-
Sub-Total	20.0	18.4	17.4	18.4	20.7	19.8	21.0	22.0	22.5
Bunkers	0.3	0.5	0.4	0.3	0.4	0.6	0.9	1.1	1.1
Total	20.3	18.9	17.7	18.7	21.2	20.4	21.9	23.1	23.6

[1] See Notes in Part I.4.

Oil Demand by Main Product Groups[1]
Demande de produits raffinés par groupes principaux

Chile

	1971	1973	1978	1984	1985	1986	1987	1988	1989
Thousand tonnes									
NGL/LPG/Ethane	347	402	461	436	426	434	470	510	545
Naphtha	45	60	51	43	33	32	35	36	38
Motor Gasoline	1 348	1 197	1 009	1 058	998	1 038	1 094	1 220	1 329
Aviation Fuels	138	123	165	161	156	196	214	217	236
Kerosene	417	512	320	149	116	147	171	188	231
Gas Diesel	752	785	1 044	1 383	1 416	1 534	1 600	1 815	2 075
Heavy Fuel Oil	1 539	1 434	1 411	1 011	943	935	880	976	1 112
Other	26	10	18	11	8	17	20	146	404
Refinery Fuel	286	320	374	285	290	288	324	349	379
Sub-Total	4 898	4 843	4 853	4 537	4 386	4 621	4 808	5 457	6 349
Bunkers	195	189	86	23	29	46	43	51	117
Total	5 093	5 032	4 939	4 560	4 415	4 667	4 851	5 508	6 466
Thousand barrels/day									
NGL/LPG/Ethane	11.0	12.8	14.7	13.8	13.5	13.8	14.9	16.2	17.3
Naphtha	1.0	1.4	1.2	1.0	0.8	0.7	0.8	0.8	0.9
Motor Gasoline	31.5	28.0	23.6	24.7	23.3	24.3	25.6	28.4	31.1
Aviation Fuels	3.1	2.7	3.6	3.5	3.4	4.3	4.7	4.7	5.1
Kerosene	8.8	10.9	6.8	3.2	2.5	3.1	3.6	4.0	4.9
Gas Diesel	15.4	16.0	21.3	28.2	28.9	31.4	32.7	37.0	42.4
Heavy Fuel Oil	28.1	26.2	25.7	18.4	17.2	17.1	16.1	17.8	20.3
Other	0.5	0.2	0.3	0.2	0.2	0.3	0.4	0.4	0.6
Refinery Fuel	5.7	6.4	7.7	5.8	5.9	5.9	6.6	7.1	7.7
Sub-Total	105.1	104.6	104.9	98.7	95.6	100.8	105.3	116.3	130.3
Bunkers	3.6	3.4	1.6	0.4	0.5	0.8	0.8	0.9	2.1
Total	108.7	108.0	106.5	99.1	96.2	101.6	106.1	117.3	132.4

	1990	1991	1992	1993	1994	1995	1996	1997	1998
Thousand tonnes									
NGL/LPG/Ethane	567	606	673	737	754	828	898	937	1 013
Naphtha	41	43	48	53	57	63	60	68	73
Motor Gasoline	1 374	1 422	1 567	1 653	1 857	2 009	2 151	2 237	2 327
Aviation Fuels	283	273	318	363	348	401	441	557	649
Kerosene	177	221	263	278	273	271	320	319	291
Gas Diesel	2 250	2 301	2 357	2 665	2 911	3 215	3 494	3 968	3 907
Heavy Fuel Oil	1 108	1 122	1 238	1 352	1 441	1 608	1 694	1 752	1 346
Other	443	324	420	409	409	395	412	644	760
Refinery Fuel	401	421	463	453	482	524	672	527	798
Sub-Total	6 644	6 733	7 347	7 963	8 532	9 314	10 142	11 009	11 164
Bunkers	183	185	227	297	316	383	254	308	439
Total	6 827	6 918	7 574	8 260	8 848	9 697	10 396	11 317	11 603
Thousand barrels/day									
NGL/LPG/Ethane	18.0	19.3	21.3	23.4	24.0	26.3	28.5	29.8	32.2
Naphtha	1.0	1.0	1.1	1.2	1.3	1.5	1.4	1.6	1.7
Motor Gasoline	32.1	33.2	36.5	38.6	43.4	47.0	50.1	52.3	54.4
Aviation Fuels	6.2	5.9	6.9	7.9	7.6	8.7	9.6	12.1	14.1
Kerosene	3.8	4.7	5.6	5.9	5.8	5.7	6.8	6.8	6.2
Gas Diesel	46.0	47.0	48.0	54.5	59.5	65.7	71.2	81.1	79.9
Heavy Fuel Oil	20.2	20.5	22.5	24.7	26.3	29.3	30.8	32.0	24.6
Other	0.6	0.6	0.6	0.6	-	-	-	-	-
Refinery Fuel	8.2	8.6	9.5	9.4	10.0	11.0	14.1	11.3	17.0
Sub-Total	136.0	140.8	152.0	166.2	177.9	195.3	212.5	226.8	229.9
Bunkers	3.3	3.4	4.1	5.4	5.8	7.0	4.6	5.6	8.0
Total	139.3	144.2	156.2	171.6	183.6	202.3	217.1	232.5	238.0

[1] See Notes in Part I.4.

Oil Demand by Main Product Groups[1]
Demande de produits raffinés par groupes principaux

People's Republic of China

	1971	1973	1978	1984	1985	1986	1987	1988	1989
Thousand tonnes									
NGL/LPG/Ethane	-	-	800	1 530	1 597	2 020	2 153	2 318	2 515
Naphtha	-	-	1 040	3 240	4 920	3 830	5 560	7 020	7 915
Motor Gasoline	4 465	6 110	9 562	12 193	13 996	15 002	16 241	17 965	18 568
Aviation Fuels	-	-	-	578	632	600	747	756	797
Kerosene	3 000	3 000	3 335	3 314	3 282	3 324	3 139	2 895	2 885
Gas Diesel	9 445	12 580	17 335	17 527	19 323	21 120	22 974	25 562	27 330
Heavy Fuel Oil	11 055	15 120	24 418	26 569	25 848	27 219	28 195	29 181	29 310
Other	9 392	12 782	30 895	18 585	17 045	18 748	18 665	18 126	17 644
Refinery Fuel	1 600	1 800	4 500	3 251	4 707	5 295	5 541	5 787	7 166
Sub-Total	38 957	51 392	91 885	86 787	91 350	97 158	103 215	109 610	114 130
Bunkers	-	-	-	-	-	-	-	-	-
Total	38 957	51 392	91 885	86 787	91 350	97 158	103 215	109 610	114 130
Thousand barrels/day									
NGL/LPG/Ethane	-	-	25.4	48.5	50.8	64.2	68.4	73.5	79.9
Naphtha	-	-	24.2	75.2	114.6	89.2	129.5	163.0	184.3
Motor Gasoline	104.3	142.8	223.5	284.2	327.1	350.6	379.6	418.7	433.9
Aviation Fuels	-	-	-	12.7	13.9	13.2	16.4	16.6	17.5
Kerosene	63.6	63.6	70.7	70.1	69.6	70.5	66.6	61.2	61.2
Gas Diesel	193.0	257.1	354.3	357.2	394.9	431.7	469.6	521.0	558.6
Heavy Fuel Oil	201.7	275.9	445.5	483.5	471.6	496.7	514.5	531.0	534.8
Other	181.6	249.0	609.8	355.7	326.9	359.3	357.5	346.4	338.0
Refinery Fuel	31.4	35.8	85.8	62.9	93.9	105.5	110.4	114.8	140.5
Sub-Total	775.7	1 024.2	1 839.3	1 749.9	1 863.2	1 980.8	2 112.3	2 246.2	2 348.7
Bunkers	-	-	-	-	-	-	-	-	-
Total	775.7	1 024.2	1 839.3	1 749.9	1 863.2	1 980.8	2 112.3	2 246.2	2 348.7

	1990	1991	1992	1993	1994	1995	1996	1997	1998
Thousand tonnes									
NGL/LPG/Ethane	2 542	3 009	3 456	4 234	4 807	6 773	8 317	8 801	11 068
Naphtha	8 432	9 631	9 874	11 487	12 674	14 447	15 892	17 439	17 412
Motor Gasoline	18 841	22 095	24 457	30 661	26 746	28 861	31 824	33 119	33 287
Aviation Fuels	1 011	1 402	1 489	1 761	2 000	2 751	3 350	4 543	4 390
Kerosene	2 574	2 509	2 507	2 479	2 518	2 693	2 546	2 512	2 704
Gas Diesel	26 785	29 782	31 875	38 034	36 566	42 867	46 594	53 940	52 485
Heavy Fuel Oil	29 336	29 305	30 956	33 965	31 518	33 114	35 514	36 019	34 859
Other	15 864	14 706	15 245	13 241	20 067	17 276	22 768	30 851	30 777
Refinery Fuel	7 152	7 134	7 880	7 045	6 322	7 763	8 143	10 027	8 564
Sub-Total	112 537	119 573	127 739	142 907	143 218	156 545	174 948	197 251	195 546
Bunkers	-	-	-	2 631	2 631	797	692	678	357
Total	112 537	119 573	127 739	145 538	145 849	157 342	175 640	197 929	195 903
Thousand barrels/day									
NGL/LPG/Ethane	80.8	95.6	109.5	134.6	152.8	215.3	263.6	279.7	351.8
Naphtha	196.4	224.3	229.3	267.5	295.1	336.4	369.1	406.1	405.5
Motor Gasoline	440.3	516.4	570.0	716.5	625.1	674.5	741.7	774.0	777.9
Aviation Fuels	22.2	30.7	32.5	38.3	43.5	60.0	72.8	98.9	95.7
Kerosene	54.6	53.2	53.0	52.6	53.4	57.1	53.8	53.3	57.3
Gas Diesel	547.4	608.7	649.7	777.4	747.3	876.1	949.7	1 102.4	1 072.7
Heavy Fuel Oil	535.3	534.7	563.3	619.7	575.1	604.2	646.2	657.2	636.1
Other	302.5	275.5	284.4	263.0	380.3	321.7	445.9	604.0	603.2
Refinery Fuel	140.8	140.4	155.0	150.5	136.0	162.2	174.1	216.6	180.8
Sub-Total	2 320.2	2 479.5	2 646.8	3 020.0	3 008.6	3 307.6	3 717.0	4 192.2	4 181.0
Bunkers	-	-	-	50.6	50.6	15.4	13.4	13.1	7.0
Total	2 320.2	2 479.5	2 646.8	3 070.6	3 059.2	3 323.0	3 730.4	4 205.3	4 187.9

[1] See Notes in Part I.4.

Oil Demand by Main Product Groups[1]
Demande de produits raffinés par groupes principaux

Chinese Taipei

	1971	1973	1978	1984	1985	1986	1987	1988	1989
Thousand tonnes									
NGL/LPG/Ethane	167	336	593	848	909	1 043	1 115	1 217	1 295
Naphtha	-	-	1 057	2 079	2 412	2 731	2 654	2 724	2 534
Motor Gasoline	488	687	1 221	1 999	2 084	2 269	2 631	3 116	3 631
Aviation Fuels	469	879	644	500	553	575	661	654	783
Kerosene	25	28	32	63	80	106	88	91	70
Gas Diesel	627	974	2 013	2 555	2 595	2 786	3 007	3 226	3 549
Heavy Fuel Oil	4 141	6 138	10 712	7 118	5 853	6 682	6 592	8 857	10 078
Other	415	470	494	672	998	1 027	1 233	1 317	1 585
Refinery Fuel	327	420	960	1 708	1 775	1 878	1 873	1 745	1 666
Sub-Total	6 659	9 932	17 726	17 542	17 259	19 097	19 854	22 947	25 191
Bunkers	123	401	554	583	548	893	983	1 259	1 660
Total	6 782	10 333	18 280	18 125	17 807	19 990	20 837	24 206	26 851
Thousand barrels/day									
NGL/LPG/Ethane	5.3	10.7	18.8	26.1	28.0	32.1	34.4	37.4	39.9
Naphtha	-	-	24.6	48.3	56.2	63.6	61.8	63.3	59.0
Motor Gasoline	11.4	16.1	28.5	46.6	48.7	53.0	61.5	72.6	84.9
Aviation Fuels	10.3	19.2	14.1	10.9	12.1	12.6	14.4	14.2	17.0
Kerosene	0.5	0.6	0.7	1.3	1.7	2.2	1.9	1.9	1.5
Gas Diesel	12.8	19.9	41.1	52.1	53.0	56.9	61.5	65.8	72.5
Heavy Fuel Oil	75.6	112.0	195.5	129.5	106.8	121.9	120.3	161.2	183.9
Other	8.0	9.0	8.6	12.1	18.2	18.8	22.8	24.2	29.2
Refinery Fuel	6.1	7.8	17.7	34.4	36.0	38.0	37.8	35.1	33.7
Sub-Total	129.9	195.2	349.6	361.2	360.7	399.2	416.3	475.6	521.6
Bunkers	2.5	8.2	11.3	10.8	10.1	16.4	18.1	23.0	30.6
Total	132.4	203.4	360.9	372.0	370.8	415.7	434.4	498.7	552.2

	1990	1991	1992	1993	1994	1995	1996	1997	1998
Thousand tonnes									
NGL/LPG/Ethane	1 331	1 312	1 388	1 401	1 461	1 497	1 526	1 556	1 533
Naphtha	2 542	2 330	2 293	2 426	3 177	3 132	3 277	3 524	3 335
Motor Gasoline	4 099	4 404	4 875	5 251	5 620	5 956	6 267	6 386	6 708
Aviation Fuels	901	916	1 147	1 151	1 462	1 704	1 840	1 874	2 002
Kerosene	39	32	29	39	43	31	40	29	31
Gas Diesel	3 664	3 856	3 927	4 243	4 639	4 659	4 513	4 147	4 380
Heavy Fuel Oil	10 355	11 229	10 727	11 488	11 423	12 673	12 153	12 430	12 385
Other	1 644	1 685	1 694	1 817	1 927	1 974	2 195	2 377	2 824
Refinery Fuel	1 517	1 101	1 075	1 203	1 405	1 494	1 544	1 589	1 557
Sub-Total	26 092	26 865	27 155	29 019	31 157	33 120	33 355	33 912	34 755
Bunkers	1 575	1 490	2 048	1 951	2 148	2 452	2 373	2 830	3 131
Total	27 667	28 355	29 203	30 970	33 305	35 572	35 728	36 742	37 886
Thousand barrels/day									
NGL/LPG/Ethane	41.0	40.4	42.7	43.2	45.4	46.5	47.2	48.3	47.6
Naphtha	59.2	54.3	53.3	56.5	74.0	72.9	76.1	82.1	77.7
Motor Gasoline	94.7	101.7	112.3	121.3	129.8	137.6	144.3	147.5	154.9
Aviation Fuels	19.6	19.9	24.9	25.0	31.8	37.0	39.9	40.7	43.5
Kerosene	0.8	0.7	0.6	0.8	0.9	0.7	0.8	0.6	0.7
Gas Diesel	76.7	80.7	81.9	88.8	97.1	97.5	94.2	86.8	91.6
Heavy Fuel Oil	187.5	203.4	193.7	208.0	206.9	229.5	219.5	225.1	224.3
Other	30.7	31.4	31.3	33.5	35.7	36.6	40.6	44.2	52.6
Refinery Fuel	31.3	22.9	22.3	25.1	29.2	30.9	32.0	33.0	32.2
Sub-Total	541.4	555.3	562.9	602.2	650.6	689.0	694.7	708.2	725.1
Bunkers	28.8	27.2	37.3	35.7	39.2	44.8	43.6	51.8	57.3
Total	570.2	582.6	600.2	637.9	689.8	733.8	738.3	760.0	782.4

[1] See Notes in Part I.4.

Oil Demand by Main Product Groups[1]
Demande de produits raffinés par groupes principaux

Colombia

	1971	1973	1978	1984	1985	1986	1987	1988	1989
Thousand tonnes									
NGL/LPG/Ethane	270	338	224	389	393	411	365	379	416
Naphtha	246	222	185	215	210	210	210	190	230
Motor Gasoline	1 987	2 254	3 065	3 732	3 861	3 925	4 159	4 448	4 734
Aviation Fuels	247	284	432	452	445	405	433	497	488
Kerosene	433	410	375	220	225	206	285	231	195
Gas Diesel	888	905	1 023	1 359	1 381	1 365	1 544	1 617	1 661
Heavy Fuel Oil	1 448	1 549	982	281	195	151	152	164	119
Other	260	227	191	613	631	684	715	871	909
Refinery Fuel	302	250	355	360	421	306	230	273	261
Sub-Total	6 081	6 439	6 832	7 621	7 762	7 663	8 093	8 670	9 013
Bunkers	300	250	85	61	70	60	96	45	48
Total	6 381	6 689	6 917	7 682	7 832	7 723	8 189	8 715	9 061
Thousand barrels/day									
NGL/LPG/Ethane	8.6	10.7	7.1	12.3	12.5	13.1	11.6	12.0	13.2
Naphtha	5.7	5.2	4.3	5.0	4.9	4.9	4.9	4.4	5.4
Motor Gasoline	46.4	52.7	71.6	87.0	90.2	91.7	97.2	103.7	110.6
Aviation Fuels	5.5	6.3	9.5	9.9	9.7	8.9	9.5	10.9	10.7
Kerosene	9.2	8.7	8.0	4.7	4.8	4.4	6.0	4.9	4.1
Gas Diesel	18.1	18.5	20.9	27.7	28.2	27.9	31.6	33.0	33.9
Heavy Fuel Oil	26.4	28.3	17.9	5.1	3.6	2.8	2.8	3.0	2.2
Other	5.0	4.4	3.7	11.8	12.2	13.2	13.8	16.8	17.6
Refinery Fuel	6.2	5.1	7.4	7.7	9.3	6.5	4.8	5.7	5.5
Sub-Total	131.3	139.8	150.4	171.2	175.4	173.3	182.2	194.2	203.3
Bunkers	5.7	4.7	1.6	1.2	1.4	1.2	1.9	0.9	1.0
Total	136.9	144.6	152.0	172.3	176.8	174.5	184.0	195.1	204.2

	1990	1991	1992	1993	1994	1995	1996	1997	1998
Thousand tonnes									
NGL/LPG/Ethane	419	438	526	556	544	545	642	680	734
Naphtha	250	-	200	-	-	-	-	-	-
Motor Gasoline	4 682	4 809	5 176	5 131	5 334	5 546	5 562	5 758	5 730
Aviation Fuels	521	517	578	567	676	708	694	729	722
Kerosene	269	242	274	284	155	163	158	161	119
Gas Diesel	1 759	1 877	1 794	1 891	2 180	2 315	2 604	2 753	2 788
Heavy Fuel Oil	128	221	251	167	117	238	159	219	170
Other	987	1 001	1 126	2 073	2 308	2 633	2 683	2 870	2 661
Refinery Fuel	258	296	391	623	563	646	680	538	590
Sub-Total	9 273	9 401	10 316	11 292	11 877	12 794	13 182	13 708	13 514
Bunkers	104	119	143	119	148	183	189	205	201
Total	9 377	9 520	10 459	11 411	12 025	12 977	13 371	13 913	13 715
Thousand barrels/day									
NGL/LPG/Ethane	13.3	13.9	16.7	17.7	17.3	17.3	20.3	21.6	23.3
Naphtha	5.8	-	4.6	-	-	-	-	-	-
Motor Gasoline	109.4	112.4	120.6	119.9	124.7	129.6	129.6	134.6	133.9
Aviation Fuels	11.4	11.3	12.6	12.4	14.7	15.4	15.1	15.9	15.7
Kerosene	5.7	5.1	5.8	6.0	3.3	3.5	3.3	3.4	2.5
Gas Diesel	36.0	38.4	36.6	38.6	44.6	47.3	53.1	56.3	57.0
Heavy Fuel Oil	2.3	4.0	4.6	3.0	2.1	4.3	2.9	4.0	3.1
Other	19.1	19.3	21.7	39.9	44.4	50.7	51.5	55.3	51.2
Refinery Fuel	5.4	6.2	7.9	13.3	12.1	14.0	14.7	11.5	12.5
Sub-Total	208.4	210.6	231.1	250.9	263.2	282.1	290.5	302.5	299.3
Bunkers	2.1	2.4	2.8	2.4	3.0	3.7	3.8	4.1	4.1
Total	210.5	213.0	233.9	253.3	266.2	285.8	294.3	306.6	303.3

[1] See Notes in Part I.4.

Oil Demand by Main Product Groups[1]
Demande de produits raffinés par groupes principaux

Congo

	1971	1973	1978	1984	1985	1986	1987	1988	1989
Thousand tonnes									
NGL/LPG/Ethane	2	2	1	8	5	8	5	4	4
Naphtha	-	-	-	-	-	-	-	-	-
Motor Gasoline	26	28	35	52	53	58	61	51	55
Aviation Fuels	20	22	27	37	32	24	13	13	61
Kerosene	8	10	18	2	2	2	2	2	34
Gas Diesel	70	75	128	129	133	134	136	126	99
Heavy Fuel Oil	32	27	3	10	10	10	-	-	1
Other	-	-	-	-	12	8	5	5	11
Refinery Fuel	-	-	-	30	5	30	40	35	42
Sub-Total	158	164	212	268	252	274	262	236	307
Bunkers	-	-	3	8	9	10	-	-	-
Total	158	164	215	276	261	284	262	236	307
Thousand barrels/day									
NGL/LPG/Ethane	0.1	0.1	0.0	0.3	0.2	0.3	0.2	0.1	0.1
Naphtha	-	-	-	-	-	-	-	-	-
Motor Gasoline	0.6	0.7	0.8	1.2	1.2	1.4	1.4	1.2	1.3
Aviation Fuels	0.4	0.5	0.6	0.8	0.7	0.5	0.3	0.3	1.5
Kerosene	0.2	0.2	0.4	0.0	0.0	0.0	0.0	0.0	0.7
Gas Diesel	1.4	1.5	2.6	2.6	2.7	2.7	2.8	2.6	2.0
Heavy Fuel Oil	0.6	0.5	0.1	0.2	0.2	0.2	-	-	0.0
Other	-	-	-	-	0.2	0.2	0.1	0.1	0.2
Refinery Fuel	-	-	-	0.5	0.1	0.5	0.7	0.6	0.8
Sub-Total	3.3	3.4	4.5	5.7	5.4	5.8	5.5	4.9	6.6
Bunkers	-	-	0.1	0.1	0.2	0.2	-	-	-
Total	3.3	3.4	4.5	5.8	5.5	6.0	5.5	4.9	6.6

	1990	1991	1992	1993	1994	1995	1996	1997	1998
Thousand tonnes									
NGL/LPG/Ethane	4	5	4	4	4	3	3	3	3
Naphtha	-	-	-	-	-	-	-	-	-
Motor Gasoline	40	41	42	45	37	52	53	53	53
Aviation Fuels	66	65	64	65	66	67	69	69	69
Kerosene	34	34	35	34	34	49	48	48	48
Gas Diesel	98	103	80	84	84	80	82	82	61
Heavy Fuel Oil	2	2	5	4	4	4	4	4	4
Other	11	15	20	22	23	20	20	20	20
Refinery Fuel	41	41	42	42	40	38	38	38	-
Sub-Total	296	306	292	300	292	313	317	317	258
Bunkers	5	6	13	8	8	8	10	10	46
Total	301	312	305	308	300	321	327	327	304
Thousand barrels/day									
NGL/LPG/Ethane	0.1	0.2	0.1	0.1	0.1	0.1	0.1	0.1	0.1
Naphtha	-	-	-	-	-	-	-	-	-
Motor Gasoline	0.9	1.0	1.0	1.1	0.9	1.2	1.2	1.2	1.2
Aviation Fuels	1.6	1.6	1.5	1.6	1.6	1.6	1.6	1.7	1.7
Kerosene	0.7	0.7	0.7	0.7	0.7	1.0	1.0	1.0	1.0
Gas Diesel	2.0	2.1	1.6	1.7	1.7	1.6	1.7	1.7	1.2
Heavy Fuel Oil	0.0	0.0	0.1	0.1	0.1	0.1	0.1	0.1	0.1
Other	0.2	0.3	0.4	0.4	0.4	0.4	0.4	0.4	0.4
Refinery Fuel	0.7	0.7	0.8	0.8	0.7	0.7	0.7	0.7	-
Sub-Total	6.4	6.6	6.2	6.4	6.2	6.7	6.8	6.8	5.7
Bunkers	0.1	0.1	0.2	0.1	0.1	0.1	0.2	0.2	0.8
Total	6.4	6.7	6.5	6.6	6.4	6.9	7.0	7.0	6.5

[1] See Notes in Part I.4.

Oil Demand by Main Product Groups[1]
Demande de produits raffinés par groupes principaux

Democratic Republic of Congo

	1971	1973	1978	1984	1985	1986	1987	1988	1989
Thousand tonnes									
NGL/LPG/Ethane	1	9	8	10	10	10	11	11	11
Naphtha	-	-	-	-	-	-	-	-	-
Motor Gasoline	134	159	168	192	175	176	182	185	185
Aviation Fuels	88	91	126	120	126	120	124	130	133
Kerosene	69	72	29	85	75	62	88	84	85
Gas Diesel	297	300	339	395	360	267	338	340	345
Heavy Fuel Oil	48	77	116	195	180	180	196	245	208
Other	-	-	-	18	24	23	31	34	49
Refinery Fuel	28	30	8	16	12	8	12	16	14
Sub-Total	665	738	794	1 031	962	846	982	1 045	1 030
Bunkers	131	120	30	32	30	35	32	31	31
Total	796	858	824	1 063	992	881	1 014	1 076	1 061
Thousand barrels/day									
NGL/LPG/Ethane	0.0	0.3	0.3	0.3	0.3	0.3	0.4	0.3	0.4
Naphtha	-	-	-	-	-	-	-	-	-
Motor Gasoline	3.1	3.7	3.9	4.5	4.1	4.1	4.3	4.3	4.3
Aviation Fuels	1.9	2.0	2.8	2.6	2.8	2.6	2.7	2.8	2.9
Kerosene	1.5	1.5	0.6	1.8	1.6	1.3	1.9	1.8	1.8
Gas Diesel	6.1	6.1	6.9	8.1	7.4	5.5	6.9	6.9	7.1
Heavy Fuel Oil	0.9	1.4	2.1	3.5	3.3	3.3	3.6	4.5	3.8
Other	-	-	-	0.3	0.5	0.4	0.6	0.7	0.9
Refinery Fuel	0.6	0.6	0.2	0.3	0.2	0.2	0.2	0.3	0.3
Sub-Total	14.1	15.7	16.8	21.5	20.1	17.7	20.5	21.6	21.5
Bunkers	2.4	2.2	0.5	0.6	0.5	0.6	0.6	0.6	0.6
Total	16.5	17.9	17.3	22.1	20.7	18.4	21.1	22.2	22.0

	1990	1991	1992	1993	1994	1995	1996	1997	1998
Thousand tonnes									
NGL/LPG/Ethane	12	13	12	13	13	13	13	13	12
Naphtha	-	-	-	-	-	-	-	-	-
Motor Gasoline	207	208	193	220	226	235	235	235	222
Aviation Fuels	142	147	134	150	153	159	159	159	150
Kerosene	77	90	83	96	97	113	113	113	107
Gas Diesel	328	332	301	335	339	341	341	341	321
Heavy Fuel Oil	283	292	274	294	295	280	280	280	264
Other	49	44	42	57	60	59	59	59	55
Refinery Fuel	15	17	16	16	16	16	16	16	15
Sub-Total	1 113	1 143	1 055	1 181	1 199	1 216	1 216	1 216	1 146
Bunkers	31	35	29	32	34	35	35	35	33
Total	1 144	1 178	1 084	1 213	1 233	1 251	1 251	1 251	1 179
Thousand barrels/day									
NGL/LPG/Ethane	0.4	0.4	0.4	0.4	0.4	0.4	0.4	0.4	0.4
Naphtha	-	-	-	-	-	-	-	-	-
Motor Gasoline	4.8	4.9	4.5	5.1	5.3	5.5	5.5	5.5	5.2
Aviation Fuels	3.1	3.2	2.9	3.3	3.4	3.5	3.5	3.5	3.3
Kerosene	1.6	1.9	1.8	2.0	2.1	2.4	2.4	2.4	2.3
Gas Diesel	6.7	6.8	6.1	6.8	6.9	7.0	7.0	7.0	6.6
Heavy Fuel Oil	5.2	5.3	5.0	5.4	5.4	5.1	5.1	5.1	4.8
Other	0.9	0.8	0.8	1.1	1.2	1.1	1.1	1.1	1.1
Refinery Fuel	0.3	0.3	0.3	0.3	0.3	0.3	0.3	0.3	0.3
Sub-Total	23.1	23.7	21.8	24.5	24.9	25.3	25.3	25.3	23.9
Bunkers	0.6	0.6	0.5	0.6	0.6	0.6	0.6	0.6	0.6
Total	23.7	24.4	22.3	25.1	25.5	26.0	25.9	26.0	24.5

[1] See Notes in Part I.4.

Oil Demand by Main Product Groups[1]
Demande de produits raffinés par groupes principaux

Costa Rica

	1971	1973	1978	1984	1985	1986	1987	1988	1989
Thousand tonnes									
NGL/LPG/Ethane	7	10	17	15	15	16	17	20	23
Naphtha	-	-	-	-	-	-	-	-	-
Motor Gasoline	97	126	155	126	128	140	168	181	206
Aviation Fuels	7	9	19	12	13	21	26	33	36
Kerosene	21	21	25	15	14	13	13	10	9
Gas Diesel	179	232	398	285	294	319	335	366	386
Heavy Fuel Oil	70	97	162	111	134	118	123	111	167
Other	-	-	19	14	25	13	18	12	14
Refinery Fuel	29	30	16	31	30	44	39	41	42
Sub-Total	410	525	811	609	653	684	739	774	883
Bunkers	-	-	-	-	-	-	-	-	-
Total	410	525	811	609	653	684	739	774	883
Thousand barrels/day									
NGL/LPG/Ethane	0.2	0.3	0.5	0.5	0.5	0.5	0.5	0.6	0.7
Naphtha	-	-	-	-	-	-	-	-	-
Motor Gasoline	2.3	2.9	3.6	2.9	3.0	3.3	3.9	4.2	4.8
Aviation Fuels	0.2	0.2	0.4	0.3	0.3	0.5	0.6	0.7	0.8
Kerosene	0.4	0.4	0.5	0.3	0.3	0.3	0.3	0.2	0.2
Gas Diesel	3.7	4.7	8.1	5.8	6.0	6.5	6.8	7.5	7.9
Heavy Fuel Oil	1.3	1.8	3.0	2.0	2.4	2.2	2.2	2.0	3.0
Other	-	-	0.4	0.3	0.5	0.2	0.3	0.2	0.3
Refinery Fuel	0.5	0.5	0.3	0.6	0.6	0.8	0.7	0.8	0.8
Sub-Total	8.6	11.0	16.9	12.7	13.5	14.3	15.5	16.3	18.5
Bunkers	-	-	-	-	-	-	-	-	-
Total	8.6	11.0	16.9	12.7	13.5	14.3	15.5	16.3	18.5

	1990	1991	1992	1993	1994	1995	1996	1997	1998
Thousand tonnes									
NGL/LPG/Ethane	23	26	28	30	35	40	46	50	58
Naphtha	-	-	12	6	5	4	2	2	2
Motor Gasoline	208	226	441	428	430	423	437	447	502
Aviation Fuels	42	41	62	73	95	100	94	103	122
Kerosene	8	9	7	3	8	8	7	8	8
Gas Diesel	386	429	533	548	724	714	627	583	790
Heavy Fuel Oil	163	166	198	194	199	206	179	192	207
Other	8	2	14	18	39	33	26	31	37
Refinery Fuel	38	20	25	20	25	31	30	30	12
Sub-Total	876	919	1 320	1 320	1 560	1 559	1 448	1 446	1 738
Bunkers	-	-	-	-	-	-	-	-	-
Total	876	919	1 320	1 320	1 560	1 559	1 448	1 446	1 738
Thousand barrels/day									
NGL/LPG/Ethane	0.7	0.8	0.9	1.0	1.1	1.3	1.5	1.6	1.8
Naphtha	-	-	0.3	0.1	0.1	0.1	0.0	0.0	0.0
Motor Gasoline	4.9	5.3	10.3	10.0	10.0	9.9	10.2	10.4	11.7
Aviation Fuels	0.9	0.9	1.4	1.6	2.1	2.2	2.0	2.2	2.7
Kerosene	0.2	0.2	0.1	0.1	0.2	0.2	0.1	0.2	0.2
Gas Diesel	7.9	8.8	10.9	11.2	14.8	14.6	12.8	11.9	16.1
Heavy Fuel Oil	3.0	3.0	3.6	3.5	3.6	3.8	3.3	3.5	3.8
Other	0.2	0.0	0.2	0.3	0.7	0.6	0.5	0.5	0.6
Refinery Fuel	0.7	0.4	0.5	0.4	0.5	0.6	0.5	0.5	0.2
Sub-Total	18.4	19.4	28.1	28.2	33.1	33.1	30.9	31.0	37.2
Bunkers	-	-	-	-	-	-	-	-	-
Total	18.4	19.4	28.1	28.2	33.1	33.1	30.9	31.0	37.2

[1] See Notes in Part I.4.

Oil Demand by Main Product Groups[1]
Demande de produits raffinés par groupes principaux

Cote d'Ivoire

	1971	1973	1978	1984	1985	1986	1987	1988	1989
Thousand tonnes									
NGL/LPG/Ethane	6	7	11	23	25	19	18	22	22
Naphtha	-	-	-	-	-	-	-	-	-
Motor Gasoline	160	177	246	200	201	201	220	187	181
Aviation Fuels	42	50	78	87	91	87	90	110	104
Kerosene	39	39	49	70	72	68	64	64	61
Gas Diesel	257	268	410	376	372	406	466	488	467
Heavy Fuel Oil	282	331	483	318	199	174	105	95	91
Other	24	32	63	168	80	157	200	149	149
Refinery Fuel	21	43	72	78	104	94	72	69	67
Sub-Total	831	947	1 412	1 320	1 144	1 206	1 235	1 184	1 142
Bunkers	20	15	2	196	233	130	130	130	130
Total	851	962	1 414	1 516	1 377	1 336	1 365	1 314	1 272
Thousand barrels/day									
NGL/LPG/Ethane	0.2	0.2	0.4	0.7	0.8	0.6	0.6	0.7	0.7
Naphtha	-	-	-	-	-	-	-	-	-
Motor Gasoline	3.7	4.1	5.7	4.7	4.7	4.7	5.1	4.4	4.2
Aviation Fuels	0.9	1.1	1.7	1.9	2.0	1.9	2.0	2.4	2.3
Kerosene	0.8	0.8	1.0	1.5	1.5	1.4	1.4	1.4	1.3
Gas Diesel	5.3	5.5	8.4	7.7	7.6	8.3	9.5	9.9	9.5
Heavy Fuel Oil	5.1	6.0	8.8	5.8	3.6	3.2	1.9	1.7	1.7
Other	0.5	0.6	1.2	3.2	1.5	3.0	3.8	2.9	2.9
Refinery Fuel	0.4	0.9	1.4	1.5	2.1	1.9	1.5	1.4	1.4
Sub-Total	16.9	19.3	28.7	27.0	23.8	25.0	25.8	24.7	23.9
Bunkers	0.4	0.3	0.0	3.7	4.5	2.4	2.4	2.4	2.4
Total	17.3	19.5	28.7	30.7	28.3	27.4	28.1	27.1	26.3

	1990	1991	1992	1993	1994	1995	1996	1997	1998
Thousand tonnes									
NGL/LPG/Ethane	22	27	27	21	41	39	63	59	75
Naphtha	-	-	-	-	-	-	-	-	-
Motor Gasoline	170	155	146	138	155	154	166	163	163
Aviation Fuels	85	87	85	82	85	82	82	82	82
Kerosene	56	59	56	51	53	59	59	59	59
Gas Diesel	453	479	461	447	467	535	461	515	564
Heavy Fuel Oil	69	76	70	67	83	94	47	42	48
Other	66	73	64	59	64	65	65	73	73
Refinery Fuel	61	63	57	66	72	77	69	67	73
Sub-Total	982	1 019	966	931	1 020	1 105	1 012	1 060	1 137
Bunkers	39	87	102	89	108	87	89	89	89
Total	1 021	1 106	1 068	1 020	1 128	1 192	1 101	1 149	1 226
Thousand barrels/day									
NGL/LPG/Ethane	0.7	0.9	0.9	0.7	1.3	1.2	2.0	1.9	2.4
Naphtha	-	-	-	-	-	-	-	-	-
Motor Gasoline	4.0	3.6	3.4	3.2	3.6	3.6	3.9	3.8	3.8
Aviation Fuels	1.8	1.9	1.8	1.8	1.8	1.8	1.8	1.8	1.8
Kerosene	1.2	1.3	1.2	1.1	1.1	1.3	1.2	1.3	1.3
Gas Diesel	9.3	9.8	9.4	9.1	9.5	10.9	9.4	10.5	11.5
Heavy Fuel Oil	1.3	1.4	1.3	1.2	1.5	1.7	0.9	0.8	0.9
Other	1.3	1.4	1.2	1.1	1.2	1.3	1.2	1.4	1.4
Refinery Fuel	1.3	1.3	1.2	1.4	1.5	1.6	1.5	1.4	1.5
Sub-Total	20.8	21.5	20.4	19.6	21.7	23.4	21.8	22.8	24.6
Bunkers	0.7	1.6	1.9	1.7	2.0	1.6	1.7	1.7	1.7
Total	21.5	23.1	22.2	21.3	23.7	25.0	23.5	24.5	26.2

[1] See Notes in Part I.4.

Oil Demand by Main Product Groups[1]
Demande de produits raffinés par groupes principaux

Croatia

	1971	1973	1978	1984	1985	1986	1987	1988	1989
Thousand tonnes									
NGL/LPG/Ethane	-	-	-	-	-	-	-	-	-
Naphtha	-	-	-	-	-	-	-	-	-
Motor Gasoline	-	-	-	-	-	-	-	-	-
Aviation Fuels	-	-	-	-	-	-	-	-	-
Kerosene	-	-	-	-	-	-	-	-	-
Gas Diesel	-	-	-	-	-	-	-	-	-
Heavy Fuel Oil	-	-	-	-	-	-	-	-	-
Other	-	-	-	-	-	-	-	-	-
Refinery Fuel	-	-	-	-	-	-	-	-	-
Sub-Total	-	-	-	-	-	-	-	-	-
Bunkers	-	-	-	-	-	-	-	-	-
Total	-	-	-	-	-	-	-	-	-
Thousand barrels/day									
NGL/LPG/Ethane	-	-	-	-	-	-	-	-	-
Naphtha	-	-	-	-	-	-	-	-	-
Motor Gasoline	-	-	-	-	-	-	-	-	-
Aviation Fuels	-	-	-	-	-	-	-	-	-
Kerosene	-	-	-	-	-	-	-	-	-
Gas Diesel	-	-	-	-	-	-	-	-	-
Heavy Fuel Oil	-	-	-	-	-	-	-	-	-
Other	-	-	-	-	-	-	-	-	-
Refinery Fuel	-	-	-	-	-	-	-	-	-
Sub-Total	-	-	-	-	-	-	-	-	-
Bunkers	-	-	-	-	-	-	-	-	-
Total	-	-	-	-	-	-	-	-	-

	1990	1991	1992	1993	1994	1995	1996	1997	1998
Thousand tonnes									
NGL/LPG/Ethane	-	184	164	190	194	206	179	213	194
Naphtha	-	115	129	142	162	153	133	112	108
Motor Gasoline	-	591	511	497	545	575	600	678	737
Aviation Fuels	-	32	25	63	90	85	78	82	88
Kerosene	-	4	2	10	1	8	7	6	2
Gas Diesel	-	953	909	928	986	1 061	1 023	1 190	1 207
Heavy Fuel Oil	-	858	954	963	804	1 032	998	1 039	1 261
Other	-	215	153	158	214	208	181	331	312
Refinery Fuel	-	498	395	476	490	558	485	479	666
Sub-Total	-	3 450	3 242	3 427	3 486	3 886	3 684	4 130	4 575
Bunkers	-	-	-	-	45	33	29	24	26
Total	-	3 450	3 242	3 427	3 531	3 919	3 713	4 154	4 601
Thousand barrels/day									
NGL/LPG/Ethane	-	6.9	6.0	7.3	7.5	8.0	6.9	8.1	7.4
Naphtha	-	2.7	3.0	3.3	3.8	3.6	3.1	2.6	2.5
Motor Gasoline	-	13.8	11.9	11.6	12.7	13.4	14.0	15.8	17.2
Aviation Fuels	-	0.7	0.5	1.4	2.0	1.8	1.7	1.8	1.9
Kerosene	-	0.1	0.0	0.2	0.0	0.2	0.1	0.1	0.0
Gas Diesel	-	19.5	18.5	19.0	20.2	21.7	20.9	24.3	24.7
Heavy Fuel Oil	-	15.7	17.4	17.6	14.7	18.8	18.2	19.0	23.0
Other	-	4.0	2.8	3.0	4.1	4.0	3.5	6.2	5.7
Refinery Fuel	-	9.8	7.6	9.3	9.9	11.3	9.8	9.5	13.2
Sub-Total	-	73.0	67.7	72.6	74.8	82.8	78.1	87.4	95.7
Bunkers	-	-	-	-	0.9	0.6	0.6	0.5	0.5
Total	-	73.0	67.7	72.6	75.7	83.5	78.7	87.8	96.2

[1] See Notes in Part I.4.

Oil Demand by Main Product Groups[1]
Demande de produits raffinés par groupes principaux

Cuba

	1971	1973	1978	1984	1985	1986	1987	1988	1989
Thousand tonnes									
NGL/LPG/Ethane	58	67	102	121	124	127	126	133	138
Naphtha	8	32	76	205	202	207	186	184	194
Motor Gasoline	951	1 093	1 333	1 371	1 241	1 263	1 250	1 319	1 393
Aviation Fuels	88	111	306	204	214	182	220	231	204
Kerosene	399	461	447	657	662	669	632	653	652
Gas Diesel	1 145	2 167	1 714	2 239	2 191	2 184	2 533	2 364	2 485
Heavy Fuel Oil	3 427	3 581	4 893	5 079	5 313	5 372	5 278	5 747	6 009
Other	100	161	240	276	476	603	722	712	683
Refinery Fuel	157	169	200	329	243	267	258	273	272
Sub-Total	6 333	7 842	9 311	10 481	10 666	10 874	11 205	11 616	12 030
Bunkers	-	-	-	190	219	228	96	122	108
Total	6 333	7 842	9 311	10 671	10 885	11 102	11 301	11 738	12 138
Thousand barrels/day									
NGL/LPG/Ethane	1.8	2.1	3.2	3.8	3.9	4.0	4.0	4.2	4.4
Naphtha	0.2	0.7	1.8	4.8	4.7	4.8	4.3	4.3	4.5
Motor Gasoline	22.0	25.3	30.9	31.6	28.7	29.2	28.8	30.4	32.2
Aviation Fuels	1.9	2.4	6.7	4.5	4.7	4.0	4.8	5.0	4.5
Kerosene	8.5	9.8	9.5	13.9	14.0	14.2	13.4	13.8	13.8
Gas Diesel	23.4	44.3	35.0	45.6	44.8	44.6	51.8	48.2	50.8
Heavy Fuel Oil	62.5	65.3	89.3	92.4	96.9	98.0	96.3	104.6	109.6
Other	1.7	2.7	4.0	4.6	8.3	10.6	12.7	12.5	12.0
Refinery Fuel	3.0	3.2	3.8	6.2	4.7	5.1	5.0	5.2	5.2
Sub-Total	125.1	156.0	184.2	207.4	210.8	214.5	221.1	228.2	237.0
Bunkers	-	-	-	3.5	4.1	4.2	1.8	2.3	2.0
Total	125.1	156.0	184.2	210.9	214.9	218.8	222.9	230.4	239.0

	1990	1991	1992	1993	1994	1995	1996	1997	1998
Thousand tonnes									
NGL/LPG/Ethane	130	114	94	64	75	89	88	88	96
Naphtha	341	329	372	367	351	428	400	380	373
Motor Gasoline	1 169	776	546	411	419	432	475	497	475
Aviation Fuels	323	285	208	199	168	176	268	303	277
Kerosene	640	620	438	310	260	237	222	193	153
Gas Diesel	2 354	1 818	1 523	1 299	1 436	1 511	1 653	1 729	1 611
Heavy Fuel Oil	5 429	4 367	3 982	3 975	4 202	4 666	4 879	5 156	5 094
Other	572	363	766	1 115	1 359	1 298	1 386	1 360	1 389
Refinery Fuel	91	75	39	27	23	38	37	41	44
Sub-Total	11 049	8 747	7 968	7 767	8 293	8 875	9 408	9 747	9 512
Bunkers	244	230	167	100	104	84	104	113	115
Total	11 293	8 977	8 135	7 867	8 397	8 959	9 512	9 860	9 627
Thousand barrels/day									
NGL/LPG/Ethane	4.1	3.6	3.0	2.0	2.4	2.8	2.8	2.8	3.1
Naphtha	7.9	7.7	8.6	8.5	8.2	10.0	9.3	8.8	8.7
Motor Gasoline	27.0	17.9	12.5	9.4	9.6	9.9	10.9	11.4	11.0
Aviation Fuels	7.0	6.2	4.5	4.3	3.7	3.8	5.8	6.6	6.0
Kerosene	13.6	13.1	9.3	6.6	5.5	5.0	4.7	4.1	3.2
Gas Diesel	48.1	37.2	31.0	26.5	29.3	30.9	33.7	35.3	32.9
Heavy Fuel Oil	99.1	79.7	72.5	72.5	76.7	85.1	88.8	94.1	92.9
Other	10.1	6.4	13.5	19.8	24.1	23.0	24.5	24.1	24.6
Refinery Fuel	2.0	1.6	0.8	0.6	0.5	0.8	0.8	0.9	0.9
Sub-Total	219.0	173.4	155.8	150.3	159.9	171.4	181.2	188.2	183.4
Bunkers	4.5	4.2	3.1	1.9	1.9	1.6	1.9	2.1	2.1
Total	223.5	177.6	158.8	152.2	161.8	172.9	183.2	190.3	185.5

[1] See Notes in Part I.4.

Oil Demand by Main Product Groups[1]
Demande de produits raffinés par groupes principaux

Cyprus

	1971	1973	1978	1984	1985	1986	1987	1988	1989
Thousand tonnes									
NGL/LPG/Ethane	24	29	31	41	42	43	46	46	45
Naphtha	-	-	-	-	-	-	-	-	-
Motor Gasoline	108	137	100	118	124	130	140	147	154
Aviation Fuels	49	75	61	135	143	148	169	170	209
Kerosene	14	13	7	8	7	7	9	10	10
Gas Diesel	133	156	131	197	204	218	239	250	274
Heavy Fuel Oil	285	365	437	460	449	523	583	626	640
Other	17	30	26	38	31	37	41	40	44
Refinery Fuel	-	37	17	21	10	17	27	30	30
Sub-Total	630	842	810	1 018	1 010	1 123	1 254	1 319	1 406
Bunkers	2	10	13	37	34	36	68	68	86
Total	632	852	823	1 055	1 044	1 159	1 322	1 387	1 492
Thousand barrels/day									
NGL/LPG/Ethane	0.8	0.9	1.0	1.3	1.3	1.4	1.5	1.5	1.4
Naphtha	-	-	-	-	-	-	-	-	-
Motor Gasoline	2.5	3.2	2.3	2.8	2.9	3.0	3.3	3.4	3.6
Aviation Fuels	1.1	1.6	1.3	2.9	3.1	3.2	3.7	3.7	4.5
Kerosene	0.3	0.3	0.1	0.2	0.1	0.1	0.2	0.2	0.2
Gas Diesel	2.7	3.2	2.7	4.0	4.2	4.5	4.9	5.1	5.6
Heavy Fuel Oil	5.2	6.7	8.0	8.4	8.2	9.5	10.6	11.4	11.7
Other	0.3	0.5	0.5	0.7	0.6	0.7	0.7	0.7	0.7
Refinery Fuel	-	0.7	0.3	0.4	0.2	0.4	0.6	0.6	0.6
Sub-Total	12.9	17.2	16.3	20.6	20.6	22.8	25.4	26.6	28.4
Bunkers	0.0	0.2	0.3	0.7	0.7	0.7	1.3	1.3	1.6
Total	12.9	17.3	16.5	21.3	21.3	23.5	26.7	27.8	30.1

	1990	1991	1992	1993	1994	1995	1996	1997	1998
Thousand tonnes									
NGL/LPG/Ethane	49	49	55	51	50	51	51	52	50
Naphtha	-	-	-	-	-	-	-	-	-
Motor Gasoline	163	170	172	169	180	183	186	191	195
Aviation Fuels	236	280	272	231	237	260	249	245	258
Kerosene	12	12	17	16	17	17	18	20	21
Gas Diesel	308	311	378	392	404	446	465	489	526
Heavy Fuel Oil	577	685	763	797	837	759	814	813	879
Other	73	121	135	182	180	191	217	227	232
Refinery Fuel	29	29	30	26	38	30	28	30	31
Sub-Total	1 447	1 657	1 822	1 864	1 943	1 937	2 028	2 067	2 192
Bunkers	58	56	59	50	62	69	91	99	99
Total	1 505	1 713	1 881	1 914	2 005	2 006	2 119	2 166	2 291
Thousand barrels/day									
NGL/LPG/Ethane	1.6	1.6	1.7	1.6	1.6	1.6	1.6	1.7	1.6
Naphtha	-	-	-	-	-	-	-	-	-
Motor Gasoline	3.8	4.0	4.0	4.0	4.2	4.3	4.3	4.5	4.6
Aviation Fuels	5.1	6.1	5.9	5.0	5.1	5.6	5.4	5.3	5.6
Kerosene	0.3	0.3	0.4	0.3	0.4	0.4	0.4	0.4	0.4
Gas Diesel	6.3	6.4	7.7	8.0	8.3	9.1	9.5	10.0	10.8
Heavy Fuel Oil	10.5	12.5	13.9	14.5	15.3	13.8	14.8	14.8	16.0
Other	1.3	1.9	2.1	2.9	2.9	3.0	3.4	3.6	3.6
Refinery Fuel	0.6	0.6	0.6	0.5	0.8	0.6	0.6	0.6	0.6
Sub-Total	29.5	33.2	36.3	36.9	38.5	38.5	40.0	40.9	43.3
Bunkers	1.1	1.1	1.1	0.9	1.2	1.3	1.7	1.9	1.9
Total	30.6	34.3	37.4	37.8	39.6	39.8	41.7	42.7	45.1

[1] See Notes in Part I.4.

Oil Demand by Main Product Groups[1]
Demande de produits raffinés par groupes principaux

Dominican Republic

	1971	1973	1978	1984	1985	1986	1987	1988	1989
Thousand tonnes									
NGL/LPG/Ethane	35	46	60	100	141	105	114	126	128
Naphtha	-	-	-	-	-	-	-	-	-
Motor Gasoline	315	385	401	416	479	487	625	500	726
Aviation Fuels	25	26	38	82	52	84	92	107	32
Kerosene	19	14	14	9	15	25	28	25	147
Gas Diesel	229	173	383	488	346	465	605	677	660
Heavy Fuel Oil	516	859	995	1 229	835	1 148	1 017	1 184	1 085
Other	-	-	-	-	-	-	-	-	-
Refinery Fuel	-	-	-	-	-	-	-	-	-
Sub-Total	1 139	1 503	1 891	2 324	1 868	2 314	2 481	2 619	2 778
Bunkers	-	-	-	-	-	-	-	-	-
Total	1 139	1 503	1 891	2 324	1 868	2 314	2 481	2 619	2 778
Thousand barrels/day									
NGL/LPG/Ethane	1.1	1.5	1.9	3.2	4.5	3.3	3.6	4.0	4.1
Naphtha	-	-	-	-	-	-	-	-	-
Motor Gasoline	7.4	9.0	9.4	9.7	11.2	11.4	14.6	11.7	17.0
Aviation Fuels	0.5	0.6	0.8	1.8	1.1	1.8	2.0	2.3	0.7
Kerosene	0.4	0.3	0.3	0.2	0.3	0.5	0.6	0.5	3.1
Gas Diesel	4.7	3.5	7.8	9.9	7.1	9.5	12.4	13.8	13.5
Heavy Fuel Oil	9.4	15.7	18.2	22.4	15.2	20.9	18.6	21.5	19.8
Other	-	-	-	-	-	-	-	-	-
Refinery Fuel	-	-	-	-	-	-	-	-	-
Sub-Total	23.5	30.5	38.4	47.1	39.4	47.5	51.7	53.8	58.1
Bunkers	-	-	-	-	-	-	-	-	-
Total	23.5	30.5	38.4	47.1	39.4	47.5	51.7	53.8	58.1

	1990	1991	1992	1993	1994	1995	1996	1997	1998
Thousand tonnes									
NGL/LPG/Ethane	109	178	222	259	302	321	340	425	435
Naphtha	-	-	-	-	-	-	-	-	-
Motor Gasoline	692	638	781	574	586	597	609	621	632
Aviation Fuels	36	34	48	46	52	55	59	62	66
Kerosene	156	144	202	192	218	233	247	263	283
Gas Diesel	747	687	882	867	898	929	960	995	1 015
Heavy Fuel Oil	1 112	1 126	1 173	1 238	1 312	1 352	1 392	1 477	1 514
Other	-	-	-	-	-	-	-	-	-
Refinery Fuel	-	-	-	-	-	-	-	-	-
Sub-Total	2 852	2 807	3 308	3 176	3 368	3 487	3 607	3 843	3 945
Bunkers	-	-	-	-	-	-	-	-	-
Total	2 852	2 807	3 308	3 176	3 368	3 487	3 607	3 843	3 945
Thousand barrels/day									
NGL/LPG/Ethane	3.5	5.7	7.0	8.2	9.6	10.2	10.8	13.5	13.8
Naphtha	-	-	-	-	-	-	-	-	-
Motor Gasoline	16.2	14.9	18.2	13.4	13.7	14.0	14.2	14.5	14.8
Aviation Fuels	0.8	0.7	1.0	1.0	1.1	1.2	1.3	1.3	1.4
Kerosene	3.3	3.1	4.3	4.1	4.6	4.9	5.2	5.6	6.0
Gas Diesel	15.3	14.0	18.0	17.7	18.4	19.0	19.6	20.3	20.7
Heavy Fuel Oil	20.3	20.5	21.3	22.6	23.9	24.7	25.3	27.0	27.6
Other	-	-	-	-	-	-	-	-	-
Refinery Fuel	-	-	-	-	-	-	-	-	-
Sub-Total	59.3	58.9	69.9	67.0	71.3	73.9	76.4	82.2	84.4
Bunkers	-	-	-	-	-	-	-	-	-
Total	59.3	58.9	69.9	67.0	71.3	73.9	76.4	82.2	84.4

[1] See Notes in Part I.4.

Oil Demand by Main Product Groups[1]
Demande de produits raffinés par groupes principaux

Ecuador

	1971	1973	1978	1984	1985	1986	1987	1988	1989
Thousand tonnes									
NGL/LPG/Ethane	7	14	63	161	185	214	250	292	322
Naphtha	-	-	-	198	201	199	173	203	200
Motor Gasoline	427	495	974	957	975	1 025	981	1 266	1 285
Aviation Fuels	86	125	104	134	145	165	162	171	166
Kerosene	50	51	360	289	291	310	179	173	203
Gas Diesel	263	333	668	805	928	968	1 042	1 048	1 018
Heavy Fuel Oil	319	317	667	1 029	1 208	1 012	1 033	946	886
Other	32	45	126	224	251	277	276	313	262
Refinery Fuel	47	58	168	281	80	231	160	140	84
Sub-Total	1 231	1 438	3 130	4 078	4 264	4 401	4 256	4 552	4 426
Bunkers	90	100	7	44	37	43	109	100	139
Total	1 321	1 538	3 137	4 122	4 301	4 444	4 365	4 652	4 565
Thousand barrels/day									
NGL/LPG/Ethane	0.2	0.4	2.0	5.1	5.9	6.8	7.9	9.3	10.2
Naphtha	-	-	-	4.6	4.7	4.6	4.0	4.7	4.7
Motor Gasoline	10.0	11.6	22.8	22.3	22.8	24.0	22.9	29.5	30.0
Aviation Fuels	1.9	2.7	2.3	2.9	3.2	3.6	3.5	3.7	3.6
Kerosene	1.1	1.1	7.6	6.1	6.2	6.6	3.8	3.7	4.3
Gas Diesel	5.4	6.8	13.7	16.4	19.0	19.8	21.3	21.4	20.8
Heavy Fuel Oil	5.8	5.8	12.2	18.7	22.0	18.5	18.8	17.2	16.2
Other	0.6	0.9	2.3	4.2	4.7	5.1	5.1	5.7	4.7
Refinery Fuel	0.9	1.1	3.2	5.4	1.6	4.3	3.1	2.6	1.7
Sub-Total	25.9	30.4	65.9	85.7	90.0	93.3	90.6	97.7	96.2
Bunkers	1.6	1.9	0.1	0.8	0.7	0.8	2.0	1.8	2.5
Total	27.5	32.2	66.1	86.5	90.7	94.1	92.6	99.6	98.7

	1990	1991	1992	1993	1994	1995	1996	1997	1998
Thousand tonnes									
NGL/LPG/Ethane	338	385	417	449	483	553	576	633	636
Naphtha	200	210	-	-	-	-	-	-	-
Motor Gasoline	1 236	1 353	1 380	1 300	1 258	1 222	1 330	1 326	1 355
Aviation Fuels	185	180	177	196	177	182	215	219	158
Kerosene	181	129	108	52	46	30	30	30	30
Gas Diesel	1 187	1 159	1 318	1 285	1 537	1 870	1 991	2 075	2 161
Heavy Fuel Oil	960	1 048	1 014	886	883	1 123	1 043	1 067	1 209
Other	241	207	165	189	278	230	254	261	302
Refinery Fuel	100	104	78	56	55	70	74	33	35
Sub-Total	4 628	4 775	4 657	4 413	4 717	5 280	5 513	5 644	5 886
Bunkers	185	240	466	454	515	340	404	445	365
Total	4 813	5 015	5 123	4 867	5 232	5 620	5 917	6 089	6 251
Thousand barrels/day									
NGL/LPG/Ethane	10.7	12.2	13.2	14.3	15.4	17.6	18.3	20.1	20.2
Naphtha	4.7	4.9	-	-	-	-	-	-	-
Motor Gasoline	28.9	31.6	32.2	30.4	29.4	28.6	31.0	31.0	31.7
Aviation Fuels	4.0	3.9	3.9	4.3	3.9	4.0	4.7	4.8	3.5
Kerosene	3.8	2.7	2.3	1.1	1.0	0.6	0.6	0.6	0.6
Gas Diesel	24.3	23.7	26.9	26.3	31.4	38.2	40.6	42.4	44.2
Heavy Fuel Oil	17.5	19.1	18.5	16.2	16.1	20.5	19.0	19.5	22.1
Other	4.4	3.8	3.2	3.6	5.3	4.4	4.9	5.0	5.8
Refinery Fuel	2.0	2.0	1.5	1.1	1.1	1.5	1.4	0.7	0.7
Sub-Total	100.3	104.0	101.5	97.2	103.5	115.4	120.4	124.1	128.7
Bunkers	3.4	4.5	8.7	8.5	9.6	6.2	7.4	8.1	6.7
Total	103.7	108.6	110.2	105.7	113.1	121.6	127.8	132.2	135.4

[1] See Notes in Part I.4.

Oil Demand by Main Product Groups[1]
Demande de produits raffinés par groupes principaux

Egypt

	1971	1973	1978	1984	1985	1986	1987	1988	1989
Thousand tonnes									
NGL/LPG/Ethane	106	118	297	580	615	666	739	722	865
Naphtha	-	-	23	-	-	-	8	53	-
Motor Gasoline	570	674	959	1 809	1 958	2 025	2 140	2 175	2 141
Aviation Fuels	103	167	193	406	447	367	553	442	469
Kerosene	977	1 112	1 389	2 118	2 220	2 328	2 450	2 412	2 386
Gas Diesel	1 222	1 200	1 859	4 178	3 804	3 675	3 856	3 764	4 147
Heavy Fuel Oil	2 639	2 622	4 325	7 637	6 247	5 836	6 606	6 647	6 443
Other	282	218	344	871	1 024	1 178	1 114	1 104	1 131
Refinery Fuel	167	228	396	658	683	724	755	749	742
Sub-Total	6 066	6 339	9 785	18 257	16 998	16 799	18 221	18 068	18 324
Bunkers	19	165	705	1 776	1 522	1 629	1 700	1 670	1 700
Total	6 085	6 504	10 490	20 033	18 520	18 428	19 921	19 738	20 024
Thousand barrels/day									
NGL/LPG/Ethane	3.4	3.8	9.4	18.4	19.5	21.2	23.5	22.9	27.5
Naphtha	-	-	0.5	-	-	-	0.2	1.2	-
Motor Gasoline	13.3	15.8	22.4	42.2	45.8	47.3	50.0	50.7	50.0
Aviation Fuels	2.3	3.7	4.2	8.8	9.7	8.0	12.0	9.6	10.2
Kerosene	20.7	23.6	29.5	44.8	47.1	49.4	52.0	51.0	50.6
Gas Diesel	25.0	24.5	38.0	85.2	77.7	75.1	78.8	76.7	84.8
Heavy Fuel Oil	48.2	47.8	78.9	139.0	114.0	106.5	120.5	121.0	117.6
Other	5.4	4.2	6.6	15.0	17.6	20.4	19.2	19.2	19.9
Refinery Fuel	3.3	4.6	7.9	13.1	13.7	14.4	15.0	14.9	14.8
Sub-Total	121.6	127.8	197.5	366.4	345.1	342.3	371.2	367.2	375.3
Bunkers	0.3	3.0	13.0	33.0	28.3	30.2	31.6	30.9	31.5
Total	121.9	130.9	210.5	399.4	373.5	372.4	402.8	398.1	406.8

	1990	1991	1992	1993	1994	1995	1996	1997	1998
Thousand tonnes									
NGL/LPG/Ethane	845	874	976	1 050	1 240	1 454	1 610	1 864	2 016
Naphtha	-	-	-	-	-	-	-	-	-
Motor Gasoline	2 172	2 082	1 957	1 877	1 878	1 951	2 013	2 080	2 150
Aviation Fuels	480	444	459	479	567	602	813	432	395
Kerosene	2 333	2 165	2 004	1 629	1 426	1 334	1 266	1 197	1 114
Gas Diesel	4 315	4 364	4 370	4 479	4 743	5 314	5 758	6 216	6 725
Heavy Fuel Oil	6 742	6 628	6 529	5 179	3 923	4 793	4 360	5 026	7 250
Other	1 202	1 183	1 186	1 232	1 242	1 397	1 340	1 411	1 562
Refinery Fuel	770	800	810	820	830	855	851	908	894
Sub-Total	18 859	18 540	18 291	16 745	15 849	17 700	18 011	19 134	22 106
Bunkers	1 700	1 710	1 410	1 500	2 444	2 501	3 076	3 032	2 242
Total	20 559	20 250	19 701	18 245	18 293	20 201	21 087	22 166	24 348
Thousand barrels/day									
NGL/LPG/Ethane	26.9	27.8	30.9	33.4	39.4	46.2	51.0	59.2	64.1
Naphtha	-	-	-	-	-	-	-	-	-
Motor Gasoline	50.8	48.7	45.6	43.9	43.9	45.6	46.9	48.6	50.2
Aviation Fuels	10.4	9.6	9.9	10.4	12.3	13.1	17.6	9.4	8.6
Kerosene	49.5	45.9	42.4	34.5	30.2	28.3	26.8	25.4	23.6
Gas Diesel	88.2	89.2	89.1	91.5	96.9	108.6	117.4	127.0	137.4
Heavy Fuel Oil	123.0	120.9	118.8	94.5	71.6	87.5	79.3	91.7	132.3
Other	21.0	20.7	20.7	21.6	21.7	24.4	23.4	24.8	27.2
Refinery Fuel	15.4	16.0	16.2	16.5	16.7	17.2	17.1	18.1	17.8
Sub-Total	385.1	378.9	373.7	346.3	332.8	370.9	379.5	404.3	461.3
Bunkers	31.5	31.6	26.3	28.0	45.3	46.3	56.6	55.9	41.5
Total	416.6	410.5	400.0	374.4	378.0	417.2	436.1	460.2	502.8

[1] See Notes in Part I.4.

Oil Demand by Main Product Groups[1]
Demande de produits raffinés par groupes principaux

El Salvador

	1971	1973	1978	1984	1985	1986	1987	1988	1989
Thousand tonnes									
NGL/LPG/Ethane	10	14	27	29	30	32	35	46	50
Naphtha	-	-	-	-	-	-	-	-	-
Motor Gasoline	98	114	159	134	141	139	153	163	168
Aviation Fuels	11	15	21	27	33	25	20	33	27
Kerosene	36	35	30	18	22	22	19	18	19
Gas Diesel	103	130	219	174	191	189	219	235	233
Heavy Fuel Oil	151	227	192	112	121	117	193	185	173
Other	-	18	24	13	11	11	9	16	16
Refinery Fuel	24	26	32	22	25	24	26	24	10
Sub-Total	433	579	704	529	574	559	674	720	696
Bunkers	-	-	-	-	-	-	-	-	-
Total	433	579	704	529	574	559	674	720	696
Thousand barrels/day									
NGL/LPG/Ethane	0.3	0.5	0.9	0.9	1.0	1.0	1.1	1.5	1.6
Naphtha	-	-	-	-	-	-	-	-	-
Motor Gasoline	2.3	2.7	3.8	3.2	3.3	3.3	3.6	3.8	4.0
Aviation Fuels	0.2	0.3	0.5	0.6	0.7	0.5	0.4	0.7	0.6
Kerosene	0.8	0.8	0.7	0.4	0.5	0.5	0.4	0.4	0.4
Gas Diesel	2.1	2.7	4.5	3.6	3.9	3.9	4.5	4.8	4.8
Heavy Fuel Oil	2.7	4.1	3.4	2.0	2.2	2.1	3.5	3.3	3.1
Other	-	0.3	0.5	0.2	0.2	0.2	0.2	0.3	0.3
Refinery Fuel	0.5	0.5	0.6	0.4	0.5	0.5	0.5	0.5	0.2
Sub-Total	9.0	11.9	14.8	11.3	12.3	12.0	14.2	15.3	15.0
Bunkers	-	-	-	-	-	-	-	-	-
Total	9.0	11.9	14.8	11.3	12.3	12.0	14.2	15.3	15.0

	1990	1991	1992	1993	1994	1995	1996	1997	1998
Thousand tonnes									
NGL/LPG/Ethane	51	56	63	76	86	99	97	105	118
Naphtha	-	-	-	-	-	-	-	-	-
Motor Gasoline	166	173	213	236	257	285	295	307	341
Aviation Fuels	34	25	41	43	44	47	48	47	60
Kerosene	17	17	18	17	17	17	18	18	24
Gas Diesel	261	374	413	513	644	730	508	609	648
Heavy Fuel Oil	173	252	295	286	300	324	343	504	524
Other	22	20	27	26	29	34	45	32	32
Refinery Fuel	7	16	10	12	16	17	18	16	16
Sub-Total	731	933	1 080	1 209	1 393	1 553	1 372	1 638	1 763
Bunkers	-	-	-	-	-	-	-	-	-
Total	731	933	1 080	1 209	1 393	1 553	1 372	1 638	1 763
Thousand barrels/day									
NGL/LPG/Ethane	1.6	1.8	2.0	2.4	2.8	3.2	3.1	3.4	3.8
Naphtha	-	-	-	-	-	-	-	-	-
Motor Gasoline	3.9	4.1	5.0	5.6	6.1	6.7	6.9	7.2	8.0
Aviation Fuels	0.7	0.5	0.9	0.9	1.0	1.0	1.1	1.0	1.3
Kerosene	0.4	0.4	0.4	0.4	0.4	0.4	0.4	0.4	0.5
Gas Diesel	5.3	7.7	8.4	10.5	13.2	14.9	10.4	12.5	13.3
Heavy Fuel Oil	3.1	4.5	5.3	5.1	5.4	5.8	6.1	9.0	9.4
Other	0.4	0.4	0.5	0.5	0.6	0.7	0.9	0.6	0.6
Refinery Fuel	0.2	0.3	0.2	0.3	0.4	0.4	0.4	0.4	0.4
Sub-Total	15.7	19.7	22.8	25.7	29.6	33.1	29.3	34.5	37.3
Bunkers	-	-	-	-	-	-	-	-	-
Total	15.7	19.7	22.8	25.7	29.6	33.1	29.3	34.5	37.3

[1] See Notes in Part I.4.

Oil Demand by Main Product Groups[1]
Demande de produits raffinés par groupes principaux

Eritrea

	1971	1973	1978	1984	1985	1986	1987	1988	1989
Thousand tonnes									
NGL/LPG/Ethane	-	-	-	-	-	-	-	-	-
Naphtha	-	-	-	-	-	-	-	-	-
Motor Gasoline	-	-	-	-	-	-	-	-	-
Aviation Fuels	-	-	-	-	-	-	-	-	-
Kerosene	-	-	-	-	-	-	-	-	-
Gas Diesel	-	-	-	-	-	-	-	-	-
Heavy Fuel Oil	-	-	-	-	-	-	-	-	-
Other	-	-	-	-	-	-	-	-	-
Refinery Fuel	-	-	-	-	-	-	-	-	-
Sub-Total	-	-	-	-	-	-	-	-	-
Bunkers	-	-	-	-	-	-	-	-	-
Total	-	-	-	-	-	-	-	-	-
Thousand barrels/day									
NGL/LPG/Ethane	-	-	-	-	-	-	-	-	-
Naphtha	-	-	-	-	-	-	-	-	-
Motor Gasoline	-	-	-	-	-	-	-	-	-
Aviation Fuels	-	-	-	-	-	-	-	-	-
Kerosene	-	-	-	-	-	-	-	-	-
Gas Diesel	-	-	-	-	-	-	-	-	-
Heavy Fuel Oil	-	-	-	-	-	-	-	-	-
Other	-	-	-	-	-	-	-	-	-
Refinery Fuel	-	-	-	-	-	-	-	-	-
Sub-Total	-	-	-	-	-	-	-	-	-
Bunkers	-	-	-	-	-	-	-	-	-
Total	-	-	-	-	-	-	-	-	-

	1990	1991	1992	1993	1994	1995	1996	1997	1998
Thousand tonnes									
NGL/LPG/Ethane	-	-	-	-	1	1	1	2	1
Naphtha	-	-	-	-	-	-	-	-	-
Motor Gasoline	-	-	6	10	12	13	14	15	16
Aviation Fuels	-	-	4	4	4	5	10	10	7
Kerosene	-	-	8	13	14	17	20	22	21
Gas Diesel	-	-	55	102	115	129	135	138	104
Heavy Fuel Oil	-	-	69	77	87	92	116	97	70
Other	-	-	25	14	13	8	11	19	6
Refinery Fuel	-	-	-	6	7	3	7	4	-
Sub-Total	-	-	167	226	253	268	314	307	225
Bunkers	-	-	228	157	212	138	149	74	-
Total	-	-	395	383	465	406	463	381	225
Thousand barrels/day									
NGL/LPG/Ethane	-	-	-	-	0.0	0.0	0.0	0.1	0.0
Naphtha	-	-	-	-	-	-	-	-	-
Motor Gasoline	-	-	0.1	0.2	0.3	0.3	0.3	0.4	0.4
Aviation Fuels	-	-	0.1	0.1	0.1	0.1	0.2	0.2	0.2
Kerosene	-	-	0.2	0.3	0.3	0.4	0.4	0.5	0.4
Gas Diesel	-	-	1.1	2.1	2.4	2.6	2.8	2.8	2.1
Heavy Fuel Oil	-	-	1.3	1.4	1.6	1.7	2.1	1.8	1.3
Other	-	-	0.5	0.3	0.3	0.2	0.2	0.4	0.1
Refinery Fuel	-	-	-	0.1	0.2	0.1	0.2	0.1	-
Sub-Total	-	-	3.2	4.5	5.0	5.3	6.2	6.1	4.5
Bunkers	-	-	4.1	2.9	3.9	2.5	2.7	1.4	-
Total	-	-	7.4	7.4	8.9	7.9	8.9	7.5	4.5

[1] See Notes in Part I.4.

Oil Demand by Main Product Groups[1]
Demande de produits raffinés par groupes principaux

Estonia

	1971	1973	1978	1984	1985	1986	1987	1988	1989
Thousand tonnes									
NGL/LPG/Ethane	-	-	-	-	-	-	-	-	-
Naphtha	-	-	-	-	-	-	-	-	-
Motor Gasoline	-	-	-	-	-	-	-	-	-
Aviation Fuels	-	-	-	-	-	-	-	-	-
Kerosene	-	-	-	-	-	-	-	-	-
Gas Diesel	-	-	-	-	-	-	-	-	-
Heavy Fuel Oil	-	-	-	-	-	-	-	-	-
Other	-	-	-	-	-	-	-	-	-
Refinery Fuel	-	-	-	-	-	-	-	-	-
Sub-Total	-	-	-	-	-	-	-	-	-
Bunkers	-	-	-	-	-	-	-	-	-
Total	-	-	-	-	-	-	-	-	-
Thousand barrels/day									
NGL/LPG/Ethane	-	-	-	-	-	-	-	-	-
Naphtha	-	-	-	-	-	-	-	-	-
Motor Gasoline	-	-	-	-	-	-	-	-	-
Aviation Fuels	-	-	-	-	-	-	-	-	-
Kerosene	-	-	-	-	-	-	-	-	-
Gas Diesel	-	-	-	-	-	-	-	-	-
Heavy Fuel Oil	-	-	-	-	-	-	-	-	-
Other	-	-	-	-	-	-	-	-	-
Refinery Fuel	-	-	-	-	-	-	-	-	-
Sub-Total	-	-	-	-	-	-	-	-	-
Bunkers	-	-	-	-	-	-	-	-	-
Total	-	-	-	-	-	-	-	-	-

	1990	1991	1992	1993	1994	1995	1996	1997	1998
Thousand tonnes									
NGL/LPG/Ethane	-	-	13	8	11	7	7	9	9
Naphtha	-	-	-	-	-	-	-	-	-
Motor Gasoline	-	-	228	235	287	248	281	305	295
Aviation Fuels	-	-	12	45	15	18	16	22	16
Kerosene	-	-	-	-	49	23	-	-	-
Gas Diesel	-	-	332	337	342	313	358	378	408
Heavy Fuel Oil	-	-	879	898	766	550	549	484	449
Other	-	-	41	44	51	43	46	42	47
Refinery Fuel	-	-	-	-	-	-	-	-	-
Sub-Total	-	-	1 505	1 567	1 521	1 202	1 257	1 240	1 224
Bunkers	-	-	157	158	101	90	93	102	108
Total	-	-	1 662	1 725	1 622	1 292	1 350	1 342	1 332
Thousand barrels/day									
NGL/LPG/Ethane	-	-	0.4	0.3	0.4	0.2	0.2	0.3	0.3
Naphtha	-	-	-	-	-	-	-	-	-
Motor Gasoline	-	-	5.3	5.5	6.7	5.8	6.5	7.1	6.9
Aviation Fuels	-	-	0.3	1.0	0.3	0.4	0.3	0.5	0.4
Kerosene	-	-	-	-	1.0	0.5	-	-	-
Gas Diesel	-	-	6.8	6.9	7.0	6.4	7.3	7.7	8.3
Heavy Fuel Oil	-	-	16.0	16.4	14.0	10.0	10.0	8.8	8.2
Other	-	-	0.7	0.8	0.9	0.8	0.8	0.7	0.8
Refinery Fuel	-	-	-	-	-	-	-	-	-
Sub-Total	-	-	29.5	30.8	30.3	24.1	25.2	25.2	24.9
Bunkers	-	-	3.0	3.1	1.9	1.7	1.8	1.9	2.0
Total	-	-	32.5	33.9	32.2	25.8	27.0	27.1	26.9

[1] See Notes in Part I.4.

Oil Demand by Main Product Groups[1]
Demande de produits raffinés par groupes principaux

Ethiopia

	1971	1973	1978	1984	1985	1986	1987	1988	1989
Thousand tonnes									
NGL/LPG/Ethane	3	4	3	5	5	6	5	5	5
Naphtha	-	-	-	-	-	-	-	-	-
Motor Gasoline	82	84	85	109	116	123	124	119	119
Aviation Fuels	45	42	31	106	108	115	156	163	169
Kerosene	4	8	10	10	7	8	13	12	12
Gas Diesel	190	188	219	186	182	242	324	349	380
Heavy Fuel Oil	103	107	55	83	95	98	84	90	91
Other	29	31	38	34	33	33	43	33	38
Refinery Fuel	35	22	26	36	37	44	39	35	35
Sub-Total	491	486	467	569	583	669	788	806	849
Bunkers	22	5	1	10	10	13	4	7	7
Total	513	491	468	579	593	682	792	813	856
Thousand barrels/day									
NGL/LPG/Ethane	0.1	0.1	0.1	0.2	0.2	0.2	0.2	0.2	0.2
Naphtha	-	-	-	-	-	-	-	-	-
Motor Gasoline	1.9	2.0	2.0	2.5	2.7	2.9	2.9	2.8	2.8
Aviation Fuels	1.0	0.9	0.7	2.3	2.4	2.5	3.4	3.5	3.7
Kerosene	0.1	0.2	0.2	0.2	0.1	0.2	0.3	0.3	0.3
Gas Diesel	3.9	3.8	4.5	3.8	3.7	4.9	6.6	7.1	7.8
Heavy Fuel Oil	1.9	2.0	1.0	1.5	1.7	1.8	1.5	1.6	1.7
Other	0.5	0.5	0.7	0.6	0.6	0.6	0.8	0.6	0.7
Refinery Fuel	0.7	0.4	0.5	0.7	0.7	0.9	0.7	0.7	0.7
Sub-Total	10.1	10.0	9.7	11.9	12.2	14.0	16.4	16.8	17.7
Bunkers	0.4	0.1	0.0	0.2	0.2	0.2	0.1	0.1	0.1
Total	10.5	10.1	9.7	12.0	12.4	14.2	16.5	16.9	17.8

	1990	1991	1992	1993	1994	1995	1996	1997	1998
Thousand tonnes									
NGL/LPG/Ethane	5	5	3	4	3	2	2	2	3
Naphtha	-	-	-	-	-	-	-	-	-
Motor Gasoline	134	145	82	98	112	125	132	130	134
Aviation Fuels	168	175	35	49	62	63	58	55	63
Kerosene	12	13	68	95	110	130	155	169	159
Gas Diesel	373	355	251	301	330	368	417	453	506
Heavy Fuel Oil	111	136	28	57	90	97	98	105	110
Other	37	43	39	43	46	59	40	40	40
Refinery Fuel	44	36	3	4	4	4	4	2	-
Sub-Total	884	908	509	651	757	848	906	956	1 015
Bunkers	14	15	147	187	186	169	153	70	-
Total	898	923	656	838	943	1 017	1 059	1 026	1 015
Thousand barrels/day									
NGL/LPG/Ethane	0.2	0.2	0.1	0.1	0.1	0.1	0.1	0.1	0.1
Naphtha	-	-	-	-	-	-	-	-	-
Motor Gasoline	3.1	3.4	1.9	2.3	2.6	2.9	3.1	3.0	3.1
Aviation Fuels	3.7	3.8	0.8	1.1	1.4	1.4	1.3	1.2	1.4
Kerosene	0.3	0.3	1.4	2.0	2.3	2.8	3.3	3.6	3.4
Gas Diesel	7.6	7.3	5.1	6.2	6.7	7.5	8.5	9.3	10.3
Heavy Fuel Oil	2.0	2.5	0.5	1.0	1.6	1.8	1.8	1.9	2.0
Other	0.7	0.8	0.7	0.8	0.9	1.1	0.8	0.8	0.8
Refinery Fuel	0.8	0.7	0.1	0.1	0.1	0.1	0.1	0.0	-
Sub-Total	18.4	18.9	10.6	13.6	15.7	17.6	18.8	19.9	21.1
Bunkers	0.3	0.3	2.7	3.4	3.4	3.1	2.8	1.3	-
Total	18.6	19.1	13.3	17.0	19.1	20.7	21.6	21.2	21.1

[1] See Notes in Part I.4.

Oil Demand by Main Product Groups[1]
Demande de produits raffinés par groupes principaux

Gabon

	1971	1973	1978	1984	1985	1986	1987	1988	1989
Thousand tonnes									
NGL/LPG/Ethane	-	-	-	5	5	5	4	4	9
Naphtha	-	-	-	78	77	78	66	54	-
Motor Gasoline	-	12	13	33	38	40	29	36	50
Aviation Fuels	8	10	22	26	27	46	47	54	68
Kerosene	-	13	19	95	98	81	69	79	25
Gas Diesel	25	35	288	143	155	203	161	164	209
Heavy Fuel Oil	127	135	231	178	207	102	106	124	30
Other	-	-	8	27	30	28	31	43	10
Refinery Fuel	-	-	-	-	-	-	-	-	40
Sub-Total	160	205	581	585	637	583	513	558	441
Bunkers	65	65	60	70	72	80	68	71	134
Total	225	270	641	655	709	663	581	629	575
Thousand barrels/day									
NGL/LPG/Ethane	-	-	-	0.2	0.2	0.2	0.1	0.1	0.3
Naphtha	-	-	-	1.8	1.8	1.8	1.5	1.3	-
Motor Gasoline	-	0.3	0.3	0.8	0.9	0.9	0.7	0.8	1.2
Aviation Fuels	0.2	0.2	0.5	0.6	0.6	1.0	1.0	1.2	1.5
Kerosene	-	0.3	0.4	2.0	2.1	1.7	1.5	1.7	0.5
Gas Diesel	0.5	0.7	5.9	2.9	3.2	4.1	3.3	3.3	4.3
Heavy Fuel Oil	2.3	2.5	4.2	3.2	3.8	1.9	1.9	2.3	0.5
Other	-	-	0.1	0.5	0.6	0.5	0.6	0.8	0.2
Refinery Fuel	-	-	-	-	-	-	-	-	0.8
Sub-Total	3.0	4.0	11.4	12.0	13.0	12.2	10.7	11.5	9.3
Bunkers	1.2	1.2	1.1	1.3	1.4	1.5	1.3	1.3	2.5
Total	4.2	5.2	12.6	13.3	14.4	13.7	12.0	12.9	11.8

	1990	1991	1992	1993	1994	1995	1996	1997	1998
Thousand tonnes									
NGL/LPG/Ethane	10	12	12	12	13	13	15	16	17
Naphtha	-	-	10	10	10	10	10	10	10
Motor Gasoline	40	39	34	33	31	35	34	41	45
Aviation Fuels	62	57	56	51	67	62	67	94	92
Kerosene	28	31	30	30	27	30	29	62	26
Gas Diesel	163	172	176	200	174	206	232	251	294
Heavy Fuel Oil	-	-	28	32	32	29	37	32	32
Other	18	56	34	54	54	65	64	59	67
Refinery Fuel	20	22	27	28	29	30	30	30	32
Sub-Total	341	389	407	450	437	480	518	595	615
Bunkers	-	-	192	188	136	64	70	-	-
Total	341	389	599	638	573	544	588	595	615
Thousand barrels/day									
NGL/LPG/Ethane	0.3	0.4	0.4	0.4	0.4	0.4	0.5	0.5	0.5
Naphtha	-	-	0.2	0.2	0.2	0.2	0.2	0.2	0.2
Motor Gasoline	0.9	0.9	0.8	0.8	0.7	0.8	0.8	1.0	1.1
Aviation Fuels	1.4	1.3	1.2	1.1	1.5	1.4	1.5	2.1	2.0
Kerosene	0.6	0.7	0.6	0.6	0.6	0.6	0.6	1.3	0.6
Gas Diesel	3.3	3.5	3.6	4.1	3.6	4.2	4.7	5.1	6.0
Heavy Fuel Oil	-	-	0.5	0.6	0.6	0.5	0.7	0.6	0.6
Other	0.3	1.1	0.6	1.0	1.0	1.2	1.2	1.1	1.2
Refinery Fuel	0.4	0.5	0.6	0.6	0.6	0.7	0.7	0.7	0.7
Sub-Total	7.3	8.3	8.6	9.5	9.2	10.1	10.8	12.6	12.9
Bunkers	-	-	3.5	3.4	2.5	1.2	1.3	-	-
Total	7.3	8.3	12.1	12.9	11.7	11.3	12.1	12.6	12.9

[1] See Notes in Part I.4.

Oil Demand by Main Product Groups[1]
Demande de produits raffinés par groupes principaux

Georgia

	1971	1973	1978	1984	1985	1986	1987	1988	1989
Thousand tonnes									
NGL/LPG/Ethane	-	-	-	-	-	-	-	-	-
Naphtha	-	-	-	-	-	-	-	-	-
Motor Gasoline	-	-	-	-	-	-	-	-	-
Aviation Fuels	-	-	-	-	-	-	-	-	-
Kerosene	-	-	-	-	-	-	-	-	-
Gas Diesel	-	-	-	-	-	-	-	-	-
Heavy Fuel Oil	-	-	-	-	-	-	-	-	-
Other	-	-	-	-	-	-	-	-	-
Refinery Fuel	-	-	-	-	-	-	-	-	-
Sub-Total	-	-	-	-	-	-	-	-	-
Bunkers	-	-	-	-	-	-	-	-	-
Total	-	-	-	-	-	-	-	-	-
Thousand barrels/day									
NGL/LPG/Ethane	-	-	-	-	-	-	-	-	-
Naphtha	-	-	-	-	-	-	-	-	-
Motor Gasoline	-	-	-	-	-	-	-	-	-
Aviation Fuels	-	-	-	-	-	-	-	-	-
Kerosene	-	-	-	-	-	-	-	-	-
Gas Diesel	-	-	-	-	-	-	-	-	-
Heavy Fuel Oil	-	-	-	-	-	-	-	-	-
Other	-	-	-	-	-	-	-	-	-
Refinery Fuel	-	-	-	-	-	-	-	-	-
Sub-Total	-	-	-	-	-	-	-	-	-
Bunkers	-	-	-	-	-	-	-	-	-
Total	-	-	-	-	-	-	-	-	-

	1990	1991	1992	1993	1994	1995	1996	1997	1998
Thousand tonnes									
NGL/LPG/Ethane	-	-	-	1	1	1	15	39	96
Naphtha	-	-	-	-	-	-	-	-	-
Motor Gasoline	-	-	392	234	69	44	438	472	418
Aviation Fuels	-	-	9	10	2	2	46	48	103
Kerosene	-	-	85	42	19	7	30	31	115
Gas Diesel	-	-	255	138	64	23	184	166	241
Heavy Fuel Oil	-	-	516	268	115	44	97	85	168
Other	-	-	60	44	20	6	18	51	22
Refinery Fuel	-	-	7	3	3	1	1	1	1
Sub-Total	-	-	1 324	740	293	128	829	893	1 164
Bunkers	-	-	-	-	-	-	-	-	-
Total	-	-	1 324	740	293	128	829	893	1 164
Thousand barrels/day									
NGL/LPG/Ethane	-	-	-	0.0	0.0	0.0	0.5	1.2	3.1
Naphtha	-	-	-	-	-	-	-	-	-
Motor Gasoline	-	-	9.1	5.5	1.6	1.0	10.2	11.0	9.8
Aviation Fuels	-	-	0.2	0.2	0.0	0.0	1.0	1.0	2.2
Kerosene	-	-	1.8	0.9	0.4	0.1	0.6	0.7	2.4
Gas Diesel	-	-	5.2	2.8	1.3	0.5	3.8	3.4	4.9
Heavy Fuel Oil	-	-	9.4	4.9	2.1	0.8	1.8	1.6	3.1
Other	-	-	1.1	0.8	0.4	0.1	0.3	1.0	0.4
Refinery Fuel	-	-	0.2	0.1	0.1	0.0	0.0	0.0	0.0
Sub-Total	-	-	27.0	15.2	5.9	2.7	18.2	19.9	25.9
Bunkers	-	-	-	-	-	-	-	-	-
Total	-	-	27.0	15.2	5.9	2.7	18.2	19.9	25.9

[1] See Notes in Part I.4.

Oil Demand by Main Product Groups[1]
Demande de produits raffinés par groupes principaux

Ghana

	1971	1973	1978	1984	1985	1986	1987	1988	1989
Thousand tonnes									
NGL/LPG/Ethane	3	4	13	8	11	12	9	10	12
Naphtha	-	-	-	-	-	-	-	-	-
Motor Gasoline	189	199	257	177	215	221	215	278	326
Aviation Fuels	42	26	33	30	32	33	50	51	51
Kerosene	80	95	135	73	78	92	112	108	136
Gas Diesel	222	236	280	243	283	266	217	333	273
Heavy Fuel Oil	99	106	86	48	27	46	40	39	69
Other	18	21	34	51	38	37	75	74	75
Refinery Fuel	17	23	36	30	32	27	-	-	-
Sub-Total	670	710	874	660	716	734	718	893	942
Bunkers	51	59	30	20	25	28	27	29	28
Total	721	769	904	680	741	762	745	922	970
Thousand barrels/day									
NGL/LPG/Ethane	0.1	0.1	0.4	0.3	0.4	0.4	0.3	0.3	0.4
Naphtha	-	-	-	-	-	-	-	-	-
Motor Gasoline	4.4	4.7	6.0	4.1	5.0	5.2	5.0	6.5	7.6
Aviation Fuels	0.9	0.6	0.7	0.7	0.7	0.7	1.1	1.1	1.1
Kerosene	1.7	2.0	2.9	1.5	1.7	2.0	2.4	2.3	2.9
Gas Diesel	4.5	4.8	5.7	5.0	5.8	5.4	4.4	6.8	5.6
Heavy Fuel Oil	1.8	1.9	1.6	0.9	0.5	0.8	0.7	0.7	1.3
Other	0.4	0.4	0.7	1.0	0.7	0.7	1.4	1.4	1.4
Refinery Fuel	0.4	0.5	0.8	0.7	0.7	0.6	-	-	-
Sub-Total	14.2	15.0	18.7	14.0	15.4	15.8	15.4	19.1	20.3
Bunkers	1.0	1.2	0.6	0.4	0.5	0.6	0.6	0.6	0.6
Total	15.2	16.2	19.3	14.4	15.9	16.4	15.9	19.7	20.8

	1990	1991	1992	1993	1994	1995	1996	1997	1998
Thousand tonnes									
NGL/LPG/Ethane	12	18	27	40	55	64	76	70	69
Naphtha	-	-	-	-	-	-	-	-	-
Motor Gasoline	325	243	306	341	324	344	371	389	416
Aviation Fuels	52	51	53	56	56	57	57	57	57
Kerosene	122	120	127	130	130	132	132	132	135
Gas Diesel	264	299	283	322	401	434	478	530	616
Heavy Fuel Oil	46	45	47	45	49	47	49	54	50
Other	79	81	76	58	59	60	60	60	60
Refinery Fuel	-	-	61	55	56	57	58	-	45
Sub-Total	900	857	980	1 047	1 130	1 195	1 281	1 292	1 448
Bunkers	21	22	22	22	24	24	24	25	25
Total	921	879	1 002	1 069	1 154	1 219	1 305	1 317	1 473
Thousand barrels/day									
NGL/LPG/Ethane	0.4	0.6	0.9	1.3	1.7	2.0	2.4	2.2	2.2
Naphtha	-	-	-	-	-	-	-	-	-
Motor Gasoline	7.6	5.7	7.1	8.0	7.6	8.0	8.6	9.1	9.7
Aviation Fuels	1.1	1.1	1.2	1.2	1.2	1.3	1.2	1.3	1.3
Kerosene	2.6	2.5	2.7	2.8	2.8	2.8	2.8	2.8	2.9
Gas Diesel	5.4	6.1	5.8	6.6	8.2	8.9	9.7	10.8	12.6
Heavy Fuel Oil	0.8	0.8	0.9	0.8	0.9	0.9	0.9	1.0	0.9
Other	1.5	1.5	1.4	1.1	1.1	1.1	1.1	1.1	1.1
Refinery Fuel	-	-	1.3	1.2	1.2	1.2	1.3	-	1.0
Sub-Total	19.4	18.4	21.2	22.9	24.7	26.2	28.1	28.3	31.7
Bunkers	0.4	0.5	0.4	0.5	0.5	0.5	0.5	0.5	0.5
Total	19.9	18.8	21.7	23.4	25.2	26.7	28.6	28.8	32.2

[1] See Notes in Part I.4.

Oil Demand by Main Product Groups[1]
Demande de produits raffinés par groupes principaux

Gibraltar

	1971	1973	1978	1984	1985	1986	1987	1988	1989
Thousand tonnes									
NGL/LPG/Ethane	-	-	-	-	-	-	-	-	-
Naphtha	-	-	-	-	-	-	-	-	-
Motor Gasoline	4	4	3	4	5	4	8	12	12
Aviation Fuels	5	7	5	3	4	3	5	8	8
Kerosene	-	-	-	-	-	-	-	-	-
Gas Diesel	10	6	4	8	9	18	16	16	12
Heavy Fuel Oil	8	9	10	15	15	15	15	15	17
Other	11	12	9	8	10	9	9	12	12
Refinery Fuel	-	-	-	-	-	-	-	-	-
Sub-Total	38	38	31	38	43	49	53	63	61
Bunkers	179	195	142	114	285	362	355	458	435
Total	217	233	173	152	328	411	408	521	496
Thousand barrels/day									
NGL/LPG/Ethane	-	-	-	-	-	-	-	-	-
Naphtha	-	-	-	-	-	-	-	-	-
Motor Gasoline	0.1	0.1	0.1	0.1	0.1	0.1	0.2	0.3	0.3
Aviation Fuels	0.1	0.2	0.1	0.1	0.1	0.1	0.1	0.2	0.2
Kerosene	-	-	-	-	-	-	-	-	-
Gas Diesel	0.2	0.1	0.1	0.2	0.2	0.4	0.3	0.3	0.2
Heavy Fuel Oil	0.1	0.2	0.2	0.3	0.3	0.3	0.3	0.3	0.3
Other	0.2	0.2	0.2	0.2	0.2	0.2	0.2	0.2	0.2
Refinery Fuel	-	-	-	-	-	-	-	-	-
Sub-Total	0.8	0.8	0.6	0.7	0.9	1.0	1.1	1.3	1.2
Bunkers	3.4	3.7	2.7	2.1	5.3	6.7	6.6	8.5	8.1
Total	4.2	4.5	3.3	2.9	6.1	7.7	7.6	9.7	9.3

	1990	1991	1992	1993	1994	1995	1996	1997	1998
Thousand tonnes									
NGL/LPG/Ethane	-	-	-	-	-	-	-	-	-
Naphtha	-	-	-	-	-	-	-	-	-
Motor Gasoline	12	12	16	16	16	16	16	16	16
Aviation Fuels	7	8	4	4	4	4	4	4	4
Kerosene	-	-	-	-	-	-	-	-	-
Gas Diesel	15	31	37	37	50	48	48	48	48
Heavy Fuel Oil	25	17	17	30	50	50	50	50	50
Other	12	11	15	15	15	15	15	15	15
Refinery Fuel	-	-	-	-	-	-	-	-	-
Sub-Total	71	79	89	102	135	133	133	133	133
Bunkers	443	855	855	858	850	854	854	854	854
Total	514	934	944	960	985	987	987	987	987
Thousand barrels/day									
NGL/LPG/Ethane	-	-	-	-	-	-	-	-	-
Naphtha	-	-	-	-	-	-	-	-	-
Motor Gasoline	0.3	0.3	0.4	0.4	0.4	0.4	0.4	0.4	0.4
Aviation Fuels	0.2	0.2	0.1	0.1	0.1	0.1	0.1	0.1	0.1
Kerosene	-	-	-	-	-	-	-	-	-
Gas Diesel	0.3	0.6	0.8	0.8	1.0	1.0	1.0	1.0	1.0
Heavy Fuel Oil	0.5	0.3	0.3	0.5	0.9	0.9	0.9	0.9	0.9
Other	0.2	0.2	0.3	0.3	0.3	0.3	0.3	0.3	0.3
Refinery Fuel	-	-	-	-	-	-	-	-	-
Sub-Total	1.4	1.6	1.8	2.1	2.7	2.6	2.6	2.6	2.6
Bunkers	8.2	15.8	15.8	15.9	15.7	15.8	15.8	15.8	15.8
Total	9.6	17.4	17.6	17.9	18.4	18.5	18.4	18.5	18.5

[1] See Notes in Part I.4.

Oil Demand by Main Product Groups[1]
Demande de produits raffinés par groupes principaux

Guatemala

	1971	1973	1978	1984	1985	1986	1987	1988	1989
Thousand tonnes									
NGL/LPG/Ethane	20	27	39	58	69	82	83	89	94
Naphtha	-	-	-	-	-	-	-	-	-
Motor Gasoline	166	210	298	247	250	235	271	271	292
Aviation Fuels	48	62	48	39	38	34	43	40	39
Kerosene	55	64	50	59	55	45	34	34	36
Gas Diesel	161	214	365	293	324	260	322	351	364
Heavy Fuel Oil	284	283	376	235	233	111	139	145	142
Other	32	36	79	123	122	48	57	74	67
Refinery Fuel	37	45	39	32	32	25	26	28	27
Sub-Total	803	941	1 294	1 086	1 123	840	975	1 032	1 061
Bunkers	57	65	118	120	120	115	120	125	120
Total	860	1 006	1 412	1 206	1 243	955	1 095	1 157	1 181
Thousand barrels/day									
NGL/LPG/Ethane	0.6	0.9	1.2	1.8	2.2	2.6	2.6	2.8	3.0
Naphtha	-	-	-	-	-	-	-	-	-
Motor Gasoline	3.9	4.9	7.0	5.8	5.8	5.5	6.3	6.3	6.8
Aviation Fuels	1.1	1.4	1.1	0.9	0.8	0.8	0.9	0.9	0.9
Kerosene	1.2	1.4	1.1	1.2	1.2	1.0	0.7	0.7	0.8
Gas Diesel	3.3	4.4	7.5	6.0	6.6	5.3	6.6	7.2	7.4
Heavy Fuel Oil	5.2	5.2	6.9	4.3	4.3	2.0	2.5	2.6	2.6
Other	0.6	0.7	1.4	2.3	2.3	0.9	1.1	1.4	1.2
Refinery Fuel	0.7	0.9	0.8	0.6	0.6	0.5	0.5	0.5	0.5
Sub-Total	16.5	19.6	26.9	22.9	23.8	18.5	21.3	22.4	23.2
Bunkers	1.1	1.3	2.4	2.4	2.5	2.4	2.5	2.5	2.5
Total	17.7	20.9	29.3	25.3	26.3	20.9	23.8	25.0	25.7

	1990	1991	1992	1993	1994	1995	1996	1997	1998
Thousand tonnes									
NGL/LPG/Ethane	100	101	110	120	123	140	153	159	168
Naphtha	-	-	-	-	-	-	-	-	-
Motor Gasoline	314	357	360	366	390	494	526	550	657
Aviation Fuels	41	37	40	40	38	43	47	50	50
Kerosene	38	35	36	39	38	43	46	49	49
Gas Diesel	383	470	651	522	573	666	630	684	802
Heavy Fuel Oil	152	162	196	393	400	494	503	515	823
Other	71	75	79	72	84	79	19	27	27
Refinery Fuel	26	26	26	17	18	19	20	21	11
Sub-Total	1 125	1 263	1 498	1 569	1 664	1 978	1 944	2 055	2 587
Bunkers	120	120	120	120	120	120	120	120	120
Total	1 245	1 383	1 618	1 689	1 784	2 098	2 064	2 175	2 707
Thousand barrels/day									
NGL/LPG/Ethane	3.2	3.2	3.5	3.8	3.9	4.4	4.8	5.1	5.3
Naphtha	-	-	-	-	-	-	-	-	-
Motor Gasoline	7.3	8.2	8.3	8.6	9.1	11.5	12.3	12.9	15.4
Aviation Fuels	0.9	0.8	0.9	0.9	0.8	0.9	1.0	1.1	1.1
Kerosene	0.8	0.7	0.8	0.8	0.8	0.9	1.0	1.0	1.0
Gas Diesel	7.8	9.6	13.3	10.7	11.7	13.6	12.8	14.0	16.4
Heavy Fuel Oil	2.8	3.0	3.6	7.2	7.3	9.0	9.2	9.4	15.0
Other	1.3	1.4	1.5	1.3	1.6	1.5	0.4	0.5	0.5
Refinery Fuel	0.5	0.5	0.5	0.4	0.4	0.4	0.4	0.5	0.2
Sub-Total	24.6	27.4	32.2	33.6	35.6	42.3	41.9	44.4	55.0
Bunkers	2.5	2.5	2.4	2.5	2.5	2.5	2.4	2.5	2.5
Total	27.1	29.9	34.7	36.1	38.1	44.8	44.3	46.8	57.4

[1] See Notes in Part I.4.

Oil Demand by Main Product Groups[1]
Demande de produits raffinés par groupes principaux

Haiti

	1971	1973	1978	1984	1985	1986	1987	1988	1989
Thousand tonnes									
NGL/LPG/Ethane	1	1	2	4	5	4	4	4	7
Naphtha	-	-	-	-	-	-	-	-	-
Motor Gasoline	25	28	48	44	43	45	50	53	59
Aviation Fuels	6	8	12	12	14	19	20	23	22
Kerosene	4	3	8	13	14	21	19	21	24
Gas Diesel	54	48	76	101	101	93	102	115	133
Heavy Fuel Oil	35	36	70	43	42	58	70	64	64
Other	-	-	-	-	-	-	-	-	-
Refinery Fuel	1	-	1	-	-	-	-	-	-
Sub-Total	126	124	217	217	219	240	265	280	309
Bunkers	-	-	-	-	-	-	-	-	-
Total	126	124	217	217	219	240	265	280	309
Thousand barrels/day									
NGL/LPG/Ethane	0.0	0.0	0.1	0.1	0.2	0.1	0.1	0.1	0.2
Naphtha	-	-	-	-	-	-	-	-	-
Motor Gasoline	0.6	0.7	1.1	1.0	1.0	1.1	1.2	1.2	1.4
Aviation Fuels	0.1	0.2	0.3	0.3	0.3	0.4	0.4	0.5	0.5
Kerosene	0.1	0.1	0.2	0.3	0.3	0.4	0.4	0.4	0.5
Gas Diesel	1.1	1.0	1.6	2.1	2.1	1.9	2.1	2.3	2.7
Heavy Fuel Oil	0.6	0.7	1.3	0.8	0.8	1.1	1.3	1.2	1.2
Other	-	-	-	-	-	-	-	-	-
Refinery Fuel	0.0	-	0.0	-	-	-	-	-	-
Sub-Total	2.6	2.6	4.5	4.5	4.6	5.0	5.5	5.8	6.5
Bunkers	-	-	-	-	-	-	-	-	-
Total	2.6	2.6	4.5	4.5	4.6	5.0	5.5	5.8	6.5

	1990	1991	1992	1993	1994	1995	1996	1997	1998
Thousand tonnes									
NGL/LPG/Ethane	8	7	6	6	6	9	6	6	7
Naphtha	-	-	-	-	-	-	-	-	-
Motor Gasoline	60	56	54	49	16	67	78	85	93
Aviation Fuels	23	20	12	11	-	22	19	19	21
Kerosene	23	18	14	16	5	25	28	35	35
Gas Diesel	128	123	119	114	31	167	198	209	221
Heavy Fuel Oil	68	60	45	21	2	17	24	68	46
Other	7	6	9	8	-	4	2	9	8
Refinery Fuel	-	-	-	-	-	-	-	-	-
Sub-Total	317	290	259	225	60	311	355	431	431
Bunkers	-	-	-	-	-	-	-	-	-
Total	317	290	259	225	60	311	355	431	431
Thousand barrels/day									
NGL/LPG/Ethane	0.3	0.2	0.2	0.2	0.2	0.3	0.2	0.2	0.2
Naphtha	-	-	-	-	-	-	-	-	-
Motor Gasoline	1.4	1.3	1.3	1.1	0.4	1.6	1.8	2.0	2.2
Aviation Fuels	0.5	0.4	0.3	0.2	-	0.5	0.4	0.4	0.5
Kerosene	0.5	0.4	0.3	0.3	0.1	0.5	0.6	0.7	0.7
Gas Diesel	2.6	2.5	2.4	2.3	0.6	3.4	4.0	4.3	4.5
Heavy Fuel Oil	1.2	1.1	0.8	0.4	0.0	0.3	0.4	1.2	0.8
Other	0.1	0.1	0.2	0.2	-	0.1	0.0	0.2	0.2
Refinery Fuel	-	-	-	-	-	-	-	-	-
Sub-Total	6.6	6.1	5.4	4.8	1.3	6.7	7.5	9.0	9.1
Bunkers	-	-	-	-	-	-	-	-	-
Total	6.6	6.1	5.4	4.8	1.3	6.7	7.5	9.0	9.1

[1] See Notes in Part I.4.

Oil Demand by Main Product Groups[1]
Demande de produits raffinés par groupes principaux

Honduras

	1971	1973	1978	1984	1985	1986	1987	1988	1989
Thousand tonnes									
NGL/LPG/Ethane	5	7	9	12	12	13	13	16	13
Naphtha	-	-	-	-	-	-	-	-	-
Motor Gasoline	87	96	103	103	104	113	123	124	136
Aviation Fuels	7	9	17	21	38	25	26	29	30
Kerosene	32	34	42	35	32	57	60	66	69
Gas Diesel	149	175	234	310	275	262	301	329	359
Heavy Fuel Oil	59	67	88	105	87	83	90	114	146
Other	-	-	-	-	-	-	-	-	-
Refinery Fuel	25	23	27	26	24	12	17	15	11
Sub-Total	364	411	520	612	572	565	630	693	764
Bunkers	-	-	-	-	-	-	-	-	-
Total	364	411	520	612	572	565	630	693	764
Thousand barrels/day									
NGL/LPG/Ethane	0.2	0.2	0.3	0.4	0.4	0.4	0.4	0.5	0.4
Naphtha	-	-	-	-	-	-	-	-	-
Motor Gasoline	2.1	2.3	2.4	2.4	2.5	2.7	2.9	2.9	3.2
Aviation Fuels	0.2	0.2	0.4	0.5	0.8	0.5	0.6	0.6	0.7
Kerosene	0.7	0.7	0.9	0.7	0.7	1.2	1.3	1.4	1.5
Gas Diesel	3.1	3.6	4.8	6.3	5.6	5.4	6.2	6.7	7.4
Heavy Fuel Oil	1.0	1.2	1.6	1.9	1.5	1.5	1.6	2.0	2.6
Other	-	-	-	-	-	-	-	-	-
Refinery Fuel	0.5	0.4	0.5	0.5	0.5	0.2	0.4	0.3	0.2
Sub-Total	7.6	8.6	10.9	12.7	12.0	11.9	13.3	14.5	15.9
Bunkers	-	-	-	-	-	-	-	-	-
Total	7.6	8.6	10.9	12.7	12.0	11.9	13.3	14.5	15.9

	1990	1991	1992	1993	1994	1995	1996	1997	1998
Thousand tonnes									
NGL/LPG/Ethane	12	13	13	13	23	36	38	43	32
Naphtha	-	-	-	-	-	-	-	-	-
Motor Gasoline	132	136	161	186	218	234	209	248	282
Aviation Fuels	29	29	39	38	24	24	19	21	24
Kerosene	67	67	85	68	57	52	45	50	58
Gas Diesel	323	312	354	428	511	614	485	498	676
Heavy Fuel Oil	142	147	160	126	138	305	350	317	318
Other	-	-	-	-	-	-	-	-	-
Refinery Fuel	7	6	7	1	1	1	1	1	2
Sub-Total	712	710	819	860	972	1 266	1 147	1 178	1 392
Bunkers	-	-	-	-	-	-	-	-	-
Total	712	710	819	860	972	1 266	1 147	1 178	1 392
Thousand barrels/day									
NGL/LPG/Ethane	0.4	0.4	0.4	0.4	0.7	1.2	1.2	1.4	1.0
Naphtha	-	-	-	-	-	-	-	-	-
Motor Gasoline	3.1	3.2	3.8	4.4	5.1	5.5	4.9	5.9	6.7
Aviation Fuels	0.6	0.6	0.8	0.8	0.5	0.5	0.4	0.5	0.5
Kerosene	1.4	1.4	1.8	1.4	1.2	1.1	1.0	1.1	1.2
Gas Diesel	6.6	6.4	7.2	8.8	10.5	12.6	9.9	10.2	13.9
Heavy Fuel Oil	2.5	2.6	2.8	2.2	2.4	5.4	6.2	5.6	5.6
Other	-	-	-	-	-	-	-	-	-
Refinery Fuel	0.1	0.1	0.1	0.0	0.0	0.0	0.0	0.0	0.0
Sub-Total	14.8	14.8	17.1	18.1	20.6	26.3	23.6	24.6	29.0
Bunkers	-	-	-	-	-	-	-	-	-
Total	14.8	14.8	17.1	18.1	20.6	26.3	23.6	24.6	29.0

[1] See Notes in Part I.4.

Oil Demand by Main Product Groups[1]
Demande de produits raffinés par groupes principaux

Hong Kong, China

	1971	1973	1978	1984	1985	1986	1987	1988	1989
Thousand tonnes									
NGL/LPG/Ethane	46	72	97	146	151	174	174	184	189
Naphtha	26	32	63	189	204	254	285	304	351
Motor Gasoline	114	132	165	198	191	189	196	229	245
Aviation Fuels	447	523	633	818	809	930	982	1 221	1 418
Kerosene	210	161	174	127	114	123	125	110	97
Gas Diesel	666	805	1 080	1 058	966	1 306	1 226	1 509	1 446
Heavy Fuel Oil	1 860	1 867	3 365	1 992	1 638	1 329	975	987	993
Other	69	84	77	108	110	117	120	103	156
Refinery Fuel	-	-	-	-	-	-	-	-	-
Sub-Total	3 438	3 676	5 654	4 636	4 183	4 422	4 083	4 647	4 895
Bunkers	642	1 179	551	317	1 003	838	911	1 075	1 367
Total	4 080	4 855	6 205	4 953	5 186	5 260	4 994	5 722	6 262
Thousand barrels/day									
NGL/LPG/Ethane	1.5	2.3	3.1	4.6	4.8	5.5	5.5	5.8	6.0
Naphtha	0.6	0.7	1.5	4.4	4.8	5.9	6.6	7.1	8.2
Motor Gasoline	2.7	3.1	3.9	4.6	4.5	4.4	4.6	5.3	5.7
Aviation Fuels	9.7	11.4	13.8	17.7	17.6	20.2	21.3	26.5	30.8
Kerosene	4.5	3.4	3.7	2.7	2.4	2.6	2.7	2.3	2.1
Gas Diesel	13.6	16.5	22.1	21.6	19.7	26.7	25.1	30.8	29.6
Heavy Fuel Oil	33.9	34.1	61.4	36.2	29.9	24.3	17.8	18.0	18.1
Other	1.3	1.6	1.4	2.0	2.0	2.2	2.2	1.9	2.9
Refinery Fuel	-	-	-	-	-	-	-	-	-
Sub-Total	67.7	73.0	110.7	93.8	85.6	91.8	85.8	97.6	103.3
Bunkers	11.7	21.5	10.1	6.1	18.6	15.6	17.1	20.0	25.7
Total	79.5	94.5	120.8	99.9	104.3	107.4	102.9	117.6	129.0

	1990	1991	1992	1993	1994	1995	1996	1997	1998
Thousand tonnes									
NGL/LPG/Ethane	203	202	214	196	188	186	162	146	140
Naphtha	392	432	437	455	510	512	508	531	552
Motor Gasoline	237	270	295	322	340	334	335	344	356
Aviation Fuels	1 781	1 581	2 092	2 373	2 730	2 919	3 023	3 202	3 010
Kerosene	86	78	82	69	64	62	55	53	41
Gas Diesel	1 661	2 053	2 352	2 659	2 952	2 808	2 604	2 654	4 611
Heavy Fuel Oil	534	441	905	548	196	232	100	85	80
Other	155	124	143	146	189	175	185	313	166
Refinery Fuel	-	-	-	-	-	-	-	-	-
Sub-Total	5 049	5 181	6 520	6 768	7 169	7 228	6 972	7 328	8 956
Bunkers	1 457	1 244	1 527	1 575	1 706	2 303	2 377	2 149	2 843
Total	6 506	6 425	8 047	8 343	8 875	9 531	9 349	9 477	11 799
Thousand barrels/day									
NGL/LPG/Ethane	6.5	6.4	6.8	6.2	6.0	5.9	5.1	4.6	4.4
Naphtha	9.1	10.1	10.1	10.6	11.9	11.9	11.8	12.4	12.9
Motor Gasoline	5.5	6.3	6.9	7.5	7.9	7.8	7.8	8.0	8.3
Aviation Fuels	38.7	34.3	45.3	51.6	59.3	63.4	65.5	69.6	65.4
Kerosene	1.8	1.7	1.7	1.5	1.4	1.3	1.2	1.1	0.9
Gas Diesel	33.9	42.0	47.9	54.3	60.3	57.4	53.1	54.2	94.2
Heavy Fuel Oil	9.7	8.0	16.5	10.0	3.6	4.2	1.8	1.6	1.5
Other	2.9	2.3	2.6	2.7	3.5	3.2	3.4	5.9	3.0
Refinery Fuel	-	-	-	-	-	-	-	-	-
Sub-Total	108.2	111.1	137.9	144.4	153.9	155.2	149.6	157.4	190.6
Bunkers	27.3	23.4	28.4	29.4	31.9	43.4	44.5	40.3	54.7
Total	135.5	134.4	166.3	173.8	185.8	198.6	194.2	197.7	245.3

[1] See Notes in Part I.4.

Oil Demand by Main Product Groups[1]
Demande de produits raffinés par groupes principaux

India

	1971	1973	1978	1984	1985	1986	1987	1988	1989
Thousand tonnes									
NGL/LPG/Ethane	202	263	408	886	1 275	1 529	1 740	1 979	2 362
Naphtha	1 107	1 454	2 377	3 153	3 107	3 194	2 875	3 258	3 421
Motor Gasoline	1 553	1 630	1 499	2 026	2 247	2 428	2 735	2 974	3 370
Aviation Fuels	824	872	1 174	1 323	1 441	1 588	1 627	1 691	1 767
Kerosene	3 595	3 461	3 952	5 838	6 168	6 522	7 088	7 595	8 089
Gas Diesel	5 385	6 510	9 464	14 485	15 858	16 852	18 385	19 704	21 820
Heavy Fuel Oil	4 770	5 729	6 260	7 786	7 955	8 016	8 150	8 360	8 700
Other	2 090	2 433	2 245	2 335	2 642	2 846	2 946	3 293	3 608
Refinery Fuel	1 077	1 128	1 409	2 005	2 314	2 606	2 504	2 454	2 678
Sub-Total	20 603	23 480	28 788	39 837	43 007	45 581	48 050	51 308	55 815
Bunkers	230	234	183	204	109	122	228	278	235
Total	20 833	23 714	28 971	40 041	43 116	45 703	48 278	51 586	56 050
Thousand barrels/day									
NGL/LPG/Ethane	6.4	8.4	13.0	28.1	40.4	48.4	55.1	62.6	74.8
Naphtha	25.8	33.9	55.4	73.2	72.4	74.4	67.0	75.7	79.7
Motor Gasoline	36.3	38.1	35.0	47.2	52.5	56.7	63.9	69.3	78.8
Aviation Fuels	18.0	19.0	25.6	28.7	31.4	34.6	35.4	36.7	38.4
Kerosene	76.2	73.4	83.8	123.5	130.8	138.3	150.3	160.6	171.5
Gas Diesel	110.1	133.1	193.4	295.2	324.1	344.4	375.8	401.6	446.0
Heavy Fuel Oil	87.0	104.5	114.2	141.7	145.2	146.3	148.7	152.1	158.7
Other	37.1	43.3	40.1	41.8	47.3	50.9	52.4	58.2	63.9
Refinery Fuel	20.9	21.9	27.3	38.7	44.8	50.4	48.7	47.7	52.1
Sub-Total	417.8	475.5	587.8	818.1	888.8	944.4	997.2	1 064.5	1 163.9
Bunkers	4.3	4.3	3.5	3.8	2.1	2.3	4.3	5.2	4.4
Total	422.0	479.8	591.3	821.9	890.9	946.7	1 001.5	1 069.7	1 168.3

	1990	1991	1992	1993	1994	1995	1996	1997	1998
Thousand tonnes									
NGL/LPG/Ethane	2 643	2 862	3 159	3 272	3 823	4 189	4 803	5 057	5 519
Naphtha	3 397	3 418	3 380	3 310	3 258	3 781	3 802	4 425	5 172
Motor Gasoline	3 583	3 587	3 592	3 761	4 037	4 037	4 527	5 005	5 511
Aviation Fuels	1 711	1 563	1 582	1 705	1 894	2 072	2 191	2 228	2 132
Kerosene	8 423	8 412	8 573	8 618	8 931	9 334	9 577	9 869	11 807
Gas Diesel	21 346	24 702	26 277	27 830	30 269	32 471	35 798	36 629	35 945
Heavy Fuel Oil	8 855	8 765	9 207	8 914	9 287	10 481	11 515	11 000	12 742
Other	3 673	3 644	3 873	3 556	3 617	3 204	3 504	3 498	3 702
Refinery Fuel	2 706	2 621	2 793	2 830	3 093	3 178	3 384	3 478	3 623
Sub-Total	56 337	59 574	62 436	63 796	68 209	72 747	79 101	81 189	86 153
Bunkers	414	393	454	513	539	254	601	1 083	940
Total	56 751	59 967	62 890	64 309	68 748	73 001	79 702	82 272	87 093
Thousand barrels/day									
NGL/LPG/Ethane	83.6	90.5	99.6	103.6	120.7	132.3	151.1	159.5	174.9
Naphtha	79.1	79.6	78.5	77.1	75.9	88.1	88.3	103.0	120.4
Motor Gasoline	83.7	83.8	83.7	87.9	94.3	94.3	105.5	117.0	128.8
Aviation Fuels	37.2	34.0	34.3	37.1	41.2	45.1	47.5	48.5	46.4
Kerosene	178.6	178.4	181.3	182.7	189.4	197.9	202.5	209.3	250.4
Gas Diesel	436.3	504.9	535.6	568.8	618.6	663.7	729.7	748.6	734.7
Heavy Fuel Oil	161.6	159.9	167.5	162.7	169.5	191.2	209.5	200.7	232.5
Other	65.3	64.8	68.6	63.4	64.2	55.7	60.6	60.6	64.1
Refinery Fuel	52.6	51.0	54.2	55.0	60.2	61.8	65.7	67.7	70.5
Sub-Total	1 178.1	1 246.9	1 303.4	1 338.2	1 434.0	1 530.2	1 660.4	1 714.8	1 822.6
Bunkers	8.0	7.6	8.7	9.8	10.3	5.1	11.9	20.8	18.4
Total	1 186.1	1 254.5	1 312.1	1 348.0	1 444.3	1 535.3	1 672.3	1 735.6	1 841.0

[1] See Notes in Part I.4.

Oil Demand by Main Product Groups[1]
Demande de produits raffinés par groupes principaux

Indonesia

	1971	1973	1978	1984	1985	1986	1987	1988	1989
Thousand tonnes									
NGL/LPG/Ethane	3	13	46	110	145	198	217	234	265
Naphtha	-	-	-	-	334	340	350	350	667
Motor Gasoline	1 236	1 425	2 393	2 967	3 029	3 295	3 578	3 833	4 184
Aviation Fuels	143	170	551	496	498	488	645	708	776
Kerosene	2 430	2 978	5 367	5 837	5 674	5 635	5 602	5 803	6 022
Gas Diesel	2 433	2 972	5 306	8 292	8 169	7 783	8 363	9 307	9 956
Heavy Fuel Oil	992	1 109	2 079	2 736	3 186	2 710	2 926	2 715	2 775
Other	439	1 233	483	258	382	415	420	633	1 541
Refinery Fuel	181	201	318	1 035	1 279	1 609	1 670	1 784	1 508
Sub-Total	7 857	10 101	16 543	21 731	22 696	22 473	23 771	25 367	27 694
Bunkers	226	428	274	268	220	160	175	237	251
Total	8 083	10 529	16 817	21 999	22 916	22 633	23 946	25 604	27 945
Thousand barrels/day									
NGL/LPG/Ethane	0.1	0.4	1.5	3.5	4.6	6.3	6.9	7.4	8.4
Naphtha	-	-	-	-	7.8	7.9	8.2	8.1	15.5
Motor Gasoline	28.9	33.3	55.9	69.1	70.8	77.0	83.6	89.3	97.8
Aviation Fuels	3.2	3.7	12.0	10.8	10.8	10.6	14.0	15.4	16.9
Kerosene	51.5	63.2	113.8	123.4	120.3	119.5	118.8	122.7	127.7
Gas Diesel	49.7	60.7	108.4	169.0	167.0	159.1	170.9	189.7	203.5
Heavy Fuel Oil	18.1	20.2	37.9	49.8	58.1	49.4	53.4	49.4	50.6
Other	8.8	24.9	9.7	5.2	7.6	8.2	8.2	12.1	29.6
Refinery Fuel	3.5	3.9	6.1	19.3	23.9	29.9	31.1	33.1	28.2
Sub-Total	163.8	210.4	345.4	450.1	470.9	468.0	495.1	527.3	578.2
Bunkers	4.2	7.9	5.1	4.9	4.0	3.1	3.2	4.4	4.7
Total	168.1	218.3	350.5	455.1	474.9	471.1	498.3	531.7	582.9

	1990	1991	1992	1993	1994	1995	1996	1997	1998
Thousand tonnes									
NGL/LPG/Ethane	318	359	349	469	553	639	721	829	829
Naphtha	409	76	104	151	409	144	298	183	80
Motor Gasoline	4 687	4 982	5 309	5 495	6 150	6 776	7 396	8 283	8 489
Aviation Fuels	904	893	1 023	1 157	1 292	1 389	1 604	933	637
Kerosene	6 380	6 561	6 811	7 072	7 250	7 518	7 948	8 113	8 257
Gas Diesel	11 419	12 538	13 688	15 866	15 294	15 986	17 353	19 635	17 689
Heavy Fuel Oil	3 740	4 332	4 271	4 656	3 553	3 465	3 723	4 830	4 947
Other	799	918	752	1 048	1 115	1 172	1 316	1 052	683
Refinery Fuel	1 993	1 984	1 988	2 035	1 952	2 053	1 958	1 858	2 372
Sub-Total	30 649	32 643	34 295	37 949	37 568	39 142	42 317	45 716	43 983
Bunkers	541	248	254	243	203	89	344	315	342
Total	31 190	32 891	34 549	38 192	37 771	39 231	42 661	46 031	44 325
Thousand barrels/day									
NGL/LPG/Ethane	10.1	11.4	11.1	14.9	17.6	20.3	22.9	26.3	26.3
Naphtha	9.5	1.8	2.4	3.5	9.5	3.4	6.9	4.3	1.9
Motor Gasoline	109.5	116.4	123.7	128.4	143.7	158.4	172.4	193.6	198.4
Aviation Fuels	19.7	19.4	22.2	25.2	28.1	30.2	34.8	20.3	13.9
Kerosene	135.3	139.1	144.0	150.0	153.7	159.4	168.1	172.0	175.1
Gas Diesel	233.4	256.3	279.0	324.3	312.6	326.7	353.7	401.3	361.5
Heavy Fuel Oil	68.2	79.0	77.7	85.0	64.8	63.2	67.7	88.1	90.3
Other	14.1	16.3	13.3	18.8	20.2	21.2	23.9	19.3	12.4
Refinery Fuel	38.6	38.1	38.4	38.9	37.3	39.3	37.9	36.2	46.0
Sub-Total	638.5	677.9	711.8	788.9	787.6	822.1	888.3	961.4	925.7
Bunkers	10.3	4.6	4.7	4.5	3.7	1.7	6.3	5.8	6.3
Total	648.8	682.5	716.5	793.4	791.3	823.8	894.6	967.2	931.9

[1] See Notes in Part I.4.

Oil Demand by Main Product Groups[1]
Demande de produits raffinés par groupes principaux

Islamic Republic of Iran

	1971	1973	1978	1984	1985	1986	1987	1988	1989
Thousand tonnes									
NGL/LPG/Ethane	402	451	494	869	850	750	741	830	1 290
Naphtha	311	346	-	-	-	-	-	10	10
Motor Gasoline	1 300	1 854	3 771	4 752	5 166	4 444	4 736	4 713	5 451
Aviation Fuels	310	840	1 187	561	595	489	524	420	410
Kerosene	1 713	2 321	4 740	6 450	6 752	4 780	5 736	5 671	5 140
Gas Diesel	2 884	3 947	7 713	12 411	11 556	13 315	14 261	14 137	14 716
Heavy Fuel Oil	1 588	2 217	4 481	9 029	10 086	12 297	11 876	12 193	12 401
Other	640	767	1 307	1 401	1 479	2 080	2 285	2 471	2 371
Refinery Fuel	1 559	1 608	2 042	1 852	1 942	2 837	2 222	2 102	2 387
Sub-Total	10 707	14 351	25 735	37 325	38 426	40 992	42 381	42 547	44 176
Bunkers	5 819	7 010	3 443	1 700	1 650	1 600	1 350	1 500	2 207
Total	16 526	21 361	29 178	39 025	40 076	42 592	43 731	44 047	46 383
Thousand barrels/day									
NGL/LPG/Ethane	12.8	14.3	15.7	27.5	27.0	23.8	23.6	26.3	41.0
Naphtha	7.2	8.1	-	-	-	-	-	0.2	0.2
Motor Gasoline	30.4	43.3	88.1	110.8	120.7	103.9	110.7	109.8	127.4
Aviation Fuels	6.8	18.3	25.9	12.3	13.1	10.8	11.6	9.3	9.1
Kerosene	36.3	49.2	100.5	136.4	143.2	101.4	121.6	119.9	109.0
Gas Diesel	58.9	80.7	157.6	253.0	236.2	272.1	291.5	288.1	300.8
Heavy Fuel Oil	29.0	40.5	81.8	164.3	184.0	224.4	216.7	221.9	226.3
Other	12.3	14.7	25.1	26.8	28.4	39.9	43.8	47.3	45.5
Refinery Fuel	28.4	29.3	37.3	33.7	35.4	51.8	40.5	38.3	43.6
Sub-Total	222.1	298.4	532.0	764.8	788.0	828.0	860.0	861.1	902.8
Bunkers	106.5	128.1	62.9	31.4	30.7	29.6	25.2	28.0	40.3
Total	328.6	426.6	594.9	796.1	818.7	857.6	885.1	889.1	943.1

	1990	1991	1992	1993	1994	1995	1996	1997	1998
Thousand tonnes									
NGL/LPG/Ethane	1 370	1 600	1 693	1 667	1 694	1 689	1 816	1 837	1 941
Naphtha	130	140	140	140	140	253	348	539	1 759
Motor Gasoline	6 076	6 800	7 050	7 912	8 254	8 438	8 886	9 414	10 096
Aviation Fuels	561	590	605	573	591	180	211	239	219
Kerosene	5 300	5 975	6 058	7 203	7 653	1 355	1 440	1 372	1 276
Gas Diesel	16 146	17 220	17 549	18 176	18 703	16 883	17 396	18 235	18 924
Heavy Fuel Oil	11 900	11 900	12 463	12 723	12 816	13 242	14 248	14 914	12 228
Other	3 000	3 300	3 350	3 655	3 665	2 932	3 196	3 224	3 172
Refinery Fuel	2 400	2 200	2 200	2 133	2 157	2 182	2 288	2 356	2 319
Sub-Total	46 883	49 725	51 108	54 182	55 673	47 154	49 829	52 130	51 934
Bunkers	2 700	2 700	1 517	851	831	761	465	345	608
Total	49 583	52 425	52 625	55 033	56 504	47 915	50 294	52 475	52 542
Thousand barrels/day									
NGL/LPG/Ethane	43.5	50.8	53.7	53.0	53.8	53.7	57.6	58.4	61.7
Naphtha	3.0	3.3	3.3	3.3	3.3	5.9	8.1	12.6	41.0
Motor Gasoline	142.0	158.9	164.3	184.9	192.9	197.2	207.1	220.0	235.9
Aviation Fuels	12.5	13.1	13.4	12.6	13.0	4.1	4.8	5.4	5.0
Kerosene	112.4	126.7	128.1	152.7	162.3	28.7	30.5	29.1	27.1
Gas Diesel	330.0	351.9	357.7	371.5	382.3	345.1	354.6	372.7	386.8
Heavy Fuel Oil	217.1	217.1	226.8	232.2	233.8	241.6	259.3	272.1	223.1
Other	57.5	63.3	64.1	70.1	70.3	50.2	54.2	54.9	53.9
Refinery Fuel	43.8	40.1	40.0	40.3	40.8	41.2	43.3	45.1	43.6
Sub-Total	961.9	1 025.3	1 051.3	1 120.6	1 152.5	967.6	1 019.3	1 070.3	1 078.1
Bunkers	49.3	49.3	27.8	15.7	15.3	13.9	8.5	6.3	11.1
Total	1 011.1	1 074.6	1 079.1	1 136.3	1 167.8	981.5	1 027.7	1 076.5	1 089.2

[1] See Notes in Part I.4.

Oil Demand by Main Product Groups[1]
Demande de produits raffinés par groupes principaux

Iraq

	1971	1973	1978	1984	1985	1986	1987	1988	1989
Thousand tonnes									
NGL/LPG/Ethane	42	57	325	528	629	740	861	1 046	1 178
Naphtha	65	32	90	-	-	-	-	-	700
Motor Gasoline	419	465	963	2 173	2 345	2 500	2 520	2 650	2 750
Aviation Fuels	75	117	284	336	354	629	745	675	716
Kerosene	649	635	889	800	760	750	790	800	800
Gas Diesel	543	700	1 990	3 950	4 200	4 115	4 195	4 350	4 600
Heavy Fuel Oil	1 343	1 580	1 719	3 013	3 425	3 000	3 400	3 400	3 400
Other	299	422	797	1 117	1 182	1 877	1 649	1 963	1 582
Refinery Fuel	149	176	191	1 387	1 509	1 434	1 436	1 434	1 420
Sub-Total	3 584	4 184	7 248	13 304	14 404	15 045	15 596	16 318	17 146
Bunkers	84	80	120	150	150	100	200	130	130
Total	3 668	4 264	7 368	13 454	14 554	15 145	15 796	16 448	17 276
Thousand barrels/day									
NGL/LPG/Ethane	1.3	1.8	10.3	16.7	20.0	23.5	27.4	33.2	37.4
Naphtha	1.5	0.7	2.1	-	-	-	-	-	16.3
Motor Gasoline	9.8	10.9	22.5	50.6	54.8	58.4	58.9	61.8	64.3
Aviation Fuels	1.6	2.6	6.2	7.3	7.7	13.7	16.2	14.6	15.6
Kerosene	13.8	13.5	18.9	16.9	16.1	15.9	16.8	16.9	17.0
Gas Diesel	11.1	14.3	40.7	80.5	85.8	84.1	85.7	88.7	94.0
Heavy Fuel Oil	24.5	28.8	31.4	54.8	62.5	54.7	62.0	61.9	62.0
Other	5.7	7.8	14.9	20.5	21.7	35.8	31.2	37.0	29.3
Refinery Fuel	2.7	3.2	3.5	26.7	29.1	27.8	27.8	27.7	27.5
Sub-Total	72.0	83.6	150.4	274.2	297.8	314.0	326.0	341.7	363.4
Bunkers	1.5	1.5	2.2	2.7	2.7	1.8	3.6	2.4	2.4
Total	73.6	85.1	152.5	276.9	300.5	315.8	329.6	344.1	365.7

	1990	1991	1992	1993	1994	1995	1996	1997	1998
Thousand tonnes									
NGL/LPG/Ethane	1 020	300	518	1 182	1 186	1 166	1 131	1 144	1 180
Naphtha	600	450	600	500	500	492	477	483	498
Motor Gasoline	2 800	2 000	2 460	2 740	2 970	2 990	2 858	2 886	2 988
Aviation Fuels	916	250	358	385	440	425	412	419	431
Kerosene	600	560	675	885	1 012	978	948	964	992
Gas Diesel	4 700	4 300	4 800	5 250	5 387	5 322	5 322	5 375	5 428
Heavy Fuel Oil	3 600	3 840	4 010	4 700	5 309	5 218	5 140	5 157	5 370
Other	1 778	1 335	2 120	2 490	2 588	2 574	2 550	2 559	2 584
Refinery Fuel	1 269	870	1 130	1 420	1 670	1 634	1 599	1 609	1 671
Sub-Total	17 283	13 905	16 671	19 552	21 062	20 799	20 437	20 596	21 142
Bunkers	130	-	-	-	-	-	-	-	-
Total	17 413	13 905	16 671	19 552	21 062	20 799	20 437	20 596	21 142
Thousand barrels/day									
NGL/LPG/Ethane	32.4	9.5	16.4	37.6	37.7	37.1	35.8	36.4	37.5
Naphtha	14.0	10.5	13.9	11.6	11.6	11.5	11.1	11.2	11.6
Motor Gasoline	65.4	46.7	57.3	64.0	69.4	69.9	66.6	67.4	69.8
Aviation Fuels	19.9	5.4	7.8	8.4	9.6	9.2	8.9	9.1	9.4
Kerosene	12.7	11.9	14.3	18.8	21.5	20.7	20.0	20.4	21.0
Gas Diesel	96.1	87.9	97.8	107.3	110.1	108.8	108.5	109.9	110.9
Heavy Fuel Oil	65.7	70.1	73.0	85.8	96.9	95.2	93.5	94.1	98.0
Other	33.3	25.7	41.3	48.9	50.6	50.4	49.8	50.1	50.6
Refinery Fuel	24.5	16.9	21.8	27.5	32.2	31.5	30.7	31.0	32.2
Sub-Total	364.1	284.6	343.6	409.8	439.6	434.2	425.1	429.7	441.0
Bunkers	2.4	-	-	-	-	-	-	-	-
Total	366.4	284.6	343.6	409.8	439.6	434.2	425.1	429.7	441.0

[1] See Notes in Part I.4.

Oil Demand by Main Product Groups[1]
Demande de produits raffinés par groupes principaux

Israel

	1971	1973	1978	1984	1985	1986	1987	1988	1989
Thousand tonnes									
NGL/LPG/Ethane	111	123	126	135	127	201	223	225	209
Naphtha	64	31	206	273	226	317	383	412	385
Motor Gasoline	613	675	748	932	1 010	1 115	1 246	1 321	1 400
Aviation Fuels	-	-	-	-	-	-	-	-	-
Kerosene	575	789	730	662	640	630	705	683	720
Gas Diesel	770	866	951	914	843	989	1 078	1 091	1 164
Heavy Fuel Oil	2 606	2 957	3 700	2 939	2 891	2 921	3 177	3 623	3 592
Other	329	416	352	171	125	146	166	198	184
Refinery Fuel	162	179	233	187	560	578	632	634	722
Sub-Total	5 230	6 036	7 046	6 213	6 422	6 897	7 610	8 187	8 376
Bunkers	-	-	-	-	113	118	136	116	130
Total	5 230	6 036	7 046	6 213	6 535	7 015	7 746	8 303	8 506
Thousand barrels/day									
NGL/LPG/Ethane	3.5	3.9	4.0	4.3	4.0	6.4	7.1	7.1	6.6
Naphtha	1.5	0.7	4.8	6.3	5.3	7.4	8.9	9.6	9.0
Motor Gasoline	14.3	15.8	17.5	21.7	23.6	26.1	29.1	30.8	32.7
Aviation Fuels	-	-	-	-	-	-	-	-	-
Kerosene	12.2	16.7	15.5	14.0	13.6	13.4	15.0	14.4	15.3
Gas Diesel	15.7	17.7	19.4	18.6	17.2	20.2	22.0	22.2	23.8
Heavy Fuel Oil	47.6	54.0	67.5	53.5	52.8	53.3	58.0	65.9	65.5
Other	6.4	8.1	6.9	3.3	2.4	2.8	3.2	3.8	3.5
Refinery Fuel	3.0	3.3	4.3	3.4	10.2	10.5	11.5	11.5	13.2
Sub-Total	104.2	120.2	139.8	125.1	129.1	140.1	154.8	165.4	169.6
Bunkers	-	-	-	-	2.1	2.2	2.6	2.2	2.4
Total	104.2	120.2	139.8	125.1	131.2	142.3	157.4	167.6	172.1

	1990	1991	1992	1993	1994	1995	1996	1997	1998
Thousand tonnes									
NGL/LPG/Ethane	228	215	215	240	264	296	315	412	413
Naphtha	403	478	482	450	550	645	730	775	887
Motor Gasoline	1 484	1 546	1 650	1 720	1 834	1 970	2 029	2 005	2 002
Aviation Fuels	-	-	-	-	-	-	-	-	-
Kerosene	735	710	725	838	852	936	993	1 012	1 070
Gas Diesel	1 291	1 351	1 527	1 514	1 739	2 049	1 996	2 269	2 586
Heavy Fuel Oil	3 512	3 336	3 564	3 588	3 834	4 089	3 953	3 810	4 020
Other	291	283	349	534	488	540	527	441	567
Refinery Fuel	712	721	745	856	794	637	500	586	508
Sub-Total	8 656	8 640	9 257	9 740	10 355	11 162	11 043	11 310	12 053
Bunkers	122	153	191	191	203	207	94	180	153
Total	8 778	8 793	9 448	9 931	10 558	11 369	11 137	11 490	12 206
Thousand barrels/day									
NGL/LPG/Ethane	7.2	6.8	6.8	7.6	8.4	9.4	9.9	13.1	13.1
Naphtha	9.4	11.1	11.2	10.5	12.8	15.0	17.0	18.0	20.7
Motor Gasoline	34.7	36.1	38.5	40.2	42.9	46.0	47.3	46.9	46.8
Aviation Fuels	-	-	-	-	-	-	-	-	-
Kerosene	15.6	15.1	15.3	17.8	18.1	19.8	21.0	21.5	22.7
Gas Diesel	26.4	27.6	31.1	30.9	35.5	41.9	40.7	46.4	52.9
Heavy Fuel Oil	64.1	60.9	64.9	65.5	70.0	74.6	71.9	69.5	73.4
Other	5.6	5.4	6.7	10.2	9.4	10.4	10.1	8.5	10.9
Refinery Fuel	13.0	13.2	13.6	15.6	14.5	11.6	9.1	10.7	9.3
Sub-Total	175.9	176.2	188.0	198.3	211.4	228.7	227.0	234.5	249.6
Bunkers	2.3	2.9	3.6	3.6	3.9	4.0	1.8	3.4	2.9
Total	178.3	179.1	191.6	201.9	215.3	232.7	228.8	237.9	252.5

[1] See Notes in Part I.4.

Oil Demand by Main Product Groups[1]
Demande de produits raffinés par groupes principaux

Jamaica

	1971	1973	1978	1984	1985	1986	1987	1988	1989
Thousand tonnes									
NGL/LPG/Ethane	24	38	39	33	32	34	36	37	40
Naphtha	-	-	-	-	-	-	-	-	-
Motor Gasoline	330	236	206	181	176	184	197	207	229
Aviation Fuels	137	150	87	99	126	118	134	135	144
Kerosene	22	22	56	38	46	52	55	63	55
Gas Diesel	254	359	185	175	169	173	171	215	291
Heavy Fuel Oil	1 066	1 847	1 745	1 232	1 060	961	1 010	998	1 343
Other	40	39	26	21	16	12	17	23	18
Refinery Fuel	74	80	14	13	13	12	33	35	15
Sub-Total	1 947	2 771	2 358	1 792	1 638	1 546	1 653	1 713	2 135
Bunkers	51	117	37	13	12	20	21	13	30
Total	1 998	2 888	2 395	1 805	1 650	1 566	1 674	1 726	2 165
Thousand barrels/day									
NGL/LPG/Ethane	0.8	1.2	1.2	1.0	1.0	1.1	1.1	1.2	1.3
Naphtha	-	-	-	-	-	-	-	-	-
Motor Gasoline	7.7	5.5	4.8	4.2	4.1	4.3	4.6	4.8	5.4
Aviation Fuels	3.0	3.3	1.9	2.2	2.7	2.6	2.9	2.9	3.1
Kerosene	0.5	0.5	1.2	0.8	1.0	1.1	1.2	1.3	1.2
Gas Diesel	5.2	7.3	3.8	3.6	3.5	3.5	3.5	4.4	5.9
Heavy Fuel Oil	19.5	33.7	31.8	22.4	19.3	17.5	18.4	18.2	24.5
Other	0.7	0.7	0.5	0.4	0.3	0.2	0.3	0.4	0.3
Refinery Fuel	1.4	1.6	0.3	0.3	0.3	0.3	0.7	0.7	0.3
Sub-Total	38.7	53.7	45.5	34.9	32.2	30.6	32.7	33.9	42.0
Bunkers	0.9	2.1	0.7	0.3	0.2	0.4	0.4	0.3	0.6
Total	39.7	55.9	46.2	35.1	32.5	31.0	33.1	34.2	42.6

	1990	1991	1992	1993	1994	1995	1996	1997	1998
Thousand tonnes									
NGL/LPG/Ethane	43	44	48	48	50	50	61	62	66
Naphtha	-	-	-	-	-	-	-	-	-
Motor Gasoline	234	247	233	275	282	353	411	437	466
Aviation Fuels	148	142	150	155	155	167	181	187	192
Kerosene	34	30	42	43	53	54	20	15	11
Gas Diesel	315	272	251	309	473	485	499	512	526
Heavy Fuel Oil	1 607	1 715	1 762	1 727	1 584	1 839	1 889	1 935	1 958
Other	19	18	18	14	23	28	33	22	16
Refinery Fuel	60	-	6	3	2	2	2	2	2
Sub-Total	2 460	2 468	2 510	2 574	2 622	2 978	3 096	3 172	3 237
Bunkers	36	42	44	37	37	37	37	37	37
Total	2 496	2 510	2 554	2 611	2 659	3 015	3 133	3 209	3 274
Thousand barrels/day									
NGL/LPG/Ethane	1.4	1.4	1.5	1.5	1.6	1.6	1.9	2.0	2.1
Naphtha	-	-	-	-	-	-	-	-	-
Motor Gasoline	5.5	5.8	5.4	6.4	6.6	8.3	9.6	10.2	10.9
Aviation Fuels	3.2	3.1	3.3	3.4	3.4	3.6	3.9	4.1	4.2
Kerosene	0.7	0.6	0.9	0.9	1.1	1.1	0.4	0.3	0.2
Gas Diesel	6.4	5.6	5.1	6.3	9.7	9.9	10.2	10.5	10.8
Heavy Fuel Oil	29.3	31.3	32.1	31.5	28.9	33.6	34.4	35.3	35.7
Other	0.3	0.3	0.3	0.2	0.4	0.5	0.6	0.4	0.3
Refinery Fuel	1.2	-	0.2	0.1	0.0	0.0	0.0	0.0	0.0
Sub-Total	48.0	48.1	48.8	50.4	51.7	58.6	61.0	62.8	64.2
Bunkers	0.7	0.9	0.9	0.8	0.8	0.8	0.8	0.8	0.8
Total	48.8	48.9	49.7	51.2	52.5	59.4	61.8	63.5	65.0

[1] See Notes in Part I.4.

Oil Demand by Main Product Groups[1]
Demande de produits raffinés par groupes principaux

Jordan

	1971	1973	1978	1984	1985	1986	1987	1988	1989
Thousand tonnes									
NGL/LPG/Ethane	14	17	42	83	77	85	90	99	100
Naphtha	-	-	-	-	-	-	-	-	-
Motor Gasoline	83	107	229	309	309	311	314	319	322
Aviation Fuels	45	50	131	232	215	175	174	174	224
Kerosene	102	92	137	141	127	136	136	152	141
Gas Diesel	122	165	407	683	737	783	784	805	809
Heavy Fuel Oil	69	116	179	919	962	1 177	1 273	1 127	1 114
Other	34	22	96	132	133	138	153	147	111
Refinery Fuel	32	42	58	102	152	113	162	170	179
Sub-Total	501	611	1 279	2 601	2 712	2 918	3 086	2 993	3 000
Bunkers	-	-	-	-	-	9	32	-	-
Total	501	611	1 279	2 601	2 712	2 927	3 118	2 993	3 000
Thousand barrels/day									
NGL/LPG/Ethane	0.4	0.5	1.3	2.6	2.4	2.7	2.9	3.1	3.2
Naphtha	-	-	-	-	-	-	-	-	-
Motor Gasoline	1.9	2.5	5.4	7.2	7.2	7.3	7.3	7.4	7.5
Aviation Fuels	1.0	1.1	2.9	5.1	4.7	3.9	3.8	3.8	4.9
Kerosene	2.2	2.0	2.9	3.0	2.7	2.9	2.9	3.2	3.0
Gas Diesel	2.5	3.4	8.3	13.9	15.1	16.0	16.0	16.4	16.5
Heavy Fuel Oil	1.3	2.1	3.3	16.7	17.6	21.5	23.2	20.5	20.3
Other	0.6	0.4	1.6	2.2	2.2	2.3	2.6	2.5	1.9
Refinery Fuel	0.6	0.8	1.1	1.9	2.9	2.2	3.1	3.3	3.4
Sub-Total	10.4	12.7	26.7	52.6	54.8	58.7	61.8	60.2	60.8
Bunkers	-	-	-	-	-	0.2	0.6	-	-
Total	10.4	12.7	26.7	52.6	54.8	58.9	62.4	60.2	60.8

	1990	1991	1992	1993	1994	1995	1996	1997	1998
Thousand tonnes									
NGL/LPG/Ethane	121	131	155	167	185	203	220	249	243
Naphtha	-	-	-	-	-	-	-	-	-
Motor Gasoline	400	384	421	435	454	487	516	533	530
Aviation Fuels	229	163	210	221	235	251	299	277	225
Kerosene	159	158	284	236	230	216	192	208	182
Gas Diesel	847	822	852	892	964	1 037	1 103	1 236	1 232
Heavy Fuel Oil	1 185	1 212	1 479	1 492	1 519	1 661	1 670	1 644	1 829
Other	132	123	149	167	140	136	142	129	143
Refinery Fuel	154	180	147	147	216	177	240	232	242
Sub-Total	3 227	3 173	3 697	3 757	3 943	4 168	4 382	4 508	4 626
Bunkers	-	-	-	-	4	9	3	6	3
Total	3 227	3 173	3 697	3 757	3 947	4 177	4 385	4 514	4 629
Thousand barrels/day									
NGL/LPG/Ethane	3.7	4.0	4.7	5.1	5.7	6.2	6.7	7.6	7.5
Naphtha	-	-	-	-	-	-	-	-	-
Motor Gasoline	9.2	8.8	9.7	10.0	10.4	11.2	11.8	12.3	12.2
Aviation Fuels	5.2	3.7	4.8	5.0	5.4	5.7	6.8	6.3	5.1
Kerosene	3.5	3.4	6.2	5.1	5.0	4.7	4.2	4.5	4.0
Gas Diesel	17.5	17.0	17.6	18.4	19.9	21.4	22.7	25.5	25.5
Heavy Fuel Oil	21.8	22.2	27.1	27.4	27.9	30.5	30.6	30.2	33.6
Other	2.2	2.1	2.5	2.8	2.3	2.3	2.4	2.1	2.4
Refinery Fuel	2.9	3.3	2.7	2.8	4.1	3.3	4.5	4.4	4.5
Sub-Total	66.0	64.7	75.3	76.7	80.7	85.3	89.7	92.9	94.7
Bunkers	-	-	-	-	0.1	0.2	0.1	0.1	0.1
Total	66.0	64.7	75.3	76.7	80.8	85.5	89.7	93.1	94.7

[1] See Notes in Part I.4.

Oil Demand by Main Product Groups[1]
Demande de produits raffinés par groupes principaux

Kazakhstan

	1971	1973	1978	1984	1985	1986	1987	1988	1989
Thousand tonnes									
NGL/LPG/Ethane	-	-	-	-	-	-	-	-	-
Naphtha	-	-	-	-	-	-	-	-	-
Motor Gasoline	-	-	-	-	-	-	-	-	-
Aviation Fuels	-	-	-	-	-	-	-	-	-
Kerosene	-	-	-	-	-	-	-	-	-
Gas Diesel	-	-	-	-	-	-	-	-	-
Heavy Fuel Oil	-	-	-	-	-	-	-	-	-
Other	-	-	-	-	-	-	-	-	-
Refinery Fuel	-	-	-	-	-	-	-	-	-
Sub-Total	-	-	-	-	-	-	-	-	-
Bunkers	-	-	-	-	-	-	-	-	-
Total	-	-	-	-	-	-	-	-	-
Thousand barrels/day									
NGL/LPG/Ethane	-	-	-	-	-	-	-	-	-
Naphtha	-	-	-	-	-	-	-	-	-
Motor Gasoline	-	-	-	-	-	-	-	-	-
Aviation Fuels	-	-	-	-	-	-	-	-	-
Kerosene	-	-	-	-	-	-	-	-	-
Gas Diesel	-	-	-	-	-	-	-	-	-
Heavy Fuel Oil	-	-	-	-	-	-	-	-	-
Other	-	-	-	-	-	-	-	-	-
Refinery Fuel	-	-	-	-	-	-	-	-	-
Sub-Total	-	-	-	-	-	-	-	-	-
Bunkers	-	-	-	-	-	-	-	-	-
Total	-	-	-	-	-	-	-	-	-

	1990	1991	1992	1993	1994	1995	1996	1997	1998
Thousand tonnes									
NGL/LPG/Ethane	-	-	700	216	157	116	89	75	62
Naphtha	-	-	-	-	-	-	-	-	-
Motor Gasoline	-	-	3 833	3 405	2 411	2 182	2 219	1 906	1 919
Aviation Fuels	-	-	982	331	485	431	358	221	278
Kerosene	-	-	1 001	528	252	342	292	237	268
Gas Diesel	-	-	5 832	4 408	3 143	3 454	2 161	2 197	2 466
Heavy Fuel Oil	-	-	5 557	4 762	4 361	3 275	2 997	2 308	2 133
Other	-	-	3 088	2 160	827	348	630	657	1 034
Refinery Fuel	-	-	730	580	566	438	465	367	343
Sub-Total	-	-	21 723	16 390	12 202	10 586	9 211	7 968	8 503
Bunkers	-	-	-	-	-	-	-	-	-
Total	-	-	21 723	16 390	12 202	10 586	9 211	7 968	8 503
Thousand barrels/day									
NGL/LPG/Ethane	-	-	22.2	6.9	5.0	3.7	2.8	2.4	2.0
Naphtha	-	-	-	-	-	-	-	-	-
Motor Gasoline	-	-	89.3	79.6	56.3	51.0	51.7	44.5	44.8
Aviation Fuels	-	-	21.3	7.2	10.5	9.4	7.8	4.8	6.0
Kerosene	-	-	21.2	11.2	5.3	7.3	6.2	5.0	5.7
Gas Diesel	-	-	118.9	90.1	64.2	70.6	44.0	44.9	50.4
Heavy Fuel Oil	-	-	101.1	86.9	79.6	59.8	54.5	42.1	38.9
Other	-	-	59.3	41.4	15.9	6.7	12.1	12.6	19.8
Refinery Fuel	-	-	13.9	11.1	10.8	8.4	8.9	7.0	6.6
Sub-Total	-	-	447.2	334.4	247.7	216.7	188.0	163.4	174.2
Bunkers	-	-	-	-	-	-	-	-	-
Total	-	-	447.2	334.4	247.7	216.7	188.0	163.4	174.2

[1] See Notes in Part I.4.

Oil Demand by Main Product Groups[1]
Demande de produits raffinés par groupes principaux

Kenya

	1971	1973	1978	1984	1985	1986	1987	1988	1989
Thousand tonnes									
NGL/LPG/Ethane	8	12	17	22	22	19	26	27	26
Naphtha	-	-	-	-	-	-	-	-	-
Motor Gasoline	197	233	291	318	348	395	421	425	477
Aviation Fuels	186	261	335	265	267	269	256	261	280
Kerosene	46	53	78	111	121	138	167	185	211
Gas Diesel	249	301	353	445	473	510	580	547	569
Heavy Fuel Oil	399	411	464	406	377	458	373	358	336
Other	34	31	67	70	71	69	87	70	56
Refinery Fuel	64	58	74	45	44	71	83	78	86
Sub-Total	1 183	1 360	1 679	1 682	1 723	1 929	1 993	1 951	2 041
Bunkers	476	458	219	160	145	158	169	183	158
Total	1 659	1 818	1 898	1 842	1 868	2 087	2 162	2 134	2 199
Thousand barrels/day									
NGL/LPG/Ethane	0.3	0.4	0.5	0.7	0.7	0.6	0.8	0.9	0.8
Naphtha	-	-	-	-	-	-	-	-	-
Motor Gasoline	4.6	5.4	6.8	7.4	8.1	9.2	9.8	9.9	11.1
Aviation Fuels	4.1	5.7	7.3	5.8	5.8	5.9	5.6	5.7	6.1
Kerosene	1.0	1.1	1.7	2.3	2.6	2.9	3.5	3.9	4.5
Gas Diesel	5.1	6.2	7.2	9.1	9.7	10.4	11.9	11.1	11.6
Heavy Fuel Oil	7.3	7.5	8.5	7.4	6.9	8.4	6.8	6.5	6.1
Other	0.6	0.5	1.2	1.3	1.3	1.3	1.6	1.3	1.0
Refinery Fuel	1.2	1.1	1.4	0.9	0.8	1.4	1.7	1.6	1.8
Sub-Total	24.1	27.9	34.6	34.8	35.9	40.1	41.7	40.9	43.1
Bunkers	8.9	8.6	4.1	3.0	2.7	3.0	3.2	3.4	2.9
Total	32.9	36.5	38.8	37.8	38.6	43.1	44.9	44.3	46.0

	1990	1991	1992	1993	1994	1995	1996	1997	1998
Thousand tonnes									
NGL/LPG/Ethane	27	25	27	25	28	31	31	31	31
Naphtha	-	-	-	-	-	-	-	-	-
Motor Gasoline	398	363	347	352	352	379	399	391	396
Aviation Fuels	272	239	316	357	481	440	450	436	422
Kerosene	184	175	175	165	173	243	254	268	318
Gas Diesel	592	590	599	577	564	627	673	664	634
Heavy Fuel Oil	343	336	341	369	379	323	394	357	367
Other	59	60	63	59	42	55	49	43	60
Refinery Fuel	89	75	80	181	187	184	173	184	185
Sub-Total	1 964	1 863	1 948	2 085	2 206	2 282	2 423	2 374	2 413
Bunkers	178	149	140	166	137	62	61	60	61
Total	2 142	2 012	2 088	2 251	2 343	2 344	2 484	2 434	2 474
Thousand barrels/day									
NGL/LPG/Ethane	0.9	0.8	0.9	0.8	0.9	1.0	1.0	1.0	1.0
Naphtha	-	-	-	-	-	-	-	-	-
Motor Gasoline	9.3	8.5	8.1	8.2	8.2	8.9	9.3	9.1	9.3
Aviation Fuels	5.9	5.2	6.9	7.8	10.5	9.6	9.8	9.5	9.2
Kerosene	3.9	3.7	3.7	3.5	3.7	5.2	5.4	5.7	6.7
Gas Diesel	12.1	12.1	12.2	11.8	11.5	12.8	13.7	13.6	13.0
Heavy Fuel Oil	6.3	6.1	6.2	6.7	6.9	5.9	7.2	6.5	6.7
Other	1.0	1.1	1.1	1.1	0.8	1.0	0.9	0.8	1.1
Refinery Fuel	1.8	1.5	1.6	3.7	3.8	3.8	3.5	3.7	3.8
Sub-Total	41.2	39.0	40.7	43.6	46.3	48.0	50.7	49.9	50.7
Bunkers	3.3	2.8	2.6	3.1	2.5	1.2	1.1	1.1	1.1
Total	44.5	41.8	43.3	46.7	48.8	49.2	51.9	51.0	51.8

[1] See Notes in Part I.4.

Oil Demand by Main Product Groups[1]
Demande de produits raffinés par groupes principaux

Dem. People's Rep. of Korea

	1971	1973	1978	1984	1985	1986	1987	1988	1989
Thousand tonnes									
NGL/LPG/Ethane	-	-	-	-	-	-	-	-	-
Naphtha	-	-	-	-	-	-	-	-	-
Motor Gasoline	300	350	700	900	950	1 400	2 400	1 940	1 975
Aviation Fuels	-	-	-	-	-	-	-	-	-
Kerosene	30	35	101	220	230	245	255	265	282
Gas Diesel	263	285	605	1 100	1 175	1 230	1 230	1 320	1 365
Heavy Fuel Oil	90	95	326	528	572	752	805	867	885
Other	-	-	-	-	-	-	-	-	-
Refinery Fuel	-	-	38	72	78	103	110	118	120
Sub-Total	683	765	1 770	2 820	3 005	3 730	4 800	4 510	4 627
Bunkers	-	-	-	-	-	-	-	-	-
Total	683	765	1 770	2 820	3 005	3 730	4 800	4 510	4 627
Thousand barrels/day									
NGL/LPG/Ethane	-	-	-	-	-	-	-	-	-
Naphtha	-	-	-	-	-	-	-	-	-
Motor Gasoline	7.0	8.2	16.4	21.0	22.2	32.7	56.1	45.2	46.2
Aviation Fuels	-	-	-	-	-	-	-	-	-
Kerosene	0.6	0.7	2.1	4.7	4.9	5.2	5.4	5.6	6.0
Gas Diesel	5.4	5.8	12.4	22.4	24.0	25.1	25.1	26.9	27.9
Heavy Fuel Oil	1.6	1.7	5.9	9.6	10.4	13.7	14.7	15.8	16.1
Other	-	-	-	-	-	-	-	-	-
Refinery Fuel	-	-	0.7	1.3	1.4	1.9	2.0	2.1	2.2
Sub-Total	14.7	16.5	37.5	59.0	63.0	78.7	103.3	95.6	98.4
Bunkers	-	-	-	-	-	-	-	-	-
Total	14.7	16.5	37.5	59.0	63.0	78.7	103.3	95.6	98.4

	1990	1991	1992	1993	1994	1995	1996	1997	1998
Thousand tonnes									
NGL/LPG/Ethane	-	-	-	-	-	-	-	-	-
Naphtha	-	-	-	-	-	-	-	-	-
Motor Gasoline	1 900 ·	1 860	1 860	1 840	1 820	1 800	1 785	1 663	1 580
Aviation Fuels	-	-	-	-	-	-	-	-	-
Kerosene	282	270	270	265	259	253	247	230	219
Gas Diesel	1 375	1 345	1 345	1 340	1 330	1 320	1 310	1 221	1 160
Heavy Fuel Oil	859	837	837	833	826	816	807	752	714
Other	-	-	-	-	-	-	-	-	-
Refinery Fuel	116	113	113	112	111	110	109	102	97
Sub-Total	4 532	4 425	4 425	4 390	4 346	4 299	4 258	3 968	3 770
Bunkers	-	-	-	-	-	-	-	-	-
Total	4 532	4 425	4 425	4 390	4 346	4 299	4 258	3 968	3 770
Thousand barrels/day									
NGL/LPG/Ethane	-	-	-	-	-	-	-	-	-
Naphtha	-	-	-	-	-	-	-	-	-
Motor Gasoline	44.4	43.5	43.3	43.0	42.5	42.1	41.6	38.9	36.9
Aviation Fuels	-	-	-	-	-	-	-	-	-
Kerosene	6.0	5.7	5.7	5.6	5.5	5.4	5.2	4.9	4.6
Gas Diesel	28.1	27.5	27.4	27.4	27.2	27.0	26.7	25.0	23.7
Heavy Fuel Oil	15.7	15.3	15.2	15.2	15.1	14.9	14.7	13.7	13.0
Other	-	-	-	-	-	-	-	-	-
Refinery Fuel	2.1	2.1	2.1	2.0	2.0	2.0	2.0	1.9	1.8
Sub-Total	96.3	94.0	93.8	93.3	92.3	91.3	90.2	84.3	80.1
Bunkers	-	-	-	-	-	-	-	-	-
Total	96.3	94.0	93.8	93.3	92.3	91.3	90.2	84.3	80.1

[1] See Notes in Part I.4.

Oil Demand by Main Product Groups[1]
Demande de produits raffinés par groupes principaux

Kuwait

	1971	1973	1978	1984	1985	1986	1987	1988	1989
Thousand tonnes									
NGL/LPG/Ethane	36	31	44	626	411	609	672	698	822
Naphtha	-	-	-	-	-	-	-	-	-
Motor Gasoline	490	555	713	1 094	1 115	1 179	1 185	1 260	1 372
Aviation Fuels	109	71	306	332	307	310	296	316	352
Kerosene	32	36	35	25	27	33	33	33	28
Gas Diesel	217	246	424	2 102	3 032	2 585	2 340	3 239	2 641
Heavy Fuel Oil	3	28	549	1 276	1 532	1 464	1 559	1 785	1 780
Other	54	37	113	280	361	391	583	627	919
Refinery Fuel	3 534	3 971	2 395	3 260	2 937	2 686	2 885	3 247	3 303
Sub-Total	**4 475**	**4 975**	**4 579**	**8 995**	**9 722**	**9 257**	**9 553**	**11 205**	**11 217**
Bunkers	1 815	2 076	1 777	651	688	639	669	773	810
Total	**6 290**	**7 051**	**6 356**	**9 646**	**10 410**	**9 896**	**10 222**	**11 978**	**12 027**
Thousand barrels/day									
NGL/LPG/Ethane	1.1	1.0	1.4	19.8	13.1	19.4	21.4	22.1	26.1
Naphtha	-	-	-	-	-	-	-	-	-
Motor Gasoline	11.5	13.0	16.7	25.5	26.1	27.6	27.7	29.4	32.1
Aviation Fuels	2.4	1.5	6.7	7.2	6.7	6.7	6.4	6.8	7.7
Kerosene	0.7	0.8	0.7	0.5	0.6	0.7	0.7	0.7	0.6
Gas Diesel	4.4	5.0	8.7	42.8	62.0	52.8	47.8	66.0	54.0
Heavy Fuel Oil	0.1	0.5	10.0	23.2	28.0	26.7	28.4	32.5	32.5
Other	0.9	0.6	1.9	5.0	6.5	7.2	10.8	11.8	17.4
Refinery Fuel	65.3	73.2	44.4	60.2	54.3	49.8	53.7	60.1	61.3
Sub-Total	**86.3**	**95.6**	**90.4**	**184.3**	**197.1**	**190.9**	**197.0**	**229.4**	**231.6**
Bunkers	33.4	38.2	32.8	12.0	12.7	11.8	12.4	14.3	15.0
Total	**119.8**	**133.9**	**123.2**	**196.3**	**209.8**	**202.7**	**209.3**	**243.7**	**246.6**

	1990	1991	1992	1993	1994	1995	1996	1997	1998
Thousand tonnes									
NGL/LPG/Ethane	456	125	659	748	820	916	936	1 017	1 309
Naphtha	-	-	-	-	-	-	-	-	-
Motor Gasoline	787	600	1 153	1 281	1 362	1 469	1 556	1 693	1 782
Aviation Fuels	162	131	280	295	325	356	377	409	481
Kerosene	15	7	28	22	23	24	22	26	30
Gas Diesel	367	695	500	419	432	469	434	493	527
Heavy Fuel Oil	141	1 145	912	890	1 044	699	1 245	1 337	1 567
Other	709	2 359	3 462	3 857	4 785	4 485	5 745	6 009	6 961
Refinery Fuel	256	1 376	2 366	1 772	2 289	3 049	2 224	2 783	3 649
Sub-Total	**2 893**	**6 438**	**9 360**	**9 284**	**11 080**	**11 467**	**12 539**	**13 767**	**16 306**
Bunkers	180	-	428	360	548	588	556	665	623
Total	**3 073**	**6 438**	**9 788**	**9 644**	**11 628**	**12 055**	**13 095**	**14 432**	**16 929**
Thousand barrels/day									
NGL/LPG/Ethane	14.5	4.0	20.9	23.8	26.1	29.1	29.7	32.3	41.6
Naphtha	-	-	-	-	-	-	-	-	-
Motor Gasoline	18.4	14.0	26.9	29.9	31.8	34.3	36.3	39.6	41.6
Aviation Fuels	3.5	2.8	6.1	6.4	7.1	7.7	8.2	8.9	10.5
Kerosene	0.3	0.1	0.6	0.5	0.5	0.5	0.5	0.6	0.6
Gas Diesel	7.5	14.2	10.2	8.6	8.8	9.6	8.8	10.1	10.8
Heavy Fuel Oil	2.6	20.9	16.6	16.2	19.0	12.8	22.7	24.4	28.6
Other	13.5	46.8	68.2	75.8	93.9	87.7	112.5	118.0	136.8
Refinery Fuel	5.4	25.2	43.2	33.2	43.0	56.9	41.7	52.0	67.8
Sub-Total	**65.7**	**128.1**	**192.5**	**194.4**	**230.2**	**238.6**	**260.2**	**285.8**	**338.3**
Bunkers	3.3	-	7.9	6.7	10.3	11.0	10.4	12.4	11.7
Total	**69.1**	**128.1**	**200.4**	**201.1**	**240.5**	**249.6**	**270.6**	**298.3**	**350.0**

[1] See Notes in Part I.4.

Oil Demand by Main Product Groups[1]
Demande de produits raffinés par groupes principaux

Kyrgyzstan

	1971	1973	1978	1984	1985	1986	1987	1988	1989
Thousand tonnes									
NGL/LPG/Ethane	-	-	-	-	-	-	-	-	-
Naphtha	-	-	-	-	-	-	-	-	-
Motor Gasoline	-	-	-	-	-	-	-	-	-
Aviation Fuels	-	-	-	-	-	-	-	-	-
Kerosene	-	-	-	-	-	-	-	-	-
Gas Diesel	-	-	-	-	-	-	-	-	-
Heavy Fuel Oil	-	-	-	-	-	-	-	-	-
Other	-	-	-	-	-	-	-	-	-
Refinery Fuel	-	-	-	-	-	-	-	-	-
Sub-Total	-	-	-	-	-	-	-	-	-
Bunkers	-	-	-	-	-	-	-	-	-
Total	-	-	-	-	-	-	-	-	-
Thousand barrels/day									
NGL/LPG/Ethane	-	-	-	-	-	-	-	-	-
Naphtha	-	-	-	-	-	-	-	-	-
Motor Gasoline	-	-	-	-	-	-	-	-	-
Aviation Fuels	-	-	-	-	-	-	-	-	-
Kerosene	-	-	-	-	-	-	-	-	-
Gas Diesel	-	-	-	-	-	-	-	-	-
Heavy Fuel Oil	-	-	-	-	-	-	-	-	-
Other	-	-	-	-	-	-	-	-	-
Refinery Fuel	-	-	-	-	-	-	-	-	-
Sub-Total	-	-	-	-	-	-	-	-	-
Bunkers	-	-	-	-	-	-	-	-	-
Total	-	-	-	-	-	-	-	-	-

	1990	1991	1992	1993	1994	1995	1996	1997	1998
Thousand tonnes									
NGL/LPG/Ethane	-	-	75	35	13	-	-	-	-
Naphtha	-	-	-	-	-	-	-	-	-
Motor Gasoline	-	-	412	273	102	212	230	115	227
Aviation Fuels	-	-	60	16	13	81	68	45	53
Kerosene	-	-	-	-	-	-	-	-	-
Gas Diesel	-	-	460	236	59	132	144	100	81
Heavy Fuel Oil	-	-	498	255	95	92	100	88	108
Other	-	-	404	350	135	82	104	122	172
Refinery Fuel	-	-	-	-	-	-	-	-	-
Sub-Total	-	-	1 909	1 165	417	599	646	470	641
Bunkers	-	-	-	-	-	-	-	-	-
Total	-	-	1 909	1 165	417	599	646	470	641
Thousand barrels/day									
NGL/LPG/Ethane	-	-	2.4	1.1	0.4	-	-	-	-
Naphtha	-	-	-	-	-	-	-	-	-
Motor Gasoline	-	-	9.6	6.4	2.4	5.0	5.4	2.7	5.3
Aviation Fuels	-	-	1.3	0.3	0.3	1.8	1.5	1.0	1.2
Kerosene	-	-	-	-	-	-	-	-	-
Gas Diesel	-	-	9.4	4.8	1.2	2.7	2.9	2.0	1.7
Heavy Fuel Oil	-	-	9.1	4.7	1.7	1.7	1.8	1.6	2.0
Other	-	-	7.7	6.7	2.6	1.6	2.0	2.3	3.3
Refinery Fuel	-	-	-	-	-	-	-	-	-
Sub-Total	-	-	39.4	24.0	8.6	12.7	13.6	9.7	13.4
Bunkers	-	-	-	-	-	-	-	-	-
Total	-	-	39.4	24.0	8.6	12.7	13.6	9.7	13.4

[1] See Notes in Part I.4.

Oil Demand by Main Product Groups[1]
Demande de produits raffinés par groupes principaux

Latvia

	1971	1973	1978	1984	1985	1986	1987	1988	1989
Thousand tonnes									
NGL/LPG/Ethane	-	-	-	-	-	-	-	-	-
Naphtha	-	-	-	-	-	-	-	-	-
Motor Gasoline	-	-	-	-	-	-	-	-	-
Aviation Fuels	-	-	-	-	-	-	-	-	-
Kerosene	-	-	-	-	-	-	-	-	-
Gas Diesel	-	-	-	-	-	-	-	-	-
Heavy Fuel Oil	-	-	-	-	-	-	-	-	-
Other	-	-	-	-	-	-	-	-	-
Refinery Fuel	-	-	-	-	-	-	-	-	-
Sub-Total	-	-	-	-	-	-	-	-	-
Bunkers	-	-	-	-	-	-	-	-	-
Total	-	-	-	-	-	-	-	-	-
Thousand barrels/day									
NGL/LPG/Ethane	-	-	-	-	-	-	-	-	-
Naphtha	-	-	-	-	-	-	-	-	-
Motor Gasoline	-	-	-	-	-	-	-	-	-
Aviation Fuels	-	-	-	-	-	-	-	-	-
Kerosene	-	-	-	-	-	-	-	-	-
Gas Diesel	-	-	-	-	-	-	-	-	-
Heavy Fuel Oil	-	-	-	-	-	-	-	-	-
Other	-	-	-	-	-	-	-	-	-
Refinery Fuel	-	-	-	-	-	-	-	-	-
Sub-Total	-	-	-	-	-	-	-	-	-
Bunkers	-	-	-	-	-	-	-	-	-
Total	-	-	-	-	-	-	-	-	-

	1990	1991	1992	1993	1994	1995	1996	1997	1998
Thousand tonnes									
NGL/LPG/Ethane	-	-	58	61	52	34	35	38	44
Naphtha	-	-	-	-	-	-	-	-	-
Motor Gasoline	-	-	493	478	457	412	434	354	350
Aviation Fuels	-	-	90	26	21	29	34	48	36
Kerosene	-	-	29	8	7	10	4	4	2
Gas Diesel	-	-	790	722	564	502	431	396	391
Heavy Fuel Oil	-	-	1 089	1 116	1 415	1 017	1 141	791	733
Other	-	-	191	28	-	15	3	41	75
Refinery Fuel	-	-	-	-	-	-	-	-	-
Sub-Total	-	-	2 740	2 439	2 516	2 019	2 082	1 672	1 631
Bunkers	-	-	-	-	-	-	-	-	-
Total	-	-	2 740	2 439	2 516	2 019	2 082	1 672	1 631
Thousand barrels/day									
NGL/LPG/Ethane	-	-	1.8	1.9	1.7	1.1	1.1	1.2	1.4
Naphtha	-	-	-	-	-	-	-	-	-
Motor Gasoline	-	-	11.5	11.2	10.7	9.6	10.1	8.3	8.2
Aviation Fuels	-	-	2.0	0.6	0.5	0.6	0.7	1.0	0.8
Kerosene	-	-	0.6	0.2	0.1	0.2	0.1	0.1	0.0
Gas Diesel	-	-	16.1	14.8	11.5	10.3	8.8	8.1	8.0
Heavy Fuel Oil	-	-	19.8	20.4	25.8	18.6	20.8	14.4	13.4
Other	-	-	3.7	0.5	-	0.3	0.1	0.8	1.4
Refinery Fuel	-	-	-	-	-	-	-	-	-
Sub-Total	-	-	55.5	49.5	50.3	40.7	41.7	33.9	33.2
Bunkers	-	-	-	-	-	-	-	-	-
Total	-	-	55.5	49.5	50.3	40.7	41.7	33.9	33.2

[1] See Notes in Part I.4.

Oil Demand by Main Product Groups[1]
Demande de produits raffinés par groupes principaux

Lebanon

	1971	1973	1978	1984	1985	1986	1987	1988	1989
Thousand tonnes									
NGL/LPG/Ethane	68	76	70	75	90	91	95	75	91
Naphtha	-	-	-	-	-	-	-	-	-
Motor Gasoline	403	517	473	640	739	666	680	479	600
Aviation Fuels	264	347	201	101	121	151	180	85	60
Kerosene	24	20	19	40	70	80	60	50	60
Gas Diesel	254	318	250	535	547	509	520	350	300
Heavy Fuel Oil	632	907	902	760	974	1 090	1 097	967	999
Other	-	-	-	-	-	-	-	-	-
Refinery Fuel	105	206	130	38	31	56	63	30	5
Sub-Total	1 750	2 391	2 045	2 189	2 572	2 643	2 695	2 036	2 115
Bunkers	230	21	10	-	-	-	-	-	-
Total	1 980	2 412	2 055	2 189	2 572	2 643	2 695	2 036	2 115
Thousand barrels/day									
NGL/LPG/Ethane	2.2	2.4	2.2	2.4	2.9	2.9	3.0	2.4	2.9
Naphtha	-	-	-	-	-	-	-	-	-
Motor Gasoline	9.4	12.1	11.1	14.9	17.3	15.6	15.9	11.2	14.0
Aviation Fuels	5.8	7.6	4.4	2.2	2.6	3.3	3.9	1.8	1.3
Kerosene	0.5	0.4	0.4	0.8	1.5	1.7	1.3	1.1	1.3
Gas Diesel	5.2	6.5	5.1	10.9	11.2	10.4	10.6	7.1	6.1
Heavy Fuel Oil	11.5	16.6	16.5	13.8	17.8	19.9	20.0	17.6	18.2
Other	-	-	-	-	-	-	-	-	-
Refinery Fuel	1.9	3.8	2.4	0.7	0.6	1.0	1.2	0.5	0.1
Sub-Total	36.5	49.3	42.0	45.8	53.8	54.7	55.9	41.7	43.9
Bunkers	4.2	0.4	0.2	-	-	-	-	-	-
Total	40.7	49.7	42.2	45.8	53.8	54.7	55.9	41.7	43.9

	1990	1991	1992	1993	1994	1995	1996	1997	1998
Thousand tonnes									
NGL/LPG/Ethane	91	95	128	130	153	116	124	141	100
Naphtha	-	-	-	-	-	-	-	-	-
Motor Gasoline	600	750	1 041	1 200	1 123	1 327	1 379	1 310	1 412
Aviation Fuels	60	75	107	120	146	208	107	109	107
Kerosene	60	50	9	4	4	4	4	4	4
Gas Diesel	300	560	430	662	795	916	879	1 387	1 515
Heavy Fuel Oil	999	1 100	870	1 125	1 297	1 443	1 724	1 870	1 474
Other	-	-	22	50	67	67	109	88	110
Refinery Fuel	5	20	20	-	-	-	-	-	-
Sub-Total	2 115	2 650	2 627	3 291	3 585	4 081	4 326	4 909	4 722
Bunkers	-	10	10	-	-	-	-	-	-
Total	2 115	2 660	2 637	3 291	3 585	4 081	4 326	4 909	4 722
Thousand barrels/day									
NGL/LPG/Ethane	2.9	3.0	4.1	4.1	4.9	3.7	3.9	4.5	3.2
Naphtha	-	-	-	-	-	-	-	-	-
Motor Gasoline	14.0	17.5	24.3	28.0	26.2	31.0	32.1	30.6	33.0
Aviation Fuels	1.3	1.6	2.3	2.6	3.2	4.5	2.3	2.4	2.3
Kerosene	1.3	1.1	0.2	0.1	0.1	0.1	0.1	0.1	0.1
Gas Diesel	6.1	11.4	8.8	13.5	16.2	18.7	17.9	28.3	31.0
Heavy Fuel Oil	18.2	20.1	15.8	20.5	23.7	26.3	31.4	34.1	26.9
Other	-	-	0.4	1.0	1.3	1.3	2.1	1.7	2.1
Refinery Fuel	0.1	0.4	0.4	-	-	-	-	-	-
Sub-Total	43.9	55.1	56.2	69.9	75.6	85.6	89.8	101.7	98.6
Bunkers	-	0.2	0.2	-	-	-	-	-	-
Total	43.9	55.3	56.4	69.9	75.6	85.6	89.8	101.7	98.6

[1] See Notes in Part I.4.

Oil Demand by Main Product Groups[1]
Demande de produits raffinés par groupes principaux

Libya

	1971	1973	1978	1984	1985	1986	1987	1988	1989
Thousand tonnes									
NGL/LPG/Ethane	37	20	63	100	110	100	240	240	208
Naphtha	-	-	-	320	450	480	410	330	632
Motor Gasoline	210	302	670	948	968	1 001	979	1 000	1 031
Aviation Fuels	87	112	177	362	335	305	366	290	302
Kerosene	65	74	110	73	94	85	64	62	150
Gas Diesel	16	578	1 304	1 864	1 567	1 416	1 494	1 600	1 620
Heavy Fuel Oil	176	217	722	1 955	1 734	1 715	1 980	2 181	2 029
Other	23	75	130	350	247	205	176	182	200
Refinery Fuel	12	14	180	286	335	462	590	630	620
Sub-Total	626	1 392	3 356	6 258	5 840	5 769	6 299	6 515	6 792
Bunkers	3	2	12	8	12	15	20	50	70
Total	629	1 394	3 368	6 266	5 852	5 784	6 319	6 565	6 862
Thousand barrels/day									
NGL/LPG/Ethane	1.2	0.6	2.0	3.2	3.5	3.2	7.6	7.6	6.6
Naphtha	-	-	-	7.4	10.5	11.2	9.5	7.7	14.7
Motor Gasoline	4.9	7.1	15.7	22.1	22.6	23.4	22.9	23.3	24.1
Aviation Fuels	1.9	2.4	3.9	7.8	7.3	6.6	8.0	6.3	6.6
Kerosene	1.4	1.6	2.3	1.5	2.0	1.8	1.4	1.3	3.2
Gas Diesel	0.3	11.8	26.7	38.0	32.0	28.9	30.5	32.6	33.1
Heavy Fuel Oil	3.2	4.0	13.2	35.6	31.6	31.3	36.1	39.7	37.0
Other	0.4	1.3	2.4	6.1	4.4	3.8	3.2	3.3	3.6
Refinery Fuel	0.2	0.3	3.6	5.9	6.9	9.6	12.4	13.1	13.0
Sub-Total	13.6	29.1	69.7	127.7	120.8	119.8	131.6	135.0	141.9
Bunkers	0.1	0.0	0.2	0.1	0.2	0.3	0.4	0.9	1.3
Total	13.6	29.1	69.9	127.8	121.1	120.1	132.0	135.9	143.2

	1990	1991	1992	1993	1994	1995	1996	1997	1998
Thousand tonnes									
NGL/LPG/Ethane	250	290	300	340	360	408	411	422	439
Naphtha	700	820	710	810	820	926	933	936	936
Motor Gasoline	1 040	1 050	1 063	1 207	1 477	1 675	1 687	1 716	1 788
Aviation Fuels	202	212	227	237	288	291	318	299	299
Kerosene	180	180	192	190	232	234	256	271	271
Gas Diesel	1 620	1 625	1 640	1 898	2 322	2 340	2 332	2 331	2 399
Heavy Fuel Oil	2 000	2 000	2 020	2 050	2 542	2 975	2 973	3 252	3 347
Other	150	130	140	130	142	173	174	225	225
Refinery Fuel	620	620	620	620	709	716	726	728	728
Sub-Total	6 762	6 927	6 912	7 482	8 892	9 738	9 810	10 180	10 432
Bunkers	80	80	70	90	80	90	90	90	90
Total	6 842	7 007	6 982	7 572	8 972	9 828	9 900	10 270	10 522
Thousand barrels/day									
NGL/LPG/Ethane	7.9	9.2	9.5	10.8	11.4	13.0	13.0	13.4	14.0
Naphtha	16.3	19.1	16.5	18.9	19.1	21.6	21.7	21.8	21.8
Motor Gasoline	24.3	24.5	24.8	28.2	34.5	39.1	39.3	40.1	41.8
Aviation Fuels	4.4	4.6	4.9	5.2	6.3	6.3	6.9	6.5	6.5
Kerosene	3.8	3.8	4.1	4.0	4.9	5.0	5.4	5.7	5.7
Gas Diesel	33.1	33.2	33.4	38.8	47.5	47.8	47.5	47.6	49.0
Heavy Fuel Oil	36.5	36.5	36.8	37.4	46.4	54.3	54.1	59.3	61.1
Other	2.7	2.3	2.4	2.3	2.5	3.0	3.0	3.9	3.9
Refinery Fuel	13.0	13.0	13.0	13.0	14.9	15.1	15.2	15.3	15.3
Sub-Total	142.0	146.3	145.3	158.5	187.5	205.1	206.2	213.7	219.1
Bunkers	1.5	1.5	1.3	1.6	1.5	1.6	1.6	1.6	1.6
Total	143.5	147.7	146.6	160.2	188.9	206.8	207.8	215.4	220.7

[1] See Notes in Part I.4.

Oil Demand by Main Product Groups[1]
Demande de produits raffinés par groupes principaux

Lithuania

	1971	1973	1978	1984	1985	1986	1987	1988	1989
Thousand tonnes									
NGL/LPG/Ethane	-	-	-	-	-	-	-	-	-
Naphtha	-	-	-	-	-	-	-	-	-
Motor Gasoline	-	-	-	-	-	-	-	-	-
Aviation Fuels	-	-	-	-	-	-	-	-	-
Kerosene	-	-	-	-	-	-	-	-	-
Gas Diesel	-	-	-	-	-	-	-	-	-
Heavy Fuel Oil	-	-	-	-	-	-	-	-	-
Other	-	-	-	-	-	-	-	-	-
Refinery Fuel	-	-	-	-	-	-	-	-	-
Sub-Total	-	-	-	-	-	-	-	-	-
Bunkers	-	-	-	-	-	-	-	-	-
Total	-	-	-	-	-	-	-	-	-
Thousand barrels/day									
NGL/LPG/Ethane	-	-	-	-	-	-	-	-	-
Naphtha	-	-	-	-	-	-	-	-	-
Motor Gasoline	-	-	-	-	-	-	-	-	-
Aviation Fuels	-	-	-	-	-	-	-	-	-
Kerosene	-	-	-	-	-	-	-	-	-
Gas Diesel	-	-	-	-	-	-	-	-	-
Heavy Fuel Oil	-	-	-	-	-	-	-	-	-
Other	-	-	-	-	-	-	-	-	-
Refinery Fuel	-	-	-	-	-	-	-	-	-
Sub-Total	-	-	-	-	-	-	-	-	-
Bunkers	-	-	-	-	-	-	-	-	-
Total	-	-	-	-	-	-	-	-	-

	1990	1991	1992	1993	1994	1995	1996	1997	1998
Thousand tonnes									
NGL/LPG/Ethane	-	-	98	96	93	93	74	94	108
Naphtha	-	-	-	-	-	-	-	-	-
Motor Gasoline	-	-	675	545	442	607	665	662	634
Aviation Fuels	-	-	66	37	39	40	33	31	28
Kerosene	-	-	-	-	-	-	-	-	-
Gas Diesel	-	-	994	773	638	621	612	673	716
Heavy Fuel Oil	-	-	2 247	1 851	1 910	1 452	1 500	1 176	1 524
Other	-	-	311	211	279	421	155	191	273
Refinery Fuel	-	-	-	388	117	92	109	392	467
Sub-Total	-	-	4 391	3 901	3 518	3 326	3 148	3 219	3 750
Bunkers	-	-	-	-	-	-	-	62	51
Total	-	-	4 391	3 901	3 518	3 326	3 148	3 281	3 801
Thousand barrels/day									
NGL/LPG/Ethane	-	-	3.1	3.1	3.0	3.0	2.3	3.0	3.4
Naphtha	-	-	-	-	-	-	-	-	-
Motor Gasoline	-	-	15.7	12.7	10.3	14.2	15.5	15.5	14.8
Aviation Fuels	-	-	1.4	0.8	0.8	0.9	0.7	0.7	0.6
Kerosene	-	-	-	-	-	-	-	-	-
Gas Diesel	-	-	20.3	15.8	13.0	12.7	12.5	13.8	14.6
Heavy Fuel Oil	-	-	40.9	33.8	34.9	26.5	27.3	21.5	27.8
Other	-	-	5.9	4.0	5.2	8.0	2.8	3.4	4.9
Refinery Fuel	-	-	-	7.1	2.1	1.7	2.0	7.5	9.0
Sub-Total	-	-	87.3	77.2	69.4	66.9	63.1	65.3	75.2
Bunkers	-	-	-	-	-	-	-	1.1	1.0
Total	-	-	87.3	77.2	69.4	66.9	63.1	66.4	76.2

[1] See Notes in Part I.4.

Oil Demand by Main Product Groups[1]
Demande de produits raffinés par groupes principaux

Former Yugoslav Republic of Macedonia

	1971	1973	1978	1984	1985	1986	1987	1988	1989
Thousand tonnes									
NGL/LPG/Ethane	-	-	-	-	-	-	-	-	-
Naphtha	-	-	-	-	-	-	-	-	-
Motor Gasoline	-	-	-	-	-	-	-	-	-
Aviation Fuels	-	-	-	-	-	-	-	-	-
Kerosene	-	-	-	-	-	-	-	-	-
Gas Diesel	-	-	-	-	-	-	-	-	-
Heavy Fuel Oil	-	-	-	-	-	-	-	-	-
Other	-	-	-	-	-	-	-	-	-
Refinery Fuel	-	-	-	-	-	-	-	-	-
Sub-Total	-	-	-	-	-	-	-	-	-
Bunkers	-	-	-	-	-	-	-	-	-
Total	-	-	-	-	-	-	-	-	-
Thousand barrels/day									
NGL/LPG/Ethane	-	-	-	-	-	-	-	-	-
Naphtha	-	-	-	-	-	-	-	-	-
Motor Gasoline	-	-	-	-	-	-	-	-	-
Aviation Fuels	-	-	-	-	-	-	-	-	-
Kerosene	-	-	-	-	-	-	-	-	-
Gas Diesel	-	-	-	-	-	-	-	-	-
Heavy Fuel Oil	-	-	-	-	-	-	-	-	-
Other	-	-	-	-	-	-	-	-	-
Refinery Fuel	-	-	-	-	-	-	-	-	-
Sub-Total	-	-	-	-	-	-	-	-	-
Bunkers	-	-	-	-	-	-	-	-	-
Total	-	-	-	-	-	-	-	-	-

	1990	1991	1992	1993	1994	1995	1996	1997	1998
Thousand tonnes									
NGL/LPG/Ethane	41	42	35	18	15	12	19	16	26
Naphtha	-	-	-	-	-	-	-	-	-
Motor Gasoline	161	163	163	182	241	207	209	214	170
Aviation Fuels	5	7	7	18	19	31	26	30	16
Kerosene	-	-	-	-	-	-	-	-	-
Gas Diesel	258	250	250	395	88	220	457	366	304
Heavy Fuel Oil	452	425	425	370	377	297	486	360	322
Other	84	45	45	37	41	48	40	41	44
Refinery Fuel	52	35	35	-	-	-	-	-	-
Sub-Total	1 053	967	960	1 020	781	815	1 237	1 027	882
Bunkers	-	-	-	-	-	-	-	-	-
Total	1 053	967	960	1 020	781	815	1 237	1 027	882
Thousand barrels/day									
NGL/LPG/Ethane	1.3	1.3	1.1	0.6	0.5	0.4	0.6	0.5	0.8
Naphtha	-	-	-	-	-	-	-	-	-
Motor Gasoline	3.8	3.8	3.8	4.3	5.6	4.8	4.9	5.0	4.0
Aviation Fuels	0.1	0.2	0.2	0.4	0.4	0.7	0.6	0.7	0.3
Kerosene	-	-	-	-	-	-	-	-	-
Gas Diesel	5.3	5.1	5.1	8.1	1.8	4.5	9.3	7.5	6.2
Heavy Fuel Oil	8.2	7.8	7.7	6.8	6.9	5.4	8.8	6.6	5.9
Other	1.5	0.8	0.8	0.6	0.7	0.8	0.7	0.7	0.8
Refinery Fuel	0.9	0.6	0.6	-	-	-	-	-	-
Sub-Total	21.2	19.6	19.3	20.7	15.9	16.6	24.9	20.9	18.0
Bunkers	-	-	-	-	-	-	-	-	-
Total	21.2	19.6	19.3	20.7	15.9	16.6	24.9	20.9	18.0

[1] See Notes in Part I.4.

Oil Demand by Main Product Groups[1]
Demande de produits raffinés par groupes principaux

Malaysia

	1971	1973	1978	1984	1985	1986	1987	1988	1989
Thousand tonnes									
NGL/LPG/Ethane	56	32	93	172	211	249	303	348	381
Naphtha	-	-	-	-	-	-	-	-	-
Motor Gasoline	663	817	961	1 832	1 988	2 073	2 186	2 332	2 460
Aviation Fuels	132	167	215	362	281	418	422	448	485
Kerosene	72	139	326	346	300	291	261	247	205
Gas Diesel	1 600	1 695	1 991	3 172	3 070	2 995	3 159	3 453	4 071
Heavy Fuel Oil	1 553	1 338	2 572	2 902	2 751	2 724	2 636	2 671	2 694
Other	1	2	185	324	394	391	365	374	323
Refinery Fuel	62	60	60	123	157	83	78	103	100
Sub-Total	4 139	4 250	6 403	9 233	9 152	9 224	9 410	9 976	10 719
Bunkers	34	23	48	60	99	74	94	63	69
Total	4 173	4 273	6 451	9 293	9 251	9 298	9 504	10 039	10 788
Thousand barrels/day									
NGL/LPG/Ethane	1.8	1.0	3.0	5.5	6.7	7.9	9.6	11.0	12.1
Naphtha	-	-	-	-	-	-	-	-	-
Motor Gasoline	15.5	19.1	22.5	42.7	46.5	48.4	51.1	54.4	57.5
Aviation Fuels	2.9	3.6	4.7	7.9	6.1	9.1	9.2	9.7	10.5
Kerosene	1.5	2.9	6.9	7.3	6.4	6.2	5.5	5.2	4.3
Gas Diesel	32.7	34.6	40.7	64.7	62.7	61.2	64.6	70.4	83.2
Heavy Fuel Oil	28.3	24.4	46.9	52.8	50.2	49.7	48.1	48.6	49.2
Other	0.0	0.0	3.1	5.4	6.6	6.6	6.1	6.3	5.4
Refinery Fuel	1.4	1.3	1.3	2.7	3.4	1.8	1.7	2.3	2.2
Sub-Total	84.1	87.1	129.1	188.9	188.6	190.9	195.9	207.8	224.5
Bunkers	0.6	0.4	1.0	1.2	2.0	1.4	1.7	1.2	1.3
Total	84.7	87.5	130.0	190.1	190.6	192.3	197.7	209.0	225.8

	1990	1991	1992	1993	1994	1995	1996	1997	1998
Thousand tonnes									
NGL/LPG/Ethane	513	568	865	1 041	1 472	2 234	1 510	1 796	1 586
Naphtha	-	-	-	-	-	-	-	-	-
Motor Gasoline	2 761	2 984	3 166	3 490	3 939	4 329	4 954	5 317	5 572
Aviation Fuels	613	670	742	850	952	1 125	1 295	1 396	1 570
Kerosene	197	175	154	143	147	172	190	163	160
Gas Diesel	4 467	4 959	5 367	5 342	5 802	5 981	6 912	7 383	6 602
Heavy Fuel Oil	3 787	3 661	3 468	3 712	3 376	3 608	4 157	4 496	3 840
Other	238	477	566	635	664	633	744	847	619
Refinery Fuel	129	141	124	83	156	404	283	178	163
Sub-Total	12 705	13 635	14 452	15 296	16 508	18 486	20 045	21 576	20 112
Bunkers	89	109	73	60	246	163	185	169	448
Total	12 794	13 744	14 525	15 356	16 754	18 649	20 230	21 745	20 560
Thousand barrels/day									
NGL/LPG/Ethane	16.3	18.1	27.1	33.5	46.1	71.6	47.7	56.7	50.3
Naphtha	-	-	-	-	-	-	-	-	-
Motor Gasoline	64.5	69.7	73.8	81.8	92.3	101.4	115.7	124.5	130.5
Aviation Fuels	13.3	14.6	16.1	18.4	20.6	24.4	28.0	30.3	34.0
Kerosene	4.2	3.7	3.3	3.1	3.2	3.7	4.1	3.5	3.5
Gas Diesel	91.3	101.4	109.4	109.8	119.2	122.9	141.6	151.7	135.7
Heavy Fuel Oil	69.1	66.8	63.1	67.1	61.0	65.2	75.0	81.3	69.4
Other	4.0	8.0	9.4	10.6	11.1	10.6	12.4	14.1	10.3
Refinery Fuel	2.8	3.1	2.7	1.8	3.4	8.9	6.2	3.9	3.6
Sub-Total	265.6	285.3	304.8	326.1	357.0	408.6	430.7	466.1	437.3
Bunkers	1.7	2.1	1.4	1.1	4.5	3.0	3.4	3.1	8.2
Total	267.3	287.4	306.2	327.3	361.5	411.6	434.1	469.2	445.5

[1] See Notes in Part I.4.

Oil Demand by Main Product Groups[1]
Demande de produits raffinés par groupes principaux

Malta

	1971	1973	1978	1984	1985	1986	1987	1988	1989
Thousand tonnes									
NGL/LPG/Ethane	5	5	25	14	8	13	15	15	15
Naphtha	-	-	-	-	-	-	-	-	-
Motor Gasoline	39	46	41	41	20	56	60	65	65
Aviation Fuels	57	63	57	82	47	67	72	70	70
Kerosene	28	35	16	18	10	12	15	15	15
Gas Diesel	27	33	23	54	26	104	100	105	105
Heavy Fuel Oil	113	138	193	228	159	313	301	310	310
Other	-	-	-	4	8	-	7	6	6
Refinery Fuel	-	-	-	-	-	-	-	-	-
Sub-Total	269	320	355	441	278	565	570	586	586
Bunkers	61	23	30	20	20	32	30	30	30
Total	330	343	385	461	298	597	600	616	616
Thousand barrels/day									
NGL/LPG/Ethane	0.2	0.2	0.8	0.4	0.3	0.4	0.5	0.5	0.5
Naphtha	-	-	-	-	-	-	-	-	-
Motor Gasoline	0.9	1.1	1.0	1.0	0.5	1.3	1.4	1.5	1.5
Aviation Fuels	1.2	1.4	1.2	1.8	1.0	1.5	1.6	1.5	1.5
Kerosene	0.6	0.7	0.3	0.4	0.2	0.3	0.3	0.3	0.3
Gas Diesel	0.6	0.7	0.5	1.1	0.5	2.1	2.0	2.1	2.1
Heavy Fuel Oil	2.1	2.5	3.5	4.1	2.9	5.7	5.5	5.6	5.7
Other	-	-	-	0.1	0.1	-	0.1	0.1	0.1
Refinery Fuel	-	-	-	-	-	-	-	-	-
Sub-Total	5.5	6.5	7.3	8.9	5.5	11.3	11.4	11.7	11.7
Bunkers	1.2	0.4	0.6	0.4	0.4	0.6	0.6	0.6	0.6
Total	6.7	7.0	7.9	9.3	5.9	11.9	12.0	12.3	12.3

	1990	1991	1992	1993	1994	1995	1996	1997	1998
Thousand tonnes									
NGL/LPG/Ethane	15	16	18	18	18	16	17	18	19
Naphtha	-	-	-	-	-	-	-	-	-
Motor Gasoline	65	67	70	85	85	73	77	77	81
Aviation Fuels	70	80	79	79	79	111	107	112	163
Kerosene	16	16	16	16	16	18	19	55	57
Gas Diesel	105	127	112	120	120	120	170	170	86
Heavy Fuel Oil	310	296	323	431	410	468	488	497	523
Other	6	6	6	6	6	-	14	11	18
Refinery Fuel	-	-	-	-	-	-	-	-	-
Sub-Total	587	608	624	755	734	806	892	940	947
Bunkers	30	40	50	30	30	45	79	80	46
Total	617	648	674	785	764	851	971	1 020	993
Thousand barrels/day									
NGL/LPG/Ethane	0.5	0.5	0.6	0.6	0.6	0.5	0.5	0.6	0.6
Naphtha	-	-	-	-	-	-	-	-	-
Motor Gasoline	1.5	1.6	1.6	2.0	2.0	1.7	1.8	1.8	1.9
Aviation Fuels	1.5	1.7	1.7	1.7	1.7	2.4	2.3	2.4	3.5
Kerosene	0.3	0.3	0.3	0.3	0.3	0.4	0.4	1.2	1.2
Gas Diesel	2.1	2.6	2.3	2.5	2.5	2.5	3.5	3.5	1.8
Heavy Fuel Oil	5.7	5.4	5.9	7.9	7.5	8.5	8.9	9.1	9.5
Other	0.1	0.1	0.1	0.1	0.1	-	0.3	0.2	0.3
Refinery Fuel	-	-	-	-	-	-	-	-	-
Sub-Total	11.8	12.3	12.5	15.0	14.7	16.0	17.7	18.7	18.9
Bunkers	0.6	0.8	1.0	0.6	0.6	0.9	1.5	1.5	0.9
Total	12.4	13.0	13.5	15.6	15.2	16.9	19.1	20.2	19.8

[1] See Notes in Part I.4.

Oil Demand by Main Product Groups[1]
Demande de produits raffinés par groupes principaux

Republic of Moldova

	1971	1973	1978	1984	1985	1986	1987	1988	1989
Thousand tonnes									
NGL/LPG/Ethane	-	-	-	-	-	-	-	-	-
Naphtha	-	-	-	-	-	-	-	-	-
Motor Gasoline	-	-	-	-	-	-	-	-	-
Aviation Fuels	-	-	-	-	-	-	-	-	-
Kerosene	-	-	-	-	-	-	-	-	-
Gas Diesel	-	-	-	-	-	-	-	-	-
Heavy Fuel Oil	-	-	-	-	-	-	-	-	-
Other	-	-	-	-	-	-	-	-	-
Refinery Fuel	-	-	-	-	-	-	-	-	-
Sub-Total	-	-	-	-	-	-	-	-	-
Bunkers	-	-	-	-	-	-	-	-	-
Total	-	-	-	-	-	-	-	-	-
Thousand barrels/day									
NGL/LPG/Ethane	-	-	-	-	-	-	-	-	-
Naphtha	-	-	-	-	-	-	-	-	-
Motor Gasoline	-	-	-	-	-	-	-	-	-
Aviation Fuels	-	-	-	-	-	-	-	-	-
Kerosene	-	-	-	-	-	-	-	-	-
Gas Diesel	-	-	-	-	-	-	-	-	-
Heavy Fuel Oil	-	-	-	-	-	-	-	-	-
Other	-	-	-	-	-	-	-	-	-
Refinery Fuel	-	-	-	-	-	-	-	-	-
Sub-Total	-	-	-	-	-	-	-	-	-
Bunkers	-	-	-	-	-	-	-	-	-
Total	-	-	-	-	-	-	-	-	-

	1990	1991	1992	1993	1994	1995	1996	1997	1998
Thousand tonnes									
NGL/LPG/Ethane	-	-	100	41	20	20	22	27	28
Naphtha	-	-	-	-	-	-	-	-	-
Motor Gasoline	-	-	405	221	213	225	218	249	211
Aviation Fuels	-	-	33	21	13	15	20	23	21
Kerosene	-	-	-	20	5	2	8	7	11
Gas Diesel	-	-	598	469	391	383	325	312	248
Heavy Fuel Oil	-	-	1 745	1 298	465	365	343	242	195
Other	-	-	22	14	17	18	17	16	11
Refinery Fuel	-	-	-	-	-	-	-	-	-
Sub-Total	-	-	2 903	2 084	1 124	1 028	953	876	725
Bunkers	-	-	-	-	-	-	-	-	-
Total	-	-	2 903	2 084	1 124	1 028	953	876	725
Thousand barrels/day									
NGL/LPG/Ethane	-	-	3.2	1.3	0.6	0.6	0.7	0.9	0.9
Naphtha	-	-	-	-	-	-	-	-	-
Motor Gasoline	-	-	9.4	5.2	5.0	5.3	5.1	5.8	4.9
Aviation Fuels	-	-	0.7	0.5	0.3	0.3	0.4	0.5	0.5
Kerosene	-	-	-	0.4	0.1	0.0	0.2	0.1	0.2
Gas Diesel	-	-	12.2	9.6	8.0	7.8	6.6	6.4	5.1
Heavy Fuel Oil	-	-	31.8	23.7	8.5	6.7	6.2	4.4	3.6
Other	-	-	0.4	0.3	0.3	0.4	0.3	0.3	0.2
Refinery Fuel	-	-	-	-	-	-	-	-	-
Sub-Total	-	-	57.7	40.9	22.8	21.1	19.6	18.4	15.4
Bunkers	-	-	-	-	-	-	-	-	-
Total	-	-	57.7	40.9	22.8	21.1	19.6	18.4	15.4

[1] See Notes in Part I.4.

Oil Demand by Main Product Groups[1]
Demande de produits raffinés par groupes principaux

Morocco

	1971	1973	1978	1984	1985	1986	1987	1988	1989
Thousand tonnes									
NGL/LPG/Ethane	92	126	244	353	383	415	454	496	548
Naphtha	-	-	-	-	-	-	-	-	-
Motor Gasoline	330	373	393	338	329	336	348	364	370
Aviation Fuels	110	154	181	207	221	207	207	212	214
Kerosene	77	74	78	59	54	45	53	47	47
Gas Diesel	521	639	1 035	1 165	1 224	1 306	1 371	1 483	1 529
Heavy Fuel Oil	643	896	1 512	2 216	2 197	1 996	1 832	1 960	2 237
Other	115	125	147	131	126	142	153	151	177
Refinery Fuel	26	42	64	100	96	96	96	96	100
Sub-Total	1 914	2 429	3 654	4 569	4 630	4 543	4 514	4 809	5 222
Bunkers	77	103	52	14	12	12	13	12	20
Total	1 991	2 532	3 706	4 583	4 642	4 555	4 527	4 821	5 242
Thousand barrels/day									
NGL/LPG/Ethane	2.9	4.0	7.8	11.2	12.2	13.2	14.4	15.7	17.4
Naphtha	-	-	-	-	-	-	-	-	-
Motor Gasoline	7.7	8.7	9.2	7.9	7.7	7.9	8.1	8.5	8.6
Aviation Fuels	2.4	3.4	3.9	4.5	4.8	4.5	4.5	4.6	4.6
Kerosene	1.6	1.6	1.7	1.2	1.1	1.0	1.1	1.0	1.0
Gas Diesel	10.6	13.1	21.2	23.7	25.0	26.7	28.0	30.2	31.3
Heavy Fuel Oil	11.7	16.3	27.6	40.3	40.1	36.4	33.4	35.7	40.8
Other	2.2	2.4	2.8	2.5	2.4	2.7	2.9	2.9	3.2
Refinery Fuel	0.6	0.9	1.4	2.2	2.1	2.1	2.1	2.1	2.2
Sub-Total	39.8	50.4	75.5	93.6	95.4	94.4	94.7	100.7	109.2
Bunkers	1.5	2.0	1.0	0.3	0.2	0.2	0.3	0.2	0.4
Total	41.3	52.3	76.5	93.8	95.7	94.7	94.9	100.9	109.6

	1990	1991	1992	1993	1994	1995	1996	1997	1998
Thousand tonnes									
NGL/LPG/Ethane	512	634	719	786	850	897	951	1 007	1 036
Naphtha	-	-	-	-	-	-	-	-	-
Motor Gasoline	381	385	400	402	402	378	375	380	385
Aviation Fuels	249	183	213	220	207	231	238	247	280
Kerosene	48	45	46	46	42	48	39	57	80
Gas Diesel	1 706	1 804	2 118	2 352	2 578	2 615	2 640	2 621	2 662
Heavy Fuel Oil	2 161	2 134	2 417	2 291	2 569	2 106	1 687	1 889	1 852
Other	213	241	238	197	232	230	212	229	236
Refinery Fuel	100	90	100	105	110	104	106	112	115
Sub-Total	5 370	5 516	6 251	6 399	6 990	6 609	6 248	6 542	6 646
Bunkers	20	13	14	14	13	13	13	13	13
Total	5 390	5 529	6 265	6 413	7 003	6 622	6 261	6 555	6 659
Thousand barrels/day									
NGL/LPG/Ethane	16.3	20.1	22.8	25.0	27.0	28.5	30.1	32.0	32.9
Naphtha	-	-	-	-	-	-	-	-	-
Motor Gasoline	8.9	9.0	9.3	9.4	9.4	8.8	8.7	8.9	9.0
Aviation Fuels	5.4	4.0	4.6	4.8	4.5	5.0	5.2	5.4	6.1
Kerosene	1.0	1.0	1.0	1.0	0.9	1.0	0.8	1.2	1.7
Gas Diesel	34.9	36.9	43.2	48.1	52.7	53.4	53.8	53.6	54.4
Heavy Fuel Oil	39.4	38.9	44.0	41.8	46.9	38.4	30.7	34.5	33.8
Other	3.8	4.4	4.3	3.5	4.1	4.1	3.8	4.1	4.2
Refinery Fuel	2.2	2.0	2.2	2.3	2.4	2.3	2.3	2.5	2.5
Sub-Total	111.9	116.2	131.3	135.8	147.9	141.6	135.5	142.0	144.6
Bunkers	0.4	0.3	0.3	0.3	0.3	0.3	0.3	0.3	0.3
Total	112.3	116.5	131.6	136.1	148.2	141.9	135.7	142.3	144.9

[1] See Notes in Part I.4.

Oil Demand by Main Product Groups[1]
Demande de produits raffinés par groupes principaux

Mozambique

	1971	1973	1978	1984	1985	1986	1987	1988	1989
Thousand tonnes									
NGL/LPG/Ethane	9	10	7	10	10	10	10	4	4
Naphtha	-	-	-	-	-	-	-	-	-
Motor Gasoline	100	98	78	66	40	33	35	36	38
Aviation Fuels	37	37	30	23	30	40	41	42	40
Kerosene	41	41	55	12	5	4	4	5	5
Gas Diesel	102	94	98	112	170	195	195	200	205
Heavy Fuel Oil	179	167	186	169	173	175	175	176	156
Other	34	40	6	5	-	-	-	-	-
Refinery Fuel	33	32	25	5	-	-	-	-	-
Sub-Total	**535**	**519**	**485**	**402**	**428**	**457**	**460**	**463**	**448**
Bunkers	244	208	68	103	30	25	30	28	25
Total	**779**	**727**	**553**	**505**	**458**	**482**	**490**	**491**	**473**
Thousand barrels/day									
NGL/LPG/Ethane	0.3	0.3	0.2	0.3	0.3	0.3	0.3	0.1	0.1
Naphtha	-	-	-	-	-	-	-	-	-
Motor Gasoline	2.3	2.3	1.8	1.5	0.9	0.8	0.8	0.8	0.9
Aviation Fuels	0.8	0.8	0.7	0.5	0.7	0.9	0.9	0.9	0.9
Kerosene	0.9	0.9	1.2	0.3	0.1	0.1	0.1	0.1	0.1
Gas Diesel	2.1	1.9	2.0	2.3	3.5	4.0	4.0	4.1	4.2
Heavy Fuel Oil	3.3	3.0	3.4	3.1	3.2	3.2	3.2	3.2	2.8
Other	0.6	0.7	0.1	0.1	-	-	-	-	-
Refinery Fuel	0.6	0.6	0.5	0.1	-	-	-	-	-
Sub-Total	**10.8**	**10.5**	**9.8**	**8.1**	**8.6**	**9.2**	**9.3**	**9.3**	**9.0**
Bunkers	4.6	3.9	1.3	1.9	0.6	0.5	0.6	0.6	0.5
Total	**15.4**	**14.4**	**11.1**	**10.1**	**9.3**	**9.7**	**9.9**	**9.8**	**9.5**

	1990	1991	1992	1993	1994	1995	1996	1997	1998
Thousand tonnes									
NGL/LPG/Ethane	4	5	4	4	4	4	4	5	6
Naphtha	-	-	-	-	-	-	-	-	-
Motor Gasoline	40	41	31	32	35	39	40	43	49
Aviation Fuels	41	36	27	35	43	18	19	25	27
Kerosene	23	5	11	12	13	16	17	24	30
Gas Diesel	210	174	234	290	230	239	230	233	263
Heavy Fuel Oil	21	12	16	19	10	21	8	15	12
Other	7	8	8	6	6	6	7	7	7
Refinery Fuel	-	-	-	-	-	-	-	-	-
Sub-Total	**346**	**281**	**331**	**398**	**341**	**343**	**325**	**352**	**394**
Bunkers	28	9	3	2	2	3	3	3	1
Total	**374**	**290**	**334**	**400**	**343**	**346**	**328**	**355**	**395**
Thousand barrels/day									
NGL/LPG/Ethane	0.1	0.2	0.1	0.1	0.1	0.1	0.1	0.2	0.2
Naphtha	-	-	-	-	-	-	-	-	-
Motor Gasoline	0.9	1.0	0.7	0.7	0.8	0.9	0.9	1.0	1.1
Aviation Fuels	0.9	0.8	0.6	0.8	0.9	0.4	0.4	0.5	0.6
Kerosene	0.5	0.1	0.2	0.3	0.3	0.3	0.4	0.5	0.6
Gas Diesel	4.3	3.6	4.8	5.9	4.7	4.9	4.7	4.8	5.4
Heavy Fuel Oil	0.4	0.2	0.3	0.3	0.2	0.4	0.1	0.3	0.2
Other	0.1	0.2	0.2	0.1	0.1	0.1	0.1	0.1	0.1
Refinery Fuel	-	-	-	-	-	-	-	-	-
Sub-Total	**7.3**	**5.9**	**6.9**	**8.3**	**7.2**	**7.2**	**6.8**	**7.4**	**8.3**
Bunkers	0.6	0.2	0.1	0.0	0.0	0.1	0.1	0.1	0.0
Total	**7.8**	**6.1**	**6.9**	**8.3**	**7.2**	**7.2**	**6.9**	**7.4**	**8.3**

[1] See Notes in Part I.4.

Oil Demand by Main Product Groups[1]
Demande de produits raffinés par groupes principaux

Myanmar

	1971	1973	1978	1984	1985	1986	1987	1988	1989
Thousand tonnes									
NGL/LPG/Ethane	5	7	4	3	4	5	4	4	9
Naphtha	-	-	-	-	-	-	-	-	-
Motor Gasoline	162	174	220	260	235	261	177	153	152
Aviation Fuels	30	34	30	37	40	40	20	20	20
Kerosene	287	241	113	18	7	8	5	5	5
Gas Diesel	324	231	326	410	397	430	269	271	262
Heavy Fuel Oil	183	77	203	213	194	215	120	115	115
Other	228	221	245	205	228	175	163	114	109
Refinery Fuel	52	40	49	51	50	53	32	29	29
Sub-Total	1 271	1 025	1 190	1 197	1 155	1 187	790	711	701
Bunkers	2	1	-	-	-	-	-	-	-
Total	1 273	1 026	1 190	1 197	1 155	1 187	790	711	701
Thousand barrels/day									
NGL/LPG/Ethane	0.2	0.2	0.1	0.1	0.1	0.2	0.1	0.1	0.3
Naphtha	-	-	-	-	-	-	-	-	-
Motor Gasoline	3.9	4.2	5.3	6.2	5.6	6.3	4.2	3.7	3.6
Aviation Fuels	0.7	0.7	0.7	0.8	0.9	0.9	0.4	0.4	0.4
Kerosene	6.2	5.2	2.5	0.4	0.2	0.2	0.1	0.1	0.1
Gas Diesel	6.7	4.8	6.8	8.5	8.2	8.9	5.6	5.6	5.4
Heavy Fuel Oil	3.6	1.5	4.0	4.2	3.8	4.2	2.4	2.3	2.3
Other	4.3	4.2	4.6	3.8	4.3	3.3	3.1	2.1	2.0
Refinery Fuel	1.1	0.9	1.1	1.1	1.1	1.2	0.7	0.6	0.6
Sub-Total	26.7	21.7	25.0	25.1	24.2	25.1	16.6	15.0	14.8
Bunkers	0.0	0.0	-	-	-	-	-	-	-
Total	26.7	21.7	25.0	25.1	24.2	25.1	16.6	15.0	14.8

	1990	1991	1992	1993	1994	1995	1996	1997	1998
Thousand tonnes									
NGL/LPG/Ethane	9	9	10	12	13	9	11	12	13
Naphtha	-	-	-	-	-	-	-	-	-
Motor Gasoline	125	123	153	169	194	209	204	248	299
Aviation Fuels	28	25	28	33	44	46	54	52	49
Kerosene	2	1	9	11	12	13	8	3	2
Gas Diesel	305	275	362	449	540	787	878	1 009	946
Heavy Fuel Oil	118	115	142	137	94	108	100	106	105
Other	68	69	133	93	135	28	35	44	41
Refinery Fuel	29	31	70	59	67	49	45	73	62
Sub-Total	684	648	907	963	1 099	1 249	1 335	1 547	1 517
Bunkers	-	-	27	1	5	3	5	4	2
Total	684	648	934	964	1 104	1 252	1 340	1 551	1 519
Thousand barrels/day									
NGL/LPG/Ethane	0.3	0.3	0.3	0.4	0.4	0.3	0.4	0.4	0.4
Naphtha	-	-	-	-	-	-	-	-	-
Motor Gasoline	3.0	3.0	3.7	4.1	4.7	5.0	4.9	6.0	7.2
Aviation Fuels	0.6	0.5	0.6	0.7	1.0	1.0	1.2	1.1	1.1
Kerosene	0.0	0.0	0.2	0.2	0.3	0.3	0.2	0.1	0.0
Gas Diesel	6.3	5.7	7.5	9.3	11.2	16.3	18.1	20.9	19.6
Heavy Fuel Oil	2.3	2.3	2.8	2.7	1.8	2.1	2.0	2.1	2.1
Other	1.3	1.3	2.7	1.9	2.7	0.5	0.6	0.7	0.7
Refinery Fuel	0.6	0.7	1.5	1.2	1.4	1.0	0.9	1.5	1.3
Sub-Total	14.5	13.7	19.2	20.5	23.4	26.5	28.2	32.8	32.3
Bunkers	-	-	0.5	0.0	0.1	0.1	0.1	0.1	0.0
Total	14.5	13.7	19.7	20.5	23.5	26.6	28.3	32.9	32.4

[1] See Notes in Part I.4.

Oil Demand by Main Product Groups[1]
Demande de produits raffinés par groupes principaux

Nepal

	1971	1973	1978	1984	1985	1986	1987	1988	1989
Thousand tonnes									
NGL/LPG/Ethane	-	-	1	1	1	2	3	3	-
Naphtha	-	-	-	-	-	-	-	-	-
Motor Gasoline	10	10	8	13	13	14	16	17	12
Aviation Fuels	3	4	11	17	18	18	18	17	10
Kerosene	16	28	26	39	46	55	47	69	97
Gas Diesel	25	27	41	67	68	79	84	79	104
Heavy Fuel Oil	-	1	1	8	10	12	6	6	-
Other	-	-	-	5	2	3	3	3	2
Refinery Fuel	-	-	-	-	-	-	-	-	-
Sub-Total	54	70	88	150	158	183	177	194	225
Bunkers	-	-	-	-	-	-	-	-	-
Total	54	70	88	150	158	183	177	194	225
Thousand barrels/day									
NGL/LPG/Ethane	-	-	0.0	0.0	0.0	0.1	0.1	0.1	-
Naphtha	-	-	-	-	-	-	-	-	-
Motor Gasoline	0.2	0.2	0.2	0.3	0.3	0.3	0.4	0.4	0.3
Aviation Fuels	0.1	0.1	0.2	0.4	0.4	0.4	0.4	0.4	0.2
Kerosene	0.3	0.6	0.6	0.9	1.0	1.2	1.0	1.5	2.2
Gas Diesel	0.5	0.6	0.8	1.4	1.4	1.7	1.8	1.7	2.2
Heavy Fuel Oil	-	0.0	0.0	0.1	0.2	0.2	0.1	0.1	-
Other	-	-	-	0.1	0.0	0.1	0.1	0.1	0.0
Refinery Fuel	-	-	-	-	-	-	-	-	-
Sub-Total	1.1	1.5	1.9	3.2	3.4	4.0	3.9	4.2	4.9
Bunkers	-	-	-	-	-	-	-	-	-
Total	1.1	1.5	1.9	3.2	3.4	4.0	3.9	4.2	4.9

	1990	1991	1992	1993	1994	1995	1996	1997	1998
Thousand tonnes									
NGL/LPG/Ethane	4	6	8	10	13	18	22	29	39
Naphtha	-	-	-	-	-	-	-	-	-
Motor Gasoline	16	27	26	22	25	32	32	35	37
Aviation Fuels	15	23	22	22	26	32	34	39	47
Kerosene	77	125	118	128	130	174	195	223	231
Gas Diesel	110	169	162	167	182	225	219	255	271
Heavy Fuel Oil	6	11	20	25	26	13	13	26	29
Other	3	4	3	5	6	7	8	10	11
Refinery Fuel	-	-	-	-	-	-	-	-	-
Sub-Total	231	365	359	379	408	501	523	617	665
Bunkers	-	-	-	-	-	-	-	-	-
Total	231	365	359	379	408	501	523	617	665
Thousand barrels/day									
NGL/LPG/Ethane	0.1	0.2	0.3	0.3	0.4	0.6	0.7	0.9	1.2
Naphtha	-	-	-	-	-	-	-	-	-
Motor Gasoline	0.4	0.7	0.6	0.5	0.6	0.8	0.8	0.9	0.9
Aviation Fuels	0.3	0.5	0.5	0.5	0.6	0.7	0.8	0.9	1.0
Kerosene	1.7	2.8	2.6	2.8	2.9	3.9	4.3	4.9	5.1
Gas Diesel	2.3	3.6	3.4	3.5	3.8	4.7	4.6	5.4	5.7
Heavy Fuel Oil	0.1	0.2	0.4	0.5	0.5	0.2	0.2	0.5	0.5
Other	0.1	0.1	0.1	0.1	0.1	0.1	0.1	0.2	0.2
Refinery Fuel	-	-	-	-	-	-	-	-	-
Sub-Total	5.0	8.0	7.8	8.2	8.9	11.0	11.5	13.6	14.7
Bunkers	-	-	-	-	-	-	-	-	-
Total	5.0	8.0	7.8	8.2	8.9	11.0	11.5	13.6	14.7

[1] See Notes in Part I.4.

Oil Demand by Main Product Groups[1]
Demande de produits raffinés par groupes principaux

Netherlands Antilles

	1971	1973	1978	1984	1985	1986	1987	1988	1989
Thousand tonnes									
NGL/LPG/Ethane	79	57	82	65	60	60	38	35	39
Naphtha	-	-	-	-	-	-	-	-	-
Motor Gasoline	67	78	179	100	100	100	99	98	85
Aviation Fuels	49	55	46	46	41	36	37	37	37
Kerosene	45	46	20	28	30	30	30	18	20
Gas Diesel	404	351	349	300	250	187	202	215	201
Heavy Fuel Oil	1 670	1 835	742	570	540	492	492	420	381
Other	24	24	27	35	35	33	33	33	33
Refinery Fuel	2 390	2 640	1 600	1 080	460	508	408	320	219
Sub-Total	**4 728**	**5 086**	**3 045**	**2 224**	**1 516**	**1 446**	**1 339**	**1 176**	**1 015**
Bunkers	2 500	2 312	2 259	1 950	1 985	1 675	1 600	1 600	1 650
Total	**7 228**	**7 398**	**5 304**	**4 174**	**3 501**	**3 121**	**2 939**	**2 776**	**2 665**
Thousand barrels/day									
NGL/LPG/Ethane	2.5	1.8	2.6	2.1	1.9	1.9	1.2	1.1	1.2
Naphtha	-	-	-	-	-	-	-	-	-
Motor Gasoline	1.5	1.8	4.1	2.3	2.3	2.3	2.3	2.2	2.0
Aviation Fuels	1.1	1.2	1.0	1.0	0.9	0.8	0.8	0.8	0.8
Kerosene	1.0	1.0	0.4	0.6	0.6	0.6	0.6	0.4	0.4
Gas Diesel	8.1	7.0	7.0	6.0	5.0	3.7	4.0	4.3	4.0
Heavy Fuel Oil	30.5	31.7	12.8	9.8	9.3	8.5	8.5	7.2	6.6
Other	0.5	0.5	0.5	0.7	0.7	0.6	0.6	0.6	0.6
Refinery Fuel	43.6	45.6	27.6	18.6	7.9	8.8	7.0	5.5	3.8
Sub-Total	**88.7**	**90.5**	**56.1**	**41.0**	**28.7**	**27.3**	**25.1**	**22.2**	**19.4**
Bunkers	45.8	40.3	39.5	34.1	34.8	29.4	28.0	28.0	29.0
Total	**134.5**	**130.8**	**95.5**	**75.1**	**63.4**	**56.6**	**53.2**	**50.1**	**48.4**

	1990	1991	1992	1993	1994	1995	1996	1997	1998
Thousand tonnes									
NGL/LPG/Ethane	45	45	45	52	51	52	55	55	55
Naphtha	-	-	-	-	-	-	-	-	-
Motor Gasoline	85	85	85	43	59	63	69	69	69
Aviation Fuels	37	49	49	59	60	61	62	62	62
Kerosene	25	25	30	101	108	108	111	111	111
Gas Diesel	197	228	253	222	297	295	293	293	293
Heavy Fuel Oil	358	360	373	349	364	371	383	383	397
Other	33	33	38	98	88	89	65	65	65
Refinery Fuel	142	140	127	44	30	21	10	10	10
Sub-Total	**922**	**965**	**1 000**	**968**	**1 057**	**1 060**	**1 048**	**1 048**	**1 062**
Bunkers	1 675	1 725	1 725	1 697	1 714	1 720	1 724	1 724	1 724
Total	**2 597**	**2 690**	**2 725**	**2 665**	**2 771**	**2 780**	**2 772**	**2 772**	**2 786**
Thousand barrels/day									
NGL/LPG/Ethane	1.4	1.4	1.4	1.7	1.6	1.7	1.7	1.7	1.7
Naphtha	-	-	-	-	-	-	-	-	-
Motor Gasoline	2.0	2.0	2.0	1.0	1.4	1.5	1.6	1.6	1.6
Aviation Fuels	0.8	1.1	1.0	1.3	1.3	1.3	1.3	1.3	1.3
Kerosene	0.5	0.5	0.6	2.1	2.3	2.3	2.3	2.4	2.4
Gas Diesel	3.9	4.6	5.0	4.4	5.9	5.9	5.8	5.9	5.9
Heavy Fuel Oil	6.2	6.2	6.4	6.0	6.3	6.4	6.6	6.6	6.9
Other	0.6	0.6	0.7	1.8	1.6	1.6	1.2	1.3	1.3
Refinery Fuel	2.5	2.4	2.2	0.8	0.5	0.4	0.2	0.2	0.2
Sub-Total	**17.9**	**18.8**	**19.4**	**19.1**	**20.9**	**21.0**	**20.9**	**20.9**	**21.2**
Bunkers	29.5	30.5	30.4	29.8	30.2	30.3	30.3	30.3	30.3
Total	**47.4**	**49.2**	**49.8**	**48.9**	**51.1**	**51.3**	**51.1**	**51.3**	**51.5**

[1] See Notes in Part I.4.

Oil Demand by Main Product Groups[1]
Demande de produits raffinés par groupes principaux

Nicaragua

	1971	1973	1978	1984	1985	1986	1987	1988	1989
Thousand tonnes									
NGL/LPG/Ethane	13	16	16	16	17	19	19	16	18
Naphtha	-	-	-	-	-	-	-	-	-
Motor Gasoline	112	140	167	110	107	118	120	107	82
Aviation Fuels	24	23	25	21	21	19	36	27	24
Kerosene	17	18	16	12	15	22	15	13	9
Gas Diesel	163	175	226	216	213	231	245	227	203
Heavy Fuel Oil	168	180	319	211	218	275	269	250	187
Other	12	42	29	14	18	17	28	23	47
Refinery Fuel	10	12	12	8	10	10	10	10	12
Sub-Total	**519**	**606**	**810**	**608**	**619**	**711**	**742**	**673**	**582**
Bunkers	-	-	-	-	-	-	-	-	-
Total	**519**	**606**	**810**	**608**	**619**	**711**	**742**	**673**	**582**
Thousand barrels/day									
NGL/LPG/Ethane	0.4	0.5	0.5	0.5	0.5	0.6	0.6	0.5	0.6
Naphtha	-	-	-	-	-	-	-	-	-
Motor Gasoline	2.6	3.3	3.9	2.6	2.5	2.8	2.8	2.5	1.9
Aviation Fuels	0.5	0.5	0.6	0.5	0.5	0.4	0.8	0.6	0.5
Kerosene	0.4	0.4	0.3	0.3	0.3	0.5	0.3	0.3	0.2
Gas Diesel	3.3	3.6	4.6	4.4	4.4	4.7	5.0	4.6	4.1
Heavy Fuel Oil	3.1	3.3	5.8	3.8	4.0	5.0	4.9	4.5	3.4
Other	0.2	0.8	0.5	0.3	0.3	0.3	0.5	0.4	0.9
Refinery Fuel	0.2	0.3	0.3	0.2	0.2	0.2	0.2	0.2	0.3
Sub-Total	**10.8**	**12.6**	**16.5**	**12.5**	**12.7**	**14.5**	**15.2**	**13.7**	**11.9**
Bunkers	-	-	-	-	-	-	-	-	-
Total	**10.8**	**12.6**	**16.5**	**12.5**	**12.7**	**14.5**	**15.2**	**13.7**	**11.9**

	1990	1991	1992	1993	1994	1995	1996	1997	1998
Thousand tonnes									
NGL/LPG/Ethane	17	20	21	23	27	28	29	29	35
Naphtha	-	-	-	-	-	-	-	-	-
Motor Gasoline	87	100	121	117	110	112	111	116	133
Aviation Fuels	27	24	16	14	18	21	23	22	28
Kerosene	9	12	12	13	14	13	13	12	12
Gas Diesel	210	197	214	226	281	306	307	357	429
Heavy Fuel Oil	241	233	288	270	285	272	377	379	427
Other	29	24	35	39	31	25	32	56	46
Refinery Fuel	12	13	14	13	14	12	13	16	18
Sub-Total	**632**	**623**	**721**	**715**	**780**	**789**	**905**	**987**	**1 128**
Bunkers	-	-	-	-	-	-	-	-	-
Total	**632**	**623**	**721**	**715**	**780**	**789**	**905**	**987**	**1 128**
Thousand barrels/day									
NGL/LPG/Ethane	0.5	0.6	0.7	0.7	0.9	0.9	0.9	0.9	1.1
Naphtha	-	-	-	-	-	-	-	-	-
Motor Gasoline	2.0	2.3	2.8	2.7	2.6	2.6	2.6	2.7	3.1
Aviation Fuels	0.6	0.5	0.3	0.3	0.4	0.5	0.5	0.5	0.6
Kerosene	0.2	0.3	0.3	0.3	0.3	0.3	0.3	0.3	0.3
Gas Diesel	4.3	4.0	4.4	4.6	5.7	6.3	6.3	7.3	8.8
Heavy Fuel Oil	4.4	4.3	5.2	4.9	5.2	5.0	6.9	6.9	7.8
Other	0.5	0.4	0.6	0.7	0.6	0.5	0.6	1.1	0.9
Refinery Fuel	0.3	0.3	0.3	0.3	0.3	0.3	0.3	0.3	0.4
Sub-Total	**12.8**	**12.7**	**14.6**	**14.6**	**15.9**	**16.2**	**18.3**	**20.0**	**22.9**
Bunkers	-	-	-	-	-	-	-	-	-
Total	**12.8**	**12.7**	**14.6**	**14.6**	**15.9**	**16.2**	**18.3**	**20.0**	**22.9**

[1] See Notes in Part I.4.

Oil Demand by Main Product Groups[1]
Demande de produits raffinés par groupes principaux

Nigeria

	1971	1973	1978	1984	1985	1986	1987	1988	1989
Thousand tonnes									
NGL/LPG/Ethane	12	17	32	67	86	88	87	84	69
Naphtha	-	-	-	-	-	-	-	-	-
Motor Gasoline	509	700	2 362	2 900	2 900	2 885	3 057	3 312	3 576
Aviation Fuels	77	121	315	411	422	300	280	280	278
Kerosene	216	257	671	1 304	1 307	1 528	1 630	1 551	1 503
Gas Diesel	450	704	1 851	2 275	2 103	1 823	1 670	1 754	1 599
Heavy Fuel Oil	274	438	753	804	1 015	762	689	938	950
Other	88	290	486	269	280	307	376	255	453
Refinery Fuel	62	69	92	426	433	481	483	533	533
Sub-Total	1 688	2 596	6 562	8 456	8 546	8 174	8 272	8 707	8 961
Bunkers	5	13	45	100	110	372	204	307	340
Total	1 693	2 609	6 607	8 556	8 656	8 546	8 476	9 014	9 301
Thousand barrels/day									
NGL/LPG/Ethane	0.4	0.5	1.0	2.1	2.7	2.8	2.8	2.7	2.2
Naphtha	-	-	-	-	-	-	-	-	-
Motor Gasoline	11.9	16.4	55.2	67.6	67.8	67.4	71.4	77.2	83.6
Aviation Fuels	1.7	2.6	6.9	8.9	9.2	6.5	6.1	6.1	6.0
Kerosene	4.6	5.5	14.2	27.6	27.7	32.4	34.6	32.8	31.9
Gas Diesel	9.2	14.4	37.8	46.4	43.0	37.3	34.1	35.8	32.7
Heavy Fuel Oil	5.0	8.0	13.7	14.6	18.5	13.9	12.6	17.1	17.3
Other	1.7	5.6	9.3	5.1	5.4	5.9	7.2	4.9	8.7
Refinery Fuel	1.2	1.4	1.8	8.4	8.6	9.4	9.5	10.7	10.8
Sub-Total	35.7	54.3	140.0	180.7	182.8	175.6	178.3	187.2	193.1
Bunkers	0.1	0.2	0.9	1.9	2.1	7.1	3.9	5.8	6.4
Total	35.7	54.5	140.9	182.7	185.0	182.8	182.1	192.9	199.5

	1990	1991	1992	1993	1994	1995	1996	1997	1998
Thousand tonnes									
NGL/LPG/Ethane	40	56	49	36	21	10	11	11	10
Naphtha	-	-	-	-	-	-	-	-	-
Motor Gasoline	2 625	3 380	3 969	3 336	2 534	3 016	3 301	3 308	3 265
Aviation Fuels	302	354	363	369	306	432	413	498	426
Kerosene	1 298	1 253	1 232	1 062	807	959	1 051	1 053	1 039
Gas Diesel	1 396	1 787	2 280	2 054	1 596	1 963	2 430	3 017	2 976
Heavy Fuel Oil	877	824	2 002	1 600	1 103	752	1 015	1 135	1 115
Other	451	766	2 108	1 025	1 494	1 679	1 582	1 751	1 592
Refinery Fuel	530	485	504	553	576	681	784	835	686
Sub-Total	7 519	8 905	12 507	10 035	8 437	9 492	10 587	11 608	11 109
Bunkers	186	354	349	354	94	344	344	344	344
Total	7 705	9 259	12 856	10 389	8 531	9 836	10 931	11 952	11 453
Thousand barrels/day									
NGL/LPG/Ethane	1.3	1.8	1.6	1.1	0.7	0.3	0.3	0.4	0.3
Naphtha	-	-	-	-	-	-	-	-	-
Motor Gasoline	61.3	79.0	92.5	78.0	59.2	70.5	76.9	77.3	76.3
Aviation Fuels	6.6	7.7	7.9	8.0	6.6	9.4	8.9	10.8	9.3
Kerosene	27.5	26.6	26.1	22.5	17.1	20.3	22.2	22.3	22.0
Gas Diesel	28.5	36.5	46.5	42.0	32.6	40.1	49.5	61.7	60.8
Heavy Fuel Oil	16.0	15.0	36.4	29.2	20.1	13.7	18.5	20.7	20.3
Other	8.6	14.5	40.2	19.4	28.6	32.1	30.2	33.5	30.5
Refinery Fuel	10.7	9.9	10.2	11.0	11.3	13.5	15.5	16.5	13.3
Sub-Total	160.6	191.0	261.3	211.2	176.3	200.0	222.1	243.2	232.8
Bunkers	3.6	6.7	6.6	6.7	1.9	6.5	6.5	6.5	6.5
Total	164.2	197.7	267.9	217.9	178.2	206.4	228.6	249.6	239.3

[1] See Notes in Part I.4.

Oil Demand by Main Product Groups[1]
Demande de produits raffinés par groupes principaux

Oman

	1971	1973	1978	1984	1985	1986	1987	1988	1989
Thousand tonnes									
NGL/LPG/Ethane	-	-	-	18	23	29	25	27	30
Naphtha	-	-	-	-	-	-	-	-	-
Motor Gasoline	30	37	165	352	397	421	402	415	438
Aviation Fuels	4	4	104	180	182	189	127	129	140
Kerosene	6	9	13	11	9	10	10	5	10
Gas Diesel	45	46	215	452	525	520	508	540	536
Heavy Fuel Oil	-	-	-	-	-	-	-	-	-
Other	-	-	72	123	106	-	4	4	20
Refinery Fuel	-	-	-	88	92	100	79	96	100
Sub-Total	85	96	569	1 224	1 334	1 269	1 155	1 216	1 274
Bunkers	1 250	1 230	637	96	114	120	150	180	200
Total	1 335	1 326	1 206	1 320	1 448	1 389	1 305	1 396	1 474
Thousand barrels/day									
NGL/LPG/Ethane	-	-	-	0.6	0.7	0.9	0.8	0.9	1.0
Naphtha	-	-	-	-	-	-	-	-	-
Motor Gasoline	0.7	0.9	3.9	8.2	9.3	9.8	9.4	9.7	10.2
Aviation Fuels	0.1	0.1	2.3	3.9	4.0	4.1	2.8	2.8	3.0
Kerosene	0.1	0.2	0.3	0.2	0.2	0.2	0.2	0.1	0.2
Gas Diesel	0.9	0.9	4.4	9.2	10.7	10.6	10.4	11.0	11.0
Heavy Fuel Oil	-	-	-	-	-	-	-	-	-
Other	-	-	1.4	2.4	2.0	-	0.1	0.1	0.4
Refinery Fuel	-	-	-	1.6	1.7	1.8	1.4	1.7	1.8
Sub-Total	1.8	2.1	12.2	26.1	28.6	27.5	25.1	26.2	27.6
Bunkers	22.8	22.4	11.6	1.7	2.1	2.2	2.7	3.3	3.6
Total	24.6	24.5	23.8	27.8	30.7	29.7	27.8	29.5	31.2

	1990	1991	1992	1993	1994	1995	1996	1997	1998
Thousand tonnes									
NGL/LPG/Ethane	66	82	73	57	78	63	63	52	66
Naphtha	-	-	-	-	-	-	-	-	-
Motor Gasoline	478	521	560	589	593	617	639	678	700
Aviation Fuels	296	256	163	192	164	147	152	172	181
Kerosene	4	5	5	5	4	4	4	4	3
Gas Diesel	433	459	513	595	641	675	656	678	745
Heavy Fuel Oil	507	1 661	1 427	965	521	954	824	442	281
Other	24	27	30	43	41	41	41	41	41
Refinery Fuel	122	130	112	94	147	150	152	137	161
Sub-Total	1 930	3 141	2 883	2 540	2 189	2 651	2 531	2 204	2 178
Bunkers	20	21	22	247	25	27	27	30	39
Total	1 950	3 162	2 905	2 787	2 214	2 678	2 558	2 234	2 217
Thousand barrels/day									
NGL/LPG/Ethane	2.1	2.6	2.3	1.8	2.5	2.0	2.0	1.7	2.1
Naphtha	-	-	-	-	-	-	-	-	-
Motor Gasoline	11.2	12.2	13.1	13.8	13.9	14.4	14.9	15.8	16.4
Aviation Fuels	6.4	5.6	3.5	4.2	3.6	3.2	3.3	3.7	3.9
Kerosene	0.1	0.1	0.1	0.1	0.1	0.1	0.1	0.1	0.1
Gas Diesel	8.9	9.4	10.5	12.2	13.1	13.8	13.4	13.9	15.2
Heavy Fuel Oil	9.3	30.3	26.0	17.6	9.5	17.4	15.0	8.1	5.1
Other	0.4	0.5	0.5	0.8	0.8	0.8	0.8	0.8	0.8
Refinery Fuel	2.2	2.4	2.0	1.7	2.7	2.7	2.8	2.5	2.9
Sub-Total	40.5	63.0	58.0	52.1	46.0	54.4	52.1	46.5	46.5
Bunkers	0.4	0.4	0.4	4.5	0.5	0.5	0.5	0.5	0.7
Total	40.9	63.4	58.4	56.6	46.5	54.9	52.6	47.0	47.2

[1] See Notes in Part I.4.

Oil Demand by Main Product Groups[1]
Demande de produits raffinés par groupes principaux

Pakistan

	1971	1973	1978	1984	1985	1986	1987	1988	1989
Thousand tonnes									
NGL/LPG/Ethane	4	5	11	70	70	75	100	130	131
Naphtha	-	-	-	-	-	-	-	-	-
Motor Gasoline	261	315	445	755	784	823	872	940	989
Aviation Fuels	368	339	440	465	459	499	464	475	466
Kerosene	422	436	644	690	760	811	877	876	989
Gas Diesel	1 122	1 248	1 636	2 653	2 784	2 980	3 499	3 975	4 286
Heavy Fuel Oil	735	711	584	1 662	1 943	2 070	2 123	2 374	2 438
Other	147	133	307	297	346	373	360	383	377
Refinery Fuel	97	97	114	148	155	166	170	168	180
Sub-Total	3 156	3 284	4 181	6 740	7 301	7 797	8 465	9 321	9 856
Bunkers	92	158	81	46	25	13	10	15	22
Total	3 248	3 442	4 262	6 786	7 326	7 810	8 475	9 336	9 878
Thousand barrels/day									
NGL/LPG/Ethane	0.1	0.2	0.4	2.2	2.2	2.4	3.2	4.1	4.2
Naphtha	-	-	-	-	-	-	-	-	-
Motor Gasoline	6.1	7.4	10.4	17.6	18.3	19.2	20.4	21.9	23.1
Aviation Fuels	8.0	7.4	9.6	10.1	10.0	10.8	10.1	10.3	10.1
Kerosene	8.9	9.2	13.7	14.6	16.1	17.2	18.6	18.5	21.0
Gas Diesel	22.9	25.5	33.4	54.1	56.9	60.9	71.5	81.0	87.6
Heavy Fuel Oil	13.4	13.0	10.7	30.2	35.5	37.8	38.7	43.2	44.5
Other	2.8	2.6	5.9	5.4	6.3	6.8	6.5	6.9	6.9
Refinery Fuel	1.9	1.9	2.3	2.9	3.1	3.3	3.4	3.3	3.6
Sub-Total	64.3	67.1	86.2	137.1	148.4	158.4	172.4	189.3	200.9
Bunkers	1.7	2.9	1.5	0.8	0.5	0.2	0.2	0.3	0.4
Total	66.0	70.0	87.7	138.0	148.8	158.7	172.6	189.6	201.3

	1990	1991	1992	1993	1994	1995	1996	1997	1998
Thousand tonnes									
NGL/LPG/Ethane	127	147	131	135	122	171	205	187	206
Naphtha	-	-	-	-	-	-	-	-	-
Motor Gasoline	821	829	899	989	992	983	1 062	1 197	1 223
Aviation Fuels	455	504	540	568	553	556	586	582	597
Kerosene	1 133	962	632	640	608	602	614	522	514
Gas Diesel	4 277	4 305	5 036	5 501	5 912	6 149	6 600	6 405	6 509
Heavy Fuel Oil	3 101	3 271	3 875	4 274	5 102	5 629	6 706	6 959	7 819
Other	212	319	334	376	338	361	341	342	370
Refinery Fuel	180	185	190	210	212	205	213	205	210
Sub-Total	10 306	10 522	11 637	12 693	13 839	14 656	16 327	16 399	17 448
Bunkers	34	18	13	15	17	15	12	18	15
Total	10 340	10 540	11 650	12 708	13 856	14 671	16 339	16 417	17 463
Thousand barrels/day									
NGL/LPG/Ethane	4.0	4.7	4.2	4.3	3.9	5.4	6.5	5.9	6.5
Naphtha	-	-	-	-	-	-	-	-	-
Motor Gasoline	19.2	19.4	21.0	23.1	23.2	23.0	24.8	28.0	28.6
Aviation Fuels	9.9	11.0	11.7	12.3	12.0	12.1	12.7	12.6	13.0
Kerosene	24.0	20.4	13.4	13.6	12.9	12.8	13.0	11.1	10.9
Gas Diesel	87.4	88.0	102.6	112.4	120.8	125.7	134.5	130.9	133.0
Heavy Fuel Oil	56.6	59.7	70.5	78.0	93.1	102.7	122.0	127.0	142.7
Other	3.7	5.7	5.9	6.6	6.0	6.4	6.0	6.0	6.6
Refinery Fuel	3.6	3.7	3.8	4.2	4.3	4.1	4.3	4.1	4.2
Sub-Total	208.4	212.4	233.0	254.6	276.2	292.1	323.8	325.7	345.5
Bunkers	0.6	0.3	0.2	0.3	0.3	0.3	0.2	0.3	0.3
Total	209.0	212.8	233.3	254.9	276.5	292.4	324.0	326.0	345.8

[1] See Notes in Part I.4.

Oil Demand by Main Product Groups[1]
Demande de produits raffinés par groupes principaux

Panama

	1971	1973	1978	1984	1985	1986	1987	1988	1989
Thousand tonnes									
NGL/LPG/Ethane	23	29	40	50	51	55	58	56	55
Naphtha	-	-	-	-	-	-	-	-	-
Motor Gasoline	234	256	259	205	203	210	198	191	192
Aviation Fuels	6	6	2	2	2	2	1	1	1
Kerosene	8	5	4	4	3	3	4	3	4
Gas Diesel	169	209	264	296	310	313	332	267	283
Heavy Fuel Oil	330	368	305	228	173	183	227	132	146
Other	13	15	9	32	12	12	6	4	6
Refinery Fuel	141	160	123	65	91	76	97	75	77
Sub-Total	**924**	**1 048**	**1 006**	**882**	**845**	**854**	**923**	**729**	**764**
Bunkers	5 808	5 264	2 220	740	745	725	845	710	860
Total	**6 732**	**6 312**	**3 226**	**1 622**	**1 590**	**1 579**	**1 768**	**1 439**	**1 624**
Thousand barrels/day									
NGL/LPG/Ethane	0.7	0.9	1.3	1.6	1.6	1.7	1.8	1.8	1.7
Naphtha	-	-	-	-	-	-	-	-	-
Motor Gasoline	5.5	6.0	6.1	4.8	4.7	4.9	4.6	4.5	4.5
Aviation Fuels	0.1	0.1	0.0	0.0	0.0	0.0	0.0	0.0	0.0
Kerosene	0.2	0.1	0.1	0.1	0.1	0.1	0.1	0.1	0.1
Gas Diesel	3.5	4.3	5.4	6.0	6.3	6.4	6.8	5.4	5.8
Heavy Fuel Oil	6.0	6.7	5.6	4.1	3.2	3.3	4.1	2.4	2.7
Other	0.2	0.3	0.2	0.6	0.2	0.2	0.1	0.1	0.1
Refinery Fuel	2.9	3.2	2.4	1.3	1.8	1.5	1.9	1.5	1.5
Sub-Total	**19.1**	**21.6**	**21.0**	**18.6**	**18.0**	**18.2**	**19.5**	**15.7**	**16.4**
Bunkers	107.1	96.8	41.3	13.6	13.7	13.3	15.5	13.0	15.8
Total	**126.2**	**118.4**	**62.3**	**32.2**	**31.7**	**31.5**	**35.1**	**28.7**	**32.2**

	1990	1991	1992	1993	1994	1995	1996	1997	1998
Thousand tonnes									
NGL/LPG/Ethane	62	65	68	68	73	79	83	85	90
Naphtha	-	-	-	-	-	-	-	-	-
Motor Gasoline	214	223	240	259	287	301	317	340	381
Aviation Fuels	2	2	2	2	3	2	3	3	4
Kerosene	5	5	6	8	8	8	8	9	11
Gas Diesel	296	380	404	402	438	450	442	549	718
Heavy Fuel Oil	158	180	209	222	273	270	279	251	329
Other	7	4	10	13	12	13	19	47	47
Refinery Fuel	77	79	97	101	104	106	133	44	44
Sub-Total	**821**	**938**	**1 036**	**1 075**	**1 198**	**1 229**	**1 284**	**1 328**	**1 624**
Bunkers	1 040	1 110	1 160	1 060	1 060	1 060	1 060	1 060	1 060
Total	**1 861**	**2 048**	**2 196**	**2 135**	**2 258**	**2 289**	**2 344**	**2 388**	**2 684**
Thousand barrels/day									
NGL/LPG/Ethane	2.0	2.1	2.2	2.2	2.3	2.5	2.6	2.7	2.9
Naphtha	-	-	-	-	-	-	-	-	-
Motor Gasoline	5.0	5.2	5.6	6.1	6.7	7.0	7.4	7.9	8.9
Aviation Fuels	0.0	0.0	0.0	0.0	0.1	0.0	0.1	0.1	0.1
Kerosene	0.1	0.1	0.1	0.2	0.2	0.2	0.2	0.2	0.2
Gas Diesel	6.1	7.8	8.2	8.2	9.0	9.2	9.0	11.2	14.7
Heavy Fuel Oil	2.9	3.3	3.8	4.1	5.0	4.9	5.1	4.6	6.0
Other	0.1	0.1	0.2	0.2	0.2	0.2	0.4	0.9	0.9
Refinery Fuel	1.5	1.5	1.9	2.0	2.0	2.1	2.6	1.0	1.0
Sub-Total	**17.7**	**20.1**	**22.0**	**22.9**	**25.5**	**26.2**	**27.3**	**28.6**	**34.6**
Bunkers	19.1	20.4	21.2	19.5	19.5	19.5	19.4	19.5	19.5
Total	**36.8**	**40.4**	**43.3**	**42.4**	**44.9**	**45.7**	**46.8**	**48.0**	**54.1**

[1] See Notes in Part I.4.

Oil Demand by Main Product Groups[1]
Demande de produits raffinés par groupes principaux

Paraguay

	1971	1973	1978	1984	1985	1986	1987	1988	1989
Thousand tonnes									
NGL/LPG/Ethane	-	8	5	29	32	30	33	28	41
Naphtha	-	-	-	-	-	-	-	-	-
Motor Gasoline	80	83	123	101	109	122	134	147	137
Aviation Fuels	6	8	19	22	23	35	32	33	35
Kerosene	18	13	19	11	18	15	13	13	14
Gas Diesel	47	59	195	301	297	301	316	322	374
Heavy Fuel Oil	33	61	56	21	20	24	34	45	40
Other	9	11	9	8	12	14	13	12	12
Refinery Fuel	11	14	19	9	12	13	10	10	10
Sub-Total	204	257	445	502	523	554	585	610	663
Bunkers	-	-	-	-	-	-	-	-	-
Total	204	257	445	502	523	554	585	610	663
Thousand barrels/day									
NGL/LPG/Ethane	-	0.2	0.2	0.9	1.0	0.9	1.0	0.9	1.3
Naphtha	-	-	-	-	-	-	-	-	-
Motor Gasoline	1.9	1.9	2.9	2.3	2.5	2.8	3.1	3.4	3.1
Aviation Fuels	0.1	0.2	0.4	0.5	0.5	0.8	0.7	0.7	0.8
Kerosene	0.4	0.3	0.4	0.2	0.4	0.3	0.3	0.3	0.3
Gas Diesel	1.0	1.2	4.0	6.1	6.0	6.1	6.4	6.5	7.6
Heavy Fuel Oil	0.6	1.1	1.0	0.4	0.4	0.4	0.6	0.8	0.7
Other	0.2	0.2	0.2	0.1	0.2	0.2	0.2	0.2	0.2
Refinery Fuel	0.2	0.3	0.3	0.2	0.2	0.2	0.2	0.2	0.2
Sub-Total	4.3	5.4	9.3	10.7	11.2	11.8	12.5	12.9	14.2
Bunkers	-	-	-	-	-	-	-	-	-
Total	4.3	5.4	9.3	10.7	11.2	11.8	12.5	12.9	14.2

	1990	1991	1992	1993	1994	1995	1996	1997	1998
Thousand tonnes									
NGL/LPG/Ethane	50	54	59	63	65	65	66	66	78
Naphtha	-	-	-	-	-	-	-	-	-
Motor Gasoline	146	155	165	189	216	246	209	187	224
Aviation Fuels	3	3	3	3	3	3	3	3	3
Kerosene	8	8	8	7	8	8	7	7	8
Gas Diesel	383	365	477	553	632	728	711	807	884
Heavy Fuel Oil	49	45	61	46	86	81	33	38	38
Other	12	12	10	10	9	12	12	12	12
Refinery Fuel	10	-	8	7	-	-	-	-	-
Sub-Total	661	642	791	878	1 019	1 143	1 041	1 120	1 247
Bunkers	-	-	-	-	-	-	-	-	-
Total	661	642	791	878	1 019	1 143	1 041	1 120	1 247
Thousand barrels/day									
NGL/LPG/Ethane	1.5	1.7	1.8	1.9	2.0	2.0	2.0	2.0	2.4
Naphtha	-	-	-	-	-	-	-	-	-
Motor Gasoline	3.4	3.6	3.8	4.4	5.0	5.7	4.8	4.3	5.2
Aviation Fuels	0.1	0.1	0.1	0.1	0.1	0.1	0.1	0.1	0.1
Kerosene	0.2	0.2	0.2	0.1	0.2	0.2	0.1	0.1	0.2
Gas Diesel	7.8	7.4	9.7	11.2	12.8	14.8	14.4	16.4	17.9
Heavy Fuel Oil	0.9	0.8	1.1	0.8	1.6	1.5	0.6	0.7	0.7
Other	0.2	0.2	0.2	0.2	0.2	0.2	0.2	0.2	0.2
Refinery Fuel	0.2	-	0.1	0.1	-	-	-	-	-
Sub-Total	14.2	13.9	16.9	18.9	21.8	24.4	22.3	23.9	26.7
Bunkers	-	-	-	-	-	-	-	-	-
Total	14.2	13.9	16.9	18.9	21.8	24.4	22.3	23.9	26.7

[1] See Notes in Part I.4.

Oil Demand by Main Product Groups[1]
Demande de produits raffinés par groupes principaux

Peru

	1971	1973	1978	1984	1985	1986	1987	1988	1989
Thousand tonnes									
NGL/LPG/Ethane	62	81	116	121	118	137	144	161	159
Naphtha	-	-	-	-	-	-	-	-	-
Motor Gasoline	1 224	1 419	1 243	1 222	1 118	1 194	1 313	1 338	1 097
Aviation Fuels	235	255	345	336	233	255	274	243	204
Kerosene	527	593	722	785	806	933	1 045	1 111	957
Gas Diesel	837	875	1 188	1 489	1 441	1 482	1 604	1 590	1 601
Heavy Fuel Oil	1 525	1 482	1 784	1 423	1 483	1 563	1 754	1 687	1 571
Other	107	106	142	116	140	172	161	103	72
Refinery Fuel	138	142	211	246	139	205	201	209	174
Sub-Total	4 655	4 953	5 751	5 738	5 478	5 941	6 496	6 442	5 835
Bunkers	29	36	67	82	77	67	63	60	60
Total	4 684	4 989	5 818	5 820	5 555	6 008	6 559	6 502	5 895
Thousand barrels/day									
NGL/LPG/Ethane	2.0	2.6	3.7	3.8	3.8	4.4	4.6	5.1	5.1
Naphtha	-	-	-	-	-	-	-	-	-
Motor Gasoline	28.6	33.2	29.0	28.5	26.1	27.9	30.7	31.2	25.6
Aviation Fuels	5.2	5.6	7.5	7.3	5.1	5.5	6.0	5.3	4.4
Kerosene	11.2	12.6	15.3	16.6	17.1	19.8	22.2	23.5	20.3
Gas Diesel	17.1	17.9	24.3	30.4	29.5	30.3	32.8	32.4	32.7
Heavy Fuel Oil	27.8	27.0	32.6	25.9	27.1	28.5	32.0	30.7	28.7
Other	2.1	2.0	2.7	2.2	2.7	3.3	3.1	2.0	1.4
Refinery Fuel	2.8	2.8	4.2	5.0	2.9	4.1	4.0	4.1	3.4
Sub-Total	96.7	103.7	119.3	119.6	114.1	123.8	135.2	134.3	121.6
Bunkers	0.6	0.7	1.3	1.6	1.5	1.3	1.2	1.2	1.2
Total	97.2	104.4	120.6	121.2	115.6	125.1	136.4	135.4	122.8

	1990	1991	1992	1993	1994	1995	1996	1997	1998
Thousand tonnes									
NGL/LPG/Ethane	175	189	201	211	239	262	268	294	347
Naphtha	-	-	-	-	-	-	-	-	-
Motor Gasoline	1 232	1 188	1 092	1 160	1 075	1 095	1 194	1 220	1 188
Aviation Fuels	266	267	269	285	308	400	451	440	418
Kerosene	743	711	723	706	683	683	718	713	693
Gas Diesel	1 580	1 545	1 967	1 946	2 105	2 770	2 886	3 099	2 854
Heavy Fuel Oil	1 562	1 392	1 015	1 189	1 224	1 679	1 967	1 759	1 726
Other	66	83	67	89	91	78	54	152	169
Refinery Fuel	123	119	117	124	122	143	180	214	178
Sub-Total	5 747	5 494	5 451	5 710	5 847	7 110	7 718	7 891	7 573
Bunkers	11	25	69	66	64	131	1	21	45
Total	5 758	5 519	5 520	5 776	5 911	7 241	7 719	7 912	7 618
Thousand barrels/day									
NGL/LPG/Ethane	5.6	6.0	6.4	6.7	7.6	8.3	8.5	9.3	11.0
Naphtha	-	-	-	-	-	-	-	-	-
Motor Gasoline	28.8	27.8	25.5	27.1	25.1	25.6	27.8	28.5	27.8
Aviation Fuels	5.8	5.8	5.8	6.2	6.7	8.7	9.8	9.6	9.1
Kerosene	15.8	15.1	15.3	15.0	14.5	14.5	15.2	15.1	14.7
Gas Diesel	32.3	31.6	40.1	39.8	43.0	56.6	58.8	63.3	58.3
Heavy Fuel Oil	28.5	25.4	18.5	21.7	22.3	30.6	35.8	32.1	31.5
Other	1.3	1.6	1.3	1.7	1.7	1.5	1.0	2.9	3.2
Refinery Fuel	2.4	2.3	2.4	2.5	2.4	2.8	3.5	4.2	3.4
Sub-Total	120.4	115.6	115.2	120.7	123.4	148.7	160.4	165.1	159.0
Bunkers	0.2	0.5	1.4	1.3	1.3	2.7	0.0	0.4	0.9
Total	120.6	116.1	116.6	122.0	124.8	151.3	160.4	165.5	160.0

[1] See Notes in Part I.4.

Oil Demand by Main Product Groups[1]
Demande de produits raffinés par groupes principaux

Philippines

	1971	1973	1978	1984	1985	1986	1987	1988	1989
Thousand tonnes									
NGL/LPG/Ethane	119	158	223	186	181	211	205	302	361
Naphtha	2	33	170	42	92	51	43	15	28
Motor Gasoline	1 864	1 937	1 791	1 017	990	1 072	1 189	1 289	1 442
Aviation Fuels	250	280	350	360	346	346	421	478	544
Kerosene	419	428	469	312	268	292	324	356	433
Gas Diesel	1 520	1 739	2 155	2 254	2 072	2 137	2 355	2 529	3 025
Heavy Fuel Oil	3 188	4 047	5 384	3 847	3 099	2 817	4 147	4 363	4 678
Other	240	311	302	193	118	156	179	199	240
Refinery Fuel	125	206	271	429	392	402	906	909	488
Sub-Total	7 727	9 139	11 115	8 640	7 558	7 484	9 769	10 440	11 239
Bunkers	412	261	187	80	-	-	-	-	-
Total	8 139	9 400	11 302	8 720	7 558	7 484	9 769	10 440	11 239
Thousand barrels/day									
NGL/LPG/Ethane	3.8	5.0	7.1	5.9	5.8	6.7	6.5	9.6	11.5
Naphtha	0.0	0.8	4.0	1.0	2.1	1.2	1.0	0.3	0.7
Motor Gasoline	43.6	45.3	41.9	23.7	23.1	25.1	27.8	30.0	33.7
Aviation Fuels	5.5	6.1	7.6	7.8	7.5	7.5	9.2	10.4	11.8
Kerosene	8.9	9.1	9.9	6.6	5.7	6.2	6.9	7.5	9.2
Gas Diesel	31.1	35.5	44.0	45.9	42.3	43.7	48.1	51.5	61.8
Heavy Fuel Oil	58.2	73.8	98.2	70.0	56.5	51.4	75.7	79.4	85.4
Other	4.6	5.8	5.7	3.6	2.2	2.9	3.3	3.7	4.5
Refinery Fuel	2.4	4.0	5.4	9.1	8.4	8.6	18.7	18.6	10.3
Sub-Total	157.9	185.4	223.8	173.6	153.7	153.3	197.2	211.2	228.8
Bunkers	7.6	4.8	3.6	1.5	-	-	-	-	-
Total	165.5	190.2	227.4	175.1	153.7	153.3	197.2	211.2	228.8

	1990	1991	1992	1993	1994	1995	1996	1997	1998
Thousand tonnes									
NGL/LPG/Ethane	396	409	469	533	615	719	800	888	944
Naphtha	-	-	-	-	-	-	-	-	-
Motor Gasoline	1 507	1 299	1 453	1 623	1 767	2 080	2 334	2 526	2 654
Aviation Fuels	506	494	564	591	633	624	774	974	679
Kerosene	462	463	506	547	552	564	590	653	647
Gas Diesel	2 677	2 830	3 852	5 196	5 592	5 697	5 931	6 406	6 236
Heavy Fuel Oil	4 614	3 304	4 256	4 527	5 151	5 672	5 602	5 818	5 657
Other	262	299	286	305	310	319	359	353	342
Refinery Fuel	491	473	512	519	538	648	715	697	597
Sub-Total	10 915	9 571	11 898	13 841	15 158	16 323	17 105	18 315	17 756
Bunkers	-	-	23	23	-	73	73	73	73
Total	10 915	9 571	11 921	13 864	15 158	16 396	17 178	18 388	17 829
Thousand barrels/day									
NGL/LPG/Ethane	12.6	13.0	14.9	16.9	19.5	22.9	25.4	28.2	30.0
Naphtha	-	-	-	-	-	-	-	-	-
Motor Gasoline	35.2	30.4	33.9	37.9	41.3	48.6	54.4	59.0	62.0
Aviation Fuels	11.0	10.7	12.2	12.9	13.8	13.6	16.8	21.2	14.8
Kerosene	9.8	9.8	10.7	11.6	11.7	12.0	12.5	13.8	13.7
Gas Diesel	54.7	57.8	78.5	106.2	114.3	116.4	120.9	130.9	127.5
Heavy Fuel Oil	84.2	60.3	77.4	82.6	94.0	103.5	101.9	106.2	103.2
Other	4.9	5.6	5.4	5.8	5.9	6.0	6.8	6.7	6.5
Refinery Fuel	10.4	10.0	10.8	11.0	11.3	13.7	15.1	14.7	12.5
Sub-Total	222.8	197.7	243.8	284.9	311.8	336.6	353.7	380.8	370.2
Bunkers	-	-	0.5	0.5	-	1.4	1.4	1.4	1.4
Total	222.8	197.7	244.3	285.3	311.8	338.0	355.1	382.2	371.6

[1] See Notes in Part I.4.

Oil Demand by Main Product Groups [1]
Demande de produits raffinés par groupes principaux

Qatar

	1971	1973	1978	1984	1985	1986	1987	1988	1989
Thousand tonnes									
NGL/LPG/Ethane	-	-	6	335	343	450	504	454	622
Naphtha	-	-	-	-	-	-	-	-	-
Motor Gasoline	39	62	126	234	235	247	255	271	260
Aviation Fuels	-	21	77	83	75	93	101	85	90
Kerosene	4	6	5	10	15	15	15	15	15
Gas Diesel	35	44	142	185	177	163	156	169	170
Heavy Fuel Oil	-	-	-	-	-	-	-	-	-
Other	17	22	84	18	4	4	2	-	-
Refinery Fuel	1	1	14	5	33	45	52	63	40
Sub-Total	96	156	454	870	882	1 017	1 085	1 057	1 197
Bunkers	-	-	-	-	-	-	-	-	-
Total	96	156	454	870	882	1 017	1 085	1 057	1 197
Thousand barrels/day									
NGL/LPG/Ethane	-	-	0.2	10.6	10.9	14.3	16.0	14.4	19.8
Naphtha	-	-	-	-	-	-	-	-	-
Motor Gasoline	0.9	1.4	2.9	5.5	5.5	5.8	6.0	6.3	6.1
Aviation Fuels	-	0.5	1.7	1.8	1.6	2.0	2.2	1.8	2.0
Kerosene	0.1	0.1	0.1	0.2	0.3	0.3	0.3	0.3	0.3
Gas Diesel	0.7	0.9	2.9	3.8	3.6	3.3	3.2	3.4	3.5
Heavy Fuel Oil	-	-	-	-	-	-	-	-	-
Other	0.4	0.5	1.7	0.4	0.1	0.1	0.0	-	-
Refinery Fuel	0.0	0.0	0.3	0.1	0.7	0.9	1.1	1.3	0.9
Sub-Total	2.1	3.4	9.8	22.3	22.7	26.8	28.8	27.6	32.5
Bunkers	-	-	-	-	-	-	-	-	-
Total	2.1	3.4	9.8	22.3	22.7	26.8	28.8	27.6	32.5

	1990	1991	1992	1993	1994	1995	1996	1997	1998
Thousand tonnes									
NGL/LPG/Ethane	627	674	726	781	815	904	974	1 129	1 184
Naphtha	-	-	-	-	-	-	-	-	-
Motor Gasoline	303	307	323	339	347	371	389	414	439
Aviation Fuels	109	86	99	98	131	135	174	197	179
Kerosene	10	4	10	13	20	20	21	21	21
Gas Diesel	166	166	181	194	220	254	277	331	344
Heavy Fuel Oil	-	-	-	-	-	-	-	-	-
Other	-	-	-	-	-	-	-	-	-
Refinery Fuel	40	40	45	52	51	47	53	59	63
Sub-Total	1 255	1 277	1 384	1 477	1 584	1 731	1 888	2 151	2 230
Bunkers	-	-	-	-	-	-	-	-	-
Total	1 255	1 277	1 384	1 477	1 584	1 731	1 888	2 151	2 230
Thousand barrels/day									
NGL/LPG/Ethane	19.9	21.4	23.0	24.8	25.9	28.7	30.9	35.9	37.6
Naphtha	-	-	-	-	-	-	-	-	-
Motor Gasoline	7.1	7.2	7.5	7.9	8.1	8.7	9.1	9.7	10.3
Aviation Fuels	2.4	1.9	2.1	2.1	2.8	2.9	3.8	4.3	3.9
Kerosene	0.2	0.1	0.2	0.3	0.4	0.4	0.4	0.4	0.4
Gas Diesel	3.4	3.4	3.7	4.0	4.5	5.2	5.6	6.8	7.0
Heavy Fuel Oil	-	-	-	-	-	-	-	-	-
Other	-	-	-	-	-	-	-	-	-
Refinery Fuel	0.9	0.9	1.0	1.1	1.1	1.0	1.2	1.3	1.4
Sub-Total	33.9	34.8	37.6	40.2	42.9	47.0	51.0	58.3	60.6
Bunkers	-	-	-	-	-	-	-	-	-
Total	33.9	34.8	37.6	40.2	42.9	47.0	51.0	58.3	60.6

[1] See Notes in Part I.4.

Oil Demand by Main Product Groups[1]
Demande de produits raffinés par groupes principaux

Romania

	1971	1973	1978	1984	1985	1986	1987	1988	1989
Thousand tonnes									
NGL/LPG/Ethane	203	245	242	210	185	207	216	211	204
Naphtha	420	481	743	-	-	-	-	-	-
Motor Gasoline	2 009	2 270	2 581	1 656	1 405	1 308	1 821	1 700	1 689
Aviation Fuels	21	21	15	-	-	-	-	-	245
Kerosene	981	1 016	1 005	570	487	571	563	625	411
Gas Diesel	2 657	3 402	4 599	4 094	4 071	3 692	4 295	3 935	4 009
Heavy Fuel Oil	2 580	3 440	6 461	4 076	5 487	6 690	7 294	6 737	6 674
Other	1 600	1 679	1 850	1 281	1 258	1 479	1 319	1 207	1 537
Refinery Fuel	761	841	1 214	1 384	1 388	1 462	1 512	1 479	1 668
Sub-Total	11 232	13 395	18 710	13 271	14 281	15 409	17 020	15 894	16 437
Bunkers	-	-	-	-	-	-	-	-	-
Total	11 232	13 395	18 710	13 271	14 281	15 409	17 020	15 894	16 437
Thousand barrels/day									
NGL/LPG/Ethane	6.5	7.8	7.7	6.7	5.9	6.6	6.9	6.7	6.5
Naphtha	9.8	11.2	17.3	-	-	-	-	-	-
Motor Gasoline	47.0	53.1	60.3	38.6	32.8	30.6	42.6	39.6	39.5
Aviation Fuels	0.5	0.5	0.4	-	-	-	-	-	5.3
Kerosene	20.8	21.5	21.3	12.1	10.3	12.1	11.9	13.2	8.7
Gas Diesel	54.3	69.5	94.0	83.4	83.2	75.5	87.8	80.2	81.9
Heavy Fuel Oil	47.1	62.8	117.9	74.2	100.1	122.1	133.1	122.6	121.8
Other	29.7	31.0	34.1	22.5	22.2	25.5	23.0	20.4	27.3
Refinery Fuel	15.8	17.4	25.1	28.9	29.0	30.6	31.7	31.0	35.2
Sub-Total	231.3	274.8	378.1	266.3	283.6	302.9	336.9	313.8	326.2
Bunkers	-	-	-	-	-	-	-	-	-
Total	231.3	274.8	378.1	266.3	283.6	302.9	336.9	313.8	326.2

	1990	1991	1992	1993	1994	1995	1996	1997	1998
Thousand tonnes									
NGL/LPG/Ethane	221	222	224	212	177	245	187	286	335
Naphtha	-	-	-	107	122	56	177	42	291
Motor Gasoline	2 083	1 744	2 150	1 125	1 232	1 052	1 444	1 561	1 548
Aviation Fuels	227	161	242	248	165	181	86	123	110
Kerosene	357	284	354	114	35	23	31	66	55
Gas Diesel	4 028	3 436	3 630	2 856	2 562	2 414	3 506	3 142	2 963
Heavy Fuel Oil	8 162	6 895	4 759	5 209	4 073	4 592	5 227	5 253	3 650
Other	1 284	939	1 490	1 688	2 164	2 440	1 974	2 547	2 269
Refinery Fuel	1 616	1 153	591	689	951	623	563	535	647
Sub-Total	17 978	14 834	13 440	12 248	11 481	11 626	13 195	13 555	11 868
Bunkers	-	-	-	-	-	-	-	-	108
Total	17 978	14 834	13 440	12 248	11 481	11 626	13 195	13 555	11 976
Thousand barrels/day									
NGL/LPG/Ethane	7.0	7.1	7.1	6.7	5.5	7.7	5.9	9.0	10.5
Naphtha	-	-	-	2.5	2.8	1.3	4.1	1.0	6.8
Motor Gasoline	48.7	40.8	50.1	26.3	28.8	24.6	33.7	36.5	36.2
Aviation Fuels	4.9	3.5	5.2	5.4	3.6	3.9	1.9	2.7	2.4
Kerosene	7.6	6.0	7.5	2.4	0.7	0.5	0.7	1.4	1.2
Gas Diesel	82.3	70.2	74.0	58.4	52.4	49.3	71.5	64.2	60.6
Heavy Fuel Oil	148.9	125.8	86.6	95.0	74.3	83.8	95.1	95.8	66.6
Other	23.0	16.5	27.7	31.9	41.0	46.3	36.5	48.1	42.0
Refinery Fuel	34.4	24.2	12.4	14.5	19.6	13.7	12.3	11.7	14.2
Sub-Total	356.9	294.0	270.7	243.1	228.7	231.1	261.6	270.5	240.4
Bunkers	-	-	-	-	-	-	-	-	2.0
Total	356.9	294.0	270.7	243.1	228.7	231.1	261.6	270.5	242.5

[1] See Notes in Part I.4.

Oil Demand by Main Product Groups[1]
Demande de produits raffinés par groupes principaux

Russia

	1971	1973	1978	1984	1985	1986	1987	1988	1989
Thousand tonnes									
NGL/LPG/Ethane	-	-	-	-	-	-	-	-	-
Naphtha	-	-	-	-	-	-	-	-	-
Motor Gasoline	-	-	-	-	-	-	-	-	-
Aviation Fuels	-	-	-	-	-	-	-	-	-
Kerosene	-	-	-	-	-	-	-	-	-
Gas Diesel	-	-	-	-	-	-	-	-	-
Heavy Fuel Oil	-	-	-	-	-	-	-	-	-
Other	-	-	-	-	-	-	-	-	-
Refinery Fuel	-	-	-	-	-	-	-	-	-
Sub-Total	-	-	-	-	-	-	-	-	-
Bunkers	-	-	-	-	-	-	-	-	-
Total	-	-	-	-	-	-	-	-	-
Thousand barrels/day									
NGL/LPG/Ethane	-	-	-	-	-	-	-	-	-
Naphtha	-	-	-	-	-	-	-	-	-
Motor Gasoline	-	-	-	-	-	-	-	-	-
Aviation Fuels	-	-	-	-	-	-	-	-	-
Kerosene	-	-	-	-	-	-	-	-	-
Gas Diesel	-	-	-	-	-	-	-	-	-
Heavy Fuel Oil	-	-	-	-	-	-	-	-	-
Other	-	-	-	-	-	-	-	-	-
Refinery Fuel	-	-	-	-	-	-	-	-	-
Sub-Total	-	-	-	-	-	-	-	-	-
Bunkers	-	-	-	-	-	-	-	-	-
Total	-	-	-	-	-	-	-	-	-

	1990	1991	1992	1993	1994	1995	1996	1997	1998
Thousand tonnes									
NGL/LPG/Ethane	-	-	3 487	4 410	3 955	4 204	4 900	4 207	3 213
Naphtha	-	-	6 379	8 240	3 577	-	-	-	-
Motor Gasoline	-	-	30 996	27 543	26 323	26 078	24 701	24 741	24 404
Aviation Fuels	-	-	14 100	11 951	9 953	9 271	9 239	9 229	9 220
Kerosene	-	-	882	616	283	143	130	107	51
Gas Diesel	-	-	52 374	41 573	30 456	29 476	24 615	25 485	23 188
Heavy Fuel Oil	-	-	70 274	71 063	54 931	46 278	41 331	36 773	35 896
Other	-	-	23 416	29 640	16 925	23 448	23 503	23 640	23 101
Refinery Fuel	-	-	9 095	1 284	-	1 809	-	-	-
Sub-Total	-	-	211 003	196 320	146 403	140 707	128 419	124 182	119 073
Bunkers	-	-	-	-	-	-	-	-	-
Total	-	-	211 003	196 320	146 403	140 707	128 419	124 182	119 073
Thousand barrels/day									
NGL/LPG/Ethane	-	-	110.5	140.2	125.7	133.6	155.3	133.7	102.1
Naphtha	-	-	148.1	191.9	83.3	-	-	-	-
Motor Gasoline	-	-	722.4	643.7	615.2	609.4	575.7	578.2	570.3
Aviation Fuels	-	-	305.9	259.9	216.5	201.6	200.3	200.6	200.4
Kerosene	-	-	18.7	13.1	6.0	3.0	2.7	2.3	1.1
Gas Diesel	-	-	1 067.5	849.7	622.5	602.4	501.7	520.9	473.9
Heavy Fuel Oil	-	-	1 278.8	1 296.7	1 002.3	844.4	752.1	671.0	655.0
Other	-	-	432.8	560.3	315.4	440.9	451.0	457.0	447.4
Refinery Fuel	-	-	230.8	28.1	-	39.6	-	-	-
Sub-Total	-	-	4 315.5	3 983.5	2 986.8	2 875.1	2 638.9	2 563.6	2 450.2
Bunkers	-	-	-	-	-	-	-	-	-
Total	-	-	4 315.5	3 983.5	2 986.8	2 875.1	2 638.9	2 563.6	2 450.2

[1] See Notes in Part I.4.

Oil Demand by Main Product Groups[1]
Demande de produits raffinés par groupes principaux

Saudi Arabia

	1971	1973	1978	1984	1985	1986	1987	1988	1989
Thousand tonnes									
NGL/LPG/Ethane	51	77	195	399	2 338	3 476	3 496	3 704	3 597
Naphtha	-	-	-	-	-	-	-	-	-
Motor Gasoline	524	700	2 480	5 830	6 327	6 354	6 199	6 213	5 766
Aviation Fuels	199	316	1 133	1 884	1 928	1 886	1 820	1 892	1 689
Kerosene	133	155	132	181	208	206	143	195	171
Gas Diesel	809	1 081	5 079	13 541	12 819	12 477	12 428	13 394	13 847
Heavy Fuel Oil	632	1 179	3 618	2 858	2 521	2 913	2 685	2 913	3 343
Other	385	559	853	5 775	6 109	6 296	6 883	6 566	6 535
Refinery Fuel	825	925	937	1 313	1 720	1 834	2 933	3 233	3 135
Sub-Total	3 558	4 992	14 427	31 781	33 970	35 442	36 587	38 110	38 083
Bunkers	13 004	13 803	7 205	5 221	9 098	7 464	6 083	5 221	1 918
Total	16 562	18 795	21 632	37 002	43 068	42 906	42 670	43 331	40 001
Thousand barrels/day									
NGL/LPG/Ethane	1.6	2.4	6.2	12.6	68.2	100.7	101.2	107.0	103.8
Naphtha	-	-	-	-	-	-	-	-	-
Motor Gasoline	12.2	16.4	58.0	135.9	147.9	148.5	144.9	144.8	134.8
Aviation Fuels	4.3	6.9	24.6	40.8	41.9	41.0	39.5	41.0	36.7
Kerosene	2.8	3.3	2.8	3.8	4.4	4.4	3.0	4.1	3.6
Gas Diesel	16.5	22.1	103.8	276.0	262.0	255.0	254.0	273.0	283.0
Heavy Fuel Oil	11.5	21.5	66.0	52.0	46.0	53.2	49.0	53.0	61.0
Other	7.6	11.0	16.7	114.0	121.1	124.9	137.3	130.5	130.5
Refinery Fuel	16.5	18.3	18.6	26.1	34.7	36.8	59.0	64.7	62.7
Sub-Total	73.2	101.9	296.7	661.3	726.2	764.4	788.0	818.1	816.1
Bunkers	237.5	252.2	132.7	95.0	166.0	136.2	111.0	95.0	35.0
Total	310.7	354.1	429.5	756.3	892.2	900.5	899.0	913.1	851.1

	1990	1991	1992	1993	1994	1995	1996	1997	1998
Thousand tonnes									
NGL/LPG/Ethane	3 857	3 851	3 964	4 654	5 342	5 342	5 564	6 808	8 523
Naphtha	-	-	-	-	-	-	-	-	-
Motor Gasoline	6 650	6 902	7 754	8 537	9 004	8 900	9 559	9 643	9 728
Aviation Fuels	2 595	3 378	1 912	2 075	2 499	2 413	2 456	2 448	2 414
Kerosene	279	434	230	236	236	212	216	215	212
Gas Diesel	13 749	14 238	15 602	16 636	16 782	14 819	15 909	16 546	16 477
Heavy Fuel Oil	3 343	2 960	3 077	2 795	3 069	3 449	4 201	3 641	4 744
Other	6 693	7 397	7 953	9 379	9 135	9 255	9 700	10 323	9 673
Refinery Fuel	3 335	3 388	3 405	3 405	3 405	3 265	3 696	3 386	3 920
Sub-Total	40 501	42 548	43 897	47 717	49 472	47 655	51 301	53 010	55 691
Bunkers	1 863	2 521	2 638	2 521	2 083	1 936	1 936	1 936	1 936
Total	42 364	45 069	46 535	50 238	51 555	49 591	53 237	54 946	57 627
Thousand barrels/day									
NGL/LPG/Ethane	111.7	111.5	114.8	135.0	154.7	154.7	160.8	196.8	245.7
Naphtha	-	-	-	-	-	-	-	-	-
Motor Gasoline	155.4	161.3	180.7	199.5	210.4	208.0	222.8	225.4	227.3
Aviation Fuels	56.4	73.4	41.4	45.1	54.3	52.4	53.2	53.2	52.4
Kerosene	5.9	9.2	4.9	5.0	5.0	4.5	4.6	4.6	4.5
Gas Diesel	281.0	291.0	318.0	340.0	343.0	302.9	324.3	338.2	336.8
Heavy Fuel Oil	61.0	54.0	56.0	51.0	56.0	62.9	76.4	66.4	86.6
Other	133.5	147.6	158.2	187.1	182.2	184.9	193.2	206.2	193.1
Refinery Fuel	66.7	68.0	68.2	68.4	68.4	65.6	74.0	68.3	78.3
Sub-Total	871.6	916.1	942.2	1 031.1	1 074.0	1 035.9	1 109.3	1 159.1	1 224.7
Bunkers	34.0	46.0	48.0	46.0	38.0	35.3	35.2	35.3	35.3
Total	905.6	962.1	990.2	1 077.1	1 112.1	1 071.2	1 144.5	1 194.4	1 260.1

[1] See Notes in Part I.4.

Oil Demand by Main Product Groups[1]
Demande de produits raffinés par groupes principaux

Senegal

	1971	1973	1978	1984	1985	1986	1987	1988	1989
Thousand tonnes									
NGL/LPG/Ethane	3	3	4	13	16	16	18	26	29
Naphtha	-	-	-	-	-	-	-	-	-
Motor Gasoline	83	92	123	102	92	89	86	84	80
Aviation Fuels	95	113	118	132	135	134	112	131	148
Kerosene	11	11	13	11	11	11	11	11	10
Gas Diesel	89	106	131	218	250	220	229	196	199
Heavy Fuel Oil	184	210	255	294	297	300	300	299	336
Other	11	28	20	15	12	9	12	12	12
Refinery Fuel	19	19	35	12	10	15	15	7	4
Sub-Total	495	582	699	797	823	794	783	766	818
Bunkers	962	983	523	124	105	88	57	33	45
Total	1 457	1 565	1 222	921	928	882	840	799	863
Thousand barrels/day									
NGL/LPG/Ethane	0.1	0.1	0.1	0.4	0.5	0.5	0.6	0.8	0.9
Naphtha	-	-	-	-	-	-	-	-	-
Motor Gasoline	1.9	2.2	2.9	2.4	2.2	2.1	2.0	2.0	1.9
Aviation Fuels	2.1	2.5	2.6	2.9	2.9	2.9	2.4	2.8	3.2
Kerosene	0.2	0.2	0.3	0.2	0.2	0.2	0.2	0.2	0.2
Gas Diesel	1.8	2.2	2.7	4.4	5.1	4.5	4.7	4.0	4.1
Heavy Fuel Oil	3.4	3.8	4.7	5.4	5.4	5.5	5.5	5.4	6.1
Other	0.2	0.5	0.4	0.3	0.2	0.2	0.2	0.2	0.2
Refinery Fuel	0.4	0.4	0.7	0.2	0.2	0.3	0.3	0.2	0.1
Sub-Total	10.1	11.8	14.2	16.2	16.8	16.2	15.9	15.7	16.7
Bunkers	18.1	18.4	10.1	2.4	2.1	1.8	1.1	0.7	0.9
Total	28.2	30.2	24.4	18.6	18.9	17.9	17.0	16.3	17.6

	1990	1991	1992	1993	1994	1995	1996	1997	1998
Thousand tonnes									
NGL/LPG/Ethane	33	38	42	45	50	57	68	76	87
Naphtha	-	-	-	-	-	-	-	-	-
Motor Gasoline	79	76	75	68	67	72	74	78	84
Aviation Fuels	144	127	128	102	126	144	162	160	179
Kerosene	10	10	10	10	10	10	10	12	14
Gas Diesel	212	220	249	250	243	246	257	322	391
Heavy Fuel Oil	336	328	357	319	360	366	383	402	418
Other	12	12	11	12	18	19	19	19	19
Refinery Fuel	7	4	6	6	2	6	6	6	6
Sub-Total	833	815	878	812	876	920	979	1 075	1 198
Bunkers	36	1	17	5	22	29	58	-	70
Total	869	816	895	817	898	949	1 037	1 075	1 268
Thousand barrels/day									
NGL/LPG/Ethane	1.0	1.2	1.3	1.4	1.6	1.8	2.2	2.4	2.8
Naphtha	-	-	-	-	-	-	-	-	-
Motor Gasoline	1.8	1.8	1.7	1.6	1.6	1.7	1.7	1.8	2.0
Aviation Fuels	3.1	2.8	2.8	2.2	2.7	3.1	3.5	3.5	3.9
Kerosene	0.2	0.2	0.2	0.2	0.2	0.2	0.2	0.3	0.3
Gas Diesel	4.3	4.5	5.1	5.1	5.0	5.0	5.2	6.6	8.0
Heavy Fuel Oil	6.1	6.0	6.5	5.8	6.6	6.7	7.0	7.3	7.6
Other	0.2	0.2	0.2	0.2	0.3	0.4	0.4	0.4	0.4
Refinery Fuel	0.2	0.1	0.1	0.1	0.0	0.1	0.1	0.1	0.1
Sub-Total	17.1	16.8	18.0	16.7	18.0	19.0	20.3	22.4	25.0
Bunkers	0.7	0.0	0.3	0.1	0.4	0.6	1.2	-	1.4
Total	17.8	16.8	18.3	16.8	18.5	19.6	21.5	22.4	26.5

[1] See Notes in Part I.4.

Oil Demand by Main Product Groups[1]
Demande de produits raffinés par groupes principaux

Singapore

	1971	1973	1978	1984	1985	1986	1987	1988	1989
Thousand tonnes									
NGL/LPG/Ethane	4	6	45	72	432	387	390	390	430
Naphtha	-	48	55	397	611	641	953	986	1 040
Motor Gasoline	213	276	300	403	370	410	436	463	483
Aviation Fuels	221	314	777	992	1 012	1 136	1 218	1 296	1 329
Kerosene	105	79	55	60	60	60	60	60	60
Gas Diesel	353	334	552	752	717	715	724	682	842
Heavy Fuel Oil	771	1 091	1 672	2 374	2 466	2 528	2 674	3 121	3 244
Other	155	335	407	321	450	412	438	416	429
Refinery Fuel	500	790	1 107	1 328	1 243	1 220	1 160	1 560	1 560
Sub-Total	2 322	3 273	4 970	6 699	7 361	7 509	8 053	8 974	9 417
Bunkers	2 877	4 383	4 882	5 057	4 897	8 360	8 346	9 884	9 700
Total	5 199	7 656	9 852	11 756	12 258	15 869	16 399	18 858	19 117
Thousand barrels/day									
NGL/LPG/Ethane	0.1	0.2	1.4	2.3	13.7	12.3	12.4	12.4	13.7
Naphtha	-	1.1	1.3	9.2	14.2	14.9	22.2	22.9	24.2
Motor Gasoline	5.0	6.5	7.0	9.4	8.6	9.6	10.2	10.8	11.3
Aviation Fuels	4.8	6.8	16.9	21.5	22.0	24.7	26.5	28.1	28.9
Kerosene	2.2	1.7	1.2	1.3	1.3	1.3	1.3	1.3	1.3
Gas Diesel	7.2	6.8	11.3	15.3	14.7	14.6	14.8	13.9	17.2
Heavy Fuel Oil	14.1	19.9	30.5	43.2	45.0	46.1	48.8	56.8	59.2
Other	2.9	6.3	7.6	5.9	8.4	7.6	8.1	7.6	7.9
Refinery Fuel	9.1	14.4	22.2	26.5	25.0	24.5	23.3	32.0	32.1
Sub-Total	45.4	63.7	99.4	134.5	152.9	155.6	167.5	185.7	195.7
Bunkers	53.1	80.9	90.4	93.4	90.7	154.3	155.2	183.0	178.1
Total	98.6	144.6	189.8	227.9	243.5	309.9	322.7	368.7	373.8

	1990	1991	1992	1993	1994	1995	1996	1997	1998
Thousand tonnes									
NGL/LPG/Ethane	393	365	189	189	189	143	237	214	214
Naphtha	1 237	1 149	1 074	1 202	1 288	1 696	1 988	2 418	3 207
Motor Gasoline	447	476	513	535	556	656	666	690	718
Aviation Fuels	1 784	1 497	1 611	2 072	2 533	2 474	2 561	2 806	2 268
Kerosene	60	35	35	40	42	42	43	43	43
Gas Diesel	972	1 081	992	959	1 130	1 206	1 156	1 213	1 225
Heavy Fuel Oil	4 500	4 474	4 605	5 365	5 467	5 575	3 840	4 234	4 428
Other	666	619	574	496	523	507	558	558	558
Refinery Fuel	2 740	3 085	3 150	3 225	3 389	3 036	3 303	3 303	3 303
Sub-Total	12 799	12 781	12 743	14 083	15 117	15 335	14 352	15 479	15 964
Bunkers	10 981	10 389	12 810	11 438	11 477	11 419	14 280	16 191	17 198
Total	23 780	23 170	25 553	25 521	26 594	26 754	28 632	31 670	33 162
Thousand barrels/day									
NGL/LPG/Ethane	12.5	11.6	6.0	6.0	6.0	4.5	7.5	6.8	6.8
Naphtha	28.8	26.8	24.9	28.0	30.0	39.5	46.2	56.3	74.7
Motor Gasoline	10.4	11.1	12.0	12.5	13.0	15.3	15.5	16.1	16.8
Aviation Fuels	38.8	32.5	34.9	45.0	55.0	53.8	55.5	61.0	49.3
Kerosene	1.3	0.7	0.7	0.8	0.9	0.9	0.9	0.9	0.9
Gas Diesel	19.9	22.1	20.2	19.6	23.1	24.6	23.6	24.8	25.0
Heavy Fuel Oil	82.1	81.6	83.8	97.9	99.8	101.7	69.9	77.3	80.8
Other	12.3	11.5	10.6	9.3	9.8	9.6	10.5	10.5	10.5
Refinery Fuel	52.6	59.0	59.3	60.8	63.9	57.3	62.2	62.4	62.4
Sub-Total	258.6	257.0	252.5	280.0	301.5	307.3	291.7	316.1	327.2
Bunkers	201.7	191.1	234.7	210.6	212.2	211.0	264.2	301.8	320.8
Total	460.3	448.2	487.2	490.6	513.7	518.3	556.0	617.8	648.0

[1] See Notes in Part I.4.

Oil Demand by Main Product Groups[1]
Demande de produits raffinés par groupes principaux

Slovak Republic

	1971	1973	1978	1984	1985	1986	1987	1988	1989
Thousand tonnes									
NGL/LPG/Ethane	6	14	17	19	18	17	21	17	16
Naphtha	-	-	-	-	-	-	-	-	1 038
Motor Gasoline	459	532	550	421	408	436	452	493	489
Aviation Fuels	-	-	-	-	-	-	-	-	-
Kerosene	7	8	11	11	12	12	20	13	11
Gas Diesel	1 049	1 205	1 168	1 164	1 171	1 169	1 199	1 225	1 208
Heavy Fuel Oil	1 106	1 201	1 278	2 495	2 487	2 316	2 195	2 011	1 824
Other	243	295	294	490	438	425	428	419	944
Refinery Fuel	281	339	446	123	123	118	118	110	138
Sub-Total	3 151	3 594	3 764	4 723	4 657	4 493	4 433	4 288	5 668
Bunkers	-	-	-	-	-	-	-	-	-
Total	3 151	3 594	3 764	4 723	4 657	4 493	4 433	4 288	5 668
Thousand barrels/day									
NGL/LPG/Ethane	0.2	0.4	0.5	0.6	0.6	0.5	0.7	0.5	0.5
Naphtha	-	-	-	-	-	-	-	-	24.2
Motor Gasoline	10.7	12.4	12.9	9.8	9.5	10.2	10.6	11.5	11.4
Aviation Fuels	-	-	-	-	-	-	-	-	-
Kerosene	0.1	0.2	0.2	0.2	0.3	0.3	0.4	0.3	0.2
Gas Diesel	21.4	24.6	23.9	23.7	23.9	23.9	24.5	25.0	24.7
Heavy Fuel Oil	20.2	21.9	23.3	45.4	45.4	42.3	40.1	36.6	33.3
Other	4.7	5.7	5.6	9.8	8.8	8.6	8.6	8.4	18.6
Refinery Fuel	5.2	6.6	8.6	2.5	2.5	2.4	2.4	2.2	2.8
Sub-Total	62.6	71.8	75.0	92.0	91.0	88.1	87.2	84.5	115.7
Bunkers	-	-	-	-	-	-	-	-	-
Total	62.6	71.8	75.0	92.0	91.0	88.1	87.2	84.5	115.7

	1990	1991	1992	1993	1994	1995	1996	1997	1998
Thousand tonnes									
NGL/LPG/Ethane	15	77	45	37	38	34	33	35	32
Naphtha	517	505	684	560	626	670	670	670	670
Motor Gasoline	469	396	453	414	482	504	459	548	570
Aviation Fuels	-	-	-	-	26	39	38	35	29
Kerosene	9	20	60	2	21	2	5	1	2
Gas Diesel	1 124	878	842	701	735	852	806	743	820
Heavy Fuel Oil	1 535	1 422	1 212	1 020	920	642	662	538	509
Other	1 188	933	506	497	487	498	726	796	618
Refinery Fuel	109	122	118	13	13	-	-	27	-
Sub-Total	4 966	4 353	3 920	3 244	3 348	3 241	3 399	3 393	3 250
Bunkers	-	-	-	-	-	-	-	-	-
Total	4 966	4 353	3 920	3 244	3 348	3 241	3 399	3 393	3 250
Thousand barrels/day									
NGL/LPG/Ethane	0.5	2.4	1.4	1.2	1.2	1.1	1.0	1.1	1.0
Naphtha	12.0	11.8	15.9	13.0	14.6	15.6	15.6	15.6	15.6
Motor Gasoline	11.0	9.3	10.6	9.7	11.3	11.8	10.7	12.8	13.3
Aviation Fuels	-	-	-	-	0.6	0.8	0.8	0.8	0.6
Kerosene	0.2	0.4	1.3	0.0	0.4	0.0	0.1	0.0	0.0
Gas Diesel	23.0	17.9	17.2	14.3	15.0	17.4	16.4	15.2	16.8
Heavy Fuel Oil	28.0	25.9	22.1	18.6	16.8	11.7	12.0	9.8	9.3
Other	23.2	18.2	9.9	9.4	9.2	9.0	13.3	14.7	11.3
Refinery Fuel	2.2	2.5	2.5	0.3	0.3	-	-	0.5	-
Sub-Total	100.1	88.5	80.7	66.6	69.4	67.4	70.0	70.5	67.9
Bunkers	-	-	-	-	-	-	-	-	-
Total	100.1	88.5	80.7	66.6	69.4	67.4	70.0	70.5	67.9

[1] See Notes in Part I.4.

Oil Demand by Main Product Groups[1]
Demande de produits raffinés par groupes principaux

Slovenia

	1971	1973	1978	1984	1985	1986	1987	1988	1989
Thousand tonnes									
NGL/LPG/Ethane	-	-	-	57	44	43	42	43	39
Naphtha	-	-	-	2	17	11	23	11	19
Motor Gasoline	-	-	-	325	363	434	482	520	530
Aviation Fuels	-	-	-	23	25	32	29	28	27
Kerosene	-	-	-	-	-	-	-	-	-
Gas Diesel	-	-	-	728	693	760	742	759	788
Heavy Fuel Oil	-	-	-	236	239	255	272	259	272
Other	-	-	-	-	-	-	1	5	3
Refinery Fuel	-	-	-	5	6	6	11	13	14
Sub-Total	-	-	-	1 376	1 387	1 541	1 602	1 638	1 692
Bunkers	-	-	-	-	-	-	-	-	-
Total	-	-	-	1 376	1 387	1 541	1 602	1 638	1 692
Thousand barrels/day									
NGL/LPG/Ethane	-	-	-	1.8	1.4	1.4	1.3	1.4	1.2
Naphtha	-	-	-	0.0	0.4	0.3	0.5	0.3	0.4
Motor Gasoline	-	-	-	7.6	8.5	10.1	11.3	12.1	12.4
Aviation Fuels	-	-	-	0.5	0.5	0.7	0.6	0.6	0.6
Kerosene	-	-	-	-	-	-	-	-	-
Gas Diesel	-	-	-	14.8	14.2	15.5	15.2	15.5	16.1
Heavy Fuel Oil	-	-	-	4.3	4.4	4.7	5.0	4.7	5.0
Other	-	-	-	-	-	-	0.0	0.1	0.1
Refinery Fuel	-	-	-	0.1	0.1	0.1	0.2	0.2	0.3
Sub-Total	-	-	-	29.1	29.5	32.8	34.1	34.9	36.0
Bunkers	-	-	-	-	-	-	-	-	-
Total	-	-	-	29.1	29.5	32.8	34.1	34.9	36.0

	1990	1991	1992	1993	1994	1995	1996	1997	1998
Thousand tonnes									
NGL/LPG/Ethane	34	40	43	39	38	35	39	74	75
Naphtha	6	21	9	-	-	28	76	11	18
Motor Gasoline	565	526	580	689	765	821	923	913	798
Aviation Fuels	26	10	11	17	18	20	18	19	18
Kerosene	-	-	-	-	-	-	-	-	1
Gas Diesel	777	831	681	870	915	1 060	1 283	1 303	1 562
Heavy Fuel Oil	262	201	219	248	280	244	257	210	158
Other	3	3	11	7	6	5	7	4	112
Refinery Fuel	14	13	18	8	9	16	14	16	16
Sub-Total	1 687	1 645	1 572	1 878	2 031	2 229	2 617	2 550	2 758
Bunkers	-	-	-	-	-	-	-	-	-
Total	1 687	1 645	1 572	1 878	2 031	2 229	2 617	2 550	2 758
Thousand barrels/day									
NGL/LPG/Ethane	1.1	1.3	1.4	1.2	1.2	1.1	1.2	2.4	2.4
Naphtha	0.1	0.5	0.2	-	-	0.7	1.8	0.3	0.4
Motor Gasoline	13.2	12.3	13.5	16.1	17.9	19.2	21.5	21.3	18.6
Aviation Fuels	0.6	0.2	0.2	0.4	0.4	0.4	0.4	0.4	0.4
Kerosene	-	-	-	-	-	-	-	-	0.0
Gas Diesel	15.9	17.0	13.9	17.8	18.7	21.7	26.2	26.6	31.9
Heavy Fuel Oil	4.8	3.7	4.0	4.5	5.1	4.5	4.7	3.8	2.9
Other	0.1	0.1	0.2	0.1	0.1	0.1	0.1	0.1	1.9
Refinery Fuel	0.3	0.2	0.3	0.1	0.2	0.3	0.3	0.3	0.3
Sub-Total	36.0	35.2	33.7	40.3	43.6	47.9	56.1	55.2	58.9
Bunkers	-	-	-	-	-	-	-	-	-
Total	36.0	35.2	33.7	40.3	43.6	47.9	56.1	55.2	58.9

[1] See Notes in Part I.4.

Oil Demand by Main Product Groups[1]
Demande de produits raffinés par groupes principaux

South Africa

	1971	1973	1978	1984	1985	1986	1987	1988	1989
Thousand tonnes									
NGL/LPG/Ethane	106	133	159	193	176	177	197	221	249
Naphtha	50	53	50	50	50	50	50	50	50
Motor Gasoline	3 459	3 975	4 159	4 867	4 678	4 764	5 251	5 687	5 921
Aviation Fuels	329	360	564	616	606	565	577	640	690
Kerosene	535	535	535	340	341	365	401	452	478
Gas Diesel	2 566	3 222	4 127	4 288	4 185	3 992	4 125	4 431	4 443
Heavy Fuel Oil	862	1 198	820	580	633	547	522	470	481
Other	746	737	1 280	2 070	1 930	1 930	1 893	2 160	2 149
Refinery Fuel	604	645	778	808	770	770	770	750	750
Sub-Total	9 257	10 858	12 472	13 812	13 369	13 160	13 786	14 861	15 211
Bunkers	3 504	2 771	1 200	1 100	1 100	1 000	1 000	1 191	1 053
Total	12 761	13 629	13 672	14 912	14 469	14 160	14 786	16 052	16 264
Thousand barrels/day									
NGL/LPG/Ethane	3.4	4.2	5.1	6.1	5.6	5.6	6.3	7.0	7.9
Naphtha	1.2	1.2	1.2	1.2	1.2	1.2	1.2	1.2	1.2
Motor Gasoline	80.8	92.9	97.2	113.4	109.3	111.3	122.7	132.5	138.4
Aviation Fuels	7.2	7.9	12.4	13.4	13.2	12.3	12.6	13.9	15.0
Kerosene	11.3	11.3	11.3	7.2	7.2	7.7	8.5	9.6	10.1
Gas Diesel	52.4	65.9	84.3	87.4	85.5	81.6	84.3	90.3	90.8
Heavy Fuel Oil	15.7	21.9	15.0	10.6	11.6	10.0	9.5	8.6	8.8
Other	13.8	13.7	24.0	38.9	36.4	36.4	35.8	40.7	40.6
Refinery Fuel	11.0	11.8	14.2	14.7	14.1	14.1	14.1	13.6	13.7
Sub-Total	196.9	230.8	264.7	292.9	284.1	280.2	294.9	317.4	326.5
Bunkers	64.4	51.0	22.3	20.5	20.5	18.7	18.7	22.1	19.7
Total	261.3	281.8	287.0	313.4	304.6	298.9	313.6	339.5	346.1

	1990	1991	1992	1993	1994	1995	1996	1997	1998
Thousand tonnes									
NGL/LPG/Ethane	240	255	253	253	268	267	256	285	284
Naphtha	50	50	50	-	-	-	-	-	-
Motor Gasoline	6 251	6 442	6 469	6 784	7 099	8 068	7 791	7 898	8 024
Aviation Fuels	706	706	819	979	965	1 122	1 288	1 429	1 507
Kerosene	559	556	512	694	733	702	739	779	842
Gas Diesel	4 387	4 261	4 026	4 195	4 332	4 622	4 721	4 804	4 842
Heavy Fuel Oil	548	476	397	559	755	720	607	534	519
Other	2 133	2 077	2 144	634	774	759	835	780	760
Refinery Fuel	750	750	750	1 427	1 615	1 733	1 634	1 279	1 525
Sub-Total	15 624	15 573	15 420	15 525	16 541	17 993	17 871	17 788	18 303
Bunkers	1 927	2 118	2 742	2 442	3 115	3 805	2 759	2 145	2 488
Total	17 551	17 691	18 162	17 967	19 656	21 798	20 630	19 933	20 791
Thousand barrels/day									
NGL/LPG/Ethane	7.6	8.1	8.0	8.0	8.5	8.5	8.1	9.1	9.0
Naphtha	1.2	1.2	1.2	-	-	-	-	-	-
Motor Gasoline	146.1	150.5	150.8	158.5	165.9	188.5	181.6	184.6	187.5
Aviation Fuels	15.4	15.4	17.8	21.3	21.0	24.4	28.0	31.1	32.8
Kerosene	11.9	11.8	10.8	14.7	15.5	14.9	15.6	16.5	17.9
Gas Diesel	89.7	87.1	82.1	85.7	88.5	94.5	96.2	98.2	99.0
Heavy Fuel Oil	10.0	8.7	7.2	10.2	13.8	13.1	11.0	9.7	9.5
Other	40.3	39.4	40.5	11.5	14.5	14.2	15.6	14.6	14.4
Refinery Fuel	13.7	13.7	13.6	33.3	37.7	40.5	38.1	29.9	35.6
Sub-Total	335.8	335.8	332.0	343.4	365.5	398.6	394.2	393.7	405.6
Bunkers	35.6	39.1	50.3	45.2	57.7	70.4	51.0	39.9	45.8
Total	371.4	374.9	382.4	388.7	423.2	469.1	445.1	433.6	451.4

[1] See Notes in Part I.4.

Oil Demand by Main Product Groups[1]
Demande de produits raffinés par groupes principaux

Sri Lanka

	1971	1973	1978	1984	1985	1986	1987	1988	1989
Thousand tonnes									
NGL/LPG/Ethane	-	-	5	11	13	16	19	20	29
Naphtha	-	-	-	-	-	-	-	-	-
Motor Gasoline	114	123	121	119	122	131	140	158	177
Aviation Fuels	18	51	92	116	119	113	115	100	49
Kerosene	270	275	238	157	165	157	153	163	162
Gas Diesel	297	355	368	560	525	524	649	569	531
Heavy Fuel Oil	112	257	184	200	110	94	129	135	144
Other	6	48	35	44	37	52	39	39	28
Refinery Fuel	65	62	50	45	48	51	49	49	44
Sub-Total	882	1 171	1 093	1 252	1 139	1 138	1 293	1 233	1 164
Bunkers	385	433	338	335	328	394	412	362	312
Total	1 267	1 604	1 431	1 587	1 467	1 532	1 705	1 595	1 476
Thousand barrels/day									
NGL/LPG/Ethane	-	-	0.2	0.3	0.4	0.5	0.6	0.6	0.9
Naphtha	-	-	-	-	-	-	-	-	-
Motor Gasoline	2.7	2.9	2.8	2.8	2.9	3.1	3.3	3.7	4.1
Aviation Fuels	0.4	1.1	2.0	2.5	2.6	2.5	2.5	2.2	1.1
Kerosene	5.7	5.8	5.0	3.3	3.5	3.3	3.2	3.4	3.4
Gas Diesel	6.1	7.3	7.5	11.4	10.7	10.7	13.3	11.6	10.9
Heavy Fuel Oil	2.0	4.7	3.4	3.6	2.0	1.7	2.4	2.5	2.6
Other	0.1	0.9	0.7	0.8	0.7	1.0	0.7	0.7	0.5
Refinery Fuel	1.2	1.2	1.0	0.9	0.9	1.0	1.0	1.0	0.8
Sub-Total	18.3	23.9	22.5	25.7	23.7	23.8	26.9	25.7	24.4
Bunkers	7.1	8.0	6.3	6.2	6.1	7.3	7.6	6.6	5.8
Total	25.4	31.9	28.9	31.9	29.8	31.0	34.5	32.3	30.2

	1990	1991	1992	1993	1994	1995	1996	1997	1998
Thousand tonnes									
NGL/LPG/Ethane	34	38	44	52	64	72	87	99	113
Naphtha	-	-	-	-	-	-	-	-	-
Motor Gasoline	179	159	165	173	184	188	198	195	204
Aviation Fuels	57	70	81	72	146	166	195	205	200
Kerosene	167	161	189	191	208	224	229	225	236
Gas Diesel	548	590	786	744	804	879	1 198	1 690	1 424
Heavy Fuel Oil	125	146	252	252	236	155	363	471	638
Other	31	40	42	45	38	42	63	55	55
Refinery Fuel	51	47	39	42	54	17	49	104	99
Sub-Total	1 192	1 251	1 598	1 571	1 734	1 743	2 382	3 044	2 969
Bunkers	391	312	304	352	345	344	385	258	-
Total	1 583	1 563	1 902	1 923	2 079	2 087	2 767	3 302	2 969
Thousand barrels/day									
NGL/LPG/Ethane	1.1	1.2	1.4	1.7	2.0	2.3	2.8	3.1	3.6
Naphtha	-	-	-	-	-	-	-	-	-
Motor Gasoline	4.2	3.7	3.8	4.0	4.3	4.4	4.6	4.6	4.8
Aviation Fuels	1.2	1.5	1.8	1.6	3.2	3.6	4.2	4.5	4.3
Kerosene	3.5	3.4	4.0	4.1	4.4	4.8	4.8	4.8	5.0
Gas Diesel	11.2	12.1	16.0	15.2	16.4	18.0	24.4	34.5	29.1
Heavy Fuel Oil	2.3	2.7	4.6	4.6	4.3	2.8	6.6	8.6	11.6
Other	0.6	0.8	0.8	0.9	0.7	0.8	1.2	1.1	1.1
Refinery Fuel	1.0	0.9	0.8	0.8	1.1	0.4	1.0	2.4	2.5
Sub-Total	25.1	26.3	33.2	32.8	36.4	37.0	49.6	63.5	62.0
Bunkers	7.2	5.8	5.6	6.5	6.4	6.4	7.1	4.8	-
Total	32.4	32.1	38.8	39.3	42.8	43.4	56.7	68.3	62.0

[1] See Notes in Part I.4.

Oil Demand by Main Product Groups[1]
Demande de produits raffinés par groupes principaux

Sudan

	1971	1973	1978	1984	1985	1986	1987	1988	1989
Thousand tonnes									
NGL/LPG/Ethane	4	5	4	7	8	10	9	11	10
Naphtha	-	-	-	16	18	17	19	19	20
Motor Gasoline	97	103	158	179	190	202	143	243	219
Aviation Fuels	107	119	54	65	68	64	61	60	63
Kerosene	107	81	24	37	31	36	31	31	31
Gas Diesel	561	682	514	559	704	640	521	804	715
Heavy Fuel Oil	242	564	270	269	322	346	246	291	272
Other	31	33	36	105	96	87	92	88	91
Refinery Fuel	4	5	5	3	3	4	2	4	3
Sub-Total	1 153	1 592	1 065	1 240	1 440	1 406	1 124	1 551	1 424
Bunkers	-	2	4	5	5	4	5	5	5
Total	1 153	1 594	1 069	1 245	1 445	1 410	1 129	1 556	1 429
Thousand barrels/day									
NGL/LPG/Ethane	0.1	0.2	0.1	0.2	0.3	0.3	0.3	0.3	0.3
Naphtha	-	-	-	0.4	0.4	0.4	0.4	0.4	0.5
Motor Gasoline	2.3	2.4	3.7	4.2	4.4	4.7	3.3	5.7	5.1
Aviation Fuels	2.4	2.7	1.2	1.4	1.5	1.4	1.3	1.3	1.4
Kerosene	2.3	1.7	0.5	0.8	0.7	0.8	0.7	0.7	0.7
Gas Diesel	11.5	13.9	10.5	11.4	14.4	13.1	10.6	16.4	14.6
Heavy Fuel Oil	4.4	10.3	4.9	4.9	5.9	6.3	4.5	5.3	5.0
Other	0.6	0.6	0.7	2.0	1.8	1.7	1.8	1.7	1.8
Refinery Fuel	0.1	0.1	0.1	0.1	0.1	0.1	0.0	0.1	0.1
Sub-Total	23.6	31.9	21.7	25.3	29.4	28.8	23.0	31.9	29.3
Bunkers	-	0.0	0.1	0.1	0.1	0.1	0.1	0.1	0.1
Total	23.6	32.0	21.8	25.4	29.5	28.8	23.1	32.0	29.4

	1990	1991	1992	1993	1994	1995	1996	1997	1998
Thousand tonnes									
NGL/LPG/Ethane	14	16	14	11	17	20	26	25	33
Naphtha	20	22	20	22	23	23	24	24	24
Motor Gasoline	218	190	230	161	206	199	212	242	223
Aviation Fuels	36	41	42	45	50	50	52	52	52
Kerosene	32	37	36	37	40	41	62	62	62
Gas Diesel	1 060	849	802	575	768	735	682	812	795
Heavy Fuel Oil	343	399	315	192	422	366	315	462	352
Other	95	97	92	101	103	102	104	104	104
Refinery Fuel	4	5	4	2	3	4	4	5	3
Sub-Total	1 822	1 656	1 555	1 146	1 632	1 540	1 481	1 788	1 648
Bunkers	7	7	7	7	7	8	8	8	8
Total	1 829	1 663	1 562	1 153	1 639	1 548	1 489	1 796	1 656
Thousand barrels/day									
NGL/LPG/Ethane	0.4	0.5	0.4	0.4	0.5	0.6	0.8	0.8	1.0
Naphtha	0.5	0.5	0.5	0.5	0.5	0.5	0.6	0.6	0.6
Motor Gasoline	5.1	4.4	5.4	3.8	4.8	4.7	4.9	5.7	5.2
Aviation Fuels	0.8	0.9	0.9	1.0	1.1	1.1	1.2	1.2	1.2
Kerosene	0.7	0.8	0.8	0.8	0.8	0.9	1.3	1.3	1.3
Gas Diesel	21.7	17.4	16.3	11.8	15.7	15.0	13.9	16.6	16.2
Heavy Fuel Oil	6.3	7.3	5.7	3.5	7.7	6.7	5.7	8.4	6.4
Other	1.8	1.9	1.8	1.9	2.0	2.0	2.0	2.0	2.0
Refinery Fuel	0.1	0.1	0.1	0.0	0.1	0.1	0.1	0.1	0.1
Sub-Total	37.3	33.8	31.9	23.7	33.3	31.6	30.5	36.6	34.1
Bunkers	0.1	0.1	0.1	0.1	0.1	0.2	0.2	0.2	0.2
Total	37.5	33.9	32.0	23.8	33.5	31.7	30.7	36.8	34.2

[1] See Notes in Part I.4.

Oil Demand by Main Product Groups[1]
Demande de produits raffinés par groupes principaux

Syria

	1971	1973	1978	1984	1985	1986	1987	1988	1989
Thousand tonnes									
NGL/LPG/Ethane	23	27	31	231	248	265	289	293	256
Naphtha	-	-	313	184	211	211	178	17	70
Motor Gasoline	212	281	533	736	788	828	943	960	1 100
Aviation Fuels	75	159	238	284	215	295	266	300	336
Kerosene	244	215	282	203	161	212	178	172	230
Gas Diesel	820	689	2 000	2 990	3 055	2 920	3 192	2 927	3 200
Heavy Fuel Oil	577	542	768	2 992	2 175	2 520	3 046	3 276	2 353
Other	223	195	220	478	552	372	346	366	400
Refinery Fuel	89	79	150	564	489	450	529	399	400
Sub-Total	2 263	2 187	4 535	8 662	7 894	8 073	8 967	8 710	8 345
Bunkers	-	-	-	-	-	-	-	-	-
Total	2 263	2 187	4 535	8 662	7 894	8 073	8 967	8 710	8 345
Thousand barrels/day									
NGL/LPG/Ethane	0.7	0.9	1.0	7.3	7.9	8.4	9.2	9.3	8.1
Naphtha	-	-	7.3	4.3	4.9	4.9	4.1	0.4	1.6
Motor Gasoline	5.0	6.6	12.5	17.2	18.4	19.4	22.0	22.4	25.7
Aviation Fuels	1.6	3.5	5.2	6.2	4.7	6.4	5.8	6.5	7.3
Kerosene	5.2	4.6	6.0	4.3	3.4	4.5	3.8	3.6	4.9
Gas Diesel	16.8	14.1	40.9	60.9	62.4	59.7	65.2	59.7	65.4
Heavy Fuel Oil	10.5	9.9	14.0	54.4	39.7	46.0	55.6	59.6	42.9
Other	4.3	3.7	4.2	9.1	10.6	7.1	6.6	7.0	7.7
Refinery Fuel	1.6	1.4	2.7	10.3	8.9	8.2	9.7	7.3	7.3
Sub-Total	45.7	44.6	93.7	174.0	160.9	164.6	182.0	175.7	171.0
Bunkers	-	-	-	-	-	-	-	-	-
Total	45.7	44.6	93.7	174.0	160.9	164.6	182.0	175.7	171.0

	1990	1991	1992	1993	1994	1995	1996	1997	1998
Thousand tonnes									
NGL/LPG/Ethane	264	270	280	288	319	356	368	392	420
Naphtha	50	50	50	52	58	101	105	111	115
Motor Gasoline	1 220	1 258	1 060	1 110	1 090	1 063	1 101	1 142	1 176
Aviation Fuels	390	324	330	214	237	156	162	235	246
Kerosene	252	260	226	231	245	239	247	284	272
Gas Diesel	3 305	3 620	4 095	4 160	4 210	4 283	4 433	5 051	5 203
Heavy Fuel Oil	3 100	3 702	3 270	3 885	4 350	4 624	4 786	4 046	4 648
Other	400	300	350	348	386	376	389	389	389
Refinery Fuel	400	400	410	420	420	447	463	391	450
Sub-Total	9 381	10 184	10 071	10 708	11 315	11 645	12 054	12 041	12 919
Bunkers	-	-	-	-	-	-	-	-	-
Total	9 381	10 184	10 071	10 708	11 315	11 645	12 054	12 041	12 919
Thousand barrels/day									
NGL/LPG/Ethane	8.4	8.6	8.9	9.2	10.1	11.3	11.7	12.5	13.3
Naphtha	1.2	1.2	1.2	1.2	1.4	2.4	2.4	2.6	2.7
Motor Gasoline	28.5	29.4	24.7	25.9	25.5	24.8	25.7	26.7	27.5
Aviation Fuels	8.5	7.0	7.2	4.6	5.1	3.4	3.5	5.1	5.3
Kerosene	5.3	5.5	4.8	4.9	5.2	5.1	5.2	6.0	5.8
Gas Diesel	67.5	74.0	83.5	85.0	86.0	87.5	90.4	103.2	106.3
Heavy Fuel Oil	56.6	67.5	59.5	70.9	79.4	84.4	87.1	73.8	84.8
Other	7.7	5.8	6.7	6.7	7.4	7.2	7.4	7.5	7.5
Refinery Fuel	7.3	7.3	7.5	7.7	7.7	8.2	8.4	7.1	8.2
Sub-Total	191.0	206.3	203.8	216.1	227.8	234.2	241.8	244.5	261.4
Bunkers	-	-	-	-	-	-	-	-	-
Total	191.0	206.3	203.8	216.1	227.8	234.2	241.8	244.5	261.4

[1] See Notes in Part I.4.

Oil Demand by Main Product Groups[1]
Demande de produits raffinés par groupes principaux

Tajikistan

	1971	1973	1978	1984	1985	1986	1987	1988	1989
Thousand tonnes									
NGL/LPG/Ethane	-	-	-	-	-	-	-	-	-
Naphtha	-	-	-	-	-	-	-	-	-
Motor Gasoline	-	-	-	-	-	-	-	-	-
Aviation Fuels	-	-	-	-	-	-	-	-	-
Kerosene	-	-	-	-	-	-	-	-	-
Gas Diesel	-	-	-	-	-	-	-	-	-
Heavy Fuel Oil	-	-	-	-	-	-	-	-	-
Other	-	-	-	-	-	-	-	-	-
Refinery Fuel	-	-	-	-	-	-	-	-	-
Sub-Total	-	-	-	-	-	-	-	-	-
Bunkers	-	-	-	-	-	-	-	-	-
Total	-	-	-	-	-	-	-	-	-
Thousand barrels/day									
NGL/LPG/Ethane	-	-	-	-	-	-	-	-	-
Naphtha	-	-	-	-	-	-	-	-	-
Motor Gasoline	-	-	-	-	-	-	-	-	-
Aviation Fuels	-	-	-	-	-	-	-	-	-
Kerosene	-	-	-	-	-	-	-	-	-
Gas Diesel	-	-	-	-	-	-	-	-	-
Heavy Fuel Oil	-	-	-	-	-	-	-	-	-
Other	-	-	-	-	-	-	-	-	-
Refinery Fuel	-	-	-	-	-	-	-	-	-
Sub-Total	-	-	-	-	-	-	-	-	-
Bunkers	-	-	-	-	-	-	-	-	-
Total	-	-	-	-	-	-	-	-	-

	1990	1991	1992	1993	1994	1995	1996	1997	1998
Thousand tonnes									
NGL/LPG/Ethane	-	-	-	22	7	7	7	7	7
Naphtha	-	-	-	-	-	-	-	-	-
Motor Gasoline	-	-	4 763	2 988	996	996	996	996	996
Aviation Fuels	-	-	15	17	5	5	5	5	5
Kerosene	-	-	-	-	-	-	-	-	-
Gas Diesel	-	-	282	226	75	75	75	75	75
Heavy Fuel Oil	-	-	309	100	33	33	33	33	33
Other	-	-	226	130	43	43	43	43	161
Refinery Fuel	-	-	-	-	-	-	-	-	-
Sub-Total	-	-	5 595	3 483	1 159	1 159	1 159	1 159	1 277
Bunkers	-	-	-	-	-	-	-	-	-
Total	-	-	5 595	3 483	1 159	1 159	1 159	1 159	1 277
Thousand barrels/day									
NGL/LPG/Ethane	-	-	-	0.7	0.2	0.2	0.2	0.2	0.2
Naphtha	-	-	-	-	-	-	-	-	-
Motor Gasoline	-	-	111.0	69.8	23.3	23.3	23.2	23.3	23.3
Aviation Fuels	-	-	0.3	0.4	0.1	0.1	0.1	0.1	0.1
Kerosene	-	-	-	-	-	-	-	-	-
Gas Diesel	-	-	5.7	4.6	1.5	1.5	1.5	1.5	1.5
Heavy Fuel Oil	-	-	5.6	1.8	0.6	0.6	0.6	0.6	0.6
Other	-	-	4.3	2.5	0.8	0.8	0.8	0.8	2.6
Refinery Fuel	-	-	-	-	-	-	-	-	-
Sub-Total	-	-	127.0	79.8	26.6	26.6	26.5	26.6	28.3
Bunkers	-	-	-	-	-	-	-	-	-
Total	-	-	127.0	79.8	26.6	26.6	26.5	26.6	28.3

[1] See Notes in Part I.4.

Oil Demand by Main Product Groups[1]
Demande de produits raffinés par groupes principaux

United Republic of Tanzania

	1971	1973	1978	1984	1985	1986	1987	1988	1989
Thousand tonnes									
NGL/LPG/Ethane	5	5	6	7	7	7	6	6	6
Naphtha	-	-	-	-	-	-	-	-	-
Motor Gasoline	109	120	100	85	87	85	90	92	85
Aviation Fuels	27	27	52	40	40	37	35	61	64
Kerosene	58	75	94	57	58	93	85	97	100
Gas Diesel	184	216	177	189	191	192	194	194	199
Heavy Fuel Oil	72	101	84	81	95	95	96	97	102
Other	44	41	47	22	25	16	19	16	16
Refinery Fuel	19	15	11	15	15	15	15	15	15
Sub-Total	518	600	571	496	518	540	540	578	587
Bunkers	17	20	30	26	26	25	23	24	24
Total	535	620	601	522	544	565	563	602	611
Thousand barrels/day									
NGL/LPG/Ethane	0.2	0.2	0.2	0.2	0.2	0.2	0.2	0.2	0.2
Naphtha	-	-	-	-	-	-	-	-	-
Motor Gasoline	2.5	2.8	2.3	2.0	2.0	2.0	2.1	2.1	2.0
Aviation Fuels	0.6	0.6	1.1	0.9	0.9	0.8	0.8	1.3	1.4
Kerosene	1.2	1.6	2.0	1.2	1.2	2.0	1.8	2.1	2.1
Gas Diesel	3.8	4.4	3.6	3.9	3.9	3.9	4.0	4.0	4.1
Heavy Fuel Oil	1.3	1.8	1.5	1.5	1.7	1.7	1.8	1.8	1.9
Other	0.8	0.8	0.9	0.4	0.5	0.3	0.4	0.3	0.3
Refinery Fuel	0.4	0.3	0.2	0.3	0.3	0.3	0.3	0.3	0.3
Sub-Total	10.9	12.5	12.0	10.4	10.8	11.3	11.3	12.1	12.3
Bunkers	0.3	0.4	0.5	0.5	0.5	0.5	0.4	0.4	0.4
Total	11.2	12.9	12.5	10.8	11.3	11.8	11.7	12.5	12.7

	1990	1991	1992	1993	1994	1995	1996	1997	1998
Thousand tonnes									
NGL/LPG/Ethane	7	7	5	6	6	6	6	6	6
Naphtha	-	-	-	-	-	-	-	-	-
Motor Gasoline	92	91	99	108	108	98	98	98	98
Aviation Fuels	70	70	67	46	49	49	36	36	36
Kerosene	102	104	96	70	70	72	72	72	72
Gas Diesel	207	194	195	195	195	197	201	281	155
Heavy Fuel Oil	105	109	109	105	105	105	105	105	105
Other	16	17	15	15	16	16	16	16	16
Refinery Fuel	15	15	14	14	14	14	14	14	14
Sub-Total	614	607	600	559	563	557	548	628	502
Bunkers	26	22	22	22	22	23	23	23	23
Total	640	629	622	581	585	580	571	651	525
Thousand barrels/day									
NGL/LPG/Ethane	0.2	0.2	0.2	0.2	0.2	0.2	0.2	0.2	0.2
Naphtha	-	-	-	-	-	-	-	-	-
Motor Gasoline	2.2	2.1	2.3	2.5	2.5	2.3	2.3	2.3	2.3
Aviation Fuels	1.5	1.5	1.5	1.0	1.1	1.1	0.8	0.8	0.8
Kerosene	2.2	2.2	2.0	1.5	1.5	1.5	1.5	1.5	1.5
Gas Diesel	4.2	4.0	4.0	4.0	4.0	4.0	4.1	5.7	3.2
Heavy Fuel Oil	1.9	2.0	2.0	1.9	1.9	1.9	1.9	1.9	1.9
Other	0.3	0.3	0.3	0.3	0.3	0.3	0.3	0.3	0.3
Refinery Fuel	0.3	0.3	0.3	0.3	0.3	0.3	0.3	0.3	0.3
Sub-Total	12.9	12.7	12.5	11.7	11.8	11.7	11.4	13.1	10.5
Bunkers	0.5	0.4	0.4	0.4	0.4	0.4	0.4	0.4	0.4
Total	13.3	13.1	12.9	12.1	12.2	12.1	11.8	13.5	10.9

[1] See Notes in Part I.4.

Oil Demand by Main Product Groups[1]
Demande de produits raffinés par groupes principaux

Thailand

	1971	1973	1978	1984	1985	1986	1987	1988	1989
Thousand tonnes									
NGL/LPG/Ethane	50	77	162	548	604	635	646	758	855
Naphtha	-	-	-	-	-	-	-	-	-
Motor Gasoline	939	1 104	1 656	1 525	1 527	1 654	1 930	2 114	2 409
Aviation Fuels	400	724	628	965	990	1 096	1 202	1 387	1 638
Kerosene	153	169	220	232	121	114	103	100	94
Gas Diesel	2 181	2 628	3 251	4 423	4 655	4 835	5 335	6 134	7 245
Heavy Fuel Oil	1 630	2 378	3 849	2 797	2 041	2 088	1 968	2 342	3 057
Other	280	255	230	260	170	200	292	166	304
Refinery Fuel	53	77	83	79	139	139	149	92	107
Sub-Total	5 686	7 412	10 079	10 829	10 247	10 761	11 625	13 093	15 709
Bunkers	65	86	123	284	202	295	375	364	491
Total	5 751	7 498	10 202	11 113	10 449	11 056	12 000	13 457	16 200
Thousand barrels/day									
NGL/LPG/Ethane	1.6	2.5	5.3	17.8	19.6	20.7	21.0	24.6	27.8
Naphtha	-	-	-	-	-	-	-	-	-
Motor Gasoline	22.2	26.1	39.2	36.0	36.1	39.1	45.6	49.8	57.0
Aviation Fuels	9.2	16.7	14.4	22.1	22.8	25.2	27.6	31.8	37.7
Kerosene	3.3	3.7	4.8	5.0	2.6	2.5	2.2	2.2	2.1
Gas Diesel	43.7	52.6	65.1	88.3	93.2	96.8	106.8	122.5	145.1
Heavy Fuel Oil	29.9	43.6	70.5	51.1	37.4	38.3	36.1	42.8	56.0
Other	4.7	4.3	3.9	4.4	2.9	3.4	5.3	2.8	5.5
Refinery Fuel	1.2	1.7	1.8	1.7	3.1	3.1	3.4	2.0	2.3
Sub-Total	115.8	151.2	205.0	226.5	217.8	229.1	248.1	278.5	333.5
Bunkers	1.2	1.6	2.3	5.3	3.8	5.5	7.0	6.8	9.1
Total	117.0	152.8	207.3	231.8	221.6	234.6	255.1	285.3	342.6

	1990	1991	1992	1993	1994	1995	1996	1997	1998
Thousand tonnes									
NGL/LPG/Ethane	928	1 008	1 124	1 219	1 845	1 892	2 684	2 467	2 494
Naphtha	-	-	-	-	-	-	-	-	-
Motor Gasoline	2 670	2 824	3 147	3 575	4 071	4 590	5 059	5 384	5 238
Aviation Fuels	1 782	1 719	2 121	2 142	2 341	2 389	2 492	2 667	2 495
Kerosene	97	88	90	86	92	80	78	70	43
Gas Diesel	8 361	8 483	8 818	10 228	11 279	13 274	15 192	14 993	13 077
Heavy Fuel Oil	4 544	5 261	6 276	6 928	7 692	8 739	8 455	7 853	6 992
Other	722	412	613	727	521	610	904	1 064	822
Refinery Fuel	112	119	147	172	200	239	318	367	346
Sub-Total	19 216	19 914	22 336	25 077	28 041	31 813	35 182	34 865	31 507
Bunkers	528	580	673	733	914	939	759	809	568
Total	19 744	20 494	23 009	25 810	28 955	32 752	35 941	35 674	32 075
Thousand barrels/day									
NGL/LPG/Ethane	30.2	32.8	36.4	39.6	64.7	65.1	96.3	85.3	86.9
Naphtha	-	-	-	-	-	-	-	-	-
Motor Gasoline	63.1	66.8	74.2	84.5	96.3	108.5	119.3	127.3	123.8
Aviation Fuels	41.0	39.5	48.6	49.2	53.8	54.9	57.1	61.3	57.4
Kerosene	2.1	1.9	2.0	1.9	2.0	1.7	1.7	1.5	0.9
Gas Diesel	167.4	169.9	176.1	204.8	225.9	265.8	303.4	300.3	261.9
Heavy Fuel Oil	83.3	96.4	114.7	127.0	141.0	160.2	154.5	143.9	128.2
Other	13.8	7.4	11.3	13.5	9.3	10.8	16.3	19.1	14.4
Refinery Fuel	2.5	2.6	3.2	3.8	4.4	5.2	7.0	8.0	7.6
Sub-Total	403.3	417.3	466.5	524.3	597.3	672.4	755.6	746.7	681.1
Bunkers	9.8	10.8	12.5	13.6	17.0	17.5	14.1	15.0	10.5
Total	413.2	428.1	479.0	537.9	614.3	689.8	769.7	761.7	691.6

[1] See Notes in Part I.4.

Oil Demand by Main Product Groups[1]
Demande de produits raffinés par groupes principaux

Togo

	1971	1973	1978	1984	1985	1986	1987	1988	1989
Thousand tonnes									
NGL/LPG/Ethane	-	-	-	-	-	-	-	-	-
Naphtha	-	-	-	-	-	-	-	-	-
Motor Gasoline	24	31	45	66	55	61	67	68	68
Aviation Fuels	-	-	-	-	-	-	-	-	-
Kerosene	20	14	14	17	11	16	14	16	19
Gas Diesel	53	44	49	45	31	68	75	69	69
Heavy Fuel Oil	-	-	-	-	-	-	-	-	-
Other	-	-	-	-	-	-	-	-	-
Refinery Fuel	-	-	-	-	-	-	-	-	-
Sub-Total	97	89	108	128	97	145	156	153	156
Bunkers	-	-	-	-	-	-	-	-	-
Total	97	89	108	128	97	145	156	153	156
Thousand barrels/day									
NGL/LPG/Ethane	-	-	-	-	-	-	-	-	-
Naphtha	-	-	-	-	-	-	-	-	-
Motor Gasoline	0.6	0.7	1.1	1.5	1.3	1.4	1.6	1.6	1.6
Aviation Fuels	-	-	-	-	-	-	-	-	-
Kerosene	0.4	0.3	0.3	0.4	0.2	0.3	0.3	0.3	0.4
Gas Diesel	1.1	0.9	1.0	0.9	0.6	1.4	1.5	1.4	1.4
Heavy Fuel Oil	-	-	-	-	-	-	-	-	-
Other	-	-	-	-	-	-	-	-	-
Refinery Fuel	-	-	-	-	-	-	-	-	-
Sub-Total	2.1	1.9	2.4	2.8	2.2	3.2	3.4	3.3	3.4
Bunkers	-	-	-	-	-	-	-	-	-
Total	2.1	1.9	2.4	2.8	2.2	3.2	3.4	3.3	3.4

	1990	1991	1992	1993	1994	1995	1996	1997	1998
Thousand tonnes									
NGL/LPG/Ethane	-	-	-	-	-	-	-	-	-
Naphtha	-	-	-	-	-	-	-	-	-
Motor Gasoline	74	63	64	23	65	77	109	87	98
Aviation Fuels	11	9	9	4	8	14	16	14	19
Kerosene	22	19	17	8	21	24	32	31	37
Gas Diesel	69	59	40	28	56	55	90	67	82
Heavy Fuel Oil	-	-	-	-	-	-	-	-	-
Other	8	19	4	-	34	40	40	64	69
Refinery Fuel	-	-	-	-	-	-	-	-	-
Sub-Total	184	169	134	63	184	210	287	263	305
Bunkers	-	-	-	-	-	-	-	-	-
Total	184	169	134	63	184	210	287	263	305
Thousand barrels/day									
NGL/LPG/Ethane	-	-	-	-	-	-	-	-	-
Naphtha	-	-	-	-	-	-	-	-	-
Motor Gasoline	1.7	1.5	1.5	0.5	1.5	1.8	2.5	2.0	2.3
Aviation Fuels	0.2	0.2	0.2	0.1	0.2	0.3	0.3	0.3	0.4
Kerosene	0.5	0.4	0.4	0.2	0.4	0.5	0.7	0.7	0.8
Gas Diesel	1.4	1.2	0.8	0.6	1.1	1.1	1.8	1.4	1.7
Heavy Fuel Oil	-	-	-	-	-	-	-	-	-
Other	0.2	0.4	0.1	-	0.6	0.8	0.8	1.2	1.3
Refinery Fuel	-	-	-	-	-	-	-	-	-
Sub-Total	4.0	3.6	2.9	1.4	3.9	4.5	6.2	5.6	6.5
Bunkers	-	-	-	-	-	-	-	-	-
Total	4.0	3.6	2.9	1.4	3.9	4.5	6.2	5.6	6.5

[1] See Notes in Part I.4.

Oil Demand by Main Product Groups[1]
Demande de produits raffinés par groupes principaux

Trinidad-and-Tobago

	1971	1973	1978	1984	1985	1986	1987	1988	1989
Thousand tonnes									
NGL/LPG/Ethane	15	30	29	56	47	47	44	77	47
Naphtha	-	-	-	-	-	-	-	-	-
Motor Gasoline	172	227	270	388	400	407	402	380	364
Aviation Fuels	67	38	40	64	69	84	50	50	40
Kerosene	30	30	30	13	11	15	8	8	7
Gas Diesel	62	134	132	162	136	143	126	121	128
Heavy Fuel Oil	38	29	45	37	10	10	26	7	11
Other	221	35	273	185	171	71	47	26	18
Refinery Fuel	544	541	417	75	102	105	123	111	114
Sub-Total	**1 149**	**1 064**	**1 236**	**980**	**946**	**882**	**826**	**780**	**729**
Bunkers	1 656	1 658	1 001	168	99	27	87	55	40
Total	**2 805**	**2 722**	**2 237**	**1 148**	**1 045**	**909**	**913**	**835**	**769**
Thousand barrels/day									
NGL/LPG/Ethane	0.5	1.0	0.9	1.8	1.5	1.5	1.4	2.4	1.5
Naphtha	-	-	-	-	-	-	-	-	-
Motor Gasoline	4.0	5.3	6.3	9.0	9.3	9.5	9.4	8.7	8.4
Aviation Fuels	1.5	0.8	0.9	1.4	1.5	1.8	1.1	1.1	0.9
Kerosene	0.6	0.6	0.6	0.3	0.2	0.3	0.2	0.2	0.1
Gas Diesel	1.3	2.7	2.7	3.3	2.8	2.9	2.6	2.5	2.6
Heavy Fuel Oil	0.7	0.5	0.8	0.7	0.2	0.2	0.5	0.1	0.2
Other	4.2	0.7	5.2	3.5	3.3	1.3	0.9	0.5	0.3
Refinery Fuel	11.3	11.2	8.4	1.6	2.2	2.3	2.7	2.4	2.5
Sub-Total	**24.1**	**22.8**	**25.9**	**21.6**	**21.0**	**19.9**	**18.7**	**17.9**	**16.6**
Bunkers	30.8	30.7	18.9	3.1	1.9	0.5	1.7	1.0	0.8
Total	**54.8**	**53.5**	**44.8**	**24.7**	**22.9**	**20.4**	**20.4**	**19.0**	**17.3**

	1990	1991	1992	1993	1994	1995	1996	1997	1998
Thousand tonnes									
NGL/LPG/Ethane	47	49	53	49	51	58	54	64	68
Naphtha	-	-	-	-	-	-	-	-	-
Motor Gasoline	351	349	335	313	286	288	293	315	325
Aviation Fuels	62	62	78	54	48	55	47	52	56
Kerosene	5	7	7	7	14	16	25	29	30
Gas Diesel	143	177	176	159	186	197	217	241	246
Heavy Fuel Oil	31	28	30	8	18	6	11	3	3
Other	9	46	14	14	22	18	22	7	9
Refinery Fuel	102	120	190	142	167	143	144	125	129
Sub-Total	**750**	**838**	**883**	**746**	**792**	**781**	**813**	**836**	**866**
Bunkers	35	30	37	53	52	52	15	17	17
Total	**785**	**868**	**920**	**799**	**844**	**833**	**828**	**853**	**883**
Thousand barrels/day									
NGL/LPG/Ethane	1.5	1.6	1.7	1.6	1.6	1.8	1.7	2.0	2.2
Naphtha	-	-	-	-	-	-	-	-	-
Motor Gasoline	8.1	8.0	7.7	7.2	6.6	6.6	6.7	7.2	7.4
Aviation Fuels	1.3	1.3	1.7	1.2	1.0	1.2	1.0	1.1	1.2
Kerosene	0.1	0.1	0.1	0.1	0.3	0.3	0.5	0.6	0.6
Gas Diesel	2.9	3.6	3.6	3.3	3.8	4.0	4.4	4.9	5.0
Heavy Fuel Oil	0.6	0.5	0.5	0.1	0.3	0.1	0.2	0.1	0.1
Other	0.2	0.9	0.3	0.3	0.4	0.3	0.4	0.1	0.2
Refinery Fuel	2.2	2.6	4.2	3.1	3.7	3.1	3.1	2.7	2.8
Sub-Total	**17.0**	**18.7**	**19.8**	**16.8**	**17.7**	**17.6**	**18.1**	**18.8**	**19.5**
Bunkers	0.7	0.6	0.7	1.0	1.0	1.0	0.3	0.3	0.3
Total	**17.6**	**19.3**	**20.5**	**17.8**	**18.7**	**18.6**	**18.4**	**19.2**	**19.8**

[1] See Notes in Part I.4.

Oil Demand by Main Product Groups[1]
Demande de produits raffinés par groupes principaux

Tunisia

	1971	1973	1978	1984	1985	1986	1987	1988	1989
Thousand tonnes									
NGL/LPG/Ethane	20	29	72	149	168	181	191	233	258
Naphtha	9	9	12	15	15	12	6	3	3
Motor Gasoline	99	118	145	203	220	220	216	238	246
Aviation Fuels	122	141	244	121	99	92	145	164	172
Kerosene	72	84	106	133	136	139	143	142	152
Gas Diesel	330	411	765	1 007	1 020	983	981	934	1 006
Heavy Fuel Oil	481	469	624	989	840	1 129	741	1 108	1 247
Other	42	46	71	81	63	4	4	4	4
Refinery Fuel	53	38	43	33	56	56	56	56	56
Sub-Total	1 228	1 345	2 082	2 731	2 617	2 816	2 483	2 882	3 144
Bunkers	19	17	12	5	4	2	-	-	-
Total	1 247	1 362	2 094	2 736	2 621	2 818	2 483	2 882	3 144
Thousand barrels/day									
NGL/LPG/Ethane	0.6	0.9	2.3	4.7	5.3	5.8	6.1	7.4	8.2
Naphtha	0.2	0.2	0.3	0.3	0.3	0.3	0.1	0.1	0.1
Motor Gasoline	2.3	2.8	3.4	4.7	5.1	5.1	5.0	5.5	5.7
Aviation Fuels	2.7	3.1	5.3	2.6	2.2	2.0	3.2	3.6	3.7
Kerosene	1.5	1.8	2.2	2.8	2.9	2.9	3.0	3.0	3.2
Gas Diesel	6.7	8.4	15.6	20.5	20.8	20.1	20.1	19.0	20.6
Heavy Fuel Oil	8.8	8.6	11.4	18.0	15.3	20.6	13.5	20.2	22.8
Other	0.8	0.9	1.4	1.5	1.2	0.1	0.1	0.1	0.1
Refinery Fuel	1.0	0.8	0.9	0.7	1.1	1.1	1.1	1.1	1.1
Sub-Total	24.7	27.3	42.7	56.0	54.4	58.0	52.2	60.0	65.5
Bunkers	0.4	0.3	0.2	0.1	0.1	0.0	-	-	-
Total	25.1	27.7	43.0	56.1	54.5	58.1	52.2	60.0	65.5

	1990	1991	1992	1993	1994	1995	1996	1997	1998
Thousand tonnes									
NGL/LPG/Ethane	287	287	297	282	286	302	323	342	352
Naphtha	2	2	1	-	-	-	-	-	-
Motor Gasoline	263	272	289	299	301	312	324	340	346
Aviation Fuels	179	132	187	212	244	242	249	283	301
Kerosene	142	144	153	164	152	165	168	178	182
Gas Diesel	1 031	1 053	1 065	1 207	1 230	1 262	1 339	1 434	1 488
Heavy Fuel Oil	1 147	1 493	1 376	1 368	989	793	809	773	778
Other	5	6	6	164	145	136	147	148	166
Refinery Fuel	65	56	54	62	64	68	68	68	68
Sub-Total	3 121	3 445	3 428	3 758	3 411	3 280	3 427	3 566	3 681
Bunkers	22	29	25	19	19	19	19	18	18
Total	3 143	3 474	3 453	3 777	3 430	3 299	3 446	3 584	3 699
Thousand barrels/day									
NGL/LPG/Ethane	9.1	9.1	9.4	9.0	9.1	9.6	10.2	10.9	11.2
Naphtha	0.0	0.0	0.0	-	-	-	-	-	-
Motor Gasoline	6.1	6.4	6.7	7.0	7.0	7.3	7.6	7.9	8.1
Aviation Fuels	3.9	2.9	4.1	4.6	5.3	5.3	5.4	6.1	6.5
Kerosene	3.0	3.1	3.2	3.5	3.2	3.5	3.6	3.8	3.9
Gas Diesel	21.1	21.5	21.7	24.7	25.1	25.8	27.3	29.3	30.4
Heavy Fuel Oil	20.9	27.2	25.0	25.0	18.0	14.5	14.7	14.1	14.2
Other	0.1	0.1	0.1	3.1	2.8	2.6	2.8	2.8	3.2
Refinery Fuel	1.3	1.1	1.1	1.2	1.3	1.4	1.4	1.4	1.4
Sub-Total	65.6	71.5	71.4	78.1	71.9	69.9	72.9	76.4	78.8
Bunkers	0.4	0.5	0.5	0.3	0.3	0.3	0.3	0.3	0.3
Total	66.0	72.0	71.9	78.4	72.2	70.2	73.3	76.7	79.2

[1] See Notes in Part I.4.

Oil Demand by Main Product Groups[1]
Demande de produits raffinés par groupes principaux

Turkmenistan

	1971	1973	1978	1984	1985	1986	1987	1988	1989
Thousand tonnes									
NGL/LPG/Ethane	-	-	-	-	-	-	-	-	-
Naphtha	-	-	-	-	-	-	-	-	-
Motor Gasoline	-	-	-	-	-	-	-	-	-
Aviation Fuels	-	-	-	-	-	-	-	-	-
Kerosene	-	-	-	-	-	-	-	-	-
Gas Diesel	-	-	-	-	-	-	-	-	-
Heavy Fuel Oil	-	-	-	-	-	-	-	-	-
Other	-	-	-	-	-	-	-	-	-
Refinery Fuel	-	-	-	-	-	-	-	-	-
Sub-Total	-	-	-	-	-	-	-	-	-
Bunkers	-	-	-	-	-	-	-	-	-
Total	-	-	-	-	-	-	-	-	-
Thousand barrels/day									
NGL/LPG/Ethane	-	-	-	-	-	-	-	-	-
Naphtha	-	-	-	-	-	-	-	-	-
Motor Gasoline	-	-	-	-	-	-	-	-	-
Aviation Fuels	-	-	-	-	-	-	-	-	-
Kerosene	-	-	-	-	-	-	-	-	-
Gas Diesel	-	-	-	-	-	-	-	-	-
Heavy Fuel Oil	-	-	-	-	-	-	-	-	-
Other	-	-	-	-	-	-	-	-	-
Refinery Fuel	-	-	-	-	-	-	-	-	-
Sub-Total	-	-	-	-	-	-	-	-	-
Bunkers	-	-	-	-	-	-	-	-	-
Total	-	-	-	-	-	-	-	-	-

	1990	1991	1992	1993	1994	1995	1996	1997	1998
Thousand tonnes									
NGL/LPG/Ethane	-	-	297	254	805	968	379	272	324
Naphtha	-	-	-	-	-	-	-	-	-
Motor Gasoline	-	-	713	527	493	451	483	347	313
Aviation Fuels	-	-	-	-	-	-	-	-	-
Kerosene	-	-	-	-	-	-	-	-	-
Gas Diesel	-	-	1 164	999	945	864	924	457	357
Heavy Fuel Oil	-	-	2 541	834	834	763	816	1 369	1 097
Other	-	-	-	-	-	-	-	-	-
Refinery Fuel	-	-	230	200	197	180	193	250	287
Sub-Total	-	-	4 945	2 814	3 274	3 226	2 795	2 695	2 378
Bunkers	-	-	-	-	-	-	-	-	-
Total	-	-	4 945	2 814	3 274	3 226	2 795	2 695	2 378
Thousand barrels/day									
NGL/LPG/Ethane	-	-	8.1	7.0	22.5	26.7	10.7	7.7	9.2
Naphtha	-	-	-	-	-	-	-	-	-
Motor Gasoline	-	-	16.6	12.3	11.5	10.5	11.3	8.1	7.3
Aviation Fuels	-	-	-	-	-	-	-	-	-
Kerosene	-	-	-	-	-	-	-	-	-
Gas Diesel	-	-	23.7	20.4	19.3	17.7	18.8	9.3	7.3
Heavy Fuel Oil	-	-	46.2	15.2	15.2	13.9	14.8	25.0	20.0
Other	-	-	-	-	-	-	-	-	-
Refinery Fuel	-	-	5.0	4.4	4.3	3.9	4.2	5.5	6.3
Sub-Total	-	-	99.7	59.3	72.9	72.8	59.8	55.6	50.1
Bunkers	-	-	-	-	-	-	-	-	-
Total	-	-	99.7	59.3	72.9	72.8	59.8	55.6	50.1

[1] See Notes in Part I.4.

Oil Demand by Main Product Groups[1]
Demande de produits raffinés par groupes principaux

Ukraine

	1971	1973	1978	1984	1985	1986	1987	1988	1989
Thousand tonnes									
NGL/LPG/Ethane	-	-	-	-	-	-	-	-	-
Naphtha	-	-	-	-	-	-	-	-	-
Motor Gasoline	-	-	-	-	-	-	-	-	-
Aviation Fuels	-	-	-	-	-	-	-	-	-
Kerosene	-	-	-	-	-	-	-	-	-
Gas Diesel	-	-	-	-	-	-	-	-	-
Heavy Fuel Oil	-	-	-	-	-	-	-	-	-
Other	-	-	-	-	-	-	-	-	-
Refinery Fuel	-	-	-	-	-	-	-	-	-
Sub-Total	-	-	-	-	-	-	-	-	-
Bunkers	-	-	-	-	-	-	-	-	-
Total	-	-	-	-	-	-	-	-	-
Thousand barrels/day									
NGL/LPG/Ethane	-	-	-	-	-	-	-	-	-
Naphtha	-	-	-	-	-	-	-	-	-
Motor Gasoline	-	-	-	-	-	-	-	-	-
Aviation Fuels	-	-	-	-	-	-	-	-	-
Kerosene	-	-	-	-	-	-	-	-	-
Gas Diesel	-	-	-	-	-	-	-	-	-
Heavy Fuel Oil	-	-	-	-	-	-	-	-	-
Other	-	-	-	-	-	-	-	-	-
Refinery Fuel	-	-	-	-	-	-	-	-	-
Sub-Total	-	-	-	-	-	-	-	-	-
Bunkers	-	-	-	-	-	-	-	-	-
Total	-	-	-	-	-	-	-	-	-

	1990	1991	1992	1993	1994	1995	1996	1997	1998
Thousand tonnes									
NGL/LPG/Ethane	-	-	2 681	1 275	1 439	2 281	1 702	1 584	1 485
Naphtha	-	-	-	-	-	-	-	-	-
Motor Gasoline	-	-	5 797	4 142	3 836	3 984	3 585	3 200	3 311
Aviation Fuels	-	-	718	1 202	1 043	891	784	674	699
Kerosene	-	-	75	16	14	12	11	10	8
Gas Diesel	-	-	8 826	7 344	6 308	6 775	5 055	4 795	4 909
Heavy Fuel Oil	-	-	17 402	11 675	7 882	8 080	6 028	5 459	5 572
Other	-	-	3 747	1 920	1 703	1 410	1 648	1 400	1 250
Refinery Fuel	-	-	929	621	579	538	409	384	397
Sub-Total	-	-	40 175	28 195	22 804	23 971	19 222	17 506	17 631
Bunkers	-	-	-	-	-	-	-	-	-
Total	-	-	40 175	28 195	22 804	23 971	19 222	17 506	17 631
Thousand barrels/day									
NGL/LPG/Ethane	-	-	85.0	40.5	45.7	72.5	53.9	50.3	47.2
Naphtha	-	-	-	-	-	-	-	-	-
Motor Gasoline	-	-	135.1	96.8	89.6	93.1	83.6	74.8	77.4
Aviation Fuels	-	-	15.6	26.1	22.7	19.4	17.1	14.7	15.3
Kerosene	-	-	1.6	0.3	0.3	0.3	0.2	0.2	0.2
Gas Diesel	-	-	179.9	150.1	128.9	138.5	103.0	98.0	100.3
Heavy Fuel Oil	-	-	316.7	213.0	143.8	147.4	109.7	99.6	101.7
Other	-	-	70.9	37.6	32.7	27.0	31.3	26.8	23.9
Refinery Fuel	-	-	18.5	12.5	11.6	10.7	8.1	7.7	7.9
Sub-Total	-	-	823.2	577.0	475.3	508.8	407.0	372.2	373.9
Bunkers	-	-	-	-	-	-	-	-	-
Total	-	-	823.2	577.0	475.3	508.8	407.0	372.2	373.9

[1] See Notes in Part I.4.

Oil Demand by Main Product Groups[1]
Demande de produits raffinés par groupes principaux

United Arab Emirates

	1971	1973	1978	1984	1985	1986	1987	1988	1989
Thousand tonnes									
NGL/LPG/Ethane	-	-	-	18	24	26	27	53	52
Naphtha	-	-	-	-	-	-	-	-	-
Motor Gasoline	79	119	391	775	805	844	887	935	948
Aviation Fuels	7	36	134	420	494	401	427	484	570
Kerosene	-	-	-	-	-	-	-	-	-
Gas Diesel	50	125	454	640	628	565	638	764	757
Heavy Fuel Oil	-	-	541	1 653	1 775	1 819	1 824	1 946	1 973
Other	-	-	4	11	21	25	26	39	43
Refinery Fuel	-	-	-	197	234	276	295	283	289
Sub-Total	136	280	1 524	3 714	3 981	3 956	4 124	4 504	4 632
Bunkers	-	-	-	-	-	-	-	-	-
Total	136	280	1 524	3 714	3 981	3 956	4 124	4 504	4 632
Thousand barrels/day									
NGL/LPG/Ethane	-	-	-	0.6	0.8	0.8	0.9	1.7	1.7
Naphtha	-	-	-	-	-	-	-	-	-
Motor Gasoline	1.8	2.8	9.1	18.1	18.8	19.7	20.7	21.8	22.2
Aviation Fuels	0.2	0.8	2.9	9.1	10.7	8.7	9.3	10.5	12.4
Kerosene	-	-	-	-	-	-	-	-	-
Gas Diesel	1.0	2.6	9.3	13.0	12.8	11.5	13.0	15.6	15.5
Heavy Fuel Oil	-	-	9.9	30.1	32.4	33.2	33.3	35.4	36.0
Other	-	-	0.1	0.2	0.4	0.5	0.5	0.8	0.9
Refinery Fuel	-	-	-	4.3	5.1	6.0	6.5	6.2	6.3
Sub-Total	3.0	6.1	31.3	75.4	81.1	80.6	84.2	91.9	94.8
Bunkers	-	-	-	-	-	-	-	-	-
Total	3.0	6.1	31.3	75.4	81.1	80.6	84.2	91.9	94.8

	1990	1991	1992	1993	1994	1995	1996	1997	1998
Thousand tonnes									
NGL/LPG/Ethane	51	48	45	43	54	70	86	87	89
Naphtha	-	-	-	-	-	-	-	-	-
Motor Gasoline	960	986	1 034	1 103	1 081	1 042	1 421	1 551	1 680
Aviation Fuels	656	630	604	706	695	686	676	669	661
Kerosene	-	-	-	-	-	-	-	-	-
Gas Diesel	749	863	977	889	972	981	991	1 001	1 172
Heavy Fuel Oil	2 001	2 056	2 155	2 298	2 282	1 929	1 516	1 527	1 516
Other	46	48	51	57	47	56	67	71	71
Refinery Fuel	296	333	345	328	361	361	357	369	357
Sub-Total	4 759	4 964	5 211	5 424	5 492	5 125	5 114	5 275	5 546
Bunkers	-	-	-	-	-	139	124	117	-
Total	4 759	4 964	5 211	5 424	5 492	5 264	5 238	5 392	5 546
Thousand barrels/day									
NGL/LPG/Ethane	1.6	1.5	1.4	1.4	1.7	2.2	2.7	2.8	2.8
Naphtha	-	-	-	-	-	-	-	-	-
Motor Gasoline	22.4	23.0	24.1	25.8	25.3	24.4	33.1	36.2	39.3
Aviation Fuels	14.3	13.7	13.1	15.3	15.1	14.9	14.6	14.5	14.4
Kerosene	-	-	-	-	-	-	-	-	-
Gas Diesel	15.3	17.6	19.9	18.2	19.9	20.1	20.2	20.5	24.0
Heavy Fuel Oil	36.5	37.5	39.2	41.9	41.6	35.2	27.6	27.9	27.7
Other	0.9	0.9	1.0	1.1	0.9	1.1	1.3	1.4	1.4
Refinery Fuel	6.5	7.3	7.5	7.2	7.9	7.9	7.8	8.1	7.8
Sub-Total	97.5	101.7	106.3	110.9	112.4	105.8	107.4	111.4	117.3
Bunkers	-	-	-	-	-	2.5	2.3	2.1	-
Total	97.5	101.7	106.3	110.9	112.4	108.3	109.7	113.5	117.3

[1] See Notes in Part I.4.

Oil Demand by Main Product Groups[1]
Demande de produits raffinés par groupes principaux

Uruguay

	1971	1973	1978	1984	1985	1986	1987	1988	1989
Thousand tonnes									
NGL/LPG/Ethane	37	35	38	44	45	47	53	64	67
Naphtha	-	-	11	11	12	13	13	14	14
Motor Gasoline	257	225	203	175	174	177	187	200	214
Aviation Fuels	23	22	30	10	10	12	15	12	12
Kerosene	200	193	144	68	55	51	51	52	50
Gas Diesel	345	364	455	409	386	378	391	412	449
Heavy Fuel Oil	710	772	822	264	246	238	292	558	616
Other	89	86	113	48	43	49	72	68	66
Refinery Fuel	57	57	65	50	45	39	70	69	70
Sub-Total	1 718	1 754	1 881	1 079	1 016	1 004	1 144	1 449	1 558
Bunkers	88	105	82	110	105	78	96	95	89
Total	1 806	1 859	1 963	1 189	1 121	1 082	1 240	1 544	1 647
Thousand barrels/day									
NGL/LPG/Ethane	1.2	1.1	1.2	1.4	1.4	1.5	1.7	2.0	2.1
Naphtha	-	-	0.3	0.3	0.3	0.3	0.3	0.3	0.3
Motor Gasoline	6.0	5.3	4.7	4.1	4.1	4.1	4.4	4.7	5.0
Aviation Fuels	0.5	0.5	0.7	0.2	0.2	0.3	0.3	0.3	0.3
Kerosene	4.2	4.1	3.1	1.4	1.2	1.1	1.1	1.1	1.1
Gas Diesel	7.1	7.4	9.3	8.3	7.9	7.7	8.0	8.4	9.2
Heavy Fuel Oil	13.0	14.1	15.0	4.8	4.5	4.3	5.3	10.2	11.2
Other	1.6	1.5	2.0	0.9	0.8	0.9	1.4	1.3	1.3
Refinery Fuel	1.1	1.1	1.3	1.0	0.9	0.8	1.4	1.3	1.4
Sub-Total	34.6	35.1	37.5	22.4	21.3	21.1	23.8	29.6	31.8
Bunkers	1.7	2.0	1.6	2.1	2.0	1.5	1.9	1.9	1.8
Total	36.3	37.1	39.1	24.5	23.3	22.6	25.7	31.4	33.6

	1990	1991	1992	1993	1994	1995	1996	1997	1998
Thousand tonnes									
NGL/LPG/Ethane	70	75	78	80	85	83	96	100	105
Naphtha	15	15	16	15	15	15	16	11	11
Motor Gasoline	215	228	243	266	291	304	315	322	350
Aviation Fuels	11	15	16	17	18	15	8	7	7
Kerosene	49	53	46	43	35	33	32	25	24
Gas Diesel	429	452	649	652	612	616	705	732	768
Heavy Fuel Oil	287	386	286	237	179	255	347	350	372
Other	56	69	65	62	61	56	81	84	91
Refinery Fuel	71	81	98	31	11	71	104	95	70
Sub-Total	1 203	1 374	1 497	1 403	1 307	1 448	1 704	1 726	1 798
Bunkers	115	139	199	306	183	368	379	307	280
Total	1 318	1 513	1 696	1 709	1 490	1 816	2 083	2 033	2 078
Thousand barrels/day									
NGL/LPG/Ethane	2.2	2.4	2.5	2.5	2.7	2.6	3.0	3.2	3.3
Naphtha	0.3	0.3	0.4	0.3	0.3	0.3	0.4	0.3	0.3
Motor Gasoline	5.0	5.3	5.7	6.2	6.8	7.1	7.3	7.5	8.2
Aviation Fuels	0.2	0.3	0.4	0.4	0.4	0.3	0.2	0.2	0.2
Kerosene	1.0	1.1	1.0	0.9	0.7	0.7	0.7	0.5	0.5
Gas Diesel	8.8	9.2	13.2	13.3	12.5	12.6	14.4	15.0	15.7
Heavy Fuel Oil	5.2	7.0	5.2	4.3	3.3	4.7	6.3	6.4	6.8
Other	1.1	1.3	1.2	1.2	1.2	1.1	1.5	1.6	1.7
Refinery Fuel	1.4	1.6	1.9	0.6	0.2	1.4	2.0	1.8	1.4
Sub-Total	25.4	28.7	31.4	29.8	28.1	30.8	35.9	36.4	38.0
Bunkers	2.3	2.7	3.8	6.1	3.6	6.9	7.3	5.9	5.3
Total	27.6	31.4	35.2	35.9	31.7	37.8	43.2	42.3	43.4

[1] See Notes in Part I.4.

Oil Demand by Main Product Groups[1]
Demande de produits raffinés par groupes principaux

Uzbekistan

	1971	1973	1978	1984	1985	1986	1987	1988	1989
Thousand tonnes									
NGL/LPG/Ethane	-	-	-	-	-	-	-	-	-
Naphtha	-	-	-	-	-	-	-	-	-
Motor Gasoline	-	-	-	-	-	-	-	-	-
Aviation Fuels	-	-	-	-	-	-	-	-	-
Kerosene	-	-	-	-	-	-	-	-	-
Gas Diesel	-	-	-	-	-	-	-	-	-
Heavy Fuel Oil	-	-	-	-	-	-	-	-	-
Other	-	-	-	-	-	-	-	-	-
Refinery Fuel	-	-	-	-	-	-	-	-	-
Sub-Total	-	-	-	-	-	-	-	-	-
Bunkers	-	-	-	-	-	-	-	-	-
Total	-	-	-	-	-	-	-	-	-
Thousand barrels/day									
NGL/LPG/Ethane	-	-	-	-	-	-	-	-	-
Naphtha	-	-	-	-	-	-	-	-	-
Motor Gasoline	-	-	-	-	-	-	-	-	-
Aviation Fuels	-	-	-	-	-	-	-	-	-
Kerosene	-	-	-	-	-	-	-	-	-
Gas Diesel	-	-	-	-	-	-	-	-	-
Heavy Fuel Oil	-	-	-	-	-	-	-	-	-
Other	-	-	-	-	-	-	-	-	-
Refinery Fuel	-	-	-	-	-	-	-	-	-
Sub-Total	-	-	-	-	-	-	-	-	-
Bunkers	-	-	-	-	-	-	-	-	-
Total	-	-	-	-	-	-	-	-	-

	1990	1991	1992	1993	1994	1995	1996	1997	1998
Thousand tonnes									
NGL/LPG/Ethane	-	-	-	-	-	53	60	57	56
Naphtha	-	-	273	-	-	-	-	-	-
Motor Gasoline	-	-	2 102	1 939	1 889	1 225	1 178	1 407	1 370
Aviation Fuels	-	-	83	56	89	317	312	277	294
Kerosene	-	-	-	-	-	66	46	69	100
Gas Diesel	-	-	2 608	2 733	2 394	2 061	1 946	2 011	2 017
Heavy Fuel Oil	-	-	2 170	1 919	1 525	1 981	2 116	2 048	1 796
Other	-	-	1 000	803	796	808	764	944	897
Refinery Fuel	-	-	246	130	117	205	198	221	219
Sub-Total	-	-	8 482	7 580	6 810	6 716	6 620	7 034	6 749
Bunkers	-	-	-	-	-	-	-	-	-
Total	-	-	8 482	7 580	6 810	6 716	6 620	7 034	6 749
Thousand barrels/day									
NGL/LPG/Ethane	-	-	-	-	-	1.6	1.8	1.7	1.7
Naphtha	-	-	6.3	-	-	-	-	-	-
Motor Gasoline	-	-	49.0	45.3	44.1	28.6	27.5	32.9	32.0
Aviation Fuels	-	-	1.8	1.2	1.9	6.9	6.8	6.0	6.4
Kerosene	-	-	-	-	-	1.4	1.0	1.5	2.1
Gas Diesel	-	-	53.2	55.9	48.9	42.1	39.7	41.1	41.2
Heavy Fuel Oil	-	-	39.5	35.0	27.8	36.1	38.5	37.4	32.8
Other	-	-	19.1	15.4	15.3	15.3	14.5	18.0	17.1
Refinery Fuel	-	-	4.6	2.8	2.6	4.5	4.3	4.8	4.8
Sub-Total	-	-	173.5	155.6	140.7	136.6	134.0	143.4	138.1
Bunkers	-	-	-	-	-	-	-	-	-
Total	-	-	173.5	155.6	140.7	136.6	134.0	143.4	138.1

[1] See Notes in Part I.4.

Oil Demand by Main Product Groups[1]
Demande de produits raffinés par groupes principaux

Venezuela

	1971	1973	1978	1984	1985	1986	1987	1988	1989
Thousand tonnes									
NGL/LPG/Ethane	158	220	451	988	1 307	1 092	1 097	1 120	1 583
Naphtha	-	-	-	-	-	-	-	-	-
Motor Gasoline	3 176	3 714	6 097	6 916	7 008	7 027	7 195	7 467	6 929
Aviation Fuels	88	111	197	418	247	259	312	332	336
Kerosene	486	474	467	393	380	375	394	398	325
Gas Diesel	1 246	1 742	3 090	3 184	2 732	3 144	2 960	3 084	2 974
Heavy Fuel Oil	745	990	1 141	2 567	2 749	2 575	2 384	2 393	2 557
Other	1 404	1 648	1 376	1 846	1 175	1 164	2 699	1 333	881
Refinery Fuel	2 873	3 240	2 394	2 214	2 188	2 083	1 740	1 972	2 682
Sub-Total	10 176	12 139	15 213	18 526	17 786	17 719	18 781	18 099	18 267
Bunkers	2 754	2 928	862	541	532	435	458	784	725
Total	12 930	15 067	16 075	19 067	18 318	18 154	19 239	18 883	18 992
Thousand barrels/day									
NGL/LPG/Ethane	5.0	7.0	14.3	31.3	41.5	34.7	34.9	35.5	50.3
Naphtha	-	-	-	-	-	-	-	-	-
Motor Gasoline	74.2	86.8	142.5	161.2	163.8	164.2	168.1	174.0	161.9
Aviation Fuels	2.0	2.5	4.3	9.1	5.4	5.7	6.8	7.2	7.3
Kerosene	10.3	10.1	9.9	8.3	8.1	8.0	8.4	8.4	6.9
Gas Diesel	25.5	35.6	63.2	64.9	55.8	64.3	60.5	62.9	60.8
Heavy Fuel Oil	13.6	18.1	20.8	46.7	50.2	47.0	43.5	43.5	46.7
Other	26.9	31.6	26.4	35.3	22.5	22.3	51.7	25.5	16.9
Refinery Fuel	61.7	69.1	51.4	47.3	47.0	44.8	37.8	42.9	56.6
Sub-Total	219.2	260.6	332.8	404.1	394.3	390.9	411.8	399.9	407.4
Bunkers	50.5	53.7	16.0	10.1	9.9	8.1	8.5	14.5	13.4
Total	269.7	314.4	348.8	414.2	404.2	399.1	420.3	414.4	420.9

	1990	1991	1992	1993	1994	1995	1996	1997	1998
Thousand tonnes									
NGL/LPG/Ethane	1 637	1 667	1 844	2 126	2 049	2 227	2 470	2 144	2 045
Naphtha	-	-	-	-	-	-	-	-	-
Motor Gasoline	7 045	7 435	7 486	8 256	8 214	8 648	8 910	8 559	8 352
Aviation Fuels	314	286	288	349	284	306	287	288	261
Kerosene	317	304	268	279	263	278	137	93	84
Gas Diesel	3 175	3 389	2 932	3 620	3 696	3 855	4 308	4 751	4 541
Heavy Fuel Oil	2 058	1 203	1 260	1 314	933	773	604	1 406	1 193
Other	899	2 275	2 332	1 029	4 756	3 167	3 406	5 278	1 870
Refinery Fuel	2 798	2 481	3 329	2 574	2 605	2 515	1 080	2 122	3 488
Sub-Total	18 243	19 040	19 739	19 547	22 800	21 769	21 202	24 641	21 834
Bunkers	754	929	844	821	762	693	682	610	507
Total	18 997	19 969	20 583	20 368	23 562	22 462	21 884	25 251	22 341
Thousand barrels/day									
NGL/LPG/Ethane	52.0	53.0	58.4	67.6	65.1	70.8	78.3	68.1	65.3
Naphtha	-	-	-	-	-	-	-	-	-
Motor Gasoline	164.6	173.8	174.5	192.9	192.0	202.1	207.7	199.3	194.5
Aviation Fuels	6.9	6.2	6.3	7.6	6.2	6.7	6.3	6.2	5.6
Kerosene	6.7	6.4	5.7	5.9	5.6	5.9	2.9	2.0	1.8
Gas Diesel	64.9	69.3	59.8	74.0	75.5	78.8	87.8	97.1	92.8
Heavy Fuel Oil	37.6	22.0	22.9	24.0	17.0	14.1	11.0	25.5	21.6
Other	17.2	43.6	41.7	18.0	89.0	56.5	60.5	96.4	31.3
Refinery Fuel	59.1	53.3	68.8	55.1	54.6	52.3	22.8	40.7	69.1
Sub-Total	409.0	427.5	438.0	445.1	505.1	487.2	477.1	535.3	482.0
Bunkers	14.0	17.3	15.6	15.3	14.1	12.8	12.6	11.2	9.3
Total	423.0	444.7	453.7	460.4	519.2	500.0	489.7	546.6	491.4

[1] See Notes in Part I.4.

Oil Demand by Main Product Groups[1]
Demande de produits raffinés par groupes principaux

Vietnam

	1971	1973	1978	1984	1985	1986	1987	1988	1989
Thousand tonnes									
NGL/LPG/Ethane	5	14	3	5	-	-	-	-	-
Naphtha	-	-	-	-	-	-	-	-	-
Motor Gasoline	922	920	185	303	330	343	401	505	431
Aviation Fuels	2 247	2 000	-	86	78	115	120	134	111
Kerosene	50	50	80	167	153	150	182	185	215
Gas Diesel	1 664	1 460	295	776	843	874	1 043	1 029	974
Heavy Fuel Oil	747	1 200	325	462	470	556	613	584	518
Other	25	10	45	40	-	61	73	74	65
Refinery Fuel	-	-	-	-	-	-	-	2	2
Sub-Total	5 660	5 654	933	1 839	1 874	2 099	2 432	2 513	2 316
Bunkers	-	-	-	-	-	-	-	-	-
Total	5 660	5 654	933	1 839	1 874	2 099	2 432	2 513	2 316
Thousand barrels/day									
NGL/LPG/Ethane	0.2	0.4	0.1	0.2	-	-	-	-	-
Naphtha	-	-	-	-	-	-	-	-	-
Motor Gasoline	21.5	21.5	4.3	7.1	7.7	8.0	9.4	11.8	10.1
Aviation Fuels	50.8	44.8	-	1.9	1.7	2.5	2.6	2.9	2.4
Kerosene	1.1	1.1	1.7	3.5	3.2	3.2	3.9	3.9	4.6
Gas Diesel	34.0	29.8	6.0	15.8	17.2	17.9	21.3	21.0	19.9
Heavy Fuel Oil	13.6	21.9	5.9	8.4	8.6	10.1	11.2	10.6	9.5
Other	0.5	0.2	0.9	0.8	-	1.2	1.4	1.4	1.2
Refinery Fuel	-	-	-	-	-	-	-	0.0	0.0
Sub-Total	121.7	119.7	18.9	37.6	38.5	42.9	49.7	51.6	47.7
Bunkers	-	-	-	-	-	-	-	-	-
Total	121.7	119.7	18.9	37.6	38.5	42.9	49.7	51.6	47.7

	1990	1991	1992	1993	1994	1995	1996	1997	1998
Thousand tonnes									
NGL/LPG/Ethane	-	-	-	-	11	47	55	125	195
Naphtha	-	-	-	-	-	-	-	-	-
Motor Gasoline	653	521	630	874	972	1 184	1 319	1 469	1 174
Aviation Fuels	102	104	173	166	177	229	262	288	230
Kerosene	214	174	159	191	201	236	250	265	264
Gas Diesel	1 242	1 094	1 311	1 923	2 005	2 174	2 425	2 745	3 067
Heavy Fuel Oil	564	601	945	747	784	1 058	1 133	1 217	1 420
Other	29	28	-	-	160	189	194	113	118
Refinery Fuel	2	2	2	-	-	-	2	-	-
Sub-Total	2 806	2 524	3 220	3 901	4 310	5 117	5 640	6 222	6 468
Bunkers	-	-	-	-	-	-	-	-	-
Total	2 806	2 524	3 220	3 901	4 310	5 117	5 640	6 222	6 468
Thousand barrels/day									
NGL/LPG/Ethane	-	-	-	-	0.4	1.5	1.7	4.0	6.2
Naphtha	-	-	-	-	-	-	-	-	-
Motor Gasoline	15.3	12.2	14.7	20.4	22.7	27.7	30.7	34.3	27.4
Aviation Fuels	2.2	2.3	3.7	3.6	3.8	5.0	5.7	6.3	5.0
Kerosene	4.5	3.7	3.4	4.1	4.3	5.0	5.3	5.6	5.6
Gas Diesel	25.4	22.4	26.7	39.3	41.0	44.4	49.4	56.1	62.7
Heavy Fuel Oil	10.3	11.0	17.2	13.6	14.3	19.3	20.6	22.2	25.9
Other	0.6	0.5	-	-	3.1	3.6	3.7	2.2	2.3
Refinery Fuel	0.0	0.0	0.0	-	-	-	0.0	-	-
Sub-Total	58.3	52.0	65.8	81.0	89.5	106.5	117.2	130.7	135.1
Bunkers	-	-	-	-	-	-	-	-	-
Total	58.3	52.0	65.8	81.0	89.5	106.5	117.2	130.7	135.1

[1] See Notes in Part I.4.

Oil Demand by Main Product Groups[1]
Demande de produits raffinés par groupes principaux

Yemen

	1971	1973	1978	1984	1985	1986	1987	1988	1989
Thousand tonnes									
NGL/LPG/Ethane	2	6	8	61	57	68	74	93	120
Naphtha	-	-	-	-	-	-	-	-	-
Motor Gasoline	145	165	206	295	319	408	472	492	490
Aviation Fuels	27	32	71	128	146	151	161	171	71
Kerosene	53	48	95	152	202	140	167	179	179
Gas Diesel	72	238	337	695	724	773	782	838	900
Heavy Fuel Oil	54	54	165	336	492	476	417	424	430
Other	-	-	-	-	-	50	100	100	68
Refinery Fuel	58	84	68	119	124	127	112	112	110
Sub-Total	411	627	950	1 786	2 064	2 193	2 285	2 409	2 368
Bunkers	364	320	660	390	400	380	400	410	420
Total	775	947	1 610	2 176	2 464	2 573	2 685	2 819	2 788
Thousand barrels/day									
NGL/LPG/Ethane	0.1	0.2	0.3	1.9	1.8	2.2	2.4	2.9	3.8
Naphtha	-	-	-	-	-	-	-	-	-
Motor Gasoline	3.4	3.9	4.8	6.9	7.5	9.5	11.0	11.5	11.5
Aviation Fuels	0.6	0.7	1.5	2.8	3.2	3.3	3.5	3.7	1.5
Kerosene	1.1	1.0	2.0	3.2	4.3	3.0	3.5	3.8	3.8
Gas Diesel	1.5	4.9	6.9	14.2	14.8	15.8	16.0	17.1	18.4
Heavy Fuel Oil	1.0	1.0	3.0	6.1	9.0	8.7	7.6	7.7	7.8
Other	-	-	-	-	-	1.0	1.9	1.9	1.3
Refinery Fuel	1.1	1.5	1.2	2.2	2.3	2.3	2.0	2.0	2.0
Sub-Total	8.7	13.1	19.8	37.2	42.8	45.7	48.0	50.7	50.2
Bunkers	6.8	6.0	12.2	7.3	7.5	7.1	7.5	7.7	7.9
Total	15.5	19.1	32.0	44.6	50.3	52.8	55.5	58.3	58.1

	1990	1991	1992	1993	1994	1995	1996	1997	1998
Thousand tonnes									
NGL/LPG/Ethane	152	179	243	292	330	356	387	428	438
Naphtha	-	-	-	-	-	-	-	-	-
Motor Gasoline	804	857	924	814	857	1 073	1 045	1 032	1 032
Aviation Fuels	55	56	83	83	88	88	83	88	88
Kerosene	74	116	121	109	116	116	114	118	118
Gas Diesel	603	832	868	672	712	713	680	709	709
Heavy Fuel Oil	228	448	528	286	305	305	343	345	345
Other	68	43	48	58	58	58	60	59	59
Refinery Fuel	110	110	115	110	115	115	129	130	130
Sub-Total	2 094	2 641	2 930	2 424	2 581	2 824	2 841	2 909	2 919
Bunkers	400	320	375	110	100	100	100	100	100
Total	2 494	2 961	3 305	2 534	2 681	2 924	2 941	3 009	3 019
Thousand barrels/day									
NGL/LPG/Ethane	4.8	5.7	7.7	9.3	10.5	11.3	12.3	13.6	13.9
Naphtha	-	-	-	-	-	-	-	-	-
Motor Gasoline	18.8	20.0	21.5	19.0	20.0	25.1	24.4	24.1	24.1
Aviation Fuels	1.2	1.2	1.8	1.8	1.9	1.9	1.8	1.9	1.9
Kerosene	1.6	2.5	2.6	2.3	2.5	2.5	2.4	2.5	2.5
Gas Diesel	12.3	17.0	17.7	13.7	14.6	14.6	13.9	14.5	14.5
Heavy Fuel Oil	4.2	8.2	9.6	5.2	5.6	5.6	6.2	6.3	6.3
Other	1.3	0.7	0.8	1.0	1.0	1.0	1.0	1.0	1.0
Refinery Fuel	2.0	2.0	2.1	2.0	2.1	2.1	2.3	2.4	2.4
Sub-Total	46.2	57.3	63.8	54.3	58.1	64.0	64.3	66.3	66.6
Bunkers	7.5	6.0	7.0	2.1	1.9	1.9	1.9	1.9	1.9
Total	53.7	63.3	70.8	56.5	60.0	65.9	66.2	68.2	68.5

[1] See Notes in Part I.4.

Oil Demand by Main Product Groups[1]
Demande de produits raffinés par groupes principaux

Federal Republic of Yugoslavia

	1971	1973	1978	1984	1985	1986	1987	1988	1989
Thousand tonnes									
NGL/LPG/Ethane	-	-	-	-	-	-	-	-	-
Naphtha	-	-	-	-	-	-	-	-	-
Motor Gasoline	-	-	-	-	-	-	-	-	-
Aviation Fuels	-	-	-	-	-	-	-	-	-
Kerosene	-	-	-	-	-	-	-	-	-
Gas Diesel	-	-	-	-	-	-	-	-	-
Heavy Fuel Oil	-	-	-	-	-	-	-	-	-
Other	-	-	-	-	-	-	-	-	-
Refinery Fuel	-	-	-	-	-	-	-	-	-
Sub-Total	-	-	-	-	-	-	-	-	-
Bunkers	-	-	-	-	-	-	-	-	-
Total	-	-	-	-	-	-	-	-	-
Thousand barrels/day									
NGL/LPG/Ethane	-	-	-	-	-	-	-	-	-
Naphtha	-	-	-	-	-	-	-	-	-
Motor Gasoline	-	-	-	-	-	-	-	-	-
Aviation Fuels	-	-	-	-	-	-	-	-	-
Kerosene	-	-	-	-	-	-	-	-	-
Gas Diesel	-	-	-	-	-	-	-	-	-
Heavy Fuel Oil	-	-	-	-	-	-	-	-	-
Other	-	-	-	-	-	-	-	-	-
Refinery Fuel	-	-	-	-	-	-	-	-	-
Sub-Total	-	-	-	-	-	-	-	-	-
Bunkers	-	-	-	-	-	-	-	-	-
Total	-	-	-	-	-	-	-	-	-

	1990	1991	1992	1993	1994	1995	1996	1997	1998
Thousand tonnes									
NGL/LPG/Ethane	-	55	25	10	10	11	27	47	45
Naphtha	-	599	200	164	155	173	226	385	385
Motor Gasoline	-	462	300	160	154	166	540	790	770
Aviation Fuels	-	100	60	40	38	42	55	95	95
Kerosene	-	36	25	27	25	28	37	54	52
Gas Diesel	-	678	513	246	308	348	423	784	600
Heavy Fuel Oil	-	1 404	1 162	565	447	465	783	1 059	684
Other	-	309	249	213	201	224	319	292	288
Refinery Fuel	-	-	-	-	-	-	-	-	-
Sub-Total	-	3 643	2 534	1 425	1 338	1 457	2 410	3 506	2 919
Bunkers	-	-	-	-	-	-	-	-	-
Total	-	3 643	2 534	1 425	1 338	1 457	2 410	3 506	2 919
Thousand barrels/day									
NGL/LPG/Ethane	-	1.7	0.8	0.3	0.3	0.4	0.9	1.5	1.4
Naphtha	-	13.9	4.6	3.8	3.6	4.0	5.2	9.0	9.0
Motor Gasoline	-	10.8	7.0	3.7	3.6	3.9	12.6	18.5	18.0
Aviation Fuels	-	2.2	1.3	0.9	0.8	0.9	1.2	2.1	2.1
Kerosene	-	0.8	0.5	0.6	0.5	0.6	0.8	1.1	1.1
Gas Diesel	-	13.9	10.5	5.0	6.3	7.1	8.6	16.0	12.3
Heavy Fuel Oil	-	25.6	21.1	10.3	8.2	8.5	14.2	19.3	12.5
Other	-	5.8	4.7	4.0	3.8	4.3	6.0	5.3	5.3
Refinery Fuel	-	-	-	-	-	-	-	-	-
Sub-Total	-	74.7	50.5	28.7	27.2	29.6	49.5	72.8	61.6
Bunkers	-	-	-	-	-	-	-	-	-
Total	-	74.7	50.5	28.7	27.2	29.6	49.5	72.8	61.6

[1] See Notes in Part I.4.

Oil Demand by Main Product Groups[1]
Demande de produits raffinés par groupes principaux

Former Yugoslavia

	1971	1973	1978	1984	1985	1986	1987	1988	1989
Thousand tonnes									
NGL/LPG/Ethane	74	203	339	408	365	377	417	434	398
Naphtha	-	399	574	668	474	647	828	1 049	1 040
Motor Gasoline	1 343	1 474	2 254	1 652	1 700	2 190	1 954	2 276	2 330
Aviation Fuels	211	234	271	331	326	356	369	391	395
Kerosene	-	-	-	-	-	-	-	-	-
Gas Diesel	2 999	2 874	3 392	3 389	3 860	4 305	4 113	4 240	3 438
Heavy Fuel Oil	2 943	3 729	5 633	4 202	4 035	4 320	4 388	5 325	5 299
Other	643	643	1 170	957	1 284	1 109	1 548	1 456	1 560
Refinery Fuel	306	340	757	1 298	1 203	1 371	1 438	1 546	1 463
Sub-Total	8 519	9 896	14 390	12 905	13 247	14 675	15 055	16 717	15 923
Bunkers	-	-	-	-	-	-	-	-	-
Total	8 519	9 896	14 390	12 905	13 247	14 675	15 055	16 717	15 923
Thousand barrels/day									
NGL/LPG/Ethane	2.4	6.5	10.8	12.9	11.6	12.0	13.3	13.8	12.6
Naphtha	-	9.3	13.4	15.5	11.0	15.1	19.3	24.4	24.2
Motor Gasoline	31.4	34.4	52.7	38.5	39.7	51.2	45.7	53.0	54.5
Aviation Fuels	4.6	5.1	5.9	7.2	7.1	7.8	8.0	8.5	8.6
Kerosene	-	-	-	-	-	-	-	-	-
Gas Diesel	61.3	58.7	69.3	69.1	78.9	88.0	84.1	86.4	70.3
Heavy Fuel Oil	53.7	68.0	102.8	76.5	73.6	78.8	80.1	96.9	96.7
Other	12.3	12.3	22.4	18.3	24.6	21.3	29.7	27.8	29.9
Refinery Fuel	5.6	6.2	14.9	24.7	22.9	26.2	27.6	29.6	28.0
Sub-Total	171.3	200.6	292.2	262.7	269.5	300.3	307.6	340.4	324.8
Bunkers	-	-	-	-	-	-	-	-	-
Total	171.3	200.6	292.2	262.7	269.5	300.3	307.6	340.4	324.8

	1990	1991	1992	1993	1994	1995	1996	1997	1998
Thousand tonnes									
NGL/LPG/Ethane	414	504	267	257	257	264	264	350	340
Naphtha	790	780	338	1 270	615	661	751	833	846
Motor Gasoline	2 299	2 616	1 654	1 624	1 820	1 887	2 394	2 721	2 605
Aviation Fuels	376	254	119	168	201	215	215	265	257
Kerosene	-	-	27	37	26	36	44	60	55
Gas Diesel	3 319	2 686	2 673	2 663	2 566	2 966	3 471	3 937	3 976
Heavy Fuel Oil	5 450	3 775	2 860	2 226	2 004	2 137	2 626	2 773	2 533
Other	1 676	1 608	458	453	507	531	595	717	807
Refinery Fuel	1 394	1 060	448	484	499	574	499	495	682
Sub-Total	15 718	13 283	8 844	9 182	8 495	9 271	10 859	12 151	12 101
Bunkers	-	-	-	-	45	33	29	24	26
Total	15 718	13 283	8 844	9 182	8 540	9 304	10 888	12 175	12 127
Thousand barrels/day									
NGL/LPG/Ethane	13.2	15.1	9.2	9.4	9.5	9.8	9.6	12.5	12.0
Naphtha	18.4	18.2	7.9	29.6	14.3	15.4	17.4	19.4	19.7
Motor Gasoline	53.7	61.1	38.5	38.0	42.5	44.1	55.8	63.6	60.9
Aviation Fuels	8.2	5.5	2.6	3.7	4.4	4.7	4.7	5.8	5.6
Kerosene	-	-	0.6	0.8	0.6	0.8	0.9	1.3	1.2
Gas Diesel	67.8	54.9	54.5	54.4	52.4	60.6	70.7	80.5	81.3
Heavy Fuel Oil	99.4	68.9	52.0	40.6	36.6	39.0	47.8	50.6	46.2
Other	32.1	30.2	8.5	8.6	9.6	10.1	11.2	13.2	14.7
Refinery Fuel	26.8	20.1	8.6	9.5	10.1	11.6	10.1	9.8	13.5
Sub-Total	319.7	274.0	182.3	194.4	180.0	196.1	228.3	256.6	255.0
Bunkers	-	-	-	-	0.9	0.6	0.6	0.5	0.5
Total	319.7	274.0	182.3	194.4	180.8	196.7	228.8	257.0	255.5

[1] See Notes in Part I.4.

Oil Demand by Main Product Groups[1]
Demande de produits raffinés par groupes principaux

Zambia

	1971	1973	1978	1984	1985	1986	1987	1988	1989
Thousand tonnes									
NGL/LPG/Ethane	1	3	3	4	4	3	3	4	4
Naphtha	5	5	-	-	-	-	-	-	-
Motor Gasoline	146	171	144	117	117	105	107	118	110
Aviation Fuels	14	21	55	52	52	46	51	54	53
Kerosene	4	8	27	34	35	25	31	38	38
Gas Diesel	271	320	269	225	191	188	185	185	192
Heavy Fuel Oil	1	85	167	102	95	90	90	77	75
Other	40	39	37	52	61	35	39	40	36
Refinery Fuel	-	21	41	35	35	32	35	35	35
Sub-Total	482	673	743	621	590	524	541	551	543
Bunkers	-	-	-	-	-	-	-	-	-
Total	482	673	743	621	590	524	541	551	543
Thousand barrels/day									
NGL/LPG/Ethane	0.0	0.1	0.1	0.1	0.1	0.1	0.1	0.1	0.1
Naphtha	0.1	0.1	-	-	-	-	-	-	-
Motor Gasoline	3.4	4.0	3.4	2.7	2.7	2.5	2.5	2.8	2.6
Aviation Fuels	0.3	0.5	1.2	1.1	1.1	1.0	1.1	1.2	1.2
Kerosene	0.1	0.2	0.6	0.7	0.7	0.5	0.7	0.8	0.8
Gas Diesel	5.5	6.5	5.5	4.6	3.9	3.8	3.8	3.8	3.9
Heavy Fuel Oil	0.0	1.6	3.0	1.9	1.7	1.6	1.6	1.4	1.4
Other	0.8	0.7	0.7	1.0	1.1	0.6	0.7	0.7	0.7
Refinery Fuel	-	0.4	0.8	0.7	0.7	0.6	0.7	0.7	0.7
Sub-Total	10.3	14.1	15.3	12.8	12.2	10.9	11.2	11.5	11.3
Bunkers	-	-	-	-	-	-	-	-	-
Total	10.3	14.1	15.3	12.8	12.2	10.9	11.2	11.5	11.3

	1990	1991	1992	1993	1994	1995	1996	1997	1998
Thousand tonnes									
NGL/LPG/Ethane	10	10	9	9	10	10	10	10	10
Naphtha	-	-	-	-	-	-	-	-	-
Motor Gasoline	106	109	99	101	105	107	110	110	110
Aviation Fuels	40	41	41	40	40	38	38	38	38
Kerosene	29	30	27	32	30	32	31	31	31
Gas Diesel	185	190	173	181	195	195	197	197	197
Heavy Fuel Oil	70	72	72	73	70	72	74	74	74
Other	33	33	31	31	32	32	32	32	32
Refinery Fuel	34	34	34	34	35	36	36	36	36
Sub-Total	507	519	486	501	517	522	528	528	528
Bunkers	-	-	-	-	-	-	-	-	-
Total	507	519	486	501	517	522	528	528	528
Thousand barrels/day									
NGL/LPG/Ethane	0.3	0.3	0.3	0.3	0.3	0.3	0.3	0.3	0.3
Naphtha	-	-	-	-	-	-	-	-	-
Motor Gasoline	2.5	2.5	2.3	2.4	2.5	2.5	2.6	2.6	2.6
Aviation Fuels	0.9	0.9	0.9	0.9	0.9	0.8	0.8	0.8	0.8
Kerosene	0.6	0.6	0.6	0.7	0.6	0.7	0.7	0.7	0.7
Gas Diesel	3.8	3.9	3.5	3.7	4.0	4.0	4.0	4.0	4.0
Heavy Fuel Oil	1.3	1.3	1.3	1.3	1.3	1.3	1.3	1.4	1.4
Other	0.6	0.6	0.6	0.6	0.6	0.6	0.6	0.6	0.6
Refinery Fuel	0.7	0.7	0.7	0.7	0.7	0.7	0.7	0.7	0.7
Sub-Total	10.6	10.9	10.1	10.5	10.8	10.9	11.0	11.1	11.1
Bunkers	-	-	-	-	-	-	-	-	-
Total	10.6	10.9	10.1	10.5	10.8	10.9	11.0	11.1	11.1

[1] See Notes in Part I.4.

Oil Demand by Main Product Groups[1]
Demande de produits raffinés par groupes principaux

Zimbabwe

	1971	1973	1978	1984	1985	1986	1987	1988	1989
Thousand tonnes									
NGL/LPG/Ethane	4	5	6	3	5	5	4	6	4
Naphtha	-	-	-	-	-	-	-	-	-
Motor Gasoline	184	234	189	171	185	198	164	188	191
Aviation Fuels	27	44	79	91	105	123	100	116	104
Kerosene	42	46	30	30	35	51	35	36	43
Gas Diesel	240	301	315	395	411	402	487	476	426
Heavy Fuel Oil	-	-	-	1	1	-	-	-	-
Other	42	50	39	16	16	18	17	17	17
Refinery Fuel	-	-	-	-	-	-	-	-	-
Sub-Total	539	680	658	707	758	797	807	839	785
Bunkers	-	-	-	-	-	-	-	-	-
Total	539	680	658	707	758	797	807	839	785
Thousand barrels/day									
NGL/LPG/Ethane	0.1	0.2	0.2	0.1	0.2	0.2	0.1	0.2	0.1
Naphtha	-	-	-	-	-	-	-	-	-
Motor Gasoline	4.3	5.5	4.4	4.0	4.3	4.6	3.8	4.4	4.5
Aviation Fuels	0.6	1.0	1.7	2.0	2.3	2.7	2.2	2.5	2.3
Kerosene	0.9	1.0	0.6	0.6	0.7	1.1	0.7	0.8	0.9
Gas Diesel	4.9	6.2	6.4	8.1	8.4	8.2	10.0	9.7	8.7
Heavy Fuel Oil	-	-	-	0.0	0.0	-	-	-	-
Other	0.8	1.0	0.7	0.3	0.3	0.3	0.3	0.3	0.3
Refinery Fuel	-	-	-	-	-	-	-	-	-
Sub-Total	11.6	14.7	14.2	15.1	16.2	17.1	17.2	17.9	16.8
Bunkers	-	-	-	-	-	-	-	-	-
Total	11.6	14.7	14.2	15.1	16.2	17.1	17.2	17.9	16.8

	1990	1991	1992	1993	1994	1995	1996	1997	1998
Thousand tonnes									
NGL/LPG/Ethane	6	6	6	6	7	12	12	12	12
Naphtha	-	-	-	-	-	-	-	-	-
Motor Gasoline	238	206	261	285	290	429	435	435	435
Aviation Fuels	79	139	79	72	98	110	118	118	118
Kerosene	51	43	38	51	54	54	42	42	42
Gas Diesel	516	592	584	598	731	645	634	634	634
Heavy Fuel Oil	-	-	-	-	-	-	-	-	-
Other	27	27	25	25	26	27	27	27	27
Refinery Fuel	-	-	-	-	-	-	-	-	-
Sub-Total	917	1 013	993	1 037	1 206	1 277	1 268	1 268	1 268
Bunkers	-	-	-	-	-	-	-	-	-
Total	917	1 013	993	1 037	1 206	1 277	1 268	1 268	1 268
Thousand barrels/day									
NGL/LPG/Ethane	0.2	0.2	0.2	0.2	0.2	0.4	0.4	0.4	0.4
Naphtha	-	-	-	-	-	-	-	-	-
Motor Gasoline	5.6	4.8	6.1	6.7	6.8	10.0	10.1	10.2	10.2
Aviation Fuels	1.7	3.0	1.7	1.6	2.1	2.4	2.6	2.6	2.6
Kerosene	1.1	0.9	0.8	1.1	1.1	1.1	0.9	0.9	0.9
Gas Diesel	10.5	12.1	11.9	12.2	14.9	13.2	12.9	13.0	13.0
Heavy Fuel Oil	-	-	-	-	-	-	-	-	-
Other	0.5	0.5	0.5	0.5	0.5	0.5	0.5	0.5	0.5
Refinery Fuel	-	-	-	-	-	-	-	-	-
Sub-Total	19.6	21.6	21.2	22.2	25.7	27.7	27.4	27.5	27.5
Bunkers	-	-	-	-	-	-	-	-	-
Total	19.6	21.6	21.2	22.2	25.7	27.7	27.4	27.5	27.5

[1] See Notes in Part I.4.

Oil Demand by Main Product Groups[1]
Demande de produits raffinés par groupes principaux

Other Africa

	1971	1973	1978	1984	1985	1986	1987	1988	1989
Thousand tonnes									
NGL/LPG/Ethane	10	12	16	39	38	39	76	87	91
Naphtha	-	-	2	4	5	5	6	7	7
Motor Gasoline	407	471	697	825	843	856	1 215	1 247	1 197
Aviation Fuels	182	255	330	425	394	416	481	493	649
Kerosene	166	166	175	264	292	290	361	377	375
Gas Diesel	782	844	982	1 373	1 463	1 459	1 733	1 682	1 649
Heavy Fuel Oil	638	672	734	703	709	755	1 097	1 129	1 035
Other	114	156	195	331	302	316	445	459	467
Refinery Fuel	33	34	35	-	-	-	-	-	-
Sub-Total	2 332	2 610	3 166	3 964	4 046	4 136	5 414	5 481	5 470
Bunkers	968	799	592	597	584	630	625	670	709
Total	3 300	3 409	3 758	4 561	4 630	4 766	6 039	6 151	6 179
Thousand barrels/day									
NGL/LPG/Ethane	0.3	0.4	0.5	1.2	1.2	1.2	2.4	2.8	2.9
Naphtha	-	-	0.0	0.1	0.1	0.1	0.1	0.2	0.2
Motor Gasoline	9.5	11.0	16.3	19.2	19.7	20.0	28.4	29.1	28.0
Aviation Fuels	4.0	5.6	7.3	9.4	8.7	9.2	10.6	10.8	14.5
Kerosene	3.5	3.5	3.7	5.6	6.2	6.2	7.7	8.0	8.0
Gas Diesel	16.0	17.3	20.1	28.0	29.9	29.8	35.4	34.3	33.7
Heavy Fuel Oil	11.6	12.3	13.4	12.8	12.9	13.8	20.0	20.5	18.9
Other	2.2	3.0	3.7	6.3	5.7	6.0	8.4	8.7	8.9
Refinery Fuel	0.6	0.6	0.6	-	-	-	-	-	-
Sub-Total	47.8	53.6	65.6	82.5	84.5	86.2	113.1	114.3	114.9
Bunkers	18.7	15.6	11.4	11.4	11.1	12.1	11.9	12.8	13.6
Total	66.5	69.2	77.0	93.9	95.6	98.3	125.0	127.1	128.5

	1990	1991	1992	1993	1994	1995	1996	1997	1998
Thousand tonnes									
NGL/LPG/Ethane	98	105	115	115	119	124	127	133	133
Naphtha	7	8	8	8	8	8	8	8	8
Motor Gasoline	1 121	1 157	1 179	1 220	1 236	1 256	1 272	1 339	1 339
Aviation Fuels	671	601	644	696	635	681	667	677	677
Kerosene	373	402	412	441	441	465	510	536	536
Gas Diesel	1 494	1 541	1 609	1 660	1 639	1 629	1 673	1 764	1 764
Heavy Fuel Oil	979	1 027	1 062	1 089	1 143	1 172	1 175	1 175	1 175
Other	424	449	462	475	490	485	490	509	509
Refinery Fuel	-	-	-	-	-	-	-	-	-
Sub-Total	5 167	5 290	5 491	5 704	5 711	5 820	5 922	6 141	6 141
Bunkers	637	621	607	601	684	708	685	705	705
Total	5 804	5 911	6 098	6 305	6 395	6 528	6 607	6 846	6 846
Thousand barrels/day									
NGL/LPG/Ethane	3.1	3.3	3.6	3.7	3.8	3.9	4.0	4.2	4.2
Naphtha	0.2	0.2	0.2	0.2	0.2	0.2	0.2	0.2	0.2
Motor Gasoline	26.2	27.0	27.5	28.5	28.9	29.4	29.6	31.3	31.3
Aviation Fuels	15.1	13.5	14.4	15.6	14.3	15.3	14.9	15.2	15.2
Kerosene	7.9	8.5	8.7	9.4	9.4	9.9	10.8	11.4	11.4
Gas Diesel	30.5	31.5	32.8	33.9	33.5	33.3	34.1	36.1	36.1
Heavy Fuel Oil	17.9	18.7	19.3	19.9	20.9	21.4	21.4	21.4	21.4
Other	8.0	8.5	8.7	9.0	9.2	9.2	9.2	9.6	9.6
Refinery Fuel	-	-	-	-	-	-	-	-	-
Sub-Total	108.9	111.3	115.2	120.1	120.1	122.5	124.3	129.4	129.4
Bunkers	12.2	11.9	11.6	11.5	13.1	13.7	13.1	13.5	13.5
Total	121.1	123.2	126.8	131.5	133.2	136.1	137.4	142.9	142.9

[1] See Notes in Part I.4.

Oil Demand by Main Product Groups[1]
Demande de produits raffinés par groupes principaux

Other Asia

	1971	1973	1978	1984	1985	1986	1987	1988	1989
Thousand tonnes									
NGL/LPG/Ethane	11	14	15	26	28	26	28	25	28
Naphtha	-	-	-	1	1	1	-	-	-
Motor Gasoline	276	257	358	333	337	287	362	344	355
Aviation Fuels	209	205	168	380	381	355	387	367	384
Kerosene	60	65	81	46	55	68	86	84	96
Gas Diesel	436	464	566	684	856	817	997	961	974
Heavy Fuel Oil	428	846	581	474	510	490	531	517	536
Other	17	31	51	38	37	38	35	39	35
Refinery Fuel	-	-	-	-	-	-	-	-	-
Sub-Total	1 437	1 882	1 820	1 982	2 205	2 082	2 426	2 337	2 408
Bunkers	181	189	148	90	63	83	76	62	66
Total	1 618	2 071	1 968	2 072	2 268	2 165	2 502	2 399	2 474
Thousand barrels/day									
NGL/LPG/Ethane	0.4	0.4	0.5	0.8	0.9	0.8	0.9	0.8	0.9
Naphtha	-	-	-	0.0	0.0	0.0	-	-	-
Motor Gasoline	6.5	6.0	8.4	7.8	7.9	6.7	8.5	8.0	8.3
Aviation Fuels	4.7	4.6	3.7	8.3	8.3	7.8	8.4	8.0	8.4
Kerosene	1.3	1.4	1.7	1.0	1.2	1.4	1.8	1.8	2.0
Gas Diesel	8.9	9.5	11.6	13.9	17.5	16.7	20.4	19.6	19.9
Heavy Fuel Oil	7.8	15.4	10.6	8.6	9.3	8.9	9.7	9.4	9.8
Other	0.3	0.6	1.0	0.7	0.7	0.7	0.7	0.7	0.7
Refinery Fuel	-	-	-	-	-	-	-	-	-
Sub-Total	29.8	37.9	37.4	41.1	45.8	43.1	50.3	48.3	49.9
Bunkers	3.5	3.6	2.9	1.8	1.3	1.7	1.5	1.2	1.3
Total	33.3	41.5	40.3	42.9	47.0	44.8	51.9	49.6	51.3

	1990	1991	1992	1993	1994	1995	1996	1997	1998
Thousand tonnes									
NGL/LPG/Ethane	32	33	30	30	31	33	34	36	36
Naphtha	-	-	-	-	-	-	-	-	-
Motor Gasoline	376	367	345	342	341	339	339	339	339
Aviation Fuels	389	358	316	308	229	234	228	228	228
Kerosene	124	69	68	66	62	64	66	71	71
Gas Diesel	1 028	1 032	868	853	838	858	864	902	902
Heavy Fuel Oil	552	584	582	585	585	584	588	614	614
Other	38	33	25	32	30	30	29	30	30
Refinery Fuel	-	-	-	-	-	-	-	-	-
Sub-Total	2 539	2 476	2 234	2 216	2 116	2 142	2 148	2 220	2 220
Bunkers	66	72	72	72	75	78	78	82	82
Total	2 605	2 548	2 306	2 288	2 191	2 220	2 226	2 302	2 302
Thousand barrels/day									
NGL/LPG/Ethane	1.0	1.0	1.0	1.0	1.0	1.0	1.1	1.1	1.1
Naphtha	-	-	-	-	-	-	-	-	-
Motor Gasoline	8.8	8.6	8.0	8.0	8.0	7.9	7.9	7.9	7.9
Aviation Fuels	8.5	7.8	6.9	6.7	5.0	5.1	5.0	5.0	5.0
Kerosene	2.6	1.5	1.4	1.4	1.3	1.4	1.4	1.5	1.5
Gas Diesel	21.0	21.1	17.7	17.4	17.1	17.5	17.6	18.4	18.4
Heavy Fuel Oil	10.1	10.7	10.6	10.7	10.7	10.7	10.7	11.2	11.2
Other	0.7	0.6	0.5	0.6	0.6	0.6	0.5	0.6	0.6
Refinery Fuel	-	-	-	-	-	-	-	-	-
Sub-Total	52.7	51.3	46.0	45.8	43.6	44.2	44.2	45.8	45.8
Bunkers	1.3	1.4	1.4	1.4	1.5	1.6	1.6	1.6	1.6
Total	54.0	52.7	47.5	47.2	45.1	45.8	45.7	47.4	47.4

[1] See Notes in Part I.4.

Oil Demand by Main Product Groups[1]
Demande de produits raffinés par groupes principaux

Other Latin America

	1971	1973	1978	1984	1985	1986	1987	1988	1989
Thousand tonnes									
NGL/LPG/Ethane	48	48	68	90	97	101	76	78	78
Naphtha	-	-	-	-	-	-	-	-	-
Motor Gasoline	347	511	477	466	445	471	541	567	622
Aviation Fuels	379	225	412	341	336	329	431	413	453
Kerosene	180	268	242	166	173	151	188	203	262
Gas Diesel	866	1 245	1 063	953	844	1 043	1 153	1 096	1 236
Heavy Fuel Oil	44	48	522	192	200	257	504	512	495
Other	8	27	73	166	173	181	192	181	161
Refinery Fuel	-	-	-	-	-	-	-	-	3
Sub-Total	1 872	2 372	2 857	2 374	2 268	2 533	3 085	3 050	3 310
Bunkers	994	1 111	872	464	578	401	322	245	208
Total	2 866	3 483	3 729	2 838	2 846	2 934	3 407	3 295	3 518
Thousand barrels/day									
NGL/LPG/Ethane	1.5	1.5	2.2	2.9	3.1	3.2	2.4	2.5	2.5
Naphtha	-	-	-	-	-	-	-	-	-
Motor Gasoline	8.1	11.9	11.1	10.9	10.4	11.0	12.6	13.2	14.5
Aviation Fuels	8.4	5.0	9.0	7.5	7.4	7.2	9.5	9.0	10.0
Kerosene	3.8	5.7	5.1	3.5	3.7	3.2	4.0	4.3	5.6
Gas Diesel	17.7	25.4	21.7	19.4	17.3	21.3	23.6	22.3	25.3
Heavy Fuel Oil	0.8	0.9	9.5	3.5	3.6	4.7	9.2	9.3	9.0
Other	0.1	0.5	0.8	1.4	1.5	1.6	1.8	1.6	1.2
Refinery Fuel	-	-	-	-	-	-	-	-	0.1
Sub-Total	40.5	51.0	59.5	49.0	46.9	52.3	63.1	62.3	68.1
Bunkers	18.6	20.6	16.2	8.6	10.6	7.4	5.9	4.5	3.9
Total	59.1	71.5	75.7	57.6	57.6	59.7	69.0	66.8	72.0

	1990	1991	1992	1993	1994	1995	1996	1997	1998
Thousand tonnes									
NGL/LPG/Ethane	105	124	129	129	138	148	150	156	156
Naphtha	-	-	-	-	-	-	-	-	-
Motor Gasoline	646	629	631	636	651	680	696	716	716
Aviation Fuels	432	431	399	373	427	435	450	465	465
Kerosene	246	257	215	220	232	235	240	250	250
Gas Diesel	1 248	1 145	1 039	1 081	1 068	1 114	1 137	1 177	1 177
Heavy Fuel Oil	487	508	499	509	515	432	444	463	463
Other	157	165	157	168	171	170	171	173	173
Refinery Fuel	3	3	3	3	3	1	3	3	3
Sub-Total	3 324	3 262	3 072	3 119	3 205	3 215	3 291	3 403	3 403
Bunkers	223	225	221	225	225	224	230	232	232
Total	3 547	3 487	3 293	3 344	3 430	3 439	3 521	3 635	3 635
Thousand barrels/day									
NGL/LPG/Ethane	3.3	3.9	4.1	4.1	4.4	4.7	4.8	5.0	5.0
Naphtha	-	-	-	-	-	-	-	-	-
Motor Gasoline	15.1	14.7	14.7	14.9	15.2	15.9	16.2	16.7	16.7
Aviation Fuels	9.5	9.5	8.8	8.3	9.5	9.6	9.9	10.3	10.3
Kerosene	5.2	5.5	4.5	4.7	4.9	5.0	5.1	5.3	5.3
Gas Diesel	25.5	23.4	21.2	22.1	21.8	22.8	23.2	24.1	24.1
Heavy Fuel Oil	8.9	9.3	9.1	9.3	9.4	7.9	8.1	8.4	8.4
Other	1.1	1.3	1.1	1.3	1.4	1.3	1.4	1.4	1.4
Refinery Fuel	0.1	0.1	0.1	0.1	0.1	0.0	0.1	0.1	0.1
Sub-Total	68.8	67.6	63.6	64.6	66.6	67.2	68.6	71.2	71.2
Bunkers	4.2	4.2	4.1	4.2	4.2	4.2	4.3	4.4	4.4
Total	73.0	71.8	67.7	68.9	70.8	71.4	73.0	75.6	75.6

[1] See Notes in Part I.4.

INTERNATIONAL ENERGY AGENCY
ENERGY STATISTICS DIVISION
POSSIBLE STAFF VACANCIES

The Division is responsible for statistical support and advice to the policy and operational Divisions of the International Energy Agency. It also produces a wide range of annual and quarterly publications complemented by a data service on microcomputer diskettes. For these purposes, the Division maintains, on a central computer and an expanding network of microcomputers, extensive international databases covering most aspects of energy supply and use.

- Vacancies for statistical assistants occur from time to time. Typically their work includes:

- Gathering and vetting data from questionnaires and publications, discussions on data issues with respondents to questionnaires in national administrations and fuel companies.

- Managing energy databases on a mainframe computer and microcomputers in order to maintain accuracy and timeliness of output.

- Preparing computer procedures for the production of tables, reports and analyses.

Seasonal adjustment of data and analysis of trends and market movements.

- Preparing studies on an ad-hoc basis as required by other Divisions of the International Energy Agency.

Nationals of any OECD Member Country are eligible for appointment. Basic salaries range from 15 600 to 20 700 French francs per month, depending on qualifications. The possibilities for advancement are good for candidates with appropriate qualifications and experience. Tentative enquiries about future vacancies are welcomed from men and women with relevant qualifications and experience. Applications in French or English, specifying the reference "ENERSTAT" and enclosing a curriculum vitae, should be sent to:

Beth Hunter
Head of Administrative Unit
IEA, 9 rue de la Fédération
75739 Paris CEDEX 15, France
beth.hunter@iea.org

AGENCE INTERNATIONALE DE L'ENERGIE
DIVISION DES STATISTIQUES DE L'ENERGIE
VACANCES D'EMPLOI EVENTUELLES

La Division est chargée de fournir une aide et des conseils dans le domaine statistique aux Divisions administratives et opérationnelles de l'Agence internationale de l'énergie. En outre, elle diffuse une large gamme de publications annuelles et trimestrielles complétées par un service de données sur disquettes pour micro-ordinateur. A cet effet, la Division tient à jour, sur un ordinateur central et un réseau de plus en plus étendu de micro-ordinateurs, de vastes bases de données internationales portant sur la plupart des aspects de l'offre et de la consommation d'énergie.

Des postes d'assistant statisticien sont susceptibles de se libérer de temps à autre. Les fonctions dévolues aux titulaires de ces postes sont notamment les suivantes :

- Rassembler et valider les données tirées de questionnaires et de publications, ainsi que d'échanges de vues sur les données avec les personnes des Administrations nationales ou des entreprises du secteur de l'énergie qui répondent aux questionnaires.

- Gérer des bases de données relatives à l'énergie sur un ordinateur central et des micro-ordinateurs en vue de s'assurer de l'exactitude et de l'actualisation des données de sortie.

- Mettre au point des procédures informatiques pour la réalisation de tableaux, rapports et analyses. Procéder à l'ajustement saisonnier des données et analyses relatives aux tendances et aux fluctuations du marché.

- Effectuer des études en fonction des besoins des autres Divisions de l'Agence internationale de l'énergie.

Ces postes sont ouverts aux ressortissants des pays Membres de l'OCDE. Les traitements de base sont compris entre 15 600 et 20 700 francs français par mois, suivant les qualifications. Les candidats possédant les qualifications et l'expérience appropriées se verront offrir des perspectives de promotion. Les demandes de renseignements sur les postes susceptibles de se libérer qui émanent de personnes dotées des qualifications et de l'expérience voulues seront les bienvenues. Les candidatures, rédigées en français ou en anglais et accompagnées d'un curriculum vitae, doivent être envoyées, sous la référence "ENERSTAT", à l'adresse suivante :

Beth Hunter
Head of Administrative Unit
IEA, 9 rue de la Fédération
75739 Paris CEDEX 15, France
beth.hunter@iea.org

MULTILINGUAL PULLOUT

Italian
Japanese

MULTILINGUAL PULLOUT

Spanish
Russian

MULTILINGUAL PULLOUT

Chinese

NINE ANNUAL PUBLICATIONS

Coal Information 2000

Issued annually since 1983, this publication provides comprehensive information on current world coal market trends and long-term prospects. Compiled in cooperation with the Coal Industry Advisory Board, it contains thorough analysis and current country-specific statistics for OECD Member countries and selected non-OECD countries on coal prices, demand, trade, production, productive capacity, emissions standards for coal-fired boilers, coal ports, coal-fired power stations and coal data for non-OECD countries. This publication is a key reference tool for all sectors of the coal industry as well as for OECD Member country governments. *Published August 2000.*

Electricity Information 2000

This publication brings together in one volume the IEA's data on electricity and heat supply and demand in the OECD. The report presents a comprehensive picture of electricity capacity and production, consumption, trade and prices for the OECD regions and individual countries in over 20 separate tables for each OECD country. Detailed data on the fuels used for electricity and heat production are also presented. *Published September 2000.*

Natural Gas Information 2000

A detailed reference work on gas supply and demand, covering not only the OECD countries but also the rest of the world. Contains essential information on LNG and pipeline trade, gas reserves, storage capacity and prices. The main part of the book, however, concentrates on OECD countries, showing a detailed gas supply and demand balance for each individual country and for the three OECD regions: North America, Europe and Asia-Pacific, as well as a breakdown of gas consumption by end-user. Import and export data are reported by source and destination. *Published September 2000.*

Oil Information 2000

A comprehensive reference book on current developments in oil supply and demand. The first part of this publication contains key data on world production, trade, prices and consumption of major oil product groups, with time series back to the early 1970s. The second part gives a more detailed and comprehensive picture of oil supply, demand, trade, production and consumption by end-user for each OECD country individually and for the OECD regions. Trade data are reported extensively by origin and destination. *Published August 2000.*

Energy Statistics of OECD Countries 1997-1998

No other publication offers such in-depth statistical coverage. It is intended for anyone involved in analytical or policy work related to energy issues. It contains data on energy supply and consumption in original units for coal, oil, natural gas, combustible renewables/wastes and products derived from these primary fuels, as well as for electricity and heat. Data are presented for the two most recent years available in detailed supply and consumption tables. Historical tables summarise data on production, trade and final consumption. Each issue includes definitions of products and flows and explanatory notes on the individual country data. *Published June 2000.*

Energy Balances of OECD Countries 1997-1998

A companion volume to *Energy Statistics of OECD Countries*, this publication presents standardised energy balances expressed in million tonnes of oil equivalent. Energy supply and consumption data are divided by main fuel: coal, oil, gas, nuclear, hydro, geothermal/solar, combustible renewables/wastes, electricity and heat. This allows for easy comparison of the contributions each fuel makes to the economy and their interrelationships through the conversion of one fuel to another. All of this is essential for estimating total energy supply, forecasting, energy conservation, and analysing the potential for interfuel substitution. Complete energy balances are presented for the two most recent years available. Historical tables summarise key energy and economic indicators as well as data on production, trade and final consumption. Each issue includes definitions of products and flows and explanatory notes on the individual country data as well as conversion factors from original units to tonnes of oil equivalent. *Published June 2000.*

Energy Statistics of Non-OECD Countries 1997-1998

This new publication offers the same in-depth statistical coverage as the homonymous publication covering OECD countries. It includes data in original units for 103 individual countries and nine main regions. The consistency of OECD and non-OECD countries' detailed statistics provides an accurate picture of the global energy situation. For a description of the content, please see *Energy Statistics of OECD Countries* above. *Published September 2000.*

Energy Balances of Non-OECD Countries 1997-1998

A companion volume to the new publication *Energy Statistics of Non-OECD Countries*, this publication presents energy balances in million tonnes of oil equivalent and key economic and energy indicators for 103 individual countries and nine main regions. It offers the same statistical coverage as the homonymous publication covering OECD Countries, and thus provides an accurate picture of the global energy situation. For a description of the content, please see *Energy Balances of OECD Countries* above. *Published September 2000.*

CO_2 Emissions from Fuel Combustion - 2000 Edition

In order for nations to tackle the problem of climate change, they need accurate greenhouse gas emissions data. This publication provides a new basis for comparative analysis of CO_2 emissions from fossil fuel combustion, a major source of anthropogenic emissions. The data in this book are designed to assist in understanding the evolution of these emissions from 1971 to 1998 on a country, regional and worldwide basis. They should help in the preparation and the follow-up to the Sixth Conference of the Parties (COP-6) meeting under the U.N. Climate Convention in the Hague in November 2000. Emissions were calculated using IEA energy databases and the default methods and emissions factors from the *Revised 1996 IPCC Guidelines for National Greenhouse Gas Inventories*. *Published October 2000.*

TWO QUARTERLIES

Oil, Gas, Coal and Electricity, Quarterly Statistics

Oil statistics cover OECD production, trade (by origin and destination), refinery intake and output, stock changes and consumption for crude oil, NGL and nine selected oil product groups. Statistics for natural gas show OECD supply, consumption and trade (by origin and destination). Coal data cover the main OECD and world-wide producers of hard and brown coal and major exporters and importers of steam and coking coal. Trade data for the main OECD countries are reported by origin and destination. Electricity statistics cover production (by major fuel category), consumption and trade for 29 OECD countries. Quarterly data on world oil and coal production are included, as well as world steam and coking coal trade.

Energy Prices and Taxes

This publication responds to the needs of the energy industry and OECD governments for up-to-date information on prices and taxes in national and international energy markets. It contains for OECD countries and certain non-OECD countries prices at all market levels: import prices, industry prices and consumer prices. The statistics cover the main petroleum products, gas, coal and electricity, giving for imported products an average price both for importing country and country of origin. Every issue includes full notes on sources and methods and a description of price mechanisms in each country.

For more information on the IEA statistics publications, please feel free to contact Ms. Sharon Michel in the Energy Statistics Division, Tel: (33 1) 40 57 66 25; Fax: (33 1) 40 57 66 49.

To complement its publications, the Energy Statistics Division produces diskettes containing the complete databases which are used for preparing the statistics publications. State-of-the-art software allows you to access and manipulate all these data in a very user-friendly manner and includes graphic and mapping facilities.

The diskette service includes:

Annual Diskettes

- Energy Statistics of OECD Countries, 1960-1998
- Energy Balances of OECD Countries, 1960-1998
- Energy Statistics of non-OECD Countries, 1971-1998
- Energy Balances of non-OECD Countries, 1971-1998
- CO_2 Emissions from Fuel Combustion

- Natural Gas Information
- Oil Information
- Coal Information
- Electricity Information

Quarterly Diskettes

- Energy Prices and Taxes

Monthly Diskettes

- The IEA Monthly Oil Data Service (see box below)

The IEA Monthly Oil Data Service

Diskettes

The IEA Monthly Oil Data Service provides the detailed database of historical and projected information which is used in preparing the IEA's monthly Oil Market Report (OMR) and includes all the information previously contained in the Monthly Oil Statistics (MOS) diskettes. The IEA Monthly Oil Data Service comprises three packages:

- Supply, Demand, Balances and Stocks;
- Trade;
- Field-by-Field Supply;

available separately or combined, either as a subscriber service on the Internet or on diskettes.

Internet

The Internet version is available two days after the official release of the Oil Market Report. Diskettes are mailed four days afterwards.

Annual Oil and Gas Statistics, Energy Statistics and Balances of OECD and Non-OECD Countries are also available on **CD Rom** in the OECD Statistical Compendium (for additional information on the CD-ROM contact the OECD Publications Service, Tel: (33 1) 45 24 82 00; Fax: (33 1) 49 10 42 76. Moreover, the IEA site on **Internet** contains key energy indicators by country, graphs on the world and OECD's energy situation evolution from 1971 to the most recent year available, as well as selected databases for demonstration. The IEA site can be accessed at: **http://www.iea.org**. For more information, please feel free to contact the Energy Statistics Division of the IEA by Fax: (33 1) 40 57 66 49 or Phone: (33 1) 40 57 66 25.

Order Form

Publications

OECD BONN OFFICE

c/o DVG mbh (OECD)
Birkenmaarstrasse 8
D-53340 Meckenheim, Germany
Tel: (+49-2225) 926 166/168
Fax: (+49-2225) 926 169
E-mail: oecd@dvg.dsb.net
Internet: www.oecd.org/bonn

OECD MEXICO CENTRE

Edificio INFOTEC
Av. Presidente Mazarik 526
Colonia: Polanco
C.P. 11560 - Mexico D.F.
Tel: (+52-5) 281 38 10
Fax: (+52-5) 280 04 80
E-mail: mexico.contact@oecd.org
Internet: www.rtn.net.mx/ocde

OECD CENTRES

*Please send your order
by mail, fax, or by e-mail
to your nearest
OECD Centre*

OECD TOKYO CENTRE

Landic Akasaka Building
2-3-4 Akasaka, Minato-ku
Tokyo 107-0052, Japan
Tel: (+81-3) 3586 2016
Fax: (+81-3) 3584 7929
E-mail: center@oecdtokyo.org
Internet: www.oecdtokyo.org

OECD WASHINGTON CENTER

2001 L Street NW, Suite 650
Washington, D.C., 20036-4922, US
Tel: (+1-202) 785-6323
Toll-free number for orders:
(+1-800) 456-6323
Fax: (+1-202) 785-0350
E-mail: washington.contact@oecd.org
Internet: www.oecdwash.org

I would like to order the following publications

ANNUAL PUBLICATIONS - 2000 Edition	PUBLICATION DATE	QTY	PRICE*	TOTAL
☐ Energy Statistics of OECD Countries 1997-1998	June 2000		$110	
☐ Energy Balances of OECD Countries 1997-1998	June 2000		$110	
☐ Energy Statistics of Non-OECD Countries 1997-1998	September 2000		$110	
☐ Energy Balances of Non-OECD Countries 1997-1998	September 2000		$110	
☐ Coal Information 2000	August 2000		$200	
☐ Electricity Information 2000	September 2000		$130	
☐ Natural Gas Information 2000	September 2000		$150	
☐ Oil Information 2000	August 2000		$150	
☐ CO_2 Emissions from Fuel Combustion 1971-1998	October 2000		$150	

Please enter my subscription as indicated below

QUARTERLY PUBLICATIONS	QTY	SINGLE COPY*	ANNUAL*	TOTAL
☐ Energy Prices and Taxes		$110	$350	
☐ Oil, Gas, Coal and Electricity Statistics		$110	$350	

*Postage and packing fees will be added to each order.

DELIVERY DETAILS

Name   Organisation

Address

Country   Postcode

Telephone   Fax

PAYMENT DETAILS

☐ I enclose a cheque payable to IEA Publications for the sum of $ _____

☐ Please debit my credit card (tick choice).   ☐ Access/Mastercard   ☐ Diners   ☐ VISA   ☐ AMEX

Card no: ⌐_␣_␣_␣_␣_␣_␣_␣_␣_␣_␣_␣_¬

Expiry date: ⌐_␣_␣_␣_¬   Signature:

Order Form

Diskettes

OECD BONN OFFICE

c/o DVG mbh (OECD)
Birkenmaarstrasse 8
D-53340 Meckenheim, Germany
Tel: (+49-2225) 926 166/168
Fax: (+49-2225) 926 169
E-mail: oecd@dvg.dsb.net
Internet: www.oecd.org/bonn

OECD MEXICO CENTRE

Edificio INFOTEC
Av. Presidente Mazarik 526
Colonia: Polanco
C.P. 11560 - Mexico D.F.
Tel: (+52-5) 281 38 10
Fax: (+52-5) 280 04 80
E-mail: mexico.contact@oecd.org
Internet: www.rtn.net.mx/ocde

OECD CENTRES

*Please send your order
by mail, fax, or by e-mail
to your nearest
OECD Centre*

OECD TOKYO CENTRE

Landic Akasaka Building
2-3-4 Akasaka, Minato-ku
Tokyo 107-0052, Japan
Tel: (+81-3) 3586 2016
Fax: (+81-3) 3584 7929
E-mail: center@oecdtokyo.org
Internet: www.oecdtokyo.org

OECD WASHINGTON CENTER

2001 L Street NW, Suite 650
Washington, D.C., 20036-4922, US
Tel: (+1-202) 785-6323
Toll-free number for orders:
(+1-800) 456-6323
Fax: (+1-202) 785-0350
E-mail: washington.contact@oecd.org
Internet: www.oecdwash.org

*I would like to order the following diskettes**

ANNUAL DISKETTES* - 2000 Edition	RELEASE DATE	QTY	PRICE**	TOTAL
☐ Energy Statistics of OECD Countries 1960-1998	June 2000		$500	
☐ Energy Balances of OECD Countries 1960-1998	June 2000		$500	
☐ Energy Statistics of Non-OECD Countries 1971-1998	September 2000		$500	
☐ Energy Balances of Non-OECD Countries 1971-1998	September 2000		$500	
☐ *Combined subscription of the above four series*	–		*$1200*	
☐ Coal Information 2000	August 2000		$500	
☐ Electricity Information 2000	September 2000		$500	
☐ Natural Gas Information 2000	September 2000		$500	
☐ Oil Information 2000	August 2000		$500	
☐ CO_2 Emissions from Fuel Combustion 1960/71-1998	October 2000		$500	

Please enter my subscription as indicated below

QUARTERLY DISKETTES*	QTY	PRICE**	TOTAL
☐ Energy Prices and Taxes (four quarters)		$800	

*Prices are for single user licence. Please contact us for pricing information on multi-user licences.
**Postage and packing fees will be added to each order.

DELIVERY DETAILS

Name   Organisation

Address

Country   Postcode

Telephone   Fax

PAYMENT DETAILS

☐ I enclose a cheque payable to IEA Publications for the sum of $ _____

☐ Please debit my credit card (tick choice).   ☐ Access/Mastercard   ☐ Diners   ☐ VISA   ☐ AMEX

Card no: ⌊_⌋_⌊_⌋_⌊_⌋_⌊_⌋_⌊_⌋_⌊_⌋_⌊_⌋_⌊_⌋_⌊_⌋_⌊_⌋_⌊_⌋

Expiry date: ⌊_⌋_⌊_⌋_⌊_⌋_⌊_⌋

Signature:

International Energy Agency, 9, rue de la Fédération, 75739 Paris CEDEX 15
PRINTED IN FRANCE BY STEDI
(61 00 15 3P) ISBN 92-64-08511-4 - 2000